MW01201936

THE OXFORD HANDBOOK OF

# THE HISTORY
# OF QUANTUM
# INTERPRETATIONS

# THE OXFORD HANDBOOK OF

# THE HISTORY
# OF QUANTUM
# INTERPRETATIONS

*Edited by*
OLIVAL FREIRE JR,
*Assistant Editors*
GUIDO BACCIAGALUPPI,
OLIVIER DARRIGOL,
THIAGO HARTZ,
CHRISTIAN JOAS,
ALEXEI KOJEVNIKOV,
*and*
OSVALDO PESSOA JR

OXFORD
UNIVERSITY PRESS

# OXFORD
## UNIVERSITY PRESS

Great Clarendon Street, Oxford, OX2 6DP,
United Kingdom

Oxford University Press is a department of the University of Oxford.
It furthers the University's objective of excellence in research, scholarship,
and education by publishing worldwide. Oxford is a registered trade mark of
Oxford University Press in the UK and in certain other countries

First Edition published in 2022

Impression: 1

Published in the United States of America by Oxford University Press
198 Madison Avenue, New York, NY 10016, United States of America

British Library Cataloguing in Publication Data
Data available

Library of Congress Control Number: 2022900685

ISBN 978-0-19-884449-5

Printed and bound by
CPI Group (UK) Ltd, Croydon, CR0 4YY

# TABLE OF CONTENTS

*List of Contributors*                                                        xi

Introduction                                                                   1
OLIVAL FREIRE JR, GUIDO BACCIAGALUPPI, OLIVIER DARRIGOL,
THIAGO HARTZ, CHRISTIAN JOAS, ALEXEI KOJEVNIKOV,
AND OSVALDO PESSOA JR

## PART I QUANTUM PHYSICS—SCIENTIFIC AND PHILOSOPHICAL ISSUES UNDER DEBATE

1. Quantum Mechanics is Routinely Used in Laboratories with Great
   Success, but No Consensus on its Interpretation has Emerged              7
   FRANCK LALOË

2. Philosophical Issues Raised by Quantum Theory and its
   Interpretations                                                         53
   WAYNE C. MYRVOLD

## PART II HISTORICAL LANDMARKS OF THE INTERPRETATIONS AND FOUNDATIONS OF QUANTUM PHYSICS

3. Quantization Conditions, 1900–1927                                      77
   ANTHONY DUNCAN AND MICHEL JANSSEN

4. Of Weighting and Counting: Statistics and Ontology in the
   Old Quantum Theory                                                      95
   MASSIMILIANO BADINO

5. Dead as a Doornail? Zero-Point Energy and Low-Temperature
   Physics in Early Quantum Theory                                        117
   HELGE KRAGH

6. The Early Debates about the Interpretation of Quantum
   Mechanics                                                     135
   MARTIN JÄHNERT AND CHRISTOPH LEHNER

7. Foundations and Applications: The Creative Tension in the
   Early Development of Quantum Mechanics                        173
   CHRISTIAN JOAS

8. The Statistical Interpretation: Born, Heisenberg, and von
   Neumann, 1926–27                                              203
   GUIDO BACCIAGALUPPI

9. A Perennially Grinning Cheshire Cat? Over A Century
   of Experiments on Light Quanta and Their Perplexing
   Interpretations                                               233
   KLAUS HENTSCHEL

10. The Evolving Understanding of Quantum Statistics            255
    DANIELA MONALDI

11. The Measurement Problem                                     281
    OSVALDO PESSOA JR

12. Einstein's Criticism of Quantum Mechanics                  303
    MICHEL PATY

13. Tackling Loopholes in Experimental Tests of Bell's Inequality  339
    DAVID I. KAISER

14. The Measuring Process in Quantum Field Theory              371
    THIAGO HARTZ

15. The Interpretation Debate and Quantum Gravity             393
    ALEXANDER S. BLUM AND BERNADETTE LESSEL

16. Quantum Information and the Quest for Reconstruction of
    Quantum Theory                                              417
    ALEXEI GRINBAUM

17. Natural Reconstructions of Quantum Mechanics               437
    OLIVIER DARRIGOL

18. The Axiomatization of Quantum Theory through Functional
    Analysis: Hilbert, von Neumann, and Beyond                 473
    KLAAS LANDSMAN

19. Tony Leggett's Challenge to Quantum Mechanics and its Path
    to Decoherence                                                          495
    FÁBIO FREITAS

## PART III PLACES AND CONTEXTS RELEVANT FOR THE INTERPRETATIONS OF QUANTUM THEORY

20. The Copenhagen Interpretation                                          521
    DON HOWARD

21. Copenhagen and Niels Bohr                                              543
    ANJA SKAAR JACOBSEN

22. Grete Hermann's Interpretation of Quantum Mechanics                    567
    ELISE CRULL

23. Instrumentation and the Foundations of Quantum Mechanics               587
    CLIMÉRIO PAULO DA SILVA NETO

24. Early Solvay Councils: Rhetorical Lenses for Quantum
    Convergence and Divergence                                             615
    JOSÉ G. PERILLÁN

25. The Foundations of Quantum Mechanics in Post-War
    Italy's Cultural Context                                               641
    FLAVIO DEL SANTO

26. Foundations of Quantum Physics in the Soviet Union                     667
    JEAN-PHILIPPE MARTINEZ

27. Early Japanese Reactions to the Interpretation of Quantum
    Mechanics, 1927–1943                                                   687
    KENJI ITO

28. Form and Meaning: Textbooks, Pedagogy, and the Canonical
    Genres of Quantum Mechanics                                            709
    JOSEP SIMON

29. Chien-Shiung Wu's Contributions to Experimental Philosophy             735
    INDIANARA SILVA

30. On How *Epistemological Letters* Changed the Foundations
of Quantum Mechanics                                                    755
SEBASTIÁN MURGUEITIO RAMÍREZ

31. Quantum Interpretations and 20th Century Philosophy
of Science                                                              777
THOMAS RYCKMAN

# PART IV  HISTORICAL AND PHILOSOPHICAL THESES

32. Bohr and the Epistemological Lesson of Quantum Mechanics            797
STEFANO OSNAGHI

33. Making Sense of the Century-Old Scientific Controversy over
the Quanta                                                              825
OLIVAL FREIRE JR

34. Orthodoxy and Heterodoxy in the Post-war Era                        847
KRISTIAN CAMILLERI

35. The Reception of the Forman Thesis in Modernity and
Postmodernity                                                           871
PAUL FORMAN

36. Quantum Historiography and Cultural History: Revisiting
the Forman Thesis                                                       887
ALEXEI KOJEVNIKOV

37. The Co-creation of Classical and Modern Physics and
the Foundations of Quantum Mechanics                                    909
RICHARD STALEY

38. Interpretation in Electrodynamics, Atomic Theory, and
Quantum Mechanics                                                       937
GIORA HON AND BERNARD R. GOLDSTEIN

# PART V  THE PROLIFERATION OF INTERPRETATIONS

39. Hidden Variables                                                           957
    JEFFREY BUB

40. Pure Wave Mechanics, Relative States, and Many Worlds                      987
    JEFFREY A. BARRETT

41. Is QBism a Possible Solution to the Conceptual Problems
    of Quantum Mechanics?                                                     1007
    HERVÉ ZWIRN

42. Agential Realism—A Relation Ontology Interpretation
    of Quantum Physics                                                        1031
    KAREN BARAD

43. The Relational Interpretation                                             1055
    CARLO ROVELLI

44. The Philosophy of Wholeness and the General and New
    Concept of Order: Bohm's and Penrose's Points of View                     1073
    JEAN-JACQUES SZCZECINIARZ AND JOSEPH KOUNEIHER

45. Spontaneous Localization Theories: Quantum Philosophy
    between History and Physics                                               1103
    VALIA ALLORI

46. The Non-Individuals Interpretation of Quantum Mechanics                   1135
    DÉCIO KRAUSE, JONAS R. B. ARENHART, AND OTÁVIO BUENO

47. Modal Interpretations of Quantum Mechanics                                1155
    DENNIS DIEKS

48. A Brief Historical Perspective on the Consistent Histories
    Interpretation of Quantum Mechanics                                       1175
    GUSTAVO RODRIGUES ROCHA, DEAN RICKLES, AND FLORIAN J. BOGE

49. Einstein, Bohm, and Bell: A Comedy of Errors                              1197
    JEAN BRICMONT

50. The Statistical (Ensemble) Interpretation of Quantum
    Mechanics                                                    1223
    ALEXANDER PECHENKIN

51. Stochastic Interpretations of Quantum Mechanics              1247
    EMILIO SANTOS

*Index*                                                          1265

# LIST OF CONTRIBUTORS

**Valia Allori,** Professor of Philosophy, Northern Illinois University

**Jonas R. B. Arenhart,** Associate Professor, Universidade Federal de Santa Catarina

**Guido Bacciagaluppi,** Associate Professor of Foundations of Physics, Freudenthal Institute, Utrecht University

**Massimiliano Badino,** Associate Professor in Logic and Philosophy of Science, University of Verona

**Karen Barad,** Professor, University of California at Santa Cruz

**Jeffrey A. Barrett,** Chancellor's Professor of Logic and Philosophy of Science, UC Irvine

**Alexander Blum,** Research group leader at the Max Planck Institute for the History of Science (MPIWG)

**Florian J. Boge,** Postdoctoral Researcher at Bergische Universität Wuppertal

**Jean Bricmont,** Professor, Catholic University of Louvain

**Jeffrey Bub,** Distinguished University Professor Emeritus in the Department of Philosophy, the Institute for Physical Science and Technology, and the Joint Center for Quantum Information and Computer Science, University of Maryland

**Otávio Bueno,** Professor of Philosophy and Cooper Senior Scholar in Arts and Sciences, University of Miami

**Kristian Camilleri,** Senior Lecturer in History and Philosophy of Science, University of Melbourne

**Elise Crull,** Associate Professor of Philosophy, The City College of New York

**Olivier Darrigol,** Research Director at Centre National de la Recherche Scientifique, UMR SPHere, Paris

**Dennis Dieks,** Professor, Utrecht University

**Anthony Duncan,** Professor Emeritus, Department of Physics and Astronomy, University of Pittsburgh

**Paul Forman,** Curator of Modern Physics, Emeritus, National Museum of American History, Smithsonian Institution

**Olival Freire Jr,** Professor, Physics Institute, Universidade Federal da Bahia

**Fábio Freitas,** Associate Professor of History, Philosophy and Physics Teaching, Physics Institute, Universidade Federal da Bahia

**Bernard R. Goldstein,** University Professor Emeritus University of Pittsburgh

**Alexei Grinbaum,** Researcher at the Commissariat à l'énergie atomique (CEA-Saclay/ Larsim), France

**Thiago Hartz,** Assistant Professor, Instituto de Matemática, Universidade Federal do Rio de Janeiro

**Klaus Hentschel,** Professor for History of Science & Technology, University of Stuttgart

**Giora Hon,** Professor Emeritus of Philosophy, University of Haifa

**Don Howard,** Professor, Department of Philosophy, University of Notre Dame

**Kenji Ito,** Associate Professor at the Graduate University for Advanced Studies, SOKENDAI

**Anja Skaar Jacobsen,** Physics teacher at the Copenhagen Adult Education Centre (KVUC), Denmark

**Martin Jähnert,** Research and Teaching Associate, Technical University Berlin & Max Planck Institute for the History of Science (MPIWG)

**Michel Janssen,** Professor, School of Physics and Astronomy, University of Minnesota

**Christian Joas,** Director, Niels Bohr Archive, and Associate Professor, Department of Science Education, University of Copenhagen

**David I. Kaiser,** Germeshausen Professor of the History of Science and Professor of Physics, Massachusetts Institute of Technology

**Alexei Kojevnikov,** Professor at the Department of History of the University of British Columbia, Vancouver

**Joseph Kouneiher,** Professor of mathematical physics and engineering sciences, Côte d'Azur University

**Helge Kragh,** Emeritus professor, Niels Bohr Institute, University of Copenhagen

**Décio Krause,** Professor of Philosophy, Universidade Federal de Santa Catarina and Graduate Program in Logic and Metaphysics, Universidade Federal do Rio de Janeiro

**Franck Laloë,** Researcher at the CNRS, Laboratoire Kastler Brossel, ENS Paris

**Klaas Landsman,** Chair of Mathematical Physics, Radboud University Nijmegen

**Christoph Lehner,** Independent researcher, Berlin

**Bernadette Lessel,** Postdoctoral Fellow at the Max Planck Institute for the History of Science (MPIWG)

**Jean-Philippe Martinez,** Postdoctoral researcher at the RWTH Aachen University

**Daniela Monaldi,** Course Director at the Department of Science & Technology Studies, York University

**Wayne C. Myrvold,** Professor, Department of Philosophy, The University of Western Ontario

**Stefano Osnaghi,** Postdoctoral researcher at the Institute for Quantum Optics and Quantum Information, Vienna; Archives Husserl, ENS, Paris

**Michel Paty,** Emeritus Research Director, CNRS - Université Paris Diderot

**Climério Paulo da Silva Neto,** Assistant Professor of History, Philosophy and Physics Teaching, Physics Institute, Universidade Federal da Bahia

**Alexander Pechenkin,** Professor, Department of philosophy and methodology of science, Lomonosov Moscow State University; Senior researcher, S. Vavilov Institute of the History of Science and Technology

**José G. Perillán,** Associate Professor of Physics and Science, Technology, and Society at Vassar College

**Osvaldo Pessoa Jr,** Associate Professor, Department of Philosophy, Universidade de São Paulo

**Sebastián Murgueitio Ramírez,** Postdoctoral Research Fellow in Philosophy of Physics, University of Oxford

**Dean Rickles,** Professor of History and Philosophy of Modern Physics, University of Sydney

**Gustavo Rodrigues Rocha,** Assistant Professor, Physics Department, Universidade Estadual de Feira de Santana

**Carlo Rovelli,** AMU Université, Université de Toulon, CNRS, CPT, Marseille; Department of Philosophy and Rotman Institute of Philosophy, University of Western Ontario, and Perimeter Institute

**Thomas Ryckman,** Professor of Philosophy, Stanford University

**Flavio Del Santo,** Researcher at the Institute of Quantum Optics and Quantum Information (IQOQI-Vienna) and the University of Vienna

**Emilio Santos,** Emeritus Professor of Physics, Universidad de Cantabria

**Indianara Silva,** Associate Professor, Physics Department, Universidade Estadual de Feira de Santana

**Josep Simon,** Ramón y Cajal Research Fellow, Institut interuniversitari López Piñero, Universitat de València

**Richard Staley,** Hans Rausing Lecturer and Reader in History and Philosophy of Science, University of Cambridge

**Jean-Jacques Szczeciniarz,** Professeur émérite, Université Paris Diderot Paris 7

**Hervé Zwirn,** Research Director, CNRS, ENS Paris-Saclay, France

# INTRODUCTION

OLIVAL FREIRE JR, GUIDO BACCIAGALUPPI,
OLIVIER DARRIGOL, THIAGO HARTZ,
CHRISTIAN JOAS, ALEXEI KOJEVNIKOV,
AND OSVALDO PESSOA JR

THIS Handbook is dedicated to the memory of Silvan Samuel Schweber (1928–2017), who was the first to note that we needed such a volume, but who unfortunately was unable to see the work to its completion and sign this introduction with us. Sam Schweber was an accomplished physicist in the domain of quantum field theories until he moved to history of sciences where he wrote influential works such as a history of quantum electrodynamics and a biography of Hans Bethe (Schweber 1994, 2012; see also, on Schweber's works, Gavroglu and Renn, 2007). More importantly, he was an exceptional human being and a source of inspiration and encouragement for all of us.

Quantum mechanics, created in 1925–1927, is approaching its centenary with an impressive record. It became the backbone of most research in physics, led to applications such as the transistor and laser, and prompted an upheaval in the philosophy of science. Its scope and its precision have been constantly growing, and it is now promising more powerful computers, safer cryptography, and more sophisticated sensing devices. This century of conquests has also been a time of ongoing debates about the foundations and interpretation of the theory, which has been referred to as the quantum controversy. The Oxford Handbook of the History of Quantum Interpretations is dedicated to these debates. Although quantum physics is not the first scientific theory to spark fierce debates about its foundations, the resulting controversy has been unusually long and remains open. Half a century ago, the historian of physics Max Jammer aptly called it 'a story without an ending' (Jammer, 1974, p. 521). There still is no ending or closure in sight.

In fact, the debates on the foundations and interpretation of this physical theory began when its first elements appeared with the works of Max Planck, Albert Einstein, and Niels Bohr at the beginning of the 20th century. They coalesced between 1925 and 1927 when the very quantum theory was still *in statu nascendi*. These early debates were championed mainly by Einstein, Bohr, Born, Schrödinger, Pauli, Heisenberg, and Jordan and expressed quite a variety of conflicting philosophical views, which can be grouped around four main issues: *Anschaulichkeit–Unanschaulichkeit* (roughly translated as visualizability–unvisualizability), continuity–discontinuity, the wave–particle dilemma, and causality–acausality. Since 1927 the debates have attracted new generations of physicists and the agenda has diversified widely, with varying emphasis through time.

The history of the quantum debates has not been smooth: the stakes, the actors, and the intensity of these debates have also varied considerably over time. They gained more traction when it became clear that they were helping to improve the shared understanding of quantum physics, regardless of the motivations of the physicists engaged in the controversy. Indeed, in large part thanks to these debates, physicists have become familiar with the now essential concepts of entanglement and decoherence, which have provided the foundations for the currently flourishing research on quantum information.

Interpretations and foundations of quantum physics have now become an appealing subject for a few mainstream physicists. The reasons for this are diverse. Steven Weinberg (2017) once surprised many by declaring: 'I'm not as sure as I once was about the future of quantum mechanics.' Reasons for this unease were related to the difficulty of conveying physical meaning to the mathematical formalism of quantum mechanics. For Weinberg, 'it is a bad sign that those physicists today who are most comfortable with quantum mechanics do not agree with one another about what it all means.' In a recent textbook, Weinberg (2013, p. 95) analysed several interpretations of quantum mechanics and arrived at the conclusion that 'today there is no interpretation of quantum mechanics that does not have serious flaws, and that we ought to take seriously the possibility of finding some more satisfactory other theory, to which quantum mechanics is merely a good approximation.' Franck Laloë, researcher at the École Normale Supérieure and a contributor to this handbook, presents the subject along the same lines, with a sense of history: 'We have a rare situation in the history of sciences: consensus exists concerning a systematic approach to physical phenomena, involving calculation methods having an extraordinary predictive power; nevertheless, almost a century after the introduction of these methods, the same consensus is far from being reached concerning the interpretation of the theory and its foundations.' (Laloë, 2012, p. xi). In a book provocatively titled *Do we really understand quantum mechanics?* he concluded with a strong metaphor: 'This is reminiscent of the colossus with feet of clay.' The Austrian physicist Anton Zeilinger, who is deeply involved in experiments regarding quantum foundations and quantum information, did something quite unusual for a respected scientist: he conducted a survey among thirty-three participants at a conference on the foundations of quantum mechanics, and asked 'What is your favorite interpretation of quantum mechanics?' Every participant could give more than one answer. The results revealed 42% support for 'Copenhagen', 24% for 'Information-based/information-theoretical', 18% for 'Everett' in its diverse versions, 9% support for 'Objective collapse' such as GRW, 6% in favour of 'Quantum

Bayesianism', 6% for 'Relational quantum mechanics', and 12% for 'Other' or 'I have no preferred interpretation'. While the authors of the survey cautiously presented their results as a 'snapshot', it is surely indicative of the current diversity of views (Schlosshauer, Kofler, and Zeilinger, 2013).

Presenting the diversity of interpretations of quantum mechanics requires a caveat. Some of the so-called 'interpretations' are not just interpretations of a fixed formalism, they are different theories as they alter the mathematical and conceptual structure of quantum mechanics. To be more precise, we should speak of the existence of a continued debate about the interpretations *and the foundations* of quantum mechanics. That said, at certain moments in this volume, including in the title, we will simply use 'history of interpretations', 'history of quantum interpretations' or even 'quantum controversy' for brevity.

While the subject of interpretations and foundations of physics appeals to physicists as well as philosophers of physics, it has also attracted intense attention from historians of science. Not only did quantum physics produce a fundamental revolution in sciences, but its dramatic history also drew together diverse factors related to ideas, values, experiments, instruments, philosophy, politics, ideology, cultural changes, and professional biases, among others. It thus presents a challenging case for the analysis of how complex the workings of contemporary science can be. Furthermore, it is not a linear history. The subject has been and continues to be successively much debated, forgotten, stigmatized, revived, and amplified. At each of these stages some of these factors have been particularly influential. As we will now see, an example that can illustrate this multifaceted story is the recognition of entanglement as a signature quantum concept.

# THE STORY OF ENTANGLEMENT

The early story of entanglement was closely connected with a conceptual issue: whether or not quantum mechanics was a complete theory in a certain sense. This issue was the matter of multiple debates among Niels Bohr, Albert Einstein, Erwin Schrödinger, and John von Neumann, among others. Later, this was forgotten or put to one side. David Bohm tried to revive it in the early 1950s only to face the hostility of his colleagues. In 1964 John Bell revisited it by deriving a crucial contradiction between quantum mechanics and local realism to be taken to the lab benches. Physicists' hostility to this type of consideration lasted even after the first experiments. John Clauser's professional career suffered as a result of it. However, the cultural and political unrest of the late 1960s and early 1970s propelled the subject and new experiments by Alain Aspect drew wider attention to it. In the early 1990s, it received a new boost from dramatic advances in techniques (for instance the use of Parametric Down Conversion as the source of a pair of entangled photons), better experiments, and technological expectations related to the field of quantum information. In 2010 John Clauser, Alain

Aspect, and Anton Zeilinger received the Physics Wolf Prize for their experimental tests of quantum correlations. In the last ten years we have seen even more compelling experiments on entanglement. More recently, experiments led by Zeilinger, with the participation of David Kaiser (who wrote the chapter on experiments on Bell's theorem for this handbook) used cosmic sources (light from distant quasars) in order to close possible loopholes in earlier experiments; a team led by the Chinese physicist Pan Jianwei used entangled photons in experiments involving satellites and earth-based laboratories; and the press has covered the competition among major corporations to design the first prototypes of quantum computers.

The story of entanglement has been marked throughout by an irony, indeed a ruse of history, which may explain physicists' current interest in research on the foundations and interpretation of quantum mechanics. Most of the first physicists to work on this subject were 'quantum dissidents', as they challenged the common view that foundational issues were already solved (Freire Jr, 2015). Furthermore, most of them expected quantum mechanics would run into problems for yet unexplored phenomena. John Bell, the father of Bell's theorem, used to say that quantum mechanics is 'rotten', Hamlet's famous line, in an oblique reference to Niels Bohr, the Danish physicist and the main authority on the standard interpretation of quantum theory. The ruse is that in trying to attack the foundations of quantum mechanics, the quantum dissidents ended up consolidating and developing this theory, which has entered the 21st century as vindicated as ever.

# The scope of this Handbook

The history of the interpretations and foundations of quantum mechanics has gained momentum and professional rigour since the project *Sources for History of Quantum Physics* was completed in the mid-1960s (Kuhn, 1967). It has also contributed to the renewal of perspectives in the history of science as a whole, (see Badino, 2016 and in the chapters on Forman's thesis in this volume). Comprehensive works dealing with the full plethora of subjects and periods are nevertheless scant, noticeable exceptions being Max Jammer's books (Jammer, 1966 and 1974) and a compendium with entries on concepts, experiments, history, and philosophy of quantum mechanics (Greenberger, Hentschel, and Weinert, 2009). A particularly influential reference has been Jammer's 1974 book *The Philosophy of Quantum Mechanics: The Interpretations of Quantum Mechanics in Historical Perspective*. This Handbook, with its wide scope and span, attempts to enrich this genre of works.

The Oxford Handbooks offer an ideal format to address the challenge of writing a huge volume, as comprehensive as possible, bringing together a diversity of authors. This multiplicity was necessary not only due to the difficulty of finding a single author able to cover all aspects of the quantum interpretation debates, but also because it was not desirable to have a single author. The interpretations and foundations of quantum mechanics is such a controversial subject that a single author would inevitably bring certain biases to the writing. A recent book illustrates this potential risk. While organizing the book *Elegance and Enigma*, Maximilian Schlosshauer dealt with a similar issue and asked, given the diversity of views and the entrenched controversy about this contemporaneous subject, 'how could a single author do the field full justice without coloring her story?' The solution he found was to produce a book strictly based on questions answered by 17 actors (most of them physicists and a few of them philosophers) in the field of foundations of quantum mechanics (Schlosshauer, 2011). In this handbook, we have brought together a number of historians of physics and also physicists and philosophers of physics who are sensitive to the task of writing texts with historically informed goals. Despite the large number of contributors, we are aware that a few topics, such as weak measurement or quantum cosmology, have not been adequately addressed. Some chapters are dedicated to synthesizing earlier studies. Many others deal with subjects not yet covered in the literature in the history of science. Although the collaboration of historians, philosophers, and physicists on a single project is not common practice, we decided early on that such collaboration was necessary to provide a comprehensive, multifaceted history of the interpretations and foundations of quantum physics. We hope the result will meet our readers' expectations and perhaps guide them in the unfinished quest for quantum understanding.

This Oxford Handbook may serve physicists, both senior and junior, and physics students who are contemplating the interpretations of quantum mechanics. We hope it may be particularly useful to those who are engaged with, or attracted by, current research in the field of quantum information. We hope it may give them a flavour of the history that led to or did the ground work for the current investigations. The book will help map the roads already taken and the issues already dealt with as well as those yet open and it should also be of interest to the many philosophers of science who work on the philosophy of quantum mechanics. Last but not least, it should interest historians of science and particularly historians of physics who deal with quantum physics, the ensuing debates, their cultural conditions, and their technological impact.

## ACKNOWLEDGEMENTS

First of all, we thank the authors who contributed to this volume. In addition, its production was possible thanks to the dialogue with many colleagues who supported us in putting together this volume. The names of a few of them appear in the following chapters. Furthermore, we would like to thank the staff at Oxford University Press who supported this work, particularly Sonke Adlung, Harriet Konishi, Katherine Ward, Manikandan Santhanam, and Giulia Lipparini, as well as the early reviewers of the proposal. We were assisted by Denise Key and Climério Paulo da Silva Neto in the final stages of the production of the manuscript. Earlier versions of the Handbook proposal and some chapters were presented at the Laboratório de História da Ciência (LAHCIC), in Bahia, Brazil; we thank the comments from colleagues and students. We are grateful to the Brazilian agencies CNPQ (408011/2018-1) and CAPES-PRINT/UFBA for supporting the research that led to this volume.

## References

Badino, M. (2016). What Have the Historians of Quantum Physics Ever Done for Us? *Centaurus*, **58**, 327–346.

Freire Jr, O. (2015). *The Quantum Dissidents – Rebuilding the Foundations of Quantum Mechanics 1950–1990*. Berlin: Springer.

Gavroglu, K., and Renn, J. (eds) (2007). *Positioning the History of Science*. Dordrecht: Springer.

Greenberger, D., Hentschel, K., and Weinert, F. (eds.) (2009). *Compendium of Quantum Physics: Concepts, Experiments, History and Philosophy*. Berlin: Springer.

Jammer, M. (1966). *The Conceptual Development of Quantum Mechanics*. New York: McGraw-Hill.

Jammer, M. (1974). *The Philosophy of Quantum Mechanics: The Interpretations of Quantum Mechanics in Historical Perspective*. New York: John Wiley & Sons.

Kuhn, T. S. (1967). *Sources for history of quantum physics; an inventory and report* [by] Thomas S. Kuhn [and others]. Philadelphia: American Philosophical Society.

Laloë, F. (2012). *Do We Really Understand Quantum Mechanics?* New York: Cambridge University Press.

Schlosshauer, M. (ed.) (2011). *Elegance and Enigma – The Quantum Interviews*. Heidelberg: Springer.

Schlosshauer, M., Kofler, J., and Zeilinger, A. (2013). A snapshot of foundational attitudes toward quantum mechanics. *Stud. Hist. Philos. Mod. Phys.*, **44**,222–230.

Schweber, S. S. (1994). *QED and the men who made it: Dyson, Feynman, Schwinger, and Tomonaga*. Princeton, NJ: Princeton University Press.

Schweber, S. S. (2012). *Nuclear forces: the making of the physicist Hans Bethe*. Cambridge, MA: Harvard University Press.

Weinberg, S. (2013). *Lectures on Quantum Mechanics*. Cambridge: Cambridge University Press.

Weinberg, S. (2017). 'The Trouble with Quantum Mechanics'. *The New York Review of Books*, January 19.

# CHAPTER 1

# QUANTUM MECHANICS IS ROUTINELY USED IN LABORATORIES WITH GREAT SUCCESS, BUT NO CONSENSUS ON ITS INTERPRETATION HAS EMERGED

FRANCK LALOË*

PHYSICS is a very robust scientific discipline. Of course, from time to time, one of its component needs to be replaced by a new element, sometimes completely different; scientific revolutions do occur in physics. It nevertheless remains a stable edifice, which has sometimes been compared to a table with many legs; if one leg is broken, or needs replacement for some reason, the table does not collapse. Even revolutions do not revolve the table upside down; a new stronger component in the theory replaces the old one and the whole edifice remains operational.

What is true, nevertheless, is that all feet of the table are not of equal importance. Two main pillars in contemporary physics are more essential than the others: quantum theory and general relativity. Both provide general frameworks inside which all other theories are supposed to fit. Quantum theory provides a formalism in terms of operators acting in a space of states, a formalism which applies in principle to all physical systems. The other pillar, general relativity, provides the structure of space-time inside which all physical events occur. It would be rewarding to claim that these

* laloe@lkb.ens.fr

two basic theories connect to each other in a seamless way, or are complementary. Alas, this is not the case, for at least two reasons. The first is that gravity is at the core of general relativity, and that it seems to be very difficult to propose a quantum field theory that includes gravity; this is still a domain of active theoretical research. The second reason is more conceptual. While the interpretation of general relativity is almost universally accepted, constant and vivid discussions concerning the interpretation of quantum mechanics still take place. Proponents of various interpretations still argue to defend their point of view against others. This is an unusual situation in physics. When new theories appear, it is perfectly normal if discussions should initially arise about how this theory should be interpreted. But, in most cases, these discussions tend to decay after some time, and a consensual interpretation emerges, in particular in textbooks. Strangely enough, quantum mechanics does not follow this scenario: one century after its appearance, active discussions and even controversies about its interpretation still constantly take place.

The word 'interpretation', when applied to a physical theory, can be understood in several ways. A minimalistic view would be to consider that an interpretation is just a set of metaphors or illustrative images facilitating the use of the theory without any real implication on its nature, such as Feynman diagrams for instance. Another definition of the word would refer to the development of correspondence rules between the mathematical formalism and the experiments. Still another view would see an interpretation as a method to eliminate internal insufficiencies or contradictions, or even to provide a consistent view of the physical world that emerges from the theory. The interpretations we will discuss in this text belong to the second and third, more ambitious, categories.

The purpose of this chapter is to give an overview of the present situation concerning quantum mechanics, without any claim to be complete or exhaustive. In fact, so many different interpretations of quantum theory have been proposed in the literature that even a whole book could not include a proper description of all of them. We will therefore limit ourselves to brief descriptions of the main classes of interpretations, their differences, but also their similarities. More detailed information can for instance be obtained from Jammer (1966), Mehra and Rechenberg (1982), Darrigol (1992), and d'Espagnat (1971). This chapter is not an introduction to quantum mechanics: we will assume that the reader is already familiar with the basic formalism of this theory. Many of the ideas presented here are discussed in more detail in Laloë (2019).

## 1.1  QUANTUM MECHANICS IN LABORATORIES

How do physicists use quantum mechanics in their everyday practice, in their laboratories? Most physicists are more busy applying the theory than interpreting it. This attitude is natural: physics is a theory rooted in experiments, and it turns out that the calculations required to make predictions concerning experiments are practically independent of the interpretation. Why then worry about interpretation? This may explain a natural tendency of many physicists to think that 'the interpretation of quantum mechanics is something to leave to philosophers', the role of physicists being to perform useful experiments and calculations. Indeed, numerous great discoveries have been made in quantum physics without worrying deeply about the foundations.

Nevertheless, the human mind is built in such a way that it is difficult, if not impossible, to use pure algebra without any mental picture of the physics involved. It seems likely that most quantum physicists, consciously or unconsciously, create in their mind an image of how things evolve in quantum physics: for instance, how the wave function evolves in time, what physical processes act on it, etc. In other words, even those who claim that they do not care about the interpretation of quantum mechanics may be using one, even in a vague sense. So, discussing the interpretations of quantum mechanics is not without interest.

We begin with a discussion of decoherence since, for some time, it was rather popular to consider that decoherence provides a satisfying answer to all conceptual difficulties of quantum mechanics. Even if this idea turned out to be delusive, the effects of quantum decoherence are worth studying in a first step.

## 1.1.1 Decoherence

Because physical systems can be described in quantum mechanics by state vectors[1] (or wave functions) belonging to a linear space of states (often called the Hilbert space), a basic tenet is the superposition postulate: quantum states can be linearly combined, and this can give rise to interference effects. Many experiments have confirmed this postulate. For instance, interference experiments have been performed with large molecules containing more than 300 atoms (Hornberger *et al.*, 2012), and more recently 2000 (Fein *et al.*, 2019); for the moment no limit has been observed on the maximum size of quantum objects that can interfere. Nevertheless, there exist practical limits on the possibilities of observing interference phenomena, and an important limit is the phenomenon of decoherence.

Decoherence takes place as soon as the interfering object interacts with its surrounding, and for instance changes the quantum state of its environment in a way that depends on the path taken by the object in the interferometer. Assume that, in a double slit experiment, the test object crossing the two slits scatters a photon (or any particle) in two different ways, depending on whether it crosses the upper or lower slit. If the two quantum states reached by the photon are orthogonal, quantum mechanics predicts that the interference effect completely disappears when the test object is observed later.

---

[1] One can also describe physical systems with density operators, which are equivalent to a set of state vectors associated with probabilities. For our present discussion, state vectors are sufficient.

The theory actually predicts that the interference property has been transferred to the full system comprising the object and the photon: an interference can only be observed if both are made to interfere, which amounts to performing what is called a 'two-particle interference experiment'. Now, if the test object scatters two photons, one has to perform a three-particle interference experiment, etc. which quickly becomes very difficult, if not impossible, for large numbers of photons. Generally speaking, since large objects have a strong tendency to scatter photons or other particles, in practice the coherence is lost and no interference can be observed. The relation between decoherence and quantum measurement have been extensively discussed in the literature, see for instance Zeh (1970), Zurek (1981, 1982, 2003a, b), and Schlosshauer (2005).

At this point, it is tempting to decide that decoherence solves the difficulties of quantum measurement: one could think that the disappearance of quantum interference effects is sufficient to restore classicality. Several authors have proposed this idea, which still reappears regularly in the literature. A closer examination nevertheless shows that this explanation is, at best, only partly correct. The so-called 'measurement problem' requires understanding why a single macroscopic result emerges from a quantum measurement process, while the dynamical equations of quantum mechanics seem to indicate that several results occur simultaneously. The real problem is therefore, not to explain why a coherent superposition of these possibilities does not occur (which is indeed explained by decoherence), but why no superposition at all subsists, coherent or not. In other words, decoherence explains what happens during the first stage of an experiment, before a single result appears in a measurement, but not how this single result emerges; further analysis is required to understand this second step.

## 1.1.2 Common Sense in Laboratories

If this explanation is not satisfactory, why do most physicists seem to have so little interest in solving the problem? Probably because, in practice, they know how to avoid these difficulties by just applying common sense (or physical intuition) even if, at a fundamental level, the logic behind the method may remain rather vague. This pragmatic attitude has proved fruitful, opening the way to many discoveries, which would not necessarily have been possible if the physicists had remained blocked at the level of the foundations.

Applying the theory as it is to concrete examples can be more productive than constantly questioning it. In addition, in many experiments, there is no need to understand the details of the quantum measurement process, since the measurement is performed over a large collection of individual quantum systems. For instance, in a nuclear magnetic resonance experiment, one measures the total magnetization of a large collection of spins, never that of a single spin. It is then sufficient to calculate the average of a macroscopic quantum operator, and this average is a number that can be treated as a classical quantity. There is no need to include the perturbating effect of the measurement process itself. This process then appears as classical, and the difficulties associated with a quantum regime can be seen as academic, with no practical impact. Still, if the problems associated with the conceptual difficulties of quantum mechanics are not given first priority, even very pragmatic physicists acknowledge that logical consistency of the theory requires more than the calculation of averages. We should also be able to somehow incorporate into the theory the detailed description of the events occuring during a single observation of single quantum systems. We now discuss a few popular strategies for reaching this objective. The first (section 1.1.2.1) is to break the von Neumann chain by hand, when it 'obviously' goes too far; the appearance of uniqueness is applied, so to say, above the formalism of the theory. Another popular strategy (section 1.1.2.2) is to use the 'correlation interpretation', a point of view where macroscopic events are indeed considered as classical and unique by definition (or by common sense). In this view, the role of the theory is just to provide probabilities relating preparation events to measurement and observation events. Other scientists prefer to emphasize the central role of information in the theory (section 1.1.2.3).

### 1.1.2.1 *Breaking the von Neumann Chain by Hand; the Projection Postulate (Collapse)*

In his famous 1932 treatise on quantum mechanics, John von Neumann (1932) discusses quantum measurement by considering the measurement apparatus as a quantum system (we come back to von Neumann's contributions in section 1.2.1.1). He observes that, if the measured system is before measurement in a superposition of several eigenstates of measurement, during its interaction with the measurement apparatus it gets entangled with it: the state vector then contains a superposition of several possible states for the system S, each associated with a different final position of the pointer of the apparatus (the device indicating the measurement result). Now, if a second measurement apparatus is used to measure the result indicated by the first, the same phenomenon takes place: now we have three entangled systems, each in states associated with the different measurement results. The chain continues ad infinitum and is often called 'the von Neumann infinite regress'.

To break this chain, and to introduce the uniqueness of the measurement results, von Neumann proposed to introduce his 'projection postulate': because the result of any measurement has to take a well-defined value, the propagation of entanglement has to stop at some point, which means that only one component of the state vector subsists (projection, or collapse, of the state vector). Invoking the 'principle of psycho-physical paralellism' (von Neumann, 1932, pp. 419–421)—according to which it must be possible to describe the subjective perception processes as if they were parts of the physical world—von Neumann incorporates the observer in the chain by discussing the impression of an image in the retina, in the optic nerve, and in the brain. He then shows that a consistent quantum theory can be obtained by merely assuming that the presence of the observer produces the projection, which puts a final end to the chain.

This postulate provides a very convenient way to apply quantum mechanics to real experiments, and this is probably why it is nowadays included in most textbooks. At a fundamental level, a difficulty remains, namely to predict at which point precisely this projection/collapse takes place. The question was left open by von Neumann, since in practice the exact position of this point does not affect the result.

Physicists often put an end to the infinite von Neumann chain in an implicit way, based on their physical intuition. One possible empirical rule is to consider that, as soon as 'significant' decoherence takes place, the von Neumann chain automatically stops: all its branches but one spontaneously disappear. In other words, one systematically associates emergence of uniqueness with macroscopic decoherence. For instance, as soon as a measuring apparatus is inserted in the experiment, and as soon as this apparatus is able to register results, one considers that it is obvious that it registers a single result (whether or not a human being observes it), instead of remaining in a quantum superposition. In any case, clearly there is no hope of ever observing the physical effects of coherent superposition when they have propagated too far in the environment; assuming that the superposition has been resolved into one of its components creates no risk of contradiction with experimental observations.

Of course, the difficulty is to define the exact meaning of the word 'significant' in this context. Questions such as 'At what degree of entanglement, precisely, is the von Neumann chain resolved into a single branch?' are left to common sense and personal judgement; in this sense, the theory is more a phenomenology than a complete construction. Theories with modified Schrödinger dynamics (section 1.2.6) are built to rationalize this approach by introducing a precise physical mechanism to stop the von Neumann chain. They provide precise answers to such questions, but the mechanism allowing the emergence of a single branch is more intrinsic to the quantum system (it may involve the masses of the particles) than induced by a classical environment.

In the so-called 'Wigner interpretation' of quantum mechanics (Wigner, 1961), the origin of state vector reduction is related to human consciousness. The idea is similar to breaking the von Neumann chain when the measurement apparatus registers a result, but here the chain stops as soon as the linear superposition involves different states of human consciousness. This is reminiscent of von Neumann's analysis mentioned above. Similar views were also discussed by London and Bauer as early as 1939 (London and Bauer, 1939). These authors emphasize that state vector reduction restores a pure state from a statistical mixture of the measured subsystem (see section 1.1.1), as well as 'the essential role played by the consciousness of the observer in this transition from a mixture to a pure state'; they then explain this special role by the faculty of introspection of conscious observers. Others prefer to invoke 'special properties' of the electrical currents that correspond to perception in a human brain. Actually, common sense only suggests that human consciousness provides at least an upper limit, a boundary beyond which the von Neumann chain certainly does not propagate; whether or not this propagation stops exactly at this limit, or before, is another question.

## 1.1.2.2 *Correlation Interpretation*

The correlation interpretation can be seen as a 'minimalistic interpretation', a common core that is shared by all other interpretations. Anyone who finds it insufficient is free to add more elements to it, for instance by importing pieces from other interpretations. We will use the words 'correlation interpretation' to describe this point of view, since it puts the emphasis on the correlations between successive results of experiments. In this interpretation, the uniqueness of macroscopic events (including acts of measurements) is postulated, as an obvious property of the macroscopic world that we perceive around us. Another postulate is that these events occur in a stochastic way; the objective of the theory is not to make deterministic predictions, but just to provide probabilities associated with all possible situations of preparation, evolution, and measurements performed on quantum systems.

This interpretation is based on the so-called Wigner formula. Assume that a system is described by the initial density operator $\rho(t_0)$:

$$\rho(t_0) = |\Psi(t_0)\rangle\langle\Psi(t_0)| \tag{1}$$

and that measurements are performed at times $t_1$ and $t_2$, with $t_2 > t_1$. Result $m$ for the first measurement is associated with a projector $P_M(m)$, and result $n$ for the second measurement with a projector $P_N(n)$. In the Heisenberg picture, these projectors become time-dependent operators $\hat{P}_M(m,t)$ and $\hat{P}_N(n,t)$. The Wigner formula expresses the probability of obtaining result $m$ at time $t_1$ followed by result $n$ at time $t_2$ as:

$$\mathcal{P}(m, t_1; n, t_2) = Tr\{\hat{P}_N(n, t_2)\hat{P}_M(m, t_1)\rho(t_0)\hat{P}_M(m, t_1)\hat{P}_N(n, t_2)\} \tag{2}$$

We note that, using circular permutation under the trace, one can in fact suppress one of the extreme projectors $\hat{P}_N(n_2; t_2)$ in formula (2), but not the others. Relation (2) can easily be generalized to more than two measurements (by inserting additional projectors on both sides in the reverse time order), and to situations where $\rho(t_0)$ describes a statistical mixture instead of a pure state.

**No need of state vector reduction** Equation (2) can be seen as a consequence of the postulate of state vector reduction. Conversely, it is also possible to take this equation as a starting point, which then becomes a postulate in itself giving the probability of any sequence of measurements in a perfectly unambiguous way. The von Neumann projection can then be derived as an effective rule. In fact, the projection rule then becomes superfluous, since the generalized (multi-time) Born rule given by (2) is sufficient to predict the probabilities of any sequence of measurements—in a sense, the state vector reduction is also contained in some way in the trace operation of (2), but no explicit reference to it is indispensable.

It is also necessary to postulate that the results of any measurements can give only one of the eigenvalues of the corresponding operator, and that the result is fundamentally random, as also assumed when using the Born rule. The advantage is that, if one just uses formula (2) and ignores state vector reduction, the problems associated with a difficult coexistence of two different evolution postulates disappear; no discontinuous jump of any mathematical quantity ever occurs in the formalism. Why not then give up entirely the other postulates and just use this single formula for all predictions of results?

Some physicists consider that this is indeed the best solution: if one accepts the idea that the purpose of physics is only to correlate the preparation of a physical system, contained mathematically in $\rho(t_0)$, with all possible sequence of results of measurements (by providing their probabilities), nothing more than (2) is needed. Why then worry about what happens in a single realization of an experiment? For instance, in his famous treatise, Paul Dirac (1930, page 7 of the first edition) writes 'The only object of theoretical physics is to calculate results that can be compared with experiments, and it is quite unnecessary that any satisfying description of the whole course of phenomena should be given'. In page 10 of the fourth edition (1958) he expresses a similar idea by writing 'The main object of physical science is not the provision of pictures, but is the formulation of laws governing the phenomena and the application of these laws to the discovery of new phenomena'.

**Discussion** The 'correlation interpretation' is based on the use of density operators, but not contradictory with the use of a state vector, which then expresses a preparation procedure rather than a physical property of the measured system. For instance, Peres (1984) writes 'a state vector is not a property of a physical system, but rather represents an experimental procedure for preparing or testing one of more physical systems'.

Within this interpretation, questions such as 'How should the physical system be described when one first measurement has already been performed, but before the second measurement is decided' are dismissed as superfluous or meaningless. This interpretation is also clearly in complete opposition with the famous EPR reasoning (Einstein, Podolsky, and Rosen, 1935), since it shows no interest whatsoever in questions related to physical reality as something 'in itself'. If one considers that the notion of the EPR elements of reality itself is completely irrelevant to physics, all potential problems related to Bell-, GHZ-, and Hardy-type considerations are automatically solved. The same is true of the emergence of a single result in a single experiment; in a sense, the Schrödinger cat paradox is eliminated from the scope of physics, because the paradox is not expressed in terms of correlations. A positive feature of this point of view is that the boundary between the measured system and the environment provided by the measuring devices is flexible; an advantage of this flexibility is that the method is well suited for successive approximations in the treatment of a measurement process, for instance the tracks left by a particle in a bubble chamber as discussed by John Bell (1981; 1987, pp. 117–138).

### 1.1.2.3 *Emphasizing Information*

Emphasizing the role of information provides another popular interpretation of quantum mechanics (Peres, 1993, 2005; Zeilinger, 1999; Brukner and Zeilinger, 1999; Fuchs, 2001, 2002; Pitowsky, 2003; Auletta, 2001, 2005; Bub, 2011). The informational point of view has become even more natural with the recent rise of the field of quantum information. Information can be about the whole experimental set-up, including preparation and measurement devices. In this case, we recover Bohr's emphasis on the relevance of the whole experimental set-up (section 1.2.1). Information can also include elements that are acquired when experimental results become known, and undergo sudden changes when the results are observed; this property can be invoked to introduce the von Neumann state vector reduction, seen as a 'purely mental process' (Englert *et al.*, 1999) arising from an update of information.

It is clear that any measurement process gives rise to a flow of information from the location where the measurement takes place to the environment (including, of course, the observer). Indeed, the interaction between the measured system and the measurement apparatus, and then the environment, initiates a von Neumann chain, during which entanglement progresses further and further in the environment (Auletta, 2001, 2005; Auletta *et al.*, 2014). Consider a fixed volume containing all the measurement apparatus; as long as the entangled chain does not leave this volume, since any Hamiltonian evolution preserves pure states as well as entropy, the amount of entropy contained in the volume remains constant; but, as soon as entanglement proceeds beyond the fixed volume, the properties of the physical system contained inside the volume must be obtained by a partial trace (over the system outside the volume), and the entropy contained in the volume increases. To the local observer, the effect of this leak of entanglement towards the outside world appears as an entropy production.

One may apply the informational point of view in various ways. One may just emphasize the information content of the state vector or, more strictly, consider that the nature of the state vector itself is mostly informational, or even take the more extreme view where it is purely informational. Here is for instance how Fuchs (2002) describes the programme of informational quantum theory:

> The quantum system represents something real and independent of us; the quantum state represents a collection of degrees of belief about *something* to do with that system ... The structure called quantum mechanics is about the interplay between these two things—the subjective and the objective. The task before us is to separate the wheat from the chaff. If the true quantum state represents subjective information, then how much of its mathematical support structure might be of the same character? Some of it, maybe most of it, but surely not all of it.

One can also relate this point of view with the correlation interpretation; it can actually be used to complement it. It offers a rather natural explanation to the discontinuities arising from state vector reduction, but some real difficulties remain. For instance, questions may be asked about the division of the world into systems providing information and those in which information is acquired, or the description of independent reality during the experiment. Wigner's friend paradox is not problematic since, as long as the friend outside the closed laboratory has less information than the other, he continues to use an unreduced state vector, while the friend inside has already reduced his.

# 1.2 SOME INTERPRETATIONS

As already mentioned in the introduction, it is impossible to list all interpretations of quantum mechanics that have been proposed, and of course even more impossible to describe all of them in detail. The situation is made even more complicated by the fact that some of these interpretations can be combined in various ways, giving rise to many possible combinations. We will therefore limit ourselves to a discussion of the main families of interpretations, without any claim of exhaustivity.

## 1.2.1 Standard and Similar Interpretations

The standard interpretation of quantum mechanics is often called the 'Copenhagen interpretation', by reference to Niels Bohr who lived in this city, but also to other famous physicists who interacted and worked with him: Heisenberg, Pauli, Jordan, etc. The words 'orthodox interpretation' are also often used. The difficulty, nevertheless, is that all these 'orthodox' physicists have often expressed views that are not identical, even sometimes divergent, and have changed over time. Fortunately, most textbooks on quantum mechanics use very similar postulates, probably for pedagogical reasons.

### 1.2.1.1 *Textbook Interpretation, von Neumann's Collapse*

We have already discussed in section 1.1.2.1 some elements of the interpretation of quantum mechanics that is used in textbooks. All interpretations of quantum mechanics contain the continuous, deterministic, evolution of the wave function/state vector due to the Schrödinger equation. This evolution applies when a physical system evolves freely, without being subject to measurement. The corresponding dynamics may take various forms, for instance the Schrödinger equation, or the von Neumann equation ruling the time evolution of the density operator. In the Heisenberg picture, it is also possible to consider that the state vector and density operator are independent of time, while the time evolution is transposed to the operators describing the physical quantities. In all cases, at this stage the dynamics is completely deterministic.

In order to introduce the nondeterministic character of quantum mechanics, one then postulates the Born rule: when an observable described by an operator $A$ is measured, the result can only be one of the eigenvalues of this operator; the probability of obtaining a particular value $a$ of $A$ is given by the modulus square of the scalar product of the corresponding eigenvector of $A$ by the state vector.[2]

---

[2] For the sake of simplicity, we assume that the eigenvalue $a$ is nondegenerate, but the Born rule can easily be generalized to degenerate eigenvalues, by a summation over an orthonormal basis of eigenvectors.

Most textbooks then include the von Neumann projection postulate as a second postulate of evolution. Assume for instance that a physical system is prepared at time $t_0$, evolves freely (without being measured) until time $t_1$, where it undergoes a first measurement, and then evolves freely (according to the Schrödinger or von Neumann equation) again until time $t_2$, where a second measurement is performed. Just after the first measurement at time $t_1$, when the corresponding result of measurement is known, it is very natural to consider that both the initial preparation and the first measurement are part of a single preparation process of the system. One then associates to this preparation a state vector that includes the information of the first result; in other words, the state vector is updated to include the interaction with the first measurement apparatus as well as the information acquired during the corresponding measurement. This is precisely what the von Neumann state vector reduction (or state collapse) postulate does. The new 'reduced' state vector can then be used as an initial state to calculate the probabilities of the different results corresponding to the second measurement, at time $t_2$.

It is easy to generalize the rule to more than two successive measurements. Each time a measurement is performed, as soon as a result is obtained, the state vector 'jumps' to a new value that includes this new information (but may also erase some previous information). Only one exception can occur: if the same measurement is performed repeatedly, at times that are sufficiently close to avoid any Schrödinger evolution of the system between the measurements, all results are then necessarily the same and, after the first measurement, the state vector reduction has no additional effect (it nevertheless becomes effective again as soon as a different observable is measured).

Different stages in the evolution of the state vector thus appear in this general scheme. Between preparation and measurements, the state vector evolves continuously according to the Schrödinger equation. Later, when the a measurement is performed, the system interacts with a measurement apparatus. The probabilities of the various outcomes can be calculated from the state vector by using the Born rule, but the Schrödinger equation itself does not select a single result. The uniqueness of the outcome in the quantum description of the system is associated with state vector reduction, a process that makes the state vector jump discontinuously (and randomly) to a new value. The emergence of a single result is then obtained (one could say 'forced') explicitly in the state vector by retaining only the component corresponding to the observed outcome; all components of the state vector associated with the other results are put to zero, hence the name 'reduction'. The reduction process is discontinuous and irreversible. In this scheme, separate rules and equations are therefore introduced, one for the 'natural' continuous evolution of the system between measurements, another for the measurements performed on it. The difficulty is then to understand precisely how to avoid possible conflicts between these two different postulates. Nevertheless, in practice, physicists know how to use common sense to avoid this difficulty (section 1.1.2), so that applying the theory creates no problem at all.

### 1.2.1.2 *Bohr*

In Bohr's point of view, the purpose of quantum mechanics is not to describe the microscopic world. It is, rather, to make correct predictions concerning the observations made by physicists using well-defined preparation and measurement procedures. For him, it is not possible to express more than this in 'ordinary language', and therefore meaningless to try and describe what really happens with a microscopic quantum system, what its properties are at every time, etc.

Bohr defines the purpose of physics in the following way (Bohr, 1960; Norris, 2000, p. 233): 'Physics is to be regarded not so much as the study of something a priori given, but rather as the development of methods of ordering and surveying human experience. In this respect our task must be to account for such experience in a manner independent of individual subjective judgement and therefore objective in the sense that it can be unambiguously communicated in ordinary human language.'

As a consequence, there is no point in trying to describe the detailed properties and evolution of quantum systems, since 'We must be clear that when it comes to atoms, language can be used only as in poetry. The poet, too, is not nearly so concerned with describing facts as with creating images and establishing mental connections' (Pranger, 1933, p. 11). He even goes further when he writes: 'There is no quantum world. There is only an abstract physical description. It is wrong to think that the task of physics is to find out how Nature is. Physics concerns what we can say about Nature' (page 204 of Jammer, 1966; Petersen, 1963). Or, similarly: 'there is no quantum concept' (Chevalley, 1994).

**Role of the state vector** Clearly, for Bohr, the state vector does not provide a direct description of reality; it is just a tool allowing us to calculate correct probabilities, with the help of the Born rule. In Bohr (1950, p. 52) he writes: 'The entire formalism is to be considered as a tool for deriving predictions of definite or statistical character, as regards information obtainable under experimental conditions' If an experiment is made of two or more successive measurements, the whole experimental set-up should be considered as a single device; one may ascribe intermediate mathematical values to the state vector of the quantum system at intermediate times, i.e. between two successive measurements, but these values are not physically relevant. The rules of quantum mechanics then provide the probabilities corresponding to all possible series of results. In this view, one should not ask what is the physical state of the measured system between the first measurement and the second: separating the system from the whole experimental apparatus has no meaning in Bohr's interpretation of quantum mechanics (nonseparability). Assume for instance that the polarization of a photon is measured nondestructively somewhere in Europe, that it propagates along a polarization-preserving optical fibre, and reaches America where another polarization measurement is performed. In practice, most physicists consider intuitively that 'something having physical properties' has propagated from one site of measurement to the other, and that physics has something to say about these properties. For Bohr, this intuition is not particularly meaningful: nothing that physics should describe separately propagates along the fibre.

In this view, the conflict between Schrödinger evolution and von Neumann projection postulates disappears. Nevertheless, another difficulty arises since, in any experiment, one has to clearly distinguish between two different parts: the observed system(s) and the measurement apparatus(es). Only the latter are directly accessible to human experience and can be described with ordinary language, as in classical physics. A consequence is that the measurement apparatuses then have a very specific role. They provide the results of measurements to observers, so that they are at the origin of our perception of the physical world; but, at the same time, they also introduce an irreducible nondeterministic component into the theory. The problem is then to decide where to put the frontier between the two different parts. If, for instance, the distinction is to be made in terms of size of the systems, one could ask from what size a physical system is sufficiently macroscopic to be considered as directly accessible to human experience, and will behave as a measurement apparatus. If the distinction is made in another way than size, then more elaborate rules should be specified to clarify this concept. For Bohr, it is not so much the size of the apparatus than its measuring function that matters; one could imagine situations where the same physical system is measured in one experiment, and then plays the role of the measurement apparatus (or some part of it) in another experiment. In Bohr (1935, p. 701), he writes: 'This necessity of discriminating in each experimental arrangement between those parts of the physical system considered which are to be treated as measuring instruments and those which constitute the objects under investigation may indeed be said to form a *principal distinction between the classical and quantum-mechanical description of physical phenomena*'.

**Resolution of paradoxes, the Schrödinger cat, etc.** Bohr's point of view is of course consistent. As an illustration, we now discuss how it answers paradoxes such a the well-known Schrödinger paradox. In an article entitled 'The Present Situation of Quantum Mechanics', Schrödinger (1935) writes (page 238 of the translation by Trimmer (1980)):

> One can even set up quite ridiculous cases. A cat is penned up in a steel chamber, along with the following device (which must be secured against direct interference by the cat): in a Geiger counter, there is a tiny bit of radioactive substance, so small that perhaps in the course of the hour, one of the atoms decays, but also, with equal probability, perhaps none; if it happens, the counter tube discharges, and through a relay releases a hammer that shatters a small flask of hydrocyanic acid. If one has left this entire system to itself for an hour, one would say that the cat still lives if meanwhile no atom has decayed. The first atomic decay would have poisoned it. The psi-function of the entire system would express this by having in it the living and dead cat (pardon the expression) mixed or smeared out in equal parts.

The essence of Schrödinger's argument is to consider a von Neumann chain starting from one (or a few) radioactive atomic nucleus. The device is designed so that, if the nucleus emits a photon, this photon is detected by a gamma ray detector, which provides an electric signal, which then undergoes an electronic amplification, and triggers a macroscopic mechanical system that automatically opens a bottle of poison and eventually kills the cat. This is what happens in the branch of the state vector where a photon has been emitted; but none of these events takes place in the branch where no photon is emitted. When the probability that a photon has been emitted is about 1/2, the system reaches a state with two components of equal weight, one where the cat is alive and one where it is dead. The equations of evolution seem to predict that the cat is at the same time alive and dead, instead of being alive or dead ('and-or question'). Schrödinger considers this coexistence of totally different states of the cat as an obvious impossibility (a 'quite ridiculous case') and concludes, therefore, that something must have happened to stop the von Neumann chain before it went so far. The challenge is then to explain macroscopic uniqueness: why, at a macroscopic level, a unique result (the alive, or dead, cat) emerges, while this does not happen within the linear Schrödinger equation.

For Bohr, there is no paradox at all. The apparent paradox occurs only because Schrödinger extrapolates the quantum formalism to a situation where it does not apply. The role of the wave function, and of its evolution, is only to provide probabilities of observing results obtained with macroscopic apparatuses; it is not to describe the real events taking place at every time. Common sense and experience show that superpositions of macroscopically different states such as dead and alive cats are never observed. The role of the wave function is only to give the probabilities of the corresponding events. Using the wave equation to predict that macroscopic superpositions may exist and be observed is just using the quantum formalism outside of its range. In other words, Schrödinger takes his own equation too seriously!

Even if this reasoning is convincing, we are still faced with a logical problem that did not exist before, in classical mechanics, when nobody thought that measurements providing information should be treated as special processes in physics. We learn from Bohr that we should not try to transpose our experience of the everyday world to microscopic systems; then, for each experiment, where exactly is the limit between these two worlds? In von Neumann's approach also, it becomes necessary to introduce a 'split' between the measured system and the measurement apparatus, or between a standard quantum evolution and a measurement process. It has often been argued that the precise position of this split is irrelevant in practice, since it does not affect the physical predictions. It remains nevertheless true that, under these conditions, the theory is not perfectly well defined. Is it really sufficient to remark that the distance between microscopic and macroscopic is so huge that the exact position of the frontier between them is unimportant? Bell (and others) complained about the shifty character of this division (Bell, 1992); in the words of Mermin (2012), 'Bell deplored a shifty split that haunts quantum mechanics'. As we will see for instance in sections 1.2.5 and 1.2.6, the purpose of some other interpretations of quantum mechanics is to suppress any necessity of a split.

### 1.2.1.3 *Variants*

**The Copenhagen interpretations** One often speaks of the 'Copenhagen interpretation', to refer to Bohr's interpretation and to the fact that Bohr lived in Copenhagen. But, in fact, the Copenhagen interpretation is not defined in the same way by all authors, depending on the importance they give to the various philosophical positions of Bohr, which evolved over time. A special mention should be made of his general notion of 'complementarity' which, according to him, may play an essential role, not only in physics, but also in all scientific disciplines. For instance, Howard (2004) writes:

> Most importantly, Bohr's complementarity interpretation makes no mention of wave packet collapse . . . or a privileged role for the subjective consciousness of the observer. Bohr was also in no way a positivist. Much of what passes for the Copenhagen interpretation is found in the writings of Werner Heisenberg, but not in Bohr. Indeed, Bohr and Heisenberg disagreed for decades in deep and important ways. The idea that there was a unitary Copenhagen point of view on interpretation was, it shall be argued, a post-war invention, for which Heisenberg was chiefly responsible.

Nevertheless, in most of these variants, the state vector (or wave function) is not defined as a direct description of a quantum physical system, but associated with a preparation procedure. When introducing the Copenhagen interpretation, Stapp (1972) writes: 'The specifications A on the manner of preparation of the physical system are first transcribed into a wave function $\Psi_A(x)$'. Peres (1984) writes something similar: 'a state vector is not a property of a physical system, but rather represents an experimental procedure for preparing or testing one or more physical systems', and: 'quantum theory is incompatible with the proposition that measurements are processes by which we discover some unknown and pre-existing property'. Seen in this way, a wave function is not subjective, but an objective representation (independent of the observer) of a preparation procedure, rather than of the isolated physical system itself. We now illustrate the variety of the family of Copenhagen interpretations with a few examples.

**Veiled reality** The 'veiled reality' interpretation of quantum mechanics introduced by Bernard d'Espagnat (1971, 1979, 1985, 1995) does not use a mathematical formalism that differs from that of the standard interpretation, but proposes a conceptual and philosophical framework that is specific. As for Bohr, the general framework is realist, but the definition of reality does not involve human perception relayed by measurement apparatuses; the existence of reality is considered fundamental, with no need to refer to humans and their perceptive structure. D'Espagnat analyses the theory of measurement, the EPR and Bell arguments, the relation between counterfactuality and realism, and the consequences of the intersubjective agreement. From this he concludes that quantum mechanics cannot lead to descriptive interpretations of individual objects. This leads him to distinguish between independent reality, of which at best the general structures are accessible, and empirical reality (perceived phenomena). His conclusion is that the ultimate reality is a 'veiled reality', only marginally accessible to discursive knowledge.

As we see later, other interpretations of quantum mechanics also distinguish between two levels of reality. This is for example the case of the de Broglie–Bohm theory (section 1.2.5.1), with a physical field that can be manipulated (the wave function) and the particle positions (which are observable, but not manipulable). But the two approaches remain very different, in particular because the two levels of reality appear in the mathematical formalism of the de Broglie–Bohm theory itself, but not in the the the veiled reality interpretation.

**Hardy's axioms for probabilities** Lucien Hardy (2001) proposed to replace the axioms of the standard formulation of quantum theory by a set of 'five reasonable axioms' concerning probabilities. The 'state' associated with a preparation of a quantum system is defined as the mathematical object that can be used to determine the probabilities associated with the outcomes of any measurement of the system. The type of any quantum system is characterized by two integers $K$ and $N$:

- $K$ is the 'number of degrees of freedom', defined as the minimum number of probability measurements required to determine the mathematical state.
- $N$ is the 'dimension', defined as the maximum number of states that can be distinguished with certainty with a single measurement operation.

Hardy assumes that $K$ and $N$ are finite or countably infinite. His five axioms are:

- When the same experiment is repeated many times on an ensemble of $n$ systems, for each of them the relative frequencies of the various outcomes tend to the same limit.
- Considered as a function of $N$, $K(N)$ takes the minimum value consistent with the axioms.
- If for some reason the state of a system is constrained to belong to a $M$-dimensional subspace, the system behaves like a system of dimension $M$.
- A composite system made up of subsystems $A$ and $B$ has dimension $N = N_A \times N_B$.
- There exists a continuous reversible transformation on a system that transforms an arbitrary state of the system into any other state.

From these postulates, one can then reconstruct the formalism of quantum mechanics. A remarkable point is that the quantum character arises only from the continuity of the last postulate: if this continuity is not postulated, one merely obtains classical probability theory.

Olivier Darrigol (2015) has analysed this axiomatic construction, proposed a simplified version of this new foundation, and discussed variants and the relation between the axioms and the correspondence arguments.

**Contextual interpretation** The CSM interpretation (for Contexts, Systems, Modalities), introduced by Alexia Auffèves and Philippe Grangier (2016a,b), proposes a notion of physical reality that is defined in terms of the relations between a quantum object and the environment provided by the ensemble of its measurement apparatuses. Three postulates are introduced:

(i) The 'modality' attached to a physical system is defined as the ensemble of values of a complete set of physical quantities that can be predicted with certainty and measured repeatedly on this system. The meaning of the word 'complete' refers to the largest possible set compatible with certainty and repeatability. This complete set of physical quantities is called a context; it corresponds to a given experimental setting. The modality is attributed jointly to the system and the context.

(ii) Different modalities exist for a given context. Their number, $N$, is the same in all relevant contexts; they are then mutually exclusive: if one set of predictions is true, the others are in general wrong. The value of $N$ is a characteristic property of a given quantum system, called the dimension.

(iii) The various contexts of a given physical system are related by transformations having the structure of a continuous group $G$.

Postulate (i) introduces the notion that is usually attached to the state vector; postulate (ii) on quantization; postulate (iii) on the relations between different contexts. One can then show that a theory that is compatible with these postulates is necessarily probabilistic; the introduction of probabilities then no longer appears as a postulate per se, but a consequence of other postulates. Moreover, one can also derive the whole quantum formalism, so that the structure of quantum mechanics then emerges as a consequence of an interplay between the quantized number of 'modalities' accessible to a quantum system, and the continuum of 'contexts' that define these modalities.

This approach is related to the Gleason theorem (Gleason, 1957; Busch, 2003). It is also rather similar to Bohr's point of view, where physical reality can only be defined in terms of the whole experimental set-up; an important difference is nevertheless the central role of quantization, as expressed by postulate (ii).

**QBism** Probabilities can be defined as objective or subjective. In the former case, probabilities relate to an event that happens in reality, independently of any observer, consciousness, or anyone's knowledge about the phenomenon. A probability then just characterizes the frequency of occurrence of a given phenomenon among an ensemble of possible occurrences. In the latter case, probabilities may define the degree of belief of an observer, often called an 'agent' in this context, about the possibility of this occurrence. Subjective probabilities belong to the family of Bayesian probabilities, which are interpreted as reasonable expectations, states of knowledge, or characterizations of personal beliefs (Cox, 1946; de Finetti, 2017).

A recent version of the interpretations based on Bayesian probabilities is QBism, an abbreviation for Quantum Bayesianism (Caves *et al.*, 2007; Fuchs, 2010; Healey, 2017; Fuchs *et al.*, 2013). Within QBism, the quantum state associated with a physical system characterizes the probabilities that an agent assigns to the results of possible future measurements. The quantum state is therefore not associated with the physical system itself only, or its physical preparation, but with a couple agent + system; as a consequence, two different agents may then assign two distinct quantum states to the same physical system. The quantum state is subjective and summarizes the knowledge of agents and their beliefs about the results of their future experience. For instance, Christopher Fuchs (2010) writes: 'For the Quantum Bayesian, quantum theory is not something outside probability theory... but rather it is an addition to probability theory itself'.

Within QBism, discontinuous jumps of the state vector become perfectly natural, as sudden jumps of classical subjective probabilities when some new information is taken into account. The agents play a central role, since the whole quantum formalism deals only with their knowledge and assignments, not with the real world. There is still a frontier between different kinds of situations: ordinary Schrödinger evolution for totally isolated systems, and situations where information is collected by observers. Fuchs (2010) notes that 'In contemplating a quantum measurement, one makes a conceptual split in the world: one part is treated as an agent, and the other as a kind of reagent or catalyst'. But this frontier is no longer objective; as David Mermin (2012) writes 'The splits reside, not in the objective world, but at the boundaries between that world and the experiences of the various agents who use quantum mechanics'. State vector reduction can only occur when an agent has the 'mental facility to use quantum mechanics to update its state assignments on the basis of its subsequent experience'.

## 1.2.2 Consistent Histories

The interpretation of 'consistent histories' can also be seen as a variant of the Bohr interpretation, but it offers a logical framework to discuss the evolution of a closed quantum system, without any reference to measurements. It is also sometimes just called 'history interpretation'. The main ideas were introduced and developed by Robert Griffiths (1984, 2002), Roland Omnès (1988, 1994, 1999), and Murray Gell-Mann and James Hartle (1993). The reader interested in more precise information than our brief introduction on the subject should go to the provided references—for a general presentation, see also an article in *Physics Today* (Griffiths and Omnès, 1999) and the references contained therein, or the introductory review article by Pierre Hohenberg (2010).

### 1.2.2.1 *Families of Histories*

Any orthogonal projector $P$ on a subspace $\mathcal{F}$ of the space of states of a system has two eigenvalues, $+1$ corresponding to all the states belonging to $\mathcal{F}$, and 0 corresponding to all states that are orthogonal to $\mathcal{F}$. A measurement process can be associated with $P$: if the state of the system belongs to $\mathcal{F}$, the result of the measurement is $+1$; if the state is orthogonal to $\mathcal{F}$, the result is 0. If this measurement is made at time $t_1$ on a system that is initially (at time $t_0$) described by a density operator $\rho(t_0)$, the probability for finding the state of the system in $\mathcal{F}$ (result $+1$) at time $t_1$ is then given by formula (2). A similar result can obviously be generalized to several subspaces $\mathcal{F}_1$, $\mathcal{F}_2$, $\mathcal{F}_3$, etc., and several measurement times $t_1$, $t_2$, $t_3$, etc. (we assume $t_1 < t_2 < t_3 < ...$). The probability for finding the state of the system at time $t_1$ in $\mathcal{F}_1$, then at time $t_2$ in $\mathcal{F}_2$, then at time $t_3$ in $\mathcal{F}_3$, etc. is, according to the Wigner formula:

$$P(\mathcal{F}_1, t_1; \mathcal{F}_2, t_2; \mathcal{F}_3, t_3 ...) = \mathrm{Tr}\{...\hat{P}_3(t_3)\hat{P}_2(t_2)\hat{P}_1(t_1)\rho(t_0)\hat{P}_1(t_1)\hat{P}_2(t_2)\hat{P}_3(t_3)...\} \quad (3)$$

where, as before, the $\hat{P}_i(t_i)$ are the projectors over subspaces $\mathcal{F}_1$, $\mathcal{F}_2$, $\mathcal{F}_3$ in the Heisenberg picture.

A 'history' of the system can then be associated with this equation: a history $\mathcal{H}$ is defined by a series of arbitrary times $t_i$, each of them associated with an orthogonal projector $P_i$ over some subspace; its probability is given by (3), which, for simplicity, we will write as $P(\mathcal{H})$. In other words, a history is the selection of a particular path, or branch, for the state vector in a von Neumann chain, defined mathematically by a series of times and projectors. In this point of view, no reference is made to measurements: a history describes inherent physical properties of the system itself, without having to invoke observations or interactions with external physical systems.

A gigantic number of different histories may be defined, so it is useful to group them into families of histories. A family is defined again by an arbitrary series of times $t_1$, $t_2$, $t_3, ...$, but now we associate to each of these times $t_i$ an ensemble of orthogonal projectors $P_{i,n}$ that, when summed over $n$, restore the whole initial space of states. For each time, we then have a series of mutually orthogonal projectors that provide a decomposition of the unity operator:

$$\sum_n P_{i,n} = 1 \quad (4)$$

This gives the system a choice, for each time $t_i$, among many projectors, which corresponds to a choice among many histories of the same family. It is actually easy to see from (4) and (3) that the sum of probabilities of all histories of a given family is equal to 1:

$$\sum_{\text{histories of a family}} P(\mathcal{H}) = 1 \quad (5)$$

This relation implies that the physical system will always follow one, and only one, of the histories of the family.

### 1.2.2.2 *Consistency Condition for Families*

A consistency condition for families can be introduced. Consider first the simplest case where two projectors only, occurring at time $t_i$, have been grouped into one single projector to build a new history. The two 'parent' histories then correspond to two exclusive possibilities (they contain orthogonal projectors), so that their probabilities add independently in the sum (5). The daughter history is exclusive of neither of its parents, and contains less information on the physical properties of the system at time $t_i$: the system may have either of the properties associated with the parents. Now, a general theorem states that the probability associated with an event that can be realized by either of two exclusive events is the sum of the individual probabilities. One then expects that the probability of the daughter history should be the sum of the parent probabilities. Nevertheless, in quantum mechanics, relation (3) shows that this is not necessarily the case; since any projector, $\hat{P}_2(t_2)$ for instance, appears twice in the formula, replacing it by a sum of projectors introduces four terms: two terms that give the sum of probabilities, as expected, but also two crossed terms (or 'interference terms') between the parent histories. Therefore, the probability of the daughter history is in general different from the sums of the parent probabilities.

It is possible to restore the additivity of probabilities by considering only families for which the crossed terms vanish. This introduces the condition:

$$\mathrm{Tr}\{...\hat{P}_{3,n_3}(t_3)\hat{P}_{2,n_2}(t_2)\hat{P}_{1,n_1}(t_1)\rho(t_0)\hat{P}_{1,n_1'}(t_1)\hat{P}_{2,n_2'}(t_2)\hat{P}_{3,n_3'}(t_3)...\}$$

$$\propto \quad \delta_{n_1,n_1'} \times \delta_{n_2,n_2'} \times \delta_{n_3,n_3'} \times ... \quad (6)$$

In this relation, the product of $\delta$ in the right-hand side forces the left-hand side of (6) to vanish if at least one pair of the indices $(n_1,n_1')$, $(n_2,n_2')$, $(n_3,n_3')$, etc., contains different values. If they are all equal, the trace merely gives the probability $\mathcal{P}(\mathcal{H})$ associated with the particular history of the family. This introduces the notion of a 'consistent family': if condition (6) is fulfilled for all projectors of a given family of histories, this family is said to be logically consistent, or consistent for short.

### 1.2.2.3 *Describing the Evolution of an Isolated System*

Assume a consistent family of histories has been chosen to describe an isolated system; any consistent family may be selected but, as soon as this choice is made, it cannot be changed. This unique choice provides a well-defined logical frame, and a series of possible histories that are accessible to the system and give information at all intermediate times $t_1$, $t_2$,... Which history will actually occur in a given realization of the physical system is impossible to predict in advance, since it depends on some fundamentally random process that selects one single history among all those of the family. The corresponding probability $\mathcal{P}(\mathcal{H})$ is given by the right-hand side of (3); since this formula is identical to that of standard quantum mechanics, this postulate ensures that the standard predictions of the theory are automatically recovered. For each realization of the experiment, the system then possesses at each time $t_i$ all physical properties associated to the particular projectors $P_{i,n}$ that occur in the selected history. We then obtain a description of the evolution of its physical properties that can be significantly more accurate than that given by its state vector. For instance, if the system is a particle and if the projector is a projector over some region of space, we will say that for a particular history the particle is in this region at the corresponding time, even if the whole Schrödinger wave function extends over a much larger region. Or, if a photon strikes a beam splitter, or enters a Mach–Zehnder interferometer, some histories of the system may include information on which trajectory is chosen by the photon.

Clearly, the smaller the subspaces associated with the projectors $P_{i,n}$, the more accuracy is gained. By contrast, no information is gained if all $P_{i,n}$ are projectors over the whole space of states, but this corresponds to a trivial case of little interest.

### 1.2.2.4 *Incompatibility: the Simultaneous Use of Several Different Consistent Families is Meaningless*

Within the history interpretation, all consistent families are equally valid, while they obviously lead to totally different descriptions of the evolution of the same physical system. The history interpretation considers that different consistent families should be regarded as mutually exclusive (except, of course, in very particular cases where the two families can be embedded into a single large consistent family). Any family may be used in a logical reasoning, but not combined together with others. We are free to choose any point of view in order to describe the evolution of the system and to ascribe properties to the system. Later, in another totally independent step, another consistent family may also be chosen in order to develop other logical considerations within this different frame; but it would be totally meaningless to combine considerations arising from the two frames. This fundamental rule, which is somewhat reminiscent of Bohr's complementarity, must be constantly kept in mind when one uses this interpretation. See Griffiths (1996) for a discussion of how to reason consistently in the presence of disparate families, and Griffiths (1998) for simple examples.

## 1.2.3 Ensemble (Statistical) Interpretations

The family of 'ensemble interpretations' of quantum mechanics, also often called 'statistical interpretation', considers that the description given by a state vector applies only to ensemble systems prepared in identical conditions. Single systems or single experiments are therefore not included in this description. Einstein is often considered as the father of these interpretations. In a letter to Schrödinger written in 1932 (Balibar *et al.*, 1989), he writes: 'The $\Psi$ function does not describe the state of a single system but (statistically) an ensemble of systems'. Within this point of view, the function $\Psi$ then contains an information that is similar to the information contained in the phase space distribution of classical statistical physics—certainly a very useful descripion, but not the most precise possible description of a given physical system within this theory.

A classic on the statistical interpretations is the review article by Leslie Ballentine (1970), who writes: 'Several arguments are advanced in favour of considering the quantum state description to apply only to an ensemble of similarly prepared systems, rather than supposing, as is often done, that it exhaustively represents an individual physical system. Most of the problems associated with the quantum theory of measurements are artifacts of the attempt to maintain the latter interpretation'. He distinguishes between two classes of interpretations:

(i) 'The statistical interpretation...according to which a pure state...need not provide a complete description of an individual system.

(ii) Interpretations which assert that a pure state provides a complete and exhaustive description of an individual system. This class contains...several versions of the Copenhagen interpretation'. He considers that 'hypothesis (ii) is unnecessary for quantum theory, and moreover leads to serious difficulties'.

Even if one adheres to the statistical interpretation (i), two different attitudes are possible:

- Either one decides that this limitation of pure state descriptions to ensembles proves that this description cannot be complete; more variables are therefore necessary to specify which particular system is considered inside the ensemble. This point of view naturally leads to theories with additional (hidden) variables.
- Or one decides that a theory describing only ensembles of systems is perfectly satisfactory, so that one may consider that it has reached its final form. When a single experiment is performed, a fundamentally random process takes place and makes a single result appear; the description of this process is not necessary within the theory.

Physicists using this interpretation are not all very explicit about which of these two possibilities they favour. With or without explicit reference to additional variables, one can find a number of authors who support the idea that the quantum state vector should be used only for the description of statistical ensembles. This general discussion is of course related to the status of the wave function. See for instance the article by Yakir Aharonov et al. (1993), with the title 'Meaning of the wave function'. These authors discuss how it is possible to determine the time evolution of a state vector by actual measurements. This is done by considering a series of measurements that last a long time, named 'protective measurements', during which the wave function is prevented from changing appreciably by means of another interaction it undergoes at the same time. To obtain a complete determination of the state vector, the method requires performing measurements under various experimental conditions; it is therefore necessary to have access to a large sample of identically prepared quantum systems. This supports the statistical interpretaton.

A recent example of this family of interpretations is discussed by Armen Allahverdyan, Roger Balian, and Theo Nieuwenhuizen (2017), who propose a theory of ideal quantum measurements relying on subensembles. The idea is that, inside the whole ensemble of all realizations of a given measurement experiment, as soon as the measured system $S$ and the pointer of the measurement apparatus $M$ have become strongly correlated according to the von Neumann model, they distinguish subensembles of realizations. The smallest subensembles are of course individual realizations. It is assumed that these subensembles can be described by the same formalism and dynamic equations as the whole ensemble. From a detailed study of the relaxation created by the interaction between $S$ and $M$, and from a series of adequate interpretative principles, Allahverdyan *et al.* (2017) propose a progressive introduction of the von Neumann postulate of projection (by successive embedded density operators), in close connexion with the properties of the dynamics of the interaction between $S$ and $M$.

## 1.2.4 Everett

In 1957, Hugh Everett introduced an interpretation that he named 'relative state interpretation' (Everett, 1957), but that is nowadays often called 'Everett interpretation'; its various forms are also sometimes called 'many-worlds interpretation' (MWI), 'many-minds interpretation', or 'branching universe interpretation' (the word 'branching' refers here to the branches appearing in the state vector of the universe).

In these family of interpretations, the Schrödinger equation is taken even more seriously than in any other interpretation. Indeed, the equation is assumed to be valid at any time: before, during and after a measurement process. When an experiment is performed, one considers that single results do not emerge, except, of course, when the Born rule yields a probability equal to unity for one result: all possibilities associated with nonzero probabilities are in fact realized at the same time! As a consequence, the von Neumann chain is never broken, but its tree is left free to develop its many branches ad infinitum.

The Schrödinger equation predicts that, after a measurement, several subsystems are entangled: the measured quantum system, the measurement apparatus, the observer, and the environment if information has leaked into it. Everett (1957, page 456) writes: 'there does not exist anything like a single state for one subsystem ... one can arbitrarily choose a state for one subsystem and be led to the relative state for the remainder'— until now, this seems to be just a description of quantum entanglement, a well-known concept. But the novelty is that Everett considers observers as purely physical systems, to be treated within the theory exactly on the same footing as the rest of the environment. He models it by an automatically functioning machine, coupled to the recording devices and registering past sensory data, as well as its own machine configurations. This leads Everett to the idea that 'current sensory data, as well as machine configuration, is immediately recorded in the memory, so that all the actions of the machine at a given instant can be considered as functions of the memory contents only'; similarly, all relevant experience that the observer retains from the past is also registered in this memory. From this Everett concludes that 'there is no single state of the observer; ... with each succeeding observation (or interaction), the observer's state branches into a number of different states ... All branches exist simultaneously in the superposition after any sequence of observations'.

For Everett, the emergence of well-defined results from experiments is not considered as a reality, but only as a delusion taking place in the mind of the observer. In fact, all branches of the state vector are equally real, meaning that the same is true for all possible outcomes of an experiment after it has been performed. What the state vector of the system + apparatus + observer + environment does is to constantly ramify into branches corresponding to all results of measurement, without ever selecting a single one. The observer is part of this ramification process, but can never bring to his consiousness more than one result of measurement. In other words, each 'component of the observer' remains completely unaware of all the others, as well as of the state vectors that are associated to them (hence the name 'relative state interpretation'). The delusion of the emergence of a single result in any experiment then appears as a consequence of the limitations of the human mind!

The Everett interpretation leads to conclusions that are identical to those of the Copenhagen interpretation; in this sense it is not falsifiable. Bryce DeWitt even considers that this interpretation is a mere consequence of the formalism: after asking the question 'Could the solution to the dilemma of indeterminism be a universe in which all possible outcomes of an experiment actually occur?' he states that 'the mathematical formalism of the quantum theory is capable of yielding its own interpretation' (DeWitt, 1970)—see also the interesting debate[3] that was stirred by the publication of this point of view.

The Everett interpretation may look like a particularly simple way to avoid difficult problems, and is therefore attractive. It is very useful in quantum cosmology, in particular because it allows one to consider the state vector of the universe. Nevertheless, on further study most realize how difficult it is to really comprehend. For instance, how do probabilities emerge in this completely deterministic theory? We refer to section 11.12 of Laloë (2019) for a more detailed discussion. We should mention that the interpretation has been criticized by several authors, for instance Dieter Zeh (1970) considers the theory as ambiguous because dynamical stability conditions are not considered. Asher Peres (1993) calls it a 'bizarre theory'. Antony Leggett (2002) discusses this interpretation in the following terms: 'The branches of the superposition which we are not conscious of are said to be "equally real", though it is not clear ... what these words,

---

[3] 'Quantum mechanics debate', *Phys. Today*, 24, 36–44 (April 1971); 'Still more quantum mechanics', *Phys. Today*, 24, 11–15 (October 1971).

ostensibly English, are supposed to mean'. One question is what we should expect from a physical theory; does it have to explain how we perceive results of experiments; and if it should, what kind of explanation would be satisfactory? Since the emphasis is put not on the physical properties of the systems themselves, but on the effects that they produce on our minds, notions such as perception (Everett (1957) speaks of 'trajectory of the memory configuration') and psychology become part of the debate. The Everett interpretation is certainly attractive aesthetically, but remains somewhat mind boggling.

## 1.2.5  Additional Variables

Quantum theories with additional variables are theories where the quantum system is described by variables that are added to the standard description by a state vector. These theories are often built mathematically in order to make exactly the same predictions as standard quantum mechanics, expressed in term of the probabilities associated to all possible measurements. It is then clear that experiments cannot be used to discriminate between these theories and standard quantum mechanics. Theories with additional variables (sometimes also called 'hidden variables') nevertheless have a real conceptual interest since they offer a description of quantum phenomena that is different from the standard description, providing an interesting point of view on the content of quantum physics. We will limit ourselves to two theories with additional variables: the de Broglie–Bohm (dBB) theory, and the so-called modal interpretation.

### 1.2.5.1  *dBB Theory*

In the dBB theory, the quantum dynamics is enriched by adding positions (often called 'Bohmian positions') to the standard state vector. While, in standard quantum mechanics, a single particle may manifest itself either as a wave or as a particle (Bohrian complementarity), in dBB theory it is simultaneously described by both, at any time. The idea was introduced in 1926–1927 with the early work of Louis de Broglie (1927, 1956, 1987), along the lines of his thesis (de Broglie, 1924). Actually, de Broglie first elaborated his 'theory of the double solution', where the same wave equation has two solutions: the usual continuous wave function $\Psi(\mathbf{r})$, and a solution with mobile singularities $u(\mathbf{r})$ representing the physical particle itself. He then simplified it by considering only the point of space where the singularity occurs, and naming it the 'position' of the particle. Here, we will discuss only this simpler version; for a derivation of the double solution theory, see Holland (2020).

  More than 20 years later, David Bohm (1952) independently elaborated a more complete version of the pilot wave theory. He completed the theory by considering systems of more than one particle, and introducing various notions, such as the notion of 'empty waves'. Nowadays, one often speaks of the 'de Broglie–Bohm' theory.

**Dynamical equations** The dynamical equations of the dBB theory provide the coupled evolution of two set of quantities:

- the wave function of the $N$ particles (which, for the sake of simplicity, we assume to be spinless):

$$\Phi(\mathbf{r}_1,\mathbf{r}_2...\mathbf{r}_N) = R(\mathbf{r}_1,\mathbf{r}_2...\mathbf{r}_N)e^{i\xi(\mathbf{r}_1,\mathbf{r}_2...\mathbf{r}_N)} \qquad (7)$$

of modulus $R(\mathbf{r}_1,\mathbf{r}_2...\mathbf{r}_N)$ and phase $\xi(\mathbf{r}_1,\mathbf{r}_2...\mathbf{r}_N)$. This function evolves over time according to the standard Schrödinger equation.

- the Bohmian positions $\mathbf{Q}_2$, $\mathbf{Q}_2...,\mathbf{Q}_N$ of the $N$ particles. Each of these positions evolves according to the guiding equation:

$$\frac{\mathrm{d}}{\mathrm{d}t}\mathbf{Q}_i = \frac{\hbar}{m}\mathbf{\nabla}_{r_i}\xi(\mathbf{r}_1,\mathbf{r}_2...\mathbf{r}_N) \qquad (8)$$

The velocity of the position then depends only on the gradient of the phase $\xi$ of the wave function, not on its modulus $R$. As a consequence, vanishingly small wave functions may have a finite influence on the position of the particles. For instance, with a Gaussian wave packet, the influence of the wave packet on the velocity of the particle is comparable near the centre of the wave packet or at large distances, where the wave function is vanishingly small. Of course, situations where the position of the particle is extremely far from the centre of the wave packet are very rare, but if they occur by chance, the position is guided exactly with the same efficiency as in any other place.

Physically, what the guiding relation (8) means is that the motion of the Bohmian position is guided by the fluid of probability. It can easily be generalized to particles with spins and described by Pauli spinors.

**Condition of quantum equilibrium** The dBB theory then postulates that the initial distribution of the position variables $\mathbf{Q}_1$, $\mathbf{Q}_2,...,$ $\mathbf{Q}_N$ is totally random, and that it reproduces exactly the initial quantum probability distribution $|\Phi(\mathbf{Q}_1,\mathbf{Q}_2,...)|^2$ of standard theory for position measurements. This condition is often called the 'quantum equilibrium condition'. For a given realization of an experiment, there is no way to select which value of the position is realized inside the distribution; from one realization to the next, a new completely random choice of position is spontaneously made by Nature. This assumption therefore conserves the fundamentally random character of the predictions of quantum mechanics.

Then, combining the Schrödinger equation with the form of the 'quantum velocity term' (8), one can show that quantum equilibrium is stable: if at some time $t$ the distribution of positions is equal to $|\Psi(\mathbf{Q}_1,\mathbf{Q}_2,...)|^2$, the equality also holds at any time $t + dt$. This automatically provides a close contact with all the predictions of quantum mechanics, since all predictions concerning the probabilities of position measurements are then identical. Under the effect of the quantum velocity term, the Bohmian positions are constantly dragged by the wave function and can never move away from it; they remain in regions of space where it does not vanish, which ensures that the guiding formula (8) never becomes indeterminate. Another important consequence of this postulate is to avoid a conflict with relativity, since arbitrary distributions would introduce the possibility of superluminal signalling (Valentini, 2002, 2009). Since the Born rule is a consequence of the quantum equilibrium, one can then consider that this rule is not an independent postulate of quantum mechanics, but merely a consequence of the relativistic impossibility of instantaneous signal transmission.

**Trajectories** The guiding relation (8) can be used to plot trajectories for the particles, which provide interesting illustration of quantum effects in various cases. Many examples of such trajectories can be found in the book by Peter Holland (1993). Generally speaking, the Bohmian trajectories are often very different from classical trajectories, which is natural, since otherwise quantum effects could not appear. For instance, in an interference experiment, the trajectories of particles in free space are not necessarily straight lines, which allows the particles to avoid the nodes of the interference pattern.

Especially interesting are the so-called 'surrealistic trajectories', called in this way by Englert *et al.* (1992), who proposed to study an interference experiment in a particular case (see also the discussion of Scully (1998)). These authors discuss situations where the Bohmian trajectory of a test particle crosses one slit of the interference device, while it interacts with a 'which way' detector. When the detector moves slowly after interacting with the test particle, it may seem to indicate that the particle crossed the other slit than the one crossed by the trajectory of the test particle. The apparent paradox can be lifted by remarking that, to understand the indications of the which way detector, it is necessary to understand its dynamics, including possible quantum effects. If the detector is microscopic, quantum nonlocal effects affecting both the particle and the detector may completely change the interpretation of the indications of the pointer. Nevertheless, if the detector is macroscopic, the surrealistic trajectories simply never occur (Tastevin and Laloë, 2018).

**Applications, cosmology** Since more variables are introduced, the dBB dynamics is richer than the standard dynamics of quantum mechanics. It nevertheless provides identical results. This adds more flexibility in the calculations and can lead to useful tools for constructing approximations, typically in many-body problems. It turns out that introducing Bohmian positions is a convenient way to study problems in molecular dynamics (Lopreore and Wyatt, 1999) and molecular adsorption on surfaces (Prezhdo and Brooksby, 2001; Gindensperger *et al.*, 2000, 2002a, b); quantum Monte Carlo methods and conditional wave functions may provide good simulations of atoms in ultra-strong laser fields (Christov, 2007, 2009); moreover, the behaviour of electrons in nano-electronics circuits can also be studied (Shifren *et al.*, 2000; Albareda *et al.*, 2013). General reviews of many possible applications are given in Oriols and Mompart (2012) and Benseny *et al.* (2014).

Applications also occur in physical cosmology, in particular in the study of the evolution of the structure of the early universe during the initial 'big bang' period.

Quantum mechanics and gravity simultaneously play an important role in this difficult problem. General relativity introduces an equation giving the metric, where the general relativistic scale factor appears as a quantum operator. Perturbation expansions can be introduced with respect to small inhomogeneous and anisotropic fluctuations of the metric, but even then the solution still remains very difficult, because of the presence of the quantum operator associated with the relativistic scale factor.

One can then develop a Bohmian theory for quantum gravity (Peter *et al.*, 2005, 2006; Pinho and Pinto-Neto, 2007), just along the same lines as for the electromagnetic field. This does not mean that quantum effects are ignored: Bohmian trajectories are sensitive to interference or tunnelling, and their use does not suppress any quantum result. For instance, the Bohmian trajectory can be used in theories where the universe collapses and expands, but where the collapse takes place only until a small (but nonzero) value of the scale factor is reached, and where the quantum effects of gravity become so strongly repulsive that the universe rebounds back out ('Big Bounce') (Acacio de Barros *et al.*, 1998). The method is particularly well suited to defining a global time in cosmology.

**Discussion** In the dBB interpretation, the positions $Q_i$ of the material particles are considered to be the real positions of the particles, and the only physical quantities that can directly be observed. In any measurement, what is observed at the end is the position of some object, for instance the pointer of the measurement apparatus. This applies to all massive particles, which have a position operator in quantum mechanics. How to interpret fields, for instance the electromagnetic field and its photon (which has no position operator) is less obvious. For massive particles, the first question that arises is the status of the wave function, which guides the real positions. Should it be considered as real, because anything that influences something real must be real also? Various opinions about this question have been expressed.

For instance, Bell wrote (chapter 18 of Bell, 1981, 1987, p. 128): 'No one can understand this theory (the dBB theory) until he is willing to think of $\Psi$ as a real objective field rather than just a probability amplitude'. Seen in this way, a 'particle' always involves a combination of both a position and the associated field, which is physically real. Since this field can extend at arbitrarily large distance from the Bohmian position, it is then not surprising if two particles should influence each other even if their positions are very far away: the influence results from the interaction between two physical objects, a position and a field, as for instance in classical electromagnetism.

A different point of view (Bricmont, 2016) considers that the wave function is not a real physical field, but a mathematical function that has the role to provide, through its partial derivatives, the velocities of the Bohmian positions. In an experiment, the wave function characterizes the preparation procedure of the quantum physical system. Its role is similar to that of a Hamiltonian or a Lagrangian in classical mechanics. Nowadays, this latter point of view seems to be more frequent among the supporters of the dBB interpretation.

It is often said that the dBB interpretation offers a deterministic version of quantum mechanics. This statement is nevertheless ambiguous, if not worse. In dBB theory, what determines the result of a quantum measurement is the ensemble of Bohmian positions of the system and measurement apparatus, but these positions cannot be known in advance for fundamental reasons (persistence of the quantum equilibrium). In other words, the Bohmian positions are the mathematical ingredients which, in the dynamics, introduce the stochasticity in a way that perfectly reproduces the Born rule. Trajectories exist in the dBB theory, but the existence of trajectories does not mean that one can predict the future behaviour of particles. Some retrodictions are indeed possible: if one observes the position of a particle in a given experimental set-up, one can calculate the past trajectory of this particle. To summarize, a deterministic dynamics does not mean that the theory itself is deterministic. More details on the dBB interpretation, a few variants, its applications, field theory, etc. can be found in section H-1 of Chapter XI of Laloë (2019).

### 1.2.5.2 *Modal Interpretations*

The modal interpretations provide another class of interpretations where additional variables are introduced in quantum mechanics. These variables are not positions, as in the dBB interpretation, but another state vector. The modal interpretations (van Fraassen, 1972, 1974, 1991; Dickson and Dieks, 2007; Lombardi and Dieks, 2012) attribute more properties to physical systems than standard quantum mechanics does, and these properties are described by an additional state vector, which is perfectly similar to the standard state vector (it belongs to the same space of states). It nevertheless plays a different role, since it directly characterizes the real physical properties of the physical system. Therefore, within these interpretations, a system may for instance have a perfectly well-defined value for a physical observable, even if its standard state vector is not an eigenstate of this observable.

Bas van Fraassen (1972, 1974, 1991) associates two distinct states (kets) with any physical system:

- The 'dynamical state', which is the usual quantum state. It describes the evolution of the system, which, for an isolated system, is given by the Schrödinger (or von Neumann) equation. In the modal interpretation, the state vector reduction never occurs for the dynamical state.
- A 'value state' representing the physical properties of the system at any time.

Assume that a system $S$ is part of a larger system $T$, which is assumed to be isolated. We now discuss how the dynamical and the value states describing $S$ may differ if the dynamical state of $S$ (density operator obtained by partial trace) is not a projector over a pure state (at least two of its eigenvalues do not vanish). Various possibilities have been suggested to define the value state. Van Fraassen initially proposed relatively general rules, assuming that the value state could be any state appearing in the decomposition of the dynamical state (it can be any linear combination of the eigenstates with nonzero eigenvalues of the partial density operator $\rho_S$ defining the dynamical state of $S$). The definition of the value state then remains rather loose. Other authors have proposed to be more specific and to use the bi-orthonormal decomposition (Schmidt decomposition) to write the entangled dynamical state of the whole system $T$ as:

$$|\Psi\rangle = \sum_n c_n |\varphi_n\rangle \otimes |\Phi'_n\rangle \qquad (9)$$

where the $|\varphi_n\rangle$ are normalized and mutually orthogonal states describing $S$, and the $|\Phi'_n\rangle$ are similar states for the system complementing $S$ in $T$. In the modal interpretation, the $|\varphi_n\rangle$ are then the possible value states describing system $S$. Of course, if all $c_n$ but one are zero, within standard quantum mechanics system $S$ is already in a pure state, and there is nothing new. But the modal interpretation postulates that, even when several $c_n$ do not vanish, system $S$ has all the properties associated with a single $|\varphi_n\rangle$. This postulate remains fundamentally nondeterministic: the only possible prediction is given by the probability $|c_n|^2$ for $S$ to be in state $|\varphi_n\rangle$, which reconstructs the Born rule (at this stage, the use of the dynamical state cannot be avoided). But, even if it is impossible in advance to predict which of the accessible states will be reached, when it is reached, all sets of propositions about $S$ that would be true if $S$ was in state $|\varphi_n\rangle$ within standard quantum mechanics are indeed true. This point of view is called 'modal' because it leads to a modal logic of quantum propositions.

As mentioned above, the value state of $S$ may contain more information about the physical properties of $S$ than its dynamical state. But the set of physical properties resulting from this more precise description should not exceed the maximum that is permitted by a standard description with any quantum state. One can then never attribute simultaneous sharp definitions to noncommuting operators, position and momentum for instance (as opposed to Bohmian mechanics, where a particle has a perfectly well-defined position and velocity at every time).

Consider a measurement process involving system $S$ and a measurement apparatus $M$. During the initial stage of interaction between $S$ and $M$, both systems develop entanglement. This is sufficient to introduce a value state for $S$ that differs from its standard description (with a density operator). From this moment, the interpretation guarantees that both subsystems have all the properties associated with the emergence of a single result from the measurement (macroscopic uniqueness).

Similar ideas were later developed by Simon Kochen (1985) with more emphasis on the relational character of properties of physical systems. Dennis Dieks (1988, 1989a, b, 1994) introduces another view where systems do have intrinsic properties, and discusses how measurement processes and the existence of macroscopic behaviour can be understood within the modal interpretation. Richard Healey (1989, 1993) has proposed an 'interactive interpretation', which has similarities with a modal interpretation.

## 1.2.6 Modified Schrödinger Dynamics

When Schrödinger introduced his equation, his main purpose was to propose an equation that would reproduce the experimental observations concerning the spectrum of the hydrogen atom. Since then, the equation has been found to correct predictions covering many domains, in all atomic and molecular physics, solid state physics, etc. It is also the essential element of modern chemistry. The purpose of the theories introducing modified Schrödinger dynamics is to find an equation that preserves these correct predictions, but also extends the domain of these predictions even further, including the observation of a single experimental result in a quantum experiment. In this perspective, the usual necessity of introducing the notion of observers, macroscopic observation apparatuses, etc. vanishes: any measurement becomes a simple interaction process between a system and an apparatus, and can be treated within a completely universal dynamics.

All the interpretations that we have discussed until now are built to reproduce the predictions of standard quantum mechanics exactly. By contrast, theories with modified Schrödinger dynamics are really different theories, since they may in some cases make different predictions, even if these cases never occur with microscopic systems.

### 1.2.6.1 GRW

Giancarlo Ghirardi, Alberto Rimini, and Tullio Weber (GRW) (1986, 1987) introduced a new version of quantum theory based on a 'unified dynamics for microscopic and macroscopic systems'. This unification is obtained by introducing a 'spontaneous localization' (SL) term into the Schrödinger equation, a term that may suddenly change the state vector by localizing its wave function. This process occurs at random times, with a time constant that is adjusted so that quantum superposition of macroscopically distinct states (QSMDS) are rapidly destroyed, while microscopic systems remain practically unaffected.

The simplest case occurs with a single particle described by a state vector $|\Psi(t)\rangle$. The effect of one of the spontaneous localization terms is to suddenly replace $|\Psi(t)\rangle$ by a ket $|\Psi'(t)\rangle$ according to:

$$|\Psi(t)\rangle \Rightarrow |\Psi'(t)\rangle = \frac{F_j|\Psi(t)\rangle}{\langle\Psi(t)|(F_j)^2|\Psi(t)\rangle} \tag{10}$$

where the denominator of this expression ensures the conservation of the norm of the state vector. In this equation, $F_j$ is a Hermitian operator, diagonal in the position representation. It localizes the particle around the point of space $\mathbf{r}_j$ with an accuracy characterized by a free parameter $\alpha$ ($\alpha^{-1/2}$ is a length):

$$F_j = ce^{-\alpha(\mathbf{R}-\mathbf{r}_j)^2/2} \tag{11}$$

where $\mathbf{R}$ is the position operator of the particle and $c$ a real normalization factor.

In fact, a whole set of operators $F_j$ is introduced, which obey the condition:

$$\sum_j (F_j)^2 = 1 \tag{12}$$

The $r_j$ may also define discrete positions of the nodes of a lattice with a unit cell that has a size much smaller than $1/\sqrt{\alpha}$; the value of $c$ then depends on this size. Here, we continue to write discrete sums with index $j$, but the transposition to continuous indices and integrals is simple.

GRW assume that the localization processes corresponding to the various $r_j$ constantly act in parallel, each with a probability per unit time given by:

$$\gamma \langle \Psi(t) | (F_j)^2 | \Psi(t) \rangle \tag{13}$$

In this relation, $\gamma$ is a free parameter of the theory. Condition (12) ensures that the total probability for any sort of hit is independent of the initial state $|\Psi(t)\rangle$. When the random effect of these localization processes is added to the usual Schrödinger evolution, one obtains a theory where, for each possible realization of the random hits at all times, the state vector follows a trajectory that is well defined; it is of course different for each realization.

During the first localization process, if $\Psi(\mathbf{r},t)$ is the wave function associated with state $|\Psi(t)\rangle$, according to (11) the probability of occurrence of the process with index $j$ is:

$$\mathcal{P}_j = \gamma c^2 \int d^3r \, e^{-\alpha(\mathbf{R}-\mathbf{r}_j)^2} |\Psi(\mathbf{r},t)|^2 \tag{14}$$

These localization processes are therefore more likely to be centred at values of $\mathbf{r}_j$ where the density of probability $|\Psi(\mathbf{r},t)|^2$ is maximal. Whatever $\mathbf{r}_j$ is selected in the first localization process, the wave function after this process is multiplied by $e^{-\alpha(\mathbf{r}-\mathbf{r}_j)^2/2}$, which tends to restrict the wave function to a neighborhood of $\mathbf{r}_j$ with spatial extension $\alpha^{-1/2}$.

Let us now assume that the wave function does not have the time to evolve before a second localization process takes place; the second is then more likely to be centred at a point not too far from $\mathbf{r}_j$. Similarly, the third localization is likely to select a point that is in the neighborhood of the two preceding localization points, and so on: after a few localization processes, the wave function ends up well localized around a point $\mathbf{r}$ that is random, but well defined. This creates the spontaneous spatial reduction process of the wave function.

Now, if the wave function evolves between the localization processes, the standard spreading of the wave packet tends to broaden it; it therefore tends to counterbalance the effects of localization. A balance between the opposite processes may occur. The succession of the points of localization reconstructs a trajectory for the particle, as the track created in a cloud or spark chamber.

This dynamics can be generalized to a system made of $N$ particles, by assuming that all the particles independently undergo localization processes. The operators $F_j$ are then replaced by Hermitian operators $F_j^i$ acting on the $i$-th particle, and their effect is summed over both indices $i$ and $j$ in the Schrödinger equation. Again, for each realization, one obtains a single (approximate) trajectory for the state vector $|\Psi(t)\rangle$ describing the physical system. The effect of the random processes is to spatially localize the wave function around a single point of configuration space with $3N$ dimensions, with a spatial extension $\alpha^{-1/2}$ in every direction.

An interesting feature of the localization process appears if a system of many particles is a superposition of two quantum states that are localized in different regions of space. This is for instance the case if the pointer of a measurement apparatus indicates two different results at the same time (or in any Schrödinger-cat-like situation). The localization of a single particle is then sufficient to cancel one of the components of the coherent superposition. Since all particles are constantly subject to the same GRW random process, the probability per unit time that one of them will be localized, i.e. the probability of the disappearance of the superposition, is proportional to the number of particles. Large quantum superpositions therefore disappear almost instantaneously, while microscopic systems can avoid any localization process for very long times.

It is of course also possible to study the evolution of the density operator $\rho$ describing an ensemble of realizations of the same physical system. The pilot equation providing the time evolution of $\rho$ contains, in addition to the standard Hamiltonian commutator, localization terms that are analogous to the Lindblad form of relaxation terms. For more details, we refer the reader to section 11.10 of Laloë (2019).

## 1.2.6.2  *CSL*

It is also possible to avoid the discontinuous character of the GRW theory, while retaining the attractive features of the GRW model. Philip Pearle (1989) showed that this result can be obtained by adding terms of 'continuous spontaneous localization' (CSL) to the usual Hamiltonian in the Schrödinger evolution. For this purpose, he introduces Markov processes depending on a set of random functions of time $w_j(t)$ having a broad frequency spectrum (white noise). They contain a time rate $y$ as well as a set of mutually commuting Hermitian operators $A_j$. A full compatibility with the standard notion of identical particles in quantum mechanics is realized with an appropriate choice of the operators. In this point of view, the state vector still evolves according to a differential equation, but this equation includes random functions of time (Itô stochastic differential equation) as well as antiHermitian terms; therefore, the norm of the state vector $|\Psi(t)\rangle$ is no longer conserved.

The statistical properties of the random functions are defined precisely from the variations of this norm: one postulates a CSL 'probability rule', which states that the probability for realizing any time dependence $w_j(t)$ that leads to a given value of $|\Psi(t)\rangle$ is equal to $\langle\Psi(t)|\Psi(t)\rangle^2$. The nonlinear character of the theory is then obvious: we have a feedback process where the evolution of $|\Psi(t)\rangle$ depends on functions $w_j(t)$, which, in turn, have statistical properties depending on the norm of $|\Psi(t)\rangle$ itself. The postulate strongly favours the realizations of the random functions that keep a large norm to the state vector, while reducing the probability of all the others that give exponentially small values to the norm (even if they correspond to many more possibilities). This choice of statistical properties remains consistent with the independent Markovian evolution of each realization of the state vector.

Let us first consider a simple case where a single operator $A$ (with eigenvalues $a_n$) is introduced; $A$ corresponds to the observable measured in a quantum measurement process. After the interaction between the measured system and the apparatus, both systems are entangled. Under the effect of antiHermitian terms controlled by the random functions $w_j(t)$, the modulus of each probability amplitude $c_n(t)$ fluctuates in time. Among the large number of possible values of $w_j(t)$, according to the CSL probability rule, only a very small proportion may occur with a nonnegligible probability: the subensemble leading to a large sum over $n$ of all $|c_n(t)|^2$. One can then show that, among these special functions, the best to provide a large norm for the state vector are those that give a large value to one $|c_n(t)|^2$ only. The reason is that $w_j(t)$ can favour one value of $n$ during its fluctuations, but not several at the same time; situations where the fluctuations of $w_j(t)$ successively favour two (or more) of these coefficients lead to a sort of dilution of the norm preservation effect, resulting in an exponentially smaller value of the total norm at the end. In fact, the fluctuations of the random functions break the symmetry between all possible measurement outcomes, in a way that reproduces state vector reduction.

In full CSL theory, $A$ is replaced by a series of position localization operators $A_j$, acting on all particles of the system, and localizing them at all possible positions in space ($j$ then becomes an index for spatial positions, and thus may be continuous). As in GRW theories, perfect localizations of the particles would transfer an infinite amount of kinetic energy to them. To avoid this problem, one postulates that the localization provided by each $A_j$ is imperfect, and takes place over a spatial range $\alpha^{-1/2}$; all $A_j$ are mutually commuting operators. Despite these changes concerning the definition of the operators, the essence of the localization process remains similar to that of GRW. It introduces a selection that, at the end, localizes every particle into a single random region of space – a spatial reduction process of the quantum state.

With microscopic quantum systems such as a single particle, an atom or molecule, etc., the probability of occurrence of any collapse process remains extremely low, even for a very long time since $\gamma$ is small. For macroscopic systems in quantum superpositions of two spatially distinct states (QSMDS), collapse is very likely to occur rapidly and to cancel all components of the QSMDS, but one. This process is fast since all the particles involved are constantly subject to localization, while the localization of a single particle is sufficient to destroy one of the two components.

The study of Markov processes and continuous spontaneous localization for identical particles was expanded by Ghirardi, Pearle, and Rimini (1990). They showed that, for an ensemble of systems, one obtains a 'Lindblad form' giving the following time evolution:

$$\frac{d\rho(t)}{dt} = \frac{1}{i\hbar}[H(t), \rho(t)] + \frac{\gamma}{2}\sum_{j=1}^{N}[2A_j\rho A_j - (A_j)^2\rho(t) - \rho(t)(A_j)^2] \qquad (15)$$

where the $A_j$ are position localization operators. Discrete Markov processes in Hilbert space can be reduced, in the limit of small hits with infinite frequency, to a continuous spontaneous localization.

Both GRW and CSL theories require including new universal constants, which appear in the modified Schrödinger dynamics, and are adjusted to introduce a collapse only when necessary, and in particular avoid any contradiction with any presently known fact. These constants may in a sense seem to be ad hoc physical quantities, introduced only for technical reasons. Actually, they have an important conceptual role: they define the border between the microscopic and macroscopic world. The corresponding border is unavoidable, but ill-defined, in the Copenhagen interpretation; within localization theories, it is introduced in a perfectly defined way.

### 1.2.6.3 *Gravity Induced Collapse*

Instead of introducing an arbitrary Gaussian localization process, Diósi (1989) proposed to introduce a collapse of the state vector involving the standard Newton gravitation law, inserted in an appropriate stochastic dynamics. This theory does not require the introduction of any dimensional parameter (while GRW and CSL require two); a single dimensionless quantity is introduced, which is assumed to be of order of unity. Another interesting feature is that this theory provides a universal mass density process, which applies in the same way to any kind of particle. Ghirardi, Grassi, and Rimini (1990) nevertheless soon showed that this theory suffers from serious problems, in particular at short distances, and for instance predicts unrealistically short lifetimes for some nuclei.

Similar ideas have been proposed by Roger Penrose within the context of general relativity, and even considerations on the nature of consciousness (Penrose, 1989, 1994). This author discusses situations involving quantum superpositions of the same massive object located in two different regions of space (Penrose, 1996). In the absence of gravity, this superposition can have a very well-defined energy but, within general relativity, because masses curve space-time, this situation implies a superposition of two different space-times. Penrose then uses the 'principle of general covariance' to study the properties of the time translation operator, and shows that the considered situation necessarily has an energy uncertainty $\Delta E$. He then conjectures that the inverse of $\Delta E$ corresponds to a finite lifetime of the initial superposition (time–energy uncertainty relation). As a consequence, the superposition is unstable, and should spontaneously decay into one of its components. This is equivalent with a state vector reduction, and ensures macroscopic uniqueness of the position of all massive bodies. Penrose considers that 'this proposal does not provide a *theory* of quantum state reduction. It merely indicates the level at which deviations from the standard linear Schrödinger (unitary) evolution are to be expected owing to gravitational effects'. Other contributions may be found in Diósi (2014); Tilloy and Diósi (2016); Adler (2016); Gasbarri *et al.* (2017). Reviews of this class of theories can be found in § II-B of Bassi *et al.* (2013) and in Singh (2015).

In Laloë (2020), the present author proposes a model of quantum collapse where gravity is also the source of collapse, but is considered as a classical field originating from the Bohmian positions of the particles (instead of the quantum density obtained from the many body wave function). The equations of the model are completely deterministic, the only source of randomness being the initial randomness of the Bohmian position, as in the de Broglie–Bohm (dBB) theory. They are based on the correlation properties of the Bohmian positions: if a QSMDS occurs, for instance if a pointer reaches a superposition of two states localized in different regions of space, all the Bohmian positions nevertheless remain grouped together in the same region. This is because the positions are driven by the wave function in configuration space, and because of the cohesion forces inside the pointer introduce an attractive part in the Hamiltonian that inhibits configurations where the particles can spread in different regions. In a particular realization of the experiment, if all Bohmian particles attract the wave function, they will localize it around one single position of the pointer. This provides a projection of the N-particle wave function that is similar to the von Neumann projection postulate. Its dynamics necessitates the introduction of one single, dimensionless constant.

Because the source of the curvature of space-time is the Bohmian positions of the particles, this model is consistent with a general view where space-time remains classical. The various quantum fields then propagate inside this classical space-time frame. Similar ideas have been discussed for instance by Ward Struyve (2015, 2017) and Antoine Tilloy (2019).

# 1.3 CONCLUSION

We still have a large variety of interpretations of quantum mechanics that are supported by different groups of physicists. It therefore seems possible that this theory, as wonderful and efficient as it is presently, may not have reached its final form, and may even evolve significantly in the future. The two pillars of physics mentioned in the introduction, relativity and quantum mechanics, are not incompatible, but they do not really belong to the same logical frame; the internal consistency of the whole edifice is not particularly clear.

## 1.3.1 An Internal Tension in Physics

Already in 1928 Bohr was aware of a tension between a quantum description and relativity. In Bohr (1928, page 580) he wrote 'We learn from the quantum theory that the appropriateness of our usual causal space-time description depends entirely upon the small value of the quantum of action as compared to the actions involved in ordinary sense descriptions'. He actually considers that a space-time description and the claims of causality are complementary, and therefore incompatible. Later (page 586), he adds 'The fundamental difficulties opposing a space-time description of a system of particles in interaction appear at once from the inevitability of the super-position principle in the description of the behaviour of individual particles'. Similar ideas are implied in his reply (Bohr, 1935) to the famous Einstein–Podolsky–Rosen article (Einstein *et al.*, 1935): because of what is often called the 'failure of ordinary space-time description' Bohr considers that the EPR argument, which is based on the independence of random events taking place in different regions of space, is not valid.

Incidentally, it is frustrating to see that so many scientific journalists, when describing the famous Einstein–Bohr debate, write that the essence of the debate was determinism, because Einstein stubbornly did not want to accept indeterminism. In fact, both had actually perfectly understood that what was at stake was the possibility of a causal description of measurements performed in remote regions of space on entangled quantum systems. For Einstein, a description in terms of space-time events should be possible for any thought experiments. For Bohr, the measurement process performed with an apparatus was to be considered as an inseparable whole, even if the apparatus consisted of two parts widely separated in space. So, determinism in itself was not the issue!

John Bell (1987, chapter 18), at the end of his essay 'Speakable and unspeakable in quantum mechanics', writes: 'We have an apparent incompatibility, at the deepest level, between the two fundamental pillars of contemporary theory' (quantum mechanics and relativity). Of course, 'apparent incompatibility' does not mean 'contradiction': the theory resting on these two pillars is not self-contradictory as long as quantum mechanics does not imply superluminal signalling. Abner Shimony (1993, p. 131) expresses this idea by writing: 'In this sense there may be a peaceful coexistence between quantum mechanics and relativity'.

At present, the tension between general relativity and quantum mechanics is still not considered as fully resolved. Standard quantum mechanics considers that the various particles and quanta describing the forces in quantum field theory propagate in a 'flat' space of special relativity. By contrast, general relativity changes the structure of space-time, since all the masses are sources of gravitation that introduce a curvature of space-time. In general relativity, gravitation is not quantized, while quantum field theory naturally tends to quantize all fields, which leads to introducing particles named 'gravitons' for the gravitational field. Moreover, as discussed for instance by Richard Feynman (see Hartfield (2002)) and Roger Penrose (1996), the occurrence of QSMDS should put space-time into quantum superpositions of different geometries. Generally speaking, within quantum mechanics the structure of space-time should have quantum fluctuations, which creates various difficulties and paradoxes.

As mentioned at the end of section 1.2.6.3, two views on gravitation are therefore possible: either the gravitational field is a field as the other fields (electromagnetic, strong forces, weak forces) and should be quantized within a more general theory that would encompass quantum mechanics and general relativity; or general relativity provides the classical frame of space-time inside which the other quantum fields propagate. Numerous attempts have been made to fulfil the first programme, and to quantize gravitation: supersymmetries, string theory, quantum loop gravity, non-commutative geometry, correlated world lines, etc. For general or more specific reviews, see for instance Julia and Zinn-Justin (1992); Polchinski (1998); Rovelli (2004); Carlip *et al.* (2015); Nath (2016); Greene (2010); Kiefer (2012); Carney *et al.* (2019).

We have already mentioned quantum cosmogenesis, where one problem is to study the evolution of the structure of the early Universe during the initial 'big bang' period. Both quantum mechanics and gravity play an important role in this problem. General relativity provides an equation for the metric, where the general relativistic scale factor should be replaced by a quantum operator. The calculations then become untractable. One can nevertheless develop a Bohmian theory for quantum gravity, with a Bohmian variable for the metric (Peter *et al.*, 2005; Pinho and Pinto-Neto, 2007) and a trajectory of the scale factor. This trajectory can be calculated to successive orders, which makes it possible to inject the corresponding values into the successive perturbation equations of the other degrees of freedom. Quantum effects are taken into account, since Bohmian trajectories are sensitive to interference or tunnelling, and their use does not suppress any quantum result.

## 1.3.2 The Influence on the Development of Physics

To non-physicists, it may seem surprising that the relative fuzziness of the interpretation of quantum mechanics has so little influence on the advances of physics in laboratories. Are all physicists adepts of what is sometimes described humorously as 'Shut up and calculate' (Mermin, 1989)? To some extent, yes. The reason is that, until now, the common sense interpretation discussed in section 1.1.2 has proved sufficient to interpret all experiments. Indeed, the 'correlation interpretation', or some variant of it, manages to handle all situations encountered in physics laboratories perfectly well. Physicists know that the progress of physics is rooted in experiments, and indeed the success in experimental physics has been spectacular for many decades. For instance, many experiments that were considered as pure thought experiments, and described as such by the founding fathers of quantum mechanics, have been performed, or are now within reach experimentally. Schrödinger once wrote that it will never be possible to measure a single electron but, in fact, a single electron undergoing quantum jumps has already been observed long ago (see for instance Peil and Gabrielse (1999)). Nowadays experiments are routinely done where single particles are observed in a quantum regime, and where 'state vector reduction is observed in real time' (see for instance Laloë (2019, chapter 10)).

It has often been claimed, including especially by the founding fathers of quantum mechanics, that this theory has a significant impact on the philosophical view we have of the universe. In retrospect, one century later, this strong impact of physics on philosophy seems to have been overestimated: various interpretations have since been proposed, supporting different philosophical views, and still no definitive logical argument has been found to select one or the other. For instance, the definition of reality in the history interpretation, dBB interpretation, or GRW, have very little in common, but choosing one or the other remains a question of personal taste. Actually, it seems that quantum mechanics is at the same time very robust and prone to philosophical indifference. Fortunately in a sense, it can indeed be used without any problems by physicists who have very different views on its interpretation.

Of course, no one knows how this theory will evolve in the future: maybe, as Einstein hoped, a universal point of view encompassing at the same time general relativity and all useful features of quantum mechanics will emerge one day. The only certitude is that this future theory will necessarily incorporate many of the useful features of the present state of quantum mechanics.

## REFERENCES

Acacio de Barros, J., Pinto-Neto, N., and Sagioro-Leal, M.A. (1998). The causal interpretation of dust and radiation fluid non-singular quantum cosmologies. *Phys. Lett. A*, **241**, 229–39.

Adler, S. L. (2016). Gravitation and the noise needed in objective reduction models. In M. Bell and S. Gao (eds), *Quantum Nonlocality and Reality: 50 Years of Bell's Theorem*, Cambridge: Cambridge University Press, pp. 390–399.

Aharonov, Y., Anandan, J., and Vaidman, L. (1993). Meaning of the wave function. *Phys. Rev. A*, **47**, 4616–26.

Albareda, G., Marian, D., Benali, A., Yaro, S., Zanghì, N., and Oriols, X. (2013). Time-resolved electron transport with quantum trajectories. *J. Comput. Electron.*, **12**, 405–19.

Allahverdyan, A. E., Balian, R., and Nieuwenhuizen, T. M. (2017). A subensemble theory of ideal quantum measurement processes. *Annals of Physics*, **376**, 324–52.

Auffèves, A. and Grangier, P. (2016a). Contexts, Systems and Modalities: A New Ontology for Quantum Mechanics. *Found. Phys.*, **46**, 121–37.

Auffèves, A. and Grangier, P. (2016b). Recovering the quantum formalism from physically realist axioms. *arXiv:1610.06164v2*.

Auletta, G. (2001). *Foundations and Interpretations of Quantum Mechanics*. Singapore: World Scientific.

Auletta, G. (2005). Quantum Information as a General Paradigm. *Found. Phys.*, **35**, 787–15.

Auletta, G., Fortunato, M., and Parisi, G. (2014). *Quantum Mechanics*. Cambridge: Cambridge University Press.

Balibar, F., Darrigol, O., and Jech, B. (1989). *Albert Einstein, Oeuvres Choisies I, Quanta*. Paris: Editions du Seuil et Editions du CNRS.

Ballentine, L. E. (1970). The Statistical Interpretation of Quantum Mechanics. *Rev. Mod. Phys.*, **42**, 358–81.

Bassi, A., Lochan, K., Satin, S., Singh, T. P., and Ulbricht, H. (2013). Models of wave-function collapse, underlying theories, and experimental tests. *Rev. Mod. Phys.*, **85**, 471–27.

Bell, J. S. (1981). Quantum mechanics for cosmologists. In C. J. Isham, R. Penrose, and D. W. Sciama (eds), Quantum Gravity 2, A Second Oxford Symposium, Oxford: Clarendon.

Bell, J. S. (1987). *Speakable and Unspeakable in Quantum Mechanics*. Cambridge: Cambridge University Press. Second augmented edition (2004) contains the complete set of J. Bell's articles on quantum mechanics.

Bell, J. S. (1992). Six possible worlds for quantum mechanics. *Found. Phys.*, **22**, 1201–15.

Benseny, A., Albareda, G., Sanz, A. S., Mompart, J., and Oriols, X. (2014). Applied Bohmian mechanics. *Eur. Phys. J.*, **68**, 286.

Bohm, D. (1952). A Suggested Interpretation of the Quantum Theory in Terms of 'Hidden' Variables. *Phys. Rev.*, **85**, 166–179 and 180–193.

Bohr, N. (1928). The Quantum Postulate and the Recent Development of Atomic Theory. *Supplements to Nature*, April, 580–590.

Bohr, N. (1935). Can Quantum-Mechanical Description of Physical Reality be Considered Complete? *Phys. Rev.*, **48**, 696–02.

Bohr, N. (1950). On the Notions of Causality and Complementarity. *Science*, **111**, 51–54.

Bohr, N. (1960). The unity of human knowledge. In N. Bohr, *Atomic Physics and Human Knowledge*, New York: Wiley.

Bricmont, J. (2016). *Making Sense of Quantum Mechanics*. Cham, Switzerland: Springer.

Brukner, C. and Zeilinger, A. (1999). Operationally Invariant Information in Quantum Measurements. *Phys. Rev. Lett.*, **83**, 3354–57.

Bub, J. (2011). Quantum probabilities: An information-theoretic interpretation. In C. Beisbart and S. Hartmann (eds), *Probabilities in Physics*, Oxford: Oxford University Press.

Busch, P. (2003). Quantum States and Generalized Observables: A Simple Proof of Gleason's Theorem. *Phys. Rev. Lett.*, **91**, 120403.

Carlip, S., Chiou, D., Ni, W., and Woodard, R. (2015). Quantum gravity: A brief history of ideas and some prospects. *Int. J. Mod. Phys. D*, **24**, 1530028.

Carney, D., Stamp, P. C. E., and Taylor, J. M. (2019). Tabletop experiments for quantum gravity: A user's manual. *Classical and Quantum Gravity*, **36**, 034001.

Caves, C. M., Fuchs, C. A., and Schack, R. (2007). Subjective probability and quantum certainty. *Stud. Hist. Phil. Mod. Phys.*, **38**, 255–74.

Chevalley, C. (1994). Niels Bohr's words and the Atlantis of Kantianism. In J. Faye and H. Folse (eds), *Niels Bohr and Contemporary Philosophy*, Dordrecht: Kluwer, pp. 33–57.

Christov, I. P. (2007). Time-dependent quantum Monte Carlo: Preparation of the ground state. *New Journal Phys.*, **9**, 70.

Christov, I. P. (2009). Polynomial-Time-Scaling Quantum Dynamics with Time-Dependent Quantum Monte Carlo. *J. Phys. Chem. A*, **113**, 6016–21.

Cox, R. T. (1946). Probability, Frequency and Reasonable Expectation. *American Journal of Physics*, **14**, 1–13.

Darrigol, O. (1992). *From C-Numbers to q-Numbers: The Classical Analogy in the History of Quantum Theory*. University of California Press.

Darrigol, O. (2015). Shut up and contemplate! Lucien Hardy's reasonable axioms for quantum theory. *Studies in Philosophy of Mod. Phys.*, **52**, 328–42.

de Broglie, L. (1924). *Recherches Sur La Théorie Des Quanta*. Ph.D. thesis, Paris.

de Broglie, L. (1927). La mécanique ondulatoire et la structure atomique de la matière et du rayonnement. *J. Physique et le Radium*, **8**, 225–41.

de Broglie, L. (1956). *Tentative d'Interprétation Causale et Non-Linéaire de La Mécanique Ondulatoire*. Paris: Gauthier-Villars.

de Broglie, L. (1987). Interpretation of quantum mechanics by the double solution theory. In *Ann. Fond. Louis de Broglie*, Volume 12, 4.

de Finetti, B. (2017). *Theory of Probability: A Critical Introductory Treatment*. New York: Wiley.

d'Espagnat, B. (1971). *Conceptual Foundations of Quantum Mechanics*. Amsterdam: Benjamins.

d'Espagnat, B. (1979). *A La Recherche Du Réel*. Paris: Gauthier-Villars Bordas.

d'Espagnat, B. (1985). *Une Incertaine Réalité, La Connaissance et La Durée*. Paris: Gauthier-Villars.

d'Espagnat, B. (1995). *Veiled Reality: An Analysis of Present Day Quantum Mechanics Concepts*. Boston, MA: Addison Wesley. Translation of B. d'Espagnat (1994), *Le réel voilé, analyse des concepts quantiques*, Paris: Fayard.

DeWitt, B. S. (1970). Quantum mechanics and reality. *Phys. Today*, **23**, 30–35.

Dickson, M. and Dieks, D. (2007). Modal interpretation of quantum mechanics. In *Stanford Encyclopedia of Philosophy*. http://plato.stanford.edu/entries/qm-modal/ (now replaced by Lombardi and Dieks (2012), but still accessible on the site of the Encyclopedia).

Dieks, D. (1988). The Formalism of Quantum Theory: An Objective Description of Reality? *Annalen der Physik*, **500**, 174–90.

Dieks, D. (1989a). Quantum mechanics without the projection postulate and its realistic interpretation. *Found. Phys.*, **19**, 1397–1423.

Dieks, D. (1989b). Resolution of the measurement problem through decoherence of the quantum state. *Phys. Lett. A*, **142**, 439–46.

Dieks, D. (1994). Modal interpretation of quantum mechanics, measurements, and macroscopic behavior. *Phys. Rev. A*, **49**, 2290–00.

Diósi, L. (1989). Models for universal reduction of macroscopic quantum fluctuations. *Phys. Rev. A*, **40**, 1165–74.

Diósi, L. (2014). Gravity-related spontaneous wave function collapse in bulk matter. *New Journal of Physics*, **16**, 105006.

Dirac, P. A. M. (1930). *The Principles of Quantum Mechanics*. Oxford: Oxford University Press.

Einstein, A., Podolsky, B., and Rosen, N. (1935). Can Quantum-Mechanical Description of Physical Reality Be Considered Complete? *Phys. Rev.*, **47**, 777–780. Reproduced in J. A. Wheeler and W. H. Zurek (eds) (1983), *Quantum Theory of Measurement*, Princeton: Princeton University Press, pp. 138–41.

Englert, B., Scully, M. O., Süssmann, G., and Walther, H. (1992). Surrealistic Bohm Trajectories. *Z. Naturforschung*, **47a**, 1175–1186.

Englert, B-G., Scully, M. O., and Walther, H. (1999). Quantum erasure in double-slit interferometers with which-way detectors. *Am. J. Phys.*, **67**, 325–329. See the first few lines of IV.

Everett, H. (1957). 'Relative State' Formulation of Quantum Mechanics. *Rev. Mod. Phys.*, **29**, 454–462. Reprinted in J.A. Wheeler and W.H. Zurek (eds) (1983), *Quantum Theory and Measurement*, Princeton: Princeton University Press, pp. 315–323.

Fein, Y. Y., Geyer, P., Zwick, P., Kiałka, F., Pedalino, S., Mayor, M., Gerlich, S., and Arndt, M. (2019). Quantum superposition of molecules beyond 25 kDa. *Nature Physics*, 15, 1242–45.

Fuchs, C. A. (2001). Quantum Foundations in the Light of Quantum Information. *arXiv:quant-ph/0106166*.

Fuchs, C. A. (2002). Quantum Mechanics as Quantum Information (and only a little more). *arXiv:quant-ph/0205039*.

Fuchs, C. A. (2010). QBism, the Perimeter of Quantum Bayesianism. *arXiv:1003.5209v1 [quant-ph]*.

Fuchs, C. A., Mermin, N. D., and Schack, R. (2013). An Introduction to QBism with an Application to the Locality of Quantum Mechanics. *arXiv:1311.5253v1 [quant-ph]*.

Gasbarri, G., Toroš, M., Donadi, S., and Bassi, A. (2017). Gravity induced wave function collapse. *Phys. Rev. D*, 96, 104013.

Gell-Mann, M. and Hartle, J. B. (1993). Classical equations for quantum systems. *Phys. Rev. D.*, 47, 3345–82.

Ghirardi, G., Grassi, R., and Rimini, A. (1990). Continuous-spontaneous-reduction model involving gravity. *Phys. Rev. A*, 42, 1057–64.

Ghirardi, G. C., Pearle, P., and Rimini, A. (1990). Markov processes in Hilbert space and continuous spontaneous localization of systems of identical particles. *Phys. Rev. A*, 42, 78–89.

Ghirardi, G. C., Rimini, A., and Weber, T. (1986). Unified dynamics for microscopic and macroscopic systems. *Phys. Rev. D*, 34, 470–91.

Ghirardi, G. C., Rimini, A., and Weber, T. (1987). Disentanglement of quantum wave functions. *Phys. Rev. D*, 36, 3287–9.

Gindensperger, E., Meier, C., and Beswick, J. A. (2000). Mixing quantum and classical dynamics using Bohmian trajectories. *J. Chem. Phys.*, 113, 9369–72.

Gindensperger, E., Meier, C., and Beswick, J. A. (2002a). Quantum-classical dynamics including continuum states using quantum trajectories. *J. Chem. Phys.*, 116, 8. https://doi.org/10.1063/1.1415452.

Gindensperger, E., Meier, C., Beswick, J. A., and Heitz, M-C. (2002b). Quantum-classical description of rotational diffractive scattering using Bohmian trajectories: Comparison with full quantum wave packet results. *J. Chem. Phys.*, 116, 23. https://doi.org/10.1063/1.1471904.

Gleason, A. M. (1957). Measures on the Closed Subspaces of a Hilbert Space. *Journal of Mathematics and Mechanics*, 6, 885–93.

Greene, B. (2010). *The Elegant Universe*. New York: W.W. Norton and company.

Griffiths, R. B. (1984). Consistent histories and the interpretation of quantum mechanics. *J. Stat. Phys.*, 36, 219–72.

Griffiths, R. B. (1996). Consistent histories and quantum reasoning. *Phys. Rev. A*, 54, 2759–74.

Griffiths, R. B. (1998). Choice of consistent family, and quantum incompatibility. *Phys. Rev. A*, 57, 1604–18.

Griffiths, R. B. (2002). *Consistent Quantum Theory*. Cambridge: Cambridge University Press.

Griffiths, R. B. and Omnès, R. (1999). Consistent Histories and Quantum Measurements. *Phys. Today*, 52, 26–31.

Hardy, L. (2001). Quantum Theory From Five Reasonable Axioms. *arXiv:quant-ph/0101012*.

Hartfield, B. (ed.) (2002). *Lectures on Gravitation*. Boulder, CO: Westview Press.

Healey, R. (1989). *The Philosophy of Quantum Mechanics: An Interactive Interpretation*. Cambridge: Cambridge University Press.

Healey, R. (1993). Measurement and quantum indeterminateness. *Found. Phys. Lett.*, **6**, 307–16.

Healey, R. (2017). Quantum-Bayesian and pragmatist views of quantum theory. In *Stanford Encyclopedia of Physics*. https://plato.stanford.edu/entries/quantum-bayesianhttps://plato.stanford.edu/entries/quantum-bayesian.

Hohenberg, P. C. (2010). Colloquium: An introduction to consistent quantum theory. *Rev. Mod. Phys.*, **82**, 2835–44.

Holland, P. (1993). *The Quantum Theory of Motion*. Cambridge: Cambridge University Press.

Holland, P. (2020). Uniting the wave and the particle in quantum mechanics. *Quantum Stud.: Math. Found.*, **7**, 155–78.

Hornberger, K., Gerlich, S., Haslinger, P., Nimmrichter, S., and Arndt, M. (2012). Colloquium: Quantum interference of clusters and molecules. *Rev. Mod. Phys.*, **84**, 157–73.

Howard, D. (2004). Who Invented the 'Copenhagen Interpretation'? A Study in Mythology. *Philosophy of Science*, **71**, 669–82.

Jammer, M. (1966). *The Conceptual Development of Quantum Mechanics*. New York: McGraw-Hill. Second edition (1989).

Julia, B. and Zinn-Justin, J. (ed.) (1992). *Gravitation and Quantizations*. Amsterdam: North Holland. Proceedings of the 57th session of the Les Houches summer school in theoretical physics.

Kiefer, C. (2012). *Quantum Gravity*. Oxford: Oxford University Press.

Kochen, S. (1985). A new interpretation of quantum mechanics. In P. Mittelstaedt and P. Lahti (eds), *Symposium on the Foundations of Modern Physics*, Singapore: World Scientific, pp. 151–169.

Laloë, F. (2019). *Do We Really Understand Quantum Mechanics?*, 2nd edn. Cambridge: Cambridge University Press.

Laloë, F. (2020). A model of quantum collapse induced by gravity. *Eur. Phys. J. D*, **74**, 25.

Leggett, A. J. (2002). Testing the limits of quantum mechanics: Motivation, state of play, prospects. *J. Phys. Condens. Matter*, **14**, R415–51.

Lombardi, O. and Dieks, D. (2012). Modal interpretations of quantum mechanics. In *Stanford Encyclopedia of Philosophy*. https://plato.stanford.edu/entries/qm-modal/.

London, F. and Bauer, E. (1939). *La théorie de l'observation en mécanique quantique*. Volume 775, Actualités Scientifiques et Industrielles, Exposés de Physique Générale. Hermann, Paris. Translated into English as 'The theory of observation in quantum mechanics' in J. A. Wheeler and W. H. Zurek (eds) (1983), *Quantum Theory of Measurement*, Princeton: Princeton University Press, pp. 217–259; see in particular 11, but also 13 and 14.

Lopreore, C. L. and Wyatt, R. E. (1999). Quantum Wave Packet Dynamics with Trajectories. *Phys. Rev. Lett.*, **82**, 5190–93.

Mehra, J. and Rechenberg, H. (1982). *The Historical Development of Quantum Theory, Vol. 4*. New York: Springer.

Mermin, N. D. (1989). What's Wrong with this Pillow? *Physics Today*, **42**, 9–11.

Mermin, N. D. (2012). Quantum mechanics: Fixing the shifty split. *Physics Today*, **65**, 8–10.

Nath, P. (2016). *Supersymmetry, Supergravity and Unification*. Cambridge: Cambridge University Press.

Norris, C. (2000). *Quantum Theory and the Flight from Realism: Philosophical Responses to Quantum Mechanics*. London: Routledge.

Omnès, R. (1988). Logical reformulation of quantum mechanics. *J. Stat. Phys.*, **53**. 'I: Foundations', 893–932; 'II: Interferences and the EPR experiments', 933–955; 'III: Classical limit and irreversibility', 957–975.

Omnès, R. (1994). *The Interpretation of Quantum Mechanics*. Princeton: Princeton University Press.

Omnès, R. (1999). *Understanding Quantum Mechanics*. Princeton: Princeton University Press.

Oriols, X. and Mompart, J. (2012). *Applied Bohmian Mechanics: From Nanoscale Systems to Cosmology*. Singapore: Pan Stanford Publishing Pte. Ltd.

Pearle, P. (1989). Combining stochastic dynamical state-vector reduction with spontaneous localization. *Phys. Rev. A*, **39**, 2277–89.

Peil, S. and Gabrielse, G. (1999). Observing the Quantum Limit of an Electron Cyclotron: QND Measurements of Quantum Jumps between Fock States. *Phys. Rev. Lett.*, **83**, 1287–90.

Penrose, R. (1989). *The Emperor's New Mind*. Oxford: Oxford University Press.

Penrose, R. (1994). *Shadows of the Mind*. Oxford: Oxford University Press.

Penrose, R. (1996). On Gravity's role in Quantum State Reduction. *General Relativity and Gravitation*, **28**, 581–00.

Peres, A. (1984). What is a state vector? *Am. J. Phys.*, **52**, 644–50.

Peres, A. (1993). *Quantum Theory: Concepts and Methods*. Dordrecht: Kluwer.

Peres, A. (2005). Einstein, Podolsky, Rosen, and Shannon. *Found. Phys.*, **35**, 511–14.

Peter, P., Pinho, E. J. C., and Pinto-Neto, N. (2006). Gravitational wave background in perfect fluid quantum cosmologies. *Phys. Rev. D*, **73**, 104017.

Peter, P., Pinho, E. J. C., and Pinto-Neto, N. (2005). Tensor perturbations in quantum cosmological backgrounds. *JCAP*, **7**, 014.

Petersen, A. (1963). The Philosophy of Niels Bohr. *Bulletin of the Atomic Scientists*, **XIX**, 8–14.

Pinho, E. J. C., and Pinto-Neto, N. (2007). Scalar and vector perturbations in quantum cosmological backgrounds. *Phys. Rev. D*, **76**, 023506.

Pitowsky, I. (2003). Betting on the outcomes of measurements: A Bayesian theory of quantum probability. *Studies in History and Philosophy of Modern Physics*, **34**, 395–14.

Polchinski, J. (1998). *String Theory*. Volume I and II. Cambridge: Cambridge University Press.

Pranger, R. J. (1933). Discussions about Language. Quoted in R. J. Pranger (1972), *Defense Implications of International Indeterminacy*, Washington, DC: American Enterprise Institute for Public Policy Research.

Prezhdo, O. V. and Brooksby, C. (2001). Quantum Backreaction through the Bohmian Particle. *Phys. Rev. Lett.*, **86**, 3215–19.

Rovelli, C. (2004). *Quantum Gravity*. Cambridge: Cambridge University Press.

Schlosshauer, M. (2005). Decoherence, the measurement problem, and interpretations of quantum mechanics. *Rev. Mod. Phys.*, **76**, 1267–1305.

Schrödinger, E. (1935). Die gegenwärtige Situation in der Quantenmechanik. *Naturwissenschaften*, **23**, 807–812, 823–828, 844–849.

Scully, M. O. (1998). Do Bohm Trajectories Always Provide a Trustworthy Physical Picture of Particle Motion? *Phys. Scripta*, T **76**, 41–46.

Shifren, L., Akis, R., and Ferry, D. K. (2000). Correspondence between quantum and classical motion: Comparing Bohmian mechanics with a smoothed effective potential approach. *Phys. Lett. A*, **274**, 75–83.

Shimony, A. (1993). *Search for a Naturalistic World View*. Volume II. Cambridge: Cambridge University Press.

Singh, T. P. (2015). Possible role of gravity in collapse of the wave-function: A brief survey of some ideas. *J. Physics: Conference Series*, **626**, 012009.

Stapp, H. P. (1972). The Copenhagen Interpretation. *American Journal of Physics*, **40**, 1098–1116.

Struyve, W. (2015). Semi-classical approximations based on Bohmian mechanics. *Int. J. Mod. Phys. A*, **35**(14). doi: 10.1142/S0217751X20500700.

Struyve, W. (2017). Towards a novel approach to semi-classical gravity. In K. Chamcham, J. Silk, J. D. Barrow, and S. Saunders (eds), *The Philosophy of Cosmology*, Cambridge: Cambridge University Press. arXiv:1902.02188 (2019).

Tastevin, G. and Laloë, F. (2018). Surrealistic Bohmian trajectories do not occur with macroscopic pointers. *Eur. Phys. J. D*, **72**, 183. arXiv:1802.03783 [quant-ph].

Tilloy, A. (2019). Does gravity have to be quantized? Lessons from non-relativistic toy models. *arXiv:1903.01823*.

Tilloy, A. and Diósi, L. (2016). Sourcing semiclassical gravity from spontaneously localized quantum matter. *Phys. Rev. D*, **93**, 024026.

Trimmer, J. D. (1980). The Present Situation in Quantum Mechanics: A Translation of Schrödinger's 'Cat Paradox' Paper. *Proc. Amer. Phil. Soc.*, **24**, 323–338. Also available in J.A. Wheeler and W.H. Zurek (eds) (1983), Quantum Theory of Measurement, Princeton: Princeton University Press, pp. 152–67.

Valentini, A. (2002). Signal-locality in hidden-variables theories. *Phys. Lett. A*, **297**, 273–78.

Valentini, A. (2009). Beyond the quantum. *Physics World*, **22**, 32–37.

van Fraassen, B. C. (1972). A formal approach to the philosophy of science. In R. Colodny (ed.), *Paradigms and Paradoxes: The Philosophical Challenge of the Quantum Domain*, Pittsburgh: University of Pittsburgh Press, pp. 303–366.

van Fraassen, B. C. (1974). The Einstein–Podolsky–Rosen paradox. *Synthese*, **29**, 291–309.

van Fraassen, B. C. (1991). *Quantum Mechanics: An Empiricist View*. Oxford: Clarendon Press.

von Neumann, J. (1932). *Mathematische Grundlagen Der Quantenmechanik*. Berlin: Springer. *Mathematical Foundations of Quantum Mechanics*. Princeton: Princeton University Press (1955), see in particular Chap. VI.

Wigner, E. P. (1961). Remarks on the mind-body question. In I. J. Good (ed.), *The Scientist Speculates*, London: Heinemann, pp. 284–302., Reprinted in E. P. Wigner (1967), *Symmetries and Reflections*, Indiana University Press, pp. 171–84.

Zeh, H. D. (1970). On the interpretation of measurement in quantum theory. *Found. Phys.*, **1**, 69–76.

Zeilinger, A. (1999). A foundational principle for quantum mechanics. *Foundations of Physics*, **29**, 631–43.

Zurek, W. H. (1981). Pointer basis of quantum apparatus: Into what mixture does the wave packet collapse? *Phys. Rev. D*, **24**, 1516–25.

Zurek, W. H. (1982). Environment-induced superselection rules. *Phys. Rev. D*, **26**, 1862–80.

Zurek, W. H. (2003a). Decoherence, einselection, and the quantum origins of the classical. *Rev. Mod. Phys.*, **75**, 715–75.

Zurek, W. H. (2003b). Environment-assisted invariance, entanglement, and probabilities in quantum physics. *Phys. Rev. Lett.*, **90**, 120404.

# CHAPTER 2

# PHILOSOPHICAL ISSUES RAISED BY QUANTUM THEORY AND ITS INTERPRETATIONS

### WAYNE C. MYRVOLD

## 2.1 INTRODUCTION

THE philosophical questions surrounding quantum theory revolve around the question: What, if anything, does the empirical success of quantum theory tell us about the physical world? Since the key papers formulating what we now call quantum mechanics were published in the years 1925–27, we are only a few years away from the centennial of the theory's inception. As we approach the centennial, there is more intense discussion than ever about its import.

Why is this? At the heart of the discussions is the following situation. We have in our textbooks, and teach to our students, what amounts to an operational recipe sufficiently precise for most applications. We learn to associate quantum states with various physical situations, and use them to calculate probabilities of outcomes of experiments. This is enormously important, as it is the basis both for the experimental testing of the theory and for its application. The success of the theory in these contexts is the reason we are taking it seriously at all. But the operational recipe does not, without further ado, yield anything like a clear description of what the physical systems to which it is applied are like, or what they are doing in between experiments.

The various approaches to the question of description of physical systems and processes (among which are those that hold that we should *refrain* from describing physical systems) are sometimes referred to as *interpretations* of quantum mechanics. This terminology is potentially misleading, for two reasons. The first is that it might suggest that the task of interpreting quantum theory is akin to supplying a model for an uninterpreted formal system. This is nothing at all like the task at hand. We don't have an uninterpreted formal system, or a mathematical theory devoid of physical significance. What we have is a formalism with an agreed-upon operational significance (or, at least, sufficiently close to agreed-upon for most applications). The question is what, if anything, is to be added to this operational core.

The second reason that the phrase 'interpretations of quantum mechanics' is potentially misleading is that some of the avenues of approach involve formulation of a physical theory distinct from standard quantum theory, in some cases differing in empirical content. These are not merely different interpretations of a common theory, but alternate physical theories.

# 2.2 What is a Quantum Theory?

In this section quantum theories are briefly described, with an emphasis on the agreed-upon operational core, which every interpretational project must take into account. Quantum theories can be expressed in a number of different mathematical forms that are equivalent as far as the operational core is concerned. To avoid the pitfall of tying interpretational matters too closely to any particular formulation, we focus on what all the formulations have in common. This means eschewing, in the first instance, talk of Hilbert spaces or wave functions. Readers who find the presentation disconcertingly unfamiliar may be reassured that these can be introduced when desired.

A *quantum-mechanical theory* is a quantum theory of a system (such as a finite number of particles) having finitely many degrees of freedom. A quantum theory of a system with infinitely many degrees of freedom is a *quantum field theory*. We use the term *quantum theory* to embrace both quantum-mechanical theories and quantum field theories.

## 2.2.1 Constructing Quantum Theories

To construct a quantum theory, one identifies a system or systems of interest, and the dynamical variables that are to be modelled. The system could, for example, be the familiar textbook example of a hydrogen atom, with the variables to be modelled being the positions and momenta of a proton and an electron, and perhaps also their spins.

It is a peculiarity of quantum theories that, in order to formulate one, we begin with a classical theory and subject it to a procedure known as *quantization*. This typically begins with a Lagrangian or Hamiltonian formulation of a classical theory. In these formulations, the configuration of a system is represented by variables $\{q_1,...,q_n\}$. These could, for example, be positions of a number of point particles, or they could specify the positions and angular orientations of a number of rigid bodies. One associates with each of these configuration variables $q_i$ its *conjugate momentum* $p_i$. The complete physical state of a system is given by a specification of the values of these *canonical variables*, $\{(q_i,p_i)|i = 1, ...,n\}$. The set of all possible states is called the *phase space* of the system. Any dynamical variable of the system is a function of the canonical variables $\{(q_i,p_i)\}$.

A quantum theory is constructed by taking the canonical variables of a system and associating with them elements of an algebra $\mathcal{A}_Q$ with a non-commutative multiplication.[1] We will refer to the elements of this algebra as *operators*. For any two operators $A$, $B$, we define the *commutator*,

$$[A, B] = AB - BA. \tag{1}$$

When $AB$ is equal to $BA$, $A$ and $B$ are said to *commute*.

The distinctive quantum relations are the *canonical commutation relations*. These specify commutators for the operators $\{(Q_i,P_i)\}$ that correspond to canonical variables $\{(q_i,p_i)\}$. The rules are:

- Operators corresponding to different degrees of freedom commute. This means that, for distinct $i$, $j$, $Q_i$ and $P_i$ commute with $Q_j$ and $P_j$.
- $[Q_i,P_i] = i\hbar\mathbb{1}$, where $\mathbb{1}$ is the identity operator and $\hbar = h/2\pi$, where $h$ is Planck's constant. The special operator $\mathbb{1}$ is the multiplicative identity; the result of multiplying any operator $A$ by $\mathbb{1}$ is just $A$ itself.

For any operator $A$, there is an operator $A^\dagger$, called the *adjoint* of $A$. These satisfy,

- $(A^\dagger)^\dagger = A$.
- For any operators $A$, $B$, and any complex numbers $a$, $b$,
  i) $(a A + b B)^\dagger = a^\star A\dagger + b^\star B^\dagger$.
  ii) $(AB)^\dagger = B^\dagger A^\dagger$.

An operator that is its own adjoint is said to be *self-adjoint*. We can associate with any operator a set of real or complex numbers called its *spectrum*. If the operator is self-adjoint, its spectrum consists of real numbers only.

We associate with any experiment a self-adjoint operator whose spectrum is the set of possible values of the outcome variable. For an experiment that, classically, would be regarded as a measurement of a given quantity that is a function of the canonical variables $\{(q_i,p_i)\}$, the associated operator is the corresponding function of the operators $\{(Q_i,P_i)\}$ (this does not yield a unique prescription, but this is not a matter we will go into in this chapter).

[1] It is an algebra over the complex numbers. This means that any operator can be multiplied by any complex number, that operators can be added and multiplied, and that multiplication distributes over addition.

We associate quantum states with preparation procedures that the system can be subjected to. A *quantum state* is an assignment $\rho$ of numbers to operators, required to satisfy the conditions,

- *Positivity.* For any $A$, $\rho(A^\dagger A)$ is a non-negative real number.
- *Normalization.* $\rho(\mathbb{1}) = 1$.
- *Linearity.* For any complex numbers $a$, $b$,

$$\rho(a\,A + b\,B) = a\,\rho(A) + b\,\rho(B).$$

For self-adjoint $A$, the value of $\rho(A)$ is to be interpreted as the expectation value, in state $\rho$, of the outcome of an experiment with which is associated the operator $A$. The positivity condition ensures that self-adjoint operators are assigned real numbers. This condition, together with the normalization condition, ensures that, if the spectrum of a self-adjoint operator $A$ is bounded, $\rho(A)$ is not above or below the bounds of the spectrum of $A$. This is required in order for these numbers to be interpreted as expectation values for the outcomes of experiments that yield results in the spectrum of $A$. The linearity condition is a nontrivial constraint, as it relates expectation values assigned to outcomes of incompatible experiments. It is a central principle of quantum mechanics, but is not something that is dictated by the operational significance of $\rho(A)$ as an expectation value of the outcomes of an experiment (one could imagine other, non-quantum theories that violate it).

For any state $\rho$ and any self-adjoint operator $A$, let $a$ be the expectation value of $A$ in state $\rho$, $\rho(A)$. We define the *variance* of $A$ in state $\rho$ as,

$$\mathrm{Var}_\rho(A) = \rho((A - a\,\mathbb{1})^2) = \rho(A^2) - \rho(A)^2. \tag{2}$$

This is one way to quantify the spread in the probability distribution of outcomes of an $A$ experiment. It is small if the distribution is tightly focused near the expectation value $\rho(A)$. It is zero only when there is a single outcome that will be obtained with probability one. If this is the case—that is, if $Var_\rho(A)$ is equal to zero—then $\rho$ is said to be an *eigenstate* of $A$, with *eigenvalue* $\rho(A)$. For such a state, one in which there is a definite value of the observable corresponding to $A$ that will with certainty be obtained as the outcome of an appropriate experiment, it is usual to ascribe the property of possessing this value to the system. This is known as the *eigenstate–eigenvalue link*. For example, a state that is an eigenstate of the operator corresponding to energy, with eigenvalue $E$, is taken to be a state in which the system has energy $E$. This has been, since the early days of quantum mechanics, a central interpretational principle of quantum theories.

Given any two states $\rho_1, \rho_2$, and any two positive numbers $p_1$ and $p_2$ that sum to one, we can always form the corresponding *mixture* of the states,

$$\bar{\rho} = p_1\,\rho_1 + p_2\,\rho_2. \tag{3}$$

An example of a preparation procedure with which a state like that would be associated is one that employs some randomizing device to choose between preparation of state $\rho_1$ and state $\rho_2$, with probabilities $p_1$ and $p_2$. Mixtures of more than two states are defined analogously. A state that is not a mixture of any two distinct states is called a *pure state*.

The content of the operational core of a quantum theory is completely encapsulated in the structure of the algebra $\mathcal{A}_Q$, the association of certain operators with experimental procedures, and of states with preparation procedures. However, for most purposes, this is not the most convenient formulation of the theory. It is often useful to construct a representation of the algebra as operators operating on vectors in a Hilbert space, in which any pure state $\psi$ can be represented by a vector $|\psi\rangle$. If we're willing to countenance Hilbert spaces that are more capacious than they need to be, we can construct a representation in which *every* state, pure or mixed, is represented by a state vector.

For any two distinct states $\rho_1, \rho_2$, represented by Hilbert space vectors $|\psi_1\rangle, |\psi_2\rangle$, and any complex numbers $a$, $b$, there is another state that is represented by the vector

$$|\phi\rangle = a|\psi_1\rangle + b|\psi_2\rangle. \tag{4}$$

Vector addition is also referred to as *superposition*, and $|\phi\rangle$ is said to be a superposition of $|\psi_1\rangle$ and $|\psi_2\rangle$. Resist the temptation to talk as if some states are superpositions, and some are not. Any vector $|\phi\rangle$ is equal to infinitely many superpositions of other vectors.

If $|\psi_1\rangle$ and $|\psi_2\rangle$ are eigenvectors of some operator $A$, with eigenvalues $a_1$ and $a_2$, respectively, then the eigenstate–eigenvalue link tells us that in states represented by those vectors, the system has the corresponding properties. If $a_1$ and $a_2$ are distinct, then $|\phi\rangle$, as defined by (4), is *not* an eigenstate of $A$, and the eigenstate–eigenvalue link is silent on whether we are to ascribe any properties corresponding to $A$ to the system in such a state. This is at the core of the so-called *measurement problem*, which will be presented in section 2.4.

In a quantum-mechanical theory of a system consisting of $n$ particles without spin, a quantum-mechanical state can be represented by a square-integrable function $\psi(x_1, x_2, \ldots, x_n)$, called a *wave function*. A wave function representing the state is not unique; any two functions that differ only on a set of measure zero represent the same state, and multiplying any wave function by any complex number yields another function that represents the same state. A wave function yields probabilities for outcomes of detection experiments as follows: the probability of finding particle 1 in a set $\Delta_1$, particle 2 in $\Delta_2$, etc., is given by the integral of $|\psi|^2$ over the region of configuration space with $x_1$ in $\Delta_1$ and $x_2$ in $\Delta_2$, etc., divided by the integral of $|\psi|^2$ over all of configuration space. For a system consisting of $n$ particles with spin, the total spin state of the system can be represented by a vector in a finite-dimensional Hilbert space $\mathcal{H}_S$. A wave function for such a system is an assignment of a vector in $\mathcal{H}_S$ to each point in configuration space.

## 2.2.2 Entangled and Unentangled States

Consider a quantum theory of two non-overlapping systems. The algebra of observables $\mathcal{A}_Q$ has two commuting subalgebras, $\mathcal{A}_A$ and $\mathcal{A}_B$, corresponding to the observables of the two subsystems. A state $\rho$ is a *product state* if any only if

$$\rho(AB) = \rho(A)\rho(B) \tag{5}$$

for all $A$ in $\mathcal{A}_A$ and $B$ in $\mathcal{A}_B$. A state that is either a product state or mixture of product states is called a *separable state*. A state (pure or mixed) that is not a separable state is an *entangled state*.

We can also characterize pure entangled states more directly. A pure state $\rho$ of $\mathcal{A}_Q$ is a product state if the restriction of $\rho$ to $\mathcal{A}_A$ is a pure state of $\mathcal{A}_A$; it is an entangled state if the restriction of $\rho$ to $\mathcal{A}_A$ is a mixed state of $\mathcal{A}_A$.

For any state that is not a product state, the state of a composite system is not uniquely determined by the states of the components, even if the state of the composite is pure. This is a striking difference between quantum and classical theories. For a classical theory, the restriction of any pure state—that is, a maximally specific state description—of a composite to one of its components is a pure state of the component, and specification of the states of the components uniquely determines the state of the composite. Following Howard (1985), this feature of classical theories has come to be known as *separability*, and the fact that it is not satisfied by quantum theories as *nonseparability*.

## 2.2.3 Temporal Evolution: Schrödinger and Heisenberg Pictures

Suppose that we have a system whose dynamical variables are $\{(q_i(t), p_i(t))\}$. To construct a quantum theory of the system, we require operators $\{(Q_i(t), P_i(t))\}$ to represent the dynamical variables.

The dynamical laws of our quantum theory specify how expectation values of variables at different times are related to each other. The basic equation of evolution is,

$$i\hbar \frac{d}{dt}\rho(A(t)) = \rho(A(t)H - HA(t)), \tag{6}$$

where $H$ is the operator corresponding to the system's Hamiltonian $H$.

Suppose, now, we want to construct a Hilbert space representation of our theory. This means assigning, to each operator $A(t)$, a Hilbert space operator $\hat{A}(t)$, and choosing, for each time $t$, a density operator $\hat{\rho}(t)$ to represent the state, in such a way that

$$\rho(A(t)) = \text{Tr}\,[\hat{\rho}(t)\hat{A}(t)]. \tag{7}$$

As we have to specify both $\hat{A}(t)$ and $\hat{\rho}(t)$, this gives us some leeway. One way to do this is to choose the same Hilbert space operators $(\hat{Q}_i, \hat{P}_i)$ to represent $(q_i(t), p_i(t))$ at all times. Then the density operators $\hat{\rho}(t)$ will have to satisfy,

$$i\hbar\frac{d}{dt}\hat{\rho}(t) = \hat{H}\hat{\rho}(t) - \hat{\rho}(t)\hat{H}. \tag{8}$$

For a pure state represented by a state vector $|\psi(t)\rangle$, we have,

$$i\hbar\frac{d}{dt}|\psi(t)\rangle = \hat{H}|\psi(t)\rangle. \tag{9}$$

This is the *Schrödinger equation*, and the choice of Hilbert space representation on which the operators representing $(q_i(t), p_i(t))$ are time-independent is called the *Schrödinger picture*.

Another choice is to choose a fixed density operator $\hat{\rho}$ to represent the state at any time. This requires the operators $\hat{A}(t)$ to satisfy

$$i\hbar\frac{d}{dt}\hat{A}(t) = \hat{A}(t)\hat{H} - \hat{H}\hat{A}(t). \tag{10}$$

This is the *Heisenberg equation of motion*, and the Hilbert space representation on which the density operator representing the state is time-independent is called the *Heisenberg picture*.

It should be emphasized that these two choices are two Hilbert space representations of *one and the same quantum theory*; the physical content is the same. It would be a mistake to take the time-independence of the Heisenberg-picture density operator to suggest that what is being modelled is an unchanging physical situation. What is changed is the location of the time-dependence in the mathematical apparatus used to model a changing physical situation.

In some cases a clear meaning can be given to the idea that quantities associated with a system at different times are values, at different times, of the 'same' dynamical variable. For example, in a theory set in Galilean spacetime one can choose a reference frame, and consider the position-coordinate of a particle, at different times, as differing values of the same dynamical variable. This, of course, requires a notion of trans-temporal identity of particles.

In other cases, it may be inconvenient to do so. Consider a classical field theory on Minkowski spacetime. A particular solution of the field equations will specify, for each spacetime point, a field value at that point. One can pick a set of timelike lines that jointly cover the spacetime (whether or not these are inertial trajectories), and consider how the field value changes with position on a line. If one is considering the response of some device that monitors the field, it makes sense to consider how the field variable changes along the worldline of the detector. But the structure of the theory may be more perspicuously represented without such considerations, and without identifying any two field values as values of the 'same' dynamical quantity at different times. For this reason, the Heisenberg picture is usually regarded as more suitable for relativistic quantum field theories.

# 2.3 THE COLLAPSE POSTULATE: SOME HISTORY

Textbook formulations of quantum mechanics usually include an additional postulate about how to assign a state vector after an experiment, according to which one replaces the quantum state with an eigenstate of the 'measured' observable, corresponding to the result obtained. Unlike the unitary evolution applied otherwise, this is a discontinuous change of the quantum state, sometimes referred to as *collapse of the state vector*, or *state vector reduction*. There are two interpretations of the postulate about collapse, corresponding to two different conceptions of quantum states. If a quantum state represents nothing more than our knowledge about the system, then the collapse of the state to one corresponding to the observed result can be thought of as representing nothing more than an updating of knowledge. If, however, quantum states represent physical reality, in such a way that distinct pure states always represent distinct physical states of affairs, then the collapse postulate entails an abrupt, perhaps discontinuous, change of the physical state of the system. Considerable confusion can arise if the two interpretations are conflated.

The collapse postulate occurs already in the general discussion at the fifth Solvay Conference in 1927 (see Bacciaguppi and Valentini, 2009, 437–450). It is also found already in Heisenberg's *The Physical Principles of the Quantum Theory*, based on lectures presented in 1929 (Heisenberg, 1930a, 27; 1930b, 36). Von Neumann, in his reformulation of quantum theory a few years later, distinguished between two types of processes: Process 1., which occurs upon performance of an experiment, and Process 2., the unitary evolution that takes place as long as no measurement is made (von Neumann, 1932; 1955, §V.I). He does not take this distinction to be a difference between two physically distinct processes. Rather, the invocation of one process or the other depends on a somewhat arbitrary division of the world into an observing part and an observed part (see von Neumann, 1932, 224; 1955, 420).

There is a persistent misconception that, for von Neumann, collapse is to be invoked only when a conscious observer becomes aware of the result. This is the opposite of his attitude; for him it is essential that the location of the boundary between the observed part of the world and the observing part is somewhat arbitrary. It may be placed between the system under study and the experimental apparatus. On the other hand, we could include the experimental apparatus in the quantum description, and place the cut at the moment when light indicating the result hits the observer's retina. Or we could go further, and include the retina and relevant parts of the observer's nervous system in the quantum system. That the cut may be pushed arbitrarily far into the perceptual apparatus of the observer is required, according to von Neumann, by the principle of psycho-physical parallelism.

The collapse postulate does not appear in the first edition (1930) of Dirac's *Principles of Quantum Mechanics*; it is introduced in the second edition (1935), which appeared subsequent to von Neumann's treatment. Dirac, in contrast to Heisenberg and von Neumann, takes the distinction between unitary and collapse evolution to be a distinction between two physical processes. Also, for Dirac it is an act of *measurement*, not *observation*, that causes a system to 'jump' into an eigenstate of the observable being measured (Dirac, 1935, 26). According to Dirac, this jump is caused by the interaction of the system with the experimental apparatus.

A formulation of a version of the collapse postulate according to which a measurement is not completed until the result is observed is found in London and Bauer (1939). For them, as for Heisenberg, this is a matter of an increase of knowledge on the part of the observer.

Wigner (1961) combined elements of the two interpretations. Like those who take the collapse to be a matter of updating of belief in light of information newly acquired by an observer, he takes collapse to take place when a conscious observer becomes aware of an experimental result. However, like Dirac, he takes it to be a real physical process. His conclusion is that consciousness has an influence on the physical world not captured by the laws of quantum mechanics. This involves a rejection of von Neumann's principle of psycho-physical parallelism, according to which it must be possible to treat the process of subjective perception as if it were a physical process like any other.[2]

## 2.4 THE SO-CALLED 'MEASUREMENT PROBLEM'

If it is possible for any quantum theory to be a comprehensive physical theory, it must be capable of treating of our experimental apparatus, and, indeed, everything else. Suppose, then, we analyse an experimental set-up quantum-mechanically.

Let $S$ be the system to be experimented on, which we call the *studied system*. We suppose that it has at least two distinguishable states, $|+\rangle_S$ and $|-\rangle_S$, and that an apparatus, $A$, can be devised that distinguishes these states. This means that there are

---

[2] Despite this, Wigner's proposal is sometimes wrongly attributed to von Neumann, and is sometimes called the 'von Neumann–Wigner interpretation.'

distinguishable sets $A^+$ and $A^-$ of states of the apparatus, which we will call *indicator states*, such that the apparatus can be coupled to the system $S$ in such a way that, if the apparatus is started out in a ready state and the studied system in the state $|+\rangle_S$, the apparatus will evolve to a state in $A^+$, and, if the apparatus is started out in a ready state and the studied system in the state $|-\rangle_S$, the apparatus will evolve to a state in $A^-$. We do not assume that the apparatus is isolated from its environment, so we include the relevant parts of the environment in our description.

The evolution of the composite system should satisfy,

$$|+\rangle_S|R\rangle_{AE} \Rightarrow |'+'\rangle_{SAE};$$
$$|-\rangle_S|R\rangle_{AE} \Rightarrow |'-'\rangle_{SAE}; \tag{11}$$

where $|R\rangle_{AE}$ is a state of the apparatus plus its environment in which the apparatus is ready to perform an experiment, and $|'+'\rangle_{SAE}$ and $|'-'\rangle_{SAE}$ are states of the composite system in which the apparatus is indicating the results $+$ and $-$, respectively.

Suppose, now, that the system $A$ is started out in a state that is a nontrivial superposition of $|+\rangle_S$ and $|-\rangle_S$,

$$a|+\rangle_S + b|-\rangle_S,$$

with $a$ and $b$ both nonzero. If the evolution of the composite system is linear, as it must be if the Schrödinger equation applies, then we must have,

$$(a|+\rangle_S + b|-\rangle_S)|R\rangle_{AE} \Rightarrow a|'+'\rangle_{SAE} + b|'-'\rangle_{SAE}. \tag{12}$$

What to make of a state like this? It is not a state in which the apparatus is in either one of the indicator sets $A^+$ or $A^-$. It is not an eigenstate of the apparatus indicator variable, but a superposition of distinct eigenstates. The eigenstate–eigenvalue link, therefore, offers no guidance. The interaction of the apparatus with its environment may result in an entangled state of the apparatus and its environment that is such that the state of the apparatus is a mixture of distinct indicator states—a process known as *decoherence*— but this does not help with the interpretational issue, as the system containing the apparatus and a sufficiently wide portion of its environment is still a superposition of macroscopically distinct states. It is often said that a state like (12) conflicts with our experience, according to which experimental apparatus is, at the end of an experiment, always indicating some determinate result. This is highly misleading, because the real issue is whether we can make sense of (12) as a possible state of a system containing the experimental apparatus. If we don't know what it would be like to find the apparatus in such a state, then it makes no sense either to affirm or to deny that we have ever found the apparatus in such a state.

Though we have chosen an experimental set-up as an illustration, situations in which linear evolution would yield superpositions of macroscopically distinct states are ubiquitous. Nonetheless, the problem of what to make of this fact—that applying linear evolution to quantum states involving macroscopic objects will lead to superpositions of macroscopically distinct states—has come to be known as the *measurement problem*.

If there is a unique outcome of the experiment, and if (12) is the correct quantum state, then the outcome fails to be represented by the quantum state, which must be supplemented by something that *does* indicate the outcome. On the other hand, it might be that neither Schrödinger evolution nor any other linear evolution applies to situations like the one envisaged, and that the correct evolution leads to a state that we *can* take to be indicating a determinate outcome. These two options were summarized by J. S. Bell in his remark, 'Either the wavefunction, as given by the Schrödinger equation, is not everything, or it is not right' (Bell, 1987a, 41, 1987b and 2004, 201). This gives us a (*prima facie*) neat way of classifying approaches to the so-called 'measurement problem'.

- There are approaches that involve a denial that a quantum wave function (or any other way of representing a quantum state) yields a complete description of a physical system.
- There are approaches that involve modification of the dynamics to produce a collapse of the quantum state in appropriate circumstances.
- There are approaches that reject both horns of Bell's dilemma, and hold that quantum states undergo unitary evolution at all times and that there is no more to be said about the physical state of a system than can be represented by a quantum state.

We include in the first category approaches that deny that a quantum state should be thought of as representing anything in physical reality at all. If quantum states do not represent anything, and if there is something rather than nothing, then quantum states do not represent everything. In this category are Bohrian approaches, according to which there are principled reasons not to seek a complete description, and Einsteinian approaches, according to which seeking a theory that need not leave anything out in its descriptions is a project worthy of pursuit.

Also included in the first category are approaches that take quantum states to represent something, but not everything, in physical reality. These include 'hidden-variables' theories, and modal interpretations (see Lombardi and Dieks, 2017). The best-known and most thoroughly worked-out theory of this sort is the de Broglie–Bohm pilot wave theory, which takes particles with definite trajectories as the basic ontology. The role of the wave function is to provide dynamics for the particles. See Bacciagaluppi and Valentini (2009) for a historical introduction, and Dürr, Goldstein, and Zanghì (1992) and Pearle and Valentini (2006) for current perspectives.

The second category embraces the dynamical collapse theory programme, which seeks a modified dynamics that approximates unitary evolution in the domains in which we have good evidence for its correctness, and approximates collapse in other situations, including, but not limited to, experimental set-ups. The best-known version of this is the Ghirardi–Rimini–Weber (GRW) theory (Ghirardi, Rimini, and Weber, 1986), referred to by its creators as *Quantum Mechanics with Spontaneous Localization* (QMSL). On this theory, Schrödinger evolution of the quantum state is punctuated by discontinuous jumps. The GRW theory has the defect that it does not respect the symmetrization/antisymmetrization requirements for states of a system containing identical particles. This is remedied in a successor theory, the *Continuous Spontaneous Localization* (CSL) theory (Pearle, 1989; Ghirardi, Pearle, and Rimini, 1990).

Approaches that reject both horns of Bell's dilemma are typified by Everettian, or 'many-worlds' interpretations. The basic idea is denial that there is a unique experimental outcome; rather, there is a splitting, and different results obtain on different branches of the multiverse. These have their roots in the work of Hugh Everett III (see Barrett and Byrne, 2012). See Saunders *et al.* (2010), Wallace (2012), and Carroll and Singh (2019) for some recent approaches along these lines.

An approach that does not fit neatly into these categories is the *relational interpretation* advocated by Carlo Rovelli. It is akin in some ways to Everett's original conception, which he called the *relative-state* interpretation. It differs from it in not taking quantum states to be representational. For more on this, see Rovelli's contribution to this volume, and also Laudisa and Rovelli (2019), and references therein.

# 2.5 BELL'S THEOREM, NONLOCALITY, AND RELATIVITY

'Bell's theorem' is a term used for a family of theorems of the following form. From a condition on probabilities that is motivated, in part, by locality considerations, an inequality is derived constraining correlations between results of spatially separated experiments, which is violated by the predictions of quantum mechanics. See the *Stanford Encyclopedia of Philosophy* entry (Myrvold, Genovese, and Shimony, 2019) for an overview more detailed than found here.

The distinctive condition needed to derive Bell-type inequalities is the condition that correlations between outcomes of spatially separated systems be *locally explicable*. This has two parts: that the correlations be explicable in a certain sense, and the explanation be local. The explicability condition was taken by Bell to be the condition that correlations between events that are not in a direct cause-effect relation with each other be attributable to a common cause (see Bell 1981, C2–55; 1987b and 2004, 152). This is a version of a principle that has been called by Reichenbach (1956, §56) the *Principle of the Common Cause*, and, though Reichenbach made no pretence to originality, has been called *Reichenbach's Common Cause Principle*. The Common Cause Principle, together with the assumption that experimental outcomes at spacelike separation are not in a direct cause-effect relation with each other, yields the condition that Shimony (1986; 1990) has called *outcome independence*. Causal locality requires that a choice of experiment made at one location does not affect the probabilities of outcomes of another experiment performed at spacelike separation. This condition is called *parameter independence*. In some of his writings, Bell combined locality and causality considerations in a principle he called 'The Principle of Local Causality' (Bell, 1976; 1990).

In addition to this condition of local explicability, there are supplementary assumptions of the sort taken for granted in all experimental science, such as the assumption that it is possible, via some randomizing procedure, to render one's choice of experiment statistically independent of the state of the system on which the experiment is done. This condition is referred to as a 'no-conspiracies' assumption, or 'measurement independence', or, in some of the recent literature, 'statistical independence'. Though the assumption has been denied by some, we will in what follows restrict our attention to views consistent with this assumption.

Violation of Bell inequalities has been abundantly confirmed by experiment. What does this tell us?

It is sometimes said that violation of Bell inequalities straightforwardly entails violation of relativistic causality. Things are not so simple, as there is no interpretation-independent answer to the question of compatibility with relativity.

The question is most straightforward in connection with hidden-variables theories such as the de Broglie–Bohm theory. Any deterministic theory that violates Bell inequalities must violate parameter independence, and thus must have cause-effect dependencies between spacelike separated events.

Because, in a multi-particle system, the velocity of each particle may depend on the positions of all the others, the de Broglie–Bohm theory requires a preferred relation of distant simultaneity for its formulation. There is a series of theorems that show that any theory of this sort, on which the quantum state is supplemented by extra variables that are required to have a probability distribution given by the Born rule, must employ a distinguished relation of distant simultaneity, as it is not possible to satisfy the postulate about probabilities on arbitrary spacelike hypersurfaces. See Berndl *et al.* (1996); Dickson and Clifton (1998); Arnztenius (1998); and Myrvold (2002, 2009). This has the consequence that such theories require a dynamically distinguished relation of distant simultaneity; see Myrvold (2021, §5.5.1) for the argument.

Dynamical collapse theories, on the other hand, do not require a preferred relation of distant simultaneity for their formulation. There is an extension of the GRW theory to a relativistic context (Dove, 1996; Dove and Squires, 1996; Tumulka, 2006), which involves a fixed, finite number of noninteracting particles. A generalized version, for a finite number of interacting distinguishable particles, is presented by Tumulka (2021). There are also extensions of the CSL theory to the context of relativistic quantum field theories (Bedingham, 2011a,b; Pearle, 2015).

These theories involve probabilistic correlations between spacelike separated events that are *not* attributable to events in their common past. That is, they involve a rejection of the Reichenbach Common Cause Principle. The question of whether a theory such as this is in violation of any restriction on causal relations that is motivated by considerations of special relativity has been a hotly debated one. Several authors have argued over the years, in different ways, in favour of the compatibility of theories like that with the requirements of special relativity; these include Shimony (1978; 1984; 1986), Jarrett (1984), Skyrms (1984), Redhead (1987), Ghirardi and Grassi (1996), and Ghirardi (2012). See Myrvold (2016) for a recent argument for compatibility of special relativity with violations of Bell inequalities.

# 2.6 Ontological Questions Concerning Quantum States

## 2.6.1 The Question of Quantum State Realism

We have introduced quantum states via their operational significance: they encode probabilities of outcomes of experiments. Should we think of them as representing some feature of the system to which they are ascribed?

Positions that deny that quantum states represent features of physical reality have a history as old as quantum theory itself. This is one thing that Bohr and Einstein agreed upon. For Bohr, all description of physical reality must be couched in classical terms, and the limits of classical physics are the limits of physical description; quantum wave functions have only 'symbolic' status (see Bohr, 1934, 17). Einstein argued, in several places (see, e.g., Einstein, 1936), that quantum states should be regarded as akin to the probability distributions of classical statistical mechanics, that is, as representing incomplete knowledge of some deeper underlying physical state. The chief locus of difference between the two had to do with the propriety of seeking a deeper level of description.

The idea that quantum states are like that is an attractive one. It faces considerable obstacles, and it should be non-controversial that quantum states are not *just* like classical probability distributions.

A useful way of sharpening the question of realism about quantum states is afforded by the framework constructed by Harrigan and Spekkens (2010). This framework makes explicit some principles that are deeply embedded in our reasoning about the world.

Suppose that Alice has a choice of two or more preparation procedures that she can subject a system to. Having made the choice, she passes the system on to Bob, who can do an experiment on the system, and, from the outcome, reliably identify the procedure Alice has chosen. We would take this as an indication that distinct choices of preparation on Alice's part result in physical differences in the system being prepared, and that the outcome of Bob's experiment is informative about these differences.

Suppose, now, that we begin to consider what a theoretical model of a set-up like this would look like. This would involve some set $\Lambda$ of possible physical states of the system (Harrigan and Spekkens call this the *ontic state space*). We associate with a preparation procedure $\psi$ a probability distribution $P_\psi$ over appropriate subsets of $\Lambda$. Suppose that Bob's experiment has potential outcomes $\{o_1,\ldots,o_k\}$. We associate with Bob's experiment *response probability functions* $f_k(\lambda)$. The function $f_k(\lambda)$ yields the probability that the $k$th outcome of the experiment is obtained, if the state of the system experimented on is $\lambda$.

We say that the preparations $\psi$ and $\phi$ are *distinguishable* if there is an experiment whose outcome discriminates between them with certainty. If $\psi$ and $\phi$ are distinguishable, then (as one would expect) there is no set of states that has nonzero probability of being realized on both preparations. Call a pair of preparations $\psi$, $\phi$ *ontically distinct* if there is no subset of the state space $\Lambda$ that has nonzero probability of being realized on both preparations.

Preparations corresponding to distinguishable quantum states must be ontically distinct. The question to be addressed is whether, for all pairs of distinct pure quantum states, including those that are not distinguishable, the corresponding preparations are ontically distinct. Harrigan and Spekkens say that a theory is $\psi$-*ontic* if, according to the theory, preparations corresponding to distinct pure quantum states are always ontically distinct. They define $\psi$-*epistemic* as the negation of $\psi$-ontic. This is potentially misleading terminology. Consider, for example, a classical system, whose ontic state is represented by a point in its phase space. Suppose that one could learn either its position, or its momentum, but not both, though it always *has* determinate position and momentum. Any position is compatible with any momentum, and hence, for any position $x$ and momentum $p$, the set of ontic states corresponding to position $x$ overlaps with the set of states corresponding to momentum $p$. That doesn't mean that there is anything epistemic about position or momentum. Furthermore, to call a model '$\psi$-epistemic' if there are distinct pure quantum states whose associated probability distributions have *some* overlap, no matter how small, is potentially misleading, as it might suggest that the goal of constructing an interpretation on which quantum states are like classical probability distributions has been achieved. This, however, would require that the model be what has been called a *maximally $\psi$-epistemic* model (Barrett *et al.*, 2014). On such a model, the indistinguishability of quantum states is fully explained by overlap of the corresponding probability distributions on ontic state space.

There are a number of theorems concerning the viability of the programme of constructing a theory that is maximally $\psi$-epistemic, or, failing that, a theory that is not $\psi$-ontic. In particular, Barrett *et al.* (2014) show that no theory that reproduces quantum probabilities for outcomes of experiments and fits into the framework just sketched can be maximally $\psi$-epistemic, or even come close to being so. Pusey, Barrett, and Rudolph (PBR) show that, provided that the theory satisfies a postulate called the *Preparation Independence Postulate*, it must be $\psi$-ontic in order to reproduce quantum probabilities for outcomes of experiments (Pusey, Barrett, and Rudolph, 2012).

The Preparation Independence Postulate is a postulate to the effect that it is possible to subject a pair of distinct systems $A$ and $B$ to preparation procedures that render their ontic states probabilistically independent of each other. This postulate involves an assumption, called the *Cartesian Product Assumption*, to the effect that, for a preparation of that sort, the state of the composite system $AB$ can be fully represented by specifying a state of $A$ and a state of $B$. This is a nontrivial restriction on the state spaces employed in the theory. A weaker assumption, called the *Preparation Uninformativeness Condition*, which makes no assumptions about the structure of the state spaces, was suggested by Myrvold (2018c, 2020). On the basis of this weaker assumption, a weaker conclusion can be derived. The conclusion that is derived from this condition is that, on any theory that satisfies it, pure quantum states $|\psi\rangle$ and $|\phi\rangle$ that are not too close to each other are ontically distinct. Here, the condition of not being too close is that the absolute value of their inner product be less than $1/\sqrt{2}$.

## 2.6.2 The Ontological Status of Quantum States

Suppose that we are realists about quantum states. This means that distinct pure quantum states represent physically distinct states of affairs. This still leaves open the question of what sorts of physical reality these states represent. In this section we briefly discuss some options.

### 2.6.2.1 *Quantum State Monism*

Could there be nothing more to the world than what is represented by a quantum state?

Recall that a quantum theory is not an uninterpreted formalism. A quantum theory involves an identification of physical quantities to be represented, and an association of operators with those quantities. The eigenstate–eigenvalue link yields property attributions in the special case of eigenstates. If we had a dynamical collapse theory that produced eigenstates of the right sorts of dynamical quantities—if, for example, it yielded definite mass or energy content for regions of space that are small on the macroscopic scale—then such a theory could, in a straightforward way, be a quantum state monist theory. Sometimes scepticism is expressed about this, but this scepticism seems aimed at a different project, a project that would involve starting with a mathematical formalism devoid of physical interpretation and attempting to interpret it physically.

Things are not so simple, because dynamical collapse theories do not produce eigenstates of appropriate physical quantities, and there are principled reasons for not expecting a dynamical collapse theory to do that. For this reason, Ghirardi and collaborators proposed a weakening of the eigenstate–eigenvalue link, according to which a system is to be ascribed a property if its quantum state is *sufficiently close* to being an eigenstate of the corresponding operator (Ghirardi, Grassi, and Pearle, 1990, 1298; see also Ghirardi, Grassi, and Benatti, 1995, 13). This modification has been dubbed, by Clifton and Monton (1999), the *fuzzy link*.[3] For a defence of quantum state monism along the lines proposed by the originators of the GRW and CSL theories, see Myrvold (2018a, 2019).

Everettian theories seem to be best interpreted along these lines. Since such theories eschew collapse, on such a theory a quantum state will not typically be anywhere near an eigenstate of familiar macroscopic variables. However, the quantum state of any bounded region of spacetime will typically be a *mixture* of states in which macroscopic variables are near-definite, and the terms of these mixtures will evolve independently, and can be taken to represent quasi-classical domains.

### 2.6.2.2 *The Project Known as 'Wave Function Realism'*

Consider again a quantum theory of $n$ distinguishable particles, and suppose that we choose to represent quantum states in this theory by wave functions on the configuration space of the particles (recall from section 2.2.1 that this is optional). The wave function representation is not unique; a wave function representing a quantum state $\rho$ is represented by a class of functions, not a single one; two functions that differ only on a set of measure zero, or differ only by a multiplicative constant, represent the same quantum state.

When de Broglie introduced the precursors of our quantum-mechanical wave functions, the thought was that these would be akin to electromagnetic fields. A stumbling block for an interpretation of this sort is that multi-particle wave functions are functions of $n$ points in space, and have to be, to encode correlations between the positions of particles. That is, they are what Belot (2012) has called 'multi-fields'.

Suppose, however, one wanted—perhaps out of a commitment to separability—to construct an alternative to standard quantum theory according to which the basic ontology consisted of a specification of local conditions at every point of some fundamental space (see Albert, 1996). This is a project that has come to be known as *wave function realism*—a misleading terminology, as every form of realism about quantum states will maintain that wave functions, as one way of representing quantum states, represent something physically real.

The basic idea is that quantum theories are to be embedded in a more encompassing framework that would include theories with no relation to quantum theories as we have characterized them. We are to imagine some class of dynamical laws for fields in this framework that are such that for some—but not all—choices of dynamics the evolution of the field on a $3n$-dimensional fundamental space will mimic the evolution of a wave

---

[3] Peter Lewis (2016, 86–90) distinguishes between a *fuzzy link*, according to which there is some precise threshold $p$ such that a system possesses the property $A = a$ if and only if the probability of finding some other value is less than $p$, and a *vague link* according to which possession of a definite property is a matter of degree. It is hard to imagine what arguments there could be (or even what it might mean) for there to be a precise threshold. Albert and Loewer (1996) argue, correctly in my opinion, that there could be no such precise threshold, and that the modified link must be somewhat vague. This is what I mean by a fuzzy link.

function defined at $n$ points in a 3-dimensional space; the theory restricted to evolutions of the sort will then be functionally equivalent to a quantum theory.

I describe this as a *project* because those engaged in it have not yet said in any detail what the field on the fundamental space is thought to be, except in the case of the nonrelativistic theory of $n$ distinguishable particles. One can perhaps dimly see how the case of particles with spin is to be handled, and even more dimly, that of a quantum field theory (see Myrvold 2015 for some options). It is also unclear what sort of structure is presumed for the fundamental space: does it have some built-in metric or causal structure, or is this to be regarded as emergent also?

The project should be regarded as a work-in-progress. See Ney and Albert (2013) for a collection of essays connected with this project. Among the matters that require clarification are the goals and motivations of the project. For some discussion of this, see Ney (2019, 2021).

### 2.6.2.3 *Primitive Ontology, and the Nomic View of Quantum States*

As mentioned above, the originators of dynamical collapse theories originally advocated a quantum state monist ontology, with a modified version of the eigenstate–eigenvalue link. In recent years this proposal has somewhat fallen out of favour, to be replaced with a 'primitive ontology' approach on which a theory must posit some basic ontology, which is the stuff of which ordinary objects are made, and the role of quantum states is to provide dynamics for that ontology. On this view, what quantum states represent physically is more like a dynamical law than 'stuff' as usually construed. See Allori *et al.* (2008, 2014) and Allori (2013) for exposition and defence of this view. This sort of conception applies most straightforwardly to the de Broglie–Bohm theory, on which the primitive ontology is particles with definite trajectories, and the wave function is a guiding wave.

One advantage of this view is that it makes transparent how quantum wave functions should transform under dynamical symmetries. If one starts with the primitive ontology, and takes the role of the wave function to provide dynamics for it, then the condition that the set of dynamically possible trajectories be invariant under a symmetry operation yields guidance as to how the mathematical representation of the wave function should change under symmetry transformations.

It should be emphasized that the nomic view of quantum states is a *realist* conception of quantum states, in the sense used in section 2.6.1. Preparation procedures have an influence on quantum states, and there is a matter of fact about *which* quantum state has been prepared. Where the nomic view differs from other conceptions of quantum states has to do with what sort of matter of fact it is.

### 2.6.2.4 *A Comment on 'Spacetime State Realism'*

Wallace and Timpson (2010) introduce the term 'Spacetime State Realism', which they characterize as 'a view which takes the states associated to spacetime regions as fundamental'. This strikes me as misleading, as the terminology suggests that what is being proposed is some alternative to other sorts of realism about quantum states.

But recall that a quantum state is an assignment of expectation values to operators representing physical quantities. If the basic quantities are taken to be ones that pertain to bounded spacetime regions (which are the ones that are relevant to outcomes of experiments, as these take place within bounded spacetime regions), then for any spacetime region there is an associated algebra, and hence an associated state. This is explicit in the algebraic formulation of quantum field theory, but any presentation of quantum theory will have to make sense of 'observables' associated with bounded spacetime regions. Spacetime state realism, then, is simply realism about quantum states, and a view that takes states associated to spacetime regions as fundamental is simply a view that takes quantum states as fundamental.

# 2.7 CONCLUSION

This has by no means been an exhaustive overview of the philosophical discussions surrounding quantum mechanics. Among topics not touched upon are interpretational and conceptual issues peculiar to quantum field theories. We have said little about the arguments offered by those who reject realism about quantum states in favour of such a position. Also omitted is the project of reconstructing quantum theory on operational or information-theoretical principles, and any detailed discussion of the classical-quantum interface. The *Stanford Encyclopedia of Philosophy* article, 'Philosophical Issues in Quantum Theory' (Myrvold, 2018b) provides some pointers to the relevant literature on these topics.

# REFERENCES

Albert, D. and Loewer, B. (1996). Tails of Schrödinger's cat. In R. K. Clifton (ed.), *Perspectives on Quantum Reality*, Dordrecht: Kluwer Academic Publishers, pp. 81–92.

Albert, D. Z. (1996). Elementary quantum metaphysics. In J. T. Cushing, A. Fine, and S. Goldstein (eds), *Bohmian Mechanics and Quantum Mechanics: An Appraisal*, Dordrecht: Kluwer, pp. 277–284.

Allori, V. (2013). Primitive ontology and the structure of fundamental physical theories. In A. Ney and D. Z. Albert (eds), *The Wave Function: Essays on the Metaphysics of Quantum Mechanics*, New York: Oxford University Press, pp. 58–75.

Allori, V., Goldstein, S., Tumulka, R., and Zanghì, N. (2008). On the common structure of Bohmian mechanics and the Ghirardi–Rimini–Weber theory. *The British Journal for the Philosophy of Science*, **59**, 353–389.

Allori, V., Goldstein, S., Tumulka, R., and Zanghì, N. (2014). Predictions and primitive ontology in quantum foundations: A study of examples. *The British Journal for the Philosophy of Science*, **65**, 323–352.

Arnztenius, F. (1998). Curiouser and curiouser: A personal evaluation of modal intepretations. In D. Dieks and P. E. Vermaas (eds), *The Modal Interpretation of Quantum*

*Mechanics*, The Western Ontario Series in Philosophy of Science, Volume 60. Dordrecht: Springer, pp. 337–377.

Bacciagaluppi, G. and Valentini, A. (2009). *Quantum Theory at the Crossroads: Reconsidering the 1927 Solvay Conference*. Cambridge: Cambridge University Press.

Barrett, J., Cavalcanti, E. G., Lal, R., and Maroney, O. J. E. (2014). No $\psi$-epistemic model can fully explain the indistinguishability of quantum states. *Physical Review Letters*, **112**, 250403.

Barrett, J. A. and Byrne, P. (eds) (2012). *The Everett Interpretation of Quantum Mechanics: Collected Works 1955–1980 with Commentary*. Princeton: Princeton University Press.

Bedingham, D. (2011a). Relativistic state reduction model. *Journal of Physics: Conference series*, **306**, 012034.

Bedingham, D. (2011b). Relativistic state reduction dynamics. *Foundations of Physics*, **41**, 686–704.

Bell, J. S. (1976). The theory of local beables. *Epistemological Letters*, **9**, 11–24. Reprinted in *Dialectica*, **39** (1985), 86–96, and in Bell (1987b, 2004), 52–62.

Bell, J. S. (1981). Bertlmann's socks and the nature of reality. *Journal de Physique*, **42**, 611–37. Reprinted in Bell (1987b, 2004), 139–158.

Bell, J. S. (1987a). Are there quantum jumps? In C. Kilmister (ed.), *Schrödinger: Centenary celebration of a polymath*, Cambridge: Cambridge University Press, pp. 41–52. Reprinted in Bell (1987b, 2004), pp. 201–212.

Bell, J. S. (1987b). *Speakable and Unspeakable in Quantum Mechanics*. Cambridge: Cambridge University Press.

Bell, J. S. (1990). La nouvelle cuisine. In A. Sarlemijn and P. Kroes (eds), *Between Science and Technology*, Amsterdam: North-Holland. Reprinted in Bell (2004), pp. 232–48.

Bell, J. S. (2004). *Speakable and Unspeakable in Quantum Mechanics*, 2nd edn. Cambridge: Cambridge University Press.

Belot, G. (2012). Quantum states for primitive ontologists: A case study. *European Journal for Philosophy of Science*, **2**, 67–83.

Berndl, K., Dürr, D., Goldstein, S., and Zanghì, N. (1996). Nonlocality, Lorentz invariance, and Bohmian quantum theory. *Physical Review A*, **53**, 2062–2073.

Bohr, N. (1934). Introductory survey (1929). In N. Bohr, *Atomic Theory and the Description of Nature*, Cambridge: Cambridge University Press, pp. 1–24.

Carroll, S. M. and Singh, A. (2019). Mad-dog Everettianism: Quantum mechanics at its most minimal. In A. Aguirre, B. Foster, and Z. Merali (eds), *What is Fundamental?*, Cham, Switerland: Springer, pp. 95–104.

Clifton, R. and Monton, B. (1999). Losing your marbles in wavefunction collapse theories. *The British Journal for the Philosophy of Science*, **50**, 697–717.

Cushing, J. T., Fine, A., and Goldstein, S. (eds) (1996). *Bohmian Mechanics and Quantum Mechanics: An Appraisal*. Dordrecht: Kluwer.

Dickson, M. and Clifton, R. (1998). Lorentz-invariance in modal interpretations. In D. Dieks and P. E. Vermaas (eds), *The Modal Interpretation of Quantum Mechanics*, The Western Ontario Series in Philosophy of Science, Volume 60. Dordrecht: Springer, pp. 9–47.

Dieks, D. and Vermaas, P. E. (eds) (1998). *The Modal Interpretation of Quantum Mechanics*. The Western Ontario Series in Philosophy of Science, Volume 60. Dordrecht: Kluwer Academic Publishers.

Dirac, P. A. M. (1935). *Principles of Quantum Mechanics*, 2nd edn. Oxford: Oxford University Press.

Dove, C. and Squires, E. J. (1996). *A local model of explicit wavefunction collapse*. arXiv:quant-ph/9605047v1.

Dove, C. J. (1996). *Explicit wavefunction collapse and quantum measurement*. Ph. D. thesis, Durham University.

Dürr, D., Goldstein, S., and Zanghì, N. (1992). Quantum equilibrium and the origin of absolute uncertainty. *Journal of Statistical Physics*, **67**, 843–907.

Einstein, A. (1936). Physik und Realität. *Journal of the Franklin Institute*, **221**, 349–382. English translation in Einstein (1954).

Einstein, A. (1954). Physics and reality. In A. Einstein, *Ideas and Opinions*, New York: Crown Publishers, Inc., pp. 290–323. Translation of Einstein (1936).

Ghirardi, G. (2012). Does quantum nonlocality irremediably conflict with special relativity? *Foundations of Physics*, **40**, 1379–1395.

Ghirardi, G. and Grassi, R. (1996). Bohm's theory versus dynamical reduction. In J. T. Cushing, A. Fine, and S. Goldstein (eds), *Bohmian Mechanics and Quantum Mechanics: An Appraisal*, Dordrecht: Kluwer, pp. 353–377.

Ghirardi, G., Grassi, R., and Pearle, P. (1990). Relativistic dynamical reduction models: General framework and examples. *Foundations of Physics*, **20**, 1271–1316.

Ghirardi, G. C., Grassi, R., and Benatti, F. (1995). Describing the macroscopic world: Closing the circle within the dynamical reduction program. *Foundations of Physics*, **25**, 5–38.

Ghirardi, G. C., Pearle, P., and Rimini, A. (1990). Markov processes in Hilbert space and continuous spontaneous localization of systems of identical particles. *Physical Review A*, **42**, 78–89.

Ghirardi, G. C., Rimini, A., and Weber, T. (1986). Unified dynamics for microscopic and macroscopic systems. *Physical Review D*, **34**, 470–491.

Harrigan, N. and Spekkens, R. W. (2010). Einstein, incompleteness, and the epistemic view of quantum states. *Foundations of Physics*, **40**, 125–157.

Heisenberg, W. (1930a). *Die Physikalische Prinzipien der Quantentheorie*. Leipzig: Verlag von S. Hirzel. English translation in Heisenberg (1930b).

Heisenberg, W. (1930b). *The Physical Principles of the Quantum Theory*. Chicago: University of Chicago Press. Translation, by Carl Eckart and Frank C. Hoyt, of Heisenberg (1930a).

Howard, D. (1985). Einstein on locality and separability. *Studies in History and Philosophy of Science*, **16**, 171–201.

Jarrett, J. (1984). On the physical significance of the locality conditions in the Bell arguments. *Noûs*, **18**, 569–589.

Laudisa, F. and Rovelli, C. (2019). Relational quantum mechanics. In E. N. Zalta (ed.), *The Stanford Encyclopedia of Philosophy* (Winter 2019 edn), Stanford, CA: Metaphysics Research Lab, Stanford University.

Lewis, P. J. (2016). *Quantum Ontology: A Guide to the Metaphysics of Quantum Mechanics*. Oxford: Oxford University Press.

Lombardi, O. and Dieks, D. (2017). Modal interpretations of quantum mechanics. In E. N. Zalta (ed.), *The Stanford Encyclopedia of Philosophy* (Spring 2017 edn), Stanford, CA: Metaphysics Research Lab, Stanford University.

Lombardi, O., Fortin, S., López, C., and Holik, F. (eds) (2019). *Quantum Worlds: Perspectives on the Ontology of Quantum Mechanics*. Cambridge: Cambridge University Press.

London, F. and Bauer, E. (1939). *La Théorie de l'Observation en Méchanique Quantique*. Paris: Hermann & Co. English translation in London and Bauer (1983).

London, F. and Bauer, E. (1983). The theory of observation in quantum mechanics. In J. A. Wheeler and W. H. Zurek (eds), *Quantum Theory and Measurement*, Princeton: Princeton University Press, pp. 217–259. Translation of London and Bauer (1983).

Myrvold, W., Genovese, M., and Shimony, A. (2019). Bell's theorem. In E. N. Zalta (ed.), *The Stanford Encyclopedia of Philosophy* (Spring 2019 edn), Stanford, CA: Metaphysics Research Lab, Stanford University.

Myrvold, W. C. (2002). Modal intepretations and relativity. *Foundations of Physics*, **32**, 1773–1784.

Myrvold, W. C. (2009). Chasing chimeras. *The British Journal for the Philosophy of Science*, **60**, 635–646.

Myrvold, W. C. (2015). What is a wave function? *Synthese*, **292**, 3247–3274.

Myrvold, W. C. (2016). Lessons of Bell's theorem: Nonlocality, yes; action at a distance, not necessarily. In M. Bell and S. Gao (eds), *Quantum Nonlocality and Reality: 50 Years of Bell's Theorem*, Cambridge: Cambridge University Press, pp. 238–260.

Myrvold, W. C. (2018a). Ontology for collapse theories. In S. Gao (ed.), *Collapse of the Wave Function: Models, Ontology, Origin, and Implications*, Cambridge: Cambridge University Press, pp. 97–123.

Myrvold, W. C. (2018b). Philosophical issues in quantum theory. In E. N. Zalta (ed.), *The Stanford Encyclopedia of Philosophy* (Fall 2018 edn), Stanford, CA: Metaphysics Research Lab, Stanford University.

Myrvold, W. C. (2018c). $\psi$ -ontology result without the Cartesian product assumption. *Physical Review A*, **97**, 052109.

Myrvold, W. C. (2019). Ontology for relativistic collapse theories. In O. Lombardi, S. Fortin, C. López, and F. Holik (eds), *Quantum Worlds: Perspectives on the Ontology of Quantum Mechanics*, Cambridge: Cambridge University Press, pp. 9–31.

Myrvold, W. C. (2020). On the status of quantum state realism. In S. French and J. Saatsi (eds), *Scientific Realism and the Quantum*, Oxford: Oxford University Press, pp. 219–251.

Myrvold, W. C. (2021). Relativistic constraints on interpretations of quantum mechanics. In E. Knox and A. Wilson (eds), *The Routledge Companion to Philosophy of Physics*, Abingdon: Routledge.

Ney, A. (2019). Locality and wave function realism. In O. Lombardi, S. Fortin, C. López, and F. Holik (eds), *Quantum Worlds: Perspectives on the Ontology of Quantum Mechanics*, Cambridge: Cambridge University Press, pp. 164–182.

Ney, A. (2021). *The World in the Wave Function: A Metaphysics for Quantum Physics*. New York: Oxford University Press.

Ney, A. and D. Z. Albert (eds) (2013). *The Wave Function: Essays on the Metaphysics of Quantum Mechanics*. New York: Oxford University Press.

Pearle, P. (1989). Combining stochastic dynamical state-vector reduction with spontaneous localization. *Physical Review A*, **39**, 913–923.

Pearle, P. (2015). Relativistic dynamical collapse model. *Physical Review D*, **91**, 105012.

Pearle, P. and Valentini, A. (2006). Generalizations of quantum mechanics. In J.-P. Françoise, G. L. Naber, and T. S. Tsun (eds), *Encyclopedia of Mathematical Physics*, Amsterdam: Elsevier, pp. 265–276.

Pusey, M. A., Barrett, J., and Rudolph, T. (2012). On the reality of the quantum state. *Nature Physics*, **8**, 475–478.

Redhead, M. (1987). *Incompleteness, Nonlocality, and Realism*. Oxford: Oxford University Press.

Reichenbach, H. (1956). *The Direction of Time*. Berkeley, CA: University of Los Angeles Press.

Saunders, S., Barrett, J., Kent, A., and Wallace, D. (eds) (2010). *Many Worlds? Everett, Quantum Theory, and Reality*. Oxford: Oxford University Press.

Shimony, A. (1978). Metaphysical problems in the foundations of quantum mechanics. *International Philosophical Quarterly*, **18**, 3–17.

Shimony, A. (1984). Controllable and uncontrollable non-locality. In S. Kamefuchi (ed.), *Foundations of Quantum Mechanics in Light of New Technology*, Tokyo: The Physical Society of Japan, pp. 225–30. Reprinted in Shimony (1993), pp. 130–138.

Shimony, A. (1986). Events and processes in the quantum world. In R. Penrose and C. J. Isham (eds), *Quantum Concepts in Space and Time*, New York: Oxford University Press, pp. 182–203. Reprinted in Shimony (1993), pp. 140–162.

Shimony, A. (1990). An exposition of Bell's theorem. In A. I. Miller (ed.), *Sixty-Two Years of Uncertainty*, New York: Plenum, pp. 33–43. Reprinted in Shimony (1993), pp. 90–103.

Shimony, A. (1993). *Search for a Naturalistic Worldview, Volume II: Natural Science and Metaphysics*. Cambridge: Cambridge University Press.

Skyrms, B. (1984). EPR: Lessons for metaphysics. *Midwest Studies in Philosophy*, **9**, 245–255.

Tumulka, R. (2006). A relativistic version of the Ghirardi–Rimini–Weber model. *Journal of Statistical Physics*, **125**, 825–844.

Tumulka, R. (2021). A Relativistic GRW Flash Process with Interaction. In V. Allori. A. Bassi, D. Dürr, and N. Zanghì (eds.), *Do Wave Functions Jump? Perspectives of the Work of GianCarlo Ghirardi*, Springer, 321–347.

von Neumann, J. (1932). *Mathematische Grundlagen der Quantenmechanik*. Berlin: Verlag von Julius Springer. English translation in von Neumann (1955).

von Neumann, J. (1955). *Mathematical Foundations of Quantum Mechanics*. Princeton: Princeton University Press. Translation, by Robert T. Beyer, of von Neumann (1932).

Wallace, D. (2012). *The Emergent Multiverse*. Oxford: Oxford University Press.

Wallace, D. and Timpson, C. G. (2010). Quantum mechanics on spacetime I: Spacetime state realism. *The British Journal for the Philosophy of Science*, **61**, 697–727.

Wigner, E. (1961). Remarks on the mind-body question. In I. J. Good (ed.), *The Scientist Speculates*, London: Heinemann, pp. 284–302.

Robertson, H. (1929), 'The Uncertainty Principle', in *Quantum Theory and Measurement*, ed. J. A. Wheeler and W. H. Zurek (1983) — first published in *Physical Review* and reprinted in Wheeler and Zurek, Oxford University Press.

Saunders, A. (1995), 'Relativism', problems in the foundations of quantum mechanics, in *Quantum Interpretation*, Chapter 18, xxxx.

Saunders, A. (2014), 'Comodels and time', in *The Everett Interpretation*, ed. S. Barentson, in *Foundations of Quantum Physics*, in the light of the Frontiers: Foundations Physics Review in Japan, pp. xxx, as discussed in Saunders (2014), pp. xxx-xxx.

Sklar, A. (1969), 'From time into space: the quantum world', in *Reality* and Its Interpretation, Quantum space and time, New York, Oxford University Press, pp. xxx-xxx, reproduced in Saunders (1995), pp. xxx-xx.

Stapp, M. (1993), 'An exposition of Bell's theorem', in *F. Selleri ed. Quantum* Mechanics New Synthesis, London, pp. xx-xx, reprinted in Stapp essays, pp. xxx-x.

Stamatescu, A. (1993), *Schwarz and Nijhuis in Machinery*, Volume *Real Natural Sciences and Humanities*, Cambridge, Cambridge University Press.

Struma, B. (1938), EPR, Lessons for metaphysics, *Midwest Studies in Philosophy*, xxx, xxx.

Tumulka, R. (2006), 'A relativistic version of the Ghirardi–Rimini–Weber model', *Journal of Statistical Physics* 125, 821–840.

Vaidman, L. (2014), A tentative by GRW flash-point with modification, in V. Allori, A. Bassi, D. Dürr, and N. Zanghi (eds.), *Do Wave-Functions Jump? Perspectives on the Work of* GianCarlo Ghirardi, pp. xx–xx.

von Neumann, J. (1932), *Mathematische Grundlagen der Quantenmechanik*, Berlin, Verlag Julius Springer, English translation as von Neumann (1955).

von Neumann, J. (1955), *Mathematical Foundations of Quantum Mechanics* (translated from German translation by Robert T. Beyer), Princeton University Press.

Wallace, D. (2012), *The Emergent Multiverse*, Oxford University Press.

Wallace, D. and Timpson, C. (2010), 'Quantum mechanics on spacetime I: Spacetime state realism', in *The British Journal for the Philosophy of Science* 61, 697–727.

Wigner, E. (1961), 'Remarks on the mind-body question', in J. Heisenberg (ed.), *The Scientist Speculates*, London: Heinemann, pp. xxx–xxx.

# CHAPTER 3

..................................................................

# QUANTIZATION CONDITIONS, 1900–1927

..................................................................

ANTHONY DUNCAN AND MICHEL JANSSEN

## 3.1 OVERVIEW

..................................................................

WE trace the evolution of quantization conditions from Max Planck's introduction of a new fundamental constant ($h$) in his treatment of blackbody radiation in 1900 to Werner Heisenberg's interpretation of the commutation relations of modern quantum mechanics in terms of his uncertainty principle in 1927.[1]

In the most general sense, quantum conditions are relations between classical theory and quantum theory that enable us to construct a quantum theory from a classical theory. We can distinguish two stages in the use of such conditions. In the first stage, the idea was to take classical mechanics and modify it with an additional quantum structure. This was done by cutting up classical phase space. This idea first arose in the period 1900–1910 in the context of new theories for blackbody radiation and specific heats that involved the statistics of large collections of simple harmonic oscillators. In this context, the structure added to classical phase space was used to select equiprobable states in phase space. With the arrival of Bohr's model of the atom in 1913, the main focus in the development of quantum theory shifted from the statistics of large numbers of oscillators or modes of the electromagnetic field to the detailed structure of individual atoms and molecules and the spectra they produce. In that context, the additional structure of phase space was used to select a discrete subset of classically possible motions. In the second stage, after the transition to modern quantum mechanics in 1925–1926, quantum theory was completely divorced from its classical substratum and quantum conditions became a means of controlling the symbolic translation of classical relations into relations between quantum-theoretical quantities,

---

[1] This chapter is based on a two-volume book on the genesis of quantum mechanics (Duncan and Janssen 2019–2022).

represented by matrices or operators and no longer referring to orbits in classical phase space.[2]

More specifically, the development we trace in this essay can be summarized as follows. In late 1900 and early 1901, Planck used discrete energy units $\varepsilon = h\nu$ in his statistical treatment of radiating charged harmonic oscillators with resonance frequency $\nu$. However, he still allowed the energy of these oscillators to take on the full continuum of values. It was not until more than five years later that Albert Einstein first showed that one only arrives at the Planck law for blackbody radiation if the energy of Planck's oscillators is restricted to integral multiples of $h\nu$.

This Planck–Einstein condition was inextricably tied to a particular mechanical system, i.e., a one-dimensional simple harmonic oscillator. In his lectures on radiation theory, published in 1906, Planck suggested a more general condition for one-dimensional bound periodic systems by carving up the phase space of such systems into areas of size $h$. In late 1915, he generalized this idea to systems with several degrees of freedom (by a slicing procedure in the multidimensional phase space). By that time, Arnold Sommerfeld had independently found a procedure to quantize phase space that turned out to be equivalent to Planck's. Using this procedure, Sommerfeld was able to generalize the circular orbits of Niels Bohr's hydrogen atom (selected through quantization of the orbital angular momentum of the electron) to a larger set of Keplerian orbits of varying size and eccentricity, selected on the basis of the quantization of phase integrals such as $\oint p\,dq = nh$, where $p$ is the momentum conjugate to some generalized coordinate $q$ and the integral is to be taken over one period of the motion. Bohr's angular momentum quantum number gave the value of just one such integral, for the angular coordinate. Quantized orbits with different eccentricities became possible once the phase integral for the radial coordinate was similarly subjected to quantization. It was soon realized by the astronomer Karl Schwarzschild that the procedures of Planck and Sommerfeld are equivalent and that they amount to treating the classical problem in action-angle variables $(j_i, w_i)$, familiar from celestial mechanics, with the action variables restricted to multiples of $h$, $J_i = \oint p_i\,dq_i = n_i h$.

The transition from the old to the new quantum theory began in 1924 with the transcription (inspired by Bohr's correspondence principle) of the classical action derivative $d/dJ$ as a discrete difference quotient, $(1/h)\Delta/\Delta n$, by Hans Kramers, John Van Vleck, Max Born, and others. This transcription procedure was critical in the development of Kramers's dispersion theory for the elastic scattering of light. It led to the introduction of complex coordinate amplitudes depending on a pair of states linked by a quantum transition. In his famous *Umdeutung* (i.e., reinterpretation) paper, Heisenberg reinterpreted these amplitudes as two-index arrays with a specific non-commutative multiplication rule. Applying the transcription procedure to the Sommerfeld phase integral itself, he arrived at a nonlinear quantization constraint on these amplitudes. He showed that this constraint is just the high-frequency limit of the

---

[2] Cf. the concise account of the emergence of quantum mechanics by Darrigol (2009).

Kramers dispersion formula, known as the Thomas–Kuhn sum rule. Born quickly recognized that Heisenberg's two-index arrays are nothing but matrices and that the multiplication rule is simply the rule for matrix multiplication. Rewriting Heisenberg's quantization condition in matrix language, Born and Pascual Jordan arrived at the familiar commutation relation $[p_k, q_l] \equiv p_k q_l - q_l p_k = (h/i)\delta_{kl}$ of modern quantum mechanics (with $\hbar \equiv h/2\pi$ and $\delta_{kl}$ the Kronecker delta). Around the same time, and independently of Born and Jordan, Paul Dirac showed that this commutation relation is the exact analogue of Poisson brackets in classical mechanics. The commutation relation for position and momentum was also found to be satisfied by the operators representing these quantities and acting on wave functions in the alternative form of quantum mechanics developed by Erwin Schrödinger in late 1925 and early 1926. This commutation relation represents the central locus for the injection of Planck's constant into the new quantum theory. In 1927, Heisenberg interpreted it in terms of his uncertainty principle.

In the balance of this chapter, we examine the developments sketched above in more detail.[3]

# 3.2 THE EARLIEST QUANTUM CONDITIONS

In December 1900, Planck introduced the relation $\varepsilon = h\nu$ to provide a derivation of the empirically successful new formula he had first presented a couple of months earlier for the spectral distribution of the energy in blackbody radiation. Unlike Einstein and, independently, Paul Ehrenfest, Planck did not apply the relation $\varepsilon = h\nu$ to the energy of the radiation itself but to the energy of tiny charged harmonic oscillators (which he called 'resonators'), spread throughout the cavity and interacting with the radiation in it. Planck used the relation to count the number of possible microstates of a collection of such resonators. He then inserted this number into Boltzmann's formula relating the entropy of a macrostate to the number of microstates realizing it. The formula Planck thus found for the entropy of a resonator leads directly to the law for blackbody radiation now named after him. Planck did not restrict the possible values of the resonator energy to integral multiples $nh\nu$ (where $n$ is an integer), he only assumed that energies between successive values of $n$ should be lumped together when counting microstates. It was not until six years later that Einstein finally showed that Planck's derivation only works if the resonator energies are, in fact, restricted to integral multiples of $h\nu$.[4]

---

[3] For much more detailed accounts, see Jammer (1966), Mehra and Rechenberg (1982–2001), Darrigol (1992), and Duncan and Janssen (2019–2022).

[4] Planck (1900a, 1900b, 1901), Einstein (1905), Ehrenfest (1906), Einstein (1906). Following the publication of Kuhn's (1978) revisionist account of the origins of quantum theory, historians have reevaluated Planck's work and its early reception. Some of the most prominent contributions to this reevaluation are (in roughly chronological order): Klein's review of Kuhn's book (Klein, Shimony, and

That same year, 1906, Planck published his lectures on radiation theory, in which he reworked his own discretization of resonator energy by dividing the phase space of a resonator, spanned by its position and momentum, into cells of size $h$. In 1908, he finally accepted that the energy of a resonator is quantized: a resonator is only allowed to be at the edges of the cells in phase space. In 1911, he once again changed his mind in what came to be known as 'Planck's second theory'. He now proposed that a resonator can absorb energy from the ambient radiation continuously but release energy only when its own energy is an integral multiple of $h\nu$ and then only in integral multiples $n h\nu$.[5]

In 1913, drawing both on the quantum theory of blackbody radiation and on the British tradition of atomic modelling, Bohr, in the first part of his famous trilogy, proposed a quantum model of the hydrogen atom. He showed that this model gives the correct formula for the Balmer lines with a value for the Rydberg constant in excellent agreement with the spectroscopic data. Bohr quantized the energy of the electron in the hydrogen atom but eventually settled on quantizing its angular momentum instead, restricting its value to integer multiples of $h/2\pi$ (or, in modern notation, $\hbar$). Although he allowed elliptical orbits, Bohr did most of his calculations for circular orbits. This did not affect his results as he only used one quantization condition. In the second and third parts of his trilogy, he used this same quantization condition for planar models of more complicated atoms and molecules.[6]

## 3.3 QUANTIZATION CONDITIONS IN THE OLD QUANTUM THEORY

Bohr's success in accounting for the most prominent features of the hydrogen spectrum led to a shift in work on quantum theory. Instead of dealing with large collections of harmonic oscillators, physicists began to focus on individual atoms and tried to account for their spectra with or without external electric or magnetic fields on the basis of models similar to Bohr's. These efforts resulted in what, after the transition to modern quantum mechanics in the mid-1920s, came to be known as the old quantum theory. Its undisputed leader, besides Bohr in Copenhagen, was Sommerfeld in Munich. The development of the old quantum theory can be followed in successive editions

Pinch 1979, pp. 430–34), Kuhn's (1984) response to his critics, Needell (1980, 1988), Darrigol (1992, 2000, 2001) and Gearhart (2002). This debate informed our discussion of the early history of quantum theory in Duncan and Janssen (2019–2022, Vol. 1, Chs. 2–3).

[5] Planck (1906), Planck to Lorentz, October 7, 1908 (Lorentz 2008, Doc. 197; for discussion of this letter, see Kuhn 1987, p. 198). For 'Planck's second theory' see the second edition of his lectures on radiation theory (Planck 1913).

[6] Bohr (1913). On the genesis and reception of the Bohr model, see Heilbron and Kuhn (1969) and Kragh (2012).

of his book *Atomic Structure and Spectral Lines* (*Atombau und Spektrallinien*), which became known as the bible of atomic theory. Ehrenfest referred to its author (though not as a compliment) as the theory's pope.[7]

In two papers presented to the Munich Academy in December 1915 and January 1916, Sommerfeld rephrased Bohr's quantization condition in terms of Planck's phase-space quantization, with the understanding that only states at the edge of Planck's cells are allowed. Sommerfeld elaborated on these ideas in a paper published in two installments in *Annalen der Physik*. In one dimension, this phase-space quantization rule restricts the values of what were called phase integrals, the integral of the conjugate momentum $p$ of some generalized coordinate $q$ over one period of the motion, to integer multiples of $h$: $\oint p \, dq = nh$. This allowed Sommerfeld to subsume the quantized oscillators of Planck and Einstein and Bohr's model of the hydrogen atom under one quantization rule. Moreover, he generalized Bohr's model by allowing elliptical as well as circular orbits. He accomplished this by applying his phase-space quantization rule both to the radial coordinate and to the angular coordinate and their conjugate momenta. He recovered Bohr's quantum number $n$ as the sum of the two quantum numbers $n_r$ and $n_\varphi$ that he had introduced to quantize the orbits in a hydrogen atom in polar coordinates. The energy levels Bohr had identified thus correspond to multiple orbits with different combinations of Sommerfeld's radial and angular quantum numbers. This degeneracy is lifted, Sommerfeld showed, when the relativistic dependence of the mass of the electron on its velocity is taken into account. Sommerfeld thus found a formula for the fine structure of the hydrogen spectrum. This formula survives to this day even though, compared to modern quantum mechanics, Sommerfeld's quantum numbers for angular momentum are all off by 1. Even more baffling, he derived it without any knowledge of electron spin.[8]

As Sommerfeld was adapting Bohr's quantization condition to Planck's phase-space ideas, Planck himself, in two presentations to the German Physical Society in November and December 1915, tried to generalize the phase-space slicing he had introduced for one-dimensional oscillators to systems of multiple degrees of freedom. In a system described by $n$ independent canonical coordinates $q_1, ..., q_n$ and $n$ conjugate momenta $p_1, ..., p_n$, the $2n$-dimensional phase-space was to be sliced into cells of equal phase-space volume $h^n$. The surfaces producing this slicing were prescribed by $n$ functions $g_i(q_j, p_k)$ (with $i, j, k = 1, ..., n$), subject to the quantization conditions $g_i = n_i h$ at the boundaries of the cells. Planck placed two requirements on these functions. First, during the completely classical motion of the system in between the quantum jumps characteristic of transitions between stationary states in the Bohr picture, the system should either stay between the boundaries of two cells or move along one of these

---

[7] Sommerfeld (1919). An English translation of the third edition appeared shortly after the publication of the German original (Sommerfeld 1923). Sommerfeld's contributions are discussed in Eckert (2013a, 2013b, 2013c, 2014).

[8] Sommerfeld (1915a, 1915b, 1916). For discussion of the fortuitous character of Sommerfeld's derivation of the fine-structure formula, see Yourgrau and Mandelstam (1979) and Biedenharn (1983).

boundaries. The simplest way to enforce this is the one adopted by Planck: choose phase-space functions $g_i$ that are constants of the motion. Second, the requirement that cells take up equal volumes of phase space entails a factorization of the phase-space measure: the functions $g_i$ had to be chosen so that the multidimensional volume $dq_1...dq_n dp_1...dp_n$ could be rewritten as $dg_1...dg_n$. Planck's procedure could only be implemented on a case-by-case basis—and rather awkwardly at that. Planck applied his method to a number of cases (two-dimensional oscillators, Coulomb problem, three-dimensional rigid body) but did not find the correct quantized energy levels in all cases. The basic idea, however, was correct. In the action-angle formalism, developed in the context of celestial mechanics and transferred to atomic physics a few months later by Schwarzschild, the two conditions that Planck imposed on the slicing functions $g_i$ are automatically satisfied by the action variables $J_i$ in cases where such variables exist. Had Planck been *au courant* with the action-angle formalism, he might well have recognized this and anticipated Schwarzschild's seminal work a few months later.[9]

The Planck–Sommerfeld rule for quantizing phase integrals was found independently by William Wilson and Jun Ishiwara. It was left to Schwarzschild, however, to make the connection between phase integrals and action variables, well-known to astronomers knowledgeable about celestial mechanics and the techniques of Hamilton–Jacobi theory. In a letter of March 1916, Schwarzschild alerted Sommerfeld to this connection. Combining these techniques from celestial mechanics with Sommerfeld's quantum condition, Schwarzschild in short order derived a formula for the Stark effect in hydrogen, the splitting of its spectral lines in an external electric field. Sommerfeld's former student Paul Epstein arrived at essentially the same result at essentially the same time.[10]

What made action variables natural candidates for quantization was that they were so-called adiabatic invariants. As early as 1913, as can be gathered from a letter to Abram Joffe of February that year, Ehrenfest had realized the importance for quantum theory of a theorem found independently by Kalman Szily, Rudolf Clausius, and Ehrenfest's teacher Ludwig Boltzmann. This theorem asserts that under slow changes of the parameters of a mechanical system undergoing periodic motion, the integral of its kinetic energy over a single period is time invariant. For a harmonic oscillator with its total energy restricted to integral multiples of $h\nu$, Ehrenfest realized, this means that the adiabatic invariant $\overline{E_{\text{kin}}}/\nu$, the ratio of its average kinetic energy and its characteristic frequency, has to be set equal to $\frac{1}{2}n h$. Around the same time and independently of Bohr, Ehrenfest used an adiabatic-invariance argument to quantize the angular momentum of diatomic molecules: $L = n\hbar$. As soon as he saw Sommerfeld's phase

---

[9] Planck (1916).

[10] Wilson (1915), Ishiwara (1915), Schwarzschild to Sommerfeld, March 1, 1916 (Sommerfeld 2000, Doc. 240), Schwarzschild (1916), Epstein (1916a, 1916b). On the history of action-angle variables, see Nakane (2015). On Schwarzschild alerting Sommerfeld to action-angle variables, see Eckert (2013a, 2014). For the explanation of the Stark effect in the old (and the new) quantum theory, see Duncan and Janssen (2014, 2015).

integral quantization, as he explained in a letter to his Munich colleague of May 1916, Ehrenfest made the connection with adiabatic invariants. He formally introduced what came to be known as the adiabatic principle, one of the pillars of the old quantum theory, in a paper published later that year, first in the Proceedings of the Amsterdam Academy and, shortly thereafter, in *Annalen der Physik*. In the latter paper, he formulated his 'adiabatic hypothesis' in a particularly concise way: 'Under reversible adiabatic transformation of a system, (quantum-theoretically) "allowed" motions are always changed into "allowed" motions.' The following year, Jan Burgers, one of Ehrenfest's students in Leiden, supplied the proof that individual action variables are adiabatic invariants.[11]

The application of the action-angle formalism by Schwarzschild and Epstein to the Stark effect was seen as a major success for the old quantum theory, on par with Sommerfeld's elucidation of the fine structure. It also illustrates, however, a fundamental problem that would become an important factor in the eventual demise of the old quantum theory. Given the basic picture of atoms as miniature solar systems and the use of techniques borrowed from celestial mechanics to calculate the allowed energy levels, it was only natural to think of electrons as orbiting the nucleus on classical orbits. The Stark effect formed one example—its magnetic counterpart, the Zeeman effect, would provide a more dramatic one—where this picture turned out to be highly problematic. As Bohr, Sommerfeld, and Epstein realized, it makes a difference in which coordinates the quantum conditions are imposed. Even though the choice of coordinates does not affect the energy levels found, it does affect the shape of the orbits.[12]

# 3.4 THE TRANSITION TO QUANTUM MECHANICS AND THE APPEARANCE OF THE MODERN COMMUTATION RELATIONS

A more serious problem for the picture of orbits arose in attempts to adapt the classical theory of optical dispersion of Hermann von Helmholtz, Hendrik Antoon Lorentz, and Paul Drude to the quantum theory of Bohr and Sommerfeld. The classical theory was developed to deal with the phenomenon of anomalous dispersion, the effect that the index of refraction decreases rather than increases with frequency in ranges around the absorption frequencies of the material under study. This phenomenon could be

---

[11] Ehrenfest to Joffe, February 20, 1913, quoted and discussed by Klein (1970, p. 261), Ehrenfest (1913a; 1913b; 1916a; 1916b, the passage we quoted from this last paper can be found on p. 328), Ehrenfest to Sommerfeld, May 1916 (Sommerfeld 2000, Doc. 254; quoted in Klein 1970, p. 286), Burgers (1917a, 1917b). For further discussion of the adiabatic principle, see Navarro and Pérez (2004, 2006), Pérez (2009), and Duncan and Pérez (2016).

[12] Duncan and Janssen (2014, 2015).

explained on the assumption that matter contains large numbers of charged oscillators with resonance frequencies at these absorption frequencies.[13]

These oscillating charges could not simply be replaced by the orbiting electrons of the Bohr–Sommerfeld model of the atom. To recover the Balmer formula, Bohr had been forced to sever the relation between the orbital frequency of the electron circling the nucleus in the hydrogen atom and the frequency of the radiation emitted or absorbed upon quantum jumps of the electron from one orbit to another, a frequency given by the Bohr frequency condition $h\nu = |E_i - E_f|$ (where the subscripts $i$ and $f$ refer to initial and final orbit). Only in the limit of high quantum numbers do these transition frequencies merge with orbital frequencies. In 1913, Bohr had actually used the requirement that these two frequencies merge in this limit to put the quantum condition he needed to recover the Balmer formula on a more secure footing. In the limit of high quantum numbers, Bohr's quantum theory thus merged with classical electrodynamics according to which the frequencies of radiation are always (overtones of) the frequencies of the oscillations generating the radiation. Over the next few years, Bohr greatly expanded the use of analogies with classical electrodynamics to develop his own quantum theory. He eventually introduced the term 'correspondence principle' to characterize this approach. Severing radiation and orbital frequencies meanwhile was widely seen as the most radical aspect of Bohr's model. It also meant that dispersion becomes anomalous at frequencies that differ sharply from the orbital frequencies of the electrons. Dispersion thus posed a serious problem for the old quantum theory.[14]

In 1921, Rudolf Ladenburg, an experimental physicist in Breslau, addressed a problem for the *classical* dispersion theory. The number of 'dispersion electrons' one found by fitting the dispersion formula to the experimental data was much lower than one would expect. Drawing on Einstein's quantum theory of radiation to replace amplitudes of radiation by probabilities of emission or absorption, Ladenburg replaced numbers of electrons by numbers of electron jumps and thus arrived at a formula that at least to some extent takes care of this problem. Ladenburg had no solution, however, for the problem that dispersion appears to be anomalous at the wrong frequencies in the old quantum theory. He just replaced orbital frequencies with transition frequencies in the classical formula without any theoretical justification because this was clearly what the experimental evidence indicated. Another limitation of Ladenburg's formula was that it only applied to the ground state of an atom.[15]

Ladenburg's work drew the attention of Bohr and, in 1924, his assistant Kramers found a generalization of Ladenburg's formula that removed its limitations. Kramers

---

[13]  On dispersion in classical theory and the old quantum theory, see Jordi Taltavull (2017).

[14]  The term 'correspondence principle' does not occur in the main body of Bohr (1918, Parts I and II). It does appear, however, in an appendix to Part III, which finally saw the light of day in November 1922, although a manuscript existed already in 1918 as the first two parts went to press. For further discussion of the correspondence principle, see, e.g., Darrigol (1992), Fedak and Prentis (2002), Bokulich (2008), Rynasiewicz (2015), and Jähnert (2019).

[15]  Ladenburg (1921), Einstein (1917).

combined the sophisticated techniques the old quantum theory had borrowed from celestial mechanics with Bohr's correspondence-principle approach. He considered the scattering of an electromagnetic wave by some generic mechanical system with one electron and derived a classical dispersion formula for such a system that has the form of a derivative with respect to an action variable of an expression containing amplitudes and frequencies of the oscillations induced in that system by an electromagnetic wave. To turn this into a quantum formula, Kramers replaced amplitudes by transition probabilities (as Ladenburg had done before him), orbital frequencies by transition frequencies, and—and this was Kramers's main innovation—derivatives by difference quotients. The construction of this quantum formula guaranteed that it merges with classical theory in the limit of high quantum numbers. In this limit, after all, transition frequencies and orbital frequencies can be used interchangeably and the difference between successive integers in the allowed values of the action variable (which, as Schwarzschild had first shown, was just Sommerfeld's phase integral) becomes so small that derivatives can be replaced by difference quotients. In the spirit of Bohr's correspondence principle, Kramers now took the leap of faith that this formula would continue to hold all the way down to the lowest quantum numbers.[16]

Kramers initially only published his formula in two short notes in *Nature*. Over the Christmas break of 1924–1925, however, he teamed up with Heisenberg, a former student of Sommerfeld's who was visiting Copenhagen, to write a paper providing a detailed exposition and further extension of the results he had found before. This paper played a central role in the train of thought that led to the famous *Umdeutung* (reinterpretation) paper with which Heisenberg laid the foundation for matrix mechanics. One of the striking features of the Kramers dispersion formula is that it only depends on transitions between the orbits used in its derivation. It no longer refers to individual orbits. This seems to have given Heisenberg the key idea of setting up a new framework for all of physics in which any quantity that used to be represented by a number connected to one particular orbit is represented instead by an array of numbers connected to all possible transitions between orbits. The reason Heisenberg referred to this as *Umdeutung* is that he retained the laws of classical mechanics relating these quantities. He only reinterpreted the nature of the quantities related by these laws. Heisenberg emphasized that all observable quantities (frequencies, intensities, and polarizations of radiation) correspond not to individual orbits but to transitions between them. He hoped to eliminate the increasingly problematic orbits altogether by focusing on transitions between stationary states.[17]

To achieve this goal, Heisenberg also needed to replace the basic Bohr–Sommerfeld quantization condition, which, after all, gave the allowed values of the action variable for individual orbits. In keeping with his *Umdeutung* programme, Heisenberg looked at

[16] Kramers (1924a, 1924b). Full derivations of the Kramers dispersion formula were first published by Born (1924) and Van Vleck (1924a,1924b).

[17] Kramers and Heisenberg (1925). On the path leading from Kramers's dispersion theory to Heisenberg's (1925) *Umdeutung* paper, see, e.g., Dresden (1987) and Duncan and Janssen (2007).

the change in values of these action variables in transitions between orbits. Transcribing the classical formula for such changes into a quantum one, in a manner analogous to what Kramers had done in dispersion theory, Heisenberg arrived at a formula he had encountered before. Right around that time and independently of one another, Werner Kuhn in Copenhagen and Willy Thomas in Breslau had derived an expression for the high-frequency limit of the Kramers dispersion formula that has come to be known as the Thomas–(Reiche–)Kuhn sum rule. This is what Heisenberg used as his quantization condition in the *Umdeutung* paper.[18]

Heisenberg had not just studied with Sommerfeld in Munich and spent time in Bohr's institute in Copenhagen, he had also co-authored a paper with Born in Göttingen. In the early 1920s and under Born's leadership, Göttingen had emerged alongside Copenhagen and Munich as a third leading centre for work on the old quantum theory. When Born read the *Umdeutung* paper, he immediately recognized that the arrays of quantities Heisenberg had introduced were nothing but matrices and that Heisenberg's peculiar non-commutative multiplication was nothing but the standard rule for matrix multiplication. He also realized that Heisenberg's quantization condition, the Thomas–Kuhn rule, is equivalent to the diagonal elements of the commutation relation $\hat{q}\hat{p} - \hat{p}\hat{q} = i\hbar$ (where hats indicate that these quantities are matrices). His student, Jordan, showed that the off-diagonal elements vanish. They reported these results in a joint paper elaborating on Heisenberg's *Umdeutung* paper. Heisenberg generalized this commutation relation from one to multiple degrees of freedom. The resulting commutation relations, $[\hat{q}_k, \hat{p}_l] = i\hbar\,\delta_{kl}$, are central to the first authoritative exposition of matrix mechanics, the *Dreimännerarbeit* of Born, Heisenberg, and Jordan.[19]

Around the same time and independently of the work of Born and Jordan, Dirac, using arguments from dispersion theory, derived a precise correspondence between classical Poisson brackets and the commutator of the corresponding quantum variables. Dirac's procedure precisely imitates the methodology of the Kramers–Heisenberg derivation of the dispersion formula for inelastic (Raman) light scattering, in which the amplitude for a transition between two quantum states $a$ and $b$ was expressed as a sum of amplitudes for transitions via two distinct intermediate states $c$ and $d$ (i.e., as a sum of the amplitudes for the sequential transitions $a \rightarrow c \rightarrow b$ and $a \rightarrow d \rightarrow b$). The transition between classical formulas involving derivatives with respect to the action and quantum ones involving discrete differences of amplitudes with varying quantum numbers was accomplished via the transcription procedure $d/dJ \rightarrow 1/h\,\Delta/\Delta n$ by now familiar from the work of Kramers, Born, Van Vleck, and Heisenberg. Dirac's beautiful derivation shows that the commutator of any two kinematical variables (divided by Planck's constant $h$), interpreted according to Heisenberg's matrix reinterpretation, corresponds precisely to the Poisson bracket of the associated classical quantities in the limit of large quantum numbers.[20]

---

[18]  Kuhn (1925), Thomas (1925), Reiche and Thomas (1925).

[19]  Born and Heisenberg (1923), Born and Jordan (1925), Born, Heisenberg, and Jordan (1926).

[20]  Dirac (1926a, 1926b). For analysis, see Darrigol (1992) and Kragh (1990).

Completely independently of the work of the Göttingen group, and of Dirac in Cambridge, a formulation of quantum theory based on a continuum wave theory was developed in late 1925 and early 1926 by Schrödinger, working in Zurich. The theory, inspired by the work of Louis de Broglie on matter waves (via Einstein's second paper on the quantum theory of the ideal gas), and by the analogies already discovered almost a century earlier by William Rowan Hamilton between geometrical optics and particle mechanics, posited the existence of a well-defined solution $\psi(\vec{r})$ to a wave equation associated in some way with the dynamics of a single particle. In his first paper on wave mechanics published in January 1926, this wave equation was obtained as the solution of a variational problem based on the classical Hamilton–Jacobi equation, and the appearance of energy quantization—for the bound states, with negative energy, of the hydrogen atom—is a consequence of the imposition of regularity and finiteness conditions on the wave function $\psi$. Once the wave function $\psi(\vec{r})$ is required to *remain finite* as $r \to \infty$, the Bohr–Balmer quantization of the bound states of the hydrogen atom follows immediately. A few months later, Schrödinger (and independently, Wolfgang Pauli in Hamburg and Carl Eckart at Caltech) established the connection between his wave functions and the matrices of Heisenberg et al., at which point it became clear that the matrices so defined would also satisfy the commutation relation which served as the point of departure of matrix mechanics.[21]

In late 1926, Jordan and Dirac independently of one another found essentially the same formalism, now known as the Dirac–Jordan statistical transformation theory, unifying matrix mechanics, wave mechanics, Dirac's $q$-number theory and yet another version of the new quantum theory, the operator calculus of Born and Norbert Wiener. A few months later, drawing on Jordan's work, Heisenberg showed that the commutation relations central to the new theory express what we now know as the uncertainty principle. Around the same time, John von Neumann introduced the Hilbert space formalism of quantum mechanics and showed that matrix mechanics and wave mechanics correspond to two different incarnations of Hilbert space, the space $l^2$ of square-summable sequences and the space $L^2$ of square-integrable functions, respectively.[22]

---

[21] De Broglie (1924, 1925), Einstein (1925), Schrödinger (1926a, 1926b), Eckart (1926), Pauli to Jordan, April 12, 1926 (Pauli 1979, pp. 315–20; translated and discussed by van der Waerden 1973). For discussion of Schrödinger's use of the optical-mechanical analogy, see Joas and Lehner (2009); for discussion of his equivalence proof, see Muller (1997–1999).

[22] Born and Wiener (1926), Jordan (1927a, 1927b), Dirac (1927), Heisenberg (1927), von Neumann (1927). For discussion of the transition from Jordan's transformation theory to von Neumann's Hilbert space formalism, see Duncan and Janssen (2013).

# 3.5 CONCLUSION

As this brief overview shows, the canonical commutation relations $\hat{q}_i\hat{p}_j - \hat{p}_j\hat{q}_i = i\hbar\delta_{ij}$ at the heart of modern quantum mechanics can be traced back to Heisenberg's use of the Thomas–Kuhn sum rule, a corollary of the Kramers dispersion formula, as the quantization condition in his *Umdeutung* paper. Heisenberg was led to this quantization condition by transcribing (in a manner analogous to how Kramers arrived at his dispersion formula) the phase-integral quantization condition $\oint p\,dq = nh$ of the old quantum theory found by Sommerfeld, Wilson, and Ishiwara, and clarified by Schwarzschild and Epstein, who identified and exploited the connection of these phase integrals to the action variables familiar from celestial mechanics. What, in turn, had inspired Sommerfeld to adopt the phase-integral quantization condition was Planck's reworking of the condition $\varepsilon = h\nu$ central to the derivation of his black-body radiation law.

## ACKNOWLEDGMENT

We thank Olivier Darrigol, Olival Freire Jr., and Alexei Kojevnikov for helpful comments on an earlier draft of this essay.

## REFERENCES

Aaserud, F., and Heilbron, J. L. (2013). *Love, Literature, and the Quantum Atom. Niels Bohr's 1913 Trilogy Revisited*. Oxford: Oxford University Press.

Aaserud, F., and Kragh, H. (eds.) (2015). *One Hundred Years of the Bohr Atom: Proceedings from a Conference* (Scientia Danica: Series M: Mathematica et Physica, 1). Copenhagen: Royal Danish Academy of Sciences and Letters.

Badino, M., and Navarro, J. (eds.), *Research and Pedagogy: A History of Quantum Physics through Its Textbooks*. Berlin: Edition Open Access, pp. 117–135.

Biedenharn, L. C. (1983). The 'Sommerfeld Puzzle' Revisited and Resolved. *Foundations of Physics*, **13**, 13–34.

Bohr, N. (1913). On the Constitution of Atoms and Molecules (Parts I–III). *Philosophical Magazine*, **26**, 1–25 (I), 476–502 (II), 857–875 (III). Reprinted in facsimile in Bohr (1972–2008, Vol. 2, pp. 161–233) and in Aaserud and Heilbron (2013, pp. 203–273).

Bohr, N. (1918). On the Quantum Theory of Line Spectra. Parts I and II. *Det Kongelige Danske Videnskabernes Selskab. Skrifter. Naturvidenskabelig og Matematisk Afdeling*, **8** (Raekke, IV.1), 1–100. Reprinted in facsimile in Bohr (1972–2008, Vol. 3, pp. 67–166). Introduction and Part I reprinted in Van der Waerden (1968, pp. 95–136).

Bohr, N. (1972–2008). *Collected Works*. Edited by L. Rosenfeld, F. Aaserud, E. Rüdiger, et al. Amsterdam: North Holland.

Bokulich, A. (2008). *Reexamining the Quantum-Classical Relation: Beyond Reductionism and Pluralism*. Cambridge: Cambridge University Press.

Born, M. (1924). Über Quantenmechanik. *Zeitschrift für Physik*, **26**, 379–395. English translation in Van der Waerden (1968, pp. 181–198).

Born, M., and Heisenberg, W. (1923). Die Elektronenbahnen im angeregten Heliumatom. *Zeitschrift für Physik*, **16**, 229–243.

Born, M., Heisenberg, W., and Jordan, P. (1926). Zur Quantenmechanik II. *Zeitschrift für Physik*, **35**, 557–615. English translation in van der Waerden (1968, 321–385).

Born, M., and Jordan, P. (1925). Zur Quantenmechanik. *Zeitschrift für Physik*, **34**, 858–888. Page references to Chs. 1–3 are to the English translation in Van der Waerden (1968, pp. 277–306). Ch. 4 is omitted in this translation.

Born, M., and Wiener, N. (1926). Eine neue Formulierung der Quantengesetze für periodische und nicht periodische Vorgänge. *Zeitschrift für Physik*, **36**, 174–187.

Broglie, L. de (1924). *Recherche sur la théorie des quanta*. PhD diss., University of Paris.

Broglie, L. de (1925). Recherche sur la théorie des quanta. *Annales de physique*, **3**, 22–128.

Burgers, J. M. (1917a). Adiabatische invarianten bij mechanische systemen. Parts 1–3: *Koninklijke Akademie van Wetenschappen te Amsterdam. Wis- en Natuurkundige Afdeeling. Verslagen van de Gewone Vergaderingen*, **25**, 849–857 (I), 918–927 (II), 1055–1061 (III). English translation: Adiabatic Invariants of Mechanical Systems. *Koninklijke Akademie van Wetenschappen te Amsterdam. Section of Sciences. Proceedings*, **20**, 149–157 (I), 158–162 (II), 163–169 (III).

Burgers, J. M. (1917b). Adiabatic Invariants of Mechanical Systems. *Philosophical Magazine*, **33**, 514–520.

Darrigol, O. (1992). *From c-Numbers to q-Numbers: The Classical Analogy in the History of Quantum Theory*. Berkeley: University of California Press.

Darrigol, O. (2000). Continuities and Discontinuities in Planck's *Akt der Verzweiflung*. *Annalen der Physik*, **9**, 951–960.

Darrigol, O. (2001). The Historians' Disagreement Over the Meaning of Planck's Quantum. *Centaurus*, **43**, 219–239.

Darrigol, O. (2009). A Simplified Genesis of Quantum Mechanics. *Studies in History and Philosophy of Modern Physics*, **40**, 151–166.

Dirac, P. A. M. (1926a). The Fundamental Equations of Quantum Mechanics. *Proceedings of the Royal Society of London*, **A109**, 642–653. Reprinted in Van der Waerden (1968, pp. 307–320).

Dirac, P. A. M. (1926b). Quantum Mechanics and a Preliminary Investigation of the Hydrogen Atom. *Proceedings of the Royal Society of London*, **A110**, 561–569. Reprinted in Van der Waerden (1968, pp. 417–427).

Dirac, P. A. M. (1927). The Physical Interpretation of the Quantum Dynamics. *Proceedings of the Royal Society of London*, **A113**, 621–641.

Dresden, M. (1987). *H. A. Kramers: Between Tradition and Revolution*. New York: Springer.

Duncan, A., and Janssen, M. (2007). On the Verge of *Umdeutung* in Minnesota: Van Vleck and the Correspondence Principle. 2 Pts. *Archive for History of Exact Sciences*, **61**, 553–624, 625–671.

Duncan, A., and Janssen, M. (2013). (Never) Mind your *p*'s and *q*'s: Von Neumann versus Jordan on the Foundations of Quantum Theory. *The European Physical Journal H*, **38**, 175–259.

Duncan, A., and Janssen, M. (2014). The Trouble with Orbits: The Stark Effect in the Old and the New Quantum Theory. *Studies In History and Philosophy of Modern Physics*, **48**, 68–83.

Duncan, A., and Janssen, M. (2015). The Stark Effect in the Bohr–Sommerfeld Theory and in Schrödinger's Wave Mechanics. In Aaserud and Kragh (2015, pp. 217–271).

Duncan, A., and Janssen, M. (2019–2022). *Constructing Quantum Mechanics*. Vol. 1. *The Scaffold, 1900–1923*. Vol. 2. *The Arch, 1923–1927*. Oxford: Oxford University Press.

Duncan, A., and Pérez, E. (2016). The Puzzle of Half-integral Quanta in the Application of the Adiabatic Hypothesis to Rotational Motion. *Studies In History and Philosophy of Modern Physics*, **54**, 1–8.

Eckart, C. (1926). Operator Calculus and the Solution of the Equations of Quantum Dynamics. *Physical Review*, **28**, 711–726.

Eckert, M. (2013a). *Historische Annäherung*. In Sommerfeld (2013, pp. 1–60).

Eckert, M. (2013b). *Arnold Sommerfeld. Atomphysiker und Kulturbote 1868–1951. Eine Biographie*. Göttingen: Wallstein. English translation (by T. Artin): *Arnold Sommerfeld. Science, Life and Turbulent Times*. New York: Springer.

Eckert, M. (2013c). Sommerfeld's *Atombau und Spektrallinien*. In Badino and Navarro (2013, pp. 117–135).

Eckert, M. (2014). How Sommerfeld Extended Bohr's Model of the Atom. *European Physical Journal H*, **39**, 141–156.

Ehrenfest, P. (1906). Zur Planckschen Strahlungstheorie. *Physikalische Zeitschrift*, **7**, 528–532. Reprinted in facsimile in Ehrenfest (1959, pp. 120–124).

Ehrenfest, P. (1913a). Bemerkung betreffs der spezifischen Wärme zweiatomiger Gase. *Verhandlungen der Deutschen Physikalischen Gesellschaft*, **15**, 451–457. Reprinted in facsimile in Ehrenfest (1959, pp. 333–339).

Ehrenfest, P. (1913b). A Mechanical Theorem of Boltzmann and Its Relation to the Theory of Energy Quanta. *Koninklijke Akademie van Wetenschappen te Amsterdam, Section of Sciences, Proceedings*, **16**, 591–597. Reprinted in facsimile in Ehrenfest (1959, 340–346).

Ehrenfest, P. (1916a). On Adiabatic Changes of a System in Connection with the Quantum Theory. *Koninklijke Akademie van Wetenschappen te Amsterdam, Section of Sciences, Proceedings*, **19**, 576–597. Reprinted in facsimile in Ehrenfest (1959, pp. 378–399).

Ehrenfest, P. (1916b). Adiabatische Invarianten und Quantentheorie. *Annalen der Physik*, **51**, 327–352. German translation of Ehrenfest (1916a).

Ehrenfest, P. (1959). *Collected Scientific Papers*. Edited by M. J. Klein. Amsterdam: North Holland.

Einstein, A. (1905). Über eine die Erzeugung und die Verwandlung des Lichtes betreffenden heuristischen Gesichtspunkts. *Annalen der Physik*, **17**, 132–148. Reprinted in Einstein (1987–2018, Vol. 2, Doc. 14).

Einstein, A. (1906). Zur Theorie der Lichterzeugung und Lichtabsorption. *Annalen der Physik*, **20**, 199–206. Reprinted in Einstein (1987–2018, Vol. 2, Doc. 34).

Einstein, A. (1917). Zur Quantentheorie der Strahlung. *Physikalische Zeitschrift*, **18**, 121–128. First published in: *Physikalische Gesellschaft Zürich. Mitteilungen*, **18** (1916): 47–62. Reprinted in Einstein (1987–2018, Vol. 6, Doc. 38). English translation in van der Waerden (1968, 63–77).

Einstein, A. (1925). Quantentheorie des einatomigen idealen Gases. Zweite Abhandlung. *Preußische Akademie der Wissenschaften* (Berlin). *Physikalisch-mathematische Klasse. Sitzungsberichte*, 3–14. Reprinted in Einstein (1987–2018, Vol. 14, Doc. 385).

Einstein, A. (1987–2018). *The Collected Papers of Albert Einstein*. 15 Vols. Edited by J. Stachel, M. J. Klein, R. Schulmann, D. Barkan Buchwald, et al. Princeton: Princeton University Press.

Epstein, P. (1916a). Zur Theorie des Starkeffektes. *Annalen der Physik*, 50, 489–521.

Epstein, P. (1916b). Zur Quantentheorie. *Annalen der Physik*, 51, 168–188.

Fedak, W. A., and Prentis, J. J. (2002). Quantum Jumps and Classical Harmonics. *American Journal of Physics*, 70, 332–344.

Gearhart, C. A. (2002). Planck, the Quantum, and the Historians. *Physics in Perspective*, 4, 170–215.

Heilbron, J. L. (1981). *Historical Studies in the Theory of Atomic Structure*. New York: Arno Press.

Heilbron, J. L., and Kuhn, T. S. (1969). The Genesis of the Bohr Atom. *Historical Studies in the Physical Sciences*, 1, 211–290. Reprinted in Heilbron (1981, pp. 149–228).

Heisenberg, W. (1925). Über die quantentheoretische Umdeutung kinematischer und mechanischer Beziehungen. *Zeitschrift für Physik*, 33, 879–893. English translation in Van der Waerden (1968, pp. 261–276).

Heisenberg, W. (1927). Über den anschaulichen Inhalte der quantentheoretischen Kinematik und Mechanik. *Zeitschrift für Physik*, 43, 172–198.

Hoffmann, D. (ed.) (2008). *Max Planck: Annalen Papers*. Weinheim: Wiley.

Ishiwara, J. (1915). Die universelle Bedeutung des Wirkungsquantums. *Proceedings of the Tokyo Mathematico-Physical Society*, 8, 106–116. Annotated translation, 'The universal meaning of the quantum of action' (by K. Pelogia and C. Alexandre) in *The European Journal of Physics H*, 42 (2017), 523–536.

Jähnert, M. (2019). *Practicing the Correspondence Principle in the Old Quantum Theory. A Transformation Through Application*. Cham, Switzerland: Springer.

Jammer, M. (1966). *The Conceptual Development of Quantum Mechanics*. New York: McGraw-Hill.

Joas, C., and Lehner, C. (2009). The Classical Roots of Wave mechanics: Schrödinger's Transformations of the Optical-mechanical Analogy. *Studies in History and Philosophy of Modern Physics*, 40, 338–351.

Jordan, P. (1927a). Über eine neue Begründung der Quantenmechanik. *Zeitschrift für Physik*, 40, 809–838.

Jordan, P. (1927b). Über eine neue Begründung der Quantenmechanik. II. *Zeitschrift für Physik*, 44, 1–25.

Jordi Taltavull, M. (2017). *Transformation of Optical Knowledge from 1870 to 1925: Optical Dispersion between Classical and Quantum Physics*. PhD diss., Humboldt University, Berlin.

Kangro, H. (ed.) (1972). *Planck's Original Papers in Quantum Physics*. London: Taylor & Francis. Translations: D. ter Haar and S. G. Brush.

Klein, M. J. (1970). *Paul Ehrenfest. Vol. 1. The Making of a Theoretical Physicist*. Amsterdam: North Holland.

Klein, M. J., Shimony, A., and Pinch, T. (1979). Paradigm Lost? A Review Symposium. *Isis*, 4, 429–440.

Kragh, H. (1990). *Dirac. A Scientific Biography*. Cambridge: Cambridge University Press.

Kragh, H. (2012). *Niels Bohr and the Quantum Atom. The Bohr model of Atomic Structure, 1913–1925*. Oxford: Oxford University Press.

Kramers, H. A. (1924a). The Law of Dispersion and Bohr's Theory of Spectra. *Nature*, 113, 673–676. Reprinted in Van der Waerden (1968, pp. 177–180).

Kramers, H. A. (1924b). The Quantum Theory of Dispersion. *Nature*, 114, 310–311. Reprinted in Van der Waerden (1968, pp. 199–201).

Kramers, H. A., and Heisenberg, W. (1925). Über die Streuung von Strahlung durch Atome. *Zeitschrift für Physik*, 31, 681–707. English translation in Van der Waerden (1968, pp. 223–252).

Kuhn, T. S. (1978). *Black-Body Theory and the Quantum Discontinuity, 1894–1912*. Oxford: Oxford University Press. Reprinted, including Kuhn (1984) as a new afterword, as Kuhn (1987).

Kuhn, T. S. (1984). Revisiting Planck. *Historical Studies in the Physical Sciences*, 14, 231–252. Reprinted in Kuhn (1987, 349–370).

Kuhn, T. S. (1987). *Black-Body Theory and the Quantum Discontinuity, 1894–1912*, 2nd edn. Chicago: University of Chicago Press.

Kuhn, W. (1925). Über die Gesamtstärke der von einem Zustande ausgehenden Asborptionslinien. *Zeitschrift für Physik*, 33, 408–412. Translated and reprinted in Van der Waerden (1968), pp. 253–257.

Ladenburg, R. (1921). Die quantentheoretische Deutung der Zahl der Dispersionselektronen. *Zeitschrift für Physik*, 4, 451–468. Translated and reprinted in Van der Waerden (1968), pp. 139–157.

Lorentz, H. A. (2008). *The Scientific Correspondence of H. A. Lorentz. Vol. 1*. Edited by A. J. Kox. New York: Springer.

Mehra, J., and Rechenberg, H. (1982–2001). *The Historical Development of Quantum Theory*. 6 Vols. New York, Heidelberg, Berlin: Springer.

Muller, F. A. (1997–1999). The Equivalence Myth of Quantum Mechanics. 2 Pts. plus addendum. *Studies in History and Philosophy of Modern Physics*, 28, 35–61, 219–247 (I–II), 30, 543–545 (addendum).

Nakane, M. (2015). The Origins of Action-Angle Variables and Bohr's Introduction of Them in a 1918 Paper. In Aaserud and Kragh (2015, pp. 290–309).

Navarro, L., and Pérez, E. (2004). Paul Ehrenfest on the Necessity of Quanta (1911): Discontinuity, Quantization, Corpuscularity, and Adiabatic Invariance. *Archive for the History of the Exact Sciences*, 58, 97–141.

Navarro, L., and Pérez, E. (2006). Paul Ehrenfest: The Genesis of the Adiabatic Hypothesis, 1911–1914. *Archive for the History of the Exact Sciences*, 60, 209–267.

Needell, A. A. (1980). *Irreversibility and the Failure of Classical Dynamics: Max Planck's Work on the Quantum Theory, 1900–1915*. PhD diss., Yale University.

Needell, A. A. (1988). 'Introduction,' In M. Planck, *The Theory of Heat Radiation*. New York: American Institute of Physics, pp. xi–xlv.

Pauli, W. (1979). *Wissenschaftlicher Briefwechsel mit Bohr, Einstein, Heisenberg u.a. Band I: 1919–1929/Scientific Correspondence with Bohr, Einstein, Heisenberg, a.o. Vol: I: 1919–1929*. Edited by A. Hermann, K. von Meyenn, and V. F. Weisskopf. New York, Heidelberg, Berlin: Springer.

Pérez, E. (2009). Ehrenfest's Adiabatic Theory and the Old Quantum Theory, 1916–1918. *Archive for the History of the Exact Sciences*, 63, 81–125.

Planck, M. (1900a). Über eine Verbesserung der Wien'schen Spectralgleichung. *Deutsche Physikalische Gesellschaft. Verhandlungen*, 2, 202–204. Reprinted in Planck (1958, Vol. 1, 687–689). English translation in Kangro (1972, pp. 35–45).

Planck, M. (1900b). Zur Theorie des Gesetzes der Energieverteilung im Normalspektrum. *Deutsche Physikalische Gesellschaft. Verhandlungen*, **2**, 237–245. Reprinted in Planck (1958, Vol. 1, pp. 698–706). English translation in Kangro (1972, pp. 35–45).

Planck, M. (1901). Über das Gesetz der Energieverteilung im Normalspektrum. *Annalen der Physik*, **4**, 553–563. Reprinted in facsimile in Planck (1958, Vol. 1, pp. 717–727) and in Hoffmann (2008, pp. 537–547). Slightly abbreviated English translation in Shamos (1959, pp. 305–313).

Planck, M. (1906). *Vorlesungen über die Theorie der Wärmestrahlung*. Leipzig: Barth. Reprinted in Planck (1988, 243–470).

Planck, M. (1913). *Vorlesungen über die Theorie der Wärmestrahlung*, 2nd edn. (1st edn.: 1906). Leipzig: Barth. Reprinted in Planck (1988, 1–239). English translation: *The Theory of Heat Radiation* (New York: Dover, 1991).

Planck, M. (1916). Die Physikalische Struktur des Phasenraumes. *Annalen der Physik*, **50**, 385–418. Reprinted in facsimile in Planck (1958, Vol. 2, pp. 386–419) and in Hoffmann (2008, pp. 654–687).

Planck, M. (1958). *Physikalische Abhandlungen und Vorträge*. 3 Vols. Braunschweig: Vieweg.

Planck, M. (1988). *The Theory of Heat Radiation*. New York: American Institute of Physics. With an introduction by A. A. Needell.

Reiche, F., and Thomas, W. (1925). Über die Zahl der Dispersionselektronen, die einem stationären Zustand zugeordnet sind. *Zeitschrift für Physik*, **34**, 510–525.

Rynasiewicz, R. (2015). The (?) Correspondence Principle. In Aaserud and Kragh (2015, pp. 175–199).

Schrödinger, E. (1926a). Quantisierung als Eigenwertproblem. (Mitteilung I–IV). *Annalen der Physik*, **79**, 361–376 (I), 489–527 (II); **80**, 437–490 (III); **81**, 109–139 (IV). Translation, 'Quantisation as a Problem of Proper Values (Parts I–IV),' in Schrödinger (1982).

Schrödinger, E. (1926b). Über das Verhältnis der Heisenberg-Born-Jordanschen Quantenmechanik zu der meinen. *Annalen der Physik*, **79**, 734–756. Reprinted in Schrödinger (1927, pp. 62–84; 1984, Vol. 3, pp. 143–165). Translation, 'On the Relation between the Quantum Mechanics of Heisenberg, Born, and Jordan, and that of Schrödinger,' in Schrödinger (1982, pp. 45–61).

Schrödinger, E. (1982). *Collected Papers on Wave Mechanics*. Third (augmented) English edition. Providence, RI: American Mathematical Society Chelsea Publishing.

Schrödinger, E. (1984). *Gesammelte Abhandlungen/Collected Papers*. 4 Vols. Vienna, Braunschweig/Wiesbaden: Verlag der Österreichischen Akademie der Wissenschaften, Vieweg.

Schwarzschild, K. (1916). Zur Quantenhypothese. *Königlich Preussische Akademie der Wissenschaften* (Berlin). Sitzungsberichte, 548–568.

Shamos, M. H. (ed.) (1959). *Great Experiments in Physics. Firsthand Accounts from Galileo to Einstein*. New York: Dover.

Sommerfeld, A. (1915a). Zur Theorie der Balmerschen Serie. *Königlich Bayerische Akademie der Wissenschaften zu München. Mathematischphysikalische Klasse. Sitzungsberichte*, 425–458. Reprinted in facsimile in Sommerfeld (2013). English translation: Sommerfeld (2014a).

Sommerfeld, A. (1915b). Die Feinstruktur der Wasserstoff- und der Wasserstoff-ähnlichen Linien. *Königlich Bayerische Akademie der Wissenschaften zu München. Mathematischphysikalische Klasse. Sitzungsberichte*, 459–500. Reprinted in facsimile in Sommerfeld (2013). English translation: Sommerfeld (2014b).

Sommerfeld, A. (1916). Zur Quantentheorie der Spektrallinien. Parts I–III. *Annalen der Physik*, **51**, 1–94 (I–II), 125–167 (III). Reprinted in Sommerfeld (1968, Vol. 3, pp. 172–308).

Sommerfeld, A. (1919). *Atombau und Spektrallinien*. 1st edn. Braunschweig: Vieweg. 2nd edn: 1921; 3rd edn: 1922; 4th edn: 1924.

Sommerfeld, A. (1923). *Atomic Structure and Spectral Lines*. London: Methuen. Translation (by H. L. Brose) of 3rd edn of Sommerfeld (1919).

Sommerfeld, A. (1968). *Gesammelte Schriften*. 4 Vols. Edited by F. Sauter. Braunschweig: Vieweg.

Sommerfeld, A. (2000). *Wissenschaftlicher Briefwechsel. Band 1: 1892–1918*. Edited by M. Eckert and K. Märker. Berlin, Diepholz, München: Deutsches Museum.

Sommerfeld, A. (2013). *Die Bohr-Sommerfeldsche Atomtheorie. Sommerfelds Erweiterung des Bohrschen Atommodells 1915/16 kommentiert von Michael Eckert*. Berlin: Springer.

Sommerfeld, A. (2014a). On the Theory of the Balmer Series. *European Physical Journal H*, **9**, 157–177. English translation of Sommerfeld (1915a) with an introduction by M. Eckert (2014).

Sommerfeld, A. (2014b). The Fine Structure of Hydrogen and Hydrogen-like Lines. *European Physical Journal H*, **39**, 179–204. English translation of Sommerfeld (1915b) with an introduction by M. Eckert (2014).

Thomas, W. (1925). Über die Zahl der Dispersionselektronen, die einem stationären Zustand zugeordnet sind. *Die Naturwissenschaften*, **13**, 627.

Van der Waerden, B. L. (ed.) (1968). *Sources of Quantum Mechanics*. New York: Dover.

Van der Waerden, B. L. (1973). From Matrix Mechanics and Wave Mechanics to Unified Quantum Mechanics. In J. Mehra (ed.), *The Physicist's Conception of Nature*, Dordrecht: Reidel, pp. 276–293.

Van Vleck, J. H. (1924a). The Absorption of Radiation by Multiply Periodic Orbits, and Its Relation to the Correspondence Principle and the Rayeigh-Jeans law. Part I. Some Extensions of the Correspondence Principle. *Physical Review*, **24**, 330–346. Reprinted in Van der Waerden (1968, pp. 203–222).

Van Vleck, J. H. (1924b). The Absorption of Radiation by Multiply Periodic Orbits, and Its Relation to the Correspondence Principle and the Rayeigh-Jeans law. Part II. Calculation of Absorption by Multiply Periodic Orbits. *Physical Review*, **24**, 347–365.

Von Neumann, J. (1927). Mathematische Begründung der Quantenmechanik. *Königliche Gesellschaft der Wissenschaften zu Göttingen. Mathematisch-physikalische Klasse. Nachrichten*, 1–57.

Wilson, W. (1915). The Quantum Theory of Radiation and Line Spectra. *Philosophical Magazine*, **29**, 795–802.

Yourgrau, W., and Mandelstam, S. (1979). *Variational Principles in Dynamics and Quantum Theory*, 3rd rev. edn. New York: Dover.

CHAPTER 4

# OF WEIGHTING AND COUNTING

## Statistics and Ontology in the Old Quantum Theory

MASSIMILIANO BADINO

## 4.1 INTRODUCTION: PHILOSOPHICAL PRELIMINARIES

'DEGREES of freedom should be weighted, not counted.'[1] With this witty paraphrase of Schiller, who was in turn paraphrasing an old Yiddish saying,[2] Arnold Sommerfeld summarized a fundamental conceptual tension of statistical mechanics. Sommerfeld was discussing the problem of specific heats in gases, one of the 'clouds of nineteenth-century physics', as William Thomson famously dubbed it. In his quip, he was putting his finger on the fact that what one counts statistically does not always automatically make sense from a dynamical point of view.

This tension traverses the entire history of quantum statistics, and this article tries to unfold the main fracture lines. It is often stated that quantum statistics deals with indistinguishable particles, or it abandons the notion of an individual particle. It is important for our story to keep these concepts separate. The complex relation between distinguishability and individual identity goes back to Leibniz's principle of indiscernibles[3] and depends essentially on which properties must be considered intrinsic. In general, objects are distinguished by properties such as mass, colour, shape, etc.

---

[1] (Sommerfeld, 1911, p. 1061).

[2] In his play *Demetrius*, Schiller refers to votes, while the Yiddish saying refers to words.

[3] For a discussion of this point, see (French, 1989; French and Krause 2006, pp. 15–17). For an overview of the problem of individuality in science, see (Dorato and Morganti, 2013).

However, even if two objects have the same intrinsic properties (say, for example, two identical coins), they are still two distinct individuals. Hence, the central ontological question is: What confers identity to physical objects? For brevity's sake, let's call the criteria to determine the individual identity an ontology. Thus, an ontology is a way to single out the furniture of the universe.

Classical statistical mechanics rests on the delicate equilibrium between two different ontologies. First, there is the ontology of dynamics or D-ontology for short. According to D-ontology, identity is the unifying element of a continuous dynamical story. Two particles, albeit similar in their intrinsic properties, are two different individuals because they belong to two uninterrupted spatio-temporal trajectories.[4] Hence, each particle is labelled by a dynamical story whose uniqueness is sanctioned by the Hamiltonian equations of motion. Moreover, the continuity of this story is a necessary condition to ensure individual identity.

In statistics, however, individuals are entities whose permutation generates a new countable event. In other words, individuals are difference-makers: if one swaps them, the ensuing state is statistically different from the original one and must be accordingly counted. While D-ontology rests on the continuity of trajectories, the ontology of statistics or S-ontology rests on elementary states' equiprobability: states generated by permuting individuals have the same probability. Thus, while the dynamic identity is conferred by persistence over a continuous spatio-temporal trajectory, the statistical identity is ensured by countability under equiprobable events.

This chapter discusses how this tension panned out in the history of quantum statistics. It is organized as follows. In section 4.2, I summarize the basics of Boltzmann's statistics, and in section 4.3, I discuss Planck's peculiar use of combinatorics in his radiation theory. I argue that Planck's reluctance to take an ontological commitment toward microscopic particles combined with his opportunistic use of statistics contributed to making the conceptual tension between D-ontology and S-ontology even more problematic. In sections 4.4 and 4.5, I analyse how the tension developed in radiation theory and gas theory in the 1910s. During these developments, it became increasingly clear that Planck's statistics introduced a form of interdependence, but opinions were split whether it had to be interpreted dynamically or statistically. In the epilogue, I briefly discuss the transition to Bose–Einstein statistics.

# 4.2 PROLOGUE: BOLTZMANN'S STATISTICS

The name of Ludwig Boltzmann is inextricably linked to the birth of modern statistical mechanics. Although he was not the first to apply statistics to thermodynamics, he was certainly the first to use combinatorial arguments, most notably in his famous 1877

---

[4] (French and Krause, 2006, pp. 40–51).

article. To be sure, he had briefly touched upon similar issues at the end of another paper, back in 1868.[5] On that occasion, Boltzmann calculates the marginal probability for a gas particle to have a certain energy. He assumes that an $n$-particle system's total energy can be divided into $p$ elements of magnitude $\varepsilon$. The probability that the energy of an arbitrary particle lies in the interval between $i\varepsilon$ and $(1 + i)\varepsilon$ depends on the number of ways in which the remaining energy can be distributed among the remaining particles or, which is the same, the number of ways in which the particles can be distributed over the corresponding energy intervals. In this model, both the particles and the energy intervals are individuals, i.e., the particles' permutation is a countable event. In the general case, the number of ways in which one can distribute $n$ individual particles over $p$ individual boxes amounts to:

$$J(n,p) = \frac{(n + p - 1)!}{(n - 1)!(p - 1)!} \tag{1}$$

Bear in mind this number because it will play a key role in our story.[6]

From an early stage, Boltzmann was convinced that statistics and mechanics were tightly interwoven and must play an equal role in explaining thermal equilibrium. In the 1870s, he explored multiple strategies and, even when he was not using statistics explicitly, one could see it lurking behind purely mechanical analyses.[7] But it is in the 1877 paper that combinatorics feature prominently.[8] While conceived to investigate the relations between the second principle of thermodynamics and probability calculus, this article is historically important because of the role it would play in Planck's blackbody theory of December 1900. To fully appreciate this role, it is necessary to dive into the intricacies of Boltzmann's combinatorial models.

In 1877 Boltzmann developed no less than three different combinatorial models whose relations reveal the connections between statistics and ontology. The first and most famous model is described through an analogy with tickets drawn from an urn.[9] Let us assume the total energy of a system of $n$ particles be divided up into $p$ elements of fixed magnitude $\varepsilon$. Let us also assume we have an urn with countless tickets, each of which carries a number between 0 and $p$. Imagine now to make $n$ draws, and at each draw, the corresponding number of elements is attributed to a certain particle and then returned to the urn. After $n$ draws, one obtains an allocation of energy elements over

---

[5] (Boltzmann 1868, 1909 I, pp. 92–96); for a discussion of this argument, see (Badino 2011 pp. 359–60).

[6] (Boltzmann, 1909 I, p. 85). For the details of Boltzmann's combinatorial calculation, see (Badino, 2009 pp. 83–85; Costantini et al., 1996 pp. 284–88; Uffink, 2007 pp. 955–56).

[7] See (Badino, 2011). It should be noted that this reading is at odds with the common wisdom according to which Boltzmann followed a purely kinetic approach to thermal equilibrium until he was convinced by his colleague Josep Loschmidt that use of statistics was inevitable. For the classical interpretation, see (Klein, 1973; Brown et al., 2009). For a general analysis of Boltzmann's works in statistical mechanics, see (Darrigol, 2018a).

[8] (Boltzmann, 1877, 1909 II, pp. 164–23).

[9] (Boltzmann, 1909 II, pp. 167–86).

the particles, which Boltzmann calls a *complexion*. Likely, the sum total of the energies will be larger or smaller than the system's total energy. In that case, we repeat the entire procedure until we have a large number of acceptable complexions.

Now, Boltzmann notices that, while a permutation of energies between two particles gives a different complexion, it leaves untouched the energy distribution. Thus, assuming that all complexions are equiprobable, the probability of the distribution is proportional to the number of complexions corresponding to it, that is, the number of permutations of a complexion. As for the normalization factor, this is the total number of ways of distributing $p + 1$ individual tickets[10] over $n$ individual particles or, which is the same, $n$ individual particles into $p + 1$ individual boxes. This is precisely number (1) above, hence the probability of distribution is:

$$ P = \frac{1}{J(n, p + 1)} \frac{n!}{n_0! \ldots n_p!} \tag{2} $$

where $n_i$ is the $i$-th occupation number, which is the number of particles with $i$ elements of energy. By maximizing the probability and letting $\varepsilon$ go to zero, Boltzmann finds Maxwell's distribution.

But Boltzmann isn't through with combinatorics. First, he toys shortly with a provisional model, which must be rejected because it does not satisfy exchangeability.[11] Much later in the paper, however, he introduces a third and more interesting urn model.[12] Contrary to the first procedure, he now supposes to make $p$ draws with reintroduction from an urn containing the $n$ particles suitably labelled. At the end of the $p$ draws, each particle is assigned a number of energy elements equal to the number of times it was drawn. This urn model automatically satisfies the energy constraint, but this is perhaps the least interesting of its features. Although the final result is again a complexion, this urn model is statistically very different. One allocates a single energy element over a certain particle at each step, but the energy elements are statistically non-individual because swapping two draws does not change the final result. As one makes $p$ draws, the model consists of distributing $p$ non-individual energy elements over $n$ individual particles. Boltzmann calculates the total number of complexions generated by this model, which turns out to be:

$$ J'(n, p) = \frac{(n + p - 1)!}{(n - 1)! p!} = J(n, p + 1) \tag{3} $$

In other words, the total number of ways of distributing $n$ individual particles over $p+1$ individual energy levels is equal to the total number of ways of distributing $p$ non-individual energy elements over $n$ individual particles.

---

[10]  Differently from the 1868 model, here 0 energy is a possible value.

[11]  (Boltzmann, 1909 II, pp. 171–72); for a discussion of this model, see (Costantini *et al.*, 1996, pp. 288–92).

[12]  (Boltzmann, 1909 II, pp. 211–14).

Boltzmann wasn't too moved by what he probably considered a formal coincidence. In hindsight, an argument can be made that Boltzmann is here switching between two different statistics. In the first urn model, individual particles are allocated over *individual energy levels*, and each such configuration is equiprobable: classical Maxwell–Boltzmann statistics rule the model. The third model behaves differently. An elementary configuration is *the number* of energy elements that go over individual particles. Hence, a countable event is the permutation of groups of energy elements, which means that they follow Bose–Einstein statistics. This difference might appear purely formal (after all, the energy levels in the first model are defined by the number of energy elements), and, certainly, Boltzmann never invites the reader to consider energy elements as distinguishable. However, some writers discussed the possibility that Boltzmann was, in fact anticipating the Bose–Einstein statistics.[13] For Boltzmann, the abundance of formal models was not a philosophical problem because his statistical mechanics was firmly rooted in the classical D-ontology. There is no doubt that he thought of particles as identifiable by their continuous spatio-temporal history. At times he compares molecules to macroscopic individuals,[14] and in the introduction to his *Lectures on the Principles of Mechanics*, he explicitly states the continuity principle and declares that 'it only allows us to recognize the same material point at different times.'[15] A robust D-ontology constrains the S-ontology and allows Boltzmann to tell apart genuine combinatorial models from formal artefacts. In the next sections, we shall see how this natural coupling between the two ontologies was disrupted by Planck's radiation theory and how physicists at the beginning of the 20th century laboriously searched to restore it in a new form.

# 4.3 ENTER THE QUANTUM

## 4.3.1 Planck's Programme

With few exceptions, Boltzmann himself never made use of his complicated combinatorial model after 1877. He probably regarded it more as an ingenious illustration of irreversibility's statistical nature than a physically workable instrument. One reason this model is so famous is that it made a surprising comeback in a completely unrelated ambit: Max Planck's theory of blackbody radiation.[16]

---

[13] See, for example, (Bach, 1987, 1990; Costantini and Garibaldi, 1997).

[14] This analogy, which was customary for Maxwell too, is stated at the beginning of his famous 1872 paper on the Boltzmann equation (Boltzmann, 1872, 1909 I, pp. 316–402).

[15] (Boltzmann, 1897, p. 9).

[16] On the connection between Boltzmann's combinatorics and Planck's theory see (Hoyer, 1980; Darrigol, 1988).

When Planck started studying the blackbody problem, in the late 1890s, it seemed to be a moribund, if not dead, horse. Introduced by Gustav Kirchhoff in 1860 as a formal tool to investigate thermal radiation, a blackbody was conceived as a physical system able to absorb all the radiation impinging on it.[17] As Kirchhoff proved by means of the second law of thermodynamics, the energy distribution of a blackbody is particularly simple because it is a universal function of temperature only. This theorem led not only to construct experimental approximations of a blackbody but also to find theoretical constraints on the energy distribution such as the Stefan–Boltzmann law and Wien's displacement law. Increasingly accurate measurements in the mid-1880s showed that the energy distribution was characteristically bell-shaped, which suggested that the radiation law contained an exponential function of the energy and temperature. By cunningly combining thermodynamic, kinetic, and electromagnetic arguments, in 1896 Wilhelm Wien reached an expression for the energy distribution that, apparently, fitted well the existing observations.

Thus, when Planck started to work on the blackbody theory, the problem in itself did not offer a theoretical challenge. Planck, however, was aiming at a much larger target. A fierce opponent of Boltzmann's statistical view of irreversibility, he was determined to show that, under certain conditions, a conservative system behaves in a strictly irreversible way. The blackbody was described by the reversible Maxwell equations, but, at the same time, it had the traits of an irreversible thermodynamic process. Besides, it was a perfect case for three reasons. First, it was theoretically simple and fairly well-known. Second, the thermal features of blackbody radiation were independent of the kind of matter–radiation interaction occurring within the cavity. Third, being a purely electromagnetic phenomenon, Planck could eschew all the quandaries related to molecular collisions.

Planck supposed a spherical cavity filled up with radiation and a Hertzian resonator at its centre.[18] It is important to notice that the resonator interacts only with the field component at nearly the same characteristic frequency; therefore, it is not able to change the energy distribution of the cavity. However, it does make the radiation more spatially uniform by absorbing a plane wave and re-emitting a spherical wave. Hence, at this stage, Planck was interested only in the isotropy of the blackbody radiation. Using Maxwell's equations, the resonator equations, and reasonable boundary conditions, Planck calculated the field-resonator interaction and its time-reversal to show that the latter does not fulfil the boundary conditions. As the time-reversal was not a physically acceptable solution to the electromagnetic problem, Planck concluded that the field-resonator interaction was a strictly irreversible process.[19]

Unsurprisingly, Boltzmann opposed this conclusion. He realized that Planck's argument was embarrassingly flawed, as he summarized in a letter to Felix Klein:

---

[17] The standard reference on the experimental background of Planck's theory is (Kangro, 1970).

[18] On the details of Planck's theory of radiation, see (Kuhn, 1978, pp. 72–91; Darrigol, 1992, pp. 29–50; Badino, 2015, pp. 41–80; Duncan and Janssen, 2019, pp. 51–64).

[19] (Planck, 1898a, 1958 I, pp. 508–31).

Herr Planck has reversed the exciting wave for a specific case, but he has completely forgotten that the wave prior emitted by the resonator must be reversed as well. From the circumstances that he has obtained a totally counterintuitive formula, he has not concluded that he was wrong, but rather that he had found out a process whose reversal is not possible. I have sent him directly my considerations, a move that will not necessarily shorten the dispute; I'm curious to hear his response.[20]

Boltzmann also proved that if the time-reversal is calculated correctly, it is necessarily a physically acceptable solution to Planck's electromagnetic problem. Planck had no choice but to abandon his original project and find an alternative route to irreversibility. His new argument, which he developed in 1898 and 1899, hinged on two main pillars.

## 4.3.2 Natural Radiation and Entropy

Although badly hit by Boltzmann's criticism, Planck did not give up his thermodynamics-inspired approach to radiation theory. This approach rested on a sharp distinction between physically meaningful macroscopic quantities and the mysterious microscopic world. The former are empirically measurable, slow-varying quantities such as the field intensity and the resonator energy, while the latter is a physically inaccessible and ontologically suspicious realm. More precisely, Planck regarded macroscopic quantities as the time averages of innumerable fast-changing monochromatic field vibrations, which have no independent physical meaning on their own. The statement that the microscopic field vibrations change so rapidly and are so disorderly that they can be safely ignored and replaced by their averages is the hypothesis of natural radiation, Planck's first pillar.[21]

To appreciate the ontological role played by natural radiation, it is instructive to contrast it with kinetic theory. Microscopic field quantities correspond to individual molecular quantities such as position and velocity, while macroscopic quantities are averages such as total energy. Natural radiation is tantamount to stating that molecular dynamics is so complicated and uncontrollable that one might as well black-box much of what is happening at that microscopic level and confine the calculations to macroscopic, measurable quantities.

The second pillar of Planck's theory is the definition of entropy.[22] After Boltzmann's radical rebuttal of his first argument, Planck resorted to his opponent's favourite strategy: to show that a certain state function existed, which increases monotonically over time. To define electromagnetic entropy, Planck worked backwards. He knew that

---

[20] Boltzmann to Felix Klein, 12 February 1898 (Höflechner, 1994, doc. 462).
[21] The hypothesis of natural radiation is first introduced in (Planck, 1898b, 1958 I, pp. 532–59). For a discussion, see (Kuhn, 1978 pp. 80–82; Badino, 2015 pp. 60–71).
[22] (Planck, 1899, 1958 I, pp. 560–600).

the maximum of that entropy must give Wien's law, which holds at equilibrium. He then worked out an entropy formula that fitted such a requirement and showed that, given natural radiation, the entropy calculated for the combination of the free field and the field-resonator interaction was a monotonically increasing function.[23]

## 4.3.3 Resorting to Combinatorics

As we have seen, Planck's general argument for irreversibility was tightly related to the fate of Wien's law. At the beginning of 1900, however, Wien's law was being challenged from several quarters. Already in March, Otto Lummer and Ernst Pringsheim argued that at a temperature as high as 1800 K and in the region of wavelengths between 12 and 18 μ, 'the Wien–Planck spectral equation does not represent the black radiation measured by us.'[24] The situation deteriorated further in October when Heinrich Rubens and Ferdinand Kurlbaum found a marked failure of Wien's law at wavelengths equal to 51.2 μ. Planck was informed in advance and could work out a new energy distribution formula by interpolation,[25] but the new law was incompatible with his previous entropy formula. It was at this point that Planck resorted to Boltzmann's 1877 model.[26] Much ink has been spilled over Planck's combinatorial argument in December 1900. Here I want merely to recapitulate the key differences with Boltzmann's original procedure.

To find the equilibrium distribution between resonators and free field, Planck assumes the energy divided into quanta of fixed magnitude $h\nu$, so that the total energy to be allocated over the resonators with natural frequency $\nu_i$ can be written as $E_i = P_i h\nu_i = P\varepsilon_i$. If there are $N_i$ resonators with that frequency, the number of ways of distributing the energy quanta is:

$$J'(N_i, P_i) = \frac{(N_i + P_i - 1)!}{(N_i - 1)!P_i!} \tag{4}$$

Using (4) as the equilibrium probability and plugging it into the so-called Boltzmann's principle that entropy is proportional to the logarithm of the state probability, Planck arrives at the correct energy distribution law.[27]

There are two important points to notice about Planck's combinatorial procedure. First, historians have been discussing for years the status that Planck ascribed to the quantum in 1900. While Martin Klein argued that Planck considered the energy

---

[23] Because of the formal similarity between Planck's electromagnetic entropy and Boltzmann's $H$-function, several authors speak of an electromagnetic $H$-theorem (Kuhn, 1978, pp. 72–91; Darrigol, 1992, pp. 45–50).

[24] (Lummer and Pringsheim, 1900, p. 171).

[25] (Planck, 1900b, 1958 I, pp. 687–89).

[26] (Planck, 1900c, 1958 I, pp. 698–06).

[27] For the details of Planck's combinatorial calculations, see (Badino, 2015, pp. 94–98).

elements as discrete indistinguishable units to be allocated individually on resonators, Thomas Kuhn countered that the energy elements could play the same role as continuous and distinguishable energy intervals.[28] To be sure, statistics, with some cautions, supports both claims. Remembering equation (3), Planck's equation (4) can be interpreted as the total number of ways of distributing $N_i$ individual resonators over $P_i + 1$ individual energy intervals or $P_i$ non-individual energy elements over $N_i$ individual resonators. The formal ambiguity that worked for Boltzmann's 1877 model works here as well. Planck was aware of this ambiguity[29] and, after all, it fitted perfectly with his thermodynamic approach. As we have seen, Planck's general strategy was to black-box the microscopic part of his theory, so it was just a fortunate coincidence that statistics allowed him to maintain a non-committal position about the ontology of resonators and quanta.

However, one could argue that, where statistics fails, physics comes to the rescue. In Planck's combinatorial model, the resonators play the role of molecules so that they might be granted individuality: the D-ontology can be applied to them. By contrast, energy elements are not localized, so they cannot be labelled. This brings me to my second point. There is a fundamental ontological fracture between the two parts of Planck's theory, and the statistical underdetermination only makes it worse. Clearly, the combinatorial part relies on the analogy between resonators and molecules: it is this analogy that enables Planck to deploy Boltzmann's procedure. But in the radiation part, the elementary entities are the field monochromatic components. In a paper written in March 1900, Planck himself warns the reader against a quick analogy between resonators and molecules, for the same reasons as above.[30] Hence, both the S-ontology and the D-ontology of Planck's theory are ultimately confused and unstable. This instability, as we shall see, is the original sin of Planck's theory.

# 4.4 EHRENFEST, EINSTEIN, AND THE RIDDLE OF RADIATION THEORY

It is difficult to tell the story of the emergence of quantum statistics. It consists of several seemingly unrelated research lines, exotic ideas, and obscure arguments. It comes as no surprise that, in describing this episode, some writers used the term 'serendipity' to signal the absence of coherent development.[31] If we focus on the tension

---

[28] (Klein, 1962; Kuhn, 1978, 1984). For a survey of the debate, see (Badino, 2009; Gearhart, 2002).

[29] Much later, in a passage of his *Lectures on the Heat Radiation*, Planck makes an explicit reference to the twofold models underlying his combinatorial calculations (Planck, 1906 pp. 151–52).

[30] (Planck, 1900a, 1958 I, pp. 668–86); see especially pp. 673–674.

[31] See, for instance (Delbruck, 1980; Bergia, 1987). Some writers have described this story as the thermodynamic route to quantum mechanics as opposed to the traditional atomic route (Darrigol, 1991, 2002; Desalvo, 1992).

between D-ontology and S-ontology, we can, with a good measure of approximation, single out three reactions to it. First, Ehrenfest and his disciples insisted that Planck's combinatorics called for a radical change in the classical S-ontology. Second, Einstein, albeit in agreement with Ehrenfest's diagnosis, thought that a way out of the impasse was to modify the D-ontology through wave–particle duality. Lastly, Planck shifted his attention from radiation theory back to thermodynamics and tried to justify the new S-ontology as the result of the application of the quantum hypothesis to the ideal gas.

From the very beginning, it was clear that the problem of Planck's theory lay in the mix of combinatorics and quanta. As for the resonator, it seemed to play only a marginal role. Lord Rayleigh, James Jeans, and Hendrick Antoon Lorentz showed that they could be effectively replaced by normal vibration modes in the free field. Paul Ehrenfest even reinforced this conclusion by showing that Planck's resonator-field interaction mechanism could not ensure an entropy increase.[32] To muddle the situation even more, in 1905, Einstein published a landmark article where he proposed a daring analogy between gas and radiation. By comparing the entropy variations of an ideal gas and cavity radiation in the Wien regime, Einstein concluded that radiation behaved as if it consisted of corpuscles of energy $E = nh\nu$.[33] Although the idea that the quanta existed in empty space was too unorthodox to be accepted, Einstein's theory had the merit of giving a statistical backdrop to Wien's law and thus rapidly became a useful term of comparison. In particular, understanding the statistical difference between Einstein's and Planck's quanta became Paul Ehrenfest's main research theme.

A loyal follower of Boltzmann and the guardian of the sacred fire of Boltzmannian statistical mechanics, Ehrenfest pursued this goal with his characteristically obsessive devotion. In 1906, he set out to clarify once and for all what made it possible to obtain the correct radiation law.[34] He considered a cavity full of radiation (without resonators) and defined a tridimensional distribution function for the normal modes in terms of their frequency, amplitude, and momentum. He then calculated the maximum entropy for the cavity radiation under the constraints of normalization and total energy. Unsurprisingly, the calculation yielded the Rayleigh–Jeans law, whose validity is limited to the low-frequency regime. Ehrenfest then investigated how the argument could be modified to get Planck's law in its stead and concluded that the energy quanta's fixed magnitude was the key additional constraint. This general analysis convinced Ehrenfest that the quantization was a purely formal device, which somehow had to play a role in the statistical part of the theory. In other words, while quantization of energy was a sufficient condition for Planck's law, Einstein's corpuscularization was

[32] (Ehrenfest, 1905). Ehrenfest's criticism was, in fact, a bit unfair. Planck's mechanism was not meant to redistribute energy among frequencies, but rather as a spatial equilibrator, a point Ehrenfest himself would later acknowledge.

[33] (Einstein, 1905).

[34] (Ehrenfest, 1906); for a discussion, see (Navarro and Pérez, 2004, pp. 101–102).

not. For Ehrenfest, it was a matter of finding the correct S-ontology implicit in Planck's combinatorics, while the D-ontology was secondary.

In the meantime, Einstein was working in a different direction. He also realized that quantization was a sufficient condition but insisted that the solution of the mystery lay in the physics. These thoughts led him to a hypothesis possibly even more destabilizing than the light quantum. In 1909, Einstein argued that if we calculate the fluctuation of the cavity radiation using Planck's law, the resulting formula is made of two parts: a classical expression for waves' interference and the fluctuation generated by a system of independent particles.[35] To Einstein, this weird cohabitation meant that classical D-ontology underlying statistical mechanics had to be supplemented with some interaction mechanism between the particles whose deep nature was hidden in their undulatory features. Convinced of the corpuscular structure of radiation, Einstein argued that quanta were physical individuals, but their individuality had to be negotiated with a wave aspect.[36]

Ehrenfest thought differently. During the late 1900s, he worked intensely on the problem of radiation, trying to understand why Planck got the right answer, although his combinatorics was incompatible with Boltzmann's.[37] The breakthrough happened in 1911 when he proved that quantization was not only sufficient for Planck's law, but it was, in fact, necessary.[38] Near the end of the article and almost in passing, Ehrenfest finally hit the nail on the head. The key difference between Einstein's and Planck's quanta was that the former were independent, while the latter manifested a non-classical statistical interdependence. Planck's peculiar twofold derivation of the radiation law (see section 4.3.3) had masked the fact that quanta are statistically very different from gas molecules. To clarify this point, Ehrenfest argues, it suffices the following consideration. In Einstein's combinatorics, if a particle has, say, $n$ quanta of energy, this is the result of receiving one quantum in $n$ independent attributions. By contrast, if a resonator has $n$ quanta, they have been allocated together so that the individual quanta making up the total energy are interdependent. Of course, this is reminiscent of Boltzmann's third model, but the fact that the quanta do not vanish makes the key difference. Ehrenfest realizes that this simple fact changes the S-ontology radically because it introduces a new countable event: a permutation of an entire bunch of quanta changes the state. This point is best formulated in an article authored with

---

[35] (Einstein, 1909). On Einstein's work in radiation, theory see (Duncan and Janssen, 2019, pp. 94–107).

[36] The commitment toward an enlarged D-ontology marked Einstein's work in radiation theory; see, for example, his 1916–17 theory of emission and absorption in which he attributes a momentum to light quanta (Einstein, 1917).

[37] Ehrenfest's research notebooks registered his strenuous efforts to cope with the statistical puzzle. Particularly in note 843, written on 21 March 1911 (of the Russian Calendar), he expresses all his frustration that Planck's procedure 'must be wrong' and still leads to the correct result ('but how', double underscore). See (Navarro and Pérez, 2004, p. 119).

[38] (Ehrenfest, 1911). For a detailed analysis, see (Klein, 1970, pp. 245–51; Navarro and Pérez, 2004, pp. 110–18).

Kamerlingh Onnes: 'The real object which is counted remains the number of all the different distributions of $N$ resonators over the energy grades 0, $\varepsilon$, 2$\varepsilon$,...with a given total $P$.'[39]

Thus, the discussion on Planck's radiation theory generated two research paths, which, albeit closely related, were often seen as competing. Some scholars such as Ehrenfest and Władysław Natanson[40] focused primarily on unearthing the S-ontology hidden in the ambiguities of Planck's combinatorial procedure. By contrast, for other physicists like Einstein, understanding the D-ontology was much more illuminating. An illustration of how detached these two paths could become is the short dispute between Mieczyslaw Wolfke and Georg Krutkow. A follower of Einstein, Wolfke published two short communications in which he proposed to treat Einstein's light quanta as 'light atoms' that can be dynamically labelled and combinatorially treated to obtain Planck's law. Krutkow, who was a student of Ehrenfest, replied that Einstein's light quanta were independent and, as such, could not be correctly treated by Planck's combinatorics. Taken by surprise, Wolfke argued that Einstein's quanta were *existentially* independent but *spatially* interdependent, that is, they, presumably, were individuals connected by some sort of physical interaction.[41] Krutkow did not pursue the discussion further because it was obvious that they saw the ontological problem from two very different perspectives.

# 4.5 THE ISSUE OF ENTROPY EXTENSIVITY

## 4.5.1 Early Attempts at Gas Quantization

During the 1910s, the quantum was applied to atoms in two circumstances. One was, famously, the construction of atomic models by Bohr and Sommerfeld. The other was the much less famous but not less important attempt at formulating a quantum theory of the ideal gas. To understand the contest of this attempt, a little detour into physical chemistry is necessary.

The most important quantity for calculating equilibrium in chemical reactions is the so-called equilibrium constant, which is the ratio between forward and backward rates. At the end of the 19th century, by applying thermodynamics to chemistry, Jacobus Henricus van 't Hoff and Josiah Willard Gibbs managed to find connections between the equilibrium constant, temperature, and maximum work but failed to give a direct method to measure it from calorimetric data.[42] The problem was that classical

[39] (Ehrenfest and Kamerlingh Onnes, 1915, p. 873).
[40] (Natanson, 1911); on Natanson's fascinating analysis see (Kokowski, 2019).
[41] (Krutkow, 1914; Wolfke, 1913a, 1913b).
[42] For a survey on the problem of chemical equilibrium and its relation with thermodynamics, see (Badino and Friedrich, 2013, pp. 299–02).

thermodynamics provided no tools to determine integration constants. A break-through happened in 1906 with Walther Nernst's heat theorem, which entailed a zero value for the integration constant of entropy. Unfortunately, Nernst's arguments cannot be applied to gases, thus a workaround had to be found. Nernst noticed that the van 't Hoff equation for a gas could be solved by exploiting a relation between its integration constant and the Clausius–Clapeyron equation, which gives the pressure of a gas in equilibrium with a condensate. The integration constant of the van 't Hoff equation was dubbed by Nernst the chemical constant.

While this further change of constant set a different experimental problem, it did not necessarily ease the experimenter's task because it was difficult to measure chemical constants at temperatures low enough to have an equilibrium with a condensate. Somewhat unexpectedly, help came from quantum theory. In the third edition of his *Lectures on Thermodynamics*, Planck stressed that, combined with Nernst's heat theorem, quantum theory entailed that it was possible to express the entropy constant in terms of universal quantities.[43] This hint was picked up by Otto Sackur, who spotted a theoretical relation between the chemical constant of a gas and its entropy constant.[44] The problem was thus reduced to finding a quantum expression of the entropy of a gas. From a theoretical standpoint, however, the application of the quantum hypothesis to the ideal gas is far from trivial, the reason being that quantization requires a natural frequency, while there is no such thing for gas molecules. Sackur circumvented the obstacle by exploiting another of Planck's hints. In *Heat Radiation*, Planck had noticed that the fixed magnitude of the energy element entailed a partition of the phase space of a resonator into 'elementary regions'. Sackur's simple but brilliant idea was to transfer this procedure to the gas phase space and then apply Boltzmann's combinatorial approach to arrive at an entropy formula.[45] It worked surprisingly well, and Sackur managed to find an expression for the entropy constant of a monatomic ideal gas, but there was a small problem with extensivity. The nature of this problem is deeply entrenched with the tension between D-ontology and S-ontology.

It is well known from thermodynamics that entropy is additive (i.e., the entropy of a system is equal to the sum of the entropies of its subsystems) and extensive (i.e., it depends on the quantity of the system, for instance, the number of gas molecules). Under certain conditions, these two properties lead to a phenomenon called the Gibbs paradox.[46] Let's assume a container divided by a partition into two equal volumes filled up with gas. Let's now remove the partition and let the two amounts of gas mix. Two things can happen. If the gases in both volumes are chemically indistinguishable, the

[43] (Planck, 1911, pp. 268–69).

[44] On Sackur's life and work, see (Badino and Friedrich, 2013).

[45] (Sackur, 1911, 1912).

[46] On the physical definition and the philosophical consequences of the Gibbs paradox, see (Pešić, 1991; Dieks, 2018; Saunders, 2018); on its history see (Darrigol, 2018b).

removal of the partition can be performed reversibly, and the final entropy is the sum of the initial entropies. By contrast, if the gases in the volumes are different, the mixing process is irreversible and leads to an additional entropy term, called mixing entropy. The paradoxical aspect of the entire process lies in the fact that the mixing entropy is nonzero even for infinitesimally different gases but disappears suddenly when they become indistinguishable, a behaviour that apparently contradicts the fact that entropy is a state function. In fact, this discontinuous behaviour is a consequence of the clear-cut distinction between reversible and irreversible processes. Gas molecules on either side of the partition are D-ontologically, and therefore S-ontologically, distinguishable regardless of their chemical similarity, and their swap counts as a different microstate. This means that some statistical adjustment is necessary to safeguard entropy extensivity in those cases in which, from a thermodynamic standpoint, no mixing entropy is produced. Gibbs solved the issue by introducing a distinction between two notions of statistical description or, in his parlance, phase.[47] Although a permutation of molecules makes a difference in *specific* (microscopic) phases, it does not change the *generic* (macroscopic) phase.[48] Sackur's method to save extensivity was more cumbersome but consequential nonetheless. He supposed to divide the volume $V$ of the gas into $N/n$ independent sub-volumes $v$, $N$ being the total number of molecules and $n$ the molecules in each sub-volume. As the sub-volumes are arbitrary, Sackur is, in fact, introducing an interdependence between the size of the elementary regions and the number of molecules.

In one of those odd coincidences that occasionally occur in the history of science, the seventeen-year-old Hugo Tetrode published, almost at the same time, an alternative approach to the same problem.[49] Tetrode's solution was much more in the spirit of Gibbs' statistical mechanics, although his language was still largely Boltzmannian. Contrary to Sackur, Tetrode worked directly with probability and noticed that, if the particles are similar and 'exchangeable', the definition of probability must be accordingly modified by cancelling out the permutations originated by exchanging similar particles. This entails that the Boltzmannian probability must be corrected by a division by $N!$, where $N$ is the number of particles. This 'corrected Boltzmann counting' as we call it today, is introduced by Tetrode as a straightforward consequence of Gibbs' distinction between specific and generic phases.

The duality in gas theory reminds us of the duality we already observed in radiation theory. Once again, physicists were split between focusing on the S-ontological side of the problem or instead looking for a D-ontological solution in terms of the definition of physical identity.

[47]  (Willard Gibbs, 1902, p. 187).
[48]  On the importance of the concept of generic phase for indistinguishability see (Saunders, 2020).
[49]  (Tetrode, 1912b); see also (Tetrode, 1912a). On Tetrode's life and work, see (Dieks and Slooten, 1986).

## 4.5.2  Planck's Quantum Theory of the Ideal Gas

In the second edition of *Heat Radiation*, Planck had stressed that the fixed size of the phase space cell and the absolute meaning of entropy were the distinctive features of quantum theory.[50] Thus, when Sackur published his idea of interdependence between elementary regions and the number of molecules, Planck received it sympathetically. In particular, he saw Sackur's idea hinging upon a critical point: the disanalogy between resonators and gas molecules. He first elaborated on this idea in a lecture delivered at the Wolfskehl conference in April 1913. Resonators, Planck argued, occupy a fixed place, and they only interact with the free field. Thus, each resonator can be easily identified in space and time. This is not the case for gas molecules, which move and interact with each other by collisions. According to Planck, this entails a difference in the structure of the elementary regions of the corresponding phase spaces. While regions representing the state of resonators are individual ellipses, the elementary regions of each molecule should somewhat depend on the other molecules' state. This holistic interdependence was the physical reason for Sackur's correction to attain extensivity.[51]

Planck's argument was indeed sketchy and, unsurprisingly, did not pass the test of Hendrik Antoon Lorentz: it seemed to invoke a mutual determination in position and momenta of the molecules, which was inconsistent both with usual statistical assumptions and with the physical nature of an ideal gas.[52] These points were well received, and Planck refined his approach. He accepted that extensivity could be reached by dividing the number of complexions by $N!$, but could not force himself to regard this as a mere formal trick. On the contrary, he was convinced that a modification in the S-ontology called for some change in the D-ontology of the gas molecules. In the ensuing years, he concentrated his efforts on studying the structure of the phase space, a research that brought him very close to the programme pursued by Arnold Sommerfeld in atomic theory.

By introducing the partition function, Planck was able to show that the calculation of the most important thermodynamic functions boils down to the computation of the accessible states, which, in turn, depends on how one partitions the phase spaces into elementary regions.[53] In classical statistical mechanics, the elementary regions are infinitesimal in size and are defined by the integrals of motion, but in quantum physics, the constant $h$ determines a finite size of the regions. So far, Planck's procedure simply generalized to an arbitrary mechanical system the argument developed for the resonator in 1906. But now came a new step. We need to distinguish, Planck argued, between two formal spaces: one describing the individual state of the system and one describing

---

[50]  See (Planck, 1912, sec. 125).

[51]  (Planck, 1914, pp. 7–8, 1958 II, pp. 320–21); on Planck's quantum theory of monatomic ideal gas see (Badino, 2010).

[52]  See (Lorentz, 1915).

[53]  See (Planck, 1916b, 1958 p. II, 420–34); see also (Planck, 1916a, 1958 p. II, 386–19).

the physically meaningful states. This difference is introduced to account for a combinatorial fact:

> [I]f groups of similar atoms are present in the body (...) a more or less large number of physically equivalent points of the phase space will be ascribed to a certain physical state of the body because a given point of the phase space of a single individual atom depends on determined coordinates and velocity. As many per-mutations of similar atoms are possible, so many phase points will correspond to a given physical state. Therefore, to clarify the expression, I will distinguish between 'phase point' [Phasenpunkt] and 'state point' [Zustandpunkt].[54]

Sackur's idea of dividing up the formal space into two different ways here makes a second and more sophisticated appearance. Planck states that while the phase space accounts for atoms' individuality in terms of their Hamiltonian properties of position and momenta, we still need to account for their physical indistinguishability. As this is combinatorially dealt with by cancelling out a corresponding number of complexions, Planck introduces a new formal space whose elementary regions—or state-regions as he calls them—correspond to an $N$-dependent cluster of phase regions. Planck's move sounds familiar. In his 1899 radiation theory, he had affirmed that individual oscilla-tions in the electromagnetic field had no independent physical meaning as opposed to macroscopic quantities. Here, again, Planck expresses a physical difference in terms of individuality and physical meaning.

The conceptual similarities between Planck's radiation theory and his quantum theory of gas did not escape the attention of Ehrenfest. In an article written with Viktor Trkal, Ehrenfest insisted that Planck's absolute entropy approach was wrongheaded. Using Boltzmann's dissociation theory, they showed that extensivity could be easily recovered as a property of entropy differences rather than entropy itself.[55] Hence, Planck's abstract distinction between phase- and state-space was unwarranted. One needs to stay with good old thermodynamics and refer entropy to corresponding reversible processes. Extensivity would then come out as a natural output of Boltz-mann's theory.

Planck replied almost immediately with an article and, very tellingly, with a brand new section of the fourth edition of his *Heat Radiation*.[56] As in the Wolfskehl lecture, he moved from statistics back to dynamics and, once again, he exploited the disanalogy between resonators and gas molecules. The former are fixed in space and cannot interact, while the latter are in continuous interaction. This generates an *Austausch-möglichkeit* (exchange possibility), i.e., molecules can be exchanged without altering the macroscopic state. It is important to stress that the physical reason for the *Austausch-möglichkeit* is the series of correlations and symmetries originated by the molecular

---

[54] (Planck, 1916b, p. 661, 1958 II, p. 428).
[55] (Ehrenfest and Trkal, 1921).
[56] (Planck, 1921a, 1958 II, pp. 527–534, 1921b, sec. V).

interaction: '[in the case of gas] we have no system of separate molecules, but a single structure arranged by symmetries and these symmetries consist in the fact that there is no physical mark that allows to single out a certain atom if one considers the gas first in one state and then in the other.'[57] Despite the obscure language, it is not difficult to grasp what Planck is after. The correction of the traditional Boltzmann counting entails a sort of D-ontological interdependence between particles that Planck interpreted, much like 1899, as a feature of the system's mysterious microdynamics. Although he is not explicit about this point, Planck arguably regarded the $N!$ division and the *Austauschmöglichkeit* on a par with the hypothesis of natural radiation: a simplifying assumption to account for the complex microscopic interactions.

## 4.6 EPILOGUE

Famously, the breakthrough eventually happened in radiation theory. In 1924, Satyendra Nath Bose, an obscure Indian physicist, sent to Einstein a short article in which he proposed a new derivation of Planck's law. Bose's simple yet effective idea was to start with a new definition of a countable event, i.e., a distribution of quanta (treated as a particle) over cells, and then recast this event in an apparently classical formula:[58]

$$W = \prod_s \frac{A_s!}{P_0^s! P_1^s! \dots} \tag{5}$$

where $W$ is the number of ways of distributing $A^i$ cells over $N^j$ quanta, and $P^s_i$ is the occupancy number, i.e., the number of cells containing $i$ quanta. In this way, Bose scaled up the statistical description to a new definition of a countable event and a new analogue of the distribution in Boltzmann's statistics. In Bose's S-ontology, the cells and not the quanta were statistically independent. Einstein, who arranged publication of it, immediately realized the potential of Bose's derivation and applied it to gas theory. He found the Sackur–Tetrode entropy as well as the behaviour predicted by Nernst's theorem at low temperature and even a new condensation phenomenon.[59]

Ehrenfest reacted immediately, pointing out that the statistical interdependence displayed by the quanta was not new: one could see it at work in Planck's radiation and gas theory all along. Einstein agreed and acknowledged that his new gas theory 'expresses indirectly an implicit hypothesis about the mutual influence of the molecules of a totally new and mysterious kind.' Now that the S-ontology was more or less clear, it remained to understand the D-ontology, and this was no easy task. Einstein came back to his idea that the mysterious influence could be due to the particles' wave features.

[57] (Planck, 1921b, p. 209).    [58] (Bose, 1924).
[59] (Einstein, 1924, 1925). On Bose's statistics and Einstein's application to the gas see (Monaldi, 2009; Pérez and Sauer, 2010).

The wave–particle duality had recently found new support with de Broglie's theory of matter waves. Eventually, the man who put all these strands together was Erwin Schrödinger. An expert on thermodynamics and statistical mechanics, in 1926, Schrödinger treated quantized matter waves like Debye's normal modes, thus inaugurating a new undulatory version of quantum mechanics. After the developments in 1924–1926, classical statistics was definitely overcome. Quantum mechanics brought in a new conceptual arsenal made of the wave function, eigenvalues, degeneracy, and even a third type of statistics. The search for a new alignment between S-ontology and D-ontology lasted for some more years, but this is another story.

## REFERENCES

Bach, A. (1987). Indistinguishability or Distinguishability of the Particles of Maxwell-Boltzmann Statistics. *Physics Letters A*, **125**(9), 447–50.

Bach, A. (1990). Boltzmann's Probability Distribution of 1877. *Archive for History of Exact Sciences*, **41**(1), 1–40.

Badino, M. (2009). The Odd Couple: Boltzmann, Planck and the Application of Statistics to Physics (1900–1913). *Annalen der Physik*, **18**(2–3), 81–101.

Badino, M. (2010). Das Verfolgen einer Idee: Plancks Theorie idealer Gase. In D. Hoffmann (ed.), *Max Planck und die moderne Physik*, Berlin: Springer, pp. 135–48.

Badino, M. (2011). Mechanistic Slumber vs. Statistical Insomnia: The Early Phase of Boltzmann's H-theorem (1868–1877). *European Physical Journal H*, **36**, 353–78.

Badino, M. (2015). *The Bumpy Road. Max Planck from Radiation Theory to the Quantum (1896–1906)*. New York: Springer.

Badino, M., and Friedrich, B. (2013). Much Polyphony but Little Harmony: Otto Sackur's Groping for a Quantum Theory of Gases. *Physics in Perspective*, **15**(3), 295–19.

Bergia, S. (1987). Who Discovered the Bose-Einstein Statistics? In M. G. Doncel, A. Hermann, L. Michel, and A. Pais (eds), *Symmetry in Physics (1600–1980)*, Bellaterra: Universitat Autònoma de Barcelona, pp. 221–48.

Boltzmann, L. (1868). Studien über das Gleichgewicht der lebendigen Kraft zwischen bewegten materiellen Punkten. *Sitzungsberichte der Akademie der Wissenschaften zu Wien*, **58**, 517–60.

Boltzmann, L. (1872). Weitere Studien über das Wärmegleichgewicht unter Gasmolekülen. *Sitzungsberichte der Akademie der Wissenschaften zu Wien*, **66**, 275–70.

Boltzmann, L. (1877). Über die Beziehung zwischen dem zweiten Hauptsatze der mechanischen Wärmetheorie und der Wahrscheinlichkeitsrechnung respective den Sätzen über das Wärmegleichgewicht. *Sitzungsberichte der Akademie der Wissenschaften zu Wien*, **76**, 373–35.

Boltzmann, L. (1897). *Vorlesungen über die Principe der Mechanik*, Vol. 1, Leipzig: Barth.

Boltzmann, L. (1909). *Wissenschaftliche Abhandlungen*, Vols. 1–3, Leipzig: Barth.

Bose, S. N. (1924). Planck's Gesetz und Lichtquantenhypothese. *Zeitschrift für Physik*, **26**, 178–81.

Brown, H. R., Myrvold, W., and Uffink, J. (2009). Boltzmann's H-theorem, its Discontents, and the Birth of Statistical Mechanics. *Studies in History and Philosophy of Modern Physics*, **40**, 174–91.

Costantini, D., and Garibaldi, U. (1997). A Probabilistic Foundation of Elementary Particle Statistics. Part I. *Studies in History and Philosophy of Modern Physics*, **28**, 483–06.

Costantini, D., Garibaldi, U., and Penco, M. A. (1996). Ludwig Boltzmann alla nascita della meccanica statistica. *Statistica*, **3**, 279–00.

Darrigol, O. (1988). Statistics and Combinatorics in Early Quantum Theory. *Historical Studies in the Physical Science*, **19**, 18–80.

Darrigol, O. (1991). Statistics and Combinatorics in Early Quantum Theory, II: Early Symptoma of Indistinguishability and Holism. *Historical Studies in the Physical Science*, **21**, 237–98.

Darrigol, O. (1992). *From c-numbers to q-numbers. The Classical Analogy in the History of Quantum Theory*. Berkeley, CA: University of California Press.

Darrigol, O. (2002). Quantum Theory and Atomic Structure, 1900–1927. In M. J. Nye (ed.), *The Cambridge History of Science: Volume 5: The Modern Physical and Mathematical Sciences*, Vol. 5, Cambridge: Cambridge University Press, pp. 329–49.

Darrigol, O. (2018a). *Atoms, Mechanics, and Probability. Ludwig Boltzmann's Statistico-Mechanical Writings: An Exegesis*. Oxford: Oxford University Press.

Darrigol, O. (2018b). The Gibbs Paradox: Early History and Solutions. *Entropy*, **20**(6), 443–97.

Delbruck, M. (1980). Was Bose–Einstein Statistics Arrived at by Serendipity? *Journal of Chemical Education*, **57**(7), 467–70.

Desalvo, A. (1992). From the Chemical Constant to Quantum Statistics: A Thermodynamic Route to Quantum Mechanics. *Physics*, **29**, 465–37.

Dieks, D. (2018). The Gibbs Paradox and Particle Individuality. *Entropy*, **20**(6), 466–81.

Dieks, D., and Slooten, W. J. (1986). Historic Papers in Physics: The Case of Hugo Martin Tetrode, 1895–1931. *Czechoslovak Journal of Physics*, **36**, 39–42.

Dorato, M., and Morganti, M. (2013). Grades of Individuality. A Pluralistic View of Identity in Quantum Mechanics and in the Sciences. *Philosophical Studies*, **163**(3), 591–10.

Duncan, A., and Janssen, M. (2019). *Constructing Quantum Mechanics: Volume 1: The Scaffold: 1900–1923*, Oxford: Oxford University Press.

Ehrenfest, P. (1905). Über die physikalischen Voraussetzungen der Planck'schen Theorie der irreversiblen Strahlungsvorgänge. *Sitzungsberichte der Akademie der Wissenschaften zu Wien*, **114**, 1301–14.

Ehrenfest, P. (1906). Zur Planckschen Strahlungstheorie. *Physikalische Zeitschrift*, **7**, 528–32.

Ehrenfest, P. (1911). Welche Züge der Lichtquantenhypothese spielen in der Theorie der Wärmstrahlung eine wesentliche Rolle? *Annalen der Physik*, **36**, 91–118.

Ehrenfest, P., and Kamerlingh Onnes, H. (1915). Vereinfachte Ableitung der kombinatorischen Formel, welche der Planckschen Strahlungstheorie zugrunde liegt. *Annalen der Physik*, **46**, 1021–24.

Ehrenfest, P., and Trkal, V. (1921). Ableitung des Dissociationgleichwichts aus der Quantentheorie und darauf beruhende Berechnung chemischer Konstanten. *Annalen der Physik*, **65**, 609–28.

Einstein, A. (1905). Über einen die Erzeugung und Verwandlung des Lichtes betreffenden heuristischen Gesichtspunkt. *Annalen der Physik*, **17**, 132–48.

Einstein, A. (1909). Zur gegenwärtigen Stand des Strahlungsproblems. *Physikalische Zeitschrift*, **10**, 185–93.

Einstein, A. (1917). Zur Quantentheorie der Strahlung. *Physikalische Zeitschrift*, **18**, 121–28.

Einstein, A. (1924). Quantentheorie des einatomigen idealen Gases I. *Sitzungsberichte der Preussischen Akademie der Wissenschaft*, 261–67.

Einstein, A. (1925). Quantentheorie des einatomigen idealen Gases II. *Sitzungsberichte der Preussischen Akademie der Wissenschaft*, 3–14.

French, S. (1989). Why the Principle of the Identity of Indiscernibles is not Contingently True Either. *Synthese*, 78(2), 141–66.

French, S., and Krause, D. (2006). *Identity in Physics: A Historical, Philosophical, and Formal Analysis*. Oxford: Oxford University Press.

Gearhart, C. (2002). Planck, the Quantum, and the Historians. *Physics in Perspective*, 4, 170–15.

Höflechner, W. (1994). *Ludwig Boltzmann. Leben und Briefe*. Graz: Akademische Druck und Verlaganstalt.

Hoyer, U. (1980). Von Boltzmann zu Planck. *Archive for History of Exact Sciences*, 23, 47–86.

Kangro, H. (1970). *Vorgeschichte des Planckschen Strahlungsgesetzes*. Wiesbaden: Steiner.

Klein, M. J. (1962). Max Planck and the Beginnings of the Quantum Theory. *Archive for History of Exact Sciences*, 1, 459–79.

Klein, M. J. (1970). *Paul Ehrenfest, Vol. 1. The Making of a Theoretical Physicist*. Amsterdam: North Holland.

Klein, M. J. (1973). The Development of Boltzmann's Statistical Ideas. *Acta Physica Austriaca, Supplementum*, 10, 53–106.

Kokowski, M. (2019). The Divergent Histories of Bose–Einstein Statistics and the Forgotten Achievements of Władysław Natanson (1864-1937). *Studia Historiae Scientiarum*, 18, 327–64.

Krutkow, G. (1914). Aus der Annahme unabhängiger Licht-quanten folgt die Wiensche Strahlungsformel. *Physikalische Zeitschrift*, 15, 133–36.

Kuhn, T. S. (1978). *Black-Body Theory and the Quantum Discontinuity, 1894-1912*. Oxford: Oxford University Press.

Kuhn, T. S. (1984). Revisiting Planck. *Historical Studies in the Physical Science*, 14, 232–52.

Lorentz, H. A. (1915). Some remarks on the theory of monoatomic gases. *Proceedings Koninklijke Akademie van Wetenschappen Te Amsterdam*, 19, 737–51.

Lummer, O., and Pringsheim, E. (1900). Über die Strahlung des schwarzen Körpers für lange Wellen. *Verhandlungen der Deutschen Physikalische Gesellschaft*, 2(12), 163–80.

Monaldi, D. (2009). A Note on the Prehistory of Indistinguishable Particles. *Studies in History and Philosophy of Modern Physics*, 40(4), 383–94.

Natanson, L. (1911). Über die statistische Theorie der Strahlung. *Physikalische Zeitschrift*, 12, 659–66.

Navarro, L., and Pérez, E. (2004). Paul Ehrenfest on the Necessity of Quanta (1911): Discontinuity, Quantization, Corpuscularity, and Adiabatic Invariance. *Archive for History of Exact Sciences*, 58(2), 97–141.

Pérez, E., and Sauer, T. (2010). Einstein's Quantum Theory of the Monatomic Ideal Gas: Non-statistical Arguments for a New Statistics. *Archive for History of Exact Sciences*, 64(5), 561–12.

Pešić, P. D. (1991). The Principle of Identicality and the Foundations of Quantum Theory. I. The Gibbs Paradox. *American Journal of Physics*, 59(11), 971–74.

Planck, M. (1898a). Über irreversible Strahlungsvorgänge. 3. Mitteilung. *Sitzungsberichte der Preussischen Akademie der Wissenschaften*, 1, 1122–45.

Planck, M. (1898b). Über irreversible Strahlungsvorgänge. 4. Mitteilung. *Sitzungsberichte der Preussischen Akademie der Wissenschaften*, 2, 449–76.

Planck, M. (1899). Über irreversible Strahlungsvorgänge. 5. Mitteilung. *Sitzungsberichte der Preussischen Akademie der Wissenschaften*, 1, 440–80.

Planck, M. (1900a). Entropie und Temperatur strahlender Wärme. *Annalen der Physik*, 4, 719–37.

Planck, M. (1900b). Über eine Verbesserung der Wienschen Spektralgleichung. *Verhandlungen der Deutschen Physikalische Gesellschaft*, 2, 202–204.

Planck, M. (1900c). Zur Theorie des Gesetzes der Energieverteilung im Normalspektrum. *Verhandlungen der Deutschen Physikalische Gesellschaft*, 2, 237–45.

Planck, M. (1906). *Vorlesungen über die Theorie der Wärmestrahlung*. Leipzig: Barth.

Planck, M. (1911). *Vorlesungen über Thermodynamik*, 3rd edn. Leipzig: Barth.

Planck, M. (1912). *Vorlesungen über die Theorie der Wärmestrahlung*, 2nd edn. Leipzig: Barth.

Planck, M. (1914). Die gegenwärtige Bedeutung der Quantenhypothese für die kinetische Gastheorie. In D. Hilbert (ed.), *Vorträge über die kinetische Theorie der Materie und der Elektrizität*, Göttingen: Teubner, pp. 3–16.

Planck, M. (1916a). Die physikalische Struktur des Phasenraumes. *Annalen der Physik*, 50, 385–18.

Planck, M. (1916b). Über die absolute Entropie einatomiger Körper. *Sitzungsberichte der Preussischen Akademie der Wissenschaft*, 653–67.

Planck, M. (1921a). Absolute Entropie und chemische Konstante. *Annalen der Physik*, 66, 365–72.

Planck, M. (1921b). *Vorlesungen über die Theorie der Wärmestrahlung*, 4th edn. Leipzig: Barth.

Planck, M. (1958). *Physikalische Abhandlungen und Vorträge*. (M. Von Laue, ed.), Vols. 1–3. Braunschweig: Vieweg u. Sohn.

Sackur, O. (1911). Die Anwendung der kinetischen Theorie der Gase auf chemische Probleme. *Annalen der Physik*, 36, 958–80.

Sackur, O. (1912). Die universelle Bedeutung des sogenannten elementaren Wirkungsquantums. *Annalen der Physik*, 40, 67–86.

Saunders, S. (2018). The Gibbs Paradox. *Entropy*, 20(8), 552–76.

Saunders, S. (2020). The Concept 'Indistinguishable.' *Studies in History and Philosophy of Modern Physics*, 71, 37–59.

Sommerfeld, A. (1911). Das Plancksche Wirkungsquantum und seine allgemeine Bedeutung für die Molekularphysik. *Physikalische Zeitschrift*, 12, 1057–69.

Tetrode, H. M. (1912a). Berichtigung zu meiner Arbeit: 'Die chemische Konstante der Gase und das elementare Wirkungsquantum.' *Annalen der Physik*, 39, 255–56.

Tetrode, H. M. (1912b). Die chemische Konstante der Gase und das elementare Wirkungsquantum. *Annalen der Physik*, 38, 434–42.

Uffink, J. (2007). Compendium of the Foundations of Classical Statistical Mechanics. In J. Butterfield and J. Earman (eds), *Philosophy of Physics*, Vol. 2, Amsterdam: North Holland, pp. 923–1074.

Willard Gibbs, J. (1902). *Elementary Principles in Statistical Mechanics*. Woodbridge: Ox Bow Press.

Wolfke, M. (1913a). Zur Quantentheorie. Zweite vorläufige Mitteilung. *Verhandlungen der Deutschen Physikalischen Gesellschaft*, 15, 1215–18.

Wolfke, M. (1913b). Zur Quantentheorie. Vorläufige Mitteilung. *Verhandlungen der Deutschen Physikalischen Gesellschaft*, 15(21), 1123–29.

....................................................

# CHAPTER 5

# DEAD AS A DOORNAIL? ZERO-POINT ENERGY AND LOW-TEMPERATURE PHYSICS IN EARLY QUANTUM THEORY

....................................................

### HELGE KRAGH

In classical physics the absolute temperature $T$ of a body is a measure of the kinetic energy of the atoms or molecules making up the body. Consequently, at $T = 0$ the particles will be at rest. The intuitively clear connection between temperature and kinetic energy, or between temperature and molecular speed, was maintained in the original formulation of quantum theory of 1900 but not in the version that Max Planck proposed eleven years later. In this version there appeared for the first time the counterintuitive notion of zero-point energy (ZPE), which was much discussed in the physics community, if generally met with scepticism. The hypothetical energy was associated with half-integral quantum numbers, which had no natural place in the accepted theory of atoms.

In 1924 spectroscopic experiments provided the first convincing evidence for molecular ZPE and the following year the concept was theoretically justified by the new quantum mechanics. However, the question of whether or not radiation in free space could be ascribed a ZPE, and if so whether it had any physical implications, remained controversial. The early history of the concept of ZPE was closely connected with the interest of quantum physicists in phenomena at very low temperature. If only by accident, the discovery of superconductivity and Planck's introduction of ZPE date from the same year.

# 5.1 NEVER AT REST

The important concept of ZPE is distinctly a result of quantum mechanics, but at the time when it appeared in Werner Heisenberg's famous *Umdeutung* paper in September 1925 it had been known for more than a decade. Heisenberg found that the energy of a one-dimensional harmonic oscillator vibrating with the frequency $v$ was not given by $hv$ but by

$$E_n = (n + \tfrac{1}{2})hv,$$

where $n = 0, 1, 2, \ldots$ and $h$ is Planck's constant. For an oscillator vibrating in three dimensions the additional energy term was $3hv/2$. In April 1926, Erwin Schrödinger arrived at the same result from his very different theory of wave mechanics. Still unaware of the formal equivalence between his own wave mechanics and the Göttingen quantum mechanics, he commented, 'Strangely, our quantum levels are precisely the same as in Heisenberg's theory!' Moreover, Schrödinger briefly pointed out that, 'From a formal point of view this is the old question of the zero-point energy [*Nullpunktse-nergie*] which was already raised in connection with the dilemma of choosing between Planck's first and second theory.'[1] Indeed, the concept of ZPE first turned up in Planck's so-called 'second theory' as he communicated it to the German Physical Society on 3 February 1911, originally using the name *Restenergie* for what he considered a quantum analogue to the rest energy $m_0c^2$ of a particle in relativity mechanics. Later the same year he presented his ideas to the first Solvay Conference, and in early 1913 the theory appeared in its canonical formulation in the second edition of Planck's *Vorlesungen über die Theorie der Wärmestrahlung*, which was published in an English translation the following year.

The fundamental notion that a quantum oscillator or some other system can never be at rest thus originated in an incorrect and short-lived theory, to be reinvented in the context of quantum mechanics more than a decade later. Although Planck's second theory was soon abandoned, the controversial concept of a ZPE continued to be discussed within the framework of the old quantum theory. By the summer of 1925, shortly before the emergence of quantum mechanics, it was supported by experiments but still lacked a convincing theoretical foundation.[2]

In his original theory of 1900, Planck assumed that for the purpose of entropy calculations, the distribution of energy over a set of oscillators was discrete rather than continuous. In his second theory, as he expounded it in works between 1911 and 1913, Planck restricted the discrete energy quanta to the emission process whereas he

---

[1] Schrödinger (1926), p. 516.
[2] Accounts of the early history of the ZPE include Milloni and Hsieh (1991), Mehra and Rechenberg (1999), and Kragh (2012a). See also https://en.wikipedia.org/wiki/Zero-point_energy#History.

supposed absorption to occur continuously in accordance with the laws of classical electrodynamics. Emission and absorption were thus treated on an unequal basis. In a letter to Paul Ehrenfest of 1913, Planck expressed his mixed feelings about the new theory and the discrete features it still retained: 'I fear that your hatred of the zero-point energy extends to the electrodynamic emission I introduced and that leads to it. But what's to be done? For my part, I hate discontinuity of energy even more than discontinuity of emission.'[3] Even in the second theory the energy of an oscillator was not quantized as it could vary continuously. At absolute zero, oscillators would possess energies randomly distributed between $E = 0$ and $E = h\nu$.

The ZPE mentioned by Planck in his letter to Ehrenfest appeared in his calculation of how the average energy of an oscillator of frequency $\nu$ varied with the temperature. With $k$ denoting Boltzmann's constant, his result was

$$\bar{E} = \frac{h\nu}{2}\frac{\exp\left(\frac{h\nu}{kT}\right) + 1}{\exp\left(\frac{h\nu}{kT}\right) - 1} = \frac{h\nu}{2} + \frac{h\nu}{\exp\left(\frac{h\nu}{kT}\right) - 1}$$

As Planck commented in his 1911 address to the German Physical Society, it follows that for $T \rightarrow 0, \bar{E}$ 'does not become 0, but equals ½$h\nu$. This rest-energy remains with the oscillator, on the average, at the absolute zero of temperature. It cannot lose it, for it does not emit energy so long as $\bar{U}$ [$\bar{E}$] is smaller than $h\nu$.'[4] In his reasoning leading to the ZPE, Planck made use of a correspondence argument which bears some resemblance to Bohr's later correspondence principle.[5]

Planck fully realized that the ZPE was a concept completely different to classical physics and he accordingly was uncertain whether or not it existed as a physically measurable quantity. Since the new energy expression for an oscillator at fixed frequency differed from the original one by only an additive constant, it would have no effect on the specific heat as given by $c = d\bar{E}/dT$. At the time Einstein's 1907 quantum theory of the specific heat of solids (or Peter Debye's 1912 improvement of it) had received experimental confirmation, but this was of no use to differentiate between the two radiation theories.

Although Planck did not think of the ZPE as a quantity with direct experimental consequences, he suggested that in a more qualitative sense it might be empirically justified. For example, it was known at the time that the half-life of radioactive substances and the energy released by them remained uninfluenced by even the most extreme cold. Like some other scientists at the time, Planck speculated that somehow radioactivity was due to intra-atomic changes independent of the temperature. He also referred to the photoelectric effect as a phenomenon that might possibly be explained in terms of the ZPE. Still in 1915 he was optimistic about his brainchild. 'I am more

[3] Quoted in Kuhn (1978), p. 253.
[4] The 1911 paper is reprinted in Planck (1958), pp. 249–259.
[5] Kuhn (1978), p. 240; Whitaker (1985).

convinced than ever that zero-point energy is an indispensable element,' he wrote in a letter to Heike Kamerlingh Onnes in Leiden.[6]

Planck's second quantum theory was important but unsuccessful. The asymmetry between emission and absorption was widely seen as unsatisfactory, and there were also problems with the theory's consistency. More important was that the theory seemed unable to explain optical spectra and squarely disagreed with Niels Bohr's new theory of atomic structure proposed in 1913. As a result, the second theory gradually disappeared from the scene of physics. Planck belatedly admitted that 'This second formulation of the quantum theory may be considered today...as finally disproved.'[7] However, one result of the theory survived. The ZPE originated within the second theory, but after about 1916 it was as if it lived its own life.

# 5.2 THE BOHR ATOM AND ZERO-POINT ENERGY

The concept of a motion which can never be brought to rest not only appeared in Planck's second theory but also, if only implicitly and in a very different way, in Bohr's atomic theory of 1913. In this theory the electron in a hydrogen atom in its ground state $n = 1$ revolved continuously around the nucleus with a mechanical frequency given by the natural constants $m$, $e$, and $h$. Although Bohr did not mention it explicitly, the motion of the electron was independent of the temperature. To some critics this was a reason to question the validity of his atomic model. One of the critics was the American physical chemist Gilbert N. Lewis, who saw an analogy between Planck's ZPE and Bohr's model since both operated with motion at zero temperature. 'Planck in his elementary oscillator which maintains its motion at the absolute zero, and Bohr in his electron moving in a fixed orbit, have invented systems containing electrons of which the motion produces no effect on external charges.'[8] To Lewis it was 'logically objectionable' to postulate a motion which produced no physical effect whatsoever.

In Bohr's seminal series of papers from the fall of 1913 he cited three of Planck's papers on the second theory but without referring to the ZPE. It is possible that the second theory inspired parts of Bohr's thinking, such as has been argued by some historians of physics. However, other historians disagree and there is no solid evidence for the suggestion.[9] In any case, Bohr thought that Planck's concept of atomic oscillators was foreign to or even inconsistent with the nuclear model on which Bohr's

[6] Quoted in Van Delft (2007), p. 491. Planck further stated that he had 'the strongest evidence' for the existence of a ZPE, but without revealing what this evidence was.
[7] Planck (1923), p. 537. On the fate of the second theory, see Kuhn (1978), pp. 252–254.
[8] Lewis (1916), p. 773. For other of Lewis' objections to Bohr's theory, see Kragh (2012b), p. 118–121.
[9] For arguments for and against the suggestion that Planck's second theory influenced Bohr or at least was relevant to it, see Whitaker (1985).

theory rested. In a lecture presented to the Danish Physical Society in December 1913 he expressed his disbelief: 'No one has ever seen a Planck's resonator, not indeed even measured its frequency of oscillation; we can observe only the period of oscillation of the radiation which is permitted.'[10]

In a long letter of December 1915 to his friend, the Swedish physicist Carl Wilhelm Oseen, Bohr objected that Planck's second theory was inapplicable to systems with more than one degree of freedom such as a diatomic molecule. He sketched his own alternative, which 'gives a simple explanation of the mysterious zero-point energy and, when applied to the hydrogen molecule an agreement with the experiments on specific heat.'[11] In an unpublished article of 1916 Bohr elaborated, now arguing that the a priori probability $P_n$ of a quantum system being in a state $n$ varied as $n^{r-1}$, where $r$ is the number of degrees of freedom. The probability of finding a rotator ($r = 2$) in the state $n$ would thus be $P_n = n$. Bohr observed that a system of $N$ harmonic oscillators of $r = 2$ and frequency $\omega$ would have non-zero energy at $T = 0$. He thus regained a ZPE, but stressed that his argument for it was quite different from the one of Planck, which he labelled a 'postulate'. According to Bohr:

> The present theory, in contrast to Planck's theory, gives the same variation with temperature of the specific heat for a vibrator of one and two degrees of freedom. [For] $T = 0$ we do not get $E = 0$ but $E = Nh\omega$. This so-called zero-point energy has here an origin quite distinct from that in Planck's theory. In the present theory it arises simply from the fact that according to [our] assumption there is no probability of a periodic system of several degrees of freedom being in the state corresponding to $n = 0$.[12]

Apart from applying his theory of periodic systems to the specific heat of molecular hydrogen at low temperatures, Bohr also applied it to his quantized hydrogen atom where the energy levels vary as $E_n \sim -n^{-2}$, where $n$ is the principal quantum number. Since the hydrogen atom has three degrees of freedom, $P_n = n^2$. Commenting on what he called this 'peculiar case of zero-point energy' he stated that the state $n = 0$ was nonsensical because it corresponds to an infinite negative energy: 'In order to obtain agreement with experiments it must be assumed that the normal state of the system corresponds to $n = 1$.'

Bohr did not deal with the ZPE in his further work on atomic theory. He only returned to the subject much later, in an important paper with Léon Rosenfeld from 1933, and then in connection with the 'paradox' of the infinite ZPE formally appearing in quantum field theory (see also Section 5.7).[13]

---

[10] English translation in Bohr (1922), p. 10.

[11] Bohr (1981), p. 567.

[12] Bohr (1981), p. 456. See also Gearhart (2010), pp. 146–147. Bohr withdrew the article shortly before it was to be published in *Philosophical Magazine*.

[13] The Bohr–Rosenfeld paper is reprinted in Bohr (1996), pp. 55–121, where the comment on ZPE appears on p. 74.

# 5.3 Einstein's Ambivalence

Is the ZPE physically real or just a theoretical artefact? Can it be detected? For more than a decade the questions remained unanswered despite many suggestions to address them by means of experiments. For example, during the second Solvay Conference in 1913 the question was raised by Walther Nernst, who in a comment to Max Laue's theory on X-ray diffraction by crystal lattices wondered whether X-ray measurements might provide evidence for the shadowy ZPE. Several others of the participants in Brussels took part in the discussion following Nernst's remarks. Kamerlingh Onnes, responding to Planck's report on his second quantum theory, objected that the ZPE made it difficult to understand the resistance of mercury in terms of oscillating atoms. It was on this basis, he said, that it was possible to predict that 'the resistance of mercury below 4° absolute is practically zero... [and] it seems difficult to explain this phenomenon by means of the new formula [of Planck's second theory].'[14]

Einstein also took part in the Solvay discussion, suggesting that somehow the issue might relate to the new phenomenon of superconductivity.[15] However, he argued that ZPE was unrelated to elastic oscillations and for this reason he was opposed to Nernst's suggestion that zero-point oscillations might influence the X-ray diffraction pattern. Einstein was at the time greatly interested in ZPE and may have been the first to come up with a definite argument for its existence that did not rely exclusively on theory.

In a paper of early 1913 written jointly with Otto Stern, Einstein examined theoretically the rotational energy of diatomic molecules and compared the corresponding specific heats ($c = dE_{rot}/dT$) with those measured experimentally. In Einstein's model of the hydrogen molecule, or the Einstein–Stern model, the molecules were assumed to rotate with the same frequency at a given temperature $T$. Since the frequency depended on $T$, and this was also the case with the ZPE, the existence of a ZPE term would affect the calculated specific heat. Einstein and Stern hoped in this way to provide evidence for or against the hypothesis. By comparing the theoretical expression for $c(T)$ with recent data obtained by Arnold Eucken for molecular hydrogen at low temperatures, they found a reasonable agreement if a ZPE of $\frac{1}{2}hv$ was taken into account. On the other hand, quantum theory without ZPE showed no agreement at all with Eucken's data. The two physicists cautiously concluded that Eucken's experiments 'make probable the existence of a zero-point energy equal to $\frac{1}{2}hv$.'[16]

However, less than a year later Einstein retracted his support of the ZPE assumption. Among the reasons for his turnaround was that he and Stern found that a derivation of Planck's radiation law necessitated a zero-point contribution of $hv$ rather than $\frac{1}{2}hv$. They thus arrived at the correct Planck spectrum using the wrong ZPE, a puzzling

[14] Eucken (1914), p. 107.
[15] Authier (2013), p. 161; Sauer (2007), p. 170.
[16] Einstein and Stern (1913), p. 560. For context and details, see Gearhart (2010), pp. 128–132 and Milloni and Hsieh (1991).

result that in retrospect can be ascribed to their disregard of the ZPE of the electro-magnetic field. Moreover, the first value $h\nu$ spoiled the agreement with Eucken's data on hydrogen's specific heat. What was intended to be a clarification of the controversial ZPE issue only resulted in further confusion. By the fall of 1913 Einstein had with-drawn his support of the concept introduced by Planck two years earlier. At the Solvay conference in October he made his stance clear: 'I no longer consider the arguments for the existence of zero-point energy that I and Mr. Stern put forward to be correct.' A few days later, in a letter to Ehrenfest, he declared ZPE to be 'dead as a doornail' (*Mausetot*). To his assistant and collaborator Ludwig Hopf he wrote: 'One hopes Debije [Debye] will soon demonstrate the incorrectness of the hypothesis of zero-point energy, the theoretical untenability of which became glaringly obvious to me soon after the publication of the article I co-authored with Mr. Stern.'[17]

This was not Einstein's final verdict, though, for he continued thinking about the question. The ZPE entered significantly in Einstein's collaboration of 1915 with the Dutch physicist Wander J. de Haas on molecular currents. This work resulted in what is called the Einstein–de Haas effect, which is basically the rotation of a freely suspended ferromagnetic body due to changes in its magnetization.[18] According to the Curie (or Curie–Langevin) law molecular magnetic moments should persist even as $T \to 0$. From this Einstein conjectured that there must exist a temperature-independent magnetic moment arising from a current of rotating intra-atomic elec-trons and that the associated energy at $T = 0$ might represent a ZPE.

The conjecture of non-radiating atomic electrons contradicted Maxwellian electro-dynamics, but on the other hand it agreed nicely with Bohr's postulate of stationary states. Bohr was aware of the unintended support of his still controversial postulate to which he called attention in a paper of 1915, albeit without referring to ZPE.[19] One might expect that Einstein and de Haas also noted the connection to Bohr's theory, but they did not refer to it in any of their publications.

Einstein's earlier optimism with respect to the ZPE had proved a mistake, but now, in his experiments with de Haas, he thought to have found fresh evidence for it. To his friend Michele Besso he expressed his new-found optimism: 'The experiment will soon be finished...[and] it will also have proved the existence of a zero-point energy.' Although Einstein was more cautious in print, he still thought that the circular motion of molecular currents 'must be the so-called zero-point energy—a concept that arouses a quite understandable resistance in many physicists.'[20] It seems that Einstein was of two minds with respect to the ZPE question. In one of his papers on molecular currents he stated that no theoretical physicist 'can at present utter the words "zero-point energy" without breaking into a half-embarrassed, half-ironic smile.' Some years later, in his correspondence with Ehrenfest, he returned to the subject, now suggesting that the

[17] Einstein (1995), Doc. 22; Einstein (1993), Doc. 480 and Doc. 481.
[18] See Frenkel (1979) and Galison (1982) on the Einstein–de Haas effect and its connection to the ZPE.
[19] Bohr (1915), p. 397.
[20] Einstein and de Haas (1915), p. 153; Einstein (1998), Doc. 56.

maximum density in liquid helium might throw light on the question of whether ZPE existed or not.[21] Apparently the suggestion was not meant to be half-ironic.

# 5.4 LOW-TEMPERATURE PHYSICS AND OTHER EXPERIMENTS

Contrary to what Einstein thought in 1913, the idea of ZPE was not dead as a doornail but continued to attract much attention. It gave rise to a large number of experiments, the aim of which was to find macroscopic manifestations of the mysterious energy. According to the British physicist Frederick Lindemann, who was a former collaborator of Nernst, separation of isotopes by means of fractional distillation might provide an answer. In a paper of 1919 he investigated isotope effects in the vapour pressure of monatomic solids, which he argued would depend on whether or not a ZPE existed. His calculations indicated that in the presence of ZPE the lighter isotope would have a higher vapour pressure than the heavier isotope, and the other way around if ZPE was absent. Lindemann generally argued that a separation of isotopes would be possible, but that 'the difference should be measurable [only] if there is no "Nullpunktsenergie".'[22] In other words, isotope separation by distillation would prove the non-existence of ZPE. Unfortunately, experiments of the kind suggested by Lindemann were at the time too unreliable to settle the question.

Theoretical arguments similar to Lindemann's were also suggested by Stern, who calculated the vapour pressure above the surface of a solid conceived as a collection of atoms oscillating harmonically in three dimensions. Stern's comparison with experimental data did not yield an answer to the ZPE puzzle any more than Lindemann's reasoning did. All he could do was to express the pious hope that the ZPE of solids, if it were real, would find its interpretation in 'the more recent works of N. Bohr [in which] this hypothesis in a somewhat modified form has acquired a very deep meaning.' He may have had Bohr's correspondence principle in mind but did not elaborate. Stern later recalled that he discussed the question with Wolfgang Pauli in the early 1920s. He tried 'to convert Pauli to the zero-point energy against which he [Pauli] had the gravest hesitations.'[23]

Most of the phenomena studied with the aim of demonstrating the ZPE were in the low-temperature region which, with the liquefaction of helium in 1908 and the discovery of superconductivity three years later, had recently acquired a new dimension. The focus on low temperatures is understandable, since one would expect a considerable zero-energy effect only in cases where $kT$ is small compared to $h\nu$. One of the phenomena was the X-ray diffraction pattern from crystals, which at low

---

[21] Einstein (1915), p. 237; Einstein (2009), Doc. 219.
[22] Lindemann (1919), p. 181.    [23] Stern (1919); Enz (2002), p. 150.

temperatures would presumably be influenced by a zero-point motion of the constituent atoms. As a result, the intensity of the diffracted radiation would be reduced even as the temperature approached absolute zero. The idea was first forwarded by Debye in his important works of 1913 on the intensity of X-rays diffracted by a crystal, but at the time it was little more than a suggestion. Only in 1928, with improved accuracy of intensity measurements and computations based on quantum mechanics, was it possible to solve the problem, which was principally due to the British physicist Reginald William James. According to James and his co-workers, the analysis of the diffraction patterns from rock-salt demonstrated convincingly a ZPE of $\frac{1}{2}h\nu$ per degree of freedom.[24] The result was reassuring but no surprise, for at the time the ZPE of material systems, such as implied by quantum mechanics, was generally accepted.

Many of the experiments related to ZPE took place at the new laboratories devoted to low-temperature physics in Berlin, Leiden, and elsewhere. Working at Nernst's laboratory in Berlin, the young physical chemists Kurt Bennewitz and Franz Simon calculated in 1923 the melting points of hydrogen, argon, and mercury, which they interpreted as possible evidence for a ZPE contribution to their latent heats. Moreover, they suggested that the ZPE might be responsible for the difficulty in solidifying helium even at the lowest temperatures attainable. According to Bennewitz and Simon, the ZPE in liquid helium acted as an internal pressure which expanded the liquid to such a low density that no rigid atomic structure could be maintained.[25] For a period, Dutch physicists associated with Kamerlingh Onnes' laboratory in Leiden were keenly interested in this and related problems. Thus, in his Nobel Prize lecture of 1913 Kamerlingh Onnes referred to the Einstein–Stern support of ZPE and generally stated that the ZPE had contributed to make low-temperature measurements a field of prime interest to physicists.

The Utrecht physicist Willem Keesom was perhaps the period's most outspoken supporter of the ZPE, which he defended at a meeting in Göttingen in 1913, arguing that the hypothesis was needed to account for helium's equation of state. Keesom also applied ZPE to the theory of ferromagnetism, to deviations from the Curie–Langevin law at low temperatures, and to absorption spectra. He even thought that in certain spectroscopic experiments, 'one observes the zero-point velocity of the molecules directly.'[26]

As Keesom saw it, the evidence in favour of ZPE was far stronger than the counter-evidence. However, this view was not shared by the majority of physicists. Evidence is not proof, and proof was still lacking.

---

[24] James, Waller, and Hartree (1928); Authier (2013), p. 201.
[25] Gavroglu and Goudaroulis (1989), p. 99 take the Bennewitz–Simon paper to be the first 'direct evidence' for ZPE, but this is not how it was conceived by contemporary physicists.
[26] Van Delft (2007), p. 490.

# 5.5  MOLECULAR SPECTROSCOPY

According to the fundamental assumptions of Bohr's atomic theory of 1913, quantum numbers had to be integers. However, the extension of Bohr's theory to cover not only line spectra but also the band spectra of rotating and vibrating molecules suggested that somehow half-integral quantum numbers had to be taken into account.[27] This kind of strange quantum numbers not only turned up in spectroscopy but also, in a different way, in Heisenberg's and Alfred Landé's early attempts to understand the anomalous Zeeman effect. However, Bohr disliked the hypothesis of half-integral quantum numbers, which he thought was irreconcilable with the fundamental principles of quantum theory.

Based on Bohr's correspondence principle, in early 1920 the German physicist Fritz Reiche argued that to explain experimental data the quantization rule for a rotating molecule had to be of the form

$$E_{rot} = (n + \tfrac{1}{2})^2 \frac{h^2}{8\pi^2 J}$$

where $J$ denotes the molecule's moment of inertia and $n = 0, 1, 2, \dots$. Reiche realized that the spectroscopic 'half-quanta' implied a rotational energy at zero temperature and thus supported the more general notion of ZPE of which he was clearly in favour. Interestingly, he stated that the idea of such half-quanta was first suggested to him in conversations with Einstein. Whether or not Einstein was the first to come up with the idea, in a letter to Ehrenfest of March 1920, Einstein, referring to the band spectra, suggested that 'in the case of rotation the allowed motions are not given by $n$ but by $n + \frac{1}{2}$.' He found this to be 'a most remarkable business.' And later the same year, in a letter to Fritz Haber, he wrote: 'If no zero-point energy of rotation existed, there would be a certain fraction of nonrotating molecules that ... would probably have to exhibit a deflection. However, the specific heat function of $H_2$ as well as the Bierrum [sic] spectrum of HCl speaks *for* a rotational zero-point energy.'[28]

Thus, although half-integral quantum numbers were controversial from a theoretical point of view, experiments strongly indicated their existence in rotational and vibrational spectra. As a result, they were soon incorporated in the theories of band spectra proposed by Adolf Kratzer and other leaders of molecular spectroscopy. If half-quanta were real, so presumably were ZPEs at least in the realm of molecules. As a consequence, within a few years the concept of ZPE became respectable among molecular physicists if not yet among physicists generally.

---

[27]  See Kragh (2012b), pp. 239–245, where references to the literature are given.

[28]  Einstein (2004), Doc. 335 and Einstein (2006), Doc. 162. The reference to the 'Bierrum spectrum' is to the Danish chemist Niels Bjerrum's work on the infrared absorption spectra of gases.

In an important study of 1924 the young American physical chemist Robert S. Mulliken, a future Nobel laureate in chemistry, provided solid evidence for the reality of ZPE. In a careful investigation of the band spectra of boron monoxide BO in its two isotopic compositions $B^{10}O^{16}$ and $B^{11}O^{16}$ Mulliken concluded that the spectra could only be understood on the assumption of half-integral quantum numbers. His preliminary report appeared in *Nature* in September 1924 and the full report only in the March 1925 issue of *Physical Review*. In the latter paper Mulliken made it clear that his work implied 'a null-point energy of ½ quantum each of vibration and rotation.'[29] Although his study did not prove the ZPE hypothesis in its general form, it was widely considered a confirmation of the hypothesis, which after that time largely ceased to be controversial.

In other words, the ZPE was confirmed experimentally before it received theoretical blessing in the form of Heisenberg's quantum mechanics. For this reason, Heisenberg's 1925 deduction of a ZPE does not count as a proper prediction, at least not in the ordinary sense of the term. Likewise, although the ZPE follows naturally from Heisenberg's uncertainty principle for position and momentum (in one dimension $\Delta p_x \Delta x \geq h/2\pi$), it is hardly correct to see it as a prediction from this principle. In his famous paper of 1927 introducing the uncertainty relations, Heisenberg did not mention the ZPE as a consequence of the position–momentum relation. Incidentally, neither Heisenberg in 1925 nor Schrödinger in 1926 referred to Mulliken's paper or to other papers in the tradition of molecular spectroscopy. Within the framework of the still existing old quantum theory the dual concepts of ZPE and half-quanta remained anomalous. One might therefore expect that Mulliken's discovery added to the crisis that soon led to the new quantum mechanics, but this was not the case. The discovery was well known among chemists but it aroused little attention among the leading quantum theorists.

# 5.6 Nernst's Vacuum

By the summer of 1926 the ZPE was no longer controversial insofar as it concerned atoms, molecules, and other material systems. After all, it was recognized as an inevitable consequence of quantum mechanics. On the other hand, the question of whether ZPE was valid also for free space, or for a space occupied by radiation alone, was seen in a very different light. The origin of the zero-point vacuum is found in a paper Nernst presented to the German Physical Society in 1916 and in which he defended the unorthodox idea that the ether —or vacuum, as most physicists would say—was filled with electromagnetic zero-point radiation: 'Even without the existence of radiating matter, that is, matter heated above absolute zero or somehow excited,

---

[29] Mulliken (1925), p. 281.

empty space...is filled with radiation.'[30] He thus claimed that the ZPE hypothesis was equally valid for material systems and the free ether.

However, Nernst did not accept Planck's version of ZPE. As an alternative he suggested that for each degree of freedom, where classical theory assigns the energy $\frac{1}{2}kT$ the ZPE becomes $\frac{1}{2}\,h\nu$. This implies that the lowest possible state of a harmonic oscillator becomes $h\nu$ and not, as in Planck's second theory, $\frac{1}{2}\,h\nu$. For the energy density of the electromagnetic ZPE radiation Nernst adopted the Rayleigh–Jeans formula for blackbody radiation in which he replaced the quantity $kT$ with $h\nu$. To avoid an infinite total energy density integrated over all frequencies he introduced a cut-off frequency corresponding to some maximum frequency $\nu_m$ and thus arrived at

$$\rho = \frac{2\pi h}{c^3}\nu_m^4$$

Assuming $\nu_m = 10^{20}$Hz, Nernst obtained the 'quite enormous' value $\rho = 1.5 \times 10^{16}$J/cm$^3$ corresponding to ca. 150g/cm$^3$. He further showed the 'truly remarkable result' that if zero-point radiation enclosed in a container is compressed, neither its energy density nor its spectral distribution will be affected. Based on his own ZPE version of quantum theory, Nernst also proposed a model of the H$_2$ molecule that differed from Bohr's earlier model. According to Nernst the two revolving electrons, each with a kinetic energy of $\frac{1}{2}\,h\nu$, were in equilibrium with the zero-point radiation and therefore did not radiate. In this way he explained the stability of the molecule without adopting Bohr's radical postulate of stationary states in which the laws of ordinary electrodynamics were invalid.

Nernst's speculations on the ZPE of the ether were connected with his growing concern with cosmological physics, such as he elaborated in a booklet of 1921 entitled *Das Weltgebäude im Lichte der neueren Forschung*. In this and other works he suggested that ether–matter transmutations might continually take place in the depths of space and in this way provide a cosmic cycle that annulled the cosmic heat death due to the continual increase in entropy. He speculated that atoms of heavy chemical elements might appear out of the ether's zero-point fluctuations, and that these atoms or their decay products would then again disappear into the ZPE of the infinite sea of ether. Nernst argued that the free ether must have a small capacity for absorbing heat rays and that this absorption of heat would eventually turn up as a ZPE in the ether and a corresponding non-zero temperature. During the 1920s and 1930s he was much occupied with questions of astrophysics and cosmology, aiming to develop a model of the universe in which matter and energy were eternally in a state of equilibrium. Whereas most astronomers accepted the expanding universe in the early 1930s, Nernst defended a static steady-state model. He strongly disagreed with

---

[30] Nernst (1916), p. 86. For a full description and references to the literature, see Kragh (2012a).

the theory of cosmic expansion and instead interpreted Hubble's law as a quantum effect in a stationary universe.[31]

The ideas that Nernst entertained with regard to ether and ZPE were known in the German physics community but without attracting much interest. Most mainstream physicists found his arguments for a vacuum zero-point radiation to be unconvincing. Reiche was aware of Nernst's theory but found it to be too 'radical' to be taken seriously. As the physics professor Siegfried Valentiner commented, 'It is much more difficult to conceive the presence of such a zero-point energy in the vacuum filled with electrical radiation than it is to assume that the existence of the zero-point energy is a peculiarity of the oscillators.' The American physicist and chemist Richard Tolman agreed:

> This 'nullpunkt energie' in the Nernst treatment is in equilibrium with radiant energy in the ether. On rise of temperature, energy is drawn not only from the surroundings but also from the reservoir of 'nullpunkt energie' and the principle of the conservation of energy becomes merely statistically true rather than true for the individual elements of the system. It is evident that the theories in question (like so much of quantum theory) are still in their birth-pangs.[32]

Only much later did Nernst's ideas attract wide attention, as it was recognized that the hypothesis of an energy-rich vacuum has elements in common with the modern concept of dark energy. Nernst has even been portrayed as the 'grandfather of dark energy.'[33]

## 5.7 THE FREE RADIATION FIELD

Nernst's proposal of assigning a ZPE to pure radiation dissociated from matter was dismissed by most contemporary physicists. In the mid-1920s it was briefly discussed by Stern and also by Wilhelm Lenz, who in a paper of 1926 considered the role of radiation in a closed universe described by Einstein's field equations. Lenz derived a formula relating the temperature of radiation to the radius of the universe but deliberately left out zero-point radiation because he found it to be a concept with no secure foundation in physics. If the zero-point radiation existed, the associated energy would act gravitationally by increasing the curvature of space and hence making the universe much smaller—'one would obtain a vacuum energy density of such a value that the world would not reach even to the Moon.'[34] To avoid such an absurdity it

[31] On Nernst's unorthodox cosmological views, see Kragh (1995) and Bartel and Huebener (2007), pp. 308–320.
[32] Tolman (1920), p. 1189.    [33] Kragh (2012c).
[34] On Lenz's paper, and on the quote and its history, see Kragh (2012a), pp. 223–224.

seemed better to forget about the zero-point radiation. Lenz discussed this and related issues with Pauli, who denied the reality of a ZPE in free space.

Shortly after the emergence of the Göttingen quantum mechanics and its justification of ZPE for material systems, Pascual Jordan attacked the problem of quantizing the electromagnetic field. Conceiving the field inside a cavity as a superposition of harmonic oscillators he concluded that there had to exist an infinite ZPE as given by $\frac{1}{2}\sum h\nu_i$, where the summation is over the total number of degrees of freedom of the field. However, Jordan thought of the ZPE of the field as merely a mathematical quantity with no direct physical meaning, such as he wrote to Einstein in late 1925. Einstein apparently agreed, for in a letter to Ehrenfest of 12 February 1926 he expressed his disbelief in the new quantum mechanics or what he called 'the Heisenberg–Born scheme.' Among the reasons for his negative attitude he mentioned that, 'a zero-point energy of cavity radiation should not exist. I believe that Heisenberg, Born, and Jordan's argument in favour of it (fluctuations) is feeble.' In discussions with Jordan he also objected that the ZPE of cavity radiation per unit volume would be infinite, but Jordan responded that this was not a problem since in experiments only energy differences turn up.[35]

As far as the physical meaning of the ZPE of space was concerned, Jordan agreed with Einstein that it was non-physical and beyond measurement. This was also the opinion of Pauli, who in a joint paper with Jordan found an easy way to get rid of the infinite zero-point term. The two authors emphasized the difference between the ZPE of a material system and the one of an electromagnetic field:

> Contrary to the eigen-oscillations in a crystal lattice (where theoretical as well as empirical reasons speak to the presence of a zero-point energy), for the eigen-oscillations of the radiation no physical reality is associated with this 'zero-point energy' of ½ $h\nu$ per degree of freedom. We are here dealing with strictly harmonic oscillators, and since this 'zero-point energy' can neither be absorbed nor reflected ... it seems to escape any possibility for detection. For this reason it is probably simpler and more satisfying to assume that for electromagnetic fields this zero-point radiation does not exist at all.[36]

Of course, the verdict of the specialists in quantum mechanics of the late 1920s was not the final answer to the question of the ZPE associated with electromagnetic radiation. Ten years later it had become part of standard physics that zero-point oscillations in empty space are inevitable and can never cease completely. An absolute vacuum in the sense of classical physics is impossible. Or to phrase it differently, the physical vacuum is entirely different from nothingness. But there still was no experimental evidence that vacuum ZPE and its related fluctuations really existed.

[35]  Mehra and Rechenberg (1982), pp. 276–277.
[36]  Jordan and Pauli (1928), p. 154.

# 5.8 Roads to Dark Energy

Physicists in the late 1920s could not possibly foresee the crucial role that ZPE came to play in quantum field theory as well as in the apparently quite different area of physical cosmology. From a modern point of view there are zero-point fluctuations associated with the ZPE of *all* quantum fields and not only the electromagnetic field. As regards cosmology the consensus view is that cosmic space is filled with vacuum energy with a density given by the cosmological constant $\Lambda$ introduced by Einstein in his original cosmological model of 1917. Moreover, Nernst's old speculation of a vacuum energy has been vindicated by experiments on the so-called Casimir effect predicted by the Dutch physicist Hendrik Casimir in 1948. The Casimir effect, which is an attraction between uncharged and perfectly conducting plates, is generally taken as evidence of zero-point fluctuations in the quantum vacuum.

Despite this consensus view, the Casimir effect does not prove the ZPE for quantum fields in any strict sense. The same effect, qualitatively as well as quantitatively, can be derived by methods that do not rely on the ZPE assumption, which has caused some physicists to doubt if there is any connection between the Casimir effect and ZPE, or if empty space needs be ascribed an energy density due to ZPE oscillations. They argue that neither the Casimir effect nor any other known phenomenon demonstrates the reality of ZPE in field physics. More generally, ZPE does not follow unambiguously from quantum field theory but depends on the chosen quantization procedure; for this and other reasons, ZPE is not a necessity but is rather to be seen as a heuristic and computational aid.[37]

Only relatively recently, with the discovery of the accelerated expansion of the universe in the late 1990s, did the concept of 'dark energy' come to play a central role in cosmology. According to the standard view, the still mysterious dark energy is a manifestation of vacuum ZPE. The process leading to the current understanding of dark energy took two separate ways, one connected with the cosmological constant and the other with the quantum vacuum.

In an address of 1933 the Belgian physicist and pioneer cosmologist Georges Lemaître interpreted the cosmological constant as a vacuum energy density given by

$$\rho_{vac} = \frac{c^2}{8\pi G}\Lambda$$

where $G$ is the constant of gravity. 'Everything happens as though the energy *in vacuo* would be different from zero,' he said.[38] He found the corresponding pressure to be negative, $p_{vac} = -\rho_{vac}c^2$, meaning that it is a tension rather than a pressure. Lemaître

---

[37] Jaffe (2005); Gründler (2017).
[38] Lemaître (1934), p. 12. For some of the steps leading to the modern concept of dark energy, see Kragh (2012a), pp. 226–233.

did not associate his interpretation with the ZPE of space or otherwise relate it to quantum physics. At about 1960 Lemaître's insight was practically forgotten and the cosmological constant generally believed to be zero. Although the Casimir effect was known, it was not yet confirmed quantitatively and it was not thought to have any cosmological significance. Only in 1967 did the Russian physicist Yakov Zeldovich suggest that the cosmological constant might be understood in terms of the quantum vacuum. Today it is generally accepted that the greater part of the energy content of the universe (about 73 per cent) is due to the dark energy or cosmological constant expressing the zero-point fluctuations in empty space. Unfortunately, the quantum-field calculations of the latter quantity differ by an enormous factor from the observed value of the cosmological constant and this profound difficulty has still not been resolved. But this is another story. What matters is that ZPE is all over.

## References

Authier, A. (2013). *Early Days of X-Ray Crystallography*. Oxford: Oxford University Press.

Bartel, H.-G., and Huebener, R. P. (2007). *Walther Nernst: Pioneer of Physics and Chemistry*. Singapore: World Scientific.

Bohr, N. (1915). On the quantum theory of radiation and the structure of the atom. *Philosophical Magazine*, 30, 394–415.

Bohr, N. (1922). *The Theory of Spectra and Atomic Constitution*. Cambridge: Cambridge University Press.

Bohr, N. (1981). *Niels Bohr Collected Works*, vol. 2, ed. Ulrich Hoyer. Amsterdam: North-Holland.

Bohr, N. (1996). *Niels Bohr Collected Works*, vol. 7, ed. Jørgen Kalckar. Amsterdam: Elsevier.

Einstein, A. (1915). Experimenteller Nachweis der Ampèreschen Molekularströme. *Naturwissenschaften*, 3, 237–38.

Einstein, A. (1993). *The Collected Papers of Albert Einstein*, vol. 5, eds. M. J. Klein, A. J. Kox, and R. Schulmann. Princeton: Princeton University Press.

Einstein, A. (1995). *The Collected Papers of Albert Einstein*, vol. 4, eds. M. J. Klein, A. J. Kox, J. Renn, and R. Schulmann. Princeton: Princeton University Press.

Einstein, A. (1998). *The Collected Papers of Albert Einstein*, vol. 8, eds. R. Schulmann, A. J. Kox, M. Janssen, and J. Illy. Princeton: Princeton University Press.

Einstein, A. (2004). *The Collected Papers of Albert Einstein*, vol. 9, eds. D. K. Buchwald, R. Schulmann, and T. Sauer. Princeton: Princeton University Press.

Einstein, A. (2006). *The Collected Papers of Albert Einstein*, vol. 10, eds. D. K. Buchwald, J. Illy, and V. I. Holmes. Princeton: Princeton University Press.

Einstein, A. (2009). *The Collected Papers of Albert Einstein*, vol. 12, eds. D. K. Buchwald, et al. Princeton: Princeton University Press.

Einstein, A., and de Haas, W. J. (1915). Experimenteller Nachweis der Ampèreschen Molekularströme. *Deutsche Physikalische Gesellschaft, Verhandlungen*, 17, 152–70.

Einstein, A., and Stern, O. (1913). Einige Argumente für die Annahme einer molekularen Agitation beim absoluten Nullpunkt. *Annalen der Physik*, 40, 551–60.

Enz, C. P. (2002). *No Time to be Brief: A Scientific Biography of Wolfgang Pauli*. Oxford: Oxford University Press.

Eucken, A. (ed.) (1914). *Die Theorie der Strahlung und der Quanten: Verhandlungen auf einer von E. Solvay einberufenen Zusammenkunft.* Halle: W. Knapp.

Frenkel, V. Y. (1979). On the history of the Einstein–de Haas effect. *Soviet Physics Uspekhi,* **22**(7), 580–87.

Galison, P. (1982). Theoretical predispositions in experimental physics: Einstein and the gyromagnetic experiments, 1915–1925. *Historical Studies in the Physical Sciences,* **12**, 285–323.

Gavroglu, K., and Goudaroulis, Y. (1989). *Methodological Aspects of the Development of Low Temperature Physics 1881–1956.* Dordrecht: Kluwer Academic.

Gearhart, C. A. (2010). 'Astonishing successes' and 'bitter disappointment': The specific heat of hydrogen in quantum theory. *Archive for History of Exact Sciences,* **64**, 113–202.

Gründler, G. (2017). The zero-point energy of elementary fields. Arxiv:1711.03877.

Jaffe, R. L. (2005). Casimir effect and the quantum vacuum. *Physical Review D,* **72**, 021301.

James, R. W., Waller, I., and Hartree, D. R. (1928). An investigation into the existence of zero-point energy in the rock-salt lattice by an X-ray diffraction method. *Proceedings of the Royal Society A,* **118**, 334–50.

Jordan, P., and Pauli, W. (1928). Zur Quantenelektrodynamik ladungsfreier Felder. *Zeitschrift für Physik,* **47**, 151–73.

Kuhn, T. (1978). *Black-Body Theory and the Quantum Discontinuity.* Oxford: Clarendon Press.

Kragh, H. (1995). Cosmology between the wars: The Nernst–Macmillan alternative. *Journal for the History of Astronomy,* **26**, 93–115.

Kragh, H. (2012a). Preludes to dark energy: Zero-point energy and vacuum speculations. *Archive for History of Exact Sciences,* **66**, 199–240.

Kragh, H. (2012b). *Niels Bohr and the Quantum Atom: The Bohr Model of Atomic Structure 1913–1925.* Oxford: Oxford University Press.

Kragh, H. (2012c). Walther Nernst: Grandfather of dark energy? *Astronomy and Geophysics,* **53**(2), 1.24–26.

Lewis, G. N. (1916). The atom and the molecule. *Journal of the American Chemical Society,* **38**, 762–85.

Lindemann, F. (1919). Note on the vapour pressure and affinity of isotopes. *Philosophical Magazine,* **38**, 173–81.

Lemaître, G. (1934). Evolution of the expanding universe. *Proceedings of the National Academy of Sciences,* **20**, 12–17.

Mehra, J., and Rechenberg, H. (1982). *The Historical Development of Quantum Theory, Vol. 4.* New York: Springer.

Mehra, J., and Rechenberg, H. (1999). Planck's half-quanta: A history of the concept of zero-point energy. *Foundations of Physics,* **29**, 91–132.

Milloni, P. W., and Hsieh, M.-L. (1991). Zero-point energy in early quantum theory. *American Journal of Physics,* **59**(8), 684–97.

Mulliken, R. S. (1925). The isotope effect in band spectra: The spectrum of boron monoxide. *Physical Review,* **25**, 259–94.

Nernst, W. (1916). Über einen Versuch, von quantentheoretischen Betrachtungen zur Annahme stetiger Energieänderungen zurückzukehren. *Verhandlungen der Deutschen Physikalischen Gesellschaft* **18**, 83–116.

Planck, M. (1923). Die Bohrsche Atomtheorie. *Naturwissenschaften,* **11**, 535–37.

Planck, M. (1958). *Physikalische Abhandlungen und Vorträge,* vol. 2. Braunschweig: Vieweg und Sohn.

Sauer, T. (2007). Einstein and the early theory of superconductivity, 1919–1922. *Archive for History of Exact Sciences*, **61**, 159–211.

Schrödinger, E. (1926). Quantisierung als Eigenwertproblem, II. *Annalen der Physik*, **79**, 489–527.

Stern, O. (1919). Die Molekulartheorie des Dampfdruckes fester Stoffe und ihre Bedeutung für die Berechnung chemischer Konstanten. *Zeitschrift für Elektrochemie*, **25**, 66–80.

Tolman, R. C. (1920). The entropy of gases. *Journal of the American Chemical Society*, **42**, 1185–93.

Van Delft, D. (2007). *Freezing Physics: Heike Kamerlingh Onnes and the Quest for Cold.* Amsterdam: Royal Dutch Academy of Science.

Whitaker, M. A. B. (1985). Planck's first and second theories and the correspondence principle. *European Journal of Physics*, **6**, 266–70.

CHAPTER 6

..................................................................

# THE EARLY DEBATES
# ABOUT THE
# INTERPRETATION OF
# QUANTUM MECHANICS

..................................................................

## MARTIN JÄHNERT AND CHRISTOPH LEHNER

QUANTUM mechanics emerged in 1925/26 via two distinct historical routes, leading to matrix and wave mechanics, respectively. These two theories seemed different in nearly every aspect. Each covered different albeit overlapping phenomena, had very different mathematical formulations, and resulted from sharply diverting ideas on physical theories. It soon became clear to the historical actors, however, that both theories were closely related. Shortly after the advent of wave mechanics, both Erwin Schrödinger and Wolfgang Pauli showed that the two theories could be translated into each other and that they therefore led to the same observational results. This made the contrast between the different methodological approaches and the physical ideas even more striking.

In this situation, Schrödinger on one side and the Göttingen and Copenhagen physicists on the other (especially Werner Heisenberg, Max Born, and Niels Bohr) entered into—at times heated—discussions on the interpretation of what they came to understand as a single theory, namely quantum mechanics. This debate centred around the competing claims of wave and matrix mechanics to offer a deeper understanding of quantum phenomena and played an important role in the further development of quantum mechanics as both parties challenged each other to extend their theories to new empirical domains and to clarify the fundamental concepts of the theory.[1]

In the historiography of quantum mechanics, the debate between Schrödinger and the Göttingen school has been prominently interpreted as a fight for hegemony over

---

[1] Beller (1999) argued that Schrödinger's call for a spatiotemporal interpretation was a necessary corrective to the positivism of Heisenberg and Jordan.

the emerging quantum mechanics and its interpretation, which was fuelled by personal motivations and the sociological circumstances of German academia at the time.[2] The present contribution takes the epistemological, methodological, and physical questions of the debate to be more than mere rhetoric. This is not to imply that the discussion was not in part fuelled by personal motivations and the sociological circumstances of German academia at the time. Nevertheless, the struggle to understand the quantum mechanical formalism was led as a rational debate that played an important role in the development of the new theory and extended its application beyond the explanation of atomic spectra to a genuinely universal theory.

The discussion focused on two questions: whether quantum phenomena had to be described in terms of continuity or discontinuity,[3] and whether quantum mechanics had to be understandable in an *anschaulich* (intuitive) way or not.[4] Whereas the question of continuity and discontinuity was intimately connected with the ontological dispute whether corpuscles or waves were the basic entities of quantum mechanics and whether processes should be described in terms of a continuous temporal evolution or in terms of quantum transitions, the question of *Anschaulichkeit* was of a methodological nature. This part of the discussion has often been taken as a debate about ontological realism defended by Schrödinger versus positivism championed by Heisenberg or Pascual Jordan. However, Schrödinger's position was not motivated by realism in the traditional sense, i.e. by whether the elements of the theory correspond to the true building blocks of nature, but rather by a methodological or even psychological demand for images in scientific reasoning.[5] The issue at stake with this question was the possibility and necessity of any interpretation of quantum mechanics beyond the mathematical formalism and its operationalist understanding in terms of measurement results. As we will argue, this question was an eminently practical matter for the contemporaries and not just or even primarily a clash of philosophical prejudices, as which it has been interpreted later. It concerned the possibility of extending quantum mechanics into new domains and of integrating the established knowledge from classical mechanics into quantum theory.

---

[2] For example, see Cassidy (1992), Beller (1999), MacKinnon (1980,1982), as well as Alexei Kojenikov's notion of the *quasi-free postdoc* (Alexei Kojevnikov 'Knabenphysik: The birth of quantum mechanics from a postdoctoral viewpoint' Talk at the HQ1-conference, Berlin, 2007), or the notion of the *physicist philosopher* proposed by Stöltzner (2003).

[3] The aspect of continuities vs discontinuities is traditionally seen as the core of the argument between Schrödinger and Heisenberg, Bohr. See for example Cassidy (1992) or Bacciagaluppi and Valentini (2009).

[4] The German word *Anschaulichkeit* literally means visualizability and has often been translated in that sense. It can also mean more generally comprehensibility in a pragmatic sense. The question of *Anschaulichkeit* is discussed by De Regt (1997, 1999), Camilleri (2009b) and Miller (1978, 1982).

[5] This idea has been discussed by Bitbol (1996) and De Regt (1997, 1999). De Regt, in particular, links Schrödinger's demand for *Anschaulichkeit* to Ludwig Boltzmann's *Bild* conception of scientific understanding and to his own contextual approach of scientific understanding (De Regt and Dieks, 2005). See also Darrigol (1992b, 272–274), who discusses Schrödinger's methodological realism in the context of his work on statistical mechanics.

# 6.1 THE NECESSITY OF SPACE-TIME PICTURES AND SCHRÖDINGER'S DEMAND FOR *ANSCHAULICHKEIT*

As mentioned in the introduction, the interpretational debate on quantum mechanics took off in the spring of 1926 as a clash of two competing theories, matrix and wave mechanics, which appeared to be divergent in terms of mathematical methods as well as concerning their normative ideas on physical theories.

The first of these theories, presented in Heisenberg's seminal paper 'On the Quantum Theoretical Reinterpretation [German '*Umdeutung*'] of Kinematical and Mechanical Relations' (Heisenberg, 1925) and the so-called 'Dreimännerarbeit' by Born, Jordan, and Heisenberg (Born *et al.*, 1926), relied on the basic concepts of Bohr's quantum theory of the atom. Quantum systems were described as physical entities by means of stationary states and transitions. A quantum system possessed a number of stationary states with a definite energy and at least initially it was assumed that the system was always in one of these states. Physical processes took place in the form of discrete transitions between these states, which were usually associated with a change in energy and hence with the emission or absorption of radiation according to the frequency condition. In the Bohr–Sommerfeld quantum theory of the atom, the main focus had been on the stationary states and their energies, which were determined using classical mechanics and a quantum condition. Matrix mechanics, by contrast, defined and constructed quantum systems through their transitions.

The heuristics and epistemic motivations underlying this approach have been subject to intense historiographical analysis and debate, with several competing attempts to explain how Heisenberg arrived at his seminal ideas.[6] While we have proposed our own version of this story elsewhere, we believe that the subtleties of this discussion are of little consequence for the present discussion. To understand the interpretational debate it is more important to characterize how Heisenberg and subsequently Born and Jordan attempted to make sense of matrix mechanics, and

---

[6] One major tradition claimed that Heisenberg's work relied on virtual oscillators and BKS theory but that he suppressed this reliance after the failure of BKS and the associated demise of virtual oscillators (MacKinnon, 1977; Serwer, 1977). Darrigol (1992a) and Duncan and Janssen (2007) argued against this that virtual oscillators and BKS theory did not play a constructive role in Heisenberg's work. Instead they traced Heisenberg's *Umdeutung* to the idea of a symbolic translation of classical into quantum expressions. This idea played a central role in the theory of dispersion and, as Darrigol (1992a) pointed out, also came into play in Heisenberg's work on multiplet intensities in the fall of 1924. Blum *et al.* (2017) as well as Jähnert (2019) put further emphasis on the role of the multiplet intensity problem. They analysed Heisenberg's work with Ralph Kronig and Wolfgang Pauli on the problem of multiplet intensities and the theoretical practices developed in this context. Based on this analysis they argued that Heisenberg's motivation for a new quantum kinematics, his choice of the anharmonic oscillator as the paradigmatic case, and the introduction of the multiplication rule can be better understood when placed in this context.

how this led to the formulation of their programmatic and normative positions on quantum mechanics and on physical theories in general.

For the purposes of this chapter it suffices to say that the development of Heisenberg's *Umdeutung* and matrix mechanics was driven primarily by calculational and formal considerations. It grew out of a highly exploratory approach by Heisenberg, leading to a new calculational scheme, and was then transformed into the equally formal mechanical framework of matrix mechanics. This is clearly visible in Heisenberg's initial work on *Umdeutung*. Following Bohr's correspondence principle, he took the transition frequencies and transition probabilities as the basic kinematic properties describing the system and sought to develop a new 'quantum kinematics', a new concept of motion, in which transition frequencies and transition probabilities/amplitudes took the place of the Fourier coefficients and frequencies in the classical description of motion. Developing this scheme in the summer of 1925 and eventually publishing his seminal paper, Heisenberg gradually reached an idea about the computational aspects of his new scheme and was able to obtain results which lent confidence in the empirical adequacy and internal consistency of his scheme. At the same time, he did not come very far in reaching a deeper understanding of this new concept of motion or what his new scheme meant as a whole. In one of his letters, still thinking primarily in terms of the calculational details, he admitted to Pauli:

> I would also like to understand what the equations of motion actually mean if one considers them as relations between transition probabilities.[7]

As the physical meaning of the equation of motion remained deeply unclear, Heisenberg neither offered a new model for the atom nor answered open interpretational questions like the one above. Instead, he explicitly rejected the classical model of the Bohr atom and came to the conclusion that it was not possible to form any spatio-temporal representation for a quantum system.[8] In the programmatic introduction to his paper, he argued that this was the most important lesson to be learned from the defects of Bohr's quantum theory:

> [Facing the failure of classical mechanics] it seems advisable to abandon the hope once and for all to observe the until now unobserved quantities (like position, frequency of the electron) [...] and to try to develop a quantumtheoretical mechanics in which only relations between observable quantities occur.
>
> (Heisenberg, 1925, p. 880)

---

[7] Heisenberg to Pauli, 24 June 1925, in: Hermann *et al.* (1979), p. 228.

[8] He thereby followed Pauli, who had advocated a radical break with the well-established ways of representing a physical theory as early as 1921 and is documented in Hendry (1984, pp. 19–23 and 63–66). Heisenberg came to agree with Pauli by 1925 and wanted to 'kill the concept of an orbit once and for all.' (Heisenberg to Pauli, 9 July 1925, in: Hermann *et al.* (1979), p. 231.)

This statement has been scrutinized at length within the historiographical debate mentioned above;[9] the historical actors, by contrast, interpreted it rather literally. For them, and this is the important point for the interpretational debate, Heisenberg's call for a quantum mechanics based 'observable quantities' given in the introduction of the paper naturally appeared as Heisenberg's programme for the emerging quantum mechanics and as a call for a positivistic approach to physics.

This programme was endorsed by Heisenberg himself, Born, Jordan, and Pauli. They proclaimed that the new quantum mechanics with its abstract nature was a 'truly discontinuous' theory that was free from any model and solely concerned with observable quantities.[10] Born and Jordan then elaborated Heisenberg's calculational scheme into a new mechanical formalism, which was modelled on the structure of Hamiltonian mechanics; the basic physical quantities, position and momentum, were given by an array of numbers (matrices). Thereby these quantities did not, however, regain their classical meaning but were emptied even further as the position matrix was no longer defined through but only associated with the transition probabilities of the system, while there was no operational definition for the momentum matrix outside the formalism. As such, the underlying physical picture was left in the shadows and the search for an interpretation of the theory was postponed for the sake of applying the formalism to problems like the Zeeman effect or the hydrogen atom, exploring the new possibilities and developing new methods for putting the theory to work.

At the same time, the abandoning of a space-time description of the atom—as Heisenberg, Born, and Jordan knew very well—was costly.

> Such a system of quantum theoretical relations between observable quantities will, however, show the deficit compared with the previous quantum theory, that it cannot be immediately interpreted geometrically intuitively (*anschaulich*), because the electron's motions cannot be described in ordinary concepts of space and time.   (Born *et al.*, 1926, p. 558)

Matrix mechanics—even for its founders—lacked the intuitiveness (*Anschaulichkeit*) of a space-time interpretation that the old quantum theory and classical physics had to offer. But as long as matrix mechanics was empirically adequate and mathematically consistent a missing interpretation of matrix mechanics in terms of models was a deficit that Heisenberg, Born, and Jordan were willing to accept.[11] For the time being,

---

[9]  Initially, it was interpreted as the methodological starting point for the construction of *Umdeutung*. This view has been rejected for a long time and instead most regarded it as a rhetorical strategy and mere window-dressing serving to justify the new calculation scheme while lacking a consistent physical model for it. See MacKinnon (1977, p. 177) and Duncan and Janssen (2007). While Heisenberg's positivistic rhetoric certainly cannot be taken at face value, recent scholarship has taken the more moderate position that it can also be seen as a first tentative attempt to circumscribe and reflect on the steps taken in the development of *Umdeutung*. See Blum *et al.* (2017).

[10]  See (Born and Jordan, 1925a, p. 879) and (Heisenberg, 1925, p. 880).

[11]  For a more elaborate discussion of matrix mechanics and its shortcomings see Beller (1983a, 1999) as well as Bonk (1994).

the creators of matrix mechanics remained largely silent on the physical content and interpretation of matrix mechanics. In their letters and papers, Heisenberg, Born, and Jordan mostly described the theory through its departure from classical physics and its modes of a spatio-temporal representation.

Other physicists, especially from the older generation, did not assent. This was particularly true for Erwin Schrödinger, who was convinced that his development of wave mechanics, which was based on a fairly straightforward picture of electron states as standing waves within the atom, was superior to matrix mechanics. From his perceived position of superiority Schrödinger described matrix mechanics thus:

> The intellectual situation of the Göttingen authors is roughly that of a person who has learned to calculate the amplitude and phase of the plucked chord but does not yet know that they can be combined into a well-known intuitive (*anschaulichen*) picture.[12]

In his elaboration of wave mechanics Schrödinger agreed with the matrix mechanists that Bohr's orbital models had to be given up (Schrödinger, 1926b, p. 509). However, he opposed Heisenberg's, Pauli's, and Jordan's call for a positivist physics and believed that abandoning all models went too far. For Schrödinger, quantum mechanics like any other physical theory needed some form of *Anschaulichkeit* or neither the phenomena nor the physical theory could be understood. In this sense matrix mechanics had given up a constitutive aim of physics and committed what Schrödinger called *Waffenstreck-ung*, the surrender of one's weapons.

Stressing this point in the following debate, Schrödinger emphasized the necessity of *Anschaulichkeit* repeatedly. Indebted to the legacy of Ludwig Boltzmann through his teacher Franz Exner, he followed the Boltzmannian tradition that demanded for concrete models underlying a physical theory. His demand for *Anschaulichkeit*, however, took a more general form, as its basis was not necessarily in classical mechanics or Euclidean geometry.[13] Schrödinger looked for such a model for his wave mechanics from the beginning and interpreted the wave equation as describing the vibrations of a physical medium in space. On this basis, he tried to understand particles as a super-position of waves forming a wave packet and their electromagnetic properties as those of a charge distribution in three-dimensional space. He took this approach as the key for understanding the theory and applying it to new physical problems.[14] However, Schrödinger encountered serious problems in the attempt to develop such an interpretation. Most importantly, the wave function was not defined in the geometrical

---

[12] Schrödinger to Thirring, 17 March 1926, in: (Meyenn, 2011, pp. 199–200).

[13] (Wessels, 1983); (Bitbol and Darrigol, 1992);(Bitbol, 1996);(Stöltzner, 2003);(De Regt, 1997, 1999) The demand for concrete models was not limited to the Viennese school of statistical mechanics. Einstein's notion of constructive theories, which explain physical phenomena on the basis of simple conceptual building blocks, describes a similar approach to physics. In his distinction between theories of principle and constructive theories, Einstein in particular attributes *Anschaulichkeit* to being a characteristic of constructive theories (Einstein, 1919).

[14] Schrödinger (1926b,a,c).

space of three dimensions but rather in a high-dimensional configuration space. This space was abstractly defined and could not be simply reduced to geometrical space. The problem of connecting these two spaces occupied Schrödinger during the whole period from 1926–1928 and was one of the major arguments against the *Anschaulichkeit* of wave mechanics that even sympathetic contemporaries like Albert Einstein and Arnold Sommerfeld brought forth. Already in his notebooks of 1926 Schrödinger began understanding this irreducibility of wave functions for which he later coined the concept of *entanglement* (*Verschränkung*), since it implied that even noninteracting systems could be coupled in their physical properties.[15]

The meta-scientific convictions that motivated Schrödinger's call for *Anschaulichkeit*, however, did not carry much weight for Heisenberg. As long as he was able to calculate and predict adequate results on the basis of matrix mechanics he could claim that *Anschaulichkeit* was not a necessity for a physical theory but merely a matter of utility. This situation changed when Schrödinger, Pauli, and others proved the equivalence between matrix and wave mechanics.[16] It was within the context of the equivalence argument that Schrödinger's call for *Anschaulichkeit* took another form which forced Heisenberg to consider the issue more seriously. Schrödinger now took up Heisenberg's criterion that judged theories solely by their success. He argued that it was the possibility of extending quantum mechanics to new problems, which would decide the question of whether a spatio-temporal model was needed or not. One of these problems was atomic collision, in which electrons are scattered by the atoms of a medium at certain angles, which can be detected on a screen. To Schrödinger these experiments showed the necessity of *Anschaulichkeit*:

> The problems, which (…) come into question for the further extension of the atomic dynamics, are posed to us in an eminently *anschaulich* manner, as for example: how do two colliding atoms or molecules bounce off each other, how do electrons or α-particles get deflected, when [...] they are shot through an atom. To approach such problems, it is necessary to oversee the continuous transition between the macroscopic *anschaulich* mechanics and the micromechanics of the atom.   (Schrödinger, 1926d, p. 753)

Matrix mechanics so far was only able to account for periodic systems like atomic orbits through the formal translation of their treatment in Hamiltonian mechanics.

---

[15] Schrödinger Notebooks Undulatorische Statistik I and II, AHQP 41-1-002, 41-1-003. Schrödinger's development of the concept of entanglement will be discussed in a forthcoming paper by C. Lehner and Jos Uffink, 'Schrödinger and the emergence of the EPR argument.'

[16] Schrödinger gave his argument in his paper 'Über das Verhältnis der Heisenberg–Born–Jordanschen Theorie zu der meinen,' Schrödinger (1926d). This equivalence was also shown by Pauli (letter to Jordan from Pauli, 9 April 1926, in: Hermann *et al.* (1979), pp. 319–320) and others. Muller (1997) shows that Schrödinger and Pauli did not provide mathematically rigorous equivalence proofs. As Perovic (2008) argued, this was of little relevance to the historical actors. For them the ability to translate matrix elements into wave functions was a sufficient pragmatic argument for the equivalence of the two theories.

How could it be made to address experiments like atomic scattering, in which a spatio-temporal process and its geometrical setup is essential? For Schrödinger this meant that the geometrical situation of the experiment had to be translated into a quantum theoretical description of the microphysical collision phenomenon and that hence there had to be some kind of conception of a spatio-temporal process within quantum mechanics. '[If] one feels obliged to suppress *Anschauung* in atomic dynamics [ . . . ] for epistemological reasons' (Schrödinger, 1926a, p. 753), such a transformation from microphysical to macrophysical descriptions can hardly be achieved as the presumably key concepts are not part of quantum mechanics.

While he did not see any possible way to tackle the collision problem within matrix mechanics, Schrödinger argued that wave mechanics could make the transition from macroscopic to microscopic dynamics understandable through the optical-mechanical analogy: classical mechanics represents the motion of a particle as a spatial trajectory, and one can consider this as a geometrical limit of the wave mechanical representation of a wave packet, like the beams of geometrical optics are a limit of wave trains in wave optics. In the wave mechanical picture a particle has to be represented as a superposition of waves which form a wave packet and stay together in a limited region of space. Schrödinger succeeded in showing the possibility of such a wave packet for the harmonic oscillator and stipulated that his approach would also be possible for more sophisticated problems.[17]

Schrödinger's new argument posed a serious challenge and set much of the agenda for Heisenberg, Bohr, Born, and Pauli in the next stage of interpretational debate. Unlike the previous call for *Anschaulichkeit*, the new argument was based on desiderata that Heisenberg shared. First there was the challenge of extending quantum mechanics to new empirical domains like scattering, which matrix mechanics also had to account for. Second there was the transition between the microscopic quantum description and the macroscopic world of the experiment. In terms of spatio-temporal models such a transition could be thought of as a gradual shift of scale. Even if Heisenberg wanted to keep avoiding such models, he nonetheless had to account for the emergence of the macroscopic world.

In his June 1926 paper 'Mehrkörperproblem und Resonanz in der Quantenmecha-nik' (Heisenberg, 1926), Heisenberg felt the need to address Schrödinger's arguments. He acknowledged that *anschauliche Bilder* had a heuristic value in order to operation-alize the otherwise abstract mathematical formalism. He even agreed with Schrödinger that these pictures could be used to extend the quantum mechanical formalism. However, he did not come up with an answer for the specific problem of scattering and kept insisting that *Anschaulichkeit* was a matter of utility. Rather, he warned that

---

[17] Schrödinger had elaborated his view on the optical-mechanical analogy (see Joas and Lehner (2009) for a detailed account) and put forward the idea of considering particles as wave packets in the second communication (Schrödinger, 1926b) and proved the possibility of constructing stable wave packets for the case of the harmonic oscillator in his paper 'Der stetige Übergang von der Mikro- zur Makromechanik' (Schrödinger, 1926a).

putting too much trust in pictures would probably modify the theory in a way that would threaten the consistency of the mathematical formalism (Heisenberg, 1926, pp. 412–413).

Heisenberg therefore liked to think of pictures as mere analogies, which could be abandoned once one got into trouble with them, as had been the case with the pictures of old quantum theory. While Heisenberg acknowledged that the demand for *Anschaulichkeit* was justified, he had reason to doubt that Schrödinger's solution could give a consistent model of atomic processes since he could not accept that quantum phenomena could be explained in purely continuous terms. Phenomena such as the photoelectric effect, and the experiments of Franck and Herz and of Stern and Gerlach, displayed the discontinuous nature of quantum theory and Schrödinger's continuous theory so far could not account for them.[18] Schrödinger could not agree with this pragmatic take on models. From his Boltzmannian point of view it was exactly the visualizable models that guarantee the consistency of the formalism. He insisted that one had to apply models consistently throughout physics and in that sense take them seriously beyond their heuristic usage.

Second, Schrödinger's idea to connect classical and quantum mechanics directly via his idea of wave packets contradicted an essential aspect of the correspondence principle, which formed the core of matrix mechanics. Where Schrödinger saw quantum theory and classical theory merging into another, the correspondence principle stated the opposite. Classical and quantum theory, though analogous in their mathematical structure, were distinct conceptual frameworks. In the limiting case, where both theories led to the same results, the conceptual gaps between them were not diminished whatsoever.[19] Hence Schrödinger's idea of wave packets that should directly transfer the spatio-temporal description of classical mechanics into quantum theory inappropriately jumped this conceptual gap. While Heisenberg would eventual start to search for a spatio-temporal interpretation and to think about the relation between the macroscopic and microscopic mechanics, these challenging problems did not force him to adopt Schrödinger's approach.

Heisenberg and Schrödinger first met in person in Munich in June 1926. Schrödinger presented his wave mechanics in two talks at Sommerfeld's colloquium. After the talks, Heisenberg challenged him that wave mechanics could not explain the discontinuity of the quantum effects.[20] This first encounter left both physicists with a puzzle.

---

[18] 'As nice as Schr[ödinger] is in person, I do find his physics very strange; you feel 26 years younger when you hear about it. Schr[ödinger] throws everything "quantum theoretical" overboard, namely the photoelectric effect, Franck's scattering, the Stern–Gerlach effect, etc.; then, it isn't difficult to produce a theory. Only, it doesn't fit experience.' Heisenberg to Pauli, 28 July 1926, in: Hermann *et al.* (1979), pp. 337–338, our translation.

[19] For Bohr's thinking on the conceptual gap between classical and quantum physics, see Darrigol (1992a), (Tanona, 2002) and Jähnert (2019).

[20] There are no records of Schrödinger's talk nor of Heisenberg's challenges. Heisenberg described the situation in his popular book 'Der Teil und das Ganze' first published in 1969 (Heisenberg, 2005, p. 91), and reported to Pauli in the above-mentioned letter from 28 July 1926.

Schrödinger of course knew of the quantum effects Heisenberg had pointed out to him. But he believed that seemingly discontinuous phenomena such as the photoelectric effect or the Franck–Hertz experiment could be given a continuous explanation, just as the discrete energy levels of atoms could be explained in terms of his wave equation.[21] Schrödinger claimed that continuity was a necessary element for the understanding of a phenomenon as a spatio-temporal process. As the discussion was going on, however, he had to acknowledge that the pictures he had to offer were far from being capable of explaining all quantum effects and would eventually have to be revised a good deal.[22]

In order to challenge Heisenberg's view that discrete transition processes were necessary for the description of the quantum effects, Schrödinger argued that quantum mechanics so far did not have a self-sufficient explanation of the intensity and width of spectral lines, which he saw as depending on a theory of the interaction of matter and the electromagnetic field. Matrix mechanics in the tradition of the correspondence principle simply postulated the identity of matrix elements with the transition probabilities and treated the radiation process as a black box.[23] This lack of a dynamical theory of the interaction between matter and radiation was not only a theoretical matter, Schrödinger argued that it led to problems in the interpretation of experiments. In particular, he pointed to phenomena such as the attenuation of radiation or the width of spectral lines which were connected to the problem of radiation damping in classical electrodynamics. The former problem had been studied quantitatively in Wilhelm Wien's decay experiments on canal rays[24] and led to the conclusion that the decay constants of the first three main lines of the hydrogen spectrum were identical. In matrix mechanics or any theory explaining radiation on the basis of discrete transitions, this could hardly be expected. Here, exponential decay would result from the statistics of a large number of individual transitions, in which the

[21] Schrödinger held that discontinuities could be understood as resonance phenomena. This conviction remained even after Born had proposed his statistical interpretation. See Schrödinger to Born, 2 November 1926 (AHQP 41.7.).

[22] Schrödinger acknowledges the deficits of his picture in a letter to Wien, 25 August 1926, in: (Meyenn, 2011, p. 305) and admitted that he could not think of a way to approach the problem either as a problem of calculation or as a problem of reasoning.

[23] Agreeing with Schrödinger, Born acknowledged that this was not yet a 'rational theory' of the transition processes and their dynamics (Born, 1926a, 827). See Joas and Lehner (2009) as well as in a forthcoming paper by Alexander Blum and Martin Jähnert entitled 'Closed theories? The Coevolution (and eventual Separation) of Quantum Mechanics and QED 1925–1927' for an account of Schrödinger's struggles to develop a theory of the interaction between radiation and matter on the basis of wave mechanics.

[24] Wien had conducted these experiments in the early 1920s and written a series of articles on it (Wien, 1919, 1921, 1924). Heisenberg—having studied in Munich and worked with Sommerfeld—knew of the experiments and had even written a paper together with Sommerfeld in 1922 (Sommerfeld and Heisenberg, 1922), in which they had connected the width of spectral lines with the correspondence principle and the damping of radiation.

decay time is set by the transition probability.[25] Because the transition probabilities vary for each line, each line decays with a different time.

As he left the meeting, Heisenberg was quite unsatisfied. In addition to the puzzle of Wien's decay experiments he had the distinct feeling that he did not convince any of the physicists around and that they were about to turn to the wave mechanical approach.[26] Heisenberg addressed both problems in letters to Pauli and Bohr. He tried to convince Pauli to work on the decay experiments from the viewpoint of matrix mechanics and informed Bohr about the situation in Munich.[27]

In response, Bohr invited Schrödinger to Copenhagen. Schrödinger accepted and gave two talks on his wave mechanics in October 1926. While Schrödinger was in Copenhagen he and Bohr discussed the issues of discontinuity and the possibilities of a space-time description.[28] Schrödinger could not come up with a consistent explanation for the various quantum effects on the basis of one continuous space-time description. For Bohr this was not due to a temporary inadequacy of wave mechanics. It was rather—Bohr argued in a line of thought that would later be an essential point in his complementarity argument—that it was impossible to describe quantum phenomena in classical physics language and the space-time pictures associated with it. To represent the formalism one had to operate with different pictures depending on the physical situation.

For Schrödinger this was unacceptable, as multiple pictures could not be used in a clear, unambiguous way. Rather he hoped that a future modification of wave mechanics would contain a consistent spatio-temporal representation even if it did not accord to classical physics. He argued again that certain phenomena such as the width of spectral lines were not accounted for in quantum mechanics and that the theory was in need of a modification in any case.

[25] This type of explanation had already been developed within the old quantum theory by Stern and Voelmer in 1919 (Stern and Voelmer, 1919) and described by Pauli in his review article (Pauli, 1925). In Schrödinger's wave mechanics the decay would normally not be exponential since it depended on the interference of two proper vibrations of the atom (Joos, 1926).

[26] The impression that wave mechanics was taking over might have been most troubling to Heisenberg, especially in light of Max Born's work on atomic collisions, published in July 1926 right after the encounter in Munich. Born's work will discussed in the next section since it opened a new line of argument in the debate. At this point it is important to note that Heisenberg, in his immediate response, saw it as endorsing wave mechanics as the more easily applicable framework (Heisenberg did not challenge this point) and as moving towards Schrödinger's interpretational position in favour of waves. See Heisenberg to Pauli, 28 July 1926, in: Hermann *et al.*, (1979), pp. 337–338.

[27] Heisenberg to Pauli, 28 July 1926, in: Hermann *et al.* (1979), pp. 337–338. Heisenberg hoped that Wien's findings could be explained by the fact that the atoms got excited again within the beam. See Schrödinger to Wien, 21 October 1926, in: (Meyenn, 2011, pp. 321–322). The letter to Bohr is not available anymore but only mentioned in Heisenberg's recollections in (Heisenberg, 2005, p. 91).

[28] The discussions between Bohr and Schrödinger were not recorded apart from Heisenberg's narrative in (Heisenberg, 2005, p. 94). However, some of the arguments can be reconstructed from Schrödinger's letters to Bohr and Wien shortly before and after he went to Copenhagen. See Schrödinger to Wien, 25 August 1926, Schrödinger to Wien, 21 October 1926, and Schrödinger to Bohr, 23 October 1926, in: (Meyenn, 2011, pp. 306 and 320–325).

Here there seems to be a contradiction and you say: here our present words and concepts simply do not suffice. I cannot content myself with this statement and I cannot deduce the right for me to operate further with contradictory assertions. One can weaken the assertions by saying for example that the atomic ensemble behaves 'in certain sense, as if…' and 'in a certain sense, as if…', but that is nothing more than a judicial remedy, which cannot be translated into clear thinking. I do not think that it is impossible to construct pictures that actually show the above behaviour. Radiation damping, width of spectral lines are not integrated in the theory. What I have in mind is only one thesis: one must not, even if one hundred attempts have failed, give up the hope to reach the solution, I do not say through classical pictures, but through logically consistent ideas of the actual nature of the spatio-temporal events. It is exceedingly probable that this is possible.[29]

Rejecting Bohr's attempt to deal with contradictions by resorting to multiple descriptions, Schrödinger kept insisting on an interpretation through a perhaps new ontology in ordinary space and time. He did so for methodological reasons arguing that 'contradiction-free pictures' were necessary for 'clear thinking'. In order to obtain it, he was willing to keep the theory open for revision. His main argument in this direction and indeed the main agenda in his research in 1926 was to search for an integration of wave mechanics and (classical) electrodynamics.[30]

By the fall of 1926, Heisenberg in particular went in a different direction and increasingly rejected the option of a revision of quantum mechanics. He came to argue that quantum mechanics was a consistent, logically closed theory and thus externalized problems such as radiation damping as the subject of a yet to be found quantum theory of electrodynamics.[31] The upshot of this emerging position for Heisenberg's stance in the interpretational debate was to consider the quantum mechanical formalism as given and to ask how classical models, as interpretational devices, could be applied consistently to account for experiments showing quantum behaviour.

In summary of this section, the debate between Schrödinger, Heisenberg, and Bohr (one could also include Pauli, Sommerfeld, Jordan, and Born) up to the summer and fall of 1926 centred on the question of what kind of interpretation was needed for quantum mechanics. This question needed to be debated after matrix mechanics had been developed as a scheme for calculations in Heisenberg's *Umdeutung* and after this calculational scheme had been recast into a mechanical formalism. Both these steps

[29] Schrödinger to Bohr, 23 October 1926, in: (Meyenn, 2011, p. 324).

[30] For Schrödinger's agenda, see Schrödinger (1926d). Schrödinger's struggles to develop a theory of the interaction between radiation and matter on the basis of wave mechanics have been described to some extent by Joas and Lehner (2009) and are part of paper currently prepared by Alexander Blum and Martin Jähnert.

[31] For Heisenberg's notion of closed theories, see Chevallier (1988) and Bokulich (2004). Heisenberg's arguments for the closure of quantum mechanics and separation from quantum electrodynamics will be discussed in the aforementioned paper by Blum and Jähnert.

had not touched upon on the constructive basis of the theory but had focused on the renunciation of classical models.

From this perspective, Heisenberg, Born, and Jordan, at least initially, argued that all physical information could already be expressed through the mathematical formalism and the operationalist definitions of its concepts. Schrödinger contested this rejection of *Anschaulichkeit* and called for an interpretation of quantum mechanics. As he saw it, it was one of the ultimate goals of physics but also a methodological necessity to have an interpretation in terms of spatio-temporal pictures or models, not necessarily identical with but certainly not categorically different from classical ones.

The basis for a debate beyond conflicting methodological approaches was given with the proof of equivalence between matrix and wave mechanics. It was in this context that Schrödinger challenged his opponents with the concrete problem of particle collisions and the unsettled relationship between classical and quantum theory. These problems were for him evidence for the necessity of a spatio-temporal understanding of quantum mechanics beyond the abstract mathematical formalism. While no one in Copenhagen or Göttingen accepted Schrödinger's interpretation, his wave mechanics undeniably offered a new and fruitful way to approach the solution of long-standing problems. The successes of wave mechanics and Schrödinger's challenges to quantum mechanics could not be denied. As we will see in the next sections, it was indeed the adaptation of formal methods of wave mechanics that allowed Max Born to treat the problem of atomic collisions and scattering and thereby eventually led Heisenberg to accept the necessity of a spatio-temporal description and fuel the search for a physical picture underlying quantum mechanics.

# 6.2 THE STATISTICAL INTERPRETATION AND TRANSFORMATION THEORY

Whereas Schrödinger's call for *Anschaulichkeit* was undisputedly of major importance for the debates in Göttingen and Copenhagen, the interpretation of quantum mechanics was also strongly interconnected with the formal development of quantum mechanics. It was through the development of transformation theory and the associated advent of the probabilistic interpretation in the work of Born, Pauli, Jordan, and Paul Dirac that matrix mechanists addressed a tension between the quantum theoretical probability concept and the framework of mechanics that was inherent in matrix mechanics from the beginning.

The beginnings of this reflective process can be found in Born's work on atomic collisions. In June 1926, he produced the quantum mechanical account of atomic scattering that Schrödinger had deemed crucial for further development but which Born approached from a perspective quite different from Schrödinger's. Born had been working on the problem of atomic collisions since 1925. He had unsuccessfully tried to account for 'aperiodic phenomena' in a pre-matrix-mechanics paper with Jordan (Born and Jordan, 1925b) and in the context of matrix mechanics (Born *et al.*, 1926; Born and Wiener, 1926). It was only with the formalism of wave mechanics that Born managed to treat the collision problem successfully[32] and explicitly acknowledged that his success depended on the possibility of giving a spatio-temporal description of the physical process in wave mechanics:

> [ ... ] quantum mechanics does not only formulate and solve the problem of the stationary states but also the problem of transition processes. Schrödinger's formulation seems to do justice to this situation by far in the simplest way. Moreover it allows keeping the ordinary conception of space and time, in which the events take place in a normal way.   (Born, 1926b, 826)

Born's wave mechanical account relied on a clear picture of the collision process: in the simplest case, he treated atomic collision as the interaction of the wave function representing a scattering particle with a potential describing the scattering medium. This meant that a spatially extended wave approached the scattering centre and was scattered in all possible directions. However, even though this treatment was based on a spatio-temporal description of the collision process, it did not square well with Schrödinger's initial wave packet interpretation of a particle. As the outgoing wave was spread out in all directions it could not represent particles in the form of Schrödinger's wave packets, which were localized in a narrow region of space. Instead Born proposed that the scattered wave did not actually represent the particle itself but that the amplitudes of the wave at a certain point determined the probability distribution for different scattering directions of the particles.

This new interpretation of the wave function was at odds with both Heisenberg's and Schrödinger's approach to the interpretation of quantum mechanics up to this point. On the one hand, Born went back to a spatio-temporal picture of the collision process and hence distanced himself from the rejection of visualizability in early matrix mechanics. On the other hand, his interpretation differed substantially from the kind of interpretation envisioned by Schrödinger, who took the wave function to be a physical object with definite energy and momentum and represented the electron as a material wave packet with a charge density. Born's newly interpreted wave function was a guiding field, determining the outcome of an experiment, and hence did not resemble a physical object in Schrödinger's sense.

The idea of the guiding field combined the notion of a field (similar to classical field theory) with the notion of probabilities within a mechanical framework. This conception was not invented by Born in 1926. Born referred to Einstein's idea from the early 1920s, that the light wave determined the trajectories of light quanta, which alone carry

---

[32] Born (1926b,a, 1927a). For a detailed reconstruction of Born's efforts to account for scattering processes see Wessels (1980). For further work on Born's interpretation see Jammer (1966), Konno (1978), Cartwright (1987), Beller (1990), Soon (1996).

energy and momentum.[33] Moreover, the connection of transition probabilities with a field intensity had played a crucial role in the development of the correspondence principle and hence in the establishment of matrix mechanics. Once the equivalence between matrix elements and the products of the eigenfunctions was established, it was quite straightforward to interpret the amplitudes of wave functions as probability amplitudes in the sense of the matrix calculus. As a major contributor to these developments, Born was in a perfect position to extend this statistical element to wave mechanics in the case of the collision problem. Interpreting the wave function as a probability field, he also came to discuss the connection between statistics and mechanics which was inherent in matrix mechanics from the beginning. While transition probabilities had almost exclusively been used to characterize the black box of the emission process, Born's statistical interpretation described matter: the wave function described the probability of finding a given value for the position or momentum of an electron. For Born, this aspect pointed to a more general insight into the foundations of quantum mechanics:

> This point of view consists in seeing quantum mechanics as a fusion (*Verschmelzung*) of mechanics and statistics in the following sense: the new mechanics does not, like the old one, answer the question: 'how does a particle move' but the question 'how probable it is, that the particle moves in given way.' Particles (point charges, electrons) are thus assumed [ ... ]. The particles are always accompanied by a wave process; these de Broglie–Schrödinger waves depend on the forces and through the square of their amplitude determine the probability of a particle being present in the state of motion characterized by the wave.[34]

Born's new view of quantum mechanics took the probability amplitudes as the basic quantities characterizing the material particles and the Schrödinger equation as the law of their dynamical evolution. In this sense Born stated further:

> Classical mechanics solves the following problem. Given the configuration (position and momenta) at time $t = 0$; what is the configuration at time $t = T$, if the forces acting in the meantime are given? Quantum mechanics, I claim, solves the analogous problem: up to the time $t = 0$ the probability of a configuration (the $|c_n|^2$) is given;

---

[33] (Born, 1926b, p. 804). See also (CPAE 7, 2002, p. 486). Similar ideas were also expressed by Louis de Broglie whose seminal ideas about matter waves were the inspiration for Schrödinger's work. De Broglie, whose work was embedded in the smaller French and Belgian community, developed his interpretational ideas in virtual isolation. A detailed discussion of de Broglie's work and his interpretation is given in Bacciagaluppi and Valentini (2009). De Broglie entered the debate only at the Solvay Conference in 1927 and did not take part actively in the debate between Schrödinger and the Göttingen physicists before that.

[34] (Born, 1927a, pp. 167–168). This paper is the third on atomic collision processes. Born also expressed this interpretation in a survey paper sent to *Die Naturwissenschaften* in March 1927 (Born, 1927b, p. 240).

what is the probability of the configuration after the time $T$, if in the meantime (and only in this time) an external given force is acting?    (Born, 1927a, 171–172)

Born's interpretation of quantum mechanics as a 'fusion of mechanics and statistics' is an explicit reflection on the aforementioned tension within matrix mechanics. This tension was not one between particle or waves but rather between the quantum theoretical probability concept and the framework of mechanics. It had been expressed already in the basic theoretical entity of matrix mechanics: matrix elements which functioned both as dynamical variables and as probabilities. They were quantum theoretical translations of the classical Fourier coefficients of orbital motion and transition probabilities at the same time. Adopting this perspective in wave mechanics implied that the Schrödinger equation, which determines the evolution of the eigen-functions and hence of the matrix elements, makes transition probabilities subject to a deterministic dynamics in close analogy to classical mechanics.

Born's quantum mechanical probability amplitudes were alien to the framework of either classical mechanics or statistical physics. In these theories, probabilities were not thought of as a physical property of a system evolving under given forces but usually thought as an epistemic tool to describe the ignorance of an observer.[35] The new quantum theoretical concept of a 'probability amplitude'[36] emerged from an integration of two distinct fields of knowledge and could not be reduced to one or the other.

In the conclusion to his paper, Born addressed the question whether the probability interpretation of the wave function implied some form of indeterminism of atomic processes. While he saw it as an indication that determinism for atomic processes should be given up, he admitted that the probability interpretation did not necessarily imply indeterminism, since one could still postulate what came to be known as hidden variables, i.e., the existence of a more detailed description of individual processes. He stressed only that these additional parameters were of no relevance to the working physicist and that the statistical predictions were all that were relevant for experimental situations.

It should be noted that neither Born's position nor the ensuing debate about the statistical interpretation fit well with the traditional reading of the interpretation debate as a struggle between determinism and indeterminism. For example, Paul Forman saw the defence of indeterminism as a central argument for the cultural influences on the development of quantum physics (Forman, 1971). While Born was one of the few physicists at the time who actually referred to indeterminism, his arguments and his main conclusion did not concern this issue. It was only in his later years that he identified indeterminism as a central element of quantum mechanics.[37]

---

[35] Cartwright (1987) has also pointed the fact that Born's interpretation is at odds with our ordinary conception of probability. From Born's statement that probabilities obey a deterministic dynamics and thus are an object, on which a force is acting, she has argued that his probabilities are real and not epistemic.

[36] The term was coined by Jordan (1927b, 3), who took up Born's and Pauli's statistical interpretation.

[37] Born (1955).

The tension between mechanics and statistics created by Born's statistical interpret-ation of wave mechanics triggered new discussions about the meaning of quantum mechanics, which we will discuss here. While the Göttingen–Copenhagen community explored the quantum probability concept, the critics of matrix mechanics, especially Schrödinger and Einstein, pointed out paradoxes that this concept entailed. In the case of Schrödinger, this new line of discussion emerged in immediate reaction to Born's collision theory, in which he argued against Born's solution for the collision problem and his new interpretation. Schrödinger did so not because he opposed indeterminism in general, but because he rejected Born's reintroduction of the discontinuities of the old quantum theory and of the particle picture. As Schrödinger saw it, Born together with the Göttingen and Copenhagen community dogmatically took these discontinu-ities as secured by experiment and thereby denied the possibility of searching for the continuous explanations he envisioned.[38] Schrödinger himself had advocated indeter-minism in atomic physics in his inaugural lecture at Zürich in 1922, which he published as (Schrödinger, 1929), even though he distanced himself from this radical indeterminism in 1926.[39] After contemplating Born's interpretation, he repeatedly stated in his letters dating from winter 1926 to 1928 that under the statistical inter-pretation his wave equation would have to be considered as analogous to the Fokker–Planck equation, which describes the global evolution of an ensemble of systems obeying probabilistic dynamics. However, Schrödinger immediately saw how prob-lematic this analogy was: the Fokker–Planck equation is not reversible in time since it describes a process of growing disorder; the Schrödinger equation on the other hand is time reversible, just like a classical wave equation. 'I would like to believe it [the probabilistic interpretation], because it is really much more comfortable; if only I could calm my conscience that it is not irresponsible to get over the difficulties through such a cheap buy.'[40] In a letter to Born from 2 November 1926, Schrödinger argued that there was a contradiction between the approach of waves spread out in space and the corpuscular concepts used by Born for the interpretation. Schrödinger based this critique on two separate arguments.

The first one involved the explanation of the particle tracks in the Wilson cloud chamber and the general conceptualization of collision processes. Schrödinger con-trasted Born's wave treatment of collision with the classical account of particle colli-sions. Classically, collisions were treated as particles passing a scattering object on a well-defined trajectory at a certain distance from the scattering centre and interacting with the scattering potential. In this treatment the distance from the centre is the key parameter in the scattering process. The closer the particle passes the scattering centre the larger the angle at which it is scattered. In Born's approach, scattering was treated without considering the distance between the particle and the atom. This quantity

[38] Schrödinger to Born, 2 November 1926, in: (Meyenn, 2011, pp. 328–332).
[39] Schrödinger to Wien, 25 August 1926, in: (Meyenn, 2011, p. 306).
[40] Schrödinger to Planck, 4 July 1927, in: (Meyenn, 2011, p. 421). See also Schrödinger to Born, 8 June 1927, in: (Meyenn, 2011, p. 409).

could not even be formulated in Born's approach, as the incoming particle was represented by a plane wave and the scattered particle by a spherical wave, which both extended throughout all space.

Schrödinger now asked how Born could meaningfully describe an individual process of scattering if he assumed that electrons had a definite position. First assume that the electron's position does not play a role for the scattering process. For example, an electron passing on the left of an atom at 10 Å could still be scattered to the right, and hence one would get the same probability distribution for the scattering directions regardless of the distance between the incoming electron and the centre. Then the talk about a point particle with a trajectory defined with a precision of 10 Å is meaningless and the description through the 'vicious wave packets'[41] contains everything that is physically relevant. Now assume that the position does play a role and the probability of different scattering angles depends on the trajectory of the incoming electron. Then Born's approach does not cover the individual process and one needs to search for a representation of the individual scattering event that is more detailed than Born's.

In his second argument, Schrödinger compared Born's corpuscular interpretation of wave mechanics with the already disproven Bohr–Kramers–Slater theory of 1924. The BKS theory had attempted to describe Compton scattering without the assumption of photons, assuming that each scattered electron was accompanied by a spherical electromagnetic wave going off in all possible directions. This wave in turn induced detection events probabilistically, so that energy and momentum were conserved only in the statistical average, not individually. However, this statistical explanation could not account for the correlation of detection events observed in the Bothe–Geiger experiment.[42]

Schrödinger argued that Born's interpretation of the wave function as a field of probabilities led to the same picture and merely added particles. This addition, he argued, would only be able to fix the problem of correlation if the position of the particles played a role in the triggering of the counter. This meant in turn that Born's description of scattering without the particles' positions could not be regarded as a complete description of the individual event. Combining his two arguments, Schrödinger thus pitted two versions of the statistical interpretation against each other: where the first took the wave function as a complete description while abandoning particle positions, the second saw the wave function as an incomplete description allowing for a more detailed underlying description with localized particles. He argued the claim of completeness in the first alternative was particularly problematic in light of available experimental knowledge and implicitly advocated further modifications of quantum mechanics.

Schrödinger's letter to Born shows the first awareness of a problem that occupies the discussions about the interpretation of quantum mechanics to this day: whether the wave function should be interpreted epistemically, i.e. as an incomplete description of

---

[41] Schrödinger to Born, 2 November 1926, in: (Meyenn, 2011, p. 331).
[42] Bohr *et al.* (1924).

particles with a definite position, or ontically, i.e. as an objectively existing spatially extended object. If it is interpreted ontically, then it cannot explain (without further assumptions) the appearance of definite positions upon measurement, such as Wilson cloud chamber tracks. If it is interpreted epistemically, this implies that there should be a more complete theory where the particle positions appear explicitly in the formalism. If the quantum mechanists wanted to maintain that the definite orbits of the old quantum theory had to be abandoned, it became a central challenge how to explain the appearance of definite particle tracks. Hence the explanation of the Wilson cloud chamber tracks became a challenge for quantum mechanics and the relation of quantum mechanical states to observed values was a central problem of Heisenberg's 1927 uncertainty paper, which will be discussed in section 6.3.

In his response to Schrödinger's letter, Born admitted that he was not able to completely dissolve the challenges to the statistical interpretation. In the case of the Bothe–Geiger experiment Born stated that it was not the picture of particles impinging on the counter but the Schrödinger equation that took care of the observed correlations through energy and momentum conservation. The particle picture was in this sense unnecessary to explain the Bothe–Geiger experiment. Its only merit was to visualize the energy-momentum conservation in classical terms. As he said, 'no one is forced to visualize [zu veranschaulichen] anything.'[43] If one still wanted to do so, Born thought, no other picture than that of a corpuscle guided by a wave field was possible.

To Schrödinger's challenge to specify his interpretation of the wave function, Born replied somewhat evasively. He rejected Schrödinger's attempts to interpret the wave function as a statement about the spatio-temporal behaviour of a system in ordinary three-dimensional space with the already mentioned argument that the wave function was a function in $3N$-dimensional configuration space and could not be reduced to one in ordinary space. Against Schrödinger's interpretation of the wave function Born also referred to a letter by Pauli, which will be discussed below. Pauli had shown that the wave function was not tied to q-space. Rather one could transform this q-function into a representation in p-space or mixed pq-space. Hence Schrödinger's preference for q-space as tied to the models in the 'Anschauungsraum' was quite unjustified.

Just as Born's treatment challenged Schrödinger's interpretation of the wave function, it also posed a problem for the matrix mechanists: Born's treatment of atomic collision was based on wave mechanics, and it was not clear how it could be translated into the formalism of matrix mechanics. The attempt to translate Born's wave-mechanical treatment of collisions into the scheme of matrix mechanics was one root of the development of the generalization of matrix mechanics, transformation theory, which became the basis of the modern Hilbert space formulation of quantum mechanics, but it also informed Heisenberg's search for a more intuitive interpretation of quantum mechanics.

---

[43] Born to Schrödinger, 6 November 1926, in: (Meyenn, 2011, p. 334).

The first thoughts about such a translation was presented by Pauli in a letter to Heisenberg, which, even though it did not contain anything like a solid argument or calculations, circulated in the Copenhagen and Göttingen community and had a wider impact.[44] The main problem of applying matrix mechanics, as formulated in the Dreimännerarbeit and its subsequent applications, to the scattering problem, was that it could only be applied to closed periodic systems. Free electrons performing aperiodic motions were out of reach of its formal technical apparatus. To remedy this situation, Pauli considered an electron on a large circular orbit as an approximation for the free electron moving in a straight line. He then introduced a local potential and calculated the perturbation induced by it. This perturbation introduced transition probabilities between the different orbits. If the radius of the orbit was sufficiently large, this could be seen as a model of scattering of a free particle in a one-dimensional setting.

Extending this approach, Pauli attempted to treat the three-dimensional case. Here he realized that specifying an energy level of the unperturbed motion did not completely specify the initial state. For each energy there are many different possible states, which can be characterized by other constants of the motion (e.g. the momentum or the angular momentum). Depending on what additional constants of the motion are chosen, Pauli found, one obtains different initial states, which he likened to different kinds of 'particle swarms', i.e., ensembles of particles with that constant of motion fixed, but a varying conjugate variable (position or angle). For example, a state of given momentum corresponded to a swarm of particles moving in parallel with a specific momentum and constant energy with the conjugate positions necessarily completely unspecified. This, so Pauli wrote, meant that if the momenta were taken as 'controlled', the positions had to be left 'uncontrolled'.

Pauli was not sure about the physical meaning of this result and described it as a 'dark point' emerging from his considerations:

> The first question is why only the p's, and certainly not the p's as well as the q's, can be prescribed both with arbitrary certainty. [ ... ] One can look at the world with the p-eye and one can look at it with the q-eye, but if one wants to open both at the same time, then one goes crazy.
>      (Pauli to Heisenberg, 19 October 1926, in: Hermann *et al.* (1979), p. 347)

Exploring this dark point further Pauli recognized that he could make 'a mathematical joke'. He could show that having set up the Schrödinger wave function as a probability distribution of the position, it was possible to Fourier transform this probability distribution into a probability distribution of the momentum. In other words, it was only possible to look at the world with either the q- or the p-eye, but both views of the world gave the same information.

---

[44] Pauli to Heisenberg, 19 October 1926, in: Hermann *et al.* (1979), pp. 340–349.

Heisenberg, who read Pauli's letter and discussed it with his colleagues in Copenhagen, replied enthusiastically to Pauli's ideas.[45] For him, and likewise for Born, Pauli had shown that the waves in p-space were as real as the waves in q-space and the possibility of a transformation between the two showed once more that the wave function was not connected with the ordinary 'Anschauungsraum' as Schrödinger was hoping. More importantly, Pauli's insight became the starting point for Heisenberg's thoughts on the uncertainty relation which we will discuss in section 6.3.

Heisenberg was not the only one for whom Born's and Pauli's ideas were of great importance. Jordan, who had developed the idea of canonical transformations in matrix mechanics as the equivalent of the canonical transformations of classical Hamiltonian mechanics (Born *et al.*, 1926; Jordan, 1926a,b), incorporated the statistical interpretation into his general theory of transformation between the various existent formulations of quantum mechanics (Jordan, 1927b).[46] In this context Pauli's transformation between the q- and the p-representation of the wave function became an example of such a canonical transformation. Based on this, Jordan generalized the statistical interpretation of the wave function into a statistical interpretation of the elements of the transformation matrix, which Jordan now called probability amplitudes and treated as the fundamental quantities of the theory. Starting from these amplitudes Jordan wanted to build up quantum mechanics as an axiomatic theory following from statistical postulates. Jordan pointed out that these amplitudes combined through a simple multiplication rule. This, however, implied that the probabilities as squares of the amplitudes did not combine like classical probabilities via multiplication but that there was what Jordan called 'interference of probabilities'. In his habilitation speech (Jordan, 1927a) Jordan concluded from this that transformation theory was not yet a satisfactory statistical theory and that it had to be developed further.[47] This fundamental theory might even be deterministic.

Nevertheless, Jordan was convinced that transformation theory already allowed drawing the fundamental lesson that any future interpretation of quantum mechanics could not be based on ontological claims but had to be radically positivistic. As he would state in 1936 in his programmatic popular account 'Physics in the 20th century':

> The atom as we know it today no longer has the tangible and visualizable properties of the atoms of Democritus. It has been stripped of all sensible qualities and can only be characterized by a system of mathematical equations. The unbridgeable opposition of materialistic philosophy and positivistic epistemology stands out especially clearly at this point. With this insight, one of the most prominent elements of the materialist world view has been liquidated once and for all. At the same time, the positivistic epistemology has been confirmed and justified decisively.    (Jordan, 1936, pp. 122–123)

---

[45] Heisenberg to Pauli, 28 October 1926, in: Hermann *et al.* (1979), pp. 349–350.
[46] Duncan and Janssen (2013).
[47] By 1927, Jordan was convinced that the formal foundation of quantum physics had to be sought in quantum field theory (Lehner, 2011).

The basis for this bold metatheoretical claim was Jordan's conviction that the symmetry of different descriptions established by transformation theory implied that there was no *one* fundamental physical description and that therefore statements about unobservable entities in quantum mechanics (which corresponded to one specific description, i.e. the wave or particle picture) were meaningless.[48]

Independently from Pauli and Jordan in Hamburg and Göttingen, Paul Dirac, who was in Copenhagen at the time, also worked on an overarching formalism and on a 'physical interpretation' of quantum mechanics (Dirac, 1927). For the overarching formalism, Dirac developed a mathematical theory of transformations of q-numbers, his more abstract characterization of matrices. Just like Jordan this enabled him to understand Schrödinger's wave functions as transformation functions from a representation in a position basis to a representation in an energy basis. Unlike Jordan's attempt to develop quantum mechanics axiomatically, Dirac formulated his theory in close parallel to classical mechanics. In analogy with classical physics the dynamics of quantum mechanics were deterministic. Accordingly, Dirac remained silent on the ontological or epistemological implications of transformation theory. Rather, he saw its function in establishing a connection between the formalism and actual observations. It was the general framework that determined which physical information could be extracted from the mathematical formalism of quantum mechanics. In contrast to classical mechanics one could not extract a complete specification of any physical quantity from the formalism. The quantum mechanical description could only specify one half of the dynamical variables. In both Dirac's approach and Pauli's swarm picture, the complete specification of one variable left the other variable completely unspecified. Unlike Pauli, who assumed that the formalism itself implied statistical predictions about these variables, Dirac stressed that any statistical prediction required an explicit statistical assumption about the distribution of the quantity conjugate to the specified dynamical variable. Therefore, statistics was not part of the dynamics of quantum mechanics:

> The notion of probabilities does not enter into the ultimate description of mechanical processes: only when one is given some information that involves a probability [ ... ] can one deduce results that involve probabilities.[49]

Dirac's and Jordan's transformation theory became important for the interpretational debate in two respects: firstly, they offered a unified theory of quantum mechanics, into which matrix and wave mechanics were embedded as two different formulations.

---

[48] See Jordan (1934) for a defence of positivism as a general epistemological principle, and Darrigol (1986, pp. 232–233), Cini (1982), Lehner (2011) for discussions of Jordan's positivism.

[49] (Dirac, 1927, p. 641). See Darrigol (1986) for a detailed historical account of Dirac's quantum mechanics.

Secondly, this unified theory essentially rested on the assumption that the transformation amplitudes gave rise to observable probabilities. Hence transformation theory generalized the probabilistic interpretations of Schrödinger's wave function by Born and Pauli into a central feature of quantum mechanics.

Even though it played a foundational role for transformation theory, the probabilistic element of quantum mechanics gave rise to different views on its physical meaning. Dirac held that the evolution of a physical system was still deterministic. Heisenberg approved of this in his letter to Pauli sketching the uncertainty paper[50] and stressed the implication that all statistics only came into the theory due to observations. For Pauli statistics entered at an earlier stage—he saw quantum mechanics as giving a statistical description of the indeterministic kinematical behaviour of an ensemble of physical systems.[51] Lastly, Jordan took the probability amplitudes as fundamental quantities in their own right and tried to base the theory on postulates about these probability amplitudes.[52]

# 6.3 Understanding Quantum Mechanics—The Principles of Uncertainty and Complementarity

Heisenberg followed the developing transformation theory and the statistical interpretation very closely as he wrestled with Schrödinger's challenges and his own interpretational attempts from October 1926 up to March 1927, when he published his famous paper 'Über den anschaulichen Inhalt der quantentheoretischen Kinematik und Mechanik' ('On the *anschaulich* content of quantum theoretical kinematics and mechanics').[53] As mentioned in section 6.2, this process took off with Heisenberg's reaction to Pauli's idea of a Fourier transformation between q- and p-space and the associated dark point. As Heisenberg saw it, there was a positive way to make sense of Pauli's idea. Whereas the idea that position and momentum cannot be specified with arbitrary certainty was hard to square with the classical particle picture, he argued, the classical wave picture provided a basic analogy for understanding the apparent mutual exclusiveness of position and momentum:

---

[50] Heisenberg to Pauli, 23 February 1927, in: Hermann *et al.* (1979), p. 377.
[51] Pauli to Heisenberg, 19 October 1926, in: Hermann *et al.* (1979), p. 348.
[52] (Jordan, 1927a).
[53] (Heisenberg, 1927). Mara Beller has argued (Beller 1985), that Heisenberg developed his thoughts in response to Jordan. His reaction to Jordan's paper is documented in two letters to Pauli (Heisenberg to Pauli, 5 February 1927, in: Hermann *et al.* (1979), pp. 373–375, and Heisenberg to Pauli, 23 February 1927, in: Hermann *et al.* (1979), pp. 376–381).

The equation $pq - qp = hi$ always corresponds to the fact in the wave picture that it does not make sense to talk about a monochromatic wave at a certain time (or in a short time period). If one does not make the line too sharp, the time interval not too short, then this makes perfect sense. Analogously it does not make sense to talk about the position of a corpuscle with a certain velocity. If one is not particular about velocities and positions, then this makes perfect sense. Thus one understands that it makes sense to talk about position and velocities of a body macroscopically.
(Heisenberg to Pauli, 28 October 1926, in: Hermann *et al.* (1979), p. 350)

A short wave train, Heisenberg recalled, is never monochromatic. This wave will always have a range of different frequencies. Only for a wave that extends over a long period of time does this range of frequencies get narrower. This is reflected in the basic equation $\Delta v \Delta T \sim 1$, which shows that a wave of $\Delta v = 0$ is only possible if the time difference is infinitely large. This was the scheme to understand Pauli's statement on quantum mechanics. As long as the position of a particle was sharply defined the momentum could not be. So one could never predict positions and momenta jointly with absolute certainty. In the macroscopic world one could take positions and momenta jointly because one did not use them too exactly.

Following this idea, Heisenberg initially hoped that one could develop a discontinuous space-time and that continuous space and time would emerge only for a large number of particles as a statistical effect. In this discontinuous world, Heisenberg wrote in two letters in November 1926, the concept of a particle velocity did not make exact sense as one needed two infinitesimally close points to define it. More generally all classical concepts that were based on continuous space-time pictures could not be transferred into the quantum world.

Heisenberg did not make much progress in his search for a discontinuous space-time and ultimately declared that he did not know what wave or particle meant anymore.[54] At some point, Heisenberg discarded his ideas on discontinuous space and time and started to adopt a short-lived operationalist approach to physics.[55] A first formulation of this operationalism can be found in a letter to Pauli on 5 February 1927. There, he reacted to Jordan's habilitation speech and complained that Jordan talked about the probability of an electron being in a certain position without giving a definition of what the 'position of an electron' meant. In the next letter, from 23 February, he developed his own answer to this question, which is an outline of the position in his uncertainty paper, which he published in March 1927.

---

[54] The idea of a discontinuous space and time resurfaces frequently in late 1926 in Heisenberg's letters. This idea can be traced back to the attempt by Born to arrive at a 'fully discontinuous physics'. For a more detailed discussion of Heisenberg's speculations on a statistical conceptualization of space-time see Camilleri (2007b, pp. 183–185).

[55] Heisenberg mentioned his abandoned project of a discontinuous space-time description in the introduction to his paper on the uncertainty relation (Heisenberg, 1927, pp. 172–173). A detailed analysis of Heisenberg's take on operationalism is given by Camilleri (2007b).

In both the letter and the paper, Heisenberg argued that the concept of position, just like any physical concept, had to be defined through the way it is measured. This operationalism is valid both in quantum theory as well as in classical physics. The main difference, which Heisenberg stressed in his letter to Pauli and later in his uncertainty paper, was that the measurement of a quantity always induces a disturbance in the system and that this disturbance is negligible in classical physics but plays a central role in quantum theory.

To illustrate this point, Heisenberg discussed a thought experiment, which became central for discussions surrounding the uncertainty relation. In it, he considered the problem of measuring the position of an electron in a gamma-ray microscope in relation to the disturbance of the momentum induced by this measurement and argued that the position of an electron and its momentum can only be measured simultaneously with finite precision: To determine the position of an electron with a certain precision $\Delta q$, he began, the gamma-rays used by the microscope have to have a wavelength smaller than $\Delta q$. Due to the de Broglie relation between wavelength and momentum, this implies that the momentum of the gamma-ray light quanta is larger than $\frac{h}{\Delta q}$. The scattering of the light quantum off the electron hence introduces a change in the momentum of at least this magnitude. The more accurate the position is determined, the less we know about the momentum of the electron.

In the second part of his paper, Heisenberg reformulated this idea in a formal and general manner showing that this indefiniteness was not due to the particular experimental situation of the gamma-ray microscope. Within the framework of Dirac's and Jordan's transformation theory he analysed the momentum distribution of Schrödinger's wave packet and showed that a wave packet with a width of $\Delta q$ had a dispersion in momentum space of $\Delta p$, where the product $\Delta q \Delta p$ was of the order of magnitude of $\frac{h}{2\pi}$:

$$\Delta q \Delta p \sim \frac{h}{2\pi}.$$

In general any matrix in quantum mechanics can be understood as being defined in a multidimensional space that represents all possible experimental situations. For each situation the matrix gives not just a value but also a probable error for the measurement of the physical quantity represented by the matrix. This led to a new intuitively appealing interpretation of Schrödinger's wave packet: it was an error curve for a measurement.

For Heisenberg, the main conclusion of these two arguments for the interpretational debate was that one could define concepts like position and momentum in quantum mechanics in analogy with those concepts in classical (particle) mechanics. But in order to do so one had to accept an indefiniteness in the value of conjugated variables. This resolved Pauli's 'dark point': even though both position and momentum can be defined just like for a classical particle, the two concepts cannot be applied jointly in a precise manner. This is the physical content of the noncommutativity of position and momentum, now expressed through the uncertainty relation.

Heisenberg saw his uncertainty paper as serving two somewhat different purposes. It was both a 'commentary' on the meaning of transformation theory[56] and an answer to Schrödinger's challenge to produce an intuitive interpretation of quantum mechanics. Addressing the latter, he considered Schrödinger's challenge of linking micromechanics and macromechanics. He still firmly rejected Schrödinger's idea that macromechanics would emerge directly from the short wavelength limit of wave mechanics and that a particle could be constructed as a material wave packet.[57] Yet, he adopted Schrödinger's wave packet for the description of physical systems and reinterpreted it as a 'probability packet' in the sense of Born's statistical interpretation. According to his argument from transformation theory, the wave packet gives the probability of finding a system in a certain position. As time evolves, the wave packet disperses so that the location of the particle is ever less specified. Only a measurement of position will locate the particle again and thus 'reduce the wave packet' (Heisenberg, 1927, p. 186) to a size given by the measurement precision.

This reduction of the probability packet, Heisenberg argued, explained the emergence of classical trajectories as observed in the Wilson cloud chamber and thus answered Schrödinger's challenge: a trajectory had to be regarded as a number of subsequent detections of the particle and correspondingly as reductions of the wave packet. In this respect, micromechanics and macromechanics were not very different at all: if the initial state was given with a certain imprecision, both gave only statistical predictions and hence the transition from micromechanics to macromechanics could be achieved without resorting to Schrödinger's interpretation.

Heisenberg's interpretation thus took up Schrödinger's wave packet idea but showed how the dispersion of the wave packet did not lead to inconsistencies. Whereas the wave packet dispersed indefinitely so that particles could not remain localized in Schrödinger's interpretation, Heisenberg's statistical reinterpretation allowed for a reduction of the wave packet to a smaller size through measurement. This in itself presented strong support for Born's statistical interpretation. However, Heisenberg did not answer Schrödinger's question how the quantum mechanical probabilities had to be understood. As Schrödinger had pointed out in the discussion with Born, it made an important difference whether the probability was to be seen as an epistemic probability, with the particle already in a specific place before the measurement, or as the ontic

---

[56] Heisenberg to Jordan, 7 March 1927 (AHQP 18.2). Heisenberg felt that such a commentary did not bring much new for Dirac, Jordan, or Pauli, but that he could clarify the physical significance of transformation theory.

[57] (Heisenberg, 1927, p. 185). A wave packet orbiting an atom, Heisenberg argued, would emit electromagnetic waves whose frequencies are integer multiples of the orbiting frequency. Because this was not the case in observed spectra, realistic wave packets have to disperse and hence cannot be used to explain the formation of a stable particle. This, Heisenberg argued, makes it impossible to construct a direct transition from micromechanical description of wave mechanics to classical particle mechanics. With this argument Heisenberg made reference to Bohr's correspondence principle, from which it was clear that the clear identity of radiation frequency with the overtones of a Fourier series was only possible in the classical limit.

probability for a physical localization of a previously delocalized particle. If the probability was epistemic, then particle trajectories did objectively exist and quantum mechanics was not a complete theory. If the probability was ontic, something new and strange happened at the moment of measurement, and the reduction of the wave packet was a novel physical process. Heisenberg did not take a stand on this question, only remarking at the end of the paper that speculations about an unobserved world behind the phenomena seemed pointless to him. It was only in the fall of 1927 at the Solvay conference that the reduction of the wave packet would be discussed as a problem for the interpretation of quantum mechanics.

Instead, Heisenberg went on to show how his new interpretation could be useful in the discussion of other thought experiments. One of these problems, which had been discussed privately by Bohr, Pauli, and Heisenberg in the context of BKS, was the interpretation of coherent fluorescent radiation.[58] In 1924 Bohr had considered atoms illuminated by radiation with a frequency that was equal to one of its transition frequencies. Bohr had assumed then that a small part of the atoms jumped to a higher state and reemitted incoherent radiation of the same frequency, while the rest of the atoms—just as in the case of dispersion—covibrated in phase with the incident radiation without changing their state and thus sent out coherent radiation. In 1925 Heisenberg described a thought experiment that separated these two radiation processes.[59] Here, a molecular beam would first pass through light causing fluorescence radiation and would then be sent through a Stern–Gerlach magnet causing the beam to split into one beam composed solely of excited molecules and another composed only of unexcited molecules. If there were indeed two radiation mechanisms at play, the different beams should emit coherent and incoherent radiation respectively. If, however, radiation took place in accordance with the Bohr model only, there would only be incoherent radiation emitted by the beam of excited molecules.

This second option, which both Heisenberg and Pauli both came to adopt, made it even more puzzling how there could be both coherent and incoherent radiation in the unresolved beam. In 1927, Heisenberg returned to this puzzle and used the uncertainty relations to explain the situation. Formulating an uncertainty relation between energy and time, Heisenberg argued that the coherent fluorescent radiation had to be produced by atomic states with a definite phase of oscillation. Such definite phases could, however, exist only as long as the energy of the system is not specified. This was apparently the case when the molecular beam passed through the light. As soon as it entered the Stern–Gerlach magnet, however, the energy was determined as the atoms were separated into two beams. The determination of the energy states destroys the initial phase relations and hence the coherence of the resulting radiation so that only the excited atoms radiate in an incoherent way, while the unexcited atoms do not radiate (Heisenberg, 1927, 192).

---

[58] Bohr *et al.* (1924, p. 789), Bohr (1924), Pauli to Bohr, 2 October 1924, in: Hermann *et al.* (1979), p. 165.
[59] Heisenberg to Pauli, 16 November 1925, in: Hermann *et al.* (1979), p. 256.

Heisenberg's resolution of this long-standing puzzle is interesting for the present discussion because it shows that his use of the uncertainty relation goes far beyond the questions of the nature and knowability of quantum mechanical states that have dominated the later philosophical debate. Rather, with the example of fluorescence radiation, Heisenberg came to tease out an insight into the nature of the new quantum mechanics which shook previously unquestioned assumptions. In early matrix mechanics, it had still been assumed in accordance with the Bohr model that an atom was always in a particular stationary state and radiation was emitted when it jumped between them. Heisenberg's analysis, by contrast, implied that atoms could be in a superposition of states.

In summary, Heisenberg's uncertainty paper first introduced the uncertainty relation on the basis of considerations about the limitations of measurement and of transformation theory. He then used the uncertainty relation as a prescription to analyse physical situations as measurements that disturb a physical system. One aim of this argument was to resolve the two challenges Schrödinger had posed to quantum mechanics: the connection between the abstract formalism of quantum mechanics and a concrete physical situation, and the problem of the transition from macroscopic to microscopic mechanics. For both problems Heisenberg accepted that one needed an understanding of quantum mechanics beyond the mathematical formalism and with it some form of *Anschaulichkeit*. In the gamma-ray microscope Heisenberg used the particle picture to analyse the physical situation. In the case of resonance radiation he pictured atoms flying through a magnetic field and being separated according to energy in space. But unlike Schrödinger, who thought that this understanding could only be achieved through a concrete, visualizable physical model, Heisenberg claimed that visualizability neither was equivalent to nor the essential property of *Anschaulichkeit*. He framed this essential property at the beginning of his paper:

> We believe we can understand a physical theory intuitively (*anschaulich*), if we can think of the experimental consequences of that theory qualitatively in all simple cases, and if we have realized at the same time that the application of the theory never contains inner contradictions.   (Heisenberg, 1927, p. 172)

For Heisenberg, *Anschaulichkeit* did not result from a model or spatio-temporal picture; rather it was a more abstract form of intuition, which was revealed in the ability to foresee the outcome of an experiment without working through detailed calculations. From this point of view every technique guiding physicists' expectations (be it a model, a principle, or a thought experiment) would be considered *anschaulich*. Hence, the deeper aim of the paper was to show that the uncertainty relation established a fit between the theoretical description and the limits of measurement, which guaranteed that inconsistencies could not arise.

While Heisenberg was using the particle mechanical picture in the cases of position-momentum uncertainties, he was not, like Schrödinger, proposing a new ontological model and then trying to show its adequacy. His interpretation aimed to give a prescription to analyse physical situations without running into inconsistencies. To do so, Heisenberg developed a new way of analysing physical systems and their inter-actions. These interactions now appeared to Heisenberg as measurements which determined one aspect of a physical system while perturbing another in accordance to the uncertainty relation. In this way, the uncertainty relation explicated the com-mutation relations: one could understand noncommuting matrices as representing quantities that do not have joint exact values. Heisenberg's new way of thinking allowed one to keep the existing classical pictures and still avoid inconsistencies in the account of the phenomena. Hence the overall conclusion of the uncertainty paper is not simply that physics has to be operationalized or that classical concepts are limited, as the standard reading has been,[60] but that the interpretation is *anschaulich* in this new sense and that it is a productive tool to analyse physical situations.

Already, when he was sketching his argument for the uncertainty paper in the letter to Pauli from 23 February 1927, Heisenberg had closed with the statement that thinking in terms of uncertainties leads to a new *anschaulich* understanding of quantum mechanics:

> If one believes, as I do, that one understands physical laws intuitively [*anschaulich*] if one can tell 'by feeling' [*gefühlsmäßig*][61] in every case, what results [from the physical situation] then such considerations as above perhaps help a little further; in my case, they alleviated my conscience.
> (Heisenberg to Pauli, 23 February 1927, in: Hermann *et al.* (1979), p.381)

Therefore it seems fair to say that Heisenberg's redefinition of *Anschaulichkeit* was not solely a rhetorical strategy, designed to justify a new interpretation that was shying away from a new ontology in Schrödinger's sense.[62]

Heisenberg's uncertainty argument was taken as a very important step in the development of quantum mechanics by Bohr and in the Copenhagen and Göttingen communities, and was debated intensely in the spring and summer of 1927. Bohr found Heisenberg's paper to be of major importance: it showed the consistency and com-pleteness of the quantum mechanical formalism and demonstrated the correct appli-cation of classical pictures to a physical situation without running into the paradoxes in which the old quantum theory got entangled in the description of quantum phenomena.[63] While Bohr thereby acknowledged that Heisenberg had indeed provided

---

[60] For example, Popper (2012), p. 140–143.

[61] Both terms would normally be translated as 'intuitively' in English, but Heisenberg here stresses the distinction between their literal meaning 'visually' versus 'by feeling'.

[62] As Kristian Camilleri has pointed out, Heisenberg's new notion of *Anschaulichkeit* fits well into a more general philosophical reflection on the notions of space and time, ranging from interest in non-Euclidean geometry in the late 19th century, to the debates about space and time in the theory of relativity and the reflections of the logical empiricists (Camilleri, 2009b, pp. 50–53; see also Stöltzner, 2003, pp. 302–303.)

[63] For the importance of the uncertainty relation see Bohr (1928a), p. 250 and Kalckar (1985), pp. 23, 62, 96.

an *anschaulich* understanding of quantum mechanics in his new sense, he had a rather different idea of what this *Anschaulichkeit* involved.

As is already apparent in his discussion with Schrödinger in October 1926, Bohr's approach to the interpretation of quantum mechanics focused on the indispensability but limited applicability of classical pictures in quantum theory. Through his work on the correspondence principle within the old quantum theory (Tanona, 2002) Bohr had come to accept this dialectical relationship and now finally concluded that there was no description of radiation in terms of either particles or waves. Rather, only a description operating with both of these mutually exclusive pictures would be adequate. In contrast to Heisenberg, however, these pictures were not affected by the imprecision implied in the uncertainty relation. Each picture was to be taken as exact, but as an abstraction since neither waves nor particles could give a full account of the various possible physical situations. From this perspective, the uncertainty relation had a different meaning for Bohr: it showed that classical concepts complement each other in the description of microphysical entities but it did not imply that these concepts were only applicable with an inevitable vagueness as it did for Heisenberg.[64]

This difference between Bohr's and Heisenberg's interpretational approach came to the fore in Bohr's response to Heisenberg's gamma-ray thought experiment. Bohr strongly disagreed with Heisenberg that the uncertainty relation was a consequence of the discontinuous change during measurements and was therefore quite upset that Heisenberg had submitted the paper without Bohr's consent. Specifically, he criticized Heisenberg's analysis of the gamma-ray microscope thought experiment. If one treated both gamma-rays and electrons only as particles—Bohr argued—it would still be possible to deduce the value of both the momentum and the position of the electron in the microscope simultaneously since the recoil momentum of the photon could be used to determine the original momentum of the electron if the original momentum of the photon was known. The uncertainty occurred only due to the wave nature of the gamma-rays, which relates the resolution of the microscope to its aperture and to the wavelength of the gamma-rays. Hence, for Bohr, the uncertainty relation could not be justified by invoking only particle ideas and Heisenberg's relation expressed the necessity of using both the wave and the particle picture.[65]

Bohr's first public exposition of his position on the duality of wave and particle pictures appeared in his lecture in Como in September 1927, which was the basis for his canonical formulation of complementarity (in Bohr, 1928b,a) and therefore has been seen as the founding event of the Copenhagen interpretation. In this lecture entitled 'Das Quantenpostulat und die neuere Entwicklung der Atomistik' Bohr tried to deliver a more general synthesis of his thoughts on the interpretation of quantum mechanics

---

[64] (Camilleri 2007a).

[65] The disagreement between Heisenberg and Bohr on the gamma-ray microscope thought experiment led to a heated discussion, which put a stress on their relationship. In the end Heisenberg appended an acknowledgement of Bohr's criticism to his already submitted paper but did not revise it as a whole. (See Kalckar, 1985; Camilleri, 2009b; Cassidy, 1992.)

and of the development that quantum mechanics had undergone since its advent in 1925. He presented this synthesis in his principle of complementarity, according to which quantum phenomena could not be interpreted in terms of either particle- or wave-pictures; one had to use both even though they were mutually exclusive.[66]

However, complementarity was not a mere reformulation of wave–particle duality, it resulted from Bohr's more general reflection on the observability of atomic systems and on the fundamental idea of Heisenberg's uncertainty paper. The central idea of Heisenberg's paper, according to Bohr, was that 'no observation of atomic phenomena is possible without their essential disturbance'. This implied that an exactly defined atomic state is unobservable, while the observation of an atomic phenomenon cannot be captured by a definite description. The incompatibility of the exact definition of a system with the possibility of its observation is the basis for Bohr's complementarity and it is this impossibility that Heisenberg's uncertainty relation expresses (Bohr, 1985, pp. 91 and 93). A consequence of this dilemma is the impossibility of giving a description that is both spatio-temporally exact and causal in the sense of fulfilling the conservation of energy and momentum. However, as mentioned above, that does not change the meaning of the concepts of physics. This meaning is fixed by the ideas of classical physics, which are necessary to interpret our experiments, even though their use leads to paradoxes in the description of quantum phenomena.

For Bohr, this complementarity between observation and definition was apparent also in Schrödinger's wave mechanics. The pure (energy) eigenstates, which are the solutions of the Schrödinger equation, describe unobserved systems and are therefore abstractions (Bohr, 1985, pp. 96–97). The interaction involved in any observation will lead to a superposition of these eigenstates, allowing for localized systems such as wave packets, but in which energy and momentum are not defined anymore. Even though Bohr uses wave mechanics as a corrective against Heisenberg's reading of the uncertainty relation, he distances himself from Schrödinger's hope that wave mechanics allows a return to a continuous spatio-temporal description of quantum phenomena.

With this take on the fundamental difficulty of quantum mechanics, Bohr took a middle ground between wave and matrix mechanics and tried to make the conflicting interpretations compatible. He shifted the discussion from competing ontological claims (such as waves versus particles or continuity versus discontinuity) to a

---

[66] Bohr's complementarity argument underwent a complicated development from 1925 until 1950 and changed considerably over time, and there has been plenty of research on editorial questions, intellectual history, and the sociological issues in the reception of Bohr's thought (Beller, 1983b; Camilleri, 2009a; Howard, 2004). In the Collected Works of Niels Bohr, Jørgen Kalckar has offered a compilation of the letters and manuscripts in the context of Bohr's work on complementarity, the Como lecture and subsequent papers in *Nature* and *Die Naturwissenschaften* titled 'The Quantum Postulate and the Recent Development of Atomic Theory' (Bohr, 1928b) and 'Das Quantenpostulat und die neuere Entwicklung der Atomistik' (Bohr, 1928a). We focus here on the manuscript version that Bohr wrote on October 12–13 between the Como lecture and the Solvay Conference (Bohr, 1985, pp. 91–98) as it contains the essential arguments Bohr presumably made in Como and is the basis of the published versions.

complementarity of conceptual schemes (wave versus particle description or causal description versus space-time description). However, Bohr's solution did not satisfy either camp. Heisenberg saw Bohr's attempts as an elevation of wave–particle duality from a problem to a principle:

> Bohr wants to write a general paper on the 'conceptual foundation' of quantum theory from the perspective: 'There are waves and corpuscles',—if one starts from that, one can of course make everything consistent.
>
> (Heisenberg to Pauli, 16 May 1927, in: Hermann *et al.* (1979), p. 394)

Schrödinger, on the other hand, saw Bohr's attempts as mired in the limited concepts of classical mechanics and argued for new physical concepts adequate for describing quantum phenomena:

> [The uncertainty relation] seems to demand authoritatively the introduction of *new* concepts in which this limitation [of classical concepts] does *not* exist *any more*. Because what is unobservable in principle should not be contained in our conceptual scheme at all, it should not be representable in it. In the *adequate* conceptual scheme it must not look as if our possibilities of experience were limited by unfortunate circumstances.
>
> (Schrödinger to Bohr, 5 May 1928, in: Meyenn (2011), p. 454)

For both Heisenberg and Schrödinger, Bohr's interpretation did not go far enough in searching for new conceptual grounds for the interpretation of quantum mechanics. Nonetheless, Heisenberg's uncertainty paper and Bohr's Como lecture provided an answer to Schrödinger's challenge to offer an *anschaulich* interpretation of quantum mechanics. Both interpretations departed from the model-based approach, with which Schrödinger was aiming for a unified description of nature, and abandoned the idea of modifying the formalism of quantum mechanics. Instead Bohr's and Heisenberg's interpretation considered the quantum mechanical formalism as given and substantially complete and tried to find a consistent application of classical pictures that was in agreement with this formalism. For them, the uncertainty relations were a means to achieve just that.

Despite their differences, Heisenberg never publicly criticized Bohr's ideas about complementarity, and Bohr's interpretation of the uncertainty paper quickly became accepted by the majority of physicists working in quantum mechanics. When the leading researchers of the community met at the Solvay conference in Brussels in the fall of 1927, de Broglie and Schrödinger still presented their positions, and the general discussion did involve some questions about the 'reduction of the wave packet'.[67] However, the rapid successes of quantum mechanics in atomic and molecular physics

---

[67] The conference is treated in detail in Bacciagaluppi and Valentini (2009), which also contains a translation of the proceedings.

convinced a majority of physicists of Bohr's and Heisenberg's position that quantum mechanics was a complete and consistent theory and that questions about its interpretation could not contribute anymore to its development, even though the theory was not relativistic and particularly could not treat light quanta. The attempt to generalize quantum mechanics to a relativistic theory of quantum fields (Dirac, 1928; Heisenberg and Pauli, 1929a,b) quickly showed serious problems that threatened to undermine the whole programme. Soon after, von Neumann's mathematical codification of transformation theory into his theory of the Hilbert space demonstrated that the reduction of the wave packet was just a particular instance of a fundamental paradox, the measurement problem (Neumann, 1932). Nevertheless, by then the physics community had widely accepted Heisenberg's claim that quantum mechanics was a closed theory and Bohr's explication of it in terms of complementarity. When Einstein *et al.* (1935) and Schrödinger (1935) finally published their now famous papers on the problems that quantum mechanical entanglement posed for our understanding of quantum mechanics, they were mostly ignored. The debate about the interpretation of quantum mechanics ceased to be a creative force in physics for decades.

We have attempted to show that this was not so in the years of the creation of quantum mechanics, that the sharp differences in the interpretation of the formalism were an important factor in its further development into the theory we now know. We have mainly limited ourselves here to the debate between Heisenberg, Schrödinger, and Bohr. As Mara Beller (1999) has already shown, this was merely a small subgroup of a much wider network of dialogues that accompanied the evolution of quantum physics, which was a creation of many, and came with a substantial amount of competitiveness and rivalry. As we hope to have made plausible in one striking example, this rivalry did not exclude a rational and fruitful debate.

# REFERENCES

Bacciagaluppi, G. and Valentini, A. (2009). *Quantum Theory at the Crossroads. Reconsidering the Solvay Conference.* Cambridge: Cambridge University Press.

Beller, M. (1983a). Matrix Theory Before Schrödinger: Philosophy, Problems, Consequences. *Isis,* **74,** 469–491.

Beller, M. (1983b). *The Genesis of Interpretations of Quantum Physics, 1925–1927.* Ann Arbor: University of Michigan Press.

Beller, M. (1985). Pascual Jordan's Influence on the Discovery of Heisenberg's Indeterminacy Principle. *Archive for History of Exact Sciences,* **330,** 387–349.

Beller, M. (1990). Born's Probabilistic Interpretation: A Case Study of Concepts in Flux. *Studies in History and Philosophy of Science,* **21,** 563–88.

Beller, M. (1999). *Quantum Dialogue. The Making of a Revolution.* Chicago: University of Chicago Press.

Bitbol, M. (1996). *Schrödinger's Philosophy of Quantum Mechanics.* Dordrecht: Kluwer.

Bitbol, M. and Darrigol, O. (eds) (1992). *Erwin Schrödinger: Philosophy and the Birth of Quantum Mechanics.* Gif-sur-Yvette: Editions Frontières.

Blum, A. S., Jähnert, M., Lehner, C., and Renn, J. (2017). Translation as Heuristics. Heisenberg's Turn to Matrix Mechanics. *Studies in History and Philosophy of Modern Physics*, **60**, 3–22.

Bohr, N. (1924). Zur Polarisation des Fluoreszenzlichtes. *Die Naturwissenschaften*, **49**, 1115–1117.

Bohr, N. (1928a). Das Quantenpostulat und die neuere Entwicklung der Atomistik. *Die Naturwissenschaften*, **160**, 245–257.

Bohr, N. (1928b). The Quantum Postulate and the Recent Development of Atomic Theory. *Nature*, **121**, 580–590.

Bohr, N. (1985). *Collected Works Part I of Volume 6: Foundations of quantum physics I: (1926–1932)*. Amsterdam: North-Holland.

Bohr, N., Kramers, H. A., and Slater, J. C. (1924). The Quantum Theory of Radiation. *Philosophical Magazine*, **470**, 785–802.

Bokulich, A. (2004). Open or Closed? Dirac, Heisenberg, and the Relation between Classical and Quantum Mechanics. *Studies in History and Philosophy of Modern Physics*, **35**, 377–396.

Bonk, T. (1994). Bemerkungen zur Interpretation, Bestätigung und Progressivität der frühen Matrizenmechanik. *Journal for General Philosophy of Science*, **25**, 1–25.

Born, M. (1926a). Quantenmechanik der Stossvorgänge. *Zeitschrift für Physik*, **38**, 803–827.

Born, M. (1926b). Zur Quantenmechanik der Stossvorgänge (Vorläufige Mitteilung). *Zeitschrift für Physik*, **37**, 863–867.

Born, M. (1927a). Das Adiabtenprinzip in der Quantenmechanik. *Zeitschrift für Physik*, **40**, 167–192.

Born, M. (1927b). Quantenmechanik und Statistik. *Die Naturwissenschaften*, **10**, 238–242.

Born, M. (1955). Statistical Interpretation of Quantum Mechanics. *Science*, **122**, 675–679.

Born, M. and Jordan, P. (1925a). Zur Quantenmechanik. *Zeitschrift für Physik*, **34**, 858–888.

Born, M. and Jordan, P. (1925b). Zur Quantentheorie aperiodischer Vorgänge. *Zeitschrift für Physik*, **330**, 479–505.

Born, M. and Wiener, N. (1926). Eine neue Formulierung der Quantengesetze für periodische und nicht periodische Vorgänge. *Zeitschrift für Physik*, **36**, 174–187.

Born, M., Heisenberg, W., and Jordan, P. (1926). Zur Quantenmechanik II. *Zeitschrift für Physik*, **35**, 557–615.

Camilleri, K. (2007a). Bohr, Heisenberg and the Divergent Viewpoints of Complementarity. *Studies in the History and Philosophy of Modern Physics*, **38**, 514–528.

Camilleri, K. (2007b). Indeterminacy and the Limits of Classical Concepts: The Turning Point in Heisenberg's Thought. *Perspectives on Science*, **150**(2), 176–199.

Camilleri, K. (2009a). Constructing the Myth of the Copenhagen Interpretation. *Perspectives on Science*, **170**(1), 26–57.

Camilleri, K. (2009b). *Heisenberg and the Interpretation of Quantum Mechanics: The Physicist as Philosopher*. Cambridge: Cambridge University Press.

Cartwright, N. (1987). Max Born and the Reality of Quantum Probabilities. In L. Krüger, G. Gigerenzer, and M. S. Morgan (eds), *The Probability Revolution Vol. 2 Ideas in the Sciences*, Cambridge, MA: MIT Press, pp. 409–417.

Cassidy, D. C. (1992). *Uncertainty: The Life and Science of Werner Heisenberg*. New York: Freeman.

Chevallier, C. (1988). Physical Reality and closed theories in Werner Heisenberg's early papers. In D. Batens and J. P. van Bendegem (eds), *Theory and experiment*, Dordrecht: Reidel.

Cini, M. (1982). Cultural Traditions and Enviromental Factors in the Development of Quantum Electrodynamics (1925–1933). *Fundamenta Scientiae*, **3**, 229–253.

CPAE 7 (2002). *The Collected Papers of Albert Einstein. Vol. 7: The Berlin Years: Writings, 1918–1921*. Princeton: Princeton University Press.

Darrigol, O. (1986). The Origin of Quantized Matter Waves. *Historical Studies in the Physical and Biological Sciences*, **16**, 197–253.

Darrigol, O. (1992a). *From c-Numbers to q-Numbers: The Classical Analogy in the History of Quantum Theory*. Berkeley, CA: University of California Press.

Darrigol, O. (1992b). Schrödinger's Statistical Physics and some Related Themes. In M. Bitbol and O. Darrigol (eds), *Erwin Schrödinger. Philosophy and the Birth of Quantum Mechanics*, Gif-sur-Yvette: Éditions Frontières, pp. 237–276.

De Regt, H. W. (1997). Erwin Schrödinger, Anschaulichkeit, and Quantum Theory. *Studies In History and Philosophy of Science Part B: Studies In History and Philosophy of Modern Physics*, **280**, 461–481.

De Regt, H. W. (1999). Ludwig Boltzmann's *Bildtheorie* and Scientific Understanding. *Synthese*, **1190**, 113–134.

De Regt, H. W. and Dieks, D. (2005). A Contextual Approach to Scientific Understanding. *Synthese*, **144**, 137–170.

Dirac, P. A. M. (1927). The Physical Interpretation of the Quantum Dynamics. *Proceedings of the Royal Society of London. Series A*, **113**, 621–641.

Dirac, P. A. M. (1928). The Quantum Theory of the Electron. *Proceedings of the Royal Society of London. Series A*, **117**, 610–624.

Duncan, A. and Janssen, M (2007). On the Verge of *Umdeutung* in Minnesota: Van Vleck and the Correspondence Principle. Parts I and II. *Archive for History of Exact Sciences*, **61**, 553–624, 625–671.

Duncan, A. and Janssen, M. (2013). (Never) Mind Your p's and q's: von Neumann versus Jordan on the Foundations of Quantum Theory. *The European Physical Journal H*, **380**, 175–259.

Einstein, A. (1919). Time, Space, and Gravitation. *The Times (London)*, 28 November, pp. 13–14.

Einstein, A., Podolsky, B., and Rosen, N. (1935). Can quantum-mechanical description of reality be considered complete? *Physical Review*, **47**, 777–780.

Forman, P. (1971). Weimar culture, causality, and quantum theory, 1918–1927: Adaptation by German physicists and mathematicians to a hostile intellectual environment. *Historical Studies in the Physical Sciences*, **3**, 1–115.

Heisenberg, W. (1925). Über quantentheoretische Umdeutung kinematischer und mechanischer Beziehungen. *Zeitschrift für Physik*, **330**, 879–893.

Heisenberg, W. (1926). Mehrkörperproblem und Resonanz in der Quantenmechanik. *Zeitschrift für Physik*, **380**, 411–426.

Heisenberg, W. (1927). Über den anschaulichen Inhalt der quantentheoretischen Kinematik und Mechanik. *Zeitschrift für Physik*, **43**, 172–198.

Heisenberg, W. (2005). *Der Teil und das Ganze*, 6th edn. München: Piper.

Heisenberg, W., and Pauli, W. (1929a). Zur Quantendynamik der Wellenfelder. *Zeitschrift für Physik*, **56**, 1–61.

Heisenberg, W., and Pauli, W. (1929b). Zur Quantentheorie der Wellenfelder. II. *Zeitschrift für Physik*, **590**(3–4), 168–190.

Hendry, J. (1984). *The Creation of Quantum Mechanics and the Bohr-Pauli Dialogue*, volume 14 of *Studies in the History of Modern Science*. Dordrecht and Boston and Hingham, MA: Springer.

Hermann, A., von Meyenn, K., and Weisskopf, V. F. (eds) (1979). *Wissenschaftlicher Briefwechsel mit Bohr, Einstein, Heisenberg u.a., Volume 1: 1919–1929*. New York, Heidelberg, Berlin: Springer.

Howard, D. (2004). Who invented the Copenhagen Interpretation? A Study in Mythodology. *Philosophy of Science*, 71, 669–682.

Jähnert, M. (2019). *Practicing the Correspondence Principle in the Old Quantum Theory: A Transformation Through Application*. Archimedes. Dordrecht: Springer.

Jammer, M. (1966). *The Conceptual Development of Quantum Mechanics*. New York: McGraw-Hill.

Joas, C., and Lehner, C. (2009). The Classical Roots of Wave Mechanics: Schrödinger's Transformation of the Optical-mechanical Analogy. *Studies in History and Philosophy of Modern Physics*, 40, 338–351.

Joos, G. (1926). Das Abklingleuchten in der Schrödinger'schen Atomtheorie. *Sitzungsberichte der Mathematisch-Physikalischen Klasse der Bayerischen Akademie der Wissenschaften zu München*, 399–404.

Jordan, P. (1926a). Über kanonische Transformationen in der Quantenmechanik. *Zeitschrift für Physik*, 37, 383–386.

Jordan, P. (1926b). Über kanonische Transformationen in der Quantenmechanik II. *Zeitschrift für Physik*, 38, 513–517.

Jordan, P. (1927a). Kausalität und Statistik in der modernen Physik. *Die Naturwissenschaften*, 50(5), 105–110.

Jordan, P. (1927b). Über eine neue Begründung der Quantenmechanik I. *Zeitschrift für Physik*, 40, 809–838.

Jordan, P. (1934). Über den positivistischen Begriff der Wirklichkeit. *Die Naturwissenschaften*, 220, 485–490.

Jordan, P. (1936). *Die Physik des 20. Jahrhunderts: Einführung in den Gedankeninhalt der modernen Physik*. Die Wissenschaft; 88. Braunschweig: Vieweg.

Kalckar, J. (1985). Introduction. Copenhagen Discussions Prior to the Establishment of the Uncertainty Relations. In J. Kalckar (ed.), *Niels Bohr Collected Works, Part I of Volume 6*, Amsterdam: North-Holland, pp. 7–51.

Konno, H. (1978). The Historical Roots of Born's Probability Interpretation. *Japanese Studies in History of Science*, 17, 129–145.

Lehner, C. (2011). Mathematical Foundations and Physical Visions: Pascual Jordan and the Field Theory Programme. In K.-H. Schlote and M. Schneider, *Mathematics Meets Physics: A Contribution to their Interaction in the 19th and the first half of the 20th century*, Franfurt-am-Main: Verlag Harri Deutsch, pp. 272–292.

MacKinnon, E. M. (1977). Heisenberg, Models and the Rise of Matrix Mechanics. *Historical Studies in the Physical Sciences*, 8, 135–188.

MacKinnon, E. M. (1980). The Rise and Fall of the Schrödinger Interpretation. In P. Suppes (ed.), Studies in the Foundations of Quantum Mechanics, East Lansing: University of Chicago Press, pp. 1–58.

MacKinnon, E. M. (1982). *Scientific Explanation and Atomic Physics*. Chicago: University of Chicago Press.

Meyenn, K. von (ed.) (2011). *Eine Entdeckung von ganz außerordentlicher Tragweite: Schrödingers Briefwechsel zur Wellenmechanik und zum Katzenparadoxon.*, Volume 1. Heidelberg: Springer.

Miller, A. I. (1978). Visualization Lost and Regained: The Genesis of the Quantum Theory in the Period 1913–27. In J. Wechsler (ed.), *On Aesthetics in Science*, Cambridge, MA: MIT Press, pp. 73–102.

Miller, A. I. (1982). Redefining Anschaulichkeit. In A. Shimony and H. Felsbach (eds), *Physics as Natural Physilosophy: Essays in Honor of Lazlo Tisza on his Seventy-fith Birthday*, Cambridge: Cambridge University Press, pp. 376–411.

Muller, F. A. (1997). The Equivalence Myth of Quantum Mechanics. *Studies in the History and Philosophy of Modern Physics*, **28**, 219–247.

Neumann, J. von (1932). *Mathematische Grundlagen der Quantenmechanik*. Berlin: Springer.

Pauli, W. (1925). Quantentheorie. In K. Scheel (ed.), *Handbuch der Physik*, volume 23, Berlin: Springer, pp. 1–279.

Perovic, S. (2008). Why Were Matrix Mechanics and Wave Mechanics Considered Equivalent? *Studies in the History and Philosophy of Modern Physics*, **39**, 444–461.

Popper, K. R. (2012). *Ausgangspunkte. Meine intellektuelle Entwicklung*. Heidelberg: Mohr Siebeck.

Schrödinger, E. (1926a). Der stetige Übergang von der Mikro- zur Makromechanik. *Die Naturwissenschaften*, **140**(28), 664–666.

Schrödinger, E. (1926b). Quantisierung als Eigenwertproblem (Zweite Mitteilung). *Annalen der Physik*, **79**, 486–527.

Schrödinger, E. (1926c). Quantisierung als Eigenwertproblem (Dritte Mitteilung). *Annalen der Physik*, **80**, 437–490.

Schrödinger, E. (1926d). Über das Verhältnis der Heisenberg–Born–Jordanschen Quantenmechanik zu der meinen. *Annalen der Physik*, **790**, 143–165.

Schrödinger, E. (1929). Was ist ein Naturgesetz? *Die Naturwissenschaften*, **170**, 9–11.

Schrödinger, E. (1935). Die gegenwärtige Situation in der Quantenmechanik. *Die Naturwissenschaften*, **23**, 807–812, 823–828, 844–849.

Serwer, D. (1977). *Unmechanischer Zwang*: Pauli, Heisenberg, and the Rejection of the Mechanical Atom. *Historical Studies in the Physical Sciences*, **8**, 189–256.

Sommerfeld, A., and Heisenberg, W. (1922). Eine Bemerkung über relativistische Röntgendubletts und Linienschärfe. *Zeitschrift für Physik*, **10**, 393–398.

Soon, G. I. (1996). Experimental Constraints on Formal Quantum Mechanics: The Emergence of Born's Quantum Theory of Collision Processes in Göttingen, 1924–1927. *Archive for History of Exact Sciences*, **50**, 73–101.

Stern, O., and Voelmer, M. (1919). Über die Abklingungszeit der Fluoreszenz. *Physikalische Zeitschrift*, **20**, 183–188.

Stöltzner, M. (2003). *Causality, Realism and the Two Strands of Boltzmann's Legacy (1896–1936)*. Bielefeld: Universität Bielefeld.

Tanona, S. D. (2002). *From Correspondence to Complementarity: The Emergence of Bohr's Copenhagen Interpretation of Quantum Mechanics*. Ph. D. thesis, Indiana University.

Wessels, L. (1980). What Was Born's Statistical Interpretation? *PSA: Proceedings of the Biennial Meeting of the Philosophy of Science Association*, **2**, 187–200.

Wessels, L. (1983). Erwin Schrödinger and the Descriptive Tradition. In R. Aris, H. E. Davis, and R. H. Stuewer (eds), *Springs of Scientific Creativity: Essays on Founders of Modern Science*, Minneapolis: University of Minnesota Press, pp. 254–278.

Wien, W. (1919). Über Messungen der Leuchtdauer der Atome und der Dämpfung der Spektrallinien. I. *Annalen der Physik*, **60**, 597–639.

Wien, W. (1921). Über Messungen der Leuchtdauer der Atome und der Dämpfung der Spektrallinien. II. *Annalen der Physik*, **66**, 229–236.

Wien, W. (1924). Über Messungen der Leuchtdauer der Atome und der Dämpfung der Spektrallinien. III. *Annalen der Physik*, **73**, 483–504.

CHAPTER 7

# FOUNDATIONS AND APPLICATIONS

*The Creative Tension in the Early Development of Quantum Mechanics*

CHRISTIAN JOAS

## 7.1 INTRODUCTION

WHEN historians and philosophers of science discuss the earliest phase of the history of the interpretation of quantum mechanics,[1] besides the now classic papers presenting the formalism,[2] a rather limited number of additional primary sources from ca. 1925–1927 get a lot of mileage: Max Born's two papers on the probability interpretation, Werner Heisenberg's uncertainty paper, Niels Bohr's Como lecture, and a few others.[3] Add to this papers that contain early explicit attempts at interpreting quantum mechanics differently than in the emerging Copenhagen way—such as those of Erwin Schrödinger, Erwin Madelung, or Louis de Broglie—that are also discussed by historians and philosophers of science, especially when they form initial points in lineages that extend to interpretations of quantum mechanics that are still being discussed today.[4]

All these papers are typically treated as contributions to the interpretational debate(s) proper and not, or at least not primarily, appreciated as papers in which the formalism

---

[1] Unless explicitly stated otherwise, throughout this chapter and somewhat anachronistically, 'quantum mechanics' stands for both Heisenberg's as well as Born's and Jordan's matrix mechanics, Schrödinger's wave mechanics, and their amalgamations into what textbooks nowadays refer to as quantum mechanics.

[2] Heisenberg (1925b); Born and Jordan (1925); Born *et al.* (1926); Schrödinger (1926a,b,d,e).

[3] Born (1926a,b); Heisenberg (1927b); Bohr (1928a,b).

[4] Jammer (1974, ch. 2), for instance focuses on Schrödinger (1926f); Madelung (1927); de Broglie (1927). See also: Cushing (1994).

of quantum mechanics was extended,[5] nor as papers in which the range of validity of quantum mechanics was expanded to bear on phenomena in specific physical systems going beyond simple, periodic, two-body systems like the hydrogen atom, such as scattering problems, more complex many-electron atoms (including helium), molecules, solids, or the atomic nucleus, even though many of them do one or the other. But does a division into (a) papers that established the formalism (interesting, but usually not crucial for the history of interpretations), (b) papers on interpretation proper (highly relevant for the history of interpretations), and (c) papers on specific applications (genuinely uninteresting for the history of interpretations) really hold up to historical scrutiny?[6] My answer to this question is no, and this chapter is an attempt to explain why, and why one should care, and also how it may affect views on the history of the interpretations of quantum mechanics if one does.

Especially philosophers of science, but also physicists themselves and to a lesser degree historians, have a tendency to draw a demarcation line between the formalism of quantum mechanics, i.e., a clearly demarcated theory or formalism, often in the mature form given to it by John von Neumann in his 1932 *Mathematische Grundlagen der Quantenmechanik*,[7] and its interpretations.[8] This dissociation of formalism and interpretation is not historically warranted for the earliest years of quantum mechanics, and it was certainly not so in the actors' minds, as will be shown later in the chapter. In the mid-to-late 1920s, quantum mechanics was neither a monolithic and inert theoretical formalism, nor did the actors see its interpretation as dissociated from its further elaboration.[9] That not all actors agreed on its interpretation is a truism, but even those who seemingly agreed with, say, Niels Bohr's views, in fact did not, or only to a much lesser extent than usually acknowledged.[10]

Therefore, it is worthwhile to give up the somewhat myopic perspective on the debates about the interpretation of quantum mechanics being fought out on relatively few pages of paper produced by a very select number of usually well-known actors—the

---

[5] On the relevance of formalism to Bohr in writings that otherwise seem to eschew formalism, see Dieks (2017). See also Tanona (2004).

[6] Throughout this paper, as is common usage among physicists, 'applications' stand for the use of a theory (here: quantum mechanics) to model a specific phenomenon or idealized system. Note in particular that following this usage, 'applications' do not refer to technological applications nor to experimental instruments, but primarily to elements of theoretical practice. On the important connections between the foundations of quantum mechanics and instrumentation as well as experimental techniques, especially in the latter half of the twentieth century, see Climério da Silva Neto's chapter in this volume.

[7] von Neumann (1932).

[8] See, e.g., Jammer (1974), esp. ch. 1.

[9] See, e.g., Beller (1999); Kojevnikov (2011); Kojevnikov (2020). See also Badino (2016) for a comprehensive review of historical scholarship regarding the history of quantum physics.

[10] See, e.g., Beller (1996); Howard (2004); Camilleri (2007); Camilleri (2009). To which extent the emergence of a rigid demarcation line between formalism and interpretation of quantum mechanics is a by-product of the post-WWII invention of the Copenhagen interpretation is an open question. As Don Howard points out, Léon Rosenfeld in a 1957 paper decries 'the false problem ("*Scheinproblem*") of "interpreting a formalism"' as a 'short-lived decay-product of the mechanistic philosophy of the nineteenth century'; see Rosenfeld (1957), p. 41; Howard (2004), p. 679.

few references mentioned above, plus maybe the correspondences of Pauli, Heisenberg, Born, and Bohr. This narrow perspective hides many interesting developments that are relevant for understanding the early interpretational debates, and much can be gained from widening it. More papers should be studied, by more actors, and in more depth, many of which have not yet been translated into English, because they have been seen as derivative and mostly elaborating the content of the formalism of quantum mechanics or pertaining to strictly derivative applications.[11] I argue that a thick reading of more literature, also off the beaten path,[12] is needed in order to fully appreciate both what was at stake for the individual actors and how open-ended their perspectives on quantum mechanics and physics as a whole were in the mid-to-late 1920s. The same holds beyond 1927, which is often—and to a large degree rightfully—taken to be the canonical end point of the initial, intensely creative phase of quantum mechanics, after which the amalgamated Bohr–Heisenberg–Copenhagen interpretation hardened and—thus goes the consensus view—for several decades buried all interpretational creativity under an ever-thickening crust of dogmatism and orthodoxy,[13] and as 'more mundane subjects' occupied the physicists' agenda.[14]

Today and at least since Bell's inequalities and their experimental tests by Alain Aspect and others,[15] both experimental and theoretical applications of quantum mechanics in contexts such as quantum optics, quantum information, or quantum computing are not merely making use of the theory's formalism or contributing to deciding between tenable and untenable interpretations, they also lead to a further articulation of the theory and thus to a better understanding of its physical content and meaning.[16] Regardless of whether one is convinced by the Kuhnian rhetoric of a 'second' quantum revolution,[17] or that of a 'third' one,[18] even sceptics might agree that quantum mechanics and its interpretation are evolving further in part due to these applications. Why should it have been any different in the volatile early years of quantum mechanics?

In what follows, I will first discuss the cut between foundations and applications (section 7.2) and explain why this cut is mostly artificial and anachronistic when

---

[11] Providing more translations into English of original papers from the old quantum theory and the early days of quantum mechanics is an important desideratum, and scholarly journals should be open to publishing such translations, ideally alongside expert commentary by historians of science. The most widely used commented collection of original papers in English, van der Waerden (1967), contains very helpful translations of many important papers, yet consciously avoids papers on wave mechanics, since they were planned to be assembled in a second volume that was never published, or on applications and extensions such as the Zeeman effect, spin, or statistics (see van der Waerden (1967), esp. vi). In a way, van der Waerden thus—unwillingly—contributed to biases regarding what the core corpus of the foundational papers on quantum mechanics consists of.

[12] Besides published articles, also the correspondence of less prominent actors and other historical sources, such as textbooks, should receive further attention. On the importance of textbooks as historical sources, see Badino and Navarro (2013); Badino (2019). See also Josep Simon's chapter in this volume.

[13] For a critical view on this narrative, see Don Howard's chapter in this volume as well as Beller (1996); Howard (2004); Camilleri (2007, 2009).

[14] Freire Jr. (2015), p. 19.    [15] Freire Jr. (2015).

[16] See, e.g., Bub (2016); Koberinski and Müller (2018); Janas et al. (forthcoming).

[17] Freire Jr. (2015).    [18] Celi et al. (2017), referring to quantum optics and cold atoms.

applied in hindsight to the early phase of quantum mechanics, which was an open-ended and fluid period of scientific development that does not at all fit the caricature of a finished formalism awaiting its interpretation (section 7.3). I will then try and classify the motivations of physicists for undertaking applications and provide examples of applications from the early years of quantum mechanics (section 7.4). Finally, by way of conclusion, I will discuss how these and other examples contributed to debates about the meaning and interpretation of quantum mechanics, which reached deeper than just to the question of how the formalism should be interpreted, and what might be gained by studying in more detail the early applications of quantum mechanics (section 7.5).

## 7.2 THE ARTIFICIAL CUT BETWEEN FOUNDATIONS AND APPLICATIONS

Applications of a physical theory are not usually acknowledged to be of considerable import to the theory's foundations. And that is, of course, true of many—almost all—applications: applying a theory can simply mean solving a problem that the theory was intended for, in the very basic sense of Thomas S. Kuhn's 'normal science'.[19] As long as one does not encounter any conceptual problems or produce any contradictory results, the theory's foundations are safe and sound. Better yet, every additional problem solved in this way serves as a consistency check of the theory and thus contributes to its acceptance. This first dimension of applying a theory—validation through problem solving—qualifies applications as *mere* applications of a theory. If anything, they add stability to the theoretical foundations, and this does not make such applications the most exciting object of study for physicists, nor for historians or philosophers of science.[20] This first dimension of applications is often acknowledged by the historical actors themselves when writing about the numerous applications of early quantum mechanics. For instance, Pascual Jordan, in his 1927 *Naturwissenschaften* piece on 'The Development of the New Quantum Mechanics,' states:

> Simultaneously with the progressive elucidation of the general laws of quantum mechanics, new evidence for the empirical correctness of this theory has been provided by a large number of applications by various authors.[21]

---

[19] Kuhn (1962).
[20] Except, for instance, when historians or philosophers are interested in science pedagogy or everyday scientific practice.
[21] Jordan (1927), p. 616. (German original: 'Gleichzeitig mit der fortschreitenden Aufklärung der allgemeinen Gesetzmäßigkeiten der Quantenmechanik sind durch eine große Fülle von Anwendungen von verschiedenen Verfassern immer neue Beweise für die empirische Richtigkeit dieser Theorie geliefert worden'.) All translations of quotes are mine unless stated otherwise.

Jordan then goes on to conclude that 'the stabilization of our quantum theoretical knowledge' had 'finally occurred'.[22]

Slightly more exciting are applications that extend the range of validity of a theory into new empirical phenomena or provide novel calculational schemes as part of the formalism. When reflecting on the early applications of quantum mechanics, some actors focus on this second dimension of applications. Friedrich Hund, for instance, in his 1967 history of quantum theory, states:

> With the probability interpretation and transformation theory, the principles of quantum mechanics were by and large known at the end of 1926. Since spring 1926, the Schrödinger equation had been a convenient method for solving the simpler problems, adapted to the mathematical knowledge of physicists at that time. Around 1927, these circumstances led to a *flood of applications* and the development of practical methods of calculation.[23]

Also historians and philosophers of physics often stress this very dimension of applications, for instance Max Jammer when he states:

> Satisfied that the theory 'works', since it provided unambiguous answers whenever invoked, physicists engaged themselves rather in solving problems which so far had defied all previous attempts or which promised to open up new avenues of research. The year 1927 thus not only became the year in which the quantum-mechanical formalism, in all its essential points, received a formal completion and a consistent interpretation; 1927 also witnessed a veritable *avalanche of elaborations and applications* of the new conceptions and led to new insights in atomic physics to an unprecedented extent.[24]

Following such accounts, and using Kuhnian terms, the quantum mechanics revolution ended quickly, little over two years after *Umdeutung*. What was left was a 'flood' or an 'avalanche' of normal science and problem solving in which a completed and canonized theory was being put to practical use, maybe in new domains of science,

---

[22] Jordan (1927), p. 616. (German original: 'die endlich eingetretene Stabilisierung unserer quantentheoretischen Erkenntnis'.)

[23] Hund (1967), p. 169. (German original: 'Die Prinzipien der Quantenmechanik waren Ende 1926 mit der Wahrscheinlichkeitsdeutung und der Transformationstheorie im Großen und Ganzen erkannt. In der Schrödinger-Gleichung lag seit dem Frühjahr 1926 ein bequemes und den damaligen Mathematikkenntnissen der Physiker angepaßtes Verfahren zur Lösung der einfacheren Aufgaben vor. Diese Umstände zeitigten um 1927 eine *Flut von Anwendungen* und die Ausbildung von praktischen Rechenverfahren.')

[24] Jammer (1966), p. 362 (emphasis added). Note that I am not claiming that Jammer (1966) discusses only this dimension of applications. As Olival Freire Jr. has pointed out to me, the very existence of two separate and each on their own very influential volumes by Jammer—Jammer (1966) on *The Conceptual Development of Quantum Mechanics* and Jammer (1974) on *The Philosophy of Quantum Mechanics: The Interpretations of Quantum Mechanics in Historical Perspective*—might have contributed to the perception of 'a gap, not intentionally, for sure, between conceptual development and interpretations of quantum mechanics'.

yet of little to no import to the foundations of quantum mechanics.[25] In such accounts, the role played by applications is thus straightforward and practical and not very relevant to the theory's foundations. The story of how applications of quantum mechanics to molecules, solids, or nuclei opened up new avenues for research in physics, therefore, would not belong in a history of quantum mechanics itself, but rather in separate histories of the new subdisciplines associated with these phenomena. If this was all that applications can be in the practice of physicists, drawing a firm line between foundations and applications would indeed be warranted.

But there is another, third dimension of applying a theory that goes beyond *mere* application and makes the cut suggested in the above quotes—between foundations and applications—appear eerily artificial, if not altogether misleading. It involves the articulation of the theory, potentially its extension or even overthrow, and the further clarification of its meaning and physical content. This fundamentally constructive role of applications, i.e., their ability to modify or add to the conceptual framework or to lead to an understanding of the formalism, is less often explicitly acknowledged by the actors, yet statements of this kind can be found. Either implicitly, for example when Werner Heisenberg in the endgame of the old quantum theory put his hope into '"sharp" applications ["scharfe" Anwendungen]' of Bohr's correspondence principle,[26] e.g., to the Zeeman effect or to optical dispersion; or also explicitly, when Niels Bohr in his 1922 Wolfskehl lectures laid out his best bid for a strategy going forward:

> When attempting to formulate the principles of the quantum theory, we encounter these difficulties, which are formidable, indeed. In such a situation, the most cautious procedure is always to stick to the applications of the principles.[27]

In a 1924 paper titled 'On Quantum Mechanics' which despite its title predates Heisenberg's *Umdeutung* paper and thus what is today known as quantum mechanics by about one year, Max Born evaluates and compares the respective promises of different classes of applications of the correspondence principle and states:

> As long as one does not know the laws of the influence of light on the atom, and thus the connection between dispersion, the structure of the atom, and the quantum jumps, one will certainly remain in the dark concerning the interactions between multiple electrons in an atom.[28]

---

[25] On subdisciplinary histories, see: Joas and Hartz (2019); James and Joas (2015).

[26] See, e.g., Heisenberg (1925a), p. 617.

[27] Niels Bohr, Wolfskehl lectures 1922, 2nd lecture (13 June 1922), in: Bohr (1977), p. 351. I am indebted to Richard Staley for pointing me to this passage. Note that when Bohr speaks of applications, he usually speaks of the 'consistent application' of certain principles or conceptions, not of applications of a theory or a formalism; also in Bohr (1928a,b) and many other writings. See also Stefano Osnaghi's chapter in this volume.

[28] Born (1924), p. 379. (German original: 'Solange man die Gesetze der Einwirkung des Lichtes auf Atome, also den Zusammenhang der Dispersion mit dem Atombau und den Quantensprüngen, nicht kennt, wird man erst recht über die Gesetze der Wechselwirkung zwischen mehreren Elektronen eines Atoms im Dunkeln sein.')

For Born, the most likely route towards achieving a consistent extension of quantum theory—a true quantum *mechanics* replacing classical mechanics—lay in recognizing that the application of quantum theory to many-body problems ('interactions between multiple electrons') was intimately connected to applications of quantum theory to the problem of the interaction between radiation and matter ('influence of light on the atom'). In his 1924 paper, Born thus took as his point of departure Kramers's 1924 theory of optical dispersion and explored how it might be used to devise perturbative schemes appropriate for studying at the same time the coupling of electrons to one another and the coupling of electrons to radiation. For Born, the narrowing of research interests to dispersion in the time immediately preceding *Umdeutung* therefore was not a turn away from trying to apply quantum theory to the structure of matter but rather the best available strategy for achieving this very goal.[29]

Scientific theories, especially while being developed but also thereafter, are not static, but highly fluid, and they cannot be disentangled from the contexts in which they are applied.[30] When faced with a novel theory for a given phenomenon—whether developed by themselves or by someone else—physicists soon tend to ask whether that theory might also hold more generally, i.e., for other closely related phenomena, or for phenomena that are not usually seen as directly related, or maybe even for all physical phenomena.[31] This move is not necessarily driven by an expectation that the extension of the range of application of the theory is successful or even possible. Rather, obtaining a negative result in itself constitutes a valuable insight about the limits of the theory and might even reveal something previously unknown about the validity of the theory in its original context. Yet also when an attempt to extend the theory's range of validity proves successful, this does not necessarily mean that the theory itself remains unchanged during this attempt. Along the way, the physicist might discover the necessity to actively alter the theory's original formulation or some of the concepts entering it, often ever so slightly, in order to remove contradictions and to accommodate the new phenomenon within the theory's range of validity. This, in turn, changes how the theory works in its original context, and it might actually yield new and unexpected predictions also in this original context. In short, it is a viable research strategy to go look for applications of a theory to new areas which are not yet contained in its original domain of application, not just in a 'colonist' move to conquer new territory for a theory at hand, but even—and this is where my argument hinges—to improve things at home.

[29] James and Joas (2015), especially p. 665.

[30] For accounts of such a view of theory, see, e.g., Darrigol (2008); Badino (2015). See also: James and Joas (2015). Throughout this chapter, I am deliberately using a rather fuzzy notion of 'theory' and avoid using the word 'model' even though sometimes probably more appropriate from the point of view of philosophical rigour.

[31] This pragmatic strategy of methodological (and not necessarily metaphysical, see Bitbol (1996), especially pp. 13–15) realism builds upon the successes of many different specific epistemological approaches physicists adopted throughout history, e.g., Newton's or Boltzmann's. For a short discussion, see: Joas and Katzir (2011). On Boltzmann, see also: de Regt (2017).

This bidirectionality of knowledge transfer and of knowledge being applied in new contexts is well-known in grand narratives of exchanges between different local contexts. It is a prominent trope in postcolonial history, which often considers how cultural or scientific exchange between the colonists and the colonized did not unidirectionally and exclusively affect the knowledge systems of the colonized, but also had effects on those of the colonists. The mechanism can also be found at work beyond knowledge transfer between geographical spaces, in processes of knowledge transfer between different fields or conceptual spaces, and in the context of application.[32] Examples abound in the history of physics: inspired by Joseph-Louis Lagrange's mechanics, Sir William Rowan Hamilton in the 1830s developed a similar scheme for optics and later realized how this scheme in turn could be applied to the mechanics of moving bodies, leading to an elegant new formulation of mechanics, which would ultimately provide a key conceptual ingredient for Schrödinger's wave mechanics (Hamilton's optical-mechanical analogy) nearly a century later.[33] Applications of Bohr's correspondence principle in the old quantum theory in the early-to-mid 1920s were hoped and expected to lead the way to a more general quantum theoretical framework.[34] And Heisenberg's *Umdeutung* as well as Schrödinger's wave mechanics can be seen as successful products of this general research programme, which uncovered—through specific applications to optical dispersion or atomic spectra in the case of Heisenberg, or to the hydrogen spectrum in Schrödinger's case—the core elements of the two new theories we now collectively refer to as quantum mechanics.[35]

The line between foundations and applications—terms regularly used by the actors themselves—therefore gets blurry. Applications are not per se irrelevant for the foundations. They rarely lead to a complete *bouleversement* of a theory in its original context, but this is by no means impossible either, as the examples of Heisenberg's *Umdeutung* and Schrödinger's wave mechanics demonstrate. And if the strategy of extending or clarifying a theory through application was as omnipresent and ultimately successful in the old quantum theory, it would be very surprising if the actors gave up on such a successful strategy in light of the new quantum mechanics, even more so because few at first saw quantum mechanics as a firm cornerstone for the future theoretical development and many, at least early on, expected another radical theoretical innovation to wait around the next corner.

[32] Carrier (2011), especially pp. 20–27, who discusses technological applications (rather than applications of a theory, as done here) argues for the 'epistemic dignity of application-oriented research' and states: 'Applied challenges may raise fundamental questions which need to be addressed if the practical task is to be mastered. Applied research *never merely* taps the system of knowledge and combines known elements of knowledge in a novel way' (Carrier, 2011, p. 24, emphasis added).

[33] Hankins (1980), especially chapter 4; Joas and Lehner (2009), especially pp. 340–42. Note that Hamilton repeatedly refers to the transfer of his idea of the characteristic function from optics to mechanics as an application of his dynamical principle; see Hamilton (1834), p. 248.

[34] Jähnert (2019).

[35] *Umdeutung* refers to Heisenberg (1925b) and is discussed in Duncan and Janssen (2007a,b); Blum *et al.* (2017). See also MacKinnon (1977); Dresden (1987); Darrigol (1992).

## 7.3 The Open-endedness of Early Quantum Mechanics

Neither matrix nor wave mechanics were full-fledged theories at the times of their original formulations in 1925 and early 1926, respectively.[36] Rather, they were tentative formalisms initially with very narrow domains of successful application. Testing and extending their range of validity through applications to new domains was a natural strategy in the hunt for clues on how to make sense of the new formalisms and on how to articulate them further.

In mid-1926, within a few months after the genesis of wave mechanics, matrix and wave mechanics were recognized as equivalent despite the diverging original motivations and ontological commitments of their originators, their different domains of validity (intensities and transition probabilities in line spectra versus the energy levels of the stationary states in the hydrogen atom), and the stark contrasts between the mathematical forms they took (infinite-matrix algebra versus differential equations).[37] While the equivalence proofs by Schrödinger, Eckart, and Pauli both grew out of attempts to apply and extend the formalism and acted as catalysts for new applications providing extensions of their range of validity or establishing fruitful connections between the two approaches, they did not make the task of finding an interpretation of the now two-headed formalism any easier.[38] To the contrary, interpretation increasingly turned into an activity destined to establish the superiority of one or the other formalism on different grounds than those of empirical accuracy, i.e., despite their empirical or 'mathematical' equivalence.[39]

Both Heisenberg and Schrödinger in 1926 and 1927 (and also beyond), while acknowledging equivalence, increasingly sought to demonstrate the epistemological superiority—i.e., an epistemological *inequivalence*—of their respective approaches, also and importantly through applications of their theories to new domains. Much of the work of both Heisenberg and Schrödinger in 1926–1927 and of the early interpretational debates can better be understood when keeping this specific angle in mind.

Already in his second communication, i.e., roughly a month before submitting his equivalence proof, Schrödinger expresses the hope that matrix and wave mechanics 'will not fight each other, but rather, precisely because of the extraordinary difference in

[36] Heisenberg (1925b); Born and Jordan (1925); Born *et al.* (1926); Schrödinger (1926a,b).

[37] It is important to stress that the various equivalence proofs did not achieve a full-fledged mathematical equivalence between the two formalisms. See: Perović (2008); Muller (1997a,b). This underlines yet again the open-endedness of the development in the early phase of quantum mechanics.

[38] Schrödinger (1926c); Eckart (1926); Wolfgang Pauli to Pascual Jordan, 12 April 1926, in Pauli (1979), pp. 315–20. See also: van der Waerden (1973).

[39] For a more in-depth discussion of the interpretational debates in 1925–1927 and of whether or not they really constituted a philosophical debate in the strict sense, see Kojevnikov (2020), especially chapter 6.

starting point and method, will complement each other' and then goes on to add—with tangible optimism for his own wave mechanics—'in that one will help where the other fails'.[40] Heisenberg makes very similar statements on several occasions, for instance in his paper on the 'Many-body problem and resonance in quantum mechanics' in which he explains the splitting between singlet and triplet term systems in ortho- and parahelium introducing the concept of resonance, i.e., exchange, (see also later in this section), submitted from Copenhagen on 11 June 1926:

> Because of the mathematical equivalence of Schrödinger's approach [wave mechanics] with quantum mechanics [read: matrix mechanics], however, the question of the physical processes underlying the equations could be regarded for the time being as a question of expediency concerning our intuition [*eine unsere Anschauung betreffende Zweckmäßigkeitsfrage*]; but *only as long as we do not try to extend the foundations of this quantum theory based on the intuitive pictures we have chosen*. For the many-body problems to be treated here, we require a direct connection to the points of view underlying quantum mechanics [read: matrix mechanics], while Schrödinger's approach might require essential modifications of the established equations.[41]

Directly following this passage, Heisenberg provides arguments for the superiority of his own intuitions over Schrödinger's, before closing on a conciliatory note by stressing that wave and matrix mechanics indeed do complement one another.

At least until the summer of 1927, it was still open how to further develop and understand quantum mechanics. Many of the actors saw the new mechanics very much as a promising, yet ultimately open-ended endeavour that still needed to demonstrate its full generality, and not as a theory nearing completion that simply lacked a few final elements (such as, say, uncertainty and complementarity) or merely a consistent interpretation. Physicists therefore continued to test the range of validity of the new

---

[40] Schrödinger (1926b), p. 513. (German original: '...daß diese beiden Vorstöße einander nicht bekämpfen, vielmehr, gerade wegen der außerordentlichen Verschiedenheit des Ausgangspunktes und der Methode, einander ergänzen werden, indem der eine weiterhilft, wo der andere versagt.') Soon thereafter, Schrödinger (1926f) attempted to obtain a physical interpretation of his wavefunction and to establish a connection of wave mechanics to ordinary macroscopic mechanics by constructing wave packets of harmonic-oscillator eigenfunctions in an attempt to model classical particles. Interestingly, this was at least in part motivated by Schrödinger's desire to demonstrate that the complex nature of the wavefunction was merely an unphysical artefact and, as insinuated in Schrödinger (1926e), that a future extension of wave mechanics would provide a real (as opposed to complex) fourth-order (instead of second-order) wave equation. See Karam (2020).

[41] Heisenberg (1926a), p. 412, emphasis added. (German original: 'Wegen der mathematischen Äquivalenz des Schrödingerschen Verfahrens mit der Quantenmechanik könnte allerdings die Frage nach dem den Gleichungen zugrunde liegenden physikalischen Geschehen einstweilen als eine unsere Anschauung betreffende Zweckmäßigkeitsfrage betrachtet werden; aber nur solange wir nicht versuchen, auf Grund der einmal gewählten anschaulichen Bilder die Grundlage dieser Quantentheorie zu erweitern. Für die hier zu behandelnden Mehrkörperprobleme möchten wir einen direkten Anschluß an die der Quantenmechanik zugrunde liegenden Gesichtspunkte verlangen, während Schrödingers Darstellungsweise wesentliche Änderungen der bisherigen Gleichungen als möglich erscheinen läßt.')

mechanics and explore its extension in all possible directions, fully aware, or even hoping, that successful applications might lead to a revision of quantum mechanics or to new fundamental insights necessitating a new theoretical framework that went beyond quantum mechanics: relativity, aperiodic phenomena (i.e., scattering), and electron spin all called for extensions of the formalism. High on the agenda was also an extension of quantum mechanics to many-body systems such as complex atoms (beginning with helium, which had resisted all attempts at explanation in the old quantum theory), molecules, solids, or nuclei. And so was the connection to electro-dynamics, i.e., quantum electrodynamics.

The available historical sources abound with discussions of applications and of lists and plans of which applications to attack next. Schrödinger's many published writings of 1926–1927 show him methodically working through numerous applications of nonrelativistic wave mechanics to many different domains, such as his development of a perturbative scheme to treat the Stark effect, while at the same time his unpub-lished notebooks reveal that he worked on a much more ambitious, private (and overall unsuccessful) research programme aimed at fulfilling the promises he saw in a theory that based itself purely on a wave (or field) ontology and would resolve the problems he was facing in finding a consistent physical interpretation of matter waves.[42] Pauli's correspondence with Heisenberg and also with Bohr shows them discussing various open problems and possible applications of quantum mechanics, e.g., connected to electron spin, such as in a letter from Pauli to Bohr dated 12 March 1926 which mentions 'Goudsmit's electron, the explanation of the fine structure, [ . . . ] the helium spectrum, the distance of singlet and triplet term systems, equivalent orbits [Pauli's exclusion principle].'[43]

The sense of openness, to be sure, was paired with a widespread optimism regarding the outlook and promises of quantum mechanics and future theoretical developments building upon it, rather than with confusion or disillusionment. Felix Bloch, a student in Zurich during the crucial early phase of quantum mechanics, who in the fall of 1927 moved to Leipzig to work with Heisenberg, half a century later described the mood in those early days of quantum mechanics as follows:

> I don't think many of us realized that we had just gone through quite a unique era; we thought that this was just the way physics was normally to be done and only wondered why clever people had not seen that earlier. Almost any problem that had been tossed around years before could now be reopened and made amenable to a consistent treatment. To be sure, there were a few minor difficulties left, such as the

---

[42] Joas and Lehner (2009), especially pp. 349–50. One of Schrödinger's unpublished notebooks of the time—probably from 1927—bears the title 'The pending questions [*Die schwebenden Fragen*]' (AHQP 41-2-002), in which he considers the relativistic extension of wave mechanics in the light of electron spin and tries to apply his findings to the Zeeman effect. This notebook and many others of the time demonstrate what Michel Bitbol has called Schrödinger's 'over-revolutionary' attitude, as opposed to the conservatism Schrödinger is usually accused of. See Bitbol (1996), especially pp. 20–24.

[43] Wolfgang Pauli to Niels Bohr, 12 March 1926, in Pauli (1979), pp. 310–12, on p. 311.

infinite self-energy of the electron and the question of how it could exist in the nucleus before beta decay; and nobody had yet derived the numerical value of the fine-structure constant. But we were sure that the solutions were just around the corner and that any new ideas that might be called for in the process would be easily supplied in the unlikely event that this should be necessary. Well, the last fifty years have taught us at least to be a little more modest in our expectations.[44]

For Bloch, as for most other actors, the heyday of quantum mechanics was over in late 1927, when agreement on what later would come to be known as *the* Copenhagen interpretation was reached by the Göttingen–Copenhagen physicists (i.e., Bohr, Heisenberg, Born, Pauli, Jordan, and also Dirac, as opposed to Einstein, Schrödinger, Lorentz, and de Broglie) in, as John Hendry has put it, a 'least common denominator of the leading quantum physicists' views'.[45] Many things were still open, especially the question of how to connect quantum mechanics with electrodynamics, but these lingering issues, which were high on everyone's agenda, were compartmentalized and declared as lying outside the scope of the foundations of quantum mechanics proper.

## 7.4 'FOUNDATIONAL' APPLICATIONS?

How to make sense of this creative tension in the early development of quantum mechanics which had the actors searching after both the interpretation and possible extensions of the new quantum mechanics via the exploration of concrete applications? Which motivations drove physicists to turn to applications of quantum mechanics in 1925–1927 and beyond, and how did what they ended up with match their initial expectations? Several motivations have already been touched upon, but I would like to anyway try and provide a more systematic overview here, and to complement this overview with concrete examples of applications that did have a foundational impact on quantum mechanics.

Before doing so, it is important to mention that physicists did not necessarily spell out their motivations for tackling a specific application, that their motivations often changed along the way, and that sometimes applications serendipitously led to insights that scientists might not have anticipated when embarking on their endeavour. Mara Beller has studied in detail Max Born's mid-1926 work on scattering, which ultimately led to the probability interpretation of Schrödinger's wavefunction.[46] In his two mid-1926 papers, Born extends wave mechanics to aperiodic phenomena. According to Beller, the probability interpretation is merely a by-product of this endeavour. She stresses that 'Born's aim in his first collision paper [Born, 1926a] was not to contribute

---

[44] Bloch (1976), p. 26.    [45] Hendry (1984), p. 127.    [46] Born (1926a,b).

to the clarification of interpretational issues, as his later recollections suggest, but to solve a particular (yet crucial) scientific problem.'[47] Instead,

> Born's probabilistic interpretation was a conceptional contribution that crystallized over a considerable period of time. It was a process during which Born's ideas and commitments underwent significant changes. In fact, during the formative stage all of Born's intellectual pronouncements were *fluid and uncommitted*.[48]

Beller's analysis also suggests that actors' later recollections on why they undertook specific applications have to be treated with utmost caution, as they often get flavoured or even downright distorted by hindsight.

What, then, were those motivations?

First and foremost, as mentioned above, few (if any) in the early days of quantum mechanics expected the theory to remain the best bid for years if not decades to come. Hopes were high for modifications to the theory that would, among other things, make it an inherently relativistic theory (think of Schrödinger's initial attempts at formulating a relativistic wave equation, Dirac's struggles with making sense of his relativistic wave equation, or the many attempts towards quantum electrodynamics) or establish the adequacy of one specific interpretation and philosophical purview and the inadequacy of others (think of Heisenberg's work on many-body systems,[49] briefly discussed above and in more detail below, or even more importantly his uncertainty paper,[50] or of Schrödinger's attempts at formulating a physical interpretation of many-body wavefunctions that would take them from higher-dimensional configuration space into real space). Applications were a vehicle towards uncovering potential avenues to an extension of quantum mechanics in the not-so-distant future and towards clarifying which overall approach was most likely to lead to this extension.

Secondly, quantum mechanics early on was not much more than a rump theory, or actually two. Many of the successes of the old quantum theory had not yet been recovered in quantum mechanics by, say, mid-1926.[51] At least initially, there could be no talk of quantum mechanics replacing or superseding the old quantum theory when looking at the narrow empirical content it covered. Bit by bit, both matrix and wave mechanics had to be demonstrated to cover familiar ground, such as in Pauli's

---

[47] Beller (1990), p. 564. See also: Beller (1999), especially pp. 41–49. Note that the opposite, i.e., a physicist undertaking an application of quantum mechanics in a hope to find clues towards its interpretation and turning up empty, might be much more common.

[48] Beller (1990), p. 564 (emphasis added). Note that 'fluid' here alludes to Yehuda Elkana's idea of 'concepts in flux', see: Elkana (1970).

[49] Heisenberg (1926a); Heisenberg (1927a).    [50] Heisenberg (1927b).

[51] Midwinter and Janssen (2013) refer to such (not necessarily permanent) situations in which the application of a newer theory fails to account for things that the application of an older theory was able to explain as 'Kuhn losses'. In the case they study—electric and magnetic susceptibilities in classical physics, the old quantum theory, and quantum mechanics through John H. van Vleck's work—quantum mechanics eventually recovered results from classical physics that the old quantum theory had never been able to come to grips with.

January 1926 application of matrix mechanics to the hydrogen atom, which was a major argument for putting continued trust into matrix mechanics.[52] In Schrödinger's case, explaining the Stark effect, a key success of the old quantum theory,[53] took serious effort and the development of a perturbative scheme for wave mechanics and was only presented in his third communication.[54]

One of the big successes of the old quantum theory, quantum statistics, was particularly resistant to being integrated into quantum mechanics during the course of 1926–1927.[55] This happened when Heisenberg, Pauli, Dirac, and others ventured to extend quantum mechanics to many-body systems. An important example in this context is Pauli's exclusion principle, about which Pauli in the conclusion to his January 1925 paper had conjectured that it 'likely can only be tackled after a *future deepening* of the fundamental principles of quantum theory.'[56] In the introduction to his 1926 *Mehrkörperproblem* paper, which was submitted on 11 June 1926 and partly delivered on Pauli's hope, Heisenberg remarked: 'Additional rules like Pauli's exclusion of equivalent orbits in their present form do not have a place in the mathematical scheme of quantum mechanics. One could thus ponder a *failure* of quantum mechanics with respect to the problem of equivalent orbits.'[57] Much later, he would reflect upon the integration of the exclusion principle into quantum mechanics as follows:

> In the summer months of that same year [1926], the connection between Pauli's exclusion principle and wave as well as quantum mechanics was clarified. On the one hand, based on the quantum mechanics of the helium atom with which I had preoccupied myself, one could prove that one only obtained the correct terms, satisfying Pauli's principle, if one required the wave function to be antisymmetric in the particle coordinates; and on the other hand, Fermi and Dirac managed to show in general that permuting the coordinates of two arbitrary electrons was equivalent to Pauli's principle and, when applied to an ideal gas, led to a new statistics. Thereby, the meaning of Pauli's exclusion principle had been clarified once and for all.[58]

That electrons obey Fermi statistics—instead of Bose statistics, as originally and errone-ously assumed—was a realization Heisenberg and Pauli reached even later, in late 1926.[59] Applications were thus a locus for the combination of different lines of successful development from the old quantum theory into what we now consider one integral

---

[52] Pauli (1926). See also: Kragh (1985), especially pp. 118–19.
[53] Duncan and Janssen (2014).
[54] Schrödinger (1926d); Epstein (1926).
[55] For a comprehensive account of this process, see Daniela Monaldi's chapter in this volume.
[56] Pauli (1925), p. 783 (emphasis added).
[57] Heisenberg (1926a), p. 413 (emphasis added).     [58] Heisenberg (1960).
[59] James and Joas (2015), especially pp. 676–79. It is worth noting here that Fermi statistics itself was not a product of quantum mechanics, but actually a late success of the old quantum theory, just like Bose–Einstein statistics. The prominent role quantum statistics plays in many later quantum mechanics textbooks makes it easy to forget this.

theory.[60] They allowed physicists to exploit as yet untapped knowledge structures, developed within the old quantum theory, in efforts to further solidify quantum mechanics.[61]

Thirdly, testing and establishing the adequacy and generality of quantum mechanics by charting and extending its domain of applicability was another driving force for applications. Despite widespread agreement that many phenomena which had escaped explanation in the old quantum theory would eventually come to be explained using quantum mechanics, the theory still had to make good on this promise in the early days of quantum mechanics.

Open problems abounded. As already mentioned, the spectrum of the hydrogen atom was explained by quantum mechanics in early 1926,[62] but explaining the helium spectrum took until mid-1926.[63] By July 1926, Heisenberg was able to triumphantly conclude that 'quantum mechanics allows for a qualitative description of the spectrum also of atoms with two electrons up to the finest details.'[64] More complex atoms, i.e., atoms with more than two electrons, or molecules, would take even longer.[65] On occasion also new concepts and new empirical knowledge had to be integrated into the theory, such as electron spin, initially foreign to quantum mechanics, so much so that its integration into the theory was a bit like changing the spinning wheels while riding a bicycle.[66] Heisenberg and Jordan achieved its integration into matrix mechanics in March 1926 through applying matrix mechanics to the anomalous Zeeman effect.[67] Pauli's 1927 integration of spin into wave mechanics (which brought about the so-called Pauli spin matrices) was as much an attempt to integrate spin as to come to grips with many-body problems in quantum mechanics.[68] Physicists' motivations for many other applications of quantum mechanics outside of atomic physics and atomic spectroscopy also fall into this category, and the establishment of quantum mechanics as *the* foundational theory for a wide range of physical phenomena would have been unthinkable without success in these applications.

Fourth, applications held a promise towards deepening the understanding of what the quantum-mechanical formalism really *meant*. In a 1928 Göttingen speech, Max Born commented on the new formalisms of matrix and wave mechanics: 'At first, however, they were only formalisms, and it was a matter of discovering their meaning *a posteriori*.'[69] Similar statements can also be found from other important actors (see below). Applications contained leads on better understanding the kinds of predictions quantum mechanics generated and on possible interpretations, in a broader sense than

---

[60] James and Joas (2015).

[61] This is reminiscent of what happened around 1900 when physicists, first and foremost Albert Einstein, began exploring the consequences of Planck's quantum of action. See: Büttner *et al.* (2003).

[62] Pauli (1926).        [63] Heisenberg (1926a).        [64] Heisenberg (1926b).

[65] James and Joas (2015).        [66] Forman (1968); Tomonaga (1997), especially pp. 43–62.

[67] Heisenberg and Jordan (1926).        [68] Pauli (1927).

[69] Born (1928), quoted after Born (1969), p. 25. (German original: 'Aber zunächst waren es eben nur Formalismen, und es kam darauf an, ihren Sinn nachträglich herauszufinden.'). Note that this quote also makes a cut between formalism and 'meaning'.

we today tend to speak of the interpretation of quantum mechanics, e.g., including the search for physically intuitive pictures and *Anschaulichkeit*.[70] And different actors had different notions of what was required to establish the physical meaning of the formalism.

J. Robert Oppenheimer is a particularly interesting case. When interviewed by Thomas S. Kuhn in 1963 and asked about Heisenberg's idea of resonance (i.e., exchange), he stated: 'That was very exciting. That there was a lot of talk about and I regarded it as a kind of discovery of the *meaning* of quantum theory.'[71] When prompted by Kuhn to expand on what he meant by 'meaning', Oppenheimer first explained how he perceived the status of early quantum mechanics:

> Well, here first in very primitive systems one had the new formalisms. In the case of matrix mechanics a rather clear idea of the connection between the formalism and the observation, but also a very limited idea because what one was talking about were these dipole amplitudes in optical phenomena. Well, that's a very small part of what you can find out about an atom. Then with the wave mechanics one had a more supple formalism which had this connection to experiment but which was already very thin and, through the probabilistic interpretation and Born himself, some greater physical content. But let us say the two equations, $pq - qp$ and the Hamiltonian equations on the one hand and Schrödinger's equation on the other, had a lot to say about the world of nature and it had to be explored. [ ... ] I thought that it was a period in which, using the formalism or its more or less natural extensions, a great deal could be discovered about the content of quantum mechanics which couldn't have been known to its inventors.[72]

Oppenheimer then went on to compare the status of quantum mechanics ca. 1926 with that of general relativity in the early 1960s, which at that point, like the interpretation of quantum mechanics, was undergoing a renaissance of its own:[73]

> This is a very rare thing; I mean, one is still in that state in general relativity. It's a formalism rather poorly warranted by any experiment or observation whose consequences are far richer than anyone knows. Well, that's for many reasons, but quantum mechanics was also a formalism which had a hell of a lot to say if you knew how to ask it some questions and I think the two things that happened in the' 26–'27 winter—but that was also when the uncertainty principle began coming up—but the two things that revealed most about phenomena which had not been adequately described before were the radiation theory and the exchange terms.[74]

---

[70] See the chapter by Martin Jähnert and Christoph Lehner in this volume.

[71] J. Robert Oppenheimer, interview by Thomas S. Kuhn, Institute for Advanced Study, Princeton, NJ, 20 November 1963, AHQP M/f no. 1419, sec. 4, on p. 14 of transcript (emphasis added). I am indebted to Alexander S. Blum for pointing me to this passage.

[72] J. Robert Oppenheimer, interview by Thomas S. Kuhn (note 71), p. 14–15 of the transcript.

[73] See, e.g., Blum *et al.* (2015).

[74] J. Robert Oppenheimer, interview by Thomas S. Kuhn (note 71), p. 14–15 of the transcript.

Kuhn then asked Oppenheimer how typical this attitude would have been at the time:

> How many people would have thought of themselves as exploring nature by trying to see what you could do unexpected with the formalism as against applying the formalism to straight problem solving?[75]

And Oppenheimer replied:

> I think when Heisenberg looked at the helium atom he was trying to solve a problem, but I think that Jordan was certainly seeing how far he could push the formalism. After all, in a small way the Stark effect was of the former kind; of seeing what was in the theory; and molecules were problem solving because one knew that there was something to get straight. So, the distinction isn't all that sharp. But I think that if Heisenberg had found that there wasn't anything new but just that the integrals of wave functions happened to give the helium spectrum right, it would have been problem solving. It was the fact that there was an element of novelty and something which had never been described before which turned it from solving a problem into *exploring the content and meaning.*[76]

For Oppenheimer, Heisenberg's application of quantum mechanics to the helium spectrum revealed something genuinely new and previously unknown about the 'content and meaning' of quantum mechanics. It is unclear whether Heisenberg would have subscribed to this reading, but it is worth noting that Heisenberg's motivation for attacking the many-body problem in quantum mechanics indeed was at least twofold, as becomes evident from his paper.[77] His motivation was not merely to derive a formalism that would enable physicists to deal with quantum many-body systems, but also to clarify the physical interpretation of matrix mechanics and to thereby demonstrate its superiority over wave mechanics.

Heisenberg's concept of resonance, today commonly referred to as exchange, was seminal for many different areas of quantum theory. This brings us to the fifth and final motivation for pursuing applications, without necessarily claiming that the list presented here is exhaustive.

Fifth, then, new insights in the form of novel concepts and interpretive devices emerged directly from applications and contributed to the further articulation of quantum mechanics. The most famous example, Born's probability interpretation, has already been discussed above. There are many more examples. Heisenberg's concept of resonant exchange, for instance, would play a crucial role in the development of physical theories of the chemical bond, particularly the 1927 work of Heitler and London on the hydrogen molecule, which lies at the onset of modern quantum chemistry.[78] It also played an important role in Heisenberg's own work. In May 1928,

---

[75] Ibid., p. 15 of the transcript.
[76] Ibid., p. 15 of the transcript (emphasis added).    [77] Heisenberg (1926a).
[78] Heitler and London (1927).

Heisenberg submitted a paper on the theory of ferromagnetism to *Zeitschrift für Physik*,[79] in which he explained ferromagnetism as resulting from exchange forces similar to the ones he had discussed in his *Mehrkörperproblem* paper.[80] Whereas Weiss in his 1907 theory of ferromagnetism had simply postulated the existence of a molecular field, Heisenberg now was able to explain it as originating from quantum-mechanical exchange forces.

In a letter to Niels Bohr, Heisenberg, now in Leipzig, on 23 July 1928 explained what had brought him to considering the application of quantum mechanics to ferromagnets. Being convinced that the 'questions of principle [regarding the foundational questions of nonrelativistic quantum mechanics] . . . are now completely resolved', he had turned to Dirac's relativistic theory, about which he was 'unhappy' due to its 'inconsequence'. He goes on: 'Thus I find the current situation [in relativistic quantum theory] quite absurd and have therefore moved on, almost out of desperation, to a completely different area, that of ferromagnetism.' This application of quantum mechanics contributed to the spelling out of what indistinguishability implied physically in many-body quantum mechanics, just like Oppenheimer's work on exchange in electron–electron scattering or Felix Bloch's work under Heisenberg's supervision, at about the same time, on electrons in metals.[81] In that work, Bloch developed the concept of what later became known as the Bloch wave, a significant step beyond Sommerfeld's theory of the electron gas in metals, which had been based on classical Drude–Lorentz theory with a dash of Fermi statistics, and, unlike Bloch's, had not been a truly quantum-mechanical theory of electrons in metals. Bloch would later apply his ideas to spin waves in ferromagnets.[82]

Another concept that is a direct product of attempts to apply and thereby extend quantum mechanics is entanglement, a key ingredient in many a debate on the interpretation of quantum mechanics. Entanglement is often said to have been recognized as a key feature of quantum mechanics only around 1935 in the context of the EPR *gedankenexperiment*. Awareness of the concept—not the name it would eventually bear, which was indeed coined in German (*Verschränkung*) by Schrödinger in 1935— is, however, older and dates to the phase immediately following the genesis of quantum mechanics, and it emerged in the context of application.

Erwin Schrödinger encountered entanglement when trying to apply wave mechanics to coupled systems, such as diatomic molecules, in his unpublished notebooks. The notebooks show that his motivation for applying wave mechanics to coupled systems was the wish to better understand many-body wavefunctions in configuration space and to make sense of what they might physically mean. In a notebook titled

---

[79] Heisenberg (1928). See also Carson (1996a,b).

[80] As pointed out by the editors of Pauli's correspondence (Pauli, 1979, p. 442), Heisenberg had considered this kind of explanation already during his stay in Copenhagen in late 1926, while working on resonance in the helium spectrum (Heisenberg, 1926a), in letters to Pauli dated 4 November 1926 (Pauli, 1979, pp. 352–53) and 15 November 1926 (Pauli, 1979, pp. 354–56).

[81] Oppenheimer (1928); Bloch (1929).    [82] Bloch (1930).

'Undulatory statistics. I. [*Undulatorische Statistik. I.*]', probably from late 1926,[83] under the heading 'Coupling of arbitrary systems [*Koppelung beliebiger Systeme*]', Schrödinger states about the oscillatory states of a coupled system:

> The joint oscillatory states, which immediately arise from the interaction even if the individual systems had been in definite eigenstates before switching on the interaction, are of such a type that they cannot anymore be resolved into states of the individual systems. Even worse: It seems that even if one slowly switches on the interaction and then removes it again, the resulting oscillatory state will be of the above-discussed 'unresolvable' [i.e., not resolvable into states of the individual systems] type.[84]

Entanglement thus is a concept that Schrödinger arrived at—at least in its embryonic form—while applying wave mechanics in a hope to better understand or *interpret* his wavefunctions.

But well beyond such prominent examples, applications harboured many further new concepts that are now part and parcel of quantum mechanics, such as quantummechanical tunnelling, the Bloch wave, spin waves, the Brillouin zone, or the Fermi surface, which since have become common elements of the language of quantum mechanics.[85] Intermediate-level concepts such as these bridge the gap between fundamental laws and empirical observations and reify novel aspects of the theory or provide modelling strategies at the boundary between microscopic and macroscopic descriptions.[86] Add to that techniques of approximation, such as the Born–Oppenheimer approximation, or methods of computation, such as Ryleigh–Schrödinger perturbation theory, Slater determinants, or Pauli spin matrices, which helped establish the generality of quantum mechanics in practice and not just in theory, and it becomes clear that applications were crucial in providing physicists with the means to articulate the new mechanics and thus to bring to life the initially barren universe of quantum mechanics.[87]

---

[83] The notebook mentions Born's probability interpretation as the 'statistical obscenity from Göttingen [*statistische Schweinerei aus Göttingen*]' and implicitly mentions Wentzel (1926), a paper by Gregor Wentzel from November 1926.

[84] Erwin Schrödinger, Notebook 'Undulatorische Statistik. I.,' AHQP 41-1-002, online: https://phaidra.univie.ac.at/detail/o:165508 (German original: '[...] die gemeinsamen Schwingungszustände, die beim Wechselwirken sofort sich einstellen, und zwar selbst dann sich einstellen, wenn die Einzelsysteme vor Beginn der Wechselwirkung nur mit je einer Eigenschwingung angeregt waren, von solcher Art sind, daß sie nicht mehr in Zustände der Einzelsysteme auflösbar sind. Schlimmer: Es scheint, selbst wenn man die Wechselw. langsam anbringt und wieder aufhält[?], der übrigbleibende Schwingungszustand von der oben [eben?] besprochenen "nichtauflösbaren" Art sein wird.'). The (partial) transcription of this notebook is due to Christoph Lehner and myself, and so is the English translation of the quote. Christoph Lehner and Jos Uffink are currently preparing an article on 'Schrödinger and the prehistory of entanglement'. See also: Uffink (2020). Bitbol (1996), especially p. 7, makes a similar point regarding Schrödinger (1927).

[85] See, e.g., Joas and Eckert (2017).    [86] James and Joas (2015).

[87] James and Joas (2015).

Applying quantum mechanics would remain a promising strategy for uncovering interesting new physics for years, if not decades, after the end of the initial heyday of quantum mechanics in 1927. They acted as drivers of conceptual and interpretational innovation and time and again provided new concepts which today often are seen as part and parcel of quantum mechanics. For most of the actors who had participated in the genesis of quantum mechanics, two areas especially were expected to yield progress: the relativistic extension of quantum mechanics, including prominently quantum electrodynamics, and nuclear physics. In his famous magnetic monopole paper, Dirac in 1931 stated:

> There are at present fundamental problems in theoretical physics awaiting solution, e.g., the relativistic formulation of quantum mechanics and the nature of atomic nuclei (to be followed by more difficult ones such as the problem of life), the solution of which problems will presumably require a more drastic revision of our fundamental concepts than any that have gone before.[88]

Besides the relativistic extension of quantum mechanics, nuclear physics was *the* area in which progress was expected that would show a way beyond (nonrelativistic) quantum mechanics. Rudolf Peierls begins his introduction to Volume 9 of the *Niels Bohr Collected Works* as follows: 'Niels Bohr followed the early development of nuclear physics with deep interest. He realised that this was a subject which was bound to touch the limitations of the existing quantum mechanics, or to extend them.'[89] To Bohr, whether quantum mechanics was even capable of describing the atomic nucleus remained a puzzle until the *annus mirabilis* of nuclear physics in 1932, and, along the way, Bohr for a while was even willing to dispense with the principle of conservation of energy in order to explain beta decay.[90]

But areas beyond relativity and the nucleus were seen as promising sources of new concepts as well, such as attempts to explain superconductivity using quantum mechanics, which was another arena in which physicists searched for clues as to how quantum mechanics could potentially be extended.[91] Fritz London's concept of macroscopic quantum states, such as in superconductors, can also be seen as part of the further articulation of quantum mechanics and as an attempt at contributing to its interpretation.[92] Applications played key constructive roles in extending quantum mechanics and providing new concepts in quantum mechanics and quantum field

---

[88] Dirac (1931), p. 60. See also Kragh (1981); Blum (2014). Note that Dirac's mathematical strategy, discussed by him immediately after the quote provided here, in essence is also a strategy of *application* of quantum mechanics, just not an empiricist one. In his paper, Dirac explores the mathematical formalism of quantum mechanics and shows that a specific class of entities—magnetic monopoles—is not precluded by the formalism and thus a new mathematical feature of quantum mechanics that can be interpreted in terms of physical entities.

[89] Peierls (1986), p. 3.

[90] Peierls (1986); Stuewer (1979, 2018).

[91] See, e.g., Pauli to Bohr, 16 January 1929, in Pauli (1979).

[92] See, e.g., Joas and Waysand (2014); Monaldi (2017).

theory, such as second quantization, antiparticles, quasiparticles and collective excitations, effective theories, renormalization, symmetry breaking, exchange forces, quantum criticality, to name just a few.[93] To be sure, not all applications of quantum mechanics to new phenomena were ultimately relevant to the foundations of quantum mechanics or led to conceptual innovation. Yet many more than are usually considered did bring about changes in how physicists think about the content and meaning of quantum mechanics.

## 7.5 THE MEANING OF QUANTUM MECHANICS—AND ITS INTERPRETATION

When discussing the interpretational debates in the early days of quantum mechanics, ca. 1925–1927, philosophers of science, physicists, and occasionally also historians of science, tend to focus on an ever-recurring, narrow corpus of primary sources and disregard a large amount of fascinating material, some of which has never been translated into English, which could at the very least further elucidate some of the main protagonists' positions as well as the evolution of their ideas. Specifically, the early applications of quantum mechanics, especially those going beyond atomic physics, are all-too-easily dismissed as uninteresting and irrelevant to the grander debates on interpretation and thus relegated to subdisciplinary histories. Yet, if interpretation was indeed the central issue for the key actors at the time, why were all those physicists—including Heisenberg, Pauli, and Schrödinger—investing so much time and effort in working out applications of the theory?

In this chapter, I have attempted to chart a course for gaining a better understanding of which roles specific problems can play in the development of a general theory like quantum mechanics, especially at the early stages of its development, during the crucial phase of its canonization into a finished theory. In many ways, quantum mechanics is a unique case, also because of its complicated intertheoretical relationship with both the old quantum theory as a direct precursor theory—which was not replaced wholesale, but reinterpreted by quantum mechanics—and with classical mechanics as another precursor theory. I hope to have shown that the actors who perceived quantum mechanics as an open-ended endeavour at least until the fall of 1927 used applications to better understand the kinds of predictions quantum mechanics generates, to widen its domain of validity and establish its generality. Rather than being derivative and subsequent, applications of quantum mechanics were drivers of conceptual innovation

[93] For examples of such concepts from, respectively, condensed matter physics, quantum chemistry, or nuclear physics, see, e.g., Hoddeson et al. (1992); Gavroglu and Simões (2012); Stuewer (1979); Stuewer (2018). On quasiparticles and collective excitations, see in particular Kojevnikov (1999); Kojevnikov (2002); Blum and Joas (2016).

and thus are of relevance to everyone interested in the conceptual development of quantum physics—philosophers, physicists, and historians alike. Quantum mechanics was being transformed through it being implemented.[94]

The actors also used applications to make sense of the meaning of quantum mechanics, both in the wide sense of the word 'meaning' and in the narrower sense of 'interpretation'. The difference between these two terms, 'meaning' and 'interpretation', which at first might appear synonymous, is subtle, yet in my eyes important. On the one hand, 'meaning' is a term frequently used by the actors (as established in some of the quotes above). It appears to be used in a broader and also more open-ended way than 'interpretation' typically is. 'Meaning' includes questions of inner consistency of the theory, of its practical application, of its extensibility and applicability to new phenomena, and it also covers the search for physically intuitive pictures and specific partially explanatory models which are not necessarily consistent with the full theory and all of its predictions.[95] It is thus a fitting term for making sense of a theory that is under construction and tentatively being applied to new domains.

On the other hand, the way the term 'interpretation of quantum mechanics' is often used today retroactively and anachronistically narrows it to a few key issues—say, the physical interpretation of the wavefunction and its collapse, as well as issues emanating from EPR and other key interpretational debates, such as reality, locality, determinism, or completeness. It boils the question of interpretation of quantum mechanics down to its bare essentials, and thus basically to the question of how to interpret the symbols and elements of the quantum mechanical formalism. This serves to reinforce the very cut between formalism and interpretation which I in this chapter attempted to deconstruct and expose as largely artificial, at least for the early phase of quantum mechanics. 'Interpretational issue' would not usually be invoked as a label for questions like 'Can quantum mechanics explain why there is ortho- and parahelium?' (the question at the outset of Heisenberg, 1926a) or 'Can quantum mechanics explain why electrons in metals have such long mean-free paths?' (one of the questions leading to the formulation of the Bloch wave). This is why I decided to use both terms—'meaning' and 'interpretation'—in this chapter, and why I suggest that they are not synonymous.

The reassessment of some of the work done by physicists in the mid-to-late 1920s and beyond also places a question mark behind the interpretational 'dark age' that allegedly began in 1927 and only ended in the postwar years. This 'dark age' might well turn out to be an artefact of the rather myopic view that many historians have taken because they followed to an unhealthy extent what both the actors and some philosophers of physics considered as truly fundamental and all too easily discarded more mundane subjects and 'mere' applications as well as their potential contributions to the always ongoing further clarification of the meaning of quantum mechanics. In turn, this reassessment could also lead to a reevaluation of subdisciplinary histories of

---

[94] James and Joas (2015). The catchphrase 'transformation through implementation' was coined by Jürgen Renn.
[95] On the last point, see Ramsey (2000), especially p. 561.

physics and of their relevance for histories of the interpretation of quantum mechanics, such as histories of solid-state physics, quantum chemistry, or nuclear physics,[96] including histories of the spread of quantum theory to geographical regions outside of central Europe.[97]

Looking at the history of quantum mechanics in this way might also help bring the discussions of the actors regarding the interpretation of the theory closer to both physics and physicists, unlike some of the sophisticated and technically refined philosophical work on quantum foundations which—unjustly of course—sometimes gets ridiculed by working physicists.[98] And it might equip historians and philosophers of physics with important tools to study what has much more recently been going on in fields such as quantum information, quantum optics, and quantum computing, where 'application' has taken on a whole new meaning—that of practical and technological application—which the story I presented here did not even attempt to take into account.

## ACKNOWLEDGEMENTS

Some of the ideas contained in this chapter have their origin in joint research about the early applications of quantum mechanics conducted with Jeremiah James (James and Joas, 2015). I am also indebted to the members of the *Project on the History and Foundations of Quantum Physics*, which I was a part of from 2007–2012 at the Max Planck Institute for the History of Science and the Fritz Haber Institute of the Max Planck Society in Berlin, and in particular to Christoph Lehner, Jürgen Renn, Michel Janssen, Shaul Katzir, Massimiliano Badino, Alexander S. Blum, Arianna Borrelli, Dieter Hoffmann, Martin Jähnert, Daniela Monaldi, Jaume Navarro, Arne Schirrmacher, and Marta Jordi Taltavull. I would also like to thank Olival Freire Jr. for suggesting to me to write a chapter on this topic, for coming up with the idea for its title, and for his valuable comments; Richard Staley for very helpful remarks; Karin Tybjerg for inviting me to give a talk at Medical Museion at the University of Copenhagen in 2018, where some of the claims made in this chapter were presented for the first time; Alexei Kojevnikov and Anja Skaar Jacobsen for detailed and very helpful comments on a previous version; and Ricardo Karam for his insightful comments on a previous version as well as for our joint teaching of a course on the history of quantum mechanics at the University of Copenhagen in the spring of 2020, which enabled me to dig deeper into many of the questions discussed here.

---

[96] Hoddeson et al. (1992); Gavroglu and Simões (2012), Stuewer (1979, 2018).
[97] See, e.g., Nakane (2019) and Kenji Ito's chapter in this volume.
[98] See, e.g., MacKinnon (2016).

## References

Badino, M. (2015). Three Dogmas on Scientific Theory. Online: https://philpapers.org/rec/BADTDO-7 (last accessed: 25 October 2020).

Badino, M. (2016). What Have the Historians of Quantum Physics Ever Done for Us? *Centaurus*, **58**, 327–46.

Badino, M. (2019). Schooling the Quantum Generations: Textbooks and Quantum Cultures from the 1910s to the 1930s. *Berichte zur Wissenschaftsgeschichte*, **42**(4), 290–306.

Badino, M., and Navarro, J. (eds) (2013). *Research and Pedagogy: A History of Quantum Physics through Its Textbooks*. Berlin: Edition Open Access.

Beller, M. (1990). Born's Probabilistic Interpretation: A Case Study of 'Concepts in Flux'. *Studies in History and Philosophy of Science*, **21**(4), 563–88.

Beller, M. (1996). The Rhetoric of Antirealism and the Copenhagen Spirit. *Philosophy of Science*, **63**(2), 183–204.

Beller, M. (1999). *Quantum Dialogue: The Making of a Revolution*. Chicago: University of Chicago Press.

Bitbol, M. (1996). *Schrödinger's Philosophy of Quantum Mechanics*. Dordrecht: Springer.

Bloch, F. (1929). Über die Quantenmechanik der Elektronen in Kristallgittern. *Zeitschrift für Physik*, **52**(7), 555–600.

Bloch, F. (1930). Zur Theorie des Ferromagnetismus. *Zeitschrift für Physik*, **61**, 206–19.

Bloch, F. (1976). Heisenberg and the Early Days of Quantum Mechanics. *Physics Today*, **29**(12), 23–27.

Blum, A. (2014). From the Necessary to the Possible: The Genesis of the Spin-Statistics Theorem. *The European Physical Journal H*, **39**(5), 543–74.

Blum, A., Lalli, R., and Renn, J. (2015). The Reinvention of General Relativity: A Historiographical Framework for Assessing One Hundred Years of Curved Space-Time. *Isis*, **106**(3), 598–620.

Blum, A. S., and Joas, C. (2016). From Dressed Electrons to Quasiparticles: the Emergence of Emergent Entities in Quantum Field Theory. *Studies in History and Philosophy of Modern Physics*, **53**, 1–8.

Blum, A., Jähnert, M., Lehner, C., and Renn, J. (2017). Translation as Heuristics: Heisenberg's Turn to Matrix Mechanics. *Studies in History and Philosophy of Modern Physics*, **60**, 3–22.

Bohr, N. (1928a). The Quantum Postulate and the Recent Development of Atomic Theory. In *Atti del Congresso Internazionale dei Fisici 11-20 Settembre 1927, Como-Pavia-Roma*, Vol. 2, Bologna: Nicola Zanichelli, pp. 565–88.

Bohr, N. (1928b). The Quantum Postulate and the Recent Development of Atomic Theory. *Nature (Suppl.)*, **121**, 580–90.

Bohr, N. (1977). *Collected Works, Vol. 4: The Periodic System (1920–1923)*, ed. J. Rud Nielsen. Amsterdam: North-Holland.

Bohr, N. (1986). *Collected Works, Vol. 9: Nuclear Physics (1929–1952)*, ed. Sir Rudolf Peierls. Amsterdam: North Holland.

Born, M. (1924). Über Quantenmechanik. *Zeitschrift für Physik*, **26**, 379–395.

Born, M. (1926a). Zur Quantenmechanik der Stoßvorgänge (Vorläufige Mitteilung). *Zeitschrift für Physik*, **37**, 863–67.

Born, M. (1926b). Quantenmechanik der Stoßvorgänge. *Zeitschrift für Physik*, **38**, 803–27.

Born, M. (1928). Über den Sinn der physikalischen Theorien [speech given on 10 November 1928]. *Nachrichten von der Gesellschaft der Wissenschaften zu Göttingen, Geschäftliche Mitteilungen aus dem Berichtsjahr 1928/29*, pp. 51–70.

Born, M. (1969). *Physics in My Generation*. New York: Springer.

Born, M., and Jordan, P. (1925). Zur Quantenmechanik. *Zeitschrift für Physik*, **34**, 858–88.

Born, M., Heisenberg, W., and Jordan, P. (1926). Zur Quantenmechanik. II. *Zeitschrift für Physik*, **35**, 557–615.

de Broglie, L. (1927). La mécanique ondulatoire et la structure atomique de la matière et du rayonnement. *Le Journal de Physique et Le Radium*, **8**(5), 225–41.

Bub, J. (2016). *Bananaworld: Quantum Mechanics for Primates*. Oxford: Oxford University Press.

Büttner, J., Renn, J., and Schemmel, M. (2003). Exploring the Limits of Classical Physics: Planck, Einstein, and the Structure of a Scientific Revolution. *Studies in History and Philosophy of Modern Physics*, **34**(1), 37–59.

Camilleri, K. (2007). Bohr, Heisenberg and the Divergent Views of Complementarity. *Studies in History and Philosophy of Modern Physics*, **38**(3), 514–28.

Camilleri, K. (2009). Constructing the Myth of the Copenhagen Interpretation. *Perspectives on Science*, **17**(1), 26–57.

Carrier, M. (2011). Knowledge, Politics, and Commerce: Science Under the Pressure of Practice. In M. Carrier and A. Nordmann (eds), *Science in the Context of Application*, Dordrecht: Springer, pp. 11–30.

Carson, C. (1996a). The Peculiar Notion of Exchange Forces—I: Origins in Quantum Mechanics, 1926–1928. *Studies in History and Philosophy of Modern Physics*, **27**(1), 23–45.

Carson, C. (1996b). The Peculiar Notion of Exchange Forces—II: From Nuclear Forces to QED, 1929–1950. *Studies in History and Philosophy of Modern Physics*, **27**(2), 99–131.

Celi, A., Sanpera, A., Ahufinger, V., and Lewenstein, M. (2017). Quantum Optics and Frontiers of Physics: the Third Quantum Revolution. *Physica Scripta*, **92**(1), 013003–16.

Cushing, J. T. (1994). *Quantum Mechanics: Historical Contingency and the Copenhagen Hegemony*. Chicago: University of Chicago Press.

Darrigol, O. (1992). *From c-Numbers to q-Numbers: The Classical Analogy in the History of Quantum Theory*. Berkeley, CA: University of California Press.

Darrigol, O. (2008). The Modular Structure of Physical Theories. *Synthese*, **162**(2), 195–23.

De Regt, H. W. (2017). *Understanding Scientific Understanding*. Oxford: Oxford University Press.

Dieks, D. (2017). Niels Bohr and the Formalism of Quantum Mechanics. In J. Faye and H. J. Folse (eds), *Niels Bohr and the Philosophy of Physics: Twenty-First-Century Perspectives*, London: Bloomsbury Academic, pp. 303–34.

Dirac, P. A. M. (1931). Quantised singularities in the electromagnetic field. *Proceedings of the Royal Society, Series A*, **133**(821), 60–72.

Dresden, M. (1987). *H. A. Kramers: Between Tradition and Revolution*. New York: Springer.

Duncan, A., and Janssen, M. (2007a). On the Verge of *Umdeutung* in Minnesota: Van Vleck and the Correspondence Principle. Part One. *Archive for History of Exact Sciences*, **61**(6), 553–624.

Duncan, A., and Janssen, M. (2007b). On the Verge of *Umdeutung* in Minnesota: Van Vleck and the Correspondence Principle. Part Two. *Archive for History of Exact Sciences*, **61**(6), 625–71.

Duncan, A., and Janssen, M. (2014). The Trouble with Orbits: The Stark Effect in the Old and the New Quantum Theory. *Studies in History and Philosophy of Modern Physics*, 48, 68–83.

Eckart, C. (1926). Operator calculus and the solution of the equations of motion of quantum dynamics. *Physical Review*, 28, 711–26.

Elkana, Y. (1970). 'Helmholtz' 'Kraft': An Illustration of Concepts in Flux. *Historical Studies in the Physical Sciences*, 2, 263–98.

Epstein, P. S. (1926). The Stark effect from the point of view of Schroedinger's quantum theory. *Physical Review*, 28(4), 695–710.

Freire Jr., O. (2015). *The Quantum Dissidents: Rebuilding the Foundations of Quantum Mechanics (1950–1990)*. Berlin: Springer.

Forman, P. (1968). The Doublet Riddle and Atomic Physics *circa* 1924. *Isis*, 59(2), 156–74.

Gavroglu, K., and Simões, A. (2012). *Neither Physics nor Chemistry: A History of Quantum Chemistry*. Cambridge, MA: MIT Press.

Hamilton, W. R. (1834). On a general method in dynamics; by which the study of the motions of all free systems of attracting or repelling points is reduced to the search and differentiation of one central relation, or characteristic function. *Philosophical Transactions of the Royal Society*, 124, 247–308.

Hankins, T. L. (1980). *Sir William Rowan Hamilton*. Baltimore: Johns Hopkins University Press.

Heisenberg, W. (1925a). Über eine Anwendung des Korrespondenzprinzips auf die Frage nach der Polarisation des Fluoreszenzlichtes. *Zeitschrift für Physik*, 31, 617–26.

Heisenberg, W. (1925b). Über quantentheoretische Umdeutung kinematischer und mechanischer Beziehungen. *Zeitschrift für Physik*, 33, 879–93.

Heisenberg, W. (1926a). Mehrkörperproblem und Resonanz in der Quantenmechanik. *Zeitschrift für Physik*, 38, 411–26.

Heisenberg, W. (1926b). Über die Spektra von Atomsystemen mit zwei Elektronen. *Zeitschrift für Physik*, 39, 499–18.

Heisenberg, W. (1927a). Mehrkörperprobleme und Resonanz in der Quantenmechanik. II. *Zeitschrift für Physik*, 41, 239–67.

Heisenberg, W. (1927b). Über den anschaulichen Inhalt der quantentheoretischen Kinematik und Mechanik. *Zeitschrift für Physik*, 43, 172–98.

Heisenberg, W. (1928). Zur Theorie des Ferromagnetismus. *Zeitschrift für Physik*, 49, 619–36.

Heisenberg, W. (1960). Erinnerungen an die Zeit der Entwicklung der Quantenmechanik. In M. Fierz and V. F. Weisskopf (eds), *Theoretical Physics in the Twentieth Century*, New York: Interscience, pp. 40–47.

Heisenberg, W., and Jordan, P. (1926). Anwendung der Quantenmechanik auf das Problem der anomalen Zeemaneffekte. *Zeitschrift für Physik*, 37, 263–77.

Heitler, W., and London, F. (1927). Wechselwirkung neutraler Atome und homöopolare Bindung nach der Quantenmechanik. *Zeitschrift für Physik*, 44(6), 455–72.

Hendry, J. (1984). *The Creation of Quantum Mechanics and the Bohr-Pauli Dialogue*. Dordrecht: D. Reidel.

Hoddeson, L., Braun, E., Teichmann, J., and Weart, S. (eds) (1992). *Out of the Crystal Maze: Chapters from the History of Solid- State Physics*. New York: Oxford University Press.

Howard, D. (2004). Who Invented the 'Copenhagen Interpretation'?: A Study in Mythology. *Philosophy of Science*, 71(5), 669–82.

Hund, F. (1967). *Geschichte der Quantentheorie*. Mannheim: BI.

Jähnert, M. (2019). *Practicing the Correspondence Principle in the Old Quantum Theory.* Cham: Springer.

Janas, M., Cuffaro, M. E., and Janssen, M. (forthcoming). *Understanding Quantum Raffles. Quantum Mechanics on an Information-Theoretic Approach: Structure and Interpretation.* Cham: Springer. Preprint version: Janas, M., Cuffaro, M. E., and Janssen, M. *Putting Probabilities First: How Hilbert Space Generates and Constrains Them.* arXiv.org preprint, online: https://arxiv.org/abs/1910.10688v1

James, J., and Joas, C. (2015). Subsequent and Subsidiary? Rethinking the Role of Applications in Establishing Quantum Mechanics. *Historical Studies in the Natural Sciences,* **45**(5), 641–702.

Jammer, M. (1966). *The Conceptual Development of Quantum Mechanics.* New York: McGraw Hill.

Jammer, M. (1974). *The Philosophy of Quantum Mechanics: The Interpretations of Quantum Mechanics in Historical Perspective.* New York: John Wiley and Sons.

Joas, C., and Eckert, M. (2017). Arnold Sommerfeld and Condensed Matter Physics. *Annual Review of Condensed Matter Physics,* **8**(1), 31–49.

Joas, C., and Hartz, T. (2019). Quantum Cultures: Historical Perspectives on the Practices of Quantum Physicists. *Berichte zur Wissenschaftsgeschichte,* **42**(4), 281–89.

Joas, C., and Katzir, S. (2011). Analogy, Extension, and Novelty: Young Schrödinger on Electric Phenomena in Solids. *Studies in History and Philosophy of Modern Physics,* **42**, 43–53.

Joas, C., and Lehner, C. (2009). The Classical Roots of Wave Mechanics: Schrödinger's Transformations of the Optical-Mechanical Analogy. *Studies in History and Philosophy of Modern Physics,* **40**(4), 338–51.

Joas, C., and Waysand, G. (2014). Superconductivity—A Challenge to Modern Physics. In K. Gavroglu (ed.), *History of Artificial Cold, Scientific, Technological and Cultural Issues,* Dordrecht: Springer, pp. 83–92.

Jordan, P. (1927). Die Entwicklung der neuen Quantenmechanik. *Die Naturwissenschaften,* **15**(3), 614–23, 636–49.

Karam, R. (2020). Schrödinger's Original Struggles with a Complex Wave Function. *American Journal of Physics,* **88**(6), 433–38.

Koberinski, A., and Müller, M. P. (2018). Quantum Theory as a Principle Theory: Insights from an Information Theoretic Reconstruction. In M. E. Cuffaro and S. C. Fletcher (eds), *Physical Perspectives on Computation, Computational Perspectives on Physics,* Cambridge: Cambridge University Press, pp. 257–80.

Kojevnikov, A. (1999). Freedom, Collectivism, and Quasiparticles: Social Metaphors in Quantum Physics. *Historical Studies in the Physical and Biological Sciences,* **29**(2), 295–31.

Kojevnikov, A. (2002). David Bohm and Collective Movement. *Historical Studies in the Physical and Biological Sciences,* **33**, 161–92.

Kojevnikov, A. (2011). Philosophical Rhetoric in Early Quantum Mechanics 1925–27: High Principles, Cultural Values and Professional Anxieties. In C. Carson, A. Kojevnikov, and H. Trischler (eds), *Weimar Culture and Quantum Mechanics: Selected Papers by Paul Forman and Contemporary Perspectives on the Forman Thesis,* London and Singapore: Imperial College Press and World Scientific, pp. 319–48.

Kojevnikov, A. (2020). *The Copenhagen Network: The Birth of Quantum Mechanics from a Postdoctoral Perspective.* Cham: Springer.

Kragh, H. (1981). The Concept of the Monopole: A Historical and Analytical Case-Study. *Studies in History and Philosophy of Science*, **12**(2), 141–72.

Kragh, H. (1985). The Fine Structure of Hydrogen and the Gross Structure of the Physics Community, 1916–26. *Historical Studies in the Physical Sciences*, **15**(2), 67–125.

Kuhn, T. S. (1962). *The Structure of Scientific Revolutions*. Chicago: University of Chicago Press.

MacKinnon, E. (1977). Heisenberg, Models, and the Rise of Matrix Mechanics. *Historical Studies in the Physical Sciences*, **8**, 137–88.

MacKinnon, E. (2016). Why Interpret Quantum Physics? *Open Journal of Philosophy*, **6**, 86–102.

Madelung, E. (1927). Quantentheorie in hydrodynamischer Form. *Zeitschrift für Physik*, **40**(3), 322–26.

Midwinter, C., and Janssen, M. (2013). Kuhn Losses Regained: Van Vleck from Spectra to Susceptibilities. In M. Badino and J. Navarro (eds), *Research and Pedagogy: A History of Quantum Physics through Its Textbooks*, Berlin: Edition Open Access, pp. 137–205.

Monaldi, D. (2017). Fritz London and the Scale of Quantum Mechanisms. *Studies in History and Philosophy of Modern Physics*, **60**, 35–45.

Muller, F. A. (1997a). The Equivalence Myth of Quantum Mechanics—Part I. *Studies in History and Philosophy of Modern Physics*, **28**(1), 35–61.

Muller, F. A. (1997b). The Equivalence Myth of Quantum Mechanics—Part II. *Studies in History and Philosophy of Modern Physics*, **28**(2), 219–47.

Nakane, M. (2019). Yoshikatsu Sugiura's Contribution to the Development of Quantum Physics in Japan. *Berichte zur Wissenschaftsgeschichte*, **42**(4), 338–56.

von Neumann, J. (1932). *Mathematische Grundlagen der Quantenmechanik*. Berlin: Springer.

Oppenheimer, J. (1928). On the Quantum Theory of Electronic Impacts. *Physical Review*, **32**(3), 361–76.

Pauli, W. (1925). Über den Zusammenhang des Abschlusses der Elektronengruppen im Atom mit der Komplexstruktur der Spektren. *Zeitschrift für Physik*, **31**, 765–83.

Pauli, W. (1926). Über das Wasserstoffspektrum vom Standpunkt der neuen Quantenmechanik. *Zeitschrift für Physik*, **36**(5), 336–63.

Pauli, W. (1927). Zur Quantenmechanik des magnetischen Elektrons. *Zeitschrift für Physik*, **43**, 601–23.

Pauli, W. (1979). *Scientific Correspondence with Bohr, Einstein, Heisenberg, a. o., Volume I: 1919–1929* (Sources in the History of Mathematics and Physical Sciences 2), ed. A. Herrmann, K. von Meyenn, V. F. Weisskopf, New York: Springer.

Peierls, Sir Rudolf (1986). Introduction. In N. Bohr, *Collected Works, Vol. 9: Nuclear Physics (1929–1952)*, ed. Sir Rudolf Peierls, Amsterdam: North Holland, pp. 3–83.

Perović, S. (2008). Why Were Matrix Mechanics and Wave Mechanics Considered Equivalent? *Studies in History and Philosophy of Modern Physics*, **39**(2), 444–61.

Ramsey, J. (2000). Of Parameters and Principles: Producing Theory in Twentieth Century Physics and Chemistry. *Studies in History and Philosophy of Modern Physics*, **31**(4), 549–67.

Rosenfeld, L. (1957). Misunderstandings About the Foundations of Quantum Theory. In S. Körner (ed.), *Observation and interpretation: A Symposium of Philosophers and Physicists (Colston Papers 9)*, New York: Academic Press, pp. 41–61.

Schrödinger, E. (1926a). Quantisierung als Eigenwertproblem (Erste Mitteilung). *Annalen der Physik*, **79**, 361–76.

Schrödinger, E. (1926b). Quantisierung als Eigenwertproblem (Zweite Mitteilung). *Annalen der Physik*, **79**, 489–527.

Schrödinger, E. (1926c). Über das Verhältnis der Heisenberg–Born–Jordanschen Quantenmechanik zu der meinen. *Annalen der Physik*, **79**, 734–56.

Schrödinger, E. (1926d). Quantisierung als Eigenwertproblem (Dritte Mitteilung: Störungstheorie mit Anwendung auf den Starkeffekt der Balmerlinien). *Annalen der Physik*, **80**, 437–90.

Schrödinger, E. (1926e). Quantisierung als Eigenwertproblem (Vierte Mitteilung). *Annalen der Physik*, **81**, 109–39.

Schrödinger, E. (1926f). Der stetige Übergang von der Mikro- zur Makromechanik. *Die Naturwissenschaften*, **14**(2), 664–66.

Schrödinger, E. (1927). Energieaustausch nach der Wellenmechanik. *Annalen der Physik*, **83**(1), 956–68.

Stuewer, R. H. (ed.) (1979). *Nuclear Physics in Retrospect: Proceedings of a Symposium on the 1930s.* Minneapolis: University of Minnesota Press.

Stuewer, R. H. (2018). *The Age of Innocence: Nuclear Physics between the First and Second World Wars.* Oxford: Oxford University Press.

Tanona, S. (2004). Idealization and Formalism in Bohr's Approach to Quantum Theory. *Philosophy of Science*, **71**(5), 683–95.

Tomonaga, S.-I. (1997). *The Story of Spin.* Chicago: University of Chicago Press.

Uffink, J. (2020). Schrödinger's Reaction to the EPR Paper. In M. Hemmo and O. Shenker (eds), *Quantum, Probability, Logic: The Work and Influence of Itamar Pitowsky* (Jerusalem Studies in the History of Science), Cham: Springer, pp. 545–66.

van der Waerden, B. L. (1967). *Sources of Quantum Mechanics.* Amsterdam: North Holland.

van der Waerden, B. L. (1973). From Matrix Mechanics and Wave Mechanics to Unified Quantum Mechanics. In J. Mehra (ed.), *The Physicist's Conception of Nature*, Dordrecht: Springer, pp. 276–93.

Wentzel, G. (1926). Zur Theorie des photoelektrischen Effekts. *Zeitschrift für Physik*, **40**, 574–89.

CHAPTER 8

.....

# THE STATISTICAL INTERPRETATION: BORN, HEISENBERG, AND VON NEUMANN, 1926–27

.....

GUIDO BACCIAGALUPPI[1]

## 8.1 INTRODUCTION

.....

IN 1954, Max Born was awarded the Nobel prize for physics 'for his fundamental research in quantum mechanics, especially for his statistical interpretation of the wavefunction'.[2] For the Nobel committee this presumably meant what by then was the standard reading of the wavefunction: a state of a quantum system that determines irreducible probabilities for the results of quantum measurements, as arguably codified in the textbooks by Dirac (1930) and von Neumann (1932). As the latter explicitly writes in his section on 'The statistical interpretation' (1932, Section III.2, p. 109):

> This conception of quantum mechanics, which recognizes its statistical statements as the true form of the laws of nature and gives up the principle of causality, is the so-called statistical interpretation. It was formulated by M. BORN. It is the only interpretation of quantum mechanics consistently implementable today [ . . . ]

Almost the same words recur in the conclusion of a little paper published by Einstein in 1953 in the *Festschrift* for Born's retirement from the University of Edinburgh: 'the only interpretation of the Schrödinger equation acceptable so far is the statistical

---

[1] Descartes Centre for the History and Philosophy of the Sciences and the Humanities, Utrecht University (email: g.bacciagaluppi@uu.nl).

[2] From https://www.nobelprize.org/prizes/physics/1954/summary/, consulted 13 August 2020.

interpretation given by Born' (Einstein, 1953, pp. 39–40). What Einstein had in mind, however, was something *completely different*. In this paper Einstein took a macroscopic ball (say 1mm in diameter) bouncing between the walls of a box. Quantum mechanically we can describe it using a stationary wave, namely an equal-weight superposition of a wave travelling from left to right and one travelling from right to left. Einstein noted that this description yields the correct statistics, namely a uniform probability distribution for the position of the particle in the box, and probability ½ each for momentum directed left or right. But it is clearly not a description of an *individual macroscopic ball*.[3] Hence (Einstein, 1953, pp. 39–40):

> [...] the only interpretation of the Schrödinger equation acceptable so far is the statistical interpretation given by Born. This, however, does not yield a real description of the individual system, but only statistical statements for ensembles of systems.

Presumably, Einstein was writing tongue-in-cheek, since Born and he heartily *disagreed* about the issue of the completeness of quantum mechanics. But Einstein's understanding of the term 'statistical interpretation' is far closer to Born's original ideas as presented in 1926 than what von Neumann and presumably the Nobel committee understood. As a matter of fact, the 'statistical interpretation' underwent radical change in a very short time (1926–27)—even though this change seems to have been subsequently largely forgotten (among others by Born himself).[4]

This chapter will sketch this development. I shall distinguish three phases, to be discussed respectively in sections 8.2–8.4:

- June–October 1926: Born's papers on collisions and on the adiabatic theorem (as well as Dirac's 'On the theory of quantum mechanics');
- October 1926–October 1927: Heisenberg's and Jordan's papers on fluctuation phenomena, Dirac's and Jordan's transformation theory, Heisenberg's uncertainty paper, and Born and Heisenberg's joint report at the 1927 Solvay conference;
- October–November 1927: von Neumann's 'Wahrscheinlichkeitstheoretischer Aufbau der Quantenmechanik' ['Probabilistic construction of quantum mechanics'].

As we shall see, it is the first phase that makes intelligible why Einstein could talk of 'the statistical interpretation given by Born' meaning that $\psi$ is an incomplete description of an individual quantum system. The second phase is closer to what we now take to be standard quantum mechanics, in particular with an increased emphasis on the

---

[3] Einstein uses this example also as the basis for his criticism of Bohm's (1952a,b) theory, because although in the latter the individual ball does have a position and a momentum, it is motionless *until* we open the box: this is not the macroscopic behaviour we wish to recover. Cf. Myrvold (2003) and Bacciagaluppi (2016a) for discussions.

[4] See e.g. Beller (1990).

role of measurements. (But there were marked differences in the understanding of the theory among the likes of Dirac, Jordan, and Born and Heisenberg.) By the time von Neumann had finished with it (with Born's blessing), the 'statistical theory' had essentially become what we know from the later textbooks. Von Neumann had also given a complete characterization of the statistical ensembles in quantum mechanics, in particular showing that there are no ensembles that are dispersion-free for all quantum-mechanical quantities.[5] This last result appears to establish that indeterminism in quantum mechanics is irreducible, and contrasts starkly with what Einstein appears to understand under 'the statistical interpretation given by Born'. The final section, 8.5, will accordingly review the relation between the statistical interpretation and indeterminism.[6]

## 8.2 Phase I: Born

The immediate background for Born's introduction of the statistical interpretation was the conflict between matrix mechanics and wave mechanics. As initially developed, the two theories shared some of the same successes, e.g. the calculation of the hydrogen spectrum (Pauli, 1926; Schrödinger, 1926). However, they were far from being 'equivalent'.[7] They also employed two fundamentally different physical pictures, suggesting in fact the potential for subsequent empirical disagreement.[8]

In today's terminology, the wave mechanical understanding of the wavefunction $\psi$ was as an 'ontic state', e.g. as literally representing a smeared-out electron. Schrödinger was able to interpret quantization of energy in terms of the discreteness of eigenoscillations, and hoped to derive other quantum phenomena as arising from his continuous and deterministic wave equation. (For instance, he developed a treatment of radiation phenomena still used today as a semiclassical approach.[9])

The matrix mechanical picture instead assumed that systems were always in stationary states, and randomly performed quantum jumps between them.[10] Matrices

---

[5] That is, for every ensemble there are quantities with a non-trivial distribution of values (non-zero dispersion).

[6] This chapter is a much updated version of a lecture given at the 5th Tübingen Summer School in the History and Philosophy of Science, August 2016. The lecture was itself based on material in Bacciagaluppi and Valentini (2009) and Bacciagaluppi (2008), to which I refer the reader for further details, and on my talk on 'Von Neumann's no-hidden-variables theorem (and Hermann's critique)', given as a joint LogiCIC/LIRa seminar, University of Amsterdam, May 2016. When not otherwise noted, all translations are by myself and all emphases are original.

[7] See the classic discussion by Muller (1997a,b, 1999).

[8] Cf. Bacciagaluppi and Valentini (2009, section 4.6) and Bacciagaluppi and Crull (2021, section 3.3).

[9] Cf. e.g. Bacciagaluppi and Valentini (2009, section 4.4).

[10] This is not in fact stated explicitly but has to be read between the lines. The closest to a smoking gun is presumably Jordan's derivation of the blackbody fluctuation formula at the end of the 'three-man paper' by Born, Heisenberg, and Jordan (1926). There Jordan uses a Boltzmannian approach, and

described collectively the possible states and the possible transitions (including selection rules), but there was no object providing a description of the state of a system. Radiation intensities (transition probabilities) were determined via correspondence arguments.[11]

Even its proponents, however, thought that there were open problems within matrix mechanics. As Born and Heisenberg retrospectively put it in their Solvay report (Bacciagaluppi and Valentini, 2009, p. 383 [translation slightly amended]):

> The most noticeable defect of the original matrix mechanics consists in that at first it appears to give information not about actual phenomena, but rather only about possible states and processes. It allows one to calculate the possible stationary states and processes. It allows one to calculate the possible stationary states of a system; further it states the nature of the harmonic oscillation that can manifest itself as a light wave in a quantum jump. But it says nothing about when a given state is present, or when a change is to be expected. The reason for this is clear: matrix mechanics deals only with closed periodic systems, and in these there are indeed no changes. In order to have true processes, as long as one remains in the domain of matrix mechanics, one must direct one's attention to a *part* of the system; this is no longer closed and enters into interaction with the rest of the system. The question is what matrix mechanics can tell us about this.

As discussed by Wessels (1980), aperiodic phenomena had been on Born's mind ever since he had started working on modifying Bohr's theory of the atom in 1921. In particular, Born and Franck (1925a,b) had written two papers together on the dynamical interaction between two atoms leading either to the atoms forming a molecule or coming apart again in a collision. This was followed by the first of Born's papers with Jordan,[12] which among other things contained a section on collisions in which they suggested investigating asymptotic motions, in particular in the context of the Ramsauer effect (Born and Jordan, 1925a). After the momentous introduction of the new quantum kinematics by Heisenberg (1925), in their following paper Born and Jordan (1925b) announced further work on aperiodic phenomena, and Born noted the difficulties of doing so—involving something like continuous matrices—in the third section of the 'three-man paper' (Born, Heisenberg, and Jordan, 1926). His first concrete contribution to the problem was in fact work with Norbert Wiener on an 'operator formalism' that generalized the matrix formalism. They applied it to the free

---

modifies the Lorentz–Ehrenfest calculation for an element of a vibrating string. Specifically, he uses matrix variables instead of classical variables, formally obtaining wave and particle terms in the fluctuation formula. In order for this to be in fact a fluctuation formula, however, Jordan must believe that each string element (a part of the system) is jumping between states of different energies. For details, see Bacciagaluppi, Crull, and Maroney (2017). Cf. also Duncan and Janssen (2008).

[11] It has recently been shown that the problem of intensities had been a driving concern in the development of matrix mechanics; see the wonderful paper by Blum *et al.* (2017).

[12] Jordan was originally one of Franck's assistants, with whom he co-authored the article on scattering for the *Handbuch der Physik* (Franck and Jordan, 1926).

particle, obtaining in a rather contrived way something that looked like classical inertial motion (Born and Wiener, 1926a,b).[13]

Born then turned to use *wave mechanics* to treat collisions in two fundamental papers (Born, 1926a,b—the first of which was a 'preliminary communication'). The scenario treated by Born is the collision between an electron initially described by a plane wave (i.e. a stationary state) and an atom initially also in a stationary state. Born determined the asymptotic behaviour of the combined system solving the time-independent Schrödinger equation by perturbative methods. For the case of an inelastic collision, the result is what one now calls an *entangled* wavefunction that is the superposition of a series of stationary states of the atom and a continuum of outgoing plane waves. Born thought that these components must clearly be the *possible* final states after the collision, and he identified the (mod-squared) expansion coefficients as the corresponding *probabilities*.[14] Since the initial state is a stationary state of the atom and a 'uniform rectilinear motion [*sic*]' of the electron, these probabilities are also equal to the transition probabilities, i.e. the probabilities for quantum jumps from the initial state to one of the possible final states.[15] As also first emphasized by Wessels (1980),[16] in these papers Born's 'statistical interpretation' concerns exclusively the distribution of the stationary states of the atom and of the free electron.

Importantly, the corresponding probabilities are not probabilities for 'finding' a system in a certain stationary state upon measurement: the atom and the electron (at least when the interaction is completed) are assumed to *be* in a stationary state. Thus, in line with Einstein's understanding, the Schrödinger wavefunction is *not* a complete description of a physical system, because in general it does not specify the actual state of a system. Note that nowhere in these papers does Born use the word 'state' to refer to anything but a stationary state. On the other hand, a description in terms of the actual stationary states need not be *causally* complete. Indeed, Born himself considers it plausible that the transition probabilities are an expression of genuine indeterminism, and he uses the picture of the wavefunction as a deterministically evolving 'guiding field'. But, as Beller (1990) emphasizes, at this stage Born is open-minded about determinism being still possible in principle.

Born's proposed interpretation appears fully explicitly in a later paper from the same year, 'Das Adiabatenprinzip in der Quantenmechanik' ['The adiabatic principle in

---

[13] See also Bacciagaluppi and Valentini (2009, section 3.4) and Bacciagaluppi (2008, section 2.2).

[14] The correct expression appears in the preliminary communication only in a footnote added in proof. Presumably independently, Dirac (1926, section 5) also identified squared amplitudes in the energy basis expansion (the sum of which he noted was conserved) as the frequencies of the various stationary states in an ensemble. The problem treated by Dirac was that of an atom subject to an external perturbation, in particular an external electromagnetic field, and he rederived the Einstein coefficients in a special case (cf. Darrigol 1992, p. 333).

[15] For details, see e.g. Wessels (1980), Beller (1990), and Bacciagaluppi and Valentini (2009, section 3.4).

[16] About earlier work she notes: 'Hund (1967, pp. 56–157) and Konno (1978, pp. 141–44) provide the only historical accounts of Born's work in which Born's original statistical interpretation is stated clearly enough to distinguish it from the interpretation now commonly attributed to Born' (Wessels, 1980, p. 197).

quantum mechanics'] (Born, 1926c), in which he investigates and discusses his approach further (much more so than in the better-known collision papers). It turns out to be quite surprising:[17]

- Particles exist (at least when evolving freely).
- They are accompanied by de Broglie–Schrödinger waves.
- During free evolution, systems are always in stationary states.
- If at time $t = 0$ the wavefunction of a system is

$$\psi(x, 0) = \sum_m c_m \psi_m(x), \tag{1}$$

where the $\psi_m(x)$ are eigenfunctions of energy, then the expressions $|c_m|^2$ are the probabilities for the occurrence of the corresponding stationary states. Born calls them 'state probabilities' (he again reserves the term 'state' for stationary states).

- Consider the application of an external force (or an interaction): no 'anschaulich' representation of what takes place may be possible, but 'quantum jumps' occur, in the sense that after the external intervention ceases, the system will generally be in a different stationary state.
- The evolution of the state probabilities is nevertheless well-defined and is determined for arbitrary times $t = T$ by the solution of the Schrödinger equation,

$$\psi(x, T) = \sum_n C_n \psi_n(x). \tag{2}$$

- For the case that $\psi(x, 0) = \psi_m(x)$ for some $m$, the solution can be determined explicitly (in terms of the external potential). Writing for this case

$$\psi(x, T) = \sum_n b_{mn} \psi_n(x), \tag{3}$$

the expressions $|b_{mn}|^2$ are then the transition probabilities for the quantum jump from $\psi_m(x)$ at $t = 0$ to $\psi_n(x)$ at $t = T$. (In the 'adiabatic limit' they tend to $\delta_{mn}$, thus arguably recovering classical behaviour.)

---

[17] For more details see Bacciagaluppi (2008, section 2.1). This paper is not discussed explicitly by Wessels (1980), but it is covered in Beller (1990). Despite Beller's excellent discussion of the wave and particle pictures in Born's work, I find it difficult to assess in what precise sense Born thinks of particles: it is quite possible that at least for the case of free particles (or even stationary orbits?) he envisages a picture of well-defined geometric trajectories, to be given up during collisions. Beller also points out how initially Born appears to have been enthusiastic about Schrödinger's idea of describing quantum jumps as a continuous process. I would like to emphasize, however, that already in the first collision paper the final (entangled) wavefunction is strictly incompatible with it. Born's views on quantum mechanics from this period are also presented informally in two near-identical papers (Born, 1927a,b), which were expanded versions of a talk given by Born in August 1926.

- In the general case $\psi(x, 0) = \sum_m c_m \psi_m(x)$, the state probabilities at time $t = T$ have the form

$$|C_n|^2 = |\sum_m c_m b_{mn}|^2 . \tag{4}$$

What Born says about this general case is intriguing (Born, 1926c, p. 174):

> The quantum jumps between two states labelled by $m$ and $n$ thus do *not* occur as independent events; for in that case the above expansion should be simply $\sum_m |c_m|^2 |b_{mn}|^2$.

This suggests an inaccurate understanding of interference as being due to *different electrons* not jumping independently (which seems only confirmed by Born's discussion of 'natural light' later in the paper).[18] Understanding interference was to play an important role in the next phase of development of the statistical interpretation.

# 8.3 PHASE II: FROM BORN TO BORN & HEISENBERG

Born's collision papers managed to make *both* Heisenberg and Schrödinger furious. Heisenberg, because Born had successfully solved an open problem in matrix mechanics using wave mechanics; Schrödinger, because at the same time Born had reinterpreted wave mechanics 'in Heisenberg's sense'.[19]

Soon, however, Heisenberg (1927a) (following an idea suggested by Pauli in a letter of 19 October 1926) as well as Jordan (1927a) translated Born's results into the

---

[18] In footnotes on pp. 174 and 180, Born references Dirac (1926) as also pointing out that the quantum jumps are not independent (and become so if one averages over the phases), and generally noting that their points of view are in agreement. Dirac writes explicitly: 'One cannot take spontaneous emission into account without a more elaborate theory involving the positions of the various atoms *and the interference of their individual emissions*' (Dirac, 1926, p. 677, emphasis added). Further evidence for an early understanding of interference as a many-particle effect can be taken from comments by Einstein at the 1927 Solvay conference, where he distinguishes between two conceptions of the wavefunction (both of which he finds problematic): one as pertaining to single electrons and one in which '[t]he de Broglie–Schrödinger waves do not correspond to a single electron, but to a cloud of electrons extended in space' (Bacciagaluppi and Valentini, 2009, p. 441).

[19] In Born's own words: 'the fundamental ideas of the matrix form of the theory initiated by Heisenberg [ ... ] have grown directly out of the natural description of atomic processes in terms of "quantum jumps" and emphasize the classically-geometrically incomprehensible nature of these phenomena. It is settled that both forms of the theory arrive at the same results for stationary states; the question is only how one should treat non-stationary processes. Here, Schrödinger's formalism turns out to be substantially more convenient, provided one interprets it in Heisenberg's sense. I therefore wish to advocate a merging of both points of view, in which each fulfils a very particular role' (Born, 1926c, p. 168).

language of matrix mechanics—precisely by looking at what matrix mechanics could say about 'parts of systems'. Specifically, Heisenberg and Jordan considered two weakly coupled atoms with energy differences in common (all or one, respectively). Because of the coupling, the energy matrices of the two atoms are no longer diagonal in the energy basis of the composite system, and thus are time-dependent, exhibiting a slow 'exchange of energy'. The main aim of the papers is to demonstrate how this result is in fact equivalent to the picture of matching quantum jumps in the two atoms. Born had calculated the transition probabilities for the quantum jumps in terms of the coefficients of the final wavefunction, and Heisenberg and Jordan now calculated them in terms of the elements of the transformation matrix between the two energy bases. This arguably provided a rigorous calculation of the transition probabilities by purely matrix mechanical means. In the same letter of 19 October Pauli also suggested that the statistical interpretation be applied to other quantities, specifically that the modulus squared of both $\psi(\mathbf{x})$ and its Fourier transform $\varphi(\mathbf{p})$ should be interpreted as probability densities for position and momentum.[20]

The next step was the introduction of transition probabilities also between quantities other than energy. The formal tools for this were developed, again independently of each other, mainly by Dirac (1927) and by Jordan (1927b,c) with the so-called transformation theory. In Dirac's treatment, one considers arbitrary conjugate quantities $\xi$ and $\eta$ and some arbitrary quantity $g(\xi,\eta)$. If one assumes an ensemble in which $\xi$ has a definite value and $\eta$ is uniformly distributed, then Dirac's results yield the frequency of the values of $g$ in the ensemble. The main tool that Dirac developed was the theory of the transformation matrices between the energy representation of $g$ and its $\xi$-representation. Heisenberg's paper on fluctuation phenomena (Heisenberg, 1927a) was explicitly cited as a special example, as was Born's work on collisions. Jordan (1927b, pp. 810–11) credits Pauli with the suggestion that there should be well-defined transition amplitudes between any two physical quantities. His theory is developed along these lines, including an axiomatization of quantum mechanics using probability amplitudes as the basic notion.[21]

---

[20] Cf. Pauli to Heisenberg, 19 October 1926 (Pauli, 1979, pp. 340–49). Heisenberg's paper was received shortly afterwards on 6 November 1926. For further details, see Bacciagaluppi and Valentini (2009, section 3.4.4). Pauli's suggestion of $|\psi(\mathbf{x})|^2$ as position density also appeared in a footnote in a paper on gas theory and paramagnetism (Pauli, 1927, p. 83), but of the further correspondence between Heisenberg and Pauli in this period only Heisenberg's letters have survived. Note that Heisenberg (1927a) talks about the two atoms exchanging a 'sound quantum', which suggests a connection with Jordan's treatment of the vibrating string in Born, Heisenberg, and Jordan (1926) and the related discussion of the analogy between the radiation field and excitations of a lattice (see Bacciagaluppi, Crull, and Maroney, 2017).

[21] For Dirac's transformation theory, see Darrigol (1992, pp. 337–45). For a detailed treatment of Jordan, including the involvement of Pauli, see Duncan and Janssen (2009). London (1926), too, made contributions to transformation theory, as also acknowledged by Jordan. On these see Lacki (2004) and again Duncan and Janssen (2009).

Dirac's paper was actually called 'The physical interpretation of the quantum dynamics', and while this was not quite spelled out, Dirac (1927, p. 641) concluded with the suggestion of:

[ ... ] a point of view for regarding quantum phenomena rather different from the usual ones. One can suppose that the initial state of a system determines definitely the state of the system at any subsequent time. If, however, one describes the state of the system at an arbitrary time by giving numerical values to the co-ordinates and momenta, then one cannot actually set up a one-to-one correspondence between the values of these co-ordinates and momenta initially and their values at a subsequent time. All the same one can obtain a good deal of information (of the nature of averages) about the values at the subsequent time considered as functions of the initial values. The notion of probabilities does not enter into the ultimate description of mechanical processes: only when one is given some information that involves a probability (e.g., that all points in $\eta$-space are equally probable for representing the system) can one deduce results that involve probabilities.

Born was delighted with Jordan's work, writing to Wentzel on 13 December 1926:[22]

Mr Jordan has now carried out a huge generalization of the idea that the amplitude of the Schrödinger function is related to the state probabilities. Pauli had considered such a generalization, but Jordan had the idea for the mathematical trick it requires. It now seems to turn out that using this conception quantum mechanics can manage with the usual space and usual time, but that a general probabilistic kinematics takes the place of the causal laws.

Similarly, Heisenberg was hugely impressed with Dirac's results, but still wondered about a satisfactory interpretation.[23] And this is when Heisenberg introduced uncertainty.

The uncertainty paper (Heisenberg, 1927b) is often quite correctly read in the context of the debate with Schrödinger about *Anschaulichkeit* (about the applicability of spatio-temporal pictures in quantum mechanics). But it is equally relevant to the development of the statistical interpretation. The idea that physical quantities are well-defined under the appropriate measurement conditions goes along with the idea that quantum probabilities refer to finding certain values *upon measurement*. In this context, Heisenberg disagrees with Jordan about the need to revise probabilistic notions to accommodate interference: no such revision is required because the different probabilities (in Born's notation above, $|\sum_m c_m b_{mn}|^2$ and $\sum_m |c_m|^2 |b_{mn}|^2$) refer to different measurements (Heisenberg, 1927b, pp. 183–84). And what the statistical interpretation allows us to do is 'conclude through certain statistical rules from one experiment to the possible results of another experiment' (p. 184). Now probabilities

---

[22] AHQP-66, section 3–044, as mentioned by Wessels (1980, p. 197).
[23] See Heisenberg to Pauli, 23 November 1926 (Pauli, 1979, pp. 357–60).

pertain to arbitrary quantum mechanical quantities and stationary states are no longer privileged (pp. 190–91). Finally—adding a twist to Dirac's suggestion about how probabilities enter quantum mechanics—Heisenberg takes uncertainty to be fundamental: the law of causality becomes inapplicable because it is impossible in principle to know the present in all its determining data (p. 197). According to Beller this is 'the beginning of a real commitment to indeterministic philosophy, as opposed to physicists' earlier tentative employment of statistical considerations' (Beller, 1990, p. 583).[24]

These various strands were then brought together at the fifth Solvay conference of October 1927 in a joint report by Born and Heisenberg, where they set out what for them was the definitive view of the statistical interpretation.[25] Born and Heisenberg largely follow Born's discussion in the adiabatic paper, but give a significantly different treatment of the 'theorem of the interference of probabilities', emphasizing the use of 'usual probabilities' along the lines of Heisenberg's uncertainty paper.[26] Much like Heisenberg's take on the law of causality in the uncertainty paper, the law of total probability is not refuted but its antecedent is not satisfied.[27]

On this reading interference applies to individual systems just as well as to ensembles. Furthermore, compared to Born's case of collisions, an electron is no longer assumed to always be in a stationary state, but only when its energy has in fact been measured. Indeed, the analysis is generalized to any observable that has been measured. In this sense, what Born and Heisenberg present as the statistical interpretation is significantly closer to what we now know as standard quantum mechanics.[28]

[24] Heisenberg's fuller argument for excluding a deeper causal picture is briefly hinted at on pp. 188–189, and was a version of the argument later made famous by Feynman using the two-slit experiment. In that familiar version, interference shows that a particle cannot have a trajectory through one slit, because it would have to know whether the other slit is open or closed. More generally for Heisenberg any 'hidden variables' would suppress interference effects. For details see e.g. Bacciagaluppi and Crull (2009, esp. section 3.2) and Bacciagaluppi and Crull (2021, Chapters 4 and 14). On the two-slit experiment, see also the remarks by Weizsäcker in his letter to Grete Hermann of 17 December 1933 (Herrmann, 2019, pp. 439–44).
[25] The report and ensuing discussion are translated in Bacciagaluppi and Valentini (2009, pp. 372–01 and 402–405). For commentary, see Bacciagaluppi and Valentini (2009, sections 3.4.6 and 6.1.2).
[26] I now see this reading of what Born and Heisenberg mean with 'usual probabilities' as more natural than the one I suggested in Bacciagaluppi (2008, p. 274). The emphasis on usual probabilities is also an implicit criticism of Jordan's choice of probability amplitudes as the basic notion for his axiomatization of quantum mechanics. Explicitly, on p. 392 of their report Born and Heisenberg note that amplitudes are not directly observable, and point out that von Neumann's recent work (Hilbert, von Neumann, and Nordheim 1928; von Neumann, 1927a) suffers from no such drawbacks (nor from problems with Dirac's δ-functions).
[27] Again, if quantities had hidden values, they would perform the transitions and interference would be destroyed. Such ideas also seem related to Heisenberg's work on the S-matrix of the 1940s, to his ideas on potentialities of the 1950s, and to other scattered remarks even later.
[28] Quirkily enough, some kind of intermediate position resurfaces in Born's Waynflete lectures of 1948. There Born states that systems are indeed in stationary states and perform quantum jumps between them, but *which* systems this applies to depends on the subjective choice of how we distinguish between a system and its environment (Born, 1949, pp. 99–101).

But close is no cigar. The physical picture in Born and Heisenberg's report seems to be that what is 'out there' are on the one hand values of measured quantities, on the other transition probabilities; several aspects in the report suggest that they still do not take $\psi$ as the 'state' of the system. Indeed, in Born and Heisenberg's eyes, the work by Heisenberg, Jordan, and Dirac shows that matrix mechanics can stand on its own feet. Schrödinger waves may be a useful tool for calculating transition probabilities, but the latter can be defined directly in terms of transformation matrices. Note that, even though the interpretational section of the report was drafted by Born, these aspects suggest Heisenberg's hand.[29]

While the report also includes the idea of uncertainty as grounding the statistical aspects of the theory, it is remarkably silent on the 'reduction of the wave packet', which had also been introduced in the uncertainty paper by Heisenberg (1927b, p. 186). Already in that paper Heisenberg (1927b, pp. 183–84) had suggested that measuring the energy in an atomic beam passing through successive Stern–Gerlach magnets would randomize the phases in a superposition of energy states. This seems to suggest an unorthodox form of the collapse postulate, in which superpositions are maintained but with randomized phases. Such a form of the collapse is discussed explicitly in Heisenberg's Chicago lectures (Heisenberg, 1930, pp. 60–61), and would seem to be inconsistent with the repeatability of measurements; but it is inconsistent with the repeatability of measurements *only if* we assume that $\psi$ is the 'state' of a quantum system and determines probabilities for measurement results. It is unobjectionable if we assume that what determines the probabilities of the results are actual values and transition probabilities, and that $\psi$ is just a book-keeping device that can be changed if and when convenient.[30]

Additional support for this reading may perhaps be found in Born's remarks on the reduction of the wave packet in the General Discussion at the Solvay conference (Bacciagaluppi and Valentini, 2009, pp. 437–39), where, prompted by Einstein, Born gave a simplified treatment of α-particle tracks in a cloud chamber. Born cites remarks by Pauli to the effect that reduction can be dispensed with if one includes a description also of the apparatus (i.e. the cloud chamber), but does not go into details, except perhaps with his remark towards the end to the effect that:

[t]o the 'reduction' of the wave packet corresponds the choice of one of the two directions of propagation $+x_0, -x_0$, which one must take as soon as it is established that one of the two points 1 and 2 is hit, that is to say, that the trajectory of the packet has received a kink.

[29] Dirac's position was very different. As emphasized by Darrigol (1992, p. 344), Dirac (1927) does not yet contain any expression for quantum states, but we have seen that in his conclusion Dirac does refer to the idea of a deterministically evolving 'state'. Possibly, by describing his point of view as 'rather different from the usual ones', he wishes to distance himself from both Heisenberg and Schrödinger. See also Dirac's unusually extensive remarks on the interpretation of the theory at the fifth Solvay conference (Bacciagaluppi and Valentini, 2009, pp. 446–48).

[30] For further details see Bacciagaluppi and Valentini (2009, sections 3.2, 6.2, 6.3, and 11.3), and Bacciagaluppi (2008, section 3).

If this is read as an alternative to the description in terms of reduction, Born is saying that *instead* of collapsing the wavefunction, one can keep the original wavefunction but take a definite value for the direction of propagation of the α-particle. In his own remarks Heisenberg also objected to Dirac's suggestion that collapse was a natural process (Bacciagaluppi and Valentini, 2009, pp. 449–50).

# 8.4 PHASE III: VON NEUMANN

In the very same General Discussion at the Solvay conference Born announced a new paper by von Neumann (Bacciagaluppi and Valentini, 2009, p. 448):[31]

> I should like to point out, with regard to the considerations of Mr Dirac, that they seem closely related to the ideas expressed in a paper by my collaborator J. von Neumann, which will appear shortly. The author of this paper shows that quantum mechanics can be built up using the ordinary probability calculus, starting from a small number of formal hypotheses; the probability amplitudes and the law of their composition do not really play a role there.

This paper is clearly 'Wahrscheinlichkeitstheoretischer Aufbau der Quantenmechanik' (von Neumann, 1927b), which was presented by Born himself in the session of 11 November 1927 of the Göttingen Academy of Sciences. It had been preceded by two other papers, one by Hilbert, von Neumann, and Nordheim (1928), which still built on Jordan's approach, and one by von Neumann alone, in which he developed the Hilbert-space formalism of quantum mechanics (von Neumann, 1927a).[32] Then, in the November paper, von Neumann went on to treat explicitly the statistical aspects of quantum mechanics. The paper is of fundamental importance in the development of the statistical interpretation and less well-known than it ought to be, so it is useful to summarize it in some detail. But the main points can be stated briefly as follows.

Von Neumann is concerned with 'statistical theories', i.e. theories about ensembles of systems. In such theories, measurements of physical quantities generally lead to non-trivial distributions of values for any given ensemble, and a theory makes predictions about values and their distributions. What characterizes a *non*-statistical theory are the algebraic relations between its physical quantities; and the crucial difference between classical mechanics and quantum mechanics as non-statistical theories is that the former has a commutative algebra of observables and the latter a non-commutative one. Von Neumann shows that the statistical aspects of quantum mechanics need not be postulated ad hoc: quantum mechanics as a statistical theory can be completely and rigorously derived purely by imposing 'usual' probabilistic axioms on expectation

---

[31] The final remark again appears to favourably contrast von Neumann's approach to Jordan's.
[32] Cf. also Lacki (2000), Duncan and Janssen (2012), and Mitsch (forthcoming).

values for ensembles of systems whose physical quantities are characterized by the non-commutative algebra of matrix mechanics. These probabilistic axioms are positivity and linearity (on the latter we shall have much more to say). The resulting theory indeed completely determines the possible values and value distributions for the physical quantities. Its object are ensembles, characterized by (pure or mixed) states, which can be specified by results of measurements. Thus von Neumann's version of the statistical interpretation now includes both the notion of a quantum state, and that of the collapse of the state upon measurement. This is essentially modern quantum mechanics. Furthermore, while in classical mechanics as a statistical theory the pure states (those that can no longer be decomposed into other states) are such that all physical quantities have 'sharp' distributions (characterized by a single value), for each pure state in quantum mechanics there are quantities that do not have a sharp distribution. Thus the incompatibility of physical quantities leads inexorably to the indeterministic character of quantum mechanics. Readers interested specifically in the relation between statistical interpretation and indeterminism may skip directly to the final section of this chapter. The rest of this section spells out the details of von Neumann's paper (followed by a few brief comments).

In Section I, von Neumann distinguishes between the 'wave theory' of quantum mechanics and the 'statistical theory' initiated by Born, Pauli, and London, by which he means the transformation theory of Jordan and Dirac. Von Neumann's topic is the statistical theory, which according to him typically answers the following questions:

  (i)   What values can a physical quantity have?
  (ii)  What are the a priori probabilities for these values?
  (iii) What are the probabilities given results of other measurements?

One can consider a statistical theory also in a classical context, in which, however, non-trivial probabilities can be in principle eliminated by performing sufficiently many measurements. This is not so in quantum mechanics, where pairs of measurements are generally mutually incompatible, as emphasized by Dirac (1926) and by Heisenberg (1927b). On the other hand, according to von Neumann, the usual presentation of the statistical quantum mechanics 'rather dogmatically' *postulates* certain quantities to be probabilities (in wave mechanical language the weights in decompositions of the wavefunction), and one then checks the empirical correctness of the predictions. Furthermore, the relation between the theory as usually presented and the theory of probability remains at best unclear. So for instance Jordan (1927b) proposes a theory based on probability amplitudes, but Heisenberg (1927b) dissents. Von Neumann's aim in this paper is to provide a systematic *derivation* of statistical quantum mechanics from the usual theory of probability with a few supplementary assumptions.

Section II describes the basic assumptions of the derivation. Von Neumann considers a physical system $\mathfrak{S}$ and quantities $a, b, \ldots$ on the system. A function $f(a)$ of a quantity $a$ is always well-defined, namely as the quantity that for each value $x$ of $a$ takes the value $f(x)$. A statistical theory concerns *ensembles* of such systems, which are interpreted epistemically (or at least, von Neumann says they represent 'knowledge' about a system—with inverted commas in the original). Measurements of physical quantities lead to a distribution of results for each given ensemble. There should be a maximally disordered ensemble in which all 'states' of the system are equally probable (no 'knowledge' about the system—yielding a priori distributions for measurement results) and ensembles arising from it through measurements (yielding distributions conditional on certain previous measurement results). Each ensemble is in fact characterized by the expectation values for all quantities. Von Neumann wishes to describe quantum mechanics as such a statistical theory, keeping as close as possible to the ordinary notions of probability and expectation value.[33]

One needs to make a crucial distinction between quantities that can or cannot be measured simultaneously (von Neumann remarks that if two quantities are compatible they are both functions of a third quantity[34]). A function $f(a,b)$ of two compatible quantities is always well-defined, while in general it is meaningless to talk of functions of incompatible quantities. On the other hand, it is always possible to define the *sum* of two incompatible quantities if one assumes expectation values to be additive: $a + b$ is the quantity with

$$E(a + b) = E(a) + E(b) \tag{5}$$

for all ensembles.[35] The definition can be trivially extended to arbitrary linear combinations of two quantities or even countably many quantities when the corresponding series converges. Therefore, even in a case like that of quantum mechanics, expectation functionals $E(a)$ can be required to satisfy:

A.  For any convergent real linear combination $\alpha a + \beta b + \gamma c + ...$,

$$E(\alpha a + \beta b + \gamma c + ...) = \alpha E(a) + \beta E(b) + \gamma E(c) + ...$$

Furthermore, it is clear that if a quantity $a$ takes only positive values, its expectation value in any ensemble will also be positive:

B.  For $a \geq 0$, $E(a) \geq 0$.

These are both features of expectation values in ordinary probability theory: linearity (extended by continuity to countable linear combinations) and positivity. However,

---

[33] Two further assumptions—not needed to characterize ensembles and states—are introduced later and make precise what von Neumann understands by 'measurement': repeatability in Section VI and non-contextuality in Section VII.

[34] Strangely, he also states that this is a feature specific to quantum mechanics (cf. his footnote 8 on p. 248).

[35] Note that von Neumann is thereby also taking physical quantities to be characterized by their expectation values in all ensembles.

von Neumann explicitly leaves out normalization (remarking that it is more important to have well-defined 'relative probabilities').[36]

After these 'conceptual' assumptions, von Neumann then introduces a 'formal' assumption: namely that physical quantities in quantum mechanics correspond bijectively to diagonalizable ('normal') linear operators on a separable Hilbert space.[37] For von Neumann's purposes the relevant aspects of this correspondence are the following:

C. Let $S, T, \ldots, \alpha S + \beta T + \ldots$ be diagonalizable. If $S, T, \ldots$ correspond to $\mathfrak{a}, \mathfrak{b} \ldots$ then $\alpha S + \beta T + \ldots$ corresponds to $\alpha \mathfrak{a} + \beta \mathfrak{b} + \ldots$

D. Let $S$ be diagonalizable, and $f(x)$ be a real function. If $S$ corresponds to $\mathfrak{a}$, then $f(S)$ corresponds to $f(\mathfrak{a})$.[38]

That is, the correspondence is such that the functional relations between the physical quantities are represented by the functional relations between the operators. In particular, the *sum* of two operators corresponds to the sum defined via (5). This, von Neumann notes, conforms to actual practice, but he is fully aware that it is a very radical feature of quantum mechanics. In his footnote 9 on p. 249 he takes the example of the mutually incompatible quantities $\mathbf{Q}^2$, $\mathbf{P}^2$, and $\mathbf{Q}^2 + \mathbf{P}^2$: even though the first two can take continuous values and $\mathbf{Q}^2 + \mathbf{P}^2$ takes only discrete values, the quantum-mechanical expectation values *are* in fact linear.

Note that C. and D. are part and parcel of matrix mechanics (in Hilbert-space formulation) seen as a *non-statistical* theory. They simply express the algebraic structure of the physical quantities of the theory. A. and B. instead are assumptions of the 'usual probability theory'. On the basis of these assumptions, von Neumann will answer questions (i), (ii), and (iii): what values physical quantities take in quantum mechanics, what the maximally disordered state is, and which other states arise from it through which measurements—in particular recovering the usual probabilistic interpretation of the wavefunction. The rest of the paper carries this out.

Sections III and IV are headed 'General form of the expectation values. States'. (They are the ones containing his characterization of quantum mechanical ensembles and what later became known as von Neumann's 'impossibility theorem' for hidden variables.) In Section III von Neumann shows from A.–D. that expectation functionals $E(S)$ on the diagonalizable operators $S$ are of the form

$$\mathrm{Tr}(SU) \qquad (6)$$

for some positive linear symmetric operator $U$, and that any such operator defines an expectation functional—a central result that is remarkably easy to prove. In Section IV

---

[36] Note that von Neumann was writing before the now standard axiomatization of probability theory by Kolmogorov.

[37] The spectral theorem for self-adjoint operators was proved in full only in von Neumann (1929), but von Neumann (1927a, Sections IX and X) already contained partial results.

[38] Such a function is to be understood along the lines of von Neumann (1927a, Section XIV).

he then defines pure ensembles as ones for which the operator $U$ has no non-trivial decomposition $\eta U^* + \vartheta U^{**}$ (with $\eta, \vartheta > 0$), and shows that (up to an overall positive factor) they are given by the one-dimensional projection operators $P_\varphi$ with $\varphi$ a normalized vector in the Hilbert space. Any two $\varphi$ related by a phase factor define the same pure ensemble; the corresponding expectation values are normalized; the operator $S$ corresponding to a quantity $\mathfrak{a}$ is simply the operator that yields $E(\mathfrak{a})$ for all pure ensembles via $(\varphi, S\varphi)$ (in von Neumann's notation: $Q(\varphi, S\varphi)$); and $E(\mathfrak{a})$ is dispersion-free if and only if $\varphi$ is an eigenvector of $S$, in which case the value of $S$ is the corresponding eigenvalue. It follows that (unlike in classical mechanics) for each pure ensemble there are physical quantities with non-zero dispersion. Note that von Neumann equates 'pure' ensembles with 'homogeneous' ones ('i.e. those in which all systems [...] are in the same state', p. 255). In this sense, while ensembles in general are arguably interpreted epistemically, pure ensembles are ones in which all systems are in the same ontic state ('Thereby we shall have found all states in which the system $\mathfrak{S}$ can be', p. 255). Von Neumann has thus recovered the *statistical interpretation of the wavefunction*, as well as introduced the notion of a *quantum state* and the *eigenstate–eigenvalue link* that defines which values observables have in which states and thus answering question (i) above.

Sections V and VI are titled 'Measurements and states'. Although von Neumann has abstractly characterized ensembles and states in Section IV, operationally they are defined in terms of results of measurements, so he needs to establish how they are related to measurements. In Section V he shows that quantities $\mathfrak{a}_1, ..., \mathfrak{a}_m$ are jointly measurable if and only if they correspond to operators $S_1, ..., S_m$, all elements of whose resolutions of the identity commute.[39] If the measurements only fix the value of some quantity $\mathfrak{a}_j$ to a certain interval $I_j$, then only the corresponding projection operator $E_j(I_j)$ in the resolution of the identity of $S_j$ needs to commute with the elements of the resolutions of the identity of the other operators. Von Neumann also shows that the product of two compatible quantities $\mathfrak{a}_1, \mathfrak{a}_2$ corresponds to the product $S_1 S_2$. Then in Section VI, he considers a measurement that fixes the values of $\mathfrak{a}_1, ..., \mathfrak{a}_m$ to within the intervals $I_1, ..., I_m$ from the (yet to be determined) 'maximally disordered' ensemble. Assuming that measurements are *repeatable*, he shows that the positive operator $U$ characterizing the resulting ensemble commutes with the projector $E := E_1(I_1)E_2(I_2)...E_m(I_m)$, and in fact $U = EU = UE$. In the special case in which $E$ is one-dimensional, $U$ is thus the corresponding projection operator $P_\varphi$. In modern terminology, von Neumann has derived that in a maximal measurement the state *collapses* to an eigenstate of the measured observable.

Finally, Sections VII and VIII are titled 'Statistical relations between different measurements'. Von Neumann returns to the question of what ensembles arise from the maximally disordered ensemble through measurements that fix values of physical quantities only to within given intervals. He now makes a *non-contextuality* assumption: that the ensemble arising from such a measurement is independent of which other

---

[39] Resolutions of the identity had been introduced in von Neumann (1927a, Section IX).

measurements are carried out compatibly with it. In other words (and in modern terminology), that such a measurement of a non-maximal observable can be implemented by measuring an arbitrary maximal observable compatible with it and then coarse-graining over the finer results.[40] Under this assumption, von Neumann shows that $U$ is a mixture of projections $P_{\varphi_p}$ for which $\{\varphi_p\}$ spans the range of $E$ and that the corresponding weights are all equal, so that (up to a proportionality factor) $U = E$.[41] In Section VIII he then uses this result to answer questions (iii) and (ii) above. Starting from an initial maximally disordered state, for two successive measurements that fix the values of two sets of quantities $a_1,...,a_m$ and $b_1,...,b_n$ to certain intervals (defining two not necessarily commuting projection operators $E,F$), he expresses the (unnormalized) *transition probability* between the corresponding results as $\mathrm{Tr}(EF)$ – an expression he had proposed and discussed in von Neumann (1927a, Sections XII–XIV) as generalizing the transition probabilities of the usual transformation theory. He also determines the 'maximally disordered ensemble' as the identity operator $U = 1$, noting that it indeed assigns the same (relative) probability 1 to all pure states (even though a notion of 'volume' on the unit sphere of a separable Hilbert space may be ill-defined[42]), and that it can be prepared also by mixing in equal proportions the elements of an arbitrary complete orthonormal system. Ensembles of the form $U = E$ can be analogously prepared by mixing in equal proportions the elements of an arbitrary orthonormal system spanning the range of $E$, showing that decompositions of mixed states are non-unique. (For von Neumann this shows that the only homogeneous ensembles, i.e. the only true 'states', are pure.) Finally, if the range of $E$ is finite-dimensional, then the statistical distributions defined by $U = E$ can be normalized by dividing by the dimension, while if it is infinite-dimensional one obtains indeed only relative distributions.

Section IX is a 'Summary'. Von Neumann states that he has uniquely derived the usual statistical quantum mechanics from the 'usual probability theory', from the fundamental assumptions that measurements in general disturb a system, that certain pairs of measurements are incompatible, and that nevertheless measurements are repeatable, and from the formal assumption that in quantum mechanics physical quantities and their functional relations are represented by operators and their

[40] From von Neumann's book (1932, Chapter V.1) it is clear that this is in fact the only way he conceptualizes the measurement of a non-maximal observable. This was later criticized by Lüders (1950), who introduced the now standard form of the collapse postulate for the case of non-maximal observables.

[41] It may be surprising that while in Section III von Neumann characterizes ensembles as arbitrary unnormalized density matrices, here in Section VII he is claiming that ensembles arising through non-maximal measurements are always (multi-dimensional) projection operators. But note that he is attempting to characterize only ensembles that arise from an initial state that is 'maximally disordered'. Von Neumann does not have the general notion of a measurement as a completely positive map, but he can certainly conceive of preparing more general ensembles by *mixing* different ensembles of the form $U = E$. (He explicitly mentions mixing in Section VIII.)

[42] Cf. the note added in proof on p. 256, where von Neumann compares his approach advantageously to that of Weyl (1927).

functional relations. He remarks that, since states always evolve deterministically according to the Schrödinger equation, the statistical and 'acausal' character of quantum mechanics arises specifically through the in-principle limitations of measurements pointed out by Heisenberg (1927b).

Von Neumann's statistical quantum mechanics is now arguably the theory as we know it, and not just because he uses a Hilbert-space formulation (albeit only infinite-dimensional). Unlike the theory as presented by Born and Heisenberg at the 1927 Solvay conference—which included recognizable transition probabilities but no recognizable notions of states or collapse—von Neumann's statistical theory includes a characterization of both pure and mixed states, respectively as wavefunctions and (unnormalized) density operators; it includes the general form of the Born rule for expectation values $\mathrm{Tr}(US)$ and a generalized form of transition probabilities $\mathrm{Tr}(EF)$; and it includes a recognizable version of the collapse postulate for the general case of maximal measurements and in a special case of non-maximal measurements.

The physical picture behind von Neumann's formulation of the theory is also fairly clear. He appears to believe that a physical system is always in some pure state. These states are represented by Schrödinger's wavefunctions, but can be equivalently characterized in terms of the values of a maximal set of commuting observables. Measurements are interventions on a system that in general irreducibly disturb it, forcing the system into one of the eigenstates of the measured observable (cf. von Neumann's footnote 33 on p. 264). Ensembles, represented by generally unnormalized density operators, are understood epistemically. A priori probabilities corresponding to no knowledge are represented by the identity operator. Homogeneous ensembles are identified with pure states. Unlike the case of classical statistical mechanics, these are not dispersion-free.

One could comment extensively on all of these points. Because of entanglement, quantum systems are not always in pure states.[43] Measurements are only given a phenomenological description.[44] The requirement of a well-defined notion of a priori probabilities resurfaces in von Neumann's later work on rings of operators.[45] One should not identify homogeneous ensembles and pure states.[46] But we want to

[43] The EPR state is an example where the subsystems are described by the maximally mixed state, but—precisely if quantum mechanics is complete—not by any specific decomposition into either position or momentum eigenstates.

[44] Not only is there no 'measurement problem' yet, but also no attempt to model the process of measurement. Von Neumann was going to discuss the latter in Chapter VI of his book, but it is arguably Pauli who did the most interesting early work on modelling the process of measurement in Section A.9 of his 1933 handbook article (Pauli, 1933). On Pauli's interest in the measurement process, cf. above Born's remarks on the cloud chamber, Pauli's letter to Bohr of 17 October 1927 (Pauli, 1979, pp. 411–13), and the discussion in Bacciagaluppi and Crull (2021, section 5.5).

[45] As discussed by Rédei (1996), it was a motivation for von Neumann in the mid-1930s to think of type-II factors as a possible replacement for Hilbert space in the formulation of quantum mechanics.

[46] If we define a homogeneous ensemble as one in which systems are all in the same state, then subsystems of an EPR pair are indeed always in the same state, even though they are described by the identity matrix which is non-uniquely decomposable into pure states.

conclude by focusing on the last point, the lack of dispersion-free states, and specifically on whether and how von Neumann can be understood as having shown that 'the statistical interpretation [ ... ] is the only interpretation of quantum mechanics consistently implementable today'.

## 8.5 STATISTICAL INTERPRETATION
## AND INDETERMINISM

The statistical interpretation of quantum mechanics was connected from the start to the issue of indeterminism. We saw how Born had opted for a middle course between the discontinuity of matrix mechanics and the continuity of wave mechanics, with the latter determining the probabilities for the former. As discussed in detail by Beller (1990), we also saw how Born nevertheless was only tentatively taking the quantum probabilities as an expression of indeterminism, and that it was maybe Heisenberg in the uncertainty paper who first claimed that the statistical interpretation was an expression of irreducible indeterminism. The explicit references to Heisenberg in von Neumann's paper indicate that von Neumann understood his 'probabilistic construction' of quantum mechanics as endorsing Heisenberg on this point. Specifically, I would like to suggest that von Neumann saw himself as providing a rigorous proof of Heisenberg's intuition that irreducible probabilities follow from the *incompatibility* of quantum mechanical quantities.[47] Indeed, von Neumann's distinction between classical and quantum mechanics is precisely that at the non-statistical level the former is commutative and the latter is not, and that at the statistical level the probabilities in classical mechanics can always be eliminated (which makes clear they are merely epistemic) but that this is impossible in quantum mechanics.

The emergence of this crucial distinction can be seen implicitly also in the previous development of the statistical interpretation. In Born's 1926 papers, the statistical distributions are over the stationary states of essentially non-interacting systems, thus over the values of a set of *commuting* quantities. In Heisenberg and Jordan's fluctuation papers, we first explicitly see probabilities arising through consideration of *non*-commuting matrices, namely the Hamiltonians with and without an interaction term. This is generalized in the transformation theory to arbitrary pairs of non-commuting quantities, and taken in the uncertainty paper as the very reason why

[47] Wigner (1970, note 1) states that von Neumann's main argument against hidden variables was based on the impossibility of preparing dispersion-free states when considering successive measurements of incompatible observables. This argument is in fact presented in von Neumann's book, but von Neumann nevertheless raises the question of whether dispersion-free states exist at all (von Neumann, 1932, Section IV.1). The argument and Wigner's remarks are briefly discussed by Jammer (1974, pp. 266–67). Thanks to Chris Mitsch for correspondence on this point, and see Mitsch (forthcoming) for further discussion.

probabilities are needed. Finally, von Neumann proves that the non-commutative algebra of matrix mechanics leads to pure states that always involve non-trivial probabilities.

Should any of this rule out the very possibility of fundamental determinism? Max Born appears to have been the first to explicitly draw stronger conclusions from von Neumann's result. In a little-known paper of 1929 he wrote:[48]

> Although the new theory seems thus well established in experience, one can still pose the question of whether *in the future*, through extension or refinement, it might not be made *deterministic again*. In this regard one must note: it can be shown in a mathematically exact way that the established formalism of quantum mechanics allows for *no* such completion. If thus one wants to retain the hope that determinism will return someday, then one must consider the present theory to be *contentually false*; specific statements of this theory would have to be refuted experimentally. Therefore, in order to convert the adherents of the statistical theory, the determinist should not protest but rather test.

But it is von Neumann himself in his 1932 book who discusses the issue most explicitly. In the 'Introduction' section he announces his purpose with these words (von Neumann, 1932, pp. 2–3):

> In analysing the fundamental questions it will be shown in particular how the statistical formulas of quantum mechanics can be derived from some qualitative fundamental assumptions. Further, the question is discussed in detail of whether it is possible to trace the statistical character of quantum mechanics back to an ambiguity (i.e. incompleteness) in our description of nature: this explanation would best fit the general principle according to which all probabilistic statements arise from the incompleteness of our knowledge. This explanation 'by hidden parameters', as well as another related one that ascribes the 'hidden parameters' to the observer and not to the observed system, has in fact been proposed several times. However, it turns out that this cannot be achieved in a satisfactory way, or more precisely: such an explanation is incompatible with certain qualitative fundamental postulates of quantum mechanics.

Von Neumann then sets up the question in Section III.2 ('The statistical interpretation'), discussing hidden parameters relating to the system in Section IV.1 ('Fundamental basis of the statistical theory') and Section IV.2 ('Proof of the statistical formulas'), and hidden parameters relating to the apparatus at the beginning of Section VI.3 ('Discussion of the measurement process'). In the later literature, these two discussions have become known respectively as von Neumann's 'impossibility theorem' for hidden variables (or some such name) and his 'insolubility theorem' for the measurement problem of quantum mechanics. The latter is the proof that measurement statistics cannot be reproduced by assuming a mixed state for the apparatus

---

[48] Born (1929, pp. 117–18), as translated in Crull and Bacciagaluppi (2017, p. 223).

and a unitary interaction between system and apparatus (cf. Brown, 1986). For von Neumann, the two are parallel cases in the sense that they attempt to explain the quantum mechanical statistics by assuming respectively that the state of the system or the state of the apparatus is not homogeneous.[49]

Section IV.2 is the one that interests us most. After presenting his results of 1927, von Neumann returns to the question of whether the quantum statistics could be explained by assuming that the homogeneous states are in fact mixtures of dispersion-free states characterized by 'hidden parameters'. He states that this is impossible, citing two reasons that at first may seem to beg the question: homogeneous states would not be homogeneous after all, and dispersion-free states have just been shown not to exist. But then he adds (von Neumann, 1932, p. 171):

> It should be noted that here we did not need to discuss the further details of the mechanism of the 'hidden parameters': the established results of quantum mechanics can in no way be rederived with their help, it is even impossible that the same physical quantities with the same functional relations exist (i.e. that [D.] and [C.] hold), if other variables ('hidden parameters') should exist in addition to the wavefunction.
>
> [...] the relations assumed by quantum mechanics [...] would have to fail already for [...] known quantities. Thus it is not at all a question of interpretation of quantum mechanics, as often assumed, rather the latter ought to be objectively false in order for a behaviour of elementary processes to be possible that is other than statistical.

Recall that the original purpose of von Neumann's 1927 paper was to provide a rigorous derivation of the statistical aspects of quantum mechanics from its algebraic structure together with general assumptions from the theory of probability. What von Neumann is saying here is that the existence of hidden variables would imply that in quantum mechanics we have *misidentified* the algebraic structure of the physical quantities. As Dieks (2017, section 4) emphasizes, von Neumann's fundamental assumption is that classical kinematics should be replaced by the new quantum kinematics. Now this assumption may turn out to be false, but for von Neumann all the evidence was in favour of it: the new kinematics had already been securely established through the matrix mechanics of Heisenberg (1925), Born and Jordan (1925b) and Born, Heisenberg, and Jordan (1926), well before Born introduced the first elements of the 'statistical interpretation' in his papers of 1926.

Compare this with Bohm's (1952a,b) theory—which does what von Neumann was reputed to have shown to be impossible: provide a deterministic underpinning ('causal interpretation') of quantum mechanics. It does so precisely by rejecting the quantum

---

[49] Mitsch (forthcoming), following Wigner (1970, note 1), suggests that Schrödinger was in fact the originator of some of the proposals von Neumann is criticizing (during the years they spent together in Berlin). On measurements in quantum mechanics, see also Schrödinger to von Neumann, 25 December 1929 (Meyenn, 2011, pp. 474–76).

kinematics. The true algebra of physical quantities of a system in Bohm's theory is the commutative algebra generated by the position operators, and for these quantities (and all 'hidden states') von Neumann's assumptions straightforwardly hold, including the assumption of linearity. Other operators simply do not correspond to single physical quantities. As Bohm himself remarked, discussing von Neumann's theorem, 'the so-called "observables" are [...] not properties belonging to the observed system alone' (1952b, p. 187). In any measurement one indeed measures some physical quantity of the system— but (except for functions of position) that quantity will depend on the details of the experimental context. Different quantities are then mistakenly identified with each other because they happen to have the same statistics in all ensembles characterized by quantum states. According to Bub (2010), the significance of von Neumann's theorem for the hidden variables debate is that it establishes this kind of contextuality.[50]

The significance of von Neumann's result for the hidden variables debate has been traditionally framed in the context of Bell's (1966) criticism. Bell pointed out that if one wishes to construct a theory in which hidden states determine without dispersion the values of quantum mechanical observables, then it is unreasonable to require of the hidden states that

$$E(\alpha S + \beta T) = \alpha E(S) + \beta E(T) \tag{7}$$

(the conjunction of von Neumann's assumptions **A.** and **D.**): dispersion-free values of quantum observables must equal eigenvalues, and eigenvalues of non-commuting operators in general do not behave additively.[51]

---

[50] Bohm himself (1952b, pp. 187–88) inaccurately suggests that the experimental results depend on the hidden variables of both system and measuring device, but no apparatus hidden variables are needed. What is crucial is that the same hidden variables for the system lead to different results depending on a measurement context—quantum observables are 'contextual observables'. For analyses of measurement contextuality in the Bohm theory, see Pagonis and Clifton (1995) and Barrett (1999, section 5.2). See also the analogy between the Bohm theory and the 'firefly box' in Bacciagaluppi (2016b, section 6).

[51] It has also been appreciated recently that Grete Hermann had already criticized von Neumann's reputed argument against hidden variables thirty years earlier (see e.g. Seevinck, 2017). Hermann's criticism, too, centres on the linearity assumption, but unlike Bell she is aware that von Neumann fully realizes how non-trivial it is (recall von Neumann's own example of $Q_2$, $P_2$, and $Q_2+P_2$). Instead she believes that von Neumann smuggles it in illicitly. But her overall assessment of the significance of von Neumann's result is precisely that the existence of hidden variables would imply that quantum mechanics has misidentified the algebraic structure of the physical quantities it describes. For details see Dieks (2017, section 5). The criticism first appears in a manuscript from 1933 (Dirac Archives, DRAC 3/11) in which Hermann defends the possibility of deterministically completing quantum mechanics (and which contains the above passage from Born (1929) as an epigraph). Hermann's criticism was then published as section 7 ('The circle in Neumann's proof') in Hermann (1935), an extended analysis of quantum mechanics in which instead Hermann provides an alternative neo-Kantian argument for the completeness of the theory closely related to Bohr's complementarity. Both Hermann's 1933 manuscript and her 1935 essay are translated in Crull and Bacciagaluppi (2017, Chapters 14 and 15)—with extensive discussions—and have been (re-)published in German in Herrmann (2019, pp. 185–203 and 205–58). See also the extremely interesting correspondence from 1968 between Hermann and Jammer (where Hermann retracts the specific claim of circularity) published in Herrmann (2019, pp. 599–10).

In recent years the criticism of von Neumann has become somewhat more strident. In a later interview, Bell even characterized assumption (7) as 'silly' and 'foolish',[52] with a number of authors following suit (e.g. Mermin, 1993). Recent work by Bub (2010), Dieks (2017), and Mitsch (forthcoming) instead counters this assessment in various ways by restoring von Neumann's work to its historical context.[53]

Let us be clear. There are *no* mistakes in von Neumann's mathematics: taking self-adjoint operators and their functional relations as representing physical quantities and their functional relations, it follows that expectation values are represented by density operators, and that no such expectation values are dispersion-free for all physical quantities. And it is equally clear that the result does *not* rule out deterministic hidden variables theories: Bohm's theory is an unequivocal example of such a theory. It is also presumably the case that many physicists mistakenly did take von Neumann's theorem to rule out hidden variables theories (as many physicists nowadays mistakenly take Bell's theorem to rule them out!).[54] The question is whether von Neumann's theorem in fact establishes something interesting about hidden variables and what von Neumann himself thought it established.[55]

In terms of the distinction standard today between the 'completeness' and 'correctness' of quantum mechanics, if one theoretically completes quantum mechanics by assuming the existence of Bohmian corpuscles, quantum mechanics does nevertheless remain empirically correct. To insist that this would make quantum mechanics 'objectively wrong' may seem disingenuous. Knowledge of the Bohm theory, however, gives us the benefit of hindsight. The historically accurate question may be too difficult ever to decide, but it is whether the likes of Born and von Neumann considered the possibility (however implausible) that a theory with a different algebraic structure

[52] *Omni* magazine (May 1988, p. 88), quoted in Mermin (1993).

[53] As mentioned, Bub (2010) argues that von Neumann's theorem shows that hidden variables theories cannot be constructed on the model of classical statistical mechanics but have to be contextual, and Dieks (2017) that von Neumann's crucial assumption is indeed the new quantum kinematics. Mitsch (forthcoming) argues that the 'uniqueness' of the statistical quantum mechanics is to be seen in the context of Hilbert's axiomatic method, which aimed in particular to define a unique formal apparatus capturing a given physical content and was thus understood as both provisional and relative.

[54] Note that (with few exceptions, most notably Born), the other 'founding fathers' of quantum mechanics rarely referred to von Neumann's theorem when arguing for the completeness of quantum mechanics. This can be seen, for instance, in their reactions to the EPR paper (see Bacciagaluppi and Crull, 2021). Bohr's arguments were based on complementarity, Heisenberg's (as mentioned) were based on interference. Pauli did not refer explicitly to von Neumann but (*pace* Bell's later criticism) he did take the example of $Q_2$, $P_2$, and $Q_2+P_2$ as showing the impossibility of hidden variables: if $Q$ and $P$ had continuously distributed dispersion-free values, so would $Q_2+P_2$. As first discussed by Fine (1994), in his first paper on entanglement Schrödinger (1935) took up this example, and presented it as part of a dilemma: the EPR argument showed that all quantum mechanical observables had (non-contextual!) values, but these could not obey any obvious functional relations.

[55] Note that replying on Bohr's behalf to Hermann's 1933 manuscript, Weizsäcker in his letter to Hermann of 17 December 1933 agrees that determinism is still viable, but wishes to argue that the 'self-imposed restrictions of quantum theory' are not arbitrary (Herrmann, 2019, p. 440). Also Jammer (1974, pp. 272–275, esp. fn 45), in discussing Hermann's criticism, remarks perceptively on what von Neumann's result does or does not show.

might be able to deliver the same empirical predictions as quantum mechanics, or whether they tacitly assumed that any such theory would automatically contradict also the empirical predictions of quantum mechanics. If the former, they would have been mistaken in discounting that possibility, but perfectly rational in judging it implausible. If the latter, they would indeed have been drawing an unwarranted conclusion from von Neumann's theorem.[56]

In later years, von Neumann does not seem to have returned $\Psi$ much to the issue of 'hidden parameters'. He discusses it in print in discussion comments on Bohr's talk at the 1938 Warsaw conference—a rare occasion where Bohr actually cites von Neumann's theorem as clarifying how 'the fundamental principle of superposition of quantum mechanics [sic] logically excludes the possibility of avoiding the non-causal character of the formalism by any kind of conceivable introduction of additional variables' (Bohr, 1939, p. 17), before going on to a more detailed discussion of complementarity as 'providing a direct generalization of the very ideal of causality' (p. 28). In his comments, von Neumann characterizes a physical system in terms of its measurable physical quantities and their algebraic relations; he then considers ensembles, and defines a 'purely' statistical theory as one in which pure states are not dispersion-free (not 'causal'). Given that measurable quantities in quantum mechanics are given by self-adjoint operators, from the 'obvious properties' of expectation values one easily derives the general trace formula (6), from which it follows that pure states in quantum mechanics are not causal. Quantum mechanics is thus the first example of a 'properly' statistical theory, unlike classical statistical mechanics (which can in fact be completed via 'hidden parameters'). Von Neumann concludes that 'no causal explanation of quantum mechanics is possible unless one sacrifices some part of the theory as it exists today' (Bohr, 1939, p. 38). Again, this short exposition can be given the nuanced readings discussed above. Dieks (2017, section 6) also reports that, after Bohm's (1952a,b) papers appeared, von Neumann seems to have agreed that Bohm's

---

[56] Note that the situation is subtle even in the Bohm theory. If one considers ensembles with arbitrary position distributions, then the theory will indeed violate the empirical predictions of quantum mechanics (for instance, it will violate the no-signalling theorem). A theory with different algebraic structure does lead to different empirical predictions. But if one considers special 'equilibrium' ensembles in which positions are distributed according to the Born distribution $|\psi(x)|^2$, the empirical predictions of the theory become the same as those of standard quantum mechanics (in particular the self-adjoint operators emerge as contextual observables). Not only that, the theoretical regime defined by these ensembles is stable in the sense that—whether they evolve freely or are subjected to any (quantum) measurements—these ensembles will always remain 'equilibrium' ensembles. De Broglie in 1927 already knew that the $|\psi(x)|^2$-distribution was stable under time evolution (Bacciagaluppi and Valentini, 2009, p. 70) but did not have Bohm's (1952b) measurement theory for quantum observables other than position. The only author before Bohm who seriously considered the distinction between physical quantities and quantum operators—and even the idea of hidden variables in 'disequilibrium' violating the empirical predictions of quantum mechanics—appears to have been Grete Hermann; cf. section 5 of her 1933 manuscript and her letter to Heisenberg of 9 February 1934 (Herrmann, 2019, Letter 7, pp. 457–61). For a detailed discussion of the Bohmian equilibrium regime see e.g. Dürr, Goldstein, and Zanghí (1992). For Bohmian disequilibrium see e.g. Valentini (2002a,b).

theory reproduced all the predictions of quantum mechanics, without objecting to it on the basis of his earlier results.

A final remark that seems revealing about what von Neumann thought his theorem showed was published after von Neumann's death by Léon Van Hove (who had worked for a few years at the Institute for Advanced Studies at Princeton) in a short paper on 'Von Neumann's contributions to quantum theory' that was part of an issue of the *Bulletin of the American Mathematical Society* in memory of von Neumann (Van Hove, 1958, p. 98):

> Von Neumann devoted in his book considerable attention to a point which had not been discussed in the 1927 papers and which was later the subject of much controversy. It is the question of the possible existence of 'hidden variables', the consideration of which would eliminate the noncausal element involved in the measuring process. Von Neumann could show that hidden parameters with this property cannot exist if the basic structure of quantum theory is retained. Although he mentioned the latter restriction explicitly, his result was often quoted without due reference to it, a fact which sometimes gave rise to unjustified criticism in the many discussions devoted through the years to the possibility of an entirely deterministic reformulation of quantum theory.

I wish to conclude by suggesting a more dispassionate assessment of von Neumann's result in the light of another theme in the foundations of quantum mechanics, namely the difference between classical and quantum probability. Von Neumann's choice of linearity of expectation values as a probabilistic axiom may strike a modern reader as relying on superfluous algebraic structure (as opposed to Boolean structure), but von Neumann's paper is a formalization of quantum mechanics as a probability theory for incompatible quantities, and his impossibility theorem is the first proof establishing the fundamental difference between classical and quantum probability.[57]

## ACKNOWLEDGEMENTS

My heartfelt thanks go to Olival Freire for support and encouragement and for the wonderful way in which he has put together this volume, and to Christian Joas for his extremely helpful and detailed comments (as Heisenberg wrote, 'Wissenschaft entsteht im Gespräch'). I am very grateful to Sonja Smets and Soroush Rafiee Rad for the opportunity of presenting parts of this material in Amsterdam, to Marco Giovanelli for that of presenting in Tübingen, and to the audiences of those presentations. I have discussed related materials with more people than I can remember, but for recent discussions and correspondence I wish to thank especially Chris Mitsch (also for a draft of his paper) and Damon Moley.

---

[57] Cf. the classic discussion by Pitowsky (1994), or for an introductory overview see again Bacciagaluppi (2016b).

## REFERENCES

Bacciagaluppi, G. (2008). The statistical interpretation according to Born and Heisenberg. In C. Joas, C. Lehner, and J. Renn (eds), *HQ-1: Conference on the History of Quantum Physics (vols I & II)*, MPIWG preprint series, vol. 350, Berlin: Max-Planck-Institut für Wissenschaftsgeschichte, Chapter 14, pp. 269–288 (vol. II).

Bacciagaluppi, G. (2016a). Einstein, Bohm, and Leggett–Garg. In E. Dzhafarov, S. Jordan, R. Zhang, and V. Cervantes (eds), *Contextuality from Quantum Physics to Psychology*, Singapore: World Scientific, pp. 63–76.

Bacciagaluppi, G. (2016b). Quantum probability: an introduction. In A. Hájek and C. Hitchcock (eds), *The Oxford Handbook of Probability and Philosophy*, Oxford: Oxford University Press, pp. 545–572.

Bacciagaluppi, G., and Crull, E. (2009). Heisenberg (and Schrödinger, and Pauli) on hidden variables. *Studies in History and Philosophy of Modern Physics*, 40, 374–382.

Bacciagaluppi, G., and Crull, E. (2021). *The Einstein Paradox: The Debate on Nonlocality and Incompleteness in 1935*. Cambridge: Cambridge University Press, forthcoming.

Bacciagaluppi, G., Crull, E., and Maroney, O. (2017). Jordan's derivation of blackbody fluctuations. *Studies in History and Philosophy of Modern Physics*, 60, 23–34.

Bacciagaluppi, G., and Valentini, A. (2009). *Quantum Theory at the Crossroads: Reconsidering the 1927 Solvay Conference*. Cambridge: Cambridge University Press.

Barrett, J. A. (1999). *The Quantum Mechanics of Minds and Worlds*. Oxford: Oxford University Press.

Bell, J. S. (1966). On the problem of hidden variables in quantum mechanics. *Reviews of Modern Physics*, 38(3), 447–475.

Beller, M. (1990). Born's probabilistic interpretation: a case study of 'concepts in flux'. *Studies in History and Philosophy of Science*, 21(4), 563–588.

Blum, A., Jähnert, M., Lehner, C., and Renn, J. (2017). Translation as heuristics: Heisenberg's turn to matrix mechanics. *Studies in History and Philosophy of Modern Physics*, 60, 3–22.

Bohm, D. (1952a). A suggested interpretation of the quantum theory in terms of 'hidden' variables. I. *Physical Review*, 85(2), 166–179.

Bohm, D. (1952b). A suggested interpretation of the quantum theory in terms of 'hidden' variables. II. *Physical Review*, 85(2), 180–193.

Bohr, N. (1939). Le problème causal en physique atomique (with discussion). In C. Białobrzeski (ed.), *Les nouvelles théories de la physique*, Paris: Institut International de Coopération Intellectuelle, pp. 11–48.

Born, M. (1926a). Zur Quantenmechanik der Stoßvorgänge. *Zeitschrift für Physik*, 37, 863–867.

Born, M. (1926b). Quantenmechanik der Stoßvorgänge. *Zeitschrift für Physik*, 38, 803–827.

Born, M. (1926c). Das Adiabatenprinzip in der Quantenmechanik. *Zeitschrift für Physik*, 40, 167–192.

Born, M. (1927a). Physical aspects of quantum mechanics. *Nature*, 119, 354–357.

Born, M. (1927b). Quantenmechanik und Statistik. *Die Naturwissenschaften*, 15(10), 238–242.

Born, M. (1929). Über den Sinn der physikalischen Theorien. *Die Naturwissenschaften*, 17(7), 109–118.

Born, M. (1949). *Natural Philosophy of Cause and Chance*. Oxford: Clarendon Press.

Born, M., and Franck, J. (1925a). Bemerkungen über Dissipation der Reaktionswärme. *Annalen der Physik*, **76**, 225–230.

Born, M., and Franck, J. (1925b). Quantentheorie und Molekülbildung. *Zeitschrift für Physik*, **31**, 411–429.

Born, M., Heisenberg, W., and Jordan, P. (1926). Zur Quantenmechanik, II. *Zeitschrift für Physik*, **35**, 557–615.

Born, M., and Jordan, P. (1925a). Zur Quantentheorie aperiodischer Vorgänge. *Zeitschrift für Physik*, **35**, 479–505.

Born, M., and Jordan, P. (1925b). Zur Quantenmechanik. *Zeitschrift für Physik*, **34**, 858–888.

Born, M., and Wiener, N. (1926a). Eine neue Formulierung der Quantengesetze für periodische und nichtperiodische Vorgänge. *Zeitschrift für Physik*, **36**, 174–187.

Born, M., and Wiener, N. (1926b). A new formulation of the laws of quantization of periodic and aperiodic phenomena. *Journal of Mathematics and Physics M.I.T.*, **5**, 84–98.

Brown, H. R. (1986). The insolubility proof of the quantum measurement problem. *Foundations of Physics*, **16**(9), 857–870.

Bub, J. (2010). Von Neumann's 'no hidden variables' proof: a re-appraisal *Foundations of Physics*, **40** (9–10), 1333–1340.

Crull, E., and Bacciagaluppi, G. (eds) (2017). *Grete Hermann: Between Physics and Philosophy.* Dordrecht: Springer.

Darrigol, O. (1992). *From c-Numbers to q-Numbers: The Classical Analogy in the History of Quantum Theory.* Berkeley, CA: University of California Press.

Dieks, D. (2017). Von Neumann's impossibility proof: mathematics in the service of rhetorics. *Studies in History and Philosophy of Modern Physics*, **60**, 136–148.

Dirac, P. A. M. (1926). On the theory of quantum mechanics. *Proceedings of the Royal Society A*, **112**, 661–677.

Dirac, P. A. M. (1927). The physical interpretation of the quantum dynamics. *Proceedings of the Royal Society A*, **113**, 621–641.

Dirac, P. A. M. (1930). *The Principles of Quantum Mechanics*, 1st edn. ord: Oxford University Press.

Duncan, A., and Janssen, M. (2008). Pascual Jordan's resolution of the conundrum of the wave–particle duality of light. *Studies in History and Philosophy of Modern Physics*, **39**(3), 634–666.

Duncan, A., and Janssen, M. (2009). From canonical transformations to transformation theory, 1926–1927: the road to Jordan's Neue Begründung. *Studies in History and Philosophy of Modern Physics*, **40**(4), 352–362.

Duncan, A., and Janssen, M. (2012). (Never) mind your P's and Q's: von Neumann versus Jordan on the foundations of quantum theory. *European Physical Journal H*, **38**(2), 175–259.

Dürr, D., Goldstein, S., and Zanghì, N. (1992). Quantum equilibrium and the origin of absolute uncertainty. *Journal of Statistical Physics*, **67**, 843–907.

Einstein, A. (1953). Elementare Überlegungen zur Interpretation der Grundlagen der Quantenmechanik. In E. Appleton *et al.* (eds), *Scientific Papers Presented to Max Born on his Retirement from the Tait Chair of Natural Philosophy in the University of Edinburgh*, Edinburgh: Oliver and Boyd, pp. 33–40.

Fine, A. (1994). Schrödinger's version of EPR, and its problems. In P. Humphreys (ed.), *Patrick Suppes: Scientific Philosopher*, vol. 2, Dordrecht: Kluwer, pp. 29–38.

Franck, J., and Jordan, P. (1926). Anregung von Quantensprüngen durch Stöße (mit Ausschluß der Erscheinungen an Korpsukularstrahlen hoher Geschwindigkeit). In H. Geiger and H. Scheel (eds), *Handbuch der Physik* (1st edn), vol. XXIII, Berlin: Springer, pp. 641–775.

Heisenberg, W. (1925). Über quantentheoretische Umdeutung kinematischer und mechanischer Beziehungen. *Zeitschrift für Physik*, 33, 879–893.

Heisenberg, W. (1927a). Schwankungserscheinungen und Quantenmechanik. *Zeitschrift für Physik*, 40, 501–506.

Heisenberg, W. (1927b). Über den anschaulichen Inhalt der quantentheoretischen Kinematik und Mechanik. *Zeitschrift für Physik*, 43, 172–198.

Hermann, G. (1935). Die naturphilosophischen Grundlagen der Quantenmechanik. *Abhandlungen der Fries'schen Schule*, 6(2), 75–152. Reprinted in Herrmann (2019), pp. 205–258. Translated in Crull and Bacciagaluppi (2017), pp. 239–278.

Herrmann, K. (ed.) (2019). *Grete Henry-Hermann: Philosophie—Mathematik—Quantenmechanik*. Wiesbaden: Springer.

Hilbert, D., Neumann, J. von, and Nordheim, L. (1928). Über die Grundlagen der Quantenmechanik. *Mathematische Annalen*, 98, 1–30.

Hund, F. (1967). *Geschichte der Quantentheorie*. Mannheim: Bibliographisches Institut.

Jammer, M. (1974). *The Philosophy of Quantum Mechanics: The Interpretations of Quantum Mechanics in Historical Perspective*. New York: John Wiley and Sons.

Jordan, P. (1927a). Über quantenmechanische Darstellung von Quantensprüngen. *Zeitschrift für Physik*, 40, 661–666.

Jordan, P. (1927b). Über eine neue Begründung der Quantenmechanik, I. *Zeitschrift für Physik*, 40, 809–838.

Jordan, P. (1927c). Über eine neue Begründung der Quantenmechanik, II. *Zeitschrift für Physik*, 44, 1–25.

Konno, H. (1978). The historical roots of Born's probabilistic interpretation. *Japanese Studies in the History of Science*, 17, 129–145.

Lacki, J. (2000). The early axiomatizations of quantum mechanics: Jordan, von Neumann and the continuation of Hilbert's program. *Archive for History of Exact Sciences*, 54, 279–318.

Lacki, J. (2004). The puzzle of canonical transformations in early quantum mechanics. *Studies in History and Philosophy of Modern Physics*, 35, 317–344.

London, F. (1926). Winkelvariable und kanonische Transformationen in der Undulationsmechanik. *Zeitschrift für Physik*, 40, 193–210.

Lüders, G. (1950). Über die Zustandsänderung durch den Meßprozeß. *Annalen der Physik*, 443(5–8), 322–328.

Mermin, N. D. (1993). Hidden variables and the two theorems of John Bell. *Reviews of Modern Physics*, 65(3), 803–815.

Meyenn, K. von (ed.) (2011). *Eine Entdeckung von ganz außerordentlicher Tragweite: Schrödingers Briefwechsel zur Wellenmechanik und zum Katzenparadoxon*. Heidelberg: Springer.

Mitsch, C. (forthcoming). The (not so) hidden contextuality of von Neumann's 'no hidden variables' proof.

Muller, F. A. (1997a). The equivalence myth of quantum mechanics. Part I. *Studies in History and Philosophy of Modern Physics*, 28, 35–61.

Muller, F. A. (1997b). The equivalence myth of quantum mechanics. Part II. *Studies in History and Philosophy of Modern Physics*, 28, 219–247.

Muller, F. A. (1999). The equivalence myth of quantum mechanics (addendum). *Studies in History and Philosophy of Modern Physics*, **30**, 543–545.

Myrvold, W. (2003). On some early objections to Bohm's theory. *International Studies in the Philosophy of Science*, **17**(1), 7–24.

Pagonis, C., and Clifton, R. (1995). Unremarkable contextualism: dispositions in the Bohm theory. *Foundations of Physics*, **25**(2), 281–296.

Pauli, W. (1926). Über das Wasserstoffspektrum vom Standpunkt der neuen Quantenmechanik. *Zeitschrift für Physik*, **36**, 336–363.

Pauli, W. (1927). Über Gasentartung und Paramagnetismus. *Zeitschrift für Physik*, **41**, 81–102.

Pauli, W. (1933). Die allgemeinen Prinzipien der Wellenmechanik. In H. Geiger and H. Scheel (eds), *Handbuch der Physik* (2nd edn), vol. XXIV(1), Berlin: Springer, pp. 83–272.

Pauli, W. (1979) *Wissenschaftlicher Briefwechsel mit Bohr, Einstein, Heisenberg u.a., Teil I: 1919–1929*, ed. by A. Hermann, K. von Meyenn, and V. F. Weisskopf. Berlin and Heidelberg: Springer.

Pitowsky, I. (1994). George Boole's 'conditions of possible experience' and the quantum puzzle. *The British Journal for the Philosophy of Science*, **45**(1), 95–125.

Rédei, M. (1996). Why John von Neumann did not like the Hilbert space formalism of quantum mechanics (and what he liked instead). *Studies in History and Philosophy of Modern Physics*, **27**(4), 493–510.

Schrödinger, E. (1926). Quantisierung als Eigenwertproblem (erste Mitteilung). *Annalen der Physik*, **79**, 361–376.

Schrödinger, E. (1935). Discussion of probability relations between separated systems. *Proceedings of the Cambridge Philosophical Society, Mathematical and Physical Sciences*, **31**(4), 555–563.

Seevinck, M. (2017). Challenging the gospel: Grete Hermann on von Neumann's no-hidden-variables proof. In Crull and Bacciagaluppi (2017), pp. 107–117.

Valentini, A. (2002a). Signal-locality in hidden-variables theories. *Physics Letters A*, **297**, 273–278.

Valentini, A. (2002b). Subquantum information and computation. *Pramana—Journal of Physics*, **59**, 269–277.

Van Hove, L. (1958). Von Neumann's contributions to quantum theory. *Bulletin of the American Mathematical Society*, **64**(3), 95–99.

von Neumann, J. (1927a). Mathematische Begründung der Quantenmechanik. *Nachrichten der Akademie der Wissenschaften in Göttingen. II., Mathematisch-Physikalische Klasse*, **1927**(20 May), 1–57.

von Neumann, J. (1927b). Wahrscheinlichkeitstheoretischer Aufbau der Quantenmechanik. *Nachrichten von der Gesellschaft der Wissenschaften zu Göttingen. Mathematisch-Physikalische Klasse*, **1927**(11 November), 245–272.

von Neumann, J. (1929). Allgemeine Eigenwerttheorie Hermitischer Funktionaloperatoren. *Mathematische Annalen*, **102**, 49–131.

von Neumann, J. (1932). *Mathematische Grundlagen der Quantenmechanik*. Berlin: Springer.

Wessels, L. (1980). What was Born's statistical interpretation? *PSA: Proceedings of the Biennial Meeting of the Philosophy of Science Association*, **1980**(2), 187–200.

Weyl, H. (1927). Quantenmechanik und Gruppentheorie. *Zeitschrift für Physik*, **46**(1–2), 1–46.

Wigner, E. P. (1970). On hidden variables and quantum mechanical probabilities. *American Journal of Physics*, **38**(8), 1005–1009.

# CHAPTER 9

# A PERENNIALLY GRINNING CHESHIRE CAT? OVER A CENTURY OF EXPERIMENTS ON LIGHT QUANTA AND THEIR PERPLEXING INTERPRETATIONS

## KLAUS HENTSCHEL

THE **terms** 'light quantum' or 'photon'—derived from the Greek word for light, φως (*phos*)—are omnipresent in modern science and technology, in the media, and even in art. Photon-based applications such as the laser and derivative devices such as the compact disc, DVD player, and the bar code scanner, define our daily lives. Scientists and engineers in materials research and technology at many 'photonics' centres collaborate to develop new applications of quantum optics. This prevalence of the term in research and production nonetheless does not necessarily imply that there is general agreement about what 'photons' actually are. The interpretations and mental models used by various groups for this concept could hardly be more different.

Why could it be useful—indeed important—for a modern reader to think about the complex conceptual history of photons, instead of just concentrating on today and tomorrow? For a deeper understanding of what we mean by light quanta, it is instructive to study the history behind the concept and the cognitive obstacles that feature in it, faced by some of the most brilliant physicists. Einstein himself was never able to come fully to grips with his own conceptual creation. In 1951 he wrote to a lifelong friend and confidante: 'All those 50 years of careful pondering have not brought me closer to the answer to the question: "What are light quanta?" Today any old scamp believes he knows, but he's deluding himself.'[1] As late as 1955, Willis Lamb (1913–2008), like Einstein a theoretical physicist and Nobel laureate, announced: 'There is no such thing as a photon. Only a comedy of errors and historical accidents led to its popularity among physicists and optical scientists. I admit that the word is short and convenient. Its use is also habit forming.' (Lamb 1995, 77) These persistent, conspicuous and deep problems confounding some of the greatest minds in the history of physics cannot be so easily dismissed.

As is described elsewhere (Klein, 1962, 1963; Kuhn, 1978; Needell, 1980; Heilbron, 1986; Gearhart, 2010; Badino, 2015; Hentschel, 2018, chap. 2), Planck's quantum hypothesis was 'somewhat of a premature birth. Essential properties of nature to which the quantity $h$ could be attached still had been scarcely researched or classified. While Planck's radiation formula rapidly gained acceptance as empirically valid, initially his theory was not explored further.' (Hund, 1984, p. 29) The first person to study more closely not only Planck's interpolation formula but also his derivation from December 1900 and the theoretical assumptions underpinning them was Albert Einstein (1879–1955), at that time a still unknown examiner at the Berne patent office. One of the 12 papers published during his *annus mirabilis* of 1905 dealt with 'a heuristic point of view concerning the production and transformation of light.' This modest title referred to what was a 'very revolutionary' introduction of light quanta. Einstein argued that quantization was indeed not limited to material resonators or to the interaction between matter and the field (as Planck had hoped). Quantization was required of the electromagnetic field energy itself:

> When a light ray is spreading from a point, the energy is not distributed continuously over ever-increasing spaces, but consists of a finite number of energy quanta that are localized in points in space, move without dividing, and can be absorbed or generated only as a whole.[2]

**What did this new 'heuristic point of view' achieve?** First and foremost, it opened up new avenues for a—hopefully consistent—theoretical new framework to understand light, in which continuous electrodynamics (modelling light as continuous electromagnetic waves) and statistical mechanics (operating with discontinuous particles) would also find their place. Empirically, the new vantage point allowed the prediction of many experimental effects as well as explanations for known empirical findings. Einstein mentioned the following topics in 1905:

- Stokes's rule on photoluminescence: The re-emission frequency is always smaller than the initiator frequency. Einstein realized that this previously

---

[1] Einstein in a letter to his former colleague at the Bernese patent office, Michele Besso, 12 Dec. 1951, in: Speziali (ed) (1972), p. 453.
[2] This is Einstein's definition of the light quantum in March 1905, CPAE vol. 2, doc. 14, p. 133, (transl. ed., p. 87).

incomprehensible rule simply follows from the law for the conservation of energy, $E = h \cdot v$.

- The photoelectric effect, first observed in 1888 by Hallwachs and experimentally analysed by Lenard in 1902: UV radiation that hits a cathode in a vacuum triggers the emission of cathode rays. Lenard attributed the energy of the released radiation to an energy within those particles prior to the emission. Einstein thought it originated from the absorbed UV light quanta, less the work of emission $W_A$. This explains why $E = h \cdot v - W_A$ and is not proportional to the square of the amplitude, as would be expected in classical electrodynamics.
- The short-wave limit of x-ray bremsstrahlung $v < E_{max}$: This, too, follows from the energy conservation law. The energy released in the form of x-rays at frequency $v$ from charged particles that are suddenly slowed down cannot exceed the maximal energy of the slowed-down particles.
- The spectral density of blackbody radiation: Einstein demonstrated the compatibility of his light-quantum hypothesis with Planck's formula for the energy density of blackbody radiation from 1900. It had already been confirmed experimentally, but this provided a deeper theoretical underpinning. (On these experiments see Kangro, 1970; cf. Dorling, 1971; Norton, 2008; Darrigol, 2014 on Einstein's argumentation.)

The harmony between all of these empirically verified predictions or conclusive new explanations of known effects that had hitherto been difficult to understand, if at all, yielded a 'consilience of inductions' as defined by the English philosopher of science William Whewell (1794–1866), one of the strongest indicators of the correctness of a proposition.

Between 1905 and 1925 Einstein himself used various verbal descriptions for his new concept of 'light quanta'. Table 9.1 displays an astonishing variety of expressions that he used to denote his object of study.

**Table 9.1  Einstein's own terminology with first mentions 1905–24, based on my own counting in the original papers Einstein (1905) and Einstein (1924/25)**

| | | |
|---|---|---|
| light quanta *Lichtquanten* | Einstein 1905 | in Einstein (1905) 6 × |
| energy quanta *Energiequanten* | Planck 1900 | in Einstein (1905) 17 × |
| light energy quanta *Lichtenergiequanten* | Planck 1900 | in Einstein (1905) 4 × |
| elementary quanta *Elementarquanten* | Planck 1900 | in Einstein (1905) 2 × |
| energy projectiles *Energieprojektile* | in Einstein (1924) | *Berliner Tageblatt* 3 × |
| light corpuscles *Lichtkorpuskeln* | in Einstein (1924/25) | 1 × |

In a popular lecture, Einstein concluded in 1927 that some experiments supported 'that light be projectile-like in character, hence be corpuscular. [ . . . ] A ray in which energy and momentum are localized in a point shape does not essentially differ from a corpuscular ray; we have revived Newton's corpuscles' (Einstein, 1927, p. 546). Of course, he also admitted that other properties, such as the capacity of light to interfere, 'are not explained by the quantum interpretation.' Faced with this abundance of synonyms, it is rather surprising that—as far as I can see—Albert Einstein never once used the expression 'photon', despite working in an English-speaking setting at the Princeton Institute for Advanced Study from 1933. The following **competing neologisms** by some of Einstein's contemporaries could be added to the same semantic field from this early period. (For specific references see Hentschel, 2018, p. 25.) The Polish physicist Mieczyslaw Wolfke (1883–1947), who defended his postdoctoral thesis under Einstein at the University of Zurich in 1913, spoke of *Lichtatome*. Terminologically this suggests a return to the Newtonian conception of particles, but that is not at all what Wolfke meant. The same applies to a later adaptation of this compound in the Anglo-Saxon variant, 'atoms of light' (used by L.T. Troland in 1917, Ornstein and Zernike (1920), as well as by Ehrenfest & Joffé in 1924. In 1926 Charles D. Ellis delivered three lectures on 'The atom of light and the atom of electricity' at the Royal Institution in London (Ellis, 1926). In France, Louis de Broglie (1922) and Fred Wolfers (1926) referred to *atomes de lumière*. Gilbert N. Lewis added the variant 'particle of light' (Lewis, 1926b), and Lewis and W. Band, 'light corpuscles'. Other formulations include the metaphorical 'bullets' from 1917 by Daniel Frost Comstock (1883–1970), who proposed 'bullets of energy, one might say', to describe 'separate units [of radiant energy] in space' (Comstock and Troland, 1917, I, § 10, 46).

Arthur Holly Compton (1892–1962) later took over this expression as 'light bullets' (Compton, 1925, p. 246: 'light bullets': 'light as consisting of streams of little particles'), and his fellow countryman Robert A. Millikan (1868–1953) also used a similarly visual rewording in a textbook from 1935: 'photon, or light-dart, theory of radiation'.[3] Other exotic oddities include Arthur Llewellyn Hughes's expression from 1914, 'that light was molecular in structure, each "molecule" or unit containing an amount of energy $h\nu$ which could not be subdivided.' He evidently overlooked that molecules are certainly *not* indivisible entities but—quite unlike 'light quanta'—can be split into atoms. This expression also found its followers, such as Wolfke, who took this metaphor literally and considered real bonds between individual light quanta to explain fluctuations in radiation density.[4]

---

[3] Millikan (1935, 2nd edn, p. 259); the term 'light dart' stems from Ludwig Silberstein (1922) and was also adopted by the British physicist Edward Neville da Costa Andrade (1887–1971) in Andrade (1930/36), p. 128 and in 1957.

[4] See Wolfke (1921); cf. further Kojevnikov (2002), p. 200, for a few attempts along these lines, all of which failed.

Such a **multiplicity of terms** is typical of the early developmental phase of a still diffuse, nascent concept. As already mentioned, in Einstein's oeuvre, the term 'photon' is nowhere to be found. Between 1917 and 1926, this term had been introduced independently in four different contexts: two of them in sensory physiology, one in photochemistry, and the last and most famous one in physical chemistry (cf. Kragh, 2014). In 1926, the American physical chemist Gilbert N. Lewis also introduced the term which is now often attributed to him, but *not* in today's meaning. Lewis wrongly presumed that the number of photons is conserved. According to the modern interpretation, however, it is not at all a conserved quantity, because photons are generated in light emission and annihilated in absorption processes.[5] Consequently, G. N. Lewis is definitely *not* the father of the modern photon as a *concept*, even though he was one of the first to use the **term** in the physicochemical context. It is thus important to distinguish between terms, concepts, and mental models. During the period 1926–35, we find in the English-speaking literature 'light quantum' and 'photon' in about equal frequency. The term 'photon' gained the upper hand in the Anglo-American context after 1945, especially after the laser came into experimental realization (see layer (iv) in the next section).

## 9.1 TWELVE LAYERS OF MEANING

As is shown *in extenso* in Hentschel (2018), the complex history of the photon concept is composed of many different layers, many of which were formed well before 1905 while other additional ones emerged only much later. The latest layers of meaning derive from QED and QFT during the second half of the twentieth century.

Altogether, I identified twelve different semantic layers that must be viewed together, in order to be able to understand properly the meaning and usage of the concept of 'light quantum' or 'photon'.

(i) The oldest of these semantic layers is the mental model of light as being similar to a stream of particles, which dates back to Antiquity. However, this earliest proto-concept was radically reshaped many times, over the course of history, especially in Isaac Newton's corpuscular theory of light (Newton, 1704; Eisenstaedt, 2005, 2007; cf. Hentschel, 2018, sections 3.1 and 4.1). We now know that this naive mental model of light quanta as point-like particles is anything but suitable because in experimental situations such as interference, light quanta prove to have considerable extension and

---

[5] See Lewis (1926c) on this supposed 'conservation of photons,' which he had called 'light corpuscles' in the same journal half a year previously, related to naive quasi-Newtonian particulate notions, see Lewis (1926a).

that, generally speaking, the localization of photons is impossible in quantum field theory (Newton and Wigner, 1949).

(ii) That **the velocity of light $c$ is finite** had already been established prior to Newton. Ole Rømer detected this in his observations of time delays in the transits of Jupiter's moons, although he did not give any concrete figure for it. Huygens, Cassini, Halley, and other astronomers later offered the first specific estimates.[6] It was not evident to Newton that **the velocity of light is exactly the same for different colours**. On the contrary, his derivation of the law of refraction seemed to indicate that the degree of refrangibility of light was correlated with its velocity. Newton himself had shown in his *New Theory about Light and Colours* from 1672 that components of light of different colours manifest different angles of refraction. This suggested that the variously coloured components of light either had different mass or that they would propagate at different velocities through the same medium. Because the red component of the spectrum was the least refractive, according to the latter assumption, it actually ought to be the most rapidly moving one. Then the particles of light at the other, blue end of the spectrum would be the slowest relative to the other colours. In 1691, Newton asked the Astronomer Royal, John Flamsteed (1646–1719), to observe Jupiter's moons. Does a terrestrial observer first perceive the red component of the light right after their transits behind the planet and the blue component only afterwards? Flamsteed's negative reply dissuaded Newton of the hypothesis that red light must be faster than blue light.[7]

It took until the nineteenth century for experimental technology to advance far enough along to be able to measure the velocity of light in terrestrial experiments. With sophisticated instrumentation, such as reliable rapidly rotating mirror systems or interferometers, it was shown that the velocity of light did not change when an emitter of light or the medium in which the light is propagating was itself in motion relative to the observer. In Einstein's special theory of relativity, the **constancy of the velocity of light in vacuum** was elevated to the status of a postulate. This is also valid for Einstein's interpretation of light as photons. The velocity of light in vacuum is thus independent of the velocity of the emitter.

(iii) The third semantic layer can also be traced back to Antiquity. It ranges at least up to Einstein's famous papers about induced emission from 1916 and 1917 and to QED. In query 5 of the *Opticks*, Newton wrote in 1704: 'Do not Bodies and Light act mutually upon one another; that is to say, Bodies upon Light in emitting, reflecting, refracting and inflecting it, and Light upon Bodies for heating them, putting their parts into a vibrating motion wherein heat consists?' (Isaac Newton: *Opticks* (1704), ed. by I.B. Cohen, New York: Dover 1952, p. 339) The mental model of light as an emission of light particles from an emitter and a re-absorption at an absorber easily

---

[6] See Wroblewski (1985) and a special issue of *Centaurus*, 54 (2012), pp. 1–102 about Rømer.

[7] Turnbull et al., eds. (1959–77), vol. 3, p. 202; Shapiro, 1993, p. 218; Eisenstaedt (2005), (2007), p. 30.

carried over into Einstein's theory of light quanta, and into Bohr's and Sommerfeld's atomic model.

(iv) As with ping-pong balls volleyed between two table-tennis players, such an **emission and absorption** of light quanta implied a **transfer of momentum**. It became measurable after 1900 in sophisticated experiments. Mirrors suspended under controlled conditions *in vacuo* and exposed on one side to intense radiation led to a small but traceable rotation of the mirrors.[8] Einstein was aware of this recent experimental confirmation of radiation pressure and mentioned it in his 1905 paper. He thus showed that his theory of light quanta was also able to cope with light pressure which had been explained classically on the basis of Maxwell's electrodynamics by Pointing (1884).

(v)–(vii) Both Einstein's photon conception as well as Maxwell's interpretation of light as electromagnetic waves implied that such a **transfer of momentum was accompanied by a transfer of energy,** which in turn became measurable around 1900 in the so-called **photoelectric effect.** Electromagnetic radiation of variable wavelength hits a cathode in an evacuated cathode-ray tube. Contrary to classical expectations, the **energy of the emitted electrons did not depend on the intensity of the incoming radiation but only on its frequency.** In 1916, Robert Millikan—against his own expectations—firmly established the linear **dependency of the energy $E$ on the frequency $\nu$:** $E = h\,\nu$ (Millikan, 1916a,b). This confirmed Planck's energy quantization as well as Einstein's photon hypothesis—even if Millikan and Compton baulked at the idea. Furthermore, this result now also allowed a precise experimental determination of Planck's constant $h$.[9]

In 1923, such a **transfer of energy and momentum** was also demonstrated in the **Compton effect,** i.e., in Arthur H. Compton's original experiment, the scattering of x-rays by electrons bound to graphite atoms (cf. Stuewer (1975) and primary literature cited there). Nevertheless, how exactly to interpret the way that light might 'transfer' momentum remained unclear for a long time. Did it mean that the light quanta attributed to light have their own non-vanishing mass at rest? At first this was occasionally supposed: around 1922 Louis de Broglie (1892–1987), for instance, estimated the mass of a light quantum as finite, but smaller than $10^{-50}$g. Nowadays, photons are usually assumed to be **strictly massless energy quanta of the electromagnetic field** (see Okun 2006 about experimental limits $< 7 \times 10^{-17}$ eV>. Nevertheless, in phenomena such as radiation pressure and the Compton effect, light transfers energy and momentum onto material particles. For many contemporaries, it seemed self-contradictory that a 'massless' light quantum should nevertheless act like a material particle in collision processes such as the Compton effect. But from the formalism of the special theory of relativity for the

---

[8] See Hentschel (2018), section 3.4 on the pioneering experiments of Lebedew (1901) and Nichols and Hull (1901, 1903), and on earlier failed attempts by Bennet (1792) and others.

[9] On the photoelectric effect cf. Hughes (1914), Wheaton (1978), Brush (2007), and Hentschel (2018), section 3.6.

squared four-vector $(\mathbf{p},E)$ of momentum $\mathbf{p}$ and energy $E$, it follows that for a vanishing mass $m = 0$:

$$E^2 = (pc)^2 + (mc^2)^2 = (pc)^2 + 0 \rightarrow E = pc.$$

This **ultra-relativistic limit valid for photons** inheres a strict proportionality between energy and momentum $p = E/c = h\nu/c$. In the photon-based explanation for the Compton effect, the same formula reappeared and yielded a mental model of a relativistic scattering of light quanta in matter. The perfect fit between theoretical prediction and Compton's experimental results provided strong support for Einstein's photon hypothesis in the mid-1920s, and disarmed Einstein's critics.

(viii) In 1909, Einstein turned to another experimentally accessible area in which quantum phenomena are manifest—namely, fluctuations in their spatial distribution and in radiation pressure. Although these fluctuations are normally far too small to be directly detectible, systems exist that significantly augment such fluctuations statistically. Under suitable circumstances, such fluctuations even become macroscopically visible. In a lecture delivered at a conference in Salzburg in 1909, Einstein presented a thought experiment concerning a mirror in a vacuum. He considered a freely suspended mirror with a surface $f$ positioned between two radiation fields $V_1$ and $V_2$. When energy fluctuations occur in the two partial volumes $V_1$, $V_2$, then the mirror at the dividing plane is bombarded from both sides with innumerable collisions of various intensities, causing it to fluctuate between the left and right volume. This trembling motion of the mirror resembles Brownian molecular motion, which Einstein had explained in his third world-famous paper from late 1905. This sophisticated thought experiment demonstrated once again that the radiation field transmits not only energy of electromagnetic waves onto the mirror but also momentum (semantic layer iv). For the changes in radiation pressure $\Delta$ on the mirror surface $f$ during the time interval $\tau$ from random fluctuations in the radiation field at energy density $\rho$, Einstein obtained:

$$\Delta^2 = f\tau d\nu/c[\mathrm{h}\,\nu\,\rho + c^3\rho^2/(8\pi\nu^2)]$$

Analogously, because of the independent motion or interference 'of narrowly extended complexes of energy $h\,\nu$' in volume $V$, the energy fluctuation $\Delta E$ is:

$$(\Delta E)^2 = V d\nu\,[\mathrm{h}\,\nu\,\rho + c^2\rho^2/(8\pi\nu^2)]$$

Both expressions in square brackets on the right-hand side have the same form. They are a sum of two terms, the first of which could be traced back to a collection of corpuscular light quanta of energy $h\,\nu$, whereas the second term could be derived under the precondition of interference between waves of frequency $\nu$, derivable from Maxwell's theory of continuous electromagnetic radiation. This curious result led Einstein to a peculiar duality between wave-like and particle-like aspects of the radiation field. The correct, complete result only came from the sum of these two terms. Einstein concluded that 'the two structural properties (the undulatory structure and the quantum structure) simultaneously displayed by radiation according to the Planck formula should not be considered as mutually incompatible.' (Einstein, 1909, 499–500, in CPAE (1989–2015), vol. 2, 393–4; cf. Klein, 1964)

This was a far-sighted anticipation of what later became known as the **wave-particle duality** (for surveys see Wheaton, 1983 and Bhatta, 2021). In retrospect, it is surprising to modern readers that another decade had to elapse before physicists were able to gain clearer insight into the dual nature of light. Thus the photon manifests the paradoxical consequences of wave–particle duality in its most extreme form: Depending on the experimental situation, the photon appears to us either as a wave or a particle, but *prior* to this measurement or the foregoing preparations associated with the quantum-mechanical state, a commitment to one or the other is impossible and futile. (Redhead, 1983 and Cao, 1998, 18-9, 170-3 speak of a new 'ephemeral ontology': photons are quasi-particles capable of being created or absorbed, without permanent existence and individuality, and likewise fields losing their continuity.)

(ix) In 1916, Einstein rederived the Planck formula for the density of energy distribution in a black body on the basis of statistical considerations on the emission and reabsorption of radiation by matter. For every quantum jump from a higher to a lower energy level there exists a probability A, but aside from this 'spontaneous emission'—a causally undetermined process—the emission of electromagnetic radiation is also increased by the presence of other radiation of the same frequency. The probability B of this second process, called 'induced emission', thus depends on the electromagnetic field intensity already present in the vicinity of the emitter (Einstein, 1916, 1917). This is made use of in lasers, in which diffuse energy pumped into an active medium is converted into coherent electromagnetic radiation, as shown schematically in fig. 9.1. The experimental realization of 'light amplification through stimulated emission of light' (laser) in 1960, by Theodore Maiman (1927–2007), who used a ruby crystal to create red laser-light, brilliantly confirmed Einstein's prediction of stimulated emission. (On the history of the laser see e.g., Bromberg, 1991; Hecht, 2005.) Today lasers are one of the most often used spin-offs of physical experimentation in all kind of scientific, technological, industrial, and commercial branches; and 'photonics' centres were created world-wide.[10]

(x) **Doublet splitting** spectra with only two component lines were quite frequent in spectroscopy, but they posed a serious problem for the semiclassical Bohr–Sommerfeld atomic model. According to its multiplicity formula $2m + 1$, which had led to excellent fits with the normal Zeeman splitting of spectral lines emitted in electric and magnetic

---

[10] On the history of the term photonics and of photonics as a research branch see Krasnodebski (2018).

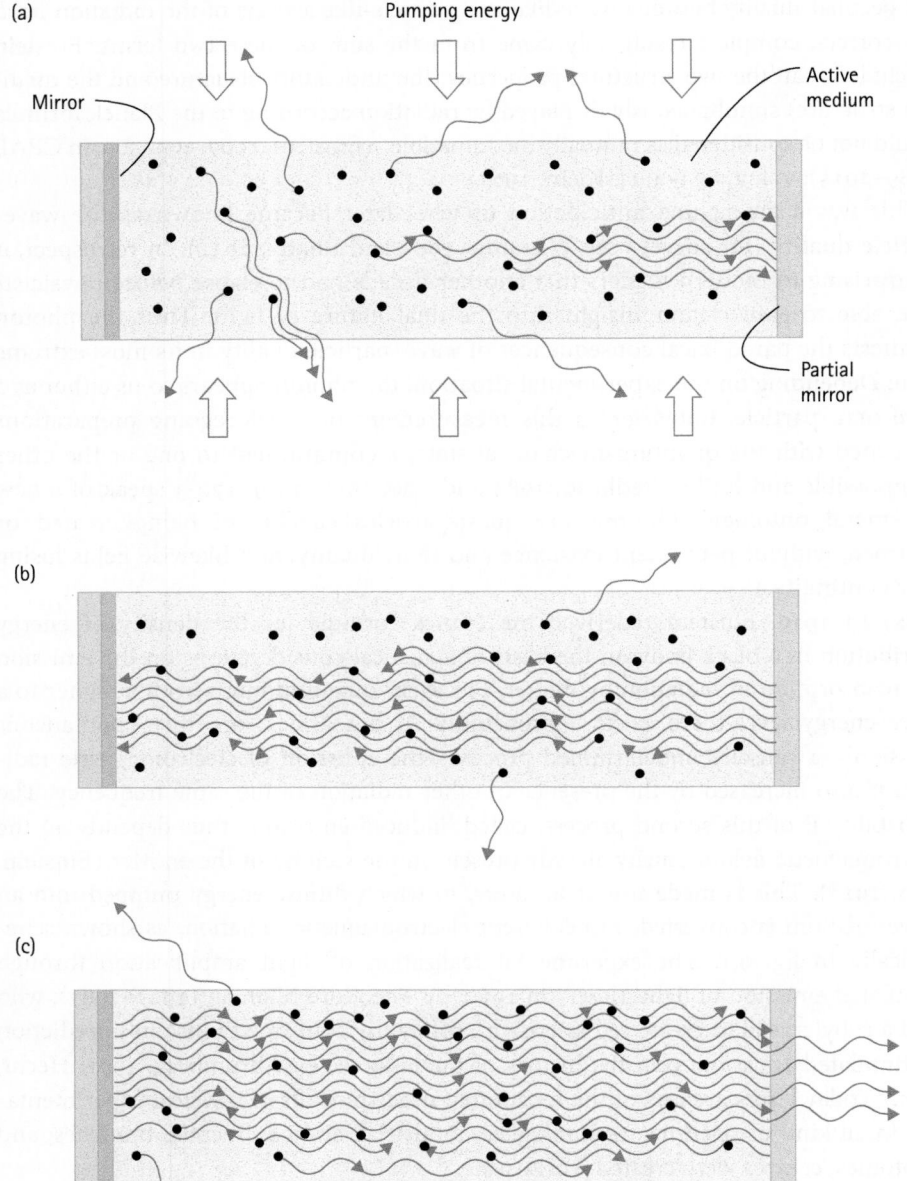

FIGURE 9.1 In lasers, light rays volleyed back and forth between two mirrors cause an amplifying cascade of induced emissions of coherent and monochromatic radiation in the laser's gain medium, which form a coherent beam of light upon exiting out of the resonance area (through a partial mirror on the right). From http://abyss.uoregon.edu/~js/images/laser_pump.gif (accessed 17 March 2016).

fields, $m$ should not be integral here, but half-integral. Furthermore, experiments by Otto Stern and Walther Gerlach in Frankfurt in 1922 on atomic beams of highly heated silver sent through an inhomogeneous magnetic field also yielded two values for the magnetic moment of the silver atoms. Werner Heisenberg (1901–1976), at that time still a young student of Sommerfeld in Munich, speculated in 1922 that a new **half-integer quantum number** was a kind of average between two actually integral quantum numbers, one half of which could be associated with the atomic core and the other half to the electronic shell. Another young student of Sommerfeld, Wolfgang Pauli (1900–1958), arrived at the even more outlandish interpretation of a half-integer quantum number: a 'mechanically indescribable ambiguity', characteristic of the outermost optically active electron (*Leuchtelektron*). In January 1925, Pauli introduced a new quantum number into a mathematical description of this ambiguity, initially denoted by the letter $\mu$, with $\mu = \pm \frac{1}{2}$, and also postulated that a new selection criterion $\Delta \mu = 0$ or $\pm 1$ must apply for physically permissible transitions. Each electron was hence described by a set of four quantum numbers $n$, $l$, $m$, and $\mu$ (sometimes also denoted as $n$, $l$, $j$, and $s$).[11]

The introduction of **spin** came at the expense of another radical, if not audacious abrogation in classical physics, though. Pauli considered classical physics fundamentally incapable of covering such a 'mechanically indescribable ambiguity'. At this point the young upstarts Heisenberg and Pauli were standing on the very threshold between the old semiclassical quantum theory and quantum mechanics, which they would begin to develop a few months later in mid-1925.

(xi) When Heisenberg, Dirac, Schrödinger, and others created quantum mechanics in 1925/26, Pauli's assumption of spin as the fourth quantum number of electrons carried over into the new framework. It soon became clear that not only electrons, but photons as well had an additional in-built property, a spin. Interpreted as indicator of an internal angular momentum, this spin could have the values of +1 or −1, depending on the orientation of the photon with respect to the exterior field (0 is excluded because of the photon's vanishing rest mass). Taken together with their **indistinguishability** (cf., e.g., Redhead & Teller 1992), a property first clearly spelled out by the Polish physicist Ladislas Natanson (1864–1937) in 1911, this led to a completely new statistics of light quanta, distinguishing them further from classical particles (obeying Boltzmann statistics) and from electrons and other fermions (with spin ½). The new **class of particles with integer spin was later redubbed bosons,** and the corresponding statistics was called **Bose-Einstein statistics.** The Indian physicist Satyendra Nath Bose (1894–1974) had written a letter to Albert Einstein initiating the publication of several papers

---

[11] On the history of spin see Meyenn (1979, vol. 1, 1980/81), as well as van der Waerden (1960) and Tomonaga (1974/97), lecture 2, Serwer (1977), and Heilbron (1983). Experimental evidence for spinning photons was provided in spectroscopy by Raman and Bhagavantam (1931), Bhagavantam (1932) in Raman spectra.

in the Proceedings of the Prussian Academy of Sciences in 1924. Bose (1924) and Einstein (1924/25)—soon followed by others—spelled out the strange implications of the new statistics soon named after them: **clumping, clustering or bunching of bosons** with similar quantum properties. Whereas electrons and other fermions obey the Pauli principle, which thus precludes the co-presence of two fermions with the same quantum numbers in the same region, bosons cluster together. The clustering of many photons in laser light discussed above is a prime example of it. **Bose–Einstein condensation** is another consequence, which became experimentally accessible in 1995. For photons in particular, photon clumping became verifiable in the Hanbury Brown and Twiss (HBT) effect and in photon bunching experiments.[12]

The **HBT effect** (1955–57) derived from interferometric experiments by Robert Hanbury Brown (1916–2002) and his collaborator Richard Twiss (1920–2005) to measure the size of Sirius, one of the nearest fixed stars, comparing correlations of light intensities from both sides of the star. They installed two concave mirrors and photodetectors a few metres apart from each other. After successfully testing the arrangement, they demonstrated that with increasing distance between the two mirrors there is a rapidly decreasing positive cross-correlation between the two detectors. This allowed them to provide an estimate of the angular size of Sirius, but the very low intensity of recorded photons also implied 'photon bunching', i.e., a tendency of photons to cluster together (Hanbury Brown and Twiss (1956a,b, 1957), Silva and Freire (2013), and further references there).

A similar **photon bunching** was later observed in the so-called **Hong–Ou–Mandel dip** of 1987, established by Leonard Mandel (1927–2001) and collaborators in Rochester. Such photon correlation experiments frequently showed that at very low intensities correlations between two photons from the same source were statistically *not* evenly distributed anymore. For intervals lesser than or equal to the coherence time $\tau = \lambda/c$, the probability of counting two or more pulses at a time was higher than the statistical Poisson distribution would have led one to expect. In 1956 Edward M. Purcell (1912–1997) initially named this phenomenon 'clumping' in connection with the experiment by Hanbury Brown and Twiss, but later he mostly referred to 'bunching of light' or 'photon bunching'. This tendency of photons to cluster in groups is derivable from Bose–Einstein statistics.

(xii) With the advent of quantum electrodynamics (QED) and quantum field theory (QFT) in the late 1920s and 1930s, yet another meaning arose for photons: their **interpretation as the massless exchange particle of electromagnetic interaction**. The most important characteristics and limiting conditions of their interaction were derived from requirements such as gauge invariance and invariance in Lorentz

---

[12] On the following, see Silva and Freire (2013), Hentschel (2018), sec. 8.1, 8.6–8, and further references there.

transformations.[13] Unlike the older semiclassical theories, QED built upon this pioneering research and from the late 1940s began to develop apace. A quantization of energy occurs not only for matter, but also for the electromagnetic field itself. The photon was thus reinterpreted as the exchange particle of electromagnetic interactions. This may be imagined as a ping-pong ball (the photon) that is volleyed back and forth between two players (the charged particles), whose actions are related to each other only by these volleys. Ping-pong balls and other material exchange particles have a non-vanishing mass of their own. This is not the case with photons, That is precisely why electromagnetic interactions have such a much longer range than other interactions mediates by massive bosons. By contrast, the strong and weak nuclear forces in the interior of the atomic nucleus, responsible for beta decay and for atomic cohesion, are relayed by ponderous exchange particles with a very limited range not exceeding the radius of an atom (Cao, 1998, 182f.).

According to QED, electromagnetic interaction is effectuated by a virtual exchange of photons of frequency $v$ for a very short interval of time $\Delta t \leq h/v$. Richard Feynman (1918–1988) hit upon an ingenious way to visualize schematically lengthy calculations of perturbation theory to higher orders in 1948. They soon became known as 'Feynman diagrams'.[14] Straight lines represent elementary particles, particularly electrons, and wavy lines represent photons, whereas the time axis in these diagrams runs from below upwards (see fig. 9.2).

The point to keep in mind, though, is that electromagnetic interaction is the effect of the summation of *all* of these virtual processes in all orders of magnitude, so that it is incorrect to (mis)interpret the wavy line in the above diagram as the real trajectory of a photon—it is just one possibility (leading to one term in lowest order) among plenty of others (in higher orders of perturbation theory).[15]

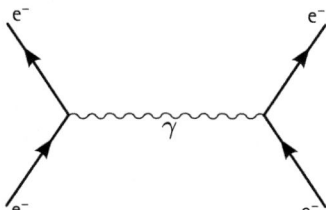

FIGURE 9.2 An elementary interaction between two electrons, with a virtual photon as mediator of electromagnetic interaction.Source: https://de.wikipedia.org/wiki/Datei:Feynmandiagram.svg (accessed 10 February 2021).

[13] On QED see, e.g., Feynman (1961) and Jauch and Rohrlich (1955), which only proceed from abstract premises. Feynman (1985) provides an intuitive introduction. Schweber (1994), Miller (1994) provide QED history.

[14] On Feynman's life and work see, e.g., Mehra (1994) and Schweber (1994), chapter 8. The history of Feynman diagrams and their application in quantum field theory is recounted by Kaiser (2004) and Wüthrich (2011).

[15] On the controversial issue of whether virtual particles are nothing but formal tools, see Cao (1998), pp. 20–23, 176ff., Bacelar Valente (2011) vs Fox (2008), and further sources mentioned there.

# 9.2 HOW TO INTERPRET PHOTONS TODAY?

We must give up the local-realist interpretation that is familiar to us from macroscopic experience with particles since the days of the Greek atomists. Newton and, from 1900 on, the quantum pioneers also could not resist the temptation to interpret light handsomely as a stream of 'atoms', 'corpuscles', 'bullets', or even 'molecules of light' or 'quanta'. An operationalist or instrumentalist interpretation would resist our habitual 'objectivizing' in everyday life, — what is referred to as 'reification of entities' in natural philosophy. The photon is merely a man-made concept, assumed in an attempt to model phenomena in the physics laboratory and in the world we live in. Our intuition and our concepts must constantly be adapted to our environment and our knowledge about it in order for this orientation to be consistent and pragmatically appropriate: Mental models of particles and waves, derived from the macroscopic world, are inappropriate to describe the microcosmos of quantum phenomena.

In particular, we have to avoid illegitimate attributions of spatiotemporal photon locality, or of their supposed smallness. Of course, experiments were performed with high-energy gamma or x-rays, in which the energy and momentum of photons was transmitted over long distances without causing any noticeable broadening. This blatantly contradicts the classical model of a spreading spherically symmetrical wave and only encourages such metaphors as 'needle rays' or 'light bullets'. That was why experimenters like J. J. Thomson were motivated early on to consider corpuscular models, which Johannes Stark or A. H. Compton, for instance, connected with Einstein's light-quantum hypothesis after 1905 (see layers (v)–(vii) in section 9.1). The scattering experiments by Compton and Simon (1922–27), as well as the later ones by Bothe and Geiger (1924–25) and by Raman in 1930, also seemed to suggest that after interacting with electrons electromagnetic radiation is emitted in very definite directions and not in the shape of spherical wave fronts. However, in 1927 Arthur Jeffrey Dempster (1886–1950) and Harold F. Batho, two coworkers of A.H. Compton in Chicago, published diffraction experiments performed with an echelon grating at the helium line 4471$\lambda$ at very low light intensities, corresponding to an average number of only 95 light quanta per second at this wavelength. A clearly recognizable interference pattern still appeared on a surface measuring 32 square millimetres set at a distance of 34 cm behind the echelon grating. The two experimenters concluded that even single light quanta must be able to interfere with themselves over larger surfaces in space 'when the quanta emitted from the volume of the source used were completely separated in time, showing that a single quantum obeys the classical laws of partial transmission and reflection at a half-silvered mirror and of subsequent combination with the phase difference required by the wave theory of light' (Dempster and Batho, 1927, p. 644). Subsequent experiments, with improved photon detectors and working increasingly clearly within the regime of lowest photon numbers and light intensities, confirmed the hypothesis that in these cases photons occupy spatially very extended regions of space-time and definitely should not be imagined as almost point-shaped. According to quantum electrodynamics, photons are quantized states of the electro-magnetic field whose energy generally belongs to the whole region of space occupied by the radiation field. In general, localized states do not exist *prior* to a measurement, which depending on the interpretation leads either to a collapse of the state function or to decoherence. Quantum field theorists have shown that in QFT in general no position operator exists for photons: 'in any case, the energy of a photon is distributed over the entire volume of the field and there is, in general, no use in attaching a coordinate to the photon' (Strnad 1986, p. 650; cf. Newton and Wigner 1949). It is true: 'a fully-quantized theory of matter-radiation interaction can lend a characteristic of spatial discreteness to the photon when it interacts with a finite-sized atom', but there are many other types of interaction, and in general 'there is no particle-creation operation that creates a photon at an exact point in space' (quotes from Muthukrishnan, Scully, and Zubairy (2003), S24–S25, thereby explicitly withdrawing earlier contrary claims in Scully and Sargent, 1972).

There is another heritage of our evolutionary origins that we must abandon in the quantum world: the classical premise of the individualizability of particles. Although peas in a pod all look so similar, in principle it is possible to number them or attach different labels to each individual one of them. In the case of elementary particles or photons, this is no longer valid. They do not fall under classical statistics, for which the possibility of individualization (in Latin, *haecceitas*) is presumed (see, e.g., French (2015) as well as Lyre in Friebe et al. (2015), chapter 3). They are subject to very different quantum statistics. Our indistinguishable light quanta (photons) are spin-1 particles, i.e. bosons, which obey Bose-Einstein statistics (cf. layer (xi) in section 9.1). This inability to individualize, this impossibility to associate an individual photon with one of the two beams, is essential to the formation of interference patterns. Such which-way interference experiments have demonstrated this, with the interference pattern vanishing as soon as we have information about the path that an individual photon has taken. Even if—on the average—there is only one photon at a time, interference can still occur, but any knowledge about the path taken will immediately destroy the interference pattern (Grangier *et al.*, 1986).

Today the light quantum or photon is accepted as a convenient conceptual model, as a useful mental crutch. Countless physicists, optics specialists, and technicians lean on this very helpful conceptual label in their daily perambulations, true to Roy J. Glauber's motto: 'I don't know anything about photons, but I know one when I see one.'[16] Taking this shirt-sleeve attitude, we can very simply define the photon operationally: 'A photon is what a photodetector detects'.[17] But it remains controversial to this day

---

[16] Bon mot by Glauber at the summer school in Les Houches in 1963, quoted as the motto of the paper by Holger Mack and Wolfgang Schleich in Roychoudhuri and Roy (2003), p. 28.

[17] quoting Roy J. Glauber from the contribution by Scully *et al.* in Roychoudhuri and Roy (2003), p. 18.

whether these photons really exist, and if so, in what sense; in other words, by which underlying interpretation. Some (such as Marlan O. Scully or Anton Zeilinger) consider the photon a fascinating, intrinsically quantum-mechanical object that is indispensable for the purposes of quantum optics or quantum field theory (Scully and Zubairy, 1997, chaps. 1 & 21; Grangier, 2005; Zeilinger et al., 2005; Zeilinger, 2010; Shadbolt et al., 2014). Others—including the Nobel prizewinner Willis Lamb—consider it an anachronism, spare ballast, or even a nuisance. But how seriously should we take such heretical conclusions? They are the result of burgeoning usages of the word 'photon'. The ability to explain the outcomes of many semiclassical experiments without necessarily assuming photons does not prove that the concept of photons is superfluous since there are, after all, many other experiments which do require such a concept. Single-photon experiments, EPR experiments with photon–photon correlations in the 1980s by Aspect and collaborators in Paris (Aspect, 1982; Grangier et al., 1986), as well as which-way experiments implementing Wheeler's delayed choice (for references see Ma et al., 2016) and quantum teleportation experiments with single photons over many miles distance (see, e.g., Zeilinger (2010), Pirandola et al. (2015), Hentschel (2018), sections 8.4 and 8.9 for recent surveys with references), all confirm the indispensability of the photon concept, as strange and mysterious as it might still appear to us.

## REFERENCES

Andrade, E. N. (1930/36). *The Mechanism of Nature.* London: Bell & Sons.

Aspect, A., Grangier, P., and Roger, G. (1982). Experimental realization of Einstein–Podolsky–Rosen–Bohm Gedankenexperiment—a new test of Bell's inequalities. *Physical Review Letters*, **49**(2), 91–94.

Bacelar Valente, M. (2011). Are virtual quanta nothing but formal tools? (Reply to Fox, 2008). *International Studies in the Philosophy of Science*, **25**, 39–53.

Badino, M. (2015). *The Bumpy Road—Max Planck from Radiation Theory to the Quantum 1896–1906.* Berlin: Springer.

Bhagavantam, S. (1932). Evidence for a spinning photon. *Indian Journal of Physics*, **7**, 107–138.

Bhatta, V. S. (2021). Critique of wave-particle duality of single photons. *Journal for General Philosophy of Science* 52.

Bose, S. N. (1924). Plancks Gesetz und Lichtquantenhypothese. *Zeitschrift für Physik*, **26**, 178–181.

Bothe, W., and Geiger, H. (1924). Ein Weg zur experimentellen Nachprüfung der Theorie von Bohr, Kramers, und Slater. *Zeitschrift für Physik*, **26**, 44–58.

Bothe, W., and Geiger, H. (1925). Über das Wesen des Comptoneffekts: Ein experimenteller Beitrag zur Theorie der Strahlung. *Zeitschrift für Physik*, **32**, 639–663.

Bromberg, J. L. (1991). *The Laser in America, 1950–1970.* Cambridge, MA: MIT Press.

Cao, T. Y. (1998). *Conceptual Development of 20th Field Theories.* Cambridge: Cambridge University Press.

Compton, A. H. (1925). Light waves or light bullets? *Scientific American*, **133** (October), 246–247.

Compton, A. H., and Simon, A. W. (1925). Directed quanta of scattered x-rays. *The Physical Review*, **26**(2), 289–299.

Comstock, D. F., and Troland, L. T. (1917). *The Nature of Matter and Electricity: An Outline of Modern Views*. New York: Van Nostrand.

Darrigol, O. (2014). The Quantum Enigma. In Michel Janssen and Christoph Lehner (eds), *The Cambridge Companion to Einstein*, Cambridge: Cambridge University Press, pp. 117–142.

de Broglie, L. (1922). Rayonnement noir et quanta de lumière. *Journal de Physique*, **6**(3), 422–428.

de Broglie, L. (1923). Radiation: Ondes es quanta. *Comptes Rendus*, **177**, 507–510.

Brush, S. G. (2007). How ideas became knowledge: The light-quantum hypothesis 1905–1935. *Historical Studies in the Physical and Biological Sciences*, **37**, 205–246.

CPAE (1989–2015). *The Collected Papers of Albert Einstein*. Princeton: Princeton University Press. Vol. 2: *The Swiss Years: Writings 1900–1909*, eds. John Stachel et al., 1989. Vol. 6: *The Berlin Years: Writings 1914–1917*, eds. A.J. Kox et al., 1996. Vol. 14: *The Berlin Years: Writings and Correspondence April 1923–May 1925*, eds. Diana Kormos Buchwald et al., 2015.

Dempster, A. J., and Batho, H. F. (1927). Light quanta and interference. *Physical Review*, **30**(2), 644–648.

Dorling, J. (1924). Das Kompton'sche Experiment. *Berliner Tageblatt*, 20 April, suppl., p. 1. Reprinted in CPAE, Vol. 14 (2015) doc. 236, pp. 364–367. English trans. by A.M. Hentschel: 'The Compton experiment. Does science exist for its own sake?', pp. 231–234.

Dorling, J. (1971). Einstein's introduction of photons—Argument by analogy or deduction from the phenomena? *British Journal for the Philosophy of Science*, **22**, 1–8.

Einstein, A. (1905). Über einen die Erzeugung und Umwandlung des Lichtes betreffenden heuristischen Standpunkt. *Annalen der Physik*, **17**(4), 132–148. Reprinted in CPAE, Vol.2 (1989) doc. 14, pp. 134–169; trans. by A. Beck: 'On a heuristic point of view concerning the production and transformation of light', pp. 86–103.

Einstein, A. (1909). Über die Entwicklung unserer Anschauungen über das Wesen und die Konstitution der Strahlung. (a) *Physikalische Zeitschrift*, **10**, 817–825, (b) *Verhandlungen der Deutschen Physikalischen Gesellschaft*, **7**, 482–500, reprint in CPAE, Vol. 2 (1989) doc. 60; trans. ed.: On the development of our views concerning the nature and constitution of radiation, pp. 379–394.

Einstein, A. (1916). Strahlungs-Emission und -Absorption nach der Quantentheorie. *Verhandlungen der Deutschen Physikalischen Gesellschaft*, **18**, 318–323. Reprinted in CPAE, Vol. 6 (1996), doc. 34, pp. 363–370. English trans.: 'Emission and absorption of radiation in quantum theory', pp. 212–217.

Einstein, A. (1917). Zur Quantentheorie der Strahlung. *Physikalische Zeitschrift*, **18**, 121–128. Reprinted in CPAE, Vol. 6 (1996), doc. 38, pp. 381–398. English trans. by A. Engel: 'On the quantum theory of radiation', pp. 220–233.

Einstein, A. (1924/25). Quantenthesorie des einatomigen idealen Gases I–III, *Sitzungsberichte der Preußischen Akademie der Wissenschaften, math.-phys. Klasse* (a) I, 1924: 261–267; (b) II & III, 1925: 3–14, 18–25. Reprinted in CPAE, Vol. 14, docs. 283, 385, 427, pp. 433–441, 580–594, 648–657. Trans. ed.: 'Quantum theory of the monatomic ideal gas', pp. 276–283, 371–383, 418–425.

Einstein, A. (1927). Theoretisches und Experimentelles zur Frage der Lichtentstehung. *Zeitschrift für angewandte Chemie*, **40**, 546.

Eisenstaedt, J. (2005). *Avant Einstein. Relativité, lumière, gravitation*. Paris: Presses Univ. de France.

Eisenstaedt, J. (2007). From Newton to Einstein. A forgotten relativistic optics of moving bodies. *American Journal of Physics*, **75**, 74–79.

Ellis, C. D. (1926). The light-quantum hypothesis. *Nature*, **117**, 895–897.

Feynman, R. P. (1961). *Quantum Electrodynamics*. New York: Benjamin.

Feynman, R. P. (1985). *QED. The Strange Theory of Light and Matter*. Princeton: Princeton University Press.

Fox, T. (2008). Haunted by the spectre of virtual particles: A philosophical reconsideration. *Journal for General Philosophy of Science*, **39**(1), 30–51.

French, S. (2015). Identity and Individuality in Quantum Theory. *Stanford Encyclopedia of Philosophy* at https://plato.stanford.edu/entries/qt-idind/ (3 August 2015).

Friebe, C. *et al.* (eds) (2015). *Philosophie der Quantenphysik*. Heidelberg: Springer.

Gearhart, C. A. (2010). 'Astonishing successes' and 'bitter disappointment': The specific heat of hydrogen in quantum theory. *Archive for History of Exact Sciences*, **64**, 113–202.

Grangier, P., Roger, G., and Aspect, A. (1986). Experimental-evidence for a photon anticorrelation effect on a beam splitter—a new light on single-photon interferences. *Europhysics Letters*, **1**(4), 173–179.

Grangier, P. (2005). Experiments with Single Photos. *Séminaire Poincaré*, **2**, 1–26.

Hanbury Brown, R., and Twiss, R. W. (1956a). A test of a new type of stellar interferometer on Sirius. *Nature*, **178**, 1046–1048.

Hanbury Brown, R., and Twiss, R. W. (1956b). Correlation between photons in two coherent beams of light. *Nature*, **177**, 27–29.

Hanbury Brown, R., and Twiss, R. W. (1957). The question of correlation between photons in coherent beams of light. *Nature*, **179**, 1128–1129.

Hecht, J. (2005). *Beam. The Race to Make the Laser*. Oxford: Oxford University Press.

Heilbron, J. L. (1986). *The Dilemmas of an Upright Man: Max Planck and the Fortunes of German Science*. Berkeley, CA: University of California Press.

Heilbron, J. L. (1983). The origin of the exclusion principle. *Historical Studies in the Physical Sciences*, **13**, 261–310.

Hentschel, K. (2018). *Photons. The History and Mental Models of Light Quanta*. Cham, Switzerland: Springer.

Hong, C. K., Ou, Z. Y., and Mandel, L. (1987). Measurement of subpicosecond time intervals between two photons by interference. *Physical Review Letters*, **59**, 2044–2046.

Hughes, A. L. (1914). *Photo-Electricity*. Cambridge: Cambridge University Press.

Hund, F. (1984). *Geschichte der Quantentheorie*. Mannheim: Bibliographisches Institut.

Jauch, J. M., and Rohrlich, F. (1955). *The Theory of Photons and Electrons*. New York, Heidelberg, Berlin: Springer.

Kaiser, D. (2004). *Drawing Theories Apart. The Dispersion of Feynman Diagrams in Postwar Physics*. Chicago: University of Chicago Press.

Kangro, H. (1970). *Vorgeschichte des Planckschen Strahlungsgesetzes*. Wiesbaden: Steiner.

Klein, M. J. (1962). Max Planck and the beginnings of the quantum theory. *Archive for History of Exact Sciences*, **1**, 459–479.

Klein, M. J. (1963). Einstein's first paper on quanta. *Natural Philosopher*, **2**, 59–86.

Klein, M. J. (1964). Einstein and the wave–particle duality. *Natural Philosopher*, **3**, 1–49.

Kojevnikov, Alexei (2002). Einstein's Fluctuation Formula and the Wave-Particle Duality. In *Einstein Studies in Russia*, ed. by Yuri Balashov & Vladimir Vizgin (Boston: Birkhäuser), pp. 181–228.

Kragh, H. S. (2014). The names of physics: plasma, fission, photon. *European Physical Journal*, H **39**, 263–281.

Krasnodębski, M. (2018). Throwing light on photonics: The genealogy of a technological paradigm. *Centaurus*, **60**(2), 3–24.

Kuhn, T. S. (1978). *Black-Body Theory and the Quantum Discontinuity 1894–1912*. Oxford: Oxford University Press.

Lamb, W. E. (1995). Anti-photon. *Applied Physics B*, **60**, 77–84.

Lewis, G. N. (1926a). The nature of light. *Proceedings of the National Academy of Sciences*, **12**, 22–29.

Lewis, G. N. (1926b). Light waves and light corpuscles. *Nature*, **117**, 236–238.

Lewis, G. N. (1926c). The conservation of photons. *Nature*, **118**, 874–875.

Ma, X.-S., Kofler, J., and Zeilinger, A. (2016). Delayed-choice gedanken experiments and their realizations. *Reviews of Modern Physics*, **88**, 015005

Mehra, J. (1994). *The Beat of a Different Drum—The Life and Science of Richard Feynman*. Oxford: Clarendon Press.

Meyenn, K. von (1980/81). Paulis Weg zum Ausschließungsprinzip. *Physikalische Blätter*, **36** (1980), 293–298; **37** (1981), 31–19.

Meyenn, K. von, (ed.) (1979). *Wolfgang Pauli: Wissenschaftlicher Briefwechsel mit Bohr, Einstein, Heisenberg u.a.* Vol. 1: *(1919–1929)*. Berlin: Springer.

Miller, A. I. (ed.) (1994). *Early Quantum Electrodynamics. A Sourcebook*. Cambridge: Cambridge University Press.

Millikan, R. A. (1916a). Einstein's photoelectric equation and contact electromotive force. *Physical Review*, **7**(2), 18–32.

Millikan, R. A. (1916b). A direct photoelectric determination of Planck's *h*. *Physical Review*, **7**(2), 355–388.

Millikan, R. A. (1935). *Electrons (+ and -), Protons, Photons, Neutrons, and Cosmic Rays*. Chicago: University of Chicago Press. 2nd expanded edn 1947.

Muthukrishnan, A., Scully, M. O., and Zubairy, M. S. (2003). The concept of the photon—revisited. *Opticas and Photonics. New Trends*, **3**(1), S18–S27.

Natanson, L. (1911). Statistische Theorie der Strahlung. *Physikalische Zeitschrift*, **12**, 659–666. In Polish in *Bulletin de l'Académie des Sciences de Cracovie*, A **1911**, 134–148.

Needell, A. A. (1980). *Irreversibility and the Failure of Classical Dynamics: Max Planck's Work on the Quantum Theory 1900–1915*. Ann Arbor, Michigan: University Microfilms.

Newton, Isaac (1672). New Theory about Light and Colours. *Philosophical Transactions of the Royal Society of London*, **6**(80), 3075–3087.

Newton, I. (1704). *Opticks*. Reprint ed. by I.B. Cohen, New York: Dover (1952).

Newton, T. D. and Wigner, E. P. (1949). Localized states for elementary systems. *Reviews of Modern Physics*, **21**, 400–406.

Norton, J. (2008). *Einstein's Miraculous Argument of 1905: The Thermodynamic Grounding of Light Quanta*. HQ1 Max-Planck-Institute Preprint, pp. 63–78.

Okun, L. B. (2006). Photon: history, mass, charge, *Acta Physica Polonica*, B **37**, 565–573.

Ornstein, L. S. and Zernike, F. (1920). Energiewisselingen der zweite Straling en light-atomen?. *Verslagen van de gewone vergaderingen der afdeling Natuurkunde. Koniklijke Nederlandse Akademie van Wetenschappen te Amsterdam* **28**, 281–292.

Pirandola, S. *et al.* (2015). Advances in Quantum Teleportation. *Nature Photonics*, **9**(10), 641–652.

Poynting, J. H. (1884). On the Transfer of Energy in the Electromagnetic Field. *Philosophical Transactions of the Royal Society of London*, **175**, 343–361.

Raman, C. V., and Bhagavantam, S. (1931) Experimental proof of the spin of the photon. *Indian Journal of Physics*, **6**, 353–366.

Redhead, M. L. G. (1983). Quantum field theory for philosophers. *PSA 1982*, Vol. **2** (1983), 57–99.

Redhead, M., and Teller, P. (1992). Particle labels and the theory of indistinguishable particles in quantum mechanics. *British Journal for the Philosophy of Science*, **43**, 201–218.

Roychoudhuri, C., and Roy, R. (eds) (2003). The Nature of Light. What is a photon? *Photonics News Suppl.*, OPN-Trends, Oct., S1–S35.

Schweber, S. S. (1994). *QED and the Men Who Made It: Dyson, Feynman, Schwinger and Tomonaga*. Princeton: Princeton University Press.

Scully, M. O., and Sargent, M. (1972). The concept of the photon. *Physics Today*, **25**(3), 38–47.

Scully, M. O., and Zubairy, M. S. (1997). *Quantum Optics*. Cambridge: Cambridge University Press.

Serwer, D. (1977). Unmechanischer Zwang: Pauli, Heisenberg and the rejection of the mechanical atom, 1923–1925. *Historical Studies in the Physical Sciences*, **8**, 189–256.

Shadbolt, P., *et al.* (2014). Testing foundations of quantum mechanics with photons. *Nature Physics*, **10**, 278–286.

Shapiro, A. (1993). *Fits, Passions and Paroxysms. Physics, Method and Chemistry and Newton's Theories of Colored Bodies and Fits of Easy Reflection.* Cambridge: Cambridge University Press.

Silberstein, L. (1922). Quantum Theory of Photographic Exposure, Philosophical Magazine 6th series, 44: 257–273 and 44: 956–968.

Silva, I., and Freire, O. (2013). The concept of the photon in question—The controversy surrounding the HBT effect 1957–58. *Historical Studies in the Natural Sciences*, **43**(4), 453–491.

Speziali, P. (ed.) (1972). *Albert Einstein—Michele Besso: Correspondence 1903–1955*. Paris: Hermann.

Strnad, J. (1986). Photons in introductory quantum physics. *American Journal of Physics*, **54**, 650–652.

Stuewer, R. H. (1975). *The Compton Effect: Turning Point in Physics*. New York: Science History Publishers.

Tomonaga, S.-I. (1974/97). *Spin wa meguru.* (1st Japanese ed. 1974); English trans: *The Story of Spin*. Chicago: University of Chicago Press, 1997.

Turnbull, H. W., Scott, J. F., Hall, A. R., and Tilling, L. (eds) (1959–77). *The Correspondence of Isaac Newton*. Cambridge: Cambridge University Press.

van der Waerden, B. L. (1960). Exclusion principle and spin. In M. Fierz and V.F. Weisskopf (eds), *Theoretical Physics in the Twentieth Century*, New York: Interscience.

Wheaton, B. (1983). *The Tiger and the Shark. Empirical Roots of Wave-Particle Dualism.* Cambridge: Cambridge University Press.

Wheaton, B. R. (1978). Philipp Lenard and the photoelectric effect, 1889–1911. *Historical Studies in the Physical Sciences*, **9**, 299–322.

Wolfers, F. (1926). Une action probable de la matière sur les quanta de radiation. *Comptes Rendus hebdomadaires de l'Académie des Sciences*, **183**, 276–277.

Wolfke, Mieczyslaw (1921). Einsteinsche Lichtquanten und räumliche Struktur der Strahlung. *Physikalische Zeitschrift* 22: 375-379.

Wroblewski, A. (1985). De mora luminis: A spectacle in the acts with a prologue and an epilogue. *American Journal of Physics*, **54**, 620–630.

Wüthrich, A. (2011). *The Genesis of Feynman Diagrams*. Heidelberg: Springer.

Zeilinger, A. (2010). *Dance of the Photons*. New York: Farrar, Straus and Giroux.

Zeilinger, A., Weihs, G., Jonnewein, T., and Aspelmeyer, M. (2005). Happy centenary, photon. *Nature*, **433**, 230–238.

CHAPTER 10

......................................................................................................

# THE EVOLVING
# UNDERSTANDING OF
# QUANTUM STATISTICS

......................................................................................................

DANIELA MONALDI

## 10.1 THE CHOICE THAT DID NOT HAPPEN

......................................................................................................

THE transition from classical to quantum physics has produced a profound change in
the concept of elementary particles. In the quantum world, particles are no longer
scaled-down versions of idealized mass-produced macroscopic objects, as for instance
perfect bullets or monochrome billiard balls, which, although perfectly equal, are
identifiable by their spatio-temporal trajectories, but are so completely non-identifiable
that all their imaginable permutations constitute one single physical state. In physicists'
everyday language, classical particles are said to be 'distinguishable', quantum particles
'indistinguishable'.

Indistinguishability is thus commonly taken to be the defining attribute of quantum
statistics, the theory that statistically describes and explains the behaviour of microphysical entities. It is indistinguishability that separates quantum statistics from its
parent theory, the so-called classical statistical mechanics.[1] But when and how, exactly,
did this transition occur? When did particles become indistinguishable? The apparent
simplicity of this question conceals several layers of complication.[2] To begin with, the
terminology of indistinguishability, though common, is neither unanimous nor
unequivocal, for even classical particles, if they belong to the same particle species

[1] On the history of the concept of 'classical' physics, see Staley (2005).
[2] This question was asked by the late Jürgen Ehlers at one of the reading-group meetings of the
History and Foundations of Quantum Physics project at the Max-Planck-Institut für Wissenschaftsgeschichte. I wish to acknowledge Professor Ehlers for the inspiration and the MPIWG group for the
initial fostering of my research on this question.

and are imagined at rest, are observationally indistinguishable. For this reason, some authors describe the conceptual transition as the change from the classical idea of indistinguishability to a quantum idea of indistinguishability. Others evade the ambiguity by stressing that quantum particles are 'absolutely' indistinguishable. Philosophers of physics have finer distinctions and a richer vocabulary for the complexities involved in different conceptions of physical and microphysical objects (Castellani, 1998; French and Krause, 2006; Saunders, 2006). When particles became indistinguishable, thus, depends on what exactly one means by 'indistinguishable'.

Another complication lies with the definition of 'particles', which also changed dramatically over the course of the twentieth century, with the development of microphysics, quantum mechanics, quantum field theory, and high-energy physics, to accommodate old but reconceptualized microphysical entities as well as a host of new ones. Thus, when particles became indistinguishable also depends on what one means by 'particles'. Finally, the concepts of particles and indistinguishability did not change independently from one another but evolved in a process of mutual adjustment, being closely related parts of a changing conceptual, formal, and material environment. A consequence of this complex history is that the precise meaning of particle indistinguishability remains one of the open questions in the ongoing construction of the conceptual foundations of quantum physics. This does not mean that the historical question is useless. On the contrary, an investigation of the entangled evolution of the concepts of particles and indistinguishability in the early history of quantum statistics can be of help in clarifying the genesis of the framework in which today's foundational research is carried out.

Modern physics textbooks typically explain the difference between the classical statistics of particles, the Maxwell–Boltzmann statistics, and the quantum statistics, the Bose–Einstein and Fermi–Dirac statistics, by stressing indistinguishability as one of the defining features of bosons and fermions, and by making it dependent on the underlying microphysical mechanics, quantum mechanics. This line of argument is found also in historical and philosophical literature. Max Jammer, for example, framed the development of quantum statistics as one of the far-reaching consequences of Heisenberg's uncertainty relations (Jammer, 1966, pp. 338–345). Jammer conceded that the question of the possibility of existence in nature of '"like" particles, that is, particles exactly equal in all their qualities or physical specifications' had 'a long history of its own'. Nonetheless, he claimed that 'Heisenberg, one of the first quantum physicists to question the doctrine of determinism, was also the first to recognize the importance of the notion of like particles for the new scheme of conceptions.' (Jammer, 1966, p. 342). Then, after a cursory mention of the Fermi–Dirac and Bose–Einstein statistics, he concluded (Jammer, 1966, p. 344):

> These results not only lent weight to the concept of like particles; they also showed that like particles may be indistinguishable, that is, may lose their identity, a conclusion which follows from the uncertainty relations or, more precisely, from the impossibility of keeping track of the individual particles in the case of the interactions of like particles.

The same conflation of historical reconstruction and rational justification was outlined by Mary Hesse in *Models and Analogies in Science* (1966, p. 50):

> We are unable to identify individual electrons, hence it is meaningless to speak of the self-identity of electrons, hence electrons are like pounds, shillings, and pence in a balance and not like indistinguishable billiard balls, and hence they conform to Fermi–Dirac statistics.

Hesse, however, presented this standard argument only to debunk it. As she pointed out, the nature and behaviour of particles cannot be deduced from the observer's inability to distinguish them. The standard argument presents the transition to particle indistinguishability as if it were a logical consequence of the principles of quantum mechanics. Hesse argued that it was, instead, a choice between two physical models. Hesse was chiefly interested in a philosophical analysis of the role of models in physical theory, not in the history of quantum statistics. Her aim was to refute the claim that quantum mechanics had shown models to be unnecessary for the construction of theories. Thus, she was arguing that the creators of quantum mechanics, under the cover of anti-modelling principles, quietly operated a choice between 'two indistinguishability models' for microphysical entities, the model of billiard balls, and that of 'pounds, shillings, and pence' in a bank balance (Hesse, 1966, pp. 49–50).

Recent historical and philosophical scholarship, turning attention from rational reconstructions to scientists' practices, has not only supported but greatly expanded Hesse's core thesis about the constitutive role of models and analogies in physicists' practices (Morgan and Morrison, 1999; Bailer-Jones, 2009). Nonetheless, an examination of historical evidence shows that both the standard argument and Hesse's model-choice scenario fare poorly as historical accounts, not least because they run against the chronological order of events. The first of the quantum statistics, Bose–Einstein statistics, preceded the earliest formulation of quantum mechanics by over a year, and the second quantum statistics, Fermi–Dirac statistics, predated by several months the indeterminacy principle, which established the impossibility of reidentifying individual electrons. Far from deriving from quantum mechanics, quantum statistics played instead an important role in its genesis. But both the standard argument and the model-choice scenario are historically inaccurate also for another reason. The birth of quantum statistics can no more be reduced to a choice between two pre-existing models of particles than to a logical inference from some methodological principle, mathematical structure, experimental result, or other foundational element. A clarification of the development of quantum statistics and the indistinguishability concept requires a more flexible description of the open-ended interplay of conceptual models, mathematical formalism, and changing historical conditions in physicists' theoretical practices.

Theoretical formalism played a primary role in the invention of quantum statistics, not only operationalizing physicists' conceptions but also taking them in unexpected directions. It is useful to characterize the formal quantum mathematical apparatus as a 'theoretical technology' or a set of 'paper tools' (Warwick, 1995; Klein, 1999; Klein, 2001; Kaiser, 2005). This characterization helps to highlight not only the necessity of interpreting the mathematical formalism, but also its constructive power and its versatility (Dieks, 2005). The work of theorists can be loosely described as the application of technical skills to develop formal tools and the simultaneous application of imaginative resources to interpret the tools. Implicit or explicit interpretations are necessary to find solutions to problems, and to produce the extensions, idealizations, and approximations through which theories are established, expanded, and connected to experimental results. Still, it is a property of tools to be relatively autonomous from the interpretations. Theoretical technologies are like material technologies in that they have meanings as well as interpretive flexibility (Bijker *et al.*, 1987; Bijker, 1995). While most of the time meanings are tacit, new applications often require the explicit articulation and the negotiation of conflicting interpretations. Technologies in the making are open to a range of possible interpretations, which are often perceived as mutually incompatible by different relevant groups of creators, promoters, and users. It is only when, through a process of stabilization and closure, one interpretation comes to prevail that a stable meaning emerges, which can appear in hindsight to have been implicit in the technology all along.

In this chapter, I aim to re-examine the early interpretations of quantum statistics to bring back to light the interpretive flexibility of the quantum statisticians' paper tools. The result of my analysis is that the work of interpreting quantum statistics was neither an inference nor a choice, but a slow and unsystematic process of modification of the classical model of particles to adapt it to the changing theoretical apparatus and theoretical context. Physicists were uncertain and tentative in their early interpretive attempts. The early interpretations can be broadly classified into three kinds, depending on the main ideas on which they relied:

- The statistical interdependence of particles in systems of equal particles, commonly expressed as the *lack* of independence, or *loss* of independence, to stress that it contravened a core feature of the only statistical model of microphysical entities known until then, the ideal gas of Maxwell's and Boltzmann's molecular-kinetic theory.
- The symmetry of the quantum formalism under permutations of equal particles.
- The lack of individuality of particles, also expressed as the *loss* of individuality, to stress the difference from the usual conception of particles as classical mass points.

## 10.2 A NEW STATISTICS OF NON-INDEPENDENCE

Quantum statistics was born from a series of events that began in June 1924, when Albert Einstein, director of the Kaiser Wilhelm Institute for Physics in Berlin and Nobel laureate, received a manuscript by Satyendranath Bose, then an unknown young

Bengali physicist in Dacca, India (today Dhaka, Bangladesh) (Bose, 1924; Pais, 1979; Darrigol, 1991; Stachel, 2002; Banerjee, 2016; Monaldi, 2019).[3] Einstein swiftly translated Bose's paper into German and had it published in *Zeitschrift für Physik*. He then proceeded to apply Bose's formal procedure to a monoatomic ideal gas, and published the first of what would become, in the span of seven months, a trilogy of articles on a new quantum statistical theory, now known as Bose–Einstein statistics (Einstein, 1924; Einstein, 1925a; Einstein, 1925c). Yet, Bose had no intention of starting a change in statistics or transforming the conception of particles. His only aim was to free the quantum derivation of Planck's law from any residual input from classical electrodynamics, and for this he thought it would be sufficient to adapt Planck's quantum treatment of the ideal gas to Einstein's light quanta.[4] He took light quanta to be nearly conventional particles, massless and not numerically conserved but otherwise analogous to the molecules of kinetic-molecular theory, and set out to follow Planck's procedure of applying Boltzmann's combinatorial calculation of entropy to an assembly of such free particles in their quantized phase space. A few physicists had already vainly attempted to arrive at Planck's radiation law by applying gas statistics to light quanta, but Bose, isolated by geographical distance, was uninformed about their trials. He did not know that Abram F. Joffé, Paul Ehrenfest, and Yuri A. Krutkow had proven repeatedly and conclusively that if light quanta were understood to be analogous to molecules, the application of Boltzmann's statistics to them did not lead to Planck's law but to the empirically incorrect Wien's law (Bergia *et al.*, 1985; Howard, 1990; Darrigol, 1991; Kojevnikov, 2002; Fick and Kant, 2013; Monaldi, 2019). Unaware of all this, Bose took the wrong path afresh. He then happened to misapply one of Boltzmann's formulas along the way, and so landed on the intended destination, namely, Planck's law (Pais, 1979; Bergia, 1987; Stachel, 2000; Stachel, 2002; Darrigol, 2014; Monaldi, 2019). This incident led the physicist Max Delbrück to wonder whether Bose–Einstein statistics was arrived at by serendipity (Delbrück, 1980). But serendipity is the faculty of discovering by accident something one was not looking for, while Bose found exactly what he wanted to find. Chance obviously did play a key role in the happy outcome, but it did so by operating in a space already mapped by conceptual models and mathematical formalism. Bose was able to get the right result applying a paper tool incorrectly to the wrong physical model. This occurrence bears testimony to the partial autonomy and constructive power of paper tools. Bose's error lay in the incoherence between his formal manipulations and his conceptual interpretation of them. He thought that he was using old tools to blaze a new trail to Planck's destination, when he was ,instead, unwittingly retracing Planck's formal steps.

When Einstein first read Bose's manuscript, he did not see that it was a case of two wrongs making a right. Einstein was understandably dazzled by Bose's result, since he

---

[3] Bose to Einstein, 4 June 1924, Doc. 261 in Buchwald *et al.* (2015a). All translations are mine unless otherwise stated.

[4] On the complicated relationship between light quanta and Planck's law, see also Badino in this collection. On Einstein's hypothesis of light quanta, see Hentschel in this collection.

was at that moment amid a contested resurgence of his nearly twenty-year-old hypothesis of light quanta. On the one hand, recent experiments with x-rays and γ-rays were providing evidence for a corpuscular structure of radiation, which could be used to support the hypothesis (Wheaton, 1983; Stuewer, 1975). On the other hand, Niels Bohr, Hendrick Kramers, and John Slater had just published a theory that aimed to shoot it down for good (Klein, 1970; Howard, 1990; Stuewer, 2014). Einstein must have welcomed Bose's unexpected support, yet he had to admit that it puzzled him. He wrote to his friend Ehrenfest,

> Bohr, Kramers, and someone else abolished 'loose' quanta. They won't be dispensable, though. The Indian, Bose, presented a nice derivation of Planck's law, and its constant, on the basis of loose light quanta. Derivation elegant, but the essence remains obscure.[5]

Despite its interpretational obscurity, Bose's method, as a formal tool, was capable of being applied to gas molecules too, and Einstein wasted no time. As he was writing to Ehrenfest, he had already submitted a paper in which he had begun to develop his new quantum theory of the ideal gas using Bose's method (Einstein, 1924).

When the paper appeared, it did not take long for a Viennese physicist, Otto Halpern, to point out in private to Herr Professor Einstein that what he and Bose had done was incorrect.[6] So prodded, Einstein came up with his interpretation. He replied that, yes, Halpern was right, Bose's method did violate one of the fundamental requirements of statistical mechanics, namely the mutual independence of the statistical entities, but that was no mistake. On the contrary, it was the defining feature of a new 'elementary statistical hypothesis'. The new statistical hypothesis was opposed and alternative to the 'the notion of *independent* atom-like quanta', and implicitly assumed 'certain statistical dependencies' between the particles.[7] Halpern was not the only physicist mystified by Einstein's adoption of Bose's procedure. The strange new method appeared unavoidably incongruent in the shared culture of statistical thermodynamics, in which the formal apparatus of the Maxwell–Boltzmann statistics thrived in symbiosis with the billiard-ball model of ideal gas molecules and its analogues (Monaldi, 2019). In early February the following year, Einstein received in the same day two similarly concerned letters from this cultural milieu, one from Vienna and the other from Zurich. The one was from Adolf Smekal, a Privatdozent at the University of Vienna who had been tasked with reviewing Bose's and Einstein's papers for the *Physikalische Berichte*. Smekal was sending Einstein a typewritten draft of his review,

---

[5] Einstein to Ehrenfest, 12 July 1924, Doc. 285 in Buchwald *et al.* (2015a), pp. 442–43. English translation in Buchwald *et al.* (2015b), pp. 284–85.

[6] Halpern to Einstein, 26 August 1924, Doc. 308, in Buchwald *et al.* (2015a), pp. 481–82. English translation in Buchwald *et al.* (2015b), pp. 311–12. On the Einstein–Halpern exchange, see also Pérez and Sauer (2010).

[7] Einstein to Halpern, after 26 August 1924, Doc. 309, in Buchwald *et al.* (2015a), pp. 483. English translation in Buchwald *et al.* (2015b), pp. 312–313, on p. 313, emphasis in the original.

to share his 'misgivings' about what he termed 'the Bose statistics'.[8] He also wrote that Hans Thirring, another Viennese physicist, had informed him of Halpern's objections. Einstein's reply is lost, but evidently it won Smekal over. Smekal fully endorsed the non-independence interpretation in the brief remarks on the new statistics that he had time to add to the essay on the foundations of statistical mechanics he wrote for the fifth instalment of the *Encyclopedia of the Mathematical Sciences*, as well as in a paper he wrote in 1925 (Smekal, 1925; Smekal, 1926, p. 1214). The other letter was from Erwin Schrödinger. Schrödinger, who was professor of theoretical physics at the University of Zurich and had long been engaged in statistical thermodynamics, respectfully inquired if Einstein was sure he was not making a blunder.[9] In reply, Einstein reassured Schrödinger that there was 'certainly no error' in his calculations.[10] He embraced the term, 'the Bose statistics', further calling it 'a special statistics' and contrasting it to 'the classical statistics'. Using the elementary example of two molecules in two cells, he explained that in the new statistics 'the quanta or molecules are not treated as *independent of one another*'. As a consequence, 'the molecules sit together relatively more frequently than according to the hypothesis of the molecules being statistically independent'.[11] The result of this novel treatment was not in contradiction with Einstein's quantum theory of radiation of 1916–1917, as Schrödinger feared, because the effects of non-independence were relevant only in conditions of great molecular density. When the molecules were sufficiently rarefied, the Maxwell–Boltzmann distribution was recovered.[12]

Einstein laid out his interpretation in print in a second, more extensive instalment of his quantum gas theory, which was written in December 1924 and published at the beginning of February 1925 (Einstein, 1925a). In this paper, Einstein reported that Ehrenfest and 'other colleagues' had criticized him and Bose for not treating the light quanta and the gas molecules 'as statistically mutually independent structures' (Einstein, 1925b, p. 373).[13] These critics were perfectly right, Einstein stated. The lack of statistical independence was indeed the distinctive characteristics of the new theory. Einstein's argumentative strategy was a gamble of remarkable audacity. Not only was he introducing into physics, for the first time, a statistical method different from that of Maxwell and Boltzmann. He was also proposing, for the first time, to model material particles on his still hypothetical and enigmatic quanta rather than the other way

---

[8] Smekal to Einstein, 5 February 1925, Doc. 434, in Buchwald *et al.* (2015a), p. 664. English translation in Buchwald *et al.* (2015b), p. 430.

[9] Schrödinger to Einstein, 5 February 1925, Doc. 433, in Buchwald *et al.* (2015a), pp. 662–63. English translation in Buchwald *et al.* (2015b), pp. 429–30.

[10] Einstein to Schrödinger, 28 February 1925, Doc. 446, in Buchwald *et al.* (2015a), pp. 677–78. English translation in Buchwald *et al.* (2015b), pp. 438–439, on p. 439.

[11] Doc. 446 in Buchwald *et al.* (2015a), emphasis in the original.

[12] Doc. 446 in Buchwald *et al.* (2015a), on p. 439.

[13] About the contacts between Einstein and Ehrenfest in this period, and Ehrenfest's role in stimulating Einstein's thinking about the quantum gas, see Pérez and Sauer (2010). On Einstein's invention of a new statistics, see Monaldi (2019).

around. Furthermore, he billed the whole idea as worthy of consideration because it completed the analogy between gas of quanta and gas of molecules by making it symmetrical, though he knew that the analogy was shaky in the asymmetric form already. To be sure, he also tried to shore up the symmetrical analogy empirically, on the one side pointing to experimental validation with Planck's law for light quanta, and on the other side linking the low-temperature behaviour of his quantum gas to Walther Nernst's theory of gas degeneracy. He had already shown in the first paper that at low temperatures Bose's method satisfied Planck's reformulation of Nernst's theorem in terms of vanishing entropy, and at high temperatures it reproduced all the known statistical-thermodynamic results. Adding to that, he now predicted the low-temperature phenomenon that would become famous as Bose–Einstein condensation (Einstein, 1925a, pp. 4–5). He also dedicated the second half of the paper to deriving other observable consequences that could be tested experimentally (Einstein, 1925a, pp. 10–14).

Einstein, Ehrenfest, and members of their circle in St. Petersburg, Zurich, Berlin, and Leiden had long known that if light quanta existed, Planck's law demanded that they were not independent, and that treating light quanta as statistically independent in analogy with ideal gas molecules led necessarily to Wien's law. In fact, it was their investigations of the possibility of deriving Planck's law from the hypothesis of light quanta that made explicit the fundamental role of statistical independence in kinetic-molecular theory (Bergia *et al.*, 1985; Howard, 1990; Darrigol 1991; Kojevnikov 2002; Fick and Kant, 2013; Monaldi, 2019). For Ehrenfest and like-minded physicists, the lack of statistical independence was sufficient reason to deny the physical existence of light quanta. Either these objects were statistically independent ('loose', as Einstein put it in his letter to Ehrenfest, evidently hinting to their long-running dialogue on this topic) or they were just mathematical constructs, not physical entities: *tertium non datur* (Ehrenfest, 1911; Ehrenfest and Kamerlingh Onnes, 1915; Krutkow, 1914). Einstein's view on the nature of light quanta was more nuanced, steeped as it was in twenty years of puzzling over the wave–particle duality of light (Klein, 1964; Howard, 1990; Darrigol 1991; Kojevnikov, 2002; Darrigol, 2014).

The only way to reconcile the non-independence interpretation with the classical conception of particles was to admit that the particles were not free, as the ideal gas model required, but interacted with one another. Thus, as Einstein put it, the new statistics expressed indirectly 'a certain hypothesis' about 'a mutual influence of the molecules,' which was, at the moment, 'completely puzzling' (Einstein, 1925b, p. 374). Likewise, in the letter to Schrödinger, Einstein wrote of an 'interaction between the molecules' that was taken into account statistically, although its physical nature remained 'puzzling'.[14] For him, the lack of independence could be made intelligible within the atomistic worldview by admitting the existence of a new form of interaction among the particles. Halpern, Ehrenfest, and Schrödinger used this point to reject the new statistics,

---

[14] Doc. 446 in Buchwald *et al.* (2015a), on p. 439.

because they could not imagine any form of interaction that could make Bose's formulas correct, and even less one that could justify Einstein's formulas, which after all were supposed to apply to the free molecules of an ideal gas. Einstein, in contrast, held out the hope that this seeming impossibility would turn into an opportunity.

Einstein had to recognize that Bose's method offered no solution to the old conundrum of the wave–particle duality of radiation. As he wrote to Halpern, 'Bose's derivation, therefore, cannot be regarded as an actual theoretical basis for Planck's law, rather only as its reduction to a certain simple, yet arbitrary statistical elementary hypothesis.'[15] Bose's success just confirmed to him that a new theory had to be invented, a fusion of wave and particle theory. But now Einstein's imagination was excited a new by the possibility that the same interdependence that existed between light quanta could exist also among gas molecules, because this would reveal a 'deep essential relationship between radiation and matter' (Einstein, 1925b, p. 376). He found further support for this idea in his calculation of energy fluctuations of the quantum gas, which turned out to be composed of two terms formally analogous to those he had derived in 1909 for radiation. The first term would be present by itself if the molecules or light quanta were mutually independent. The second term suggested an underlying wave-like behaviour for the gas because in the case of radiation it corresponded to fluctuations due to wave interference. For this reason, Einstein was thrilled when, just as he was completing the second paper on the quantum gas, he read the Ph.D. thesis of a student of Paul Langevin, Louis de Broglie. In a letter dated December 16th, Einstein gushed to Langevin that the young de Broglie had 'lifted a corner of the great veil.'[16] And on the same day he wrote to Lorentz that de Broglie's theory was 'a feeble ray of light on this worst of our physics enigmas'.[17] The young French theorist had independently endeavoured to give full shape precisely to the symmetrical gas-radiation analogy that Einstein envisioned. Thus, Einstein cited de Broglie's theory of matter waves in support of his interpretation of the gas energy fluctuations in terms of a dual nature, corpuscular and undulatory, of the gas. Einstein's endorsement of de Broglie's theory was the instigation for Schrödinger's wave mechanics and for Pascual Jordan's early version of quantum field theory (Hanle, 1977a,b; Bernardini and Donini, 1980; Darrigol, 1986; Darrigol, 1992a; Duncan and Janssen, 2007). For Einstein, however, the wave–particle duality of matter and radiation was to remain forever an enticing but unsolved mystery.

The first interpretation of the new statistics, then, was the product of the small subculture within quantum statistical physics that focused on the question of the relation between the light quanta and Planck's radiation law. Einstein's confident

[15] Einstein to Halpern, after 26 August 1924, Doc. 309, in Buchwald *et al.* (2015b), pp. 312–313, on p. 313.

[16] Einstein to Langevin, 16 December 1924, Doc. 398, in Buchwald *et al.* (2015a), pp. 608–609, on p. 608. English translation in Buchwald *et al.* (2015b), pp. 393–394, on p. 393.

[17] Einstein to Lorentz, 16 December 1924, Doc. 399, in Buchwald *et al.* (2015a), pp. 609–610, on p. 609. English translation in Buchwald *et al.* (2015b), pp. 394–395, on p. 395.

framing of Bose's methods as a new statistics of non-independence is easily under-standable against the background of the history of light quanta. But a hand-written note that Halpern added to his letter raises the intriguing possibility that Einstein briefly toyed also with another possibility. Referring to a previous postcard by Einstein that is now unfortunately lost, Halpern wrote that the 'interchangeability of quanta inside the same cell' mentioned by 'Herr Professor' (Einstein, presumably) could not serve the purpose of distinguishing between Wien's law and Planck's law.[18] The reason was that interchangeability applied to both statistical hypotheses. As Halpern sharply observed in the letter, the real difference lay in the designation of microstates of equal probability, which could not be decided by interchangeability considerations. The interchangeability or exchange-symmetry argument would play a large role in inter-pretations of quantum statistics after it was linked to the quantum mechanics of systems of multiple particles in the second half of 1926. It also played a role in the quantum theory of ideal gases before 1925.[19] It is therefore necessary to briefly review it to inquire what role it played in Einstein's first interpretation of the new statistics and its immediate aftermaths.

## 10.3  THE EXCHANGE-SYMMETRY ARGUMENT BEFORE 1926

Until then, the quantum theory of radiation and the quantum theory of the ideal gas had grown separately though contiguously, as separated branches of the growing quantum tree, divided because they were aimed at different physical systems, but stemming from the common quantum hypothesis and linked by an organic web of partial analogies (Klein, 1964; Darrigol, 1991; Desalvo, 1992; Monaldi, 2009). Now, Einstein's new statistics promised to merge the two theories completely. The quantum theory of the ideal gas had developed around Planck's definition of absolute entropy, and had hitherto consisted of the application of the combinatorial version of Boltz-mann's statistics to an assembly of material particles in a quantized phase space (Badino, 2009). Bose's method was meant to mimic it, and therefore started on the same premise, but Einstein showed that it departed from it because it adopted a different way of counting the microstates of equal probability, or 'complexions' as they were called following Boltzmann. To underline the divergence, Einstein called the older count 'the hypothesis of the mutual statistical independence of the gas molecules' (Einstein, 1925a, pp. 5–6). He then highlighted one consequence that followed from the different count. Gas theorists commonly assumed that entropy had to be an extensive

---

[18]  Halpern to Einstein, 26 August 1924, Doc. 308, in Buchwald *et al.* (2015a), pp. 481–482, on p. 482. English translation in Buchwald *et al.* (2015b), pp. 311–312, on p. 312.

[19]  On the quantum theory of the ideal gas before 1925, see also Badino in this collection.

quantity, and to ensure extensivity they routinely subtracted by hand a term that depended only on the number of molecules in the gas. The entropy formula so produced was the Sackur–Tetrode formula, which was known to be empirically correct at high temperature. Thanks to the Boltzmann–Planck equation between entropy and thermodynamic probability, the subtraction was mathematically equivalent to dividing the thermodynamic probability of a gas state by the factorial of the number of molecules, an operation that could be regarded as the quantum translation of J. Willard Gibbs' formal definition of generic phases in 'systems composed of a great number of entirely similar particles' (Gibbs, 1902, p. 220). This mathematical operation was physically insignificant to Gibbs, and remained so in quantum theory as long as only entropy differences were regarded as physically significant. But if one insisted, as did Planck, that entropy was an absolute quantity, then the subtraction had to be justified with an adequate physical interpretation. Planck and his followers had struggled to find a convincing justification for this formal artifice since its first introduction in 1912. Planck's argument revolved around the idea that since the molecules of the same gas were perfectly exchangeable with one another, microstates that differed only by the exchange of equal molecules ought not to be counted as distinct states. In Planck's words, in the gas of equal molecules (Planck, 1921, p. 209),

> [...] we do not have a system of mutually separated molecules, but a single entity endowed with symmetries, and these symmetries consist of the fact that there is no feature that allows one to recognize a given atom if one considers the gas first in one state and then in another state. Therefore, two states of the gas that differ only by the exchange of two atoms do not produce a new complexion.

Planck had had to defend his argument repeatedly against stiff criticism, notably from Ehrenfest, among others (Ehrenfest and Trkal, 1921; Planck, 1922; Desalvo, 1992; Darrigol, 1992a). Einstein, too, considered Planck's faith in absolute entropy ungrounded, and the factorial division a matter of convenience and 'basically insignificant' (Einstein, 1916, p. 126). Now, having been persuaded by Bose's procedure to finally pay close attention to Planck's combinatorial calculation, he anatomized it to find out where exactly Bose had diverged from it. He was thus able to spot a fatal flaw of the factorial division: it had the unfortunate effect of making the entropy negative at low temperatures. Since, then, the factorial division was only valid at sufficiently high temperature, exchange symmetry could not be a valid justification for it. As far as Einstein was concerned, it was a distinct merit of Bose's method that it had no use for the factorial artifice and thus no need for the associated exchange-symmetry argument.

Planck agreed completely with Einstein's interpretation of the difference between the two gas theories. The new theory derived from the hypothesis of statistical non-independence, and was therefore fundamentally different from his theory, which derived from statistical independence supplemented by exchange symmetry. There were now two gas statistics to choose from, and the choice could only be made empirically. But if experiment confirmed the non-independence hypothesis, Planck noted, it would entail 'a fundamental modification of the ordinary conception of the nature and mode of interaction of the molecules' (Planck, 1925, p. 57). It is remarkable that it never occurred to Planck that the exchange-symmetry argument, too, would require a fundamental modification of the ordinary conception of molecules.

The first to associate exchange symmetry to the new statistics was Erwin Schrödinger. After receiving Einstein's explanation of the meaning of Bose's method, Schrödinger undertook a comparative analysis of the competing definitions of entropy found in the literature (Schrödinger, 1925). He found that the factorial division was an incoherent attempt to implement the exchange-symmetry argument. The premise of exchangeability in the quantized phase space of a large number of equal molecules at all temperatures did not lead to Planck's absolute entropy, but rather, Schrödinger argued, to Einstein's new statistics. Far from warming him up to Bose's method, this finding confirmed his dislike of it. For him, as for Einstein, the 'elimination of permutations from the statistical calculus' that distinguished it only made sense as the formal consequence of non-independence, but he could not see 'for the time being any possibility of understanding the peculiar kind of interaction among molecules' that it implied (Schrödinger, 1925, pp. 440). Schrödinger was much more attracted to the idea 'of proceeding not from the quantization of the individual molecules but from that of the whole gas'. But after exploring this 'beautiful thought' for two dense pages, he came to the conclusion that it was 'impossible to develop without arbitrariness' (Schrödinger, 1925, p. 440).

Einstein found Schrödinger's analysis 'illuminating', and wrote to him with a suggestion on how to obtain, perhaps, a theory based on treating the gas as a single entity, if one wanted to. This led to a series of letters and a publication by Schrödinger on the quantum levels of a gas of fully exchangeable molecules. On 3 November 1925 Schrödinger informed Einstein that he had finally got hold of de Broglie's thesis, and that it had completely clarified to him the 'undulatory theory of the gas' that Einstein referenced in the second quantum gas paper.[20] Schrödinger was now busy expanding on de Broglie's theory. He planned to explain away the Bose statistics by switching around, as he put it, 'the concepts "material substrate" and "energy content".'[21] He detailed the formal structure of this switch in a paper titled 'On Einstein's Gas Theory' (Schrödinger, 1926). He noticed that the phase space cells of Bose's procedure formally corresponded to the quantized field vibrations of a model of radiation developed by James Jeans in 1905 and Peter Debye in 1910, in which the field vibrations were analogues of ideal gas molecules and followed the old statistics. Schrödinger then doubled the analogy back onto itself so he could shift the weight of physical interpretation onto the old-statistics side: he re-interpreted the phase space as an imaginary quantized field, which in his view was able to provide meaning to the formalism precisely because it obeyed the Boltzmann statistics. The gas molecules became energy quanta carried by the waves, and were

---

[20] Schrödinger to Einstein, 3 November 1925, Doc. 101 in Buchwald *et al.* (2015a), pp. 181–183, on p. 182. English translation in Buchwald *et al.* (2015b), pp. 120–121, on p. 121.

[21] Schrödinger to Einstein, 4 Dec 1925, Doc. 123 in Buchwald *et al.* (2015a), pp. 215–217, on p. 216. English translation in Buchwald *et al.* (2015b), pp. 143–145, on p. 144.

therefore allowed to display, in their ephemeral existence, the odd statistics of Bose. By Schrödinger's own account, it was the dialogue with Einstein on the interpretation of Bose's formalism that guided him to de Broglie's matter-wave theory and hence to the formulation of wave mechanics.[22] His motivation, as he vividly expressed it, was the desire to restore the old statistics, which he called 'natural', and to avoid the 'sacrificium intellectus' that the new statistics inflicted (Schrödinger, 1926, p. 95).

Schrödinger's feeling that the new statistics demanded intellectual sacrifice stemmed from the shared background of atomistic and mechanistic modes of understanding, which were built not only into the formal apparatus of kinetic-molecular theory, but also into the expanding material culture of microphysics, and thus permeated the evolution of statistical mechanics and quantum physics. The alliance of molecular mechanics and classical statistical formalism was well known to be incomplete, notably on the front of irreversibility (Uffink, 2007). Yet, its promise had become so entrenched as to be perceived as natural. Not by accident, Schrödinger was also the first to suggest that the new statistics entailed a loss of individuality, appealing to a notion of individuality rooted in the idealization of a material body in classical mechanics. From his in-depth analysis of the new statistics, he concluded (Schrödinger, 1926, p. 101),

> If in a specific case experimental facts force us to apply the Bose statistics to a class of objects (and this of course is by no means established in the case of the gas), in my point of view we have to draw the conclusion that this class of objects are not real individuals but energy excitation states.

The emphasis Schrödinger placed on the hypothetical status of this prospect indicates that, despite his theoretical elaboration of the inverted gas-radiation analogy, he was not ready to let go of the conception of gas molecules as 'real individuals'. From there, in a surprising and deeply ambivalent move, he went right on to pursue the particle–wave analogy, only no longer from the perspective of a gas as a whole but from that of an individual electron.

## 10.4 QUANTUM STATISTICS, QUANTUM MECHANICS, AND THE NEW FORM OF EXCHANGE SYMMETRY

The year 1926 saw a series of interrelated developments that led to the linkage of quantum statistics to quantum mechanics, mainly at the hands of Werner Heisenberg and Paul Dirac (Heisenberg, 1926a; Dirac, 1926). Heisenberg and Dirac were both

[22] Schrödinger to Einstein, 21 January 1926, Doc. 174 in Buchwald *et al.* (2018a), pp. 314–315, on p. 315. English translation in Buchwald *et al.* (2018b), pp. 192–193, on. 193. Schrödinger to Einstein, 23 April 1926, Doc. 264 in Buchwald *et al.* (2018a), pp. 442–43.

working to extend the newly created quantum mechanics from single-particle to multiple-particle systems. They worked independently, Heisenberg in Copenhagen and Dirac in Cambridge, but were interacting through correspondence and the quantum physics network. Despite their different trainings and personal styles, they found common ground in the anti-modelling arguments they used to rationalize their methods of theory construction (Camilleri, 2009; Darrigol, 1992b). Yet, they did rely on a conceptual model, namely, the classical corpuscular model, even as they were radically restructuring its formal foundations.[23] The papers in which they broached multiple-particle quantum mechanics were their first published responses to Schrödinger's wave mechanics; they belong in the context of the debates that erupted about the interpretation of the quantum mechanical formalisms.

In his seminal paper on the helium atom, Heisenberg manufactured a clever but fraught argument to weave together the quantum mechanics of equal-particle systems, Pauli's exclusion principle, and Bose–Einstein statistics. He mistakenly believed that the Bose–Einstein statistics was a modification of classical statistics required by the exclusion principle. He did not realize until months later that they were formally incompatible. His declared motivation was to oppose Schrödinger's interpretation of wave mechanics as a theory of matter waves, and reaffirm instead the corpuscular interpretation of quantum mechanics, while appreciating and using the formal equivalence between wave mechanics and matrix mechanics. His argument hinged on the notion that, for systems composed of two or more equal particles, every total energy state was multifold in principle because it was realized by all possible particle permutations. Generalizing his solution of the helium spectrum to any number of equal particles, he conflated the permutation number with two other numbers, the statistical weight of each energy state and the number of non-combining sets of states that solved the quantum-mechanical equations of a system of interacting electrons. He then assigned to the exclusion principle the role of selecting the correct set of solutions while also reducing the statistical weights to produce the quantum statistics (Heisenberg, 1926a, pp. 423–426). In this way, Heisenberg appropriated the exchange-symmetry argument as part of a contrived interpretation that tied the quantum-mechanical formalism to the new statistics. He insisted that this tangle confirmed the validity of the corpuscular model, even though he warned that in quantum mechanics, for interacting systems, it no longer made 'physical sense to talk of' the motion of single electrons because any non-symmetrical function of the electrons did not represent an observable quantity (Heisenberg, 1926a, p. 423). As a concession to physical modelling, he suggested that if one wanted to form a picture of the processes in question, one could imagine the electrons to 'exchange place periodically in a continuous way' (Heisenberg, 1926a, p. 421 and p. 423). This suggestion would live on and give rise to the notion of exchange forces (Carson, 1996).

---

[23] See Monaldi (2013) for a detailed analysis of Heisenberg (1926a) and Dirac (1926).

Dirac (who was informed about Heisenberg's results but probably had not read the paper) established the rule that the wave function of a system of equal particles must be either symmetrical or anti-symmetrical under particle permutations. He purported to derive this rule from Heisenberg's principle of observability, invoking, that is, the methodological prescription that a theory should 'enable one to calculate only observable quantities' to justify the choice of representing states differing only by particle exchanges as single states. Without this choice, he claimed, the theory would 'enable one to calculate the intensities' of transitions to particle-permuted states separately, while the transitions were 'physically indistinguishable' (Dirac, 1926, p. 667). He then developed a new form of statistics that correctly incorporated Pauli's principle (which we now know as the Fermi–Dirac statistics), and associated it to the anti-symmetric wave functions solely on the basis of this principle.[24] Dirac's version of the exchange-symmetry argument might seem, in retrospect, to exemplify historically Hesse's model-choice scenario. But in 1926 Dirac was only making a choice between two possible variants of the formalism, not between the classical model of particles, the billiard-ball model, and a new quantum one, the "pounds, shillings, and pence" model that unified material particles and light quanta. On the contrary, in the only explicitly interpretative comment he made, he used his new statistics to firmly reject the radiation-matter analogy. He asserted, instead, a fundamental ontological distinction between material particles, which obeyed the exclusion principle and the individualistic statistics linked to it, and light quanta, which obeyed instead the communitarian Bose–Einstein statistics. Dirac's superselection rule sealed the integration of the two quantum statistics, the Bose–Einstein and Fermi–Dirac statistics, with multi-particle quantum mechanics, and the problem of interpreting the statistics was soon engulfed by that of interpreting quantum mechanics.

## 10.5 A GENERAL FORMALISM OF QUANTUM STATISTICS

Einstein's non-independence interpretation was ignored by Heisenberg and Dirac, and generally lost prominence to exchange symmetry in the new theoretical context of quantum mechanics. It continued to be relevant, however, for theorists working in the tradition of statistical mechanics. Dirac's mentor, Ralph H. Fowler, for example, responded to the new developments by producing a general form of statistics that yielded the three statistics, 'classical', 'Einstein's', and 'Dirac's', as special cases by means of specific assignations of statistical weights (Fowler, 1926, p. 433). Fowler

---

[24] Dirac was not the first to develop the quantum statistics based on Pauli's principle. Pascual Jordan wrote it down in December 1925, but his paper never got published (Schücking, 1999). Enrico Fermi independently developed it in early 1926 in the framework of the old quantum theory (Fermi 1926a,b).

deployed a technique that he had co-created with Charles G. Darwin in 1922 to generalize statistical mechanics to a wide range of physical systems in a mathematically rigorous way. Darwin and Fowler had made explicit the conditions of applicability of their general method. They were the defining conditions of the kinetic-molecular theory of the ideal gas, namely, the large number, perfect equality, and statistical independence of the elements composing the statistical assembly (Fowler and Darwin, 1922). Fowler was an early supporter of quantum mechanics; yet, he dismissed outright the interpretation of any statistics as a statistics of 'indistinguishable' elements, arguing in squarely classical terms that it 'must be possible ideally to recognize differences between particles moving with the same energy in different directions' (Fowler, 1926, p. 440). His general formulation had the virtue of eliminating the need for the unacceptable assumption of indistinguishability, replacing it with specific assignations of statistical weights. Since in classical theory the assignation was justified by the Liouville theorem, he was confident that it would soon be provided by 'the analogue of Liouville's theorem' in the new mechanics (Fowler, 1926, p. 445.) But how were the correct statistics to be assigned to different physical systems? Siding with Dirac, Fowler asserted that, while the Bose–Einstein choice was correct for radiation, it was 'almost certain' that the Fermi–Dirac choice was 'the true form for gaseous assemblies' (Fowler, 1926, p. 445). He expressed the same view in his comprehensive textbook, *Statistical Mechanics*, which appeared in 1929. The statistical treatment of atoms, molecules, and electrons must be grounded in quantum mechanics, which was primarily 'a purely "atomic" theory', concerned with 'the properties of individual atoms and molecules, and the interactions of pairs of such[.]' Quantum mechanics required the Fermi–Dirac statistics for material particles, and the statistical weights had to be regarded as atomic constants determined by the laws of atomic mechanics (Fowler, 1929, pp.2–3).

In contrast, Ehrenfest and his doctoral student, George Uhlenbeck, tried vainly to stem the tide that construed quantum statistics as a formal and physical consequence of quantum mechanics, and used the generalizability of the statistical-mechanical formalism to reclaim the primacy of modelling in theory construction. They argued that all three statistics were compatible with quantum mechanics. Only when symmetry requirements were imposed on the formalism was the old statistics ruled out in favour of the new ones (Ehrenfest and Uhlenbeck, 1927). Evidently unconvinced by Dirac's appeal to observational symmetry, they argued that the appropriate statistics was to be chosen on the basis of the appropriate physical model for the system in question. They further warned that it was premature to assign models on the basis of the theoretical formalism when an interpretation of multi-dimensional wave functions was still wanting. In particular, they noted that although the Pauli principle in its original form could be visualized as a 'prohibition' of equivalent orbits for electrons in an atom, in the generalized form of a restriction to anti-symmetric wave functions, it did not admit a corpuscular interpretation.

Uhlenbeck expanded on this argument and conducted a systematic comparison of the three statistics in his doctoral dissertation (Uhlenbeck, 1927.) Starting from the foundational analysis conducted by the Ehrenfests in 1911 (Ehrenfest and Ehrenfest-Afanassjewa, 1911), he used a combination of Gibbs' method and the Darwin–Fowler method to construct a general form of statistical mechanics similar to Fowler's. He tackled the question of the assignation of the different statistics to different physical systems systematically, distinguishing two paradigmatic systems, thermal radiation and ideal gas, and two interpretive models, the corpuscular model and the quantized wave model. The Boltzmann case followed from the assumption of statistical independence, and hence it applied to a system 'constituted of $n$ little corpuscles' interacting only through mechanical collisions (Uhlenbeck, 1927, p. 25.) Uhlenbeck pointed out that, despite appearing obvious, this assumption was unproven. The Bose–Einstein and Fermi–Dirac cases, instead, followed from the assumption of non-independence, and applied to models consisting of superpositions of eigenwaves. They differed from one another only because the Bose–Einstein statistics placed no restriction on the number of quanta per wave, while the Fermi–Dirac statistics limited the quanta to no more than one. The empirical correctness of Bose's method spoke in favour of the undulatory model for radiation, but the question of which model was appropriate for the ideal gas was now open. If empirical evidence were to show that particles obeyed the statistics of non-independence, Uhlenbeck concluded (inverting the argument he and Ehrenfest had presented earlier), an undulatory conception of matter would become necessary (Uhlenbeck, 1927, pp. 73–79).

# 10.6 Photons, Exchange Symmetry, and the Secret Life of Electrons

The same uncertainty reigned at the 1927 Solvay Congress. At this meeting, Arthur H. Compton successfully proposed the name 'photons' for light quanta, which was a signal that these entities were being accepted as having physical existence and were starting to be classified as particles, but with all the ambiguities that the emerging issue of wave–particle duality entailed for the concept of particles (Bacciagaluppi and Valentini, 2009, pp. 332–333). There was broad consensus that photons followed the Bose–Einstein statistics while electrons and protons, the only material particles known at the time, followed the Fermi–Dirac statistics. But opinions diverged as to whether this difference in statistics confirmed a fundamental distinction between radiation and matter, or was simply due to different particle properties such as charge and angular momentum (Bacciagaluppi and Valentini, 2009, pp. 501–520). Max Born and Heisenberg pointed out that the rule of symmetry for the wave function was an arbitrary addition to quantum mechanics. Nonetheless, they also claimed that the new statistics had 'a perfectly legitimate place' in quantum mechanics but not in classical mechanics (Bacciagaluppi and Valentini, 2009, p. 436).

Initially, quantum theorists working on multi-particle systems were focused on electrons and on the symmetry properties of the quantum formalism, and hence their interpretation of quantum statistics hinged on exchange symmetry. This was especially the case for those applying group theory. Hermann Weyl, for example, in his book, *Gruppentheorie und Quantenmechanik*, evoked the peculiar consequences of exchange symmetry in quantum mechanics by personifying a system of two electrons as a pair of identical twins named Hans and Karl (Weyl, 1928, p. 188):

> It is impossible to keep the identical individuals of the same nature, Hans and Karl, each by himself, in his persisting identity with himself. From electrons, one cannot in principle demand a proof of their alibi. So prevails Leibniz's principle of the *coincidentia indiscernibilium* in the modern quantum theory.

Weyl's quirky metaphor and his elliptic mention of Leibniz's principle were suggestive but, alas, not particularly clarifying. At any rate, the exclusive reference to electrons indicates that he was not trying to articulate a unified interpretation of both quantum statistics or of all microphysical entities.

The successful applications of multi-particle quantum mechanics reinforced the impression that exchange symmetry represented, in some enigmatic quantum-mechanical way, physical particle exchanges. Dirac returned to the physical meaning of the quantum formalism in 1929, in a paper known for the reductionist declaration that the 'underlying physical laws' of most of physics and chemistry were now 'completely known' (Dirac, 1929, p. 714). The object of the paper was to translate the 'exchange phenomena' described by group theory into 'the language of quantum mechanics' (Dirac, 1929, p. 716). By 'exchange phenomena' Dirac meant the large 'exchange energies' due to the '*exchange* (austausch) interaction of the electrons, which arises owing to the electrons being indistinguishable one from another.' By way of explanation, he added, 'Two electrons may change places without our knowing it, and the proper allowance for the possibility of quantum jumps of this nature, which can be made in a treatment of the problem by quantum mechanics, gives rise to the new kind of interaction' (Dirac, 1929, p. 715). He conjured up the notion of secret quantum exchanges also in his textbook, *The Principles of Quantum Mechanics*, and popularized with it the terminology of 'absolutely indistinguishable' particles (Dirac, 1930, p. 198):

> If a system in atomic physics contains a number of particles of the same kind, e.g. a number of electrons, the particles are absolutely indistinguishable one from another. No observable change is made when two of them are interchanged. This circumstance gives rise to some curious phenomena in quantum mechanics having no analogue in the classical theory, which arise from the fact that in quantum mechanics a transition may occur resulting in merely the interchange of two similar particles, which transition then could not be detected by any observational means. A satisfactory theory ought, of course, to count two observationally indistinguishable states as the same state and to deny that any transition does occur when two similar particles exchange place. We shall find that such a theory can be developed in agreement with the principles of quantum mechanics.

# 10.7 THE LOSS OF INDIVIDUALITY

Physicists' interpretations were understandably incomplete and unstable in those years, and this was also true for Heisenberg (Beller, 1999; Camilleri, 2009.) In a *Naturwissenschaften* article in November 1926, after some intense debating with Schrödinger, Heisenberg relaxed the positivistic posture, and argued that the object of quantum mechanics was to investigate the 'kind of physical reality' possessed by the fundamental components of matter, electrons and atoms, to which it was not possible to grant 'the same degree of immediate reality as the objects of everyday experience'. He even opened up to the radiation-matter analogy, conceding that '*electrons* possess[ed] the same degree of reality as the *light quanta*' and that the same 'remarkable dualism' that existed for light might be admitted for material particles as well (Heisenberg, 1926b, pp. 989 and 992). Then, referring to Schrödinger's analysis of the Bose–Einstein statistics, he picked up the hint that Bose–Einstein particles were not real individuals. Unlike Schrödinger, however, Heisenberg embraced the new statistics precisely because it placed restrictions on the reality of particles, indeed, the same restrictions implied by quantum mechanics. Just as in quantum mechanics it was not possible to assign to a particle a specific position as a function of time, in quantum statistics it was not possible to reidentify a specific particle in a collection of equal particles. This meant that 'the individuality of a corpuscle [could] be lost' (Heisenberg, 1926b, p. 993).

The loss of individuality was also floated by Paul Langevin at the 1927 Solvay Congress. During the general discussion, Langevin noted that while in the old statistics one attributed an individuality to each component of a system, in the new statistics one had to suppress the individuality of the constituents and substitute it with the individuality of the states of motion. To Langevin, the suppression of individuality seemed appropriate to the nature of microphysical components in general because of their 'complete identity of nature' (Bacciagaluppi and Valentini, 2009, p. 501). Still, he agreed with the general consensus that photons and material particles were fundamentally different and followed different statistics. He classified composite systems like molecules together with photons, and conjectured that material particles were distinguished from photons and molecules by their impenetrability (Bacciagaluppi and Valentini, 2009, p. 502).

Commenting on Heisenberg's utterance about the loss of individuality, Paul Forman remarked, 'never again did this simple truth flow from Heisenberg's pen' (Forman, 1984, p. 211). Forman took the loss of individuality of particles to be 'the most direct and least arguable consequence' and 'the one inarguable "lesson" of quantum mechanics, and the statistics integral to it', and argued that this lesson was oddly neglected (Forman, 1984, pp. 210 and 213). In fact, Heisenberg soon reclaimed the notion of individual particles as foundational to quantum mechanics, and formulated the uncertainty relations on that basis. Niels Bohr, on his part, starting with his formulation of complementarity in the summer of 1927, propounded the thesis that quantum physics demonstrated the individuality of atomic processes. Interpreting Bohr's emphasis on individuality as applying to material particles as well, Forman argued that this denial of the clearest lesson of quantum mechanics demonstrated the physicists' adaptation to the dominant cultural values of the Weimar period. *Individualität*, the 'ideal of the autonomous individual personality', was in fact a cultural theme of the time, and Niels Bohr and others associated it to irrationality and to the failure of causality (Forman, 1984, p. 212). Forman's analysis is a healthy reminder that the work of physicists, whether theoretical or experimental, is not isolated from broader cultural, social, and material contexts. A concept like individuality has deep cultural resonances, and scientists' imaginative resources are not sealed within aseptic professional cultures. Yet, the concept of individuality was not univocal, and the lesson of quantum statistics was not as simple and clear as Forman made it to be. It took very long to discern anything like a meaningful lesson amidst incomplete, confusing, and conflicting signals, and the lesson is, to a significant extent, still a work in progress (Castellani, 1998; French and Krause, 2006; Emch, 2007).

## 10.8 CONCLUSION

The work that was needed to interpret quantum statistics cannot be described as a logical inference or as a choice between two pre-formed models. It was, instead, a gradual and long-drawn process of modifying the familiar pictures of particles and of radiation to adapt them to the innovations in theoretical technology, and to the changing theoretical and experimental contexts. Physicists undertook this process using the conceptual resources, skills, and expertise they had acquired with their training and their practice within their professional communities and cultures, which were embedded in larger communities and culture. This process illustrates the constructive power and interpretive flexibility of theoretical apparatus. Today's explanations of quantum statistics typically combine the three ideas of non-independence, exchange symmetry, and lack of individuality. All three of these ideas were discussed in the early period, but the early interpreters discussed them for the most part separately, and even, in some cases, as antagonistic. It was only in the years following WWII, when particle physics emerged as an independent field, that these ideas began to congeal, at least on the surface, into a unified interpretation, the idea of quantum indistinguishability as taught by modern physics textbooks, described by Jammer, and encapsulated by Hesse in the analogy between particles and currency units in a bank balance.

The evolution of the understanding of quantum statistics can be epitomized by Dirac's arc. In 1926, Dirac forged the link between quantum mechanics and quantum statistics while sternly rejecting the matter-radiation analogy. Nineteen eventful years later, he officiated what can perhaps be regarded as the inauguration of the new field of elementary particle physics by making that analogy the foundation of the new edifice. At a lecture in Paris at the end of 1945, before an audience that was eagerly waiting to hear about the latest developments in nuclear physics, he spoke of an entirely new species of particles, the mesons, and proposed a new classification of all microphysical entities, no longer based on the old distinction between light and matter but on their statistics, subdividing the entities in two classes, to which he gave the telling names of 'bosons' and 'fermions' (Dirac, 1945, p. 1252; Kragh, 1990, p. 36 and note 73 on p. 322; Farmelo, 2009, p. 331). This sea change in the conception of particles did not occur at once. It had started with the invention of Bose–Einstein statistics and had been propelled by the formulation of quantum mechanics, but was also driven by a host of other theoretical, experimental, and instrumental developments, which had occurred since the late 1920s and were still ongoing, such as the growth of relativistic quantum field theory, the invention of high-energy physics apparatus, the discovery of new particles, and the detection of their instability.

# References

Badino, M. (2009). The Odd Couple: Boltzmann, Planck, and the Application of Statistics to Physics (1900–1913). *Annalen der Physik*, **18** (2–3), 81–101.

Bailer-Jones, D. M. (2009). *Scientific Models in Philosophy of Science*. Pittsburgh: University of Pittsburgh Press.

Bacciagaluppi, G. and Valentini, A. (2009). *Quantum Theory at the Crossroads: Reconsidering the 1927 Solvay Congress*. Cambridge: Cambridge University Press.

Banerjee, S. (2016). Transnational Quantum: Quantum Physics in India Through the Lens of Satyendranath Bose. *Physics in Perspective*, **18**, 157–181.

Beller, M. (1999). *Quantum Dialogue. The Making of a Revolution*. Chicago and London: The University of Chicago Press.

Bergia, S. (1987). Who Discovered Bose-Einstein Statistics? In M. G. Doncel *et al.* (eds), *Symmetries in Physics (1600–1980)*, Bellaterra: Seminario de Historia de las Ciencias de la Universidad Autónoma de Barcelona, pp. 221–248.

Bergia, S. *et al.* (1985). Side Paths in the History of Physics: The Idea of Light Molecule from Ishiwara to De Broglie. *Rivista di Storia della Scienza*, **2**, 71–97.

Bernardini, P., and Donini, E. (1980). Perché fu Einstein il tramite tra de Broglie e Schrödinger? In G. Battimelli *et al.* (eds), *La ristrutturazione della scienza fra le due guerre mondiali*, Vol. I, Rome: La Goliardica, pp. 149–187.

Bijker, W. E. (1995). *Of Bicycles, Bakelites, and Bulbs. Toward a Theory of Sociotechnical Change*. Cambridge, MA: The MIT Press.

Bijker, W. E., *et al.* (eds) (1987). *The Social Construction of Technological Systems. New Directions in the Sociology and History of Technology*. Cambridge, MA: The MIT Press.

Bose, S. (1924). Wärmegleichgewicht im Strahlungsfeld bei Anwesenheit von Materie. *Zeitschrift für Physik*, **27**, 384–387.

Buchwald, D. K. *et al.* (eds) (2015a). *The Collected Papers of Albert Einstein. Volume 14: The Berlin Years: Writings and Correspondence, April 1923–May 1925*. Princeton: Princeton University Press.

Buchwald, D. K. *et al.* (eds) (2015b). *The Collected Papers of Albert Einstein. Volume 14: The Berlin Years: Writings and Correspondence, April 1923–May 1925*. English Translation. Princeton: Princeton University Press.

Buchwald, D. K. *et al.* (eds) (2018a). *The Collected Papers of Albert Einstein. Volume 15: The Berlin Years: Writings and Correspondence, June 1925–May 1927*. Princeton: Princeton University Press.

Buchwald, D. K. *et al.* (eds) (2018b). *The Collected Papers of Albert Einstein. Volume 15: The Berlin Years: Writings and Correspondence, June 1925–May 1927*. English Translation. Princeton: Princeton University Press.

Camilleri, K. (2009). *Heisenberg and the Interpretation of Quantum Mechanics. The Physicist as Philosopher*. Cambridge, UK: Cambridge University Press.

Carson, C. (1996). The Peculiar Notion of Exchange Forces—I. Origins in Quantum Mechanics, 1926–1928. *Studies in History and Philosophy of Quantum Physics*, **27**, 23–45.

Castellani, E. (ed.) (1998). *Interpreting Bodies. Classical and Quantum Objects in Modern Physics*. Princeton, NJ: Princeton University Press.

Darrigol, O. (1986). The Origin of Quantized Matter Waves. *Historical Studies in the Physical and Biological Sciences*, **16**, 197–253.

Darrigol, O. (1991). Statistics and Combinatorics in Early Quantum Theory, II: Early Symptoma of Indistinguishability and Holism. *Historical Studies in the Physical Sciences*, **21**, 237–298.

Darrigol, O. (1992a). Schrödinger's Statistical Physics and Some Related Themes. In M. Bitbol and O. Darrigol (eds), *Erwin Schrödinger. Philosophy and the Birth of Quantum Mechanics*, Gif-sur-Yvette: Editions Frontières, pp. 197–253.

Darrigol, O. (1992b). *From C-Numbers to Q-Numbers. The Classical Analogy in the History of Quantum Physics*. Berkeley, CA: University of California Press.

Darrigol, O. (2014). The Quantum Enigma. In M. Janssen and C. Lehner (eds), *The Cambridge Companion to Einstein*, New York: Cambridge University Press, pp. 118–142.

Delbrück, M. (1980). Was Bose-Einstein Statistics Arrived at by Serendipity? *Journal of Chemical Education*, **57**, 467–470.

Desalvo, A. (1992). From the Chemical Constant to Quantum Statistics: A Thermodynamic Route to Quantum Mechanics. *Physis*, **29**, 465–537.

Dieks, D. (2005). The Flexibility of Mathematics. In G. Boniolo *et al.* (eds), *The Role of Mathematics in Physical Sciences*, Dordrecht: Springer, pp. 115–129.

Dirac, P. A. M. (1926). On the Theory of Quantum Mechanics. *Proceedings of the Royal Society of London A*, **112**, 661–677.

Dirac, P. A. M. (1929). Quantum Mechanics of Many-Electron Systems. *Proceedings of the Royal Society of London A*, **123**, 714–733.

Dirac, P. A. M. (1930). *The Principles of Quantum Mechanics*. Oxford: Clarendon Press.

Dirac, P. A. M. (1945). Quelques développements sur la théorie atomique. Conférence faite au Palais de la Découverte le 6 Décembre 1945. In R. H. Dalitz (ed.), *The Collected Works of P. A. M. Dirac*, Cambridge and New York: Cambridge University Press, 1995, pp. 1246–266.

Duncan, A. and Janssen, M. (2007). Pascual Jordan's Resolution of the Conundrum of the Wave-Particle Duality of Light. *Studies in History and Philosophy of Modern Physics*, **39**, 634–666.

Ehrenfest, P. (1911). Welche Züge der Lichtquantenhypothese spielen in der Theorie der Wärmestrahlung eine wesentliche Rolle? *Annalen der Physik*, **36**, 91–118.

Ehrenfest, P. and Ehrenfest-Afanassjewa, T. (1911). Begriffliche Grundlagen der statistischen Auffassung in der Mechanik. In F. Klein and C. Müller (eds), *Encyklopädie der Mathematischen Wissenschaften mit Einschluss ihrer Anwendungen* IV, 2, IV, 32. Leipzig: Teubner, pp. 3–90.

Ehrenfest, P. and Kamerlingh Onnes, H. (1915). Vereinfachte Ableitung der kombinator-ischen Formel, welche der Planckschen Strahlungstheorie zugrunde liegt. *Annalen der Physik*, **46**, 1021–1024.

Ehrenfest, P. and Trkal, V. (1921). Ableitung des Dissoziationsgleichgewichts aus der Quan-tentheorie und darauf beruhende Berechnung der chemischen Konstanten. *Annalen der Physik*, **370**, 609–628.

Ehrenfest, P. and Uhlenbeck, G. E. (1927). Die wellenmechanische Interpretation der Boltz-mannschen Statistik neben der der neuren Statistiken. *Zeitschrift für Physik*, **41**, 24–26.

Einstein, A. (1916). On the Theory of Tetrode and Sackur for the Entropy Constant. Translated manuscript, Doc. 26 in Klein *et al.* (eds) (1997), *The Collected Papers of Albert Einstein. Volume 6: The Berlin Years: Writings, 1914–1917*. English Translation Supple-ment. Princeton: Princeton University Press, pp. 121–131.

Einstein, A. (1924). Quantentheorie des einatomigen idealen Gases. *Sitzungsberichte der Preussischen Akademie der Wissenschaften. Physikalisch-mathematische Klasse*, pp. 261–267. Reprinted as Doc. 283 in Buchwald *et al.* (2015a), pp. 434–440.

Einstein, A. (1925a). Quantentheorie des einatomigen idealen Gases. Zweite Abhandlung. *Sitzungsberichte der Preussischen Akademie der Wissenschaften. Physikalisch-mathematische Klasse*, pp. 3–14. Reprinted as Doc. 385 in Buchwald *et al.* (2015a), pp. 581–592.

Einstein, A. (1925b). Quantum Theory of the Monatomic Ideal Gas. Second Paper. Reprinted as Doc. 385 in Buchwald *et al.* (2015b), pp. 371–383.

Einstein, A. (1925c). Zur Quantentheorie des idealen Gases. *Sitzungsberichte der Preussischen Akademie der Wissenschaften. Physikalisch-mathematische Klasse*, pp. 18–25. Reprinted as Doc. 427 in Buchwald *et al.* (2015a), pp. 649–56.

Emch, G. G. (2007). Quantum Statistical Physics. In J. Butterfield and J. Earman (eds), *Philosophy of Physics*, Amsterdam and Boston: Elsevier/North-Holland, pp. 1075–1182.

Farmelo, G. (2009). *The Strangest Man. The Hidden Life of Paul Dirac, Mystic of the Atom.* New York: Basic Books.

Fermi, E. (1926a). Sulla quantizzazione del gas perfetto monoatomico. *Atti dell'Accademia dei Lincei*, **3**, 145–49.

Fermi, E. (1926b). Zur Quantelung des idealen einatomigen Gases. *Zeitschrift für Physik*, **36**, 902–12.

Fick, D. and Kant, H. (2013). The Concepts of Light Atoms and Light Molecules and Their Final Interpretation. In S. Katzir *et al.* (eds), *Traditions and Transformations in the History of Quantum Physics*, Edition Open Access, Max Planck Institute for the History of Science, pp. 89–124.

Forman, P. (1984). *Kausalität, Anschaulichkeit*, and *Individualität*, or, How Cultural Values Prescribed the Character and the Lessons Ascribed to Quantum Mechanics. Reprinted in C. Carson *et al.* (eds), *Weimar Culture and Quantum Mechanics. Selected Papers by Paul Forman and Contemporary Perspectives on the Forman Thesis*, London and Singapore: Imperial College Press/World Scientific, 2011, pp. 203–219.

Fowler, R. H. (1926). General Forms of Statistical Mechanics with Special Reference to the Requirements of the New Quantum Mechanics. *Proceedings of the Royal Society of London A*, **113**, 432–49.

Fowler, R. H. (1929). *Statistical Mechanics. The Theory of the Properties of Matter in Equilibrium*. Cambridge: Cambridge University Press.

Fowler, R. H. and Darwin, C. G. (1922). On the Partition of Energy. *Philosophical Magazine*, Series 6, **44**(261), 450–79.

French, S. and Krause, D. (2006). *Identity in Physics: A Historical, Philosophical, and Formal Analysis*. Oxford: Clarendon Press.

Gibbs, J. W. (1902). *Elementary Principles in Statistical Mechanics*. New York: Charles Scribner's Sons.

Hanle, P. (1977a). The Coming of Age of Erwin Schrödinger: His Quantum Statistics of Ideal Gases. *Archive for History of Exact Sciences*, 17, 165–192.

Hanle, P. (1977b). The Schrödinger–Einstein Correspondence and the Sources of Wave Mechanics. *American Journal of Physics*, 47, 644–48.

Heisenberg, W. (1926a). Mehrkörperproblem und Resonanz in der Quantenmechanik. *Zeitschrift für Physik*, 38, 411–426.

Heisenberg, W. (1926b). Quantenmechanik. *Die Naturwissenschaften*, 45, 989–94.

Heitler, W. and London, F. (1927). Wechselwirkung neutraler Atome und homöopolare Bindung nach der Quantenmechanik. *Zeitschrift für Physik*, 44, 455–72.

Hesse, M. B. (1966). *Models and Analogies in Science*. Notre Dame, IN: University of Notre Dame Press.

Howard, D. (1990). 'Nicht sein kann was nicht sein darf,' or the Prehistory of EPR, 1909–1935: Einstein's Early Worries about the Quantum Mechanics of Composite Systems. In A. I. Miller (ed.), *Sixty-Two Years of Uncertainty*, Boston: Springer, pp. 61–111.

Jammer, M. (1966). *The Conceptual Development of Quantum Mechanics*. New York: McGraw-Hill.

Kaiser, D. (2005). *Drawing Theories Apart. The Dispersion of Feynman Diagrams in Postwar Physics*. Chicago: The University of Chicago Press.

Klein, M. (1964). Einstein and the Wave–Particle Duality. *The Natural Philosopher*, 3, 3–49.

Klein, M. (1970). The First Phase of the Bohr–Einstein Dialogue. *Historical Studies in the Physical Sciences*, 2, 1–39.

Klein, M. J. et al. (eds.) (1997). *The Collected Papers of Albert Einstein. Volume 6: The Berlin Years: Writings, 1914–1917*. English Translation Supplement. Princeton: Princeton University Press.

Klein, U. (1999). Techniques of Modelling and Paper-Tools in Classical Chemistry. In M. Morrison and M. S. Morgan (eds), *Models as Mediators. Perspectives on Natural and Social Science*, Cambridge, UK: Cambridge University Press, pp. 146–67.

Klein, U. (2001). Paper Tools in Experimental Cultures. *Studies in History and Philosophy of Science*, 32(2), 265–302.

Kojevnikov, A. (2002). Einstein's Fluctuation Formula and the Wave-Particle Duality. In Y. Balashov and V. Vizgin (eds), *Einstein Studies in Russia. Einstein Studies, Vol. 10*, Boston: Birkhäuser, pp. 181–228.

Kragh, H. (1990). *Paul Dirac. A Scientific Biography*. Cambridge, UK: Cambridge University Press.

Krutkow, Y. A. (1914). Aus der Annahme unabhängiger Lichtquanten folgt die Wiensche Strahlungsformel. *Physikalische Zeitschrift*, 15, 133–36.

Monaldi, D. (2009). A Note on the Prehistory of Indistinguishable Particles. *Studies in History and Philosophy of Modern Physics*, 40, 383–94.

Monaldi, D. (2013). Early Interactions of Quantum Statistics and Quantum Mechanics. In S. Katzir, C. Lehner, and J. Renn (eds), *Traditions and Transformations in the History of Quantum Physics*, Edition Open Access, Max Planck Institute for the History of Science, pp. 125–147.

Monaldi, D. (2019). The Statistical Style of Reasoning and the Invention of Bose–Einstein Statistics. *Berichte zur Wissenschaftsgeschichte*, **42**, 307–37.

Morgan, M. S. and Morrison, M. (1999). *Models as Mediators. Perspectives on Natural and Social Science*. Cambridge UK: Cambridge University Press.

Pais, A. (1979). Einstein and the Quantum Theory. *Reviews of Modern Physics*, **51**, 863–914.

Pérez, E. and Sauer, T. (2010). Einstein's Quantum Theory of the Monatomic Ideal Gas: Non-statistical Arguments for a New Statistics. *Archive for History of Exact Sciences*, **64**, 561–612.

Planck, M. (1921). *Vorlesungen über die Theorie der Wärmestrahlung*, 4th revised edn. Leipzig: Ambrosius Barth Verlag.

Planck, M. (1922). Absolute Entropie und chemische Konstante. *Annalen der Physik*, **371**, 365–72.

Planck, M. (1925). Zur Frage der Quantelung einatomiger Gase. *Sitzungsberichte der Königlich Preussischen Akademie der Wissenschaften*, pp. 49–57.

Saunders, S. (2006). Are quantum particles objects? *Analysis*, **66**(1), 52–63.

Schrödinger, E. (1925). Bemerkungen über die statistische Entropiedefinition beim idealen Gas. *Sitzungsberichte der Preussischen Akademie der Wissenschaften*, pp. 434–41.

Schrödinger, E. (1926). Zur Einsteinsche Gastheorie. *Physikalische Zeitschrift*, **27**, 95–101.

Schücking, E. L. (1999). Jordan, Pauli, Politics, Brecht, and a Variable Gravitational Constant. *Physics Today*, October 1999, pp. 26–31.

Smekal, A. (1925). Zwei Beiträge zur Bose-Einsteinschen Statistik. *Zeitschrift für Physik*, **33**, 613–22.

Smekal, A. (1926). Allgemeine Grundlagen der Quantenstatistik und Quantentheorie. In A. Sommerfeld (ed.), *Encyklopädie der Mathematischen Wissenschaften mit Einschluss ihrer Anwendungen* V, 3, 28, Wiesbaden: Vieweg+Teubner Verlag, pp. 861–1214.

Stachel, J. (2000). Einstein's Light-Quantum Hypothesis, or Why Didn't Einstein Propose a Quantum Gas a Decade-and-a-Half Earlier? In D. Howard and J. Stachel (eds), *Einstein: The Formative Years, 1879–1909*, Boston: Birkhäuser, pp. 231–251.

Stachel, J. (2002). Einstein and Bose. In *Einstein from 'B' to 'Z'*, Boston: Birkhäuser, pp. 519–538.

Staley, R. (2005). On the Co-Creation of Classical and Modern Physics. *Isis*, **96**, 530–58.

Stuewer, R. H. (1975). *The Compton Effect: Turning Point in Physics*. New York: Science History Publications.

Stuewer, R. H. (2014). The Experimental Challenge of Light Quanta. In M. Janssen and C. Lehner (eds), *The Cambridge Companion to Einstein*, New York: Cambridge University Press, pp. 143–166.

Uffink, J. (2007). Compendium of the Foundations of Classical Statistical Physics. In J. Butterfield and J. Earman (eds), *Philosophy of Physics*, Amsterdam and Boston: Elsevier/North-Holland, pp. 923–1074.

Uhlenbeck, G. E. (1927). *Over Statistische Methoden in de Theorie der Quanta*. 's-Gravenhage: Martinus Nijhoff.

Warwick, A. (1995). The Sturdy Protestants of Science: Larmor, Trouton, and the Earth's Motion through the Ether. In J. Z. Buchwald (ed.), *Scientific Practice. Theories and Stories of Doing Physics*, Chicago and London: The University of Chicago Press, pp. 300–343.

Weyl, H. (1928). *Gruppentheorie und Quantenmechanik*. Leipzig: Hirzel.

Wheaton, B. R. (1983). *The Tiger and the Shark. Empirical Roots of Wave–Particle Dualism*. Cambridge UK: Cambridge University Press.

# CHAPTER 11

..........................................................................................

# THE MEASUREMENT
# PROBLEM

..........................................................................................

OSVALDO PESSOA JR

## 11.1 INTRODUCTION

..........................................................................................

QUANTUM mechanics (QM) is the physical theory that describes atoms and radiation below the scale of a nanometre. The theory may be structured as follows. A closed system, such as an atom that does not interact significantly with its environment (but only with classically described components such as analysers), is described by a state that changes over time according to the Schrödinger wave equation (or an equivalent one). Such a state evolution occurs in a unique way, so one can say that the evolution is 'deterministic', in the sense that given the Hamiltonian operator that describes the system and the initial conditions, the quantum-mechanical state at any given time is determined in a unique way. The operator that describes this evolution obeys the mathematical property of 'unitarity', which expresses its reversible character. This is the case as long as the system remains closed, a condition that is broken when a measurement takes place.

In QM, the result of a measurement on an individual system is usually unpredictable, so the theory can only predict the relative frequencies of measurement outcomes on an ensemble of similarly prepared individual systems. In other words, for an individual measurement the theory can only give the probabilities for the different possible results. After an individual measurement, if the system is not destroyed (as in the absorption of a photon by an electron), it usually changes to a new state, a state that depends on the result obtained. This unpredictable transition was initially called 'collapse of the wave packet' or 'state reduction', being described in a formal way by John von Neumann ([1932] 1955), later generalized for the degenerate case by Gerhart Lüders ([1951] 2006), and called the 'projection postulate' by Henry Margenau (1958, p. 29).

The so-called 'measurement problem' arises from the opposition between the predictable unitary evolution of a quantum-mechanical state of a closed system, described by the Schrödinger equation (in the non-relativistic case), and the unpredictable transition that is associated with the measurement process, described by the projection postulate (or some equivalent rule for the collapse of the wave function). This opposition becomes a problem when two hypotheses are assumed: (1) the measurement apparatus (which might include the conscious observer) may be adequately described as a quantum system; (2) the 'composite system' (which includes the quantum object and the measurement apparatus) may be considered a closed system, in relation to the external environment. In this case, the composite system should evolve unitarily (since it is a closed system), in a predictable way, but at the same time unpredictable state reductions should be taking place during the measurements involving apparatus and quantum object. How can these two apparently contradictory possibilities be reconciled?

In this chapter, we will examine the history of this problem until around 1990. After briefly examining the 'pre-von Neumann' period and von Neumann's formulation of the problem, the proposed solutions will be separated into four main classes: (i) 'objectivist' solutions, which consider that the correct explanation does not involve a conscious observer who causes collapses or that enters into superpositions; (ii) the 'subjectivist' views, that attribute to the human observer a causal role in the process of state reduction; (iii) the 'many worlds' interpretation, which postulates that the human observer enters into a quantum superposition, so that the apparent reduction is an expression of a 'relative state'; (iv) views that consider that the measurement problem is a pseudoproblem.

Objectivist solutions include the view that the thermodynamic process describing the interaction between the quantum object and a macroscopic amplifier in a meta-stable state leads to the collapse of the wave function, even when no actual amplification occurs (in null-result measurements), and is associated with the proposals of Ludwig (1953), Daneri, Loinger, and Prosperi (1962), and many others. The subjectivist views include the proposal of London and Bauer ([1939] 1983) that conscious observation entails the objective collapse of the quantum-mechanical state, and also views that consider that different observers might have different judgements on whether a collapse has taken place, based on Wigner's friend thought-experiment (Wigner, 1962). The many-worlds views, inaugurated by Hugh Everett (1957), differ from the subjective views in that state reductions never actually occur (the state evolution is always unitary), but may be classed together with these due to the essential role played by the human observer in the solution of the measurement problem.

The debate on the measurement problem blossomed in the 1960s, with new versions of von Neumann's insolubility proofs of the measurement problem, the consideration of null-result measurements, and the proposal of non-linear corrections to the Schrödinger equation. In the 1970s and 1980s, new objectivist ideas were put forth, such as the notion that all systems are open, leading to environmentally induced decoherence or to the idea that gravity plays a role in wave-function collapse. The theory of stochastic localizations was proposed, and merged with the non-linear approach. Many solutions also assumed that the apparatus should be described with infinite degrees of freedom. We also explore briefly the points of view that dismiss the issue as a pseudoproblem.

## 11.2 THE 'PRE-VON NEUMANN' PERIOD

The first formulation of the measurement problem of quantum mechanics is usually associated with the work of John von Neumann ([1932] 1955), but its roots lie at the beginning of quantum theory. The measurement problem is a descendent of the 'wave–particle paradox' that troubled radiation physics during the first quarter of the century.

As early as 1897, J.J. Thomson expressed what Wheaton (1983, p. 76) has called the 'paradox of quantity': why is it that only a vanishingly small fraction of gas molecules is ionized by x-rays, when one should expect that all molecules would be uniformly affected by the wave pulse? This problem was later met by W.H. Bragg in 1906, who also voiced a 'paradox of quality', relative to the energy content of a pulse (Wheaton, 1983, p. 86). When an electron of specific energy decelerates and generates a wave front of an x-ray impulse which is spread over a large area, it was found that the x-ray energy absorbed by a single molecule of the gas is equal to that of the initial electron. One would expect that the energy of the x-ray pulse is spread over the sphere of propagation, so only a small fraction of its energy should be locally absorbed by a gas molecule, so how can it be that the full energy of the impulse is concentrated in a single molecule?

Corpuscular models of light were formulated qualitatively by Thomson (1904) and quantitatively by Einstein (in 1905, using Planck's constant), and corpuscular views of x-rays were advanced by Bragg and Stark, but the latter were generally ignored and later were unable to explain the phenomena of interference, observed for x-rays by 1912. Millikan's (1916) experimental confirmation of Einstein's photoelectric law, the observation of the photoelectric effect in x-rays and γ-rays in 1921, and the successful explanation of the Compton effect using the corpuscular model of the quantum of light (1922) opened the way for Louis de Broglie's concept of *wave–particle duality*, and later on for quantum mechanics (1926).

The experimental evidence that all matter exhibits undulatory aspects (electron diffraction experiments were reported in 1927) extended the paradoxes encountered for electromagnetic radiation to all of matter. How did the new quantum mechanics 'solve' the wave–particle paradoxes?

The answer that developed in the years 1927–1928 introduced a special role for *measurement* in affecting changes in atomic systems. It was possibly Heisenberg's ([1927] 1983, pp. 64, 83) justification of his indeterminacy relations by means of the γ-ray microscope that first emphasized the inevitable disturbance of the observer on the atomic system being observed. As Heisenberg stated at the 5th Solvay Congress, which took place in Brussels in October 1927:

I do not agree with Mr Dirac when he says that, in the described experiment, nature makes a choice. [...] I should rather say, as I did in my last paper, that the observer himself makes the choice, because it is only at the moment when the observation is made that the 'choice' has become a physical reality and that the phase relationship in the waves, the power of interference, is destroyed.

(Solvay [1928] 2009, pp. 449–50)

A little later, Niels Bohr published his views on the inevitable 'interaction' of the 'agency of observation' with the atomic system. While Heisenberg tended to give priority to the corpuscular view, Bohr (1928, pp. 54–55) professed a clearly dualistic interpretation which put wave and particle aspects of matter on an equal footing, as exclusive but 'complementary' aspects of nature.

Before this dualistic view took hold as the 'orthodox' interpretation of QM, and parallel to the recognition of the role of the measuring apparatus in disturbing the object system, a new view of microscopic phenomena that became known as the *wave interpretation* was suggested by Schrödinger's version of QM. Since Schrödinger's wave function for one particle, $\Psi(r, t)$, could be identified with the intuition we have of a three-dimensional wave, one could be tempted to associate it with a real wave propagating in space. However, for $N$ interacting particles, the composite wave function $\Psi(r_1, r_2, \dots r_N, t)$ could only be adequately represented in the $3N$-dimensional 'configuration space', which does not correspond to our common intuition about reality (Solvay [1928] 2009, pp. 437–439). At any rate, while the wave-like properties observed for matter follow directly from the wave function in configuration space, how could one account for the observed particle-like properties of matter?

Two interpretative schemes connected this corpuscular aspect with the act of measurement: Max Born ([1926] 1983, p. 54) interpreted the wave function of scattered radiation as a 'probability wave' whose squared amplitude gives the probabilities for detection; and Heisenberg introduced the notion that position measurements '*reduce the wave packet*' which represents a particle (Heisenberg [1927] 1983, p. 74), thus accounting for the linear trajectories observed in a cloud chamber. At the 5th Solvay Congress, Born was able to answer Einstein's version of the paradox of quantity: 'Now, if one associates a spherical wave with each emission process, how can one understand that the track of each α particle appears as a (very nearly) straight line?' Born explained how 'the corpuscular character of the phenomenon [can] be reconciled here with the representation by waves' by making use of the 'reduction of the probability packet' which does not occur 'for as long as one does not observe ionization'. In order to describe what happens after the observation, one must '"reduce" the wave packet in the immediate vicinity of the drops' (Solvay [1928] 2009, p. 437).

Within this 'probability wave' interpretation of QM, the wave–particle paradox was solved by attributing to the act of observation the power to reduce a spatially extended probability wave to a narrow wave packet. One problem with this explanation of wave–particle duality was that the concepts of 'observation', 'measurement', and 'apparatus' were not defined in an unambiguous way. In the example of the cloud chamber, is the mere ionization of a gas molecule to be considered an observation, or must it include the condensation around the molecule, or even the illumination of the chamber and perception by the scientist?

The measurement problem did not arise for the complementarity interpretation of Bohr ([1928] 1934, p. 53) because he took the detection of discrete quanta as the starting point of QM, as the 'quantum postulate, which attributes to any atomic process an essential discontinuity, or rather individuality [ ... ]', not to be further explained. In addition, he excluded the measuring apparatus from the quantum description, although occasionally the 'double nature' (classical and quantum) of the macroscopic apparatus was invoked by Bohr (see Jammer 1974, pp. 473–474). The measurement problem tends not to arise within antirealist interpretations, for which it makes no sense to ask questions the answers to which are not verifiable, such as whether the collapse of $\Psi(r,t)$ could occur without a human observation being made, or which stage of the measurement process would be responsible for the collapse of $\Psi(r,t)$. We see, therefore, that to deal with the measurement problem, we cannot fully adopt Bohr's version of the Copenhagen interpretation (nor the ensemble interpretation, as we will see in section 11.4), although it does arise in the so-called Princeton interpretation of von Neumann and Wigner, which is the more mathematical version of the orthodox interpretation that describes the apparatus as a quantum system. By way of discussion, we will treat the wave function as an objective entity, although the wave-probabilistic interpretation has problems, one of which we have already mentioned (for further discussion, see Gibbins, 1987, pp. 43–45).

## 11.3 JOHN VON NEUMANN

Wave–particle duality was 'solved' within a probability wave interpretation of QM by the claim that an observation (or a measurement) leads to a collapse of the wave packet. The question of *how* this collapse occurs may be called the *general measurement problem*, or 'objectification problem' (Busch *et al.*, 1996, pp. 73, 91). For example, considering measurements of *position* of an electron, assumed to have a discrete spectrum, the general measurement problem may be stated as the question of how, during measurement, a quantum superposition of eigenstates of position $\sum_i a_i|\varphi_i\rangle$ can be transformed into a single eigenstate of position $|\varphi_k\rangle$, that corresponds to the macroscopic point-like signal obtained in the measurement. The mathematical description of this state reduction is given by the projection postulate, but can this process be further analysed?

Von Neumann's approach was to take into account the interaction between the quantum object and the measurement apparatus (initially in a generic state $|\xi_0\rangle$), and he supposed that the macroscopic measuring device can be described as a quantum system (von Neumann, 1955, p. 422). Object and apparatus would thus form a 'composite' quantum system, initially in state $\Sigma_i a_i |\varphi_i\rangle \cdot |\xi_0\rangle$. The interaction between the quantum object and the measuring apparatus would lead to a unitary evolution in time, leading to a correlated composite system: $\Sigma_i a_i |\varphi_i\rangle \cdot |\xi_i\rangle$. At this point, the projection postulate must be invoked to describe the discontinuous and irreversible transition to the post-measurement collapsed state $|\varphi_k\rangle \cdot |\xi_k\rangle$.

Von Neumann ([1932] 1955, pp. 419–421) follows Heisenberg (1930, p. 64) in claiming that the 'cut' between the observed quantum system and the classical apparatus can be drawn anywhere between the object and the observer. He considers that the chain of observation goes all the way into the brain of the scientist, but there will always be an observer or 'abstract "ego"' which cannot be described physically. He arrives at this conclusion invoking 'the principle of psycho-physical parallelism'.

Adding the quantum-mechanical description of the apparatus didn't bring real progress in elucidating the projection postulate, but von Neumann then considers the possibility of describing the apparatus not as a pure quantum state, but as a statistical mixture of such pure states. The justification for this assumption is that the unpredictability of measurement outcomes could be due to the *limited knowledge* that one has of the apparatus states (or of our own conscious states) (von Neumann, 1955, p. 438). Could this more general description of the apparatus state lead to the correct statistical distribution of the measurement outcomes, using only unitary evolution (without the projection postulate)? If this were the case, the laws of QM would be 'complete', without having to add the projection postulate. Because of this, Fine (1973, pp. 568–570) called this mathematical problem the 'completeness problem'.

If the unitary evolution of the composite system could explain the projection postulate, and given the deterministic nature of unitary evolutions, one would have a *cryptodeterministic* (Whittaker, 1943, p. 461) interpretation of QM, in the sense that the determinism of natural laws would be preserved, although in a hidden way, and the unpredictability of the results of individual measurements would arise from the limited knowledge about the initial state of the systems in interaction with the quantum object.

After arguing that the thesis of limited knowledge is *prima facie* plausible, von Neumann provided an argument against the solubility of the completeness problem, by showing that the probabilities that must appear in the statistical mixture of the apparatus states would have to depend on what is the state of the object system (and so would not furnish the desired behaviour for all object states). Similar 'insolubility proofs' would be developed in the 1960s, as we will see in section 11.6.

After being ruled out by von Neumann's insolubility argument, the cryptodeterministic view became plausible again with the rise of the thermodynamic amplification programme (section 11.5) and the solutions offered to the completeness problem (see for example Green, 1958, p. 889). Another defence of the thesis of limited knowledge would be given by Heisenberg (1958, p. 53–54; see Stein and Shimony, 1971, p. 64).

# 11.4 THE SUBJECTIVIST APPROACH AND EARLY REACTIONS

As we have seen, the general measurement problem is the question of how the collapse of the wave packet occurs. Von Neumann's answer was this process is described by the projection postulate, which accompanies any act of measurement. Two problems unfold from this solution to the general measurement problem.

(i) *'Characterization' problem* (a term adapted from Cartwright, 1983, p. 196): under what conditions should the projection postulate be applied, and if it always accompanies a measurement, what characterizes a measurement or an observation? At what stage of the measurement, understood as a physical process, does the collapse of the state vector occur? During the coupling with a detection plate, or with the amplifier? During conscious observation?

(ii) *'Completeness' problem* (seen in section 11.3): can the apparently indeterministic state transition described by the projection postulate be explained by the unitary evolution which applies to closed systems, plus some ingenious model of the measurement process?

The statuses of these two problems are different. The characterization problem requires some positive answer, while the completeness problem might just be answered with a negation or shown to be 'insoluble'. The questions, however, are not independent: if the completeness problem receives a positive answer, then the characterization problem will also be solved.

The characterization problem is not explicitly addressed by von Neumann, but it is implicit in his work that what characterizes an 'observation' is the presence of an intelligent observer or consciousness. The projection postulate should therefore be applied only when an intelligent being makes an observation. Von Neumann (1955, p. 421) acknowledged the influence of Leo Szilard ([1929] 1983, pp. 539–540, 544), who had explored Maxwell's demon and concluded that the 'intervention of intelligent beings' in a thermodynamic system during a measurement (resulting in an increase in knowledge about the system) necessarily involves an increase in entropy (energy dissipation).

This *subjectivist* or 'mentalist' solution to the characterization problem was stated more clearly by London and Bauer ([1939] 1983, pp. 249, 252), for whom the 'irreversible transformation of the state of the measured object' is due to the 'faculty of introspection' or the 'immanent knowledge' that the conscious observer has of his own state, allowing him to 'cut the chain of statistical correlations' in a 'creative act of "making objective"'. Adopting this view, Heitler (1949, pp. 193–195) argued that if an electron, with a spatially distributed wave function, falls in succession on two thin parallel photographic plates, the collapse will occur on the second plate if it is observed first by a conscious being.

A new era of philosophical discussions on the foundations of QM arose from the famous paper by Einstein, Podolsky, and Rosen on the measurement of position and momentum of two particles that are non-interacting but correlated (i.e., have interacted in the past). In a study of their argument for the 'incompleteness' of quantum theory, Schrödinger ([1935] 1983, p. 157) presented his 'cat paradox' thought-experiment, which is the most famous statement of the characterization problem. A cat is enclosed in a steel chamber together with a tiny bit of radioactive substance, which has a probability ½ for triggering a detector. To this detector a 'diabolical device' is hooked so that if a triggering occurs, the cat is killed, while the cat remains alive if no radiation is detected. QM describes the state of the radioactive atom as a superposition of emission and non-emission states, but what would the state of the entire macroscopic system be? According to the subjectivist view, it would be a superposition of living and dead cats until an observation were made, and the state were reduced to either dead or live cat. This solution, however, sounds absurd since our intuitive idea of a classical object is that it does not exist in such superpositions, and furthermore, that its macroscopic states are not affected by the act of observation. Schrödinger ([1935] 1983, pp. 166–167) hoped that such 'antinomies of entanglement' could be solved by appropriately defining a time operator.

We have thus seen how the measurement problem was associated in the 1930s with a subjectivist argument for the inseparability of object and observer. This view coexisted in peace with the complementarity interpretation, although the latter downplayed the importance of the measurement problem. More realist views such as that of Schrödinger used the measurement problem as an argument for the incompleteness of QM.

# 11.5  THE THERMODYNAMIC AMPLIFICATION PROGRAMME

After the war, the subjectivist approach fell into disfavour, and in the 1950s the characterization problem generated a multitude of *objectivist* attempts at formulating a mechanism for reduction in a macroscopic measuring apparatus, in a way that would also fit in with the complementarity interpretation.

Back in 1942, the Japanese physicist Taketani (1971, pp. 68–71) had given a talk that emphasized that in the irreversible process of measurement, the cut in the chain of observation would take place during amplification. But the publication that launched the new objectivist research programme was written in 1949 by none other than Pascual Jordan, who in the 1930s had put forth an 'idealistic' position which stated that 'we ourselves produce the results of measurement' (see Jammer, 1974, p. 161). Jordan (1949, pp. 270–271) took a measurement to be a real 'macrophysical' process which makes two waves 'incoherent'. The 'decision' (reduction) taken by a photon has been completed when a macrophysical record arises from an 'avalanche process', so that the conscious observer is not necessary for reduction to occur.

Regarding the *completeness problem*, a new axiom would have to be introduced in QM, connecting large numbers of particles to states without coherence, which would give an objectivist foundation for the projection postulate (Jordan, 1949, pp. 274–275). This new principle could be a 'hypothesis of elementary disorder' involving 'random a priori phases', like the one introduced into quantum statistical mechanics by Pauli in 1928 and used to justify the irreversibility of macroscopic processes from reversible micro-scopic laws (by means of an H-theorem; see for example ter Haar, 1955, pp. 318–323).

Concerning the *characterization problem*, Jordan made two suggestions. He pro-posed, as we saw, that reduction would occur with the formation of any macroscopic register. In another passage, however, he suggested that the thermal motion of the molecules in a birefringent prism would be sufficient for the destruction of interfer-ence: 'Generally we can regard Brownian movement as that factor which is suited to create incoherence and to destroy every possibility of interference' (Jordan,1949, p. 273). This specific suggestion is clearly wrong, as can be shown by recombining the beams and passing them through another birefringent prism, which reinstates the initial beam, as shown classically by Jules Jamin, in 1868.

The idea that the *random phases* introduced by the apparatus on the object wave function during measurement are responsible for reduction was treated mathematic-ally by Bohm (1951, pp. 120–124, 600–602), who did not go into details about the physical origin of this effect. Bohm was inspired by a footnote by Feynman (1948, p. 369), who was himself inspired by von Neumann. Feynman's footnote indicates that attempts at an objectivist explanation were already circulating within the physics community.

Bohm (1951, pp. 604–608) was also the first to describe a particular thought-experiment, later analysed by Ludwig (1954, pp. 136–138) and Wigner (1963, p. 10), the 'reversible Stern–Gerlach set-up', in which a recombination is made of the separ-ated beams. Like the Jamin experiment mentioned above, this set-up indicates that no collapse occurs when the components are separated by the magnet. Although such an experiment has not been implemented, for technical reasons, an equivalent one with neutron interferometry has been put in practice (see Badurek *et al.*, 1986).

Jordan's emphasis on the 'avalanche process' as a sufficient condition for reduction was made independently by Ludwig (1953, pp. 486, 504–506), for whom the measure-ment process involves a coupling of the microscopic object with a macroscopic apparatus in a 'metastable' state (such as a cloud chamber), leading to a thermodynam-ically irreversible process whose final state fixes the outcome of the measurement. He presented a version of the ergodic theorem that accounted for this process for meas-urements of the 1st kind (to be defined in section 11.6). The stated aim of this programme was to ensure the consistency of QM, and it amounted to a search for a thermodynamic mechanism that would solve the completeness problem, explaining how macroscopic equilibrium states can arise from a reversible equation of motion.

This paradigm of *thermodynamic amplification (TDA)* was shared by most of the physicists concerned with the process of measurement during the 1950s. Bohr ([1955] 1958, p. 73) could now characterize more precisely the observation of an atomic phenomenon as 'based on registrations obtained by means of suitable amplification devices with irreversible functioning'. The year of 1957, with the Symposium on *Observation and interpretation in the philosophy of physics* (Körner, 1962), showed a marked increase in studies on the 'measurement problem', a term apparently first used by Georg Süssmann (1958, p. 5). Within the TDA programme, the problem was considered already solved, with the theoretical reduction of the measurement problem to the problem of the origin of irreversibility in statistical mechanics. All that was left was a derivation of the ergodic theorem or the H-theorem that would apply to the measurement problem in a completely satisfactory way, justifying the approximations being used. In the words of Paul Feyerabend (1962, p. 128), 'all the processes which happen during measurement can be understood on the basis of the [microscopic] equations of motion only'.

The culmination of the TDA programme was the theory of Daneri, Loinger, and Prosperi (1962), based on ergodicity conditions defined a few years earlier by Léon van Hove. The macroscopic apparatus is initially in a state of metastable equilibrium. When an interaction with a microscopic object takes place, the apparatus shifts to an out-of-equilibrium state, characterized by a change in certain constants of motion (such as energy). As the apparatus evolves towards a new equilibrium state within this channel, it tends in the limit of infinite times towards a state with a very high number of complexions in relation to other states, leading to the desired loss of coherence of the macroscopic state. The state reduction of the microscopic object would not occur during the interaction with the macroscopic apparatus, but is associated with the process that occurs after the end of the interaction (Rosenfeld, 1965, pp. 225, 230), and that can be identified with the process of *amplification*.

Another model of TDA that was simpler and quite influential (see Blokhintsev [1960] 1968, pp. 91–98) was elaborated by H.S. Green (1958), a former collaborator of Born. Describing an ingenious model for detectors, Green argued that the interference terms between two detectors *tends to zero* as the number of oscillators that compose the detectors increases. Years later, Furry (1966, pp. 56–59) would show that Green's conclusions were based on an illegitimate averaging procedure (see d'Espagnat, 1976, pp. 217–219).

It is interesting that the thermodynamic amplification approach to the measurement problem was well received by physicists close to the complementarity interpretation, although it describes the macroscopic apparatus as a quantum system. The TDA programme was seen as an extension or 'completion' of the Copenhagen view (see Jammer, 1974, pp. 492–493). The same can be said of von Neumann's description of the apparatus as a quantum system.

Von Neumann's insolubility argument did not hinder the development of the TDA approach. This is probably because the proof did not explicitly refer to macroscopic systems, which are subject to additional constraints such as the ergodicity conditions. Another way of describing macroscopic systems not encompassed by von Neumann's proof would be to consider them as having infinite degrees of freedom, or evolving in time spans that are infinitely longer than the durations of microscopic interactions. It is only under these limits that the preceding solutions apply exactly.

Although the thermodynamic amplification programme was strong in the 1950s, it was not free from criticisms. This can be exemplified by the debate between Süssman and Feyerabend at the 1957 Colston Symposium (Körner, 1962, pp. 140–147). There was an agreement that quantum theory could not explain exactly the transition from a pure composite state to a classical-like mixture of such states. The debate was whether it was justified to *approximate* the pure state of a macroscopic system by a mixture on the grounds that both are indistinguishable for a macroscopic observer (Feyerabend, 1962, p. 127). This 'justification for approximation' is necessary when using models with infinities to describe systems of finite size evolving in finite times.

Through the 1960s the strength of the TDA programme would wane, in part due to additional criticisms, and in part due to stronger versions of von Neumann's insolubility proof. The most important argument used against the thesis that amplification is a necessary condition for state reduction was that this thesis could not explain the existence of 'null-result experiments'. This is exemplified by a Stern–Gerlach experiment for a single atom (of known time of arrival) in which one of the two detectors is removed, and the remaining detector (assumed to have perfect efficiency) does not trigger: one can infer that the atom, previously described in a superposed state of position, is now in a well-defined path. In this case one has state reduction without amplification. This experiment, discussed in 1959 by Mauritius Renninger as a criticism of the complementarity interpretation (see Jammer, 1974, pp. 495–496), was used by Klaus Tausk, in a preprint of 1966, as a criticism of the Daneri–Loinger–Prosperi proposal (see Pessoa, Freire, and De Greiff, 2008), an idea repeated in passing by Jauch, Wigner, and Yanase (1967, p. 150). A response to this criticism was given by Loinger (1968), arguing that his theory did not require an amplification to take place, but only a coupling between the quantum object and the detector (an issue which was later explored by Dicke, 1981).

Another factor that contributed to the loss of popularity of the Daneri–Loinger–Prosperi model was Wigner's (1963, pp. 333–335) proof that the completeness problem is insoluble, reaffirming that the thermodynamic models could not furnish an 'exact' solution for the measurement problem. Furthermore, with the rise of the decoherence approach in the 1980s (section 11.8), it became clearer that the description of the measurement process in terms of statistical ensembles can attempt to explain the disappearance of interference terms and to give insight concerning the characterization problem, but cannot fully answer the question of how individual collapses occur (see Bell, 1990, p. 25). Still, this opinion has not deterred the tradition of proposing solutions to the measurement problem in the framework of quantum statistical mechanics (see Allahverdyan *et al.*, 2013).

# 11.6 INSOLUBILITY PROOFS

As we saw in section 11.3, if the completeness problem were 'soluble', then one could explain the unpredictability of the results of quantum measurement as being due to our ignorance concerning the initial state of the measuring apparatus (or the environment), and not to some intrinsic randomness of nature or to the existence of hidden variables in the object system. The projection postulate would then be reducible to the unitary evolution described by the Schrödinger equation.

To show that the completeness problem is soluble, one would have to construct a composite quantum system that satisfies certain conditions. These would include the two hypotheses presented in section 11.1. The first is: (1) the macroscopic apparatus may be adequately described as a quantum system. Since an apparatus is composed of a finite number of particles, the state of the apparatus could be represented in a Hilbert space of finite dimension. Here, it is also necessary to specify whether this state is pure or mixed. For greater generality a mixture is assumed, although this assumption brings with it the problem of interpreting mixtures.

The other hypothesis presented in section 11.1 is as follows: (2) the composite quantum system (object and apparatus) can be considered closed in relation to the environment (the rest of the universe). A corollary of this hypothesis is that the temporal evolution of the composite system is *unitary*.

In addition, a 'pointer assumption' is introduced: (3) the different measurement results correspond to distinct final states of the apparatus (pointer states). (4) A more precise specification of this pointer assumption is a condition of *solubility*. For an individual measurement in which the apparatus is considered a pure system, this condition is very restrictive: one would require that at the end of the measurement the apparatus be in one of the pointer states (which are generally degenerate), and never in a superposition of them. For mixtures, the solubility condition can be relaxed, requiring only that the density matrix corresponding to the final mixture of the apparatus be diagonal (in the representation of the pointer states).

(5) Finally, one must define precisely what is meant by 'measurement' (or 'premeasurement', according to the more refined analysis of Busch *et al.*, 1996). A measurement is a subclass of the interactions governing the evolution of a composite system, and must satisfy a 'distinguishability' criterion. The usual criterion requires that the measurement be able to distinguish classes of states of the object for which the mean values of a self-adjoint operator are different (Fine, 1969, p. 112).

In addition to the criterion of distinguishability, it is common to introduce a restriction on the class of measurements. The classes of measurement usually considered have been called 'repeatable' (of the 1st kind, or ideal) and 'predictable' (of the 2nd kind) measurements. A measurement of the 1st kind is such that if, immediately after being performed, it is repeated, the result obtained will certainly be the same. In other words, the final state of the object after the measurement is the eigenstate corresponding to the eigenvalue obtained as a result. If the object starts off in one of these eigenstates, of course, the measurement outcome will have a unique value with probability 1. This is the type of measurement considered by Dirac and von Neumann in their treatises on QM, and by Wigner in his insolubility proof. Predictable measurements were defined by Landau and Peierls, and are those that for each possible measurement outcome, there is an object eigenstate which gives that outcome with probability 1 (as for measurements of the 1st kind), but after the measurement the object is not in the initial eigenstate anymore (for instance, by losing energy in the interaction with the measurement apparatus). Even more general measurement classes can be defined (see Fine, 1969, p. 112; Busch *et al.*, 1996, p. 7).

A measurement that satisfies the conditions above (1–5) would constitute a solution to the completeness problem. One proves, however, that such a measurement cannot be defined, therefore proving the 'insolubility' of the completeness problem. Denoting the condition for unitarity (2) by U, that of measurement (5) by M, that of solubility (4) by S, and the additional hypotheses (1,3) by H, one can express the general scheme of the insolubility proofs in the following form: $H, U, S \Rightarrow \neg M$, where '$\Rightarrow$' is the sign of implication and '$\neg$' of negation. Accepting H and U, this implies that there are no measurements (of the specified type) that fulfil the solubility condition: $H, U, M \Rightarrow \neg S$.

In the 1960s, the first insolubility proof was that of Wigner (1963, pp. 333–335) for repeatable measurements without degeneracy. This result brought a new wave of interest in the 'measurement problem', favouring the aforementioned position of Süssmann, according to which the results of the TDA programme could not be valid in an exact way. This therefore weakened the thesis that state reduction occurred with the coupling of the quantum object with an amplification device in a metastable state. Several other proofs followed, such as that of d'Espagnat (1966) for predictable measurements, that of Fine (1970) for more general measurements, and that of Busch and Shimony (1996) for positive operator-valued measures.

## 11.7 OTHER APPROACHES
### BEFORE THE 1970s

Dozens of proposals have been made to address the measurement problem, including arguments that it is nothing more than a pseudoproblem. Here some trends before the 1970s will be mentioned.

A view that tended to dissolve the enigmatic character of the measurement problem was the so-called ensemble interpretation of QM, according to which the state vector does not describe an individual system but a statistical ensemble of identically prepared systems (see Jammer 1974, pp. 440–447). According to Kemble (1937, pp. 326–329), the reduction of the wave packet simply reflects a mental process in which the object system is transferred from an initial ensemble to one of its subensembles, during the 'selective process' of measurement. Such a view is connected to an instrumentalist or *epistemic* interpretation of the wave function, which in the words of Kemble is 'merely a subjective computational tool and not in any sense a description of objective reality'.

Kemble's version of the ensemble interpretation was criticized by Henry Margenau, who also rejected the subjectivist view and the role attributed to 'conscious cognizance'. He used the general measurement problem and von Neumann's insolubility proof as an argument *against* the projection postulate (Margenau, 1936, pp. 241–242). His analysis of the measurement process was focused on position determinations, distinguishing the operation of state preparation from that of measurement, which usually involves absorption and destruction of the object system (Margenau, 1937, pp. 356–359).

Proponents of the ensemble interpretation tend to consider that there is no 'measurement problem', or that it is a pseudoproblem. This view is especially true in instrumentalist versions of the interpretation, such as that of James Park, a former student of Margenau. According to this view, QM is a theory only about the statistics of measurement results, and not about the intrinsic properties of individual physical objects. This position leads to a rejection of the concept of state reduction, and the measurement problem is considered meaningless (Park, 1973).

From another perspective, Unruh (1986, pp. 244–249) has stressed that there is no measurement problem in the Heisenberg picture. In this picture, the continuous change of the system's attributes is expressed by the unitary evolution of the operators, while any change in our knowledge of the system due to measurement is encoded as a change in the state vector.

In the 1950s, in parallel with the rise of the thermodynamic amplification programme, David Bohm (1952) elaborated his famous interpretation of QM in terms of 'hidden variables'. Any particle would have a well-defined position and momentum, and these variables would evolve continuously and deterministically in time. The interaction between the object and the particles of the apparatus would be ruled by a non-local 'quantum mechanical potential' that undergoes violent fluctuations which are very sensitive to the precise initial conditions. Louis de Broglie (1957, p. 84) also advanced a theory of measurement based on hidden variables, where the determination of position plays a fundamental role.

Another radical solution to the general problem of measurement was Hugh Everett's (1957) 'relative state' approach, which postulated that the universe as a whole should be described by a single wave function that evolves deterministically according to the Schrödinger equation. According to this view, the collapse of the wave function doesn't really occur, but is an illusion that arises after the quantum state of the conscious observer gets entangled with the superposition of the composite system (object plus apparatus). Our conscious perception of the world is defined by a single 'trajectory of memory configurations', but as a whole the universe branches out into many parallel universes during each act of measurement.

Another way out of the completeness problem would be that the linear Schrödinger equation should include an additional non-linear term which becomes important in certain circumstances, such as in living beings (Wigner 1962, pp. 180–181) or in macroscopic systems in general (Ludwig, 1961, pp. 156–160). Such non-linearity may be postulated as 'intrinsic' to the evolution of any system or as arising 'extrinsically' due to an interaction with the environment. To date, however, there has been no clear evidence of violation of the superposition principle (i.e., linearity) in QM (see section 11.9).

An outgrowth of the discussions on a 'justification for approximation' (see section 11.5, with Feyerabend and Süssman) which attracted much attention in the 1960s was a pragmatic solution to the measurement problem called the 'classical properties approach', associated with Josef-Maria Jauch (1964). Even though the evolved state of a composite object and apparatus system is pure and entangled, a third component system such as an amplifier could be coupled solely to the apparatus system, and the final state of this amplifier could reflect a mixture of composite states (determined by a partial trace, as done in the environmentally induced decoherence programme) which is indistinguishable from the actual pure state (see discussion in d'Espagnat, 1976, pp. 173–185).

# 11.8 OBJECTIVIST APPROACHES IN THE 1970s AND 1980s

The 1970s were marked by a series of new objectivist approaches that can be divided into three groups, according to their main idea: infinite apparatus, open systems, or stochasticity.

The idea of treating the apparatus as a system with infinite degrees of freedom was already present in the TDA programme, and led to exact solutions in this limit. Hepp (1972) constructed a model that involves an apparatus consisting of an infinite array of spin ½ particles, and which leads to state reduction in the limit of infinite times. These two limit conditions were criticized by Bell (1975) as being unphysical. But the idea of building infinite models of the measurement apparatus led to many papers.

Many other models of the measurement apparatus using statistical quantum mechanics have been proposed ever since (see the review by Allahverdyan et al., 2013), continuing the TDA programme and incorporating the decoherence approach. One of these early models, worth singling out, is the model of Machida and Namiki (1980), which represents the apparatus system not in a single Hilbert space, but in a direct sum of an infinite number of individual spaces, each with a different number of particles (the 'Fock space'). An interesting feature of their model is that state reduction occurs during detection (e.g., during the interaction with a metallic plate), before amplification, which is an interesting hypothesis still not amenable to experimental testing. As with the TDA programme, the null-result experiment would be explained by the object's interaction with the detector, without resulting in signal amplification.

A second guiding idea used for solving the measurement problem in the 1970s involved the thesis that all macrosocopic systems are open to interactions with the environment, so that the composite system consisting of object plus apparatus is not governed by a unitary evolution operator, as in Schrödinger's wave mechanics. This idea had already been explored by Bohm (1951, pp. 138–140) and Heisenberg (1955, pp. 22–23, 26–27) and had an important role in Everett's relative state interpretation. The fact that the composite system is open thus violates one of the hypotheses of the insolubility proofs (thesis 2 of section 11.6), so that the insolubility of the completeness problem does not follow.

A physical argument for the impossibility of closing any macroscopic system is that its discrete energy levels are extremely dense, so that it is always strongly correlated with other macroscopic bodies in the environment (Zeh 1970, p. 73). To clarify this, consider that a single spin ½ particle in a magnetic field has two reasonably separate energy levels, so that a small external effect such as the presence of weak radio waves would not significantly affect the probability of transition between the levels. As other particles are added to the system, the number of levels multiplies, and the energetic distance between them decreases. For a macroscopic body, then, a faint external effect like a fluctuation in the background radiation of the universe could significantly affect the transition probabilities. Therefore, there are no macroscopic closed systems. In statistical mechanics, this principle was used by Burbury (1894) to solve the problem of irreversibility, that is, to explain the Second Law of thermodynamics in microscopic terms, and is now known as the 'interventionist hypothesis' (Sklar, 1993, pp. 250–254).

Within the open systems paradigm, two important approaches to the measurement problem were developed. The *environmentally induced decoherence programme* performs a statistical analysis of the time-dependent interaction between a quantum system and its environment, and then applies a 'partial trace' that averages out the state of the environment. In this way, it describes the process by which the pure state of a quantum system evolves in time into an improper mixture, with increase in entropy. In this way, a quantum system described as a coherent superposition, say of position eigenstates, loses its interference terms and becomes a classical-like set of incoherent position eigenstates. Such decoherence arises from fluctuations in the environment, and follows an exponential decay law like that derived by Caldeira and Leggett (1985) for a specific model of the environment. The decoherence time is inversely proportional to the temperature, to the mass and to the square of the spatial separation between different parts of the wave packet, being extremely short for usual macroscopic objects (see Zurek, 1991).

When such an approach is applied to the measurement process, an environment is added to the composite system (object plus apparatus), an interaction with the environment is handpicked, and then the mathematical operation of taking the partial trace results in an improper mixture which is diagonal in the basis of the composite pointer states $\{|\varphi_i \cdot |\xi_i\rangle\}$, thus 'explaining' why $\{|\varphi_i\rangle\}$ are the eigenstates of the operator being measured. As mentioned above, the key feature of this approach is to consider the unitary evolution of system plus environment until a certain moment, and then to apply a partial trace that averages out the state of the environment. One can argue that the application of such a partial trace in the measurement process amounts to a statistical version of the projection postulate (Pessoa, 1998, pp. 332–333).

The decoherence approach does not solve the measurement problem for the individual case, because it does not explain why a superposition of position eigenstates collapses to a specific position eigenstate. But it in principle offers a quantum-mechanical explanation of why a certain measurement apparatus leads to the measurement of a specific observable, i.e. why a certain set of pointer states is selected from a specific interaction between the composite system and the environment surrounding the system, if only such interactions could be determined. One consequence of this approach is the suggestion that an environment can 'make measurements' or 'monitor' continuously a quantum-mechanical system, leading to its decoherence without the presence of a human observer or even a measurement apparatus. This *suggests* that collapses can occur in noisy environments, without measurement apparatuses, but there is no detailed description of this process for an *individual* quantum system: decoherence is a process described for a statistical ensemble of systems (see analysis by Schlosshauer, 2005).

The other approach that accepts the open systems paradigm is the thesis that gravity is responsible for the collapse of the quantum state. Károlyházy (1966) postulated that space-time has a 'slightly smeared metric' that destroys the coherence between parts of a macroscopic body. Roger Penrose (1986) also investigated a possible connection between gravity and state reduction, corresponding to a rise in 'gravitational entropy'.

A third guiding idea that also accepts that there is an objective process of reduction assumes that such a process is intrinsically stochastic, and not cryptodeterministic. A theory developed by Ghirardi, Rimini, and Weber (GRW) assumes that any system has a very small probability of undergoing a spontaneous localization (Benatti *et al.*, 1987). For a system with a small number of particles, such a localization would occur very rarely, and would not violate the Schrödinger equation. For a macroscopic system, however, composed of a large number of entangled particles, such a collapse would happen quite frequently. That would explain why we only observe reduction when a macroscopic apparatus couples to the quantum object. A general criticism against this model is that it is 'ad hoc', only explaining the phenomenon for which it was developed. In this line, one can also criticize the hypothesis of spontaneous localization because it introduces an additional universal constant (the frequency of collapses) to the small set of universal physics constants.

Another approach which sees reduction as a stochastic process uses dynamic models of reduction which introduce a stochastic correction to the Schrödinger equation (in line with Ludwig's 1961 proposal), without worrying about the physical origin of such white noise (Pearle, 1986).

# 11.9 Conclusions

Twenty years into the 21st century, the measurement problem continues to be an unsolved challenge in the foundations of physics. In a poll conducted by Schlosshauer *et al.* (2013) with 33 researchers into the foundations of quantum mechanics, one of the questions asked about the measurement problem. To their surprise, only 24 per cent considered it 'a severe difficulty threatening quantum mechanics'. In addition, 27 per cent took it as 'a pseudoproblem', 15 per cent chose the answer 'solved by decoherence', and 39 per cent 'solved/will be solved in another way' (another 27 per cent chose 'none of the above'—multiple answers were allowed). In another question concerning interpretations, 42 per cent chose 'Copenhagen', 18 per cent picked 'Everett (many worlds and/or many minds)', 24 per cent 'information-based/information-theoretical', and 9 per cent 'objective collapse (e.g. GRW, Penrose)'. In another question about the observer, only 6 per cent answered 'plays a distinguished physical role (e.g. wave-function collapse by consciousness)'.

With this, we can estimate how the four general classes of answers to the measurement problem, mentioned in section 11.1, fare today: (i) objectivist views, considering 'decoherence' plus 'solved in another way' minus 'many worlds' and minus 'subjectivism': around 30 per cent; (ii) subjectivist interpretations, around 6 per cent; (iii) many worlds approach, around 18 per cent; (iv) pseudoproblem, around 27 per cent.

The most noteworthy tendency seems to be a decline in interest or in the attribution of importance to the measurement problem, possibly correlated with the rise of information theoretical approaches. This decline in interest is to be expected for problems for which few theoretical ideas or novel experiments have occurred.

Still, the objectivist programme has been developing with more detailed models, as surveyed by Allahverdyan *et al.* (2013) and Bassi *et al.* (2013). And the advance of technology has brought about experiments of ever increasing precision in the foundations of quantum mechanics, marked by the superposition experiments led by Marcus Arndt with fullerene molecules in 1999, which twenty years later have attained superpositions of molecules with 2000 atoms (Fein *et al.*, 2019). These experiments have extended the known limits of quantum superpositions, addressing the question of whether macroscopic superpositions are possible.

With the increasing sophistication of experimental and theoretical tools, are we to expect that the measurement problem will be solved in our century?

This paper benefitted from the comments by Linda Wessels, Harvey Brown, Olival Freire Jr., and Alexei Kojevnikov.

# References

Allahverdyan, A. E., Balian, R., and Nieuwenhuizenc, T. M. (2013). Understanding quantum measurement from the solution of dynamical models. *Physics Reports*, **525**, 1–166.

Badurek, G., Rauch, H., and Tuppinger, D. (1986). Polarized neutron interferometry. In D. M. Greenberger (ed.), *New techniques and ideas in quantum measurement theory. Annals of the New York Academy of Sciences*, **480**, 133–46.

Bassi, A., Lochan, K., Satin, S., Singh, T. P., and Ulbricht, H. (2013). Models of wave-function collapse, underlying theories, and experimental tests. *Reviews of Modern Physics*, **8**, 471–527.

Bell, J. S. (1975). On wave packet reduction in the Coleman–Hepp model. *Helvetica Physica Acta*, **48**, 93–98. Reprinted in Bell, J. S. (1987), *Speakable and unspeakable in quantum mechanics*, Cambridge: Cambridge University Press, pp. 45–51.

Bell, J. S. (1990). Against 'measurement'. In A. I. Miller (ed.), *Sixty-two years of uncertainty*, New York: Plenum, pp. 17–31.

Benatti, F., Ghirardi, G. C., Rimini, A., and Weber, T. (1987). Quantum mechanics with spontaneous localization and the quantum theory of measurement. *Il Nuovo Cimento B*, **100**, 27–41.

Blokhintsev, D. I. (1968). *The philosophy of quantum mechanics*. Dordrecht: Reidel. [Russian original: 1960.]

Bohm, D. (1951). *Quantum theory*. Englewood Cliffs: Prentice-Hall.

Bohm, D. (1952). A suggested interpretation of the quantum theory in terms of 'hidden' variables. I. II. *Physical Review*, **85**, 166–193. Reprinted in J. A. Wheeler and W. H. Zurek (eds.) (1983), *Quantum theory and measurement*, Princeton: Princeton University Press, pp. 369–96.

Bohr, N. (1928). The quantum postulate and the recent development of atomic theory. *Nature*, **121**, 580–90. Reprinted in Bohr, N. (1931), *Atomic physics and the description of nature*, Cambridge: Cambridge University Press, pp. 52–91. Reprinted in J. A. Wheeler and W. H. Zurek (eds.) (1983), *Quantum theory and measurement*, Princeton: Princeton University Press, pp. 87–126.

Bohr, N. (1958). Unity of knowledge. In N. Bohr, *Atomic physics and human knowledge*, New York: Wiley, pp. 67–82. First published in 1955, in L. G. Leary (ed.), *The unity of knowledge*, New York: Doubleday, pp. 47–62.

Born, M. (1983). On the quantum mechanics of collisions. Transl. J. A. Wheeler and W. H. Zurek, in J. A. Wheeler and W. H. Zurek (eds.), *Quantum theory and measurement*, Princeton: Princeton University Press, pp. 52–55. [German original: 1926].

Burbury, S. H. (1894). Boltzmann's minimum function. *Nature*, **51**, 78–78.

Busch, P., Lahti, P. J., and Mittelstaedt, P. (1996). *The quantum theory of measurement*, 2nd ed. Berlin: Springer.

Busch, P. and Shimony, A. (1996). Insolubility of the quantum measurement problem for unsharp observables. *Studies in History and Philosophy of Modern Physics B*, **27**, 397–404.

Caldeira, A. O. and Leggett, A. J. (1985). Influence of damping on quantum interference: an exactly soluble model. *Physical Review A*, **31**, 1059–66.

Cartwright, N. (1983). *How the laws of physics lie*. Oxford: Clarendon.

Daneri, A., Loinger, A., and Prosperi, G. M. (1962). Quantum theory of measurement and ergodicity conditions. *Nuclear Physics*, **33**, 297–319. Reprinted in J. A. Wheeler and W. H. Zurek (eds.) (1983), *Quantum theory and measurement*, Princeton: Princeton University Press, pp. 657–79.

Dicke, R. H. (1981). Interaction-free quantum measurements: a paradox? *American Journal of Physics*, **49**, 925–30.

de Broglie, L. (1957). *La théorie de la mesure en mécanique quantique*. Paris: Gauthiers-Villars.

d'Espagnat, B. (1966). Two remarks on the theory of measurement. *Supplemento al Nuovo Cimento*, **4**, 828–38.

d'Espagnat, B. (1976). *Conceptual foundations of quantum mechanics*, 2nd edn. Reading, MA: Benjamin.

Everett, H., III. (1957). Relative state formulation of quantum mechanics. *Reviews of Modern Physics*, **29**, 454–62. Reprinted in J. A. Wheeler and W. H. Zurek (eds.) (1983), *Quantum theory and measurement*, Princeton: Princeton University Press, pp. 315–23.

Fein, Y.Y., Geyer, P., Zwick, P., Kiałka, F., Pedalino, S., Mayor, M., Gerlich, S., and Arndt, M. (2019). Quantum superposition of molecules beyond 25 kDa. *Nature Physics*, **15**, 1242–45.

Feyerabend, P. K. (1962). On the quantum-theory of measurement. In S. Körner (ed.), *Observation and interpretation in the philosophy of physics: with special reference to quantum mechanics*, New York: Dover, pp. 121–30.

Feynman, R. P. (1948). Space-time approach to non-relativistic quantum mechanics. *Reviews of Modern Physics*, **20**, 367–87.

Fine, A. I. (1969). On the general quantum theory of measurement. *Proceedings of the Cambridge Philosophical Society*, **65**, 111–22.

Fine, A. I. (1970). Insolubility of the quantum measurement problem. *Physical Review D*, **2**, 2783–87.

Fine, A. I. (1973). The two problems of quantum measurement. In P. Suppes, L. Henkin, A. Joja and G. C. Moisil (eds), *Logic, Methodology and Philosophy of Science IV*, Amsterdam: North-Holland, pp. 567–81.

Furry, W. H. (1966). Some aspects of the quantum theory of measurement. In W. E. Brittin (ed.), *Lectures in Theoretical Physics*, vol. VIII A, Boulder: University of Colorado Press, pp. 1–64.

Gibbins, P. (1987). *Particles and paradoxes*. Cambridge: Cambridge University Press.

Green, H. S. (1958). Observation in quantum mechanics. *Il Nuovo Cimento*, **9**, 880–89.

Heisenberg, W. (1983). The physical content of quantum kinematics and mechanics. Transl. J. A. Wheeler and W. H. Zurek, in J. A. Wheeler and W. H. Zurek (eds.), *Quantum theory and measurement*, Princeton: Princeton University Press, 62–84. [German original: 1927.]

Heisenberg, W. (1930). *The physical principles of quantum theory*. Chicago: Chicago University Press.

Heisenberg, W. (1955). The development of the interpretation of the quantum theory. In W. Pauli (ed.), *Niels Bohr and the development of physics*, London: Pergamon, pp. 12–29.

Heisenberg, W. (1958). *Physics and philosophy*. New York: Harper.

Heitler, W. (1949). The departure from classical thought in modern physics. In P. A. Schilpp (ed.), *Albert Einstein, philosopher-scientist*, Urbana: Open Court, pp. 181–98.

Hepp, K. (1972). Quantum theory of measurement and macroscopic observables. *Helvetica Physica Acta*, **45**, 237–48.

Jammer, M. (1974). *The philosophy of quantum mechanics*. New York: Wiley.

Jauch, J.-M. (1964). The problem of measurement in quantum mechanics. *Helvetica Physica Acta*, **37**, 293–316.

Jauch, J.-M., Wigner, E. P., and Yanase, M. M. (1967). Some comments concerning measurements in quantum mechanics. *Il Nuovo Cimento*, **48 B**, 144–51.

Jordan, P. (1949). On the process of measurement in quantum mechanics. *Philosophy of Science*, **16**, 269–78.

Károlyházy, F. (1966). Gravitation and quantum mechanics of macroscopic objects. *Il Nuovo Cimento*, **42**, 390–402.

Kemble, E. C. (1937). *The fundamental principles of quantum mechanics*. New York: McGraw-Hill.

Körner, S. (ed.), in collaboration with M. H. L. Pryce (1962). *Observation and interpretation in the philosophy of physics: with special reference to quantum mechanics*. Proceedings of the 9th Symposium of the Colston Research Society held in the University of Bristol, April 1st–April 4th, 1957. New York: Dover.

Loinger, A. (1968). Comments on a recent paper concerning measurements in quantum mechanics. *Nuclear Physics*, **108**, 245–49.

London, F. W. and Bauer, E. (1983). The theory of observation in quantum mechanics. Transl. A. Shimony *et al.*, in J. A. Wheeler and W. H. Zurek (eds.), *Quantum theory and measurement*, Princeton: Princeton University Press, pp. 217–59. [French original: 1939.]

Lüders, G. (2006). Concerning the state-change due to the measurement process. Transl. K. A. Kirkpatrick, *Annalen der Physik (Leipzig)*, **15**, 663–70. [German original: 1951.]

Ludwig, G. (1953). Der Messprozess. *Zeitschrift für Physik*, **135**, 483–511.

Ludwig, G. (1954). *Die Grundlagen der Quantenmechanik*. Berlin: Springer.

Ludwig, G. (1961). Gelöste und ungelöste Probleme des Meßprozesses in der Quantenmechanik. In F. Bopp (ed.), *Werner Heisenberg und die Physik unserer Zeit*, Braunschweig: Vieweg, pp. 150–81.

Machida, S. and Namiki, M. (1980). Theory of measurement in quantum mechanics: mechanism of reduction of wave packet. I. II. *Progress in Theoretical Physics*, **63**, 1457–73, 1833–47.

Margenau, H. (1936). Quantum mechanical description. *Physical Review*, **49**, 240–42.

Margenau, H. (1937). Critical points in modern physical theory. *Philosophy of Science*, **4**, 337–370.

Margenau, H. (1958). Philosophical problems concerning the meaning of measurement in physics. *Philosophy of Science*, **25**, 23–33.

Park, J. L. (1973). The self-contradictory foundations of formalistic quantum measurement theories. *International Journal of Theoretical Physics*, **8**, 211–18.

Pearle, P. (1986). Models for reduction. In R. Penrose and C. J. Isham (eds), *Quantum concepts in space and time*, Oxford: Clarendon, pp. 84–108.

Penrose, R. (1986). Gravity and state vector reduction. In R. Penrose and C. J. Isham (eds), *Quantum concepts in space and time*, Oxford: Clarendon, pp. 129–46.

Pessoa Jr., O. (1998). Can the decoherence approach help to solve the measurement problem? *Synthese*, **113**, 323–46.

Pessoa Jr., O., Freire Jr., O., and De Greiff, A. (2008). The Tausk controversy on the foundations of quantum mechanics: physics, philosophy, and politics. *Physics in Perspective*, **10**, 138–62.

Rosenfeld, L. (1965). The measuring process in quantum mechanics. *Supplement of the Progress of Theoretical Physics*, extra number: 222–31.

Schlosshauer, M. (2005). Decoherence, the measurement problem, and interpretations of quantum mechanics. *Reviews of Modern Physics*, **76**, 1267–1305.

Schlosshauer, M., Kofler, J. and Zeilinger, A. (2013). A snapshot of foundational attitudes toward quantum mechanics. *Studies in History and Philosophy of Modern Physics*, **44**, 222–230.

Schrödinger, E. (1983). The present situation in quantum mechanics. Transl. J. D. Trimmer, in J. A. Wheeler and W. H. Zurek (eds.), *Quantum theory and measurement*, Princeton: Princeton University Press, pp. 152–67. [German original: 1935].

Sklar, L. (1993). *Physics and chance: philosophical issues in the foundations of statistical mechanics*. Cambridge: Cambridge University Press.

Solvay, L'Institut International de Physique (2009). General discussion of the new ideas presented. In G. Bacciagaluppi and A. Valentini (eds), *Quantum theory at the crossroads: reconsidering the 1927 Solvay Conference*, Cambridge: Cambridge University Press, pp. 432–70. [French original: 1928.]

Stein, H. and Shimony, A. (1971). Limitations on measurement. In B. d'Espagnat (ed.), *Foundations of quantum mechanics*, New York: Academic, pp. 56–76.

Süssmann, G. (1958). Über den Meßvorgang. *Abhandlungen der Bayerischen Akademie der Wissenschaften, Mathematisch-Naturwissenschaftliche Klasse*, Neue Folge, **88**, 3–41.

Szilard, L. (1983). On the decrease of entropy in a thermodynamic system by the intervention of intelligent beings. Transl. A. Rapoport and M. Knoller, in J. A. Wheeler and W. H. Zurek (eds.), *Quantum theory and measurement*, Princeton: Princeton University Press, pp. 539–48. [German original: 1929].

Taketani, M. (1971). Observation problem of quantum mechanics. *Progress in Theoretical Physics*, **50**, 65–72.

ter Haar, D. (1955). Foundations of statistical mechanics. *Reviews of Modern Physics*, **27**, 289–338.

Unruh, W. G. (1986). Quantum measurement. In D. M. Greenberger (ed.), *New techniques and ideas in quantum measurement theory. Annals of the New York Academy of Sciences*, **480**, 242–49.

von Neumann, J. (1955). *Mathematical Foundations of Quantum Mechanics*. Transl. R. T. Beyer. Princeton: Princeton University Press. [German original: 1932.]

Wheaton, B. S. (1983). *The tiger and the shark: empirical roots of the wave-particle dualism*. Cambridge: Cambridge University Press.

Wheeler, J. A., and Zurek, W. H. (eds.) (1983). *Quantum theory and measurement*. Princeton: Princeton University Press.

Whittaker, E. T. (1943). Chance, freewill and necessity in the scientific conception of the universe. *Proceedings of the Physical Society*, **55**, 459–71.

Wigner, E. P. (1962). Remarks on the mind-body question. In I. J. Good (ed.), *The scientist speculates*, New York: Basic Books, pp. 284–302. Reprinted in J. A. Wheeler and W. H. Zurek (eds.) (1983), *Quantum theory and measurement*, Princeton: Princeton University Press, pp. 168–81.

Wigner, E. P. (1963). The problem of measurement. *American Journal of Physics*, **31**, 6–15. Reprinted in J. A. Wheeler and W. H. Zurek (eds.) (1983), *Quantum theory and measurement*, Princeton: Princeton University Press, pp. 324–41.

Zeh, H.-D. (1970). On the interpretation of measurement in quantum theory. *Foundations of Physics*, **1**, 69–76. Reprinted in J. A. Wheeler and W. H. Zurek (eds.) (1983), *Quantum theory and measurement*, Princeton: Princeton University Press, pp. 342–49.

Zurek, W. H. (1991). Decoherence and the transition from quantum to classical. *Physics Today*, **44**(10), 36–44.

..............................................................................................

# CHAPTER 12

# EINSTEIN'S CRITICISM OF QUANTUM MECHANICS

..............................................................................................

### MICHEL PATY

## 12.1 INTRODUCTION[1]

..............................................................................................

AFTER having played a fundamental pioneering role in the birth of quantum physics, unveiling important non-classical properties and formulating a first coherent theoretical approach, Einstein was, as it is well known, far less enthusiastic about its constitution as *quantum mechanics*. He himself, together with Max Born in the years 1920–1924,[2] coined the expression that was to be adopted, from 1925–1927 on, to designate the proposed new theory. This forerunner use must be understood as referring already, not to limited mechanical models or explanations, but indeed to *analytical mechanics* that was, to the eyes of all physicists and mathematicians, the classical standard of a fundamental mathematized physical theory. To formulate such a new theory, that would no more be classical but proper to the quantum domain, was the aim for Einstein as well as for Born and their colleagues. And when the founders

---

[1] This present text is based on a previous article by the author, entitled 'The Nature of Einstein's objections to the Copenhagen interpretation of quantum mechanics' and published in 1995 (Paty (1995)). It has been partly modified (without changing the main argument), and updated for relevant bibliography.

[2] Einstein had already used the expression '*quantum mechanics*' in a letter to Max Born dated 4 June 1919, in Einstein and Born (1969), and later in a letter to Paul Ehrenfest dated 31 May 1924 (Einstein Archives, and Einstein (1987–2018), vol. 15) where he underlines it. Max Born used it the same year (1924) to entitle his article 'Über Quantenmechanik' (Born (1924)), before using it as the definitive label for the theory in the founding papers by himself alone (Born, 1926, 1927) and in collaboration with his disciples Werner Heisenberg and Pascual Jordan (Born and Jordan, 1925; Born, Heisenberg, and Jordan 1926; Born and Heisenberg 1928), and by Paul Dirac as well (Dirac, 1926 a,b). On the side of the founders of *wave mechanics*, they called their theory either quantum mechanics (Schrödinger, 1926, 1928), or 'the quantum theory' or 'the dynamics of quanta' (de Broglie, 1924, 1928) in the titles of their respective expositions.

and proponents of the theory presented at the 1927 Solvay Conference baptized it with these words, it was in the perspective of having achieved that very goal. But Einstein did not share this opinion and, from then on and over the years up to his death, he constantly argued against their pretension to have settled a definitive and complete quantum theory, without denying the successes and advances allowed by the new theory, which he readily and often warmly acknowledged.

Anyhow, it seemed to many physicists—actually to most of them as time went by and quantum physics experienced considerable development—that a gap was widening between Einstein's reservations and the main stream of research, in which they themselves were engaged with enthusiasm, and that impression has been amplified and propagated over the years, even up to the present day. The width of the gap may even seem, in the eyes of some, to be extending further with the considerable developments that quantum physics has been experiencing, both in terms of the field of knowledge and in terms of correlated technological advances.

The fight may indeed look unequal between, on the one hand, objections which were clearly of a fundamental nature but appeared ultimately more epistemological than physical in the proper sense and, on the other, very effective advances being acquired continuously since that time and up to now in the knowledge of **both** theoretical and practical physics, and even in its technological applications. Think of the mastery obtained on the quantum properties of atomic, nuclear, and subnuclear (or 'high energy') matter, and for the latter, of the quantum theory of the interaction fields which achieved a partial unification in the form of the 'Standard Model', and of the possibility, acquired in this way, to conceive, formulate, and explore a quantum cosmology, brought together with relativistic cosmology under the aegis of the 'Standard Cosmological Model'. Or, closer to elementary and fundamental quantum processes, think of the implications of specific properties such as *quantum entanglement*, a direct consequence of *the superposition principle* of state functions on which the quantum theory is based, which reveals a whole range of previously unsuspected properties of quantum matter: from 'non local-separability' of correlated subsystems to the transmission of *quantum information* (opening the possibility of new technologies such as *quantum computing* and *quantum cryptology*), as well as to the '*decoherence* phenomenon', that makes it possible to conceive an objective approach to the *problem of the measurement of quantum systems*, and to better master the *problem of the connection between the quantum and the classical domains* of the physical world.

Do these—impressive to an unprecedented degree—advances of quantum physics, achieved through the present quantum theory structured by quantum mechanics, henceforth make obsolete Einstein's objections, these having now at best purely historical interest? This is what many physicists thought at the most intense moments of controversy, and such is possibly the opinion of more than one today. At the same time, a few others, among the most creative actors of these developments, considered, like Richard Feynman—one of the founders of quantum electrodynamics—that 'no one really understands quantum mechanics'. By this apparently paradoxical statement, not devoid of irony, he was alluding—so I suppose—to the exceptionally 'formal' character of the theory and its decidedly indirect relationship to the sensitive knowledge given to us in experiments. There is no doubt that a full deep understanding is not necessarily required in order to carry out theoretical thinking and obtain results. But it remains indispensable to *understand* it up to a certain point in order to advance, and the *critical mind* remains awakened at one level or another on the path of knowledge. And in this last point stand the lessons we can receive from Einstein's criticisms of the quantum theory of his time.

The gain in perspective that all the further developments of quantum physics just mentioned above offers nowadays, far from abolishing the questions of foundations and interpretation that had marked the beginnings of this field of physics and its now so powerful theory, allows us to look at them anew nowadays, for further understanding in the light of these lessons. Clearly, these preoccupations or concerns do not oppose the historical viewpoint which we shall follow now. On the contrary, the historical-epistemological approach for recent contemporary science enlightens the ways by which scientific knowledge is progressively built by human minds anxious on the whole to understand the world and to criticize their understanding so as to refine it.

In what follows, a presentation of Einstein's objections to quantum mechanics and to its dominant interpretation is proposed, situating them in their time and scientific context, examining their arguments and trying to estimate their eventual impact on the further evolution of quantum physics, in its advances as well as in its deeper understanding.

First the importance of the 'philosophical climate' in the milieu of quantum physicists at that time is emphasized, a climate dominated by the so-called 'Copenhagen orthodoxy' and Bohr's 'idea (or philosophy) of complementarity'. What Einstein was primarily reluctant to do was to accept without question the fundamental character of the theory of quantum mechanics, presented as an abstract mathematical formalism related indirectly (through 'interpretation') to results of experiments, and to modify for it the basic principles of knowledge, by adopting an operationalist view on science that gives up the presupposition of an objective physical reality, independent of any observer and of the ways of its observation.

Then the main lines of Einstein's own programme with respect to quantum physics are stressed, a programme which is to be considered in relation with other previous and contemporary attempts and achievements of his own and of other physicists as well, in that and other domains. This programme is related to Einstein's epistemological views (on physical reality, causality, and determinism, and on the physical meaning of theoretical relations and conceptual coherence) and reflects aspects of his *scientific style*[3] as a searcher.

The critical arguments he opposed to quantum mechanics are followed in their evolution from 1927 (the year of the presentation of the theory at the Solvay Conference) to 1935 (publication of the 'EPR' article with a decisive argument, the so-called

---

[3] On this notion, see for example Paty (1996) and references there quoted.

'EPR thought experiment'), and later on over the years in his endeavours to refine the formulation of his critiques by continuing to explore the conceptual implications of the EPR thought experiment in a more and more precise and significant way that deepened what was truly at stake. Through this evolution one sees the nature of his reservations move from an objection to quantum mechanics towards an objection to its favoured interpretation, i.e. the pretension of its proponents to consider it as a complete theory.[4]

These elaborations and correlated debates contributed highly to the formulation and clarification of quantum concepts such as that of *non local-separability* that was to be experimentally evidenced years later. Other related further clarifications and advances have been made possible thanks to or inspired by Einstein and the critical questioning of others (about *entanglement, quantum transmission of information, decoherence,* distinction and relations between probability and statistics, etc.).

Some remaining fundamental questions will be evoked in the concluding comments.

# 12.2 THE CONTEXT OF THE 'COMPLEMENTARITY INTERPRETATION'

As said at the beginning, the true nature of Einstein's well-known opposition to the standard interpretation of quantum mechanics has often been underestimated or misunderstood. As a matter of fact, most of the quantum physicists, i.e. physicists working in the domain of quantum phenomena (in quantum optics, condensed matter, atomic, nuclear, and particle physics) seem to admit that in the Einstein–Bohr dialogue or controversy it was indeed Bohr who was right and that Einstein's position was 'an effect of fossilization', as he himself expressed it, with ironic self-derision, in a letter to his friend Max Born,[5] who himself was not far from thinking so. In effect, Max Born's commentaries to the edition of his correspondence with Einstein (published 14 years after Einstein's death) witness this incomprehension that has lasted since the actual

---

[4] See further on, in particular, the next section. May we anticipate somewhat at this point by offering already a brief summary of the fundamental status of what illustrated this incompleteness to Einstein? *Non local-separability* between the two correlated subsystems of the EPR thought experiment, inherent to the quantum theory formulation (one owes to Einstein to have extracted it explicitly as a physical concept, be it negatively), would (to him) imply non-physical instantaneous action-at-a-distance, which cannot be, and thus would impeach the theory to describe individual physical systems. The theory would then be valid, but could only provide a statistical description of ensembles of systems. Such is actually the meaning of Einstein's formula 'God does not play dice', thereby more complex than it may look at first sight. But further investigations both theoretical and experimental, inspired and guided by this state of things, have shown that *non local-separability* is confirmed as a physical fact (without action-at-a-distance) and corresponds well to a *fundamental quantum physical concept* that one has henceforward 'to live with'. Indeed, its implications have recently considerably enriched quantum physics, as we comment on in section 12.7.

[5] A. Einstein, letter to Max Born, 7 November 1944, in Einstein and Born (1969).

establishment of quantum mechanics.[6] 'The Einstein of today is changed', Born wrote in Schilpp's volume of homages to Einstein, and he contrasted Einstein's late saying in 1944 that, contrary to Born believing in a God playing dice, he (Einstein) tried 'to grasp in a wildly speculative way' the 'perfect laws in the world of things existing as real objects',[7] to the young Einstein who 'used probability as a tool for dealing with nature just like any scientific device',[8] and who was an empiricist as much as Born and his quantum colleagues were now. Obviously there was here a deep misunderstanding about the true nature of Einstein's 'constructive' contributions to quantum theory.[9] Other outstanding physicists who contributed to the edification of quantum mechanics, such as Heisenberg, for example, had similar conceptions (Wolfgang Pauli, although he adhered to the orthodox position, had a more nuanced understanding of Einstein's dissatisfaction with quantum mechanics).

At best, they did not understand why one of the most prominent pioneers of quantum physics did not accept the theory when it established itself on what they thought were firm bases and manifested itself as maybe the most powerful theory in the history of physics. Among the physicists of the next generation, who shared this conviction received from their masters, Pais' account of the subject is typical. His book, *Subtle is the Lord*, which is probably the most complete and authoritative scientific biography of Einstein, and which deals with great detail with Einstein's contributions to early quantum theory, is rather succinct when he considers the case of Einstein's unease with quantum mechanics, and undervalues and even denies the interest of Einstein's objections as regards improvements in our comprehension of that theory. Speaking of the Einstein–Podolsky–Rosen paper of 1935, Pais concludes that it 'contains neither a paradox nor any flaw of logic' (I (M.P.) agree with that part of the statement), and he adds: 'It (the paper) simply concludes that objective reality is incompatible with the assumption that quantum mechanics is complete. This conclusion has not affected subsequent development in physics, and it is doubtful that it ever will.' And Pais adds the following: 'He (Bohr) did not believe that the

---

[6] *Wave* and *quantum mechanics* were established in the years 1924–1927, with the works of Louis de Broglie, Erwin Schrödinger, Max Born, Werner Heisenberg, Pascual Jordan, Paul Dirac and others. The key date of the advent of these theories (actually they were shown to be practically equivalent by Schrödinger by the same time) can be taken as the Solvay Conference held in Brussels in November 1927, where its presentation to physicists was so to speak godfathered by Niels Bohr. Bohr had been asked to develop epistemological-philosophical considerations that could tame the theory's uncommon peculiarities, namely the distance between its mathematical unusual abstract form and the observable phenomena. Bohr presented the ideas that were to be called afterwards the '*Copenhagen interpretation* or *philosophy of complementarity*', which he had developed a few months previously in the Como Conference of Physics (see for instance, Jammer (1966, 1974), Holton (1988)). The title of his Solvay presentation was simply 'The postulate of quanta and the new development of the atomistic' (Bohr, 1928).

[7] Einstein, letter to Max Born, 7 November 1944, *in* Einstein and Born (1969).

[8] Born (1949), in Schilpp (1949), p. 176.

[9] Born's assertion of Einstein having been empiricist in his youth, shared by other illustrious commentators, does not stand up to the analysis of Einstein's works of the corresponding period. For a more extensive discussion on this point, cf. Paty (1993a, 1996, forthcoming a and b).

Einstein–Podolsky–Rosen paper called for any change in the interpretation of quantum mechanics. Most physicists (myself included) agree with this opinion.'[10] Pais was here a bad prophet, just as Bohr was, if one considers the posterity that the EPR argument would later know.[11]

Our aim in the present contribution is, contrary to these 'dominant' conceptions, to propose a different understanding of Einstein's objections to quantum mechanics, in giving Einstein the credit that his position can be different from a mere adherence to a conservative and obsolete view on physics. This aim is twofold. It has first a historical dimension, because we need, in order to grasp the very essence of Einstein's objections, to place them in the context in which they have been formulated, which can be characterized grossly as the 'complementarity context'; we also need to regard Einstein's argumentation on quantum mechanics in the light of his own research programme in physics at the same time. The other aim of this chapter is more epistemological, or even physical, although it is also connected with historical matters: contrary to Pais' opinion, our preferred view is that Einstein's objections did have an effect on our understanding of quantum mechanics.

We shall not enter into too many details about the context of complementarity, to which Einstein reacted by his objections to quantum mechanics and its interpretation, as it is dealt with elsewhere by others in this Handbook.[12] Let us merely briefly summarize here a few elements of this context.

The edification of quantum (or wave) mechanics as a mathematically formalized theory was very fast, in the years 1924–1927. Its starting point was Einstein's so-called 'first theory of quanta' or 'semi-classical theory of quanta' of 1916, together with more recent results, and it was affected already by the two features Einstein was dissatisfied with in his theory (while at the same time registering them while waiting further deeper theoretical understanding), as he had expressed it after exposing his results: the *wave–particle duality* character of light, and the only up to then *statistical* character of the predictions (from the probabilistic expression).[13] Einstein had hoped that, in this last respect, the theory, in progressing, would provide a causal prediction for the amplitudes of emission and absorption related to individual events; this was not really the case, at least from a fundamental point of view, as the quantum theory of radiation, which was being developed at the same time, based itself on the quantum mechanical formalism. And that state of things was indeed theorized and formalized by quantum mechanics with *Born's (physical) interpretation* of the state function ($\Psi$) of the considered physical system, which relates it to the probability ($P = |\psi|^2$) for the system to be in that state, and thus providing theoretically the probabilistic-statistical distribution given by observation.[14] Max Born, as well as Einstein, and most physicists of the time thought equivalently of probability and statistics.

---

[10] Pais (1982), p. 456.    [11] See further on.
[12] In this volume, see Howard (2022) and Osnaghi (2022). See also: Paty (1985, 2021).
[13] Einstein (1916b).    [14] Born (1927).

(Incidentally, let us make a comment here which might be of some help to our actual understanding of *physical meaning or content* versus *quantum formal theoretical statements*. Further elucidations have led later on to make a distinction between a *probabilistic* statement (of a *theoretical* nature) and a *statistical* one (related to *experimental* results).[15] The expression '*probability amplitude*', that has been adopted in quantum physics to characterize the *physically meaningful use of the state function*, happens happily to illustrate such a distinction: there is no meaning attached to 'probability amplitude' in the pure mathematical theory of probability and it only can be meaningful from a physical (physically structurant) point of view, namely as a *physical-theoretical concept*. As such, endowed with its relational theoretical properties—superposition principle, etc.—far from being reduced to its practical function of providing statistical distributions, it sums up and integrates at least potentially all the physical properties of the considered physical system.)

Now, the theory, presented in a very formal mathematical way, was considered by its proponents to be *definitive* and *complete*, 'not subject to any further modification', in the words of Born and Heisenberg.[16] It included the two previous 'weak points', so qualified by Einstein, as cornerstones of the formal edifice, which Bohr's conception of complementarity (presented at the same conference), justified from a fundamental point of view, calling for a new philosophy of knowledge. Indeed, one also meets at this stage with the problem of 'interpretation', which appears, in the conception of the Copenhagen–Göttingen school, as necessary from the physical point of view to complement an abstract mathematical formalism. But such 'interpretation' appears indeed as a mixture of, on one side, physical interpretation of axiomatic statements and mathematical symbols and relations and, on the other side, of philosophical propositions such as the 'principle of complementarity' itself (and also the starting consideration that physical reality, in Heisenberg's own words, is only defined 'at the very moment of observation').[17]

One can then follow the series of objections put forward by Einstein, which seem to evolve according to the periodization suggested by Pais.[18] From 1927 to 1931, Einstein tried to show some inconsistency in quantum mechanics, taking the Heisenberg 'uncertainty relations' in order to demonstrate, by physical examples, that they cannot be as absolute as claimed. Then, from 1931 on, Einstein did not look any more for inconsistencies but tried to show that quantum mechanics was 'incomplete'. In fact, throughout all these years, it is the problem of the completeness of the theory that Einstein had essentially in his mind, and one can say that, even when he had been looking previously for inconsistencies, it was above all *inconsistencies in view of the claim of completeness* that he contemplated.

These arguments culminate with the 'EPR' article, which subsequently became famous under this name (with variants: EPR paper, argument, paradox, mental

---

[15] See, for example Paty (2001).    [16] Born and Heisenberg (1928).
[17] Heisenberg, in the discussion, in Solvay (1928).    [18] Pais (1982).

experiment, etc.), supposedly refuted shortly afterwards by Bohr.[19] By completeness, the authors of the paper meant that 'any element of the physical reality which can be predicted with certainty can be put in correspondence with an element of the physical theory'. What they had in mind was the pretension of quantum mechanics to deal thoroughly with the object it aimed at. We shall come back to the EPR reasoning, and show that Einstein and his collaborators, by letting arise in that manner the question of completeness, actually, rightly and explicitly, put their finger on the non-local, or non-locally separable, character of quantum mechanics, that was to them its fundamental weakness. (In fact, it is Einstein himself alone who, in his own and further writings, made this concept fully explicit, as we shall see.) The EPR collaboration, and unambiguously Einstein later on, demanded *local separability* as a requirement for interaction-independent 'elements of physical reality', or physical systems, although, as we know now, recent developments have shown that states systems in the quantum domain may be not locally separable—and thus, in this respect, quantum mechanics does not hold the defect of incompleteness. On the contrary, it holds this—up to then unnoticed—physical property. (Let us remark here that the question of completeness of quantum mechanics was affected also by the problem of *quantum measurement*, at least up to the developments that happened many years later, near the turn of the century, of the theory of decoherence and its experimental corroborations).[20] But as we noted above, Einstein's non-adherence to a property of non-locality (non local-separability) for physical systems does not preclude his priority in having clearly pointed and extracted it from the quantum theoretical description.[21]

## 12.3  EINSTEIN'S PROGRAMME FOR THE QUANTUM THEORY

In his scientific biography of Einstein, Abraham Pais quotes a sentence of the latter praising in 1926 two of Schrödinger's papers on quantum rules, and he adds: 'It was the last time he would write approvingly about quantum mechanics.'[22] But this is not really true, for Einstein wrote on many occasions positive statements about that theory. For example, in 'Physics and reality' (1936): 'Probably never before has a theory been evolved which has given a key to the interpretation and calculation of such a

---

[19] Einstein, Podolsky, and Rosen (1935); Bohr (1935). Actually Bohr was possibly somewhat disturbed at first reading by the EPR argument, and needed one month to polish his reply.

[20] On decoherence, see for instance: Zeh (1970), Joos and Zeh (1985), Zurek (1991, 2002, 2003, 2004), Joos *et al.* (2003); on the experimental evidence for it: Haroche, Brune, and Raimond (1997); for further considerations and epistemological comments: Schlosshauer (2005, 2007), Paty (2009).

[21] Its importance and novelty (although it was already contained in the quantum theoretical formulation) shall be underlined further on. See also Paty (1986).

[22] Pais (1982), p. 442. These papers have been reprinted together in Schrödinger (1926). (Cf. Letter of Einstein to M. Besso, 1 May 1926, in Einstein and Besso (1972)).

heterogeneous group of phenomena of experience as has the quantum theory.' In many other texts he gave similar appreciations on the heuristic importance of quantum mechanics and its strength as a physical theory. In his 'Autobiographical notes', written in 1946 and published in 1949, he refers to it as 'the most successful physical theory of our period, viz., the statistical quantum theory which, about twenty-five years ago, took on a consistent logical form (Schrödinger, Heisenberg, Dirac, Born). This is the only theory at present which permits a unitary grasp of experiences concerning the quantum character of micro-mechanical events'. And he adds, significantly (because this sets quantum mechanics on a par with the theory of general relativity): 'This theory, on the one hand, and the theory of relativity, on the other, are both considered correct in a certain sense, although their combination has resisted all efforts up to now.'[23]

In 1948, in his article 'QuantenMechanik und Wirklichkeit' ('Quantum Mechanics and reality'), published in the Swiss philosophical journal *Dialectica*, he states that this theory constitutes an important advance, which can be considered in a certain sense as definitive, in the knowledge of the physical world.[24] And in his 'Reply to criticisms', in the Schilpp volume (1949) already referred to: 'I fully recognize the very important progress which the statistical quantum theory has brought to theoretical physics. In the mechanical problems ( . . . ), (this theory) even now presents a system which, in its closed character, correctly describes the empirical relations between stable phenomena as they were theoretically to be expected. This theory is until now the only one which unites the corpuscular and ondulatory dual character of matter in a logically satisfactory fashion; and the (testable) relations, which are contained in it, are, within the natural limits fixed by the indeterminacy relations, complete. *The formal relations which are given in this theory—i.e., its entire mathematical formalism—will probably have to be contained, in the form of logical inferences, in every future theory.*'[25] And again, in 1953: 'I have no doubt that the present quantum theory (or better, "quantum mechanics") is the most perfect theory compatible with experience, in so far as one bases the description on the concepts of material point and potential energy as elementary concepts'.[26]

All these utterances (one could add other ones, taken for example from his correspondence) clearly state that quantum mechanics is a powerful tool, which contains important elements of the 'ultimate truth'. It is therefore obviously unfair to say repeatedly that he did not accept the theory. He did not accept it, indeed, as the definitive answer which its proponents claimed it was; that is why it is important to situate Einstein's views on quantum mechanics in relation to his own programme of research.

The positive appreciations we just quoted above were indeed always mitigated with a restrictive comment, always the same, about the fundamental *incompleteness* of quantum mechanics, which is mostly expressed as a feature, not at all of invalidation

---

[23]  Einstein (1946), in Schilpp (1949), p. 81.        [24]  Einstein (1948).
[25]  Einstein (1949), in Schilpp (1949), pp. 666–67. My emphasis (M.P.).
[26]  Einstein (1953), pp. 6–7.

(quantum conditions must be taken into account in any future theory of atomic processes), but of the impossibility of taking it (its system of concepts and propositions) as a starting point for further fundamental developments of a theory which would unify the microcospic and the macroscopic worlds. The quotation we have taken above from Einstein's 1948 *Dialectica* article is followed by this comment: 'In spite of this, however, I believe that the theory is apt to beguile us into error in our search for a uniform basis for physics, because, in my belief, it is an incomplete representation of real things, although it is the only one which we can build out of the fundamental concepts of force and material points (*quantum corrections to classical mechanics*). The incompleteness of the representation is the outcome of the statistical nature (incompleteness) of the laws.'[27] This type of consideration is constant through the texts from the EPR time up to his last writings. Einstein's worries about quantum mechanics do not concern its validation as general law with respect to phenomena, but they have to do with *the fundamental problem of its physical principles*.

We should recall here what Einstein's conceptions about the fundamental characters of a physical theory (*theories of principles*, and *constructive theories*)[28] were. We would then encounter the most important trait of his own programme regarding physical theory in general and of his concern with progresses in the quantum domain in particular. He considered his task to be one of investigation in the fundamental matters, leaving aside any other aspects of research—not because he considered them as devoid of value. His personal concern was not to find a theory that 'works', but to find a theory deeply rooted in fundamental principles, and this definition of his own research programme does not preclude other types of research work. Already in 1924, in a letter to Maurice Solovine, Einstein said the following about his work: 'I had always an interest towards philosophy, but only as a second concern. My interest towards science has in fact always been limited to the study of principles, and this provides the best explanation of all my behaviour. The same reason explains why I published so little, for the strong desire to grasp principles resulted as an effect in fruitless efforts'.[29]

In his 'Preliminary remarks on the fundamental concepts'[30] (a rather significant title) in the book of homage to Louis de Broglie published in 1953, Einstein gives the following precision to show how his own approach is at variance with that of de Broglie (which was to restore causal determinism through hidden supplementary variables):[31] 'I nevertheless unceasingly looked for another way of solving the quanta riddle or at least to help preparing the solution. These researches were based on a deep discomfort, of a principle nature ("prinzipieller Natur"), which the bases of the statistical quantum theory inspired me...'

---

[27] My emphasis (M.P.). But this does not mean that incompleteness is to be identified merely with indeterminism (more on this later on).

[28] See, in particular, Einstein (1936), pp. 96–97, and other texts ('The foundation of physical theory', Einstein (1940), for instance). About this notion (theory of principles), see Paty (1993a).

[29] Letter to Maurice Solovine, 30 October 1924, in Einstein (1956), pp. 48–49. Note the modesty of this confession.

[30] '*Grundbegriff*'. Einstein (1953), p. 5.

[31] On this programme and Einstein's position, cf. Paty (1993b).

Preoccupied essentially at this stage with fundamental features of physical theories, Einstein was not satisfied by dealing with the existing ones, i.e. quantum theory and even general relativity, but sought to proceed further towards an integrated 'theory of principles'. Quantum mechanics and (general) relativity theory, although they represent, each for its side, something of the truth about nature, cannot be combined:[32] here lies, in fact, the essence of the problem, and the variety of approaches and beliefs among physicists begins. Einstein adds to his statement: 'This is probably the reason why among contemporary theoretical physicists there exist entirely differing opinions concerning the question as to how the theoretical foundations of the physics of the future will appear. Will it be a field theory; will it be in essence a statistical theory?'[33]

Thus, Einstein expressed dissatisfaction with quantum mechanics in the form of a programme which had, in his eyes, to be performed towards a satisfactory quantum theory. He was not looking for the past but to a further stage of our physical knowledge. 'What does not satisfy me in that theory' (quantum mechanics), he wrote in his 'Reply to criticisms', 'is its attitude towards that which appears to me to be the *programmative aim of all physics: the complete description of any (individual) real situation (as it supposedly exists irrespective of any act of observation or substantiation)*'.[34]

Here stands the question of the nature of Einstein's programme. It is, generally speaking, a realistic, objective, and unifying one, with the permanent claim that 'there is something like the "real state" of a physical system, which exists objectively, independently of any observation or measurement, and which can in principle be described by the means of expression of physics.'[35] This, for the epistemological side. More precisely, as to the properly quantum theoretical side, Einstein had his own direction in mind, which he delineated in the following terms: 'Continuous functions in the four-dimensional (continuum) as basic concepts of the theory'. 'Rigid adherence to this programme can rightfully be asserted to me', he did add (and correct) in answering the accusation of Bohr and Pauli of a 'rigid adherence to classical theory', and after having noticed that 'there is, strictly speaking, today no such a thing as a classical-field theory' (i.e., a field theory which would be 'complete' in a specific sense, i.e. a theory of the field-creating masses, which was, for the physics of fields, Einstein's long-range problem and goal).[36]

---

[32] Einstein (1946), pp. 80–81.

[33] More recently, since the seventies, the striking successes of the quantum field theory in the subnuclear domain have deeply modified the perspectives, with the specification of the dynamics of the fields of interaction and of the principles of symmetry *and invariance* that rule them. One could say that on the quantum side also, it is the field which comes first (not continuous indeed, but quantized) and the fundamental symmetry principles to which it is submitted. And unification is ultimately the target.

[34] Einstein (1949), in Schilpp (1949), p. 667. My emphasis (M.P.).

[35] Einstein (1953), p. 7.        [36] Einstein (1949), in Schilpp (1949), p. 675.

Einstein's programme for quantum physics was thus strongly linked with his programme for field theory: for him, quantum as well as (continuous) field physics were unachieved, and his concern was about how to achieve them, or better, to accomplish them together. Such an aspect of 'completeness', which appears as completeness endowed with a different meaning from the one which is strictly involved in the EPR argument,[37] is nevertheless to be kept in mind when we deal with the restricted sense it has in the EPR paper and, more widely, in consideration of the problem of completeness of quantum mechanics: but it is not to be superimposed on it (contrary to what Karl Popper states in the third volume of his *Postcript*, on *The quantum schism*).[38] There are early indications, from Einstein himself, that he was partly motivated in his unitary field research programme by the preoccupation of resolving the quantum riddle[39] (see, in this respect, his 1946 'Autobiographical notes').

From 1920, Einstein had been looking for an overdetermined field theory, i.e. a theory in which the field variables would be overdetermined by appropriate equations whose number would be larger than the number of field variables.[40] He proposed the following reasoning: take general covariance; require agreement of the equations with Maxwell's (i.e electromagnetism) and gravitation (i.e. general relativity) theories; require that the equations which overdetermine the fields have symmetrical static solutions which describe the proton and electron: it then should be so that one obtains conditions of overdetermination which restrict the choice of the initial conditions of the fields. Instead of an under-causality, one would obtain in such a way an overcausality ('Überkausalität').[41] Such a programme was praised up to 1925 by Born and others.

To recall in detail Einstein's views about the relations between quantum mechanics and the theory of relativity would exceed the limits of this contribution. Let us merely mention that he considered that the two theories were foreign to each other and were not to be fused together, due to the strong difference of their epistemological features and their physical meanings. He had many worries about the problems related with quantum field theory (he saw in it a kind of a monster, importing (special-) relativistic features into quantum mechanics, and exhibiting infinities). What interested him, however, in this respect, was not so much the special but the general relativity theory, as the problem he eventually considered was that of a unification of the dynamics of atomic matter with the dynamics of gravitation.

What is certain is that, even when dealing with the apparently quite different problem of the continuous field theory, in the line of general relativity towards a unified theory of gravitation and electromagnetism, Einstein was keeping in his mind the preoccupation with the problem of quantum physics, and this was one of his

---

[37] On the various kinds of completeness invoked by Einstein, see: Paty (1988b).

[38] Popper (1983).

[39] Einstein (1923). Cf. Pais (1982), p. 464; Stachel (1986).

[40] Letter to M. Besso, 5 January 1924, in Einstein and Besso (1972). John Stachel has given a profound analysis of this expectation by Einstein (Stachel, 1986).

[41] Einstein (1936, 1940). He stated: 'Field theory programme, but not yet performed'.

motivations for pursuing his research.[42] Up to the end of his life, his programme was 'to find the equations of the total field'. However, interestingly enough, he never mixed, in his scientific contributions, considerations on quantum theory and on the continuous field. Such an attitude, he had had since his first articles on these subjects, in 1905, and it appears to be a characteristic feature of his scientific method or, rather, *style* in physical research.[43]

It is only recently that such a programme, namely looking for a total field of matter, has become the main concern of quantum physicists; indeed they are looking for a *quantum field* (formulated on a quantum conceptual basis), but they struggle to encounter, although—so to speak—from the other side of the hill, the problem with which Einstein had an endless concern and secret or hidden struggle, on his own continuum physics side: the unification of the quantum domain and of general relativity, or gravitation.

## 12.4 THE EVOLUTION OF EINSTEIN'S OBJECTIONS AND THE CLARIFICATION OF QUANTUM CONCEPTS

Einstein's argumentation concerning the alleged deficiencies of quantum mechanics, throughout the years, appears as centred around two problems: the *probabilistic-statistical* character of quantum theory and its inherent *non-locality*. Such were, to him, the essential features of its incompleteness. Einstein attained very soon after the constitution of quantum mechanics these essential traits of his criticism of it or, better, of the interpretation of quantum mechanics which considered it as a fundamental and complete theory, although his formulation of the main arguments evolved somewhat with time, as we shall now show.

It is obvious for probabilities, or 'statistical character'—a feature of quantum theory which, if we recall, he noted already in 1916, seeing in it a weakness for the theory then still in progress, one which he hoped at that time that a further theoretical advance would get rid of.

As we know, quantum mechanics did not alter this character and, on the contrary, it reinforced the role of probability in the theoretical description, changing its status from a mere tool (through the use of statistics) to a fundamental ingredient of the theoretical formulation, through the physical interpretation of the state function as an 'amplitude

---

[42] Pais rightly notices in his book (Pais (1982), pp. 465–66), founding himself on Einstein's correspondence, that the problems of the unified field theory and of quantum physics were simultaneously present to his thought.
[43] On this, see Paty (1993a, 1996, forthcoming c).

of probability'.[44] Einstein did not deny the utility and importance of probability and statistics for theoretical elaboration, and had used it constantly in his own 'constructive elaborations' since the earliest ones. His way of introducing probability in quantum physics, already in 1903, by means of a reinterpretation in terms of frequency (for a given physical state) of Boltzmann entropy formula and of calculating fluctuations around it, had led him from his 1905 paper on radiation energy quantization to his 1916–1917 synthesis, with the demonstration of the radiation momentum, and to his 1925 calculation of monoatomic gas ('Bose–Einstein') statistics. Along all of this pathway, Einstein considered probability as a useful (and somewhat efficient) tool to point out some characteristic features of quantum phenomena, that could be attained in that indirect way. Also, he constantly praised the so-called 'Born interpretation' as one of the most powerful achievements of quantum mechanics. But he always considered that a fundamental quantum theory should go further than this *probabilistic* (or, equivalently for him and others at that time, *statistical*) aspect which was, in his view, only provisional and waiting for further and deeper characterizations.

But his demand for some kind of causal determinism or, more exactly, for a non-purely statistical character of the theory, is more refined than what is generally thought, as one can see by following his argument of incompleteness. He wanted to show (such is in essence the argument) that the (quantum) description cannot in any case be referred to individual systems. To demonstrate this, he did put forward the assertion of a natural (for him) *local-separability principle*. The *non-separability* (or *entanglement* in Schrödinger's parallel approach[45]) of the correlated subsystems in a EPR-type situation (of correlated given subsystems initially forming one overall system endowed with conservation laws), which resulted theoretically from the non-factorizability of their $\Psi$-functions in the mathematical formulation of quantum mechanics (derived from the *principle of superposition*), entailed as a consequence the non-local character of these $\Psi$-functions, from which one must infer the statistical character of the description. For only in this way could non-locality for individual (sub)systems be avoided, according to him. The reason for it being the identification he considered between, on the one side, *non-locality* for the two subsystems having interacted and then each going on its way and, on the other side (as the subsystems supposedly maintain their correlation as a whole), *some kind of instantaneous action-at-a-distance between them*, although the latter could not be admitted physically, according to the well-established theory of special relativity.

As a matter of fact, Einstein had demonstrated—so he thought—with his EPR argument yielding such a consequence (action-at-a-distance in the case of individual events), that the 'probability amplitude' or state-function could only lead to a statistical description, thus *leaving open the problem of the representation of an individual system*

[44] Born (1927). As we have commented previously.
[45] Schrödinger (1935).

to which, so he thought, the present quantum theory could not pretend, being for that reason an incomplete theory. True, the required *separability or locality principle for individual quantum physical systems*, which is a decisive point in his argumentation, is not present in the formulation of quantum mechanics, as he recognized, but he considered it highly desirable—and indeed indispensable.

Thus *non-locality* plays a role of prime importance in Einstein's rejection of the alleged *fundamental* character *of quantum mechanical probabilistic assertions*, which he expressed in the famous sentence 'God does not play dice'.

Let us now come to this concept (*non-locality*, or *non local-separability*), which took time before being completely evidenced and better understood. It is only recently that it has been fully appreciated as a fundamental feature of quantum mechanics, and commentators have seldom realized that it was one of Einstein's main points in his criticisms. Anyhow, it consequently inspired further elucidatons (theoretical as well as experimental) which would be decisive in its further evidencing. Even Bohr seems to have been unaware that non-locality was at the root of Einstein's argumentation, and did not even explicitly realize that it was also an inherent feature of quantum mechanics (his reference to the implication of an observer complicated the picture). Notwithstanding, it is usually thought that it was Bohr who made non local-separability explicit in his answer to the EPR argument. But, as a matter of fact, it was not so, strictly speaking: non local-separability has been pointed out at first, although somewhat clumsily, in the Einstein–Podolsky–Rosen paper itself, and later very clearly in Einstein's further elucidations (true, as an undesirable formal property to be set aside).

Let us try to clarify that point. We said that Einstein paid attention to it very early on: we can trace it back to a remark Einstein made about the indistinguishability of quantum particles in his 1925 work on the theory of perfect monoatomic gases. In it, he evidenced the 'Bose–Einstein statistics' for quantum particles with integer spins. Answering a criticism made by Paul Ehrenfest and others to his and Bose's papers of 1924–1925 (Bose, 1924; Einstein 1924b), Einstein emphasized the fact that (quantum) particles are not dealt with, in these papers, as mutually independent, and that their treatment, i.e. the 'Bose–Einstein' (as it would be called afterward) counting of particles as indistinguishable, 'express(es) indirectly a certain hypothesis on a mutual influence of the molecules which for the time being is of a quite mysterious nature'.[46] This comment could be extended as well to the case of half-integer spin particles obeying the 'Fermi–Dirac statistics' and thus subject to the Pauli exclusion principle, that were formulated in the same period.

But *non-locality* as such was to be explicitly pointed out when macrocospic distances are involved. Hence it would manifest itself in EPR-type situations (in the theoretical

---

[46] Einstein (1925), p. 3. See also Pais (1982), p. 430. The Bose–Einstein condensation has been put experimentally in evidence near the end of the 20th century, a long time after Einstein's theoretical prediction of it. See: Griffin, Snoke, and Stringari (1995), Cornell and Wieman (1998).

formulation[47] and in phenomena, we would say today, in the light of our further knowledge); as for Einstein, as we shall see, it was only in the formalism—if one would extend its interpretation to the description of individual systems. Before the EPR paper, however, and as early as the Solvay Conference of 1927, something which sounds like what would later be called *non-locality* is present in Einstein's objections. On that occasion, in the general discussion opened by Bohr who presented his idea (soon to become a 'principle', of a philosophical nature) of complementarity, Einstein expressed his dissatisfaction with the alleged fundamental character of quantum mechanics, by taking the example of electrons being diffracted by a slit and hitting a screen covered by a photographic plate. He argued that a purely statistical theory was incomplete, because each electron has a trajectory which could be detected at least approximately, and that if one admitted such an individual description of each electron, then one would have to face a difficulty : if $\Psi$ is related to an individual particle, expressing 'the probability that this particle is located at a given place', such interpretation of the function 'implies a quite peculiar mechanism of action at-a-distance, which forbids the wave, continuously distributed in space, from producing an action in two places on the screen'. This last sentence refers to a *non-locality* which was unacceptable to Einstein, for the reason he indicated (it would involve an (instantaneous) action-at-a-distance, contrary to the special relativity theory).[48] In this first thought about non-locality, that arose from the wave character of the physical system under consideration, Einstein saw it as contradictory with the particle concept which is nevertheless utilized together with the wave concept.

Once again (such a contradictory character he had pointed out as early as 1909, stating that it would demand an approach towards quantum phenomena which would break with classical theories), Einstein found himself with a feature which was not to be solved simply by asserting that we can deal, from a fundamental point of view, only with the wave and/or with the particle concepts as claimed by Bohr's complementarity conception: there was still something more fundamental to find. We deal provisionally with these classical concepts, so he thought, at the cost of using a theory of a purely statistical nature. With the statistical interpretation of the $\Psi$-function, one is free from any action-at-a-distance that he saw as inherent to particle non-locality. Such was Einstein's view in 1927, and the essence of his further reasoning is present in this simple argument, which was already aimed at refuting the pretension of quantum mechanics to be a fundamental complete theory. Einstein would even try to do more

---

[47] One used to (and still today very often does), in quantum physics, speak of 'formalism', which strictly speaking would refer to the mathematical form only, even when this form is directly submitted to the requirement of physical representation; the expression 'theoretical formulation', which we use here, seems more appropriate as it refers to the considered *physical theory*. Actually, such a received way of speaking reflects the circumstances of the formation of quantum mechanics that were guided by the use of mathematical relations very adequate to those of the phenomena under study, and mediated by a 'translation' or 'interpretation' (a physical one) between the two that was not obvious at first glance.

[48] Einstein's intervention in Solvay (1928), pp. 255–56.

for some time, and show that quantum mechanics has logical inconsistencies in giving absolute credit to Heisenberg relations. Such is his radiation-in-a-box experiment of 1930, reported by Bohr,[49] who refuted it by invoking the theory of general relativity. (Let us mention *en passant* that, despite the fact that Bohr's answer has unanimously been considered to be the right one for a long time, it has recently been shown to be inexact, the argument to keep the fourth indeterminacy relation owing nothing to general relativity, and being of a much more general character).[50] Such is also Einstein's article with Tolman and Podolsky of 1931 about 'Knowledge of past and future in quantum mechanics',[51] where he presents for the first time, without more details, the idea of correlation of distant particles.

Anyhow, he would henceforth admit that quantum mechanics is free of inner contradictions but consider it as incomplete, in the sense he then specified, as we have already indicated. It is in trying to show this incompleteness by physical arguments that he made explicit the non-local character of such physical systems as described by quantum mechanics. We notice, in these attempts, that incompleteness is not, in Einstein's view, merely to be identified a priori with indeterminism and the statistical character of the description.[52] Einstein's motivation was not, even in his first objections, primarily to restore determinism against probabilistic description but to point out non-locality as an insufficiency of the quantum theoretical formalism, which prevents the description of individual events and systems and therefore leaves us with only a statistical description.

Before the EPR paper, in 1933, Einstein put forward a reasoning (unpublished and presented only privately) that has been reported later by Léon Rosenfeld. This early consideration, interestingly enough, shows itself as a close prefiguration of the EPR argument.[53] Rosenfeld tells that, at that time, he thought that Einstein meant merely to illustrate unfamiliar features of quantum phenomena by this example. But actually, emphasis was given in it to the non-local character of quantum phenomena (probability being only present in the background if we keep in mind the 1927 Solvay Conference commentary).

The next objection is the EPR reasoning of 1935. The example of two quantum particles having interacted in the past is fully developed in the EPR article as regards its quantum mechanical description. We have already commented about some of its main lines, and now we shall restrict ourselves to analysing in some more detail its demonstration of the non-local character of the quantum mechanical description, if one considers it as applying to individual events or physical systems.

---

[49] Bohr (1949).    [50] Borzeskowski and Treder (1988).
[51] Einstein, Tolman, and Podolsky (1931). See a description in Paty (forthcoming c).
[52] Cf. Paty (1993b).    [53] Rosenfeld (1967). Cf. Paty (1985).

# 12.5 CLASSICAL AND QUANTUM CORRELATIONS BETWEEN DISTANT SUBSYSTEMS

To understand more clearly what is at stake here, let us recall that the EPR argument is based on the idea of particle correlation in physics and that the nature of these correlations is quite different in the classical and in the quantum cases. The difference is shown, in the EPR paper,[54] through the question of the determination of conjugate physical quantities such as coordinates and momenta. It becomes eventually immediately clearer if we consider the EPR argument as reformulated by David Bohm in his book of 1951, *Quantum theory*,[55] by taking as 'conjugate' or 'incompatible' variables (non-commuting operators) two components of the angular momentum of the subsystems $A$ and $B$, instead of considering (as with EPR) the coordinate and momentum. Spin conservation holds as in the classical case ($J_x(A) = -J_x(B)$, $J_y(A) = -J_y(B)$, $J_z(A) = -J_z(B)$), but the quantum correlation is restrictive when compared with a classical one in that it is not possible to have a simultaneous determination of the various components of a unique spin.

At this stage, the specificity of quantum correlations comes from (and shows itself in) the probabilistic definition of the wave function, the non-commutation of the operators which represent the physical quantities considered, the choice made by the act of measurement of one spin component (if we consider this case), at the exclusion of the other ones, i.e., the reduction of the wave packet or reduction process. In summary, it is the fundamental statements (or axioms) of quantum mechanics that make the difference, and the EPR argument in effect took as its target these statements or axioms, when one confronts their consequences with the requirements of 'local realism', which in fact would correspond to classical type correlations. (Let us note, however, that it appeared explicitly as 'local realism' only in Einstein's further rehearsals of the argument. This local character was implicit in the requirements for 'elements of the physical reality'.) For the EPR authors, the inability of quantum theory to predict correlated physical quantities was evidence for the incompleteness of that theory. Such was the function of the thought experiment which they invoked: to investigate the deepness and the range of the fundamental statements of quantum theory. But this inquiry was done from outside, as the criterion considered (local realism in the fully explicit version) was foreign to these quantum theoretical statements and could not be determinant if one considered only these statements as pertinent. Bohr did not express it clearly in his refutation of the EPR argument, but Einstein himself did, for instance in his 1948 article in *Dialectica*, in which he emphasizes that the 'separability principle' is not part of the formalism of quantum

---

[54] Einstein, Podolsky, and Rosen (1935).    [55] Bohm (1951).

theory, and states that such a requirement is nevertheless reasonable when one deals with individual physical systems considered in space.

Bohr's answer to the EPR argument did not consider non-separability (such as it was to be understood afterwards) explicitly, and it did not go further than the EPR description of the phenomenon. Bohr's answer was so impregnated with his philosophical view of complementarity and observation (linking the description of the physical system and the observer's means), that he did not identify unambiguously the quantum concept at stake here, *non local-separability*. It was again Einstein who 'translated' in such terms Bohr's refutation of the EPR argument. He did it in a direct and clear way some years later, when replying to Bohr's criticism of his own conception in Schilpp's volume of homage *Albert Einstein philosopher-scientist*.[56] 'Of the "orthodox" quantum theoreticians whose position I know', he wrote, 'Niels Bohr's seems to me to come nearest to doing justice to the problem. *Translated into my own way of putting it*,[57] he argues as follows: "If the partial systems A and B form a total system which is described by its Ψ-function $\Psi$ (AB), there is no reason why any mutually independent existence (state of reality) should be ascribed to the partial systems A and B viewed separately, *not even if the partial systems are spatially separated*[58] *from each other at the particular time under consideration*"'. And Einstein continues 'translating' in his own words Bohr's position: '"The assertion that, in this latter case, the real situation of B could not be (directly) influenced by any measurement taken on A is, therefore, within the framework of quantum theory, unfounded and (as the paradox shows) unacceptable"'. By this last sentence, Einstein makes explicit (although attributing it to Bohr, through his own 'translation'), that non-separability is a non-locality, in the sense that, in such an event, the exact respective localizations of A and B do not matter.

Returning to the original Einstein–Bohr dialogue (or controversy) of 1935, we can say that what Bohr added to Einstein's previous description in the EPR paper (put in Einstein's way of viewing, i.e. considering that physical states must be *conceived independently of their measurement*) is merely that there is no reason for any mutual independent existence of substates. But it is only much later on that the *acceptance as a fact* of these specific quantum correlations among such subsystems, justified by the theoretical formulation, submitted to experimental verification through Bell's inequalities, would led to the full adoption of the specific quantum concept of *non local-separability*.

Nevertheless, if we dare add, the presence of non-separability in the quantum theoretical formulation, as inherent to the quantum description before any experiment would show it directly, explains why many physicists did not immediately recognize the importance of Bell's theorem. Also why, when the latter had been shown to be experimentally testable, most physicists considered—this time, with good reasons— that quantum correlation predictions would win against locality ones, before the

---

[56] Einstein (1949), pp. 681–82.    [57] My emphasis (M.P.).

[58] *'Spatially distant'* might be more exact (my comment, and again my emphasis, M.P.).

realization of convincing experiments which effectively concluded in that way:[59] they favoured the overall evidence given to quantum theory by so many phenomena mainly in subatomic matter. Strictly speaking, however, one could admit that the domain of most of these phenomena (the subatomic one) is silent about locality in macroscopic conditions.

It is indeed this non-separability that expresses the quantum character of the correlation under consideration. Compared with it, the classical correlation is rather light or weak: despite the relation between physical quantities due to the conservation law, the two subsystems of a classical individual system are completely separable one from the other, then separated and, each one moving on its side, getting more and more remote from each other: a modification of one of them has no effect on the other as soon as they are sufficiently far from each other to be free of any mutual interaction (local separation). On the contrary, the description of the quantum correlation is not so simple and cannot be done by using images of that intuitive and classical kind (billiard balls . . . ).

It is only when one thinks of quantum correlations in imaged or intuitive terms of the above classical (-macroscopic) type, that one eventually considers—as many people do—the quantum situation for an individual system as being the exact opposite of the classical situation: in such a view the quantum correlation (with no interaction) would express the (instantaneous) modification of one of the subsystems as the effect of a modification of the other despite their remoteness (this latter being supposed to leave them autonomous). One could say, in a way, that Einstein made such a translation.[60] Therefore, in his view, the quantum theoretical description does not hold for individual quantum systems. That is why he sent back quantum mechanics to a purely statistical interpretation. But such a statement is only a translation as, if one stands on the quantum description side, considering the quantum quantities involved, there is nothing physical (defined as such) that interacts.

Strictly speaking, the quantum correlation is nothing but what is expressed by the theoretical relationship ('formalism') of the quantum theory: it is described by quantum mechanics and is implied by its fundamental statements. It is, precisely, the *non-separability* of the state function describing the overall system with regard to the two state functions representing each of the subsystems: from the point of view of the theoretical formulation, this property of the state functions comes from the non-factorizability of those of the two subsystems (in this simple case) due to the superposition principle. Unless one modifies the initial conditions (which is not possible here in a causal way once they are given), we cannot separate (split) the wave functions of the subsystems into independent eigenfunctions—if we restrict ourselves to the description of physical systems by quantum mechanics. If one of the subsystems (electron, photon, etc.) were submitted to a measurement in order to determine its

---

[59] Cf. Paty (1977).       [60] Cf. Einstein (1949), p. 682.

spin state (in Bohm's 1951 retaking of the EPR example),[61] this would correspond to a change of the initial condition (and subsequently to a change in the problem): the further initial condition would henceforth be the electron (photon, etc.) singled out, with the exactly determined (by the measurement) value of its measured spin on a given direction.

It is useful to clarify here a trait of Einstein's reasoning with the EPR-type thought experiment which has led to some misunderstanding. It has to deal with a difference in the expression of the argument between the 'three-men' EPR redaction and Einstein's own way of putting it. Einstein expressed in a letter to Schrödinger, a short while after its publication, his dissatisfaction with the writing-up of the EPR paper—done in fact by Podolsky, because of his better familiarity with the English language—which he thought clumsy in some respects, saying that 'the main point has been buried under erudition'.[62] Although some commentators have taken profit from this avowal to claim that the EPR argument was not Einstein's, it seems clear that its essence is fully of his own: from the idea of quantum particles' theoretical correlation (through conservation laws) to avoid perturbation (argument generally invoked to justify the uncertainty relations), up to the conclusion about the incompleteness of the quantum description.[63] But there is a difference between the emphasis given in the two expositions (EPR's and Einstein's) to the fundamental reason for incompleteness.

The EPR paper insists on the determination of the conjugate quantum variables, such as coordinate and momentum, which ultimately provide the state function. Without measuring those of one of the subsystems, one can afford to them 'reality', by making use of the measurement on the other subsystem together with the relation of conservation entailed from the initial state. Thus, the EPR paper stated, as quantum mechanics says nothing of such elements of reality, it is therefore incomplete—from which many people have thought, afterwards, that *hidden variables* were implicit in such a reasoning to recover these lost elements of reality. Actually, Einstein was not sympathetic to the 'hidden variables' alternative searches (he thought that such solutions were 'too cheap'), although he did not want to discourage its proponents such as

---

[61] Bohm (1951). David Bohm developed his theoretical model of quantum theory supplemented by hidden variables to restore determinism, these supplementary variables being non-local, in conformity with quantum mechanics (Bohm (1952). See also: Bohm and Aharonov (1957), Bohm and Hiley (1975)). This model was useful to John Bell in his later developments that led to his theorem of non-locality. A generalization of Bell's theorem has been given afterwards that was no more tributary to hidden variables. See Bell (1964, 1966, 1976, 1987), Stapp (1980), Greenberger, Horne, and Zeilinger (1989). High precision EPR-type experiments based on Bell's theorem have some time later definitively concluded in favour of non-locality: Aspect, Dalibar, and Roger (1982), Aspect, Grangier and Roger (1981, 1982), Aspect (1983). See also d'Espagnat (1984), Freedman and Clauser (1972), and Paty (1977, 1982).

[62] Einstein, letter to E. Schrödinger, August 1935 (Einstein's Archives).

[63] Cf. Howard (1986).

Louis de Broglie and David Bohm, in the name of pluralism of conceptions and of support to orthodoxy opponents, and valued their contributions.[64]

When Einstein expressed the argument in his own terms, he emphasized not what happened to the variables but what the description of the state was—supposed to represent the real system. This consideration was also present in the EPR paper, but did not seem—'buried under erudition'—so central as it is in Einstein's own explicit thought. A complete measurement of a complete commuting set $A$ of quantities of the first subsystem ($I$), he stated, defines not only the $\Psi$-function of it ($\Psi_I^A$), but also, through the correlation coming from the initial state, the $\Psi$-function of the second subsystem ($II$), ($\Psi_{II}^A$), without any measurement having been performed on this last one ($II$). But the same can be said with a set $B$, different from and incompatible with $A$ (in the sense of quantum mechanics, that is through anticommutation relations and thereof Heisenberg inequalities), leading to another determination, without direct measurement, of the state of the second system ($\Psi_{II}^B$).

As these two states, obtained by two different and independent ways, have no reason to be the same, there is no unique, non-ambiguous correspondence between the state of a system and its theoretical description, unless one admits an (unacceptable, for Einstein and special relativity) instantaneous influence at-a-distance of what happened to the first system ($I$) on to the second ($II$), i.e. Einstein's transcription of non-locality. As this reasoning was done for the consideration of individual systems, one must admit, according to Einstein, that the $\Psi$-function does not describe an individual system: it is only related to ensembles of systems. It is for that reason that, to him, quantum mechanics is incomplete. *Separability* and *locality* (independence of systems differently located in space) are essential in such a statement. That is to say, *apparent non-separability or non-locality* is, for him, the quantum theoretical reason of incompleteness: it is not indeterminism, which is merely a consequence.[65]

Non-separability corresponds to the theoretical property of 'entanglement', pointed out by Schrödinger and Einstein in 1935, which relates according to the superposition principle two quantum systems, once they have been connected through an interaction (and the two irreversibly linked sub-subsystems constitute thereafter only one proper quantum single system). We are referring here to the thought experiments of Einstein and his 'EPR' colleagues and of Schrödinger with a cat, to which we may add, previous to both of them, that one of Einstein's powder barrel, standing in a linear superposition of non-exploded and exploded states, which he suggested in a letter to Schrödinger dated August 1935, and which might have inspired in part the latter in proposing his imaginary situation of a dead and alive cat. Einstein wanted to indicate to Schrödinger that the $\Psi$-function could not be a description of a 'real state of things', as it seemed to him that such was his correspondent's interpretation. But Schrödinger explained in his answer to Einstein that his present interpretation of $\Psi$ was no more exactly in terms of

---

[64] And he helped David Bohm a lot when persecuted by McCarthyism. See Paty (1993b), Freire (1999).

[65] For a fuller description and analysis of Einstein's argumentation, see Paty (2021).

real state. And in fact, Schrödinger wanted to illustrate the difficulty of conceiving the linking up between the quantum description with the macroscopic-classical one.[66] Actually, both Einstein's barrel of gunpowder and Schrödinger's cat, although macroscopic, are supposedly directly coupled as such with a genuine quantum system in a quantum superposition, which looks obviously unrealistic. On the contrary, the EPR case couples two quantum (sub)states in an easily imaginable way. To enter into a quantum treatment, the barrel of gunpowder as well as the cat should have been taken from their quantum constituents, being henceforth (as we know now) submitted to the process leading to decoherence... Indeed, their meaning, for both Einstein and Schrödinger, was a symbolic one, as they themselves intended (in the sense indicated above).

## 12.6 Non-locality as a Criterion and as a Physical Fact

It has been argued by Don Howard[67] that Einstein's position evolved towards an acceptance of 'non-separability' although he always maintained his opposition to 'non-locality'. For sure, the use of each of the two terms can be attributed differently, the first one expressing a theoretical relationship, that of entanglement, while the second one expresses its consequence regarding a spatial property.[68] But as a result non-separability and non-locality melt into one and the same concept of quantum physics, which can be called unambiguously '*non local-separability*'. Although it should be argued in greater detail, not possible here, a sharp distinction like Howard's cannot be found in Einstein's writings, even in his *Dialectica* article of 1948,[69] which gives the clearest definition of his understanding of these concepts.

Our view here is that Einstein considered that non-separability and non-locality are one and the same concept, because quantum mechanics commonly also deals with the problem of measurement, in its statements when they refer to physical situations that can be experimented. Non-locality, as it stands in the definition used by Howard, corresponds to the *criterion of non-locality* proposed by John S. Bell following Einstein's considerations. The *concept* in itself is more general than the *criterion*, which refers specifically to observational situations, and *non-locality* is now admitted as a *quantum-mechanical concept*, rooted in the theoretical description, which has been established through appropriate experiments to the status of a physical fact,[70] without any reference to a transmission of a physical influence between two spatially separated subsystems. On the contrary, Howard's definition seems to equate non-locality with

---

[66] Schrödinger (1935); Einstein, letter to Schrödinger dated 8 August 1935, in Einstein archives; Schrödinger, letter to Einstein dated 19 August 1935, in Einstein archives. Cf., for example, Paty (2021), chapter 5.

[67] Howard (1986).      [68] Paty (1982).

[69] Einstein (1948).      [70] See, for example, Paty (1986).

such an action-at-a-distance, which leads to an oversimplification of the problem and also to a distortion of Einstein's thought, which is read in an anachronistic way, through subsequent developments.

The quantum theoretical definition of *non local-separability*, actually present from the start in the theoretical formulation of quantum mechanics, although unnoticed as to its specificity initially, became explicit as an effect of the discussion between Einstein together with Podolsky and Rosen, on one side, and Bohr on the other. As for Schrödinger, he considered it as *entanglement*, exemplified by the superposition of the two states, dead and alive, of his poor 'enquantumed' cat, as a difficulty to conceive the junction of the quantum and the classical-macroscopic regimes in the description of matter. But *entanglement* or *non local-separability* remained optional as a property: it was possible to imagine in principle a modification of the theory that would eliminate it. As such, it was, for those who were not content with the theory of quantum mechanics (because, in the case of Einstein, of its non-local property consequence), a manifestation of its incompleteness (for it prevented the $\Psi$-function to represent a single system and restricted its physical meaning to a statistical ensemble only).

As for those who would admit that the standard quantum mechanics describes through the $\Psi$-function individual systems as well as ensembles of systems, the consideration of non-separability or non-locality was, in the context of the time, hidden under the apparently more generally considered problem of causal determination, a problem also in debate in the interpretation controversies.

The possibility or not to restore determinism in quantum physics had been considered far before the concept of *non local-separability* came to the forefront. And it still was the aim of all attempts at alternative descriptions of microsystems even after the EPR argument, until rather recently; indeed the EPR so-called 'paradox' strengthened the motivation for those attempts, but the very essence of the question remained unperceived, up to the outstanding work of John Stuart Bell in 1964–1966.[71] A theorem demonstrated by John von Neumann in 1932 had established the incompatibility between the strict quantum mechanics formulation and deterministic hidden variables,[72] which had mostly discouraged further searches in the latter direction (de Broglie had abandoned his 'double solution' and 'pilot wave' theories and rejoined the Copenhagen interpretation throughout the thirties and the forties, and he retook them with his collaborators through the impulse of the David Bohm's own theory of hidden variables published in 1952).[73] A decade later, looking for the proof of incompatibility of alternative deterministic theories and quantum mechanics, Bell realized that von Neumann's theorem suffered exceptions, as it was based on restrictive hypotheses, and

---

[71] Bell (1964, 1966).

[72] von Neumann (1932). The proposed causal theories with hidden variables are thoroughly described by Olival Freire in his book *The Quantum Dissidents*, Freire (2015). See also other contributions in the present volume.

[73] Bohm (1952). Bohm conceived his theoretical idea independently of that of de Broglie's, published in 1927, which he came to know by Einstein's comment in his letter acknowledging Bohm for sending his work.

that there was an entire class of causal hidden variables which was not a priori incompatible with quantum mechanics: the 'non-local' class, to which those of Bohm pertained. Bell had been put on this track by noticing that the hidden variables considered in Bohm's theory were in effect *both causal and non-local* and ensured compatibility with ordinary quantum mechanics.[74] He had the idea of his own criterion of locality, by which he intended to demarcate quantum mechanics from deterministic supplementary parameters,[75] while he meditated on the EPR thought-experiment as reformulated by Bohm and by Bohm and Aharonov,[76] and also particularly on the following sentence by Einstein, commenting on the EPR situation in his 'Autobiographical notes': 'But on one supposition we should, in my opinion, absolutely hold fast: the real factual situation of the system *A* is independent of what is done with the system *B*, which is spatially separated from the former'.[77] We must notice also that the measurement problem is indeed present in these formulations of locality.

When we retrospectively consider the very content of Einstein's argumentation, we cannot omit to take into account that, nowadays (and very recently), *non-locality* (or *non-local separability*) in quantum physics has been given a fundamental, and probably definitive, status: it has been confirmed as a physical fact,[78] theoretically justified before having been experimentally verified, and it is founded on a conceptual ground which is more fundamental than the one considered through the question of determinism and hidden supplementary variables. Furthermore it has been shown, after its unveiling, to entail highly important consequences that will be summarily evoked in the concluding remarks. The origin of the concept is to be found nowhere else than in Einstein's argumentation when he objected to the quantum mechanics claim for completeness. True, Einstein disliked the concept, for the reasons exposed above, seeing in it evidence for the incomplete character of the theory.[79] But this does not preclude in any way that he indeed pointed out the property corresponding to the concept (*non local-separability*), which itself emerged from the inquiry into the physical meaning of some fundamental theoretical implications of quantum mechanics. And this reminds us of one of Einstein's privileged attitudes in exploring yet unknown domains of physics, and particularly the quantum one: *critical questioning as a heuristic theoretical tool.*

[74] This he did in his 1966 paper, written before the 1964 one. On these circumstances, see Jammer (1974). Bohm was aware of this non-locality, but Bell clarified it fully. De Broglie's hidden variables were, on the contrary, local ones.

[75] Bell (1964).    [76] Bohm (1952), Bohm and Aharonov (1957).

[77] Einstein (1949), pp. 84–85.

[78] See Paty (1982, 1986, and 1988a, chapter 6), for argumentation and bibliography.

[79] See for example the continuation of Einstein's remark on non-locality already quoted from his 'Autobiographical notes' (Einstein (1949), p. 682): 'By this way of looking at the matter it becomes evident that the paradox forces us to relinquish one of the following two assertions: 1) the description by means of the Ψ-function is complete; 2) the real states of spatially separated objects are independent of each other'.

From all these developments one must admit that non local-separability of correlated subsystems is part of the quantum theory, fundamentally inherent to it. This does not close absolutely the question whether quantum mechanics is a complete theory in the sense here considered. What can be said is that insofar as non local-separability is a physical fact, theoretically founded, it cannot be seen anymore as a weak point in the theory and as a criterion for its incompleteness. The quantum theory, as it stands, is coherent and powerful, although as we shall comment in the conclusion, it still holds some obscurity, at least to our intuitive intellectual grasp.

## 12.7 CONCLUDING COMMENTS. FURTHER ADVANCES MADE POSSIBLE IN THE LINE OF EINSTEIN AND OTHERS' CRITICAL QUESTIONING, AND REMAINING FUNDAMENTAL ISSUES

It is time to conclude this presentation and analysis of Einstein's discomfort about quantum mechanics, the beginnings of which he contributed so fruitfully to. We may say that his critiques have not been sterile and have contributed also to further advances that have appeared effective a somewhat long time afterwards. We shall now briefly sketch various advances in quantum physics that have occurred since Einstein's time and that are, in one way or another, related to his questions and critiques. On the whole, Einstein's fundamental questioning appears to have partially inspired the progressive unveiling of various new important and fundamental states of things.

The question of whether one can consider individual quantum physical systems, and of whether the quantum theory describes them as such, so decisive for Einstein (it had been one of his worries about the wave–particle duality already before the advent of quantum mechanics,[80] it was central in his requirements about *physical reality* and *theoretical completeness*), has been settled positively both theoretically and experimentally. In particular it has been possible, in quantum optics and in atomic, nuclear, and particle physics, to produce one by one and study *individual* systems, be they

---

[80] See his earlier comments on the experiment of W. Bothe and H. Geiger (1924, 1925) that refuted, as he had anticipated, the theoretical model of N. Bohr, H. A. Kramers, and J. C. Slater (1924) proposing a pure statistical relationship in a Compton-type experiment (scattering of an electromagnetic radiation endowed with a momentum on an atomic electron). Bothe and Geiger's result was a correlation between the outgoing radiation and electron, that evidenced the conservation of energy and momentum in each individual interaction. See Paty (2020).

radiations, atoms, or particles, and to obtain evidence that the theory describes them adequately as such.[81]

One must therefore admit that *non local-separability* and *individuality* have definitively to be taken together in the description of physical systems by the quantum theory. And consequently, if one follows Einstein's requirements, quantum theory cannot be said to be incomplete in this respect (contrary to what he thought in his time). One must take account of their conceptually necessary conciliation, as long as one considers correlated systems that remain in a characterizable quantum regime, i.e. before their decoherence (to which we come). The remaining fundamental question would henceforth be that of *the status of the concept of space in a quantum regime*, which is met also with respect to other situations (for example, Bose–Einstein condensates,[82] neutron stars, transitions, and frontiers in the cosmology of the early quantum Universe when the latter gives rise to relativistic cosmology, etc.). The question might be of decisive importance, as to the last situation, in the search for a quantum theory of gravitation.

The genuine quantum physical property of 'entanglement' or non (local-)separability, whose explicit knowledge we owe initially to Einstein and Schrödinger, has been duely explored further and has led to major consequences for quantum physics. One is the phenomenon of 'decoherence', which corresponds to the loss of the 'quantum regime' resulting from the multiplication of successive correlating interactions suffered by an entangled physical system in its path through matter. (We come back to it next.) The other one is (in the opposite direction) the *transmission of specific (quantum) information* between quantum systems (related to the entanglement of these) as long as they can be maintained in a quantum regime before decoherence occurs. The mastery of quantum information and its transmission would be of upmost importance for technological applications : *quantum computing* and *quantum cryptography*, incomparably more powerful than the ordinary ones.

As to the *decoherence* process, it sheds more light on the still present problem of *measurement of quantum systems* with *classical devices* (an unsurpassable limitation due to our human constraints). A few more words on that last issue might be suitable. The rather recent studies on *decoherence phenomena* have shown that measurement of a physical quantity operates on systems that are no longer in simple quantum states,

---

[81] See for example: Pflegor and Mandel (1967), Grangier (1986), Haroche, Brune, and Raymond (1997), Haroche (2012), Wineland (2013). The Nobel prize for 2012 was awarded to Serge Haroche and David Wineland *'for ground-breaking experimental methods that enable measuring and manipulation of individual quantum systems'*.

[82] See for example: Cornell and Wieman (1998), Griffin, Snoke, and Stringari (1995). The individual indistinguishable identical bosons aggregated in the same state that constitute a Bose–Einstein condensate are countable (and counted); the space they occupy together grows with their accumulation going on, up eventually to become visible (in a microscopic tube for instance) as it is observed in the phenomena. Non-locality of the constituents of such a *collective quantum state* which occupies a delimited space volume, manifests itself at a quasi-macroscopic level. In such a situation one is brought to the question of what could be said, physically, about such space?

but embedded in already decohered ones, that is with classical-like behaviour, entailing statistical distribution of the results. The theorical description of the decoherent process was first proposed by Dieter Zeh and developed by him and others, and further high-precision experiments have since been carried out and confirmed the picture.[83] The theoretical calculation describes in all details how the considered initial (quantum) particle (or system) gets entangled through successive interactions with elementary quantum components of the material milieu met in its path towards and inside the detector, giving rise to further higher superpositions of states. Following such a process illustrates as a kind of 'visualization' through the play of the physical magnitude relationships, and so to speak lets us intellectually 'see' a state of quantum superposition as propagating and, through this, makes it easier to conceive the possibility and the physical reality of such intricate quantum states, more and more complex, by interiorizing them 'as in themselves'. The overall system, *in fine* (finally), decoheres, which can be followed in time with precision. And it appears that when measurement occurs, decoherence has already happened.

The statistical results are thereafter equated to the square of the (theoretical) amplitude of probability (or state function), which indeed, according to the quantum theory, is the *proper quantity endowed with physical meaning*, and which has thus received its *definite value* by this *indirect measurement*. Statistics as such lose henceforth their apparent primacy in the effective quantum description to the benefit of the state function, which is theoretically unequivocal. This new state of things would reinforce the claim for completeness of quantum mechanics (in Einstein's meaning of the word completeness, and contrary to his estimation), but at the same time it liberates the latter from the common usual inhibition with respect to a realist assertion, inhibition that was maintained by the measurement problem as conceived according to the operationalist attitude. And this, at least, Einstein would have most probably highly appreciated.

All these developments tend—so it seems, at least—if we think of what they mean from the point of view of knowledge, to reinforce the idea of *physical reality* and *objectivity* for quantum physics, which was Einstein's main demand in his elaborations and criticisms. To speak of a quantum domain, which exists independently from any observer, henceforth appears to make sense for the present *physical description of matter* as well as for the *transformations of the early Universe*, which we shall never see but which we can think of and try to describe.

In this progress, still unsolved fundamental questions are remaining. They refer ultimately to *our intellectual means to represent the world symbolically*, and to their increasingly indirect and abstract character compared to the information given by our senses.

---

[83] On the theory and experiments of decoherence and their consequences, see: Zeh (1970), Zurek (1991, 2002), Joos and Zeh (1985), d'Espagnat (1984), Omnès (1994, 1999), Haroche, Brune, and Raimond (1997), Joos *et al.* (2003), Paty (2003), Schlosshauer (2005).

# BIBLIOGRAPHY

Aspect, A. (1983). *Trois tests expérimentaux des inégalités de Bell par mesure de polarisation de photons*, Thèse de doctorat ès-sciences physiques, Université Paris-Sud, Orsay.

Aspect, A., Dalibar, J. and Roger, G. (1982). Experimental tests of Bell's inequalities using time-varying analyzers. *Physical Review Letters*, **49**, 1804–07.

Aspect, A., Grangier, P. and Roger, G. (1981). Experimental tests of realistic local theories via Bell's theorem. *Physical Review Letters*, **47**, 1981, 460–63.

Aspect, A., Grangier, P., and Roger, G. (1982). Experimental realization of Einstein–Podolsky–Rosen–Bohm *Gedankenexperiment*: a new violation of Bell's inequalities. *Physical Review Letters*, **49**, 91–94.

Bacciagaluppi, G. and Valentini, A. (2009). *Quantum theory at the Crossroads: Reconsidering the 1927 Solvay Conference*. Cambridge: Cambridge University Press.

Bell, J. S. (1964). On the Einstein–Podolsky–Rosen paradox. *Physics*, **1**, 195–200. Repr. in Bell (1987), pp. 14–21.

Bell, J. S. (1966). On the problem of hidden variables in quantum mechanics. *Review of Modern Physics*, **38**, 447–52. Repr. in Bell (1987), pp. 1–13.

Bell, J. S. (1976). Einstein–Podolsky–Rosen experiments. In *Proceedings of the Symposium on Frontier problems in high energy physics, in honour of Gilberto Bernardini on his 70th birthday, Pisa, 1976*, Pisa: Scuola Normale Superiore, pp. 33–45. Repr. in Bell (1987), pp. 81–92.

Bell, J. S. (1987). *Speakable and Unspeakable in Quantum Mechanics*. Cambridge: Cambridge University Press.

Bohm, D. (1951). *Quantum theory*. Englewood Cliffs: Prentice Hall.

Bohm, D. (1952). A suggested interpretation of the Quantum theory in terms of 'hidden variables', I and II. *Physical Review*, **85**, 166–79; 180–193.

Bohm, D. and Aharonov, Y. (1957). Discussion of experimental proof for the paradox of Einstein, Rosen and Podolsky. *Physical Review*, **108**, 1070.

Bohm, D. and Hiley, D. (1975). On the Intuitive Understanding of Non-Locality as Implied by Quantum theory, *Foundations of Physics*, **5**, 93–109; also in J. Leite Lopes and M. Paty (eds.) (1977), *Quantum Mechanics, a half century later*, Dordrecht: Reidel, pp. 206–225.

Bohr, N. (1928). Le postulat des quanta et le nouveau développement de l'atomistique. In Solvay (1928), pp. 215–47.

Bohr, N. (1935). Can quantum mechanical description of physical reality be considered complete? *Physical Review*, **48**, 696–702.

Bohr, N. (1949). Discussion with Einstein on epistemological problems in atomic physics. In P.-A. Schilpp (ed.), *Albert Einstein, philosopher-scientist*, Evanston, IL: The Library of Living Philosophers, pp. 199–242.

Bohr, N. (1972). *Collected works*, vols. 1 to 5, edited by L. Rosenfeld and J. R. Nielsen. Amsterdam/New York: North Holland/Elsevier.

Bohr, N., Kramers, H. A., and Slater, J. C. (1924). The quantum theory of radiation. *The Philosophical Magazine*, **47**, 1924, 785–822.

Borzeskowski (von), H.-H., and Treder, H.-J. (1988). *The meaning of quantum gravity*. Dordrecht: Reidel.

Born, M. (1924). Uber Quantenmechanik. *Zeitschrift für Physik*, **26**, 379–95. Repr. in Born (1963), vol. 2, pp. 61–77.

Born, M. (1926). Quantenmechanik der Stossvorgänge. *Zeitschrift für Physik*, **38**, 803–827. Repr. in Born (1963), vol. 2, pp. 233–58.

Born, M. (1927). Quantenmechanik und Statistik. *Naturwissenschaftlich*, **15**, 238–42.

Born, M. (1949). Einstein's statistical theories. In P.-A. Schilpp (ed.), *Albert Einstein, philosopher-scientist*, Evanston, IL: The Library of Living Philosophers, pp. 161–78.

Born, M. (1963). *Ausgewählte Abhandlungen*. Göttingen: Vandenhoeck and Ruprecht.

Born, M., and Heisenberg, W. (1928). La mécanique des quanta. In Solvay (1928), pp. 143–83.

Born, M., Heisenberg, W., and Jordan, P. (1926). Zur Quantenmechanik II. *Zeitschrift für Physik*, **35**, 557–615. Repr. in Born (1963), vol. 2, pp. 155–213.

Born, M., and Jordan, P. (1925). Zur Quantenmechanik. *Zeitschrift für Physik*, **34**, 858–88. Repr. in Born (1963), vol. 2, pp. 124–154.

Bose, S. N. (1924). Planck's Gesetz und Lichtquantenhypothese. *Zeitschrift für Physik*, **26**, 1924, 178–81. (Transl. from English by A. Einstein.) English original version, 'Planck's law and the hypothesis of light quanta', repr. in Theimer, H., and Ram, B. (1976), The beginning of quantum statistics, *American Journal of Physics*, **44**, 1056–1057.

Bothe, W., and Geiger, H. (1924). Ein Weg zu experimentellen Nachprüfung der Theorie von Bohr, Kramers und Slater. *Zeitschrift für Physik*, **26**, 1924, 44. Engl. transl.: Experimental test of the theory of Bohr, Kramers, and Slater, in R. Lindsay, (ed.). *Early concepts of energy in atomic physics*, Dowden, Hutchinson and Ross, Stroutsburgh, Penns., 1979, pp. 230–31.

Bothe, W., and Geiger, H. (1925). Über das Wesen des Comptoneffekts ; eine experimentelles Beitrag zur Theorie des Strahlung. *Naturwissenschaft*, **13**, 440; *Zeitschrift für Physik*, **32**, 639–63.

Broglie, L. de (1924). *Recherches sur la théorie des quanta*, Thèse, Paris, 1924; *Annales de physique*, 10 ème série, 3, 1925, 22–128; ré-éd., Masson, Paris, 1963.

Broglie, L. de (1928). La nouvelle dynamique des quanta. In Solvay (1928), pp. 105–32.

Cornell, E. A., and Wieman, C. E. (1998). Bose–Einstein condensate. *Scientific American*, **278**, 40–45.

d'Espagnat, B. (1984). Nonseparability and the Tentative Descriptions of Reality. *Physics Report*, **110**, 201–64.

Dirac, P. A. M. (1926a). On the theory of quantum mechanics. *Proceedings of the Royal society of London A*, **112**, 661–77.

Dirac, P. A. M. (1926b). The physical interpretation of the quantum dynamics. *Proceedings of the Royal society of London A*, **113**, 621–41.

Einstein, A. (1905). Ueber einen die Erzeugung und Verwandlung des Lichtes betreffenden heuristischen Gesischtspunkt. *Annalen der Physik*, **17**, 132–148; also in *Collected Papers*, vol. 2, pp. 150–166.

Einstein, A. (1916a). Strahlung-emission und -absorption nach der Quantentheorie. *Deutsche Physikalische Gesellschaft, Verhandlungen*, XVIII, 318–23.

Einstein, A. (1916b). Zur Quantentheorie der Strahlung. *Physikalische Gesellschaft Mitteilungen* (Zürich), 1916, 47–62; *Physikalische Zeitschrift*, **18**, 121–128. Engl. transl., 'On the quantum theory of radiation', in van der Waerden (1967), pp. 63–78.

Einstein, A. (1921). Ueber ein den Elementarprozess der Lichtemission betreffendes Experiment. *Preussische Akademie der Wissenschaften, Sitzungsberichte*, 882–83.

Einstein, A. (1922a). Quantentheoretische Bemerkungen zur Supraleitung der Metalle. *Gedenboek Kammerling Onnes* (11. 3.1922), Leiden, 429–35.

Einstein, A. (1922b). Theorie der Lichtfortpflanzung in dispergierenden Medien. *Preussische Akademie der Wissenschaften, Phys. Math. Klasse, Sitzungsberichte*, 18–22.

Einstein, A. (1923). Bietet die feldtheorie Möglichkeiten für die Lösung des Quantenproblems? *Preussische Akademie der Wissenschaften, Phys. Math. Klasse, Sitzungsberichte*, 1923, pp. 359–64.

Einstein, A. (1924a). Das Komptonsche Experiment. *Berliner Tageblatt*, 20 April 1924, Supplt.

Einstein, A. (1924b). Wärmegleichgewicht im Strahlungsfeld bei Anwesenheit von Materie. *Zeitschrift für Physik*, **27**, 392–93. [Joint note to N. Bose's article, cf. Bose (1924)].

Einstein, A. (1924–1925). Quantentheorie des einatomigen idealen Gases, I and II. *Preussische Akademie Wissenschaften, Phys. Math. Klasse, Sitzungsberichte*, 1924, 261–67; 1925, pp. 3–14.

Einstein, A. (1925). Quantentheorie des idealen Gases, *Preussische Akademie Wissenschaften, Phys. Math. Klasse, Sitzungsberichte*, 1925, pp. 18–25.

Einstein, A. (1936). Physik und Realität. [English transl.: Physics and reality.] In Einstein (1954) (ed. 1981), pp. 283–314.

Einstein, A. (1940). The fundaments of theoretical physics. In Einstein (1954) (ed. 1981), pp. 315–26.

Einstein, A. (1946). Autobiographishes [Autobiographical notes]. In P.-A. Schilpp (ed.) (1949), *Albert Einstein, philosopher-scientist*, Evanston, IL: The Library of Living Philosophers, pp. 1–94.

Einstein, A. (1948). Quantenmechanik und Wirklichkeit. *Dialectica*, **2**, 320–24.

Einstein, A. (1949). Reply to criticism. Remarks concerning the essays brought together in this co-operative volume. In P.-A. Schilpp (ed.), *Albert Einstein, philosopher-scientist*, Evanston, IL: The Library of Living Philosophers, pp. 665–88.

Einstein, A. (1953). Einleitende Bemerkungen über Grundbegriffe. Remarques préliminaires sur les principes fondamentaux (Fr. transl. by M.-A. Tonnelat). In *Louis de Broglie, physicien et penseur*, Paris: Albin Michel, pp. 4–15.

Einstein, A. (1954). *Ideas and opinions.* New transl. and revision by Sonja Bargmann. New York: Crown.

Einstein, A. (1956). *Briefe an Maurice Solovine. Lettres à Maurice Solovine.* Paris: Gauthier-Villars.

Einstein, A. (1987–2018). *The Collected Papers of Albert Einstein*, vols. 1–15. Edited by J. Stachel, M. Klein *et al.*, Princeton: Princeton University Press. (Writings and Correspondence up to May 1927. Further ones are in the course of publication).

Einstein, A. (1989–1993). *Oeuvres choisies*, 5 vols. French transl. Paris: Seuil/éd. du CNRS.

Einstein, A., and Besso, M. (1972). *Correspondance, 1903–1955*, published by P. Speziali (original texts and French transl.). Paris: Hermann.

Einstein, A., and Born, M. (1969). *Briefwechsel 1916–1955.* Munich: Nymphenburger Verlagshandlung. Engl. transl., *The Born-Einstein letters*, ed. by M. Born, New York: Walker, 1971.

Einstein, A., and Ehrenfest, P. (1922). Quantentheoretische Bemerkungen zum experiment von Stern und Gerlach. *Zeitschrift für Physik*, **11**, 31–34.

Einstein, A., Podolsky, B., and Rosen, N. (1935). Can quantum mechanical description of physical reality be considered complete? *Physical Review*, **47**, 777–80.

Einstein, A., Tolman, R.C., and Podolsky, B. (1931). Knowledge of past and future in quantum mechanics. *Physical Review*, **37**, 780–81.

Freedman, S. J. and Clauser, J. F. (1972). Experimental test of local-hidden variable theories. *Physical Review Letters*, **28**, 938–41.

Freire Jr, O. (1999). *David Bohm e a controversia dos quanta*. Campinas (Brazil): Coleção CLE.

Freire Jr, O. (2015). *The Quantum Dissidents. Rebuilding the Foundations of Quantum Mechanics (1950–1990)*. Berlin, Heidelberg: Springer.

Grangier, P. (1986). *étude expérimentale de propriétés non-classiques de la lumière ; interférences à un seul photon*. Thèse de doctorat ès-sciences physiques, Université Paris-Sud, Orsay, 1986.

Greenberger, D. M., Horne, M.A., and Zeilinger, A. (1989). Going beyond Bell's theorem. In M. Kafatos (ed.), *Bell's Theorem, Quantum Theory and Conceptions of the Universe*, Dordrecht: Kluwer, pp. 69–72.

Griffin, A, Snoke, D.W., and Stringari, S. (eds) (1995). *Bose–Einstein condensation*. Cambridge: Cambridge University Press.

Haroche, S. (2012). Controlling Photons in a Box and Exploring the Quantum to Classical Boundary, *Nobel Lecture*, Stockholm, 8 December 2012, pp. 63–107.

Haroche, S., Brune, M., and Raimond, J.-M. (1997). Experiments with Single Atoms in a Cavity : Entanglement, Schrödinger's Cats and Decoherence. *Philosophical Transactions of the Royal Society of London A*, **355**, 2367–80.

Holton, G. (1988). The Roots of Complementarity. *Daedalus*, **117**, 151–97.

Howard, D. (1986). Einstein on locality and separability. *Studies in History and Philosophy of Science*, **16**, 171–201.

Howard, D. (2022, this volume). The Copenhagen Interpretation.

Jammer, M. (1966). *The Conceptual Development of Quantum Mechanics*. New York: McGraw-Hill.

Jammer, M. (1974). *The Philosophy of Quantum Mechanics. The interpretations of Quantum Mechanics in Historical Perspective*. New York: Wiley and Sons.

Joos, E., and Zeh, H. D. (1985). The emergence of classical properties through interactions with the environment. *Zeitschrift für Physik*, B **59**(2), 223–43.

Joos, E., Zeh, H. D., Kiefer, C., Giulini, D., Kupsch, J., and Stamatescu, I.-O. (2003). *Decoherence and the appearance of a classical world in Quantum theory*, 2nd edn. Berlin: Springer,.

Leite Lopes, J. and Paty, M. (eds) (1977). *Quantum Mechanics, a half century later*. Dordrecht: Reidel.

Neumann, J. von (1932). *Mathematische Gründlagen der Quanten-Mechanic*. Berlin: Springer Verlag.

Omnès, R. (1994). *The Interpretation of Quantum Mechanics*. Princeton: Princeton University Press.

Omnès, R. (1999). *Understanding Quantum Mechanics*. Princeton: Princeton University Press.

Osnaghi, S. (2022, this volume). Bohr and the epistemological lesson of quantum mechanics,

Pais, A. (1982). *Subtle is the Lord. The science and life of Albert Einstein*. Oxford: Oxford University Press.

Paty, M. (1977). The recent attempts to verify quantum mechanics. In J. Leite Lopes and M. Paty (eds.), Quantum Mechanics, a half century later, Dordrecht: Reidel, pp. 261–89.

Paty, M. (1982). L'inséparabilité quantique en perspective. *Fundamenta Scientiae*, **3**, pp. 79–92.

Paty, M. (1985). Einstein et la complémentarité au sens de Bohr. Du retrait dans le tumulte aux arguments d'incomplétude. *Revue d'histoire des sciences*, **38**, 325–51.

Paty, M. (1986). La non-séparabilité locale et l'objet de la théorie physique. *Fundamenta Scientiae*, 7, 47–87.

Paty, M. (1988a). *La Matière dérobée. L'appropriation critique de l'objet de la physique contemporaine*. Paris: Edition des Archives contemporaines.

Paty, M. (1988b). Sur la notion de complétude d'une théorie physique. In N. Fleury, S. Joffily, J. A. Martins Simões, and A. Troper (eds), *Leite Lopes Festchrift. A pioneer physicist in the third world (dedicated to J. Leite Lopes on the occasion of his seventieth birthday)*, Singapore: World Scientific, pp. 143–64.

Paty, M. (1989). Einstein, Popper et le débat quantique aujourd'hui. In R. Bouveresse (dir.), *Karl Popper et la science d'aujourd'hui*, Colloque de Cerisy, 1–11 juillet 1981, Paris: Aubier, pp. 255–72.

Paty, M. (1990). Reality and Probability in Mario Bunge's Treatise. In G. Dorn and P. Weingartner (eds), *Studies on Mario Bunge's Treatise*, Series 'Poznan Studies in the Philosophy of the Sciences and Humanities', Amsterdam-Atlanta: Rodopi, pp. 301–22.

Paty, M. (1993a). *Einstein philosophe. La physique comme pratique philosophique*. Coll. 'Philosophie d'aujourd'hui'. Paris: Presses Universitaires de France.

Paty, M. (1993b). Sur les variables cachées de la mécanique quantique: Albert Einstein, David Bohm et Louis de Broglie. *La Pensée*, **292**, 93–116.

Paty, M. (1995). The Nature of Einstein's objections to the Copenhagen Interpretation of Quantum Mechanics. *Foundations of Physics*, **25**, 183–204.

Paty, M. (1996). Le Style d'Einstein, la nature du travail scientifique et le problème de la découverte. *Revue philosophique de Louvain*, **94**, 447–70.

Paty, M. (1999). Are quantum systems physical objects with physical properties? *European Journal of Physics*, **20**, 373–88. (Special issue on 'Unsolved problems of physics'.)

Paty, M. (2000a). Interprétations et significations en physique quantique. *Revue Internationale de philosophie* (Brussels), **212**(2), 1999–2042.

Paty, M. (2000b). The quantum and the classical domains as provisional parallel coexistents. In D. Krause, S. French, and F. Doria (eds), *In honor of Newton da Costa), Synthese* special issue, Dordrecht: Kluwer, pp. 179–200.

Paty, M. (2001). Physical quantum states and the meaning of probability. In M. C. Galavotti, P. Suppes, and D. Costantini (eds), *Stochastic Causality*, Stanford, CA: CSLI Publications (Center for Studies on Language and Information), Chapter 14, pp. 235–55.

Paty, M. (2002). La Physique quantique ou l'entraînement de la forme mathématique sur la pensée physique. In C. Mataix and A. Rivadulla (eds), *Física cuantica y realidad. Quantum physics and reality*, Madrid: Editorial Complutense, pp. 97–134.

Paty, M. (2003). The concept of quantum state: new views on old phenomena. In A. Ashtekar, R. S. Cohen, D. Howard, J. Renn, S. Sarkar, and A. Shimony (eds), *Revisiting the Foundations of Relativistic Physics: Festschrift in Honour of John Stachel*, Boston Studies in the Philosophy and History of Science, Dordrecht: Kluwer Academic Publishers, pp. 451–78.

Paty, M. (2009). On Today's Understanding of the Concepts and Theory of Quantum Physics. A Philosophical Reflection. In M. S. D. Cattani, L. C. B. Crispino, M. O. C. Gomes, and A. F. S. Santoro (eds), *Trends in Physics: Festschrift in Homage to Prof. José Maria Filardo Bassalo*, São Paulo: Livraria da Física, pp. 209–35.

Paty, M. (2015). Le rôle des mathématiques en physique et la place de l'expérience dans la pensée d'Albert Einstein. In E. Barbin, and P. Cléro (eds), *Les mathématiques et l'expérience. Ce qu'en ont dit les philosophes et les mathématiciens*, Collection 'La République des Lettres. Symposium', Paris: Hermann, pp. 313–36.

Paty, M. (forthcoming a). Einstein's scientific style in the exploration of the quantum domain (A view on the relationship between theory and its object), Proceedings of Prof. Robert Cohen Symposium organized at the Institute of Science, Technology and Society and Tsinghua University Library, Tsinghua University, Beijing, Beijing (China), 26–27 april 2010.

Paty, M. (forthcoming b). Sur le cheminement d'Einstein vers une théorie quantique du rayonnement, in Mayrargue, Arnaud & Fauque, Danielle (éds.), Séminaire "Histoire de la Lumière", Université Paris-Diderot, in press.

Paty, M. (forthcoming c). *Einstein, les quanta et le reel,* in completion to be published.

Pflegor, R. L., and Mandel, L. (1967). Interference of Independent Photon Beams. *Physical Review,* **159,** 1084.

Popper, K. (1983). *The Quantum Schism,* volume 3 of *The Postcript to the Logic of Scientific Discovery.* London: Hutchinson.

Rosenfeld, L. (1967). Niels Bohr in the thirties: consolidation and extension of the conception of complementarity. In S. Rozental (ed.), *Niels Bohr, his life and work,* New York: Wiley and Sons, pp. 114–136.

Schilpp, P.-A. (ed.) (1949). *Albert Einstein, philosopher-scientist.* Evanston, IL: The Library of Living Philosophers.

Schlosshauer, M. (2005). Decoherence, the measurement problem and interpretation of quantum mechanics. *Reviews of Modern Physics,* **76**(4), 1267–1305.

Schlosshauer, M. (2007). *Decoherence and the Quantum-to-Classical Transition,* 1st edn. Berlin/Heidelberg: Springer.

Schrödinger, E. (1926). *Abhandlungen zur Wellemechanik.* Leipzig: Barth. 2nd edn, 1928, *Mémoires sur la mécanique ondulatoire* (1933), French transl. by Alexandre Proca, with unedited notes by E. Schrödinger, Paris: Alcan.

Schrödinger, E. (1928). La mécanique des ondes. In Solvay (1928), pp. 185–206.

Schrödinger, E. (1935). Die Gegenwärtige Situation in der Quantenmechanik. *Die Naturwissenschaften,* **23,** 807–12; 824–828; 844–849. Also in Schrödinger (1984), vol. 3, pp. 484–501.

Schrödinger, E. (1984). *Gesammelte Abhandlungen. Collected papers,* 4 vols. Braunschweig/ Wien: Verlag der Oesterreichischen Akademie der Wissenschaften/Vieweg und Sohn.

Solvay (1928). *Rapports et discussions du cinquième Conseil de physique tenu à Bruxelles du 24 au 29 octobre 1927, sous les auspices de l'Institut International de Physique Solvay.* Paris: Gauthier-Villars. (An English transl. is provided in Bacciagaluppi and Valentini (2009).)

Stachel, J. (1986). Einstein and the quantum: fifty years of struggle. In R. Colodny (ed.), *From quarks to quasars,* Pittsburgh: University of Pittsburgh Press, pp. 349–381.

Stapp, H. (1980). Locality and reality. *Foundations of Physics,* **10,** 767–95.

van der Waerden, B. L. (ed.) (1967). *Sources of Quantum Mechanics.* Amsterdam: North Holland.

Wineland, D. (2013). Nobel lecture : superposition, entanglement, and raising Schrödonger's cat. *Review of Modern Physics,* **85,** 1103–14.

Zeh, H. D. (1970). On the Interpretation of Measurement in Quantum theory. *Foundations of Physics,* **1,** 69.

Zurek, W. H. (1991). Decoherence and the Transition from Quantum to Classical. *Physics Today,* **44,** 36–44.

Zurek, W. H. (2002). Decoherence and the transition from Quantum to classical—Revisited. *Los Alamos Science,* **27,** 2–25.

Zurek, W. H. (2003). Decoherence, einselection and the quantum origin of the classical. *Review of Modern Physics*, **75**(3), 715.

Zurek, W. H. (2004). Quantum Darwinism and Envariance. In J. D. Barrow, P. C. W. Davies, and C. H. Harper (eds), *Science and Ultimate Reality*, Cambridge: Cambridge University Press, pp. 121–137.

# CHAPTER 13

# TACKLING LOOPHOLES IN EXPERIMENTAL TESTS OF BELL'S INEQUALITY

## DAVID I. KAISER[1]

## 13.1 INTRODUCTION

BELL'S inequality (1964, 1990) remains a hallmark achievement of modern physics, and a touchstone for efforts to distinguish between quantum mechanics and various alternatives. In particular, Bell's inequality sets a strict threshold for how strongly correlated the outcomes of measurements on two or more particles can be, if the underlying theory of nature that describes those particles' behaviour satisfies certain criteria, often labeled 'local realism' and associated with the famous paper by Albert Einstein, Boris Podolsky, and Nathan Rosen (1935). In local-realist theories, the outcome of a measurement performed on a particle at one location cannot depend on actions undertaken at an arbitrarily distant location.[2] Quantum mechanics is not compatible with local realism and, as Bell demonstrated, quantum mechanics predicts

[1] Program in Science, Technology, and Society and Department of Physics, Massachusetts Institute of Technology, Cambridge, Massachusetts 02139 USA. Email: dikaiser@mit.edu

[2] Early work on Bell's inequality, including Bell's first derivation (1964), was deeply influenced by the EPR paper (Einstein, Podolsky, and Rosen 1935), in which the authors argued that particles should be considered to have definite properties on their own, prior to and independent of physicists' efforts to measure them ('realism'), and that distant events should not influence local ones arbitrarily quickly ('locality'). More recent work has clarified the minimal requirements for Bell's inequality to hold: the measurement outcome at one detector should not depend on either the detector setting or the measurement outcome at a distant detector, and the selection of detector settings on each experimental run should be independent of the properties of the particles to be measured. For recent, succinct discussions of 'local realism' in the context of Bell's inequality, see the Appendix of Clauser (2017) and Section 3.1 of Myrvold, Genovese, and Shimony (2019). Note that Bell's inequality does not apply to formulations such as Bohmian mechanics (Bohm 1952a, 1952b), which, as Bell (1964) noted, has a 'grossly non-local structure'.

that measurements on pairs of particles in so-called 'entangled' states can be *more strongly correlated* than the local-realist bound would allow.[3]

Virtually every published experimental test of Bell's inequality, stretching over half a century, has found results compatible with the predictions of quantum mechanics, and (hence) in violation of Bell's inequality.[4] Yet since the earliest efforts to subject Bell's inequality to experimental test, physicists have recognized that several 'loopholes' must be addressed before one may conclude that local-realist alternatives to quantum mechanics really have been ruled out. The loopholes consist of logical possibilities—however seemingly remote or implausible—by which a local-realist theory could give rise to correlated measurements that mimic the expectations from quantum theory, exceeding Bell's bound.[5]

In this chapter, I discuss the three major loopholes that have been identified for experimental tests of Bell's inequality. In section 13.2, I briefly review the form of Bell's inequality on which most experimental efforts have been focused. This form, which was introduced by John Clauser, Michael Horne, Abner Shimony, and Richard Holt (1969) soon after Bell published his original paper on the topic, is usually referred to as the 'Bell–CHSH inequality'. Several of those physicists, often in close dialogue with Bell himself, were also among the first to identify various loopholes. The first of these, known as the 'locality loophole', is the subject of section 13.3. In section 13.4, I discuss the 'fair-sampling loophole', while in section 13.5 I turn to the 'freedom-of-choice loophole'. Brief concluding remarks follow in section 13.6. As described below, tests of Bell's inequality have been performed on many different physical systems, subjecting different types of particles to measurements with different types of detectors. In this chapter I focus primarily on conceptual analysis of the various loopholes, more than on the details of particular experimental implementations.

# 13.2 THE BELL–CHSH INEQUALITY AND THE FIRST EXPERIMENTAL TESTS

Most experimental tests of Bell's inequality have concerned correlations among measurements on pairs of particles. Such tests can be pictured as in Fig. 13.1: a source ($\sigma$) in the centre of the experiment emits a pair of particles which travel away from the

[3] For historical treatments, see Freire (2006, 2015), Bromberg (2006), Gilder (2008), Kaiser (2011, 2020), and Whitaker (2016). For a range of philosophical responses, see d'Espagnat (1976), Redhead (1987), Cushing and McMullin (1989), Albert (1992), Bub (1997, 2016), Bokulich and Jaeger (2010), Maudlin (2011), Cramer (2016), Bertlmann and Zeilinger (2017), and Myrvold, Genovese, and Shimony (2019). Recent popular accounts include Orzel (2009), Zeilinger (2010), Musser (2015), Becker (2018), Ball (2018), Bub and Bub (2018), Greenstein (2019), and Brody (2020).

[4] The only published experimental test of Bell's inequality that appeared to contradict the predictions from quantum theory was published in (Faraci *et al.*, 1974), though that experiment was criticized in Kasday, Ullman, and Wu (1975) and Clauser and Shimony (1978).

[5] For reviews, see Clauser and Shimony (1978), Zeilinger (1999), Weihs (2009), Brunner *et al.* (2014), Larsson (2014), and Giustina (2017).

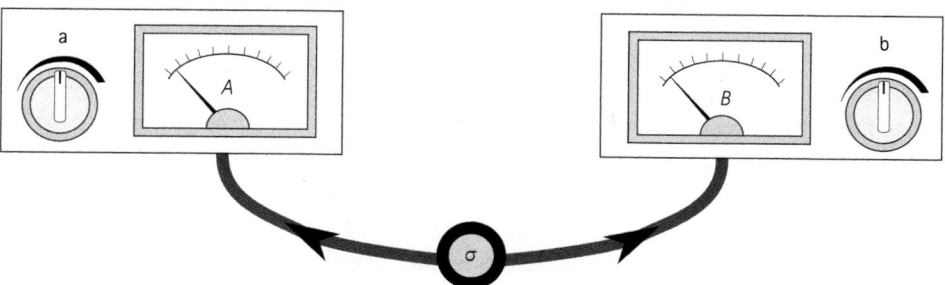

**FIGURE 13.1** Schematic illustration of a typical Bell test. A source σ emits a pair of particles, which travel in opposite directions. At each detector, a physicist selects a particular measurement to be performed by adjusting the detector settings (**a**, **b**); each detector then yields a measurement outcome $(A, B)$. (Adapted from Gallicchio, Friedman, and Kaiser (2014).)

source in opposite directions. At each detector, a physicist selects a particular measurement to be performed by adjusting the detector settings (**a**,**b**); each detector then yields a measurement outcome $(A,B)$. For example, if the particles emitted from the source consist of pairs of electrons, a physicist at the left detector might choose to measure the spin of the left-moving electron along the **x**-axis, or along the **y**-axis, or along some intermediate angle; her choice of basis in which to measure the electron's spin is labeled **a**. The physicist at the right detector chooses to measure the spin of the right-moving electron along a particular orientation in space by adjusting the detector setting **b**. In this example, for any pair of detector settings (**a**,**b**) that have been selected, the measurement outcomes $(A, B)$ at each detector can only be spin-up or spin-down. If we label the measurement outcome spin-up as $+1$ and spin-down as $-1$, then we have $A(\mathbf{a}), B(\mathbf{b}) \in \{+1, -1\}$.[6]

For any pair of detector settings (**a**,**b**), we may construct the correlation function

$$E(\mathbf{a}, \mathbf{b}) \equiv \langle A(\mathbf{a}) \, B(\mathbf{b}) \rangle, \tag{1}$$

where the angular brackets indicate averages over the many experimental runs in which pairs of particles were subjected to measurements with detector settings (**a**, **b**). For measurements of a property such as spin, the outcomes $A(\mathbf{a})$ and $B(\mathbf{b})$ can only

---

[6] David Bohm first suggested that EPR-type experiments could be conducted using measurements of observables such as spin, which have discrete sets of possible measurement outcomes, in his influential textbook on quantum mechanics: Bohm (1951), pp. 614–622. Bell was inspired by Bohm's variation while working on Bell (1964). (See Kaiser (2011), pp. 31–37.) On the wider impact of Bohm's textbook, see Kaiser (2020), chap. 8. A beautiful variation on Bell's original argument—which (in principle) can force an empirical contradiction between predictions from local realism and quantum mechanics with a single set of measurements rather than statistical averages over many experimental runs—concerns measurements of a discrete observable such as spin on $N$-particle entangled states, with $N \geq 3$. See Greenberger, Horne, and Zeilinger (1989) and Greenberger, Horne, Shimony, and Zeilinger (1990).

ever be $\pm 1$, so on any given experimental run, the product $A(\mathbf{a})\,B(\mathbf{b})$ can only ever be $\pm 1$. Upon averaging over many runs in which the detector settings were $(\mathbf{a}, \mathbf{b})$, the correlation function $E(\mathbf{a}, \mathbf{b})$ therefore satisfies $-1 \leq E(\mathbf{a,b}) \leq 1$.

One might try to account for the behaviour of such correlation functions $E(\mathbf{a}, \mathbf{b})$ by constructing a local-realist theory and using it to calculate $p(A, B|\mathbf{a}, \mathbf{b})$, the conditional probability that physicists would find measurement outcomes $A$ and $B$ upon selecting detector settings $\mathbf{a}$ and $\mathbf{b}$. Bell (1964, 1971) argued that within any local-realist formulation, such conditional probabilities would take the form

$$
\begin{aligned}
p(A, B|\mathbf{a}, \mathbf{b}) &= \int d\lambda\, p(\lambda)\, p(A, B|\mathbf{a}, \mathbf{b}, \lambda) \\
&= \int d\lambda\, p(\lambda)\, p(A|\mathbf{a},\lambda)\, p(B|\mathbf{b}, \lambda) \,.
\end{aligned}
\tag{2}
$$

Here $\lambda$ represents all the properties of the particles prepared at the source $\sigma$ that could affect the measurement outcomes $A$ and $B$. Bell imagined that whatever specific form the variables $\lambda$ took, their values on a given experimental run would be governed by some probability distribution $p(\lambda)$. Given detector setting $\mathbf{a}$ at the left detector, there would be some probability $p(A|\mathbf{a}, \lambda)$ to find measurement outcome $A$ at that detector, and likewise some probability $p(B|\mathbf{b},\lambda)$ to find outcome $B$ at the right detector given detector setting $\mathbf{b}$.[7] Note that these expressions encode 'locality': nothing about the probability to find outcome $B$ at the right detector depends on either the setting $(\mathbf{a})$ or the outcome $(A)$ at the distant detector, and vice versa (Bell, 1964, 1971, 1990). (For a helpful and succinct discussion, see Myrvold, Genovese, and Shimony, 2019.) In his original derivation, Bell (1964) quoted from Einstein's 'Autobiographical Notes'. As Einstein had written, 'But on one supposition we should, in my opinion, absolutely hold fast: the real factual situation of the system $S_2$ [the particle being measured at the right detector] is independent of what is done with the system $S_1$ [the particle at the left detector], which is spatially separated from the former.' (Einstein (1949), p. 85. See also Howard (1985), Fine (1986), and Gutfreund and Renn (2020).) See Fig. 13.2.

Bell's inequality can be cast in a particularly simple form if we consider experiments in which each particle is subjected to measurement in one of two detector settings: either $\mathbf{a}$ or $\mathbf{a}'$ at the left detector, and $\mathbf{b}$ or $\mathbf{b}'$ at the right detector. Then one may consider a particular combination of correlation functions, as one varies the pairs of detector settings:

$$
S \equiv |E(\mathbf{a,b}) + E(\mathbf{a',b}) - E(\mathbf{a,b'}) + E(\mathbf{a',b'})| \,.
\tag{3}
$$

---

[7] In his original derivation, Bell (1964) assumed that the measurement outcomes were governed by deterministic functions $A(\mathbf{a}, \lambda)$ and $B(\mathbf{b}, \lambda)$. He generalized his derivation to stochastic models, with conditional probabilities $p(A|\mathbf{a}, \lambda)$ and $p(B|\mathbf{b}, \lambda)$, in Bell (1971).

**FIGURE 13.2** John S. Bell in his office at CERN, 1982. (Courtesy CERN.)

The quantity $S$, first derived by Clauser, Horne, Shimony, and Holt (1969), is known as the Bell–CHSH parameter. Closely following Bell's original reasoning (1964), the CHSH authors demonstrated that for any model in which conditional probabilities $p(A, B|\mathbf{a}, \mathbf{b})$ took the form of Eq. (2), the parameter $S$ obeys the inequality (Clauser *et al.*, 1969)

$$S \leq 2 \,. \tag{4}$$

Eq. (4) is known as the Bell–CHSH inequality.[8] A straightforward calculation (see, e.g., Sakurai (1994), pp. 223–232) suffices to show that for pairs of particles prepared in a maximally entangled state, such as

$$|\Psi^{(\pm)}\rangle = \frac{1}{\sqrt{2}}\{|+1\rangle_A \otimes |-1\rangle_B \pm |-1\rangle_A \otimes |+1\rangle_B\} \,, \tag{5}$$

---

[8] As in Bell (1964), the original CHSH derivation (1969) applied to local-realist models in which the measurement outcomes were given by deterministic functions $A(\mathbf{a}, \lambda)$ and $B(\mathbf{b}, \lambda)$. The CHSH inequality in Eq. (4) also applies to stochastic models in which $A(\mathbf{a}, \lambda) \rightarrow p(A|\mathbf{a}, \lambda)$ and $B(\mathbf{b}, \lambda) \rightarrow p(B|\mathbf{b}, \lambda)$. See, e.g., Appendix A of Hall (2011) and Myrvold, Genovese, and Shimony (2019).

quantum mechanics predicts that $S$ could *violate* the bound in Eq. (4), achieving a maximum value

$$S_{QM}^{\max} = 2\sqrt{2}. \tag{6}$$

According to quantum mechanics, the value $S_{QM}^{\max}$ should arise for particular choices of settings $(\mathbf{a}, \mathbf{a}')$ and $(\mathbf{b}, \mathbf{b}')$. The value $S_{QM}^{\max} = 2\sqrt{2}$ is known as the 'Tsirelson bound' (Cirel'Son, 1980).

Consider, for example, an experiment involving pairs of linearly polarized photons prepared in the state $|\Psi^{(\pm)}\rangle$ of Eq. (5), with $|+1\rangle \rightarrow |H\rangle$ (horizontal polarization) and $|-1\rangle \rightarrow |V\rangle$ (vertical polarization) with respect to some orientation in space. If the photons travel along the $z$-axis toward each detector, then the detector settings $(\mathbf{a}, \mathbf{a}', \mathbf{b}, \mathbf{b}')$ are simply unit vectors pointing at various angles within the $xy$ plane, along which polarizing filters could be oriented. A photon in state $|H\rangle_A$ with respect to orientation $\mathbf{a}$ would yield measurement outcome $A(\mathbf{a}) = +1$, whereas a photon in state $|V\rangle_A$ along $\mathbf{a}$ would yield $A(\mathbf{a}) = -1$.[9] For this set-up, the quantum-mechanical prediction for the correlation function is simply $E(\mathbf{a}, \mathbf{b}) = -\cos(2\theta_{ab})$, where $\cos\theta_{ab} \equiv \mathbf{a} \cdot \mathbf{b}$.[10] In that case, the Tsirelson bound corresponds to the choice of settings $(\mathbf{a}, \mathbf{a}') = (0°, 45°)$ and $(\mathbf{b}, \mathbf{b}') = (22.5°, 67.5°)$. On the other hand, both local-realist theories and quantum mechanics predict that $S \leq 2$ for choices such that $\mathbf{a} \cdot \mathbf{b} = \mathbf{a}' \cdot \mathbf{b}' = 0$ or 1, regardless of the angle between $\mathbf{a}$ and $\mathbf{a}'$.

The size of the difference between predictions from local-realist theories and from quantum mechanics—$S \leq 2$ versus $S \leq 2\sqrt{2}$ for clever choices of detector settings— is sufficiently large that some physicists quickly began to imagine conducting experimental tests of Bell's inequality. One of the first to highlight this possibility was Henry Stapp, a research scientist at the Lawrence Berkeley Laboratory in California. Stapp had been trained in particle physics at Berkeley in the early

---

[9] The earliest experimental tests involving polarized photons used single-channel measuring devices at each detector. Hence (most) photons whose polarization aligned with the orientation of the polarizing filter would pass through the filter and be registered by a device such as a photomultiplier tube, yielding $A(\mathbf{a}) = +1$, whereas photons whose polarization was perpendicular to the orientation of the polarizing filter would yield no registered detection, coded as $A(\mathbf{a}) = -1$. In practice, such an approach combined data on perpendicular polarization with all other reasons that the detector might have failed to register a photon on a given experimental run. Later experiments adopted two-channel measuring devices at each detector, taking advantage of the fact that a photon in state $|H\rangle_A$ with respect to orientation $\mathbf{a}$ will pass through the polarizing filter along a distinct trajectory, different from that of a photon in state $|V\rangle_A$. See, e.g., Aspect (2002) and Myrvold, Genovese, and Shimony (2019).

[10] The form of $E(\mathbf{a}, \mathbf{b})$ in this case is easy to understand. When measured along the same orientation in space, $\mathbf{a} = \mathbf{b}$ and hence $\theta_{ab} = 0°$, pairs of polarized photons prepared in the state $|\Psi^{(+)}\rangle$ of Eq. (5) should be perfectly anti-correlated, with $E(\mathbf{a}, \mathbf{b}) = -1$. If the photons in that state are measured along perpendicular orientations in space, with $\theta_{ab} = 90°$, then the polarization measurements should be perfectly correlated, with $E(\mathbf{a}, \mathbf{b}) = +1$. The variation of $E(\mathbf{a}, \mathbf{b})$ with $\theta_{ab}$ follows from considering measurements in rotated bases. For example, if the polarizer at the left detector is rotated by angle $\varphi$, the eigenstates of the new detector setting $\tilde{\mathbf{a}}$ will be given by $|\tilde{H}(\varphi)\rangle_A = \cos\varphi |H\rangle_A + \sin\varphi |V\rangle_A$ and $|\tilde{V}(\varphi)\rangle_A = -\sin\varphi |H\rangle_A + \cos\varphi |V\rangle_A$.

1950s; for his Ph.D. dissertation, he had studied spin correlations in proton–proton scattering experiments. During the summer of 1968—even before the CHSH version of Bell's inequality had appeared—Stapp wrote a preprint noting that Bell's inequality could be tested in experiments much like the proton-scattering ones on which he had previously focused. The critical update that would be required, compared to previous experiments, would be to vary the angles along which the protons' spins were measured, to avoid only measuring in bases such that $\mathbf{a} \cdot \mathbf{b} = \mathbf{a}' \cdot \mathbf{b}' = 0$ or $1$. As Stapp wrote, 'The precise experiments considered here have not all actually been performed. But they are only slight variations of experiments that have been performed' (Stapp, 1976).[11]

Around the same time, other physicists hit upon a similar idea. Abner Shimony, at the time a young professor in both the Physics and Philosophy Departments at Boston University, wondered whether data from previous correlation experiments, which had been conducted for other reasons, could be used to test Bell's inequality. Together with then-graduate student Michael Horne, he delved into the published literature, conducting what Horne playfully dubbed 'quantum archaeology'. Much like Stapp, however, Horne and Shimony realized that previous correlation experiments had failed to consider the range of angles among the bases $(\mathbf{a}, \mathbf{a}', \mathbf{b}, \mathbf{b}')$ for which quantum mechanics predicts $S > 2$.[12]

Independent of Stapp, Shimony, and Horne, John Clauser also became intrigued by the possibility of testing Bell's inequality in a laboratory experiment, while still in graduate school. He wrote directly to Bell in February 1969, asking if anyone had conducted such an experiment during the years since Bell's (1964) paper had appeared. Upon hearing back from Bell that no one had as yet performed such an experiment— and with Bell's additional encouragement that if Clauser were to measure something different from what quantum mechanics predicts, that would 'shake the world!'— Clauser began thinking about how to perform such a test. In the midst of that work, he learned of Shimony's and Horne's interest in Bell's inequality, and soon they began to collaborate together (Clauser, 2002). (See also Freire (2006), pp. 590–91, and Kaiser (2011), pp. 43–45.)

---

[11] Stapp originally composed and circulated (Stapp, 1976) during the summer of 1968, in advance of a conference on 'Quantum Theory and Beyond', which was held in Cambridge, England. The paper was not published in the conference proceedings, and Stapp later released the paper as a technical report from the Lawrence Berkeley Laboratory 'to fill continuing requests' (Stapp, 1976). On Stapp's training and his early interest in Bell's inequality, see Kaiser (2011), pp. 55–56. In 1976, around the time that Stapp circulated his technical report, M. Lamehi-Rachti and W. Mittig (1976) reported results of their analysis of violations of Bell's inequality using low-energy proton–proton scattering data. They reported good agreement with the quantum-mechanical predictions, though given experimental limitations they needed to make several additional assumptions in order to put the proton scattering data into a form with which to test Bell's inequality, beyond those typically required of Bell tests (Clauser and Shimony, 1978).

[12] Abner Shimony, interview with Joan Bromberg, 9 September 2002, transcript available in the Niels Bohr Library of the Center for History of Physics, American Institute of Physics, College Park, Maryland. See also Kaiser (2011), pp. 45–46.

Soon after Clauser began a postdoctoral fellowship at the Lawrence Berkeley Laboratory, he asked his supervisor, quantum-electronics pioneer Charles Townes, if he could design and conduct an experimental test of Bell's inequality as a side project, separate from the main research project for which Townes had hired him. Townes agreed and arranged for Clauser to work with Stuart Freedman, at the time a graduate student at the laboratory. Freedman and Clauser used pairs of linearly polarized photons in a maximally entangled state, emitted by atomic cascades within excited calcium atoms. They mounted the polarizers in such a way that their orientations at the left and right detectors could be adjusted. The distance between the two detectors was approximately four metres. To collect their data, Freedman and Clauser first fixed the polarizer orientations at each detector, beginning with $(\mathbf{a}, \mathbf{b}) = (0°, 22.5°)$. Then they recorded the measurement outcomes $A(\mathbf{a})$ and $B(\mathbf{b})$ within brief coincidence windows ($\Delta t = 8.1$ ns), to ensure that the pairs of measurements $(A, B)$ on a given experimental run pertained to a single pair of entangled photons that had been emitted from the source. Upon collecting many thousands of measurements on pairs of photons with the polarizers set to these orientations and averaging those results, they constructed the correlation function $E(\mathbf{a}, \mathbf{b})$. Then they paused the experiment, rotated the polarizer on the left side from orientation $\mathbf{a} = 0°$ to $\mathbf{a}' = 45°$ while keeping the polarizer on the right side fixed at $\mathbf{b} = 22.5°$, and conducted new measurements with which to construct $E(\mathbf{a}', \mathbf{b})$—and so on, until they had collected sufficient data with the various joint settings $(\mathbf{a}, \mathbf{b})$, $(\mathbf{a}', \mathbf{b})$, $(\mathbf{a}, \mathbf{b}')$, and $(\mathbf{a}', \mathbf{b}')$ to construct the combination of correlation functions needed for the Bell–CHSH parameter $S$ in Eq. (3). Their findings were equivalent to $S = 2.388 \pm 0.072$, violating the Bell–CHSH inequality of Eq. (4) by more than five standard deviations (Freedman and Clauser, 1972).[13] (See also Clauser and Shimony (1978), Freire (2006), and Kaiser (2011).) See Fig. 13.3.

Around the same time, Richard Holt and his graduate-school supervisor Francis Pipkin performed their own experimental test of Bell's inequality at Harvard. Like Freedman and Clauser, they conducted measurements on pairs of linearly polarized photons in a maximally entangled state, in this case using photons emitted from a particular cascade in excited mercury atoms. They used the combination of detector settings $(\mathbf{a}, \mathbf{a}', \mathbf{b}, \mathbf{b}')$ predicted by quantum mechanics to yield the maximum violation of Eq. (4). Yet unlike Freedman and Clauser's experiment, Holt and Pipkin measured

---

[13] Freedman and Clauser (1972) reported their results in terms of a quantity $\Delta(\varphi)$, closely related to the Bell–CHSH parameter $S$ of Eq. (3). For choices of detector settings such that $\theta_{ab} = \theta_{a'b} = \theta_{a'b'} = \varphi$ and $\theta_{ab'} = 3\varphi$, the quantities are related by $S = |4\Delta(\varphi) + 2|$. They measured $\Delta(22.5°) = 0.104 \pm 0.026$ and $\Delta(67.5°) = -1.097 \pm 0.018$, which represent violations of $S \leq 2$ by 4.0 and 5.4 standard deviations, respectively. None of the early experiments found results close to saturating the (theoretical) Tsirelson bound of quantum mechanics, $S_{QM}^{max} = 2\sqrt{2} = 2.83$, largely because of limited detector efficiencies, even though they were able to measure $S > 2$ to high statistical significance. On subtleties of normalization for various Bell-like inequalities, which can complicate direct comparisons between parameters like the Bell–CHSH parameter $S$ and related quantities, see Clauser (2017).

FIGURE 13.3 John Clauser working on the instrumentation with which he and Stuart Freedman conducted the first experimental test of Bell's inequality, in 1972. (Courtesy Lawrence Berkeley National Laboratory.)

results suggesting a strong *compatibility* with local-realist theories, equivalent to $S = 1.728 \pm 0.104$, easily consistent with the Bell–CHSH bound $S \leq 2$ and in strong disagreement with the prediction from quantum mechanics. They circulated a preprint of their results but never pursued formal publication, given their own lingering doubts about possible systematic errors. (See Freire (2006), p. 595.) Three years later, Clauser repeated their experiment, using the same cascade within excited mercury atoms to produce the entangled photons, and measured the equivalent of $S = 2.308 \pm 0.0744$, a violation of the Bell–CHSH inequality by more than 4 standard deviations. (Around the same time, Edward Fry and Randall Thompson independently performed a Bell test at Texas A & M University using entangled photons from excited mercury atoms, and, like Clauser, measured a strong violation of Bell's inequality: Fry and Thompson (1976), Clauser and Shimony (1978), and Fry and Walther (2002).) In the course of repeating the Holt–Pipkin experiment, Clauser found that the measured correlations depended sensitively upon stresses in the walls of the glass bulb containing the mercury vapour as well as in the lenses used to focus the emitted photons toward their respective detectors. Although Holt and Pipkin had reported observing similar stresses in the mercury bulb during their experiment, they had not attempted to repeat their measurements.[14] (Clauser (1976), Clauser and Shimony (1978), Freire (2006).)

Since Freedman and Clauser's original experiment (1972), virtually every published test has measured violations of Bell's inequality, consistent with the quantum-mechanical predictions (Clauser and Shimony, 1978; Brunner *et al.*, 2014). One might therefore ask: following the 1976 repetitions of the Holt–Pipkin experiment (Clauser, 1976; Fry and Thompson, 1976), alongside similar experiments completed during the mid-1970s (Clauser and Shimony, 1978; Freire, 2006), why have physicists continued to subject Bell's inequality to experimental test? The answer is that each of the early experiments was subject to multiple loopholes: explanations consistent with local realism that could (in principle) account for the experimental results.

## 13.3  THE LOCALITY LOOPHOLE

The first loophole that physicists identified for tests of Bell's inequality is often referred to as the 'locality loophole'. It concerns the flow of information during a given experimental run. In particular, could any communication among elements of the experiment—with information travelling at or below the speed of light—account for the strong correlation among measurement outcomes, even if the particles being subjected to test really did obey local realism?

Each experimental run in a test of the Bell–CHSH inequality involves (at least) five relevant events, whose space-time arrangement we may depict as in Fig. 13.4: experimenters must select the detector settings **a** and **b**, emit the entangled particles from the source $\sigma$, and perform a measurement on each particle, yielding outcomes $A$ and $B$. (As usual, we adopt coordinates such that light travels one unit of space in one unit of time, so that light-like trajectories follow $45°$ diagonals.) If the experimenters *first* select the detector settings and *later* emit particles from the source (as shown on the left side of Fig. 13.4), then a local-realist description could readily account for the observed correlations between $A$ and $B$. In such a case, the measurement outcome $A$ could depend on information about detector setting **b** and/or

---

[14] Holt and Pipkin actually measured a closely related quantity to the Bell–CHSH parameter $S$, which had been introduced in Freedman and Clauser (1972): $|R(\varphi) - R(3\varphi)|/R_0$, where $R(\varphi)$ is the number of coincidence count rates when the relative orientation of the polarizers at the two detectors is equal to $\theta_{ab} = \varphi$, and $R_0$ is the coincidence count rate when both polarizers are removed from their respective detectors. According to quantum mechanics, for ideal measurements the parameter $R(\varphi) = \frac{1}{4}[1 + cos(2\varphi)]$. For $\varphi = 22.5°$, the quantum-mechanical prediction is therefore $|R(\varphi) - R(3\varphi)|/R_0 = \sqrt{2}/4$, whereas local-realist models predict $|R(\varphi) - R(3\varphi)|/R_0 \leq 1/4$. Holt and Pipkin reported $|R(22.5°) - R(67.5°)|/R_0 = 0.216 \pm 0.013$ considerably below the local-realist threshold of 0.25, much less the quantum-mechanical prediction of 0.35 (Clauser and Shimony, 1978). When Clauser repeated their experiment (1976), he measured $|R(22.5°) - R(67.5°)|/R_0 = 0.2885 \pm 0.0093$, violating the local-realist bound by 4.1 standard deviations.

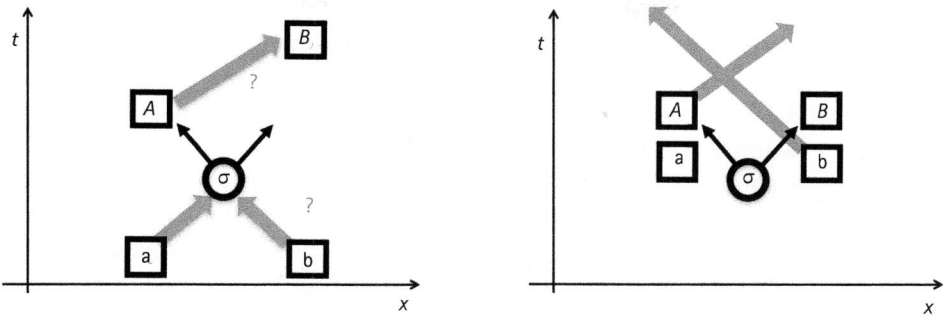

FIGURE 13.4 During a test of the Bell–CHSH inequality, experimenters must select the detector settings **a** and **b**; emit the pair of entangled particles from the source σ; and perform a measurement on each particle, yielding outcomes $A$ and $B$. In the space-time arrangement shown on the left, information travelling at or below the speed of light (thick grey lines) could account for the correlations among measurement outcomes even if the particles obeyed local realism. In the space-time arrangement shown on the right, the relevant events have been spacelike separated, so that such information exchange would not suffice to account for the measured correlations.

outcome $B$ could depend on setting **a**, such that the expression in the integrand of the top line of Eq. (2) would no longer factorize: $p(A, B|\mathbf{a},\mathbf{b},\lambda) \neq p(A|\mathbf{a}, \lambda)\,p(B|\mathbf{b},\lambda)$. Likewise, if the measurement on the left-moving particle were completed well before the measurement on the right-moving particle, then the detector on the right side could exploit information about the distant measurement to arrange a correlated outcome. Such explanations would be compatible with local realism. On the other hand, if special care were taken with the space-time arrangement of these five crucial events, as shown on the right side of Fig. 13.4, then the locality assumptions under which the bottom line of Eq. (2) had been derived would hold, and none of the scenarios depicted on the left side of Fig. 13.4 would be available for a local-realist account of the measured correlations.[15]

John Bell articulated a version of the locality loophole in his original article on Bell's inequality (1964). He closed his now-famous paper by writing that the quantum-mechanical predictions might apply 'only to experiments in which the settings of the instruments are made sufficiently in advance to allow them to reach some mutual rapport by exchange of signals with velocity less than or equal to that of light'. It would therefore be 'crucial', he concluded, to conduct experiments 'in which the settings are changed during the flight of the particles'.[16] John Clauser agreed. In his first letter to

---

[15] Several authors have considered *retrocausal* models, in which information flows backwards in time within the experiment. In such models, the retrocausal information flow yields a correlation between the detector settings and hidden variables, akin to the freedom-of-choice loophole discussed below. See Costa de Beauregard (1978), Argaman (2010), and Price and Wharton (2015).

[16] Bell (1964) cited an article by David Bohm and Yakir Aharonov (1957), in which they had argued that a proper test of the original Einstein–Podolsky–Rosen scenario should include the provision to change the detector settings while the particles were 'still in flight'. Bohm had made a similar observation in his discussion of EPR in his 1951 textbook: Bohm (1951), p. 622.

Bell (written in February 1969), Clauser noted that 'it might also be possible to "rotate" the polarizers by means of magneto-optic effects while the photons are in flight' (quoted in Freire (2006), p. 591). Achieving such fast switching among polarizer orientations, however, proved to be quite a technical challenge. As noted in section 13.2, when Freedman and Clauser conducted their experiment in 1972, they first manually set **a** and **b** for a given run by adjusting their polarizers to particular orientations before emitting the entangled photons. (Holt and Pipkin followed the same approach in their 1973 test, as did the other experiments conducted during the 1970s; see Clauser and Shimony (1978) and Freire (2006).) Bell returned to the point when summarizing discussions at a 1976 summer school devoted to the foundations of quantum mechanics, declaring that the experiments that had been conducted to date 'have nothing to do with Einstein locality', and that a test 'of the very highest interest' would be one in which 'the polarization analysers are in effect re-set while the photons are in flight' (quoted in Freire (2006), p. 606).

A participant in that 1976 summer school, Alain Aspect, was already hard at work on just such an experiment (Aspect, 1976). Together with colleagues Jean Dalibard and Gérard Roger at the Institut d'Optique Théorique et Appliquée in Orsay, near Paris, Aspect designed and built an experiment with fast-changing acoustico-optical switches inserted in the photons' paths from the source σ to the left and right detectors. Depending on which orientation the optical switch on the left side happened to be in when a photon arrived, that photon would be directed toward one of two polarizing filters, oriented at different angles; and likewise for the optical switch on the right side. The polarizer orientations $(\mathbf{a}, \mathbf{a}')$ on the left and $(\mathbf{b}, \mathbf{b}')$ on the right were fixed in advance, while the optical switches on the left and right sides changed every 10 ns. The detectors in Aspect's experiment were each about 6 metres from the source of entangled particles, which meant that photons emitted from the source would require at least 20 ns to travel to their respective detectors. Hence the optical switches on each side changed one or more times while each pair of photons was in flight. The particular detector settings $(\mathbf{a}, \mathbf{a}')$ and $(\mathbf{b}, \mathbf{b}')$ that each photon encountered therefore had not been fixed at the time of the photons' emission, and the measurements at each detector were completed such that no signal travelling at light speed could inform the distant detector about the settings or outcome at the local detector before each side had completed its measurement. Aspect and his colleagues thus performed the first Bell test, in 1982, in which the critical events were arranged as in the right side of Fig. 13.4. Even with this improved space-time arrangement, they measured a violation of the Bell–CHSH inequality by five standard deviations (Aspect, Dalibard, and Roger 1982). (See also Aspect (2002), Freire (2006), and Kaiser (2011), pp. 175–178.)

Aspect and his colleagues noted in their original article that their acoustico-optical switches did not determine the detector settings $(\mathbf{a}, \mathbf{a}')$ or $(\mathbf{b}, \mathbf{b}')$ in a truly random manner. Instead, the switches operated quasi-periodically, which suggested—at least in principle— that information would have been available at the photons' source σ, in advance of each emission event, that could have sufficed to predict the detector settings that each photon would ultimately encounter (Aspect, Dalibard, and Roger, 1982; Zeilinger, 1986).

In addition, the detectors on the left and right sides in Aspect's experiment were linked in real time via electronic coincidence circuits; results from each side were not recorded independently (Aspect, Dalibard, and Roger, 1982). Each of these points suggested that although Aspect's 1982 experiment clearly represented an enormous milestone in tests of Bell's inequality, the 'locality' loophole had not yet been closed conclusively.

More than fifteen years later, Anton Zeilinger and his group, at the time based at the University of Innsbruck in Austria, completed a new Bell test that addressed the locality loophole head on. Led by Gregor Weihs, the group set up two detector stations 400 metres apart, making it easier to determine the measurement outcomes $A$ and $B$ at spacelike-separated events. (For this experiment, the light travel time between detectors was $1.3\,\mu$ s rather than the 40 ns in Aspect's experiment in Orsay.) Each of the detector stations was equipped with its own quantum random number generator (QRNG), a device that could output a fresh bit (either a 0 or a 1) at a rate of 500 MHz.[17] The output from each random-number generator was linked to an electro-optical modulator, a device that could quickly rotate the basis in which a photon's polarization would be measured by an angle proportional to the applied voltage, changing bases at a frequency up to 30 MHz. Each detector station also had its own atomic clock with an accuracy of 0.5 ns, with which the time of each detection event at each detector could be recorded. Using this scheme, information about the detector settings $(\mathbf{a}, \mathbf{a}')$ and $(\mathbf{b}, \mathbf{b}')$ for a given run should not have been available at either the emission event $\sigma$ or at the distant measurement events that yielded $A$ or $B$, and no direct link connected the two detector stations, more conclusively achieving the space-time arrangement depicted on the right side of Fig. 13.4. The group measured $S = 2.73 \pm 0.02$, a violation of the Bell–CHSH inequality by more than 35 standard deviations (Weihs *et al.*, 1998). (See also Weihs (2002), Tittel *et al.*, (1998), and Aspect (1999).)

# 13.4 THE FAIR-SAMPLING LOOPHOLE

Near the conclusion of their article, Weihs, Zeilinger, and their colleagues noted that 'while our results confirm the quantum theoretical predictions, we admit that, however unlikely, local realistic or semiclassical interpretations [of their experimental results] are still possible', if one invoked a different loophole than locality: 'we would then have to assume that the sample of pairs registered is not a faithful representative of the whole ensemble emitted' (Weihs *et al.*, 1998). After all, in their experiment, they had successfully completed measurements on only about 5% of all the photon pairs that

---

[17] The quantum random number generator (QRNG) used in (Weihs *et al.*, 1998) produced a rapid bitstream of 0's and 1's by shining the output from a light-emitting diode onto a beam splitter. In principle, each photon encountering the beam splitter had a 50–50 chance to be transmitted or reflected. Each path (transmission and reflection) was monitored by a photomultiplier capable of detecting single photons. Depending on which detector recorded a photon within a very brief time interval ($\Delta t \simeq 2\,\text{ns}$), the device would output a 0 or a 1.

had been emitted from the source. Tests of the Bell–CHSH inequality require performing statistical averages over measurements on many pairs of particles. What if the
subset of particles that was successfully detected had been drawn from some biased
sample, skewing the statistical results? This second loophole has been dubbed the
'detector-efficiency loophole' or 'fair-sampling loophole'.

Physicists typically define the 'efficiency', $\eta$, of a given detector as the probability that
for any particle impinging upon the device, the detector will register a definite
measurement outcome. In any real experiment, the efficiency will be less than 100%,
with $0 < \eta < 1$. If one assumes that the detector efficiencies are identical for the two
detector stations in a test of the Bell–CHSH inequality, and that the detectors operate
independently of each other, then only in a fraction $\eta^2$ of experimental runs will each
detector successfully perform a measurement on its member of an entangled pair. In a
fraction $\eta(1 - \eta)$ of experimental runs, only the left detector will register a measurement of its particle while the right detector registers nothing, and in a separate fraction
$\eta(1 - \eta)$ of runs, only the right detector will complete a measurement. Finally, in a
fraction $(1 - \eta)^2$ of runs, neither detector will register a measurement. If one only
considers experimental runs in which at least one detector completes a measurement,
then the Bell–CHSH inequality of Eq. (4) is modified to read

$$S \leq \frac{4}{\eta} - 2 \,, \tag{7}$$

with the Bell–CHSH parameter $S$ defined in Eq. (3).[18] (See Clauser and Horne (1974),
Garg and Mermin (1987), Eberhard (1993), Weihs (2009).) In the limiting case of
detectors with perfect efficiency, $\eta \to 1$, Eq. (7) reverts to the original form of the Bell–
CHSH inequality, $S \leq 2$. On the other hand, if the efficiency of each detector is below a
critical threshold, $\eta \leq \eta_*$ with

$$\eta_* \equiv 2(\sqrt{2} - 1) = 0.828 \,, \tag{8}$$

---

[18] If one assumes perfect detector efficiencies, $\eta \to 1$, then the correlation function $E(\mathbf{a}, \mathbf{b})$ in Eq. (1)
may be evaluated as $E(\mathbf{a}, \mathbf{b}) = \sum_{A,B=\pm 1} (ABN_{ab}^{AB})/N_{ab}^{\text{tot}}$, where $N_{ab}^{\text{tot}}$ is the total number of entangled
pairs that are emitted when the detector settings are $(\mathbf{a},\mathbf{b})$, and $N_{ab}^{AB}$ is the number of double-coincidence
measurements in which the left and right detectors yield $\{A, B\} \in \{+1, -1\}$. However, if one takes into
account imperfect detector efficiencies, one may define $E'(\mathbf{a}, \mathbf{b}) = \sum_{A,B=\pm,0} (ABN_{ab}^{AB})/N_{ab}^{AB}$, in which
$A = 0 \ (B = 0)$ indicates the lack of a successful measurement at the left (right) detector. If one neglects
the (unobservable) runs in which neither detector completes a measurement, then one finds
$E'(\mathbf{a}, \mathbf{b}) = [N_{ab}^{\text{double}}/(N_{ab}^{\text{double}} + N_{ab}^{\text{single}})] \, E(\mathbf{a}, \mathbf{b}) = [\eta/(2 - \eta)]E(\mathbf{a}, \mathbf{b})$, where the number of double-
coincidence measurements is $N_{ab}^{\text{double}} = \eta^2 N_{ab}^{\text{tot}}$ and the number of single-sided measurements is
$N_{ab}^{\text{single}} = 2\eta(1 - \eta)N_{ab}^{\text{tot}}$. One may then construct $S' = |E'(\mathbf{a}, \mathbf{b}) + E'(\mathbf{a}', \mathbf{b}) - E'(\mathbf{a}, \mathbf{b}') + E'(\mathbf{a}', \mathbf{b}')|$, and,
using the same arguments as in Bell (1971), derive that $S' \leq 2$ for any local-realist model in which
$p(A, B|\mathbf{a}, \mathbf{b})$ takes the form of Eq. (2). Taking into account the different normalizations of $E'(\mathbf{a}, \mathbf{b})$ and
$E(\mathbf{a}, \mathbf{b})$, one then arrives at the updated inequality for the original Bell–CHSH parameter $S$, defined in
terms of $E(\mathbf{a}, \mathbf{b})$, as in Eq. (7).

then the inequality in Eq. (7) becomes $S \leq 2\sqrt{2}$, indistinguishable from the Tsirelson bound for quantum mechanics, $S_{QM}^{max}$ of Eq. (6). In other words, if the detector efficiencies are less than 82.8%, then a local-realist explanation could account for any experimental result that found $2 \leq S \leq 2\sqrt{2}$ simply by invoking the argument that the pairs of particles that happened to be detected during the experiment represented a biased (rather than 'fair') sample of all the pairs that had been emitted. Several physicists developed explicit local-realist models, with conditional probabilities $p(A, B|\mathbf{a}, \mathbf{b})$ satisfying Bell's form in Eq. (2), that could exploit non-detection events at either detector in order to mimic the predictions from quantum mechanics. (See Pearle (1970), Clauser and Horne (1974), Fine (1982), Marshall, Santons, and Selleri (1983), Garg and Mermin (1987), and Eberhard (1993).)

Although the fair-sampling loophole was identified as early as 1970 (Pearle, 1970)—and Clauser, Horne, and Shimony focused on it in various papers during the 1970s (Clauser and Horne, 1974; Clauser and Shimony, 1978)—addressing this loophole in a real experiment proved to be quite challenging, given technological limitations on available instrumentation. In fact, more than thirty years elapsed between the identification of the loophole and the earliest experiments to address it. The first groups to attempt Bell tests that closed the fair-sampling loophole used pairs of slow-moving, entangled ions in high-fidelity magnetic traps, rather than entangled photons. Since the traps kept the ions accessible for long periods of time, the teams achieved very high efficiencies, $\eta \simeq 0.98$, easily above the critical threshold $\eta_*$, and managed to measure violations of the Bell–CHSH inequality. A group at the U.S. National Institute of Standards and Technology (NIST) in Boulder, Colorado performed such a test on pairs of beryllium ions in 2001, finding $S = 2.25 \pm 0.03$ (Rowe *et al.*, 2001), and a separate group, based at the University of Maryland, measured $S = 2.22 \pm 0.07$ with pairs of ytterbium ions in 2008 (Matsukevich *et al.*, 2008). (See also Ansmann *et al.* (2009).)

A few years later, in 2013, two groups exploited advances in highly efficient single-photon detectors to conduct Bell tests with polarization-entangled photons that closed the fair-sampling loophole. By using cryogenically cooled transition-edge sensors (TES) operating at the superconducting transition, the teams achieved detector efficiencies $\eta \geq 0.90$. By conducting short-distance tests in which the entangled photons travelled to their respective detectors via carefully shielded optical fibres, the total losses between emission at the source and measurements at the detectors remained sufficiently low to enable tests of a Bell-type inequality while closing the fair-sampling loophole.[19]

---

[19] The photon experiments in (Christensen *et al.*, 2013; Giustina *et al.*, 2013) each measured violations of a Bell-type inequality first derived by Philippe Eberhard (1993), who demonstrated that the fair-sampling loophole could be closed with overall efficiencies *below* $\eta_*$ in Eq. (8) if one performed measurements on pairs of particles in the *non*-maximally entangled state $|\Psi\rangle = [1 + r^2]^{-1/2}(|H\rangle_A \otimes |V\rangle_B + r|V\rangle_A \otimes |H\rangle_B)$, with $r$ a real parameter within the range $0 < r < 1$. In addition, these experiments exploited recent advances to efficiently produce polarization-entangled photons via spontaneous parametric down conversion (Kwiat *et al.*, 1995): photons of a particular frequency are directed from a pump laser onto a special nonlinear crystal, which absorbs the incoming photons and emits pairs of photons that conserve overall energy, linear momentum, and angular momentum.

One group measured violations of the inequality by nearly 8 standard deviations (Christensen *et al.*, 2013), and the other group by more than 65 standard deviations (Giustina *et al.*, 2013). As an indication of the technical challenges posed by these experiments, note that each of them—those using trapped ions in the early and mid-2000s (Rowe *et al.*, 2001; Matsukevich *et al.*, 2008) and those using entangled photons in 2013 (Christensen *et al.*, 2013; Giustina *et al.*, 2013)—focused *only* on closing the fair-sampling loophole, and did not even attempt to address the locality loophole.

An enormous milestone was achieved late in 2015, when three groups performed Bell tests that closed both the locality and fair-sampling loopholes in the same experiments. The first group to accomplish this feat was directed by Ronald Hanson at the Delft University of Technology in the Netherlands. Hanson's group set up three stations across the university campus. Stations $A$ and $B$, which were separated by 1.3 km, each included a single electron spin degree of freedom, associated with a single nitrogen-vacancy (NV) defect centre in a diamond chip. Each spin was entangled with a single photon; the photons from stations $A$ and $B$ were then transmitted to a central station $C$ via optical fibres. Upon arrival at station $C$, the photons from stations $A$ and $B$ were subjected to a Bell-state measurement, which (as in entanglement-swapping protocols (Żukowski *et al.*, 1993)) projected the associated electron spins at stations $A$ and $B$ into a maximally entangled state. After the heralding photons left stations $A$ and $B$ but before they arrived at station $C$, quantum random number generators (QRNGs) at stations $A$ and $B$ selected bases in which the electron spins would be measured. As in Weihs's experiment, the selections of detector settings at stations $A$ and $B$ were spacelike-separated from the preparation of the entangled state, and the measurements of each spin at stations $A$ and $B$ were spacelike-separated from each other, thereby closing the locality loophole. Mean-while, by using an 'event-ready' protocol, in which the joint detection of photons at station $C$ indicated the successful preparation of an entangled state, the group could carefully monitor the total number of entangled states produced, and thereby verify that their series of spin measurements closed the fair-sampling loophole. For their first experiment, the group conducted 245 experimental runs and measured $S = 2.42 \pm 0.20$ (Hensen *et al.*, 2015). A few months later, the group collected data on an additional 300 runs. With the combined datasets, they measured $S = 2.38 \pm 0.14$, a violation of the Bell–CHSH inequality by 2.7 standard deviations (Hensen *et al.*, 2016).

Soon after Hanson's group completed its first experiment, two other groups conducted experiments that likewise closed both the locality and fair-sampling loopholes (Shalm *et al.*, 2015; Giustina *et al.*, 2015). Building directly upon the 2013 experiments with high-fidelity single-photon detectors (Christensen *et al.*, 2013; Giustina *et al.*, 2013), each of these groups performed Bell tests with polarization-entangled photons. To close the locality loophole, the groups needed to increase the distances between the source of entangled photons and the detector stations, so that the relevant space-time events could be arranged as in the right side of Fig. 13.4. For the experiment led by Krister Shalm at the National Institute of Standards and Technology (NIST) in Boulder, Colorado, the team set up a triangular arrangement. The entangled source sat at the vertex of a (nearly) right triangle; photons were transmitted to detector stations $A$ and $B$ via optical fibres. Stations $A$ and $B$ were each about 130 m from the entangled source, and 185 m from each

**FIGURE 13.5** Marissa Giustina (centre) describes her 2015 experiment to colleagues in the sub-basement of the Hofburg Palace in Vienna. The source of entangled photons is within the kiosk, visible in the middle of the hallway. Optical fibres (beneath the shielding in the centre of the floor) transmitted the photons to detector stations on opposite ends of the long hallway, each 29 metres from the source. (Photo by the author.)

other. The light travel time from the source to a detector station was thus about 0.43 μs. Detector settings at stations $A$ and $B$ were determined by QRNGs co-located at each detector, which produced fresh, random bits about every 5 ns, thereby changing multiple times while the entangled photons were in flight. The group measured clear violation of a Bell-type inequality; the probability that their experimental results could have arisen from a local-realist model, in which $p(A, B|\mathbf{a}, \mathbf{b})$ took the form of Eq. (2), was $p = 2.3 \times 10^{-7}$ (Shalm *et al.*, 2015). If one makes some simplifying assumptions and adopts Gaussian statistics, this $p$ value corresponds to a violation of the relevant inequality by more than 5 standard deviations.[20]

At the same time, Anton Zeilinger's group in Vienna completed its own Bell test using polarization-entangled photons while closing the locality and fair-sampling loopholes. Led by Marissa Giustina, the group adopted a collinear spatial arrangement for the various experimental stations, rather than the triangular set-up of the NIST group. Optical fibres transmitted entangled photons from the central source toward detector stations on opposite ends of a long, narrow hallway in a sub-basement of the fabled Hofburg Palace in central Vienna. See Fig. 13.5. (The project constituted a

---

[20] As I discuss in section 13.6, there has been substantial progress and innovation in the statistical analysis of recent Bell tests, as well as in experimental designs. In particular, the use of Gaussian statistics implies several simplifying assumptions, and hence many experimental groups now report their results in terms of $p$-values rather than (or in addition to) standard deviations.

major part of Giustina's doctoral dissertation. Many graduate students *feel* as if they are trapped in a castle dungeon while working on their theses...) Each detector station was 29 m from the source, yielding a light travel time of just 96.7 ns from source to detector station. Within that brief time period, QRNGs at each detector station implemented fresh detector settings within 26 ns windows, ensuring spacelike-separation of the relevant events, as in the right side of Fig. 13.4. Like the NIST group, the Vienna experiment measured a substantial violation of a Bell-like inequality. The probability that their measured correlations would arise in a local-realist theory of the form in Eq. (2) was just $p = 3.74 \times 10^{-31}$; if one again (naively) adopts Gaussian statistics, this result corresponds to a violation of the relevant Bell inequality by nearly 12 standard deviations (Giustina *et al.*, 2015). (See also Rosenfeld *et al.* (2017).)

## 13.5 THE FREEDOM-OF-CHOICE LOOPHOLE

In 1976 Shimony, Horne, and Clauser identified a third significant loophole in tests of Bell's inequality, which had eluded Bell himself (Shimony, Horne, and Clauser, 1976).[21] (A few years later, Richard Feynman independently articulated a version of this loophole: Feynman (1982), p. 485.) The third loophole is often denoted the 'measurement-dependence loophole', the 'settings-dependence loophole', or the 'freedom-of-choice loophole', and is conceptually distinct from the locality loophole. The locality loophole concerns the flow of information *during* a given experimental run, and relies upon direct communication between parts of the apparatus to account for the strong correlations; locality, in other words, concerns whether $p(A, B|\mathbf{a}, \mathbf{b}, \lambda)$ factorizes as $p(A|\mathbf{a},\lambda)\,p(B|\mathbf{b},\lambda)$. The freedom-of-choice loophole, on the other hand, concerns whether any *common cause* could have established statistical correlations between the parameters $\lambda$ that affect measurement outcomes and the selection of detector settings $(\mathbf{a}, \mathbf{b})$, such that $p(\lambda|\mathbf{a}, \mathbf{b}) \neq p(\lambda)$. Such statistical correlations could arise even in the absence of direct communication between parts of the experimental apparatus.

Shimony, Horne, and Clauser (1976) pointed out that in Bell's original derivation of his inequality, he had relied upon expressions of the form in Eq. (2) for the conditional probabilities $p(A, B|\mathbf{a}, \mathbf{b})$. Yet the law of total probability requires that one write such expressions as

---

[21] Shimony, Horne, and Clauser originally circulated their analysis (1976) in the informal newsletter *Epistemological Letters*, which served as a forum for discussions of Bell's inequality and other issues in the foundations of quantum mechanics throughout the 1970s, at a time when many physics journals, such as the *Physical Review*, downplayed the topic. Several years later, their exchange with Bell was republished in the philosophical journal *Dialectica* (Bell *et al.*, 1985). Bell later included his own contributions to the exchange (1976, 1977) as chapters in his well-known book, *Speakable and Unspeakable in Quantum Mechanics* (1990), where they appear as chapters 7 and 12. On the role of *Epistemological Letters*, see Freire (2006), pp. 602–603, and Kaiser (2011), p. 122.

$$p(A,B|\mathbf{a},\mathbf{b}) = \int d\lambda \, p(A, B|\mathbf{a},\mathbf{b},\lambda) \, p(\lambda|\mathbf{a},\mathbf{b}). \qquad (9)$$

(On the law of total probability, see, e.g., Blitzstein and Hwang (2014), section 2.3.) In general, when calculating $p(A, B|\mathbf{a}, \mathbf{b})$ one must take into account possible correlations between $\lambda$ and the detector settings $(\mathbf{a}, \mathbf{b})$, represented by the term $p(\lambda|\mathbf{a}, \mathbf{b})$, *regardless* of whether $p(A, B|\mathbf{a}, \mathbf{b}, \lambda)$ factorizes as $p(A|\mathbf{a},\lambda) \, p(B|\mathbf{b},\lambda)$. In his original derivation, however, Bell had tacitly neglected any possible correlation between $\lambda$ and $(\mathbf{a}, \mathbf{b})$, writing simply $p(\lambda)$ in place of $p(\lambda|\mathbf{a}, \mathbf{b})$. Via Bayes's theorem, that was equivalent to replacing $p(\mathbf{a}, \mathbf{b}|\lambda) \to p(\mathbf{a}, \mathbf{b})$, that is, to assuming (by fiat) that the selection of detector settings $(\mathbf{a}, \mathbf{b})$ was entirely independent of the parameters $\lambda$ that could affect the behaviour of the entangled particles. As Bell (1976) wrote in his exchange with Shimony, Horne, and Clauser, 'It has been assumed that the settings of instruments are in some sense free variables—say at the whim of the experimenters—or in any case not determined in the overlap of the backward light cones'.

A dozen years after Shimony, Horne, and Clauser identified the freedom-of-choice loophole, Carl Brans (1988) developed an explicit local-realist model that exploited nontrivial correlations $p(\lambda|\mathbf{a}, \mathbf{b}) \neq p(\lambda)$ in order to mimic the predictions from quantum mechanics for Bell tests. More recent theoretical work has clarified that the freedom-of-choice loophole offers by far the most efficient means by which local-realist models can produce correlations that exceed Bell's inequality. Only a *minuscule* amount of statistical correlation between the selection of detector settings $(\mathbf{a}, \mathbf{b})$ and the parameters $\lambda$ is required for local-realist models to mimic the correlations of a maximally entangled quantum state like $|\Psi^{(\pm)}\rangle$ in Eq. (5), for example—over twenty times less coordination (or 'mutual information') than required for local-realist models that exploit the locality loophole in order to reproduce the quantum-mechanical predictions.[22] (See Hall (2010, 2011, 2016), Barrett and Gisin (2011), Banik *et al.* (2012), and Friedman *et al.* (2019).) In other words, despite claims that have occasionally been made in the literature, local-realist models that exploit the freedom-of-choice loophole certainly do *not* require the strong assumption of 'superdeterminism', in which experimenters' every single action would be determined by initial conditions (set, for example, at the time of the Big Bang). Rather, the freedom-of-choice loophole merely requires that in a small fraction of experimental runs, the source of entangled particles could predict (better than chance) at least one of the detector settings, $\mathbf{a}$ or $\mathbf{b}$, that would be used on a given run (Friedman *et al.*, 2019).

---

[22] One may quantify the amount of correlation required for a local-realist model to mimic the quantum-mechanical predictions by exploiting the freedom-of-choice loophole in terms of the mutual information, $I = \sum_{\lambda,\mathbf{a},\mathbf{b}} p(\lambda|\mathbf{a}, \mathbf{b}) p(\mathbf{a}, \mathbf{b}) \log_2[p(\lambda|\mathbf{a}, \mathbf{b})/p(\lambda)]$. The most efficient local-realist models that can reproduce predictions for correlations in a maximally entangled two-particle state by exploiting the freedom-of-choice loophole require merely $I = 0.046 \simeq 1/22$ of a bit of mutual information (Friedman *et al.*, 2019). Local-realist models that exploit the locality loophole in order to mimic the quantum-mechanical predictions for Bell tests, on the other hand, require at least 1 full bit of mutual information (Hall, 2011).

The freedom-of-choice loophole thus appears to be quite robust, theoretically: the basic rules for calculating conditional probabilities require that the term $p(\lambda|\mathbf{a}, \mathbf{b})$ be included in expressions like Eq. (9), and the possibility that $p(\lambda|\mathbf{a}, \mathbf{b}) \neq p(\lambda)$ offers the most efficient mechanism by which a local-realist model could yield strong correlations among measurement outcomes. So how might one address the freedom-of-choice loophole experimentally? The remarkable experiments described in section 13.4, which managed to close both the locality and fair-sampling loopholes, relied upon quantum random number generators (QRNGs) to select detector settings $(\mathbf{a}, \mathbf{b})$ for each experimental run. According to quantum mechanics, the outputs of such devices should be intrinsically random, and hence unpredictable. But the purported intrinsic randomness of quantum mechanics is part of what is at *stake* in tests of Bell's inequality, so the use of QRNGs in Bell tests raises the spectre of circularity. Put another way: if the world were in fact governed by a local-realist theory rather than by quantum mechanics, then the behaviour of QRNGs should—in principle—be susceptible to a description of the form in Eq. (9), including the critical term $p(\lambda|\mathbf{a}, \mathbf{b})$. If that conditional probability incorporated even modest statistical correlations between $\lambda$ and the selection of various detector settings $(\mathbf{a}, \mathbf{b})$, then the outputs of the QRNGs might very well *appear* to be random—that is, they might pass the usual suite of randomness tests, such that knowledge of the previous $N$ bits would not suffice to predict bit $N + 1$ at greater than chance levels. Yet the outputs could nonetheless have been sufficiently correlated with the (unseen) parameters $\lambda$ to produce measurement outcomes that yield $S \to S_{QM}^{max} = 2\sqrt{2}$ in tests of the Bell–CHSH inequality. (See also Pironio (2015).)

Physicists have pursued two distinct approaches to address the freedom-of-choice loophole in recent experiments. One approach has been to crowd-source seemingly random bits in real time. During the course of a single day—30 November 2016—about 100,000 volunteers around the world, dubbed 'Bellsters', played a specially designed video game. Their task was to try to produce an unpredictable sequence of 0's and 1's; while they played, a sophisticated machine-learning algorithm analysed each Bellster's first few entries and tried to predict what the next one would be. With real-time feedback from the algorithm, players could improve their scores by making their next selections less predictable. The outputs from all those volunteers—which totalled nearly $10^8$ (quasi-)random bits—were directed via high-speed networks to twelve laboratories distributed across five continents: from Australia to Shanghai, Vienna to Barcelona, Buenos Aires to Boulder, Colorado. In each of those laboratories that day, the real-time Bellster bits determined which detector settings $(\mathbf{a}, \mathbf{b})$ would be used, run by run, in independent Bell tests. Every participating laboratory measured statistically significant violations of Bell's inequality: Abellán *et al.* (2018).

A complementary approach to addressing the freedom-of-choice loophole has been to isolate the events that determine detector settings $(\mathbf{a}, \mathbf{b})$ as much as possible, in space and time, from the rest of the experiment. In place of QRNGs or large numbers of Earthbound volunteers, such 'Cosmic Bell' experiments make use of astronomical random number generators (ARNGs) for selecting detector settings: devices that perform real-time astronomical measurements of light from distant objects, and rapidly convert some aspect of those measurements into a (quasi-)random bitstream (Gallicchio, Friedman, and Kaiser, 2014). For example, in the experiments reported in Handsteiner *et al.* (2017), Leung *et al.* (2018), and Rauch *et al.* (2018), the ARNGs used dichroic filters to rapidly distinguish light from astronomical sources that was more red or more blue than some reference wavelength.

The first Cosmic Bell test was performed in Vienna in April 2016, by a collaboration including Anton Zeilinger and his group together with astrophysicists at MIT, Harvey Mudd College, and NASA's Jet Propulsion Laboratory. Polarization-entangled photons were emitted from the roof of the Institute for Quantum Optics and Quantum Information (IQOQI). One detector station, on the top floor of the Austrian National Bank (about 0.5 km from IQOQI), included an ARNG that performed real-time measurements of light from a bright Milky Way star that was about 600 lightyears from Earth. The other detector station, in a university building (more than 1 km from IQOQI, in the opposite direction), included its own ARNG trained on a distinct Milky Way star about 1930 lightyears from Earth (Handsteiner *et al.*, 2017).

In keeping with the space-time arrangement shown on the right in Fig. 13.4, each ARNG implemented a fresh detector setting at its station every few microseconds, while the entangled photons were in flight, and the measurement outcomes $(A, B)$ were determined at spacelike-separated events. In addition, the causal alignment of the three experimental stations and the two astronomical sources was carefully analysed, to ensure that the causal wave front from the stellar emission event that was intended for the ARNG at the Austrian National Bank arrived at its intended location before any information about that astronomical photon could have arrived at either the source of entangled photons or the distant detector station (and vice versa). In this way, the experiment closed the locality loophole. In addition, by selecting the detector settings on each run based on events that had occurred hundreds of years ago, quadrillions of miles from Earth, the experiment pushed back to 600 years ago the most recent time by which any local-realist mechanism could have exploited the freedom-of-choice loophole to engineer the necessary correlations between detector settings and properties of the entangled photons. The experiment measured $S = 2.502 \pm 0.042$, violating the Bell–CHSH inequality by nearly 12 standard deviations (Handsteiner *et al.*, 2017).

In January 2018, the group performed a second Cosmic Bell test, this time using a pair of 4 m telescopes at the Roque de los Muchachos Observatory atop La Palma, in the Canary Islands. See Fig. 13.6. With the larger telescopes, the ARNGs could measure light from cosmologically distant sources: high-redshift quasars rather than Milky Way stars. The group performed measurements on pairs of polarization-entangled photons while detector settings for each experimental run were determined by emission events that had occurred 7.78 billion years ago for one detector station and 12.21 billion years ago for the other. (For reference, the Big Bang occurred 13.80 billion years ago: Aghanim *et al.* (2018).) As with the first Cosmic Bell experiment, causal alignment was carefully analysed to ensure that no information about a given cosmic emission event could have arrived at either the source of entangled particles or at the distant detector before that cosmic photon was measured by its intended ARNG. Fresh

**FIGURE 13.6** (*Top left*) Andrew Friedman, Jason Gallicchio, Anton Zeilinger, and David Kaiser discuss early plans for Cosmic Bell tests at MIT, October 2014. (From the author's collection.) (*Top right*) Johannes Handsteiner sets up the astronomical random number generator (ARNG) in the Austrian National Bank in Vienna for the first Cosmic Bell test, April 2016. (Courtesy Sören Wengerovsky.) (*Bottom left*) Anton Zeilinger (back to the camera) discusses observing options in the control room of the William Herschel Telescope on La Palma, January 2018. Others shown (from left to right) are Christopher Benn (leaning), Thomas Scheidl, Armin Hochrainer, and Dominik Rauch. (Photo by the author.) (*Bottom right*) Two of the large telescopes at the Roque de los Muchachos Observatory in La Palma; on the left is the Telescopio Nazionale Galileo, which the Cosmic Bell group used during its January 2018 experiment. (Courtesy Calvin Leung.)

detector settings were implemented within brief windows (of order $1\,\mu s$) while the entangled photons were in flight, again closing the locality loophole. The experiment measured $S = 2.646 \pm 0.070$, violating the Bell–CHSH inequality by more than 9 standard deviations. By deploying ARNGs focused on cosmologically distant quasars, the experiment pushed back to nearly 8 billion years ago the most recent time by which any local-realist influences could have exploited the freedom-of-choice loophole to engineer the observed violation of the Bell–CHSH inequality. Given the space-time arrangement of the particular quasars used for the experiment, the past light cones of each emission event, and the expansion history of the universe since the Big Bang, this second Cosmic Bell experiment excluded such local-realist, freedom-of-choice scenarios from 96.0% of the space-time volume of the past light cone of the experiment, extending from the Big Bang to the present time (Rauch *et al.*, 2018). (See also Kaiser (2020), chap. 4.)

The two Cosmic Bell experiments thus managed to close the locality loophole while constraining the freedom-of-choice loophole by dozens of orders of magnitude, compared to earlier, pioneering efforts to address this third loophole (Scheidl *et al.*, 2010; Aktas *et al.*, 2015). Because the Cosmic Bell experiments relied upon free-space transmission of the entangled photons rather than transmitting the photons via low-loss optical fibres, however, they were not able to close the fair-sampling loophole. In a stunning accomplishment, a separate group in Shanghai led by Jian-Wei Pan conducted a Bell test in which they distributed entangled photons via optical fibres across relatively short distances: about 90 m between the source and each detector station. The (vacuum) light travel time between source and each detector station was therefore about 300 ns, requiring a correspondingly faster rate of generating and implementing fresh detector settings for each experimental run. Pan's group used ARNGs focused on very bright, nearby stars, with which they could implement fresh detector settings within windows as brief as 250 ns.[23] The group measured violations of a Bell-like inequality with $p = 7.87 \times 10^{-4}$ (roughly equivalent to 3.4 standard deviations), while closing the locality and fair-sampling loopholes and constraining any local-realist, freedom-of-choice scenario to have been set in motion no more recently than 11.5 years prior to the experiment (Li *et al.*, 2018).

# 13.6 CONCLUSIONS

John Bell first formulated his now-famous inequality in 1964, and physicists have subjected Bell's inequality to experimental tests since 1972. Virtually every published test has measured violations of Bell's inequality, consistent with predictions from quantum mechanics. Over that period, physicists have identified several significant loopholes and devised clever, updated experimental designs, all with the goal of producing the most compelling evidence possible with which to evaluate the core question that had animated Bell's work: is the universe governed by a theory compatible with local realism, or not? Bolstered by a recent slew of experiments that have addressed various combinations of the major loopholes, the evidence against local realism is stronger than ever.[24]

---

[23] The brighter the astronomical object, the greater the flux of astronomical photons that an ARNG can collect per unit time, and hence the quicker an ARNG can output a bitstream of (quasi-)random numbers.

[24] See Hensen *et al.* (2015; 2016), Shalm *et al.* (2015), Giustina *et al.* (2015), Rosenfeld *et al.* (2017), Abellán *et al.* (2018), Handsteiner *et al.* (2017), Rauch *et al.* (2018), and Li *et al.* (2018).

Recent efforts to test Bell's inequality have featured advances in statistical analysis as well as experimental design. It was common in early experiments, for example, to adopt Gaussian statistics when analysing the statistical significance of a measured violation of (say) the Bell–CHSH inequality. Yet Gaussian statistics rely on several simplifying assumptions. First, and most obvious, the Gaussian distribution is an idealized form that holds in the limit of an infinite number of measurements, $N \rightarrow \infty$; when applied to any finite series of measurements, one must adopt an additional assumption about the convergence of the actual statistical distribution to the idealized form. Several recent Bell tests have involved large numbers of measurements on pairs of particles, with $N \sim 10^4 - 10^7$ (Christensen et al., 2013; Giustina et al., 2013; Shalm et al., 2015; Giustina et al., 2015; Handsteiner et al., 2017; Rauch et al., 2018; Li et al., 2018), which might plausibly justify the approximation $N \rightarrow \infty$, though other recent tests (especially those involving event-ready protocols) have included as few as $N \sim 10^2 - 10^3$ measurements (Matsukevich et al., 2008; Hensen et al., 2015; Hensen et al., 2016; Rosenfeld et al., 2017). In other words, successful tests of Bell-type inequalities do not always approximate the domain $N \rightarrow \infty$ for which Gaussian statistics might be appropriate.

More important, use of the Gaussian distribution is predicated on the assumption that relevant variables for each experimental run are independent and identically distributed (often abbreviated as 'i.i.d.'). This assumption is usually violated in real experiments. For example, whatever processes are used to select detector settings $(\mathbf{a}, \mathbf{b})$ on a given experimental run usually do not yield precisely equal numbers of trials with the various joint-settings pairs $(\mathbf{a}, \mathbf{b})$, $(\mathbf{a}', \mathbf{b})$, $(\mathbf{a}, \mathbf{b}')$, and $(\mathbf{a}', \mathbf{b}')$—hence a local-realist mechanism could exploit the 'excess predictability' that certain combinations of detector settings can be expected to arise more frequently than others (Kofler et al., 2016). Even more subtle, the i.i.d. assumption neglects what have come to be called 'memory' effects. Like a seasoned poker player who carefully tracks the cards that have been played and updates her strategy over the course of a game, a local-realist mechanism could make use of information about the previous detector settings and measurement outcomes and adjust its strategy over the course of an experiment. Exploiting such locally available information would remain compatible with local-realist conditional probabilities of the form in Eq. (2), let alone Eq. (9). Taking into account both excess predictabilities and 'memory' effects therefore requires more sophisticated calculations of the statistical significance with which measured correlations in a Bell test exceed what could be accounted for by a local-realist scenario.[25]

In recent years, several groups of physicists have found an additional motivation for performing tests of Bell's inequality, beyond the enduring question about local realism. Quantum entanglement is now at the core of new devices, including quantum computation and quantum encryption. Such real-world technologies will only function as expected if entanglement, as described by quantum mechanics, is a robust fact of nature rather than an illusion that arises from some local-realist underpinning. In

---

[25] See especially Kofler et al. (2016), Gill (2003, 2014), Bierhorst (2015), Elkouss and Wehner (2016), Handsteiner et al. (2017), and Rauch et al. (2018).

particular, many quantum encryption protocols rely upon embedded Bell tests to verify the security of a communication channel. If some local-realist mechanism could exploit loopholes like locality, fair sampling, and/or freedom of choice to produce the expected results in a Bell test, then (in principle) such mechanisms would also be available to eavesdroppers or hackers, intent on gaining access to unauthorized information.[26] Some of the most ambitious and audacious recent tests of Bell's inequality—including the breathtaking experiment by Jian-Wei Pan and his group, involving pairs of polarization-entangled photons emitted from the specially built *Micius* satellite, in low-Earth orbit, and measured at detector stations 1200 km apart from each other on Earth (Yin *et al.*, 2017)—have been key components in building and testing real-world quantum encryption infrastructure (Liao *et al.*, 2017; Liao *et al.*, 2018). In our new era of quantum information science, the stakes for tests of local realism have only grown, even beyond the deep questions that drove Bell, Clauser, Horne, Shimony, and their early colleagues to pursue tests of Bell's inequality.

# ACKNOWLEDGEMENTS

I am grateful to Olival Freire for inviting me to contribute to this *Handbook* and for many years of engaging discussion about the history of Bell's theorem, and to Guido Bacciagaluppi and Jason Gallicchio for helpful comments on an early draft of this chapter. I am also grateful to my colleagues from the 'Cosmic Bell' collaboration, in particular Andrew Friedman, Jason Gallicchio, Alan Guth, Johannes Handsteiner, Calvin Leung, Dominik Rauch, Thomas Scheidl, and Anton Zeilinger, for years of close collaboration and camaraderie. This chapter is dedicated to the memory of Andrew S. Friedman (1979–2020). This work was supported in part by the U.S. National Science Foundation INSPIRE grant No. PHY-1541160. Part of this work was conducted in MIT's Center for Theoretical Physics, and supported in part by the U.S. Department of Energy under Contract No. DE-SC0012567.

# REFERENCES

Abellán, C., *et al.* (2018). Challenging local realism with human choices. *Nature*, 557(7704), 212–16. doi:10.1038/s41586-25618-0085-3. arXiv: 1805.04431 [quant-ph].

Aghanim, N., *et al.* (2018). Planck 2018 results, VI: Cosmological parameters. arXiv: 1807.06209 [astro-ph.CO].

Aktas, D., Tanzilli, S., Martin, A., Pütz, G., Thew, R., and Gisin, N. (2015). Demonstration of quantum nonlocality in the presence of measurement dependence. *Phys. Rev. Lett.*, 114(22), 220404. doi:10.1103/PhysRevLett.114.220404. arXiv: 1504.08332 [quant-ph].

---

[26] See especially Ekert (1991), Gisin *et al.* (2002), Scarani *et al.* (2009), Alléaume *et al.* (2014), Zhang *et al.* (2018), Xu *et al.* (2019), and Pirandola *et al.* (2019).

Albert, D. Z. (1992). *Quantum Mechanics and Experience*. Cambridge, MA: Harvard University Press.

Alléaume, R., Branciard, C., Bouda, J., Debuisschert, T., Dianati, M., Gisin, N., Godfrey, M., et al. (2014). Using quantum key distribution for cryptographic purposes: A survey. *Theo. Computer Science*, **560**, 62–81. doi:10.1016/j.tcs.2014.09.018. arXiv: quant-ph/0701168.

Ansmann, M., Wang, H., Bialczak, R. C., Hofheinz, M., Lucero, E., Neeley, M., O'Connell, A. D., et al. (2009). Violation of Bell's inequality in Josephson phase qubits. *Nature*, **461**(7263), 504–506. doi:10.1038/nature08363.

Argaman, N. (2010). Bell's theorem and the causal arrow of time. *Am. J. Phys.*, **78**, 1007–13. doi:10.1119/1.3456564. arXiv: 0807.2041 [quant-ph].

Aspect, A. (1976). Proposed experiment to test the nonseparability of quantum mechanics. *Phys. Rev. D*, **14**, 1944–51. doi:10.1103/PhysRevD.14.1944.

Aspect, A. (1999). Bell's inequality test: More ideal than ever. *Nature*, **398**, 189–90. doi:10.1038/18296.

Aspect, A. (2002). Bell's theorem: The naive view of an experimentalist. In R. Bertlmann and A. Zeilinger (eds), *Quantum [Un]Speakables: From Bell to Quantum Information*, Berlin: Springer, pp. 119–53.

Aspect, A., Dalibard, J., and Roger, G. (1982). Experimental test of Bell's inequalities using time varying analyzers. *Phys. Rev. Lett.*, **49**, 1804–807. doi:10.1103/PhysRevLett.49.1804.

Ball, P. (2018). *Beyond Weird: Why Everything You Thought You Knew about Quantum Physics is Different*. Chicago: University of Chicago Press.

Banik, M., Gazi, M. D. R., Das, S., Rai, A., and Kunkri, S. (2012). Optimal free will on one side in reproducing the singlet correlation. *J. Phys. A*, **45**(20), 205301. doi:10.1088/1751-8113/45/20/205301. arXiv: 1204.3835 [quant-ph].

Barrett, J., and Gisin, N. (2011). How much measurement independence is needed to demonstrate nonlocality? *Phys. Rev. Lett.*, **106**, 100406. doi:10.1103/PhysRevLett.106.100406. arXiv: 1008.3612 [quant-ph].

Becker, A. (2018). *What is Real: The Unfinished Quest for the Meaning of Quantum Physics*. New York: Basic Books.

Bell, J. S. (1964). On the Einstein-Podolsky-Rosen paradox. *Physics Physique Fizika*, **1**, 195–200. doi:10.1103/PhysicsPhysiqueFizika.1.195.

Bell, J. S. (1971). Introduction to the hidden-variable question. In B. d'Espagnat (ed.), *Proceedings of the International School of Physics 'Enrico Fermi,' Foundations of Quantum Mechanics*, New York: Academic Press, pp. 171–81.

Bell, J. S. (1976). The theory of local beables. *Epistemological Letters*, **9**, 11–24.

Bell, J. S. (1977). Free variables and local causality. *Epistemological Letters*, **15**, 79–84.

Bell, J. S. (1990). *Speakable and Unspeakable in Quantum Mechanics*. New York: Cambridge University Press.

Bell, J. S., Shimony, A., Horne, M. A., and Clauser, J. F. (1985). An exchange on local beables. *Dialectica*, **39**(2), 85–110. doi:10.1111/j.1746-8361.1985.tb01249.x.

Bertlmann, R., and Zeilinger, A. (eds) (2017). *Quantum [Un]Speakables II: Half a Century of Bell's Theorem*. Berlin: Springer.

Bierhorst, P. (2015). A robust mathematical model for a loophole-free Clauser–Horne experiment. *J. Phys. A*, **48**(19), 195302. doi:10.1088/1751-8113/48/19/195302. arXiv: 1312.2999 [quant-ph].

Blitzstein, J. K., and Hwang, J. (2014). *Introduction to Probability*. New York: CRC Press.

Bohm, D. (1951). *Quantum Mechanics*. Englewood Cliffs, NJ: Prentice-Hall.

Bohm, D. (1952a). A suggested interpretation of the quantum theory in terms of hidden variables. 1. *Phys. Rev.*, **85**, 166–79. doi:10.1103/PhysRev.85.166.

Bohm, D. (1952b). A suggested interpretation of the quantum theory in terms of hidden variables. 2. *Phys. Rev.*, **85**, 180–93. doi:10.1103/PhysRev.85.180.

Bohm, D., and Aharonov, Y. (1957). Discussion of experimental proof for the paradox of Einstein, Rosen, and Podolsky. *Phys. Rev.*, **108**(4), 1070–76. doi:10.1103/PhysRev.108. 1070.

Bokulich, A., and Jaeger, G. (eds) (2010). *Quantum Information and Entanglement*. New York: Cambridge University Press.

Brans, C. H. (1988). Bell's theorem does not eliminate fully causal hidden variables. *Int. J. Theo. Phys.*, **27**(2), 219–26. doi:10.1007/BF00670750.

Brody, J. (2020). *Quantum Entanglement*. Cambridge, MA: MIT Press.

Bromberg, J. L. (2006). Device physics vis-à-vis fundamental physics in Cold War America: The case of quantum optics. *Isis*, **97**, 237–59. doi:10.1086/504733.

Brunner, N., Cavalcanti, D., Pironio, S., Scarani, V., and Wehner, S. (2014). Bell nonlocality. *Rev. Mod. Phys.*, **86**(2), 419–78. doi:10.1103/RevMod Phys.86.419. arXiv: 1303.2849 [quant-ph].

Bub, J. (1997). *Interpreting the Quantum World*. New York: Cambridge University Press.

Bub, J. (2016). *Bananaworld: Quantum Mechanics for Primates*. New York: Oxford University Press.

Bub, T., and Bub, J. (2018). *Totally Random: Why Nobody Understands Quantum Mechanics*. Princeton, NJ: Princeton University Press.

Christensen, B. G., McCusker, K. T., Altepeter, J. B., Calkins, B., Gerrits, T., Lita, A. E., Miller, A., *et al.* (2013). Detection-loophole-free test of quantum nonlocality, and applications. *Phys. Rev. Lett.*, **111**(13), 130406. doi:10.1103/PhysRevLett.111.130406. arXiv: 1306.5772 [quant-ph].

Cirel'Son, B. S. (1980). Quantum generalizations of Bell's inequality. *Lett. Math. Phys.*, **4**(2), 93–100. doi:10.1007/BF00417500.

Clauser, J. F. (1976). Experimental investigation of a polarization correlation anomaly. *Phys. Rev. Lett.*, **36**(21), 1223–26. doi:10.1103/PhysRevLett.36.1223.

Clauser, J. F. (2002). Early history of Bell's theorem. In R. Bertlmann and A. Zeilinger (eds), *Quantum [Un]Speakables: From Bell to Quantum Information*, Berlin: Springer, pp. 61–98.

Clauser, J. F. (2017). Bell's theorem, Bell inequalities, and the 'probability normalization loophole'. In R. Bertlmann and A. Zeilinger (eds), *Quantum [Un]Speakables II: Half a Century of Bell's Theorem*, Berlin: Springer, pp. 451–84.

Clauser, J. F., and Horne, M. A. (1974). Experimental consequences of objective local theories. *Phys. Rev. D*, **10**(2), 526–35. doi:10.1103/PhysRevD.10.526.

Clauser, J. F., Horne, M. A., Shimony, A., and Holt, R. A. (1969). Proposed experiment to test local hidden variable theories. *Phys. Rev. Lett.*, **23**, 880–84. doi:10.1103/PhysRevLett.23.880.

Clauser, J. F., and Shimony, A. (1978). Bell's theorem: Experimental tests and implications. *Rept. Prog. Phys.*, **41**, 1881–927. doi:10.1088/0034-4885/41/12/002.

Costa de Beauregard, O. (1978). S-matrix, Feynman zigzag and Einstein correlation. *Phys. Lett. A*, **67**(3), 171–74. doi:10.1016/0375-9601(78)90480-2.

Cramer, J. G. (2016). *The Quantum Handshake: Entanglement, Nonlocality, and Transactions*. Berlin: Springer.

Cushing, J. T., and McMullin, E. (eds) (1989). *Philosophical Consequences of Quantum Theory: Reflections on Bell's Theorem*. Notre Dame, IN: University of Notre Dame Press.

d'Espagnat, B. (1976). *Conceptual Foundations of Quantum Mechanics*, 2nd edn. Reading, MA: W. A. Benjamin.

Eberhard, P. H. (1993). Background level and counter efficiencies required for a loophole-free Einstein–Podolsky–Rosen experiment. *Phys. Rev. A*, **47**(2), R747–50. doi:10.1103/PhysRevA.47.R747.

Einstein, A. (1949). Autobiographical notes. In P. A. Schilpp (ed.),*Albert Einstein: Philosopher-Scientist*, LaSalle, IL: Open Court, pp. 3–94.

Einstein, A., Podolsky, B., and Rosen, N. (1935). Can quantum mechanical description of physical reality be considered complete? *Phys. Rev.*, **47**, 777–80. doi:10.1103/PhysRev.47.777.

Ekert, A. K. (1991). Quantum cryptography based on Bell's theorem. *Phys. Rev. Lett.*, **67**(6), 661–63. doi:10.1103/PhysRevLett.67.661.

Elkouss, D., and Wehner, S. (2016). (Nearly) optimal $p$ values for all Bell inequalities. *npj Quantum Information*, **2**, 16026. doi:10.1038/npjqi.2016.26. arXiv:1510.07233 [quant-ph].

Faraci, G., Gutkowski, D., Notarrigo, S., and Pennisi, A. R. (1974). An experimental test of the EPR paradox. *Lett. Nuovo Cimento*, **9**, 607–11. doi:10.1007/BF02763124.

Feynman, R. P. (1982). Simulating physics with computers. *Int. J. Theo. Phys.*, **21**(6–7), 467–88. doi:10.1007/BF02650179.

Fine, A. (1982). Some local models for correlation experiments. *Synthese*, **50**, 279–94. doi:10.1007/BF00416904.

Fine, A. (1986). *The Shaky Game: Einstein, Realism, and the Quantum Theory*. Chicago: University of Chicago Press.

Freedman, S. J., and Clauser, J. F. (1972). Experimental test of local hidden-variable theories. *Phys. Rev. Lett.*, **28**, 938–41. doi:10.1103/PhysRevLett.28.938.

Freire, O., Jr. (2006). Philosophy enters the optics laboratory: Bell's theorem and its first experimental tests (1965–1982). *Stud. Hist. Phil. Mod. Phys.*, **37**, 577–616. doi:10.1016/j.shpsb.2005.12.003.

Freire, O., Jr. (2015). *The Quantum Dissidents: Rebuilding the Foundations of Quantum Mechanics, 1950–1990*. Berlin: Springer.

Friedman, A. S., Guth, A. H., Hall, M. J. W., Kaiser, D. I., and Gallicchio, J. (2019). Relaxed Bell inequalities with arbitrary measurement dependence for each observer. *Phys. Rev. A*, **99**(1), 012121. doi:10.1103/PhysRevA.99.012121. arXiv: 1809.01307 [quant-ph].

Fry, E. S., and Thompson, R. C. (1976). Experimental test of local hidden-variable theories. *Phys. Rev. Lett.*, **37**(8), 465–68. doi:10.1103/PhysRevLett.37.465.

Fry, E. S., and Walther, T. (2002). Atom based tests of the Bell inequalities: The legacy of John Bell continues. In R. Bertlmann and A. Zeilinger (eds), *Quantum [Un]Speakables: From Bell to Quantum Information*, Berlin: Springer, pp. 103–18.

Gallicchio, J., Friedman, A. S., and Kaiser, D. I. (2014). Testing Bell's inequality with cosmic photons: Closing the setting-independence loophole. *Phys. Rev. Lett.*, **112**(11), 110405. doi:10.1103/PhysRevLett.112.110405. arXiv: 1310.3288 [quant-ph].

Garg, A., and Mermin, N. D. (1987). Detector inefficiencies in the Einstein-Podolsky-Rosen experiment. *Phys. Rev. D*, **35**(12), 3831–35. doi:10.1103/PhysRevD.35.3831.

Gilder, L. (2008). *The Age of Entanglement: When Quantum Physics was Reborn*. New York: Knopf.

Gill, R. D. (2003). Time, finite statistics, and Bell's fifth position. In *Proceedings of Foundations of Probability and Physics*, Växjö, Sweden: Växjö University Press, pp. 179–206. arXiv: quant-ph/0301059 [quant-ph].

Gill, R. D. (2014). Statistics, causality and Bell's theorem. *Statist. Sci.*, **29**, 512–28. doi:10.1214/14-STS490. arXiv: 1207.5103 [stat.AP].

Gisin, N., Ribordy, G., Tittel, W., and Zbinden, H. (2002). Quantum cryptography. *Rev. Mod. Phys.*, 74(145), 146–95. doi:10.1103/RevModPhys.74.145. arXiv: quant-ph/0101098.

Giustina, M. (2017). On loopholes and experiments. In R. Bertlmann and A. Zeilinger (eds), *Quantum [Un]Speakables II: Half a Century of Bell's Theorem*, Berlin: Springer, pp. 485–501.

Giustina, M., Mech, A., Ramelow, S., Wittmann, B., Kofler, J., Beyer, J., Lita, A., *et al.* (2013). Bell violation using entangled photons without the fair-sampling assumption. *Nature*, 497(7448), 227–30. doi:10.1038/nature12012. arXiv: 1212.0533 [quant-ph].

Giustina, M., Versteegh, M. A. M., Wengerowsky, S., Handsteiner, J., Hochrainer, A., Phelan, K., Steinlechner, F., *et al.* (2015). Significant-loophole-free test of Bell's theorem with entangled photons. *Phys. Rev. Lett.*, 115(25), 250401. doi:10.1103/PhysRevLett.115.250401. arXiv: 1511.03190 [quant-ph].

Greenberger, D. M., Horne, M. A., Shimony, A., and Zeilinger, A. (1990). Bell's theorem without inequalities. *Am. J. Phys.*, 58, 1131–43. doi:10.1119/1.16243.

Greenberger, D. M., Horne, M. A., and Zeilinger, A. (1989). Going beyond Bell's theorem. In M. Kafatos (ed.),*Bell's Theorem, Quantum Theory, and Conceptions of the Universe*, Dordrecht: Kluwer, pp. 69–72. arXiv: 0712.0921 [quant-ph].

Greenstein, G. (2019). *Quantum Strangeness: Wrestling with Bell's Theorem and the Ultimate Nature of Reality*. Cambridge, MA: MIT Press.

Gutfreund, H., and Renn, J. (2020). *Einstein on Einstein: Autobiographical and Scientific Reflections*. Princeton, NJ: Princeton University Press.

Hall, M. J. W. (2010). Local deterministic model of singlet state correlations based on relaxing measurement independence. *Phys. Rev. Lett.*, 105(25), 250404. doi:10.1103/PhysRevLett.105.250404. arXiv: 1007.5518 [quant-ph].

Hall, M. J. W. (2011). Relaxed Bell inequalities and Kochen–Specker theorems. *Phys. Rev. A*, 84(2), 022102. doi:10.1103/PhysRevA.84.022102. arXiv: 1102.4467 [quant-ph].

Hall, M. J. W. (2016). The significance of measurement independence for Bell inequalities and locality. In T. Asselmeyer-Maluga (ed.),*At the Frontier of Spacetime*, Berlin: Springer, pp. 189–204. doi:10.1007/978-3-319-31299-6_11. arXiv: 1511.00729 [quant-ph].

Handsteiner, J., Friedman, A. S., Rauch, D., Gallicchio, J., Liu, B., Hosp, H., Kofler, J., *et al.* (2017). Cosmic Bell test: Measurement settings from Milky Way stars. *Phys. Rev. Lett.*, 118(6), 060401. doi:10.1103/PhysRevLett.118.060401. arXiv: 1611.06985 [quant-ph].

Hensen, B., Bernien, H., Dréau, A. E., Reiserer, A., Kalb, N., Blok, M. S., Ruitenberg, J. *et al.* (2015). Loophole-free Bell inequality violation using electron spins separated by 1.3 kilometres. *Nature*, 526(7575), 682–86. doi:10.1038/nature15759. arXiv: 1508.05949 [quant-ph].

Hensen, B., Kalb, N., Blok, M. S., Dréau, A. E., Reiserer, A., Vermeulen, R. F. L., Schouten, R. N., *et al.* (2016). Loophole-free Bell test using electron spins in diamond: Second experiment and additional analysis. *Scientific Reports*, 6, 30289. doi:10.1038/srep30289. arXiv: 1603.05705 [quant-ph].

Howard, D. (1985). Einstein on locality and separability. *Stud. Hist. Phil. Sci.*, 16(3), 171–201. doi:10.1016/0039-3681(85)90001-9.

Kaiser, D. (2011). *How the Hippies Saved Physics: Science, Counterculture, and the Quantum Revival*. New York: W. W. Norton.

Kaiser, D. (2020). *Quantum Legacies: Dispatches from an Uncertain World*. Chicago: University of Chicago Press.

Kasday, L. R., Ullman, J. D., and Wu, C. S. (1975). Angular correlation of Compton-scattered annihilation photons and hidden variables. *Nuovo Cimento B*, 25(2), 633–61. doi:10.1007/BF02724742.

Kofler, J., Giustina, M., Larsson, J.-Å., and Mitchell, M. W. (2016). Requirements for a loophole-free photonic Bell test using imperfect setting generators. *Phys. Rev. A*, **93**, 032115. doi:10.1103/PhysRevA.93.032115. arXiv: 1411.4787 [quant-ph].

Kwiat, P. G., Mattle, K., Weinfurter, H., Zeilinger, A., Sergienko, A. V., and Shih, Y. (1995). New high-intensity source of polarization-entangled photon pairs. *Phys. Rev. Lett.*, 75(24), 4337–41. doi:10.1103/PhysRevLett.75.4337. https://link.aps.org/doi/10.1103/PhysRevLett.75.4337.

Lamehi-Rachti, M., and Mittig, W. (1976). Quantum mechanics and hidden variables: A test of Bell's inequality by the measurement of the spin correlation in low-energy proton-proton scattering. *Phys. Rev. D*, 14(10), 2543–55. doi:10.1103/PhysRevD.14.2543.

Larsson, J.-Å. (2014). Loopholes in Bell inequality tests of local realism. *J. Phys. A*, 47(42), 424003. doi:10.1088/1751-8113/47/42/424003. arXiv: 1407.0363 [quant-ph].

Leung, C., Brown, A., Nguyen, H., Friedman, A. S., Kaiser, D. I., and Gallicchio, J. (2018). Astronomical random numbers for quantum foundations experiments. *Phys. Rev. A*, 97(4), 042120. doi:10.1103/PhysRevA.97.042120. arXiv: 1706.02276 [quant-ph].

Li, M.-H., Wu, C., Zhang, Y., Liu, W.-Z., Bai, B., Liu, Y., Zhang, W., *et al.* (2018). Test of local realism into the past without detection and locality loopholes. *Phys. Rev. Lett.*, 121(8), 080404. doi:10.1103/PhysRevLett.121.080404. arXiv: 1808.07653 [quant-ph].

Liao, S.-K., Cai, W.-Q., Liu, W.-Y., Zhang, L., Li, Y., Ren, J.-G., Yin, J., *et al.* (2017). Satellite-to-ground quantum key distribution. *Nature*, **549**, 43–47. doi:10.1038/nature23655.

Liao, S.-K., Cai, W.-Q., Handsteiner, J., Liu, B., Yin, J., Zhang, L., Rauch, D., *et al.* (2018). Satellite-relayed intercontinental quantum network. *Phys. Rev. Lett.*, **120**, 030501. doi:10.1103/PhysRevLett.120.030501. arXiv: 1801.04418 [quant-ph].

Marshall, T. W., Santos, E., and Selleri, F. (1983). Local realism has not been refuted by atomic cascade experiments. *Phys. Lett. A*, 98(1–2), 5–9. doi:10.1016/0375-9601(8 3)90531-5.

Matsukevich, D. N., Maunz, P., Moehring, D. L., Olmschenk, S., and Monroe, C. (2008). Bell inequality violation with two remote atomic qubits. *Phys. Rev. Lett.*, **100**(15), 150404. doi:10.1103/PhysRevLett.100.150404. https://link.aps.org/doi/10.1103/PhysRevLett.100.150404.

Maudlin, T. (2011). *Quantum Non-Locality and Relativity*, 3rd edn. Oxford: Blackwell.

Musser, G. (2015). *Spooky Action at a Distance*. New York: Macmillan.

Myrvold, W., Genovese, M., and Shimony, A. (2019). Bell's theorem. In E. N. Zalta (ed.), *The Stanford Encyclopedia of Philosophy*, Metaphysics Research Lab, Stanford University. https://plato.stanford.edu/archives/spr2019/entries/bell-theorem/.

Orzel, C. (2009). *How to Teach Quantum Physics to Your Dog*. New York: Scribner.

Pearle, P. M. (1970). Hidden-variable example based upon data rejection. *Phys. Rev. D*, 2(8), 1418–25. doi:10.1103/PhysRevD.2.1418.

Pirandola, S., Andersen, U. L., Banchi, L., Berta, M., Bunandar, D., Colbeck, R., Englund, D., *et al.* (2019). Advances in quantum cryptography. arXiv: 1906.01645 [quant-ph].

Pironio, S. (2015). Random 'choices' and the locality loophole. arXiv: 1510.00248 [quant-ph].

Price, H., and Wharton, K. (2015). Disentangling the quantum world. *Entropy*, 17(11), 7752–67. doi:10.3390/e17117752. arXiv: 1508.01140 [quant-ph].

Rauch, D., Handsteiner, J., Hochrainer, A., Gallicchio, J., Friedman, A. S., Leung, C., Liu, B., *et al.* (2018). Cosmic Bell test using random measurement settings from high-redshift quasars. *Phys. Rev. Lett.*, 121(8), 080403. doi:10.1103/PhysRevLett.121.080403. arXiv: 1808.05966 [quant-ph].

Redhead, M. (1987). *Incompleteness, Nonlocality, and Realism: A Prolegomenon to the Philosophy of Quantum Mechanics*. New York: Oxford University Press.

Rosenfeld, W., Burchardt, D., Garthoff, R., Redeker, K., Ortegel, N., Rau, M., and Weinfurter, H. (2017). Event-ready Bell test using entangled atoms simultaneously closing detection and locality loopholes. *Phys. Rev. Lett.*, **119**(1), 010402. doi:10.1103/PhysRevLett.119.010402. arXiv: 1611.04604 [quant-ph].

Rowe, M. A., Kielpinski, D., Meyer, V., Sackett, C. A., Itano, W. M., Monroe, C., and Wineland, D. J. (2001). Experimental violation of a Bell's inequality with efficient detection. *Nature*, 409(6822), 791–94. doi:10.1038/35057215.

Sakurai, J. J. (1994). *Modern Quantum Mechanics*. Reading, MA: Addison-Wesley.

Scarani, V., Bechmann-Pasquinucci, H., Cerf, N. J., Dušek, M., Lütkenhaus, N., and Peev, M. (2009). The security of practical quantum key distribution. *Rev. Mod. Phys.*, **81**, 1301–50. doi:10.1103/RevModPhys.81.1301. arXiv: 0802 4155 [quant-ph].

Scheidl, T., Ursin, R., Kofler, J., Ramelow, S., Ma, X.-S., Herbst, T., Ratschbacher, L., *et al.* (2010). Violation of local realism with freedom of choice. *Proc. Nat. Acad. Sci. (USA)*, 107(46), 19708–13. doi:10.1073/pnas.1002780107. arXiv: 0811.3129 [quant-ph].

Shalm, L. K., Meyer-Scott, E., Christensen, B. G., Bierhorst, P., Wayne, M. A., Stevens, M. J., Gerrits, T., *et al.* (2015). Strong loophole-free test of local realism. *Phys. Rev. Lett.*, **115**(25), 250402. doi:10.1103/PhysRevLett.115.250402. arXiv: 1511.03189 [quant-ph].

Shimony, A., Horne, M. A., and Clauser, J. F. (1976). Comment on 'The theory of local beables'. *Epistemological Letters*, **13**, 1–9.

Stapp, H. P. (1976). *Correlation experiments and the nonvalidity of ordinary ideas about the physical world*. Lawrence Berkeley Laboratory report LBL-5333.

Tittel, W., Brendel, J., Zbinden, H., and Gisin, N. (1998). Violation of Bell inequalities by photons more than 10 km apart. *Phys. Rev. Lett.*, **81**(17), 3563–66. doi:10.1103/PhysRevLett.81.3563. arXiv: quant-ph/9806043 [quant-ph]

Weihs, G. (2002). Bell's theorem for space-like separation. In R. Bertlmann and A. Zeilinger (eds), *Quantum [Un]Speakables: From Bell to Quantum Information*, Berlin: Springer, pp. 155–62.

Weihs, G. (2009). Loopholes in experiments. In F. Weinert, K. Hentschel, and D. Greenberger (eds), *Compendium of Quantum Physics: Concepts, Experiments, History, and Philosophy*, Berlin: Springer, pp. 348–55.

Weihs, G., Jennewein, T., Simon, C., Weinfurter, H., and Zeilinger, A. (1998). Violation of Bell's inequality under strict Einstein locality conditions. *Phys. Rev. Lett.*, **81**(23), 5039–43. doi:10.1103/PhysRevLett.81.5039. arXiv: quantph/9810080 [quant-ph].

Whitaker, A. (2016). *John Stewart Bell and Twentieth-Century Physics: Vision and Integrity*. Oxford: Oxford University Press.

Xu, F., Ma, X., Zhang, Q., Lo, H.-K., and Pan, J.-W. (2019). Quantum cryptography with realistic devices. arXiv: 1903.09051 [quant-ph].

Yin, J., Cao, Y., Li, Y.-H., Liao, S.-K., Zhang, L., Ren, J.-G., Cai, W.-Q., *et al.* (2017). Satellite-based entanglement distribution over 1200 kilometers. *Science*, 356(6343), 1140–44. doi:10.1126/science.aan3211. arXiv: 1707.01339 [quant-ph].

Zeilinger, A. (1986). Testing Bell's inequalities with periodic switching. *Phys. Lett. A*, **118**, 1–2. doi:10.1016/0375-9601(86)90520-7.

Zeilinger, A. (1999). Experiment and the foundations of quantum physics. *Rev. Mod. Phys.*, **71**, S288–97. doi:10.1103/RevModPhys.71.S288.

Zeilinger, A. (2010). *Dance of the Photons: From Einstein to Quantum Teleportation*. New York: Farrar, Straus & Giroux.

Zhang, Q., Xu, F., Chen, Y.-A., Peng, C.-Z., and Pan, J.-W. (2018). Large scale quantum key distribution: Challenges and solutions. *Optics Express*, **26**, 24260. doi:10.1364/ OE.26.024260. arXiv: 1809.02291 [quant-ph].

Żukowski, M., Zeilinger, A., Horne, M. A., and Ekert, A. K. (1993). 'Event-ready-detectors' Bell experiment via entanglement swapping. *Phys. Rev. Lett.*, **71**(26), 4287–90. doi:10.1103/ PhysRevLett.71.4287.

# CHAPTER 14

# THE MEASURING PROCESS IN QUANTUM FIELD THEORY

## THIAGO HARTZ

## 14.1 INTRODUCTION

THE natural phenomena analysed by quantum theory imposed complex experimental situations in which the ways of thinking inherited from 19th century physics no longer seemed to work. Fundamental ideas such as causality and determinism were challenged. The weirdness of the quantum world was encapsulated in ideas such as non-locality, entanglement, complementarity, quantum jumps, wave-function collapse, the exclusion principle, the uncertainty principle, and the role of the observer in quantum measurement.

These ideas appeared in the non-relativistic domain. However, when quantum theory was applied in the relativistic domain, new oddities emerged, such as antimatter, vacuum polarization, virtual particles, spontaneous symmetry breaking, and violation of parity. It is natural to ask whether relativistic quantum phenomena can be understood in the same framework as the non-relativistic ones.

In a practical sense, quantum field theory (QFT) may be defined as the current physical theory that accounts for the behaviour of relativistic quantum systems. In a more formal sense, it is defined as the physical theory that provides the quantization of classical systems with infinitely many degrees of freedom, such as a vibrating string, the electromagnetic field, or the gravitational field. Although the former definition is more intuitive, I will adopt the latter here as it subsumes the former and is more adequate for historical and conceptual purposes.

Does QFT, thus defined, bring new interpretational issues on stage? James Cushing argued that 'as a source to illustrate how theories are constructed and established ( . . . ), quantum field theory does provide a wide range of useful examples', but that 'at the foundational level there is essentially nothing new in quantum field theory that is not already present in non-relativistic quantum mechanics' (Cushing, 1988, p. 25). According to him, the interpretational issues of QFT already existed in ordinary quantum mechanics. Michael Redhead opposed this view, arguing that QFT brings at least eight new philosophical problems (Redhead, 1988). His concerns are more of a metaphysical nature. He stated, quoting Howard Stein, that 'the quantum theory of fields is the contemporary locus of metaphysical research' (Redhead, 1982, p. 57).

In this chapter I will discuss some interpretational issues of QFT from a historical perspective. Among quantum field theories, I will pay particular attention to the quantum theory that describes the interaction between matter and the electromagnetic field, namely, quantum electrodynamics. Physicists often praise it as 'the most stringently tested—and the most dramatically successful—of all physical theories' (Peskin and Schroeder, 2007, p. 198), due to the outstanding agreement between experimental measurements and theoretical predictions. For instance, for the anomalous magnetic moment $g$ of the electron:[1]

$$\left(\frac{g-2}{2}\right)_{\text{exp.}} = 0.00115965218073(28)$$

$$\left(\frac{g-2}{2}\right)_{\text{theo.}} = 0.001159652181643(764)$$

(1)

The theory, nevertheless, was not so successful in its early days. The first attempts at developing quantum electrodynamics, in the period 1927–1946, faced enormous difficulties. In particular, almost every calculation led to divergent integrals. Physicists began to ask themselves whether they were dealing with the correct theory. Some of them expected a new scientific revolution. In the late 1940s, however, the problem was circumvented using the already existing formalism with the aid of a technique called renormalization. Strong suspicion about the renormalization programme persisted at least until the 1970s, when the conceptual understanding of effective field theories and of the renormalization group finally placed the theory on a safer ground.

Since many physicists believed for almost fifty years that they had the wrong theory, they were forced to scrutinize the formalism, in order to find where the mistake lay. This led to a number of articles about the foundations of QFT, particularly in the critical periods 1930–1933 and 1952–1967. Nevertheless, the debates over foundations never achieved in QFT the same prominence as they had in the history of non-relativistic quantum mechanics.

The goal of this chapter is to provide a historical analysis of the interpretational issues encountered by quantum field theorists mainly in the period 1925–1960. By interpretational issue in quantum theory I mean a situation in which a classical picture

---

[1] In order to arrive at this agreement, the theoretical prediction must take into account not only the electromagnetic interaction between the electron and the electromagnetic field, but also corrections from the electroweak theory.

does not provide the proper understanding of phenomena. The issues may be related to the description of phenomena, to the role of mathematics, and/or to the predictive power of theory. I will not dwell on ontological issues. Instead, I will pay special attention to the measuring process in QFT, and discuss how it is related to quantum field uncertainties, to vacuum energy, and to the divergent integrals.

I will rely on both primary and secondary literature. Max Jammer, Olivier Darrigol, Olival Freire Jr., Anthony Duncan, and Michel Janssen, among others, have provided comprehensive—technical, conceptual, and contextual—accounts of the history of non-relativistic quantum mechanics and its interpretation. No similar account exists for the history of QFT. Two brilliant first attempts in this direction were made almost forty years ago by Darrigol (1982) and Schweber (1984, 1994). There are also notable contributions by Wentzel (1960), Pais (1986), Miller (1994, pp. 1–118), and Cao (1997, pp. 123–267). Other studies about the history of QFT discuss specific authors—often within biographies—or ideas. For instance, there are analyses of the works of Pascual Jordan, Paul Dirac, Léon Rosenfeld, Richard Feynman, Julian Schwinger, and Hans Bethe. There are also some studies about the history of quantum field measurement, vacuum polarization, divergent calculations, S-matrix, quantum gravity, and several other aspects of QFT. Despite these works, we know far less about the history of QFT than we know about the history of non-relativistic quantum mechanics.

From the 1982 meeting of the Philosophy of Science Association onwards, philosophers of physics began to pay attention to QFT. A large number of articles, PhD theses, and books have been written about the philosophical (mostly ontological) aspects of the theory. Since the measuring process has no central role in most of these works,[2] I will not discuss them here.

This chapter is structured as follows. In order to understand the debates on the interpretation of QFT, a big picture of the development of that theory is required. Accordingly, section 14.2 presents a history of some episodes of QFT. I chose episodes that are related to discussions of the following sections without intending to be comprehensive in my historical account. Section 14.3 explains the debate over quantum field measurement and how this is connected to some interpretative problems of QFT. Finally, section 14.4 discusses how pragmatism and mathematics helped to shape physicists' view of QFT.

# 14.2  SOME EPISODES FROM THE HISTORY OF QUANTUM FIELD THEORY

From the early days of quantum theory, physicists including Max Planck, Paul Ehrenfest, Albert Einstein, Peter Debye, and Niels Bohr were concerned about the

[2] A notable exception is Bokulich (2003).

emission and absorption of electromagnetic radiation by matter. Blackbody radiation, the photoelectric effect, wave–particle duality, Einstein's A and B coefficients, the Compton effect, and the quantum atom exhibited new features of light. The formula $\Delta E = h\nu$ correctly accounted for the emission and absorption of radiation by atoms and molecules, but the existence of spontaneous emission, which was related to Einstein's A coefficient, was yet to be explained. It was not clear at the time whether the free electromagnetic field had, as Einstein alleged, a quantum nature.

In the 1910s, as physicists struggled to model atoms with several electrons and gathered a large amount of empirical data related to specific heats at low temperatures, the emphasis shifted from the emission and absorption of radiation to the internal motion of atoms and molecules. Which were the allowed orbital motions of electrons in atoms? How do molecules vibrate and rotate? Until the mid-1920s, physicists answered these questions using the old quantum theory, a set of ill-defined rules, methods, and principles. In 1925–1927, a systematic approach to atomic and molecular motion finally appeared, namely, quantum mechanics.

Looking back at that time from our current perspective, it may sound reasonable, and even natural, that after dealing with atomic motion and structure, physicists would turn to electromagnetic radiation with the same systematic approach, developing a quantum theory of the electromagnetic field. However, that was far from clear in the 1920s. In 1926, Bohr, Werner Heisenberg, Erwin Schrödinger, and others believed that the electromagnetic field should remain classical (Darrigol, 1984, pp. 433–435; Duncan and Janssen, 2008, pp. 640–643). Pascual Jordan and Paul Dirac dissented.

## 14.2.1 The First Quantizations of the Electromagnetic Field

As early as 1925, Jordan suggested that the components of the electromagnetic field should be matrices, since the position and the linear momentum of the electric charges were, according to the new quantum mechanics, also matrices (Born and Jordan, 1925, pp. 883–888). It was then common knowledge that the motion of a classical string fixed at both ends is the superposition of normal modes, which behave like harmonic oscillators. In the famous *Dreimännerarbeit* (Born, Heisenberg, and Jordan, 1926), Jordan quantized the motion of such a string and obtained the total energy:

$$\mathcal{E} = \sum_{i=1}^{\infty} \left( n_i + \frac{1}{2} \right) h\nu_i, \tag{2}$$

where $\nu_i$ is the frequency of the $i$-th normal mode and $n_i$ is the number of quanta with frequency $\nu_i$. The sum $\sum_{i=1}^{\infty} n_i h\nu_i$ Jordan called 'thermal energy', while the sum $\sum_{i=1}^{\infty} h\nu_i/2$ was the so-called 'zero-point energy', a residual energy that exists even when all $n_i$ are zero, i.e., in a vacuum. That residual energy is constant and infinite. Since energy is defined up to a constant, the zero-point energy could be discarded.

That was the first divergent calculation of QFT, a recurrent pattern that would soon plague the entire theory. Jordan also calculated the mean squared energy fluctuation, and thus retrieved Einstein's famous 1909 formula. This formula encapsulates wave–particle duality for it consists of two terms, the first corresponding to a particle behaviour of light and the second to a wave behaviour (Pais, 1982, p. 403; Duncan and Janssen, 2019, p. 124).

Soon after, Paul Dirac (1927) invented a new method, called second quantization, that enabled him to develop a quantum theory of a photon assembly in interaction with an atom. In parallel, Dirac quantized the electromagnetic field after decomposing it in its normal modes. Jordan and Pauli (1928) provided a relativistically covariant quantization of the free electromagnetic field. Jordan and Wigner (1928) showed how to represent an assembly of fermions with a quantized matter field. Based on these developments Heisenberg and Pauli (1929) put forward a much improved and more general mathematical formalism for QFT. Through a gauge fixing strategy, they developed a Hamiltonian approach to quantum electrodynamics. The equivalence of the different formulations of quantum electrodynamics was established in the early 1930s.[3]

## 14.2.2 Attempts at Eliminating the Infinities of Quantum Field Theory

As soon as physicists began to apply quantum electrodynamics to specific problems, it became clear that the theory was somehow incorrect because most calculations led to infinite results. The zero-point energy of the electromagnetic field was just the first and the least troublesome of a series of divergencies.[4] In general, the interaction between quantized fields yielded, beyond the first order of perturbation, a number of divergent integrals.

The physicists' attitudes towards the problem can be sorted into two groups. On one side, a group of physicists wanted to change the classical theory before quantization. As early as 1929, Jordan claimed that the infinities of QFT were inherited from the classical theory of the electron (Lehner, 2011, p. 283). The classical theory should therefore be adjusted before being quantized. Born and Leopold Infeld's 1934 non-linear electrodynamics, Arthur March's 1936 discrete space-time, Dirac's 1938 covariant electrodynamics, and John Wheeler and Richard Feynman's 1941 field-less electrodynamics were attempts in that direction (Schweber, 1994, p. xxv; Kragh, 1995; Darrigol, 2019). On the other side, another group wanted to change quantum theory. They expected a new quantum revolution, similar to the one that had happened in

---

[3] An analysis of the works of Jordan, Dirac, Pauli, and Heisenberg can be found in Darrigol (1982, pp. 40–124), Schweber (1994, pp. 1–75), and Enz (2002, pp. 175–93).

[4] A careful discussion of those divergences can be found in Darrigol (1982, pp. 125–48), Pais (1986, pp. 360–96), and Rueger (1992).

1925–1927 (Kojevnikov, 2004, p. 86). Two members of that group were Lev Landau and Rudolf Peierls who I will discuss in section 14.3. Wheeler's 1937 and Heisenberg's 1943 S-matrix theories were also representatives of that second group (Blum, 2017).

The difficulties with quantum electrodynamics did not prevent it from serving as a model for other QFTs in the 1930s, which presented, as expected, the same trouble with divergent quantities. So when Hideki Yukawa and Enrico Fermi proposed a theory of short-range nuclear forces and an explanation for β-decay respectively, they found their inspiration in quantum electrodynamics, and argued based on analogy (Darrigol, 1988).

## 14.2.3  The Physical Meaning of the Zero-Point Energy

The pragmatic strategy was to simply discard the divergent quantities, subtracting them from the calculations as unphysical constants. During the 1930s, physicists developed several methods of this so-called 'subtraction physics', albeit not very successfully. The zero-point energy was a paradigmatic example of this.

Walter Nernst, who had envisioned the existence of the zero-point energy of the electromagnetic field as early as 1916, had already doubted the existence of any physical effect related to that energy: 'only with the use of mirrors that are efficient also at very short-wavelength radiation would the zero-point radiation manifest itself' (quoted from Enz, 2002, p. 182). Echoing that attitude, Heisenberg and Pauli (1929, p. 154) observed that 'no physical reality is associated to that "zero-point energy" $h\nu/2$ per degree of freedom (...) because that "zero-point radiation" cannot be absorbed nor scattered nor reflected' (translation from Enz, 2002, p. 181). Pauli repeatedly expressed the same belief:

> [It] is more consistent not to introduce here a zero-point energy of $[(1/2)h\nu]$ per degree of freedom, in contrast to the material oscillator. For, on the one hand, because of the infinite number of degrees of freedom this would lead to an infinitely large energy per unit volume on the other hand it would be unobservable in principle since it is neither emitted, absorbed or scattered, hence cannot be enclosed inside walls and, as is evident from experience, it also does not produce a gravitational field.    (Pauli, 1933, p. 250; translation from Enz, 2002, p. 151)

Niels Bohr also opposed the existence of a vacuum infinite energy density in 1946, in particular in the context of the Dirac sea, since it would imply 'the existence in free space of an energy density and electric density [that] would be far too great to conform to the basis of general relativity theory' (quoted from Kragh and Overduin, 2014, p. 58).

Frustrating those beliefs, Hendrik Casimir found a physical meaning for the zero-point energy in 1947. He was trying to find the long-range correction to the van der Waals force between two neutral, polarizable atoms. The short-range expression had been given by Fritz London and Robert Eisenschitz in 1930. Casimir, in collaboration with Dirk Polder, took into account the finite speed of light and, after a tedious calculation in fourth-order perturbative quantum electrodynamics, obtained that in the long range the force would decrease faster than in the short range, in agreement with then recent experiments in colloidal chemistry (Rechenberg, 1999; Farina, 2006). As Casimir recollects (see also Farina, 2006; Kragh and Overduin, 2014, p. 58):

> During a visit I paid to Copenhagen, it must have been in 1946 or 1947, Bohr asked me what I had been doing and I explained our work on van der Waals forces. (...) I then explained I should like to find a simple and elegant derivation of my results. Bohr thought this over, then mumbled something like 'must have something to do with the zero-point energy'. That was all, but in retrospect I have to admit that I owe much to this remark.   (Casimir, 1999, p. 6)

Following Bohr's advice, Casimir imagined that the presence of atoms would impose a boundary condition on the electromagnetic field that surrounds them, changing the field zero-point energy. What has a physical meaning, he claimed, is the variation of that energy, which occurs when the allowed frequencies change, that is, when the position of the atoms change (Casimir, 1949). He therefore defined a new energy, nowadays called Casimir energy:

$$\mathcal{E}_C(r) = \lim_{\beta \to 0} \left[ \left( \sum_{k,\sigma} \frac{1}{2} h\nu_k e^{-\beta\nu_k} \right)_A - \left( \sum_{k,\sigma} \frac{1}{2} h\nu_k e^{-\beta\nu_k} \right)_B \right], \tag{3}$$

where the sum, which can be continuous or discrete, is over the wave vector $\mathbf{k}$ and the polarization $\sigma$. In the $A$ term there is a boundary condition that depends on the distance $r$ between the atoms, while in the $B$ term there is no boundary condition. Casimir introduced cutoff functions (that I represented in the above expression by the exponential factors) as a regularization technique, but they have a physical justification, which he fully appreciated: real physical boundaries are not capable of imposing boundary conditions on high-frequency electromagnetic waves, that is, the boundaries are transparent to high-frequency radiation. One must remove the regularization parameter $\beta$ at the end of calculation.

   The energy $\mathcal{E}_C(r)$, which is finite, is the relevant energy of the system. As Casimir phrased it, 'the variation [of the zero-point] energy due to the presence of a particle (...) can be interpreted as the electromagnetic energy of that particle' (Casimir, 1949). Whenever the distance between those atoms is altered, there is a correspondent alteration in the allowed frequencies and, therefore, in the Casimir energy. Therefore, the force on the atom (or, in the general case, on the boundary) is given by $F(r) = -\mathcal{E}'_C(r)$. Using such a method, Casimir was able to recalculate his previous derivation of the long-range van der Waals force in a much easier way. This result was presented at the *Colloque sur la theorie de la liaison chimique*, which took place in Paris in April 1948. Casimir's method was so successful that nowadays all forces that come from variations of the zero-point energy carry his name (see Farina, 2006).

Casimir's work is an example of the general strategy used to deal with divergent quantities in QFT in the late 1940s, which consisted of first making a *regularization* (for instance, inserting a cutoff function that makes the expression finite and that must be removed at the end of calculations) and then a *renormalization* (a redefinition of the physically relevant quantity; in Casimir's case, a redefinition of the energy).[5] That implementation of subtraction physics, of which Casimir's work is one example, came to be known as the renormalization programme.


## 14.2.4  The Renormalization Programme

An unrelated development of the renormalization programme happened in the United States and in Japan at the same time. Whereas direct verification of Casimir's predictions had to wait until the late 1950s, the American and Japanese theoretical achievements went hand-in-hand with the experimental results, and therefore made a much stronger case in defence of the renormalization programme.

With the development of microwave techniques during World War II, Willis Lamb Jr. and Robert Retherford were able to measure a tiny difference between the energy levels $2S_{1/2}$ and $2P_{1/2}$ of the hydrogen atom, a result that became known as the Lamb shift (Schweber, 1994, pp. 206–219). If one takes into account solely the interaction between the electron and the proton (via Dirac's equation), those levels should be degenerate; if one couples the electron with the surrounding quantum electromagnetic field, the energy difference between those levels becomes infinite. Hans Bethe was able to tackle the problem in 1947 using mass renormalization, that is, a redefinition of the mass of the electron, and obtained a theoretical estimate in good agreement with the experimental value (Schweber, 1994, pp. 228–231). A similar explanation of the Lamb shift was found in Japan at about the same time by Sin-itiro Tomonaga (Schweber, 1994, pp. 268–272).

In 1947–1952 Tomonaga, Julian Schwinger, Richard Feynman, and Freeman Dyson developed systematic methods of implementing the renormalization programme in perturbative expansions (Schweber, 2005; Wüthrich, 2011; Darrigol, 2019). Green functions, effective actions, path integrals, and Feynman diagrams became the core of QFT. Their approaches were conservative, preserving the existing theory and focusing on new calculational techniques (Schwinger, 1948, p. 1440):

> [At] moderate energies the known divergences of [quantum electrodynamics] can be isolated as unobservable quantities, which can be regarded as renormalization constants. ( . . . ) [The Lamb shift] can be accounted for unambiguously without new concepts or appreciable modifications of the theory.

---

[5] For a discussion of these two terms, see Schweber (1994, pp. 595–605).

Freeman Dyson showed the equivalence between the existing formalisms and, most importantly, formulated the concept of renormalizability. A theory is said to be renormalizable when all infinite quantities can be absorbed in a redefinition of a finite number of physical parameters (such as the electron mass and charge). Renormalizability became a criterion for theory selection (Schweber, 1994, p. xii).

The renormalization programme was embraced by some of the most prominent physicists of the previous generation, including Pauli, Heisenberg, and Léon Rosenfeld. However, there were some opponents, the most famous being Dirac and Lev Landau, as well as several members of the new generation, as I will discuss in section 14.4.2. Despite these critics, a new generation was educated in the United States and in Japan in the early 1950s learning that quantum electrodynamics was no longer an open problem, and that they should then approach the other fundamental interactions, namely, the gravitational and nuclear ones, with similar methods. However, perturbative renormalization methods could not be easily extended to those interactions. That remained the major concern of quantum field theorists until the mid-1970s, when nuclear forces were finally understood through a renormalizable gauge field theory, including ideas such as asymptotic freedom and confinement. The quantum theory of the gravitational field remains, to the present day, an open problem.

After this overview of the history of QFT, I turn to its interpretational debates, with emphasis on the measuring process. My goal is to explain what a quantum field measurement is and to discuss the role it played in the understanding of the basic features of QFT.

# 14.3  THE QUANTUM FIELD MEASUREMENT

In the early 20th century, a growing number of physicists came to the conclusion that quantum processes, e.g. atomic jumps, the photoelectric effect, and the molecular zero-point energy, could not be visualized in classical terms. Bohr's famous rejection of the space-time description of the motion of atomic particles was an important step in that direction. It was in this context that the well-known dispute between Heisenberg and Schrödinger about the *Anschaulichkeit* (visualizability) of quantum phenomena took place, a major outcome of which was Heisenberg's 1927 uncertainty principle. Heisenberg propounded the gamma-ray microscope as a thought experiment that could explain in simple terms the impossibility of determining the position and the linear momentum of an electron (Darrigol, 1991, pp. 153–156).

A couple of years later, while lecturing in Chicago, Heisenberg devised two thought experiments to investigate the uncertainty relations for components of the quantized electromagnetic field. He expressed doubts about them in a letter to Bohr written on 16 June 1929 (Kalckar, 1996, pp. 5–7). This letter started a long controversy about the measurability of quantum fields, a debate that involved a number of physicists and that had a lasting impact on the development of QFT and on the development of the interpretation of quantum theory as a whole.[6]

## 14.3.1 The 'Small War' over Quantum Fields

Heisenberg started from a basic assumption: one can only measure an electromagnetic field by observing its action on a charged test body. He considered a beam of charged particles (perhaps inspired by cloud chambers) and from its deflection, considering the wave–particle duality, arrived at the following uncertainties for the average, over a volume $V = \ell^3$, of the electric field component $E_x$ and magnetic field component $H_y$,

$$\Delta E_x \, \Delta H_y \gtrsim \frac{hc}{\ell^4}. \tag{4}$$

His letter was discussed in Copenhagen and Zurich, and two young physicists, Lev Landau and Rudolf Peierls, decided to develop Heisenberg's ideas further. They considered a single electron as the test body. The uncertainties in the position and momentum of the electron impose limitations in the determination of the field components—a point that became the source of many misunderstandings, as will be discussed in section 14.3.3. They soon noticed that the field uncertainties were much larger than expected by Heisenberg due to the test body's radiation reaction. They arrived at single component uncertainties of the electric field,

$$\Delta E_x \gtrsim \frac{\sqrt{\hbar c}}{(cT)^2}, \tag{5}$$

where $T$ is the time interval used to perform the measurement. This single component uncertainty has no counterpart in the mathematical formalism. Thus, there was no agreement between the mathematical formalism (the possibilities of definition of the electromagnetic field) and the measurement analysis (the possibilities of observation of the electromagnetic field). Following Bohr's statement that 'the consistency [of quantum theory] can be judged only by weighing the possibilities of definition and observation' (Bohr, 1928, p. 580), they claimed that the quantum theory of the electromagnetic field was not a consistent theory. According to them, this was the origin of all the trouble that such a theory was facing with its divergent integrals. They also predicted that in 'the correct relativistic quantum theory (which does not yet exist), there will therefore be no physical quantities and no measurements in the sense of [the non-relativistic quantum] mechanics' (Landau and Peierls, 1931, p. 68; translated by

---

[6] A careful analysis of that controversy can be found in Darrigol (1991), Miller (1991), Kalckar (1996), and Jacobsen (2011). I follow these four works here but with a slightly different emphasis.

ter Haar, 1965, p. 50). Landau 'frequently complained of having come a little too late to fully partake in the quantum revolution' (Kojevnikov, 2004, p. 85), and expected that his work with Peierls would lead to a new scientific revolution.

That claim deeply disturbed Bohr, who was then the leading figure of the quantum revolution. However, he could not initially find a mistake in their argument. As Heisenberg explained in a letter to Pauli on 12 March 1931 (translated by Kalckar, 1996, pp. 441–442), Bohr's concerns were initially methodological:

> Bohr agrees with the uncertainty relations of Landau and Peierls. He considers their derivations to be sloppy in some places, but this point is not essential. The main criticism is rather directed at the conclusions that Landau and Peierls draw from the uncertainty relations. Bohr thinks that the uncertainty relations do not at all mean that relativistic wave mechanics is too narrow and has to give way for a more general formalism.

Bohr sketched some thoughts on the problem with the help of the 21-year-old Casimir (Kalckar, 1996, p. 520); however, it was with Rosenfeld, an older and more experienced physicist, that he wrote his final answer. Bohr and Rosenfeld argued that Landau and Peierls did not present the best possible measurement. The test body should not be an electron, but a body with arbitrary mass and with the charge smeared on a region, so the charge density is finite. Then, with the aid of compensation devices, the radiation reaction may be completely eliminated (Bohr and Rosenfeld, 1933). A finite charge density means that the atomic constitution of the test body is neglected, i.e., the test body is macroscopic. That was, apparently, the very first moment that Bohr stated that a measuring apparatus must be macroscopic.

The article runs to 63 pages and demonstrated a sophisticated reasoning, full of convoluted sentences and epistemological intricacies. Nevertheless, the final result is quite simple: the mathematical formalism agrees with the measurement analysis. Therefore, quantum electrodynamics is a consistent theory. Considered to be the winner of that fierce dispute, Bohr and Rosenfeld's article soon became a classic. A friend of Bohr once said: 'It is a very good paper that one does not have to read. You just have to know it exists' (Pais, 1991, p. 362). The important message was that there is nothing inconsistent in quantum electrodynamics. Landau's expected revolution was frustrated. As Bohr confessed in a letter to Pauli written on 15 February 1934 (translated by Kalckar, 1996, p. 33): 'I am afraid that such an attitude may perhaps at first appear much too reactionary; but if I am not quite mistaken, it is really the only sober view on atomic problems that is possible at the moment.'[7] Bohr's collaborator Oskar Klein called this controversy 'our small war' (Jacobsen, 2011, p. 384).

In the next sections, I will raise three questions about the interpretation of QFT and explain the Bohrian perspective (and other perspectives not so Bohrian) about them.

---

[7] This statement refers specifically to the macroscopic aspect of the test body, but can be applied to the article as a whole.

## 14.3.2  What is the Role of Measurement in Quantum Field Theory?

Jørgen Kalckar, one of Bohr's last students, stated (1971, p. 127): 'Measurability problems in the sense of quantum theory are consistency problems. The discussion of imaginary experiments provides a testing ground for the compatibility of concepts in situations which are simple enough to allow a comprehensive analysis.' But what should one do if the theory is not consistent? Landau and Peierls claimed that it should be discarded. Quantum measurement was, according to them, a criterion for theory selection. Heisenberg's letter to Pauli quoted in section 14.3.1 shows that Bohr's first reaction was to say that the measurement analysis should not have this function. The measurement analysis in QFT should be, according to him, of the same kind as in quantum mechanics: it should be a reflection about the use of classical concepts.

> In order to realise what words like position and momentum actually imply, the only method is to go back to the imaginary experiments, the idealised manipulations of measuring apparatus, which allow us to assign a physical object a definite position and momentum. This is the *raison d'être* of Bohr's famous discussion of measuring processes.   (Rosenfeld, 1963a, p. 27)

The readers of the 1933 papers attributed other roles to field measurements, almost never in the original sense suggested by Bohr.

## 14.3.3  What is the Source of Field Uncertainties?

Perhaps the subtlest point in Bohr's writing about quantum mechanics is the difference between *classical* and *macroscopical*. This was a major source of confusion during the 20th century. As can be seen, for instance, in a letter from Hugh Everett to Aage Petersen (Bohr's last assistant), from 31 May 1957 (quoted by Osnaghi, Freitas, and Freire, 2009, p. 106):

> You talk of the massiveness of macrosystems allowing one to neglect further quantum effects (in discussions of breaking the measuring chain), but never give any justification for this flatly asserted dogma. ( . . . ) You vigorously state that when apparatus can be used as measuring apparatus then one cannot simultaneously give consideration to quantum effects—but proceed blithely to apply [the uncertainty relations] to such devices, tacitly admitting quantum effects.

In a Bohrian sense, macroscopical does not necessarily mean classical. A macroscopic system is a system in which the atomic constitution is neglected. That system can be—and often is, in Bohr's writing—treated quantum mechanically. Indeed, in the discussions about Einstein's box and in the analysis of quantum field measurement, for instance, Bohr writes down uncertainty relations to the measuring apparatuses. In the 1933 paper, the test body is macroscopic (and, therefore, has a finite charge density), but has uncertainties due to the quantal aspect of the interaction. Rosenfeld stressed this point several times:

> [The measurement] analysis shows that the reciprocal limitations of measurability of field components predicted by the quantized theory arise as a consequence of the impossibility of controlling the number of photons contained in the interaction of the test-bodies in the course of the measuring process; if, however, this interaction were entirely classical, it could be completely compensated, and there would be no limit to the measurability of the field even when due account is taken of the mechanical uncertainty relations to which the test bodies are subjected. I insist on this because the view was expressed in conversation that, according to the analysis, the electromagnetic field quantization is necessarily entailed by the quantization of the motion of the test bodies: there is no such logical necessity in either case.
>
> (Rosenfeld in Infeld, 1964, p. 220)

Rosenfeld was opposing Bryce DeWitt's assertion that the quantization of the test bodies's motion would imply the quantization of the field to which they are coupled (Hartz and Freire, 2015, pp. 411–412; Blum and Hartz, 2017, pp. 123–128). This assertion can be found in Heisenberg's and in Landau and Peierls' analyses of field measurements. DeWitt, however, believed his result to be a mathematical theorem: '[The] quantization of a given system implies also the quantization of any other system to which it can be coupled. By a principle of induction, therefore, the quantum theory must be extended to all physical systems.' He added: 'The demonstration is essentially a paraphrase, applicable to completely arbitrary systems, of the Bohr–Rosenfeld paper on electromagnetic commutators' (DeWitt, 1962, p. 619). Rosenfeld reacted to DeWitt and to other physicists who made similar claims. He recalled the proper role of measurement in quantum theory:

> The ultimate necessity of quantizing the electromagnetic field (or any other field) can only be founded on experience, and all that considerations of measurability of field components can do is to illustrate the consistency of the way in which the mathematical formalism of a theory embodying such quantization is linked with the classical concepts on which its use in analysing the phenomena rests.
>
> (Rosenfeld, 1963b, pp. 353–354)

His words had little impact. In the period 1952–1970, as the foundations of quantum mechanics emerged as an area of research, several physicists revisited quantum field measurement (Hartz and Freire, 2015). The measuring process became a theoretical tool, which was used to develop, constrain, and select approaches to quantum theory.

### 14.3.4 Why Do Calculations Diverge?

Landau and Peierls argued that QFT calculations diverged because the theory was not consistent. They used an electron as a test body and therefore measured the field at a precise point. Bohr and Rosenfeld considered a macroscopic test body and measured the field average over a space-time region. This choice allowed them to eliminate the test body radiation reaction. They never claimed that using field averages would also eliminate the divergencies of the theory; however, that was the lesson that many physicists extracted from the article, perhaps because Bohr and Rosenfeld, in their derivation of the commutator of field averages, smoothed out the delta functions that existed in the article by Jordan and Pauli (1928).

This does not mean that Bohr and Rosenfeld were not interested in the divergence problem, quite the opposite. In the 1933 paper, they made a systematic effort to understand the sources of difficulties. The use of macroscopic test bodies, whose atomic constitution could be neglected, separated the field-fluctuation divergences from the classical electron theory divergences. In Landau and Peierls' argument both problems were entangled, a belief that was shared by most physicists at the time (see Jordan's opinion in section 14.2.2).

The 1933 article also shed some light on the zero-point energy problem. From an epistemological perspective, one cannot say that there are field fluctuations unless there is a charge to measure the field. What could be said then about the vacuum? Bohr mentioned this subtle point in his letter to Pauli written on 15 February 1934 (translated by Kalckar, 1996, p. 34):

> The idea that the field concept has to be used with great care lies also close at hand when we remember that all field effects in the last resort can only be observed through their effects on matter. Thus, as Rosenfeld and I showed, it is quite impossible to decide whether the field fluctuations are already present in empty space or merely created by the test bodies.

In the end, the field concept was not used with as much care by other physicists. The method that eventually solved the divergence problems—renormalization—was not concerned with epistemological subtleties. It was a pragmatic solution, as I will explain in the next section. David Kaiser remarked (2007, p. 4): 'Ironically, Bohr and Rosenfeld might have won the "battle" [against Landau and Peierls] over quantum fields, but they lost the "war" over the proper role of the theorist.'

## 14.4 POSTWAR PRAGMATISM AND ITS OPPONENTS

Silvan Schweber stated that 'the solution advanced by Feynman, Schwinger, and Dyson was pragmatic and conservative' (Schweber, 1986, pp. 97–98). This characterization can be traced back to Schwinger and Dyson themselves (Dyson, 1965; about Schwinger, see section 14.2.4). Why was such a solution accepted by the physics community? Schweber proposed a strong historiographical thesis in order to answer that question, to which I now turn.

## 14.4.1  The Acceptance of the Renormalization Programme

Schweber assumes that 'not merely theoretical physicists, but also theoretical physics is subject to pervasive and consequential vogues of style and substance' (Schweber, 1986, p. 59). Inspired by Paul Forman's thesis, he argued that the acceptance of the renormalization programme was a consequence of the pragmatism of the postwar physics in the United States, a 'war-forged pragmatism' (Kaiser, 2014, p. 155). As Kaiser explains, the 'oft-discussed pragmatic character of American science, and of postwar American theoretical physics in particular, must be understood in terms of these changes in physicists' infrastructure' (Kaiser, 2004, p. 16).

Therefore, according to Schweber, the acceptance of renormalization did not happen simply because of the outstanding agreement between theory and experiment, but also (and most importantly) because the 'the pragmatic, utilitarian outlook—which had been reinforced by the wartime experiences—(. . .) gave the philosophical and ideological underpinning' of postwar physics (Schweber, 1994, p. 150). The renormalization methods 'dispelled doubts about the adequacy of quantum electrodynamics and gave renewed faith in the possibility of a field theoretic explanation for nuclear phenomena' (Schweber, 1986, p. 98).

Schweber's characterization of postwar pragmatism seems quite correct, and his explanation of the acceptance of the renormalization programme is appealing. However, as I hope to show in the following section, he perhaps overestimated the impact of the renormalization programme.

## 14.4.2  The New Mathematical Physics of the 1950s

The renormalization method was a major achievement of postwar physics. However, it was far from being the final word on the divergencies of QFT, since renormalization was hardly accepted in the early 1950s as the final solution outside a small circle of physicists, situated mostly in the United States, in Japan, and in Switzerland. As a consequence of the dissatisfaction, a new community emerged, whose approach was called the new mathematical physics. The physicists and mathematicians gathered in such a group—led by Arthur Wightman, Rudolf Haag, Irving Segal, Karl Friedrichs, and Res Jost—believed that the renormalization methods had questionable foundations. There were also intermediary figures, such as Gunnar Källén, who was a supporter of the renormalization programme and, nevertheless, an opponent of

Schwinger's methods. All of them claimed that quantum electrodynamics was far from being a solved problem. Joseph M. Cook, in his PhD thesis, written under Segal at the University of Chicago in 1951, aptly expressed the group's agenda:

> Although few nonspecialists have had opportunity to become familiar with the language of modern pure mathematics, quantum theory seems to have reached a point where it must use that language if it is to find a genuine escape from the divergence difficulties. Divergence can not be properly coped with when convergence itself has never been rigorously defined.    (Cook, 1953, p. 222)

This community gathered on several occasions during the 1950s, for instance at a conference in Lille in 1957, at the Varenna summer school in 1958 and 1968, and at the Solvay conference in 1967. Of course, its members disagreed about the best possible approach. They even argued about what the new approach should be called. Several names were suggested, for instance, axiomatic QFT (Wightman), algebraic QFT (Haag), and general theory of quantized fields (Jost). Källén mockingly called Wightman's approach 'epsilontics', and complained in a letter to Haag in 1958: 'If this kind of mathematics ever becomes a fashion in physics I am going to abandon the subject' (Jarlskog, 2014, p. 403). Haag (1970, pp. 5–7) identified three attitudes in the new community:

> 1) (...) We need a radical change of our concepts, a radical new idea. It is just as futile to approach elementary particles with the conceptual structure of 1930 as it was to attack atomic physics within the frame of classical mechanics. (...)
>
> 2) (...) It is worthwhile to develop a framework which incorporates the old principles, formulating them precisely, separating the essential and the peripheral features of traditional Quantum Field Theory, recognizing the numerous mathematical pitfalls. (...)
>
> 3) (...) The most fruitful task for the theoretician at present is to analyse experiments, looking for regularities and for phenomenological models which describe the essential features.
>
>    It is unfortunately in the nature of ideologies that they tend to crystallize. One has to make a determined effort to keep the channels of communication between different camps open.

Those aligned with the second 'ideology' believed that the cause of the divergent calculations was the point-like aspect of field interaction. Therefore, in order to eliminate the infinities from theory, one should use space-time average of field quantities. As I mentioned in section 14.3.4, there was a widespread belief that such an idea could be traced back to Bohr:

> It was recognized very early, however, by Bohr and Rosenfeld that, even in the case of a free field, no physical meaning could be attached to the values of the field at a particular point—only the suitably smoothed averages over finite space-time regions had such a meaning.    (Segal, 1961, p. 1)

It was not clear, however, how those smoothed averages should be mathematically implemented. Wightman, in collaboration with the mathematician Lars Gårding, favored Laurent Schwartz's theory of distributions, Haag favored the algebra of local observables, while Segal favored $C^*$-algebras. There was, nonetheless, a consensus about the importance of guiding principles. QFT should respect, for instance, causality, unitarity, and relativistic covariance (Schweber, 2002, p. 385). Paradoxically, they found inspiration in Dyson's claim that renormalizability could be used as a criterion for theory selection. There was strong hope that a clear statement of theoretical requirements would constrain the theoretical development, allowing physicists to arrive at the correct theory from mathematical considerations, with scarce input from experimental results. Therefore, it comes as no surprise that the hero of that generation was Eugene Wigner, who in 1939 had provided a classification of the irreducible representations of the Poincaré group, relating the space-time symmetries to the possible mass and spin of elementary particles (Haag, 2010, pp. 265–266).

These theoretical values soon found supporters worldwide, including Nikolai Bogoliubov, Huzihiro Araki, Walter Thirring, Klaus Hepp, Guido Bolini, Juan José Giambiagi, Jorge André Swieca, Bert Schroer, André Lichnerowicz, and Ray Streater, to mention just a few. New journals were created, in particular Haag's *Communications in Mathematical Physics* (Haag, 2010, p. 284). Despite a tremendous amount of work, mathematical physicists have not been able, until now, to obtain through rigorous methods all the theoretical predictions obtained using the pragmatic renormalization methods. As Haag recollected (1970, p. 5):

> The following story was reported to me. A few years ago, Klaus Hepp gave some lectures in the Brandeis summer school. At some stage he praised the beauty of axiomatic field theory. Next day he found the note on the blackboard:
>
> 'Axiom 1: Axiomatic Field Theory is beautiful in an empty sort of way.'
>
> Presumably this note expresses also pretty accurately the feelings of the majority of today's audience and indeed there is an element of truth in it.

Nowadays, virtually all physicists believe that the divergent calculations can be traced back to the point-like aspect of field interactions and that renormalization is an acceptable solution, although many of them still feel slightly uncomfortable about it. They often claim that if one wants to be more rigorous, it is just a matter of using space-time field averages and all calculations become finite. Bohr and Rosenfeld's analysis of quantum field measurement, which once allegedly gave support to this claim, is almost forgotten. It is now clear that even if the theory is finite, renormalization is necessary. Physicists' current attitudes towards the interpretation of QFT are a syncretism of several research agendas, as often happens in science.

## ACKNOWLEDGEMENTS

I am very grateful to Olivier Darrigol, Carlos Farina, Olival Freire, Reinaldo de Melo e Souza, and Denise Key for their comments on a previous version of this chapter.

## REFERENCES

Blum, A., and Hartz, T. (2017). The 1957 quantum gravity meeting in Copenhagen: An analysis of Bryce S. DeWitt's report. *The European Physical Journal H*, **42**(2), 107–157.

Blum, A. (2017). The state is not abolished, it withers away: How quantum field theory became a theory of scattering. *Studies in History and Philosophy of Modern Physics*, **60**, 46–80.

Bohr, N., and Rosenfeld, L. (1933). Zur Frage der Messbarkeit der elektromagnetischen Feldgrössen. *Kgl. Danske Vidensk. Selskab. Math.-Fys. Medd*, **12**(8), 1–65.

Bohr, N. (1928). The Quantum Postulate and the Recent Development of Atomic Theory. *Nature*, **121**(3050), 580–590.

Bokulich, P. (2003). *Horizons of Description: Black Holes and Complementarity*. PhD Thesis, Graduate Program in Philosophy, University of Notre Dame.

Born, M., and Jordan, P. (1925). Zur Quantenmechanik. *Zeitschrift für Physik*, **34**(1), 858–888.

Born, M., Heisenberg, W., and Jordan, P. (1926). Zur Quantenmechanik. II. *Zeitschrift für Physik*, **35**(8), 557–615.

Cao, T. Y. (1997). *Conceptual Developments of 20th Century Field Theories*. Cambridge: Cambridge University Press.

Casimir, H. B. G. (1949). Sur les forces van der Waals-London. *Journal de Chimie Physique*, **46**, 407–410.

Casimir, H. B. G. (1999). Some remarks on the history of the so-called Casimir effect. In M. Bordag (ed.), *Casimir Effect 50 Years Later*. Singapore: World Scientific, pp. 3–9.

Cook, J. M. (1953). The mathematics of second quantization. *Transactions of the American Mathematical Society*, **74**(2), 222–245.

Cushing, J. (1988). Foundational Problems in and Methodological Lessons from Quantum Field Theory. In H. R. Brown and R. Harré (eds.), *Philosophical Foundations of Quantum Field Theory*. Oxford: Clarendon Press, pp. 25–39.

Darrigol, O. (1982). *Les débuts de la théorie quantique des champs (1925–1948)*. Thèse pour le doctorat de troisième cycle, Université de Paris 1 (Panthéon-Sorbonne).

Darrigol, O. (1984). La genèse du concept de champ quantique. *Annales de Physique*, **9**(3), 433–501.

Darrigol, O. (1988). The Quantum Electrodynamical Analogy in Early Nuclear Theory or the Roots of Yukawa's Theory. *Revue d'histoire des sciences*, **41**(3), 225–297.

Darrigol, O. (1991). Cohérence et complétude de la mécanique quantique : l'exemple de « Bohr-Rosenfeld ». *Revue d'histoire des sciences*, **44**(2), 137–179.

Darrigol, O. (2019). The magic of Feynman's QED: from field-less electrodynamics to the Feynman diagrams. *The European Physical Journal H*, **44**(4), 349–369.

DeWitt, B. S. (1962). Definition of Commutators via the Uncertainty Principle. *Journal of Mathematical Physics*, **3**(4), 619–624.

Dirac, P. A. M. (1927). The quantum theory of the emission and absorption of radiation. *Proceedings of the Royal Society A*, **114**(767), 243–265.

Duncan, A., and Janssen, M. (2008). Pascual Jordan's resolution of the conundrum of the wave-particle duality of light. *Studies in History and Philosophy of Modern Physics*, **39**(3), 634–666.

Duncan, A., and Janssen, M. (2019). *Constructing Quantum Mechanics: Volume 1—The Scaffold: 1900–1923*. Oxford: Oxford University Press.

Dyson, F. J. (1965). Tomonaga, Schwinger, and Feynman Awarded the Nobel Prize for Physics. *Science*, **150**, 588–589.

Enz, C. P. (2002). *No time to be brief: A scientific biography of Wolfgang Pauli*. Oxford: Oxford University Press.

Farina, C. (2006). The Casimir Effect: Some Aspects. *Brazilian Journal of Physics*, **36**(4A), 1137–1149.

Haag, R. (1970). Observables and Fields. in S. Deser, M. Grisaru, and H. Pendleton (eds.), *Lectures on Elementary Particle and Quantum Field Theory*. Cambridge, MA: MIT Press, pp. 1–89.

Haag, R. (2010). Some people and some problems met in half a century of commitment to mathematical physics. *The European Physical Journal H*, **35**(3), 263–307.

Hartz, T., and Freire, O. (2015). Uses and appropriations of Niels Bohr's ideas about quantum field measurement, 1930–1965. In F. Aaserud and H. Kragh (eds.), *One hundred years of the Bohr atom: Proceedings from a conference*. Copenhagen: Det Kongelige Danske Videnskabernes Selskab.

Heisenberg, W., and Pauli, W. (1929). Zur Quantendynamik der Wellenfelder. *Zeitschrift für Physik*, **56**(1–2), 1–61.

Infeld, L. (ed.) (1964). *Conférence internationale sur les théories relativistes de la gravitation: Sous la direction de L. Infeld*. Paris: Gauthier-Villars.

Jacobsen, A. (2011). Crisis, Measurement Problems and Controversy in Early Quantum Electrodynamics: The Failed Appropriation of Epistemology in the Second Quantum Generation. In C. Carson, A. Kojevnikov, and H. Trischler (eds.), *Weimar Culture and Quantum Mechanics: Selected Papers by Paul Forman and Contemporary Perspectives on the Forman Thesis*. London: Imperial College Press, pp. 375–396.

Jarlskog, C. (2014). *Portrait of Gunnar Källén: A Physics Shooting Star and Poet of Early Quantum Field Theory*. Cham: Springer.

Jordan, P., and Pauli, W. (1928). Zur Quantenelektrodynamik ladungsfreier Felder. *Zeitschrift für Physik*, **47**, 151–173.

Jordan, P., and Wigner, E. (1928). Über das Paulische Äquivalenzverbot. *Zeitschrift für Physik*, **47**, 631–651.

Kaiser, D. (2004). *Drawing Theories Apart: The Dispersion of Feynman Diagrams in Postwar Physics*. Chicago: University of Chicago Press.

Kaiser, D. (2007). *Comments on 'Interpreting Quantum Mechanics: A Century of Debate'*. HSS Session, November 2007, unpublished.

Kaiser, D. (2014). Shut Up and Calculate! *Nature*, **505**(7482), 153–155.

Kalckar, J. (1971). Measurability Problems in the Quantum Theory of Fields. In B. d'Espagnat (ed.), *Foundations of Quantum Mechanics*, New York: Academic Press, pp. 127–169.

Kalckar, J. (ed.) (1996). *Niels Bohr Collected Works—Volume 7: Foundations of Quantum Physics II (1933–1958)*. Amsterdam: Elsevier.

Kojevnikov, A. (2004). *Stalin's Great Science: The Times and Adventures of Soviet Physicists*. London: Imperial College Press.

Kragh, H. S., and Overduin, J. M. (2014). *The Weight of the Vacuum: A Scientific History of Dark Energy*. Heidelberg: Springer.

Kragh, H. (1995). Arthur March, Werner Heisenberg, and the search for a smallest length. *Revue d'histoire des sciences*, 48(4), 401–434.

Landau, L., and Peierls, R. (1931). Erweiterung des Unbestimmtheitsprinzips für die relativistische Quantentheorie. *Zeitschrift für Physik*, 69(1–2), 56–69.

Lehner, C. (2011). Mathematical foundations and physical visions: Pascual Jordan and the field theory program. In K.-H. Sclote and M. Schneider (eds.), *Mathematics meets physics: A contribution to their interaction in the 19th and the first half of the 20th century*. Frankfurt: Harri Deutsch, pp. 271–293.

Miller, A. I. (1991). Measurement Problems in Quantum Field Theory. In A. I. Miller (ed.), *Sixty-Two Years of Uncertainty*. New York: Plenum Press, pp. 139–152.

Miller, A. I. (ed.) (1994). *Early Quantum Electrodynamics*. Cambridge: Cambridge University Press.

Osnaghi, S., Freitas, F., and Freire Jr, O. (2009). The origin of the Everettian heresy. *Studies in History and Philosophy of Modern Physics*, 40(2), 97–123.

Pais, A. (1982). *Subtle is the Lord: The Science and Life of Albert Einstein*. Oxford: Oxford University Press.

Pais, A. (1986). *Inward Bound: Of Matter and Forces in the Physical World*. Oxford: Clarendon Press.

Pais, A. (1991). *Niels Bohr's Times: In Physics, Philosophy, and Polity*. Oxford: Oxford University Press.

Pauli, W. (1933). Die allgemeinen Prinzipien der Wellenmechanik. In H.Geiger and K. Scheel (eds.), *Handbuch der Physik*—Band V, Teil 1, Berlin: Springer, pp. 83–272.

Peskin, M. E., and Schroeder, D. V. (2007). *Introduction to Quantum Field Theory*, 2nd edn. Cambridge, MA: Perseus Books.

Rechenberg, H. (1999). Historical Remarks on Zero-Point Energy and the Casimir Effect. In M. Bordag (ed.), *Casimir Effect 50 Years Later*, Singapore: World Scientific, pp. 10–19.

Redhead, M. (1982). Quantum Field Theory for Philosophers. In *Proceedings of the 1982 Biennial Meeting of the Philosophy of Science Association, Volume Two: Symposia and Invited Papers*, Chicago: University of Chicago Press, pp. 57–99.

Redhead, M. (1988). A Philosopher Looks at Quantum Field Theory. In H. R. Brown and R. Harré (eds.), *Philosophical Foundations of Quantum Field Theory*, Oxford: Clarendon Press, pp. 9–23.

Rosenfeld, L. (1963a). Matter and Force After Fifty Years of Quantum Theory. In S. K. Runcorn (ed.), *Physics in the Sixties*. London: Oliver & Boyd, pp. 9–30.

Rosenfeld, L. (1963b). On quantization of fields. *Nuclear Physics*, 40, 353–356.

Rueger, A. (1992). Attitudes towards infinities: Responses to anomalies in quantum electrodynamics, 1927–1947. *Historical Studies in the Physical and Biological Sciences*, 22(2), 309–337.

Schweber, S. S. (1984). Some Chapters for a History of Quantum Field Theory. In B. S. DeWitt and R. Stora (eds.), *Relativité, Groupes et Topologie*, Amsterdam: Elsevier, pp. 37–220.

Schweber, S. S. (1986). The empiricist temper regnant: Theoretical physics in the United States 1920-1950. *Historical Studies in the Physical and Biological Sciences*, 17(1), 55–98.

Schweber, S. S. (1994). *QED and the Men Who Made It: Dyson, Feynman, Schwinger, and Tomonaga.* Princeton: Princeton University Press.

Schweber, S. S. (2002). Quantum Field Theory: From QED to the Standard Model. In M. J. Nye (ed.), *The Cambridge History of Science—Volume 5: The Modern Physical and Mathematical Sciences.* Cambridge: Cambridge University Press, pp. 375–393.

Schweber, S. S. (2005). The sources of Schwinger's Green's functions. *Proceedings of the National Academy of Sciences,* **102**(22), 7783–7788.

Schwinger, J. (1948). Quantum Electrodynamics. I. A Covariant Formulation. *Physical Review,* **74**(10), 1439–1461.

Segal, I. E. (1961). Foundations of the theory of dynamical systems of infinitely many degrees of freedom, II. *Canadian Journal of Mathematics,* **13**, 1–18.

ter Haar, D. (ed.) (1965). *Collected papers of L. D. Landau.* New York: Gordon and Breach.

Wentzel, G. (1960). Quantum Theory of Fields (until 1947). In M. Fierz and V. F. Weisskopf (eds.), *Theoretical Physics in the Twentieth Century.* New York: Interscience Publishers, pp. 48–77.

Wüthrich, A. (2011). *The Genesis of Feynman Diagrams.* Dordrecht: Springer.

# CHAPTER 15

........................................................................................

# THE INTERPRETATION
# DEBATE AND QUANTUM
# GRAVITY

........................................................................................

## ALEXANDER S. BLUM AND BERNADETTE LESSEL

THE histories of quantum gravity (QG) and of the interpretation debate are strangely parallel: both began as marginal issues in the decades after the establishment of quantum mechanics, but by the end of the 20th century they were recognized as perhaps the two major open questions in the foundations of theoretical physics. This move into the limelight happened around the same time for both quantum gravity and the interpretation debate, in the early 1980s. For the interpretation debate that is well documented by other chapters in this handbook; for quantum gravity it is marked by the 'first superstring revolution' and the establishment of loop quantum gravity (Rickles, 2014; Rovelli, 2004).

Around the time of this simultaneous move to the mainstream we can observe, in the mid-1980s, a surge of research into the connection between quantum gravity and the interpretation of quantum mechanics, which continues to this day. Mapping the manifold developments of this research over the past three decades would be a fascinating subject, both for historical analysis and technical review, but is beyond the scope of this short chapter. Instead we describe the developments leading up to this surge and the emergence of central points of connection.

One such point of connection is the role of the 'classical'; clearly prominent in the interpretation debate, it also played a central role in the quantization of gravity, given that this programme implied a quantization of the bedrock of classical physics: space and time. In section 15.1, we will look at Léon Rosenfeld's opposition to the quantization of the gravitational field, which was importantly based on his reading Niels Bohr's philosophy and its implications for the relation between the quantum and the classical. Another point of connection is cosmology, and, as we shall discuss in section 15.2, it was concerns of quantum cosmology that led to John Wheeler's short-lived but enthusiastic support for the Everett interpretation, which might otherwise never have received even the little attention it did. We will focus here on the detailed relation between specific (technical) problems in quantum cosmology and the interpretation of quantum mechanics. In section 15.3, we will look at how the interpretation debate and quantum gravity research came together, in DeWitt's promotion of the Everett interpretation beyond the confines of quantum cosmology, in John Bell's first attempts at bringing other interpretations (Bohmian mechanics) to bear on cosmological issues, and in Roger Penrose's programme of obtaining an interpretation of quantum mechanics from a theory of quantum gravity.

# 15.1 LÉON ROSENFELD

From at least 1932, Rosenfeld was highly critical towards the idea of an empirically unjustified formal quantization of the gravitational field—an attitude that we will trace back to the influence of Bohr's interpretation of quantum mechanics on Rosenfeld, which began in 1931. However, the formation of this attitude cannot be understood without taking into account Rosenfeld's early work. For he started his career with attempts to formally unite general relativity and wave mechanics, followed by pioneering work on the quantization of the gravitational field.

In 1927, after having left university, Rosenfeld combined his freshly acquired knowledge in relativity and current problems of wave mechanics to develop a wave equation in five dimensions (Rosenfeld, 1927) which was similar to a formulation that Oskar Klein published later in the same year (Peruzzi and Rocci, 2018). Being in Paris at that time, Rosenfeld stood under the influence of Louis de Broglie whose opinion about the interpretation of quantum mechanics was that the notion of wave functions on configuration space was not sufficiently physical. Considering the wave function as something real, he was instead seeking for a description of quantum phenomena through wave functions in ordinary space, which he hoped that Rosenfeld's unified formalism could provide. However, after leaving Paris for Göttingen in October 1927, Rosenfeld did not continue to work on the five-dimensional formalism. Furthermore, Max Born and Pascual Jordan quickly managed to convince Rosenfeld to abandon the realist interpretation of quantum mechanics.[1]

But when he became Wolfgang Pauli's assistant in Zürich in 1929, bringing together concepts from quantum mechanics and general relativity came round a second time for Rosenfeld. This time, he got provoked by Pauli to tackle the problem of the quantization of gravitation, particularly the gravitational effects of light quanta. This endeavour was not set in the grand metaphysical scheme of uniting the two big theories of nature, quantum physics and general relativity. The motivation here was very pragmatic,

---

[1] Interview of Léon Rosenfeld conducted by Thomas S. Kuhn and John L. Heilbron. (Oral History Interviews by the American Institute of Physics), 1 July 1963, https://www.aip.org/history-programs/niels-bohr-library/oral-histories/4847-1.

namely to understand where the infinite self-energy of the electron in QED came from. Unfortunately, Rosenfeld's results were not very pleasant; he calculated the self-energy of the photon in its gravitational field to be infinite, which showed that the divergences of the self-energies in QED were not the result of a classical singularity, but genuinely quantum, which in turn created the impression that there was something deeply wrong with the whole theory. But Rosenfeld's work was much more significant than its contribution to the comprehension of the origins of the infinities in QED. In fact, he was the first person who attempted a direct quantization of the gravitational field, both in the linearized (Rosenfeld, 1930a) and in the full non-linear case (Rosenfeld, 1930b), thereby developing several methods that are still of relevance today (Blum and Rickles, 2018; Salisbury and Sundermeyer, 2017).

However, even though it was not just Pauli, but also Werner Heisenberg and Paul Dirac, who knew of Rosenfeld's work, it seems that it remained largely unnoticed at that time. Salisbury and Sundermeyer (2017) suppose that Rosenfeld may have felt discouraged by Pauli to advertise his work, or to claim ownership when his methods were gradually rediscovered by other people.[2] But in 1931, Rosenfeld began working with Niels Bohr on a reply to Lev Landau and Rudolf Peierls's article 'Erweiterung des Unbestimmtheitsprinzips für die relativistische Quantentheorie' (Landau and Peierls, 1931). Landau and Peierls claimed that there are inconsistencies in the measuring process of the electromagnetic field, in the sense that there are uncertainties in the measurement of the field components that are unrelated to the uncertainties predicted by quantum theory, and that these inconsistencies are responsible for the divergences in QED.

These kinds of problems fell exactly under Bohr's purview. And Rosenfeld, who had just arrived in Copenhagen to be Bohr's assistant when Landau and Peierls came out with their paper, was exactly the right person arriving at the right time to explain to Bohr, who apparently had not previously been interested in QED,[3] the necessary fundamentals. What Bohr taught Rosenfeld in return was a better understanding of the interpretation of quantum mechanics and a greater appreciation of the physical concepts behind the formalism. This, in particular, seems to have led Rosenfeld to re-evaluate his earlier quantization attempts and their consequences. When in 1932 he published the lecture notes to a course on QFT that he gave at the Institut Henri Poincaré in February 1931[4] (Rosenfeld, 1932), he added to those notes a final paragraph that was entitled 'Critique de la théorie de l'électron et du rayonnement'. In this paragraph he concluded:

[2] Rosenfeld preceded Peter Bergmann's work on non-linear quantized gravity by 20 years.

[3] Interview between Thomas S. Kuhn, John L. Heilbron, and Rosenfeld (Oral History Interviews by the American Institute of Physics), 19 July 1963, https://www.aip.org/history-programs/niels-bohr-library/oral-histories/4847-2. On the Landau–Peierls–Bohr–Rosenfeld debate see also (Darrigol, 1991).

[4] Rosenfeld moved to Copenhagen in June 1931.

> The first successes of the theory of quantization of the radiation field made it possible to suppose for a moment that within the framework of quantum mechanics, this method provided a rigorous mathematical expression of the correspondence principle. But in reality, [ ... ], further approximation leads immediately to meaningless results, so that the new formalism can only be conceived as a symbolic computational procedure [ ... ].[5]

Consequently,

> [T]here is no reason to think that this same method is likely to provide us with an adequate description of the interactions between material particles, or, to express the same idea in other terms, that the concept of the photon (which immediately results from the quantization of the electromagnetic field) can be legitimately applied to the analysis of fields other than pure radiation fields.

With this analysis, Rosenfeld clearly turned his back on his Zürich work on QFT and quantization of gravity, deploring its lack of physical grounding. For quantization 'has no other role than to translate in a condensed and elegant way the fundamental statistical laws relating to quantum systems'.[6] The formal quantization procedure therefore needs to be seen as an end product of theory development, as opposed to a starting point. It does not have any inherent physical power to turn classical systems into quantum ones if physical facts are not demanding it. And at its core, this would remain Rosenfeld's attitude towards field quantization for the rest of his life.

The many hours in discussion with Bohr finally led in 1933 to the celebrated 'Zur Frage der Messbarkeit von elektromagnetischen Feldgrössen' (Bohr and Rosenfeld, 1933), in which Bohr and Rosenfeld demonstrate that measurements of the free electromagnetic field can be performed consistently after all, as long as one settles on measuring spatio-temporal averages instead of the value at a specific point.

After this period, Rosenfeld mainly focused on nuclear physics and made a name for himself as a fierce and proactive defender and spokesperson of Bohr's interpretation of quantum mechanics (Jacobsen, 2012). He was, however, brought back to the debate on quantization of gravity many years later when he was invited to participate in the conference on 'The role of gravitation in physics' at Chapel Hill in 1957. This conference was organized by Cécile and Bryce DeWitt and explicitly dedicated three out of eight sessions to the problem of the quantization of the gravitational field. Problems of measurements of the gravitational field were 'placed first on the agenda, in an attempt to keep physical concepts as much as possible in the foreground in a subject [ ... ] which suffers from lack of experimental guideposts' (DeWitt and Rickles, 2011, p. 167). Very early in these sessions, Rosenfeld remarked that difficulties appear if one tries to extend his and Bohr's theory of measurement straightforwardly to the

---

[5] English translation. The French original can be found in (Rosenfeld, 1932, p. 89).
[6] English translation. The French original can be found in (Rosenfeld, 1932, p. 89).

gravitational field. Features of the electromagnetic field that were crucial in their argumentation don't find their analogue in the gravitational field. Consequently, Rosenfeld assumed that there will be a limit in the accuracy with which the gravitational field can be measured. At the conference, no agreement was reached as to whether this is actually the case and how far such a limit would delegitimize formal attempts to quantize gravity. Rosenfeld himself withdrew from his previous firm statement a bit later in the event, uttering a few possibilities about how the case for the gravitational field could be rescued after all. Generally, it is fair to say that in Chapel Hill, Rosenfeld discussed open-mindedly about the problem of quantization of gravitation, though being sceptical and constantly bringing to attention the necessity of a physical grounding for the whole endeavour. For example, in the session in which 'The Necessity of Gravitational Quantization' was discussed, he finally said (DeWitt and Rickles, 2011):

> It is difficult for me to imagine a quantized metric unless, of course, this quantization of the metric is related to the deep-seated limitations of the definitions of space and time in very small domains corresponding to internal structures of particles. That is one prospect we may consider. The whole trouble, of course, which raises all these doubts, is that we have too few experiments to decide things one way or the other.

Rosenfeld gave an elaboration of his position 'that the case for quantization of the gravitation field is perhaps not as obvious as it is sometimes made out to be' (Infeld, 1964, p. 219) in the form of a long-delayed, but then unequivocal, reaction to Bryce DeWitt's work of the early 1960s. At that time, DeWitt was attempting 'to develop the formalism [of quantized gravity] itself with the aid of the ideas of the theory of measurability' of Bohr and Rosenfeld (DeWitt, 1962, p. 270).[7] In January 1961, DeWitt sent a copy of his new article, (DeWitt, 1962), to Rosenfeld, asking, as politely as probably possible, for written comments.[8] However, it seems that Rosenfeld could not have cared less about these matters, as no direct reply from Rosenfeld to DeWitt is known.[9] Also, to a request of Hermann Bondi to give a report on quantization, Rosenfeld replied in November 1961:[10]

> I think I could make a report on quantization of gravitation much shorter than 30 minutes. The full text of it would be: Et ego censeo gravitatem non quantificandam esse, or vernacular words to that effect.

---

[7] More on the 'Uses and appropriations of Niels Bohr's ideas about quantum field measurement' can be found in (Hartz and Freire, 2015).

[8] Letter by DeWitt to Rosenfeld, 11 January 1961 (Léon Rosenfeld Papers, Box 8, Folder 11; Niels Bohr Archive, Copenhagen).

[9] See also Thiago Hartz's talk on 'Bryce DeWitt's road to the Many Worlds' at the HQ-4 meeting in San Sebastián in 2015: https://youtu.be/8VE9Pbn46DE.

[10] Letter by Rosenfeld to Bondi, Leiden, 7 November 1961 (Léon Rosenfeld Papers, Box 8, Folder 11; Niels Bohr Archive, Copenhagen).

But for some reason he still accepted an invitation to attend the third edition of the International Conference on Relativity and Gravitation held in Jabłonna in July 1962, even though he knew that it would 'mainly, but not only, deal with waves and radiation in general relativity and the problem of quantizing the gravitational field'.[11] And at this conference, DeWitt gave a talk on precisely the article he had sent to Rosenfeld (Infeld, 1964, p. 131), who attended this talk, but in this moment again doesn't seem to have shown particular interest in DeWitt's findings.[12] It was only two days later, during the General Discussion, when Rosenfeld finally made a statement about DeWitt's work (Infeld, 1964, p. 219). It was, however, not a commentary on DeWitt's work per se, but in fact a thorough critique of DeWitt's claim that the gravitational field must be quantized.

Rosenfeld began his criticism by admitting sympathetically that the universality of the quantum of action can tempt one to believe that any classical theory is a limiting case of some quantal theory. However, he continued:

> One must not lose sight of the fact that the formulation of any theory in its application to given physical situations involves the specification of the system under consideration and of the external conditions under which it is investigated: such specifications, which represent the essential link between the theoretical description and the physical observation, are necessarily expressed in terms of classical concepts [ . . . ].

To Rosenfeld, the metrical tensor 'appears as such a c-number specification of conditions of observation, and there is no logical imperfection in regarding as fundamental the classical, unquantized, form of the equations expressing the connexion of the metrical or gravitation field with the other fields.' This was actually the first time that Rosenfeld had explicitly used ideas from the Copenhagen interpretation of quantum mechanics to argue against the logical necessity of a quantization of the metric. This same point has also consequences for the application of any interpretation of quantum theory that includes the measurement apparatus in the quantal description, as '[i]n the particular case of the measurement of gravitation quantities, it is unavoidable to have some classical metrical substratum for the localization of the testbodies.'

---

[11] Letter by Andrzej Trautman and Leopold Infeld to Rosenfeld, Warszawa, January 1962 (Léon Rosenfeld Papers, Box 8, Folder 8; Niels Bohr Archive, Copenhagen).

[12] In the subsequent discussion, Rosenfeld merely responded to criticism by DeWitt that in the Bohr–Rosenfeld paper 'they recognized that there were experimental limitations due to the atomic structure of matter', but, according to DeWitt, 'ignored this completely'. Rosenfeld then explained that there is a logical difference between the limitation in the use of classical concepts for practical purposes, referring to a case that DeWitt mentioned in his talk and which the latter referred to as a 'fundamental limitation', and 'the question of consistent use of the classical concepts in interpreting the theory', which is was Bohr and Rosenfeld had set out to investigate.

To give an idea of how a reconciliation of quantum theory and gravitation could look like that is not at odds with the Bohrian interpretation of quantum mechanics, Rosenfeld afterwards proposed a semi-classical approach,[13] where in the general relativistic field equations the geometrical Einstein tensor is set equal to the expectation value of a q-valued energy-momentum tensor. However, after elaborating briefly why quantization of gravitation is probably meaningless anyways since the quantal effects would be too small to become appreciable, Rosenfeld went on to make his final point, namely that, in any case, 'all such an analysis [of the limits of measurability of gravitational quantities, as carried out by DeWitt] can tell us is whether an assumed quantization of the gravitation field is consistent with the quantization of the other fields.' But not whether the gravitational field has to be quantized or not, as DeWitt had claimed.

In 1963, Rosenfeld published the short article 'On quantization of fields' (Rosenfeld, 1963), repeating almost all of the points he had raised in his monologue at the General Discussion in Jabłonna, sometimes even up to the precise wording. However, this article was apparently provoked by Ernest M. Henley's and Wolfgang Thirring's textbook on Elementary Quantum Field Theory (Henley and Thirring, 1962). In this book, Rosenfeld states, the authors 'instead of plainly acknowledging the empirical origin of the connexion between quantized motion and quantized radiation [...], endeavour to "prove" it as if it were a logical necessity'. For this 'proof' they even '[sought] support' in the Bohr–Rosenfeld paper.[14] So, their delinquency, in Rosenfeld's eyes, was analogous to that of DeWitt and the programme of formal quantization of gravitation, and in both cases he felt the need to react and set things straight, stating, just as he did in his 'Critique de la théorie de l'électron et du rayonnement':

> It is nice to have at one's disposal such exquisite mathematical tools as the present methods of quantum field theory, but one should not forget that these methods have been elaborated in order to describe definite empirical situations, in which they find their only justification. Any question as to their range of application can only be answered by experience, not by formal argumentation.

[13] This semi-classical approach to reconcile quantum theory and gravitation must have been haunting the quantization community since its beginning; however, it was only put on a thorough grounding by Christian Møller in 1959 (Infeld, 1964) (see also section 15.2).

[14] Thirring and Henley corresponded about Rosenfeld's criticism (Walter Thirring Estate, Austrian Central Library for Physics, Folder Henley Correspondence), complaining that Rosenfeld had misunderstood their book. Henley even composed a letter to Rosenfeld to defend their claims against his criticism. But this letter was probably never sent. In their book, Henley and Thirring eventually changed the reference to the Bohr–Rosenfeld article into references to other works, by Werner Heisenberg and by Julian Schwinger. This resulted in the curious fact that there are now two circulating versions of the first edition: one with a reference to Bohr–Rosenfeld and one with a reference to Heisenberg and to Schwinger.

# 15.2 JOHN WHEELER AND THE EVERETT INTERPRETATION

Rosenfeld's rejection of the programme of quantizing gravity was, as we have seen, partly motivated by (his reading of) the Copenhagen interpretation of quantum theory. In turn, we shall see in the following how those actively pursuing this programme would tend to gravitate (pardon the pun) toward alternative interpretations, more suited to their needs. It is in this way that we can understand the great interest that John Wheeler, who began to pursue the programme of quantizing gravity in the early 1950s (Blum and Brill, 2019, section 5), initially showed in the 'relative state' interpretation of quantum mechanics given in the thesis of his PhD student Hugh Everett III.

The story of the origins of Everett's interpretation has been discussed in detail by Osnaghi *et al.* (2009), and there is no need to retell that story here. We will instead focus specifically on the connection between Everett's interpretation and the work on quantum gravity and cosmology in the group of John Wheeler at Princeton in the mid-1950s. On this issue, Freire, Freitas, and Osnaghi did not say much. What little they said, however, remains entirely correct also in the light of our more detailed analysis, so that we take the relevant passage from their paper [p. 103] as our starting point:

> Even though Everett denied having received any external input for undertaking his work [ ... ] he and [Charles] Misner allude to the influence that Wheeler's charac-teristic approach to theoretical physics might have exerted on the development of the relative state formulation. Misner says: 'He [Wheeler] was preaching this idea that you just look at the equations and there were the fundamentals of physics [ ... ] you followed their conclusions and gave them a serious hearing. He was doing that on these solutions of Einstein's equations like Wormholes and Geons.' And Everett replies: 'I've got to admit that that is right, and might very well have been totally instrumental in what happened.'

> The analysis of Everett's early writings does not indicate that his search for an original approach to quantum mechanics was inspired by issues of cosmology. Yet, there is little doubt that Wheeler's interest in Everett's ideas was enhanced by his recent involvement in that area of research. This is mostly apparent from the final version of the dissertation, in the drafting of which Wheeler took an important part.

What do we have to add to this? On the one hand, a rather surprising common-origins story that is not made explicit by Freire, Freitas, and Osnaghi. Wheeler's methodology described by Misner was referred to by Wheeler himself as 'daring conservatism'. And as shown by one of the authors (Blum and Brill, 2019), it was this methodology that led Wheeler to pursue quantum gravity, as a combination of the established fundamental truths of physics (general relativity, quantum theory), in contrast to the rampant speculation and incessant postulation of new fields and particles he saw at play in contemporary high-energy physics. What thus joins quantum gravity and the Everett interpretation is the attitude of working with what one has got (Einstein equations and quantization rules on the one hand, the Schrödinger equation on the other), rather than bringing in extraneous concepts (new particles or classical observers).

This is hardly more than an intriguing observation. Our main focus will thus be on a different point: How exactly did Wheeler (and then other physicists after him) connect the Everett interpretation to quantum gravity and quantum cosmology? The obvious and fundamental connection is of course through Everett's concept of a 'Universal Wave Function'. A universal wave function provided, as Wheeler put it in his published commentary to Everett's thesis, the only 'self-consistent system of ideas [ . . . ] to explain what one shall mean by quantizing a closed system like the universe of general relativity' (Wheeler, 1957a, p. 465). Such a closed system could not provide an outside, classical observer and was thus anathema to the Copenhagen interpretation, as we have discussed in the previous section.

But we need to look somewhat more in detail at the connection between the Everett interpretation and relativistic cosmology, beyond the mere fact that Everett's theory formally contained a wave function that depended on the variables describing the spatio-temporal structure of the entire universe. For the more specific implications of this connection, we find various hints in Wheeler's notebooks and letters, which are preserved at the American Philosophical Society Library in Philadelphia. In particular we will be drawing on his series of Relativity notebooks, specifically notebooks 2 (abbreviated WRII) and 3 (WRIII).

Let us first state what the Everett interpretation did not (need to) do in Wheeler's view: provide a motivation for the quantization of gravity. Starting from the Everett interpretation, one might indeed argue that every physical system needs to be quantized to be included in the universal wave function. But for Wheeler the logic was the other way around. He strongly believed in the need to quantize gravity for independent reasons. Twice in the mid-1950s he had extended discussions with experts in the field on the formal difficulties involved in quantizing general relativity—with Bryce DeWitt (who had written his PhD thesis on perturbative quantum gravity with Julian Schwinger) and with Jim Anderson (who had written his PhD thesis on canonical quantum gravity with Peter Bergmann). In both cases, Wheeler insisted that the formal difficulties had to be surmountable, because general relativity had to be quantizable for physical reasons. In his notes on the conversation with Anderson (15 March 1954, WRII, p. 137), Wheeler remarked '*Have* to be able to quantize Bohr–Rosenfeld argument', referring to the argument for quantization that Rosenfeld would later so vehemently reject. And similarly in his conversation with DeWitt (25 January 1955, WRIII, p. 113): 'Get into contradiction JAW [John Archibald Wheeler] claims strongly but de W. not so certain that one will get into great trouble if other fields quantized, gravitation not.'[15]

---

[15] DeWitt then apparently went on to explicate how such a theory without a quantized gravitational field would work, pitching a semi-classical version of the Einstein equations, where the source is the expectation value of the energy-momentum tensor. This is to our knowledge the first recorded instance of these equations, now also known as the Møller–Rosenfeld Equations.

So for Wheeler, the need for the Everett interpretation arose from the need to quantize gravity and the concomitant need to introduce something like a wave function of the universe. But was the Everett interpretation relevant to the programme of quantizing gravity beyond this simple statement? Wheeler was not very explicit on this in his published papers of the time. We have already cited his very vague remarks in his comment on Everett's paper. A similarly vague and equivalent statement is to be found in footnote 1 of (Wheeler, 1957b). More explicit remarks can be found in Wheeler's notebooks. Here we can identify two specific hopes regarding the impact of Everett's interpretation on quantum cosmology. We shall discuss them in turn.

The first one concerned dealing with a universal wave function that depends on (and thus, in some sense, must give a probability distribution over) global properties of space-time, not just local, field-like ones. In particular, Wheeler was concerned with a wave function that depended on the dimensionality of space. This idea seems to have arisen in the first months of 1954. Wheeler had primarily been working on the classical description of geons, including 'neutrino geons', i.e., stable, particle-like field configurations composed of a classical spinorial field and the gravitational field (the original geons used the electromagnetic instead of a spinorial field). The various kinds of geons seemed to offer rich possibilities for a classical description of particle-like entities in terms of a pure field theory. But after calculations on 'doughnut geons' on 30 March 1954, we find some reflective remarks, dated 8 April 1954 (WRII, p.146), written on a 'train nearing Jackson, SC':

> Last night reviewed status of geons + elementary particles with Dale Babcock [a physical chemist, whom Wheeler knew from the Manhattan Project]. He asked about fundamentality of grav-e.m.-ν description of elem. particles, even if it can be worked out. I agree that his question has a strong point. Can therefore at this stage *either* stop at a description which works down to geons of size several multiples that of elem. particles, and say qualitatively, we can only speak roughly about how things go from here on in from here on out renormalizations, etc; *or* try to go whole hog. Remember Bohr: 'Pit your difficulties against each other.' If then there is to be any understanding of the great issue of the unification of g-em-ν, do we not need some other great difficulty to play off against it? What about the *dimensionality of space-time* as this issue? Make a first try just to see the nature of the problem. How do we know that there is any dimensionality at bottom? Suppose number of dimensions indefinite.

Invoking a Bohrian methodology,[16] Wheeler thus felt that it would not be sufficient to simply see how a quantum theory of gravity would modify his classical geon solutions, but that he would have to further complicate matters by making the dimensionality of space-time indefinite in quantum gravity in order to get the necessary creative dynamics going—Wheeler was certainly not one to shy away from hard problems.[17] The difficulty

---

[16] On which see the forthcoming PhD thesis of Stefano Furlan.
[17] The epigraph to WRIII was: 'The human brain needs to be put [sic] tasks beyond its power'.

with an indefinite dimensionality was of course how to interpret a cosmological wave function that gives a probability distribution over the dimensionality of space-time. In his notebooks, Wheeler frequently reflected on his research programme, and in one such set of notes, dated 21 July 1954 (i.e., several months after initially considering an indefinite dimensionality, WRIII, p. 25), we find him pondering:

> Idea is, that just as q[uantum] mech[anics] sum over paths, a q[uantum] mech[anical] sum over 'ylem'[18] gives probability concentrated about 4-dim spacetime. [...] Have also the question what we will mean by probability amplitude what possibilities exist to determine the probability amplitude for the whole universe!

Wheeler primarily thought in terms of Feynman's path integral formulation of quantum mechanics when considering these questions, as witnessed, e.g., by a letter he wrote to John von Neumann on 12 October 1954 (WRIII, p. 79). In the letter he sought von Neumann's mathematical advice on dealing with spaces with indefinite dimensionality:

> This circumstance invited one to consider the Feynman sum to be extended, not only over all signatures and topologies for a given dimensionality, but also over all dimensionalities. One knows that general relativity is rather a wash-out in 2 and 3 dimensions, and one might suppose that 5 and higher dimensionalities for some built-in weighting reason don't cut much ice, if one wants to understand along these lines why four is the important dimensionality in a classical world.

There are, however, hints that Wheeler also considered the relevance of Everett's interpretation to this problem, which he discussed with several of his students in the spring of 1955, among them Charles Misner, Hale Trotter, Pierre Conner, and, indeed, Hugh Everett. In notes on these discussions (dated 14 April 1955, and thus at a time when Everett presumably had the basic ideas of his thesis in place, WRIII p. 141), we find among the conclusions: 'Hugh Everett - might not ask as much as Feynman scheme.'[19]

The second hope that Wheeler attached to the Everett interpretation was that it would help with integrating Mach's principle into quantum gravity. Mach's principle had played a central role in Wheeler's thinking about general relativity from the outset. In the classical theory, a realization of Mach's principle meant (to Wheeler) that a given mass distribution would uniquely define the metric. This in turn, he conjectured

---

[18] This refers to the primordial matter in big bang nucleosynthesis, as introduced in (Alpher, 1948).
[19] It should be kept in mind when interpreting this short note that Wheeler considered both Feynman's path integral and Everett's relative state-formulation as reformulations of quantum mechanics, which also brought with them new interpretations. So, on 19 March 1972, he remarked in a letter to Max Jammer (Barrett and Byrne, 2012, Chapter 22), when speaking about Everett: 'Bohr did not take to his way of describing q. mechanics, as he also earlier had not accepted Feynman's way of describing q. mechanics.' Wheeler thus did not perform a distinction that we nowadays consider natural, namely to take path integrals as a reformulation of quantum mechanics, and Everett's work as a reinterpretation.

following Einstein, was only the case if space was closed.[20, 21] As for time, Wheeler was on the fence for a long time whether the universe should also be temporally closed.[22] In any case, we find on p. 153 of WRIII, in 'Reflections on Plan of Physics' dated 6 May 1955, Wheeler posing these questions in the context of quantum gravity:

> If one sticks to compact manifolds, should system be closed in time as well as in space. If so, does idea of wave function have use only over limited portions of manifold, in sense that trans[ition] prob[ability] has δ-f[u]n[ction] character or identity form over whole manifold. Important to explore Everett's ideas + see if one can describe all of nature by a single wave f[u]n[ction].

The precise line of thought here is somewhat sketchy. But given the invocation of Machian themes and the problematization of a trivial transition probability, it appears that Wheeler is here addressing for the first time a question that will reappear later on in the work of DeWitt: what if the wave function of the universe is unique (as the metric for a given mass distribution would be in Machian classical general relativity)? The world would be stuck in a single energy eigenstate with no non-trivial time evolution on a global scale. In such a case, the Feynman approach, which was concerned precisely with the calculation of transition probabilities, would be of no use. Indeed, the entire notion of a propagator, time-evolving a state from the asymptotic past to the asymptotic future, would lose its relevance, as Wheeler would remark in January 1957 at the Chapel Hill conference, after a talk by Stanley Deser on the possibility that quantum gravity effects might serve to regularize quantum field theoretical propagators:

> [T]he whole nature of quantum mechanics may be different in general relativity since in every other part of quantum theory, the space in which physics is going on is thought of as being divorced from the physics. General relativity, however, includes the space as an integral part of the physics and it is impossible to get outside of space to observe the physics. Another important thought is that the concept of eigenstates of the total energy is meaningless for a closed universe. However, there exists the proposal that there is one 'universal wave function'. This function has already been discussed by Everett, and it might be easier to look for this 'universal wave function' than to look for all the propagators.[23]

---

[20] (Blum and Brill, 2019, p. 18) This was still his assessment in 1955, see WRIII, page 230, where Wheeler cites Einstein as considering 'compactness in space, cyl[indrical] universe, as essence of Mach's principle.'

[21] By 'closed' we mean here specifically finite and without boundary, i.e., a closed manifold. Both Wheeler and DeWitt would at times also speak of 'compactness' in this context, which might also refer to a finite universe with boundary; however there is no indication that either ever considered a finite universe with boundary (and boundary conditions), and thus we will consider the terms 'compact' and 'closed' to be interchangeable in the context of our historical narrative.

[22] See, e.g., p. 238 of WRIII, where Wheeler ponders the '3 possibilities? Time closed [ . . . ]; time infinite; or time ends sharply'.

[23] (DeWitt and Rickles, 2011, pp. 269–70).

What we can glean from these remarks is that Wheeler hoped that the Everett interpretation might help analyse the physics of a closed, quantum universe in which the notions of energy eigenstates and time evolutions became problematic.

Both of these points (probability amplitudes for global, topological variables and trivial time evolution for a unique universal wave function) are only briefly touched in Wheeler's notebooks. In the spring of 1956, his interest in the Everett interpretation would take a decisive blow, when Copenhagen refused to give Everett's ideas the imprimatur. This episode is told in detail in (Osnaghi et al., 2009). For our context, we need only look a little more closely at one exchange (Barrett and Byrne, 2012, Chapter 12). After a visit by Wheeler in April 1956, a seminar was held at the Bohr Institute in Copenhagen to discuss a draft of Everett's thesis, led by Alexander Stern, an American physicist in residence at the Institute. On 20 May, Stern reported to Wheeler the (largely negative) verdict reached at this seminar. With regards to the concept of the universal wave function, Stern did not criticize it because it was in conflict with the Copenhagen interpretation and with the need for an external classical domain; rather he remarked that it was an insufficient concept in and of itself, that it seemed to 'lack meaningful content'. And in his reply (25 May 1956), Wheeler largely agreed; rather than cite the ideas from his notebook, which were very sketchy but tied to concrete theoretical problems, he mainly interpreted Everett's work as showing how the concept of a wave function of the universe was a generalization of, and thus compatible with, the wave function of the Copenhagen interpretation:

> Everett traces out a correspondence between the 'correlations' in his model universe on the one hand, and on the other hand what we observe when we go about making measurement.

But as to why the universal wave function should be introduced in the first place, Wheeler did not offer any specific reasons, describing Everett's postulate as a 'creative act, beyond any step by step pre-justification' (and thus, one should add, disavowing its origins in daring conservatism), vaguely hinting at the 'convenience' of introducing it:

> If the universal wave function is not subject to external observation, is it not as you put it 'a matter of theology'? To this question I should be frank in saying I have no complete answer, nor am I sure that it is necessary to give one. Is there any difference on this score between Everett's universe and Laplace's universe? No one seriously believed that it would be a practical possibility ever to know at one moment the position and velocity of every particle, but it was convenient to *postulate* that these quantities nevertheless had well defined values. Likewise Everett *postulates* that the 'universal wave function' had at one moment a well defined dependence upon the state variables, and therefore also a well defined dependence upon these variable at every other moment.

But no mention of, for example, what those variables might be. As we have seen, Wheeler's remarks in his published work were then similarly vague, and he soon stopped citing Everett, all through the 1960s. Most significantly perhaps, in a philo-sophical article published in the relaunched journal *The Monist*, Wheeler (1962, pp. 63–64) capped the following passage:

> But what does it mean to 'observe' the geometry of a closed universe? There is no platform outside the system on which one can stand to conduct the measurement! Evidently one must seek new insight into the meaning of the probability amplitude—most probably from the mathematical formalism itself.

by citing not Everett but, of all people, Rosenfeld (1955), and did not even mention Everett (or the universal wave function) when considering the question whether the 'geometrodynamical model universe also [has] a multitude of possible energy states' (p. 65). In a uniquely Wheelerian twist, Wheeler ultimately returned to the Copen-hagen interpretation in the 1970s (now elevated to the status of 'the quantum prin-ciple'), arguing that it actually provided the optimal conceptual framework for (quantum) cosmology. Rather than having to postulate an ideal observer, who is able to see the entire universe (i.e., the universal wave function) from outside, it embraced a holistic view where we cannot think ourselves outside of the universe:

> [B]eyond the rules of quantum mechanics for calculating answers from a Hamil-tonian stands the quantum principle. It tells what question it makes sense for the observer to ask. It promotes observer to participator. It joins participator with system in a 'wholeness' (Niels Bohr) [...] quite foreign to classical physics. It demolishes the view we once had that the universe sits safely 'out there', that we can observe what goes on in it from behind a foot-thick slab of plate glass without ourselves being involved in what goes on. We have learned that to observe even so minuscule an object as an electron we have to shatter that slab of glass. We have to reach out and insert a measuring device. We can put in a device to measure position or we can insert a device to measure momentum. But the installation of one prevents the insertion of the other. We ourselves have to decide which it is that we will do. Whichever it is, it has an unpredictable effect on the future of that electron. To that degree the future of the universe is changed. We changed it. We have to cross out that old word 'observer' and replace it by the new word 'participator'.[24]

While this re-assertion of the Copenhagen interpretation is certainly not unreasonable and would lead Wheeler to his formulation of the anthropic principle, it did not, to our knowledge, have much of an impact on the debates on the interpretation of quantum mechanics. Instead of following Wheeler's trajectory, we thus now turn to the further development of the Everett interpretation. And with Wheeler abandoning the Everett interpretation and Everett himself leaving physics altogether,[25] it was a different

---

[24] (Patton and Wheeler, 1975, pp. 560–62).
[25] On Everett's biography, see (Byrne, 2010).

physicist who kept the Everett interpretation and the connection to quantum gravity alive: Bryce DeWitt.

# 15.3 FROM DEWITT'S ENDORSEMENT TO THE OXFORD WORKSHOPS

DeWitt was one of Everett's earliest readers, as he edited the special issue of *Reviews of Modern Physics* in which Everett's and Wheeler's papers were published. DeWitt was immediately intrigued and mentioned Everett on several occasions, such as his talk at Jablonna, mentioned in section 15.1. But he only became a convert to and a proselytizer of Everett's interpretation about a decade later.

The exact cause of this conversion is somewhat murky, in particular the question whether the main driving force was questions of interpretation or quantum gravity. Thiago Hartz has clearly advocated the former,[26] primarily citing Rosenfeld's harsh rejection of DeWitt's attempts at using a Bohr–Rosenfeld analysis of quantum gravity. DeWitt himself, unsurprisingly, gave a different story, though it also takes interpretational questions as its starting point (DeWitt, 2011, p. 95). DeWitt invoked a meeting with Max Jammer, probably around 1966, after the completion of Jammer's work on the early history of quantum mechanics (Jammer, 1966); Jammer was preparing his successor volume on the philosophy of quantum mechanics, and DeWitt pointed him to Everett, whom Jammer had not heard of. DeWitt further invoked discussions with his own graduate student, Neill Graham, who had approached DeWitt wanting to write a thesis on 'foundations of quantum mechanics'. And we have definite evidence that— at least originally—DeWitt did not expect the Everett interpretation to be of much relevance for quantum gravity, as he wrote to Wheeler (7 May 1957, (Barrett and Byrne, 2012, Chapter 16)) after first reading Everett's manuscript:

> [I]t seems extremely unlikely that current physics (including quantum gravidy-namics!) will be much affected by the new point of view.

But the mid-1960s also saw an unusual visitor at DeWitt's North Carolina Institute of Field Physics, the itinerant Hungarian emigré Frigyes Károlyházy, who was investigating how quantum uncertainties in the space-time metric might lead to decoherence of quantum superpositions for macroscopic objects (Károlyházy, 1966), implying that the connection between quantum gravity and quantum foundations was already being investigated in Chapel Hill before DeWitt's official endorsement of Everett.

---

[26] See Thiago Hartz's talk on 'Bryce DeWitt's road to the Many Worlds' at the HQ-4 meeting in San Sebastian in 2015: https://youtu.be/8VE9Pbn46DE.

Whatever the original motivation may have been for DeWitt to return to the Everett interpretation, he immediately (re-)established its connection with quantum gravity. Already when first presenting the Everett interpretation, at the Battelle Rencontres meeting in Seattle in 1967, DeWitt (1968) framed his presentation in terms of quantum cosmology, in greater detail than Wheeler had ever done, thereby also addressing some new aspects. He began:

> Among the lectures to which we have listened in these Rencontres have been several by Penrose which called attention to the inadequacy of classical general relativity for dealing with the phenomenon of gravitational collapse. It has frequently been suggested that this inadequacy might be overcome by building a theory which takes quantum effects into account.

By 1967, the status of quantum gravity had thus changed: with the singularity theorems of Hawking (1965) and Penrose (1965), the quantization of gravity had not just received a new motivation, but also, for the first time, potential empirical domains of applicability, namely the early universe and black holes. DeWitt then went on:

> Since space-time serves as the common arena for both classical and quantum processes it is evident that one faces unusual difficulties in this theory in attempting to define Bohr's classical realm. The difficulties are particularly acute in a cosmology which assumes that space is compact. Wheeler has shown us that in this case the *state* of the world is very naturally represented as a *function over 3-geometries*. [ ... ] [T]here can be no doubt that this function is quite literally a *wave function of the universe*.
>
> To anyone educated in the Copenhagen school such a function is meaningless, since it leaves no room for a classical realm. The external observer has no place to stand. In order to give meaning to a universal wave function one must find a radically new way of looking at quantum mechanics. This is what Everett and Wheeler have done.

Now, in addition to the explicitness, which is conspicuously lacking from Wheeler's and Everett's extant considerations, DeWitt's framing is noteworthy for introducing the compactness of space as a mere possibility. Due to Wheeler's Machian predilections, this had always been assumed, more or less tacitly. For DeWitt it was merely one option and in the trilogy on quantum gravity he published that same year, he devoted the first part (DeWitt, 1967a) (which features the Everett interpretation and which we will consequently discuss in the following) to 'closed, finite worlds' (p. 1115) and the second and third part (DeWitt, 1967b, c) to infinite worlds. And these latter parts make no reference to Everett (or the interpretation of quantum mechanics in general) whatsoever.

This did not affect the role of the many-worlds interpretation (as DeWitt began calling it) as an interpretation of quantum mechanics more generally; but for DeWitt it was cosmologically relevant only for a closed universe. In this specific context, however, DeWitt really engaged with the impact of the Everett interpretation on the nitty-gritty technical details. In doing so, he in fact addressed the two points that we have already identified in Wheeler's notes, viz. the probability interpretation for the universal wave function as a whole and the problem of the possible uniqueness of the universal wave function. We will discuss these two points in turn.

Regarding the probability interpretation for the universal wave function, this was of course closely connected to a central issue in the Everett theory as an interpretation of non-relativistic quantum mechanics, namely how the notion of probability arose in a theory that was purportedly deterministic. Everett's answer (Everett III, 1957) was the following: in order to make statements about relative frequencies, we need to be able to identify a 'typical' branch. Just as in statistical mechanics, the notion of typicality rests on assigning a measure, to branches in the case of quantum mechanics, to trajectories in phase space for statistical mechanics. Finding a suitable measure, e.g., by imposing desired additivity properties, is essentially a mathematical problem, and Everett then went on to show that the only suitable measure on branches, i.e., on the elements of a superposition of orthogonal states written as $\Sigma_i a_i \phi_i$, is the modulus squared of the amplitudes, i.e., $|a_i|^2$, giving the probabilities of the usual Born interpretation.

Everett had applied this interpretation to states where the $\phi_i$ were product states composed of the eigenstate of a system and of an observer registering the corresponding eigenvalue. But nothing in the derivation of the probabilities relied on this factorization, and on the face of it Everett's derivation appeared to allow the assignment of probabilities also to branchings of the universal wave function as a whole, e.g., (to take up Wheeler's hopes) into states with different dimensionality. Independently of the cosmological implications (at least explicitly), DeWitt had objected to this derivation from the start. In the already-mentioned letter to Wheeler of 7 May 1957, DeWitt remarked:

> The point at issue is quintessentially summed up in a single phrase of Everett's—in the section headed 'Quantitative Interpretation, Measure . . . . etc.'—: 'Probability theory is equivalent to measure theory mathematically . . . .'. Yes, but *not* epistemologically. There is a vast difference between the two.

It was thus hardly a coincidence that a central result of Neil Graham's thesis (Graham, 1971), which DeWitt also presented prominently in his Battelle lecture (with due attribution of credit), was a new derivation of the Born rule for probability in an Everettian context. This derivation relied not on the analogy between measure theory and probability, but actually introduced the relative frequency $W_j^A$ of measuring a certain eigenvalue $A_j$ for an observable $A$ as an Hermitian operator (and thus an observable) in its own right. This operator was defined to act on a Hilbert space composed of a large number of identical systems for which the observable $A$ could be measured independently. The derivation of the Born rule thus relied not merely on the assignment of a measure to elements of an orthogonal decomposition of a state, but also on the association of that orthogonal decomposition with an actual measurement. As Graham emphasized in his thesis (p. 45):

> There are an infinity of ways in which its state $|\psi\rangle$ can be expanded in an orthonormal basis. But as long as we consider the quantum system *by itself* all these possible bases will be on equal footing. An isolated quantum system *cannot* be given a parallel world interpretation, since there is no unique prescription for resolving the state of the system into states of parallel worlds.

To some extent this obviated the advantage, emphasized by Wheeler, that the Everett interpretation provided a superior conceptual framework for dealing with isolated systems, such as in particular the universe as a whole. It certainly meant that there really was no straightforward way, as Wheeler had hoped, to interpret a probability distribution over global properties of the universe. Such a wave function could, in what Jammer would call the Everett–Wheeler–Graham interpretation, only be understood in two ways. For one, it could be peaked sharply around a certain value of that global property, in which case Graham and DeWitt assumed that this peak value was indeed the actual value of the quantity in question:

> Note that we are *not* assuming a probability interpretation for $|\psi(x)|^2$. We are just saying this: Look at the mathematical function $|\psi(x)|^2$. If it has a single sharp peak around a particular $x_0$, then $X$ has a definite value, which is $x_0$. If $|\psi(x)|^2$ does not have a single sharp peak then the observable $X$ does not have a definite numerical value and some other method of interpretation must be used.   (Graham, 1971, p. 8)

The other possibility was to decompose the universal wave function into different branches, corresponding to different values of the global property, and then identify one of those branches as our world (and thus the associated value of the global property as our value), without being able to make any quantitative statements regarding the 'probability' of the specific branch we were living in.

In fact, DeWitt would make use of both possibilities in his work on canonical quantum gravity in the mid-1960s, where he introduced for the first time a differential equation for the wave function of the universe, the so-called Wheeler–DeWitt (WdW) equation (DeWitt, 1967a). The Wheeler–DeWitt equation is of the form

$$(\mathbf{H} + \mathrm{H})\psi = 0 \tag{1}$$

where $\mathbf{H}$ is the Hamiltonian for the metric variables and $\mathrm{H}$ is the Hamiltonian for the matter content of the universe. The WdW equation is not a Schrödinger equation, because it singles out a specific eigenvalue of the Hamiltonian operator (zero) and there is thus no corresponding time-dependent equation and no straightforward way to include a time parameter. This, in its simplest form, is what is known as the problem of time.

DeWitt addressed this issue for the simplified model of a quantum Friedmann universe, where the universal wave function depends just on one metric variable, namely the radius of the universe $R$. DeWitt could quite easily obtain a solution to the thus simplified WdW equation in the WKB approximation under the assumption that H was just a constant energy, but this wave function was necessarily static; without a time parameter it could not express any form of expansion or contraction of the universe. In order to have something that could eventually be used as a time parameter, DeWitt included a further dynamical variable $q$, describing a collective motion of the matter content of the universe. But even if, in a universal wave function $\psi(R,q)$, $q$ was interpreted as a time component, one would still be hard pressed to interpret the temporal development of this wave function as the expansion or contraction of the universe, precisely because there was no clear way how to interpret such a wave function as a probability distribution over the possible values of $R$.

DeWitt did not raise this concern explicitly in his paper, but he most certainly and straightforwardly circumvented it. He first constructed a general WKB solution for the Wheeler–DeWitt equation. The existence of such a solution required that the eigenvalues of H and H were precisely matched, so that one was the negative of the other. DeWitt then further assumed that there was not just one such matched pair, but a large number, i.e., that the total Hamiltonian H + H had a highly degenerate zero eigenvalue. He then used this large number of solutions to construct sharply peaked wave packets, whose peaks traced out a quasi-classical trajectory in the $R - q$-plane. Due to the commensurability condition on the gravitational and matter energy eigenvalues, these trajectories were actually closed Lissajous figures of finite length. According to the interpretation of such a sharply peaked wave packet as giving the actual value of the physical quantity in question, these trajectories could then be interpreted (when adequately parameterized) as describing a cyclic universe, where phases of expansion (increasing $R$) alternated with phases of contraction (decreasing $R$), while avoiding the singular behaviour at the turning point (big bang/crunch) which one obtained in classical general relativity.

This construction of a non-singular cyclic universe from quantum cosmology of course relied in an essential manner on the WdW equation having a large number of solutions; without this large number of solutions, one would not be able to construct sharply peaked wave packets. This brings us to the second issue we have already seen in Wheeler's notes: what if the wave function of the universe is unique, i.e., what if the WdW equation (for a given matter Hamiltonian) has just one solution? DeWitt was well aware of this possibility, indeed even considered it the more plausible, not because of Mach's principle, but due to the striking alignment between matter and gravitational Hamiltonians necessary to construct even one solution. In order to at least sketch a possible resolution of this potential difficulty, DeWitt drew on the other possible interpretation of the wave function of the universe that his and Graham's re-interpretation of Everett allowed for, namely, that we are living just in one component of this wave function, utterly oblivious to the other components (and unable to in any way quantify the degree to which they exist or are probable). For while no wave packet tracing out a classical time evolution can be constructed from just one eigenfunction, a component of that eigenfunction, DeWitt argued, might very well do just that, thereby providing an explanation for the observed temporal evolution of the universe even for the case of a unique solution of the WdW equation.

While this was certainly the most sophisticated use of the Everett interpretation for quantum cosmology at the time, it played little role in the reception of that interpretation in the debate on the interpretation of quantum mechanics that followed DeWitt's endorsement. DeWitt wrote an article on the many-worlds interpretation for *Physics Today* (DeWitt, 1970), briefly mentioning also its cosmological relevance. In the following year, *Physics Today* published a number of letters in reaction to DeWitt's article (Ballentine *et al.*, 1971); cosmology was not mentioned. Jammer (1974) also discussed the Everett interpretation in his book on *The Interpretation of Quantum Mechanics in Historical Perspective*, but only very briefly mentioned its relevance for 'the quantum theory of general relativity' (p. 508). DeWitt had brought Everett to the interpretation debate, but it would take a few more years to bring the interpretation debate to quantum cosmology.

Indeed, we can trace the first attempts to (a) bring other interpretations (besides Everett) to bear on the problems of quantum cosmology and (b) to find novel interpretations that not only are compatible with quantum gravity (like Everett), but actually rely on it, to a symposium on quantum gravity in Oxford in 1980. The symposium was the second of its kind, organized by Chris Isham, Roger Penrose, and Dennis Sciama. The first one had been held in 1974 and saw essentially no mention of the foundations of quantum theory, with the exception of Wheeler's idiosyncratic take on the Copenhagen interpretation, mentioned earlier (Isham *et al.*, 1975). Things were slightly different the next time around.

On the one hand, Isham had got interested in the interpretation debate and the philosophy of quantum mechanics, had met John Bell, one of the protagonists of that debate, during a visit to CERN, and had invited him to the second Oxford workshop.[27] Bell in turn had been one of the first in the emerging quantum foundations community to engage with the Everett interpretation, albeit critically. Indeed, he had developed a reading of the Everett interpretation in which it was essentially the same as the de Broglie–Bohm interpretation, with the sole difference that the Bohm interpretation selected one of the branches as real (the one containing the actual particle), while the Everett interpretation considered them all to be equally real (Bell, 1976). In Oxford, Bell (1981) consequently gave a talk entitled 'Quantum Mechanics for Cosmologists' in which he argued that the de Broglie–Bohm interpretation shared Everett's conception of a 'world wave function' and could thus equally well be brought to bear on problems of cosmology. Bell's analysis did not go much deeper than that and the talk was primarily concerned with interpretative issues; Isham remembers that the talk was received 'politely' by the audience. But Bell had demonstrated that quantum gravity

---

[27] E-mail from Chris Isham to one of the authors (AB), 28 July 2020.

(and cosmology in particular) was no longer the sole domain of the Everett interpretation and was now an arena in which different interpretations could test their mettle.

On the other hand, Penrose had been led to the question of quantum measurement through his interest in the emerging black hole information paradox. In his talk at the Oxford symposium (Penrose, 1981), Penrose proposed to identify the non-unitarity involved in the creation and subsequent Hawking evaporation of a black hole with the non-unitarity involved in the collapse of the wave function upon measurement: objects that were 'sufficiently "macroscopic" that they significantly affect the space-time geometry' (p. 267) were supposed to be responsible for wave function collapse. These ideas were not yet fleshed out. It was also not the first time that gravitation was invoked in explaining the measurement process; we briefly mentioned Károlyházy's work on decoherence through quantum uncertainties in the metric. But by explicitly identifying non-unitary wave function collapse as a result of gravitation, Penrose was arguing that quantum gravitation played an essential role in the interpretation of quantum mechanics, not just as an arbiter between different interpretations, but as a physical principle underlying the interpretation itself.

That these new roles for quantum gravity in the interpretation debate, being proposed by Penrose and Bell, were not just flashes in the pan, but significantly changed the relation between the two fields, can be clearly gauged from the programme of the third instalment of the Oxford Symposium held in 1984. This meeting was, as Penrose and Isham remarked in the preface, 'more concerned with quantum theory and its foundations than with gravitational theory proper' and consequently carried the title 'Quantum Concepts in Space and Time'. It contained papers on neutron interferometry and the possibility of using it to test the effects of gravitation (Anton Zeilinger), the Aharonov–Bohm effect in the presence of a gravitational field (Jeeva Anandan), gravitational state reduction (by both Károlyházy and Penrose), Everettian cosmology (Frank Tipler), Everett and quantum computing (David Deutsch), and an argument that quantum gravity considerations favoured Nelson's stochastic interpretation (Lee Smolin) (Penrose and Isham, 1986). Clearly, a new era in the relation between quantum gravity and the interpretation debate had begun, an era that we are still in today.

## References

Alpher, R. (1948). A neutron-capture theory of the formation and relative abundance of the elements. *Physical Review*, **74**, 1577–1589.

Ballentine, L. E., Pearle, P., Walker, E. H., Sachs, M., Koga, T., Gerver, J., and DeWitt, B. (1971). Quantum-mechanics debate. *Physics Today*, **24**(4), 36–44.

Barrett, J. A. and Byrne, P. (eds) (2012). *The Everett Interpretation of Quantum Mechanics: Collected Works 1955–1980 with Commentary*. Princeton: Princeton University Press.

Bell, J. (1976). The Measurement Theory of Everett and de Broglie's Pilot Wave. In M. Flato, Z. Maric, A. Milojevic, D. Sternheimer, and J. P. Vigier (eds), *Quantum Mechanics*,

*Determinism, Causality, and Particles*, Dordrecht: D. Reidel Publishing Company, pp. 11–17.

Bell, J. (1981). Quantum mechanics for cosmologists. In C. Isham, R. Penrose, and D. Sciama (eds), *Quantum Gravity 2: A Second Oxford Symposium*, Oxford: Clarendon Press, pp. 611–637.

Blum, A. S. and Brill, D. (2019). Tokyo Wheeler or the epistemic preconditions of the renaissance of relativity. arXiv:1905.05988 [physics.hist-ph].

Blum, A. S. and Rickles, D. (eds) (2018). *Quantum Gravity in the First Half of the XXth Century: A Sourcebook*. Berlin: Edition Open Access.

Bohr, N. and Rosenfeld, L. (1933). Zur Frage der Messbarkeit der elektromagnetischen Feldgrössen. *Mathematisk-fysiske Meddelelser Det Kgl. Danske Videnskabernes Selskab.*, **12**(8).

Byrne, P. (2010). *The Many Worlds of Hugh Everett III.* Oxford: Oxford University Press.

Darrigol, O. (1991). Cohérence et complétude de la mécanique quantique: l'exemple de 'Bohr-Rosenfeld'. *Revue d'histoire des sciences*, **44**(2), 137–179.

DeWitt, B. S. (1962). The Quantization of Geometry. In *Gravitation: an Introduction to Current Research*, Wiley, pages 266–381.

DeWitt, B. S. (1967a). Quantum theory of gravity. I. The canonical theory. *Physical Review*, **160**, 1113–1148.

DeWitt, B. S. (1967b). Quantum theory of gravity. II. The manifestly covariant theory. *Physical Review*, **162**, 1195–1239.

DeWitt, B. S. (1967c). Quantum theory of gravity. III. Applications of the covariant theory. *Physical Review*, **162**, 1239–1256.

DeWitt, B. S. (1968). The Everett-Wheeler Interpretation of Quantum Mechanics. In C. M. DeWitt and J. A. Wheeler (eds), *Battelle Rencontres: 1967 Lectures in Mathematics and Physics*, New York: W. A. Benjamin, pp. 318–332.

DeWitt, B. S. (1970). Quantum mechanics and reality. *Physics Today*, **23**(9), 30–35.

DeWitt, C. M. (2011). *The Pursuit of Quantum Gravity: Memoirs of Bryce DeWitt from 1946 to 2004.* Berlin: Springer.

DeWitt, C. M. and Rickles, D. (eds) (2011). *The Role of Gravitation in Physics: Report from the 1957 Chapel Hill Conference.* Berlin: Edition Open Access.

Everett III, H. (1957). 'Relative state' formulation of quantum mechanics. *Reviews of Modern Physics*, **29**(3), 454–462.

Graham, R. N. (1971). *The Everett Interpretation of Quantum Mechanics.* PhD thesis, University of North Carolina at Chapel Hill.

Hartz, T. and Freire, O. (2015). Uses and appropriations of Niels Bohr's ideas about quantum field measurement, 1930–1965. In F. Aaserud and H. Kragh (eds), *One hundred years of the Bohr atom: Proceedings from a conference*, volume 1 of *Scientia Danica. Series M, Mathematica et physica*, Copenhagen: Det Kongelige Danske Videnskabernes Selskab., pp. 397–418.

Hawking, S. W. (1965). Occurrence of singularities in open universes. *Physical Review Letters*, **15**, 689–690.

Henley, E. M. and Thirring, W. (1962). *Elementary Quantum Field Theory.* New York: McGraw-Hill.

Infeld, L. (ed.) (1964). *Proceedings, Relativistic Theories of Gravitation: Warsaw and Jablonna, Poland, July 25–31, 1962.* Oxford: Pergamon Press.

Isham, C., Penrose, R., and Sciama, D. (eds) (1975). *Quantum Gravity: An Oxford Symposium.* Oxford: Clarendon Press.

Jacobsen, A. (2012). *Léon Rosenfeld: Physics, philosophy, and politics in the twentieth century.* Singapore: World Scientific Publishing.

Jammer, M. (1966). *The conceptual development of quantum mechanics.* New York: McGraw-Hill.

Jammer, M. (1974). *The Philosophy of Quantum Mechanics: The Interpretations of Quantum Mechanics in Historical Perspective.* New York: Wiley.

Károlyházy, F. (1966). Gravitation and quantum mechanics of macroscopic objects. *Il Nuovo Cimento,* 42, 390–402.

Landau, L. and Peierls, R. (1931). Erweiterung des Unbestimmtheitsprinzips für die relativistische Quantentheorie. *Zeitschrift für Physik,* 69(1–2), 56–69.

Osnaghi, S., Freitas, F., and Freire Jr., O. (2009). The origin of the Everettian heresy. *Studies in History and Philosophy of Modern Physics,* 40, 97–123.

Patton, G. M. and Wheeler, J. A. (1975). Is physics legislated by cosmogony? In C. Isham, R. Penrose, and D. Sciama (eds), *Quantum Gravity: An Oxford Symposium,* Oxford: Clarendon Press, pp. 538–591.

Penrose, R. (1965). Gravitational collapse and space-time singularities. *Physical Review Letters,* 14, 57–59.

Penrose, R. (1981). Time-asymmetry and quantum gravity. In C. Isham, R. Penrose, and D. Sciama (eds), *Quantum Gravity: An Oxford Symposium,* Oxford: Clarendon Press, pp. 244–272.

Penrose, R. and Isham, C. (eds) (1986). *Quantum Concepts in Space and Time.* Oxford: Clarendon Press.

Peruzzi, G. and Rocci, A. (2018). Tales from the prehistory of Quantum Gravity. *The European Physical Journal H,* 43(2), 185–241.

Rickles, D. (2014). *A brief history of string theory.* Berlin, Heidelberg: Springer.

Rosenfeld, L. (1927). L'univers à cinq dimensions et la mécanique ondulatoire. *Bull. Acad. Roy. Belgique,* 13(5), 326–328.

Rosenfeld, L. (1930a). Über die Gravitationswirkungen des Lichtes. *Zeitschrift für Physik,* 65(9–10), 589–599.

Rosenfeld, L. (1930b). Zur Quantelung der Wellenfelder. *Annalen der Physik,* 397(1), 113–152.

Rosenfeld, L. (1932). La théorie quantique des champs. *Annales de l'Institut H. Poincaré,* 2, 25–91.

Rosenfeld, L. (1955). On quantum electrodynamics. In L. Rosenfeld and V. Weisskopf (eds), *Niels Bohr and the Development of Physics,* Oxford: Pergamon Press, pp. 70–95.

Rosenfeld, L. (1963). On quantization of fields. *Nuclear Physics,* 40, 353–356.

Rovelli, C. (2004). *Quantum Gravity.* Cambridge: Cambridge University Press.

Salisbury, D. and Sundermeyer, K. (2017). Léon Rosenfeld's general theory of constrained Hamiltonian dynamics. *The European Physical Journal H,* 42, 23–61.

Wheeler, J. A. (1957a). Assessment of Everett's 'Relative State' Formulation of Quantum Theory. *Reviews of Modern Physics,* 29, 463–465.

Wheeler, J. A. (1957b). On the nature of quantum geometrodynamics. *Annals of Physics,* 2, 604–614.

Wheeler, J. A. (1962). The universe in the light of general relativity. *The Monist,* 47(1), 40–76.

CHAPTER 16

.................................................................................................

# QUANTUM INFORMATION AND THE QUEST FOR RECONSTRUCTION OF QUANTUM THEORY

.................................................................................................

ALEXEI GRINBAUM

## 16.1 An Overview of the Problem

.................................................................................................

THE development of quantum information has revolutionized our understanding of quantum foundations by opening new lines of questioning. Unthinkable before the 'second quantum revolution' initiated by John Bell's inequalities, they arise from a mathematical approach that anchors fundamental issues, e.g. causality or compositionality, in the formalism of finite-dimensional Hilbert spaces.

Bell's inequalities and the Kochen–Specker theorem provided first tangible assessments of non-classicality. John Wheeler's dictum 'It from Bit' marked the birth of an approach using such quantitative measures to perform classically impossible tasks. It then became possible to take such tasks as fundamental properties of quantum theory as well as explanatory elements for its interpretation. On this view, quantum theory employs classically unavailable protocols or resources. Quantumness, previously credited to the superposition of states, stems from features such as quantum entanglement, discord, contextuality, fundamental randomness, no-cloning, teleportation, compositional structure, various cryptographic protocols, or time and postselection paradoxes. Many of these non-classical features can be seen as resources, and quantum theory as a 'resource theory'. The non-classical resources, when they are exploited by quantum technologies, enable real-world non-classical devices.

Such non-classical concepts have opened a new vista at quantum theory, by placing it at a particular measurable distance from other 'post-quantum' models, which involve either modified quantitative bounds or algebraic or computational properties that are different from the usual quantum ones. The question 'why the quantum?' can now be assessed against the background of such models, greatly deepening our knowledge of the interrelations between various non-classical properties of the theory, their fundamental or derivative character, and their import for experimental protocols. This new knowledge of the structure of quantum theory, however precious it may be for theorists and experimentalists, has not yet yielded a generally satisfactory explanation of 'why the quantum?'.

The philosophical interpretation that '[quantum] physics is about information' has emerged as a logical consequence of the new mathematical results. The concept of information can be understood in various ways, based on the Shannon, von Neumann, or Kolmogorov entropies, but also incorporating such quantum features as the possibility to continuously vary one's information state. Philosophers have been actively debating the pros and cons of these 'informational' interpretations. The question 'information about what?', usually dismissed as irrelevant by their proponents, remains a stumbling block for those opposed to taking information as a basic concept. This debate has produced a new wave of instrumental, or operational, accounts, while also contributing to the emergence of a particular informational form of structural realism. At the same time, quantum cryptography developed an array of device-independent protocols, which have found a broader use in quantum foundations. Device-independent models erase the implicit trust in the description of quantum systems involved in a given setting. The spread of this view from quantum cryptography to general quantum physics raises a new interpretative challenge: device-independent approaches exacerbate the need for a minimal conceptual reading to support their mathematical formalism based on formal languages. The view that a physical theory may be 'about' language emerges in the contemporary debate as a philosophical lesson of quantum information for interpreting quantum theory.

## 16.2 FROM QUANTUM LOGIC TO RESOURCE THEORIES

A feeling of unease with the conceptual status of quantum theory may be obtained by following many avenues. Two are of particular concern here. First, a historic avenue: from Niels Bohr to modern philosophical approaches, the task of interpreting quantum theory implied a task of interpreting the notion of quantum observer. This concept, hotly contested in quantum foundations, has been significantly clarified in the past three decades by an array of insights from quantum information. Second, an axiomatic avenue. Axiomatic reconstructions of quantum mechanics began in the 1930s with the advent of quantum logic (Birkhoff and von Neumann, 1936) in line with the axiomatic programme inherited from Hilbert (Hilbert et al., 1927). By the end of the 20th century, this logical vision was largely replaced by an operational approach using a palette of generalized probabilistic or device-independent models. While the old approach stressed the complete—and often obscure—rigour of mathematical physics, the latter favours clarity of meaning and a direct link between the formalism and the underlying principles that purport to explain physical reality.

In the modern sense, an axiomatic reconstruction of quantum theory is a derivation of its formalism from a set of principles with a clear meaning (see, e.g., Grinbaum, 2007). Contrary to the interpretations of quantum mechanics, the reconstructions possess supplementary persuasive power provided by mathematical derivation. Theoretic results are established as theorems: 'Why is it so?'—'Because we derived it.' The question of meaning, previously asked with regard to the formalism, now bears on the selection of first principles. It is now routine to explain the significance of a principle or a postulate in natural language: e.g., quantum theory is *about* probabilities, a particular composition rule, and a principle of continuity (Hardy, 2000). This state of affairs stands in stark contrast with the heavily mathematical axiomatic systems of quantum logic (Mackey, 1957; Piron, 1964, 1972; Ludwig, 1985). It has enabled a new view of quantum theory leading to significant advances in understanding its foundations.

Axiomatic reconstructions are a primary way to single out the most important features of quantum theory. As we shall revisit in section 16.3, von Neumann himself was dissatisfied with the Hilbert space formalism. To understand the significance of different components of the formalism in isolation from others, i.e. taken one by one, one needs a mathematical setting in which such philosophical questions could be asked and investigated. Such models have been created with an intention to reproduce important, although not all of the, features usually qualified as being 'uniquely' quantum. This informs a central and yet ongoing search in quantum foundations: what is the key property that would convincingly distinguish quantum theory from the alternatives? The hope is that if a new theory comes to replace quantum mechanics, then all essentially quantum characteristics will have to be emergent in the new theory, perhaps only in some approximation. This would greatly enhance our understanding of the strangeness of quantum mechanics, while also providing for the very accurate empirical predictions obtained via the standard Hilbert space formalism.

One essentially quantum mechanical property is a peculiar composition rule for systems, which gives rise to non-locality. The older quantum logical reconstructions devoted much effort to the derivation of properties pertaining to individual systems, leaving the composition as a problematic issue. The operational approaches reverse this situation. The composition is now seen as a fundamental ingredient: various alternative composition rules have been imagined and their consequences explored with the goal of clarifying the meaning of the tensor product rule in the standard quantum theory (Barrett, 2007a; Barnum *et al.*, 2007).

This emphasis on the rules of composition has resulted in an entire body of work purporting to derive quantitative bounds on the amount of correlations between subsystems, most often the Tsirelson bound of the bipartite Bell–CHSH inequalities (Bell, 1964; Cirel'son, 1980; Clauser *et al.*, 1969). To this end, a number of intentionally non-quantum ('postquantum') models were set up to reach a different, typically higher

**Table 16.1 Entanglement and contextuality as non-classical resources. Adopted with modifications from (Dourdent, 2018)**

| Property | Non-locality | Contextuality |
|---|---|---|
| Valid for a multipartite system | Yes | Yes |
| Valid for a unique system | No | Yes |
| State-independent | No | Yes |
| Minimal dimension of the Hilbert space | 2 | 3 |
| Witnessing requires measurement | Yes | No |
| Exists in device-independent models | Yes | Postquantum |

than quantum bound on bipartite correlations (Popescu and Rohrlich, 1994; van Dam, 2000; Pawlowski *et al.*, 2009; Oppenheim and Wehner, 2010; Popescu, 2014). A systematic comparison between these models and quantum theory sheds some light on the origin of quantum correlations. For example, it was shown that no bipartite composition principle is sufficient to fully characterize the quantum correlation bound (Gallego *et al.*, 2011). It is an open question whether quantum correlations can be captured through a set of principles involving a finite number of observers; recent results suggest a negative answer (Ji *et al.*, 2020). Furthermore, many tentative principles proposed in the operational reconstructions are satisfied by a set of correlations larger than the quantum set (Navascués and Wunderlich, 2010). The search for a conceptually satisfactory and mathematically rigorous limiting principle for quantum non-locality is still ongoing.

A lot of research in quantum foundations has been devoted to the notion of quantum contextuality, a non-classical trait of quantum theory that is not equivalent to entanglement (Table 16.1). Many years after its publication, the Kochen–Specker theorem, which shows that the concept of entanglement does not capture all the non-classical features of quantum theory, is again enjoying increased attention (Kochen and Specker, 1965, 1967). In view of this research, the problem of characterizing non-classical theories can be reformulated as a search for the ultimate non-classical resource.

Like in the case of bipartite correlations, postquantum models exhibiting generalized contextuality provide a helpful comparison between the principles that give rise to either specifically quantum or generic properties (Spekkens, 2007; Abramsky *et al.*, 2015). This is illustrated in Table 16.2. Many candidate resources as well as many candidate postulates are now available for reconstructing the composition rule as one unique feature of quantum theory (Cabello, 2012; Masanes and Müller, 2011). On the best current understanding, this problem bears on the decomposition of a global system rather than the composition of subsystems:

> It is our point of view that the operation of forming a composite system $\mathcal{H}_A \otimes \mathcal{H}_B$ from its subsystems $\mathcal{H}_A$ and $\mathcal{H}_B$ should not be a fundamental structure in a physical theory. The point is that nature presents us with a huge quantum system which we

Table 16.2 Notions used by various authors to express non–classicality. The first group was used by the 'founding fathers' of quantum theory, while the second group illustrates a modern understanding of quantum resources. The selection is adopted from (Grinbaum, 2017b)

| Non-classical feature | Example of author |
| --- | --- |
| Uncertainty | W. Heisenberg |
| Complementarity | N. Bohr |
| Quantum superposition | E. Schrödinger |
| Entanglement and non-locality | J. Bell |
| Tensor product structure | (Coecke and Kissinger, 2017) |
| Amount of non-locality | (Bub, 2016) |
| Cloning and teleportation | (Zeilinger, 2010) |
| Quantum discord | (Vedral, 2012) |
| Quantum contextuality | (Mermin, 1990), (Cabello, 2012) |
| Quantum randomness | (Gisin, 2014) |
| Cryptographic protocols | (Scarani, 2006, 2010) |
| Time and postselection paradoxes | (Aharonov and Rohrlich, 2005) |

observe and conduct experiments on, and in some ways this total system behaves as if it were composed of smaller parts. [ ... ] Hence it seems that the correct question would be 'When does a physical system behave like it were composed of smaller parts?' rather than 'How do physical systems compose to composite systems?'. [ ... ] Note that this is in stark contrast to many other approaches to the foundations of quantum theory, in which the operation of forming a composite system from subsystems is a fundamental structure.   (Fritz, 2012; Fritz *et al.*, 2013)

Two philosophical lessons are central to the reconstruction programme of quantum theory: an increased explanatory value arising from the use of mathematical derivation and a focus on clear meaning of the principles. Both lessons have been successfully implemented in the modern information-theoretic view of quantum theory. This view begins with choosing a resource. In one or several steps, the resource is shown to lead to a classically unavailable result or advantage in a particular communication or computational task. Quantum theory is then interpreted as a toolbox, or a 'resource theory', that can be put to use in order to achieve such unusual performance (Coecke *et al.*, 2016; Brandão *et al.*, 2013; Veitch *et al.*, 2014; Streltsov *et al.*, 2017). This is a pragmatic interpretation inspired by the growing application of quantum information to the development of quantum technologies. Instead of wondering what the world is like according to quantum mechanics, one searches for its meaning by asking how this theory can be employed beyond what can be done classically. Actively harvesting quantum resources, rather than contemplating the strangeness of quantum reality, is a modern avenue that leads to a better understanding of quantum theory.

# 16.3 Operational Reconstructions

An interpretation of quantum mechanics attempts to give a clear meaning to this physical theory. The main difficulty has to do with the measurement problem: a reversible unitary evolution is replaced, at the moment of measurement, by an irreversible wave function collapse. The interpretations made sense of it in many ways, most often either by postulating projective measurement to be a fundamental ingredient of the theory or by completely denying the collapse. The reconstruction programme begins with a realization that a meaning cannot be satisfactory if it is merely heaped over and above the mathematics of quantum mechanics, instead of coming all the way along with the formalism as it arises in a derivation of the theory. One of the most vocal expressions of this idea belongs to Rovelli:

> Quantum mechanics will cease to look puzzling only when we will be able to *derive* the formalism of the theory from a set of simple physical assertions ('postulates', 'principles') about the world. Therefore, we should not try to append a reasonable interpretation to the quantum mechanical formalism, but rather to *derive* the formalism from a set of experimentally motivated postulates.    (Rovelli, 1996)

A *reconstruction* is the following schema: theorems and major results of physical theory are formally derived from simpler mathematical assumptions. These assumptions or axioms, in turn, appear as a formal translation of a set of physical principles. Thus a reconstruction consists of three stages: a set of physical principles, their mathematical representation, and a derivation of the formalism of the theory. Established as valid results, theorems and equations of the theory lose most of their philosophical ambiguity; in particular, within an operational reconstruction the measurement problem arguably becomes a 'pseudo-problem' (Bub, 2016, p. 223).

Any reconstruction starts with a choice of physical principles, e.g. the well-known 'five simple axioms' operational approach (Hardy, 2000) begins with a selection of measurement settings and results. They form a convex set, which is further characterized by two natural numbers, $K$ and $N$. $K$ is the number of degrees of freedom of the system and is defined as the minimum number of probability measurements needed to determine the state. Dimension $N$ is defined as the maximum number of states that can be reliably distinguished from one another in a single measurement. The axioms are:

**H1** *Probabilities.* In the limit as $n$ becomes infinite, relative frequencies (measured by taking the proportion of times a particular outcome is observed) tend to the same value for any case where a given measurement is performed on an ensemble of $n$ systems prepared by some given preparation.

**H2** *Simplicity.* $K$ is determined by a function of $N$ where $N = 1, 2, \ldots$ and where, for each given $N$, $K$ takes the minimum value consistent with the axioms.

**H3** *Subspaces.* A system whose state is constrained to belong to an $M$ dimensional subspace behaves like a system of dimension $M$.

**H4** *Composite systems.* A composite system consisting of subsystems $A$ and $B$ satisfies $N = N_A N_B$ and $K = K_A K_B$.

**H5** *Continuity.* There exists a continuous reversible transformation on a system between any two pure states of that system.

The point of this approach is that the operational meaning of the axioms can be grasped relatively easily, as already suggested by the axioms' short names. It seems natural to couch the approach of this kind in the instrumentalist philosophy. However, one's philosophical stance plays no particular role in the derivation: Hardy acknowledges that the axioms can be adopted by a realist, a hidden variable theorist, or a partisan of collapse interpretations, while Timpson points out that 'an operationalist black-box formulation of a theory might be most appropriate...but there is still an ontological story to be told too' (Timpson, 2013, p. 187). An operational reconstruction should not be seen a commitment to anti-realism or instrumentalism (Letertre, 2021): it is operational in the minimal philosophical sense, i.e. it gives a clearer understanding of quantum theory irrespectively of the ontological justification of the axioms.

Even if the operational approach brings additional clarity to understanding quantum mechanics, it has its own limits: as exemplified by Hardy's axiom H5, any complete reconstruction of quantum theory contains an element that does not stem from purely empirical considerations. The recourse to mathematical abstraction cannot be eliminated and appears compulsory despite many historic and recent attempts to challenge the role of mathematics in physical theory (Wigner, 1960; Grinbaum, 2019; Rédei, 2020; Gisin, 2020). Intentionally incomplete reconstructions (see section 16.5) show that this apparent failure to provide an entirely operational grounding for an axiomatic reconstruction has deep roots: most attempts to replace mathematical abstraction with constructive arguments lead to models that are close to, but not fully coincident with, quantum theory.

To delve deeper into history, the story of mathematical abstraction in quantum mechanics has its roots in the debates of the 1920s. Max Born happened to know matrix multiplication; he told Werner Heisenberg that there existed a mathematical theory for his rules of calculation of energy levels. John von Neumann helped to replace matrix mechanics by the theory of operators on a Hilbert space, which then formed a basis of his canonical textbook (von Neumann, 1932). This successful construction of a new theory had two reasons. First, Heisenberg's empirically motivated rules were brought into a general mathematical framework. Second, this framework produced new predictions that were later confirmed experimentally. Henceforth the position of the Hilbert spaces, imported by von Neumann from mathematics into the foundations of quantum theory, became unshakeable. How unavoidable was this course of events? Von Neumann famously did not think much of the inevitability of using Hilbert spaces in the foundations of quantum theory. In 1935, he wrote to Garrett Birkhoff: 'I would like to make a confession which may seem immoral. I do not believe absolutely in

Hilbert space any more' (von Neumann, 2005). Obviously, von Neumann did not think that the Hilbert space formalism was wrong; instead he believed that vector spaces were not essential. Quantum logic sought to replace this description of quantum theory by a different approach based on orthomodular lattices and algebras. Even if Heisenberg's empirical calculation rules came first historically, this was no reason to believe that they provided good indication about what was truly important in quantum theory. An exploration of what the theory tells us, as von Neumann realized, had to be carried out mathematically; quantum logic was but the first among many attempts at establishing a novel mathematical formalism for the already existing quantum theory.

# 16.4 HISTORIC MILESTONES

This section presents several major historic milestones in the philosophy of physics on the road leading to modern information-theoretic reconstructions. In the 1980s John Wheeler emphasized the difficulty of stating sharply where 'the community of observer-participators' begins and ends (Wheeler, 1983). Quantum theory says nothing about the physical composition of the observer: this term has no further theoretical description. One cannot infer from a set of quantum measurements if the observer is a conscious human being, a computing machine, or the whole Universe. Whatever it is materially, the observer has some information about the system. This is indeed the only condition: an observer must somehow register information coming from measurement. This viewpoint has been slowly distilled in many debates prior to Bell's 1964 'second quantum revolution' (Aspect, 2004), from Bohr to Everett, and has continued and developed in novel ways since the advent of quantum information.

Bohr's Como lecture in 1927 became a foundation of what came to be known as the Copenhagen interpretation of quantum mechanics. Its main point: 'Only with the help of classical ideas is it possible to ascribe an unambiguous meaning to the results of observation... It lies in the nature of physical observation, that all experience must ultimately be expressed in terms of classical concepts' (Bohr, 1934, p. 94). One can classify different readings of this statement into two groups based on the meaning of the term 'classical'. The first is a straightforward inference that quantum mechanics requires classical mechanics:

> It is in principle impossible to formulate the basic concepts of quantum mechanics without using classical mechanics.   (Landau and Lifshitz, 1977, p. 2)

This is a mechanistic, even a materialist, view that implies that the world consists of physical objects, quantum or classical, described by respective mechanical theories, and the special status of measurement is entirely due to the interplay between them. The second interpretation of Bohr's dictum is that the quantum mechanical experiments can only be described by classical *language*:

Bohr went on to say that the terms of discussion of the experimental conditions and of the experimental results are *necessarily* those of 'everyday language', suitably 'refined' where necessary, so as to take the form of classical dynamics. It was apparently Bohr's belief that this was the only possible language for the *unambiguous communication* of the results of an experiment.    (Bohm, 1971, p. 38)

Linguistic readings like this one invoke 'classical language', 'classical concepts', or 'classical terms', and seem to require someone who is in command of such concepts: the observer. When Bohr put forward these ideas in the late 1920s and early 1930s, he repeatedly introduced such formulations; however, in the next decade he began to move away from them. Without fully reneging on the importance of classical language, in the 1940s and 1950s Bohr focused his philosophical analysis on the property of communicability of the measurement results between observers. Short of an a priori requirement, classical language has become a solution to achieving communicability. Bohr insisted that the problem of communicability be treated mathematically, rather than by giving a special role to human consciousness.

Writing in 1939, Fritz London and Edmond Bauer noted that the quantum mechanical observer must be a human person. As subjectivists, they ruled out an objective description of reality: 'It seems that the result of measurement is intimately linked to the consciousness of the person making it' (London and Bauer, 1939, p. 48). Consciousness was an ultimate characteristic of the observer. They solved the problem of objective description by postulating, in a 'community of scientific consciousness, an agreement on what constitutes the object of the investigation' (London and Bauer, 1939, p. 49). The nature of this agreement remains mysterious. London and Bauer's work was among the early examples of subjectivist, deeply first-person approaches to interpreting quantum mechanics.

The interpretation given by London and Bauer was taken on by Eugene Wigner. The consciousness of the observer 'enters the theory unavoidably and unalterably' and corresponds to the impression produced by the measured system on the observer (Wigner, 1976). Faithful to Bohr in this respect, Wigner notes that the wave function 'exists' only in the sense that 'the information given by the wave function is communicable'. Observers agree because they communicate measurement results to one another; however, Wigner's understanding of communicability fell significantly behind Bohr's insistence on expressing it in mathematical terms. Indeed, Wigner merely interpreted it as a mainstream application of the eigenvalue problem:

The communicability of information means that if someone else looks at time $t$ and tells us whether he saw a flash, we can look at time $t + 1$ and observe a flash with the same probabilities as if we had seen or not seen the flash at time $t$ ourselves.

(Wigner, 1961)

The nature of objectivity, or at least of intersubjective agreement, remains an open issue in the 'Wigner's friend' Gedankenexperiment introduced in the same publication. This thought experiment has recently become a subject of renewed attention, even of experimental tests, in a modern setting with two or more 'friends' as well as two or more 'Wigners'. Such analysis aims at understanding the foundational and philosophical implications of the ability to communicate measurement results between observers, most notably the relativity of facts (Frauchiger and Renner, 2018; Brukner, 2018; Bong et al., 2020).

Back in the middle of the 20th century, Léon Brillouin insisted that information, including the information possessed by the observer, must be defined with the exclusion of all human elements (Brillouin, 1964). Satosi Watanabe believed that objectivity arose from the direction of time common to all observers: 'The past-to-future directions of all observers coincide. This statement has a well-defined physical meaning, for "positive time direction" is a Lorentz-invariant concept' (Watanabe, 1953, p. 387). As we shall see below, many years after Watanabe's work, his intuition became a topic of theoretical investigation in the studies of indefinite causal orders in quantum theory.

Hugh Everett thought differently. Although he is better known for his interpretation of quantum mechanics, which was subsequently called the 'many-worlds interpretation' (DeWitt and Graham, 1973), Everett's insights were not limited to it. Observers, according to Everett, are systems that possess memory, i.e. 'parts . . . whose states are in correspondence with past experience' (Everett, 1957). They do not have to be human but include 'automatically functioning machines, possessing a sensory apparatus and coupled to recording devices'. Observation relies on the necessary assumption that records be distinguishable: 'If we are to able to call the interaction an observation at all, the requirement that the observer's state change in a manner which is different for each eigenfunction is necessary'. It seems reasonable to assume that differently constituted observers with the same memory size will be equally able to register measurement results obtained on a quantum system. Material 'hardware' does not matter, because quantum mechanics uses abstract mathematics. Everett's heavy use of the concept of information in his thesis, published in full only in 1973 (Everett, 1973), leaves us on the threshold of an information-theoretic treatment of the quantum observer.

Several decades after Everett, Carlo Rovelli proposed another relational interpretation (Rovelli, 1996). According to Rovelli, all physical systems have the capacity to act as observers, i.e. possess information about other systems. Information should be seen as an observer-dependent, rather than objective, notion,[1] Quantum mechanical states are therefore relational: they are indexed, first, by the observed system, and second, by the observing system that has obtained information about the observed system. Trying to remove the second index—to 'liberate' the notion of information from its relational character—is meaningless in any relational view (Di Biagio and Rovelli, 2021). Conceptually speaking, 'information' is nothing more than a correlation between the degrees of freedom of the observed and observing systems; in particular, it has no semantics and no everyday meaning in natural language.

---

[1] The same conclusion is currently much debated in the literature on Wigner's Gedankenexperiment, see (Brukner, 2020).

Echoing Rovelli, Christopher Fuchs presented an ambitious programme at the turn of the century: 'The task is not to make sense of the quantum axioms by heaping more structure, more definitions...on top of them, but to throw them away wholesale and start afresh. We should be relentless in asking ourselves: From what deep *physical* principles might we *derive* this exquisite mathematical structure? [...] I myself see no alternative but to contemplate deep and hard the tasks, the techniques, and the implications of quantum information theory' (Fuchs, 2002). Following Peres (1993), Fuchs insisted that a view of measurement as a positive operator-valued measure (POVM), rather than a von Neumann projective measurement, is key to understanding quantum mechanics. His was a decision-theoretic Bayesian approach (Fuchs, 2001), which later evolved, under the influence of the work in quantum information and of the subjective view on probabilities, into a mathematical and philosophical paradigm of 'QBism' (Fuchs and Schack, 2013). Today, the programme of QBism can be seen as one of the most ambitious information-theoretic interpretations of quantum theory (Table 16.3).

Originally QBists believed that reality is what 'remains' when the axioms of quantum theory—as many of them as possible—are reformulated in the form of information-theoretic statements. In the early 2000s, one such candidate 'element of reality'—to use an expression dating back to the EPR article (Einstein *et al.*, 1935)—was the dimension of the Hilbert space. This position has later evolved into a particular variety of

**Table 16.3 QBism as a culmination of information–theoretic approach to quantum theory. Reproduced with permission from (Fuchs, 2002)**

Quantum Mechanics:

*The Axioms and Our Imperative!*

| | |
|---|---|
| States correspond to density operators $\rho$ over a Hilbert space $H$. | *Give an information theoretic reason if possible!* |
| Measurements correspond to positive operator-valued measures (POVMs) $\{E_d\}$ on $H$. | *Give an information theoretic reason if possible!* |
| $H$ is a complex vector space, not a real vector space, not a quaternionic module. | *Give an information theoretic reason if possible!* |
| Systems combine according to the tensor product of their separate vector spaces, $H_{AB} = H_A \otimes H_B$. | *Give an information theoretic reason if possible!* |
| Between measurements, states evolve according to trace-preserving completely positive linear maps. | *Give an information theoretic reason if possible!* |
| By way of measurement, states evolve (up to normalization) via outcome-dependent completely positive linear maps. | *Give an information theoretic reason if possible!* |
| Probabilities for the outcomes of a measurement obey the Born rule for POVMs $\mathrm{tr}(\rho E_d)$. | *Give an information theoretic reason if possible!* |

The distillate that remains—the piece of quantum theory with no information theoretic significance—will be our first unadorned glimpse of 'quantum reality'. Far from being the end of the journey, placing this conception of nature in open view will be the start of a great adventure.

'participatory realism': rather than relinquishing the idea of reality, QBism submits that 'reality is more than any single third-person, objective perspective can capture' (Fuchs, 2016). At long last, certain elements of Bohr's worldview are conjugated in this philosophy with an emphasis on the subjective first-person approach typical of the 20th-century phenomenology. This is not without resemblance with the work by London and Bauer; however, Fuchs's view is more detailed and informed by the philosophical research stemming from Husserl's critique of objectivity (Husserl, 1954), e.g. (d'Espagnat, 1994; Bitbol, 2000; Petitot, 1997). Without expounding on these latest shifts in QBist philosophy, it will be sufficient to note that the information-theoretic viewpoints on quantum theory also have an affinity with the neo-Copenhagen approaches, even though they explicitly depart from Bohr's original views by putting forward a formal mathematical notion of information, free of any semantics, that has no connection with natural language.

# 16.5 INFORMATION-THEORETIC POSTQUANTUM MODELS

A complete reconstruction aims at deriving the full structure of quantum theory. A new type of information-theoretic reconstructions renounces this ambition. Such toy models began to appear in the early 2000s, most notably in the work of Spekkens and Barrett (Aaronson, 2004, 2005; Barrett, 2007a; Hardy, 1999; Popescu and Rohrlich, 1994; Smolin, 2005; Spekkens, 2007). While they were derogatively called 'fantasy quantum mechanics' or 'quantum mechanics lite', they managed to produce a signifi-cant impact on research in the foundations of physics. This was due to their method-ology: a reconstruction may contain important insights by targeting only a certain element of quantum theory. One builds a model which, from the very beginning, is not intended to be complete; nevertheless, it allows for a better understanding of the structure of quantum theory. This piecemeal approach tries to solve one aspect of the puzzle at a time, rather than attempt a reconstruction of the entire edifice. The models that addressed the Tsirelson bound soon started to be called postquantum models.

Most postquantum models are based on information-theoretic principles. To com-pare them with quantum theory, one checks whether the former reproduces quantum computational phenomena available in the latter:

- Does the model allow for superluminal signalling?
- Does the model allow for bit commitment?
- Does the model allow for teleportation of states, dense coding, or remote steering?
- Does the model allow exponential speed-up relative to classical computation or solving NP-complete problems in polynomial time?

Other points of comparison do not necessarily relate to computation:

- Does the model allow non-locality? If yes, what is the measure of non-locality present in the model?
- Is the model contextual? If yes, is there a difference between quantum contextuality and the form of contextuality present in the model? Can this difference be evaluated quantitatively?
- Does the model possess a continuum of states, measurements, and transformations between states?

For example, Spekkens's toy model, while it contains only a small number of states and just one fundamental principle, accommodates such quantum phenomena as non-commutativity, interference, the multiplicity of convex decompositions of a mixed state, no-cloning, and teleportation (Spekkens, 2007; Bartlett *et al.*, 2012; Pusey, 2012). It does not include a continuous state space, the existence of a Bell theorem, or contextuality. All of the latter are therefore rigorously unconnected with the appearance of the former.

Generalized probabilistic theories form another major avenue in the analysis of quantum foundations (Barrett, 2007a,b; Barrett *et al.*, 2005). Models of this type are nowadays often studied jointly with a class of models aimed at exploring non-standard causality with the tools of quantum theory. In spacetime, events $A$ and $B$ can be in three causal relations: either $A$ is before $B$, $B$ is before $A$, or $A$ and $B$ are causally separated, i.e. they lie on a spacelike interval. Quantum mechanics admits causal structures that do not correspond to any of these cases. Heuristically, this can be pictured as putting the order between $A$ and $B$ in a quantum superposition. More precisely, several approaches to indefinite causal orders were proposed using the mathematical frameworks aimed at capturing causal relations, e.g. the general 'process matrix' paradigm or the more intuitive, albeit less rigorous, 'quantum switch' (Oreshkov *et al.*, 2012; Chiribella *et al.*, 2013; Hardy, 2007; Araújo *et al.*, 2015; Chiribella and Kristjánsson, 2019; Ibnouhsein and Grinbaum, 2015). While such frameworks are not always mathematically equivalent, all of them enable the exploration of one fundamental idea: an indefinite causal order. This is a new kind of an inherently quantum phenomenon shedding light on a notion hitherto explored mainly in spacetime theories. Its reality has recently been demonstrated in several experimental implementations of the quantum switch (Procopio *et al.*, 2015; Rubino *et al.*, 2016; Goswami *et al.*, 2018b,a; Wei *et al.*, 2018; Procopio *et al.*, 2019; Rubino *et al.*, 2020).

Much like the debate on non-classical resources has not yielded one consensual answer to the question about the key quantum resource, these models of generalized causality contain a new element, the precise meaning of which remains yet to be distilled. To gauge what is brought by quantum theory into the study of causality, quantum control on causal order can be treated as a resource that provides non-classical communication advantage, i.e., two noisy channels in a quantum switch can transmit more information than any of these channels individually (Abbott *et al.*, 2016;

Ebler *et al.*, 2018; Loizeau and Grinbaum, 2020). More generally, causal orders become yet another resource that can be harnessed in the protocols of quantum information. Taken as a resource, the quantum switch exhibits an intricate interplay between two non-classical factors, both of which are capable of providing a communication advantage: quantum superposition and non-commutativity. It remains a genuine combination of these contributing properties without being reducible to any one of them.

To achieve a better theoretical understanding of the advantage demonstrated in such set-ups, one needs in particular to explore its origin. While the notion of controlling quantum channels via a quantum subsystem is a useful heuristic, its rigorous mathematical formulation demonstrates that the action of observers involved in such control and measurement operations is hardly ever local in time and space (Guérin *et al.*, 2019). While this approach has the benefit of immediately clarifying the physical interest of quantum causality, it relies on a currently unresolved question as to whether any local party can operationally exercise quantum control (Oreshkov, 2019). The very notion of 'local observer' in quantum theory becomes an issue open to mathematical and conceptual investigation.

To take another example, a toy model called non-local or Popescu–Rohrlich boxes (Popescu and Rohrlich, 1994) allows for non-local correlations that are strictly stronger than those allowed by quantum mechanics. One learns that non-locality is not an exclusively quantum feature: it is present in other possible theories, in which its amount can exceed that of quantum theory. This development was among the earliest examples of so called device-independent models in quantum information (Grinbaum, 2017a). An information-theoretic axiom of no-signalling describes processes in a model connecting the inputs $x, y \in \{0, 1\}$ and the outputs $a, b \in \{0, 1\}$ of two parties according to the joint probability distribution:

$$P(ab \,|\, xy) = \begin{cases} 1/2 : & a + b = xy \bmod 2 \\ 0 : & \text{otherwise.} \end{cases} \tag{1}$$

While the Popescu–Rohrlich box is a general algebraic framework designed to go beyond quantum theory, the application of the no-signalling principle implies that this box will respect the laws of special relativity. Its device-independent non-local structure accommodates a violation of the Tsirelson bound (Cirel'son, 1980) by reaching the maximum amount of correlations in the CHSH inequality (Bell, 1964; Clauser *et al.*, 1969). Hailed as a 'very important recent development' (Popescu, 2014), device-independent models like the Popescu–Rohrlich box are characterized by the absence of assumptions about the internal workings of the box. Its 'interior' is not described by a particular physical theory. The box is unknown territory which, since it is assumed to be of interest for physical theory, is also a territory of science. The entire set-up belongs within the boundaries of physics; at the same time, it opens up new possibilities to redefine these very boundaries, as device-independent methods erase one of the main dogmas of quantum theory. In quantum mechanics, it is assumed that a measurement setting is chosen in earnest, i.e., the observer trusts the system to be constituted of precisely the degrees of freedom described by the theory. This trust in preparation devices is usually not subject to theoretical scrutiny. Device-independent methods convert it into a theoretical problem.

The absence of trust is also a concern that quantum cryptography is designed to address. Cryptographic tests do not rely on the degrees of freedom pertinent to the system; they only involve inputs and outputs in a device-independent protocol. Quantum cryptography has an array of methods for dealing with adversaries which, via action upon sources, effectively turn systems into untrusted entities, including randomness generation (Colbeck, 2006; Pironio et al., 2010), quantum key distribution (Barrett et al., 2005), estimation of the states of unknown systems (Bardyn et al., 2009), certification of multipartite entanglement (Bancal et al., 2011), and distrustful cryptography (Aharon et al., 2016). Some of these cryptographic protocols have found a broader use in quantum information and quantum foundations. When quantum theory is viewed in a such a device-independent way, only some of its aspects come into play, e.g. the amount of non-locality or contextuality. What is new is that these approaches to essentially quantum features do not require that the notion of system be present in the theory. They are based on languages that are formal, departing from Bohr's view about the necessity of natural language for interpreting quantum theory. Yet such models describing the inputs and the outputs remain informative for the philosophy of science, for a theory based on formal languages still maintains clear physical implications. Jointly with the results on indefinite causal orders, which convert the presence of local observers into a theoretic problem, these findings initiate—perhaps only with minor exaggeration—a new conceptual revolution in quantum foundations: systems and observers, two of the most problematic metatheoretical notions in orthodox quantum mechanics, can now be given a rigorous treatment within mathematical frameworks.

Postquantum models, as well as the entire domain of device-independent approaches, demonstrate how one can achieve in practice what has been the initial promise of the programme of intentionally incomplete information-theoretic reconstructions: gaining a deeper insight into the structure of quantum theory. The lessons learned in the last decades would be unthinkable without the tools from quantum information theory. The key challenge that lies before us is to bring these lessons to theoretical fruition as well as to introduce them into mainstream teaching and pedagogy of quantum physics.

# REFERENCES

Aaronson, S. (2004). Is quantum mechanics an island in theoryspace? In A. Khrennikov (ed.), *Proceedings of the Växjö Conference 'Quantum Theory: Reconsideration of Foundations'*, Växjö, Sweden: Växjö University Press.

Aaronson, S. (2005). Quantum computing, postselection, and probabilistic polynomial-time. *Proc. Roy. Soc. Lond. A*, **461**, 3473–3482.

Abbott, A., Giarmatzi, C., Costa, F., and Branciard, C. (2016). Multipartite causal correlations: Polytopes and inequalities. *Phys. Rev. A*, **94**, 032131.

Abramsky, S., Barbosa, R. S., Kishida, K., Lal, R., and Mansfield, S. (2015). Contextuality, cohomology and paradox. In S. Kreutzer (ed.), *24th EACSL Annual Conference on Computer Science Logic (CSL 2015)*, volume 41, Schloss Dagstuhl–Leibniz–Zentrum für Informatik: Dagstuhl, Germany, pp. 211–228.

Aharon, N., Massar, S., Pironio, S., and Silman, J. (2016). Device-independent bit commitment based on the CHSH inequality. *New J. of Physics*, **18**, 025014.

Aharonov, Y., and Rohrlich, D. (2005). *Quantum Paradoxes: Quantum Theory for the Perplexed*. New York: Wiley.

Araújo, M., Branciard, C., Costa, F., Feix, A., Giarmatzi, C., and Brukner, Č. (2015). Witnessing causal nonseparability. *New Journal of Physics*, **17**(10), 102001.

Aspect, A. (2004). John Bell and the second quantum revolution. In J. Bell, *Speakable and Unspeakable in Quantum Mechanics*, revised edition, Cambridge: Cambridge University Press, pp. xvii–xl.

Bancal, J.-D., Gisin, N., Liang, Y.-C., and Pironio, S. (2011). Device-independent witnesses of genuine multipartite entanglement. *Phys. Rev. Lett.*, **106**, 250404.

Bardyn, C.-E., Liew, T. C. H., Massar, S., McKague, M., and Scarani, V. (2009). Device-independent state estimation based on Bell's inequalities. *Phys. Rev. A*, **80**, 062327.

Barnum, H., Barrett, J., Leifer, M., and Wilce, A. (2007). Generalized no-broadcasting theorem. *Physical Review Letters*, **99**(24), 240501.

Barrett, J. (2007a). Information processing in non-signalling theories. *Phys. Rev. A*, **75**, 032304.

Barrett, J. (2007b). Information processing in generalized probabilistic theories. *Phys. Rev. A*, **75**, 032304.

Barrett, J., Hardy, L., and Kent, A. (2005). No signaling and quantum key distribution. *Physical Review Letters*, **95**(1), 010503.

Bartlett, S., Rudolph, T., and Spekkens, R. (2012). Reconstruction of Gaussian quantum mechanics from Liouville mechanics with an epistemic restriction. *Phys. Rev. A*, **86**, 012103.

Bell, J. (1964). On the Einstein–Podolsky–Rosen paradox. *Physica*, **1**, 195–200.

Birkhoff, G. and von Neumann, J. (1936). The logic of quantum mechanics. *Ann. Math. Phys.*, **37**, 823–843. Reprinted in: J. von Neumann (1961), *Collected Works*, Vol. IV, Oxford: Pergamon Press, pp. 105–125.

Bitbol, M. (2000). Physique quantique et cognition. *Revue Internationale de Philosophie*, **212**(2), 299–328.

Bohm, D. (1971). On Bohr's views concerning the quantum theory. In T. Bastin (ed.), *Quantum Theory and Beyond*, Cambridge: Cambridge University Press, pp. 33–40.

Bohr, N. (1934). *Atomic Theory and the Description of Nature*. Cambridge: Cambridge University Press.

Bong, K.-W., Utreras-Alarcón, A., Ghafari, F., Liang, Y.-C., Tischler, N., Cavalcanti, E. G., Pryde, G. J., and Wiseman, H. M. (2020). A strong no-go theorem on the Wigner's friend paradox. *Nature Physics*. doi: 10.1038/s41567-020-0990-x.

Brandão, F. G. S. L., Horodecki, M., Oppenheim, J., Renes, J. M., and Spekkens, R. W. (2013). Resource theory of quantum states out of thermal equilibrium. *Phys. Rev. Lett.*, **111**, 250404.

Brillouin, L. (1964). *Scientific Uncertainty and Information*. New York: Academic Press.

Brukner, Č. (2018). A no-go theorem for observer-independent facts. *Entropy*, **20**, 350.

Brukner, Č. (2020). Facts are relative. *Nature Physics*. doi: 10.1038/s41567-020-0984-8.

Bub, J. (2016). *Bananaworld: Quantum Mechanics for Primates*. Oxford: Oxford University Press.

Cabello, A. (2012). The contextual computer. In H. Zenil (ed.), *A Computable Universe: Understanding and Exploring Nature as Computation*, Singapore: World Scientific, pp. 595–604.

Chiribella, G., and Kristjánsson, H. (2019). Quantum Shannon theory with superpositions of trajectories. *Proceedings of the Royal Society A: Mathematical, Physical and Engineering Sciences*, **475**(2225), 20180903.

Chiribella, G., D'Ariano, G. M., Perinotti, P., and Valiron, B. (2013). Quantum computations without definite causal structure. *Physical Review A*, **88**(2), 022318.

Cirel'son, B. S. (1980). Quantum generalizations of Bell's inequality. *Lett. Math. Phys.*, **4**(2), 93–100.

Clauser, J., Holt, R., Horne, M., and Shimony, A. (1969). Proposed experiment to test local hidden-variable theories. *Phys. Rev. Lett.*, **23**, 880–884.

Coecke, B., and Kissinger, A. (2017). *Picturing Quantum Processes: A First Course in Quantum Theory and Diagrammatic Reasoning*. Cambridge: Cambridge University Press.

Coecke, B., Fritz, T., and Spekkens, R. W. (2016). A mathematical theory of resources. *Information and Computation*, **250**, 59–86.

Colbeck, R. (2006). *Quantum And Relativistic Protocols For Secure Multi-Party Computation*. PhD thesis, University of Cambridge.

d'Espagnat, B. (1994). *Le réel voilé*. Paris: Fayard. English translation: *Veiled Reality: An Analysis of Present-Day Quantum Mechanical Concepts*, Boulder: Westview Press, 2003.

DeWitt, B., and Graham, N. (eds) (1973). *The Many-Worlds Interpretation of Quantum Mechanics*. Princeton: Princeton University Press.

Di Biagio, A., and Rovelli, C. (2021). Stable Facts, Relative Facts. *Found. Phys.*, **51**, 30.

Dourdent, H. (2018). Contextuality, witness of quantum weirdness. arXiv:1801.09768.

Ebler, D., Salek, S., and Chiribella, G. (2018). Enhanced communication with the assistance of indefinite causal order. *Phys. Rev. Lett.*, **120**, 120502.

Einstein, A., Podolsky, B., and Rosen, N. (1935). Can quantum-mechanical description of physical reality be considered complete? *Phys. Rev.*, **47**, 777.

Everett, H. (1957). 'Relative state' formulation of quantum mechanics. *Rev. Mod. Phys.*, **29**, 454–462.

Everett, H. (1973). The theory of the universal wave function. In B. DeWitt and N. Graham (eds), *The Many-Worlds Interpretation of Quantum Mechanics*, Princeton: Princeton University Press, pp. 3–140.

Frauchiger, D., and Renner, R. (2018). Quantum theory cannot consistently describe the use of itself. *Nature Communications*, **9**, 3711.

Fritz, T. (2012). Tsirelson's problem and Kirchberg's conjecture. *Reviews in Mathematical Physics*, **24**(05), 1250012.

Fritz, T., Sainz, A. B., Augusiak, R., Brask, J. B., Chaves, R., Leverrier, A., and Acín, A. (2013). Local orthogonality as a multipartite principle for quantum correlations. *Nature Communications*, **4**, 2263.

Fuchs, C. (2001). Quantum foundations in the light of quantum information. In A. Gonis and P. Turchi (eds), *Decoherence and its Implications in Quantum Computation and Information Transfer: Proceedings of the NATO Advanced Research Workshop, Mykonos, Greece, June 25–30, 2000*, Amsterdam: IOS Press, pp. 39–82.

Fuchs, C. (2002). Quantum mechanics (and only a little more). In A. Khrennikov (ed.), *Quantum Theory: Reconsideration of foundations*, Växjo, Sweden: Växjo University Press, pp. 463–543.

Fuchs, C. A. (2016). On participatory realism. In I. T. Durham and D. Rickles (eds), *Information and Interaction: Eddington, Wheeler, and the Limits of Knowledge*, Berlin: Springer, pp. 113–134.

Fuchs, C. A. and Schack, R. (2013). Quantum-Bayesian coherence. *Rev. Mod. Phys.*, **85**, 1693–1715.

Gallego, R., Würflinger, L. E., Acín, A., and Navascués, M. (2011). Quantum correlations require multipartite information principles. *Phys. Rev. Lett.*, **107**, 210403.

Gisin, N. (2014). *Quantum Chance: Nonlocality, Teleportation and Other Quantum Marvels*. Cham: Springer.

Gisin, N. (2020). Mathematical languages shape our understanding of time in physics. *Nat. Phys.*, **16**, 114–116.

Goswami, K., Cao, Y., Paz-Silva, G., Romero, J., and White, A. (2018a). Communicating via ignorance: Increasing communication capacity via superposition of order. arXiv:1807.07383.

Goswami, K., Giarmatzi, C., Kewming, M., Costa, F., Branciard, C., Romero, J., and White, A. G. (2018b). Indefinite causal order in a quantum switch. *Phys. Rev. Lett.*, **121**, 090503.

Grinbaum, A. (2007). Reconstruction of quantum theory. *British Journal for the Philosophy of Science*, **58**, 387–408.

Grinbaum, A. (2017a). How device-independent approaches change the meaning of physical theory. *Studies in History and Philosophy of Science Part B*, **58**, 22–30.

Grinbaum, A. (2017b). Narratives of quantum theory in the age of quantum technologies. *Ethics and Information Technology*, **19**, 295–306.

Grinbaum, A. (2019). The effectiveness of mathematics in physics of the unknown. *Synthese*, **196**, 973–989.

Guérin, P. A., Rubino, G., and Brukner, Č. (2019). Communication through quantum-controlled noise. *Phys. Rev. A*, **99**, 062317.

Hardy, L. (1999). Disentangling nonlocality and teleportation. arXiv:quant-ph/9906123.

Hardy, L. (2000). Quantum theory from five reasonable axioms. arXiv:quant-ph/00101012.

Hardy, L. (2007). Towards quantum gravity: a framework for probabilistic theories with non-fixed causal structure. *Journal of Physics A: Mathematical and Theoretical*, **40**(12), 3081–3099.

Hilbert, D., von Neumann, J., and Nordheim, L. (1927). Über die Grundlagen der Quanten-mechanik. *Math. Ann.*, **98**, 1–30. (Reprinted in J. von Neumann (1961), *Collected Works*, Vol. I, Oxford: Pergamon Press, pp. 104–133).

Husserl, E. (1954). *Die Krisis der europäischen Wissenschaften und die transzendentale Phänomenologie: Eine Einleitung in die phänomenologische Philosophie*. Leiden: Martinus Nijhof. English translation: *The Crisis of European Sciences and Transcendental Phenomenology*, Evanston, IL: Northwestern University Press, 1970.

Ibnouhsein, I., and Grinbaum, A. (2015). Information-theoretic constraints on correlations with indefinite causal order. *Phys. Rev. A*, **92**, 042124.

Ji, Z., Natarajan, A., Vidick, T., Wright, J., and Yuen, H. (2020). MIP*=RE. https://arxiv.org/abs/2001.04383.

Kochen, S. and Specker, E. (1965). Logical structures arising in quantum theory. In J. Addison, L. Henkin, and A. Tarski (eds), *The Theory of Models*, Amsterdam: North-Holland, pp. 177–189.

Kochen, S., and Specker, E. (1967). The problem of hidden variables in quantum mechanics. *Journal of Mathematics and Mechanics*, **17**, 59–87.

Landau, L., and Lifshitz, E. (1977). *Quantum mechanics*. Oxford: Pergamon Press. First Russian edition: Leningrad: State RSFSR Publishers, 1948.

Letertre, L. (2021). The operational framework for quantum theories is both epistemologically and ontologically neutral. *Studies in History and Philosophy of Science Part A*, **89**, 129–137.

Loizeau, N., and Grinbaum, A. (2020). Channel capacity enhancement with indefinite causal order. *Phys. Rev. A*, **101**, 012340.

London, F., and Bauer, E. (1939). *La théorie de l'observation en mécanique quantique*. Paris: Hermann. English translation in (Wheeler and Zurek, 1983, pp. 218–259).

Ludwig, G. (1985). *An Axiomatic Basis for Quantum Mechanics*. Berlin, Heidelberg: Springer.

Mackey, G. (1957). Quantum mechanics and Hilbert space. *Amer. Math. Monthly*, **64**, 45–57.

Masanes, L., and Müller, M. (2011). A derivation of quantum theory from physical requirements. *New Journal of Physics*, **13**, 063001.

Mermin, N. D. (1990). *Boojums All the Way through: Communicating Science in a Prosaic Age*. Cambridge: Cambridge University Press.

Navascués, M., and Wunderlich, H. (2010). A glance beyond the quantum model. *Proceedings of the Royal Society A: Mathematical, Physical and Engineering Science*, **466**, 881–890.

Oppenheim, J., and Wehner, S. (2010). The uncertainty principle determines the nonlocality of quantum mechanics. *Science*, **330**(6007), 1072–1074.

Oreshkov, O. (2019). Time-delocalized quantum subsystems and operations: on the existence of processes with indefinite causal structure in quantum mechanics. *Quantum*, **3**, 206.

Oreshkov, O., Costa, F., and Brukner, Č. (2012). Quantum correlations with no causal order. *Nature Communications*, **3**, 1092.

Pawlowski, M., Paterek, T., Kaszlikowski, D., Scarani, V., Winter, A., and Zukowski, M. (2009). Information causality as a physical principle. *Nature*, **461**, 1101–1104.

Peres, A. (1993). *Quantum Theory: Concepts and Methods*. Dordrecht: Kluwer Academic Publishers.

Petitot, J. (1997). Philosophie transcendantale et objectivité physique. *Philosophiques*, **24**(2), 367–388.

Piron, C. (1964). Axiomatique quantique. *Helvetica Physica Acta*, **36**, 439–468.

Piron, C. (1972). Survey of general quantum physics. *Found. Phys.*, **2**, 287–314.

Pironio, S., *et al.* (2010). Random numbers certified by Bell's theorem. *Nature*, **464**, 1021–1024.

Popescu, S. (2014). Nonlocality beyond quantum mechanics. *Nature Physics*, **10**, 264–270.

Popescu, S., and Rohrlich, D. (1994). Nonlocality as an axiom for quantum theory. *Foundations of Physics*, **24**, 379.

Procopio, L. M., Moqanaki, A., Araújo, M., Costa, F. M., Calafell, I. A., Dowd, E. G., Hamel, D. R., Rozema, L. A., Brukner, Č., and Walther, P. (2015). Experimental superposition of orders of quantum gates. *Nature Communications*, **6**, 7913.

Procopio, L. M., Delgado, F., Enriquez, M., Belabas, N., and Levenson, J. A. (2019). Communication through quantum coherent control of $N$ channels in a multi-partite causal-order scenario. *Entropy*, **21**, 1012.

Pusey, M. (2012). Stabilizer notation for Spekken's toy theory. *Found. Phys.*, **42**, 688.

Rédei, M. (2020). On the tension between physics and mathematics. *Journal for General Philosophy of Science*. doi: 10.1007/s10838-019-09496-0.

Rovelli, C. (1996). Relational quantum mechanics. *Int. J. of Theor. Phys.*, **35**, 1637.

Rubino, G., Rozema, L. A., Feix, A., Araújo, M., Zeuner, J. M., Procopio, L. M., Brukner, Č., and Walther, P. (2016). Experimental verification of an indefinite causal order. arXiv:1608.01683.

Rubino, G., Rozema, L. A., Ebler, D., Kristjánsson, H., Salek, S., Guérin, P. A., Abbott, A. A., Branciard, C., Brukner, Č., Chiribella, G., and Walther, P. (2020). Experimental quantum communication enhancement by superposing trajectories. arXiv:2007.05005v2 [quant-ph].

Scarani, V. (2006). *Quantum Physics: A First Encounter: Interference, Entanglement, and Reality.* Oxford: Oxford University Press.

Scarani, V. (2010). *Six Quantum Pieces: A First Course in Quantum Physics.* Singapore: World Scientific.

Smolin, J. (2005). Can quantum cryptography imply quantum mechanics? *Quantum Information and Computation*, 5, 161–169.

Spekkens, R. (2007). Evidence for the epistemic view of quantum states: A toy theory. *Phys. Rev. A*, 75, 032110.

Streltsov, A., Adesso, G., and Plenio, M. B. (2017). Colloquium: Quantum coherence as a resource. *Rev. Mod. Phys.*, 89, 041003.

Timpson, C. G. (2013). *Quantum Information Theory and the Foundations of Quantum Mechanics.* Oxford: Oxford University Press.

van Dam, W. (2000). *Nonlocality and communication complexity.* PhD thesis, Faculty of Physical Sciences, University of Oxford.

Vedral, V. (2012). *Decoding Reality.* Oxford: Oxford University Press.

Veitch, V., Mousavian, S. A. H., Gottesman, D., and Emerson, J. (2014). The resource theory of stabilizer quantum computation. *New Journal of Physics*, 16, 013009.

von Neumann, J. (1932). *Mathematische Grundlagen der Quantenmechanik.* Berlin: Springer. English translation: *Mathematical Foundations of Quantum Mechanics*, Princeton: Princeton University Press, 1955.

von Neumann, J. (2005). *Selected letters.* Edited by M. Rédei. American Mathematical Society, London Mathematical Society.

Watanabe, S. (1953). Réversibilité contre irréversibilité en physique quantique. In A. George (ed.), *Louis de Broglie, physicien et penseur*, Paris: Albin Michel, pp. 385–400.

Wei, K., Tischler, N., Zhao, S.-R., Li, Y.-H., Arrazola, J. M., Liu, Y., Zhang, W., Li, H., You, L., Wang, Z., Chen, Y.-A., Sanders, B. C., Zhang, Q., Pryde, G. J., Xu, F., and Pan, J.-W. (2018). Experimental quantum switching for exponentially superior quantum communication complexity. *Phys. Rev. Lett.*, 122, 120504.

Wheeler, J. (1983). Law without law. In J. Wheeler and W. Zurek (eds), *Quantum Theory and Measurement*, Princeton: Princeton University Press, pp. 182–213.

Wheeler, J., and Zurek, W. (eds) (1983). *Quantum Theory and Measurement.* Princeton: Princeton University Press.

Wigner, E. (1960). The unreasonable effectiveness of mathematics in the natural sciences. *Communications in Pure and Applied Mathematics*, 13, 1–14.

Wigner, E. (1961). Remarks on the mind-body question. In I. Good (ed.), *The Scientist Speculates*, London: Heinemann, pp. 284–302. Reprinted in Wheeler and Zurek (1983), pp. 168–181.

Wigner, E. (1976). Interpretation of quantum mechanics. Lectures given in the Physics Department of Princeton University. Published in Wheeler and Zurek (1983), pp. 260–314.

Zeilinger, A. (2010). *Dance of the Photons.* New York: Farrar Straus Giroux.

CHAPTER 17

..................................................................................................

# NATURAL RECONSTRUCTIONS OF QUANTUM MECHANICS

..................................................................................................

OLIVIER DARRIGOL

## 17.1 INTRODUCTION

..................................................................................................

In the first quarter of the 20th century, studies of blackbody radiation, low-temperature degeneracy, and atomic and molecular constitution led to the strangest of all theories, quantum mechanics, in a gradual and intricate way that does not help much understanding its nature. The basic mathematical structure of this theory, involving Hilbert spaces and operators acting on this space, defies intuitive understanding. It is therefore desirable to reconstruct quantum mechanics from assumptions or axioms more natural than those of a Hilbert space.

Since the early years of quantum mechanics, there have been many attempts to reformulate quantum mechanics, with various motivations including increased mathematical rigour and power, more direct insight into its empirical consequences, conceptual homogenization, and reduction. Some reconstructions yield broad extensions of quantum mechanics, with the purpose of easing its more problematic applications to quantum fields and gravitation. Other reconstructions are only partial: they reproduce only some central features of the theory and they purport to deepen our understanding of these features by partial reduction and comparison.[1]

The purpose of this essay is not to review the entire spectrum of reconstructions. The focus is on reconstructions yielding the full kinematic apparatus of quantum mechanics (the Hilbert space structure and the associated probability distributions) in a natural manner. Among those, only three of the most influential ones are retained: quantum

---

[1] Cf. Grinbaum (2007).

logic, deformation of the Lie group of classical evolutions, and Lucien Hardy's five reasonable axioms.

What should we mean by *natural* reconstruction? The answer is not that the implied deductions should rely on simple, transparent mathematics. This would be too much to ask and too contingent on one's level of mathematical competence. We should only require the naturalness of the primitive assumptions or axioms on which the deduction is based. Naturalness may refer to empirical immediacy, to empirical veracity, to mathematical simplicity, to fittingness in a familiar conceptual framework, or to correspondence à la Bohr.

Empirical immediacy stipulates the direct operational significance of a basic assumption. This may be achieved by focusing on devices for preparation, transformation, and measurement. Empirical veracity further requires the axiom to be a generalization of commonly accepted experimental facts. We would of course be happy if all the axioms met this criterion: we would thus be able to deduce quantum mechanics from a few empirically obvious principles just as we can, for instance, derive thermodynamics from the impossibility of two forms of perpetual motion (with some background knowledge). The mathematical simplicity of a basic assumption is a less convincing criterion of naturalness as long as there is no philosophical reason to presuppose that nature chooses the simplest mathematical options. The third notion of naturalness, fittingness in a familiar conceptual framework, is equally unconvincing. We may for instance seek fitness with probability theory or fitness with information theory. In both cases, the naturalness is debatable because what is natural from the point of view of probability or information theory need not be natural from a physical point of view. This is why in the following I will exclude reconstructions whose naturalness reduces to mathematical simplicity or information-theoretic homeliness. We are left with reconstructions based on empirical immediacy, empirical adequacy, and correspondence arguments à la Bohr.

As correspondence arguments play a central role in quantization by deformation and also (implicitly) in Hardy's reconstruction, a few words are in order. These arguments have often been criticized for their lack of rigour, along the lines: they do not by themselves lead to any definite theory and they must be combined with postulates that seem to contradict the classicality assumed by correspondence. This criticism, which Bohr already had to endure, is based on a confusion between conditions imposed on the possibilities of measurement with conditions on the deeper nature of the measured systems. As Bohr strove to explain, the language in which we express the conditions of experimentation need not be applicable to the objects on which we experiment. And a common observational language is needed in order to compare the old and new theories. At the very least we must assume that every homogeneous macroscopic system has (or appears to have) well-defined macroscopic properties. As we will see, this limited kind of correspondence suffices to justify a few of Hardy's axioms.

Correspondence arguments are also criticized for their making a more fundamental theory depend on a less fundamental one. This is true for the arguments that presuppose asymptotic agreement of the new theory with classical theory. This is much less the case for correspondence requirements that only appeal to a more limited classicality. At any rate, whether or not we regard a smaller-scale theory as more fundamental than a larger-scale theory, it remains true that the corroborated predictions of the latter theory should be included in the former. This is one of the most severe constraints we can legitimately impose on a new theory. Historically, it played a central role in the construction of the major theories of modern physics, including special relativity, general relativity, and quantum mechanics. Similarly, weaker correspondence requirements may play a central role in a natural reconstruction of quantum mechanics.

Our first example of natural reconstruction will be by deformation of classical mechanics. The essential results are the equivalence of all one-parameter deformations of the Poisson algebra of infinitesimal evolutions in Hamiltonian mechanics (for a sufficiently simple topology of phase-space), proven by André Lichnerowicz and Simone Gutt in 1979, and the agreeing of one of these deformations with Joseph Moyal's phase-space formulation of quantum mechanics. Roughly, there is only one way to deform classical mechanics and this leads to the formal apparatus of quantum mechanics. Our second example will be the quantum logic initiated by John von Neumann in the mid-1930s. The central idea is to base quantum mechanics on simple, natural assumptions for the lattice of experimental Yes–No questions that may be asked of a physical system. It turns out that the lattice properties assumed by von Neumann for finite dimension and later by Constantin Piron for infinite dimension lead to a generalized Hilbert space. To some extent, these properties are operationally justified. Our third example is Lucien Hardy's reconstruction of quantum mechanics as a generalized probability theory from 'five reasonable axioms' in 2001. In this approach, statistical correlation between discrete measurements is the most basic notion. The states of a system are defined through measurement probability distributions, which may be seen as the expression of information content. The axioms have simple empirical grounding, or they may be justified through correspondence arguments. They lead to the Hilbert-space structure of quantum mechanics in a relatively effortless manner.

For the sake of brevity, I only describe the most seminal works for each kind of reconstruction and I will not describe their abundant ulterior developments. I enunciate the various theorems needed in these three reconstructions without giving the proofs, which the reader can find in the cited original works. As these proofs often rely on unnecessarily difficult and powerful mathematics, the reader may also consult the simplified proofs given in earlier work of mine on which the present essay is largely based.[2]

---

[2] Darrigol (2014), chap. 8; Darrigol (2015).

# 17.2 The Deformation of Classical Mechanics

## 17.2.1 Quantum Mechanics in Phase Space

Consider a classical Hamiltonian system with the Hamiltonian $H(q,p)$. To every quantity $g(q,p)$ we may associate its average value

$$\bar{g} = \int \rho(q,p)g(q,p)\mathrm{d}q\mathrm{d}p \qquad (1)$$

for a given value of the probability distribution $\rho(q,p)$ in phase space. The evolution of this distribution is given by

$$\dot{\rho} = \{H,\rho\}, \qquad (2)$$

in which the Poisson bracket is defined by

$$\{f,g\} = \frac{\partial f}{\partial q}\frac{\partial g}{\partial p} - \frac{\partial f}{\partial p}\frac{\partial g}{\partial q}. \qquad (3)$$

In quantum mechanics, the physical quantity $g$ is represented by a Hermitian operator $\mathbf{g}$ and its average value is given by

$$\langle \mathbf{g} \rangle = \mathrm{Tr}(\boldsymbol{\rho}\mathbf{g}) \qquad (4)$$

in a state represented by the density matrix $\boldsymbol{\rho}$. The evolution of this operator is given by

$$i\hbar\dot{\boldsymbol{\rho}} = [\mathbf{H},\boldsymbol{\rho}]. \qquad (5)$$

We may naturally wonder whether quantum mechanics can be brought closer to classical mechanics through a one-to-one correspondence between a Hermitic operator $\mathbf{f}$ and a real distribution in phase space $f(q,p)$. During World War II, José Moyal in England and Hilbrand Groenewold in the Netherlands independently discovered that Hermann Weyl's quantization formula of 1932,

$$\mathbf{f} = \frac{1}{4\pi^2}\int f(q,p)e^{i[\alpha(\mathbf{q}-q)+\beta(\mathbf{p}-p)]}\mathrm{d}q\mathrm{d}p\mathrm{d}\alpha\mathrm{d}\beta, \qquad (6)$$

did the job. It could indeed be inverted to get

$$f(q,p) = 2 \int dq' e^{2ipq'/\hbar} \langle q - q' | \mathbf{f} | q + q' \rangle, \tag{7}$$

an expression already used by Eugene Wigner in 1932 while studying quasi-classical approximations of quantum statistical mechanics.[3]

With this reciprocal correspondence, we have

$$\mathrm{Tr}(\boldsymbol{\rho} \mathbf{g}) = \int \rho(q,p) g(q,p) dq dp / h, \tag{8}$$

which means that any quantum-mechanical average can be replaced with a classical phase-space average. To the quantum product of two quantities $\mathbf{f}$ and $\mathbf{g}$ corresponds the *Moyal product* of the associated phase-space densities $f(q,p)$ and $g(q,p)$:

$$f * g = \sum_{n'n''} \left( \frac{i\hbar}{2} \right)^{n'+n''} \frac{(-1)^{n''}}{n'!n''!} (\partial_q^{n'} \partial_p^{n''} f)(\partial_q^{n''} \partial_p^{n'} g), \tag{9}$$

or, in the symbolic notation used by Groenewold and Moyal,

$$f * g = f e^{(i\hbar/2)(\overleftarrow{\partial}_q \overrightarrow{\partial}_p - \overleftarrow{\partial}_p \overrightarrow{\partial}_q)} g,$$

in which the arrows above the derivatives indicate on which side they are operating.

From this product we may form the *Moyal bracket*

$$\{\{f,g\}\} = \frac{1}{i\hbar} (f * g - g * f), \tag{10}$$

which corresponds to $[\mathbf{f}, \mathbf{g}]/i\hbar$. This bracket differs from the Poisson bracket $\{f,g\}$ by terms involving powers of $\hbar$ and higher derivatives of the functions $f$ and $g$. By means of these new brackets, Eq. (5) for quantum-mechanical equation translates into the phase-space equation

$$\dot{\rho} = \{\{H, \rho\}\}. \tag{11}$$

When $\hbar$ reaches zero, this equation degenerates into the classical equation (2).

In order that the distribution $\rho(q,p)$ yields a density matrix (Hermitian, positive, and of trace *one*) by Weyl quantization, it must meet certain conditions. In particular, it must be real, it must satisfy

---

[3] Moyal (1949); Groenewold (1946). Cf. Curtright and Zachos (2011); Zachos, Fairlie, and Curtright (2005). In Eq. (7), I have multiplied the original Wigner transform by $h$, so that $f$ and $\mathbf{f}$ have the same dimension and so that it becomes the exact inverse of the Weyl transform.

$$\iint \rho(q,p)\mathrm{d}q\mathrm{d}p/h = 1 \tag{12}$$

as if $\rho/h$ were a probability distribution, and it must satisfy the inequality

$$|\rho(q,p)| \leq 2. \tag{13}$$

Unlike a true probability, it can take negative values.


## 17.2.2 Deforming the Poisson Algebra

Is there any advantage in reformulating the equations of quantum mechanics in classical phase-space? Evidently, distributions in phase-space are easier to visualize than Hermitian operators. This is why they are frequently used in modern quantum optics. However, the non-positive character of the distribution forbids its interpretation as a probability in phase space. There is something irremediably formal in the phase-space formulation of quantum mechanics.

Yet there is something mathematically remarkable about the Moyal bracket: it is a deformation of the Poisson bracket in the precise sense given by Murray Gerstenhaber in 1964. Namely, it is a $C^\infty$ bilinear alternate function $\{f,g\}_\hbar$ of the phase-space functions $f$ and $g$ and of the parameter $\hbar$ defining a Lie algebra (satisfying the Jacobi identity) in the space of phase-space functions, varying continuously with $\hbar$ and coinciding with the usual Poisson bracket for $\hbar = 0$. This is so true that in 1975 the French mathematician Jacques Vey rediscovered the Moyal bracket by deforming the Poisson bracket, without prior knowledge of Moyal's work. Four years later, in a systematic study of all possible deformations of the Poisson brackets, André Lichnerowicz in Paris and Simone Gutt in Brussels both discovered that all deformations of the Poisson bracket were equivalent on manifolds for which the second Betti number vanishes. This topological condition is trivially met in the flat $\mathbf{R}^{2n}$ case which is most commonly met in physics.[4]

The equivalence of two deformed brackets $\{f,g\}_\hbar$ and $\{f,g\}_\hbar'$ is defined as the existence of a differential operator

$$T = \mathrm{Id} + \sum_{s=1}^{\infty} \hbar^s T_s \text{ such that } T\{f,g\}_\hbar' = \{Tf, Tg\}_\hbar \tag{14}$$

for any pair $(f,g)$ of phase functions. This equivalence evidently preserves the equation of motion $\dot\rho = \{H,\rho\}_\hbar$. In addition, the equivalence $T(f*'g) = Tf * Tg$ for the associated star products preserves the possibility of translating this equation in operator language by replacing the Wigner transform $W$ through the transform $W'$ such that

---

[4]  Vey (1975); Gutt (1979); Lichnerowicz (1979).

$W'(g) = W(Tg)$. The original equivalence proof involves the Chevalley cohomology and it is not for the amateur mathematician.

## 17.2.3 Morals

This highly remarkable result means that in a deep mathematical sense, the Lie algebra of quantum-mechanical evolutions is a unique natural extension of the Lie algebra of classical Hamiltonian evolution. More recently, in 2011, Maurice de Gosson and Basil Hiley discovered a more direct correspondence between the two algebras. These results mean that quantum mechanics could have been discovered by mathematicians studying the deformations of Lie algebras. Indeed, it can be said that Jacques Vey independently rediscovered quantum mechanics with some delay.[5]

To be true, without already knowing about quantum physics, there would be no reason to think that the result of a purely mathematical deformation would have physical meaning. On the contrary, the non-positive character of the relevant distributions would be a serious obstacle to physical interpretation; and the Hilbert-space operator version of the Moyal algebra would not jump to the eyes. The chief protagonists of quantization by deformation were well aware of this shortcoming. For instance, Lichnerowicz's collaborator David Sternheimer wrote:[6]

> A word of caution may be needed here. It is possible to intellectually imagine new physical theories by deforming existing ones... Nevertheless, such intellectual constructs, even if they are beautiful mathematical theories, need to be somehow confronted with physical reality in order to be taken seriously in physics. So some physical intuition is still needed when using deformation theory in physics.

If we forget about the difficulty of physically interpreting the result of the deformation, there are reasons to think that deformation quantization is not a merely mathematical game and has import in judging the necessity of quantum mechanics qua physical theory. Firstly, we should expect any physically meaningful generalization of classical mechanics to contain classical mechanics in the limiting case in which the evolutions are macroscopic. Secondly, as long as the concept of time remains valid, we should expect the new theory to be about a group of continuous evolutions in time. Combining these two conditions, and knowing about Planck's constant, we should expect the new group of evolutions to be a one-parameter deformation of the old group. This should be true independently of the particular way in which we represent the action of the group. For Lichnerowicz and Gutt, this action is on phase-space distributions. Although the deformation destroys the classical interpretation of these distributions, the existence of an alternative, meaningful interpretation should not come as a surprise if we believe the deformed algebra to represent true evolutions in this world.

[5] Gosson and Hiley (2009, 2011).    [6] Sternheimer (1998), p. 2.

# 17.3  QUANTUM LOGIC

## 17.3.1  From the Spectral Theorem to Quantum Logic

A basic difficulty in the rigorous mathematical formulation of quantum mechanics is the unbounded character of its operators, which implies the possibility of non-normalizable eigenstates in the continuous part of the spectrum. For instance, in the Schrödinger representation, the eigenfunction of the momentum operator $-i\hbar\partial/\partial q$ with the momentum value $p$ is the non-normalizable function $\psi(q) = e^{ipq/\hbar}$. In order to avoid this difficulty without enlarging the Hilbert space of states, in 1930 John von Neumann introduced his spectral theorem in which the eigenvectors are replaced with appropriate projectors.[7]

More exactly, von Neumann defines a *spectral measure* as an operator-valued measure $P_\lambda d\lambda$ such that for any measurable subset $\Lambda$ of $\mathbf{R}$ the integral $P_\Lambda = \int_{\lambda\in\Lambda} P_\lambda d\lambda$

is an orthogonal projector and such that the subspaces associated with the projectors $P_\Lambda$ and $P_{\Lambda'}$ are orthogonal whenever the subsets $\Lambda$ and $\Lambda'$ are disjoint. An operator $X$ is said to be *self-adjoint* if and only if its definition domain is dense in the Hilbert space (as is the case for the position and momentum operators, even though they are unbounded) and the usual relation of conjugation $\langle\phi|X|\psi\rangle = \langle\psi|X|\phi\rangle^*$ holds. According to von Neumann's spectral theorem, such operators admit the spectral decomposition

$$X = \int_{-\infty}^{+\infty} \lambda P_\lambda d\lambda. \tag{15}$$

Von Neumann based his statistical interpretation of quantum mechanics on the projectors $P_\Lambda$. The eigenvalue 1 of this projector corresponds to a quantum state in which the quantity $X$ takes a value belonging to the set $\Lambda$, and the eigenvalue 0 corresponds to a quantum state in which the quantity $X$ takes a value belonging to the complementary set. Thus, the projectors for any possible quantity symbolize Yes–No questions from which the result of any measurement can be inferred. As von Neumann noted in 1932, the properties of the projectors define 'a sort of logical calculus' in which different questions may not be simultaneously decidable.[8]

Von Neumann and Garrett Birkhoff developed this calculus or quantum logic in an influential memoir of 1936. A basic proposition $a$ on a physical system is there defined by the invariant subspace $A$ of the associated projector. The relation '$a$ implies $b$' is identified with the inclusion of the associated subspaces, the generalized conjunction 'meet of $a$ and $b$' with the intersection of the associated subspaces, the general disjunction 'join of $a$ and $b$' with the closed linear sum of the associated subspaces,

---

[7] von Neumann (1930). Cf. Lacki (2000).    [8] von Neumann (1932), p. 253.

the negation of $a$ with the orthogonal complement of the associated subspace, the always false proposition '0' with the trivial subspace $\{0\}$, and the always true proposition '1' with the entire Hilbert space H. In symbols, we have

| $a \leq b$ | $a \wedge b$ | $a \vee b$ | $\bar{a}$ | 0 | 1 |
|---|---|---|---|---|---|
| $A \subset B$ | $A \cap B$ | $A + B$ | $A^{\perp}$ | $\{0\}$ | H |

The relation $a \leq b$ is a relation of partial order; the meet $a \wedge b$ is the greatest lower bound of $a$ and $b$ with respect to this relation; the join $a \vee b$ is the least upper bound of $a$ and $b$. The set of propositions thus forms what mathematicians call a *lattice*. This lattice is *orthocomplemented*, namely: it has a minimal element 0 and a maximal element 1; to every $a$ corresponds an orthocomplement $\bar{a}$ satisfying $\bar{\bar{a}} = a$, $a \wedge \bar{a} = 0$, $a \vee \bar{a} = 1$; if $a \leq b$ then $\bar{b} \leq \bar{a}$.[9]

Ordinary logic shares this orthocomplemented lattice structure. In addition, it enjoys the distributivity

$$a \wedge (b \vee c) = (a \wedge b) \vee (a \wedge c) \quad \text{and} \quad a \vee (b \wedge c) = (a \vee b) \wedge (a \vee c). \tag{16}$$

Quantum logic does not have this Boolean property, except when the projectors associated with the three propositions $a$, $b$, $c$ commute. If the Hilbert space is of finite dimension, it enjoys the weaker property

$$\text{If } a \leq c \quad \text{then} \quad a \vee (b \wedge c) = (a \vee b) \wedge c. \tag{17}$$

which is called *modularity*. To sum up, the subspaces of a Hilbert space of finite dimension define a modular orthocomplemented lattice of propositions.

## 17.3.2 From the Axioms of Quantum Logic to the Hilbert-Space Structure

Conversely, von Neumann and Birkhoff proved that any irreducible orthocomplemented modular lattice of finite dimension is isomorphic to the lattice of subspaces of a generalized Hilbert space. Roughly, quantum logic implicitly contains something like the Hilbert space structure.

In order to prove this remarkable result, Birkhoff and von Neumann relied on two successive isomorphisms, firstly between a lattice and a projective geometry, then between a projective geometry and the set of subspaces to a vector space. According to the first isomorphism, *any irreducible complemented modular lattice of finite*

---

[9] Birkhoff and von Neumann (1936). On lattice theory, cf. Birkhoff (1940). On the history of quantum logic, cf. Jammer (1974), chap. 8; Rédei (2007); Dalla Chiara, Giuntini, and Rédei (2007).

*dimension (higher than two) is isomorphic to a projective geometry of finite definition.* This statement appeals to the following definitions:

- A lattice is said to be *irreducible* if there is a third minimal non-zero element smaller than the join of any two distinct minimal non-zero elements of the lattice (in the contrary case, the lattice would be the direct product of several irreducible lattices).
- A lattice is said to be *complemented* if there exist a 0 and a 1 and if every element $a$ admits a complement $a'$ such that $a \vee a' = 1$ and $a \wedge a' = 0$.
- The dimension of a lattice is said to be finite if there is a maximal value to the number of elements in a chain $0 < a_1 < a_2 < \ldots < 1$.
- An abstract *projective geometry of finite dimension* is defined by a set of elements of increasing but bounded 'dimension' called points, lines, planes, etc. and satisfying the four following axioms:

  $P_1$: Two distinct points are contained in one and only one line.

  $P_2$: If A, B, C are points not all on the same line, and D and E are two distinct points such that B, C, D are on a line and C, A, E are on a line, then there is a point F such that A, B, F are on a line and also D, E, F are on a line.

  $P_3$: Every line contains at least three points.

  $P_4$: The set of points on lines through any $k$-dimensional element and a fixed point not on the element is a $(k + 1)$-dimensional element, and every $(k + 1)$-dimensional element can be defined in this way.

The isomorphism is obtained by assimilating the various points, lines, etc. with elements of the lattice of increasing dimension, and by identifying the implication $a \leq b$ with the geometric statement '$a$ is on $b$.' Consequently, the meet of two elements corresponds to their geometric intersection, and their join to the smallest geometric element that contains both.

According to the second isomorphism, for a geometric dimension $N - 1$ higher than two, any projective geometry is isomorphic to the set of subspaces of a vector space of dimension $N$. A point of the projective geometry corresponds to a subspace of dimension 1, a line to a subspace of dimension 2, etc. The inclusion of a geometric element in another corresponds to inclusion for the corresponding subspaces. The division ring on which the vector space is built is obtained by a projective construction for the sum and product of two points on a line. This isomorphism had been a well-known result of projective geometry since the 19th century.[10]

Combining the two isomorphisms, we may conclude that every irreducible complemented modular lattice of finite dimension higher than three has the same structure as the set of subspaces of a vector space. For quantum logic, we also want the lattice to be orthocomplemented. This brings a further restriction on the vector space: according to a theorem by von Neumann, this vector space must be what I will call a **K\***-space.

---

[10] See, e.g., Garner (1981).

In such a space, the division ring is equipped with a 'star conjugation' with the properties

$$(x+y)^* = x^* + y^*, \ (xy)^* = y^* x^*, \quad \text{and} \quad x^{**} = x \qquad (18)$$

for any two elements $x$ and $y$ of the division ring, and the vector space is equipped with the definite form $<\ ,\ >$ satisfying

$$\begin{aligned}
<a, b+c> &= <a,b> + <a,c>, \ <a+b, c> = <a,b> + <a,c> \\
<b, ax> &= <b,a>x, \ <bx, a> = x^* <b,a>, \ <a,b> = <b,a>^*
\end{aligned} \qquad (19)$$

for any three vectors $a$, $b$, $c$ and for any element $x$ of the division ring. Clearly, a $K^*$-space is a generalization of a Hilbert space based on the ring $C$ of complex numbers. In the present case of finite dimension, there is no further restriction on the choice of the division ring $K$. It could be a finite (necessarily) commutative ring, and it could be a non-commutative ring such as the ring of quaternions. It is still remarkable that the axioms of quantum logic would lead to a structure closely resembling the state-space of quantum mechanics.

## 17.3.3 Interpreting and Justifying the Axioms of Quantum Logic in Finite Dimension

From a purely mathematical point of view, the generalization of the Boolean lattice of ordinary logic to a modular orthocomplemented lattice is fairly natural. It is based on a simple weakening of the mutual distributivity of the meet and the join, leading to the modularity already contemplated by Richard Dedekind in the late 19th century for purely mathematical reasons. This sort of naturalness is not sufficient from a physical point of view. We want to know whether, in the present case of finite dimension, the Yes–No experimental questions on a physical system should naturally be expected to share the structure of a modular orthocomplemented lattice.

Firstly, the basic axioms of a lattice have the desired naturalness in the case of finite dimension. The partial ordering $a \leq b$ has an evident operational meaning: the answer to the question $b$ is Yes if the answer to the question $a$ is Yes. This interpretation is compatible with the basic properties of the lattice implication, because the corresponding tests are repeatable and because the operational implication is transitive. In the present case of finite dimension, the existence and the properties of the meet and the join simply derive from the partial order. Orthocomplementation obtains simply by permuting the Yes and No of an experimental question. In an orthocomplemented modular lattice, irreducibility is equivalent with the requirement that 0 and 1 are the only elements compatible with all the elements of the lattice. Operationally, this means that there are no non-trivial questions that do not interfere with any other question.

In the contrary case, the non-trivial question could be used to further specify the system (as is done in quantum mechanics with super-selection rules).[11]

There remains modularity, which is not so easy to justify from a physical point of view. For this purpose, it is useful to replace modularity with two properties introduced by Constantin Piron in his Geneva dissertation in 1964: weak modularity and atomicity. An orthocomplemented lattice is said to be *weakly modular* if and only if for any two elements $a$ and $b$, $a \leq b$ implies $a$ is compatible with $b$ (meaning that $a, \bar{a}, b, \bar{b}$ engender a Boolean sublattice). An orthocomplemented weakly modular lattice is called an *orthomodular lattice*. An *atom* (also called a point in the geometric interpretation) is a minimal non-zero element of the lattice. An *atomic lattice* is a lattice satisfying the two following axioms:

$A_1$: Every element contains an atom.

$A_2$ (*covering law*): If $a$ and $b$ are elements of the lattice and $e$ an atom, one can never have $a < b < a \vee e$ ($a \vee e$ at most covers $a$).

According to a theorem by Piron, *any orthomodular atomic lattice of finite dimension is modular.*[12]

Operationally, weak modularity means that if a test $b$ has a well-defined result when $a$ has just been tested, then the testing sequence $a, b, a$ always yields the same result for the two tests of $a$ (and so, too, do the two tests of $b$ in the sequence $b, a, b$). This seems reasonable, because the premise $a \leq b$ intuitively implies that the test $b$ refines our knowledge of the system without destroying knowledge acquired by the test $a$. For finite dimension, the atomic axiom $A_1$ holds necessarily since a chain below any given element of the lattice cannot be indefinitely lengthened by inserting a non-zero element under its least element. We are left with the hardest part: justifying the covering law.

It is easy to see that whenever an atom $e$ is compatible with $a$, then $a \vee e$ at most covers $a$. When the atom $e$ is not compatible with $a$, we may rely on the identity $a \vee e = a \vee d$, with $d = (a \vee e) \wedge \bar{a}$. Since $d$ is compatible with $a$, the covering law will be justified if we can find a physical reason for $d$ being an atom. This is what Piron managed to do in 1969. An atom $e$ of the lattice is associated with a maximal repeatable test, and can therefore represent the state of the system produced by this test (a pure state in quantum-mechanical language). Suppose that the system was originally in this state and that a test $a$ then gives a No answer. After this test, we expect the system to still be in a maximally known state, say $f$. This state ought to be determinable by the set of tests that do not interfere with it. This set comprises a second test of $a$, and the test of every proposition $x$ containing $e$ and compatible with $a$. Indeed the latter kind of test may be indifferently performed before or after the (first) test of $a$ (owing to

[11] Cf. Jauch (1968), pp. 124–26.
[12] Piron (1964), pp. 446 (weak modularity), 448 (atomicity), 460 (proof of the theorem).

compatibility interpreted as non-interference)[13] and it obviously does not alter the state $e$ in the former case. Compatibility implies

$$x = (x \vee a) \wedge (x \vee \bar{a}) \geq (e \vee a) \wedge (e \vee \bar{a}) \equiv b. \tag{20}$$

The element $b$ is itself a possible $x$, since it contains $e$ and is compatible with $a$. Consequently, the conditions $f \leq b$ and $f \leq \bar{a}$ should fully determine the atom $f$. Equivalently, there should be only one atom $f$ such that $f \leq \bar{a} \wedge b = (a \vee e) \wedge \bar{a} = d$. This can be true only if $d$ is an atom and $f = d$. Piron thus determined the final state and justified the covering law:[14]

> It is important to remark that without this axiom [the covering law] we cannot determine the final state of the system; and although the measurement may be ideal, [without this axiom] the perturbation results in a loss of information, even if we take the response of the system into account.

Altogether the axioms of quantum logic of finite dimension, namely, lattice structure, orthocomplementarity, and modularity receive natural justifications in an operational context in which the elements of the lattice refer to repeatable, binary tests on a physical system. It is easy to imagine with Piron that the tests may or may not interfere, and that they may produce maximal knowledge of the state of the system. These intuitions justify modularity in a subtle but fairly convincing manner. The main shortcoming of the quantum-logical axioms in finite dimension is that they do not exactly lead to the desired Hilbert-space structure. They define a broad generalization of this structure, the K*-space structure, in which the ring of complex numbers of the usual Hilbert space of quantum mechanics can be replaced by any division ring K equipped with a star conjugation.[15]

## 17.3.4 Infinite Dimension

The quantization of a classical Hamiltonian system typically leads to a Hilbert space of infinite dimension. It may therefore seem artificial to focus on lattices of fine dimension, as we have done so far. Yet this is not so absurd from a physical point of view. In a fruitful approximation, physicists can concretely isolate a finite-dimensional subspace of the global Hilbert space, for instance when they deal with resonant transitions between two levels of an atoms, or when they deal with the internal degrees of freedom

---

[13] The test is 'ideal' in Piron's sense, namely, it does not modify the outcome of a compatible test.

[14] Jauch and Piron (1969), pp. 847–48; Piron (1976), p. 69 (quotation); Stachow (1984). On other justifications of the covering law, see Wilce (2009), section 5.

[15] On later developments of quantum logic and on improvements of its operational grounding, cf. Coecke, Moore, and Wilce (2000); Gabbay, Lehmann, and Engesser (2009); Wilce (2009).

(spin) of a particle. It therefore seems reasonable to first establish the relevant lattice structure in finite dimension, from which the K*-space emerges, and then generalize this space to infinite dimension.

The other route, favoured by Piron and his followers, consists in directly generalizing the finite-dimensional lattice structure to infinite dimension. Remember that for finite dimension the lattice must be orthocomplemented, irreducible, and modular. Unfortunately, the lattice of subspaces of a Hilbert space of infinite dimension is easily seen not to be modular. Piron circumvented this difficulty by replacing modularity with the conjunction of atomicity and weak modularity. He then proved his fundamental theorem according to which *any irreducible complete orthomodular atomic lattice can be represented by the lattice of subspaces of a K*-space.*[16]

Again the choice of the division ring is free, as long as it can support a star conjugation. However, in her Konstanz University dissertation of 1995 Maria Pia Solèr proved that the choice of K is restricted to R, C, or H (quaternions) if and only if there exists an infinite orthonormal sequence of vectors in the K*-space. This is a remarkable result because the quantum-logical axioms do not make any reference to a measurable continuum: they imply simple binary tests, not any measurement in the strict sense. There are physical arguments against a quantum mechanics based in a real Hilbert space, and the little-studied quaternionic case is likely to be equivalent to the usual quantum mechanics for particles with a proper internal symmetry group. We are left with the usual Hilbert space as the only plausible candidate.[17]

This advantage of the infinite-dimensional approach is unfortunately counterbalanced by the difficulty of finding physical justifications for some of the axioms in the infinite case.[18] In particular, the existence of the meet (greatest lower bound) and the join (least upper bound) is no longer an automatic consequence of the partial order. It must be postulated, and it lacks any simple physical interpretation. Also, atomicity is no longer warranted. One must postulate that the indefinite refining of a test is impossible, which is tantamount to assuming the basic discreteness of the results of measurement. Lastly, Solèr's assumption of an infinite orthonormal sequence is in need of a simple physical justification. One could for instance argue that for an unbounded physical quantity, there are infinitely many compatible binary tests for the value of this quantity (corresponding to the infinite sequence of diagonalizing state vectors in quantum mechanics). For the rest, it seems that the properties that are operationally meaningful for finite dimension remain valid for infinite dimension even when their operational meaning is lost. There is no rigorous justification of this felicitous circumstance.

---

[16] Piron (1964). A student of Mackey's, Malcom Donald MacLaren, and two Japanese mathematicians, Ichiro Amemiya and Huzihiro Araki, perfected the proof: cf. Primas (1981), p. 212.

[17] Solèr (1995). Cf. Holland (1995).

[18] For a lucid discussion of these difficulties, cf. Primas (1981), pp. 214–19.

## 17.3.5 Probabilities, States, and Evolution

In the context of quantum logic, a natural way to define the state of a system is by giving the probability of a positive answer for every possible Yes–No question. This is indeed the definition proposed by the Harvard mathematician George Mackey in 1957 in a rigorous axiomatization of quantum mechanics based on quantum logic and probability. The correspondence between these questions and the subspaces of a Hilbert space being established, the state is defined through a probability measure on the set of subspaces of a Hilbert space. According to a theorem published in the same year (1957) by Mackey's colleague Andrew Gleason, *any probability measure on the subspaces of a Hilbert space can be represented by a density matrix.* We thus reach the usual quantum mechanical representation of a state.[19]

We may then follow Mackey in defining the evolution of the system through a probability-conserving evolution of the density matrix. In symbols, $(\alpha\rho + \beta\sigma)' = \alpha\rho' + \beta\sigma'$ for any two density matrices $\rho$ and $\sigma$ and for any two weights $\alpha$ and $\beta$ such that $\alpha \geq 0, \beta \geq 0$ and $\alpha + \beta = 1$. Using a theorem by Richard Kadison on automorphisms in C*-algebras, Mackey proved that *any mixture-preserving one-to-one mapping of the set of density matrices onto itself could be expressed as $\rho' = U\rho U^{-1}$, U being a unitary or anti-unitary operator U unique up to a phase factor.* Mackey then used the uniformity of time to argue that $U(\tau)$ and $U^2(\tau/2)$ only differed by a phase factor. Since the square of an anti-linear operator is linear, the anti-unitary option is excluded. We are left with the usual unitary evolution of quantum mechanics.[20]

## 17.3.6 Morals

There is something exhilarating in the capacity of quantum logic to generate a generalization of the Hilbert-space structure of quantum mechanics from very simple and fairly natural axioms about the Yes–No questions we may ask on a physical system. For a long time, this success seemed limited because the choice of the division ring in the generalized Hilbert space structure was very free. Solèr's theorem greatly enhanced it by limiting this choice to a Hilbert space based on the field **R**, **C**, or **H**. However, the fundamental theorems by von Neumann, Piron, and Solèr linking the Hilbert-space structure to the axioms of quantum logic are not easy to prove and they involve a kind of mathematics (lattice theory, projective geometry, C*-algebras) rarely used in theoretical physics. Another difficulty resides in the operational justification of the axioms.

---

[19] Mackey (1957); Gleason (1957).

[20] Mackey (1963), pp. 81–82; Kadison (1951); Piron and Jauch favour another approach in which the evolution is assumed to be an automorphism of the lattice of propositions. This leads to the same result by Wigner's theorem, which states that all automorphisms of the lattice of subspaces of a Hilbert spaces are generated by unitary or anti-unitary operators. Cf. Beltrametti and Cassinelli (1981), pp. 252–54.

Although the operational meaning of the axioms is clear in the case of finite dimension, an act of faith is needed to extend them to the case of infinite dimension.[21]

Another difficulty of quantum logic resides in its ignoring composite systems. Nowadays, the tensor-product definition of the state-space of a composite system is regarded as an essential feature of quantum mechanics, leading to entangled states and all associated quantum strangeness. Unfortunately, there is no simple lattice-theoretical counterpart to the tensor product of standard quantum mechanics.[22] For this reason, the more recent reconstructions of quantum mechanics tend to include axioms regarding composite systems. This newer trend does not annihilate the merits of quantum logic as an operational elucidation of the Hilbert-space structure of quantum mechanics. What it shows is the need to diversify our reconstructions of quantum mechanics.

## 17.4 Reconstructions à la Hardy

As was already mentioned, in 1957 George Mackey defined the states of a quantum system through the probabilities of measurement outcomes. So did too Günther Ludwig in the highly elaborate axiomatics he began to develop in the 1950s and perfected in the 1980s. Parts of Mackey's and Ludwig's axioms were mathematically motivated, the rest inspired by quantum logic. In a never refereed ArXiv paper written in 2001 in Oxford, the British theoretical physicist Lucien Hardy retained the probabilistic, operational definition of states, but ignored quantum logic and relied on 'five reasonable axioms' for these probabilities. In preparation for looking at this new kind of reconstruction, it will first be shown that for a two-level quantum system (a spin), the quantum-mechanical probability formulas result from very simple assumptions on the relative frequencies of measurement outcomes.[23]

### 17.4.1 The One-Half Spin System

There is a continuous infinity of possible measurements of a spin, as there are infinitely many directions of space in which an angular momentum can be measured. In contrast, there are only two possible outcomes for each of these measurements: $+\hbar/2$ and $-\hbar/2$. If the system is found to have the momentum $+\hbar/2$ in a given direction, a subsequent measurement performed in a direction making an angle $\theta$ with the former direction will give either $+\hbar/2$ or $-\hbar/2$. Let us repeat the same preparation

---

[21] See the discussions in Dalla Chiara, Giuntini, and Rédei (2007), pp. 228–31; Mittelstaedt (2011), p. 64.

[22] Cf. Wilce (2009), section 7, and reference there to the no-go theorems by David Foulis and Charles Randall and by Diederik Aerts.

[23] Mackey (1957); Ludwig (1983, 1985); Hardy (2001).

and the same measurement a large number of times. If $p_+$ and $p_-$ denote the frequencies of the two possible outcomes, we must have

$$p_+ - p_- = \cos\theta \tag{21}$$

in order that the average angular momentum in the direction $\theta$ be equal to the projection of the initial angular momentum on this direction. This is so because by a correspondence argument we expect the total angular momentum (or magnetic moment) of a large number of identically prepared, non-interacting spin-particles to behave as the angular momentum of a macroscopic object under measurement. Since $p_+ + p_- = 1$, we have[24]

$$p_+ = \cos^2(\theta/2), \ p_- = \sin^2(\theta/2) \tag{22}$$

To sum up, the double-valuedness of spin, the spatial character of spin measurement, and a correspondence requirement together imply the well-known quantum-mechanical expression for the correlations between spin measurements in two different directions. In a different notation, the correlation probability for $+\hbar/2$ spin components in the directions (unit vectors) $\mathbf{u}$ and $\mathbf{u}'$ is

$$p(\mathbf{u}, \mathbf{u}') = \frac{1 + \mathbf{u} \cdot \mathbf{u}'}{2}. \tag{23}$$

In polar coordinates for which $\mathbf{u} = (\cos\theta, \sin\theta\cos\phi, \sin\theta\sin\phi)$, we have

$$p(\mathbf{u}, \mathbf{u}') = \frac{1}{2}[1 + \cos\theta\cos\theta' + \sin\theta\sin\theta'\cos(\phi - \phi')] \tag{24}$$
$$= |\cos(\theta/2)\cos(\theta'/2) + \sin(\theta/2)\sin(\theta'/2)e^{i(\phi-\phi')}|^2.$$

Introducing a bidimensional Hilbert space with two orthogonal state vectors $|+\rangle$ and $|-\rangle$ corresponding to the spins $+\hbar/2$ and $-\hbar/2$ in the polar direction, the vectors

$$|+_\mathbf{u}\rangle = \cos(\theta/2)|+\rangle + e^{i\phi}\sin(\theta/2)|-\rangle \text{ and } |+_{\mathbf{u}'}\rangle = \cos(\theta'/2)|+\rangle + e^{i\phi'}\sin(\theta'/2)|-\rangle \tag{25}$$

are such that

$$p(\mathbf{u}, \mathbf{u}') = |\langle +_\mathbf{u}|+_{\mathbf{u}'}\rangle|^2. \tag{26}$$

---

[24] A similar reasoning is found in Comte (1996), although Comte uses 'homogeneity' (see the axiom $D_1''$ below) instead of correspondence.

We thus see that the full quantum kinematics of a two-level system derives from a very simple combination of discreteness, symmetry, and correspondence.

Now consider a system of two particles with spin one-half, prepared so that its total angular momentum vanishes. Suppose that the spin of the first particle is found to be $+\hbar/2$ in a given direction. Then the spin of the second particle must be $-\hbar/2$ in the same direction by conservation. By a similar correspondence argument, the probabilities $P_{\varepsilon\varepsilon'}(\theta)$ of finding $\varepsilon\hbar/2$ and $\varepsilon'\hbar/2$ (with $\varepsilon,\ \varepsilon' = -1,\ 1$) for the spins of the two particles in two directions making the angle $\theta$ must verify

$$P_{++} - P_{+-} = -\frac{1}{2}\cos\theta \tag{27}$$

because for a large ensemble of identically prepared pairs for which the spin measurement in a given direction has given the outcome $\hbar/2$ for the first particle, the average angular momentum of all the second particles in a direction making the angle $\theta$ with the former direction should be the projection of $-\hbar/2$ in this direction.

Taking into account the normalization

$$P_{++} + P_{+-} = P_{-+} + P_{--} = \frac{1}{2}, \tag{29}$$

we get

$$P_{++}(\theta) = \frac{1}{2}\sin^2(\theta/2) \quad \text{and} \quad P_{+-}(\theta) = \frac{1}{2}\cos^2(\theta/2). \tag{30}$$

Again, this is the result given by quantum mechanics, with

$$|0\rangle = \frac{1}{\sqrt{2}}(|+\rangle|-\rangle - |-\rangle|+\rangle) \tag{31}$$

for the prepared state, and

$$\langle+|\langle+_\theta| = \langle+|\otimes\Big(\cos(\theta/2)\langle+| + \sin(\theta/2)\langle-|\Big) \tag{32}$$

for the projecting measurement state in the ++ case. As is well known, these correlations violate the Bell inequalities. Thus, a singular consequence of quantum mechanics, the violation of EPR locality, can be derived from a simple combination of discreteness, conservation, and correspondence.[25]

---

[25] I remember hearing this argument or a similar one from Claude Comte in the early 1990s.

These two arguments cannot really pass for an a priori derivation of quantum mechanical laws, for they involve two empirical facts: the existence of a two-level system for which possible measurements are mapped by unit vectors in geometrical space, and the existence of combined spin states for which the total angular momentum vanishes. The simplicity of the underlying assumptions is nonetheless striking. We could try to abstract them from the peculiarities of the 1/2 spin and to generalize them to all quantum systems. This would lead, for instance, to the following variant of Hardy's reconstruction.

## 17.4.2 Quantum Mechanics from Five Assumptions

The following reconstruction of quantum mechanics is based on a series of definitions ($\Delta$) and axioms (A). The axioms are heuristic assumptions rather than axioms in a strict mathematical sense.

$\Delta_1$: A system can be concretely and generically prepared in different *states*. The state can be determined by performing *measurements* on identically prepared copies of the system. The implied measurements are ideal, in the sense that they yield exactly the same outcome after immediate repetition.

$\Delta_2$: Two successive measurements are said to be *compatible* if their outcome does not depend on the order in which they are performed.

$\Delta_3$: A *maximal measurement* is a set of compatible measurements such that the outcome of any further compatible measurement is completely determined.

$\Delta_4$: A given outcome of a maximal measurement is called a *determination*.

$\Delta_5$: A system is said to be in a *pure state* after been maximally measured.

$\Delta_6$: A *transformation* of a system is any physical action on the system that causes a change of state.

$A_1$ (*Discreteness*): Measurement outcomes are discrete. For an *N-level system*, by definition the number of distinct outcomes of a maximal measurement has the finite value $N$, called the *dimension* of the system.

$A_2'$ (*Probabilities*): For copies of the same system generically produced in the same state, the indefinite repetition of a given maximal measurement yields a definite probability for each outcome.

$A_2''$ (*Degrees of freedom*): A finite number of such probabilities is sufficient to characterize the *state* of the system. The minimal value of this number defines the *number of degrees of freedom K* of the system.

$A_3$ (*Subspaces*): An $N$-level system whose states are constrained so that the probability of $N - 2$ of the distinct outcomes of a given maximal measurement is zero behaves like a two-level system.

$A_4'$ (*Composition of systems*): The state of a composite system is entirely determined by the probability of joint measurements performed on the components of this system.

$A_4''$ (*Pure conditioning*): For a bipartite system globally in a pure state, a maximal measurement performed on one part of the system produces a pure state of the other part.

$A_5'$ (*Reversibility*): Between any two pure states of a system there exists a reversible transformation.

$A_5''$ (*Continuity*): This transformation can be achieved in a continuous, gradual manner.

$A_5'''$: For a 2-level system, if the two measurement outcomes are equiprobable for a given determination, the gradual transformation can be done in a way that preserves this equiprobability.

These definitions and axioms are easily seen to be compatible with a standard quantum mechanics in which the state of the system is represented by the matrix density $\rho$, a pure state by a vector in Hilbert space, and a measurement probability by $\mathrm{tr}\rho P$ if $P$ denotes the projector associated with the measurement. We will now see that, reciprocally, our five axioms imply the matrix density representation of states.[26]

By the probability axioms $A_2$, a state of the system is characterized by the vector $p = (p_1, p_2, \ldots, p_K)$ defined by the probabilities $p_r$ of $K$ different determinations. A determination may then be characterized by the vector $p'$ of the associated pure state. First consider a two-level system for which $N = 2$, and replace the vector $p$ with the more convenient vector $x$ of components $x_r = 2p_r - 1$ (so that $x = 0$ corresponds to the completely mixed state). Through fairly simple reasoning, the discreteness axiom $A_1$, the probability axioms $A_2$, and the transformation axioms $A_5$ together imply that the set of pure states of a two-level system ($N = 2$) is a unit sphere of dimension $K - 1$ in $x$-space. In addition, the probability of the determination $x'$ when the system is in the state $x$ is simply given by

$$P(x, x') = \frac{1}{2}(1 + x \cdot x').  \tag{33}$$

The number $K$ of degrees of freedom remains to be determined. The continuity axiom $A_5''$ excludes the value $K = 1$ (ordinary theory of probabilities). Then the axioms $A_3$ and $A_4$ regarding subspaces and composite systems leave $K = 3$ as the only possibility. The idea of the demonstration is first to prove the existence of entangled states for a composite system made of two two-level systems, and then to use the axiom $A_4''$ of pure conditioning and quantitative conditions of purity. For $K = 3$, the former

---

[26] A full derivation of this implication is given in Darrigol (2015), pp. 332–336, mostly based on ideas by Hardy (2001); Dakić and Brukner (2009); Masanes and Müller (2010); Chiribella, D'Ariano, and Perinotti (2011)—except for a simpler derivation of $K = 3$ for $N = 2$.

expression of measurement probabilities in a two-level system matches the formula (23) for a spin 1/2, and therefore agrees with quantum mechanics.

For arbitrary $N$, the combination axiom $A_4'$ is easily seen to imply

$$K(N) + 1 = N^\kappa, \tag{34}$$

wherein $\kappa$ is an integer. Since $K = 3$ for $N = 2$, this integer is two, and

$$K(N) = N^2 - 1. \tag{35}$$

This number agrees with the number of independent real coefficients in a density matrix.

It remains to be proved that the probability of a determination in a given state can generally be expressed under the form $\langle\psi|\rho|\psi\rangle$, where $\rho$ is a positive Hermitian matrix of trace 1 and $|\psi\rangle$ is a vector representing the state of the system in a Hilbert space of dimension $N$. This can be done by repeated application of the subspace axiom $A_3$ to subspaces of dimension 2 for which the density matrix representation has already been established.

Granted that our five axioms imply the basic space-structure of quantum mechanics, it remains to be seen how natural these axioms are. Two kinds of correspondence argument à la Bohr will here play a significant role:

$K_1$: *It should be possible to formulate the predictions of quantum theory as relations or correlations between quantities that share the classical attributes of definiteness* (non-ambiguity, as Bohr says), *measurability, and distributivity.* (By distributivity, I mean that measurements on a composite system are reducible to measurements on the parts of the system).

$K_2$: *The average value of a measured micro-property over a macroscopic ensemble of identically prepared copies of the same micro-system should have a definite value and respect the symmetries of the associated macroscopic measurement.*

The principle $K_1$ justifies the definition $\Delta_1$ of states through measurements, as well as the probability axiom $A_2'$. The distributivity in $K_1$ justifies the composition axiom $A_4'$. It also leads to the pure conditioning axiom $A_4''$ if we further assume that whenever we have maximal information on a system, we still have maximal information on this system after performing an ideal (yet incomplete) measurement.[27] The principle $K_2$ justifies the axiom $A_2''$ for the number of degrees of freedom, because without it the macroscopic ensemble would have more degrees of freedom than required classically (for instance, for a spin, three independent measurements should suffice to determine the state, because the associated macroscopic moment has three independent

---

[27] This supposition is made to justify the covering law in quantum logic: see Piron (1976), p. 69.

components). It also justifies the transformation axioms $A_5$ because the transformation properties of a macroscopic ensemble of microsystems in the same pure state directly translate into transformation properties for the pure state itself.[28]

We are left with $A_1$, $A_3$. The discreteness axiom $A_1$ has no justification from a classical point of view. On the contrary, it should be regarded as a basic empirical characteristic of measurement results in quantum physics. The subspace axiom $A_3$ should similarly be regarded as the expression of our empirical ability to extract subsystems of lower dimension from systems of larger dimension.

Altogether we see that the definition of states through probabilities of measurement outcomes, a few correspondence requirements for these probabilities, and a basic assumption of discreteness of measurement results together lead to the state-space of quantum mechanics. In conformity with Bohr's intuition that quantum theory results from a harmonious melding of continuity and discontinuity, the derivation involves the discreteness axiom $A_1$ and the continuity axiom $A_5''$. In conformity with the contemporary emphasis on entangled states, the derivation also involves composite systems and subsystems.

## 17.4.3  Hardy's Five Reasonable Axioms

The previous reconstruction is largely based on Hardy's 'five reasonable axioms' of 2001, with a few improvements suggested by later developments. The basic ingredients of Hardy's approach are devices for preparing, transforming, and measuring a system, and states defined by the probabilities of measurement outcomes. Hardy introduces the 'dimension' $N$ and the 'number of degrees of freedom' $K'$ of the system, which correspond to my $N$ and my $K + 1$ respectively.[29] His axioms read:

$H_1$ *Probabilities*: Relative frequencies. . .tend to the same value (which we call the probability) for any case where a given measurement is performed on an ensemble of $n$ systems prepared by some given preparation in the limit as $n$ becomes infinite.

$H_2$ *Simplicity*: $K'$ is determined by a function of $N$. . ., and for each given $N$, $K'$ takes the minimum value consistent with the axioms.

$H_3$ *Subspaces*: A system whose state is constrained to belong to an $M$ dimensional subspace. . .behaves like a system of dimension $M$.

$H_4$ *Composite systems*: A composite system consisting of subsystems A and B satisfies $N = N_A N_B$, $K' = K'_A K'_B$.

---

[28] Hardy and his followers do not explicitly rely on correspondence arguments. I do not know their opinion on such arguments.

[29] Hardy's preference for $K'$ over $K$ comes from his including a probability for the system not to be detected by the measuring device.

$H_5$ *Continuity*: There exists a continuous reversible transformation on a system between any two pure states of that system.

This list of axioms has much in common with the axioms $A_1, \ldots, A_5$. Axiom $H_1$ is the same as $A_2'$; The subspace axiom $H_3$ is a generalization of $A_3$. Hardy justifies it as follows:

> This axiom is motivated by the intuition that any collection of distinguishable states should be on an equal footing with any other collection of the same number of distinguishable states. In logical terms, we can think of distinguishable states as corresponding to propositions. We expect a probability theory pertaining to $M$ propositions to be independent of whether these propositions are a subset or some larger set or not.

Axiom $H_4$ can be derived from $A_4'$, as Hardy himself shows. Hardy does not regard the discreteness of measurement results (my $A_1$) as an axiom. Nor does he regard the finiteness of $K$ (my $A_2'$) as an axiom; he introduces this property in a casual remark: 'Since most physical theories have some structure, a smaller set of probabilities pertaining to a set of carefully chosen measurements may be sufficient to determine the state.' The simplicity axiom $H_2$ later turned out to be unnecessary (more on this in a moment).[30]

Hardy regards all his axioms (except the simplicity axiom) as 'natural' from the point of view of the theory of probabilities. The axiom $H_5$ of continuity is the one that excludes classical probability and forces us to adopt quantum probability theory if we accept the simplicity axiom. Hardy's justification of the continuity axiom reads:

> Given the intuition that pure states represent definite states of a system we expect to be able to transform the state of a system from any pure state to any other pure state. It should be possible to do this in a way that does not extract information about the state and so we expect this can be done by a reversible transformation.

Implicitly, this is a correspondence argument because the idea of a definite state that can be transformed continuously is a classical idea. Arguably, empirical arguments and correspondence arguments are the true justification of Hardy's axioms. Indeed the natural character of an axiom from a probability-theory point of view or from an information-theory point of view is not equivalent to its natural character from a physical point of view. For example, the discontinuity of measurement outcomes is unproblematic from a probability-theory point of view, and yet physicists had difficulty admitting it for dynamical quantities and they did so under much empirical pressure; the continuity of measurement possibilities may perhaps be justified by an information-theoretic argument (as Hardy tries to do), but it should be rather puzzling

---

[30]  Hardy (2001), p. 10 for $H_1$, p. 14 for $H_3$, p. 2 for the casual remark.

for a physicist who has accepted quantum discontinuity unless he or she evokes a correspondence argument.[31]

In a subsequent summary of his new reconstruction of quantum theory, Hardy explains his motivations:

> There are various reasons for developing reasonable axioms. Firstly, physics is primarily about explanation and we can be said to have explained quantum theory more deeply if we give reasonable axioms. Secondly, by having a deeper understanding of the origin of quantum theory we are more likely to be able to extend or adapt the theory to new domains of applicability (such as quantum gravity). Thirdly, the fact that we put quantum theory and classical probability theory on such a similar footing may point the way to a deeper appreciation of the relationship between classical and quantum information. And finally, these new axioms may shed some light on the interpretation of quantum theory.

Hardy himself illustrated some of these potentialities in subsequent works. In an essay published in 2004 in memory of Rob Clifton, he used his fiducial-measurement approach to prove that realist interpretations of quantum mechanics necessarily implied infinite 'ontological excess baggage' (as is indeed the case in the Bohm–Vigier interpretation). In a series of memoirs beginning in 2005, he modified his original assumptions to make them independent from any pre-assumed causal structure. This reconstruction relies on the concept of information and on a notion of circuits similar to that of Chiribella, D'Ariano, and Perinotti (see the later subsection on their work). It is the basis of Hardy's ongoing attempt to conciliate gravitation theory (in which the causal structure is dynamical) with quantum theory (in which the causal structure is usually fixed). In general, Hardy condemns the 'Shut up and calculate!' attitude of the average quantum physicist and propounds, together with Robert Spekkens, the alternative slogan: 'Shut up and contemplate!'[32]

## 17.4.4 Dakić and Brukner

The clarity and elegance of Hardy's derivation of quantum-mechanics state-space and the simplicity of his axioms have attracted much legitimate attention. An evident defect of this derivation is the artificial character of the simplicity axiom $H_2$. Hardy himself wondered about the possibility of more complicated theories involving higher $\kappa$ exponents in the relation $K = N^\kappa - 1$.[33] In 2009, the Vienna-based theorists Borivoje Dakić and Časlav Brukner answered this question negatively by showing that $K$ could

---

[31] Hardy (2001), p. 15.
[32] Hardy (2002), p. 74; Hardy (2004, 2005, 2007, 2011); Hardy and Spekkens (2010), p. 4. The 'Shut up and calculate!' characterization belongs to Mermin (1989), p. 9.
[33] Hardy (2001), p. 13.

not exceed three in the two-level case $N = 2$. Their argument being based on the consideration of entangled states for composite systems, they published it under the title 'Quantum theory and beyond: Is entanglement special?'. They relied on the following system of axioms:[34]

$D_1$ (Information capacity): An elementary system has the information carrying capacity of at most one bit. All systems of the same information carrying capacity are equivalent.

$D_2$ (Locality): The state of a composite system is completely determined by local measurements on its subsystems and their correlations.

$D_3$ (Reversibility): Between any two pure states there exists a reversible transformation.

The axiom $D_2$ is the same as axiom $A_4'$ and it is directly related to Hardy's axiom $H_4$. The axiom $D_3$ is the same as the axiom $A_5'$, which is a much weakened form of $H_5$, for it does not assume $A_5''$, namely the existence of a continuous sequence of transformations gradually bringing the first pure state to coincide with the second. This axiom warrants that a compact group acts transitively on the subspace of pure states. Its weakness is compensated by the strength of axiom $D_1$, which in fact contains two subaxioms:

$D_1'$: An elementary system has the information carrying capacity of at most one bit.

$D_1''$: All systems of the same information carrying capacity are equivalent.

Since the information carrying capacity is nothing but the number $N$ of distinct outcomes of a maximal measurement, axiom $D_1''$ is an information-theoretic rephrasing of Hardy's subspace axiom $H_3$.[35] Dakić and Brukner translate their information-theoretic axiom $D_1'$ into 'any state of a two dimensional system can be prepared by mixing at most two basis (i.e. perfectly distinguishable in a measurement) states.' In the mid-1990s, Claude Comte and Daniel Fivel had already taken the $N$-dimensional generalization of this property as an axiom.[36] Apparently unaware of these works, Brukner borrowed axiom $D_1'$ from his Viennese colleague and collaborator Anton Zeilinger.

In a celebration of Daniel Greenberger's sixty-fifth birthday, Zeilinger suggested basing quantum theory on the principle that '*An elementary system represents the truth value of one proposition*' or, equivalently, that '*An elementary system carries 1 bit of information.*' From this principle he derived randomness and entanglement:

[34] Dakić and Brukner (2009, 2011).
[35] The authors credit Grinbaum (2007) for the rephrasing.
[36] Comte (1996); Fivel (1994).

An elementary system can only give a definite result in one specific measurement. The irreducible randomness in other measurements is then a necessary consequence. For composite systems entanglement results if all possible information is exhausted in specifying joint properties of the constituents.

For instance, there can only be one direction of spin measurement for which the spin of a particle can have a definite value because the spin state can only contain the reply to a single Yes–No question. In any other direction, the result of the measurement must be random. For a composite system of two spins, the joint property that the two particles have the same spin in one direction and the other joint property that the two particles have the same spin in another direction exhaust all possible information since the global system has two bits. The answer to other questions, for instance about the spin of one of the particles in a given direction, should be random: this is the signature of an entangled state.[37]

Seduced by this reasoning, Dakić and Brukner turned Zeilinger's information-theoretic principle into the most potent axiom of their theory, which is $D_1$. Their first remarkable result is that the axioms $D_1'$ and $D_3$ are sufficient to determine the probability theory for $N = 2$ and for any given value of $K$. Their proof is similar to the reasoning given above, except that the axiom $D_1'$ replaces the transformation axiom $A_5$ in the derivation of the fact that every point of the sphere $S^{K-1}$ defines a pure state.[38]

Next comes the proof that $K \leq 3$ for a two-level system. Dakić and Brukner first prove that the axioms $D_1$ and $D_2$ (which contain $A_3$ and $A_4'$) imply that the system obtained by combining two two-level systems has entangled states. The rest of their ingenious argument relies on rotations (isometries of determinant *one*) in $\mathbf{R}^K$. $K$ being inferior to 3 and the even value $K = 2$ being excluded by $K = N^\kappa - 1$, the options $K = 1$ (classical probabilities) and $K = 3$ (quantum probabilities) are the only ones left. Dakić and Brukner then evoke Hardy's consideration of subspaces in order to justify that for $\kappa = 2$ the states of any $N$-level system can be represented by a quantum-mechanical density matrix.

Dakić and Brukner's proof that $K \leq 3$ presupposes that every rotation in $\mathbf{R}^K$ is a physical transformation. As Hardy first noticed, this is true only for odd values of $K$ except seven, and the proof requires advanced group theory. Even values of $K$ need not be considered since they are excluded by $K = N^\kappa - 1$. There remains the exceptional case $K = 7$.[39] Another weakness of Dakić and Brukner's approach is their reliance on the qubit axiom $D_1'$, which can hardly be said to be natural. Why should every physical two-level system be assimilated to a one-bit information facility? Is it not highly unnatural to assume, when there is a continuum of possibilities of measurement, that every state of the system can be obtained as a mixture of the outcomes of a single measurement? Does not the axiom in itself contain all the quantum weirdness that we would want to derive rather than assume?

---

[37] Zeilinger (1999), pp. 631, 635.      [38] Dakić and Brukner (2009), p. 5.
[39] Hardy's remark is reported in Masanes and Müller (2010), p. 22.

## 17.4.5 Masanes and Müller

In 2010, Lluís Masanes and Markus Müller published 'A derivation of quantum theory from physical requirements' in which they addressed some of the shortcomings of Dakić and Brukner's approach. The informal version of their axioms or 'requirements' read:

$M_1$: In systems that carry one bit of information, each state is characterized by a finite set of outcome probabilities.

$M_2$: The state of a composite system is characterized by the statistics of measurements on the individual components.

$M_3$: All systems that effectively carry the same amount of information have equivalent state spaces.

$M_4$: Any pure state of a system can be reversibly transformed into any other.

$M_5$: In systems that carry one bit of information, all mathematically well-defined measurements are allowed by the theory.

Axiom $M_1$ is related to Hardy's first axiom, and it corresponds to the axiom $A_2$ restricted to two-level systems. Axiom $M_2$ on composite systems is the same as $A_4$ or $H_4$. Axiom $M_3$ is the subspace axiom $A_3$ or $H_3$. Axiom $M_4$ coincides with $A_5'$ or $D_3$. The last axiom, $M_5$, is new. It replaces Hardy's continuity $H_5$ or the Comte–Fivel–Zeilinger axiom $D'_1$ in the derivation of the set of pure states for a two-level system. Its more precise expression reads: 'All tight effects correspond to allowed measurements'. Translated in the $x$-vector language, a 'tight effect' is any affine function

$$F(x) = \frac{1}{2}(1 + a \cdot x)$$

through which the image of the state space is the whole interval $[0, 1]$. An 'allowed measurement' is a $F$ function for which the vector $a$ defines a determination. By simple reasoning the identification of tight effects with allowed measurements implies that every point of the sphere $S^{K-1}$ defines a permitted state.[40]

Masanes and Müller regard tight effects as innocent mathematical idealizations of concrete measurements ('mathematically well-defined measurements'). This is questionable. Do not they confuse mathematical simplicity with physical plausibility? The main advantage of their axiom is that it leads to the desired result (the Bloch sphere of pure states) in the most direct manner. It may be a little more natural than axiom $D'_1$, but it is still mysterious.

A clearer advantage of Masanes and Müller's approach over Dakić and Brukner's is a rigorous derivation that the existence of entangled states (for a composite system made of two two-level systems) leads to $K = 3$ for two-level systems. Unfortunately, the

[40] Masanes and Müller (2010), pp. 1, 5.

argument requires advanced group theory. Masanes and Müller also provide their own group-theoretical proof, based on axiom $M_5$, that states can be represented by density matrices for any dimension.

## 17.4.6 Chiribella, D'Ariano, and Perinotti

From the times of Hardy's seminal reconstruction of quantum theory, there has been a growing tendency to formulate and justify the axioms of quantum mechanics by information-theoretical means. This is a natural evolution, considering the present importance of researches on quantum-mechanical information processing and quantum computing. In 1990, John Archibald Wheeler famously defined the 'It from bit' programme for reducing physics to the processing of information. Although this sort of reductionism has often been criticized, it has inspired a few arguments for the information-theoretic necessity of quantum theory. Zeilinger's qubit considerations and their exploitation by Brukner and Dakić have already been mentioned. Another interesting information-theoretic deduction of quantum mechanics, sketched by Carlo Rovelli in 1996 and fully developed by Alexei Grinbaum in 2003, relies on an information-theoretic reframing of quantum logic.[41]

Still another information-theoretic deduction is found in a memoir of 2003 by three philosophers of physics: Rob Clifton, Jeffrey Bub, and Hans Halvorson (CBH). The gist of their argument is a proof of the three following assertions:

1) The impossibility of supraluminal communication between two systems entails the commutativity of the associated algebras of observables.

2) The impossibility of perfectly broadcasting the information contained in an unknown physical state entails the non-commutativity of the algebra of observables of an individual system.

3) The impossibility of unconditionally secure bit commitment entails the existence of entangled states.

The first impossibility (micro-causality) is a mere consequence of relativity theory; the second and third impossibilities are well-known consequences of quantum mechanics applied to quantum cryptography. CBH regard these three impossibilities as given information-theoretic facts and study their consequences in the powerful language of

---

[41] Wheeler (1990); Rovelli (1996); Grinbaum (2003, 2004). Philip Goyal derives complex probability amplitudes through the symmetries of a generalized, information-based probability theory (see his website: https://www.philipgoyal.org/). On information-theoretic reconstructions, cf. Fuchs (2003); Jaeger (2018).

$C^*$-algebras. This language generalizes the operator algebra on Hilbert spaces and it is meant to encompass every past and future physical theory.[42]

No matter how interesting the CBH result may be as an information-theoretic characterization of quantum theory, it cannot pass for an argument for the naturalness of quantum mechanics. There are three reasons for that. Firstly, the $C^*$-algebraic framework is much too abstractly mathematical to pass for an a priori natural frame in which to formulate physical theories.[43] For a rationalist exploitation of the CBH result, one would first need to derive this framework from simple operational considerations, which does not seem easier than deriving the Hilbert space structure of quantum mechanics. A second shortcoming has to do with the contents of CBH's information-theoretic principles. Even if they could be shown to be natural from an information-theoretic point of view, this would not make their physical realization in elementary systems more natural. Thirdly, CBH do not prove that quantum mechanics results from their principles. What they construct is a generalized quantum theory defined by a $C^*$-algebra satisfying the algebraic constraints that derive from their information-theoretic principles. In fact, they want this generality because they have in mind situations (quantum field theory and quantum gravity) in which it might be needed.

Very recently Giulio Chiribella, Giacomo Mauro D'Ariano, and Paolo Perinotti (CDP) have offered an 'informational derivation of quantum theory' that does not have the first and third of the defects of CBH's derivation. CDP arrive at the state-space of quantum mechanics, and they do so in a purely operational framework based on probability distributions for 'circuits' resulting from the connection of physical devices:

> Our principles do not refer to abstract properties of the mathematical structures that we use to represent states, transformations, or measurements, but only to the way in which states, transformations, and measurements combine with each other.

CDP give the following informal statement of their axioms:[44]

$C_1$ *Causality*: The probability of a measurement outcome at a certain time does not depend on the choice of measurements that will be performed later.

$C_2$ *Perfect distinguishability*: If a state is not completely mixed (i.e., if it cannot be obtained as a mixture from any other state), then there exists at least one state that can be perfectly distinguished from it.

$C_3$ *Ideal compression*: Every source of information can be encoded in a suitable physical system in a lossless and maximally efficient fashion. Here *lossless* means that

---

[42] Clifton, Bub, and Halvorson (2003), pp. 2–3. The no-broadcasting theorem is an extension to mixed states of the no-cloning theorem given by William Wootters, Wojciech Zurek, and Dennis Dieks in 1982. Bit commitment is a cryptographic notion defined by Gilles Brassard, David Chaum, and Claude Crépeau in 1988. On the early history of quantum information, cf. Kaiser (2011). For a criticism of information-theoretic reductionism, cf. Deutsch (2003).

[43] The same criticism is found in Grinbaum (2007), p. 402.

[44] Chiribella, D'Ariano, and Perinotti (2011), pp. 2, 3.

the information can be decoded without errors and *maximally efficient* means that every state of the encoding system represents a state in the information source.

$C_4$ *Local distinguishability*: If two states of a composite system are different, then we can distinguish between them from the statistics of local measurements on the component systems.

$C_5$ *Pure conditioning*: If a pure state of system AB undergoes an atomic measurement on system A, then each outcome of the measurement induces a pure state on system B.

$C_6$ *Purification postulate*: Every state has a purification. For fixed purifying system, every two purifications of the same state are connected by a reversible transformation on the purifying system.

The causality axiom $C_1$ is so evident that all other axiomatizers assumed it without stating it. The local distinguishability axiom $C_4$ is a rewording of $A_4$ or $D_2$. The other axioms are more original. The pure conditioning postulate $C_5$ was not used in earlier axiomatizations. It coincides with the axiom $A_4''$ introduced and discussed in section 17.2 of this essay. The purification postulate $C_6$ means that every state of a system may be regarded as the marginal state of a subsystem of a larger system that is in a pure state. CDP note the affinity with Schrödinger's remark of 1935:

> An optimal knowledge of the whole does not imply an optimal knowledge of its parts—that is the whole mystery. I would not call that *one* but rather *the* characteristic trait of quantum mechanics, the one that enforces its entire departure from classical lines of thought.

Somewhat artificially, CDP include the existence of reversible transformations between any two pure states in the purification postulate. The axioms $C_1$, $C_2$, and $C_3$ serve to derive the duality between pure states and pure determinations as well as the representation of any state as a convex mixture of pure states; whereas in my simplified approach these facts are trivial consequence of the definition of general states as mixtures of states produced by maximal measuring devices. Axioms $C_5$ and $C_6$ sustain a proof that $K = N^2 - 1$ and allow the derivation of the density-matrix representation of states. Interestingly, CDP do not need the subspace axiom. They deduce the equivalence (up to a reversible transformation) of all systems with the same dimension from their own axioms.[45]

Unfortunately, the purification postulate is hard to swallow.[46] To assume this postulate is to admit in the very basis of the theory the quantum oddities deplored

---

[45] Chiribella, D'Ariano, and Perinotti (2011), p. 2 (on Schrödinger), p. 29 (no subspace axiom); Schrödinger (1935), p. 555. The purification postulate directly implies the no-cloning and the impossibility of bit commitment assumed by CBH.

[46] Chiribella, D'Ariano, and Perinotti (2011), p. 2, try to justify the purification principle by having it express the possibility of reducing thermodynamic irreversibility to reversible interaction with an uncontrolled environment.

by Schrödinger. As explained by CDP, the true advantage of their approach is its providing direct, instructive links between new information-theoretic axioms for quantum mechanics and the various quantum-information theorems to which physicists have lately devoted much attention.[47]

CDP criticize earlier reconstructions of quantum theory for involving uninterpreted mathematical assumptions.[48] This charge certainly applies to Ludwig's old axiomatics, despite Ludwig's intention to provide physically justified axioms; it also applies to Masanes and Müller's 'tight effect' axiom; but it does not truly apply to Hardy's approach because his only uninterpreted axiom, the simplicity axiom, is now known to be superfluous; and the charge has no grip on Dakić and Brukner's reconstruction.

## 17.4.7 Morals

Natural axioms à la Hardy permit a convincing demonstration that the consistent melding of the discontinuity of measurement results with the continuity of measurement possibilities necessarily leads to the density-matrix representation of physical states. Are the implied axioms on measurement probabilities natural enough to suggest some rational necessity of quantum mechanics? Hardy flirted with this idea when he posted his 'five reasonable axioms':[49]

> Quantum theory is simply a new type of probability theory. Like classical probability theory it can be applied to a wide range of phenomena. However, the rules of classical probability theory can be determined by pure thought alone without any particular appeal to experiment (though, of course, to develop classical probability theory, we do employ some basic intuitions about the nature of the world). Is the same true of quantum theory? Put another way, could a 19th century theorist have developed quantum theory without access to the empirical data that later became available to his 20th century descendants? In this paper it will be shown that quantum theory follows from five very reasonable axioms which might well have been posited without any particular access to empirical data.

Could we truly have reached quantum mechanics by brooding over a generalized probability theory? A first difficulty in this rationalist fiction concerns the evolution of systems. So to say, the reconstructions of Hardy and his followers provide only the kinematics of quantum mechanics, that is, its representation of physical states. They do not tell us how the states evolve, except that probability should be conserved. In the continuous case in which no measurement is performed, the latter property implies the existence of a Hamiltonian operator from which the evolution derives. Hardy regards the precise expression of the Hamiltonian as a contingent element to be drawn from experience.

---

[47] Chiribella, D'Ariano, and Perinotti (2011), p. 38.
[48] Chiribella, D'Ariano, and Perinotti (2011), p. 2.      [49] Hardy (2001), p. 1.

A more fundamental limitation to a rationalist exploitation of Hardy's reconstruction is inherent in the assumption of strictly discrete (ideal) measurement outcomes (axiom $A_1$). As was already said, this assumption is unproblematic from the point of view of probability theory but it is unnatural from a 19th-century physics point of view. Historically, quantum discontinuity was a difficult conquest by a few daring young physicists in the early 20th century, against the prejudices of the older generation. Its origins are sometimes regarded as empirical, for instance in the discreteness and universality of atomic spectra; and sometimes as intertheoretical, in the paradoxes regarding the interaction between radiation and a large assembly of atoms (ultraviolet catastrophe). The associated arguments for the discreteness of measurement outcomes are not as compelling as the deduction of quantum mechanics from this discreteness in a reconstruction à la Hardy. In the present state of this approach, we should probably content ourselves with the insight that quantum discontinuity, if it is admitted as a fundamental feature of the micro-world and if it is complemented with natural axioms concerning the relation between micro- and macro-world, necessarily leads to quantum mechanics as we know it.

## 17.5 CONCLUSIONS

A first striking feature of the three reconstructions presented is the contrast between the naturalness of their basic assumptions and the mathematical abstractness of the derived Hilbert-space structure. This is what we wanted in order to domesticate this highly abstract structure and the resulting oddities of quantum physics.

The second striking feature is how much the three reconstructions differ from each other. A first superficial difference concerns the rigour of the employed mathematics. The two first approaches, by deformation of the Hamiltonian structure and by non-Boolean logic, rest on rigorous mathematics done by professional mathematicians and they are strictly axiomatic. In contrast, Hardy's 'axioms' are not axioms in a strict mathematical sense. But they are sharp enough to lead, in an idealized form and with the help of a few easily accepted additional assumptions, to well-defined mathematical statements whose consequences can be pursued rigorously. It would not be difficult to provide a list of genuine mathematical axioms for Hardy's reconstruction. Not much would be gained in this process. What truly matters, in the comparison between the three reconstructions, is the physical nature of the basic assumptions and the domains of mathematics needed to derive the Hilbert-space structure from them.

Let us consider more closely the differences between the three reconstructions. The approach by deformation of Hamiltonian mechanics essentially rests on the full correspondence principle, implying the correspondence between the classical dynamics and the new, deformed dynamics, whereas quantum logic does not need any correspondence argument (except perhaps to justify Solèr's condition), and Hardy's assumptions depend on more restricted forms of correspondence. Accordingly, the deformation-based reconstruction is the only one in which Planck's constant plays a role, as the parameter of the deformation. The other reconstructions lead to the bare algebra of quantum magnitudes and their evolution (*quantum kinematics*), but they do not tell us more on the nature of the associated operators. To be true, they could be completed to include position and momentum operators, commutation rules, a Hamiltonian operator etc. (*quantum dynamics*), and this has indeed been done for quantum logic by Constantin Piron and Josef-Maria Jauch. It would not be difficult to do the same for Hardy's reconstruction by supplementing its kinematic-kind of correspondence with a dynamic-kind of correspondence.

Quantum logic is singular and attractive by the extreme simplicity of its basic concepts: Yes–No questions and their combinations. No one would deny that the account of any experiment can be principally reduced to the answer of a sufficiently large set of such binary questions. Without presupposing anything on the nature of the questions, and just making simple assumptions on the associated lattice, quantum logic leads to a generalized Hilbert-space structure. It does so at the cost of some not too easy mathematics, and the operational justification of some of its axioms is far from obvious.

Hardy's axiomatics involves from the start more refined concepts of state and measurement than quantum logic: it defines states through probabilities of discrete measurement results. Its basic assumptions can all be justified by a combination of simple empirical and correspondence arguments. The mathematics needed to deduce the Hilbert-space structure from these assumptions is relatively simple and the deductions are much shorter than in the two earlier reconstructions. This advantage probably results from the fact that Hardy's axioms directly assume a few essential features of any experimentation in physics: quantitative measurement, statistical evaluation, combination of systems, and extraction of subsystems.

To summarize, the deformation strategy wins for its ability to yield the full quantum dynamics. But it does so in a formal manner only, in a non-interpreted phase-space with negative probabilities, and at the cost of highly advanced mathematics. Quantum logic wins for the utmost simplicity of its basic concepts. But the axioms it posits for the lattice of propositions are not all easy to justify in an operationally convincing manner; it requires ample mathematical knowledge; and it does not easily lead to the specific Hilbert space of quantum mechanics. Hardy's reconstruction is the overall winner because it reaches quantum kinematics at the lowest mathematical cost from assumptions that would seem natural to any educated physicist. Also, it properly reflects the contemporary concern of quantum physicists with the composition of systems and with the informational content of quantum states.

We are left with the following puzzle: Why is it that the same Hilbert-space kinematics derives from so utterly different premises? At first glance, there is nothing in common between deforming Poisson brackets, generalizing the Boolean lattice of ordinary logic, and regulating the probabilistic correlations between discrete measurements. The three reconstructions are rooted in three different grounds: Hamiltonian mechanics, logical lattices, probability theory. The first and third reconstructions assume some correspondence with classical concepts while the second does not; the second and third assume discreteness, while the first does not; the third relies on the composition of systems while the first and second do not; the second assumes an operational form of non-commutativity, while the first and third do not. In the end we get three different intuitions of the origins of quantum kinematics: 1) it is an ineluctable consequence of a generalization of the algebra of classical infinitesimal evolutions; 2) it emerges from a natural extension of the lattice of Yes–No empirical questions when the questions do not commute; 3) it results from the harmonious blending of the discontinuity of measurement results with the continuity of the possibilities of measurement. Although none of these justifications is perfect, their convergence strongly suggests some inevitability of quantum kinematics. No matter how we try to consistently depart from classicality, we arrive at the Hilbert-space structure of quantum mechanics. *Omnes viae Romam ducunt.*

## REFERENCES

Beltrametti, Enrico, and Gianni Cassinelli (1981). *The logic of quantum mechanics.* Reading, MA: Addison-Wesley.

Birkhoff, Garrett (1940). *Lattice theory.* New York: American Mathematical Society.

Birkhoff, Garrett, and John von Neumann (1936). The logic of quantum mechanics. *Annals of Mathematics*, **37**, 823–843.

Chiribella, Giulio, Giacomo Mauro D'Ariano, and Paolo Perinotti (2011). Informational derivation of quantum theory. *Physical Review A*, **84**, 012311.

Clifton, Rob, Jeffrey Bub, and Hans Halvorson (2003). Characterizing quantum theory in terms of information-theoretic constraints. *Foundations of Physics*, **33**, 1561–1591.

Comte, Claude (1996). Symmetry, relativity and quantum mechanics. *Il nuovo cimento*, **111**B, 937–956.

Curtwright, Thomas, and Cosmas Zachos (2011). *Quantum mechanics in phase space.* arXiv:1104.5269v2 [physics.hist-ph]

Dakić, Borivoje, and Časlav Brukner (2009). Quantum theory and beyond: Is entanglement special? arXiv:0911.0695v1 [quant-ph]

Dakić, Borivoje, and Časlav Brukner (2011). Quantum theory and beyond: Is entanglement special? In Hans Halvorson (ed.), *Deep beauty: Understanding the quantum world through mathematical innovation*, Cambridge: Cambridge University Press, pp. 365–392.

Dalla Chiara, Maria Luisa, Roberto Giuntini, and Miklós Rédei (2007). The history of quantum logic. In Dov M. Gabbay and John Woods (eds), *The many valued and non-monotonic turn in logic*, vol. 8 of *Handbook of the history of logic* (11 vols.), Amsterdam: Elsevier, pp. 205–283.

Darrigol, Olivier (2014). *Physics and necessity: Rationalist pursuits from the Cartesian past to the quantum present.* Oxford: Oxford University Press.

Darrigol, Olivier (2015). 'Shut up and contemplate!': Lucien Hardy's reasonable axioms for quantum theory. *Studies in History and Philosophy of Modern Physics*, **52**, 328–342.

Deutsch, David (2003). It from qubit. In J. Barrow, P. Davies, and C. Harper (eds), *Science and ultimate reality*, Cambridge: Cambridge University Press, pp. 90–102.

Garner, Lynn (1981). *An outline of projective geometry.* New York: North Holland.

Fivel, Daniel (1994). How interference effects in mixtures determine the rules of quantum mechanics. *Physical Review A*, **50**, 2108–2119.

Fuchs, Christopher (2003). Quantum mechanics as quantum information, mostly. *Journal of Modern Optics*, **50**, 987–1023.

Gabbay, Dov, Daniel Lehmann, and Kurt Engesser (eds) (2009). *Handbook of quantum logic and quantum structures: Quantum logic*. Amsterdam: Elsevier.

Gleason, Andrew (1957). Measures on the closed subspaces of a Hilbert space. *Journal of Mathematics and Mechanics*, **6**, 885–893.

Gosson, Maurice de, and Basil Hiley (2009). The symplectic camel and the uncertainty principle: The tip of an iceberg? *Foundations of Physics*, **99**, 194–204.

Gosson, Maurice de, and Basil Hiley (2011). Imprints of the quantum world in classical physics. *Foundations of Physics*, **11**, 1415–1436.

Grinbaum, Alexei (2003). Elements of information-theoretic derivation of the formalism of quantum theory. *International Journal of Quantum Information*, **1**, 289–300.

Grinbaum, Alexei (2004). *Le rôle de l'information dans la théorie quantique*. PhD diss. Ecole Polytechnique.

Grinbaum, Alexei (2007). Reconstruction of quantum theory. *British Journal for the Philosophy of Science*, **58**, 387–408.

Groenewold, Hilbrand (1946). On the principles of elementary quantum mechanics. *Physica*, **12**, 405–460.

Gutt, Simone (1979). Equivalence of deformations and associated *-products. *Letters in Mathematical Physics*, **3**, 297–309.

Hardy, Lucien (2001). Quantum theory from five reasonable axioms. arXiv:quant-ph/0101012 [25 Sep. 2001; I have mostly used version 4, which dates from 1 Feb 2008].

Hardy, Lucien (2002). Why quantum theory? In Jeremy Butterfield and Tomasz Placek (eds), *Proceedings of the NATO advanced research workshop on modality, probability, and Bell's theorem*, Amsterdam, IOS Press, pp. 61–74. Also in arXiv:quant-ph/0111068.

Hardy, Lucien (2004). Quantum ontological excess baggage. *Studies in History and Philosophy of Modern Physics*, **35**, 267–276.

Hardy, Lucien (2005). Probability theories with dynamic causal structure: A new framework for quantum gravity. arXiv:gr-qc/0509120

Hardy, Lucien (2007). Towards quantum gravity: A framework for probabilistic theories with non-fixed causal structure. *Journal of Physics A*, **40**, 3081–3099.

Hardy, Lucien (2011). Reformulating and reconstructing quantum theory. arXiv:1104.2066

Hardy, Lucien, and Robert Spekkens (2010). Why physics needs quantum foundations. arXiv:1003.5008

Holland, Samuel (1995). Orthomodularity in infinite dimensions; a theorem of M. Solèr. *Bulletin of the American Mathematical Society*, **32**, 205–234.

Jaeger (2018). Information and the reconstruction of quantum physics. *Annalen der Physik*, **531**. https://doi.org/10.1002/andp.201800097

Jammer, Max (1974). *The philosophy of quantum mechanics: The interpretations of quantum mechanics in historical perspective*. New York: Wiley.

Jauch, Josef (1968). *Foundations of quantum mechanics*. Reading, MA: Addison-Wesley.

Jauch, Josef, and Constantin Piron (1969). On the structure of quantal proposition systems. *Helvetica Physica Acta*, **42**, 842–848.

Kadison, Richard (1951). Isometries of operator algebras. *Annals of Mathematics*, **54**, 325–338.

Kaiser, David (2011). *How the hippies saved physics: Science, counterculture, and the quantum revival*. New York: Norton & Co.

Lichnerowicz, André (1979). Équivalence et existence des $*_v$-produits sur une variété symplectique. Académie des sciences, *Comptes rendus A*, **289**, 349–353.

Lacki, Jan (2000). The early axiomatizations of quantum mechanics: Jordan, von Neumann and the continuation of Hilbert program. *Archive for History of Exact Sciences*, **54**, 279–318.

Ludwig, Günther (1983). *Foundations of quantum mechanics*. 2 vols. Berlin: Springer.

Ludwig, Günther (1985). *An axiomatic basis for quantum mechanics*. 2 vols. Berlin: Springer.

Mackey, George (1957). Quantum mechanics and Hilbert space. *American Mathematical Monthly, Supplement*, **64**(8), 45–57.

Mackey, George (1963). *The mathematical foundations of quantum mechanics*. Reading, MA: Benjamin/Cummings.

Masanes, Lluís, and Markus Müller (2010). A derivation of quantum theory from physical requirements. arXiv:1004.1483v2 [quant-ph].

Mermin, David (1989). What's wrong with this pillow? *Physics Today*, **42**(4), 9–11.

Mittelstaedt, Peter (2011). *Rational reconstructions of modern physics*. Heidelberg: Springer.

Moyal, José Enriques (1949). Quantum mechanics as a statistical theory. *Proceedings of the Cambridge Philosophical Society*, **45**, 99–124.

Piron, Constantin (1964). Axiomatique quantique. *Helvetica Physica Acta*, **37**, 439–468.

Piron, Constantin (1976). *Foundations of quantum mechanics*. Reading, MA: Benjamin.

Primas, Hans (1981). *Chemistry, quantum mechanics and reductionism*. Berlin: Springer.

Rédei, Miklós (2007). The birth of quantum logic. *History and Philosophy of Logic*, **28**, 107–122.

Rovelli, Carlo (1996). Relational quantum mechanics. *International Journal of Theoretical Physics*, **35**, 1637–78.

Solèr, Maria Pia (1995). Characterization of Hilbert spaces with orthomodular spaces. *Communications in Algebra*, **23**, 219–243.

Stachow, Ernst-Walther (1984). Structures of quantum languages for individual systems. In P. Mittelstaedt and E.-W. Stachow (eds), *Recent developments in quantum logic*, Zürich: Bibliographisches Institut, pp. 129–146.

Sternheimer, Daniel (1998). Deformation quantization: Ten years after. arXiv:math/9809056v1 [math.QA].

Vey, Jacques (1975). Déformation du crochet de Poisson sur une variété symplectique. *Commentarii Mathematici Helvetici*, **50**, 421–454.

von Neumann, John (1930). Allgemeine Eigenwerttheorie Hermitescher Funktionaloperatoren. *Mathematische Annalen*, **102**, 49–131.

von Neumann, John (1932). *Mathematische Grundlagen der Quantenmechanik*. Berlin: Springer.

Wheeler, John. Archibald (1990). Information, physics, quantum: The search for links. In W. Zurek (ed.), *Complexity, entropy, and the physics of information*. Redwood City: Addison-Wesley, pp. 3–28.

Wilce, Alexander (2009). Quantum logic and probability theory. In Edward N. Zalta (ed.), *The Stanford Encyclopedia of Philosophy* (Spring 2009 Edition), http://plato.stanford.edu/archives/spr2009/entries/qt-quantlog/ (last accessed Jan. 2019).

Zachos, Cosmas, David Fairlie, and Thomas Curtright (2005). *Quantum mechanics in phase space: An overview with selected papers*. Singapore: World Scientific.

Zeilinger, Anton (1999). A foundational principle for quantum theory. *Foundations of Physics*, **29**, 631–643.

CHAPTER 18

..........

# THE AXIOMATIZATION OF QUANTUM THEORY THROUGH FUNCTIONAL ANALYSIS

## Hilbert, von Neumann, and Beyond

..........

KLAAS LANDSMAN

# 18.1 INTRODUCTION[1]

..........

DIJKSTERHUIS (1961) concludes his masterpiece *The Mechanization of the World Picture* (which ends with Newton) with the statement that the process described in the title consisted of the mathematization of the natural sciences, adding that this process had been completed by 20th-century physics. As such, the topic of this chapter seems a perfect illustration of Dijksterhuis's claim, perhaps even *the* most perfect illustration.[2]

However, there is an important difference between the application of calculus to classical mechanics and the application of functional analysis to quantum mechanics: Newton invented calculus in the context of classical mechanics,[3] whereas functional

---

[1] This chapter has a strict word limit, as a consequence of which the discussion is often terse. For example, instead of explaining the technical details, for which I refer to books like Landsman (2017), I have tried to sketch the relevant history at an almost sociological level. I am deeply indebted to Michel Janssen, Miklos Rédei, and a referee for comments.

[2] Einstein's theory of general relativity is an equally deep and significant example of the process in question, but I would suggest that his application of Riemannian geometry to physics (already contemplated by Riemann himself) was less unexpected than the application of functional analysis to quantum theory.

[3] The fact that Newton subsequently erased his own calculus from the *Principia* does not change this; see Whiteside (1970) and Guicciardini (2009) for an analysis of this bizarre historical conundrum.

analysis was certainly *not* created with quantum theory in mind. In fact, the interaction between the two fields only started in 1927, when quantum mechanics was almost finished at least from a physical point of view, and also functional analysis had most of its history behind it.[4] Functional analysis *did* have its roots in *classical* physics. Monna (1973), Dieudonné (1981), and Siegmund-Schultze (2003) trace functional analysis back to various sources:

1. *The Calculus of Variations*, which by itself has a distinguished history involving (Johann) Bernoulli, Euler, Lagrange, Legendre, Jacobi, and others. This was one of the sources of the idea of studying (not necessarily *linear*) spaces of functions and functions on such spaces, i.e. functionals (*idem dito*). This was picked up in the 1880s by the so-called Italian school of functional analysis, involving Ascoli and Arzelà (whose theorem was the first rigorous result in the subject), Volterra, Pincherle, and to some extent Peano (1888), who first axiomatized vector spaces.
2. *Infinite systems of linear equations with an infinite number of unknowns*, initially coming from Fourier's work on heat transfer in the 1820s; his idea of what we now call 'Fourier analysis' links certain linear partial differential equations (PDEs) with such systems. This link was almost immediately generalized by the *Sturm–Liouville theory* of linear second-order differential equations from the 1830s, in which the crucial idea of eigenfunctions and eigenvalues originates. Around 1890, the analysis of Hill, Poincaré, and von Koch on the motion of the Moon provided further inspiration.
3. *Integral equations*, first studied by Abel in the 1820s and independently by Liouville in the 1830s in connection with problems in mechanics. From the 1860s onwards integral equations were used by Beer, Neumann, and others as a tool in the study of harmonic functions and the closely related *Dirichlet problem* (which asks for a function satisfying Laplace's equation on a domain with prescribed boundary value, and may also be seen as a variational problem). This problem, in turn, came from the study of vibrating membranes via PDEs from the 18th century onwards; it is hardly a coincidence that the Dirichlet problem was eventually solved rigorously in 1901 by Hilbert (Monna, 1975), whose role in functional analysis is described below.

However, functional analysis equally benefited from—and was eventually even one of the highlights of—the abstract or 'modernist' turn that mathematics took in the 19th century,[5]

[4] See Monna (1973), Dorier (1995), and Moore (1995) for the history of linear structures, and Bernkopf (1966), Monna (1973), Steen (1973), Dieudonné (1981), Birkhoff and Kreyszig (1984), Pier (2001), and Siegmund-Schultze (1982, 2003) for the historical development of functional analysis. According to most authors this development occupied a period of about 50 years, starting in the 1880s and ending in 1932 with the books by Banach, Stone, and von Neumann published in that year (see section 18.5).

[5] See Mehrtens (1990) and Gray (2008) in general, and Siegmund-Schultze (1982) for functional analysis.

in shaking off its roots in the real (physical) world and becoming an independent intellectual and abstract activity. In fact, it was exactly this turn that made the completely unexpected application of functional analysis to quantum theory possible, and hence it seems no accident that Hilbert was a crucial player both in the decisive phase of the modernist turn *and* in the said application. Thus Hilbert played a double role in this:

1. Through his general views on mathematics (which of course he instilled in his pupils such as Weyl and von Neumann) and the ensuing scientific atmosphere he had created in Göttingen.[6] Hilbert's views branched off in two closely related directions:
   - His relentless emphasis on *axiomatization*, which started (at least in public) with his famous memoir *Grundlagen der Geometry* from 1899 (Volkert, 2015).
   - His promotion of the interplay between mathematics and physics (Corry, 2004a).

These came together in his *Sixth Problem* (from the famous list of 23 in 1900):[7]

> Mathematical Treatment of the Axioms of Physics. The investigations on the foundations of geometry suggest the problem: To treat in the same manner, by means of axioms, those physical sciences in which already today mathematics plays an important part; in the first rank are the theory of probabilities and mechanics.   (Hilbert, 1902)

2. Through his contributions to functional analysis, of which he was one of the founders:

> By the depth and novelty of its ideas, [Hilbert (1906)] is a turning point in the history of Functional Analysis, and indeed deserves to be considered the very *first* paper published in that discipline.   (Dieudonné, 1981, p. 110)

The connection between quantum theory and functional analysis could be made so quickly because in the 1920s Göttingen did not only have the best mathematical institute in the world (going back to Gauß, Riemann, and now Hilbert), but, due to the presence of Born, Heisenberg, and Jordan, and others, was also one of the main centres in the creation of quantum mechanics in the crucial years 1925–1927. It was this combination that enabled the decisive contributions of von Neumann (who spent 1926–1927 in Göttingen, see section 18.4).

It cannot be overemphasized how remarkable the link between quantum theory and functional analysis is. The former is the physical theory of the atomic world that was developed between 1900 and 1930, written down for the first time in systematic form in Dirac's celebrated textbook *The Principles of Quantum Mechanics* from 1930 (Jammer,

---

[6] '*One cannot overstate the significance of the influence exerted by Hilbert's thought and personality on all who came out of [the Mathematical Institute at Göttingen]*' (Corry, 2018). See also Rowe (2018).

[7] See e.g. Corry (1997, 2004a, 2018) and references therein. It is puzzling that Hilbert did not mention Newton's *Principia* in this light, which was surely the first explicit and successful axiomatization of physics.

1989). The latter is a mathematical theory of infinite-dimensional vector spaces (and linear maps between these), endowed with some notion of convergence (i.e. a topology), either in abstract form or in concrete examples where the 'points' of the space are often functions.[8] These two topics appear to have nothing to do with each other whatsoever, and hence the work of von Neumann (1932) in which they are related seems nothing short of a miracle. The aim of my paper is to put this miracle in some historical perspective.

## 18.2 HILBERT: AXIOMATIC METHOD

I believe this: as soon as it is ripe for theory building, anything that can be the subject of scientific thought at all falls under the scope of the axiomatic method and hence indirectly of mathematics. By penetrating into ever deeper layers of axioms in the sense outlined earlier we also gain insight into the nature of scientific thought by itself and become steadily more aware of the unity of our knowledge. Under the header of the axiomatic method mathematics appears to be called into a leading role in science in general.   (Hilbert, 1918, p. 415)

Hilbert (1918) begins his essay on axiomatic thought (of which the above text is the end) by stressing the importance of the connection between mathematics and neigbouring fields, especially physics and epistemology, and then says that the essence of this connection lies in the *axiomatic method*. By this, he simply means the identification of certain sentences (playing the role of axioms) that form the foundation of a specific field in the sense that its theoretical structure (Hilbert uses the German word *Fachwerk*) can be (re)constructed from the axioms via logical principles. Axioms typically state relations between 'things' (*Dinge*), like 'points' or 'lines', which are defined implicitly through the axioms and hence may change their meaning if the axiom systems in which they occur change, as is the case in e.g. non-Euclidean geometry.[9] The epistemological status of the axioms differs between fields. For example, Hilbert considered geometry initially a natural science:

Geometry also emerges from the observation of nature, from experience. To this extent, it is an *experimental science*. [ . . . ] All that is needed is to derive [its] foundations from a minimal set of *independent axioms* and thus to construct the whole edifice of geometry by *purely logical means*. In this way geometry is turned into a *purely mathematical science*.
(Hilbert in 1898–99, quoted in Corry, 2004a, p. 90)[10]

[8] For example, in Hilbert spaces convergence is defined via the norm coming from the inner product.
[9] The revolutionary nature of this view may be traced from Hilbert's correspondence with Frege, who apparently never accepted (or even grasped) this point (Gabriel *et al.*, 1980; Blanchette, 2018).
[10] From unpublished lecture notes by Hilbert, emphasis in original. Translation: Corry (2004a).

This does not mean that he treated the axioms of geometry as 'true' (as Euclid had done): Hilbert often stressed the tentative and malleable nature of axiom systems,[11] and acknowledged that axioms for physics might even be inconsistent, in which case finding new, consistent axioms is an important source of progress (Corry, 2004a; Majer, 2014).[12]

Hilbert is famous for his purely formal treatment of axioms,[13] which indeed was striking all the way from the *Grundlagen der Geometrie* in 1899 to his swan song *Grundlagen der Mathematik* (Hilbert and Bernays, 1934, 1939), but in fact such formality is always strictly limited to the logical analysis of axiom systems (notably his relentless emphasis on consistency and to a lesser extent on completeness) and the validation of proofs. Indeed, except for logic Hilbert made almost no contribution to the axiomatization of mathematical structures, although, starting already in the 19th century with e.g. Dedekind, Peano, and Weber, this became a central driving force of 20th-century mathematics (Corry, 2004b).

# 18.3 HILBERT: FUNCTIONAL ANALYSIS

Inspired by Fredholm's theory of integral operators,[14] in the paper mentioned by Dieudonné in the Introduction (section 18.1) above, Hilbert (1906) introduced many of the key tools of functional analysis, such as bounded and compact operators and spectral theory, culminating in his discovery of continuous spectra.[15] However, what we now see as *the* central aspect of functional analysis, namely its linear structure, is absent! Hilbert's analysis is entirely given in terms of quadratic forms $K(x) = \Sigma_{p,q} k_{pq} x_p x_q$, where the sequence $(x)$ satisfies $\Sigma_k |x_k|^2 \leq 1$, so that he works on what we would now call the closed unit ball of the Hilbert space $\ell^2$ of square-summable

---

[11] As exemplified by the seven editions of *Grundlagen der Geometrie* Hilbert published during his lifetime!

[12] Though Einstein would speak of 'principles' rather than 'axioms', many of his key contributions to physics, such as special relativity, general relativity, and the EPR argument are examples of this strategy.

[13] This has earned Hilbert the undeserved reputation of being a 'formalist', which is remote from his actual views on mathematics. The purely symbolic treatment of axioms and proofs was not even new with Hilbert; he apparently took it from Russell (Mancosu, 2003; Ewald and Sieg, 2013) and hence indirectly from Frege and Peano. But unlike Russell, until the last decade of his career dedicated to proof theory, Hilbert stated axioms quite informally, using a combination of mathematical and natural language.

[14] [The day (in 1901) on which Holmgren spoke on Fredholm's work in Hilbert's seminar] '*was decisive for a long period in Hilbert's life and for a considerable part of his fame*' (Blumenthal, 1935, p. 410).

[15] Since he lacked the concept of a linear operator Hilbert used a somewhat cumbersome definition of a spectrum; the modern definition is due to Riesz (1913) and was also adopted by von Neumann, see section 18.4.

sequences, which is compact in what we now call the weak topology, which Hilbert also introduced himself to great effect (Dieudonné, 1981).

As pointed out at the end of section 18.2, though at first sight odd,[16] it seems typical for Hilbert not to rely on axiomatized mathematical structures, let alone that he would care to refer to Peano (1888), in which the concepts of a vector space and a linear map had been axiomatized. Perhaps Peano's axiomatization was really unknown in Göttingen, where Hilbert's former student Weyl (1918) rediscovered the axioms for a vector space in the context of Einstein's theory of general relativity (at least, he did not cite Peano either).[17] However, the essentially linear nature of Hilbert's constructions was soon noted and developed by various mathematicians, notably Hilbert's own student Schmidt (1908), who introduced the space $\ell^2$ including its inner product and even norm in modern form,[18] and a bit later by F. Riesz (1913), who rewrote most of Hilbert's results in the modern way via bounded or compact linear operators on $\ell^2$ (the notion of a linear operator as such had already appeared before, notably in the Italian school of functional analysis).

Around 1905, Hadamard and his student Fréchet (partly inspired by the Italian school) emphasized the idea of looking at functions as points in some (infinite-dimensional) vector space, including an early use of topology, then also a new field— it was Fréchet (1906) who in his thesis introduced metric spaces. This idea, with its convenient notion of convergence, crossed with the Hilbert school through the introduction of $L^2$-spaces, including the spectacular and unexpected isomorphism $L^2([a, b]) \cong \ell^2$ due to Riesz (1907) and Fischer (1907). A truly geometric or spatial view of functional analysis was subsequently developed especially by Riesz (1913, 1918), culminating in the axiomatic development of Banach spaces in the 1920s by Helly, Wiener, and Banach (Monna, 1973; Pietsch, 2007).[19]

[16] Dieudonné (1981) explains the 19th-century emphasis on matrices and quadratic forms at the expense of vectors and linear maps, so that Hilbert had one foot in the 19th century and the other in the 20th.

[17] See, however, Corry (2004a, section 9.2) on the culture of 'nostrification' in Hilbert's Göttingen: 'It was widely understood, among German mathematicians at least, that "nostrification" encapsulated the peculiar style of creating and developing scientific ideas in Göttingen, and not least because of the pervasive influence of Hilbert. Of course, "nostrification" should not be understood as mere plagiarism.' (loc. cit. p. 419).

[18] The first author to use the term 'Hilbert space' (Hilbertscher Raum) was Schönflies (1908), but he meant the closed unit ball in $\ell^2$ (which was historically spot on, since that is what Hilbert analysed!). Riesz (1913) used l'espace hilbertien for what we now call $\ell^2$, and both the notion and the name Hilbert space for the general abstract concept we now take it to mean was introduced by von Neumann (1927a).

[19] As an intermediate step from Hilbert to Banach spaces, $L^p$ spaces were introduced by F. Riesz (1909). Frigyes Riesz (whose younger brother Marcel was also a well-known mathematician) was almost unique at the time in being familiar with the Italian, French, and German schools of functional analysis, whose ideas he combined. Dieudonné (1981, p. 145) calls Riesz (1918), which develops the (spectral) theory of compact operators on Banach spaces avant la lettre, 'one of the most beautiful papers ever written'.

# 18.4 VON NEUMANN: FOUNDATIONS OF QUANTUM THEORY

*Methoden der mathematischen Physik* by Courant and Hilbert (1924) put the lid on the Göttingen school in functional analysis. It was meant to describe *classical* physics.

> And now, one of those events happened, unforeseeable by the wildest imagination, the like of which could tempt one to believe in a pre-established harmony between physical nature and mathematical mind:[20] Twenty years after Hilbert's investigations *quantum mechanics* found that the observables of a physical system are represented by the linear symmetric operators on a Hilbert space and that the eigenvalues and eigenvectors of that operator which represent *energy* are the energy levels and corresponding stationary quantum states of the system. Of course, this quantum-physical interpretation added greatly to the interest in the theory and led to a more scrupulous investigation of it, resulting in various simplifications and extensions.    (Weyl, 1951, p. 541)

Although Weyl (who frequently fell into lyrical overstatements in his philosophical writings) may have been right about the events in question being unforeseeable, the historical record shows considerable continuity, too. Perhaps a slightly more balanced judgement is:

> This revolution was made possible by *combining* a concern for rigorous foundations with an interest in physical applications, *and* by coordinating the relevant literature in depth.    (Birkhoff and Kreyszig, 1984, pp. 306–307)

At least the first two aspects were exemplified by Hilbert, who had lectured on the mathematical foundations of physics since 1898 (Sauer and Majer, 2009; Majer and Sauer, 2021), and, helped by various assistants,[21] organized a regular research seminar on the latest developments in physics (Reid, 1970; Schirrmacher, 2019). With Born, Heisenberg, and Jordan all at Göttingen at the time, during the Winter Semester of 1926–1927 Hilbert lectured on quantum theory, with book-length lecture notes by Nordheim.[22]

---

[20] 'Pre-established harmony', a philosophical concept originally going back to Leibniz, was a popular concept in the Göttingen of Hilbert, where it referred to the relationship between mathematics and physics, or more generally between the human mind and nature (Pyenson, 1982; Corry, 2004a). Minkowski, Hilbert, Born, and Weyl himself all used it as appropriate, and Corry (2004a, pp. 393–94) even claims that it was '*one of the most basic concepts that underlay the whole scientific enterprise in Göttingen*', adding that '*Hilbert, like all his colleagues in Göttingen, was never really able to explain, in coherent philosophical terms, its meaning and the possible basis of its putative pervasiveness, except by alluding to "a miracle".*'

[21] Hilbert's first assistant (at the time unpaid) had been Born in 1904; from 1922–1926 it was Nordheim.

[22] These may be found in Sauer and Majer (2009), pp. 507–706. Half of the course, on the 'old quantum theory' was practically reproduced from Hilbert's earlier lectures on quantum theory during 1922–1923.

These lectures are very impressive and cover almost everything, from Hamilton–Jacobi theory and the 'old quantum theory' to Heisenberg's matrix mechanics, Schrödinger's wave mechanics, Born's probability interpretation, and finally Jordan's *Neue Begründung*, i.e. his attempt (simultaneous with Dirac's) to unify the last three ingredients into a single formalism.[23]

'Coordinating the relevant literature in depth' was therefore certainly taken care of on the physics side, but on the mathematical side Hilbert's surprising lack of interest in the axiomatization of new mathematical structures except logic (cf. section 18.2) still played a role:

> The German school [in functional analysis, i.e. Hilbert's school] remained reserved with respect to the more abstract concepts of set theory and axiomatics until well into the 1920s.   (Siegmund-Schultze, 2003, p. 385)

In 1926 Hilbert (1862–1943) attracted the internationally acknowledged young genius von Neumann (1903–1957) to spend the academic year 1926–1927 in Göttingen in order to work on his Proof Theory,[24] but in actual fact the latter mostly worked on the mathematical foundations of quantum theory and added his insights into axiomatic set theory to this.[25]

The paper by Hilbert, von Neumann, and Nordheim (1927), now obsolete, is actually a summary of Hilbert's (inconclusive) views, based on his lectures; the decisive establishment of the interaction between quantum theory and functional analysis is entirely due to von Neumann (1927a,b), with mathematical details further elaborated in von Neumann (1930a,b) and a unified exposition in his famous book from 1932, which remains a classic. Apart from his discussion of quantum statistical mechanics and of the measurement problem, which were ground-breaking contributions to physics but are less relevant for our topic, the main accomplishments of von Neumann (1932) in the light of functional analysis are:[26]

1. *Axiomatization of the notion of a Hilbert space (previously known only in examples).* Although Schmidt, Riesz, and others had used spaces like $\ell^2$ and $L^2$ in a geometric way 20 years earlier, including their inner products and norms, the

[23] See Duncan and Janssen (2009, 2013) for a detailed survey of Jordan's *Neue Begründung*. Older and somewhat complementary histories of this period include Jammer (1989) and Mehra and Rechenberg (2000).

[24] Hilbert got a fellowship for von Neumann from the International Education Board (a subsidiary of the Rockefeller Foundation). In the fall of 1927 von Neumann moved to Berlin as a *Privatdozent*, where Hilbert's former student Schmidt provided him with the concept of self-adjointness he had initially missed in setting up his spectral theory for closed unbounded operators (Birkhoff and Kreyszig, 1984, p. 309).

[25] Von Neumann had made brilliant contributions to set theory already in his late teens. For further information about von Neumann see Oxtoby *et al.* (1958), Heims (1980), Glimm *et al.* (1990), Macrae (1992), Bródy and Vámos (1995), Rédei (2005b), and the rare but insightful manuscript Vonneumann (1987).

[26] Even finding *one* of these would have been impressive, especially for someone who was only 23 years old.

abstract concept of a Hilbert space (unlike that of a Banach space), with an axiomatically defined inner product and ensuing completeness requirement in the associated norm, was still lacking before 1927. The novelty of von Neumann's coordinate-free approach to Hilbert spaces, which in particular saw matrices as coordinate-dependent expressions of operators, is illustrated by the fatherly advice he got in Berlin from Schmidt:[27]

No! No! You shouldn't say operator, say matrix!    (Bernkopf, 1968, p. 346)

2. *Establishment of a spectral theorem for (possibly unbounded) self-adjoint operators.* This was a vast abstraction and generalization of practically all of the spectral theory done in Hilbert's school, including the work of Weyl and Carleman on what (since von Neumann) are called unbounded self-adjoint operators and their deficiency indices.[28]

3. *Axiomatization of quantum mechanics in terms of Hilbert spaces (and operators):*
   (a) *Identification of observables with (possibly unbounded) self-adjoint operators.* One of von Neumann's main goals was a rigorous proof of the equivalence between matrix mechanics (which almost deliberately lacked states) and wave mechanics (which initially lacked observables). As a first ingredient, Heisenberg's matrices (which were 'quantum-mechanical reinterpretations of classical observables', as in the title of Heisenberg (1925)) were reinterpreted once again, now as self-adjoint operators on some Hilbert space like $\ell^2$. The need for *unbounded* operators, e.g. for position, momentum, and energy, emerged at once.[29]
   (b) *Identification of pure states with one-dimensional projections (or rays).* And this was the second ingredient of the equivalence proof. Identifying Schrödinger's wave-function $\Psi$ with a unit vector in the Hilbert space $L^2(\mathbb{R}^3)$ was an accomplishment by itself, but on top of this, von Neumann (and Weyl) quickly recognized the importance of the fact that such vectors only define (pure) states *up to a phase*. Thus the cleanest way to define (pure) states is to identify them with one-dimensional projections rather than vectors.[30]
   (c) *Identification of transition amplitudes with inner products.* Next to the goal just stated, *the* point of von Neumann's axiomatization was to provide a home to the mysterious transition amplitudes $\langle \varphi | \psi \rangle$ that were at the heart of

---

[27] Apart from the fact that it wasn't mathematically rigorous, Dirac's (1930) approach was not so much coordinate-free, but invariant under coordinate changes. See also Kronz and Lupher (2019).

[28] This history is very nicely explained by Dieudonné (1981), Chapter VII. See also Stone (1932).

[29] This was clear to von Neumann from Schrödinger's realization of the position and momentum operators, cf. von Neumann (1932), p. 48. Later also more abstract arguments from the canonical commutation relations were given, which force these operators to be unbounded (Wintner, 1947; Wielandt, 1949).

[30] Following Minkowski's example from his *Geometry of Numbers*, in the completely different context of quantum mechanics von Neumann defined pure states to be extreme points of the convex set of all states.

Jordan's *Neue Begründung* (Duncan and Janssen, 2013), which Born, Dirac, and Pauli also regarded as the essence of quantum mechanics. If $|\psi\rangle$ and $|\varphi\rangle$ are unit vectors in some Hilbert space, von Neumann took the amplitude $\langle\varphi|\psi\rangle$ to be their inner product, with corresponding transition probability $P(\varphi,\psi) = |\langle\varphi|\psi\rangle|^2$. In terms of the one-dimensional projections $e = |\varphi\rangle\langle\varphi|$ and $f = |\psi\rangle\langle\psi|$, this gives $P(e,f) = \mathrm{Tr}(ef)$, where Tr is the trace.[31]

(d) *A formula for the Born rule stating the probability of measurement outcomes.* Von Neumann's Born probability to find a result $\lambda \in I$ in a measurement of some self-adjoint operators $A$ in a state $\rho$ is given by $\mathrm{Tr}(\rho E(I))$, where $E(I)$ is the spectral projection for a subset $I$ in the spectrum of $A$ (and generalizations thereof to commuting observables). For $\rho = |\psi\rangle\langle\psi|$, $I = \{\lambda\}$, and $E(I) = |\varphi\rangle\langle\varphi|$, assuming $A\varphi = \lambda\varphi$ and the eigenvalue $\lambda$ is nondegenerate, this recovers the transition probability in the previous item.

(e) *Identification of general states with density operators.* Von Neumann *proved* this identification by showing that if Exp is a *linear* map from the (real) vector space of all bounded self-adjoint operators $A$ on some Hilbert space $H$ to $\mathbb{R}$ that is *normalized* ($\mathrm{Exp}(\mathbb{I}) = 1$), *dispersion-free* ($\mathrm{Exp}(A^2) = \mathrm{Exp}(A)^2$), and satisfies a continuity condition (which is automatic if $H$ is finite-dimensional), then $\mathrm{Exp}(A) = \mathrm{Tr}(\rho A)$ for some density operator $\rho$ on $H$. Confusingly, he took this correct, noncircular, and interesting result for a proof that no hidden variables can underly quantum mechanics. See Bub (2011), Dieks (2016), and Mitsch (2021) for fair accounts that do justice to von Neumann's own intentions.

(f) *Identification of propositions with closed subspaces (or the projections thereon).*[32] This was further developed by Birkhoff and von Neumann (1936), whose calculus of such propositions initiated the field of quantum logic (Rédei, 1998).

[31] It should be admitted that this did not clinch the issue. Dirac and Jordan used probability amplitudes like $\langle x|p\rangle = exp(-ixp/\hbar)$, where $x,p \in \mathbb{R}$, but 'eigenvectors' like $|x\rangle$ and $|p\rangle$ for the continuous spectrum of some operator (like position and momentum here) are undefined in Hilbert space (since their norm would be infinite, i.e. $\sqrt{\delta(0)}$). Von Neumann circumvented this problem by first practically starting his book with a tirade against Dirac's mathematics, and second, by writing down expressions like $\mathrm{Tr}(E(I)F(J))$, where $E(I)$ and $F(J)$ are the spectral projections for subsets $I$ and $J$ in the spectra of some self-adjoint operators $A$ and $B$, respectively. Unfortunately, these 'transition probabilities' are not only unnormalized; they may even be infinite. This was one of the reasons why von Neumann felt uncomfortable with his formalism right from the start, and later sought a way out of this problem through a combination of lattice theory à la Birkhoff and von Neumann (1936) and the theory of operator algebras he had also created himself, cf. section 18.5.2. The key is the existence of type ii$_1$ factors, which admit a *normalized* trace tr, i.e. $\mathrm{tr}(\mathbb{I}) = 1$. Replacing $\mathrm{Tr}(E(I)F(J))$ by $\mathrm{tr}(E(I)F(J))$ then makes the transition probabilities finite as well as normalized, but this only works if the spectral projections lie in the said factor. See Araki (1990) and Rédei (1996, 2001).

[32] Propositions in this sense are yes–no questions. In classical physics these would correspond to subsets of phase space, which von Neumann, then, 'quantized' into closed linear subspaces of Hilbert space $H$. These bijectively correspond to (orthogonal) projections (i.e. bounded operators $p$ on $H$ such that $p^2 = p^* = p$).

In particular, von Neumann provided *two* separate (but closely related) axiomatizations:[33]

- of Hilbert space as an abstract mathematical structure (almost *contra* Hilbert, as should be clear from the quote by Siegmund-Schultze above and the end of section 18.2);
- of quantum mechanics as a physical theory (entirely in the spirit of Hilbert).

# 18.5  Quantum Theory and Functional Analysis since 1932

In this section we give a brief overview of three areas of functional analysis that have had fruitful interactions with quantum theory since the initial breakthrough during 1927–1932.

## 18.5.1  Unbounded Operators

As already mentioned, motivated by quantum mechanics, von Neumann (1930a,b, 1932) developed an abstract theory of self-adjoint operators, culminating in his spectral theorem. Some of this theory was constructed independently and simultaneously in the US by Stone (1932), who also found a result that von Neumann strangely missed and which is extremely important for the mathematical foundations of quantum theory: *Stone's Theorem* (to be distinguished from the closely related and equally famous *Stone–von Neumann Theorem*),[34] shows that (possibly unbounded) self-adjoint operators and (continuous) unitary representations of the additive group $\mathbb{R}$ on a Hilbert space are equivalent; this is a rigorous version of the link between a Hamiltonian $h$ and a unitary time-evolution $u_t = exp(-ith)$.

The step from abstract theory to concrete examples that were actually useful for quantum mechanics turned out to be highly nontrivial.[35] The first (and still most important) results in applying the abstract theory to atomic Hamiltonians are

---

[33] See Lacki (2000), Rédei (2005a), and Rédei and Stöltzner (2006) on von Neumann's methodology.

[34] Found independently by Stone (1931) and von Neumann (1931), this theorem establishes the uniqueness of irreducible representations of the canonical commutation relations that are integrable to unitary representations of the Heisenberg group; the link between these notions was, prior to Stone's Theorem, first described by Weyl (1928). See Summers (2001) and Rosenberg (2004) for history and later developments.

[35] Simon (2018, p. 176) mentions that around 1948 von Neumann told Bargmann that '*self-adjointness for atomic Hamiltonians was an impossibly hard problem and that even for the hydrogen atom, the problem was difficult and open*', adding that von Neumann's attitude may have discouraged work on the problem.

due to Kato (1951), who thereby established the study of *Schrödinger operators* as a mathematical discipline.[36]

## 18.5.2 Operator Algebras and Noncommutative Geometry

Von Neumann's contributions to the interplay between quantum theory and functional analysis did not end with his axiomatization from the period 1927–1932. Parallel to (but apparently not inspired by) the development of quantum field theory,[37] in the 1930s he initiated the study of *rings of operators*, now called *von Neumann algebras*. This theory was supplemented and refined by the work of Gelfand and Naimark (1943), who (inspired by von Neumann algebras as well as by Gelfand's earlier work on commutative Banach algebras, but apparently not by physics directly) founded the field of *C\*-algebras*.[38]

Jointly, von Neumann algebras and C\*-algebras are called *operator algebras*, which may be studied both abstractly and as algebras of concrete (bounded) operators on some Hilbert space. This flexibility allows a huge generalization of the pure Hilbert space formalism of von Neumann (1932), in which the algebra of all bounded operators on a given Hilbert space is replaced by an arbitrary (abstract) operator algebra. As such, one may continue to work with states, observables, and expectation values (Segal, 1947). From the 1960s onwards operator algebras have become an important tool in mathematical physics, initially applied to quantum systems with infinitely many degrees of freedom, as in quantum statistical mechanics (Ruelle, 1969; Bratteli and Robinson, 1981, 1987; Haag, 1992; Simon, 1993) and quantum field theory (Haag, 1992; Araki, 1999; Brunetti, Dappiaggi, and Fredenhagen, 2015).[39] In turn, these physical applications have also given rise to various new mathematical ideas.[40] Furthermore, since the 1980s the field of operator algebras has been greatly refined and

---

[36] See also his textbook Kato (1966), followed by the four-volume series Reed and Simon (1972–1978), many other books, and more briefly Simon (2000). Kato's work is described in detail in Simon (2018).

[37] In various places von Neumann mentioned quantum mechanics, ergodic theory, lattice theory, projective geometry (which he turned into *continuous geometry*), and group representation theory as inspirations for operator algebras. Oddly, his direct attempts to describe quantum theory in a more algebraic fashion (Jordan, von Neumann, and Wigner 1934; von Neumann, 1936), have had little impact on physics so far.

[38] See Petz and Rédei (1995) for the history of von Neumann algebras, Doran and Belfi (1986) for C\*-algebras, and Kadison (1982) for both. A brief survey also appears in Kronz and Lupher (2019). The founding papers 'On rings of operators' I–V are collected in von Neumann (1961); nos. I, II, and IV are co-authored by von Neumann's assistant Murray. The paper Gelfand and Naimark (1943) that started C\*-algebras is reprinted (with notes by Kadison) in Doran (1994).

[39] The recollections of Haag (2010), arguably the main player in this field, are a valuable historical source.

[40] The work on the classification of von Neumann algebras for which Connes received the Fields Medal in 1982 is a good example: this relied on ideas originating in quantum statistical mechanics (Connes, 1994).

expanded by the toolkit of *noncommutative geometry* (Connes, 1994), which so far has been applied to many areas of physics, ranging from particle physics (Connes and Marcolli, 2008; van Suijlekom, 2015) to deformation quantization and the classical limit of quantum mechanics (Rieffel, 1994; Landsman, 1998).

## 18.5.3 Distributions

Another major development in functional analysis that is relevant for quantum physics was the theory of *distributions* due to Schwartz (1950–1951).[41] Though closely related to Hilbert and Banach spaces through all kinds of natural embeddings and dualities, spaces of distributions belong to the wider class of *locally convex topological vector spaces*, which incidentally were introduced by von Neumann (1935). The facts that the Dirac delta-function, which had annoyed von Neumann (1932, p. 2) so much, becomes a well-defined object in distribution theory, and that the 'rigged Hilbert space' approach to distributions (Gelfand and Vilenkin, 1964) even gives a rigorous and satisfactory version of Dirac's continuous eigenfunction expansions (Maurin, 1968), have had surprisingly little impact on quantum mechanics, even when it is seen in the light of mathematical physics.[42]

Instead, the main applications of distribution theory to quantum physics have been to *quantum field theory*, in at least three originally different but closely related ways:

1. Through Wightman's *Axiomatic Quantum Field Theory*, where quantum fields are defined as unbounded operator-valued distributions (Streater and Wightman, 1964).
2. Through *causal perturbation theory*, an approach to renormalization based on the splitting of distributions with causal support into retarded and advanced parts.[43]
3. Through *microlocal analysis*, a phase space approach to distributions due to Hörmander (1990), which has become a key tool in quantum field theory on curved space-time.[44]

---

[41] In Chapter VI of his autobiography, Schwartz (2001) gives some history. For example, on pp. 227–28 he writes that he was inspired by the Dirac delta-function, PDEs, divergent integrals, de Rham currents, and duality in topological vector spaces, but at the time of discovery (1944–1945) was unaware of previous relevant work by Heaviside, Bochner, Carleman, and Sobolev. See also Dieudonné (1981), Chapter VIII, Lützen (1982), Barany, Paumier, and Lützen (2017), and Kronz and Lupher (2019).

[42] See e.g. van Eijndhoven and Graaf (1986) and Bohm (1994). The rigged Hilbert space approach is not needed for spectral theory, though transition amplitudes like $\langle x|p \rangle$ in footnote 31 now become well defined.

[43] This mathematically rigorous approach to renormalization has a long pedigree, but Epstein and Glaser (1973) is generally regarded as a key paper. For later work see e.g. Scharf (1995, 2001) and Rejzner (2016).

[44] See e.g. Bär and Fredenhagen (2009), Brunetti, Dappiaggi, and Fredenhagen (2015), and Gérard (2019).

In fact, these three areas can no longer be separated, neither from each other nor from the operator-algebraic approach to quantum field theory mentioned in the previous subsection: their coalescence is one of the frontiers of contemporary research in mathematical physics.

# 18.6 EPILOGUE

> Functional analysis arose in the early 20th century and gradually, conquering one stronghold after another, became a nearly universal mathematical doctrine, not merely a new area of mathematics, but a new mathematical world view. Its appearance was the inevitable consequence of the evolution of all of 19th-century mathematics, in particular classical analysis and mathematical physics. [ . . . ] Its existence answered the question of how to state general principles of a broadly interpreted analysis in a way suitable for the most diverse situations.
>
> (Vershik, 2006, p. 438, quoted by MacCluer, 2009, p. vii)

This passage explains to some extent why the spectacular and unexpected application of functional analysis to quantum theory was possible: though originating in problems from classical physics, the 'modernist' turn of mathematics towards abstraction and axiomatization that brought the subject into the 20th century made almost every field of mathematics universally applicable. Moreover, much as quantum theory was originally meant to merely describe the atomic domain but subsequently, through its extension to quantum field theory, in fact turned out to be a theory of all of physics (except perhaps gravity), through von Neumann's invention of operator algebras as well as Schwartz's theory of distributions (both partly inspired by quantum mechanics), functional analysis continued to provide an appropriate mathematical language also for quantum field theory.

Having said this, the question *why* functional analysis (here taken to be the original *linear* theory) and especially Hilbert spaces (or operator algebras) underlie quantum physics remains unanswered. Perhaps starting with Birkhoff and von Neumann (1936), many people have tried to derive the mathematical formalism from plausible physical principles, but I believe that every such derivation so far contains a contingent or even incomprehensible part in order to derive the (complex) Hilbert space formalism.[45] In

---

[45] In Birkhoff and von Neumann (1936) the modular law is already problematic; in refinements of their lattice-theoretic approach based on the reconstruction theorem of Solèr (1995) one has to assume orthomodularity *and* the existence of an infinite orthonormal set (and still needs further arguments to single out $\mathbb{C}$ over $\mathbb{R}$ or $\mathbb{H}$), etc. Mackey (1963) himself admits defeat by simply postulating that the lattice of propositions of a quantum system, for which he first gives many promising axioms, is isomorphic to the projection lattice $\mathcal{P}(H)$ of some Hilbert space $H$. In my own approach based on the axiomatization of transition probability spaces, axiom $C^*2$ on page 104 of Landsman (1998), which prescribes the transition probabilities of a two-level system, seems to lack any physical justification. See also Grinbaum (2007).

this respect, the connection between quantum theory and functional analysis remains mysterious.

Finally, let me note that this was a *winner's* (or 'whig') history, full of hero-worship: following in the footsteps of Hilbert, von Neumann established the link between quantum theory and functional analysis that has lasted. Moreover, partly through von Neumann's own contributions (which are on a par with those of Bohr, Einstein, and Schrödinger), the precision that functional analysis has brought to quantum theory has greatly benefited the foundational debate. However, it is simultaneously a *loser's* history: starting with Dirac, and continuing with Feynman, up until the present day, physicists have managed to bring quantum theory forward in utter (and, in my view, arrogant) disregard for the relevant mathematical literature. As such, functional analysis has so far failed to make any real contribution to quantum theory as a branch of physics (as opposed to mathematics), and in this respect its role seems to have been limited to something like classical music or other parts of human culture that adorn life but do not change the economy or save the planet. On the other hand, like general relativity, perhaps the intellectual development reviewed in this paper is one of those human achievements that make the planet worth saving.

# REFERENCES

Araki, H. (1990). Some of the legacy of John von Neumann in physics: Theory of measurement, quantum logic and von Neumann algebras in physics. In J. Glimm, J. Impagliazzo, and I. Singer (eds), *The Legacy of John von Neumann. Proceedings of Symposia in Pure Mathematics* 50. Providence: American Mathematical Society, pp. 119–36.

Araki, H. (1999). *Mathematical Theory of Quantum Fields*. Oxford: Oxford University Press.

Banach, S. (1932). *Théorie des Opérations Linéaires*. Warszawa: Instytut Matematyczny Polskiej Akademii Nauk, New York: Chelsea.

Bär, C., and Fredenhagen, K. (eds) (2009). *Quantum Field Theory on Curved Spacetimes: Concepts and Mathematical Foundations*. Berlin: Springer.

Barany, M., Paumier, A.-S., and Lützen, J. (2017). From Nancy to Copenhagen to the World: The internationalization of Laurent Schwartz and his theory of distributions. *Historia Mathematica*, 44, 367–94.

Bernkopf, M. (1966). The development of function spaces with particular reference to their origins in integral equation theory. *Archive for History of Exact Sciences*, 3, 1–96.

Birkhoff, G., and Kreyszig, E. (1984). The establishment of functional analysis. *Historia Mathematica*, 11, 258–321.

Bernkopf, M. (1968). A history of infinite matrices: A study of denumerably infinite linear systems as the first step in the history of operators defined on function spaces. Archive for History of Exact Sciences 4, 308–58.

Birkhoff, G., and von Neumann, J. (1936). The logic of quantum mechanics. *Annals of Mathematics*, 37, 823–43.

Blanchette, P. (2018). The Frege–Hilbert Controversy. In E. N. Zalta (ed.), *The Stanford Encyclopedia of Philosophy* (Fall 2018 edition), https://plato.stanford.edu/archives/fall2018/entries/frege-hilbert/.

Blumenthal, O. (1935). Lebensgeschichte. *David Hilbert: Gesammelte Abhandlungen Vol. III*, pp. 388–429. Berlin: Springer.

Bohm, A. (1994). *Quantum Mechanics: Foundations and Applications*. New York: Springer.

Bratteli, O., and Robinson, D. W. (1981). *Operator Algebras and Quantum Statistical Mechanics. Vol. II: Equilibrium States, Models in Statistical Mechanics*. Berlin: Springer.

Bratteli, O., and Robinson, D. W. (1987). *Operator Algebras and Quantum Statistical Mechanics. Vol. I: C\*- and W\*-Algebras, Symmetry Groups, Decomposition of States*, 2nd edn. Berlin: Springer.

Bródy, F., and Vámos, T. (eds) (1995). *The Neumann Compendium*. Singapore: World Scientific.

Brunetti, R., Dappiaggi, C., and Fredenhagen, K. (eds) (2015). *Advances in Algebraic Quantum Field Theory*. Dordrecht: Springer.

Bub, J. (2011). Is von Neumann's 'no hidden variables' proof silly? In H. Halvorson (ed.), *Deep Beauty: Mathematical Innovation and the Search for Underlying Intelligibility in the Quantum World*, Cambridge: Cambridge University Press, pp. 393–408.

Connes, A. (1994). *Noncommutative Geometry*. San Diego: Academic Press.

Connes, A., and Marcolli, M. (2008). *Noncommutative Geometry, Quantum Fields, and Motives*. New Delhi: Hindustan Book Agency.

Corry, L. (1997). David Hilbert and the axiomatization of physics (1894–1905). *Archive for History of Exact Sciences*, **51**, 83–198.

Corry, L. (2004a). *David Hilbert and the Axiomatization of Physics (1898–1918): From Grundlagen der Geometrie to Grundlagen der Physik*. Dordrecht: Kluwer.

Corry, L. (2004b). *Modern Algebra and the Rise of Mathematical Structures*, 2nd revised edn. Basel: Springer.

Corry, L. (2018). Hilbert's sixth problem: between the foundations of geometry and the axiomatization of physics. *Philosophical Transactions of the Royal Society A*. DOI: 10.1098/rsta.2017.0221.

Courant, R., and Hilbert, D. (1924). *Methoden der mathematischen Physik I*. Berlin: Springer.

Dieks, D. (2016). Von Neumann's impossibility proof: Mathematics in the service of rhetorics. *Studies in History and Philosophy of Modern Physics*, **60**, 136–48.

Dieudonné, J. (1981). *History of Functional Analysis*. Amsterdam: North-Holland.

Dijksterhuis, E. J. (1961). *The Mechanization of the World Picture*. Oxford: Oxford University Press. Translation of *De Mechanisering van het Wereldbeeld* (Meulenhoff, Amsterdam, 1950).

Dirac, P. A. M. (1930). *The Principles of Quantum Mechanics*. Oxford: Clarendon Press.

Doran, R. S. (ed.) (1994). *C\*-algebras: 1943–1993. Contemporary Mathematics* Vol. 167. Providence: American Mathematical Society.

Doran, R. S., and Belfi, V. (1986). *Characterization of C\*-algebras*. New York: Marcel Dekker.

Dorier, J.-L. (1995). A general outline of the genesis of vector space theory. *Historia Mathematica*, **22**, 227–61.

Duncan, A., and Janssen, M. (2009). From canonical transformations to transformation theory, 1926–1927: The road to Jordan's *Neue Begründung*. *Studies in History and Philosophy of Modern Physics*, **40**, 352–62.

Duncan, A., and Janssen, M. (2013). (Never) Mind your *p*'s and *q*'s: Von Neumann versus Jordan on the foundations of quantum theory. *The European Physical Journal H*, **38**, 175–259.

Eijndhoven, S. J. L. van, and Graaf, J. de (1986). *A Mathematical Introduction to Dirac's Formalism*. Amsterdam: North-Holland (Elsevier).

Epstein, H., and Glaser, V. (1973). The role of locality in perturbation theory. *Annales de l'Institute Henri Poincaré A (Physique théorique)*, **19**, 211–95.

Ewald, W., and Sieg, W. (2013). *David Hilbert's Lectures on the Foundations of Arithmetic and Logic, 1917–1933*. Heidelberg: Springer.

Fischer, E. (1907). Sur la convergence en moyenne. *Comptes Rendus de l'Académie des Sciences*, **144**, 1022–24.

Fréchet, N. (1906). Sur quelques points du calcul functionel. *Rendiconti del Circolo Matematico di Palermo*, **22**, 1–74.

Gabriel, G., Kambartel, F., and Thiel, C. (1980). *Gottlob Frege's Briefwechsel mit D. Hilbert, E. Husserl, B. Russell, sowie ausgewählte Einzelbriefe Freges*. Hamburg: Felix Meiner Verlag.

Gelfand, I. M., and Naimark, M. A. (1943). On the imbedding of normed rings into the ring of operators in Hilbert space. *Sbornik: Mathematics*, **12**, 197–213.

Gelfand, I. M., and Vilenkin, N. J. (1964). *Generalized Functions. Vol. 4: Some Applications of Harmonic Analysis. Rigged Hilbert Spaces*. New York: Academic Press.

Gérard, C. (2019). Microlocal analysis of quantum fields on curved spacetimes. https://arxiv.org/abs/1901.10175.

Glimm, J., Impagliazzo, J., and Singer, I. (eds) (1990). *The Legacy of John von Neumann. Proceedings of Symposia in Pure Mathematics 50*. Providence: American Mathematical Society.

Gray, J. D. (2008). *Plato's Ghost: The Modernist Transformation of Mathematics*. Princeton: Princeton University Press.

Grinbaum, A. (2007). Reconstruction of quantum theory. *British Journal for the Philosophy of Science*, **58**, 387–408.

Guicciardini, N. (2009). *Isaac Newton on Mathematical Certainty and Method*. Cambridge, MA: The MIT Press.

Haag, R. (1992). *Local Quantum Physics: Fields, Particles, Algebras*. Heidelberg: Springer.

Haag, R. (2010). Some people and some problems met in half a century of commitment to mathematical physics. *The European Physics Journal H*, **35**, 263–307.

Heims, S. J. (1980). *John von Neumann and Norbert Wiener: From Mathematics to the Technologies of Life and Death*. Cambridge, MA: MIT Press.

Heisenberg, W. (1925). Über quantentheoretische Umdeutung kinematischer und mechanischer Beziehungen. *Zeitschrift für Physik*, **33**, 879–93.

Hilbert, D. (1902). Mathematical Problems. Lecture delivered before the International Congress of Mathematicians at Paris in 1900. *Bulletin of the American Mathematical Society*, **8**, 437–79. Translated from *Göttinger Nachrichten*, 1900, pp. 253–97.

Hilbert, D. (1906). Grundzüge einer allgemeinen Theorie der linearen Integralgleichungen. Vierte Mitteilung. *Nachrichten von der Gesellschaft der Wissenschaften zu Göttingen, Mathematisch-Physikalische Klasse*, 157–227. Reprinted in Hilbert (1912).

Hilbert, D. (1912). *Grundzüge einer allgemeinen Theorie der linearen Integralgleichungen*. Leipiz und Berlin: Teubner. https://archive.org/details/grundzugeallgoohilbrich.

Hilbert, D. (1918). Axiomatisches Denken. *Mathematische Annalen*, **78**, 405–415. Reprinted in *David Hilbert: Gesammelte Abhandlungen Vol. III*, pp. 146–56 (1935). Berlin: Springer.

Hilbert, D., and Bernays, P. (1934, 1939). *Grundlagen der Mathematik, Bd. I, II*. Berlin: Springer.

Hilbert, D., von Neumann, J., and Nordheim, L. (1927). Über die Grundlagen der Quantenmechanik. *Mathematische Annalen*, **98**, 1–30.

Hörmander, L. (1990). *The Analysis of Linear Partial Differential Operators I*, 2nd edn. Berlin: Springer.

Jammer, M. (1989). *The Conceptual Development of Quantum Mechanics*, 2nd edn. New York: American Institute of Physics.

Jordan, P., von Neumann, J., and Wigner, E. P. (1934). On an algebraic generalization of the quantum mechanical formalism. *Annals of Mathematics*, **35**, 29–64.

Kadison, R. V. (1982). Operator algebras: The first forty years. *Proceedings of Symposia in Pure Mathematics* 38(1), pp. 1–18. Providence: American Mathematical Society.

Kato, T. (1951). Fundamental properties of Hamiltonian operators of Schrödinger type. *Transactions of the American Mathematics Society*, **70**, 195–211.

Kato, T. (1966). *Perturbation Theory for Linear Operators*. Berlin: Springer.

Kronz, F., and Lupher, T. (2019). Quantum Theory and Mathematical Rigor. In E. N. Zalta (ed.), *The Stanford Encyclopedia of Philosophy* (Fall 2019 edition), https://plato.stanford.edu/archives/fall2019/entries/qt-nvd/.

Lacki, J. (2000). The early axiomatizations of quantum mechanics: Jordan, von Neumann and the continuation of Hilbert's program. *Archive for History of Exact Sciences*, **54**, 279–318.

Lützen, J. (1982). *The Prehistory of the Theory of Distributions*. New York: Springer.

Landsman, K. (1998). *Mathematical Topics Between Classical and Quantum Mechanics*. New York: Springer.

Landsman, K. (2017). *Foundations of Quantum Theory: From Classical Concepts to Operator Algebras*. Cham: Springer Open. https://www.springer.com/gp/book/9783319517766.

MacCluer, B. D. (2009). *Elementary Functional Analysis*. New York: Springer.

Mackey, G. W. (1963). *The Mathematical Foundations of Quantum Mechanics*. New York: Benjamin.

Macrae, N. (1992). *John von Neumann: The Scientific Genius Who Pioneered the Modern Computer, Game Theory, Nuclear Deterrence, and Much More*. Providence: American Mathematical Society.

Majer, U. (2014). The 'axiomatic method' and its constitutive role in physics. *Perspectives on Science*, **22**, 56–79.

Majer, U., and Sauer, T. (eds) (2021). *David Hilbert's Lectures on the Foundations of Physics, 1898–1914*. Berlin, Heidelberg: Springer.

Mancosu, P. (2003). The Russellian influence on Hilbert and his school. *Synthese*, **137**, 59–101.

Maurin, K. (1968). *Generalized Eigenfunction Expansions and Unitary Representations of Topological Groups*. Warsaw: Polish Scientific Publishers.

Mehra, J., and Rechenberg, H. (2000). *The Historical Development of Quantum Theory. Vol. 6: The Completion of Quantum Mechanics 1926–1941. Part 1: The Probabilistic Interpretation and the Empirical and Mathematical Foundation of Quantum Mechanics, 1926–1936*. New York: Springer-Verlag.

Mehrtens, H. (1990). *Moderne Sprache Mathematik*. Frankfurt a/M: Suhrkamp.

Mitsch, C. (2021). The (not so) hidden contextuality of von Neumann's 'No Hidden Variables' proof. Forthcoming.

Monna, A. F. (1973). *Functional Analysis in Historical Perspective*. Utrecht: Oosthoek.

Monna, A. F. (1975). *Dirichlet's Principle: A Mathematical Comedy of Errors and its Influence on the Development of Analysis*. Utrecht: Oosthoek, Scheltema and Holkema.

Moore, G. H. (1995). The axiomatization of linear algebra, 1875–1940. *Historia Mathematica*, **22**, 262–303.

Neumann, J. von (1927a). Mathematische Begründung der Quantenmechanik. *Nachrichten von der Gesellschaft der Wissenschaften zu Göttingen, Mathematisch-Physikalische Klasse*, 1–57. https://eudml.org/doc/59215.

Neumann, J. von (1927b). Wahrscheinlichkeitstheoretischer Aufbau der Quantenmechanik. *Nachrichten von der Gesellschaft der Wissenschaften zu Göttingen, Mathematisch-Physikalische Klasse*, 245–72. https://eudml.org/doc/59230.

Neumann, J. von (1930a). Allgemeine Eigenwerttheorie hermitischer Funktionaloperatoren. *Mathematische Annalen*, **102**, 49–131.

Neumann, J. von (1930b). Zur Algebra der Funktionaloperationen und Theorie der normalen Operatoren. *Mathematische Annalen*, **102**, 370–427.

Neumann, J. von (1931). Die Eindeutigkeit der Schröderingerschen Operatoren. *Mathematische Annalen*, **104**, 570–78.

Neumann, J. von (1932). *Mathematische Grundlagen der Quantenmechanik*. Berlin: Springer–Verlag. English translation: *Mathematical Foundations of Quantum Mechanics*. Princeton: Princeton University Press (1955).

Neumann, J. von (1935). On complete topological spaces. *Transactions of the American Mathematical Society*, **37**, 1–20.

Neumann, J. von (1936). On an algebraic generalization of the quantum mechanical formalism (Part i). *Sbornik: Mathematics*, **1**, 415–484.

Neumann, J. von (1961). *Collected Works, Vol. III: Rings of Operators*. A. H. Taub, (ed.). Oxford: Pergamon Press.

Oxtoby, J. C., Pettis, B. J., and Price, G. B. (1958). John von Neumann: 1903–1957. *Bulletin of the American Mathematical Society*, **64**, No. 3, Part 2.

Peano, G. (1888). *Calcolo geometrico secondo l'Ausdehnungslehre di H. Grassmann: preceduto dalla operazioni della logica deduttiva*. Torino: Fratelli Bocca Editori.

Petz, D., and Rédei, M. (1995). John von Neumann and the theory of operator algebras. In F. Bródy and T. Vámos (eds), *The Neumann Compendium*, Singapore: World Scientific, pp. 163–81.

Pier, J. (2001). *Mathematical Analysis During the 20th Century*. New York: Oxford University Press.

Pietsch, W. (2007). *History of Banach Spaces and Linear Operators*. Basel: Birkhäuser.

Pyenson, L. R. (1982). Relativity in late-Wilhelminian Germany: The appeal to a pre-established harmony between mathematics and physics. *Archive for History of Exact Sciences*, **27**, 137–55.

Rédei, M. (1996). Why John von Neumann did not like the Hilbert space formalism of quantum mechanics (and what he liked instead). *Studies in History and Philosophy of Modern Physics*, **27**, 493–510.

Rédei, M. (1998). *Quantum Logic in Algebraic Approach*. Dordrecht: Kluwer Academic Publishers.

Rédei, M. (2001). Von Neumann's concept of quantum logic and quantum probability. In M. Rédei and M. Stöltzner (eds), *John von Neumann and the Foundations of Quantum Physics*, Dordrecht: Kluwer, pp. 153–72.

Rédei, M. (2005a). John von Neumann on mathematical and axiomatic physics. In G. Boniolo *et al.* (eds), *The Role of Mathematics in the Physical Sciences*, Dordrecht: Springer, pp. 43–54.

Rédei, M. (2005b). *John von Neumann: Selected Letters*. Providence: American Mathememmatical Society.

Rédei, M., and Stöltzner, M. (eds) (2001). *John von Neumann and the Foundations of Quantum Physics*. Dordrecht: Kluwer.

Rédei, M., and Stöltzner, M. (2006). Soft axiomatisation: John von Neumann on method and von Neumann's method in the physical sciences. In E. Carson and R. Huber (eds), *Intuition and the Axiomatic Method*, Dordrecht: Springer, pp. 235–50.

Reed, M., and Simon, B. (1972–1978). *Methods of Modern Mathematical Physics. Volume I: Functional Analysis. Volume II: Fourier Analysis, Self-Adjointness. Volume III: Scattering Theory. Volume IV: Analysis of Operators*. New York: Academic Press.

Reid, C. (1970). *Hilbert*. Berlin: Springer.

Rejzner, K. (2016). *Perturbative Algebraic Quantum Field Theory: An Introduction for Mathematicians*. Cham: Springer.

Rieffel, M. A. (1994). Quantization and $C^*$-algebras. *Contemporary Mathematics*, **167**, 66–97.

Riesz, F. (1907). Sur les systèmes orthogonaux des functions. *Comptes Rendus de l'Académie des Sciences*, **144**, 615–19.

Riesz, F. (1909). Untersuchungen über Systeme integrierbarer Funktionen. *Mathematische Annalen*, **69**, 449–97.

Riesz, F. (1913). *Les Systèmes d'Équations Linéaires à une Infinité d'Inconnues*. Paris: Gauthiers–Villars.

Riesz, F. (1918). Über lineare Funktionalgleichungen. *Acta Mathematica*, **41**, 71–98.

Rosenberg, J. (2004). A selective history of the Stone–von Neumann Theorem. *Contemporary Mathematics*, **365**, 331–53.

Rowe, D. E. (2018). *A Richer Picture of Mathematics: The Göttingen Tradition and Beyond*. Cham: Springer.

Ruelle, D. (1969). *Statistical Mechanics: Rigorous Results*. New York: Benjamin.

Sauer, T., and Majer, U. (eds) (2009). *David Hilbert's Lectures on the Foundations of Physics, 1915–1927*. Dordrecht: Springer.

Scharf, G. (1995). *Finite Electrodynamics: The Causal Approach*, 2nd edn. Berlin: Springer.

Scharf, G. (2001). *Quantum Gauge Theories: A True Ghost Story*. New York: Wiley.

Schirrmacher, A. (2019). *Establishing Quantum Physics in Göttingen: David Hilbert, Max Born, and Peter Debye in Context, 1900–1926*. Cham: Springer.

Schmidt, E. (1908). Über die Auflösung linearer Gleichungen mit unendlich vielen Unbekannten. *Rendiconti del Circolo Matematico di Palermo*, **xxv**, 53–77.

Schönflies, A. (1908). Die Entwicklung der Lehre von den Punktmannigfaltigkeiten, Zweiter Teil. *Jahresberichte der deutschen Mathematiker-Vereinigung, Ergänzungsband*. Leipzig: Teubner.

Schwartz, L. (1950–1951). *Théorie des Distributions. Vols. 1 & 2*. Paris: Hermann.

Schwartz, L. (2001). *A Mathematician Grappling with his Century*. Basel: Birkhäuser.

Segal, I. E. (1947). Postulates for general quantum mechanics. *Annals of Mathematics*, **48**, 930–48.

Sieg, W. (2013). *Hilbert's Programs and Beyond*. Oxford: Oxford University Press.

Siegmund-Schultze, R. (1982). Die Anfänge der Funktionalanalysis und ihr Platz im Unwandlungsprozeß der Mathematik um 1900. *Archive for History of Exact Sciences*, **26**, 13–71.

Siegmund-Schultze, R. (2003). The origins of functional analysis. In H. N. Jahnke (ed.), *A History of Analysis*, Providence: American Mathematical Society, pp. 385–408.

Simon, B. (1993). *The Statistical Mechanics of Lattice Gases. Vol. I.* Princeton: Princeton University Press.

Simon, B. (2000). Schrödinger operators in the twentieth century. *Journal of Mathematical Physics*, **41**, 3523–55.

Simon, B. (2018). Tosio Kato's work on non-relativistic quantum mechanics, part 1 and part 2. *Bulletin of Mathematical Sciences*, **8**, 121–232 and **9**, 1–99.

Solèr, M. P. (1995). Characterization of Hilbert spaces by orthomodular spaces. *Communications in Algebra*, **23**, 219–43

Steen, L. A. (1973). Highlights in the history of spectral theory. *American Mathematical Monthly*, **80**, 359–81.

Stone, M. H. (1931). Linear transformations in Hilbert space, III: Operational methods and group theory. *Proceedings of the National Academy of Sciences of the United States of America*, **16**, 172–75.

Stone, M. H. (1932). *Linear Transformations in Hilbert space and their Applications to Analysis.* Providence: American Mathematical Society.

Streater, R. F., and Wightman, A. S. (1964). *PCT, Spin and Statistics, and All That.* New York: Benjamin.

Suijlekom, W. D. van (2015). *Noncommutative Geometry and Particle Physics.* Dordrecht: Springer.

Summers, S. J. (2001). On the Stone–von Neumann uniqueness theorem and its ramifications. In M. Rédei and M. Stöltzner (eds), *John von Neumann and the Foundations of Quantum Physics*, Dordrecht: Kluwer, pp. 135–52.

Vershik, A. M. (2006). The life and fate of functional analysis in the twentieth century. In A. A. Bolibruch *et al.* (eds), *Mathematical Events of the Twentieth Century*, Berlin: Springer, pp. 437–47.

Volkert, K. (2015). *David Hilbert: Grundlagen der Geometrie.* Berlin: Springer.

Vonneumann, N. (1987). *John von Neumann as Seen by his Brother.* Typescript.

Weyl, H. (1918). *Raum—Zeit—Materie.* Berlin: Springer.

Weyl, H. (1928). *Gruppentheorie und Quantenmechanik.* Leipzig: Hirzel.

Weyl, H. (1951). A half-century of mathematics. *The American Mathematical Monthly*, **58**, 523–53.

Whiteside, D. T. (1970). The Mathematical Principles Underlying Newton's Principia Mathematica. *Journal for the History of Astronomy*, **1**, 116–38.

Wielandt, H. (1949). Über die Unbeschränktheit der Operatoren der Quantenmechanik. *Mathematische Annalen*, **121**, 21.

Wintner, A. (1947). The unboundedness of quantum-mechanical matrices. *Physical Review*, **71**, 738–39.

........................................................................................

# CHAPTER 19

# TONY LEGGETT'S CHALLENGE TO QUANTUM MECHANICS AND ITS PATH TO DECOHERENCE

........................................................................................

FÁBIO FREITAS

## 19.1 INTRODUCTION

........................................................................................

IT would be complete nonsense to even consider the idea of a Nobel Prize winner as someone unknown. However, the way one understands the career of others can be quite selective, separating what is most valuable in a specific context from what is undesirable, that needs to be forgotten. In some sense, this is exactly the case of Sir Anthony Leggett. He has virtually all the recognition a physicist can have, such as the Nobel Prize, the Wolf Prize, becoming a fellow of the Royal Society, and a knighthood from the Queen of the United Kingdom. Yet, most of this recognition arose because of his technical solutions to very difficult problems, more specifically understanding the superfluidity phase of helium 3 and how macroscopically large quantum systems behave upon interaction with the environment. However, while these problems did indeed contribute to the development of physics, there was also great philosophical insight involved in these solutions that, in some senses, remains ignored by most physicists. Furthermore, it is important to emphasize that this kind of reasoning, namely unifying physics and philosophy, came way before his recognition as a major physicist. While Anthony Leggett started his research programme that led to decoherence only at the end of the 1970s, the decision that drove him to foundational studies was taken long before. We could say that it happened before he even became a physicist.

Decoherence is the brand name for the coupling between a quantum system and its environment. Through this coupling the system loses the superposition of its eigenstates, which is the singular signature of a quantum system. Technically, the system's mathematical description evolves from a pure state to a mixture, and conceptually decoherence concerns the transition from the quantum to the classical description. Modelling such evolution was the goal pursued by physicists who worked on it during the 1980s. Eventually, it was taken to the lab and these experiments were the rationale for the 2012 Physics Nobel Prize awarded to Serge Haroche and David Wineland. There were several distinct and independent roads to conceptualizing and calculating decoherence and the physicists involved in this endeavour included others such as H. Dieter Zeh, Erich Joos, and Wojciech H. Zurek (Camilleri, 2009). In this paper we focus on the road taken by Leggett and pay some attention to the contributions of Caldeira, who was his PhD student. By focusing on Leggett's trajectory, we are interested not only in his contribution to the establishment of decoherence, but also in his singular approach to the foundations of quantum mechanics.

The paper is organized as follows: sections 19.2 and 19.3 are dedicated to Leggett's path from philosophy to physics and his own singular way towards research on foundational issues in quantum physics, while sections 19.4 and 19.5 deal with his conceptual and technical approach to the foundations of quantum mechanics. Section 19.4's heading—How to put Schrödinger's Cat in a Lab?—refers to his attempts to devise systems which could be used to describe the existence of linear superposition of macroscopically distinguishable states and still be apt to become a real experiment instead of a Gedankenexperiment. Macroscopic quantum tunnelling with Squids was for Leggett the best candidate for it. Section 19.5 presents the work of Caldeira, under Leggett's supervision, on how to model such systems and its main conclusion, namely, that damping always tends to suppress quantum tunnelling. Section 19.6 is where we discuss how Leggett challenged the validity of linear superpositions, that is quantum mechanics' validity, at the macroscopic level. We suggest an explanation for why he did not suffer professional obstacles related to his point of view. To conclude, in section 19.6, we present another explanation for a different but related issue, namely, the undervaluation and even the scant acknowledgement of his views on the foundations of quantum physics among physicists and philosophers who work in this field.

# 19.2 PHYSICS AND PHILOSOPHY

When we look at the history of foundations of quantum mechanics, we tend to separate it into two different periods regarding the background of the physicists involved. If we look at the founding fathers, mostly in a prewar European context, we usually consider that the protagonists had a solid knowledge of philosophy. Einstein, Bohr, Heisenberg,

Pauli, and several others, having studied in the late 19th and early 20th centuries, had a strong philosophical background, as was shown in the debates around the newborn quantum theory. As we move to the second wave of debates, mostly after World War II, and more specifically during the early context of the Cold War in the United States, the training of physicists was quite different. With early roots in the pragmatic character of American physics and the new needs of the Cold War for scientific training, philosophy was not considered an important topic.[1] One interesting point of comparison is the different receptions to both newborn quantum theory and QED renormalization techniques. The fathers of quantum theory engaged in long lasting debates over the meaning of the new theory during the 1920s and 1930s, yet when Feynman, Schwinger, Tomonaga, and Dyson reformed QED during the 1950s, applying very efficient 'patches' to solve infinity problems, there were no deep philosophical questions involved. The important thing was that these 'patches' worked and solved the problems they were intended to solve. Physics in its practice and the training of physicists had changed. As Schweber suggests, 'The workers of the 1930s, particularly Bohr and Dirac, and also Heisenberg, had sought solutions in terms of *revolutionary* departures. Special relativity and quantum mechanics had been created by revolutionary steps. The solution advanced by Feynman, Schwinger, and Dyson, was *pragmatic* and *conservative*' (Schweber, 1986, pp. 97–8).

Leggett (Figure 19.1) would, somehow, be misplaced in this new context, belonging to an older era. We are used to physicists embarking on a science career while still very young. Leggett, born to a couple of physics and mathematics teachers, actually wanted to follow the more prestigious path at the time, the humanities, more specifically the Greats.[2] Going to Oxford in 1955, he had the opportunity to study the classics, Greek and Latin, classical literature, and a considerable amount of philosophy, the field he actually considered following in his later career. In fact this decision had already been made quite early, when he was 13. Such an early decision also had another impact; he studied hardly any science before college, nor during it. The joke he tells about the influence of such studies in his later career is that, unlike his fellow physicists, he knows the actual meaning of Greek letters. As we shall see, this influence goes far deeper than that.[3]

However, at some point, Leggett decided that he did not really want to become a philosopher, as he 'was very dissatisfied with the fact that there seemed to be no hard subject criteria in philosophy as to whether what you're doing was right or wrong'.[4] This also had to do with the way philosophy was being practised in the English context,

[1] See Kaiser (Kaiser, 2002) and (Kaiser, Forthcoming) for the training of physicists in quantum mechanics and (Kevles, 1977) for a broader context about physicists in the United States. See also (Schweber, 1986) for a general tendency for pragmatism among United States theoretical physicists.

[2] Also known as classics or by its official name, *Literae Humaniores*.

[3] For his biographical data, see mainly his Nobel Prize biography (Leggett, 2003a). Additional material used in this dissertation includes interviews by Babak Ashrafi on 25 March 2005, and by me, on 3 August 2011. I would like to acknowledge the Niels Bohr Library Archives and Babak Ashrafi for allowing me to consult this first version of the interview.

[4] Interviews by Babak Ashrafi on 25 March 2005 and by the author on 3 August 2011.

**FIGURE 19.1** Anthony Leggett, 1983. Credits: University of Illinois, courtesy AIP Emilio Segrè Visual Archives, W. F. Meggers Gallery of Nobel Laureates Collection.

being much more around analytical philosophy and less focused on ontological and epistemological questions that would later be present in his professional career. Furthermore, with such limitations in mind, science became more appealing, more objective. 'I kind of felt that I wanted to work in an intellectual discipline in which there were, in some sense, hard objective criteria on whether your ideas are right or not'.[5] There is a bit of irony when we realize that later he would be doing research that has been classified as 'Experimental Metaphysics'.[6]

For the change to be possible, it would be necessary to get a second degree. It was usually very hard to obtain funding for second degrees, but he benefited from the Sputnik effect. With the launch of the first artificial satellite by the Soviet Union, the West became extremely worried about the shortage of scientific manpower. It became urgent to 'produce' as many scientists and engineers as possible. As he has described, 'I

---

[5]  Interviews by Babak Ashrafi on 25 March 2005 and by the author on 3 August 2011.

[6]  As coined by Abner Shimony to describe mainly those experiments to test more specifically Bell's inequalities, and as a byproduct to test ontological and epistemological questions about the nature of space-time and of our knowledge of such. See, for instance, the volume dedicated to him, 'Experimental metaphysics: Quantum mechanical studies for Abner Shimony' (Cohen, Horne, and Stachel, 1997).

only note the debt I owe in this context to the former Soviet general Sergei Korolev' (Leggett, 2020).

Even so, the transition was not easy for Leggett. First, he had no scientific background, and, second, he wanted to get a classified degree, which meant that he had to finish both his degrees in under six years. Now, he only had two years as he had already spent four on his first degree, which meant that he had to finish his degree in physics in half of the regular time and with no background training in science. Of course this was no simple task, and the fact that indeed he was able to finish in time was evidence of the talent he would later show throughout his career. In 1961, he earned his second degree, and a few years later, in 1964, under the supervision of Dirk ter Haar, he obtained his Ph.D. in physics, with a dissertation on condensed matter, a field that would mark his whole career.

## 19.3 THE PATH TO FOUNDATIONS

The 1950s and 1960s marked a change in the foundations of quantum mechanics (Freire, 2003, 2004, 2015). From the whole debate that took place during these years, two main themes would mark Leggett's path. The first one was David Bohm. In 1951, while still in Princeton but on the verge of going to Brazil, he developed his new formulation of quantum theory, the so called hidden variable programme. David Bohm proposed that we could use additional hidden variables, in the form of a quantum potential, in order to fully describe the dynamics of quantum systems, which would allow us to calculate trajectories for quantum particles. However, more importantly, his proposal marked a return to a classical determinism. The Heisenberg relations would still remain valid, but not as a limitation from nature, emerging from the uncontrollable interaction of the quantum system with the measuring apparatus. If not for that, quantum particle movement would follow a pretty regular path with trajectories predicted by the theory. While its development lacked more concrete results, with its predictions remaining in a very limited set of results, with no relativistic generalization, this concept touched a whole generation of physicists interested in foundational questions by showing that a new conceptual scheme could, at least in principle, be developed and used to replace 'regular' quantum mechanics.

The second concerns Bell inequalities. Extremely influenced by David Bohm, in 1964 John Bell questioned if it would be possible to construct a model with hidden variables that could yield different results from ordinary quantum theory, as Bohmian mechanics does not. Although using hidden variables that were different from Bohm's (Bell's model used local hidden variables, Bohm's were non-local), he could show that, in some very specific experimental contexts, no theory using local hidden variables could predict exactly the same results as ordinary quantum theory. Even though he did not have a 'theory' in the same sense as Bohm's, his result opened up the possibility of testing if quantum mechanics could possibly be wrong, and if so, whether there would exist local hidden variables. Yet, his proposal still needed to be developed before reaching the laboratories, and it was mainly John Clauser and Abner Shimony who were responsible for this from 1969, with several experiments being performed in the following decades.

As Freire has indicated, these debates around the foundations of quantum mechanics received the stigma of being non-scientific, philosophical (Freire, 2009). While this would certainly be a problem for many physicists, Leggett, with such a unique background, became quite interested in them. Despite carrying out typical technical research on low-temperature physics, he frequently paid attention to foundational debates. The actual turning point was in 1972, after a series of lectures given by Brian Easlea, at the time also at Sussex. Easlea had first developed a career in physics, but then moved to other topics such as history and philosophy of physics and later social sciences. During the late 1960s, he lectured on a classic foundational problem, the so-called measurement problem. This contact made Leggett rethink his entire career and take a drastic position: he would no longer do the kind of physics that was published in the *Physical Review B*, the main journal for low-temperature and solid-state physics.

This was not a trivial decision. To stop doing mainstream physics and begin focusing on research that could easily be identified as at best philosophy, or at worst mumbo-jumbo, had been a problem for most scientists who chose this path. Fortunately, Leggett had an advantage that few taking the same decision as him also had, namely a permanent position. He was conscious of the problems such decisions could entail for his career, yet he knew that at least he would still keep his job. As we will see, it is interesting that despite committing himself almost full time to foundational research, this would never pose a problem for him.

While the regular path of 'dissidents' (Freire, 2009, 2015) was to focus on very specific conceptual problems of quantum theory, Leggett took a different approach. With his training in low-temperature physics, a field that had been blooming since the late 1950s, with several new problems and theoretical challenges, he chose to face one that was quite unique: the superfluidity of helium 3.[7]

Helium 3 is one of the isotopes of helium. It is extremely rare when compared to the most common isotope, helium 4.[8] Its abundance is so low that it has become one of the most expensive materials on Earth. At the same time, its applications are quite vast, including atomic bombs and nuclear fusion reactors, but more commonly it serves to refrigerate systems below 1 Kelvin, temperatures needed in almost every particle accelerator. The situation is so extreme that mining it on the Moon has even been

---

[7] For a portrait of the field at the time, see 'Solid State and temperature and low temperature physics in the USSR', organized and half written by ter Haar, Leggett's Ph.D. advisor; and 'Key problems of Physics and Astrophysics', written by Vitaly Ginzburg, who shared the 2003 Physics Nobel prize with Leggett (and also Alexei Abrikosov). See (Organization for Economic Co-Operation and Development (OECD), 1964) and (Ginzburg, 1975).

[8] 'The abundance of He4 is $10^7$ times that of He3' (Ginzburg, 1975).

considered (Wittenberg *et al.*, 1992). Recently, a shortage of He3 affected the working of several physics experimental centres.[9] However, if helium 3 is quite hard to find, its theoretical and experimental studies were abundant. By 1962, it was not clear yet whether it would have a superfluid phase or not. Dirk ter Haar described it in 1964 as follows: 'Gor'kov and Pitaevskii (1962) have studied the possibility of a transition of He3 into a superfluid state. This might happen through the formation of so-called Cooper pairs as in superconductors. ( . . . ) They estimate this transition to happen between $2 \cdot 10^{-4}$ and $8 \cdot 10^{-3}$ K.' Ginzburg also examined these, portraying them as extremely difficult problems: 'L. D. Landau told me once that his attempts to solve the problem of the second-order phase transitions had demanded greater effort than any other problem he had worked upon'. More specifically, 'It has been discussed for ten years already that the atoms of $^{3}$He may "adhere" to each other forming pairs with an integral spin and undergoing Bose–Einstein condensation ( . . . ) transform to some superfluid state. Such a state is analogous to a superconducting state but, as $^{3}$He atoms are neutral, the atom in this state must be superfluid rather than superconducting; however, superconductivity may also be called superfluidity, but in a system of charged particles. ( . . . ) Meanwhile, it was found in 1972 and 1973 that not one but two phase transitions occur in the liquid $^{3}$He under very low but yet attainable temperatures of about $2.7 \cdot 10^{-3}$ and $2.0 \cdot 10^{-3}$K (under the pressure of about 34 atm, though).' These were the works of Douglas Osheroff, Robert Richardson, and David Lee, on the experimental part, and Leggett, on the theoretical dimension, and these phase transitions are precisely what Ginzburg called 'exotic transitions'. The first three were awarded the 1996 Nobel Prize in Physics and Leggett the 2003 one. Ginzburg concluded that 'studies into the superfluidity of $^{3}$He will, undoubtedly, make up a whole new chapter in the physics of low, or better to say ultralow, temperatures'.[10]

Leggett realized that the superfluidity of He3 was one of those phenomena that was indeed unique, and that it could reveal deeper aspects of nature. In fact, in such extreme conditions, it might even be possible to show that quantum mechanics would no longer hold. For him, this was the opportunity to show that QM would break down. But how could he show this? The answer was, in some sense, quite simple. He just had to apply quantum mechanics to the problem. Well, it was far from a simple problem. The quantum explanation of superfluidity was quite new, as we have seen. So, Leggett set himself the task of 'solving' this problem, i.e. applying quantum mechanics to model and describe it. The catch to showing that QM would break down is indeed to be able to apply it 'correctly'. Then, if it was well applied to the problem and in fact the theory was not able to handle such an extreme situation, the experimental results, already available, would differ from the best theoretical models. 'And, I had actually, I got so interested in the foundations of quantum mechanics over the last few years that I had actually been intending to go off and do that. And, I thought, in fact, I actually said to myself, "When I come back from this holiday in Scotland, I'm going to sit down

[9] When it happened, the price of one litre of He3 rose to over US$2000 (Adee, 2007).
[10] All the quotations above from Ginzburg are from (Ginzburg, 1975).

and really start reading quantum measurement literature and so forth and really go into this in a big way." But, this result of Bob's quite literally struck me so surprisingly that I seriously began to consider the possibility that it was evident, the first evidence that quantum mechanics was breaking down under these very extreme—because you have to remember, you're dealing with a very dense system at very low temperatures where almost no one had been before. These were conditions which were really quite anomalous by ordinary terrestrial standards. And so, was it conceivable that quantum mechanics was actually breaking down?"[11] Everything was on track until something unexpected, at least to Leggett, happened. In the end, he failed to show that QM would not work; rather, he showed that it worked perfectly!

From the point of view of foundational research, nothing interesting happened here, but from a wider perspective, Leggett solved an extremely challenging problem, one that saw him awarded several prizes. Since then, he has always been recognized as an extremely talented physicist, and was able to secure a high-flying career, crowned in 2003 with the Nobel Prize in Physics. While not central to the later development of research around superfluidity, there is a point which is important to highlight. Together with the hope that QM would break, there is also his perception that solid-state physics and theories were just as fundamental as microphysics, in the sense that they were not just a mere application of QM, but a fundamental theory in their own right. The idea was that the properties of large-scale matter, in the context of solid-state physics, would not be just a consequence of the properties of the individual atoms, and the theories also not deducible from QM. For him, this would guarantee a consistent view. There are, of course, deeper meanings involved in this view, but with it he could both keep going with his research on superfluidity and approach a different problem, to show QM wrong.[12]

His following step as a foundational researcher was down a different path. Going to the African continent to work a non-consecutive year at the Kwame Nkrumah University of Science and Technology, in the city of Kumasi, Ghana, Leggett had much more free time than he was used to. Because of this, in addition to teaching, actually the only time he presented a course on quantum mechanics per se, he was also doing research. However, with very few resources and without being able to use the current literature, he decided to approach a topic for which a lack of availability of literature wouldn't make a difference: developing a new type of Bell inequalities. This is both an indication of his interest in the debates about the foundation of quantum mechanics, and also an indication of how this field was seen. The fact that a professional physicist

[11]  Interviews by Babak Ashrafi on 25 March 2005 and by me on 3 August 2011.

[12]  For his ideas about the fundamental aspect of solid-state physics, see mainly his 1992 article. Also, he presented those ideas in his popularization of physics book, *The Problems of Physics*, in 1987. Similar lines of thinking have been presented by both Philip Anderson in 1972 and, before, in 1961, by Brian Pippard. Recently, Joe Martin has been studying how the debates around this problem affected solid-state physics in the United States. See (Leggett, 1992), Leggett, 1987a), (Anderson, 1972), (Pippard, 1961), (Martin, 2015), and (Martin, 2018). We would like to thank Christian Joas for bringing my attention to these debates and Joas and Joe Martin for valuable discussion on this.

understood that he was able to do research in such a field without literature shows how low the perceived prestige was and how little research was being done around it. Leggett did write a paper, but only published it this century.[13]

## 19.4 HOW TO PUT SCHRÖDINGER'S CAT IN A LAB?

Upon his return to Sussex, his new research programme emerged. This time, he was not alone but accompanied by Amir Caldeira, his new Ph.D. student. Originally from Brazil[14], where later he returned to follow his career, he had been a student at the Pontifical Catholic University of Rio de Janeiro (PUC), then one of the most prestigious Departments of Physics in Brazil. In 1964, Brazil underwent a military coup, in which the democratically elected government was overthrown. The dictatorship, with direct support from the United States of America, initiated a policy of suppressing civil rights and persecuting civilians who had any sympathy for socialism and left wing parties, and later, more broadly, anyone who criticized the government. With this policy, several physicists had to flee Brazil, while others were arrested and tortured, including students. Despite not being directly affected by this climate, Caldeira was raised in this context. One year before he went to university, in 1969, Luiz Davidovich, today one of the most important Brazilian physicists, was expelled from the same university.[15]

Despite the political turmoil, PUC was a distinguished university. Apart from Nicim Zagury, who supervised Caldeira's Master's degree, he also had classes with Andre Swieca, Luciano Videira, Moyses Nussenzveig, then at Rochester, and Jayme Tiomno, former student of John Wheeler, who also had been exiled from the country a few years earlier for political reasons. After starting engineering, he later switched to physics, graduating in 1974. He then chose to join the Masters Programme in Physics at the same university, with Nicim Zagury[16] as supervisor.

In 1976, Caldeira presented his thesis, 'A study on relaxation and parametric excitation in two coupled bosonic systems'. The problem, proposed by Zagury, involved studying the effect of dissipation in a coupled bosonic system interacting

[13] See (Leggett, 2003b). For the experimental tests, see (Gröblacher, *et al.*, 2007). See also the editorial comment by Aspect (Aspect, 2007).

[14] All the biographical data comes from an interview by Olival Freire Jr. and Fabio Freitas, 12 January 2009. As the interview was conducted in Portuguese, all translations are by the author.

[15] Despite not being directly connected to Leggett and Caldeira, Davidovich would be a key member doing the theoretical part of Serge Haroche's experiments on decoherence in the 1990s, which led to the Nobel Prize for Haroche (Freire, 2015).

[16] Zagury would also be a part of the theoretical team for Haroche's experiments on decoherence.

with a reservoir, using quantum mechanics. They were trying to develop 'a systematic treatment for the study of relaxation and excitation of two coupled bosonic systems that interact with a reservoir'.[17] This research took six months longer than he expected, delaying his plans. Like most skilled students in Brazil at the time, he was eager to do a doctorate abroad. The obvious choice would be the United States of America, as the majority of Brazilian physicists had been trained there. Yet, as the USA had been directly involved in the Brazilian coup, Caldeira's generation was not so keen to study there. Also, for them, together with the political contempt, the USA did not seem to offer the same kind of personal experience as Europe could provide. Yet, the American influence over Brazil would still help lead his fate, as he spoke English. The other choice of an English speaking country that had a tradition in physics was, obviously, England. 'Why England? For a simple reason, and it was a political reason, because the United States was that thing, that prejudice against Americans. Some colleagues may say they hadn't [prejudice], but at the time this was quite common. Also, living in Europe was more interesting.'[18]

Caldeira was accepted at Sussex and received a scholarship from the Brazilian federal agency CAPES. While not his first choice, it soon became clear that his natural supervisor was Leggett. As Leggett was still in Ghana, and then in the USA for a year, Gabriel Barton became a provisional advisor. Upon his return, Leggett posed a problem about nucleation in helium 3, which should be caused by a false vacuum decay. By this time, Leggett had lost interest in it, but Caldeira was looking for a problem on phase transitions, so this would be something interesting. Leggett left Caldeira working alone on this very difficult problem: the system had 18 degrees of freedom. Despite liking it, it seemed more difficult than what was usually required for a Ph.D. Luckily, he soon got to know Terry Clark and Squids.

Squids, an acronym for Superconducting Quantum Interference Devices, are particularly sensitive magnetometers. They have several practical applications, most notably in biological systems, because of their extremely high sensitivity. Of their several applications, one possibility is as q-bits in quantum computers. While this is recent, its history goes back a little. Brian Josephson, as a doctorate student in 1962 at Trinity College, Cambridge University, at just 22 years old, began to be interested in the newly proposed concept of 'broken symmetry'. Seeking ways to observe it experimentally at the Cavendish Laboratory, Josephson realized that he could set two superconducting devices, separated by a thin insulating layer, and focus on understanding the effect that the phase had on the supercurrents. Until then, the phase of the associated wave was not regarded as having a physical meaning, being just a mathematical artefact. Josephson was able to show that in such a set-up the currents would emerge even if you had zero voltage applied and that it also would be very sensitive to the magnetic

---

[17]  For the Master's thesis, see (Caldeira, 1976).
[18]  Interview by Olival Freire Jr. and Fabio Freitas, 12 January 2009.

field (in the case of zero voltages, the remaining term of such effect is $j_z = j_i sin\varphi$[19]). The explanation is that a current emerges from the interference terms related to the currents on both sides, and the tunnelling that could occur would be a function of the phase difference among the wave functions associated with each current. However, since all electrons on each side (actually all the Cooper pairs) would behave collectively, you end up with wave functions with a single degree of freedom describing a very large number of particles and, in this sense, describing a macroscopic entity. Therefore, since this entity, namely the current, can tunnel through the barrier, you end up with macroscopic quantum tunnelling.[20]

This is the kind of problem that could have deeper implications. While the idea of a kind of macroscopic quantum phenomena had been around for some time as a way to describe and explain both superconductivity and superfluidity, this might be different. Terry Clark, an experimental physicist from Sussex, understood just that in 1976, and thought that Squids 'given suitable experimental conditions ( . . . ) should display manifestly quantum mechanical behaviour over macroscopic length scales' (Clark, 1991). This was not the first time that someone was contemplating quantum mechanical behaviour over macroscopic lengths. Erwin Schrödinger had also contemplated it in 1935, not with some kind of electronic device, but with a living–dead cat. Clark was establishing the grounds for the 'cat' to go into the lab. As was already clear to him, such an extremely speculative research programme might have been just madness, so he sought advice from an expert on foundational issues. He looked to David Bohm, who was at Bristol University, and Bohm gave him enough encouragement to pursue it further, which he did.

A couple of years later, in 1978, cats became even more afraid of physicists, as this programme received support, this time on the theoretical part. Clarke presented a course about semiconductors and, at some point, began discussing all the potentialities he had been envisioning around Squids for foundational purposes. One of the problems he presented had special appeal for our Ph.D student, Amir Caldeira. It was to understand thermal fluctuations in a Squid, and how it could affect the tunnelling effect at the Josephson junction. Clark also had talked to Leggett, who approached Caldeira suggesting it to him. This time, both would be happy about it. First, it only had one degree of freedom, as it was in principle much simpler than those helium 3 problems. Second, because Caldeira immediately saw how he could tackle this problem. Just as he did in his Master's thesis, he could use dissipation in order to understand how Squids would behave in a thermal bath. This new problem would be a breakthrough for Leggett's research programme. Together with Caldeira, they could set a new challenge to quantum mechanics, to see whether the macroscopic quantum behaviour, with

---

[19] This is known as one of the Josephson equations (1st Josephson relation), where $j$ is the current, $j_i$ is a constant known as the critical current, and $\varphi$ is the phase difference of the wave function for each side of the junction.

[20] For a broad discussion on this topic, see (Takagi, 2002).

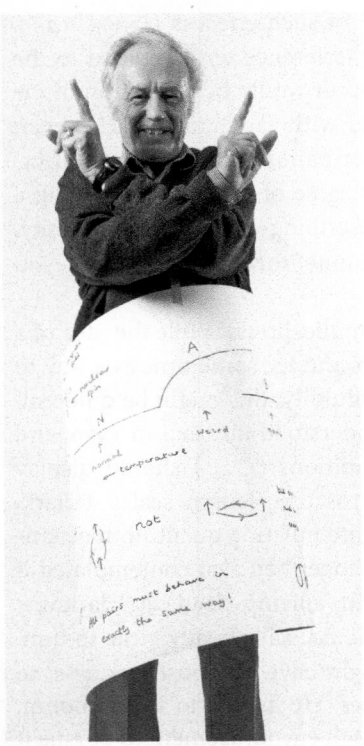

**FIGURE 19.2** Anthony Leggett in the exhibition 'Accelerating Nobels', under the project 'Nobel Drawings: Conceptual photography project with Nobel laureates', by Volker Steger, for the inauguration of the LHC. Credits: Volker Stegert. Courtesy of CERN/Volker Steger.

indeed macroscopic dimension, could be correctly described by quantum theory, and hopefully to show where it would break (Figure 19.2).

By 1980 Caldeira had finished his doctorate, and presented his dissertation on 'Macroscopic Quantum Tunnelling and Related Topics' (Caldeira, 1980). While completely aware of the foundational implications of this work, he himself was not so keen on it. In the dissertation, there are no foundational considerations whatsoever, just the application of formal techniques to semi-ideal problems. Leggett, on the other hand, took a decisive step in his career, publishing his first two articles directly related to the foundations of quantum mechanics. In the first, published in 1978, he timidly gave directions regarding research on low-temperature physics, but also indicated that Squids might pose some deep questions on the measurement problem. In the second one, published in 1980, however, he was more explicit about his intentions: he wanted to test the hypothesis of whether or not linear quantum mechanics could be applied to macroscopic systems.[21] A few months later, Caldeira's dissertation would be ready.

[21] See (Leggett, 1978) and (Leggett, 1980).

## 19.5 THE BIRTH OF DECOHERENCE

As we have mentioned, Caldeira's dissertation did not have any foundational discussion as a major topic. Yet Leggett's programme was contained within it. The dissertation revolved around two major questions. The first one, 'Is there any physical system which may exhibit quantum tunnelling on a macroscopic scale?' (Caldeira, 1980) is in a broad sense Leggett's programme, just changing quantum tunnelling for quantum behaviour, and he always believed that Squids with quantum tunnelling was the best candidate so far to study this type of quantum behaviour. From then on, Leggett would be fully dedicated to it. The second problem, 'Once a macroscopic closed system shows quantum tunnelling, would the coupling to a reservoir exert any sort of influence on the tunnelling rate?' (Caldeira, 1980) was the actual birth of decoherence for physics. While today even posing this question might seem weird, as we are so used to understanding that quantum-like properties tend to disappear because of such interactions, back then this was really an open problem.[22] Caldeira himself, after emphasizing that this was, in fact, the main problem of their work, answered 'At the end we concluded that damping always tends to suppress quantum tunnelling', adding that 'Although our last result was proved only for a specific model interaction with the reservoir we believe it can be generalized ( . . . ) however, this is a subject for more careful investigation' (Caldeira, 1980). Indeed, it would be possible to generalize it.

One year later, together they published their first letter, but the first mention of their results appeared in 1980. Roger Koch, Dale van Harlingen, and John Clarke, through preprints, mentioned that macroscopic quantum tunnelling should decrease with higher damping. Yet, about the same time, Allan Widom, with Terry Clark as co-author, in some letters argued in the opposite direction. In their second letter, in 1982, they claimed that 'In the quantum tunnelling regime, dissipation increases the barrier transmission probability', and they 'attribute the difference between the results here reported and those of Caldeira and Leggett to a divergent renormalization'. After an exchange of comments in *Physical Review Letters*, Widom and Clark concluded that they 'look forward to reading the forthcoming article by Caldeira and Leggett and remain open minded to the possibility that our statements might be in error'. They went on to add: 'However, we do not see such an error at the present time'. The 1983 article would solve this.[23]

---

[22] In some sense, even today it still is an open problem (but in a different sense). Leggett has argued that despite the extremely fast coupling of the quantum system with the environment, this coupling can be adiabatic in an extremely large number of cases, so not only in principle but also in specific experimental contexts it should be possible to deal with this level of interaction without losing the quantum-like properties of macroscopic systems. Yet, despite his position, it remains true that most believe that all quantum-like properties disappear because of such interactions and this is the base of what is called decoherence.

[23] See (Koch, van Harlingen, and Clarke, 1980), (Clark and Widom, 1981), (Widom and Clark, 1982a), (Caldeira and Leggett, 1981), (Caldeira and Leggett, 1982) and (Widom and Clark, 1982b).

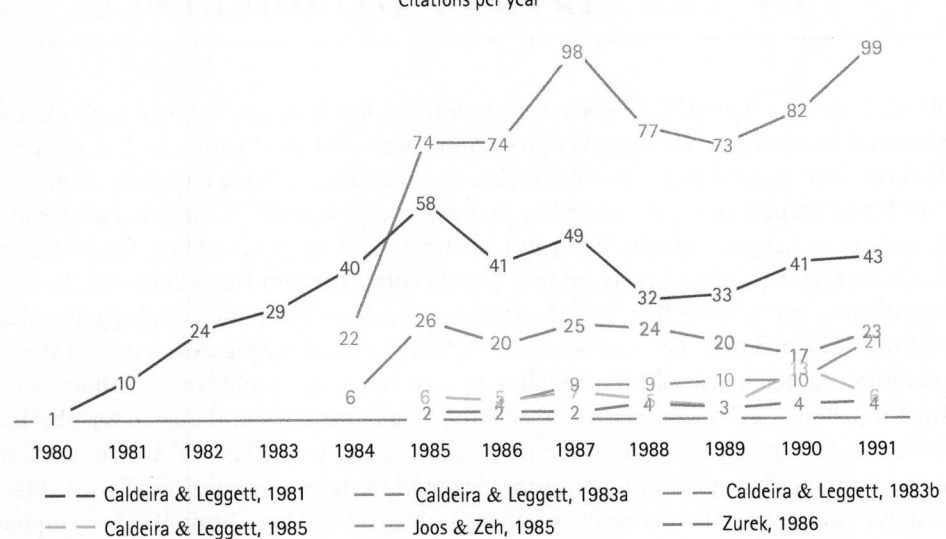

**FIGURE 19.3** Citations per year from 1980 to 1991, representing the immediate reception to the works.

*Source*: ISI/Web of Science.

Right at the beginning, Caldeira and Leggett said: '[we] attempt to motivate, define, and resolve the question "what is the effect of dissipation on quantum tunnelling"'.[24] However, while this was the general topic, the main interest was slightly wider, as we shall see in the following section. Right now it is important to emphasize how important their work was being perceived as. Leggett's new research programme was able to convince the physics community of the importance that the environment had upon quantum-like properties, and, in the same vein, that there were enough problems around it for students and researchers to dedicate their time to it.

Still in 1983, the third article arising from Caldeira's dissertation would appear in print.[25] In it, the Feynman–Vernon path integral approach was applied to study a Brownian particle motion, but as in the general programme, under the influence of the environment. Finally, their last article together would be published in 1985.[26] In this one, the theme of decoherence would become even more explicit and closer to what would be done later. This was calculating how long it would take for a quantum system to lose its quantum-like properties. They argued that the environment would serve as a sort of quantum apparatus, claims that had been presented before both by Dieter Zeh and Wojciech Zurek, but without a concrete example of how this would work and

---

[24] See (Caldeira and Leggett, 1983a).   [25] See (Caldeira and Leggett, 1983b).
[26] See (Caldeira and Leggett, 1985).

without a more complete development of the physics around it.[27] In fact, both Zeh, with his Ph.D. student Erich Joos, and Zurek would develop more technical works during the 1980s, but mostly they focused on simpler systems than the ones being modelled by Caldeira and Leggett and quite far away from any experimental tests. Joos and Zeh's main technical work during this period was (Joos and Zeh, 1985) and Zurek's was (Zurek, 1986). Joos and Zeh's dealt with a simpler system than those used by Caldeira and Leggett, and Zurek's was largely based on the previous work of Caldeira and Leggett. It is possible to infer their impacts from 1980 to 1991 from their citations dynamics presented in Figure 19.3, with Caldeira and Leggett's works receiving way more attention. With this, in no sense are we saying that these works by Joos, Zeh, and Zurek, and their other works, were not important both for the development of physics and, in particular, to the development of decoherence studies. What we mean is *that for the development of what would later be called decoherence, even more during the decade of the 1980s, Caldeira and Leggett's work set the physical basis from which other works would follow, not only those works that would later be connected to foundations of quantum mechanics, but also those connected with more practical applications of QM* and those that were seeking experimental evidence.

As it is possible to infer from the graph above, these works were extremely well received. Apart from the 1985 article, which had smaller generality, their first three articles received over 1000 citations each. Their 1983 article from the *Annals of Physics* was soon receiving over 50 citations each year, an article that talks in its introduction about Schrödinger's cat and the inapplicability of QM on the macroscopic domain. As we may see, this did not matter for the physics community, because apart from that, there was enough physics to be done around it, not just on foundational and philosophical physics. As such, the physics of decoherence indeed began here. From now on, the transition from quantum to classical was a true research programme with defined physical problems, physical methods, and also a philosophical background dispute regarding the future of QM.

# 19.6 PROVING QUANTUM MECHANICS WRONG

How come several physicists working on foundations of quantum mechanics, with ideas far more orthodox than Leggett's, had so many problems in their careers, while Leggett had none? Furthermore, his foundational work was extremely well recognized! For instance, he had a key role regarding the recognition of the foundation of physics as an autonomous field. As he tells, 'In 1984, motivated by what seemed to us a

---

[27] Camilleri presents an overview of both Zeh's and Zurek's arguments regarding how they view the philosophical implications of what would later be known as decoherence (Camilleri, 2009).

particularly foolish paper on Bell's theorem that had appeared in *PRL*, Anupam and I had written an indignant letter to the then editor of *PRL* admonishing him to apply the same standards to manuscripts in the area of quantum foundations as those used in other areas of physics. The result (which in retrospect I should have anticipated!) was that I was asked, and agreed, to become the first divisional associate editor (DAE) of *PRL* for the newly created division of quantum foundations, a post which I held until 1996.' Not only was he publicly seen as a researcher on foundations of physics (but not only as such), he also had the role of contributing to the establishment of the field as a part of mainstream physics, at least for the prestigious journal *Physical Review Letters* (*PRL*).[28]

The idea that Leggett faced no problems is even more interesting when we understand the situation a little better. The context in which he was involved was post-Bell's inequality tests. After a first period of tests during the 1970s, with almost undeniable confirmation of QM in the early 1980s, more specifically in 1982, Alan Aspect had in some sense solved Bell's conundrum. Quantum mechanics had been verified to an extent that very few loopholes remained, none of them serious enough to put in danger the meaning of the experimental results. Yet, even during the earlier decade, it was quite unfashionable to think of quantum mechanics as either wrong or incomplete. The hot topic was to understand the true meaning of nonlocality, and, as Everett's interpretation was becoming important, to understand quantum formalism with universal validity.[29]

Yet, for Leggett, none of this was particularly important. He would always recognize that QM had a very strong domain of validity, as he was still applying it to solve problems. However, much in the same way as Albert Einstein in the EPR, he was also applying it to find problems in its applications. As we have seen, he had done it before, failing to prove it wrong about the superfluidity of helium 3 (and being awarded the Nobel for this). Now he chose the more unstudied field of the applicability of QM to describe macroscopic superpositions. The outline of this challenge appeared in his very first article about foundations, in 1980. In it, he asked: 'What experimental evidence do we have that quantum mechanics is valid at the macroscopic level?'[30] Yet, this question, as posed, did not do full justice to what he was arguing. As a former researcher on superfluidity, he of course knew that QM was applicable and in fact worked at the macroscopic level, something that was already widely recognized at the time, even at textbook level. Feynman's chapter on superconductivity in 1965, for instance, was called 'The Schrödinger Equation in a Classic Context: A Seminar on Superconductivity'.[31] So, then, the question is a little more subtle. That atoms collectively behaved according to QM, even if sometimes this would happen on a (relatively) large scale, was very well known, but did true macroscopic systems behave according to QM in a linear superposition?

We first have to define what a *true* macroscopic quantum behaviour would be. Leggett, then, defined disconnectivity. In simple terms, disconnectivity describes the

---

[28] (Leggett, 2020).
[29] For this context, see Freire (2006) and Osnaghi, Freitas, and Freire (2009).
[30] (Leggett, 1980).      [31] (Feynman, Leighton, and Sands, 1965).

quantity of particles effectively interacting collectively to produce the macroscopic quantum effect.[32] For instance, in the He3 superfluidity phenomenon, despite there being many particles involved, the true interaction would be in a Cooper pair, which leads to $D = 2$. Whereas in a cat, despite the fact that we are unable to write a density matrix for all the particles in it, all of them (or at least many them) indeed interact together to form a living (or dead) cat, leading to a higher $D$. So a cat would be a great candidate on which to perform experiments to test the validity of QM on *true* macroscopic quantum systems, as Schrödinger had already realized. Yet, apart from several other problems, it can be quite hard to get a cat inside a box, as they only do what they want to, so the quest for simpler systems, but also with higher $D$, would be central for Leggett from now on.[33]

Macroscopic quantum tunnelling fitted the bill and Leggett focused on it in the following years. What made it, and more specifically coherence in Squids, so special is that the time the system takes to be damped and to lose its coherence would be much greater than other candidates, even more because of the extremely low temperatures required for the superconducting device. This would, in principle, allow one to observe quantum coherence at the macroscopic level, or at least infer it from the tunnelling effect. If we assume the universal validity of quantum mechanics, this would present no problem at all. The point that Leggett makes is that this is not a trivial assumption. To put it more precisely, there was no evidence that linear quantum mechanics would be applicable to macroscopic systems, and every test so far would not be able to differentiate a pure system from an ensemble. In his words:

> Clearly the argument as to whether the pure state $(2.7)$[34] is or is not distinguishable from a mixture is only of relevance if one believes that $(2.7)$ is the correct description in the first place. But this description only follows under the assumption that the linear laws of quantum mechanics can be applied strictly to any physical system, however macroscopic and complex. This assumption is not a trivially obvious one; it would, for example, not necessarily be a priori absurd to postulate that, at a certain level of complexity, nonlinear terms begin to play a role and cause superpositions of the form $(2.7)$ to evolve continuously into one of their branches.    (Leggett, 1980)

While not directly advocating a non-linear approach, Leggett was not at all hiding the fact that we have no secure bases to assume the universal validity of QM and, even more strongly, that it would probably break down once we could perform experiments about macroscopic tunnelling on Squids. While it was possible to use several examples

---

[32] In his book, Takagi, who collaborated with Leggett, presented a definition as 'the maximum number of those democratically counted degrees of freedom which are involved in an irreducible linear combination'. See (Leggett, 1980) for this and the rest of the paragraph, and (Takagi, 2002).

[33] In sum, to test QM macroscopically, one needs a system that is a superposition of states describing $n$ particles instead of $n$ superpositions of states describing single particles.

[34] The state 2.7 is $|apparatus\rangle = \Sigma_i c_i |X_i\rangle$, where $X_i$ is a macroscopically distinct state of the apparatus.

during the 1980s, two of them are a little more striking. The first is from their 1983 article in the *Annals of Physics*. It is rather long, but very revealing:

> Finally, it should of course be emphasized that all the calculations of this paper have been carried out within the conventional framework of quantum mechanics, that is, under the assumption that this framework can indeed be extrapolated to the macroscopic scale in the sense discussed in the Introduction. Should it eventually turn out that for a particular type of physical system quantum tunnelling is not observed under the conditions the theory predicts it should be, no doubt the most obvious inference would be the calculations, or the model on which they are based, are wrong; however, an alternative inference, which it would unwise to exclude totally a priori, would be that quantum mechanics cannot in fact be extrapolated in this way.'   (Caldeira and Leggett, 1983)

So, by no means was this a secret. It was actually quite clear, as was remarked at the end of the article. But, if his motivation was to prove QM wrong, also known as the most successful physical theory that we have ever had, how come he became so important?

This leads us to the second example. In 1983, Leggett was invited to give a course at the Nato Advanced Study Institute. That year, the theme was Percolation, Localization, and Superconductivity, and, as the editors Allen Goldman and Stuart Wolf, stated, 'the study of MQT [Macroscopic Quantum Tunnelling] is the newest subject in this grouping and is of fundamental significance for the quantum theory of measurement', but, at the same time, 'the macroscopic quantum tunnelling which is closely associated with the concept of quantum noise may determine the ultimate sensitivity of Josephson devices to electromagnetic signals'. So while this certainly could be important for foundational studies, it was clear that there could be practical and even technological applications and 'a complete understanding of the role of damping in these systems appears to require more experimental and theoretical work', making it clear that this was something rather open for both researchers and students.[35] The first series of lectures in print are from Leggett. In the first line of the introduction, he says that 'In discussions of the quantum theory of measurement, a crucial question is whether the usual laws of quantum mechanics can be applied to macroscopic bodies, and in particular, whether it is legitimate to assume the occurrence in nature of linear superpositions of states with macroscopically different properties', adding 'that this is not a matter of "quantum theology" but can be tested, at least indirectly, by experiment'. Again, he argues that his 'general approach will be to assume that the linear laws of quantum mechanics do apply without modification to macroscopic bodies and to explore the consequence of this assumption. Naturally, if the experiments were to fail to show the predicted results, the assumption might have to be re-examined' (Leggett, 1984b).

Yet, while this was at the core of his programme, other physicists did not think along the same lines. To exemplify, we may use the following lecture by Vinay Ambegoakar,

---

[35] See (Goldman and Wolf, 1984) for the quotations above.

'Quantum Dynamics of Superconductors and Tunneling between Superconductors'. Initially he wanted to make clear that both he and Leggett were doing, in some sense, the same thing: 'Since A.J. Leggett's lectures at this institute take a very much phenomenological point of view, mine should complete his rather well'. Yet, 'one matter I leave entirely to Leggett. That is the general question of whether ordinary quantum mechanics describes transitions between macroscopically distinct quantum states in superconducting devices. ( . . . ) I would be <u>most</u> surprised if it does not, and it would never occur to me to doubt that it does.' He continues pledging loyalty: 'What follows is a technical but straightforward application of the quantum mechanical machinery which—basically mysterious though it may be—we have all learned to operate with instructions from Copenhagen. As for Schrödinger's cat, my way out of that conundrum is to remark that, as a reluctant co-owner of one, I know that cats are more devious—for which read complex—than superconductors' (Ambegoakar, 1984). By no means does this indicate any kind of misunderstanding, either personal or cognitive. Both thanked each other for their respective lectures and both understood clearly what the other was doing. The fact portrayed here is that the physics community chose to separate Leggett's physics from his 'theology'. What he was doing was so important that instead of passing it strictly to the foundational domain, at the margins, they embraced it, just *pretending* that his deep philosophical insights did not exist.

Briefly, we may use one more example. In its January edition of 1984, *Physics Today* presented a report 'Physics News in 1983'. It had been organized by the American Institute of Physics for the previous 15 years, but for the first time it was presented in *Physics Today*. It was divided by fields, and each field had its contents selected by members of the American Physical Society, i.e., its subdivisions as astrophysics division, fluid dynamics, education, electron and atomic physics, etc. In the condensed matter part, Leggett's work had been chosen as noteworthy, so he wrote a short piece describing his programme, presenting similar ideas to those described above and explaining his challenge of QM. However, it is more interesting to see what the editor of this session, Miles Klein, said, dedicating one of four paragraphs to it: 'Tunneling is an important manifestation of quantum mechanical behavior and is found in nuclei, molecules, crystals, and many-electron systems such as superconducting junctions and, perhaps, in nonlinear one-dimensional conductors. Tunneling on a macroscopic scale presents, on the one hand conceptual problems associated with the foundations of quantum mechanics and on the other hand useful behavior that may be incorporated into devices such as superconducting transistors. At a finer level, tunneling now allows the production of images of surfaces on an atomic scale'. It is therefore clear that the editor was aware of foundational implications that this research might have and, while mentioning them, felt it necessary to emphasize its practical applications and technological improvements.[36]

---

[36] See (Klein, 1984) and (Leggett, 1984a).

But if we can account for how a researcher so attached to foundations of QM could become so important and recognized, the issue of why he was ignored by the 'mainstream' milieu of foundational researchers remains. It was not that he wanted to keep his distance from them. He published in the *Foundations of Physics*, the 'official' journal of this community, and he was, as we have mentioned, the first divisional associate editor for the area of foundations of physics at *Physical Review Letters*. He joined virtually every single conference on the theme, always presenting his ideas, he wrote an article for David Bohm's Festschrift, and he even discussed it in an acclaimed science popularization book published by Oxford University Press, *The Problems of Physics*, dedicating a full chapter to it, 'Skeletons in the Cupboard'.[37]

While harder to evaluate, there are a few indications of why it was like this. In *The Problems of Physics*,[38] a book dedicated to discussing in layman terms the present situation of physics and the prospects for future research, Leggett argued that 'some of the views to be explored in this chapter, particularly towards the end, would probably be characterized by the more charitable of my colleagues as heterodox, and by the less charitable quite possibly as crackpot'.[39] So, if his fellow mainstream physicists had no problem with his ideas (maybe they were the more charitable), we can only imagine why his fellow foundational researchers had problems with his thoughts. First, after the most definitive tests of Bell inequalities, 'nobody' was going against QM. The more general feeling, even during the 1970s, was that QM was strongly confirmed and those out of synchrony were being left behind. However, it is not completely true that QM was so confirmed. Quite a few names were trying to find loopholes in the experimental tests, but Leggett did not seek support among them: 'such loopholes can indeed be found, but however many have been closed ( . . . ) a sufficiently ingenious objector will almost certainly find yet more' and 'All one can say is that most of these objections seem to most physicists so contrived and *ad hoc* that in any other context they would be dismissed out of hand'.[40] Furthermore, another group becoming important were supporters of the Everett interpretation, also known as 'many-worlds'. Besides the fact that they assumed the universal validity of the linearity of quantum mechanics equations, Leggett thought of it as an 'exotic solution'. To make it clear, 'it seems to me that the many-worlds is nothing more than a verbal placebo, which gives the superficial impression of solving the problem at the cost of totally devaluing the concepts central

---

[37] See (Leggett, 1987a) for the book and (Leggett, 1987b) for the article in Bohm's volume.

[38] The book is part of a series called The Problems of Science. 'This group of OPUS books describes the current state of key scientific subjects, with special emphasis on the questions now at the forefront of research'. Aside from physics, the other volumes were on biology, chemistry, evolution, and mathematics, and the whole series has been reissued as Oxford Classic texts. So, despite the name, this was not a provocative piece, instead just a portrait of current physics as seen by one of the main theorists of the field.

[39] For this citation and the following ones in this paragraph, see (Leggett, 1987a).

[40] He went on to say that 'Whether one believes that the a priori arguments in favour of local objectivity are so compelling that it is legitimate to grasp even at such straws to save it must of course remain a matter of taste' (Leggett, 1987a).

to it, in particular, the concept of "reality"'. And, finally, 'I believe that our descendants two hundred years from now will have difficulty understanding how a distinguished group of scientists of the late twentieth century, albeit still a minority, could ever for a moment have embraced a solution which is such manifest philosophical nonsense'. Given the fact that a quite large part of the debate about foundational issues happened in informal forums, such as popularization books, as still happens nowadays, these words were of very special importance, and even more so when we realize that such non-technical texts are a very important connection among scientists and philosophers.

Finally, it is also worth mentioning the emergence of information. While Leggett never paid any attention to this, information and decoherence became intimately connected during the 1990s and later, as one can see from the name of the field itself: Quantum information. As Leggett himself claimed recently, 'Of course, in retrospect what I was seeing was the first stirrings of the quantum-information revolution that was to sweep through physics at the end of the twentieth century—certainly this was one of the most profound developments in my time, though one in which I did not really participate directly'.[41] Furthermore, the way the field developed, around practical applications of old quantum foundational challenges, strongly supported the universal validity of quantum theory. Not only were earlier experiments that supported quantum mechanics being performed in several new ways, like the same Bell inequalities but with more than two entangled particles, larger and larger entangled systems, measurements on individual systems, and so on, every single one of them further confirming quantum mechanics, each time with a greater degree of precision. Proving quantum mechanics wrong became the pursuit of few and from the 1990s onwards it was driven off the agenda altogether.

As the quantum information field practically swallowed the foundations of quantum mechanics, almost no one was still betting that quantum mechanics could be wrong.[42]

---

[41] (Leggett, 2020).

[42] After all the experimental results, this would also affect Leggett's thinking. In 2013, he claimed that 'When I first started thinking seriously about this, way back around 1980, I quite seriously hoped that when you got to the level of the so-called "flux qubit" ( . . . ) by that time something else might have happened', something else than the confirmation of quantum mechanics on yet another level. He then adds, 'Right now, it looks as if quantum mechanics is working fine at that level' (Burton, 2020). The general feeling was affected by the new zeitgeist, and so was Leggett. Still, his hopes would not change, just the feeling of when it would happen. In 1999, he concluded an article in *Physics Today* asking: 'Whither quantum mechanics in the next millennium? We do not know, of course, but here are two reasonable guesses for the short term. First, irrespective of whether or not' quantum computation' becomes a reality, the exploitation of the weird properties of entangled states is only in its infancy. Second, experimental work related to the measurement paradox will become progressively more sophisticated and eventually advance into the areas of the brain and of consciousness. This, of course, assumes that physicists will maintain their current faith in quantum mechanics as a complete description of physical reality. This is something on which I would personally bet only at even odds for the year 2100, and bet heavily against as regards the year 3000!' (Leggett, 1999). Again, in 2014, he claimed that 'In 50 years, I think there will have been a major revolution in cosmology and I think there's a small but non-zero chance that we will have pushed quantum mechanics in the direction of macroscopic world to the point it will fail and break down'. When pushed a little further as to whether or not, ultimately, he

And if this was not enough, several names that had also been involved in the decoherence approach became quite important in the information field, such as Zurek, Zeh, and David Deutsch. This led people to forget about the importance Leggett had in the development of decoherence and also defined how the earlier proponents of this approach were to become the recognized fathers of decoherence.

Philosophers themselves also seemed to have problems with Leggett's quite unique style on foundations of quantum mechanics. From the very beginning of the debates, most discussions and examples centred on extremely simple systems and many idealized situations. And, ever since, when we look at foundational debates, most cases are extremely simple. We may take several examples. David Bohm's proposal, certainly one of the most technical of all of them, was actually closer to classical mechanics, and its applications were somewhat basic. Everett's formulation was just plain linear quantum mechanics, without any more complicated applications.[43] Bell's inequalities, albeit slightly more complicated when it got to more specific models that in fact became experiments, in its general idea were quite simple, not much more complicated than EPR. Schrödinger's cat and almost every other example could ultimately fit in the general scheme of $|\Psi\rangle = c(|\uparrow\rangle + |\downarrow\rangle)$, i.e. very simple two-level systems.[44] At the same time, Leggett's foundational physics involved understanding extremely technical aspects of low-temperature physics and detailed applications of QM in not so simplified systems, as Leggett was eager to get these models into the laboratory. What we are trying to argue here is not that all philosophers did not have enough training and knowledge in physics to understand his work, as some certainly did, but that this, among the other reasons presented, would be one more reason not to pay enough attention to him. It is possible to illustrate the philosopher's attitude with one strong example. In 1987, Leslie Ballentine published one of the most comprehensive resource letters about the foundations of QM after Bell's inequalities, i.e. for the previous 20 years (Ballentine, 1987). In it, he covered virtually everything that was being debated at the time, except for Leggett's programme. There was just a single mention of him, about a very small letter that Leggett (with A. Garg[45]) contributed to a debate over an article claiming that Bell's experiments did not test local hidden variable

believed it will definitely break down, he still answered positively (Burton, 2020). So, as we argue, if even Leggett could not see QM breaking in the short term, others would naturally doubt this possibility to an even greater degree.

[43] Or, in some sense, any applications at all.

[44] This is in fact so strong that a book dedicated to teaching foundational matters was written almost entirely about two-level systems, reducing many problems in Hilbert space to regular two-dimensional vector space (Hughes, 1992).

[45] Anupam Garg collaborated with Leggett in the derivation of the so-called Leggett–Garg inequalities. These inequalities bear some similarity with Bell's inequalities, but instead of photons, they were derived with Squids in order to test whether two conditions can still be maintained alongside quantum mechanics: 1) Macroscopic realism and 2) noninvasive measurability at the macroscopic level. Since their development, in 1985, there have been several experiments ruling out both conditions. Yet, it still remains an important source of foundational experiments, just like Bell's inequalities. See (Leggett and Garg, 1985) and (Formaggio *et al.*, 2016) for one extreme case of violation of the inequalities.

theories, with no mention at all of his own programme. Ballentine knew therefore, at least at some level, that Leggett was paying attention to the foundational field. He, like several others, was just not paying any attention at all to Leggett.

## References

Adee, S. (2007). *Physics Projects Deflate for Lack of Helium-3*. Retrieved November 5, 2011, from IEEE Spectrum: http://spectrum.ieee.org/biomedical/diagnostics/physics-projects-deflate-for-lack-of-helium3

Ambegoakar, V. (1984). Quantum Dynamics of Superconductors and Tunneling between Superconductors. In A. Goldman, and S. Wolf (eds), *Localization, Percolation, and Superconductivity*, New York: Plenum Press, pp. 43–64.

Anderson, P. A. (1972). More is different. *Science*, **177**(4047), 393–96.

Aspect, A. (2007). Quantum mechanics: To be or not to be local. *Nature*, **446**, 866–67.

Ballentine, L. E. (1987). Resource Letter IQM-2: Foundations of Quantum Mechanics since the Bell inequalities. *American Journal of Physics*, **55**(9), 785–92.

Burton, H. (2020). *The Problems of Physics, Reconsidered: A Conversation with Tony Leggett*. Newcastle: Open Agenda Publishing.

Caldeira, A. (1976). *Um estudo sobre a relaxação e excitação paramétrica em dois sistemas de bosons acoplados*. Master's dissertation, Pontifícia Universidade Católica, Rio de Janeiro.

Caldeira, A. (1980). *Macroscopic Quantum Tunnelling and Related Topics*. (unpublished Doctoral dissertation or Master's thesis). University of Sussex, England.

Caldeira, A., and Leggett, A. (1981). Influence of Dissipation on Quantum Tunneling in Macroscopic Systems. *Physical Review Letters*, **46**(4), 211–14.

Caldeira, A., and Leggett, A. (1982). Comment on 'Probabilities for quantum tunneling through a barrier with linear dissipative system'. *Physical Review Letters*, **48**(22), 1571.

Caldeira, A., and Leggett, A. (1983a). Quantum tunneling in a dissipative system. *Annals of Physics*, **149**, 374–56.

Caldeira, A., and Leggett, A. (1983b). Path integral approach to quantum Brownian movement. *Physica*, **121A**, 587–16.

Caldeira, A., and Leggett, A. (1985). Influence of damping on quantum interference: an exactly soluble model. *Physical Review A*, **31**(2), 1059–66.

Camilleri, K. (2009). A history of entanglement: Decoherence and the interpretation problem. *Studies in History and Philosophy of Modern Physics*, **40** (4), 290–02.

Cohen, R., Horne, M., and Stachel, J. (eds.) (1997). *Experimental Metaphysics: Quantum Mechanical Studies for Abner Shimony, Volume One*. Dordrecht: Kluwer Academic Publishers.

Clark, T. (1991). Macroscopic Quantum Objects. In B. Hiley and F. D. Peat (eds), *Quantum Implications: Essays in honor of David Bohm*, London: Routledge, pp. 121–50.

Clark, T. D., and Widom, A. (1981). Quantum tunneling paths in superconducting weak paths. *Physical Review Letters*, **46**(26), 1704.

Daniel J. Kevles (1995 [1977]). *The Physicists: The History of a Scientific Community in Modern America*, Cambridge, MA: Harvard University Press, 489 pp

Feynman, R., Leighton, R., and Sands, M. (1965). *The Feynman Lectures on Physics—Volume III*. Redwood City: Addison-Wesley.

Formaggio, J. A., Kaiser, D. I., Murskyj, M. M., and Weiss, T. E. (2016). Violation of the Leggett–Garg Inequality in Neutrino Oscillations. *Physical Review Letters*, **117**(5), 050402.

Freire Jr., O. (1999). *David Bohm e a controvérsia dos quanta*. Campinas: CLE–Unicamp.

Freire Jr., O. (2003). A story without an ending: the quantum physics controversy 1950–1970. *Science and Education*, **12**, 573–86.

Freire Jr., O. (2004). The historical roots of 'foundations of quantum physics' as a field of research (1950–1970). *Foundations of Physics*, **34**(11), 1741–60.

Freire Jr., O. (2005). Science and exile—David Bohm, the Cold War, and a new interpretation of quantum mechanics. *Historical Studies in the Physical and Biological Sciences*, **36**(1), 1–34.

Freire Jr., O. (2006). Philosophy enters the optical laboratory: Bell's theorem and its first experimental tests (1965–1982). *Studies in History and Philosophy of Modern Physics*, **37**, 577–16.

Freire Jr., O. (2007). Orthodoxy and Heterodoxy in the Research on the Foundations of Quantum Physics: E.P. Wigner's Case. In B. de Sousa Santos (ed.), *Cognitive Justice in a Global World: Prudent Knowledges for a Decent Life*, Lanham: Lexington Books, pp. 203–24.

Freire Jr., O. (2009). Quantum dissidents: research on foundations of quantum theory *circa* 1970. *Studies in History and Philosophy of Modern Physics*, **40**, 280–89.

Freire Jr., O. (2011). Continuity and Change: Charting David Bohm's Evolving Ideas on Quantum Mechanics. In D. Krause, and A. Videira (eds), *Brazilian Studies in Philosophy and History of Science: An account of recent works*, Boston: Springer, pp. 291–99.

Freire Jr., O. (2015). *The Quantum Dissidents*. Berlin, Heidelberg: Springer.

Ginzburg, V. (1975). *Key Problems of Physics and Astrophysics*. Moscow: MIR Publishers.

Goldman, A., and Wolf, S. (1984). Preface. In A. Goldman, and S. Wolf, *Percolation, Localization, and Superconductivity*, New York: Plenum Press, pp. V–VII.

Gröblacher, S., Paterek, T., Kaltenbaek, R., Brukner, C., Zukowski, M., Aspelmeyer, M., *et al.* (2007). An experimental test of non-local realism. *Nature*, **446**, 871–75.

Hughes, R. (1992). *The Structure and Interpretation of Quantum Mechanics*. Boston: Harvard University Press.

Joos, E., and Zeh, H. D. (1985). The emergence of classical properties through interaction with the environment. *Zeitschrift für Physik B: Condensed Matter*, **59**, 223–43.

Kaiser, D. (2002). Cold war requisitions, scientific manpower, and the production of American physics after World War II. *Historical Studies on Physical Sciences*, **33**(1), 131–59.

Kaiser, D. (2005). *Pedagogy and the practice of science: historical and contemporary perspectives*. Cambridge, MA: MIT Press.

Kaiser, D. (2007). Turning physicists into quantum mechanics. *Physics World*, May, 28–33.

Kaiser, D. (Forthcoming). Training Quantum Mechanics: Enrollments and Epistemology in Modern Physics. In D. Kaiser, *American Physics and the Cold War Bubble*. Chicago: Chicago University Press.

Klein, M. (1984). Condensed Matter. *Physics Today*, January, S14.

Koch, R., van Harlingen, D. J., and Clarke, J. (1980). Quantum-noise theory for resistively shunted Josephson junction. *Physical Review Letters*, **45**(26), 2132–35.

Leggett, A. (1978). Prospects in ultralow temperature physics. *Journal de Physique*, **39**(8), 1264–1269.

Leggett, A. (1980). Macroscopic quantum systems and the quantum theory of measurement. *Progress of Theoretical Physics*, **69**, 80–100.

Leggett, A. (1984a). Macroscopic Quantum Tunneling. *Physics Today*, January, pp. S15–16.

Leggett, A. (1984b). Macroscopic quantum tunneling and related effects in Josephson systems. In A. Goldman, and S. Wolf (eds), *Percolation, Localization, and Superconductivity*, New York: Plenum Press, pp. 1–42.

Leggett, A. (1987a). *The problems of physics*. Oxford: Oxford University Press.

Leggett, A. (1987b). Reflections on the quantum measurement paradox. In B. Hiley and F. D. Peat (eds), *Quantum Implications: Essays in Honour of David Bohm*, London: Routledge, pp. 85–104.

Leggett, A. (1992). On the nature of research in condensed-state physics. *Foundations of Physics*, **22**(2), 221–33.

Leggett, A. (1999). Quantum theory: weird and wonderful, *Physics World*, **12**(12), 73–77.

Leggett, A. (2003a). *Anthony J. Leggett*. Nobel prize biographical. https://www.nobelprize.org/prizes/physics/2003/leggett/biographical/ Retrieved November 7, 2011.

Leggett, A. (2003b). Nonlocal Hidden-Variable Theories and Quantum Mechanics: An Incompatibility Theorem. *Foundations of Physics*, **33**(10), 1469–93.

Leggett, A. J. (2020). Matchmaking between Condensed Matter and Quantum Foundations, and Other Stories: My Six Decades in Physics. *Annual Review of Condensed Matter Physics*, **11**(1), 1–16.

Leggett, A. J., and Garg, A. (1985). Quantum mechanics versus macroscopic realism: Is the flux there when nobody looks? *Physical Review Letters*, **54**(9), 857–60

Martin, J. D. (2015). Fundamental Disputations. *Historical Studies in the Natural Sciences*, **45**(5), 703–57.

Martin, J. D. (2018). *Solid State Insurrection: How the science of substance made American physics matter*. Pittsburgh: University of Pittsburgh Press.

Organization for Economic Co-Operation and Development (OECD) (1964). *Solid State and Low Temperature Physics in the USSR*. Paris: OECD Publications.

Osnaghi, S., Freitas, F., and Freire Jr., O. (2009). The origin of Everettian heresy. *Studies in History and Philosophy of Modern Physics*, **40**, 97–123.

Pippard, A. B. (1961). The cat and the cream. *Physics Today*, **14**, 38–41.

Schweber S. S. (1986). The empiricist temper regnant: Theoretical physics in the United States 1920–1950, *Historical Studies in the Physical and Biological Sciences* **17**(1): 55–98.

Takagi, S. (2002). *Macroscopic Quantum Tunneling*. Cambridge: Cambridge University Press.

Widom, A., and Clark, T. (1982a). Probabilities for quantum tunneling through a barrier with linear passive dissipation. *Physical Review Letters*, **48**(2), 63–65.

Widom, A., and Clark, T. (1982b). Widom and Clark Respond. *Physical Review Letters*, **48**(22), 1572.

Wittenberg, L. J., Cameron, E. N., Kulcinski, G. L., Ott, S. H., Santarius, J. F., Sviatoslavsky, G. I., Sviatoslavsky, I. N., and Thompson, H. E. (1992). A Review of 3He Resources and Acquisition for Use as Fusion Fuel. *Fusion Technology*, **21**(4), 2230–53.

Zurek, W. H. (1986). Reduction of the wave packet: How long does it take? In G. T. More and M. O. Scully (eds), *Frontiers of nonequilibrium statistical mechanics* New York: Plenum Press, pp. 145–49.

# CHAPTER 20

## THE COPENHAGEN INTERPRETATION

### DON HOWARD

## 20.1 INTRODUCTION: THERE IS NO UNITARY COPENHAGEN INTERPRETATION

THE term 'Copenhagen interpretation' became entrenched in the later twentieth century physics and philosophy of physics literature to designate a cluster of ideas associated, rightly or wrongly, with leading thinkers centred around Niels Bohr and the Bohr Institute in Copenhagen. Foremost among those views were:

(1) complementarity
(2) a necessary role for classical modes of description in accounts of measurements
(3) the completeness of quantum mechanics
(4) objective indeterminacy
(5) wave–particle duality
(6) measurement-induced wave-packet collapse
(7) a disturbance analysis of quantum indeterminacy
(8) a privileged role for the subjective observer in the quantum realm.

Widely but wrongly assumed to represent the position of Bohr himself, several of these ideas, most notably (6), (7), and (8), were to be found more commonly in the writings of Werner Heisenberg, and core members of the Copenhagen community were badly divided over their status and validity.

How the label 'Copenhagen interpretation' came to be attached to this cluster of views is a story that has been told elsewhere (Howard, 2004; see also Camilleri, 2007, 2009). The term was first used only rather late, in a 1955 essay by Heisenberg, 'The Development of the Interpretation of the Quantum Theory', which appeared in a volume in honour of Bohr, edited by Wolfgang Pauli (Pauli, 1955).[1] Heisenberg announces in the opening paragraph that his aim is to review the history of the emergence of the modern quantum theory from 1924 to 1927, essentially the formulation of the matrix mechanics and wave mechanics formalisms, and then to examine 'the criticisms which have recently been made against the Copenhagen interpretation of the quantum theory' (Heisenberg, 1955, p. 12). The use of the definite article is clearly intended to suggest the existence of a unitary Copenhagen interpretation championed by Bohr, Heisenberg, and the other core members of the extended Copenhagen community. A few pages later he dates its birth precisely: 'What was born in Copenhagen in 1927 was not only an unambiguous prescription for the interpretation of experiments, but also a language in which one spoke about Nature on the atomic scale, and in so far a part of philosophy' (Heisenberg, 1955, p. 16). In the fall of 1955, Heisenberg still more prominently canonized the expression 'The Copenhagen Interpretation of Quantum Theory' by making it the title of the third of his Gifford Lectures, *Physics and Philosophy* (Heisenberg, 1958).

A central feature of Heisenberg's version of a Copenhagen interpretation was its premising a privileged role for the subjective observer in the quantum realm, it being the observer's intervention in the form of measurement or observation that induces wave-packet collapse. This had long been a theme in Heisenberg's writings on the interpretation of quantum mechanics, but, as we shall see, it was most definitely not Bohr's view. Indeed, from the very beginning, in 1927, this was a fundamental point of difference between Bohr and Heisenberg. Nonetheless, Heisenberg's invention of 1955 quickly became entrenched, with a wide array of authors, including, notably, David Bohm, Paul Feyerabend, and Norwood Russell Hanson, deploying the notion of a unitary Copenhagen interpretation, under that name, for a variety of rhetorical purposes (see Howard, 2004, pp. 677–680 for details).[2] More than anyone else, however, it was Karl Popper who did most to give the invention currency, as he repeatedly derided the subjectivism of the 'Copenhagen interpretation' for its having undermined the proud objectivity of physical science. Thus, in one of his most famous attacks, Popper wrote:

> This is an attempt to exorcize the ghost called 'consciousness' or 'the observer' from quantum mechanics, and to show that quantum mechanics is as 'objective' a theory as, say, classical statistical mechanics.... The opposite view, usually called the *Copenhagen interpretation* of quantum mechanics, is almost universally accepted.

---

[1] The German original of Heisenberg's essay was published the following year, using the expression, 'die Kopenhagener Deutung' (Heisenberg, 1956, p. 289).

[2] Hanson does distinguish a 'Bohr Interpretation' from the 'Copenhagen Interpretation' in a 1959 paper, but in a manner importantly different from that presented here. What distinguishes the Bohr interpretation for Hanson is, first, Bohr's insistence that all quantum mechanical descriptions concern 'phenomena'—the system of interest in a specific experimental context—and, second, 'the *excathedra* utterances of the melancholy Dane' and 'the science-fiction excesses of some of Bohr's prose' (Hanson, 1959, pp. 327–28). He does not dissociate Bohr's views from collapse interpretations.

In brief it says that *'objective reality has evaporated', and that quantum mechanics does not represent particles, but rather our knowledge, our observations, or our consciousness of particles.*    (Popper, 1967, p. 7)

Given that Bohr never endorsed a subjectivist interpretation of the quantum theory, it is a curious fact that when Popper sought to document the target of such polemics, his footnotes always led to Heisenberg and not Bohr, even though he made Bohr, by name, the chief malefactor (see Howard, 2012).

Since so much historical confusion has descended from Heisenberg's invention of the idea of a unitary Copenhagen interpretation, care will be taken in what follows to disentangle the various ideas too often conflated under that one, misleading name. Clarity and fairness of attribution oblige one to distinguish Bohr's view on interpretation from Heisenberg's. But while there was not a unitary Copenhagen interpretation, it is fair to say that there was what Heisenberg had characterized in the preface to his 1929 University of Chicago lectures, *The Physical Principles of the Quantum Theory*, as the 'Copenhagen spirit of the quantum theory' (Heisenberg, 1930, [iv]), a spirit shared among a number of physicists with strong connections to Bohr, Heisenberg, and the Bohr Institute, including, at its core, Pauli, Leon Rosenfeld, John von Neumann, and Pascual Jordan.[3] They all shared a commitment to principles (1)–(4), however much they might have interpreted them differently. But most important was their commitment to the view that quantum mechanics in its standard formulation neither allowed of nor needed any emendation or supplementation, especially not a hidden variables interpretation (see Cushing, 1994 and Beller, 1999).

Consider, now, each of the above-listed principles in turn.

## 20.2  COMPLEMENTARITY

The year 1927 was, indeed, crucial in the shaping of the Copenhagen spirit, for it was the year in which Heisenberg introduced indeterminacy (Heisenberg, 1927) and Bohr introduced complementarity, the latter in his famous 'Como' lecture, 'The Quantum Postulate and the Recent Development of Atomic Theory' (Bohr, 1928). Here is how Bohr first explained complementarity:

Now, the quantum postulate implies that any observation of atomic phenomena will involve an interaction with the agency of observation not to be neglected. Accordingly, an independent reality in the ordinary physical sense can neither be ascribed to the phenomena nor to the agencies of observation. . . .

---

[3] Other authors use the expression 'Copenhagen School' as a more all-embracing designation for the community around Bohr and Heisenberg and, thus, derivatively for the somewhat heterogeneous cluster of viewpoints espoused by them. See Camilleri (2009).

This situation has far-reaching consequences. On one hand, the definition of the state of a physical system, as ordinarily understood, claims the elimination of all external disturbances. But in that case, according to the quantum postulate, any observation will be impossible, and, above all, the concepts of space and time lose their immediate sense. On the other hand, if in order to make observation possible we permit certain interactions with suitable agencies of measurement, not belonging to the system, an unambiguous definition of the state of the system is naturally no longer possible, and there can be no question of causality in the ordinary sense of the word. The very nature of the quantum theory thus forces us to regard the space-time coordination and the claim of causality, the union of which characterizes the classical theories, as complementary but exclusive features of the description, symbolizing the idealization of observation and definition respectively.

(Bohr, 1928, p. 580)

Many have puzzled over what Bohr means here by a complementary relationship between 'space-time coordination' and 'the claims of causality', all the more so because, in later years, it became customary for Bohr and others to characterize complementarity as a relationship between observables represented by non-commuting operators, such as position and momentum, two such observables being both jointly necessary for a complete account of quantum systems and mutually exclusive. But already in the Como lecture Bohr made it clear that the latter is more or less exactly what he had in mind, writing:

In the language of the relativity theory, the content of the relations (2) [the Heisenberg position–momentum and energy–time indeterminacy relations] may be summarised in the statement that according to the quantum theory a general reciprocal relation exists between the maximum sharpness of definition of the space-time and energy-momentum vectors associated with the individuals. This circumstance may be regarded as a simple symbolical expression for the complementary nature of the space-time description and the claims of causality. At the same time, however, the general character of this relation makes it possible to a certain extent to reconcile the conservation laws with the space-time coordination of observations, the idea of a coincidence of well-defined events in a space-time point being replaced by that of unsharply defined individuals within finite space-time regions.   (Bohr, 1928, p. 582)

'Space-time coordination', then, refers simply to fixing by observation the variables—position and time—that, in classical mechanics, characterize a point on a system's trajectory, while 'the claims of causality' point towards the dynamical variables—momentum and energy—that, by satisfying a conservation law, secure the deterministic behaviour of the system.

More interesting, if even more widely misunderstood, is the concise, one-step argument that Bohr gives for complementarity:

The quantum postulate implies that any observation of atomic phenomena will involve an interaction with the agency of observation not to be neglected.

Accordingly, an independent reality in the ordinary physical sense can neither be ascribed to the phenomena nor to the agencies of observation.

Too many have read this as implying that Bohr meant to deny the reality of the quantum world, thereby endorsing an instrumentalist interpretation of the quantum theory. But that was clearly not his point, because what he writes is that neither the phenomena nor the agencies of observation possess an 'independent reality'. Instead, what Bohr is here asserting is that complementarity derives from quantum entanglement, the measurement interaction putting the measuring instruments and the observed object into a non-factorizable, entangled joint state. That might appear to be a curious reading if one thinks that entanglement was introduced only in the wake of the Einstein–Podolsky–Rosen (EPR) argument in 1935 (Einstein, Podolsky, and Rosen, 1935). In fact, already by 1927 it was widely understood that entanglement was one of the chief novelties of the quantum theory. The term 'entanglement' was first introduced by Erwin Schödinger in 1935 (Schrödinger, 1935, p. 827), but the physical fact of entanglement was well entrenched in the literature in 1927 (see Howard, 1990, 2021; Uffink, 2020, p. 246). Understanding why it is that entanglement entails complementarity requires our turning to the next of the core ideas associated with the Copenhagen way of regarding the quantum theory, Bohr's doctrine of classical concepts.

## 20.3 CLASSICAL CONCEPTS

Bohr was well known for his insistence that, in order for one to give an unambiguous and, hence, objective account of experiments in the quantum domain, one must employ classical modes of description:

> It is decisive to recognize that, *however far the phenomena transcend the scope of classical physical explanation, the account of all evidence must be expressed in classical terms.* The argument is simply that by the word 'experiment' we refer to a situation where we can tell others what we have done and what we have learned and that, therefore, the account of the experimental arrangement and of the results of the observation must be expressed in unambiguous language with suitable application of the terminology of classical physics.   (Bohr, 1949, p. 209)

There has long been confusion about what, exactly, Bohr meant by this. It might seem that he meant merely the employment of classical particle mechanics and electrodynamics. But it turns out that his point was a more subtle one. For Bohr, a classical description is one in which it is assumed that the object and the measuring instruments are not described as being in an entangled, post-measurement joint state. This is Bohr's clearest explanation of the point:

> The elucidation of the paradoxes of atomic physics has disclosed the fact that the unavoidable interaction between the objects and the measuring instruments sets an absolute limit to the possibility of speaking of a behaviour of atomic objects which is independent of the means of observation.
>
> We are here faced with an epistemological problem quite new in natural philosophy, where all description of experience has so far been based on the assumption, already inherent in ordinary conventions of language, that it is possible to distinguish sharply between the behaviour of objects and the means of observation. This assumption is not only fully justified by all everyday experience but even constitutes the whole basis of classical physics.... As soon as we are dealing, however, with phenomena like individual atomic processes which, due to their very nature, are essentially determined by the interaction between the objects in question and the measuring instruments necessary for the definition of the experimental arrangement, we are, therefore, forced to examine more closely the question of what kind of knowledge can be obtained concerning the objects. In this respect, we must, on the one hand, realize that the aim of every physical experiment—to gain knowledge under reproducible and communicable conditions—leaves us no choice but to use everyday concepts, perhaps refined by the terminology of classical physics, not only in all accounts of the construction and manipulation of the measuring instruments but also in the description of the actual experimental results. On the other hand, it is equally important to understand that just this circumstance implies that no result of an experiment concerning a phenomenon which, in principle, lies outside the range of classical physics can be interpreted as giving information about independent properties of the objects.   (Bohr, 1939a, p. 269)

On the face of it, there seems to be a contradiction here. We must describe our experiments in classical terms, meaning by means of a description that treats the object and the measuring instruments as not being entangled even though the correct, fundamental quantum description implies that they are entangled, so that observation does not yield information about the independent properties of the objects. How is this tension to be resolved?

A clue is provided by another well-known thesis of Bohr's, namely, his assertion that one has a well-defined 'phenomenon', that being a technical term in Bohr's vocabulary, only if one includes in one's description of the phenomenon an account of the specific experimental arrangements involved:

> The essential lesson of the analysis of measurements in quantum theory is thus the emphasis on the necessity, in the account of the phenomena, of taking the whole experimental arrangement into consideration, in complete conformity with the fact that all unambiguous interpretation of the quantum mechanical formalism involves the fixation of the external conditions, defining the initial state of the atomic system concerned and the character of the possible predictions as regards subsequent observable properties of that system. Any measurement in quantum theory can in fact only refer either to a fixation of the initial state or to the test of such predictions, and it is first the combination of measurements of both kinds which constitutes a

well-defined phenomenon....The conditions, which include the account of the properties and manipulation of all measuring instruments essentially concerned, constitute in fact the only basis for the definition of the concepts by which the phenomenon is described.   (Bohr, 1939b, p. 20)

The key point here is that a specification of the experimental context is necessary for defining the concepts—presumably the 'classical' concepts—whereby the phenomenon is to be described. But in what way does a determination of the experimental context define those concepts?

Recall that a classical description is, for Bohr, one that assumes that the object and the instruments are not entangled. It follows that the context defines the classical concepts by somehow fixing the manner in which the object and the instruments are to be described as being independent of one another. How that works was explained by Bohr in his reply to the EPR paper, where he carefully analysed the two different kinds of measurements that could be performed with a two-slit diffraction apparatus. Bohr noted that, if the first diaphragm, the slit in which collimates the particle beam, is firmly fixed to the lab bench, then one has an experimental arrangement suitable for a measurement of the vertical position of a particle passing through the slit, whereas if that diaphragm hangs freely on a spring, then one is, in effect, performing a measurement of the particle's momentum along the vertical axis. He then explained that in the first case, that of a position measurement, the vertical position degrees of freedom of both the diaphragm and the particle are being treated classically, hence, as if the particle and the diaphragm were not entangled, while the momentum degrees of freedom are being treated quantum mechanically, meaning as entangled. The opposite is the case when the movable diaphragm gives us a momentum measurement (see Bohr, 1935, p. 698, and for a fuller discussion, see Howard, 2021).

We now have the tools to resolve the above-mentioned seeming contradiction. An entangled, post-measurement joint state provides the correct, fundamental, quantum mechanical account of the measurement. But the degrees of freedom that must be correlated for the measurement to be the kind of measurement that it is, and only those degrees of freedom, are described as not entangled. While Bohr did not unpack this idea in more formal terms, that can be done by means of a theorem first proved in 1979 for observables with discrete spectra (Howard, 1979) and later for observables with continuous spectra (Clifton and Halvorson, 1999, 2002). Consider any pair of interacting systems and the entangled, pure, joint state that describes them exactly. Then specify a context in the form of a set of compatible, joint observables represented by commuting joint operators. The theorem asserts that there will exist a mixture over joint eigenstates of those joint operators that gives for those joint observables, but only those, exactly the same statistical predictions as the pure state. But a description in terms of a mixture can be read as assuming no entanglement with respect to the degrees of freedom that are singled out by the stipulated context and, thus, as a classical description (see Howard, 1994, 2021).

## 20.4 COMPLETENESS OF QUANTUM MECHANICS

That quantum mechanics was a complete theory, one not requiring supplementation by a hidden variables interpretation was, as mentioned above, the most central commitment uniting the core members of the Copenhagen community. Three important moments highlight this fact. The first was Louis de Broglie's presentation of a pilot-wave, hidden variables interpretation—essentially the same as David Bohm's later resurrection of the idea (Bohm, 1952a, 1952b)—at the 1927 Solvay conference (de Broglie, 1928) and the ensuing discussion in which Pauli presented what, at the time, seemed a devastating critique involving inelastic scattering (Solvay Congress, 1928, pp. 280–282), leading de Broglie to abandon for many years his interest in deterministic alternatives to standard quantum mechanics (see Cushing, 1994, pp. 118–123).

The second was von Neumann's publication in 1932 of what was widely taken to be an impossibility proof for hidden variables interpretations of quantum mechanics (von Neumann, 1932, pp. 167–173). Later criticized by Grete Hermann (Hermann, 1935) and, more famously, by John Bell (Bell, 1966) for begging the question against hidden variable theories that do not assume that the sum of the eigenvalues of two non-commuting operators should be the eigenvalue of the sum of those operators, the proof does, nonetheless, demonstrate the impossibility of dispersion-free ensembles in standard quantum mechanics and so rules out a mere hidden variables supplementation of the theory (see Bacciagaluppi and Crull, 2009; Dieks, 2017; and Seevinck, 2016). Moreover, von Neumann's prestige was such that the proof did seem to have the effect of dampening interest in hidden variable theories for two decades.

By far the most significant moment demonstrating the Copenhagen community's firm commitment to the completeness of quantum mechanics was its intense reaction to the 1935 attempted proof of the theory's incompleteness by Einstein, Podolsky, and Rosen and the more refined and sophisticated version of the proof later offered by Einstein. It would not be an overstatement to describe the reaction as mild panic. Pauli gave it vivid expression in a letter to Heisenberg of 15 June 1935, where he wrote:

> *Einstein* has once again expressed himself publicly on quantum mechanics. . . . It is well known that that is a catastrophe every time it happens. . . .
>
> At least I will concede to him that, if a student in earlier semesters would make such objections to me, I would take him for being quite intelligent and full of promise. Since as a result of this publication there exists something of a danger of confusion in public opinion—especially in America—it might be advisable to send a reply to the *Physical Review*, which I would very much like to encourage *You* to do.
>
> (Pauli, 1985, p. 402)

Heisenberg did write a reply (Heisenberg, 1935), which he shared with at least Pauli, Einstein, and Schrödinger. It is an interesting paper, containing not only an answer to

EPR but very sophisticated reflections on the possibility of a hidden variables interpretation, distinguished, among other things, by Heisenberg drawing there, for the first time, a distinction between contextual and non-contextual hidden variable theories (see Bacciagaluppi and Crull, 2021, pp. 119–125). But, unfortunately, Heisenberg's reply was never published, as he deferred to Bohr's published reply (Bohr, 1935).

The EPR argument runs as follows (Einstein, Podolsky, and Rosen, 1935). Consider a pair of previously interacting particles, I and II, that are now no longer interacting and spatially separated. Consider, as well, any pair of observables, $A$ and $B$, represented by non-commuting operators, such as position and momentum. After the interaction has ceased, one could choose to measure either $A$ or $B$ on system I, and, depending on the choice of the measurement, one can predict with certainty the value of that same observable for system II. But then, according to EPR, since the measurement on system I cannot disturb system II, there must exist an element of physical reality corresponding to the value of both $A$ and $B$ for system II. On the other hand, quantum mechanics implies that, since $A$ and $B$ are represented by non-commuting operators, a relationship between them analogous to the Heisenberg indeterminacy principle must obtain, meaning that system II cannot possess simultaneously definite values of $A$ and $B$. But the thought experiment proves that there are elements of reality corresponding to definite values of both $A$ and $B$ for system II. Therefore, since quantum mechanics does not contain a counterpart for each of these elements of physical reality, it is incomplete.

Bohr's reply (Bohr, 1935) focuses on the physical relationship between the two particles after the interaction has ceased and the question of whether a measurement performed on system I does or does not 'disturb' system II. He writes that, while there is no direct physical influence upon II from the measurement carried out on I, 'there is essentially the question of *an influence on the very conditions which define the possible types of predictions regarding the future behaviour of the system*' (Bohr, 1935, p. 700). What kind of influence is this? As Bohr had explained in the previous paragraph where he was discussing the two-slit diffraction experiment, 'we are, in the "freedom of choice" offered [the choice of an observable to measure on I] just concerned with a *discrimination between different experimental procedures which allow of the unambiguous use of complementary classical concepts*' (Bohr, 1935, p. 699). In effect, Bohr is just rehearsing here the point discussed above about the manner in which the fixing of the total experimental context by the choice of the observable to measure on system I determines which 'classical' description obtains, that in which the degrees of freedom associated with $A$ are treated classically, when it is $A$ that is measured on system I, or the degrees of freedom associated with $B$, if we measure $B$ on system I. And all of this is derivative from the fact that the interaction between I and II leaves them in an entangled joint state. As Bohr writes near the beginning of his reply:

Indeed the *finite interaction between object and measuring agencies* conditioned by the very existence of the quantum of action entails—because of the impossibility of controlling the reaction of the object on the measuring instruments if these are to serve their purpose—the necessity of a final renunciation of the classical ideal of causality and a radical revision of our attitude toward the problem of physical reality.   (Bohr, 1935, p. 697)

Or, as he had written in the Como paper, 'an independent reality in the ordinary physical sense can neither be ascribed to the phenomena nor to the agencies of observation' (Bohr, 1928, p. 580). The measurement performed on system I might not have caused a direct physical disturbance of II, but neither I nor II possessed an independent reality subsequent to their interaction, thanks to entanglement.

That entanglement was, in fact, the key point at issue in Bohr's debate with Einstein over the completeness of quantum mechanics became much clearer after Einstein explained, first in correspondence with Schrödinger in the summer of 1935 and later in print, that he had really intended a much simpler and more direct argument than that presented in the EPR paper. His preferred argument again considers the relationship between two previously interacting systems. It is based on two assumptions: (1) locality, or the absence of any subsequent sub-luminal interactions between the two; and (2) separability, or the existence of independent physical states for both. If those two conditions hold, then there exists one and only one real physical state of system II. But, depending on the choice of the observable to measure on system I, quantum mechanics describes the pair by means of two different joint state functions and, therefore, attributes two different quantum states to II. The mere fact that two different state functions for II are associated with what has to be one and only one real physical state of II alone suffices to demonstrate the incompleteness of quantum mechanics. Einstein provided what read like transcendental arguments for the necessity of both locality and separability. But, of course, the separability principle is incompatible with entanglement, and so quantum mechanics does not satisfy both principles, and Einstein's argument becomes moot (see Howard, 1985).

Einstein and Bohr continued to argue the point for many years, and, late in life, Einstein, himself, made it clear that quantum entanglement was the heart of the matter. Writing about the larger debate over the completeness of quantum mechanics, Einstein says:

> Of the 'orthodox' quantum theoreticians whose position I know, Niels Bohr's seems to me to come nearest to doing justice to the problem. Translated into my own way of putting it, he argues as follows:
> If the partial systems $A$ and $B$ form a total system which is described by its $\psi$-function $\psi(AB)$, there is no reason why any mutually independent existence (state of reality) should be ascribed to the partial systems $A$ and $B$ viewed separately, *not even if the partial systems are spatially separated from each other at the particular time under consideration.* The assertion that, in this latter case, the real situation of $B$ could not be (directly) influenced by any measurement taken on $A$ is, therefore, within the framework of quantum theory, unfounded and (as the paradox shows) unacceptable.
> By this way of looking at the matter it becomes evident that the [EPR] paradox forces us to relinquish one of the following two assertions:

(1) the description by means of the ψ-function is *complete*
(2) the real states of spatially separated objects are independent of one another.

(Einstein 1949, 681–682)

## 20.5  OBJECTIVE INDETERMINACY

It was widely assumed among the members of the Copenhagen community that a straightforward corollary of the thesis that quantum mechanics is complete is that quantum mechanical indeterminacy is an objective feature of the physical world as described by quantum mechanics and not a reflection of our ignorance of yet-to-be-discerned properties or hidden variables, a knowledge of which would fix univocally the objectively possessed values of all observables. But understanding the precise nature of objective, quantum indeterminacy proved to be a challenge.

That a quantum description of nature suggested the existence of objective randomness was clear from the time of the original Bohr model of the atom, which postulated the existence of spontaneous, random jumps of orbital electrons between different energy levels (Bohr, 1913). That this randomness was something fundamental was further suggested by Einstein's discovery in 1916 of an elegant new derivation of the Planck formula for blackbody radiation based on the introduction of probability amplitudes for spontaneous emission and induced emission and absorption of photons (Einstein, 1916). It was only after the development of the wave mechanics and matrix mechanics formalisms by 1926 that the deeper significance of the Einstein transition amplitudes began to emerge with Born's introduction of the idea that the wave function represents a probability wave in configuration space, with the square modulus of the wave function representing the probability of finding the system in a specific eigenstate (Born, 1926a, 1926b). This was followed the next year by Heisenberg's derivation of the indeterminacy principle (Heisenberg, 1927), about which he later wrote: 'It was now assumed in quantum mechanics that real states can always be represented as vectors in Hilbert space (or as "mixtures" of such vectors). The indeterminacy principle was the simple expression for this assumption' (Heisenberg, 1955, p. 15).

Since the wave function for a system of $n$ particles lives in an abstract, $3n$-dimensional configuration space rather than three-dimensional physical space and directly represents only probability amplitudes, not probabilities, themselves, questions remain about the proper understanding of quantum probabilities. If one thinks that only that which lives in three-space is objectively real, then one might incline to the view that quantum probabilities, though a necessary and unavoidable feature of the quantum description of nature, represent not real physical properties of systems but only our knowledge of those systems. This would not be in the sense of incomplete knowledge, since the completeness of quantum mechanics is taken for granted by the members of the Copenhagen community, but in the sense of our not yet knowing for sure, prior to a measurement being performed, what the outcome of the measurement will be.

Heisenberg, himself, inclined toward this view. On the other hand, Heisenberg also held that what the wave function represents between two measurements is a 'tendency', an expression that some want to read as akin to the Aristotelian notion of 'potentiality', with a measurement then representing a transition from potentiality to actuality (Heisenberg, 1958, pp. 46, 50–51; Shimony, 1983).

Bohr did not write as much as did Heisenberg about the nature of quantum probabilities. One rare occasion was a late recollection about a discussion at the 1927 Solvay meeting about how to understand the merely statistical nature of quantum predictions. Dirac had suggested that we speak of 'nature' making a choice. Heisenberg replied that it is the experimenter who makes the choice in deciding what observable to measure. Bohr comments:

> Any such terminology would, however, appear dubious since, on the one hand, it is hardly reasonable to endow nature with volition . . . while, on the other hand, it is certainly not possible for the observer to influence the events which may appear under the conditions he has arranged. To my mind, there is no alternative than to admit that, in this field of experience, we are dealing with individual phenomena and that our possibilities of handling the measuring instruments allow us only to make a choice between the different complementary types of phenomena we want to study.   (Bohr, 1949, p. 223)

One recalls Bohr's often-made point that entanglement, or 'an interaction with the agency of observation not to be neglected', entails that instrument and object lack independent physical reality and, further, that there will be a complementarity between 'space-time description' and the 'claims of causality'. It is precisely that complementarity that forces us to locate a multi-particle wave function not in three-space, but in configuration space. One could well imagine, then, that, for Bohr, the criterion for objectively real physical descriptions would not be that they are formulated in physical space-time. If so, Bohr might well have been comfortable with taking quantum probability amplitudes that live in configuration space as being as real as we can get in quantum mechanics.

# 20.6 WAVE–PARTICLE DUALITY

'Wave–particle duality' is the expression adopted in the 1920s to name the fact that, in some sense, quantum systems exhibit both wave-like and particle-like behaviours in different circumstances, as when, in the two-slit diffraction experiment, one gets a diffraction pattern when one does not monitor through which slit in the second diaphragm the systems pass and no diffraction pattern when one does. It is often wrongly if understandably conflated with complementarity, which latter asserts a complementary relationship between any pair of observables represented by non-commuting operators.

In the form in which it is most commonly referenced, the idea was first introduced by de Broglie in his 1924 doctoral dissertation, where he postulated that it was not only electromagnetic radiation that exhibited such a dual nature but that all massive particles had associated with them a real wave with a wavelength related to the particle's mass by $\lambda = (h/mv)(1 - v^2/c^2)^{1/2}$ (de Broglie, 1925). Davisson and Germer's demonstration of electron diffraction in 1927 was taken as experimental confirmation of de Broglie's hypothesis (Davisson and Germer, 1927a, 1927b). Schrödinger's development of the wave mechanics formalism was partially inspired by the work of de Broglie, but the fact that the wave function is defined in configuration space and not in three-space and that, on the Born interpretation, it represents only a probability amplitude implies that what is responsible for the wave-like behaviour is not, contrary to de Broglie's original hypothesis, a real wave in three-space.

While de Broglie fully deserved the 1929 Nobel prize for discovering the wave nature of electrons, the idea that something wave-like was associated with material particles had been in the air for a considerable time. It is a crucial part of the Bohr–Kramers–Slater theory of 1924, which postulated that virtual radiation fields associated with two atoms determine the emission and absorption probabilities for the exchange of radiation between them (Bohr, Kramers, Slater, 1924). And, while he never published the idea, Einstein was known at the time to have been interested in the idea of 'guiding fields'. Recalling discussions in Einstein's Berlin colloquium, Wigner writes:

> He was very early aware of the wave–particle duality of the behaviour of light (and also of particles); in their propagation they show a wave character and show, in particular, interference effects. Their emission and absorption are instantaneous, they behave at these events like particles. In order to explain this duality of their behaviour, Einstein proposed the idea of a 'guiding field' (*Führungsfeld*). This field obeys the field equation for light, that is, Maxwell's equation. However, the field only serves to guide the light quanta or particles; they move into the regions where the intensity of the field is high.   (Wigner, 1980, p. 463)

As it happens, Einstein had first speculated about something like wave–particle duality as early as 1909 (see Klein, 1964 and Howard, 1990).

It is noteworthy that Bohr placed little emphasis on wave–particle duality and discussed it only infrequently. When he did, as in the Como paper, it was mainly to emphasize the limited, merely 'symbolic' character of either the wave or particle picture taken alone, since each, in its own way, is a purely classical, causal, space-time mode of description (Bohr, 1928, p. 581), and to emphasize that a complete account of quantum phenomena requires the full quantum theory, which implies a 'renunciation' of a simultaneously causal and space-time description and embraces the complementarity between these two ways of regarding nature (Bohr, 1928, p. 580).

# 20.7 MEASUREMENT-INDUCED
# WAVE-PACKET COLLAPSE

After complementarity, no idea is more strongly associated with what came to be regarded as the Copenhagen interpretation than measurement-induced wave-packet collapse. The notion that measurement occasioned a discontinuous, nonlinear transformation of the wave function from a pure case to a mixture, one component of which manifested itself as the measurement result, had been around since at least 1927, when Heisenberg introduced it in his paper on the indeterminacy principle. It was assumed for the obvious reason that, when we look, we seem nearly always to find a definite measurement outcome. Linear, Schrödinger dynamics cannot explain that, because it evolves superpositions always and only into superpositions. The idea was first given a more rigorous, formal expression by von Neumann in 1932 in his axiomatization of quantum mechanics, this in the form of what he terms 'Process 1', which is contrasted with ordinary Schrödinger evolution, or 'Process 2' (von Neumann 1932, 186). Wave-packet collapse was thus canonized as an essential part of standard quantum mechanics, a status further reinforced by Fritz London and Edmund Bauer's detailed analysis of the measurement process in 1939 (London and Bauer, 1939).

But even though collapse became, in this way, a central feature of textbook quantum mechanics, one had to acknowledge its somewhat ad hoc character. After all, a measurement is just a physical interaction, and the historical accident of the momentary presence of conscious, human observers in the universe should make no difference in the operation of fundamental, physical laws. So why should there be a different law governing this one subclass of physical interactions that we designate as measurements? That puzzle is what gave rise in later years to so-called 'no-collapse' interpretations of quantum mechanics, including, most importantly, Bohm-type hidden variable theories (Bohm, 1952a, 1952b) and Everettian many-worlds or relative-state interpretations (Everett, 1957). Others sought to explain collapse by means of an ordinary physical process, such as gravitation (Károlyházy, 1966; Penrose, 1986). Some even suggested that collapse was triggered in the consciousness of the subjective observer (Wigner, 1962).

However central wave-packet collapse might have been in what Heisenberg called the Copenhagen interpretation, Bohr never endorsed it. As was discussed earlier, Bohr had a very different way of understanding the quantum physics of measurement. Bohr starts from the fact that measurement, being a physical interaction governed by linear, Schrödinger dynamics, yields an entangled, pure, post-measurement joint state for the object and the measuring instruments. Our having a well-defined 'phenomenon' requires a specification of the experimental circumstances, namely, which object and instrument degrees of freedom must be correlated for it to be the kind of measurement that it is. That specification then permits a description in which, while the entangled joint state is, in principle, the correct description, one can, in an 'as-if' sense, describe those correlated degrees of freedom 'classically' by means of a mixture over corresponding, joint, object–instrument eigenstates. For the object and instrument degrees of freedom correlated in the measurement, and only for those, that mixture gives exactly the same statistical predictions as the entangled, pure, post-measurement joint state. Thus, in Bohr's view, the only dynamical law is the Schrödinger equation and there is no collapse. Instead, one describes the post-measurement joint state as a mixture, but that mixture is a kind of fiction that works only for the degrees of freedom correlated in that specific measurement.

# 20.8 DISTURBANCE ANALYSIS OF QUANTUM INDETERMINACY

What is the origin of quantum indeterminacy? Heisenberg often sought to explain it in terms of an observation's physical disturbance of the observed object. In his 1929 University of Chicago lectures, Heisenberg illustrated the idea with the example of trying to measure an electron's momentum with a microscope (Heisenberg, 1930, pp. 20–23). Let $\lambda$ be the wavelength of the light illuminating the electron, with $\varepsilon$ being the angular spread of the scattered photons as determined by the size of the microscope's aperture. Classical optical laws for the resolving power of a microscope yield an uncertainty in the position measurement of $\Delta x = \lambda/\sin \varepsilon$. Assume that one photon is scattered from the electron which receives a Compton recoil of roughly $h/\lambda$. Since the direction of the scattered photon is uncertain, there will be an uncertainty in the recoil of $\Delta p \sim (h/\lambda)\sin \varepsilon$. From this it follows that $\Delta x \, \Delta p \sim h$ (Heisenberg, 1930, p. 21).

This illustration has, itself, become deeply entrenched in the textbook literature. But there is what one might think to be an obvious conceptual problem with it, in that it tacitly assumes that the collision of the photon and the electron changes, in a way that cannot be precisely fixed, what had been a precise, pre-measurement value of the electron's momentum. The problem is that the very indeterminacy principle that Heisenberg seeks to explain in this way rules out the existence of a sharp, pre-measurement value of the electron's momentum. Thus, as an explanation of indeterminacy, the microscope thought experiment is incoherent, as are many other handwaving arguments of a similar type.

Bohr recognized the problem with the disturbance analysis of indeterminacy from the very start. In the Como paper, which was first given as a lecture in September 1927, just six months after Heisenberg submitted his classic paper on indeterminacy (Heisenberg, 1927), Bohr wrote the following about the microscope thought experiment and a thought experiment for determining the electron's momentum by measuring the Doppler effect on the scattered radiation:

The essence of this consideration is the inevitability of the quantum postulate in the estimation of the possibilities of measurement. A closer investigation of the possibilities of definition would still seem necessary in order to bring out the general complementary character of the description. Indeed, a discontinuous change of energy and momentum during observation could not prevent us from ascribing accurate values to the space-time coordinates, as well as to the momentum-energy components before and after the process. The reciprocal uncertainty which always affects the values of these quantities is, as will be clear from the preceding analysis, essentially an outcome of the limited accuracy with which changes in energy and momentum can be defined.   (Bohr, 1928, pp. 582–583)

Bohr here explains the conceptual error in the thought experiment and emphasizes that a deeper understanding of the indeterminacy principle requires our examining not only the 'possibilities of measurement' but also the 'possibilities of definition', meaning the complementary relationship between position and momentum.

## 20.9 Privileged Role for the Subjective Observer

A final hallmark of what Heisenberg called the Copenhagen interpretation is the claim that, unlike in classical physics, the subjective observer plays a privileged role in the quantum realm. But different thinkers characterized that privileged role in different ways. An extreme version of this view is the suggestion from Wigner that wave-packet collapse is initiated by the registration of a measurement result in the subjective consciousness of the observer. That was not Heisenberg's view.

That the observer might play a crucial role in quantum mechanics is suggested by the concern mentioned above that, from a purely physical point of view, measurement interactions differ in no essential way from other interactions, raising the question why a different, nonlinear, process would be required to describe measurement, one leading to collapse of the wave packet. The presence of a sentient observer might seem to be the only circumstance implying a unique status for measurement interactions. How, then, properly, to incorporate the observer into the measurement process?

In von Neumann's treatment of the measurement process a key point is the invocation of the principle of psycho-physical parallelism, which asserts that subjective mental processes are strictly correlated with purely physical processes (von Neumann, 1932, p. 223). That means that, when the observer is incorporated into the description of the measurement process, it is only the observer's physical aspect that plays a role, not the observer's subjective consciousness. Another key point is that, when the observer is brought into the picture, there must be considerable latitude in where one draws the boundary between what we designate as the 'observer' and the 'measuring instruments', since, were this not so, the principle of psycho-physical parallelism might be violated (von Neumann, 1932, p. 224). The bulk of the formal work in von Neumann's treatment of the measurement process is devoted to proving that such is, indeed, the case (von Neumann, 1932, p. 225–232).

But this still does not give us an answer to the question why the presence of a human observer should make such a difference. Heisenberg attempted to answer that question in the very essay in which he first introduced the name 'the Copenhagen interpretation'. He starts by considering a classical description of a measurement, using the example of the emission of an electron from a piece of metal being registered by the darkening of a photographic plate. He notes that, for the registration to take place, which is a thermodynamically irreversible process, the metal (the observed object) and the plate (the measuring instrument), must be coupled with the environment. But then, given our ignorance of the detailed microstate of the environment, we have to describe the combination of object plus instrument using a Gibbsian canonical ensemble. Over time, that ensemble will contain a growing proportion of states that include a blackened plate, but it will never evolve into an ensemble in which all states correspond to a blackened plate. So far this description of the measurement is 'objective'. But, says Heisenberg, it also contains a 'subjective element'.

> Namely, in the absence of an observer, the mathematical representation of the system would go on changing continuously, in the way we have outlined. If, however, the observer is present, he will suddenly register the fact that the plate is blackened. The transition from the possible to the actual is thereby completed as far as he is concerned; he correspondingly alters the mathematical representation discontinuously, and the new ensemble contains only the blackened photographic plate. This discontinuous change is naturally *not* contained in the mechanical equations of the system; it corresponds exactly to the 'reduction of wave-packets' in the quantum theory.... We see from this that the characterization of a system by an ensemble not only specifies the properties of this system, but also contains information about the extent of the observer's knowledge of the system.
> (Heisenberg, 1955, p. 26)

The situation in the quantum theory is analogous, where we use mixtures to describe the object plus instrument in interaction with the environment.

> The compound system of system and measuring apparatus is therefore now described mathematically by a mixture, and thus the description contains, besides its objective features, also the previously discussed statements about the observer's knowledge. If the observer later registers a certain behaviour of the measuring apparatus as actual, he thereby alters the mathematical representation discontinuously, because a certain one among the various possibilities has proved to be the real one. The discontinuous 'reduction of wave-packets', which cannot be derived from Schrödinger's equation, is thus, exactly as in Gibbs thermodynamics, a consequence of the transition from the possible to the actual.   (Heisenberg, 1955, p. 27)

Heisenberg's analysis of the measurement process is curious in two respects. The first is that it posits a role for the observer in quantum theory that is basically the same as in classical mechanics, so we still do not have an explanation of what might be distinctive about measurements in the quantum world. The second is that Heisenberg explicitly attributes this analysis to Bohr (Heisenberg, 1955, pp. 25, 27), when, in fact, it is importantly different from Bohr's account of quantum measurements in several crucial respects. The most salient difference is that Bohr always emphasized the radical novelty in what the quantum theory implies regarding measurements, namely, that, unlike in classical physics, the object and the measuring instrument form an entangled pair, meaning that it is no longer possible to speak of measurement as revealing independent properties of the observed object. Bohr was quoted above as saying:

> We are here faced with an epistemological problem quite new in natural philosophy, where all description of experience has so far been based on the assumption, *already inherent in ordinary conventions of language*, that it is possible to distinguish sharply between the behaviour of objects and the means of observation. This assumption is not only fully justified by all everyday experience *but even constitutes the whole basis of classical physics*.   (Bohr, 1939a, p. 269)

Thus, for Bohr, the role of the observer in quantum theory must be radically different from that in classical physics. For Bohr, that difference is not that the presence of the observer induces wave-packet collapse. For Bohr, the difference is that the human observer, regarded as a physical system, becomes deeply entangled with the observed object.

But then it is also noteworthy that Bohr, in fact, never wrote about the role of a subjective observer in quantum theory. Here, as in every discussion of the topic, Bohr spoke only about the measuring instruments or the 'means' or 'agency' of observation. For Bohr, the subjective observer plays no special role at all. Thus, in one 1958 essay he wrote:

> Far from involving any special intricacy, the irreversible amplification effects on which the recording of the presence of atomic objects rests rather remind us of the essential irreversibility inherent in the very concept of observation. The description of atomic phenomena has in these respects a perfectly objective character, in the sense that no explicit reference is made to any individual observer and that therefore, with proper regard to relativistic exigencies, no ambiguity is involved in the communication of information.   (Bohr, 1958a, pp. 310–311)

And in another essay from the same year he explained:

> In view of the influence of the mechanical conception of nature on philosophical thinking, it is understandable that one has sometimes seen in the notion of complementarity a reference to the subjective observer, incompatible with the objectivity of scientific description. Of course, in every field of experience we

must retain a sharp distinction between the observer and the content of the observations, but we must realize that the discovery of the quantum of action has thrown new light on the very foundation of the description of nature, and revealed hitherto unnoticed presuppositions to the rational use of the concepts on which the communication of experience rests. In quantum physics, as we have seen, an account of the functioning of the measuring instruments is indispensable to the definition of phenomena and we must, so-to-say, distinguish between subject and object in such a way that each single case secures the unambiguous application of the elementary physical concepts used in the description. Far from containing any mysticism foreign to the spirit of science, the notion of complementarity points to the logical conditions for description and comprehension of experience in atomic physics. (Bohr, 1958b, p. 172)

Bohr could hardly have been more explicit about the subjective observer playing no role in quantum theory.

## 20.10 CONCLUSION: WHAT'S IN A NAME?

Contrary to what Heisenberg insinuated, there is, as we have seen, no unitary Copenhagen interpretation.[4] Though Heisenberg and Bohr agreed on such points as the completeness of quantum mechanics, they disagreed on other crucial points, such as measurement-induced wave-packet collapse and a privileged role for the subjective observer. However deeply entrenched the name 'Copenhagen interpretation' has become, the causes of clarity of thought and historical accuracy would perhaps best be served by our simply not using that expression and speaking, instead, about the Bohr interpretation and the Heisenberg interpretation. Respecting their disagreements in no way diminishes our appreciation of the way in which both gave expression to the 'Copenhagen spirit'.

## ACKNOWLEDGEMENT

My sincere thanks to Christian Joas for his many, very helpful comments and suggestions on an early draft of this article.

---

[4] Heisenberg's possible motivations for the invention of a unitary Copenhagen interpretation are discussed in Howard (2004).

## REFERENCES

Bacciagaluppi, G., and Crull, E. (2009). Heisenberg (and Schrödinger and Pauli) on Hidden Variables. *Studies in History and Philosophy of Modern Physics*, 40, 374–382.

Bacciagaluppi, G., and Crull, E. (2021). *The Einstein Paradox: The Debate on Nonlocality and Incompleteness in 1935*. Cambridge: Cambridge University Press.

Bell, J. S. (1966). On the Problem of Hidden-Variables in Quantum Mechanics. *Reviews of Modern Physics*, 38, 447–452.

Beller, M. (1999). *Quantum Dialogue: The Making of a Revolution*. Chicago: University of Chicago Press.

Bohm, D. (1952a). A Suggested Interpretation of the Quantum Theory in Terms of 'Hidden Variables. I. *Physical Review*, 85, 166–179.

Bohm, D. (1952b). A Suggested Interpretation of the Quantum Theory in Terms of 'Hidden Variables. II. *Physical Review*, 85, 180–193.

Bohr, N. (1913). On the Constitution of Atoms and Molecules, Part I. *Philosophical Magazine*, 26(151), 1–24.

Bohr, N. (1928). The Quantum Postulate and the Recent Development of Atomic Theory. *Nature* (Suppl.), 121, 580–590.

Bohr, N. (1935). Can Quantum-Mechanical Description of Physical Reality Be Considered Complete? *Physical Review*, 48, 696–702.

Bohr, N. (1939a). Natural Philosophy and Human Cultures. In *Comptes Rendus du Congrès International de Science, Anthropologie et Ethnologie*, Copenhagen: Einar Munksgaard, pp. 86–95. Reprinted in *Nature*, 143(1939), 268–272.

Bohr, N. (1939b). The Causality Problem in Atomic Physics. In *New Theories in Physics*, Paris: International Institute of Intellectual Co-operation, pp. 11–30.

Bohr, N. (1949). Discussion with Einstein on Epistemological Problems in Atomic Physics. In Schilpp (1949), pp. 199–241.

Bohr, N. (1958a). Quantum Physics and Philosophy – Casuality and Complementarity. In R. Klibansky (ed.), *Philosophy in the Mid-Century, A Survey*, Florence, Italy: La nuova Italia editrice, pp. 308–314.

Bohr, N. (1958b). On Atoms and Human Knowledge. *Daedalus*, 87, 164–175.

Bohr, N., Kramers, H. A., and Slater, J. C. (1924). The Quantum Theory of Radiation. *Philosophical Magazine*, 47, 785–802.

Born, M. (1926a). Zur Quantenmechanik der Stossvorgänge. *Zeitschrift für Physik*, 37, 863–867.

Born, M. (1926b). Quantenmechanik der Stossvorgänge. *Zeitschrift für Physik*, 38, 803–827.

Camilleri, K. (2007). Bohr, Heisenberg and the Divergent Views of Complementarity. *Studies in History and Philosophy of Modern Physics*, 38, 514–528.

Camilleri, K. (2009). Constructing the Myth of the Copenhagen Interpretation. *Perspectives on Science*, 17, 26–57.

Clifton, R., and Halvorson, H. (1999). Maximal Beable Subalgebras of Quantum Mechanical Observables. *International Journal of Theoretical Physics*, 38, 2441–2484.

Clifton, R., and Halvorson, H. (2002). Reconsidering Bohr's reply to EPR. In T. Placek and J. Butterfield (eds), *Non-locality and Modality*, Dordrecht and Boston: Kluwer, pp. 3–18.

Cushing, J. T. (1994). *Quantum Mechanics: Historical Contingency and the Copenhagen Hegemony*. Chicago: University of Chicago Press.

de Broglie, L. (1925). Recherches sur la Théorie des Quanta. *Annales de Physique*, **10**(3), 22–128.

de Broglie, L. (1928). La Nouvelle Dynamique des Quanta. In Solvay Congress (1928), 105–132.

Davisson, C., and Germer, L. (1927a). The Scattering of Electrons by a Single Crystal of Nickel. *Nature*, **119**, 558–560.

Davisson, C., and Germer, L. (1927b). Diffraction of Electrons by a Crystal of Nickel. *Physical Review*, **30**, 705–740.

Dieks, D. (2017). Von Neumann's Impossibility Proof: Mathematics in the Service of Rhetorics. *Studies in History and Philosophy of Modern Physics*, **60**, 136–148.

Einstein, A. (1916). Strahlungs-Emission und -Absorption nach der Quantentheorie. *Deutsche Physikalische Gesellschaft. Verhandlungen*, **18**, 318–323.

Einstein, A. (1949). Remarks Concerning the Essays Brought together in this Co-operative Volume. In Schilpp (1949), pp. 665–688.

Einstein, A., Podolsky, B., and Rosen, N. (1935). Can Quantum-Mechanical Description of Physical Reality Be Considered Complete? *Physical Review*, **47**, 777–780.

Everett, H. (1957). 'Relative State' Formulation of Quantum Mechanics. *Reviews of Modern Physics*, **29**, 454–462.

Hanson, N. R. (1959). Five Cautions for the Copenhagen Interpretation's Critics. *Philosophy of Science*, **26**, 325–337.

Heisenberg, W. (1927). Über den anschaulichen Inhalt der quantentheoretischen Kinematik und Mechanik. *Zeitschrift für Physik*, **43**, 172–198.

Heisenberg, W. (1930). *The Physical Principles of the Quantum Theory*. Chicago: University of Chicago Press.

Heisenberg, W. (1935). Ist eine deterministische Ergänzung der Quantenmechanik möglich? In Pauli (1985), pp. 409–418. English translation: 'Is a Deterministic Completion of Quantum Mechanics Possible?' In Bacciagaluppi and Crull (2021), pp. 257–270.

Heisenberg, W. (1955). The Development of the Interpretation of the Quantum Theory. In Pauli (1955), pp. 12–29.

Heisenberg, W. (1956). Die Entwicklung der Deutung der Quantentheorie. *Physikalische Blätter*, **12**, 289–304,

Heisenberg, W. (1958). The Copenhagen Interpretation of Quantum Theory. In *Physics and Philosophy. The Revolution in Modern Science*, New York: Harper and Row, The Gifford Lectures at St. Andrews, winter term, 1955–1956.

Hermann, G. (1935). Die naturphilosophischen Grundlagen der Quantenmechanik. *Abhandlungen der Fries'schen Schule*, **6**, 75–152.

Howard, D. (1979). *Complementarity and Ontology: Niels Bohr and the Problem of Scientific Realism in Quantum Physics*. Dissertation. Boston University.

Howard, D. (1985). Einstein on Locality and Separability. *Studies in History and Philosophy of Science*, **16**(3), 171–201.

Howard, D. (1990). 'Nicht sein kann was nicht sein darf,' or the Prehistory of EPR, 1909–1935: Einstein's Early Worries about the Quantum Mechanics of Composite Systems. In A. Miller (ed.), *Sixty-Two Years of Uncertainty: Historical, Philosophical, and Physical Inquiries into the Foundations of Quantum Mechanics*, New York: Plenum, pp. 61–111.

Howard, D. (1994). What Makes a Classical Concept Classical? Toward a Reconstruction of Niels Bohr's Philosophy of Physics. In J. Faye and H. Folse (eds), *Niels Bohr and Contemporary Philosophy*, Dordrecht: Kluwer, pp. 201–229.

Howard, D. (2004). Who Invented the Copenhagen Interpretation? A Study in Mythology. *Philosophy of Science*, **71**, 669–682.

Howard, D. (2012). Popper and Bohr on Realism in Quantum Mechanics. *QUANTA*, **1**, 33–57. (http://dx.doi.org/10.12743 per cent2Fquanta.v1i1.9)

Howard, D. (2021). Complementarity and Decoherence. In G. Jaeger *et al.* (eds), *Quantum Arrangements*, New York: Springer. (Forthcoming.)

Károlyházy, F. (1966). Gravitation and Quantum Mechanics of Macroscopic Objects. *Il Nuovo Cimento A*, **42**, 390–402.

Klein, M. J. (1964). Einstein and the Wave-Particle Duality. *The Natural Philosopher*, **3**, 3–49.

London, F., and Bauer, E. (1939). *La Théorie de l'observation en mécanique quantique*. Paris: Hermann & Cie.

Pauli, W. (ed.) (1955). *Niels Bohr and the Development of Physics*. London: Pergamon.

Pauli, W. (1985). *Wissenschaftlicher Briefwechsel mit Bohr, Einstein, Heisenberg u.a./Scientific Correspondence with Bohr, Einstein, Heisenberg a.o.* Vol. 2, *1930–1939*. Karl von Meyenn, ed. Berlin: Springer-Verlag.

Penrose, R. (1986). Gravity and State Vector Reduction. In R. Penrose and C. J. Isham (eds), *Quantum Concepts of Space and Time*, Oxford: Oxford University Press, pp. 129–146.

Popper, K. (1967). Quantum Mechanics Without 'the Observer'. In M. Bunge (ed.), *Quantum Theory and Reality*, New York: Springer, pp. 1–12.

Schilpp, P. A. (ed.) (1949). *Albert Einstein: Philosopher-Scientist*. Evanston, IL: The Library of Living Philosophers.

Schrödinger, E. (1935). Die gegenwärtige Situation in der Quantenmechanik. *Die Naturwissenschaften*, **23**, 807–812, 823–828, 844–849.

Shimony, A. (1983). Reflections on the Philosophy of Bohr, Heisenberg and Schrödinger. In R. S. Cohen and L. Laudan (eds), *Physics, Philosophy and Psychoanalysis*, Dordrecht: D. Reidel, pp. 209–221.

Seevinck, M. (2016). Challenging the Gospel: Grete Hermann on von Neumann's No-Hidden-Variables Proof. In E. Crull and G. Bacciagaluppi (eds), *Grete Hermann—Between Physics and Philosophy*, Dordrecht: Springer, pp. 107–117.

Solvay Conference (1928). *Electrons et Photons: Rapports et Discussions du Cinquième Conseil de Physique tenu à Bruxelles du 24 au 29 Octobre 1927 sous les Auspices de l'Institut International de Physique Solvay*. Paris: Gauthier-Villars.

Uffink, J. (2020). Schrödinger's Reaction to the EPR Paper. In M. Hemmo and O. Shenker (eds), *Quantum, Probability, Logic: The Work and Influence of Itamar Pitowsky*, Cham: Springer, pp. 545–566.

von Neumann, J. (1932). *Mathematische Grundlagen der Quantenmechanik*. Berlin: Julius Springer.

Wigner, E. (1962). Remarks on the Mind-Body Question. In I. J. Good (ed.), *The Scientist Speculates*, London: Heinemann, pp. 284–301.

Wigner, E. (1980). Thirty Years of Knowing Einstein. In H. Woolf (ed.), *Some Strangeness in the Proportion: A Centennial Symposium to Celebrate the Achievements of Albert Einstein*, Reading, MA: Addison-Wesley, pp. 461–468.

# CHAPTER 21

# COPENHAGEN AND NIELS BOHR

ANJA SKAAR JACOBSEN

THE Danish capital Copenhagen is a particularly important place in the history of the interpretation of quantum physics for at least three reasons. First, this is where discussions about the interpretation of quantum mechanics started and where they were initially located. The first animated discussions about how to understand Erwin Schrödinger's wave mechanics took place between Schrödinger, Niels Bohr, and Werner Heisenberg during a visit by Schrödinger to Copenhagen in the fall of 1926. Subsequently epistemological ideas were formed among some of the founding fathers of quantum physics, notably Bohr, Heisenberg, and Wolfgang Pauli. Soon Heisenberg derived the uncertainty relations, and with Bohr's publications on the subject concepts like discontinuity, indeterminacy, acausality, irrationality, indivisibility, individuality, and complementarity entered the literature to characterize the new quantum physics. Thought experiments were imagined in order to elucidate and discuss what the physical meaning of the new theory was. These discussions continued vividly in Copenhagen throughout the late 1920s and well into the 1930s, not least prompted by Albert Einstein's objections to the new ideas. With reference to these pioneering ideas we have the origin of the so-called Copenhagen Interpretation. That is the second reason for Copenhagen being an important place; the only broadly accepted interpretation of quantum mechanics is named after it.

Finally, after World War II physicists with alternative interpretations always looked to Copenhagen for a response to their ideas, and the Copenhagen Interpretation became a reference point by which all other interpretations are compared and measured or opposed. Physicists in Copenhagen, notably Léon Rosenfeld, often met the new interpretations as challenges which Bohr's ideas had to be defended against. Else the new ideas were conceived as misunderstandings of quantum theory and therefore as ideas that should be corrected, outright dismissed, or defeated.

This leaves us with questions as to why the creation of much of quantum mechanics and its interpretation could and did take place in Copenhagen, and I will attempt to answer these questions in the following. I begin by briefly outlining the Danish research policy in the interwar years, and then I focus on Niels Bohr, the Institute for Theoretical Physics, and Bohr's interpretation of quantum physics. I will finally give a brief review of how the physicists in Copenhagen responded to the new critique and development in the interpretation of quantum physics after World War II.

## 21.1 INTERWAR SCIENTIFIC INTERNATIONALISM

Great puzzlement has sometimes been expressed about the unlikely event that Copenhagen became the world centre of the revolutionary development in quantum mechanics—and for good reasons; after all, Denmark did not have an impressive history as a great scientific nation prior to the 20th century. Alexei Kojevnikov has conjectured that had it not been for World War I, probably the development would have taken place in Germany rather than in Denmark. Indeed, the dire situation in Germany after the War and German hyperinflation took their toll on science in that country.[1]

Meanwhile, according to Henrik Knudsen and Henry Nielsen, in the last year of the World War I the idea took hold in parliamentary and scientific circles in the small, neutral Nordic countries that the initiative to resume international scientific relations after the War should come from them, and that at the same time they ought to play an important role in this reestablishment of scientific liaisons. Because Denmark and the other Nordic countries had profited economically from the Great War through their neutrality, they saw it as their duty to act as mediators for rebuilding European cultural society once peace was reestablished, and that included securing good working conditions for scientific research. At the same time there was also a widespread expectation, probably paired with some optimistic opportunism, that the neutral countries had a particular mission to fulfil in linking scientific circles in the great powers, and that it could in actual fact benefit and even boost science in the small, neutral countries.[2]

In Denmark the economic and geopolitical situation was particularly favourable. The Danish state had a trunk of money from the sale of the Danish West Indies to the US in 1916, which it decided to spend on science. A visionary and innovative science

---

[1] Kojevnikov (2020).
[2] Knudsen and Nielsen (2012), pp. 117–18, Robertson (1979), p. 50.

policy emphasizing internationalism was initiated. A manifest result of the new science policy was the establishment of the Rask-Ørsted Foundation, proudly named after the two prominent 19th century scientists, linguist Rasmus Rask and physicist Hans Christian Ørsted. This foundation had the twofold goal to support Danish science and international research at the same time. Thus, it would support international cooperation which included participation of Danish scientists. Danish researchers' participation in the reestablishment of international scientific cooperation was seen as an integral part of a policy to further European relaxation and in that way stabilize the political landscape surrounding Denmark.[3]

The new policy made Denmark suitable as a meeting ground for scientists coming from countries which had been at war with each other only a few years before. Copenhagen played an important role as a venue in which also German physicists, who were otherwise largely excluded from the scientific community, could participate in meetings after World War I. Later on, in the 1930s after Hitler took power in Germany, Copenhagen and Bohr's institute served in another capacity, viz., to provide shelter for shorter or longer durations for many physicists who fled from Germany because of race or politics until they found a more permanent position abroad, often through Bohr's intervention.[4]

The Rask-Ørsted Foundation and the new science policy reflected a view on science promoted by the Danish state according to which cooperation was a central element in scientific practice, and according to which scientific collaboration preceded competition. Bohr drew heavily on the new policy in the establishment and building up of his Institute. As a result of the new science policy a new social role was attributed to the scientific elite, namely as informal international ambassadors and bridgebuilders. Bohr possessed formidable political and diplomatic skills, and he turned out to be an eminent fundraiser and organizer of science. With time he established himself and his institute as a principal and exemplary actor in the promotion of the new international mission of science, which at the same time was meant to secure Denmark as a great scientific nation. Throughout his life Bohr believed in the importance of internationalism in science, and it pervaded all areas of his work.[5]

Between the wars Copenhagen gained the reputation as one of Europe's leading scientific centres. The International Education Board (IEB), which was founded in 1923 by the American Rockefeller Foundation in order to support education internationally, considered the Danish capital as a centre of rare importance. For its size it was remarkable that four Danish scientists received Nobel Prizes in the period 1900–1945, Niels Bohr for physics in 1922, the other three in medicine.[6]

[3] Knudsen and Nielsen (2012), p. 122.

[4] Knudsen and Nielsen (2012), p. 118–120, 129, Robertson (1979), pp. 126–29, Kojevnikov (2020), Pais (1991), pp. 381–86, Aaserud (1990), BCW Vol. 11, p. 6.

[5] Knudsen and Nielsen (2012), pp. 120, 123, 134, BCW Vol. 11.

[6] Aaserud (1990), pp. 21–36, Söderqvist et al. (1998), p. 10, Nielsen and Nielsen (2001).

## 21.2 NIELS BOHR AND THE INSTITUTE
## FOR THEORETICAL PHYSICS

No matter how favourable the Danish science policy and funding system were, it is safe to say that the development of quantum physics would not have taken place in Copenhagen had it not been for Niels Bohr. Not only did he have outstanding skills as a fundraiser and in scientific leadership, but he was also exceptionally enthusiastic about the new quantum physics, which he had pioneered. Bohr was Mr. Quantum per se. In Sweden, for example, there might also have been good conditions for scientific work, but in Sweden the physicists were initially rather sceptical about the new quantum theory.[7]

Bohr had achieved international fame for his radical new contributions to atomic theory as expressed in his 1913 atomic model. Planck and Einstein had suggested the existence of the quantum, and Bohr fully embraced the quantum idea in his atomic theory. He was determined to promote and develop his atomic theory, and he did it with great success. Although his atomic postulates lacked theoretical foundations, the majority of physicists were impressed with the theory's empirical success. They recognized that Bohr's theory was an important advance and accepted that it defined the course for future research. Within a short span of years, the theory had achieved a near paradigmatic status, and as mentioned Bohr received the Nobel Prize in 1922 precisely for his work on the structure of atoms.[8]

Following a postdoc position with Ernest Rutherford in Manchester, Bohr returned to Copenhagen in 1916 in the midst of the catastrophic world war to become professor of physics at the University of Copenhagen. He was devoted to strengthening the science of physics in his native country. He was initially accommodated at the Polytechnical School with very poor means to carry out research. The path to the establishment of a new independent institute for physics proved long and intricate. However, the need for a decent institute was commonly acknowledged and by help from his family, friends, private sources, and the ministry of education, Bohr finally secured approval of a proposal for a new institute in October 1918. The new three-storey building of the Institute for Theoretical Physics at the University of Copenhagen was finished in March 1921. It included a small experimental laboratory and had, in addition to the director, three staff positions for an assistant, a mechanic, and a secretary.[9]

Bohr's speech at the Dedication of the Institute for Theoretical Physics laid out his vision for the future of this place. His institute was to be a place where physicists could plunge into new unknown theoretical territory. He foresaw that new discoveries and

[7] Grandin (2013).    [8] Kragh (2012), p. 90.
[9] Robertson (1979), Pais (1991), p. 170.

bold ideas could be fostered here primarily by the younger generation of physicists. Therefore, he considered it:

> of the greatest significance not just to depend on the abilities and powers of a limited circle of researchers; but the task of having to introduce a constantly renewed number of younger people into the results and methods of science contributes in the highest degree to continually taking up questions for discussion from the new sides; and, not least from the contributions of the younger people themselves, new blood and new ideas are constantly introduced into the work.[10]

After Bohr's speech the Rector of the University addressed Bohr with the following words:

> The University is pleased with the work begun here and that we have found such an excellent leader. I thank you for your efforts and congratulate you. We are proud of you, and we expect much of your work. You have succeeded in gathering around you both Danish and foreign scientists and have thereby in the finest way continued the international collaboration that was broken off by the World War. It is with pleasure and expectation that I declare the Institute open.[11]

Five years later, the institute had turned into what has been called a Mecca for scientific pilgrimage of theoretical physics from all over the world, and an unprecedented international cooperation in science unfolded. More than sixty visiting researchers from 17 countries plus a larger number of short-term visitors worked there during the 1920s, making crucial contributions to the new physics. Bohr fully grasped how to use the favourable political winds of scientific neutralism and internationalism in combination with the relatively easy access to Danish funding to the advantage of his institute, which he skilfully turned into a main centre of the scientific revolution of quantum mechanics.[12]

Bohr soon managed to get support for an extension of the institute, this time not from Danish funding, but from the Rockefeller-funded International Education Board (IEB). In fact, Bohr's Institute was the first to receive support from the IEB. This organization had an elitist funding policy toward basic science, providing funds for the best institutions and scientists and setting no specific conditions on the kind of work done. As one of the most prestigious and enterprising physicists at the time, Bohr profited substantially from this policy. With additional funding for instruments from the Carlsberg Foundation, the expansion of the institute was completed by the end of 1926.[13]

---

[10] *BCW* Vol. 3, p. 301.     [11] Robertson (1979), p. 9.
[12] For a more detailed history of the institute's development during this period see Kojevnikov (2020), Robertson (1979), pp. 156–59, Aaserud (1990), Knudsen and Nielsen (2012), pp. 123–26.
[13] Aaserud (1990), pp. 21–25, Robertson (1979).

The IEB also provided a fellowship programme from which Bohr's institute bene-fited greatly. The majority of the fellows came from Göttingen in Germany, where professors had a close relationship with Copenhagen. Famously, Heisenberg was the first young fellow to arrive in September 1924. He was followed by Friedrich Hund, Pascual Jordan, Lothar Nordheim, and Walter Heitler. From Holland came Samuel Goudsmit and from the Soviet Union George Gamow. The Rask-Ørsted Foundation provided support for an average of three visitors per year to the institute during the 1920s, including Oskar Klein from Sweden, Wolfgang Pauli from Austria, Yoshio Nishina from Japan, and others.[14]

Probably inspired by his mentor Rutherford, Bohr orchestrated this large group of exceptionally gifted foreign physicists effectively by providing freedom of research, discussion, and cooperation among them as well as by nurturing their individual talents and creativity. This paternal role suited Bohr's personality and general outlook on life. He was charismatic and at the same time had great empathy for the young physicists. Apart from providing advice and guidance in their research work and arranging grants for them, Bohr also saw to all kinds of details related to their stay. Indeed, fundraising for institute extensions and for travel and accommodation for the many physicists who spent time at the institute was a colossal administrative task for Bohr, and it took up a good portion of his work schedule and threatened his health. There were three main languages, Danish, German, and English. As a kind of tribute to Denmark's and Bohr's hospitality, many of the visiting physicists had learned Danish by the time they left the institute. The visiting physicists were introduced to Copen-hagen, when Bohr took them to the city's iconic buildings and cinemas.[15]

With respect to scientific research, Bohr also preferred to work in collaboration with others rather than on his own. Bohr's preferred way of working was through personal conversations or dialogues, which could take place on walks in the city's parks, while biking, on sailing trips, at his home—from 1932 the honorary residence at Carlsberg—or at his summer house in Tisvilde north of Copenhagen. For his own papers, he usually had an assistant to type or write down by hand the text that Bohr dictated, while at the same time discussing the ideas and wordings with Bohr. In other words, Bohr needed a sounding board to produce his writings. This assisting role had been performed by his brother, his mother, his wife, and later it was the young physicists who spent time in Copenhagen who worked with him. With respect to the interpret-ation of quantum physics, Bohr found an excellent collaborator in Rosenfeld.[16]

Seminars brought together most of the physicists and research students in the auditorium. These seminars had no time limit and they often lasted several hours. They were informal gatherings with spontaneous interruptions through questions being asked for the speaker to clarify a point or in order to put forward an opposing point of view, or simply humorous remarks that would lead to a general outburst of

---

[14] Aaserud (1990), pp. 25–27, Robertson (1979).
[15] Robertson (1979), pp. 16, 135, 152–53. Aaserud (1990).
[16] Aaserud (1990), pp. 11–13, Jacobsen (2012).

laughter. This informal tone often came as a surprise to visitors from universities abroad where far more formal procedures for conducting seminars were practised. The informal tone was also reflected in the fact that several of the young physicists addressed each other with 'du' at a time when the German language and the term 'Sie' were otherwise the norm. The informality and the humour were integral parts of the atmosphere at the institute. It is exemplified by the now famous Faust parody performed in connection with the Copenhagen Conference at Easter in 1932 on the hundredth anniversary of Goethe's death. The humour was even institutionalized in a kind of festschrift called *Jocular Physics*, which appeared on the occasions of Bohr's 50th, 60th, and 70th birthdays.[17]

The research practice, the cooperative spirit and network, the atmosphere, and the daily life at Bohr's institute has been described as the *Copenhagen spirit* by many of the visiting physicists after Heisenberg first coined the concept in his Chicago lectures in 1929. From 1929 and through the 1930s a series of conferences were held in Copenhagen annually around Easter. These so-called 'Copenhagen conferences' constituted a venue much more informal than the Solvay meetings; no programme had been planned in advance, for example. New theories were presented and discussed by physicists who were working or had previously worked in Copenhagen, who were Bohr's friends as well as some newcomers. Thus, the atmosphere at the meetings was in favour of the latest developments in quantum theory.[18]

In the period 1918–1927 Bohr and his collaborators became famous for the use of the so-called *correspondence principle*, a sort of heuristic theoretical tool that proved empirically fruitful and the use of which culminated with Heisenberg's construction of matrix mechanics. This principle was mainly used by Bohr and his collaborators and was sometimes referred to by physicists not attached to Copenhagen as waving a magic wand.[19]

Why did Copenhagen become the embodiment of *the philosophical implications* of the new revolutionary quantum physics? Again, it is safe to say that this could not have happened if it had not been for Bohr's preoccupation with the epistemology of quantum theory and his unique and idiosyncratic take on it, paired with his status in the physics community. Unlike in the later post-war era, between the wars many physicists, including Max Born, Pauli, and Paul Ehrenfest, were of the opinion that knowledge about the philosophical implications of quantum theory was important for the quantum physicist, and that this could be achieved through a stay with Bohr in Copenhagen. Born suggested in 1928 that if

---

[17] Robertson (1979), pp. 134–35, Jacobsen (2012), pp. 51, 54. About *Jocular Physics* see Segré (2007), Halpern (2012), Beller (1999a), Jacobsen (2012), pp. 122–23.

[18] Jacobsen (2012), pp. 31–32, Heisenberg (1930), preface, Aaserud (1990), pp. 6–15.

[19] Kragh (2012), pp. 189–20, Darrigol (1992), Jammer (1966), pp. 199–200, Bokulich and Bokulich (2020).

several of [Arnold Sommerfeld's] most recent and most significant students have achieved great things in the study of foundations, this must...be due to later influences, particularly to contact with Bohr.[20]

Indeed, for the young postdocs and other visitors, a stay in Copenhagen involved, if they were receptive to it, discussions with Bohr about the epistemological aspects of quantum theory. Naturally, Bohr's close collaborators Klein and Rosenfeld were well-versed in Bohr's thoughts. Meanwhile, the discussions could be rather heated and intense if the physicist in question disagreed with Bohr. Schrödinger had that experience when he visited Copenhagen in 1926, and Bohr famously kept the discussion going even after Schrödinger fell sick. Heisenberg experienced Bohr's relentless insistence in 1927 when he and Bohr disagreed about the meaning of the uncertainty relations. A few years later, what has jokingly been referred to by Klein as a 'small war' broke out between the Russian physicist Lev Landau and the German Rudolf Peierls about the measurability of the electromagnetic field, see below.[21]

However, far from all of the young postdocs spent much time on interpretational issues. According to the Danish physicist Christian Møller,

> Although we listened to hundreds and hundreds of talks about these things, and we were interested in it, I don't think, except Rosenfeld perhaps, that any of us were spending so much time with this thing.[22]

## 21.3 INTERPRETING QUANTUM MECHANICS IN COPENHAGEN

Why were the physicists preoccupied with interpretation in the first place? The answer is that the mathematical foundations of matrix mechanics and wave mechanics were constructed in an empirical or exploratory process through a long series of steps of trial-and-error, guided by some principles of theory construction such as the correspondence principle, Heisenberg's principle of observability, and Schrödinger's use of Hamilton's optical-mechanical analogy. Like all empirical equations, however, these mathematical systems do not in themselves carry clues to their physical meaning. When he constructed matrix mechanics, Heisenberg retained the classical intuitive notions of position and momentum, but restricted their applicability, and he stressed the importance of observable quantities. On the basis of his successful wave mechanics, Schrödinger attacked Bohr's views on discontinuity and the conception of quantum

---

[20] Born (1928), p. 1036. English translation in Seth (2010), p. 184.
[21] BSC 23, O. Klein to L. D. Landau and R. Peierls, 16 February 1931. Jacobsen (2012), pp. 52–55, 61–75.
[22] Kuhn (1963), p. 21.

jumps. In September 1926 Bohr invited Schrödinger to Copenhagen for discussions on the matter. In the end Schrödinger did not abide to Bohr's strenuous efforts to convince him about the existence of quantum jumps, but Schrödinger's objections to Bohr's interpretation stimulated further discussions in Copenhagen.[23]

Heisenberg saw the root of the conflict as a lack of a definite interpretation of the quantum-mechanical formalism and, as mentioned, thought the step forward was to stress the importance of observables in the theory. Bohr, on the other hand, was convinced that this cure was not enough. He was concerned that the mathematical formalism hid the gist of the physical problem at stake in quantum theory. He distinguished between the presence of observables in the theory—this was merely a matter of *definition*, that is, the theory's statements—and the epistemological conditions for actual *observation*, that is, measurability. In order to find out whether observable quantities could be measured in practice or in principle, Bohr suggested analysing idealized measurements of them, that is using thought experiments. He thus insisted that the limitations of the classical concepts should be founded directly on physical principles, and that physical understanding should precede further development of the formalism. Bohr was far from being alone in raising this concern; it was a general worry among such physicists as Born, Ehrenfest, Schrödinger, Louis de Broglie, and Einstein, that the new quantum theory needed physical interpretation in order to ensure its fruitfulness and status as a physical theory, so that it was not completely detached from 'reality', whatever each of them meant by that. The interpretation of quantum mechanics became from then on of utmost importance in Copenhagen.[24]

The first or original 'Copenhagen Interpretation' of quantum theory developed over several years through discussions among Bohr, Heisenberg, and Pauli among others and motivated by objections from Schrödinger, Einstein, and others. The discussions can be followed in the bulky correspondence between the protagonists. Here I will limit myself to presenting some of the results of these discussions, and, since Bohr's ideas have often been criticized or even dismissed as obscure, I will deal mainly with Bohr's views in order to clarify some of them a little.[25]

Bohr was a serious and independent thinker. He had a unique, idiosyncratic take on the interpretation of quantum physics, and his ideas were profound and not trivial. Regrettably, Bohr's papers are notoriously difficult to read and grasp, and much ink has been spilled on criticizing Bohr's writing style (if possible, his lectures are said to have been even worse). His sentences are unreasonably long for todays' standard and loaded

[23] Heisenberg (1925), Joas and Lehner (2009), Jammer (1966), pp. 324–25, Jaynes (1989), p. 8. For Bohr's letter of invitation to Schrödinger and Bohr's and Schrödinger's review of their discussions in Copenhagen, see Meyenn (2011), pp. 307–309, 323–25.

[24] Heisenberg (1964), p. 94, Jacobsen (2012), pp. 24–25, 52–53, 59–60, Joas and Lehner (2009), p. 349, Jammer (1966), p. 324. See also Osnaghi (2021).

[25] In her highly critical book Mara Beller made use of what she called a dialogic approach to the history of the quantum revolution. She focused on dialogues emerging from the correspondence and conversations among the quantum physicists and used it to ridicule Bohr and to construct what in my view is a completely distorted picture of the Copenhagen Interpretation. Beller (1999b).

with quirky concepts that were then and now alien to many physicists when describing physical phenomena. These were concepts such as discontinuity, irrationality, indivisibility, individuality, indeterminacy, and complementarity, which I shall return to in the following. Perhaps for that reason, many a philosopher of science has attempted to trace Bohr's views back to some of the philosophers he had studied during his education and upbringing. It seems to have been Max Jammer's otherwise impressive book *The Conceptual Development of Quantum Mechanics* from 1966 that sparked the speculations about Bohr's philosophical background. In the book, Jammer claimed that Bohr was much influenced by the philosophers Søren Kierkegaard, Harald Høffding, William James, and P. M. Møller. Jammer's book has had tremendous influence on how Bohr has been perceived among historians and philosophers of science. It is only natural that Bohr should have been inspired by the philosophical teachings he had been introduced to in his youth as well as the views of his father, Christian Bohr, who was an eminent physiologist. However, I would lean towards Bohr scholars who argue that Bohr's thinking cannot simply be reduced to being based on one or more of the mentioned philosophers. Anyhow, in my view Bohr was foremost a physicist whose physics it is quite possible to make sense of without this approach.[26]

Another much more critical, less anecdotic, and less hagiographic approach to explain what has been described as Bohr's monocracy and the emergence of the new vocabulary and values in quantum physics has been to seek explanation in the cultural and social environment in which the quantum physicists found themselves after World War I. Paul Forman pioneered this approach. He has suggested that the German-speaking physicists' willingness to accept such notions as acausality, indeterminacy, and individuality in quantum physics, reflects their adaptation to a hostile intellectual environment with anti-scientific sentiments in Weimar culture. Whether Bohr's vocabulary reflected his anxiety at this time of general crisis will not be analysed here. It was generally acknowledged that the new quantum mechanics was built on the ruins of and was a response to a great crisis in quantum theory during 1922–1925. However, my perception of Bohr is that he was an enthusiastic and optimistic physicist who saw the glass as half full rather than half empty, and his writings do not strike me as particularly gloomy. Furthermore, this approach seems to lose sight of the physical content of the physicists' writings or explain it away. My aim here is the opposite. Rather than following in the slipstream of this critical exposition I will offer my own analysis of Bohr's ideas.[27]

---

[26] Jammer (1966), pp. 172–176, 349. Jan Faye sees Bohr's thinking as owing much to Høffding, and he sees a strong link through Høffding's philosophy to Immanuel Kant's philosophy and his notion of *Das Ding an Sich* in his treatise *Kritik der reinen Vernunft*. Faye (2010), p. 55. David Farvholdt was in opposition to that view in his rather polemical book Farvholdt (2009). Historians who opposes a reductionistic view include Heilbron (2013), p. 106, Rosenfeld (1945), Rosenfeld (1963). About Rosenfeld's response to Jammer's book see Jacobsen (2012), p. 267 footnote.

[27] Forman (1971), Forman (1984). My list of Bohr's ideas is not exhaustive. For Bohr's focus on the use of classical concepts, for example, see Howard (2021).

The development of Bohr's thoughts reached some sort of culmination with his famous lecture at the Volta celebrations in Como in September 1927 and through the informal discussions with Einstein at the Solvay congress a month after. Bohr spent several years subsequently clarifying, consolidating, and communicating his views on interpretation. Discussions on interpretation in Copenhagen continued well into the 1930s, then also in the context of relativistic quantum mechanics. The following is based mainly on Bohr's Como lecture.[28]

## 21.3.1  The Quantum Postulate

According to Bohr, the need for a new physics and a new interpretation rested on Planck's discovery of the universal quantum of action $h$. $h$ relates the *particle* properties energy $E$ and momentum $p$ to the *wave* properties frequency $f$ and wavelength $\lambda$ in the two formulae $E = h \cdot f$ and $\lambda = h/p$. Bohr called this *the quantum postulate* and for him this was the essence of the new theory. Firstly, the quantum postulate tells us that the values of physical properties, such as for instance energy of radiation, attributed to atomic systems comes in portions, i.e. it is discrete or quantized, and properties such as radiation energy cannot exist in infinitely small amounts. This is the *discontinuous* aspect of the description. In addition, since Planck's quantum of action sets the limit for how small an energy and momentum exchange can be between systems, Planck's quantum of action constitutes the truly *indivisible* element in the theory of the *atom*. The Greek root of the word atom, *atomos*, means indivisible.[29]

Secondly, the quantum postulate had far-reaching consequences for the usual way of making idealizations in physics, according to Bohr. When defining the state of a physical system, say a free electron, physicists usually make an abstraction such as eliminating all external disturbances in order to explore how the electron will ideally behave, or where it is located, if it is isolated from its disturbing surroundings. However, if in a corresponding measurement situation, we were to isolate the electron in question, then, true enough, it would not be disturbed by external surroundings, but that would also make the measurement of its position impossible. The reason is that the properties of the electron, such as its position or its charge, are only accessible for both definition and observation when the electron interacts with another system, say a photon or another charged particle. Observation of the properties of the electron entails an interaction with it through some means of measurement. Because of the finite size of the quantum of action, the measurement interaction can never be neglected compared with the size of the usual interactions in physics, because the measurement interaction is at best of the same order of magnitude as the interactions which it is meant to measure. Therefore, the information that can be deduced from the measurement is not about the electron in itself, but it is about the electron in

---

[28] Jacobsen (2012), pp. 61–94.    [29] Bohr (1928), Bohr (1929).

interaction with the measurement device. 'What we observe is not nature in itself, but nature exposed to our method of questioning', as Heisenberg later phrased it. The implication of this new situation in quantum physics, Bohr later stated, is that we are both actors and spectators when exploring nature. This implication of quantum theory for the description of *individual* quantum systems was thoroughly discussed, analysed, and elucidated by Bohr and Heisenberg in connection with Heisenberg's thought experiment to localize an electron by means of a gamma-ray microscope. Several physicists, including de Broglie, Einstein, and Schrödinger, questioned whether quantum physics should at all be understood as saying anything about individual quantum systems or whether the theory only puts forward statements of general statistical character. They thought it would never be possible to make experiments with single particles. Still, Bohr maintained that quantum mechanics reflected *individuality*, and he explored its consequences through idealized thought experiments.[30]

Bohr referred to the act of measurements as making a space-time description whereas he denoted the theoretical definition of the system's state a causal description of the system. The fact that the two forms of description can be combined unproblematically in classical physics relies on the assumption that a physical phenomenon can be observed while still be considered to be isolated from its surroundings. However, this is not the case in quantum physics, as elucidated above, due to the finite size of the quantum of action. Therefore, according to Bohr, a measurement will also affect the definition of the system's state; it becomes necessary to take the interaction with the measurement device into account in the theoretical description. But then the system is no longer looked at in reasonable isolation. Hence, the act of observation of the system disrupts the causal connection between the preceding and the future development of the system's defined state. In other words, a measurement on an atom, for example, changes its defined state, so that its prehistory is basically eliminated (the example of stationary states is mentioned in more detail below). This is contrary to successive measurements in classical physics, the performance of which will provide us with more precise knowledge of the system's state and lead to more precise future predictions of the object's behaviour. Thus, there is a fundamental difference between measurements in classical and in quantum physics. Bohr referred to this strange implication of the limited size of the quantum of action as an *irrational* element that was brought in by the act of observation.[31]

To Bohr it therefore seemed to be impossible to introduce a sharp division between object and measurement device in quantum theory. As a result of this correlation between object and measurement apparatus, we can decide *either* to make a measurement *or* to define the system—we cannot do both at the same time. Bohr referred to this as *complementary* descriptions of the atomic system, i.e., both descriptions are necessary to give us complete information about the system—and that is because the

---

[30] Bohr (1928), Heisenberg (1958), p. 58, Hüttemann (2002), Bohr (1949), p. 236, Schrödinger (1952). For an excellent discussion of the gamma-ray microscope thought experiment see Petruccioli (2011).

[31] Bohr (1928).

theoretical description is only a statistical or probabilistic description—however, they are at the same time mutually incompatible descriptions. There is more about complementary descriptions below.

The quantum of action also appears in Heisenberg's uncertainty or indeterminacy relations, which he derived between position and momentum, and energy and time duration in 1927. They express a fundamental limitation in the application of the classical mechanical concepts both in the definition of the individual quantum systems and in the description of the means of observation. Therefore, the extent to which the intervention by a measurement will affect the description is unknown or uncontrollable in principle, according to Bohr. Hence, according to Bohr, Heisenberg's relations should not be interpreted as uncertainty in the usual sense experienced in any measurement situation, as was Heisenberg's first suggestion. Bohr and Heisenberg had a serious confrontation about the interpretation of the uncertainty relations. I refer to Jähnert and Lehner in this volume for the treatment of the disagreement between Bohr and Heisenberg.[32]

## 21.3.2 Stationary States

In classical physics there is no distinction between the description of free electrons and electrons bound in atoms. However, in order to account for spectral lines and the mechanical stability of matter, Bohr postulated in his atomic model of 1913 that atoms can only exist in certain stationary states. The concept of stationary state was meant as an abstraction pertaining to an atom isolated from all external interactions, i.e. the atom was seen as a closed system. According to the quantum postulate this means that the atom evades any possibility of being observed in this state. Furthermore, any observation of a bound electron, for example by means of collision or radiation interactions, that would allow distinguishing between the stationary states and determining the orbit of the electron, will be accompanied by a change in the state of the atom. Thus, the observation will destroy our knowledge of the prehistory of the atom. In other words, the knowledge we get from the observation cannot be linked to the knowledge we had with the originally defined state. For this reason, the concept of stationary state entails a limit to the knowledge we can achieve about the individual electrons' whereabouts in the atom.[33]

Schrödinger's objections to the idea of stationary states and the quantum jumps between them as well as his hope for the wave function to represent the real atomic state and its development in time, prompted Bohr to analyse anew his conception of stationary states, which he had struggled with since the publication of his 1913 atomic model, now in the light of Heisenberg's indeterminacy relations concerning energy and

---

[32] Heisenberg (1927), Petruccioli (2011), Jähnert and Lehner (2021).
[33] Bohr (1928).

time development. In fact, Bohr seems to have anticipated the inverse relationship between energy uncertainty and time duration of a stationary state expressed in Heisenberg's uncertainty relation, when in the Bohr–Kramer–Slater paper in 1924, he argued that there is a 'limit of definition of the motion and of the energy in the stationary states' which manifests itself in a finite line width. The stationary state, he said, 'imposes an a priori limit to the accuracy with which the motion in these states can be described by means of classical electrodynamics.' Anyhow, according to Heisenberg's relation, a well-defined energy of the stationary state eliminates the possibility for a well-defined time description. This means that not even processes of spontaneous emission of radiation, which happen in the absence of any external influences, can be accounted for when considering the atom as a closed system, since spontaneous emission sets a limit to the lifetime of the stationary states.[34]

### 21.3.3 Complementarity

Bohr's use of the concept of complementarity to characterize the relation between the definition of the atomic system and its measurement was briefly mentioned above. If we want complete knowledge of the quantum system, it does not suffice to have only one of these forms of description; we need both descriptions, but they exclude each other simultaneously. According to Bohr, the new and growing wealth of empirical evidence was in need not only of a new mathematical formalism and a new physical interpretation accompanying the empirical formulas; the new physics *also* needed a new logical frame, i.e. a new leading physical principle, if you like, akin to what characterizes a Kuhnian paradigm. Complementarity was to be the way of seeing the world that provided the epistemological precondition for making interpretations at all.[35]

Bohr came to this insight because the classical concepts such as waves and particles, for instance, clearly fell short of a comprehensive account without inconsistencies appearing. Wave–particle duality fell outside the realm of classical physics, since the wave and particle models are mutually exclusive descriptions in classical physics, but they nevertheless spring from the same identities in quantum theory. In the mathematical expression of the quantum postulate mentioned above: $E = h \cdot f$ and $\lambda = h/p$, both equations relate features of radiation which are at the same time mutually exclusive in terms of wave and particle descriptions. In order to frame the new knowledge of wave–particle duality, Bohr thought it was necessary to introduce a new kind of logic, which he called complementarity, and which could accommodate both features without contradictions. Contrary to Schrödinger, who dreamt of

---

[34] Bohr (1928), p. 588, Bohr *et al.* (1924), p. 795, Kragh (2012), pp. 331–32. That Bohr anticipated the reciprocal relationship in the Heisenberg relations is also suggested in Petruccioli (2011), p. 622, with reference to a Bohr manuscript that is very likely to have been misdated in *BCW* Vol. 6. *BCW* Vol. 6, pp. 57–65.

[35] Bohr (1949), pp. 93–94, Ladyman (2002), pp. 98–100.

reducing quantum theory to wave mechanics, and other physicists such as David Bohm later in the century who wished to reduce quantum theory to a particle theory, Bohr held that both aspects were important for an exhaustive account, even though the two aspects were simultaneously mutually exclusive. Complementarity could account for both aspects without reducing the one to the other. Bohr also conceived of Heisenberg's indeterminacy relations as providing examples of complementary descriptions, and he referred to matrix and wave mechanics as complementary forms of descriptions. Soon Bohr began to suggest that the conditions for observation and description in other scientific fields such as biology and psychology may also be termed complementary. It is likely that Bohr's conception of complementarity had inspired him since his youth.[36]

## 21.3.4  Epistemology versus Ontology

Bohr distinguished between epistemology, that is *our description* of nature—in terms of the empirical theory of quantum mechanics—and ontology, that is *the essence* of that nature. He saw quantum physics as an epistemological theory. The implication of the quantum postulate for a measurement in atomic physics was, according to Bohr, that it is no longer possible to attribute to physical quantities an existence independent of our measurement of them. On the other hand, the measurement should not be understood as a kind of mechanical disturbance of the system's 'real' properties prior to the measurement. We simply cannot talk about the properties of the system in an absolute sense if it is left alone. When discussing Bohr's ideas, it is therefore important to emphasize and keep in mind that Bohr considered physics to be *only* about what *we can say about nature*: 'In our description of nature, the purpose is not to disclose the real essence of the phenomena but only to track down, so far as it is possible, relations between the manifold aspects of our experience.' Or as Rosenfeld later phrased it: 'the task of science is not to picture the world as a spectacle watched from outside,..., but rather to give us the means of communicating, in a rational and objective way, the experience derived from our interaction with the world around us, of which we ourselves are a part.'[37]

In the late 1920s Bohr did not specifically address the probabilistic nature of quantum mechanics, and his thoughts on this aspect of the theory seem not yet sufficiently matured and articulated. Consequently, he lacked a clear and transparent vocabulary to tackle the distinction between epistemology and ontology. Bohr was aware that a new understanding of probability was needed in quantum physics and in 1949 he addressed the issue of the statistical description more directly, but still in a vague manner. Only late in Bohr's life did a new conception of subjective probability gain ground which captures the ideas he was attempting to put forward.[38]

---

[36] Bohr (1928), Rosenfeld (1945), p. 318, 321, Jammer (1966), pp. 345–53, *BCW* Vol. 10.
[37] Bohr (1929), p. 18, Rosenfeld (1945), p. 330, Faye (2010), pp. 52–53.
[38] Bohr (1949), pp. 205–206, 223, Jaynes (1989), p. 9.

However, Bohr could not escape inconsistencies in his viewpoints. His interpretation of Heisenberg's uncertainty relations seems to be of an ontological nature, because either Bohr says that it is the measurement situation that makes the limitation on the quantities position and velocity expressed in the indeterminacy relations, or it is an inherent condition for describing the measurement situation, so that it is by nature given that the two properties cannot be determined with equal precision. However, in the latter case we seem to have an ontological interpretation of the Heisenberg relations, as pointed out by Jan Faye.[39]

Einstein raised objections to Bohr's and Heisenberg's ideas at the Solvay meetings in 1927 and 1930, notably to the interpretation of Heisenberg's relations and the quantum correlation between object and measurement apparatus. The epistemology–ontology distinction was a central point separating Einstein and Bohr in their discussions. It came to the fore with the so-called EPR paper in 1935, published by Einstein, Boris Podolsky, and Nathan Rosen, and Bohr's reply, which is discussed by Osnaghi in this volume. Einstein found quantum mechanics incomplete in an ontological sense whereas Bohr found quantum mechanics complete in an epistemological sense.[40]

## 21.3.5 Thought Experiments

In the work on interpretation of quantum mechanics in Copenhagen idealized thought experiments played an important role. Heisenberg and Bohr discussed the gamma-ray microscope in 1927, and Einstein and Bohr's discussion also centred around thought experiments, for example the photon box experiment in 1930. Bohr suggested analysing idealized measurements in order to explore and specify the conditions under which agreement between theoretical predictions and experiments devised to check them could be achieved. Idealized experiments would serve to explore on the one hand the meaning of concepts of a strictly quantum character such as stationary state, uncertainty relation, wave–particle duality, and the correlation between quantum system and measurement apparatus that would all vanish from the theory in the classical limit. On the other hand, it was explored how classical concepts such as the electromagnetic field should be understood in quantum theory. To Bohr, it was not important if these idealized experiments could be realized in practice in the laboratory. The purpose of his analyses was to pair the theory's statements with the possibility of measuring these properties in an idealized experimental set-up in order to arrive at an interpretation of them. In this way, his sole purpose when examining measurability was to give meaning to the formalism. This, however, was never a trivial point in Bohr's papers and talks and has led to many misunderstandings. Many physicists had difficulty understanding

---

[39] Faye (2010), pp. 44–45.
[40] Einstein *et al.* (1935), Bohr (1935). The best discussion I have read about the Bohr–Einstein debate is Howard (2020). See also Howard (2021) and Osnaghi (2021). Jaynes (1989), p. 8.

the use or purpose of Bohr's thought experiments. It was never to provide a 'theory of measurement' like John von Neumann's in his axiomatic elucidation of quantum measurements in his book *Mathematische Grundlagen der Quantenmechanik* (1932). Bohr never recognized or showed interest in von Neumann's axiomatic approach, but it was von Neumann's mathematical exposition of the so-called measurement problem that has been taken as the starting point for most later discussions on this topic.[41]

Bohr's work on measurability culminated with his joint work with Rosenfeld on the measurability of the electromagnetic field resulting in the famous Bohr–Rosenfeld paper published in 1933. This collaborative work was prompted by a joint paper on the measurability of electric fields by the 'young Turks' Landau and Peierls in 1931. Landau and Peierls derived uncertainty relations for the electric field components and then analysed their measurability. On the basis of their idealized thought experiment in which they took into account field fluctuations and the field arising from the point charge meant to probe the field, they boldly concluded that electric fields could not be measured as accurately as the theory stated. Moreover, Landau and Peierls came to believe that the predictions of singularities and negative energy states in Dirac's electron theory were also part of the problem with measurability. Against this background they claimed that fields were not observables and suggested abandoning quantum electrodynamics altogether. Bohr considered the arguments by Landau and Peierls sloppy and naïve and insisted that they had misunderstood what measurability entailed in quantum theory. Among other things, Landau and Peierls based their arguments on a premise about repeatability of successive measurements which, however, has to be abandoned in quantum theory. Bohr later attempted to clarify that 'we can hardly, from considerations of measurements, find arguments against the theory of the type of Dirac, and of the field theory. The importance of the discussion of measurement is more that we get to know a number of things about the character of the restrictions in the use of classical pictures.' Hence, Bohr's message to Landau and Peierls was that the kind of logical investigation undertaken when analysing measurability was a way to interpret the use of the classical concept of field in quantum theory, it was not an assessment of the quantum theory of electrodynamics as such.[42]

Bohr's unrestrained reaction and his harsh critique of Landau and Peierls can in an odd way be interpreted as a compliment to them for drawing attention to an epistemological issue, which Bohr considered extremely important. This was a recurring feature of the discussions about interpretation in Copenhagen and should be seen as a distinctive trait in Bohr's personality. He reacted with equal emotion and relentlessness in discussions about interpretation with Schrödinger, Heisenberg, and Einstein. Bohr wrote to Gamow in 1933: 'I hope it will be a comfort to Landau and Peierls that the stupidities they have committed in this respect are no worse than those which we all,

---

[41] Rosenfeld (1963), p. 530, Jacobsen (2012), pp. 64–65. For a new edition of the English translation of von Neumann's book, see von Neumann (2018).

[42] Landau and Peierls (1983), Bohr and Rosenfeld (1933), Bohr MSS 'Solvay Conference' October 1930 [sic 1933] Morning October 28, microfilm 12, AHQP. Jacobsen (2012), pp. 81–84.

including Heisenberg and Pauli, have been guilty of in this controversial subject.' In proposing Landau for the Nobel Prize in 1962, Bohr wrote that Landau and Peierls' work illuminated 'the known paradoxes of the radiation phenomena in a new way' and thus 'Landau and Peierls' work gave rise to a resumed examination and clarification of the problems with measuring electromagnetic field quantities,' the closure of which of course Bohr and Rosenfeld were responsible for. However, Landau and Peierls never reconciled with Bohr's 'clarification' of their paper. In particular, Peierls never acknowledged Bohr's distinction between idealized thought experiments and real experiments.[43]

Bohr felt it was his duty to communicate the new revolutionary findings in modern physics to a broader lay audience. His younger collaborators, Klein and Rosenfeld, were also involved in this activity of popularization of complementarity, in which Bohr even extended the lessons of quantum physics, as he saw them, into areas such as biology and psychology. Probably this spreading of the gospel of complementarity to areas other than physics did more harm than good to the reception of Bohr's interpretation in the physics community. It is, however, beyond the scope of this chapter to deal with that aspect.[44]

# 21.4 POSTWAR COPENHAGEN RESPONSE TO NEW IDEAS ON THE INTERPRETATION OF QUANTUM MECHANICS

The driving force in the development of the interpretation of quantum mechanics was probably always criticism of Bohr's ideas. Bohr would have testified to this development being fruitful when the criticism came from Schrödinger, Einstein, and Landau and Peierls during the early years. When criticism arose anew in the post-war era, however, Bohr did not consider it fruitful anymore. According to Rosenfeld, at this time Bohr considered new discussions about the interpretation of quantum physics obsolete, or at least a finished chapter, if not trivial and even a waste of physicists' time.[45] Nor did Bohr show any sympathy towards new interpretations. Bohr did not change his own position much, but he attempted to spell it out once again first of all in the important paper, 'Discussion with Einstein on Epistemological Problems in Atomic Physics', which appeared in the book *Albert Einstein: Philosopher—Scientist* (Schilpp, 1949). This paper attracted huge attention. It was of course addressed to Einstein, whom Bohr admired more than any other physicist. Bohr gave his own viewpoint once

---

[43] N. Bohr to G. Gamow, 21 January 1933, in *BCW* Vol. 9, p. 571. BSC, N. Bohr, A. Bohr, B. Mottelson, C. Møller, and L. Rosenfeld to Kungliga Vetenskapsakademiens Nobelkommitté, 30 Jan 1962, carbon copy. Jacobsen 2012, pp. 85–90.

[44] Jacobsen (2012), pp. 124–41, Heilbron (1985), *BCW* Vol. 10.

[45] Jacobsen (2012), p. 256.

again on how to interpret quantum physics. In addition, he gave his version of his discussions with Einstein about quantum mechanics in the late 1920s and early 1930s. It was perhaps one last attempt to win Einstein over, but Bohr probably knew his attempt was in vain. Thus, the paper included an apology to Einstein for disagreeing with him on this issue. Bohr seems to have had enormous regrets that the two of them could not agree on the questions about how to understand quantum mechanics.[46]

Referring to the Soviet Union, Loren Graham has called the decade 1949–1958 'the age of the banishment of complementarity'. Following the new Soviet ideological line in science, art, and literature, *Zhdanovschina*, Party ideologists banned Bohr's and Heisenberg's ideas about the interpretation of quantum physics, because they were seen as being both positivistic and idealistic and thus bourgeois, and therefore not compliant with proletarian, that is, materialistic, science proper. This development in the Soviet Union partly sparked the renewed interest in the interpretation of quantum physics among left-wing physicists in the West. It also seems to have been during the ensuing debates that the term 'Copenhagen Interpretation' emerged.[47]

The criticism from a communist perspective also reached the Danish press and was promoted by the young communist physicist and politician Ib Nørlund (Bohr's wife Margrethe's nephew). At the Institute for Theoretical Physics, the situation was carefully monitored. In the following years the dispute between Nørlund, his supporters, and his opponents, gained momentum in the Danish press. Bohr did not involve himself directly in the dispute with Nørlund, but a dispute with the professor of philosophy at the University of Copenhagen, Jørgen Jørgensen, was another matter. Jørgensen was attracted to communism and during the 1940s he also followed the new line of thought dictated from Moscow. He thus echoed the communist critique of Bohr's and Heisenberg's ideas in his philosophy textbook for first-year students. All students at the University took his introductory course in philosophy and they were therefore exposed to Jørgensen's criticism of Bohr's ideas. A meeting was set up between Bohr and Jørgensen in 1953 in the club of physics students called *Parentesen* (the parenthesis), in order for them to discuss their disagreements, but they could not find common ground.[48]

During the thaw in East–West intellectual relations after Joseph Stalin's death, a visit to Copenhagen was arranged for the Soviet physicist Vladimir Fock, primarily in order for him to discuss the interpretation of quantum physics with Bohr. The visit took place in early 1957. Their meeting involved a give and take on both sides. On the one hand, Bohr acknowledged that he had used a vocabulary that could be misunderstood as positivistic in some of his texts, and he would be more careful in the future. On his part Fock returned to the Soviet Union and translated many of Bohr's central papers

[46] Bohr (1949), p. 240.
[47] Graham (1987), p. 328, Cross (1991), pp. 738–42, Jammer (1974), pp. 250–51, Freire Jr. (1997), Jacobsen (2007, 2011, 2012), Camilleri (2009), Howard (2004), Howard (2021).
[48] Jacobsen (2012), pp. 268–70.

into Russian, with the result that the opposition towards Bohr's writings in the Soviet Union eased off.[49]

The opinions by Bohr and the old quantum guard had an enormous effect and influence on the question of interpretation of quantum mechanics in the American context in the 1950s. For as long as Bohr lived, the quantum dissidents, as Freire Jr. has named the physicists who proposed new interpretations, approached Bohr to get his approval or at least his viewpoint on their new proposals. This included, among others, David Bohm and Hugh Everett. However, they never succeeded. Bohr retreated from the foreground in the discussions and left it to his collaborators, mainly Pauli and Rosenfeld, to defend complementarity and other features of the original interpretation from misunderstandings. New interpretations were relentlessly dismissed, sometimes rather arrogantly by Rosenfeld. Rosenfeld's acting in the quantum controversy suggests that he sometimes considered the best defence to be a patronizing attack. This attitude, which Freire Jr. has referred to with a strong political metaphor as a Non-Proliferation Treaty came to mark the response in Copenhagen to criticism and new interpretations in the post-war era.[50]

According to Franck Laloë, '[u]ntil now, no new fact whatsoever (or new reasoning) has appeared that has made the Copenhagen Interpretation obsolete in any sense'. Indeed, many physicists still consider the original ideas, or some variation of them, the best interpretation in order to avoid paradoxes. However, in merely accepting Bohr's epistemological ideas, there may not have been the same incentive to develop something like John Stewart Bell's inequality, nor to test experimentally whether it is violated. Many physicists are not content with physics being just about finding correlations between phenomena or calculations that ends up in numbers to be tested experimentally, which is what the Copenhagen Interpretation offers. They are at least as interested in finding physical explanations as to why the correlations that can be discovered are the way they are. History suggests that post-war scientific development was driven by the physicists, young and rebellious, in the 1950s and 1960s, who dared to challenge the exclusively epistemological interpretation of quantum physics seated in Copenhagen. This seems to have had a positive effect on scientific development leading to where we are today with, among other things, the possibility of performing entanglement experiments on huge as well as on minute scales.[51]

## ACKNOWLEDGEMENTS

I wish to thank Christian Joas, Ricardo Avalar Sotomaior Karam, and Thiago Hartz for critical comments and valuable suggestions for improvements on previous versions of this chapter.

[49] Jacobsen (2011) and (2012), pp. 290–301. For a fresh and welcome view on Fock's contributions to the interpretation of quantum mechanics, see Martinez (2019).

[50] Freire Jr. (2015), pp. 45–46, 75–77, 82, 107–108, 114–115, 345. See also the discussions between Eugene Wigner and Rosenfeld on pp. 153–61. Jacobsen (2012), pp. 277–289, 303–308.

[51] Harré (1972), p. 47, Laloë (2012), p. 2, Fuchs *et al.* (2014), Jaynes (1989), p. 9, Freire Jr. (2015).

# References

## Abbreviations

AHQP, Archive for the History of Quantum Physics
   *BCW, Niels Bohr Collected Works*, 12 Vols., eds. L. Rosenfeld, E. Rüdinger, and
F. Aaserud, Amsterdam: Elsevier, 1972–2007.
   BSC, Niels Bohr Scientific Correspondence, The Niels Bohr Archive, Copenhagen.
   *SP, Selected Papers of Léon Rosenfeld*, Boston Studies in the Philosophy of Science,
Vol. 21, eds. R. S. Cohen and J. J. Stachel, Dordrecht: D. Reidel, 1979.

## Books, articles etc.

Aaserud, F. (1990). *Redirecting Science: Niels Bohr, Philanthropy and the Rise of Nuclear Physics*. Cambridge: Cambridge University Press.
Beller, M. (1999a). Jocular Commemorations: The Copenhagen Spirit. *Osiris*, Vol. 14, *Commemorative Practices in Science: Historical Perspectives on the Politics of Collective Memory*, Chicago: University of Chicago Press, pp. 252–273.
Beller, M. (1999b). *Quantum Dialogue: The Making of a Revolution*. Chicago: The University of Chicago Press.
Bohr, N. (1928). The Quantum Postulate and the Recent Development of Atomic Theory. *Nature*, **121**, 580–590.
Bohr, N. (1929). Introductory Survey. In *Atomic Theory and the Description of Nature*, Cambridge: Cambridge University Press, 1961). Originally published as a contribution to a Festschrift in *Die Naturwissenschaften* in connection with Planck's 50 years' jubilee in 1929. *BCW* Vol. 6, pp. 279–302.
Bohr, N. (1935). Can Quantum-Mechanical Description of Physical Reality be Considered Complete? *Physical Review*, **48**, 696–702.
Bohr, N. (1949). Discussion with Einstein on Epistemological Problems in Atomic Physics. In P. A. Schilpp (ed.), *Albert Einstein: Philosopher-Scientist*, Evanston, IL: The Library of Living Philosophers, Vol. VII, pp. 201–241. *BCW* Vol. 7, pp. 339–381.
Bohr, N., Kramers, H. A., and Slater, J. C. (1924). The Quantum Theory of Radiation. *Philosophical Magazine*, **47**, 785–802.
Bohr, N., and Rosenfeld, L. (1933). On the Question of Measurability of Electromagnetic Field Quantities. *SP*, pp. 357–400. Originally published in German in *Det Kongelige Danske Videnskabernes Selskabs Mathematisk-fysiske Meddelelser*, **12**(8), 3–65. *BCW* Vol. 7, pp. 123–166.
Bokulich, A., and Bokulich, P. (2020). Bohr's Correspondence Principle. *Stanford Encyclopedia of Philosophy*, https://plato.stanford.edu/entries/bohr-correspondence/ (accessed on 26 December 2020).
Born, M. (1928). Sommerfeld als Begründer einer Schule. *Naturwissenschaften*, **16**, 1035–1036.
Camilleri, K. (2009). Constructing the Myth of the Copenhagen Interpretation. *Perspectives on Science*, **17**(1), 26–57.
Carson, C., Kojevnikov, A., and Trischler, H. (2011). *Weimar Culture and Quantum Mechanics: Selected Papers by Paul Forman and Contemporary Perspectives on the Forman Thesis*. London and Singapore: Imperial College Press and World Scientific.

Cross, A. (1991). The Crisis in Physics: Dialectical Materialism and Quantum Theory. *Social Studies of Science*, **21**, 735–759.

Darrigol, O. (1992). *From c-numbers to q-numbers: The classical analogy in the history of quantum theory*. Berkeley, CA: University of California Press.

Einstein, A., Podolsky, B., and Rosen, N. (1935). Can Quantum-Mechanical Description of Physical Reality Be Considered Complete? *Physical Review*, **47**, 777–780.

Favrholdt, D. (2009). *Filosoffen Niels Bohr*. Copenhagen: Informations Forlag.

Faye, J. (2010). *Kvantefilosofi: Ved erkendelsens grænser?* Aarhus: Aarhus Universitetsforlag.

Forman, P. (1971). Weimar Culture, Causality, and Quantum Theory, 1918–1927: Adaptation by German Physicists and Mathematicians to a Hostile Intellectual Environment. *Historical Studies in the Physical Sciences*, **3**, 1–115. Reprinted in Carson *et al.* (2011), pp. 85–201.

Forman, P. (1984). *Kausalität, Anschaulichkeit*, and *Individualität*, or How Cultural Values Prescribed the Character and the Lessons Ascribed to Quantum Mechanics. In N. Stehr and V. Meja (eds), *Society and Knowledge: Contemporary Perspectives in the Sociology of Knowledge*, New Brunswick: Transaction Books, pp. 333–347. Reprinted in Carson *et al.* (2011), pp. 203–219.

Freire Jr., O. (1997). Quantum Controversy and Marxism. *Historia Scientiarum*, 7(2), 137–152.

Freire Jr., O. (2015). *The Quantum Dissidents: Rebuilding the Foundations of Quantum Mechanics (1950–1990)*. Berlin: Springer.

Fuchs, C. A., Mermin, N. D., and Schack, R. (2014). An introduction to QBism with an application to the locality of quantum mechanics. *Am. J. Phys.*, **82**(8), 749–754.

Graham, L. (1987). *Science, Philosophy, and Human Behaviour in the Soviet Union*. New York: Columbia University Press.

Grandin, K. (2013). Niels Bohr frem til 1930 set fra et svensk perspektiv. In L. Bruun, F. Aaserud, and H. Kragh (eds), *Bohr på ny*, Copenhagen: Epsilon, pp. 71–85.

Halpern, P. (2012). Quantum Humor: The Playful Side of Physics at Bohr's Institute for Theoretical Physics. *Physics in Perspective (PIP)*, **14**(3), 279–99. doi: 10.1007/s00016-011-0071-8.

Harré, R. (1972). The Forms of Reasoning in Science. In *The Philosophies of Science*, Oxford: Oxford University Press, pp. 35–48.

Heilbron, J. L. (1985). The earliest missionaries of the Copenhagen spirit. *Revue d'Histoire des Sciences*, **38**(3–4), 195–230.

Heilbron, J. L. (2013). Nascent Science: The Scientific and Psychological Background to Bohr's Trilogy. In F. Aaserud and J. L. Heilbron (eds), *Love, Literature, and the Quantum Atom: Niels Bohr's 1913 Trilogy Revisited*, Oxford: Oxford University Press, pp. 103–200.

Heisenberg, W. (1925). Über quantentheoretische Umdeutung kinematischer und mechanischer Beziehungen. *Zeitschrift für Physik*, **33**, 879–893.

Heisenberg, W. (1927). Über de anschaulichen Inhalt der quantentheoretischen Kinematik und Mechanik. *Zeitschrift für Physik*, **43**, 172–198.

Heisenberg, W. (1930). *The physical Principles of the Quantum Theory*. Transl. C. Eckart and F.C. Hoyt. Chicago: Dover.

Heisenberg, W. (1958). *Physics and Philosophy*. New York: Harper Torchbooks.

Heisenberg, W. (1964). Kvanteteorien og dens fortolkning. In *Niels Bohr: Hans liv og virke fortalt af en kreds af venner og medarbejdere*, Copenhagen: J. H. Schultz Forlag, pp. 90–103.

Howard, D. (2004). Who Invented the Copenhagen Interpretation? A Study in Mythology. *Philosophy of Science*, **71**, 669–682.

Howard, D. (2020). Revisiting the Einstein-Bohr Dialogue. https://www3.nd.edu/~dhoward1/ (accessed on 28 June 2020).

Howard, D. (2021). The Copenhagen Interpretation. This volume.

Hüttemann, A. (2002). Idealizations in Physics. In M. Ferrari and I. Stamatescu (eds), *Symbol and Physical Knowledge: On the Conceptual Structure of Physics*, Heidelberg/Berlin: Springer, pp. 177–92.

Jacobsen, A. S. (2007). Léon Rosenfeld's Marxist defense of Complementarity. *Historical Studies in the Physical and Biological Sciences*, **37 Supplement**, 3–34.

Jacobsen, A. S. (2011). 'Strife about complementarity' and Léon Rosenfeld's conception of Marxism. *Yearbook for European Culture of Science*, **6**, 237–251.

Jacobsen, A. S. (2012). *Léon Rosenfeld: Physics, Philosophy, and Politics in the Twentieth Century*. Singapore: World Scientific.

Jähnert, M., and Lehner, C. (2021). The Early Debates about the Interpretation of Quantum Mechanics. This volume.

Jammer, M. (1966). *The conceptual development of quantum mechanics*. New York: McGraw-Hill.

Jammer, M. (1974). *The Philosophy of Quantum Mechanics: The Interpretations of Quantum Mechanics in Historical Perspective*. New York: Wiley.

Jaynes, E. T. (1989). Clearing Up the Mysteries—The Original Goal. In J. Skilling (ed.), *Maximum Entropy and Bayesian Methods* (Proceedings Volume), Dordrecht: Kluwer, pp. 1–27.

Joas, C., and Lehner, C. (2009). The classical roots of wave mechanics: Schrödinger's transformations of the optical-mechanical analogy. *Studies in History and Philosophy of Modern Physics*, **40**, 338–351.

Knudsen, H., and Nielsen, H. (2012). Pursuing Common Cultural Ideals: Niels Bohr, Neutrality, and International Scientific Collaboration during the Interwar Period. In R. Letteval, G. Somsen, and S. Widmalm (eds), *Neutrality in Twentieth-Century Europe: Intersections of Science, Culture and Politics after the First World War*, London: Routledge.

Kojevnikov, A. (2020). *The Copenhagen Network: The Birth of Quantum Mechanics from a Postdoctoral Perspective*. Cham: Springer.

Kragh, H. (2012). *Niels Bohr and the Quantum Atom: The Bohr Model of Atomic Structure 1913–1925*. Oxford: Oxford University Press.

Kuhn, T. S. (1963). Interview with C. Møller, 29 July, AHQP.

Ladyman, J. (2002). *Understanding Philosophy of Science*. New York: Routledge.

Laloë, F. (2012). *Do We Really Understand Quantum Mechanics?* Cambridge: Cambridge University Press

Landau, L. and Peierls, R. (1983). Extension of the Uncertainty Principle to Relativistic Quantum Theory. In J. A. Wheeler and W. H. Zurek (eds), *Quantum Theory and Measurement*, Princeton: Princeton University Press, pp. 465–476. Originally published in German in *Zeitschrift für Physik*, **69**(56), (1931).

Martinez, J.-P. (2019). Beyond Ideology: Epistemological Foundations of Vladimir Fock's Approach to Quantum Theory. *Berichte zur Wissenschaftsgeschichte*, **42**(4), 400–423. doi: 10.1002/bewi.201900008. Also published in this volume.

Meyenn, K. von (2011). *Eine Entdeckung von ganz au?erordentlicher Tragweite, Band 1, Schrödingers Briefwechsel zur Wellenmechanik un zum Katzenparadoxon*. Heidelberg: Springer.

Nielsen, H., and Nielsen, K. (2001). *Neighbouring Nobel: The History of Thirteen Danish Nobel Prizes*. Aarhus: Aarhus University Press.

Osnaghi, S. (2021). Bohr and the epistemological lesson of quantum mechanics. This volume.

Pais, A. (1991). *Niels Bohr's Times: In Physics, Philosophy, and Polity*. Oxford: Clarendon Press.

Petruccioli, S. (2011). Complementarity before uncertainty. *Archive for History of Exact Sciences*, **65**, 591–624. doi: 10.1007/s00407-011-0087-0

Robertson, P. (1979). *The Early Years: The Niels Bohr Institute 1921–1930*. Copenhagen: Akademisk Forlag.

Rosenfeld, L. (1945). Niels Bohr. An Essay Dedicated to Him on the Occasion of His Sixtieth Birthday, October 7, 1945. Reprinted in *SP*, pp. 313–326. Originally published by North-Holland, Amsterdam, 2nd edn, 1961. Nordita reprint No. 57.

Rosenfeld, L. (1963). Niels Bohr's Contribution to Epistemology. Reprinted in *SP*, pp. 522–535. Originally published in *Physics Today*, **16**(10), 47–52, 54. Nordita reprint No. 110.

Schrödinger, E. (1952). Are there quantum jumps? Part II. *The British Journal for the Philosophy of Science*, 3(11), 233–242.

Segrè, G. (2007). *Faust in Copenhagen: A Struggle for the Soul of Physics*. London: Jonathan Cape.

Seth, S. (2010). *Crafting the Quantum: Arnold Sommerfeld and the Practice of Theory, (1890–1926)*. Cambridge, MA: MIT Press.

Söderqvist, T., Faye, J., Kragh, H., and Rasmussen, F. A. (eds) (1998). *Videnskabernes København*. Roskilde: Roskilde Universitetsforlag.

von Neumann, J. (2018). *Mathematical Foundations of Quantum Mechanics: New Edition*. N. A. Wheeler (ed.), R. T. Beyer (transl). Princeton: Princeton University Press.

# CHAPTER 22

# GRETE HERMANN'S INTERPRETATION OF QUANTUM MECHANICS

ELISE CRULL

## 22.1 INTRODUCTION

IN the 1930s, a time when familiar figures like Einstein, Schrödinger, Bohr, Born, and Heisenberg grappled intensely with the question of how to interpret quantum mechanics, a young doctoral student of Emmy Noether's named Grete Hermann became interested in defending Kant's principle of causality in the face of this new and apparently indeterministic theory. In 1933 Hermann composed a manuscript on determinism in quantum mechanics (Hermann, 1933) which she sent to Dirac and also to Copenhagen, where it was read with interest by Bohr, Heisenberg, and von Weizsäcker. Based on the promise shown in this essay, Hermann was invited to join Heisenberg's colloquium in Leipzig in the winter term 1934–1935. She accepted this invitation. Her visit culminated in two substantial essays: one published in March of 1935 concerning the natural-philosophical foundations of quantum mechanics (Hermann, 1935a), and a more general essay on the significance of modern physics for epistemology, submitted for—and winning—the 1934 Avenarius Prize (Hermann, 1937a).

Hermann's 1935 essay is one of the first, and finest, philosophical treatments of quantum mechanics. Although her aim was to demonstrate consilience between the principle of causality and quantum theory, she far exceeds this goal: she in fact provides the contours of a neo-Kantian interpretation of quantum mechanics. Due to its author's rigorous dual training in mathematics and natural philosophy, this interpretation does particular justice to the intricacies of the theory and offers a view unlike others forming the canon of early interpretations. Along the way she uncovers or makes more perspicuous several key aspects of the theory, including the role of Bohr's twin principles of correspondence and complementarity, as well as the necessity of

referencing observational context when interpreting measurement results. She is also
the first to clearly articulate in print the uniquely quantum-mechanical phenomenon
christened 'entanglement' by Schrödinger later that same year.

In her later philosophical writings she expands on the discussion begun in 1935 of
these characteristic features of quantum mechanics, arguing that the central significance
of this new physics is not, contrary to popular thought, the failure of the law of causality,
but rather the failure of the classical assumption that measurements and observations of
physical systems can be obtained and interpreted independently of context.

In this chapter I fill out the interpretation of quantum mechanics suggested by
Hermann in her 1935 essay, drawing also on aspects of later published contributions to
her larger natural philosophy project. In so doing I examine how Hermann was led to
appreciate the import of this theory for radically reconceptualizing not just natural
science, but all domains of human inquiry. We begin with an overview of Hermann's
background context in section 22.2, and then move to a discussion of her interpretation of
quantum mechanics in section 22.3. The final section, 22.4, offers some reflections on
Hermann's interpretation when placed rightfully within the established narrative of
foundations of quantum mechanics, and when considered alongside contemporary views.

## 22.2 HERMANN'S PHILOSOPHICAL, MATHEMATICAL, AND SCIENTIFIC CONTEXT

Grete Hermann was born in 1901 in Bremen, and by age 24 was one of few women in
all of Germany who had earned a doctoral degree. She matriculated at the University of
Göttingen in 1921 to study mathematics, philosophy, and physics—a fortuitous time to
study these subjects as Göttingen was then in its halcyon days, not only with respect to
political and social advancement (which in part accounts for Hermann's being permitted to pursue a doctorate there as a woman and under the advisement of a woman:
Emmy Noether), but also boasted renowned faculty and students in mathematics and
physics. In addition to Noether, the mathematics department was then home to Hilbert
and his student Paul Bernays (who along with Reichenbach would become a teacher of
Carl Hempel), visiting fellow J. von Neumann, Felix Klein, and Richard Courant. Physics
at Göttingen boasted not only theoretical prowess with the likes of Born (with students
Robert Oppenheimer and future Nobel laureate Maria Goeppert Mayer), Heisenberg,
Jordan, and Hund—but also notable experimentalists like James Franck, who was in
daily conversation with Born's circle (van der Waerden, 1968, p. 19).

After completing a mathematics doctorate under Noether (and thereby gaining the
dual distinction of being Noether's first doctoral student and only female doctoral
student), Hermann became deeply involved in philosophy and joined the circle of

Leonard Nelson. Nelson was a prominent philosopher and close friend of Hilbert's who spent his too-brief career establishing and evangelizing for the New Friesian School, which followed the natural-philosophical tradition of neo-Kantian Jakob Friedrich Fries (Paparo, 2017, p. 40).[1] This tradition included, among many other interesting tenets, the call to engage in critical philosophy not only at the theoretical level but in addition deducing—and then implementing—practical aspects of these claims. The former project was accomplished through application of Kant's transcendental schema in the specific manner advocated by Fries, who argued that the twelve categories of pure concepts of understanding—i.e. the conditions which make knowledge possible, including natural knowledge (which is of course the aim of science)—are not, contra Kant, a priori concepts to be adhered to strictly. They are to be interpreted instead as empirical concepts which form a loose pattern qua analogies for organizing sensory experience. In this way Fries attempted to reframe critical philosophy as an explicitly scientific method for obtaining, ordering, clarifying, and ultimately unifying all knowledge.[2]

Because of Fries' relocation of the categories of understanding from the domain of the a priori to that of the empirical, his interpretation was frequently accused of psychologism: it rendered Kant's schema overly dependent upon the particularities of human cognition. Nelson presumably wished to avoid these accusations by approaching Fries' framework using a strict method of working backwards from the data of experience to analytic principles, then trimming away propositions which played no role in this process. Paparo writes of Nelson's rigorous philosophical method: '[I]t works regressively from the consequences back to the reason and discards all other unnecessary characteristics from the original judgement. This process does not bring new knowledge (since it is deductive) but causes a transformation; through reflection, vague, confused judgements are transformed into clear concepts' (Paparo, 2017, p. 42).

Thus was the Nelsonian approach to the theoretical task of Friesian critical philosophy. For the practical task, Nelson believed he could use the precise, rational judgments obtained from the theoretical process to construct *scientifically grounded* educational, legal, and ethical systems. I note here a point of departure for Nelson from Fries which shall later become important in our analysis of Hermann: despite Fries' reinterpretation of the categories as analogies, Nelson maintained the hope of constructing new systems of knowledge by applying at least the category of causality in a strict, absolute way. For example, Nelson used his natural-philosophical method to deduce a rigorous causal procedure for making *moral* decisions (Leal, 2017, sections 2.3–2.4).

---

[1] For the following overview of Fries, Nelson, and the natural-philosophical tradition I have drawn on the just-cited work of Paparo as well as Leal (2017) and Soler (2017).

[2] This arguably places Friesian neo-Kantians in closer step with the Marburg neo-Kantians than the Baden school, in that the Friesians and the Marburg school were primarily concerned to show that the transcendental method was not only consilient with advances in natural science but provided the very conditions for such knowledge (see the editor's introduction to Luft (2015)).

It was through her intimate work alongside Nelson and her enduring interest in physics that Hermann would be inspired to address the perceived conflicts arising between the worldview of Nelsonian natural philosophy and that of modern physics. Of course Nelson and his students weren't the first neo-Kantians to appreciate that modern physics, in particular the theory of relativity and quantum mechanics, would pose serious threats to central aspects of Kant's philosophy. One such threat concerned Kant's insistence on the a priori status of space and time qua pure intuitions. Another threat was Kant's inclusion of causality among the categories of understanding, indicating that causal relations were essential for the project of ordering sensory experiences. While the introduction of non-Euclidean geometries via general relativity was an affront to Kant's thoroughly Newtonian conceptions of space and time, the apparent indeterminism of the new quantum theory challenged to the core the notion of Kantian causality.

The fathers of quantum physics, while philosophical in their own way, did not have significant exposure to the nuances of neo-Kantian scholarship: they were, rightly so, more interested in understanding the strange theoretical implications and empirical results of the theory. And for their part, neo-Kantian philosophers attempting to reconcile modern physics with transcendental idealism lacked the mathematical training necessary for comprehending the terse operator calculus of Dirac and the obscure matrix mechanics favoured by Göttingen physicists. Indeed, the more scientifically minded neo-Kantians of the time (e.g. Schlick, Cassirer, Reichenbach) centred their efforts on reconciling non-Euclidean geometry with relativity. What little they said concerning quantum mechanics and causality indicated their acceptance of the belief that this physics posed a nontrivial threat to the principle of causality.

Enter Hermann: the unique climate she participated and thrived within in 1920s Göttingen, coupled with her intensive dual training in abstract mathematics under Noether and in Friesian neo-Kantianism and natural philosophy under Nelson, ideally situated her to wrest from quantum mechanics whatever vestiges of causality could be found therein. And this she did, answering the familiar challenge with a bold and novel response: she claimed, quite in contradiction with public opinion on the matter, not only that quantum mechanics gave no proof of genuinely acausal events, but that the physics was in fact already causally complete (albeit *retroactively*). Furthermore, she insisted that the truly significant result for natural philosophy from quantum mechanics wasn't to do with the survival of the causal principle but about a different aspect of transcendental idealism altogether: quantum mechanics shows that the Kantian ideal of applying the categories of understanding to the data of experience in an absolute, objective manner was quite impossible—thus Fries' reconceptualization of Kant stands justified. Indeed: quantum mechanics shows that one has to deal with irrevocably context-dependent observations and inferences in order to obtain *any* knowledge, even beyond the domain of physics.

Hermann did not come to this radical thesis directly. Her first essay on philosophy and quantum mechanics is a 1933 manuscript titled 'Determinism in Quantum Mechanics' (Hermann, 1933), in which one finds Hermann attempting a more traditional approach to the question of causality. There she lays down a two-pronged argument to the effect that the way of determinism is not yet shut: on the one hand, no one had yet succeeded in showing that such a deterministic completion was *impossible*, while on the other hand, employing Dirac's formalism one could remove all and only those aspects of the theory which render it 'indeterministic' without loss of coherence and empirical adequacy. Though she would change her mind regarding the incompleteness question, this early manuscript makes important headway in terms of the philosophy of quantum mechanics, not least by the following. In arguing for the first prong, Hermann provides the first known proof of the inadequacy of von Neumann's 1932 no-go theorem for hidden variables. She also takes an interesting stab at interpreting Heisenberg's uncertainty relations. In arguing for the second prong, she engages in a very interesting discussion on how—and how not—to interpret the wavefunction.[3]

Because Hermann had relied on Dirac's formalism for her arguments in the 1933 essay, she sent a copy to Dirac along with a cover letter asking for feedback.[4] More importantly, perhaps, she also sent a copy of her essay to the physicists at Copenhagen. We learn from a 17 December 1933 letter to Hermann from her life-long friend Gustav Heckmann—who'd been a student of Born's in Göttingen, was an avid Nelsonian, and was at that time visiting Copenhagen—that Bohr, Heisenberg, and Heisenberg's star pupil von Weizsäcker had all read her work with seriousness, and intended to compose a response.[5] Weizsäcker also writes to Hermann on that same day, thanking her on behalf of Bohr for sending the manuscript. He then engages with her essay in some detail, agreeing that indeterminism is not logically entailed by the theory. Weizsäcker discusses the gamma-ray microscope experiment, and from notes Hermann has scribbled on this letter it is clear she has started to think more deeply about the relationship between macroscopic and microscopic phenomena, and the different role measurement apparently plays in investigations of the latter (K. Herrmann, 2019, pp. 439–444). This letter marks the beginning of a long and fruitful exchange—both in person and via frequent correspondence—between Hermann and Weizsäcker concerning philosophy and physics.[6]

---

[3] For more on Hermann's 1933 manuscript, especially as it relates to her 1935 essay, see Crull and Bacciagaluppi (2017b). Detailed analyses of Hermann's argument for the circularity in von Neumann's proof can be found in Seevinck (2017). To read the 1933 manuscript in the original German, see K. Herrmann (2019, pp. 185–203); to read the first English translation of the same, see Hermann (2017a).

[4] The letter is transcribed in K. Herrmann (2019, p. 435); no evidence of a response from Dirac has been discovered.

[5] This letter is transcribed in K. Herrmann (2019, pp. 437–38), translated into English in Crull and Bacciagaluppi (2017a, pp. 221–22), and discussed more thoroughly in Crull and Bacciagaluppi (2017b).

[6] Indeed: they trade several letters concerning quantum mechanics in the short period following the letter of 17 December 1933. I do not discuss the content of this early correspondence here, though not for lack of great material. The reason for this is that, as we shall see, Hermann comes to a rather different understanding of quantum mechanics by the time she publishes her 1935 essay and it is this 'mature' interpretation of the physics which undergirds her subsequent philosophical work.

Heisenberg was so impressed with the potential exhibited in Hermann's 1933 manuscript that he invited her to attend his quantum mechanics lectures in Leipzig, which she did as early as February 1934 (K. Herrmann, 2019, p. 457). While she was there, she had many deep conversations with Heisenberg and his *Leipzigkreis* (including Weizsäcker) about natural philosophy and transcendental idealism.[7] It is after these travels and conversations that Hermann sits down to compose her chief published works in the philosophy of quantum mechanics: the 1935 essay titled 'The Natural-Philosophical Foundations of Quantum Mechanics', and her prize-winning entry in the 1934 Saxon Academy of Science essay competition on the topic 'The Significance of Modern Physics for the Theory of Knowledge'.[8]

Despite constant political upheaval and imminent threat of danger in these years due to her activism against the Nazi state, Hermann continued to pursue the more general natural-philosophical task put before her by her investigations of quantum mechanics. In particular, in the period after her participation in Heisenberg's seminars at Leipzig in 1935 until 1948 she publishes five further pieces concerning physics—three brief review articles and two original essays—which will inform my analysis here, though to a lesser degree than her earlier essays focusing on quantum mechanics. She is also in touch with many prominent thinkers on the Continent as is evident from her correspondence (which I also draw on in my analysis, but again to a lesser degree than the essay of 1935). From correspondence we may infer that her 1935 treatise was either read, or copies of it were requested, by Heisenberg, Weizsäcker, Bernays, Laue, Scholz, Kratzer, Bohr, and Pauli. There is also evidence that she sent or intended to send copies of the essay to others of the European intelligentsia, among them Einstein, Schrödinger, Courant, Born, de Broglie, Kraft, and Titeica (K. Herrmann, 2019, Part III). Thus although prima facie Hermann's work in the philosophy of physics may appear a rather minimal contribution—just one published essay and one short article from 1935!—upon further investigation it becomes clear she was active in various thriving intellectual circles in this period, and by all visible accounts deeply respected therein. All the more reason to recover her story in full detail; all the more tragedy for our late coming to this task.[9]

---

[7] One can read in K. Herrmann (2019, Part III) the wonderful missives Hermann sends her mother describing heated disputes with Heisenberg and others during her time in Leipzig.

[8] The 1935 essay was first published in March (which one will note pre-dates the EPR paper) in the journal of the New Friesian School (Hermann, 1935a); it is reprinted in German in K. Herrmann (2019, pp. 205–58) and translated into English as Hermann (2017b). Also in 1935 Hermann published in *Die Naturwissenschaften* excerpts distilled from her longer essay (Hermann, 1935c); this article was first translated into English as Hermann (1999) and is given a new English translation in Bacciagaluppi and Crull (2022). Her prize-winning essay was not published until 1937 (Hermann, 1937a). It is reprinted in K. Herrmann (2019, pp. 335–77) and an English translation is in the works.

[9] I would be remiss not to make special mention here of those precious few but crucial early works on Hermann's philosophy of quantum mechanics. First to my knowledge was Jammer, who in his canonical text (Jammer, 1974) is cued by Heisenberg's memoirs (which included a chapter about philosophical conversations he'd had with Hermann and Weizsäcker in Leipzig) to briefly introduce Hermann and

The first short piece Hermann publishes following her 1935 essay is a review of Popper's newly published *Logik der Forschung* (Hermann, 1935b). The second short piece is a review published in *Erkenntnis* of a talk by Schlick (but delivered by Philipp Frank) at the 2nd International Congress for Unity of Science in Copenhagen in June of 1936 (Hermann, 1936).[10] Hermann finishes her summary of the talk with a few tantalizing comments hinting that a Friesian is able to provide satisfactory explanations of precisely those aspects of quantum mechanics causing trouble for the positivist (e.g. Schlick and his compatriots of Vienna Circle fame). She does not go into details but instead refers the reader to her long essay (Hermann, 1935a). The final short piece is a concise statement of her natural-philosophical position prepared on the occasion of the Congrès Descartes held in Paris during August of 1937 (Hermann, 1937c).[11]

As for her substantive post-1935 contributions to the philosophy of physics, in 1937 Hermann publishes, again in the journal of the New Friesian School, an essay titled 'On the foundations of physical statements in older and modern theories' (Hermann, 1937b).[12] There she provides further insight into her interpretation of quantum mechanics by setting it alongside a Nelsonian–Friesian interpretation of various other theories of modern physics: special and general relativity and Maxwellian electromagnetism.

That Hermann remained convinced of the applicability of natural-philosophical lessons deduced from quantum mechanics to all fields (including—of special importance to her and fellow Nelsonians—pedagogy, psychology, and ethics) is evident in that despite the horrors of the War and the ensuing chaos of restructuring Germany's educational system, Hermann continued to correspond, debate, and publish regarding this topic. The last essay I shall consider appeared in the interdisciplinary journal

---

describe her thesis that quantum mechanics was causally complete via retrodiction. He also includes mention of her no-go proof of *von Neumann's* no-go proof. From K. Herrmann's volume (2019, pp. 599–610) we learn that when Jammer wrote to Hermann in 1968 to ask if the portions of his forthcoming book concerning her were accurate, he took the occasion to express several criticisms of her retrodictive causality thesis—nevertheless concluding that 'none of this detracts from the importance of your work, and in particular does not affect your emphasis on the relative nature of the state description' (K. Herrmann, 2019, p. 603; my translation). Hermann responds graciously that while his description of her work is accurate, his criticisms of it are not! Despite Jammer's inclusion of Hermann in this foundational book, no further work on her philosophy of physics appeared until the late 1990s when Dirk Lemma translated into English for the first time the abridged version of Hermann's 1935 essay as published in *Die Naturwissenschaften* (Lumma, 1999). Last of the first Hermann Evangelists is Léna Soler, who in 1996 authored an introduction and substantive postface to accompany the first French translation of Hermann's 1935 essay (Hermann, 1996); since then Soler has been actively contributing to scholarship on Hermann regarding her philosophy of science and her neo-Kantianism.

[10] In a comprehensive review of the conference proceedings given by Werkmeister, the author writes that the Congress began 'under the shadow of death', as news of Schlick's murder reached them during the first morning session (Werkmeister, 1936).

[11] These three short pieces are all reprinted in K. Herrmann (2019, Part II); they are translated into English as Hermann (2020a,b,c) with an introduction by Bacciagaluppi (2020).

[12] Reprinted in K. Herrmann (2019, pp. 275–334). An English translation of this essay is in the pipeline.

*Studium Generale* in 1948 and was simply titled 'Causality in Physics' (Henry-Hermann, 1948). There Hermann holds the line she had maintained since 1935: what philosophers must give up in order to maintain consilience with modern physics is not the sacrosanct assumption that continuous causal chains underlie all natural processes, but rather the hope—dear to neo-Kantians and positivists alike—of unifying all science within a single, absolute, objective framework.

## 22.3 HERMANN'S INTERPRETATION OF QUANTUM MECHANICS[13]

In good keeping with the natural-philosophical tradition to which she subscribed, Hermann's interpretation of quantum mechanics begins with an examination of the facts of experience. From grounds of empirical evidence she will apply the logically coherent and (she argues) logically complete Dirac formalism of quantum mechanics in order to pare away from the central concepts of the theory any aspect not strictly deducible from experimental data or the mathematical formalism. What remains, however improbable, must be taken into consideration for any future project in natural philosophy. Recall that for the Nelsonian–Friesian, one's philosophical propositions are to be revised in light of scientific developments, not vice versa. Because her methodology for interpreting quantum mechanics is philosophically rigorous, it is important to present her conclusions in the order they are originally given rather than as a list of decontextualized propositions. Indeed, Hermann finishes her introductory remarks in the 1935 essay with a comment about the necessity of interpreting quantum mechanics which makes clear her particular reasons for proceeding as she does. She warns that 'portrayals' of quantum theory given by physicists should not 'be taken on without scrutiny':

> For these portrayals—understandably—far exceed what has directly proved itself reliable through successful predictions or the correct rendering of observed data. This direct empirical corroboration pertains strictly speaking only to the formalism of the theory and its correspondence to the data of observation. The conceptual interpretation of the formalism that must necessarily accompany it in every physical theory only received direct physical corroboration in as far as the corresponding formalism proves to be an appropriate means of describing observed events. Where this interpretation is ambiguous, then, one can only make a physical-empirical decision between the different views if they give rise to different extensions of the formalism.   (Hermann, 2017b, p. 241)

---

[13] For reasons described above, I consider Hermann (1935a) to be the best representation of her mature view of quantum mechanics. Thus this text will serve as my *locus classicus* in this section.

Unsurprisingly, she then immediately turns to a discussion of experiments that demonstrate wave–particle duality. These data irrefutably demonstrate the breakdown of classical assumptions about one's ability to provide a full catalogue of determinate (or at least, determinable) properties of an isolated system. Enter the Heisenberg uncertainty relations: despite the fact that in quantum mechanics one experiences a distinctly nonclassical interdependence of measurable quantities, the uncertainty relations provide the precise domain within which we can safely apply classical concepts to quantum phenomena. As Hermann will argue, however, while one may apply the usual classical concepts (e.g. position, momentum, trajectory) within the limits prescribed by the uncertainty relations and through use of Bohr's correspondence principle, one must never forget that these concepts, though familiar, are being applied to quantum phenomena as mere analogies. More on this philosophical point comes only after Hermann has laid down some further principles about the (in classical physics, pedestrian) task of relating data to the formalism. As this task crucially relies on one's ability to make causal inferences, let us begin with an exploration of Hermann's thesis of retrodictive causality.

## 22.3.1  The Retrodictive Causality Thesis

What exactly is the nature of the indeterminacy necessarily introduced into quantum mechanics by the Heisenberg relations? If this indeterminacy represents anything deeper than a circumscription of knowledge, then the Kantian principle of causality would indeed be dealt a crippling blow. As a practicing neo-Kantian, Hermann understands that her entire project hangs on a careful analysis of this point of departure from classical, causally complete, theories. She writes: 'Whereas classically the state of a system can be expressed through a mere enumeration of the values of all occurring physical quantities, the quantum mechanical formalism employs novel symbols in the description of the state that express the mutual dependence of the determination of different quantities' (Hermann, 2017b, p. 244). These novel symbols are the uncertainty relations along with 'the probabilistic interpretation of the Schrödinger [wave] functions based on them', and Hermann's task is to examine 'whether they already show more than the merely provisional limitation of future predictions' (Hermann, 2017b, p. 246).

The examination carried out with respect to the Heisenberg uncertainty relations in her 1935 essay is repeated with greater clarity in her 1948 article; there she concludes that the position–momentum uncertainty relation can be interpreted neither as an epistemic limitation—where 'we think of it subjectively as the statement of an unavoidable ignorance [Unwissenheit] about the position and motion of the relevant corpuscle'— nor can we in a naive, ontological way interpret this relation 'objectively, and ascribe the quantities $\Delta q$ and $\Delta p$ to the examined physical process itself as objective features' (K. Herrmann, 2019, p. 387; my translation). Under the epistemic interpretation one represents the electron 'as a little flying corpuscle, for which—due to the disturbance of its motion by any measurement—the exact determination of position and velocity simply cannot be verified' (K. Herrmann, 2019, p. 387; my translation). However, as she writes a little farther on:

> The subjective interpretation fails already with dualism experiments. They force the realization that the elementary particle is not absolutely a corpuscle, else interference phenomena would be impossible. The uncertainty relation therefore cannot mean that position and momentum values objectively pertain to the electron, of which the physicist just cannot have arbitrarily exact knowledge. Experience shows instead that the position and momentum variables of the electron can only be known within the limits defined by the uncertainty relations.
>
> (K. Herrmann, 2019, p. 387; my translation)

But neither can one interpret the uncertainty relation in an objective way:

> [W]e get into difficulty immediately if we ask what happens when, say, we perform an exact measurement of position of an electron with quantity $\Delta q$. Such a measurement is possible, according to the principles of quantum mechanics. Its result can only be obtained as one of the results predicted by the rules of probability. It leads to a new determinate state of the electron solely in terms of position, fixed within a small uncertainty limit $\Delta q$, whereas the momentum, which was previously sharply determined, exhibits a large uncertainty interval. If we stick to the objective interpretation of the indeterminacy relation, we get a picture wherein a widely spread wave-peak has experienced an influence at the instant of measurement, and suddenly has contracted into a narrow—but for that reason tall, and so rapidly diverging—wave-peak. Correspondingly, an exact measurement of velocity must be associated with a sudden flattening of an initially steep and narrow wave-peak. Because of the uncertainty of their results, these processes must pertain to the allegedly 'acausal' occurrences.
>
> Now admittedly quantum mechanics has established that any measurement of the atomic is associated [verbunden] with an interaction between measuring instrument and measured physical object, through which the state of the object is changed. But these changes are presented to us here as something like wave-peaks undergoing very strange processes of contraction and propagation. For example: the idea, inadmissible in modern physics, of a wave train of astronomical proportions that, when need be, can contract from any spatial dimension to the size of an atom in an instant, and in so far as it does this, with an excessive velocity. Even aside from peculiarities of this type, it can be shown that the quantum mechanical formalism has no place for the postulation of such events.
>
> (K. Herrmann, 2019, pp. 387–388; my translation)

Clearly, then, because an objective interpretation of the uncertainty relations necessitates the postulation of the superluminal collapse of a wave packet—a physical process for which she sees no evidence in nature—such an interpretation is inadmissible. She continues her argument against this interpretation, saying that even aside from strange collapse events and the like, such a reading of the uncertainty relations must succumb to Heisenberg's cut argument: no matter at which place (the quantum mechanical or classical mechanical side of the cut) or to which system (measuring device or thing measured) one attempts to supplement the formalism with 'real' properties (read: hidden variables), the result is over-determination with the formalism (K. Herrmann, 2019, pp. 388–389).

In the 1948 essay Hermann is in strong agreement with Heisenberg's presentation of the cut argument in his unpublished reply to EPR in the summer of 1935 (we note: which he wrote after, and with explicit reference to, Hermann's 1935 essay). There he uses the cut argument to demonstrate that any attempt to supplement quantum mechanics with hidden variables will result in over-determination of the statements of quantum mechanics, ergo quantum mechanics cannot in this way be made causally complete.[14] Of course in Hermann's 1935 essay her central task is to demonstrate that quantum mechanics poses no threat to the concept of causality, and so her argument is pitched slightly differently. There she emphasizes two points: (i) all a Kantian needs to assume in terms of causation—indeed all she should expect, even for classical theories—is that for every event there exists some cause which has determined it. And (ii) although quantum mechanics is not deterministic in a way that permits the prediction of measurement outcomes, it nevertheless still satisfies the requirement of a continuous causal chain leading to events. This causal chain is not available before a measurement takes place (and this is due to the relative context of observation—more on which anon), but it is available after measurement via retrodiction.

Ad (i): The concept of predictability is, Hermann points out, often bound up with causality. Understandably so, for classical mechanics has trained us to think this association is more than fortuitous. Yet one must not conflate association with equivalence, for conceptual analysis reveals that the concept of causality and the concept of predictability are inequivalent. Although certainly Hermann was primed to endorse a relaxing of Kant's principle of causality through her agreement with Fries' interpretation of the twelve categories of understanding as mere analogies not to be applied strictly to the data of experience, she makes an independently strong case for the disentangling of these concepts. Although it is indeed difficult to see how predictability is *not* implied by an assumption that to each effect there exists a cause, this difficulty (so argues Hermann) is due to our expectations being so powerfully shaped by everyday experience. Hermann presses in here, stating that quantum mechanics has certainly shown us a new feature of the world—but it is not the existence of truly indeterministic events. Quantum mechanics not only leaves causality intact but even helps to sharpen our understanding of it. As Nelson's natural-philosophical method promises, analysis allows us to shave away from the concept of causality its unnecessary (but undoubtedly useful) fat: association with predictability. And we see this is so for all

---

[14] See Bacciagaluppi and Crull (2022) for the first full English translation of Heisenberg's manuscript response to EPR.

instances in which we apply the concept of causality, even beyond the scope of quantum mechanics.

Ad (ii): Though the uncertainty relations (grounded in experiment ergo insurmountable) imply the inability to predict the exact outcome of a measurement, after the measurement has taken place one can trace a continuous causal chain for the system of interest using its wavefunction (which evolves univocally *between* interactions) from the measurement outcome back to its initial state. So for example, in the gamma-ray microscope thought experiment one measures the desired property of the electron by placing, say, a photographic plate in either the image plane or focal plane of the microscope (respective of which property one wishes to measure, position or momentum). After the photon interacts with the electron and passes into the microscope, it records on the photographic plate certain information about the electron's state at the time of their collision. Hermann argues that a continuous causal chain can be inferred by tracing the state of the photon backwards in time from the darkening of the photographic plate to the moment of its collision with the electron.

Hermann's thesis that causality is recoverable retrodictively within quantum mechanics seems to have caused a stir among the neo-Kantians, who did not think a causal chain thus identified was sufficiently causal. Fellow Nelsonian and mathematician Paul Bernays exchanged several letters with Hermann on this point. In response to Bernays' criticisms, Hermann stresses what she'd already written in her essay: what is required of the Kantian is even less strict than the ability to pick out the correct, determinate sequence of cause and effect (K. Herrmann, 2019, pp. 497–500). After walking her readers through the gamma-ray microscope thought experiment (including her fascinating introduction of a third scenario where the measurement takes place at infinity and one is left with a linear combination of the electron and photon wavefunctions in which individual states are no longer separable (see Hermann, 2017b, p. 258, and Crull, 2017), Hermann admits that the existence of the causal chain available only after the measurement is, in this case, only indirect, and relative to the particular context of the given observation.

From the perspective of modern measurement theory, Bernays is correct to argue that such a process does not in general allow us to specify a unique causal chain, this due to the failure of one-to-one mapping between states of the system and of the apparatus.[15] As we shall see below, Hermann herself is clearly aware—in spirit, if not in these words—that retrodictive causality is not a satisfactory answer to the measurement problem if it is understood as the problem of how a *particular* definite outcome came to be measured. It is, however, a satisfactory answer to the measurement problem if by it one means the puzzle of how *some* definite outcome or other came to be measured. And, as Hermann argues, answering this latter question is sufficient for demonstrating causal continuity in quantum mechanical processes.

---

[15] A detailed analysis of measurement theory and Hermann's 1935 essay is in Bacciagaluppi (2017).

Thus Hermann's answer to the completeness question at the heart of 1930s interpretation debates is as follows: attempts to find causal completions to quantum mechanics are futile, not because (say) the insertion of hidden variables will result in over-determination on one side of the cut or the other, but rather because quantum mechanics is already complete. Indeed: she seems to indicate by her reasoning in the 1935 essay that one begins from the experiments, proceeds *to* the analytically cleaned-up concept of causality (with the help of the uncertainty relations and the correspondence principle uniting the formalism to the data), and from this concludes causal completeness. The cut argument, invoked as it is only late in the 1935 essay, merely reinforces this earlier conclusion derived from application of the natural-philosophical method.[16]

## 22.3.2 The Relative Context of Observation Thesis

Although Hermann believes she has resolved the causality issue satisfactorily, she must yet explain why, if this causal chain really exists, it cannot be found out beforehand. In other words: why, despite the causal completeness of quantum mechanics, do we find ourselves in the following situation? 'For any quantum mechanically characterized state of a physical system there are measurements whose results cannot be predicted on the basis of knowledge of this state; in such measurements one tests sometimes this, sometimes that result' (Hermann, 2017b, p. 253). In modern parlance, she here admits that her retrodictive causality thesis does not yet do what interpretations of quantum mechanics are supposed to do: explain whence this *specific* measurement outcome.

Hermann responds to the problem she has just posed in the section immediately following, unambiguously titled 'The Solution: The Relative Character of Quantum Mechanics'. It is perhaps a matter for later discussion whether or not she satisfactorily responds to the problem of specific measurement outcomes according to modern interpretations of quantum mechanics, but I suspect her interpretation holds its own in that arena. I limit my analysis in this chapter, however, to the way in which the relative character of quantum mechanics is meant to explain why the measurement problem persists despite the causal completeness of the theory. It is also in her reasoning on this point that she gives, one might suggest, a better description of Bohr's interpretation—of what really should have been the 'Copenhagen interpretation'—than Bohr ever did himself.[17]

---

[16] For a more detailed comparison of the use of the cut argument by Heisenberg and Hermann, see Bacciagaluppi and Crull (2022).

[17] That Hermann is 'more Bohr than Bohr' was already pointed out by Jammer (1974). In previous written work and presentations, Bacciagaluppi and I have attempted to further substantiate this claim of Jammer's: see for example Crull (2017), Bacciagaluppi (2017), and chapter 4 of Bacciagaluppi and Crull (2022). Bacciagaluppi has an additional article in preparation tentatively titled 'Better than Bohr', in which this claim forms the central thesis.

Earlier I mentioned that for Hermann the duality experiments necessitate Heisenberg's uncertainty relations, and that these equations coupled with Bohr's correspondence principle provided the technical apparatus for prescribing precisely the limits within which one can apply classical, intuitive concepts. It is only within these boundaries that such concepts have meaning. Thus far we know the 'why' of it: why we, in quantum mechanics, are obligated to bring otherwise mutually contradictory concepts, images, and even descriptions alongside one another to account for observed physical processes. Yet the 'how' question remains outstanding: how do we unite mutually contradictory concepts, images, and descriptions? It is to answer this question that Hermann introduces Bohr's principle of complementarity. Her take, however, is that this principle involves three different, yet non-separable, notions: each provides a slightly different answer to this 'how' question, and taken altogether they explain the different aspects of indeterminacy arising in quantum mechanics.[18]

The first aspect of complementarity is that which emerges directly out of the dualism experiments: the wave picture verses the particle picture. The second occurs *within* a single picture, where there exists complementarity among certain relevant degrees of freedom: e.g. in the particle picture one encounters complementary between position and momentum. Lastly and most importantly is arguably the 'aspect' of complementarity Bohr seems to have leaned on most heavily: in order to provide a satisfactorily coherent description of natural processes, whether microphysical or macrophysical or somewhere between, there necessarily exists a complementary relationship between quantum modes of description and classical modes of description. Within the quantum mechanical mode of description, constructed (as explained below) using unintuitive symbols, one gives up intuitiveness yet retains completeness (as classical descriptions ultimately fail with respect to quantum phenomena). Within the classical mode of description one maintains intuition but at the expense of completeness. It is in this context that Hermann first introduces Heisenberg's cut argument—to further illustrate the complementary relationship which exists between the classical and the quantum modes of description.

The story regarding intuition within these modes of description draws heavily on Kantian terminology. I have written in some depth about this aspect of Hermann's interpretation in Crull (2017, section 10.3), but here I sketch the essential ideas. Crucial to Kant's transcendental method is a division between intuitive (*anschaulich*) concepts and unintuitive ones. For Kant the data of experience are transformed into objects within the scientific understanding by applying the concepts of pure intuition (space and time) to the sensible data of experience. In the case of classical theories this might be depicted as follows: the empirical data are directly applied or inserted into the classical equations using intuitive, classical concepts (in what Kant calls 'schematic' mode of representation). But in cases where the data are not directly sensible—to wit: the unobservable processes described by quantum mechanics—a gap opens up

---

[18] On Bohr's twin principles of correspondence and complementarity, see the beautiful section 13 of Hermann (2017b).

separating the direct application of data to theory: between the unintuitive, non-sensible empirical data on one hand and the formalism on the other, one now requires a bridge. Since for Kant this bridge can only be constructed from the available forms of pure intuition and the categories of understanding—and yet we know that these categories of understanding cannot apply in the usual sense of direct representation—we must rely on a *symbolic* mode of representation. Certain symbols act as analogies for relevant intuitive concepts, providing a relationship for the unintuitive relata.

Hermann applies this Kantian apparatus to quantum mechanics as follows: the classical, intuitive concept of position in three-dimensional space cannot apply within quantum mechanics, as the state of a quantum system is doubly unintuitive: (i) it cannot be assigned fully determinate values for all of its degrees of freedom in a given context, and (ii) it is represented in the formalism by a symbol—the wavefunction—which is unintuitive because it exists in a configuration space which is generally larger than three dimensions. We form our bridge between the unintuitive empirical results of quantum mechanical experiments and the unintuitive symbols in the quantum formalism by using intuitive concepts—in this example, the classical concept of position—but remember that this bridge only links the relata with the strength of *analogy*.

One notices that the use of analogy in Hermann's interpretation is as yet merely playing its usual Kantian role. What her particular Friesian perspective further yields (as described beginning in section 15 of the 1935 treatise) is the radical claim that due to the relative context of observation, *no* connection between the data of experience and a formalism, whether in classical or in quantum mechanics, can be simply interpreted as absolute or objective. All such connections, in a way, must thereby be interpreted using analogous concepts. Furthermore, due to the constraints of the uncertainty relations, this bridge can only exist and can only meaningfully refer to its relata when the whole contraption—bridge plus relata—are applied *within a single observational context*.

This perspective is what will motivate her interpretation of the other novel symbol of quantum mechanics: the wavefunction. In a letter of 17 October 1935 in which Hermann is responding to Bernays' lengthy critique, she reminds him that a probabilistic interpretation of the wavefunction is only half of the picture: a correct interpretation of the wavefunction must retain the complementary 'double character' of being fully deterministic when applied within a single observational context (that is, between interactions) *and* being probabilistic when applied across multiple observational contexts (K. Herrmann, 2019, pp. 498–499).

And now she is prepared to make her grand conclusion: there are, it is true, revisions forced upon natural philosophy by quantum mechanics. But these revisions have nothing to do with causality (which she has shown can be maintained) but rather with this: a Friesian interpretation of the categories of understanding coupled with the necessity of describing natural processes only ever with respect to a particular observational context compel us to abandon all hope of ever developing a unified scientific framework which is objective and absolute.

As she argues in the final sections of her 1935 essay dedicated to natural-philosophical conclusions (Hermann, 2017b, p. 271 ff.), Kant's categories without the help of Fries cannot adequately capture the process by which we gain scientific understanding. Returning to her first major point about causality, Hermann claims that if all the physicist had was Kant's principle that each effect has a necessary cause then the physicist is left 'with only a strangely empty formal schema in his hands' whereby one would need to locate precisely defined causes for each event for increasingly infinitesimal time intervals. This is clearly impossible, and also not what one really does when one does science. What one really does in the pursuit of scientific understanding is couple the causal relationship with various other spatial *and* temporal observations, and this goes beyond any 'pure' application of the cause–effect relationship described by Kant. Fries, asserts Hermann, does better. Arguing this point in true Nelsonian fashion, she starts from physics itself. There the formalism makes particular use of differential equations to express law-like relationships between systems, and differential equations depict a richer connection between concepts than mere cause and effect. In particular, they 'put the time derivative of these [individual state] variables into functional relationships with the variables themselves' (Hermann, 2017b, p. 272). Since differential equations make individual causes and effects impossible to isolate, a strict application of the causal principle in Kant's sense would be impossible. Must we then forsake the concept of causality for any theory whose dynamical laws are differential equations? Not if we turn to Fries: applying causality in his looser, analogous sense allows the transcendental schema to accommodate differential equations as central features of our scientific understanding.[19]

The final piece of Hermann's interpretation takes the lessons of quantum mechanics for natural philosophy and applies them more generally. She has argued forcibly that due to the necessity of ultimately entering a quantum mechanical mode of description at the atomic level, *any* scientific model must be limited to depicting a network of relationships only relative to a given context. This point established, she begins the final section of her 1935 essay—provocatively titled 'The Splitting of Truth'—with these words:

> These considerations lead to an ever deeper connection between the results of quantum mechanics and the reflections of the critical philosophy. The proof in transcendental idealism that natural knowledge is inadequate for capturing reality but rather only picks out, in an incomplete way, a relational network whose foundations remain indeterminate within the scope of this knowledge, opens the way for the possibility of different mutually independent yet mutually compatible modes of confronting reality through perception.   (Hermann, 2017b, p. 276)

---

[19] Bernays, being a philosopher and a mathematician, would complain in particular about Hermann's interpretation of differential equations here, saying that these equations can only, qua mathematical relationships, depict 'virtual' causal relationships and not 'real ones' in nature. Hermann, unsurprisingly, disagrees, arguing that since on the Kantian picture one cannot directly access the *Ding an sich*, it is out of place to demand more direct access to 'nature' in the guise of causal relationships than what is available to us indirectly via differential equations (K. Herrmann, 2019, pp. 497–500).

# 22.4  CONCLUDING REMARKS

Because her project is one of precious few attempts to provide an explicitly Kantian interpretation of quantum mechanics,[20] and due to the fact that her view savours of transcendental idealism, her interpretation should only with suitable caution be brought alongside contemporary approaches to quantum mechanics. Her view, for instance, substantiates neither the stark realism entailed by approaches with primitive ontologies nor those espousing global determinism. She finds the idea of wavefunction collapse repellent, and would, I suspect, have nontrivial qualms with relationalist and perspectivalist views (though there is, no doubt, some temptation to read especially her points about the splitting of truth in exactly this way). Furthermore, the fact that she considers wrong-headed any 'forward in time' posing of the problem of specific measurement outcomes (why *this* definite value rather than *that* one) makes her interpretation an altogether different creature than contemporary views, which take this problem as central. For instance, her defence of causal determinism existing only relative to a particular context might bring to mind contextual hidden variables theories like that of de Broglie–Bohm. But then, she explicitly rejects hidden variables. In short, it seems fair to say that with one foot squarely on Kant's turf and the other squarely on Bohr's and Heisenberg's, her interpretation truly covers unique ground.

## REFERENCES

Bacciagaluppi, G. (2017). 'Bohr's Slit and Hermann's Microscope'. In E. Crull and G. Bacciagaluppi (eds), *Grete Hermann: Between Physics and Philosophy*, Berlin: Springer, pp. 135–147.

Bacciagaluppi, G. (2020). Translation of Three Short Papers by Grete Hermann. *General Philosophy of Science*, 51(4), 615–619.

Bacciagaluppi, G. (n.d.). 'Better than Bohr: Grete Hermann and the Copenhagen Interpretation'. In preparation.

Bacciagaluppi, G., and Crull, E. (2022). *The Einstein Paradox: The Debate on Nonlocality and Incompleteness in 1935*. Cambridge: Cambridge University Press.

Crull, E. (2017). Hermann and the Relative Context of Observation. In E. Crull and G. Bacciagaluppi (eds), *Grete Hermann: Between Physics and Philosophy*, Berlin: Springer, pp. 149–169.

Crull, E., and Bacciagaluppi, G. (eds) (2017a). *Grete Hermann: Between Physics and Philosophy*. Berlin: Springer.

---

[20] Soler mentions only two other neo-Kantians contemporary to Hermann who write at any length about quantum mechanics: Kojève and Cassirer. To these I add a minor third (minor because his *Philosophic Foundations of Quantum Mechanics* appears only in 1944): Reichenbach. I explore Hermann's interpretation alongside other neo-Kantian works on modern physics in detail in a forthcoming paper.

Crull, E. and Bacciagaluppi, G. (2017b). Grete Hermann's Lost Manuscript on Quantum Mechanics. In E. Crull and G. Bacciagaluppi (eds), *Grete Hermann: Between Physics and Philosophy*. Berlin: Springer, pp. 119–134.

Henry-Hermann, G. (1948). Die Kausalität in der Physik. *Studium Generale*, 1(6), 375–383.

Hermann, G. (1933). Determinismus und Quantenmechanik. Dirac Archives DRAC 3/11.

Hermann, G. (1935a). Die naturphilosophischen Grundlagen der Quantenmechanik. *Abhandlungen der Fries'schen Schule (Neue Folge)*, 6(2), 69–152.

Hermann, G. (1935b). Besprechung: K. Popper, *Logik der Forschung. Zur Erkenntnistheorie der modernen Naturwissenschaft. Physikalische Zeitschrift*, 36(13), 481–482.

Hermann, G. (1935c). Die naturphilosophischen Grundlagen der Quantenmechanik (Auszug). *Die Naturwissenschaften*, 23(42), 718–721.

Hermann, G. (1936). Zum Vortrag Schlicks. *Erkenntnis*, 6(5/6), 342–343.

Hermann, G. (1937a). Die Bedeutung der Modernen Physik für die Theorie der Erkenntnis. In *Drei mit dem Richard Avenarius-Preis ausgezeichnete Arbeiten von Dr G. Hermann, Dr E. May, Dr Th. Vogel*, Leipzig: S. Hirzel, pp. 3–44.

Hermann, G. (1937b). Über die Grundlagen physikalischer Aussagen in den älteren und den modernen Theorien. *Abhandlungen der Fries'schen Schule (Neue Folge)*, 6(3/4), 309–398.

Hermann, G. (1937c). Die naturphilosophischen Bedeutung des Übergangs von der klassischen zur modernen Physik. In R. Bayer (ed.), *Travaux du IX$^e$ Congrès International de Philosophie—Congrès Descartes. Vol. VII, Causalité et Déterminisme*, Actualités Scientifiques et Industrielles 536, Paris: Hermann et C$^{ie}$, pp. 99–101.

Hermann, G. (1996). *Les fondements philosophiques de la mécanique quantique*. Translation of Hermann (1935a) by A. Schnell. Edited, with introduction and postface by L. Soler. Paris: Vrin.

Hermann, G. (1999). The Foundations of Quantum Mechanics in the Philosophy of Nature. *The Harvard Review of Philosophy*, 7(1), 35–44. Translation of Hermann (1935c) with introduction by D. Lumma.

Hermann, G. (2017a). Determinism and Quantum Mechanics. In E. Crull and G. Bacciagaluppi (eds), *Grete Hermann: Between Physics and Philosophy*, Berlin: Springer, pp. 223–237. Translation of Hermann (1933) by E. Crull and G. Bacciagaluppi.

Hermann, G. (2017b). 'Natural-Philosophical Foundations of Quantum Mechanics'. In: E. Crull and G. Bacciagaluppi (eds), *Grete Hermann: Between Physics and Philosophy*, Berlin: Springer, pp. 239–278. Translation of Hermann (1935a) by E. Crull and G. Bacciagaluppi.

Hermann, G. (2020a). Review of: K. Popper, *Logik der Forschung. Zur Erkenntnistheorie der modernen Naturwissenschaft*. *General Philosophy of Science*, 51(4), 621–623. Translation of Hermann (1935b) by G. Bacciagaluppi.

Hermann, G. (2020b). Comment on Schlick. *General Philosophy of Science*, 51(4), 625–626. Translation of Hermann (1936) by G. Bacciagaluppi.

Hermann, G. (2020c). The Significance for Natural Philosophy of the Move from Classical to Modern Physics. *General Philosophy of Science*, 51(4), 627–629. Translation of Hermann (1937c) by G. Bacciagaluppi.

Herrmann, K. (ed.) (2019). *Grete Henry-Hermann: Philosophie–Mathematik–Quantenmechanik*. Wiesbaden: Springer.

Jammer, M. (1974). *The Philosophy of Quantum Mechanics: The Interpretations of Quantum Mechanics in Historical Perspective*. Hoboken, NJ: John Wiley.

Leal, F. (2017). Grete Hermann as a Philosopher. In E. Crull and G. Bacciagaluppi (eds), *Grete Hermann: Between Physics and Philosophy*, Berlin: Springer, pp. 17–34.

Luft, S. (ed.) (2015). *The Neo-Kantian Reader*. London/New York: Routledge.

Lumma, D. (1999). The Foundations of Quantum Mechanics in the Philosophy of Nature. *The Harvard Review of Philosophy*, VII, 35–44. Translation of Hermann (1935c) with introduction.

Paparo, G. (2017). Understanding Hermann's Philosophy of Nature'. In E. Crull and G. Bacciagaluppi (eds), *Grete Hermann: Between Physics and Philosophy*, Berlin: Springer, pp. 35–51.

Reichenbach, H. (1998). *Philosophic Foundations of Quantum Mechanics*. New York: Dover. The original version of this text was published in 1944 by University of California Press.

Seevinck, M. (2017). Challenging the Gospel: Grete Hermann on von Neumann's No-Hidden-Variables Proof. In E. Crull and G. Bacciagaluppi (eds), *Grete Hermann: Between Physics and Philosophy*, Berlin: Springer, pp. 107–117.

Soler, L. (2017). The Convergence of Transcendental Philosophy and Quantum Physics: Grete Henry-Hermann's 1935 Pioneering Proposal. In E. Crull and G. Bacciagaluppi (eds), *Grete Hermann: Between Physics and Philosophy*, Berlin: Springer, pp. 55–69.

van der Waerden, B. L. (ed.) (1968). *Sources of Quantum Mechanics*. New York: Dover.

Werkmeister, W. H. (1936). The Second International Congress for the Unity of Science. *The Philosophical Review*, 45(6), 593–600.

# CHAPTER 23

......................................................................

# INSTRUMENTATION AND THE FOUNDATIONS OF QUANTUM MECHANICS

......................................................................

## CLIMÉRIO PAULO DA SILVA NETO

## 23.1 INTRODUCTION

......................................................................

In the second half of the 20th century, various factors whose roles varied over time and contexts drove quantum foundations (QF) from the margins to the mainstream of physics. No doubt, a major factor was the materialization of thought experiments conceived in the early debates among the founding fathers of quantum mechanics. The preconditions for this rested not only on theoretical developments such as the famous Bell's theorem, which figures prominently in the narratives on the history of quantum foundations, but also on the impressive evolution of instruments and experimental techniques since World War II. Physicists working in QF have long stressed the importance of new instruments and technologies that enable them to produce, manipulate, and detect individual particles to consolidate their field. Historians, however, have only begun to investigate what instruments, technologies, and techniques were the most important, how they came about, and what they reveal about the historical development of QF.[1]

In pioneering studies on the material culture of QF, Joan Lisa Bromberg (2006, 2008) showed how physicists used new devices and techniques to test the complementarity principle in the early 1980s. Wilson Bispo et al. (2011, 2013), in turn, discussed the instrumentation of the early experimental tests of Bell's theorem in the 1970s. The

---

[1] A large body of literature, including chapters in this volume, has shown how philosophical and political views and trends, generational and contextual changes, and discipline-building efforts were crucial to explain the rise and consolidation of the field. Olival Freire (2015) offers a comprehensive narrative of this complex process. Joan Bromberg (2008) discussed the perspective of physicists and called for historical studies on the role of new instruments and techniques in the development of the quantum foundations.

common issue these studies addressed is the moment when it became possible to transform the thought experiments, devised in the pre-war debates among the founding fathers of quantum mechanics, into real experiments. For Bromberg, this is crucial because physicists cherish the possibility of testing their claims in real experiments. After all, 'experimentation is one way that physicists define their field and draw the border between it and philosophy' (Bromberg, 2008, p. 351). This chapter goes further to discuss other experiments and traces the historical development of the critical instrumentation to a larger extent. Its central claim is that, except for tuneable lasers, the instruments used in the foundational experiments were available about a decade before physicists applied them to quantum foundations. Additionally, it provides some novel insights into how the strategies physicists used to choose their instruments and experimental set-up influenced their results and the overall history of experimental quantum foundations.

## 23.2 BELL'S INEQUALITY EXPERIMENTS

The first and most significant foundational experiments were Bell's inequality experiments (BIE). They were the realization of the thought experiment proposed in 1935 by Albert Einstein, Boris Podolsky, and Natan Rosen (hereafter the 'EPR experiment'). It meant to show that the quantum mechanical description of reality is incomplete, but ended up showing that QM is incompatible with the conjunction of the premises of locality and realism. In the 1950s, David Bohm gave visibility to the EPR experiment when he reformulated it in terms of a binary spin system (EPRB experiment) and created his famous hidden-variable theory that reproduced all the experimental results of QM. Inspired by Bohm's accomplishments, in 1964, John Bell demonstrated with a mathematical theorem that any hidden-variable theory which satisfies the premises of locality and realism yields experimental predictions for the EPRB experiment that conflict with quantum mechanics.

Bell's profound theorem made it possible, in principle, to confront the predictions of QM with those of local hidden-variable theories (LHVT) experimentally. However, the practical realization of such tests became feasible with the extension of the theorem by John Clauser, Michael Horne, Abner Shimony, and Richard Holt (alias CHSH). They extended the theorem to realizable systems and showed that existing experimental techniques and instruments could test Bell's theorem for photon polarization correlations. CHSH proposed a straightforward experimental set-up comprising a source of entangled photons and two symmetrically placed analysers and detectors (Fig. 23.1). In the proposal, photons created simultaneously leave the source with correlated linear polarizations. The analysers select a given component of polarization, allowing only photons with that component to reach the detectors. Depending on the orientation of the analysers, there may be simultaneous or single detections. Assuming the sample of photons detected is representative of all the photons emitted, the predictions of QM and LHVT for the coincidence rates are functions ($\delta$) of the orientations of the analysers. Any LHVT

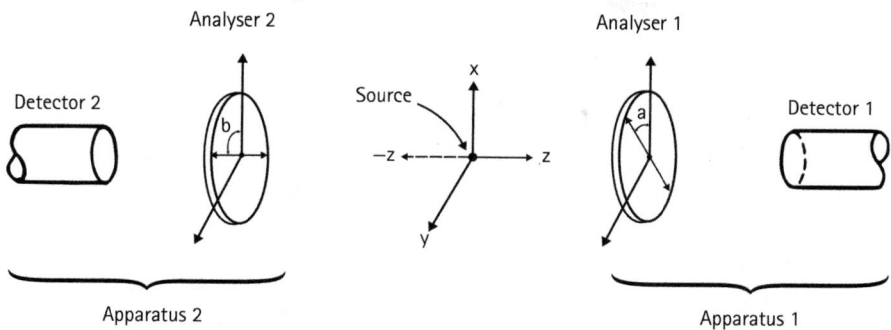

**FIGURE 23.1** Apparatus configuration used in the proof by CHSH. Drawing by the author, based on Figure 2 of Clauser and Shimony (1978).

predicts values for δ limited by an inequality, called Bell's inequality (e.g., δ ≤ 0), while QM violates that inequality for some orientations of the analysers (Clauser *et al.*, 1969).

Before writing their paper, CHSH surveyed the literature in search of experimental results that could test Bell's theorem, a procedure they later dubbed 'quantum archaeology' (Horne *et al.*, 1990, p. 361). They found two experiments that had measured the polarization correlations in pairs of entangled photons and could test the inequality they obtained with the generalization of Bell's theorem. The first, performed by Chien-Shiung Wu and her student Irving Shaknov in 1950, measured the correlations between photons emitted in electron–positron annihilation. The second, performed by Carl Kocher and Eugene Commins in 1967, examined photon pairs emitted in a photon cascade of calcium atoms. Their results could not test the theorem, because they had measured the correlations at angles for which the predictions of QM and LHVT coincide, but their experiments could be modified to do so. The following sections address the instruments and techniques that made these experiments possible, discussing the source of entangled photons, analysers, and detection sets respectively.[2]

## 23.2.1 Sources of Entangled Photons

### 23.2.1.1 *Positron Annihilation*

Used in four BIE *circa* 1975, positron annihilation is the oldest experimental source of entangled photons, its development beginning in the mid-1930s.[3] In this phenomenon, a positron, the positively charged antiparticle of the electron, fuses with an electron

---

[2] Many authors have discussed these experiments. In this volume, see Kaiser (2021). Bispo *et al.* (2011, 2013) concentrate on the material culture of the first experiments. For historical narratives see, for instance, Freire (2015) and Kaiser (2011). For more technical presentations, see Clauser and Shimony (1978), Whitaker (2000), and Duarte (2019).

[3] These experiments were variants of Wu and Shaknov (1950). For a description and appraisal of them, see Clauser and Shimony (1978).

producing a pair of γ-ray photons flying in diametrically opposite directions. Two developments were particularly important in the process of harnessing positron annihilation for practical purposes. The first one was the realization that the annihilation radiation was a key component of the radiations observed in experiments on the absorption and scattering of gamma rays that were part of a heated controversy around 1933. These experiments yielded the first instances of production and annihilation of positrons in laboratories and helped to convince sceptics that the positive particle with electronic mass observed by Carl Anderson in 1932 and by Patrick Blackett and Giuseppe Occhialine in 1933 was the anti-electron predicted by Paul Dirac in papers published between 1928 and 1931 (Roqué, 1997). The second important development was the discovery of artificial radioactivity in 1934 and the subsequent production of artificial, positron-emitting radioisotopes in newly built particle accelerators. Although the realization that annihilation produces entangled photons would come much later, by the end of the 1930s production and annihilation of positrons were stable experimental techniques.

The early instances of the production of annihilation radiation consisted of irradiating metals with traditional radioactive sources. This was what some physicists were doing around 1933, when they observed unexpected secondary radiation with energy equivalent to the rest mass of electrons. Competing explanations appeared, but a consensus had emerged by the beginning of 1934. Young theoreticians, such as Rudolf Peierls, Max Delbrük, Robert Oppenheimer, and George Uhlenbeck, successfully used Dirac's theory to account for the results of γ-ray absorption experiments. Moreover, new, independent experiments by Theodore Heiting, by Irène and Frédéric Joliot-Curie, and by Jean Thibaud made the interpretation of the secondary radiation produced in the γ-ray absorption experiments as resulting from electron-positron annihilation more compelling (Roqué, 1997).

The breakthrough towards the production of positron annihilation sources came with the discovery of artificial radioactivity. In one of their experiments the Joliot-Curies reported that by bombarding $^{27}$Al atoms with α particles they had produced $^{30}$P, an artificial and short-lived positron-emitting element. When the news of this experiment reached Berkeley, in February 1934, the team led by Ernest Lawrence used their brand-new cyclotron to repeat the Joliot-Curies' experiment using deuterons as projectiles. They discovered that almost anything they bombarded became radioactive. Once electrons and positrons were major products of the decay of artificial isotopes, subsequent studies catalogued the positron-emitting activity of several radioisotopes and led to the discovery of efficient positron emitters such as $^{64}$Cu (with a half-life of 12.8 hours) and $^{22}$Na (a singularly long-lived such isotope with a half-life of about 2.6 years), commonly used in positron sources nowadays (Heilbron and Seidel, 1990, pp. 373–386).

Only after World War II, however, did it become clear that positron annihilation produced pairs of entangled photons. In the latter years of the 1940s, several experimentalists, following a suggestion by John A. Wheeler, tested the prediction that the γ-rays created in positron annihilations have perpendicular polarizations. As positron sources, they used artificial radioisotopes produced in cyclotrons, by this time widely available in North America and Europe. Wu and Shaknov (1950), for instance, used a $^{64}$Cu source activated at the Columbia cyclotron. The unequivocal association between the photon correlations measured in those experiments and entanglement would be made only in 1957 when David Bohm and Yakir Aharonov pointed out that the Wu–Shaknov experiment had measured correlations similar to the one discussed in the EPR paper (Maia Filho and Silva, 2019; Silva, 2021). Hence, positron annihilation sources have been around since the 1930s, but only much later would they be unequivocally linked to the foundational issues around entanglement.[4]

### 23.2.1.2 *Photon Cascade*

The photon cascade source of photon pairs, developed in the 1950s, was the predominant source of entangled photons in Bell tests up to the 1980s. To create the cascade, atoms in a beam are excited to an energy level from which the decay to the ground state occurs in a two-step process (Fig. 23.2b). In Bell tests, experimentalists used two techniques to achieve this: *optical pumping* and *electron bombardment*. In optical pumping, the atoms are excited by the absorption of light, while in electron bombardment they are excited by electron–atom collisions. The quality of the cascade source depends on the excitation technique and the atomic beam density at the intersection region, two aspects which were developed extraordinarily in the first post-war decades.

Although the fundamentals of the experimental techniques for producing and exciting atomic beams date back to the first decades of the 20th century, the photon-cascade source as a tool to produce pairs of entangled photons was a product of the post-war expansion of atomic and molecular beam (AMB) methods. Atomic beams

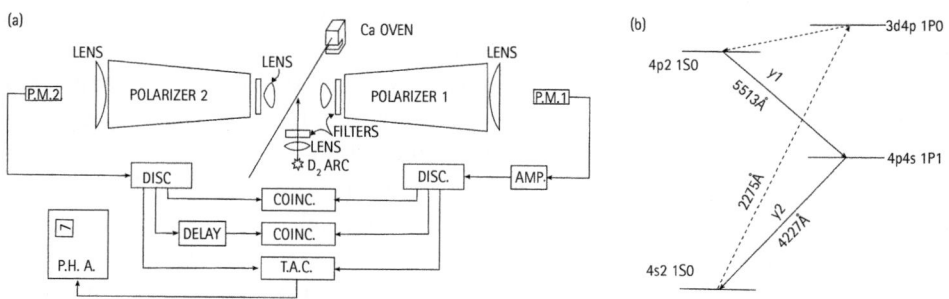

FIGURE 23.2 (a) Diagram of a photon cascade experiment (Freedman and Clauser, 1972); (b) energy levels involved in the cascade transitions. The photons observed were by $\gamma_1$ and $\gamma_2$. Adapted from the paper by the author.

---

[4] This was arguably even before entanglement was widely recognized as a phenomenon. Although Schrödinger was aware of the phenomenon in the late 1920s, as evidenced in his notebooks, he coined the term entanglement in the aftermath of the EPR paper (Joas, 2021).

were first produced by the French physicist Louis Dunoyer circa 1911. Allowing gas in a high-pressure oven to leak through a small orifice into an evacuated chamber, he produced a linear beam of sodium atoms (a unidimensional gas, as he saw it), which could be observed by the fluorescence produced after illumination with light from a sodium flame. Discontinued by the outbreak of the war, the use of atomic beams to study the interaction between matter and radiation was resumed in the 1920s when it was used in the famous Stern-Gerlach experiment and by some spectroscopists to circumvent the limitations of existing spectral lamps. However, the systematic application of atomic beams in spectroscopy, which led to the development of the beam excitation techniques, began only circa 1934 as part of an effort by spectroscopists to partake in the exciting developments of nuclear physics with high-resolution measurement of hyperfine spectra. In England, D. A. Jackson and H. Kuhn built upon the apparatus designed by the French physicist A. Bogros in the mid-1920s to excite the beams with radiation (optical pumping) and study emission and absorption spectroscopy. In Germany, on the other hand, Karl W. Meissner and K. F. Luft and R. Minkowski and H. Bruck independently opted to develop the electron impact method, which, despite its complexity, offered a broader range of tunability. By the time of the outbreak of the second world war, only a handful of groups were using atomic beams in spectroscopy, but the possibility of using the technique for low-cost and low-energy nuclear physics lured many other researchers into taking up and developing the technique.[5]

For our purpose here, it is essential to notice that the basics of the excitation technique used by Freedman and Clauser are as old as the atomic beam technique itself. However, it was a long way until this technique was perfected and standardized to become a source of entangled photons. The photon cascade sources required more than the ability to produce and excite atomic beams. They demanded knowledge on the inner structure of atoms and matter-radiation interactions which would emerge only in the upcoming decades in the process of, and as a result of, the creation and application of quantum mechanics. In particular, they required the habit of thinking of light in terms of photons, or quanta, which was consolidated only after, in the middle of the 1950s, physicists began to use detection sets comprising photomultiplier tubes and coincidence circuits (discussed below) to detect faint light.[6] In 1955, for instance, a group of Canadian physicists led by Eric Brannen applied "nuclear coincidence methods" to study atomic transitions (Brannen et al., 1955). This experiment was

---

[5] For the pre-war application of atomic beam in spectroscopy, see Meissner (1942). For high-resolution spectroscopy and an alternative for those who did not have a particle accelerator to do nuclear physics, see Schawlow (1982), p. 9.

[6] As the photomultipliers and coincidence circuits travel beyond nuclear physics, they began to raise fundamental questions that optics physicists did not face before. The controversy around the experiment of Hanbury Brown and Twiss, discussed by Silva and Freire Jr. (2013), is an example of this. This migration of instrumentation brought optics to the quantum domain and set the stage for the rise of the new discipline of quantum optics in the 1960s, which would yield significant contributions to quantum foundations. See Hentschel (2018) for a history of the mental models of photons.

strikingly similar to the first tests of Bell's inequality and stimulated several physicists to apply coincidence techniques to study atomic transitions in photon cascades (Kaul, 1966).

Thus, when CHSH began to conceive the first Bell tests, optical pumping and electron bombardment were well-known beam-excitation techniques, and both were used in the first experiments. While Freedman and Clauser (1972) used optical pumping by an ultraviolet deuterium arc lamp, Holt and Pipkin (1974) and Clauser (1976) used electron bombardment. There was no notable difference in source performance in these experiments. Both techniques had pros and cons, yielding relatively low coincidence rates that required experimental runs of more than a week to collect the data for reliable tests.

Yet, further developments stimulated by the pursuit of different types of lasers, led to extraordinary improvements in the subsequent decades. One particularly notable development was the advent of the continuous wave (CW) tuneable laser devised in 1970 by Otis Peterson and collaborators at Eastman Kodak (Peterson *et al.*, 1970). Their Rhodamine 6G-dye laser became the wand of spectroscopists. 'No other optical source can provide a comparable combination of tunability, resolution, and power' (Hollberg, 1990, p. 186).

The CW dye laser was too recent to be used in the first Bell tests, but soon Fry and Thompson (1976) mobilized it to design a Bell test with an improved excitation technique. As the energy gap between the Hg cascade levels was beyond the reach of tuneable lasers, they resorted to a combination of electron bombardment with optical pumping. They excited Hg atoms to an intermediary level with electron bombardment and then to the final level with optical pumping. With this technique, they obtained a high data accumulation rate and ran the experiment in only 80 minutes.

Impressive as that mark may have been, electron-beam fluctuations still limited the efficiency of Fry and Thompson's excitation technique. Aspect and colleagues circumvented this limitation resorting to the nonlinear phenomenon of two-photon absorption, which occurs when, under intense radiation provided by lasers, the atom absorbs two photons as if they were a single one of frequency equal to the sum of the frequencies of the absorbed photons.[7] With this technique, Aspect *et al.* (1981, p. 460) obtained an excitation rate 'more than ten times greater than that of Fry and Thompson'.

The power of the technique used in Aspect's experiment can be appreciated by comparing it to the one used in Freedman and Clauser (1972). Whereas in the experiment by Freedman and Clauser only about 7 per cent of the excited atoms underwent the cascade transition, and the typical coincidence rates with polarizers removed ranged from 0.3 to 0.1 counts per second (Freedman and Clauser, 1972, p 940), in Aspect's experiments 100 per cent of the excited atoms underwent the cascade transitions, and the typical coincidence rate was 240 counts per second. They

---

[7] Maria Goeppert-Mayer predicted this phenomenon in the 1930s, but its observation had to wait for the laser (Kaiser and Garrett, 1961).

achieved much higher statistical accuracy in only 17 minutes. That achievement, however, was due to a combination of the efficient source with novel analysers that we shall consider in a later section.

### 23.2.1.3 *Spontaneous Parametric Down Conversion*

Aspect's experiments left little room for improvement in photon-cascade tests. Nevertheless, the realization that extrinsic variables such as position and momentum or energy and time could be objects of BIE and a new kind of source of photon pairs, based on the nonlinear phenomenon known as Spontaneous Parametric Down Conversion (SPDC), led to new rounds of experiments. In SPDC, photons from an intense laser beam scatter inside a nonlinear crystal and split into two photons—signal and idler (Fig. 23.3a). The photon pairs created in this process are correlated by energy, momentum, and polarization. Respecting conservation laws, the energies of the signal and idler photons add up to the energy of the pump photon, and the same is valid for momentum (Fig. 23.3b). Polarization correlations change according to the type of SPDC, defined by the dispersion surfaces inside the crystal and the angle of incidence of the pump beam. In type-I SPDC, the output of the crystal is a cone-shaped light beam where all photons have the same polarization. In type-II, the output comprises two cone-shaped beams with orthogonal polarizations (Fig. 23.3c) (Zeilinger, 1999).

The history of SPDC dates back to 1966 when David Klyshko of Moscow State University presented the first theoretical prediction of the phenomenon and

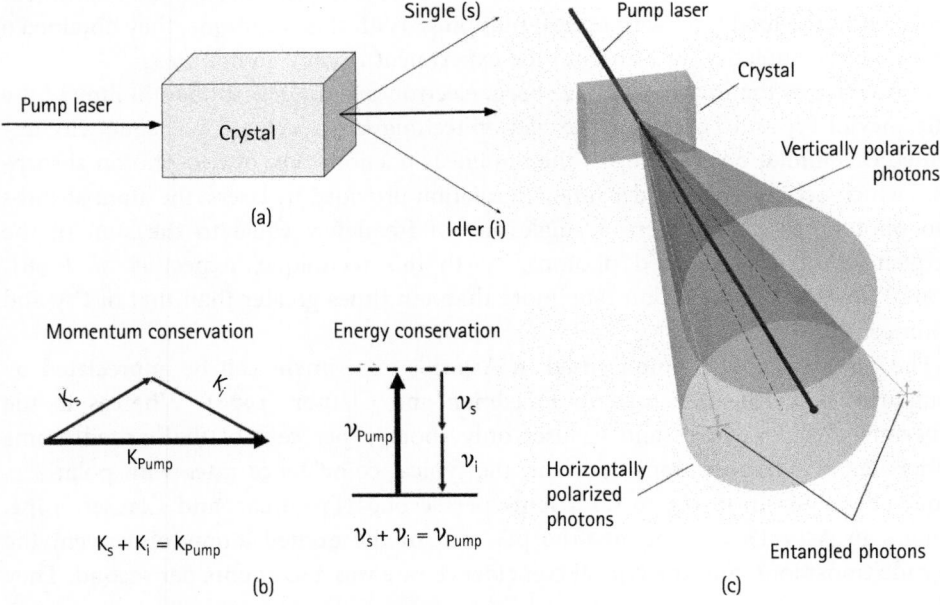

**FIGURE 23.3** Photon correlations in SPDC. Figure by the author based on images under the GNU Free Documentation License.

demonstrated the possibility of observing it by pumping a lithium niobate crystal with a CW laser (Klyshko, 1967). In the following year, three research groups observed the phenomenon (Magde *et al.*, 1967; Akhmanov *et al.*, 1967; Harris *et al.*, 1967). In 1969, Zel'dovich and Klyshko (1969) predicted photon correlations as the ones used in Bell tests, which were measured in 1970 by David Burnham and Donald Weinberg (1970) from NASA. Since then, many physicists in nonlinear and quantum optics have investigated SPDC as a source of non-classical photon states. Despite these early developments, the physicists involved with SPDC throughout the 1970s were not aware of, or not interested in, the experiments in foundations of quantum mechanics. Physicists involved in foundational experiments would discover the source only around the mid-1980s (Freire Jr, 2015, pp. 291–293).

Compared to the photon-cascade sources, SPDC is remarkable. First, because of its simplicity. As the pioneer of BIE with SPDC Yanhua Shih summarized: 'You have a laser, you have a crystal, and you generate a pair immediately. You don't need to spend millions of dollars like [ . . . ] Aspect did.'[8] This meant a dramatic decrease in the cost and size of the experiments. Second, because of its collection efficiency. Whereas photon-cascade sources emit photons in all directions, and only the pairs emitted in the direction of the symmetrically placed detection sets were collected, SPDC sources emit photons in well-defined directions, allowing highly efficient collection. As physicists began to scrutinize the effect, they mastered it to the point that they produced violations of Bell's inequality by over 100 standard deviations in less than 5 min, significantly exceeding the previous mark (Kwiat *et al.*, 1995). This made SPDC the supreme source of entangled photons (Freire Jr, 2015, pp. 300–301).

## 23.2.2 Analysers

### 23.2.2.1 *Analysers for low-energy photons*

The function of the analysers in BIE was to select photons according to their polarization. This process was trivial for low-energy photons such as the ones created in photon cascades, thanks to the existence of highly efficient polarizers for visible light. Most of the photon cascade tests in the 1970s used the pile-of-plate polarizers. The only exception was the calcite prism polarizer used in Holt and Pipkin's experiment, but its low efficiency discouraged further uses. The pile-of-plate polarizers comprise a series of plates inclined so that the incidence angle of the unpolarized radiation is equal to the Brewster angle (Fig. 23.4a). In this case, the reflected ray has only photons with

---

[8] Interview of Yanhua Shih and Morton Rubin by Joan Bromberg on May 14 2001, Niels Bohr Library & Archives, American Institute of Physics, College Park, MD, USA, www.aip.org/history-programs/niels-bohr-library/oral-histories/24558. Shih's interest in SPDC led to a collaboration with Russian physicists which drew them into the experimental quantum foundations *circa* 1990. Prior to that the interest in Bell's inequalities in Russia was restricted to a circle of mathematicians interested in quantum information theory (Martinez, 2021).

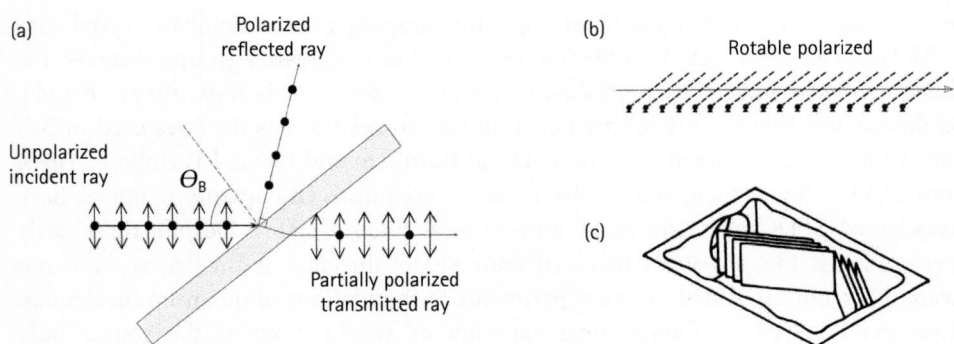

**FIGURE 23.4** Pile-of-plates polarizers. (a) shows what happens to unpolarized light when the incident light matches the Brewster condition. (b) Polarizer used in Clauser (1976). (c) Polarizers used by Fry and Thompson (1976). Figures rendered by the author based on the original figures presented in the papers.

horizontal linear polarization, with the electric field perpendicular to the plane of incidence (represented by the dots in the figure). Therefore, the transmitted ray is partially polarized. After passing through several plates, the transmitted light is all but fully polarized. In the first experiment, with ten plates, Freedman and Clauser filtered the horizontally polarized light to less than 4 per cent. In his next experiment, with fifteen plates, Clauser reduced that number to about 1 per cent. Hence, in terms of polarization efficiency, there was practically no room for improvement.

However, Alain Aspect and collaborators improved polarization analysis significantly with new kinds of analysers. In their careful preparation of the experiments, they created two different experimental schemes. In the first, they designed a four-channel analyser using a polarizing cube beamsplitter, which was able to detect both components of polarization simultaneously (Aspect, Grangier, and Roger, 1982). In the second, they designed a time-varying analyser that changed the orientation of the polarizers with the photons in flight (Aspect, Dalibard, and Roger, 1982).

The experiment with the four-channel analysers was analogous to a Stern–Gerlach experiment. As shown in Fig. 23.5, in this arrangement, the photons leaving the source are directed to polarizing cube beamsplitters (CBS) that split them according to their polarization components. The photomultipliers on the top of the figure detect the transmitted photons, and the photomultipliers in the middle detect the reflected photons. The new ingredient in this set-up, the polarizing cube beamsplitter, was one of the many remarkable results of thin-film optics, first built by Mary Banning during WWII. Cementing several layers of thin films between two prisms in such a way that the incidence angle of light on the layers matches the Brewster condition, Banning obtained a very compact polarizer with efficiency over 98 per cent (Banning, 1947, pp. 796–797; Michaud and Wills, 2016, p. 43).

It is easy to gauge the impact of the CBS because Aspect's group used the same source and detectors with pile-of-plate polarizers in one experiment and with the CBS

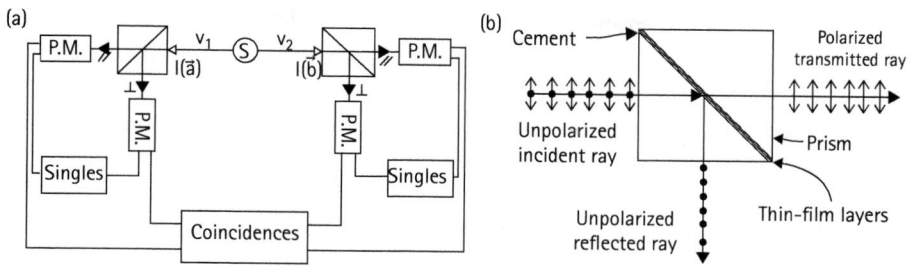

**FIGURE 23.5** (a) Version of the EPRB experiment by Aspect, Grangier, and Roger (1982, p. 92). Reprinted figure with permission. Copyright (2021) by the American Physical Society. (b) polarizing cube beamsplitter.

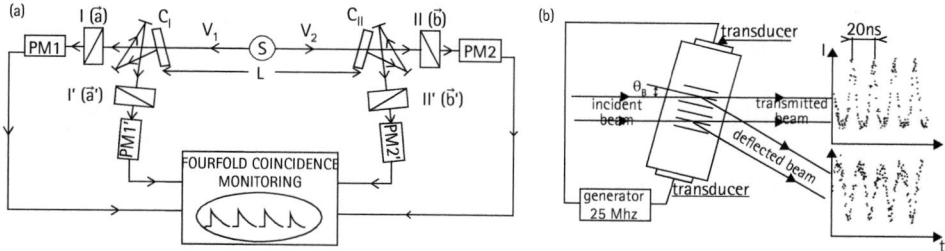

**FIGURE 23.6** (a) Configuration of the experiment with time-varying analysers. The switches are $C_I$ and $C_{II}$. (b) The mechanism used to change the orientation of the analysers. Reprinted figure with permission from Aspect, Dalibard, and Roger (1982, pp. 1805–06). Copyright (2021) by the American Physical Society.

in another. Whereas with the pile-of-plate polarizer Aspect *et al.* (1981) obtained a violation by 13 standard deviations in approximately 27 minutes, with the CBS Aspect, Grangier, and Roger (1982) obtained a violation by 46 standard deviations in only 17 minutes. Hence, by detecting both components of polarization, the CBS yielded more accurate measurements in less time. Furthermore, it had the practical advantages of being much smaller and easier to handle.

The time-varying analyser devised by Aspect, Dalibard, and Roger (1982) consisted of a mechanism that switched the direction of propagation of the light beams towards two polarizers with different orientations (Fig. 23.6a). The mechanism was ultrasonic standing waves generated by two electro-acoustic transducers of 25 MHz in vessels filled with water (Fig. 23.6b). It performed the switches in about 10 ns, a time much shorter than the time necessary for any signal exchange between detectors at the speed of light (40 ns), ruling out the possibility of communication between the detectors within the speed of light which could influence the outcome of the experiment.

These kinds of acousto-optic modulators began to be developed in the 1930s when Peter Debye and Francis Sears showed that ultrasonic standing waves in water worked as a diffraction grating (Sears and Debye, 1932). In the 1960s, several acousto-optic modulators were already standard tools in experimental physics (Williams, 1962). Like the cube beamsplitter, the acousto-optic modulators used to devise the time-varying analysers were part of the instrumentation of experimental physics before Bell's theorem. In principle, they could have been part of the first Bell tests. However, in practice it was difficult to combine them with sources of lower efficiency and more cumbersome polarizers.

### 23.2.2.2 *Analysers for γ-rays*

The lack of efficient polarizers for high-energy photons such as the γ-rays was the main challenge for experimentalists who endeavoured to turn Wheeler's proposal of detecting the correlations between the annihilation photons into a real experiment. They had to infer the γ-ray polarizations from Compton scattering, in which the incident radiation, after collision with a target, is scattered with lower energy. The British physicists Maurice Pryce and John Ward showed that the quantum-mechanical Compton scattering distribution of low-energy photons depends on the polarization of the γ-rays (Pryce and Ward, 1947). Concisely, the scattered photons from γ-rays with orthogonal polarizations would be scattered orthogonally. Although it may sound easy, some attempts to measure the scattering distribution in the late 1940s yielded poor and conflicting results. Only in 1950 did Wu and Shaknov settle the matter (Maia Filho and Silva, 2019; Silva, 2021). The analysis of the polarization of the annihilation photons rested on the two following assumptions: (i) one may, in principle, construct polarizers for high-energy photons; and (ii) the results obtained in experiments with these polarizers will be equal to the results obtained in a Compton scattering experiment (Kasday *et al.*, 1975).

This analysis of photon correlations rendered all positron annihilation tests unfit to test Bell's inequality. To infer the polarization of the γ-rays produced in the annihilation, they had to rely on the Compton scattering distribution formula obtained by a quantum mechanical description of the photons. Thus, the test of quantum mechanics depended on quantum mechanics. Furthermore, the assumption that the photons were in quantum mechanical states conflicted with Bell's theorem because the state of the system postulated by Bell was not a quantum mechanical state, but a more general one. These considerations led Shimony and Horne to conclude that 'no variant of the Wu–Shaknov experiment can provide a test of the predictions based on Bell's theorem' (Clauser and Shimony, 1978, p. 1915). Thus, the several tests of Bell's inequality with positron annihilation performed around the mid-1970s, by the end of that decade, were considered inherently flawed because of an instrumental limitation.

## 23.2.3  Detection Sets: Photomultipliers and Electronic Circuits

The detection sets used to test Bell's theorem comprised photomultipliers strung to coincidence circuits and other electronic devices that analysed the photomultipliers' output. The main requirements for those sets were the capacity to detect single photons of a given wavelength, low noise, and a temporal resolution that was good enough to discriminate between photons arriving at tiny intervals from each other.

The development of photomultipliers capable of detecting single photons began in the 1930s. They rely on a combination of the effects of photoelectric and secondary emission, both known in the beginning of the century. While in the photoelectric effect, an electron might be ejected from a photosensitive cathode by a sufficiently energetic photon, in secondary emission, several electrons might be ejected by the collision of a single energetic electron with the surface. Chiefly motivated by the blossoming television industry, engineers and scientists worked hard to combine these two effects to develop sensitive light detectors. The Soviet physicist Leonid Kubetsky achieved some of the first outstanding results. In 1934, he demonstrated a vacuum tube comprising a photocathode made of a compound of silver, oxygen and caesium (Ag-O-Cs), followed by a series of electrodes, called dynodes, submitted to an increasing electric potential. At each dynode the number of electrons is multiplied, creating an avalanche-like effect. In his own account, Kubetsky's multistage tube demonstrated in 1934 amplified the signal about one thousand times and by 1936 he had developed several models of tubes with amplification up to $10^6$ (Fig. 23.7) (Kubetsky, 1937). This latter level of amplification turns a single electron into a current pulse of the order of 0.1 mA. However, besides amplifying, a photomultiplier should have high peak quantum efficiency (QE), which translates in the percentage of the photons incident on the photocathode that yields a signal output, and low thermoionic dark current, the current in the absence of light. The Ag-O-Cs photocathode used in Kubetsky's tubes had broad spectral sensitivity, but low peak QE (0.5 per cent) and high dark current, which made it unsuitable for photon counting.

When the Radio Corporation of America (RCA) started developing multistage photomultipliers in *circa* 1935, it lagged behind Kubetsky's laboratory, but it went on to take the lead in the development of commercial photomultipliers in the subsequent decades. In 1941, RCA released its first commercially successful photomultiplier, which used the Cs-Sb photocathode discovered in 1936 by the German physicist Paul Görlich. With nine stages, the 931A had a maximum amplification of $3 \times 10^6$, with peak QE of 12.5 per cent around 400 nm. That photomultiplier found application in radar jamming and scintillation counting and was employed in both the radar and Manhattan projects. During the war, its production rose from hundreds per year to thousands per month, which stimulated further development (Engstrom, 1980, pp. 4–5).

Sensitive to blue light and fast-responding, the RCA photomultiplier worked hand in glove with the fast detection developed in nuclear physics. The technique of detecting

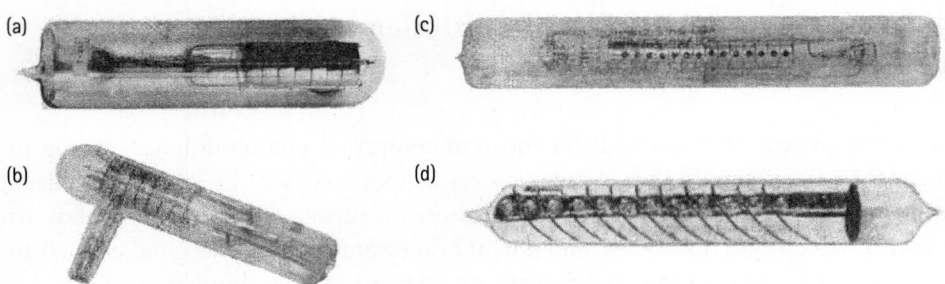

**FIGURE 23.7** Photomultiplier tubes by Kubetsky (1937): (a) and (b) first practical devices tested in 1934 with gains up to $10^3$, (c) and (d) were tested in 1935 and had gains of about $10^6$. Adapted from photographs presented in the paper.

radioactivity by the scintillation that high-energy particles produce in some materials, called scintillators, had been replaced in the 1930s by the electronic Geiger–Müller counters. These were more sensitive and objective than detecting scintillations by the human eye with the help of microscopes. In combination with the coincidence circuit first devised by Bruno Rossi in 1930, the Geiger–Müller counter found broad application and developed actively in the following decades. However, the scintillation counting method re-emerged in the 1940s thanks to the use of photomultipliers to detect the scintillations and the electronics revolution started during the war. The coupling of scintillators with photomultiplier tubes connected to the array of amplifier, high-resolution coincidence circuit, time-to-amplitude converter, and pulse-height analyser resulted in powerful detection sets that became central elements of the material culture of post-war physics (Rheinberger, 2001; Galison, 1997, pp. 454–455). The Wu–Shaknov experiment is one example of the successful application of this combination of photomultipliers as scintillation counters and coincidence circuits. The main reason for the superiority of their experiment, in comparison with the previous ones, was the use of the new scintillation counters developed during the war. The detector used in their experiment was about ten times more efficient than the Geiger–Müller counters (Maia Filho and Silva, 2019, p. 5; Silva, 2021). It was this combination of photomultipliers with coincidence circuits, amplifiers, and pulse-height analysers that opened the way to single photon counting.

The high-resolution coincidence circuit developed by Sergio DeBenedetti and Howard Richings in 1952 gives a good idea of the state of the art in the early 1950s. Associated with the scintillation counters used by Wu and Shaknov, it had a time resolution of about one nanosecond and, at the expense of efficiency, could 'measure time intervals of the order of $10^{-10}$s between radiations to an accuracy of 10 per cent' (Minton, 1956, p. 129). This might suggest that early in the 1950s the sensitivity and time resolution of photomultipliers and coincidence circuits matched the requirements for single photon counting. However, the QE and dark current of photomultipliers available in the early 1950s, most of which used the Cs-Sb photocathode, produced unsatisfactory photon-counting statistics and scientists were only beginning to under-stand how to exploit the full potential of photomultiplier to measure exceedingly faint light. A Google Ngram search for the phrase 'single photon counting' shows that it was used timidly from 1945 to 1954, when it began to grow exponentially. The first wave of works peaked around 1960 and receded until 1964 when a second and much more powerful wave began. This pattern matches well the discovery of new photocathodes such as the bialkali Na-K-Sb and the multialkali Na-K-Sb-Cs discovered respectively in 1953 and 1963 by Alfred Sommer, the German emigré who became RCA's top specialist in photoemissive materials. These new photocathodes led to further increase in peak QE in the visible region (near 30 per cent) and a remarkable decrease in dark current (Sommer, 1983). By the mid-1960s, for a 10-min counting period, researchers could measure 'an illumination level as low as three photons per second [...] with small error' (Morton, 1968, p 5). The intended pedagogical demonstration of the EPR argument by Kocher and Commins (1967) was one of the pioneering works in this second wave of photon counting.

## 23.3 COMPLEMENTARITY TESTS

As the experiments in QF diversified in the 1980s, there appeared different versions of experiments designed to examine Bohr's claim that complementary wave-like and particle-like properties cannot be observed simultaneously in a given quantum system. We have already discussed many of the instruments and techniques that enabled these experiments: lasers, thin-film beamsplitters, photomultipliers, and high-resolution coincidence circuits. In this section, we will see how they were employed to test the principle of complementarity and the dual nature of light and matter in the delayed choice experiment. Furthermore, we will discuss the new technique of neutron inter-ferometry used for a similar purpose.

### 23.3.1 Delayed-choice Experiments

In the late 1970s, as Wheeler revisited a thought experiment conceived by Bohr to illustrate the complementarity between wave and particle descriptions of microscopic systems, he proposed seven possible versions of delayed-choice experiments, and actively promoted them among experimentalists. Two teams accepted the challenge: one led by Carroll Alley at the University of Maryland, and another led by Herbert Walther at the Max Planck Institute for Quantum Optics in Garching, West Germany. These teams had the instruments and expertise to materialize Wheeler's delayed-choice experiment and were 'driven by the newly achieved possibility that [Bohr's wave–particle complementarity] could be tested in the laboratory' (Bromberg, 2008, p. 328).

Alley, who had not previously been involved with foundational issues, was among the principal investigators of two ambitious, large-scale, cold-war style research programmes. One made use of state-of-the-art technologies to measure distances between points on the Earth and on the Moon with accuracy within ten centimetres; the other aimed to detect relativistic effects on time measurements at different velocities and distances from the centre of the Earth. Alley's team drew resources from both programmes to set up their delayed-choice experiment (Bromberg, 2008, pp. 335–336).

However, the demands of the delayed-choice experiment were more modest than the origins of its instruments. In their experiment, Alley and colleagues (Fig. 23.8) split a weak laser beam into two sub-beams that were reflected by mirrors towards a second beamsplitter where they could be recombined. On the path of one of the sub-beams, they inserted a switch which consisted of a Pockels cell—a voltage-controlled waveplate—followed by a polarizer. When activated, the Pockels cell rotated the polarization of the photons by 90°, and they were deflected off the path by the polarizer, towards a detector placed to collect the deflected light. This arrangement enabled the

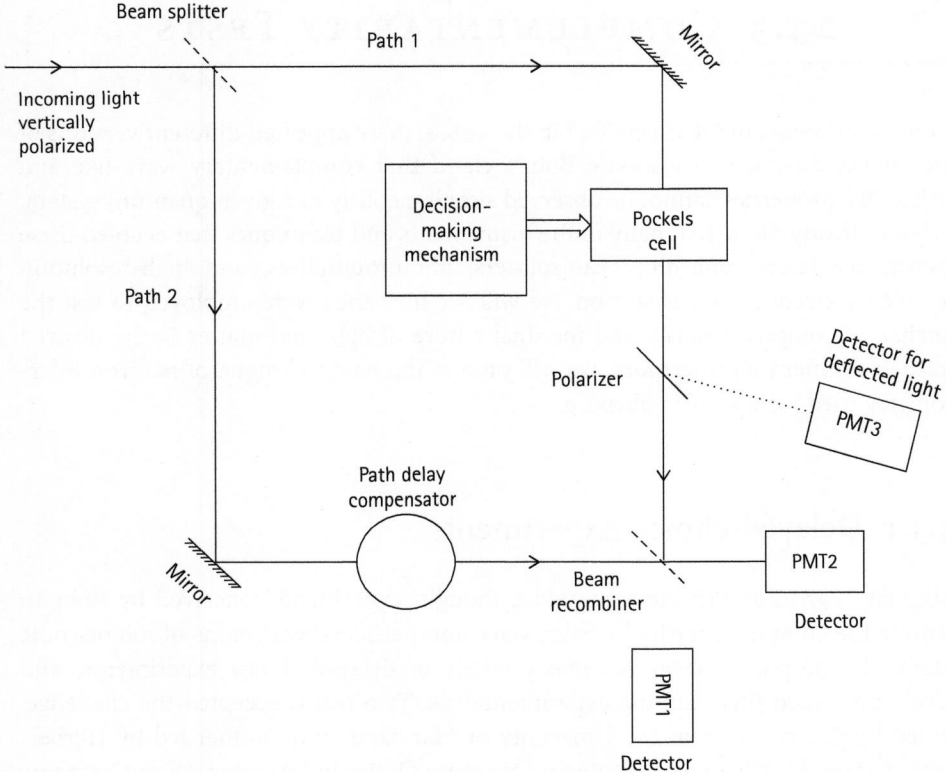

**FIGURE 23.8** Delayed-choice experiment performed at the University of Maryland. The Germans used a very similar arrangement (Hellmuth *et al.*, 1987). Author's rendering based on Figure 2 in Bromberg (2008).

identification of the path the photons took, and no interference would arise. If the Pockels cell was not activated, the photons travelled through to be recombined in the second beamsplitter, giving rise to an interference pattern. The decision-making mechanism, which activated the Pockels cell, was a photomultiplier rigged up so that it would emit, or fail to emit, an electron with 50–50 probability (Bromberg, 2008).

So far, we have covered most of the instruments used in Alley's experiment, which, at the time of the experiment, had been around for decades. The only instrument we have yet to discuss is the Pockels cell. Its working principle is based on the 19th-century discovery by Friedrich Pockels that the birefringence of some transparent crystals varies linearly with electric tension. Attaching electrodes to such crystals, one may thus modulate their birefringence. Physicists had been using electro-optic modulators such as Pockels and Kerr cells for several decades when their laser-related applications (e.g., Q-switching and modulation of laser radiation) led to a surge of interest in such devices (Kaminow, 1974, p. xiii). In the early 1970s, fast Pockels cells used to produce sub-nanosecond laser pulses had a rise time shorter than 0.5 ns (Johnson and Steinmetz, 1972). In Alley et al.'s experiment, the cell could switch between the experimental set-ups about six times while the photon was in flight, considering a distance of 1 m between the beamsplitter and the Pockels cell. The Pockels cell used in the German delayed-choice experiment was a more modest one, with a rise time of 4 ns (Hellmuth et al., 1987).

Hence, if the possibility of testing complementarity in the early 1980s was 'newly achieved' it was not because of the new instruments. Rather, it was due to Wheeler's rebranding and advertising of Bohr's thought experiment among experimentalists who had the instruments and skills to translate it into real experiments, but were not familiar with the foundational debates.

## 23.3.2 Matter Waves

In the 1980s, there were also tests and illustrations of complementarity using massive particles such as neutrons and electrons. No one questioned that these particles behave like waves in some settings—this had been well established since the 1920s. Experiments such as Akira Tonomura's, in which single electrons arrived one by one at a screen to form an interference pattern over time, were more of a demonstration than a test of a new principle. The question that led to tests of complementarity with matter waves was whether the new instruments and techniques could exhibit simultaneously the wave and particle features of single quantum systems, or whether nature precludes such a possibility.

As with the delayed-choice experiment, this test was suggested by a theorist interested in quantum foundations, the French physicist Jean Pierre Vigier, a champion of the Bohm–de Broglie wave–particle description of quantum phenomena. The theories of Bohm and de Broglie had their differences, but they both diverged from QM in that they presumed the simultaneous existence of matter and waves. As Vigier adhered to the Bohm–de Broglie description of quantum phenomena for epistemological reasons, he hoped that new experiments could cast light on what happens throughout the experiment, not only the outcomes, thus revealing what lies beneath quantum phenomena (Bromberg, 2008).

In 1983, Vigier came across a paper describing an experiment of neutron interferometry that seemed to contradict complementarity. In the experiment, an Austrian team led by Helmut Rauch demonstrated the quantum mechanical prediction that the superposition of a beam of spin-down neutrons with another of spin-up neutrons yields a beam of neutrons with horizontal spin. To do so, they sent a neutron beam into a perfect-crystal interferometer, divided it into two sub-beams, and recombined them after a radio-frequency alternating current coil had inverted the spin of neutrons in one sub-beam. They then measured the direction of the spin of the resultant beam, which conformed to the quantum-mechanical prediction (Bromberg, 2008, p. 342). Vigier saw in the fact that the radio-frequency coil exchanged photons with neutrons in an interferometry experiment the possibility of extracting the information of the path taken by the particle while preserving the interference pattern. An exchange between Vigier and Rauch led to an adaptation of the interferometry experiment suggested by Vigier in 1984.

As with the instruments discussed above, the history of the instrumentation of the neutron interferometry experiments involves emblematic developments in the history of cold war physics. The basic technique employed was the perfect-crystal neutron interferometry, first tested in 1974, in Vienna's TRIGA reactor, by Rauch, Ulrich Bonse, and Wolfgang Teimer. The key elements here were the wartime methods of growing perfect semiconductor crystals, used in radar microwave detectors, and the post-war development of neutron physics.

In the post-war period, neutrons were studied not only as inducers of nuclear reactions but, thanks to their wave properties, also as probes for matter in solid-state physics. This latter application fostered the development of a new field called neutron optics, which dealt with the wave properties of neutrons (Klein and Werner, 1983). When, in 1965, Ulrich Bonse and Michael Hart manufactured an x-ray interferometer out of a perfect silicon crystal, they pointed out that one could use perfect crystals to design a neutron interferometer, given that at room temperature neutrons have a wavelength close to x-ray. As Rauch turned to neutron optics at the beginning of the 1970s, he collaborated with Bonse to design a perfect single-crystal neutron interferometer, carrying out the first successful tests on 11 January 1974 (Rauch *et al.*, 1974).

In the following years, Rauch established a prolific research group that used the single-crystal neutron interferometer in a wide range of investigations. Their research programmes from the beginning are illustrative of the mix of fundamental and applied investigations typical of the period (Bromberg, 2008). Some experiments were tests of 'fundamental propositions of quantum mechanics' (Klein and Werner, 1983, p. 259), but not tests of quantum foundations. Namely, they were not within the field's agenda, which was shaped by the early debates among the founding fathers of quantum mechanics.[9] Only in 1984 did they mobilize their instruments for the test of complementarity. Again, the instrument was available about a decade before it was used to probe quantum foundations.

## 23.4 DECOHERENCE EXPERIMENTS

The last group of foundational experiments, which appeared in the 1990s, addressed the old problem of the transition from the quantum to the classical world, illustrated by Erwin Schrödinger with his famous cat thought experiment, and framed by John von Neumann as the measurement problem. This addressed the following questions: how and why a system found in a superposition of quantum mechanical eigenstates, evolving linearly according to Schrödinger's equation, upon measurement, collapses into a single eigenstate? The approaches that enabled the investigation of these issues experimentally emerged in the 1960s and 1980s, with studies into the interaction of quantum-mechanical systems with their surrounding environment and the understanding of how the environment destroys the quantum superposition, a process called environment-induced decoherence, or simply decoherence.

The approaches to decoherence adopt two basic postulates: (1) one can study the interaction of a measuring apparatus with a microscopic system describing the apparatus quantum mechanically, and (2) as a macroscopic system, a measuring apparatus is inevitably an open system coupled with its surrounding environment. The consequences of these assumptions were derived independently in three distinct research paths. The first was published by Heinz Dieter Zeh in 1970, and the other two in papers of the early 1980s by Wojciech Zurek and by Amir Caldeira in collaboration with Anthony Leggett. Using the standard formalism of quantum mechanics, they showed that the system–environment interaction destroys the quantum superposition, leading the system into a mixture that can be predicted by analysing the dynamic evolution of the apparatus–environment wavefunction. One of the first results obtained independently by Zeh and Zurek was the derivation of superselection rules previously postulated to exclude states of quantum superpositions that are in principle possible but never observed. Studying dissipation in quantum systems, Caldeira and Leggett produced a model that formed the basis for much of the work on decoherence and enabled calculations with experimental implications such as, for instance, the time it takes for the quantum superposition to vanish.[10]

---

[9] We might question how meaningful this distinction was for physicists in the period, for, in 1976, Rauch was invited to the Varenna school on foundations of quantum mechanics because of one such test of quantum-mechanical predictions (Freire, 2015).

[10] For the works of Zeh and Zurek, see Camilleri (2009). For the works of Caldeira and Leggett, see Freiras (2021).

The predictions of decoherence programmes were brought to the laboratory in 1996 by two leading teams in quantum optics. In Boulder, Colorado, a team of the National Institute of Standards and Technology (NIST), led by David Wineland, used laser pulses to cool down and manipulate trapped beryllium ions. In Paris, a team of the Collège de France and École Normale Supérieure, led by Serge Haroche, used Rydberg atoms to manipulate photons trapped in a superconducting cavity. Wineland and Haroche received the 2012 Nobel Prize for their 'methods of manipulating quantum systems in ways that were previously thought unattainable' (Smart, 2012, p. 16). At the core of those experiments were the techniques for ion trapping, laser cooling, super-conducting cavities, and the preparation of Rydberg atoms. As we shall see below, these techniques, developed in the 1970s and early 1980s, reflect the impressive progress of quantum electronics since the 1950s.

## 23.4.1  Ion Traps and Laser Cooling

Ion traps stemmed from the German tradition of research on atomic and molecular beams and mass spectroscopy, which developed at the University of Göttingen in the post-war period. Wolfgang Paul had just moved from Göttingen to Bonn, in 1952, when he reformulated the quadrupole lenses used to focus atomic beams so that, instead of focusing the beam on a given path, they confined the ions to a closed space (Paul, 1990). Paul's trap was taken up by Hans Dehmelt, who had learnt the craft of high-precision spectroscopy in post-war Göttigen and subsequently moved to the USA, where he settled at the University of Washington. Dehmelt and his collab-orators perfected the ion trap to the point that, when Wineland joined his laboratory in the early 1970s for a postdoc, they trapped a single electron for nearly a year (Wineland et al., 1973; Smart, 2012).

Then, tuneable lasers had just appeared with their unique combination of tuneability and extremely narrow linewidth. However, the Doppler broadening caused by the oscillation of the ions inside the trap limited spectroscopic resolution. To increase the resolution, Dehmelt and Wineland realized that they had to freeze the ions to a near standstill. Thinking the problem through, they envisaged a straightforward technique to dampen the ion oscillations using a tuneable laser.

Given that the most significant photon–atom momentum transfer occurs when the atom absorbs a photon, one can slow atoms down by making only the atoms moving against the laser beam absorb radiation. To this end, Dehmelt and Wineland turned the Doppler effect into a stepping stone. From the reference frame of atoms moving against the laser beam, the laser radiation is Doppler-shifted towards higher frequencies, while from the reference frame of atoms moving away from the laser source, the radiation is Doppler-shifted towards lower frequencies. Thus, by tuning the laser to a frequency slightly lower than the frequency of an electronic transition, for the atoms moving towards the laser, the radiation frequency Doppler-shifts towards the transition frequency. The atoms may then absorb a photon and undergo electronic transition. After the excited-state lifetime, the atom spontaneously emits a photon in a random direction and returns to the ground state. While the predominant momentum transfers from photons to atoms are always in the same direction, the recoil momentum due to spontaneous emission is random. After successive iterations, the atoms may be brought to a near standstill (Wineland and Itano, 1987; Smart, 2012).

Before Wineland and Dehmelt could demonstrate the technique they called side-band cooling, Wineland joined the time and frequency division of the National Bureau of Standards to work on the development of atomic clocks, and Dehmelt went on a sabbatical to the University of Heidelberg. Working independently on sideband cooling, they announced their successes almost simultaneously in 1978 (Neuhauser et al., 1978; Wineland et al., 1978). More refined successive experiments reached lower temperatures with single ions. In 1980 the Heidelberg team published the first photographs of a single ion nearly at rest in a trap, with a temperature below 36 mK (Wineland and Itano, 1987). The technology was ripe for the use of ions to test quantum mechanics. As Smart (2012, p. 16) put it, 'the ability to cool ions to the vibrational ground state also confers the ability to reintroduce vibrational quanta in a controlled manner. And with that, the stage was set for spectacular tests of quantum mechanics.'

## 23.4.2 Cavity Quantum Electrodynamics (QED)

Upon completing his doctoral studies, Serge Haroche decided to master the laser as a research tool for atomic physics and signed up for a postdoc at Stanford University under Arthur Schawlow, a pioneer of laser spectroscopy. During his time in California, Haroche understood that tuneable lasers enabled the production of precise states of Rydberg atoms—excited atoms in which one or more electrons have very high principal quantum number (n) and, consequently, large orbits. These atoms can be a thousand times larger than their ground-state counterparts and span about 100 nm. More significant for Haroche, the transitions between two adjacent Rydberg states are in the microwave region, making Rydberg atoms excellent tools to test the dressed atom formalism he had developed with Claude Cohen-Tannoudji and the predictions of quantum electrodynamics in confined spaces (cavity QED).[11]

By 1980, a few physicists had realized that Rydberg atoms were outstanding probes for testing cavity QED predictions. Some of the most exciting predictions were that cavities could stimulate or inhibit electronic transitions in the atoms they confined. The transition only occurs if its frequency corresponds to a cavity mode. For a cavity consisting of flat mirrors, for instance, if the distance between the mirrors is smaller

---

[11] Serge Haroche—Biographical. NobelPrize.org. Nobel Media AB 2020. Friday 24 April 2020. <https://www.nobelprize.org/prizes/physics/2012/haroche/biographical/>

than half of the transition wavelength ($d < \lambda/2$), the cavity has no mode for the radiation emitted in that transition. Hence, the confined atoms remain excited. However, if a cavity with high quality factor (Q) is precisely tuned to the resonance radiation, it sharply increases the rate of stimulated emission, and the excited atoms may emit at rates hundreds of times the rate of free-space emission (Haroche and Raimond, 1985; Haroche and Kleppner, 1989).

The first years of the 1980s were thrilling as success abounded in the preparation of Rydberg atoms and cavities (Haroche and Raimond, 1985).[12] Haroche's team announced the first breakthrough. They sent a beam of Rydberg atoms of sodium through a Fabry–Perot high-Q superconducting cavity tuned to a millimetre-wave transition and observed that the cavity magnified about 500 times the spontaneous emission rate (Goy et al., 1983). Shortly after, Herbert Walther's group at the Max Planck Institute for Quantum Optics announced a closed superconducting cavity with Q so high that they observed maser action with one Rydberg atom at a time inside the cavity (Meschede et al., 1985), demonstrating the first micromaser. Next, Daniel Kleppner's team at MIT sent Rydberg atoms through a cavity made of narrowly spaced aluminium plates and confirmed the prediction that confinement could 'turn off' spontaneous emission (Hulet et al., 1985). Besides confirming the predictions of cavity QED, these tour de forces of atomic, molecular, and optical physics opened up the prospects of new investigations into the transition between the microscopic and the macroscopic worlds.

Comparing the instruments used by Haroche's and Walther's teams in the 1980s (Goy et al., 1983; Meschede et al., 1985) with the experimental scheme proposed by Haroche's team to probe the decoherence effect in 1996 (Brune et al., 1996), we may conclude that by 1985 the technology was ripe for the test of decoherence. All the major elements were already available. The Fabry–Perot superconducting cavity designed by Haroche's team (Fig. 23.9a) did not have a quality factor as high as that of the German group, but its open design allowed the preparation of atoms in some delicate superpositions which were unattainable in closed cavities. They had also devised a method to control the Rydberg atoms' speed so that they could control the interaction time between the atoms and the cavity (Smart, 2012).

# 23.5 CONCLUSIONS

The material culture of the foundational experiments, as presented above, allows us to gauge when the technology and experimental techniques offered the possibility of translating foundational thought experiments into real experiments. For the most part, the technology was ripe about a decade before physicists turned to it to carry

---

[12] Haroche and Raimond's technical report conveys their excitement. Some outstanding results, such as those from Walther's and Kleppner's laboratories, were reported from private communications.

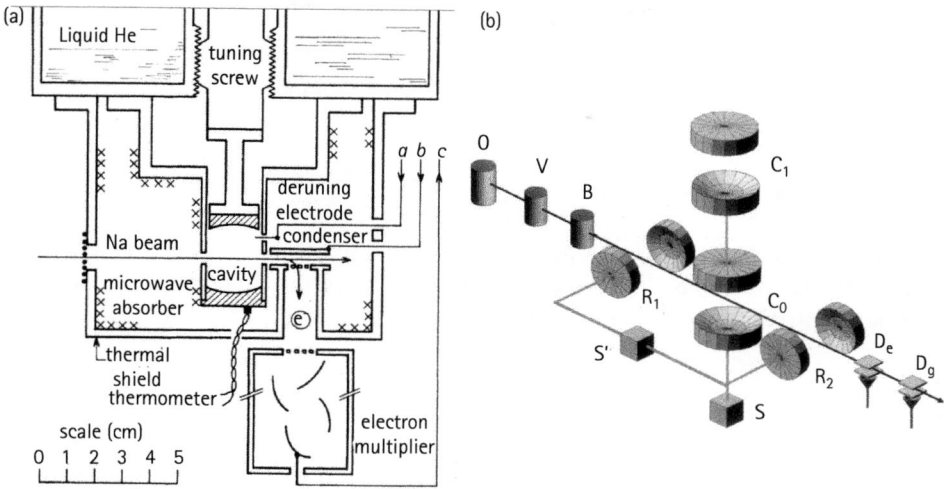

**FIGURE 23.9** (a) Experimental arrangement in Goy *et al.* (1983, p. 1904). (b) Scheme used in Raimond, *et al.* (1997, p. 1965). O is an atomic-beam source, L1 are diode lasers, L2 a tuneable laser, which prepares the Rydberg state in B, C is a superconducting cavity, R are zones with microwave fields created by S, and D are ionization detectors. Reprinted figure with permission. Copyright (2021) by the American Physical Society.

out foundational experiments. If the possibility of experimental tests had been newly achieved in the late 1960s for Bell's theorem, in the early 1980s for complementarity tests, and in the 1990s for decoherence, it was not due to the invention of new instruments, but rather to the efforts of physicists who recrafted the thought experiments to match the existing experimental possibilities and promoted them among experimentalists, or even performed the experiments themselves.

The 'quantum archaeology' adopted by CHSH turned out to be a double-edged sword. Instead of searching for instruments to design experiments resembling as close as possible the EPRB experiment, they searched for experiments that had already been performed in the hope that the available data could be used to test Bell's theorem. Although this strategy might have shortened the route towards the realization of the EPRB experiment, it led to not-so-adequate instruments and set-ups. We saw that the polarizing beamsplitters and the acousto-optic modulator used by Aspect and his colleagues were around before Bell's theorem appeared. What allowed Aspect's team to obtain their grand violations of Bell's inequality was their strategy of searching for the best instruments for an EPRB experiment.

Instruments and techniques such as lasers, SPDC, molecular beams, neutron interferometers, and laser cooling were products of wartime or cold war physics. While this study shows that they emerged with, and continuously spurred, fundamental research questions, it also reveals that it took several years for these instruments to be applied to clarify fundamental questions asked outside of the communities of physicists directly involved with their development. For them to be applied to the experimental quantum foundations, it took not only time but also broader cultural, generational, and disciplinary changes, as well as the cognitive efforts and persuasion strategies of the physicists who inhabited the social spheres of quantum foundations.

## ACKNOWLEDGEMENTS

Thanks to Olival Freire Jr for the invitation to write this chapter, Osvaldo Pessoa Jr and Eckhard Wallis for helpful criticisms, and my colleagues of the *Laboratório de História da Ciência* (LAHCIC) for stimulating discussions.

## REFERENCES

Akhmanov, S. A., *et al.* (1967). Quantum noise in parametric light amplifiers. *ZhETF Pis'ma*, **6**(4), 575–578.

Aspect, A., Dalibard, J., and Roger, G. (1982). Experimental test of Bell's inequalities using time-varying analyzers. *Physical Review Letters*, **49**(25), 1804–1807.

Aspect, A., Grangier, P., and Roger, G. (1981). Experimental tests of realistic local theories via Bell's theorem. *Physical Review Letters*, **47**(7): 460–463.

Aspect, A., Grangier, P., and Roger, G. (1982). Experimental realization of Einstein–Podolsky–Rosen–Bohm Gedankenexperiment: A new violation of Bell's inequalities. *Physical Review Letters*, **49**(2), 91–94.

Banning, M. (1947). Practical Methods of Making and Using Multilayer Filters. *Journal of the Optical Society of America*, **37**(10), 792–797.

Bispo, W. F. de O., and David, D. F. G. (2011). Sobre a cultura material dos primeiros testes experimentais do teorema de Bell: uma análise das técnicas e dos instrumentos (1972–1976). In O. Freire Jr., J. L. Bromberg, and O. Pessoa Jr. (eds) *Teoria Quântica: estudos históricos e implicações culturais*, São Paulo: Livraria da Física, pp. 97–107.

Bispo, W. F. de O., David, D. F. G., and Freire Jr., O. (2013). As contribuições de John Clauser para o primeiro teste experimental do teorema de Bell: uma análise das técnicas e da cultura material. *Revista Brasileira de Ensino de Física*, **35**(3).

Brannen, Eric, Hunt, F. R., Adlington, R. H., and Nicholls, R. W. (1955). Application of nuclear coincidence methods to atomic transitions in the wave-length range Δλ 2000-6000 Å. *Nature*, **175**(4462), 810–811.

Bromberg, J. L. (2006). Device Physics vis-à-vis Fundamental Physics in Cold War America. *Isis*, **97**(2), 237–259.

Bromberg, J. L. (2008). New Instruments and the Meaning of Quantum Mechanics. *Historical Studies in the Natural Sciences*, **38**(3), 325–352.

Brune, M., *et al.* (1996). Observing the Progressive Decoherence of the 'Meter' in a Quantum Measurement. *Physical Review Letters*, **77**(4), 4887–4890.

Burnham, D. C., and Weinberg, D. L. (1970). Observation of simultaneity in parametric production of optical photon pairs. *Physical Review Letters*, **25**(2), 84–87.

Camilleri, K. (2009). A history of entanglement: Decoherence and the interpretation problem. *Studies in History and Philosophy of Science Part B: Studies in History and Philosophy of Modern Physics*, **40**(4), 290–302. Elsevier.

Clauser, J. F., *et al.* (1969). Proposed experiment to test separable hidden-variable theories. *Physical Review Letters*, **23**(15), 880–884.

Clauser, J. F. (1976). Experimental Investigation of a Polarization Correlation Anomaly. *Physical Review Letters*, **36**(21), 1223–1226.

Clauser, J. F., and Shimony, A. (1978). Bell's theorem. Experimental tests and implications. *Reports on Progress in Physics*, **41**(12), 1881–1927.

Duarte, F. J. (2019). *Fundamentals of Quantum Entanglement.* Bristol, UK: IOP Publishing.

Engstrom, R. (1980). *Photomultiplier Handbook: Theory, Design, Application.* New York: RCA Corporation.

Freedman, S. J., and Clauser, J. F. (1972). Experimental test of local hidden-variable theories. *Physical Review Letters*, **28**(14),: 398–491.

Freire Jr., O. (2015). *The Quantum Dissidents: Rebuilding the Foundations of Quantum Mechanics (1950–1990).* Berlin, Heidelberg: Springer-Verlag.

Freitas, F. H. A. (2021). Tony Leggett's Challenge to Quantum Mechanics and its Path to Decoherence. This volume.

Fry, E. S., and Thompson, R. C. (1976). Experimental Test of Local Hidden-Variable Theories. *Physical Review Letters*, **37**(8) 465–468.

Galison, P. (1997). *Image and Logic: a Material Culture of Microphysics.* Chicago and London: The University of Chicago Press.

Goy, P., *et al.* (1983). Observation of cavity-enhanced single-atom spontaneous emission. *Physical Review Letters*, **50**(24), 1903–1906.

Haroche, S., and Kleppner, D. (1989). Cavity quantum electrodynamics. *Physics Today*, **42**(1), 24–30.

Haroche, S., and Raimond, J. M. (1985). Radiative Properties of Rydberg States in Resonant Cavities. In B. Bederson and D. R. Bates (eds), *Advances in Atomic and Molecular Physics*, New York: Academic Press, pp. 347–411.

Harris, S. E., Oshman, M. K., and Byer, R. L. (1967). Observation of Tunable Optical Parametric Fluorescence. *Physical Review Letters*, **18**(18), 732–734.

Heilbron, J. L., and Seidel, R. W. (1990). *Lawrence and His Laboratory: A History of the Lawrence Berkeley Laboratory*, Vol. I. Berkeley, Los Angeles, Oxford: University of California Press.

Hentschel, Klaus (2018). *Photons: The History and Mental Models of Light Quanta.* Cham: Springer.

Hellmuth, T., *et al.* (1987). Delayed-choice experiments in quantum interference. *Physical Review A*, **35**(6), 2532–2541.

Hollberg, L. (1990). CW Dye Lasers. In F. J. Duarte and L. W. Hillman (eds), *Dye Laser Principles with Applications,* New York: Academic Press, pp. 185–238.

Holt, R. A., and Pipkin, F. M. (1974). Precision measurement of the lifetime of the $7\,^3S_1$ state of mercury. *Physical Review A*, **9**(2), 581–584.

Horne, M. A., Shimony, A., and Zeilinger, A. (1990). Down-conversion Photon Pairs: A New Chapter In the History of Quantum Mechanical Entanglement. In *Proceedings of the International Conference on Fundamental Aspects of Quantum Theory,* Singapore: World Scientific, pp. 356–372.

Hulet, R. G., Hilfer, E. S., and Kleppner, D. (1985). Inhibited spontaneous emission by a Rydberg atom. *Physical Review Letters*, **55**(20), 2137–2140.

Joas, C. (2021). Foundations and Applications—The Creative Tension in the Early Development of Quantum Mechanics. This volume.

Johnson, B. C., and Steinmetz, L. (1972). Increased output energy for Bandwidth-Limited Nanosecond Oscillator. In G. I. Kachen Jr., W Clements, and M Genin (eds), *Laser-fusion program semiannual report—January–June 1972*, Livermore, CA: Lawrence Livermore Laboratory, pp. 6–9. Available at: https://www.osti.gov/servlets/purl/4570253/.

Kaiser, D. (2011). *How the Hippies Saved Physics: Science, Counterculture, and the Quantum Revival.* New York: W. W. Norton.

Kaiser, D. (2021). Tackling Loopholes in Experimental Tests of Bell's Inequality. This volume.

Kaiser, W., and Garrett, C. G. B. (1961). Two-Photon Excitation in CaF$_2$:Eu$^{2+}$. *Physical Review Letters*, 7(6), 229–231.

Kaminow, I. P. (1974). *An introduction to Electrooptic Devices: Selected Reprints and Introductory Text.* New York and London: Academic Press.

Kasday, L. R., Ullman, J. D., and Wu, C. S. (1975). Angular correlation of Compton-scattered annihilation photons and hidden variables. *Il Nuovo Cimento B Series 11*, 25(2), 633–661.

Kaul, R. D. (1966). Observation of optical photons in cascade. *Journal of the Optical Society of America*, 56(9), 1262–1263.

Klein, A. G., and Werner, S. A. (1983). Neutron optics. *Reports on Progress in Physics*, 46(3), 259–335.

Klyshko, D. N. (1967). Coherent Photon Decay in a Nonlinear Medium. *ZhETF Pis'ma*, 6(1), 23–25.

Kocher, C. A., and Commins, E. D. (1967). Polarization correlation of photons emitted in an atomic cascade. *Physical Review Letters*, 18(15): 575–577.

Kubetsky, L. A. (1937). Multiple Amplifier. *Proceedings of the Institute of Radio Engineers*, 25(4), 421–433.

Kwiat, P. G., et al. (1995). New high-intensity source of polarization-entangled photon pairs. *Physical Review Letters*, 75(24), 4337–4341.

Magde, D., Scarlet, R., and Mahr, H. (1967). Noncollinear parametric scattering of visible light. *Applied Physics Letters*, 11(12), 381–383.

Maia Filho, A. M., and Silva, I. (2019). O experimento WS de 1950 e as suas implicações para a segunda revolução da mecânica quântica. *Revista Brasileira de Ensino de Física*, 41(2), e20180182.

Martinez, J. P. (2021). Foundation of Quantum Physics in the Soviet Union. This volume.

Meschede, D., Walther, H., and Müller, G. (1985). One-atom maser. *Physical Review Letters*, 54(6), 551–554.

Meissner, K. W. (1942). Application of atomic beams in spectroscopy. *Reviews of Modern Physics*, 14(2–3), 68–78.

Michaud, S., and Wills, S. (2016). OSA Centennial Snapshots: Global Conflict, Thin Films, and Mary Banning. *Optics and Photonics News*, 27(3), 38–45.

Minton, G. H. (1956). Techniques in high-resolution coincidence counting. *Journal of Research of the National Bureau of Standards*, 57(3), 119–129.

Morton, G. A. (1968). Photon Counting. *Applied Optics*, 7(1), 1–10.

Neuhauser, W., et al. (1978). Optical-sideband cooling of visible atom cloud confined in parabolic well. *Physical Review Letters*, 41(4), 233–236.

Paul, W. (1990). Electromagnetic Traps for Charged and Neutral Particles (Nobel Lecture). *Angewandte Chemie International Edition in English*, 29(7), 739–748.

Peterson, O. G., Tuccio, S. A., and Snavely, B. B. (1970). CW operation of an organic dye solution laser. *Applied Physics Letters*, 17(6), 245–247.

Pryce, M. H. L., and Ward, J. C. (1947). Angular Correlation Effects with Annihilation Radiation. *Nature*, **160**(4065), 435.

Raimond, J. M., *et al.* (1997). Reversible decoherence of a mesoscopic superposition of field states. *Physical Review Letters*, **79**(11), 1964–1967.

Rauch, H., Treimer, W., and Bonse, U. (1974). Test of a single crystal neutron interferometer. *Physics Letters A*, **47**(5), 369–371.

Rheinberger, H.-J. (2001). Putting Isotopes to Work: Liquid Scintillation Counters, 1950–1970. In B. Joerges and T. Shinn (eds), *Instrumentation Between Science, State and Industry*, Dordrecht: Springer Netherlands, pp. 143–174.

Roqué, X. (1997). 'The Manufacture of the Positron. *Studies in History and Philosophy of Science Part B: Studies in History and Philosophy of Modern Physics*, **28**(1), 73–129.

Schawlow, Arthur L. (1982). Spectroscopy in a New Light. *Science*, **217**(4554), 9–16.

Sears, F. W., and Debye, P. (1932). On the Scattering of Light by Supersonic Waves. *Proceedings of the National Academy of Sciences of the United States of America*, **18**(6), 409–414.

Silva, I. L. (2021). Chien-Shiung Wu's Contributions to Experimental Philosophy. This volume.

Silva, I., and Freire Jr., O. (2013). The Concept of the Photon in Question: The Controversy Surrounding the HBT Effect circa 1956–1958. *Historical Studies in the Natural Sciences*, **43**(4), 453–491.

Smart, A. G. (2012). Physics Nobel honors pioneers in quantum optics. *Physics Today*, **65**(12), 16–18.

Sommer, A. H. (1983). The Element of Luck in Research—Photocathodes 1930 to 1980. *Journal of Vacuum Science & Technology A: Vacuum, Surfaces, and Films*, **1**(2), 119–124.

Whitaker, M. A. B. (2000). Theory and experiment in the foundations of quantum theory. *Progress in Quantum Electronics*, **24**(1), 1–106.

Williams, D. (ed.) (1962). *Molecular Physics, Methods in Experimental Physics*. New York: Academic Press.

Wineland, D. J., Drullinger, R. E., and Walls, F. L. (1978). Radiation-pressure cooling of bound resonant absorbers. *Physical Review Letters*, **40**(25), 1639–1642.

Wineland, D. J., Ekstrom, P., and Dehmelt, H. (1973). Monoelectron oscillator. *Physical Review Letters*, **31**(21), 1279–1282.

Wineland, D. J., and Itano, W. M. (1987). Laser cooling. *Physics Today*, **40**(6), 34–40.

Wu, C. S., and Shaknov, I. (1950). The angular correlation of scattered annihilation radiation. *Physical Review*, **77**(1), 136.

Zeilinger, A. (1999). Experiment and the foundations of quantum physics. *Reviews of Modern Physics*, **71**(2), 288–297.

Zel'dovich, B. Y., and Klyshko, D. N. (1969). Field Statistics in Parametric Luminescence. *ZhETF Pis'ma*, **9**(1), 40–43.

# CHAPTER 24

EARLY SOLVAY COUNCILS

*Rhetorical Lenses for Quantum Convergence and Divergence*

JOSÉ G. PERILLÁN

## 24.1 INTRODUCTION

ON 9 June 1911, twenty-three letters went out to many of the world's most prominent physicists, inviting them to a '*Conseil scientifique international*' (Lambert, 2015, p. 2024). It was an unusual invitation, from an unlikely source. The wealthy Belgian industrialist and philanthropist Ernest Solvay issued these confidential invitations to join him in Brussels later that fall for a prestigious conference. His invitation read like that of a leading physicist petitioning his specially selected colleagues to do something unprecedented. Solvay was asking them to join him in forming an elite international scientific Council in order to tackle what seemed a particularly urgent scientific problem. 'To all appearances' he began, 'we find ourselves at this moment in the middle of a new evolution of the principles on which the classical molecular and kinetic theory of matter is based.' The letter pointed explicitly to 'radiation' and 'specific heats' as two fundamental points of dissonance between classical theories and experimental data. According to Solvay's letter, 'Mr. Planck and Mr. Einstein' had shown how these 'contradictions' between theory and experiment 'disappear,' by employing the 'doctrine of energy steps'. Yet, as Solvay clearly noted, an acceptance of this new quantum interpretation 'would necessarily and undeniably entail a vast reform of our present fundamental theories...' (Solvay, 9 June 1911; as quoted in Lambert, 2015, p. 2025).

Although Solvay himself is not remembered as a quantum revolutionary, his letter profoundly influenced 20th century physics, especially the quantum revolution. The fact that he did not write the letter himself is critical, because the story behind the Solvay Councils and their revolutionary role is important to the history of quantum theory. The initial architect of the unprecedented '*Konzil*' of 1911 (see Fig. 24.1), and Solvay's ghost

**FIGURE 24.1** Official portrait of the 1911 Solvay Council (Brussels, Belgium). Left to right, standing: R. Goldschmidt, M. Planck, H. Rubens, A. Sommerfeld, F. Lindemann, L. de Broglie, M. Knudsen, F. Hasenöhrl, G. Hostelet, E. Herzen, J. Jeans, E. Rutherford, H. Kamerlingh Onnes, A. Einstein, and P. Langevin. Left to right, seated: W. Nernst, M. Brillouin, E. Solvay, H. Lorentz, E. Warburg, J. Perrin, W. Wien, M. Curie, and H. Poincaré.

Photograph by Benjamin Couprie. Courtesy of the Solvay Institutes, Brussels.

letter-writer, was the prominent German physical chemist Walther Nernst. For more than a year Nernst had been regularly talking and corresponding with Solvay, as well as with two of the most celebrated theorists of the day: Max Planck in Berlin and Hendrik Antoon Lorentz in Leiden. These exchanges helped Nernst refine his vision of a Council of elite physicists that could collaborate to tackle a foundational problem of such profound difficulty that it required an international gathering.

The first translations of Nernst's *Konzil* to French had resulted in *Concile*, rather than the more secular form *Conseil* on which they eventually settled. The idea of a scientific Council with an etymology coloured by religious overtones led Einstein to sarcastically call it the 'witches' Sabbath in Brussels' (Einstein to Besso, Prague, 21 October 1911; as quoted in Barkan, 1993, p. 67). However, unlike at religious conclaves, no votes would be cast and no election results announced. Before Solvay convened the 1911 Council, there had been few international scientific conferences. With 750 participants from 24 countries, the 1900 Conference in Paris convened by the French Physical Society was certainly international, but it was sprawling and lacked the defined objective of Solvay's Council. Nernst's vision for the 1911 Council was more in line with the first international chemistry congress held in Karlsruhe, Germany, in 1860. With the state of chemistry then in complete disarray, over one hundred leading chemists met to discuss the divergence in systems of atomic weights and molecular formulas. They were looking to stabilize and standardize the foundations of chemistry.

As Nernst told his colleagues at the Hôtel Métropole in Brussels on 30 October 1911, although no resolution was reached at Karlsruhe in 1860, the congress itself focused much-needed attention on the problem of theoretical divergence and 'soon afterward complete clarity was achieved' (Eucken, 1914, pp. 5–6; as quoted in Barkan, 1993, p. 73). Although Nernst's claim of 'complete clarity' may be somewhat hyperbolic, Karlsruhe did have a lasting 'catalytic effect on the evolution of chemistry' (Everts, 2010). In the aftermath of the 1860 meeting, it 'set the stage for the periodic system' of elements upon which all modern chemistry sits (Gordin, 2019, p. 17). In much the same way, Nernst hoped that his unique physics Council convened by Solvay in 1911 would either solve the quantum dilemma or mark a clear path towards a solution. Ultimately, conversations between Nernst and Solvay during the summer of 1910 about foundational problems in physics were transformational, not because of their scientific content but because they heralded a new era in the patronage, institutionalization, and internationalization of science.

The regularly convened meetings in Brussels that have come to be known as the celebrated 'Solvay Conferences' are ubiquitous in quantum narratives. In particular, the 1911 Council is mythologized for being the first of its kind and for the rarified level of its participants—eleven out of the eighteen were, or would eventually become, Nobel laureates (Devriese and Wallenborn, 1999, p. 15). In addition, the 1927 (see Fig. 24.2) and 1930 meetings seem to capture physicists' mythological imaginations about the great quantum interpretation debates between Einstein and the great Danish physicist Niels Bohr. Again and again, in popular historical narratives we are told about scientific consensus with regard to the interpretation of quantum theory being forged in the furnace of rational discourse at these legendary Solvay Councils.[1] Two of the most celebrated scientific heroes of all time facing off in a battle of wits and *Gedanken-experiments*.[2] The wild-haired rebel genius continually challenging the stoic, unwavering Dane. Both heroes examining every implication of this counter-intuitive theory, leaving no alternative but to accept the inevitable quantum orthodoxy (see Fig. 24.3). These myth-historical portrayals of the development of quantum theory are rhetorically powerful narratives, yet careful study of the context surrounding these Councils paints a different picture.[3]

---

[1] There is no shortage of historical narratives about the quantum revolution that rely on this particular theme. For three popular examples see: (Gribbin, 1984), (Jones, 2008), and (Kumar, 2011).

[2] The term *Gedankenexperiment* is German for 'thought experiment'. This form of inquiry and analysis is used to think through difficult physical problems that otherwise might not be testable in a real laboratory setting. Although the legacy of thought experiments goes back hundreds of years, in the 20th century Einstein became famous for his persistent use of this modality.

[3] Myth-history was a term used by physicist Leon Lederman to distinguish storytelling from scholarly history. For more on the use and impact scientific myth-histories see (Perillán, 2021).

**FIGURE 24.2** Official portrait of the 1927 Solvay Council (Brussels, Belgium). Left to right, back row: A. Piccard, E. Henriot, P. Ehrenfest, E. Herzen, T. de Donder, E. Schrödinger, J. E. Verschaffelt, W. Pauli, W. Heisenberg, R. H. Fowler, L. Brillouin. Middle row: P. Debye, M. Knudsen, W. L. Bragg, H. A. Kramers, P. A. M. Dirac, A. H. Compton, L. de Broglie, M. Born, N. Bohr. Front row: I. Langmuir, M. Planck, M. Curie, H. A. Lorentz, A. Einstein, P. Langevin, C. Guye, C. T. R. Wilson, O. W. Richardson. ABSENT: Sir W. H. Bragg, H. Deslandres, E. Van Aubel.

Photograph by Benjamin Couprie. Courtesy of the Solvay Institutes, Brussels.

Although the particulars of the science discussed at these celebrated Brussels meetings are important to a rigorous understanding of the development of quantum theory, the form of rhetoric employed by scientists and the context of how the Councils were planned and executed is equally relevant to this history. In most cases, the scientific insights traditionally associated with these Councils were actually hashed out before or after the prominent scientists met in Brussels. The presentations and subsequent discussion sessions that made up these invitation-only conferences were opportunities for scientists to employ rhetorical strategies, while leveraging their social networks, to influence scientific consensus and their own standing in the scientific community. Studying the content as well as the context of the early Solvay Councils gives us a unique lens into the development of quantum theory, its interpretations, and the dynamics of the international physics community during the first decades of the 20th century.

FIGURE 24.3 Niels Bohr and Albert Einstein deep in conversation.

Photograph by Paul Ehrenfest, courtesy
AIP Emilio Segrè Visual Archives.

## 24.2  BIRTH OF A COUNCIL: NERNST'S AND SOLVAY'S HIDDEN AGENDAS

When he moved to Berlin in 1905, Nernst brought with him the 'Göttingen spirit of Felix Klein'. His attention focused more on 'mathematics, physics, and industry', than on pure chemistry (Barkan, 1999, p. 110). Trained as a physicist, Nernst had carved out a career as one of the world's leading physical chemists. Known for his 'exceptional experimental dexterity and ingenuity' in designing micro-scale instruments, Nernst also pushed the frontier of experimental investigations on the properties of materials under extreme physical conditions, including high temperatures and pressures (Barkan, 1999, pp. 111 and 127). Moving to Berlin, Nernst began to shift his attention from studying mostly applied practical problems such as the effects of high temperatures on electrolytic conduction in solids, to more fundamental questions concerning thermodynamic laws and their applications to chemistry.

Nernst's famous heat theorem, published in 1906, led to predictions of the strange behaviour of specific heats at very low temperatures. This theorem, later interpreted by Planck as the Third Law of Thermodynamics, suggested that specific heats would tend to vanish as temperatures approached absolute zero. These predictions diverged from those made using classical statistical mechanics. Einstein's 1907 paper on 'Planck's theory of radiation and the theory of specific heat' inspired Nernst to engage with the emerging idea of energy quantization while trying experimentally to confirm the predictions of his heat theorem at very low temperatures (Barkan, 1999, pp. 164–67). After visiting Einstein in March 1910, he became convinced that seriously engaging with the 'quantum hypothesis' would in all likelihood prove critical to understanding the theoretical underpinnings of his heat theorem and could be an integral part of any future fundamental theory of solids (Barkan, 1999, pp. 182–83).

Writing to a friend shortly after his visit to Zürich, Nernst praised Einstein as a 'Boltzmann redivivus' declaring his quantum hypothesis as 'among the most remarkable thought [constructions] ever' (Nernst to Schuster, Lausanne, 17 March 1910; as quoted in Barkan, 1993, p. 62). Nernst envisioned the emerging quantum theory as a potentially revolutionary and disruptive force in physics and chemistry. There seemed to be fundamental contradictions between theory and experiment at the smallest scales. As such, Nernst thought it was imperative that the most accomplished scientists come together in one place and time to collectively hammer out a coherent understanding of the emerging quantum revolution.

That was Nernst's stated intent outlined in the invitation letter he ghost-wrote for Solvay. Yet, in addition to this overt intent, Nernst had a personal, more hidden agenda. As one of the few physical chemists engaging in this line of research, Nernst saw an opportunity to champion the importance of his heat theorem within the emerging quantum domain. Bringing together some of the most influential physicists in the world to focus on the implications of the quantum hypothesis could give him the chance to ensure his scientific legacy. Although it seems counter to the Mertonian norm of disinterestedness, Nernst's concern for his scientific legacy should not be brushed aside, or overlooked, as motivation to create an international scientific Council.

From 1901 on, he had engaged in a polemical interpersonal exchange with the Swedish physical chemist Svante Arrhenius. Once close friends, the two had become estranged, personal animosity spilling over into their professional conduct. By 1910, Arrhenius had won the Nobel Prize for Chemistry and secured a position of great influence within the Nobel Foundation in awarding prizes for both chemistry and physics. That same year, in response to a nomination of Nernst, Arrhenius categorically opposed the German's candidacy for this and all future Nobel prizes based on ill-defined 'ethical and moral failures', declaring that 'as long as this stain remains unwashed, [Nernst's] chances are minimal' (Arrhenius to Tammann, 22 December 1910; as quoted in Barkan, 1999, pp. 226–27). It is no coincidence that by the time Nernst was awarded the Nobel Prize for Chemistry in 1921, he had the unfortunate distinction of being the longest-tenured candidate, having received an unprecedented 93 nominations. In each instance, Arrhenius was responsible for evaluating the nominations and rejecting Nernst's candidacy (Crawford, 1999, p. 52). The growing clout and success of his former friend, and current rival, fuelled Nernst's desire to establish himself and his legacy in the international scientific community. Since his candidacy for the Nobel Prize was continuously undermined by Arrhenius' social and political capital, Nernst began actively looking for ways to enlarge his own social and scientific capital.

**FIGURE 24.4**  Portrait of Ernest Solvay.

Photograph by Benjamin Couprie.
Courtesy of the Solvay Institutes, Brussels.

Throughout his career in Göttingen, and later in Berlin, Nernst had been very successful in establishing strong and lasting relationships with industry. Always looking to recruit new benefactors and patrons, in 1909 Nernst and a colleague proposed the Belgian industrialist Ernest Solvay (see Fig. 24.4) for the Leibnitz Medal of the Prussian Academy of Sciences. The medal was awarded for Solvay's 'services in, and rich donations for, the promotion of the sciences' (Fisher, 1922, p. 156; as quoted in Barkan, 1999, p. 187). Solvay had made his fortune inventing, refining, and then commercializing a low-cost manufacturing process to produce soda ash from ammonia (Coupain, 2015, pp. 2076–78). As soda ash was employed widely in various industrial settings, including the manufacture of textiles, soap, glass, iron, and steel, the new Solvay process became indispensable. Solvay has been described by biographers as an 'heir of the industrial revolution' and as a fervent believer in scientific progress (Devriese and Wallenborn, 1999, pp. 1–2). He was 'imbued with the scientism and optimism of the *Belle Époque*', and believed in a 'general scientific order' that would eventually result in one 'great universal law' capable of explaining 'the movement of planets, the mechanics of living beings, and the functioning of societies' (Coupain, 2015, p. 2082).

Forced to give up his pursuit of scientific studies as a young man, Solvay's 'life's great dream' was to be 'in contact with scientists, [and if possible,] to become in some small way a scientist' himself (Devriese and Wallenborn, 1999, p. 5). In order to pursue this 'great dream' of being a valuable part of scientific progress, Solvay leveraged his wealth, social capital, and business acumen to build and manage an extensive network of scientists and engineers. Many of these consultants were either employed directly by

Solvay, or were associated with the local university in Brussels—*Université libre de Bruxelles* (ULB) (Coupain, 2015, pp. 2084–86). Through these relationships, beginning in the early 1890s, Solvay generously funded multiple institutes. The first was the Institute of Physiology (1893), followed by one in Sociology (1902), and a dedicated School of Business (1904) (Devriese and Wallenborn, 1999, pp. 11–14).

By 1900 Solvay had become a powerful philanthropist, throwing his weight behind diverse social, political, and intellectual causes. Yet physics was always at the heart of his sweeping scientific and social unification theories. In 1887, when pluralism in theoretical physics modelling had alternatively placed the ether, energy, entropy, molecular dynamics, and electrodynamics at the centre of various 'unified' theoretical frameworks, Solvay tried his hand at creating his own universal theory—*Gravitique* with gravity as its focal point (Galison, 2007, p. 2). This original work would evolve into a universal 'law of energetics' that he described as a theory of *Gravito-Matérialitique* (Devriese and Wallenborn, 1999, p. 10). Through conversations with Robert Goldschmidt, a professor of physical chemistry at the ULB (Devriese and Wallenborn, 1999, p. 14), about his universal theories and dreams about contributing to scientific progress, Solvay connected with Nernst and his grand idea for a *Conseil scientifique international*. Although Solvay well knew that he couldn't directly contribute to the highly specialized discourse at the 1911 Solvay Council, he still took advantage of a captive audience to discuss his system of *Gravito-Matérialitique* in his opening address. He also made a point of circulating a 100-page summary of his latest universal theory among the participants before the Council drew to a close (Heilbron, 2013, p. 9).

In both cases, the personal, covert motivations of Nernst and Solvay helped drive their vision for the Council yet failed to produce the exact results they desired. Nernst's heat theorem was hardly the centrepiece of the Council's discourse. The physicists seemed much more interested in discussing the quantum hypothesis with regard to radiation than its connection to specific heats. As for Solvay, although he did briefly present his *Gravito-Matérialitique* theory and circulate his paper on the topic, nobody seems to have taken his ideas seriously. Ultimately, Solvay's 'scientific, social, and political ideas have left few lasting traces' (Devriese and Wallenborn, 1999, p. 22), yet the international networks and alliances he forged with Nernst and other scientists laid the foundations for his ground-breaking Institutes for Physics (1912) and Chemistry (1913) (Devriese and Wallenborn, 1999, p. 11).

# 24.3 IMPACTS OF THE FIRST SOLVAY COUNCIL

Just after the first Solvay Council Einstein wrote a letter to Solvay thanking him profusely for providing an 'extremely beautiful week [ ... ] in Brussels', claiming that the 'Solvay Congress shall always remain one of the most beautiful memories of my life'

(Einstein to Solvay, Prague, 22 November 1911; as quoted in Mehra, 1975, chapter XIV). Yet one month later Einstein confessed a distinctly different, and private, reflection to his friend Michele Besso: 'This Congress had an aspect similar to the wailing at the ruins of Jerusalem. Nothing positive came out of it. [ . . . ] I did not benefit much, as I did not hear anything which was not known to me already' (Einstein to Besso, 26 December 1911; as quoted in Mehra, 1975, chapter XIV). Although Einstein's second account leans to the dramatic, most modern analyses of the 1911 Solvay Council side with this more critical assessment. At the first Solvay Council, not much was accomplished towards galvanizing physicists' understanding of the fledgling quantum theory (Barkan, 1993, p. 66).

In fact, as Einstein noted in the report he delivered to the Council, the 'so-called quantum theory' was more a 'useful device' than a coherent theory; and yet, classical mechanics alone could 'no longer be considered a sufficient schema for the theoretical representation of all physical phenomena'. This muddled, dissonant characterization of the current state of physics seemed disastrous and incoherent to many established physicists of the day (Einstein, Solvay, 1911, p. 436; as quoted in Galison, 2007, p. 6). Surveying the thirteen reports presented at the Council, one outstanding theme is the degree to which the participants struggled deeply with the essential tension between divergence and convergence. This tension showed in the plurality of views on how to incorporate the quantum hypothesis while still subscribing to the dominant classical paradigm (Straumann, 2011, pp. 379–99).

Shortly after the Solvay Council of 1911, the eminent French physicist Henri Poincaré wrote a summary of one of the sessions and pointed directly to the schism emerging from this quantum hypothesis: 'what struck me in the discussions that we just heard was seeing one same theory based in one place on the principles of the old mechanics, and in another on the new hypotheses that are their negation' (Poincaré, Solvay, 1911, p. 451; as quoted in Galison, 2007, p. 7). In two telling post-Council letters Einstein summed up much of this ambivalent dissonance. In a letter to a friend, he noted that 'nobody really knows anything. The whole story would have been a delight for the diabolical Jesuit fathers' (Einstein to Zangger, 15 November 1911; as quoted in Barkan, 1993, p. 68). Then, eight days later, Einstein wrote despondently to Lorentz admitting that '[t]he $h$-disease looks ever more hopeless' (Einstein to Lorentz, 23 November 1911; as quoted in Barkan, 1993, p. 68).

Although the 1911 meeting in Brussels failed to be the galvanizing force for consensus that Nernst and Solvay hoped, it did bring questions about the quantum hypothesis to a broader subset of the international physics community. More importantly, the organizational success of that first Council was formative. It resulted in a long tradition of Councils as elite international gatherings focused on interrogating the latest experimental and theoretical discourse on critical scientific topics. As such, that first Solvay Council served as a ground-breaking template for future Councils and a broader transformational force in the international organization of science.

As Solvay finished his opening remarks at the first day of the Council, he stepped aside and ensured the success not only of this first Council but of the future Solvay

Institutes and associated Councils to follow. In the lead-up to Brussels, Solvay and Nernst had the self-awareness and vision to know that, although they had set the wheels in motion, they could not effectively preside over the Council. At the outset of the meetings, Solvay recognized as much: 'I am now happy to cede my place to our eminent President, M[onsieur] Lorentz' (Solvay, 30 October 1911; as quoted in Galison, 2007, p. 3).

After early discussions, Nernst had first approached Planck with the offer to preside over the Council. This seemed like a natural choice—after all, he was a highly regarded and established theoretical physicist whose pioneering work on blackbody radiation had first sparked the community's interest in the quantization of energy. However, during initial correspondence and deliberations over how to organize the meeting, Planck made clear that he believed it was premature to convene an international Council dedicated to exploring the quantum hypothesis (Devriese and Wallenborn, 1999, p. 14). Planck was, after all, still very much the 'reluctant revolutionary', not quite ready to accept the quantum hypothesis as anything more than a crude heuristic tool (Kragh, 2000, p. 31).

In many ways Lorentz was the ideal person to preside as Chair of the 1911 Solvay Council. He was already a Nobel Laureate, having shared the 1902 prize for his pioneering work on electron theory and his analysis of the Zeeman effect. Then, in 1904, Lorentz showed his courage in extending classical electrodynamic theory at the cost of accepting counter-intuitive physical implications. In doing so he would develop his eponymous transformations and pave the way for Einstein's Special Theory of Relativity. By 1910, Lorentz was seen as a classically grounded elder statesman of science who also had the intellectual integrity and courage to openly examine orthogonal positions to his own. In many ways, Lorentz was a natural bridge for a physics community in flux. He had 'completed what was left unfinished by his predecessors and prepared the ground for the fruitful reception of the new ideas based on the quantum theory' (The Nobel Prize, 1967).

# 24.4 LORENTZ TRANSFORMATIONS: SCIENTIFIC INTERNATIONALIZATION AND DIPLOMACY

One might ask, why didn't Einstein preside over the 1911 Solvay Council? Hard as it may be to believe, at thirty-two years old Einstein was still relatively unknown in 1911. Although it had been six years since his *annus mirabilis*, he had only been a university-affiliated academic for two years. It was thanks in large part to his role during this first Solvay Council that his reputation became established in the international physics community. In addition, Einstein lacked the ideal temperament for such scientific diplomacy. 'He was driven, impatient, biting in his sarcasm, and wouldn't or couldn't hide his disdain for bad or wrong-headed approaches.' On the other hand, Lorentz, with his scientific standing, natural acumen for languages, even-keeled temperament, and an approach to physics that might best be described as 'radical conservatism', was ideally suited to take the reins at Solvay in 1911 (Galison, 2007, p. 4).

Reflecting on the 1911 Solvay Council, Einstein recounted Lorentz's 'incomparable tact and incredible virtuosity. He speaks all three languages [French, German, and English] equally well and possesses unique scientific acumen' (Einstein to Zangger, 15 November 1911; as quoted in Mehra, 1975, chapter XIV). In subsequent correspondence Einstein noted that 'Lorentz is a miracle of intelligence and subtle tact—a living work of art. In my opinion he was the most intelligent of all the theoreticians present' (Einstein to Zangger, 16 November 1911; as quoted in Mehra, 1975, chapter XIV). This assessment of Lorentz was not unique to Einstein. Kamerlingh Onnes wrote that Lorentz had 'excelled in [his] task' as Chair of the 1911 Solvay Council and had ultimately 'won everyone's heart'. Onnes went on to note that Lorentz 'was able to lead us' with his 'exceptional clarity, facility, and friendliness,' so that 'whenever there [was] any difference of opinion, [he was able] to create and maintain such a pleasant, amicable, and cheerful yet serious tone' (Onnes to Crommelin, 3 November 1911; as quoted in Berends, 2015, p. 2096).

On the last day of the Council, Solvay met privately with some senior participants in the meeting. He wished to continue his support of, and direct involvement in, resolving the physics problems associated with the quantum hypothesis that had been discussed in Brussels. Seeing Lorentz's masterful turn in presiding over the Council, Solvay asked the Dutchman to develop a plan for a foundation that would support experiments and future Councils dedicated to the progress of physics research (Berends, 2015, p. 2096). Upon his return to Leiden, Lorentz began transforming his career and professional affiliations so that he might take up his new role as venerated scientific diplomat (Berends, 2015, pp. 2100–01). On 1 May 1912, the International Solvay Institute for Physics (ISIP) was officially founded by Solvay with Lorentz as its chief architect and first President of its *Comité Scientifique* (CS). In addition to Lorentz, this scientific committee, charged with organizing future Solvay Councils and dictating the scientific agenda of ISIP-funded projects, was made up of internationally renowned physicists and physical chemists. The French were represented by Marie Curie and Marcel Brillouin, the Belgians by Goldschmidt, the Dutch by Kamerlingh Onnes, the Danes by Martin Knudsen, the British by Ernest Rutherford, and the Germans by Emil Warburg and Nernst (Berends, 2015, p. 2098).

The Nobel Foundation stimulated scientific research primarily through the awarding of prizes, but the ISIP was quite different. As a private system of patronage within science, it relied on an international committee of scientists to set scientific agendas, dictate funding priorities, and organize regular elite scientific Councils on specific topics of interest. The model, orchestrated by Lorentz and supported by Solvay, was set up with the intent to be as free of bureaucratic and political interference as possible.

There was an Administrative Commission (CA) of three members selected by Solvay, the UBL, and the King of Belgium. This administrative body managed the finances and decided on fellowships, but all scientific decisions were left to the CS. Apart from the legendary Councils, the ISIP was responsible for funding numerous critical research projects around the world and supporting young Belgian scientists through its fellowship programme. In the first three years of ISIP's operation (1912–1914) it funded 40 research projects, 20 of which were based in Germany (Berends, 2015, pp. 2099–2100).

In January 1913, as Lorentz was preparing for the second Solvay Council, he published a short article highlighting his optimism for a future of international peace through scientific collaboration. In 'International science promotes peace' he argued that the process of scientific and technological innovation had fundamentally changed. Now 'the sheer size of the accomplished progress [ . . . ] requires scientific internationalism'. He ended the article with a question: 'Who would deny that cooperation and the pursuit of the same goal will eventually have to create precious feelings of appreciation, solidarity, and good fellowship and thereby will stimulate peace?' Eighteen months later, Lorentz's vision of attaining lasting peace through scientific internationalism sustained a serious blow when World War I broke out and Germany invaded Solvay's home country of Belgium (Lorentz, January, 1913; as quoted in Berends, 2015, p. 2101).

To make things worse, on 4 October 1914, ninety-three of the most respected German intellectuals co-signed the now infamous 'Appeal to the civilized world', defending Germany's position in the war and denying many accusations of war crimes perpetrated by their military. Many of the most prominent German scientists signed the Appeal, including Nernst, Planck, Philipp Lenard, Wilhelm Röntgen, Wilhelm Ostwald, and Wilhelm Wien. In defiance of a growing narrative of unwarranted German militarism and aggression, the 'Appeal' claimed: 'It is not true that Germany is to blame for this war. [ . . . ] Without this German militarism, German culture would have been wiped off the face of the earth. It was to protect our culture that militarism in our country was born' (Heilbron, 1987, appendix; as quoted in Marage and Wallenborn, 1999, p. 113).

Scientists and other intellectuals in England and France responded immediately. The British noted that it was 'German armies alone' that had 'deliberately destroyed or bombarded such monuments of human culture as the Library at Louvain and the Cathedrals at Reims and Malines'. The French response protested assertions in the 'Appeal' that the 'intellectual future of Europe' should be solely linked to the 'future of German science' (as quoted in Berends, 2015, p. 2103). As a citizen of the neutral Netherlands, Lorentz found himself caught in the middle of escalating military and rhetorical hostilities. Although the summary reports of the 1913 Solvay Council were sent to press before the outbreak of the war, their publication would be severely delayed. The proceedings of the 1913 Solvay Council were eventually published in 1921 with a note indicating that during the 'occupation' of Belgium, it had been impossible for the ISIP to get the reports 'passed by the German censors' (as quoted in Marage and Wallenborn, 1999, p. 112).

Lorentz's day-to-day management of the ISIP was becoming nearly impossible, as the CS had quickly become uncommunicative and highly siloed. Throughout the war, the Dutch scientist and diplomat made a point of staying in constant communication with his colleagues on both sides of the divide (see Fig. 24.5). He implored them to re-engage and find a way to lessen hostilities. In addition, he always found ways to remind them of the efforts of Solvay and the ISIP to advance science (Berends, 2015, pp. 2105–08). However, the fact that Solvay had been disproportionately generous with German scientists in the months leading up to the war, only to have his homeland invaded by the German army, and then insulted by the 'disgraceful manifesto of the 93' made a de-escalation of rhetoric extremely difficult (Lorentz to Solvay, 10 January 1919; as quoted in Marage and Wallenborn, 1999, p. 114). Considering all that had happened, many saw their 'Solvay family' as irrevocably broken (Berends, 2015, p. 2106). Yet, despite intensely vitriolic rhetoric, Lorentz continued his quest for a lasting peace through the internationalization of science.

**FIGURE 24.5** Hendrik Lorentz in Leiden *circa* 1923 with Arthur Eddington (seated); standing left to right: Albert Einstein, Paul Ehrenfest, and Willem de Sitter.

# 24.5 SOLVAY COUNCILS DURING AN ERA OF INTERNATIONAL BOYCOTTS

Although Arthur Eddington had warned his colleagues that using the pursuit of truth 'as a barrier fortifying national feuds is a degradation of the fair name of science' (Eddington, 1916, p. 271; as quoted in Stanley, 2007, pp. 88–89), the hostilities of World War I, and the 'Appeal' in particular, had a lasting effect on the international physics community. Bowing to Lorentz's diplomatic efforts, Planck tried to assuage the bitter feelings associated with his signing of the 'Appeal'. In April 1916 he and Lorentz arranged for the publication of an 'open letter' penned by Planck. In it, he attempted to explain the context around the 'Appeal'. Planck admitted that 'its formulation' had given the 'wrong perception of the views of its signatories'. He also conceded that there were 'spheres of the spiritual and moral world which are beyond the fights of peoples and that an honest cooperation in the care of these international cultural goods' is possible. Planck had certainly taken a step toward reconciliation, but his open letter was unevenly received (Planck, 1916; as quoted in Berends, 2015, p. 2107).

Any goodwill garnered from efforts like Lorentz and Planck's were seriously undercut at the end of World War I. In the wake of hostilities, Germany was left with a severely crippled economy and significant social and political unrest. The Versailles Treaty later cemented more ill will. Germans came away determined to show the world that they were still a cultural and scientific force. In November 1918, Planck made this point clear in his address to the Prussian Academy of Sciences: 'If our enemy has taken from our fatherland all defence and power [ ... ] there is one thing which no foreign or domestic enemy has yet taken from us: that is the position which German science occupies in the world.' He went on to pledge that the academy under his leadership would 'defend [its position] with every available means' (Planck, 1918; as quoted in Forman, 1973, p. 163). Planck was unequivocal in his address; the post-war international scientific community was still deeply embroiled in elevating nationalism above the scientific ideals of cooperation and universalism.

Throughout the 1920s important centres of quantum physics research and teaching thrived at universities in Germany (Munich and Göttingen) and in neutral nations like Denmark (Copenhagen) and Holland (Leiden), allowing a free exchange of ideas and personnel among them (Eckert, 2001, pp. 151–61). But impediments to broader international scientific collaboration persisted, for German and Austrian scientists were barred from attending international conferences during the early and mid-1920s (Kevles, 1995, pp. 185–221; Canales, 2015, pp. 96 and 121–27). By one account, from 1919 to 1925 German scientists were excluded from 165 international scientific conferences including the third and fourth Solvay Councils of 1921 and 1924 (Kragh, 2002, pp. 143–45).

As early as 1919, in anticipation of this post-war international schism, Lorentz had delicately lobbied Solvay not to 'formally exclude the Germans' from future Solvay Councils. Lorentz recognized that 'for the time being, Belgian and French scientists want[ed] nothing more to do with' the Germans, yet he argued against a blanket ban. After all, not all German scientists had signed the 'disgraceful manifesto of the 93'. In particular, Lorentz was interested in keeping the door open to German pacifists like Einstein and his vice-president on the ISIP's CS, Emil Warburg (Lorentz to Solvay, 10 January 1919; as quoted in Marage and Wallenborn, 1999, pp. 113–14). On the other hand, the French physicist Marcel Brillouin wanted a blanket ban on all German scientists including Einstein, as well as any other nationals, like the Dutchman Peter Debye, who had spent time in Germany. Eventually the ISIP's CS and CA, in consultation with Solvay, settled on a full exclusion of German scientists except for Einstein, whom they considered 'of some ill-defined nationality' (Tassel, 1919; as quoted in Marage and Wallenborn, 1999, p. 115). Einstein initially accepted the invitation to the 1921 Solvay Council but ultimately decided against attending (Marage and Wallenborn, 1999, p. 115).

In the lead-up to the 1924 Solvay Council, the exclusion policy still stood. Einstein wrote to Lorentz in protest. 'I am of the opinion that politics should have no place in scientific matters. It leads to a situation where individuals are judged responsible for acts of government in their native countries. If I took part in the Council, I would be actively supporting [an exclusion] which I consider to be deeply unjust' (Einstein to Lorentz, 16 August 1923; as quoted in Marage and Wallenborn, 1999, p. 115). Based on the boycotts of German scientists, one might surmise that German science would have suffered tremendously both during and after World War I; in fact, within the field of quantum theory, the opposite is true. While international scientific institutions like Solvay's ISIP excluded German scientists, private American industrial powers such as General Electric recognized the larger importance of German innovation and began actively funding German science (Cassidy, 2010, pp. 121–23). These sources of private funding, along with the strong networks of exchange established among the leading 'quantum schools' in Germany and neutral countries like Denmark and the Netherlands, became critical to mitigating effects of the boycotts (Eckert, 2001, pp. 151–61).

In addition to the political turmoil of the 1920s, the planning period for the 1927 Solvay Council was contemporaneous with a period of intense, dynamic scientific exchanges concerning unsettled foundational issues within the evolving quantum mechanical picture. This scientific fluidity represented a period of high divergence in which many varied attempts were deployed to bring concepts like French physicist Louis de Broglie's wave–particle duality and Bohr's 'damn quantum jumps' into some recognizable order (Heisenberg, 1969; as quoted in Moore, 1992, p. 228). Among contending formulations were the now ubiquitously recognizable matrix and wave mechanics, but also lesser-known alternatives abandoned along the way, including de Broglie's pilot wave theory and Einstein's unpublished attempt at his own hidden variables theory.[4] This divergence was further inflamed by rising tension

---

[4] Mara Beller refers to this period in the development of quantum theory as being in 'creative dialogical flux' as she tracks the polyphony of contradictory voices and ideas, many times coming from the same physicist (Beller, 2001, p. 3). For other examples of this divergence see: (Jammer, 1966, pp. 291–93; Belousek, 1996; Brillouin, 1926; Klein, 1926).

between pragmatists and philosophically minded physicists. Just before the Council was set to meet in the fall of 1927, the British physicist Oliver Lodge encapsulated this tension; he feared that in developing the new quantum theory, 'absurdities will be retained by philosophically minded interpreters, and will be foisted into their scheme of reality' (Lodge, 1927, pp. 423–24).

Within this pluralistic, fragmented context Lorentz managed to adroitly navigate the difficulties of international scientific diplomacy and shepherd the historic encounter at the fifth Solvay Council in Brussels during Fall 1927. As chairman of the ISIP's CS, Lorentz had been trying unsuccessfully to end the Institute's exclusionary policy for years, but momentum began to swing his way with the imminent inclusion of Germany into the League of Nations. After meeting with the members of both the CS and CA and pleading his case directly before the Belgian King, Lorentz offered Einstein a seat on the CS and extended invitations to other carefully selected German physicists to participate in the upcoming fifth Solvay Council (Berends, 2015, p. 2109). During the initial planning, Lorentz had proposed the Council focus on 'the conflict and the possible reconciliation between the classical theories and the theory of quanta' (Lorentz to Ehrenfest, 29 March 1926; as quoted in Bacciagaluppi and Valentini, 2009, p. 9). Although Lorentz eventually settled on 'Electrons and Photons' as the theme of the Council, the wording of the original proposal is telling, as it reveals just how fractured he saw the international physics community, both politically and scientifically (Bacciagaluppi and Valentini, 2009, pp. 6–9 and 253).

## 24.6 FORCED CLOSURE—1927 SOLVAY COUNCIL AS CONVERGENT LENS

The fifth Solvay Council of 1927 is traditionally assigned central importance in quantum narratives as a critical salvo in the Einstein–Bohr interpretational debates— a battleground between scientific traditionalists and a new quantum ontology that threatened to override physicists' classical interpretations of the natural world. While scientific debates certainly took place in Brussels during the fall of 1927, the conference itself served more as a converging lens than a pitched battlefield. In many ways, the outcome of this Solvay Council was neither the beginning nor the end of a great philosophical battle. Instead, it became an important rhetorical moment for a scientific community embroiled in turbulent political and scientific divergence (Perillán, 2018, pp. 34–35).

As in previous Solvay Councils, the presentations began by surveying the current state of experimental techniques and results associated with the topic of discussion. The Solvay presentations on experimental techniques reflect several important points. First, the foundations of quantum theory were thoroughly grounded in experimental observations. The rising power of Erwin Schrödinger's wave equation and the matrix mechanical formalisms championed by Werner Heisenberg and Max Born were due to their success in applications, not to an arbitrary decision to make them canonical. Second, within the community of experimental physicists working on atomic physics, there were interpretational differences that mirrored those in the theoretical camp. Physicists like William L. Bragg preferred to interpret the results of their experiments using classical understandings of electricity and magnetism (Bragg, 1928, pp. 259–88), while Arthur H. Compton was beginning to accept Bohr's notion of complementary and incommensurable pictures of wave–particle duality (Compton, 1928, pp. 301–23). Finally, theorists such as Bohr, Born, Heisenberg, and Wolfgang Pauli felt so emboldened by success in applying their quantum formalisms to explain experimental observations that they proceeded to make overly dogmatic declarations (Bacciagaluppi and Valentini, 2009, pp. 291–98 and 324–38).

In the sessions that followed, presentations on the theoretical foundations of quantum mechanics were read. Lorentz had made a point of inviting three distinct theoretical views of the 'new mechanics'. First, de Broglie presented his deterministic pilot wave theory that sought to synthesize the wave and particle pictures of quantum systems by assuming underlying hidden variables. This was followed by Heisenberg and Born's joint presentation of their matrix mechanics, Heisenberg's principle of indeterminacy, and Born's probabilistic interpretation of the wave function, psi. Finally, Schrödinger presented his wave mechanical formulation and his physical interpretation of psi.

With all the theoretical divergence leading up to the Council, one would have expected fireworks from participants. Instead, the tenor seemed more directed toward leveraging social networks and pushing aggressively for theoretical consolidation by a powerful bloc of scientists. Bohr was not invited to present in Brussels, but one month earlier he had set the stage for this forced consolidation by presenting his ideas on complementarity at a conference in Como, Italy. His goal at both Como and Solvay was to use complementarity as a way to 'harmonize the apparently conflicting views taken by different scientists [on quantum theory]' (Bohr, 1928; as quoted in Jammer, 1974, p. 86). Bohr came to the Solvay Council aiming to push his colleagues toward an interpretation of the new mechanics that rested on ideas congealing in the quantum schools of Copenhagen and Göttingen. Bohr, Born, Heisenberg, and Pauli were ready to break with classical norms and embrace a new mechanics that had no trajectories, was not visualizable, was fundamentally indeterministic, and distinctly contingent on observation. According to their interpretation, wave and particle pictures could never be synthesized, and no hidden variables were necessary to make sense of the microworld.[5]

After de Broglie's paper summarized his deterministic hidden variables formulation of quantum mechanics, there was a brief discussion where some participants like Lorentz asked clarifying questions and others like Born and Pauli raised more critical

---

[5] They were not alone. To this short list, one should add others like Pascual Jordan and P.A.M. Dirac.

objections.[6] In each case, de Broglie was able to stand his ground and articulate reasonable, if not convincing, responses (Bacciagaluppi and Valentini, 2009, pp. 364–70). At one point, Léon Brillouin launched an impassioned defence of his compatriot's deterministic wave mechanical approach: 'It seems to me that no serious objection can be made to the point of view of L. de Broglie. [ ... ] There is no contradiction between the point of view of L. de Broglie and that of other authors' (as quoted in Bacciagaluppi and Valentini, 2009, p. 365).

In the immediate discussion following de Broglie's paper, there were no clear disqualifying objections to its validity as a contending formulation and interpretation of quantum theory. In addition, Schrödinger had yet to give his paper with his own formulation and interpretation of quantum theory; so, as Born and Heisenberg began to give their report presenting the core principles of matrix mechanics, the principle of indeterminacy, and the non-physical probabilistic interpretation of psi, there was no reason to think that the quantum interpretation debate had been settled.

In that context, it seemed to some premature when Born and Heisenberg wielded their rhetorical axe to pre-emptively cut down quantum formulation debates. Without offering convincing arguments against established alternative formulations, they simply asserted that 'quantum mechanics [was now] a closed theory, whose fundamental physical and mathematical assumptions [were] no longer susceptible of any modification'. In addition, they declared that indeterminism was unavoidable, a 'fundamental' principle that simply and unambiguously 'agrees with experience' (Born and Heisenberg, 1928, p. 398). This forced closure was clearly not a result of vigorous scientific debate at the fifth Solvay Council. It was a dramatic rhetorical ploy intended to end arguments that had been raging for some time. Amid all the political and scientific schisms of the 1920s, an alliance had been forged between Copenhagen and Göttingen that became explicit with Bohr's address in Como and Born and Heisenberg's presentation at the fifth Solvay Council.

Retorting to this sudden insistence that quantum mechanics represented a closed theory, not susceptible to alteration in its physical assumptions, Lorentz made explicit a growing point of tension among some participants. During the general discussion section of the Council, Lorentz commented that raising indeterminism to a fundamental principle seemed arbitrary and unjustified. The great Dutch scientist complained that the newest interpretation of quantum theory seemed to be 'a priori ... forbid[ding]' him to visualize or even ponder particle trajectories. In one of his last great scientific acts, Lorentz warned his colleagues that the reconciliation of quantum theory with classical principles of determinism should not be abandoned (as quoted in Bacciagaluppi and Valentini, 2009, pp. 432–33; Galison, 2007, pp. 10–13). He died three months later. It is clear that Einstein, Schrödinger, and de Broglie agreed with Lorentz's general critique of

---

[6] Pauli's objection to de Broglie's pilot wave theory was a rather mundane one, not the famous critique he later levelled during the general discussion involving the inelastic scattering of electrons by a rigid rotator.

the vision of quantum theory presented by Born and Heisenberg at Solvay, yet the lack of a coordinated alternate position weakened their opposition.

## 24.7 WHAT HAPPENED TO THE 1927 SOLVAY DEBATES?

While, undoubtedly, forced rhetorical arguments played a critical role in the push to close debates around deterministic formulations of quantum theory, it would be misleading to interpret these Solvay discussions without considering the concrete physical arguments in play. For example, during the general discussion, Pauli raised an empirically based critique of de Broglie's pilot wave theory that has become ubiquitous in the standard historiography of this period. In the exchange, Pauli describes a scenario involving the inelastic scattering of electrons by a rigid rotator. He claims that de Broglie's pilot wave picture is problematic because it fails to account for all expected scattering outcomes. Whether or not Pauli's critique legitimately undercut the pilot wave theory, its perceived relevance certainly contributed to de Broglie's eventual abandonment of his alternate deterministic interpretation of quantum theory.

In addition to the Pauli–de Broglie exchange, the general discussion at 1927 Solvay contained what has come to be seen as the first salvo in the now-famous series of Bohr–Einstein Solvay Council *Gedankenexperiment* debates. In Einstein's thought experiment, he offered two interpretations of a single-slit diffraction experiment. The first was an ensemble picture in which 'the de Broglie–Schrödinger waves' correspond to a 'cloud of electrons spread out in space'. He claimed that the second interpretation, akin to that presented by Born and Heisenberg, was 'a complete theory of single processes', in which each individual electron has a corresponding wave function. Einstein objected vigorously to this second interpretation on multiple grounds. Most egregiously, the second interpretation required a 'mechanism of action at a distance', which would contradict the 'postulate of relativity'. Einstein concluded that Born and Heisenberg's stated formulation could not be considered complete. According to Einstein, one had to use 'Schrödinger's wave, but at the same time [localize] the particle during propagation'. Here he gave credit to de Broglie, 'I think that [he] is right to look in this direction' (as quoted in Marage and Wallenborn, 1999, pp. 167–68).

However, in the fall of 1927 Einstein remained unconvinced that the pilot wave theory was the answer. Earlier that year he had attempted, then abandoned, a hidden variables approach to quantum mechanics (Belousek, 1996). Einstein had been struggling with the deep implications of quantum statistics for years, complaining to Born in December 1926 that 'the Old One [ . . . ] does not throw dice' (Einstein to Born, December 1926; as quoted in Marage and Wallenborn, 1999, p. 164). As a result, Einstein set out to circumvent the fundamental indeterminacy that seemed to be the root of the emerging quantum mechanics, but was forced to abandon his approach when he realized it would require him to accept non-local interactions of 'energetically independent subsystems' (as quoted in Howard, 2007, p. 66).[7] Although Einstein agreed with de Broglie's objections to quantum indeterminism, ultimately he did not think that introducing hidden variables via the pilot wave theory was an acceptable solution.

Much has been written about Einstein's *Gedankenexperiment* at the 1927 Solvay Council and Bohr's response. Unfortunately, all that survives in the published proceedings is confusion from Bohr about Einstein's objections and a restatement of his presentation of complementarity from the Como address (Bacciagaluppi and Valentini, 2009, pp. 14 and 21). Much of the lively Bohr–Einstein debate attributed to the 1927 Solvay Council is based on later recollections by Bohr and others. Of these, most prominent is Bohr's 1949 memoir account of the exchange and the unofficial discussions that continued late into the evening (Bohr, 1949; Howard, 2007, pp. 72–74).[8] The following point is critical: although both the de Broglie–Pauli and Einstein–Bohr exchanges have been cited as decisive moments in the history of quantum interpretations, they came well after Born, Heisenberg, and Bohr had unilaterally declared quantum mechanics a complete theory and the debate closed.

Ultimately, what won the day at the 1927 Solvay Council was positional asymmetry. On one side, a coordinated front pushed for consolidation and closure, while on the other side sat physicists who were uncomfortable with the rising tide of consensus but mounted no unified opposition. Without clear, coordinated opposition, Bohr, Born, Heisenberg, and many of their colleagues left Solvay insisting that quantum mechanics was now a closed theory with indeterminism an unavoidable ontological principle of the new quantum reality. There seemed to be a growing sense that, in order to maximize the community's efforts at extending and applying quantum mechanics, the epistemological pluralism that had saturated the community in the years leading up to this fifth Solvay Council must be reduced to some semblance of congealed consensus (Perillán, 2018, pp. 37–39).

## 24.8 THE EINSTEIN–BOHR DEBATES: OUTLIERS WITHIN AN INTERPRETATIONAL LULL

In the wake of Lorentz's death, many physicists failed to heed his warning not to abandon the reconciliation of quantum and classical theories. Einstein noticed this trend of interpretational complacency in the physics community. In a letter to

---

[7] This would later be termed quantum entanglement by Schrödinger and become the thrust of Einstein's critique of the orthodox interpretation of quantum theory in the famous EPR paper of 1935.

[8] Bohr's 1949 memoir is clearly a rational reconstruction, coloured by the intervening debates and the evolution of theoretical understandings of quantum theory.

Schrödinger in May 1928, he complained that the 'Heisenberg–Bohr tranquilizing philosophy—or religion?—[...] provides a gentle pillow for the true believer from which he cannot very easily be aroused' (Einstein, 1928, p. 31).

If the pragmatic majority remained somnambulant on issues of quantum interpretation, a vocal minority continued the work started by Bohr, Born, Heisenberg, and Pauli at Como and Solvay in Fall 1927. This group has been termed the 'The earliest missionaries of the Copenhagen spirit' (Heilbron, 1985). The use of the term 'missionaries' here alludes to the fervour and zealotry with which some adherents proselytized the core principles, which eventually became synonymous with the Copenhagen interpretation of quantum theory (Heilbron, 1985, p. 196). Although Heisenberg was not Bohr's most ardent 'disciple' on notions of interpretation, he was a leading propagator of this early Copenhagen spirit. His reactions to Bohr's complementarity principle in the fall of 1927 were tentatively supportive, but by the summer of 1928 Heisenberg was boldly declaring that the 'fundamental questions [in quantum theory] had been completely solved' and that Bohr's complementarity was the capstone of the whole interpretation (as quoted in Heilbron, 1985, p. 201).

To underscore his view, while touring the United States during the spring of 1929, Heisenberg began talking of a '*Kopenhagener Geist der Quantentheorie*' or a 'Copenhagen spirit of quantum theory' (as quoted in Heilbron, 1985, p. 201). Dissemination of this Copenhagen spirit came as a result of physicists, like Heisenberg, who found themselves and their ideas reaching beyond continental Europe through lecture tours, voluntary and forced immigration, and publication of new quantum textbooks. One must be careful not to characterize this Copenhagen spirit as a unitary interpretation, hegemonic power, or a singular front. Even though Heisenberg and others subscribed to it in general terms, they certainly did not see eye-to-eye on all interpretational matters (Howard, 2004; Camilleri, 2009). More accurately, the dissemination of the Copenhagen spirit via 'missionaries' like Heisenberg represented a loose 'resonance' among this core set of quantum theorists (Ito, 2005, pp. 151–83). Yet this Copenhagen spirit, illusory as it may have been, became a powerful agent of 'paradigmatic consensus' and orthodoxy within the international physics community (Beller, 2001, p. xiii).

In 1929 Paul Dirac echoed Heisenberg, observing that '[t]he underlying physical laws necessary for the mathematical theory of a large part of physics and the whole of chemistry are [...] completely known', and the sole focus should be on clarifying the 'application of these laws' (Dirac, 1929, as quoted in Schweber, 1994, p. xxii). By the time of the sixth Solvay Council of 1930, orthodoxy was undoubtedly congealing around the Copenhagen spirit. Surprisingly, even de Broglie and his pilot wave were not immune to this rising consensus. Although Einstein had somewhat encouraged his deterministic pilot wave research programme at the fifth Solvay Council, de Broglie published a book in 1930 on wave mechanics in which he explicitly abandoned his own formulation of quantum theory, claiming that in light of the prevailing interpretation, it was no longer possible to consider his own as 'satisfactory' (de Broglie, 1930, p. 7).

In most historical accounts, the high point of the celebrated Einstein–Bohr Solvay debates over quantum *Gedankenexperiments* took place at the sixth Solvay Council in

Brussels during the fall of 1930. A particularly famous exchange has been told so often that one can almost vividly imagine the back-and-forth between these two scientific legends. According to Léon Rosenfeld's account, one evening, after official sessions had ended, Einstein proposed his now famous light box *Gedankenexperiment*. The box was a device that had a controlled light source and a high-precision slit that could be opened and closed, almost instantaneously, thereby allowing an experimenter to release a single photon at a precise time. If the experimenter were to weigh the box before and after the photon's release, he or she would obtain measurements of arbitrary precision of a system's energy and time, thereby violating Heisenberg's principle of indeterminacy.

Bohr was shocked by both the simplicity and the deep implications of Einstein's light box *Gedankenexperiment*. Throughout the evening Bohr and his colleagues struggled to find a way out of this conundrum. The great Dane, in existential turmoil, warned everyone that if 'Einstein were right,' then it 'would be the end of physics'. The following morning at breakfast, Bohr announced his solution to Einstein's apparent paradox and the 'salvation of physics'. Einstein had forgotten to account for the effects of the box's displacement within a gravitational field on the frequency of the clock (as quoted in Rosenfeld, 1968, p. 232). If one were able to precisely measure the energy of the emitted photon by weighing the box, the physical displacement of the box would produce a gravitational red-shift that would result in an uncertainty in the time measured by the clock. Bohr's application of Einstein's general theory of relativity and the concept of gravitational red-shift produced just enough uncertainty to rescue the prevailing interpretation of quantum theory.

In this cat-and-mouse game, Einstein had once again challenged the completeness of the prevailing formulation of quantum theory on the biggest scientific stage, using a *Gedankenexperiment* that attempted to show the logical shortcomings of the current conceptualization—only to have his challenge summarily dismissed by Bohr. It's as if Einstein was pleading with his colleagues to echo Lorentz's radical conservatism and remain interpretationally nimble—'*ignoramus* not *ignorabimus*' (Galison, 2007, p. 13). Yet, for the most part, these debates failed to represent the broader contemporaneous discourse. By the close of the 1930 Solvay Council, most physicists working on quantum mechanics had abandoned research programmes on alternate formulations and interpretations, preferring to work on pragmatic applications of quantum mechanics.

This pragmatic shift coincided with a lull in interpretational debates and the codification of quantum mechanics in early textbooks. Throughout the 1930s and 1940s, in the quantum physics community we see an open, vibrant debate about applications and extensions of quantum formulations. Yet around core issues of interpretation, there seemed little interest in publicly challenging a Copenhagen spirit that was becoming entrenched quantum orthodoxy (Freire Jr., 2015, Chapter 2.1). Einstein stands as one of the last stalwarts who refused to accept the elevation of epistemological questions of indeterminism to a priori statements of ontological principle with regard to the microworld.

From 1911 to 1930, the first six Solvay Councils played a dynamic role in the development of quantum theory. They were important stages upon which formative scientific debates took place and helped to form the modern conception of quantum theory. Yet, it's critical that we think about the context in which these Councils took place. For the most part, physicists were not convinced by rational, evidence-based arguments in these meetings. The Councils were performative stages upon which elite physicists leveraged their networks, and all forms of rhetorical strategies, to make their ideas stick. In the political and scientific turbulence of the 1920s, they were used to segregate scientific communities, and then subsequently to heal those same scientific and political divisions. When studied in context, the early Solvay Councils serve as historical markers—rhetorical lenses of convergence and divergence. In doing so, they remind us of the contingent nature of history as well as the human and social dimensions that affect the heart of science.

# REFERENCES

Bacciagaluppi, G., and Valentini, A. (2009). *Quantum Theory at the Crossroads: Reconsidering the 1927 Solvay Conference.* Cambridge, UK: Cambridge University Press.

Barkan, D. K. (1993). The Witches' Sabbath: The First International Solvay Congress in Physics. *Science in Context*, **6**(1), 59–82.

Barkan, D. K. (1999). *Walther Nernst and the Transition to Modern Physical Science.* Cambridge, UK: Cambridge University Press.

Beller, M. (2001). *Quantum Dialogue: The Making of a Revolution.* Chicago: University of Chicago Press.

Belousek, D. (1996). Einstein's 1927 Unpublished Hidden Variables Theory: Its Background, Context and Significance. *Stud. Hist. Phil. Mod. Phys.*, **21**(4), 437–461.

Berends, F. A. (2015). Lorentz, the Solvay Councils and the Physics Institute. *The European Physical Journal. Special Topics*, **224**(10), 2091–2111.

Bohr, N. (1949). Discussion with Einstein on Epistemological Problems in Atomic Physics. In P. A. Schilpp (ed.), *Albert Einstein: Philosopher-Scientist*, Evanston, IL: The Library of Living Philosophers, pp. 199–241.

Born, M., and Heisenberg, W. (1928). Quantum mechanics. Published in the Fifth Solvay Council proceedings and reprinted in G. Bacciagaluppi and A. Valentini (2009), *Quantum Theory at the Crossroads: Reconsidering the 1927 Solvay Conference*, Cambridge, UK: Cambridge University Press.

Bragg, W. L. (1928). The intensity of X-ray reflection. Published in the Fifth Solvay Council proceedings and reprinted in G. Bacciagaluppi and A. Valentini (2009), *Quantum Theory at the Crossroads: Reconsidering the 1927 Solvay Conference*, Cambridge, UK: Cambridge University Press.

Brillouin, L. (1926). The New Atomic Mechanics. *Journal de Physique et le Radium*, **7**, 135. Reprinted in: *Selected Papers on Wave Mechanics*, translated by Winifred Deans, London: Blackie & Son Limited, 1928.

Camilleri, K. (2009). Constructing the Myth of the Copenhagen Interpretation. *Perspectives on Science*, **17**(1), 26–57.

Canales, J. (2015). *The physicist and the philosopher: Einstein, Bergson, and the debate that changed our understanding of time.* Princeton, NJ: Princeton University Press.

Cassidy, D. C. (2010). *Beyond Uncertainty: Heisenberg, Quantum Physics, and The Bomb.* New York: Bellevue Literary Press.

Compton, A. H. (1928). Disagreements between experiment and the electromagnetic theory of radiation. Published in the Fifth Solvay Council proceedings and reprinted in G. Bacciagaluppi and A. Valentini (2009), *Quantum Theory at the Crossroads: Reconsidering the 1927 Solvay Conference,* Cambridge, UK: Cambridge University Press.

Coupain, N. (2015). Ernest Solvay's Scientific Networks. From Personal Research to Academic Patronage. *Eur. Phys. J. Special Topics,* **224,** 2075–2089.

Crawford, E. (1999). The Solvay Councils and the Nobel Institution. In P. Marage and G. Wallenborn (eds), *The Solvay Councils and the Birth of Modern Physics,* Basel: Birkhäuser, pp. 48–54.

De Broglie, L. (1930). *An Introduction to the Study of Wave Mechanics.* Translated by H.T. Flint. London, UK: Methuen & Co. Ltd.

Devriese, D. and Wallenborn, G. (1999). Ernest Solvay: The System, The Law and the Council. In P. Marage and G. Wallenborn (eds), *The Solvay Councils and the Birth of Modern Physics,* Basel: Birkhäuser, pp. 1–23.

Dirac, P. A. M. (1929). Quantum mechanics of many-electron systems. *Proceedings of the Royal Society London A,* **123,** 714–733.

Eckert, M. (2001). The Emergence of Quantum Schools: Munich, Göttingen and Copenhagen as new Centers of Atomic Theory. *Ann. Phys.,* **10,** 152–153.

Einstein, A. (1928). In K. Przibram (ed.) (1967), *Letters on wave mechanics: Schrödinger, Planck, Einstein, Lorentz,* New York: Philosophical Library.

Everts, S. (2010). When Science Went International. *American Chemical Society: Chemical & Engineering News* (3 September 2010) Online: http://pubsapp.acs.org/cen/science/88/8836sci1.html?

Forman, P. (1973). Scientific Internationalism and the Weimar Physicists: The Ideology and its Manipulation in Germany after World War I. *Isis,* **64,** 150–180.

Freire Jr., O. (2015). *The Quantum Dissidents: Rebuilding the Foundations of Quantum Mechanics (1950–1990).* Berlin: Springer.

Galison, P. (2007). Solvay Redivivus. In D. Gross, M. Henneaux, and A. Sevrin (eds), *The Quantum Structure of Space and Time: Proceedings of the 23rd Solvay Conference on Physics—Brussels, Belgium, 1–3 December 2005,* Hackensack, NJ: World Scientific Publishing Co.

Gordin, M. D. (2019). *A Well-Ordered Thing: Dmitrii Mendeleev and the Shadow of the Periodic Table.* Princeton, NJ: Princeton University Press.

Gribbin, J. (1984). *In Search of Schrodinger's Cat.* London: Bantam Press.

Heilbron, J. (1985). The earliest missionaries of the Copenhagen spirit. *Revue d'histoire des sciences,* **38**(3), 195–230.

Heilbron, J. (2013). The First Solvay Council: 'A Sort of Private Conference'. In D. Gross, M. Henneaux, and A. Sevrin (eds), *The Theory of the Quantum World: Proceedings of the 25th Solvay Conference on Physics—Brussels, Belgium, 19–22 October 2011,* Hackensack, NJ: World Scientific Publishing Co.

Howard, D. (2004). Who invented the 'Copenhagen Interpretation'? A Study in Mythology. *Philosophy of Science,* **71**(5), 669–682.

Howard, D. (2007). Revisiting the Einstein–Bohr Dialogue. *Iyyun: The Jerusalem Philosophical Quarterly/* עיון: רבעון פילוסופי, **56**, 57–90.

Ito, K. (2005). The *Geist* in the Institute: The Production of Quantum Physics in 1930s Japan. In D. Kaiser (ed.), *Pedagogy and the Practice of Science*, Cambridge, MA: The MIT Press, pp. 151–183.

Jammer, M. (1966). *The Conceptual Development of Quantum Mechanics (History of Modern Physics, 1800–1950)*. New York: McGraw-Hill.

Jammer, M. (1974). *The Philosophy of Quantum Mechanics: The Interpretations of Quantum Mechanics in Historical Perspective*. New York: John Wiley & Sons.

Jones, S. (2008). *The Quantum Ten: A Story of Passion, Tragedy, Ambition, and Science*. Oxford: Oxford University Press.

Kevles, D. (1995). *The Physicists: The History of a Scientific Community in Modern America*. Cambridge, MA: Harvard University Press.

Klein O. (1926). The Atomicity of Electricity as a Quantum Theory Law. *Nature*, **118**(2971), 516.

Kragh, H. (2000). Max Planck: the reluctant revolutionary. *Physics World*, **13**(12), 31.

Kragh, H. (2002). *Quantum Generations: A History of Physics in the Twentieth Century*. Princeton, NJ: Princeton University Press.

Kumar, M. (2011). *Quantum: Einstein, Bohr, and the Great Debate About the Nature of Reality*. New York: W.W. Norton & Co.

Lambert, F. J. (2015). Einstein's Witches' Sabbath in Brussels: The Legend and the Facts. *Eur. Phys. J. Special Topics*, **224**, 2023–2040.

Lodge, O. (1927). Truth or Convenience. *Nature*, **119**(2994), 423–424.

Marage, P. and Wallenborn, G. (1999). 1913–1921: From the Second to the Third Council. In P. Marage and G. Wallenborn (eds), *The Solvay Councils and the Birth of Modern Physics*, Basel: Birkhäuser, pp. 112–133.

Mehra, J. (1975). *The Solvay Conferences on Physics: Aspects of the Development of Physics Since 1911*. Boston, MA: D. Reidel Publishing Company.

Moore, W. J. (1992). *Schrödinger: Life and Thought*. Cambridge, UK: Cambridge University Press.

The Nobel Prize (1967). *Nobel Lectures. Physics 1901–1921*. Amsterdam: Elsevier Publishing Co. Accessed via Lorentz, H.A.—Biographical. NobelPrize.org. Nobel Media AB 2019. Thu. 19 Sep 2019. <https://www.nobelprize.org/prizes/physics/1902/lorentz/biographical/>

Perillán, J. G. (2018). Quantum Narratives and the Power of Rhetorical Omission: An Early History of the Pilot Wave Interpretation of Quantum Theory. *Historical Studies in the Natural Sciences*, **48**(1), 24–55.

Perillán, J.G. (2021). *Science Between Myth and History: The Quest for Common Ground and Its Importance for Scientific Practice*. Oxford, UK: Oxford University Press.

Rosenfeld, L. (1968). Some Concluding Remarks and Reminiscences. In *Fundamental Problems in Elementary Particle Physics: Proceedings of the 14th Council on Physics, held at the University of Brussels, October 1967*. New York, NY: John Wiley Interscience.

Schweber, S. (1994). *QED and the Men Who Made It*. Princeton, NJ: Princeton University Press.

Stanley, M. (2007). *Practical Mystic: Religion, Science, and A.S. Eddington*. Chicago: University of Chicago Press.

Straumann, N. (2011). On the first Solvay Congress in 1911. *The European Physical Journal H*, **36**(3), 379–399.

# CHAPTER 25

# THE FOUNDATIONS OF QUANTUM MECHANICS IN POST-WAR ITALY'S CULTURAL CONTEXT

## FLAVIO DEL SANTO

## 25.1 THE CULTURAL ROOTS OF THE RENEWED INTEREST IN THE FOUNDATIONS OF QUANTUM MECHANICS

FROM a historiographical point of view, the change of fortune that the foundations of quantum mechanics (FQM) underwent over time is a fascinating case study of the interplay between science and the social-political environment within which it develops. Indeed, while FQM was highly debated in the early days of this theory and today are part of the mainstream research in physics, they have experienced a long period of near-oblivion. One of the reasons for the disappearance of FQM from physicists' agendas was the change of structures and priorities of scientific research during and after World War II, when physics became a tremendous driving force for new and wondrous applications, especially in the military sector. This period went down in history alongside the notorious expression 'shut up and calculate!' (see Kaiser, 2011).

In recent years, several works emphasized the role played by different cultural contexts in creating the conditions that led the field of FQM to experience a renaissance (Jammer, 1974; Trischler and Kojevnikov, 2011; Kaiser, 2011; Freire, 2014; Baracca *et al.*, 2017; Baracca and Del Santo, 2017; Freire, 2019). This revival was a complex and discontinuous process prompted by a few pockets of resistance that arose in several different countries, often under the impulse of the social-political context. To illustrate the struggles of the physicists who pioneered the renaissance of FQM, Olival Freire Jr. has called them 'the quantum dissidents' (Freire, 2014). The latter encompass physicists like John Bell—who cultivated his interest in FQM as a mere 'hobby', yet he put forward one of the most striking results of FQM, Bell's theorem (Bell, 1964)—or David Bohm, who was led by his commitment to Marxism to formulate a non-standard realistic interpretation of quantum theory (Bohm, 1952).

Certain research environments reached a critical mass of physicists working on FQM already in the 1960s and 1970s, sometimes featuring very unconventional aims. It is the case, recently reconstructed by David Kaiser (2011), of the *Fundamental Fysiks Group* in Berkeley (California) which—in the late 1970s, in the context of a non-academic *New Age* mood—'planted the seeds that would eventually flower into today's field of quantum information science' (Kaiser, 2011).

Another paradigmatic example occurred almost a decade earlier in Italy, where young, radically politicized physicists also revived interest in FQM. Being active at several Italian Universities (Bari, Bologna, Catania, Firenze, Rome, and Trieste), they channelled the general political atmosphere of the 1968 left-wing struggles for dismantling the 'capitalistic society', yet also within their field of expertise, namely physics. These radical physicists not only criticized the involvement of their colleagues in the military sector, but challenged the very contents of modern science, regarded as a product of the 'capitalistic society'. In such a way, the prevalent field of high-energy physics—its methods and practices (*Big Science*)—became the target of harsh criticisms, too.

Quantum theory, instead, became the whipping post of these unsatisfied young physicists who saw in the newly introduced Bell inequalities the tool to show its limits of validity, thus legitimizing their overall critique of scientific practice. These very ideological aims provided the motivation for the beginning of a systematic study of the state-of-the-art results in FQM that could have provided the opportunity for the introduction of a totally new physical framework. Although most of these attempts eventually turned out to be unsuccessful—the experiments on Bell's theorem confirmed the predictions of quantum mechanics—they opened a lasting period of intense research on FQM that involved dozens of Italian physicists and helped pave the way towards a recognition of FQM as a full-fledged field of physics. In fact, some of the Italian results acquired an international reputation, such as the celebrated 'objective collapse model' developed by Giancarlo Ghirardi and his collaborators (Ghirardi *et al.*, 1986) or the probabilistic solution to the measurement problem proposed by Marcello Cini (1983).

This chapter aims at reconstructing the cultural, political, and ideological context that led to a revival of the foundations of quantum mechanics in Italy and at providing an overview of the main lines of research carried out from the post-war period until the end of the 1980s.

## 25.2 THE 1960S: THE SEEDS OF THE ITALIAN INTEREST TOWARDS FQM

Like a great deal of intellectual activity in Europe, Italian physics was jeopardized by the scattering out of intellectuals from Europe after the advent of various forms of fascism and the consequent imposition of racial laws (in Italy and in Nazi Germany). The outbreak of World War II, with the great involvement of physics in war technologies, completed the job of nearly annihilating any fundamental debate on FQM, not only in Europe but generally.

In the immediate post-war period, Edoardo Amaldi (1908–1989) contributed a great deal to rebuilding modern physics in Italy, with a strong focus on nuclear and high energy physics. Yet, it was especially Piero Caldirola (1914–1984)—professor of theoretical physics in Milan—that triggered new interest on FQM. Caldirola authored the entry 'Quantistica, Meccanica' of the *Enciclopedia Italiana* (Caldirola, 1961), that resulted in a quite comprehensive review of the main open conceptual issues in quantum physics.[1] Therein he acknowledged a proliferation of interpretational concerns, affirming: 'the last decade [...] is characterized by an intense renewal of the attempts aimed at providing quantum mechanics with an interpretation different from that of the school of Copenhagen' (Caldirola, 1961). Such an entry circulated as a booklet among the students of Caldirola's courses at the University of Milan and was to have a remarkable impact on some of those who were to play important roles in the following decade (such as Angelo Baracca and, likely, Ghirardi).[2]

Caldirola formed a school of theoretical physics, known as the 'School of Milan', whose research was mainly focused on ergodic methods in statistical mechanics but, under his influence, some of his pupils brought together these statistical concepts with interest towards FQM. Indeed, in 1962, Adriana Daneri, Angelo Loinger, and Giovanni Maria Prosperi developed a mathematical model that formalized—making use of the ergodic theorem from statistical physics—Bohr's explanation of the 'collapse' of the quantum wave function due to an objective interaction between the macroscopic measurement apparatus and a microscopic (quantum) system (Daneri *et al.*, 1962). This paper was timely and triggered a tremendous international debate, ultimately providing a concrete basis for the ongoing quarrel between Eugene Wigner and Léon Rosenfeld about the objectivity of the wave-function collapse.[3] Rosenfeld, who arguably

---

[1] In fact, the first work by Caldirola on FQM had appeared as early as 1957 in the philosophy journal 'Il Pensiero' (Caldirola and Loinger, 1957), but this likely had no resonance among the physicists.

[2] The fact that this book circulated among Caldirola's students in Milan is recalled by A. Baracca (see also Baracca *et al.*, 2017).

[3] It would not be possible to enter here the details of this debate, which has, however, been thoroughly discussed in (Freire, 2014, chapters 4.4, 5.2). Nevertheless, it is worth mentioning that this involved some of the most pre-eminent physicists concerned with FQM in those years; among them—besides the aforementioned Rosenfeld and Wigner—Jauch, Shimony, Yanase, and Bohm.

was the staunchest amongst the supporters of Bohr's view, stated that 'the Italian physicists have conclusively established the full consistency of the algorithm (of quantum mechanics), leaving no loop-hole for extravagant speculations' (Rosenfeld, 1965).[4]

As a further witness of the early Italian interest in quantum foundations, it is worth mentioning that, in 1967, Bruno Ferretti (1913–2010), a former collaborator of Enrico Fermi, then professor of theoretical physics in Bologna, organized a cycle of seminars on FQM.[5] This initiative involved the whole theoretical group in Bologna, where each physicist was requested to present a seminal paper on recent developments in FQM. Among the young physicists involved were Baracca and Franco Selleri, who were to play a pivotal role in the promotion of FQM and of other critical scientific activities.

Indeed, the following years saw a definitive break with the 'old guard', and an unprecedented blossoming of the research on FQM in Italy. In the next section, we will show how this revival was, in Italy, deeply intertwined with the political radicalization of the young Italian physicists and their cry for an alternative science.

## 25.3 THE 1968 EFFECT: A REBIRTH OF FQM IN THE CONTEXT OF LEFT-WING POLITICAL ACTIVISM

### 25.3.1 The General Social-Political Context

Before getting to the heart of the renewed interest that FQM underwent at the turn of the 1960s, we deem it necessary to quickly recall the general atmosphere of social and political struggles that characterized those years.

The student protests of Berkeley (1964), the 'Prague Spring' (1968), the 'French May' (1968), as well as the radicalization of the labour struggles of the Italian 'Hot Autumn' (1969) had a tremendous impact on society as a whole, and in particular on the themes that were set as a priority by the intellectuals. The active involvement of scientists in the dreadful Vietnam war, coming only a few decades after the awareness of physicists' responsibility in the creation of the atomic bomb, sensitized many young scholars (see Moore, 2013). This caused, throughout the 1960s, several outstanding incidents, such as the student strike at MIT (Boston) against military research, the intrusion of

---

[4]  In the following years, Caldirola, too, helped a great deal to popularize in Italy the solution to the measurement problem proposed by his pupils. See, e.g., the paper (Caldirola, 1965), and the textbook (Caldirola, 1974), which devoted a whole chapter to the solutions of the measurement problem.

[5]  The series of seminars is recalled by A. Baracca, V. Monzini, E. Verondini, and S. Graffi who all participated in the initiative; communications to the author on 16 June 2014.

protesters into the meetings of the American Physical Society, and the boycotting of some talks given by eminent physicists, all members of the JASON advisory group.[6] This was the case for the Nobel laureate Murray Gell-Mann, who was prevented from giving a talk at the Collège de France (Paris) in 1972 and again in CERN (Geneva) by young activist physicists. The same happened to Sydney Drell in Rome and to John Wheeler in Erice (Italy). Also the Nobel Prize winner Wigner was openly criticized at the Varenna School in 1970 (see below), and in Trieste in 1972, where he reacted by displaying a banner with the words: 'I am flattered by your accusations. They are compliments for me.'[7]

Moreover, on the occasion of the first Moon landing in 1969, Marcello Cini (1923–2012) published a controversial essay, Il Satellite della Luna (Cini, 1969), wherein he denounced the deep economic and military interests, disguised as scientific curiosity, behind the space race. This had a profound impact on the young critical physicists who became sensitive to the problems of scientists' responsibility (see Baracca et al., 2017).

Like in several other countries where left-wing political movements were gaining momentum, in Italy students and young researchers did not limit themselves to renouncing physicists' involvement in military research, but initiated a critical analysis of the role of scientists in the (capitalistic) society, proposing concrete initiatives to change it. They sought in the methods and the contents of modern (capital-intensive) science some of the roots of the problematic issues of society. This atmosphere of unrest also seeped into the Italian Society of Physics (SIF) in October 1968, when exponents of the student movements interrupted the meeting of the Steering Committee (Baracca et al., 2017). Moreover, thanks to the initiative of Franco Selleri—who, as we will see in the following sections, was the real initiator of the renewal of the research on FQM in this ideological context—the Steering Committee of SIF voted in favour of renouncing a substantial funding grant from NATO for the year 1970.[8]

Following these premises, a new generation of radical, Marxist, Italian physicists started searching for alternative ideas and practices. It is precisely in this politicized context that the new wave of Italian research on FQM came about. It ought to be stressed that one aspect that renders the Italian case remarkably different from most of its international counterparts seems to be that in Italy the political uneasiness that

[6] JASON is a group of distinguished physicists (today encompassing also other scientists), recruited (originally by the American Institute for Defense Analysis, IDA) to advise the US Government about science and technology, especially on classified military applications. It was established on the initiative of the distinguished physicists John Wheeler and Eugene Wigner and the mathematician Oskar Morgenstern in 1960 (see e.g., Finkbeiner, 2006). JASON members (who, '[d]epending on how they're counted, they number between thirty and sixty at any given time', Finkbeiner, 2006) played a significant role in the Vietnam war, when they promoted the 'McNamara Line', an electronic barrier installed in South Vietnam to prevent infiltrations, and thus became the target of the new anti-war movements of activist scientists (see Vitale, 1976).

[7] A full reconstruction of these episodes can be found in (Freire, 2014, chapter 6) and (Vitale, 1976).

[8] Eventually, despite the vote, the funding was accepted due to technical problems with the withdrawal of the application (Baracca et al., 2017).

spread among physicists enjoyed to a large extent the support of the institutions, in particular of SIF. Thus the usual clash between the academic establishment and the bottom-up movements of renewal was in Italy quite blurred. As we shall see, some of the protagonists of the Italian revival of FQM in those years played, on the one hand, the role of young, radical critics while, at the same time, being appointed to influential positions, such as members of the Steering Committee of SIF or the editorial board of its journal, *Il Nuovo Cimento*. It was not by chance that that journal became a pivotal international forum for the studies on FQM throughout the 1970s and 1980s.

## 25.3.2  Franco Selleri's Early Dissatisfaction with Science and His Pioneering Initiatives

The undisputed protagonist of the revival of FQM in Italy—besides other political initiatives intertwined with physics—was Franco Selleri (1936–2013). After his graduation at the University of Bologna in 1958, Selleri made remarkable contributions to particle physics, which at that time was by far the most widespread field of physics, e.g., the introduction of the 'peripheral one-pion model' (Bonsignori and Selleri, 1960).

However, as early as 1965, Selleri found it was 'evident that there were problems, fundamental problems in Physics'.[9] But it was during a visiting professorship at the University of Gothenburg (Sweden), in 1966, that the major shift in Selleri's ideas took place, when he read the book *Conceptions de la Physique Contemporaine* by Bernard d'Espagnat (1965), about which he would later recall:[10]

> It was a revelation. It was something fantastic to see how many problems were open in quantum mechanics. [ ... ] Reading that book was a great discovery. There was a new field possible to create and a lot of research to do. [ ... ] It was fascinating to see that so many possibilities were open. So it was clear that the Copenhagen approach was not unique, was not obligatory.[11]

Inspired by d'Espagnat's book, in 1969 Selleri published his first work on FQM: a short note entitled 'On the wave function in quantum mechanics' (Selleri, 1969a), whose preprint he had sent to Louis de Broglie, who, 'manifest[ed] his great interest towards the research of the Italian physicist and the intent to intensify the contacts' (Nutricati, 1998). That work is more the manifesto of a programme than an actual research paper, but it is of great historical interest because it marks the moment in which realistic

---

[9] Franco Selleri interviewed by Olival Freire Jr. on 24 June 2003. Niels Bohr Library and Archives, American Institute of Physics (AIP), College Park, MD, USA, www.aip.org/history-programs/niels-bohr-library/oral-histories/28003–1.

[10] Selleri's shift towards FQM is discussed in greater detail in (Baracca *et al.*, 2017) and in a doctoral thesis entirely devoted to the figure of Franco Selleri (Romano, 2020).

[11] Franco Selleri interviewed by Olival Freire Jr. on 24 June 2003.

interpretations (in terms of hidden variables) entered the Italian physics landscape. Therein, Selleri maintained:

> Even though the theories of hidden variables are not completely developed, an important shift of philosophical attitude can be noticed: particles and waves are now objectively existing entities.

These positions were further developed by Selleri in the course of a series of lectures on *Quantum Theory and Hidden Variables* (Selleri, 1969b) that he gave in Frascati (Italy) in June–July 1969. Within these lectures, Selleri formulated a 'realistic postulate', which advocates a double ontological solution for particles and waves (i.e., they are both taken to objectively exist) and that would characterize his philosophical position ever after:

> An elementary particle is always associated to a wave objectively existing. [ ... ] 'Objectively' is taken to mean: independently on [*sic*] all observers. [ ... ] The wave has to be thought of as a real entity in some kind of postulated medium. Thus wave and particle are reminiscent of a boat in a lake. Boat and wave are both objectively existing and are found to be associated, in the sense that you cannot find a boat without a wave; the opposite is, however, possible.

Selleri was then able to put forward his ideas in a very institutional context, using—at least at first—the reputation of SIF as a stepping stone to launch his unconventional views and reach a broad audience of Italian physicists. In fact, as a member of the Steering Committee of SIF between late 1968 and 1970, Selleri had the opportunity to promote critical initiatives aimed at rethinking the current scientific system. Despite the overall reluctance towards discussing fundamental problems of physics at that time, Selleri proposed, in March 1969, to devote to FQM a one-week course of the celebrated 'Enrico Fermi' Summer School in Varenna, organized annually by SIF since 1953. In the minutes of one of their meetings, one can read that 'after a wide discussion, the Steering Committee expresse[d] favourable opinion to the above-mentioned course and decide[d] to propose Prof. d'Espagnat as a director, highlighting and considering of interest a comparison of the opinions of Daneri, Prosperi, Loinger and Wigner, Bohm, de Broglie. As secretary of the course [was] proposed Selleri' (Baracca *et al.*, 2017).

The acceptance, at that time, of such an official course on FQM by a national physical society appears to be remarkable and characteristic of the Italian case. In fact, as already mentioned, at that time there were just a few main spots of activities on FQM: in France the pupils of Louis de Broglie, in particular Jean-Pierre Vigier (see below), were active in reviving the idea of the 'double-solution' (i.e. the corpuscular and undulatory natures of quantum particles). This had gathered new momentum after the proposed hidden variable model of David Bohm (1952), who, based in UK, was also one of the protagonists of the research on FQM. Vigier and Bohm were dissidents both in physics and in politics (being committed Marxists; see Freire, 2011, 2014, 2019).

However, despite their renown and the personal support of eminent physicists (de Broglie and Einstein, respectively) it does not seem that Vigier in France and Bohm in UK managed to breach the institutions as Italian physicists did. Although it is not clear what are the factors that prepared the ground in Italy for this open-mindedness at the institutional level, this may again be related to the fact that quite radical political views were present and accepted in Italy, which, as a matter of fact, had the biggest national communist party in Western Europe. This political acceptance at the institutional level could thus have been reflected in the structure of scientific societies, in particular of SIF, at that time.

In fact, Selleri's success in this endeavour was surely assisted by the presence on the board of SIF of yet another radical left-wing critical physicist, the above-mentioned Cini, who was to have a major impact on the following critique of science and later on FQM (see section 25.4.4 below). Moreover, the President of SIF at that time, Giuliano Toraldo di Francia (1916–2011), was an open-minded physicist, sympathetic towards left-wing ideas (though more institutional and less radical than Selleri and Cini), and very sensitive to ethical and philosophical problems of science.[12]

## 25.3.3  The Varenna School on Foundations of Quantum Mechanics of 1970

The Varenna summer course on FQM opened on 29 June 1970,[13] and it constituted a meeting with a unique concomitance of features: the high profile of its teachers, the interest and the political unrest of its students, the open debates about interpretational topics—which had remained almost completely quiescent in public physics events since before World War II—and the first discussions about Bell's theorem. All together this makes clear why this School has been called 'the Woodstock of the quantum dissidents' (Freire, 2014).

The unconventional character of the meeting was already evident from the letter that the director of the School, d'Espagnat, sent to each participant beforehand (d'Espagnat, 1971):

> Let me suggest to you the following agreement: that we should not take as goals the conversion of the heretic but rather a better understanding of his standpoint; that we should not suggest that we consider as a stupid fool anybody in the audience (lest the stupid fools should in the end appear clearly to be ourselves!); that we should try to cling to facts; and that nevertheless we should be prepared to hear without indignation very nonconformist views which have no immediate bearing on facts.

[12]  See Toraldo di Francia's obituary by SIF: https://www.sif.it/riviste/sif/sag/ricordo/toraldo.

[13]  The origin, development, and aftermath of the Varenna School of 1970 are discussed in (Freire, 2014, chapter 6.3) and in (Baracca et al., 2017, sections 5–7).

The School brought together 84 participants, of which the teachers were J. Andrade e Silva, J. Bell, B. d'Espagnat, B. S. De Witt, J. Ehlers, A. Frenkel, K.-E. Hellwig, F. Herbut, M. Jauch, J. Kalckar, L. Kasday, G. Ludwig, H. Neumann, G. M. Prosperi, C. Piron, F. Selleri, A. Shimony, H. Stein, M. Vigičić, E. Wigner, M. Yanase, and H. D. Zeh. (The proceedings also contained chapters from Louis de Broglie and David Bohm, who, however, did not participate.) The topics tackled in the School were organized into three main themes: (i) measurement and basic concepts, (ii) hidden variables and non-locality, and (iii) interpretation and proposals. The only Italians among the teachers were Prosperi—who again proposed his solution to the measurement problem (see section 25.2)—and Selleri, who presented a lecture to reaffirm his realistic position and made use of Bell's inequalities (at that time virtually unknown), applying them to particle physics. In this work, Selleri also made evident his ideological motivations, by stating that:

> practically nobody in the society of which the physicist is a part has any doubt about the actual existence of an objective reality outside of the observer. [ ... ] In this time where the social responsibility of the scientist is so strong, where the destruction or the survival of the world depends also on him, it is important to develop a science not in basic contradiction with the social reality.    (In d'Espagnat, 1971)

Overall, Varenna was a unique opportunity for those 'dissident' physicists who, scattered around the world and working mostly independently, were timidly facing the main research directions of physics (at that time largely oriented towards practical applications) by exploring its neglected fundamental and philosophical aspects. At the genuine scientific level, 'the school helped network these scientists, bringing together most of the physicists who would go on to contribute to the blossoming of this research in the 1970s' (Freire, 2014). It is worth mentioning that Bryce de Witt announced in Varenna his conversion to Everett's *many-worlds interpretation* (as de Witt had named it and as it has been popularly referred to ever since).

But there is more to it than that: the School was also a melting pot of ideas for the young physicists who, already sensitized by the general political climate, were eager to revolutionize their field of expertise both in its structures and scientific agendas. Among them were, besides Selleri: Angelo Baracca, Vincenzo Capasso, Gianni De Franceschi, Carlo De Marzo, Donato Fortunato, Gianni Mattioli, Alessandro Pascolini, Marta Restignoli, Luigi Solombrino, Tito Tonietti (a mathematician), and Livio Triolo. Many of them were to contribute to the research into quantum foundations in the following years. These attendants of the School, together with some of their international colleagues, gathered every evening after the lectures and engaged in critical discussions. Starting from the content of the courses, such as the paradoxes of quantum theory, they managed to relate these highly specialized and abstract concepts, such as the interpretation of the quantum formalism, with contemporary societal and political issues. A theme that became dominant was, in fact, the social responsibility of scientists in society. As a result, these politicized students produced a collective 12-page document, *Notes on the connection between science and society* (see Baracca *et al.*, 2017), which was distributed to all the teachers and participants. The document aimed to raise awareness among the physicists about the oft-omitted relation between science and economical-political interests, and voiced the concept, which was to become influential in those years, of the *non-neutrality of science*:

> It is instead extremely important to realize that science is certainly not neutral [ ... ]. The structures of a scientific theory reproduce the categories of the culture of the dominant classes. [ ... ] The limitation of the individual consciousness to laboratory activity, disregarding any judgement of the social application of research, results in indiscriminate support for all applications of science (e.g., atomic bomb, chemical and bacteriological warfare). [ ... ] The scientist's incapacity to control the product of his research facilitates its cultural manipulations and the creation of consensus. [ ... ] This mystification is formalized in the powerful theory ('scientism') which assumes the intrinsic ability of science to solve all the human problems. [ ... ] Now we conclude that a pre-decision is strongly needed concerning the social structure in which men live and act. That is, a pre-decision on the historical and social role of the scientist, on his responsibility, on the fact that no concept or activity is neutral.

In conclusion, the Varenna School of 1970 represented a gathering of physicists who would soon recognize themselves as a new community of 'dissidents'. Simultaneously, the *zeitgeist*, with an impetus of political renewal, entered the scientific debate and stimulated several young physicists to criticize the traditional values of scientific practice, as well as the content of research. Quantum foundations thus became the starting point for concrete change.

## 25.3.4 Further Critical Activities about Science in the 'Capitalistic Society'

In the overall struggle for changing scientific practice, the turn towards research in FQM was intertwined with a number of other critical activities.[14] Besides the mentioned critique of the practices in high-energy physics, again triggered by Selleri, a major endeavour was carried out by the same young radical physicists who engaged in activities on the history of physics.

This came about once more within SIF: On the proposal of Selleri, SIF organized, in June 1972, a *Study Day on Science in the Capitalistic Society* in Florence. Whereas the Varenna School of 1972 devoted a course to *The History of Twentieth Century Physics* (Weiner, 1977), where prominent historians and physicists (such as Edoardo Amaldi, Paul Dirac, and Hendrik Casimir) held lectures before an audience which included

---

[14] It lies beyond the scope of this chapter to thoroughly discuss the critique of science that involved the young Italian physicists in the 1970s. The interested reader is referred to (Freire, 2014, Chapter 6.3), (Baracca *et al.*, 2017, sections 9–12, 14) and (Baracca and Del Santo, 2017).

several of the participants from Varenna 1970. Prompted by this, many of them engaged in professional research on the history of physics in the following years (such as A. Baracca, S. Bergia, G. Ciccotti, M. Cini, C. De Marzo, and E. Donini). History was regarded as a means to understanding the historical roots of modern physics and its positioning in the 'capitalistic society', taking evident inspiration from Marx's analysis of the political economy (*historical materialism*).[15]

This new approach soon entered a controversy with the established school of history and philosophy of science—also from a Marxist perspective—born around the figure of Ludovico Geymonat (1908–1991). This school, which had mostly a humanistic tradition and encompassed scholars such as Giulio Giorello, Enrico Bellone, and Silvano Tagliagambe, voiced an analysis of the history of science based on Marx's *dialectical materialism* (see e.g. Bellone *et al.*, 1974). This approach was criticized by the physicists (turned historians), insofar as it seemed to imply absolute objectivity of science and therefore its legitimation as a neutral activity while neglecting the relationship between scientific development and societal needs.

## 25.4 THE 1970S AND 1980S: THE BLOSSOMING OF ACTIVITIES ON FQM IN ITALY

The many seeds of uneasiness towards established scientific practice in the aftermath of the Varenna School on FQM led several Italian physicists to forever abandon the mainstream topics of research and pursue professional research in quantum foundations (as well as on the history of physics in many cases). Some of them created schools in their universities and established international collaborations.

The Italian journals of the group *Il Nuovo Cimento*, edited by SIF, played a pivotal role in disseminating work on FQM, also at the international level. Between 1969 and 1985, ca. 360 research papers on FQM authored by Italian groups appeared in these journals (Benzi, 1988).[16] We provide here an overview of the main research programmes on FQM in Italy in the decades of the 1970s and 1980s.

---

[15] These activities led to many degree theses, research papers, dissemination articles, and books. See (Baracca *et al.*, 2017) for a bibliography.

[16] It ought to be remarked that in 1969—when the first Italian papers on FQM started to appear—the editorial board of *Il Nuovo Cimento* was composed only of members of the Steering Committee of SIF, including Selleri and Cini, while the editor in chief was SIF's president Toraldo di Francia. From the following year, while Toraldo di Francia remained the editor, the editorial board was vastly enlarged to include physicists of international renown, some of whom were also concerned with FQM, such as Hendrik Casimir, Raul Gatto, Josef-Maria Jauch, Giuseppe Occhialini, Bruno Pontecorvo, and Emilio Segré.

## 25.4.1  Selleri's group in Bari and its Collaborations

As already mentioned, Selleri had been the main catalyst of the political unrest and its intertwinement with physics that led to the Italian revival of FQM. In 1968, Selleri settled at the University of Bari, in the south of Italy, as its first-ever professor of theoretical physics— a deliberate choice that allowed him full freedom for his research. In Bari, Selleri started teaching Quantum Theory in Winter 1968 and supervised the first thesis on quantum physics, and soon after some of his students fully embraced his ideas, making Bari a centre for research on FQM.[17]

Selleri started prolific collaborations with some of his pupils, among whom were Vincenzo Capasso, Donato Fortunato, Augusto Garuccio, and Nicola Cufaro-Petroni. As acknowledged also by Kaiser (2011), 'Selleri and his group in Bari, Italy, had been among the earliest and most active researchers on Bell's theorem anywhere in the world'. However, it ought to be remarked that their aim was at first to disprove quantum mechanics by proposing experimental tests of Bell's inequalities (Selleri, 1971, 1972). But afterwards—once the first experimental results corroborated quantum mechanics—Selleri and collaborators tried to show that the standard interpretation of Bell's inequalities had always been erroneous, insofar as it required additional assumptions.[18] A conviction that Selleri upheld until the end of his career: 'Bell's inequality has never been checked experimentally. They have checked something else. That is to say another inequality based on local realism plus additional assumptions'.[19]

Between 1979 and 1981, Selleri tried to stress the paradoxical aspects of quantum entanglement—again with the aim of proving a breakdown of quantum mechanics—by proposing protocols for superluminal signalling, which he presented in the course of scientific meetings, without however ever publishing his ideas.[20] These proposals were similar to those put forward by Nick Herbert from the aforementioned Fundamental Fysiks Group in the USA: the so-called 'QUICK' experiment in 1979 and the 'FLASH' experiment in 1981 (Herbert, 1982; see also Kaiser, 2011). Selleri became very

---

[17] The first thesis supervised by Selleri on quantum mechanics was: Giglietto Antonio (1968), 'Meccanica quantistica e processi markoviani', University of Bari, Italy. I am indebted to Luigi Romano for this reference (communication to the author on 28 December 2019).

[18] Some works on this topic are (Capasso, Fortunato, and Selleri, 1973); (Fortunato and Selleri, 1976); (Fortunato, Gariuccio, and Selleri, 1977) and (Selleri and Tarozzi, 1978). We limit ourselves to this partial list, but the works of Selleri's group on Bell's inequalities continued until 1990. Comprehensive bibliographies of Selleri's work on FQM can be found in (Romano, 2020) and (Nutricati, 1998).

[19] Franco Selleri interviewed by Olival Freire Jr. on 24 June 2003. Niels Bohr Library and Archives, American Institute of Physics (AIP), College Park, MD, USA, www.aip.org/history-programs/niels-bohr-library/oral-histories/28003-1.

[20] Ghirardi recalled two preprints by Selleri on this topic: 'Einstein locality and the quantum-mechanical long-distance effects', 1979, based on his presentation at a meeting in Udine, and a second preprint by Cufaro-Petroni, Garruccio, Selleri, and Vigier, 1980 (see Kaiser, 2011). Herbert Pietschmann recalls that Selleri presented before him and Walther Thirring a proposal for superluminal communication during the leave of absence that Selleri spent in Vienna in 1980–1981 (Pietschmann interviewed by the author on 15 November 2016).

sympathetic to these proposals, which were however conceived with the different spirit of actually achieving superluminal communication. After some correspondence, Selleri and Herbert met at a conference in Perugia and the former 'was besotted with Herbert's latest proposal [ ... ]. More important, he pushed copies of Herbert's preprint on several other colleagues, and helped convince an experimental physicist from Pisa to mount a real test of Herbert's design' (Kaiser, 2011).

All these proposed experiments assumed that it was possible to prepare a large number of identical copies of unknown polarized modes of a laser, which turned out to be a fundamental flaw. Yet, to show the infeasibility of these proposals, physicists had to come up with a solution that today goes under the name of 'quantum no-cloning theorem': it is fundamentally impossible to perfectly copy an unknown quantum state (Wootters and Zurek, 1982). Modern quantum cryptography relies directly on this theorem to guarantee communication security. However, as stressed by Kaiser (2011), the formulation of such a fundamental result was prompted by these idiosyncratic proposals.[21]

At the beginning of the 1980s, Selleri's group moved its focus toward experimental proposals for the detection of the hypothesized *empty waves* (Selleri, 1982a, 1982b), (Andrade e Silva *et al.*, 1983). In fact, as has already been mentioned, Selleri upheld a 'dual' ontological solution for particles and waves (ideas very similar to those of de Broglie and the early Bohm). However, while in Selleri's view particles are always associated with a wave, there might also be waves that do not carry particles (empty waves). The proposed set-up for their detection featured single photons sent through a beamsplitter. On the left branch a photodetector was located, and on the right one a laser-amplifying tube. Selleri's idea was that after postselecting the cases when the photon gets detected on the left, the 'ghost' photon on the right would still cause a (detectable) stimulated emission in the laser-amplifying tube. These proposals continued until 1990, but no experiments seem to have been carried out and the very feasibility of these proposals was questionable (Mückenheim *et al.*, 1988).

Moreover, Selleri spent a leave of absence at the University of Vienna in 1980–1981. There, he lectured on FQM: this plausibly had an impact on the Viennese students, and, as a matter of fact, on the young but influential physicist Roman U. Sexl who 'followed [his] lectures in Vienna and invited [him] to write a book' on this topic.[22] The book (Selleri, 1983) was indeed published (in German) in 1983 and became quite well known in Austria and Germany.

---

[21] Priority in the discovery of the no-cloning theorem is slightly contested. It is usually attributed to Wootters and Zurek (1982) and independently to Dieks, but Ghirardi, in his referee report of Herbert's FLASH experiment for the journal *Foundations of Physics*, in 1981, had already provided a proof of the theorem that, however, remained unpublished. Moreover, it was recently shown that this theorem had already been proven in 1970 (Ortigoso, 2018).

[22] Franco Selleri interviewed by Olival Freire Jr. on 24 June 2003. Niels Bohr Library and Archives, American Institute of Physics (AIP), College Park, MD, USA, www.aip.org/history-programs/niels-bohr-library/oral-histories/28003-1.

Another important aspect of Selleri's impact on FQM was his ability in networking with international collaborators. In 1969, Selleri established contact with French Nobel laureate Louis de Broglie, with whom he had regular correspondence and several meetings; the whole Bari group had intense exchanges with the *Fondation de Broglie* in Paris, in particular with Georges Lochak, Pierre Claverie, Simon Diner, and Olivier Costa de Beauregard.

Another lasting collaboration was established in 1978 with Vigier, who had had a prime role in rebuilding the foundations of quantum mechanics since the 1950s.[23] This fully involved Cufaro-Petroni, who spent the years 1978–1979 in Paris with Vigier, and Garuccio who worked in Paris with him in 1980. As a matter of fact, between 1979 and 1985, they published with Vigier 23 and 13 papers, respectively (see Baracca *et al.*, 2017).

It is noteworthy that Vigier could count among his allies in the fight for reclaiming realism in quantum physics the influential philosopher Karl Popper, who had been opposing the Copenhagen interpretation since 1934 (see Del Santo, 2019, 2020). Popper and Vigier together published a paper on quantum foundations that also involved Garuccio (Garuccio *et al.*, 1981). In 1983, Vigier introduced Popper to Selleri (see Del Santo, 2018), who promptly organized the conference *Open Questions in Quantum Physics*, to attract Popper to Bari. At the conference, Popper had the opportunity to present his variant of the EPR *Gedankenexperiment*, usually referred to as just 'Popper's experiment', and this marked his entry into the community of physicists concerned with FQM.[24] Allegedly, Popper's experiment was capable of empirically discriminating between a realistic interpretation of quantum physics and the standard Copenhagen interpretation, by violating Heisenberg's uncertainty principle. Also due to Selleri's effort, Popper's experiment became known internationally and triggered an intense, decade-long debate, also involving several Italian physicists. The experimentalist Francesco de Martini, in Rome, proposed to implement a variant that could be realized in his lab, but that did not persuade Popper. Selleri himself became convinced that the experiment was in principle infeasible due to the impossibility of preparing a suitable source of entangled particles (Bedford and Selleri, 1985). Ghirardi rebutted Popper's proposal at a conference co-organized by Selleri in Cesena in 1985, calling it a 'misunderstanding about the EPR analysis' (Ghirardi, 1988).

Eventually, Popper's experiment was realized in 1999 at the University of Maryland, on Garuccio's suggestion, by Yanua Shih and Yoon-Ho Kim, who surprisingly found that Popper's predictions were right. They published this result in a controversial paper entitled 'Experimental realization of Popper's experiment: Violation of the uncertainty principle?' (Kim and Shih, 1999). It was only recently that a formal analysis of Popper's

---

[23]  Vigier was the assistant of de Broglie who convinced the latter to come back to his ideas of the pilot wave, after introducing to him the work of Bohm. Vigier has been one of the major influences in creating an international network between the quantum dissidents in the post-war era (see Besson, 2018).

[24]  The genesis of Popper's experiment and its reception, as well as the involvement of Selleri's group in its popularization, is discussed in (Del Santo, 2018) and (Freire, 2004).

experiment showed that his proposal was in principle not able to test the Copenhagen interpretation (see Del Santo, 2018).

Finally, in 1979 Selleri's group began a collaboration with the experimental group of Vittorio Rapisarda, from the University of Catania, for the design of an experimental test of Bell's inequality (Garuccio and Rapisarda, 1981; see also Nutricati, 1998). However, during the preparation of the experiment 'Foca-2' (Falciglia *et al.*, 1982), Rapisarda prematurely died in a car accident on a visit to Bari in 1982. Furthermore, the philosopher Gino Tarozzi co-organized with Selleri a series of international conferences on FQM that had a significant resonance: in Bari in 1983 (Tarozzi and van der Merwe, 1985) and, in 1985, in Cesena (Tarozzi and van der Merwe, 1988) and Urbino (van der Merwe *et al.*, 1988).

From the beginning of the 1990s, however, Selleri's interest towards FQM progressively faded, also due to the more and more evident empirical confirmations of quantum mechanical predictions in Bell-type experiments. However, some of his pupils kept working on quantum foundations and, in particular, Garuccio became specialized in quantum optics.

## 25.4.2 Giancarlo Ghirardi in Trieste

Giancarlo Ghirardi (1935–2018), who arrived a little later to the interest towards FQM, was not part of those radical physicists who questioned the validity of quantum mechanics, but was directly influenced by them. In hindsight, his contributions have perhaps been the most outstanding among all the Italian studies on quantum foundations.[25]

As a student in Milan in the late 1950s, Ghirardi became aware of the fundamental issues of quantum physics quite early on, in a conference talk given by Prosperi. However, like many of his generation (including Bell), he pursued a safer career in mainstream physics until he got a permanent position in Trieste in the mid-1970s (see Kaiser, 2011). It was again Selleri who stimulated him to enter a professional activity on FQM, as recalled by Ghirardi himself. However, Ghirardi could not fully endorse Selleri's radical ideological programme: 'I attended the lectures of Selleri and with a certain sympathy because I was left-wing. However, when I heard him say that Quantum Mechanics is a bourgeois science and ought to be rejected because it is unacceptable for a worker, then I felt very, very far away.'[26]

After a few minor publications, such as on the implications of the Aharonov–Bohm effect (Ghirardi *et al.*, 1976) and the stochastic interpretation of quantum mechanics (Ghirardi *et al.*, 1978), it was in 1979 that Ghirardi started playing an important role in the international landscape. Indeed, Ghirardi and Weber (1979) found a flaw in

---

[25] A personal recollection of Ghirardi's activities on FQM can be found in (Ghirardi, 2007).
[26] Email from Ghirardi to the author on 30 September 2016.

Herbert's 'QUICK' proposal for faster-than-light communication. Moreover, in his referee report of Herbert's FLASH experiment, dated 22 April 1981, Ghirardi, while recommending a rejection of the paper, proved the 'quantum no-cloning theorem' (see footnote 21). Ghirardi also acknowledged the indirect influence of Selleri in developing the model that gave him international fame:

> [A]fter a seminar of Selleri it appeared clear to me that the core of the [measurement] problem was the superposition of the microstates that generates by reduction of the [wave] packet a statistical mixture. Since then I always had crystal clear in my mind that without breaking the linearity of the theory, we cannot get out of the contradiction.[27]

Indeed, Ghirardi and his collaborators Alberto Rimini and Tullio Weber developed a modification of the standard quantum formalism, adding non-linear terms to the standard Schrödinger equation that account for the collapse of the wave function (Ghirardi et al., 1984, 1986), called the 'GRW theory'.[28] This was the first instance of an 'objective collapse' model, in which the wave function spontaneously undergoes a collapse at a random instant of time, according to an average time-rate proportional to the number of constituents of the system under study. As such, this provides a quantitative measure of the macroscopicity of a system and thus a neat solution to Schrödinger's cat paradox. The GRW theory was praised and popularized by Bell (1987), who connected Ghirardi's group with Philip Pearle. That collaboration led to the formulation of the more sophisticated model of 'continuous spontaneous localization' (e.g., Ghirardi et al., 1990).[29]

## 25.4.3  Angelo Baracca in Florence and Silvio Bergia in Bologna

Among the young physicists sensitized by the atmosphere of protest against the military-industrial complex practices in high-energy physics and galvanized by the new trends in FQM, also thanks to the Varenna School of 1970, were Angelo Baracca and Silvio Bergia. They held professorships of theoretical physics at the Universities of Florence and Bologna, respectively. They soon started a collaboration which led to a first paper in 1974 (Baracca et al., 1974), wherein they attempted to operationally characterize the differences between entanglement and statistical mixtures. Following

---

[27]  Email from Ghirardi to the author on 30 September 2016.
[28]  The paper (Ghirardi, Rimini, and Weber, 1986) gathered an impressive resonance, having been cited ca. 2700 times according to Google Scholar (last accessed 11 July 2020).
[29]  Continuous spontaneous localization had been independently formulated by Nicolas Gisin (1989). Moreover, the ideas of Ghirardi and co-workers have been further developed by some of their pupils, notably by Angelo Bassi, who recently put forward an experimental proposal to test collapse models against standard quantum mechanics (see https://www.nytimes.com/2020/06/25/magazine/angelo-bassi-quantum-mechanic.html?smid=em-share).

the Italian line of research initiated by Selleri, however, their work was also carried out with the expectation that an experimental test of Bell's inequalities could prove quantum mechanics wrong. In Summer 1974, they organized a national conference in Frascati, which brought together most of the Italian scholars that were entering this field, thus providing a first opportunity for this community to form an identity and exchange ideas.

Baracca, who became interested in Bohm's views at the Varenna School of 1970, visited Bohm at Birkbeck College in London for two months in 1974. This collaboration led to a common paper (Baracca et al., 1975), wherein they put forward the idea that Bell's theorem 'has no essential relationship to hidden variables, but rather that it is mainly significant as a test for whether or not the laws of quantum mechanics have to be extended in certain new ways'. This again demonstrates the hope of employing Bell's inequalities to prove the limits of validity of quantum mechanics. Baracca also started supervising theses on FQM, of which the first was authored by Roberto Livi on Bell-type inequalities for multivalued observables and proposed experimental tests using molecules (Livi, 1977).

The collaboration between Baracca's and Bergia's groups continued until 1980—extending the research on Bell's inequalities, and generalizing it to higher-dimensional variables (Baracca et al., 1974; Baracca et al., 1977; Baracca et al., 1978; Bergia et al., 1980). Later, Baracca moved his interest away from research in FQM, whereas Bergia and collaborators kept working on proposed tests of Bell's inequalities (Bergia and Cannata, 1982; Bergia et al., 1985). They then proceeded to extend the ideas of Edward Nelson, according to which the evolution of a quantum mechanical system can be described in terms of stochastic processes (Bergia et al., 1988, 1989).

## 25.4.4  Marcello Cini in Rome

Marcello Cini, professor of theoretical physics in Rome, was one of the central figures of the critiques of science in the 1970s (see Gagliasso et al., 2015; Aronova and Turchetti, 2016). Cini was a dynamical intellectual, and, like Selleri, a radical leftist: a member of the Italian Communist Party since the 1940s, he became a dissident thereof when he co-founded the alternative communist newspaper Il Manifesto. In 1967, he also visited Vietnam during the war, as a member of the International War Crimes Tribunal (Russell Tribunal).

At the turn of the 1970s, Cini was a member of the steering committee and vice-president of SIF, and became a reference point for the Italian critique of science. He thus started working on the history of physics (with a Marxist approach) and on the social responsibility of scientists. His aforementioned work concerning military interests in the space programmes (Cini, 1969) and his book against the neutrality of science, written in a Marxist spirit, L'ape e l'architetto (Ciccotti et al., 1976) had a tremendous impact on the critical scientists (see Aronova and Turchetti, 2016).

Moreover, from the late 1970s, the focus of Cini's research in physics also moved towards FQM; an interest he kept until the end of his career. Similarly to the other Italians, Cini developed a critique against the Copenhagen interpretation of quantum theory, proposing a formal model that includes the measurement apparatus in the description of quantum mechanics (Cini *et al.*, 1979; Cini, 1983). Cini claimed that the fundamental problems of quantum mechanics stem from the idealization of isolated systems and the assumption that the measurement apparatus should lie outside of the domain of the theory (as a classical object). Indeed, he claimed that 'the postulate of wave packet collapse, introduced as an extra assumption in quantum mechanics [ ... ], can be dropped and replaced by the Schrödinger time evolution of the total system object + apparatus'. Although this approach was limited to particular cases only, it garnered a certain international interest, and even the prestigious journal *Nature* devoted a commentary in its 'News and Views' section to it.[30]

## 25.4.5  Silvano Tagliagambe and Quantum Physics in the USSR

As a pupil of the aforementioned Geymonat school, Silvano Tagliagambe completed his studies in philosophy in 1968 with a thesis on Hans Reichenbach's interpretation of quantum physics. Between 1971 and 1974, he was in Moscow for a specialization in the philosophy of quantum mechanics under the supervision of the physicists Ya. P. Terletsky (1912–1993) and Vladimir Fock (1898–1974) and later with the philosopher Mikhail Omelayanovskij (1904–1979).

In the USSR, where Marxist ideology had pervaded all fields of knowledge, FQM had a two-sided tradition: the one voiced by Terletsky saw in the Copenhagen interpretation a despicable form of idealism, incompatible with materialism, while the opposite position—of which Fock was the most illustrious exponent—regarded Bohr's principle of complementarity as an expression of Marx's dialectical materialism (see Freire, 2011). Thanks to these interactions, Tagliagambe soon became an expert on the philosophy of quantum mechanics in the Soviet context. In 1972, he edited the book *L'interpretazione materialistica della meccanica quantistica: Fisica e filosofia in URSS* (Tagliagambe, 1972), which represented an exceptional testimony of the philosophical debate on FQM in the USSR. In his introduction, Tagliagambe stressed that 'the volume [ ... ] has first and foremost an informational aim, being in its scope to give an idea to the Italian reader of the multitude of works that the Soviet scholars are carrying out [ ... ] on the most difficult problems of philosophy of science, and in particular of philosophy of physics' (Tagliagambe, 1972). Moreover, the volume also aimed at voicing the applicability of the methods of dialectical materialism (explicitly referring to Lenin's understanding thereof) to modern science. The volume, after a

---

[30] 'Uncertainties about uncertainty principle'. *Nature*, **302**, 377 (1983).

foreword by Geymonat and a historical preface by Tagliagambe, collected essays (translated into Italian by Tagliagambe himself) authored by seven physicists and sixteen philosophers on the occasion of various conferences held in the USSR between 1966 and 1971. Remarkably, the book also contains the first published version, in a language different from Russian, of the memoir that Fock wrote about the discussions he had with Bohr in Copenhagen in 1957. Therein, he supports his 'compatibilist' view between the Copenhagen interpretation and dialectical materialism, stating: 'Bohr's thought was always deeply dialectical [...]. Such dialecticism was not "spontaneous": Bohr told me that he studied dialectics in his youth and he always had held it in high esteem.' (Fock, in Tagliagambe, 1972).[31]

However, this book came about within the already mentioned diatribe between historical and dialectical materialism in the context of the history of physics in Italy and presumably did not have an impact on the Italian physicists concerned with FQM.[32]

## 25.4.6 The Erice School of 1976

Among the academic events devoted to FQM organized in Italy in the late 1970s and 1980s stands the international workshop 'Thinkshop on Physics', held in Erice in April 1976. The School was directed by Bell and d'Espagnat; Selleri and his pupils also participated in the event. If Varenna 1970 had represented a milestone in the legitimation of FQM, Erice provided the first opportunity for a new research community concerned with FQM to recognize itself. On this note, John Clauser, who was among the first to work on Bell's inequalities and a pioneer of modern foundations of quantum physics, later recalled that 'the sociology of the conference was as interesting as was its physics. The quantum subculture finally had come out of the closet' (Clauser, 1992).

It is also remarkable that Anton Zeilinger—who was to become one of the highest authorities in quantum foundations and quantum optics—became aware of the topics of FQM in Erice; in his words, 'There, I heard for the first time about Bell's theorem, about the Einstein–Podolsky–Rosen paradox, about entanglement, and the like' (Bertlmann and Zeilinger, 2013).

---

[31] For a recent analysis of Fock's stance on FQM also in relation to Bohr's ideas see [that is correct] Martinez, 2019).

[32] The school of Geymonat, to which Tagliagambe belonged, had a manifest Marxist approach and was to have a lasting influence on the philosophy of science in Italy. However, the new generation of physicists who revived the FQM in the 1970s harshly criticized their approach (see section 25.3.4). This likely prevented the work of Tagliagambe (1972) from becoming influential among the physicists sensitized about FQM.

## 25.4.7 Other Activities

We have until here focused on those lines of research which have been more evidently inspired by the politicized spirit of the time. As a matter of fact, FQM flourished in Italy throughout the 1970s and 1980s and it would be impossible to provide here a complete list of all the Italian research programmes on the subject matter.[33]

A number of other physicists, who were not active in the political struggles that inspired this revival, also performed research on FQM in the following years, often acquiring an international reputation. This was the case—just to mention a few notable ones—for Enrico Beltrametti and Gianni Cassinelli, in Genova, and Maria Luisa Dalla Chiara (a philosopher) and Toraldo di Francia, in Florence. They all contributed a great deal, eventually becoming international authorities, to an approach to FQM called 'quantum logic'. This subfield of FQM was pioneered by Garrett Birkhoff and John von Neumann (1936), and was experiencing a renewal in the 1960s and 1970s. They attracted international attention, being praised by the eminent philosopher Bas van Fraassen. The latter even co-organized, with Beltrametti, a conference in Erice on 'Current issues in quantum logic' (Beltrametti and van Fraassen, 1981).

It ought to be remarked that the 'School of Milan', mostly under the leadership of Prosperi, continued a prolific production of publications on FQM, after the aforementioned initial period in the 1960s, with new scholars such as Alberto Barchielli, Ludovico Lanz, and Pietro Bocchieri, some of whom are still active in this or closely related fields.

## 25.5 CONCLUDING REMARKS

In this chapter, we have analysed the contingent conditions that led to a revival of research into FQM, in Italy at the turn of the 1970s. We drew a connection between its origins and the social-political struggles for change of the 1968 left-wing movements, which also involved a young generation of physicists. What was remarkable in terms of the Italian case study is that prime academic institutions, such as the Italian Physical Society, represented a stepping stone for these young radical physicists to revolutionize the sensitivity towards an unconventional research field like FQM. These physicists criticized the structures of science (such as the military and industrial practices of the so-called *Big Science* in high-energy physics) and regarded science as yet another manifestation of the capitalistic character of modern society. They thus saw in FQM a natural starting point to dismantling the certainties of contemporary physics and thus opening new room for a radical change in the practices and contents of physics and

---

[33] A comprehensive bibliography (encompassing 362 references) of the Italian studies on FQM up until 1985 can be found in (Benzi, 1988), and in the rest of this section we refer the reader to that work for the missing references (see https://link.springer.com/chapter/10.1007/978-94-009-2947-0_20).

science. While most of the initial goals of the Italian researchers turned out to be unattainable—such as seeking the breakdown of quantum mechanics by means of Bell's theorem—dozens of Italian physicists became sensitized towards the fundamental issues of quantum physics. Between them they published hundreds of papers on quantum foundations between the end of the 1960s and the 1980s and this arguably helped a great deal in creating the conditions that made FQM an established research field in physics.

## ACKNOWLEDGEMENTS

I wish to express my gratitude to Christian Joas, Stefano Osnaghi, Martin Renner, and Joshua Morris for their comments which greatly improved this chapter. I am moreover thankful to Luigi Romano for having shared with me some of his findings on archival material about Franco Selleri.

A considerable part of the research underlying this chapter was conducted with Angelo Baracca and Silvio Bergia, and published by Baracca *et al.* (2017).

## REFERENCES

Andrade e Silva, J., Selleri, F., and Vigier, J. P. (1983). Some Possible Experiments on Quantum Waves. *Lettere al Nuovo Cimento*, **36**, 503–508.

Aronova, E., and Turchetti, S. (2016). *Science Studies during the Cold War and Beyond.* New York: Palgrave Macmillan.

Baracca, A., Bergia, S., Bigoni, R., and Cecchini, A. (1974). Statistics of observations for proper and improper mixtures in quantum mechanics. *Rivista del Nuovo Cimento*, **4**, 169.

Baracca, A., Bergia, S., and Restignoli, M. (1974). On the comparison between quantum mechanics and local hidden variable theories: Bell's type inequality for multi-valued observables. In N. Mitra, I. Slaus, V. Bhasin, and V. Gupta (eds), *Proceedings of the International Conference on Few Body Problems in Nuclear and Particle Physics*, Amsterdam: North-Holland.

Baracca, A., Bergia, S., Cannata, F., Ruffo, S., and Savoia, M. (1977). Is a Bell-Type Inequality for Nondicotomic Observables a Good Test of Quantum Mechanics? *International Journal of Theoretical Physics*, **16**, 491.

Baracca, A., Bergia, S., and Del Santo, F. (2017). The origins of the research on the foundations of quantum mechanics (and other critical activities) in Italy during the 1970s. *Studies in History and Philosophy of Modern Physics*, **57**, 66–79.

Baracca, A., Bohm, D., Hiley, B., and Stuart, A. (1975). On some notions concerning locality and nonlocality in the quantum theory. *Il Nuovo Cimento*, **28B**, 435.

Baracca, A., Cornia, A., Livi, R., and Ruffo, S. (1978). Quantum mechanics, first kind states and local hidden variables: three experimentally distinguishable situations. *Il Nuovo Cimento B*, **43**, 65.

Baracca, A., and Del Santo, F. (2017). La giovane Generazione dei fisici e il rinnovamento delle scienze in Italia negli anni Settanta. *Altronovecento: Ambiente, Tecnica, Società*, **34**.

Bedford, D., and Selleri, F. (1985). On Popper's new EPR-experiment. *Lettere al Nuovo Cimento*, **42**(7), 325–328.

Bell, J. S. (1964). On the Einstein–Podolsky–Rosen paradox. *Physics Physique Fizika*, **1**(3), 195.

Bell, J. S. (1987). Are there quantum jumps? In C. W. Kilmister (ed.), *Schrödinger, Centenary of a Polymath*, Cambridge: Cambridge University Press, pp. 41–52.

Bellone, E., Geymonat, L., Giorello, G., and Tagliagambe, S. (1974). *Attualità del Materialismo Dialettico*. Rome: Editori Riuniti.

Beltrametti, E. G., and Van Fraassen, B. C. (1981). *Current issues in quantum logic*. Boston: Springer Science & Business Media.

Benzi, M. (1988). Italian Studies in the Foundations of Quantum Physics. A Bibliography (1965–1985). In G. Tarozzi and A. van der Merwe (eds), *The Nature of Quantum Paradoxes: Italian Studies in the Foundations and Philosophy of Modern Physics*, Dordrecht: Kluwer Academic Publishers, pp. 403–425.

Bergia, S., and Cannata, F. (1982). Higher-Order Tensors and Tests of Quantum Mechanics. *Foundations of Physics*, **12**, 843.

Bergia, S., Cannata, F., Cornia, A., and Livi, R. (1980). On the Actual Measurability of the Density Matrix of a Decaying System by Means of Measurements on the Decay Products. *Foundations of Physics*, **10**, 723.

Bergia, S., Cannata, F., and Monzoni, V. (1985). Explicit Examples of Theories Satisfying Bell's Inequality: Do They Miss Their Goal Prior to Contradicting Experiments? *Foundations of Physics*, **15**, 145.

Bergia, S., Cannata, F., and Pasini, A. (1988). Space Time Fluctuations and Stochastic Mechanics: Problems and perspectives. In L. Kostro, A. Posiewnik, J. Pycacz, and M. Zukowski (eds), *Problems in Quantum Physics*, Gdansk: World Scientific.

Bergia, S., Cannata, F., and Pasini, A. (1989). On the possibility of interpreting quantum mechanics in terms of stochastic metric fluctuations. *Physics Letters*, **137**A, 21.

Besson, V. (2018). *L'interprétation causale de la mécanique quantique: biographie d'un programme de recherche minoritaire (1951–1964)*. Doctoral dissertation, Education, University of Lyon.

Bertlmann, R., and Zeilinger, A. (eds.) (2013). *Quantum (un)speakables: from Bell to quantum information*. Vienna: Springer.

Birkhoff, G., and von Neumann, J. (1936). The logic of quantum mechanics. *Annals of Mathematics*, **37**(4), 823–843.

Bonsignori, F., and Selleri, F. (1960). Pion cloud effects in pion production experiment. *Il Nuovo Cimento*, **15**(3), 465–478.

Bohm, D. (1952). A suggested interpretation of the quantum theory in terms of 'hidden variables' I. *Physical Review*, **85**(2), 166.

Caldirola, P. (1961). Quantistica, Meccanica, entry in the *Enciclopedia Italiana*, III Appendix. Rome: Treccani.

Caldirola, P. (1965). Teoria della misurazione e teoremi ergodici nella meccanica quantistica. *Giornale di Fisica*, **6**, 228–237.

Caldirola, P. (1974). *Dalla microfisica alla macrofisica.* Milan: Mondadori.

Caldirola, P., and Loinger, A. (1957). *L'interpretazione della teoria quantistica*. Milan: Il Pensiero.

Capasso, V., Fortunato, D., and Selleri, F. (1973). Sensitive Observables of Quantum Mechanics. *International Journal of Theoretical Physics*, **7**, 319–326.

Ciccotti, G., Cini, M., De Maria, M., Jona-Lasinio, G., Donini, E., *et al.* (1976). *L'Ape e l'Architetto: Paradigmi Scientifici e Materialismo Storico*. Milan: Feltrinelli

Cini, M. (1969). Il Satellite della Luna. *Il Manifesto (Rivista)*, September.

Cini, M. (1983). Quantum theory of measurement without wave packet collapse. *Il Nuovo Cimento B*, **73**(1), 27–56.

Cini, M., De Maria, M., Mattioli, G., and Nicolò, F. (1979). Wave packet reduction in quantum mechanics: a model of a measuring apparatus. *Foundations of Physics*, **9**(7), 479–500.

Clauser, J. F. (1992). Early history of Bell's theory and experiment. In T. D. Black (ed.), *Foundations of Quantum Mechanics*, Singapore: World Scientific.

Daneri, A., Loinger, A., and Prosperi, G. M. (1962). Quantum theory of measurement and ergodicity conditions. *Nuclear Physics*, **33**, 297–319.

Del Santo, F. (2018). Genesis of Karl Popper's EPR-like experiment and its resonance amongst the physics community in the 1980s *Studies in History and Philosophy of Modern Physics*, **62**, 56–70.

Del Santo, F. (2019). Karl Popper's forgotten role in the quantum debate at the edge between philosophy and physics in 1950s and 1960s. *Studies in History and Philosophy of Modern Physics*, **67**, 78–88.

Del Santo, F. (2020). An Unpublished Debate Brought to Light: Karl Popper's Enterprise against the Logic of Quantum Mechanics. *arXiv* preprint, 1910.06450.

d'Espagnat, B. (1965). *Conceptions de la physique contemporaine: les interprétations de la mécanique quantique et de la mesure*. Paris: Editions Hermann.

d'Espagnat, B. (ed.) (1971). *Proceedings of the International School of Physics 'Enrico Fermi', Foundations of Quantum Mechanics 1970*. NewYork, London: Academic Press.

Falciglia, F., Garuccio, A., and Pappalardo, L. (1982). Rapisarda's experiment: on the four-coincidence equipment 'FOCA-2', a test for nonlocality propagation. *Lettere al Nuovo Cimento*, **34**, 1–4.

Finkbeiner, A. (2006). *The Jasons: The secret history of science's postwar elite*. New York: Penguin.

Fortunato, D., and Selleri, F. (1976). Sensitive Observables in Infinite-Dimensional Hilbert Spaces. *International Journal of Theoretical Physics*, **15**, 333–338.

Fortunato, D., Garuccio, A., and Selleri, F. (1977). Observable Consequences from Second-Type State Vectors of Quantum Mechanics. *International Journal of Theoretical Physics*, **16**, 1–6.

Freire Jr, O. (2004). Popper, 'Probabilidade e mecânica quântica'. *Episteme*, **18**, 103–127.

Freire Jr, O. (2011). On the connections between the dialectical materialism and the controversy on the quanta. *Jahrbuch Für Europäische Wissenschaftskultur*, **6**, 195–210.

Freire Jr, O. (2014). *The Quantum Dissidents: Rebuilding the Foundations of Quantum Mechanics (1950–1990)*. Berlin: Springer.

Freire Jr, O. (2019). *David Bohm: A Life Dedicated to Understanding the Quantum World*. Cham, Switzerland: Springer Nature.

Gagliasso, E., Della Rocca, M., and Memoli, R. (2015). *Per una scienza critica, Marcello Cini e il presente: filosofia, storia e politiche della ricerca*. Pisa: Edizioni ETS.

Garuccio, A., Popper, K. R., and Vigier, J. P. (1981). Possible direct physical detection of de Broglie waves. *Physics Letters A*, **86**(8), 397–400.

Garuccio, A., and Rapisarda, V. (1981). Bell's inequalities and the four-coincidence experiment. *Nuovo Cimento*, **65**, 289.

Ghirardi, G. C. (1988). Some Critical Considerations on the Present Epistemological and Scientific Debate on Quantum Mechanics. In G. Tarozzi and A. van der Merwe (eds), *The Nature of Quantum Paradoxes: Italian Studies in the Foundations and Philosophy of Modern Physics*. Dordrecht: Kluwer Academic Publishers, pp. 89–105.

Ghirardi, G. C. (2007). Some reflections inspired by my research activity in quantum mechanics. *Journal of Physics A: Mathematical and Theoretical*, 40(12), 2891.

Ghirardi, G. C., Omero, C., Rimini, A., and Weber, T. (1978). The Stochastic Interpretation of Quantum Mechanics: a Critical Review. *Rivista del Nuovo Cimento*, 1(3), 1–34.

Ghirardi, G. C., Pearle, P., and Rimini, A. (1990). Markov processes in Hilbert space and continuous spontaneous localization of systems of identical particles. *Physical Review A*, 42, 78.

Ghirardi, G. C., Rimini, A., and Weber, T. (1976). Implications of the Bohm–Aharonov Hypothesis. *Il Nuovo Cimento*, 31B, 177.

Ghirardi, G. C., Rimini, A., and Weber, T. (1984). A Model for a Unified Quantum Description of Macroscopic and Microscopic Systems, Quantum Probability and Applications. In L. Accardi, *et al.* (eds), *Quantum Probability and Applications II*. Berlin: Springer.

Ghirardi, G. C., Rimini, A., and Weber, T. (1986). Unified dynamics for microscopic and macroscopic systems. *Physical Review D*, 34(2), 470.

Ghirardi, G. C., and Weber, T. (1979). On Some Recent Suggestions of Superluminal Communication through the Collapse of the Wave Function. *Lettere Nuovo Cimento*, 26, 599.

Gisin, N. (1989). Stochastic Quantum Dynamics and Relativity. *Helvetica Physica Acta*, 62(4), 363–371.

Herbert, N. (1982). FLASH-A superluminal communicator based upon a new kind of quantum measurement. *Foundations of Physics*, 12, 1171–1179.

Jammer, M. (1974). *The Philosophy of Quantum Mechanics: The Interpretations of Quantum Mechanics in Historical Perspective*. New York: Wiley.

Kaiser, D. (2011). *How the hippies saved physics: science, counterculture, and the quantum revival*. New York: WW Norton & Co.

Kim, Y. H., and Shih, Y. (1999). Experimental realization of Popper's experiment: Violation of the uncertainty principle? *Foundations of Physics*, 29(12), 1849–1861.

Livi, R. (1977). New Tests of Quantum Mechanics for Multivalued Observables. *Lettere Nuovo Cimento*, 19, 272.

Martinez, J.-P. (2019). Beyond Ideology: Epistemological Foundations of Vladimir Fock's Approach to Quantum Theory. *Berichte zur Wissenschaftsgeschichte*, 42(4), 400–423.

Moore, K. (2013). *Disrupting science: Social movements, American scientists, and the politics of the military, 1945–1975*. Princeton, NJ: Princeton University Press.

Mückenheim, W., Lokai, P., and Burghardt, B. (1988). Empty waves do not induce stimulated emission in laser media. *Physics Letters A*, 127(8), 387–390.

Nutricati, P. (1998). *Oltre I Paradossi della Fisica Moderna: I Fisici Italiani per il Rinnovamento di Teoria Quantistica e Relatività*. Bari: Dedalo.

Ortigoso, J. (2018). Twelve years before the quantum no-cloning theorem. *American Journal of Physics*, 86(3), 201–205.

Romano, L. (2020). *Franco Selleri and his contribution to the debate on Particle Physics, Foundations of Quantum Mechanics and Foundations of Relativity Theory*. Doctoral dissertation, Università degli Studi di Bari.

Rosenfeld, L. (1965). The measuring process in quantum mechanics. *Suppl. Progr. Theor. Phys.*, E65, 222–231.

Selleri, F. (1969a). On the wave function in quantum mechanics. *Lettere al Nuovo Cimento*, 1(17), 908–910.

Selleri, F. (1969b). 'Quantum Theory and Hidden Variables', lectures held in Frascati (June–July), LNF—69/75 CNEM-Laboratori Nazionali di Frascati (unpublished).

Selleri, F. (1971). Realism and the Wave-Function of Quantum Mechanics. In B. d'Espagnat (ed.), *Proceedings of the International School of Physics 'Enrico Fermi', Foundations of Quantum Mechanics 1970*. New York, London: Academic Press.

Selleri, F. (1972). A Stronger Form of Bell's Inequality. *Lettere al Nuovo Cimento*, 3(14), 581–582.

Selleri, F. (1982a). Can an Actual Existence be Granted to Quantum Waves? *Annales de la Fondation Louis de Broglie*, 7, 45–73.

Selleri, F. (1982b). On the direct observability of quantum waves. *Foundations of Physics*, 12, 1087–1112.

Selleri, F. (1983). *Die Debatte um die Quantentheorie*. Berlin: Springer-Verlag.

Selleri, F., and Tarozzi, G. (1978). Nonlocal Theories Satisfying Bell's Inequality. *Nuovo Cimento*, 48B(1), 120–130.

Tagliagambe, S. (ed.) (1972). *L'Interpretazione materialistica della meccanica quantistica. Fisica e filosofia in URSS*. Milan: Feltrinelli.

Tarozzi, G., and van der Merwe, A. (eds) (1985). *Open Questions in Quantum Physics*. Dordrecht: Reidel Publishing Co.

Tarozzi, G., and van der Merwe, A. (eds) (1988). *The Nature of Quantum Paradoxes: Italian Studies in the Foundations and Philosophy of Modern Physics*. Dordrecht: Kluwer Academic Publishers.

Trischler, C., and Kojevnikov, A. (eds) (2011). *Weimar culture and quantum mechanics: Selected papers by Paul Forman and contemporary perspectives on the Forman thesis*. Singapore: World Scientific.

van der Merwe, A., Selleri, F., and Tarozzi, G. (eds) (1988). *Microphysical Reality and Quantum Formalism (Vol. 1 and 2)*. Dordrecht: Kluwer Academic Publishers.

Vitale, B. (ed.) (1976). *The War Physicists*. Napoli: Istituto de F. T.

Weiner, C. (ed.) (1977). *Proceedings of the International School of Physics 'Enrico Fermi', History of Twentieth Century Physics 1972* New York, London: Academic Press.

Wootters, W., and Zurek, W. (1982). A single quantum cannot be cloned. *Nature*, 299, 802–803.

CHAPTER 26

........................................................

# FOUNDATIONS OF QUANTUM PHYSICS IN THE SOVIET UNION

........................................................

JEAN-PHILIPPE MARTINEZ

## 26.1 INTRODUCTION

........................................................

SOVIET physics underwent revolutionary changes at the start of the 20th century. Some were scientific, based on the new quantum and relativity theories, others socio-cultural as the October Revolution of 1917 deeply impacted its trajectory. In the 1920s, the Bolsheviks, who embraced the ideals of progress, rationality, and scientism, significantly contributed to the reorganization of the various sciences, which had suffered from isolation from the rest of the world from the start of World War I to the end of the Russian civil war. A brilliant generation of theoretical physicists thus emerged and had the opportunity to jump on the bandwagon of the quantum revolution. By the 1930s they had established practices and schools that propelled the Soviet Union to the rank of one of the leading centres of world theoretical physics. However, Soviet physicists also had to deal with the authoritarian dimension of the Bolshevik regime, mainly characterized by strict political control over science and the imposition of a single ideological line, that of dialectical materialism. By overturning the classical conceptions of matter and the processes of measurement and observation, but also by raising the questions of causality, action at a distance, and conservation of energy, quantum theory challenged many ideals of Marxist philosophy and became the object of particular attention. This therefore made the USSR, despite its strict framework, one of the most conducive environments for the continuing discussion of the foundations and inter-pretation of quantum physics. In the following, different episodes from the history of quantum theory in the Soviet Union are highlighted, discussing not only its scientific richness but also how the singularity of the socio-cultural and political context of the USSR played a role in the scientific creativity of its physicists.

## 26.2 COLLECTIVISM IN QUANTUM PHYSICS

In the fall of 1926, the very first course in quantum mechanics in the Soviet Union was given by Yakov I. Frenkel, who in 1929 was also the first Soviet physicist to publish a textbook on the subject, in German.[1] He had just returned to Leningrad from a Rockefeller fellowship in Göttingen, where, for a year, he had been a privileged witness to the elaboration of the theory. Frenkel did not participate in its development but contributed significantly to the application of quantum ideas in his own field, condensed matter physics. In particular, he was the pioneer of the collective approach, which gave birth to several new physical models, today united under the concept of quasiparticles. The use of metaphors and the transfer of ideas between scientific and political discourse characterized Frenkel's style, and he borrowed the term 'collectivist' from the language of the revolutionary era to discuss the degrees of freedom of atoms and particles. As a form of social organization based on the pooling of the means of production, collectivism was considered to be an alternative to liberal individualism by socialists, including non-Marxists, like Frenkel. When the Soviet physicist attempted in 1924 to replace the classic electronic Drude–Lorentz metal model with a theory based on quantum ideas and Bohr's atomic theory, the collectivist metaphor took on meaning. In this approach, atoms are forced so close to each other that their outermost orbits overlap, and the electric current is represented by electrons gliding from one another in a chain. Electrons are not absolutely free, nor do they belong to individual atoms.[2]

To develop the idea of one of his mentors, Abram F. Ioffe, who suggested that ions in a crystal could leave their place and wander in interatomic space, Frenkel extended his model in 1926. He introduced the concept of the hole, known today as a 'Frenkel defect', by noting that ions in motion would liberate an 'empty space' in the lattice. As it would behave like a particle and travel through the lattice by jumping from one atomic position to another, similar to the collectivist electrons, Frenkel characterized such empty space as a 'negative atom', an 'ion of the opposite sign'. This concept had much in common with Paul Dirac's hole theory in quantum electrodynamics. It could even have inspired it, the Briton having met his Soviet colleague in 1928 during the Sixth Congress of Russian Physicists. Nevertheless, Frenkel's work received a lukewarm reaction from the international scientific community, especially because in 1928 Felix Bloch proposed a rigorous application of quantum ideas to metals. Convinced of the relevance of the collective approach in condensed matter physics, the Soviet physicist

---

[1] Frenkel (1929). Frenkel's textbook was not translated in the Soviet Union until 1933. The first quantum mechanics textbook in Russian was by Fock (1932).

[2] Frenkel (1924). For in-depth consideration of the birth and the development of the collective approach in condensed matter physics, see Kojevnikov (1999; 2004, chapters 3 and 10).

was then inspired by the work of one of his compatriots, Igor E. Tamm, to develop its application.[3]

Tamm was the first Moscow theorist to use the new quantum mechanics. The recipient of a fellowship from the Lorentz fund in 1928, during his stay in Leiden he became a close friend of Paul Dirac, whose second quantization method he later used to study elastic waves in crystals. This resulted in the introduction in 1930 of one of the first quasiparticles, the 'elastic quantum'—named phonon in 1932 by Frenkel—to interpret the change in light frequency during scattering in a solid. Tamm understood phonons as the phenomenological description of collectively oscillating atoms with individuality but without much freedom. Frenkel, who was particularly sensitive to such an approach, then generalized it in 1931 to discuss the excitation of atoms in a lattice of insulators. He introduced another hypothetical particle, the 'quantum of excitation', or exciton, as an intermediary in the process of absorption of light by solids and its further transformation into heat.[4]

Meanwhile, it was the band theory of free electrons, developed by Alan H. Wilson, which quickly imposed itself in Western Europe and Northern America as the mainstream approach to condensed matter physics. However, while most Soviet physicists accepted at least parts of it, they also had reservations as to one of its basic hypotheses, that of the freedom it granted to electrons. The collective approach then offered them an interesting substitute. In the mid-1930s, Semyon P. Shubin and Sergei V. Vonsovsky developed a collectivist alternative to Heisenberg's theory of ferromagnetism. In the 1950s, Ilya M. Lifshitz and his collaborators in Kharkov treated electrons in metals as quasiparticles. But it was Lev D. Landau who mainly contributed to the enrichment of the collectivist picture of solid-state physics. In 1933 he introduced the concept of the polaron to discuss the case of trapped electrons in a crystal lattice deformed by the displacement of atoms from their equilibrium position, i.e., by phonons. In the late 1930s, with various collaborators in Moscow, he relied on Tamm's model of phonons to establish a more general method of elementary or collective excitation. Also, in the 1940s, he refined the mathematical technique of Anatoly A. Vlasov to describe collective interactions among electrons in plasma.[5]

Because Stalin had implemented the forced collectivization of agriculture in 1928, any physicist from the USSR was conscious of the particular socio-cultural roots on which the collectivist model had been based. Nevertheless, not everyone shared the same taste for Frenkel's scientific style, and new contributions left less space for political metaphors. For Soviet physicists the collectivist approach remained a powerful tool to tackle the thorny question of the degrees of freedom of atoms and electrons in solids. They thus helped bring the collectivist approach in solid-state physics to

---

[3] Frenkel (1926); Bloch (1929). On Ioffe and the institutional background of Soviet physics in the interwar period, see Josephson (1991).

[4] Tamm (1930); Frenkel (1931a, 1931b).

[5] Kojevnikov (1999; 2004, chapters 3 and 10) pays particular attention to Soviet physics and its many contributions.

worldwide recognition in the 1950s, but they also established what would become
during the Cold War one of the strongest traditions of Soviet theoretical physics, that of
condensed matter physics.

# 26.3 UNCERTAINTY RELATIONS: THE RUSSIAN BRANCH OF THE COPENHAGEN SCHOOL

Besides the application of quantum ideas to the study of solids, some of the young
generation of Soviet theoretical physicists who graduated in the 1920s were also keen to
become fully involved in the developments of quantum field theory. For Landau, who
often complained of having come a little too late to participate fully in the quantum
revolution, it even represented an opportunity to attempt a new revolutionary breakthrough in quantum physics. Thanks to fellowships which enabled him from October
1929 to March 1931 to visit some of the most renowned places in Western European
physics, Landau started to work in this direction with Rudolf Peierls in Zurich. The duo
set out to find a corpuscular analogue of the Heisenberg–Pauli quantum electrodynamics, as did another Soviet physicist, Vladimir A. Fock. The latter, a brilliant Leningrad
theorist who had spent a year in Göttingen in 1927–1928 working with Max Born, met
with Peierls in August 1930 at a conference in Odessa. A short-lived collaboration that
included Landau then began.[6]

A discussion of quantum field measurement was not central to their first draft. Fock
considered this with Pascual Jordan in an article prepared in October during a meeting
in Kharkov. They proposed new uncertainty relations for quantum field measurements, which received a mixed reception because they involved the elementary charge
$e$, making the concept of an electromagnetic field dependent on the atomicity of matter.
This criticism was notably formulated by Bohr, who discussed the matter with Landau
in Copenhagen in November. It motivated the Soviet theorist to address the issue upon
his return to Zurich, giving a whole new impulse to his work with Peierls. The
collaboration with Fock, further complicated by distance, was ended. The initial draft
was redesigned and submitted for publication in March 1931. From an analysis of the
Heisenberg relations and their relativistic generalization, Landau and Peierls then
asserted the impossibility of performing quantum field measurements. Niels Bohr,
who could not accept that the notion of the quantum field was excluded from the scope
of quantum mechanics, forcefully argued with Landau but also carried out long and
gruelling research to save the theory. It was only in 1933, in an article with Léon

---

[6] On Landau's trajectory, see Kojevnikov (2004, chapter 4). The collaboration between Landau,
Peierls, and Fock is discussed in Kojevnikov (1988).

Rosenfeld, that he finally managed, in specific conditions, to make sense of the applicability of the concept of field in the quantum domain.[7]

Apart from the importance of this episode for the foundations of quantum mechanics,[8] the construction of its historical narrative for the early 1930s highlights a golden age of Soviet theoretical physics. Indeed, it shows that the latter was fully included in the positive dynamics of worldwide science and participated significantly in the developments of quantum theory. In the USSR, Fock and Landau, as well as other physicists such as Frenkel and Tamm, drew on their international experience to establish themselves in the 1930s as pillars of Soviet theoretical physics. Fock took the chair of theoretical physics at the Leningrad State University, while Landau founded his own school in Kharkov—before he moved to Moscow in 1937—known worldwide for the excellence of its training. In the field of quantum theory, they brought to the Soviet Union the dynamism and modernity of Western European ideas and practices. This is notably illustrated by the work of their friend and close collaborator, Matvei P. Bronstein, whose attempt to quantize gravitational waves was a pioneering contribution to quantum gravity. His analysis of field measurability for the latter—inspired by the above-mentioned contributions by Landau–Peierls and Bohr–Rosenfeld—revealed an essential difference with the results obtained in quantum electrodynamics and predicted a profound modification of our classical notions of space and time.[9]

These Soviet physicists also showed deep respect for Niels Bohr and his visit in 1934 to Leningrad, Moscow, and Kharkov—the three main centres of Soviet physics—marked a great period of scientific exchange, as well as a form of recognition of the quality of theoretical practices in the USSR. Commentators and physicists then began to speak of a 'Russian branch' of the Copenhagen school.[10] Nevertheless, this label would wrongly be considered a form of subordination. Soviet physicists cultivated a strong sense of independence, and although they were aware of the value of the legacy they had received, they were also determined to make it their own and to question it with the revolutionary ardour of a young Landau at the dawn of the 1930s. Regarding the problem of measurement in quantum theory, the latter actually never recognized Bohr and Rosenfeld's conclusions. Also, Tamm and his mentor in Moscow, Leonid I. Mandelstam, opened a related debate when they proposed in 1945 a time–energy uncertainty relation based on the correlation between energy dispersion and time variation of dynamical variables. This remains widely discussed nowadays as an alternative to the traditional Bohr–Heisenberg perspective. In the Soviet Union, the latter was then chiefly defended by Fock. In 1947, with Nikolai S. Krylov, he opposed

---

[7] Jordan and Fock (1930); Landau and Peierls (1931); Bohr and Rosenfeld (1933). On the scientific debates over quantum field measurement, see Darrigol (1991), Gorelik (2005).

[8] Hartz (2022).

[9] Bronstein (1936a, 1936b). On Landau's school and its creation in the 1930s, see Hall (2008). On Bronstein, see Gorelik (2005).

[10] Jammer (1974), p. 248; Graham (1987), p. 322.

the sole statistical significance of the Mandelstam–Tamm relationship. In the early 1960s, he also criticized the proposal by Yakir Aharonov and David Bohm for a precise measurement of energy in an arbitrarily short time interval. The introduction of discontinuous functions of time into the Hamiltonian operator led, for Fock, to a *petitio principii*, as it constituted in itself a violation of the Heisenberg relations. Notwithstanding, it must be underlined that the Leningrad theorist did not act blindly in this defence of the Copenhagen approach to quantum theory, as will be shown later. Over time, he developed a fine critical analysis of Bohr's complementarity interpretation.[11]

# 26.4 Materialism and Antireductionism: The Copenhagen Interpretation Revisited

Fock was one of the Soviet physicists most concerned with problems related to the foundations of quantum theory, and from the mid-1930s he devoted a good part of his career to discussing its interpretation. He was strongly influenced by Niels Bohr—who personally explained to him complementarity when they met in 1934 in the Soviet Union—and also by the Marxist philosophy of dialectical materialism, which gave him tools to rethink the quantum revolution. Fock saw dialectical materialism as a philosophy with valuable epistemological content and incorporated this influence into his daily practice of science in the form of two main epistemological principles that were not limited to Marxist circles: scientific realism—or materialism—and antireductionism. On the one hand, Fock's materialism manifested itself in his approach to the mathematical formalism that, for him, was a direct reflection of material reality. He considered it the physicist's duty to make sense of the formalism of a theory and to work towards its conversion into purely physical concepts that would provide an exact description of the outside world. On the other hand, Fock's antireductionism consisted of understanding science as governed by a broad combination of theories. If relations of approximations linked the latter, their field of application could not wholly overlap—in the sense of a theory of everything—since each theory proved to be defined by its own conceptual apparatus. As a consequence, for the Soviet physicist, the formulation of a physical theory must be accompanied by a strict evaluation of its scope as well as that of the various concepts used.[12]

---

[11] Mandelstam and Tamm (1945); Fock and Krylov (1947); Aharonov and Bohm (1961); Fock (1962). On Landau, see Gorelik (2005), p. 1099. More on the time–energy uncertainty relation is in Jammer (1974), pp. 136–56.

[12] On Fock's epistemology of science, see Martinez (2019a, 2020).

Fock discussed quantum mechanics along these lines, and his rigorous conceptual analysis resulted in what Jammer characterized in 1974 as '[o]ne of the most trenchant and acclamatory formulations of [Bohr's] relational version of complementarity.'[13] Indeed, Fock interpreted the wave function as a catalogue of potential interactions but also extended the concept of relativity from the reference system to the means of observation. Nevertheless, it must be underlined that his interpretation resulted from attempts to creatively resolve the contradictions between Bohr's approach to quantum theory and dialectical materialism. Fock's disagreements can be summarized as follows: '[Bohr] diminishes the role of quantum mechanics (leaving it only a symbolic meaning) and exaggerates the importance of the uncertainty relation.'[14] The Soviet physicist could not accept that his Danish colleague only gave symbolic value to the mathematical formalism of the theory and that he consequently overestimated the role of the measuring device in observational situations and underestimated the properties of atomic objects.

As a materialist, Fock was determined to reveal the true nature of quantum phenomena, and, as an antireductionist, to accept their fundamental distinction from classical physics. He thus paid particular attention to the physical meaning of the wave function and associated the concept of 'state' it describes with that of 'potential possibility', which represents the probability a quantum object has of having one or another behaviour according to each possible external influence. It is, therefore, the act of measurement that transforms the virtually possible for each physical quantity, the potential possibilities, into a fait accompli. Fock understood probabilities as a real physical property of a physical system. As a consequence, by introducing the concept of potential possibility, he affirmed that the wave function contains not only a description of the behaviour of physical objects with respect to the means of observation but also of their properties. In this sense, Fock's interpretation of quantum mechanics went beyond that of Copenhagen.[15]

The theorist's adherence to Marxist philosophy was at the basis of the originality of his point of view, and his was not an isolated case. After the Bolshevik Revolution the basic ideas of dialectical materialism spread in the USSR. And with the imposition of a single ideological line in the 1930s, they became not only known to most physicists but also part of their regular education and practices. Therefore, the reception of quantum physics as well as the nature of interpretational debates in the Soviet Union were greatly influenced by dialectical materialism. Of course, the Bohr–Einstein dispute over realism in quantum physics transcended borders and was widely discussed by the Soviet physicists (see sections 26.5 and 26.6). But in Soviet conditions, as seen from the example of Fock, antireductionism also played a significant role, in addition to realism. It helped Soviet physicists accept the radical break with classical physics that quantum

[13] Jammer (1974), p. 202.
[14] Fock (1951a), p. 13.
[15] Fock (1957, 1965). For historical consideration of Fock's interpretation of quantum mechanics, see Graham (1987), pp. 320–43; Freire Jr. (1994); Martinez (2019a).

theory entailed. Indeed, dialectical materialism taught them that the laws and the conceptual apparatus of quantum physics have to be fundamentally different from those of classical physics, because each level of material reality has its own qualitative characteristics. The search for the latter became a dominant feature of the contributions of Soviet physics to the foundations of quantum mechanics.[16]

# 26.5 QUANTUM MECHANICS AS A THEORY OF ENSEMBLES

The Bohr–Einstein debates in quantum mechanics were notably introduced into the Soviet Union by Fock, who undertook the translation and publication of the original articles dealing with the EPR paradox in 1936. They were accompanied by a long introduction where the Leningrad theorist took a clear stand on the side of his Danish colleague, arguing for the compatibility of complementarity with a realistic approach to quantum mechanics. Nonetheless, part of the Soviet community was also receptive to Einstein's viewpoint on quantum mechanics and his suggestion that a statistical understanding of the theory would avoid conceptual difficulties related to its interpretation. In that frame, the $\psi$ function would not describe the condition of *one* single system but that of *an ensemble* of systems. Such an approach, called statistical—or ensemble—interpretation, was then regarded as a solution to the problems with realism and causality, on which Bohr's discourse was sometimes considered ambiguous and incompatible with the teachings of dialectical materialism.[17]

Konstantin V. Nikolsky, a researcher at the Lebedev Physical Institute of the Academy of Sciences in Moscow, was the first to work along these lines. As early as 1936, he published an article which described quantum mechanics as a non-classical statistical theory and analysed mathematically the differences with classical statistics resulting from the introduction of the quantum of action. Similar to Fock, his former professor in Leningrad, Nikolsky understood probabilities as a property of quantum systems. Therefore, he saw as a virtue of his approach the possibility of carrying out statistical predictions that, for him, objectively characterized quantum processes. Nevertheless, Fock immediately criticized Nikolsky's proposal, arguing that quantum mechanics allows the treatment of individual processes and that if its predictions are generally statistical, it cannot be reduced to statistics. Fock's authority in Soviet physics then

---

[16] Kojevnikov (2011).

[17] Fock (1936). It is worth highlighting that Fock summed up the debates in a dispute over the meaning of the wave function, shifting the focus from experimental and observational situations to that of mathematical formalism. It marked the starting point for Fock's independence in his approach to the interpretation of quantum mechanics, as discussed in section 26.4. On the history of statistical interpretation, see Jammer (1974), pp. 439–69; Pechenkin (2022). More on this interpretation in the Soviet context is in Graham (1987), pp. 320–43; Kojevnikov (2011).

contributed to marginalization of Nikolsky's point of view. However, it did not prevent the latter from further developing his ideas in a monograph he published in 1940, at the dawn of World War II. This publication greatly influenced Dmitrii I. Blokhintsev, also a theorist from Moscow, who further developed the Soviet ensemble interpretation and, after the war, brought it to national and international recognition.[18]

Blokhintsev began from the early 1940s to investigate mathematical and technical questions related to a statistical approach to quantum mechanics. It led to the publication in 1949 of an influential textbook that took up Nikolsky's frame of thought: it affirmed the qualitatively non-classical status of atomic phenomena, and the statistical—but different from classical statistics—nature of quantum theory, i.e. its inability to deal with individual processes. However, Blokhintsev went further in his reasoning and proposed a careful analysis of the main concepts of quantum mechanics, thus providing the first coherent development of the ensemble interpretation. In particular, he defined the eigenfunction as a description of a 'pure ensemble', as it was related to identically prepared atomic systems. The act of measurement hence turned a pure ensemble into a 'mixed' one, the different particles affected by the measuring device being in a new state, described by a new wave function. The measuring device, accordingly, was understood to be like a spectral analyser, since the newly created mixed ensembles were considered as a spectral resolution of the initial ensemble into components or sub-ensembles. As a dialectical materialist, Blokhintsev argued that his position preserved the necessary objectivity for a materialist approach to modern physics. His argumentation in this direction mostly relied on the concept of ensemble and the supposed inapplicability of quantum mechanics to individual particles. Indeed, while performing a measurement would change the state of a particular particle and assign it to a new ensemble, it would not affect the other particles belonging to the initial ensemble. They would remain in their previous state.[19]

On that specific point, Blokhintsev was notably criticized by Heisenberg, who underlined in 1955 that to assign a particle to an ensemble necessarily requires knowledge of the system by the observer. In the Soviet Union, the ensemble interpretation was criticized in 1952 by Fock, who kept defending the idea that quantum mechanics describes individual atomic objects, and perceived ensembles as 'speculative constructions'. He identified a circularity in Blokhintsev's definition of an ensemble, because it was characterized by means of a wave function, which was itself determined through the ensemble. Also, Fock stated the impossibility of treating the ensembles as statistical collectives, resulting from sorting entities with a common characteristic. Indeed, in quantum mechanics, atomic objects do not have defined values. The controversy between Fock and Blokhintsev subsequently determined the scientific content of most of the Soviet debates on the interpretation of quantum mechanics in the 1950s (section 26.7).[20]

---

[18] Nikolsky (1936, 1940); Fock (1937).
[19] Blokhintsev (1949).
[20] Heisenberg (1955); Fock (1952).

# 26.6 THE THEORY OF INDIRECT MEASUREMENT AS A THIRD WAY

Probability theory and its understanding in the Soviet context played a special role in debates on the interpretation of quantum mechanics. Mathematics was also the object of meticulous analysis in a dialectical materialist framework. This notably resulted, in the 1930s, in the criticism by Aleksandr Ya. Khinchin—the founder of the Moscow probability school—of Richard von Mises' positivistic understanding of probability as the frequency of a particular result during a series of experimental tests. Although the Soviet mathematician agreed with his Austrian colleague that the methods of probability theory can only deal with a large number of similar constituent entities, he sought to restore the objectivity of the theory by claiming its status as a mathematical rather than experimental science. Consequently, he approached probability as referring to the relative proportions of elements sharing a common characteristic in statistical collectives. It was, for Khinchin, a real mathematical property, the application of which was left to statistics. These debates had a significant impact on the Soviet physics community. It is therefore not surprising that the ensemble interpretation appeared in Moscow in circles close to Khinchin. Nevertheless, another voice was also heard, that of Mandelstam, one of the founders of Moscow theoretical physics, who preferred von Mises' approach to probability. He proposed a third way between the Copenhagen and ensemble interpretations.[21]

Mandelstam initially developed his views in a series of lectures given at Moscow State University in 1939—and published only posthumously in 1950—on 'The Foundations of Quantum Mechanics (The Theory of Indirect Measurement)'. In these lectures, his discussion of the EPR paradox revealed the main specificities of his interpretation. Mandelstam had ambivalent feelings about the arguments that Bohr and Einstein exchanged in their debates. He agreed with Bohr on the completeness of the theory, but could not accept a potential action at a distance, on the grounds that it would undermine the principle of causality. Also, even though he rejected the EPR argument, he was nonetheless seduced by Einstein's suggestion that a statistical understanding of quantum mechanics could resolve conceptual contradictions. The Moscow physicist was also influenced by his reading of von Neumann's 1932 axiomatic approach to quantum mechanics—which resulted in his use of the term 'ensemble' instead of 'collective'—which defended the completeness of quantum mechanics in a mathematical language conducive to the analysis of statistical ensembles.[22]

Nevertheless, Mandelstam rejected von Neumann's belief that the theory describes individual processes. He therefore followed Einstein's idea that physical parameters and the uncertainty relations refer to ensembles, and that the act of measurement leads to the reduction of these into narrower sub-ensembles. In the EPR thought experiment,

---

[21] Khinchin's criticism of von Mises is discussed by Siegmund-Schultze (2004). On Mandelstam's interpretation, see Pechenkin (2000, 2002, 2022); Kojevnikov (2011).
[22] Mandelstam (1972); von Neumann (1932).

such an approach would mean that the two spatially separated systems I and II do not interact simultaneously. If they both originate from the same ensemble, it is an operation of measuring an observable which leads to the selection of a first sub-ensemble, considered as system I. Consequently, while the uncertainty relations hold for this system, Mandelstam believed that nothing prevents the conjugate observable from being measured on another sub-ensemble, identified as system II. This reasoning subsequently influenced the position taken by Blokhintsev, who praised and used it in his 1949 textbook on quantum mechanics.[23]

However, in their entirety, the respective positions of Mandelstam and Blokhintsev rested on divergent philosophical foundations. The former was an operationalist, which explains his sympathy for the ideas of von Mises on probability. For him, physicists had to coordinate mathematical symbols with atomic objects by providing prescriptions according to which the numerical values of physical quantities could be extracted. In quantum mechanics, this prompted him to emphasize a discussion of the measurement processes rather than the meaning of the wave function—unlike dialectical materialists. This concern led him to his considerations on the EPR paradox, his collaboration with Tamm on the time–energy uncertainty relation (section 26.3), but also to be the first to underline the exceptional nature of situations when measurement is direct, in the sense that it has a macroscopic character and is accessible to the observer. In a large majority of cases, however, the quantum system that we want to measure interacts with another micro-system on which direct measurements are possible. It is only then, in a second step, that the values of the quantities relevant to the first system are provided by theoretical calculations. This explains why Mandelstam characterized quantum mechanics as a 'theory of indirect measurement'.[24]

Despite its positivistic features that hindered its diffusion in the Soviet context (see section 26.7), the novelty of his approach resonated in the community of Soviet physicists, especially in Moscow where Mandelstam had established an influential school of theoretical physics. Besides Blokhintsev, others like Tamm and Vitaly L. Ginzburg praised the profundity and clearness of his 1939 lectures. Even Fock, who strongly condemned any statistical approach to quantum theory, proved receptive to the idea of indirect measurement.[25]

# 26.7 QUANTUM PHYSICS AND SOVIET IDEOLOGY

The previous sections have highlighted various characteristics of Soviet epistemology with respect to the interpretation of quantum mechanics. On the subjects of

---

[23] Mandelstam (1972); Blokhintsev (1949).
[24] Mandelstam (1972).
[25] Tamm (1979), p. 135; Ginzburg (1999), p. 435; Fock (1951b).

measurement, objectivity, or causality, the above-mentioned contributions were responding to questions raised and debated internationally. At the same time, as mentioned earlier, the official status of dialectical materialism as an essential component of state ideology in the USSR also contributed from the 1930s on to the establishment of a socio-political context which played an essential role in the emergence, diffusion, and reception of the different points of view. Not only did Soviet physicists pay more attention to the epistemological content of the Marxist philosophy, but their scientific criticism was also often accompanied by a 'social' criticism that denounced or exposed scientific deviations and perversions. With respect to quantum mechanics, various Soviet philosophers but also several conservative physicists criticized the positions of the founding members of quantum theory for their flirtations with positivism and idealism and wanted to reject the Copenhagen interpretation. This explains why Nikolsky, in 1937, answered Fock's criticism of his statistical interpretation of quantum mechanics by denouncing the idealism that he ascribed to his former professor and other members of the Russian branch of the Copenhagen school. Later, such ideological attacks multiplied, and the latter had to react. As illustrated by Fock's 1938 defence of complementarity in the philosophical journal *Pod znamenem marksizma* [Under the banner of Marxism], he and other quantum physicists started to make explicit references to dialectical materialism in their writings to strengthen their position in polemics with philosophers.[26]

This context foreshadowed a delicate situation for Soviet physics in the post-war period. From 1946, Andrei A. Zhdanov, secretary of the Central Committee of the Soviet Communist Party, developed an ideological doctrine which affirmed the division of the world into two camps: the 'imperialist', led by the United States, and the 'democratic', led by the Soviet Union. This brought back various controversies from the 1930s and initiated the most intense ideological campaign in the history of Soviet scholarship. Zhdanov's mention, during a speech in June 1947, of the 'Kantian vagaries of modern bourgeois atomic physicists' mobilized the community of physicists. In February 1948, Mosey A. Markov, a Moscow theorist, published in *Voprosy Filosofii* [Questions of philosophy] an article that defended Bohr's complementarity within a dialectical materialist framework. The publication triggered a strong reactionary response. The methodological seminar of the physics department of the Moscow State University evaluated the article negatively, and Markov, who taught nuclear physics there, was dismissed. Aleksandr A. Maksimov, one of the ideological watchdogs of the Soviet philosophy of science, branded the theorist as the leader of ideological perversions of modern physics and pushed for the replacement of the editor-in-chief of *Voprossy Filosofii*. Soviet journals began to regularly publish devastating criticisms of the Copenhagen interpretation, such as that of Yakov P. Terletsky, a professor of theoretical physics at Moscow State University, who regularly took part in post-war debates over quantum mechanics. In the same spirit, the publication in 1950

---

[26] Nikolsky (1937); Fock (1938). On the 1930s ideological context of Soviet quantum mechanics, see Graham (1987), pp. 320–43.

of Mandelstam's lectures gave rise to tense discussions in view of the lectures' positiv-istic tendencies.[27]

While quantum theory was never in danger of being condemned in the USSR, this succession of events marked a definite increase in ideological pressure on physicists and on their philosophical interpretations of the theory. More than ever, they became aware that questions of interpretation, if not carefully handled, could lead them to precarious situations. On the one hand, this explains why complementarity was often omitted in their works, textbooks, and public statements in the late 1940s and early 1950s and, if mentioned, then only to be criticized. For example, the volume dealing with quantum mechanics from the famous *Course of Theoretical Physics* by Landau and Evgeny M. Lifshitz covered Heisenberg's uncertainty relations at length but avoided any reference to their interpretation by Bohr. On the other hand, the ideological situation encouraged the development and favourable reception of Blokhintsev's ensemble interpretation. His 1949 textbook, which not only criticized idealistic approaches to quantum theory but also provided an alternative that could be compat-ible with dialectical materialism, was seen as an excellent response to the Soviet ideological authorities and received much attention.[28]

For his part, Fock adopted an original position towards ideological pressure in the landscape of Soviet physics: he understood that the defence of modern theories had to include a critical revision of some of their concepts in order to avoid a head-on collision with dialectical materialism. Following this line, he developed a criticism of Bohr's interpretation. In an article published in 1951, the Leningrad theorist explained for the first time the essence of his epistemological divergences with his Danish colleague. But in the spirit of the times, he also condemned the positivistic features of the latter's scientific discourse. As far as complementarity was concerned, Fock made direct mention of it only to reject Bohr's attempts to apply it in other fields beyond physics. At the same time, Fock continued to express his opposition to the ensemble interpret-ation. In 1952, he formulated his most extensive criticism of Blokhintsev. Again, his epistemological discourse was coupled with a philosophical discourse in which he pointed out various inconsistencies between the ensemble interpretation and dialectical materialism.[29]

Such ideological markers strongly accompanied debates on the interpretation of quantum mechanics in the Soviet Union for about a decade. It was only after Stalin's death in 1953 that the situation could be normalized. Physicists, who notably benefited from their successes in the race for nuclear weapons, won in terms of autonomy, and the ideological pressure faded. In 1958, a conference held in Moscow on philosophical

---

[27] Markov (1947). A rich literature is available to discuss quantum mechanics in relation to the ideological context in the post-war Soviet Union. See, among others: Müller-Markus (1966); Graham (1987), pp. 320–43; Cross (1991); Gorelik (1991); Andreev (2000); Kojevnikov (2004), pp. 217–44; Freire Jr. (2011); Sonin (2017). On Terletsky, see Ichikawa (2019).

[28] Landau and Lifshitz (1948); Blokhintsev (1949).

[29] Fock (1951a, 1952). On Fock's attitude in the early 1950s, see Martinez (2019b).

aspects of natural science brought together Blokhintsev, Fock, and Terletsky, as well as many influential philosophers. As a sign of reconciliation, it sealed a form of public consensus on the peaceful coexistence of different possible Marxist interpretations of quantum mechanics, including the view of Fock, which retained complementarity. Subsequent debates gradually lost their degree of controversy.[30]

# 26.8 POST-WAR SOVIET QUANTUM MECHANICS AND INTERNATIONAL CONTEXT

From the 1930s, Stalinist policies gradually isolated the scientific community by imposing a form of economic and cultural self-sufficiency. The second half of the 1950s saw the international reopening of the Soviet Union. The restart of international relations then allowed the introduction into the USSR of the causal interpretation—explicitly deterministic and non-local—which had been proposed in 1952 by David Bohm. As a dialectical materialist, he had received the support of various Marxist physicists in the West, such as Jean-Pierre Vigier, in France, who contributed to the development of the causal interpretation and became one of its main defenders. In the Soviet Union, Terletsky, after he had initially sided with Blokhintsev, gradually embraced this position in the debates. In 1955, Terletsky edited the Russian translation of various articles by Bohm. At the 1958 Moscow conference on the philosophical aspects of the natural sciences, he clearly expressed his sympathy for the causal interpretation, offering a new option within the Soviet spectrum of Marxist approaches to quantum mechanics. Once again, Fock objected to this approach from 1956 onwards, along with his pupil Yuri N. Demkov. Remarkably, their voices also directly fed the French debates, where the causal interpretation was particularly influential. In this sense, Fock joined Rosenfeld, who, as the faithful right-hand man of Bohr, led from Copenhagen a real fight against the causal interpretation and the defence of complementarity. After the criticism levelled in 1951 by the Soviet physicist against his Danish colleague, their common stance against Bohm offered a possibility for compromise, which later led to the return of complementarity into favour in the USSR.[31]

A decisive step in this direction occurred during a 1957 meeting between Fock and Bohr in Copenhagen. The former was invited to give lectures on the theory of relativity, but lively discussions on the interpretation of quantum mechanics also took place. Fock not only explained his epistemological differences with Bohr but also some of the ideological grievances held against the latter in the Soviet Union. A year later, the

---

[30] Fedoseev (1959), pp. 655–56. On the changes experienced by Soviet physics after Stalin's death, see Ivanov (2002).

[31] Bohm (1952); Demkov (1957); Fock (1960). On the French debates over the interpretation of quantum mechanics see Besson (2018). On Fock and Rosenfeld's opposition to the causal interpretation, see Skaar Jacobsen (2012).

Danish physicist published an article which clarified his position towards causality—making it clear that it was specifically determinism that had to be abandoned in quantum mechanics—and praised the objective character of his complementarity interpretation. Fock perceived it as a direct response and acceptance of his criticism, and the Leningrad theorist started using this episode to publicly strengthen his defence of complementarity and to favour the diffusion of Bohr's ideas in the Soviet context. In 1961, the publication of the Russian version of Bohr's *Atomic Physics and Human Knowledge*, as well as the Danish physicist's visit to Moscow, contributed to the wide acceptance of complementarity in the scientific community. From this point on, Fock's slightly independent approach to quantum mechanics became the main reference in the USSR and was praised for its ability to free the Copenhagen interpretation from inclinations towards positivism. Even Mikhail E. Omelianovsky, one of the most influential Soviet philosophers of science, took Fock's side after having supported Blokhintsev in the early 1950s. Later, the contributions of Igor S. Alekseev, who from 1963 started to vigorously defend the compatibility of Bohr's complementarity with Marxism, also played an important role in legitimizing the discourse of the Copenhagen interpretation among Soviet philosophers.[32]

In the physics community, alternative views continued to be discussed, but had relatively little resonance. Among others, Georgy V. Ryazanov relied on the Gibbs distribution and Feynman's idea that there may exist in quantum mechanics a general expression for the probability amplitude of any event to describe quantum-mechanical probabilities as sums over particle paths. Yuri P. Rybakov, a former student of Terletsky, kept arguing for the causal interpretation of quantum mechanics. Also, Aleksei A. Tyapkin supported the possibility of establishing a broader theory that would provide a description of the movement of micro-objects between the moments of measurement. But in the end, the most influential work actually came from Blokhintsev, whose book *The Philosophy of Quantum Mechanics*, published in Russian in 1966, was later widely distributed in Europe and North America.[33]

One could have expected the publication in 1964 of Bell's inequalities to forcefully revive the Soviet debates on the foundations of quantum theory, but this did not happen. Possibly fatigued by decades of ideological tensions, Soviet physicists remained relatively indifferent to and unproductive on this topic. Besides sporadic interventions related to hidden variables or non-locality, the main contributions were experimental. In particular, David N. Klyshko, professor at Moscow State University, played a significant role in the development of Bell test experiments.[34] The renewed global interest in the foundations of quantum mechanics had, in fact, an outstanding impact on the community of Soviet mathematicians. In parallel with scientists from the United

[32] Bohr (1958, 1961); Alekseev (1995). On the episode of Fock's meeting with Bohr, see Freire Jr. (1994); Martinez (2019a). On Omelianovsky, see Graham (1987), pp. 343–47.

[33] Ryazanov (1958); Rybakov (1974); Blokhintsev (1968). On Tyapkin, as well as on debates over the interpretation of quantum mechanics in the post-war period, see Graham (1987), pp. 347–53.

[34] Silva Neto (2022).

States, from the 1960s they contributed significantly to the establishment of quantum information theory. Alexander S. Holevo, a pioneer in the field, was still a student at the Moscow Institute of Physics and Technology when John S. Bell published his inequalities and when Ruslan L. Stratonovich, one of the founders of the theory of stochastic differential equations, proposed a formulation of optical communications using quantum mechanics. This context led Holevo to discover a link between quantum entropy and Shannon's information, and to establish in the early 1970s the upper limit of the speed of communication in the transmission of a classic message via a quantum channel. Also, in 1980, Boris S. Tsirelson, a Leningrad mathematician, generalized the Bell inequalities to evidence an upper limit to quantum mechanical correlations between distant events. Nowadays known as Tsirelson's bound, it remained relatively unknown until the end of the USSR, but later made an important contribution to the development of quantum information and computer science.[35]

# 26.9 CONCLUSION

In a study echoing Forman's thesis, Alexei Kojevnikov argued that unlike Germany, where quantum mechanics was notably built on values of *Anschaulich*, *Individualität*, and abandonment of strict *Kausalität*, the case of the Soviet Union suggests an entirely different distribution of ideological preferences at the basis of its scientists' approach to the theory. The broad panorama established here supports this observation. From the beginning of the century, revolutionary ideas had spread through Soviet minds and it had become easier for physicists to accept the *Unanschaulich* (non-visual) character of quantum theory, especially since the antireductionist dimension of dialectical materialism taught them to expect a radical break between the classical and the quantum worlds. Also, as illustrated by the field of condensed matter physics and the development of statistical interpretations of quantum mechanics, the Soviet context, supported by a political ideology which gave more value to the group than to the individual, was conducive to the rejection of *Individualität* in favour of collectivist approaches. Finally, since the abandonment of causality was considered unacceptable by dialectical materialism, most physicists had to cautiously revise, rather than reject, the role of causality in quantum physics.[36]

Although not shared by everyone, these values, which have their sources in the socio-cultural context of the Soviet Union, have accompanied the intellectual trajectories of scientists who had to deal with a unique ideological environment for decades. The authoritarian character of this milieu was its greatest weakness, but it also proved to be one of its strengths, because it forced quantum mechanics specialists to thoroughly examine the established truths and to look for alternatives, to widen the

[35] Holevo (1973, 1979); Tsirelson (1980). On quantum information theory, see Hayashi (2006).
[36] Kojevnikov (2011, 2022).

spectrum of possibilities. Soviet scientists thus seized this opportunity to contribute significantly throughout the 20th century to a better understanding of the foundations of quantum theory.

# REFERENCES

Aharonov, Y., and Bohm, D. (1961). Time in the quantum theory and the uncertainty relation for time and energy. *Physical Review*, **122**, 1649–58.

Alekseev, I. S. (1995). *Deyatel'nostnaya kontseptsiya poznaniya i real'nosti*. Moscow: Russo.

Andreev, A. V. (2000). *Fiziki ne Zhutyat: Stranitsy Sotsial'noy Istorii Nauchno-Issledovatel'-skogo Instituta Fiziki pri MGU (1922–1954)*. Moscow: Progress-Traditsiya.

Besson, V. (2018). *L'interprétation causale de la mécanique quantique : biographie d'un programme de recherche minoritaire (1951–1964)*. Ph.D. thesis, Université de Lyon.

Bloch, F. (1929). Über die Quantenmechanik der Elektronen in Kristallgittern. *Zeitschrift für Physik*, **52**, 555–600.

Blokhintsev, D. I. (1949). *Osnovy kvantovoi mekhaniki. Izdanie vtoroe, pererabotannoe*. Moscow, Leningrad: GITTL.

Blokhintsev, D. I. (1968). *The philosophy of quantum mechanics*. New York: Humanities Press.

Bohm, D. (1952). A suggested interpretation of Quantum Theory in terms of Hidden Variables, Part 1 and 2. *Physical Review*, **85**(2), 166–93.

Bohr, N. (1958). Quantum Physics and Philosophy. Causality and Complementarity. In R. Klibansky, *La philosophie au milieu du vingtième siècle*, Chroniques, Firenze: La Nuova Italia Editrice, pp. 308–14.

Bohr, N. (1961). *Atomnaya fizika i chelovecheskoye poznaniye*. Moscow: Izdatel'stvo inostrannoy literatury.

Bohr, N., and Rosenfeld, L. (1933). Zur Frage der Messbarkeit der elektromagnetischen Feldgrössen. *Det Kongelige Danske Videnskabernes Selskabs, Mathematisk-fysiske Meddelelser*, **12**(8), 1–65.

Bronstein, M. P. (1936a). Quantentheorie schwacher Gravitationsfelder. *Physikalische Zeitschrift der Sowjetunion*, **9**(2–3), 140–57.

Bronstein, M. P. (1936b). Kvantovanie gravitatsionnykh voln. *Zhurnal eksperimental'noy i teoreticheskoy fiziki*, **6**(3), 195–236.

Cross, A. (1991). The Crisis in Physics: Dialectical Materialism and Quantum Theory. *Social Studies of Science*, **21**(4), 735–59.

Darrigol, O. (1991). Cohérence et complétude de la mécanique quantique : l'exemple de « Bohr-Rosenfeld ». *Revue d'histoire des sciences*, **44**(2), 137–79.

Demkov, Yu. (1957). Sur les tentatives de reconsidérer l'interprétation statistique de la mécanique quantique. *La Nouvelle Critique*, **93**, 107–23.

Fedoseev, P. (ed.) (1959). *Filosofskiye voprosy sovremennogo yestestvoznaniya. Trudy vsesoyuznogo soveshchaniya*. Moscow: Izdatel'stvo Akademii Nauk SSSR.

Fock, V. (1932). *Nachala kvantovoy mekhaniki*. Leningrad: Izdatel'stvo KUBUCH.

Fock, V. (1936). Mozhno li schitat', chto kvantovo-mekhanicheskoye opisaniye fizicheskoy real'nosti yavlyayetsya polnym? (Vstupitel'naya stat'ya k odnoimennym stat'yam Eynshteyna i Bora). *Uspekhi fizicheskikh nauk*, **16**(4), 436–57.

Fock, V. (1937). K stat'e Nikol'skogo 'Printsipy kvantovoi mekhaniki.' *Uspekhi fizicheskikh nauk*, 17(4), 552–54.

Fock, V. (1938). K diskussii po voprosam fiziki. *Pod znamenem marksizma*, 1, 149–59.

Fock, V. (1951a). Kritika vzglyadov Bora na kvantovuyu mekhaniku. *Uspekhi fizicheskikh nauk*, 45(9), 3–14.

Fock, V. (1951b). L. I. Mandel'shtam, Polnoe sobranie trydov, t. V. *Uspekhi fizicheskikh nauk*, 45(9), 160–63.

Fock, V. (1952). O tak nazyvayemykh ansamblyakh v kvantovoy mekhanike. *Voprosy filosofii*, 4, 170–74.

Fock, V. (1957). On the Interpretation of Quantum Mechanics. *Czechoslovak Journal of Physics*, 7, 643–56.

Fock, V. (1960). Critique épistémologique des théories récentes. *La Pensée*, 91, 8–15.

Fock, V. (1962). Criticism of an attempt to disprove the uncertainty relation between time and energy. *Soviet Physics JETP*, 15, 784–86.

Fock, V. (1965). *Kvantovaya fizika i stroyeniye materii*. Leningrad: Izdatel'stvo Leningradskiy Gosudarstvennyy Universitet.

Fock, V., and Krylov, N. (1947). On the uncertainty relation between time and energy. *Journal of Physics (USSR)*, 11, 112–20.

Freire Jr., O. (1994). Sobre o diálogo Fock-Bohr. *Perspicillum*, 8(1), 63–84.

Freire Jr., O. (2011). On the connections between the dialectical materialism and the controversy on the quanta. *Jahrbuch für Europäische Wissenschaftskultur/Yearbook for European Culture of Science*, 6, 195–210.

Frenkel, Ya. I. (1924). Beitrag zur Theorie der Metalle. *Zeitschrift für Physik*, 29, 214–40.

Frenkel, Ya. I. (1926). Über die Wärmebewegung in festen and flüssigen Körpern. *Zeitschrift für Physik*, 35, 652–69.

Frenkel, Ya. I. (1929). *Einführung in die Wellenmechanik*. Berlin: Springer.

Frenkel, Ya. I. (1931a). On the transformation of light into heat in solids. I. *Physical Review*, 37, 17–44.

Frenkel, Ya. I. (1931b). On the transformation of light into heat in solids. II. *Physical Review*, 37, 1276–94.

Ginzburg, V. (1999). Kakiye problemy fiziki i astrofiziki predstavlyayutsya seychas osobenno vazhnymi i interesnymi (tridtsat' let spustya, prichem uzhe na poroge XXI veka)? *Uspekhi fizicheskikh nauk*, 169(4), 419–41.

Gorelik, G. (1991). Fizika universitetskaya i akademicheskaya. *Voprosy istorii estestvoznaniya I tekhniki*, 2, 31–46.

Gorelik, G. (2005). Matvei Bronstein and quantum gravity: 70th anniversary of the unsolved problem. *Physics-Uspekhi*, 48(10), 1039–53.

Graham, L. R. (1987). *Science, Philosophy, and Human Behavior in the Soviet Union*. New York: Columbia University Press.

Hall, K. (2008). The Schooling of Lev Landau: The European Context of Postrevolutionary Soviet Theoretical Physics. *Osiris*, 23(1), 230–59.

Hartz, T. (2022). The measuring process in quantum field theory. In O. Freire Jr. (ed.), *The Oxford Handbook of the History of Quantum Interpretations* (this volume), Oxford: Oxford University Press.

Hayashi, M. (2006). *Quantum Information: An Introduction*. Heidelberg: Springer.

Heisenberg, W. (1955). The development of the interpretation of the quantum theory. In W. Pauli (ed.), *Niels Bohr and the development of physics*, London: Pergamon Press, pp. 12–29.

Holevo, A. S. (1973). Bounds for the quantity of information transmitted by a quantum communication channel. *Problems of Information Transmission*, **9**(3), 177–83.

Holevo, A. S. (1979). On capacity of a quantum communications channel. *Problems of Information Transmission*, **15**(4), 247–53.

Ichikawa, H. (2019). Materialist Perestroika of Quantum Dynamics and Soviet Ideology: Yakov Petrovich Terletsky (1912–1993). *Historia Scientiarum*, **28**(2), 134–51.

Ivanov, K. (2002). Science after Stalin: Forging a New Image of Soviet Science. *Science in Context*, **15**(2), 317–38.

Jammer, M. (1974). *The Philosophy of Quantum Mechanics. The Interpretations of Quantum Mechanics in Historical Perspective*. New York: John Wiley & Sons.

Jordan, P., and Fock, V. A. (1930). Neue Unbestimmtheitseigenschaften des electromagnetischen Feldes. *Zeitschrift für Physik*, **66**(3–4), 206–209.

Josephson, P. R. (1991). *Physics and Politics in Revolutionary Russia*. Berkeley, CA: University of California Press.

Kojevnikov, A. (1988). V. A. Fock i metod vtorichnogo kvantovaniya. In *Issledovaniya po istorii fiziki i mekhaniki*, Moscow: Nauka, pp. 113–38.

Kojevnikov, A. (1999). Freedom, Collectivism, and Quasiparticles: Social Metaphors in Quantum Physics. *Historical Studies in the Physical and Biological Sciences*, **29**(2), 295–31.

Kojevnikov, A. (2004). *Stalin's Great Science: The Times and Adventures of Soviet Physicists*. London: Imperial College Press.

Kojevnikov, A. (2011). Probability, Marxism, and Quantum Ensembles. *Jahrbuch für Europäische Wissenschaftskultur/Yearbook for European Culture of Science*, **6**, 211–35.

Kojevnikov, A. (2022). Quantum Historiography and Cultural History: Revisiting the Forman Thesis. In O. Freire Jr. (ed.), *The Oxford Handbook of the History of Quantum Interpretations* (this volume), Oxford: Oxford University Press.

Landau, L., and Lifshitz, E. (1948). *Kvantovaya mekhanika (nerelyativistskaya teoriya)*. Leningrad: Gosudarstvennoye izdatel'stvo RSFSR.

Landau, L., and Peierls, R. (1931). Erweiterung des Unbestimmtheitsprinzips für die relativistische Quantentheorie. *Zeitschrift für Physik*, **69**(1), 56–69.

Mandelstam, L. (1972). Lektsii po osnovam kvantovoy mekhaniki (teoriya kosvennykh izmereniy) (1939). In *Lektsii po optike, teorii otnositel'nosti i kvantovoy mekhanike*, Moscow: Nauka, pp. 325–88.

Mandelstam, L., and Tamm, I. (1945). The uncertainty relation between energy and time in non-relativistic quantum mechanics. *Journal of Physics (USSR)*, **9**, 249–54.

Markov, M. (1947). O prirode fizicheskogo znaniia. *Voprosy filosofii*, **2**, 140–76.

Martinez, J.-P. (2019a). Beyond ideology: epistemological foundations of Vladimir Fock's approach to quantum theory. *Berichte zur Wissenschaftsgeschichte/History of Science and Humanities*, **42**(4), 400–23.

Martinez, J.-P. (2019b). The 'Mach argument' and its use by Vladimir Fock to criticize Einstein in the Soviet Union. In F. Stadler (ed.), *Ernst Mach—Life, Work, Influence*, Vienna Circle Institute Yearbook 22, Dordrecht: Springer, pp. 259–70.

Martinez, J.-P. (2020). « La signification fondamentale des méthodes d'approximation en physique théorique »: Vladimir Fock épistémologue. *Philosophia Scientiae*, **24**(2), 171–90.

Müller-Markus, S. (1966). Niels Bohr in the darkness and light of Soviet philosophy. *Inquiry*, **9**(1–4), 73–93.

Nikolsky, K. (1936). Printsipy kvantovoi mekhaniki. I. *Uspekhi fizicheskikh nauk*, **16**(5), 537–65.

Nikolsky, K. (1937). Otvet V. A. Foku. *Uspekhi fizicheskikh nauk*, 17(4), 554–60.

Nikolsky, K. (1940). *Kvantovye protsessy*. Moscow: GTTI.

Pechenkin, A. A. (2000). Operationalism as the Philosophy of Soviet Physics: The Philosophical Backgrounds of L. I. Mandelstam and His School. *Synthese*, 124(3), 407–32.

Pechenkin, A. A. (2002). Mandelstam's interpretation of quantum mechanics in comparative perspective. *International Studies in the Philosophy of Science*, 16(3), 265–84.

Pechenkin, A. A. (2022). The Statistical (Ensemble) Interpretation of Quantum Mechanics. In O. Freire Jr. (ed.), *The Oxford Handbook of the History of Quantum Interpretations* (this volume), Oxford: Oxford University Press.

Ryazanov, G. V. (1958). Kvantovomekhanicheskie veroianotsi kak summy po putiam. *Zhurnal eksperimental'noy i teoreticheskoy fiziki*, 35, 121–31.

Rybakov, Yu. (1974). On the Causal Interpretation of Quantum Mechanics. *Foundations of Physics*, 4(2), 149–61.

Siegmund-Schultze, R. (2004). Mathematicians Forced to Philosophize: An Introduction to Khinchin's Paper on von Mises' Theory of Probability. *Science in Context*, 17(3), 373–90.

Silva Neto, C. P. (2022). Instrumentation and the Foundations of Quantum Mechanics. In O. Freire Jr. (ed.), *The Oxford Handbook of the History of Quantum Interpretations* (this volume), Oxford: Oxford University Press.

Skaar Jacobsen, A. (2012). *Léon Rosenfeld. Physics, Philosophy, and Politics in the Twentieth Century*. Singapore: World Scientific.

Sonin, S. (2017). *Fizicheskiy idealizm: Dramaticheskiy put' vnedreniya revolyutsionnykh idey fiziki nachala XX veka (na primere istorii protivostoyaniya v sovetskoy fizike)*, 2nd edn. Moscow: Lenand.

Tamm, I. (1930). Über die Quantentheorie der molekularen Lichtzerstreuung in festen Köpern. *Zeitschrift für Physik*, 60, 345–63.

Tamm, I. (1979). O rabotakh Mandel'shtama v teoreticheskoy fizike. In S. Rytov (ed.), *Akademik Mandel'shtam. K 100 letiyu so dnya rozhdeniya*, Moscow: Nauka, pp. 131–37.

Tsirelson, B. (1980). Quantum generalizations of Bell's inequality. *Letters in Mathematical Physics*, 4, 93–100.

von Neumann, J. (1932). *Mathematische Grundlagen der Quantenmechanik*. Berlin: Springer.

CHAPTER 27

# EARLY JAPANESE REACTIONS TO THE INTERPRETATION OF QUANTUM MECHANICS, 1927–1943

KENJI ITO

## 27.1 INTRODUCTION: QUANTUM MECHANICS AND 'EASTERN' THOUGHT

IN the history of interpretation and foundation of quantum mechanics in Japan, the place and the context mattered, but not in the way often assumed. There is a myth or a perception that philosophical concepts of quantum mechanics are somehow akin to 'Eastern thought', and therefore Asian or Japanese cultures were receptive to foundational ideas of the quantum (Holton, 1973; Katz, 1986, pp. 325–331; Kothari, 1985). Even a prominent Japanese physicist promoted this myth. Léon Rosenfeld once asked Yukawa Hideki,[1] whether the Japanese physicists had experienced the same difficulty as their Western counterparts in assimilating the idea of complementarity. Yukawa answered, 'No, Bohr's argumentation has always appeared quite evident to us.' Seeing Rosenfeld surprised, Yukawa added 'You see, we in Japan have not been corrupted by Aristotle' (Rosenfeld, 1960).

What Yukawa meant by Aristotle is unclear, and quantum mechanics might not be so antithetical to Aristotle after all (Jaeger, 2017; Kožnjak, 2020). Whatever Yukawa

---

[1] Japanese personal names in this chapter are in the Japanese traditional order (the family name first, the given name second) except when they appear as authors of European language publications.

had in mind, or whether Yukawa actually said it or not, the proposition that Bohr's complementarity was obvious to the Japanese does not do justice to what actually happened when Japanese physicists and intellectuals confronted this idea from the late 1920s to the early 1940s. As we see in this chapter, some found it difficult to understand, while others attempted to interpret it using various conceptual tools, mainly borrowed from European philosophy. A culturalist approach to paint the Japanese intellectual milieu with exotic flavour might produce cheesy publications and interest a lay audience but distorts history.[2]

## 27.2 Nishina Yoshio and Complementarity

One of the first Japanese physicists seriously engaged in the interpretation of quantum mechanics was Nishina Yoshio. Nishina formed himself as a quantum physicist more in Europe than in Japan, and his interest in the interpretation of quantum mechanics also originated from his experience in Europe. He had been in Europe since 1921, mostly in Copenhagen. Originally an experimentalist and trained in electrical engineering, he had been conducting x-ray spectroscopy with Dirk Coster and others. In the summer of 1927, he decided to attend the meeting in Como and Bohr's lecture on 16 September, which was the first public presentation on the idea of complementarity.[3] Nishina moved to Hamburg in early November and stayed there until February 1928.

In Hamburg, Nishina expanded his interest to quantum mechanics and its conceptual issues. Probably stimulated by recent developments in quantum mechanics, he began studying quantum mechanics more seriously by attending Pauli's lectures and seminars. He conducted his first theoretical work with Isidor I. Rabi, who moved together with him from Copenhagen (Rabi and Nishina, 1928). Nishina's seminar notes indicate that the conceptual issues concerning quantum mechanics, in particular Heisenberg's uncertainty relations and Bohr's complementarity, were major themes of Pauli's seminar, about which Pauli had an intense discussion with Bohr after the Como lecture (Kalckar, 1985, p. 29). The entry in Nishina's notes on 8 November was titled 'Question of determinism'. According to this note, Pauli derived the uncertainty relation of frequency and time for a Gaussian wave packet. In the following week, he discussed issues concerning the measurement of position and momentum, using the

---

[2] For more historiographical discussion on this issue, see Ito (2017).

[3] A typescript dated October 12 and 13 1927 is believed to be the closest to Bohr's talk (Bohr, 1985; Kalckar, 1985, p. 29). As for Nishina, there is a biography of Nishina in English: Kim (2007). For a critical review of this biography, see Ito (2008).

gamma-ray microscope. Then he followed Bohr's argument, and concluded that space-time description and causal description were complementary.[4]

Around 20 December, Nishina returned to Copenhagen where he spent the Christmas holidays. He started his collaboration with Oskar Klein on their famous formula (Ito, 2013; Klein and Nishina, 1929; Yazaki, 2017). At the same time, he became much more deeply involved in Bohr's complementarity. Bohr asked Nishina and Klein to translate his paper from German to English (Tamaki, 1991). Nishina and Klein spent long Scandinavian nights discussing Bohr's idea of complementarity.[5] Nishina produced 18 pages of notes on the German version of Bohr's paper. One of the sheets is titled 'Inscrutable (*fushin*)', indicating that Bohr's idea was not easy to grasp for Nishina. Through his effort, however, Nishina eventually came to appreciate Bohr's idea of complementarity. In January 1928, Nishina wrote to his mentor and superior Nagaoka Hantarō:[6]

> Bohr's new theory is extremely profound. It shows that particle theory and wave theory of radiation and matter are not contradictory but complementary, and it emphasizes that each of them is only an abstract theory that represents just one side of the matter.[7]

## 27.3 THEORETICAL PHYSICS IN 1920S JAPAN AND THE INTRODUCTION OF QUANTUM MECHANICS

Before any discussion of the philosophy of quantum mechanics, quantum mechanics had to be known in Japan. Before the advent of quantum mechanics, quantum theory was introduced into Japanese publications in 1911 at latest. Nagaoka Hantarō wrote an article for a philosophy magazine in this year on what he called the 'quantum hypothesis' and translated 'quantum' as '*ryōshi*', which became the standard (Nagaoka, 1911). Although Nagaoka presented it as a hypothesis, he soon started teaching a course on quantum theory at Tokyo Imperial University. Moreover, during the 1910s, Ishiwara Jun

---

[4] Yoshio Nishina, 'Pauli Seminar', Nishina Yoshio's notes at Pauli's seminar, Sangōkan shiryō, RIKEN, Wakōshi, Japan, 1927.

[5] Nishina Yoshio, 'Discussion with Klein', manuscript, Sangōkan shiryō, RIKEN, Wakōshi, Japan.

[6] As for Nagaoka, see Itakura *et al.* (1973).

[7] All the translations from Japanese texts are mine unless otherwise noted. Nishina's notes of Pauli's lecture show that he was in Hamburg until 19 December. His note entitled 'Discussion with Klein' was dated 21 December. Nishina returned to Hamburg by 9 January, when his lecture notes of 1928 started. Sangōkan shiryō, RIKEN, Wakōshi, Japan, 1927. Correspondence from Nishina Yoshio to Nagaoka Hantarō, 28 January 1928, Nagaoka Hantarō Collection, National Museum of Nature and Science, Tsukuba, Japan.

published original papers on quantum theory (for example, Ishiwara, 1915). In the late 1920s, after the emergence of quantum mechanical theories, some young physicists started eagerly absorbing the new theories of physics more or less independently. However, only after Nishina Yoshio's return to Japan in December 1928 did Japanese physicists establish a research tradition of quantum physics. Nishina differed from earlier Japanese physicists in his familiarity with not only the new theory but also the actual collective practices of physicists in Copenhagen (Ito, 2005).[8] In July 1931, Nishina became a 'Chief Scientist (*shunin kenkyūin*)' of the Institute for Physical and Chemical Research, or RIKEN. There, he built and expanded a research group, which included theoretical studies in nuclear and elementary particle physics and solid-state physics, and experiments with cosmic rays, radioisotopes, and accelerators, as well as biological research in genetics and radiobiology. (Ito, 2002, 2005; Kim, 2007).

Despite his earlier interests, probably because of the wide range of roles he had to play in Japan, Nishina did not talk much about the philosophy of quantum mechanics after his return. One opportunity was a talk on Bohr's theory of measurement at the colloquium of RIKEN, probably in 1933. He prepared for his talk by reading Bohr's article with his young collaborators, such as Tomonaga Sin-itiro and Tamaki Hidehiko.[9] The young physicists, however, found Bohr's article uninteresting and had a hard time staying awake (Tomonaga *et al.*, 1982, p. 61). Nishina also delivered lectures on Bohr's ideas on a few occasions outside the institute, such as a meeting of electrical engineers, who probably became aware of the relevance of atomic physics to their discipline (Ito, 2018a). Relying on the Como lecture, he discussed the necessity of classical concepts, wave–particle duality, the derivation of the uncertainty relation, and the gamma-ray experiment. Nishina never used the word 'complementarity' explicitly, but he evidently had it in mind when he said: 'There is a definite boundary between [particle and wave natures], and by complementing each other, these two constitute a complete system . . .'(Nishina, 1929).[10] Nishina's lecture on wave–particle duality and complementarity was published in the magazine *Rika kyōiku* (Science Education) (Nishina, 1931a, 1931b). His next article on this topic appeared in 1935, when he used the word *hanmensei* (one-sidedness) for complementarity (Nishina, 1935).

Probably the first Japanese word for 'complementarity' appeared in 1930 in a textbook, *Ryōshiron* (Quantum Theory), written by Sakai Takuzō (1930). Although his speciality was not quantum physics, he had been teaching the course on 'quantum theory' at Tokyo Imperial University since 1928 as an associate professor (*jokyōju*) after Nagaoka's retirement. His textbook later was based on this lecture according to the

---

[8] Sugiura Yoshikatsu was the first Japanese physicist who contributed to quantum mechanics (Sugiura, 1927). He was, however, far less instrumental in creating a research tradition in Japan. Nakane (2019) claims otherwise, but Sugiura's connection to the next generation of researchers in quantum chemistry and solid state physics appears 'indirect' at best (p. 352).

[9] Probably Bohr and Rosenfeld (1933). As for the historical significance of this paper, see Darrigol (1991).

[10] Another lecture by Nishina related to complementarity was also published.

notes taken by students.[11] It was the first textbook on quantum mechanics in Japanese, one of the earliest in any language.[12] Sakai mostly used original papers on quantum mechanics to prepare his lectures and the textbook. He devoted the last section of his book to the interpretation of quantum mechanics, discussing both uncertainty relations and complementarity. He translated complementarity as *hosoku kankei* or *hosokusei*, which later authors generally followed until 1938. His explanation of complementarity seems to be based on Bohr's Como Lecture:

> In quantum theory...we cannot neglect the effect of measurement. In other words, when we try to define the system spatio-temporally, there is always room for arbitrariness. On the other hand, if we try to retain causality, then the system should remain closed, unaffected by outside influence. In that case, however, we cannot know anything about such a system by observation (this is complementarity [*hosokusei*] of spatio-temporal description and causality).    (Sakai, 1930, pp. 179–180)

Sakai's interest in the foundation of quantum mechanics probably originated from his pedagogical, not philosophical, interest. As a teacher, Sakai tried to present quantum mechanics in a way conceptually understandable to students and probably to himself.

Thus, from the very beginning a few Japanese physicists had considerable familiarity with the conceptual issues of quantum mechanics, including complementarity. Yet, at this point, Nishina's and Sakai's work and teaching seem to have inspired none of their students and younger colleagues to pay any serious attention to the interpretation of quantum mechanics. They were not particularly receptive to complementarity. Indeed, a different issue of quantum mechanics attracted much more attention from Japanese intellectuals at this point.

# 27.4 ACAUSALITY AND SCIENCE JOURNALISM

Time and place mattered in the reception of conceptual issues of quantum mechanics in Japan, because for discussions on those topics to happen, there had to be a specific ecology, which cannot be reduced to any single cultural tradition. The principal venus for discussion of conceptual issues in science were not seminar rooms or textbooks, but science magazines and related periodicals. From the 1910s onward, science journalism developed steadily in Japan. Many popular science magazines appeared. Printed materials were the primary media for science communication before the advent of

---

[11] Mizuno Zen'uemon shiryō, Nihon Butsuri Gakkai.

[12] Sakai lists several textbooks at the end (Born and Jordan, 1930; Dirac, 1930; Frenkel, 1929; Heisenberg, 1930; Sommerfeld, 1929; Weyl, 1928). Besides Bohr's Como Lecture, Sakai seems to have used Heisenberg (1930) in his text on interpretations.

television. In addition, many magazines dealing with broader topics, especially journals for intellectuals, often carried articles on natural sciences. These periodicals gave Japanese intellectuals opportunities to discuss a wide range of topics including issues related to recent developments in the natural sciences and to debate with people from various backgrounds (Takata, 1996). For Japanese scholars, writing for such magazines was a substantial source of income. Physics occupied an important place in science journalism in Japan, especially after Albert Einstein's visit in 1922. This visit itself was funded and organized by the publisher of one of the major general magazines called *Kaizō*. It had a tremendous cultural impact on Japan, creating more opportunities for scientists and science writers to publish articles on physics (Ito, 2019; Jansen, 1989; Kaneko, 1981, 1987; Rosenkranz, 2018).

In this environment, the first conceptual issue that caught the attention of Japanese physicists and intellectuals was the problem of acausality and uncertainty. In 1927, Kurihara Kaname (1927) gave a popular account of quantum mechanical acausality. Mentioning Pascual Jordan's 'Philosophical Foundations of Quantum Theory' (1927), Kurihara characterized quantum theory as a probabilistic theory. In the following year, another young physicist, Ochiai Kiichirō (1928) discussed the statistical interpretation of quantum mechanics and Heisenberg's uncertainty relations. It was, however, only in the 1930s that Japanese intellectuals started paying serious attention to the issue of causality in quantum mechanics, probably inspired by a European trend in the early 1930s. Popular writings by Arthur Eddington (1932a), James Jeans (1930), and Pascual Jordan (1932a) were translated, often in abridged form (Eddington, 1932b; Jeans, 1932, 1938; Jordan, 1932b).

Besides translations, science magazines in Japan carried numerous articles on the breakdown of causality in quantum mechanics. The most prominent science journalist in Japan was Ishiwara Jun, a former professor of theoretical physics at Tohoku Imperial University and one of the few Japanese contributors to quantum theory, who turned to science journalism when he resigned his professorship due to a scandal and after Einstein visited Japan (Nisio, 2011). He published on this subject from 1927 onwards (Ishiwara, 1927, 1931, 1932, 1933). In his 1931 article, 'On the Uncertainty Principle', Ishiwara wrote that the uncertainty principle was 'loudly' asserted as forcing us to reject causality. Although causality did not hold for electrons, nonetheless, Ishiwara claimed, it remained valid in the physical laws in the domain of everyday phenomena (p. 314). Sugai Jun'ichi, a physicist, science writer, historian of science, and later technocrat, stated a similar view. Sugai (1934) admitted that uncertainty relations would imply the rejection of mechanical causality, but he claimed that they included the old causality as a special case, hence that quantum mechanics posed a 'causality of higher order', which, Sugai claimed, encompassed the two initially contradictory physical laws, causality and uncertainty.[13]

---

[13] Hoda (1932) also expressed a similar view.

Takeuchi Tokio was another important science journalist active in these discussions (Ito, 2018b, 2019). Often inaccurate but prolific, Takeuchi was quick to jump on the seemingly trendy issue of acausality, making it sensational news. Takeuchi (1932), titled 'Physics in a Storm', for example, contained a hodgepodge of ideas from various people, including Heisenberg, Planck, Bohr, and Schrödinger. He claims that 'Today physics has entered a period of *Sturm und Drang*. Once we overcome this crisis, wonderful new phenomena will be discovered, and the deepest secrets of our theoretical knowledge will be found' (p. 56). In a later article (T. Takeuchi, 1933) he stated more explicitly that natural science faced a 'crisis', because physics, which was, according to him, the 'monarch of all the natural sciences' was undergoing a vast conceptual change. The prime reason for that change was the denial of a strict causality. We do not know when a radium atom decays; we only know its probability. Takeuchi explained that this was not due to lack of knowledge, but due to the essentially probabilistic nature of the process. Takeuchi, then, gave a short account of Heisenberg's uncertainty relations, including his gamma-ray microscope experiment, and went on to claim that the issue of acausality in physics illuminated the problem of free will, roughly repeating Jordan's argument.

Science magazines and writers preferred the topic of acausality and uncertainty to complementarity. The former was related to classical philosophical questions and much more familiar to Japanese intellectuals. It attracted the Japanese in a similar way that Eddington's or Jordan's writings appealed to European readers. Through these writings, Heisenberg's uncertainty became widely popularized. Yet, science journalists like Ishiwara and Sugai fashioned themselves as popularizers of science, not critics. Total rejection of causality or emphasis on a 'crisis of science' would undermine the authority of science and therefore their own position as writers of scientific topics. If they were to reject causality, they needed to create a substitute immediately.

# 27.5 DIALECTICAL MATERIALISM AND QUANTUM MECHANICS: MARXIST INTELLECTUALS AND KYOTO SCHOOL PHILOSOPHERS

Beyond science journalists, conceptual issues of quantum mechanics started to attract a wider audience with more philosophical interests. The intellectual milieu of 1930s Japan was in no way simply 'Eastern'. It had two major, often conflicting, streams of thought: the Kyoto School philosophy and Marxism. Marxists were actively engaged in a broad range of intellectual endeavours, including the philosophy of science. Friedrich Engels's *Dialektik der Natur* and Vladimir Lenin's *Materialism and Empirico-criticism* were translated into Japanese around 1930, providing Japanese Marxist philosophers with a foundation for their philosophy of science (Engels, 1929; Lenin, 1930). In the history of science, *Science at the Cross Roads* (International Congress of the History of Science and Technology, 1931), a collection of papers by Soviet historians and philosophers of science, which included a classic article by Boris Hessen on the social roots of Newtonian mechanics, was translated into Japanese immediately after publication of the original book in 1931 (Puroretaria Kagaku Kenkyūjo, 1932). In 1935, a prominent Marxist philosopher, Tosaka Jun, published *Kagakuron* (Theory of science), the first systematic Marxist theory of science by a Japanese author (Tosaka, 1935).

Kyoto School philosophers were also influential in pre-war Japan. This school of philosophy includes Nishida Kitarō, Tanabe Hajime, and many others who learned from them. A wide range of intellectuals, even left-wing scholars such as Tosaka, Miki Kiyoshi, and Nakai Masakazu, could be included in this group, but the mainstream Kyoto School philosophers were generally politically conservative and had a stronger affinity to Japanese or Asian traditions. The most representative work of this school is Nishida's *Zen no kenkyū* (*An Inquiry into the Good*), which synthesized European philosophy with Asian thoughts (Nishida, 1990). In the 1940s, some of them, especially Kōsaka Masaaki and Kōyama Iwao, cooperated with the Japanese government's war efforts and tried to formulate an ideological justification. Their political alignment and philosophical positions made them the natural enemies of Marxist intellectuals, who were usually materialist and much less sympathetic to the Japanese monarchy's invented traditions and ideologies.[14]

The debate between the Marxists and Kyoto School philosophers concerning quantum mechanics began with Tanabe Hajime. Tanabe started his career as a philosopher with neo-Kantian investigations of physics and mathematics. He followed the development of the natural sciences and was quick to introduce new ideas. His first two books (Tanabe, 1915; Tanabe, 1922) were philosophical introductions to the natural sciences, which inspired many philosophy and science students, including Yukawa and Taketani Mituo. Elite institutions of higher education, especially 'higher schools' developed a student culture of appreciating high culture, where students favoured European philosophical writings in addition to European literature, music, and paintings (Roden, 1980). In such an environment, learning science through Tanabe's philosophical texts was a natural practice, even among science students. From the mid-1920s, Tanabe was increasingly inclined toward Hegelian dialectics (Nagai, 1991), then he turned to Buddhist philosophy.

Tanabe's 1933 article 'The New Physics and the Dialectics of Nature' (*Shin butsurigaku to shizen benshōhō*) is important here partly because it discussed Bohr's ideas on quantum mechanics. Tanabe, borrowing Bohr's analogy of a stick, discussed the question of subject and object:

---

[14] A vast literature exists on the Kyoto School of Philosophy, both in Japanese and English. Here are two general accounts: Carter (2013); Ōhashi (2004).

When a stick is grasped normally, we sense and feel it as an object. When we hold it tightly, and use it to touch things, our sense of feeling is located at the point where the stick touches those things; hence the stick now belongs to the subject. Similarly, [in the gamma ray experiment], although the radiation of the ray is an object, it comes to belong to the subject when one uses it as a means of measurement. This duality causes the uncertainty... This reciprocity [of subject and object] requires us to revise the view of classical mechanics, in which the means of measurement was idealized, and it causes reciprocal uncertainty.   (Tanabe, 1933, p. 90)

This 'reciprocity' (kōgosei) came from the German 'reziprok' that Bohr (1929) proposed as a replacement for 'complementary'. Tanabe further argued in this paper that the emergence of new physical theories, such as relativity theory and quantum mechanics, contradicted the (Marxist) dialectics of nature (shizen benshōhō). Tanabe regarded the reciprocity of subject and object in the measurement as a dialectical process in which subject and object were synthesized in a 'higher dimension'. He called this 'nature's dialectics' (shizen no benshōhō). Since the Marxist dialectic of nature rested on materialism, it did not accommodate such a synthesis of subject and object as demanded by nature's dialectics. Therefore, Tanabe concluded, the Marxist dialectics of nature was wrong in light of the new physics.

Tanabe was not the only person in the Kyoto School interested in quantum mechanics. Nishida Kitarō, the central figure of the Kyoto School of Philosophy, learned quantum mechanics from his son Sotohiko, one of the first Japanese physicists who studied quantum mechanics. Nishida later published two articles to interpret complementarity as an idea suggesting interaction and synthesis of subject and object (Nishida, 1943a, 1943b). Quantum mechanics appeared to them to give a clue to their project to overcome subject–object duality, along with European modern rationality.

Since Japanese Marxists generally took the position of scientism, they could hardly tolerate criticism that their philosophy was scientifically obsolete. Their response was swift. Nagai Kazuo, a member of the Marxist-dominated intellectual circle called Yuibutsuron kenkyūkai (Society for Materialism Studies) countered Tanabe's argument under his pseudonym, 'Gō Hajime'.[15] Nagai denounced Tanabe as a bourgeois scholar who, limited by his class origin, could not properly understand new developments in physics and concocted philosophical idealism out of them. Against Tanabe's argument that the interaction between the subject and the object caused indeterminacy, Nagai insisted that the duality of the electron and light was an objective quality of the electron itself, independent of human cognition. Similarly, Nagai argued that uncertainty relations were objective laws of nature, not caused by anything. While criticizing Tanabe, Nagai also attacked Bohr and Heisenberg for their 'subjectivist' interpretation of quantum mechanics, chiding Tanabe for following their 'idealistic explanation'. Nagai concluded that Tanabe's dialectics of nature isolated and exaggerated only one side of dialectics, which he claimed a 'thought style' common among all kinds of idealism (Gō, 1933, p. 57).

---

[15] Left-wing intellectuals often used a pseudonym to avoid prosecution.

Taketani Mituo, a physicist and left-wing philosopher, responded as well. In 1935, Taketani was an unpaid assistant at Kyoto Imperial University, working with Yukawa and Sakata Shōichi. During the late 1930s, he was working on his 'three-stage theory' on the development of science, to be published in 1941 (Taketani, 1971b). According to his account, while Taketani was a student, Tanabe's books fascinated him, but he later claimed to have realized that Tanabe's ideas were 'irrelevant and rather misleading'. Very disappointed, he started his quest for a 'philosophy useful for quantum mechanics'. For him, Tanabe's argument was a 'fascist mysticism', which 'tried to deceive people by using quantum mechanics' (Taketani, 1968). Taketani's view of quantum mechanics with his criticism against Tanabe appeared in a paper in 1936: 'Nature's Dialectics: On quantum mechanics' (*Shizen no benshōhō: ryōshirikigaku nitsuite*) under his pseudonym, Tani Kazuo (1936b).[16] He insisted that the concept of 'state' was essential in quantum mechanics. It unified two conflicting forms of phenomena, particles and waves. He claimed:

> So-called acausal reduction of states by measurement does not imply 'agency of subject' or 'denial of causality'. Rather, measurement is based on material laws, namely the dialectics of quantum mechanical superposition laws. There is no basis for the view that the action of the subject causes the uncertainty principle.
>
> (Tani Kazuo, 1936b, p. 9).

Because of the central place occupied by Kyoto School philosophers and Marxists in Japan's intellectual scene at that time, the philosophy of quantum mechanics, especially Bohr's notion of complementarity came to acquire more visibility among Japanese intellectuals.

# 27.6 JAPANESE RESPONSES TO THE EPR PAPER

While Marxists and Kyoto School philosophers were debating over quantum mechanics in Japan, Einstein, with Boris Podolsky and Nathan Rosen (1935), launched his most famous attack on quantum mechanics, and Bohr (1935) quickly responded. This historic exchange appeared in the widely circulated *Physical Review* and could not escape attention from Japanese physicists and science writers. Science journalists saw the exchange between two giants in physics as a big event. Ishiwara, among others, quickly responded to this debate, and reported it (Ishiwara, 1935a, 1935b). In his 'Probability Theory and Natural Sciences', however, he curiously did not show much

---

[16] Taketani published an English version of this article later (Taketani, 1971a). Here I use my own translation from the Japanese version.

interest in the apparent main issue of this debate, the question of the completeness of quantum mechanics and reality. After having outlined the debate, although admitting that it might be worth physicists' consideration whether a theory was complete in Einstein's sense or not, he wrote,

> [W]e have a much more important and interesting problem, namely, whether we can recognize causal relations in quantum phenomena in general.
>
> (Ishiwara, 1935b, p. 45)

A more intense interest arose from among the Marxists. Taketani was also quick to write about the debate (Tani, 1935). For him, the exchange between EPR and Bohr marked a memorable occasion, in which 'Bohr teaches Einstein' (*Bōa Ainshutain o oshiu*) (Tani, 1936a). At this point, Taketani presented Bohr as a person who 'accomplished brilliant achievements by introducing quantum theory into the atom, and later contributed greatly to the development of atomic physics and educated many excellent physicists'. Einstein, on the other hand, was a 'former Machian', then a 'mechanical materialist', who was 'no longer able to keep up with the development of physics'(Tani, 1935). Yet Taketani expressed some reservations in his support of Bohr. He admitted that it was fair to criticize Bohr for his subjectivism and 'confusion' of subject and object (Tani, 1936a). The problem, according to Taketani, was Bohr's use of words, which was different from that of 'philosophers'. Taketani suggested that it would be necessary to reformulate Bohr's ideas in the more precise language of philosophy, by which he apparently meant dialectic materialism. Taketani remained ambivalent and inconsistent about Bohr. In 1937, he translated the EPR paper and Bohr's response under his own name (Bohr, 1937b; Einstein, Podolsky, and Rosen, 1937). In the introduction to the latter, he described Bohr's as a 'valuable and original contribution to quantum theory' that explains 'masterfully and intelligibly the esoteric measurement problem' (p. 808). Taketani criticized Bohr's 'subjective' or 'idealistic' interpretation because of his own ideological standpoint as a Marxist but supported Bohr against Einstein, who he considered as attacking quantum mechanics, to which he and his fellow physicists were deeply committed. For the younger generation of Japanese theoretical physicists like Taketani, quantum physics was at least one of the obvious future directions of research. As Fujioka (1936) shows, Japanese physicists generally accepted Bohr's response and did not take the EPR argument seriously at this point.

# 27.7 NIELS BOHR'S 1937 VISIT TO JAPAN AND ITS AFTERMATH

Against the backdrop laid down by the debates between Marxists and Tanabe as well as the responses to EPR, an even broader readership began paying attention to conceptual questions concerning quantum mechanics after 1937. The incentive was Bohr's visit to Japan, not any Japanese or Asian cultural tradition. When Nishina wrote to Nagaoka in 1928, he suggested then inviting Bohr to Japan,[17] and kept trying to materialize the plan after his return. The correspondence between Nishina and the Bohrs from 1929 to the early 1930s reveals that the trip was repeatedly postponed for various reasons until 1937.[18] Compared to Einstein's, however, Bohr's visit to Japan did not cause as much fanfare. Unfortunately for Bohr and complementarity, the luxury steamer brought the Bohrs to Japan alongside a far more famous figure: Helen A. Keller, a writer and activist already well-known worldwide for her autobiography, *The Story of My Life*, and its various adaptations. Major newspapers extensively covered her voyage to Japan and planned activities, overshadowing a modest report on Bohr's arrival.[19] Moreover, Bohr's scientific achievements did not arouse public curiosity as much as Einstein's relativity theory. Worse, Bohr was much less photogenic than Einstein. A reporter from Asahi Newspaper, who obviously expected to see someone like Einstein, reported, 'Professor Bohr has a mediocre appearance, and it is difficult to imagine that he is a world-renowned authority of quantum mechanics and atomic theory'.[20]

Within academia, however, Bohr's visit caused a small sensation. *Teikoku Daigaku Shimbun*, a weekly newspaper at Tokyo Imperial University, published articles about Bohr in every issue while he was in Japan, telling readers who he was, where he visited, and what he did. Many science magazines carried stories about Bohr's achievements, his trips in Japan, and especially his lectures.

Bohr delivered lectures on the subjects which interested him at that time, nuclear physics and the foundations of quantum mechanics. He talked much about complementarity, the uncertainty relation, and his debate with Einstein, including the *Gedankenexperiment* of the photon box and EPR's argument. Several science magazines and one newspaper printed Bohr's lectures in full. Fujioka (1937) published in *Kagaku* a summary of the whole lecture series at Tokyo Imperial University. Part of this lecture series was also reported by Takeuchi (1937) and Q. L. Z. (a pseudonym) (Q. L. Z., 1937). Nishina translated Bohr's lecture at Kagaku Chishiki Fukyū Kyōkai (Association of the Promulgation of Scientific Knowledge) and published the text in *Kagaku chishiki* ('Scientific knowledge') (Bohr, 1937a). We have no precise record of the lectures in Kyoto, but according to Takeuchi (1937), their title was 'Atomic theory and Causality (Genshiron to ingaritsu)' and the contents of the Kyoto lectures were almost the same as Bohr's reply to EPR's paper.

Bohr's visit and his lectures drastically enhanced the visibility of complementarity in Japan. Nishina now began writing about it more often and at greater length. When a

[17] Nishina, letter to Nagaoka Hantarō, 27 January 1927. Nagaoka Papers, National Museum of Science and Nature.

[18] Niels Bohr to Nishina Yoshio, 26 December 1929; 4 August 1930; Nishina Yoshio to Niels Bohr, 21 March 1934; Betty Schultz to Nishina Yoshio, 5 July 1934 (Nishina, 1984). Also translated in Nakane *et al.* (2006).

[19] *Asahi Shinbun*, morning edition, 16 April 1937, p. 11.

[20] *Asahi Shimbun*, morning edition, 16 April 1937, p. 11.

journalist asked about Bohr before his arrival, Nishina said that complementarity would exert as much influence on philosophy as relativity theory did (Nishina *et al.*, 1937). Soon after Bohr's departure, Nishina started thinking about translating Bohr's *Atomic Theory and the Description of Nature* (1934).[21] In November 1938, *Kagaku chishiki* published a special issue devoted to 'the theory of complementarity'(Hayashi, 1938; Itagaki, 1938; Kagawa, 1938; Nishina, 1938; Saegusa, 1938). Nishina guest-edited this issue and contributed an article with the same title. In his article, Nishina (1938) contentedly declared, 'Today complementarity receives wide attention in science and philosophy in general, which is, I suppose, a matter of course'. The sudden rise of interest in complementarity among Japanese intellectuals was, therefore, triggered by Bohr's visit to Japan, not by its inherent appeal to Japanese intellectuals.

While Nishina did not go farther than just explaining Bohr's ideas at this point, Tanabe, who was quick to respond to Bohr's visit, freely applied the idea of complementarity to problems of his own concern. Just before Bohr's expected arrival, Tanabe had already written an article on complementarity entitled 'The Worldview and World Picture' (*Sekaikan to sekaizō*). Without mentioning Bohr, he relied on Jordan's interpretation of complementarity and his application of this concept to the problem of free will (Jordan, 1936, chapter 5). Here Tanabe translated 'complementarity', as '*haitateki sōgo hosokusei*', which meant 'exclusive mutual complementarity', and discussed his favourite philosophical topic, inseparability of subject and object (Tanabe, 1937b, p. 184).

Tanabe's direct response to Bohr's visit and talks in Japan appeared in his paper entitled 'Philosophical significance of quantum theory' published in July (Tanabe, 1937a). This article consisted of disorganized fragments of thought, attempting to discuss the ideas of the correspondence principle and of complementarity, relying on Bohr's *Atomic Theory and the Description of Nature* (1934) and his lecture in Kyoto, which Tanabe attended, as the main sources.[22] Tanabe introduced unfamiliar terms that he coined. One of them was 'sōhosei', which became the standard translation of 'complementarity'. He also used concepts from Buddhist philosophy, with which he had become increasingly familiar in this period. One of them was the notion of sōsoku (inseparability), a concept originating from Avatamasaka Sutra (Flower Ornament Scripture), one of the most important Buddhist scriptures that describes various stages toward the enlightenment. This article is probably the best example of how Japanese intellectuals resorted to Japanese or Asian intellectual resources in order to understand conceptual aspects of quantum mechanics. He translated the relation between object and subject in quantum mechanics into the inseparability of matter and mind. Yet, most of the philosophical framework of this article relied on Hegelian dialectics, within which he tried to grasp complementarity and the correspondence principle. Indeed, he claims, '[m]utually exclusive complementarity is, indeed, nothing but the unification of

---

[21] Nishina Yoshio to Niels Bohr, 28 August 1937, pp. 616–18 in (R. Nakane *et al.*, 2006).

[22] Uchida Yōichi, a physicist in Kyoto, witnessed Tanabe at Bohr's lecture for physicists on 11 May in the physics library (Uchida, 1990).

dialectical opposites' (Tanabe, 1937a). Either way, Tanabe found complementarity difficult to understand. He described Bohr's argumentation as 'concise but deep, and hard to understand'. Tanabe confessed that, although he thought he had some idea of what Bohr meant from attending his recent lecture, his understanding was still insufficient to grasp Bohr's 'profound thought' (Tanabe, 1937a).

Tanabe was not the only one who tried to understand complementarity within the framework of dialectics. In fact, it was probably the easiest way of reducing complementarity to a familiar philosophical theory. Saigusa Hiroto, a philosopher and historian of science, for example, stated that there was no other way to find the problem of complementarity in philosophy than to look for it in Hegel's 'unmatched' theory (Saegusa, 1938). Others resorted to Kantian philosophy. Amano Kiyoshi had started his research on the history of the interpretation of quantum mechanics slightly earlier (1936). In his book on the history of quantum mechanics, published posthumously (1948), he explained complementarity in a Kantian framework.[23]

The welcoming mood to Bohr did not make all Japanese intellectuals receptive to complementarity. Taketani remained adamantly critical, arguing that complementarity was not a physical or a philosophical concept, and that it was a baseless, and merely phenomenological word, which Bohr used in order to concoct his own interpretation of quantum mechanics (Tani, 1937, p. 39). According to Taketani, Bohr's argument inevitably fell into the debris of classical theory and phenomenological theories, because he did not have a correct methodology. Taketani concluded, 'Seeing this, we keenly realize how powerful the dialectic of nature is', suggesting that the 'correct methodology' was natural dialectic and his own three-stage theory, which he considered as being derived from the former. 'Phenomenological' is only the initial stage of the development of scientific theory according to his three-stage theory (Taketani, 1971b).

Not all objections along this line were motivated by ideology. A physicist, Tomiyama Kotarō, criticized complementarity, arguing that it was unnecessary to retain classical concepts such as position and momentum. In 1939, he published a paper, 'Classical elements in quantum theory' (Ryōshiron ni okeru kotentekinaru mono), which became the first of his critical responses to Bohr's interpretation of quantum mechanics. Tomiyama believed that a new physics should have its own suitable language, so that its new contents could be most adequately expressed. Since quantum mechanics could be expressed in the simplest way by wave functions and operators, he claimed that 'wave functions' and 'operators', not classical concepts, were the building blocks for an adequate language of quantum mechanics (Tomiyama, 1939). Tomiyama's criticism was made from the viewpoint of a physicist, who, having been trained to deal in the highly formalized theory of quantum mechanics, adopted Max Born's probabilistic interpretation of wave functions and no longer had to worry about its philosophical

---

[23] See also Takata's introduction to Amano (1973). Along with Tomonaga's (1948) pseudo-historical treatment of quantum mechanics, Amano's book is the first book-length work on the history of quantum mechanics in Japanese.

problems. If one learned quantum mechanics through mathematical formalism, the elements in the formalism—wave functions, operators, and commutation relations— appeared as real as position and momentum in classical physics. It seems probable that many physicists shared Tomiyama's view but did not bother to express it.

There are exceptions. Yukawa seems to have remained interested enough in foundational questions of quantum mechanics to write a series of review articles on the theory of observation after the war (Yukawa, 1947; Yukawa, 1948a; Yukawa, 1948b). As mentioned in the last instalment of Yukawa's reviews, probably more important in this generation is Watanabe Satosi, who attempted to synthesize quantum mechanics and thermodynamics in the 1930s. He continued working on foundational questions of quantum mechanics from this perspective and could be considered a precursor of the information theoretical understanding of quantum mechanics (Watanabe, 1935; Watanabe, 1939; Watanabe, 1948; Watanabe, 1969; Toyota, 1997; Fukuda, 2015).

# 27.8 CONCLUSION: SCIENTIFIC CULTURES AND PHILOSOPHY OF QUANTUM MECHANICS

When Japanese physicists and other intellectuals encountered conceptual issues in quantum mechanics, they had various difficulties and troubles similar to those their European colleagues had. Even Nishina initially had difficulty in following some of Bohr's arguments of complementarity. This is not at all surprising. The intellectual tools that Japanese physicists and other scholars had were nurtured by an education and culture where European learning dominated. Moreover, as in the case of Tanabe, Asian intellectual resources, such as Avatamasaka Sutra, did not prove to be particularly helpful to understand conceptual issues of quantum mechanics.

Still, the place and the context mattered. The rise of science journalism and the education of the Japanese elite created a space for discussion of philosophical issues of science. Marxism and the Kyoto School philosophy, although originally imported from Europe, developed in peculiar ways in Japan and created a unique intellectual environment there. In addition, the timing mattered. External and contingent events, such as EPR or Bohr's visit to Japan, gave additional stimuli.

From the beginning, however, many younger physicists preferred to understand quantum mechanics in terms of its mathematical and physical results and focused on research, ignoring philosophical and conceptual implications. This tendency became increasingly solidified in the late 1930s and early 1940s, as a younger generation of physicists were more engaged in competing in the forefront of physics. As mentioned, among the Kyoto School philosophers, Nishida Kitarō himself published articles on quantum mechanics before the end of World War II. After the war, some physicists published on the conceptual issues of quantum mechanics, and Amano's posthumously published book on the history of quantum mechanics (Amano 1948, 1973) also included foundational discussions. These publications, however, did not change the general trend. During the second half of the 20th century, when much more eyecatching scientific events such as Japanese Nobel laureates, nuclear power, Antarctic expeditions, space exploration, and the establishment of high energy physics and other big science laboratories happened, foundational issues of quantum mechanics never received as much attention as complementarity did in 1937 from scientific journalism in Japan, although there were some conspicuous studies in this area, such as the Machida–Namiki theory (Machida and Namiki, 1980a; Machida and Namiki, 1980b).

## References

Amano, K. (1936). Ryōshiron kaishaku no hensen to sono bunken (1) (History of the interpretation of quantum mechanics and its literature). *Nippon Sūgaku Butsuri Gakkwai shi*, 10(6), 445–55.

Amano, K. (1948). *Ryōshirikigakushi (History of quantum mechanics)*. Tokyo: Nihon Kagakusha.

Amano, K. (1973). *Ryōshirikigakushi (History of quantum mechanics)*. Tokyo: Chūōkōronsha.

Bohr, N. (1929). Wirkungsquantum und Naturbeschreibung. *Naturwissenschaften*, 17, 483–86.

Bohr, N. (1934). *Atomic Theory and the Description of Nature*. Cambridge: Cambridge University Press.

Bohr, N. (1935). Can quantum-mechanical description of physical reality be considered complete? *Physical Review*, 48, 696–702.

Bohr, N. (1937a). Genshi (Atom). (Y. Nishina, Trans.), *Kagaku chishiki*, 17, 814–21.

Bohr, N. (1937b). Genshiron ni okeru ingaritsu: Butsuriteki jitsuzai no ryōshirikigakuteki kijutsu wa kanzen to kangae rareruka? (Causality in atomic theory: Can quantum mechanical description of reality be considered complete?) (M. Taketani, Trans.), *Tetsugaku kenkyū*, 22(257), 807–19.

Bohr, N. (1985). The quantum postulate and the recent development of atomic theory [1]: Unpublished manuscript dated 12–13 October 1927. In J. Kalckar (ed.), *Niels Bohr Collected Works, Vol. 6*, Amsterdam: North-Holland, pp. 89–98.

Bohr, N., and Rosenfeld, L. (1933). Zur Frage der Messbarkeit der elektromagnetischen Feldgrössen. *Det Kongelige Danske Videnskabernes Selskabs, Mathematisk-Fysiske Meddelser*, 12(8), 1–65.

Born, M., and Jordan, P. (1930). *Elementare Quantenmechanik*. Berlin: Springer.

Carter, R. E. (2013). *The Kyoto School: An Introduction*. Albany: State University of New York Press.

Darrigol, O. (1991). Cohérence et complétude de la mécanique quantique: l'exemple de 'Bohr–Rosenfeld'. *Revue d'histoire des sciences*, 44(2), 137–79.

Dirac, P. A. M. (1930). *The Principles of Quantum Mechanics*. Oxford: Clarendon Press.

Eddington, A. (1932a). Decline of determinism. *Nature*, 129, 233–40.

Eddington, A. (1932b). Ketteiron no chōraku (Decline of determinism). (J. Ishiwara, Trans.), *Kagaku*, 3, 231–234, 276–80.

Einstein, A., Podolsky, B., and Rosen, N. (1935). Can quantum mechanical description of the reality be considered complete? *Physical Review*, 47, 777–90.

Einstein, A., Podolsky, B., and Rosen, N. (1937). Ryōshirikigaku ni okeru kansoku ni tsuite (On observation in quantum mechanics). (M. Taketani, Trans.), *Tetsugaku kenkyū*, 22(251), 168–88.

Engels, F. (1929). *Shizen benshōhō (Dialectics of nature)*. (T. Katō and Y. Kako, Trans.). Tokyo: Iwanami Shoten.

Frenkel, I. (1929). *Einführung in die Wellenmechanik*. Berlin: Springer.

Fujioka, Y. (1936). Ryōshirikigaku nitaisuru Einstein-ra no gigi to koreni taisuru Bohr no kaisetsu (Questions to quantum mechanics by Einstein and others and Bohr's explanation). *Kagaku*, 6(2), 73–76.

Fujioka, Y. (1937). Bōa kyōju no kōen (Professor Bohr's lecture). *Kagaku*, 7, 278–91, 323–27.

Fukuda, K. (2015). On the Significance of Quasi-Probability in Quantum Mechanics. Ph.D. dissertation, SOKENDAI.

Gō, H. (1933). Shin butsurigaku to shizen benshōhō: Tanabe hakase no dōdai ronbun no hihan (New physics and dialectics of nature: Criticism of Dr. Tanabe's paper with the same title). *Yuibutsuron kenkyū*, 12, 33–58.

Hayashi, T. (1938). Seirigaku no sōhosei (Complementarity in physiology). *Kagaku chishiki*, 18(11), 18–21.

Heisenberg, W. (1930). *Die physikalischen Prinzipien der Quantentheorie*. Leipzig: S. Hirzel.

Hoda, S. (1932). Ingaritsu ni kansuru ichi kōsatsu (A discussion on causality). *Nihon Gakujutsu Kyōkai hōkoku*, 7(1), 7–11.

Holton, G. (1973). Roots of Complementarity. In *Thematic Origins of Scientific Thought: Kepler to Einstein* (revised edn), Cambridge, MA: Harvard University Press, pp. 99–146.

International Congress of the History of Science and Technology (ed.) (1931). *Science at the Cross Roads: Papers Presented to the International Congress of the History of Science and Technology Held in London from June 20th to July 3rd, 1931*. London: Kniga.

Ishiwara, J. (1915). Die universelle Bedeutung des Wirkungsquantums. *Proceedings of Tokyo Mathematico-Physical Society*, 8, 106–16.

Ishiwara, J. (1927). Ryōshiron no honshitsu (The essence of quantum theory). *Taiyō*, 3(11), 69–86.

Ishiwara, J. (1931). Fukakuteisei genri nitsuite (On the uncertainty principle). *Kagaku*, 1, 313–15.

Ishiwara, J. (1932). Ingaritsu no mondai (The question of causality). *Kagaku*, 3, 231–34.

Ishiwara, J. (1933). Haizenberuku no fukakuteisei kankei (Heisenberg's uncertainty relations). In *Kagaku tokubetsu daimoku*, Tokyo: Iwanami Shoten, pp. 631–46.

Ishiwara, J. (1935a). Gendai kagakukai no saikō ronsō: Butsuriteki jitsuzai no honshitsu (Top debate of modern science: Essence of physical reality). *Kagaku chishiki*, 15, 559–61.

Ishiwara, J. (1935b). Gūzenron to shizen kagaku (Probability theory and natural science). *Nihon hyōron*, 10(10), 42–54.

Itagaki, T. (1938). Geijutsugaku no ichi tokushitsu: sōhosei ni kanrenshite (A characteristic of art studies: In relation to complementarity). *Kagaku chishiki*, 18(11), 32–35.

Itakura, K., Kimura, T., and Yagi, E. (1973). *Nagaoka Hantarō den (A biography of Nagaoka Hantarō)*. Tokyo: Asahi Shimbunsha.

Ito, K. (2002). *Making Sense of Ryoshiron (Quantum Theory): Introduction of Quantum Mechanics into Japan, 1920–1940*. Cambridge, MA: Harvard University Press.

Ito, K. (2005). The *Geist* in the Institute: Production of Quantum Theorists in Prewar Japan. In D. Kaiser (ed.), *Pedagogy and the Practice of Science: Historical and Contemporary Perspectives*, Cambridge, MA: MIT Press, pp. 151–84.

Ito, K. (2008). Kim Dong-Won, Yoshio Nishina: Father of Modern Physics in Japan (review). *Physics Today*, **61**(10), 57–58. https://doi.org/10.1063/1.3001870

Ito, K. (2013). Superposing Dynamos and Electrons: Electrical Engineering and Quantum Physics in the Case of Nishina Yoshio. In S. Katzir, C. Lehner, and J. Renn (eds), *Traditions and Transformations in the History of Quantum Physics*, Berlin: Edition Open Access, pp. 183–208.

Ito, K. (2017). Cultural difference and sameness: Historiographic reflections on histories of modern physics in Japan. In K. Chemla and E. F. Keller (eds), *Cultures without Culturalism: The Making of Scientific Knowledge*, Durham, NC: Duke University Press, pp. 49–68.

Ito, K. (2018a). 'Electron Theory' and the Emergence of Atomic Physics in Japan. *Science in Context*, **31**(3), 293–320.

Ito, K. (2018b). Takeuchi Tokio to jinkōhōshasei shokuen jiken: 1940nendai hajime no kagaku sukyandaru (Takeuchi Tokio's 'Artificial Radium': A Scientific Scandal in Early 1940s Japan). *Kagakushi kenkyū*, **57**(288), 266–83.

Ito, K. (2019). Takuchi Tokio to Ainshutain (Takeuchi Tokio and Einstein). *Gendai shisō*, 192–206.

Jaeger, G. (2017). Quantum potentiality revisited. *Philosophical Transactions of the Royal Society A: Mathematical, Physical and Engineering Sciences*, **375**(2106), 20160390.

Jansen, M. B. (1989). Einstein in Japan. *The Princeton University Library Chronicle*, **50**(2), 145–54.

Jeans, J. H. (1930). *The Mysterious Universe*. Cambridge: Cambridge University Press.

Jeans, J. H. (1932). *Shin butsurigaku no uchūzō (View of the universe in new physics)*. (K. Yamamura, Trans.). Tokyo: Kōseisha.

Jeans, J. H. (1938). *Shimpi na uchū (Mysterious universe)*. (T. Suzuki, Trans.). Tokyo: Iwanami Shoten.

Jordan, P. (1927). Philosophical foundations of quantum theory. *Nature*, **119**, 566–69.

Jordan, P. (1932a). Quantenmechanik und Grundprobleme der Biologie und Psychologie. *Die Naturwissenschaften*, **20**, 815–21.

Jordan, P. (1932b). Ryōshirikigaku to seibutsugaku oyobi shinrigaku no konpon mondai (Quantum mechanics and fundamental problems of biology and psychology). (J. Ishiwara, Trans.) *Kagaku*, **3**, 9–11.

Jordan, P. (1936). *Anschauliche Quantentheorie: Eine Einführung in die moderne Auffassung der Quantenerscheinungen*. Berlin: Springer.

Kagawa, T. (1938). Shingaku no sōhosei (Complementarity in theology). *Kagaku chishiki*, **18**(11), 28–31.

Kalckar, J. (1985). Introduction. In *Niels Bohr Collected Works, Vol. 6*. Amsterdam: North-Holland, pp. 7–53.

Kaneko, T. (1981). *Ainshutain shokku (Einstein shock), 2 vols*. Tokyo: Kawade Shobō Shinsha.

Kaneko, T. (1987). Einstein's impact on Japanese intellectuals. In T. F. Glick (ed.), *The Comparative Reception of Relativity*, Dordrecht: Springer, pp. 351–79.

Katz, E. (1986). *Niels Bohr: Philosopher-Physicist*. Ph.D. dissertation, New York University.

Kim, D.-W. (2007). *Yoshio Nishina: Father of Modern Physics in Japan*. New York: Taylor & Francis.

Klein, O., and Nishina, T. (1929). Über die Streuung von Strahlung durch freie Elektronen nach der neuen relativistischen Quantendynamik von Dirac. *Zeitschrift für Physik*, **52**(11–12), 853–68.

Kothari, D. S. (1985). The complementarity principle and Eastern philosophy. In A. P. French and P. J. Kennedy (eds), *Niels Bohr: A Centenary Volume*, Cambridge, MA: Harvard University Press, pp. 325–31.

Kožnjak, B. (2020). Aristotle and Quantum Mechanics: Potentiality and Actuality, Spontaneous Events and Final Causes. *Journal for General Philosophy of Science*, 51(2). doi: 10.1007/s10838-020-09500-y.

Kurihara, K. (1927). Saikin no butsurigaku to ingaritsu (Recent physics and causality). *Tōyō gakugei zasshi*, 43(533), 501–506.

Lenin, V. I. (1930). *Yiubutsuron to keikenhihanron (Materialism and empirico-criticism)*. (F. Sano, Trans.). Tokyo: Iwanami Shoten.

Machida, S. and Namiki, M. (1980a). Theory of measurement in quantum mechanics: Mechanism of reduction of wave packet. I. *Progress of Theoretical Physics*, 63(5), 1457–73.

Machida, S and Namiki, M. (1980b). Theory of measurement in quantum mechanics: Mechanism of reduction of wave packet. II. *Progress of Theoretical Physics*, 63(6), 1833–47.

Nagai, H. (1991). Tanabe benshōhō to shizenkagaku (Tanabe dialectics and natural sciences). In Y. Takeuchi, K. Mutō, and K. Tsujimura (eds), *Tanabe Hajime: Shisō to kaisō*, Tokyo: Chikuma Shobō, pp. 18–47.

Nagaoka, H. (1911). Kuwanten kasetsu (quantum theory) nitsuite (On quantum theory). *Tetsugaku zasshi*, 26, 885–93.

Nakane, M. (2019). Yoshikatsu Sugiura's Contribution to the Development of Quantum Physics in Japan. *Berichte zur Wissenschaftsgeschichte*, 42(4), 338–56.

Nakane, R., Nishina, Y., Nishina, K., Yasaki, Y., and Ezawa, H. (eds) (2006). *Nishina Yoshio ōfuku shokanshu (Nishina Yoshio collected correspondence)* (Vol. 3). Tokyo: Misuzu Shobo.

Nishida, K. (1943a). Chishiki no kyakkansei ni tsuite (On objectivity of knowledge) (1). *Shisō*, 248, 1–43.

Nishida, K. (1943b). Chishiki no kyakkansei ni tsuite (On objectivity of knowledge) (2). *Shisō*, 249, 1–50.

Nishida, K. (1990). *An Inquiry into the Good*. (M. Abe and C. Ives, trans). New Haven: Yale University Press.

Nishina, Y. (1929). Ryōshiron to ingasei ni tsuite (Quantum theory and causality). *Denki-gakkai zasshi*, 50, 133–45.

Nishina, Y. (1931a). Hikari to busshitsu no sōji to sōi (Similarity and difference between light and matter) (1). *Rika kyōiku*, April, 18–26.

Nishina, Y. (1931b). Hikari to busshitsu no sōji to sōi (Similarity and difference between light and matter) (2). *Rika kyōiku*, May, 22–31.

Nishina, Y. (1935). Ryōshiron ni okeru kyakkan to ingaritsu (Object and causality in quantum theory). *Shisō*, 161, 57–80.

Nishina, Y. (1938). Sōhoseiriron (Theory of complementarity). *Kagaku chishiki*, 18(11), 14–17.

Nishina, Y. (1984). *Y. Nishina's Correspondence with N. Bohr and Copenhageners*. Tokyo: Nishina Kinen Zaidan.

Nishina, Y., Ishiwara, J., Uramoto, M., Kumabe, K., and Yamamoto, T. (1937). Kagaku oyobi kagakubunmei o kataru (Discussion on science and scientific civilization). *Nihon hyōron*, May, 191–223.

Nisio, S. (2011). *Kagaku jānarizumu no senkusha: Hyōden Ishiwara Jun (A pioneer of science journalism: A biography of Ishiwara Jun)*. Tokyo: Iwanami Shoten.

Ochiai, K. (1928). Shin genshi rikigaku (New atomic physics). *Rika kyōiku*, June, 19–25.

Ōhashi, R. (ed.) (2004). *Kyoto gakuha no shisō: Shuju no zō to shisō no potensharu (Thoughts of the Kyoto School: Various perspectives and potentials of its ideas)*. Kyoto: Jinmonshoin.

Q. L. Z. (1937). Nīrusu Bōa Kyōju no kōen. *Tokyo Butsurigakkō zasshi*, 46, 301–308.

Puroretaria Kagaku Kenkyūjo (ed.) (1932). *Shinkō shizenkagaku ronshū (Collected papers on new natural sciences)*. Tokyo: Sangyō rōdō chōshasho.

Rabi, I. I., and Nishina, Y. (1928). Der wahre Absorptionskoeffizient der Röntgenstrahlen nach der Quantentheorie. *Verhandlungen der Deutschen Physikalischen Gesellschaft*, 9, 6–9.

Roden, D. T. (1980). *Schooldays in Imperial Japan: A Study in the Culture of a Student Elite*. Berkeley, CA: University of California Press.

Rosenfeld, L. (1960). Niels Bohr's contribution to epistemology. *Physics Today*, 16, 47–54.

Rosenkranz, Z. (ed.) (2018). *The Travel Diaries of Albert Einstein: The Far East, Palestine & Spain*. Princeton, NJ: Princeton University Press.

Saegusa, H. (1938). Tetsugaku no sōhosei (Complementarity in philosophy). *Kagaku chishiki*, 18(11), 22–26.

Sakai, T. (1930). *Ryōshiron (Quantum theory)*. Tokyo: Iwanami Shoten.

Sommerfeld, A. (1929). *Atombau und Spektrallinien, wellenmechanischer Erganzungsband*. Braunschweig: Friedrich Vieweg und Sohn.

Sugai, J. (1934). Ingaritsu ni kansuru sho mondai (Problems concerning causality). *Risō*, 46, 68–79.

Sugiura, Y. (1927). Über die Eigenschaften des Wasserstoffmoleküls im Grundzustande. *Zeitschrift für Physik*, 45(7–8), 484–92.

Takata, S. (1996). Kagakuzasshi no senzen to sengo (Science magazines in Japan: Before and After the World War). *Butsuri*, 51(3), 189–93.

Taketani, M. (1968). Hashigaki (Introduction). In *Benshōhō no shomondai (Problems of dialectics)* (Reprint), Tokyo: Keisō Shobō, pp. 3–8.

Taketani, M. (1971a). Dialectics of nature: On quantum mechanics. *Progress of Theoretical Physics, Supplement*, 50, 27–36.

Taketani, M. (1971b). On formation of the Newton [sic] mechanics. *Progress of Theoretical Physics, Supplement*, 50, 53–64.

Takeuchi, T. (1932). Arashi no nakanaru shin butsurigaku (New physics in a storm). *Kagaku chishiki*, 12, 54–56.

Takeuchi, T. (1933). Shizenkagaku no hijōji (A crisis in natural science). *Kaizō*, 15(1), 246–50.

Takeuchi, T. (1937). Bōa Hakase no kōen (Dr. Bohr's lecture). *Tōyō gakugei zasshi*, 46, 301–308.

Tamaki, H. (1991). Nishina Yoshio no yōroppa ryūgaku kōhan (1926–1928) no riron kenkyū ni kansuru shiryō (Historical materials concerning Nishina Yoshio's theoretical studies during the second half of his stay in Europe (1926–1928)). *Butsurigaku shi nōto*, 1, 18–23.

Tanabe, H. (1915). *Saikin no shizen kagaku (Recent natural sciences)*. Tokyo: Iwanami Shoten.

Tanabe, H. (1922). *Kagaku gairon (Introduction to science)*. Tokyo: Iwanami Shoten.

Tanabe, H. (1933). Shin butsurigaku to shizen benshōhō (New physics and dialectics of nature). *Kaizō*, 15(4), 80–97.

Tanabe, H. (1937a). Ryōshiron no tetsugakuteki imi (Philosophical implications of quantum theory). *Shisō*, 181, 1–29.

Tanabe, H. (1937b). Sekaikan to sekaizō (worldview and world picture). *Kagaku*, 7, 181–86.

Tani, K. (1935). Butsurigakkai no wadai (Topics in physics). *Sekai bunka*, 1(11), 24–28.

Tani, K. (1936a). Bōa Ainshutain o oshiu (Bohr teaches Einstein). *Sekai bunka*, 2(15), 53–54.

Tani, K. (1936b). Shizen no benshōhō: Ryōshirikigaku nitsuite (Dialectics of nature: On quantum mechanics). *Sekai bunka*, **2**(15), 2–11.

Tani, K. (1937). Nīrusu Bōa kyōju no shō gyōseki ni tsuite (Professor Niels Bohr's achivements). *Sekai bunka*, **3**(29), 38–40.

Tomiyama, K. (1939). Ryōshiron ni okeru kotenteki naru mono (Classical elements in quantum theory). *Kagaku*, **9**, 156–8.

Tomonaga, S. (1948). *Ryōshirikigaku 1 (Quantum mechanics 1)*. Tokyo: Tōzai Shuppansha.

Tomonaga, S., Yamazaki, F., Takeuchi, M., Sakata, S., Nakayama, H., and Tamaki, H. (1982). Zadan Nishina sensei o shinonde (Roundtable: Memories of Nishina-senesi). In *Tomonaga Sin-itiro chosakushū, Vol. 6*, pp. 57–93.

Tosaka, J. (1935). *Kagakuron (Theory of science)*. Tokyo: Mikasa Shobō.

Toyota, T. (1997). Jōhōriron tanjō no rekishi: Watanabe Satosi senesi no gyōseki (ryōshi kakuritsuron to entoropī kaiseki). (The history of the birth of information theory: Achivements of Professor Watanabe Satosi (theory of quantum probability and entropy analysis)). *Sūrikaiseki Kenkyūjo kōkyūroku*, **1013**, 112–16.

Uchida, Y. (1990). Kyōto no jitsujō (The situation in Kyoto). *Butsuri*, **44**, 760.

Watanabe, S. (1935). *Le deuxième théorème de la thermodynamique et la mécanique ondulatoire*. Paris: Herman.

Watanabe, S. (1939). Über die Anwendung thermodynamischer Begriffe auf den Normalzustand des Atomkerns. *Zeitschrift für Physik*, **113**, 482–513.

Watanabe, S. (1948). *Jikan (Time)*. Tokyo: Hakujitsushoin.

Watanabe, S. (1969). *Knowing and Guessing: A Quantitative Study of Inference and Information*. New York: John Wiley & Sons.

Weyl, H. (1928). *Gruppentheorie und Quantenmechanik*. Leipzig: S. Hirzel.

Yazaki, Y. (2017). How the Klein–Nishina formula was derived: Based on the Sangokan Nishina Source Materials. *Proceedings of the Japan Academy, Series B*, **93**(6), 399–421.

Yukawa, H. (1947). Kansoku no riron I. (Theory of observation I.). *Shizen*, **2**(11), 11–15.

Yukawa, H. (1948a). Kansoku no riron II. (Theory of observation II.). *Shizen*, **3**(3), 31–39.

Yukawa, H. (1948b). Kansoku no riron III. (Theory of observation III.). *Shizen*, **3**(7), 34–42.

# CHAPTER 28

# FORM AND MEANING

## Textbooks, Pedagogy, and the Canonical Genres of Quantum Mechanics

### JOSEP SIMON

'One can find a "correct" meaning in textbooks, or in some philosophical writings on the quantum theory—in short, in the graveyards of science. On the research frontier nothing is immune to reappraisal...'

(Beller, 1999, p. 6)

A decade ago, the editors of a *Compendium of Quantum Mechanics* stressed the need for such a work, and its ability to highlight connections between physics, philosophy, and history for the benefit of the study of the atom (Greenberger *et al.*, 2009, p. v). It was arranged as a series of alphabetically ordered entries, just like the *Oxford Companion to the History of Modern Science.* Years later, the *Oxford Handbook of the History of Physics* introduced a compendium of longer essays and a thematic arrangement, but recognized yet again the same goals: presenting the complexity of a grown-up and rich discipline in a compact form (Heilbron, 2003; Buchwald and Fox, 2013, p. 1).

Compendia, handbooks, and companions are types of academic publication per se, which share many features with another genre: the *textbook*. A major common feature is that of being introductory reference works; another is their fundamental aim of synthesizing a whole field of knowledge in a single volume. All four can be used both in teaching and research, although the term 'textbook' has evolved more clearly than others to mean a work explicitly conceived and used for teaching purposes. All of them are challenging forms of knowledge production whether created by single or collective authorship (Simon, 2013 and 2016).

To follow a rigorous and well-established method of scholarship, when planning a chapter on the history of quantum physics textbooks, it would be wise to read the major human output on this topic, in the genre of companions, encyclopedias, articles, handbooks, treatises and monographs, research reports, laboratory notebooks, conference proceedings, grant proposals, correspondence collections, source inventories, manuscript and printed lectures, textbooks, and compendia. But, where to start? Experience advises us to check first the last reference work in the relevant field (Heilbron, 2016). Since the history of quantum physics textbooks is as much the history of quantum physics (textbooks) as the history of (quantum) physics textbooks, this is arguably the *Oxford Handbook of the History of Physics* (2013). In that loyal companion to the historian of physics, one can find a chapter on physics textbooks that focuses a great deal on the 19th-century rise of the physics textbook genre and proposes a historiographical agenda, but deals only in passing with quantum physics (Simon, 2013). There is also another chapter that provides an overview and periodization of the making of quantum mechanics in the early 20th century, and gives some mention to textbooks (Seth, 2013).

Seth's big picture of quantum mechanics is supported by primary sources such as research papers, lecture series arranged into books, correspondence, collected works, and society and academy proceedings. He also introduces three textbooks that are crucial for his account: Arnold Sommerfeld's *Atombau und Spektrallinien* (Friedrich Vieweg & Sohn, 1919, 1st ed.), as a reference for the early quantum theory; P. A. M. Dirac's *The Principles of Quantum Mechanics* (Oxford University Press, 1930) and Werner Heisenberg's *The Physical Principles of the Quantum Theory* (University of Chicago Press, 1930), as the works marking the consolidation of quantum mechanics within physics. The latter came to substitute Sommerfeld's *Atombau* as canonical works and represented two different strands within the new quantum mechanics, one more mathematical, the other more philosophical. Later on, Seth clarifies that Heisenberg's book was based on lectures given in Chicago the year before its publication, and that Dirac's book arose from lectures at Cambridge from 1927 on. Seth implies but never expresses quite explicitly that books such as those were of special interest to college students and professors, and that these lectures took place in universities. This absence of reflection on the relevance of form and audience applies not only to textbooks, but to most sources used by Seth in his chapter.

Seth's proceeding is not an exception but a general pattern in most of the literature. The history of quantum physics is one of the pioneering fields within the history of science in the use of sources such as lectures, correspondence, oral interviews, and teaching and research notebooks. Its impulse has been to follow in minute detail the evolving thoughts of physicists in a field in formation, to establish priorities on the throne of modern physics, and to clarify the concepts of contemporary physics through research on its foundations. Paradoxically, methodological reflection on the use of such sources is scarce, and the meaning of quantum mechanics has been elucidated independently (and in certain aspects, in spite) of the form of its historical sources.[1]

---

[1] There are of course exceptions to this trend, such as the innovative focus of Beller (1999) on dialogue as a tool of knowledge creation in analysing correspondence. This is noted by recent historiographical reviews such as Badino (2016). In spite of this, the field could benefit from more reflexive and broad-minded pictures, especially on the historiographical and methodological side. Reflection on

In delineating the rationale of the Sources for History of Quantum Physics (SHQP) project, John Heilbron emphasized that: '[letters and manuscripts] often permit the historian to follow the development of ideas, techniques and research projects from week to week, sometimes even from day to day. [...] When supplemented by the interviews, they give a lively impression of the nature and practice of physics two and three generations ago'. In the same guise, he characterized the interest in successive editions of textbooks such as *Atombau* (4 editions, 1919–1924): 'The individual volumes are still shots, so to speak, of a given moment in quantum physics; when compared they afford an animated picture of its development' (Heilbron, 1968, pp. 91, 100). The impressive historical potential of such sources, and their contemporary significance, led Heilbron to encourage historians of science to devote their research to this field. As he predicted, the field has grown considerably in the last half century and has made good use of the richness of sources he contributed to gather.

Seth's *Crafting the Quantum* (which follows some of Heilbron's recommendations) provides a detailed picture of the scholarly life of Sommerfeld and his research school, and emphasizes the interaction and intersection of research and pedagogy in the making of quantum mechanics. In passing, it also mentions that the first edition of Sommerfeld's *Atombau* emerged from a lecture given between 1916 and 1917 to university chemists and medical doctors and that its author defined it sometime as 'a popular book about atomic models'. With this readership in mind, through successive editions the complexity of mathematical language was kept out as much as possible, and deferred to a closing appendix (Seth, 2010 and 2008).

What strikes the historian and philosopher alike is how such contextual knowledge of the craft, form, and status of a source is not brought to the fore for a more thorough discussion, since the conclusions derived from the analysis of such a source would—for obvious methodological reasons—depend on this too. What were Sommerfeld's motivations to write such a textbook, how did he write it, what decisions did he take both about its contents and form, how did the printed form of his physics thought-style relate to his lecturing practice, what was his intended vs actual readership, what did the diversity of its readers think about the book and what were the various uses they gave to it, how was it used in teaching and in research, how did readers' reactions contribute to reshape successive editions of the book? Too many questions. They apply in fact to any quantum physics textbook. They bring together the tension between form and meaning inherent to the craft of making physics into the history of physics.

In this chapter, I call attention to the methodological importance of problematizing sources when writing the history of science, to the place of textbooks and pedagogy in the historiography of quantum physics, and to the promiscuous relationships between different genres of scientific literature shaping the interaction of research, teaching, and

sources, historical goals, and engagement with other areas of the history of physics and of science could be particularly helpful. For further evidence, compare Badino (2016) (diachronically) with Heilbron (1968) and (synchronically) with Chang (2016) or Bensaude-Vincent and Simon (2008).

popularization. I end up by discussing what allows us to qualify certain quantum physics textbooks as classic or canonical works, and suggesting future avenues for research.

## 28.1 HISTORICAL STRATA OF OUR RECENT TEXTBOOK PAST

When in 1961 Thomas Kuhn and his collaborators launched the SHQP project, their preparatory research included a survey of general and quantum physics textbooks (by authors such as Chwolson, Reiche, Van Vleck, and Sommerfeld), and handbooks such as the one edited by Geiger. The project had a major focus on interviews, but also retrieved personal and institutional papers, correspondence, memoirs, photographs, and, last but not least, lecture notes by Bohr, Born, Hilbert, Debye, Langevin, Rutherford, Hevesy and J. J. Thomson, Fowler, and Larmor. In parallel, Kuhn produced his canonical work *The Structure of Scientific Revolutions*, which promoted the salience in history and philosophy of science of research concepts such as *structure*, *pedagogy*, and *textbooks* (Kuhn *et al.*, 1967; Kuhn, 1962; Simon, 2013).

In the previous section, we saw that Heilbron held a rather appreciative (and Kuhnian) view on textbooks, as sources for historical research. It clearly contrasts with Beller's, presented at the head of this chapter. Beller is known for her innovative approach to some of the now traditional sources for quantum physics history. While her major aim was to show how the orthodox interpretation of quantum mechanics emerged in the late 1920s, she brought a new hermeneutic to correspondence, by positing *dialogue* as a driving process in the making of quantum mechanics (Beller, 1999, pp. 1–14; Badino, 2016, p. 332). She did not apply analogous interpretative sophistication to textbooks. For her, if research-frontier science is the matter of life (thus changing constantly), textbooks are dead (for their immobility and immutability). Beller did not count—as opposed to Heilbron—on the ontology of multi-edition textbooks, nor considered that geological metaphors seeing textbooks as historical strata can be as pertinent as biological metaphors depicting scientific research as a living organism.[2]

A good methodological contrast that helps us to value the historical sophistication of the SHQP project arises when comparing it to some contemporary efforts. For instance, the source compilation coordinated by mathematician and historian of science Bartel Leendert van der Waerden (a colleague of Heisenberg in 1930s Leipzig), with advice from several quantum physicists (and an original idea by Max Born). This was a compilation of quantum mechanics historical papers (preceded by short summaries), which was a useful tool (in the 1960s), but providing just one type of source, a

---

[2] In fact, Beller's perspective is quite common among physicists and some historians of physics. See e.g., Schwinger (1973), p. 414, and Kragh (1999), p. 10.

rough taxonomy, and a parsimonious justification of its selection. The same applies to Dirk ter Haar's compilation of papers in *The Old Quantum Theory*, and the personal account of Friedrich Hund (assistant to Born in the 1920s). All were volumes published in 1967 (van der Waerden, 1967; ter Haar, 1967; Hund, 1967; Weizsäcker *et al.*, 1963). Heilbron qualified them as 'a sequence of often misleading epitomes of the important papers [ . . . ] and intended for advanced students of physics'. Kuhn, however, found van der Waerden's publication 'an invaluable anthology'. In that period, historians and philosophers of science (like Kuhn himself) developed history of quantum mechanics graduate courses (in which they confronted students with original papers), for which these compilations were surely handy (Heilbron, 1968, n. 15, p. 106; Kuhn, 1967, p. 415; WennersHerron, 2011).

In a review of the SHQP project, Tetu Hirosige reminded us that 'The historian of science starts his research with facts as all other scientists do. But the historian draws his "facts" from various forms of written source materials which, in general, have existed long before he begins his research, whereas a scientist produces his scientific facts by intentionally designed experiments. This is certainly one of the notable differences in the mode of research between the history of science and science itself'. He also characterized as candid and detailed the way that Kuhn *et al.* described in the first chapter of their report 'a basic process in studying the history of science—the gathering of historical facts from various sources'.

Hirosige was not fully aware of the innovative drive of such project, not that different from those scientific experiments he imagined, since building an archive like this meant not only a passive gathering of existing materials, but an active conceptual endeavour of creating a new source platform for historians of contemporary physics. Its plan of action contained a sophisticated evaluation of the field, based on a methodical survey of available knowledge in reference works and contemporary sources. Its rationale gave priority to the recovery of different types of documents considered pertinent to write the history of quantum physics from a historian's perspective, including for instance not only research but also pedagogical questions. Moreover, since Kuhn foresaw at the time that 'at least for the last few decades, there may be special problems about manuscript sources, for scientists have in recent years increasingly substituted the telephone and personal contact for the letter as a means of informal communication', the project did not limit itself to manuscript sources. It was particularly innovative in its endeavour to create an archive of oral history sources—something unusual for the history of science up until then. This effort included a methodological explanation of procedures, which might have been candid—as noted by Hirosige—since the project members were embarked for the first time on an enterprise of this nature, and they did not appear to have many previous examples to refer to. But they displayed a major quality of the historian's craft: methodological and historiographical transparency that leads to historical accuracy (Hirosige, 1968; Cook, 1971; Kuhn, 1967, pp. 418–419).

In the same period, Max Jammer worked on a comprehensive account of the conceptual foundations of quantum mechanics. His investigation proceeded implicitly through not only classic journal papers but also an international set of influential textbooks, as main sources to elucidate the conceptual, logical, epistemological and mathematical structure of quantum theory. Although often buried in footnotes, it also showed the importance of textbook writing for contemporary physicists to clarify or solve complex scientific and philosophical implications in the making of quantum mechanics as an evolving research field. In the process of writing his book, Jammer discussed parts of it with physicists involved in the facts he dealt with (e.g. de Broglie, Born, Heisenberg, Dirac, Hund, Heitler, van der Waerden, Fierz, Jost, Andrade, Slater, Fues, and Tank) and with philosophers of science (Holton, Feyerabend, and Kuhn). In its preface, he indicated that his book was 'neither a textbook nor a collection of biographical notes nor even a study of priority questions', but in fact, it has become a classic textbook for any student of quantum physics history.

Heilbron considered that 'this important book is not quite history', and classified it in the 'historico-critical' genre. Its focus was on 'what now seem the central conceptual innovations in quantum physics, and not necessarily upon what were historically the most important steps in its development'. Analogously, Kuhn considered that it was an impressive contribution dealing with a vast literature. But, in questions of analytical depth Jammer had often acted more as a physicist than as a historian. He had fallen prey to historical presentism by attributing 'to particular experiments or theories the significance they are given in the contemporary curriculum rather than the one they had in their own time', and neglecting 'lines of development displaced by subsequent events but of vast importance in their own time'. Beller puts it another way: Jammer and other authors representing the orthodox way in the history of science are histor-ically inaccurate because they rarely question the narrative of the winners. The impact of Jammer's contribution to our field has been continuous though, through a subse-quent volume—focusing on interpretations proper—and a second edition of his first book, published in the 1980s as volume 12 of 'The History of Modern Physics, 1800–1950' series by the American Institute of Physics (Jammer, 1966, 1989, and 1974; Heilbron, 1968, p. 92; Kuhn, 1967, pp. 416–417; Beller, 1999, p. 11).

A wide range of sources and conceptual frameworks was therefore available to write the history of quantum physics. This was a mine for doctoral research. The 1970s saw the publication and up to four reprints of Daniel Kevles' *The Physicists* (based on his Ph.D. dissertation), conceived as a big picture of a scientific community or generation of successful American physicists. In the 1980s, a strand of research in the history of science took special interest in building the big picture of the community of practi-tioners of physics and its subdisciplines in Europe and the United States. Contributions typically paid equal attention to conceptual, institutional, research, and teaching aspects. They were equally inspired by works such as Kevles' and the more socially laden side of Kuhn's *Structure of Scientific Revolutions*.

Stanley Goldberg's overview of special relativity is a good example. Another one is Katherine Sopka's survey of quantum mechanics in the United States, published as volume 10 of 'The History of Modern Physics 1800–1950' series. Books like those by Kevles, Goldberg, and Sopka were considerably descriptive, but they are still major sources to understand how a modern community of physicists was configured in the

US, and the production of textbooks was an important activity for them. They proposed analytical focuses such as the role of doctoral and postdoctoral training and fellowships, seminars and symposia, European physicists' lecturing in the US, and American physicists' training in Europe, that became standard in the historiography of physics. Analogous efforts to those of Goldberg and Sopka to put the US on the map of quantum mechanics history were subsequently made by scholars from other national contexts, such as Kenji Ito for Japan. Their goal of characterizing whole communities of physics practitioners would prevail later on in big pictures such as Kragh's *Quantum Generations* and Staley's *Einstein's Generation* (Kevles, 1971; Forman *et al.*, 1975; Goldberg, 1984; Kragh, 1985 and 1999; Sopka, 1988; Ito, 2002 and 2005; Staley, 2008).

In the same period, another line of research, well-illustrated by doctoral students such as John Hendry (Ph.D., 1978) and Mara Beller (Ph.D., 1983), took off from Jammer's previous work to provide well-balanced research on historical and philosophical grounds. Hendry's book was arguably one of the first to have the term 'dialogue' in its title. He pinpointed that dialogue between historians and physicists was much required and that dialogue between Bohr and Pauli had been essential to the making of quantum mechanics. But 'dialogue' was not a central concept in his analysis of the philosophical implications of Bohr's research. Beller's doctoral dissertation and some of her early publications hinted at the term but did not develop from it. She was undoubtedly inspired by Hendry's work and more than a decade of thought maturation, before she published *Quantum Dialogue*. While proposing *dialogue* as a central analytical tool, her book preserved the ambition of her doctoral dissertation to understand both quantum physics and quantum physics history interpretations (Hendry, 1984; Beller, 1983 and 1985).[3]

With Beller, correspondence was not only a (Bohr) principle, but also a historical source, and as such an object that required historical problematization. Moreover, she crossed genres and connected them, by showing that physicists' conversations that might have started face-to-face or through correspondence, would continue through papers. Thus, letter writing was connected to paper writing—these were just different instances of the overall process of knowledge creation. From sources to conceptual interpretations, this meant that overall the making and communicating of science were intimately related—an argument developed by other historians of science with regard to other type of sources (Holmes, 1987; Waquet, 2003; Secord, 2004). In one way, Beller swam against the current: she considered that participant interviews such as those produced by the SHQP project were 'often unreliable' as historical sources—a question well known to historians: the difference between memory and history (Nora, 1984; Portelli, 1999). In other aspects, in spite of the innovative character of her methodological and historiographical approach, Beller was typically Kuhnian, from the subtitle of her book to her reluctance to consider textbooks also as sites of knowledge creation (Beller, 1999).

---

[3] Klein (1970) also used that term but did not make it central to his analysis. Beller did not use the term explicitly in her publications until 1992.

For Beller, as for many historians of science wrapped in the Kuhnian mantle, textbooks were dogmatic and static sources. For this reason, they were useless in unveiling the actual movement of a scientist's creative thoughts. Paradoxically, Frederick L. Holmes' discussion of the difference between the contexts of discovery and justification, and its concurrence in the writing of research papers shows (*pace* Kuhn), that the 'falsification' of the research practice process through writing is as characteristic of pedagogical literature as of investigative pathways (Holmes, 1987; Simon, 2013). But it was time to focus on other types of sources.

The new *fin de siècle* would see the advent of a 'science-notebook historiographical revolution'. Starting with Kathryn Olesko and following with Andrew Warwick and David Kaiser, historians of physics produced some of the most innovative work in the history of science, by a thorough analysis of student and teacher notebooks, lecture notes, and examinations. Their methodologies and focus on science pedagogy brought a breath of fresh air to our discipline. The work of Seth on Sommerfeld mentioned at the start of this chapter can be placed in this framework. However, this 'revolution' partly relied on a classic paradigm, where authors such as Warwick and Kaiser revived the work of classics such as Kuhn and Foucault, and henceforth proposed narrow concepts of discipline and pedagogy—which once again relegated textbooks to the end of the line. There were, however, other competing perspectives that have furthered new approaches to the study of science education and textbooks (Olesko, 1991 and 2006; Warwick, 2003; Kaiser, 2005a and 2005b; Simon, 2011, 2013, and 2015).

In this context, some new contributions have started to develop more thorough case studies on quantum mechanics pedagogy and textbooks, but there is still a long way to go to define the role of pedagogy and textbooks in the conceptual, epistemological, and institutional foundation of this field of physics (Kaiser, 2007 and 2020; Badino and Navarro, 2013; Badino, 2019). Unfortunately, it is still a common trope to consider that 'scientific revolutions are rendered invisible by subsequent textbook treatments written from the perspective of the new paradigm', where history is constantly falsified by physicists and their textbooks (Gooday and Mitchell, 2013, p. 752).

This and other historiographical 'revolutions' are a sign of the vitality of the history of physics as a discipline. But perhaps a more accurate and objective measure of the temperature of our field can be obtained by taking a look at one of the oldest, more abundant, and arguably more methodologically and historiographically discussed genres in the history of science, that is, biography (Söderqvist, 2011; Nye, 2015; Forstner and Walker, 2020). Let's cite a number of representative examples (regarding quantum physics) across three decades: Heilbron's (1986) and Hoffman's (2008) Planck, Walter Moore's (1989) Schrödinger, Helge Kragh's (1990) Dirac, David Cassidy's (1992 and 2009) Heisenberg, Kostas Gavroglu's (1995) London, Silvan Schweber's (2000 and 2012) Oppenheimer and Bethe, Maria Rentzi's (2007) Meitner and Blau, Michael Eckert's (2013a) Sommerfeld, and Olival Freire's (2019) Bohm. All of them give an important share in their narratives to education and pedagogy, but, most importantly, most of them consider it relevant to discuss the role of textbook reading and writing in the intellectual and professional career of their historical characters.

## 28.2 QUANTUM TEXTBOOK GENERATIONS

The year 1900 has been dubbed the 'turning point' in physics history where natural philosophy muted into modern physics, as we know it today (Forman *et al.*, 1975; Buchwald and Hong, 2003; Kragh, 1999). However, the early-1900s textbook world looked much more as its 19th-century counterpart than as a new era. When Max Planck created the equation that bore his name, it soon appeared in two textbooks that set the standard. The first was Orest Chwolson's *Lehrbuch der Physik*—a late 19th-century multi-volume work originally written in Russian, and successively updated and translated into German, French, and Spanish. Its second volume in German appeared in 1904. The second, Adolf August Winkelmann's *Handbuch der Physik*, a multi-authored encyclopedia whose first edition appeared during the last decade of the previous century. Its second edition (1903–1909) included a volume devoted to electricity and magnetism (published in 1906) which covered literature up to 1904 and had a contribution on radiation by Leo Graetz (a former student of Planck) referring to Planck's equation as well. These were extensions of the 19th-century textbook physics tradition, which had in Germany representatives such as Johann Müller, Adolf Ferdinand Weinhold, and Adolf Wüllner. Planck's youth, and his turn-of-the-century research on the radiation of the black body, was inspired by reading works such as Müller's physics treatise and John Tyndall's textbook *Heat Considered as a Mode of Motion* (German translation, 1876). Sommerfeld contributed himself to an *Enzyklopädie der mathematischen Wissenschaften* (managed by Felix Klein, 1898–1933) by editing its international multi-authored volume on mathematical physics, and contributions to the organization of its volume on mechanics. This was the textbook physics world in which physicists like Planck and Sommerfeld (who had their doctorates and *habilitation* before the new century) grew up. In their ascent from students to *privatdozenten* and assistants, and from there to professors, research institute directors, and research school leaders, they had also to comply with tasks such as the weekly taking of detailed notes of lectures (for themselves or their professors) (Kangro, 1976, pp. 8, 230; Wilson, 1912; Simon, 2013; Eckert, 2013a, pp. 49, 60, 99–108; Gispert, 1999).

Many of the aforementioned textbooks originated between the 1850s and 1870s; by successive additions and editions, they typically grew into five-volume works, such as contemporary handbooks, and lived up to the first decades of the 20th century. Many of them survived their original authors, and were run by successive teams of editors. However, between the 1910s and 1920s, their capacity to absorb new matter in relation to the fast-growing field of physics was proving ever more unfeasible. Their unstoppable growth had made them monstrous, both due to size and lack of coherence, and it was as problematic for physicists as for publishers to keep such works in shape and at a reasonable price (Simon, 2013). This meant disruption and transition in the communication system shaping the disciplinary substance of modern physics (in which textbook physics had a relevant agency). Consequently, a new subfield or specialization emerged through the production of new textbooks specifically devoted to quantum mechanics. In its turn, as has been suggested by Jammer, the making of quantum mechanics as a new tradition or subdiscipline of physics required completeness (and coherence), and textbooks had a major role in this process. As we saw in the introduction to this chapter, Seth has proposed a similar pattern, although less explicitly, in using Sommerfeld's, Heisenberg's, and Dirac's textbooks as scaffolds for his *Handbook* chapter narrative (Simon, 2013; Stichweh, 1996 and 2001; Weisz, 2006; Jammer, 1966, pp. 366–370; Seth, 2013).

A combination of several communication genres (lectures, encyclopedia articles, journal papers, and textbooks) constituted the standard system of communication of physics research at the time. Not surprisingly, there was a crossbreeding of these genres that contributed to the invigoration of discipline building. Thus, in 1916, after submitting his paper 'Zur Quantentheorie der Spektrallinien' to *Annalen der Physik* and resuming his lecturing duties, Sommerfeld began to think about the mission of writing a textbook on the topic.[4] With its almost 140 pages and its publication over two issues of the journal, Sommerfeld's paper was by far the largest, most prominent work in a journal which had in 1916 published only slightly more than twenty issues, and whose average article length was between 10 and 30 pages that year. Needless to say, publishing in the *Annalen* was not easy, and it surely had a budget related to the length of each issue and its printing production costs.

As indicated by Sommerfeld, his paper was based on two communications submitted between the previous year and early 1916 to the Akademie der Wissenschaften zu München. He had also presented this research at the weekly colloquium of his research institute. Munich had several options for discussion on advanced physics topics, of which Sommerfeld made the most: the traditional monthly Sohncke Colloquium (a joint effort of physicists at the University and the Technical University), a newer Colloquium which eventually was called after Sommerfeld, and more informal student seminars, regular and special lectures, and guest lectures in other university departments. Sommerfeld's training in this communication system was shaped since his days as a student in Klein's research centre at Göttingen. During the first years of his professorship at Munich, there were innovations such as the aforementioned weekly colloquia and seminars. Starting in this period, Sommerfeld's publishing endeavours began to be more tightly connected to his lecturing schedule, and participation in a varied array of research presentation venues addressed to physicists as well as colleagues in other faculties.

As has been worked out in detail by Michael Eckert, it was in this overall context and not just sparked by a lecture to an audience of 80 people (including 12 physicists, the rest being mainly chemists, medical professors, and philosophers), that the goal of producing a textbook crystallized in Sommerfeld's mind. Historians of physics have

---

[4] On Sommerfeld, most of the time I follow Michael Eckert's narrative, but the interpretation is mine.

often literally followed Sommerfeld's remark on that specific audience, on his intention to produce a book representing the printed form of the lectures he was giving 'popularly, i.e., without mathematics, only conceptually presented', and on his repeated qualification of the work he was writing as a 'popular book'. Accordingly, Eckert has stated that Sommerfeld's *Atombau und Spektrallinien* was in fact not a textbook. However, Sommerfeld also indicated that his book was 'in its main part for chemists, in the appendices also for physicists', depicting a mixed but clear readership (Eckert, 1993, pp. 59–60; Eckert, 2013a, pp. 155, 164–166, 205–214, and 2013b; Sommerfeld, 1916).

In fact, in spite of increasing specialization, between the 19th and mid-20th century most physics textbooks were explicitly addressed to an assorted spectrum of readers. This was a reflection of a diverse and fragmented educational framework, a highly competitive publishing market, and the publishers' and authors' intention to capture the largest body of customers, in a readership pool in expansion but still small compared to that of other fields of knowledge. We should also not forget that *Annalen der Physik und Chemie* had only relatively recently dropped the 'chemistry' in its name (by 1900), that there were numerous research areas in the new physics (including quantum mechanics) at the intersection of physics and chemistry, and that for a long time the history of contemporary physics has unfairly been biased towards theoretical and mathematical physics. In addition, Sommerfeld is well known for being more preoccupied in his lectures and pedagogical writing with the physical problems than with mathematical foundations. Finally, in order to understand Sommerfeld's authorial intentions and the genre of his work, we should first understand what 'popular' and 'popularization' actually meant at the time (Simon, 2011, 2013, and 2009; Seth, 2010).

For this purpose, it is particularly appropriate to use the work of Ludwik Fleck, who was a medical student at the time, and through his experience as a researcher proposed in the 1930s concepts such as *Denkstil* and *Denkkollektiv*. Fleck's conceptualization of scientific practice through 'thought collectives' configuring and constrained by a 'thought style', and their subdivision into small 'esoteric circles' and large 'exoteric circles', stressed the transformative and multidirectional role of communication in the making of science. In addition, he offered a useful characterization of the carriers of scientific knowledge and agents of scientific communication through the definition of three major genres. According to him, a 'thought style' is represented by '*vademecum science*', as the carrier of common expert knowledge and the tool binding a 'thought collective'. It is opposed to '*journal science*' in that it is comprehensive and consensus-based. It differs from '*popular science*' in that it is critical and organized. However, the character of the *vademecum* is also determined by the fact that every communicative action—including those leading to its configuration—makes knowledge more exoteric. Thus, communication always transformed knowledge, and it acted towards the con-stitution of 'thought styles' based on social and intellectual consensus. In this frame-work of intensive communication processes between exoteric and esoteric circles, for Fleck the quality of the 'standard' applied to a work is in fact one of the qualities of the 'popular'. Popularizing thus did not consist uniquely in communicating to lay audiences, but also to specialized scientists belonging to other disciplines and sub-disciplines within the sciences—a characteristic of the modern system of scientific disciplines, according to Rudolf Stichweh. In spite of their differences, *vademecum*, textbook, or handbook science met *journal* and *popular* science in the making of those works that we can consider as standard, canonical, or classic by the quantitative and qualitative testimony of its readers. In this framework, due both to its conception and its documented uses, Sommerfeld's *Atombau* was obviously a textbook, and as we are going to see in the following, it represented the quantum physics canon and soon became a classic (Fleck, 1979, pp. 39–41, 51, 98–9, 109–113; Simon, 2009; Fyfe, 2002; Olesko, 2005; Stichweh, 2001).

Intensive lecturing was closely related to the university profession. A major source of income for both *privatdozenten* and professors, their prestige was not only shaped by research production, but also by pedagogical and lecturing skills. Disdain of teaching over research, and a restrictive and mystifying association of the latter with genius, have been retrospective narrative tools often used by physicists themselves, but hardly corresponding to how physical knowledge was made (Ben-David, 1971, pp. 108–138; Busch, 1963; Jungnickel and McCormmach 1986, vol. 2; Waquet, 2003; Sopka, 1988, Appendix B; Kapitza, 1973, p. 755). Other major examples of textbooks arising from their authors' activity and experiences as lecturers, and essential both to research and pedagogy are for instance: Max Born's *Vorlesungen über Atommechanik* (Julius Springer, 1925), its second volume by Born and Pascual Jordan, *Elementare Quanten-mechanik* (Julius Springer, 1930) and Born's *Probleme der Atomdynamik (Dreissig Vorlesungen gehalten im Wintersemester 1925/26 am Massachusetts Institute of Technology)* (Julius Springer, 1926; trans. in English: *Problems of Atomic Dynamics*, MIT Press, 1926), Erwin Schrödinger's, *Abhandlungen zur Wellenmechanik* (J. A. Barth, 1926), and Paul Dirac's *The Principles of Quantum Mechanics* (Oxford University Press, 1930). All clearly belonged to the textbook genre, but there are differences between them. For instance, while Schrödinger's textbook is simply a compilation of papers, and Born's *Problems* arose from a series of just ten lectures held on an American tour, his *Elementare Quantenmechanik* was a much elaborated text aiming to endow the field with a matrix mechanics foundation. Even more can be said of Dirac's *Principles*, as one of the major textbooks providing completeness and coherence to quantum mechanics. It is a matter of further research determining how quantum physics lecturing fed into textbooks and vice versa. Nonetheless, writing a quantum physics textbook, or being mentioned in one of them, had undoubtedly a major role in furthering physicists' academic status and professional career. Consequently, textbooks were a matter not only of collaboration, but also of fierce competition (Mehra and Rechenberg, 1982, pp. 252–253, 281–282; Giulini, 2013; Jammer, 1966, p. 366; Eckert, 1993, pp. 94–95, 260–261).

An important part of the meaning and status of these textbooks depended on the perspectives of students, colleagues, reviewers, and other types of readers. Thus, for instance, in spite of the relevance of Born and Pascual's *Elementare Quantenmechanik*, Wolfgang Pauli criticized the difficult reading and one-sided nature of the mathematical foundation proposed by that textbook for quantum mechanics. According to him, it limited the use and interest of the book to a restricted readership. In contrast, some quantum physics textbooks were soon characterized as classic or canonical works. A common designation for those textbooks was referring to them as the 'Bible' of modern physics. Sommerfeld's textbook was called like that by fellow physicists and chemists such as Max Born, Hermann Weyl, Friedrich Paschen, and Exum Percival Lewis. It was highly praised by numerous colleagues (e.g. Lorentz and Röntgen), who emphasized Sommerfeld's textbook wealth of data, together with its clear and precise structure, arrangement, language, and conceptual exposition. In contrast to his review of Born and Pascual's book, Pauli valued the fourth edition of Sommerfeld's textbook especially because it did not rely on any particular atomic model, thus making it useful for a comprehensive range of quantum physicists. Alfred Landé and Friedrich Hund considered that it was one of the 'great standard works' and as such was known to any theoretical physicist—certainly, *Atombau* was a major agent in drawing young students towards the study of quantum mechanics. By reading *Atombau*, a generation of students and young researchers complied with a rite of passage to become part of the international community of theoretical physicists. In the early 1920s, during a six-month lecturing tour across the US, Sommerfeld was considered an 'oracle' by some American colleagues, in relation to the knowledge contained in his textbook and conferences, and Paul Ehrenfest called him—not without irony—'St. Sommerfeldus, the quantum Pope' (Mehra and Rechenberg, 1982, p. 282; Eckert 1993, pp. 59–60, 85, 94–96, 139, 260–261; Eckert, 2013b, pp. 118, 127–128, 131; Eckert, 2013a, p. 256).

The Bible is a canonical work in the Christian religion. It can be interpreted variously and contains a wealth of different voices, but it is a fundamental and standardized repository of Christian creed, and a universal classic of literature. The use of this designation for Sommerfeld's textbook expresses its canonical and standard character, as testified by a wide and international range of its readers (it was translated into at least English, French, and Russian) and by successive editions (1919, 1921, 1922, 1924, 1931, 1944, and a second volume of the textbook: 1929, 1939, 1944). It also shows that *Atombau und Spektrallinien* neither belonged to Sommerfeld as its author, nor was it just a manifesto of his research school: it was a collective work belonging to all its highly appreciative readers, and to the world community of physics and science practitioners at large. Further cases of major collective textbook writing, even across national frontiers, can be found in other examples of physics textbooks, handbooks, and encyclopedias mentioned in this chapter, or in other canonical works in the history of science such as Berzelius' *Lärbok i Kemien* (1808–1818). The 'bible' epithet was subsequently used to characterize some other quantum physics textbooks, for instance Dirac's *Principles*. It is a value that does not apply to any type of publication, but exclusively to the classics—among which science textbooks have a larger presence than is usually acknowledged (Eckert, 1993, p. 127; Eckert, 2013b, p. 117; Jammer, 1966, p. 366; Simon, 2009 and 2013; Blondel-Mégrelis, 2000).

# 28.3 TEXTBOOK FOUNDATIONS

In the closing sentence of his Ph.D. thesis, John Heilbron confided that: 'Most quantum physicists appear to think in terms of space-time pictures when working on everyday problems and only trot out the "official" interpretation when writing text-books or philosophy. But that is another story'—where 'official' stands for 'Bohr's complementarity' (Heilbron, 1964, p. 419).

In his survey of quantum physics teaching in the United States, David Kaiser has pinpointed that between the 1940s and 1950s—the period in which a major historian of science such as Heilbron, and his 12-year-elder adviser Thomas Kuhn, were studying physics—textbook narratives abandoned the historical or genealogical mould in favour of a toolkit approach. Kaiser contrasts the structure and contents of the first textbooks and some of the early lectures (which between the late 1920s and 1940s paid equal attention to mathematical formalism and philosophical interpretation), to the emerging trend a decade later.[5] By the mid-1950s, across the country's major universities, physics graduate students saw examination questions that previously dealt typically with matters of interpretation, being replaced by questions testing competence in standard calculations. Kaiser further reinforces his argument by comparing the unequal fate of two of the major American textbooks of the period: Leonard Schiff's *Quantum Mechanics* (1949) (representing the toolkit approach) and David Bohm's *Quantum Theory* (1951) (representing the philosophically laden approach). In spite of the illustrative power of such comparison, it is only partially relevant, as it explicitly circumvents more rigorous arguments acknowledging the historical contingencies of Bohm's career. Kaiser extends the development of this trend at least up until the 1970s, although he admits that in that decade, due to lower pressure of enrolments, some textbooks started to combine a larger number of calculation questions with some qualitative essay-type ones (Kaiser, 2020).

Heilbron's witness perception, riveted by Kaiser's historical sketch, is in many ways a historian of physics' favourite. Helge Kragh's textbook *Quantum Generations* frequently refers in its pages to the idea (outlined in Beller's citation at the head of this chapter) that textbooks do not reflect the change in worldview discussed in frontier theoretical physics, because this is how they 'usually are: They are by nature conservative and cautious in their attitude toward modern ideas' (Kragh, 1999, p. 10).[6] This view is common, and new adherents follow successively a similar Kuhnian route to

---

[5] Textbooks by Condon and Morse, Ruark and Urey, Landé, and Kemble. Lectures at Berkeley and Caltech, by Oppenheimer and Bloch.

[6] It is Kragh himself who defines his book as a 'textbook'. Kragh (1999), pp. xi–xiii. See also Purrington (2018).

textbooks (Kaiser, 2005b, pp. 393–409; Badino and Navarro, 2013; Badino, 2019, p. 6).[7] If such is the reality of textbooks, how would they engage in open and not consensual questions of philosophical interpretation, instead of focusing on more direct matters of quasi-mechanical mathematical calculation, as sketched by Kaiser?

According to Hendry, 'Between 1928 and 1933 the dominant position of the Copenhagen interpretation was confirmed and consolidated through the publication of a long series of textbooks and review papers, written by its adherents. Among the authors were Heisenberg, Dirac, Weyl, Born and Jordan, Kemble, Pauli, and von Neumann'. Von Neumann's *Mathematische Grundlagen der Quantenmechanik* (1932) contributed to cement this consensual framework by providing the Copenhagen interpretation with a solid mathematical foundation and endorsing it too with an explicit philosophical approach (Hendry, 1978, p. 169). Analogous attributions are given by other authors to other textbooks, although these lay especially on the mathematical and theoretical formalism side, since philosophical interpretation was rather implicit than explicit in most textbooks. Major examples are the textbooks by Dirac, Heisenberg, Born and Jordan, and even de Broglie, as major agents both in the establishment of a predominance of the Copenhagen spirit, and inauguration of a new era in which pedagogical efficiency concealed the public exposition of philosophical quarrels or the promotion of interpretative pluralism among quantum physicists. Both aspects came together, because—as it has been noted—many textbooks did not care about introducing Bohr's complementary principle while providing space for Heisenberg's uncertainty principle and especially for everyday quantum physics practice (Gandarias Perillán, 2011, pp. 243–248; Howard, 2013; Kragh, 2013 and 1999, pp. 211–212). However, a closer look at textbooks such as von Neumann's shows that it differed in important ways from Bohr's views. It introduced, for instance, a different interpretation of the measurement process (a distinction that would become clear decades later). It contributed thus to the sustained defiance of quantum orthodoxy, and indirectly, to the constitution of a research field devoted to the foundations of quantum physics (Freire, 2015, pp. 141–149).

We cannot ignore that frontier quantum physicists wrote textbooks, but most often they interest us only when they were written in times of *crisis*. In those times, the standard historian thinks that textbooks displayed disagreement, pluralistic views, confusion, and even periodical changes in perspective (see for instance successive editions of Sommerfeld's textbook and its airing of disputes with Bohr on the foundations of quantum theory). These textbooks contributed to shape *normal science*. The *normal* quantum mechanics package included a basic conceptual framework coined 'the Copenhagen interpretation'. Subsequently, new textbooks became standard and somewhat dogmatic pieces of work exclusively aimed at this type of disciplinary

---

[7] On the limitations and historical contingency of Kuhn's perspective on textbooks, see Simon (2013) and (2016). Kaiser and Warwick have non-casually shown their strictly Kuhnian take on textbooks, and a pragmatic Kuhnian reformism on other pedagogical matters. Badino and Navarro are in practice— *pace* Kuhn—essentially inspired by Kaiser and Warwick.

indoctrination that we call scientific training. With more or fewer nuances, this framework conforms to the views of all the aforementioned authors, and in practice, it is perfectly Kuhnian. It works for certain cases, but this chapter's aim is to show that nevertheless, it is insufficient. Furthermore, does not it sound too good to be true, as an allegedly 'simple textbook narrative'? In several ways, claiming some diversity and epistemological agency for quantum physics textbooks is more akin to admitting the relevance for quantum physics of conceptual dissidence and foundations research than most scholars have traditionally been willing to accept.

In contrast, historians of physics are more open minded and sympathetic to pedagogy and textbooks when producing proposals to improve current quantum physics teaching through updated history and philosophy of science contents and approaches. There is a general agreement about the fact that it is usual to find well-packed but naive (and especially inaccurate) historical content in current physics textbooks (in particular on quantum mechanics). However, researchers typically work with a few illustrative textbook examples, and although this is likely to be a dominant trend, we are still in need of more general surveys. Furthermore, we could perhaps proceed inversely, and look for those exemplary textbooks that have already made proper use of history and philosophy of science. Finally, historians and philosophers of science in dialogue with science education researchers have come to accept that, very often, historical complexity does not easily match with pedagogical effectiveness. In spite of this, a considerable number of scholars agree that science education and (quantum) physics learning would benefit from interdisciplinary collaboration in the production of teaching materials as rigorous on pedagogical as on historical and philosophical grounds. It is thus possible to shape the ontology and epistemology of quantum mechanics through textbook science (Brush, 1974; Kragh, 1992; Lautesse *et al.*, 2015; Franklin, 2016; Greca and Freire, 2014; Hentschel, 2018; Mohan, 2020). If this is possible for current teaching and textbooks, in a time of no particular crisis for quantum mechanics, how would it be—if we keep to a strict Kuhnian vision—that it might not have happened before in the already long history of quantum textbook physics?

If we stay Kuhnian, the case of David Bohm's textbook, previously mentioned, would then feature as an exception (although obviously not the only one) in a textbook landscape apparently dominated by normative aphilosophical quantum mechanics. If we pay attention to the inspiring dialogue between historians, philosophers, and educators, it can be instead a major example, among others, showing us that there is historical life beyond the Kuhnian paradigm. To understand the neglect of Bohm's alternative approach over the dominant Copenhagen interpretation, James T. Cushing appealed to historical contingency. Much in the way of Paul Forman's famous thesis, he expounded the social, political, and cultural factors that shaped both Bohm's philosophical views, and their failure to become more widely adopted among physicists at the time of their publication (Cushing, 1994; Hiley, 1997).

In his classic paper, Forman revealed the changing thoughts of physicists in the Weimar Republic, through a wide range of literature genres that accounted for what physicists heard, read, talked about, and believed: written accounts of public lectures, academic addresses, correspondence, articles in handbooks, general science periodicals, newspapers, and specialized scientific journals, and popular philosophy books. In many ways, Forman's article could be seen as a history of reading and lecturing. It has neither this historiographical approach, nor that of book history though. It focuses on ideas and intellectuals. Thus, it considers publication genres as insignificant containers and diverts from a thorough analysis of readerships and audiences. Forman's meticulous reconstruction of the intensive lecturing schedule of the leading German physicists, on broad scientific and philosophical subjects, has fairly won him an eponymous thesis. His approach has generously showed how philosophical, artistic, educational, and scientific debates were related, and how cultural and political circles (if separable at all) interacted and intersected. Forman's thesis was a ground-breaking and courageous stance in a time in which historical, philosophical, and sociological perspectives on science were at war (Rossi, 1986). Its valuable historical finesse has lived up to the present day. Nonetheless, with the present nuances of our discipline, Forman's thesis might be hampered by exceptionalism in several ways. The most important is in fact the downside of Forman's sophisticated historical contingency: the communication practices that he describes in minute detail only happen in (particular) hostile environments, as a way for scientists to regain (previously lost) social prestige. It is therefore a result of crisis, and therefore would not apply in (Kuhnian) *normal* times. In comparison, theses such as Fleck's see in analogous communication processes as those outlined in Forman's paper the standard rule that characterizes how scientific knowledge is produced.

Thus, for instance, Forman tells us that 'Heisenberg published a popular article retailing his conclusions even before his "technical" paper was printed', because of 'the physicists' general anxiousness to carry the good news to the educated public' (Forman, 1971, p. 105). As if this publication practice was exceptional or rare. Conversely, if we bring in Fleck's historical epistemology (outlined in section 28.2): a characteristic Weimar physics style was possible because of the intense communicative interaction between esoteric and exoteric circles of knowledge and between several types of science genres, including specialized journal papers (*journal science*), but also reference works (*vademecum science*), educational works (*textbook science*), and general and popular science periodicals, newspapers, and best-selling philosophy books (*popular science*) (Simon, 2009; Olesko, 2020). Furthermore, saying that quantum mechanics knowledge became popular equates to saying that it became standardized, since both qualities arise from the same processes driving the making of quantum meaning. In this framework, Heisenberg's intersectional reading and writing can be seen as common practices in the making of scientific knowledge, and not just peculiar or extraordinary concessions to the general public. Again, we can see here the relationship between form and meaning at play, but also the tension between historical, sociological, and philosophical approaches, as connected but distinct perspectives on science.

Regaining contingency through Forman's example, we can go back to that new classic of modern quantum mechanics, American dominance of the field during the second half of the 20th century: in fact, Kaiser's central argument in his overview of US quantum physics textbooks is excessively focused on the role of decaying university enrolments (a central argument in his Ph.D. thesis). He brings in one of Forman's lines of thought (also exploited by subsequent historians, such as Olesko), but neglects other major cultural and conceptual sides of the problem. For instance, in the 1960s and 1970s many influential American physicists and textbook authors were against the integration of historical and philosophical perspectives in textbooks, because they thought—among other things—that they were *tout court* not useful to physics learning. In any event, as with the debate over the applicability of Forman's thesis beyond Weimar, we should discuss if Kaiser's US picture would apply to other cases. It might have been a general trend, but it is arguably not a universal Kuhnian truth that applies to all countries or authors of quantum physics and general physics textbooks (Stambler, 1971, pp. 236–250; Rudolph, 2002; Simon, 2019). As historians, we are also left with the question of deciding if Forman's or Fleck's theses can be applied anytime and anywhere, which includes an ever zealous methodological effort to distinguish exceptional from standard, exemplary, or comparable cases, and to tune our claims to the nature of the historical sources and interpretations we use to build them. It is clear though, that in choosing effective ways of communicating quantum mechanics, pedagogical, historical, and philosophical priorities have to be negotiated.

## 28.4 CONCLUSIONS

Dealing with quantum mechanics textbooks in a national or international perspective would require a more precise genre ontology than we currently have: for instance, are we only talking about books having 'quantum mechanics' in their title, or any book designed for teaching having adjectives 'quantum', 'atomic' or 'nuclear' in their title? Taking into account the powerful disciplinary force of textbooks, a characterization of quantum mechanics as the theoretical framework underlying atomic, molecular, and nuclear physics is too overtly gross, as it would be to think that engineering is just applied science. We know too well that each of these disciplinary fields embodies a distinct and partly sovereign way of making knowledge, while holding intense interactions and intersections with the others. We should also bear in mind that there is a century-long pedagogical tradition in textbook physics going back to the mid-19th century which predominantly favoured experimental and technological perspectives, while theoretical physics was gaining power in its impetuous march towards the mid-20th century. These might be relevant arguments to consider in discussing the absence or nature of foundational and interpretative questions in quantum physics textbooks. We surely still require more research and more of this pluralistic historical contingency, to produce more accurate analyses of the role of textbooks in quantum mechanics. It is this methodological and historiographical urge that this chapter has insistently been asking for in this field of research.

In this chapter, I have provided an overview of the place of textbooks in the history and historiography of quantum physics. This is just a first glimpse into a field in need of serious development. Recent works such as Badino and Navarro's edited volume and Kaiser's brief incursions into American physics textbooks offer a range of materials on major case studies of quantum physics textbooks by a range of experienced scholars in the history of contemporary physics. Through volumes like these and relevant sections in the wealth of available biographies on quantum physicists, the interested reader can already find relevant insights on textbooks such as those by Max Planck, Fritz Reich, George Birtwistle, Arnold Sommerfeld, Max Born, Paul Dirac, John Van Vleck, Pascual Jordan, Otto Sackur, and Paul Drude, among others.

In this chapter, I have taken a different approach, emphasizing the importance of refining our historiographical and methodological approaches on science textbooks—in particular, quantum physics textbooks—in the framework of a rigorous problematization of historical sources and scientific literature genres. In this context, I have reclaimed a more central and higher status for textbooks within the history of quantum physics, by demonstrating the relevance of their historical and historiographical agency. By using a selection of case studies, I have discussed the nature of textbooks, their major role for quantum physics research, teaching, and popularizing, the ways in which standard sources for quantum physics history such as journal papers, conference proceedings, letters, lecture notes, or interviews relate to textbooks, and how they interact between them. I have advocated for the study of how scientific literary genres cross over, and how their dynamics relates to the enterprise of making and unmaking scientific disciplines. Furthermore, I have shown how we can attribute the quality of standard, classic, or canonical to a number of science works—especially textbooks—and not just elite journal papers or esoteric treatises.

Future histories of quantum textbook physics will have to look more deeply at the epistemological role of textbooks in the making of the quantum physics discipline—a topic that, to be fair, I have scarcely dealt with in this chapter. They will also have to connect pedagogical philosophy with the economies of writing and publishing, and their socio-political parameters. Moreover, in this chapter I have focused on the disruption in the communication system of physics that drove the discipline to division and specialization, and hence the emergence of a new genre: quantum textbook physics. I have not dealt with the subsequent process of disciplinary integration that by the mid-20th century pushed for the assimilation of modern physics (mainly quantum mechanics and relativity) into general physics textbooks, to supersede the outdated nature of the introductory textbooks from the 1930s and 1940s.

Finally, the history of quantum physics textbooks has hitherto exclusively concentrated on textbooks belonging to the heroic or discipline-shaping period of quantum physics, and to the historically, historiographically, and geopolitically dominant schools. In addition to those already cited in this chapter, the reader will surely be familiar with names such as Haas, Tomonaga, Rojansky, Gurney, Groenewold, Mandl, Kramers, Pohl, Houston, Dicke and Wittke, Merzbacher, Powell and Crasemann, Matthews, Trigg, Mott, White, Landau and Lifshitz, French and Taylor, Saxon, Sakurai, Jauch, or

Rubinowicz. Most of them do not have monograph studies on their textbook work, yet, but the old historiography of quantum physics textbooks will eventually fulfil this goal.

Undoubtedly, the new historiography of quantum textbook physics will also have to include in its research horizon the study of translations, cross-national textbook writing, and different criteria of relevance for analysing the making of quantum textbook physics, leading to names such as Dushman, Persico, Cini, Bandini, Messiah, Cohen-Tannoudji and Laloë, Blokhintsev, Leite Lopes, Kogan and Galitsky, Kompaneyetz, Davydov, Sokolov, Loskutov and Ternov, Goldman and Krivchenkov, Abers, Ortiz Fornaguera, Garrido, Aréjula, de la Peña, Ghoshal, Rojo Asenjo, Galindo and Pascual, Gasiorowicz, Taketani, Kotani, . . . Is there any historian there, who might have heard of them? Will historians of quantum physics textbooks depart from neoclassic intellectual history? Where is the place where the form and meaning of quantum ideas promiscuously dialogue to produce new historiographies?

# REFERENCES

Badino, M. (2016). What Have the Historians of Quantum Physics Ever Done for Us? *Centaurus*, **58**(4), 327–46.

Badino, M. (2019). 'Schooling the Quantum Generations: Textbooks and Quantum Cultures from the 1910s to the 1930s'. *Berichte zur Wissenschaftgeschichte*, **42**(4), 1–17.

Badino, M. and Navarro, J. (eds) (2013). *Research and Pedagogy: A History of Quantum Physics through the Textbooks*. Berlin: Edition Open Access.

Beller, M. (1983). *The Genesis of Interpretations of Quantum Physics, 1925–1927*. Ph.D. dissertation, University of Maryland.

Beller, M. (1985). Pascual Jordan's Influence on the Discovery of Heisenberg's Indeterminacy Principle. *Archive for History of Exact Sciences*, **33**(4), 337–49.

Beller, M. (1999). *Quantum Dialogue: The Making of a Revolution*. Chicago: The University of Chicago Press.

Ben-David, J. (1971). *The Scientist's Role in Society: A Comparative Study*. Englewood Cliffs, NJ: Prentice-Hall.

Bensaude-Vincent, B., and Simon, J. (2008). *Chemistry: The Impure Science*. London: Imperial College Press.

Blondel-Mégrelis, M. (2000). Berzelius's Textbook: In Translation and Multiple Editions, as Seen through his Correspondence. In A. Lundgren and B. Bensaude-Vincent (eds), *Communicating Chemistry: Textbooks and Their Audiences, 1789–1939*, Canton, MA: Science History Publications, pp. 233–54.

Brush, S. G. (1974). Should the History of Science Be Rated X? *Science*, **183**(4130), 1164–70.

Buchwald, J. Z. and Fox, R. (eds) (2013). *The Oxford Handbook of the History of Physics*. Oxford: Oxford University Press.

Buchwald, J. Z. and Hong, S. (2003). Physics. In D. Cahan (ed.), *From Natural Philosophy to the Sciences: Writing the History of Nineteenth-Century Science*, Chicago: University of Chicago Press, pp. 163–95.

Busch, A. (1963). The Vicissitudes of the Privatdozent: Breakdown and Adaptation in the Recruitment of the German University Teacher. *Minerva*, **1**(3), 319–41.

Cassidy, D. C. (1992). *Uncertainty: The Life and Science of Werner Heisenberg*. New York: W. H. Freeman and Co.

Cassidy, D. C. (2009). *Beyond Uncertainty: Heisenberg, Quantum Physics and the Bomb*. New York: Bellevue Literary Press.

Chang, H. (2016). Who Cares about the History of Science? *Notes and Records of the Royal Society of London*, **71**(1), 91–107.

Cook, J. F. (1971). The Archivist: Link between Scientist and Historian. *The American Archivist*, **34**(4), 377–81.

Cushing, J. T. (1994). *Quantum Mechanics: Historical Contingency and the Copenhagen Hegemony*. Chicago: Chicago University press.

Eckert, M. (1993). *Die Atomphysiker. Eine Geschichte der theoretischen Physik am Beispiel der Sommerfeldschule*. Braunschweig: Vieweg.

Eckert, M. (2013a). *Arnold Sommerfeld: Science, Life and Turbulent Times, 1868–1951*. New York: Springer.

Eckert, M. (2013b). Sommerfeld's *Atombau und Spektrallinien*. In M. Badino and J. Navarro (eds), *Research and Pedagogy: A History of Quantum Physics through Its Textbooks*, Berlin: Edition Open Access, pp. 117–35.

Fleck, L. (1979). *Genesis and Development of a Scientific Fact*. Chicago: University of Chicago Press.

Forman, P. (1971). Weimar Culture, Causality, and Quantum Theory, 1918–1927: Adaptation by German Physicists and Mathematicians to a Hostile Intellectual Environment. *Historical Studies in the Physical Sciences*, **3**, 1–115.

Forman, P., Heilbron, J. L., and Weart, S. (1975). Physics circa 1900: Personnel, Funding, and Productivity of the Academic Establishments. *Historical Studies in the Physical Sciences*, **5**, 1–185.

Forstner, C., and Walker, M. (eds) (2020). *Biographies in the History of Physics: Actors, Objects, Institutions*. Cham: Springer.

Franklin, A. (2016). Physics Textbooks Don't Always Tell the Truth. *Physics in Perspective*, **18**(1), 3–57.

Freire Jr, O. (2015). *The Quantum Dissidents: Rebuilding the Foundations of Quantum Mechanics (1950–1990)*. Berlin: Springer-Verlag.

Freire Jr, O. (2019). *David Bohm: A Life Dedicated to Understanding the Quantum World*. Cham: Springer.

Fyfe, A. (2002). Publishing and the Classics: Paley's Natural Theology and the Nineteenth-Century Scientific Canon. *Studies in History and Philosophy of Science*, **33**(4), 729–51.

Gandarias Perillán, J. (2011). *A Reexamination of Early Debates on the Interpretation of Quantum Theory: Louis de Broglie to David Bohm*. Ph.D. dissertation, University of Rochester.

Gavroglu, K. (1995). *Fritz London: A Scientific Biography*. Cambridge: Cambridge University Press.

Gispert, H. (1999). Les débuts de l'histoire des mathématiques sur les scènes internationales et le cas de l'entreprise encyclopédique de Felix Klein et Jules Molk. *Historia Mathematica*, **26**(4), 344–60.

Giulini, D. (2013). Max Born's Vorlesungen über Atommechanik, Erster Band. In M. Badino and J. Navarro (eds), *Research and Pedagogy: A History of Quantum Physics through Its Textbooks*, Berlin: Edition Open Access, pp. 207–29.

Goldberg, S. (1984). *Understanding Relativity: Origin and Impact of a Scientific Revolution*. Birkhäuser: Boston.

Gooday, G. and Mitchell D. J. (2013). Rethinking Classical Physics In J. Z. Buchwald, and R.Fox (eds), *The Oxford Handbook of the History of Physics*, Oxford: Oxford University Press, pp. 721–64.

Greca, I. M. and Freire Jr, O. (2014). Meeting the Challenge: Quantum Physics in Introductory Physics Courses. In M. Matthews (ed.), *International Handbook of Historical and Philosophical Research in Science Education*, Dordrecht: Springer, pp. 183–210.

Greenberger, D., Hentschel, K., and Weinert, F. (2009). *Compendium of Quantum Physics: Concepts, Experiments, History and Philosophy*. Heidelberg: Springer Verlag.

Haar, D. ter (1967). *The Old Quantum Theory*. Oxford: Pergamon Press.

Heilbron, J. L. (1964). *A History of the Problem of Atomic Structure from the Discovery of the Electron to the Beginning of Quantum Mechanics*. Ph.D., University of California.

Heilbron, J. L. (1968). Quantum Historiography and the Archive for History of Quantum Physics. *History of Science*, 7(1), 90–111.

Heilbron, J. L. (1986). *The Dilemmas of an Upright Man: Max Planck as Spokesman for German Science*. Berkeley, CA: University of California Press.

Heilbron, J. L. (ed.) (2003). *The Oxford Companion to the History of Modern Science*. Oxford: Oxford University Press.

Heilbron, J. L. (2016). Where to Start. In A. Blum, K. Gavroglu, C. Joas, and J. Renn (eds), *Shifting Paradigms: Thomas S. Kuhn and the History of Science*, Berlin: MPIWG, pp. 3–13.

Hendry, J. L. (1978). *An Investigation of the Mathematical Formulation of Quantum Theory and its Physical Interpretation, 1900–1927*. Ph.D. thesis, University of London.

Hendry, J. L. (1984). *The Creation of Quantum Mechanics and the Bohr–Pauli Dialogue*. Dordrecht: Reidel.

Hentschel, K. (2018). Light Quanta Reflected in Textbooks and Science Teaching. In K. Hentschel, *Photons: The History and Mental Models of Light Quanta*, Cham: Springer, pp. 133–39.

Hiley, B. J. (1997). Essay Review. Quantum Mechanics: Historical Contingency and the Copenhagen Hegemony by James T. Cushing. *Studies in the History and Philosophy of Modern Physics*, 28(2), 299–305.

Hirosige, T. (1968). [Review of *Sources for History of Quantum Physics. An Inventory and Report* by Thomas S. Kuhn; John L. Heilbron; Paul Forman; Lini Allen]. *Isis*, 59(1), 120–21.

Hoffman, D. (2008). *Max Planck: Die Enstehung der modernen Physik*. München: C. H. Beck.

Holmes, F. L. (1987). Scientific Writing and Scientific Discovery. *Isis*, 78(2), 220–35.

Howard, D. (2013). Quantum Mechanics in Context: Pascual Jordan's 1936 *Anschauliche Quantentheorie*. In M. Badino and J. Navarro (eds), *Research and Pedagogy: A History of Quantum Physics through Its Textbooks*, Berlin: Edition Open Access, pp. 267–85.

Hund, F. (1967). *Geschichte der Quantentheorie*. Mannheim: Bibliographisches Institut.

Ito, K. (2002). *Making Sense of Ryoshiron (Quantum Theory): Introduction of Quantum Mechanics into Japan, 1920–1940*. Ph.D. dissertation, Harvard University.

Ito, K. (2005). Geist in the Institute: The Production of Quantum Physicists in 1930s Japan. In D. Kaiser (ed.), *Pedagogy and the Practice of Science: Historical and Contemporary Perspectives*, Cambridge, MA: MIT Press, pp. 151–83.

Jammer, M. (1966). *The Conceptual Development of Quantum Mechanics*. New York: McGraw-Hill (New York: Tomash Publishers, American Institute of Physics, 1989, 2nd edn).

Jammer, M. (1974). *The Philosophy of Quantum Mechanics: The Interpretations of Quantum Mechanics in Historical Perspective*. New York: John Wiley and Sons.

Jammer, M. (1989). *The Conceptual Development of Quantum Mechanics*, 2nd edn. New York: Tomash Publishers, American Institute of Physics.

Jungnickel, C. and McCormmach, R. (1986). *Intellectual Mastery of Nature: Theoretical Physics from Ohm to Einstein*, Vol. 2. Chicago: University of Chicago Press.

Kaiser, D. (2005a). *Drawing Theories Apart: The Dispersion of Feynman Diagrams in Postwar Physics*. Chicago: The University of Chicago Press.

Kaiser, D. (ed.) (2005b). *Pedagogy and the Practice of Science: Historical and Contemporary Perspectives*. Cambridge MA: MIT Press.

Kaiser, D. (2007). Turning Physicists into Quantum Mechanics. *Physics World*, **20**(5), 28–33.

Kaiser, D. (2020). Training Quantum Mechanics. In D. Kaiser, *Quantum Legacies: Dispatches from an Uncertain World*, Chicago: University of Chicago Press, pp. 116–35.

Kangro, H. (1976). *Early History of Planck's Radiation Law*. London: Taylor and Francis.

Kapitza, P. L. (1973). Recollections of Lord Rutherford. In J. Mehra (ed.), *The Physicists' Conception of Nature*, Dordrecht: D. Reidel, pp. 749–65.

Kevles, D. J. (1971). *The Physicists: The History of a Scientific Community in Modern America*. New York: Alfred Knopf.

Klein, M. J. (1970). The First Phase of the Bohr–Einstein Dialogue. *Historical Studies in the Physical Sciences*, **2**, 1–39.

Kragh, H. (1985). The Fine Structure of Hydrogen and the Gross Structure of the Physics Community, 1916–26. *Historical Studies in the Physical Sciences*, **15**(2), 67–125.

Kragh, H. (1990). *Dirac: A Scientific Biography*. Cambridge: Cambridge University Press.

Kragh, H. (1992). A Sense of History: History of Science and the Teaching of Introductory Quantum Theory. *Science & Education*, **1**(4), 349–63.

Kragh, H. (1999). *Quantum Generations: A History of Physics in the Twentieth Century*. Princeton: Princeton University Press.

Kragh, H. (2013). Paul Dirac and the *Principles of Quantum Mechanics*. In M. Badino and J. Navarro (eds), *Research and Pedagogy: A History of Quantum Physics through Its Textbooks*, Berlin: Edition Open Access, pp. 249–65.

Kuhn, T. S. (1962). *The Structure of Scientific Revolutions*. Chicago: University of Chicago Press.

Kuhn, T. S. (1967). The Turn to Recent Science. *Isis*, **58**(3), 409–19.

Kuhn, T. S., Heilbron, J. L., Forman, P. L., and Allen, L. (eds) (1967). *Sources for History of Quantum Physics: An Inventory and Report*. Philadelphia: The American Philosophical Society. [https://amphilsoc.org/guides/ahqp/] (accessed 14/08/2019)

Lautesse, P., Vila Valls, A., Ferlin, F., Héraud, J.-L., Chabot, H. (2015). Teaching Quantum Physics in Upper Secondary School in France: 'Quanton' Versus 'Wave–Particle' Duality, Two Approaches of the Problem of Reference. *Science & Education*, **24**(7–8), 937–55.

Mehra, J., and Rechenberg, H. (1982). *The Historical Development of Quantum Theory*, Vol. 4, Part 2. New York: Springer.

Mohan, A. K. (2020). Philosophical Standpoints of Textbooks in Quantum Mechanics. *Science & Education*, **29**(3), 549–69.

Moore, W. (1989). *Schrödinger: Life and Thought*. Cambridge: Cambridge University Press.

Nora, P. (1984). Entre mémoire et histoire. In *Les Lieux de mémoire, I. La République*, Paris: Gallimard, pp. xvi–xlii.

Nye, M. J. (2015). Biography and the History of Science. In T. Arabatzis, J. Renn, and A. Simões (eds), *Relocating the History of Science: Essays in Honor of Kostas Gavroglu*, Cham: Springer, pp. 281–96.

Olesko, K. (1991). *Physics as a Calling: Discipline and Practice in the Königsberg Seminar for Physics*. Ithaca: Cornell University Press.

Olesko, K. (2005). The Foundations of a Canon: Kohlrausch's Practical Physics. In D. Kaiser (ed.), *Pedagogy and the Practice of Science: Historical and Contemporary Perspectives*, Cambridge, MA: MIT Press, pp. 323–55.

Olesko, K. (2006). Science Pedagogy as a Category of Historical Analysis: Past, Present, and Future. *Science & Education*, 15(7-8), 863–80.

Olesko, K. (2020). Ludwik Fleck, Alfred Schutz, and Trust in Science: The Public Responsibility of Science Education in Challenging Times. *HoST-Journal of History of Science and Technology*, 14(2), 50–72.

Portelli, A. (1999). *L'ordine è già stato eseguito. Roma, le Fosse Ardeatine, la memoria*. Roma: Donzelli.

Purrington, R. D. (2018). *The Heroic Age: The Creation of Quantum Mechanics, 1925–1940*. Oxford: Oxford University Press.

Rentetzi, M. (2007). *Trafficking Materials and Gendered Experimental Practices: Radium Research in Early 20th Century Vienna*. New York: Columbia University Press.

Rossi, P. (1986). *I ragni e le formiche. Un' apologia della storia della scienza*. Bologna: Il Mulino.

Rudolph, J. (2002). *Scientists in the Classroom: The Cold War Reconstruction of American Science Education*. New York: Palgrave.

Schweber, S. S. (2000). *In the Shadow of the Bomb: Oppenheimer, Bethe, and the Moral Responsibility of the Scientist*. Princeton: Princeton University Press.

Schweber, S. S. (2012). *Nuclear forces: The Making of the Physicist Hans Bethe*. Cambridge, MA: Harvard University Press.

Schwinger, J. (1973). A Report on Quantum Electrodynamics. In J. Mehra (ed.), *The Physicists' Conception of Nature*, Dordrecht: D. Reidel, pp. 413–29.

Secord, J. (2004). Knowledge in Transit. *Isis*, 95(4), 654–73.

Seth, S. (2008). Mystik and Technik: Arnold Sommerfeld and Early-Weimar Quantum Theory. *Berichte zur Wissenschaftgeschichte*, 31(4), 331–52.

Seth, S. (2010). *Crafting the Quantum: Arnold Sommerfeld and the Practice of Theory, 1890–1926*. Cambridge, MA: MIT Press.

Seth, S. (2013). Quantum Physics. In J. Z. Buchwald, and R.Fox (eds), *The Oxford Handbook of the History of Physics*, Oxford: Oxford University Press, pp. 814–58.

Simon, J. (2009). Circumventing the 'elusive quarries' of Popular Science: the Communication and Appropriation of Ganot's Physics in Nineteenth-century Britain. In F. Papanelopoulou, A. Nieto-Galan, and E. Perdiguero (eds), *Popularising Science and Technology in the European Periphery, 1800–2000*, Aldershot: Ashgate, pp. 89–114.

Simon, J. (2011). *Communicating Physics: The Production, Circulation and Appropriation of Ganot's Textbooks in France and England, 1851–1887*. London: Pickering and Chatto.

Simon, J. (2013). Physics Textbooks and Textbook Physics in the Nineteenth and Twentieth Centuries. In J. Z. Buchwald, and R.Fox (eds), *The Oxford Handbook of the History of Physics*, Oxford: Oxford University Press, pp. 651–78.

Simon, J. (2015). History of Science. In R. Gunstone (ed.), *Encyclopedia of Science Education*, Berlin: Springer, pp. 456–59.

Simon, J. (2016). Textbooks. In B. Lightman (ed.), *A Companion to the History of Science*, Oxford: Wiley Blackwell, pp. 400–13.

Simon, J. (2019). The Transnational Physical Science Study Committee: The Evolving Nation in the World of Science and Education (1945–1975). In J. Krige (ed.), *How Knowledge Moves: Writing the Transnational History of Science and Technology*, Chicago: University of Chicago Press, pp. 308–42.

Söderqvist, T. (2011). The Seven Sisters: Subgenres of Bioi of Contemporary Life Scientists. *Journal of the History of Biology*, **44**(4), 633–50.

Sommerfeld, A. (1916). Zur Quantentheorie der Spektrallinien. *Annalen der Physik*, **356**(17–18), 1–94 and 125–167.

Sopka, K. R. (1988). *Quantum Physics in America: The Years through 1935*. New York: Tomash Publishers, American Institute of Physics.

Staley, R. (2008). *Einstein's Generation: The Origins of the Relativity Revolution*. Chicago: University of Chicago Press.

Stambler, S. (1971). *The Impact of the Philosophical Implications of the Relativity and Quantum Theories on the Teaching of College Physics*. Ph.D. dissertation, New York University.

Stichweh, R. (1996). *Zur Entstehung des modernen Systems wissenschaftlicher Disziplinen: Physik in Deutschland, 1740–1890*. Frankfurt: Suhrkamp.

Stichweh, R. (2001). History of Scientific Disciplines. In N. J. Smelser and P. B. Baltes (eds), *International Encyclopedia of the Social & Behavioral Sciences*, Vol. 20, Amsterdam: Pergamon, pp. 13727–31.

Waerden, B. L. van der (1967). *Sources of Quantum Mechanics*. Amsterdam: North-Holland.

Waquet, F. (2003). *Parler comme un livre: L'oralité et le savoir (XVIe—XXe siècle)*. Paris: Albin Michel.

Warwick, A. (2003). *Masters of Theory: Cambridge and the Rise of Mathematical Physics*. Chicago: Chicago University Press.

Weisz, G. (2006). *Divide and Conquer: A Comparative History of Medical Specialization*. Oxford: Oxford University Press.

Weizsäcker, C. F., Kuhn, T. S., and Heilbron, J. L. (1963). Oral history interview with Carl Friedrich Weizsäcker, 1963 July 9. In *Archives for History of Quantum Physics*. College Park, MD USA: Niels Bohr Library & Archives, American Institute of Physics [www.aip.org/history-programs/niels-bohr-library/oral-histories/4947] (Accessed 01/04/20).

WennersHerron, A. (2011). Fermilab Historian Lillian Hoddeson wins APS prize. *Fermilab Today*, Oct. 24 [https://www.fnal.gov/pub/today/archive/archive_2011/today11-10-24_Lillian%20HoddesonReadMore.html] (accessed 28/03/20).

Wilson, E. B. (1912). A Treatise on Physics. *Bulletin of the American Mathematical Society*, **18**(10), 497–508.

# CHAPTER 29

# CHIEN-SHIUNG WU'S CONTRIBUTIONS TO EXPERIMENTAL PHILOSOPHY

## INDIANARA SILVA

## 29.1 INTRODUCTION

CHIEN-SHIUNG Wu moved in 1936 from the Republic of China to the US to pursue her PhD studies and in 1940 completed her doctoral thesis in nuclear physics at the University of California, Berkeley. As a woman and a foreigner, she experienced discrimination in the job market and found her first teaching position at the women's Smith College. In 1942, Wu went to Princeton University, becoming the first female instructor in its Department of Physics, and joined the Manhattan Project at Columbia University, where she studied the gaseous diffusion of uranium. After the war, she stayed on as a faculty member at Columbia, researching nuclear physics and promoting women's rights, mainly through the example of her own scientific career. Wu is renowned for her elegant 1956 experiment, which violated parity conservation for weak nuclear interactions. She received numerous awards and honours, including the National Medal of Science, the Bonner Prize, and the Wolf Prize in Physics (Rossiter, 1984, 1998; Hammond, 2009; Chiang, 2013; Maia Filho and Silva, 2019b).

This chapter focuses on Wu's lesser-known contributions to experimental philosophy (Maia Filho and Silva, 2019a, 2019b). We analyse the Wu–Shaknov (WS) experiment of 1950 and how their data helped reopen discussions about quantum interpretations through the work of David Bohm and Yakir Aharonov. We then discuss the 1975 Kasday–Ullman–Wu (KUW) test, which attempted to break down hidden-variable models, and present concluding reflections. Both contributions are important historical cases that teach us about the different ways physicists, experiments, and data can enrich scientific culture.

## 29.2 THE WS EXPERIMENT AND ITS ROLE IN QUANTUM FOUNDATIONS

The WS experiment was rooted in American post-war physics. It was partially supported by the Atomic Energy Commission (AEC), established in 1946 as a military agency to fund and control research into nuclear and later high-energy physics during the Cold War (Kevles, 1995; Cassidy, 2011; Koizumi, 2020). At first sight, their experiment was fundamental physics of annihilation radiation. However, it was also of interest to the military. The nuclear race was on, and basic research would lead to military, industrial, and civilian applications, as history has demonstrated (Forman, 1987; Kevles, 1995; Cassidy, 2011).

Positron–electron studies did exactly this. Sponsored by the US Energy Research and Development Administration, Richard Lambrecht, at Brookhaven National Laboratory, wrote a bibliographical survey about antimatter–matter interactions. And noted that '[t]he compilation is intended to be of particular value to chemists, physicists, and researchers in applied disciplines who have only in recent years begun to use positrons as a non-destructive nuclear probe' (Lambrecht, 1975, p. iii). The WS experiment was included, as were other works considered here. Annihilation Spectroscopy is now a cutting-edge technique employed in solid-state physics, the life sciences, and military technologies (Siegel, 1980; Süvegh and Marek, 2011; Slaughter, 2010).

By the time they studied electron–positron annihilation, Wu (Fig. 29.1) was an Associate Research Professor at Columbia University and Shaknov her doctoral student. Born in Boston in 1923, Shaknov was the son of a Jewish couple, William and Frances Shaknov, who emigrated from the Russian Empire in 1913 and 1915 respectively (US Census). It was a time of significant Jewish immigration to the US, which has been explained by primarily economic and demographic internal conditions and the spread of migration networks (Spitzer, 2013, 2014, 2018). In William's case, networks certainly played a role. In 1920, he obtained his first job as a Press Operator in Boston. This city received a large number of Jewish immigrants and became a highly influential Jewish community (Friedman, 1915; Sarna, Smith, and Kosofsky, 1995). William, therefore, chose a place with a supportive network. The manner of Frances' immigration also reflected this system. 'The networks of migration were mainly based on personal encounters or small close-knitted [sic] communities and in most cases close family relations,' claimed Spitzer (2014, p. 15). The Shaknovs built their life in the Boston area, where William worked as an insurance agent and later as a senior clerk (US Census, 1930; 1940a,b).

FIGURE 29.1 Chien-Shiung Wu. © AIP Emilio Segrè Visual Archives.

Irving's historical education at the English Boston High School paved the way for the 1939 scholarship he won to pursue his undergraduate studies at MIT (The Tech, 1939). There, he became Vice-President of the MIT Society of Physics Students, literary associate and literary editor of Voo Doo, the comic campus monthly, and was a member of the regatta rowing crew (The Tech, 1941a, 1942a, 1942c). In 1941, Irving received a journalism award for his article 'Showman' (The Tech, 1941b), and in the following year, he joined the Army ROTC (Reserve Officer Training Corps), MIT's military training programme, as a first lieutenant (The Tech, 1942b). He was awarded a bronze star for his services during World War II (CNA, 2019).

After the war, Shaknov went to Columbia University and in 1951 completed his doctorate in Physics. In 1953, he died tragically while serving as a civilian specialist for the US Navy's Operational Evaluation Group during the Korean War. As Tidman wrote, 'Dr. Irving Shaknov was killed when the plane in which he was collecting data was shot down during an interdiction mission.' His death sparked reflections about appropriate policies for civilian scientists (Tidman, 1984, p. 146).

Shaknov was therefore unable to witness the way in which the WS experiment would flourish in the physical sciences. It stimulated discussions in quantum foundations, provided the first entangled particles (whose effect would be central to quantum information in the second half of the 20th century), and became a non-destructive spectroscopy technique applied to different fields.

This experiment was designed to verify John Wheeler's 1946 pair theory prediction (Wu and Shaknov, 1950)—theories about interactions between electrons, positrons, and photons. As Wheeler and Ford recollected, '[t]he paper on polyelectrons that I submitted to the New York Academy of Sciences around the end of World War II was much more modest in its scope than some of my earlier dreaming about a world made of electrons' (Wheeler and Ford, 1998, p. 119). To build this world, Wheeler studied the fundamental properties of the annihilation process. When a positron and electron combined, they annihilated, and two photons were emitted in opposite directions with zero angular momentum. He predicted that these photons were orthogonally polarized and described an experimental arrangement to test this through coincidence measurements (Wheeler, 1946).

Following Wheeler's suggestion, Wu and Shaknov observed the angular correlation between the two photons in the annihilation process. In the WS set-up, the Columbia cyclotron produced isotopes of $^{64}Cu$ by bombarding a copper target with deuterons which, by being involved in an aluminium capsule S, gave off annihilation radiation. Photons were emitted in all directions, and the drilled lead block collimator selected only γ-rays in vertically opposite directions. The radiation was then scattered by aluminium and detected through scintillation counters (Wu and Shaknov, 1950). Counter-1 recorded a selected polarization, say one photon polarized in the $x$-direction, while detector-2 recorded a second polarized in the $y$-direction (Gerjuoy, 1973; Pipkin, 1979). Anthracene crystals, coupled with RCA photomultipliers, absorbed the annihilation radiation and transformed it into scintillation light, which was converted to electrical signals (Fig. 29.2). Electronic circuits received output signals for amplification, selection, and counting (Johnson, 1956; Birks, 1964; Horrocks, 1974).

In order to calculate the theoretical counting rate, Wu and Shaknov's work was based on earlier developments from Pryce and Ward at Clarendon Laboratory, which demonstrated that the coincidence rates between the scattered photons are higher for perpendicular azimuths than coplanar azimuths and have a peak scattered angle of 90° (Pryce and Ward, 1947). Snyder, Pasternack, and Hornbostel, from the Brookhaven National Laboratory, also determined the correlation effect as:

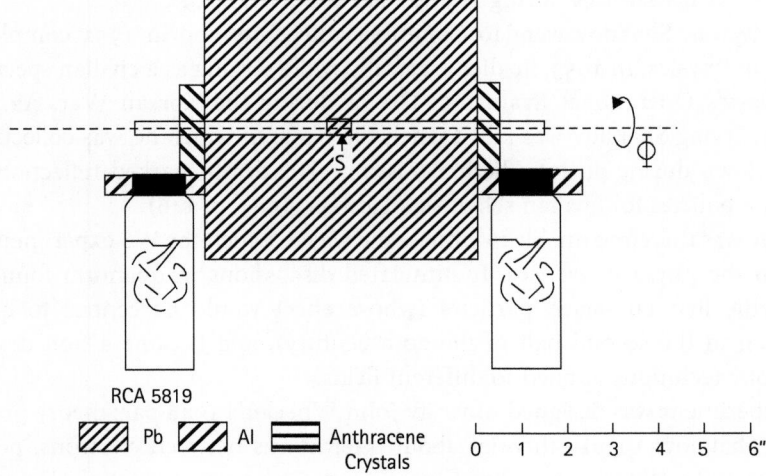

**FIGURE 29.2** The WS experimental set-up (Wu and Shaknov, 1950, p. 136). © American Physical Society.

$$\rho = \frac{N_{\varphi=\frac{\pi}{2}}}{N_{\varphi=0}} = 1 + \frac{2sin^4\theta}{\gamma^2 - 2\gamma sin^2\theta}$$

$$\gamma = 2 - cos\theta + \frac{1}{2 - cos\theta},$$

(1)

when photons were scattered through the same angle $\theta$. They derived Eq. (1) from the Klein–Nishina formula for Compton scattering, which described interactions between a free electron and a linearly polarized photon, and provided a relation between the initial and scattered momenta, as well as the angle between the polarization direction of the incident quantum and the scattered one. Assuming ideal geometry, they found the maximum theoretical asymmetry ratio $\rho$ to be 2.85 for a scattering angle of 82°, and 2.60 for 90° (Snyder, Pasternack, and Hornbostel, 1948).

In the WS experiment, the two photons were scattered at angles around $\theta = 82°$, while the azimuthal angle differences between the counters' axes $\varphi$ were 0, 90°, 180°, and 270°. The experiment registered coincidences when one of the counters was fixed while the second rotated, and vice versa. The results showed that the rates for scattered photons in a perpendicular scattering plane were higher than for those in a parallel scattering plane. After being scattered at the same angle with the counters at right angles, both annihilation radiation photons were thus detected simultaneously (Table 1, Wu and Shaknov, 1950).

The first two experiments based on Wheeler's prediction used the Geiger–Müller counter, while Wu and Shaknov's innovative arrangement used a more reliable and highly efficient scintillation counter (Bleuler and Bradt, 1948; Hanna, 1948; Wu and Shaknov, 1950). 'With this improved efficiency,' emphasized Wu and Shaknov (1950, p. 136), 'there will be an increase in the coincidence counting rate of one hundred times.' Apart from the detection process, these experiments used virtually the same experimental design, source, and scatterer material, although the WS experiment verified Wheeler's prediction more accurately (Table 29.1).

Physicists interested in Wheeler's prediction cited the same 1948 theoretical article, which assumed that 'the observation of one scattered photon gives information about the initial state of polarization of the other photon' (Snyder, Pasternack, and Hornbostel, 1948, p. 441). Interestingly, they did not raise philosophical questions, nor did they relate the correlation effect to quantum entanglement. These attitudes illustrated the power of the Copenhagen interpretation (Jammer, 1974) and Freire's thesis that foundations remained a marginalized field until the 1980s (Freire, 2014). Since this interpretation was considered mainstream and rarely questioned at the time, authors (Pryce and Ward, 1947; Snyder, Pasternack, and Hornbostel, 1948; Bleuler and Bradt, 1948; Hanna, 1948; Wu, and Shaknov, 1950) took the orthodox interpretation of quantum mechanics for granted.

Following the debates between Albert Einstein and Niels Bohr in the 1930s, there was a long period of relative silence during which the Copenhagen interpretation continued to advance without challenge. In 1952, Bohm became one of the first

Table 29.1 Instruments, techniques, and data.[1]

|  | Bleuler and Bradt(1948) | Hanna(1948) | Wu and Shaknov(1950) |
| --- | --- | --- | --- |
| Positron source | $Cu^{64}$ | $Cu^{64}$ | $Cu^{64}$ |
| Scatterer material | Aluminum | Aluminum[2] | Aluminum |
| Counter | Geiger | Geiger | Scintillation |
| Theoretical $c_\perp / c_\parallel$ | 1.7 | 1.86 | 2.00 |
| Experimental $c_\perp / c_\parallel$ | $1.9 \pm 0.3$ | $1.51 \pm 0.10$ | $2.04 \pm 0.08$ |

physicists to break this consensus and challenge the orthodox philosophy of quantum mechanics by defending a causal interpretation (Jammer, 1974; Freire, 2014, 2019; Whitaker, 2016). He developed the concept of quantum potential by manipulating the Schrödinger and Hamilton–Jacobi equations, to which particles were connected when spatially apart (Hiley and Peat, 1987) and of which he said '[t]hat new potential, they would have to explain all the new properties of matter' (Bohm, 1986). His approach rescued classical determinism, assuming that microscopic quantum entities could simultaneously have both position and momentum (Jammer, 1974; Freire, 2014, 2019; Whitaker, 2016). Bohm's heterodox interpretation of quantum mechanics not only reopened the discussion about quantum foundations (Freire, 2014, 2019), but also influenced the first generation of quantum dissidents and connected the subcultures of physics and quantum philosophy.

In 1957, at the Technion, Bohm and his graduate student Aharonov published the Discussion of Experimental Proof for the Paradox of Einstein, Rosen, and Podolsky (EPR), innovating an experimental approach to the foundations of quantum mechanics. Previously, philosophers and scientists had only applied thought experiments to discuss foundations (Brown and Fehige, 2019; Peacock, 2017).

Bohm and Aharonov considered the WS experiment to illustrate the EPR paradox and explored its implications for quantum philosophy. Unlike the original version, they assumed a molecule composed of two particles. The first had a spin $\frac{+\hbar}{2}$, while the second had a spin $\frac{-\hbar}{2}$ and the spin total remained zero. The two particles were then spatially separated to a point where no interaction occurred. When they used usual quantum mechanics to describe this system's wave function, they found

$$\psi = \frac{1}{\sqrt{2}} [\psi_+(1)\psi_-(2) - \psi_-(1)\psi_+(2)] \qquad (2)$$

in which $\psi_+(1)$ described the state of particle 1 with spin $\frac{+\hbar}{2}$, and $\psi_-(2)$, and particle 2 with spin $\frac{-\hbar}{2}$. Eq. (2) suggested that the state vectors of the two particles were correlated

---

[1] In 1949, Vlasov and Dzehelepov also conducted coincidence experiments using Geiger tubes (Belinfante, 1973; Pipkin, 1979).

[2] Hanna (1948) also used brass material in his experiment, although this was not included in our comparison.

even when apart. If you measured the spin component of particle 1 as $\frac{+\hbar}{2}$, for example, you would know that the spin of particle 2 was $\frac{-\hbar}{2}$. No matter how many times you measured particle 1, you would always have information about particle 2 without disturbing it. The explanation for this correlation was either that canonical quantum mechanics was incomplete, unable to describe a situation when particles were spatially separated, or that instantaneous interaction was possible. Either way, according to Bohm and Aharonov's spin example, when applied to non-interacting systems Copenhagen's formalism would break down (Bohm and Aharonov, 1957).

However, for Bohr, the EPR paradox was never a paradox. In 1935, he introduced the relational concept of quantum states, through which objects and measuring apparatus formed an integrated system (Jammer, 1974), and consequently, 'the questions of how correlations come about simply has no meaning'. Unlike Bohr, Bohm and Aharonov aimed to explore the correlation genesis of EPR and even felt tempted to consider 'a hidden interaction', through quantum potential between spatially separated systems. However, special relativity restricted their interpretation (Bohm and Aharonov, 1957, p. 1072).

They turned their attention to Furry's hypothesis of 1936. In this interpretation, Eq. (2) was valid only when particles 1 and 2 were interacting and would change from a pure state to a mixture

$$\psi = \psi_{+\theta,\varphi}(1)\psi_{-\theta,\varphi}(2) \tag{3}$$

when spatially separated, while $\psi_{-\theta,\varphi}(1)$ was the positive spin component of particle 1 in $\theta$, $\varphi$ directions. This implied no EPR paradox, since Furry's wave function was a product of two independent states (Selleri and Zeilinger, 1988). As each particle had a definite spin state uncorrelated to the other, when spatially separated, measuring particle 1 did not interfere with particle 2. Furry's interpretation rescued the locality principle and used uniform probability distribution to predict what quantum states would appear after the measurement process (Bohm and Aharonov, 1957). Furry insisted, '[i]t does not mean that quantum mechanics is not to be regarded as a satisfactory way of correlating and describing experience; it does illustrate the difficulty, often remarked upon by Bohr, which is inherent in the problem of the distinction between subject and object' (Furry, 1936, p. 399).

Bohm and Aharonov then compared the usual interpretation of quantum mechanics and Furry's hypothesis with the WS experimental results. Calculating the corresponding WS wave function, they found a linear combination similar to Eq. (2),

$$\phi_1 = \frac{1}{\sqrt{2}}(\psi_1 - \psi_2) = \frac{1}{\sqrt{2}}(C_1^x C_2^y - C_1^y C_2^x)\psi_0 \tag{4}$$

with $C_1^x$ and $C_1^y$ describing the creation operators for photons polarized in the $+k$ direction, while $C_2^y, C_2^x$ described those for the opposite one. The pairs of photons resulting from the annihilation process were orthogonally correlated, as in the case of EPR. '[W]e have essentially the same puzzling kind of correlations in the properties of distant particles, in which the property of any one photon that is definite is determined by a measurement on a far-away photon' (Bohm and Aharonov, 1957, p. 1074). Prior to measurement, the wave function described a superposition of polarization states, but after it, the wave function collapsed, resulting in the product $C_1^x C_2^y$ or $C_1^y C_2^x$.

Furthermore, they determined scattering probabilities from standard quantum mechanics and Furry's hypothesis, and compared these to the WS experiment. '[T]his experiment is explained adequately by the current quantum theory', they concluded, 'which implies distant correlations, of the type leading to the paradox of EPR, but not by any reasonable hypotheses implying a breakdown of the quantum theory that could avoid the paradox of EPR' (Bohm and Aharonov, 1957, p. 1075). This paradox, initially designed as a thought experiment, thus became a physical effect, as produced by Wu and Shaknov. Heterodox interpretations had no other choice but to accept correlations between non-interacting systems.

Later discussions showed that the WS result was not experimental proof of the EPR argument (Jammer, 1974; Freire, 2019). The WS detectors were designed to record photons scattered in the $x$- and $y$-planes before the measuring process. In the EPR situation, however, one should be able to measure any component of one of the photons and then predict the properties of the other (Yoshihuku, 1972). Wu and Shaknov demonstrated the polarization-dependent joint distribution of Compton scattering, rather than polarization correlations. Experiments testing the EPR argument required scattering techniques and Compton polarimeters or, as with Kocher and Commins in 1967, the use of efficient polarization filters to observe polarization correlations of lower energy photons (Jammer, 1974).

Bohm and Aharonov's work opened the door for a discussion of real experimental tests of the foundations of quantum mechanics. Using Web of Science and Google Scholar data to trace back the citations of their 1957 paper, we find that the most important breakthroughs in experimental philosophy refer to Bohm and Aharonov's paper (Table 29.2). Their work contributed to a revival of attention aimed at the 1935 EPR article, with the help of the WS data, and inspired physicists to challenge the Copenhagen interpretation.

The WS experiment, therefore, contributed to the culture of quantum mechanics, provided a background for Bohm and Aharonov's call for renewed attention to quantum foundations, and encouraged further experimental and philosophical endeavours (Table 29.2). It was also the first experiment to exhibit the entanglement effect (Bouwmeester et al., 1999; Gröblacher et al., 2007; Duarte, 2012; Whitaker, 2012, 2016). As Zeilinger confirmed, '[a]n earlier experiment by Wu and Shaknov (1950) had demonstrated the existence of spatially separated entangled states' (Zeilinger, 1999, p. S293). The term entangled was coined by Schrödinger while interpreting the EPR argument and its quantum-mechanical formalism (Jammer, 1974; Gilder, 2008; Whitaker, 2016); nowadays quantum entanglement studies include quantum cryptography, quantum teleportation, and quantum computing (Prevedel et al., 2007; Duarte, 2019).

Table 29.2 Principal citations of Bohm and Aharonov (1957) (Web of Science and Google Scholar databases, 2020).

| ARTICLE | AUTHOR | JOURNAL | YEAR |
|---|---|---|---|
| On the Einstein Podolsky Rosen paradox | Bell, J. S. | Physics Physique Fizika | 1964 |
| On the problem of hidden variables in quantum mechanics | Bell, J. S. | Reviews of Modern Physics | 1966 |
| Proposed experiment to test local hidden-variable theories | Clauser, J. F.; Horne, M. A.; Shimony, A.; Holt, R. A. | Physical Review Letters | 1969 |
| The Einstein-Podolsky-Rosen argument: Positron annihilation experiment | Kasday, L.; Ullman, J.; Wu, C. S. | Bulletin of the American Physical Society | 1970 |
| Experimental test of local hidden-variable theories | Freedman, S. J.; Clauser, J. F. | Physical Review Letters | 1972 |
| Experimental limitations to validity of semiclassical radiation theories | Clauser, J. F. | Physical Review A | 1972 |
| Experimental consequences of objective local theories | Clauser, J. F.; Horne, M. A. | Physical Review D | 1974 |
| Angular Correlation of Compton-Scattered Annihilation Photonsand Hidden Variables | Kasday, L.; Ullman, J.; Wu, C. S. | Nuovo Cimento | 1975 |
| Proposed experiment to test separable hidden-variable theories | Aspect, A. | Physics Letters A | 1975 |
| Measurement of the circular-polarization correlation in photons from an atomic cascade | Clauser, J. F. | Nuovo Cimento B | 1975 |
| Nonlocality and polarization correlations of annihilation quanta | Bohm, D. J.; Hiley, B. J. | Nuovo Cimento B | 1976 |
| Bells theorem - experimental tests and implications | Clauser, J. F.; Shimony, A. | Reports on Progress in Physics | 1978 |
| Quantum-theory and time asymmetry | Zeh, H. D. | Foundations of Physics | 1979 |
| Experimental tests of realistic local theories via Bells theorem | Aspect, A.; Grangier, P.; Roger, G. | Physical Review Letters | 1981 |
| Experimental test of Bell inequalities using time-varying analyzers | Aspect, A.; Dalibard, J.; Roger, G. | Physical Review Letters | 1982 |
| Reflections on the philosophy of Bohr, Heisenberg and Schrödinger | Shimony, A. | Journal de Physique | 1983 |

## 29.3 THE KASDAY–ULLMAN–WU (KUW) EXPERIMENT IN QUANTUM PHILOSOPHY

The Copenhagen interpretation was so successful in teaching canonical quantum mechanics that physicists often considered the research on foundations to be 'armchair philosophy', 'stigma', and 'junk science', imposing professional and personal constraints on its proponents. To survive in a hostile community, quantum rebels designed strategies to maintain their jobs and fellowships, circulate ideas and prove they were also undertaking real physics. For example, John Clauser was discredited for performing experiments on experimental philosophy and never held a permanent US university position (Freire, 2014; Whitaker, 2012, 2016).

This hostility towards foundations did not intimidate Wu. By then, she was already a distinguished experimental physicist and, in 1975, became the first female President of the American Physical Society (APS). In 1974, she was awarded the Scientist of the Year by Industrial Research, '[a] nuclear physicist, she is best known for leading the group that performed the 1956 experiment proving that parity is not conserved in weak interactions' (Anon., 1974). Her prestigious position facilitated her autonomous research agenda: unlike many other physicists, she did not have to worry about her research being tagged as junk science.

She did not go unnoticed in foundation research. For example, Bell invited her to a meeting of physicists interested in undertaking experimental philosophy, although she was unable to attend (Fig. 29.3). He clearly acknowledged the need to create a quantum philosophy community, as well as recognizing the symbolic value her image would lend to a struggling field (Freire, 2014).

The Austrian-born physicist Otto Frisch also contacted her for further information about the relative polarization of annihilation radiation. He noted, '[y]ou may wonder why I should wish to bring up that subject again. Well, I have a kind of itch to poke my fellow physicists from time to time, to remind them that physics is still conceptually quite a tricky subject' (Frisch to Wu, 5 December 1973, Wu Papers, Columbia University Archives). These words may have made Wu sufficiently comfortable to confess her own trouble with quantum philosophy:

> About three years ago, the two photon experiment was repeated by one of my students in our laboratory with great care and improvement. The results were beautiful if you forgive my saying so. However, when it came to writing up the paper and trying to explain the hidden variable theories and how to compare the predictions of Bell's hidden variable theories to our experimental results, it was a struggle.
>
> (Wu to Otto Frisch, 11 January 1974, Wu Papers,
> Columbia University Archives)

ORGANISATION EUROPÉENNE POUR LA RECHERCHE NUCLÉAIRE

**CERN** EUROPEAN ORGANIZATION FOR NUCLEAR RESEARCH

Téléphone :    (022) 41 98 11

Télex :    GENÈVE - 2 36 98

Télégramme :    CERNLAB-GENÈVE

**1211 GENÈVE 23**

Genève,    30 May 1975

Professor C.S. Wu
Pupin Physics Laboratories
Columbia University
<u>NEW YORK</u>, N.Y. 10027

Dear Professor Wu,

      Bernard d'Espagnat and I are organizing a small meeting at Erice, in Sicily, under the auspices of the Ettore Majorana Centre for Scientific Culture, during April 19–23, 1976. The intention of the meeting is to bring together people interested in experiments on the basic notions of quantum mechanics, and more particularly, people interested in <u>doing</u> such experiments. We would very much like to have your participation in this meeting.

      Your personal living expenses at Erice will be taken care of by the Centre. But unfortunately we are unable to reimburse travel expenses.

      We very much hope to hear from you soon that you will come. We would also be glad to have suggestions from you about other people to be invited.

                  Sincerely,

                  *John Bell*

                  John Bell

*Dear Dr. Bell*

*Many thanks for your kind invitation to a small meeting on the basic notions of quantum mechanics at Erice in Sicily. I am very sorry my teaching and societal duties prevent me from attending this meeting. My best wishes for the success of the meeting. Sincerely C.S. Wu*

**FIGURE 29.3** John Bell to Wu, 30 May 1975 (Wu Papers, Columbia University Archives).

The graduate student Wu is referring to was Leonard Ralph Kasday at Columbia University. Wu emphasized his scientific qualities in a recommendation letter, 'Mr. Kasday is well balanced in both theory and experimental technique. He has talent, good sense and taste, as well as perseverance. He is a promising young scientist in every sense of the word' (undated, Wu Papers, Columbia University Archives). Kasday received support from the US Atomic Energy and National Science Foundation, completing his Ph.D. in experimental philosophy in 1972.

The struggle Wu described may explain her and Kasday's subsequent collaboration with John Ullman, at Herbert Lehman College, where they used the same WS set-up, but with improved geometrical corrections to address philosophical questions. They explored what Bohm and Aharonov had proposed when they discussed the WS data, taking account of new developments in quantum foundations (Kasday, Ullman, and Wu, 1975).

In 1971, Kasday presented their results at the Varenna Summer School. This was an excellent opportunity to report on his thesis and learn from other leaders in quantum interpretations. Among these, we find Eugene Wigner, Abner Shimony, Louis de Broglie, Josef-Maria Jauch, Franco Selleri, Bryce DeWitt, Bohm, and Bell (d'Espagnat, 1971). This school contributed to the creation of a community interested in foundations and a research agenda for subsequent decades (Freire, 2014).

Kasday published the KUW results in the Varenna proceedings and thanked the German physicist Mauritius Renninger for proposing the KUW experiment. Four years later, Kasday, Ullman, and Wu published a more extended paper in *Il Nuevo Cimento*. They seem to have used different strategies for each article. The paper published in *Il Nuovo Cimento* discussed technical and instrumental details and left the discussion of foundations to the very end (Kasday, Ullman, and Wu, 1975), while the Varenna paper started with philosophical questions and compared them to experimental results (Kasday, 1971). It is hard to say with confidence, but it appears that they first selected a strategy to demonstrate real experimental physics in action, and only then related it to quantum philosophy. This psychological approach in dealing with referees may have been a strategy to avoid hostile attitudes.

They really thought that the KUW results would help to settle the battle between local hidden variables models and quantum mechanics. So they speeded up their publication in *Il Nuevo Cimento* in the hope that, as Wu noted, since '[n]o deviation from that predicted from the quantum mechanics has been detected, [ ... ] this should certainly quiet those proponents of the hidden variables in quantum mechanics' (undated, Wu Papers). Contrary to her expectations, however, the battle lasted until the 1980s. In addition, she circulated their results and interpretations prior to publication, as her letter to Stuart J. Freedman revealed, who jointly with John Clauser had performed the first test on Bell's theorem a few years before:

> In the recent past, we carried out the investigation on the Angular Correlation of Compton-scattered Annihilation Photons with much improved precision and accuracy and also made an attempt to relate its interpretation with hidden variables.

I am taking the liberty of sending you a copy of our paper (just being sent out for publication now) in thinking that you might be interested in our interpretation.
(Wu to Freedman, 16 May 1974, Wu Papers)

The KUW experiment was driven by the EPR argument, Bell's inequalities, and Furry's hypothesis.[3] In order to address these ideas, Kasday, Ullman, and Wu looked at polarization measurements of annihilation photons. In this system, they considered parity and angular momentum conservation and described its wave function as

$$\psi = \frac{|RR\rangle - |LL\rangle}{\sqrt{2}} \tag{5}$$

in which these vectors represented two photons circularly polarized to the right (R) and left (L) in $\pm \hat{z}$ direction. When the two photons were linearly polarized, they found the system state was

$$\psi = \frac{|XY\rangle - |YX\rangle}{\sqrt{2}} i \tag{6}$$

where these vectors described one photon linearly polarized along the $\hat{x}$ axis in the $+\hat{z}$ direction, the other along the $\hat{y}$ axis in the $-\hat{z}$ direction. If you measured photon 1 with circular polarization R, then you assumed that photon 2 would have R. Now, if you measured X linear polarization for photon 1, for example, you assumed photon 2 would have Y. It would therefore be possible to predict the measurements of particle 2 without disturbing it, only by measuring particle 1, even when the particles no longer interact. But the usual quantum interpretation prohibited particle 2 from having definite linear and circular polarizations simultaneously. Through this example, they revealed that the EPR argument demonstrated the incompleteness of standard quantum mechanics (Kasday, 1971; Kasday, Ullman, and Wu, 1975).

With regard to Bell's inequalities, they explained that these assumed the particle 2 measurement, since measuring particle 1, 'is determined in advance by variables over which we have no control, but which are sufficiently revealed by the first measurement so we can anticipate the result of the second' (Kasday, 1971, p. 197). This set of parameters, with linear polarizer lever arms, would measure the outputs with certainty, thereby rescuing a deterministic viewpoint. Moreover, these inequalities were rooted in the locality hypothesis that described outputs A and B as functions $A(a\lambda)$, $B(b\lambda)$, meaning that each measurement of particle 1 did not interfere in the measurement of particle 2, where $a$ and $b$ were the settings for the lever arms and $\lambda$ the set of variables (Kasday, 1971, p. 197).

---

[3] In the KUW paper, they called Furry's hypothesis the Bohm–Aharonov hypothesis. Here, we have decided to maintain the former, in order to remain consistent with Bohm and Aharonov's work.

They realized that these interpretations would lead to different values for instrumental parameters when calculating the counting rates provided by

$$R(\varphi_1 \varphi_2) = A - B cos2(\varphi_2 - \varphi_1) \qquad (7)$$

where A and B were the parameters, and $\varphi_1$ and $\varphi_2$ were the azimuthal angles. The A and B values depended on the angular width of the slits, multiple scattering in the scatterers, and the detectors' efficiency. To investigate Bell's inequality in the Compton scattering process, they had to assume an ideal linear-polarization analyser, and that quantum theory was valid for the description of this analyser, as well as for Compton scattering experiments. They derived the value of B from Bell's inequality, the Furry hypothesis, and the quantum theory prediction, then compared this to their experimental results (Kasday, 1971; Kasday, Ullman, and Wu, 1975).

$A = 1; \ B = M_1 M_2$ if quantum mechanics is valid,
$A = 1; \ B \leq M_1 M_2 / \sqrt{2}$ if a local hidden variable theory is valid,
$A = 1; \ B \leq M_1 M_2 / 2$ if Furry's hypothesis is valid.

In the KUW experiment, the annihilation photons were scattered at $S_1$, $S_2$, their energies were detected in $D_1$, $D_2$, and the lead slits in the detectors selected the azimuthal angles $\varphi_1$ and $\varphi_2$. Unlike in the WS experiment, this used plastic scintillator scatterers and a four-fold coincidence detection to avoid accidental events (Kasday, 1971; Kasday, Ullman, and Wu, 1975). 'By devising an ingenious method to use thin plastic scintillators as scatterers as well as detectors for Compton recoil electrons', noted Wu, 'he [Kasday] set a stringent criterion in selecting the events and thus dramatically reduce the background counts' (undated, Wu Papers).

Although the quantum prediction for B agreed with KUW experimental data, this experiment did not provide a conclusive argument against local-hidden models because, without the additional assumptions, it agreed with them (Kasday, 1971; Kasday, Ullman, and Wu, 1975). The KUW experimental philosophy was faced with technical and instrumental constraints. Since ideal polarization detectors for high-energy photons were unavailable, Compton polarimeters registered outputs for *either* photon was detected, *or* no photon detected, while, to test Bell's inequality, they needed to register numeral values for all outputs (Kasday, Ullman, and Wu, 1975). For example, by assigning the value (+1) to the spin-up triggered by particle 1, (−1) to the spin-down by particle 2, (0) for particle 1 not detected, and the same for particle 2 (Clauser and Shimony, 1978).

They compared their results to the 1972 Freedman and Clauser (FC) experiment, which tested Bell's inequalities through atomic cascade coincidence measurements by assuming that the probability of detecting a photon was independent of whether or not it had passed through a polarizer (Freedman and Clauser, 1972; Clauser and Shimony, 1978). Despite different assumptions, they believed the two experiments enriched one another and contributed to testing local hidden-variable theories (Kasday, Ullman, and

Wu, 1975). However, Abner Shimony, writing to Clauser, said 'I suggest that at this point you make reference to the experiment of Kasday, Ullman, and Wu, and point out that they had to make a much stronger assumption. In this way you will do justice to their work without in any way diminishing the significance of your own experiment' (Shimony to Clauser, 14 January 1972, Shimony Papers, Archives of Scientific Philosophy). Clauser followed Simony's suggestion and added a brief note to the Freedman and Clauser paper (Freedman and Clauser, 1972).

For different reasons, both KUW and Clauser were frustrated by their experimental results. The former for not breaking the local model, '[i]t would be pleasing to be able to say that the results of this experiment rule out local hidden-variable theories' (Kasday, Ullman, and Wu, 1975, p. 654). The latter for performing an experiment that confirmed the Copenhagen interpretation, despite dreams to the contrary, as Shimony noted, '[y]our paper finally arrived today. It is a classic, but unfortunately a classic tragedy, since the hero dies, and dies nobly' (Shimony to Clauser, 14 January 1972, Shimony Papers, Archives of Scientific Philosophy).

Although KUW's experiment received little attention among quantum historians (Whitaker, 2012), it contributed to the culture of experimental philosophy and encouraged outsiders to study foundations. Subsequently, positronium annihilation experiments in England and Italy began to test Bell's inequality (Wilson, Lowe, and Butt, 1976; Bruno, d'Agostino, and Maroni, 1977; Faraci, Notarrigo, and Pennisi, 1979).

## 29.4 CONCLUDING REMARKS

The WS and KUW experiments contributed to the culture of quantum philosophy. Through Bohm and Aharonov's innovative approach, the WS data served to reopen discussions about the EPR argument. Using the WS experiment to address interpretational questions, Bohm and Aharonov drew attention to two subcultures which had never previously intersected, real tests and philosophy, and anticipated future breakthroughs in quantum foundations. Although the WS experiment turned out not to be the material equivalent of the EPR thought experiment, it contributed to the Bohm and Aharonov paper and resonated further among the first generation of quantum dissidents, including Bell and Clauser. The WS experiment became the first to demonstrate quantum entanglement, which revolutionized the era of quantum information. By doing what Shimony called quantum archaeology, Bohm and Aharonov demonstrated that the WS data had a life of their own and continued importance for the problem of quantum foundations.

Unlike the WS one, the KUW experiment was consciously aimed at addressing the foundations of quantum mechanics. This case illustrated how prestigious physicists, even when seeking to defend standard quantum mechanics, could create strategies to circulate ideas from a marginalized research field. Although the KUW philosophical test did not refute hidden-variable theories, except when supported by other assumptions, it contributed to experimental philosophy research. Despite facing instrumental difficulties, including non-ideal polarizers for high-energy photons, KUW's work helped to spread research into foundations, broaden knowledge, and inspire further studies of positronium annihilation.

Wu entered the field of experimental philosophy with the KUW experiment. When she agreed to supervise Kasday's research, quantum foundations were still stigmatized as junk science. Unlike early-career physicists, such as Clauser, Wu was already an influential physicist based at Columbia University. She was interested in taking up experimental challenges and the hostile attitudes she might experience by breaking the constraints to researching quantum foundations were not the first she had to face in her research. Prior to becoming the Lady of Physics, Wu suffered gender and race segregation. She also saw her sophisticated 1956 experiment omitted from the Nobel prize, which was awarded to theoretical physicists who discovered parity non-conservation in weak interactions (Maia Filho and Silva, 2019a). It is clear that Wu knew better than most how to survive in a hostile community.

## ACKNOWLEDGEMENTS

I am grateful to Alexei Kojevnikov and Olival Freire Jr for their valuable contributions to this manuscript. My thanks also to Lucas Carvalho and Sasha Masiello, for guiding me through Ashtanga Yoga over the pandemic period.

## REFERENCES

Anon. (1974). Chien-Shiung Wu is scientist of year. *Physics Today*, **27**(11), 75.

Aspect, A. (1975). Proposed experiment to test separable hidden-variable theories. *Physics Letters A*, **54**(2), 117–18.

Aspect, A., Dalibard, J., and Roger, G. (1982). Experimental test of Bell's inequalities using time-varying analyzers. *Physical Review Letters*, **49**(25), 1804–07.

Aspect, A., Grangier, P., and Roger, G. (1981). Experimental tests of realistic local theories via Bell's theorem. *Physical Review Letters*, **47**(7), 460–63.

Belinfante, F. J. (1973). *A Survey of Hidden-Variables Theories*. International Series of Monographs in Natural Philosophy, Vol. 55. Amsterdam: Elsevier.

Bell, J. S. (1964). On the Einstein–Podolsky–Rosen paradox. *Physics Physique Fizika*, **1**(3), 195–200.

Bell, J. S. (1966). On the problem of hidden variables in quantum mechanics. *Reviews of Modern Physics*, **38**(3), 447–52.

Birks, J. B. (1964). *The theory and practice of scintillation counting*. International series of monographs in electronics and instrumentation, Vol. 27. Amsterdam: Elsevier.

Bleuler, E., and Bradt, H. L. (1948). Correlation between the states of polarization of the two quanta of annihilation radiation. *Physical Review*, **73**(11), 1398.

Bohm, D., and Aharonov, Y. (1957). Discussion of experimental proof for the paradox of Einstein, Rosen, and Podolsky. *Physical Review*, **108**(4), 1070–76.

Bohm, D. J., and Hiley, B. J. (1976). Nonlocality and polarization correlations of annihilation quanta. *Il Nuovo Cimento B*, **35**(1), 137–44.

Bohm, D. (1986). Interview by Maurice Wilkins, 6 June 1986, Niels Bohr Library & Archives, American Institute of Physics, College Park, MD, USA. www.aip.org/history-programs/niels-bohr-library/oral-histories/32977-1.

Bouwmeester, D., Pan, J.-W., Daniell, M., Weinfurter, H., and Zeilinger, A. (1999). Observation of three-photon Greenberger–Horne–Zeilinger entanglement. *Physical Review Letters*, **82**(7), 1345–49.

Brown, J. R., and Fehige, Y. (2019). Thought Experiments. *Stanford Encyclopedia of Philosophy*, Winter 2019 edition. https://plato.stanford.edu/archives/win2019/entries/thought-experiment/.

Bruno, M., d'Agostino, M., and Maroni, C. (1977). Measurement of linear polarization of positron annihilation photons. *Il Nuovo Cimento B*, **40**(1), 143–52.

Cassidy, D. C. (2011). *A Short History of Physics in the American Century*. Cambridge, MA: Harvard University Press.

Chiang, T.-C. (2013). *Madame Wu Chien-Shiung: The First Lady of Physics Research*. Singapore: World Scientific.

Clauser, J. F. (1972). Experimental limitations to the validity of semiclassical radiation theories. *Physical Review A*, **6**(1), 49–54.

Clauser, J. F. (1976). Measurement of the circular-polarization correlation in photons from an atomic cascade. *Il Nuovo Cimento B*, **33**(2), 740–46.

Clauser, J. F., and Horne, M. A. (1974). Experimental consequences of objective local theories. *Physical Review D*, **10**(2), 526–35.

Clauser, J. F., Horne, M. A., Shimony, A., and Holt, R. A. (1969). Proposed experiment to test local hidden-variable theories. *Physical Review Letters*, **23**(15), 880–84.

Clauser, J. F., and Shimony, A. (1978). Bell's theorem. Experimental tests and implications. *Reports on Progress in Physics*, **41**(12), 1881–27.

CNA (2019). Twitter Post, 8 November 2019. https://twitter.com/CNA_org/status/1192811377645232128.

d'Espagnat, B. (1971). *Proceedings of the International School of Physics Enrico Fermi. Course IL. Foundations of quantum mechanics. Course held at Varenna, Italy, 29 June 29–11 July 1970*. New York: Academic Press.

Duarte, F. J. (2012). The origin of quantum entanglement experiments based on polarization measurements.' *The European Physical Journal H*, **37**(2), 311–18.

Duarte, F. J. (2019). *Fundamentals of Quantum Entanglement*. Bristol: IOP Publishing.

Faraci, G., Notarrigo, S., and Pennisi, A. R. (1979). Simultaneous counting of true and random events in a four-fold coincidence system using two time-to-pulse-height converters. *Nuclear Instruments and Methods*, **164**(1), 157–62.

Forman, P. (1987). Behind quantum electronics: National security as basis for physical research in the United States, 1940–1960. *Historical Studies in the Physical and Biological Sciences*, **18**(1), 149–29.

Furry, W. H. (1936). Note on the quantum-mechanical theory of measurement. *Physical Review*, **49**(5), 393–99.

Freire Jr, O. (2014). *The Quantum Dissidents: Rebuilding the Foundations of Quantum Mechanics (1950–1990)*. Berlin: Springer.

Freire Jr, O. (2019). *David Bohm: A life dedicated to understanding the quantum world*. Berlin: Springer Nature.

Freedman, S. J., and Clauser, J. F. (1972). Experimental test of local hidden-variable theories. *Physical Review Letters*, 28(14), 938–41.

Friedman, L. M. (1915). Early Jewish Residents in Massachusetts. *Publications of the American Jewish Historical Society*, 23, 79–90.

Gerjuoy, E. (1973). Is the Principle of Superposition Really Necessary? In C. A. Hooker (ed.), *Contemporary Research in the Foundations and Philosophy of Quantum Theory*, Berlin: Springer, pp. 114–42.

Gilder, L. (2008). *The age of entanglement: when quantum physics was reborn*. New York: Vintage.

Gröblacher, S., Paterek, T., Kaltenbaek, R., Brukner, Č., Żukowski, M., Aspelmeyer, M., and Zeilinger, A. (2007). An experimental test of non-local realism. *Nature*, 446(7138), 871–75.

Hammond, R. (2009). *Chien-Shiung Wu: Pioneering Nuclear Physicist*. New York: Chelsea House Publications.

Hanna, R. C. (1948). Polarization of annihilation radiation. *Nature*, 162(4113), 332.

Hiley, B. J. and Peat, F. D. (eds) (1987). *Quantum implications: Essays in honour of David Bohm*. London: Routledge.

Horrocks, D. (ed.) (1974). *Applications of liquid scintillation counting*. Amsterdam: Elsevier.

Jammer, M. (1974). *The Philosophy of Quantum Mechanics; the Interpretations of Quantum Mechanics in Historical Perspective*. New York: Wiley.

Johnson, R. N. (1956). The Eisenhower Personnel Security Program. *The Journal of Politics*, 18(4), 625–50.

Kasday, L. R. (1971). Experimental test of quantum predictions for widely separated photons. In B. d'Espagnat (ed.), *Foundations of quantum mechanics, Proceedings of the International School of Physics Enrico Fermi*, New York: Academic Press, pp. 195–210.

Kasday, L., Ullman, J., and Wu, C. S. (1970). Einstein–Podolsky–Rosen argument-positron annihilation experiment. *Bulletin of the American Physical Society*, 15(4), 586.

Kasday, L. R., Ullman, J. D., and Wu, C. S. (1975). Angular correlation of Compton-scattered annihilation photons and hidden variables. *Il Nuovo Cimento B*, 25(2), 633–61.

Kevles, D. J. (1995). *The Physicists: The History of a Scientific Community in Modern America*. Cambridge, MA: Harvard University Press.

Koizumi, K. (2020). The Evolution of Public Funding of Science in the United States From World War II to the Present. *Oxford Research Encyclopedia of Physics*. https://oxfordre.com/physics/view/10.1093/acrefore/9780190871994.001.0001/acrefore-9780190871994-e-25.

Lambrecht, R. M. (1975). *Antimatter–matter interactions. I. Positrons and positronium. A bibliography, 1930–1974*. No. BNL-50510. Upton, NY: Brookhaven National Lab.

Maia Filho, A. M., and Silva, I. L. (2019a). O experimento WS de 1950 e as suas implicações para a segunda revolução da mecânica quântica. *Revista Brasileira de Ensino de Física*, 41(2), e20180182-10.

Maia Filho, A. M., and Silva, I. L. (2019b). A trajetória de Chien Shiung Wu e a sua contribuição à Física. *Caderno Brasileiro de Ensino de Física*, 36(1), 135–57.

Peacock, K. A. (2017). Happiest Thoughts: Great thought experiments of modern physics. In M. T. Stuart, Y. Fehige, J. R. Brown (eds), The Routledge Companion to Thought Experiments, Abingdon: Routledge, pp. 211–42.

Pipkin, F. M. (1979). Atomic physics tests of the basic concepts in quantum mechanics. *Advances in Atomic and Molecular Physics*, 14, 281–40.

Prevedel, R., Aspelmeyer, M., Brukner, Č., Zeilinger, A., and Jennewein, T. D. (2007). Photonic entanglement as a resource in quantum computation and quantum communication. *Journal of the Optical Society of America B*, **24**(2), 241–48.

Pryce, M. H. L., and Ward, J. C. (1947). Angular correlation effects with annihilation radiation. *Nature*, **160**(4605), 435.

Rossiter, M. W. (1984). *Women Scientists in America: Struggles and Strategies to 1940.* Baltimore, MD: Johns Hopkins University Press.

Rossiter, M. W. (1998). *Women Scientists in America: Before Affirmative Action, 1940–1972.* Baltimore, MD: Johns Hopkins University Press.

Sarna, J. D., Smith, E., and Kosofsky, S.-M. (eds) (1995). *The Jews of Boston.* Yale: Yale University Press and Jewish Philanthropies of Greater Boston.

Selleri, F., and Zeilinger, A. (1988). Local deterministic description of Einstein–Podolsky–Rosen experiments. *Foundations of Physics*, **18**(12), 1141–58.

Shimony, A. (1972). Letter from Abner Shimony to John Clauser, 14 January 1972. Correspondence, 1958–2009; Professional Correspondence—Scientific Topics, 1958–2006; Box 1, Folder 4; Archives of Scientific Philosophy, University of Pittsburgh.

Shimony, A. (1983). Reflections on the Philosophy of Bohr, Heisenberg, and Schrödinger. In R. S. Cohen and M. W. Wartofsky (eds), *A Portrait of Twenty-five Years*, Boston Studies in the Philosophy of Science, Dordrecht: Springer.

Siegel, R. W. (1980). Positron annihilation spectroscopy. *Annual Review of Materials Science*, **10**(1), 393–425.

Slaughter, R. C. (2010). *Positron Annihilation Ratio Spectroscopy (PsARS) Applied to Positronium Formation Studies.* PhD thesis, Air Force Institute of Technology. https://scholar.afit.edu/etd/2182/.

Snyder, H. S., Pasternack, S., and Hornbostel, J. (1948). Angular correlation of scattered annihilation radiation. *Physical Review*, **73**(5), 440–48.

Spitzer, Y. (2013). *Pogroms, Networks, and Migration: The Jewish Migration from the Russian Empire to the United States, 1881–1914.* September 17, 2013. https://eh.net/eha/wp-content/uploads/2013/11/Spitzer.pdf.

Spitzer, Y. (2014). *The Dynamics of Mass Migration: Estimating the Effect of Income Differences on Migration in a Dynamic Model of Discrete Choice with Diffusion.* November 25, 2014. https://economics.yale.edu/sites/default/files/spitzer_-_dynamics_of_mass_migration_141125.pdf.

Spitzer, Y. (2018). *Pogroms, Networks, and Migration: The Jewish Migration from the Russian Empire to the United States 1881–1914.* May 25, 2018. https://yannayspitzer.files.wordpress.com/2019/03/pogromsnetworksmigration_182505.pdf.

Süvegh, K., and Marek, T. (2011). Positron annihilation spectroscopies. In A. Vértes, S. Nagy, Z. Klencsár, R. G. Lovas, and F. Rösch (eds), *Handbook of Nuclear Chemistry*, Boston: Springer, pp. 1461–84.

*The Tech* (1939. September 26, http://tech.mit.edu/V59/PDF/V59-N33.pdf.

*The Tech* (1941a). January 29, http://tech.mit.edu/V60/PDF/V60-N60.pdf.

*The Tech* (1941b). May 9, http://tech.mit.edu/V61/PDF/V61-N25.pdf.

*The Tech* (1942a). April 3, http://tech.mit.edu/V62/PDF/V62-N16.pdf.

*The Tech* (1942b). April 28, http://tech.mit.edu/V62/PDF/V62-N22.pdf.

*The Tech* (1942c). May 22, http://tech.mit.edu/V62/PDF/V62-N29.pdf.

Tidman, K. R. (1984). *The Operations Evaluation Group: A History of Naval Operations Analysis.* Annapolis: Naval Institute Press.

U.S. Federal Population Census (1930). 'William Shaknov.' 1930 Census Record. Accessed 15 April 2020. https://www.ancestry.com/search/categories/35/?name=_Shaknov&name_x=1_1.

U.S. Federal Population Census (1940a). William Shaknov.' 1940 Census Record. Accessed 15 April 2020, https://www.ancestry.com/search/categories/35/?name=_Shaknov&name_x=1_1.

U.S. Federal Population Census (1940b). Frances F Shaknov. 1940 Census Record. Accessed 15 April 2020, https://www.ancestry.com/search/categories/35/?name=_Shaknov&name_x=1_1.

Wheeler, J. A. (1946). Polyelectrons. *Annals of the New York Academy of Sciences*, 48(3), 219–38.

Wheeler, J. A., and Ford, K. W. (1998). *Black Holes, Geons, and Quantum Foam; a Life in Physics*. New York: Norton.

Whitaker, A. (2012). *The New Quantum Age: From Bell's Theorem to Quantum Computation and Teleportation*. New York: Oxford University Press.

Whitaker, A. (2016). *John Stewart Bell and Twentieth-Century Physics: Vision and Integrity*. Oxford: Oxford University Press.

Wilson, A. R., Lowe, J., and Butt, D. K. (1976). Measurement of the relative planes of polarization of annihilation quanta as a function of separation distance. *Journal of Physics G: Nuclear Physics*, 2(9), 613–24.

Wu, C.-S. (undated, a). Letter draft, Chien-Shiung Wu to John Bell. C.S. (Chien-Shiung) Wu Papers 1945–1994 bulk 1960–1979; Box 1 and Folder 6; University Archives, Rare Book & Manuscript Library, Columbia University in the City of New York.

Wu, C.-S. (undated, b). Recommendation Letter. C.S. (Chien-Shiung) Wu Papers 1945–1994 bulk 1960–1979; Box 2 and Folder 27; University Archives, Rare Book & Manuscript Library, Columbia University in the City of New York.

Wu, C.-S. (1973). Letter, Otto Frisch to Chien-Shiung Wu, 5 December 1973. C.S. (Chien-Shiung) Wu Papers 1945–1994 bulk 1960–1979; Box 1 and Folder 3; University Archives, Rare Book & Manuscript Library, Columbia University in the City of New York.

Wu, C.-S. (1974a). Letter, Chien-Shiung Wu to Otto Frisch, 11 January 1974. C.S. (Chien-Shiung) Wu Papers 1945–1994 bulk 1960–1979; Box 1 and Folder 3; University Archives, Rare Book & Manuscript Library, Columbia University in the City of New York.

Wu, C.-S. (1974b). Letter, Chien-Shiung Wu to Stuart Freedman, 16 May 1974. C.S. (Chien-Shiung) Wu Papers 1945–1994 bulk 1960–1979; Box 1 and Folder 3; University Archives, Rare Book & Manuscript Library, Columbia University in the City of New York.

Wu, C.-S. (1975). Letter, John Bell to Chien-Shiung Wu, 30 May 1975. C.S. (Chien-Shiung) Wu Papers 1945–1994 bulk 1960–1979; Box 1 and Folder 6; University Archives, Rare Book & Manuscript Library, Columbia University in the City of New York.

Wu, C.-S., and Shaknov, I. (1950). The angular correlation of scattered annihilation radiation. *Physical Review*, 77(1), 136.

Yoshihuku, Y. (1972). Is the Experiment of Wu and Shaknov an Empirical Illustration of the Paradox of Einstein, Podolsky, and Rosen? *Progress of Theoretical Physics*, 48(6), 2445–9.

Zeilinger, A. (1999). Experiment and the foundations of quantum physics. *Reviews of Modern Physics*, 71(2), S288–S297.

Zeh, H.-D. (1979). Quantum theory and time asymmetry. *Foundations of Physics*, 9(11–12), 803–18.

..............................................................................................................

# CHAPTER 30

# ON HOW *EPISTEMOLOGICAL LETTERS* CHANGED THE FOUNDATIONS OF QUANTUM MECHANICS

..............................................................................................................

## SEBASTIÁN MURGUEITIO RAMÍREZ

IT is not an exaggeration to say that the 1970s marked the beginning of a new era in the foundation of quantum mechanics. These were the years when physicists and philosophers interested in the foundations of physics started to actively engage with the conceptual and empirical implications of the Bell inequality, published in 1964. Among other things, the first experiments designed to test whether quantum mechanics violated the inequality in question were carried out during these years, two journals—*Foundations of Physics* and *Epistemological Letters*—dedicated to the foundations of physics were created, and for the first time since its creation, the 'International School of Physics "Enrico Fermi"' dedicated its theme to the foundations of quantum mechanics (see Freire, 2004). Although all this is well-known, not much has been said about the exact ways that *Epistemological Letters*, in particular, helped the foundations of quantum mechanics consolidate as an important and respectable scientific discipline. And this is precisely the subject of the present chapter.

As I will argue here, at least five features of *Epistemological Letters* encouraged the foundations of quantum mechanics to flourish during a time when the discipline itself was not very well respected by the broader physics community; first, the subject matter, as this was the only journal completely dedicated to the foundations of quantum mechanics; second, the kind of methodology encouraged by the journal and the institution behind it, strongly promoting an interdisciplinary approach where philosophical and speculative discussions related to quantum physics were encouraged; third, their efforts to reach out to a wide group of scholars interested in the foundations of quantum mechanics, including both the main figures in the field and also various scholars in the early stages of their careers who would come to dominate the philosophy of physics in the decades to come; fourth, the informal style of the journal, which facilitated an environment for open, frequent, and productive debates; fifth, and more obviously, its very high quality—some of the most important papers on the foundations of quantum mechanics were first published in *Epistemological Letters*.

The structure of this chapter goes as follows. First, in section 30.1, I will offer a brief historical overview of physics as a discipline in the 1960s and 1970s, mostly focused on the American context. Then, in section 30.2, I will discuss the creation of the Association Ferdinand Gonseth and the Institut de la Méthode, which sponsored the publication of *Epistemological Letters*. Then, in section 30.3, I show exactly why *Epistemological Letters*, much more than any other publication of the time (including *Foundations of Physics*), so significantly impacted the foundations of quantum mechanics, followed by some concluding remarks.

# 30.1 Historical Context

I will approach this brief historical context from three different but complementary perspectives: from the point of view of the physics textbooks and classrooms of the time, from the first-person perspective of Clauser who was an important figure in the foundations of quantum mechanics during the 1970s, and from the position of physics journals that were published during this period.

Throughout World War II and the Cold War, physics in America was greatly shaped by military demands. Physics was the field that received the most federal funding allocated to war needs by recently founded agencies such as N.D.C.R and N.S.F. (Kevles, 1995). And of course, physics would play a crucial role in the war itself, with, for example, the development of radar and the atomic bomb (atomic bombs would continue to be developed throughout the Cold War). In part because of its rapidly expanding budget and because of the widespread public recognition of its contribution to the war effort, physics became a very attractive career choice in American society. Consequently, after World War II graduate-level enrolments in physics increased twice as quickly in the United States as all other disciplines combined (Kaiser, 2011, p. 17). The United States was producing three times as many Ph.D.s in physics as it did in pre-war times, and this number would become six times as many after Sputnik (Kaiser, 2011, p. 17). As Kaiser (2007) shows, this massive increase in enrolment significantly influenced the way quantum mechanics was taught.[1] As advanced physics classrooms started to hold more than a hundred students, professors teaching quantum mechanics had less time to deal with interpretative questions.

---

[1] See Kaiser (2020, chapter 8) for a more recent and extensive discussion on this topic.

They had less time to read (and hence assign) student essays on issues such as the meaning of the uncertainty principle or the notion of probability in the theory. By comparing lecture notes from graduate-level courses on quantum mechanics across the United States in the 1950s, Kaiser (2007, p. 30) shows that classes with an average of 13 students included five times this kind of philosophical material as compared to classes with an average class size of 40 students. In short, as one professor says in 1956, 'the philosophical issues raised by quantum mechanics...the student never has a chance to gauge their depths' (Kaiser, 2011, p. 18).

The decrease in the time dedicated towards interpretative issues was also reflected in the textbooks that professors and students used during this period. For example, one of the most popular textbooks of the time, *Quantum Mechanics* by Leonard Schiff, placed a particularly strong emphasis on calculations, which kept the students busy while avoiding discussions centred on more philosophical issues. This was just one instance of a more general pattern. Whereas textbooks on quantum mechanics before World War II used to have a fair amount of essay prompts (in some cases near a quarter of all the problems), by 1960, questions dealing with interpretative issues had shrunk to just about 10 per cent of all problems in most textbooks (Kaiser, 2007, p. 32).

In short, after World War II and during the Cold War, the number of physicists in the United States increased exponentially and this fact, when combined with the specific military needs, made teaching and research in quantum mechanics much more pragmatically orientated than before—as evidenced by the materials covered in textbooks and lectures. This rather pragmatic orientation of the discipline—the commonly deemed 'shut up and calculate' attitude—meant that students learning quantum mechanics had little to no exposure to the more philosophical questions and debates on interpretations that some of the main fathers of quantum mechanics had intensely discussed just a few decades earlier. It also meant, as we will see now, that those physicists in the early moments of their careers who were interested in pursuing more foundational questions faced strong opposition by the dominant academic physics culture of the time.

In his paper 'Notes on Early History of Bell's Theorem', John Clauser offers a first-person perspective on what the academic life of quantum physics was like in 1960s America. To better appreciate Clauser's remarks, it is pertinent to say a couple of things about what was going on with the foundations of quantum mechanics at that particular moment. In 1964, when Clauser was a graduate student at Columbia, John Bell published his paper 'On the Einstein–Podolsky–Rosen paradox'. In this work, Bell proved that, for a certain class of correlation measurements between previously interacting elementary particles, such as the polarization correlation between two photons from a two-stage atomic decay, the predictions of any local hidden variable theory would have to satisfy an inequality—now known as the Bell inequality—that is violated by the predictions made by standard quantum mechanics (Bell, 1964). Bell was motivated by his interest in non-local hidden variable theories, such as the one that David Bohm proposed in 1952 (Bohm, 1952). Bell believed that (non-local) hidden variable theories were a more conceptually satisfying theoretical framework than that afforded by orthodox quantum mechanics, mainly because of that curious loss of separate real physical states of previously interacting particles that Albert Einstein once famously derogated as 'spooky action-at-a-distance' ['spukhafte Fernwirkung'] (Einstein to Max Born, 3 March 1947; Einstein, Born, and Born, 1969, p. 215). But Bell's motivations notwithstanding, Bell's theorem pointed the way to direct experimental tests of entanglement for the first time.

In its original form, Bell's theorem was not ready for the laboratory because it made several idealized, unphysical assumptions about such crucial details as detector and polarizer efficiencies. That shortcoming was remedied when, in 1969, Abner Shimony (more on him later) along with John Clauser, Michael Horne, and Richard Holt, rederived the theorem in a form more apt to actual experiments (Clauser, Horne, Shimony, and Holt, 1969). The CHSH theorem, as it is now often dubbed, opened the door to the decades-long, still ongoing series of experimental tests of Bell's theorem that have proven, time and again, results consistent with quantum mechanics and against local-hidden variable theories.[2] By bringing an apparently abstract result from the foundations of quantum mechanics to the lab, Clauser and Shimony not only helped the foundations of quantum mechanics take a big step closer towards 'acceptance' as a legitimate scientific discipline, but they would also initiate a new era in quantum optics (for an excellent review of the origins of this collaboration between quantum optics and foundations of quantum mechanics, see Freire, 2006).

Clauser, who had been interested in the foundations of quantum mechanics since he was a graduate student, recalls this period with great disappointment:

> Any open inquiry into the wonders and peculiarities of quantum mechanics and quantum entanglement that went outside of a rigorous 'party line' [the party line of 'shut up and calculate'] was then virtually prohibited by the existence of various religious stigmas and social pressures, that taken together, amounted to an evangelical crusade against such thinking. As a result of this evangelism, *much of the early important work on Bell's Theorem was published only in an 'underground' newspaper, whose circulation was limited to members of a 'quantum-subculture', and that probably cannot be found in most physics libraries.*
>
> (Clauser, 2002, p. 62, my emphasis)

The 'underground' newspaper Clauser is referring to here is *Epistemological Letters* (we will come back to it in section 30.3). Clauser offers different examples to illustrate this so-called 'party line' that heavily hindered any work on the foundations of physics from being produced. For the sake of brevity, I will focus on two examples. First, when he was a graduate student at Columbia, Clauser asked Prof. Bob Serber for his opinion of Bell's theorem and on whether or not it was important to conduct an experiment to test if quantum mechanics violated the inequality. Serber responded, 'Well that [Bell's

---

[2] For a thorough and recent discussion of some of the main 'loopholes' in the experiments testing the violation of the Bell inequalities, see Kaiser's contribution in this volume (chapter 13).

Theorem prediction] might be worth pointing out in a letter to the editor, but no decent experimentalist would ever go to the effort of actually trying to measure it with that in mind!' (Clauser 2002, p. 71). Thankfully, Clauser did not follow this advice too closely.

Second, Clauser recalls his professors advising him that if he wanted to have a successful career, he needed to avoid all interpretative questions:

> Any physicist who openly criticized or even seriously questioned these foundations (or predictions) was immediately branded as a 'quack'. Quacks naturally found it difficult to find decent jobs within the profession... Any student who questioned the theory's foundations, or, God forbid, considered studying the associated problems as a legitimate pursuit in physics was sternly advised that he would ruin his career by doing so. I was given this advice as a student on many occasions by many famous physicists on my faculty at Columbia.   (Clauser 2002, pp. 72–73)

Obviously, Clauser does not have the fondest memories of the physics academic culture of the time in question, and we can easily imagine that he would have had a much more positive experience had he been lucky enough to work with physicists such as Eugene Wigner, David Bohm, or Bell himself, who were some of the rare examples of prominent physicists interested in the foundations of quantum mechanics. It is also worth pointing out that the popularity of the 'shut up and calculate' attitude that Clauser alludes to in these passages was not the only reason that physicists interested in foundational questions faced challenges during this time period. As the Cold War was cooling down in the late 1960s, federal funding designated for physics research was dramatically cut, causing the enrolment rates to halve in just five years (Kaiser, 2011, p. 22). This sudden loss of funds, together with the surge in physics graduates from previous years, created a rather dire job market environment. In the mid-1960s there were more physics jobs than graduates, but in 1971 there were 1053 applicants competing for just 53 jobs (Kaiser, 2011, p. 23). This would only exacerbate the pressure felt by young physicists such as Clauser who were interested in studying the foundations of quantum mechanics. Naturally, the scarce job market and reduced federal funding were not going to be conducive towards investigating questions that most of the physics community of the time did not take very seriously.

Besides textbooks, lectures, and the testimonials of figures like Clauser, there is another angle through which we can explore how the pragmatic character of the physics of the time limited research in the foundations of quantum mechanics. After World War II, because of the pragmatism of the period as well as a natural consequence of the increasing specialization of physics and academic disciplines in general, it became more difficult to find physics journals willing to publish works on this topic. There is an obvious reason for this: the foundations of quantum mechanics has always been an example of an interdisciplinary area that calls for a tight collaboration between physics and philosophy. Einstein's 'Quantum Mechanics and Reality' (1948), where he puts forward one of his clearest versions of the incompleteness argument, is an illustrative example of why an interplay between philosophy and physics is essential when addressing foundational questions in quantum mechanics. These more philosophical reflections on quantum mechanical issues were at the same time too removed and 'too interdisciplinary' for the pragmatic style of the physics in the post-WWII period.

It is not a coincidence that it was in *Dialectica*, a prestigious philosophy journal initially created by Ferdinand Gonseth, Paul Bernays, and Gaston Bachelard in 1946, and still active, where Einstein's 'Quantum Mechanics and Reality' paper was published, in a 1948 issue dedicated to Bohr's complementarity, and whose participants included figures such as Bohr, de Broglie, and Heisenberg. To this very day, that special issue of *Dialectica* stands as one of the most important published symposia on the foundations of quantum mechanics, the contributions of Bohr (1948) and Einstein (1948), especially, having become classics. Somewhat ironically, a philosophy journal was more open to continuing the debates by the founding fathers of quantum mechanics than the physics journals of the time, and this would become even more the case during the 1950s and 1960s. For instance, Clauser recalls that Samuel Goudsmit, the editor of the prestigious *Physical Review* from 1958 to 1974, enclosed a policy statement recommending the rejection of any article on the foundations of quantum mechanics unless they were 'mathematically based *and* gave new quantitative experimental predictions' (Clauser, 2002, p. 72). Surely, none of the pieces by Bohr, Einstein, Heisenberg, or de Broglie published in the special issue of *Dialectica*, and probably not even the Einstein-Podolsky-Rosen paper from 1935, would have been accepted in *Physical Review* under this policy.

As a response to this mixture of pragmatism and hyper-specialization, three physics journals relevant to our story were created in the 1960s and early 1970s.[3] On the one hand, we find the short-lived journal *Physics,* which only ran for four volumes between 1964 and 1968. The most important article published here was Bell's 1964 paper, where he presents his now very famous inequality for the first time. In the Editorial Preface, Philip Warren Anderson and Bernd Theodor Matthias stated:

> ... in our opinion physics has reached the point at which there is far more good physics written than any physicist can read, especially if he hopes to cover more than his own special field. On the other hand, it would be too bad if most physicists were to have to give up reading original material in other fields even fairly close to their own, as perforce most of them have long since given up reading articles in the other sciences. Therefore, we believe it is a good idea to institute a selective journal in which the editors try their very best to present a selection of papers which are worth the attention of all physicists.   (Anderson and Matthias, 1964)

---

[3] The historical literature taking up scientific journals as their subject is rather scant, and as far as I can tell, this work is the first one to examine the importance of these three journals within the context of the increasing specialization of physics.

So, the editors thought that Bell's paper 'was worth the attention of all physicists no matter their area'. Ironically, however, and as a clear sign of the lack of popularity of the foundations of quantum mechanics, nobody but Bell himself cited this piece in the three years after its publication (this happened in his 1966 paper, where he discussed the problems with von Neumann's proof against hidden variable theories). In contrast, all of the other papers appearing in the same issue as Bell's were cited at least twice and, in one case, seven times throughout the following three years (to this date, Bell's paper has now been cited more than six thousand times).

Then, in 1970, a journal explicitly dedicated to the foundations of physics was created. This was *Foundations of Physics*, created by Henry Margenau and Wolfgang Yourgrau, both very respectable physicists who also actively worked on the philosophy of science and the foundations of quantum mechanics. David Bohm, Louis de Broglie, V.A. Fock, Karl Popper, and Eugene Wigner, all well-known physicists (except Popper, who was a philosopher) interested in the foundations of quantum mechanics, were part of the editorial board (Freire, 2004, p. 1754). After explaining what they mean by 'foundations' and why it is a worthy area of investigation, at the end of the Editorial Preface Margenau and Yourgrau also make a reference to the fact that speculative ideas are especially encouraged in their journal (this sets a startling contrast to the policy of *Physical Review* mentioned earlier):

> Very few scientific journals today encourage speculation not tied to hard and demonstrable facts. One wonders whether brilliant ideas are not lost by this restrictive attitude. *Foundations of Physics* will publish with suitable frequency disciplined speculations suggestive of new basic approaches in physics.
>
> (Margenau and Yourgrau, 1970, p. 3)

It is interesting that the editors also viewed this kind of focus on foundations as something that would foster interdisciplinary collaborations between different sciences:

> For the most basic theories provide the approaches to other areas of science. Thus, the fundamental concepts of physics find natural applications in chemistry, astronomy, and, we hopefully observe, biology. One of the fervent desires of the editors is that this journal, committed to this view of foundations, shall be instrumental in sponsoring work that brings large physical principles to bear upon adjacent fields.
>
> (Margenau and Yourgrau, 1970, p. 1)

In the same preface, they say that the journal will focus on topics such as the equivalence of matrix mechanics and wave mechanics, the nature of observables, the strange role of variational principles, time and space, axiomatization of statistical mechanics, unified field theory, and others. Notice that foundational issues of quantum mechanics are just a subset of the many areas that the journal aimed to explore.

## 30.2 THE ASSOCIATION FERDINAND GONSETH

The same year *Foundations of Physics* was created, the well-renowned Swiss philosopher of science and mathematics, Ferdinand Gonseth, turned 80. As I noted earlier, Gonseth (1890–1975) was one of the founders of *Dialectica*. Trained in physics and mathematics at the Swiss Federal Polytechnic (EPF/ETH) in Zurich, he taught mathematics and philosophy of science at the University of Zurich, the University of Bern, and, from 1929 onwards, the EPF/ETH (Fuchs, 2007). He was the author of several important books, including *Les fondements des mathématiques* (Gonseth, 1926), *Les mathématiques et la réalité* (Gonseth, 1936), *Déterminisme et libre arbitre* (Gonseth, 1944), *La géométrie et le problème de l'espace* (Gonseth, 1945), *Le problème du temps* (Gonseth, l964). And in 1944, two years before creating *Dialectica,* he founded, together with Paul Bernays, Karl Dörr, and Karl Popper, the International Society for Logic and Philosophy of Science (Lauener, 1977, p. 113). In 1970, he expressed the desire to create a centre for 'methodological studies' in the spirit of his own work (see Bertholet, 1971). His wish was quickly satisfied when, in 1971, the Association Ferdinand Gonseth was created at Bienne with the purpose of pursuing and developing Gonseth's oeuvre. Among the different values promoted by the Association, it is noteworthy citing the following one:

> [The Association] promotes openness to the philosophy of science, to an interdisciplinary dialogue with science. Science and philosophy form one body, and all that happens again in science, whether in its methods or in its results, may resound on philosophy even in its most fundamental principles.
>
> (Bertholet, 1971; my translation)

Note how one of the core goals of the association was to strengthen the dialogue between philosophy and science, which is of course very reflective of Gonseth's own legacy. The same year as its foundation, the association also created the Institut de la Méthode to serve as a home for activities as varied as the interests of Gonseth himself:

> They range from a reflection on the teaching of mathematics to discussions on relatively technical problems from the philosophy of physics to the promotion of a dialogue between science and philosophy.    (Bertholet, 1971; my translation)

In contrast to *Foundations of Physics* and *Physics*—journals that were more interested in addressing the specializations within physics itself—this institute was very explicit in their desire to establish a dialogue between physics and philosophy. Notice also the use of the term 'philosophy of physics', which is not found in the preface to the other journals.

The secretary of the Institut was François Bonsack (1926–2006). Born in Bienne in 1926, he took a doctorate in philosophy at Geneva in 1961 with a dissertation on 'Information, thermodynamique, vie et pensée' (Bonsack, 1961). A devoted student of the work of Gonseth, Bonsack was also a lecturer in the philosophy of science at the nearby University of Neuchâtel (Prabook, 2020). From 1972 up until 2003, the Institut de la Méthode created several 'written symposia' on topics ranging from the pedagogy of mathematics to the philosophy of physics to political philosophy. These so-called 'written symposia' were not really journals or newsletters but instead functioned more like today's preprint servers. Of the eleven symposia published by the Institut de la Méthode, the one with the most issues, by far, was on the topic of hidden variables and quantum uncertainty. Its print title was *Epistemological Letters*, and it ran from November 1973 until October 1984, after publishing thirty-six issues. In 1978 and 1979 the Institut also hosted two conferences on the same topic as *Epistemological Letters*, 'Indéterminisme quantique et variables cachées', among whose participants we include d'Espagnat, Shimony, Vigier, K. Popper, and O. Costa de Beauregard (one finds the announcement of the first conference in the 17th issue of *Epistemological Letters* published in December 1977, and a list of participants in the 18th issue published in June 1978).

Before we focus our attention on *Epistemological Letters*, it is useful to consider the following question: one might ask why such a modest, private foundation such as the Association Ferdinand Gonseth, located in a small city in northwest Switzerland, successfully created and kept up a journal dedicated to the foundations of quantum mechanics. Of course, part of the answer is that Gonseth himself was interested in the foundations of quantum mechanics, as was clear from the fact that while acting as the editor of *Dialectica* he organized a whole issue on the concept of complementarity, as I mentioned earlier. But another important part of the answer involves geography. At the time *Epistemological Letters* was established, Josef-Maria Jauch and Constantin Piron were pursuing what would prove to be some of the most important work on the foundations of quantum mechanics and quantum logic while they were at the University of Geneva (see Jauch and Morrow, 1968; Jauch, 1973; Piron, 1976), which was only an hour and forty-five minutes southwest of Bienne by either car or train. And Bell himself was then working at CERN in Geneva, where he had been since the mid-1950s. Expanding our view a bit wider, we find a very influential pioneer of quantum foundations, Bernard d'Espagnat, at the Sorbonne in Paris, where he had worked since 1959, after having been Bell's colleague in the theory group at CERN for a few years in the late 1950s (see d'Espagnat ,1976, 1983). And, looking eastward, there was the remarkable physical chemist and student of the foundations of quantum mechanics, Hans Primas, at the ETH in Zurich (see Primas, 1983). By some measure, in the early 1970s the triangle comprising Paris, Geneva, and Zurich had the world's densest concentration of physicists working on the foundations of physics. Bienne and the Association Ferdinand Gonseth lay at the heart of that region.

But of course geography alone would not explain the success of *Epistemological Letters*. One must also mention Shimony's role in the production of the journal.

The 'official' editor was Bonsack, but Shimony was very much involved for the whole eleven years that *Epistemological Letters* ran. In 1970, because of the publication of the CHSH paper, Shimony had been invited by d'Espagnat to the 'International School of Physics "Enrico Fermi"', where he 'began a lasting friendship with Bell and d'Espagnat, and became still more involved with issues related both to Bell's inequalities and the measurement problem' (Freire, 2006, p. 592). So, by the time he began working on *Epistemological Letters* in 1973, Shimony was already a highly regarded figure in the community of quantum foundations in Europe, and this was going to be key in building *Epistemological Letters* into a very successful publication. It is also important to stress that because he obtained a Ph.D. in philosophy from Yale in 1953 and another one in physics from Princeton in 1963 under the supervision of Wigner, Shimony was a perfect fit for establishing the kind of dialogue between science and philosophy that the Association Ferdinand Gonseth fervently advocated.

## 30.3 *EPISTEMOLOGICAL LETTERS*

From the very start, *Epistemological Letters* emerged as a unique kind of publication. The first thing readers encounter when opening the first and second issues is a list of the people to whom *Epistemological Letters* was distributed. As the reader can appreciate in Table 30.1, nearly all of the key protagonists who helped the foundations of quantum mechanics become a more established discipline during the 1970s are listed as recipients. This includes figures who were already well renowned in the discipline, such as Wigner, Bohm, Bell, DeWitt, Shimony, Clauser, d'Espagnat, Vigier, de Broglie, Margenau, Yanase, Jauch, and Piron. But, of perhaps more importance, it also included scholars in the early stages of their careers, many of whom would become the leading philosophers of physics and science of our time. I am referring, in particular, to figures such as Howard Stein, Arthur Fine, John Earman, Bas van Fraassen, and Jeff Bub. It seems natural to speculate that the direct contact with the discussions by Bell, d'Espagnat, Shimony, and others that *Epistemological Letters* provided them would shape their own professional paths considerably. For instance, it is interesting that although by 1973 Fine had already published some papers on the foundations of quantum mechanics, including a paper on the interpretations of quantum mechanics (Fine, 1973), the first time he addressed Bell's 1964 paper was in 1974, precisely one year after *Epistemological Letters* was founded. And for the next ten years, during the time *Epistemological Letters* ran, some of Fine's most important work on the philosophy of physics would be about the Bell inequalities and their philosophical and physical significance. Indeed, in the early 1970s, philosophers working on the foundations of quantum mechanics did not really take Bell's paper to be that significant. For example, in 1974, Patrick Suppes edited a *Synthese* volume completely dedicated to the foundations of quantum mechanics with contributions by philosophers like van Fraassen, Putnam, Cartwright, and Fine. Only three out of seventeen papers in that volume

Table 30.1 On the left (under 'First participants'), we see the list of participants as found in the first issue of Epistemological Letters (I have typed the names and organized the table as found in the issue). For the second issue (under 'New participants'), new members are added, as seen in the list on the right (this is not the last time new members were announced; in the third issue a couple of new names are added as well). Notice, in the second issue, the addition of philosophers of science and physicists like van Fraassen (which is misspelled as 'Fraasen'). Fine, Earman, and Freedman, all of whom would become prominent figures in the years to come.

### First participants

| | | |
|---|---|---|
| Austria | Dürr | Chevalier |
| March | Falck | Cboquard |
| H.Thirring | Flügge | Emch |
| W.Thirring | Haken | Enz |
| Canada | Heisenberg | Fierz |
| Bunge | Hund | Beitler |
| Denmark | Jordan | Hepp |
| A.Bohr | Ludwig | Huguenin |
| France | Mittelstaedt | Jauch |
| Blanché | v. Weizsäcker | Jost |
| de Broglie | Great-Britain | König |
| Canguilhem | Bohm | Loeffel |
| Chambadal | Dirac | Mercier |
| Costa de Beauregard | Josephson | Piron |
| Des touches | Mott | Rivier |
| Dubarle | Reece | Rossel |
| d'Espagnat | Rosenfeld | Scheurer |
| P. Février | Temple | Stückelberg |
| Flato | Ziman | USA |
| Lochak | Hungary | Clauser |
| Malcor | Janossy | Cochran |
| Merleau-Ponty | Israel | Dyson |
| Mugur-Schächter | Bub | Elsaesser |
| O'Neil | Jammer | Feynman |

### New participants

| | | |
|---|---|---|
| Belgium | Japan | Gardner |
| Dockx | Yanase | Giere |
| Dopp | Yukawa | Grünbaum |
| Ladrière | Netherlands | Gudder |
| Manneback | Freudenthal | Hammer |
| Paulus | Heyting | Havas |
| Prigogine | Raven | Hempel |
| Stengers | Wolvekamp | Hoffmann |
| Canada | Poland | Hooker |
| Salman | Kotarbinski | Kleene |
| Denmark | Mostowski | Kreisel |
| Rosenfeld | Switzerland | Komar |
| Finland | Bochenski | Margenau |
| von Wright | Gonseth | Merzbacher |
| France | Piaget | Nagel |
| Lichnerowicz | Portman | Papliolios |
| Vandel | van der Waerden | Petersen |
| Western Germany | USA | Pipkin |
| Büchel | Band | Polya |
| Drieschner | Belinfante | Putnam |
| Great Britain | Church | Quine |
| Braithwaite | Cohen | Roman |
| Heine | Cooper | Salmon |
| Körner | Curry | Scott |

(Continued)

**Table 30.1  Continued**

| First participants | | |
| --- | --- | --- |
| **USA** (*Contd*) | Rosen | Holt |
| Poirier | **Japan** | Horne |
| Russo | Tomonaga | Landé |
| Thuilier | **Netherlands** | Lamb. |
| Tonnelat | Casimir | Lenzen |
| Ullmo | **Portugal** | Oldauer |
| Vigier | Andrade e Silva | Park |
| Vuillemin | **Switzerland** | Schwinger |
| **Eastern-Germany** | Amiet | Shimony |
| Treder | Bell | Watanabe |
| **Western Germany** | Bernays | Weisskopf |
| Bopp | | Wigner |
| | | **USSR** |
| | | Abrikosov |
| | | Fock |

| New participants | | |
| --- | --- | --- |
| **USA** (*Contd*) | Earman | Sperti |
| Needham | Feigl | Stachel |
| Pippard | Feinberg | Stein |
| Polanyi | Fine | Tisza |
| Popper | Finkelstein | Tutsch |
| **Italy** | van Fraasen | Wheeler |
| Montalenti | Freedman | Wightman |
| Toraldo di Francia | Friedberg | de Witt |
| Tonini | Furry | Yourgrau |

addressed Bell's papers (one paper that did so was Fine, 1974). Actually, van Fraassen's paper, 'The Einstein–Podolsky–Rosen Paradox', did not even discuss Bell's results, despite the fact that it was about hidden variables. Obviously, in the second half of the 1970s it would have been virtually impossible to find a philosophy volume on the foundations of quantum mechanics where most of the contributions ignored Bell's results (how much this changed due to *Epistemological Letters* is hard to quantify, but the fact that scholars such as Fine and van Fraassen were getting direct access to this publication probably played a significant role in this shift).

Following the list of recipients, the first issue includes a quick introduction to the symposium, which is written in French. The introduction gives a very rough overview of the most important events regarding the foundations of quantum mechanics. It mentions the double-slit experiment, Born's probability rule, Einstein's view on the incompleteness of quantum mechanics, the Bell inequalities, the CHSH paper, and the experiment by Freedman and Clauser (Freedman and Clauser, 1972) that shows the violations of the inequalities. The last line reads, 'The discussion is open for evaluating the exact importance of these results and the consequences that one ought to make'. The issue then ends with an entry by Shimony and Horne where the authors go into the details behind Einstein's incompleteness argument, the Bell inequalities, and the CHSH experiment (Shimony and Horne, 1973). In summary, the very first issue, though relatively short, sets clear expectations about what to expect of *Epistemological Letters*: open discussions about the Bell inequalities and recent experimental results favouring standard quantum mechanics. This is very important because *Epistemological Letters* was the only academic journal at the time (and since) *completely* dedicated to the foundations of quantum mechanics. The closest 'competitor' was the *Foundations of Physics*, but the latter would publish on many other foundational areas, including the foundations of thermodynamics, quantum field theory, and special and general relativity. Actually, it is surprising that, following its creation in 1970, it was 1973, after more than 50 papers had been published already in *Foundations of Physics*, when an article addressing Bell's inequalities appeared for the first time (by Jeff Bub, who was included in the list of people who received copies of *Epistemological Letters*). Furthermore, between 1970 and 1976, less than a third of the papers in *Foundations of Physics* focused on the foundations of (non-relativistic) quantum mechanics, and of these, only 9 papers (less than 4 per cent of all the papers published in the journal until that point) cited Bell's 1964 work! My purpose is not to undermine the role that *Foundations of Physics* played in helping the foundations of quantum mechanics become a more established discipline, but rather to note that even for this journal, questions on the physical and philosophical implications of Bell's theorem were marginalized during the first half of the 1970s. In contrast, *Epistemological Letters* was explicitly conceived as a publication dedicated to this topic, and the majority of its more than 170 entries explicitly took up this mission.

*Epistemological Letters* was also different from standard academic journals because of its informality, and that informality actually turned out to be one of the reasons this publication so effectively energized the foundations of quantum mechanics at the time.

This feature of the publication was intimately tied to its overall goals, as the declaration printed on the back cover of every one of its issues attests: '"Epistemological Letters" are not a scientific journal in the ordinary sense. They want to create a basis for an open and informal discussion allowing confrontation and ripening of ideas before publishing in some adequate journal.' This invitation for 'open and informal discussions' was taken very seriously by both the editors and the participants. Throughout its 36 issues, one finds more than 70 'original' contributions and more than 120 'response' articles. When the creators of *Epistemological Letters* said that they encouraged debates and confrontations of ideas, they really did mean it! Needless to say, by encouraging and facilitating an environment for these kinds of informal discussions, the journal truly enriched debates surrounding the foundations of quantum mechanics.[4]

For example, in the seventh issue of *Epistemological Letters*, published in November 1975, the editor invited Bell to submit a response to different objections to the derivation of his inequalities, objections that had all been published in previous issues of *Epistemological Letters*. Bell's response, titled 'Locality in Quantum Mechanics: Reply to critics', was an important piece in its own right because in it Bell is clearer than he was in his 1964 paper about how we should think of locality and of the hidden variables used in his theorem. Actually, one easily recognizes that this response is somewhat of a close cousin to his well-known 'Theory of local beables' paper, which would appear one year later in *Epistemological Letters* (1976). It thus seems that Bell was likely led to his theory of local beables, by a good extent, as a result of the 'confrontation and ripening of ideas' that *Epistemological Letters* so strongly promoted. Interestingly, the 'Theory of local beables' itself led to other rich debates and to the creation of very important papers that appeared first in *Epistemological Letters*. For instance, that paper led to a debate around the so-called 'free will assumption'—the assumption that the hidden variables of the particles are statistically independent from the experimenter's decision to measure one property or the other. The main participants in this conversation as it played out in *Epistemological Letters* were Shimony, Horn, and Clauser (1976), Bell (1977), and Shimony (1978). These papers have now become classics in this particular debate. In total, one finds sixteen responses to the 'Theory of local beables' scattered over five years' worth of issues. As this discussion illustrates, *Epistemological Letters* did in fact facilitate an environment that promoted open and rigorous debate amongst scholars.

As a side note, one might wonder how exactly the journal managed to keep track of all this back and forth conversation, which often involved the participation of several authors throughout different issues. By the time the third issue was published, the editors had come up with a simple but clever system to address this challenge.

---

[4] Of course, there was a more trivial sense in which the journal was informal. For one, it was printed and assembled in a rudimentary way (as can be seen from the often uneven margins). Also, at some points the equations would be typed, but at other points they would be handwritten (sometimes a single paper combined both styles). And with the goal of optimizing space, new entries could start immediately below the last entry, even if it meant starting a new article in the lower fourth of a page.

In the table of contents, to the left of each author's name, readers will find a number. If that number is an integer, say 4.0, then one knows that this entry is the fourth original contribution (that is, not a *response*) in *Epistemological Letters* since the very first issue. But if the number to the left of the author's name has some (non zero) decimal digits, say 5.3, then one knows that this is the third response to the fifth original contribution. So by just looking at the table of contents of each issue and the numbers printed on the left, one could quickly figure out if an entry is a response to an ongoing debate or an entirely new idea. Now, this system was far from perfect, as there were instances of repeated numbers for different entries. And the system was confusing at times, since in some cases the response pieces were no longer about the 'original' entry but about another response. Hence, 5.3 could indicate that this piece was a response to 5.2, as opposed to indicating that the piece is a response to 5.0. Fortunately, most of the time, the authors would specify who exactly they were responding to in the title of the paper.

Let me now focus on another aspect related to the informality of the journal, which has to do with the following passage printed on the back of each issue: 'allowing confrontation and ripening of ideas *before publishing in some adequate journal*' [my emphasis]. This is interesting because it was a way of signalling to authors that they could find here a space where speculation and half-baked ideas about quantum mechanics were well received, in startling contrast with the policies or standards of other physics journals of the time (such as *Physical Review*) where speculative papers would not be well received, or even in contrast with any other academic journal were only very polished pieces are considered for publication (as far as I can tell, *Epistemological Letters* was only peer reviewed—if ever—by the editors). Because of this, the purpose of publishing in *Epistemological Letters* (at least according to the editors) was primarily that of getting feedback from other contributors 'before publishing in some adequate journal', functioning similarly to preprint servers today such as arXiv, Phil Science Archives, or Academia.

Having said this, as a sign of the importance that this publication had in the foundations of quantum mechanics community, one does not really find cases of an author first publishing here and then moving on to a 'formal' journal but of the opposite phenomenon: authors would send papers to *Epistemological Letters* even if they had already published it in or at least submitted the paper to other journals.[5] Let me give three examples. By the time he already knew one of his papers (1974b) was accepted into *Foundations of Physics*, P.A. Moldauer sent a short version of that paper to the second issue of *Epistemological Letters* (he points out there that a more complete discussion is given in the *Foundations of Physics* version (1974a, p. 24)). Second, in the sixth issue of *Epistemological Letters* (which was published in September 1975), George Lochak published a piece in French criticizing Bell's derivation of his inequalities

---

[5] There are several cases of papers by Bell found in *Epistemological Letters* that were reprinted later in other places (more on this below), but this was not a case of an author using *Epistemological Letters* to polish a paper before sending the paper somewhere else. Rather, this happened only after several years had passed between the publication in *Epistemological Letters* and the reprinted version.

(this is one of the papers Bell responds to in the November 1975 issue I mentioned earlier). However, by October 1974 an English version of this paper had already been received by *Foundations of Physics* (as indicated on the first page of its publication as Lochak, 1976). Third, Popper, Garuccio, and Vigier published a paper in *Epistemological Letters* in July 1981 where they outlined a new experiment designed to detect de Broglie waves (Garuccio *et al.*, 1981). Although the same paper was published later in December of the same year in *Physics Letters A*, the paper was received by that journal on 18 April 1981, from which we can infer that the authors were not sending the paper to *Epistemological Letters* with the goal of getting feedback before sending it to *Physics Letters A*. At the very least, cases like this indicate that authors recognized the value in publishing in *Epistemological Letters* for its own sake (and not just for getting some preliminary feedback), even if they had already published in (or submitted to) a 'more adequate journal'.

I want to mention one more, perhaps subtler, reason that the informal style of *Epistemological Letters* benefited the foundations of quantum mechanics. Scattered at random places, one can find short notes calling the attention of readers to new conferences, books, papers published elsewhere on the foundations of quantum mechanics, and even correspondence sent to them by other scholars. For instance, in the fourth issue published in December 1974, a very short note by C. A. Hooker is printed announcing the publication of two book contributions, 'Physics and Metaphysics: A Prolegomena for the Riddles of Quantum Theory' (Hooker, 1973) and 'The Nature of Quantum Mechanical Reality: Einstein versus Bohr' (Hooker, 1972). In the sixth issue published in September 1975, the editors decided to print two contributions based on presentations from the conference 'Un demi-siècle de mécanique quantique' organized in Strasbourg in May 1974 (papers from this conference were later published as Lopes and Paty, 1977).[6] In the seventh issue, from November 1975, the editors inform others about a new paper appearing in *Helvetica Physica Acta* criticizing the derivations of the Bell inequalities. In the 17th issue published in December 1977, the editors note that they plan to organize a small conference to take place in March 1978, in Geneva, to take advantage of the fact that Shimony would be visiting them. But perhaps the best example of all is provided in issue 19 published in June 1978, where the authors made a conscious effort to report the main exchanges, in form of short fragments, between the participants of a small colloquium that included authors like Costa de Beauregard, d'Espagnat, and Shimony. Clearly, these informal announcements about other publications and about what had been discussed at conferences on the subject kept the community very alive during these crucial years.

I hope it is clear, then, that a great part of *Epistemological Letters'* success originated in its informal style; however, let me close this section by pointing out that this feature of the journal did hurt it in some ways. Since it was self-labelled as an informal journal, many scholars publishing in more standard venues do not cite papers that appeared in

---

[6] Thanks to Olival Freire for this reference.

*Epistemological Letters*, perhaps because they thought it was risky for their own careers or simply because they did not think it was necessary to do so. This meant that the impact of the journal is not adequately reflected in the number of citations received, despite the fact that many of the most significant papers on the subject were first published here. This, together with the fact that many of these important contributions were reprinted in other journals (such as in a *Dialectica* issue from 1985) or in books (such as Bell's 'Speakable and Unspeakable in Quantum Mechanics'), means that scholars today have a hard time measuring the actual influence of *Epistemological Letters* by looking at citations or by using tools such as Web of Science or Google Scholar.[7] From the perspective of actual citations, *Epistemological Letters* seems like it was a marginal player in the foundations of quantum mechanics, and it does not help that, up until 2020, there was no library in the US where one could find the entire collection (when Clauser said that it 'probably cannot be found in most physics libraries' he was understating it!). This prompted me to digitize the collection, together with Don Howard, and we are very happy to say that as of January 2020, all the issues are openly available through www.curate.nd.edu.[8] The fact that Howard had the entire collection, except for the first issue (which Howard Stein kindly gifted to us), sitting in his basement was both amusing and unsurprising since he was a student of Shimony during the years that *Epistemological Letters* was published. For an informal story of the digitization process, see https://www.nd.edu/stories/quantum-interest/.

## 30.4 CONCLUDING REMARKS

Overall, *Epistemological Letters'* strong dedication to creating a platform conducive to open debates between the most prominent scholars of the field, its evident effort to reach as many participants interested in the philosophy of quantum mechanics as possible including some raising philosophers of physics who would soon lead the discipline in the years to come, its clearly delineated subject matter, its constant effort to remind participants of conferences, books, and papers on its topic of focus, and the fact that a figure like Shimony closely worked on its production are all important features that explain why the journal had a significant impact on establishing the foundations of quantum mechanics as a respectable discipline. Who better to sum-marize the impact of *Epistemological Letters* than the editors themselves? In the preface of a special issue of *Dialectica* (1985) dedicated precisely to highlighting some of the

---

[7] However, if one uses Web of Science to analyse the citation of Bell's 1964 paper from 1973 up until the end of *Epistemological Letters*, one sees a clear upward trend, which could be a sign that *Epistemological Letters* was considerably increasing awareness of Bell's paper in the broader academic community.

[8] The website https://www.informationphilosopher.com/solutions/scientists/bell/Epistemological_Letters/ (accessed 7 December 2020), created by Bob Doyle, presents the table of contents of every issue of *Epistemological Letters*.

most important papers that had appeared in *Epistemological Letters* (including the exchange on 'The theory of local beables' between Bell, Clauser, Horne, and Shimony mentioned before), Shimony says:

> The variety of the contributions and the vigor of the debates showed that the purpose was very well accomplished. Because of the brief time interval between issues and the absence of customary refereeing procedures, it was possible to carry on a debate more rapidly than in standard journals, and speculative ideas could be more easily made public. It is remarkable that in spite of the informality of *Epistemological Letters*, the typing of the articles, including mathematical formulae, was very accurate. The reputation of the written symposium spread rapidly, and many people throughout the world wrote to be added to the list of recipients.
>
> (Shimony, 1985, p. 83)

Just as *Dialectica* did one year after it first emerged in 1947, *Epistemological Letters* stepped forward as the vehicle facilitating precisely the sort of dialogue between physics and philosophy that has rested at the heart of the foundations of quantum mechanics since its origins.

But let me end this piece by presenting something Bonsack himself said in a pamphlet found in the penultimate issue, announcing, in both French and English, the end of the symposium:

> The editors can consider, without immodesty, this written symposium as a success: 36 issues spread over more than ten years (to which two meetings in Geneva, 1978 and 1979 must be added), the contribution (with original papers) of the most distinguished students of this field, . . . , the numerous references in physical literature, all that have proved enough its utility and its contribution to the ripening of the debate.
>
> For my part—and I hope I am not the only one—I have learned very much: the ideas I first had have evolved, others became clear, especially Bell's Theorem, the conclusivity of which I questioned at the start and that finally appeared to be out of any doubt capable if calling into questions some fundamental features of our world view, particularly the non-separability between subsystems, about which we meant we could admit that they evolve independently, since they were locally disconnected and didn't show any known physical interaction.
>
> But the best things have an end. Everything—or almost everything—has been said . . . It seemed thus to the Institute for Methodology that time has come to stop.
>
> (Bonsack, 1984)

Bonsack was of course right that *Epistemological Letters* had been a success, but how wrong he was about the fact that 'almost everything' had been said! On the contrary, the early 1980s were just the beginning of a rich literature on the subject, both in physics and in philosophy. In the early 1980s, a new generation of experiments by Aspect and collaborators (1982a, 1982b) were performed, further confirming the violation of the Bell inequalities. Importantly, these new experiments (which introduced things like rapidly changing polarizers) further legitimized the foundations of quantum mechanics as a serious field of research within the physics community (see Freire, 2006, pp. 608–10). And very soon after *Epistemological Letters* stopped, philosophers like Jon Jarrett (1984), Don Howard (1985), and Fine (1986) were making long-lasting contributions regarding the logical and conceptual implications of the violation of the Bell inequalities. Thus, the end of *Epistemological Letters* was only the beginning of an even more active era on the subject.

## ACKNOWLEDGEMENTS

I want to thank Chris Hamlin and Don Howard for encouraging me to pursue this topic. I learned a lot about the journal itself and the broader historical context from personal conversations with Don, who I also worked with during the digitization process. I also want to thank Olival Freire for his very helpful comments on this piece. I also appreciate the assistance of my partner, Stacy Sivinski, who helped me edit the piece. Finally, I want to thank Natalie Meyers, from the Navari Family Center for Digital Scholarship of the University of Notre Dame, who guided me through the multiple stages involved in the digitization of the files.

## REFERENCES

Anderson, P. W., and Matthias, B. T. (1964). Editorial foreword. *Physics Physique Fizika*, **1**, i.

Aspect, A., Dalibard, J., and Roger, G. (1982a). Experimental test of Bell's inequalities using time-varying analyzers. *Physical Review Letters*, **49**(25), 1804.

Aspect, A., Grangier, P., and Roger, G. (1982b). Experimental realization of Einstein–Podolsky–Rosen–Bohm Gedankenexperiment: a new violation of Bell's inequalities. *Physical Review Letters*, **49**(2), 91.

Association Ferdinand Gonseth (1971). *L'Association Ferdinand Gonseth en quelques mots*. http://afg.logma.ch/ afg.htm (accessed 1 August 2020).

Bell, J. (1964). On the Einstein–Podolsky–Rosen paradox. *Physics Physique Fizika*, **1**, 195–200.

Bell, J. (1966). On the problem of hidden variables in quantum mechanics. *Reviews of Modern Physics*, **38**(3), 447–52.

Bell, J. (1975). Locality in Quantum Mechanics: Reply to critics. *Epistemological Letters*, (7).

Bell, J. (1976). The Theory of Local Beables. *Epistemological Letters*, (9).

Bell, J. (1977). Free variables and local causality. *Epistemological Letters*, (15).

Bell, J. and Aspect, A. (2004). *Speakable and Unspeakable in Quantum Mechanics: Collected Papers on Quantum Philosophy*. Cambridge: Cambridge University Press.

Bertholet, E. (1971). *Manifeste de l'Association Ferdinand Gonseth pour la crèation d'un institut de la méthode à Bienne*. http://afg.logma.ch/mnfst.htm (accessed 1 August 2020).

Bohm, D. (1952). A suggested interpretation of the quantum theory in terms of 'hidden' variables. I. *Physical Review*, **85**(2), 166–79.

Bohr, N. (1948). On the notions of causality and complementarity 1. *Dialectica*, 2(3–4):312–19.

Bonsack, F. (1961). *Information, thermodynamique, vie et pensée*, volume 2. Paris: Gauthier-Villars.

Bonsack, F. (1984). End of the written symposium Hidden Variables and Quantum Uncertainty. Insert. Institut de la Méthode (Association F. Gonseth).

Bub, J. (1973). On the possibility of a phase-space reconstruction of quantum statistics: A refutation of the Bell–Wigner locality argument. *Foundations of Physics*, 3(1), 29–44.

Clauser, J. (2002). Notes on Early History of Bell's Theorem. In J. Bell, R. Bertlmann, and A. Zeilinger (eds), *Quantum (Un)speakables: From Bell to Quantum Information*, Berlin: Springer.

Clauser, J. F., Horne, M. A., Shimony, A., and Holt, R. A. (1969). Proposed experiment to test local hidden-variable theories. *Physical Review Letters*, 23(15), 880.

d'Espagnat, B. (1976). *Conceptual foundations of quantum mechanics*, 2nd edn. Reading, MA: W. A. Benjamin, Inc.

d'Espagnat, B. (1983). Veiled reality. In B. d'Espagnat, *In Search of Reality*, New York: Springer, pp. 82–104.

de Broglie, L. (1948). Sur la complémentarité système. *Dialectica*, 2(3/4), 325–30.

Einstein, A. (1948). Quanten-Mechanik und Wirklichkeit. *Dialectica*, 2(3/4), 320–24.

Einstein, A., Born, M., and Born, H. (1969). Briefwechsel 1916–1955 [zwischen] Albert Einstein, Hedwig Born, und Max Born.

Einstein, A., Podolsky, B., and Rosen, N. (1935). Can quantum-mechanical description of physical reality be considered complete? *Physical Review*, 47(10), 777–80.

Fine, A. (1973). Probability and the intepretation of quantum mechanics. *The British Journal for the Philosophy of Science*, 24, 1.

Fine, A. (1974). On the completeness of quantum theory. *Synthese*, 29(1), 257–89.

Fine, A. (1986). *The shaky game: Einstein, realism, and the quantum theory*. Science and its conceptual foundations. Chicago: University of Chicago Press.

Fraassen, B. C. V. (1974). The Einstein–Podolsky–Rosen Paradox. *Synthese*, 29(1), 291.

Freedman, S., and Clauser, J. (1972). Experimental test of local hidden-variable theories. *Phys. Rev. Lett.*, 28(14), 938–41.

Freire, O. (2004). The Historical Roots of the 'Foundations of Quantum Physics' as a Field of Research (1950–1970). *Foundations of Physics*, 34(11),1741–60.

Freire, O. (2006). Philosophy enters the optics laboratory: Bell's theorem and its first experimental tests (1965–1982). *Studies in History and Philosophy of Modern Physics*, 37(4), 577–616.

Fuchs, T. (2007). *Ferdinand Gonseth.* https://hls-dhs-dss.ch/fr/articles/031363/2007-01-04/ (accessed 1 August 1, 2020).

Garuccio, A., Popper, K., and Vigier, J. P. (1981). Possible direct physical detection of de Broglie waves. *Physics Letters A*, 86(8), 397–400.

Gonseth, F. (1926). *Les fondements des mathématiques*. Paris: A. Blanchard.

Gonseth, F. (1936). *Les mathématiques et la réalité*. Paris: F. Alcan.

Gonseth, F. (1944). *Déterminisme et libre arbitre*. Neuchâtel: Editions du Griffon.

Gonseth, F. (1945). *La géométrie et le problème de l'espace*. Neuchâtel: Editions du Griffon.

Gonseth, F. (1964). *Le probleme du temps. Essai sur la méthodologie de la recherche*. Bibliotheque scientifique. Neuchâtel: Editions du Griffon.

Hooker, C. A. (1972). The Nature of Quantum Mechanical Reality: Einstein versus Bohr. In R. G. Colodny (ed.), *Paradigms and Paradoxes*, Pittsburgh: University of Pittsburgh Press, pp. 67–302.

Hooker, C. A. (1973). Physics and metaphysics: A prolegomena for the riddles of quantum theory. In C. A. Hooker (ed.), *Contemporary Research in the Foundations and Philosophy of Quantum Theory*, Dordrecht: Reidel, pp. 174–304.

Howard, D. (1985). Einstein on locality and separability. *Studies in History and Philosophy of Science Part A*, **16**(3), 171–201.

Jarrett, J. P. (1984). On the physical significance of the locality conditions in the Bell arguments. *Noûs*, **18**(4), 569–89.

Jauch, J. M. (1973). *Are quanta real?: A Galilean dialogue*. Bloomington, IN: Indiana University Press.

Jauch, J. M., and Morrow, R. A. (1968). Foundations of quantum mechanics. *Am J Ph*, **36**(8), 771.

Kaiser, D. (2007). Turning physicists into quantum mechanics. *Physics World*, **20**(5), 28.

Kaiser, D. (2011). *How the Hippies Saved Physics: Science, Counterculture, and the Quantum Revival*. New York: W. W. Norton.

Kaiser, D. (2020). *Quantum Legacies: Dispatches from an Uncertain World*. Chicago: University of Chicago Press.

Kaiser, D. (2021). Experiments on Bell's Theorem. In O. Freire (ed.), *Oxford Handbook of the History of Quantum Interpretations*, Oxford: Oxford University Press (this volume).

Kevles, D. (1995). *The Physicists: The History of a Scientific Community in Modern America*. Cambridge, MA: Harvard University Press.

Lauener, H. (1977). Ferdinand Gonseth 1890–1975. *Dialectica*, **31**(1–2), 113–18.

Lochak, G. (1975). Paramètres cachés et probabilités cachées. *Epistemological Letters*, (**6**).

Lochak, G. (1976). Has Bell's inequality a general meaning for hidden-variable theories? *Foundations of Physics*, **6**(2), 173–84.

Lopes, J. L., and Paty, M. (1977). *Quantum Mechanics: A Half Century Later*. Dordrecht: D. Reidel.

Margenau, H., and Yourgrau, W. (1970). Editorial preface. *Foundations of Physics*, **1**(1), 1–3.

Moldauer, P. (1974a). A new critique of EPR. *Epistemological Letters*, (**2**).

Moldauer, P. (1974b). Reexamination of the arguments of Einstein, Podolsky, and Rosen. *Foundations of Physics*, **4**(2), 195–205.

Myrvold, W., Genovese, M., and Shimony, A. (2019). Bell's Theorem. In E. N. Zalta (ed.), *The Stanford Encyclopedia of Philosophy*, Spring 2019 edition. Stanford, CA: Metaphysics Research Lab, Stanford University.

Pauli, W. (1948). Editorial. *Dialectica*, **2**(3/4), 307–11.

Piron, C. (1976). *Foundations of quantum physics*. Reading, MA: W. A. Benjamin, Inc.

Prabook (2020). *François Bonsack*. https://prabook.com/web/francois.bonsack/488192 (accessed 1 August 2020).

Primas, H. (1983). *Chemistry, quantum mechanics and reductionism: perspectives in theoretical chemistry*. Lecture notes in chemistry 24, 2nd corr. edn. Berlin/New York: Springer-Verlag.

Shimony, A. (1978). Reply to Bell 17.3. *Epistemological Letters*, (**18**).

Shimony, A. (1985). Introduction. *Dialectica*, **39**(2), 83–84.

Shimony, A., and Horne, M. (1973). Local Hidden-variable Theories. *Epistemological Letters*, (**1**).

Shimony, A., Horne, M., and Clauser, J. (1976). Comment on 'The Theory of Local Beables'. *Epistemological Letters*, (**18**).

Suppes, P. (1974). Introduction. *Synthese*, **29**(1), 3.

CHAPTER 31

# QUANTUM INTERPRETATIONS AND 20TH CENTURY PHILOSOPHY OF SCIENCE

THOMAS RYCKMAN

## 31.1 INTRODUCTION

PHILOSOPHICAL reflections on, and developments from, the twin revolutions of 20th century physical theory, relativity and quantum mechanics, differed significantly in depth of influence and immediacy of engagement; the course of philosophy of science in the 20th century reflects this difference. As is well-known, logical positivism (in Vienna, under M. Schlick, R. Carnap, O. Neurath) and logical empiricism (in Berlin, under H. Reichenbach) emerged in the early 1920s under the twin stars of general relativity and Whitehead and Russell's *Principia Mathematica*, together with the latter's thesis of logicism, that mathematics is logic. Despite non-trivial philosophical differences between these two movements, their ensuing philosophy of science will be, with some terminological liberty, denominated 'neopositivist'; consideration here will be limited to that philosophy and its descendants in the US and the UK.[1] Einstein's period philosophical remarks on general relativity played a major role in neopositivism's formation, portrayed by both Schlick and Reichenbach as empiricist and conventionalist in spirit (Ryckman, 2005). To the neopositivists, the main philosophical lesson of general relativity was methodological and explicitly spelled out in Einstein's 1921 lecture 'Geometry and Experience'. Reichenbach (1928) in particular

---

[1] For reasons of space, I must pass over G. Bachelard's (1934, 1951) philosophy of 'applied rationalism', a highly innovative response to quantum mechanics and the quantum formalism unfortunately not widely known outside of France.

pushed this line in a widely read philosophical monograph on spacetime theory.[2] Neopositivist 'scientific philosophy' was therefore well underway when quantum mechanics appeared in 1925/26, followed by complementarity in 1927. In accord with positivism-inflected pronouncements by the leading quantum physicists, the nascent quantum mechanics was widely deemed congenial with, even a triumph of, neopositivist philosophy of science.

The neopositivist hegemony was uprooted in the post-war period. Bohm's hidden variable theory of 1952, especially the dismissive response to it by quantum orthodoxy (proponents of what soon would be termed the 'Copenhagen interpretation'), unleashed a complex philosophical dialectic that eventually extinguished neopositivist philosophy of science. On one hand, the results of Bohm (and L. de Broglie and J.-P. Vigier) flew in the face of the orthodox contention that realism is untenable as a philosophical point of view with respect to microphysics. To realist-oriented philosophers, Bohm's hidden variable theory was a gift on a platter: an explicitly ontological and determinist theory in which particles possess definite trajectories at all times, without the subjectivity introduced into measurement by an 'observer', yet designed to coincide with standard quantum mechanics in its empirical predictions. Following more than two decades of neopositivist caricature as 'school philosophy' or 'metaphysics', realist views of physical theory revived in the period 1955–1965; currents that would transform into a specifically 'scientific realism', holding that the reasons for accepting a theory as *explanatory* (i.e., purporting to describe entities and processes of nature underlying and producing the phenomena of observation) are also reasons for accepting the entities and processes the theory postulates as existentially real.

On the other hand, quantum orthodoxy's strident repudiation of, and resistance to, Bohm's theory can be directly linked to the emergence of an *anti*-realist post-positivist philosophy of science in the 1960s. In the late 1950s N.R. Hanson had adopted the term 'paradigm' to refer to the Copenhagen interpretation, a notion thereafter found in Kuhn's 1962 account of scientific revolutions. Kuhn's philosophy of science was expressly anti-realist, while the realist tendencies of P. K. Feyerabend were to become pluralist and then increasingly relativist. Already in 1960, Feyerabend, astonished that prominent defenders of the Copenhagen interpretation (in particular, L. Rosenfeld, Heisenberg, and Bohr himself) refused to recognize even the possibility of an alternative to complementarity, rejected Kuhn's account of scientific progress through 'paradigms' as inherently conservative as well as historically inaccurate. In a series of influential papers in the early 1960s, he argued that scientific progress, rationality, objectivity, and even empiricism itself required comparison of distinct theories and their different descriptions of the

---

[2] To wit: geometry is a purely mathematical theory whose connection to observation is established by legislating correspondence rules or 'coordinative definitions' between expressions of the pseudo-Riemannian mathematical formalism (viz., the line element $ds$) and concrete instruments of measurement, rods and clocks, in this way bringing empirical content (i.e., the anomalous orbit of Mercury, the observed deflection of star light by the solar gravitational field) into a previously contentless mathematical formalism.

phenomena of observation. By the late 1960s, Feyerabend's strident advocacy of theor-etical pluralism, eventually methodological anarchism ('anything goes'), engendered or inspired many subsequent pluralisms, relativisms, and disunity of science currents. In conjunction with related developments, the result was a fundamental transformation of the entire discipline of philosophy of science.

Years later, J. S. Bell pointed to Bohm's theory as a stimulus for his own results that would furnish further grist for the conflict between realist and nonrealist philosophies of science. Experimentally demonstrated violations of Bell's inequality as well as 'Bell's theorem', that any local realist theory will empirically disagree with quantum mech-anics, led empiricist-leaning philosophers to conclude that realism is incompatible with quantum physics, but others to pronounce upon the metaphysical novelty of a non-local nature. By the 1980s philosophers of physics had embarked in pursuit of an 'experimental metaphysics' (Shimony, 1989), a project inconceivable to a previous generation of philosophers of science. That quest, and the dialectic of realism and nonrealism about quantum theory, continues into the 21st century.

## 31.2 NEOPOSITIVIST PHILOSOPHY OF SCIENCE AND QUANTUM MECHANICS 1925–1950

Fashioned upon empiricist views associated with Hume, Mach, and loosely with Einstein, while fortified (in Vienna) with the new tool of Russellian logic and type theory, Schlick, Carnap, H. Feigl, P. Frank, and a bit later K. R. Popper (who must be distinguished from neopositivism), initiated treatment of problems (the relation of theory to observation, empirical meaning of theoretical terms and concepts, methods of theory testing, and so on) to become standard fare in 20th century philosophy of science. Neopositivism's defining work, Carnap (1928), directly addressed neither relativity nor quantum mechanics but rather introduced 'scientific philosophy' as an anti-metaphysical crusade, reviving Russell's 1914 'external world programme' to reconstruct all meaningful statements of science (in the sense of '*Wissenschaft*') in terms of pure logic and reports of direct experience. These developments were under-way as quantum mechanics appeared in 1925/26, followed by complementarity in 1927. Not surprisingly, Heisenberg's programmatic call (1925) to 'discard all hope of observing hitherto unobservable quantities, such as the position and period of the electron' by attempting 'to establish a theoretical quantum mechanics ... in which only relations between observable quantities occur' was often quoted and widely celebrated.[3]

---

[3] E.g., Reichenbach (1931): '(What) has been carried out in quantum mechanics, particularly under Herr Heisenberg's direct influence, is therefore a confirmation of older considerations that previously were connected to classical physics by those of positivist-empiricist persuasion.'

Subsequent conceptual analysis of quantum mechanics under Bohr's complementarity, as well as positivism-leaning remarks of Born, Jordan, Pauli, and others, confirmed the neopositivist view of quantum mechanics as a theory without metaphysical elements, whose postulates were based on empirical fact, and whose statements concerned only observable phenomena. As a result, in its early years, quantum mechanics was widely deemed a triumph of neopositivist philosophy and even, in its general orientation, to have been anticipated by the new direction in philosophy of science.

Undoubtedly, a rhetoric of inevitability assisted this confluence with neopositivist philosophy; Born and Heisenberg's (1927, p. 398) attempt to portray quantum mechanics as a 'closed theory whose fundamental physical and mathematical assumptions are no longer susceptible to any modification' was largely successful. By 1930 (or 1932 with von Neumann's axiomatization) a widespread conviction emerged that the quantum formalism and complementarity had acquired the character of uniqueness as the only way to avoid ambiguities and pseudo-problems stemming from misleading ('metaphysical') classical intuitions and analogies. The Bohr–Einstein debate and the 1935 EPR 'paradox' were largely ignored, on the assumption that Bohr's reply to EPR achieved complete victory, rendering further debate pointless. Certainly, the neopositivist philosophers were familiar with Einstein's (and Planck's) objections to quantum mechanics, as well as their conception of physical theory as aspiring to completely describe an observer-independent reality. But these views and Einstein's programmatic papers in the 1930s on the nature of physical theory were quietly ignored (Ryckman, 2017).

Of the leading neopositivists, only Reichenbach is recognizable today as a philosopher of physics. Schlick had a 1904 Ph.D. in physical optics under Max von Laue in Berlin, a substantial correspondence with Einstein, and had published insightfully on special and general relativity. By 1925, however, he had come under the influence of L. Wittgenstein, from then on regarding traditional philosophical questions (e.g., the relation of theory to external reality, determinism, indeterminism, etc.) as 'pseudo-problems'. Following Wittgenstein, the Vienna Circle sharply distinguished between science and philosophy; the task allotted to the philosopher of science was that of logical analysis of 'the language of science'. Reichenbach, less inclined towards positivism, indeed viewed the job of philosophy differently, i.e., to assist scientists in epistemological clarification of the foundations of their theories. But his broader philosophical agenda aimed to show that quantum mechanics exemplified an antecedently formulated probabilistic theory of knowledge.[4] In general, neopositivism's interest, or intervention, in philosophical issues raised by quantum mechanics

---

[4] For a number of years, Reichenbach (e.g., Reichenbach, 1930) maintained his probabilistic critique of the principle of causality that had prefigured quantum mechanics, and that the Heisenberg indeterminacy principle was an 'exact instance' of his generalization of the principle of causality as a continuous 'probability chain', the latter notion exposing the only legitimate understanding of physical causality. A largely forgotten book (Reichenbach, 1944) advocated the use of a three-valued logic for quantum mechanical statements. Reichenbach's student H. Putnam (1957, 1969) attempted to revive the idea several times with little success.

remained quite restricted. In the 1930s, Schlick, Reichenbach, Frank, and Popper wrote on the broader philosophical significance of the uncertainty relations for the law of causality, or on the proper empiricist interpretation of Bohr's diverse remarks on complementarity. In the rapidly deteriorating European political and intellectual climate these discussions found little philosophical traction.[5]

In 1930, philosophy of science in the US existed principally in homegrown forms of evolutionary naturalism and pragmatism (J. Dewey, C.S. Peirce) or in Harvard physicist P. Bridgman's (1927) 'operationalism'. But the novel 'scientific philosophy' of Vienna and Berlin quickly became *à la mode* and in the early 1930s a younger generation of scholars (E. Nagel, C. Morris, W. V. O. Quine) either embraced or were influenced by it. The 1934 launch of the journal *Philosophy of Science* created momentum that enabled philosophy of science to become a recognized subdiscipline in US universities. Emigration of neopositivism's leading figures (Feigl, Carnap, Frank, G. Bergmann, Reichenbach, C. G. Hempel) in the 1930s shaped North American philosophy of science into the 1960s. A somewhat different situation existed in the UK in the first decades of quantum mechanics. Philosophy of science was then, with few exceptions (Cambridge philosophers C. D. Broad, B. Russell), the purview of scientists (A. S. Eddington, J. Jeans, H. Dingle, J. D. Bernal) addressing the educated laity on supposed philosophical implications of the new physical theories. It became a recognized specialization within philosophy only in the post-WWII period; the British Society for Philosophy of Science was founded in 1948 and its journal, *The British Journal of the Philosophy of Science*, in 1950. Popper, who began teaching at the LSE in 1946, was then known among English-speaking readers principally as the anti-totalitarian author of *The Open Society and its Enemies* (1945); his 1935 Viennese treatise on scientific method was not translated until 1959.

By mid-century, neopositivist philosophy of science was regnant in the US.[6] The disciplinary state of play centred upon 'rational reconstruction of the language of science' patterned upon a two-language (observational, theoretical) model linked by 'correspondence rules' while progress was marked in ever-more liberalized versions of 'criteria of cognitive significance' for the empirical interpretation of theoretical terms and statements. It continued to be broadly accepted that quantum mechanics, understood through the lens of the orthodoxy soon to be termed the Copenhagen interpretation, was congenial to, if not exemplary of, the neopositivist view of scientific theories. Popper's considerable influence in the UK, based upon his falsificationist account of scientific methodology, was growing and would be praised by leading UK scientists.

---

[5] As also was the case with G. Hermann's largely neglected monograph (1935) critiquing von Neumann's proof against hidden variables while defending a neo-Kantian 'law of causality'. Though noted in an early paper of Feyerabend (1954), Jammer (1974) is generally credited with restoring philosophical interest in Hermann.

[6] The first US anthology of readings in philosophy of science (Feigl and Brodbeck, 1953, p. 5) defined 'modern philosophy of science... [to be] a specialized branch of analytical philosophy', and the task assigned to the philosopher of science that of 'logical analysis of scientific concepts, laws, and theories'.

Already a critic of Copenhagen 'instrumentalism', Popper's 'propensity' interpretation of quantum probabilities was yet in its early stages.[7]

## 31.3 DIALECTICAL MATERIALISM AND BOHM'S THEORY

This irenic state of affairs in philosophy of science rather quickly changed following a series of events in the immediate post-war period. From an unexpected quarter came a novel demand for a 'materialist' and non-positivist interpretation of quantum theory. From 1946, especially following a widely publicized speech in June 1947, A. A. Zhdanov, Stalin's newly appointed director of Soviet cultural policy, orchestrated a vast propaganda campaign directed against Western 'bourgeois ideology', the Soviet expression for cultural ideas in the 'superstructure' reflecting the economic basis of Western capitalism (Pollock, 2006). The resulting *Zhadanovshchina* would shape the USSR's cultural and scientific life until Stalin's death in 1953. Artists, scientists, philosophers, and other intellectuals were enjoined to firmly oppose 'reactionary bourgeois idealism', to denounce 'anti-Soviet' and 'anti-socialist' ideas, and to adapt their thought, work, research, and studies in accordance with the tenets of dialectical materialism outlined by Stalin in 1938.[8] Whereas Marxist ideology in the pre-war Soviet Union was mostly tolerant of Bohr's views, in the post-war period, following the lead of Zhdanov, Soviet quantum physicists such as Blokhintsev (1951) attacked the 'Copenhagen school' and complementarity for denying causality thus promoting 'the liquidation of materialism', and for holding an idealistic view of the wave function as 'merely an expression of the information possessed by the observer' (as translated in Graham, 1966). Similar attacks on Heisenberg's and Bohr's 'idealistic figments of the imagination' were made by Soviet philosophers such as A. A. Maksimov, prompting Rosenfeld in 1949 to brief an astonished Bohr.[9] Marxist physicists and philosophers outside the Soviet Union were certainly cognizant of the Zhdanov ideological campaign.[10] Though evidence is not conclusive,

---

[7] Popper's writings on quantum mechanics and propensity interpretation of quantum probabilities were largely neglected both by physicists and philosophers; see del Santo (2018, 2019).

[8] Stalin (1938) summed up 'the principal features of the Marxist dialectical method' in four main points: 1) 'Nature Connected and Determined', 2) 'Nature is a State of Continuous Motion and Change', 3) 'Natural Quantitative Change Leads to Qualitative Change', and 4) 'Contradictions Inherent in Nature'.

[9] Jacobsen (2012, p. 268): Maksimov made an 'extremely radical speech', translated and published in a Danish journal, praising Lenin's *Materialism and Empirio-Criticism* while attacking ideas of Bohr and Heisenberg as 'idealistic figments of the imagination'. Freire, Jr. (2019, p. 69) reports that Rosenfeld warned an uncomprehending Bohr of these developments in 1949.

[10] E.g., Maurice Cornforth, former student of C. D. Broad and Wittgenstein at Cambridge, at the time the leading ideologist of the British Communist Party (1949, pp. 49–50): 'The task of dialectics is not to accept the contradictory proposition that an electron is both a wave and a particle. Its task is to disclose the real dialectical contradiction in physical processes – the objective contradiction in the physical world,

the Soviet directive to find a materialist view of quantum theory may well have prompted D. Bohm to seek a causal and ontological interpretation of quantum mechanics. At the time a committed Marxist and member of the US Communist Party, an association that would later lead to persecution and forced emigration from the US, Bohm cannot have been unaware of the Soviet ideological campaign. In fact, Blokhintsev's virulent attack appeared in Russian, which Bohm did not read.[11] But Bohm later recalled that conversations with Einstein in Princeton, as well as a paper by either Blokhintsev or Terletsky (another Soviet quantum physicist), had provided encouragement.[12] His own report (1957, p. 110) is instructive:

> [Blokhintsev and Terletsky] made it clear that it is not necessary to adopt the interpretation of Bohr and Heisenberg, and showed that instead, one may consistently regard the current quantum theory as an essentially statistical treatment, which would eventually be supplemented by a more detailed theory permitting a more nearly complete treatment of the behaviour of the individual system. They did not, however, actually propose any specific theories or models for the treatment of the individual systems. Then in 1951, partly as a result of the stimulus of discussions with Dr. Einstein, the author began to seek such a model; and indeed shortly thereafter he found a simply causal explanation of the quantum mechanics which, as he later learned, had already been proposed by de Broglie in 1927.

The very month his (1952) theory appeared, in a letter to a friend he asked, 'Why in 25 years didn't someone in the USSR find a materialistic interpretation of quantum theory?' (Freire Jr., 2005, p. 15, note 51). After decades of denial that an empirically adequate realist and deterministic completion of quantum theory was even conceivable, Bohm's 'existence theorem' (Cushing, 1994, p. 158) became a matter of fact.

## 31.4 BOHM'S THEORY AND THE EMERGENCE OF SCIENTIFIC REALISM

In a 1953 Cambridge lecture published in 1956, apparently without knowledge of John Dewey's much earlier use of the term, Popper denominated the reigning neopositivist

not a formal contradiction between propositions – and to show how the wave-like and particle-like properties manifested by electrons come into being on the basis of that real contradiction. This has not been done, but remains to be done. It is a question of physical research.'

[11] Blokhintsev's paper (1951) has not been translated into English.
[12] Jammer (1974, p. 279) reported, 'Stimulated by his discussions with Einstein and influenced by an essay which, as he told the present author, was "written in English" and "probably by Blokhintsev or some other Russian theorist like Terletsky" (footnote 63) and which criticized Bohr's approach, Bohm began to study the possibility of introducing hidden variables.' Note 63 stated: 'Bohm has forgotten the exact title and author of this paper.' See also Freire Jr. (2019, chapter 3).

conception of physical theory as 'instrumentalism' (Popper, 1956). According to Popper, instrumentalism is based on the *philosophical* idea, articulated by Cardinal Bellarmino (*contra* Galileo) and Bishop Berkeley (*contra* Newton), that theories can be *nothing but* computational tools or inference rules for prediction of observations. Popper deemed the idea exemplified in the 20th century in Bohr's complementarity; moreover, the empirical success of quantum mechanics under complementarity had resulted in instrumentalism's triumph 'without a shot being fired' over a 'Galilean' realist tradition of physical theory as a conjectured true description of the world.[13] Popper did not mention Bohm or his 1952 theory; his own later attempts (Popper, 1967) to promote realism were based on a propensity interpretation of quantum probability; these proved idiosyncratic and were not widely adopted. An early paper by Feyerabend (1954), written upon his return to Vienna after spending the academic year 1952–53 as an assistant to Popper at the LSE, does contain a concise technical summary of Bohm's 1952 theory. But like Popper, Feyerabend's interest at the time lay in the question of whether quantum mechanics demonstrates that nature, i.e., the 'real outside world' [a term Feyerabend associated with Einstein] without regard to measurement or observation, is indeterministic and the paper defended, *à la* Popper, an indeterminist metaphysics.[14] Nonetheless, already inclined to realism, Feyerabend was impressed that Bohm had produced a realist alternative to quantum orthodoxy. His writings over the next two decades would feature allegorical episodes in the history of science to illustrate the progressive nature of realist accounts of physical theory (see further in Section 31.5 below).

Inspired by the results of Bohm (and L. de Broglie and J.-P. Vigier), philosophical challenges to what by then was known as 'the Copenhagen interpretation' and its presumptuous denial of any realist characterization of the nature of microphysical inquiry were launched in the period 1955–1965 by M. Bunge, Feyerabend, Popper, and H. Putnam. Perceptible cracks in the heretofore monolithic Copenhagen hegemony went hand in hand with internal critiques of neopositivism by G. Maxwell, H. Feigl, W. Sellars, J. J. C. Smart, and others that distinguished realist accounts of scientific explanation from the neopositivist's 'deductive-nomological (D-N) model' assimilating explanation with prediction. Resuscitating the traditional realist goal of scientific inquiry as finding out how nature *really* is, these philosophers formulated within philosophy of science an explanatory (and so, *scientific*) account of realism. They essentially argued that if a physical theory can be accepted as providing an explanatory account of phenomena, then one can have good reasons for accepting the underlying

---

[13] Years later, in 1968 letters to I. Lakatos, Feyerabend (Motterlini, 1999, pp. 135, 138) strongly repudiated Popper's insinuation of realist surrender: 'Bohr makes in clear, *in every paper between 1913 and 1925*, that a new "theory" of atoms (by which he means a realistic account) is not yet available, that all he was able to provide by his own work were *instruments* of prediction . . . but that he hopes by an improvement of these instruments finally to arrive at a real theory. . . . "Without a shot?" The whole period between 1913 and 1925 is a discussion of the issue between realism and instrumentalism.'

[14] Conflating predictability and determinism, Feyerabend argued that Bohm's association of continuous trajectories to particles did not amount to a deterministic interpretation of quantum mechanics.

entities and processes postulated by the theory as literally real. An argument to this effect referencing Bohm's theory is found in an influential 1963 book *Philosophy and Scientific Realism* by Australian philosopher J. J. C. Smart advocating a 'realistic philosophy of theoretical entities'. Stating that 'it is very unlikely that quantum mechanics is in its final form, and it may be drastically revised, with some of its fundamental assumptions altered', Smart identified 'the development of deterministic theories of microphysics on the lines foreshadowed by such writers as D. Bohm and J.-P. Vigier' as among the 'weighty reasons against giving up a realistic view of the entities postulated by physical theory' (Smart, 1963, pp. 43–44, p. 46) Ironically, the realist explanatory challenges to neopositivism were countered by an assault from the opposite direction: historical accounts of 'incommensurable' concepts and theories in the course of scientific advance, and historicist theories of scientific rationality. Philosophers such as Kuhn, Feyerabend, and I. Lakatos argued that the rationality of decisions and actions of scientists could best be understood in relation to the scientific eras and contexts in which they occurred. Whether intended or not, the historicist accounts of scientific rationality led to relativist and sceptical accounts of science leading to the 'Science Wars' of the 1990s. The ascendency of realism proved to be but short-lived.[15] In 1974 Putnam (1975, p. viii) could report that 'while realists in the philosophy of science were comparatively scarce in the 1950s, they are today, happily, a flourishing species'; that species lost a flourishing member when Putnam renounced scientific realism around 1980.

# 31.5 COMPLEMENTARITY AS *PARADIGM*: HANSON, KUHN, AND FEYERABEND

Kuhn's 'breakthrough' idea (viz., the concept of *paradigm*, and related notions of *normal science, incommensurability*) that led to *The Structure of Scientific Revolutions* came in 1958–9 when he held a Fellowship at Stanford's Center for Advanced Studies in the Behavioral Sciences. Years later, Kuhn reported that at that time he had been having 'great trouble' developing his ideas and he credited an 'encounter with incommensurability' as 'the first step on the road to *Structure*' (Beller, 1999, p. 293). Historian of science Mara Beller conjectured the 'encounter' likely came from reading N. R. Hanson's *Patterns of Discovery*, a 1958 book in which the term *paradigm* occurs on page 1. The overall message of Hanson's book stressed 'how much was at work when physicists of the past disagreed' and how much is 'missed also by anyone who thinks of the history of physics as just a march of better observations and more accurate experiments' (Hanson, 1958, p. 175), themes now largely associated with Kuhn. Long out of print and mostly neglected since Hanson's death in an air crash in 1967, *Patterns*

---

[15] Witness the first sentence of A. Fine's influential 1984 article, 'Realism is dead'.

*of Discovery* built upon an illustration of theory-ladenness of observation ('Do Kepler and Tycho see the same thing to the east at dawn?') to introduce topics familiar now to philosophers of science from reading Kuhn: the Gestalt-character of perception, the distinction between 'seeing that' and 'seeing as', the 'theory-laden' character of causal description, and the absence of a neutral language in which to express observational evidence.[16] But little notice has been taken of Hanson's admission (1958, p. 2) that the first five chapters on the above themes were written with final chapter six ('Elementary Particle Physics') in mind. This chapter, with its two appendices, were to be 'the lens through which... perennial philosophical problems will be viewed'. It is therefore instructive that after a brief discussion of wave–particle duality, chapter six is largely a presentation of complementarity as the result of a conceptual struggle to fit novel observations into a productive pattern of explanation. Hanson presented complementarity as a pragmatically successful way of seeing acquired in the investigation of quantum phenomena.[17] That message is reinforced in Appendix II wherein a technical *aperçu* of quantum mechanics is followed by a brief survey of objections (de Broglie, Madelung, Bohm) to the statistical interpretation of the wave function as a description of individual systems. Hanson was not yet the dogmatic proponent of Copenhagen he would become in papers published the next year, for he allowed (Hanson, 1958, p. 174) that 'the ultimate interpretation of the $\psi$-function is not yet settled' while praising Bohm's 1952 theory as 'an important contribution to philosophy of physics and to discussions of the interpretation of $\psi$' (Hanson, 1958, p. 172). The book concludes with the pointed reminder that disputes in 'natural philosophy' are not easily resolvable since 'rarely can a man observe what does not yet exist for him as a conceptual possibility' (Hanson, 1958, p. 175). We see here seeds of the idea of 'incommensurability' as a consequence of adopting complementarity as an exclusive or *paradigmatic* 'way of seeing'. The term *incommensurability* appears to have been introduced independently by Kuhn and Feyerabend around 1961.[18]

Reading a draft of Feyerabend's largely favourable review of Bohm's *Causality and Chance in Modern Physics* apparently prompted Hanson to publish in vigorous support of the Copenhagen interpretation in 1959.[19] The intent of Feyerabend's review was to counter a scathing treatment given the book by Rosenfeld in the *Manchester Guardian*

---

[16] Kuhn's papers from the early 1960s, collected in Kuhn (1977), display considerable evidence of Hanson's influence, both in citations to Hanson, and acknowledgments of the latter's 'revolutionary reconceptualization' permitting data 'to be seen in a new way'.

[17] E.g., Hanson (1958, p. 149): 'we cannot see the micro-world as we now do without accepting the uncertainty relations as inextricable in the organization of what we encounter.' Beller (1999, p. 295) summarized: 'The bulk of the chapter is devoted to description and approbation of the Copenhagen dogma and argues the impossibility of an alternative to the orthodox interpretation.'

[18] Hoyningen-Huene (1993, p. 207, note 57): 'According to Kuhn's admittedly uncertain recollections in 1982, he and Feyerabend introduced the *term* "incommensurability" independently.'

[19] Beller, (1999, p. 299): 'Hanson's defence of Copenhagen was triggered by a draft of a favourable review of Bohm's theory by Paul Feyerabend (1960).'

(see Section 31.6).[20] In any event, Hanson now portrayed Copenhagen not merely as 'a triumphant invention of an orderly interpretative pattern for a cluster of facts which before had been simply chaotic' (Hanson, 1959, p. 326), but also the sole such 'pattern', since 'there is now no alternative to the Copenhagen Interpretation', not that there *could not* be one, but 'only that *here and now* we cannot even say what an alternative would be like' (Hanson, 1959, p. 334). Bohm's theory was no longer an alternative.

While Hanson might have first employed the term *paradigm* as a label for complementarity and the Copenhagen interpretation as a progressive 'way of seeing' in quantum physics, that notion is justly associated to Kuhn, for whom, at least in the early 1960s, the Copenhagen interpretation *was* the quantum mechanical paradigm.[21] In two letters of lengthy comments on a late draft of *The Structure of Scientific Revolutions* in 1960 or 1961, Feyerabend, then a colleague of Kuhn at Berkeley, expressly raised the case of Bohm's theory as a specific reason for rejecting the Kuhnian notion of paradigm (Hoyningen-Huene, 1995, p. 365):

> Your insistence upon faithfulness *to one and only one paradigm* is bound to result in the elimination of otherwise very important tests and it is bound in this way to reduce the empirical content of the paradigm you want to be accepted. It may well be—and Bohm and Vigier are definitely of this opinion—that the situation is the same in the present quantum theory. The 'orthodox' refuse considering alternatives and their argument is that the present point of view has not yet encountered anomalies which would necessitate reconsideration of it in its entirety. Bohm points out that the limitations of the present point of view will become evident only if one has first introduced an alternative and shown that it is preferable. Hence, if the absence of limitations is taken as a reason for not considering alternatives, then trouble will never be discovered, simply because it could be discovered only with the help of alternatives.

Rhetorical insinuation of absence of any alternative to Copenhagen was already an old theme. Feyerabend, in general an admirer and defender of Bohr's 'positivism of a higher order' against the positivist philosophy of the Copenhagen *epigoni* (e.g., Feyerabend, 1958), nonetheless recognized (Feyerabend, 1962a, p. 88) that even Bohr insisted that in the description of atomic phenomena physics can *never* go beyond the classical framework and that all future microscopic theories will have to use the notion of complementarity as a fundamental notion. Heisenberg (1955, 1958) would

---

[20]  Feyerabend (1960, p. 330, note 1) and in an undated letter to Kuhn from 1960 or 1961, stated that Rosenfeld's review of Bohm (1957) appeared in *The Manchester Guardian*. I have been unable to locate it there but Feyerabend enclosed a clipping of the review in the letter to Kuhn, asking for its return. The clipping is quoted *verbatim* in Hoyningen-Huene (1995, pp. 379–80). Except for the title, it is identical to Rosenfeld (1958).

[21]  In a 1965 London conference confronting Popper's account of scientific knowledge with that of Kuhn, Feyerabend recalled (1970, p. 206), 'I remember very well how Kuhn criticized Bohm for disturbing the uniformity of the contemporary quantum theory. [According to Kuhn], Bohm's theory is *not* permitted to change the argumentative style.'

argue along similar lines, drawing Feyerabend's criticism. But the latter's animus against Copenhagen centred mainly on L. Rosenfeld's dogmatic defence, beginning with the remarkable Rosenfeld (1953a). That paper provoked the stone that, thrown by Feyerabend, unleashed an avalanche to become known as theoretical pluralism.

## 31.6 Rosenfeld contra Bohm and the Origins of Theoretical Pluralism

Feyerabend began his philosophical career as assistant to Popper at the LSE in 1952–53; he would turn against Popper and falsificationist methodology in the mid-1960s. Long before this he had confronted the dogmatic defence of the Copenhagen point of view, especially as framed in polemics against Bohm's 1952 hidden variable theory written by Bohr's long-time amanuensis and collaborator, Belgian physicist Leon Rosenfeld. Feyerabend repeatedly quotes from three screeds Rosenfeld directed against Bohm. The first in 1953, appearing in French and English, apparently intended to deal only a glancing blow since complementarity is deemed a 'necessity' hence 'there is no need to dwell upon the medley of so-called deterministic interpretations of quantum theory... now shooting up like mushrooms after the rain' (Rosenfeld, 1953b, p. 403). Still it produced an intransigent statement Feyerabend would return to repeatedly over the next three decades; in Feyerabend's translation of the French version:[22]

> We are here not presented with a point of view which we may adopt, or reject, according to whether it agrees, or does not agree, with some philosophical criterion. It is the *unique* result of an adaptation of our ideas to a new experimental situation in the domain of atomic physics. It is therefore completely on the plane of experience... that we have to judge whether the new conceptions work in a satisfactory way.[23]

To an extreme, Rosenfeld amplified views probably held to one degree or another among many of the Copenhagen school. But here, before a scientific (French version) and lay (English version) public, Rosenfeld underscored a conviction that Bohm's

---

[22] Rosenfeld (1953a, p. 44): '*Le premier point à mettre en relief, c'est la **nécessité** avec laquelle le concept de complémentarité s'impose à nous. Nous ne sommes pas en présence d'une vue de l'espirit, qu'il nous serait loisible d'accepter ou de repousser selon qu'elle est ou non conforme à tel ou tel critère philosophique. Il s'agit uniquement d'une adaptation de nos idées à une situation expérimentale nouvelle, dans le domaine de la physique atomique. C'est donc entièrement sur le plan de l'expérience et de son interprétation immédiate qu'il convient de juger si les conceptions nouvelles remplissent adéquatement leur fonction*' (original emphasis).

[23] Cited in Feyerabend (1958, p. 89; 1961, p. 385; 1962b, pp. 191–2, and pp. 220–21; and 1963b, p. 193). Cf. Rosenfeld's revised English version of this article, (1953b, p. 394) which speaks of complementarity as a 'logical necessity'.

theory was not merely wrong, but utterly irrelevant. Complementarity was effectively a necessity of thought mandated by the facts of quantum phenomena thus *there could be no alternative interpretation* of quantum theory. This uncompromising dismissal of Bohm's 1952 'existence theorem' of a logically possible alternative to Copenhagen made a deep impression on Feyerabend, fundamentally influencing his subsequent writings on philosophy of science.

Both Feyerabend and Bohm, colleagues in Bristol at the time, were present for Rosenfeld's second attack in the first week of April 1957, at the Ninth Symposium of the Colston Research Society. The conference was the first international meeting on the foundations of quantum mechanics after World War II; the principal topic of discussion was the causal quantum theory programme of Bohm (Koznjak, 2018). The direct confrontation between Bohm and Rosenfeld proved a milestone for Feyerabend, again yielding Rosenfeld quotations to serve as rhetorical ammunition over several decades. As late as *Against Method* (1975), by which time Feyerabend had taken a much more radical turn, the same passage from Rosenfeld's 1957 contribution in Bristol is cited no less than three times as an exemplary confession both of belief in the uniqueness of the Copenhagen interpretation, and of conservative resistance to alternative theories transformed into philosophical dogma.[24] Rosenfeld's emphatic defence of what Mara Beller termed the Copenhagen 'rhetoric of finality' proved to be a continually provocative stimulus for Feyerabend. The infamous methodological anarchism would be but a late derivative of adoption and advocacy of a theoretical, cum methodological, pluralism rooted in his instinctual repudiation of Rosenfeld's dismissal of Bohm. If Rosenfeld had not existed, Feyerabend might have had to invent him.

In a number of influential papers in the early 1960s, in several of which he declared prominently that 'my general outlook derives from the work of David Bohm and K. R. Popper and from my discussions with both' (Feyerabend, 1965, p. 153; 1963a, p. 81; see also 1962a, p. 32), Feyerabend systematically developed an extensive case for pluralism in science. Stating that he has found '(t)he idea that a theoretical pluralism should be the basis of knowledge both in the dialectical philosophy of Bohm and in Popper's critical rationalism', he would seek to develop the idea within a falsificationist framework though without what yet remained of Popper's reliance upon an 'empirical core' for that basis (Feyerabend, 1965, p. 153). Feyerabend, for now, would follow Popper's deductivist account of methodology rather than the objectionable 'inductivism' he found in Bohm's philosophy of nature (see below). But many, if not all, of the epistemological advantages attributable to theoretical pluralism Feyerabend had already pinpointed in a 1960 review of Bohm (1957).

Ostensibly 'a belated review of a highly interesting and thought-provoking book', this paper in reality is Feyerabend's retort to Rosenfeld's third, and possibly most virulent critique, a review of Bohm (1957). To be sure, in the latter part of the paper,

---

[24] See Feyerabend (1975, p. 43, p. 61, and pp. 202–3). The quote from Rosenfeld (1957, p. 44) is: 'all evidence points with merciless definiteness in the...direction...that all the processes involving... unknown interactions conform to the fundamental quantum law'.

Feyerabend voiced mild criticisms of Bohm's philosophy of nature, with its dialectical materialist presupposition of an 'inexhaustible *depth* in the properties and qualities of matter' and crude materialist epistemology in which nature forces itself upon mind, compelling adaptation of concepts to experience thus enabling theories to acquire objective reference (Feyerabend, 1960, pp. 332, 334, 336). These views were not in conformity with the spirit and law of the falsificationist methodology of conjecture and refutation. Nonetheless, Feyerabend fulsomely praised Bohm, stating right away that the book gives 'an explicit refutation of the idea that complementarity, and complementarity alone, solves all the ontological and conceptual problems of microphysics' (Feyerabend, 1960, p. 321). He likened the defenders of Copenhagen to 'the more dogmatic of the medieval scholars', the sole exception being the latter followed Aristotle, not Bohr. Bohm had demonstrated both the logical and empirical possibility of alternative points of view, on the one hand by arguing that neither experience nor mathematics can establish the absolute validity of the uncertainty principle, and on the other through his 1952 theory. But Bohm's achievement was still more significant. In the first half of the paper Feyerabend identifies pluralist themes in Bohm's book, subsequently to be employed and developed at much greater length in works Feyerabend would soon write. These include: i) the fruitfulness of pre-mature theories, even metaphysical speculation, in promoting scientific advance (1962b, pp. 232–36); ii) a science purportedly free of all metaphysics risks becoming a dogmatic metaphysical system (1965, p. 150); (iii) empiricism requires clash of alternative theories and their possibly distinct interpretations of phenomena (1962a, p. 66; 1963a, p. 80; 1965, p. 150); iv) recognition of variability of meaning of scientific concepts and denial of absolute validity of one meaning (1962a, p. 81; 1962b, p. 228); v) use of alternative, or opposing, concepts is preferred to 'theoretical monism' (1963b, pp. 196–7; 1965, pp. 164–5); vii) criticism of positivism and 'radical empiricism' as a restriction on the forms of possible future experience (1963a, 1965 *passim*); viii) a broadened empiricism should recognize experimental results have only approximate validity and admit of different, even incommensurable interpretations of facts (1962b, pp. 233–4); ix) critique of a 'pragmatic' rejection of novel hypotheses or 'unfounded conjectures' on grounds of inconsistency with known facts and accepted theories, a point illuminated thorough a recurring historical parable of Aristotelian dynamics (Copenhagen) vs Copernicus and Galilean physics (Bohm) (1962a, p. 89; 1962b, pp. 226, 232; 1964a, p. 195 *ff.*). Bohm's book is also praised for following a realist philosophical tradition that proceeds 'along completely different lines' from positivism (1960, p. 327). These themes will be integrated within a methodology of falsification; criticism of accepted ideas is greatly sharpened through comparison of different empirical contents of distinct, perhaps incompatible, theories.

Feyerabend did not stop there but went on to achieve infamy for provocations and *épater la bourgeoisie* send-ups of philosophical shibboleths (rationality, objectivity, methodology, truth) of science and philosophy more generally. His contemporary reputation stems largely from his *Against Method: Outline of an Anarchist Theory of Knowledge* (1975), which became something of a countercultural icon, translated into at least 17 languages by the time he died in 1994 at age 70 (Hoyningen-Huene, 1994). Consistent with its subtitle, the book was alternately celebrated and damned for advancing the mantra that in the sciences 'anything goes'. Despite Feyerabend's own (and later) rhetorical excesses promoting relativism (e.g., 'customs, beliefs, cosmologies are not simply holy, or right, or true; they are useful, valid, true *for* some societies, useless, even dangerous, not valid, untrue *for* others' (Feyerabend, 1978, p. 7)), he understood the slogan to be a valid generalization extracted from the history of science. Its intent was also ironical: a pointed *riposte* to philosophers convinced there *is* a method of inquiry with fixed rules and procedures steering scientific progress. His enduring legacy, one linked at ever-more remote connection to its source in Copenhagen's obstreperous resistance to Bohm, may lie in the late avowal (Feyerabend, 1988, p. 21) that 'knowledge...[is] an ever-increasing *ocean of mutually incompatible alternatives*'. That statement is very much in evidence from a glance at contemporary mainstream philosophy of science where one finds that pluralism, as is befitting, has many competing voices, from anti-reductionist and disunity of science metaphysical currents, to feminist philosophy of science and heightened interest in values in science, to anti-hierarchical calls for social relevance and political engagement in the burgeoning field of science and technology studies.

## 31.7 Experimental Metaphysics

J. S. Bell's (1982, p. 990) statement of his reaction, as a student in 1952, on learning of Bohm's theory is as well-known as it is refreshing:

> In 1952 I saw the impossible done...in papers by David Bohm. Bohm showed explicitly how parameters could indeed be introduced, into nonrelativistic wave mechanics, with the help of which the indeterministic description could be transformed into a deterministic one. More importantly, in my opinion, the subjectivity of the orthodox version, the necessary reference to the 'observer' could be eliminated.... Should it not be taught, not as the only way, but as an antidote to the prevailing complacency? To show us that vagueness, subjectivity, and indeterminism, are not forced on us by experimental facts, but by deliberate theoretical choice?

Bell's results did not have appreciable impact in philosophy of science until after the 1979–1980 experiments of Aspect *et al.* showing quantum mechanical violations of the Bell inequalities and, more generally, that any local realist theory will empirically disagree with the statistical predictions of quantum mechanics (Shimony, 2002). Yet the results soon furnished further grist for the ever-present conflict between realist and non-realist philosophies of science. To empiricist philosophers, e.g., van Fraassen (1982), demonstration of significant but causally inexplicable correlations between measurement outcomes in the separated wings of Bell-type experiments furnished further evidence that realist views of physical theory could not be sustained. Taking quantum mechanics to be the epitome of an empirically successful theory, empirical success alone justified expectations about predicted correlations in Bell-type experiments. These expectations are reasonable, but are not, and conceivably cannot be, based on any understanding of possible causal mechanisms producing the correlations. Realist-inclined philosophers to the contrary accepted non-local correlations as a metaphysically deep novelty: nature, at least in some aspects or situations, has a non-local character. Resultant speculations about non-locality, superluminal signalling, the compatibility of quantum mechanics and relativity theory and so on, prompted renewed interest in, and reformulations of, Bohm's theory as well as in other alternatives to quantum orthodoxy, in particular Everett's relative state or many-worlds interpretation. The 1980s and 1990s were to be the era of 'experimental metaphysics', a project inconceivable to previous generations of philosophers of science. By the end of the 20th century, quantum realists (re: the wave function, or many worlds, or the Bohmian guidance condition expressing the velocities of particles in terms of the wave function, etc.) encountered a new opponent in a largely instrumentalist quantum information theory. These trends, and the dialectic of realism and nonrealism regarding quantum theory, continue well into the 21st century.

## References

Bachelard, G. (1934). *Le Nouvel Esprit Scientifique*. Paris: Félix Alcan.

Bachelard, G. (1951). *L'activité rationaliste de la physique contemporaine*. Paris: PUF.

Bell, J. S. (1982). On the Impossible Pilot Wave. *Foundations of Physics*, **12**(10), 989–99.

Beller, M. (1999). *Quantum Dialogue: The Making of a Revolution*. Chicago: University of Chicago Press.

Blokhintsev, D. I. (1951). *Kritika idealisticheskogo ponimaniia kvantovoi teorii.*, as reprinted in A. A. Maksimov, (ed.), *Filosofskie voprosy sovremennoi fiziki*, Moscow: Izd-Vo Akademii Nauk Sssr, pp. 358–95.

Bohm, D. (1952). A Suggested Interpretation of the Quantum Theory in Terms of 'Hidden' Variables I & II. *Physical Review*, **85**(2), 166–179, 180–193.

Bohm, D. (1957). *Causality and Chance in Modern Physics*. London: Routledge and Kegan Paul.

Born, M., and Heisenberg, W. (1927). Quantum Mechanics. As translated in G. Bacciagaluppi and A. Valentini (eds), *Quantum Theory at the Crossroads: Reconsidering the 1927 Solvay Conference*, New York: Cambridge University Press, pp. 372–401.

Bridgman, P. (1927). *The Logic of Modern Physics*. New York: Macmillan.

Carnap, R. (1928). *Der Logische Aufbau der Welt*. Berlin: Weltkreis Verlag.

Cornforth, M. (1949). *Dialectical Materialism and Science*. London: Lawrence and Wishart.

Cushing, J. T. (1994). *Quantum Mechanics: Historical Contingency and the Copenhagen Hegemony*. Chicago: University of Chicago Press.

Del Santo, F. (2018). Genesis of Karl Popper's EPR-Like Experiment and its Resonance amongst the Physics Community in the 1980s. *Studies in History and Philosophy of Modern Physics*, **62**, 56–70.

Del Santo, F. (2019). Karl Popper's forgotten role in the quantum debate at the edge between philosophy and physics in 1950s and 1960s. *Studies in History and Philosophy of Modern Physics*, **67**, 78–88.

Feigl, H., and Brodbeck, M. (eds) (1953). *Readings in The Philosophy of Science*. New York: Appleton-Century-Crofts, Inc.

Feyerabend, P. K. (1954). Determinism and Quantum Mechanics. As translated from the German original, *Philosophical Papers*, **4**, 25–48.

Feyerabend, P. K. (1958). Complementarity. *Proceedings of the Aristotelian Society, Supplementary Volumes*, **32**, 75–122.

Feyerabend, P. K. (1960). Professor Bohm's Philosophy of Nature. *British Journal for the Philosophy of Science*, **10**, 321–38.

Feyerabend, P. K. (1961). Niels Bohr's Interpretation of the Quantum Theory. In H. Feigl and G. Maxwell (eds), *Current Issues in the Philosophy of Science*, New York: Holt, Reinhart, and Winston, pp. 371–90.

Feyerabend, P. K. (1962a). Explanation, Reduction, and Empiricism. In H. Feigl and G. Maxwell (eds), *Minnesota Studies in the Philosophy of Science* 3, Minneapolis: University of Minnesota, pp. 28–97.

Feyerabend, P. K. (1962b). Problems of Microphysics. In R. Colodny (ed.), *Frontiers of Science and Philosophy*, Pittsburgh: University of Pittsburgh Press, pp. 189–283.

Feyerabend, P. K. (1963a). How to be a Good Empiricist: A Plea for Tolerance in Matters Epistemological. In B. Baurim (ed.), *Philosophy of Science: The Delaware Seminar* 2, as reprinted in *Philosophical Papers*, **3**, 78–103.

Feyerabend, P. K. (1963b). About Conservative Traits in the Sciences, and Especially in Quantum Theory, and Their Eradication. As translated from the German original, *Philosophical Papers*, **4**, 188–200.

Feyerabend, P. K. (1964). Realism and Instrumentalism: Comments on the Logic of Factual Support. In M. Bunge (ed.), *The Critical Approach to Science and Philosophy*, New York: The Free Press. As reprinted in *Philosophical Papers*, **1**, 176–202.

Feyerabend, P. K. (1965). Problems of Empiricism. In R. Colodny (ed.), *Beyond the Edge of Certainty: Essays in Contemporary Science and Philosophy*, Englewood Cliffs, NJ: Prentice Hall, Inc., pp. 145–260.

Feyerabend, P. K. (1970). Consolations for the Specialist. In I. Lakatos and A. Musgrave (eds), *Criticism and the Growth of Knowledge*, Cambridge: Cambridge University Press, pp. 197–230.

Feyerabend, P. K. (1975). *Against Method: Outline of an Anarchistic Theory of Knowledge*. London: New Left Review Press.

Feyerabend, P. K. (1978). *Science in a Free Society*. London: New Left Books.

Feyerabend, P. K. (1988). *Against Method*. Revised edition. London: Verso.

Fine, A. (1984). The Natural Ontological Attitude. In J. Leplin (ed.), *Scientific Realism*, Berkeley, CA: University of California Press, pp. 83–107.

Frank, P. (1936). Philosophische Deutungen und Mißdeutungen der Quantentheorie. *Erkenntnis*, **6**, 303–17.

Freire, Jr., O. (2005). Science and Exile: David Bohm, the Cold War, and a New Interpretation of Quantum Mechanics. *Historical Studies in the Physical and Biological Sciences*, **36**(1), 1–34.

Freire, Jr., O. (2019). *David Bohm*. (Springer Biographies.) Switzerland: Springer Nature.

Graham, L. (1966). Quantum Mechanics and Dialectical Materialism. *Slavic Review*, **25**(3), 381–410.

Hanson, N. R. (1958). *Patterns of Discovery*. New York: Cambridge University Press.

Hanson, N. R. (1959). Five Cautions for the Copenhagen Interpretation's Critics. *Philosophy of Science*, **26**(4), 325–37.

Heisenberg, W. (1925). Über quantentheoretische Umdeutung kinematischer und mechanischer Beziehungen. *Zeitschrift der Physik*, **33**, 879–93. Translated as 'Quantum-mechanical re-interpretation of kinematic and mechanical relations' in B. L. van der Waerden (ed.), *Sources of Quantum Mechanics*, Amsterdam: North Holland, 1967, pp. 261–76.

Heisenberg, W. (1955). The Development of the Interpretation of the Quantum Theory. In W. Pauli (ed.), *Niels Bohr and the Development of Physics*, New York: McGraw-Hill Book Co., Inc., pp. 12–29.

Heisenberg, W. (1958). *Physics and Philosophy: The Revolution in Modern Science*. New York: Harper and Row.

Hermann, G. (1935). *Die naturphilosophischen Grundlagen der Quantenmechanik*. Berlin: Öffentliches Leben. Translated in E. Crull and G. Bacciagaluppi (eds), *Grete Hermann— Between Physics and Philosophy*, Berlin, Heidelberg, New York: Springer, 2017, pp. 255–98.

Hoyningen-Huene, P. (1993). *Reconstructing Scientific Revolutions: Thomas S. Kuhn's Philosophy of Science*. Chicago: University of Chicago Press.

Hoyningen-Huene, P. (1994). Obituary of Paul K. Feyerabend (1924–1994). *Erkenntnis*, **40**, 289–92.

Hoyningen-Huene, P. (1995). Two Letters of Paul Feyerabend to Thomas S. Kuhn on a Draft of *The Structure of Scientific Revolutions*. *Studies in History and Philosophy of Science*, **26**(3), 353–387.

Jacobsen, A. S. (2012). *Léon Rosenfeld: Physics, Philosophy, and Politics in the Twentieth Century*. New Jersey, London, Singapore: World Scientific.

Jammer, M. (1974). *The Philosophy of Quantum Mechanics*. New York: Wiley-Interscience.

Kožnjak, B. (2018). The Missing History of Bohm's Hidden Variables Theory: The Ninth Symposium of the Colston Research Society, Bristol, 1957. *Studies in History and Philosophy of Modern Physics*, **62**, 85–97.

Kuhn, T. S. (1962) *The Structure of Scientific Revolutions*. Chicago: University of Chicago Press.

Kuhn, T. S. (1977). *The Essential Tension: Selected Studies in Scientific Tradition and Change*. Chicago: University of Chicago Press.

Motterlini, M. (ed.) (1999). *For and Against Method: Imre Lakatos and Paul Feyerabend*. Chicago: University of Chicago Press.

Pollock, E. (2006). *Stalin and the Soviet Science Wars*. Princeton, NJ: Princeton University Press.

Popper, K. R. (1934). Zur Kritik der Ungenauigkeitsrelationen. *Die Naturwissenschaften*, **48**, 807–8.

Popper, K. R. (1935). *Logik der Forschuung. Zur Erkenntnistheorie der modernen Naturwissenschaften*. Wien: Springer Verlag.

Popper, K. R. (1945). *The Open Society and its Enemies*. 2 volumes. London: Routledge and Keegan Paul.

Popper, K. R. (1956). Three Views Concerning Human Knowledge. Reprinted in H. D. Lewis (ed.), *Contemporary British Philosophy*, Third Series, London: Allen and Unwin, 2003.

Popper, K. R. (1959). *The Logic of Scientific Discovery*. London: Hutchinson and Co.

Popper, K. R. (1967). Quantum Mechanics Without 'the Observer'. In M. Bunge (ed.), *Quantum Theory and Reality*, Berlin/Heidelberg: Springer, pp. 7–44.

Putnam, H. (1957). Three-Valued Logic. *Philosophical Studies*, 8(5), 73–80.

Putnam, H. (1969). Is Logic Empirical? In R. Cohen and M. Wartofsky (eds), *Boston Studies in the Philosophy of Science* 5, Dordrecht: North Holland, D. Reidel, pp. 216–41.

Putnam, H. (1975). Introduction: Science as Approximation to Truth, dated September 1974, in *Hilary Putnam Philosophical Papers* 1, New York: Cambridge University Press, pp. viii–xiv.

Reichenbach, H. (1928). *Philosophie der Raum-Zeit-Lehre*. Berlin/Leipzig: Verlag Walter de Gruyter. Translation by Reichenbach, M. and Freund, J. as *The Philosophy of Space and Time*. New York: Dover Publications, 1958.

Reichenbach, H. (1930). Kausalität und Wahrscheinlichkeit. *Erkenntnis*, 1, 158–88.

Reichenbach, H. (1931). Diskussion über Kausalität und Quantenmechanik. *Erkenntnis*, 2, 188.

Reichenbach, H. (1944). *Philosophic Foundations of Quantum Mechanics*. Berkeley and Los Angeles, CA: University of California Press.

Rosenfeld, L. (1953a). L'évidence de la complémentarité. In *Louis de Broglie, Physicien et Penseur*. Paris: Éditions Albin Michel, pp. 43–65.

Rosenfeld, L. (1953b). Strife About Complementarity. *Science Progress*, 41(163), 393–410.

Rosenfeld, L. (1957). Misunderstandings About the Foundations of Quantum Theory. In S. Körner (ed.), *Observation and Interpretation in the Philosophy of Physics* (Proceedings of the Colston Research Society, University of Bristol, 1–4April 1957), London: Constable and Co., pp. 41–5.

Rosenfeld, L. (1958). Physics and Metaphysics, review of *Causality and Chance in Modern Physics* by Prof. David Bohm, *Nature*, 181 (8 March 1958), 658.

Ryckman, T. (2005). *The Reign of Relativity: Philosophy in Physics 1915–1925*. New York: Oxford University Press.

Ryckman, T. (2017). *Einstein*. (Routledge Philosophers series.) New York: Routledge.

Schlick, M. (1931). Die Kausalität in der gegenwärtigen Physik. *Die Naturwissenschaften*, 19, 145–62.

Shimony, A. (1989). Search for a World View which can Accommodate our Knowledge of Microphysics. In J. Cushing and E. McMullin (eds), *Philosophical Consequences of Quantum Theory: Reflections on Bell's Theorem*, Notre Dame: University of Notre Dame Press, pp. 25–39.

Shimony, A. (2002). AIP Oral History Interview of Abner Shimony by Joan Bromberg, 9 and 10 September 2002,, accessed 13March 2020: http://www.aip.org/history-programs/niels-bohr-library/oral-histories/25643

Smart, J. J. C. (1963). *Philosophy and Scientific Realism*. London: Routledge.

Stalin, J. V. (1938). *Dialectical and Historical Materialism*. New York: International Publishers.

van Fraassen, B. (1982). The Charybdis of Realism: Epistemological Implications of Bell's Inequality. *Synthese*, 52(1), 25–38.

# CHAPTER 32

# BOHR AND THE EPISTEMOLOGICAL LESSON OF QUANTUM MECHANICS

## STEFANO OSNAGHI

Planck...said that a God-like eye could certainly know what was the energy and the momentum [the position being known]...And then...I said to him: You have spoken about such an eye; but it is not a question of what an eye can see; it is a question of what you mean by knowing.

Niels Bohr (1962, Session V)

## 32.1 INTRODUCTION

ALTHOUGH Bohr had only indirectly contributed to the mathematical formalization of quantum mechanics, the foundational debate of the 1920s anointed him as the champion of orthodoxy in matters of interpretation. To be sure, granting the existence of such a thing as an *orthodox* school inspired by Bohr was not the most innocent way of acknowledging the prominence of his views, inasmuch as it clearly insinuated the idea that taking sides with Bohr implied a leap of faith.[1] Indeed, many of Bohr's colleagues were puzzled by the doctrine of complementarity that underpinned his fervent defence of the completeness of quantum theory. 'There must be quite definite

---

[1] See Camilleri (2009), Freire Jr. (2014), and references therein. In his correspondence with Schrödinger, for example, Einstein referred to Bohr as the 'Talmudic philosopher' and to his philosophy as a 'religion' (A. Einstein, letter to E. Schrödinger, 19 June 1935, quoted in Fine (1996, p. 125); A. Einstein, letter to E. Schrödinger, 31 May 1928, quoted in Murdoch (1987, p. 101)).

and clear grounds—Schrödinger once wrote to him—why you repeatedly declare that one must interpret observations classically, which lie absolutely in their essence...It must belong to your deepest conviction—and I cannot understand on what you base it.'[2] As a result, for all the many 'true believers',[3] who more or less dogmatically endorsed Bohr's maxims, there were others who equally uncritically rejected them at the outset as unintelligible.

While sociological and psychological factors may have contributed to this state of affairs,[4] the aura of obscurity which surrounded Bohr's philosophical claims should not be dismissed as anecdotal when attempting a reconstruction of his thought. On the contrary, I think that the difficulties experienced by Bohr in making others understand *the point* of his analysis of quantum mechanics are revealing of the radical—and in many respects precursory—character of his programme. Untangling the conceptual roots of such difficulties—while at the same time bringing out the remarkable affinities between Bohr's concerns and those of the philosophical avant-garde of his time (particularly with regard to the relationship between objectivity and meaning)—is important to avoid misunderstanding the scope of the intellectual enterprise to which Bohr was committed. This, in turn, is essential so as not to be misled in assessing his achievements.

Bohr considered it essential to draw a general *epistemological* lesson from quantum mechanics—a lesson, that is, that went beyond the specific *empirical* content of the theory. His struggle to make a *non-epistemic* interpretation of the uncertainty relations viable was central to this programme. In reconstructing Bohr's efforts to provide such an interpretation, I will therefore be concerned with bringing out the concrete sense in which his reflection can be understood as an inquiry into the general preconditions of knowledge—and the sense in which such an inquiry overturned received conceptions of how objectivity is constituted. Several relevant questions will remain in the background of my analysis. In particular, I will not try to determine the extent to which Bohr's epistemological insight and the corresponding methodological posture *resulted from* his gradual realization of the nature of the uncertainty relations—and were not, instead, instrumental in *guiding* him towards it.

# 32.2 FROM UNCERTAINTY TO COMPLEMENTARITY

The uncertainty relations, introduced by Heisenberg in 1927, formalized what appeared to be an intrinsic limitation on the information made available by quantum mechanics. In particular, Heisenberg's relations implied that the more precisely the

---

[2]  E. Schrödinger, letter to N. Bohr, 13 October 1935, quoted in Moore (1989, p. 313).
[3]  A. Einstein, letter to E. Schrödinger, 31 May 1928, quoted in Murdoch (1987, p. 101).
[4]  See Beller (1999), Freire Jr. (2014).

position of a particle at a given time can be predicted, the less determinate is the prediction concerning its momentum. Did this apparent limitation reflect a structural feature of atomic phenomena (as the 'orthodox' school held)? Or did it rather just show that the account provided by quantum mechanics was missing something (as Einstein and others contended)? Against the latter alternative, Heisenberg had argued that, since the 'quantum of action' makes it impossible to measure a physical variable without 'disturbing' the value of any conjugate variable, questions concerning the values simultaneously possessed by pairs of conjugate variables were empirically empty. The need to improve this argument led Bohr to introduce (in the famous Como lecture of 1927) and progressively refine (under the pressure of Einstein's attacks) his doctrine of complementarity.[5]

Bohr's starting point was the observation that the position of a particle can only be determined by means of a 'meter' which interacts with it. Thus, any measurement of position implies an exchange of momentum. Bohr argued that this precluded the possibility of determining the momentum of a particle when its position is measured. For how can the momentum of the particle be measured in practice? One would have to measure the momentum of the meter apparatus before the interaction with the particle and then again at an instant $t$ following the interaction. Assuming that the momentum of the particle before the impact on the apparatus is known, the momentum of the particle at $t$ can be deduced from these measurements by exploiting the conservation of total momentum. The particle's position at $t$ can then be computed based on this result.

Notice that, in order to ensure the sensitivity required by the implementation of the preceding scheme, the apparatus must be *loosely* connected to the support which serves as a reference frame. This means that, according to quantum mechanics, whenever the apparatus's momentum is determined, a corresponding uncertainty in its position (relative to the support) is introduced. As a result, the apparatus can no longer be used as a means to determine the position of the particle. Indeed, in order to function as a device for measuring the position, the apparatus must be *rigidly* fixed to the support in such a way that its quantum state is not significantly altered by the momentum exchanged during the interaction (the effects of the impact of the particle are irretrievably lost in the support's internal degrees of freedom). This is why Bohr regards the measurement *interaction* as 'indeterminable': to the extent that an

---

[5] Bohr (1934, pp. 52–91), Bohr (1958a, pp. 32–66). Bohr's notion of complementarity emerged from his long struggle with wave–particle duality and has its roots in the 'correspondence principle' of which he made extensive use in his pioneering work on atomic structure (Darrigol, 1992). He applied the concept to a wide range of phenomena and situations, well beyond physics. Yet he never gave a formal definition of complementarity. He used the term indifferently to refer to the relation existing between incompatible observables (see, e.g., Bohr (1963, p. 4)), the relation existing between the 'causal' and spatio-temporal descriptions of a phenomenon (see, e.g., Bohr (1934, pp. 54–55)), and the relation existing between wavelike and corpuscular behaviours of a system (Bohr (1934, p. 56)). For detailed analyses of these aspects, see Murdoch (1987), Faye (1991), Katsumori (2011), Plotnitsky (2012), Faye and Folse (1994, 2017) and references therein.

instrument works as a precise meter for position, the momentum exchanged in the measurement interaction cannot be determined.

Because the choice of the measuring apparatus influences the kind of attributes that can be ascribed to an atomic object, Bohr says that the measured object and the measuring device form a 'whole': *the finite magnitude of the quantum of action prevents altogether a sharp distinction being made between a phenomenon and the agency by which it is observed*.[6] As we shall see, Bohr will go as far as claiming that 'an independent reality in the ordinary physical sense can neither be ascribed to the phenomena nor to the agencies of observation' (Bohr, 1934, p. 54).

In view of Bohr's argument, if we *assume* the universal validity of the constraint placed by the uncertainty relations upon the predictions delivered by the theory, we should give up the idea that conjugate variables can simultaneously be *measured*. Thus, claims that imply the ascription of joint values to conjugate variables are not empirically decidable, which led Bohr to dismiss as futile the hypothesis that such values must nonetheless exist.[7] As a matter of fact, Bohr's opponents had plenty of good reasons to find this conclusion puzzling. A widely accepted epistemological view—systematized by the Kantian philosophical tradition—considered the possibility of accommodating a set of data within a causal spatio-temporal framework a precondition for them to become part of an objective account of experience.[8] If that condition could not be fulfilled, what was the objective basis for relying on the predictive prescriptions of quantum mechanics?

While Bohr acknowledged the tension between his interpretation of the uncertainty relations and Kantian epistemology, he did not dismiss the latter altogether. Instead, he tried to delineate a sort of third way which attempted to go beyond the Kantian framework without entirely losing sight of Kant's methodological lesson.[9] He urged the replacement of 'the ideal of causality by a more general viewpoint', one which did not presuppose a necessary link between objectivity and causal spatio-temporal

---

[6] Bohr (1934, p. 11). Bohr insists again and again on the 'impossibility of a strict separation of phenomena and means of observation' (Bohr, 1934, p. 96), explaining that the interaction with the measuring apparatus 'forms an inseparable part of the phenomena' (Bohr, 1963, p. 4).

[7] A concrete illustration of Bohr's stance is provided by his and Rosenfeld's work on the measurability of the quantum electromagnetic field (Darrigol, 1991).

[8] A causal description is, for Bohr (1931, p. 369), a 'description where we can use the laws of conservation of energy and momentum'. The way in which Kantian themes and conceptions are intertwined with Bohr's reflection on objectivity is far more complex and subtle than I can discuss here. See Folse (1978), von Weizsäcker (1980), Faye (1991), Kaiser (1992), Chevalley (1995), Brock (2003, 2009), Camilleri (2010), Bitbol and Osnaghi (2015), and Kauark-Leite (2017).

[9] Cf. Chevalley (1994, pp. 52–53). Catherine Chevalley (1991, 1994) has analysed Bohr's complementarity against the background of the rich philosophical tradition originated by Kant's theory of 'symbolic presentation' (by which term Kant (2000, section 59, p. 225) refers to the process that provides concepts, 'to which no sensible intuition can be adequate', with content). In particular, Chevalley has pointed out the relevance of Goethe's and Helmholtz's treatments of the 'symbolism' of scientific concepts. (See also Brock (2003).)

representability (Bohr, 1937, p. 84). The required 'consistent generalization of the ideal of causality' (Bohr, 1958a, p. 27) was provided, according to Bohr, by complementarity:

> [T]he fundamental postulate of the indivisibility of the quantum of action forces us to adopt a new mode of description designated as *complementary* in the sense that any given application of classical concepts precludes the simultaneous use of other classical concepts which in a different connection are equally necessary for the elucidation of the phenomena.   (Bohr 1934, p. 10)

Bohr's complementary mode of objectification seemed to require a novel theory of objectivity. One might expect that this theory would take the form of a new metaphysical doctrine—either in the realist sense of delineating some underlying ontology or in the Kantian sense of identifying the conditions of possibility of objectivity through the analysis of the subjective faculties (or suitable generalizations thereof) purportedly responsible for its 'constitution'. My thesis, however, is that the intuition that Bohr was trying to capture by means of complementarity was in fact deeper and more radical. For him, the 'essential inadequacy of the customary viewpoint of natural philosophy for a rational account of physical phenomena of the type with which we are concerned in quantum mechanics' (Bohr, 1935, p. 75) did not result from the endorsement of a particular metaphysical theory (be it realist or transcendental). Rather, such inadequacy had to be identified with the 'representationalist' preconception according to which the effectiveness of a predictive scheme demands an explanation capable of tracing effectiveness to some primitive relation of 'correspondence' between symbols and objects (in the same way as, in ordinary logic, the correctness of an inference is traced to a specific set-theoretic relationship between the truth conditions of the conclusion and those of the premise).[10] Complementarity was an attempt to make room for a pragmatist view of rationality which reversed this paradigm, by taking the normative structure of quantum probabilities (i.e., the predictive prescriptions associated with the 'quantum postulate') as given and *deriving* from it the constraints to which a pictorial account of the phenomena had to obey.

From today's perspective, a problem like that just described—in which the issue at stake is whether representational models can warrant the legitimacy of the norms that govern a discursive practice—can readily be identified as one of metasemantics. Things were far less clear in Bohr's times, when such issues were only beginning to be thematized and discussed by philosophers. While Bohr's discourse continued to employ metaphysical categories (not least because of the kind of objections that were put forward by his opponents), it became more and more evident to him that the task he had set for himself required a fresh analysis of how *language* works. Pivotal in this realization was the debate generated by the EPR argument.

---

[10] See Brandom (2000, pp. 1–77).

# 32.3  EPR: SEMANTICS BROUGHT
## TO THE FORE

Bohr's claim that two conjugate variables cannot simultaneously be measured lends itself to be interpreted as implying that measuring a variable inevitably *disturbs* the value of the other.[11] If this were so, a possible counterexample could be exhibited by considering measurements that involve no direct mechanical interaction with the measured object. This line of attack was developed by Einstein, Podolsky, and Rosen in their seminal paper of 1935. Before we focus on the EPR argument itself, however, it is useful to recall the concerns which lay behind Einstein's dissatisfaction with quantum mechanics and his lifelong struggle to recover 'a theory whose objects, connected by laws, are not probabilities but considered facts...'[12]

A central tenet of Einstein's was that it must be possible to establish a correspondence between the state vector *qua* set of predictive prescriptions and the putative 'physical' state of quantum systems.[13] As he wrote to Schrödinger:

> In the quantum theory, one describes a real state of a system through a normalized function, $\psi$, of the coordinates (of the configuration space)...Now one would like to say the following: $\psi$ is correlated one-to-one with the real state of the real system...If this works, then I speak of a complete description of reality by the theory. But if such an interpretation is not feasible, I call the theoretical description 'incomplete'.[14]

As for Schrödinger, he held that:

> [The condition for] a complete description of the material world in space and time [is that] it ought to be possible...to form in our mind of the physical object an idea [Vorstellung] that contains in some way everything that could be observed in some way or other by any observer, and not only the record of what has been observed simultaneously in a particular case.   (Schrödinger, 1958, p. 169)

---

[11] For example, Bohr (1934, pp. 100, 115) speaks of the 'unavoidable influence on atomic phenomena caused by observing them', and he claims that, in quantum mechanics, 'any observation necessitates an interference with the course of the phenomena'.

[12] A. Einstein, letter to M. Born, 3 March 1947, quoted in Born (1971, p. 158). Einstein (1948, p. 320) was 'inclined to believe that the description of quantum mechanics...ha[d] to be regarded as an incomplete and indirect description of reality, to be replaced at some later date by a more complete and direct one'.

[13] 'Il y a quelque chose comme "l'état réel" d'un système physique, qui existe objectivement, indépendamment de toute observation ou mesure, et qui peut en principe se décrire par les moyens d'expression de la physique.' ['There is something like the "real state" of a physical system, which exists objectively, independently of any observation or measurement, and can in principle be described by the expressive means of physics.'] (Einstein, 1953, p. 7, my translation.)

[14] A. Einstein, letter to E. Schrödinger, 19 June 1935, quoted in Howard (1985, p. 179).

While one might be tempted to trace Einstein's and Schrödinger's concerns with completeness to some dogmatic metaphysical commitment, I think that what lay behind the endeavour to 'complete' quantum mechanics was, in the first place, a *pragmatic* worry, namely the worry that basing our predictions on the Born rule required a *rational* justification. By assuming that it must be possible to translate such a worry into one of *representational* adequacy, the critics of the quantum orthodoxy were only following a venerable and largely accepted philosophical tradition (Rorty, 1979).

The problem was that—in so far as conjugate variables could be regarded as independent (that is, not related to each other via functional dependency on the same 'hidden' variable)—such an assumption could hardly be made compatible with the *necessary* character of the link that the uncertainty relations established between the probability distributions to be assigned to conjugate pairs. If the uncertainty relations are understood as a structural constraint, which applies in all circumstances regardless of any dynamical consideration, they must hold, in particular, when the probabilities are updated in view of the results of a measurement. It follows that the measurement of a variable induces a discontinuous change in the probability to be assigned to its conjugate variable. In so far as probability assignments must reflect the distribution of physical states within a given ensemble, a measurement entails therefore a suitable adjustment of such a distribution.

Needless to say, the disturbance argument thus understood is highly problematic. Especially because, as Einstein emphasized, it poses a threat to locality.[15] If, for example, we consider two systems described by an entangled state, the 'individual' density matrix that quantum mechanics assigns to one of them (that is, the state that predicts the results of measurements performed on that system alone) depends on the measurement performed on *the other* system, regardless of the latter's location in space-time. Assuming that each subsystem of a compound system must have a unique physical state of its own, the kind of updates that a measurement performed on one system enforces upon the statistical expectations concerning its distant entangled partner can hardly be understood without either postulating undesirable non-local effects or supposing that the information, which determines the good probability assignment, was, in fact, 'already there' (which would entail the incompleteness of the quantum mechanical description).

The EPR argument is an elaboration of this idea.[16] To begin with, Einstein, Podolsky, and Rosen suggested a formal criterion of completeness, according to which '*every element of the physical reality must have a counterpart in the physical theory*' if the

---

[15] Referring to Born's 'statistical approach', Einstein wrote: 'I cannot seriously believe in it because the theory cannot be reconciled with the idea that physics should be represented in time and space, free from spooky action at a distance', A. Einstein, letter to M. Born, 3 March 1947, quoted in Born (1971, p. 158).

[16] See Howard (1985, pp. 178–80), Fine (1996, pp. 36–39). As a matter of fact, Einstein was not entirely satisfied with the argument outlined in the EPR paper and thought that it did not fully express his concerns (Howard, 1990).

theory is to be deemed complete (Einstein *et al.*, 1935, p. 777). Interestingly, what should count as real is determined by the explanatory task that 'reality' is expected to perform vis-à-vis predictive norms: '*if, without in any way disturbing a system, we can predict with certainty (i.e., with probability equal to unity) the value of a physical quantity, then there exists an element of reality corresponding to that quantity*' (Einstein *et al.*, 1935, p. 777). Having provided a criterion for deciding in which circumstances we should regard a physical attribute as real, Einstein, Podolsky, and Rosen proceed to show that not all elements of reality thus defined have a counterpart in quantum mechanics.

The gist of the EPR argument can be grasped by considering the correlations existing between two spin 1/2 particles that form a maximally entangled pair (Bohm, 1951, sections 22.15–22.18). Such correlations can be exploited to predict with certainty the value of any spin observable pertaining to particle 1 by measuring the corresponding observable of particle 2. Moreover, since the particles can be arbitrarily distant from each other when the measurement on particle 2 takes place, no disturbance on particle 1 is to be expected if locality is assumed. We have thus exhibited a situation in which, at a given time, *all* incompatible spin observables can be assigned a corresponding 'element of reality'. The resulting set of elements of reality, however, has no counterpart in quantum mechanics.[17]

In a retrospective assessment of the EPR debate (Schilpp, 1949, p. 682), Einstein summarized the dilemma posed by the argument thus:

[T]he paradox forces us to relinquish one of the following two assertions:
(1) the description by means of the $\psi$-function is complete
(2) the real states of spatially separated objects are independent of each other.

Given these premises, insisting on the completeness of quantum mechanics, as Bohr did, required the rejection of (2). (Einstein remarked that this move could be avoided 'if one regard[ed] the $\psi$-function as the description of a (statistical) ensemble of systems (and therefore relinquishe[d] (1))'.)

Einstein's analysis, however, rested on substantial assumptions. In particular, because of his view of physical states, Einstein was inclined to conflate *separability* (according to which any two spatially separate systems must have independent real states) and *locality* (Einstein's *separation principle*, which is to the effect that the real state of a system cannot depend upon the kind of measurement which is performed on another, spatially separate system). According to Einstein's reconstruction, however, Bohr did make the distinction and his response to the EPR paper was intended to preserve the latter while rejecting the former (Murdoch 1987, pp. 173–5). Indeed, Bohr's reply was an attempt to dispel the threat of non-locality while explicitly rejecting the statistical interpretation of quantum probabilities suggested by Einstein. However,

---

[17] See Hooker (1972), Howard (1985), Fine (1996) for critical appraisals of the argument.

the real target of Bohr's analysis was the assumption which enforced the dilemma, namely the existence of physical states in Einstein's sense. 'The extent to which an unambiguous meaning can be attributed to such an expression as "physical reality" cannot of course be deduced from *a priori* philosophical conceptions, but...must be founded on a direct appeal to experiments and measurements' (Bohr 1935, p. 74). And conceivable experiments, Bohr thought, did not corroborate Einstein's conception of 'reality'.

The apparatus by means of which Bohr proposed to analyse the EPR situation is a straightforward extension of those employed in his previous discussions with Einstein, namely a diaphragm with two parallel slits, which act as a filter by 'preparing' a pair of quantum particles whose quantum state is such that both their relative position and the sum of their respective momenta are determined. (This is the quantum state actually discussed in the EPR paper.) Bohr's thesis was that in no actual situation are we entitled to *infer* with probability 1 *both* the position and the momentum of one particle from measurements performed on the other particle. The reason is that the measurement of the position of either particle involves an undeterminable transfer of momentum from that particle to the support of the diaphragm, which is the body defining the common spatio-temporal reference. This prevents the determination of the *total* momentum of the two particles, thereby undermining the possibility of relying on the conservation of momentum in order to infer the momentum of one particle *given* the momentum of the other.[18]

Bohr was prepared to acknowledge that, in the situation just described, there is 'no question of a mechanical disturbance of the system under investigation during the last critical stage of the measuring procedure'.[19] Yet, he pointed out that, depending on the kind of observation that is performed on the 'auxiliary body', such a body qualifies as a valid measuring instrument for determining the value of only *one* of the 'system's' conjugate variables. In this sense, Bohr speaks of '*an influence on the very conditions which define the possible types of predictions regarding the future behaviour of the system*'.[20]

[18]  See Hooker (1972), Dickson (2002), Halvorson and Clifton (2004), Whitaker (2004), Fine (2007) for a critical assessment of this argument.

[19]  Bohr (1935, p. 80). Although the notion of disturbance was not abandoned in 1935, the warnings which qualify Bohr's formulation thenceforth indicate that he became aware that his 'mechanical' argument against the simultaneous measurability of conjugate variables did not resolve—indeed could even increase—the ambiguity concerning the meaning of complementarity. (In a text of 1938, he says for example that '[s]peaking, as is often done, of disturbing a phenomenon by observation, or even of creating physical attributes to objects by measuring processes, is in fact, liable to be confusing' (Bohr 1939, p. 104). See also, e.g., Bohr (1958a, p. 73).)

[20]  Bohr (1935, p. 80). This 'influence' can be contrasted to the potential perturbations that are routinely neglected when a measurement is performed. For example, when we infer the position of a particle from the spot that it leaves on the detection screen, we are assuming that no hidden factor is changing the position of the detection screen relative to the reference frame. What Bohr's analysis is meant to show is that we are not allowed to deal with the effects of possible incompatible measurements in the same way.

Based on the preceding considerations, Bohr rejected the conclusion of Einstein, Podolsky, and Rosen. Since, when the conditions for determining the position of one particle are realized, it is *not* true that we can predict with certainty the momentum of the other, the EPR criterion of reality excludes that there can be a situation in which *both* the position and momentum of either particle *must* be real. That it should be possible to 'assign to one and the same state of the object two well-defined physical attributes in a way incompatible with the uncertainty relations' (Bohr, 1939, p. 102) is a stronger metaphysical assumption than required by the EPR criterion.

In the EPR proof, the criterion of reality provided the essential connection between pragmatic and ontological commitments. Bohr used this very connection to turn the argument on its head. From the observation that there exist no concrete experimental situations in which the probabilistic assignment prescribed by quantum mechanics *demands* the ascription of 'real' values to pairs of incompatible observables, he concluded that there is no practical *reason* to license such an ascription. If, in a given experimental context, a property need not correspond to an 'element of reality', we can just conclude that the property is not *defined* in that experimental context. This means that the information purportedly overlooked by quantum mechanics is, in effect, information about nothing. What one should give up in view of EPR, then, is not the completeness of quantum mechanics, but Einstein's notion of physical state!

> In fact, the paradox finds its complete solution within the frame of the quantum mechanical formalism, according to which no well-defined use of the concept of 'state' can be made as referring to the object separate from the body with which it has been in contact, until the external conditions involved in the definition of this concept are unambiguously fixed by a further suitable control of the auxiliary body.
>
> (Bohr 1939, p. 102)

The 'inseparability' of the measured object from the measuring instrument, thus, takes on a distinctively semantic connotation, inasmuch as the very possibility of ascribing a property to a system depends on the realization of the appropriate experimental context.[21] As Bohr would summarize a few years later: 'The elucidation of the paradoxes of atomic physics has disclosed the fact that the unavoidable interaction between the objects and the measuring instruments sets an absolute limit to the possibility of speaking of a behaviour of atomic objects which is independent of the means of observation' (Bohr, 1958a, p. 25).

This point was constantly reiterated. Bohr (1958a, p. 40) stressed the existence of 'an essential element of ambiguity in ascribing conventional physical attributes to atomic objects', arguing that 'no result of an experiment concerning a phenomenon which, in principle, lies outside the range of classical physics can be interpreted as giving information about independent properties of the objects' (Bohr 1958a, p. 26) and that these 'difficulties in talking about the properties of objects independent of the

---

[21] Murdoch (1987, pp. 151–2, 177). See also Faye (1991, p. 135).

conditions of observation' (Bohr 1958a, p. 98) imply the 'unavoidable renunciation as regards the absolute significance of ordinary attributes of objects' (Bohr, 1939, p. 105).

After 1935, Bohr starts explicitly contrasting the 'indefinability thesis', viz. the thesis according to which 'not only... the exact simultaneous position and momentum of an object cannot be measured, but also... an object cannot meaningfully be said to possess exact simultaneous values of these observables' (Murdoch, 1987, p. 139), to the statistical interpretation of the uncertainty relation: 'no unambiguous interpretation of such a relation—he says—can be given in words suited to describe a situation in which physical attributes are objectified in a classical way' (Bohr, 1948, p. 144).

> [T]he statistical character of the uncertainty relations in no way originates from any failure of measurements to discriminate within a certain latitude between classically describable states of the object, but rather expresses an essential limitation of the applicability of classical ideas to the analysis of quantum phenomena.
>
> (Bohr, 1939, p. 100)
>
> Indeed we have in each experimental arrangement suited for the study of proper quantum phenomena not merely to do with an ignorance of the value of certain physical quantities, but with the impossibility of defining these quantities in an unambiguous way.    (Bohr, 1935, p. 78)

Bohr (1937, p. 86) came even to regard a statement like 'the position and momentum of a particle cannot simultaneously be measured with arbitrary accuracy' as misleading, inasmuch as 'according to such a formulation it would appear as though we had to do with some arbitrary renunciation of the measurement of either the one or the other of the two well-defined attributes of the object, which would not preclude the possibility of a future theory taking both attributes into account on the lines of the classical physics.'

The EPR debate had therefore the effect of pushing Bohr to endorse the indefinability thesis much more explicitly than he had done before. According to Faye's reconstruction:

> It was not until Einstein, Podolsky, and Rosen challenged the completeness of quantum mechanics that Bohr was forced to take such an approach seriously, and gradually he began to recognize that his previously held epistemic argument in support of complementarity might be inadequate to meet the arguments constituting the challenge of such a position. From then on he attempted to strengthen the arguments for his own point of view by appealing more and more to... the semantic argument for the indefinability thesis... Thus it is not enough for Bohr to argue that since we cannot measure a pair of canonically conjugate parameters simultaneously we have no cognitive grounds for holding that an atomic object possesses well-defined properties... In fact he had to turn the argument upside down: which is to say that since we cannot simultaneously define exact values of conjugate variables..., we cannot measure these exact values simultaneously.
>
> (Faye, 1991, p. 185)

In brief: 'That two conjugate variables are not simultaneously measurable follows from the indefinability thesis, not the other way around...it is the indeterminacy of concepts that entails the uncertainty of knowledge' (Faye, 1991, pp. 188–9).

According to Murdoch (1987, p. 145), before 1935, the indefinability thesis was not yet clearly formulated 'in what may be called "semantic" as distinct from "ontic" terms, or in "formal" rather than "material" terms'. In the post-EPR period, however, it becomes clear that 'Bohr's interest is not so much ontological as semantical; he is concerned not so much with the nature of physical objects as with the character of the concepts under which we subsume objects'.[22] Murdoch concludes that 'thus construed, complementarity does not so much express the ontic mutual exclusiveness of the properties of exact position and exact momentum [for quantum objects] as convey a limitation on the meaningful applicability of the respective concepts [in the quantum domain]'.

While Bohr's peculiar philosophical background undoubtedly played a role in making this change of perspective possible, it is crucial to realize that the explicit endorsement of an approach focused on semantics was, to a large extent, 'enforced' upon him by the effort to provide a rational account of the fundamental character of the uncertainty relations. As long as Einstein's representationalist train of thought is not recast in semantic terms, it would be hard to escape the conclusion that quantum mechanics is incomplete. Freeing value ascriptions of their ontological burden, and treating them (at least provisionally) as pure linguistic expressions, whose assertibility is governed by the norms implicit in experimental practice, puts one in a position to question the very 'order of semantic explanation' which the representationalist story presupposes. Thus, shifting the focus towards semantics gave Bohr the conceptual latitude he needed to dismiss Einstein's *prima facie* compelling concerns, by showing that such concerns arose from assumptions whose rational necessity could itself be put into question.

# 32.4 THE CONSTITUTIVE ROLE OF CONTEXTS

According to Bohr's diagnosis, the fact that 'the whole situation in atomic physics deprives of all meaning such inherent attributes as the idealizations of classical physics would ascribe to the object' (Bohr, 1937, p. 86) required a 'radical revision of the

---

[22] In this respect, Faye (1991, pp. 188–90) has pointed out the revealing shift that occurs in Bohr's technical terminology. Thus, for example, the term 'uncertainty' tends to be replaced by 'indeterminacy', particularly in connection with Heisenberg's relations. Also, in describing atomic objects, Bohr often speaks of their 'behaviour' rather than of their 'properties', and (as disclosed by the study of the original manuscripts) he deliberately chooses to do so (Murdoch 1987, pp. 134–5, 146–47).

foundation for the unambiguous use of our most elementary physical concepts' (Bohr, 1963, p. 18). Bohr appeared to be convinced that the key to achieving such a revision was to focus upon experimental practice: the 'conditions, which include the account of the properties and manipulation of all measuring instruments essentially concerned, constitute in fact the only basis for the definition of the concepts by which the phenomenon is described' (Bohr, 1939, p. 104). In order to be 'unambiguous', the formalism must refer to a 'well-defined' phenomenon.

> [This] involves the fixation of the experimental conditions, defining the initial state of the atomic system concerned and the character of the possible predictions as regards subsequent observable properties of that system. Any measurement in quantum theory can in fact only refer to a fixation of the initial state or to the test of such predictions, and it is first the combination of measurements of both kinds which constitutes a well-defined phenomenon.[23]

As Faye (1991, pp. 185–6) suggests, the fact that Bohr 'considers a reference to the entire experimental arrangement as being what determines the conditions for the correct use of complementary concepts' can be rephrased in a modern jargon by saying that he 'now argues that a reference to the entire experimental set-up enters into the specification of the truth conditions for any statement involving the Heisenberg indeterminacy relations, which express the scope of the ascription of an exact momentum and an exact position to a quantum system'. More formally, Bub (1979, p. 118) argues that:

> Bohr regards the notion of *truth* as meaningful only in the context of a Boolean possibility structure, i.e., to ascribe a property to a system only makes sense with respect to a structure of possible properties which form a Boolean algebra. In the case of a quantum mechanical system this possibility structure is non-Boolean. The application of the classical notion of truth, or the attribution of physical properties to such a system, requires reference to a classical measuring system, which fixes a particular Boolean algebra in the non-Boolean possibility structure.

Bohr offered no formalization of such ideas.[24] We know, however, that he was not sympathetic towards the introduction of non-standard logics for describing atomic phenomena. During the final discussion of a conference of 1938, in which von Neumann had sketched his logical approach to quantum mechanics, Bohr remarked that 'he compelled himself to keep the logical forms of daily life to which actual experiments are necessarily confined', explaining that the 'aim of the idea of complementarity was to allow of keeping the usual logical forms while procuring the extension

---

[23] Bohr (1939, p. 101). See also, e.g., Bohr (1958a, pp. 7, 64, 71), Bohr (1939, p. 104).
[24] See van Fraassen and Hooker (1976), Murdoch (1987, p. 153), Howard (1994) for further analysis and formal developments.

necessary for including the new situation relative to the problem of observation in atomic physics'.[25]

That the reference to the experimental conditions may help make this programme viable is, however, far from evident. First, in so far as we deal with the *ordinary* concepts *position* and *momentum*, there seems to be nothing in their content which prevents their respective measurement contexts from being simultaneously realized. Second, if momentum is not defined when position is, how is it that, in a situation in which we ascribe a definite position to a system, we can still have reliable (if only statistical) expectations concerning measurements of momentum?

Notice that both the preceding concerns presuppose an extensional model of conceptual relations. According to such a model, the fact that the probability distribution to be assigned to an observable depends on the distribution to be assigned to another observable can only be understood by assuming that the two observables are functionally related to each other—hence, in particular, the truth conditions of their respective value ascriptions are connected via some trivial set-theoretic relation. Suppose, by contrast, that the uncertainty relations (and, more generally, the probabilistic structure of quantum mechanics) are *identified* with the *conceptual* framework of physics. Then the probability assignments derived from the Born rule in a given situation determine not only if a particular position or momentum ascription can be taken to be true, but also, more generally, whether the position or momentum ascriptions *lack* a truth value (i.e., *cannot* unambiguously be said to be either true or false). Moreover, one no longer needs to assume that an observable has a (possibly unknown) value in order to be entitled to assign probabilities to (i.e., have legitimate expectations about) the possible outcomes of its measurement. Within this perspective, probabilities are not parasitic on a given algebraic structure of propositions, which allegedly reflects some underlying ontological order. Rather, probabilities express a normative structure implicit in experimental practice, which tells us how we ought to connect our expectations about the results of possible measurements.[26]

The preceding analysis goes hand in hand with the reversal of the traditional order of semantic explanation. Truth is no longer granted a privileged status as a fundamental semantic tool; in particular, it need not enter the justification of the peculiar structure of quantum probabilities. On the contrary, truth-valuedness, and more generally representational content, appear to be *constituted* starting from—and in compliance with—such a structure. This is the sense in which complementarity 'is called for to

---

[25] Bohr (1939, pp. 115–16). See also Rosenfeld (1965).

[26] Cf. Cassirer (1956, pp. 195–96). Despite the similarities between Bohr's 'structuralism' and the views developed in the same years (particularly under the influence of Einstein's relativity) by philosophers such as Schlick, Reichenbach, and Cassirer (Coffa, 1991, Ch. 10), Bohr's approach involves a reflection on the aporias of semantic holism that those views lack. In particular, as I will argue in the last section, Bohr's attitude towards the quantum measurement problem delineates an original pragmatic way out from those aporias which goes beyond, or in any case departs from, both bare conventionalism and transcendental idealism.

provide a frame wide enough to embrace the account of fundamental regularities of nature which cannot be comprehended within a single picture' (Bohr 1963, p. 12).

To portray Bohr as engaged in such a philosophical enterprise may seem exaggerated. Not so, however, if one considers how Bohr himself conceived of his inquiry. As a physicist, Bohr was, of course, concerned with analysing certain specific physical concepts and discovering new effective ways of combining them. However, he also had a remarkable interest in understanding what determines the proper use of a concept *in general* (in 'investigating as accurately as [he could] the conditions for the use of our words' (Bohr, 1958b, p. 13)). In this sense, he regarded physics as a means to reflect on linguistic practice ('[b]ecause of the relative simplicity of physical problems, they are especially suited to investigate the use of our means of communication'), which, in turn, was the key to gaining insight into the nature of knowledge ('every analysis of the conditions of human knowledge must rest on considerations of the character and scope of our means of communication').[27] As his assistant Aage Petersen once wrote, the aim of Bohr's analysis was 'only to *make explicit* what the [quantum] formalism implies about the application of the elementary physical concepts'.[28]

Given the emphasis that Bohr placed upon the constitutive role of experimental contexts, one might be tempted to identify his semantic views with a positivist theory of meaning (in which case, his endorsement of the indefinability thesis would reflect a commitment to verificationism[29]). The kind of concerns which I have just pointed out, however, suggest other hypotheses. Murdoch (1987, p. 224) has convincingly argued that:

> The basis of Bohr's theory of meaning is pragmatism: the ascription of an exact position and an exact momentum to an object at the same time is meaningless, not principally because a property is unobservable, but because the ascription has no practical consequences whatever: it has no explanatory or predicative power.

Revealingly, the textual evidence supporting this claim predates the EPR debate:

> [As a consequence of the indivisibility of the quantum of action] a subsequent measurement to a certain degree deprives the information given by a previous measurement of its significance for predicting the future course of the phenomena. Obviously, these facts not only set a limit to the *extent* of the information obtainable by measurements, but they also set a limit to the *meaning* which we may attribute to such information.    (Bohr 1934, p. 18)

---

[27] Bohr (1958a, p. 88). Elsewhere Bohr (1937, pp. 83–84) remarks that the 'analysis of new experiences is liable to disclose again and again the unrecognized presuppositions for an unambiguous use of our most simple concepts, such as space-time description and causal connection'. See also Bohr (1958a, p. 98).

[28] Aa. Petersen, letter to H. Everett, 31 May 1957, quoted in Osnaghi *et al.* (2009, p. 117). My emphasis.

[29] To the extent that contingent statements are taken to be cognitively meaningful if and only if their truth value can be determined or confirmed *by observation*, a statement which ascribes an exact simultaneous position and momentum to an object has to be regarded as cognitively meaningless (given that it cannot be conclusively verified, or confirmed, by sensory experience). See Murdoch (1987, p. 139).

Likewise, in the Como lecture, Bohr (1934, p. 66) pointed out that, while there exist experimental situations in which one is not logically debarred from assigning both a definite position and momentum to a particle at a given time, such a double assignment adds nothing to what can be predicted based on the assignment of a position or a momentum alone.[30]

> Indeed, the position of an individual at two given moments can be measured with any desired degree of accuracy; but if, from such measurements, we would calculate the velocity of the individual in the ordinary way, it must be clearly realized that we are dealing with an abstraction from which no unambiguous information concerning the previous or future behaviour of the individual can be obtained.

The pragmatic strain in Bohr's philosophy has its roots in the neo-Kantian tradition, to which Bohr was exposed through his acquaintance with the Danish philosopher Harald Høffding.[31] In Bohr's original conception of the relationship between epistemology and language, however, one can discern a distinctively *Wittgensteinian* overtone.[32] In particular, the insight that semantics must somehow 'answer to pragmatics' (Brandom, 2000, p. 185) is, in my view, essential to properly interpret Bohr's *prima facie* 'instrumentalist' contention that 'the entire [quantum] formalism is to be considered as a tool for deriving predictions, of definite or statistical character, as regards information obtainable under experimental conditions described . . . in common language suitably refined by the vocabulary of classical physics'.[33]

Bohr's instrumentalism is often interpreted along traditional lines as a means of avoiding ontological commitments by advocating an agnostic attitude with regard to *what* the theoretical terms refer to. If this reading were correct, Bohr's critics would be right in pointing out that the paradoxes faced by the realist interpretations are not really dissolved. The appraisal changes, however, if Bohr's alleged instrumentalism is framed within a non-representational semantic approach: one then sees that Bohr is not so much suspending judgement on the kind of objects and properties that could be used to explain, via referential relations, the rules for connecting claims about quantum observables as he is implying that such rules are precisely what the claims' content boils down to.

---

[30] See Heisenberg (1930, p. 15), Park and Margenau (1968).

[31] The relation between Bohr and pragmatism was first pointed out by Henry Stapp (1972) and has been thoroughly analysed by Murdoch (1987, Ch. 10) and Faye (1991). See also Petersen (1985), Maleeh (2015), Folse (2017).

[32] Interestingly, in a letter appended to Stapp (1972), Heisenberg compared Bohr's philosophy to that of the later Wittgenstein. Some aspects of this connection are explored in Bitbol (1996, pp. 124–25), Osnaghi (2017), and Bitbol (2018). See also the discussion in Faye (1991, p. 221) and Murdoch (1987, pp. 222–25), where Bohr's implicit theory of meaning is analysed in the light of Michael Dummett's notion of objective anti-realism.

[33] Bohr (1948, p. 144). See also, e.g., Bohr (1958a, p. 64).

The central insight of *inferentialist* theories of concept use is that knowing how to use a concept requires knowing *both* the circumstances under which the concept is correctly applied (which is the aspect on which verificationists focus) *and* the consequences of applying it (which is the aspect emphasized by classical pragmatists).[34] This is precisely the kind of information provided by the quantum theory if we take it to express the norms that govern a linguistic practice whose basic assertions are probability assignments conditioned upon suitable measurements. In this sense, quantum mechanics fully qualifies as a rational practice involving the use of concepts.

Admittedly, since Bohr never formulated anything akin to a theory of meaning, any reconstruction of complementarity along these lines is doomed to rely upon conjecture and extrapolation. Nevertheless, if we consider the 'general epistemological lesson' implied in 'the fact that in atomic physics...objective description can be achieved only by including in the account of the phenomena explicit reference to the experimental conditions' (Bohr, 1963, p. 12), there is sufficient evidence to conclude that, for Bohr, such a lesson went beyond the mere impossibility of assigning a truth value to all empirical claims in every circumstance. It was the whole semantic pattern of explanation based on truth that needed to be put into question.

# 32.5 CLASSICAL BACKGROUND: A NON-FOUNDATIONALIST PERSPECTIVE

On the face of it, Bohr's contextual way of dealing with 'the peculiar feature of indivisibility, or "individuality", characterizing the elementary processes' (Bohr, 1958a, p. 34) presents an apparent shortcoming. For it seems to presuppose that, unlike the description of the atomic systems, that of the measuring apparatuses *is* free from 'ambiguity'. Is this presupposition really necessary and, if yes, is it legitimate? Bohr seemed to have no doubts. As he did not tire of repeating, '*however far the phenomena transcend the scope of classical physical explanation, the account of all evidence must be expressed in classical terms*'.[35]

No other aspect of Bohr's doctrine has given rise to a comparable stream of conflicting readings and appalled reactions. What is missing here is something analogous to the EPR paper—something that might have pushed Bohr to sharpen and clarify his views. The issue was nonetheless addressed by Bohr and his collaborators during

---

[34] The inspiration for this approach is provided by the proof-theoretic definition of logical connectives (Dummett, 1973). See Brandom (1994, 2000).

[35] Bohr (1958a, p. 39). See also, e.g., Bohr (1948, p. 144), Bohr (1958a, p. 64).

the 1950s, and the existing documents (particularly those referring to the discussions that the Copenhagen group had with John Wheeler and Hugh Everett) provide us with some useful clues. As in the case of EPR, in analysing those documents, one is faced with the task of bringing out the semantic dimension of the issue (which was largely overlooked by Bohr's interlocutors) starting from arguments that appear to revolve entirely around *physical* questions (Osnaghi *et al.*, 2009). Bohr's position is outlined in the following passage:

> In the system to which the quantum mechanical formalism is applied, it is of course possible to include any intermediate auxiliary agency employed in the measuring process. Since, however, all those properties of such agencies which, according to the aim of the measurements, have to be compared with the corresponding properties of the object, must be described on classical lines, their quantum mechanical treatment will for this purpose be essentially equivalent with a classical description. The question of eventually including such agencies within the system under investigation is thus purely a matter of practical convenience, just as in classical physical measurements; and such displacements of the section between object and measuring instruments can therefore never involve any arbitrariness in the description of a phenomenon and its quantum mechanical treatment. The only significant point is that in each case some ultimate measuring instruments, like the scales and clocks which determine the frame of space-time coordination—on which, in the last resort, even the definitions of momentum and energy quantities rest—must always be described entirely on classical lines, and consequently kept outside the system subject to quantum mechanical treatment.    (Bohr 1939, p. 104)

One may well wonder whether Bohr is here delineating a solution or he is not, instead, merely pointing to a problem. For, unless one is prepared to renounce the universal validity of quantum mechanics, the assertion that the 'ultimate measuring instruments... must always be described entirely on classical lines' must be proved to be consistent with their behaviour *qua* physical systems. While providing such a proof is no trivial task, the alternative (namely, to postulate that quantum mechanics only applies to certain classes of systems) is at least as problematic. To further complicate the situation, and much to the frustration of his critics, Bohr refrained from unequivocally endorsing either solution.

To begin with, Bohr acknowledged that any apparatus, *qua* physical system, can be described by quantum mechanics.[36] As Petersen put it in a letter, however, 'the large mass of the apparatus compared with that of the individual atomic object permits that neglect of quantum effects which is demanded for the account of the experimental

---

[36] (Bohr 1939, p. 104). '[T]he existence of the quantum of action is ultimately responsible for the properties of the materials of which the measuring instruments are built and of which the functioning of the recording devices depends' (Bohr 1958a, p. 51). The sentence is followed by the qualification that 'this circumstance is not relevant for the problems of the adequacy and completeness of the quantum-mechanical description in its aspects here discussed'.

arrangement'.[37] While Bohr often mentioned, yet never developed this point (which is essential to his proof that position and momentum cannot simultaneously be measured), it is well known that, under certain conditions, the state of a measuring apparatus evolves towards a 'statistical mixture' of macroscopically localized states, whose observable properties are not affected by exchanges of momentum taking place on an atomic scale.[38] Of course, as long as the evolution follows the rules of standard quantum mechanics, the process which blurs the phase relations between different 'pointer states' can be taken to be irreversible *only for practical purposes*. Therefore no theorem along these lines can enable the formal assignment of a truth value to propositions that ascribe a definite value to the measured observable.

If one's purpose is to change 'a theory which speaks *only* of the results of external interventions on the quantum system' into 'one in which that system is attributed *intrinsic properties*' (Bell, 1990, p. 38), a further step must be taken. Namely, ordinary quantum mechanics must be supplemented with some *ad hoc* formal mechanism capable of accounting for the transition from a coherent superposition of pointer states to a full-blown statistical mixture. The available documentation indicates that Bohr was reluctant to pursue such a programme and he was sceptical about the existing proposals (including von Neumann's 'standard' account of measurement).[39]

---

[37] Aa. Petersen, letter to H. Everett, 24 April 1957, quoted in Osnaghi *et al.* (2009, p. 119).

[38] See (Murdoch 1987, pp. 114–18). In his presentation of the 'orthodox' view, Bohm (1951, Ch. 22) discussed this point at length, offering a proof that 'a measurement process is irreversible in the sense that, after it has occurred, re-establishment of definite phase relations between the eigenfunctions of the measured variable is overwhelmingly unlikely'. Bohm remarked that this irreversibility 'greatly resembles that which appears in thermodynamic processes, where a decrease of entropy is also an overwhelmingly unlikely possibility' (Bohm, 1951, p. 608). In 1962 (the year of Bohr's death), A. Daneri, A. Loinger, and G. M. Prosperi (Daneri *et al.*, 1962) published a theory of measurement inspired by Günter Ludwig's 'thermodynamic approach' (Jammer, 1974, pp. 488–90). Such a theory was endorsed by Bohr's longtime collaborator Léon Rosenfeld (1965), who, in a letter written some years later, argued that the fact that the 'the reduction rule [for the state vector] is not an independent axiom, but essentially a thermodynamic effect, and accordingly, only valid to the thermodynamic approximation' was 'of course well known to Bohr'. L. Rosenfeld, letter to F. Belinfante, 24 July 1972, *Rosenfeld Papers*, Niels Bohr Archive, Copenhagen, quoted in Osnaghi *et al.* (2009, p. 116).

[39] Thus, for example, in a confidential report written in 1957, Rosenfeld complained that von Neumann's 'Foundations of Quantum Mechanics', 'though excellent in other respects, ha[d] contributed by its unhappy presentation of the question of measurement in quantum theory to create unnecessary confusion and raise spurious problems'. And he added that: 'Bohr's considerations were never intended to give a "theory of measurement in quantum theory", and to describe them in this way is misleading, since a proper theory of measurement would be the same in classical and quantal physics, the peculiar features of measurements on quantal systems arising not from the measuring process as such, but from the limitations imposed upon the use of classical concepts in quantum theory.' Rosenfeld concluded: 'By wrongly shifting the emphasis on the measuring process, one obscures the true significance of the argument and runs into difficulties, which have their source not in the actual situation, but merely in the inadequacy of the point of view from which one attempts to describe it. This error of method has its origin in v. Neumann's book "Foundations of Quantum Mechanics"...', Léon Rosenfeld. *Report on: Louis de Broglie, La théorie de la mesure en mécanique ondulatoire (Paris: Gauthier-Villars)*, 1957, RP, quoted in Osnaghi *et al.* (2009, pp. 99, 118). In the same report, Rosenfeld criticized those physicists who attempted 'to develop their own "theory of measurement" in opposition to what they believed to be the "orthodox" theory of measurement,

But what are the alternatives? Should Bohr's emphasis on the necessity of keeping 'all ultimate measuring instruments... outside the system for the treatment of which the quantum of action is to be taken essentially into account' and his warning that, in the interpretation of measurements, 'quantum effects... have on principle to be neglected'[40] be taken to imply that we must 'at the outset *postulate* a classical level in terms of which the definite results of a measurement can be realized' (Bohm, 1951, p. 626, my emphasis)? Is Bohr, in other words, proposing to 'draw a border between the quantum and the classical and... keep certain objects—particularly the measuring devices as well as the observers—on the classical side' (Zurek, 2003, p. 716)? According to this reading, complementarity would imply that 'the principle of superposition' has to be 'suspended "by decree" in the classical domain'.[41]

This is not a conclusion that Bohr would have endorsed. Indeed, Bohr would have found the entire argument hardly compelling, given that he did not accept one of its crucial premises, namely that the update of the predictive prescriptions, which takes place in a measurement, need be explained in terms of the evolution of some underlying physical states. If one focuses on practice, what makes a probability assignment legitimate is always another probability assignment. Thus, for example, if two systems $S_1$ and $S_2$ have interacted in such a way that their respective observables $O_1$ and $O_2$ have become perfectly correlated, the ascription of a value to $O_2$ (and the assignment of the corresponding eigenvector to $S_2$) licenses the ascription of a value to $O_1$ (and the assignment of the corresponding eigenvector to $S_1$). Clearly, this way of 'grounding' the update generates a regress. For, in principle, the ascription of a value to $O_2$ ought itself to be sanctioned by other updates concerning systems having interacted with $S_2$ in the past.[42] The regress can nevertheless be dealt with without succumbing to foundationalist concerns.[43] What happens *in practice* is that the entitlement to ordinary assertions

as presented by v. Neumann' (which Rosenfeld considered 'a distorted and largely irrelevant rendering of Bohr's argument'). Similar views were expressed in a series of conversations between Wheeler and Petersen, which took place in Copenhagen in 1956. In the notes taken on that occasion, Wheeler attributes to Petersen the following remarks: 'Von N[eumann]+Wig[ner] all nonsense; their stuff beside the point;... Von N[eumann]+Wig[ner] mess up by including [the] meas[uring] tool in [the observed] system... Silly to say apparatus has Ψ-function.' John A. Wheeler, *Notes*, 1956, quoted in Osnaghi *et al.* (2009, p. 118). For a reconstruction of von Neumann's actual views, see Becker (2004).

[40] Bohr (1939, pp. 107, 105). See also, e.g., Bohr (1958a, p. 50), Bohr (1963, pp. 3–4).

[41] Zurek (2003, p. 716). See also Rovelli (1996, p. 1671), Weinberg (2005), Omnès (1992, pp. 340–41).

[42] One way to treat the regress would be to adopt a sort of Tarskian strategy, based on the observation that 'no formalization can be complete, but must leave undefined some "primitive" concepts and take for granted without further analysis certain relations between these concepts, which are adopted as "axioms": the concrete meaning of these primitive concepts and axioms can only be conveyed in a "metalanguage" foreign to the formalism of the theory' (Rosenfeld, 1965, p. 222). (See also Mittelstaedt (1995).) However, that such a strategy can be made formally compatible with the universality of quantum mechanics is far from obvious.

[43] Bohr was famously keen on the tale *The adventures of a Danish student* by Poul Martin Møller. In a passage quoted *in extenso* in *The unity of human knowledge*, the student says: 'My endless enquiries make it impossible for me to achieve anything. Furthermore, I get to think about my own thoughts of the situation in which I find myself. I even think that I think of it, and divide myself into an infinite

concerning our immediate surroundings is granted *by default* to any competent observer. That is, there exist empirical claims that need not be *proved*: the norms implicit in the practice warrant that they can (and ought to) be accepted as long as contrary evidence is not displayed. This simple remark endows macroscopic pointer apparatuses with their privileged pragmatic status, inasmuch as assigning a particular pointer state to a macroscopic apparatus provides defeasible, yet entirely legitimate, grounds for asserting that a certain result was recorded (and for ascribing the corresponding value to the measured observable).

Truth—including the truth of assertions concerning macroscopic occurrences—plays no role in the preceding account. What is essential for the 'linguistic game' associated with experimental practice to get off the ground is that there be a class of empirical claims which practitioners are allowed to take for granted—to treat as 'innocent until indicted on the basis of reasonable suspicion' (Brandom, 1994, p. 206). These are precisely Bohr's infamous classical statements, which describe the 'ultimate measuring instruments'.

One merit of this reconstruction of Bohr's 'radical revision of the foundation for the description and explanation of physical phenomena' (Bohr, 1958a, p. 39) is that it naturally accommodates the emphasis that Bohr placed on the *direct* link between objectivity and 'unambiguous communication' in his later writings:[44]

> The argument is simply that by the word 'experiment' we refer to a situation where we can tell others what we have done and what we have learned and that, therefore, the account of the experimental arrangement and of the results of the observations must be expressed in unambiguous language with suitable application of the terminology of classical physics.   (Bohr, 1958a, p. 39)

> In actual experimental arrangements, the fulfilment of such requirements is secured by the use, as measuring instruments, of rigid bodies sufficiently heavy to allow a completely classical account of their relative positions and velocities. In this connection, it is also essential to remember that all unambiguous information concerning atomic objects is derived from the permanent marks—such as a spot on a photographic plate, caused by the impact of an electron—left on the bodies which define the experimental conditions.   (Bohr 1963, p. 3)

What justifies the use of macroscopic devices as measuring instruments is the fact that they fulfil the pragmatic-transcendental requirement about communication and *therefore*

retrogressive sequence of "I"s who consider each other. I do not know at which "I" to stop as the actual, and in the moment I stop at one, there is indeed again an "I" which stops at it. I become confused and feel a dizziness as if I were looking down into a bottomless abyss, and my ponderings result finally in a terrible headache.' To which his cousin replies: 'I cannot in any way help you in sorting out your many "I"s. It is quite outside my sphere of action, and I should either be or become as mad as you if I let myself in for your superhuman reveries. My line is to stick to palpable things and walk along the broad highway of common sense; therefore my "I"s never get tangled up.' (Bohr, 1963, p. 13.)

[44] See also Bohr (1958a, p. 67), Chevalley (1995).

enable 'truth-talk' as regards their properties—not the other way round! We can now see why Bohr appeared so reticent vis-à-vis the dilemma as to whether endorsing a fully quantum description of measurement (at the cost of remaining caught in a regress) or postulating some mechanism capable of 'reducing' the quantum state of macroscopic systems (at the cost of giving up the universality of quantum mechanics). His non-foundationalist approach had nothing to fear from the residual coherences of the states resulting from unitary models of the measurement process. Yet, he could see no reason (apart from confusion about the metaphysics of quantum states) not to assign a *mixed* state to measured systems. Quantum states are not there to *ground* empirical claims; their task is to *express* the content of such claims—by stating, in particular, which predictions ought to be endorsed by someone who asserts that a result was recorded.

This analysis sheds light, in particular, on Bohr's controversial remarks on irreversibility. Whereas Bohr observes somewhere that the possibility of describing 'both experimental conditions and observations by the same means of communication as one used in classical physics' depends on 'irreversible amplification effects such as a spot on a photographic plate left by the impact of an electron' (Bohr, 1958a, p. 88), what he typically says with regard to irreversibility is that, 'far from involving any special intricacy', amplification *only emphasizes* or *reminds us* 'of the essential irreversibility inherent in the very concept of observation'.[45] The former quotation might be taken to imply that, as long as it has not been shown that an irreversible physical process has occurred, one is not entitled to assert that a result was recorded. Since this requirement cannot be satisfied within the framework of unitary quantum mechanics, insisting on the irreversibility condition would be incompatible with maintaining (as Bohr does) that quantum mechanics is universally valid.[46] The latter quotations suggest, however, a different interpretation. In so far as irreversibility is implied in the very concept of observation, asserting that a result was recorded *is* endorsing the predictions of a statistical mixture of eigenstates of the measured observable. The legitimacy of such a probability assignment does not rest on the possibility of exhibiting a dynamical model that connects the measured values to some macroscopic 'objective' properties. Rather, the fact that the 'amplification effects' are 'practically irreversible' (Bohr, 1958a, p. 51) implies that the norm, according to which macroscopic observables ought to be treated as if they had a value, may be taken to apply to the correlated microscopic observables.

# 32.6  CONCLUSION

John Honner (1987, pp. 18–22) has contrasted Bohr's 'moderate holism' to Richard Rorty's 'thoroughgoing holism', claiming that what Bohr is ultimately concerned with is the relationship between 'picturing' reality and reality itself. Indeed, one might well

---

[45]  Bohr (1963, p. 3), Bohr (1958a, p. 89). See also, e.g., Bohr (1963, p. 25, 61, 92).
[46]  See, e.g., H. Everett, letter to Aa. Petersen, 31 May 1957, quoted in Osnaghi *et al.* (2009, p. 106).

suspect that the conceptual tensions inherent in Bohr's arguments (Bitbol, 1996, pp. 263–9), if not the very notion of complementarity itself, resulted from a residual commitment to patterns of reasoning characteristic of the representational paradigm.

These doubts are squarely rejected by Faye (1991, p. 136), who, based on an exhaustive analysis of Bohr's philosophical sources, argues that 'Bohr, like Høffding, was keenly opposed to all the ingredients' of the representational theory of knowledge.[47] Along these lines, I have attempted to show that some controversial aspects of Bohr's approach can be illuminated by the hypothesis that he was *not* committed to what Robert Brandom (2000, p. 9), referring to broadly representational theories of concepts, calls 'a platonist order of explanation'.

Bohr regarded the idea that quantum phenomena demand new concepts as essentially misleading. This seemingly conservative attitude may appear puzzling, but it is in fact quite natural for someone who identifies the content of a concept with the way in which sentences that contain it relate to other sentences within a given practice.[48] On this view, if the empirical claims involving certain concepts fail to satisfy the formal constraints imposed by some metaphysical principles, that only means that such principles fail adequately to capture the concepts' content as displayed in experimental practice.

If Bohr's effort to elucidate quantum mechanics is read through a representationalist lens—which is how Bohr's opponents typically read it—his case appears hardly convincing. On the one hand, it is difficult not to see the disturbance argument as a surreptitious (and manifestly circular) attempt to *justify* the normative character of the uncertainty relations. On the other hand, it would seem that complementarity can only replace causal spatio-temporal representation as a framework for *grounding* the objectivity of empirical claims at the price of restricting the applicability of quantum mechanics to microscopic systems.

The appraisal changes if, rather than understand Bohr's proposals as an endeavour to secure the foundations of quantum experimental practice by suitably extending the representationalist framework, we see them as an attempt to do precisely the opposite, namely, to overturn the representationalist paradigm by showing its inadequacy to account for quantum probabilities. If one takes this stance, the disturbance argument is naturally understood as a *reductio* (showing that no consistent causal spatio-temporal model can explain the predictive rules of quantum theory). Also, the task that Bohr assigns to complementarity appears to be entirely different from that of providing a generalized truth-conditional semantic framework for quantum probabilities: what complementarity does is instantiate a pragmatic approach to semantic explanation, in which truth and representational content perform no foundational task.

---

[47] According to Faye, Bohr '[e]xplicitly or implicitly rejected...a correspondence theory of truth, a picture theory of knowledge [and] strong objectivism' and he instead adhered to 'a coherence theory of truth, a non-picturing theory of knowledge [and] weak objectivism'.

[48] 'Every thought, every word, is only suited to underline a connection, which can never be fully described, but always reflected deeper.' Niels Bohr, speech at the Student Jubilee, 21 September 1928 (Manuscript Nr. 222, Niels Bohr Archive, Copenhagen). Quoted and translated from Danish by Brock (2009, p. 310).

A famous aphorism of Bohr's reads: 'It is wrong to think that the task of physics is to find out how nature is. Quantum physics is about what we can say about nature.'[49] I think the second sentence in this quotation should be taken to refer not so much to the empirical *truths* that we may hope to discover as to the inferential structure, which fixes the *sense* of empirical claims (what Bohr (1958a, pp. 67–8) calls the 'conceptual framework'). What quantum mechanics has taught us, then, is that objectivity is possible within a linguistic game whose structure is far more general than could be expected based on a straightforward extensional theory of concept use. It is because this insight applies to language *in general*, covering a good deal of what we can say, that Bohr considers the lesson of quantum mechanics so important for elucidating what we 'mean by knowing'.[50]

## ACKNOWLEDGEMENTS

I am grateful to Olivier Darrigol for his critical reading of the manuscript and helpful comments. This work was supported by grant number ANR-16-CE91-0005-01 from the *Agence nationale de la recherche* and grant number I 2906-G24 from the *Förderung der wissenschaftlichen Forschung*.

## REFERENCES

Becker, L. (2004). That von Neumann did not believe in a physical collapse. *British Journal for the Philosophy of Science*, 55(1), 121–135.

Bell, J. S. (1990). Against 'measurement'. *Physics World*, 8, 33–40.

Beller, M. (1999). *Quantum dialogue. The making of a revolution.* Chicago: University of Chicago Press.

Bitbol, M. (1996). *Mécanique quantique, une introduction philosophique.* Paris: Flammarion.

Bitbol, M. (2018). Mathematical demonstration and experimental activity: a Wittgensteinian philosophy of physics. *Philosophical Investigations*, 41(2), 188–203.

Bitbol, M., and Osnaghi, S. (2015). Bohr's complementarity and Kant's epistemology. In O. Darrigol, B. Duplantier, J.-M. Raimond, and V. Rivasseau (eds), *Niels Bohr, 1913–2013*, volume 68 of *Progress in Mathematical Physics. Poincaré Seminar 2013*, Dordrecht: Birkhäuser.

Bohm, D. (1951). *Quantum theory.* New York: Prentice–Hall.

Bohr, N. (1931). *Space-time-continuity and atomic physics. H. H. Wills Memorial Lecture, given at the University of Bristol on 5 October 1931.* In J. Kalckar (ed.), *Niels Bohr collected works: foundations of quantum physics I (1926–1932)*, volume 6, 1985, Amsterdam: Elsevier, pp. 363–370.

---

[49] Bohr quoted by Petersen (1985, p. 305).
[50] 'The epistemological lesson of atomic physics has naturally, just as have earlier advances in physical science, given rise to renewed consideration of the use of our means of communication for objective description in other fields of knowledge' (Bohr, 1958a, p. 91).

Bohr, N. (1934). *Atomic theory and the description of nature*. Cambridge: Cambridge University Press.

Bohr, N. (1935). Can quantum-mechanical description of physical reality be considered complete? *Physical Review*, 48, 696–702. Reprinted in Bohr, N., *Causality and complementarity. Supplementary papers*. The philosophical writings of Niels Bohr, Vol. 4 (pages 73–82). Edited by J. Faye and H. Folse. Woodbridge, CT: Ox Bow Press, 1998. Page numbers refer to the reprint.

Bohr, N. (1937). Causality and complementarity. *Philosophy of Science*, 4, 289–298. Reprinted in Bohr, N., *Causality and complementarity. Supplementary papers*. The philosophical writings of Niels Bohr, Vol. 4 (pages 83–91). Edited by J. Faye and H. Folse. Woodbridge, CT: Ox Bow Press, 1998. Page numbers refer to the reprint.

Bohr, N. (1939). The causality problem in atomic physics. In *New Theories in Physics*, pp. 11–45. International Institute of Intellectual Cooperation, Paris. Reprinted in Bohr, N., *Causality and complementarity. Supplementary papers*. The philosophical writings of Niels Bohr, Vol. 4 (pages 94–121). Edited by J. Faye and H. Folse. Woodbridge, CT: Ox Bow Press, 1998. Page numbers refer to the reprint.

Bohr, N. (1948). On the notions of causality and complementarity. *Dialectica*, 2, 141–148. Reprinted in Bohr, N., *Causality and complementarity. Supplementary papers*. The philosophical writings of Niels Bohr, Vol. 4 (pages 73–82). Edited by J. Faye and H. Folse. Woodbridge, CT: Ox Bow Press, 1998. Page numbers refer to the reprint.

Bohr, N. (1958a). *Atomic physics and human knowledge*. New York: John Wiley.

Bohr, N. (1958b). The unity of knowledge. Revised manuscript (January 1958) of the John Franklin Carlson Lecture delivered at Iowa State University on 5 December 1957. Niels Bohr Archive, Copenhagen.

Bohr, N. (1963). *Essays 1958–1962 on atomic physics and human knowledge*. New York: John Wiley.

Born, M. (1971). *The Born–Einstein letters*. Translated by I. Born. London: Macmillan.

Brandom, R. (1994). *Making it explicit*. Cambridge, MA: Harvard University Press.

Brandom, R. (2000). *Articulating reasons. An introduction to inferentialism*. Cambridge, MA: Harvard University Press.

Brock, S. (2003). *Niels Bohr's philosophy of quantum physics*. Berlin: Logos.

Brock, S. (2009). Old wine enriched in new bottles: Kantian flavors in Bohr's viewpoint of complementarity. In M. Bitbol, J. Petitot, and P. Kerszberg, (eds), *Constituting objectivity. Transcendental perspectives on modern physics*, volume 74 of *The Western Ontario Series in Philosophy of Science*, Dordrecht: Springer, pp. 301–316.

Bub, J. (1979). The measurement problem of quantum mechanics. In G. d. F. Toraldo (ed.), *Problems in the foundations of physics. Proceedings of the International School of Physics Enrico Fermi. Course LXXII*. Dordrecht: Reidel.

Camilleri, K. (2009). Constructing the myth of the Copenhagen interpretation. *Perspectives on Science*, 17(1), 26–57.

Camilleri, K. (2010). The Kantian framework of complementarity. *Studies in History and Philosophy of Modern Physics*, 41, 309–317.

Cassirer, E. (1956). *Determinism and indeterminism in modern physics*. New Haven, CT: Yale University Press. Translated by O. T. Benfey. Originally published as 'Determinismus und Indeterminismus in der modernen Physik' in *Göteborgs Högskolas Årsskrift*, 42, Part 3, 1936.

Chevalley, C. (1991). *Le dessin et la couleur*. Introduction to Bohr, N., *Physique atomique et connaissance humaine*. (French translation of Bohr, N., *Atomic physics and human knowledge*. New York: John Wiley, 1958.) Translated by E. Bauer, R. Omnès, and C. Chevalley. Edited by C. Chevalley. Paris: Gallimard.

Chevalley, C. (1994). Niels Bohr's words and the Atlantis of Kantianism. In J. Faye and H. J. Folse (eds), *Niels Bohr and contemporary philosophy*, volume 153 of *Boston Studies in the Philosophy of Science*, Dordrecht: Kluwer Academic Publishers, pp. 33–55.

Chevalley, C. (1995). On objectivity as intersubjective agreement. In L. Krüger and B. Falkenburg (eds), *Physik, Philosophie und die Einheit der Wissenschaften. Für Erhard Scheibe*, Heidelberg: Spektrum Akademischer Verlag, pp. 332–346.

Coffa, A. (1991). *The semantic tradition from Kant to Carnap. To the Vienna station*. Edited by L. Wessels. Cambridge: Cambridge University Press.

Daneri, A., Loinger, A., and Prosperi, G. M. (1962). Quantum theory of measurement and ergodicity conditions. *Nuclear Physics*, **33**, 297–319. Reprinted in J. A. Wheeler and W. H. Zurek (eds), *Quantum theory and measurement*, 1983, Princeton, NJ: Princeton University Press, pp. 657–679.

Darrigol, O. (1991). Cohérence et complétude de la mécanique quantique: l'exemple de 'Bohr-Rosenfeld'. *Revue d'histoire des sciences*, **44**(2), 137–179.

Darrigol, O. (1992). *From c-numbers to q-numbers*. Berkeley, CA: University of California Press.

Dickson, M. (2002). The EPR experiment: a prelude to Bohr's reply to EPR. In M. Heidelberger and F. Stadler (eds), *History of philosophy of science—new trends and perspectives*, volume 9 of *Vienna Circle Institute Yearbook*, Dordrecht: Kluwer Academic Publishers, pp. 263–275.

Dummett, M. (1973). *Frege: philosophy of language*. London: Duckworth.

Einstein, A. (1948). Quantum mechanics and reality. *Dialectica*, **2**, 320–324.

Einstein, A. (1953). Remarques préliminaires sur les concepts fondamentaux. In A. George (ed.), *Louis de Broglie: physicien et penseur*, Paris: Albin Michel, pp. 5–15.

Einstein, A., Podolsky, B., and Rosen, N. (1935). Can quantum-mechanical description of physical reality be considered complete? *Physical Review*, **47**, 777–780.

Faye, J. (1991). *Niels Bohr: his heritage and legacy. An anti-realist view of quantum mechanics*. Dordrecht: Kluwer Academic Publishers.

Faye, J., and Folse, H. J. (eds) (1994). *Niels Bohr and contemporary philosophy*, volume 153 of *Boston Studies in the Philosophy of Science*. Dordrecht: Kluwer Academic Publishers.

Faye, J., and Folse, H. J. (eds) (2017). *Niels Bohr and the philosophy of physics. Twenty-first-century perspectives*. London: Bloomsbury.

Fine, A. (1996). *The shaky game: Einstein, realism and the quantum theory*, 2nd edn. Chicago: University of Chicago Press.

Fine, A. (2007). Bohr's response to EPR: criticism and defense. *Iyyun, The Jerusalem Philosophical Quarterly*, **56**, 31–56.

Folse, H. (1978). Kantian aspects of complementarity. *Kant-Studien*, **69**, 58–66.

Folse, H. J. (2017). Complementarity and pragmatic epistemology: a comparison of Bohr and C. I. Lewis. In J. Faye and H. J. Folse (eds), *Niels Bohr and the philosophy of physics. Twenty-first-century perspectives*, London: Bloomsbury, pp. 91–114.

Freire Jr., O. (2014). *The quantum dissidents: rebuilding the foundations of quantum mechanics (1950–1990)*. Berlin: Springer.

Halvorson, H., and Clifton, R. K. (2004). Reconsidering Bohr's reply to EPR. In J. Butterfield and H. Halvorson (eds), *Quantum entanglements: selected papers of Rob Clifton*, Oxford: Oxford University Press, pp. 369–393.

Heisenberg, W. (1930). *The physical principles of the quantum theory*. Chicago: University of Chicago Press.

Honner, J. (1987). *The description of nature: Niels Bohr and the philosophy of quantum physics*. Oxford: Oxford University Press.

Hooker, C. A. (1972). The nature of quantum mechanical reality. In R. G. Colodny (ed.), *Paradigms and paradoxes: the philosophical challenge of the quantum domain*, Pittsburgh: University of Pittsburgh Press, pp. 135–172.

Howard, D. (1985). Einstein on locality and separability. *Studies in History and Philosophy of Science*, **16**, 171–201.

Howard, D. (1990). 'Nicht sein kann was nicht sein darf,' or the pre-history of EPR. In A. I. Miller (ed.), *Sixty-Two Years of Uncertainty*, New York: Plenum, pp. 61–111.

Howard, D. (1994). What makes a classical concept classical. In J. Faye and J. H. Folse (eds), *Niels Bohr and contemporary philosophy*, volume 153 of *Boston Studies in the Philosophy of Science*, Dordrecht: Kluwer Academic Publishers, pp. 201–229.

Jammer, M. (1974). *The philosophy of quantum mechanics*. New York: John Wiley.

Kaiser, D. (1992). More roots of complementarity: Kantian aspects and influences. *Studies in History and Philosophy of Science*, **23**(2), 213–239.

Kant, I. (2000). *Critique of the Power of Judgement*. Translated by P. Guyer and E. Matthews. Edited by P. Guyer. Cambridge: Cambridge University Press. Originally published 1790 as *Critik der Urtheilskraft*. Berlin and Libau: Bey Lagarde und Friederich.

Katsumori, M. (2011). *Niels Bohr's complementarity: its structure, history, and intersections with hermeneutics and deconstruction*, Volume 286 of *Boston Studies in the Philosophy of Science*. Dordrecht: Springer.

Kauark-Leite, P. (2017). Transcendental versus quantitative meanings of Bohr's complementarity principle. In J. Faye and H. J. Folse (eds), *Niels Bohr and the philosophy of physics. Twenty-first-century perspectives*, London: Bloomsbury, pp. 67–89.

Maleeh, R. (2015). Bohr's philosophy in the light of Peircean pragmatism. *Journal for General Philosophy of Science*, **46**, 3–21.

Mittelstaedt, P. (1995). Die wechselseitigen Beziehungen zwischen der Quantentheorie und ihrer Interpretation. In L. Krüger and B. Falkenburg (eds), *Physik, Philosophie und die Einheit der Wissenschaften. Für Erhard Scheibe*, Heidelberg: Spektrum Akademischer Verlag, pp. 97–117.

Moore, W. J. (1989). *Schrödinger: life and thought*. Cambridge: Cambridge University Press.

Murdoch, D. (1987). *Niels Bohr's philosophy of physics*. Cambridge: Cambridge University Press.

Omnès, R. (1992). Consistent interpretations of quantum mechanics. *Reviews of Modern Physics*, **64**, 339–382.

Osnaghi, S. (2017). Complementarity as a route to inferentialism. In J. Faye and H. J. Folse (eds), *Niels Bohr and the philosophy of physics. Twenty-first-century perspectives*, London: Bloomsbury, pp. 155–178.

Osnaghi, S., Freitas, F., and Freire Jr, O. (2009). The origin of the Everettian heresy. *Studies in History and Philosophy of Modern Physics*, **40**(2), 97–123.

Park, J. L., and Margenau, H. (1968). Simultaneous measurability in quantum theory. *International Journal of Theoretical Physics*, **1**(3), 211–283.

Petersen, A. (1985). The philosophy of Niels Bohr. In A. P. French and P. J. Kennedy (eds), *Niels Bohr, a centenary volume*, Cambridge, MA: Harvard University Press, pp. 299–310.

Plotnitsky, A. (2012). *Niels Bohr and complementarity*. Berlin: Springer.

Rovelli, C. (1996). Relational quantum mechanics. *International Journal of Theoretical Physics*, **35**(8), 1637–1678.

Rorty, R. (1979). *Philosophy and the mirror of nature*. Princeton, NJ: Princeton University Press.

Rosenfeld, L. (1965). The measuring process in quantum mechanics. *Progress of Theoretical Physics Supplements*, **1965**, 222–231.

Schilpp, P. A. (ed.) (1949). *Albert Einstein, philosopher-scientist*. Evanston, IL: The Library of Living Philosophers.

Schrödinger, E. (1958). Might perhaps energy be a merely statistical concept? *Nuovo Cimento*, **9**(1), 162–170.

Stapp, H. P. (1972). The Copenhagen interpretation. *American Journal of Physics*, **40**(8), 1098–1116.

van Fraassen, B. C., and Hooker, C. A. (1976). A semantic analysis of Niels Bohr's philosophy of quantum theory. In W. L. Harper and C. A. Hooker (eds), *Foundations of probability theory, statistical inference, and statistical theories of science—III*, volume 5c of *The Western Ontario Series in Philosophy of Science*, Dordrecht: Reidel, pp. 221–241.

von Weizsäcker, C. F. (1980). *The unity of nature*. New York: Ferrar, Straus & Giroux.

Weinberg, S. (2005). Einstein's mistakes. *Physics Today*, **58**(11), 31–35.

Whitaker, M. A. B. (2004). The EPR paper and Bohr's response: a re-assessment. *Foundations of Physics*, **34**, 1305–1340.

Zurek, W. H. (2003). Decoherence, einselection, and the quantum origins of the classical. *Reviews of Modern Physics*, **75**, 715–775.

# CHAPTER 33

# MAKING SENSE OF THE CENTURY-OLD SCIENTIFIC CONTROVERSY OVER THE QUANTA

OLIVAL FREIRE JR

## 33.1 INTRODUCTION

THE controversy about the foundations and the interpretations of quantum physics, the quantum controversy for short, stretches back over practically the last hundred years, at least since the inception of its mathematical and physical formalism between 1925 and 1927. In parallel with this long controversy, quantum physics has been considered by many as the most successful physical theory because of its ever growing scope as well as its accurate predictions. The controversy has been about a number of conceptual issues, ranging from the theory's putative completeness to its completion through additional variables and changes in its mathematical formalism or in its logic, and even about the object of study in quantum mechanics. It also concerns the role of measurement in physical theories and the compatibility between different theories, such as quantum and gravity. The physicist Franck Laloë (2012, p. xi) encapsulated this, with some sense of history, in a textbook provocatively entitled 'Do we really understand quantum mechanics?' stating: 'We have a rare situation in the history of sciences: consensus exists concerning a systematic approach to physical phenomena, involving calculation methods having an extraordinary predictive power; nevertheless, almost a century after the introduction of these methods, the same consensus is far from being reached concerning the interpretation of the theory and its foundations.' Laloë concluded with the metaphor, 'this is reminiscent of the colossus with feet of clay'. The issues at stake in the controversy are not only conceptual ones. Most of the controversy has been grounded on philosophical cleavages with different brands of realism and instrumentalism playing their roles. Furthermore, diverse contextual factors have played major roles either hindering or stimulating the development of the controversy; such contexts varying from political, cultural, and local environments to the role of individuals and generations, and experiments, instruments, and techniques.

So far the quantum controversy remains an unsolved controversy; it was so in 1974 when the historian of physics Max Jammer (1974, p. 521) called it a 'story without an ending', and it still is if we consider the number of physicists and philosophers who are engaged with it and the available plurality of choices of interpretations and models. This does not mean the whole physics community acknowledges the very existence of this controversy. In fact, it has been an inconvenient truth, a source of some discomfort for most physicists, although few have openly acknowledged this. It remains an open controversy in spite of being part of the current technological promises related to quantum information. When brought together, these threads are part of a fabric of a major episode in the history of science, or at least in the history of theoretical physics in the 20th and 21st centuries. It may be considered on the same scale as the creation of great theories or as experimental breakthroughs. Thus it comes as no surprise that a number of historians and philosophers have been paying attention to its development.

This old and persisting controversy deserves to be a subject of history of science both due to its intrinsic historical value and because some of the features of the practice of this scientific activity seem to present avenues of research in which distinct choices are still open. As once remarked by Bruno Latour (1987, p. 258), in his first rule of method, 'we study science *in action* and not ready made science or technology; to do so, we either arrive before the facts and machines are blackboxed or we follow the controversies that reopen them'. In studying this case, one has the advantage that science's black box does not need to be reopened, it has been open for a century. However, there are intellectual obstacles to framing the quantum controversy among the major scientific controversies in the history of science. Firstly, its technicalities alienate or confuse those who are not experts in physics. Secondly, some influential trends in 20th-century philosophy of science did not give room to make this controversy intelligible and even failed to acknowledge its legitimacy and existence. In addition, during most of the 20th century many physicists denied its relevance. We may thus conclude that the quantum controversy needs, first of all, to gain legitimacy as a scientific controversy in circles extending beyond the few physicists, philosophers, and historians who have engaged with it in various ways. On the one hand, as we will see in the second part of this chapter, some scholars either deny the legitimacy of this controversy or are inattentive to its evolving history. On the other hand, several historians and philosophers of science have helped to understand it. As a result, we now have a meaningful number of historical studies analysing cases, themes, and periods of the controversy.

Some of the features and themes of the quantum controversy are of interest well beyond the case of quantum mechanics; among them the tension between consensus and plurality in scientific communities and the elusive role of experiments for the resolution of scientific controversies. Furthermore, the quantum controversy brings to the forefront the possibility of the coexistence for a long time of more than one interpretation or theory and all compatible with known experimental results. This reminds us of the Duhem–Quine thesis on the underdetermination of theories by empirical data. This chapter intends to present a brief panorama of the controversy and its evolving history as well as to discuss some of its features. In the first part I present a brief overview of the chronology and cartography of the factors which have been instrumental in the history of the controversy. The analysis of these features in the concrete practice of science is the second part of this work. I conclude with the irony of the history of this controversy.

# 33.2 First Part—The Chronology and Cartography of the Controversy

We can break down this 100-year old controversy into four distinct stages regarding issues, actors, and professional dynamics. The first stage ran from the inception of the theory, in 1925, until World War II and was marked by the debate on the consistency and completeness of the theory, developments of its mathematical structure, and wider extension of the scope of the theory, which included its relativistic generalization, quantization of electromagnetic fields, and applications to molecules and solids as well as to the atomic nucleus. The first salvos were among the supporters of the matrix formulation, based on discontinuities and transition probabilities, and the enthusiastic adherents of the wave formulation, which was based on a continuous space-time arena. It was the *Anschaulichkeit* debate. Events rapidly evolved and the realization that wave functions were based on a configuration space of higher dimensions, jointly with Max Born's statistical rule and Werner Heisenberg's relations, brought visualizable alternative solutions to a standstill. In 1927, Niels Bohr presented the complementarity view not only as one possible interpretation of the theory, accommodating wave and particle descriptions and enthroning acausality in physics, but also as its epistemological lesson. John von Neumann showed that the precise arena for the quantum theory was the abstract Hilbert space and produced the quantum theory's enduring mathematical formalism. Von Neumann, however, immediately noted that this formalism could not be the end of the story and opened the way towards either the adoption of a new logic or the development of new mathematical structures, such as the operator algebras.[1]

Around the same time Albert Einstein and Erwin Schrödinger, who did not accept complementarity as the correct interpretation for the theory, proposed two critical

---

[1] The relevant literature on these early debates is extensive. In this volume, see Jähnert and Lehner (2022) and Bacciagaluppi (2022). A list of references, not comprehensive, should include: Jammer (1974, 1989), Bacciagaluppi and Valentini (2009), Darrigol (2003) and references therein, and Kojevnikov (2020).

thought experiments (EPR and the infamous cat, respectively) that would later permeate the entire development of the controversy. Einstein and Bohr's debates were a crucial chapter of this history, running from 1927 till Einstein's last days, concerning the role of determinism, completeness, realism, and locality. The debates had the active participation of the influential physicists who created the theory, or developed it, paramount among them being Bohr, Heisenberg, von Neumann, Einstein, and Schrödinger, and for a while Louis de Broglie; and the controversy was highly regarded from the intellectual point of view. It was a debate among the great founders with few younger participants engaging with it, Lev Landau and Grete Hermann being notable exceptions. It was the time when the complementarity view, articulated by Bohr, had become the most influential interpretation of the theory. However, historians have noted that this overall result, that is the acceptance of indeterminism and complementarity as foundations of the new theory, was not an inexorable result of the interaction between theory and experiments as other factors also played a role. The historian Paul Forman (1971) considered the acceptance of acausality a tenet of quantum theory resulting more from the accommodation of the German scientists to the hostile intellectual climate of Weimar republic than from an ineluctable development of the theory. While the philosopher Mara Beller (1999) argued that the rhetoric of the inevitability of the quantum lessons, just after 1927, resulted from the break from the previously existent dialogue. The pictures of Einstein and Bohr clashing over the meaning of the quantum became iconic images of the times. When physics moved its centre of gravity towards the US, interest in these foundational issues waned as the intellectual climate was ruled by the 'empiricist temper', to use Sam Schweber's (1986) words. World War II put a hold on these foundational concerns.[2]

The second stage ran from the late 1940s to the early 1980s and encompassed a plethora of subjects. The possibility of alternative interpretations was affirmed, with the successive formulation of the hidden-variables, relative-states, and statistical interpretations, created or supported by David Bohm, de Broglie, Jean-Pierre Vigier, Hugh Everett, Bryce DeWitt, Einstein, Karl Popper, and Leslie Ballentine. Measurement became a problem in the foundations of the theory, particularly as a result of debates among physicists such as Eugene Wigner, Abner Shimony, Josef-Maria Jauch, Michael Yanase, Léon Rosenfeld, Adriana Daneri, Angelo Loinger, Giovanni Maria Prosperi, Jeffrey Bub, and Klaus Tausk. Incompatibility between quantum mechanics and gravitation occupied the forefront of the research in gravitation and cosmology, particularly due to the work of DeWitt. However, the most influential development of the quantum controversy turned back to an old subject: the possibility of supplementing quantum mechanics with additional variables and the EPR experiment. In the mid-1960s John Bell brought together the possibility of introducing supplementary variables into quantum mechanics, a criticism of von Neumann's proof against this possibility, and the EPR thought experiment to obtain a theorem. Any theory with

---

[2] See, in this volume, Paty (2022), Landsman (2022), Crull (2022), Howard (2022), and Perillán (2022).

additional variables, if based on local realism, would lead to predictions conflicting with quantum mechanics. Bell's theorem moved the controversy to the lab benches, mainly those of optics. The series of experiments on Bell's theorem, from the first, led by John Clauser and Stuart Freedman and published in 1972, to the most influential, led by Alain Aspect and colleagues and published ten years later, confirming quantum predictions, brought quantum entanglement to the core of quantum mechanics.[3]

The relevance of these events was immediately recognized by Max Jammer, who put aside the planned follow-up of his *The Conceptual Development of Quantum Mechanics* and launched himself into the research and writing of his *The Philosophy of Quantum Mechanics*. According to Jammer,[4]

> I had hoped to continue this line of research with a sequel volume on the conceptual development of relativistic quantum mechanics and quantum field theory. However, John Stewart Bell's paper on hidden variables which appeared in the July 1966 issue of the *Reviews of Modern Physics*, together with his paper on the Einstein–Podolsky–Rosen paradox, threw new light on the interpretations of quantum mechanics. (...) Prompted by these developments, I wrote *The Philosophy of Quantum Mechanics*.

During this stage a new generation of physicists interested in the controversy came of age; they included Bohm, Vigier, Everett, Shimony, DeWitt, Bernard d'Espagnat, Franco Selleri, Hans-Dieter Zeh, Bell, Clauser, and Aspect.[5] In the 1950s most of the older generation remained active in the controversy and two of them, Léon Rosenfeld and Eugene Wigner, followed it into the 1960s, with opposite motivations; the former defended the citadel of complementarity against the barbarians from the new generation, and the latter moved from orthodoxy to heterodoxy and came to support the younger generation. At the beginning of this stage, the dominant view among the physicists was that all foundational issues had already been solved by the founding fathers of the discipline and the remaining questions were more a matter of philosophy than physics. This intellectual climate was perceived by two contemporaries, John Bell and Michael Nauenberg who wrote in 1966: 'We emphasize [...] that current interest in such questions is small. The typical physicist feels that they have long been answered, and that he will fully understand just how if ever he can spare 20 minutes to think about it' (Bell and Nauenberg, 1966). Physicists such as Bohm, Everett, Clauser, Tausk, and Zeh paid a high professional price for entering the controversy to criticize the dominant received view.

---

[3] See Freire Jr (2015a), Kaiser (2012, 2020, and 2022), Gilder (2008), and Silva (2022).

[4] Max Jammer, Preface to the Second Edition, July 1988, in Jammer (1989).

[5] There are now a few biographical works on these physicists; see on Selleri, (Romano, 2020), on Vigier, (Besson, 2018), on Bell, (Whitaker, 2016), on Everett, (Byrne, 2010), on Rosenfeld, (Jacobsen, 2012), and on Bohm, (Freire Jr, 2019). For a description of most of these actors as quantum dissidents, see (Freire Jr, 2015a).

Animated by the debates about Bell's theorem and supported by the accommodation of the professional milieu given the cultural and political unrest both in Europe and the US, in the late 1960s and 1970s the winds began to change and the physics community started to warm to the quantum controversy. In Italy foundations of quantum mechanics became a well-accepted topic for physics research thanks to the political movements of the times (Freire Jr, 2015a; Baracca, Bergia, and Del Santo, 2017), while in the US the counter-culture, the hippies, and the protests against the Vietnam War favoured this research (Kaiser, 2012; Freire Jr, 2015a). A physicist who witnessed this change was Aspect, for he began to look for new experiments on Bell's theorem in the unfavourable mid-1970s and at the end of his experiments became an acclaimed experimental physicist precisely because of this research. I have coined the expression 'quantum dissidents' to designate those physicists who contributed to change the professional mood related to the quantum controversy (Freire Jr, 2015a).[6]

The third stage began in 1982, after the publication of the influential experiments carried out by Alain Aspect and his team, and lasted till the mid-1990s, when the rising wave of research dedicated to quantum information gained momentum. The professional dynamics of the quantum controversy at this time was simpler than before because the controversy had become part of the regular agenda of research in physics. No longer a philosophical controversy, as it had been considered in the 1950s, it was now considered a scientific controversy, albeit with philosophical implications. It became a research topic among others, neither more nor less valued. This professional coming of age of the debates about the foundations and interpretations of quantum mechanics was the result of distinct but interacting factors. On the one hand, there were the professional activities of the quantum dissidents and the institutional accommodations. On the other hand, there was the impact of Bell's theorem, the relevant experiments, and the acknowledgement of a new physical effect, entanglement, all brought to the forefront of the physics agenda by the research related to the quantum controversy.

The third stage brought novelties in theory as well as in experiment. The conceptual transition from the quantum description of systems to the classical description was better grasped thanks to the introduction of the concept of decoherence. The concept departs from a rather simple assumption: the quantum signature description of a system, that is the superposition of at least two pure states, cannot be preserved for a long time as the inevitable interactions between such a system and its environment will dissipate the coherence represented by that superposition. The concept was developed, theoretically and mathematically, by many hands, including those of Zeh, Wojciech Zurek, Anthony Leggett, and Amir Caldeira. This theoretical breakthrough was

---

[6] Historians differ in their emphasis on the factors that were influential in the professional changes towards the quantum controversy. Kaiser (2012), Freire Jr (2015a), and Baracca, Bergia, and Del Santo (2017) value contextual changes in addition to the conceptual issues, while Bromberg (2006, p. 258; 2008) valued the availability of new instruments in the 1980s. On the new instruments, see Silva Neto (2022). About the last part of this stage, Wüthrich (2016, p. 447) disputed that foundations of quantum mechanics had become an established research field in physics.

corroborated by Serge Haroche and his collaborators, in the mid-1990s, who measured the time of loss of coherence. Decoherence did not solve the measurement problem but did shed light on the transition from the quantum description to the classical one. The spectrum of interpretations was enlarged when Giancarlo Ghirardi developed the idea of introducing nonlinearity in the Schrödinger equation in order to cope with the measurement problem. What we now call the continuous spontaneous localization theories are indeed new theories and not new interpretations. Most impressive in these times were, however, the technical improvements that enabled a variety of experiments, some previously only considered thought experiments, to check quantum mechanics predictions in the most extreme situations. One of these new techniques was a better production of pairs of entangled photons through parametric down-conversion processes, a nonlinear interaction of a photon with a crystal producing a pair of entangled photons. These new techniques may be grouped under the umbrella of techniques enabling the manipulation of single quantum systems. As identified by M. A. B. Whitaker, while reviewing experiments in the foundations of quantum theory, 'only by such experiments studying single particles, and single correlated systems, may one move beyond the rather bland information obtainable from ensembles, and study the adequacy of standard approaches to quantum theory in these more theoretically challenging circumstances' (Whitaker, 2000, p. 3). These experimental achievements justified the 2012 Nobel Prize being awarded to Serge Haroche and David Wineland 'for ground-breaking experimental methods that enable measuring and manipulation of individual quantum systems'.[7]

The conceptual seeds for the next and current stage were launched by David Deutsch, who in 1985 tried to harness quantum entanglement to improve computing power and suggested a type of Turing machine compatible with quantum mechanics to run on a future quantum computer (Deutsch, 1985); and by Charles Bennett and Gilles Brassard (1984) who produced the first cryptography protocol based on the use of quantum entanglement to encode information. A plot of the number of papers by year (Figure 33.1) with the term 'quantum information' as their topics demonstrates that while Deutsch's and Bennett and Brassard's papers attracted little attention for a few years after their publications, from the mid-1990s on there was a surge of interest in such subjects.

The plot also pinpoints the beginnings of the fourth and current stage of the quantum controversy, the mid-1990s, and its main feature, the dominance of the interest in quantum information. This research blossomed first of all driven by the technological promise of better computers and safer cryptographic protocols; all of this happening in a world experiencing the conspicuous presence of information technologies. Evidence of interest in this promise has been a surge in research funding from government bodies, including the military and private corporations. According to the journalist Jason Palmer (2017), in 2015 alone, 'about 7,000 people worldwide, with a combined budget of about $1.5bn, were working on quantum-technology research', a

[7] On decoherence, see (Freitas, 2021); on Ghirardi's work, see Allori (2021). The Nobel Prize in Physics 2012. NobelPrize.org. Nobel Media AB 2020. Sun. 26 Jul 2020. <https://www.nobelprize.org/prizes/physics/2012/summary/>

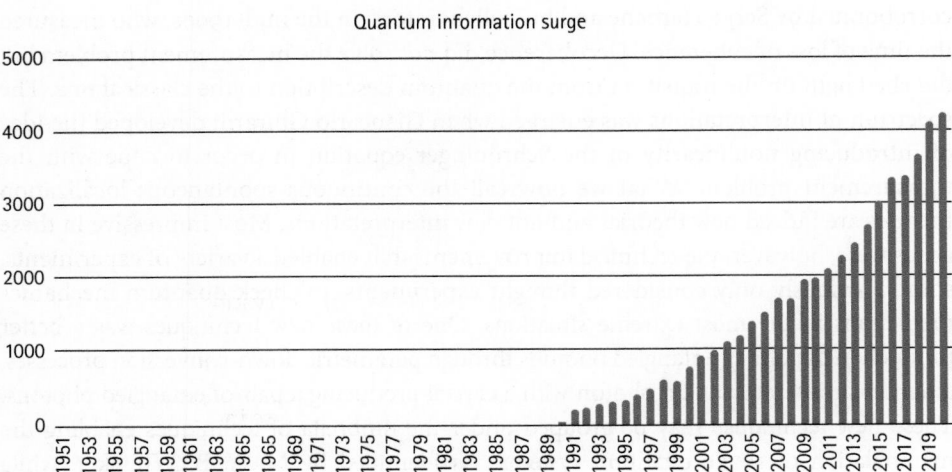

**FIGURE 33.1** Number of papers using 'quantum information' as topic.

*Source*: Web of Science, accessed 31 March 2021.

substantial part of this pool dedicated to quantum information. Quantum information research has legitimized the controversy over the interpretation of quantum mechanics because the physical effect supporting such research—quantum entanglement—emerged in the debates about this controversy. An acknowledgment of this was the 2010 Physics Wolf Prize, which was awarded to John Clauser, Alain Aspect, and Anton Zeilinger 'for their fundamental conceptual and experimental contributions to the foundations of quantum physics, specifically an increasingly sophisticated series of tests of Bell's inequalities, or extensions thereof, using entangled quantum states'.[8] However, the impact of quantum information on the research on the foundations of quantum theory has been ambiguous, to say the least. Sam Schweber echoed these concerns writing, 'people working in [ . . . ] quantum computers, . . . are principally concerned with the creation of novelty—of entities or effects that did not previously exist in the world—[ . . . ] and are no longer concerned with establishing the foundational theory that governs the interactions and determines the evolution of the structures that populate that domain' (Schweber, 2014).

Concerns with foundational theories and experiments therefore compete with harnessing quantum features for practical purposes. As our interest in this short chronological overview focuses on the ongoing quantum controversy and not on the potential applications of quantum physics, we may conclude this overview by highlighting two dominant trends. On the one hand, there have been several attempts to theoretically reconstruct quantum mechanics in order to overcome deficiencies in its foundations or to reach more general formalisms, and many of these attempts take information itself

---

[8] https://wolffund.org.il/the-wolf-prize/#Laureates.

as a central piece.[9] On the other hand, experiments on entanglement have reached phenomenological domains unthinkable a few decades ago. These include the detection of pairs of entangled photons when one of the photons travelled space scale distances between the Earth and a satellite or even, in order to control experiments in optics laboratories, photons come from cosmological distances.

# 33.3 SECOND PART

The arguments and events reported above demonstrate the existence of a long-lasting and evolving quantum controversy. But is it a legitimate scientific controversy or just a diversion in the history of quantum mechanics? Or is there a controversy at all? The answer to these questions depends on our philosophical outlook. As argued by Oscar Nudler (2011b), philosophers from Kantianism to logical positivist did not see controversies as an integral part of the production of new scientific knowledge. Thus we should begin with the post-positivists. We consider Thomas Kuhn's views on normal science and scientific revolutions not only because Kuhn's ideas were so influential but also because he was well placed to comment on the debates concerning quantum mechanics. Kuhn was trained as a theoretical physicist before entering the history and philosophy of science and acted as the leader of the project which collected documents and oral histories from physicists who acted during the creation of quantum mechanics until the early 1930s, the *Archives for the History of Quantum Physics* (Kuhn, 1967). He therefore interviewed and engaged with many of the actors of the first chronological stage of the controversy. We also took into consideration two other philosophers, Karl Popper and Paul Feyerabend, who took a very different position about the controversy. Next, we move on to take into consideration more recent studies of controversies to discuss the role of pluralism versus monocracy in science. Furthermore, we examine the role of experiments to ultimately decide among different interpretations and the possibility of a long coexistence of different empirically equivalent interpretations, which brings to the fore the Duhem–Quine thesis of underdetermination. We go on to analyse the cleavage among different philosophical assumptions and the possible uniqueness of the quantum controversy in the history of science.

## 33.3.1 Making Sense of the Quantum Controversy

In his influential book, *The Structure of Scientific Revolutions*, Kuhn (1996) suggested that the development of science alternates between periods of normal science and scientific revolutions. For Kuhn, different scientific disciplines come of age when they

---

[9] Darrigol (2022), Grinbaum (2022).

enter the stage he called normal science, when science develops through a strong consensus among its practitioners, such a consensus concerning shared paradigms or disciplinary matrices. While most of the examples taken by Kuhn to construe his theoretical framework came from the inception of modern science, he did not ignore the quantum case. Indeed, one of his most iconic citations was Max Planck meaningfully saying a new theory is accepted only when the supporters of the old one die (Kuhn, 1996, p. 151). As the creation of quantum theory in the mid-1920s was presented by Kuhn as an example of a scientific revolution, it was permissible to infer that research on quantum mechanics, after 1927, entered a period of normal science, more directed to solving problems left unsolved in the application of the theory than to challenge the foundations of the new theory. Thus, the ongoing quantum controversy cannot be framed in Kuhn's doctrine about the development of science.

The inadequacy of Kuhn's views for the case of the quantum controversy and for philosophy of science was perceived in due time by Jeffrey Bub and Mara Beller. In an important but latterly little known debate, at a symposium in Illinois organized by the philosopher Frederick Suppe, in 1969, Bohm presented a paper dealing with Kuhn's views. Bohm could not attend the symposium but his paper was presented by Jeffrey Bub, his former doctoral student. Bohm criticized the notion of incommensurability and the distinction between scientific revolutions and normal science as this implied no conceptual advancement during the stages Kuhn would call normal science. Bohm illustrated his criticisms with a case from the 19th century, contrasting Newtonian dynamics and Hamilton–Jacobi wave theory of dynamics, and another from the 20th century, arguing for the differences between Bohr and Heisenberg's views, on one hand, and von Neumann's views, on the other. Kuhn acknowledged he was not comfortable with the notion of incommensurability but sharply reacted against Bohm's views on the development of quantum mechanics and his criticisms to normal science: 'It is my impression that one of the greatest difficulties faced by people who are concerned to revise the Bohr interpretation is that none of the problems that emerge for them makes any contact whatsoever with the technical problems that physics has faced in recent years, and that has created a profound crisis for the profession'. Bub reacted by claiming Kuhn was acting as a physicist, part interested in the quantum controversy, not as a philosopher:[10]

> Something about what Professor Kuhn just said really disturbed me: it seemed to me that he was suggesting that one should really only care about what most physicists are concerned about today. Now I really find this very puzzling! What significance for the philosophy of science is it that today or in the last thirty years most physicists have been concerned with the particular sorts of problems and that certain problems have been dropped as irrelevant—namely, the problem of the interpretation of the quantum theory or the problem of vocabulary (as I think he

---

[10] Bohm's paper and Kuhn's and Bub's reactions are in Suppe (1974, pp. 374–91 and 409–13). See also Freire Jr (2019, p. 149).

referred to it)? I think this is really a very peculiar sort of way to look at the history of physics or the way physics is carried out and then to take what most physicists are doing as some sort of standard on which to base a methodology in the prescriptive sense.

Beller, at the end of the century, went on to argue that Kuhn's idea of paradigm mirrored the dominance among the physicists of the orthodox views on the interpretation of quantum mechanics. In Beller's words, 'the notion of paradigm has not only clear totalitarian implications but also dogmatic ideological roots'. Beller founded this strong claim on her identification of the 'close historical links [ . . . ] between the notion of incommensurable paradigms and the ideology of the Copenhagen dogma'.[11] With the benefit of hindsight we may say these Kuhnian lenses were challenged by the history of the controversy over the foundations of quantum theory. Indeed, it has not been a history of a shared paradigm; instead it has been a matter of permanent dispute among practitioners. Furthermore, the very existence of this controversy and its diversity of interpretations have fostered the development of physics. Bell's theorem was inspired by Bohm's interpretation, this theorem led to entanglement, and Everett's interpretation has been influential, at least from a heuristic point of view, in works on the inception of decoherence, quantum computation, and quantum gravity.

Other contemporary philosophers of science were far more tolerant of the disputes around the interpretation of quantum mechanics. Imre Lakatos (1978), for instance, with the permanent competition among scientific research programmes; Paul Feyerabend (2010), with his criticism of the existence of a universal scientific method; and Karl Popper (1959), with his view of permanent dispute among rival theories as an integral part of scientific activities, are more congenial to the existence of the quantum controversy. Furthermore, they participated in the debates about the interpretation of quantum mechanics, thus were actors in the controversy. Popper, in particular from the 1960s on, actively participated in the debates including co-authoring papers with some of the quantum dissidents and himself suggesting an experiment to test the validity of quantum mechanics (Del Santo, 2018, 2019, 2020; Del Santo and Freire Jr, 2021). The post-scriptum of his *Logic of Scientific Discovery* dedicated to quantum physics was meaningfully titled the *Quantum Schism*, which meant to make the quantum controversy a legitimate controversy in the history of 20th-century physics (Popper, 1982).[12]

In recent decades there has been a renewed interest in studies over scientific controversies, which has particularly mobilized philosophers of science as well as historians and sociologists. Examples of works entirely dedicated to the studies of controversies are Machamer, Pera, and Baltas, (2000); Engelhardt and Caplan (1987); Dascal and Boantza (2011); Gil (1990); Nudler (2011a); Raynaud (2015); and the special issue of *Science in Context* (11(2), 1998). This study has benefited from the

---

[11] Beller (1999, pp. 287–306). On Kuhn and Beller, see also Freire Jr (2016).
[12] For a discussion on 20th-century philosophers and quantum mechanics, see Ryckman (2022).

wider interest in the study of controversies at large, as testified by the book series on *Controversies* principally edited by Marcelo Dascal. However, these publications have expressed conflicting views when the controversy on the quanta is at stake, which is evidence of the difficulties of dealing with this particular one.

These volumes have a few but important traits in common which are useful for our analysis. Controversies are recognized not as exceptional moments in the development of scientific practice but as historical phenomena more widespread than previously acknowledged by scientists and positivist philosophers. The opening, development, and resolution of controversies not only encompass theoretical and observational elements; indeed, they are more complex phenomena and concern the behaviour of scientific communities, not only of individuals. As noted by Ernan McMullin (1987), scientific controversies involve much more than logical problems concerning hypotheses and evidence; they are social conflicts involving personality traits and other historical contingencies. Peter Machamer and his co-editors summarized their views stating that 'choice of theories depends upon the education and training as well as on the interests and values, in varyingly broad senses, of the scientists—individually or in groups. Even more abstractly, theories and traditions often reflect, sometimes unconsciously, higher level philosophical, ideological, or architectonic principles' (Machamer *et al.*, 2000, p. 9). They call for the use of argumentation, dialectical and rhetorical argumentation, as an analytical tool to analyse the development of controversies. Yet they are cautious about the possibility of grasping the historical development of controversies using only philosophical resources. In fact, after suggesting a philosophical pattern based on the resources of argumentation to cope with controversies, they conclude (Machamer *et al.*, 2000, p. 16):

> Although this may be considered a general pattern, nothing specific can be drawn from it. Neither the phenomenon of controversies nor its actual ways of practice can be inferred from a general philosophy of the history of science. Philosophy can analyse the phenomenon; help understand its several dimensions; make clear its sources, origins, and forms; and reflect upon the tools used, but it is up to the history of science to examine, case by case, how controversies in actual practice go, and why they go the way they do. These examinations need to use tools and concepts from sociology and other empirical disciplines, for example, discourse analysis, economics, and so on. Philosophy, history, and the sociology of science work better if the one does not neglect the results of the others.

Despite these converging views, divergence appears when the quantum controversy is called to the forefront or even just cited. Indeed, the controversy over the interpretation of the quanta has so far received scant attention from such studies; but this attention has led to conflicting views. While discussing the resolution of controversies, and to illustrate when a party in dispute is defeated, Machamer *et al.* (2000, p. 14) suggest that Einstein, who had challenged 'the logical and philosophical feasibility of the "orthodox" interpretation of quantum mechanics', eventually 'implicitly recognized that, in terms of arguments, he was defeated'. Machamer cites a letter from Einstein to Max Born, on 3 March 1947, which says 'I cannot, however, base this conviction on logical reasons, but can only produce my little finger as witness'. Machamer means Einstein was evoking here a sort of psychological resistance on his part, implicitly admitting that he had lost at the level of arguments. Marcelo Pera (2000, p. 62), in the same volume, attributed Einstein's enduring criticism of the complementarity principle to 'psychological resistance', patronizingly stating 'although it does not matter—only arguments do—it is a rich source of devices for those who are in trouble in the course of a controversy'. Such remarks, however, are not supported by historical records and studies and by philosophical analysis of Einstein's attitudes. After the cited letter, in early 1949, when the volume 'Philosopher Scientist,' honouring Einstein and edited by P. A. Schilpp (1982) was first produced, Einstein replied to the criticisms he had received from Pauli, Born, and Bohr about his interpretation of quantum mechanics. This text is part of the corpus of the quantum controversy because Einstein clearly stated his criticisms both as a philosophical criticism of positivism as well as an affirmation of a research programme for physics. Einstein (1982, p. 675) argued that he was accused of 'rigid adherence to classical theory', but showed that such terms in physics were fraught with problems. Then he concludes: 'There is, strictly speaking, today no such thing as a classical field-theory; one can, therefore, also not rigidly adhere to it'. Nevertheless, field-theory does exist as a programme: 'Continuous functions in the four-dimensional [continuum] as basic concepts of the theory'. A person who suggests a programme for the unification of physics, even acknowledging its enormous difficulties, is neither accepting defeat nor being psychologically resistant. Furthermore, Einstein's stance found an audience. According to Jammer (1974, p. 250), this volume, including Einstein's paper, 'was widely read by philosophizing physicists' and 'contributed considerably to the creation of a more critical atmosphere toward the complementarity philosophy'. Furthermore, Machamer, Pera, and Baltas missed the subtlety and irony present in the letters from Einstein to Born when scientific matters were concerned. Thus, on 15 September 1950, three years after the letter where Einstein putatively conceded defeat, he wrote to Born (2005, p. 185), 'this is where our attitudes really differ. For the time being, I am alone in my views—as Leibniz was with respect to the absolute space of Newton's theory'. It took almost 300 years for physics to resume, first with Ernst Mach and eventually with Einstein, the criticism of absolute space, which led to the abandonment of this concept. What Einstein was saying in his peculiar ironic way was that in scientific controversies he might be a minority but he was in good company.

Another illustration came when the same authors (Machamer *et al.*, 2000, pp. 15–16) suggested criteria for when there is 'no room for controversies'. These criteria appear when 'scientists do not want to change or cannot change [their views], for example, because of their merits in many fields, the claim is rejected as absurd, or neglected as unimportant, or maintained as a strange anomaly'. The list of illustrative cases was: 'the anomalous perihelion of Mercury in astronomy, [ . . . ] the lack of fossil records or of intermediate stages of species in evolutionary biology', and 'the paradoxes in quantum mechanics'. Thus, as far as the quantum controversy is concerned, with Machamer's analysis of controversies, the circle is closed back to the implication of Kuhn's views. According to these studies, the controversy over the interpretation and foundations of quantum mechanics does not exist because it should not exist, it has been in the best case a wrong direction in the history of physics. However, paraphrasing the words of the legendary Galileo, 'And Yet, It Is Heard'.[13] Its voices have been listened to by other historians and philosophers, as we have seen, but it seems to have eluded some of the philosophical literature on controversies.

It would be unfair, however, to portray all these studies on controversies as mute or in denial of the quantum controversy. Everett Mendelsohn and Ernan McMullin, in the volume organized by Engelhardt and Caplan (1987), while in divergence about the role of political factors in the study of controversies, both cited and acknowledged the quantum as a genuine scientific controversy. Gil (1990) included one chapter entirely dedicated to the debates between Einstein and Bohr and their later implications. McMullin, for instance, while exemplifying what he called controversies among principles as opposed to those among theories, cited the case—from Newton to Mach—about the concept of force and the quantum debates (McMullin, 1987, p. 73):

> The vigorous debates surrounding the quantum theory of matter were of a similar character. There was no question about the predictive successes of the new theory. Nor was there a viable alternative. But critics such as Einstein, and later Bohm, Vigier, and others urged that because of its departure from the deterministic pattern of classical mechanics, it could at best be regarded only as a stopgap. It could not, in their view, be reconciled with a larger philosophical view of nature, whose sources are wider than physics alone.

The impossibility of easily framing the controversy over the quanta in this plethora of philosophical studies of scientific controversy leaves us with an unsolved problem. Furthermore, some of the periods of this controversy, particularly the second and the third, were marked by a lack of tolerance towards the controversy among most physicists. We have recorded a number of cases of scientists whose professional careers were damaged by that lack of tolerance. This was the case with Everett, Clauser, Klaus Tausk, and Zeh; furthermore, contemporary participants, such as Shimony, Bell, and Aspect, noticed in due time the existence of a stigma against the research on the foundations (Freire Jr, 2015a). Tolerance is rather important as, according to McMullin (1987, p. 74), controversies of principles may last and 'the consequences of premature closure are much too grave'. This leads us to the value of pluralism in the practice of science. From the vantage point of hindsight, as we now know that the quantum controversy has been fecund for the development of our understanding of quantum physics, the quantum controversy reminds us of the role of pluralism in science,

---

[13] I am borrowing the title of Tonietti's book (Tonietti, 2014), which is, of course, a paraphrase of Galileo's 'Eppur si muove'.

pluralism that had been defended earlier and in other contexts than quantum mechanics by physicists such as Maxwell, Boltzmann, and Poincaré. Indeed, tolerance towards diversity has been more helpful to science than strict adhesion to the dominant views. Half a century ago, Jammer (1974, p. 521) reiterated such a claim to pluralism. After concluding that the quantum controversy remained without an ending, he stated 'let us recall the advice once given by the French moralist Joseph Joubert: "It is better to debate a question without settling it than to settle a question without debating it".' I have appealed to political metaphors to argue that, at least in science and in this case, a Hundred Flowers policy fared better than a Nonproliferation treaty (Freire Jr, 2015a, p. 326).

## 33.3.2 The Elusive Role of Experiments

The hundred-year old controversy over the quantum interpretation was marked by breakthrough experiments, the most impressive being those concerning Bell's theorem and the decoherence effect, which began in the second stage. Indeed, experiments did not play any role in the first stage. The precursors of experiments concerning the foundations of quantum mechanics were Bohm and Yakir Aharonov, in 1957, who analysed results from previous experiments by C. S. Wu and I. Sakhnov in order to check predictions from the EPR thought experiment (Silva, 2022). The first experiments on Bell's theorem were conducted by Clauser and Freedman and by Richard Holt, in 1972 (Freire Jr, 2015a). In the ongoing series of these experiments the contrast was between quantum mechanics predictions and predictions from the class of theories which used local realism as an assumption. Local realism was a reasonable expectation nurtured by many physicists, including Einstein. However, so far, no concrete alternative interpretation had shared this assumption. Bohm's 1952 hidden variable interpretation, for instance, had nonlocality embedded in its quantum potential. The relevant conclusion is that despite the number of experimental achievements in foundations of quantum mechanics, we do not have experiments that may contrast, for instance, the usual interpretations of quantum physics against the major alternative interpretations or theories, such as hidden variables, relative states, and statistical interpretations. In addition, such experiments may not be feasible as these interpretations were built to reproduce quantum predictions while departing from distinct conceptual foundations. The case is different with the continuous spontaneous localization theories as they were conceived to be different physical theories. It is not by chance that currently Angelo Bassi is engaged in testing the GRW collapse theory (Bassi et al., 2013; Henderson, 2020). It is also true that nowadays Basil Hiley expects to submit the quantum potential formulated by Bohm in 1952 to an experimental test (Flack and Hiley, 2018). As a matter of fact, a century of controversy has not been able to discard any major alternative interpretation, with experiments playing an elusive role in settling the controversy. This situation is not comfortable for physics as an experimental science and almost all of those engaged in the debates over the quantum think that future experiments may end the controversy.

However, it is enticing to think of experiments playing a further more elusive role in settling the controversy. In such a futuristic and dystopian scenario, the controversy could either fade away without resolution or enter a stage of peaceful coexistence of distinct interpretations. This conjecture reminds us of the thesis on the underdetermination of theories by empirical data (Stanford, 2016), also called the Duhem–Quine thesis in reference to the scientist, historian, and philosopher of science Pierre Duhem and the philosopher and logician Willard Van Orman Quine who formulated it. James Cushing appealed to the underdetermination thesis to explain and criticize the wider acceptance of the complementarity interpretation. Furthermore, Cushing (1994, p. 199) criticized the received view about the inevitability of theories in experimental sciences, stating 'we are not particularly uncomfortable with a lack of inevitability in other areas of history, but for science such a possibility may strike many as bordering on the sacrilegious, or at least as teetering on the brink of an abyss of rationality skepticism'. More recently, Jeffrey Bub (2005) built an analogy between the history of Einstein's principle of relativity and the quantum mechanics case to argue that if the information-theoretical constraints are considered, hidden variables and no-collapse theories are doomed to have no excess empirical content over quantum mechanics. I am not arguing that the quantum controversy vindicates the underdetermination thesis; after all, a historical case cannot corroborate a philosophical thesis. However, the Quine–Duhem thesis helps us to think how the controversy over interpretation may shoulder the impressive predictive power of quantum physics. Even with such caveats, it is challenging to think that some of our cherished interpretations of such a solid physical theory may not be the only possible description of phenomena.

## 33.3.3 The Cleavage between Different Philosophical Assumptions

Philosophical issues have also led physicists to reconsider the foundations of quantum physics and the quantum controversy as a crystal clear case in which philosophical assumptions are deeply entrenched in physics practice, many times driving it. Furthermore, at least in the case of Marxist philosophers and physicists, philosophical assumptions were deeply mixed with ideological commitments (Freire Jr, 2011; Jammer, 1974, pp. 250–1; Besson, 2018). While there has been a quest for recovering determinism and more intuitive pictures of quantum phenomena, the enduring and more relevant philosophical issue at stake has been the tension between realism and instrumentalism, in their distinct variations. The traditional interpretations, both the complementarity view and von Neumann's approach, have an instrumentalist flavour, and they were clearly the dominant views in the first stage. In the case of complementarity, this was presented as a huge epistemological lesson. In contrast, most of realistic views required alternative interpretations of quantum mechanics and the increasing acceptance of realistic views in the second half of the 20th century has been noted by different scholars (Brush, 1980; Elkana, 1984). Thus the growth of realism among physicists and philosophers involved with foundations favoured the appearance of realistic alternative interpretations, and gained traction in the second and third stage.

In the fourth and current stage of the controversy it seems there is once again a swing in philosophical trends. In parallel to the increasing research on quantum information, there have been a number of attempts to rebuild the theory attributing a central role to information, with some approaches suggesting information to be the object of the quantum theory. Quantum Bayesianism, relational interpretation, and information-theoretical approaches are examples in this direction. Some of these attempts to rebuild or reinterpret the quantum theory also imply a recovery of Bohr's interpretation, now stripped from the previous subjectivist flavour and from the expectation of being the new philosophical holy grail valid for almost all domains of experience. As recently synthesized by Jeffrey Bub (2019, p. 236), 'for the Copenhagenists, quantum mechanics is a new sort of non-representational theory for an irreducibly indeterministic universe, with a new type of nonlocal probabilistic correlation for "entangled" quantum states of separated systems, where the correlated events are intrinsically random, not merely apparently random like coin tosses'.

## 33.3.4 The Possible Uniqueness of the Quantum Controversy in the History of Science

In which sense should we speak about the uniqueness of the quantum controversy in the history of science? It is neither the very existence of a controversy, nor its bearing on foundations, for we have many other examples, some of which imply foundations. Although intellectual controversies are more common in philosophy and humanities at large, they are not strange to the history of the natural sciences. Each time we deal with issues concerning the foundations or the interpretations of scientific theories, or, with the reception of such theories, we observe analogous divergences. Canonical examples are the reception of Newton's mechanics and gravitation; the reception and later developments of Darwin's theory of natural selection; the reception of Einstein's principle of relativity; Mach's criticism of Newton's mechanics; the reception and interpretation of Wegener's continental drift theory; controversies over geological eras; debates over action at a distance and action by continuity in electromagnetism; and the debates on 20th-century cosmological models.

If the quantum controversy has any singularity, it resides in being a melting pot of at least four meaningful features. It is a controversy over the foundations of an extremely successful and widely accepted scientific theory. Secondly, it has been a long controversy. While writing on the history of dynamic systems and chaos, historians David Aubin and Amy Dahan Dalmedico (2002) mobilized the historiographical concept of *longue-durée* to deal with a time span of 90 years. The quantum controversy is, by the same token, a *longue-durée* controversy. Third, it is a controversy that even without resolution has been fecund for the development of science. We now know about entanglement and decoherence better thanks to these debates about the foundation of quantum mechanics and these quantum phenomena are essential parts of the current research on quantum information. Lastly, it is an evolving controversy and it remains open.

This last feature, an open controversy, poses particular challenges to the history of science as it is a history of the recent past, a kind of history historians have learned to deal with only very recently, indeed in the last 40 years (Garcia, 2010). Against this backdrop we may better value the wisdom of Max Jammer, who opened his 1974 book with a promise to deliver his work with a feature 'rarely, if ever', present in the scholarship about the interpretations of quantum mechanics: '*sine ira et studio*' (without anger and passion) (Jammer, 1974, p. v).

# 33.4 CODA—A RUSE OF HISTORY

The history of the quantum controversy closed the 20th century with an ironical twist. The controversy helped to produce a better understood, consolidated quantum mechanics even though some who were deeply engaged in this controversy meant to challenge this theory. One major episode in the controversy was the elaboration of Bell's theorem and the performing of relevant experiments, which led to the breaking of local realism and the wide acceptance of entanglement as a physical property of quantum systems. However, many of its main protagonists, quantum dissidents such as Bell and Clauser, expected experiments would not corroborate the quantum mechanics predictions. Another major milestone is Tony Leggett's work on decoherence. As shown by Fabio Freitas (2022) in this volume, Leggett's expectation was to reveal limitations of quantum mechanics while dealing with macroscopic systems, which remains uncorroborated in the labs to date.

The quantum dissidents won the day insofar as foundations of quantum mechanics became good physics. However, they had higher expectations. The irony is that in trying to attack the foundations of quantum mechanics, they ended up consolidating and developing quantum mechanics itself and quantum theory entered the 21st century as vindicated as ever (Freire Jr, 2015b, A51; Freitas, 2022).

## REFERENCES

Allori, V. (2022, this volume). Spontaneous Localization Theories: Quantum Philosophy between History and Physics.

Aubin, D., and Dalmedico, A. D. (2002). Writing the History of Dynamical Systems and Chaos: *Longue Durée* and Revolution, Disciplines and Cultures. *Historia Mathematica*, **29**, 273–339.

Bacciagaluppi, G. and Valentini, A. (eds) (2009). *Quantum Theory at the Crossroads: Reconsidering the 1927 Solvay Conference.* Cambridge: Cambridge University Press.

Bacciagaluppi, G. (2022, this volume). The Statistical Interpretation: Born, Heisenberg, and von Neumann, 1926–27.

Baracca, A., Bergia, S., and Del Santo, F. (2017). The origins of the research on the foundations of quantum mechanics (and other critical activities) in Italy during the 1970s. *Studies in History and Philosophy of Modern Physics*, 57, 66–79.

Bassi, A., Lochan, K., Satin, S., Singh, T. P., and Ulbricht, H. (2013). Models of wave-function collapse, underlying theories, and experimental tests. *Rev. Mod. Phys.*, 85, 471–527.

Bell, J. S., and Nauenberg, M. (1966). The moral aspect of quantum mechanics. In: A. De Shalit, H. Feshbach, and L. Van Hove (eds), *Preludes in Theoretical Physics*, Amsterdam: North Holland, pp. 279–86.

Beller, M. (1999). *Quantum Dialogue—The Making of a Revolution.* Chicago: The University of Chicago Press.

Bennett, C. H., and Brassard, G. (1984). Quantum cryptography: public key distribution and coin tossing. In: International Conference on Computers, Systems, and Signal Processing, pp. 175–79.

Besson, V. (2018). L'interprétation causale de la mécanique quantique: biographie d'un programme de recherche minoritaire (1951–1964). Ph.D. dissertation, Université Claude Bernard Lyon 1 and Universidade Federal da Bahia.

Born, M. (2005). *The Born–Einstein Letters 1916–1955.* New York: Macmillan.

Bromberg, J. L. (2006). Device physics vis-à-vis fundamental physics in Cold War America: the case of quantum optics. *Isis*, 97, 237–59.

Bromberg, J. L. (2008). New instruments and the meaning of quantum mechanics. *Hist. Stud. Nat. Sci.*, 38, 325–52.

Brush, S. G. (1980). The chimerical cat: philosophy of quantum mechanics in historical perspective. *Soc. Stud. Sci.*, (4), 393–447.

Bub, J. (2005). Quantum mechanics is about quantum information. *Found. Phys.*, 35(4), 541–60.

Bub, J. (2019). Interpreting the Quantum World: Old Questions and New Answers. *Historical Studies in the Natural Sciences*, 49(2), 226–39.

Byrne, P. (2010). *The Many Worlds of Hugh Everett III: Multiple Universes, Mutual Assured Destruction, and the Meltdown of a Nuclear Family.* New York: Oxford University Press.

Crull, E. (2022, this volume). Grete Hermann's Interpretation of Quantum Mechanics.

Cushing, J. (1994). *Quantum Mechanics—Historical Contingency and the Copenhagen Hegemony.* Chicago: The University of Chicago Press.

Darrigol, O. (2022, this volume). Natural Reconstructions of Quantum Mechanics.

Darrigol, O. (2003). Quantum Theory and Atomic Structure, 1900–1927. In M. J. Nye (ed.), *The Cambridge History of Science, Volume 5 – The Modern Physical and Mathematical Sciences*, Cambridge: Cambridge University Press, pp. 331–49.

Dascal, M., and Boantza, V. (eds) (2011). *Controversies Within the Scientific Revolution.* Amsterdam: John Benjamins.

Del Santo, F. (2018). Genesis of Karl Popper's EPR-like experiment and its resonance amongst the physics community in the 1980s. *Studies in History and Philosophy of Modern Physics*, 62, 56–70.

Del Santo, F. (2019). Karl Popper's forgotten role in the quantum debate at the edge between philosophy and physics in 1950s and 1960s. *Studies in History and Philosophy of Modern Physics*, 67, 78–88.

Del Santo, F. (2020). An unpublished debate brought to light: Karl Popper's enterprise against the logic of quantum mechanics. *Studies in History and Philosophy of Modern Physics*, **70**, 65–78.

Del Santo, F., and Freire Jr, O. (2021). Popper and the Quantum Controversy. In Z. Parusniková and D. Merritt (eds), *Karl Popper: His Science and His Philosophy*, Cham, Switzerland: Springer, pp. 17–35.

Deutsch, D. (1985). Quantum-theory, the Church–Turing principle and the universal quantum computer. *Proc. R. Soc. Lond. Ser. A Math. Phys. Eng. Sci.*, **400**(1818), 97–117.

Einstein, A. (1982). Remarks Concerning the Essays Brought Together in this Co-operative Volume. In P. A. Schilpp (ed.), *Albert Einstein Philosopher-Scientist*, La Salle, IL: Open Court, pp. 665–88.

Elkana, Y. (1984). Transformations in realist philosophy of science from Victorian Baconianism to the present day. In E. Mendelsohn (ed.), *Transformation and Tradition in the Sciences: Essays in Honor of I. Bernard Cohen*, Cambridge: Cambridge University Press, pp. 487–511.

Engelhardt, H.T., and Caplan, A.L. (eds) (1987). *Scientific Controversies: Case Studies in the Resolution and Closure of Disputes in Science and Technology*. Cambridge: Cambridge University Press.

Feyerabend, P. (2010). *Against Method*, 4th edn. New York: Verso.

Flack, R., and Hiley, B. J. (2018). Feynman Paths and Weak Values. *Entropy*, **20**, 367. doi:10.3390/e20050367.

Forman, P. (1971). Weimar culture, causality, and quantum theory, 1918–27: adaptation by German physicists and mathematicians to a hostile intellectual environment. *Hist. Stud. Phys. Sci.*, **3**, 1–115.

Freire Jr, O. (2011). On the connections between the dialectical materialism and the controversy on the quanta. *Jahrbuch für Europäische Wissenschaftskultur*, **6**, 195–210.

Freire Jr, O. (2015a). *The Quantum Dissidents—Rebuilding the Foundations of Quantum Mechanics (1950–1990)*. Berlin: Springer.

Freire Jr, O. (2015b). From the margins to the mainstream: Foundations of quantum mechanics, 1950–1990. *Ann. Phys.*, **527**(5–6), A47–A51.

Freire Jr, O. (2016). Contemporary science and the history and philosophy of science. In: A. Blum, K. Gavroglu, C. Joas, and J. Renn (eds), *Shifting Paradigms—Thomas S. Kuhn and the History of Science*, Berlin: Edition Open Access, pp. 105–14.

Freire Jr, O. (2019). *David Bohm—A Life Dedicated to Understanding the Quantum World*. Cham, Switzerland: Springer.

Freitas, F. (2022, this volume). Tony Leggett's Challenge to Quantum Mechanics and its Path to Decoherence.

Garcia, P. (2010). Histoire du temps présent. In C. Delacroix, F. Dosse, P. Garcia, and N. Offenstadt (eds), *Historiographies: concepts et débats*, volume 1, Paris: Gallimard, pp. 282–94.

Gil, F. (ed.) (1990). *Controvérsias científicas e filosóficas, Controverses scientifiques et philosophiques, Scientific and philosophical controversies*. Lisbon: Fragmentos.

Gilder, L. (2008). *The Age of Entanglement—When Quantum Physics Was Reborn*. New York: Knopf.

Grinbaum, A. (2022, this volume). Quantum Information and the Quest for Reconstruction of Quantum Theory.

Henderson, B. (2020). The Rebel Physicist on the Hunt for a Better Story than Quantum Mechanics. *The New York Times Magazine*, 25 June.

Howard, D. (2022, this volume). The Copenhagen Interpretation.

Jacobsen, A. S. (2012). *Léon Rosenfeld—Physics, Philosophy, and Politics in the Twentieth Century*. Singapore: World Scientific.

Jähnert, M., and Lehner, C. (2022, this volume). The Early Debates about the Interpretation of Quantum Mechanics.

Jammer, M. (1974). *The Philosophy of Quantum Mechanics—The Interpretations of Quantum Mechanics in Historical Perspective*. New York: Wiley.

Jammer, M. (1989). *The Conceptual Development of Quantum Mechanics*, 2nd edn. Los Angeles: Tomash Publishers.

Kaiser, D. (2012). *How the Hippies Saved Physics: Science, Counterculture, and the Quantum Revival*. New York: W. W. Norton.

Kaiser, D. (2020). *Quantum Legacies—Dispatches from an Uncertain World*, Chicago: Chicago University Press.

Kaiser, D. (2022, this volume). Tackling Loopholes in Experimental Tests of Bell's Inequality.

Kojevnikov, A. (2020). *The Copenhagen Network—The Birth of Quantum Mechanics from a Postdoctoral Perspective*. Cham, Switzerland: Springer.

Kuhn, T. S. (1967). *Sources for history of quantum physics; an inventory and report [by] Thomas S. Kuhn [and others]*. Philadelphia: American Philosophical Society.

Kuhn, T. S. (1996). *The Structure of Scientific Revolutions*, 3rd edn. Chicago: The University of Chicago Press.

Lakatos, I. (1978). *The Methodology of Scientific Research Programmes—Philosophical Papers*, volume I, edited by J. Worral and G. Currie. Cambridge: Cambridge University Press.

Laloë, F. (2012). *Do We Really Understand Quantum Mechanics?* New York: Cambridge University Press.

Landsman, K. (2022, this volume). The Axiomatization of Quantum Theory through Functional Analysis: Hilbert, von Neumann, and Beyond.

Latour, B. (1987). *Science in Action: How to Follow Scientists and Engineers Through Society*. Cambridge, MA: Harvard University Press.

Machamer, P., Pera, M., and Baltas, A. (eds) (2000). *Scientific Controversies—Philosophical and Historical Perspectives*. New York: Oxford University Press.

McMullin, E. (1987). Scientific controversy and its termination. In: H. T. Engelhardt and A. L. Caplan (eds), *Scientific Controversies: Case Studies in the Resolution and Closure of Disputes in Science and Technology*, Cambridge: Cambridge University Press, pp. 49–91.

Nudler, O. (ed.) (2011a). *Controversy Spaces. A model of scientific and philosophical change*. Amsterdam: John Benjamins.

Nudler, O. (2011b). Controversy spaces—The dialectical nature of change in the sciences and philosophy. In O. Nudler (ed.), *Controversy Spaces. A model of scientific and philosophical change*, Amsterdam: John Benjamins, pp. 9–25.

Palmer, J. (2017). Quantum technology is beginning to come into its own. *The Economist*, 9 March.

Paty, M. (2022, this volume). Einstein's Criticisms of Quantum Mechanics.

Pera, M. (2000). Rhetoric and Scientific Controversies. In P. Machamer, M. Pera, and A. Baltas (eds.), *Scientific Controversies—Philosophical and Historical Perspectives*, New York: Oxford University Press, pp. 50–66.

Perillán, J. (2022, this volume). Early Solvay Councils: Rhetorical Lenses for Quantum Convergence and Divergence.

Popper, K. R. (1959). *Logic of Scientific Discovery*. New York: Basic Books.

Popper, K. R. (1982). *Quantum Theory and the Schism in Physics. Vol. III of the Postscript to the Logic of Scientific Discovery*, edited by W. W. Bartley III. London: Hutchinson.

Raynaud, D. (2015). *Scientific Controversies—A Socio-Historical Perspective on the Advancement of Science*. New Brunswick, NJ: Transaction Publishers.

Romano, L. (2020). Franco Selleri and his contribution to the debate on particle physics, foundations of quantum mechanics, and foundations of relativity theory. Ph.D. dissertation, University of Bari Aldo Moro.

Ryckman, T. (2022, this volume). Quantum Interpretations and 20th Century Philosophy of Science.

Schweber, S. S. (1986). The empiricist temper regnant—theoretical physics in the United States, 1920–1950. *Hist. Stud. Phys. Biol. Sci.*, **17**, 55–98.

Schweber, S. S. (2014). Writing the biography of Hans Bethe: contextual history and Paul Forman. *Phys. Perspect.*, **6**, 179–217.

Silva. I. (2022, this volume). Chien-Shiung Wu's Contributions to Experimental Philosophy.

Silva Neto, C. P. (2022, this volume). Instrumentation and the Foundations of Quantum Mechanics.

Stanford, K. (2016). Underdetermination of Scientific Theory. *Stanford Encyclopedia of Philosophy*, http://plato.stanford.edu/archives/spr2016/entries/scientific-underdetermination/

Suppe, F. (ed.) (1974). *The Structure of Scientific Theories*. Champaign, IL: The University of Illinois Press.

Tonietti, T. (2014). *And Yet It Is Heard—Musical, Multilingual and Multicultural History of the Mathematical Sciences*, 2 vols, Heidelberg: Birkhäuser.

Whitaker, A. (2016). *John Stewart Bell and Twentieth-Century Physics*. Oxford: Oxford University Press.

Whitaker, M. A. B. (2000). Theory and experiment in the foundations of quantum theory. *Prog Quantum Electron*, **24**, 1–106.

Wüthrich, A. (2016). Review of O. Freire, 'The Quantum Dissidents'. *Isis*, **107**(2), 446–7.

# CHAPTER 34

## ORTHODOXY AND HETERODOXY IN THE POST-WAR ERA

### KRISTIAN CAMILLERI

## 34.1 INTRODUCTION

THE period after World War II witnessed a revival of interest in the interpretation and foundations of quantum mechanics. Albert Einstein and Erwin Schrödinger re-emerged from their wartime hiatus to resume their criticisms of the theory. In 1948, in a special issue of *Dialectica*, Einstein set out what was perhaps the clearest expression of his philosophical objections to quantum mechanics and the concept of reality that he felt must underpin a complete theoretical description. Einstein was afforded the opportunity to again present his views in the volume for the Library of Living Philosophers the following year (Einstein, 1948, 1949). In 1952 Schrödinger wrote to Niels Bohr that he had 'decided to take a firm stand against' the 'current views in quantum mechanics', particularly 'Born's probability interpretation', which he had 'disliked from the first moment on' (Schrödinger to Bohr, 3 June 1952, BSC, 32.3, AHQP). Other members of the old guard joined the chorus of dissenting voices. After having abandoned his attempt to develop his deterministic pilot-wave theory some twenty-five years earlier, a rejuvenated Louis de Broglie now declared, 'I am convinced that the whole question must be reopened' (de Broglie, 1953, p. 135).

This revival of opposition was further strengthened by events in the Soviet Union. The beginning of the Cold War saw the intensification of the ideological campaign against the enemies of dialectical materialism, following Andrei Zhdanov's speech on 24 June 1947. The subsequent discussions over quantum mechanics at the 1947 Meeting of the Academy of Sciences precipitated a critical attack on the 'Copenhagen school' from a number of Soviet physicists, notably Dimitri Blokhintsev (1952, 1953) and Ya. P. Terletsky (1952). A new generation of quantum dissidents also emerged in the early years of the Cold War, intent on pursuing alternative interpretations. David Bohm's ground-breaking paper on hidden variables in 1952 sparked a revival of interest in the subject, particularly among Marxist physicists both in Brazil and France (Bohm, 1952; Bohm and Viger, 1954; Bohm, Schiller, and Tiomno, 1955; Bohm and Schiller, 1955). At the same time the Hungarian physicist Lajos Jánossy proposed a non-linear modification of the Schrödinger equation (Jánossy, 1952). In 1957, Hugh Everett III submitted his doctoral thesis on the 'relative-state' formulation of quantum mechanics, which would later be dubbed the 'many-worlds interpretation' (Everett, 1957).

In response, the founders of quantum mechanics launched a spirited counter-offensive. Werner Heisenberg, Wolfgang Pauli, Max Born, and Léon Rosenfeld were among the leading physicists to defend the 'orthodox view' against what they saw as fundamental misunderstandings and 'philosophical prejudices' of their critics (Heisenberg, 1955, 1958; Pauli, 1953; Born, 1953, [1953] 1956a, [1953] 1956b; Rosenfeld, [1953] 1979b, [1957] 1979a). It was in this context of debate that the myth of the 'Copenhagen interpretation' was born. When Heisenberg coined the term in 1955, there was in fact no widely shared interpretation of quantum mechanics. The various attempts to defend the 'Copenhagen interpretation' took the form of a series of retrospective reconstructions that often went beyond anything we can find in the writings of the late 1920s or 30s. Various interpretational commitments were appropriated, assembled, reinterpreted, and, in some cases, even revised. As Paul Feyerabend astutely observed, defenders of the Copenhagen viewpoint were often able 'to take care of objections by development rather than reformulation', thereby creating 'the impression that the correct answer has been there all the time and that it was overlooked by the critic' (Feyerabend, [1962] 2015b, p. 104). While historical narratives over the foundations of quantum mechanics have largely revolved around notions of orthodoxy and heterodoxy, little attention has been paid to the *diachronic* history of orthodoxy.

This chapter focuses on the post-war responses of physicists such as Heisenberg, Pauli, von Neumann, Born, Dirac, and Jordan, in an attempt to cast new light on the notion of orthodoxy that has structured much of our historical thinking about the foundations of quantum mechanics. As John Henderson explains, orthodoxy has a 'dynamical character' insofar as it 'is never made, fixed, or closed', but is continually transformed in an antagonistic relationship 'with its silent collaborator and public antagonist, heresy'. It is only in response to heresy, or what might be better termed 'protoheresy', that orthodoxies are framed and articulated. 'Heretical terms and concerns, if not heretical ideas, lay at the heart of orthodoxy' (Henderson, 1998, p. 39). Unlike religious orthodoxies, however, the Copenhagen orthodoxy was never officially codified, nor was it explicitly associated with a set of doctrines by a recognized authority. Henderson's remarks point to a more nuanced understanding of the dialectical history of orthodoxy and heterodoxy. It was largely in response to the challenges presented by Einstein, Bohm, and a new generation of physicists that many defenders of quantum mechanics framed their own distinctive positions. To this extent

I shall argue that the post-war orthodoxy was a dialectical response to the new challenges it faced in the early 1950s.

Yet this dialectic did not lead at any stage to a uniform 'orthodox' position. It was therefore never an orthodoxy in the true sense of the word. Far from creating a unitary Copenhagen interpretation, the post-war debates had the effect of dramatically expanding the range of interpretations that bore the label 'orthodox' or 'Copenhagen'. As early as 1961 Paul Feyerabend noted that while many physicists 'profess to follow either Heisenberg or Niels Bohr' or 'call themselves adherents of the Copenhagen point of view', it is difficult, if not impossible, to find 'some common element in their beliefs'. For this reason, Feyerabend went on to claim, 'there is no such thing as the "Copenhagen interpretation"' (Feyerabend, [1961] 2015a, pp. 74–5). This was evident in the wide range of 'orthodox views' on the measurement problem, the ontological or epistemological meaning of the formalism, and Bohr's doctrine of classical concepts. While Bohr's idea of complementarity is still commonly seen to have formed the central plank in a unified and widely shared 'orthodox view' of quantum mechanics, which emerged in the late 1920s, extensive historical scholarship over the past thirty years has challenged, if not seriously undermined, the notion that any such consensus among the founders of quantum mechanics ever existed. In some cases, these differences amounted to fundamentally contradictory positions (Howard, 2004; Camilleri, 2009).

It might be argued that whatever their differences on the finer shades of interpretation, the defenders of the 'Copenhagen interpretation' were united in the conviction that quantum mechanics was a complete theory, or at least as complete as it could possibly be. However, even this claim does not stand up to close historical scrutiny. While the commitment to completeness has traditionally served as a convenient way to demarcate orthodoxy from heterodoxy, such a clear-cut distinction obscures the subtle but important differences in the ways that physicists such as Bohm, Einstein, Heisenberg, Born, Pauli, Dirac, Jordan, and von Neumann understood completeness in the post-war era. As I shall argue, by the mid-1950s the battle lines between orthodoxy and heterodoxy would become increasingly blurred.

This raises an important question for the historiography of quantum physics. To what extent do the range of disparate views and their transformation over time threaten to dissolve the very notion of 'the orthodoxy' in quantum mechanics? In addressing this question, I suggest that focusing attention on the *satisfaction* that physicists expressed towards the standard view of quantum mechanics (and conversely, the dissatisfaction others expressed), rather than any particular interpretation or defence of it, offers a far more promising way to understand the battle lines that were drawn after the war. However, satisfaction was not an all or nothing affair. While the debates have often been presented as a conflict between two diametrically opposed viewpoints, epitomized in the epic clashes between Bohr and Einstein, in reality we find a spectrum of attitudes ranging from a deep conviction that the new theory represented the final word, to cautious optimism, to tentative scepticism, to uneasiness and outright rejection. Orthodoxy and heterodoxy, on this view, constitute two ends of a continuous spectrum.

Yet, even this fails to capture the subtle dynamics of orthodoxy and heterodoxy. Some physicists did privately express concerns about quantum mechanics in the 1930s, but were not prepared to make these public. Given the variety of views that circulated, it might be the case that we would be better served by speaking of ortho*praxis*, rather than ortho*doxy*. Framing things in this way suggests that what we have traditionally understood as the 'orthodoxy' is not to be found in a specific set of interpretive commitments or even the completeness of quantum mechanics. Instead, it is best understood as a *resistance* on the part of physicists to any attempt to pursue alternative formulations of quantum mechanics. This resistance in some cases extended to an outright denial of the very possibility of alternative views, but in less extreme cases, it was expressed as mere indifference to foundational questions. In some instances, physicists were content to adopt a pragmatic stance, though they privately nurtured the belief that quantum mechanics was only a provisional theory.

## 34.2 Responses to Bohm: Pauli, Heisenberg, von Neumann

Bohm's papers on hidden variables in 1952 marked a significant turning point in the history of debates over the interpretation of quantum mechanics. Bohm showed, contrary to all expectations, that it was indeed possible to reformulate the theory of quantum mechanics as a deterministic theory, in which particles like electrons moved in continuous trajectories in space. Moreover, such a theory was in complete agreement with the empirical predictions of standard quantum mechanics. This is something that von Neumann, Heisenberg, and Pauli had all deemed impossible in the 1930s. Indeed, all three had devised arguments in support of this view. While Bohm's papers sparked a revival of interest in hidden variables in the 1950s, notably from Jean-Pierre Vigier and de Broglie, they were received with 'a conspiracy of silence'. Max Dresden recalled that Bohm's papers were regarded as 'juvenile deviationism', not worth wasting one's time over.[1] At the Colston symposium in Bristol in 1957, Fritz Bopp expressed a view shared by many physicists: 'Bohm's theory cannot be refuted', but at the same time, 'we don't believe in it' (Körner, 1957, p. 51).

Yet, some physicists did respond to Bohm. While the appearance of his hidden variables theory did not shake their faith in the standard quantum mechanics, it did force several leading physicists to rethink their basic philosophical commitments. Prior to the appearance of Bohm's papers, it had generally been believed that hidden variables theories were impossible. The most famous argument for this conclusion

---

[1] According to David Peat, after Max Dresden presented a seminar on Bohm's work in Princeton, Oppenheimer stood up and declared, 'If we cannot disprove Bohm, then we must agree to ignore him' (Peat, 1997, p. 133).

was of course von Neumann's impossibility proof of 1932. But Pauli and Heisenberg had also defended this view, basing their conclusions on somewhat different arguments in the 1930s (Bacciagaluppi and Crull, 2009). In a lecture delivered to the Philosophical Society in Zurich in November 1934, Pauli declared that 'no supplementation of the assertions of quantum mechanics by other assertions in the sense of determinacy is possible' without departing from the statistical predictions of quantum mechanics (Pauli, [1936] 1994a, p. 102). Yet, in the early 1950s, Pauli was forced to revise this view. After initially dismissing Bohm's work, on closer inspection he was forced to admit 'the consistency of the causal interpretation' (Bohm to Phillips, December 1951, in Talbot, 2017, p. 143). Writing to Bohm on receiving a revised manuscript of his paper, Pauli stated: 'I do not see any longer the possibility of any logical contradiction as long as your results agree completely with those of the usual wave mechanics and as long as no means is given to measure the values of your hidden parameters' (Pauli to Bohm, 3 December 1951, in von Meyenn, 1996, pp. 436–441).

However, Pauli now raised new objections. While Bohm had managed to circumvent von Neumann's proof, his theory no longer treated position and momentum on an equal footing. In the standard formulation of quantum mechanics, the wave function in position-space is treated on a par with the wave function in momentum-space, but in Bohm's theory, this was not the case. According to Bohm's theory, any attempt to measure the particle's momentum would invariably affect the measured value. It therefore could not be treated as an observable of the theory in the same sense as position. Pauli described this *'artificial asymmetry'* between position and momentum in Bohm's theory as a form of 'artificial metaphysics'. To this end, Pauli argued, 'the interpretation of quantum mechanics based on the idea of complementarity' was 'the only admissible one' (Pauli, 1953, p. 42).

A full account of the objections levelled at Bohm's theory and its general reception among physicists in the 1950s would take us beyond the scope of this chapter. However, it is worth noting that Heisenberg took a similar approach in denouncing Bohm's hidden variables theory, both in his contribution to the *Festschrift* for Bohr's seventieth birthday and his 1955–56 Gifford lectures published as *Physics and Philosophy*. While conceding that 'Bohm's interpretation cannot be refuted by experiment', he dismissed the incorporation of particle trajectories into quantum mechanics as 'a superfluous ideological superstructure', which in his view, had only served to obscure the true nature of the theory. But it was the lack of symmetry that Heisenberg highlighted as the most serious weakness of Bohm's interpretation:

> Bohm's language destroys the symmetry between position and velocity which is implicit in quantum theory; for the measurements of position Bohm adopts the usual interpretation; for measurements of velocity or momentum, he rejects it. Since the symmetry properties always constitute the most essential features of a theory; it is difficult to see what could be gained by omitting them in the corresponding language. Therefore, one cannot consider Bohm's counterproposal to the Copenhagen interpretation as an improvement.   (Heisenberg, 1958, p. 118)

Some twenty years earlier Heisenberg had argued that 'a deterministic completion of quantum mechanics is *impossible*' (von Meyenn, 1985, p. 410).[2] He now argued that any such a reformulation of the theory would *not* constitute an *improvement*. Here Heisenberg stressed that all attempts thus far to formulate alternative interpretations of quantum mechanics 'have found themselves compelled to sacrifice the symmetry properties of quantum theory'. This, for Heisenberg, was evidently too high a price to be paid for the retention of the space-time causal description of the electron's motion. To this end, Heisenberg concluded, 'the Copenhagen interpretation cannot be avoided *if these symmetry properties*—like Lorentz invariance in the theory of relativity—are held to be *a genuine feature of nature*; and every experiment yet performed supports this view' (Heisenberg, 1958, p. 128).

Whether the invariance under transformation of position to momentum can be regarded a physical symmetry in the same sense as Lorentz invariance is highly debatable (Myrvold, 2003, pp. 17–8). But from a historical point of view, this should be seen as a significant departure from the earlier defence of quantum mechanics in the 1930s. In an ironic twist, Heisenberg now appealed to an *aesthetic-realist* argument. While Heisenberg conceded that Bohm's theory could successfully reproduce all the empirical predictions of standard quantum mechanics, it failed to exhibit the all-important symmetry properties, which, for Heisenberg, constituted 'a genuine feature of nature'. Von Weizsäcker recalled that Bohm's papers were discussed in a seminar course organized by Heisenberg in the winter term 1953/54. While the 'course strengthened our conviction that all these attempts were false, we could not hide the fact that the deepest reason for our conviction was a quasi-"aesthetic" one. Quantum mechanics surpassed all competitors by its simple beauty that characterizes a complete theory' (von Weizsäcker, 1985, p. 321).

This argument must be seen in historical context. As the physicist Louis Michel later recalled, 'there was an irresistible ascent of the role of symmetries in the fifties... Most of us then shared the enthusiasm that Heisenberg had at that time' (Michel, 1989, p. 377). In his Nobel lecture in December 1957, Chen Ying Yang declared, 'it [is] scarcely possible to overemphasize the role of symmetry principles in quantum mechanics'. When confronted with 'the elegance and beautiful perfection of the mathematical reasoning involved' together with its 'complex and far-reaching consequences', one could not fail to develop 'a deep sense of respect for the power of symmetry laws' (Yang, 1958, p. 565). As Arianna Borrelli has argued, the aesthetic appeal of symmetry principles was, for many physicists, a sign that 'symmetric theories have a better chance of reflecting inner principles of nature' (Borrelli, 2017, p. 22). These realist associations were evident in a letter Heisenberg wrote to his sister-in-law Edith Kuby in 1958, about the 'incredible degree of simplicity' of the symmetries that guided his unified theory of

---

[2] The argument was contained in an unpublished manuscript entitled 'Ist eine deterministische Ergänzung der Qauntenmechanik Möglich?', which Heisenberg sent to Pauli on 2 July 1935 (von Meyenn, 1985, pp. 409–18). The original manuscript is reproduced in the Archive for the History of Quantum Physics (microfilm 45, section 11).

elementary particles. 'Not even Plato could have believed them so beautiful. For these interrelationships cannot be invented. They have been there since the creation of the world' (E. Heisenberg, 1984, p. 144). The mathematical structure of quantum mechanics, Heisenberg later declared, 'must be part of reality itself, not just our thoughts about reality' (Heisenberg, 1971, p. 68). These were hardly the utterances of a committed positivist.

We can also discern a shift in von Neumann's thinking during this time, though he adopted a more pragmatic stance. In a letter to Pauli in October 1952 Bohm reported, 'von Neumann has agreed that my interpretation is logically consistent and leads to all results of the usual interpretation' (von Meyenn, 1996, p. 392). Indeed Bohm's sources informed him that von Neumann had even thought it quite 'elegant' (Bohm to Yevick, 16 February 1952, in Talbot, 2017, p. 247). This was quite a concession from the man who twenty years earlier had 'proved' that quantum mechanics was 'in compelling logical contradiction with causality' (von Neumann [1955] 1961b, p. 327). Indeed in two important papers that appeared in 1954 and 1955, von Neumann made no mention of his proof. Instead he now claimed that the 'best description one can give today' is that we 'do not have complete determination, and that the state of the system does not determine at all what it will be immediately afterwards or later'. But, he now acknowledged, this 'may not be the ultimate one (the ultimate one may even revert to the causal form, although most physicists don't think this is likely)' (von Neumann, [1954] 1961a, p. 486). While the 'prevalent taste' up until now been 'in favour of one of the two interpretations, namely the statistical one', he pointed out that 'there have been in the last few years some interesting attempts to revive the other [causal] interpretation' (von Neumann, [1955] 1961b, p. 497). In 1932, von Neumann had argued that 'quantum mechanics would have to be objectively false, in order that another description of the elementary processes than the statistical one be possible' (von Neumann, 1955, p. 325).[3] By 1955, he saw quantum mechanics as 'an example where alternative interpretations of the same theory are possible' (von Neumann, [1955] 1961b, p. 496; See also Stöltzner, 1999, p. 257).

This raises the intriguing question of why the English translation of von Neumann's *Mathematische Grundlagen der Quantenmechanik*, published in 1955 some three years *after* von Neumann had come to learn of Bohm's theory, included the famous proof he had in formulated in the original German edition in 1932. The answer becomes clear once we realize that the translation was actually completed in October 1949, before Bohm had begun working on hidden variables theory. In a letter to the president of Dover Publications on 3 October 1949, von Neumann explained that he had already

---

[3] There has been much recent debate about how exactly von Neumann understood the significance of his theorem for the question of hidden variables. Jeffrey Bub and Dennis Dieks have defended the proof against the claim that it is "silly" (Bub, 2010, 2011; Dieks, 2017), while David Mermin and Rüdiger Schack have argued that it contains a major oversight (Mermin and Schack, 2018). See the chapters in this volume on 'Hidden Variables' (Bub, Chapter 39) and on the development of the statistical interpretation.

completed a revised translation, having deemed it necessary 'to rewrite Dr Beyer's translation'. The task had taken him about six months in all, and according to his letter, the task had been completed in May 1949 (Neumann to Cirker, 3 October 1949, in Rédei, 2005, pp. 91–2). Owing to problems with copyrights (which were vested by the United States during the war) and finding adequate mathematical types, Dover gave up on the book due to lengthy delays. It would be six years before Princeton University Press would publish the English translation.

We may surmise that by the early 1950s, von Neumann's views had shifted. Eugene Wigner later recalled that von Neumann had always given him the impression that his belief in 'the inadequacy of hidden variables theories' was *not* in fact based on the reasoning in his celebrated proof (Wigner, 1970, p. 1009). The 'true reason for his conviction of the inadequacy of the theories of hidden variables' was that 'all schemes of hidden parameters which either von Neumann himself, or anyone else whom he knew, could think of,... had some feature which made it unattractive, in fact unreasonable' (Wigner, 1971, pp. 1097–8). This seems to have been a widely shared view among physicists in the 1930s. While references to the proof were commonplace, the proof itself was seldom examined in any detail.

For von Neumann, the pressing question now was not completeness but theory choice. The 'reason for preferring one version of quantum theory' over the other, he argued, is which interpretation gives 'better heuristic guidance in extending the theory into those areas which are not yet properly explained'. While acknowledging that 'physicists certainly had definite *subjective preferences* for one description or the other', in the end, von Neumann felt the judgment of the scientific community would depend not on philosophical arguments, but on 'which succeeds in pointing the way to explaining wider areas with greater [explanatory] power' (Neumann, [1955] 1961b, pp. 497–8). Here it is worth quoting von Neumann in full:

> while there appears to be a serious philosophical controversy... it is quite likely that the controversy will be settled in quite an unphilosophical way. The decision is likely to be opportunistic in the end. The theory that lends itself better to formalistic extension towards valid new theories will overcome the other, no matter what our preference up to that point might have been. It must be emphasized that this is not a question of accepting the correct theory and rejecting the false one. It is a matter of accepting that theory which shows greater formal adaptability for a correct extension. This is a formalistic, aesthetic criterion, with a highly opportunistic flavour.
> (Neumann, [1955] 1961b, p. 498)

In this remarkable passage, von Neumann argued that heuristic considerations, not matters of philosophical principle, would eventually decide between competing interpretations of quantum mechanics. Bohm's correspondence from the time suggests that while von Neumann was convinced of the logical consistency of his theory, he was sceptical about whether it could effectively deal with spin (Bohm to Phillips, early 1952, in Talbot, 2017, p. 147). It is worth noting that 'in spite of all its successes', von

Neumann felt that standard quantum mechanics had not yet led to a satisfactory theory of quantum electrodynamics, nor to a quantum theory of elementary particles. 'About these', he said in 1955, 'we know a great deal less than about the original quantum mechanics, and we are here in the midst of grave difficulties' (von Neumann, [1955] 1961b, p. 498). Here von Neumann left the door ajar for determinism, but most physicists remained sceptical of the heuristic value of Bohm's theory.[4]

# 34.3 Is Quantum Mechanics Final? Born and Dirac

If Bohm's papers persuaded some physicists that that a deterministic completion of quantum mechanics was at least *possible*, it was not a solution that appealed to the older generation of quantum dissidents. Neither Einstein nor Schrödinger responded with much enthusiasm to Bohm's work. Einstein dismissed it as 'too cheap', while Schrödinger was not much impressed either (Einstein to Born, 12 May 1952, in Born, 1971, p. 192).[5] While it is true Einstein did briefly toy with the idea of hidden variables, and gave his public endorsement to de Broglie's search for pilot-wave theory at the 1927 Solvay conference, it was not a path he found promising and certainly not one he followed himself in his later years. This shows that the claim that quantum mechanics was an 'incomplete theory' was understood in very different ways—both by its critics and apologists.

After the war, Einstein attempted to clarify the sense in which he understood quantum mechanics to be an incomplete theory of individual process. A 'more complete theory' could not be achieved simply by carrying out a *completion* of the existing theory of quantum mechanics. Responding to the suggestion that it might be possible to devise a hidden variables theory, Einstein told one correspondent, 'I do not think that one can arrive at a description of the individual system through a simple completion of the present statistical quantum theory'. He was adamant, 'it is not possible to get rid of the statistical character of the present quantum theory by merely adding something to the latter' (Einstein to Kupperman, 14 November 1953, in Fine, 1993, p. 269). As he explained in a letter in 1954:

---

[4] At a conference, *New Research Techniques in Physics*, in Rio de Janeiro and São Paulo, on 15–29 July 1952, I. I. Rabi also remarked: 'I do not see how the causal interpretation gives us any line to work on other than the use of the concepts of quantum theory' (Freire Jr, 2015, p. 40).

[5] Einstein presented a detailed criticism of Bohm's theory in his contribution to the *Born Festschrift* (Einstein, 1953). In an exchange of letters between Einstein and Schrödinger in January/February 1953, this criticism was discussed. Schrödinger also contributed a criticism of his own in this correspondence, which Einstein did not find compelling (AHQP, 37, 5–12, 13, 14 and 15). In a letter to Miriam Yevick, Bohm related his disappointment that Schrödinger had objected to Bohm's theory, on the grounds that the transformation theory was the real core of quantum mechanics (Bohm to Yevick, 16 February 1952, in Talbot, 2017, p. 247).

The present quantum theory is in a certain sense a magnificent, self-contained system that, at least in my opinion, *cannot be made into an individual-theory by supplementing it*, e.g. any more, e.g., than Newtonian gravitational theory can be made into general relativity by supplementation. Somehow one must start from scratch, hard though that obviously is.
(Einstein to Hosemann, 9 August 1954, in Fine, 1993, pp. 269–70)

Einstein's position here was actually close to the one that von Neumann, Pauli, and Heisenberg had defended in the 1930s. Quantum mechanics was deemed a 'self-contained' or 'closed theory', which was no longer susceptible to modification.[6] What was urgently needed, in Einstein's view, was not so much a 'reinterpretation' or a 'completion' of the existing theory, as the construction of an entirely new one. This is where he parted company with Pauli and Heisenberg. Here Einstein was prepared to entertain the possibility that one might have to modify the concepts of space and time in the construction of a new theory, though he admitted, 'I have not the slightest idea what kind of elementary concepts could be used in such a theory' (Einstein to Bohm, 28 October 1954, in Fine, 1993, p. 270).

Writing to Michele Besso in 1952, Einstein conceded that quantum mechanics is 'the *most complete possible theory* compatible with experience, as long as one bases the description on the concepts of the material point and potential energy as fundamental concepts' (quoted in Stachel, 1986, p. 375). Of course, for Bohr, the hope that it might be possible to replace 'the concepts of classical physics by new conceptual forms' rested on a fundamental misunderstanding (Bohr, [1929] 1987, p. 16). Bohr remained convinced, on philosophical grounds, that 'the language of Newton and Maxwell will remain the language of physics for all time' (Bohr, 1931, p. 692). In this sense, quantum mechanics was as complete as it would ever be.

Yet, this was a view of completeness that few of Bohr's contemporaries shared. The question of whether quantum mechanics would ultimately be superseded by some deeper ontological theory was not one that many physicists felt could be answered categorically, or with any degree of certainty. While Max Born dismissed Schrödinger's

---

[6]  Heisenberg developed his notion of a 'closed theory' in later years (Heisenberg, [1948] 1974). As Alisa Bokulich explains, for Heisenberg, 'a closed theory is a tightly knit system of axioms, definitions, and laws that provides a perfectly accurate and final description of a certain limited domain of phenomena' (Bokulich, 2006, p. 91). At a 'Discussion on Determinism and Indeterminism' at the 1965 International Colloquium *Science and Synthesis*, Heisenberg attempted to clarify the sense in which he took quantum mechanics to be complete. 'Let us first speak about the old Newtonian mechanics. Is Newtonian mechanics complete, is it a closed scheme or is it not? I would say—and this may seem to you paradoxical—*it is a complete theory*, and it is absolutely impossible to improve it, in the following sense. If you can describe parts of nature with those concepts which are applied in Newtonian mechanics—namely coordinates, velocities, masses, and so on—then the equations of Newton are exact equations and every attempt to improve these equations is simply nonsense. But of course there are other parts of nature in which these concepts do not apply—this is already so in relativity, was already so in Maxwell's theory, where we had the concept of a field, and is certainly true in quantum mechanics, and so on. And in the same sense, too, I feel that quantum mechanics is complete' (Maheu *et al.*, 1971, pp. 144–45).

suggestion to develop a fully consistent wave theory as 'impracticable', he was careful to qualify his stance. '*I do not want to create the impression that I believe the present interpretation of quantum theory to be final*' (Born, [1953] 1956b, p. 131). In a paper expanding on his remarks at a discussion in December 1952, devoted to Schrödinger's recent papers on quantum jumps, Born took the opportunity to again clarify his position:

> I am far from saying that the present interpretation is perfect and final. I welcome Schrodinger's attack against the complacency of many physicists who are accepting the current interpretation because it works, without worrying about the soundness of the foundations. Yet I do not think Schrödinger has made a positive contribution to the philosophical problems.    (Born, [1953] 1956a, p. 149)

In his Waynflete Lectures in Oxford in 1949, Born had made much the same point:

> It would be silly and arrogant to deny any possibility of a return to determinism. For *no physical theory is that final*; new experiences may force us to alternatives and even revisions . . . *I expect that our present theory will be profoundly modified* . . . But I should never expect that these difficulties could be solved by a return to classical concepts. I expect just the opposite, that we shall have to sacrifice some current ideas and use still more abstract methods.    (Born, 1964, p. 109)

Far from prohibiting any possible change to the theory, Born left this an open question. Indeed, as early as 1929, Born had conceded in a letter to Einstein: 'the possible future acceptance or rejection of determinism *cannot be logically justified*. For there can always be an interpretation which lies one layer deeper than the one we know' (Born to Einstein, 13 January 1929, in Born, 1971, p. 103).[7] Born did however insist that 'if a future theory should be deterministic, it *cannot* be a modification of the present one, but must be essentially different'. In saying as much, Born was actually closer to Einstein than Bohr. Just how one could rebuild quantum theory anew, 'without

---

[7] This remark was actually a response to Einstein's criticism of the view Born had expressed in his lecture 'Über den Sinn der physikalischen Theorien', which he had presented at the public session of the *Gesellschaft der Wissenschaften zu Göttingen* on 10 November 1928 (Born, 1929). There Born had addressed 'the question of whether in the future, through extension or refinement', quantum mechanics 'might not be made deterministic again'. In answering this question, Born argued: 'it can be shown *in a mathematically exact way* that the established formalism of quantum mechanics allows for no such completion. If thus one wants to retain the hope that determinism will return someday, then one must consider the present theory to be contentually *false*; specific statements of this theory would have to be refuted experimentally. Therefore, in order to convert the adherents of the statistical theory, the determinist should not protest but rather test' (I have used the English translation in Crull and Bacciagaluppi, 2017, p. 223). In the letter to Einstein, Born wrote that he and Jordan were 'very grateful for your criticism . . . You are, of course, right that an assertion about the possible future acceptance or rejection of determinism cannot be logically justified' and 'of course we should not claim anything for which we have no rigorous proof' (Born to Einstein, 13 January 1929, in Born, 1971, p. 103).

sacrificing a whole treasure of well-established results', Born was happy to 'leave to the determinists to worry about' (Born, 1964, p. 109).

The issue that divided Born and Einstein in the 1940s and 1950s was thus not whether quantum mechanics was complete or not, but whether this mattered. As Born put it, in a letter to Einstein in 1949, 'I am inclined to make use of the formalism, and even to "believe" it in a certain sense, until something decidedly "better" turns up' (Born to Einstein, 9 May 1948, in Born, 1971, p. 175). Recognition of this point raises further questions about what in fact was the dominant view. Even if it was the view of most physicists that quantum mechanics in its current form was perfectly *satisfactory*, this did not necessarily imply a belief in its *finality*. Most physicists were content to use quantum mechanics, without concerning themselves with such questions. They simply did not share Einstein's *anxieties* about quantum mechanics. Yet, such a lack of anxiety should not be confused with a belief in completeness.

The debates over quantum mechanics brought to light questions about the future of physics. As Schrödinger put it, any attempt to draw philosophical conclusions 'from a "supposedly final" physical theory' such as quantum mechanics, as Bohr is wont to do, 'is highly suspect', for the simple reason that 'it is in the nature of any physical theory *not* to be final' (Schrödinger to Bertotti, 24 January 1960, in Bertotti, 1985, p. 85). To judge from the writings after the war, a physicist's attitude to such questions often reflected a mixture of homespun philosophy and idiosyncratic views about the nature of scientific progress. In a lecture entitled 'Phenomenon and Physical Reality', presented at the International Congress of Philosophers in Zurich in 1954, Pauli attempted to clarify the sense in which he understood quantum mechanics to be 'final'.

> The question is never: will the present theory remain as it is or not? It is always *in what* direction will it change? The answer to these invariably controversial questions can never be more than conjecture, even after all the circumstances have been weighed, among which the mathematical and logical structure of the known laws plays at least as great a part as empirical results.    (Pauli, [1957] 1994b, p. 134)

Here we find none of the earlier rhetoric of 'proof'. Nevertheless, Pauli remained adamant that whatever surprises the future of physics held, Bohr's principle of complementarity would not be eliminated. In a letter to Born, Pauli wrote: 'I am certain that the statistical character of the $\psi$-function, and thus of the laws of nature, will determine the style of the laws for at least some centuries', though Born was more equivocal (Born, 1953a, p. 150). Even here Pauli did not categorically rule out the possibility of a new kind of physics at some point in the very distant future. In discussions on this question that took place at the Colston conference in 1957, Fritz Bopp remarked, 'what we have done today was predicting the possible development of physics—we were not doing physics but metaphysics' (Körner, 1957, p. 51).

Some physicists, however, did hazard a guess as to the direction that physics might take in the future. The problems of relativistic quantum electrodynamics, to which von Neumann alluded, were generally not regarded as bearing on the completeness of quantum mechanics. But some physicists did take this view. In June 1936, Dirac wrote to Bohr arguing that 'the beauty and self-consistency of the present scheme of quantum mechanics' did *not* preclude the possibility 'of a still more beautiful scheme, in which, perhaps, the conservation laws play an entirely different role' (Dirac to Bohr, 9 June 1936, BSC, 18, AHQP). Several years later, Dirac made clear the techniques of renormalization developed in the 1940s had not altered his views on the finality of quantum mechanics:

> It seems clear that the present quantum mechanics is not in its final form. Some further changes will be needed, just about as drastic as the changes made in passing from Bohr's orbit theory to quantum mechanics ... It might very well be that the new quantum mechanics will have determinism in the way Einstein wanted ... I think it is very likely, or at any rate quite possible, that in the long run Einstein will turn out to be correct, even though for the time being physicists have to accept the Bohr probability interpretation.   (Dirac, 1982, pp. 85–6)

This was part of Dirac's more general views concerning the nature of scientific progress. In a lecture at the Canadian mathematical congress in 1949, Dirac asserted that the basic structure of quantum mechanics was 'almost certain to change with future development'. As he explained, it is 'a general feature in the progress of science that however good any theory may be, we must always be prepared to have it superseded later on by a still better theory' (Dirac, 1951, p. 11). In sharp contrast to the views expressed by Heisenberg and Pauli, in 1963 Dirac would claim: 'I think one can make a safe guess that [the] uncertainty relations in their present form will *not* survive in the physics of the future' (Dirac, 1963, p. 49). On most accounts, these views would qualify Dirac as an *opponent* of the Copenhagen orthodoxy. Yet, historians and philosophers have been reluctant, for reasons that are not altogether clear, to locate Dirac in the heterodox camp.

## 34.4 PHILOSOPHICAL ANXIETIES OVER QUANTUM MECHANICS: JORDAN AND WIGNER

In some cases, we can discern signs of discontent over quantum mechanics in the founding fathers after the war. Perhaps the most striking example of this can be seen in a little known paper presented for a symposium on the philosophical foundations of quantum theory in 1949 by Pascual Jordan. Jordan took the opportunity to reflect more deeply on the problem of measurement in quantum mechanics. In his earlier book *Anschauliche Quantentheorie*, published in 1936, Jordan had given a fairly standard account of the 'orthodox view' of measurement, in arguing that 'the *act of observation* is what first *creates* the definiteness' in an observed quantity. But he offered no clues as to what exactly occurred during this mysterious 'act of observation' (Jordan, 1936, p. 308). This has generally been considered Jordan's last word on the matter. Yet in his 1949 paper, Jordan admitted, 'there remain some questions about the process of observation itself—questions for which we do not get unambiguous answers because orthodox quantum mechanics treats the concept of "measurement" as a fundamental one which ought not to be analysed'. Here Jordan stressed that, contrary to the impression von Neumann had left, the act of observation 'must *not* be interpreted as any mental process, but as a purely physical one' (Jordan, 1949, pp. 269–70).

The measurement problem was the subject of vigorous debate among physicists in the 1950s and 60s. In a report on de Broglie's book *La Théorie de la Mesure en Mécanique Ondulatoire* written in 1957, Léon Rosenfeld took the opportunity to respond.to de Broglie's charge of subjectivism, which he saw as typical of the misguided efforts of a number of physicists in recent years in attacking 'what they believed to be the "orthodox" theory of measurement'. In reality, physicists had simply taken 'a distorted and largely irrelevant rendering of Bohr's argument by v. Neumann'. Here Rosenfeld lamented that von Neumann's work, 'though excellent in other respects, has contributed by its unfortunate presentation of the question of measurement in quantum theory to create unnecessary confusion and raise spurious problems'. Here Rosenfeld bemoaned, 'there is not a single textbook of quantum mechanics in any language in which the principles of this fundamental discipline are adequately treated, with proper consideration of the role of measurements to define the use of classical concepts in the quantal description' (LRP, Box 4 Epistemology, correspondance générale, NBA). The failure of most textbooks on quantum mechanics to deal adequately with the foundational questions was, in Rosenfeld's view, partly to blame for persistent misunderstandings, in particular the suggestion that the consciousness of the observer might play a crucial role in the collapse of the wave function.

While most physicists felt these were issues that had been dealt with adequately in the 1930s, Jordan argued that the measurement problem had *not yet* found a satisfactory resolution, either in Bohr's philosophical writings or in von Neumann's formal treatment. Jordan's anxieties over the measurement problem preceded the wave of criticisms that appeared in the early 1950s. Prior to this point, one is hard pressed to find an orthodox physicist acknowledging that measurement posed a serious problem for quantum mechanics. After completing a thorough analysis of the process of observation involving the absorption and emission of photons by atoms and in experiments concerning the polarization of photons in quantum theory, Jordan could not see how one could avoid the assumption that 'a new axiom or a new physical supposition—not already contained in the Schrödinger equation—is involved':

> Therefore I conclude ... that the notion of 'decision', 'quantum jump' or some other concept *not contained in the Schrödinger equation* is indeed necessary and unavoidable. It is then apparent that the situation—though it is clear in certain respects—does not allow a complete and final analysis; there remain open certain questions ... It seems to me that entirely new conceptions are necessary ... perhaps

the real problem is to synthesize the two fundamental notions of quantum mechanics [waves and probabilities] and unite quantum mechanics still more intimately with thermodynamics. Unable to do so myself, I should like to emphasize the urgency of further thought upon these questions.   (Jordan, 1949, pp. 275, 277)

One might read these remarks as consistent with later attempts to modify the dynamics of the Schrödinger equation. In emphasizing 'the urgency of further thought' and the necessity of 'new conceptions', Jordan here seems to have verged dangerously close to what many would regard as outright heterodoxy. However, exactly in what sense Jordan saw the orthodox formulation of quantum mechanics as 'incomplete' is difficult to say. Jordan suggested that treating entropy as a fundamental quantum concept might serve as 'a point where in the future *some generalization of the present theory might start*' (Jordan, 1949, p. 278). But these ideas were never pursued in systematic fashion. Jordan was among the first 'orthodox' physicists to publicly admit that the measurement problem in quantum mechanics constituted a *genuine problem*—and one that in his view was in urgent need of solution.

Though a number of physicists attempted to develop a quantum theory of measurement based on thermodynamic considerations in the 1950s and 60s, no consensus on measurement was ever reached. In the early 1960s Eugene Wigner publicly defended what he took to be the 'orthodox view', according to which 'it was not possible to formulate the laws of quantum mechanics in a fully consistent way without reference to the consciousness' of the observer (Wigner, [1961] 1983a, p. 169; Wigner, [1963] 1983b). By the 1970s Wigner remained open to the possibility of a fundamental revision of quantum mechanics, and he became increasingly convinced that 'far more fundamental changes will be necessary' (Wigner to Shimony, 12 October 1977, in Freire Jr, 2015, p. 167). He encouraged other physicists to pursue a range of alternative solutions to the measurement problem in the 1960s and 70s, and in doing so, 'helped to legitimize heterodoxy on this subject' (Freire Jr, 2015, p. 167).

Wigner's growing interest in foundational questions in the decades after the war might be portrayed as a gradual conversion from orthodoxy to heterodoxy. However, Wigner claimed he had *always* been troubled by certain aspects of quantum mechanics, and had several discussions with von Neumann on these questions over the years. In an interview with Kuhn in 1963, he explained 'I presented many puzzles to Johnny [von Neumann], which are still not solved and which still bother me on the theory of measurement and interpretation'. While Wigner felt 'there is some mystery here not completely cleared up', he was reluctant to make his views public. As he explained, 'during Johnny's lifetime I somehow did not want to write any paper on this. I don't know why not. As a matter of fact, I did write one, but I felt—well, I don't know' (Interview with Kuhn, 3 December 1963, AHQP). We can only speculate as to why Wigner was reluctant to publish earlier. But this suggests that theoretical physics was a cultural practice with its own socially accepted norms and conventions. An appreciation of this point serves to further complicate the standard historical narrative of orthodoxy and heterodoxy.

## 34.5 RETHINKING ORTHODOXY
## AND HETERODOXY

In what sense then can we speak of the 'orthodox view' of quantum mechanics? If Bohr, Born, Heisenberg, Jordan, Dirac, von Neumann, and Wigner offered such different, and in some cases conflicting, views on quantum mechanics, what entitles us to classify them as 'orthodox'? Scholars have typically attempted to answer this question by identifying a set of 'common commitments' that characterize the orthodox view. Yet such reconstructions are deeply problematic in failing to capture the wide variety of philosophical views held by physicists who professed to defend the Copenhagen or orthodox view (Jammer, 1974, p. 87). Both Dirac and Born expressed doubts over whether quantum mechanics was a complete or final theory. And both explicitly raised the possibility of a return to determinism at some time in the future. This suggests that what really divided Dirac and Born from Einstein and Schrödinger was not the issue of 'completeness', but rather that the former regarded the statistical formulation of quantum mechanics as a perfectly *satisfactory* theory, while the latter did not.

As Dirac put it, if we can find an interpretation of quantum mechanics 'that is satisfying to our philosophical ideas, we can count ourselves lucky. But if we cannot find such a way, *it is nothing to be really disturbed about*. We simply have to take into account that we are in a transitional stage'. In short, Dirac saw little reason to be 'disturbed' or 'bothered' with such philosophical problems, 'because they are difficulties that refer to the present stage in the development of our physical picture and are almost certain to change with future development' (Dirac, 1963, pp. 48–9). Here physicists tended to take a long-term historical view. Born too conceded it was *possible*, and even likely, that in time quantum mechanics would be superseded by a better theory. But he saw no reason to be dissatisfied with the current interpretation. By contrast, Schrödinger made no secret of the fact that he had always 'disliked the probability interpretation of wave mechanics'. In the absence of a viable alternative, he reluctantly conceded 'one had to give up opposing it and to *accept it* as an expedient interim solution' (Schrödinger, 1953, p. 20). This highlights the sense in which the 'acceptance of a theory' is by no means straightforward. Casting the attitudes of physicists in terms of simple binaries like 'acceptance' or 'rejection', as Robert Westman has argued, 'all too often masks interesting differences in the meaning of "acceptance"' (Westman, 1975, p. 165). Thus, while we might say that the vast majority of physicists 'accepted' quantum mechanics, insofar as they continued to work with the theory, this in fact tells us little about their views on completeness.

One might then argue that what really separates the 'orthodox' Born from the 'heterodox' Schrödinger was not whether or not they held quantum mechanics to be a complete theory, or even how they interpreted that theory in any deep philosophical sense, but the extent to which they saw quantum mechanics as a *satisfactory* theory. Born can be regarded as orthodox, not because of his adherence to a prescribed set of widely shared ontological or epistemological commitments, but because he was favourably disposed to the statistical formulation of quantum mechanics. On the other hand, Schrödinger's begrudging acceptance of quantum mechanics as a provisional expedient reflected his dissatisfaction. This makes the orthodox–heterodox divide less about a set of objective criteria and more about subjective and personal attitudes based on individual epistemic criteria for a physical theory. These judgments were often based on idiosyncratic views about the aim and structure of physical theory, and thus went beyond the specific question of how to interpret quantum mechanics.

But of course, satisfaction was not an all or nothing affair, and often could be expressed to varying degrees. To this extent, it is perhaps more helpful to see orthodoxy and heterodoxy, not as two polarized attitudes, but as two ends of a continuous spectrum. One could therefore 'accept' the theory either enthusiastically or reluctantly, with many shades of grey in between. In canvassing the possibility of a return to determinism, Dirac and Born both expressed views that might be regarded as heterodox, but at no stage they did they voice a deep sense of dissatisfaction with quantum mechanics. Jordan, on the other hand, did express concerns about the measurement problem. What emerges from a careful examination of the different attitudes of physicists to quantum mechanics is a range of nuanced positions, which are not adequately grasped in terms of the simple dichotomy of orthodoxy and heterodoxy. Some physicists did take the view that quantum mechanics was a complete and perfectly satisfactory theory. Others did not regard it as 'complete', but were not particularly troubled by this state of affairs.

But quite aside from what beliefs physicists may have held about quantum mechanics, there were those who kept their opinions to themselves. The distinction between what one was prepared to say publicly, or in print, and what one thought privately adds a further layer of complexity to standard accounts of the orthodoxy. In a rare moment of candour, Arnold Sommerfeld confessed that he found it difficult to resign himself to certain aspects of quantum mechanics in a letter to Carl Oseen in 1931. 'I am not very happy with "indeterminate [*unbestimmt*] physics"', he wrote, 'especially when young enthusiasts or formalists talk about it in the department for hours'. While Sommerfeld felt compelled to 'acknowledge the legitimacy of the whole way of looking at it', he allowed himself to wonder whether '*perhaps it can still be overcome by some "metaphysics"* (all physics is metaphysics according to Einstein)'. The ingenious thought experiments that Bohr had devised to demonstrate the indeterminacy in measuring a particle's position and momentum did not really strike Sommerfeld as getting to the heart of the matter. 'How inelegant, for example, the general theory of relativity would become, if one were to take into account the precision of measurement there too!' (Sommerfeld to Oseen, 22 February 1931, Eckert and Märker, 2004, p. 322).

There are intimations that other physicists harboured private reservations about other aspects of quantum mechanics. In his interview with Kuhn in 1963, I. I. Rabi recalled that when he arrived in Europe in 1927, Schrödinger's interpretational aspiration for a wave theory of matter was not taken very seriously. There were, as he recalled, simply 'no consequences of it that we could see that were useful'. Physicists accepted the statistical interpretation simply because it worked. Nevertheless, Rabi indicated that like Schrödinger, he had always regarded the probabilistic interpretation as a temporary expedient. Schrodinger 'was always unhappy about the whole thing and *I am, too*, to the very present day, in the sense that *I can't get myself to regard quantum theory as other than provisional* in some way' (Rabi, Interview with Kuhn, 8 December 1963, AHQP).[8] We find a similar view in fellow American physicist, Earle H. Kennard. In an interview in 1970, Robert Marshak recalled that there had been considerable friction in the 1930s between Kennard and Hans Bethe at Cornell, because 'Kennard did not believe in quantum mechanics'. While Kennard made a number of important 'original contributions to the new field of quantum mechanics, he never fully believed in quantum mechanics and used to constantly argue about it with Bethe' (Marshak, Interview with Wiener, 15 June 1970, AIP).[9]

Kennard's views have received little attention, in large part because his published contributions to quantum mechanics in the late 1920s give the impression that he was untroubled by interpretational issues. After spending his sabbatical in Göttingen in 1926, Kennard published a number of important papers over the next few years, building on the probabilistic interpretation. His work greatly extended the understanding of the dynamics of wave packets and he predicted what is now commonly known as the 'Kennard phase' (Kennard, 1927, 1928). Yet, as late as 1929, Kennard would claim that quantum mechanics 'cannot yet be considered as a coherent and completed theory' (Kennard, 1929, p. 78). As Joseph Rouse has argued: 'Scientists can hold heterodox beliefs about fundamental issues in their disciplines as long as their research can be taken into account and used by others' (Rouse, 2003, p. 110).

While silence on such matters has typically been taken as implying assent to Bohr's view, it is not altogether clear what physicists may have thought privately on this question. It is entirely possible that many physicists were happy to *use* quantum mechanics, without committing themselves either way on the question of whether the wave function was the most complete possible description of the state of a system. *Using* a theory does not entail accepting that theory as true or even complete. Anthony Leggett expressed the point beautifully, on the occasion of the Niels Bohr Centenary Symposium in October 1985:

> I start with an awful confession: If you were to watch me day by day, you would see me sitting at my desk solving Schrödinger's equation and calculating Green's functions and cross-sections exactly like my colleagues. But occasionally at night, when the full moon is bright, I do what in the physics community is the intellectual equivalent of turning into a werewolf: I question whether quantum mechanics is the complete and ultimate truth about the physical universe.   (Leggett, 1986, p. 53)

---

[8] American Institute of Physics, Oral History Interviews. https://www.aip.org/history-programs/niels-bohr-library/oral-histories/4836.

[9] American Institute of Physics, Oral History Interviews. https://www.aip.org/history-programs/niels-bohr-library/oral-histories/4760–1.

When Leggett uttered these words in 1985, the landscape of physics had changed appreciably. Fifty years earlier, such views were considered 'high treason' (Schrödinger to Einstein, 23 March 1936, AHQP, 37). 'If there were people in opposition', Alfred Landé later remarked, 'they didn't make their opposition public' (Interview with Kuhn, 8 March 1962, AHQP).[10] But given the views we have presented from such physicists as Jordan, Wigner, Kennard, Rabi, Sommerfeld, Born, and Dirac, one wonders how many other 'werewolves' there might have been who were not prepared to make such a public confession. Perhaps others too entertained such night thoughts. Doubts about quantum mechanics in the early years might well have been more prevalent than we tend to think.

This brings us finally to the distinction between belief and action. Most accounts of the orthodoxy have focused on commitments, beliefs, or doctrines. But it may well be that *practice* is a far more relevant historical category. As Philip Pearle would put it, 'social deviance' in quantum mechanics comes in two forms: 'Closet deviance' is 'the *belief* that standard quantum theory', in spite of its enormous success, 'has conceptual flaws. Outright deviance is the temerity to try and *do* something about it' (Pearle, 2009, pp. 257–8). Few physicists who harboured reservations about quantum mechanics were prepared to go on the attack in public, or pursue alternative lines of research. While Schrödinger, Einstein, von Laue, and Planck remained critical voices in the 1930s, their criticisms were more symbolic gestures of defiance and critical analyses of the existing theory, rather than concerted efforts to develop new research programmes. It was only in the 1950s that new interpretations began to appear, and only in the late 1970s and 80s that these formed the basis of ongoing programmes of research. Reflecting on this history, perhaps we should say that action rather than belief was the true mark of the quantum dissident.

## ABBREVIATIONS

| | |
|---|---|
| BSC | Bohr Scientific Correspondence |
| AHQP | Archive for History of Quantum Physics |
| LRP | Léon Rosenfeld Papers |
| NBA | Niels Bohr Archive, Copenhagen |
| AIP | American Institute of Physics. |

## REFERENCES

Bacciagaluppi, G., and Crull, E. (eds) (2009). Heisenberg (and Schrödinger and Pauli) on Hidden Variables. *Studies in History and Philosophy of Modern Physics*, **40**, 372–82.

Bertotti, B. (1985). The Later Work of E. Schrodinger. *Studies in History and Philosophy of Modern Physics*, **16**, 83–100.

[10] American Institute of Physics, Oral History Interviews. https://www.aip.org/history-programs/niels-bohr-library/oral-histories/4728-4.

Blokhintsev, D. (1952). Critique de la Conception Idéaliste de la Théorie Quantique. In *Questions Scientifiques*, vol. 1 *Physique*. Trans. F. Lurçat, Paris: Les Éditions de la Nouvelle Critique, pp. 95–129.

Blokhintsev, D. (1953). Kritik der philosophischen Anschauungen der sogenannten 'Kopenhagener Schule' in der Physik. *Sowjetwissenschaft*, 6(4), 545–74.

Bohm, D. (1952). A Suggested Interpretation of the Quantum Theory in Terms of 'Hidden' Variables. *Physical Review*, 85, 166–179, 180–93.

Bohm, D., and Schiller, R. (1955). A Causal Interpretation of the Pauli Equation B, *Nuovo Cimento, Suppl.*, 1, 67–91.

Bohm, D., Schiller, R., and Tiomno, J. (1955). A Causal Interpretation of the Pauli Equation A', *Nuovo Cimento, Suppl.*, 1, 48–66.

Bohm, D., and Vigier, J.-P. (1954). Model of the Causal Interpretation of Quantum Theory in Terms of a Fluid with Irregular Fluctuations. *Physical Review*, 96, 208–16.

Bohr, N. (1931). Maxwell and Modern Theoretical Physics. *Nature*, 128, 691–92.

Bohr, N. (1987). Introductory Survey [1929]. In *Atomic Theory and the Description of Nature*. The Philosophical Writings of Niels Bohr, vol. 1, Woodbridge, CT: Ox Bow Press, pp. 1–24.

Bokulich, A. (2006). Heisenberg Meets Kuhn: Closed Theories and Paradigms. *Philosophy of Science*, 73, 90–107.

Born, M. (1929). Über den Sinn der physikalischen Theorien. *Die Naturwissenschaften*, 17(7), 109–18.

Born, M. (1953). Physical Reality. *The Philosophical Quarterly*, 3, 139–49.

Born, M. (1956a). The Interpretation of Quantum Mechanics [1953]. In M. Born, *Physics in my Generation. A Selection of Papers*, London: Pergamon Press, pp. 140–150.

Born, M. (1956b). The Conceptual Situation in Physics and its Prospects for Future Development [1953]. In M. Born, *Physics in My Generation. A Selection of Papers*, London: Pergamon Press, pp. 123–39.

Born, M. (1964). *Natural Philosophy of Cause and Chance*. London: Dover Press.

Born, M. (1971). *The Born–Einstein Letters: Correspondence between Albert Einstein and Max and Hedwig Born from 1916 to 1955*. Trans. I. Born. New York: Walker.

Borrelli, A. (2017). Symmetry, Beauty and Belief in High-Energy Physics. *Approaching Religion*, 7, 22–36.

de Broglie, L. (1953). La Physique Quantique Restera-t-elle Indéterministe?. *Bulletin Société Française de Philosophie*, 46, 135–73.

Bub, J. (2010). Von Neumann's 'No Hidden Variables' Proof: A Re-appraisal. *Foundations of Physics*, 40, 1333–40.

Bub, J, (2011). Is Von Neumann's 'no hidden variables' proof silly? In H. Halvorson (ed.), *Deep Beauty—Understanding the Quantum World Through Mathematical Innovation*, Princeton, NJ: Princeton University Press, pp. 393–407.

Camilleri, K. (2009). Constructing the Myth of the Copenhagen interpretation. *Perspectives on Science*, 17, 26–57.

Crull, E., and Bacciagaluppi G. (eds) (2017). *Grete Hermann—Between Physics and Philosophy*. Dordrecht: Springer.

Dieks, D. (2017). Von Neumann's Impossibility Proof: Mathematics in the Service of Rhetorics. *Studies in the History and Philosophy of Modern Physics*, 60, 136–48.

Dirac, P. A. M. (1951). The Relation of Classical to Quantum Mechanics. In *Proceedings of the Second Canadian Mathematical Congress, Vancouver 1949*, Toronto: Toronto University Press, pp. 10–31.

Dirac, P. A. M. (1963). The Evolution of the Physicists' Picture of Nature. *Scientific American*, 208(5), 45–53.

Dirac, P. A. M. (1982). The Early Years of Relativity. In G. Holton and Y. Elkana (eds), *Albert Einstein: Historical and Cultural Perspectives: The Centennial Symposium in Jerusalem*, Princeton, NJ: Princeton University Press, pp. 79–90.

Eckert, M., and Märker, K. (eds) (2004). *Arnold Sommerfeld: Wissenschaftlicher Briefswechsel*, Band 2: 1919–1951. München: Deutsches Museum.

Einstein, A. (1948). Quanten-Mechanik und Wirklichkeit. *Dialectica*, 2, 320–24.

Einstein, A. (1949). Reply to Criticisms. In P. A. Schilpp (ed)., *Albert Einstein: Philosopher-Scientist*, La Salle: Open Court, pp. 663–88.

Einstein, A. (1953). Elementäre Überlegungen zur Interpretation der Grundlagen der Quanten-Mechanik. In E. Appleton *et. al.* (eds), *Scientific Papers Presented to Max Born*, New York: Hafner, pp. 33–40.

Everett, H. (1957). Relative State Formulation of Quantum Mechanics. *Reviews of Modern Physics*, 29, 454–62.

Feyerabend, P. (2015a). Niels Bohr's Interpretation of the Quantum Theory [1961]. In S. Gattei and J. Agassi (eds), *Physics and Philosophy: Philosophical Papers*, vol. 4, Cambridge: Cambridge University Press, pp. 74–98.

Feyerabend, P. (2015b). Problems of Microphysics [1962]. In S. Gattei and J. Agassi (eds), *Physics and Philosophy: Philosophical Papers*, vol. 4, Cambridge: Cambridge University Press, pp. 99–187.

Fine, A. (1993). Einstein's Interpretations of Quantum Theory. *Science in Context*, 6, 257–73.

Freire Jr, O. (2015). *The Quantum Dissidents: Rebuilding the Foundations of Quantum Mechanics (1950–1990)*. Heidelberg: Springer.

Heisenberg, E. (1984). *Inner Exile: Recollections of a Life with Werner Heisenberg*. Trans. S. Capellari & S. Morris. Boston: Birkhäuser.

Heisenberg, W. (1955). The Development of the Interpretation of the Quantum Theory. In W. Pauli (ed.), *Niels Bohr and the Development of Physics: Essays Dedicated to Niels Bohr on the Occasion of his Seventieth Birthday*, New York, McGraw-Hill, pp. 12–29.

Heisenberg, W. (1958). *Physics and Philosophy*. London: George Allen & Unwin.

Heisenberg, W. (1971). *Physics and Beyond: Encounters and Conversations*. Trans. A. J. Pomerans. London: George Allen & Unwin.

Heisenberg, W. (1974). The Notion of a 'Closed Theory' in Modern Science [1948]. In *Across the Frontiers*, New York: Harper & Row, pp. 39–46.

Henderson, J. B. (1998). *The Construction of Orthodoxy and Heresy: Neo-Confucian, Islamic, Jewish and Early Christian Patterns*. New York: SUNY Press.

Howard, D. (2004). Who Invented the 'Copenhagen Interpretation'? A Study in Mythology. *Philosophy of Science*, 71, 669–82.

Jánnosy, L. (1952). The Physical Aspects of the Wave-Particle Problem. *Acta Physica Hungaria*, 1, 423–67.

Jammer, M. (1974). *The Philosophy of Quantum Mechanics*. New York: John Wiley & Sons.

Jordan, P. (1936). *Anschauliche Quantentheorie: eine Einführung in die moderne Auffassung der Quantenerscheinungen*. Berlin: Springer.

Jordan, P. (1949). On the Process of Measurement in Quantum Mechanics. *Philosophy of Science*, 16, 269–78.

Kennard, E. H. (1927). Zur Quantenmechanik einfacher Bewegungstypen. *Zeitschrift für Physik*, 44, 326–52.

Kennard, E. H. (1928). On the Quantum Mechanics of a System of Particles. *Physical Review*, **31**, 876–90.

Kennard, E. H. (1929). The Quantum Mechanics of an Electron or Other Particle. *Journal of the Franklin Institute*, **207**, 47–78.

Körner S. (ed.) (1957). *Observation and Interpretation: A Symposium of Philosophers and Physicists*. London: Butterworth.

Leggett, A. (1986). Quantum Mechanics at the Macroscopic Level. In J. de Boer, E. Dal, and O. Ulfbeck (eds), *The Lesson of Quantum Theory*, Amsterdam: North-Holland, pp. 35–57.

Maheu, R. *et al.* (eds) (1971). *Science and Synthesis. An International Colloquium organized by UNESCO on the Tenth Anniversary of the Death of Albert Einstein and Teilhard de Chardin*. Berlin: Springer-Verlag.

Mermin, N. D., and Schack, R. (2018). Homer nodded: von Neumann's surprising oversight. *Foundations of Physics*, **48**, 1007–20.

Michel, L. (1989). Symmetry and Conservation Laws in Particle Physics in the Fifties. In L. Brown, M. Dresden, and L. Hoddeson (eds), *Pions to Quarks: Particle Physics in the 1950s*, Cambridge: Cambridge University Press, pp. 373–83.

Myrvold, W. C. (2003). On Some Early Objections to Bohm's Theory. *International Studies in the Philosophy of Science*, **17**, 7–24.

Pauli, W. (1953). Remarques sur le problème des paramètres cachés dans la mécanique quantique et sur la théorie de l'onde pilote. In A. George (ed.), *Louis de Broglie: Physicien et Pensuer*, Paris: Éditions Albin Michel, pp. 33–42.

Pauli, W. (1994a). Space, Time and Causality in Modern Physics [1936]. In C. P. Enz and K. von Meyenn (eds), *Wolfgang Pauli: Writings on Physics and Philosophy*, Trans. R. Schlapp, Berlin: Springer-Velag, pp. 95–105.

Pauli, W. (1994b). Phenomenon and Physical Reality [1957]. In C. P. Enz and K. von Meyenn (eds), *Wolfgang Pauli: Writings on Physics and Philosophy*, Trans. R. Schlapp, Berlin: Springer-Velag, pp. 127–35.

Pearle, P. (2009). How Stands Collapse II. In W. C. Myrvold and J. Christian (eds), *Quantum Reality, Relativistic Causality, and Closing the Epistemic Circle*, Berlin: Springer, pp. 257–92.

Peat, D. (1997). *Infinite Potential: The Life and Times of David Bohm*. Reading, MA: Addison-Wesley, Helix Books.

Rédei, M. (ed.) (2005). *John von Neumann: Selected Letters*. Providence, RI: American Mathematical Society.

Rosenfeld, L. (1979a). Misunderstandings about the Foundations of Quantum Mechanics [1957]. In R. S. Cohen and J. Stachel (eds), *Selected Papers of Léon Rosenfeld*, Dordrecht: D. Reidel, pp. 495–502.

Rosenfeld, L. (1979b). Strife about Complementarity [1953]. In R. S. Cohen and J. Stachel (eds), *Selected Papers of Léon Rosenfeld*, Dordrecht: D. Reidel, pp. 465–83.

Rouse, J. (2003). Thomas Kuhn's Philosophy of Scientific Practice. In T.Nickles (ed.), *Thomas Kuhn*, Cambridge: Cambridge University Press, pp. 101–21.

Schrödinger, E. (1953). The Meaning of Wave Mechanics. In A. George (ed.), *Louis de Broglie: Physicien et Penseur*, Paris: Édition Alben Michel, pp. 1–32.

Stachel, J. (1986). Einstein and the Quantum: Fifty Years of Struggle. In R. G. Colodny (ed.), *From Quarks to Quasars: Philosophical Problems of Modern Physics*, Pittsburgh: University of Pittsburgh Press, pp. 349–85.

Stöltzner, M. (1999). What John von Neumann Thought of the Bohm Interpretation. In D. Greenberger, D. W. Reiter, and A. Zeilinger (eds), *Epistemological and Experimental Perspectives on Quantum Physics*, Dordrecht: Springer, pp. 257–62.

Talbot, C. (ed.) (2017). *David Bohm: Causality and Chance, Letters to Three Women*. Cham, Switzerland: Springer.

Terletsky, Ya. P. (1952). Problèmes du Développement de la Théorie Quantique. In *Questions Scientifiques*, vol. 1 *Physique*, Trans. F. Lurçat, Paris: Les Éditions de la Nouvelle Critique, pp. 131–46.

von Meyenn, K. (ed.) (1985). Wolfgang Pauli: Wissenschaftlicher Briefwechsel mit Bohr, Einstein, Heisenberg u.a., vol. 2, 1930–1939. Berlin: Springer.

von Meyenn, K. (ed.) (1996). Wolfgang Pauli, Wissenschaftlicher Briefwechsel mit Bohr, Einstein, Heisenberg u.a., vol. 4. Part I 1950–1952. New York: Springer.

von Neumann, J. (1955). Mathematical Foundations of Quantum Mechanics, trans. O. Benfrey. Princeton, NJ: Princeton University Press.

von Neumann, J. (1961a). The Role of Mathematics in the Sciences and in Society [1954]. In A. H. Taub (ed.), John von Neumann. Collected Works, vol. 6, Oxford: Pergamon Press, pp. 477–90.

von Neumann, J. (1961b). Method in the Physical Sciences [1955]. In A. H. Taub (ed.), John von Neumann. Collected Works, vol. 6, Oxford: Pergamon Press, pp. 491–98.

von Weizsäcker, C. F. (1985). *Aufbau der Physik*. München: Hanser Verlag.

Westman, R. S. (1975). The Melanchthon Circle, Rheticus, and the Wittenberg Interpretation of the Copernican Theory. *Isis*, **66**, 164–93.

Wigner, E. (1970). On Hidden Variables and Quantum Mechanical Probabilities. *American Journal of Physics*, **38**, 1005–09.

Wigner, E. (1971). Rejoinder. *American Journal of Physics*, **39**, 1097–98.

Wigner, E. (1983a). Remarks on the Mind-Body Question [1961]. In J. A. Wheeler and W. H. Zurek (eds), *Quantum Theory and Measurement*, Princeton, NJ: Princeton University Press, pp. 168–81.

Wigner, E. (1983b). The Problem of Measurement [1963]. In J. A. Wheeler and W. H. Zurek (eds), *Quantum Theory and Measurement*, Princeton, NJ: Princeton University Press, pp. 324–41.

Yang, C. N. (1958). The Law of Parity Conservation and Other Symmetry Laws of Physics. *Science, New Series*, **127**(3298), 565–69.

# CHAPTER 35

# THE RECEPTION OF THE FORMAN THESIS IN MODERNITY AND POSTMODERNITY

### PAUL FORMAN

## 35.1 INTRODUCTION

IN March 2012 Polity Press published Jon Agar's 600-page history of science in the 20th century.[1] Suited, *The Economist* judged, to 'anyone who thinks seriously about science', it is distinguished, to a historian's eye, both by emphasis on the problematics of historical interpretation and by the predominance of recently published scholarship in its endnotes. Yet these two emphases are rather at odds in the one chapter advertising an interpretive issue in its title: 'Crisis: Quantum Theories and Other Weimar Sciences'. At issue is, of course, whether in the years following Germany's defeat in World War I, a widespread sense of crisis, marked by anti-scientific, and more especially Spengler-inspired anti-causal, attitudes among educated Germans, 'led physicists to ardently hope for, actively search for, and willingly embrace an acausal quantum mechanics'.[2] Agar's emphases on historical interpretation and recent scholarship are here at odds in that the controversy over this 'Forman thesis' is notable for

---

[1] Jon Agar, *Science in the 20th Century and Beyond* (Cambridge UK, Malden, MA, USA: Polity Press, 2012).

[2] Paul Forman, 'Weimar culture, causality, and quantum theory, 1918–1927: Adaptation by German physicists and mathematicians to a hostile intellectual environment', *Historical Studies in the Physical Sciences*, 3 (1971), 1–115, on p. 3. This paper, and four follow-up papers, are photographically reprinted in C. Carson, A. Kojevnikov, and H. Trischler (eds), *Weimar Culture and Quantum Mechanics: Selected Papers by Paul Forman and Contemporary Perspectives on the Forman Thesis* (London and Singapore: Imperial College Press and World Scientific, 2011), namely, 'Scientific internationalism and the Weimar

being exceptionally long-standing: as of 2012 'Weimar Culture, Causality, and Quantum Theory' (WCC) was already forty years old, and John Hendry's polemic against it, which Agar quotes extensively and exclusively for 'the other side', was over thirty.[3]

physicists: The ideology and its manipulation in Germany after World War I' (1973), 'The financial support and political alignment of physicists in Weimar Germany' (1974), 'The reception of an acausal quantum mechanics in Germany and Britain' (1979), and *Kausalität, Anschaulichkeit,* and *Individualität,* or, How cultural values prescribed the character and the lessons ascribed to quantum mechanics' (1984). A German translation of WCC, and of *'Kausalität,* etc.', incorporating the German originals of all quotations, is provided in Karl von Meyenn (ed.), *Quantenmechanik und Weimarer Republik* (Wiesbaden: Vieweg, 1994; reprinted 2013).

[3] John Hendry, 'Weimar culture and quantum causality', *History of Science,* 18 (1980), 155–78. A German translation is included in von Meyenn (ed.), *Quantenmechanik* (n. 2). Also mentioned by Agar, but wisely not quoted, is Peter Kraft and Peter Kroes, 'Adaptation of scientific knowledge to an intellectual environment. Paul Forman's Weimar culture, causality and quantum theory, 1918–1927: Analysis and criticism', *Centaurus,* 27 (1984), 76–99. The special urgency in Britain of slaying Forman's dragon arose from, on the one side, its being cited approvingly by so distinguished a philosopher-historian of science as Mary Hesse, 'Models and method in the natural and social sciences', *Methodology and Science,* 8 (1975), 163–178, on her opening page, as an outstanding example of the fact that 'social history of science is increasingly, and most interestingly, taken to mean study of the social conditioning of the theoretical belief systems of science'. And on the other side, from the circumstance that such study was beginning to be pushed aggressively by British sociologists of science, especially early and notably by Barry Barnes, *Scientific Knowledge and Sociological Theory* (London, Boston: Routledge & K. Paul, 1974), citing WCC repeatedly throughout. The first to enter the lists against WCC was the somewhat older, abler, Jon Dorling, who delivered a diatribe at the July 1976 meeting of the British Society for the History of Science. Dorling declined to publish, but the three-page summary that he circulated privately at the time (not to me, of course) makes clear that Hendry drew his broader criticisms from Dorling.
    As an indication of authoritative refusals of assent to WCC citing Hendry and/or Kraft and Kroes in support of their judgement: Michael Redhead, 'Quantum Theory', in Robert Cecil Olby, Geoffrey N. Cantor, J. R. R. Christie, and M. J. S. Hodge (eds), *Companion to the History of Modern Science* (London: Routledge, 1996), pp. 458–478, on p. 458; Gerald Holton, 'Einstein and the cultural roots of modern science', in Peter Galison, Stephen R. Graubard, Everett Mendelsohn (eds), *Science in Culture* (New Brunswick, NJ: Transaction Publishers, 2001), pp. 1–39, on p. 2; Olivier Darrigol, 'Quantum theory and atomic structure, 1900–1927', in Mary Jo Nye (ed.), *The Cambridge History of Science,* Vol. 5: *The Modern Physical and Mathematical Sciences* (Cambridge: Cambridge University Press, 2002), pp. 331–349, on p. 332. Indeed, any *pro forma* footnote reference by a historian to WCC is almost certain also to contain a citation of Hendry—in order to forestall the supposition that the writer affirms the Forman thesis.
    Hendry urged that 'in respect of the general attitude of the milieu to mathematics, physics, and causality we should distinguish a subtlety that Forman has not. For while there were indeed many attacks upon mathematics and physics from outside these disciplines, these were in all cases attacks upon their *value,* rather than upon their *content*' (p. 157, Hendry's italics). Anyone who has read WCC will recognize not only that Hendry's assertion is patently false—the attacks were obviously equally upon the content of physics and of mathematics—but also that the 'subtlety' for which Hendry pats himself on the back is just the distinction between the ideology and the content of physics, which distinction I had headlined in presenting my argument. This alone should have prevented anyone who has read WCC from conceiving confidence in Hendry's criticism of it.
    Kraft and Kroes, although categorically damning, found it impossible to deny that 'in *some* cases *certain* physicists may have started to question the validity of the principle of causality, or even to reject it, because of external factors' (p.96; the italics are theirs). While Hendry (1980) concluded by acknowledging that 'Forman's work has clearly demonstrated the poverty of a wholly internal treatment of issues such as that of causality. Physicists *were* [H's italics] influenced by the crisis-consciousness of

Hendry himself, whose historical investigations *circa* 1980 were in the service of philosophy (and correspondingly restricted and slanted)[4], did not claim to have slain the 115-page dragon against which he sallied forth. But from that time to Agar's, indeed to 2020, and probably far beyond, Hendry's twenty-four pages remain the fullest critique addressing WCC on its own turf, atomic physics in Weimar culture.[5]

Neither Hendry then, nor anyone else in the four decades since publication of Hendry's opposition, undertook the task of going thoroughly and carefully over the ground covered and the argument made in WCC to search out its authentic, demonstrable defects.[6] Recovering that ground, and more, would have been an entirely feasible doctoral dissertation project. I, having no graduate students, and being

---

post-war Europe and by the attitudes characteristic of the Weimar milieu' (p. 171). In *The Creation of Quantum Mechanics and the Bohr–Pauli Dialogue* (Dordrecht, Boston: D. Reidel, 1984), pp. 16, 18, 35, Hendry conceded, *i.a.*, that Schrödinger's 'rejection of causality was explicitly a philosophically based decision rather than one dictated by the requirements of physics'. Mara Beller, *Isis*, 77 (1986), 107–109, one of WCC's most categorical rejecters, complained in her review of this book that 'Hendry's opinion about Forman's thesis is confused', but she was careful not to confuse *Isis*'s readers by telling them what concessions Hendry had made.

[4] Hendry, introducing *The Creation of Quantum Mechanics and the Bohr–Pauli Dialogue* (n.3), p. 2, declares his aim 'is to approach the history of the theory of quantum mechanics as a means of exploring its philosophy'. Indeed his title trumpets his correspondingly limited coverage, and source material, viz., von Meyenn's then recently published first volume of Pauli's *Wissenschaftlicher Briefwechsel*. Beller, *Isis*, 77 (1986), 107–109, rightly judged that Hendry's 'account often seems to be closer to a "rational reconstruction" than to the actual historical development'.

[5] Mention must be made of one other critic of WCC, Hans [Johannes Arie] Radder, whose publications in the early 1980s are more original and contributive than Hendry's (and certainly than that by his countrymen Kraft and Kroes) but are far less frequently cited: 'Between Bohr's atomic theory and Heisenberg's matrix mechanics: A study of the role of the Dutch physicist H. A. Kramers', *Janus*, 69 (1982), pp. 223–52; 'Kramers and the Forman theses', *History of Science*, 21 (1983), pp. 165–82. Radder, then still a doctoral student in philosophy and rather taken with Lakatos, proposes that WCC be entirely redone on the basis of a 'problem shift' (I'd say, 'monster barring') that gets causality out of the sphere of scientific knowledge and quarantines it in the sphere of epistemology (read 'ideology'). Essential for this—and Kramers is central to this—is Radder's showing that Dorling, Hendry (and now we must add a long list of would-be refuters of WCC down to Staley and Stöltzner) had failed to understand what Forman (citing S.G. Brush) had pointed out, viz., that there is no incompatibility between being epistemologically positivist and being *lebensphilosophisch*. On some other points, however, Radder, whose stance was, basically, hostile to WCC, failed to emphasize support for WCC that his sources put into his hands, facts that would have strengthened his own case as well as mine—notably, the presence of Spengler on Kramers' mind, and the fact that Hendry had quashed evidence in *his* hands of the wide support in 1924 among German atomic theorists for the acausal Bohr–Kramers–Slater theory. Radder was not, however, hostile enough toward WCC for theoretical physicist Max Dresden—Kramers' student and biographer—who, as I pointed out indirectly in 'Independence, not transcendence, for the historian of science', *Isis*, 82 (1991), pp. 71–86, on p. 82, avoided referring to WCC and its author by plagiarizing Radder. (*Cf.* note 18, below.)

[6] There is an important exception: von Meyenn, in preparing his 50-page introduction to *Quantenmechanik und Weimarer Republik* (n.2), did look rather carefully over the ground that I had covered, and brought to his introduction some further supportive documentation. Moreover von Meyenn acknowledged what, to this day, all rejecters of WCC have found convenient to ignore, namely that the evidence of cultural conditioning of physicists' attitudes toward the acausal aspect of quantum

restrained by the conditions of my employment from continuing my concentrated efforts in that direction, chose not to reply to what were, after all, just niggles and quibbles, snipes and gripes. Beyond these there has been—beginning in 1988 with the first of Norton Wise's critiques—only a series of unrealized, indeed unrealizable, proposals for thoroughly redoing WCC on more ideologically acceptable grounds.[7]

mechanics, post-1925, as well as physicists' catering to cultural preferences in advertising their new quantum mechanics—evidence which I brought in the subsequent publications listed in note 2, above—are important supports for WCC's contentions. Consequently, both early and late in his lengthy introduction (pp. 12, 13, 56), von Meyenn strongly affirmed the necessity and the success of what WCC undertakes. But oddly—perhaps because his unusual career depended so heavily upon the favour of powerful physicists—at the outset of his introduction (p. 7) von Meyenn set as epigraph a quotation from Edoardo Amaldi's 1979 anti-WCC paper claiming radioactivity researchers had revealed sub-atomic indeterminism before World War I, and in his last two pages (pp. 57–58) von Meyenn expressly recanted, adopting the internalist position of this especially powerful physicist: Amaldi, 'Radioactivity, a pragmatic pillar of probabilistic conceptions', in Giuliano Toraldo di Francia (ed.), *Problems in the Foundations of Physics*, 'Scuola Internazionale di Fisica Enrico Fermi, 72, 1977, Varenna' (Bologna: Società Italiana di Fisica; Amsterdam: North Holland, 1979), 1–28. As for Amaldi's anti-WWC thesis, Helge Kragh's opinion *should be* dispositive: "it is obvious that it is anachronistic, hence unhistorical, to interpret the early history of radioactivity in accordance with the knowledge of a later generation, namely that radioactive decay is an acausal process." Kragh, *Subatomic Determinism and Causal Models of Radioactive Decay, 1903–1923*, RePoSS: Research Publications on Science Studies, #5 (University of Aarhus: Centre for Science Studies, 2009), 16.

[7] Joining, even displacing, Hendry and Kraft and Kroes as 'disprovers' in more recent references by historians and philosophers to WCC—so, notably, by Norton Wise in 2011 in *Weimar Culture and Quantum Mechanics* (n. 2), pp. 424, 429—are publications by Michael Stöltzner, and by Deborah Coen, advancing their 'Vienna indeterminism' thesis. First propounded (1999) and chiefly propagated by Stöltzner, in a dozen publications, their thesis is that Forman is wrong because there existed in Vienna, from the early years of the 20th century, an anti-determinist circle, school, tradition, and physical research programme inspired by Boltzmann and, more especially and effectively, by Franz (Serafin) Exner, Vienna's institutionally most central experimental physicist from the 1890s to *circa* 1920. As have Hendry's and Kraft and Kroes' disproofs of WCC, Stöltzner's and Coen's 'Vienna indeterminism' has been received as the proverbial 'gift horse'. On it many clamber to escape WCC, but those clamberers have preferred to leave its 'teeth' unexamined. As I emphasized at the outset of WCC, given its claims and the facts presented, any pre-World War I 'anticausality current', to the extent that such was 'subterranean', does not disprove, but supports, the thesis of WCC. Thus the question is, what did 'Vienna indeterminism' really amount to prior to 1919?

I intend to prepare critical discussions of the claims and evidence for a pre-war 'Vienna indetermin-ism'. I anticipate that its conclusions will include Kragh's opinion of Amaldi's earlier version, quoted in note 6. Here I limit myself to pointing out that its proponents know of only one publication prior to 1919 from the 'Exner circle' that advances an anti-determinist thesis—Exner's 1908 address as *Rektor* of the University of Vienna (to which I drew attention in WCC). And they know only one pre-1919 publication explicitly referring to that address—it is by Exner's nephew, and even his reference merely *pro forma*, not explicitly affirmative. Furthermore, had there indeed been such a pre-1919 'Viennese' anti-determinist perspective, persuasion, or programme, then the two leading theoretical physicists with Vienna–Boltzmann–Exner backgrounds—Ehrenfest and Schrödinger—should have been aware of it, if not themselves bearers of it. Ehrenfest, born and raised in Vienna, had attended Boltzmann's lectures and, more than any other theorist, had espoused and advanced Boltzmann's programme in statistical physics. But, as emphasized by Frans van Lunteren and M. J. Hollestelle, 'Paul Ehrenfest and the Dilemmas of Modernity', *Isis*, 104 (2013), 504–536, he evinced no awareness of such a 'Vienna indeterminism' and was himself, until the late-1920s, strongly resistant to indeterminism. Schrödinger, whose education and

Naturally, I feel aggrieved. But however revealing of the true spirit of scholarship a bare compilation of such mishandlings of WCC and its author would be, it would not provide insight into the historicity of our conceptions of scientific life and change—and consequently also into the historicity of what we historians and philosophers think proper for us, and what impermissible for us, to allege about scientists as actors and claimants to knowledge. And such historicizing is my intent here.[8]

The framework within which I here historicize the stances taken toward WCC by historians and historically minded philosophers of science is the radical transformation of Western consciousness during the last third of the 20th century—the transformation from a modernity whose principal characteristics began to take hold in the 17th century and were well established by the middle of the 19th century, to a postmodernity that is modernity's antithesis in its valuation and pursuit of knowledge and in its conception of the self in relation to society.[9]

I myself, having come of age in the 1950s, was formed as a modern, which is to say, I believed in a unique reality, knowable by, and only by, disciplined inquiry—which belief necessarily included confidence in a supposed self-correcting character of disciplinarily organized inquiry. In 1960, when I began graduate study in history, this field of humanistic scholarship was still largely scientistic in the aforesaid sense. Observing, however, over the course of the 1980s and 1990s, how historians were renouncing those affinities with science which they had touted from the later 1800s to the 1960s, I came gradually to recognize that American, Western, indeed Global, consciousness was undergoing a 'phase transition' far more general and consequential than that which Henry Adams conceived at the start of the 20th century.[10]

---

first positions were at the University of Vienna, 1906–1920, and who was, formally at least, Exner's *Assistent* from 1911 to 1920, was unware of Exner's 1908 address, and seemingly of any other Viennese anti-determinist pronouncement antedating 1919, when, in 1921, he himself first declared for indeterminism—as shown long ago by Paul A. Hanle, 'Indeterminacy before Heisenberg: The Case of Franz Exner and Erwin Schrödinger', *Hist. Stud. Phys. Sci.* 10 (1979), 225–69.

[8] I recommend to every reader of this unapologetically polemical account of the reception history of the Forman thesis that they also read Alexei Kojevnikov's complementary account, Chapter 36 in this *Oxford Handbook*. For the stimulus and opportunity to publish my account, I am grateful to Olival Freire, Jr, who had also drawn attention to its desirability in *The Quantum Dissidents: Rebuilding the Foundations of Quantum Mechanics (1950-1990)*, (Berlin: Springer-Verlag, 2015), p. 230, note 54.

[9] Paul Forman, 'The primacy of science in modernity, of technology in postmodernity, and of ideology in the history of technology', *History and Technology*, 23 (2007), 1–152; 'From the social to the moral to the spiritual: The postmodern exaltation of the history of science', in Jürgen Renn and Kostas Gavroglu (eds), *Positioning the History of Science* [Festschrift for S. S. Schweber], Boston Studies in the Philosophy of Science, vol. 248 (Dordrecht: Springer Verlag, 2007), pp. 49–55; '(Re)cognizing postmodernity: helps for historians—of science especially', *Berichte zur Wissenschaftsgeschichte*, 33 (2010), 157–75; 'On the historical forms of knowledge production and curation: modernity entailed disciplinarity, postmodernity entails antidisciplinarity', *Osiris*, 27 (2012), 56–97.

[10] Peter Novick, *That Noble Dream: The 'Objectivity Question' and the American Historical Profession* (Cambridge, UK and New York: Cambridge University Press, 1988); Henry Adams, *The Degradation of the Democratic Dogma* (New York: Macmillan, 1920); and my publications cited above, note 9.

Seeing this, I was led to a more fully historicized perspective on the reception of WCC. I had long recognized that the principled basis for rejection of WCC was the incompatibility of the Forman thesis with the modern epistemic ideology, particularly the modernist axiom of disciplinary closure and autonomy. But only in the early 1990s did I begin to understand that epistemic ideologies are integral with culturally created and authorized self-conceptions. This forced me to recognize such self-conceptions as 'principled' basis for the then emerging grounds for rejection of WCC. Specifically, whereas in modernity the self was regarded as a product of society and of social relations generally, postmoderns insisted that the only permissible perspective is an anti-disciplinary voluntarism that denies the social, its constraints, boundaries, and 'influences', and conceives the individual as an autonomous agent, freely and creatively drawing upon 'the resources available'. WCC, indigestible in modernity because inconsistent with modernity's epistemic ideology of disciplinarity, remains indigestible in anti-disciplinary postmodernity because it is inconsistent with postmodernity's epistemic ideology of anti-causal voluntarism.

# 35.2 IN MODERNITY

## 35.2.1 Disciplinarity was the Context of Composition

In 1964, when I began work toward a doctoral dissertation on scientific life in physics in German-speaking Central Europe in the aftermath of the Great War, I was by no means looking for, and still less expecting to find, substantively orientating effects of the extra-disciplinary social-cultural milieu upon fundamental concepts within the disciplines of physics and mathematics.[11] To expect, even to look for, such effects I would have had to transgress the idea, the ideal, the myth, of disciplinarity—whose central axiom is that the conceptual development of a science is determined by, and only by, what transpires inside its disciplinary boundary. And disciplinarity was taken for granted by every respected historian of modern science in those high modern decades.

Back then, self-consciously forward-looking professors and students had at least a nodding acquaintance with the Mertonian sociologists' endeavours to construct an 'axiom set' of norms for disciplinary science as social system. But such acquaintance was by no means essential, for we all knew quite as well as they—i.e., we all shared the presupposition from which they proceeded—that modern, disciplinary science

---

[11] Forman, 'The environment and practice of atomic physics in Weimar Germany: a study in the history of science' (Ph.D. dissertation, University of California, Berkeley, 1967; Ann Arbor: University Microfilms, 1968). Typical for the spirit of Kraft and Kroes's (n. 3) argumentation is their proof (p. 96) of the malicious factitiousness of WCC's thesis by pointing out that in my dissertation—written 4 years earlier, when I knew much less about the Weimar intellectual milieu—I had surmised retrenchment, not adaptation, as response to a hostile environment.

functioned as a self-regulating social system within the wider society and culture. What was most essential to this shared preconception—beyond the meritocratic presuppositions for the internal functioning of scientific disciplines—was its implicit 'skin and innards' topology, its presumption of a boundary, of an envelope, effectively isolating the discipline as a knowledge-producing system from the surrounding culture-society-economy on which the discipline depends for moral and material support, and to which, reciprocally, it supplies goods and services.[12]

It was with the intention of promoting this topological conception that, in the late 1960s, Russell McCormmach assumed the editorship of the annual, *Chymia*, renaming it *Historical Studies in the Physical Sciences*, and repurposing it as a vehicle for such 'modern' history of physics—including papers too long for quarterlies. In the forewords that McCormmach drafted for his first volumes he took the topological axiom for granted—'The discipline mediates between the conditions of the non-scientific world and the terms of reference of the practicing physicist'—and concerned himself instead with envisioning the many directions in which historians could explore its consequences.[13] Apart from provision for exceptional length, McCormmach did not foresee, but nonetheless welcomed, WCC. It headed up his third (1971) volume, published early in 1972.

McCormmach's preconception that 'the discipline mediates' was inevitably mine as well. In 1964, when I began my research, and even three years later when I completed my dissertation, the contentions of WCC were still inconceivable by me. But three more years of research made WCC, and my following publication on scientific-political alignments of German physicists, inescapable: in that one particular disciplinary population, in that one particular political-cultural context, in that one particular period—for culturally German physicists in the early post-Great War period—the

---

[12] The image 'skin and innards of science' I take from Thomas F. Gieryn, *Cultural boundaries of science: credibility on the line* (Chicago and London: University of Chicago Press, 1999), p. 21. Gieryn, who started out in the 1970s as Merton's disciple, in the 1980s discovered the importance for science of what lay outside science, leading to his stress on how largely science is socially and culturally constructed by 'boundary work'.

[13] McCormmach was there declaring the power of the discipline over 'Einstein or any other individual scientist': McCormmach, 'Editor's foreword', *Historical Studies in the Physical Sciences*, 2 (1970), ix-xx, on p. xvii. Further, Lewis Pyenson, 'Editor's Foreword', *Historical Studies in the Physical and Biological Sciences*, 37, nr 2 (2007), 189-204, on pp. 198-99, a highly sympathetic account of McCormmach's aims as editor of the journal in its early years. We historians of science should, of all scholars, have been the least susceptible to this ahistorical conception of disciplinarity as the *telos*, the perfect form, to which science, in its history, had finally and permanently attained. But then, we were ourselves, as McCormmach exemplifies, blinkered by fixation upon reshaping history of science/physics into a discipline—a discipline separate from science/physics, on the one side, and from philosophy, on the other side. That said, had that *not* been the case, WCC would not, could not, have come about. Nor could it come about *de novo* now, in postmodernity, when history of science—due especially to science's drastic demotion in cultural value (see my publications cited in note 9)—has found existence as autonomous discipline unsustainable, and realled itself with the several sciences and/or with philosophy, as it had been for centuries before the middle of the 20th. Exemplary for history's returning home to philosophy and science, as of so much else typically postmodern, is Norton Wise (see notes 24-30, below).

boundary separating physicists from the wider intellectual environment was highly permeable, virtually non-existent. Rather than the discipline mediating German physicists' interactions with their political and cultural milieu, many drew their 'terms of reference' directly from 'the conditions of the non-scientific world', and they carried back into their discipline attitudes and interests, both scientific and political, originating in that wider political-cultural milieu. As a consequence, in that place, at that time, the disciplinary institutions of physics, as well as the disciplinary controls on knowledge production within physics, came close to breaking down—indeed did partially break down.[14]

That a failure or an abandonment of the 'disciplinary orientation' became thinkable to me, that I could get an inkling of disciplinarity not as a simple truth about science in its truest form, but rather as a normative ideal or prescriptive myth intended to shape the minds and guide the actions of those inside and outside the postulated boundary, was due in no small part to my being then a regular reader of *Science*. This practice, following the example of Hunter Dupree, who in my graduate student years at Berkeley taught the history of American science and technology—he began his lectures by carrying history into the present with clippings from that day's newspaper or that week's *Science*—made me aware of the 'far-reaching accommodation of scientific ideology to a hostile intellectual milieu' that was then taking place in the United States.[15]

Some—but only some—of the propositions being thrust up so rebelliously from 'below' beginning in the mid-1960s, and finding circa 1970 astonishingly far-reaching accommodation 'above', can now be seen as proto-postmodern. But at that time I was able to understand them only as I then understood Weimar physicists' and mathematicians' accommodations, namely as aberration within, temporary deviation from, an ongoing scientific modernity—a scientific modernity whose essence I too presumed to be disciplinarity.

---

[14] The resulting congruence between the political alignments and the scientific alignments of Weimar physicists that I sketched in 'The financial support and political alignment of physicists in Weimar Germany', *Minerva*, 12 (1974), 39–66, has been substantially confirmed by the extensive archival research of Stefan L. Wolff, 'The establishment of a network of reactionary physicists in the Weimar Republic', *Weimar Culture and Quantum Mechanics* (n. 2), 293–318, thereby, in my view, providing substantial, albeit indirect, support for WCC's argument.

[15] WCC, p. 5. Forman, 'Sarton medal citation [for A.H. Dupree]', *Isis*, 82 (1991), 281–83. More than any work of my nominal mentor, T. S. Kuhn, I admired Dupree's *Science in the Federal Government: A history of policies and activities* (Cambridge, MA: Harvard University Press, 1957) and his biography of botanist *Asa Gray, 1810–1888* (Cambridge, MA: Belknap Press of Harvard University Press, 1959). Though I never enrolled in Dupree's lecture course, I audited it more than once. When Kuhn left Berkeley for Princeton in the summer of 1964—and glad I was to see him go—Dupree graciously consented to supervise my dissertation. As for scholarship in the history of physics, that I learned mainly by following the track of my more advanced fellow graduate student, John Heilbron, through books in the 'QC' class in the library stacks, keeping an eye out for the small, right-angle brackets lightly pencilled in margins, indicating what he had judged to be noteworthy passages.

## 35.2.2  Reception of WCC by Moderns

Circa 1970 I had found myself compelled by an abundance of evidence to maintain that in the first few years following World War I the proclivities of many culturally German physicists regarding the character and fundaments of their science changed very suddenly and radically, and that their motivations for these changes could not reasonably be found within the boundary of their discipline. Given the unanimity of belief in disciplinarity in those post-World War II decades, I should have expected rejection. But in those years I was still very naïve, and I idealized the academic world much as Merton did. As it was patent on WCC's pages that I had searched about more widely than anyone had before me, that fact, I thought, would gain me entrance to open minds. But the Mertonian norm of disinterestedness was a weak reed when the implications of unsealing the envelope around physics were so consequential for philosophers and for my fellow historians, not to mention physicists: to challenge the presupposition of their works was to deny their persuasiveness, and that, of course, was, and has remained, intolerable. Pursuit of these all-too-human considerations I leave to another occasion.

More to the purpose of the conceptual framework offered here—and in some degree extenuating self-interested reactions—are indications that the overwhelmingly 'reserved', when not emphatically hostile, reactions to WCC in the first decade or so following its publication were 'allergic reactions' of the critics' modernist ideological system. Such allergic reactions were, presumably, the cause of readers' failure to recognize—notwithstanding that my text was unconventionally explicit—that I was, when writing WCC, *not* happy with what I had found, that I did *not* welcome the implications for the nature of modern science that followed from my empirical researches into the Weimar physicists' and mathematicians' ideological and conceptual accommodation to their cultural milieu.[16]

Strikingly indicative of such an ideological allergy is the circumstance that a substantial fraction of the earliest and most intemperate rejections of WCC came from scholars outside the history and philosophy of physics—again showing that what WCC appeared to call in question was not a parochial disciplinary interest but an axiom of the modern mind-set.[17] Thus in *Weimar: A Cultural History* (1974), Walter Laqueur,

---

[16] Dorling and Hendry (n.3), p. 179, n. 111, like Kraft and Kroes (n.3), pp. 92, 96–97, not only failed to recognize that I was not happy about what I had found, but took it as a matter of course that findings like mine necessarily implied an intention to find such. Conversely, the proponents of the emerging radical sociology of knowledge, who read me with more open minds, could not fail to recognize my unhappiness, and thus that I was not their ally.

[17] E.g., Frans Gregersen (a distinguished Danish linguist) and Simo Koeppe, 'Against epistemological relativism', *Stud. Hist. Phil. Sci.*, 19 (1988), 447–487, p. 452, n. 5: 'Among the more well known specific analyses is Paul Forman's (1971). His hypothesis is that there is a causal relation between the anti-materialist ideology of post 1st World War Germany and the creation of quantum mechanics, especially the thesis of acausality. It has been proved that Forman's thesis is wrong—e.g. quantum mechanics was not created by Germans only and a version of acausality was introduced in physics in the last years of the 1890s (cf. Kraft and Kroes, 1984; Kragh, 1985)'. Re Kragh, see n.6, above, and n.21, below. Re 1890s, see n. 7,

whose history of the German youth movement I had quoted for support in WCC, assailed me and my account of the division of German physicists on the question of causality, asking, rhetorically, 'Was this, as some have argued, a clash of two different worldviews, the impact perhaps of Spengler and irrationalism?' His answer was an emphatically repeated 'no': 'There was no Expressionism or *Neue Sachlichkeit*, no social protest or revanchism in scientific research; there was only good science and bad science'. Albeit, 'There was a German literature, a German theatre, German schools in the visual arts, even in history and philosophy. But only a fool or a fanatic would talk about German mathematics or German physics'.[18]

Cultural historian Laqueur's passionate attachment, so clearly here expressed, to the topological conception of scientific disciplines, as the necessary condition for their production of true knowledge, underlay the first decade of rejections of WCC. It continues to this day to underlie a large fraction of them, because to this day the conceptual history of modern physics continues to be a refuge for those who cling to the ancient aspiration to transcendence through abstract knowledge. Such 'premodern moderns' and their 'premodern postmodern' successors ignore all the extraordinary circumstances obtaining in German-speaking Central Europe after the Great War, and charge me with making a universal claim about the production of knowledge—a

above. Re Germans: this has been the commonest proof 'that Forman's thesis is wrong', stressed by Dorling, Hendry, Radder, Kraft and Kroes, and on and on to Stöltzner. It is based partly on misplaced nationalism—Austrian or Danish or Dutch or Swiss or ... —and partly on ignorance of the history of scientific culture in countries bordering on Germany and having linguistic affinities with German. (Norton Wise, who knows that history, did not raise this objection.)

[18] Walter Laqueur, *Weimar: a cultural history 1918–1933* (London: Weidenfeld & Nicholson, 1974; reprinted 2000), pp. 195, 217, 219. Another clear indication of an ideological allergy: Laqueur's phrase 'as some have argued' places him (temporally possibly first) in the long line of high-status opponents of WCC who, in their indignation, refuse to dignify its author by naming him. Those with more self-control pretend that Forman and his writings do not exist—so Dresden, referred to in note 5, above. The exceptions to this overwhelmingly rejective response to WCC were usually also, but more respectably, ideologically grounded, grounded in life and education within Eastern European communism—e.g., Alexander Vucinich, 'Soviet physicists and philosophers in the 1930s: dynamics of a conflict', *Isis*, 71 (1980), 236–250, on p. 239—or grounded in reaction against Western European and American 'party line' communism. Thus the only enthusiastic response to WCC that I encountered directly within ten years of its publication was from young Italian physicists, all '68ers, all dissident Marxists. Early on, one of these, Tito Tonietti, then a postdoc high energy theorist (subsequently an accomplished historian of mathematics and music theory), sought me out at the Smithsonian Institution and requested permission to translate WCC into Italian. (Cf. Ch.25 of this Oxford Handbook: F.Del Santo, "...Italy's Cultural Context".) So my personal experience—as also the practice of *Totschweigens*, of 'killing by keeping silent', i.e., by not citing, exemplified by Laquer and Dresden—is fully consistent with Olival Freire's 2011 report that 'so far as I was able to ascertain, no physicist who supports the standard interpretation of quantum mechanics cites Forman's paper, according to the almost 200 citations registered on the *ISI-Web of Science* database (accessed 8 March 2009). In contrast, one may find a meaningful number of what I have called quantum dissidents among the citations. [ ... ] Thus, at least among practitioners of physics, the story about Forman's thesis seems to be linked either to quantum dissidents or Bohm's causal interpretation supporters'. Freire Jr, 'Causality in physics and in the history of physics: A comparison of Bohm's and Forman's papers', in *Weimar Culture and Quantum Mechanics* (n. 2), pp. 397–411, on pp. 410–11.

misrepresentation proceeding directly from their own aspiration to universally valid knowledge and their desperate need for guarantees that they can get it. Though they were wrong about my claims, they were right about the fragility of the 'crystalline sphere' of transcendence—that it would shatter, or should shatter, if struck by one hard counter example.[19]

Laqueur's invective is obviously indicative of a rejection of WCC that arises not from critical reflection but from an ideologically conditioned reflex. A more subtle indication of a primarily ideological response to WCC is imperviousness to the logic of its argument, a refusal to see the unavoidability of WCC's conclusion once one has taken the steps leading up to it. That balking, that refusal to swallow the bitter pill, is especially striking when it is exhibited—and indeed it is nearly universally exhibited— by those historians of physics who largely, even wholly, accept the accuracy of WCC's portrayal of the Weimar cultural environment. For indeed those who have made any effort to inform themselves regarding the character and contents of that intellectual milieu, no matter how ill-disposed they may be to WCC's principal contention, nearly always find themselves, as did Hendry, 'in fundamental agreement' with my characterization of it.[20]

The rare historian of quantum physics who both remains modern and who engages with WCC's argument may even take the next step, allowing, as has Helge Kragh, that Weimar physicists adapted their ideology, that is, their views about the meaning and value of physics, to that dominating their intellectual milieu. They may even acknowledge, as did Hans Radder, that some of those physicists produced 'strong exhortations and sweeping suggestions about how to get rid of causality in physics (or in science generally)', and that such statements were of the nature of an adaptation to the cultural milieu. What, however, they, as moderns, refuse is the conclusion that inevitably follows from their acceptance of the previous steps, namely that all which they admit

[19] Kraft and Kroes (n. 2) are exemplary for the appropriateness of my seemingly oxymoronic terms 'premodern moderns' and 'premodern postmoderns'. Peter Kraft, at that time professor of sociology at the Katholieke Universiteit, Nijmegen, and his collaborator, Peter A. Kroes, who had recently completed a philosophical doctorate at that university with a dissertation on time in modern physical theories, betray by their epistemic axioms, and even more by their mode of argumentation, their rootedness in Roman Catholic scholastic philosophy, even as they presume the modern, disciplinary, topologically closed, character of scientific knowledge. Scholastic echoes are still evident in Kroes's recent writings on the philosophy of technology, to which subject, moving with the times, he subsequently turned. See, notes 13 and 24–30, and, generally, Forman, '...not transcendence...', (n. 5), and Forman, '...technology in postmodernity...', (n. 9).

[20] Hendry (n.3), pp. 157, 172, n. 10. Specifically, 'Lebensphilosophie [...] was, as he [Forman] suggests, transformed by Germany's defeat [...] into a dominant cultural force'. More notably, as Norton Wise stated forthrightly in 'Forman Reformed' [1988], pp. 1–2 (see n. 24, below), 'no critic has dealt seriously with the cultural context which Forman so forcefully portrayed', and more particularly as regards its hostility toward causality, 'Since no one has seriously challenged it, I will take it as basically true'—repeated verbatim twenty years later in Wise's contribution to *Weimar Culture and Quantum Mechanics* (n. 2), p. 421.

to be the case, being the case, it is impossible that no significant consequences for the Weimar physicists' aspirations and activities as knowledge producers resulted.[21]

## 35.3 IN POSTMODERNITY

In 1992, in a paper pertinently titled 'Discipline and Bounding', Steven Shapin declared that 'We are different. The generation which began coming into the history and sociology of science from the mid-1970s has been relatively free of any such commitments'—free, that is, from commitment to a conception of science based on or incorporating an internal-external distinction, free from the topological conception, the 'skin and innards' conception, that modernity took as norm. Shapin was right— right to see this as a revolutionary change in outlook, right to see this change not as rational but as generational. This new outlook, this new sense of personhood, and consequently of all social institutions, indeed of the very idea of social institution, was *not* acquired by that cohort of historians and sociologists from their instructors in graduate school, but through the existential fact of growing up in the 1960s.[22]

Shapin is a bit older than, and came into the history and sociology of science a bit ahead of, that mid-70s-through-mid-80s cohort of Ph.D.s. So also is Norton Wise. It took them both and, to a lesser degree, that cohort too, a little while to disengage their conceptions of the process of scientific knowledge production from the modern perspectives that had been presented to them during their years of formal higher education, and to reconceive those conceptions in conformance with their postmodern proclivities. Thus for some while the disquiet WCC aroused in that cohort remained inchoate. Only in the mid-1980s, inspired in part by French theorizing, and, I imagine,

---

[21] Radder (n. 5), pp. 168–69. Kragh is, to my knowledge, the only scholar explicitly affirming WCC's underscoring of the abrupt reversal, in the winter of 1918–19, of the German physicists' ideological posture regarding the relations of physics and technology: Kragh, 'The fine structure of hydrogen and the gross structure of the physics community, 1916–26', *Hist. Stud. Phys. Sci.*, 15 (1985), 67–126, on p. 109, and Kragh, *Quantum Generations: A History of Physics in the Twentieth Century* (Princeton, NJ: Princeton University Press, 1999), pp. 140, 150, 153, noting repeatedly that 'applications were out of tune with the Weimar Zeitgeist', and that only the reactionaries had the resolve to be so. By emphasizing this buttress of WCC's argument, Kragh is stepping up onto the threshold of WCC's conclusion, but he balked at stepping across it (ibid., pp. 153–4).

[22] Steven Shapin, 'Discipline and bounding: history and sociology of science as seen through the externalism-internalism debate', *History of Science*, 30 (1992), 333–69. The anti-disciplinary animus of that generation, and the fact that it 'had absolutely nothing to do with the influence of professional historians or with the authorized rules and practices of history as an already formed discipline', is categorically affirmed of himself by historian Geoff Eley, who went up to Oxford in autumn 1967: *A Crooked Line: From Cultural History to the History of Society* (Ann Arbor: University of Michigan Press, 2005), pp. 19–20, and n.18 on pp. 211–12.

responding in part to the students whom they were then teaching, did that cohort find their postmodern footing.[23]

Only then did the main themes of the postmodern rejection of WCC appear—appear all together in a text, 'Forman Reformed', that Norton Wise drafted in 1988.[24] 'The capitulation model', Wise declared, 'is not social enough; it does not portray physicists as social beings'. Rather, said Wise, 'I begin from the premise that, in general, the social and political context of science acts as a productive, not a destructive, agent'. Thus what WCC took as extraordinary—namely, the consonance between what the cultural milieu demanded of the Weimar physicists and what nature demanded of them in order that the atom be understood—Wise took to be not at all extraordinary, but the way, 'in general', it really is and ought to be: whatever the milieu wants is right—right not merely socially and culturally, but also epistemologically, i.e., will work as a description of nature.[25]

Consistent therewith, Wise brushed aside disinterestedness, the quality of mind (or, at the very least, institutionalized incentives) central to the ideal of disciplinarity—as it was also to the premodern conception of truth-finding—and articulated the characteristically postmodern position that 'there is nothing necessarily crass about pursuing one's personal and political goals simultaneously with and even through the search for valid scientific explanations of the world'. This position, which for millennia past was considered obviously oxymoronic, can now be stated without argument because it corresponds so exactly to the presuppositions of postmodernity.[26]

---

[23] John H. Zammito, *A Nice Derangement of Epistemes: Post-Positivism in the Study of Science from Quine to Latour* (Chicago: University of Chicago Press, 2004), pp. 165–68. While not himself taking note of it, Zammito makes evident that only in the mid-1980s did hostility to sociology as science, to all generalization, and to any form of causal explanation, arise *within* science studies.

[24] Wise first circulated 'Forman Reformulated' [1988], an 11-page augmented outline of an oral presentation, which he soon reworked into the 21-page 'Forman Reformed' [1988], with the extension in length being a sketch of a notional Bohr/holism/'psychic causality'-centric alternative conceptual history of the creation of quantum mechanics. This 21-page draft, citing just ten authors apart from himself and Forman, and no primary sources, circulated as mimeographed typescript for more than twenty years, during which it was frequently cited as an adequate antidote to WCC. Wise presented his contentions as 'Forman Reformed, Again' at the 2007 conference on the Forman thesis in Vancouver, BC, and published yet another version in the volume resulting from that conference: *Weimar Culture and Quantum Mechanics* (n. 2), pp. 415–31. This published version, which now serves as that adequate antidote, contains even less attention to the evidence and argument of WCC than did 'Forman Reformed', leaning rather for support on references to Stöltzner and Coen's 'Vienna indeterminism' (see note 7, above).

[25] Wise, 'Forman Reformed' [1988], pp. 6, 7; 'Forman Reformed, Again' (2011, n. 24), p. 427: '... acts productively'. 'Forman Reformed' [1988], strictly considered, is confused and self-contradictory. Wise wants *both* to 'avoid succumbing to some form of social determination of scientific ideas' *and* to repudiate the internal-external distinction, advocating a view of the physicist, *qua* physicist, as participant, without qualification, in the wider culture and society. But when one takes account of the postmodern 'life feeling' from which Wise is proceeding, namely, untrammelled voluntarism, the contradictions in Wise's exposition may be chalked up to 'motivated' thinking—and those in Galison's too, quoted in note 27, below.

[26] Wise, 'Forman Reformed' [1988], p. 10. In the earlier version, 'Forman Reformulated' [1988], p. 6, Wise had said 'There is nothing crass'—the truly postmodern position—softening it to 'nothing

Wise's dismissal of WCC on the grounds that its perspective is 'not social enough'—
that is, on the grounds that WCC still presumes *some* distinction, *some* sort of
membrane, between the collegial worlds in which physicists work and the cultural
environments in which they live—was Wise's leading criticism. It echoed the 'everything is political' radicalism of the sixties, and the 'social construction of reality'
radicalism of the seventies, and it was the way Wise was immediately read and affirmed
(as, e.g., by Galison).[27] But then 'the science wars' broke out, and suddenly such radical
chic became very dangerous to spout.[28]

Consequently, 'not social enough' soon gave way to Wise's second, equally ideological, basis for dismissal of WCC, namely, that it is marred by allegations of
'influence', a word that betrays Forman's modern, causalist, preconceptions: 'Weimar
physicists [ ... ] are better understood as drawing on their culture, rather than being
influenced by it'—and 'resource' rhetoric runs all through Wise's paper from that point
on. This voluntarist 'resource perspective', being free of radical political taint and being
consonant with postmodernity's authentic 'life feeling'—namely, possession of autonomous, free-agent personhood—became in postmodernity the most commonly invoked
grounds for rejecting WCC.[29]

In postmodernity, not-social-enough and resource-not-influence could well be
grounds for rejection of historical writing on any conceptual topic. But WCC is not

necessarily crass' in 'Forman Reformed' and repeating that softened assertion verbatim in 'Forman
Reformed, Again' (2011), p. 428. So much for disinterestedness, the keystone in CUDOS, the set of
Mertonian norms. Compare John Dewey's *reductio ad absurdum* proof of the incompatibility between
being personally interested and being scientific, which he presented at Harvard's Tercentenary Conference (1936), as quoted by Forman, 'The primacy of science...' (n. 9), pp. 26–27. But now, in
postmodernity, this modern meaning of disinterested is almost incomprehensible, and has given way
in common parlance to 'without interest or concern'.

[27] Peter Galison, 'Multiple Constraints, Simultaneous Solutions', *PSA: Proceedings of the Biennial
Meeting of the Philosophy of Science Association*, Vol. 1988, Volume Two: *Symposia and Invited Papers*
(1988), pp. 157–163, on p. 159: 'According to Forman [ ... ] the physicists *capitulated* (Forman's term)
and introduced indeterminism into their subject. Norton Wise (forthcoming) criticizes the Forman
thesis for its reliance on an external force (Weimar pessimism) to drive internal changes within the
practice of physics (indeterminacy). Rightly, I think, Wise argues that the quantum mechanicians ought
to be considered part of their culture not as the pawns of outside powers'.

[28] Zammito, ... *Post-Positivism in the Study of Science* (n. 23), Ch. 8, 'Radical reflexivity and the
science wars', pp. 232–70.

[29] Wise, 'Forman Reformed' [1988], p.8 (quotation); resource rhetoric: pp. 6, 9 (three times), 10, 11,
16, 18. Galison's 'pawn' metaphor quoted above (n. 27) suggests much the same aversion. Mara Beller,
*Quantum dialogue: the making of a revolution* (Chicago: University of Chicago Press, 1999), p. 58,
dismissed WCC with the dictum that 'these contexts served more as resources and less as influences'. For
the sources and functions of this resource-fullness, see Forman, '(Re)cognizing postmodernity' (n. 9),
pp. 166, 174, 175. Interestingly, indeed indicatively for scholarship in postmodernity, although Wise, in
his contribution to *Weimar Culture and Quantum Mechanics* (n. 2), got to 'resource' only after spending
the first two thirds of his text inveighing against 'influence', esp. as 'causal influence', the published
reviews of that volume that I have seen—all singling out Wise's contribution for highly favourable
attention—describe and endorse it as emphasizing 'resource', so: Kristian Camilleri, *Isis*, 103 (2012),
794–96 ('we should instead see physicists as agents who were able to select certain cultural resources
available to them'); Suman Seth, *Metascience*, 22 (2013), 567–74 ('Forman's thesis functions not as a

about just 'any' topic. It is quite specifically about postmodernity's foundational issue, causality. As voluntarists, postmoderns are categorical anti-causalists, while the author of WCC not only sided explicitly, even indignantly, with Weimar's causalists, but offered a (censorious) causal account of how and why so many German physicists of that era became anti-causalists. It is thus fully appropriate that Wise devotes the first two thirds of his 2011 critique of WCC to an emphatic stance against causal history: 'It is the causal form, I argue, rather than the cultural context, that constitutes the problem [with WCC]'.[30]

What not-social-enough, resource-not-influence, and the-causal-form have in common as grounds for dismissing WCC is that they treat the factual evidence as beside the point: postmodern postulations about human nature contradict Forman's modern presuppositions and aspirations, and that suffices to vitiate WCC as contribution to historical scholarship. Putting aside the professedly ideology-*über-alles* aberrations of Communist and National Socialist regimes, the like could never have happened in modernity: in modernity fact had precedence over preference; in modernity it was necessary to allege that WCC was factually wrong, even where the appeal to fact in modern dismissals of WCC—as by Laqueur—was manifestly a fig leaf for ideological commitments regarding knowledge production. Thus the reception history of WCC exposes an essential difference between the modern and postmodern value systems: in modernity our scientific knowledge, and the means by which it is attained, was our highest cultural good, while in postmodernity self-conceptions are valued more highly than scientific conceptions, or scientific facts, or any process for establishing them.

---

positive, but as a negative model', for history should instead 'draw the physics out of the society in terms of the resources available to its practitioners'); Martin Jähnert, *Berichte zur Wissenschaftsgeschichte*, 36 (2013), 384–385, highlighting 'Kultur als Ressource'.

[30] Wise, 'Forman Reformed, Again' (n. 24), on p. 421. Wise continued, 'To put the point polemically, why in the world, if we cannot give a causal account of the formation of snowflakes, would we imagine that we could give a causal account of the formation of thoughts?' I expect that the fallaciousness of this protest will be evident to every reader of this *Handbook*. Wise, however, was so pleased with the analogy that he offered it again, for the benefit of philosophers, in 'Science as (historical) narrative', *Erkenntnis*, 75 (2011), 349–376, on p. 373. Wise has received the History of Science Society's highest marks of recognition (this last publication originated in an honorific lectureship). Nor is he alone among the field's most applauded scholars in declaring against causality. Frans van Lunteren rightly singled out Lorraine Daston and Peter Galison's *Objectivity* (New York: Zone Books, 2007) when protesting against this postmodern eschewing of 'all talk of causal factors, influences, and explanations': van Lunteren, 'Historical explanation and causality', *Isis*, 110 (2019), 321–24. Likewise, in greater detail and at greater risk to her career, Katherina Kinzel, 'Geschichte ohne Kausalität. Abgrenzungsstrategien gegen die Wissenschaftssoziologie in zeitgenössischen Ansätzen historischer Epistemologie', *Berichte zur Wissenschaftsgeschichte*, 35 (2012), 147 – 162. Here, again, the history of science was slow in catching up to 'the historical discipline', where rejection of causal explanation had already become orthodox by the mid-1990s: Forman, '(Re)cognizing postmodernity' (n. 9), p. 168, n. 11, and 'From the social to the moral to the spiritual' (n. 9).

HISTORY OF QUANTUM INTERPRETATIONS

## 35.4 CONCLUSION

How can it be that the thesis of WCC, which few historical scholars, and even fewer philosophers, unequivocally affirm, continues to hold—for half a century, now—a prominent place as stumbling block in the literature and pedagogy of the 'disciplines' of the history of science and the philosophy of science? The answer given by this account of the reception history of WCC is that in modernity, even though disciplinarity was far from fully realized in scientific and scholarly lives, adherence to the myth of disciplinarity was central to the conduct of such lives. Hence WCC's empirically well-grounded challenge to the myth of disciplinarity became the proverbial monkey on the backs of ideologically modern (and premodern) historians and philosophers of science.

Why, then, was the WCC monkey not shaken off by historians and philosophers of science in anti-disciplinary, freedom-to-make-believe postmodernity? Here the situation becomes more complicated: although, in postmodernity, the myth of disciplinarity is deprived of its ideological supports, in practice it remains indispensable to the functioning of the academic system. Whereas in modernity disciplinarity was a truly believed, constitutive myth, in postmodernity disciplinarity survives as a more-or-less conscious pretence. Consequently, the WCC monkey still clings to historians' and philosophers' backs, for it is as inseparable from that pretence as it was from the truly believed, constitutive myth.

Worse, whereas in modernity a programme of historical research that could, in principle, get the WCC monkey off all backs was easily conceived, such is not the case in postmodernity. In postmodernity the task has become far more difficult, because any replacement of WCC that *really* replaced WCC would have to satisfy the postmodern demands for boundaryless-society, resource-not-influence, and no-causal-form, while persuasively explaining away the many facts joined together by WCC's several reinforcing lines of argument. Various participants in the 2007 Vancouver conference on the Forman thesis and contributors to the subsequent publication, *Weimar Culture and Quantum Mechanics* (2011), claimed they were doing that, or would do that, but none has. It is very unlikely that anyone ever will.

....................................................................................

# CHAPTER 36

# QUANTUM HISTORIOGRAPHY AND CULTURAL HISTORY

## *Revisiting the Forman Thesis*

....................................................................................

### ALEXEI KOJEVNIKOV

THE historiography of science and technology knows several bifurcation points, when the introduction of a radically novel type of argument occurred at just the right moment to touch a sensitive nerve, spark a fundamental, often prolonged controversy, and irreversibly change the direction of the field at large. In retrospect, the list of such landmark intellectual breakthroughs would have to include Hessen's 'social and economic roots' of 1931, the Merton thesis of 1937, Koyré's *Études Galiléennes* of 1939, Kuhn's *Structure* of 1962, the Forman Thesis of 1971, Shapin and Schaffer's *Leviathan and the Air-Pump* of 1986, Haraway's *Primate Visions* of 1989, and possibly a couple of other programmatic texts. As with the rest of the list, the Forman Thesis's methodological influences stretch far beyond its original focus, but it did emerge out of the history of quantum physics and needs to be understood from its actual roots.

This chapter draws in parts from the earlier published, co-authored introduction (Carson, Kojevnikov, and Trischler, 2011). Its first section describes the founding period of quantum historiography during the 1960s. The second summarizes the main ideas of the Forman Thesis and the third examines the controversy it inspired and its subsequent influence. The fourth section discusses several key examples—some well-known, others from the more recent literature—that help test and establish the boundaries of Forman's approach, and that leads to the conclusions about its current status.

# 36.1 EARLY QUANTUM HISTORIOGRAPHY: INTELLECTUAL AND DISCIPLINARY HISTORY

While still in its formative stage during the 1960s, the professional historiography of the quantum revolution stood then at the forefront of methodological innovations in the history of science writ large. As he reviewed the discipline's major trends, Thomas Kuhn identified a shift of attention away from the ancient and old classics and towards the history of recent, i.e., 20th-century science (Kuhn, 1967). An encouragement for this move came from scientists themselves, especially physicists, who were then at the height of their Cold War power, prestige, and financial largesse. No less importantly, they could also boast a very recent fundamental revolution—relativity and quantum— conceptually as profound and awe-inspiring as any of the greatest scientific achievements of the past. Quantum mechanics by 1960 was considered complete, accomplished, and interpreted with the reigning Copenhagen philosophy. Some of the main heroes of this revolution were still alive and able to share their stories, but the deaths of Albert Einstein (1955), John von Neumann (1957), Wolfgang Pauli (1958), and Erwin Schrödinger (1961) raised the alarm among the physics community and prompted calls to record the history of the passing giants.

Responding to a physicists' initiative, with support from the American Institute of Physics and the American Philosophical Society, a team of first-generation professional historians of physics led by Kuhn used this unprecedented opportunity to create a treasure-trove of primary sources for the history of quantum physics. The US National Science Foundation granted funding for a three-year project, which enabled its almost 'big science' dimensions, by history's disciplinary standards. While empowering historians—most of whom at the time had an educational background and sometimes also research experience as scientists—this support also exposed them to the challenge of whether they could emancipate themselves, professionally and intellectually, from the authority of the scientists they studied. Physicists developed and cherished their own historical mythology, telling and retelling it on numerous celebratory and educational occasions. Quantum mechanics, too, had produced a rich quasihistorical narrative and a canon of autobiographical memoirs, which historians inevitably came to contradict and correct in multiple ways as they started careful investigations with primary documents. In his 1967 review, Kuhn took critical aim at one such immensely influential text, George Gamow's *Thirty Years that Shook Physics: The Story of Quantum Theory* (1966), as an example of unreliable popular history dominated by participants' own accounts, anecdotal memories, and disciplinary myths.

The ambitious project of the *Archive for the History of Quantum Physics* (AHQP) was not an archive in the usual sense, but a major pioneering undertaking in oral history within the field of history of science.[1] Over the course of three years, 1961–1964, Kuhn, together with graduate students John L. Heilbron and Paul Forman, and with bibliographical assistance from Lini Allen, recorded detailed and lengthy interviews with over a hundred surviving scientists who made important contributions to quantum physics and chemistry between approximately 1900 and 1935, among them Niels Bohr, Max Born, P. A. M. Dirac, and Werner Heisenberg. In conjunction with those interviews, the team sought to locate, organize, catalogue, and microfilm relevant primary documents and correspondence held by the interviewees, as well as libraries and archives. The interview transcripts and the microfilms with 'about 100,000 frames of material' of letters and manuscripts were deposited in the library of the American Philosophical Society and several other (currently more than twenty) convenient locations, thus bringing the sources much closer to potential researchers (Kuhn et al., 1967; Heilbron, 1968, p. 98). The resulting primary source collection established the empirical foundation for practically every historian working in the field since, including AHQP team members' own research into the history of quantum ideas.

The quantum revolution in physics was a large collective enterprise, but not exactly a coordinated effort. No other great scientific innovation of the period, including relativity theory, had so many crucial and chronologically overlapping contributions from dozens of prominent authors with often conflicting agendas, preferences, and aspirations. The fast reproduction rate of journal publications made it possible for a submitted paper to be published sometimes within two to three months, and about one month later already be cited in another paper submitted for publication by a different author. This explosive pattern of knowledge production also profited from close personal contacts, global geographical spread, and an unusual mobility of converts, rich correspondence networks, and informal exchanges of proof sheets of as-yet unpublished articles (Kojevnikov, 2020). Unlike a typical 19th-century model of discipline formation, no single major centre or institution of graduate training could accommodate this large community of researchers. Its members often pushed the work in diverging, sometimes contradictory directions, so that no individual leader could stay effectively in charge or claim ultimate credit for the enterprise. In the 1920s, they also produced scientific and philosophical infighting of such intensity and inconsistency of competing views that was almost unprecedented in the history of science.

Such enormous density of recorded details, thoughts, and arguments inspired a hope for a much more invasive history of ideas that would reconstruct and uncover the ways of scientific creativity—a historical strategy commensurable with Kuhn's *Structure of Scientific Revolutions* (conveniently published in 1962). Oral histories, in the long run,

---

[1] During its execution, the formal name of the project, at least the part of it funded by the NSF and its published report, was *Sources for the History of Quantum Physics*. The resulting collection of oral histories and microfilms is called officially the *Archive for the History of Quantum Physics*, the name under which the entire undertaking has become more commonly known during the subsequent decades. Heilbron (1968) provides a critical review of the early period of quantum historiography before the field's major expansion in the late 1960s and a detailed overview of the AHQP project. For a recent historiographic analysis, especially with regard to oral history, see (te Heesen, 2020).

proved a mixed blessing: while rich in personal and otherwise unavailable details, they were also partially unreliable and self-serving, because long-term memories of participants adapted to post hoc rationalizations and often contradicted many of the often-confused thoughts recorded in primary archival sources. But thousands of extant letters and manuscripts formed the basis of many detailed historical investigations, in particular in the flagship journal *Historical Studies in the Physical Sciences* (*HSPS*, 1969–1979), and allowed historians to question and challenge many historical myths and physicists' disciplinary folklore.[2] For example, Kuhn's *Black-Body Theory and the Quantum Discontinuity, 1894–1912* (1978) contradicted the cult of the founding father in quantum theory that traditionally attributed the introduction of fundamental discontinuity to Max Planck in 1900. According to Kuhn's unceremonious analysis, Planck accepted the conclusion that elementary quanta were discontinuous only reluctantly and noticeably later than other physicists, in particular Einstein and Paul Ehrenfest. The riches of available sources and the emerging power of historians to correct the scientists' disciplinary beliefs also supported the programme of 'rational reconstruction of scientific creativity', attempts to 'follow the thinking' as painstakingly as possible and recapitulate discoveries *in statu nascendi* in scientists' heads. This exaggerated hubris of the history of modern physics in its puberty, during the process of professional and intellectual emancipation from the parental authority of scientists, subsequently came under a harsh critical reassessment in (Forman, 1991).

As explained by Heilbron (2011) in his recollections about that *Sturm und Drang* era, the American present—the politically and socially turbulent 1960s—interfered with thinking about the past and prompted historians of physics to ask questions beyond the traditional repertoire of the history of ideas. The Vietnam War and the Cold War made scientists not only objects of veneration, as before, but also of criticism and suspicions, as the younger generation of Americans focused their attention on physicists' roles as weapon-makers and lobbyists. No longer seen merely as champions of ideal truth, science and its spokesmen increasingly appeared as servants and agents of the ruling political establishment and militarism. Meanwhile, the accelerated world-wide growth of R&D and higher education in the wake of the shocking launch of Sputnik by the Soviet Union, and the resulting transformation of science into a mass profession with supersized infrastructure and budgets, shifted scholarly interests towards investigating the larger scientific community, its structures, institutions, jobs, social relations, patrons, and the sources of funding, with the 'follow the money' method. These notions formed the conceptual vocabulary of the disciplinary-institutional approach to history of science, promoted in *HSPS* under the editorship of Russell McCormmach.

While the AHQP project focused primarily on individuals, the subsequent team effort by Forman, Heilbron, and Spencer Weart produced a survey of the international physics community *circa* 1900 that used a wealth of statistical data to evaluate the

---

[2] (Heilbron and Kuhn, 1969; Forman, 1969), and many other examples.

entire discipline, its demographics, institutions, positions, and social practices (Forman *et al.*, 1975). Forman's own Ph.D. thesis (1967) analysed the finances, structure, and modes of operation of the German-speaking physics community during the difficult political, economic, and social situation following the Central Powers' defeat in World War I. Forman's dissertation remained unpublished (unjustifiably so), but it influenced the approaches in many a later investigation by other historians in this new area of research that it opened up. Its historical analysis described the inner workings of the academic community and the social background in which the ideas of quantum mechanics brewed. The dissertation and two later articles–one on the post-war international boycott of German science and the other on its attempts to overcome international isolation and secure funding and political alliances (Forman, 1973, 1974)—together, served as the foundation for the seminal Forman Thesis.

# 36.2 THE FORMAN THESIS: SOCIAL AND CULTURAL HISTORY

What is usually referred to as the Forman Thesis is, more precisely, an argument, a logical sequence of theses that combine the approaches from social and cultural history to the developments in quantum physics following World War I.[3] It started with Forman's doctoral dissertation of 1967 and continued with a half-dozen articles, both preceding and following the most famous one of 1971—'Weimar Culture, Causality, and Quantum Theory, 1918–1927: Adaptation by German Physicists and Mathematicians to a Hostile Intellectual Environment'—so long, thorough, rich, and fundamental, that it effectively works as a monograph, even if published by a journal.

Forman described the intellectual climate in the economically and psychologically traumatized German cultural space after the defeat in World War I, along with the loss of the Empire, its colonies and territories, its military might and industrial prosperity. The general perception of overwhelming crisis—political, economic, and social— affected all aspects of life, including science. Subjected to harsh treatment by the victorious allies following the Versailles Treaty, the German Empire lost its sense of superiority, self-confidence, and global importance. The internationally isolated and frequently humiliated government of the Weimar Republic was also internally unstable and weak, threatened by revolutions and putsches, from both the radical right and the radical left. The German economy suffered several major blows, especially the 1922/23 hyperinflation and the 1929 stock collapse. The situation in former Austria-Hungary

---

[3] In addition to the Forman Thesis discussed here, Silvan S. Schweber also defined the 'second Forman thesis' on the symbiotic relationship between military funding agencies and the character and directions of physics research in Cold War America, analysed in Forman's papers of the 1980s, and the 'third Forman thesis', on the post-modern relationship between science and technology, studied by Forman in the 1990s (Schweber, 2014, p. 180).

was even worse, by a significant margin, as the once mighty empire split into a half-dozen small nations, each much more vulnerable than Germany to post-war political and economic insecurity. In many of these territories, the war did not end in 1918, but devolved into the continuing violence of ethnic and civil wars between various nationalistic, proto-fascist, and revolutionary forces (Gerwarth, 2016).

The German *Bürger* who lost their bank savings and prosperity—the middling classes including their intellectual subsection, the academics—felt the contrast between pre-war confidence and post-war troubles especially painfully. Their sense of a general crisis translated into, on the one hand, nostalgic feelings about the imperial past and political alienation from the left-leaning Weimar government, and on the other, disillusionment and disenchantment with many of the key ideological values of the pre-war era, such as belief in progress, positivism, modernity, and rationality. The new postwar *Zeitgeist* promoted instead a much more conservative, romantic, and pessimistic outlook and blamed the disaster from the war and the country's misfortunes on the earlier infatuation with shallow materialism and technological optimism. The fashionable philosophical and ideological treatises, including *Lebensphilosophie*, Nietzsche, and Oswald Spengler's *Decline of the West*, promoted irrational, organicist, intuitive, and anti-materialist lines of thought. In a cultural environment hostile towards the values associated with the exact sciences, physics came under severe critiques as too rational, abstract, mechanistic, and causal: 'I show that in the aftermath of Germany's defeat the dominant intellectual tendency in the Weimar academic world was a neo-romantic, existentialist "philosophy of life", revelling in crises and characterized by antagonism toward analytic rationality generally and toward the exact sciences and their technical applications particularly. Implicitly or explicitly, the scientist was the whipping boy of the incessant exhortations to spiritual renewal, while the concept—or the mere word—'causality' symbolized all that was odious in the scientific enterprise', explained Forman.

He then immediately drew attention to a 'remarkable [historical] paradox: this place and period of deep hostility to physics and mathematics was also one of the most creative in the entire history of these enterprises...I had myself previously assumed that in the face of antiscientific currents the predominant response in these highly professional sciences would be retrenchment...and reaffirmation of the discipline's traditional ideology...Yet the historian who takes even the most casual notice of the valuations of physical sciences in contemporary American society...[is] witnessing...a widespread and far-reaching accommodation of scientific ideology to a hostile intellectual environment' (Forman, 1971, pp. 4–5). In the similarly inhospitable intellectual climate of Weimar Germany, Forman observed, many academics started wavering in their attachment to rationalist values that had heretofore been central to the business of science as such.

Economically, as a profession, science was hit earlier and particularly hard in the former Habsburg lands, to the point that in the 1920s even in the capital Vienna, the famous Institute for Radium Research resorted to hiring women as regular research staff, since it could not afford to pay liveable salaries to male scientists (Rentetzi, 2004). Many Austrian and Hungarian scholars, including Erwin Schrödinger and Wolfgang

Pauli, moved to Germany, where the inflation and general economic troubles did not damage research nearly as severely. In fact, German science retained a remarkable vitality: its spokesmen and the Weimar government often saw and used it as *Macht-Ersatz*, the country's one remaining strength, a substitute for power that had been lost in most other domains: political, military, diplomatic, and economic. Under the attempts by the victorious powers to boycott and isolate it, German science was banned from many international meetings, conferences, and exchanges, and its global institutional dominance decreased. It also lost many foreign students, but redirected its main international connections and cultural imperialism to countries that had remained neutral during the war and to Soviet Russia and Japan (Kevles, 1971; Forman, 1973). Meanwhile, in its home base, research and publishing continued as actively as before, even if a prohibitive exchange rate prevented subscriptions to foreign publications and undermined opportunities for international travel. Scientific infrastructure—institutes and laboratories, built and equipped during the imperial period before the war—were still far better and richer than anywhere else in Europe. The government bureaucracies, at least in Prussia, continued to value and support science materially throughout the difficult times: professors and other salaried academics maintained liveable incomes adjusted for inflation, while grants from the newly created emergency fund *Notgemeinschaft der Deutschen Wissenschaft* partially compensated for losses in research support (Forman, 1974). Yet the erosion of prestige, security, and of their previously high social status made the majority of German academics, with a few notable exceptions such as Einstein, significantly more right-wing than their post-war governments and nostalgic about the pre-war *Kaiserreich*.

In a period when revolutions, military coups, and crises threatened German society, science and its individual disciplines were also often declared to be in a state of deep crisis and ripe for radical conceptual changes. The widespread discourse about the 'crisis of science' not merely acknowledged the economic difficulties of the profession or a disciplinary 'crisis' as in Kuhn's model of scientific revolutions, but was understood in a much more general and profound sense. Post-war doubts not only encouraged and made it easier for scientists to question the conceptual foundations of the existing knowledge; they also undermined general values that heretofore had been associated with the very essence of the exact sciences. In particular, mechanical determinism, or the principle of causality, came under severe criticism as too rationalistic and, indeed, mechanical. 'In the vocabulary of *Lebensphilosophie* there were two characteristic words: one—*Anschaulichkeit*, [visual] intuitiveness—had strongly positive connotations; the other—*Kausalität*, causality—was emphatically pejorative. And the epitome of the abstract, unintuitive, and causal mode of apprehending reality was that of the theoretical physicist', observed Forman (1979, p. 13).[4]

---

[4] Some participants in the debates proposed to draw a distinction between determinism, a more rigid, mechanical, 18th-century concept, and causality as a general, philosophical, and potentially more inclusive and amendable principle. For the purposes of this essay, these two notions will be considered roughly synonymous.

University administrators and ministry officials wanted 'to do new things', and even some older professors felt the need for new agendas and to hire representatives of the new physics (Born, 1978, p. 200). By the latter they meant first and foremost research in the atomic and quantum domains, both theoretical and experimental. Physicists who represented these new lines of research became much more willing, in comparison with more stable times, to revise or entirely abandon the fundamental principles and foundational concepts of classical physics. They also often made more than just rhetorical concessions to the fashionable philosophical critiques of the time and to the hostile intellectual environment, which they faced, most directly, as academics in the public eye at their own universities. Typically, this occurred when an exact scientist delivered a public address, a common genre in the German academic world. In these *Reden*, scholars explained and commented on the developments in their particular discipline to a gathering of university colleagues from all fields or to academically educated general audiences, in language and terms (mostly philosophical) that would be comprehensible to non-specialists. By using these sources, Forman discovered that several prominent physicists and mathematicians declared their readiness to abandon or restrict the principle of causality in physics even before the invention of quantum mechanics itself, i.e., several years prior to 1925. Once that revolutionary theory appeared, acausality was quickly ascribed to it and proclaimed the fundamental scientific principle of the new quantum mechanics of atoms and electrons.

According to Forman's analysis, outside pressure thus contributed to both ingenuity and opportunism of Weimar scientists:

> I am convinced ... that the movement to dispense with causality in physics, which sprang up so suddenly and blossomed so luxuriantly in Germany after 1918, was primarily an effort by German physicists to adapt the content of their science to the values of their intellectual environment. The explanation of the creativity of this place and period must therefore be sought, in part at least, in the very hostility of the Weimar intellectual milieu. The readiness, the anxiousness of the German physicists to reconstruct the foundations of their science is thus to be construed as a reaction to their negative prestige. Moreover the nature of that reconstruction was itself virtually dictated by the general intellectual environment: if the physicist were to improve his public image he had first and foremost to dispense with causality, with rigorous determinism, that most universally abhorred feature of the physical world picture. And this, of course, turned out to be precisely what was required for the solution of those problems in atomic physics which were then at the focus of the physicists' interest.   (Forman. 1971, pp. 7–8)

In two subsequent papers, Forman further developed and extended his original thesis. The first one looked at the reception of quantum mechanics outside of the German cultural sphere, in particular in Great Britain, where hostility towards the values of science and, correspondingly, commitment to acausality were much less pronounced at the time, and in America, where the issue of (in)determinism was largely treated with indifference (Forman, 1979). The other paper took the argument beyond the question of causality by analysing two additional and important culturally sensitive notions of the Weimar milieu: *Anschaulichkeit* (the word combining the meanings of 'visualizable' and 'intuitively grasped', depending on the context) and *Individualität* (or individuality). Forman drew the distinguishing line between the actual character of quantum physics and some philosophical features frequently ascribed to it as accommodation to the popular value-laden concepts of the time. Thus, despite the highly abstract and counterintuitive nature of the quantum description, physicists often presented its results with the label *Anschaulich*, or intuitive. Quantum mechanics' formalism abandoned the absolute individuality and distinguishability of elementary particles (Monaldi, 2009), but was publicly proclaimed to represent the opposite, their 'indestructible individuality' in the world of atoms, sometimes with an analogy to individuality in the organic world. The laws of the new theory were probabilistic and statistical, but its authors were more than willing to make much stronger ontological claims. Max Born in 1926 declared himself 'inclined to abandon determinedness in the atomic world', and a few months later Werner Heisenberg proclaimed categorically that 'quantum mechanics established definitively that the law of causality is not valid' (Forman, 1984, p. 336).

The cultural values that appealed to the predominantly conservative, anti-rationalist intellectual milieu within which the German physicists operated thus became written into the prevailing philosophical interpretation of quantum theory. 'My conclusion is that there was little connection between quantum mechanics and the philosophic constructions placed on it, or the world-view implications drawn from it. The physicists allowed themselves...to make the theory out to be whatever they wanted it to be—better, whatever their cultural milieu obliged them to want it to be. This conclusion is admittedly radical. But it does not touch the question of the social construction of reality so directly as one might at first be inclined to suppose. It is neither a statement about the physicists' practice in their laboratories nor about the physicists' theories as descriptions of reality. It is rather a meta-meta statement, a statement about the physicists' statements about their description of reality', summarized Forman (1984, pp. 342–343).

## 36.3 IMPACT AND CONTROVERSIES

The shock and uproar created by the Forman Thesis at the time of its introduction were practically guaranteed, as his landmark study explicitly contradicted then generally accepted and cherished beliefs about science. It put forward and placed in the centre of a broader discussion the argument that culture and cultural values prevalent in a given place and time condition the results of scientific research, i.e., the very content of scientific knowledge. Heilbron, whom Forman had asked to deliver the first public presentation of his 'thesis' at the Christmas 1970 meeting of the History of Science Society in Chicago, described the reaction as a 'maelstrom'. As had been anticipated.

After all, acausality at the time was generally accepted as the very core, fundamental concept of quantum mechanics, and to ascribe this glorious scientific discovery to the influence of reactionary Spenglerian philosophy seemed like a blasphemous offence against the truthfulness of physics and the purity of its spirit. For Forman, of course, the idea that famous physicists could come under the corrupting spell of a hostile public environment and reactionary ideology was not a thoughtcrime, but a lamentable reality of the Cold War (Heilbron, 2011).[5]

For many of the outraged, Forman's study represented another incarnation of the abhorrent 'externalist' approach to scientific content. Indeed, there had probably been only one even more influential and more controversial article ever published in the history of science, the 1931 analysis of classical mechanics in Boris Hessen's 'The Social and Economic Roots of Newton's *Principia*'. The two classic works did have something in common: they both enormously upset, each in its own ways, the essentially Platonic ideology of science as a pure intellectual activity, a noble search for abstract truth, supposedly in control of its intrinsic scientific method and of the criteria of true knowledge. Instead, both approached science as an essentially human, and thus also earthly, social and cultural activity, and accepted the necessary epistemological consequences of such an assumption. (Freudenthal and McLaughlin, 2009).

Yet the differences between these two papers, separated by forty years, were no less important than their similarities. Hessen developed a deliberately Marxist argument that proclaimed the influence of the economic and technological basis of the time period upon its scientific superstructure. In Forman's analysis, culture played a key role, mediating and channelling the impact of economic and social conditions. Hessen, writing at the time of the revolutionary industrialization of the Soviet Union, promoted an unabatedly progressivist view on science, without reservation counting it among the major forces of social and political progress. For Forman, in the era of DDT, napalm, and Agent Orange, the question of science's and scientists' political associations became less optimistic and more ambivalent. His study found some leading proponents of the quantum revolution entering a pact with anti-rationalist conservative ideological currents, whereas those physicists who upheld the values of causality and reason and

---

[5] I am grateful to John L. Heilbron for his letter of 21 January 2021, with a description of the meeting and the following clarifications: 'The reason Paul did not attend the Chicago meeting was that he had had an operation from which he did not feel sufficiently recovered to face what he called the "maelstrom". But he had wanted to attend and regretted the loss of conversations he had anticipated with other quantum historians. The paper as delivered was a condensed version of the third part of Paul's long Weimar article in *HSPS*. In sending it to me, he wrote that he regarded it "as tending to show that the philosophers of science were more right than we have allowed—more right about the necessity for a conventionalized, stylized, 'idealized', picture of science as the basis for a history of science." (Letter of 22 Nov) 'Maelstrom' was his word. I do not think that the reaction to the paper was as negative or noisy as your text may suggest. There were friendly people in attendance: Russ McCormmach, Stan Goldberg (I think), myself, and I do not remember how many others; Erwin Hiebert was the chairman (Paul did not think he would be favorable!) and Gerry Holton the commentator. As usual with him, Holton did not write out his comments; but you might be able to reconstruct his and Hiebert's reactions from their publications of the period.'

often adhered to more progressive politics, were nevertheless rhetorically dismissed at the time as scientifically 'conservative'. Last but not least, Hessen's essay was largely declarative and programmatic. It inspired and required further empirical justification, including Robert Merton's *Science, Technology, and Society in 17th century England* (1938). Forman's 'Weimar Culture' relied on an enormous body of primary sources, many heretofore unused, and came out of a vast empirical—archival and historical—project.

The main ideological and methodological conflict of the Cold War history of science was by that time already old, tired, and entrenched, but still arousing strong passions. The Marxist-inspired approach, imported from the Soviet Union by Hessen in the 1930s and dubbed 'externalism' as disambiguation by its ideological opponents, by the late 1960s had been largely abandoned back in the USSR but still promoted by Western Marxists such as J.D. Bernal (Bernal, 1954). It interpreted the Scientific Revolution as a product of social revolutions in early modern Europe, and science as a progressive force that was intimately linked and responsive to the economic and technological needs of the rising capitalism. To exclude subversive Marxism from major academic programmes in history of science, the Western establishment picked a different approach, also with intellectual roots in revolutionary Russia, developed by an émigré and avowedly anti-Marxist revolutionary, Alexandre Koyré (Mayer, 2004). His Anglo-American epigons mistakenly labelled it 'internalism', but Koyré did not try to isolate and restrict science to its own internal logic. His original interpretation defined the Scientific Revolution as a destruction of the Cosmos and a change in the worldview—cosmological, metaphysical, and religious all at once.[6] Koyré analysed the new European science as intimately entangled within the broadly defined Platonic world of cultural ideas and philosophical introspection of the time period, but emphatically proclaimed it above and irreducible to the profanity of material concerns and technological artisanship. As a methodological model, Forman's study did not satisfy either of the Cold War camps. To the predominantly anti-communist 'internalist' side, it was as much an anathema as the Marxist programme, and similarly subversive. For the 'old left' tradition, however, it sounded too anti-science. To them, a true science worthy of its good name was supposed to be an ally of progressive politics and the major force of social development, whereas reactionary ideological influences could only corrupt scientific knowledge, not contribute to its revolutionary advancement.

Despite the wide outrage provoked by the Forman Thesis, only a few explicitly negative rebuttals appeared in print (Hendry, 1980; Kraft and Kroes, 1984). The main counter argument relied on drawing or assuming a boundary between the outwardly oriented 'ideology' or 'rhetoric' of scientists and their supposedly 'autonomous', 'internal' knowledge. If the former, maintained the opponents, could be influenced or insincerely adjusted to hostile pressures from the outside, at the latter, internal level,

---

[6] Although teaching the concept of the Scientific Revolution continues to provide bread and butter for historians of science at many universities, debates regarding its meaning and applicability continue to this day, with shifting foci and reinterpretations. See, for example (Osler, 2000).

one should expect from scientists an entrenchment and professional autonomy rather than adaptation. Critics claimed that Forman had underestimated the internal reasons from the mounting problems in physics that around the same time pushed scientists towards accepting acausality as a fundamental feature of the atomic world (although the chronological coincidence still remained an unexplained puzzle as long as one continued to insists that 'internal' arguments did not interact with outside pressures).

Opponents also questioned the interpretations and the limited number of individual cases of physicists and mathematicians whose conversion to acausality was publicly recorded prior to 1925 and pointed out counterexamples of resistance by well-known scientists who refused to adapt even at the level of rhetoric. Hendry still had to concede that '[d]espite the criticisms that may be levelled against his analysis, Forman has succeeded in demonstrating that physicists and mathematicians were generally aware of the values of the milieu, and that this milieu did incorporate a marked hostility toward the causality principle. But when we come to the crucial claims, that there was a widespread rejection of causality in physics, and that there were no internal reasons for this rejection, then the weaknesses in his argument also become crucial. For there were strong internal reasons for the rejection of causality, and when these are taken into account, and Forman's supposed "converts to acausality" critically re-examined, it would appear that the reaction of physicists to the causality challenge was far from being accommodation, and that there may even have been a tendency to isolation.' (Hendry, 1980, p. 160).[7]

Yet, general perspectives on the nature and practice of science were already changing at the time, and the Forman Thesis both reflected and influenced these tectonic shifts. It did so precisely by undermining the boundary, or the very assumption that the internal content of science can be isolated and unequivocally separated from cultural impact and social context. In hindsight, Julia Menzel and David Kaiser observe: 'Since the article's publication, it has become a matter of principle within the history of science to insist always on the embeddedness of science in society—and there can be no doubt that even the abstruse concepts of quantum physics are worldly things produced by particular people in specific cultural contexts, toward interested ends' (Menzel and Kaiser, 2020, p. 34). Despite some temperamental objections to its findings, Forman's work has fundamentally changed directions of research and established itself as a classic in science studies, including history, sociology, and philosophy of science. In subsequent decades it became required reading in practically every graduate pro-gramme that trained students in the above fields.

Forman's analysis of Weimar physics furnished a paradigmatic example for the sociology of scientific knowledge that developed by the 1980s and made the idea of

---

[7] The known cases of resistance by Einstein, Planck, Schrödinger, and others did not contradict Forman's argument and thus were not really counterexamples. They were discussed in his paper and were necessary for the very formulation of the argument: 'My sympathies have consequently been with the conservatives in their defence of reason rather than with the "progressives" in their denigration of it.' (Forman, 1971, p. 113).

social construction widely acceptable (Hacking, 1999). The special importance of this example derived from the fact that it focused on physics and mathematics, the so-called 'hard sciences' that were and still are considered a much more challenging target for social constructivism than the life sciences and social sciences. It also dealt with a very recent breakthrough and with concepts considered true and fundamentally valid by living scientists, rather than with some outdated knowledge or antiquated theory from the era of Newton, for which it is psychologically easier to invent a historical deconstruction. Thus, some pioneers of the new sociological approaches to science could cite the case of quantum acausality as one of the most powerful demonstrations of the far-reaching influence of social factors all the way down to the hard theoretical core of scientific knowledge (Barnes, 1974; Bloor, 1981; Shapin, 1982). Eventually, another classic breakthrough by Shapin and Schaffer inquired into the very process of how the boundary between the scientific and the social, and their respective definitions, could be constructed and contested historically (Shapin and Schaffer, 1986; Shapin, 1992).

Forman's work also became one of those rare historical studies which, by relativizing the existing scientific dogmas, helped contemporary physicists such as John Bell to critically reassess them, something that Ernst Mach had also achieved heuristically for Einstein at the end of the 19th century with his critical historical exposé of absolute space and time in Newtonian mechanics. Physicists' attitudes towards the philosophical interpretation of quantum theory have changed dramatically since the 1960s. When the Forman Thesis was published, acausality of the quantum laws was still generally seen as part of the core scientific formalism, according to the then-dominant Copenhagen interpretation (Howard, 2004, 2021). Due largely to the work of Dmitry Blokhintsev, David Bohm, and John Bell, physicists' views shifted in the direction of philosophical pluralism within which different interpretations, including causal ones, are possible. Bell was aware of Forman's historical critique and used it as additional encouragement in his efforts to challenge the Copenhagen orthodoxy from within physics (Bell, 1982). In subsequent decades, ever more historians and philosophers of science also turned in their analyses to those 'conservative' physicists who had disagreed with the prevailing opinions of their colleagues and defended rationality and the causality principle in quantum mechanics, to whose previously neglected views Forman had called sympathetic attention.[8]

Since the 1960s, the history of quantum physics has matured as a field, with detailed studies of the technical formalism, philosophical questions, institutional settings, biographies, and collected editions of major contributors (Staley, 2013; Badino, 2016). Forman's work exerted profound influence on subsequent generations of researchers: its methodology, problematic, conceptual vocabulary, and questions continue to inspire further inquiries and generate controversies. Forman defined his approach as 'sociological' rather than 'psychological'. His description of scientists' ideology, or self-serving, idealized, and public representation of their activities, did

---

[8] See, in particular, (Cushing, 1994; Beller, 1999; Freire Jr, 2014).

not need to inquire whether these beliefs were individually sincere—or not. Most often, the critics questioned Forman's explanatory model as too rigid, for its assumption that the adaptation to social pressures was itself a causal, practically deterministic one-way street. 'With the information available, Forman has succeeded in demonstrating an influence of the milieu upon physicists' attitudes to causality, and were he to adopt a suitable concept of historical causation he could even assert quite reasonably that the attitudes were in some (weak) sense "caused" by the milieu... Physicists *were* influenced by the crisis-consciousness of post-war Europe and by the attitudes characteristic of the Weimar milieu. On the other hand, Forman's work has also demonstrated the dangers of a purely external treatment and the poverty of any naive social reductionism,' insisted Hendry (1980, pp. 170–171).

The above quote shows that even those who disagreed profoundly still could not deny the main discovery of a meaningful connection between Weimar cultural ideology and the quantum mechanical revolution. Many others who worked in the field could agree with Forman or disagree on the details, and yet continued to grapple with the problem of how exactly to characterize and describe the social causation's *modus operandi*. The problem presents itself as theoretical, possibly unsolvable, not just for this case but for social studies of science in general, with scholars taking different stances, from more straightforward and deterministic to indirect and variable, in a stronger or weaker sense. In the broader social and cultural history of science, into which the sociology of scientific knowledge has partly been folded, a vast range of new methodologies has been advanced since 'Weimar Culture'. In particular, the strongly causal models of interest characteristic of the early years of the sociological programme have been supplemented by (or watered down to) more modest accounts of resonances. Thus Norton Wise, in his critique of Forman, has proposed a 'model in which resources and participation replace influences and capitulation' (Wise, 2011, p. 430).

In the meantime, historians working in the genre of cultural history of science applied and extended Forman's argument further, adapting its conceptual approach to other cases and situations, and checking its applicability to different cultural milieus. To mention only a few examples of important investigations, and only those belonging to quantum historiography, Heilbron (1985) described the post-1927 spread of the Copenhagen mystical philosophy with its characteristic 'combination of imperialism and resignation', whereas Stephen G. Brush developed a longue-durée model of cultural affinities in physics, in which periods of realism and positivism alternated with more conservative and irrational (neo)romanticism (Brush, 1976, 1980). Silvan S. Schweber applied Forman-inspired analysis to the cases of Arnold Sommerfeld and Hans Bethe (Schweber, 2009, 2014) and Richard Beyler to the case of Pascual Jordan (Beyler, 2011). Dealing with historical contexts outside of Weimar Germany, Richard Staley analysed the early 20th century cultural debates about mechanics that contributed to Spengler's views on causality, Alexei Kojevnikov revealed the impact of Soviet collectivism on the development of conceptual language in solid-state and condensed-matter physics, and David Kaiser described the application of democratic ideals to models of subnuclear particles in post-war America (Staley, 2011; Kojevnikov, 1999; Kaiser, 2002).

# 36.4 REVEALING EXAMPLES: EHRENFEST, PAULI, SCHRÖDINGER, JOFFE

A June 1919 letter to Bohr from Paul Ehrenfest, professor of theoretical physics at Leiden University, almost literally confirms the core claim of the Forman Thesis: '[I]t is remarkable that precisely here, in the circles of men having much to do with technology, production, industry, patents etc., opinions develop so uniformly about perspectives of culture. Overall there is building up an uncannily intensive reaction *against rationalism* ... If I am not entirely mistaken, in the next 5–10 years we will see the following happening at the institutes of higher learning (including technical!). Professors raised as relatively *rational* and disciplined individuals will despairingly and uncomprehendingly face the complaints and demands of a relatively *"mystical"* student body. At the same time, scientifically less clear but personally warmer teachers will gain the main influence over students ... As I write this, it suddenly became so much clearer to me why, in the opinion of the young, I am so much more strongly associated with the older.'[9]

The document illustrates that, in the immediate wake of World War I, a tidal reaction against rationalism and favouring more mystical lines of thought swept through not just the intellectual public in general, but also such professionals as engineers and exact scientists previously expected to strongly resist such trends. The communication from a theoretical physicist in the Netherlands to his colleague in Denmark also signified that the mood did not remain confined to Germany and Austria, but also affected at least the neighbouring neutral countries. The letter revealed a striking admission and expectation that professors would adapt self-consciously, rather than unreflectively, to the direction of the prevailing intellectual wind. Although not abandoning his personal rationalistic convictions, Ehrenfest appeared to defer to the opinions of the younger students, pointing at an additional effective milieu capable of extracting concession from physicists. Indeed, many professors probably cared more about pleasing students in their auditoriums than academics from other disciplines. In his own field of theoretical physics, Ehrenfest regarded Bohr as precisely the kind of professor whose thoughts were too profound to be understood or even expressed clearly, which only helped him to be tremendously inspiring and resonate with the younger generation of students. Whether or not Ehrenfest's letter contained implicit advice to Bohr, and whether or not Bohr accepted the hint or arrived at similar ideas on his own, around the same time he was already inclined 'to take the most radical *or rather mystical* views imaginable' regarding the daunting problem of the quantum

---

[9] Paul Ehrenfest to Niels Bohr, 4 June 1919, Niels Bohr Scientific Correspondence, Niels Bohr Archive, Copenhagen (emphasis in the original).

interaction of matter and radiation and did not consider a modicum of such mysticism inconsistent with the practice of natural science.[10]

Five years later, in a desperate attempt to ward off the concept of light quantum, Bohr resorted to one such idea in the famous, or infamous, Bohr–Kramers–Slater theory of 1924. He proposed that mysteries of the quantum could be resolved if one assumed that the conservation laws for energy and momentum are valid only statistically, if averaged over a great many atomic interactions, but violated in individual processes at the microscopic level. The quantum community split in reaction to this radical proposal. Schrödinger, who had already two years earlier publicly declared himself opposed to causality in physics, welcomed it enthusiastically (Forman, 1971, pp. 87–88; Hendry, 1980, pp. 164–167). Others were much more reserved or sceptical: even if willing to abandon causality in principle, they were reluctant to sacrifice for this purpose the revered law of energy conservation. Pauli (in letters) vehemently opposed the 'reactionary Copenhagen Putsch'. After Bohr's proposal had been refuted in experiment, he went even further in proclaiming: 'I definitely believe that *the probability concept should not be allowed in the fundamental laws of a satisfying physical theory*. I am prepared to pay any price for the fulfilment of this desire, but unfortunately I still do not know the price for which it is to be had'.[11]

Interestingly, both Schrödinger and Pauli would reverse themselves during the subsequent process of creating quantum mechanics. The former, once he had authored wave mechanics early in 1926, traded his philosophical stance from acausality to *Anschaulichkeit*. Both concepts appealed to the ideological milieu, but *Anschaulichkeit* corresponded better to Schrödinger's latest fundamental breakthrough and his ambitious hopes for wave mechanics. Pauli changed his attitude towards fundamental probabilities once, in the wake of Max Born's probabilistic treatment of scattering, he made his own important contribution to wave mechanics in the fall of 1926. Having shown that Schrödinger's psi-function could be interpreted as the probability of the electron's position, Pauli would forever remain a staunch proponent of the statistical interpretation of quantum mechanics.[12] Schrödinger formulated his philosophical dilemma in August 1926: 'Today I no longer like to assume with Born that an individual process of this kind is "absolutely random", i.e., completely undetermined. I no longer believe today that this conception (which I championed so enthusiastically four years ago) accomplishes much. From an offprint of Born's last work in the *Zeitschr. f. Phys.* I know more or less how he thinks of things: the *waves* must be strictly causally determined through field laws; the wave functions, on the other hand, have only the meaning of probabilities for the *actual* motions of light or material particles. I believe that Born overlooks that—provided one could have this view worked

---

[10] 'or rather mystical' is inserted into the sentence above the line. Bohr to Charles Galton Darwin, July 1919, Niels Bohr Scientific Correspondence, Niels Bohr Archive, Copenhagen, draft of a presumably unsent letter. On Ehrenfest's philosophical struggles, see (Lunteren and Hollestelle, 2013).

[11] Pauli to Bohr, 17 November 1925, emphasis added (Pauli, 1979, p. 260).

[12] (Born, 1926); Pauli to Heisenberg, 19 October 1926 (Pauli, 1979, pp. 340–49).

out completely—it would depend on the taste of the observer *which* he now wishes to regard as *real*, the particle or the guiding field.'[13] Schrödinger's and Pauli's dramatic and opportunistic flip-flops on fundamental philosophical principles reveal that while the chief authors of quantum mechanics did feel compelled to relate their work to the *Zeitgeist* of the time, the latter was still rich enough to allow quantum physicists some flexibility and choices, to be better able to advance various personal agendas and interests (Kojevnikov, 2011).

By the end of 1927, the winning parties of the philosophical battle over quantum mechanics defined their choices if not completely identically, at least with sufficient overlap. Born and his Göttingen student Jordan proclaimed the abandonment of classical causality to be the main fundamental lesson of quantum mechanics. Bohr in his Como address (supported by Heisenberg, who held somewhat deviating views on the matter) stressed *Individualität*, limitation on causality, and also restricted, but still possible *Anschaulichkeit*, or spatio-temporal description of microscopic phenomena (Kojevnikov, 2020). All of these notions resonated in the Central European, German-focused academic milieu, but, as Forman has shown, they did not arouse equally strong feelings in the Anglophone culture of the time (Forman, 1979). We can find an even more striking contrast by looking at how these philosophical problems played out in a milieu with very different cultural values—in the ideological atmosphere of Soviet Russia.

From the similarly tragic experiences of World War I, the Russian revolutionaries and the German conservative intellectuals drew lessons that went in almost exactly opposite ideological directions. The ideals of progress, rationality, modernity, and scientism rose to unprecedented cultural authority in Soviet Russia following the War and the Revolution, not only among the Marxists, but among the educated public in general, especially scientists. Their emphatically pro-science general stance, however, did not prevent Soviet Marxists from feeling suspicious of certain irrational tendencies in 'bourgeois science', including quantum mechanics, or rather, its philosophy (Kojevnikov, 2012). The 'dean' and top manager of Soviet physics, Abram Joffe, attempted to assuage these concerns in 1934 when speaking to a political gathering on the occasion of the 25th anniversary of Lenin's *Materialism and Empiriocriticism*. Addressing this, 'hostile' in its own way, audience on 'The Development of Atomistic Views in the 20th Century', Joffe emphasized those features of quantum mechanics that could provide grounds for cooperation between Soviet physicists and Marxist philosophers. His choices, not surprisingly, were almost polar opposites to those preferred by his German colleagues. According to Joffe, quantum mechanics was *unanschaulich* (the corresponding Russian term is *nenagliadnyi*, or non-visual, non-pictorial), statistical but causal, and most importantly, it signified a fundamental 'loss of individuality' for quantum particles (Joffe, 1934, p. 60).

In Joffe's interpretation, *Unanschaulichkeit* stood for the truly revolutionary character of quantum mechanics: the theory appeared counterintuitive and non-visual,

---

[13] Schrödinger to Wilhelm Wien, 25 August 1926, archive of the Deutsches Museum, Munich.

904    HISTORY OF QUANTUM INTERPRETATIONS

because scientists' existing pictorial intuitions had been formed by the traditional, classical theories. The abandonment of the old ways of visual representation, including the notion of the electron's trajectory, meant that the physical laws in the microscopic world were radically new and qualitatively different from the familiar laws operating at the macroscopic level. Dispensing with *Anschaulichkeit* could thus be easily explained and even turned into an advantage in the Soviet context, in conformity with the anti-reductionism of the official Marxist philosophy of dialectical materialism (Martinez, 2021). But the principle of causality, or *Kausalität*, was ideologically sacrosanct for the Soviet Marxists and could not be questioned. Joffe and other Soviet physicists therefore carefully avoided mentioning 'acausality' or attaching this label to quantum mechanics. Instead, they proclaimed that the old crude version of mechanical determinism, dating from the 18th century, had been superseded by a more refined and sophisticated causality in the quantum world. To Soviet authors, the validity of statistical laws and probabilistic formulae did not necessitate the abandonment of causality as the funda-mental principle of science: the former could be used in quantum mechanics without sacrificing the latter (Kojevnikov, 2012).

The Soviet political ideology promoted collectivism instead of individualism, and its representatives were certainly happy to hear from Joffe that quantum statistics proved that atomic particles no longer possessed absolute individualities, and that the laws of quantum mechanics described collective behaviour and processes. Joffe reassured his Marxist audience that the existing quantum theory was still quite young and not necessarily complete, given continuing disagreements among its main contributors, but both quantum mechanics and relativity, the other profound revolutionary devel-opment in physics, were definitely confirming the philosophy of dialectical material-ism. Had it been just this talk alone, Joffe's philosophical interpretation could have been dismissed as merely rhetoric, necessary to please the authorities and protect quantum mechanics from ideological criticism in the Soviet Union. But around the same time, Soviet theoretical physicists were already designing new physical models—quasiparticles and collective excitations—that would transform the socialist philosophy of collectivism into the conceptual language and mathematical apparatus of the quantum theory of condensed matter (Kojevnikov, 1999).

# 36.5 CONCLUSIONS

The Forman Thesis *was* controversial fifty years ago, when first introduced, but it is anachronistic to continue branding it this way now. What made it scandalous back then has since become generally accepted. Most of the initial outrage came from the modernist rejection of the possibility that local and idiosyncratic culture could influ-ence the supposedly universal scientific knowledge. As the paradigmatic example demonstrating such interaction, the Forman Thesis was instrumental in the rise of new scholarly understandings of science during the 1970s and 1980s. As the number of other cases involving various cultures and scientific disciplines grew, the scholars who described them met with significantly less opposition than Forman had initially. Today the understanding that science is produced locally, in social settings, and conditioned by culture is fully accepted in cultural studies and histories of science, almost to the point of carrying no burden of proof. Current assumptions about science, however, make it harder to explain how such locally produced knowledge manages to travel across cultures and establish itself internationally; hence the importance of comparative studies related to the case described by Forman.

While its central methodological lesson became de facto commonly accepted and went to the masses, stereotypes of perception continued, by inertia, to ascribe the label 'controversial' to the original Forman Thesis. Although for somewhat different reasons, Forman's case can be compared to that of one of the physicists he studied, Erwin Schrödinger. Schrödinger's equation and the psi-function he introduced became universally known and enormously successful tools of the discipline, not requiring a reference or any explicit allocation of credit. At the same time, their author's figure and standing continued to be seen as somewhat marginal, by reputation, in large part because he did not control the field institutionally, or the interpretation of quantum mechanics. Also, as in Schrödinger's case, some aspects of the initial version of Forman's study, which is now fifty years old, are open for update, revisions, modifications, and debates about the historically evolving relationship between science and culture.

Despite its by now classic status, the Forman Thesis generates both inspiration for new studies and criticism in the field, providing a reference point for the shifting approaches and methodological changes underway in science studies. His sociological and scientistic explanatory model of causation, in particular, continues to cause disagreements and discomfort for contemporary post-modernist sensibilities. His description of physicists succumbing to ideological currents of the time flies in the face of today's currents insisting that individuals are free, even when they shop as prescribed by the latest advertisement in social media. Forman's moralism can be unsettling. He simultaneously rebuked scientists for their betrayal of disciplinary values and demonstrated that the idea of the autonomous disciplinary community is an ideological fiction, in direct opposition to Kuhn's then popular model of scientific revolutions within the self-contained scientific community. Kuhn's model also used to be somewhat controversial back in the day when it was influential, but has since become simply outdated, so that now it can be safely praised and glorified as a dead classic. This is not the case with the Forman Thesis.

## Acknowledgements

I am grateful to Olival Freire Jr, Christian Joas, John L. Heilbron, Climério Paulo da Silva Neto, Jean-Philippe Martinez, Jessica Wang, and participants at the December 2019 workshop 'Fundamentos e Interpretações da Mecânica Quântica: Aspectos Históricos e Conceituais' at Instituto de Física, Universidade Federal da Bahia, for critical and productive discussions.

## REFERENCES

Badino, M. (2016). What Have the Historians of Quantum Physics Ever Done for Us? *Centaurus*, **58**, 327–46.

Barnes, B. (1974). *Scientific Knowledge and Sociological Theory*. London: Routledge.

Bell, J. S. (1982). On the Impossible Pilot Wave. *Foundations of Physics*, **12**, 989–99.

Beller, M. (1999). *Quantum Dialogue: The Making of a Revolution*. Chicago: University of Chicago Press.

Bernal, J. D. (1954). *Science in History*. London: Watts.

Beyler, R. (2011). Jordan alias Domeier: Science and Cultural Politics in Late Weimar Conservatism. In Carson *et al.* (2011), pp. 487–503.

Bloor, D. (1981). The Strengths of the Strong Programme. *Philosophy of the Social Sciences*, **11**, 199–213.

Born, M. (1926). Zur Quantenmechanik der Stoßvorgänge. *Zeitschrift für Physik*, **37**, 863–67.

Born, M. (1978). *My Life: Recollections of a Nobel Laureate*. New York: Scribner.

Brush, S. G. (1976). Irreversibility and Determinism: Fourier to Heisenberg. *Journal of the History of Ideas*, **37**, 603–30.

Brush, S. G. (1980). The Chimerical Cat: Philosophy of Quantum Mechanics in Historical Perspective. *Social Studies of Science*, **10**, 393–447.

Carson, C., Kojevnikov, A., and Trischler, H. (2011). *The Forman Thesis: 40 Years After. Introduction to Weimar Culture and Quantum Mechanics: Selected Papers by Paul Forman and Contemporary Perspectives on the Forman Thesis*. Singapore: World Scientific, pp. 1–6.

Cushing, J. T. (1994). *Quantum Mechanics: Historical Contingency and the Copenhagen Hegemony*. Chicago: University of Chicago Press.

Forman, P. (1967). *The Environment and Practice of Atomic Physics in Weimar Germany: A Study in the History of Science*. Ph.D. dissertation, University of California, Berkeley.

Forman, P. (1969). The Discovery of the Diffraction of X-rays by Crystals: A Critique of the Myth. *Archive for History of Exact Sciences*, **6**, 38–71.

Forman, P. (1971). Weimar Culture, Causality, and Quantum Theory, 1918–1927: Adaptation by German Physicists and Mathematicians to a Hostile Intellectual Environment. *Historical Studies in the Physical Sciences*, **3**, 1–115.

Forman, P. (1973). Scientific Internationalism and the Weimar Physicists: The Ideology and its Manipulation in Germany after World War I. *Isis*, **64**, 151–80.

Forman, P. (1974). The Financial Support and Political Alignment of Physicists in Weimar Germany. *Minerva*, **12**, 39–66.

Forman, P. (1979). The Reception of an Acausal Quantum Mechanics in Germany and Britain. In S. H. Mauskopf (ed.),*The Reception of Unconventional Science*, Boulder, CO: Westview Press, pp. 11–50.

Forman, P. (1984). *Kausalität, Anschaulichkeit*, and *Individualität*, or How Cultural Values Prescribed the Character and the Lessons Ascribed to Quantum Mechanics. In N. Stehr and V. Meja (ed.), *Society and Knowledge: Contemporary Perspectives in the Sociology of Knowledge*, New Brunswick: Transaction Books, pp. 333–47.

Forman, P. (1991). Independence, not Transcendence, for the Historian of Science. *Isis*, **82**, 71–86.

Forman P., Heilbron, J. L., and Weart, S. (1975). Physics circa 1900: Personnel, Funding, and Productivity of the Academic Establishments. *Historical Studies in the Physical Sciences*, **5**, 1–185.

Freire Jr, O. (2014). *The Quantum Dissidents: Rebuilding the Foundations of Quantum Mechanics (1950–1990)*. Berlin: Springer.

Freudenthal, G. and McLaughlin, P. (eds.) (2009). The Social and Economic Roots of the Scientific Revolution: Texts by Boris Hessen and Henryk Grossmann. Dordrecht: Springer.

Gamow, G. (1966). *Thirty Years that Shook Physics: The Story of Quantum Theory*. New York: Anchor.

Gerwarth, R. (2016). *The Vanquished: Why the First World War Failed to End*. New York: Farrar, Straus and Giroux.

Hacking, I. (1999). *The Social Construction of What?* Cambridge, MA: Harvard University Press.

Heilbron, J. L. (1968). Quantum Historiography and the Archive for History of Quantum Physics. *History of Science*, **7**, 90–111.

Heilbron, J. L. (1985). The Earliest Missionaries of the Copenhagen Spirit. *Revue d'histoire des sciences*, **38**, 195–230.

Heilbron, J. L. (2011). Cold War Culture, History of Science and Postmodernity: Engagement of an Intellectual in a Hostile Academic Environment. In Carson *et al.* (2011), pp. 7–20.

Heilbron, J. L., and Kuhn, T. S. (1969). The Genesis of the Bohr Atom. *Historical Studies in the Physical Sciences*, **1**, 211–90.

Hendry, J. (1980). Weimar Culture and Quantum Causality. *History of Science*, 18: 155–80.

Howard, D. (2004). Who Invented the 'Copenhagen Interpretation'? A Study in Mythology. *Philosophy of Science*, **71**, 669–82.

Howard, D. (2021). The Copenhagen Interpretation. (This volume.)

Joffe, A. F. (1934). Razvitie Atomisticheskikh Vozzrenii v XX v. *Pod Znamenem Marksizma*, **4**, 52–68.

Kaiser, D. (2002). Nuclear Democracy: Political Engagement, Pedagogical Reform, and Particle Physics in Postwar America. *Isis*, **93**, 229–68.

Kevles, D. (1971). 'Into Hostile Political Camps': The Reorganization of International Science in World War I. *Isis*, **62**, 47–60.

Kojevnikov, A. (1999). Freedom, Collectivism, and Quasiparticles: Social Metaphors in Quantum Physics. *Historical Studies in the Physical and Biological Sciences*, **29**, 295–331.

Kojevnikov, A. (2011). Philosophical Rhetoric in Early Quantum Mechanics 1925–1927: High Principles, Cultural Values, and Professional Anxieties. In Carson *et al.* (2011), pp. 319–48.

Kojevnikov, A. (2012). Probability, Marxism, and Quantum Ensembles. *Jahrbuch für Europäische Wissenschaftskultur, 2011*, **6**, 211–35.

Kojevnikov, A. (2020). *The Copenhagen Network: The Birth of Quantum Mechanics from a Postdoctoral Perspective*. Berlin: Springer.

Kraft, P., and Kroes, P. (1984). Adaptation of Scientific Knowledge to an Intellectual Environment. Paul Forman's 'Weimar Culture, Causality, and Quantum Theory, 1918–1927': Analysis and Criticism. *Centaurus*, **27**, 76–99.

Kuhn, T. S. (1962). *The Structure of Scientific Revolutions*. Chicago: University of Chicago Press.

Kuhn, T. S. (1967). Review: The Turn to Recent Science. *Isis*, **58**, 409–19.

Kuhn, T. S. (1978). *Black-Body Theory and the Quantum Discontinuity, 1894–1912*. Oxford: Oxford University Press.

Kuhn, T. S., Heilbron, J. L., Forman, P., and Allen, L. (1967). *Sources for History of Quantum Physics: An Inventory and Report*. Philadelphia: American Philosophical Society.

Lunteren, F. H. van, and Hollestelle, M. J. (2013). Paul Ehrenfest and the Dilemmas of Modernity. *Isis*, **104**, 504–36.

Martinez, J.-P. (2021). Foundations of Quantum Physics in the Soviet Union. (This volume.)

Mayer, A.-K. (2004). Setting up a Discipline, II: British History of Science and 'The End of Ideology,' 1931–1948. *Studies in History and Philosophy of Science*, **35**, 41–72.

Menzel, J. H., and Kaiser, D. (2020). Weimar, Cold War, and Historical Explanation. *Historical Studies in the Natural Sciences*, **50**, 31–40.

Merton, R. K. (1938). Science, Technology, and Society in Seventeenth Century England. *Osiris*, **4**, 360–632.

Monaldi, D. (2009). A Note on the Prehistory of Indistinguishable Particles. *Studies in History and Philosophy of Modern Physics*, **40**, 383–94.

Osler, M. J. (ed.) (2000). *Rethinking the Scientific Revolution*. Cambridge: Cambridge University Press.

Pauli, W. (1979). *Wissenschaftlicher Briefwechsel mit Bohr, Einstein, Heisenberg u. a.* Vol 1: 1919–1929. Edited by A. Hermann, K. von Meyenn, and V. F. Weisskopf. New York: Springer.

Rentetzi, M. (2004). Gender, Politics, and Radioactivity Research in Interwar Vienna: The Case of the Institute for Radium Research. *Isis*, **95**, 359–93.

Schweber, S. S. (2009). Weimar Physics: Sommerfeld Seminar and the Causality Principle. *Physics in Perspective*, **11**, 261–301.

Schweber, S. S. (2014). Writing the Biography of Hans Bethe: Contextual History and Paul Forman. *Physics in Perspective*, **16**, 179–217.

Shapin, S. (1982). History of Science and its Sociological Reconstructions. *History of Science*, **20**, 157–211.

Shapin, S. (1992). Discipline and Bounding: The History and Sociology of Science as Seen through the Externalism-Internalism Debate. *History of Science*, **30**, 333–69.

Shapin, S., and Schaffer, S. (1986). *Leviathan and the Air-Pump: Hobbes, Boyle, and the Experimental Life*. Princeton, NJ: Princeton University Press.

Staley, R. (2011). Culture and Mechanics in Germany. 1869–1918: A Sketch. In Carson *et al.* (2011), pp. 277–92.

Staley, R. (2013). Trajectories in the History and Historiography of Physics in the Twentieth Century. *History of Science*, **51**, 151–77.

te Heesen, A. (2020). Thomas S. Kuhn, Earwitness: Interviewing and the Making of a New History of Science. *Isis*, **111**, 86–96.

Wise, M. N. (2011). Forman Reformed, Again. In Carson *et al.* (2011), pp. 415–31.

..........................................................................................................

# CHAPTER 37

# THE CO-CREATION OF CLASSICAL AND MODERN PHYSICS AND THE FOUNDATIONS OF QUANTUM MECHANICS

..........................................................................................................

## RICHARD STALEY

UNTIL relatively recently physicists and historians of science have routinely described the early 20th century as hosting a transition from classical to modern physics, and depicted physicists of the earlier period as in thrall to classical physics. Albert Einstein's 'Autobiographical Notes', for example, described Ernst Mach as upsetting the dogmatic faith that physicists of the 19th century held in classical mechanics as a foundation for all physics.[1] Yet following Allan Needell and Olivier Darrigol's argument that 'classical physics' has been applied retrospectively to a period in which physicists' views were in fact much more open and fluid, in 2005 and 2008 I showed that concepts of classical theory—such as 'classical mechanics' and 'classical thermodynamics'—had only begun to be introduced in the 1890s, and often expressed diverse, even competing theoretical approaches.[2] They were, then, created at the same time as the new physics. Further, rather than being developed by proponents (like the contemporaneous terms 'energetics' and 'the electromagnetic world view'), the versions of 'classical mechanics' and 'classical theory' that began to be accepted more widely in the early 20th century were in fact articulated by the architects of relativity and quantum theory. In the hands of Einstein and later Planck such terms were used to express limiting cases to the new theories, or theoretical approaches now surpassed. Supporting this argument in 2013,

[1] (Einstein, 1949), p. 21.
[2] (Needell, 1988, pp. xi–xliii); (Darrigol, 2001, p. 224). My original argument was in (Staley, 2005). For a focus on worldviews and a fuller account see (Staley, 2008a, 2008b).

Graeme Gooday and Daniel Mitchell queried my emphasis on the 1911 Solvay Council as creating conditions that allowed a more general concept of classical theory to be widely promulgated, and highlighted distinctions between 'classical theory' and the still more general term 'classical physics'. Arguing that much work remains to be done to understand how concepts of 'classical physics' arose, became part of physics orthodoxy and were linked to concepts of 'modern physics', Gooday and Mitchell pointed to what they regarded as the first British attempt to articulate a concept of 'classical physics' in Arthur Eddington's 1927 Gifford Lectures on the relations between science and religion.[3] They are right to think the British context illustrates the complexity of the issues involved, but did little to address quantum theory, the primary field in which the concept was most significant. This chapter takes up the challenge they have posed in an examination of the use of concepts of the classical in the emergence of quantum mechanics. An examination of British perspectives before and after the Solvay Council will highlight critical issues.

## 37.1 THE 1911 SOLVAY COUNCIL AND THE LANGUAGE OF THE CLASSICAL IN BRITAIN

In a short period from early 1911 to 1913, the mathematical physicist Samuel Bruce McLaren published a series of four papers grappling with the equipartition of energy and Planck's theory of blackbody radiation, subjects first discussed around the turn of the century. McLaren had graduated 3rd Wrangler in the Cambridge Mathematical Tripos in 1899 and lectured at the University of Birmingham from 1906. His first paper followed his teacher Joseph Larmor's 1909 Bakerian lecture in addressing the gap between matter and radiation exposed by the work of James Jeans and H.A. Lorentz (referring also to Wien's law).[4] Revealingly, McLaren addressed a field that had been rendered urgent by Planck's and Einstein's work, without considering their approach directly. McLaren's discussion of the 'old paradoxical' conclusion of the equipartition theorem also indicates then common perspectives, commenting 'To many it has always seemed that the method of statistics builds much, and to little purpose, on a very unsure foundation'.[5] In February 1911 McLaren extended Lorentz's treatment of emission and absorption of radiation across all wavelengths, showing it necessarily yielded the expression Lorentz (and Rayleigh and Jeans) had derived. Since it was only empirically confirmed for long wavelengths, he thought it must be replaced 'by some such formula as that of Planck'. Yet in his view Planck's arguments had no clear basis,

---

[3] (Gooday and Mitchell, 2013), pp. 727–28.

[4] Larmor emphasized the long-standing role of discontinuities in physics and generalized Planck's approach to natural radiation, in a highly empiricist framework. On Larmor see (Warwick, 2003).

[5] (McLaren, 1911a, p. 16).

simply stipulating that not all distributions are equally likely and deducing his law 'by reasoning in which ordinary dynamics is ignored'.[6]

McLaren's language underlines an important point about early discussions of quantum theory. Without exception—until 1911—the critical published papers from Planck, Einstein, Jeans, and others did not incorporate the language of the classical, even when recognizing sharp contrasts. Rather than reflecting the settled understanding of 19th century physicists, the equipartition theorem had to be *made* classical.[7] This occurred long after Planck's law had been accepted as empirically justified, in the 1911 Solvay Council when, quite suddenly, six participants refer to concepts of 'classical mechanics' or 'classical theory' in pre-circulated papers or discussion. The most important invocation came from Planck himself, who articulated an extension of classical mechanics to kinetic theory, thereby rendering equipartition classical—despite the controversies that had attended its application and despite the fact that his early papers simply did not consider the theorem. In 1911 Planck described the principles of classical dynamics, even extended by relativity theory, as being inadequate to meet microphysical phenomena and argued the first incontrovertible proof of this had come in the contradictions between 'classical theory' and observation opened up by the universal law of blackbody radiation—for which he had given the definitive formulation in 1900.[8]

In 1912 McLaren directly addressed the question of the emission and absorption of radiation that had been at the heart of Einstein's 1905 arguments for light quanta. Now McLaren argued that surmounting the problem posed by the equipartition theorem was beyond dynamics in its present state. He wrote:

> Some such revolution of our physical ideas as is foreshadowed in the work of Planck and Einstein is, it would seem, inevitable; at the most it may be hoped that the future will adapt and not destroy the work of the past. In this paper 'aether' still retains its classical traditional value, it is the aether of Maxwell's equations.[9]

The aether was thus a saving strategy. McLaren argued it should be no surprise that only aether, free of matter, was completely mechanical. Spectroscopic phenomena showed matter existed 'in distinct species without continuous change from one species to another', something almost impossible to explain on any purely mechanical theory. The impossibility of rigid bodies in relativity likewise indicated the 'mechanics of matter in bulk' could not deal with its ultimate parts.[10] While the main text was dated 14 March 1911, a concluding section added on 29 January 1912 commented that the 'classical theory of the aether' did not require the equipartition

---

[6] (McLaren, 1911b).
[7] (Staley, 2005); (Staley, 2008a) at Part 4, chs 9 and 10.
[8] (Planck, 1913b), p. 77; (Staley, 2005); (Staley, 2008a), pp. 409–17 on p. 412.
[9] (McLaren, 1912), p. 516.    [10] (McLaren, 1912), p. 516.

of radiant energy in the way Rayleigh's formula showed. That only followed if matter and radiation formed a single dynamical system with laws of motion deduced from a single action formula—and Einstein's statistical treatment of thermodynamics showed much less than this could serve as a mechanical foundation for thermodynamics.[11]

Gooday and Mitchell have discussed and dismissed the possibility that ether should be understood as the essence of classical physics, following Andrew Warwick's study of the Cambridge Mathematical Tripos to note major distinctions between British electricians (who had no particular allegiance to understandings of the ether) and those trained in the Tripos. Having successfully incorporated the challenge of the Michelson–Morley experiment into an electronic theory of matter closely linked to a dynamical ether, Tripos mathematicians generally interpreted Einstein's work as a limited, somewhat philosophical gloss that did not trouble their understanding of the ether.[12] Considering this stance to relativity, Gooday and Mitchell disabuse historians of simplistically linking allegiance to the ether with classical physics, but themselves neglected to examine practitioners' language in any detail. McLaren represents a fascinating example of someone absorbing Planck's and Einstein's work on several fronts and taking up the language of the classical, primarily in response to quantum theory.

Between the early draft and final version of McLaren's paper, in late March 1911 Larmor proposed an Adams Prize on the theory of radiation—possibly in order to encourage McLaren, for submissions were advised to address both spectroscopy and the constitution of radiant energy in statistical equilibrium.[13] Yet despite stimulating such research, in June Larmor turned down a confidential invitation to attend the Solvay Council—devoted to new principles in radiation theory—stating he had not been able to keep up with recent progress. His Cambridge colleague James Jeans accepted, along with Ernest Rutherford. Jeans was one of the few invited who already knew quantum theory well, having long argued that although the theorem of equipartition held it could not be expected to be realized except over millions of years (a radical stance few would have described as traditional). On their return from Brussels in early November, Rutherford published a brief account stating 'The interchange of views on many problems of modern physics was a feature of the occasion, and led to a much clearer understanding of the points at issue'.[14]

Rutherford subsequently discussed the meeting with a young Danish physicist he met in Cambridge—and Niels Bohr soon followed him to Manchester—but in the short run Henri Poincaré was still more important for British understandings of quantum theory. Stunned by the implications of work he was learning about for the

---

[11] (McLaren, 1912), p. 540.
[12] (Warwick, 2003); (Gooday and Mitchell, 2013), pp. 730–36.
[13] (McLaren and Hassé, 1925), p. v. McLaren and Nicholson shared the Prize.
[14] (Rutherford, 1911), p. 83.

first time, in January 1912 Poincaré published a proof that the quantum hypothesis was both sufficient and necessary for the derivation of Planck's law; discontinuity was an essential feature.[15] In his widely read 1902 book *La science et l'hypothèse*, Poincaré had limited 'classical mechanics' to Newtonian understandings, using 'Mécanique classique' as the title of a chapter criticizing the principle of mechanics in Newtonian form, in opposition to the 'energetic approach' discussed in the next chapter and based on the energy principle and the principle of least action.[16] After Brussels he extended its ambit to encompass Hamiltonian dynamics, and in 1912 announced that classical mechanics could not comprehend the change Planck's theory wrought. He thought quantum discontinuity the greatest and most profound revolution natural philosophy had experienced since Newton.[17] With forensic care Poincaré noted those elements of Planck's work that relied on what now appeared the uncertain foundation of 'classical electrodynamics', re-evaluating the fruits of still recent work on the basis of new imperatives.[18] Like McLaren's reference to the 'classical' value of the aether, this was a relatively new referent for the adjective. Thinking of Lorentz's development of Maxwell's theory, in 1902 Poincaré had written of the 'classic system of electrodynamics' which 'is perhaps even now not quite definitive'. Now, pairing the term 'classical' with electrodynamics offered a different emphasis on past rather than current standards.[19] In May 1912, Poincaré gave four lectures at the University of London. *Nature* reported that, discussing relativity, he described the research of modern physicists, especially Lorentz, as having brought about a revolution with the insistence on the propagation of effects over time rather than instantaneously. His final lecture on 'The Theory of Radiation' went still further. Like several other accounts indebted to the Solvay Council, Poincaré retained the counter-chronological argumentative structure of the papers presented in Brussels, discussing the equipartition theorem before considering Planck's work.[20] *Nature* reported his view that 'Planck had enunciated some ideas, which, if they were accepted, would bring about in the science of physics the most profound revolution that had occurred since the time of Newton'.[21]

Poincaré's work clearly helped British physicists accept such Planckian ideas, and Jeans in particular played an important role introducing others to quantum theory— first in Cambridge, and fourteen months later reporting to the 1913 British Association meeting in Birmingham.[22] There Jeans announced that Poincaré's proof had convinced him quantization was necessary, and like Poincaré described his past endeavours in new terms:

---

[15] (Poincaré, 1912); (McCormmach, 1967), pp. 44–49.

[16] (Poincaré, 1902).    [17] (Poincaré, 1912), p. 5.

[18] (Poincaré, 1912), pp. 29–30.

[19] (Poincaré, 1905), p. 225; (Poincaré, 1902), p. 225. Like Poincaré, Planck also referred to 'classical electrodynamics' in 1912 (Planck, 1912, p. 643).

[20] See for example (Reiche, 1913), pp. 549–53.    [21] (Anonymous, 1912).

[22] (Jeans, 1913). A more extensive report followed in (Jeans, 1914).

I have devoted years of work to an attempt, quite unfruitful as it turned out, to reconcile the laws of radiation with the classical mechanics by assuming that [its] formulae do *not* represent a real final state [of thermodynamic equilibrium; his emphasis]....the classical mechanics...really show no power of explaining the facts of radiation; the new mechanics, based on the quantum hypothesis, show just that power of explaining and predicting the facts to be expected of a new truth in its infancy.[23]

Jeans's lecture is also important for his favourable description of Bohr's recently published atomic model as an ingenious, suggestive, and convincing account of spectral lines.[24] The Cambridge mathematical physicist J. W. Nicholson probably learned of Poincaré's views from Jeans, and before Bohr wrote he had similarly introduced quantization into a model of the atom designed to yield an understanding of spectral lines. Without articulating any specific contrasts with earlier theory, Nicholson introduced Planck's constant and used it to generate new results, but preferred Arnold Sommerfeld's suggestion that the angular momentum of the atom might be quantized without requiring quantization of energy itself. He thought this presented less difficulties to the mind, and also noted the need to give Planck's abstract resonators a foundation in atomic theory.[25]

In marked contrast to Nicholson's matter of fact introduction of new approaches into a speculative model, the last paper we shall consider from McLaren related the new physics to the tenor of the period in dramatic terms:

The unrest of our time has invaded even the world of Physics, where scarcely one of the principles long accepted as fundamental passes unchallenged by all. The spirit of revolution is seen at its boldest in the theory of radiation. It is not only that Einstein's idea of the quantum is destructive of the continuous medium and all that was built upon it in the nineteenth century. His form of atomism excludes what has been fundamental in Physical science, the ideal of mechanical explanation.[26]

Einstein's theory was governed by laws of change that did not depend on motion (like chemistry) and thereby sacrificed mechanism. In response McLaren offered 'an attempt to save the classical view of radiation as a continuous wave motion. If that can be done, it seems to me a small thing to sacrifice the ordinary mechanical notions of matter.'[27]

This account shows that a small group of British physicists linked to Cambridge engaged closely with quantum theory between 1911 and 1913, in ways shaped increasingly strongly by the earlier contributions of Planck and Einstein. In large part (but not

[23] (Jeans, 1913), p. 378.
[24] (Jeans, 1913), pp. 379–80; (McCormmach, 1967), pp. 53–54.
[25] (Nicholson, 1912), p. 679.    [26] (McLaren, 1913), p. 43.    [27] (McLaren, 1913), p. 43.

entirely) in response to the Solvay Council and subsequent work, McLaren and Jeans both incorporated the language of the classical in their commentary and depicted a need to go beyond classical mechanics in significant research papers and addresses. Furthermore, their work deployed the argumentative and narrative structure of the Solvay proceedings, representing the failure of the equipartition theorem as a failure of classical mechanics or dynamics. Although this language and historical perspective had not been typical of anyone involved before the Solvay meeting, it was now becoming increasingly widespread, promulgated in public forums like *Nature*. Their purported scope and significance is surely wide enough to describe these changes as epochal and potentially relevant for physics generally, beyond radiation theory alone, but we should also note Poincaré's cautious awareness that one could question whether they would be accepted widely.

## 37.2 NIELS BOHR'S EARLY WORK AND THE CONTRAST BETWEEN CLASSICAL ELECTRODYNAMICS AND THE QUANTUM ATOM

As we shall soon see, in 1913 Niels Bohr also incorporated the language of the classical in describing the nature of his new atomic model, its relations to previous theoretical frameworks, and its implications—although another language was available and at times he wrote of 'ordinary' theory in preference. Still more interestingly, rather than simply using the term as a descriptive label for work of an earlier period, elements of that earlier work became integral to his own research generating new approaches. This process is well recognized, and it has usually been discussed using the language of the classical. For example, Olivier Darrigol's valuable 1992 study *From c-Numbers to q-Numbers* bears the subtitle *The Classical Analogy in the History of Quantum Theory*. Of course Bohr's work has been the subject of continual historical and philosophical analysis, and this has sometimes addressed his concept of the classical in particular; but apart from the valuable work of Darrigol and Alisa Bokulich (which pay little attention to the language Bohr used) these studies have typically focused on writings from the 1930s onwards.[28] In this section I will focus rather narrowly on Bohr's descriptions of his work through to the early 1920s, showing that these changed significantly over time—something that has previously gone unnoticed. Later sections will consider how

---

[28] On the role of classical concepts in Bohr see in particular: (Darrigol, 1992); (Howard, 1994); (Bokulich, 2008); (Schlosshauser and Camilleri, 2017). Valuable recent biographies have not focused on this issue: (Kragh, 2012); (Aaserud and Heilbron, 2013).

physicists in Germany and elsewhere used concepts of classical, to help us understand why Bohr's language changed when it did.

Bohr used the term 'classical' only six times in his major three-part paper on the structure of atoms and multi-nuclei systems, and the word 'modern' only once, but these uses highlight the significance of the Solvay Council for the introduction of the former concept. Bohr begins with differences between Rutherford's and J. J. Thomson's atomic models, noting that questions concerning the dimensions of the atom had been changed thoroughly by recent work on energy radiation. Footnoting the Solvay proceedings in considering its implications he wrote 'The result of the discussion of these questions seems to be a general acknowledgment of the inadequacy of the classical electrodynamics in describing the behaviour of systems of atomic size', so that it now seems necessary to introduce Planck's constant—'a quantity foreign to the classical electrodynamics'.[29] Similarly, Bohr motivated the introduction of a new concept of stability in the stationary states of electron orbits by first describing the 'inadequacy of the classical electrodynamics' to give an account of the structure of the atom. Detailing how an accelerating electron would radiate away energy and approach the nucleus of the atom according to 'the ordinary laws of electrodynamics', indicated why the discontinuity integral to Planck's theory might offer a better match with observations. This shift from the adjective 'classical' to 'ordinary' is characteristic of the rest of the paper and indicates how much more common it was to invoke a contrast with customary or ordinary approaches early in the development of a contrast of this kind. Thus, when Bohr first articulated and then restated the critical assumptions underlying the approach he would go on to develop at length, he wrote of 'ordinary electrodynamics' and 'ordinary mechanics', both in the first and third parts.[30] In publications in the 1920s these assumptions would be reframed instead in terms of the implications of classical electrodynamics and classical theory.[31]

In the rest of the paper, Bohr used the term classical only three times (on the same page), twice discussing the laws governing the collision of free and bound electrons, and then concluding this section. Although the preliminary and hypothetical nature of his considerations would be obvious, these were developed to explain facts not explained by ordinary electrodynamics, and the assumptions made 'do not seem to be inconsistent with experiments on phenomena for which a satisfactory explanation has been given by the classical dynamics and the wave theory of light'.[32] The latter can remind us of an important point: while insufficient in some respects, classical laws also offered powerful constraints or guidelines where they had proved satisfactory. Bohr subsequently attempted to use this dual-edged character as a methodological tool in developing new theory. I noted above that Bohr used the word modern just once. That

---

[29] (Bohr, 1913a), on p. 2. While his footnote states 'see for instance', in fact it is only in the Council and publications stemming from it like Poincaré's that its implications had been stated in these terms.
[30] (Bohr, 1913a), on pp. 5, 7; (Bohr, 1913b), pp. 874–75.
[31] See, for example, Bohr, Kramers, and Slater (1924), p. 787; (Bohr, 1925), p. 847.
[32] (Bohr, 1913a), p. 19.

came in conclusion, where he wrote the 'intimate connexion' between his theory and 'the modern theories of the radiation from a black body and of specific heat' would be evident.[33]

That Bohr's references to ordinary electrodynamics vastly outnumber those to classical theory offers a clear indication that this was early days in the articulation of a new approach. In the next few years, working in Copenhagen and Manchester (from October 1914 to July 1916), Bohr responded to weak points in the formulation, basis, and applications of his atomic model. For example, in a 1913 address on the structure of hydrogen published in Danish in 1914 (later translated into English), Bohr discussed the explanatory status of his work and Planck's theory, which used 'ordinary electro-dynamics' in certain calculations but introduced variant assumptions in others without attempting to give a consistent explanation.[34] Bohr concluded by saying he hoped he'd been clear enough in showing the extent to which his work was in conflict with previous theory—'the admirably coherent group of conceptions which have been rightly termed the classical theory of electrodynamics'.[35]

It is interesting that straight after justifying his use of a relatively new term in this way, Bohr largely stopped using it. Indeed, in several papers in the early years of World War I that drew on recent experimental work to shore up his approach in the face of criticism, Bohr referred in preference to 'ordinary electrodynamics', 'ordinary mech-anics', and 'the ordinary theory of dispersion', writing just once of 'the classical electron theory' in a paper on the Stark effect.[36] This hitherto unnoticed hiatus in Bohr's use of the term classical is worth probing. It was likely a response to the somewhat contro-versial status of his own theory, and the predominantly cautious stance of British colleagues. For example, while Rutherford pointed to the denial of classical mechanics in his first published comment on Bohr's theory early in 1914, later that year he spoke only of 'ordinary theory' and Nicholson described Bohr's work in similar terms when asked to offer a perspective.[37] Even after he returned to Copenhagen to a professorship in theoretical physics, Bohr continued to publish in English—although the majority of the research he drew on and cited was German.

Bohr's most important work during the war was long in gestation. In November 1917 he sent two sections of a long paper to the Royal Danish Academy of Sciences and Letters, articulating a line of thought he pursued for several years, soon giving it the name 'the correspondence principle'. Historians have often noted this re-conceptualization, but usually interpreted as a study in the relations between classical and quantum theory what was initially described in other terms.[38] Bohr aimed to address the polarization and intensity of spectral lines as well as their frequencies from a uniform perspective, and to consider especially 'the underlying assumptions in their

---

[33] (Bohr, 1913b), p. 875.    [34] (Bohr, 1922 [1914]), p. 6.    [35] (Bohr, 1922 [1914]), p. 19.
[36] (Bohr, 1914), p. 518 (for 'the classical electron theory'). (Bohr, 1915b), (Bohr, 1915a), p. 585 (for 'the ordinary theory of dispersion').
[37] (Rutherford, 1914a), p. 498; (Rutherford, 1914b), pp. 5, 12–15.
[38] For one example in an excellent study, see (Kragh, 2012), pp. 198–99.

relations to ordinary mechanics and electrodynamics'. He argued some of the difficulties facing the theory could be exposed by 'tracing the analogy between the quantum theory and the ordinary theory of radiation as closely as possible'.[39] Bohr's most concrete arguments concerned the possibility of drawing conclusions about the allowed transitions and characteristics of the radiation (even intensity and polarization) emitted from atoms in the limit of slow vibrations and large quantum numbers, where the motions of electrons in successive stationary states differ little from each other. There, by analogy with the ordinary theory of radiation he found a relationship between the harmonics present in the ordinary theory and the allowed quantum transitions.[40]

Seeds of this approach could be traced to 1913, and it reflected a conversation with Peter Debye in Göttingen in 1914 in which Bohr framed his investigation as one into the relations between quantum and classical theory.[41] Yet when Debye and Sommerfeld considered dispersion in papers published in 1915 and 1917–18, respectively, like Bohr they focused on the relations between ordinary mechanics and electrodynamics and quantum theory.[42] Marta Jordi Taltavull has offered a valuable account of the continuities between diverse accounts of dispersion between classical and quantum treatments, and highlights the fact that in 1915 Sommerfeld described dispersion as corroborating the 'peaceful coexistence' between 'classical' electrodynamics and mechanics and quantum theory—without recognizing that Sommerfeld did not use that language in this later work.[43] But others took a different approach. In two conceptual, statistical studies published in 1916 and 1917, Einstein considered relations between the 'electromagnetic-mechanical' considerations which led to Planck's law and the incompatible quantum theory that emerged from it, describing Planck's oscillators as 'a limiting case of classical electrodynamics and mechanics'.[44] Similarly Paul Ehrenfest used Wien's law (which he described as derived from 'classical foundations') with its investigation of adiabatically invariant changes as a probe to investigate 'the boundaries between the "classical" and "quantal"' (publishing in German and English from the neutral Netherlands).[45] Given the variety of models available and the ready use both these authors made of the language of the classical, Bohr's decision to use the term

---

[39] (Bohr, 1918a, 1918b). Both Darrigol and Bokulich simply read this in terms of the contrast between classical and quantum, see: (Darrigol, 1992), p. 26; (Bokulich, 2008), p. 87.

[40] (Bohr, 1918a), pp. 7, 15–16.    [41] (Bohr, 1981), p. 563.

[42] (Debye, 1915); (Sommerfeld, 1918a). Sommerfeld wrote once of the classical rules of calculation but referred far more commonly to ordinary mechanics and electrodynamics, see especially (Sommerfeld, 1918a), pp. 502–505 at p. 504.

[43] (Jordi Taltavull, 2018), p. 2. See also (Jordi Taltavull, 2016).

[44] (Einstein, 1916a), p. 322; (Einstein, 1917a), p. 124.

[45] (Ehrenfest, 1916), pp. 327–28, 330; (Ehrenfest, 1917), pp. 500–13 at pp. 500–02. My translation preserves the verbal ingenuity of the German: in the two versions Ehrenfest wrote, respectively, of the desire to find a more general point of view for the distinction between 'klassisch' and 'quantös' (p. 327); and 'to trace the boundary between the "classical region" and the "region of the quantum"' (p. 500). For Ehrenfest's early attitudes to quantum theory see (Staley, 2008a), pp. 391–92.

'ordinary' in 1918 might be as deliberate as his choice to rename Ehrenfest's adiabatic hypothesis the 'principle of mechanical transformability' in the specific context in which he used it to fix the stationary states of the atom amongst all those mechanically possible.[46]

Two and a half years later, however, in April 1920 Bohr returned to the language of the classical in an invited address to the German Physical Society on the series spectra of the elements. A year earlier he had lectured to Ehrenfest's group in Leiden and in September 1919 welcomed Sommerfeld as the first international visitor to Copenhagen, but Bohr continued to refer to ordinary electrodynamics and mechanics or the usual conceptions of physical theory until travelling to Berlin.[47] The first of his many publications in German, this paper was subsequently translated into English and featured the first full articulation of his 'correspondence principle', now postulating 'a formal correspondence between the fundamentally different conceptions of the classical electrodynamics and of the quantum theory' and arguing that 'in a certain sense' the laws of spectral series derived could be understood as 'a rational generalization of the ordinary theory of radiation'.[48] Introducing the paper, Bohr spoke of the difficulty explaining the correlation between the number of electrons and atomic number of an element on the basis of 'the classical laws of mechanics and electrodynamics', but also noted the success of 'the classical theory of radiation' in explaining interference phenomena.[49] Why did Bohr's language change in this way, retaining references to ordinary theory but now framed as an aspect of the relations between classical and quantum theory?

## 37.3 THE CLASSICAL IN GERMAN PHYSICS, 1911–1920: 'THE TREATMENT AND DEEPENING OF CLASSICAL THEORY'

It is likely that both the nature of Bohr's Berlin audience and the attention relativity had brought to a similar contrast between new and customary approaches provided incentives to return to the language of the classical. This section will chart how German physicists incorporated the classical in discussions of quantum theory after 1911, arguing this changed in character and extent through the war, and indicating reasons for this in the work of Planck, Einstein, and Sommerfeld in particular. In research papers and books published on quantum theory after the Solvay Council,

---

[46] (Bohr, 1918a), p. 8.
[47] (Bohr, 1976b [1919]), pp. 201–16; (Bohr, 1976c [1919]), pp. 217–19; (Bohr, 1976a [1919]); (Bohr, 1976d [1920]), pp. 227–40.
[48] (Bohr, 1920), pp. 423–69; (Bohr, 1922), p. v; (Bohr, 1922 [1920]), pp. 20–60 at pp. 23–24.
[49] (Bohr, 1922 [1920]), pp. 21–22.

it was soon relatively common to refer to different versions of classical theory amongst the German physics community. We can see the immediate effect of the council's language in reports on its proceedings from Fritz Reiche and Max Born, and Arnold Eucken's 1913 German translation.[50] The publication of the second edition of Max Planck's lectures on the theory of heat radiation also provided an important impetus. Planck's foreword described the need to tie the quantum hypothesis as closely as possible to 'classical dynamics' and only go further where the facts of experience give no other possibility, while also arguing that a really new principle couldn't be determined by a model functioning according to the old.[51] With its continual references to concepts of 'classical thermodynamics', 'classical electro-dynamics', 'classical dynamics', and 'classical statistical mechanics', this was the first textbook to embody a newly comprehensive understanding of classical theory.[52] However, research papers in the field were as likely to refer instead to the 'customary' mechanics, as Sommerfeld and his former student Paul Epstein did in their major 1916 papers in the *Annalen der Physik* incorporating Keplerian orbits, a relativistic treatment, and an explanation of the Stark effect on the basis of the Bohr atom.[53] Sommerfeld's work was particularly important and Bohr's terminology may have followed his example.[54] In contrast, when Einstein began publishing on quantum theory again in 1916 after completing general relativity he adopted Planck's language, referring variously to 'classical mechanics', 'classical theory', and 'classical electrodynamics'.[55] He was still more expansive in his use of the adjective in later papers, writing also of 'classical thermodynamics' and 'the classical theory of heat' in an obituary of Marian von Smoluchowski, and of the 'classical theory of dispersion' in a paper on the experimental determination of the X-ray refractive indices of solids.[56]

Einstein's expansive use of the term in this period stands in some contrast to his earlier discussions, and most other physicists. If Planck's contribution to the Solvay Council unlocked the possibility for a more general and encompassing use of the term

---

[50] (Reiche, 1913); (Born, 1914); (Eucken, 1913).    [51] (Planck, 1913a), pp. viii–ix.

[52] (Planck, 1913a). References to the classical occur at pp. vii, 118, 132 (klassischen Thermodynamik); pp. viii, 149, 159, 164, 179 (klassische/n Dynamik); pp. viii, 147, 148, 181 (klassischen Electrodynamik); p. viii (klassischen elektrodynamischen Gesetzen der Emission und Absorption), p. ix (klassische Auffassung die Physik der Atome); pp. 124, 141 (klassischen statistische Mechanik); p. 130 (klassischen Theorie).

[53] (Sommerfeld, 1916), pp. 1–94, 125–67; (Epstein, 1916), pp. 168–88. In this case an important contrast was with customary mechanics and a relativistic treatment. Michael Eckert describes Sommerfeld as having become familiar with the difficulties of a classical treatment of electron oscillations through earlier studies of Voigt's quasi-elastic approach, but does not discuss Sommerfeld's language: (Eckert, 2013), pp. 261, 268.

[54] Seth offers a discussion but re-describes what Bohr termed ordinary as 'classical'; (Seth, 2010), pp. 194–97 on p. 195.

[55] (Einstein, 1916a), p. 47 (for references to 'klassische Mechanik und Elektrodynamik' and 'klassische Theorie'), p. 49 ('klassischen Elektrodynamik'), and pp. 50, 52, 53, 60 ('klassischen Theorie'); (Einstein, 1916b), pp. 47–62; (Einstein, 1917a).

[56] (Einstein,1917b), pp. 737–38; (Einstein, 1918), pp. 86–87.

in 1911, and his 1913 textbook consolidated it, the terms in which he brought Einstein to Berlin—with an approach in 1913 leading to Einstein's move from Zürich in early 1914—surely reinforced Einstein's use of this language by giving it an intimate, personal dimension. The proposal to make Einstein a member of the Prussian Academy of Sciences that Planck, Walther Nernst, Heinrich Rubens, and Emil Warburg signed indicated their appreciation for the breadth of Einstein's contributions to 'modern physics', but also noted that as well as founding and critiquing new hypotheses, Einstein had proved a master of 'the treatment and deepening of classical theory'.[57] In turn, Einstein anthropomorphized the struggle to develop and understand quantum theory and associated it with Planck in particular, as a tribute to Planck in November 1913, and his inaugural address to the Prussian Academy of Science both show. Einstein wrote 'classical mechanics also fell victim' to the search for a universal heat radiation law, and it was too early to say whether Maxwell's electrodynamic equations would survive. On his account, Planck alone had been able to determine the radiation law theoretically, and understand it.[58]

Sixtieth birthday celebrations for Planck in April 1918 also helped further such views and language. In a special issue of *Die Naturwissenschaften*, neither Sommerfeld's overview nor Wien's or Laue's short contributions made particular use of concepts of the classical, but two substantive articles on quantum theory from Reiche and Epstein framed their subject matter in terms of classical theory.[59] As we might expect, Nernst and Warburg wrote in similar terms, with Nernst referring to classical thermodynamics and classical kinetic theory, and Warburg writing still more generally that, 'Since its classical period, especially in Germany, physics has taken on a new face'. Warburg wrote that in the beginning of his scientific career, Planck had stood completely on the grounds of classical theory, in that he had deliberately avoided atomism; and described quantum theory as having laid theoretical foundations for an abundance of fields, highlighting the work of Einstein and Bohr.[60]

Perhaps keeping this company tipped the linguistic and discursive balance for Sommerfeld. Replying to a letter Bohr sent with part I of his 1918 paper on the quantum theory of line spectra, Sommerfeld discussed his 'interesting comparison between classical and quantum-theoretical emission for large quantum numbers'. Bohr had not used that term in the original paper; nor had Sommerfeld in earlier published discussions of Bohr.[61] Nevertheless, Sommerfeld adopted it fully in his

---

[57] (Planck et al., 1993 [1913]), pp. 526–29 at pp. 527–28.

[58] (Einstein, 1913), pp. 1077–79 at p. 1078; (Einstein, 1914), pp. 739–42 at p. 740.

[59] Reiche devoted a section to the denial of classical statistics and Epstein described the customary mechanics and Maxwell–Lorentzian electrodynamics as the 'classical foundations' that had proved insufficient to ground a theory of blackbody radiation. Sommerfeld made one reference to classical optics. (Sommerfeld, 1918b), pp. 195–99 at p. 197; (Wien, 1918), pp. 203–06; (Laue, 1918), pp. 207–13; (Reiche, 1918), pp. 213–30 at pp. 214–15; (Epstein, 1918), pp. 230–53 at p. 230.

[60] (Warburg, 1918), p. 202 (for the quote); (Nernst, 1918), pp. 206–07.

[61] See N. Bohr to A. Sommerfeld, 7 May 1918, and A. Sommerfeld to N. Bohr, 18 May 1918, in (Sommerfeld, 2000).

textbook *Atombau and Spektrallinien*, which proved enormously important for further research. Introducing quantum theory, Sommerfeld described it as a child of the 20th century with its birthday in Planck's work in December 1900, focusing on the bold intervention this had forced in the previous conceptions of wave theory and outlining its violations of 'classical electrodynamics' and 'classical statistics'.[62] When Sommerfeld considered Bohr's 'principle of the analogy between wave theory and quantum theory' (contrasting it with his own methods), he re-described Bohr's ideas in terms of a contrast between the classical theory of radiation and quantum theory.[63] In 1921 the second edition was expanded to deal better with Bohr's correspondence principle (still disapproving of its mixture of classical and quantum theoretical perspectives); an English translation of the third edition was published in 1923.

While discussions of quantum theory were pursued in mixed terms through World War I, I have shown here that the promotion of classical language by Planck, Ehrenfest, and Einstein in particular was consolidated by the celebration of Planck in Germany, and joined early in the post-war period by Sommerfeld's significant textbook, as well as by Bohr's return to the language of the classical. I will now show that a strong focus on quantum theory in early post-war Nobel physics prizes led to the more widespread adoption of this language amongst broader circles internationally, together with a heightened sense of the historical and cultural significance of the conceptual innovations physicists were considering—and a generalized concept of classical physics.

## 37.4 Celebrating New Physics and Concepts of the Classical after the Great War

In June 1920 three days of prizes, banqueting, and speech-giving helped the Nobel Foundation catch up with business left over since 1914. Although the subject matter of the physics prizes ranged widely, each of the four physics Nobel Laureates taking their awards referred to the significance of quantum theory. Celebrated for x-ray diffraction, Laue described the solution of 'the great quantum mystery' as 'the most important objective currently confronting the entire field of physics'.[64] The Edinburgh physicist Charles Glover Barkla and Johannes Stark (of Greifswald and Würzburg) both devoted significant attention to the bearing of their research (in x-ray spectroscopy, and the splitting of spectral lines in electric fields) on quantum theory, expressing limitations to its scope and a distrust of Bohr's hypotheses, respectively.[65] In presenting the award to Planck the President of the Swedish Academy of Sciences described his radiation

---

[62] (Sommerfeld, 1919), pp. 213–14.    [63] (Sommerfeld, 1919), pp. 401–02.
[64] (Laue, 1967 [1920]), pp. 347–55.
[65] (Barkla, 1967 [1920]), pp. 392–99; (Stark, 1967 [1920]).

theory as 'the lodestar of modern physical research'.[66] Planck himself offered a much more personal account of its early stages than he had previously. Before the Solvay Council in 1911, Planck had described 'the principles of classical mechanics' as so well confirmed (even extended through electrodynamics and electron theory), that they had seemed a foundation for the unification of the worldview, before reaching towards the inclusive plural to argue 'Today we have to say that that hope has proven deceptive'.[67] In contrast, when looking back on the 1890s in 1920, he wrote 'I was filled at the time' with 'what would today be thought the naively charming and agreeable expectations, that the laws of classical electrodynamics' (suitably extended) would be sufficient.[68] He depicted the dramatic changes quantum theory required most clearly when discussing Bohr's work, which he said rested on hypotheses that would have been flatly rejected a generation ago, found monstrous and impractical by theoretical physicists 'brought up in the classical school'. Now, rather than finding an uncertain place, that conceptual intruder, the quantum of action was bursting asunder the old framework—although Planck thought that following Ehrenfest's adiabatic hypothesis and Bohr's recent ideas, the 'main principles of thermodynamics from the classical theory will not only rule unchallenged but will more probably become correspondingly extended'.[69]

Even in their variety of stances these addresses underline the burgeoning significance of quantum theory. As Planck admitted, the introduction of the quantum of action 'still has not yet produced a genuine quantum theory'—but that made it a vital question.[70] *Nature* recommended the extraordinary interest of his address, commenting that much of Planck's account was concerned with 'the later history of the quantum, and it is scarcely too much to say that this is simply the history of modern physics'.[71] Planck's richly personal account made the early and ongoing history of the quantum a dramatic revision but also a potential extension of classical theory; and this offers a second sense in which we might understand classical and modern physics as being co-created, with Bohr's correspondence principle furthering 'classical theory' in dialogue with quantum theory (something we will examine further in the following section).

Thus, the post-war emergence of quantum theory came with kudos, but also with an explicit, broadly framed account of historical significance. That combination seems to have been highly effective in attracting new researchers, and changing the terms in which even some of those long familiar with the theory discussed it. The British mathematical physicists H. Stanley Allen and Edmund Whittaker (both students in Cambridge in the late 1890s) provide good examples. Allen lectured at King's College London from 1905. His 1913 book on photo-electricity and a series of papers on atomic models show his familiarity with the work of Einstein, Bohr, Nicholson, and many

---

[66] (Ekstrand, 1918).    [67] (Planck, 1913b), p. 77.
[68] (Planck, 1967 [1920]). Note, however, that some of the tensions in this retrospective view are indicated in Planck's admission that he dealt with the general laws of electrodynamics in the vacuum rather than with Lorentz's electron theory because he did not trust the latter—indeed both were so recent it would have been truly unusual to have understood the former as 'classical' in the 1890s.
[69] (Planck, 1967 [1920]).    [70] (Planck, 1967 [1920]).    [71] (Anonymous, 1920), p. 509.

others, adopting and refining Bohr's model rather selectively, focusing especially on the nucleus, without particular comment.[72] In 1919 Allen joined Barkla lecturing in Edinburgh, and in 1921 published a paper on the aether and quantum theory in which he offered a new sense of the significance of his work. Rather than following the most radical adherents of relativity he would use the concept of the aether as a model, presently inadequate, to interpret physical phenomena. Drawing on McLaren to offer an account of the quantum based on the magneton, Allen concluded that rather than 'reconciling' quantum theory and classical dynamics he was concerned with understanding the nature of the quantum, accepting it demanded some modification of the old theories.[73] Whittaker taught astronomy and mathematical physics in Dublin from 1906 before taking up a chair in mathematics in Edinburgh. Having written on theories of aether and electricity in 1910, he followed the success of Einstein's general theory of relativity and the quantum theory to publish on both in 1922, first offering an account of Faraday tubes appropriate to relativity, and then developing a model for quantization which drew significant attention from British colleagues, with Allen giving an account of the extent to which it (unwittingly) betrayed classical theory.[74]

These accounts also show that relativity and quantum theory were regularly coupled, something that underlined the breadth of challenges to familiar theory, and likely reinforced use of the term classical. Key papers and text books on relativity published in the 1910s from Einstein and others typically introduced the principle of relativity as it had been used in ordinary, customary, or 'classical mechanics' before outlining the changes required for its generalization firstly to electrodynamics—introducing the finite constant $c$ for the velocity of light—and secondly to accelerated motion and gravitation. Expressing a relation of historical succession and incorporation of the old into the new rather than rival methods, these authors treated 'classical mechanics' as a foil to a new, more general dynamics that involved modifications or transformations of key concepts.[75] In research papers and textbooks, but also in more public contexts like newspaper articles and popular texts, the term 'classical' began to be used more often, without ever fully displacing other terms like 'old', 'customary', or 'Newtonian'. For example, in 1916 and 1917 Einstein introduced both his fundamental papers and popular account of relativity with discussions of classical mechanics (with the term used much more pervasively in the latter). However, in November 1919 his account of relativity for the London *Times* combatted the view that relativity overthrew or superseded Newton's work.[76]

---

[72] Allen wrote that the Planck–Einstein theory of the photo effect was not accepted because it could not be reconciled with the fundamental dynamical equations of Maxwell and Hertz, see (Allen, 1913), p. 146. Similarly, early in a series of nine papers on atomic models published between 1915 and 1920 he wrote simply that Bohr's theory is 'confessedly not dependent on the usual dynamical laws' (Allen, 1915, p. 41). See also (Kragh, 2012), pp. 102–03.

[73] (Allen, 1921b), pp. 34, 43; (Allen, 1921a).

[74] (Whittaker, 1921); (Whittaker, 1922); (Allen, 1922), pp. 218–20.

[75] (Staley, 2008a), pp. 367–75 on p. 374.

[76] (Einstein, 1974 [1916]); (Einstein, 1917c); (Einstein, 1919).

Given Einstein's popularity, the Nobel Foundation risked a scandal each year they did not acknowledge him. After navigating internal committee politics in 1922 they eventually awarded the 1921 prize to Einstein for the photoelectric effect, rather than relativity, while bestowing that year's prize on Bohr. Presenting their prizes (with Einstein absent due to the danger of returning to Europe soon after the assassination of Walther Rathenau), Svante Arrhenius referred to classical mechanics in discussing Einstein and built on earlier narratives of Planck's work to explain why Bohr was so ready to diverge substantially from Maxwell's 'classical doctrines'.[77] Bohr's banquet speech made a point of noting the internationalism the Nobel represented (something different Laureates approached quite differently). Offering an overview of recent research, his Nobel Lecture both drew on the diverse communities with whom he had worked, and became a vehicle for their further integration. Although far less personal than Planck's address, it also brought a new element to uses of the classical.

Bohr's internationalism was direct and inclusive, even if framed in racial terms. In banquet, Bohr described himself as a connecting link between the imagination and acumen of British investigators like Thomson and Rutherford, and the systematic, abstract investigations of the Germans Planck and Einstein, but also noted the intellectual solidarity and racial affinity binding Scandinavians.[78] In his lecture, discussing the recent work of many researchers across diverse traditions represented a distinct effort to bridge research traditions. In the post-war period Bohr had begun publishing key papers and books in German, English, and French (sometimes translating his Danish papers), as well as in both physics and chemical journals. Similarly, Bohr's Nobel Lecture helped consolidate international research around common problems and similar attitudes towards the past. Both R.H. Fowler's review of a 1922 collection of Bohr's papers on *The Structure of the Atom*, and *Nature*'s July 1923 introduction to a supplement publishing a translation of his Nobel Lecture illustrate these points. Fowler wrote that with the atomic theory being 'expounded semi-historically by its principal creator' his review could not function as a critical account, for that would involve 'nothing less than an exhaustive survey of the whole tendencies of modern physics'.[79] Tellingly, *Nature* titled their introduction 'Modern Physics and the Atom'. While they centred their history on research on the atom (and the significance of moving from statistical deductions and properties in bulk to individual atoms), the turn of the century formed a critical juncture and they explained that the impotence of classical electrodynamics had led Bohr to invoke quantum theory (as now all agreed was necessary), while also emphasizing the importance of Bohr's work for chemistry.[80]

Even as Bohr's work helped render the claims of quantum theory to represent modern physics newly comprehensive and not just critically fundamental, Bohr's writings (and Nobel Lecture in particular) subtly expanded the frame of the classical

[77] (Arrhenius, 1922).    [78] (Bohr, 1923 [1922]).
[79] (Fowler, 1923), pp. 523–24. Fowler had begun his work in quantum theory by filling a gap in Poincaré's proof of Planck's quantum hypothesis in (Fowler, 1921).
[80] (Anonymous, 1923).

from theories to explanations and concepts, with a newly frequent use of the term classical both concentrating its rhetorical significance and rendering it more general. Revealingly, Fowler's review focused on the explanations that current atomic theory provided, commenting 'the theory is non-mechanical—in fact, is nowadays identical with the quantum theory—and "explanation" by the theory cannot mean explanation in the classical sense'. Casually introducing this new coinage, Fowler argued correlation and coordinating facts under common general principles was enough for explanation (and the next stage would bring classical mechanics, electro-mechanics, and quantum theory together as a homogenous whole).[81] His discussion highlights something the unprecedented extent and range of references to classical in Bohr's Nobel Lecture enhanced still further (concentrating messages also conveyed in his seven Wolfskehl lectures that summer). There Bohr spoke not only of classical mechanics, electro-dynamics, and theory but of 'classical physics' twice (referring to Planck's work), and perhaps even more significantly, of classical concepts, conceptions, and ideas. Bohr wrote that according to 'classical concepts' the frequency of oscillation of a particle was also supposed to be the frequency of radiation emitted; that Rutherford's discovery of the atomic nucleus made it clear 'classical conceptions alone' could not comprehend the atom; and of the 'classical ideas' of the origin of electromagnetic radiation.[82] Bohr's discussion thus generalized the concept by associating classical not just with particular theories, laws, or fields, but with ways of thinking. This built on a long-standing tendency to write in terms of conceptions. Whereas in 1914 he referred to the 'theoretical conceptions hitherto employed', 'ordinary conceptions', or once (as mentioned earlier) 'the admirably coherent group of conceptions we can rightly term the classical theory of electrodynamics', after the war Bohr's discussions moved from writing in terms of 'usual' or 'ordinary conceptions' in 1920 and 1921 to 'classical concepts' or 'classical ideas' in 1922.[83] In June 1922 his Wolfskehl lectures opened by stating the inability to make any progress in atomic theory with classical electrodynamics, particularly in the theory of radiation—before noting also the conceptual incoherence of the quantum theory of light, which gave a definition of the frequency through its relation to Planck's constant and energy, but could yield no understanding of wavelength.[84] The second lecture began with the striking point that 'So far, only concepts developed in the classical theories, such as those of the electron and electric and magnetic forces, are available to us for describing the natural phenomena'. But with the classical picture assumed to be invalid it was a real question whether the classical concepts could be united with quantum theory without contradiction—and

[81] (Fowler, 1923), pp. 523–24.
[82] (Bohr, 1923a), with 24 references to variants of 'classical', including classical electrodynamic theory/electromagnetic theory on pp. 31, 33 (2), 34 (2), 37, 38, and 40; classical physics on p. 31 (2); classical theory on pp. 31, 32, 33, and 39 (4); classical concepts/conceptions/ideas on pp. 32 (2) and 39; and classical mechanics on pp. 32, 33, and 34.
[83] (Bohr, 1922 [1914]), pp. 10, 13, 19; (Bohr, 1922 [1920]), p. 20; (Bohr, 1922 [1921]), p. 81; (Bohr, 1922 [1921]), pp. 32, 39. (Bohr, 1977 [1922]), pp. 344–45.
[84] (Bohr, 1977 [1922]), pp. 344–45.

Bohr advocated approaching this through the conservative strategy of focusing on the applications of the principles of the quantum theory.[85] He was articulating a problematic in terms that many physicists took up in the next seven years.

# 37.5 CODA

Our study of physicists pursuing quantum theory after 1911 in Britain and Germany in particular has shown that by the early 1920s elite scientists and major journals had begun to present quantum theory as the central and encompassing feature of 'modern physics' at the same time as presenting newly general concepts of first 'classical electrodynamics', 'classical mechanics', and 'classical theory', and then 'classical physics' and 'classical concepts'. While this certainly depended primarily on the technical achievements of research papers from Planck, Einstein, Ehrenfest, Poincaré, Bohr, Sommerfeld, and many others, I have shown that after the Solvay Council had allowed a new coinage of classical theory to be voiced and echoed, the celebratory cultures of physics had given it a strongly personal dimension, most clearly for Planck, but also for Einstein. Examining the changing stances of physicists both on the edge of work in quantum theory, like McLaren and Allen, and at its heart, like Bohr and Sommerfeld, has illustrated the emotive, persuasive power of this language, as well as its contingency. But for a wartime celebration of Planck's birthday, for example, Sommerfeld and Bohr might not have incorporated this language in their key post-war textbooks, research papers, and addresses. The cultural achievement of changing the language of physics in this way is also part of its work. In this instance, it helped shape the terms in which practitioners understood the historical significance of new theory. Without that work we may have won a modern physics that was winnowed from ordinary radiation theory and dispersion theory rather than from classical physics. But even in this general form, this new concept of classical physics was strongly shaped by the specific dialogues in which it was invoked—by its relations to new theory as much as by reference to the past. Diverse relationships with new theory were to render it increasingly part of technical practice in the 1920s.

The combination of technical and cultural work that I have disclosed so far made the incorporation of the language of the classical inevitable in the further development of quantum theory. It set the stage for a remarkable period in the 1920s in which Bohr, Hendrik Kramers, John C. Slater, Born, John Van Vleck, Werner Heisenberg, Pascual Jordan, Paul Dirac, and Erwin Schrödinger (to name key figures working in Denmark, Germany, the US, and Britain) would all make central contributions to the articulation of a new quantum mechanics of the atom, and each would invoke concepts of classical theory in the process. However, the specific concepts they invoked were rather

---

[85] (Bohr, 1977 [1922]), p. 351.

diverse—in many ways characteristically different. This coda offers a brief overview of some of the most important of them in order to give an overarching perspective on the central role that concepts of classical physics played in the formation of quantum mechanics.

I begin by noting that in 1923, Bohr offered a revealingly more precise formulation of the central conceptual problematic he described in 1922, writing 'From the present point of view of physics, however, every description of natural processes must be based on ideas which have been introduced and defined by the classical theory'.[86] As we will see, by 1929 the opening clause's qualification had been removed to refer simply to *every* description, and Bohr had tied the introduction and definition of ideas more closely to experiment. The following year Bohr published an important paper co-written with Kramers and the American physicist John Slater which questioned the 'classical claim' of conservation of angular momentum, while defending the use of virtual oscillators, which already departed so far from 'the classical space-time description' as to licence such formal interpretations.[87] In 1924 Born built on Kramers's dispersion theory and Heisenberg's approach to the anomalous Zeeman effect to argue that a formal transition from classical mechanics to a 'quantum mechanics' lay very near, beginning his account with sections on 'classical perturbation theory' and 'classical dispersion theory', before articulating the transition from 'classical' to quantum 'formulae'.[88] The following year Heisenberg provided an analogue to classical mechanics in which only the connections between observable quantities appeared; which Born and Jordan would describe as an attempt to develop a conceptual system actually appropriate to these facts rather than a more or less forced accommodation of customary concepts.[89] Heisenberg's paper articulated it through a constant comparison between 'classical formulae' and 'classical magnitudes', and symbolic quantum theoretical analogues. Born and Jordan's paper showed that a 'closed mathematical theory' of quantum mechanics was possible 'in remarkably close analogy to classical mechanics', and indicated how rapidly a new perspective had been won by describing new versions of the equations of what they could now call 'the "classical" quantum theory', which had been provided by Thomas and Kuhn earlier that year.[90] Also following Heisenberg but minimizing the contrast between classical and new theory, Dirac showed that the equations of classical mechanics were not at fault, but modifying the mathematical operations by which results were obtained allowed the new theory to make use of '*All* the information supplied by the classical theory' (his emphasis).[91] He introduced 'classical rules', products, operations, and variables in articulating how to develop a quantum algebra, and later quantum variables or q-numbers in contrast to 'classical numbers'.[92] Similarly, in 1926 Born, Heisenberg, and Jordan described the

---

[86] (Bohr, 1923b), p. 117. See also (Bohr, 1924 [1923]), p. 35.

[87] (Bohr, Kramers, and Slater, 1924), pp. 792, 799.      [88] (Born, 1924), pp. 380, 386.

[89] (Heisenberg, 1925); (Born and Jordan, 1925), p. 858.

[90] (Born and Jordan, 1925), pp. 858, 869–70.

[91] (Dirac, 1925), p. 642.      [92] (Dirac, 1926).

transformation to a symbolic quantum geometry from the 'intuitive classical geom-etry'.[93] Schrödinger did not use the term classical nearly so frequently in developing his work, but did outline the respects in which he and Heisenberg, Born and Jordan had taken different departures from classical theory.[94] Meanwhile, in 1925 Bohr had described Einstein's 'fundamental theory of relativity' as perhaps 'the natural comple-tion of the classical theories', while arguing that in quantum theory, 'one is faced not with a modification of the mechanical and electrodynamical theories describable in terms of the usual physical concepts, but with an essential failure of the pictures of space and time on which the description of natural phenomena has hitherto been based'.[95]

The significance of specific analogies and the varieties of classical theory evident in this brief overview of new nomenclature indicates the degree to which classical physics was articulated in contrast to the newly developed quantum mechanics. That variety was mirrored by a manifest diversity of interpretations of the new mechanics—which often also turned significantly on physicists' different perspectives on its relations with classical physics, as Bokulich's study of Dirac, Heisenberg, and Bohr shows brilliantly.[96] Heisenberg's uncertainty principle in 1927 and the publication of Bohr's Como Lecture in 1928 both aimed to clarify this situation, by specifying the extent to which quantum mechanics showed significant limitations in the use of classical concepts, if not in their definition. In 1927 Heisenberg argued that 'all concepts used in classical mechanics are also well-defined in the realm of atomic processes. But, as a pure fact of experience, experiments that serve to provide such a definition for one quantity are subject to particular indeterminacies if we seek to provide a simultaneous definition of two canonically conjugate quantities'.[97] Bohr's strategy was more complex. He clearly took the articulation of the new mechanics to resolve features of the conceptual problematic he described in 1922 and 1923, but in two rather different respects, which on the one hand tightened its focus on phenomena rather than laws, and on the other hand generalized its significance to language and experience rather than physics alone. The first element of this elaboration was phrased in terms of the renunciation of the causal space-time coordination in examining atomic phenomena, since observation of phenomena involves an interaction of the agency of observation: quantum mechanics now showed that space-time coordination and causality were complementary and symbolized the idealization of observation and definition in classical theories, which had assumed their union.[98] The second element was articu-lated in more general discussions. In a celebration of Planck's birthday in 1929, Bohr argued that 'It lies in the nature of physical observation, nevertheless, that all experi-ence must ultimately be expressed in terms of classical concepts, neglecting the quantum of action'. The generality of this view led Bohr to state his view:

---

[93] (Born, Heisenberg, and Jordan, 1926).    [94] (Schrödinger, 1926).
[95] (Bohr, 1925), pp. 846, 848.    [96] (Bokulich, 2008).
[97] (Heisenberg, 1927), p. 179.    [98] (Bohr, 1928).

it would be a misconception to believe that the difficulties of the atomic theory may be evaded by eventually replacing the concepts of classical physics by new conceptual forms. Indeed, as already emphasized, the recognition of the limitation of our forms of perception by no means implies that we can dispense with our customary ideas or their direct verbal expressions when reducing our sense impressions to order. No more is it likely that the fundamental concepts of the classical theories will ever become superfluous for the description of physical experience.[99]

Thus, if physicists had now reached the point of defining modern physics in part through its contrast with classical physics, this was only apparently the physics of the past.

## ACKNOWLEDGEMENT

I would like to acknowledge my gratefulness for Olivier Darrigol's careful reading and helpful suggestions.

## REFERENCES

Aaserud, F., and Heilbron, J. L. (2013). *Love, Literature and the Quantum Atom: Niels Bohr's 1913 Trilogy Revisited*. Oxford: Oxford University Press.

Allen, H. S. (1913). *Photo-Electricity: The Liberation of Electrons by Light, with Chapters on Fluorescence & Phosphorescence, and Photo-Chemical Actions & Photography*. London/New York: Longmans, Green and Co.

Allen, H. S. (1915). The magnetic field of an atom in relation to theories of spectral series. *The London, Edinburgh, and Dublin Philosophical Magazine and Journal of Science*, **29**(169), 40–49.

Allen, H. S. (1921a). Faraday's 'Magnetic lines' as quanta.—Part I. *The London, Edinburgh, and Dublin Philosophical Magazine and Journal of Science*, **42**(250), 523–37.

Allen, H. S. (1921b). Æther and the Quantum Theory. *Proceedings of the Royal Society of Edinburgh*, **41**, 34–43.

Allen, H. S. (1922). The Magnetic Character of the Quantum. *Proceedings of the Royal Society of Edinburgh*, **42**, 213–20.

Anonymous (1912). M. Poincaré's Lectures at the University of London. *Nature*, **89**(2220), 279.

Anonymous (1920). The Quantum Theory. *Nature*, **106**(2668), 508–09.

Anonymous (1923). Modern Physics and the Atom. *Nature*, **112**(2801), 1–3.

Arrhenius, S. (1922). Award Ceremony Speech. https://www.nobelprize.org/prizes/physics/1922/ceremony-speech/, accessed 24 July 2020.

Barkla, C. G. (1967 [1920]). Characteristic Röntgen Radiation. Nobel Lecture, 3 June 1920. In *Nobel Lectures, Physics 1901–1921*, Amsterdam: Elsevier Publishing Company, pp. 392–99.

---

[99] (Bohr, 1934 [1929]), pp. 15–16.

Bohr, N. (1913a). On the Constitution of Atoms and Molecules. *The London, Edinburgh, and Dublin Philosophical Magazine and Journal of Science*, **26**(151), 1–25.

Bohr, N. (1913b). On the Constitution of Atoms and Molecules. Part III. Systems containing Several Nuclei. *The London, Edinburgh, and Dublin Philosophical Magazine and Journal of Science*, **26**(155), 857–75.

Bohr, N. (1914). On the Effect of Electric and Magnetic Fields on Spectral Lines. *The London, Edinburgh, and Dublin Philosophical Magazine and Journal of Science*, **27**(159), 506–24.

Bohr, N. (1915a). On the Decrease of Velocity of Swiftly Moving Electrified Particles in Passing through Matter. *The London, Edinburgh, and Dublin Philosophical Magazine and Journal of Science*, **30**(178), 581–612.

Bohr, N. (1915b). On the Quantum Theory of Radiation and the Structure of the Atom. *The London, Edinburgh, and Dublin Philosophical Magazine and Journal of Science*, **30**(177), 394–415.

Bohr, N. (1918a). On the quantum theory of line spectra, part I: On the general theory. *Det Kongelige Danske Videnskabernes Selskab, Matematisk-Fysiske Meddelser*, **4**(1), 1–36.

Bohr, N. (1918b). On the quantum theory of line spectra, part II: On the hydrogen spectrum. *Det Kongelige Danske Videnskabernes Selskab, Matematisk-Fysiske Meddelser*, **4**(1), 36–100.

Bohr, N. (1920). Über die Serienspektra der Elemente. *Zeitschrift für Physik*, **2**(5), 423–69.

Bohr, N. (1922). *The Theory of Spectra and Atomic Constitution: Three Essays*. Cambridge/London: Cambridge University Press.

Bohr, N. (1922 [1914]). On the Spectrum of Hydrogen. In *The Theory of Spectra and Atomic Constitution: Three Essays*, Cambridge/London: Cambridge University Press, pp. 1–19.

Bohr, N. (1922 [1920]). On the Series Spectra of the Elements. In *The Theory of Spectra and Atomic Constitution: Three Essays*, Cambridge/London: Cambridge University Press, pp. 20–60.

Bohr, N. (1922 [1921]). The Structure of the Atom and the Physical and Chemical Properties of the Elements. In *The Theory of Spectra and Atomic Constitution: Three Essays*, Cambridge/London: Cambridge University Press, pp. 61–126.

Bohr, N. (1923a). The Structure of the Atom. *Nature*, **112**(2801), 29–44.

Bohr, N. (1923b). Über die Anwendung der Quantentheorie auf den Atombau. *Zeitschrift für Physik*, **13**(1), 117–65.

Bohr, N. (1923 [1922]). Banquet Speech. In C. G. Santesson (ed.), *Les Prix Nobel en 1921–1922*, Stockholm: Nobel Foundation.

Bohr, N. (1924 [1923]). On the application of the quantum theory to atomic structure. *Proceedings of the Cambridge Philosophical Society (Supplement)*, 1–42.

Bohr, N. (1925). Atomic Theory and Mechanics. *Nature*, **116**, 845–52.

Bohr, N. (1928). The Quantum Postulate and the Recent Development of Atomic Theory. *Nature*, **121**, 580–90.

Bohr, N. (1934 [1929]). Introductory Survey. In *Atomic Theory and the Description of Nature*, Cambridge: Cambridge Univ. Press, pp. 1–24.

Bohr, N. (1976a [1919]). On The Program of The Newer Atomic Physics: Translation of Fragments. In L. Rosenfeld and J. Rud Nielsen (eds), *Niels Bohr Collected Works*, vol. 3, Amsterdam: North-Holland, pp. 221–26.

Bohr, N. (1976b [1919]). Problems of The Atom and The Molecule. In L. Rosenfeld and J. Rud Nielsen (eds), *Niels Bohr Collected Works*, vol. 3, Amsterdam: North-Holland, pp. 201–16.

Bohr, N. (1976c [1919]). Speech of Appreciation to Sommerfeld. In L. Rosenfeld and J. Rud Nielsen (eds), *Niels Bohr Collected Works*, vol. 3, Amsterdam: North-Holland, pp. 217–19.

Bohr, N. (1976d [1920]). On The Interaction between Light and Matter: Translation. In L. Rosenfeld and J. Rud Nielsen (eds), *Niels Bohr Collected Works*, vol. 3, Amsterdam: North-Holland, pp. 227–40.

Bohr, N. (1977 [1922]). Seven Lectures on the Theory of Atomic Structure. In J. Rud Nielsen (ed.), *Niels Bohr Collected Works*, vol. 4, Amsterdam: Elsevier, pp. 341–18.

Bohr, N. (1981). Selected correspondence. In *Niels Bohr's Collected Works*, Vol. 2, Amsterdam: North-Holland, pp. 491–611.

Bohr, N., Kramers, H. A., and Slater, J. C. (1924). The Quantum Theory of Radiation. *The London, Edinburgh, and Dublin Philosophical Magazine and Journal of Science*, 47(281), 785–02.

Bokulich, A. (2008). *Reexamining the Quantum-Classical Relation: Beyond Reductionism and Pluralism*. Cambridge/New York: Cambridge University Press.

Born, M. (1914). P. Langevin u. M. de Broglie, La théorie du rayonnement et les quanta. *Physikalische Zeitschrift*, 15, 166–67.

Born, M. (1924). Über Quantenmechanik. *Zeitschrift für Physik*, 26(1), 379–95.

Born, M., and Jordan, P. (1925). Zur Quantenmechanik. *Zeitschrift für Physik*, 34(1), 858–88.

Born, M., Heisenberg, W., and Jordan, P. (1926). Zur Quantenmechanik. II. *Zeitschrift für Physik*, 35(8), 557–15.

Darrigol, O. (1992). *From c-Numbers to q-Numbers: The Classical Analogy in the History of Quantum Theory*. California Studies in the History of Science. Berkeley, CA: University of California Press.

Darrigol, O. (2001). The Historians' Disagreements over the Meaning of Planck's Quantum. *Centaurus*, 43, 219–39.

Debye, P. (1915). Die Konstitution des Wasserstoff-Moleküls. *Sitzungsberichte der mathematisch-physikalischen Klasse der Königlich Bayerischen Akademie zu München*, 1–26.

Dirac, P. A. M. (1925). The fundamental equations of quantum mechanics. *Proceedings of the Royal Society of London Series A*, 109(752), 642–53.

Dirac, P. A. M. (1926). Quantum mechanics and a preliminary investigation of the hydrogen atom. *Proceedings of the Royal Society of London Series A*, 110(755), 561–79.

Eckert, M. (2013). *Arnold Sommerfeld: Atomphysiker und Kulturbote, 1868–1951: eine Biografie*. Abhandlungen und Berichte (Deutsches Museum (Germany)). Göttingen: Wallstein.

Ehrenfest, P. (1916). Adiabatische Invarianten und Quantentheorie. *Annalen der Physik*, 356(19), 327–52.

Ehrenfest, P. (1917). Adiabatic Invariants and the Theory of Quanta. *Philosophical Magazine*, 33, 500–13.

Einstein, A. (1913). Max Planck als Forscher. *Die Naturwissenschaften*, 1, 1077–79.

Einstein, A. (1914). Antrittsrede. *Sitzungsberichte der Preussischen Akademie der Wissenschaften*, 2, 739–42.

Einstein, A. (1916a). Strahlungs- Emission und -Absorption nach der Quantentheorie. *Deutsche Physikalische Gesellschaft. Verhandlungen*, 18, 318–23.

Einstein, A. (1916b). Zur Quantentheorie der Strahlung. *Physikalische Gesellschaft Zürich. Mitteilungen*, 18, 47–62.

Einstein, A. (1917a). Zur Quantentheorie der Strahlung. *Physikalische Zeitschrift*, 18, 121–28.

Einstein, A. (1917b). Marian von Smoluchowski. *Die Naturwissenschaften*, 5, 737–38.

Einstein, A. (1917c). *Über die spezielle und die allgemeine Relativitätstheorie: Gemeinverständ-lich*. Braunschweig: Vieweg.

Einstein, A. (1918). Lassen sich Brechungsexponente der Körper für Röntgenstrahlen experi-mentell ermitteln? *Deutsche Physikalische Gesellschaft. Verhandlungen*, **20**, 86–87.

Einstein, A. (1919). Einstein on his Theory. Time, Space and Gravitation. The Newtonian System. *The London Times*, 28 November, pp. 13–14.

Einstein, A. (1949). Autobiographical Notes. In P. A. Schilpp (ed.), *Albert Einstein: Philosopher-Scientist*, Evanston, IL: The Library of Living Philosophers, pp. 2–94.

Einstein, A. (1974 [1916]). Die Grundlage der allgemeinen Relativitätstheorie. In H. A. Lorentz, *et al.* (eds), *Das Relativitätsprinzip: Eine Sammlung von Abhandlungen*, 7th edn, Leipzig/Berlin: Teubner, pp. 81–124.

Ekstrand, A. G. (1918). Award Ceremony Speech. https://www.nobelprize.org/prizes/physics/1918/ceremony-speech/, accessed 21 July 2020.

Epstein, P. S. (1916). Zur Quantentheorie. *Annalen der Physik*, **51**, 168–88.

Epstein, P. S. (1918). Anwendungen der Quantenlehre in der Theorie der Serienspektren. *Die Naturwissenschaften*, **6**, 230–53.

Eucken, A. (ed.) (1913). *Die Theorie der Strahlung und der Quanten. Verhandlungen auf einer von E. Solvay einberufenen Zusammenkunft (30. Oktober bis 3. November 1911)*. Berlin: Verlag Chemie Gmbh.

Fowler, R. H. (1921). A simple extension of Fourier's integral theorem and some physical applications, in particular to the theory of quanta. *Proceedings of the Royal Society of London Series A*, **99**(701), 462–71.

Fowler, R. H. (1923). The Structure of the Atom. *Nature*, **111**(2790), 523–26.

Gooday, G., and Mitchell, D. (2013). Rethinking 'Classical Physics'. In J. Z. Buchwald and R. Fox (eds), *Oxford Handbook of the History of Physics*, Oxford: Oxford University Press, pp. 721–64.

Heisenberg, W. (1925). Über quantentheoretische Umdeutung kinematischer und mechan-ischer Beziehungen. *Zeitschrift für Physik*, **33**(1), 879–93.

Heisenberg, W. (1927). Über den anschaulichen Inhalt der quantentheoretischen Kinematik und Mechanik. *Zeitschrift für Physik*, **43**(3), 172–98.

Howard, D. (1994). What Makes a Classical Concept Classical? Toward a Reconstruction of Niels Bohr's Philosophy of Physics. In J. Faye and H. J. Folse (eds), *Niels Bohr and Contemporary Philosophy*, Dordrecht: Kluwer Academic, pp. 201–230.

Jeans, J. H. (1913). Discussion on Radiation. *Reports of the British Association for the Advancement of Science*, 376–86.

Jeans, J. H. (1914). *Report on Radiation and Quantum-Theory*. London: The Electrician.

Jordi Taltavull, M. (2016). Transmitting Knowledge across Divides: Optical Dispersion from Classical to Quantum Physics. *Historical Studies in the Natural Sciences*, **46**(3), 313–59.

Jordi Taltavull, M. (2018). The Uncertain Limits Between Classical and Quantum Physics: Optical Dispersion and Bohr's Atomic Model. *Annalen der Physik*, **530**(8), 1,800,104.

Kragh, H. (2012). *Niels Bohr and the Quantum Atom: The Bohr Model of Atomic Structure 1913–1925*. Oxford: Oxford University Press.

Laue, M. v. (1918). Thermodynamik und Kohärenz. *Die Naturwissenschaften*, **6**, 207–13.

Laue, M. v. (1967 [1920]). Concerning the Detection of X-ray Interferences. Nobel Lecture, June 3, 1920.In *Nobel Lectures, Physics 1901–1921*, Amsterdam: Elsevier, pp. 347–55.

McCormmach, R. (1967). Henri Poincaré and the Quantum Theory. *Isis*, **58**(1), 37–55.

McLaren, S. B. (1911a). Hamilton's equations and the partition of energy between matter and radiation. *Philosophical Magazine Series 6*, 21(121), 15–26.

McLaren, S. B. (1911b). The emission and absorption of energy by electrons. *Philosophical Magazine Series 6*, 22(127), 66–83.

McLaren, S. B. (1912). The emission and absorption of radiation in any material system and complete radiation. *Philosophical Magazine Series 6*, 23(136), 513–42.

McLaren, S. B. (1913). The theory of radiation. *Philosophical Magazine Series 6*, 25(145), 43–56.

McLaren, S. B., and Hassé, H. R. (1925). *Scientific Papers, Mainly on Electrodynamics and Natural Radiation: Including the Substance of an Adams Prize Essay in the University of Cambridge.* Cambridge: Cambridge University Press.

Needell, A. A. (1988). Introduction. In *Max Planck, The Theory of Heat Radiation* (The History of Modern Physics, 1800–1950, 11; Los Angeles/New York: Tomash and American Institute of Physics), pp. xi–xlv.

Nernst, W. (1918). Quantentheorie und neuer Wärmesatz. *Die Naturwissenschaften*, 6, 206–07.

Nicholson, J. W. (1912). The Constitution of the Solar Corona. II. *Monthly Notices of the Royal Astronomical Society*, 72(8), 677–93.

Planck, M. (1912). Über die Begründung des Gesetzes der schwarzen Strahlung. *Annalen der Physik*, 37, 642–56.

Planck, M. (1913a). *Vorlesungen über die Theorie der Wärmestrahlung* (2., teilweise umgearbeitete edn). Leipzig: J.A. Barth.

Planck, M. (1913b). Die Gesetze der Wärmestrahlung und die Hypothese der elementaren Wirkungsquanten. In A. Eucken (ed.), *Die Theorie der Strahlung und der Quanten. Verhandlungen auf einer von E. Solvay einberufenen Zusammenkunft (30. Oktober bis 3. November 1911)*, Berlin: Verlag Chemie Gmbh, pp. 77–94.

Planck, M. (1967 [1920]). The Genesis and Present State of Development of the Quantum Theory. Nobel Lecture, 2 June 1920. In *Nobel Lectures, Physics 1901–1921*, Amsterdam: Elsevier, pp. 407–20.

Planck, M., et al. (1993 [1913]). Proposal for Einstein's Membership in the Prussian Academy of Sciences. In M. J. Klein, A. J. Kox, and R. Schulmann (eds), *The Collected Papers of Albert Einstein, Vol. 5: The Swiss Years: Correspondence, 1902–1914*, Princeton, NJ: Princeton University Press, pp. 526–29.

Poincaré, H. (1902). *La Science et l'hypothèse*. Paris: Flammarion.

Poincaré, H. (1905). *Science and Hypothesis*, trans. William John Greenstreet. London: Walter Scott.

Poincaré, H. (1912). Sur la théorie des quanta. *Journal de Physique théorique et appliquée*, 2, 5–34.

Reiche, F. (1913). Die Quantentheorie. *Die Naturwissenschaften*, 1, 549–53; 568–72.

Reiche, F. (1918). Die Quantentheorie. Ihr Ursprung und ihre Entwiklung. *Die Naturwissenschaften*, 6, 213–30.

Rutherford, E. (1911). Conference on the Theory of Radiation. *Nature*, 88(2194), 82–83.

Rutherford, E. (1914a). The Structure of the Atom. *The London, Edinburgh, and Dublin Philosophical Magazine and Journal of Science*, 27(159), 488–98.

Rutherford, E. (1914b). Discussion on the Structure of the Atom. *Proceedings of the Royal Society of London Series A*, 90(615), 1–19.

Schlosshauer, M., and Camilleri, K. (2017). Bohr and the Problem of the Quantum-to-Classical Transition. In J. Faye and H. Folse (eds), *Niels Bohr and the Philosophy of Physics: Twenty-First Century Perspectives*, London: Bloomsbury, pp. 223–33.

Schrödinger, E. (1926). Über das Verhältnis der Heisenberg–Born–Jordanschen Quantenmechanik zu der meinen. *Annalen der Physik*, **79**, 734–56.

Seth, S. (2010). *Crafting the Quantum: Arnold Sommerfeld and the Practice of Theory, 1890–1926*. Cambridge, MA: MIT Press.

Sommerfeld, A. (1916). Zur Quantentheorie der Spektrallinien. *Annalen der Physik*, **51**, 1–94, 125–67.

Sommerfeld, A. (1918a). Die Drudesche Dispersionstheorie vom Standpunkte des Bohrschen Modelles und die Konstitution von $H_2$, $O_2$ und $N_2$. *Annalen der Physik*, **358**(15), 497–50.

Sommerfeld, A. (1918b). Max Planck zum sechzigsten Geburtstage. *Die Naturwissenschaften*, **6**, 195–99.

Sommerfeld, A. (1919). *Atombau und Spektrallinien*. Braunschweig: Friedrich Vieweg und Sohn.

Sommerfeld, A. (2000). *Wissenschaftlicher Briefwechsel. Band 1: 1892–1918* (edited by M. Eckert and K. Märker). Berlin/Diepholz/München: Deutsches Museum and Verlag für Geschichte der Naturwissenschaften und der Technik.

Staley, R. (2005). On the Co-Creation of Classical and Modern Physics. *Isis*, **96**, 530–58.

Staley, R. (2008a). *Einstein's Generation: The Origins of the Relativity Revolution*. Chicago: University of Chicago Press.

Staley, R. (2008b). Worldviews and Physicists' Experience of Disciplinary Change: On the Uses of 'Classical' Physics. *Studies in the History and Philosophy of Science*, **39**, 298–311.

Stark, J. (1967 [1920]). Structural and Spectral Changes of Chemical Atoms. Nobel Lecture, 3 June 1920. In *Nobel Lectures, Physics 1901–1921*, Amsterdam: Elsevier.

Warburg, E. (1918). Über Plancks Verdienste um die Experimentalphysik. *Die Naturwissenschaften*, **6**, 202–03.

Warwick, A. (2003). *Masters of Theory: Cambridge and the Rise of Mathematical Physics*. Chicago/London: University of Chicago Press.

Whittaker, E. T. (1921). On Tubes of Electromagnetic Force. *Proceedings of the Royal Society of Edinburgh*, **42**, 1–23.

Whittaker, E. T. (1922). On the Quantum Mechanism in the Atom. *Proceedings of the Royal Society of Edinburgh*, **42**, 129–42.

Wien, W. (1918). Die Entwicklung von Max Plancks Strahlungstheorie. *Die Naturwissenschaften*, **6**, 203–06.

# CHAPTER 38

# INTERPRETATION IN ELECTRODYNAMICS, ATOMIC THEORY, AND QUANTUM MECHANICS

GIORA HON AND BERNARD R. GOLDSTEIN

## 38.1 THE CLAIM

THE interpretation of quantum physics has posed a longstanding problem in the philosophy of the foundations of physics. There has never been an agreement on a single interpretation; indeed, a multitude of interpretations have proliferated. Yet, the debate presupposes that there is one correct, or true, interpretation, and this interpretation is a description of the underlying physical reality. We are persuaded that the problem of interpretation is not unique to quantum mechanics. Indeed, the problem is apparent in classical electrodynamics and was, in fact, acknowledged. In the original quantum theory, an agreed interpretation emerged, similar to that in classical mechanics, but it then became amply clear that this interpretation led to insurmountable difficulties. The breakdown of the quantum theory was addressed, as is well known, by the creators of matrix mechanics—notably Werner Heisenberg (1901–1976)—who found it necessary to replace the quantum theory with a fundamentally different one. The formal solution of the quandaries in the quantum theory provided by the discrete theory of matrix mechanics exposed the latent foundational problem of interpretation, for there arose another theory, wave mechanics, which commenced with a continuous physical picture. Indeed, the algebraic (matrix) character without any physical picture of the former theory contrasted with the latter which enhanced 'wave mechanics' for depicting physical processes and nurturing intuition. Matrix mechanics and wave mechanics are different from previous physical theories in that the problem of interpretation is evident and recalcitrant. In brief, we claim that this problem is inherent to physics, indeed, to any science which is expressed in symbolic language, that is, formulated mathematically. Put differently, the problem of interpretation may be recast in terms of the loss of equivalence between physical and mathematical descriptions. In this chapter we develop an argument to justify this claim. We contrast two cases in the history of modern physics and then analyse in some detail the position taken by Erwin Schrödinger (1887–1961). The two cases, namely, electrodynamics and the physics of the atom, serve to highlight the relation between symbolic language and verbal expressions where the latter, ultimately, reflect mental images of the physics at stake, and the former its mathematical structure.

## 38.2  THE ARGUMENT

### 38.2.1  Background

In 1931 Albert Einstein (1879–1955) succinctly captured the changing scene in physics. Writing on the occasion of the centenary of the birth of James Clerk Maxwell (1831–1879), he made two critical points. He considered Michael Faraday (1791–1867) and Maxwell jointly and then posed the problem of mechanical interpretation, claiming that no such interpretation is possible.[1] According to Einstein, a fundamental change took place in the conception of physical reality as a consequence of the study of electromagnetism.[2]

Indeed, as Einstein observed, Maxwell was committed to Faraday's conception of science.[3] In late 1857, upon receipt of a preprint of Maxwell's paper (published in 1858), Faraday sharply criticized Maxwell for not presenting his mathematical results in ordinary language. 'Would it not be a great boon to such as I to express them so?—translating them out of their hieroglyphics, that we also might work upon them by experiment. I think it must be so', Faraday exclaimed. Faraday needed the language, as he put it, to be 'popular, useful', and in a 'working state'.[4] Maxwell took this request seriously and extended the usage of the term 'translation'. For example, in 1873 he invoked this demand in a technical context, where the transformation of a physical problem in one domain to another was associated with a translation from one natural language to another.[5] This is a strong plea for recasting the symbolic into the verbal; Maxwell—most likely, following Faraday—called it 'translation'. Maxwell's demand for

---

[1] Einstein (1931), pp. 66–71.

[2] Note that Einstein's view of physical reality differed from that of Maxwell.

[3] For a recent study of Faraday, see Steinle ([2005] 2016). On Maxwell's commitment to Faraday, see Hon and Goldstein (2020).

[4] For Faraday's letter to Maxwell, dated 13 November 1857, see James (2008), pp. 304–06 as well as Campbell and Garnett (1882), pp. 145, 290.

[5] See Maxwell ([1871/1890] 1965), p. 258, where the example of translation from one natural language to another is from French to Italian.

mental imagery had to do with the expectation that the symbolic formalism could be translated back into verbal language, mental imagery being the vehicle for carrying out the translation.

In the wake of the quantum mechanical revolution, Einstein offered a perspective that Faraday and Maxwell did not foresee: a continuous field is not susceptible to any mechanical interpretation. In 1873 Maxwell was satisfied with mental images of a mechanical nature, although—as he acknowledged—the real mechanism at the micro-level had not been determined.[6] However, in 1931, more than half a century later, Einstein was in a position to declare that the search for mechanical interpretations is simply misguided. Thus, for Einstein the profound change in the conception of physical reality was not the whole story; what was novel is that the new conception was not open to any mechanical interpretation. In fact, a few years later, in 1935—as is well known—he and his co-authors of the EPR paper, Boris Podolsky (1896–1966) and Nathan Rosen (1909–1995), went even further. Their point of departure was the philosophical demand that:

> Any serious consideration of a physical theory must take into account the distinction between the objective reality, which is independent of any theory, and the physical concepts with which the theory operates. These concepts are intended to correspond with the objective reality, and by means of these concepts we picture this reality ourselves.[7]

This fundamental demand of correspondence is, in effect, the condition of completeness as defined in the EPR paper: 'every element of the physical reality must have a counterpart in the physical theory'.[8] This much is in the spirit of Faraday and Maxwell and, indeed, in the spirit of physics toward the end of the 19th century. Thus, for example, Heinrich Hertz (1857–1894) opened the Preface to his *Principles of Mechanics Presented in a New Form* (1894) in this vein: 'All physicists agree that the problem of physics consists in tracing the phenomena of nature back to the simple laws of mechanics'.[9] Indeed, in classical mechanics one is able to picture processes and to think of oneself as part of the connecting machinery. The spatial constitution of the living human body, its disposition and motion due to muscular force, all contribute to an intuitive grasp of mechanics that requires no mediation. We have here a two-way street—from physical reality to physical theory, and back, exactly as Maxwell demanded.

---

[6] Maxwell (1873), vol. 2, pp. 416–17, § 831. For a discussion, see Hon and Goldstein (2020), pp. 182–83.

[7] Einstein, Podolsky, and Rosen (1935), p. 777. Note the appeal to 'objective reality'. After these two instances of this expression in the opening paragraph of the paper, there are no more such occurrences; rather, the authors change terminology, invoking instead the expression 'physical reality'.

[8] Einstein, Podolsky, and Rosen (1935), p. 777.

[9] Hertz ([1894] 1899), Preface, p. xxi. For the expectation that light will eventually be understood mechanically see, e.g., Poynting ([1893] 1920), p. 264.

This is all reasonably clear in classical mechanics, but how does it apply in electrodynamics, let alone in quantum mechanics? And recall that, for Einstein, Maxwell's theory 'changed' physical reality. That is, in contrast to classical mechanics, completeness cannot be taken for granted in electrodynamics. However, the issue is far more acute in quantum mechanics, and the authors of the EPR paper concluded that:

> the quantum-mechanical description of physical reality given by wave functions is not complete.[10]

One of the immediate consequences of the EPR argument is that, in general, no correspondence can be obtained between the concepts of the quantum mechanical theory and elements of physical reality, and thus no mental image can be constructed that relates the concepts of quantum mechanics to elements of physical reality consistently and coherently. This lack of correspondence is commonly characterized as the problem of interpretation.

## 38.2.2  Framework: Two cases

This is the framework of our argument: we focus on two cases where the issue of interpretation has to be confronted head on. In the first case we address the different conceptions of the theory of electrodynamics by Maxwell and one of his successors, Hertz. This case concerns the separation of formalism from physical content in the context of electromagnetism. We argue that the issue of correspondence between formalism and physical content (or the lack thereof) had already arisen in electrodynamics before the revolution of quantum mechanics that took place in 1925. In the second case we recount the approach which Niels Bohr (1885–1962) and Arnold Sommerfeld (1868–1951) took vis-à-vis the physics of the atom as they developed the quantum theory. This theory preceded the quantum revolution and in effect led to it. Against the background of these two cases, we discuss Schrödinger's position, bringing to the fore the problem of interpretation.

### 38.2.2.1  *Case 1: Electromagnetism*

The move from mechanics to electromagnetism, from a domain of physics where one could think of oneself, indeed, feel oneself, as part of the connecting machinery, to a novel domain where intuition fails, posed a challenge. Maxwell's study of electromagnetism should be seen as emerging 'organically' from Faraday's work, both experimental and theoretical. In 1873 Maxwell concluded his magnum opus, *A Treatise on Electricity and Magnetism*, emphasizing that it was critically important for the development of electromagnetism to construct a mental representation of the details of the

---

[10] Einstein, Podolsky, and Rosen (1935), p. 780.

action of the medium, whose existence had been assumed, despite the difficulties in conceiving it.[11] Maxwell believed that the advancement of science depends on developing exact ideas which both facilitate mental representation of the physics at stake and warrant deductions by mathematical reasoning.[12] He adopted the formalism of Joseph-Louis Lagrange (1736–1813) as well as that of William Rowan Hamilton (1805–1865) to apply the abstract dynamics of William Thomson (1824–1907) and Peter G. Tait (1831–1901) to electromagnetism.[13] According to Maxwell, Lagrange and his followers considered only 'pure quantity' represented by symbols devoid of concepts such as velocity, momentum, and energy. While appealing to this mathematical technique, Maxwell was keen 'to retranslate the principal equations of the method into language which may be intelligible without the use of symbols'.[14]

Maxwell explicitly expressed the need for bidirectionality—from physical content to physical theory, and back. He specified his aim, namely, to cultivate dynamical ideas.

> We therefore avail ourselves of the labours of the mathematicians [Lagrange and Hamilton], and retranslate their results from the language of the calculus into the language of dynamics, so that our words may call up the mental image, not of some algebraical process, but of some property of moving bodies.[15]

Evidently, for Maxwell the theory was not just the formalism; the theory had to offer mental imagery of the physics it dealt with. Thus, in constructing his theory Maxwell sought to combine purely symbolic language with physical content. The expression 'Maxwell's theory' or, to be precise, 'Maxwell's electrodynamics', does not merely refer to the equations; rather, it refers to the equations together with a host of commitments, assumptions, derivations, and mental images. For Maxwell symbolic relations by themselves are devoid of physical content; a proper theory, indeed, a viable theory, needs to have both: it must convey physical meaning with its formalism. To use the EPR formulation, there must be a correspondence between the formal terms of the theory and elements of physical reality. In this sense, Maxwell's theory was intended to be as complete as mechanics.

It is then not surprising that Maxwell sought a mathematical formulation of electromagnetic phenomena which could be captured by mental imagery of a mechanical nature. He found that 'the medium must be in a state of mechanical stress'.[16] Maxwell's discussion of his attempts to specify this 'mechanical stress' is most illuminating. He claimed that a mechanism may be conceived which produces the effects of the electromagnetic field. However, he openly acknowledged that such a mechanically equivalent model cannot be unique and that, in principle, there are infinitely many

---

[11] Maxwell (1873), vol. 2, p. 438, § 866.    [12] Maxwell ([1873/1890] 1965), p. 360.
[13] See Hon and Goldstein (2020), chapter 7.    [14] Maxwell (1873), vol. 2, p. 194, § 567.
[15] Maxwell (1873), vol. 2, pp. 184–85, § 554. Maxwell invoked 'mental image' and 'mental representation' interchangeably; we adhere to 'mental image' since Maxwell frequently used the expression 'mechanical representation' which has a different methodological purpose.
[16] Maxwell (1873), vol. 1, p. 59, § 59.

such models of different degrees of efficiency in reproducing the electromagnetic phenomena.[17] Thus, the precise mechanism at the micro-level cannot be determined; the theory complete with mental imagery is unproven, but the set of equations holds. Here we see a clear separation between the valid formal structure and the unproven physical interpretation.

The separation between formalism and physical content was well established in physics ever since the strong analogy between heat conduction and electricity was discovered by Thomson (later known as Lord Kelvin) in 1842.[18] Given a formal correspondence between two distinct physical domains, the problems are interchangeable, and so are the solutions.

We note that in this view what is critical for the advancement of physics is the mathematical structure which the symbolic formulation of the laws exhibits and not the physical theory. Put differently, the formalism stands, while the physical content may vary. To be sure, this was not how Maxwell treated electromagnetism but, as we show below, this was the approach taken by the subsequent generations of physicists. In sum, Maxwell introduced a new approach in electromagnetism in which the formalism—cast in the form of mathematical equations—could be separated from the theory, even though this was not how Maxwell understood his theory. Indeed, Hertz capitalized on this separation and began to work with formalism without being constrained by physical meaning. Thus, he treated electricity and magnetism interchangeably and reformulated the equations for the two sets of phenomena symmetrically.

In 1884 Hertz embarked on a detailed and critical analysis of Maxwell's set of equations. This major undertaking preceded his famous *tour de force*: proving experimentally that electric waves exist. As Hertz noted, he attempted 'to demonstrate the truth [*die Gültigkeit*] of Maxwell's equations by starting from premises which are generally admitted in the opposing system of electromagnetics, and by using propositions which are familiar in it'.[19] Hertz sought to show the validity of Maxwell's set of equations even if one starts with the premises of opposing theories, perhaps alluding to the viewpoint of his mentor, Hermann von Helmholtz (1821–1894).[20] At stake then was the issue of derivation and the goal of constructing a 'bridge' between the opposing views.

Hertz was particularly sensitive to laboratory demands: his interests were theoretical but also and, perhaps even more so, experimental, much in line with Faraday's practice.

---

[17] Maxwell (1873), vol. 2, pp. 416–17, § 831.

[18] Thomson (1842). On 'strong' and 'weak' analogy, see Hon and Goldstein (2020), chapter 1, section 1.5.

[19] Hertz (1896), p. 289; Hertz ([1884] 1895), p. 313: cf. Buchwald (1994), p. 198. By 'opposing system' Hertz meant any electrodynamics which could accommodate the potential law of Franz E. Neumann (1798–1895): see Hertz (1896), p. 276, n. 1, and Hertz ([1884] 1895), p. 298, n. 2. On Neumann's potential law, see Archibald (1989).

[20] Hertz was a student in Berlin where his teachers included Helmholtz and Gustav Kirchhoff (1824–1887), both of whom served as examiners of his dissertation in 1880. See Buchwald (1994), pp. 59, 97.

He stated that his motivation was to 'make the magnetic and electrostatic systems change places [*Platz vertauschen*]'.[21] He sought to establish, first theoretically and then experimentally, that the laws governing electrostatic force and electromagnetic induction are interchangeable. Thus, 'in the laws of electric induction we need only interchange the words "electric" and "magnetic" throughout in order to obtain the inductive actions in magnetic circuits'.[22] Notice that Hertz found laws in which the forces are interchangeable, although they keep their identity as magnetic and electric.

Against this background, Hertz remarked that the two forces, namely, electric and magnetic, had usually been deduced in an asymmetrical manner and, given the goal of his paper, it is not surprising that Hertz was dissatisfied with the way Maxwell's equations had been derived: the distinct phenomena of electricity and magnetism are analogous and the formalism ought to exhibit a correspondence between them. Hence the equations should be symmetrical in the sense that analogous elements have to correspond and be interchangeable. He proceeded to eliminate the asymmetry between the forces and rendered them interchangeable by purely formal, mathematical means. The 'trick' was to differentiate Maxwell's original equations with respect to time.[23] As Hertz stated, the result of this mathematical move was that 'the electric and magnetic forces are now interchangeable [*vertauschbar*]'.[24] Hertz aimed at the following symmetrical result: 'magnetic currents act on each other according to the same laws as electric currents'.[25] Symmetry for Hertz embodied the notion of interchangeability, and Maxwell's equations were recast to exhibit this feature.

In 1889, in an essay on light and electricity, Hertz revealed the extent to which Maxwell's set of equations had impressed him:

> It is impossible to study this wonderful theory without feeling as if the mathematical equations had an independent life and an intelligence of their own, as if they were wiser than ourselves, indeed wiser than their discoverer, as if they gave forth more than he had put into them . . .[26]

Notice how Hertz shifted his admiration from the theory to the equations. This is the background to Hertz's celebrated view of Maxwell's theory. In 1892 Hertz reported in the theoretical part of the Introduction to his *Electric Waves* that he wished to simplify Maxwell's theory as much as he could by stripping it of all physical conceptions which could be dispensed with, without affecting the account of the phenomena. And he famously concluded:

---

[21] Hertz (1896), p. 292; Hertz ([1885] 1895), p. 316.
[22] Hertz (1896), p. 277; Hertz ([1884] 1895), p. 300.
[23] Hertz (1896), pp. 286–87; Hertz ([1884] 1895), pp. 310–11.
[24] Hertz (1896), p. 287 (slightly modified); Hertz ([1884] 1895), p. 311.
[25] Hertz (1896), p. 276; Hertz ([1884] 1895), p. 299.
[26] Hertz (1896), p. 318; Hertz ([1889] 1895), p. 344.

To the question, 'What is Maxwell's theory?' I know of no shorter or more definite answer than the following:—Maxwell's theory is Maxwell's system of equations.[27]

Hertz stated his belief that the formalism of the equations is independent of the theory. He then proceeded to spell out the different interpretations given to the formalism, indicating that he did not share the view that the difficulty was of a mathematical nature.[28] And he cautioned the reader that:

> scientific accuracy [*die Strenge der Wissenschaft*] requires of us that we should in no wise confuse the simple and homely figure, as it is presented to us by nature, with the gay garment [*bunte Gewand*] which we use to clothe it. Of our own free will we can make no change whatever in the form of the one, but the cut and colour of the other we can choose as we please.[29]

Hertz wished to strip from the theory everything (which he called 'garments') except the equations. The theory, with its interpretative baggage, could, and should, be separated from its purely formal-mathematical representation.[30]

To conclude Case 1: Maxwell was committed to some version of realism in electro-dynamics, but it was indeterminate; indeed, he acknowledged that there were many, in fact infinite, mechanical possibilities that can represent the correspondence between formalism and physical content in this domain of physics. Hertz, we argue, abandoned 'physical reality' in the sense that the set of equations, namely, the formalism, stands alone without the theory. By contrast, Maxwell insisted that the theory must include a mental image of the action of the medium and, thereby, physical interpretation. It is noteworthy that Hertz did not challenge Maxwell's theory; rather, he modified the formalism without raising any objections to the theory as Maxwell had presented it. He sought to reformulate it to satisfy the principle of symmetry. And we recall that in 1905 Einstein launched his special theory of relativity on the basis of the Maxwell–Hertz equations (subsequently, in 1907, he changed the reference to the Maxwell–Lorentz equations).[31]

---

[27] Hertz ([1893] 1962), pp. 20–21; Hertz (1892), pp. 22–23.

[28] Hertz ([1893] 1962), p. 22; Hertz (1892), p. 23: 'Die oft gehörte Ansicht, dass diese Schwierigkeit mathematischer Natur sei, kann ich nicht theilen.'

[29] Hertz ([1893] 1962), p. 28; Hertz (1892), p. 31. See also Hertz ([1893] 1962), p. 195, and Hertz (1892), p. 208.

[30] Hon and Goldstein (2005), pp. 492–03 and Hon and Goldstein (2006), pp. 637–38, 643–44. Oliver Heaviside (1850–1925) was also not constrained by physical considerations when he modified the formal presentation of Maxwell's theory. See, e.g., Heaviside (1893), p. vii: '[Maxwell's theory] may be, and has been, differently interpreted by different men, which is a sign that it is not set forth in a perfectly clear and unmistakeable form. There are many obscurities and some inconsistencies. Speaking for myself, it was only by changing its form of presentation that I was able to see it clearly, and so as to avoid the inconsistencies. Now there is no finality in a growing science. It is, therefore, impossible to adhere strictly to Maxwell's theory as he gave it to the world, if only on account of its inconvenient form'. Cf. Hon and Goldstein (2005), pp. 503–10. See also Hunt (1991), Darrigol ([2000] 2002), and Buchwald (1985).

[31] Einstein (1905): for Maxwell's theory, see p. 891; for the equations, see p. 907. Einstein (1907): for the equations, see p. 427.

## 38.2.2.2  Case 2: The physics of the atom

Given the belief in the universality of Newtonian mechanics, it is hardly surprising that the structure of the atom was compared to that of a planetary system. Confidence in the planetary model was greatly enhanced by its success in explaining and predicting phenomena of simple atomic structures. This success made it seem likely that a fully satisfactory theory of this kind would be found, although later on it became clear that this initial optimism was ill-founded.

Bohr's quantum theory of 1913 was slow in making an impact on the world of physics; its reception, however, accelerated in 1915 with the work of Sommerfeld who later, in his influential book, *Atombau und Spektrallinien* (first edition, 1919)—the 'bible' of atomic physics in the 1920s and 1930s[32]—argued for an atomic analogue to Kepler's three laws within the framework of Bohr's theory.[33] In this view, the planetary analogy became quantitative with experimental consequences, and took advantage of recent developments in relativistic physics. Bohr accepted Sommerfeld's planetary atom-model and lent his authority to support it, most prominently in his Nobel Lecture of 1922. That year probably marks the high point of the planetary model.

At the outset of Sommerfeld's contribution to atomic physics in 1915, he referred to 'Keplerschen Bewegung', 'Keplerschen Ellipsen', and 'Keplerschen Bahnsystems'.[34] Bohr's appeal in 1913 to orbits—definite trajectories for the circulating electrons—was taken up by Sommerfeld who then further elaborated the concept of orbit. The references to the Keplerian scheme meant that Sommerfeld could presuppose a planetary model in which the electron rotates around the nucleus in a well-defined elliptical orbit—a striking example of the search for physical content accessible to mental imagery.

Sommerfeld complemented Bohr's theory with two additional quantum numbers corresponding to the Stark and Zeeman effects, and introduced the concept of the fine structure of the spectral line configuration. In Sommerfeld's view, with his extension of Bohr's theory, the phenomena of spectral lines acquired '*a deepened theoretical significance and its origin has now multiple roots*'.[35] Sommerfeld analysed the motion of the electron within the atom and included an appeal to Einstein's special and general relativity: in its elliptical path around the nucleus the electron's velocity increases and decreases and so does its mass as predicted by special relativity; the perihelion of the electron's elliptical orbit precesses in analogy with the precession of the perihelion of the planet Mercury as predicted by general relativity. These are among the earliest applications of special and general relativity.

[32] Kragh and Nielsen (2013), p. 264.

[33] Sommerfeld ([1919] 1922) and ([1919/1922] 1923). For the reception of Bohr's theory, see Kramers and Holst ([1922] 1923), p. 145; Andrade (1957), p. 446, and Kragh and Nielsen (2013), pp. 260–62. For the contribution of Sommerfeld, see Forman and Hermann (1975), p. 529.

[34] Sommerfeld (1915).

[35] Sommerfeld ([1919/1922] 1923), p. 237, italics in the original.

A dominant feature of Sommerfeld's analysis is the appeal to planetary imagery in the analysis of atomic radiation phenomena:

> When, in our planetary system, an electron jumps into an orbit nearer the nucleus, the potential energy of the planetary system certainly becomes diminished... Hence energy is liberated. We assume that this appears in the form of energy of radiation, and that it is emitted as monochromatic radiation, that is, as radiation of *one* wavelength, in each case.[36]

Evidently, Sommerfeld had great confidence in his generalization of Bohr's theory. For our argument it is important to note that this generalization is based on Sommerfeld's deep commitment to the planetary model.

Sommerfeld transferred the problem of stability from the atom to the solar system; in other words, the stability of the atom is likened to the stability of the planetary system:

> The earth fails to fall into the sun for the reason that it develops centrifugal forces owing to its motion in its own orbit, and these forces are in equilibrium with the sun's attraction. If we transpose these ideas to our atomic model we arrive at the following view. **The atom is a planetary system in which the planets are electrons. They circulate about the central body, the nucleus.** The atom of which the atomic number is Z is composed of Z planets each charged with a single negative charge, and of a sun charged with Z positive units. The *gravitational attraction*, as expressed in Newton's law, is represented by the *electrical attraction* as given by Coulomb's law; these laws are alike in form.[37]

The imagery is quite extraordinary. Sommerfeld, in so many words, identified the atom as a planetary system: the stability of the atom is due to the same conditions that maintain the stability of the solar system, so his argument goes. Moreover, for Sommerfeld it was not merely an analogy, for he stated an identity: 'the atom is a planetary system'. This is, of course, an exaggeration, expressing enthusiasm (surely a sign of his confidence in the model) rather than objective analysis. And Sommerfeld retreated immediately to the analogy: the gravitational force is 'represented' by the Coulomb force, the two laws being alike in form. Sommerfeld was aware of the differences too. Although the laws are formally similar,

> There is a difference in that the planets repel one another in our atomic microcosm—likewise according to Coulomb's law—whereas, in the case of the solar macrocosm they undergo attraction not only from the sun but also from themselves.[38]

[36] Sommerfeld ([1919/1922] 1923), p. 144, italics in the original; for the German, see Sommerfeld ([1919] 1922), p. 178.

[37] Sommerfeld ([1919/1922] 1923), p. 65, boldface and italics in the original; for the German, see Sommerfeld ([1919] 1922), p. 79.

[38] Sommerfeld ([1919/1922] 1923), p. 65.

Still, the claim that the dynamical laws hold in the 'atomic microcosm' exactly as in the 'solar macrocosm' runs like a thread in Sommerfeld's study.

No doubt under the influence of Sommerfeld, Bohr explicitly described the atom as a planetary system.[39] All the same, the strong image of the planetary system, indeed the iconic imagery of planets orbiting the Sun,[40] dominated Bohr's analysis even of atoms of greater complexity than that of hydrogen. In effect, Bohr followed closely the planetary model that Sommerfeld had introduced.[41] Nevertheless, he was aware of the difficulties it entailed. In his Nobel lecture of 1922, Bohr noted that the analogy 'provide[s] us with an explanation', but it has its limitations:

> As soon as we try to trace a more intimate connexion between the properties of the elements and atomic structure, we encounter profound difficulties, in that essential differences between an atom and a planetary system show themselves here in spite of the analogy we have mentioned.[42]

And Bohr surmised:

> On the basis of our picture of the constitution of the atom it is thus impossible, so long as we restrict ourselves to the ordinary mechanical laws, to account for the characteristic atomic stability which is required for an explanation of the properties of elements.[43]

Notice that Bohr spoke of a picture of the constitution of the atom, a picture which, given his analysis, cannot be maintained.

Bohr acknowledged that 'it is not possible to follow the emission in detail by means of the usual conceptions';[44] indeed, the traditional, mechanical-electrodynamical approach does not even 'offer us the means of calculating the frequency of the emitted radiations'. But Bohr was not discouraged: 'Notwithstanding the fundamental departure from the ordinary mechanical and electrodynamical conceptions, . . . it is possible to give a rational interpretation of the evidence provided by the spectra'.[45] The motivating question in probing the constitution of the atom was one of construction: 'How may an atom be formed by the successive capture and binding of the electrons one by one in the field of force surrounding the nucleus?'[46] It is no surprise that 'the difficulty of the mechanical problem . . . increases with the number of the particles in the atom'.[47]

Bohr and Sommerfeld imagined a specific physical reality in a different domain from the one to which Faraday and Maxwell contributed. Unlike the case of electromagnetism, this appeal to physical content in the description of the atom led—as is well known—to a crisis. In 1925 Heisenberg responded to the breakdown of the quantum

---

[39] Bohr ([1922] 1965), p. 8.    [40] See, e.g., Goldstein and Hon (2013–2014).
[41] Bohr (1922), p. 23.    [42] Bohr ([1922] 1965), pp. 10–11.
[43] Bohr ([1922] 1965), p. 11.    [44] Bohr (1922), p. 23.    [45] Bohr (1922), p. 23.
[46] Bohr (1922), p. 75; see also Bohr ([1922] 1965), pp. 31–32.
[47] Bohr (1922), p. 57.

theory by laying the foundations for quantum mechanics: he produced equations that are independent of any physical interpretation. Heisenberg began his ground-breaking paper with a succinct description of the crisis and a call for a new approach:

> It is well known that the formal rules which are used in quantum theory for calculating observable quantities such as the energy of the hydrogen atom may be seriously criticized on the grounds that they contain, as basic elements, relationships between quantities that are apparently unobservable in principle, e.g., position and period of revolution of the electron. Thus these rules lack an evident physical foundation..., [and] the extension of the quantum rules to the treatment of atoms having several electrons has proved unfeasible.[48]

An example of a quantity to be excluded from the new theory was the time-dependent position coordinate. For an observable quantity to replace it, Heisenberg turned to probabilities for transitions between stationary states.[49]

To conclude Case 2: Bohr and Sommerfeld developed an intricate planetary model for the atom complete with Keplerian laws which retained a correspondence between formalism and physical content. In effect, we would say that Bohr and Sommerfeld tried to recover what had been abandoned by Hertz and others in the generation following Maxwell. Ultimately, these efforts of Bohr and Sommerfeld came to naught. Heisenberg's programmatic call to base the physics of the atom *only* on observable quantities resulted in a set of new mathematical equations in quantum mechanics which lacked meaningful physical content. The problem of interpretation now became evident, and more critical than ever.

## 38.3 SCHRÖDINGER'S WAVE MECHANICS: CONTINUOUS SCALE TRANSITION AND NURTURING INTUITION

Like Heisenberg, Schrödinger also responded to the crisis but, on the face of it, his approach was different. Unlike Heisenberg, Schrödinger sought to retain a semblance of physical content in the spirit of Bohr and Sommerfeld. Indeed, much like Sommerfeld, Schrödinger spoke of 'der Keplerbewegung',[50] 'das relativistische Keplerproblem',[51] 'im Falle des Keplerproblems',[52] 'Keplerbahnen von atomaren Dimensionen',[53] 'der

---

[48] Heisenberg ([1925] 1967, p. 262). Other key players in 1925-26 were Max Born (1882–1970), Pascual Jordan (1902–1980), and Wolfgang Pauli (1900–1958).

[49] See, e.g., Aitchison *et al.* (2004), pp. 1370–71.

[50] Schrödinger (1926a), p. 362.

[51] Schrödinger (1926a), p. 372; Schrödinger (1926b), p. 519 n.

[52] Schrödinger (1926b), pp. 489, 514.    [53] Schrödinger (1926b), p. 498.

ungestörten Keplerbewegung',[54] and 'hochquantigen Keplerellipsen'.[55] However, it soon became clear to Schrödinger that the two approaches, namely, his and that of Heisenberg, yield the same results in significant cases.[56]

Schrödinger's approach serves our argument well.[57] We consider two complementary perspectives that Schrödinger himself introduced: (1) the micro-macro relation, and (2) nurturing intuition.

(1)  Consider the following assertion:

> The strength of Heisenberg's programme lies in the fact that it promises to give the *line-intensities*, a question that we have not approached as yet. The strength of the present attempt ... lies in the guiding, physical point of view, which creates a bridge between the macroscopic and microscopic mechanical processes [*die Brücke schlägt zwischen dem makroskopischen und dem mikroskopischen mechanischen Geschehen*], and which makes intelligible the outwardly different modes of treatment which they demand.[58]

Shortly after the publication of Heisenberg's paper, Schrödinger acknowledged the strength of Heisenberg's approach. In juxtaposing his own theory against that of Heisenberg, he drew attention to an advantage of his own approach, namely, the apparent linkage between the macroscopic mechanical processes and those at the microscopic scale. Schrödinger went so far as to envision this linkage as a bridge between two kinds of mechanical processes at two different scales. This imagery offers, according to Schrödinger, intelligible guidance on how to address the new physics at the atomic scale. In a subsequent paper published in the same year, Schrödinger rehearsed this point. When addressing issues such as two colliding atoms, he found it necessary to survey the transition (*Übergang*) from macroscopic, perceptual mechanics to the micro-mechanics of the atom. And he explained how he understood this transition:

> Micro-mechanics appears as a refinement of macro-mechanics [*Die Mikromechanik stellt sich als eine Verfeinerung der Makromechanik dar*], which is necessitated by the geometrical and mechanical smallness of the objects, and the transition is of the same nature as that from geometrical to physical optics ...[59]

Schrödinger was clearly motivated by the possible reductive macro-micro relation. Here 'transition' replaces the image of the 'bridge', but then he added a very strong claim: micro-mechanics is a refinement of macro-mechanics. Schrödinger was keen on imagining a transition between these physical scales despite Heisenberg's insistence on

---

[54]  Schrödinger (1926b), p. 519.     [55]  Schrödinger (1926c), p. 666.
[56]  Schrödinger ([1926d] 1928).      [57]  See, e.g., Dorling (1987) and Bitbol (1996).
[58]  Schrödinger (1926b), p. 514, trans. in Schrödinger (1928), p. 30, italics in the original.
[59]  Schrödinger (1926d), p. 753, trans. in Schrödinger (1928), p. 59.

abandoning such a physical intuition as misleading and indeed futile. Evidently, Schrödinger sought to resist the epistemological objection which lies at the core of Heisenberg's approach that only observables such as 'transition probabilities' can play a role in the construction of the new physics of the atom. Schrödinger remarked:

> To me it seems extraordinarily difficult to tackle problems of the above kind [e.g., two colliding atoms], as long as we feel obliged on epistemological grounds to repress intuition in atomic dynamics [*aus erkenntnistheoretischen Gründen verpflictet fühlt, in der Atomdynamik die Anschauung zu unterdrücken*], and to operate only with such abstract ideas as transition probabilities, energy levels, etc.[60]

The transition from 'geometrical optics' to 'physical optics' was a prime example for Schrödinger that in this move from the macro- to the microscopic realm, intuition need not be lost and that physical meaning can be retained.

This brings us to the second perspective, namely, nurturing intuition, rather than repressing it, in the dynamics of the atom.

(2)  Consider the following claim:

> Leaving aside the special optical questions, the problems which the course of development of atomic dynamics brings up for consideration are presented to us by experimental physics in an eminently intuitive form [*in eminent Anschaulicher Form*]; as, for example, how two colliding atoms or molecules rebound from one another, or how an electron or α-particle is diverted, when it is shot through an atom with a given velocity and with the initial path at a given perpendicular distance from the nucleus.[61]

In the move from the macro- to the micro-scale, Schrödinger took his cue from the issues involved in a fairly intuitive way; similarly, when he conceived problems in atomic dynamics presented by experimentation, he approached the issues intuitively (his term). Thus, the following assertion does not come as a surprise:

> For me, personally, there is a special charm in the conception, . . . of the emitted frequencies as 'beats', which I believe will lead to an intuitive understanding of the intensity formulae.[62]

Evidently, Schrödinger was keen to retain an intuitive conception of physics. To be sure, he was fully aware of the difficulty in nurturing intuition when it came to the physics of the atom. He remarked:

---

[60] Schrödinger (1926d), p. 753, trans. in Schrödinger (1928), p. 59.
[61] Schrödinger (1926d), p. 753, trans. in Schrödinger (1928), p. 59.
[62] Schrödinger (1926b), p. 514, trans. in Schrödinger (1928), p. 30.

All these assertions systematically contribute to the relinquishing of the ideas of 'place of the electron' and 'path of the electron'. If these are not given up, contradictions remain. This contradiction has been so strongly felt that it has even been doubted whether what goes on in the atom could ever be described within the scheme of space and time.[63]

But for Schrödinger this meant giving up too much; he considered it an epistemic issue:

> From the philosophical standpoint, I would consider a conclusive decision in this sense as equivalent to a complete surrender. For we cannot really alter our manner of thinking in space and time, and what we cannot comprehend within it we cannot understand at all. There are such things—but I do not believe that atomic structure is one of them.[64]

This is a striking confession, for the claim that the dynamics of the atom can be understood is indeed just a belief which Schrödinger held, hoping to share it with his readers. As we have seen, he acknowledged that there are, to be sure, incomprehensible things, but the physics of the atom is not one of them. Schrödinger was then required to offer a methodological move to facilitate a conception of physics that is not alien to our intuition and, at the same time, faithful to the evidence of the physics in question—the dynamics of the atom. To illustrate this move he referred to the transition from geometrical optics to physical optics and, to be specific, to the undulatory theory:

> From our standpoint, however, there is no reason for such doubt, although or rather *because* its appearance is extraordinarily comprehensible. So might a person versed in geometrical optics, after many attempts to explain diffraction phenomena by means of the idea of the ray (trustworthy for his macroscopic optics), which always came to nothing, at last think that the *Laws of Geometry* are not applicable to diffraction, since he continually finds that light rays, which he imagines as *rectilinear* and *independent* of each other, now suddenly show, even in homogeneous media, the most remarkable *curvatures*, and obviously *mutually influence* one another. I consider this analogy as *very* strict. Even for the unexplained *curvatures*, the analogy in the atom is not lacking—think of the 'non-mechanical force', devised for the explanation of anomalous Zeeman effects.[65]

The analogy to optics, both geometrical and physical, yields a straightforward consequence, namely, one has to change physical conception when one moves from the macro-scale to the micro-scale; but does this change of conceptual framework lead to unintelligible physics? No. According to Schrödinger, our epistemic resources are

---

[63] Schrödinger (1926b), p. 508, trans. in Schrödinger (1928), p. 26.
[64] Schrödinger (1926b), pp. 508–09, trans. in Schrödinger (1928), pp. 26–27.
[65] Schrödinger (1926b), p. 509, trans. in Schrödinger (1928), p. 27, italics in the original.

sufficient to deal with the required conceptual change so that we can retain a meaningful physics.

These two perspectives, namely, (1) change of physical framework in correspondence to continuous scale transition and, at the same time, (2) retaining intuition to represent the physics with mental images, are at the root of Schrödinger's optimism for avoiding Heisenberg's formalism that is divorced from intuition. He expressed his hope—with surprise—when he compared the two theories:

> Considering the extraordinary differences between the starting points and the concepts of Heisenberg's quantum mechanics and of the theory which has been designated 'undulatory' or 'physical' mechanics, and has lately been described here, it is very strange that these two new theories agree *with one another* with regard to the known facts, where they differ from the old quantum theory.[66]

Schrödinger was probably in the best position to describe the difference between the two theories:

> That is really very remarkable, because starting points, presentations, methods, and in fact the whole mathematical apparatus, seem fundamentally different. Above all, however, the departure from classical mechanics in the two theories seems to occur in diametrically opposed directions. In Heisenberg's work the classical continuous variables are replaced by systems of discrete numerical quantities (matrices), which depend on a pair of integral indices, and are defined by *algebraic equations*. The authors themselves describe the theory as a 'true theory of a discontinuum'. On the other hand, wave mechanics shows just the reverse tendency; it is a step from classical point-mechanics towards a *continuum-theory*.[67]

The theories are equivalent in terms of their results but, by their different mathematical structures, they convey different conceptions of physics and, hence, different physical reality. Schrödinger contrasted this paradigm with recent cases in the history of physics, namely, Gustav Kirchhoff (1824–1887) and Ernst Mach (1838–1916), who both sought a theory which is merely a mathematical description of the empirical connections among phenomena. In these classical cases, the mathematical equivalence between the two theories guarantees that the physics too is equivalent. However, this does not seem to be the case in quantum mechanics. Schrödinger noted that Heisenberg's theory depended on discrete quantities whereas his own theory invoked a continuum, that is, the underlying mathematics of the two theories is entirely different. Moreover, Heisenberg's theory is intentionally devoid of any spatio-temporal 'picture' whereas Schrödinger insisted on the relevance of intuition in constructing a physical theory. It is then surprising, as Schrödinger remarked, that the two theories agree in

---

[66] Schrödinger (1926d), p. 734, trans. in Schrödinger (1928), p. 45, italics in the original.
[67] Schrödinger (1926d), p. 734, trans. in Schrödinger (1928), p. 45, italics in the original. Cf. Madrid Casado (2008), p. 153.

accounting for many phenomena.[68] Indeed, it was later shown that the two theories are formally equivalent.[69]

## 38.4 CONCLUSION

The goal of this chapter is to demonstrate the claim that the problem of interpretation in quantum mechanics is not unique, for it had already arisen in classical physics. We have detailed two cases in modern physics in which formalism is divorced from physical content. We note a parallel between the success of Maxwell's set of equations—as recast by Hertz—that was not tied to any physical interpretation, and the success of the formalism of Heisenberg which is also independent of any physical interpretation. The essential difference between the two cases is that in electrodynamics no crisis had occurred, whereas in the case of quantum mechanics the new formalism was constructed in response to the failure of the quantum theory as elaborated by Bohr and Sommerfeld. To be specific, Hertz accepted Maxwell's set of equations but noticed a lack of symmetry in the treatment of the two forces, electricity and magnetism. He then proceeded to modify the equations algebraically to make them symmetrical. In so doing, he undermined the original intention of Maxwell 'to construct a mental representation of all the details of…[the medium's] action'.[70] By contrast, Heisenberg, together with a few other physicists, discarded the quantum theory, complete with planetary images, and developed a novel formalism based solely on observable quantities. As he remarked, 'it seems more reasonable to try to establish a theoretical quantum mechanics, analogous to classical mechanics, but in which only relations between observable quantities occur'.[71] To be sure, Heisenberg thought in analogical terms but presupposed *only* observables. Schrödinger, for his part, explicitly sought to retain not only some analogical semblance between quantum mechanics and classical mechanics,[72] but the very intuitive way of thinking as well. Schrödinger's 'wave mechanics' is a *tour de force* of intuitive thinking against the trend to separate mathematical structure from physical content, where the latter can be captured by mental images (as Maxwell had demanded). In 1893, with the 'calm' separation in electrodynamics between mathematical structure and physical content in the background, the English physicist John Henry Poynting (1852–1914) was optimistic that one day light will 'once more become mechanical'.[73] Little did he know that this was not to be.

---

[68] See Schrödinger (1926d), p. 751, trans. in Schrödinger (1928), p. 58.

[69] See von Neumann ([1932] 1955); cf. Madrid Casado (2008), p. 155.

[70] Maxwell (1873), vol. 2, p. 438, § 866. Heaviside found himself in the same situation as Hertz, namely, he accepted the equations but modified them algebraically: see footnote 30, above.

[71] Heisenberg ([1925] 1967), p. 262.       [72] See, e.g., Joas and Katzir (2011).

[73] Poynting ([1893] 1920), p. 264.

## REFERENCES

Aitchison, I., MacManus, D., and Snyder, T. (2004). Understanding Heisenberg's 'magical' paper of July 1925: A new look at the calculational details. *American Journal of Physics*, 72(11), 1370–79.

Andrade, E. N. da C. (1957). The Rutherford Memorial Lecture: The Birth of the Nuclear Atom. *Proceedings of the Royal Society of London, Series A, Mathematical and Physical Sciences*, 244(1239), 437–55.

Archibald, T. (1989). Physics as a Constraint on Mathematical Research: The Case of Potential Theory and Electrodynamics. In D. E. Rowe and J. McCleary, *The history of modern mathematics. Vol. 2: Institutions and application*, Boston, New York: Academic Press, pp. 29–75.

Bitbol, M. (1996). *Schrödinger's philosophy of quantum mechanics*. Dordrecht: Kluwer.

Bohr, N. (1922). *The Theory of Spectra and Atomic Constitution*. Three essays. Cambridge: Cambridge University Press.

Bohr, N. ([1922] 1965). The structure of the atom. In *Nobel Lectures, Physics 1922–1941*. Amsterdam: Elsevier, pp. 7–43.

Buchwald, J. Z. (1985). *From Maxwell to microphysics: Aspects of electromagnetic theory in the last quarter of the nineteenth century*. Chicago and London: University of Chicago Press.

Buchwald, J. Z. (1994). *The creation of scientific effects: Heinrich Hertz and electric waves*. Chicago: University of Chicago Press.

Campbell, L., and Garnett, W. (1882). *The life of James Clerk Maxwell*. London: Macmillan.

Darrigol, O. ([2000] 2002). *Electrodynamics from Ampère to Einstein*. Oxford: Oxford University Press.

Dorling, J. (1987). Schrödinger's original interpretation of the Schrödinger equation: a rescue attempt. In C. W. Kilmister (ed.), *Schrödinger, centenary celebration of a polymath*, Cambridge: Cambridge University Press, pp. 16–40.

Einstein, A. (1905). Zur Elektrodynamik bewegter Körper. *Annalen der Physik*, 17, 891–21.

Einstein, A. (1907). Über das Relativitätsprinzip und die aus demselben gezogenen Folgerungen. *Jahrbuch der Radioaktivität und Elektronik*, 4, 411–62.

Einstein, A. (1931). Maxwell's influence on the development of the conception of physical reality. In J. J. Thomson et al. (eds), *James Clerk Maxwell; a commemoration volume, 1831–1931*, New York: `Macmillan, pp. 66–73.

Einstein, A., Podolsky, B., and Rosen, N. (1935). Can quantum-mechanical description of physical reality be considered complete? *Physical Review*, 47, 777–80.

Forman, P., and Hermann, A. (1975). Sommerfeld, Arnold. *Dictionary of Scientific Biography*, 12, 525–32.

Goldstein, B. R., and Hon, G. (2013–2014). The Image that Became the Icon for Atomic Energy. *Physis*, 49, 259–72.

Heaviside, O. (1893). *Electromagnetic theory*. Vol. 1. London: 'The Electrician' Printing and Publishing Co.

Heisenberg, W. ([1925] 1967). Über quantentheoretische Umdeutung kinematischer und mechanischer Beziehungen. *Zeitschrift für Physik*, 33, 879–93. For an English translation, see B. L. van der Waerden, *Sources of quantum mechanics*, Amsterdam: North-Holland, pp. 261–76.

Hertz, H. ([1884] 1895). Ueber die Beziehungen zwischen den Maxwell'schen electrodynamischen Grundgleichungen und den Grundgleichungen der gegnerischen Electrodynamik.

*Annalen der Physik und Chemie*, **23**, 84–103. Reprinted in Hertz (1895), 295–314. Translated in Hertz (1896), 273–290, On the Relations between Maxwell's Fundamental Electromagnetic Equations and the Fundamental Equations of the Opposing Electromagnetics.

Hertz, H. ([1885] 1895). Über die Dimensionen des magnetichen Poles in verschiedenen Maßsystemen. *Annalen der Physik und Chemie*, **24**, 114–18. Reprinted in Hertz (1895), 315–319. Translated in Hertz (1896), 291–295, On the Dimensions of Magnetic Pole in Different Systems of Units.

Hertz, H. ([1889] 1895). *Über die Beziehungen zwischen Licht und Elektricität: ein Vortrag gehalten bei der 62. Versammlung deutscher Naturforscher und Aertze in Heidelberg.* Bonn: Strauss. Reprinted in Hertz (1895), 339–354. Translated in Hertz (1896), 313–327, On the Relations between Light and Electricity.

Hertz, H. (1892). *Untersuchungen über die Ausbreitung der elektrischen Kraft.* Leipzig: Barth. Translated in Hertz ([1893] 1962).

Hertz, H. ([1893] 1962). *Electric waves.* Translated by D. E. Jones. New York: Dover.

Hertz, H. ([1894] 1899). *The principles of mechanics presented in a new form.* New York: MacMillan.

Hertz, H. (1895). *Gesammelte Werke.* Vol. 1: *Schriften vermischten Inhalts.* Ph. Lenard (ed.). Leipzig: Barth.

Hertz, H. (1896). *Miscellaneous Papers.* Translated by D. E. Jones and G. A. Schott. London and New York: Macmillan.

Hon, G., and Goldstein, B. R. (2005). How Einstein Made Asymmetry Disappear: Symmetry and Relativity in 1905. *Archive for History of Exact Sciences*, **59**, 437–44.

Hon, G., and Goldstein, B. R. (2006). Symmetry and Asymmetry in Electrodynamics from Rowland to Einstein. *Studies in History and Philosophy of Modern Physics*, **37**, 635–60.

Hon, G., and Goldstein, B. R. (2020). *Reflections on the practice of physics: James Clerk Maxwell's methodological odyssey in electromagnetism.* London and New York: Routledge.

Hunt, B. J. (1991). *The Maxwellians.* Ithaca and London: Cornell University Press.

James, F. A. J. L. (ed.) (2008). *The Correspondence of Michael Faraday, Volume 5: 1855–1860.* London: Institution of Engineering and Technology.

Joas, C., and Katzir, S. (2011). Analogy, extension, and novelty: Young Schrödinger on electric phenomena in solids. *Studies in History and Philosophy of Modern Physics*, **42**, 43–53.

Kragh, H., and Nielsen, K. H. (2013). Spreading the Gospel: A Popular Book on the Bohr Atom in its Historical Context. *Annals of Science*, **70**, 257–83.

Kramers, H. A., and Holst, H. ([1922] 1923). *The atom and the Bohr theory of its structure: An elementary presentation.* With a Foreward by E. Rutherford. Translated from Danish by Mr. and Mrs. Lindsay. London and Copenhagen: Gyldendal.

Madrid Casado, C. M. (2008). A brief history of the mathematical equivalence between the two quantum mechanics. *Latin American Journal of Physics Education*, **2**(2), 152–55.

Maxwell, J. C. ([1871/1890] 1965). Remarks on the mathematical classification of physical quantities. *Proceedings of the London Mathematical Society*, **3**, 224–32. Reprinted in Maxwell ([1890] 1965), 2: 257–266.

Maxwell, J. C. ([1873/1890] 1965). Faraday. *Nature*, **8**(11 September 1873), 397–99. Reprinted in Maxwell ([1890] 1965), vol. 2, pp. 355–360.

Maxwell, J. C. (1873). *A treatise on electricity and magnetism.* 2 vols. Oxford: Clarendon Press.

Maxwell, J. C. ([1890] 1965). *The scientific papers of James Clerk Maxwell.* W. D. Niven (ed.), 2 vols. Cambridge: University Press. Reprinted, two volumes bound as one. New York: Dover.

Poynting, J. H. ([1893] 1920). An examination of Prof. Lodge's electromagnetic hypothesis. *Electrician*, 31(1893), 575–77, 606–608, 635–636. Reprinted in Poynting (1920). *Collected scientific papers.* Cambridge: Cambridge University Press, pp. 250–268.

Schrödinger, E. (1926a). Quantisierung als Eigenwertproblem (Erste Mitteilung). *Annalen der Physik*, **79**, 361–76. Quantisation as a problem of proper values (Part I). English trans. in Schrödinger (1928), 1–12.

Schrödinger, E. (1926b). Quantisierung als Eigenwertproblem (Zweite Mitteilung). *Annalen der Physik*, **79**, 489–27. Quantisation as a problem of proper values (Part II). English trans. in Schrödinger (1928), 13–40.

Schrödinger, E. (1926c). Der stetige Übergang von der Mikro- zur Makromechanik. *Die Naturwissenschaften*, **14**(28), 664–66. The continuous transition from micro- to macro-mechanics. English trans. in Schrödinger (1928), 41–44.

Schrödinger, E. (1926d). Über das Verhältnis der Heisenberg-Born-Jordanschen Quantenmechanik zu der meinen. *Annalen der Physik*, **79**, 734–56. On the relation between the quantum mechanics of Heisenberg, Born, and Jordan, and that of Schrödinger. English trans. in Schrödinger (1928), 45–61.

Schrödinger, E. (1928). *Collected papers on wave mechanics.* Translated from the 2nd German ed. London and Glasgow: Blackie & Son.

Sommerfeld, A. (1915). Zur Theorie der Balmerschen Serie. *Sitzungsberichte der mathematisch-physikalischen Klasse der königl. bayerischen Akademie der Wissenschaften,* Jahrgang 1915, 425–58.

Sommerfeld, A. ([1919] 1922). *Atombau und Spektrallinien*, 3rd edn. Braunschweig: Vieweg & Sohn.

Sommerfeld, A. ([1919/1922] 1923). *Atomic structure and spectral lines.* Translated from the 3rd German ed. by Henry L. Brose. New York: Dutton.

Steinle, F. ([2005] 2016). *Exploratory experiments: Ampère, Faraday, and the origins of electrodynamics.* Pittsburgh: University of Pittsburgh Press.

Thomson, W. (1842). On the uniform motion of heat in homogeneous solid bodies, and its connexion with the mathematical theory of electricity. *Cambridge Mathematical Journal*, **3**, 71–84. Reprinted in Thomson (1854).

Thomson, W. (1854). On the uniform motion of heat in homogeneous solid bodies, and its connexion with the mathematical theory of electricity. *Philosophical Magazine*, Series 4, **7**, 502–15. Reprinted in Thomson (1872), 1–14.

Thomson, W. (1872). *Reprint of papers on electrostatics and magnetism.* London: Macmillan.

von Neumann, J. ([1932] 1955). *Mathematical Foundations of Quantum Mechanics*, trans. by R. T. Beyer. Princeton, NJ: Princeton University Press.

# CHAPTER 39

.....................................................................................

# HIDDEN VARIABLES

.....................................................................................

JEFFREY BUB

## 39.1 INTRODUCTION

.....................................................................................

IN 1926, the year Schrödinger demonstrated the equivalence of Heisenberg's matrix mechanics and his own wave mechanics (Schrödinger, 1926), Max Born proposed what is now known as the 'Born rule', the probabilistic interpretation of the Schrödinger wave function, in a paper on the quantum description of collisions (Born, 1926, p. 54):

> If one translates this into particles, only one interpretation is possible. $\Phi_{nm}(\alpha,\beta,\gamma)$ gives the probability for the electron, arriving from the $z$-direction, to be thrown out into the direction designated by the angles $\alpha,\beta,\gamma$, with the phase change $\delta$.

Remarkably, this is not quite right, but the mistake is corrected in a footnote added in the proof:

> More careful consideration shows that the probability is proportional to the square of the quantity $\Phi_{nm}$.

Born adds:

> From the standpoint of our quantum mechanics there is no quantity which in any individual case causally fixes the consequence of the collision; but also experimentally we have so far no reason to believe that there are some inner properties of the atom which condition a definite outcome for the collision.

The question of whether there are 'inner properties' or 'hidden variables' of a quantum system that 'condition a definite outcome' in a quantum mechanical interaction, in particular the outcome of a measurement, or whether quantum probabilities are ultimately 'irreducible', has been a recurring issue in the foundations of quantum mechanics since Born's seminal paper.

The first hidden variable theory was proposed by de Broglie in a series of papers in 1926 and 1927 (de Broglie, 1926, 1927a,b), but de Broglie stopped working on the theory after criticism by Pauli, or rather, after the 1927 Solvay conference; see the discussion by Bacciagaluppi and Valentini (2009, Chapters 10, 11). Einstein presented a hidden variable theory at the 1927 meeting of the Prussian Academy of Sciences in Berlin, but withdrew the paper before publication after realizing that, for two independent non-interacting systems, measurements performed independently on the separate systems would yield a combined outcome that differed from the predicted outcome of a measurement on the composite system. See Belousek (1996) for a detailed account. The most well-known hidden variable theory, still the topic of an ongoing research programme, is Bohm's 1952 theory (Bohm, 1952), sometimes referred to as the de Broglie–Bohm theory because of the similarity to de Broglie's earlier theory. In Bohm's theory, the difficulties with de Broglie's theory are resolved in a novel treatment of quantum measurement.

Bohm's theory is discussed in some detail elsewhere in this book. This chapter will focus on several seminal impossibility proofs, so-called 'no go' theorems, and Bohm's other hidden variable theory, the Bohm–Bub 1966 theory (Bohm and Bub, 1966).

## 39.2 IMPOSSIBILITY PROOFS

### 39.2.1 Von Neumann

Von Neumann's proof first appeared in a 1927 paper (von Neumann, 1927), and a later version in the 1932 German edition of his book *Mathematical Foundations of Quantum Mechanics* (von Neumann, 1955), where the hidden variable question is discussed in Chapter 4. The question von Neumann poses is whether a quantum system can be characterized by certain parameters in addition to the quantum state given by the Schrödinger wave function, such that a specification of the values of these 'hidden' parameters together with the wave function would 'give the values of all physical quantities exactly and with certainty', as he puts it in a preliminary discussion in Chapter 3 (von Neumann, 1955, p. 209). He begins the investigation with the following preamble (von Neumann, 1955, pp. 297–298):

> Let us forget the whole of quantum mechanics but retain the following. Suppose a system S is given, which is characterized for the experimenter by the enumeration of all the effectively measurable quantities in it and their functional relations with one another. With each quantity we include the directions as to how it is to be measured—and how its value is to be read or calculated from the indicator positions on the measuring instruments. If $\mathcal{R}$ is a quantity and $f(x)$ any function,

then the quantity $f(\mathcal{R})$ is defined as follows: To measure $f(\mathcal{R})$, we measure $\mathcal{R}$ and find the value $a$ (for $\mathcal{R}$). Then $f(\mathcal{R})$ has the value $f(a)$. As we see, all quantities $f(\mathcal{R})$ ($\mathcal{R}$ fixed, $f(x)$ an arbitrary function) are measured simultaneously with $\mathcal{R}$ … But it should be realized that it is completely meaningless to try to form $f(\mathcal{R},\mathcal{S})$ if $\mathcal{R},\mathcal{S}$ are not simultaneously measurable: there is no way of giving the corresponding measuring arrangement.

As a first step, von Neumann (1955, p. 311) imposes some structure on these quantities, which are introduced initially in terms of the statistics of measurement outcomes by two assumptions labelled **A′,B′**:

**A′**: If the quantity $\mathcal{R}$ is by nature non-negative, for example, if it is the square of another quantity $\mathcal{S}$, then Exp ($\mathcal{R}$) $\geq 0$ (where Exp represents the expectation value)

**B′**: If $\mathcal{R},\mathcal{S},\dots$ are arbitrary quantities and a, b, … real numbers, then Exp (a$\mathcal{R}$ + b$\mathcal{S}$ + ⋯) = aExp ($\mathcal{R}$) + bExp($\mathcal{S}$) + ⋯

Now von Neumann (1955, p. 309) remarks that if the quantities $\mathcal{R},\mathcal{S},\cdots$ are simultaneously measurable, then $a\mathcal{R} + b\mathcal{S} + \cdots$ is the ordinary sum, but if they are not simultaneously measurable, then 'the sum is characterized … only in an implicit way'. As an example, he refers in footnote 164 (von Neumann, 1955, pp. 309–310) to the kinetic energy of an electron in an atom, $\mathcal{E} = \mathcal{P}^2/2m + V(\mathcal{Q})$, where $\mathcal{P}^2/2m$ as a function of momentum and $V(\mathcal{Q})$ as a function of position are not simultaneously measurable.

Clearly, **B′** is proposed as a *definition* of the quantity $a\mathcal{R} + b\mathcal{S} + \cdots$ when $\mathcal{R},\mathcal{S},\dots$ are not simultaneously measurable, i.e., **B′** denotes a quantity, $\mathcal{X} \equiv a\mathcal{R} + b\mathcal{S} + \cdots$, whose expectation value is the linear sum of expectation values $a\text{Exp}(\mathcal{R}) + b\text{Exp}(\mathcal{S}) + \cdots$. With $\mathcal{X}$ defined through the observed measurement statistics of $\mathcal{R},\mathcal{S},\cdots$ in this way, it is not necesssarily the case that the value of $\mathcal{X}$,val($\mathcal{X}$), for specific values of the hidden variables and a given wave function, is equal to $a\text{val}(\mathcal{R}) + b\text{val}(\mathcal{S}) + \cdots$ if $\mathcal{R},\mathcal{S},\cdots$ are not simultaneously measurable, and it is a further question how $\mathcal{X}$ is to be measured so as to satisfy **B′** for the observed statistics. For the kinetic energy of an electron in an atom, it turns out that **B′** is satisfied if we measure the energy by measuring the frequency of the spectral lines in the radiation emitted by the electron.

Von Neumann proves that *if* the quantities $\mathcal{R}$ defined via the measurement statistics are related to corresponding quantities represented by Hilbert space operators according to assumptions (von Neumann, 1955, pp. 313–314):

I: If the quantity $\mathcal{R}$ has the operator $R$, then the quantity $f(\mathcal{R})$ has the operator $f(R)$.

II: If the quantities $\mathcal{R},\mathcal{S},\cdots$ have the operators $R,S,\cdots$, then the quantity $\mathcal{R} + \mathcal{S} + \cdots$ has the operator $R + S + \cdots$.

*then* the expectation value of a quantity is uniquely defined by the trace function:

$$\text{Exp}(\mathcal{R}) = \text{Tr}(WR) \tag{1}$$

where $W$ is a Hermitian operator independent of $\mathcal{R}$ that characterizes the ensemble.

There are no $W$ operators for dispersion-free ensembles, which von Neumann (1955, p. 306) defines as ensembles for which $\text{Exp}(\mathcal{R}^2) = (\text{Exp}(\mathcal{R}))^2$ for all $\mathcal{R}$, so no ensembles for which an $\mathcal{R}$-measurement would produce the same outcome for all members of the ensemble—an outcome that would be specified by particular values of the hidden variables. In other words, if we assume that the measurable quantities $\mathcal{R}$ are related to Hilbert space operators according to I,II, then the only possible expectation values are those that correspond to quantum states according to (1) as a generalization of the Born rule, with $W$ representing a pure or mixed quantum state.

Von Neumann concludes (von Neumann, 1955, pp. 324–325):

> It should be noted that we need not go any further into the mechanism of the 'hidden parameters', since we now know that the established results of quantum mechanics can never be re-derived with their help. In fact, we have even ascertained that it is impossible that the same physical quantities exist with the same function connections (i.e., that I,II hold) if other variables (i.e., 'hidden parameters') should exist in addition to the wave function.

> Nor would it help if there existed other, as yet undiscovered, physical quantities, in addition to those represented by the operators in quantum mechanics, because the relations assumed by quantum mechanics (i.e., I,II) would have to fail already for the by now known quantities, those that we discussed above. It is therefore not, as is often assumed, a question of a reinterpretation of quantum mechanics—the present system of quantum mechanics would have to be objectively false, in order that another description of the elementary processes than the statistical one be possible.

Von Neumann's argument has been criticized by Hermann (1935) and by Bell (1966), who focus on von Neumann's condition $\mathbf{B}'$ for the quantities $\mathcal{R}, \mathcal{S}, \cdots$. They argue that if these quantities are identified with Hilbert space operators, as in II, then for specific values of the hidden variables and a given wave function the expectation values in $\mathbf{B}'$ should be eigenvalues of the operators, but the eigenvalues of an operator defined as a sum of noncommuting operators are not generally equal to the sums of the eigenvalues of the operators in the sum. Bell (1966, p. 449) points out that for a spin-1/2 particle, the eigenvalues of $\sigma_x$ and $\sigma_y$ are both $\pm 1$, while the eigenvalues of $\sigma_x + \sigma_y$ are $\pm\sqrt{2}$. He concludes:

> It was not the objective measurable predictions of quantum mechanics which ruled out hidden variables. It was the arbitrary assumption of a particular (and impossible) relation between the results of incompatible measurements either of which might be made on a given occasion but only one of which can in fact be made.

In an interview with *Omni* magazine Bell (1988) remarks:

Yet the von Neumann proof, if you actually come to grips with it, falls apart in your hands! There is *nothing* to it. It's not just flawed, it's *silly*! ... When you translate [his assumptions] into terms of physical disposition, they're nonsense. You may quote me on that: The proof of von Neumann is not merely false but *foolish*!

Mermin and Schack (2018) support the critique by Hermann and Bell against a defence of von Neumann's argument by Bub (2010) and Dieks (2017). They argue that, even if $\mathbf{B}'$ is understood as a definition for a sum of quantities that are not simultaneously measurable, there is an alternative definition: simply define the quantity $a\mathcal{R} + b\mathcal{S} + \cdots$ as the quantity that corresponds to the Hermitian operator $aR + bS + \cdots$, where $\mathcal{R}, \mathcal{S}$ are the quantities that correspond to the operators $R, S, \cdots$.

But this, to paraphrase Bell, is silly. The whole point of von Neumann's argument is to begin with quantities defined via *measurement statistics* and then to show that if such quantities can be represented by Hilbert space operators, as in quantum mechanics, the statistics of measurement outcomes cannot be derived by averaging over values of the quantities as functions of dispersion-free states (deterministic hidden variables). If, at the start, one identifies a sum of quantities that are not simultaneously measurable with a quantity associated with the operator sum, then the possible values of this quantity are the eigenvalues of the operator sum. It follows immediately that $\mathbf{B}'$ is false for dispersion-free states, as in Bell's example of the spin eigenvalues of a spin-1/2 particle. So this is not an appropriate assumption if the aim is to exclude hidden variables—all this shows is that the proposed strategy does not lead to a 'no hidden variables' proof. It is not an objection to von Neumann's strategy.

Von Neumann begins by considering abstractly a theory characterized by a set of measurable quantities and their functional relationships, where these quantities are specified by measurement instructions—how the measurement is to be performed and the outcomes calculated from the instrument readings. A structure is imposed on these measurable quantities via the conditions $\mathbf{A}', \mathbf{B}'$, in the obvious sense for simultaneously measurable quantities and in the sense outlined above for quantities that are not simultaneously measurable. There is no mention of Hilbert space up to this point. The entire analysis is in terms of the measurable quantities of an arbitrary theory. The argument about the failure of $\mathbf{B}'$ for specific values of the hidden variables would apply equally to von Neumann's example of kinetic energy as the sum of two quantities that are not simultaneously measurable, so Bell's observation could not have escaped von Neumann.

What von Neumann demonstrates is that if these quantities are represented by Hilbert space operators according to conditions I,II, the observed statistics cannot be derived as measures over a space of dispersion-free (deterministic) hidden variable states, each of which labels pre-existing values of the quantities, 'what's there before you look'. As he says, 'the relations assumed by quantum mechanics (i.e., I,II) would have to fail already for the by now known quantities, those that we discussed above', like position, momentum, energy, so in *this sense* 'the present system of quantum mechanics would have to be objectively false'.

Von Neumann's argument does not exclude the possibility of deriving the measurement statistics of quantum observables in some other way as, for example, in Bohm's 1952 theory. Then some explanation is required for why the measurement statistics of a measurable quantity like $S_x + S_y$, implicitly defined through the measurement statistics of the quantities $S_x$ and $S_y$, can be derived by applying the trace formula to the Hilbert space operator $\sigma_x + \sigma_y$. This is what Bohm's theory does on the basis of a measurement theory, where a quantum measurement is treated as disturbing the hidden variables. See the discussion in Section 39.2.3.

In Bohm's hidden variable theory, physical quantities are introduced as classical dynamical quantities, functions of positions and momenta, and the quantum statistics defined by the trace rule for quantum observables is an artefact of a dynamical process that is not in fact a measurement of any physical quantity of the system. For von Neumann, the physical quantities of quantum systems and their functional relations are abstracted from the measurement statistics and appropriately represented by Hermitian operators, and the quantum theory, including the quantum theory of measurement, is a theory about these physical quantities.

The pre-dynamic structure of quantum mechanics, given by the Hilbert space representation of physical quantities as Hermitian operators and the trace formula for expectation values, excludes hidden variables in a similar sense to which the pre-dynamic structure of special relativity, given by the non-Euclidean geometry of Minkowski space-time, excludes hidden forces as responsible for Lorentz contraction. Lorentz aimed to maintain Euclidean geometry and Newtonian kinematics and explain the anomalous behaviour of light in terms of a dynamical theory about how rods contract as they move through the ether. His dynamical theory plays a similar role, in showing how Euclidean geometry can be preserved, as Bohm's dynamical theory of measurement in explaining how the quantum statistics can be generated in a classical or Boolean theory of probability.

## 39.2.2 Kochen and Specker

The Kochen–Specker result was first announced in a paper by Specker (Specker, 1960). Bell (1966, p. 450) essentially proved the result in a direct proof of what he presented as a corollary to Gleason's theorem (Gleason, 1957) relevant to the hidden variable question (see Brown (1992) for an illuminating discussion).

A common view of von Neumann's result, following Bell's critique (Bell, 1966), is that von Neumann proved that no noncontextual hidden variable theory can recover the quantum statistics, but from stronger assumptions than those of Kochen and Specker—in fact so strong as to reduce the result to a triviality. But the way in which Kochen and Specker (1967) pose the hidden variable question is quite different from von Neumann's approach. A hidden variable theory of quantum mechanics is characterized as introducing a space $\Omega$ of hidden variables and a measure $\mu_\psi$ on $\Omega$ so that the probability assigned to a range of values $U$ of an observable $A$ by a quantum state $\psi$ can be recovered as the measure of the set of hidden variables for which $A$, as a function $f_A$ on $\Omega$, has a value in $U$. Formally:

$$p_{A\psi}(U) = \mu_\psi(f_A^{-1}(U)) \tag{2}$$

They note that, without further constraints, this is trivially possible. To show this, they define $\Omega$ as the set of mappings from the quantum observables, $\mathcal{O}$, to the real numbers, $\Omega = \{\omega | \omega : \mathcal{O} \to \mathbf{R}\}$. For any $A \in \Omega$ they define the map $f_A : \Omega \to \mathbf{R}$ as $f_A(\omega) = \omega(A)$. Finally, they define $\mu_\psi$ as the product measure $\mu_\psi = \prod_{A \in \mathcal{O}} p_{A\psi}$. Then:

$$\mu_\psi(f_A^{-1}(U)) = \mu_\psi(\{\omega | \omega(A) \in U\}) = p_{A\psi}(U) \tag{3}$$

In this construction, the observables of a quantum system are represented as independent random variables on $\Omega$ for each quantum state $\psi$. As Kochen and Specker point out, the quantum observables are not independent: there are functional relations among observables that define an algebraic structure, and a hidden variable theory should preserve this structure. Specifically, in addition to $p_{A\psi}(U) = \mu_\psi(f_A^{-1}(U))$, a hidden variable theory should satisfy the condition that, for every Borel function $g$ and observable $A$:

$$f_{g(A)} = g(f_A) \tag{4}$$

The idempotent observables of a classical system, those for which $A^2 = A$, are represented by two-valued functions that take the value 1 on a Borel subset of phase space and 0 elsewhere. As such, they are in 1-1 correspondence with the Borel subsets of phase space representing classical properties or propositions (with 0 representing 'false' and 1 representing 'true') and form a Boolean algebra. In quantum mechanics, the idempotent observables are represented by projection operators, which have eigenvalues 0 or 1 and correspond to quantum properties or propositions. A set of mutually compatible idempotent observables represented by commuting projection operators is associated with a Boolean algebra, but not all projection operators commute. With sums and products defined for mutually compatible observables, the idempotent observables form what Kochen and Specker call a 'partial Boolean algebra'. This is a union of a family of Boolean algebras, 'pasted together' by identifying the maximum and minimum elements of all the Boolean algebras and the elements that are in common to two or more Boolean algebras, such that for every $n$-tuple of pairwise compatible elements (elements belonging to a common Boolean algebra) there is a Boolean algebra in the family containing these elements.

Kochen and Specker show that a hidden variable theory satisfying the functional relationship condition requires the partial Boolean algebra to be embeddable into a Boolean algebra. A necessary and sufficient condition for an embedding is that, for every pair of distinct elements $p,q$ in the partial Boolean algebra, there exists a homomorphism $h$ onto the two-element Boolean algebra $\{0, 1\}$ such that $h(p) \neq h(q)$, i.e., for every pair of distinct elements, considered as quantum propositions, there is a structure-preserving map that separates the elements by assigning one element the value 'false' (0) and the other element the value 'true' (1).

While two-valued homomorphisms can be defined on the partial Boolean algebra of projection operators of a two-dimensional Hilbert space, there are no such maps for a Hilbert space of three or more dimensions, much less any embeddings. The homomorphism condition implies that for every orthogonal $n$-tuple of one-dimensional projection operators or corresponding rays in an $n$-dimensional Hilbert space $\mathcal{H}_n$, one projection operator or ray is mapped onto 1 and the others onto 0. Kochen and Specker prove that this is impossible for the finite set of orthogonal triples of rays that can be constructed from 117 appropriately chosen rays in $\mathcal{H}_3$: any assignment of 0s and 1s to this set of orthogonal triples satisfying the homomorphism condition involves a contradiction. Later, Kochen and Conway reduced the number to 31 in an unpublished proof (Peres, 1993, p. 197), and several authors, including Peres (1991), Penrose (1994), and Bub (1996), published 33-ray proofs.

By contrast with von Neumann, Kochen and Specker consider the problem of hidden variables as the question of whether the observables of a quantum system can be represented as functions on a space of hidden variables satisfying the functional relation condition (4), which entails the possibility of embedding the partial Boolean algebra of quantum observables into a Boolean algebra. Shimony (1984) contrasts hidden variable theories of the type considered by Kochen and Specker with 'contextual' hidden variable theories. In a contextual hidden variable theory, the probability measure $\mu_\psi$ in (2) depends on a context, $C$, so (2) becomes

$$p_{A\psi}(U) = \mu_\psi(f_A^{-1}(U), C) \tag{5}$$

Gudder (1970), taking $C$ to be a Boolean subalgebra corresponding to a maximal observable and $U$ as belonging to the subalgebra, proved an existence theorem for contextual hidden variable theories.

A nonmaximal observable $A$ can be compatible with a maximal observable $X$ (so the operators representing $A$ and $X$ commute) and also with a maximal observable $Y$ (so the operators representing $A$ and $Y$ commute), but the operators representing $X$ and $Y$ don't commute. The observable $A$ could then be represented as a function of $X$, $f(X)$, and measured in the context of $X$ via an $X$-measurement, and also as a function of $Y$, $g(Y)$, and measured in the context of $Y$ as $Y$-measurement. The functional relationship condition requires that the value of $A$ as a function of $X$ should be the same as the value of $A$ as a function of $Y$, i.e., that the value of $A$ should not be dependent on the measurement context. Bell (1966, p. 451) argued that this constraint on a hidden variable theory (expressed as an additivity requirement for expectation values of compatible observables) cannot be justified on physical grounds:

> It was tacitly assumed that measurement of an observable must yield the same value independently of what other measurements may be made simultaneously ... These different possibilities require different experimental arrangements; there is no

*a priori* reason to believe that the results ... should be the same. The result of an observation may reasonably depend not only on the state of the system (including hidden variables) but also on the complete experimental arrangement.

Shimony calls hidden variable theories of the Gudder type contextual in an 'algebraic' sense. What Kochen and Specker proved is the impossibility of an algebraically noncontextual hidden variable extension of quantum mechanics for a system associated with a Hilbert space of three or more dimensions. The two-dimensional case is an exception, and for such systems a hidden variable theory *is* possible, as in the construction by Kochen and Specker at the beginning of this section, or Bell's toy model of spin measurements on a single particle in (Bell, 1966).

For actual measuring instruments, which are non-ideal and do not single out maximal observables, Shimony suggests that a second type of 'environmental' contextuality is relevant, in which $C$ is a specification of some aspect of the physical environment of a quantum system, in principle the whole environment, including the measuring instrument. (Note that the term is sometimes used in a slightly different sense, e.g., by Heywood and Redhead (1983).) Shimony presents Bell's result, discussed in the following section, as a proof that an environmentally contextual hidden variable cannot recover the quantum statistics without introducing the physically undesirable feature of nonlocality.

## 39.2.3  Bell

Bell (1966, p. 452) observed that the equations of motion in Bohm's 1952 hidden variable theory are nonlocal, so that 'an explicit causal mechanism exists whereby the disposition of one piece of apparatus affects the results obtained with a distant piece', and he asked whether it is possible to prove 'that *any* hidden variable account of quantum mechanics *must* have this extraordinary character'. His 1964 paper with the proof of what is now known as Bell's theorem (published two years prior to the long-delayed publication of the paper that raised the issue of locality as a constraint on hidden variables) is his response to this question. Bell (1964, p. 199) concludes that a hidden variable theory reproducing the quantum statistics must introduce 'a mechanism whereby the setting on one measuring device can influence the reading of another instrument, however remote'. More recent versions of Bell's result do not require statistical arguments (Greenberger *et al.*, 1989), or minimize the statistics needed (Hardy, 1992, 1993).

If $A$ denotes an observable of a system $S$, and $X$ and $Y$ denote incompatible observables of a system $S'$, space-like separated from $S$, the functional relationship constraint becomes a physically motivated locality condition. It follows from the result in Bell (1964) that this weaker condition cannot be satisfied in general in a Hilbert space of four or more dimensions: there are sets of observables for which value assignments satisfying the locality condition are inconsistent with the probabilities of entangled quantum states. But the scope of Bell's result is much broader than this application to algebraically contextual hidden variable theories. As Shimony points out in (Shimony, 1984), Bell's result is more properly understood as showing that a contextual hidden variable theory of the environmental type cannot recover the quantum statistics without introducing 'physically undesirable features', like nonlocality.

Bell (1964, p. 196) derived an inequality for hidden variable theories satisfying a locality constraint: 'if two measurements are made at places remote from one another the orientation of one [polarizer] does not influence the result obtained with the other'. That is, no information should be available in region **A** about alternative measurement choices made by Bob in region **B**—Alice in region **A** shouldn't be able to tell what Bob measured in region **B**, or whether Bob performed any measurement at all, by looking at the statistics of her measurement outcomes, and conversely.

Denoting the observables measured by $A$ and $B$, measurement outcomes by $a$ and $b$, and the hidden variables by $\lambda$, this means that, conditional on the hidden variables, the probabilities of particular outcomes $a$ and $b$ on the separate systems should be independent of the observable measured on the remote system:

$$p(a,b|A,B,\lambda) = p(a|A,\lambda)p(b|B,\lambda) \tag{6}$$

even though the observed probabilities, which are averaged over the hidden variables, will not be independent if there are nonlocal correlations, as there are for entangled quantum states:

$$p(a,b|A,B) \neq p(a|A)p(b|B) \tag{7}$$

For example, for a spin-1/2 system in the singlet state, if the orientations of the polarizers measuring $A$ and $B$ are the same, a measurement yields opposite outcomes, so $p(+1, -1|A,B) = 1$, while $p(+1|A) = p(-1|B) = 1/2$.

Bell's locality condition amounts to conditional statistical independence. Equivalently, the marginal probability at **A** or **B** is independent of the measurement procedure as well as the measurement outcome at the remote location. Note that in a deterministic hidden variable theory, the probabilities $p(a|A,\lambda)$ and $p(b|B,\lambda)$ would all be 0 or 1, but this restriction need not be assumed here.

There are a variety of inequalities satisfying a locality condition related to Bell's original inequality. The Clauser–Horne–Shimony–Holt (CHSH) inequality (Clauser *et al.*, 1969) is a particularly useful version, providing the possibility of an experimental test that excludes local hidden variables. This is an inequality for two pairs of binary-valued quantities, $A, A'$ with values $\pm 1$ and B, B' with values $\pm 1$, associated with two separated systems, say two separated photons **A** and **B**.

Consider the expression

$$K = \langle A|B \rangle + \langle A|B' \rangle + \langle A'|B \rangle - \langle A'|B' \rangle \tag{8}$$

where $\langle \,|\, \rangle$ denotes the expectation value averaged over $\lambda$. Assuming locality, the joint expectation values factorize for particular values of $\lambda$:

$$
\begin{aligned}
K_\lambda &= \langle A \rangle_\lambda \langle B \rangle_\lambda + \langle A \rangle_\lambda \langle B' \rangle_\lambda + \langle A' \rangle_\lambda \langle B \rangle_\lambda - \langle A' \rangle_\lambda \langle B' \rangle_\lambda \\
&= \langle A \rangle_\lambda [\langle B \rangle_\lambda + \langle B' \rangle_\lambda] + \langle A' \rangle_\lambda [\langle B \rangle_\lambda - \langle B' \rangle_\lambda]
\end{aligned}
\tag{9}
$$

Since all the quantities are $\pm 1$, the maximum and minimum values of the bracketed expressions are $\pm 2$, and when one bracketed expression takes either of these values the other bracketed expression is 0. It follows that $-2 \le K_\lambda \le 2$, and since averaging over $\lambda$ won't change the inequality:

$$
-2 \le K \le 2
\tag{10}
$$

This is the CHSH inequality.

There is an implicit assumption here, that the distribution of $\lambda$, i.e., the hidden variables, is independent of the observables $A,B$ we choose to measure. One could consider 'superdeterministic' or 'retrocausal' hidden variable theories for which this assumption is not satisfied. Putting such exotic possibilities aside, any local hidden variable theory must satisfy the CHSH inequality, but the inequality is violated by the probabilistic correlations of quantum mechanics—with maximally entangled systems it is possible to achieve a value $K = 2\sqrt{2}$, the quantum Tsirelson bound.

There were several early experimental confirmations of Bell's result, notably the experiment by Freedman and Clauser in 1972 (Freedman and Clauser, 1972) and the experiments in the 1980s by Aspect and associates. The first successful test of Bell's result in the form of the CHSH inequality was in 1981 by Aspect et al. (1981) on entangled polarization states of photons. Later in the same year Aspect et al. (1982) showed that the CHSH inequality is violated in an experiment in which the settings of the measurement devices, the orientations of the polarizers, change rapidly during the time the photons travel from the source to the polarizers, so that no information about a device setting is able to reach the source without travelling superluminally.

Shimony regards Bell's result as showing that an environmentally contextual hidden variable would have to be nonlocal to recover the correlations of entangled quantum states. Shimony remarks:

> If a contextual hidden variables theory of the environmental type had been viable, then a world view different from that of quantum mechanics, and yet different in some respects from that of classical physics, would have been suggested. In this world view a system $S$ has a definite complete state, and so does its environment. The outcome of an experiment performed upon $S$ is determined cooperatively by these two complete states. Hence this kind of theory does not have the fundamental stochastic character of quantum mechanics. ... Instead of entanglement, which is a kinematical feature of a composite system in quantum mechanics, there would be causal interaction between components, which is a dynamical matter. The outcome

of an experiment is a cooperative effect, just as it is conceived to be in classical physics. Classically, however, the system $S$ possesses a definite array of properties, and the interaction of $S$ with measuring apparatus produces an effect which is indexical of one or more of these properties—perfectly so if the apparatus is ideal, imperfectly with actual apparatus. A contextual hidden variables theory of the environmental type differs from classical physics by regarding as relational many features which in classical physics, and also in noncontextual hidden variables theories, are thought to be intrinsic to $S$.    (Shimony, 1984, p. 34, 35)

This is, in fact, the way Bohm's 1952 hidden variable theory recovers the quantum measurement statistics. In Bohm's theory, particles always have well-defined positions and momenta. The momentum of a Bohmian particle is the rate of change of its position, but the expectation value of momentum in a quantum ensemble is not derived by averaging over the particle momenta in an ensemble. Rather, the expectation value is derived via a theory of measurement, which yields the correct trace formula involving the momentum *operator* if we assume, as a contingent fact, that the probability distribution of hidden variables—particle positions—is an equilibrium distribution (in which case, the Bohmian equations of motion ensure that equilibrium is preserved). Bohm (1952, p. 387) gives an example of a free particle in a box of length $L$ with perfectly reflecting walls. Because the wave function is real, the particle is at rest. The kinetic energy of the particle is $E = P^2/2m = (nh/L)^2/2m$. How can a particle with high energy be at rest in the empty space of the box? Bohm's solution to the puzzle is that a measurement of the particle's momentum changes the wave function, which plays the role of a guiding field for the particle's motion, in such a way that the *measured* momentum values will be $\pm nh/L$ with equal probability. Bohm (1952, pp. 386–387) comments:

> This means that the measurement of an 'observable' is not really a measurement of any physical property belonging to the observed system alone. Instead, the value of an 'observable' measures only an incompletely predictable and controllable potentiality belonging just as much to the measuring apparatus as to the observed system itself.... We conclude then that this measurement of the momentum 'observable' leads to the same result as is predicted in the usual interpretation. However, the actual particle momentum existing before the measurement took place is quite different from the numerical value obtained for the momentum 'observable', which, in the usual interpretation, is called the 'momentum'.

In this respect, as indicated in Section 39.2.1, the relation between Bohm's theory and standard quantum mechanics is like the relation between Lorentz's ether theory and Einstein's special theory of relativity. Lorentz's theory explains the constancy of the velocity of light in all inertial frames as an artefact of a dynamical process in which rods contract as they move through the ether, while in Einstein's theory it is an assumption, ultimately explained by the geometry of Minkowski space (even though what we mean by space and time, and so what we mean by velocity, is rather different in relativity than in Newtonian mechanics). Bohm's theory explains the measurement statistics of electron spin as an artefact of the dynamics, while in standard quantum mechanics spin is a property of an electron that we measure in a spin measurement (even though what we mean by measurement is rather different in a noncommutative or non-Boolean theory like quantum mechanics from that in a classical or Boolean theory).

Note that Bohm's theory is contextual in both the algebraic and environmental sense, and also nonlocal. Consider a spin-1 particle and suppose $S_x^2, S_{x'}^2, S_y^2, S_{y'}^2, S_z^2$ represent the squared components of spin in the $x, x', y, y'$, and $z$ directions, respectively, where $x, y, z$ and $x', y', z$ form two orthogonal triples of directions with the $z$-direction in common. Each of these observables has eigenvalues 0 and 1 (taking units in which $\hbar = 1$ and a spin component has eigenvalues $-1, 0,$ and $+1$), corresponding respectively to a plane and a ray in a three-dimensional Hilbert space $\mathcal{H}_3$. The three 0-eigenrays of $S_x^2, S_y^2, S_z^2$ form an orthogonal triple in $\mathcal{H}_3$, and the three 0-eigenrays of $S_{x'}^2, S_{y'}^2, S_z^2$ form another orthogonal triple, with the 0-eigenray of $S_z^2$ in common. The observables $H = S_x^2 - S_y^2$ and $H' = S_{x'}^2 - S_{y'}^2$ are maximal and incompatible, each with three eigenvalues, $-1, 0,$ and $+1$, corresponding to the orthogonal triple of eigenrays defined by the 0-eigenrays of $S_x^2, S_y^2, S_z^2$, and to the orthogonal triple of eigenrays defined by the 0-eigenrays of $S_{x'}^2, S_{y'}^2, S_z^2$. The non-maximal observable $S_z^2$ can be represented as a function of $H$ and also as a function of $H'$ ($S_z^2 = H^2 = H'^2$) and so can be measured via a measurement of $H$ or via a measurement of $H'$. But, as shown by Pagonis and Clifton (1995) (see also Bub and Clifton (1996, pp. 208–211)), a measurement of $S_z^2$ via an $H$-measurement need not yield the same result as a measurement of $S_z^2$ via an $H'$-measurement, for the same initial position of the particle (the hidden variable) and the same quantum state, so Bohm's theory is contextual in the algebraic sense.

The theory is nonlocal because for two spin-1/2 particles, $S_1$ and $S_2$, initially in the singlet spin state, the outcome of a spin measurement on $S_1$ can depend on the type of spin measurement carried out on $S_2$, i.e., on the orientation of the Stern–Gerlach magnet at $S_2$. The quantum state of the composite system will evolve differently for a measurement of $x$-spin rather than $z$-spin at $S_2$ and so will affect the evolution of the position of $S_1$, the hidden variable, differently in a subsequent spin measurement at $S_1$, as in the measurement of $H$ and $H'$ on the spin-1 system. Since the outcome of a spin measurement can depend on further details of the measurement procedure, say on differences in the strengths of the magnetic field at the north and south poles of the Stern–Gerlach magnet, the theory is also contextual in the environmental sense.

## 39.3 STOCHASTIC HIDDEN VARIABLES

Bohm's 1952 hidden variable theory (Bohm, 1952) is a deterministic theory. Bohm and Vigier (1954) developed a stochastic version of the theory in which the Born distribution is produced asymptotically as an equilibrium distribution from an arbitrary probability density via stochastic field fluctuations that behave like random fluctuations in a fluid.

In the original derivation of Bell's inequality (Bell, 1964), the hidden variable is assumed to be deterministic. Bell discusses locality in the general context of stochastic hidden variables in his theory of 'local beables' (Bell, 1985), where 'the assignment of values to some beables $\Lambda$ implies, not necessarily a particular value, but a probability distribution, for another beable $A$'. The distinction he has in mind is, in the first place, between what he refers to as classically describable 'beables' like measurement settings and measurement outcomes, by contrast with quantum mechanical 'observables' represented by Hilbert space operators. In the second place, he refers to the distinction between physical quantities like the electric and magnetic fields in Maxwell's theory, which he regards as the 'beables' of the theory, and the scalar and vector potentials, which are 'non-physical' mathematical conveniences. A 'locally causal theory' is a theory of local beables in which correlations between space-like separated regions can only arise because of common causes, defined probabilistically via conditional statistical independence as in Section 39.2.3 above. So, as Bell shows in the paper, such theories satisfy the CHSH inequality (the version he derives is slightly different from the version in Section 39.2.3).

Following Bell (1987, pp. 76–177), Vink (1993) has formulated Bohm's 1952 theory as a stochastic theory. Bell constructs stochastic trajectories for fermion number density, regarded as a discrete beable for quantum field theory. The equation of motion generalizes the deterministic evolution of position in configuration space in Bohm's theory to the stochastic evolution of a discrete variable. Vink applies this idea to the stochastic evolution of a discrete variable representing the values of a maximal quantum observable $R$ for a system with an $N$-dimensional Hilbert space. The discontinuous jumps in the values $r_n$ of $R$ are governed by transition probabilities $T_{mn}dt$, where $T_{mn}dt$ denotes the probability of a jump from the value $r_n$ to $r_m$ in the time interval $dt$, giving rise to a time-dependent probability distribution of $R$-values. Vink shows that Bell's solution for the transition matrix in terms of the probability density and the probability current matrix leads to Bohm's deterministic theory in the continuum limit, when $R$ is position in configuration space.

For a single particle of mass $M$ on a one-dimensional lattice, $R = x = n\varepsilon$, $n = 1, 2 \cdots N$, where $\varepsilon$ is the distance between neighbouring lattice sites. The potential function $V(x)$ drops out of the expression for the current matrix $J_{mn}$ if $J_{mn}$ is a function of position only, so the Hamiltonian induces only nearest-neighbour transitions. As the distance between lattice sites tends to zero, the particle either remains at a given site or jumps to the next site in a particular direction. The probability of a jump depends on the absolute value of the gradient of the phase of the Schrödinger wave function at the lattice position, and the direction in which the particle jumps depends on whether the gradient of the phase is positive or negative at that lattice position. In other words, whether the particle jumps to the next lattice site, and which direction it jumps, depends stochastically only on where the particle is on the lattice through a feature of the wave function at the lattice position of the particle. As the distance between lattice sites tends to zero, the stochastic trajectories become smoother and smoother for nearest-neighbour transitions. Vink shows that the dispersion in the average displacement vanishes in the limit as $\varepsilon \to 0$, and so the discontinuous stochastic trajectories approach the continuous deterministic Bohmian trajectories in the continuum limit. Note that if we take $R$ as discretized momentum, $J_{mn}$ and hence $T_{mn}$ can be non-zero between arbitrarily distant lattice sites and will depend on the specific form of the potential function.

Vink shows that not only can any observable be elevated to the status of a beable in this sense, but *all* observables can be given this status (if it is assumed that all physical quantities are discrete at a fundamental level), i.e., we can regard all observables as simultaneously having definite values evolving in time. The price to pay is that these beables will not generally satisfy the functional relations of quantum mechanical observables for arbitrary functions, not even for beables associated with commuting observables, as Kochen and Specker showed. The Kochen–Specker functional relationship constraint is satisfied in Vink's theory for the values of measured observables, specifically for all observables that commute with a measured observable.

Any stochastic hidden variable theory can be extended to a deterministic theory (see, e.g., (Werner and Wolf, 2001, section 2.5)), so there would seem to be little to be gained by considering the more general notion with respect to the question of hidden variables in quantum mechanics. There is, nevertheless, an issue for which the distinction between deterministic and stochastic hidden variables is relevant, and the topic has been extensively discussed in the literature.

Bell's locality condition as conditional statistical independence is equivalent to the conjunction of two conditions, originally identified by Jarrett (1984) as 'completeness' and 'locality', but more commonly, following Shimony (1984), termed 'outcome independence' and 'parameter independence' (for a proof see (Bub, 1997) or (Howard, 1989, p. 231)). Outcome independence is the assumption that the outcome $a$ in (6) is independent of the remote outcome $b$ (and conversely for the outcome $b$ with respect to the outcome $a$). Parameter independence is the assumption that the outcome $a$ is independent of the remote choice of measurement setting or 'parameter' $B$ (and conversely for the outcome $b$ with respect to the measurement setting $A$). What motivates the distinction is that measurement outcomes are random and not under our control, while the choice of the measurement setting or what observable to measure *is* under our control (putting aside the possibility of superdeterminism or retrocausality). Similar conditions were introduced earlier by Suppes and Zanotti (Suppes and Zanotti, 1976), and by van Fraassen (van Fraassen, 1982). Van Fraassen's 'causality' condition is equivalent to outcome independence and his 'hidden locality' condition is equivalent to parameter independence.

From the failure of conditional statistical independence for (some) quantum correlations, it follows that either outcome independence or parameter independence fails, or both. If quantum correlations are outcome dependent but not parameter dependent, then we appear to have 'peaceful coexistence' between special relativity and quantum mechanics, as Shimony put it (Shimony 1984)—no instantaneous signalling or action at a distance from parameter dependence, but only what Shimony called 'passion at a distance' from outcome dependence. Note that in a deterministic hidden variable theory, where all the $\lambda$-probabilities are 0 or 1, outcome independence is automatically satisfied (because conditionalizing a probability that is either 0 or 1 leaves the probability unchanged), so this possibility does not arise.

Jarrett's completeness and locality conditions are not quite the same as Shimony's outcome independence and parameter independence. Jarrett explicitly considers the possible probabilistic dependence of measurement outcomes on causal factors that can be considered to be hidden variables of the measuring instruments $M_A$ and $M_B$, in addition to the measurement settings.

Jones and Clifton (Jones and Clifton, 1993) have shown that, under certain circumstances, violations of completeness can lead to superluminal signalling, no less than violations of locality. They note that the probability of $M_A$ producing a particular measurement outcome, for particular values of $\lambda$, measurement settings $A$ and $B$, and the hidden variables of $M_B$, might depend in part on the hidden variables of $M_A$ (and similarly for the probability of a particular outcome produced by $M_B$). It might also be the case that the probability of $M_A$ producing a particular measurement outcome, for particular values of $\lambda$, measurement settings $A$ and $B$, and the hidden variables of $M_A$, conditional on $M_B$ producing a particular measurement outcome, is independent of variations in the hidden variables of $M_B$ that leave the outcome produced by $M_B$ unchanged (and similarly for outcomes produced by $M_B$). They refer to the first condition as 'measurement contextualism' and the second as 'constrained locality' to distinguish this locality condition from Jarrett's locality condition, which requires measurement outcomes to be probabilistically independent of both the observable measured by a distant instrument and the hidden variables of the distant instrument.

It would then follow from measurement contextualism and constrained locality that manipulating the hidden variables of $M_B$ could affect certain measurement outcomes at $M_B$ probabilistically, and so, if completeness (outcome independence) fails, could influence the probabilities of measurement outcomes produced by $M_A$ (and similarly for manipulations of the hidden variables of $M_A$ and measurement outcomes produced by $M_B$). Constrained locality ensures that manipulating the hidden variables of the instrument $M_B$ as a stochastic 'trigger' for certain measurement outcomes produced by $M_B$ does not have a causal effect on the outcome produced by the instrument $M_A$, by some route other than the causal link between outcomes at $M_B$ and outcomes at $M_A$ associated with the failure of completeness, that cancels the effect via the outcome produced by $M_B$.

The Jones–Clifton argument shows that a failure of completeness or outcome independence under these circumstances would entail a failure of locality (because measurement outcomes at one measuring instrument would depend probabilistically on variations in the hidden variables of the distant instrument), and hence the possibility of superluminal signalling, extinguishing the hope that one could construct outcome-dependent stochastic hidden variable theories with no parameter dependence.

All actual examples of stochastic hidden variable theories involve a dependence on distant measurement procedures. A result by Colbeck and Renner (2011) shows that a stochastic hidden variable theory without dependence on distant measurement procedures is not possible, as in the deterministic case. Specifically, for two photons in an entangled state like $|\phi\rangle = \frac{1}{\sqrt{2}}|0\rangle|0\rangle + \frac{1}{\sqrt{2}}|1\rangle|1\rangle$, Colbeck and Renner show that there can't be a variable, $z$, associated with the history of the photons before the preparation of the entangled state in the reference frame of any inertial observer, that provides information about the outcomes of polarization measurements on the photons, so that the marginal probabilities, conditional on $z$, for measurement outcomes on the separate photons are closer to 1 than the probabilities of the entangled quantum state $|\phi\rangle$. In other words, the information in $|\phi\rangle$ about the probabilities of measurement outcomes is as complete as it could be. The same conclusion follows for any entangled state, not necessarily a maximally entangled state like $|\phi\rangle$ with equal coefficients for the product states in the sum, and the argument can be extended to any quantum state, for measurement outcome probabilities between 0 and 1, by exploiting the fact that a measurement interaction described by the dynamics of quantum mechanics leads to an entangled state of the measured system and the measuring instrument.

## 39.4 THE BOHM–BUB THEORY

The Bohm and Bub (1966) hidden variable theory (see also Bohm (1996)) developed as an extension and modification of the differential-space theory of Wiener and Siegel (1953, 1955); also Siegel and Wiener (1956) and Wiener *et al.* (1966, Chapter 5). For a detailed account of the relation between the two theories see Belinfante (1973).

Bohm and Bub add a nonlinear term to the linear unitary evolution of the quantum state $|\psi\rangle$ in an $n$-dimensional Hilbert space $\mathcal{H}_n$ described by the Schrödinger equation:

$$\frac{d|\psi\rangle}{dt} = -\frac{i}{\hbar}H|\psi\rangle + \gamma\sum_{j,k}p_k\left(\frac{p_j}{h_j} - \frac{p_k}{h_k}\right)P_j|\psi\rangle \tag{11}$$

The additional term involves a set of projection operators $\{P_j\}$ onto the $n$ eigenvectors $|a_j\rangle$ of some maximal (non-degenerate) observable $A$ in $\mathcal{H}_n$, as well as hidden variables represented by a unit vector $|\lambda\rangle$, which Bohm and Bub locate in a dual Hilbert space but could just as well belong to $\mathcal{H}_n$. The quantum state $|\psi\rangle$ is a linear superposition $\sum_j c_j|a_j\rangle$ with coefficients $c_j$ in the $A$-basis, and similarly $|\lambda\rangle$ is a linear superposition with coefficients $d_j$ in the same basis. The $p_j = |c_j|^2$ are the Born probabilities for the outcomes of an $A$-measurement, and $h_j = |d_j|^2$ represent the hidden variables.

The idea is that the set of projection operators $\{P_j\}$ that define the $A$-basis is determined by the large-scale environment of a system, in particular by what Bohr (1949) had in mind by 'the whole experimental arrangement' during what we call a 'measurement' in quantum mechanics. The coefficient $\gamma$ is assumed to be negligible between measurements, but large and positive during a measurement so that the nonlinear evolution dominates the unitary Schrödinger evolution. The effect is that the quantum state $|\psi\rangle$ undergoes a deterministic evolution during an $A$-measurement that projects or 'collapses' $|\psi\rangle$ onto one of the $n$ eigenstates $|a_j\rangle$ of $A$, depending on the value of the hidden variables represented by the vector $|\lambda\rangle$, which are assumed to remain essentially constant during a measurement process, however else they might change between measurements.

It is convenient to express the evolution in terms of the coefficients $c_j$ of $|\psi\rangle$ in the $A$-representation, which are functions of time and correspond to the Schrödinger wave function $\psi(x)$ in the position representation, where $x$ is position in configuration space. Writing $R_j = p_j/h_j$, the evolution is described by the equation

$$\frac{dc_j}{dt} = -\frac{i}{\hbar}\sum_k H_{jk}c_k + \gamma c_j \sum_j p_k(R_j - R_k), k = 1, 2, \cdots n \qquad (12)$$

The nonlinear term describes an evolution during which the probability $p_s$ associated with the largest ratio $p_s/h_s$ increases to 1 and the remaining $p_j$ decrease to 0, so $|\psi\rangle \rightarrow |a_s\rangle$.

To see this, consider the nonlinear evolution:

$$\frac{dc_j}{dt} = \gamma c_j \sum_k p_k(R_j - R_k), j = 1, 2, \cdots n \qquad (13)$$

It follows that $d(\sum_i p_j)/dt = 0$ because

$$\frac{dp_j}{dt} = 2\gamma p_j \sum_j p_k(R_j - R_k), j = 1, 2, \cdots n \qquad (14)$$

So the normalization of $|\psi\rangle$ is unchanged during the measurement process, even though the evolution is non-unitary. Also,

$$\frac{d(\ln p_j)}{dt} = 2\gamma \sum_k p_k(R_j - R_k), i = 1, 2, \cdots n \qquad (15)$$

If $R_s > R_j$, and $p_j \neq 0$ for $j \neq s$ (in which case $p_s = 1$), $d(\ln p_s)/dt$ will be positive and $d(\ln p_j)/dt$ negative, so the differences $R_s - R_j$ will increase for $j \neq s$ and $p_s$ will increase to 1 while $p_j$ will decrease to 0. In other words, the maximum $R_s$ will increase until $p_s = 1$ and $p_j = 0$ for $j \neq s$, which is to say $|\psi\rangle \rightarrow |a_s\rangle$ (except for a set of measure zero when there is no maximum), independently of other details in the initial conditions.

Assuming that $|\lambda\rangle$ is randomly distributed, integrating over a uniform distribution on the surface of the unit sphere yields the Born probability for the measure of hidden variable states for which $R_s > R_j$ and so for the transition $|\psi\rangle \rightarrow |a_s\rangle$. This is proved in

Bohm and Bub (1966) for a spin measurement in a two-dimensional Hilbert space. For a general proof in $\mathcal{H}_n$, see Bub (1969).

The Bohm–Bub theory is both contextual and nonlocal. If $R_1$ is the maximum ratio for a particular hidden variable vector $|\lambda\rangle$ and a basis defined by the projection operators $P_1, P_2, \cdots P_n$, it will not necessarily be the maximum ratio for a basis defined by the operators $P_1, P_2', \cdots P_n'$, where $P_j'$ do not commute with $P_j$ for $j \neq 1$. As formulated, the theory treats the measurement of a non-maximal observable as the measurement of some maximal observable of which it is a function, in conformity with Bohr's requirement to specify 'the whole experimental arrangement'. For example, in a three-dimensional Hilbert space, one could measure the non-maximal two-valued observable represented by the projection operator $P$ as a function of a maximal observable $X$ with eigenstates associated with the projection operators $P, Q, R$, or as a function of a maximal observable $Y$ with eigenstates associated with the projection operators $P, Q', R'$, where $Q'$ and $R'$ do not commute with $Q$ and $R$. For a given quantum state and a particular hidden variable vector $|\chi\rangle$, the outcome of a $P$-measurement will in general depend on the context, $X$ or $Y$, even though integrating over the hidden variables will yield the Born probabilities (which are independent of context) for the two possible outcomes.

If the change in context is nonlocal, the outcome of a measurement could depend on the nonlocal context for a particular $|\lambda\rangle$, while locality is recovered when averaging over $|\lambda\rangle$. In terms of Bell locality as conditional statistical independence, the probabilities of particular outcomes, $a$ of $A$ and $b$ of $B$ on separated systems, will not be independent of the observable measured on the distant system:

$$p(a,b|A,B,|\lambda\rangle) \neq p(a|A,|\lambda\rangle a)p(b|B,|\lambda\rangle) \qquad (16)$$

Since the outcome of an $A$-measurement for a specific value of the hidden variable is only specified in the theory with respect to a maximal observable $X$ of which $A$ and an observable $B$ of the distant system are functions, it will not in general be the case that $p(a|X,|\lambda\rangle) = p(a|Y,|\lambda\rangle)$, where $Y$ is another maximal observable of which $A$ and $B'$ are functions and $B'$ does not commute with $B$. So the probability $p(a|A,|\lambda\rangle)$ is not well-defined without specifying some observable of the distant system.

The above is a sketch of the theory. For further details, including how the theory might be elaborated for measurements of non-maximal observables, see Belinfante (1973, pp. 237–239), Tutsch (1969), and Christensen and Mattuck (1982). Tutsch (1970)—see also Gudder (1970)—proves under very general considerations that the Bohm–Bub equations are the simplest collapse equations for a hidden variable theory of this sort. See Tutsch (1968) on the collapse time.

Now notice that after a measurement of an observable $A$ that produces the transition $|\psi\rangle \rightarrow |a_s\rangle$, the hidden vectors $|\lambda\rangle$ will be in a region of the Hilbert space for which $R_s$ is a maximum, and a subsequent measurement of $A$ will produce the same outcome (because $R_j = 0$ for $j \neq s$). But a subsequent measurement of an observable $B$

incompatible with $A$ will further restrict the domain of $|\lambda\rangle$ to the intersection of the domains for the $A$-measurement and the $B$-measurement. This means that after a sequence of measurements integrating over the restricted region of $|\lambda\rangle$ in $\mathcal{H}_n$, the probabilities for a subsequent measurement will not be the Born probabilities. To avoid a contradiction with quantum mechanics, Bohm and Bub assume that the hidden variables undergo a stochastic process between measurements with a short relaxation time $\tau$ so that they rapidly randomize to the uniform distribution over the unit sphere. A sequence of measurements of incompatible observables associated with noncommuting operators, with the time interval between successive measurements shorter than $\tau$, should yield a statistics that differs from the quantum statistics. In this sense the Bohm–Bub hidden variable theory is different from Bohm's 1952 theory, which is empirically equivalent to quantum mechanics for the equilibrium distribution of hidden variables, which the equations of motion for the hidden variables maintain during and after measurements.

Papaliolios (1967) (see also Freedman *et al.* (1976)) performed an experiment in which photons were sent through a stack of three thin linear polarizers, with the transmission axes of the first two polarizers nearly crossed. Photons that get through the first two polarizers have a definite quantum state and values of $|\lambda\rangle$ in a restricted region of the Hilbert space. Rotating the third polarizer, Papaliolios found that transmission varied with the angle $\theta$ to within 1 per cent of the $\cos^2\theta$ prediction of quantum mechanics, for a transit time between the polarizers of $7.5 \times 10^{-14}$ seconds. So in this experiment the hidden variables would have to randomize in a time shorter than $10^{-14}$ seconds to produce the quantum statistics.

Bohm and Bub conjectured that $\tau \approx h/kT \approx 10^{-13}$ seconds at room temperature, but this conjecture was hardly more than a speculation, and they also suggested that $\tau$ might be related to the lifetime of a quantum state. Folman (1994) proposed that decay processes involving particles with mass might provide a more promising test of the theory than photons. Specifically, he looked at data for the decay of tau particles in high energy experiments at CERN. The tau particle is a negatively charged spin-1/2 particle with several decay modes and a mean lifetime of the order of $10^{-13}$ seconds. Folman's idea was to consider the branching of a decay into different modes at the decay vertex as analogous to the transition into one of two polarization states when a photon passes through a polarizer in the Papaliolios experiment. Intriguingly, he found that several experimental data sets showed deviations in the number of decay events over the expected distribution, as much as a 29% enhancement for short lifetime decays where one would expect unrelaxed hidden variables to play a role. Folman cautioned, however, that the decay distribution is constructed via a data analysis that involves factors such as calibration of the devices that produce and monitor tau decay, as well as procedures that discard certain measured events that might introduce a bias. So it is difficult to distinguish observed deviations that could be attributed to artefacts of the data analysis from deviations due to unrelaxed hidden variables.

# 39.5  WHY HIDDEN VARIABLES?

Bohm (1952, pp. 167–168) motivates the hidden variable theory with a critique of the Copenhagen interpretation, specifically Bohr's version of this interpretation:

> In formulating [the principle of complementarity], Bohr suggests that at the atomic level we must renounce our hitherto successful practice of conceiving of an individual system as a unified and precisely definable whole, all of whose aspects are, in a manner of speaking, simultaneously and unambiguously accessible to our conceptual gaze. Such a system of concepts, which is sometimes called a 'model', need not be restricted to pictures, but may also include, for example, mathematical concepts, as long as these are supposed to be in a precise (i.e., one-to-one) correspondence with the objects that are being described. The principle of complementarity requires us, however, to renounce even mathematical models. Thus, in Bohr's point of view, the wave function is in no sense a conceptual model of an individual system, since it is not in a precise (one-to-one) correspondence with the behaviour of this system, but only in a statistical correspondence.

Bohm and Bub (1966, pp. 456–457) begin their discussion of hidden variables with similar remarks:

> It may perhaps clarify the situation to point out here that, according to the usual interpretation of quantum mechanics, the process of collapse is equivalent to assuming that a given wave function is replaced by a second wave function after measurement, but for no assignable reason... Of course, this means the renunciation of a deterministic treatment of physical process, so that the statistics of quantum mechanics becomes *irreducible* (whereas in classical mechanics it is a simplification—in principle more detailed predictions are possible with more information).... At this point, the idea seems quite naturally to suggest itself that the differences in the second wave function (after measurement) could be referred to differences, before the measurement, in the values of some new kinds of variables, at present 'hidden' (but in principle ultimately observable with the aid of suitable new methods of observation [that] may later be suggested by thinking in terms of such variables).

The Copenhagen interpretation is a somewhat amorphous blend of various related views, often misrepresented in easily dismissible 'straw man' versions in the literature on the foundations of quantum mechanics, leaving the impression that we need to come up with a viable alternative to make sense of quantum mechanics. Here is what I think Bohr was getting at.

The essential step in going from classical to quantum mechanics is noncommutativity, or equivalently non-Booleanity for the idempotent elements, the projection operators. This is the salient feature of the theory as originally formulated by Heisenberg, Born, and Jordan, somewhat obscured by Schrödinger's version of quantum mechanics which makes the theory look like a new sort of wave theory. The non-

Boolean structure—the subspace structure of Hilbert space—is fundamentally a new sort of probability theory. As von Nemann put it, the probabilities are 'there from the start' (von Neumann, 2001, p. 245) and 'sui generis' (von Neumann, 1961). They are there from the start in the sense that they are part of the Hilbert space geometry, represented by the angles between different bases or Boolean 'frames', rather than something that is added afterwards, as probability is introduced in classical statistical mechanics as a measure over the Boolean algebra of subsets of the state space (where a state, a point $s$ in phase space, is a maximal specification of what is true and what is false, represented by the subsets containing $s$ and those not containing $s$). The non-Boolean probability theory allows new sorts of probabilistic correlations associated with nonlocal entangled quantum states, correlations without a common cause, as Bell showed, that are not possible in a commutative or Boolean theory.

In the case of relativity, we were wrong in thinking of space and time as a fixed Euclidean background with respect to which events happen and change, as Newton thought. In the case of quantum mechanics, we were wrong in thinking that probability ultimately represents a measure of ignorance about what is the case, i.e., ignorance about the pre-measurement values of observables or the properties of a system. What's really novel about quantum mechanics as a noncommutative or non-Boolean theory is not superposition and interference, which is also exhibited by waves in a fluid like water, but nonlocal entanglement.

The fundamental interpretation problem of quantum mechanics is how to understand these 'sui generis' structural probabilities. You can't assign truth or falsity to everything in a consistent way in this non-Boolean theory, so quantum probabilities don't represent ignorance about what is the case. Rather, the probabilities are relevant for a Boolean frame with respect to which it is possible to assign truth values consistently. The different possible truth value assignments to the elements in the frame refer to the different possibilities with respect to the frame. These are the possible outcomes of what we call a 'measurement' in quantum mechanics. The outcomes are intrinsically random, and the theory tells you what the probabilities for these outcomes are, given the outcome that occurred with respect to a prior, in general incompatible Boolean frame, referred to as 'preparing a quantum state'. If time passes between preparation and measurement, then the unitary dynamics evolves the whole Hilbert space, so all the Boolean frames, to provide temporally adjusted probabilities, taking account of whatever might have happened in the way of possible interactions in the time between.

Bohr did not use the term 'Boolean', but the concept is simply a precise way of codifying a significant aspect of what Bohr meant by a description 'in classical terms' in his constant insistence that (Bohr 1949, p. 209, his emphasis):

*however far the phenomena transcend the scope of classical physical explanation, the account of all evidence must be expressed in classical terms.*

By 'classical', as he goes on to explain, he meant 'unambiguous language with suitable application of the terminology of classical physics'—for the simple reason, as he put it, that we need to be able 'to tell others what we have done and what we have learned'. Formally speaking, the significance of 'classical' here as being able 'to tell others what we have done and what we have learned' is that the events in question should fit together as a Boolean algebra. The essential thing about 'classicality' for Bohr is its Booleanity, not the specific kinematic and dynamical quantities of classical physics. Boole (1998, p. 229) characterized this structure as capturing 'the conditions of possible experience'. As he put it: 'When satisfied they indicate that the data *may* have, when not satisfied they indicate that the data *cannot* have, resulted from actual observation'.

The measurement problem is how a pure quantum state 'collapses' or gets projected onto one of its eigenstates in a measurement, the eigenstate associated with the eigenvalue that is selected as the outcome of the measurement. There are two separate problems here. Since the eigenvalues and eigenstates depend on the basis in which the state is represented, the first problem is how a basis is selected, i.e., how one set of commuting observables rather than another gets assigned definite values, given that all observables can't be assigned values simultaneously. It is the problem of explaining how a particular Boolean algebra or 'frame' defined by an effectively classical probability space of macroscopic measurement outcomes can emerge in a quantum measurement process. This is the 'preferred basis' problem. The second problem is how or why an observable, once selected, gets to take on a definite value. Call this the 'definiteness' problem.

With respect to the preferred basis problem, the current view is that a particular basis is selected through the process of environmental decoherence. For macrosystems consisting of a great many ($\sim 10^{23}$) elementary quantum systems, decoherence explains why it is in principle possible but not remotely practicable to distinguish a pure state, expressed with respect to a particular basis, from a corresponding mixture over the elements of this basis. In this sense, decoherence explains why 'quantumness' gets washed out for macrosystems, and why a macroscopic arrangement of matter in a laboratory defines an effective Boolean frame associated with a particular basis. What quantum mechanics is all about is the probabilistic relationship between such Boolean frames.

All systems, including electrons and rocks, are quantum systems. We know how to implement a Boolean frame in which an electron behaves randomly, and we describe the behaviour of an electron with respect to different incompatible Boolean frames in terms of quantum mechanics. That's to say, we know how to get an electron to show its quantum properties, like interference or entanglement, which means exploiting the possibilities inherent in different Boolean frames for an electron. The difference between an electron and a rock is that we know how to do this for an electron, but we also know that, while it would be possible in principle to do this for a rock, such a feat would require a level of control over the environment that is not possible in any remotely practical sense.

On the Copenhagen interpretation, quantum mechanics as a noncommutative or non-Boolean theory is a new sort of *non-representational* theory for an indeterministic universe, in which the quantum state is a bookkeeping device for keeping track of probabilities and probabilistic correlations between intrinsically random events. Probabilities are defined with respect to a single Boolean frame, the Boolean algebra generated by the 'pointer-readings' of what Bohr (1939, pp. 23–24) referred to as the 'ultimate measuring instruments', which are 'kept outside the system subject to quantum mechanical treatment':

> In the system to which the quantum mechanical formalism is applied, it is of course possible to include any intermediate auxiliary agency employed in the measuring processes.... The only significant point is that in each case some ultimate measuring instruments, like the scales and clocks which determine the frame of space-time coordination—on which, in the last resort, even the definition of momentum and energy quantities rest—must always be described entirely on classical lines, and consequently be kept outside the system subject to quantum mechanical treatment.

Any system, of any complexity, is fundamentally a non-Boolean quantum system and can be treated as such, in principle, which is to say that a unitary dynamical analysis can be applied to whatever level of precision you like. But we have evolved to experience the occurrence of events with respect to a Boolean algebra of possibilities. At the end of a chain of instruments and recording devices, some particular system, $M$, functions as the 'ultimate measuring instrument' that defines an associated Boolean algebra, with respect to which an event corresponding to a definite measurement outcome occurs. The system $M$, or any part of $M$, can be treated quantum mechanically, but then some other system, $M'$, treated as classical or commutative or Boolean, plays the role of the ultimate measuring instrument in any application of the theory. The outcome of a measurement is an intrinsically random event at the macrolevel, *something that actually happens*, not described by the deterministic unitary dynamics, so outside the theory. Putting it differently, the 'collapse' is something you put in by hand after recording the actual outcome. The physics doesn't give it to you.

As Pauli (1994, p. 46) put it, somewhat dramatically:

> Observation thereby takes on the character of *irrational, unique actuality* with unpredictable outcome.... Contrasted with this *irrational* aspect of concrete phenomena which are determined in their *actuality*, there stands the *rational* aspect of an abstract ordering of the possibilities of statements by means of the mathematical concept of probability and the $\psi$-function.

Noncomutativity or non-Booleanity makes quantum mechanics quite unlike any theory we have dealt with before, and there is no reason, apart from tradition, to assume that a non-Boolean theory should provide the sort of explanation we are familiar with in a theory that is commutative or Boolean at the fundamental level. A representational theory proposes a primitive ontology, perhaps of particles or fields of a certain sort, and a dynamics that describes change over time. The non-Boolean physics of quantum mechanics does not provide a representational explanation of phenomena. If a 'world' in which truth and falsity, 'this' rather than 'that', makes sense is Boolean, then there is no quantum 'world' and it would be misleading to attempt to picture it, as Petersen (1963, p. 12) reports Bohr to have said:

> When asked whether the algorithm of quantum mechanics could be considered as somehow mirroring an underlying quantum world, Bohr would answer, 'There is no quantum world. There is only an abstract quantum physical description. It is wrong to think that the task of physics is to find out how nature *is*. Physics concerns what we can *say* about nature.'

In the critical comments about the Copenhagen interpretation quoted above by Bohm, and by Bohm and Bub, the sticking point in both cases is the supposed 'irreducibility' of the quantum statistics, the intrinsic randomness of measurement outcomes in quantum mechanics. What does a hidden variable programme offer in place of the Copenhagen interpretation with respect to this issue?

In Bohm's 1952 hidden variable theory, there is a fixed Boolean frame, the frame defined by position in configuration space. The point $s$ in configuration space representing the hidden variables moves under the influence of a guiding wave. A measurement of a quantum observable is understood as a particular wave motion in which the wave evolves into a superposition of relatively localized peaks in configuration space associated with different values of the observable. The outcome of the measurement is determined by the peak containing $s$. This is the explanation in the theory of how an observable gets to take on a definite value. In the Bohm–Bub theory, a maximal basis defined by a set of projection operators is selected by the large-scale environment in some way that is not further specified. The equations of motion explain how an observable associated with this basis takes on a definite value. The result in both cases is a theory that is contextual and nonlocal at the level of the hidden variables, and the impossibility proofs show that this is a general feature of hidden variable theories.

Heisenberg, Born, and Jordan gave us a non-Boolean conceptual framework for irreducibly statistical events that allows the possibility of nonlocal probabilistic correlations without a common cause, i.e., without hidden variables. There is no answer to the definiteness question in such a non-Boolean theory. Insisting on an answer means rejecting the possibility of a world in which events are intrinsically random, or simply insisting on a preference for a deterministic theory, or a statistical theory in which probabilities are introduced for practical reasons, as in classical statistical mechanics, with some sort of restriction on our ability to reduce the probabilities by getting to the underlying real states.

Quantum mechanics is nonlocal in the sense that there are probabilistic correlations produced by nonlocal entangled states that violate Bell's inequality. But any hidden variable theory that mimics this feature of quantum mechanics would have to introduce what Einstein referred to as 'spooky actions at a distance' (Born, 1971, p. 158), 'an explicit causal mechanism... whereby the disposition of one piece of apparatus affects the results obtained with a distant piece', as Bell put it. Considering stochastic hidden variables does not change this conclusion, as indicated by the results of Jones and Clifton (Jones and Clifton, 1993) and Colbeck and Renner (Colbeck and Renner, 2011). What the non-Booleanity of quantum mechanics achieves, through the irreducible structural probabilities that are 'there from the start' in the geometry of Hilbert space, is the possibility of new sorts of nonlocal probabilistic correlations that are impossible in a classical or Boolean theory, and it does so without introducing unphysical 'spooky actions at a distance'. The price for extending quantum mechanics with hidden variables to resolve the definiteness problem is a physics that requires instantaneous action at a distance at the fundamental level.

## ACKNOWLEDGEMENTS

Thanks to Guido Bacciagaluppi for a very thorough review of this chapter, and some detailed suggestions incorporated in the final version. Thanks also to Michael Dascal for many discussions on these topics and for suggesting the term 'frame' introduced in Section 39.5.

## REFERENCES

Aspect, A., Dalibard, J., and Roger, G. (1982). Experimental test of Bell's inequalities using time varying analyzers. *Physical Review Letters*, **49**, 1804–1807.

Aspect, A., Grangier, P., and Roger, G. (1981). Experimental tests of realistic local theories via Bell's theorem. *Physical Review Letters*, **47**, 460–467.

Bacciagaluppi, G., and Valentini, A. (2009). *Quantum Theory at the Crossroads: Reconsidering the 1927 Solvay Conference*. Cambridge: Cambridge University Press.

Belinfante, F. J. (1973). *A Survey of Hidden-Variables Theories*. New York: Pergamon Press.

Bell, J. S. (1964). On the Einstein–Podolsky–Rosen paradox. *Physics*, **1**, 195–200. Reprinted in J. S. Bell, *Speakable and Unspeakable in Quantum Mechanics*, Cambridge: Cambridge University Press (1987).

Bell, J. S. (1966). On the problem of hidden variables in quantum mechanics. *Reviews in Modern Physics*, **38**, 447–452. Reprinted in J. S. Bell, *Speakable and Unspeakable in Quantum Mechanics*, Cambridge: Cambridge University Press (1987).

Bell, J. S. (1985). The theory of local beables. *Dialectica*, **39**, 86–96. Reprinted in J. S. Bell, *Speakable and Unspeakable in Quantum Mechanics*, Cambridge: Cambridge University Press (1987).

Bell, J. S. (1987). Beables for quantum field theory. In J. S. Bell, Speakable and Unspeakable in Quantum Mechanics, Cambridge: Cambridge University Press, pp. 173–180.

Bell, J. S. (1988). Interview with John Bell, by Robert Crease and Charles Mann. *Omni*, May, 85.

Belousek, D. W. (1996). Einstein's 1927 unpublished hidden-variable theory: its background, context and significance. *Studies in History and Philosophy of Modern Physics*, **27**, 437–461.

Bohm, D. (1952). A suggested interpretation of quantum theory in terms of 'hidden' variables. I and II. *Physical Review*, **85**, 166–193.

Bohm, D. (1996). On the role of hidden variables in the fundamental structure of physics. *Foundations of Physics*, **26**, 719–786.

Bohm, D., and Bub, J. (1966). A proposed solution of the measurement problem in quantum mechanics by a hidden variable theory. *Reviews of Modern Physics*, **28**, 453–469.

Bohm, D., and Vigier, J. (1954). Model of the causal interpretation of quantum theory in terms of a fluid with irregular fluctuations. *Physical Review*, **96**, 208–216.

Bohr, N. (1939). The causality problem in atomic physics. In *New Theories in Physics*, Warsaw: International Institute of Intellectual Cooperation, pp. 11–30.

Bohr, N. (1949). Discussion with Einstein on epistemological problems in modern physics. In P. A. Schilpp (ed.), *Albert Einstein: Philosopher-Scientist*, Vol. VII, La Salle, IL: The Library of Living Philosophers, pp. 201–241.

Boole, G. (1998). On the theory of probabilities. *Proceedings of the Royal Society of London*, **12**, 225–252.

Born, M. (1926). Zur Quantenmechanik der Stossvorgänge. *Zeitschrift für Physik*, **37**, 863–867. Translated as 'On the quantum mechanics of collisions,' in J. A. Wheeler and W. Zurek (eds), *Quantum Theory and Measurement*, Princeton, NJ: Princeton University Press (1983), pp. 52–55.

Born, M. (1971). *The Born–Einstein Letters*. London: Walker and Co..

Brown, H. (1992). Bell's other theorem and its connection with locality, Part I. In A. van der Merwe, F. Selleri, and G. Tarozzi (eds), *Bell's Theorem and the Foundations of Modern Physics*, Singapore: World Scientific, pp. 104–116.

Bub, J. (1969). What is a hidden variable theory of quantum phenomena? *International Journal of Theoretical Physics*, **2**, 101–123.

Bub, J. (1996). Schütte's tautology and the Kochen–Specker theorem. *Foundations of Physics*, **26**, 787–806.

Bub, J. (1997). *Interpreting the Quantum World*. Cambridge: Cambridge University Press.

Bub, J. (2010). Von Neumann's 'no hidden variables' proof: a re-appraisal. *Foundations of Physics*, **40**, 1333–1340.

Bub, J., and Clifton, R. (1996). A uniqueness theorem for 'no collapse' interpretations of quantum mechanics. *Studies in History and Philosophy of Modern Physics*, **27**, 181–219.

Christensen, J. P., and Mattuck, R. D. (1982). Partial measurement in the Bohm–Bub hidden-variable theory. *Foundations of Physics*, **12**, 347–361.

Clauser, J., Horne, M., Shimony, A., and Holt, R. (1969). Proposed experiment to test hidden variable theories. *Physical Review Letters*, **23**, 880–883.

Colbeck, R., and Renner, R. (2011). No extension of quantum theory can have improved predictive power. *Nature Communications*, **2**, 411–415.

de Broglie, L. (1926). Sur la possibilité de relier les phénomènes d'interférence et de diffraction à la théorie des quanta de lumière. *Comptes Rendus Hebdomadaires des Séances de l'Académie des Sciences*, **183**, 447.

de Broglie, L. (1927a). La structure atomique de la matière et du rayonnement et la mècanique ondulatoire. *Comptes Rendus Hebdomadaires des Séances de l'Académie des Sciences*, **184**, 273.

de Broglie, L. (1927b). Sur le rôle des ondes continues en mécanique ondulatoire. *Comptes Rendus Hebdomadaires des Séances de l'Académie des Sciences*, **185**, 380.

Dieks, D. (2017). Von Neumann's impossibility proof: mathematics in the service of rhetorics. *Studies in History and Philosophy of Modern Physics*, **60**, 136–148.

Folman, R. (1994). A search for hidden variables in the domain of high energy physics. *Foundations of Physics Letters*, **7**, 191–200.

Freedman, S. J., and Clauser, J. F. (1972). Experimental test of local hidden-variable theories. *Physical Review Letters*, **28**, 938–941.

Freedman, S. J., Holt, R. A., and Papaliolios, C. (1976). Experimental status of hidden variable theories. In M. Flato, Z. Maric, A. Milojevic, D. Sternheimer, and J.P. Vigier (eds), *Quantum Mechanics, Determinism, Causality, and Particles*, Dordrecht: Reidel, pp. 1–10.

Gleason, A. (1957). Measures on the closed sub-spaces of Hilbert spaces. *Journal of Mathematics and Mechanics*, **6**, 885–893.

Greenberger, D., Horne, M., and Zeilinger, A. (1989). Going beyond Bell's theorem. In M. Kafatos (ed.), *Bell's Theorem, Quantum Theory, and Conceptions of the Universe*, Dordrecht: Kluwer Academic, pp. 69–72.

Gudder, S. P. (1970). On hidden-variable theories. *Journal of Mathematical Physics*, **11**, 431–436.

Hardy, L. (1992). Quantum mechanics, local realistic theories, and Lorentz-invariant realistic theories. *Physical Review Letters*, **68**, 2981–2984.

Hardy, L. (1993). Nonlocality for two particles without inequalities for almost all entangled states. *Physical Review Letters*, **71**, 1665–1668.

Hermann, G. (1935). Die naturphilosophischen Grundlagen der Quantenmechanik. *Abhandlungen der Fries'schen Schule*, **6**, 75–152.

Heywood, P., and Redhead, M. (1983). Nonlocality and the Kochen–Specker paradox. *Foundations of Physics*, **13**, 481–499.

Howard, D. (1989). Holism, separability, and the metaphysical implications of Bell's inequality. In J. Cushing and E. McMullin (eds), Philosophical Consequences of Quantum Theory: Reflections on Bell's Inequality, Notre Dame: University of Notre Dame Press, pp. 224–253.

Jarrett, J. (1984). On the physical significance of the locality conditions in the Bell arguments. *Noûs*, **18**, 569–589.

Jones, M., and Clifton, R. (1993). Against experimental metaphysics. In P. French, J. T. E. Euling, and H. Wettstein (eds), *Mid-West Studies in Philosophy, vol. XVII: Philosophy of Science*, Notre Dame: University of Notre Dame Press, pp. 297–316.

Kochen, S., and Specker, E. (1967). On the problem of hidden variables in quantum mechanics. *Journal of Mathematics and Mechanics*, **17**, 59–87.

Mermin, N. D., and Schack, R. (2018). Homer nodded: von Neumann's surprising oversight. *Foundations of Physics*, **48**, 1007–1020.

Pagonis, C., and Clifton, R. (1995). Unremarkable contextualism: dispositions in the Bohm theory. *Foundations of Physics*, **25**, 281–296.

Papaliolios, C. (1967). Experimental test of a hidden-variable quantum theory. *Physical Review Letters*, **18**, 622–625.

Pauli, W. (1994). Probability and physics. In *Wolfgang Pauli: Writings on Physics and Philosophy*, Berlin: Springer.

Penrose, R. (1994). On Bell non-locality without probabilities: some curious geometry. In *Quantum Reflections (in honour of J. S. Bell)*, Cambridge: Cambridge University Press.

Peres, A. (1991). Two simple proofs of the Kochen–Specker theorem. *Journal of Physics A: Mathematical and General*, **24**, L175–L178.

Peres, A. (1993). *Quantum Theory: Concepts and Methods*. Dordrecht: Kluwer.

Petersen, A. (1963). The philosophy of Niels Bohr. *Bulletin of the Atomic Scientists*, **19**, 8–14.

Schrödinger, E. (1926). Uber das Verhältnis der Heisenberg–Born–Jordanschen Quantenmechanik zu der meinem. *Annalen der Physik*, **79**, 734–756. English translation 'On the Relation between the Quantum Mechanics of Heisenberg, Born, and Jordan, and that of Schrödinger', *Collected Papers on Wave Mechanics*, New York: Chelsea Publishing Company (2003), pp. 45–61.

Shimony, A. (1984). Contextual hidden variables and Bell's inequalities. *British Journal for the Philosophy of Science*, **35**, 24–45.

Siegel, A., and Wiener, E. (1956). 'Theory of measurement' in differential space quantum mechanics. *Physical Review*, **101**, 429–432.

Specker, E. (1960). Die Logik nicht gleichzeitig entscheidbarer Aussagen. *Dialectica*, **14**, 239–246.

Suppes, P., and Zanotti, M. (1976). On the determinism of hidden variable theories with strict correlation and conditional statistical independence of observables. In P. Suppes (ed.), Logic and Probability in Quantum Mechanics, Dordrecht: Reidel, pp. 445–455.

Tutsch, J. H. (1968). Collapse time for the Bohm–Bub hidden variable theory. *Reviews of Modern Physics*, **40**, 232–234.

Tutsch, J. H. (1969). Simultaneous measurement in the Bohm–Bub hidden-variable theory. *Physical Review*, **183**, 1116–1131.

Tutsch, J. H. (1970). Mathematics of the measurement problem in quantum mechanics. *Journal of Mathematical Physics*, **12**, 1711–1718.

van Fraassen, B. (1982). The Charybdis of realism: Epistemological implications of Bell's inequality. *Synthese*, **52**, 25–38.

Vink, J. C. (1993). Quantum mechanics in terms of discrete beables. *Physical Review A*, **48**, 1808–1818.

von Neumann, J. (1927). Wahrscheinlichkeitstheoretischer Aufbau der Quantenmechanik. *Nachrichten von der Gesellschaft der Wissenschaften zu Göttingen, Mathematisch-Physikalische Klasse*, pp. 245–272.

von Neumann, J. (1955). *Mathematical foundations of quantum mechanics*. Princeton, NJ: Princeton University Press. First published as *Mathematische Grundlagen der Quantenmechanik*, Berlin: Springer (1932).

von Neumann, J. (1961). Quantum logics: strict- and probability-logics. In A. Taub (ed.), *Collected Works of John von Neumann*, Oxford and New York: Pergamon Press, pp. 195–197. A 1937 unfinished manuscript.

von Neumann, J. (2001). Unsolved problems in mathematics. In M. Rédei and M. Stöltzner (eds), *John von Neumann and the Foundations of Quantum Physics*, Dordrecht: Kluwer Academic Publishers, pp. 231–245. An address to the International Mathematical Congress, Amsterdam, 2 September 1954.

Werner, R. F., and Wolf, M. M. (2001). Bell inequalities and entanglement. *Quantum Information and Computation*, **1–25**, 517–540.

Wiener, N., and Siegel, A. (1953). A new form for the statistical postulate of quantum mechanics. *Physical Review*, **9**, 1551–1560.

Wiener, N., and Siegel, A. (1955). The differential-space theory of quantum systems. *Nuovo Cimento*, **2**, 982–1003.

Wiener, N., Siegel, A., Rankin, B., and Martin, W. T. (1966). *Differential Space, Quantum Systems, and Prediction*. Boston, MA: MIT Press.

CHAPTER 40

..........................................................................................................................

# PURE WAVE MECHANICS, RELATIVE STATES, AND MANY WORLDS

..........................................................................................................................

JEFFREY A. BARRETT

## 40.1 INTRODUCTION

..........................................................................................................................

HUGH Everett III presented *pure wave mechanics* as a solution to the quantum measurement problem. His interpretation of pure wave mechanics is the *relative-state formulation of quantum mechanics*. Subsequent commentators have, at least somewhat misleadingly, referred to Everett's formulation of quantum mechanics as the *many-worlds interpretation*.

Everett developed his formulation of quantum mechanics while a graduate student at Princeton University. We have two versions of his PhD thesis. He wrote the long version for his advisor John Wheeler in January 1956. This version was ultimately entitled 'Wave Mechanics without Probability'. Niels Bohr and his students took Everett's work to be critical of Bohr's Copenhagen interpretation. Since Wheeler did not want Everett to incur Bohr's disapproval, especially on the topic of quantum mechanics, he insisted that the thesis be revised. Wheeler worked with Everett to shorten it and make it less threatening to Bohr's views. The result was a version of the thesis approximately one third as long. The short version of the thesis was accepted in March 1957 with the title 'On the Foundations of Quantum Mechanics'. A nearly identical paper was published in *Reviews of Modern Physics* in July 1957 as '"Relative State" Formulation of Quantum Mechanics' (Everett, 1957a). Everett later edited the long version of his thesis for publication in the Bryce S. DeWitt and Neill Graham (1973) volume as 'The Theory of the Universal Wave Function'.[1]

While Wheeler was deeply concerned about Bohr's reaction, Everett did not think of his work as being closely related to the Copenhagen interpretation. Rather, as he explained in a 31 May 1957 letter to Aage Petersen (Everett, 1957b), one of Bohr's colleagues whom Everett had met at Princeton, Everett took the target of his criticism to be the standard von Neumann formulation of quantum mechanics. This is not because he thought the Copenhagen interpretation was better. Indeed, as he explained to Petersen, Everett much preferred the von Neumann formulation to Bohr's approach:

> [T]he particular difficulties with quantum mechanics that are discussed in my paper have mostly to do with the more common (at least in this country) form of quantum theory, as expressed for example by von Neumann, and not so much with the Bohr (Copenhagen) interpretation. The Bohr interpretation is to me even more unsatisfactory, and on quite different grounds. Primarily my main objections are the complete reliance on classical physics from the outset (which precludes *even in principle* any deduction at all of classical physics from quantum mechanics, as well as any adequate study of measuring processes), and the strange duality of adhering to a 'reality' concept for macroscopic physics and denying the same for the microcosm.   (Barrett and Byrne, 2012, p. 239)

At least von Neumann's formulation of quantum mechanics made a passing attempt to characterize the behaviour of all physical systems.

We will consider how Everett understood the quantum measurement problem then consider his proposal for solving it. Along the way, we will distinguish between pure wave mechanics, the relative-state formulation of quantum mechanics, and the many-worlds approach more generally. We will pay particular attention to the sense in which Everett believed that one can derive the standard statistical predictions of quantum mechanics as 'subjective appearances' from pure wave mechanics. To this end, we will consider the role that his internal model of observation, his distinction between total and relative states, and his choice of a special typicality measure play in the argument.

In brief, Everett's strategy was to characterize a sense in which the relative-state formulation of pure wave mechanics explains the standard quantum statistics for the relative measurement records of a *typical modelled observer*. Since pure wave mechanics is a deterministic theory that says nothing whatsoever about observers, relative records, typicality, or probability, he needed to add an assortment of definitions and auxiliary assumptions to the theory to get the subjective appearance of quantum probabilities. We will consider precisely what probabilities as subjective appearances are in Everett's formulation of quantum mechanics and how he derived them.

---

[1] Everett's revisions for the DeWitt and Graham (1973) volume moved the long thesis in the direction of presenting his arguments in a less tentative form. See Byrne (2010) and Barrett and Byrne (2012, p. 4) for more details regarding the history of Everett's thesis. See Osnaghi, Freitas, and Freire (2009) and Byrne (2010) for histories of Everett's thought and influence more generally.

## 40.2 THE MEASUREMENT PROBLEM

Everett used von Neumann's (1955) formulation of quantum mechanics to characterize the quantum measurement problem and his solution. He understood the standard
formulation of quantum mechanics to be the following:[2]

1. Representation of States: The state of a physical system $S$ is represented by a
   vector $|\psi\rangle_S$ of unit length in a Hilbert space $H$.
2. Representation of Observables: A physical observable $O$ is represented by a
   complete set of orthogonal eigenvectors $O$. These vectors represent the eigenstates of the observable, each corresponding to a different value.
3. Dynamical Laws:
   I. Process 1 (nonlinear collapse dynamics): If a *measurement* is made, the
      system $S$ randomly, instantaneously, and nonlinearly jumps to an eigenstate
      of the observable being measured: the probability of jumping to $|\phi\rangle_S$ when $O$
      is measured is $|\langle\psi|\phi\rangle|^2$.
   II. Process 2 (linear dynamics): If *no measurement* is made, the system $S$ evolves
      in a deterministic linear way: $|\psi(t_1)\rangle_S = \hat{U}(t_0, t_1)|\psi(t_0)\rangle_S$.

In addition to these principles, Everett began by assuming the standard interpretation
of states: a system $S$ has a definite value for observable $O$ if and only if $|\psi\rangle_S$ is an
eigenstate of $O$, and the value is given by the eigenvalue corresponding to $|\psi\rangle_S$. He used
an example of a cannonball in a superposition of positions to explain the standard
interpretation.

> Suppose... that we coupled a spin measuring device to a cannonball, so that if the
> spin is up the cannonball will be shifted one foot to the left, while if the spin is down it
> will be shifted an equal distance to the right. If we now perform a measurement with
> this arrangement upon a particle whose spin is a superposition of up and down, then
> the resulting total state will also be a superposition of two states, one in which the
> cannonball is to the left, and one in which it is to the right. There is no definite
> position for our macroscopic cannonball!   (Barrett and Byrne, 2012, p. 117)

Everett noted that this behaviour seems to be at variance with our observations since
macroscopic objects always appear to us to have definite positions. The standard
formulation of quantum mechanics explains the results of our observations by means
of the nonlinear collapse dynamics (Process 1). Specifically, a cannonball that begins in
a superposition of different locations will instantaneously and randomly jump to a state
where it has a perfectly determinate position whenever one looks for it. But Everett
took this explanation to be entirely unacceptable. As he characterized the quantum
measurement problem, '[t]he question of the consistency of the scheme arises if one

---

[2] See Barrett and Byrne (2012, pp. 73 and 175).

contemplates regarding the observer and his object-system as a single (composite) physical system. Indeed, the situation becomes quite paradoxical if we allow for the existence of more than one observer' (Barrett and Byrne, 2012, p. 73). Ultimately, he took the standard theory to be logically inconsistent.

At root, Everett took the quantum measurement problem to be a problem in consistently describing *nested measurements* in the theory. He used what he called an 'amusing, but extremely hypothetical drama' to explain the the problem. Everett's hypothetical drama was perhaps the earliest description of what is now known as the Wigner's Friend story.[3]

Everett describes a room containing an observer $A$ who measures (say) the $x$-spin of a spin-1/2 system $S$ that begins in the state

$$\alpha|\!\uparrow_x\rangle_S + \beta|\!\downarrow_x\rangle_S \tag{1}$$

then records his result in a notebook. Let the state $|\,\text{``ready''}\,\rangle_A$ represent the state where observer $A$ is ready to observe the $x$-spin of $S$ and record the result in his notebook.

Being an orthodox quantum theorist, observer $A$ predicts that the result of his measurement of $S$ will be described by Process 1 and, hence, result in either

$$|\,\text{``}\!\uparrow_x\!\text{''}\,\rangle_A|\!\uparrow_x\rangle_S \tag{2}$$

where $A$ has recorded the result "$\uparrow_x$" in the notebook or

$$|\,\text{``}\!\downarrow_x\!\text{''}\,\rangle_A|\!\downarrow_x\rangle_S \tag{3}$$

where $A$ has recorded the result "$\downarrow_x$" in the notebook, with probabilities $|\alpha|^2$ and $|\beta|^2$ respectively.

In the meantime, in Everett's story, there is an equally orthodox observer $B$ outside the room, who is in possession of the initial state of the composite system consisting of $A$ and $S$ just prior to $A$'s measurement of $S$:

$$|\,\text{``ready''}\,\rangle_A(\alpha|\!\uparrow_x\rangle_S + \beta|\!\downarrow_x\rangle_S). \tag{4}$$

Assuming that $A$ is a good observer, and hence would record "$\uparrow_x$" if the initial state of $S$ were $\uparrow_x$ and would record "$\downarrow_x$" if the initial state of $S$ were $\downarrow_x$, observer $B$ calculates, by the linearity of Process 2, that the state of $A$ and $S$ after $A$'s interaction with $S$ will be

$$\alpha|\,\text{``}\!\uparrow_x\!\text{''}\,\rangle_A|\!\uparrow_x\rangle_S + \beta|\,\text{``}\!\downarrow_x\!\text{''}\,\rangle_A|\!\downarrow_x\rangle_S. \tag{5}$$

---

[3] Eugene Wigner (1961) later retold this story without attribution to Everett. Everett may have encountered the setup in a course he took from Wigner at Princeton in the fall of 1954. See Barrett and Byrne (2012, pp. 12–14 and the footnote on p. 29).

After all, from $B$'s perspective, the interaction between $A$ and $S$ is a physical interaction between two physical systems just like any other.

The problem with all this, Everett argues, is that observer $A$'s description of the result of the interaction between $A$ and $S$, indicating that the result state must be given by either expression (2) or (3), is flatly incompatible with observer $B$'s description of the result of the interaction between $A$ and $S$, indicating that the result must be given by expression (5). Observer $B$ believes that observer $A$'s calculation must be incorrect. On the other hand, supposing that $B$ is right to calculate the resultant state using the linear Process 2, observer $A$ challenges $B$ to explain how either of them ever get determinate measurement outcomes at all. Given the manifest inconsistency of the state attributions of the two orthodox observers after the interaction between $A$ and $S$, Everett concludes: 'It is now clear that the interpretation of quantum mechanics with which we began is untenable if we are to consider a universe containing more than one observer. We must therefore seek a suitable modification of this scheme, or an entirely different system of interpretation' (Barrett and Byrne, 2012, pp. 74–5).

The reason Everett characterized the story as 'extremely hypothetical' is because it would be virtually impossible to isolate and control the state of a real observer $A$ with sufficient precision that an observer $B$ might calculate the composite evolution of system $A$ and $S$. And it would be yet more difficult for $B$ to perform a measurement that would empirically distinguish between the state that he calculates (given by (5)) and one of the states calculated by $A$ (given by (2) or (3)) but Everett took such interference measurements to always be in principle possible.[4] Observer $B$ might for example distinguish between these states by measuring an observable $Q$ of the composite system consisting of $A$ and $S$ that has state (5) as an eigenstate corresponding to eigenvalue $+1$ and

$$\alpha|\text{``}\uparrow_x\text{''}\rangle_A|\uparrow_x\rangle_S - \beta|\text{``}\downarrow_x\text{''}\rangle_A|\downarrow_x\rangle_S. \tag{6}$$

as an eigenstate corresponding to eigenvalue $-1$. If $Q$ is measured when the composite system is in state (5), $B$ would certainly get the result $+1$. But if $Q$ is measured when the composite system is in state (2) or state (3), then $B$ would sometimes get the result $+1$ and sometimes get the result $-1$. The upshot is that state (5) is not only logically incompatible with states (2) and (3), it is in principle *empirically distinguishable* from each of these states.[5]

The practical impossibility of performing such an experiment was irrelevant to Everett. What mattered to him was that the standard formulation of quantum mechanics entails a logical inconsistency if one supposes that observers are physical systems

---

[4] See for example the long quotation in the last section of this article and the discussion in Barrett and Byrne (2012, p. 150) more generally.

[5] Wigner (1961) explicitly recognized and discussed the possibility of $Q$-type measurements. In contrast with Everett, Wigner thought that such measurements would show that observer had caused a collapse.

like any other, and he took there to be no good reason to suppose otherwise. Everett held that one only has a satisfactory formulation of quantum mechanics if one can consistently assign states in the context of his hypothetical drama, providing a consistent and compelling account of nested measurements like the sequence ending with the measurement of $Q$ above and explaining the subjective experiences of the two observers.

For his part, Everett believed that $B$'s calculations were right. Specifically, he held that, contrary to what $A$ believes, the composite system of $A$ and $S$ ends up in the state:

$$\alpha | \text{“}\uparrow_x\text{”} \rangle_A |\uparrow_x\rangle_S + \beta | \text{“}\downarrow_x\text{”} \rangle_A |\downarrow_x\rangle_S. \tag{7}$$

As a result, he was committed to the conclusion that $B$ would with certainly get the result $+1$ if he were ever able to measure the observable $Q$. The task that Everett set for himself, then, was to explain why it *seems* to $A$ that he gets a determinate, repeatable, intersubjectively agreed-upon measurement result when he measures the $x$-spin of $S$ and why the determinate result *appears* to be randomly produced with probability $|\alpha|^2$ for "$\uparrow_x$" and $|\beta|^2$ for "$\downarrow_x$".

To accomplish this Everett proposed modifications to both the standard formulation of quantum mechanics and to the associated standard interpretation of states. Specifically, he proposed dropping Process 1 (the random collapse dynamics) from the theory entirely, then adding the distinction between total and relative states to explain determinate measurement outcomes in the entangled superpositions like (5) that are predicted by Process 2 (the deterministic linear dynamics).

# 40.3 THE RELATIVE STATE FORMULATION OF PURE WAVE MECHANICS

Everett believed that one might deduce the standard quantum probabilities as subjective appearances from *pure wave mechanics* together with a small handful of auxiliary assumptions necessary to interpret the theory, where pure wave mechanics was the standard von Neumann–Dirac formulation without Process 1. Pure wave mechanics then replaces the hybrid dynamics of the standard orthodox formulation of quantum mechanics with just:

3. Linear dynamics: A system $S$ *always* evolves in a deterministic, linear way: $|\psi(t_1)\rangle_S = \hat{U}(t_0,t_1)|\psi(t_0)\rangle_S$.

Everett kept part of the standard interpretation of states by assuming that a system $S$ has a determinate value for an observable $O$ if its *total state* is an eigenstate of $O$. Then to get his relative-state formulation of the theory, he added (1) an internal model of measurement where observers are treated as physical systems described by the theory, (2) a distinction between the *total* (or *universal*) *state* and the *relative state* of a system, and (3) a special typicality measure over relative states.

Everett's project then was 'to deduce the probabilistic assertions of Process 1 as subjective appearances' to observers who are themselves treated as perfectly ordinary physical systems always subject to the linear dynamics 'thus placing the theory in correspondence with experience'. He took the moral to be that '[w]e are then led to the novel situation in which the formal theory is objectively continuous and causal, while subjectively discontinuous and probabilistic' (Barrett and Byrne, 2012, p. 77). If successful, this reconciles observer *B*'s state attributions with observer *A*'s determinate experience in his hypothetical drama.

## 40.3.1 The Internal Model of Measurement

Dropping Process 1 from the standard theory resolves the conflict between the states predicted by Process 1 and Process 2, but it also immediately leads to two new problems: one we might call *the determinate record problem* and the other *the probability problem*. To address the determinate record problem one needs to explain how an observer gets a determinate measurement record at all. And to address the probability problem one needs to provide some way of understanding the standard quantum probabilities in one's new and fully deterministic theory.

Everett's strategy was to get empirical predictions in terms of the relative records of observers who were situated within the physical world described by the theory. To this end, he introduced an internal model of measurement where observers are treated as physical systems described by the theory. Specifically, observers are understood as physical automata who simply correlate the state of their physical memory with the quantum property being measured.

We have already seen the basic setup with observer *A* in the hypothetical drama. If observer *A* measures the *x*-spin of a system *S* that begins in an *x*-spin up eigenstate, the composite system will evolve from $| \text{"ready"} \rangle_A |{\uparrow_x}\rangle_S$ to $|\text{"}{\uparrow_x}\text{"}\rangle_A|{\uparrow_x}\rangle_S$. Here the final state represents the observer's physical memory recording "$\uparrow_x$" and hence explains the observer's subjective experience of getting the result '*x*-spin up'. Similarly, if *A* measures the *x*-spin of a system *S* that begins in an *x*-spin down eigenstate, the composite system will evolve from $| \text{"ready"} \rangle_A |{\downarrow_x}\rangle_S$ to $|\text{"}{\downarrow_x}\text{"}\rangle_A|{\downarrow_x}\rangle_S$. Here the final state represents the observer's physical memory recording "$\downarrow_x$" and hence explains the observer's subjective experience of getting the result '*x*-spin down'.

Everett believed that he had explained an observer's *subjective experience* if he could account for such physical measurement records. But his explanation of observer *A*'s experience in the hypothetical drama needs to be more subtle than suggested by these two cases. In the hypothetical drama the total post-measurement state of the composite system of observer *A* and system *S* is not one that describes *A* as recording the result

'$x$-spin up' and it is not one that describes $A$ as recording the result '$x$-spin down' either. In order to associate the observer's state with his subjective experience in this case, Everett appealed to a distinction between total states and relative states.

## 40.3.2 Total States and Relative States

Everett distinguished between the *total state* of a composite system and the *relative states* of its subsystems as follows. When systems like $A$ and $S$ interact linearly, this results in an entangled state for the composite system where 'to any arbitrary choice of state for one subsystem there will correspond a *relative state* for the other subsystem, which will generally be dependent upon the choice of state for the first subsystem, so that the state of one subsystem is not independent, but correlated to the state of the remaining subsystem' (Barrett and Byrne, 2012, p. 78). Since 'from our point of view all measurement and observation processes are to be regarded simply as interactions between observer and object-system which produce strong correlations' the result of a measurement interaction resulting from the linear interaction between an observer $A$ and an object system $S$ will look something like the state that observer $B$ calculates in the hypothetical drama:

$$\alpha| \text{``}\uparrow_x\text{''} \rangle_A |\uparrow_x\rangle_S + \beta| \text{``}\downarrow_x\text{''} \rangle_A |\downarrow_x\rangle_S. \tag{8}$$

In this total state, the state of the observer $A$ *relative* to the object system $S$ being $x$-spin up is $| \text{``}\uparrow_x\text{''} \rangle_A$ and the state of the observer $A$ *relative* to the object system $S$ being $x$-spin down is $| \text{``}\downarrow_x\text{''} \rangle_A$.

Everett's key interpretive move was to take the relative states of an observer's records to describe the observer's experience. He took each of the relative states of the observer $A$ to represent a 'branch' (or 'element') of the total state. Here $A$ has the relative record "$\uparrow_x$", and hence has the *relative experience* of getting the result '$x$-spin up' in the first branch of the total state, and $A$ has the relative record "$\downarrow_x$", and hence the *relative experience* of getting the result '$x$-spin down' in the second branch of the total state. The idea is that in each element of the superposition the observer-system state describes the observer as determinately recording the particular system state described by that element and hence having the corresponding subjective experience.[6]

Everett added the following footnote to the short version of his thesis when it was published to explain the point further:

---

[6] Given how the explanation is supposed to work here, Everett arguably encounters no *preferred basis problem*. He explicitly says that one can calculate relative states with respect to any subsystem or basis. But the relative states one calculates will only be relevant to his account of subjective experience if they result from a correlating interaction that produces relative observer *records*. In contrast, if branches were supposed to represent physically distinct worlds, one would have the problem of having to specify a preferred basis that describes how and when worlds split.

At this point we encounter a language difficulty. Whereas before the observation we had a single observer state afterwards there were a number of different states for the observer, all occurring in a superposition. Each of these separate states is a state for an observer, so that we can speak of the different observers described by the different states. On the other hand, the same physical system is involved, and from this viewpoint it is the *same* observer, which is in different states for different elements of the superposition (i.e., has had different experiences in the separate elements of the superposition). In this situation we shall use the singular when we wish to emphasize that a single physical system is involved, and the plural when we wish to emphasize the different experiences for the separate elements of the superposition. (E.g., the observer performs an observation of the quantity A, after which each of the observers of the resulting superposition has perceived an eigenvalue.)    (Barrett and Byrne, 2012, p. 172)

The post-measurement total state then describes a single physical observer who is associated with different relative records and hence different perceptions. In the hypothetical drama one relative observer perceives the result 'x-spin up' and the other perceives the result 'x-spin down'.

Identifying an observer's experiences with his relative records is a significant interpretational assumption, but Everett had a number of reasons to believe that he was justified in making it. Some of these had to do with a sort of diachronic consistency that results from the linear dynamics when applied to physical systems like measurement records. Others had to do with how he understood the purpose of a physical theory. We will consider the issue of diachronic consistency here, then return to the issue of how Everett understood the purpose of a physical theory later.

Suppose that observer $A$ repeats his $x$-spin measurement. If $|\text{``}\uparrow_x\text{''}\rangle_A|\uparrow_x\rangle_S$ were the state after his first measurement, then he would get '$x$-spin up' again and end up in the state $|\text{``}\uparrow_x\text{''},\text{``}\uparrow_x\text{''}\rangle_A|\uparrow_x\rangle_S$ after repeating the measurement. And if $|\text{``}\downarrow_x\text{''}\rangle_A|\downarrow_x\rangle_S$ were the state after his first measurement, then he would get '$x$-spin down' again and end up in the state $|\text{``}\downarrow_x\text{''},\text{``}\downarrow_x\text{''}\rangle_A|\downarrow_x\rangle_S$ after repeating the measurement. So, by the linearity of Process 2, if the state is

$$\alpha|\text{``}\uparrow_x\text{''}\rangle_A|\uparrow_x\rangle_S + \beta|\text{``}\downarrow_x\text{''}\rangle_A|\downarrow_x\rangle_S \tag{9}$$

after his first measurement, he will end up in the state

$$\alpha|\text{``}z\uparrow_x\text{''},\text{``}\uparrow_x\text{''}\rangle_A|\uparrow_x\rangle_S + \beta|\text{``}\downarrow_x\text{''},\text{``}\downarrow_x\text{''}\rangle_A|\downarrow_x\rangle_S \tag{10}$$

after repeating the measurement. Everett explains that this final total state 'describes the observer as having obtained the *same result* for each of the two observations'. 'Thus', he concludes, 'for every separate state of the observer in the final superposition, the result of the observation was repeatable, even though different for different states. This repeatability is, of course, a consequence of the fact that after an observation the *relative* system state for a particular observer state is the corresponding eigenstate'

(Barrett and Byrne, 2012, p. 122). The fact that the result is repeatable will make it appear to each relative observer that a collapse has occurred.

Everett made a more general point regarding the diachronic consistency of records in a 31 May 1957 letter to Bryce DeWitt defending his theory (Everett, 1957c):

> From the viewpoint of the theory, *all* elements of a superposition (all 'branches') are 'actual', none any more 'real' than another. It is completely unnecessary to suppose that after an observation somehow one element of the final superposition is selected to be awarded with a mysterious quality called reality and the others condemned to oblivion. We can be more charitable and allow the others to coexist—they won't cause any trouble anyway because all the separate elements of the superposition (branches) individually obey the wave equation with complete indifference to the presence or absence (actuality or not) of any other elements.    (Barrett and Byrne, 2012, p. 254)

What Everett had in mind here is that since Process 2 is linear, the action of the unitary time-evolution operator on the total state is precisely the same as the superposition of its action on each element of the superposition. While this is true, it is misleading to suggest that branches *never* interact with each other. Everett knew that it was always in principle possible to detect what branches there are by means of a suitable interference measurement like the Q-measurement we discussed in the hypothetical drama. But while it is *in principle* possible for different elements of the total superposition to interfere, one would expect it to be extremely unlikely that this would happen by chance or design for a macroscopic system where there are many physical degrees of freedom. To get interference between branches for the composite system $A$ and $S$, the linear dynamics would have to match up each of the relevant degrees of freedom involving $A$'s record and $S$'s $x$-spin. The upshot is that one should typically expect the branches associated with a macroscopic system to evolve more or less independently of each other under the linear dynamics.[7]

Everett further argued in the long thesis that, because of the correlations between the positions of their parts, macroscopic objects will tend to maintain their structural properties over time as described by their relative states (he used the cannonball example again to explain this point) (Barrett and Byrne, 2012, p. 135). And he argued that, because macroscopic objects are relatively massive, their total states can be understood as a superposition of quasi-classical states where they have approximately determinate positions and momenta in each branch and where 'the centers of the square amplitude distributions for the objects will move in a manner approximately obeying the laws of motion of classical mechanics, with the degree of approximation depending upon the masses and the length of time considered'. Hence one's relative records of macroscopic objects will create the appearance of quasi-classical behaviour (Barrett and Byrne, 2012, p. 136).

---

[7] In addition to what he said in his published work, we know from Everett's handwritten notes for the long thesis that he clearly understood that the more degrees of freedom a system has, the harder it is to observe interference effects. Everett's notes are collected in Barrett, Byrne, and Weatherall (2011).

## 40.3.3  Typicality and Probability

Everett clearly and repeatedly insisted that there were no probabilities in pure wave mechanics. That he took this to be a central point is reflected in the fact that he ultimately entitled the long version of his PhD thesis 'Wave Mechanics without Probability'. As he explained in setting up his project:

> [W]e shall use the language of probability theory to facilitate the exposition, and because it enables us to introduce in a unified manner a number of concepts that will be of later use. We shall nevertheless subsequently apply the mathematical definitions directly to state functions, by replacing probabilities by square amplitudes, without, however, making any reference to probability models.
>
> (Barrett and Byrne, 2012, p. 79)

Instead of probabilities, pure wave mechanics concerns a special *typicality measure* over branches. Everett's typicality measure over branches is given by the norm-squared of the amplitude associated with each branch. Formally, this measure satisfies the definition of a probability measure, but it is explicitly not to be understood as representing probabilities in any way. Rather, Everett sought to show that the sequence of an observer's relative measurement records in a *typical branch* of the total state will be randomly distributed with the standard quantum statistics. On the assumption that an observer's relative records describe her experience, it will therefore *appear* to a typical relative observer that the standard quantum probabilities are obtained for the results of her measurements. This is how he sought to deduce the probabilitistic predictions of Process 1 in pure wave mechanics.

To show that typical branches exhibit the standard quantum statistics Everett needed a typicality measure over branches—something that is not provided by pure wave mechanics itself. He introduced his typicality measure by imposing a sequence of formal constraints until a particular measure was uniquely determined. Specifically, he stipulated that the typicality measure $m$ (1) must satisfy the standard axioms for being a probability measure, (2) must be a positive function of the amplitudes of the complex-valued coefficients $a_i$ associated with the branches of the superposition $\sum a_i |\chi_i\rangle$, and (3) must satisfy a sub-branch additivity condition so that the sum of the measures associated with the branches resulting from a measurement interaction equal the measure associated with the observer before the measurement was performed. Together, Everett showed that these constraints determine the typicality measure $m$ as the norm-squared amplitude measure over branches.[8]

Consider the post-measurement state in the hypothetical drama again:

---

[8] See Barrett and Byrne (2012, pp. 123–26) for Everett's discussion of why he selected this measure.

$$\alpha|\text{``}\!\uparrow_x\text{''}\rangle_A|\!\uparrow_x\rangle_S + \beta|\text{``}\!\downarrow_x\text{''}\rangle_A|\!\downarrow_x\rangle_S. \tag{11}$$

Importantly, these two branches typically do not get the same measure—the first is assigned measure $|\alpha|^2$ and the second is assigned measure $|\beta|^2$. These measures are the probabilities that would be assigned to each outcome by Process 1 in the standard collapse theory, but again here the measure does not represent anything like the probability of each branch being realized at the expense of the other branches. Inasmuch as Everett took all branches to be equally actual, he took the probability of each branch to be *one*. For Everett, the norm-squared typicality measure was a measure *across* branches where the appearance of quantum probabilities were to be explained *within* a typical branch.

Everett used the typicality measure to argue for his two main statistical results. Consider a measuring device $M$ that is ready to make and record an infinite series of measurements on each of an infinite series of systems $S_k$ in state

$$\alpha|\!\uparrow_x\rangle_{S_k} + \beta|\!\downarrow_x\rangle_{S_k}. \tag{12}$$

Suppose that $M$ interacts with each system in turn and perfectly correlates its $k$th memory register with the $x$-spin of system $S_k$ by the linear dynamics. This will produce an increasingly complicated entangled superposition of different sequences of measurement outcomes. After one measurement, the state of $M$ and $S_1$ will be

$$\alpha|\text{``}\!\uparrow\text{''}\rangle_M|\!\uparrow\rangle_{S_1} + \beta|\text{``}\!\downarrow\text{''}\rangle_M|\!\downarrow\rangle_{S_2}. \tag{13}$$

After two measurements, the state of $M$, $S_1$, and $S_2$ will be

$$\begin{aligned}\alpha^2|\text{``}\!\uparrow\uparrow\text{''}\rangle_M|\!\uparrow\rangle_{S_1}|\!\uparrow\rangle_{S_2} &\ +\alpha\beta|\text{``}\!\uparrow\downarrow\text{''}\rangle_M|\!\uparrow\rangle_{S_1}|\!\downarrow\rangle_{S_2}\\ +\beta\alpha|\text{``}\!\downarrow\downarrow\text{''}\rangle_M|\!\downarrow\rangle_{S_1}|\!\uparrow\rangle_{S_2}&+\beta^2|\text{``}\!\downarrow\downarrow\text{''}\rangle_M|\!\downarrow\rangle_{S_1}|\!\downarrow_x\rangle_{S_2}.\end{aligned} \tag{14}$$

The first three steps of the branching process and the amplitudes associated with each branch are given in the figure:

The following two limiting properties hold for such sequences of measurements in pure wave mechanics:

**Relative Frequency:** For any $\delta$ and $\varepsilon$, there exists a $k$ such that after $k$ measurements the sum of the norm-squared of the amplitude associated with each branch where the distribution of spin-up results is within $\varepsilon$ of $|\alpha|^2$ and the distribution of spin-down results is within $\varepsilon$ of $|\beta|^2$ is within $\delta$ of one.

**Randomness:** The sum of the norm-squared of the amplitude associated with each branch where the sequence of relative records is algorithmically incompressible, or satisfies any other of the standard criteria for being random, goes to one as the number of measurements gets large.

Everett briefly sketched why one should expect these theorems to be true, but they can be proven rigorously.[9] It follows from these that in the limit as the number of measurements gets large, *almost all* branches in measure $m$ will describe sequences of measurement records that are randomly distributed with the standard quantum statistics. Hence in the limit *the records of a typical relative observer* will exhibit the standard quantum statistics. And from this Everett concludes that '[w]e have therefore shown that the statistical assertions of Process 1 will appear to be valid to almost all observers described by separate elements of the superposition... in the limit as the number of observations goes to infinity' (Barrett and Byrne, 2012, p. 127). This was Everett's deduction of quantum probabilities as subjective appearances.

## 40.4 THE CONTENT OF EVERETT'S DEDUCTION

It is important to be clear regarding what Everett did and did not do in his deduction of the standard quantum statistics as subjective appearances. He showed that a *typical relative observer* (in his special stipulated sense of typical) will end up with relative records that appear to have been randomly determined by the collapse dynamics Process 1. Note, however, that there is nothing whatsoever here that entails that one should *expect* the standard quantum probabilities to hold for the results of one's measurements nor that such statistics are *probable*. Put another way, Everett's relative-state formulation of pure wave mechanics does not make any of the standard *probabilistic predictions* of quantum mechanics. Probabilities for Everett are appearances *within* a typical branch, not a measure of the relative likelihood of each branch in fact being realized. On the assumption that a relative observer's experience is given by his relative records, the theory says that the standard quantum statistics will appear to hold for a typical relative observer, but it does not tell one that one should *expect* one's experience to be typical.

If one wanted the theory to predict the standard quantum probabilities, rather than just to say that the quantum statistics are branch-typical in a specified measure, then one would need an additional assumption that connects the typicality measure to one's probabilistic expectations. Let's call an auxiliary assumption that connects typicalities to probabilities a ★-*principle*. A ★-principle might be thought of as a sort of generalized principle of indifference.[10] A ★-principle that would work to get probabilities for Everett would be something like: one should *expect* one's experience to be given by a

---

[9] See Barrett and Byrne (2012, pp. 126–27) for Everett's argument. See chapter 4 of Barrett (1999) for proofs of the two theorems. See also Hartle (1968) and Farhi, Goldstone, and Gutmann (1989) for closely related results.

[10] The standard principle of indifference says that if one has possibilities and no special evidence, then one should assign a prior probability of to each possibility.

typical branch in the norm-squared coefficient measure sense of typical. Armed with such an auxiliary principle and a diachronic account of personal identity that makes appropriate sense of an agent's expectations concerning future observations in the context of splitting branches, one would then be able to deduce that one should *expect* the standard quantum statistics. A ★-principle would allow one to infer both probabilities and expectations *across* alternative branches and hence understand the various quantum-mechanical possibilities as more or less probable.

One is perfectly free to supplement pure wave mechanics with a measure of typicality and an associated ★-principle should one wish to do so, but (1) this is a further nontrivial addition to pure wave mechanics, (2) there is no general principle of reason that determines the choice, and (3) this is not what Everett himself did. We will discuss these in turn.

To the first point, that the ★-principle one would need to get probabilities or expectations here is nontrivial is reflected in the fact that it is usually not the case that most branches *by count* exhibit the standard quantum statistics—indeed, this is only the case for the $x$-spin measurements we have been considering if one is measuring a sequence of identically prepared systems where $|\alpha|^2 = |\beta|^2 = 1/2$. Hence, a standard principle of indifference, even if one (mistakenly) thought that it was a general principle of reason, would not work here. Rather, one needs a just-right ★-principle that works with Everett's specified typicality measure over branches to get the usual quantum expectations.

To the second point, there is good reason to suppose that reason itself places only relatively weak constraints on how one assigns prior probabilities. Specifically, one wants one's priors to be *probabilistically coherent* (that is, to satisfy the standard axioms of probability—else one is open to a Dutch book) and *non-dogmatic* (that is, not equal to zero or one—else one cannot revise one's degrees of belief by conditioning on new evidence). But there is in general no good reason at all for one to assign priors that satisfy any particular ★-principle. If there were purely rational grounds for a stipulated canonical assignment of priors, then something would go wrong if an agent did not abide by it. But a good Bayesian agent might assign any set of coherent, non-dogmatic priors, then condition on new evidence without any negative consequences whatsoever. Hence it is exceedingly difficult to provide a compelling argument for a particular ★-principle as a general principle of reason. The challenge, then, is this: if one wants to argue that a particular ★-principle is required by reason, then one needs to say precisely what goes wrong if a good Bayesian agent does not subject her beliefs to the principle in general.

The moral here is that deducing the standard quantum *probabilities* from pure wave mechanics requires special background assumptions, like the choice of a particular typicality measure and an associated ★-principle, assumptions that tie quantum amplitudes over branches to probabilities over branches. One would need to add these assumptions to pure wave mechanics explicitly. They would be, after all, a critically important part of the predictive apparatus of the theory—the part that ultimately tells one what to *expect*. This would involve adding significantly more to pure wave mechanics than what Everett himself added.[11]

This brings us to the last of the three points. Everett himself did not seek to provide probabilities over alternative branches or account for the future expectations of a rational agent—rather, he just wanted to show that the quantum statistics hold for the measurement records in a typical branch. And this has everything to do with what he considered to be the purpose of a physical theory.

# 40.5 WHAT EVERETT WANTED FROM A SATISFACTORY PHYSICAL THEORY

How Everett understood the purpose of a physical theory goes some way in explaining how he thought of the account of experience provided by the relative-state formulation of pure wave mechanics. In his 31 May 1957 letter to Bryce DeWitt, Everett explained his 'conception of the nature and purpose of physical theories in general':

> To me, any physical theory is a logical construct (model), consisting of symbols and rules for their manipulation, *some* of whose elements are associated with elements of the perceived world. If this association is an isomorphism (or at least a homomorphism) we can speak of the theory as correct, or as faithful. The fundamental requirements of any theory are logical consistency and correctness in this sense.
>
> (Barrett and Byrne, 2012, p. 253)

More specifically, Everett took a physical theory to be satisfactory if it is logically consistent and empirically faithful in the sense that one can find an observer's perceptions represented as records associated with the observer as represented in the model of the theory. Pure wave mechanics is a satisfactory physical theory because it is logically consistent and one can find an observer's perceptions represented in the relative records of a typical modelled observer. This is the only sort of empirical adequacy that mattered to Everett. Being able to find one's experience represented in the model of the theory in this way renders the theory empirically acceptable even when it predicts that records that satisfy the standard quantum statistics are neither probable nor to be expected.[12]

---

[11] See Barrett (2017) and (2018) for more details regarding Everett's deduction. See Wallace (2010b) for a careful description of a set of auxiliary assumptions that are sufficient to derive something like a probability over branches in a decohering-histories interpretation of pure wave mechanics and Barrett (2019) for a critical discussion.

[12] See Barrett (2011a) and (2015) for discussions of empirical faithfulness and how this relates to other notions of empirical adequacy.

While Everett did not take pure wave mechanics to make probabilistic predictions, he did describe how the theory might be taken to predict particularly salient aspects of our experience. An example will illustrate the point.

In his letter to DeWitt, Everett quoted an earlier letter from DeWitt to Wheeler, 'Now in your letter you say, "the trajectory of the memory configuration of a real physical observer, on the other hand, does *not* branch. I can testify to this from personal introspection, as can you. I simply do *not* branch".' To this Everett replied that just as Copernican theory, when supplemented with Newtonian mechanics, predicts that one would not feel the earth moving, pure wave mechanics predicts that one would not feel oneself branching. As Everett put it:

> I must confess that I do not see this 'branching process' as the 'vast contradiction' that you do. The theory is in full accord with our experience (at least insofar as ordinary quantum mechanics is). It is in full accord just because it *is* possible to show that no observer would ever be aware of any 'branching', which is alien to our experience as you point out. The whole issue of the 'transition from the possible to the actual' is taken care of in a very simple way—there is no such transition, *nor is such a transition necessary for the theory to be in accord with our experience.*
>
> (Barrett and Byrne, 2012, p. 254)

The prediction that an observer would not be aware of the branching process results from the fact that, inasmuch as it would require her to perform a reliable Q-type measurement on herself and her object system to show that multiple incompatible measurement results were in fact realized in alternative branches, it would be virtually impossible for a macroscopic observer to have a record of the branching process. The predictions of the theory, then, are in accord with our experience, and DeWitt's empirical worries are empty.

But that said, it is not *impossible* for an observer to detect branching. As Everett readily acknowledged, an observer might *in principle* have a record of the branching process. In the hypothetical drama, if observer B were to measure the interference observable Q of the composite system A and S and repeatedly get the result +1, this would be evidence of the branching of A into two relative observers with different relative records. This information might then be shared with observer A. But again, measuring anything like Q would be extraordinarily difficult for a system containing a macroscopic observer. Everett knew this, and it is closely connected to how he understood different branches of the total state.

# 40.6 MANY WORLDS

For Everett, different branches of the total state were equally real and they represented different subjective experiences, but they did not represent different physical systems.

As he explained in the language-difficulty footnote, after the measurement interaction, there is a single physical observer 'in different states for different elements of the superposition' (Barrett and Byrne, 2012, p. 172). Different branches were not real because they were physical copies of the original system or because they represented physically real worlds—they were real because they might in principle be *detected*. As Everett put the point:

> It is therefore improper to attribute any less validity or reality to any element of a superposition than any other element, due to this ever present possibility of obtaining interference effects between the elements. All elements of a superposition must be regarded as simultaneously existing.   (Barrett and Byrne, 2012, p. 150)

Note that since post-measurement branches might in principle be detected, they are in potential causal contact with each other and hence not at all different physical *worlds*.[13]

While Everett himself understood alternative branches as representing alternative subjective experiences in a single physical world, it quickly became the practice of his supporters to understand alternative branches as alternative physical worlds.[14] After being convinced that the relative-state formulation predicts that an observer would not detect the splitting process, DeWitt became an energetic proponent and popularizer of Everett's theory. On DeWitt's telling, the post-measurement branches simply represent physically distinct worlds each containing a physical copy of the pre-measurement observer.[15]

This view, however, has shifted somewhat. On the most sophisticated current formulations of the many-worlds interpretation, worlds are taken to be decoherence-induced emergent structures. While this picture is rather different from Everett's original view, it is in many ways closer to Everett's view than to DeWitt's. Inasmuch as decohering worlds do not involve a physical process that in any way produces new physical copies of the pre-measurement systems, the decoherence-induced branching process might be understood as producing decohering records that explain the *subjective appearance* of different physical worlds much as Everett himself might have put it. This is reflected, for example, in David Wallace's view that there is not even a simple matter of fact regarding *how many* post-measurement worlds there are—such facts, rather, depend on one's level of description.[16]

One point on which there is significant distance between Everett's views and the decohering-worlds interpretation is on the topic of probability. Proponents of decohering worlds want to recapture the standard probabilistic predictions of quantum mechanics in a stronger sense than Everett did. Among other options, a popular view is

[13] See Albert (1986) and Albert and Barrett (1995) for discussions of this point.
[14] See Barrett (2011b) for a detailed discussion of worlds and Everett's metaphysical commitments.
[15] See DeWitt's and Graham's descriptions of the theory in their contributions to DeWitt and Graham (1973).
[16] See Wallace (2010a) and (2012) for a description of the decohering-worlds account and explanations of how the emergent metaphysics and levels of description are supposed to work.

that quantum probabilities are the subjective future self-location probabilities that would be inferred by a rational agent who understood the decoherence-induced branching process.[17] For his part, Everett took it to be sufficient to show that the relative measurement records in a typical branch, in a sense of *typical* that he took to be canonical for the purpose, will exhibit the standard quantum statistics.

## References

Albert, D. Z. (1986). How to Take a Photograph of Another Everett World. *Annals of the New York Academy of Sciences: New Techniques and Ideas in Quantum Measurement Theory*, 480, 498–502.

Albert, D. Z., and Barrett, J. A. (1995). On What It Takes to Be a World. *Topoi*, 14(1), 35–37.

Barrett, J. A. (1999). *The Quantum Mechanics of Minds and Worlds*. Oxford: Oxford University Press.

Barrett, J. A. (2011a). On the Faithful Interpretation of Pure Wave Mechanics. *British Journal for the Philosophy of Science*, 62(4), 693–709.

Barrett, J. A. (2011b). Everett's Pure Wave Mechanics and the Notion of Worlds. *European Journal for Philosophy of Science*, 1(2), 277–302.

Barrett, J. A. (2015). Pure Wave Mechanics and the Very Idea of Empirical Adequacy. *Synthese*, 192(10), 3071–3104.

Barrett, J. A. (2017). Typical Worlds. *Studies in the History and Philosophy of Modern Physics*, 58, 31–40.

Barrett, J. A. (2018). Everetts Relative-State Formulation of Quantum Mechanics. In E. N. Zalta (ed.), *Stanford Encyclopedia of Philosophy*. https://plato.stanford.edu/entries/qm-everett/

Barrett, J. A. (2019). *The Conceptual Foundations of Quantum Mechanics*. Oxford: Oxford University Press.

Barrett, J. A., and Byrne, P. (eds) (2012). *The Everett Interpretation of Quantum Mechanics: Collected Works 1955–1980 with Commentary*. Princeton, NJ: Princeton University Press.

Barrett, J. A., Byrne, P., and Weatherall, J. (eds) (2011). *Hugh Everett III Manuscripts*, UCIspace @ the Libraries. Permanent url: http://hdl.handle.net/10575/1060.

Byrne, P. (2010). *The Many Worlds of Hugh Everett III: Multiple Universes, Mutual Assured Destruction, and the Meltdown of a Nuclear Family*. Oxford: Oxford University Press.

DeWitt, B. S., and Graham, N. (eds) (1973). *The Many-Worlds Interpretation of Quantum Mechanics*. Princeton, NJ: Princeton University Press.

Deutsch, D. (1999). Quantum Theory of Probability and Decisions. *Proceedings of the Royal Society of London A*, 455, 3129–3137.

Everett III, H. (1956). The Theory of the Universal Wave Function. In Barrett and Byrne (eds) (2012, pp. 72–172).

Everett III, H. (1957a). 'Relative State' Formulation of Quantum Mechanics. *Reviews of Modern Physics*, 29, 454–462. In Barrett and Byrne (eds) (2012, pp. 173–96).

---

[17] For recent discussions of probability in Everett-style formulations of quantum mechanics, see Deutsch (1999); Greaves (2007); Vaidman (2012) and (2014); Wallace (2010a), (2010b), and (2012); and Sebens and Carroll (2016). See Kent (2010) and Barrett (2019) for critical discussions.

Everett III, H. (1957b). Letter to Aage Petersen, 31 May 1957. In Barrett and Byrne (2012, pp. 238–240).

Everett III, H. (1957c). Letter to Bryce DeWitt, 31 May 1957. In Barrett and Byrne (2012, pp. 251–256).

Farhi, E., Goldstone, J., Gutmann, S. (1989). How Probability Arises in Quantum Mechanics. *Annals of Physics*, **192**(2), 368–382.

Greaves, H. (2007). Probability in the Everett Interpretation. *Philosophy Compass*, 2/1, 109–128. doi: 10.1111/j.1747–9991.2006.00054.x

Hartle, J. B. (1968). Quantum Mechanics of Individual Systems. *American Journal of Physics*, **36**(8), 704–12.

Kent, A. (2010). One World Versus Many: The Inadequacy of Everettian Accounts of Evolution, Probability, and Scientific Confirmation. In Saunders *et al.* (eds), (2010).

Osnaghi, D., Freitas, F., and Freire Jr, O. (2009). The Origin of the Everettian Heresy. *Studies in History and Philosophy of Modern Physics*, **40**, 97–123.

Saunders, S., Barrett, J., Kent, A., and Wallace, D. (eds) (2010). *Many Worlds?: Everett, Quantum Theory, and Reality*. Oxford: Oxford University Press.

Sebens, C. T., and Carroll, S. M. (2016). Self-Locating Uncertainty and the Origin of Probability in Everettian Quantum Mechanics. *British Journal for the Philosophy of Science*. doi: 10.1093/bjps/axw004. First published online: 5 July 2016.

Vaidman, L. (2012). Probability in the Many-Worlds Interpretation of Quantum Mechanics. In Y. Ben-Menahem and M. Hemmo (eds), *Probability in Physics (The Frontiers Collection XII)*, Berlin, Heidelberg: Springer, pp. 299–311.

Vaidman, L. (2014). Many-Worlds Interpretation of Quantum Mechanics. In E. N. Zalta (ed.), *The Stanford Encyclopedia of Philosophy*. https://plato.stanford.edu/entries/qm-manyworlds/. Accessed 29 July 2020.

von Neumann, J. (1955). *Mathematical Foundations of Quantum Mechanics*. Princeton, NJ: Princeton University Press. Translated by R. Beyer from *Mathematische Grundlagen der Quantenmechanik*, Berlin: Springer (1932).

Wallace, D. (2010a). Decoherence and Ontology. In Saunders *et al.* (2010), pp. 53–72.

Wallace, D. (2010b). How to Prove the Born Rule. In Saunders *et al.* (2010), pp. 227–63.

Wallace, D. (2012). *The Emergent Multiverse: Quantum Theory according to the Everett Interpretation*. Oxford: Oxford University Press.

Wigner, E. (1961). Remarks on the Mind-Body Problem. In I. J. Good (ed.), *The Scientist Speculates*, New York: Basic Books, pp. 284–302.

# CHAPTER 41

# IS QBISM A POSSIBLE SOLUTION TO THE CONCEPTUAL PROBLEMS OF QUANTUM MECHANICS?

HERVÉ ZWIRN

## 41.1 INTRODUCTION

QUANTUM Bayesianism is a type of interpretation that applies the Bayesian interpretation of probability to Quantum Mechanics. For the Bayesian interpretation everything starts by assuming some prior probabilities. Then some recipes are provided to update these probabilities on new information. For the subjective Bayesians (de Finetti, 1990; Ramsey, 1931; Savage, 1954), probabilities are related to an epistemic (hence personal and subjective) uncertainty and represent the degree of belief of an agent for the happening of an event.

There are several versions of Quantum Bayesianism (Pitowsky, 2002; Baez, 2003; Bub and Pitowsky, 2010; Bub, 2019; Youssef, 1994; Leifer and Spekkens, 2013; Caticha, 2007).

QBism is the current state of the version of Quantum Bayesianism that was initially developed by Caves, Fuchs, and Schack (2002, 2007) but has evolved in such a way that Caves now '*does not subscribe to it*' (Stacey, 2019b). It has been more recently adopted by Mermin (Mermin, 2013, 2014, 2017; Fuchs, Mermin, and Schack, 2014, 2015). The denomination QBism appeared for the first time in a 2009 arXiv article which has been published in 2013 (Fuchs and Schack, 2013). As Fuchs (2016) says:

> *The term QBism, invented in 2009, initially stood for Quantum Bayesianism, a view of quantum theory a few of us had been developing since 1993. Eventually, however, I. J. Good's warning that there are 46,656 varieties of Bayesianism came to bite us, with some Bayesians feeling their good name had been hijacked. David Mermin suggested that the B in QBism should more accurately stand for 'Bruno' as in Bruno de Finetti, so that we would at least get the variety of (subjective) Bayesianism right. The trouble is QBism incorporates a kind metaphysics that even Bruno de Finetti might have rejected! So, trying to be as true to our story as possible, we momentarily toyed with the idea of associating the B with what the early 20th-century United State Supreme Court Justice Oliver Wendell Homes Jr called 'bettabilitarianism'. It is the idea that the world is loose at the joints, that indeterminism plays a real role in the world. [ . . . ] But what an ugly, ugly word, bettabilitarianism! Therefore, maybe one should just think of the B as standing for no word in particular, but a deep idea instead: That the world is so wired that our actions as active agents actually matter.*

So, QBists think that the central subject of science is the primitive concept of experience. This seems similar to Bohr's position who insisted on saying that experiments were the basic subject of quantum mechanics, but what Bohr had in mind was the classical reading of a macroscopic apparatus, while for QBism 'experience' is much more than that. The word 'experience' must be understood here as 'personal experience', meaning for each agent what the world has induced in her throughout the course of her life. Experience is the way in which the world impinges on any agent and how the agent impinges on the world.

QBists adopt the subjective interpretation of probability according to which probabilities represent the degrees of belief of an agent and, hence, are particular to that agent. This is an important point to notice: as they say, it is a 'single user theory'. This means that:

> *Probability assignments express the beliefs of the agent who makes them, and refer to that same agent's expectations for her subsequent experiences.*
> (Fuchs, Mermin, and Schack, 2014)
>
> *When I—the agent—write down a quantum state it is my quantum state, no one else's. When I contemplate a measurement, I contemplate its results, outcomes, consequences for me, no one else—it is my experience.*   (Fuchs, 2018)

According to QBism the entities of the formalism (wave functions, observables, Hamiltonians, probabilities) are not objective and attached to an external reality. They are subjective in nature and have no absolute value but are relative to a particular agent. So they are not necessarily the same for all observers. QBists refuse the idea that the quantum state of a system is an objective property of that system (Fuchs and Schack, 2015). It is only a tool for assigning a subjective probability to the agent's future experience. It only represents the degree of belief of the agent about the possible outcomes of future measurements. So quantum mechanics does not directly say something about the 'external world'.

*But quantum mechanics itself does not deal directly with the objective world; it deals with the experiences of that objective world that belong to whatever particular agent is making use of the quantum theory.*    (Fuchs, Mermin, and Schack, 2014)

This means that QBism does not consider quantum mechanics as a theory describing the external world that we could gradually discover more and more intimately through our measurements, but as a theory that describes the agent's experience. A measurement (in the usual sense) is just a special case of what QBism calls experience, and that is any action taken by any agent on her external world. The result of the measurement is the experience the agent gets from her action on her personal world. In this sense, measurements are made continuously by every agent. A measurement does not reveal a pre-existing state of affairs but creates a result for the agent.

*A measurement does not, as the term unfortunately suggests, reveal a pre-existing state of affairs. It is an action on the world by an agent that results in the creation of an outcome—a new experience for that agent.*    (Fuchs, Mermin, and Schack, 2014)

So, QBism could be seen as an instrumentalist interpretation since it presents quantum mechanics as nothing but a tool allowing any agent to compute her probabilistic expectations for her future experience from the knowledge of the results of her past experience. Then the goal of quantum formalism is only to give recipes to allow agents to compute their personal degree of belief about what will happen if they do such and such experiment. But QBists refuse to be called instrumentalist:

*QBism has also breathed new life into instrumentalism (a term originally coined in by John Dewey to describe his brand of pragmatism). Instrumentalism, it seems, has a bad reputation. It is perceived as unable to explain the successes of our best theories, and as unwilling to reach for nature's essence. QBism tells us that this assessment may well be premature. What QBism shows is that an enriched instrumentalist attitude—in this case, seeing quantum theory as something that 'deals only with the object–subject relation', as Schrödinger once put it—may in fact be our best bet of arriving at explanations more profound and radical than anything we could have imagined.*    (Schlosshauer's foreword in Fuchs, 2015)

QBists claim that they endorse a special kind of realism they call 'participatory realism' where reality goes further than what can be said in a third-person account (Fuchs, 2016).

Perhaps a good way to introduce the basics of QBism is to quote what Fuchs (2018) says about three important tenets of this interpretation in the paper where he both acknowledges what QBism owes to Bohr and makes explicit the main divergences with the Copenhagen interpretation:

*Along the way, we lay out three tenets of QBism in some detail: 1) The Born Rule— the foundation of what quantum theory means for QBism—is a normative statement. It is about the decision-making behaviour any individual agent should strive for; it is not a descriptive 'law of nature' in the usual sense. 2) All probabilities, including all quantum probabilities, are so subjective they never tell nature what to do. This includes probability-1 assignments. Quantum states thus have no 'ontic hold' on the world. 3) Quantum measurement outcomes just are personal experiences for the agent gambling upon them. Particularly, quantum measurement outcomes are not, to paraphrase Bohr, instances of 'irreversible amplification in devices whose design is communicable in common language suitably refined by the terminology of classical physics'.*

# 41.2 Personal Experience and Subjective Probabilities

Strongly influenced by 'Wheeler's vision of a world built upon "law without law", a world where the big bang is here and the observer is a participator in the process' (Fuchs, 2018), QBism proposes a picture where the agent and the external world interact in a creative way to give birth to an experience for the agent. This allows them to separate the subjective from the objective, which in turn is the way to unscramble the omelette made by Heisenberg and Bohr according to Jaynes (1990), another great influencer for QBists:

*Our present QM formalism is not purely epistemological; it is a peculiar mixture describing in part realities of Nature, in part incomplete human information about Nature—all scrambled up by Heisenberg and Bohr into an omelette that nobody has seen how to unscramble. Yet we think that the unscrambling is a prerequisite for any further advance in basic physical theory. For, if we cannot separate the subjective and objective aspects of the formalism, we cannot know what we are talking about; it is just that simple.*

QBism then proceeds to sort what is subjective and what is objective. The quantum system is something real and independent of the agent. The quantum state represents a collection of subjective degrees of belief about it and the probabilities given by the formalism for a measurement are nothing but degrees of belief concerning the results that the agent could get if she were interacting with the quantum system. Since a measurement is any action taken by any agent on her external world and since this action creates something new for the agent, probabilities are a numerical way to express the degree of belief the agent has on such and such result and are a guide for determining the amount she would bet on a particular result.

Hence, quantum theory does not describe an objective reality existing independently of any observer and where events, that any observer can passively witness, happen by themselves. On the contrary, quantum theory is a personal guide that has the same general form and the same rules for all the agents but that each agent specifies to adapt it to her experience. Hence, quantum states, observables, Hamiltonians, density matrices etc. . . . are not absolute concepts but tools relative to each specific agent. There is no reason for two agents to share the same quantum state when considering a quantum system, nor to have the same probability for getting a particular result during a measurement. For QBists, probabilities cannot tell Nature what to do. Gambling on events is one thing, the events of the world are another. As they say, a quantum state has no 'ontic hold' on the world. They go so far as saying this even for probability 1, i.e. for an outcome that is certain; there is nothing intrinsic to the quantum system, i.e., no objectively real property of the system that guarantees this particular outcome.

> *The statement that an outcome is certain to occur is always a statement relative to a scientist's (necessarily subjective) state of belief. 'It is certain' is a state of belief, not a fact.* (Caves, Fuchs, and Schack, 2007)

Hence even if the probability of a 'yes' outcome is exactly 1 for the measurement of an agent that *'does not mean that the world is forbidden to give her a "no" outcome for this measurement'* because *'My probabilities cannot tell nature what to do'* (Fuchs, 2018).

We will see that this is a key point in the way QBism refuses the criterion of reality in the EPR argument.

## 41.3 THE MEASUREMENT PROBLEM

The measurement problem in quantum mechanics comes from the fact that there are two postulates describing the dynamics of the state of a system. The first one says that the state evolves deterministically according to the Schrödinger equation which is linear and unitary, and hence transforms a superposed state into another superposed state. The second one, the reduction postulate, says that under a measurement the wave function collapses, and the state is projected onto one of the eigenstates of the observable that is measured. So, in general, these two laws do not lead to the same results. The difficulty is that it is not possible to give a non-ambiguous description of what a measurement is. Hence, there is no clear rule to decide when each one of these two postulates should be used. The question is to explain why only one result among many possible others is obtained during a measurement. The way QBists solve the problem is very simple. First, for QBists, unitary evolution concerns the change between the belief the agent has at time to for the outcomes of a measurement she could perform at to and the belief she has at time to about a measurement she would perform at t1 > to. Second, for QBists a measurement is the experience of an agent. They assume from the beginning that the direct internal awareness of her private experience is the only phenomenon accessible to an agent which she does not model with quantum mechanics, and that the agent's awareness is the result of the experiment. Hence, there is no longer any ambiguity about when to use the reduction postulate, since it does not say anything about the 'real state' of the system that is measured and is nothing else than the updating of the agent's state assignment on the basis of her experience. QBists like to quote Asher Peres' (1978) famous sentence: '*Unperformed experiments have no results*'.

In particular, there is no measurement when there is no agent: a Stern and Gerlach apparatus cannot measure by itself the spin of a particle. QBists want to stress the fact that they depart from the standard Copenhagen interpretation in that a measurement is not '*an interaction between classical and quantum objects occurring apart from and independently of any observer*' (Fuchs and Schack, 2013). They quote here the Landau and Lifshitz formulation of the Copenhagen interpretation. To be fair, this quotation is not really faithful to Bohr's position which is more subtle than that. But it is true that even when it is taken in all its subtlety, Bohr's position is not able to solve the measurement problem in a satisfying way. Indeed it does not make any explicit resort to the observer to define a measurement while simultaneously it appeals to the use that is done (by her) of the macroscopic apparatus. For QBists an apparatus should be seen as an extension of the agent like another sense organ or a prosthesis. As such by itself it cannot create a result.

For QBism, any agent can use the quantum formalism to model any system external to herself whether they be atoms, apparatuses, or even other agents. For any agent, the personal internal awareness of other agents is inaccessible and not something she can apply quantum mechanics to. But communication with other agents (verbal exchanges for example) is. This means that an agent can use a description of another agent through a superposed state encoding her probabilities for the possible answers to any question she might ask before getting a definite answer. A consequence is that it seems that reality can differ from one agent to another. But a reservation should be made about the meaning of the comparison between the perceptions of two agents since this comparison could only be done from a meta-observer point of view, which does not exist. No third person point of view is allowed.

So an important feature of QBism is the fundamental role that the observer plays in the measurement process and the relinquishment of any absolute description of the world.

# 41.4 DECOHERENCE

Decoherence takes the interactions between the system and its environment into account (Zeh, 1970; Zurek, 1981, 1982, 1991; Caldeira and Leggett, 1981). It is usually held that it plays an important role both to solve the preferred basis problem through the selection of a 'pointer basis' and to show how the reduced density matrix of the system remains very nearly diagonal in this basis for a time exceeding the age of the universe. The Schrödinger equation is applied to a system and its environment so they become entangled and the reduced density matrix of the system made by tracing off the degrees of freedom of the environment is nearly diagonal. Decoherence is not a solution to the measurement problem (Zwirn, 1997, 2015, 2016) but it helps understand the classical appearance of the world (Joos *et al.*, 2003). In particular it explains why macroscopic objects do not exhibit interferences which the standard view of quantum mechanics predicts we should see with quantum superpositions. But for QBists the Schrödinger equation is not a law governing the behaviour of an objective quantum state. It is only a tool for describing how the agent's epistemic state evolves. Hence decoherence plays no role in their framework:

> *For a Quantum Bayesian, the only physical process in a quantum measurement is what was previously seen as 'the selection step'—i.e., the agent's action on the external world and its unpredictable consequence for her, the data that leads to a new state of belief about the system. Thus, it would seem there is no foundational place for decoherence in the Quantum Bayesian program. And this is true.*
>
> (Fuchs and Schack, 2012)

It is true that the main advantage of decoherence is to explain the classical appearance of the world if we take quantum states as representing real states of the systems. But for QBists, this does not need to be explained since the wave function is merely a probability distribution:

> *'Why does the world look classical if it actually operates according to quantum mechanics?' The touted mystery is that we never 'see' quantum superposition and entanglement in our everyday experience. But have you ever seen a probability distribution sitting in front of you? Probabilities in personalist Bayesianism are not the sorts of things that can be seen; they are the things that are thought. It is events that are seen.* (Fuchs and Stacey, 2019)

# 41.5 RECONTRUCTING THE QUANTUM FORMALISM FROM PROBABILITIES

Probabilities are at the heart of QBism. So it is natural to ask:

> *If quantum theory is so closely allied with probability theory, then why is it not written in a language that starts with probability, rather than a language that ends with it? Why does quantum theory invoke the mathematical apparatus of complex amplitudes, Hilbert spaces, and linear operators?* (Fuchs and Stacey, 2019)

One goal QBists try to achieve is to reconstruct the quantum formalism from probabilities, using tools coming from the growing field of quantum information theory (Fuchs and Stacey, 2019). One key ingredient is a hypothetical structure called a 'symmetric informationally complete positive-operator valued measure', a SIC POVM. This is a set of d2 rank-one projection operator $\Pi_i = |\varphi_i\rangle\langle\varphi_i|$ on a d-dimensional Hilbert space such that $|\langle\varphi_i||\varphi_j\rangle|^2 = \frac{1}{d+1}$ whenever i≠j. What is interesting is that the operators $H_i = \frac{1}{d}\Pi_i$ are positive semi-definite and sum to the identity so they can be considered as representing a generalization of a standard von Neumann measurement. The link between the density matrix ρ and the probabilities P(Hi) = tr(ρ Hi) is given by:

$$\rho = \sum_{i=1}^{d^2}\left[(d+1)P(H_i) - \frac{1}{d}\right]\Pi_i$$

This reconstruction is an ambitious goal that is far from being completed because it raises some very difficult mathematical problems. In particular, nobody knows today if SIC POVMs exist in general dimension (Fuchs, Hoang, and Stacey, 2017). This is not the place to analyse this programme here and, moreover, it is not necessary to understand the main conceptual points of QBism.

## 41.6 EPR and Non Locality

The Einstein, Podolsky, and Rosen (1935) paper (EPR) is probably one of the most influential papers in the field of quantum foundations, and the famous criterion of reality has been discussed thousands of times:

> If, without in any way disturbing a system, we can predict with certainty (i.e., with probability equal to unity) the value of a physical quantity, then there exists an element of reality corresponding to that quantity.

As is well known, this argument was initially intended to show that quantum mechanics is incomplete. The authors consider a system of two particles A and B which separate. By choosing freely to measure either the position or the momentum of A, it is possible to predict with certainty either the position or the momentum of B without 'disturbing' it. According to the criterion, the conclusion follows that B must have a definite position and momentum. Hence quantum mechanics is incomplete.

QBists deny this conclusion on the basis that, as we saw before, for them a probability of 1 says nothing about the system itself but only means that the agent is certain that if she does a measurement she will get the expected result. So they reject the criterion of reality.

Now the standard reasoning about EPR is that either quantum mechanics is not complete and that leads to assuming hidden variables or nature is not local through what Einstein called 'a spooky action at a distance'. As Bell's inequalities (Bell, 1964) show that local hidden variables are forbidden, the usual conclusion is that nature is not local. Indeed, since correlations between spatially separated (even by a space-like interval) measurements on A and B cannot be explained by local hidden variables, they must come from some sort of action at a distance of one measurement on the other. QBists deny this simply by noticing that (Fuchs, Mermin, and Schack, 2014):

> *Quantum correlations, by their very nature, refer only to time-like separated events: the acquisition of experiences by any single agent. Quantum mechanics, in the QBist interpretation, cannot assign correlations, spooky or otherwise, to space-like separated events, since they cannot be experienced by any single agent. Quantum mechanics is thus explicitly local in the QBist interpretation.*

# 41.7 Is QBism an Epistemic Interpretation?

It has become common to sort interpretations of quantum mechanics according to whether they treat quantum states as corresponding directly to reality or instead represent only our knowledge of some aspect of reality. This is the now classical distinction between the ontic and epistemic interpretations of the wave function (Spekkens, 2007). The ontic view posits that quantum states refer directly to reality. The epistemic view on the contrary assumes that quantum states refer only to our knowledge of part of reality. Moreover, the quantum state can be considered as a complete description of reality—which is called the $\psi$-complete view—or may require to be supplemented with additional variables—this is the $\psi$-supplemented view (Harrigan and Spekkens, 2010). The $\psi$-complete view is ontic and constitutes the orthodox interpretation of quantum mechanics. However this classification relies on the assumption that there is at least an underlying ontic state. As it is not the case for QBism, it does not fit into this classification because quantum states do not represent incomplete knowledge of some underlying reality. In this sense, QBism is not an epistemic interpretation but can be called a doxastic interpretation where quantum states are states of belief of an agent. This is a point that characterizes the difference between the initial position supported by Caves, Fuchs, and Schack before 2009 (Caves, Fuchs, and Schack, 2002, 2007) that can be called 'Quantum Bayesianism' and the current QBist position. As Stacey (2019b) says, quoting a letter that Fuchs sent to Mabuchi in 2000, the initial slogan 'maximal information is not complete' was locked into an epistemic mindset, rather than a doxastic one.

From this point of view QBism is similar (though different in many other aspects) to some other interpretations such as Leifer and Spekkens (2013), Bub and Pitowsky (2010), Brukner and Zeilinger (2009), Brukner (2017), and to pragmatist positions (Healey, 2012, 2016, 2017) which also suppose that there is no underlying ontic state.

## 41.8 QBism and the No-Go Theorems

A certain number of no-go theorems have been published recently. Pusey, Barrett, and Rudolph (2012) wanted to prove '*that any model in which a quantum state represents mere information about an underlying physical state of the system, and in which systems that are prepared independently have independent physical states, must make predictions which contradict those of quantum theory*'. Frauchiger and Renner (2018) argued that no single-world interpretation can be logically consistent, and that quantum theory cannot be extrapolated to complex systems and in particular to observers themselves. If true these no-go theorems would be a real issue for QBism. But actually they rely on precise prior assumptions that are not correct as far as QBism is concerned (Stacey, 2019a).

## 41.9 Criticisms

QBism has given rise to many criticisms since its inception. It seems that even the Wikipedia treatment of QBism has been wrong for a long time (Stacey, 2016). Many criticisms were caused by earlier ambiguous or even contradictory formulations (attributed rightly or not to QBism) that have been corrected in the recent version of QBism. As Stacey (2019b) says:

> Fuchs and Peres present a version of the Wigner's Friend thought-experiment. In their portrayal, Erwin applies quantum mechanics to his colleague Cathy, who in turn applies it to a piece of cake. The Fuchs–Peres discussion is, from a QBist standpoint, rather unforgivably sloppy about the distinctions between ontic degrees of freedom, epistemic statements about ontic quantities, and doxastic statements about future personal experiences.

This is true when we read:

> As soon as one detector was activated, her wave function collapsed. Of course nothing dramatic happened to her. She just acquired the knowledge of the kind of food she could eat.    (Fuchs and Peres, 2000)

But, beyond these early misleading presentations, it is true that the counter-intuitive picture QBism provides has often been a source of misunderstanding, when it was not of a mere expression of reluctance to consider a position which is not compatible with standard realism. Some of these criticisms missed the point and do not need to be detailed here. See for example Nauenberg's comment (Nauenberg, 2015) and the reply by Fuchs, Mermin, and Schack (2015). Norsen (2014) denied that the solution given by QBism to avoid non-locality was correct since it amounts to a quantum solipsistic solution. Khrennikov (2015) made a lot of critical comments but corrected them afterwards (Khrennikov, 2018). Stairs (2011) challenges the way probability-one assignments are dealt with by QBists. Mohrhoff (2014) agrees with some features of QBism such as the rejection of the EPR's reality criterion, the fact that a measurement is an act of creation, and that there is an external reality that cannot be fully represented by a quantum state. But he contests the QBist claims that a probability judgment is neither right nor wrong, that quantum correlations refer only to time-like separated events, and that all external systems, whether they be atoms or other agents, are to be treated quantum-mechanically. Marchildon (2015) gives a fairly good description of the main tenets of QBism and says that, once the notion of 'agent' is precisely defined (for instance, a mentally sane *Homo sapiens*), QBism is a consistent and well-defined theory. The definition of an agent is something on which we will come back in the next section. Marchildon's conclusion is nevertheless that he is not convinced, but he admits that it is mainly for personal preferences. Marchildon has the honesty to recognize that the picture of the world we get from QBism is so counter-intuitive that he does not like it. Many criticisms attempting to present more logical arguments against QBism are actually mere disguise for the same implicit personal feeling. Objections from Timpson (2013) are worth analysing more carefully, and we will come back to them in the next section.

## 41.10 DISCUSSION OF SOME ISSUES

We want to examine here questions coming neither from a bad understanding of the interpretation nor from a negative reaction due to a personal reluctance to the fact that QBism does not provide an image in line with a standard realist description of the world. Actually QBism leaves open some important questions whose answer would be necessary if one wants to get a precise detailed picture of what is going on. As Timpson (2013) says: '*Now, as a formal proposal, quantum Bayesianism is relatively clear and well developed. But it is rather less transparent philosophically*'.

I analyse below a list of questions that naturally arise when one tries to go beyond a superficial understanding, which at first sight might seem satisfying but raises many issues at second sight when one tries to understand QBism more precisely. These questions do not necessarily attack the internal consistency of QBism but are intended to better understand what it would mean for one to consider it not merely an instrumentalist position, but something that aims to give a description, even if partial, of the world. Some of them will nevertheless need clear answers to be sure that QBism avoids a real inconsistency. Depending on the way these necessary clarifications are made, the global image given by QBism can vary from pure instrumentalism to interpretations making the observer play a much more radical role (as in Convivial Solipsism that we will present in Section 41.11) than in the usual presentation made by the founders of QBism.

The first one is to clarify what it is that makes 'something' an agent. It would have been clearer for QBists to use the term 'observer' which is standard in quantum mechanics instead of 'agent' more often used in social science. In the context of physics, this gives rise to an ambiguity that may allow one to believe that something else than a human observer could be an agent. They argue that they do not have to define 'agent':

> *No, for the same reason you don't have to in any standard decision theoretic situation. [ . . . ] Thinking of probability theory in the personalist Bayesian way, as an extension of formal logic, would one ever imagine that the motion of an agent, the user of the theory, could be derived out [of] its conceptual apparatus?*
>                                                (DeBrota and Stacey, 2019)

But this way of escaping the problem is totally illegitimate here. It is true that the standard probability theory (whether Bayesian or not) does not have to define the user of the theory because the user does not play any role in the events of which probabilities quantify the occurrence. With or without any agent, the classical events occur and the agent is a passive witness to what has happened. The user is totally external. But it is not the case in QBism where, as largely emphasized by QBists, the agent is at the core of the fact that a result is obtained. It is the interaction between an agent and the external world that creates a result. Without agent, there is no result. A macroscopic apparatus is not able to create a result. So, the agent is not an entity external to the formalism as in the classical theory of probability, but an internal entity. Hence, they must define what an agent, who is necessary for creating a result, is. The way out they use is unacceptable.

Now, according to QBists, an agent must have experience. Does that mean that an agent must be conscious or that she must have perceptions? That is something that they are reluctant to accept. They don't want to be involved in the 'consciousness question'. Nevertheless, they admit that experience and consciousness are difficult to disentangle (private communication with David Mermin). But, they say that consciousness is not necessary for QBism and that experience is all that is needed. However they admit that a computer programmed to use quantum mechanics would not be an agent and that the only agents known today are human beings . . . Acknowledging that an agent must be conscious to have experience would help understanding what happens when an agent has an experience and what the status of the result she gets is.

QBism is not an idealist position and accepts the existence of an external world, even if this world is not describable in usual terms. According to QBism, it is through her interaction with the external world that the agent has an experience which is the result of the measurement.

> *An experiment is an active intervention into the course of Nature: We set up this or that experiment to see how Nature reacts. We have learned something new when we can distill from the accumulated data a compact description of all that was seen and an indication of which further experiments will corroborate that description. This is what science is about. If, from such a description, we can further distill a model of a free-standing 'reality' independent of our interventions, then so much the better.*

> *Classical physics is the ultimate example of such a model. However, there is no logical necessity for a realistic worldview to always be obtainable. If the world is such that we can never identify a reality independent of our experimental activity, then we must be prepared for that too.*    (Fuchs and Peres, 2000)

So far so good, but the following statement is much more problematic:

> *In the case of a gamble with consequences beyond an individual's expectations for their own longevity, the agent making the bet may be a community, rather than a single human being.*    (DeBrota and Stacey, 2019)

Having experience is perfectly understandable for a single human being but what does it mean for a community?

The kind of realism that QBism endorses is called 'participatory realism', wherein reality consists of more than can be captured by any putative third-person account of it (Fuchs, 2016).

So, there seems to remain no more ambiguity about what a measurement is since there is no measurement without an agent. If one admits that an agent creates a result through her experience, it becomes acceptable to claim that the role of quantum mechanics is to give rules allowing the agent to update her old beliefs with the result she got and from that, to compute her new beliefs about future experience. But once acknowledged that the quantum formalism concerns only the modelling of the agent's belief and that quantum states, operators, and Hamiltonians are subjective and relative to each agent, it remains necessary to be precise on what, independently of the formalism, bears her belief. What is the ontology? QBists are not very clear on it. Timpson (2013) tries painfully to give a description and warns the reader that, perhaps, the founders of QBism will not endorse it. The simple fact that it is necessary for Timpson to guess the ontology of QBism without even being sure that his guess will be faithful to what QBists have in mind shows that there is a genuine obscurity on this point. QBism endorses the existence of an external world independent of any agent, but it is not clear if the external world is unique and shared by all agents or if each agent has her own external world.

> *. . . the real world, the one both are embedded in—with its objects and events—is taken for granted. What is not taken for granted is each agent's access to the part of it he has not touched.*    (Fuchs, 2010a, 2010b)

QBism claims that the micro level of this world is something unspeakable. There is no law ruling the behaviour of micro-objects. Any event that occurs when a system and a measuring device interact is not determined by anything, not even probabilistically: 'Something new really does come into the world when two bits of it [system and apparatus] are united' (Fuchs, 2006). Two questions then arise. The first one is: how does this result appear from the interaction between a system at the micro-level which is unspeakable and without law, and a measuring device? Timpson proposes that systems have dispositions to give rise to events when they interact with one another.

He says that we can think of the world at each time as pregnant with possibilities, while yet, what will happen is completely undetermined:

> *The micro-objects are seats of causal powers and there is nothing to explain why they give rise to the particular manifestations of their powers that they do.*
> (Timpson, 2013)

This is reminiscent of the 'opium dormitive virtue' that Molière ridiculed. It is the kind of explanation that explains nothing and, I think, it would be more acceptable to say simply that we cannot talk of the process of giving rise to an event. The second question is the status of the result the agent gets. I think that it would be unfaithful to QBism to say that the result appears as a simple interaction between a quantum system and an apparatus and that the agent simply records it. It is probable that QBists would not agree today with Fuchs' sentence above where he said that something new does come into the world when a system and an apparatus are united, since it does not make any mention of an agent in the measurement process. That would come back to the Copenhagen interpretation and this is not what QBism says. A result is something happening in the agent's experience and without the agent, there is no result. QBism says on the one hand that the external world is independent of any agent and that a result comes from the interaction between an agent and the external world, and on the other hand that the result is only for the agent herself:

> *It says that the outcomes of quantum measurements are single-user as well! That is to say, when an agent writes down her degrees of belief for the outcomes of a quantum measurement, what she is writing down are her degrees of belief about her potential personal experiences arising in consequence of her action on the external world.*
> (Fuchs and Schack, 2014)

So the question is: is the result in the agent's mind or is it in the external world, which would imply that each agent owns her personal external world? If it is only in the mind of the agent, then the result is non-physical, and it is not true that 'something new comes into the world'. If the result is in the external world, how is it possible that the agent creates this result (which is not created simply by the interaction between the system and the apparatus)? It seems that we are back to the position of London and Bauer (1939) with a mental process that can have a physical impact on the world while resting outside of the scope of quantum mechanics. And if the result is in the agent's personal external world, is it shared with other agents or is it only available for the agent having created it? According to Fuchs and Schack (2014):

> *But by that turn, the consequences of our actions on physical systems must be egocentric as well. Measurement outcomes come about for the agent herself. Quantum mechanics is a single-user theory through and through—first in the usual Bayesian sense with regard to personal beliefs, and second in that quantum measurement outcomes are wholly personal experiences.*

So it seems that the result of a measurement made by an agent is not necessarily shared by another agent. But in this case, is this result a modification of something in the external world or not?

> *One often thinks of events as involving the modification of properties of objects, so don't we need some occurrent properties of the systems to be around to be modified if there are to be any events? Perhaps. But again we can (and should) be agnostic on the details.* (Timpson, 2013)

The fuzziness of that sentence clearly shows that many questions remain obscure. As we will see in Section 41.11, these questions find natural answers in Convivial Solipsism.

A second issue is related to the troubles QBism seems to have with explanation. If QBism considers that the quantum formalism concerns exclusively agents' beliefs, then physics does not speak of the world. As Timpson (2013) says '*Ultimately we are just not interested in agents' expectation that matter structured like sodium would conduct; we are interested in why in fact it does so*'. QBists' answers, relying on historical examples like Kepler's wrong explanation of the number of planets or on the abandonment of the idea of phlogiston (DeBrota and Stacey, 2019) to clear one's name, are not really convincing. It is true that the status of what is a good explanation has strongly changed from the Greeks to our time, but that does not mean that we don't feel when something provides no explanation at all, and that seems to be the case with QBism, especially with its emphasis on the indescribable aspect of the external world and the total absence of any law.

Now, there are some issues coming from the QBists' claim that all probabilities are purely subjective in the quantum framework. De Finetti, Ramsey, and Savage did not have quantum mechanics in mind when they designed their position. The subjective concept of probabilities makes sense in a deterministic world where the subjectivity comes from our ignorance of the details allowing us in principle to know the future result. But QBism does not say that quantum mechanics is ultimately deterministic. On the contrary, it even explicitly says that the result an agent gets during a measurement is fundamentally at random, and that there is an objective indeterminism. This makes a big difference with the context de Finetti, Ramsey, and Savage had in mind. So, it is difficult to deny that quantum probabilities, linked with the occurring of an objectively random event, do not contain at least a small, objective part. Mentioning a micro-level which is unspeakable, without law, and saying that there are no facts that determine the measurement outcome is not enough to support the claim that all probabilities are subjective. On the contrary, this seems to prove that something fundamentally random is at work and that objective probabilities are there, even if we cannot know them. Objective probabilities do not need to be computable to exist. QBism can coherently claim that these objective probabilities are not knowable, that they are out of the scope of the quantum formalism and that all probabilities taken into account in the formalism are subjective, but it cannot claim that only subjective probabilities exist. Moreover, the problem of explaining why the subjective probabilities provided by the formalism are better than anything else remains open. One answer could be that these subjective probabilities are based on an attempt to guess the underlying objective probabilities. But QBists refuse this answer and adopt an even more radical position. After a measurement having given the result d, the post measurement state is updated to |d> but that does not mean that the system is in the state |d> since this assignation only concerns the agent's state of belief. Of course, the probability for the agent to get the result d if she performs again the same measurement is 1, so the agent is sure to get d. But as, in QBism, probability 1 does not imply anything objective for the system, it does not need to be 'really' in the state |d>. The question is then: since the objective existence of the system is admitted by QBists, what can be said about the system? We come back to the questions we raised above. Has the system changed? If the answer is no, then what does it mean that the agent has created a non-pre-existing result in her interaction with the system? Is the result she saw only in her mind? And if the answer is yes, why not consider that the new state |d> objectively represents something (perhaps incomplete) about the system, be it only true for the agent herself? As Timpson (2013) says:

> Of course, this does not mean that whether or not an apparatus provides a particular read-out or any read-out at all, is a subjective matter; on measurement one of the outcomes d will, objectively, obtain.

So there is a link between the objective result d and the update of belief. Not recognizing that and maintaining at all costs that the state only represents the agent's subjective degree of belief seems a little bit artificial. That means that an agent can be certain that some particular outcome will obtain and simultaneously claim that there is no fact about what that outcome will be. That is fundamental for preserving the status QBism gives to probability 1, which is in turn necessary for avoiding the EPR criterion of reality. As Timpson (2013) rightly notices, this amounts to what he calls the 'Quantum Bayesian Moore's paradox': an agent is committed to say at the same time 'I am certain that p and it is not certain that p'. This is certainly problematic and the solution Timpson charitably proposes is not very convincing:

> The two halves of the pertinent sentences concern different levels of discussion; that the second claim—'it is not certain'—is made at a meta level of philosophical analysis; is said, perhaps, in a different tone of voice, from the ground level—'I am certain'; and thus the tension is diffused.

That seems more like a joke than a real solution . . .

At this stage, I do not claim that there is any definitive contradiction in QBism but only that all these points should be made clear in order to be able to get a global coherent picture. All the questions above may well find answers if things are made clearer about the status of the external world and what it means to be the result that an observer gets from a measurement. These issues find a natural explanation in Convivial Solipsism which is much more explicit on all these points. In the next section, I will show that answering these questions in a natural way and diminishing the emphasis put on the subjective probabilities, would lead QBism to be very near to Convivial Solipsism.

# 41.11 COMPARISON WITH CONVIVIAL SOLIPSISM (ConSol)

Convivial Solipsism (ConSol), which in spite of its name is not a solipsist position but a kind of realist one, was exposed first in a French book (Zwirn, 2000) and further developed in (Zwirn, 2015, 2016, 2017, 2020). It starts from the desire to get rid of the physical collapse of the wave function exactly as in Everett's interpretation (Everett, 1956). But the initial motivation for Convivial Solipsism comes from a remark by d'Espagnat (1971), who was totally unsatisfied with the astronomical number of worlds (or of minds) that is assumed in the various presentations of Everett's interpretation. ConSol is an interpretation inside which the physical dynamics of the universe is described by the Schrödinger equation and which states that a measurement is nothing else than the awareness of a result by a conscious observer. ConSol assumes that there is an external world that each observer models with her own wave function. This external world is not directly describable in terms of classical properties. Contrary to QBism, the wave function is considered as modelling (in a very limited way) the external world but a system represented by such a superposed wave function cannot be perceived by a human observer in all its richness. Consequently, an observer watching a superposed physical thing will perceive it through some mental filters allowing her to see only a definite result and giving her the impression that a definite result is 'really' obtained and that the system is 'really' in the state she perceives.

> Convivial Solipsism is situated in a neo-Kantian framework and assumes that there is 'something' else than consciousness, something that (according to the famous Wittgenstein sentence) it is not appropriate to talk of. This is close to what Kant calls 'thing in itself' or 'noumenal world'. Consciousness and this 'something' give rise to what each observer thinks is her reality, following Putnam's famous statement 'the mind and the world jointly make up the mind and the world'. So perception is not a passive affair: perceiving is not simply witnessing what is in front of us but is creating (independently for each us) what we perceive through a co-construction from the world and the mind. The hanging-on mechanism takes part in this co-construction and helps (very partially) understanding it through the selection it does.[1]    (Zwirn, 2016)

---

[1] Of course, it is out of the question to be able to explain how awareness happens. This is probably one of the most difficult problems of contemporary science. Neuroscientists are hardly beginning to understand some mechanisms that show how consciousness works and how different it is from how our own consciousness itself thinks it is working.

In QBism probabilities are agent's degrees of belief about the result she will get if she performs an experiment. ConSol puts less emphasis on the subjective aspect of probabilities, which is not the important point. Probabilities are simply about the observer's states of awareness, because, between many possible ones, only one appears. Indeed actually, what she sees is just the result of the limitation of her perception by these filters, and the system remains in a superposed state. This awareness has no physical impact on the systems that are measured and whose states are unchanged after the measurement. The reduction is only a way to describe what appears to the observer and does not affect the external world which remains superposed. Hence there is no physical collapse but, contrary to what happens in Everett's interpretation, the world does not split in as many versions as there are possible results. If we compare this with QBism, we see that the result of a measurement is also created in the interaction of the observer with the external world. Unlike QBism, ConSol relies on the decoherence mechanism to solve the problem of the preferred basis (Zwirn 1997, 2015, 2016). But a major common point between QBism and ConSol is that both say that quantum mechanics is about the interaction of an agent or an observer and an external world knowable at best very partially and imperfectly. But ConSol is clear about the fact that this result exists only in the observer's perception, and the reason why an observer gets a unique result is clear: that comes from the limited perceptive capacity of the observer who is unable to see a superposed state in its whole richness. QBism and ConSol both claim that it is not possible to give any familiar complete description of the external world:

> *Perhaps it is just the case that once one seeks to go beyond a certain level of detail, the world simply does not admit of any straightforward description or capturing by theory, and so our best attempts at providing such a theory do not deliver us with what we had anticipated, or with what we wanted. [ . . . ] The world, perhaps, to borrow Bell's felicitous phrase, is unspeakable below a certain level.*
>
> (Timpson, 2013)

ConSol rests on two main assumptions. The first one is the hanging-on mechanism which describes what a measurement is and how an observer must update her wave function after getting a result:

> *A measurement is the awareness of a result by a conscious observer whose consciousness selects at random (according to the Born rule) one branch of the entangled state vector written in the preferred basis and hangs-on to it. Once the consciousness is hung-on to one branch, it will hang-on only to branches that are daughters of this branch for all the following observations.*    (Zwirn, 2016)

The consequence is that even if the external world is not impacted by the measurement, everything goes for the observer as if a real result has been created and she must update her wave function by restricting it to the branch corresponding to the result she got. There is both a similarity and a difference in the way QBism and ConSol deal with measurements. The similarity is that a measurement is necessarily an interaction between an agent or an observer and the external world, and that a measurement is an act of creation. It is not a simple record of a pre-existing state of affairs. The difference is that in QBism, it seems (but that is not so clear) that the result is 'really created' in the external world of the agent, while in ConSol nothing changes in the external world, the result being entirely inside the observer's perception. As QBism, ConSol assumes that the quantum formalism is universal and can be used to model not only macroscopic systems but also other observers. That means that an observer can use a description of another observer through a superposed state encoding her probabilities for the possible answers to any question she might ask before getting a definite answer. The hanging-on mechanism guarantees that two observers will never disagree on the result of a measurement because any communication between the two is like a measurement of the one by the other (and vice versa). This is the reason why it is called Convivial. Another similarity between QBism and ConSol is that quantum states (and Hamiltonians and operators) are considered relative to the observer. QBism is clear about the personal use of quantum mechanics but is fuzzy about when a result created by an agent is shared with other agents. In ConSol, this question has no meaning because it is forbidden to speak simultaneously of the perceptions of two observers. Each sentence about a result has to be stated only from the point of view of a unique observer. Sentences like 'Alice saw "up" and Bob saw "down"' are not allowed. It seems that this is also the case for QBism since, otherwise, QBism's solution to non-locality would not work. Unlike QBism, ConSol does not consider that probability 1 does not give any information on the system, but nevertheless it avoids non-locality for the same reason as QBism: no single observer can make space-like separated measurements. Hence, if QBists are ready to go as far as to accept the fact that it is impossible to compare the results gotten by two agents, the picture given by QBism and ConSol would be similar regarding the fact that 'what is real' is only valid for one observer and that a third-person point of view is forbidden. Each observer lives in her world and QBism would merit also to be called 'a kind of solipsism' (in the weak sense adopted in ConSol) even though this is something QBists hate.

## 41.12 CONCLUSION

Many criticisms against QBism or ConSol come from the same source:

> Other confusions typically occur when the interlocutor has unwittingly switched from subjective probability to objective, or from a first-person perspective to third-person, midway through a thought process. These are habits which take discipline to avoid, at least at first.  (Stacey, 2019b)

In the case of ConSol, the confusion does not come from switching from subjective probability to objective, but from switching from a first-person perspective to third person. These two interpretations give a picture of the world which is radically different from what the majority of physicists are accustomed to, and even are ready to accept. In the list of the many interpretations that have been proposed, it is first necessary to eliminate those which are inconsistent (and there are many of them). Then the choice is largely a matter of taste since, at the moment, no reasonable and doable experiments exist to decide between those remaining.

## Acknowledgments

I want to thank Chris Fuchs for the vivid exchanges we have had in the colloquium at 'Les Treilles' and in Paris, and through many emails allowing me to better understand QBism. I want also to thank Ruediger Schack and David Mermin for former enlightening exchanges.

## References

Baez, J. C. (2003). Bayesian Probability Theory and Quantum Mechanics. http://math.ucr.edu/home/baez/bayes.html.

Bell, J. S. (1964). On the Einstein-Podolski-Rosen Paradox. *Physics*, 1, 195.

Brukner, Č. (2017). On the Quantum Measurement Problem. In R. Bertlmann and A. Zeilinger (eds), *Quantum [Un]Speakables II*. The Frontiers Collection. Cham: Springer International Publishing, pp. 95–117. *arXiv:1507.05255*.

Brukner, Č., and Zeilinger, A. (2009). Information Invariance and Quantum Probabilities. *Foundations of Physics*, 39(7), 677–89. *arXiv:0905.0653*.

Bub, J. (2019). Two dogmas redux. *arXiv:1907.06240*.

Bub, J., and Pitowsky, I. (2010). Two dogmas about quantum mechanics. In S. Saunders, J. Barrett, A. Kent, and D. Wallace (eds), *Many Worlds?: Everett, Quantum Theory and Reality*, Oxford: Oxford University Press, pp. 433–59. *arXiv:0712.4258*.

Caves, C. M., Fuchs, C. A., and Schack, R. (2002). Quantum Probabilities as Bayesian Probabilities. *Phys. Rev. A*, 65, 022305.

Caves, C. M., Fuchs, C. A., and Schack, R. (2007). Subjective Probability and Quantum Certainty. *Studies in History and Philosophy of Modern Physics*, 38(2), 255–74.

Caldeira, A. O., and Leggett, A. J. (1981). Influence of Dissipation on Quantum Tunneling in Macroscopic Systems. *Physical Review Letters*, 46(4), 211–14.

Caticha, A. (2007). From objective amplitudes to Bayesian probabilities. *AIP Conference Proceedings*, 889, 62. *arXiv:quant-ph/0610076*.

DeBrota, J. B., and Stacey, B. C. (2019). FAQBism. *arXiv:1810.13401*.

Einstein, A., Podolsky, B., and Rosen, N. (1935). Can quantum-mechanical description of physical reality be considered complete? *Phys. Rev.*, 47, 777–80.

d'Espagnat, B. (1971). *Conceptual Foundations of Quantum Mechanics*. New York: Benjamin.

Everett, H. (1956). The Theory of the Universal Wave Function. First printed in B. S. DeWitt and N. Graham (eds) (1973), *The Many-Worlds Interpretation of Quantum Mechanics*, Princeton, NJ: Princeton University Press.

de Finetti, B. (1990). *Theory of Probability*. New York: Wiley.

Frauchiger, D., and Renner, R. (2018). Quantum theory cannot consistently describe the use of itself. *Nature Communications*, **9**(1), 3711.

Fuchs, C. A. (2006). Quantum states: What the hell are they? And Darwinism all the way down (Probabilism all the way back up), http://www.perimeterinstitute.ca/personal/cfuchs/nSamizdat-2.pdf.

Fuchs, C. A. (2010a). Quantum Bayesianism at the perimeter. *arXiv:1003:5182*.

Fuchs, C. A. (2010b). QBism, the perimeter of Quantum Bayesianism. *arXiv:1003:5209*.

Fuchs, C. A. (2015). My Struggles with the Block Universe. *arXiv:1405:2390*.

Fuchs, C. A. (2016). On participatory Realism. *arXiv.org/abs/1601.04360*.

Fuchs, C. A. (2018). Notwithstanding Bohr, the Reasons for QBism. *arXiv:1705.03483*.

Fuchs, C. A., Hoang, M. C., and Stacey, B. C. (2017). The SIC Question: History and State of Play. *Axioms 6*, **21**. *arXiv:1703.07901*.

Fuchs, C. A., Mermin, D. N., and Schack, R. (2014). An Introduction to QBism with an application to the locality of quantum mechanics. *American Journal of Physics*, **82**, 749. *arXiv:1311:5253*.

Fuchs, C. A., Mermin, D. N., and Schack, R. (2015). Reading QBism: A Reply to Nauenberg. *American Journal of Physics*, **83**, 198. *arXiv:1502.02841*.

Fuchs, C. A., and Peres, A. (2000). Quantum Theory Needs no Interpretation. *Physics Today*, **53**, 70.

Fuchs, C. A., and Schack, R. (2012). Bayesian Conditioning, the Reflection Principle, and Quantum Decoherence. In Y. Ben-Menahem and M. Hemmo (eds), Probability in Physics, Berlin, Heidelberg: Springer-Verlag, pp. 233–47. *arXiv:1103.5950*.

Fuchs, C. A., and Schack, R. (2013). Quantum-Bayesian coherence. *Reviews of Modern Physics*, **85**, 1693. *arXiv:0906.2187* (2009).

Fuchs, C. A., and Schack, R. (2014). Quantum Measurement and the Paulian Idea. *arXiv:1412.4209*.

Fuchs, C. A., and Schack, R. (2015). QBism and the Greeks: why a quantum state does not represent an element of physical reality. *Physica Scripta*, **90**, 1. *arXiv:1412.4211*.

Fuchs, C. A., and Stacey, B. C. (2019). QBism: Quantum Theory as a Hero's Handbook. In E. M. Rasel, W. P. Schleich, and S. Wölk (eds), *Proceedings of the International School of Physics 'Enrico Fermi,' Course 197 – Foundations of Quantum Physics*, Italian Physical Society. *arXiv:1612.07308*.

Harrigan, N., and Spekkens, R. W. (2010). *Found. Phys.*, **40**, 125. *arXiv:0706.2661*.

Healey, R. (2012). Quantum Theory: A Pragmatist Approach. *British Journal for the Philosophy of Science*, **63**, 729–71.

Healey, R. (2016). Quantum-Bayesian and pragmatist views of quantum theory. In E. N. Zalta (ed.), *The Stanford Encyclopedia of Philosophy* (winter 2016 edn), Metaphysics Research Lab, Stanford University. https://plato.stanford.edu/archives/win2016/entries/quantum-bayesian/.

Healey, R. (2017). *The Quantum Revolution in Philosophy*. Oxford: Oxford University Press.

Jaynes, E. T. (1990). Probability in quantum theory. In W. Zurek (ed.), *Complexity, Entropy and the Physics of Information*, Redwood City, CA: Addison-Wesley, pp. 381–403.

Joos, E., Zeh, H. D., Kiefer, C., Giulini, D., Kupsh, J., and Stamatescu, I.-O. (2003). *Decoherence and the Appearance of a Classical World in Quantum Theory*. Berlin, Heidelberg: Springer.

Khrennikov, A. Y. (2015). External observer reflections on QBism. *arXiv:1512.07195*.

Khrennikov, A. Y. (2018). Towards Better Understanding QBism. *Foundations of Science*, **23**(1), 181–95.

Leifer, M. S., and Spekkens, R. W. (2013). Towards a Formulation of Quantum Theory as a Causally Neutral Theory of Bayesian Inference. *Phys. Rev. A.*, **88**(5), 052130. *arXiv:1107.5849*.

London, F., and Bauer, E. (1939). *La théorie de l'observation en mécanique quantique.* Paris: Hermann.

Marchildon, L. (2015). Why I am not a QBist. *Found. Phys.*, **45**, 754–61.

Mermin, N. D. (2013). QBism as CBism: Solving the problem of 'the Now'. *arXiv:1312.7825*.

Mermin, N. D. (2014). QBism puts the scientist back into science. *Nature*, **507**, 421–23.

Mermin, N. D. (2017). Why QBism is not the Copenhagen interpretation and what John Bell might have thought of it. In R. Bertlmann and A. Zeilinger (eds), *Quantum [Un]Speakables II*. The Frontiers Collection. Cham: Springer International Publishing. *arXiv:1409.2454*.

Mohrhoff, U. (2014). QBism: A Critical Appraisal. *arXiv:1409.3312*.

Nauenberg, M. (2015). Comment on QBism and locality in quantum mechanics. *American Journal of Physics*, **83**, 197–98. *arXiv:1502.00123*.

Norsen, T. (2014). Quantum Solipsism and Non-Locality. *Int. J. Quant. Found.*, John Bell Workshop.

Peres, A. (1978). Unperformed Experiments Have No Results. *Am. J. Phys.*, **46**, 745.

Pitowsky, I. (2002). Betting on the outcomes of measurements: A Bayesian theory of quantum probability. *arXiv:quant-ph/0208121*.

Pusey, M. F., Barrett, J., and Rudolph, T. (2012). On the reality of the quantum state. *Nature Physics*, **8**, 475. *arXiv:1111.3328*.

Ramsey, F. P. (1931). Truth and Probability. In F. P. Ramsey, *The Foundations of Mathematics and other Logical Essays*, edited by R. B. Braithwaite, New York: Harcourt, Brace and Company, p. 156.

Savage, L. J. (1954). *The Foundations of Statistics.* New York: John Wiley and Sons.

Spekkens, R. W. (2007). *Phys. Rev. A.*, **75**, 032110. *arXiv:quant-ph/0401052*.

Stacey, B. C. (2016). Von Neumann was not a Quantum Bayesian. *arXiv:1412.2409*.

Stacey, B. C. (2019a). On QBism and Assumption (Q). *arXiv:1907.03805*.

Stacey, B. C. (2019b). Ideas Abandoned on Route to QBism. *arXiv:1911.07386*.

Stairs, A. (2011). A loose and separate certainty: Caves, Fuchs and Schack on quantum probability one. *Studies in History and Philosophy of Modern Physics*.

Timpson, C. G. (2013). *Quantum information theory and the foundations of quantum mechanics.* Oxford: Oxford University Press.

Youssef, S. (1994). Quantum mechanics as Bayesian complex probability theory. *Modern Physics Letters A*, **9**(28), 2571–86. *arXiv:hep-th/9307019*.

Zeh, H. (1970). *Foundations of Physics*, **1**, 69–76.

Zurek, W. (1981). *Phys. Rev. D*, **24**, 1516.

Zurek, W. (1982). *Phys. Rev. D*, **26**, 1862.

Zurek, W. (1991). Decoherence and the Transition from Quantum to Classical. *Physics Today*.

Zwirn, H. (1997). La décohérence est-elle la solution du problème de la mesure? In B. d'Espagnat, M. Bitbol, S. Laugier (eds), *Physique et Réalité, un débat avec Bernard d'Espagnat*, Paris: Frontières/Diderot Editeur, pp. 165–76.

Zwirn, H. (2000). *Les limites de la connaissance.* Paris: Odile Jacob.

Zwirn, H. (2015). Decoherence and the Measurement Problem. *Proceedings of 'Frontiers of Fundamental Physics' 14*, PoS(FFP14), p. 223.

Zwirn, H. (2016). The Measurement Problem: Decoherence and Convivial Solipsism. *Found. Phys.*, **46**, 635. *arXiv:1505.05029.*

Zwirn, H. (2017). Delayed Choice, Complementarity, Entanglement and Measurement. *Physics Essays*, **30**, 3. *arXiv:1607.02364.*

Zwirn, H. (2020). Non Locality versus Modified Realism. *Found. Phys.*, **50**, 1–26, doi: 10.1007/s10701-019-00314-7. *arXiv:11812.06451.*

CHAPTER 42

..........................................................................................................................

# AGENTIAL REALISM—A RELATION ONTOLOGY INTERPRETATION OF QUANTUM PHYSICS

..........................................................................................................................

## KAREN BARAD

AGENTIAL Realism, a theory developed by physicist Karen Barad, offers a *relational ontology interpretation* of quantum physics. The theory is detailed in Barad's *Meeting the Universe Halfway: Quantum Physics and the Entanglement of Matter and Meaning* and other publications.[1] In these texts, Barad reconstructs Niels Bohr's philosophy-physics, takes into account more recent theoretical and experimental results, and elaborates Bohr's insights regarding what human knowledge can grasp of laboratory experiments into a more comprehensive ontological understanding of the nature of nature.[2] Barad thereby provides an interpretation of quantum physics that reworks core concepts of Western philosophy, including objectivity, causality, agency, subject, object, space, time, and matter.

A core issue in quantum physics, argues Barad, is ontological indeterminacy, not merely epistemological uncertainty. While Newtonian physics supports an ontology consisting of individual entities that precede their interactions (relata precede

[1] Karen Barad (2007), *Meeting the Universe Halfway: Quantum Physics and the Entanglement of Matter and Meaning* (Duke University Press) (hereafter '*Meeting*'). Barad began speaking and writing about the philosophical implications of Bohr's epistemology in 1987 (invited lecture delivered at Barnard College Conference on Women in the Natural Sciences, published in *The Barnard Occasional Papers*). See also K. Barad (1995), 'A Feminist Approach to Teaching Quantum Physics', in *Teaching the Majority*, edited by Sue V. Rosser (Teachers College Press), and K. Barad (1996), 'Meeting the Universe Halfway: Realism and Social Constructivism Without Contradiction', in *Feminism, Science, and the Philosophy of Science*, edited by Lynn Hankinson Nelson and Jack Nelson (Kluwer Press), originally submitted for publication in 1990.

[2] 'Philosophy-physics' denotes Bohr's commitment of the inseparability of physics and philosophy (see *Meeting*, p. 24).

relations), Barad argues that quantum physics resists such a metaphysics of individualism, and instead supports the understanding that *reality is made of relations*: that relata come into being only through and as constitutive parts of their **intra-actions** (as opposed to the usual 'interactions', the neologism marking a relational ontology rethinking of causal relations—see 42.1.5)—that is, relata only exist within relations. Intra-actions partially resolve the ontological indeterminacy in specific ways, to the exclusion of others, depending on the specific nature of the intra-action. Whereas Bohr's primary concern was the epistemological understanding of laboratory experiments, and in particular, the determination of a cut between the measuring apparatus and the object of investigation, agential realism goes further, giving an ontological account of worldly, and not just laboratory, intra-actions.

In establishing connections with core concepts of Western philosophy, Barad's work engages with a wide-ranging set of fields, including physics, philosophy of science, science studies, and social and political theories. In this article the focus will be on quantum theory, especially foundational issues in quantum physics. This chapter is divided into three sections: (42.1) reconstruction of Bohr's philosophy-physics; (42.2) the theory of agential realism—further elaboration of Bohr's philosophy-physics; and (42.3) an agential realist interpretation of quantum physics.

# 42.1 RECONSTRUCTION OF BOHR'S PHILOSOPHY-PHYSICS

## 42.1.1 Bohr's Epistemological Framework: Measurement Matters

Barad's starting point is a careful analysis and reconstruction of Bohr's philosophy-physics, offering a nuanced and unique understanding of his approach. The main points summarized here are directed towards Bohr's analysis of the nature of measurement.

Bohr questions key foundational elements of Newtonian physics that are consistent with ontological and epistemological assumptions of Western philosophy—such as the existence of individual objects with determinate properties independent of our experimental investigations of them. In particular, measurements are assumed to reveal an object's pre-existing properties (e.g., its position and momentum). On this account, the process of measurement is understood to be transparent and is therefore given scant attention in Newtonian physics. By contrast, Bohr sees the key issues in quantum physics as having to do with the question of the nature of measurement.

Barad characterizes Bohr's core argument as one of showing the fallacy of the Newtonian notion of measurement transparency, which is based on three fundamental assumptions: (1) *the metaphysics of individualism*: the world is composed of individual objects with determinate boundaries and properties; (2) *representationalism*: the property's determinate values can be represented by abstract universal concepts that have determinate meaning independent of experimental practice; and (3) *separability of knower and known*: measurements involve continuous determinable interactions enabling the values of the properties measured to be attributed to pre-measurement properties of objects separate from the measuring agencies. On Bohr's account, each of these fundamental assumptions is overturned.

While Heisenberg argued for an uncertainty principle placing a limit on what we can know, namely, the limits to simultaneously revealing a given particle's position and momentum, Bohr was critical of Heisenberg's analysis, saying it didn't go far enough. The fact that entities are disturbed when we measure them is, by itself, not a startling novelty of quantum physics—indeed, this point is already present in classical physics. For some measurements the disturbance imparted upon the object is negligible. And when the disturbance is not negligible, Newtonian physics trusts that measurement-independent values of the object's position and momentum can nonetheless be found since the disturbance can always be determined and subtracted out. It was this latter point, Bohr argued, that is no longer tenable. The question is why subtraction of the disturbance is not fully possible.

Barad argues that Bohr's analysis is based on his original and ground-breaking insight that *concepts are defined by the circumstances required for their measurements*. That is, *theoretical concepts are specific physical arrangements*, not mere ideas existing in human minds. For Bohr, measurement and description (the physical and the conceptual) entail each other. Bohr argues, for example, that the measurement of position requires an apparatus with measuring rods (i.e., rigid parts) used to define "position", whereas the measurement of momentum requires an apparatus that can respond to the transfer of momentum (i.e., one with movable parts) used to define "momentum"; the same part cannot be both immovable and movable at the same time, and therein lies the crux of the limits of measurement: it is not something mysterious but rather the incompatibility of complementary (mutually exclusive) material conditions of possibility.[3] Put differently, *position and momentum, or more generally, spatial-temporal and momentum-energy representations, are not simultaneously possible because they require mutually exclusive experimental circumstances for their determination/definition*. This crucial insight is in fact the basis on which Bohr concludes that measurement interactions are inherently indeterminable. Bohr argues that because concepts are defined by the circumstances required for their measurement, it is impossible to fully determine the nature of the measurement interaction—that is, the effect of the measurement interaction on (assumed) pre-existing values—since this would require the measuring apparatus to have complementary—mutually exclusive—features simultaneously. Therefore, *observation is only possible on the condition that the effect of the measurement is indeterminable*. Or to put it another way, Bohr argues that

---

[3] For details, see especially *Meeting*, Chapter 3, 'Niels Bohr's Philosophy-Physics: Quantum Physics and the Nature of Knowledge and Reality'.

*the measurement interaction can be accounted for only if the measuring device is itself treated as an object, defying its purpose as a measuring instrument.*

Bohr's argument concerning the indeterminacy of the measurement interaction is of profound consequence: since observations involve an indeterminable discontinuous interaction, *as a matter of principle, there is no unambiguous way to differentiate between the 'object' and the 'agencies of observation'. No inherent Cartesian subject-object distinction exists.* The question of what constitutes the object of measurement is not fixed. The boundary between the object of observation and the agencies of observation is indeterminate in the absence of an apparatus defined by its specific physical configuration. What constitutes the object of observation and co-constitutes the agencies of observation is determinable only on the condition that the measurement apparatus is specified. *The apparatus enacts a cut delineating the object from the agencies of observation. Therefore, observations do not refer to properties of observation-independent objects since they don't pre-exist as such.*

If the distinction between object and agencies of observation is not inherent, what sense, if any, should we attribute to the notion of observation? Bohr states that 'by an experiment we simply understand an event about which we are able in an unambiguous way to state the conditions necessary for the reproduction of the phenomena'.[4] Although no inherent distinction exists, every measurement involves a particular choice of apparatus, providing the conditions necessary to give meaning to a particular set of variables, at the exclusion of other essential variables, thereby placing a particular cut delineating the object from the agencies of observation. If one important consequence of this analysis is that we are not entitled to ascribe the value that we obtained for the position, for example, to some abstract notion of a measurement-independent object (i.e., the object as it presumably would have been before the measurement), what, then, does the value correspond to? If for Bohr, in the absence of a Cartesian subject/object distinction, objectivity is a matter of reproducibility, then what is the objective referent for a given measurement? In order to answer the question of the nature of the objective referent, Bohr, Barad argues, takes his analysis in an even more radical direction: *ontological indeterminacy.*

## 42.1.2 Indeterminacy, not Uncertainty

Although it is often said that complementarity and uncertainty are the cornerstones of the Copenhagen interpretation, Barad carefully distinguishes Bohr's and Heisenberg's positions regarding the nature of measurement, and argues that these respective

---

[4] Niels Bohr (1998) [1946 essay: 'Newton's Principles and Modern Atomic Mechanics'] *The Philosophical Writings of Niels Bohr, Vol. IV, Causality and Complementarity: Supplementary Papers*, edited by Jan Faye and Henry Folse, p. 130.

contributions constitute fundamentally different, indeed arguably incompatible, interpretive positions. (See Section 42.3.1 on the quantum eraser experiment.)

In Heisenberg's famous 1927 paper on the uncertainty relations, he considers the
measurement of the position of an electron using a gamma-ray (i.e., high-energy
photon) microscope.[5] In Heisenberg's analysis, the key issue is that the electron's
momentum is *disturbed* by the photon in the attempt to determine the electron's
position. This analysis, based on the notion of *disturbance*, leads Heisenberg to
conclude that the *uncertainty* relation is an *epistemic* principle—it says that there is a
limitation to what we can *know*. In other words, determinate values of the electron's
position and momentum are assumed to exist independently of measurement, however, we can't know both simultaneously; to the degree that we know the value of one
variable we remain *uncertain* about the value of the other, owing to the unavoidable
*disturbance* caused by the measurement interaction.

Barad points out that Heisenberg's analysis stops just at the point where Bohr's
begins. While Heisenberg's focus is on the disturbance entailed in measurement
interactions, Bohr introduces a second, arguably more fundamental, issue: that of the
conditions of possibility for determining the effect of the measurement interaction. In
this regard Bohr makes the following crucial point:

> Thus, a sentence like 'we cannot know both the momentum and position of an
> atomic object' raises at once questions as to the physical reality of two such
> attributes of the object, which can be answered only by referring to the conditions
> for the unambiguous use of space-time concepts, on the one hand, and dynamical
> conservation laws, on the other hand.[6]

Crucially, Bohr argues that one is not entitled to ascribe an independent physical reality
to these properties, or, for that matter, the object itself. *Heisenberg's analysis thus misses
the crucial question of Bohr's analysis: the existence of a fundamental indeterminacy,
and how the cut gets made and the indeterminacy is (partially) resolved.* Ultimately,
Heisenberg acquiesces to Bohr's point of view and acknowledges this major flaw in his
argument in an often overlooked postscript to his 1927 paper on uncertainty.

Barad argues that *for Bohr, the core issue is* not that of disturbance, but *one of the
conditions for the possibility of measurement, that is, for (partially) resolving the
inherent indeterminacy.* In other words, *for Bohr, the core issue is one of ontological
indeterminacy, not epistemological uncertainty.* Bohr understands the reciprocal relation between position and momentum, for example, in semantic and ontic terms, and
only derivatively in epistemic terms (i.e., we can't know something definite about
something for which there is nothing definite to know). Bohr's *indeterminacy principle*

---

[5] Werner Heisenberg (1927), 'The Physical Content of Quantum Kinematics and Mechanics',
*Zeitschrift fur Physik*, 43, 172–98.

[6] Niels Bohr (1963) [1949 essay: 'Discussion with Einstein on Epistemological Problems in Atomic
Physics'], *The Philosophical Writings of Niels Bohr, Vol.II, Essays, 1933–1957 On Atomic Physics and
Human Knowledge*, Woodbridge, CT: Ox Bow Press, pp. 40–41.

can be stated as follows: *complementary variables such as position and momentum are not simultaneously determinate.* The issue is not one of unknowability per se; rather, it is a question of *what can be said to simultaneously exist* – and this is conditioned by the specific physical configuration of the apparatus. Barad argues that Bohr introduced his indeterminacy principle as part of his complementarity framework that can be usefully contrasted with Heisenberg's uncertainty principle (see below). These contrasting and incompatible views of Bohr and Heisenberg call into question the coherence of the so-called Copenhagen interpretation.[7]

## 42.1.3  Bohrian Phenomenon

Barad points out that in Bohr's account a key point is *'quantum wholeness'*, or *the lack of an independent existence of 'object' and the 'agencies of observation'.* In the absence of a given apparatus there is no unambiguous way to differentiate between the object and agencies of observation; an apparatus must be introduced to (partially) resolve the ambiguity, and crucially, *the apparatus must be understood as an inseparable part of what is being described/measured.*

This is a central notion in Bohr's philosophy-physics and he uses the term *'phenomenon'* to designate *a particular instance of wholeness*:

> While, within the scope of classical physics, the interaction between object and apparatus can be neglected or, if necessary, compensated for, in quantum physics *this interaction thus forms an inseparable part of the phenomenon.* Accordingly, *the unambiguous account of proper quantum phenomena must, in principle, include a description of all relevant features of the experimental arrangement.*[8]

Barad importantly argues that for Bohr, a **phenomenon** is *the fundamental inseparability of 'observed object' and 'agencies of observation' (as delineated through the cut enacted by the specific physical configuration of the apparatus).*

## 42.1.4  Bohr's Account of Objectivity

Another crucial aspect of Bohr's philosophy-physics is the question of objectivity. Bohr insists that quantum mechanical measurements are objective. Since he also emphasizes the essential wholeness of phenomena, Bohr cannot possibly mean that measurements

---

[7] See also Mara Beller (1999), *Quantum Dialogue: The Making of a Revolution*, Chicago: University of Chicago Press.

[8] Niels Bohr (1963) [1958 essay: 'Quantum Physics and Philosophy: Causality and Complementarity'], *The Philosophical Writings of Niels Bohr, Vol.III, Essays, 1958–1962 On Atomic Physics and Human Knowledge*, Woodbridge, CT: Ox Bow Press, p. 4; italics mine.

reveal objective properties of independent objects. Rather, Bohr's use of the term *objectivity* is tied to the fact that 'no explicit reference is made to any individual observer'. For Bohr, *objective means reproducible and unambiguously communicable*—in the sense that 'permanent marks ... [are] left on bodies which define the experimental conditions'.[9] Bohr's notion of objectivity, which is not predicated on an inherent or Cartesian distinction between objects and agencies of observation, stands in stark contrast to a Newtonian sense of 'objectivity' denoting observation-independence.

According to Barad's reading of Bohr, a crucial point in his analysis is that the physical apparatus, embodying a particular concept to the exclusion of others, (contingently) marks the subject-object distinction thereby enabling Bohr to hold onto a particular conception of objectivity despite the fact that there is no inherent Cartesian cut. In short, concepts obtain their meaning in relation to a particular physical apparatus, which marks the placement of *a Bohrian cut* between the object and the agencies of observation, (partially) resolving the ontological indeterminacy. *This resolution of the ontological indeterminacy provides the condition for the possibility of objectivity.* In Bohr's account, objectivity requires accountability to 'permanent marks ... left on the bodies which define the experimental conditions'. Therefore, *'bodies which define the experimental conditions' serve as both the endpoint and the starting point for meaningful and objective scientific practice.*

The question remains: What is the objective referent for the determinate value of the measured property? Since there is no inherent distinction between object and instrument, the measured property cannot meaningfully be attributed to either an abstract object or an abstract measuring instrument. That is, the measured value is neither attributable to an observation-independent object, nor is it a property created by the active measurement (which would belie any sensible meaning of the word 'measurement'). Barad's reading is that, for Bohr: *measured properties refer to phenomena*, remembering that the crucial identifying feature of phenomena is that they include 'all relevant features of the experimental arrangement'. In other words, *the objective referent is not an observation-independent object, but rather a phenomenon.* Or to put it the other way around, *a condition for objective knowledge is that the referent is a phenomenon (and not an observation-independent object).*

Bohr's notion of objectivity—based on reproducibility and communicability of results—is epistemological in nature. Notably, this is insufficient for Einstein who insisted on the necessity of the absolute separation of observer and observed—an ontological condition—as the basis for objectivity. However, for Bohr, objectivity is not predicated on an inherent or Cartesian cut between the two; rather, what is required is an unambiguous and reproducible account of marks of bodies. For Bohr, the delineation of observer and observed is determined by the physics, and not by philosophical preconception. Einstein rejected Bohr's conception of objectivity in part because it seemed too weak and grounded not in reality but on intersubjective agreement.

[9] Niels Bohr (1963), p. 3.

## 42.1.5  A Bohrian Ontology: Phenomena and Intra-Actions

Bohr focuses on crucial epistemological questions, such as the conditions for the possibility of objective knowledge, but he does not give an account of how he understands the nature of reality. Barad attends to the implicit ontological dimensions of Bohr's philosophy-physics and proposes an ontology that coheres with Bohr's account. Barad then goes on to offer a more comprehensive account (see Section 42.2 on Agential Realism).

An essential component of Barad's reconstruction of Bohr's philosophy-physics is that *phenomena are constitutive of reality*.[10] That is, unlike in classical Newtonian physics, reality is not composed of individual entities with determinate properties; rather, *reality is made up of phenomena*. The smallest material unit of analysis is a phenomenon, which, in particular, includes the measurement apparatus that gives a determinate sense to particular features of phenomena while excluding others. Measurement apparatuses provide the conditions for the possibility of determinate boundaries and properties of 'objects' within phenomena, where *phenomena are the ontological inseparability of objects and agencies of observation*. (In their further elaboration, Barad suitably revises this understanding which is restricted to what can be said about laboratory measurements. See Section 42.2.1 below for Barad's agential realist conception of phenomena.)

Barad also argues that implicit in Bohr's account is a radically new understanding of causality. Since individually determinate entities do not exist, measurements do not entail an interaction between separate entities; rather, *determinate entities emerge from and within (that is, as co-constitutive parts of) their intra-action*. The term **intra-action** is introduced in recognition of *the ontological inseparability (entanglement) of entities that become co-constituted as (partially) determinate in particular ways (and not others) through their intra-action*. This is in contrast to the usual 'interaction', which presupposes a metaphysics of individualism (in particular, the prior existence of separately determinate entities). That is, *a Bohrian phenomenon is a specific intra-action of an 'object' and the 'measuring agencies'; the object and the measuring agencies emerge from, rather than precede, the intra-action that produces them*. That is, the measurement intra-action enacts a cut that co-constitutes and co-delineates 'object' and 'agencies of observation'.

*Intra-action is a unique notion of causality*, neither strict determinism nor total randomness (noncausal). As Bohr points out, the inseparability of the object from the apparatus 'entails...the necessity of a final renunciation of the classical ideal of causality and a radical revision of our attitude towards the problem of physical reality'.[11] Bohr rejects both poles of the usual dualistic thinking about causality—absolute freedom and strict determinism. Instead, Barad argues, for Bohr *that which*

---

[10]  For details see *Meeting*, especially pp. 125–8, 'A Bohrian Ontology: Phenomena and Intra-actions'.

[11]  Niels Bohr (1963) [1949 essay: 'Discussion with Einstein on Epistemological Problems in Atomic Physics'], *The Philosophical Writings of Niels Bohr, Vol. II, Essays, 1933–1957 On Atomic Physics and Human Knowledge*, Woodbridge, CT: Ox Bow Press, pp. 59–60.

constitutes 'cause' and that which constitutes 'effect' are co-constituted through their mutual intra-action.[12]

Crucially, then, we should understand *phenomena* not as objects-in-themselves, or as perceived objects (in the Kantian phenomenological sense), but as *specific intra-actions*. Not only is this *relational ontology understanding of phenomena* consistent with Bohr's insights, it is also consistent with recent experimental and theoretical developments in quantum physics (see Section 42.3 below).

## 42.1.6 Limitations of Bohr's Epistemological Account

Bohr's conception of apparatuses is limited in two significant ways: (1) growing out of the desire to provide an objective epistemological account of scientific measurements, Bohr crucially understands the apparatus as having a semantic role to play (giving definition to some concepts to the exclusion of others), and thereby ties it to the human-centred notion of linguistic concepts (making it difficult to understand how his insights might apply outside of the laboratory and to the larger world), and (2) in an effort to highlight complementary configurations of apparatuses he equates them with static configurations of laboratory instruments. That is, while Bohr's sketches of apparatuses show his exacting attention to detail (true to his insight that the material configuration of the apparatus matters), none of them are operable or alterable. Indeed, they are abstract images of laboratory instruments crafted by humans rather than real-world apparatuses that allow for the complexities of laboratory practices (e.g., everything it takes to make an experiment work, from turning dials, making modifications to an apparatus, applying for grants, etc.), or perhaps more to the point when it comes to proposing an ontology, practices beyond those of the laboratory that might have determinative effects.

Barad thus argues that Bohr's account of an apparatus requires major revisions: (1) a shift from a focus on human knowledge (epistemology) to that of a relational ontology; (2) a shift from linguistic concepts to discursive practices (crucially, Barad develops an understanding of *discursive practices* as material practices of mattering that are not human-based, but on the contrary, enable the notion of the 'human' to be understood as itself an emergent and contingent phenomenon); and (3) a shift from apparatuses as static laboratory set-ups to an understanding of apparatuses as ubiquitous and open-ended material-discursive practices through which all distinctions, including the core distinctions between subject and object, nature and culture, human and nonhuman, are constituted.[13]

---

[12] For details see *Meeting*, especially pp. 129–31, 'Causality'.
[13] For details see below and *Meeting*, especially Chapter 4, 'Agential Realism: How Material-Discursive Practices Matter'.

Significantly, Barad's agential realist approach offers a unique relational ontology interpretation of quantum physics. And it also provides an ontological condition for objectivity that does not require absolute separation of object and agencies of observation (as per Einstein).

## 42.2 AGENTIAL REALISM

Barad's *agential realism* approach provides an important alternative to representationalism— the idea that entities pre-exist and are independent of their representation. Representationalism is a metaphysical presupposition that has permeated much of Western thought including how scientists think about the relationship between the nature of scientific theories and reality. Upending one of the foundational beliefs of Western philosophy that words and things are independently definable/determinate, and that understanding constitutes the correct mapping of concepts onto independently existing things, agential realism rethinks the nature of theory and world and the relationship between them. It begins with insights from Bohr's understanding of quantum physics and further elaborates them, providing the basis for a unique interpretation of quantum theory.[14]

While Bohr's philosophy-physics constitutes a profound shift in the epistemological understanding of the nature of science it is limited in important and fundamental ways. Bohr's singular focus on laboratory measurements and questions of epistemology limits the implications of his account and places fundamental reliance on the role of the (unquestioned) human knowing subject. Referencing the limitations of the (so-called) Copenhagen interpretation,[15] the physicist John Bell summarizes his objections thusly: 'to restrict quantum mechanics to be exclusively about piddling laboratory operations is to betray the great enterprise. A serious formulation will not exclude the big world outside the laboratory'.[16]

This restriction is a serious matter and raises some crucial questions. Is it possible to build on Bohr's profound insights concerning the nature of measurement intra-actions without inserting the human into the foundations of the theory? Is it possible to understand this fundamental relationship ontologically and to understand the theory as saying something meaningful about the physical world rather than merely about

---

[14] Representationalism underlies both traditional forms of realism and social constructivism. In reworking core concepts of Western philosophy, agential realism engages with a wide-ranging set of fields, including physics, philosophy of science, science studies, and critical social and political theories. The unique form of realism proposed avoids the pitfalls that Rovelli mentions in cautioning against ontological and realist approaches, since ontology and realism take on new meanings.

[15] Barad argues that the so-called Copenhagen interpretation is an incoherent pastiche of various and contradictory points made by Bohr, Heisenberg, and others. See *Meeting* for more details.

[16] John Bell (1990), 'Against "Measurement"', *Physics World* (August), p. 34.

laboratory exercises or other human contrivances? Is it possible to secure objectivity on ontological, rather than merely epistemological, grounds?

In *Meeting the Universe Halfway*, Barad provides a detailed and unique explication of Bohr's account and its limitations, makes use of his insights to elaborate essential ontological dimensions of the account, and ultimately proposes an *agential realist interpretation of quantum physics* (see Section 42.3). The remainder of this section summarizes some key points of *agential realism*.[17]

## 42.2.1 Agential Realism: Understanding Phenomena Ontologically, or A Posthumanist Relational Ontology Approach to Quantum Physics

According to Barad's agential realist approach, phenomena are not mere laboratory relations between observer and observed (as per Bohr); rather, *phenomena are the ontological inseparability—entanglements—of intra-acting agencies.*[18] Phenomena are the basic units of existence: relations without pre-existing relata. Relations are not formed through interactions between or among independently existing things; rather, *intra-actions* enact *agential cuts* that co-constitute relata-within-relations. It is only through intra-actions that the differentiation of particular boundaries and properties of intra-acting agencies become determinate *within* (as co-constitutive parts of) phenomena. The very notion of intra-action—through which agencies are differentiated as they are entangled—provides a unique understanding of relationality and causality.

In performing this elaboration, Barad honours Bohr's crucial insight about the physicality of concepts while moving away from his singular focus on linguistic and epistemological concerns. Importantly, Barad proposes a crucial shift in understanding the nature of apparatuses: *apparatuses are to be understood not as mere laboratory instruments, but as open-ended and dynamic material practices through which boundaries and properties of agencies-within-phenomena are articulated.* Not only does this shift provide a much more dynamic and realistic conception of laboratory endeavours, it removes the restriction placed on quantum physics by interpretations that figure it as merely an account of laboratory or other human-based practices.

---

[17] The task of unearthing and removing humanist assumptions in Bohr's account, suitably generalizing his insights in order to obtain a robust and coherent account that can speak to the nature of reality and not merely that of human experience of reality, is a challenging undertaking. In order to accomplish this, Barad draws on philosophical notions that are generally not well-known by physicists. The specialized terms Barad uses (e.g., discursive practices) can be a stumbling block for those unfamiliar with them. See *Meeting* for more details of Barad's exposition.

[18] The term 'agencies' marks the dynamism of this ontology that replaces the more familiar notion of thingness that is the basis of the Newtonian ontology. For more details, see the agential realist understanding of agency later in this section and in *Meeting*.

Barad's elaboration begins by drawing on (and further elaborating) the philosophical notion of *discursive practices*, where *discourse* is not a synonym for language, but rather, *the material conditions of the possibility of meaning-making*. In this reconceptualization, sense-making or meaning-making (mattering), is not about signification. Rather, it is a matter of articulation, that is, of *how things differentially articulate with one another*. On Barad's account, **discursive practices** *are ongoing agential intra-actions of the world through which specific determinacies (along with complementary—materially excluded—indeterminacies) are enacted within the phenomenon thus produced. Discursive practices are causal intra-actions*—they enact causal structures through which one 'component' (the 'effect') of the phenomenon is marked by another 'component' (the 'cause') in their differential articulation. In its causal intra-activity, 'part' of the world becomes determinately bounded and propertied in its emergent articulation with another 'part' of the world. It is this iterative enactment of contingent and partial ontological determinacy that it is at the core of what 'discursivity' entails. *Discursive practices are material and causative boundary-drawing processes of iterative intra-activity.*

In this way, apparatuses are no longer strictly identified with laboratory instruments or even laboratory practices. Rather, apparatuses (in this more general sense) are material-discursive processes that are productive of and part of the phenomena they help produce. That is, **apparatuses** *are specific material configurations, or rather, dynamic reconfigurings, of the world that play a role in the production of phenomena.*[19] No longer tied to the 'human' who operates instruments from outside of nature, this formulation renders humans—understood now as specific phenomena—among the specific material configurations of the world. To the extent that human concepts have a role to play, they do so as part of the material configuration of the world. Importantly, reality does not depend on the prior or foundational existence of human beings; on the contrary, the point is to understand that humans are themselves natural phenomena that are contingently articulated, and scientific practices are simply a special case or subset of all possible processes. *Agential realism thus offers an account of the larger ontological issues of the nature of nature, rather than merely what humans can say about nature as explored under highly controlled conditions.* A key point is that specific material processes, or more precisely, specific dynamic material configurings of the world, *causally* produce specific material phenomena, as part of the ongoing differential performance of the world. *The world articulates itself differently through a dynamics of iterative intra-activity.* Put differently, *the world is made up of phenomena, that is, of specific material performances/articulations of the world.*[20]

---

[19] This shift from a metaphysics of individualism to a dynamics of relational becomings—material reconfigurings in their iterative intra-activity—suggests a need for grammatical shifts from nouns to verbs, which although it may come across as unnecessary is arguably more precise.

[20] As such phenomena are doings (processes/enactments/articulations), not things.

In Barad's agential realist account, matter does not refer to a property of independently existing things; rather, *matter* refers to phenomena (doings/articulations) in their ongoing iterative materialization. *Matter is substance in its interactive becoming—not a thing but a doing, a congealing of agency. Matter is a stabilizing and destabilizing process of iterative intra-activity.* Phenomena come to matter through this process of ongoing intra-activity. Materiality is reconceptualized in terms of the shift in the ontological objective referent from things to phenomena. Materiality and discursivity are mutually implicated in the dynamics of intra-activity. This is an outgrowth of the agential realist reconceptualizations of causality and agency.

Causal relations cannot be thought of as specific relations between pre-existing objects; rather causal relations emerge as a result of an agential cut delineating 'cause' and 'effect' in their co-constitution as specific relata within the phenomenon in question. *Causality is a matter of intra-activity.* And agency is not aligned with human intentionality or subjectivity; it is not a feature that someone or something has. Rather, *agency* is an enactment, a doing in its intra-activity. It is an enactment of iterative changes to particular processes. *Agency is a matter of the ongoing reconfiguring of the material world, including the material conditions of possibility of what comes to matter.* The notion of *intra-action*, therefore, represents a profound conceptual shift.

Crucially, *intra-actions enact* **agential separability**—*the condition of (contingent and relational) exteriority-within-phenomena, that is, differentiation within entanglements (differentiation without separability).* The notion of *agential separability* is of fundamental importance, for in the absence of a classical ontological condition of absolute exteriority between observer and observed (absolutely fundamental to Einstein's conception of objectivity and the basis for his rejection of Bohr's interpretation), it *provides an ontological condition for the possibility of objectivity.*

Importantly, then, while Bohr's conception of objectivity as reproducibility and communicability (reducing objectivity to intersubjective agreement) is limited by his epistemological focus on the objective outcome of laboratory experiments, the agential realist conception of objectivity is an ontological condition which honours Einstein's concerns without giving into his prejudice in favour of the separation of states for which there is now strong empirical and theoretical support against. According to agential realism, **objectivity** is understood in terms of *agential separability*—contingent and relational separability *within* the phenomenon—thereby offering an ontological account while respecting the nonseparability of entangled states. Crucially, on this account of objectivity, the objective referent is a phenomenon and not some alleged independently existing thing; correctly identifying the objective referent entails providing a comprehensive account of the phenomenon (the entire process).[21]

---

[21] Although in practice it is never possible to take account of the 'entire' phenomenon, not every aspect is equally important. A comprehensive account would mean not only identifying and taking account of the most important apparatuses that produce the phenomenon, but also taking account of what gets excluded.

# 42.3 AN AGENTIAL REALIST INTERPRETATION OF QUANTUM PHYSICS

In *Meeting the Universe Halfway: Quantum Physics and the Entanglement of Matter and Meaning,* Barad addresses foundational issues in quantum physics, drawing out an agential realist interpretation informed by new developments in theoretical and experimental physics. The remainder of this chapter focuses on (only) two key points from that discussion: an agential realist understanding of the quantum eraser experiment and a solution to the measurement problem.[22]

## 42.3.1 Empirical Evidence: The Quantum Eraser Experiment and Experimental Meta/physics

Barad offers a detailed discussion of the Einstein–Podolsky–Rosen (EPR) paradox and related tests of Bell's inequalities, together with the Bell–Kochen–Specker theorem, and draws out their philosophical implications: namely, the ineliminable contextuality of measurement; or to put it another way, the downfall of the metaphysics of individualism (the assumption that there are pre-existing individuals with a full set of determinate properties) This then entails the necessity of taking the full set of experimental arrangements into consideration (including changes to the apparatus) in the (full) identification of the phenomenon. Barad then proceeds to carefully analyse which-path experiments, including the quantum eraser experiment, and demonstrates both the consistency of the empirical results with a relational ontology interpretation of quantum physics, and also points to particular ways in which the assumption of metaphysical individualism continues to haunt and trouble the very interpretation of results that directly challenge and indeed outrightly deny this assumption.

The famous *gedanken* (thought) experiments that occupied Einstein and Bohr were about nothing less than the nature of reality. Einstein remained committed to a metaphysics of individualism and set out to prove that quantum mechanics, which

---

[22] Bohr is not alone in assigning epistemological concerns, and, in particular, the observer, a privileged role in quantum theory. Some interpretations go much further than Bohr and take mind or consciousness to be a primary and ineliminable ingredient. By contrast, Barad's agential realist approach takes Bohr's naturalist stance—that 'we are a part of that nature that we seek to understand'—further than Bohr, and offers an ontological account that decentres the human without situating the human as an external observer of the system of interest. Barad's *post-humanist* account is a thoroughgoing naturalism that understands humans as part of nature, and practices of knowing as natural processes of engagement with and as part of the world, rather than external impositions on the natural world. Barad demonstrates how an agential realist understanding of quantum physics provides a solution to the measurement problem without relying on a special or external system. For a more complete account, see especially Chapter 7 of *Meeting*.

seemed disinclined, if not adverse, to this metaphysics, was incomplete. By contrast, Bohr argued that quantum physics is not only complete but provides a coherent understanding of experimental results if one did not insist on imposing one's prejudice that favoured an individualist ontology. For example, with regard to the famous which-path experiment the focus of the debate between the two famous physicists was whether or not it was possible to determine which-path information without disturbing the interference pattern. In this case, Bohr argued that there is a complementary relationship between deploying an experimental arrangement to determine which-path information—that is, information about which particular path a particle travels along, or more precisely, which slit of a two-slit interference device it goes through—and the existence of an interference pattern on the detecting screen. According to Bohr, '[w]e are presented here with a choice of *either* tracing the path of a particle *or* observing interference effects',[23] but in any case quantum physics gives a complete account. In the 1990s, which-path experiments—for Bohr and Einstein occupying the realm of imagination only—became realizable in the laboratory.[24]

Experiments by Zou *et al.* (1991), for example, not only confirm Bohr's prediction that there is a complementary relationship between which-path determination and the demonstration of interference, the experiments even go further in lending support to Barad's reconstruction of Bohr's philosophy-physics where the question is one of *ability*—that is, the existence of the very conditions of possibility for distinguish*ability*—rather than one of disturbance. That is, the experiments confirm that *it is the material conditions of possibility that matter.*

> The disappearance of the interference pattern here is not the result of a large uncontrollable *disturbance*...in the spirit of the Heisenberg γ-ray microscope, but simply as a consequence of the fact that the two possible photon paths...have become distinguish*able*...whether or not this auxiliary measurement...is *actually* made... appears to make no difference. It is sufficient that it *could* be made, and that the photon path will then be identifiable, *in principle*, for the interference to be wiped out.[25]

In other words, all that is required to degrade the interference pattern is the *possibility* of distinguishing paths. That is, as Bohr had emphasized in his criticism of the EPR paper and Heisenberg's derivation and interpretation of the uncertainty relations, what

---

[23] Niels Bohr (1963) [1949 essay: 'Discussion with Einstein on Epistemological Problems in Atomic Physics'], *The Philosophical Writings of Niels Bohr, Vol. II, Essays, 1933–1957 On Atomic Physics and Human Knowledge*, Woodbridge, CT: Ox Bow Press, p. 46.

[24] See for example Scully and Drühl (1982), 'Quantum Eraser: A proposed photon correlation experiment concerning observation and "delayed choice" in quantum mechanics', *Physical Review A*, **25**(4), 2208–13; Scully, Englert, and Walther (1991), 'Quantum Optical Tests of Complementarity', *Nature*, **351**, 111–15; Dürr, Nonn, and Rempe (1998), 'Origin of quantum-mechanical complementarity approved by a "which-way" experiment in an atom interferometer', *Nature*, **395**, 33–7; among others. See also citations in footnotes 25 and 27.

[25] Zou, Wang, and Mandel (1991), 'Induced Coherence and Indistinguishability in Optical Interference', *Physical Review Letters*, **67**(3), 321; my emphasis.

is at issue is not the question of a disturbance but rather the '*possibilities of definition*' of the variables in question. Since it has been confirmed experimentally that the interference pattern disappears without any which-path measurement having actually been performed—but *just by the mere possibility of distinguishing paths*—these findings offer a clear challenge to any explanation of the destruction of the interference pattern that relies on a mechanical disturbance as its cause.

The famous quantum eraser experiment of Scully *et al.* (1982) is also highly significant. In a remarkable realization of the gedanken experiment discussed by Bohr and Einstein, their proposed experiment uses an apparatus that shoots rubidium atoms at a double-slit grating and adds a which-slit detector, built in part with micromaser cavities, placed in front of the two slits. On its way to the slits, each Rb atom encounters a laser tuned to a particular excitation; upon entering one of the micromaser cavities the atom de-excites leaving behind a tell-tale photon indicating which slit the atom went through on its way to the screen. The which-slit determination works by manipulating the internal degrees of freedom of the atom only such that it does not affect its forward momentum.[26] Hence, if the interference pattern no longer obtains, the explanation cannot be one of disturbance, which is the crux of Heisenberg's explanation. In this way, the Heisenberg disturbance-based explanation is hard-wired out of the experiment, which thus becomes a direct test of Bohr's account.

Experimental realizations of the proposal of Scully *et al.* confirm the validity of Bohr's analysis over those by Einstein and Heisenberg. Despite the lack of disturbance, they find a trade-off (complementarity) between which-path information and interference, as Bohr predicted. What then is the source of complementarity? The authors conclude that it is 'due to the system-detector correlations and not due to a random-ization of phase'.[27] That is, it is due to the correlation between the observed system and the measuring device (per Bohr) and not a disturbance (per Heisenberg). Barad argues that what this experiment shows is that *phenomena* (intra-acting agencies)—in this case, entanglements of the 'object' and the 'agencies of observation'—are in fact the root of complementarity (indeterminacy) and not uncertainty/disturbance.[28]

Barad furthermore points out the peculiarity of the framing by Scully *et al.*, who attribute complementarity to the uncertainty principle, rather than recognizing these as distinctive and contradictory positions. Their understanding is the familiar yet mistaken one, and yet it is precisely the significant distinction between Bohr's and Heisenberg's positions that *this very experiment* confirms, namely the validity of Bohr's philosophy-physics over Heisenberg's uncertainty-based epistemic interpretation. In fact, Barad argues that Bohr derives an 'indeterminacy principle'—a quantitative expression of complementarity—that has the same mathematical form but differs in derivation and

---

[26] See *Meeting*, Appendix C, 'Controversy Concerning the Relationship between Bohr's Principle of Complementarity and Heisenberg's Uncertainty Principle', pp. 402–403 for a more detailed discussion of this point.

[27] Scully and Walther (1989), 'Quantum Optical Tests of Observation and Complementarity in Quantum Mechanics', *Physical Review A*, 39(10), 5229.

[28] See also *Meeting*, Appendix B, 'The Uncertainty Principle is Not the Basis of Complementarity'.

interpretation from Heisenberg's uncertainty principle.[29] That is, the equations look identical; both can be represented as

$$\Delta x \, \Delta p \geq \frac{\hbar}{2},$$

but their interpretations and derivations are different: for Bohr, $\Delta$ stands for onto-logical indeterminacy rather than epistemic uncertainty as per Heisenberg.

Given the fact that the loss of interference is due to the entanglement of the object and the apparatus, rather than a disturbance, Scully *et al.* hypothesize that the erasure of which-path information, making the paths once again indistinguishable, might once again produce an interference pattern. This is the idea behind the **quantum eraser experiment**, which can in fact be done in **delayed-choice mode**—that is, the decision to make available conditions for erasure of which-path information (e.g., opening the shutters that separates the micromaser cavities) can be made *after the atom has gone through the one slit or the other and hit the screen.*

Remarkably, the experimenters find that it is in fact possible to find the interference pattern extant in the data set: crucially, this shows that *the entanglement is not destroyed.* This is a profound finding because, as Barad argues, it confirms the fact that which-path vs interference complementarity, following Bohr, is *not* derivative of the uncertainty principle which entails a disturbance per Heisenberg, and further-more, it is crucial to resolving the measurement problem (see Section 42.3.2). *The quantum eraser experiment thus confirms the core of Bohr's philosophy-physics, includ-ing the crucial point that the objects and the agencies of observation are inseparable co-constitutive aspects of a single phenomenon—that is, they form a material entangle-ment* (not merely a correlation of knowledge per Schrödinger[30]), *thus providing further evidence for a relational ontology interpretation.*

As such, for Bohr, the entire apparatus, or rather, following Barad's elaboration, *the entire material process must be taken into account* (in order to correctly identify the phenomenon in question). In this specific case, that would mean taking account of the entire apparatus (or set(s) of apparatuses), including the full set(s) of measurements enacted: for example, the fact that a which-path experiment was conducted, the which-path detector was modified, the modified which-path detector was used to erase which-path information for particular runs of the experiment, to name just a few elements to give of sense of what this entails. *This entire process, in its materiality, constitutes one phenomenon.*[31]

---

[29] For more details see *Meeting*, pp. 294–310, 'Complementarity II: Which-Path Experiments'.

[30] On Schrödinger's epistemic interpretation of entanglements, collapse is not a physical occurrence but rather simply a reduction in our ignorance, which is resolved by making measurements. For more discussion of Schrödinger's interpretation of entanglement, and in particular its epistemic nature, in contrast to what Bohr says, see *Meeting*, esp. p. 344.

[31] In fact, even finding that the interference pattern was evident after the erasure requires a detailed accounting of the entire process, including the series of apparatuses used and the details of their precise operation. See *Meeting*, pp. 310–17, 'Complementarity III: Quantum Erasers—Entanglements Rule!'

To hold to a metaphysics of individualism and focus on individual particles is to invite paradox. This kind of thinking has led some physicists to mistakenly claim that the quantum eraser experiment provides evidence of retrocausality: that is, the possibility of changing the past.[32] However, this interpretation is not consistent with the findings.[33] It is not the case that the experimenter can change a past that had already been present, or that atoms (somehow) fall in line with a new future simply by erasing information. Rather, the crucial point is that what was assumed to have taken place in the past, or more precisely, a taken-for-granted conception of the past itself, is the very thing that is being called into question here. Specifically, what is at issue is the very assumption that there is a pre-existing past (that includes all moments prior to a present one) where particles (have already) take(n) their place in space and time. In other words, an assumption that undergirds this paradoxical way of thinking is the notion of a past (i.e., events already passed and no longer present) that pre-exists as such. This way of thinking is inherent to the Newtonian conception of time: the absolute and equable passage of moments independent of all that happens. The quantum eraser experiment provides evidence contrary to this very conception of time. Contra Newton, the past doesn't pre-exist as such. Rather, 'past' acquires meaning and can be made determinate only in the course of specific measurements. For example, *it is only when the material conditions exist for determining if an atom passed through one slit or the other that one can give meaning to the claim that an atom will have gone through the top slit; in the absence of these conditions such a claim cannot be made.* The use of the future perfect tense here—"will have gone"—is a grammatical clue that time is phenomenal, rather than Newtonian, in nature. To summarize: The 'past' doesn't pre-exist as such; rather, it can only be given meaning within and as part of a specific phenomenon. Indeed, neither space nor time—neither where nor when— exist as determinate givens outside of phenomena.

While it has become common knowledge that entangled states seem to instantaneously communicate across large distances, in seeming contradiction to special relativity, Barad attributes to Bohr the understanding that this is an illusion created by the false assumption that there are individual particles separated in space. Bohr's objection to EPR is precisely about this point: there are not individual particles with independent existences that take their place in space and time prior to measurement; rather, they are inseparable from one another—they share the same existence, and the constituent components of the entangled state take their place in and through the measurement process.

To conclude, it is the metaphysics of individualism that compels Newton to postulate space and time as universal background features (existing independently of all that happens and) against which individuals are located and events take place. Given the downfall of individualism, as evidenced by this experiment, there is arguably a need to

---

[32] For example, U. Morhoff, 'Objectivity, retrocausation, and the experiment of Englert, Scully, and Walther,' *Am. J. Phys.* 67, 330–335 (1999).

[33] See also B.-G. Englert, M. O. Scully, and H. Walther, 'Quantum erasure in double-slit interferometers with *which*-way detectors,' *Am. J. Phys.* 67, 325–329 (1999) for a related critique of Morhoff.

rework the Newtonian conception of space and time. More in line with Einstein's insistence that time is what you measure with a clock (and space is what you measure with a ruler) albeit taken up in a Bohrian sense where the *clock (i.e., any apparatus that measures time or temporal sequence) is a material-discursive apparatus/process inseparable from that which it measures: time as it is defined and made manifest is part of a phenomenon in its specificity.* Time is not the given. Rather, *time is intra-actively constituted, that is, materialized in particular ways.* (One can make a similar argument for rethinking the nature of space.) In the case of the quantum eraser, for example, *time traces the material histories of intra-actions recorded in the data (iterative materializations).* That is, space and time (or, indeed, spacetime) need(s) to be understood in terms of phenomena, that is, in terms of iterative intra-activity: *spacetime is an emergent property of quantum entanglements.*[34] Spacetime (like position and momentum) is not a given universal, but *is itself a relational property that is defined and determinate only as a constitutive aspect of phenomena in their specificity.*

## 42.3.2 On the Problem of Measurement

> In the first place, we must recognize that a measurement can mean nothing else than the unambiguous comparison of some property of the object under investigation with a corresponding property of another system, serving as a measuring instrument, and for which this property is directly determinable according to its definition in everyday language or in the terminology of classical physics.    (Bohr)[35]

Although the question of measurement is central to Bohr's understanding of quantum physics, he does not directly discuss 'the measurement problem' which has become a focal point of discussions in the foundations of quantum mechanics. Perhaps the closest he comes to saying anything that might be seen as remotely or implicitly engaging this issue is the following:

> In the system to which the quantum mechanical formalism is applied, it is of course possible to include any intermediate auxiliary agency employed in the measuring processes. Since, however, all those properties of such agencies which, according to the aim of the measurement, have to be compared with corresponding properties of the object, *must be described on classical lines, their quantum mechanical treatment will for this purpose be essentially equivalent with a classical description.* The

---

[34] It's noteworthy that Barad offers this significant reconceptualization of the nature of space and time (spacetime) based on a critical examination of an experiment that has implications for foundational issues in quantum physics, and not from a consideration of general relativity in relation to quantum theory. Note that "emergence" cannot be thought in the usual sense of coming into being or becoming in time since time itself doesn't simply flow.

[35] Niels Bohr (1998) [1938 essay, 'The Causality Problem in Quantum Physics'], *The Philosophical Writings of Niels Bohr, Vol. 4*, edited by Jan Faye and Henry Folse (Ox Bow Press), p. 100.

question of eventually including such agencies within the system under investigation is thus purely a matter of practical convenience, just as in classical physical
measurements; and such displacements of the section [i.e., cuts] between object and
measuring instruments can therefore never involve any arbitrariness in the description of a phenomenon and its quantum mechanical treatment. *The only significant
point is that in each case some ultimate measuring instruments*, like the scales and
clocks which determine the frame of space-time coordination—on which, in the last
resort, even the definitions of momentum and energy quantities rest—*must always
be described entirely on classical lines, and consequently kept outside the system
subject to quantum mechanical treatment.*[36]

As this passage indicates, from Bohr's perspective there is no measurement problem,
per se: there is no indication that measurements entail any kind of collapse (physical or
otherwise), only cuts. What is at issue when it comes to measurements is just the set of
concerns which constitute the primary emphasis of Bohr's analysis: namely, the
unambiguous objective description of a quantum phenomenon which depends upon
a cut enacted by the larger apparatus that gives definition to particular concepts (i.e.,
descriptive terms familiar in classical physics such as the direction of a pointer) to the
exclusion of others and delineates 'object' from 'measuring instrument' within and as
co-constitutive parts of a phenomenon. Significantly, for Bohr, *the question is one of
description*, in particular, *a description that would ensure objectivity* (in Bohr's terms).

As was often the case, Bohr's attempts to clarify only spawned (further) misunderstandings. For one thing there has been a great deal of confusion about Bohr's
(seemingly unwarranted) insistence on the need to use classical concepts and to treat
the instrument classically. Furthermore, these points do not seem to address the
question that is most pertinent to those who insist that there is a measurement
problem: how is it possible to account for the fact that the measurement of any quantity
produces a determinate value even though the quantum formalism permits states with
indeterminate values? What is it about the nature of the correlation between 'object'
and 'observing instrument' such that a **superposition** of states (expressing ontological
*indeterminacy*) reduces, upon measurement, to a single determinate value for the
quantity in question?[37] Barad's explication and elaboration of Bohr's philosophy-
physics brings clarity to these issues.

According to Bohr, 'to measure the position [for example] of one of the particles can
mean nothing else than to establish a correlation between its behaviour and some

---

[36] Niels Bohr (1998) [1938 essay, 'The Causality Problem in Quantum Physics'], p. 104; my emphasis.
The question of Bohr's insistence on classical concepts has been widely misunderstood; see *Meeting*,
pp. 327–31, 'Are Classical Concepts Necessary for the Description of Quantum Phenomena?' for an
extended discussion and clarification of the issues.

[37] A *mixture* refers to a collection or ensemble of particles, each with a determinate value of the property
in question, such that the state of any given particle is determinate, although it may be unknown.
A *superposition* (of particle states) refers to an ontologically indeterminate state—a state with no determinate fact of the matter concerning the property in question. Superpositions embody quantum indeterminacy.
Superpositions and mixtures are physically distinguishable: they leave different traces.

instrument rigidly fixed to the support which defines the space frame of reference'.[38] Barad's agential realist elaboration of Bohr's framework opens up the possibility of a more explicit understanding of the nature of the correlation that measurements entail. Schrödinger, for example, introduces the notion of an entanglement between the object and the measuring system in order to try to understand the 'collapse'. However, Schrödinger construes the notion of entanglements in an epistemic sense: the shift in the wave function of the object is due to *the entanglement of our knowledge* of the 'object' with our knowledge of the 'measuring instrument'. Schrödinger then invokes the cognizing agent, a conscious subject, as the vehicle for the reduction of the wave function. By contrast, Barad argues that it is important to understand Bohr's notion of a **phenomenon** as an *ontological entanglement* of agentially intra-acting 'components' (e.g., 'object' and 'agencies' of observation). While at first glance, ontologizing the notion of entanglement might seem to make matters worse, Barad argues that not only does the resolution of the measurement quandary not require the introduction of a cognizing agent, but that the crucial point is in fact paying attention to the role of agential cuts, and in particular, agential separability. That is, this elaboration importantly resolves the difficulty without the need for a supplementary agent, like consciousness, as will be explained.

In Barad's discussion of recent experiments that address key foundational issues, evidence is presented to support the idea that *phenomena are ontological entanglements*. In their analysis of the quantum eraser experiment (see 42.3.1), Scully *et al.* show that *the entanglement of the object and the measuring apparatus accounts for the fact that an interference pattern results from the erasure of which-path information*. Indeed, the fact that an interference pattern can be found when the which-slit information is erased indicates that *the quantum coherence was not destroyed by the which-slit measurement, that is, the wave function superposition was not physically collapsed into a mixture*. In fact, the authors explain that the entanglement even accounts for the very fact that the which-slit detection 'destroys' the interference pattern. That is, *the wave function is not collapsed in the sense of physically destroying the superposition or entanglement so much as establishing/reconfiguring it (and even extending it) in such a way as to account for the observed phenomenon*. Hence, *there is an important sense in which measurements create, reconfigure, and indeed, further extend entanglements* (in the sense that measurements entail material configurations that articulate and produce correlations between 'instruments' and 'objects').

But how can we then explain the resolution of the ontological indeterminacy (as symbolized by a *superposition*) through measurement intra-actions? This paradox, whether quantum coherence is destroyed or reconfigured/extended through measurement intra-actions, goes to the core of the measurement problem.

*In an agential realist account, the resolution of ontological indeterminacy is understood in terms of the enactment of agential separability.* As discussed above, boundaries

[38] Niels Bohr (1998) [1935 essay, 'Quantum-Mechanical Description of Physical Reality'], p. 79.

and properties are only determinate *within* a given phenomenon through the enactment of an agential cut. The agential cut is determined by the materiality of the larger experimental arrangement, which delineates 'measured object' from 'measuring agency', while providing the material conditions of possibility for the articulation of particular physical variables to the exclusion of others. The body of the 'measuring instrument' is marked in its correlation with the particular agentially determinate property of the 'measured object'. *Agential separability—differentiation without separability—thereby provides an unambiguous resolution of the ontological indeterminacy within the phenomenon*; it is the cornerstone of the resolution of the measurement problem, as will be further explained.

A question that arises in this regard is Bohr's insistence on macroscopic measuring instruments that are to be described classically. This issue has been widely misunderstood. Barad helpfully clarifies Bohr's stakes, showing that it has everything to do with cuts, classical descriptions needed for objectivity, and ultimately Bohr's correspondence principle.[39] In Bohr's account, *all* systems (whether microscopic or macroscopic) evolve according to the Schrödinger equation, that is, the laws of quantum mechanics—period. So what does he mean by a 'classical description'? Bohr's insistence that classical descriptions are necessary to objectively (unambiguously) describe the outcome of a measurement is *not* an injunction to use the laws of Newtonian physics. What is at issue is a directive to use *a description appropriate for separately determinate systems* (as classical physics assumes). Separable systems can properly be described as mixtures. A *mixture* is a combination of *individual states with separately determinate values* of the property in question. Unlike the situation of an entangled state, a mixture can be expressed as the product of separate individual states. According to Barad's explication and elaboration of Bohr's account, *measurements do not produce absolutely separate states; rather, the states are agentially separable within the phenomenon*. What Bohr is essentially saying is that *within a phenomenon—where we have agential separability—*the mark on the measuring instrument (e.g., the direction of a pointer) is *describable* as a mixture (even though it is not strictly speaking a mixture). That is, it appears as a mixture if *the degrees of freedom of the instruments are bracketed, which is just what is done in describing the instrument classically*. Indeed, it is in this sense that a '**classical description**' is given. If, however, we are talking about a property that is not made determinate by the particular experimental arrangement in use (e.g., a complementary property), then we must use a quantum mechanical treatment—meaning that we must allow for *superpositions* that take the indeterminacy into account.

Now, as Bohr tells us, we are free to reposition the agential cut by using some further auxiliary apparatus such that the 'measuring agencies' are now included as part of the system under investigation (the 'object'). This constitutes the formation of some new phenomenon, and everything works in the usual way (except now, of course, the original measuring agencies no longer serve their purpose). Bohr tells us that 'the

---

[39] For more details see *Meeting*, Chapter 7 'Quantum Entanglements: Experimental Metaphysics and the Nature of Nature'.

only significant point is that in each case [no matter where the cut is made] some ultimate measuring instruments . . . must always be described entirely on classical lines, and consequently kept outside the system subject to quantum mechanical treatment'. That is, *agential separability is enacted in measurement intra-actions and it is precisely what is required for the possibility of an objective account.* Consequently, for Bohr, there is no measurement problem. *While superpositions must always be treated quantum-mechanically, the measurement question is addressed precisely by the need in securing objectivity to describe the agentially separable 'object' within a phenomenon classically (i.e., essentially as a mixture bracketing 'the agencies of observation').*

Bohr's argument, as per Barad's elaboration, is completely in line with the result obtained by Asher Peres in a paper entitled 'Quantum Measurements Are Reversible'.[40] Using a simple and elegant analysis, Peres finds that 'the state of the measured system alone, which was pure before its interaction with the instrument, *appears as a mixture as a consequence of the measurement, if the degrees of freedom of the instrument are ignored'.* But ignoring the instrument is precisely what Bohr has in mind when he says that the measuring instrument, described entirely on classical lines, is kept outside the system subject to quantum mechanical treatment. 'The state of the complete system (including the instrument) is always pure', writes Peres. The measurement process is indeed described by a 'continuous unitary transformation of the states vector of the observed system and the apparatus'; *there is no 'collapse' involved.*

*The crucial point is agential separability.* If there were an absolute separation we'd have a real mixture—independently existing entities—but that is not the case. Instead agential separability provides the possibility of giving a 'classical description'—that is, it can be described as if it were 'essentially' a mixture. That is, if we 'look inside' (within) the phenomenon, we would find that the mark the 'object'-within-the-phenomenon makes indicates a specific definite value of the property being measured. In other words, the 'measuring instrument'-within-the-phenomenon would indicate a definite value corresponding to one eigenvalue or another (e.g., there would be a pointer pointing in a specific direction). But this is precisely what we do when we perform a measurement—we are 'looking inside' a phenomenon. We don't have an outside view of the (entire) phenomenon itself, which is what is needed to observe the (extant) entanglement. To get such an outside view, we'd have to enlist a further auxiliary apparatus that can be used to measure the (original) phenomenon in question. Of course, such an attempt entails a further entanglement of the new auxiliary apparatus and the original phenomenon, constituting a new phenomenon. By some clever design we may in fact by this method be able to detect the entanglement (as in the ingenious quantum eraser experiments), but we still never see pointers indicating indeterminate values, since with every measurement we are 'looking inside a phenomenon', only now the phenomenon in question is the new extended one.

That is, *the agential cut provides for the registering of a determinate value of the property being measured, and hence, agential separability enables an objective*

---

[40] Asher Peres (1974), 'Quantum Measurements are Reversible', *American Journal of Physics*, 42, 886–91.

*description in terms of mixtures, <u>without destroying quantum coherence</u>.*[41] *What the agential cut provides is a contingent resolution of the ontological inseparability within the phenomenon and hence the conditions for objective description*: that is, it enables an unambiguous account of marks on bodies, but *only within the particular phenomenon. Strictly speaking, there is only a single entity—the phenomenon—and hence the proper objective referent for descriptive terms is the phenomenon.*

In summary, on this relational ontology account, there is no collapse. Rather, what is at issue is the enactment of an agential cut providing the material condition of *agential separability—contingent relational separability—within a phenomenon* (entanglement). Agential cuts (partially) resolve the ontological indeterminacy within the phenomenon. *The crucial point is that agential separability is enacted only <u>within</u> a particular phenomenon, and within a given phenomenon a determinate value of the property being measured is found as indicated by a mark on the 'measuring agencies' -within-the-phenomenon, which are bracketed for the purposes of providing a classical— that is, objective—description.* In this way, the value of the property in question is conditionally attributable to the object-within-the-phenomenon (not an independently existing object) although rigorously speaking the objective referent is the entire phenomenon. This point goes to the heart of the issue of objectivity: *in the absence of absolute separability, objectivity is secured through agential separability*, thus providing insight into Bohr's nearly singular focus on the objective nature of measurement. Hence, an agential realist elaboration of Bohr's epistemological account in terms of a more comprehensive ontological one provides a deeper understanding of the process of measurement and without restricting quantum physics to making claims about 'piddling' laboratory experiments. *The nature of nature is an ongoing dynamism of intra-activity through which the world is differentially articulated, and some of those articulations are what we call measurements.*

## ACKNOWLEDGEMENTS

I would like to thank Olival Freire, Jr for inviting me to contribute to this project and for his kind patience. For their support, encouragement, and interest in agential realism from early on, I thank physicists Herb Bernstein, Chris Fuchs, and Arthur Zajonc; thanks also to physicists Brian Josephson and Carlo Rovelli for reaching out more recently to express interest in agential realism. Thanks to Stephen David Engel and Alexei Kojevnikov for helpful copy-editing feedback.

---

[41] That is, decoherence actually entails the further extension/reconfiguration of the entanglement, not its dissolution. As Greenstein and Zajonc put it, "we might be tempted to argue that...the full experiment makes use of...macroscopic devices, subject to all the fluctuations which decoherence treats (i.e., the quantum behavior is destroyed because of the thermodynamic fluctuations of the larger environment). The quantum eraser experiment reveals that this, however, is not the case. In reality the full quantum behavior is still present, but it can only be seen by performing a different experiment" (*The Quantum Challenge: Modern Research on the Foundations of Quantum Mechanics*, 1997, 209).

# CHAPTER 43

·····································································································

# THE RELATIONAL
# INTERPRETATION

·····································································································

## CARLO ROVELLI

THE relational interpretation (or RQM, for Relational Quantum Mechanics) solves the measurement problem by considering an ontology of sparse *relative* events, or facts. Facts are realized in interactions between *any* two physical systems and are relative to these systems. RQM's technical core is the realization that quantum transition amplitudes determine physical probabilities only when their arguments are facts relative to the same system. The relativity of facts can be neglected in the approximation where decoherence hides interference, thus making facts approximately stable.

## 43.1 HISTORICAL ROOTS

·····································································································

In his celebrated 1926 paper (Schrödinger, 1926), Erwin Schrödinger introduced the wave function $\psi$ and computed the hydrogen's spectrum from first principles. This spectrum, however, had already been computed from first principles by Pauli four month *earlier* (Pauli, 1926), using the theory developed from Werner Heisenberg's 1925 breakthrough (Heisenberg, 1925), based on the equation

$$qp - pq = i\hbar, \tag{1}$$

with no reference to $\psi$. The theory we call 'quantum mechanics', in fact, had already evolved into its current full set of equations in the series of articles by Born, Jordan, and Heisenberg himself (Born and Jordan, 1925; Born *et al.*, 1926). Dirac, equally inspired by Heisenberg's breakthrough, got to the same structure independently in 1925, the year *before* Schrödinger's work, in a work titled 'The fundamental equations of quantum mechanics' (Dirac, 1925). (See (Fedak and Prentis, 2009; van der Waerden,

1967) for a detailed historical account.) Properly, the only Nobel Prize with the motivation 'for the creation of quantum mechanics' was assigned to Heisenberg.

So, what did Schrödinger achieve in 1926? With hindsight, he took a technical and a conceptual step. The technical step was to translate the unfamiliar algebraic language of quantum theory into a familiar one: differential equations. This brought the novel ethereal quantum theory down to the level of the average theoretical physicist. The conceptual step was to introduce the notion of 'wave function', soon evolved into the general notion of 'quantum state', $\psi$, endowing it with ontological weight.

The Relational Interpretation of Quantum Mechanics, or RQM, is based on the idea that this conceptual step, *which doesn't add anything to the predictive power of the theory*, was misleading: we are paying the price for the confusion it generated.

The mistaken idea is that the quantum state $\psi$ represents the 'actual stuff' described by quantum mechanics. This idea has pervaded later thinking about the theory, fostered by the toxic habit of introducing students to quantum theory in the form of Schrödinger's 'wave mechanics', thus betraying history, logic, and reasonableness.

The true founders of quantum mechanics saw immediately the mistakes in this conceptual step. Heisenberg was vocal in pointing them out (Kumar, 2008): first, Schrödinger's ground for considering $\psi$ to be 'real' was the claim that quantum theory is a theory of waves *in physical space*. This of course is wrong: the state of two particles cannot be expressed as two functions in physical space. Second, a pure-wave formulation misses the essential feature of quantum theory: discreteness, which must be recovered by additional assumptions as there is no reason for a physical wave to have energy related to frequency. Nobody expressed this point clearer than Schrödinger himself, who much later recognized: 'There was a moment when the creators of wave mechanics [that is, himself] nurtured the illusion of having eliminated the discontinuities in quantum theory. But the discontinuities eliminated from the equations of the theory reappear the moment the theory is confronted with what is observed'. (Schrödinger, 1996). Third, most importantly, if we take $\psi$ to be real we fall into the infamous 'measurement problem'. In its most vivid form (due to Einstein): a wave spreads over a region of space, but it suddenly concentrates into the single spot where the particle manifests itself. Schrödinger understood the difficulty with his early interpretation, changed his mind repeatedly about the interpretation of the theory (Bitbol, 1996), and became one of the most insightful contributors to the debate on the interpretation; but the badly misleading idea of taking the 'quantum state' as a faithful picture of reality stuck.

Heisenberg lost the political battle against wave mechanics for a number of reasons: differential equations are easier to work with than non-commutative algebras. 'Interpretation' wasn't so interesting for many physicists, when the equations of quantum mechanics begun producing wonders. Dirac found it easier to give the algebra a linear representation, and von Neumann followed: his robust math brilliantly focused on the algebras, but gave weight to their representation on Hilbert spaces of 'states'. And finally Niels Bohr—fatherly figure of the community—tried to mediate between his two extraordinarily brilliant bickering children, Heisenberg and Schrödinger, by obscurely agitating hands about a shamanic 'wave/particle duality'.

A way to get clarity about quantum mechanics is to undo the conceptual mess raised by Schrödinger's introduction of the 'quantum state'. This is what the Relational Interpretation does.

## 43.2 Quantum Theory is about Physical Facts, not Quantum States

RQM was proposed in the late Nineties (Rovelli, 1996), and has acquired increased clarity over time (Laudisa, 2001; Smerlak and Rovelli, 2007; Rovelli, 2016; Laudisa and Rovelli, 2017; Rovelli, 2017; Di Biagio and Rovelli, 2006). Interest in RQM has grown slowly but steadily, attracting attention during the last decade particularly from philosophers (van Fraassen, 2010; Bitbol, 2007; Bitbol, 2010; Dorato, 2016; Candiotto, 2017; Brown, 2009; Dorato, 2020).

RQM interprets quantum mechanics as a theory about physical *events of facts*. The theory provides transition amplitudes of the form $W(b, a)$ that determine the probability $P(b,a) = |W(b, a)|^2$ for a fact (or a collection of facts) $b$ to occur, given that a fact (or a collection of facts) $a$ has occurred. Facts are the independent variables of the quantum transition amplitudes.

A fact is for instance a particle having a certain value of a spin component at a certain time, or being at a certain place at a certain time. We perceive and describe the world in terms of such facts. An example of transition probability is the probability $P(L_\phi = \hbar/2, L_z = \hbar/2) = \cos^2(\phi/2)$ of having spin $\hbar/2$ in a direction at angle $\phi$ from the $z$-axis (a fact) if the spin in the direction $z$ was $\hbar/2$ (a fact). A fact is quantitatively described by the value of a variable or a set of variables.

Classical mechanics can equally be interpreted as a theory about physical facts, described by values of physical variables (points in phase space). But there are three differences between quantum facts and the corresponding facts of classical mechanics. First, their dynamical evolution laws are genuinely probabilistic. Second, the spectrum of possible facts is limited by quantum discreteness (for instance: energy or spin can have only certain values). Third, crucially, facts are *sparse* and *relative*.

Facts are *sparse*: they are realized *only* at the interactions between (any) two physical systems. This is the key physical insight that we can take from Heisenberg's seminal papers, and is a basic assumption of RQM.

Facts are *relative* to the systems that interact. That is, they are *labelled* by the interacting systems. This is the core idea of RQM. It gives a general and precise formulation to the central feature of quantum theory, on which Bohr has correctly long insisted: contextuality.

The insight of RQM is that the transition amplitudes $W(b, a)$ must be interpreted as determining physical probability amplitudes *only* if the physical facts $a$ and $b$ are *relative to the same systems*.

If all facts are relative to (or labelled by) the systems involved in the interactions, how come that we can describe a macroscopic world disregarding the labels? The reason is decoherence (Zeh, 1970; Zurek, 1981; Zurek, 1982): because of decoherence, a subset of all relative facts become *stable* (Di Biagio and Rovelli, 2006). This means that if we disregard their labelling we only miss interference effects that are anyway practically inaccessible relative to us, because of our limited access to the large number of degrees of freedom of the world. The conventional laboratory 'measurement outcomes' are a particular case of stable facts (Di Biagio and Rovelli, 2006); they are relative facts (realized in the pre-measurements) that can be considered stable because of the decoherence due to the interaction of the pointer variable with the environment.

# 43.3 The Relational Resolution of the Measurement Problem

## 43.3.1 The Problem

The measurement problem can be understood as the apparent contradiction between two postulates of the textbook formulation of quantum theory. On the one hand, the unitary evolution postulate states that the amplitude $W(b(t),a)$ for a fact $b$ to happen at time $t$ changes with $t$ according to the *linear* Schrödinger evolution equation $i\hbar\,\partial_t W(b(t), a) = H\,W(b, a)$, where $H$ is a linear operator. On the other hand, the projection postulate states that probabilities change when a fact occurs. The contradiction appears because of quantum interference: interference effects in the unitary evolution are cancelled by the projection.

Explicitly, suppose we know that a fact $a$ has happened and one of $N$ mutually exclusive facts $b_i$ ($i = 1...N$) can later happen. By composition of probabilities, we expect the probability $P(c)$ for a further fact $c$ to happen to be given by

$$P_{collapse}(c|a) = \sum_i P(c|b_i)P(b_i|a), \tag{2}$$

where $P(b|a)$ is the probability for $b$ to happen, given $a$. From the relation between probability and amplitude

$$P_{collapse}(c|a) = \sum_i |W(c, b_i)|^2 |W(b_i,a)|^2. \tag{3}$$

But since quantum probabilities are squares of amplitudes and amplitudes sum, linear evolution requires

$$P_{unitary}(c|a) = |W(c,a)|^2 = |\sum_i W(c,b_i)W(b_i,a)|^2$$
$$\neq \sum_i |W(c,b_i)|^2|W(b_i,a)|^2 = P_{collapse}(c|a). \tag{4}$$

So, what is the probability for $c$ to happen? The projection postulate demands it to be $P_{collapse}(c|a)$ but the unitary evolution postulate requires it to be $P_{unitary}(c|a)$, and the two are different because of interference.

The textbook answer is that the first holds 'if a measurement has happened', while the second holds if it hasn't. But what counts as a measurement? Does Wigner's friend's (Wigner, 1961) observation count as a measurement? Does Schrödinger's cat's (Schrödinger, 1935) observation of the releasing of the poison count as a measurement?

A positive answer to these two questions violates the universality of the evolution postulate. But a negative answer (as in the Many Worlds interpretation) obscures the relation between the theory and the facts in terms of which we account for the world.

## 43.3.2 The RQM Solution

The solution offered by RQM is that facts are labelled by systems. They are labelled by the systems involved in the interaction where the fact happens. In general, equation (3) does not hold if the $b_i$ have different labels from $c$, because the amplitudes $W(b,a)$ determine probabilities *only* if $a$ and $b$ are relative to the same system. This solves the apparent contradiction.

For instance, in the Wigner's friend scenario, the friend interacts with a system and a fact is realized *with respect to the friend*. But this fact is not realized with respect to Wigner, who was not involved in the interaction, and the probability for facts *with respect to Wigner* (realized in interactions with Wigner) still includes interference effects. A fact with respect to the friend does not count as a fact with respect to Wigner. With respect to Wigner, it only determines the establishment of an 'entanglement', namely the expectation of a correlation, between the friend and the system.

In the case of Schrödinger's cat, the release or not of the poison is a fact with respect to the cat. An external observer with superior measuring capacities can still detect interference effects (which would not have been possible if the release –or not– of the poison had become a fact in an interaction with her).

Notice that in an ontology based on facts rather than quantum states, the phrase 'Schrödinger's cat is in a quantum superposition' means *only* that we cannot use either the cat being dead or the cat being alive as inputs for transition amplitudes. This is precisely what RQM clarifies: facts are labelled by the systems involved in the interactions and the transition amplitudes $W(b,a)$ have physical meaning only if $a$ and $b$ are relative to a same system.

## 43.4 MEASUREMENT OUTCOMES AND RELATION WITH COPENHAGEN INTERPRETATION

If sufficient decoherence intervenes, the difference between $P_{collapse}(c|a)$ and $P_{unitary}(c|a)$ becomes negligible. In this situation we can safely consider the $b_i$ facts as realized, independently of their labelling. When we can disregard the labels of a fact, we call it 'stable' (Di Biagio and Rovelli, 2006). This is the case for the textbook laboratory quantum measurement outcomes (Di Biagio and Rovelli, 2006), which assume a macroscopic apparatus, and hence decoherence. The macroscopic world is entirely described by stable facts.

Could we base the ontology on the sole stable facts? This is what is done in the Copenhagen interpretation and in QBism (Fuchs and Stacey, 2019). The difficulty with this choice is that decoherence is generally approximate only, and only relative to the lack of access to environmental degrees of freedom. Therefore the stability of the stable facts is always only approximate. It requires the observer system to have properties that real systems have only approximately. RQM, on the other hand, is based on the observation that enlarging the ontology from stable facts to all relative facts resolves this difficulty, and therefore the difficulty of characterizing what is an observer and what is a measurement.

For relative facts, every interaction can be seen as a 'Copenhagen measurement', but only for the systems involved. Any physical system can play the role of the 'Copenhagen observer', but only for the facts defined with respect to itself. From this perspective, RQM is nothing else than a minimal extension of the textbook Copenhagen interpretation, based on the realization that any physical system can play the role of the 'observer' and any interaction can play the role of a 'measurement': this is not in contradiction with the permanence of interference through interactions because the 'measured' values are only relative to the interacting systems themselves and do not affect third physical systems.

In the absence of interactions, there are no facts and variables can be genuinely non-determinate, as in the Copenhagen interpretation (Brown, 2009; Wood, 2014; Calosi and Mariani, 2020).

## 43.5 MEANING OF THE QUANTUM STATE

What is then a 'quantum state'? In RQM, it is a bookkeeping of known facts, and a tool for predicting the probability of unknown facts, on the basis of the available knowledge. Since it summarizes knowledge about relative facts, the quantum state $\psi$ of a system (and a fortiori its density matrix $\rho$) does not pertain solely to the system. It pertains also to the other system involved in the interactions that gave rise to the facts considered known. Hence it is always a relative state. The idea that quantum states are relative states, namely states of a physical system relative to a second physical system, is Everett's lasting contribution to the understanding of quantum theory (Everett, 1957). Since $\psi$ is just a theoretical device we use for bookkeeping information and computing probabilities, it is not a surprise that it jumps if we learn that a fact happens.

A moment of reflection shows that the quantum states used in real laboratories where scientists use quantum mechanics concretely is *always* a *relative* state. The $\psi$ that physicists use in their laboratories to describe a quantum system is not the hypothetical universal wave function: it is the relative state, in the sense of Everett, that describes the properties of the system, relative to the apparatuses it is interacting with.

In a suitable semi-classical approximation we can write $\psi \sim e^{iS}$ where $S$ is a Hamilton–Jacobi function. This shows that the physical nature of $\psi$ is the same as the physical nature of a Hamilton–Jacobi function. Obviously giving $S$ ontological weight in classical mechanics is a mistake: $S$ is not a picture of the 'actual stuff' out there: it is a calculation device used to predict an outcome on the basis of an input.

Quantum mechanics does not need to be interpreted as the theory of the dynamics of a mysterious $\psi$ entity, from which the world of our experience emerges through some laborious argument. It can simply be interpreted as a theory for computing the probability of facts occurring given that other facts have occurred. There is no contradiction with the Pusey–Barrett–Rudolph (PBR) theorem (Pusey et al., 2012) because an implicit assumption of that theorem is not assumed in RQM (see Section 43.10).

There is a simple observation that confirms that it is a mistake to charge the quantum state with ontological weight (Rovelli, 2016). Consider particles with spin that undergo sequences of laboratory measurements of their spin components along different axes. Say that at time $t$ the spin in the $z$ direction of a particle is positive. We can predict that (if nothing else happens in-between) the spin has probability $cos^2(\phi/2)$ to be up in a direction at angle $\phi$ with the $z$ direction. *This is true irrespective of which comes earlier between $t$ and $t'$* (Rovelli, 2016): quantum probabilistic predictions are the same forth and back in time (Einstein *et al.*, 1931). Now, what is the state of the particle during the time interval between $t$ and $t'$? Answer: it depends on what we consider to be known: if I know the past (respectively, future) value, I can use the state to predict the future (respectively, past) value. This shows manifestly that the state is a coding of our information; not something the particle 'has'. It is reasonable to be realist about the values of the spin, which we observe, but not about the $\psi$ in-between, *because $\psi$ depends on a time orientation, while the observable physics does not.*

# 43.6 DISCRETENESS

Several interpretations of quantum theory neglect discreteness. Discreteness is not an accessory consequence of quantum theory, it is at its core. Quantum theory is characterized by the Planck constant $h = 2\pi\hbar$. This constant sets the scale of the discreteness of the world and determines how bad the approximation provided by classical mechanics's continuity is.

A fact is quantitively described by the values of an ensemble of variables, that got determined, or 'measured', in the interaction. The space of the values of these variables is the *phase space* of the system. For each degree of freedom, the phase space is two-dimensional. A measurement has finite precision: it determines a region $R$ of phase space. Classical mechanics assumes that $R$ can be taken to be arbitrarily small.

The volume $V(R)$ of a phase space region $R$ has dimensions $Length^2 \times Mass/Time$ per degree of freedom, namely action. The Planck constant, which has dimensions of an action, fixes the size of the smallest region that a measurement can determine:

$$V(R) \leq 2\pi\hbar \tag{5}$$

per degree of freedom. This is a most general and important physical fact at the core of quantum theory.

It follows that the number of possible values of a variable distinguishing points within a finite region $R$ of phase space is at most

$$N \leq \frac{V(R)}{2\pi\hbar} \tag{6}$$

Hence such a variable can take *discrete* values only. Any variable separating finite regions of phase space is discrete.

Quantum mechanics gives the values that a physical quantity can take. Variables are represented by self-adjoint elements $A$ of the non-commutative ($C^*$) algebra defined by $qp - pq = i\hbar$. The values $a$ that a variable $A(q, p)$ can take are the spectral values of its algebra element $A$ (those for which $(A - a1)$ has no inverse).

## 43.7 INFORMATION

The 1996 seminal RQM paper (Rovelli, 1996) indicated *information* as a key concept to understand quantum theory (under the influence of John Wheeler (Wheeler, 1990; Wheeler, 1991)) and suggested the programme of understanding quantum theory by deriving its peculiar formalism from a transparent set of elementary 'postulates' formulated in terms of information theory. The hope was that these would clarify the meaning of the formalism in the manner Einstein's two Special Relativity postulates had clarified the physical content of the Lorentz transformations.

The two postulates proposed in (Rovelli, 1996) are:

**P1**  There is a finite maximal amount of relevant information that can be obtained from a physical system.

**P2**    It is always possible to obtain new relevant information from a system by interacting with it.

Here a compact classical phase space is assumed. (Any classical system can be approximated by a system with a compact phase space.) 'Relevant' information means information that contributes to the possibility of predicting the outcome of future interactions. The two postulates are not in contradiction with one another because when new information is acquired part of the old information becomes irrelevant. For instance, measuring the spin of a spin-1/2 particle along a given direction makes the result of any previous spin measurement irrelevant for the probability distribution of future spin measurements. Related ideas were independently considered by Zeilinger and Brukner (Zeilinger, 1999; Brukner and Zeilinger, 2003).

The first postulate captures the characteristic quantum discreteness. It corresponds to (5). 'Information' means here nothing else than 'number of possible distinct alternatives'.

The second postulate captures the probabilistic aspects of the theory, because in a deterministic theory there is no way of adding new information once the full information about a system is achieved.

Historically, the paper (Rovelli, 1996) was followed by the development of epistemic interpretations like QBism and other interpretations based on quantum information (Fuchs, 1998; Fuchs, 2001; Spekkens, 2014). It has also given the impetus to the programme of reconstruction of the formalism of quantum mechanics from physically transparent postulates based on information theory (as a 'theory of principle') (Hartle, 1968; Hardy, 2001; Grinbaum, 2005; Spekkens, 2007; Dakic and Brukner, 2009; Masanes and Müller, 2011; Kochen, 2013; Oeckl, 2019), which has grown in a number of different directions. See also (Auffèves and Grangier, 2018; Auffèves and Grangier, 2019) where similar ideas have appeared. A particularly successful realization of the reconstruction programme is in the work of Höhn (Höhn, 2017; Höhn and Wever, 2017), which uses postulates directly based on P1 and P2 above.

'Information' is understood in this context in its purely physical sense, namely as correlation: a system has information about another system if the number of possible states of the two systems is less that the product of the number of possible states of each. For instance a measuring apparatus that has information about a system after the measure because its pointer variable is correlated to the variable of the system.

The term 'information' is ambiguous, with a wide spectrum of meanings ranging from epistemic states of conscious observers all the way to simply counting alternatives, à la Shannon. As pointed out by Dorato (Dorato, 2017), even in its weakest sense information cannot be taken as a primary notion from which all others can be derived, since it is always information about something. Nevertheless, information can be a powerful organizational principle in the sense of Einstein's distinction between 'principle theories' (like thermodynamics) versus 'constructive theories' (like electromagnetism) (Spekkens, 2014). The role of the general theory of mechanics is not to list the ingredients of the world—this is done by the individual mechanical theories, like the Standard Model of particle physics, general relativity, of the harmonic oscillator. The role of the general theory of mechanics (like classical mechanics or quantum mechanics) is to provide a general framework within which specific constructive theories are realized. It is in this sense that the notion of information as a number of possible alternatives plays a role in accounting for the general structure of the correlations in the physical world.

It is in this sense that the two postulates can be understood. They are limitations on the structure of the values that variables can take. The list of relevant variables, which define a physical system, and their algebraic relations, are provided by specific quantum theories.

## 43.8 THE RELATION BETWEEN FORMALISM AND INTERPRETATION

Heisenberg's breakthrough has been the idea of keeping the same equations as in the classical theory, but replacing commuting variables with non-commuting ones, satisfying (as soon later realized by Born and Jordan) equation (1). One can say that quantum theory has the same equations as the classical theory, plus this single equation. The formalism of quantum theory is condensed in a single equation. The entire quantum phenomenology follows from this equation.

As is well known, the Heisenberg uncertainty relations

$$\Delta q \, \Delta p \leq \hbar/2 \tag{7}$$

can be derived from (1) in a few lines. Hence (1) leads immediately to (6), hence to discreteness, establishing the quantitative aspect of discreteness (via $\hbar$). This is directly related to the first postulate.

On the other hand, the non-commutativity expressed in (1) reflects the fact that the result of the measurements of $q$ and $p$ depends on the order in which measurements are made: this is what blocks the possibility of a complete specification of all the variables of a system: it is directly related to the second postulate. Both postulates are made concrete in the maths by equation (1).

In other words, the non-commutativity of the algebra of the variables (measured by $\hbar$) is the mathematical expression of the physical fact that variables cannot be simultaneously sharp, hence there is a ($\hbar$-size) minimal volume attainable in phase space, and predictions are probabilistic.

The fact that values of variables can be predicted only probabilistically raises immediately the key interpretational question of quantum mechanics: when and how is a probabilistic disposition resolved into an actual value? RQM has a simple answer:

when any two systems interact, provided that we label the resulting facts with the interacting systems themselves.

## 43.9 OTHER CONSIDERATIONS

*No-go theorems for non-relative facts.* There are a number of no-go theorems for non-relative (absolute) facts (Frauchiger and Renner, 2018; Brukner, 2018), and some experimental confirmations of them (Bong *et al.*, 2020; Proietti *et al.*, 2019). The existence of facts not labelled by systems is among the inputs of these theorems, hence these no-go theorems can be taken as direct evidence in favour of RQM. See a detailed discussion in (Di Biagio and Rovelli, 2006) and also (Waaijer and van Neerven, 2021; Cavalcanti, 2019).

*Locality.* The way in which quantum nonlocality is realized in RQM (Smerlak and Rovelli, 2007; Laudisa, 2017) has been clarified recently in (Martin-Dussaud *et al.*, 2019; Pienaar, 2019), showing how the tension with relativity is alleviated by the fact that a measurement in one location cannot be a fact relative to a distantly located observer.

*Quantum Gravity.* RQM is a suitable interpretational framework for quantum gravity (Rovelli, 2004; Rovelli and Vidotto, 2014) (this was its historical motivation). In quantum gravity, we do not have a background spacetime in which to locate things. RQM works because the quantum relationalism combines in a surprisingly natural manner with the relationalism of general relativity. Locality is what makes this work (Vidotto, 2013): the quantum mechanical notion of 'physical system' can be identified with the general relativistic notion of 'spacetime region'. The quantum mechanical notion of 'interaction' between systems is identified with the general relativistic notion of 'adjacency' between spacetime regions. Locality assures that interaction requires (and defines) adjacency. Thus quantum 'states' can be associated to three-dimensional surfaces bounding spacetime regions and quantum mechanical transition amplitudes are associated to 'processes' identified with the spacetime regions themselves. In other words, variables actualize at three-dimensional boundaries, with respect to (arbitrary) spacetime partitions. The theory can then be used locally, without necessarily assuming anything about the global aspects of the universe.

## 43.10 PHILOSOPHICAL IMPLICATIONS

The beauty of the problem of the interpretation of quantum mechanics is the fact that the spectacular and unmatched empirical success of the theory forces us to give up at least *some* cherished philosophical assumption. Which one is convenient to give up is the open question.

The relational interpretation offers an alternative to the quantum state realism of Many-Worlds-like interpretations and to the strong instrumentalism of the strictly epistemic interpretations (Zeilinger, 1999; Hardy, 2001; Brukner and Zeilinger, 2003; Spekkens, 2007; Höhn and Wever, 2017; Fuchs and Stacey, 2019). It avoids introducing 'many worlds', hidden variables, physical collapse, and also avoids the instrumentalism of other epistemic interpretations. But, like any other consistent interpretations of quantum theory, it comes at a price.

It is compatible with diverse philosophical perspectives (see below). But not all. Its main cost is a challenge to a strong version of realism, which is implied by its radical relational stance.

Relationality is no surprise in physics. In classical mechanics the velocity of an object has no meaning by itself: it is only defined with respect to another object. The colour of a quark in strong-interaction theory has no meaning by itself: only the relative colour of two quarks has meaning. In electromagnetism, the potential at a point has no meaning, unless another point is taken as reference; that is, only relative potentials have meanings. In general relativity, the location of something is only defined with respect to the gravitational field, or with respect to other physical entities; and so on. But quantum theory takes this ubiquitous relationalism to a new level: the actual value of *all* physical quantities of *all* systems is only meaningful in relation to another system. Value actualization is a relational notion like velocity.

Hence the conceptual cost of RQM is giving up a strong form of realism: not only to give up the assumption that physical variables take values at all times, but also to accept that they take values at different times for different systems.

Strong realism is ingrained in our common-sense view of the world, and is often given for granted. For instance it is among the hidden hypotheses of the Pusey–Barrett–Rudolph theorem (Pusey *et al.*, 2012). The relational interpretation circumvents theorems like these (Oldofredi and Calosi, 2021) because these *assume* that at every moment of time all properties are well defined. (For a review, see (Leifer, 2014).) This assumption is explicitly denied in relational QM: properties do not exist at all times—they are properties of events and the events happen at interactions. In the same vein, in (Laudisa, 2017) Laudisa criticizes relational QM because it does not provide a 'deeper justification' for the 'state reduction process'. It is a stance based on a too strongly realist (in the narrow sense specified above) philosophical assumption.

A second element in RQM that challenges strong realism is that values taken with respect to different systems can be compared (Rovelli, 1996) (hence there is no solipsism in any sense), but the comparison amounts to a physical interaction, and its sharpness is limited by $\hbar$. Therefore we cannot escape from the limitation to partial views: there is no coherent global view available. Matthew Brown has discussed this point in (Brown, 2009).

The third element in RQM that challenges strong realism, emphasized by Dorato (Dorato, 2020), is the 'anti-monistic' stance implicit in relational QM. Since the state of a system is a bookkeeping device of interactions with *something else*, it follows immediately that there is no meaning in 'the quantum state of the full universe'. There is no 'something else' with respect to the entire universe. Everett's relative states are the only quantum states we can meaningfully talk about. Every quantum state is an Everett quantum state. (This does not prevent conventional quantum cosmology from being studied, since physical cosmology is not the science of everything: it is the science of the largest-scale degrees of freedom (Vidotto, 2017).)

In the philosophical literature RQM has been extensively discussed by Bas van Fraassen (van Fraassen, 2010) from a marked empiricist perspective, by Michel Bitbol (Bitbol, 2007; Bitbol, 2010) who has given a neo-Kantian version of the interpretation, by Mauro Dorato (Dorato, 2016) who has defended it against a number of potential objections and discussed its philosophical implication for monism and dispositionalism, and recently by Laura Candiotto (Candiotto, 2017) who has given it an intriguing reading in terms of (Ontic) Structural Realism (French and Ladyman, 2011). Metaphysical and epistemological implications of relational QM have also been discussed by Matthew Brown (Brown, 2009) and Daniel Wolf (né Wood) (Wood, 2014).

RQM has aspects in common with QBism (Fuchs *et al.*, 2014), with Healey's pragmatist approach (Healey, 2010) and is close in spirit to the view of quantum theory discussed by Zeilinger and Bruckner (Zeilinger, 1999; Brukner and Zeilinger, 2003). There are similarities with recent ideas by Auffèves and Grangier (Auffèves and Grangier, 2018; Auffèves and Grangier, 2019).

## 43.11 PERSPECTIVE

Relational QM is a radical attempt to cash out the breakthrough that originated the theory: the world is described by events or facts described by values of variables that obey the equations of classical mechanics, *but* products of these variables have a tiny non-commutativity that generically prevents sharp value assignment, leading to *discreteness*, *probability*, and to the contextual, *relational* character of value assignment.

The founders expressed this contextual character of Nature in the 'observer-measurement' language. This language requires that special systems (the observer, the classical world, macroscopic objects...) escape the quantum limitations. But nothing of that sort (and in particular no 'subjective states of conscious observers') is needed in the interpretation of QM. We can remove this exception, and realize that *any* physical system can play the role of a Copenhagen 'observer'. Relational QM is Copenhagen quantum mechanics made democratic by bringing all systems onto the same footing. Macroscopic observers, who lose information to decoherence, can forget the labelling of facts.

In the history of physics progress has often happened by realizing that some naively realist expectations were ill-founded, and therefore by dropping this kind of question: How are the spheres governing the orbits of planets arranged? What is the mechanical underpinning of the electric and magnetic fields? Into where is the universe expanding?

To some extent, one can say that modern science itself was born in Newton's celebrated 'hypotheses non fingo', which is precisely the recognition that questions of this sort might be misleading.

When everybody else was trying to find dynamical laws accounting for atoms, Heisenberg's breakthrough was to realize that the known laws were already good enough, but the question of the actual continuous orbit of the electron was ill-posed: the world is better comprehensible in terms of a sparse relational ontology. RQM is the realization that this is what we have learned about the world with quantum physics.

## ACKNOWLEDGEMENT

A special thanks to Andrea Di Biagio and Guido Baciagaluppi for corrections, suggestions, and guidance.

## REFERENCES

Auffèves, A., and Grangier, P. (2018). What is quantum in quantum randomness? *Philosophical Transactions of the Royal Society A*, **376**. arXiv:1804.04807

Auffèves, A., and Grangier, P. (2019). A Generic Model for Quantum Measurements. *Entropy*, **21**, 904. arXiv:1907.11261

Bitbol, M. (1996). *Schrödinger's Philosophy of Quantum Mechanics*. Boston Studies in the Philosophy and History of Science. Berlin:Springer.

Bitbol, M. (2007). Physical Relations or Functional Relations? A non-metaphysical construal of Rovelli's Relational Quantum Mechanics. September 2007. http://philsci-archive.pitt.edu/3506/.

Bitbol, M. (2010). *De l'intérieur du monde*. Paris: Flammarion. [Relational Quantum Mechanics is extensively discussed in Chapter 2.]

Bong, K. W., Utreras-Alarcón, A., Ghafari, F., Liang, Y. C., Tischler, N., Cavalcanti, E. G., Pryde, G. J., and Wiseman, H. M. (2020). A strong no-go theorem on the Wigner's friend paradox. *Nature Physics*. arXiv:1907.05607

Born, M., Heisenberg, W., and Jordan, P. (1926). Zur Quantenmechanik II. *Zeitschrift für Physik*, **35**, 557–615.

Born, M., and Jordan, P. (1925). Zur Quantenmechanik. *Zeitschrift für Physik*, **34**, 854–888.

Brown, M. J. (2009). Relational Quantum Mechanics and the Determinacy Problem. *British Journal for the Philosophy of Science*, **60**(4), 679–695.

Brukner, Č. (2018). A No-Go theorem for observer-independent facts. *Entropy*, **20**.

Brukner, Č., and Zeilinger, A. (2003). Information and fundamental elements of the structure of quantum theory. In L. Castell and O. Ischebeck (eds), *Time, Quantum, Information*, Berlin, Heidelberg: Springer, pp. 323–354. arXiv:0212084 [quant-ph]

Calosi, C., and Mariani, C. (2020). Quantum relational indeterminacy. Studies in History and Philosophy of Science Part B: Studies in History and Philosophy of Modern Physics, **71**, 158–169.

Candiotto, L. (2017). The reality of relations. *Giornale di Metafisica*, **2**, 537–551.

Cavalcanti, E. (2019). Implications of Local Friendliness violation to quantum causality. https://www.quantumlab.it/wp-content/uploads/2019/09/Cavalcanti_E.pdf.

Dakic, B., and Brukner, Č. (2009). Quantum Theory and Beyond: Is Entanglement Special? arXiv:0911.0695 [quant-ph] http://arxiv.org/abs/0911.0695

Di Biagio, A., and Rovelli, C. (2006). Stable Facts, Relative Facts. arXiv:2006.15543.

Dirac, P. A. M. (1925). The fundamental equations of quantum mechanics. *Proc. Roy. Soc.*, **109**, 645–653.

Dorato, M. (2016). Rovelli's relational quantum mechanics, monism and quantum becoming. In A. Marmodoro and A. Yates (eds), *The Metaphysics of Relations*, Oxford: Oxford University Press, pp. 290–324. arXiv:1309.0132

Dorato, M. (2017). Dynamical versus structural explanations in scientific revolutions. *Synthese*, **194**, 2307–2327. http://philsci-archive.pitt.edu/10982/

Dorato, M. (2020). Bohr meets Rovelli: a dispositionalist account of the quantum state. *Quantum Studies: Mathematics and Foundations*, 7, 233–245. arXiv:2001.08626

Einstein, A., Tolman, R., and Podolsky, B. (1931). Knowledge of past and future in quantum mechanics. *Phys. Rev.*, **37**, 780.

Everett, H. (1957). Relative state formulation of quantum mechanics. *Rev. Mod. Phys.*, **29**, 454–462.

Fedak, W. A., and Prentis, J. J. (2009). The 1925 Born and Jordan paper 'On quantum mechanics'. *American Journal of Physics*, 77, 128.

Frauchiger, D., and Renner, R. (2018). Quantum theory cannot consistently describe the use of itself. *Nature Communications*, **9**. arXiv:1604.07422

French, S., and Ladyman, J. (2011). In Defence of Ontic Structural Realism. In A. Bokulich and P. Bokulich (eds), *Scientific Structuralism*, Berlin: Springer, pp. 25–42.

Fuchs, C. A. (1998). Information Gain vs. State Disturbance in Quantum Theory. *Fortschritte der Physik*, **46**, 535–565.

Fuchs, C. A. (2001). Quantum Foundations in the Light of Quantum Information. In A. Gonis and P. E. A. Turchi (eds), *Decoherence and Its Implications in Quantum Computation and Information Transfer: Proceedings of the NATO Advanced Research Workshop*, Amsterdam: IOS Press. arXiv:0106166 [quant-ph].

Fuchs, C. A., Mermin, N., and Schack, R. (2014). An introduction to QBism with an application to the locality of quantum mechanics. *Am. J. Phys.*, **82**, 749–754.

Fuchs, C. A., and Stacey, B. C. (2019). QBism: Quantum theory as a hero's handbook. In E. M. Rasel, W. P. Schleich, S. Wölk (eds), *Foundations of Quantum Theory*, Proceedings of the International School of Physics 'Enrico Fermi', vol. 197, Amsterdam: IOS Press, pp. 133–202. arXiv:1612.07308

Grinbaum, A. (2005). Information-theoretic principle entails orthomodularity of a lattice. *Foundations of Physics Letters*, **18**, 563–572. arXiv:0509106 [quant-ph]

Hardy, L. (2001). Quantum Theory from Five Reasonable Axioms. arXiv:0101012 [quant-ph]

Hartle, J. B. (1968). Quantum Mechanics of Individual Systems. *American Journal of Physics*, **36**, 704. arXiv:1907.02953

Healey, R. (2010). Quantum Theory: a Pragmatist Approach. arXiv:1008.3896

Heisenberg, W. (1925). Uber quantentheoretische Umdeutung kinematischer und mechanischer Beziehungen. *Zeitschrift für Physik*, **33**(1), 879–893.

Höhn, P. A. (2017). Toolbox for reconstructing quantum theory from rules on information acquisition. *Quantum*, **1**, 38. arXiv:1412.8323

Höhn, P. A., and Wever, C. S. (2017). Quantum theory from questions. *Physical Review A*, **95**. arXiv:1511.01130

Kochen, S. (2013). A Reconstruction of Quantum Mechanics. arXiv:1306.3951 [quant-ph]. http://arxiv.org/abs/1306.3951

Kumar, M. (2008). *Quantum: Einstein, Bohr and the Great Debate About the Nature of Reality*. London: Icon Books Ltd.

Laudisa, F. (2001). The EPR argument in a relational interpretation of quantum mechanics. *Foundations of Physics Letters*, **14**, 119–132. arXiv:0011016v1 [quant-ph]

Laudisa, F. (2017). Open Problems in Relational Quantum Mechanics. arXiv:1710.07556

Laudisa F. and Rovelli, C. (2017). Relational Quantum Mechanics. In E. N. Zalta (ed.), *Stanford Encyclopedia of Philosophy*. https://plato.stanford.edu/entries/qm-relational/.

Leifer, M. S. (2014). Is the quantum state real? An extended review of $\psi$-ontology theorems. *Quanta*, **3**, 67–155. arXiv:1409.1570

Martin-Dussaud, P., Rovelli, C., and Zalamea, F. (2019). The Notion of Locality in Relational Quantum Mechanics. *Foundations of Physics*, **49**(2), 96–106. Pauli, W. (1926). Über das Wasserstoffspektrum vom Standpunkt der neuen Quantenmechanik [On the hydrogen spectrum from the standpoint of the new quantum mechanics]. *Zeitschrift für Physik*, **36**, 336–363.

Masanes, L., and Müller, M. P. (2011). A Derivation of Quantum Theory from Physical Requirements. *New J. Phys.*, **13**, 063001. https://doi.org/10.1088/1367-2630/13/6/063001

Oeckl, R. (2019). A Local and Operational Framework for the Foundations of Physics. *Adv. Theor. Math. Phys.*, **23**, 437–592. arXiv:1610.09052. http://arxiv.org/abs/1610.09052

Oldofredi, A, and Calosi, C. Relational Quantum Mechanics and the PBR Theorem: A Peaceful Coexistence. Foundations of Physics 51(4):1–21 (2021).

Pienaar, J. (2019). Comment on 'The Notion of Locality in Relational Quantum Mechanics'. *Foundations of Physics*, **49**, 1404–1414. arXiv:1807.06457

Proietti, M., Pickston, A., Graffitti, F., Barrow, P., Kundys, D., Branciard, C., Ringbauer, M., and Fedrizzi, A. (2019). Experimental test of local observer independence. *Science Advances*, **5**, 9. arXiv:1902.05080

Pusey, M., Barrett, J., and Rudolph, T. (2012). On the reality of the quantum state. *Nature Physics*, **8**, 475–478. arXiv:1111.3328

Rovelli, C. (1996). Relational Quantum Mechanics. *Int. J. Theor. Phys.*, **35**, 1637. arXiv:9609002 [quant-ph]

Rovelli, C. (2004). *Quantum Gravity*. Cambridge: Cambridge University Press.

Rovelli, C. (2016). An Argument Against the Realistic Interpretation of the Wave Function. *Foundations of Physics*, **46**(10), 1229–1237. arXiv:1508.05533

Rovelli, C. (2017). 'Space is blue and birds fly through it'. *Philosophical Transactions of the Royal Society A*, **376**, arXiv:1712.02894

Rovelli, C., and Vidotto, F. (2014). *Covariant Loop Quantum Gravity*. Cambridge: Cambridge University Press.

Schrödinger, E. (1926). Quantisierung als Eigenwertproblem (Erste Mitteilung). *Annalen der Physik*, **79**, 361–76.

Schrödinger, E. (1935). Die gegenwärtige Situation in der Quantenmechanik. *Die Naturwissenschaften*, **23**, 807–812.

Schrödinger, E. (1996). *Nature and the Greeks and Science and Humanism*. Cambridge: Cambridge University Press.

Smerlak, M., and Rovelli, C. (2007). Relational EPR. *Found. Phys.*, 37, 427–445. arXiv:0604064 [quant-ph]

Spekkens, R. W. (2007). Evidence for the epistemic view of quantum states: A toy theory. *Physical Review A*, **75**, 032110

Spekkens, R. (2014). The invasion of Physics by Information Theory. Talk at Perimeter Institute, 26 March 2014. http://pirsa.org/14030085/

van der Waerden, B. (1967). *Sources of quantum mechanics*. Amsterdam: North Holland.

van Fraassen, B. C. (2010). Rovelli's World. *Foundations of Physics*, **40**(4), 390–417. https://www.princeton.edu/~fraassen/abstract/Rovelli_sWorld-FIN.pdf.

Vidotto, F. (2013). Atomism and Relationism as guiding principles for Quantum Gravity. In 14th International Symposium Frontiers of Fundamental Physics (FFP14) : Marseille, France, 15–18 July 2014, vol. FFP14, SISSA, p. 222. doi: 10.22323/1.224.0222. https://inspirehep.net/record/1487207

Vidotto, F. (2017). Relational quantum cosmology. In K. Chamcham, J. Silk, J. D. Barrow, S. Saunders (eds), *The Philosophy of Cosmology*, Cambridge: Cambridge University Press, pp. 297–316.

Waaijer, M., and van Neerven, J. (2021). Relational analysis of the Frauchiger–Renner paradox and interaction-free detection of records from the past. *Foundations of Physics*, **51**, 45. arXiv:1902.07139

Wheeler, J. A. (1990). Information, physics, quantum: The search for links. In W. H. Zurek (ed.), *Complexity, Entropy, and the Physics of Information*, Redwood City, CA: Addison-Wesley, pp. 3–28.

Wheeler, J. A. (1991). It from bit. In L. V. Keldysh and V. Y. Fainberg (eds), *Sakharov Memorial Lectures in Physics: Proceedings of the First International Sakharov Conference on Physics, May 21–31, 1991, Moscow*, vol. 2, Hauppauge, NY: Nova Science Publishers, p. 751.

Wigner, E. (1961). Remarks on the mind-body question. In I. Good (ed.), *The Scientist Speculates*, London: Heinemann, pp. 248–302. http://www.informationphilosopher.com/solutions/scientists/wigner/Wigner_Remarks.pdf.

Wood, D. (2014). Everything Is Relative: Has Rovelli Found the Way out of the Woods? http://dwolf.eu/uploads/2/7/1/3/27138059/rqm_essay_dwood.pdf.

Zeh, H. D. (1970). On the Interpretation of Measurement in QuantumTheory. *Foundations of Physics*, **1**, 79.

Zeilinger, A. (1999). A Foundational Principle for Quantum Mechanics. *Found. Phys.*, **29**, 631–643.

Zurek, W. H. (1981). Pointer Basis of Quantum Apparatus: Into What Mixture Does the Wave Packet Collapse? *Phys. Rev. D*, **24**, 1516–1525.

Zurek, W. H. (1982). Environment induced superselection rules. *Phys. Rev. D*, **26**, 1862–1880.

CHAPTER 44

......................................................................................

# THE PHILOSOPHY OF WHOLENESS AND THE GENERAL AND NEW CONCEPT OF ORDER: BOHM'S AND PENROSE'S POINTS OF VIEW

......................................................................................

JEAN-JACQUES SZCZECINIARZ
AND JOSEPH KOUNEIHER

## 44.1 INTRODUCTION[1]

......................................................................................

FROM history we know that breakthroughs can be the result of conflict between observations and predictions (and consequently a new theory must be explored) or as the result of a resolution of some theoretical problem with new predictions which are confirmed by experiment (Freire, 2021). In both cases it is the inconsistencies which gives rise to the breakthroughs: either theory does not agree with data (experiment-led), or the theories have internal disagreements that require resolution (theory-led). Often we present the theory-led breakthroughs as a success of beautiful mathematics. However, scientists sometimes work on the right problem for the wrong reason while our conception of what counts as a beautiful theory is often based on what worked in the past. Though, in certain circumstances a new breakthrough requires new notions of beauty. Penrose, like Dirac, often refers to the idea of the beauty of physical theory as a guide:

---

[1] This paper results from the meditation of David Bohm as a deep reflection on the inseparability of philosophy and physics.

> Two spheres surround Truth: first Beauty then Morality, *'Beauty and Truth are* *intertwined, the beauty of a physical theory acting as a guide to its correctness in* *relation to the Physical World, whereas the whole issue of Morality is ultimately* *dependent upon the World of Mentality.'*    (Penrose, 2004, p. 1029)

Penrose addresses afterwards the two 'driving forces' of much physics research: *'beauty* *and miracles', 'mysterious aspect of mathematics . . . that underlies physical theory at its* *deepest level'* (Penrose, 2004, p. 1038). He finds that many highly successful physical theories including Newtonian mechanics, the equations of Maxwell's electromagnetic theory, Dirac's wave equations, and Einstein's special and general relativity are expressed by beautiful mathematical structures.

One of the first apparent problems in physics is that of interpretation. To devise an experiment, we first have to be able to say what the theory means. Interpretation is something physicists agree upon, and thus it is not objective. If an experiment fails, it can either mean that the mathematical theory behind it is flawed (i.e. the experiment falsifies the theory) or that the way we interpret the theory is wrong, although the theory itself is essentially correct. The need for interpretation undermines the sense of absolute objectiveness of scientific theories with which they are often presented, but generally this causes little difficulty in practice.

Fundamental problems in quantum mechanics which have a deep philosophical aspect concern *measurement* and *interpretation* problems. The measurement problem can be stated as follows: If the orthodox formulation of quantum theory—which in general allows attributions of only objectively indefinite properties or potentialities to physical objects (Heisenberg, 1958)—is interpreted in compliance with what is usually referred to as scientific realism, then one is faced with an irreconcilable incompatibility between the nonnegotiable linearity of quantum dynamics—which governs the evolution of the network of potentialities—and the apparent definite or actual properties of the physical objects of our 'macroscopic' world. Moreover, to date no epistemic explanation of these potentialities has been completely successful. Thus, on the one hand there is overwhelming experimental evidence in favour of the quantum mechanical potentialities, supporting the view that they comprise a novel metaphysical modality of nature situated between logical possibility and actuality, and on the other hand, there is a phenomenologically compelling proliferation of actualities in our everyday world, including even in the microbiological domain. The problem then is that a universally agreeable mechanism for transition between these two ontologically very different modalities, i.e., transition from the multiplicity of potentialities to various specific actualities, is completely missing. As delineated, this is clearly a very serious physical problem. What is more, the lack of a clear understanding of this apparent transition in the world is also quite an obstacle for any reasonable programme of scientific realism.

Quantum theory is entirely unclear about what constitutes a 'measurement'. It simply postulates that the measuring device must be classical, without defining where such a boundary between the classical and quantum lies, thus leaving the door open for those who think that human consciousness needs to be invoked for collapse. However, these experiments do not constitute empirical evidence for such claims. In the double-slit experiment carried out with single photons, all one can do is verify the probabilistic predictions of the mathematics. If the probabilities are borne out over the course of sending tens of thousands of identical photons through the double slit, the theory claims that each photon's wave function collapses, thanks to an ill-defined process called measurement.

Werner Heisenberg, among others, interpreted the mathematics to mean that reality does not exist until observed. He wrote *'The idea of an objective real world whose smallest parts exist objectively in the same sense as stones or trees exist, independently of whether or not we observe them ... is impossible'* (Heisenberg, 1958). John Wheeler, too, used a variant of the double-slit experiment to argue that *'no elementary quantum phenomenon is a phenomenon until it is a registered or observed phenomenon'* (Wheeler, 1983).

However, Penrose insisted on the fact that the question of 'reality' *must* be addressed in quantum mechanics especially if one takes the view (as many physicists tend to) that the quantum formalism applies universally to the whole of physics.[2] For then, if there is no quantum reality, there can be no reality at any level. We need, said Penrose, 'a notion of physical reality, even if only a provisional or approximate one, for without it our objective universe, and hence the whole of science, simply evaporates before our contemplative gaze!'

Another approach, called Collapse theories, argues that wave functions collapse randomly: the more the number of particles in the quantum system, the more likely the collapse. Observers merely discover the outcome. Collapse theories predict that when particles of matter become more massive than some threshold, they cannot remain in a quantum superposition of going through both slits at once, and this will destroy the interference pattern.

Roger Penrose has his own version of a collapse theory (Penrose, 1989, pp. 475–481), in which the more massive the mass of the object in superposition, the faster it will collapse to one state or the other, because of gravitational instabilities. Again, it is an observer-independent theory. No consciousness needed. Conceptually, the idea is to not just put a photon into a superposition of going through two slits at once, but to also put one of the slits in a superposition of being in two locations at once. According to

---

[2] Einstein's claim to the reality of locally separable physical systems should not be confused with attempts elsewhere to restore determinism and causality. The probabilistic nature of quantum mechanics seems to imply that it is only from statistical data that one can reconstruct the properties of individual systems; this situation is the opposite of that of classical physics, where the behaviours of statistical sets are, on the contrary, deduced from individual behaviours, as noted by both Einstein and supporters of the orthodox interpretation, in terms of finding a state of affairs. It is precisely in the perspective of returning to a more traditional situation in this respect, and of restoring, in the quantum domain, determinism and causality in the classical sense, considering the possibility of describing physical systems in a way that is not only probabilistic, that 'hidden variable theories' have been proposed by various authors as alternatives to quantum mechanics.

Penrose, the displaced slit will either stay in superposition or collapse while the photon is in flight, leading to different types of interference patterns. The collapse will depend on the mass of the slits.

In the modern quantum form, Young's experiment (or the two-slit experiment) is a conundrum to behold, raising fundamental questions about the very nature of reality. Bohm theory interprets the double-slit experiment differently, it says that reality is both wave and particle. A photon heads towards the double slit with a definite position at all times and goes through one slit or the other; so each photon has a trajectory. It is riding a pilot wave, which goes through both slits, interferes and then guides the photon to a location of constructive interference. Crucially, the theory does not need observers or measurements or a non-material consciousness. Bohm was not the only one dismayed by the answer of the Copenhagen people (Howard, 2021). Albert Einstein did not like it either. He did not believe that the probabilistic nature of quantum mechanics is fundamental (Paty, 2021; Laloë, 2021). In contrast to what is often claimed, Einstein did not think quantum mechanics was wrong. He just thought it is probabilistic the same way that classical physics is probabilistic, namely, that our inability to predict the outcome of a measurement in quantum mechanics comes from our lack of information, which is why it appears random. This missing information in quantum mechanics is usually called 'hidden variables'. If we knew the hidden variables,[3] we would be able to predict the outcome of a measurement. But the fact that the variables are 'hidden' implies the use of the 'probability' to get a particular outcome.

This article is based on a reading of David Bohm's and B. Hiley's book '*The Undivided Universe*' (Bohm and Hiley, 1995) and particularly on the topic of *Wholeness and the implicate order*. The book is a true philosophical meditation that lays the (open) foundations of physics but also of what he calls *reality*. The main thing remains the upheaval introduced by quantum physics. David Bohm forges a number of fundamental concepts intended to disintegrate the problems opened by this upheaval.

The ontological interpretation in the subtitle of the book 'An ontological interpretation of quantum theory', refers to the causal interpretation of quantum mechanics, a refinement and completion of de Broglie's 1926 pilot-wave model of non-relativistic quantum theory proposed by Bohm in 1952. This theory refers explicitly to the non-locality common to all the formulations and interpretations of quantum theory. It is to this aspect of wholeness that the title refers.

In Bohm's mechanics the reason why we can only make probabilistic predictions is linked to the fact that we do not know exactly where the particle initially was. If we measured we would find out where it was. Bohm's theory therefore considers that probability in quantum mechanics is the same type as in classical mechanics. The

---

[3] Note that the use of the term 'hidden variables' represents a terminological weakness. It would be preferable to speak of 'uncontrolled variables', insofar as one cannot, by hypothesis, say, in the current state, more about them. The term 'hidden variables' might imply that this is a matter of taste. For Bell, if these variables have a physical meaning, they are likely to manifest themselves by an effect (an effect that we do not yet know), and to take a certain value (Paty, 1982; Bub, 2021).

reason we can only predict probability for outcomes is because we have missing information. Bohm's mechanics is a hidden variables theory, and these hidden variables are the positions of the particles. This is therefore the basic assumption of Bohm's mechanics.

We also draw on Roger Penrose's deep mathematical and more generally conceptual understanding of these problems. Penrose's interest in physics began with quantum mechanics, and he formulated his ideas in geometric and topological terms. He was particularly interested in one of the fundamental properties of quantum particles called 'spin'. Spin is about 'angular momentum', which in the case of a ball basically means how fast the ball is spinning. When we try to think about quantum spin the same way, however, we find that the particle seems to rotate twice to return to where it started from: it is almost as if the object is spinning in twisted space. This motion can be described using mathematical objects called spinors. Since the late 1960s Penrose has been developing a theory, based on positing objects called twistors (Penrose, 2018; Kouneiher, 2018), adding a new twist to spinors, that reveal deep connections between quantum theory and the shapes of spacetime, and which he thinks might point the way to a theory of 'quantum gravity', achieving the long-sought goal of reconciling quantum theory and general relativity.

## 44.1.1  Quantum Entanglement and EPR experiment

Entanglement between particles, a notion first made explicit by Schrödinger (1935), is what lead to the extremely puzzling but actually observed phenomena known as *Einstein–Podolsky–Rosen* effects. They are rather subtle features of the quantum world.

The simplest EPR situation is that considered by David Bohm in 1951. We here envisage a pair of spin $\frac{1}{2}$ particles, let us say particle $P_1$ and particle $P_2$ which start together in a combined spin 0 state, and then travel away from each other. The outcome of the repeated experiment is that if we measure the spin on the particle on one side, then the spin of the particle on the other side has the opposite value.

To better understand this we should measure the spin in two different directions and not in the same direction on both sides. In quantum mechanics, the spin in two orthogonal directions has the same type of mutual uncertainty as the position and momentum. Therefore, if we measure the spin in one direction, then we do not know what is the result of the other direction. This means if we, on the left side, measure the spin in an up-down direction and on the right side measure in a horizontal direction, then there is no correlation between the measurements. If we measure them in the same direction, then the measurements are maximally correlated. Where quantum mechanics becomes important is for what happens in between, if we dial the difference in directions of the measurements from orthogonal to parallel. For this case we can calculate how strongly correlated the measurement outcomes are if the spins had been determined already at the time the original particle decayed.

More precisely, when measuring entangled particles the two detectors can both be vertical. Then there is a perfect correlation in the spin measurement: when the first particle is measured the second particle has the opposing spin. There is no randomness to this. When the angle is changed between the two detectors this correlation drops off in a sine wave shape. So when the two detectors are orthogonal this correlation drops off over 90 degrees in the sine wave shape. If a vertical spin is detected in the first particle it has no correlation to a horizontal spin measurement in the second detector—it is completely random. When there is an intermediate angle between the two, say 30 degrees as $\theta$, the correlation has dropped off by $sin^2(\theta)$ between them. Conversely there are more random measurements of spins in the first and second detector the greater this angle $\theta$ becomes. This shows there cannot be local hidden variables, if there is a correlation then a measurement in the first detector gives the opposing spin in the second detector. But often there is no correlation according to $sin^2(\theta)$. A hidden variable from when the two particles separated could not 'know' when to give a correlation and when not to. Because the detectors are spaced widely apart a first particle cannot know the angle between the two detectors. So the second particle cannot know from the start how often it should be correlated with an opposing spin to the first photon. This can also be tested by changing the angle between the detectors while the photons are moving away from each other.

This correlation has an upper bound, which is known as Bell's inequality.[4] However, and this is an important point: many experiments have shown that this bound can be violated.[5] In addition, this creates the key conundrum of quantum mechanics. If the outcome of the measurement had been determined at the time that the entangled state was created, then we cannot explain the observed correlations. However, if the spins were not already determined before the measurement, then they suddenly become determined on both sides the moment we measure at least one of them. So this is why quantum mechanics is said to be non-local. The correlations between separated particles, in quantum mechanics, when the state had been determined before measurement, forces us to give up locality.[6]

---

[4] What Bell's theorem, together with the experimental results, proves to be impossible is not determinism or hidden variables or realism but locality, in a perfectly clear sense. Bell obtained his theorem, which is thus a general theorem on non-locality, by mathematically transcribing Einstein's local separability hypothesis in terms of hidden parameters, and by comparing the predictions thus obtained to those of quantum mechanics considered for the same systems. The hidden variables which intervene in the intermediate calculations disappear in the expression of the finally comparable physical quantities. Bell insisted that neither determinism nor 'hidden variables' were presupposed in the derivation of his theorem, and therefore simply renouncing determinism or 'hidden variables' in physics would not change the impact of the theorem in any way.

[5] In other words Bell's inequality only works with observations if we do not pre-determine spin states and fails if we do; this is why we cannot assume spin states are pre-determined at and before the particles' separation, assuming statistical independence is valid.

[6] Nowadays, non-locality in quantum physics has been given a fundamental, and probably well definitive, status: it has been raised to the state of a physical fact, and it is founded on a conceptual ground which is more general than the question of determinism and hidden supplementary variables (see (Paty, 1982; Paty, 1986) for argumentation and bibliography).

## 44.1.2  Quantum Entanglement, a Geometric Point of View

First, some remarks are in order about the 'observables' in quantum mechanics. Some 'measurable' quality of a quantum system would be represented by a certain kind of operator $Q$ called *observable*. The dynamical variables (say position or momentum) would be examples of observables. Here we will follow Penrose's point of view and presentation of the principles of the quantum theory. An observable $Q$, like the position or momentum operators, has to be represented as a linear operator so that its action on the space $H$ is a linear transformation of $H$. A state $\psi$ has a definite value, the eigenvalue $q$, with respect to the observable $Q$ if $\psi$ is an eigenstate of $Q$. An essential requirement of an observable $Q$ is that its eigenvectors, corresponding to distinct eigenvalues, are orthogonal. This is a characteristic property of a *normal* operator. A *normal* operator $Q$ is one that commutes with its adjoint:

$$Q^*Q = QQ^*.$$

According to the rules of quantum mechanics, the result of a measurement, corresponding to some operator $Q$, will always be one of the eigenstates of $Q$: this correspond to a 'jumping' of the quantum state that occurs in the process of measurement, the $R$ process according to Penrose. One implication of the rules for calculating probabilities will be that the probability that a state jumps, as a result of a measurement, to an orthogonal state is always zero. The rule is that the probability of the state jumping from $|\psi\rangle$ to the eigenstate $|\phi\rangle$, of $Q$, is given by

$$|\langle\psi|\phi\rangle|^2,$$

assuming that $|\psi\rangle$ and $|\phi\rangle$ are normalized.

In the case of operators such as position or momentum, where the eigenstates are not normalizable, we get zero for the probability of finding particle in such a state. This is actually the 'correct' answer, because the probability of the position or momentum having any particular value would indeed be zero (because position and momentum are continuous parameters). We might prefer, as Penrose says, to use another kind of observable, such as that which poses a question: 'is the position within such-and-such a range of values' and a similar question for the momentum. YES/NO questions such as these can be incorporated into the quantum formalism by assigning the eigenvalue 1 to the YES answer and the eigenvalue 0 to NO. An observable of this kind is described by what is called a *projector*. And we know that a projector $E$ is self-adjoint and square to itself,

$$E^2 = E = E^*$$

Notice that when such a YES/NO is performed these operators degenerate. $Q$ is said degenerate with respect to some eigenvalue $q$, if the space of eigenvectors corresponding to $q$ is more than one-dimensional, i.e. there are non-proportional eigenvectors of $\mathbf{Q}$ corresponding to the same eigenvalue. The entire linear subspace of $\mathbf{H}$ consisting of all eigenvectors corresponding to the same eigenvalue $q$ is referred to as the *eigenspace* of $\mathbf{Q}$ corresponding to $q$. In such cases, obtaining the 'result' of the measurement, does not, in itself, tell us which state the state vector is supposed to 'jump' to.

The issue is resolved by the so-called *projection postulate* which asserts that the state $|\psi\rangle$ being subjected to measurement is orthogonally projected to the eigenspace of $\mathbf{Q}$ corresponding to $q$. (The term 'projection postulate' is often used simply for the standard quantum mechanical procedure.)

One of the best ways, according to Penrose, of expressing this projection is by use of an appropriate projector $E$, the one whose eigenspace corresponding to its YES eigenvalue 1 is identical to $\mathbf{Q}$ eigenspace corresponding to $q$. ($E$ is simply asking a more basic question than that posed by $\mathbf{Q}$, namely: is $q$ the result of the $\mathbf{Q}$ measurement?). The projection postulate tells us that the result of the measurement is that $|\psi\rangle$ jumps to $E|\psi\rangle$.

Consider the example of an entangled state, with the particular aim of seeing what implication this entangled state has. The experiment is the following. We are going to imagine that we are sitting at the left, at L, and we are going to perform a measurement of the spin of the left-hand particle $P_L$ in the 'up' direction $\uparrow$ (YES if $\uparrow$, NO if $\downarrow$). This would project the entire state $|\psi\rangle$ into $|\uparrow\rangle\langle\downarrow|$ if we got the answer YES, and it would project it into $(-)|\downarrow\rangle|\uparrow\rangle$ if we got NO.

The result would now be unentangled, except that standard U-evolution would tell us that $P_L$ is now likely to be hopelessly entangled with our own measuring apparatus L. Penrose suggests that if we get the answer YES, then our colleague, situated by the right-hand detector R, will be presented with $P_R$ having spin state $|\downarrow\rangle$, whereas if we measure NO, then our colleague will be presented with $|\downarrow\rangle$. Upon subsequently performing an 'up' measurement on $P_R$, our colleague will necessarily obtain the opposite result to our own.

For whatever direction we choose to measure, say $\swarrow$, then if our colleague chooses the same direction $\swarrow$, the result will be the opposite. Penrose presents the result in terms of an algebraic calculation. The result is clear from the rotational invariance of the spin 0, but he performs a direct algebraic calculation to verify that (where $\propto$ means 'equals' up to a non-zero overall factor).

$$|\omega\rangle \propto |\swarrow\rangle|\nearrow\rangle - |\nearrow\rangle|\swarrow\rangle,$$

where $\nearrow$ is the direction opposite to $\swarrow$. If $|\swarrow\rangle = a|\uparrow\rangle + b|\downarrow\rangle$ then $|\nearrow\rangle \propto \bar{b}|\uparrow\rangle - \bar{a}|\downarrow\rangle$. We also conclude from all this what the joint probabilities for YY, YN, NY, NN would be, if we and our colleague choose *different* directions in which to measure the spin. Suppose we choose $\nwarrow$ and our colleague chooses $\nearrow$, where the angle between $\nwarrow$ and $\nearrow$ is $\theta$, we would find the joint probabilities: agree $\frac{1}{2}(1 - \cos\theta)$, disagree $\frac{1}{2}(1 + \cos\theta)$ (where 'agree' means YY or NN and 'disagree' means YN or NY). In the case of spin 12 we can show that there is an immediate association between ordinary *spatial directions*

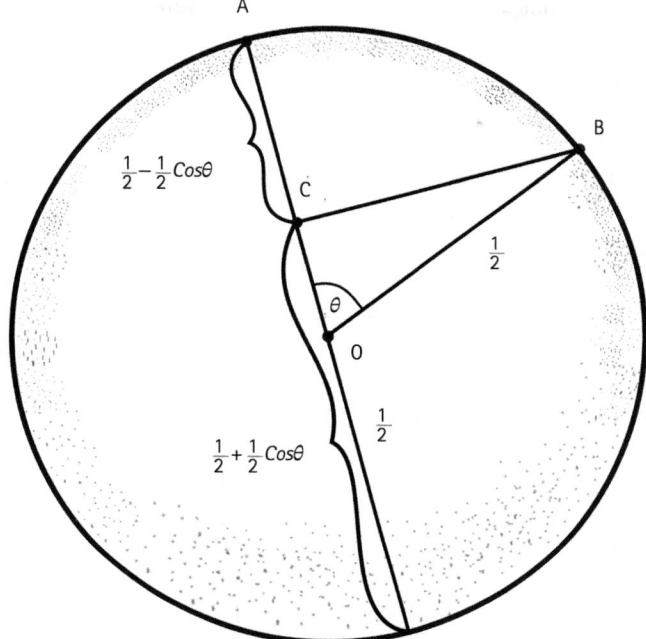

**FIGURE 44.1** Riemann sphere.

and the points of the Riemann sphere (Figure 44.1). So that the probabilities that arise in the quantum measurement, in this case, are expressed by the geometry of $\mathbb{P}H^2$.

Suppose that the initial state of a two-state system is represented by the point B on the Riemann sphere and we want to perform a YES/NO measurement corresponding to some other point on the sphere, where YES would find the state at the point A and NO would find it at the point A$'$, antipodal to A. Taking the sphere to have the radius 12, and projecting B orthogonally to C on the axis AA$'$, we find that the probability of YES is the length A$'$C which is AA$'$ and the probability of NO is the length CA, which is $12(1 - cos\theta)$ where $\theta$ is the angle between OB and OA, the sphere's centre being O. This use of the Riemann sphere clearly exhibits—once again—an explicit connection between the complex number ratios that arise in quantum mechanics and ordinary directions in space. Penrose stressed the fact that the entire projective Hilbert space $\mathbb{P}H$ for a system *is* physically meaningful.

## 44.2 BOHM'S APPROACH

During the 1940s and 1950s, it was assumed that a delayed-choice experiment, in which it is possible to show that the effect can precede the cause, would eventually be built. It was noticed that if the physics of quantum mechanics was correct and complete, such a paradoxical result would be manifest. Indeed, in 1926 Lewis proposed a delayed-choice thought experiment which appeared to show retrocausation in the Conventional Formulation of quantum mechanics (Lewis, 1926, 1930; von Weizsäcker, 1941). Retrocausation, also known as future input dependence, is when a model parameter associated with time $t$ depends on model inputs associated with times greater than $t$.

We suspect that the possibility of causality violation in the framework of quantum mechanics was one of the points that preoccupied David Bohm: '*It would appear, therefore, that the conclusions concerning the need to give up the concepts of causality, continuity of motion, and the objective reality of individual micro-objects were too hasty*' (Bohm, 1957, p. 95). Indeed, without a clear definition of cause and effect, no time ordering of events is possible, and so no entropy, which makes it difficult to consider developing the usual framework of theoretical physics (Robbin, 2014). In the usual experience of conversion between potential energy and kinetic energy, the conversion is a one-way conversion because of the loss due to friction and an arrow of time is defined. It seems that the time order of events, the arrow of time, is unavoidably bound up in the most basic principle of physics.

In his textbook *Quantum Theory*, which appeared in 1951, Bohm wanted to put quantum mechanics on a firm material basis. For him the necessity of the quantum mechanics is dictated by the properties of matter at an atomic level.[7] Bohm considered the indeterminism of quantum mechanics as a fundamental property of matter (Bohm, 1951, p. 100). This was in contrast to the supporters of the Copenhagen interpretation (Howard, 2021), who saw indeterminism as a consequence of a restriction of our knowledge of the incompatible properties of systems.

In his book, Bohm essentially adopted Bohr's point of view.[8] But, unlike Bohr, Bohm did not insist on the impossibility of a new quantum language and called for the creation of new concepts and adopted a dialectical materialism approach as far as the wave–particle dualism is concerned. He considered the prominence of the interaction with the environment:

> The most important new concept to which we are led is that any given piece of matter (for instance, an electron) is not completely identical with either a particle or a wave, but that, instead, it is something potentially capable of developing either one of these aspects of its behaviour at the expense of the other. Which of the electron's opposing potentialities will actually be realized in a given case depends as much on

---

[7] Letter from David Bohm to Miriam Yevick, 7 January 1952, DBP, uncatalogued.

[8] The central idea of Bohr's thought is the narrowness of the classical language in quantum mechanics. On the one hand we need the classical language to describe the quantum phenomena, but on the other hand the classical language is too limited to describe all qualities of the quantum phenomena. Therefore, it is necessary to use classical concepts in quantum mechanics, which contradict each other, e.g. position and momentum of a particle, or energy and time, which we call complementary concepts. Another central point in Bohr's early work is the indivisibility of quantum processes.

the nature of the systems with which the electron interacts as on the electron itself. Because the electron interacts continually with many different kinds of systems, each of which develops different potentialities, the electron will undergo continual transformations between its different possible forms of behaviour (i.e., wave or particle).   (Bohm, 1951, pp. 138–140)

However, even though Bohm's book remained within the scope of the standard interpretation, and even though he first denied the hidden variable theories, Bohm started to turn away from the standard quantum mechanics and formulated his own positions.

In 1952, Bohm published two papers in which he proposed a realistic interpretation of quantum mechanics and laid out his idea for how to make sense of the quantum mechanics (Bohm, 1952). According to Bohm, the wave equation in quantum mechanics is not what we actually observe. Instead, what we observe are particles, which are guided by the wave function. One can arrive at this interpretation in a few lines of calculation. We usually take the wave function apart into an absolute value and a phase, insert it into the Schrödinger equation, and then separate the resulting equation into its real and imaginary parts. The result is that in Bohmian mechanics the Schrödinger equation breaks down into two equations. One describes the conservation of probability and determines what the guiding field does:

$$\partial_t \sigma^2 = \frac{-1}{m} \nabla(\sigma^2 \nabla \theta). \tag{1}$$

The other determines the position of the particle with momenta $p = \nabla\theta$, and it depends on the guiding field. This second equation is usually called the *guiding equation*[9]

$$\partial_t \theta = \frac{\hbar^2}{2m} \frac{\nabla^2 \sigma}{\sigma} - \frac{(\nabla\theta)^2}{2m} - V. \tag{4}$$

Bohm in his theory puts forward an ontology of point-like particles that are located in three-dimensional space. The quantum mechanical wave function does not contain the information about the position of the particles, that is, the information about the actual

---

[9] In some references these equations can be written:

$$i\hbar \frac{\partial \Psi_t}{\partial t} = \hat{H}\Psi_t, \tag{2}$$

$$\frac{dQ}{dt} = v\Psi_t(Q). \tag{3}$$

In (3), $Q$ denotes the spatial configuration of $N$ particles in three-dimensional space and $\Psi_t$ the wave function of that configuration at time $t$. More precisely, $Q$ stands for the configuration of *all* the particles at time $t$, and $\Psi_t$ is the *universal* wave function.

particle configuration. The role of the wave function and its temporal development according to the Schrödinger equation is to fix the velocity of the particles given their position. Consequently, Bohm's theory is characterized by the Schrödinger equation for the temporal development of the wave function and an equation, known as the guiding equation, that applies the wave function to particle positions as initial conditions in order to obtain a velocity field for the particles.

The ontology of Bohm's theory thus is the same as the ontology of classical mechanics, particles moving on definite trajectories in three-dimensional space. In contrast to classical mechanics, however, Bohm's theory is a first-order theory: given the position of the particles, the law fixes their velocity. The quantum mechanical features of the world are taken into account by applying to this ontology of classical objects a *non-local* law: the guiding equation (3) makes the velocity of any particle depend on, strictly speaking, the positions of *all* the other particles.

The dynamics of Bohm's theory has indeed a remarkable property called equivariance. This means that, if the initial distribution coincides with that of Born, then it will coincide with it at all subsequent times—for the new wave function obtained by applying the Schrödinger evolution to the initial wave function. One can obviously question the justification of the choice of the initial distribution, which is a complex question. Note, however, that here the apparently 'random' character of the experimental results can be explained, as for classical dynamic systems, for example 'chaotic', by means of a hypothesis on the initial conditions, whereas, in the traditional interpretation, this randomness is not explained at all. In some way Bohm's theory gives the ontological significance of the entanglement of the wave function in configuration space. But there are no superpositions of values of properties in the world. The only property that the particles have is position.

So this is how Bohmian mechanics works: we have a particle, and it is guided by a field which in turn depends on the particle (Figure 44.2).

Bohm wrote: '*an interesting and suggestive possibility is then that of a sub-quantum mechanical level containing hidden variables*' (Bohm 1952). Such hidden variables would bring back a more classic concept of causality. Such hidden variables could be present yet are impossible to directly detect as by definition they would be below the scale of Heisenberg indeterminacy. '*For even if a sub-quantum level containing*

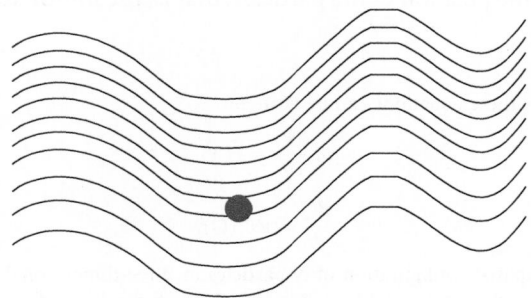

**FIGURE 44.2** Particle in guiding field.

*"hidden" variables of the type described previously should exist, these variables would then never play any real role in the prediction of any possible kind of experimental results'* (p. 85).

To use Bohm's theory we need one further assumption, one that tells us what the probability is for the particle to be at a certain place in the guiding field. This adds another equation, usually called the 'quantum equilibrium hypothesis'. It is basically equivalent to Born's rule and says that the probability of finding the particle in a particular place in the guiding field is given by the absolute square of the wave function at that place. Taken together, these equations—the conservation of probability, the guiding equation, and the quantum equilibrium hypothesis—give the exact same prediction as quantum mechanics. The important difference is that in Bohmian mechanics, the particle is really always in only one place, which is not the case in quantum mechanics.

Bohm's theory has another important feature: non-locality. The basic observation is that the wave function is a function defined on configuration space and not, like the electromagnetic field, on physical space. As John Bell (Bell, 1987, p.115) stressed,

> That the guiding wave, in the general case, propagates not in ordinary three-space but in a multidimensional-configuration space is the origin of the notorious 'non-locality' of quantum mechanics. It is a merit of the de Broglie–Bohm version to bring this out so explicitly that it cannot be ignored.

Consider two particles and assume that there is a potential located in the vicinity of the origin, and corresponding to the introduction of a measuring device acting on the first particle. The evolution of the wave function will be affected by the potential via the Schrödinger equation; however, the wave function determines the trajectories of the two particles via the Bohm equation. So the trajectory of the second particle will also (indirectly) be affected by the potential (i.e. the measuring device), even if it happens to be very far from the origin. This helps in understanding what happens when polarization or spin 'measurements' are performed on (anti-) correlated pairs. Results, as Bell shows, are not determined until after interaction with a measuring device. And that is why the perfect correlations are due to a subtle form of 'communication' between the two sides of the experience. The latter is made possible because the wave function connects distant parts of the universe through the Bohm equation.[10] Einstein, in particular, assumed that Bohm's theory shares the same characteristics as quantum mechanics, including non-locality.[11]

---

[10] This means if we have a system of ten particles then the guiding equation implies that the velocity of one particle depends on the velocities of the other particles regardless how far they are from each other.

[11] We find Einstein's exact appreciation of David Bohm's conceptions at this time in a letter he addressed to Aron Kuppermann on 14 November 1953 (Einstein Archives). A. Kuppermann, of the University of Notre Dame (Indiana), planning to organize a seminar on the ideas of David Bohm, sought the advice of Einstein. Otherwise, we know that Einstein had hopes of arriving at a complete theory of quantum phenomena. But the way he envisioned was not to start from the existing theoretical scheme

This non-local feature of Bohm mechanics is a problem because we know that quantum mechanics is, strictly speaking, only an approximation (Crull, 2021). The correct theory is really a more complicated version of quantum mechanics known as Quantum Field Theory. In QFT locality and the speed of light limit are essential. The problem now is that since Bohm mechanics is not local, it becomes difficult to make QFT out of it.[12] Some have made attempts, but currently there is simply no pilot wave alternative for the standard particle physics.

Bohmian mechanics has another odd feature that seems to have perplexed Einstein and Bell in particular (Bricmont, 2021). It is that depending on the exact initial position of the particle, the guiding field tells the particle to go one way or another, but the guiding field has a lot of valleys where particles could be going. So what happens with the empty valleys if we make a measurement?

In principle these empty valleys continue to exist. In some sense pilot wave theories are a kind of parallel-universe theories. Bohm himself, interestingly enough, seems to have changed his attitude towards his own theory. He originally thought it would in some cases give rise to predictions different from quantum mechanics.

In his 1957 book *Causality and Chance in Modern Physics*, Bohm reviewed the well-known Young's double-split experiment of 1801 (Bohm, 1957; Figure 44.3).

The thin black curves are the possible ways that the particle could go from the double slits to the screen where it is measured by following the guiding field. Just which way

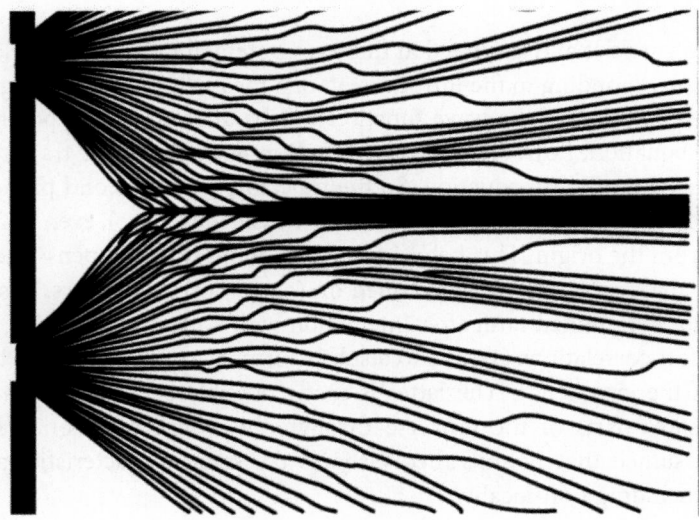

**FIGURE 44.3** Double-slit experiment.

and add to it elements which would be lacking, by simply completing the existing theory: it was necessary, in his view, to found it on other bases (of other principles, more fundamental, other concepts, less classic and 'mechanical').

[12] Bohm's theory is not Lorentz invariant and it cannot be modified easily to accommodate Lorentz invariance.

the particle goes is determined by the place it started from. The randomness in the observed outcome is simply due to not knowing exactly where the particle came from. The great thing about Bohm mechanics is that it explains what happened in quantum measurement. In Bohm mechanics, there is no real indeterminism regarding the possible outcome of a measurement. By taking into account the initial positions of the particles, the law of motion singles out the final configuration of the system plus measuring apparatus, and hence a definite outcome.

In 1959, Bohm published with Aharonov (Aharonov and Bohm, 1959) a new non-local effect.[13] In a region of space isolated from electric and magnetic fields (inside a solenoid, for example), a charged particle could still feel the effects of a force from outside the barrier. Such a force was previously described only as a mathematic potential, yet it was now observed to have a physical effect at a distance.

Bohm continued to probe the concept of non-locality and discovered a way to test EPR directly by measuring the spins of the entangled pairs along one of three axes, chosen at random (Bohm and Aharonov, 1957). In 1964 Bell devised a way to test EPR as modified by Bohm (Bell, 1964). He showed through his famous theorem that if local hidden variables exist, certain experiments could be performed involving quantum entanglement where the result would satisfy a Bell inequality. If, on the other hand, statistical correlations resulting from quantum entanglement could not be explained by local hidden variables, the Bell's inequality would be violated. As the Bell's inequality experiments had been repeated by different teams, Bohm already knew that the entangled particles are both real and non-local and that there is no hidden connection between the pairs. But he could not give up causality altogether. So he began a new definition of space, which he called a new order in physics that challenges the very idea of time and location. Bohm wanted to dispense with the problem of the time order of events by demoting the entire concept of time to something of a superficial construct (Robbin, 2014): *if an order is "enfolded" throughout all of space and time, it cannot coherently be regarded as constituting a signal that would propagate information from one place to another over a period of time* (Bohm, 1973, p. 164).

In contrast to the Copenhagen interpretation which was concerned with the nature of matter, Bohm was concerned with the nature of space and he introduced the idea of a hologram:

> Thus, the word 'electron' should be regarded as no more than a name by which we call attention to a certain aspect of the holomovement, an aspect that can be discussed only by taking into account the entire experimental situation and that cannot be specified in terms of localized objects moving autonomously through space. And, of course, every kind of 'particle' which in current physics is said to be a basic constituent of matter will have to be discussed in the same sort of terms (so that such 'particles' are no longer considered as autonomous and separately

[13] The Aharonov–Bohm effect is a quantum phenomenon described in 1949 by Werner Ehrenberg and Raymond Eldred Siday and rediscovered in 1959 by David Bohm and Yakir Aharonov.

existent). Thus, we come to a new general physical description in which 'everything implicates everything' in an order of undivided *wholeness*.   (Bohm, 1973, p. 153)

So, all points in space contain information enfolded in it about all other points of space.

There is the germ of a new notion of order here. This order is not to be understood solely in terms of regular arrangements of objects (e.g., in rows) or as a regular arrangement of events (e.g., in a series). Rather, a total order is contained, in some implicit sense, in each region of space and time.   (Bohm, 1973, p. 147)

# 44.3  BOHM'S PHILOSOPHICAL POINT OF VIEW

Here we will consider Bohm's reflections which underlie all his work as a physicist. Recall the key features of the contrast between relativistic and quantum theories. The theory of relativity requires continuity, strict causality (or determinism), and locality. In contrast, quantum theory requires non-continuity, non-causality, and non-locality. So the basic concepts of relativity and quantum theory directly contradict each other. It is therefore hardly surprising that these two theories have never been unified in a consistent way. What is needed is a qualitatively new theory, from which both relativity and quantum theory are to be derived. The best place to begin is with what they fundamentally have in common.

More generally, David Bohm's approach considers the physical laws of relativity and quantum theory underlying the basic concepts of explicate and implicate order and at another level the concept of wholeness. More importantly, the concept of undivided wholeness: 'All implicates all' even to the extent that 'we ourselves' are implicated together with 'all what we see and think about'. We are present everywhere and at all times, though only implicitly. In the case of quantum theory, the complete autonomy of the 'quantum state' of a system is supposed to hold only when it is not being observed. During an observation, it is assumed that we have to consider two initially autonomous systems that have come into interaction. One of these is described by the 'state vector of the observed object' and the other by the 'state vector of the observing apparatus'... It ultimately implies that such 'objects' have to be understood as merging with each other... to make one indivisible whole (Bohm, 1980; Bohm, 1973, p. 196).

David Bohm presents the following hypothesis: What we are proposing here is that the quantum property of a non-local, non-causal relationship of distant elements may be understood through an extension of the following notion. Each of the particles constituting a system as a projection of a 'higher -dimensional' reality, rather than a separate particle, existing together with all others in a common three-dimensional space. For example, in the experiment of Einstein–Podolsky–Rosen, each two atoms that initially combine to form a single molecule are to be regarded as a three-dimensional projection of a six-dimensional reality.

David Bohm proposes another notion of reality for which old orders of thought may cease to be relevant.[14] 'Reality' has to be considered as a process of *enfoldment* and *unfoldment* in a higher-dimensional space. The implicate order has to be extended into a multidimensional reality. This reality is one unbroken whole, including the entire universe with all its 'fields' and 'particles'. One can conceive this concept of reality by means of the principle of relative autonomy of sub-totalities.

## 44.3.1 Explicate Order and Implicate Order

The new notion of order is not to be understood solely in terms of a regular arrangement of *objects* (e. g. in rows), or as a regular arrangement of *events* (e. g. in a series), rather, a *total order* is contained, in some *implicit* sense, in each region of space and time. And as Bohm says we may be led to exploring the notion that in some sense each region contains a total structure 'enfolded' within it. This is the implicate order.

It is important to highlight the relevance of a new distinction between implicate order and explicate order. Generally speaking the laws of physics have thus far referred mainly to the explicate order. For example, the principal function of Cartesian coordinates is just to give a clear and precise description of explicate order. David Bohm proposes that in the formulation of the laws of physics, primary relevance is to be given to the implicate order, while the explicate order is to have a secondary kind of significance. A new kind of description will indeed have to be developed for discussing the laws of physics.

To generalize so as to emphasize undivided wholeness, we shall say that what 'carries' an implicate order is *the holomovement* which is an unbroken and undivided totality. In certain cases, we can abstract particular aspects of the holomovement (e. g. light, electrons, sound, etc.), but more generally, all forms of the holomovement merge and are inseparable. Thus in its totality, the holomovement is not limited in any specifiable way at all. It is not required to conform to any particular order, or to be bounded by any particular measure. Thus, says David Bohm, '*the holomovement is undefinable and immeasurable*'. He proposes a new vision of physical theories. Each theory will abstract a certain aspect that is *relevant* only in some limited context, which is indicated by some appropriate measure.

---

[14] 'We want to stress that this conception implies a conception of history of science. For example, by using the notion of order prevailing in ancient times, men were led to 'make' the fact about planetary motions by describing and measuring in terms of epicycles. In classical physics, the fact was 'made' in terms of the order of planetary orbits, measured through positions and times. In general relativity, the fact was 'made' in terms of the order of Riemannian geometry, and of measure implied by concepts such as 'curvature of space'. In quantum theory, the fact was made in terms of the order of energy levels, quantum numbers, symmetry groups, etc., along with appropriate measures' (Bohm 1973, p. 180).

What is common to the functioning of instruments generally used in physical research is that the sensitive perceptible content is ultimately describable in terms of a Euclidean system of order and measure, i. e., one that can adequately be understood in terms of ordinary Euclidean geometry. Bohm adopts the well known view of the mathematician Klein, who considers the general transformation to be the essential determining features of a geometry. In a Euclidean space of three dimensions, there are three displacement operators $D_i$. Each of these operators defines a set of parallel lines which transform into themselves under the operation in question. Together, they define concentric spheres that transform into themselves under the whole set of $R_i$. Finally, there is the dilation operator $R_0$, which transforms a sphere of a given radius into one of a different radius. Under this operation, the radial lines through the origin transform into themselves.

From any one operator $R_1, R_0$ we obtain another set $R'_1, R'_0$, corresponding to a different centre, by means of a displacement

$$(R'_1, R'_0) = D_1(R'_1, R'_0)D_1^{-1}.$$

From the $D_i$, we obtain a set of displacements $D'_i$, in new directions by the rotation

$$D'_1 = R_1 D_2 R_1^{-1}.$$

Now, if $D_1$ is a certain displacement, $(D_1)^s$ will be a displacement of a similar step. This means that displacement can be ordered naturally in an order similar to that of the integers. We can describe displacements on a *numerical* scale. This gives not only an order, but also a *measure* insofar as we treat successive displacements as equivalent in size.

In this framework, a transformation is a simple geometric change within a given explicate order. This idea is linked with that of metamorphosis. We shall use the symbol $T$ for transformation, and for a metamorphosis $M$, while $E$ denotes a whole set of transformations that are relevant in a given explicate order $(D_1, R_1, R_0)$:

$$E' = MEM^{-1}.$$

As an essential feature of a similarity metamorphosis, let us consider the example of the hologram. In this case, the appropriate metamorphosis $M$ is determined by the Green's function relating amplitudes at the illuminated structure to those at the photographic plate. For waves of definite frequency $\omega$ the Green's function is

$$G(x - y) \approx \{exp(i\omega/c)|x - x|\}/|x - y|.$$

where $\mathbf{x}$ is a coordinate relevant to the illuminated structure and $\mathbf{y}$ is the one relevant to the plate. Thus, if $A(\mathbf{x})$ is the amplitude of the wave at illuminated structure, then the amplitude $B(\mathbf{x})$ at the plate is

$$B(y) \approx \int \{[exp(i\omega)/c]|\mathbf{x} - \mathbf{y}|\}/|\mathbf{x} - \mathbf{y}|A(\mathbf{x})dx.$$

The entire illuminated structure is seen from the above equation to be 'carried' and 'enfolded' in each region of the plate in a way that evidently cannot be described in terms of a point-to-point transformation or correspondence between $\mathbf{x}$ and $\mathbf{y}$. The matrix $M(\mathbf{x}, \mathbf{y})$ which is essentially $G(\mathbf{xy})$ can thus be called a metamorphosis of the amplitudes at the illuminated structure into the amplitude at the hologram. Point-to-point transformations of space and time that preserve locality are 'enfolded' into more general operations that are similar and nevertheless are not locality-preserving point-to-point transformations.

## 44.3.2  Back to Our Reality

The *folding-unfolding* idea can be depicted in mathematical form using a Green's function approach to Schrödinger's equation (Hiley and Peat, 1987). Indeed, in Feynman's path formulation, Huygens' principle can be used to determine the wave function at some point $y$ from the wave function at the set of points $x$ on a surface at a previous time. Thus:

$$\psi(y, t_2) = \int_S G(x, y, t_1, t_2)\psi(x, t_1)dx \tag{5}$$

where $G(x, y, t_1, t_2)$ is Green's function. The wave function at all points of the surface $S$ contributes to the wave function at $y$. Thus, the information on the surface $S$ is enfolded into $\psi(y)$ which determines the quantum potential acting on the particle at $y$ so that the particle reacts to the enfolded information of a set of earlier wave functions. In turn $\psi(y)$ itself gets 'unfolded' into a series of points on a later surface $S'$. In this way we see that the quantum potential itself is determined by an enfolding-unfolding process. Consider $E$ and $E'$ as two successive explicate orders in this scheme. By continuity $EG_1 = G_2E'$, i.e. the enfolded $E$ is equal to the unfolded $E'$. $G_1$ and $G_2$ are the enfolding and unfolding Green's functions.

Bohm and his group were led to this framework when they noticed that the basic mathematics required to describe the implicate order will involve the use of matrix algebras. More precisely a direct product of a suitable Clifford algebra and a symplectic algebra. The Clifford algebras carrying the rotational symmetries and the symplectic algebra carrying the translational symmetries. Thus they have within their structures all the required symmetry properties for abstracting the spacetime continuum (Bohm, 1980).

## 44.4 PENROSE'S APPROACH

Penrose considered the reduction of the wave function as an essential conundrum of the quantum mechanics. He noted that in all expressions of quantum theory, the wave function reduction "$R$" element (ending the unobserved wave function, the "$\psi$" state) depends upon an observer or measurement. In fact, for the Copenhagen interpretation, the wave function does not exist independently, only in 'the observer's mind'.

Penrose considers that another $R$ interpretation, quantum 'environmental decoherence', leads us to an Everett-like multiverse with each state, at collapse, splitting off as a separate universe. For Penrose, this model is incomplete because it lacks 'an adequate theory of observers'. According to Penrose, environmental decoherence does not provide a consistent ontology for the reality of the world, and provides merely a pragmatic procedure. Moreover, it does not address the issue of how $R$ might arise in isolated systems, nor the nature of isolation, in which an external 'environment' would not be involved, nor does it tell us which part of a system is to be regarded as the 'environment' part, and it provides no limit to the size of that part which can remain subject to quantum superposition. He observes that the philosophical 'environmental decoherence' approach is common among theoretical physicists.

Otherwise, he finds that one model of quantum mechanics greatly diminishes or eliminates the role of observation: de Broglie–Bohmian mechanics which changes $R$ and $\psi$ factors from observer-dependent to real entities. In Bohmian mechanics, this change is partly affected by the existence of deterministic 'hidden variables' ('pilot wave' or 'guiding equation') (Penrose, 2004, pp. 1031–1033.). Penrose's own gravitational model also converts the $R$ into an objective entity. In his model, the presence of gravity replaces the observer (or measurement) as the reduction agent. He switches the causal roles of the $R$ process and consciousness. Instead of the classical quantum perspective in which an observer causes $R$, the existence of $R$ partly causes the conscious observer to exist. Although he reverses these roles, like his predecessors, Penrose continues to include conscious observation in the model of objective reality.

Penrose notes that entanglement is involved in quantum measurement. In his interpretation, 'the state reduction R [wave function collapse] can be understood as coming about because the quantum system under consideration becomes inextricably entangled with its environment' (Penrose, 2004, p. 528). Penrose believes that the quantum measurement paradox will be resolved when Einstein's theory of gravity is brought into the picture; the standard rules of quantum theory must change.

### 44.4.1 Wave-Packet Reduction according to Penrose

Penrose suspects that the reason we do not see the counter-intuitive properties of quantum particles—most notably the way they seem to exist simultaneously in two or more states, or *superpositions*—in the everyday world is that when objects get big enough to 'feel' significant gravitational force, quantum mechanics needs to be modified if it is to describe them. Because general relativity says that gravity is caused by masses bending spacetime, a quantum superposition of a large object—crudely, seeming to put it in two places at once—would have to superimpose two simultaneous structures of spacetime. That cannot be countenanced, Penrose says.

Penrose's idea is that there is some real, physical 'collapse' of a quantum superposition as objects grow big enough for gravity to become a significant part of the picture. This is still a minority view, although Penrose says experimentalists are very close to being able to test it, and he confidently predicts they will find discrepancies with the standard theory.

Penrose argues that a valid formulation of quantum state reduction replacing von Neumann's projection postulate must faithfully describe an objective physical process that he calls objective reduction. As such a physical process remains empirically unconfirmed so far, Penrose proposes that effects not currently covered by quantum theory could play a role in state reduction. Ideal candidates for him are gravitational effects since gravitation is the only fundamental interaction which is not integrated into quantum theory so far. Rather than modifying elements of the theory of gravitation (i.e., general relativity) to achieve such an integration, Penrose discusses the reverse: that novel features have to be incorporated in quantum theory for this purpose. In this way, he arrives at the proposal of gravitation-induced objective state reduction.

The judiciously employed tool in practice, the infamous postulate often referred to as the reduction of quantum state,[15] may then be understood as an objective physical phenomenon; i.e., one affording an ontological as opposed to epistemological understanding. From the physical viewpoint such a resolution of the measurement problem would be quite satisfactory. Since it would render the proliferation of diverse philosophical opinions on the matter to nothing more than a curious episode in the history of physics.

If one seeks to replace the Schrödinger evolution by a nonlinear or stochastic operation, this one should have the following properties: to be sufficiently well approximated by the evolution of Schrödinger, when discussing a small number of particles, so that the predictions of quantum mechanics remain true for the new theory, and describe the interactions, when we are interested in a large number of particles— for example, in a measuring device—essentially to the reduction. Considering the mathematical difficulty inherent in dealing with nonlinear equations, the fact that there are no such satisfactory theories today cannot be taken as a very strong argument against this suggestion. Moreover, since most physicists consider ordinary quantum mechanics to be acceptable, little testing has ultimately been done. Penrose encourages this approach with the following analogy:

---

[15] Which in orthodox formulations of the theory is taken as one of the unexplained basic postulates to resolve the tension between the linearity of quantum dynamics and the plethora of physical objects with apparent definite properties.

Nonetheless, I think it would be surprising if quantum theory were not to undergo a profound change in the future—towards something that this linearity is only an approximation of. There is certainly a history of these kinds of changes. The power and elegance of Newton's theory of universal gravitation is largely due to the fact that the forces in this theory add up linearly. But, with Einstein's general relativity, we see that this linearity is only an approximation—and the elegance of Einstein's theory exceeds even that of Newton's theory.    (Penrose, 1989)

Note that in the absence of a nonlinear theory, there are different proposals to resolve the measurement problem. One of them is a stochastic theory called 'GRW'[16] (Ghirardi *et al.*, 1986; Bassi and Ghirardi, 2003), where the wave function is reduced randomly. This theory features a modified Schrödinger evolution assigning to quantum systems a probability to undergo spontaneous localizations ('collapses') at random times. This probability is very small for microscopic systems, but it approaches 1 for macroscopic systems. The GRW theory therefore promises to ensure what the unmodified Schrödinger evolution cannot achieve, namely the unique localization of macroscopic bodies and, more specifically, unique measurement outcomes. However, it has become increasingly clear in the last two decades that modifying the Schrödinger equation is, by itself, not sufficient to solve the measurement problem. The reason is that the GRW equation is an equation that describes the temporal development of the wave function in configuration space, in contrast to an equation that describes the temporal development of matter in physical space. The fundamental dynamics of the GRW approach is free of inauspicious references to consciousness or measurement. The theory's portrayal of the ordinary macroscopic world tells a seemingly adequate story about stochastic succession of sharply concentrated wave functions in coordinate space. The intended ontology of the theory is clearly centred on the wave function as a physical field. This proposal does not tally well with ordinary intuition, and this has been regarded by some critics as a fault (see Bassi and Ghirardi, 2003). One obvious concern also is that it would be imprudent to accept GRW as a correct description, given the theory's present lack of experimental force. This is a fair methodological worry—the same point applies to all current attempts to improve the standard theory. A more serious concern focuses on the 'external' coherence of GRW, more specifically on the generalizability of the proposal into a proper Lorentz-invariant theory.

Unlike some other approaches to the philosophical problems of quantum theory, Penrose's approach implicitly takes temporal transience in the world—the incessant fading away of the specious present into the indubitable past—not as a merely phenomenological appearance, but as a bona fide ontological attribute of the world. For, clearly, any gravity-induced or other intrinsic mechanism, which purports to actualize—as a real physical process—a genuine multiplicity of quantum mechanical potentialities to a specific choice among them, evidently captures transiency, and thereby not only goes beyond the symmetric temporality of quantum theory, but also

---

[16] After the names of its authors: Ghirardi, Rimini, and Weber.

acknowledges the temporal transience as a fundamental and objective attribute of the physical world. A possibility of an empirical test confirming the objectivity of this facet of the world via Penrose's approach is by itself sufficient. It is generally believed that the classical general relativistic notion of spacetime is meaningful only at scales well above the Planck regime, and that near the Planck scale the usual classical structure of spacetime emerges purely phenomenologically via a phase transition or symmetry-breaking phenomenon.

## 44.4.2 Penrose's Approach to Non-Locality

The problem of the relationship between the existing quantum theory and objective reality at the atomic and subatomic levels can be tackled in essentially two ways:

i) One may focus attention on the formalism of the theory and attempt to deduce from it a coherent description of our measuring processes and a deeper understanding of the microworld.
ii) Alternatively, one may start from the experimental evidence and/or from models of the objective reality compatible with it and go on to investigate whether or not formalization of this knowledge can be accommodated within the broad confines of existing quantum theory.

From this perspective Penrose's epistemological position can be considered as a third option between the computational/determinist and the embodied/evolutionary approaches to scientific knowledge. For Penrose, knowledge is not a collection of data that can be stored in computer memories and that can be used to perform calculations; rather, knowledge is a state of a rational subject who can explain the reason why the process of discovery he is carrying out is likely to reveal new aspects of reality (and not only the future states of a given system). What is at stake here is the method of production of the mathematical structures that are involved in the genesis of our representation of reality, a method which aims to enlarge the possibilities of experience rather than to enclose it in a predictable totality.

Penrose's epistemological interest concerns essentially the connectivity problem and the acceptance of any form of Platonism. How can an external Platonic form (often referred to as a 'universal') be connected to or associated with a material object, or how can it cause or be involved in an instantiation in the physical world?

Penrose suggests a cyclically interacting Platonic-like world: Truth pushes (shape or form) to Physicality which pushes in turn (material of concepts) to Mentality which pushes (observational limits) to Truth. In *The Road to Reality* (Penrose, 2004), Penrose expresses his concern with the fact that if contemporary physical theory wants to reach the '*true road to reality... [Therefore] some new insights are needed.*' (Penrose, 2004, p. 1027).

Modern physicists invariably describe things in terms of mathematical models... as though they seek to find 'reality' within the Platonic world of mathematical ideals... then physical reality would appear merely as a reflection of purely mathematical laws... [but] we are a long way from any such theory.... [However] the more deeply we probe Nature's secrets, the more profoundly we are driven into Plato's world of mathematical ideals as we seek our understanding.... At present, we can only see that as a mystery.... But are mathematical notions things that really inhabit a 'world' of their own? If so, we seem to have found our ultimate reality to have its home within that entirely abstract world. Some... have difficulties with accepting Plato's mathematical world as being in any sense 'real' and would gain no comfort from a view that physical reality itself is constructed merely from abstract notions. Our own position on this matter is that we should... take Plato's world as providing a kind of 'reality' to mathematical notions... but I might balk at actually attempting to identify physical reality within the abstract reality of Plato's world.    (Penrose, 2004, pp. 1027–1029)

This connectivity or identity between physical reality and abstract reality is underlying intrinsically his geometric work on the light rays. It was during his work on light rays that Penrose discovered that his formulation of twistors (as the basic element of physics) do have non-local properties. As he later wrote:

Another guiding principle behind twistor theory is quantum non-locality. We recall from the strange EPR effects... and more specifically from the role of *quanglement* [quantum entanglement], as manifest particularly in the phenomenon of quantum teleportation... that physical behaviour cannot be fully understood in terms of entirely local influences of the normal 'causal' character. This suggests that some theory is needed in which such non-local features are incorporated.
(Penrose, 2004, p. 963)

In other words, Penrose understood that the projective space is inherently non-local and time-independent, and idealized light rays (or their generalization, with spin) appear to be, in a sense, the carriers of *quanglement* (Penrose, 2002, p. 114). Therefore, by virtue of projective structure, empty space has Minkowski metrics, spin, and non-locality.

Penrose was concerned by the paradox of time order and entropy. On a small scale, many events in physics are time-symmetric,[17] and Penrose puzzled over the 'splitting'

---

[17] The problem can be understood by imagining an open bottle containing a gas in a closed room. After the dispersion of the gas, the room is filled with the molecules and no gas remains in the bottle. The paradox is that if a very short film were made focusing on just a few molecules of gas as they crash into one another and bounce around, a physicist could not tell if the film were running forward or backward; the molecules can bounce down just as easily as they bounce up. There is, then, a disconnect between the global picture, which shows entropy (the potential energy stored in the bottle of gas is converted to the kinetic energy of the evaporation—a one-way conversion that constitutes an arrow of time), and the local picture in which actions are time-symmetric. It will likely be the case that a few molecules of gas will bounce back into the bottle, but overall no one would wait for all of the gas to be back in the bottle.

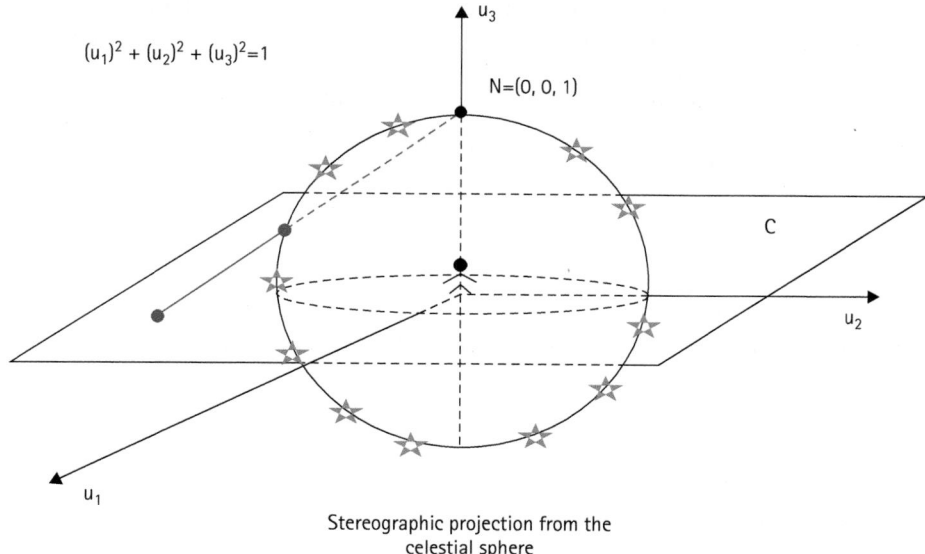

$(u_1)^2 + (u_2)^2 + (u_3)^2 = 1$

$u_3$

$N = (0, 0, 1)$

C

$u_2$

$u_1$

Stereographic projection from the
celestial sphere

**FIGURE 44.4** The two-dimensional sphere is the simplest example of a non-trivial complex manifold.

and sought a natural way to mathematically model time reversibility (Robbin, 2014). He found that natural way in an alternate formalism he discovered to model the Lorentz transformations, which describe the effects of special relativity. Penrose showed that special relativity transformations could be considered to be projective transformations on the Riemann sphere (specifically Möbius transformations), a geometric model that represents all the complex numbers together with ∞. Around the equator of this sphere are all the real numbers from 0 to infinity, both the positive numbers to infinity and the negative numbers to infinity. A longitudinal cut through the poles shows all the positive and negative imaginary numbers, $i$ (the square root of minus one). If homogeneous coordinates are used to describe rays from the centre of this sphere, then an arbitrary chunk of spacetime, with its $(+ + + -)$ Minkowski signature, can be represented as a region of this Riemann sphere (Kouneiher, 2018). Here is the example Penrose often gives:

Imagine an observer situated at a point in spacetime, out in space looking at the stars. Suppose she plots the angular position of these stars on a sphere. Now, if a second observer were to pass through the same point at the same time, but with a velocity relative to the first observer then, owing to aberration effects, he would map the stars in different positions on the sphere. What is remarkable is that the different positions of the points on the sphere are related by a special transformation called a Möbius transformation. Such transformations form precisely the group that preserves the complex structure of the Riemann sphere. Thus, the space of light rays through a spacetime point is, in a natural sense, a Riemann sphere. We find it very beautiful, moreover, that the fundamental symmetry group of physics relating observers with different velocities, the (restricted) Lorentz group, can be realized as the automorphism group of the simplest one- (complex-)dimensional manifold, the Riemann sphere.

Two–dimensional sphere $S^2 \subset \mathbb{R}^3$ is a one-dimensional complex manifold with local coordinates defined by stereographic projection (Figure 44.4). Let $(u_1, u_2, u_3) \in S^2$. Define two open subsets covering $S^2$

$$U_0 = S^2 - \{(0, 0, 1)\}, \quad U_1 = S^2 - \{(0, 0, -1)\},$$

Stereographic projection from the north pole $(0, 0, -1)$ gives a complex coordinate

$$U = \frac{u_1 + iu_2}{1 - u_3}.$$

Projecting from the south pole $(0, 0 - 1)$ gives another coordinate

$$U' = \frac{u_1 + iu_2}{1 - u_3}.$$

The domain of $U$ is the whole sphere less the north pole; the domain of $U'$ is the whole sphere less the south pole.

On the overlap $U_0 \cap U_1$, we have: $U' = 1/U$, which is a holomorphic function; this makes $S^2$ into a complex manifold $\mathbb{CP}^1$ (Riemann sphere). This auxiliary space can be thought of as the space of light rays at each point in spacetime. Given an observer in a four-dimensional spacetime at a point $O$, his *celestial sphere*, i.e., the image of planets, suns, and galaxies he sees around him, is the backward light cone at $O$ given by the two-sphere

$$t = -1 \text{ and } x^2 + y^2 + z^2 = 1. \tag{6}$$

Projections from the north pole are considered positive, pointing toward the future. Projections from the south pole are considered negative, pointing to the past. Thus a simple geometric interpretation is given to the splitting of time direction in a sophisticated model of physical spacetime. Consistently throughout his writings, Penrose has emphasized that light rays should be thought of as projective elements. The very good physical reason for this is that a light ray has a direction but no fixed length: from its point of view, it is wherever it was and where it will ever be at the same instant. Of course, slower-moving observers see things differently. We see a photon that was created in a nuclear explosion on Alpha Centauri that travelled 4.37 years until it is absorbed, and destroyed, in our eye, but for the photon itself no time has passed, no distance was travelled. Light rays are projective points and make the space of special relativity a projective space.

# 44.5  A PRELIMINARY FOR A CONCLUSION

Bohm and Penrose share the non-local but causal point of view of physics. Bohm for his part gives priority to the implicate order, while Penrose chooses the empty space provided with the Minkowskian metric. Both stress the close relationship between the local and the global and the importance of a one-way arrow of time into the future. While giving entropy and causality a predominant place (Robbin, 2014).

Bohm has an almost fractal and hologrammatic point of view of the space: the whole is included in each part of the space. A Bohmian universe is not the usual sort of physical field on physical space to which we are accustomed, but a field on the abstract space of all possible configurations, a space of enormous dimension, a space constructed, it would seem, by physicists as a matter of convenience. Moreover, the fact that Bohmian mechanics is a first-order theory, allows us to achieve Galilean invariance despite the no-action of the configuration upon the wave function. In other words, Bohmian mechanics is based on a fundamentally different sort of structure than classical mechanics, one that does not require the action–reaction principle to achieve the desired underlying symmetry. Concerning the ontological aspect of Bohmian mechanics, Bell said:

> No one can understand this theory until he is willing to think of as a real objective field rather than just a 'probability amplitude'. Even though it propagates not in 3-space but in $3N$-space.   (Bell, 1987, p. 128)

As we said before, Bohm noticed that the basic mathematics required to describe the implicate order will involve the use of a suitable Clifford algebra and a symplectic algebra. Thus they have within their structures all the required symmetry properties for abstracting the spacetime continuum (Bohm, 1980).

As for Penrose, he emphasizes the projective structure underlying the rays of light by reflecting the cones of light into spheres and the rays of light into projective points. Thus, the structured space of these light rays represents a fundamental entity in it. It is an elegant model, it accommodates the splitting of light rays into past or future, in account of their different chirality, and models the space of special relativity—what Penrose considers the starting point of all physics. In Penrose's twistor formulation there is an elusive and highly technical but still geometric reason for understanding why the future is more likely to occur next rather than the past.

So, Penrose's point of view is that the notion that the spacetime continuum should not be taken as a basic notion.[18] In his essay *On the Origins of the Twistor Theory* (Penrose, 1987), Penrose wrote:

> we think that we had very much come around to the view that massless particles and fields were to be regarded as more fundamental than massive ones.
>
> (Penrose, 1987, section 6)

---

[18]  See (Rovelli, 2021) for the relational aspect.

Penrose's approach through the spin network and twistors can be given an algebraic flavour if one recognizes that the non-relativistic spinor is simply an ideal in the Clifford algebra ($C_2$) while the twistor is a similar ideal in the conformal Clifford algebra ($C_6$) (Penrose, 1972).

To round up, the Clifford algebra structure underlies the two programmes and we can see that 'from [a] methodological point of view, Penrose's system is self-consciously a projection method, and Bohm's is an implicit projection method' (Robbin, 2014).

'Nowadays, non-locality in quantum physics has been given a fundamental and probably definitive status and it has been raised to the state of a physical fact. It is founded on a conceptual ground which is more general than the question of determinism and hidden supplementary variables' (Paty, 1982).

Finally, we would like to take into account some consideration by Abner Shimony. He stresses Bohm's metaphysical programme: 'Bohm is willing to accept non-local hidden variables theories. He seems to believe that some of the features of present-day quantum theory—especially holism and non-locality—will remain permanently in the *Weltanschauung* of physics even after the development of a sub quantum physics.' We agree with Abner Shimony that Bohm's philosophical position raises difficult issues to resolve, in particular its metaphysical holism. The order involved and the order explained here are intended to address this issue. But above all the philosophical form that the author deploys, with the introduction of the ideas of *Wholeness and implicate order*, is essentially a mode of understanding and a general support for its reflection as a physicist.

# REFERENCES

Aharonov, Y., and Bohm, D. (1959). Significance of electromagnetic potentials in quantum theory. *Phys. Rev.*, **115**, 485–491.

Bassi, A., and Ghirardi, G. C. (2003). Dynamical reduction models. *Physics Reports*, **379**, 257–426.

Bell, J. S. (1964). On the Einstein, Podolsky, Rosen paradox. *Physics*, **1**, 195–200.

Bell, J. S. (1987). *Speakable and Unspeakable in Quantum Mechanics*. Cambridge: Cambridge University Press.

Bohm, D. (1951). *Quantum theory*. Englewood Cliffs: Prentice-Hall.

Bohm, D. (1952). A suggested interpretation of the quantum theory in terms of 'hidden variables' I & II. *Phys. Rev.*, **85**, 180.

Bohm, D. (ed.) 1957). *Causality and Chance in Modern Physics*. London: Routledge.

Bohm, D. (1973). Quantum Theory as an Indication of a New Order in Physics: B. Implicate and Explicate Order in Physical Law. *Foundations of Physics*, **3**(2), 139–155.

Bohm, D. (1980). *Wholeness and the implicate order*. London: Routledge and KeganPaul.

Bohm, D. (1983). The paradox of Einstein, Rosen, and Podolsky. In J. A. Wheeler and W. H. Zurek (eds), *Quantum Theory and Measurement*, Princeton, NJ: Princeton University Press.

Bohm, D., and Aharonov, Y. (1957). Discussion of Experimental Proof for the Paradox of Einstein, Rosen, and Podolsky. *Phys. Rev.*, **108**, 1070.

Bohm, D., and Hiley, B. (1995). *The Undivided Universe*. London: Routledge.

Bricmont, J. (2021). Einstein, Bohm, and Bell: a comedy of errors. This volume.

Bub, J. (2021). Hidden variables. This volume.

Crull, E. (2021). Grete Hermann's Interpretation of Quantum Mechanics. This volume.

Freire, O. (2021). Making sense of the century-old scientific controversy over the quanta. This volume.

Ghirardi, G. C., Rimini A., and Weber, T. (1986). Unified dynamics for microscopic and macroscopic systems. *Phys. Rev. D*, **34**, 470–491.

Heisenberg, W. (1958). *Physics and Philosophy: The Revolution in Modern Science*. New York: Harper and Row.

Hiley, B. J., and Peat, F. D. (1987). *Quantum implications. Essays in Honour of David Bohm*. London: Routledge.

Howard, D. (2021). The Copenhagen Interpretation. This volume.

Kouneiher, J. (2018). Where we stand today. In J. Kouneiher (ed.), *Foundations of Mathematics and Physics, One Century after Hilbert*, Cham, Switzerland: Springer International, pp. 1–73.

Laloë, F. (2021). Quantum mechanics is routinely used in laboratories with great success, but no consensus on its interpretation has emerged. This volume.

Lewis, G. N. (1926). The nature of light. *Proc. Natl. Acad. Sci. USA*, **12**, 22–29.

Lewis, G. N. (1930). The symmetry of time in physics. *Science*, **71**, 569–577.

Paty, M. (1982). L'inséparabilité quantique en perspective. *Fundamenta Scientiae*, **3**, 79–92.

Paty, M. (1986). La non-séparabilité locale et l'objet de la théorie physique. *Fundamenta Scientiae*, **7**, 47–87.

Paty, M. (2021). Einstein's criticisms of quantum mechanics. This volume.

Penrose, R. (1972). On the Nature of Quantum Geometry. In J. R. Klauder (ed.), *Magic without Magic: John Archibald Wheeler*, San Francisco: Freeman, pp. 333–354.

Penrose, R. (1987). *On the Origins of the Twistor Theory: Gravitation and Geometry*. Naples: Bibliopolis.

Penrose, R. (1989). *The Emperor's New Mind*. Oxford: Oxford University Press.

Penrose, R. (2002). John Bell, State Reduction, and Quanglement. In R. A. Bertlmann and A. Zeilinger (eds), *Quantum [Un]speakables: From Bell to Quantum Information*, Berlin: Springer Verlag, pp. 319–331.

Penrose, R. (ed.) (2004). *The Road to Reality, A Complete Guide to the Laws of the Universe*. London: Jonathan Cape.

Penrose, R. (2018). Twistor theory as an approach to fundamental physics. In J. Kouneiher (ed.), *Foundations of Mathematics and Physics, One Century after Hilbert*, Cham, Switzerland: Springer International, pp. 253–285.

Robbin, T. (2014). *David Bohm, Roger Penrose, and the Search f or Non-local Causality*. Preprint.

Rovelli, C. (2021). The Relational Interpretation. This volume.

von Weizsäcker, C. F. (1941). On the interpretation of quantum mechanics. *Z. Phys.*, **118**, 489–509.

Wheeler, J. A. (1983). Law without law. In J. A. Wheeler and W. H. Zurek (eds), *Quantum Theory and Measurement*, Princeton, NJ: Princeton University Press, pp. 182–213.

# CHAPTER 45

# SPONTANEOUS LOCALIZATION THEORIES: QUANTUM PHILOSOPHY BETWEEN HISTORY AND PHYSICS

VALIA ALLORI

## 45.1 INTRODUCTION

SPONTANEOUS localization theories are a set of quantum theories that do not suffer from the ontological vagueness and imprecision of the so-called orthodox quantum theory. In orthodox quantum mechanics, the theory regularly taught in schools all around the globe, the notion of observer or measurement has a fundamental role: its basic evolution equation changes when a measurement is performed, or an observation is made. This theory, however, manages to account for all the empirical data collected so far by providing a successful recipe for generating experimental outcomes. This is so even if the recipe is vague, namely it is unclear what a measurement or an observation is, and it is left unspecified whether there is an underlying microscopic reality whose behaviour is able to explain the emergence of these macroscopic data. Spontaneous localization theories aim at making the quantum recipe precise: the main idea driving this type of theories is to provide a single equation which would appropriately describe both the microscopic and the macroscopic world. In other words, in spontaneous localization theories the fundamental law suitably reproduces the quantum predictions without invoking the notion of measurement or observer.

In this chapter, I review spontaneous localization theories. In doing that, I give emphasis to their conceptual foundations, rather than to their technical and mathematical aspects. After briefly discussing the problems of orthodox quantum theory (Section 45.2), I present the main ideas driving the construction of spontaneous localization theories (Section 45.3) as well as their development (Section 45.4). Then, I discuss how this type of theories is empirically adequate, how its predictions may be experimentally tested, and how a first version of spontaneous localization theories was arguably not ontologically satisfactory and needed to be supplemented (Section 45.5). I conclude with proposed relativistic extensions (Section 45.6).

# 45.2 THE TROUBLE WITH QUANTUM THEORY

Quantum theory marks one of the greatest successes in physics: no theory before quantum mechanics has been confirmed so widely and accurately. However, quantum theory also represents one of the scandals of modern science: the theory is hopelessly vague and imprecise, and disappointing in its surrender and passive renunciation to even attempt to provide a coherent picture of the microscopic reality behind the phenomena. How is it possible?

## 45.2.1 Macroscopic Superpositions

The standard story is that the theory provides no intelligible microscopic picture of the world, and physics should not even inquire, as such a description is impossible. The reasons for these claims are multiple, the most famous of which is the so-called *measurement problem*. This problem was put forward by Schrödinger in his famous cat paper to underline the unsatisfactory nature of the theory and goes roughly like this.[1] In quantum theory the complete description of any physical system, its *state*, is given by a mathematical object sometimes called the *wavefunction*. The wavefunction evolves in time according to a linear equation called the Schrödinger equation, because he first proposed it. A linear equation is an equation that typically describes waves, and waves superimpose, in contrast with particles, which do not.[2] This is the *superposition principle*. It is possible therefore that a superposition wavefunction, represented as a sum of various components, describes a state of affairs of a given physical system. Superpositions seem necessary to explain some experiments involving microscopic

[1] Schrödinger (1935). However, the main idea expressed in this paper was already circulating in a less pictorial manner in the 1920s. See Bacciagaluppi and Valentini (2009), Bricmont (2016), and Norsen (2017) for interesting historical notes.
[2] When two waves come across one another, their amplitudes combine, positively or negatively, respectively, when two crests or two troughs merge or a crest meets a through.

objects, such as the famous two-slit experiment.[3] However, these microscopic super-positions would readily *amplify to the macroscopic, observable level*. Consider a pair of particles whose state is given by the sum of a state describing a particle going on the right and of a state describing the particle going on the left, each towards a screen. This is a superposition state. However, at the end either the left screen or the right screen will show a detection spot. A superposition state instead would be more like a combination between the left screen and the right screen having detected something, and we never observe anything like that. So, if the wavefunction describes the complete state of the system, and it evolves according to the linear Schrödinger equation, then quantum theory would be empirically inadequate because it would predict unobserved macroscopic superpositions.

## 45.2.2  The Collapse Postulate

Instead of caving in in the face of empirical falsification and revising their theory, the physics community endorsed a questionable way out: *von Neumann's collapse postulate*, which eliminates macroscopic superpositions by 'measuring them out'.[4] The wavefunction evolves according to Schrödinger's equation unless a measurement is performed, after which the wavefunction rapidly and randomly collapses into one of the terms of the superposition. The result of the measurement is given by one of the terms of the superposition wavefunction, and the probability of each result is given by the square modulus of the component of the wavefunction corresponding to that result. With now two evolution equations, one which is linear and deterministic (the Schrödinger equation) and the other which is random and stochastic (the von Neumann collapse postulate), quantum theory becomes empirically adequate.

However, one needs to *precisely define what a measurement is*, because we need to know when to apply each rule. One may say that it's the action of observing that triggers the collapse. Nonetheless, this moves the mystery one step further: what is an observation? Instead of attempting to provide an answer to this question, most physicists started to ignore it, and kept making more and more predictions, which quite surprisingly came out as confirmed, apparently justifying the physicists in their dismissive attitude toward foundational questions. There were exceptions, of course: in the 1920s, de Broglie, Schrödinger, and Einstein all expressed their scepticism towards the theory so formulated.[5] Then in the 1950s, Bohm[6] and Everett[7] each proposed their solution to the measurement problem, alternative to the one proposed by von

---

[3] A beam of objects previously identified as particles (e.g., electrons) is focused towards a screen with two slits. The result is that an interference and diffraction pattern is displayed on a detection screen behind the two slits. This is unexpected with 'regular', classical particles, as one should instead see an image reproducing the two slits on the back screen: interference and diffraction are wave-like, rather than particle-like, phenomena.

[4] von Neumann (1932).        [5] See Bacciagaluppi and Valentini (2009) and references therein.
[6] Bohm (1952).        [7] Everett (1957).

Neumann: Bohm refined de Broglie's original pilot-wave proposal,[8] and Everett laid the foundations of the many-worlds theory. Traditionally the pilot-wave theory solves the problem by 'adding' particles' positions to the description provided by the wavefunction: when the wavefunction is in the superposition state of 'the particle being here' and 'the particle being there', the particle is instead either 'here' or 'there', and the information coming from both the wavefunction and the particles' locations effectively 'collapses' the wavefunction to one of the two spatial regions, thereby solving the measurement problem.[9] The many-worlds theory instead embraces the idea that superpositions exist at all scales, but we do not observe macroscopic ones because the universe suitably 'branches out' into as many copies as there are terms of the superpositions.[10]

In the late 1960s through the 1980s a different type of approach emerged: *spontaneous localization theories*, in which the evolution of the wavefunction automatically suppresses the macroscopic superpositions without the need of a measurement or an observer. More precisely, the idea is that the wavefunction evolves according to a non-linear stochastic equation which most of the time resembles the Schrödinger evolution, but at suitable times reproduces the von Neumann collapse at the relevant macroscopic scale. This is the sense in which the wavefunction spontaneously collapses, or reduces, or localizes in a small spatial region. The first successful theory in this direction is the so-called QMSL, from Quantum Mechanics with Spontaneous Localization, developed by Ghirardi, Rimini, and Weber, and thus often dubbed the GRW theory.[11] I present the history, the details, and some of the modifications of this theory in Section 45.4. Now, in the next section, let us discuss the main features that spontaneous collapse theories need to have in order to solve the measurement problem.

# 45.3 THE IDEAS DRIVING SPONTANEOUS COLLAPSE THEORIES

In this paper I focus on the conceptual, rather than the technical and mathematical aspects of spontaneous localization theories.[12] Spontaneous collapse or spontaneous localization theories are theories aimed at solving the measurement problem by eliminating macroscopic superpositions dynamically, and not with an additional postulate, hence they are also known as dynamical reduction theories. In this type of theories, in contrast with the pilot-wave theory and like the many-worlds theory, there

---

[8] de Broglie (1924).
[9] See Dürr, Goldstein, and Zanghì (2011) for recent developments of the theory.
[10] For a contemporary presentation of the theory, see Wallace (2012).
[11] Ghirardi, Rimini, and Weber (1986).
[12] For a more detailed review of collapse theories, see notably Bassi and Ghirardi (2003); Bassi *et al.* (2013); Bassi and Ghirardi (2020), and references therein.

is only the wavefunction,[13] but the Schrödinger evolution for the wavefunction is modified with the sole purpose of making the macroscopic superpositions disappear, while keeping the microscopic ones. That is, the wavefunction 'spontaneously' (in the sense as a matter of law, not because of a measurement) localizes in one or the other term of the superpositions very quickly for macroscopic objects, but not for microscopic systems. So, if the aim of the theory is to suitably tackle superpositions to solve the measurement problem dynamically, let us ask ourselves what exactly the theory needs to do.

## 45.3.1  Pure and Mixed States

The state of the system is the complete description of that system. Sometimes the state is known, in which case one talks about a *pure state*. Such a state can be a superposition of other states, such as $a + b$. Sometimes instead we do not know what the state is: e.g. it can be either $a$ or $b$ but we do not know which. In this case we say that the state is in a *statistical mixture* of states, or *mixed state*. It is important not to confuse a statistical mixture with a superposition state. In fact, a superposition of states, say $a + b$, is a state in which the system is in an indefinite state which is neither $a$ nor $b$. As such, it expresses a claim about the ontology of the system: it does not have a definite property, being an $a$ or a $b$. Instead, a statistical mixture expresses the idea that we do not know which state the system is in. Thus, it is an epistemic claim. To describe these states, it is useful to use the density matrix or statistical operator, where the diagonal terms correspond to the pure states $a$ and $b$, while the off-diagonal terms represent interference effects between them. The von Neumann collapse postulate, and therefore any measurement process, effectively transforms a pure superposition state (in which the system is in an indefinite state) into a statistical mixture (in which now the system is in a definite but unknown state). There is a vast literature on the effects of the interactions of systems with their environment. The system loses coherence (or, it decoheres): the terms of the superposition can no longer interfere with one another. In other words, the interference terms in the density matrix are damped, and effectively the *superposition state becomes a statistical mixture* as if the system has been 'measured' by its environment. Some may think that this suppression is enough to solve the measurement problem. However, it is not: to solve the measurement problem (without hidden variables or many worlds) one needs to show that the (macroscopic) system is no longer in a superposition state. However, the same density matrix with suppressed off-diagonal terms can be associated both to a mixture of states with definite properties (half the system is in state $a$ and half in state $b$) and to a mixture of superposition states (half the system is in state $a + b$, and half the state in $a - b$). Therefore, for the proposed theory *one must provide a fundamental equation for the pure state, rather than for the mixed state*, in which the (pure) superpositions are appropriately damped.

---

[13] However, see Section 45.5.2.

## 45.3.2 The Preferred Basis

In addition, notice that the so-called *quantum state*, which provides the complete description of any quantum system, can be written in any basis: the position basis, the momentum basis, the energy basis, and so on. In simpler words, one can write the function representing the quantum state as a function of different variables: as a function of position, as a function of momentum, and so on. *The wavefunction is the quantum state written in the position basis*, or in more layman terms, it is the quantum state written as a function of the position, $\psi(r)$. But there is also, for instance, $\phi(p)$, which represents the same state as a function of the momentum $p$. Regardless, there will be superpositions in all bases: for $\psi(r)$, for $\psi(p)$, and so on. And suppressing a superposition in one basis will not guarantee suppression in another one. So, a first problem emerges, called the problem of the *preferred basis*: in which basis is it important to suppress the macroscopic superpositions? Assume that, as we usually do in classical (non-relativistic, non-quantum) physics, the fundamental space in which matter is located can be represented by a three-dimensional space, and that macroscopic objects are made of microscopic entities located in this space. Then it seems plausible that one would give priority to superpositions in physical space (the space in which matter is located), rather than in the space of momenta (which is more abstract and derivative, as momentum is defined in terms of position) or to anything else. In other words, *the preferred basis is the position basis because macroscopic objects need to have a definite location* over and above any other definite property, like for instance momentum.[14]

## 45.3.3 The Trigger Problem

As mentioned, the main purpose of these theories is to solve the measurement problem by suppressing macroscopic superpositions while keeping the microscopic ones. This leads to the so-called *trigger problem*:[15] what triggers the wavefunction localization only for macroscopic objects? This type of theories can accomplish that by tying the localization mechanism to the *number of particles* the system has. In this way, the localization mechanism would not do much for systems of few particles, but for macroscopic systems, thus with many particles, it would simulate well the action of the von Neumann collapse.

To avoid unnecessary confusions, one may ask how we can even talk about particles in this theory. Strictly speaking, in fact, there are no particles in spontaneous

---

[14] Some models induce the collapse in the energy basis (see Adler, 2002, 2004, Adler *et al.*, 2001, and references therein), others in the momentum basis (Benatti *et al.*, 1988), or the spin basis (Bassi and Ippoliti, 2004, Pearle, 2012). However, there is an additional burden of proof for how these models are supposed to solve the measurement problem.

[15] Pearle (1989).

localization theories. These theories are just about the behaviour of the wavefunction, which is supposed to provide the complete description of any physical system. So, *the descriptions 'a system with many particles' or 'a system with few particles' should not be read literally*. Instead, they should be understood respectively as 'a system with many degrees of freedom' and 'a system with few degrees of freedom', thereby removing the unnecessary connection to 'particles'. However, as we will see in Section 45.5.2, more needs to be said about the ontology of these theories.

## 45.3.4 No Superluminal Signalling

Mathematically, how can we implement these requirements? Namely, which mathematical features should the new equation of motion for the wavefunction have to have in order to solve the measurement problem along the lines just discussed? Notice that the linearity of the Schrödinger equation, according to which sums of solutions are also solutions, is responsible for the unobserved macroscopic superpositions. So, suppressing macroscopic superpositions makes the equation *non-linear*. The Schrödinger equation is also deterministic, namely the initial conditions and the law completely determine the state of the system at any other time. Or, given the law and the initial conditions, there is only one possible outcome, which happens with probability 1. Alternatively, there are stochastic theories, in which the initial conditions and the law merely provide a set of probabilities with which a set of various possible outcomes will happen. Therefore, to avoid unnecessary complications, it seems that in building a theory whose law simulates the von Neumann collapse one should try first a *non-linear deterministic* modification of the (linear deterministic) Schrödinger equation: why introduce stochasticity if we can use a non-linear deterministic dynamics? Indeed, even if for other reasons, deterministic non-linear modifications of the Schrödinger equation have been proposed, but soon discarded.[16] Interestingly, the reason for this has to do with the fact that a theory should *not allow superluminal signalling* (transmission of information at a velocity greater than the speed of light). So, the spontaneous localization dynamics is non-linear and stochastic because, while non-linearity suppresses macroscopic superpositions, it can prohibit superluminal signalling *only when combined with stochasticity*. To understand where this last claim comes from, one needs to take a step back. In a classical understanding, matter is made of individual components that clump together and they interact with one another through local interactions: only things which are sufficiently close to each other manage to make a difference in one another's behaviour. This locality condition is even more important in relativity, as in this theory there is a physical limit to the highest possible velocity with which stuff can travel: the velocity of light. However, in quantum theory with the

---

[16] See, for instance, Weinberg (1989a, b, c). See also Doebner and Goldin (1992), and Białynicki-Birula and Mycielski (1976). See also the Schrödinger–Newton semiclassical theory of quantum gravity discussed in Section 45.4.

*collapse postulate* the wavefunction instantaneously and randomly collapses into one of the terms of the superpositions of the wavefunction, no matter how far away they are. This is *in tension with the locality condition of relativity*.[17] Indeed, Einstein, Podolsky, and Rosen (EPR)[18] proved that if the Schrödinger evolving wavefunction provides the complete description of every physical system, then the world must be nonlocal. This was deemed unacceptable by EPR because it would be in contrast with relativity. They concluded therefore that quantum theory is incomplete. However, later Bell started from the EPR conclusion and went to compute some measurable consequences of theories which complete quantum theory in such a local way.[19] It turned out that the empirical predictions of these theories are at odds with the ones of quantum mechanics. Therefore, one could perform a sort of crucial test. This was in fact done, and it falsified the alternatives to quantum theory as envisioned by EPR.[20] That is, any empirically adequate theory would have to be nonlocal, as locality was the only assumption made in the EPR argument.[21] So the problem was then how to reconcile quantum nonlocality with relativity.[22] Be that as it may, spontaneous localization theories are in the same condition as orthodox quantum theory: *one needs to show that the dynamical collapse mechanism is not in contrast with relativity*. This is the further constraint mentioned earlier. Gisin has shown that non-linear deterministic modifications of the Schrödinger equation to produce a dynamical collapse would permit superluminal signalling.[23] So, the only possible ways of avoiding superluminal signalling is either having a linear deterministic evolution equation or a nonlinear stochastic one. Since we need nonlinearity to solve the measurement problem (in the case of this strategy), we also need stochasticity. The model developed by Ghirardi,

---

[17] See the correspondence between Einstein, Heisenberg, and others reported e.g. in Bacciagaluppi and Valentini (2009), Bricmont (2016), Norsen (2017).

[18] Einstein, Podolsky, and Rosen (1935).    [19] Bell (1964).

[20] Freedman and Clauser (1972), Aspect *et al.* (1981, 1982).

[21] See Goldstein *et al.* (2011) and references therein. Notice that controversies surrounding the reading of Bell's theorem do not influence the discussion here.

[22] At the time, it was even more problematic because some had argued that quantum mechanics could permit superluminal signalling: not only there are nonlocal influences, but they could be used to transmit information. In 1980, however the so-called no-signalling theorem was proven, stating that quantum nonlocality cannot be used to send information faster than light (Ghirardi, Rimini, and Weber, 1980).

[23] Gisin (1989). See also Ghirardi and Grassi (1991); Polchinski (1991). Here is a brief summary of Gisin's proof. It has been shown (Davies, 1976) that only a linear deterministic evolution would transform the original statistical mixture into an equivalent one (one with the same density matrix). This is important because if the evolved density matrix is not the same as the original one, there would be superluminal signalling. In fact, take a scenario like the one considered before, with two observers measuring the properties of two particles in an entangled state. If the initial and the evolved density matrix were not the same, then the density matrix would have evolved into physically distinguishable situations. In this way one observer can let the other know what measurement he has decided to perform on his system, no matter how distant, and this allows faster than light signalling (see Bassi and Hejazi, 2015, for details). Thus, a non-linear deterministic theory would allow for superluminal signalling. The only option for a non-linear dynamics is therefore to be stochastic.

Rimini, and Weber (1986), in contrast with the other alternatives proposed so far, was the first successful localization theory of this kind. Let's discuss some of its details in the next section, as well as the ones of other spontaneous collapse theories.

# 45.4 A SHORT OVERVIEW OF SPONTANEOUS LOCALIZATION THEORIES

As we have seen, deterministic non-linear modifications of the Schrödinger equation result in superluminal propagation, so let's just discuss here the progress in non-linear stochastic modifications. Starting from the late 1960s, and into the 1980s, people started to look for ways of *formalizing measurement processes into the Schrödinger equation*. Notably, Barchielli, Lanz, and Prosperi, following the work of Ludwig,[24] studied the dynamics of a macroscopic particle when subjected to appropriate, approximate position measurements at equally spaced instants which localize the system.[25] On another front, Ghirardi, Rimini, and Fonda, attempting to *reconcile the nuclear decay law with quantum mechanics*, proposed a model for unstable systems such as nuclei in which the wavefunction of the decay fragments underwent dynamical random localization processes at random times.[26] In these papers the localization processes were phenomenological and not taken to be something fundamental, as they were due to the interaction of the system with its environment. In addition, Pearle, Gisin, Diósi, and others,[27] with the aim of solving the measurement problem, developed models to account for the wavefunction collapse in terms of a *non-linear, stochastic modification of the Schrödinger equation*. In particular, Pearle proposed a non-linear equation according to which the phases of the state vectors take random values after the interaction with the measurement apparatus. However, none of these theories could solve the preferred basis problem and could provide a general account, independent of the type of measurement performed. Also, there was the trigger problem: it was not clear how to make the localization effective for macroscopic objects but not for microscopic ones.

## 45.4.1 Quantum Mechanics with Spontaneous Localization

These papers were precursors of the first successful spontaneous localization theory, published by Ghirardi, Rimini, and Weber (1986), in which the localization mechanics is spontaneous and fundamental. This theory was able to successfully suppress only macroscopic superpositions, while keeping microscopic ones, without relying on any

---

[24] Ludwig (1967, 1968, 1970).    [25] Barchielli, Lanz, and Prosperi (1982, 1983a,b).
[26] Fonda, Ghirardi, and Rimini (1978).
[27] Pearle (1976, 1979); Gisin (1984); Diósi (1984).

measurement, avoiding the preferred basis problem and superluminal signalling. Technically, Ghirardi, Rimini, and Weber (GRW) wanted to build the collapse rule into the evolution equation. GRW chose *position measurements as fundamental.* They followed the work of Barchielli, Lanz, and Prosperi (BLP), but in contrast to them GRW assumed that the localization process occurs at random times. In the original GRW paper they proposed a *non-linear stochastic equation for the density matrix and not the wavefunction,* in contrast to the attempts of Pearle, Gisin, and Diósi mentioned above, even if the suppression mechanism effectively dampens the macroscopic superpositions of the wavefunction.[28] Importantly, Bell contributed to making the theory well known in the community when he presented the theory in terms of what this theory does to the wavefunction: the wavefunction evolves according to the Schrödinger equation for a certain amount of time, then it stochastically localizes at random times in random places, and then continues to evolve according to the Schrödinger equation, and so on.[29] Here are some more details about the *dynamical collapse mechanism* of the GRW theory, which the authors call QMSL, *Quantum Mechanics with Spontaneous Localizations,* expressed in terms of the behaviour of the wavefunction, rather than as an evolution equation for the density matrix. Every physical system, macroscopic or microscopic, is subject to random and spontaneous localization processes, which GRW called *hittings.* The nature of these hittings is left unspecified: there is no mechanism or underlying explanation of their origin or cause (see Section 45.4.2.3 for gravity induced collapse). This is, roughly, how this mechanism works. A hitting is just like a position measurement, which collapses the wavefunction in one of the terms of the superposition. A hitting is mathematically implemented by the wavefunction being instantaneously multiplied by an appropriately normalized Gaussian function. The width of the function $d$ represents the *localization accuracy,* which is a measure of how spread out the wavefunction is after the localization happens. The spontaneous collapse effectively localizes the wavefunction in a region of width $d$, suppressing it outside of that region. The localization centre is random at a point in three-dimensional space with probability density given by the squared norm of the localized wavefunction so as to reproduce the quantum predictions: there will be more hittings where the probability (in orthodox quantum theory) of finding the particle if a measurement is performed is greater. The hittings occur at random times, distributed

---

[28] I think the reason for this was mainly historical: as noted, GRW were following the work of BLP, who also developed an equation for the density matrix. Also, since GRW's idea was to build the collapse postulate into a new equation for the state, one can think of the collapse in a way which naturally leads to think of density matrices: while the collapse gives definiteness to the state, one loses information about what the state is. In fact, the indefinite pure superposition state is transformed by the collapse rule into a definite but unknown mixture: because of the measurement the superposition 'collapses' into one of its terms, but we do not know which. In this way, the collapse can be seen as transforming a pure superposition state into a statistical mixture. So, one way to formalize this transition is to write an evolution equation for the density matrix which transforms, for the appropriate macroscopic systems, the pure state into a statistical mixture.

[29] Bell (1987).

according to a Poisson distribution, with mean *collapse frequency f.* In between hittings, the wavefunction evolves linearly according to the Schrödinger dynamics.

This theory automatically takes care of *the trigger problem* because the localization of one of the constituents of a macroscopic object amounts to the localization of the object itself. In fact, the wavefunction of a macroscopic object can be thought as the product of the wavefunctions of its microscopic constituents, which are not zero only in one of the terms of the superposition. So that if one of the microscopic systems undergoes localization near a given point, all the macroscopic superpositions will also localize around the same point. This guarantees that the higher the number of particles in the object, the fastest the localization, so that while microscopic objects do not quickly localize, macroscopic objects do, as desired. Numerically, the localization accuracy $d$ and the localization frequency $f$ are new constants of nature. Their values, $d = 10^{-7}$ m and the frequency $f = 10^{-16} \, s^{-1}$ were chosen so that microscopic systems would undergo a localization on average every hundred million years, while for macroscopic systems that would be every $10^{-7}$ seconds.

A notable feature of spontaneous localization theories is that, since the evolution equation for the wavefunction is no longer the Schrödinger equation, *they do not make the same predictions as orthodox quantum theory*, in contrast with the pilot-wave theory and the many-worlds theory. In particular, as a consequence of the non-Hamiltonian character of the evolution, energy is not conserved. A rough estimate of the predicted yearly energy increase for a monoatomic gas ($N \sim 10^{23}$) is $10^{-15}$K.[30] The parameters $d$ and $f$ were chosen to make these experimental discrepancies between spontaneous localization theories and orthodox quantum mechanics so small to be currently undetectable. Nonetheless, experiments to test spontaneous localization theory are underway, as presented in Section 45.5.1.

## 45.4.2  Continuous Collapse Theories

After the first successful proposal, new spontaneous localization models were proposed, most notably Quantum Mechanics with Universal Position Localization or QMUPL, the Continuous Spontaneous Localization theory or CSL, and a gravity induced spontaneous collapse theory, also known as the Diósi–Penrose (DP) theory. All these theories, including the original GRW proposal, can be written in terms of a *non-linear stochastic evolution equation for the wavefunction* in which one can identify three components. First, there is a *Hamiltonian term* corresponding to the usual Schrödinger evolution. Then there is a *stochastic component* which forces the wavefunction to collapse toward one of the possible position measurement results (eigenstates of the position operator). The stochastic component is multiplied by a constant $\gamma$ which expresses the *strength of the collapse* process. Then there is a third term

---

[30] Bassi and Ghirardi (2020).

introduced for consistency reasons.[31] GRW-type models can be rewritten along these lines.[32] These collapse theories are qualitatively equivalent: they all induce the localization of the wavefunction in space (as opposed to e.g. energy, or momentum), and the collapse is faster, the larger the system. The difference with GRW-type theories, in which the localization mechanism is discrete, is that in all these other theories the collapse is continuous. In these theories the continuous localization which provides the localization mechanism is mathematically implemented by a stochastic equation representing a diffusion process, so it is as if the system constantly undergoes small, random fluctuations proper of Brownian motion, and one says that the body is subject to a universal force 'noise'.

The simplest continuous spontaneous localization theory is *Quantum Mechanics with Universal Position Localization*, or QMUPL, proposed by Diósi.[33] QMUPL can be seen as a simplified version of CSL,[34] and it has the advantage of allowing for a rigorous mathematical analysis. However, just as in the case of the original QMSL (the GRW theory), it is built for systems of distinguishable particles.

### 45.4.2.1 *Indistinguishable Particles*

The idea of spontaneous collapse can be generalized to identical particles of different types in the framework of the *Continuous Spontaneous Localization*, or CSL, theories in which the discontinuous jumps are replaced by a continuous stochastic evolution in a way that is more general than the QMUPL.[35] While QMSL and QMUPL use the so-called first-quantization language, CSL theories are expressed in the *second-quantization formalism*, which allows the generalization of the theory to a varying number of indistinguishable particles. Using creation and annihilation operators, the collapse mechanism makes it the case that superpositions containing different numbers of particles in different points of space are suppressed, which is equivalent to collapsing the wavefunction in space, in a second quantized language. CSL introduces a three-dimensional *mass density field M*, defined as the mass of the particles multiplied by their number density and weighted through a correlation function, which involves a

---

[31] Adler (2004) has argued that if one modifies the Hamiltonian evolution only adding a stochastic term, the norm of the wavefunction would not be conserved. This can be avoided by normalizing, which however makes the equation non-linear in a way which leads to superluminal signalling. The problem can be avoided if one adds a suitable extra term.

[32] Girardi, Rimini, and Weber (1986); Diósi (1988a, 1988b, 1990); Gatarek and Gisin (1991); Bassi (2005); Bassi, Ippoliti, and Vacchini (2005); Halliwell and Zoupas (1995).

[33] Diósi (1989).

[34] QMUPL is equivalent to CSL for distances which are small with respect of the localization length *d* (Dürr, Hinrichs, and Kolb, 2011).

[35] Pearle (1989); Ghirardi, Pearle, and Rimini (1990). Also, Tumulka (2006b,c) has developed a discrete spontaneous localization theory for a variable number of identical particles, following the work of Dove and Squires (1995) and questioning the claim that one needs to go to a continuous localization process to handle identical particles; see Ghirardi (1999) and Bassi and Ghirardi (2003). Other, however unsuccessful, proposals are Ghirardi *et al.* (1988) and Kent (1989).

correlation length $r_C$ (analogue of $d$ in QMSL).[36] That is, the collapse mechanism acts on $M$, which defines a mass field in three-dimensional space on which the stochastic terms acts. We will discuss this mass density CSL again in Section 45.5.2. *The correlation length $r_C$ and the collapse strength $\gamma$, which is the constant in front of the stochastic term, are the two parameters of CSL.* Their value is set at $\gamma = 10^{-36} \mathrm{m}^3 \mathrm{s}^{-1}$ and $r_C = 10^{-7}$ m. The collapse frequency $f$ is related to the above constant by the expression: $f = \frac{\gamma}{8\pi^{3/2} r_C^3} \cong 2.2 \times 10^{-17} \mathrm{s}^{-1}$.[37]

## 45.4.2.2 *New Physical Field: Colour of the Noise and Dissipation*

Another problem for QMUPL is that the stochastic term (sometimes also dubbed 'noise') depends only on time, while in CSL it also depends on space. This is interesting because it allows for a different reading of the evolution equation of these theories. The original attitude towards the meaning of these equations was to think that nature is intrinsically stochastic. However, it is also possible to think of the stochastic term as representing *a new physical field* filling space, which couples with (quantum) matter in such a way to suppress macroscopic superpositions. The new terms in the modified Schrödinger equation are meant to describe such a coupling. This field is essentially non-quantum (otherwise we still have the measurement problem) and couples to (quantum) matter through a non-linear coupling. As discussed in Section 45.4.2.3, some have proposed that the collapse field is connected with gravity, as gravity is non-linear, and it has not been successfully quantized yet. Because of its lack of spatial dependence in QMUPL, the stochastic field cannot be immediately identified with a physical collapse field (as physical fields usually extend in space), while this seems possible in CSL, as the field also depends on space.

Another thing to notice is that in both QMUPL and CSL the collapse field, when thought about in terms of noise, is said to be white (all frequencies of the noise contribute to the collapse with the same weight) and the evolution is Markovian (it has no memory: the probability of the state at one time depends only on the present state and not the past). However, more realistic theories would use *coloured noises*, as a physical field of gravitational cosmological origin would not be white, as the frequency spectrum needs to be bounded (it cannot have an infinite number of frequencies). This, therefore, requires a cut-off frequency which, unfortunately, makes the theory non-Markovian, and thus difficult to investigate.[38]

Even more realistic theories include *dissipative effects*. In the original theories (QMSL, QMUPL, CSL, with or without coloured noises) the collapse noise keeps exchanging energy without changing temperature: the wavefunction localizes, but at

---

[36] Ghirardi, Benatti, and Grassi (1995). See also Pearle and Squires (1994).

[37] Different values have been suggested by Adler (2007) and falsified: see Bassi, Deckert, and Ferialdi (2010); Curceanu, Hiesmayr, and Piscicchia (2015); Vinante *et al.* (2016); and Toroš and Bassi (2018).

[38] A theory of this kind is the non-Markovian QMUPL theory (Bassi and Ferialdi, 2009a, b), while a non-Markovian CSL theory is still under development (Adler and Bassi, 2007, 2008). General non-Markovian collapse theories have been discussed in Bassi and Ghirardi (2002); Diósi *et al.* (1998); Pearle (1993, 1996); Adler and Bassi (2007, 2008).

the same time the energy of the system increases steadily. Dissipative terms have been included through a position and momentum coupling between the wavefunction and the noise, and this has helped contain the energy increase predicted by the theory.[39]

### 45.4.2.3 *The Diósi–Penrose Theory*

As we have seen, the stochastic term in CSL acts on the mass density field $M$. One motivation for considering such a field in CSL has to do with the idea of explaining where the collapse is coming from: gravity.[40] In fact, gravity is universal, and its strength increases with the mass of the system. Also, the collapse needs to be non-quantum and non-linearly coupled with the (quantum) matter field. Gravity is neither quantized nor linear. Indeed, some spontaneous localization theories other than CSL have been proposed to *connect the collapse with gravity*. There are two approaches to this.[41] First, the Schrödinger–Newton (SN) equation which can be derived from a semiclassical theory of gravity, and then the Diósi–Penrose, or DP, approach. The semiclassical theory of quantum gravity is a proposed fundamental theory in which the classical (not quantized) gravitational field couples with the quantum material fields into a non-linear deterministic equation. The coupling with the gravitational field allows for dampening of the various terms of a superposition, but in an empirically inadequate way.[42] As we have seen, introducing some kind of collapse preserving determinism would lead to superluminal signalling, so the only option is to transform the SN equation into a non-linear stochastic equation. This is what has been done by Diósi and independently by Penrose, hence the name, *Diósi–Penrose (DP) theory*.[43] Penrose argued that the quantum superposition principle is in fundamental contradiction with the general covariance principle of general relativity, and accordingly he suggested that a macroscopic superposition would spontaneously decay, due to gravity, after a finite and sufficiently short time. On the other hand, Diósi wanted to connect the collapse in terms of gravitational interaction because in this way he would also *eliminate the arbitrary parameters* which enter the other spontaneous localization theories. So, he proposed a non-linear stochastic theory similar in structure to CSL but in which the stochastic field is identified with the gravitational field. Technically, this is implemented by making the *stochastic term proportional to the gravitational potential* and acting on the system through a three-dimensional mass density field. Unfortunately, this parameter-free approach fails because the gravitational potential diverges for small distances and one has to introduce an effective radius (cut-off) below which particles are considered point-like. Diósi proposed the natural choice, namely the radius of the nucleon. Notably, the collapse time scale predicted by Diósi coincides

---

[39] Bassi and Ferialdi (2012a, b).    [40] See also Ghirardi, Grassi, and Rimini (1990).

[41] Diósi (1984, 1989, 2007); Penrose (1986, 1996). For a review, see Gasbarri *et al.* (2017).

[42] A superposition of two spatial locations would always collapse into the middle, rather than 50 per cent of the time on the left and 50 per cent of the time on the right. See Bahrami *et al.* (2014a).

[43] A similar idea was first proposed by Károlyházy (Károlyházy 1976, Károlyházy *et al.* 1986). An important early study comparing the Károlyházy theory and GRW was made in Frenkel (1990).

with the one heuristically suggested by Penrose. However, with this cut-off the theory predicts an unacceptable total energy increase. Ghirardi, Grassi, and Rimini showed that this can be reduced to an acceptable level with a much larger cut-off, which however seems to lack a proper physical justification.[44] Accordingly, others have explored the possibility of limiting the energy increase by considering dissipation. This could work at the cost of adding another free parameter (a mass connected with the energy increase) but only in the mesoscopic regime (masses not too small, not too large).[45] This limitation reinforces the idea that these theories are actually phenomenological rather than fundamental descriptions of the world. Moreover, even if this theory (and its modifications) provides an attempt to explain the origin of the collapse in terms of a gravitational interaction, it does not seem to succeed at doing so, as the gravitational field does not couple to matter like one would expect (in contrast with what happens in the SN equation).[46] Also, recent experiments have put very strong constraints on these theories (see Section 45.5.1).

# 45.5 THE EMPIRICAL AND ONTOLOGICAL ADEQUACY OF SPONTANEOUS LOCALIZATION THEORIES

As we have seen, if the wavefunction evolves according to the Schrödinger equation, then without a collapse rule the theory predicts unobserved macroscopic superpositions and thus it is not empirically adequate. One way to make quantum theory empirically adequate is the von Neumann collapse rule but this would introduce a vague and double dynamics. Spontaneous localization theories are empirically adequate theories with a precise and unified quantum dynamics: the wavefunction evolves according to a non-linear equation such that microscopic systems may interfere while macroscopic ones cannot.

Objects at all scales are represented by the wavefunction (however, see Section 45.5.2). An electron, for instance, is not a point-like object with a space-time trajectory. Rather it is a field represented by the wavefunction and in a double-slit experiment it suitably crosses both slits at the same time. As a direct consequence of the modified Schrödinger dynamics, when we try to measure the electron location by

---

[44] Ghirardi Grassi, and Rimini (1990). See also Bahrami, Smirne, and Bassi (2014).

[45] Bahrami *et al.* (2014a).

[46] Another spontaneous localization theory based on quantum gravity considerations has been proposed by Ellis, Hagelinm, and Nanopoulos (1984); Ellis, Nanopoulos, and Mohanty (1989). In this theory the wavefunction of a system is effectively localized by the interaction with a bath of quantum wormholes which characterize the space-time structure at the Planck length scale. Compared to CSL and DP, this theory is parameter-free, which allows for unambiguous experimental falsifiability. See Bassi *et al.* (2013).

making it interact with a measurement apparatus like a photographic plate, the wavefunction gets localized at a point, which is the experimental outcome. Macroscopic objects, also represented by the wavefunction, are almost always narrowly localized in space so that we can think of them as moving according to the semiclassical laws for all practical purposes. For most macroscopic systems one can separate the classical centre-of-mass motion from the internal GRW-like motion.[47]

As we have seen, spontaneous localization theories introduce two new parameters, the collapse frequency and the localization length (or equivalently the collapse strength and the correlation length). The original choice for these parameters was guided by the desire for consistency with observations. However, there is room for more general considerations. In particular, since *spontaneous localization theories are only in practice empirically equivalent to quantum theory, not in principle*, there is room for exploring the possibility of *empirically falsifying quantum theory* in favour of spontaneous localization theories, as well as to put bounds on the collapse parameters due to the absence of predicted deviations from quantum theory. In other words, while for instance QMSL sets a precise value for the parameters, instead experiments can allow us to plot the physically viable regions in parameter space: the collapse frequency and the localization distance can only assume these values otherwise we would see an energy increase, or a temperature increase, or a sound emission, and so on, that we do not see.

## 45.5.1 Experiments

So, let me briefly mention some of the possible experimental tests relevant for spontaneous localization theories. First, as we have mentioned already, since the wavefunction evolves according to a modified Schrödinger equation, *energy is no longer conserved*. Moreover, one could observe the *diffusion* induced by the collapse. These two effects can be measured in cold atoms.[48] Because the collapses tend to add energy to every system, temperatures of all things increase as well, leading to some *universal warming*. One can compute the temperature increase as a function of the two parameters for various objects and obtain bounds for the value of the parameters by observing the absence of warming in these things.[49]

Also, it is possible that the collapse would create so much energy as to create a small explosion, with consequent *sound emission*. Since no sound emitted by macroscopic matter is observed, this leads to bounds on the parameters.[50]

---

[47] Ghirardi, Rimini, and Weber (1986).
[48] Laloë, Mullin, and Pearle (2014) and Bilardello *et al.* (2016).
[49] Adler (2007). However, experiments of this type seem to be inconclusive, as it is difficult to account for all the possible ways for the body to cool off. See Feldmann and Tumulka (2012).
[50] Feldmann and Tumulka (2012).

People have studied the *loss of coherence* in macromolecules[51] and in other devices of mesoscopic dimensions.[52] These devices and similar ones,[53] as well as gravitational wave detectors such as LIGO, AURIGA, and LISA Pathfinder[54] could also be used to detect the noise induced by the collapse.[55]

Another way is to constrain the parameters by determining the scale at which a given system can still interfere. Since collapse can destroy interference, every successful *diffraction and interference* experiment puts bounds on the parameters.

Because of the collapse, some otherwise stable systems may emit radiation, contrary to what is predicted by quantum theory. Experiments measuring *the spontaneous radiation emission* from atoms provide the strongest upper bound on the parameters.[56] Experiments of this type have also recently been used to test the DP theory. In fact, the gravity induced collapse depends on the mass of the system and is random. This results in a diffusion of the motion of the particles, which if charged begin to radiate. The radiation emission rate has been computed, experiments were performed, and they rule out the DP theory.[57]

Spontaneous localization theories also have empirical consequences for supercurrents in a *superconducting ring*[58] as the collapse would break the Cooper pairs, and thus, the supercurrent would spontaneously decay (unless the Cooper pairs get re-created) at a given rate.[59]

## 45.5.2  Ontology

Scientific antirealists believe that a satisfactory theory only needs to account for the experimental outcomes. However, one thing is to say that the theory has to be satisfactory at the macroscopic level in this way, and another is for the theory to provide a sensible microscopic description of reality giving rise to the results the theory accurately reproduces. This second requirement is something desirable for a *scientific*

---

[51] Arndt *et al.* (1999); Hackermueller *et al.* (2004); Eibenberger *et al.* (2013); Fein *et al.* (2019).

[52] Marshall *et al.* (2003); Bassi, Ippoliti and Adler (2005).

[53] Collett and Pearle (2003); Bahrami *et al.* (2014b); Nimmrichter *et al.* (2014); Diósi (2015); Vinante *et al.* (2016); Vinante *et al.* (2017); Zheng *et al.* (2020); Pontin *et al.* (2020).

[54] Carlesso *et al.* (2016); Helou *et al.* (2017).

[55] For recent developments see Carlesso, Vinante, and Bassi (2018); Carlesso *et al.* (2018); Schrinski, Stickler, and Hornberger (2017); Komori *et al.* (2020).

[56] Fu (1997); Adler and Bassi (2007); Adler and Ramazanoglu (2009); Adler, Bassi, and Donadi (2013); Bassi and Donadi (2014); Donadi, Deckert, and Bassi (2014); Donadi and Bassi (2014). Curceanu and colleagues (Curceanu, Hiesmayr, and Piscicchia (2015), Curceanu *et al.* (2016), Piscicchia *et al.* (2017)) have rejected Adler's proposal (2007) for a change of the frequency of the localizations, unless CSL is modified by taking a non-white noise (which is actually a reasonable assumption, if the noise is physical).

[57] See Donadi *et al.* (2020) for details.

[58] Rae (1990); Buffa *et al.* (1995); Leggett (2002); Adler (2007). However, the current estimates do not include the possibility of re-creation of the Cooper pairs.

[59] Feldmann and Tumulka (2012).

*realist*, who believes that spontaneous localization theories are able to provide approximately true descriptions of reality. By suppressing macroscopic superpositions, spontaneous localization theories will satisfy the first requirement, and thus antirealists would be happy. But what about the second? Can these theories be made satisfactory to the scientific realist too? In other words, now the question is about ontology: what is matter made of in spontaneous localization theories? If one starts from the formalism, since in spontaneous localization theories one only has an evolution equation for the wavefunction, it seems natural to think that the wavefunction represents the (microscopic and macroscopic) description of the world. This is similar to what happens in classical physics, where the fundamental equation is an evolution equation for mathematical objects that could be taken to represent point particles. This view is called *wavefunction realism.*[60] However, there are arguments against this view. The possibility of considering the wavefunction as a physical field was initially entertained by Schrödinger when he proposed his own equation.[61] Still, he dismissed it because *the wavefunction is not a field in three-dimensional space*, like electromagnetic fields, but it is on configuration space: the space of the configurations of all particles. That is, the wavefunction lives in a space with dimensions equal to three times the number of particles in the system. Schrödinger found it unacceptable to think that such an object could represent something truly vibrating.[62] This problem is sometimes called the *configuration space problem.* People who accept this as a problem are drawn to the so-called *primitive ontology approach*, according to which only mathematical objects in three-dimensional space (or four-dimensional space-time) can satisfactorily represent physical systems: particles have three-dimensional trajectories, fields vibrate in three-dimensional space. Mathematical objects in the theory which are more abstract (like for instance the wavefunction) are not the right kind of objects to represent matter.[63]

This attitude is consistent with the way in which these theories have been formulated. In fact, taking position (in three-dimensional space) as the *preferred basis* and solving the *trigger problem* by giving preference to position measurements would be arbitrary choices if we did not already think of three-dimensional space as privileged and fundamental, namely as the space in which material stuff lives.

Also, Ghirardi, Benatti, and Grassi (GBG) argued that CSL would not be ontologically satisfactory unless it is seen as a theory of something else, other than the wavefunction.[64] In fact, if the wavefunction correctly represented physical systems, *then the distance between two wavefunctions* would capture how states (as represented by wavefunctions) are physically different. However, this is not the case. Take, for instance the following three states: $h$, $h^*$, and $t$, where $h$ and $t$ represent different macroscopic properties of an object, such as being localized 'here' and 'there', while $h^*$ is a state identical to $h$, but for one particle being in a state orthogonal to the corresponding

---

[60] See Albert and Ney (2013) and references therein.     [61] Schrödinger (1928).
[62] See Bacciagaluppi and Valentini (2009); Bricmont (2016); Norsen (2017), and references therein.
[63] Allori *et al.* (2008); Allori (2013, 2015, 2019).
[64] Ghirardi, Benatti, and Grassi (1995).

particle in $h$. Then, macroscopically, $h$, and $h^*$ are indistinguishable and different from $t$. Despite this, however, the distance between $h$ and $h^*$ is equal to that between $h$ and $t$. As a consequence, GBG concluded that macroscopic systems would be better represented by something other than the wavefunction. In other words, one needs to supplement, also in spontaneous localization theories, the description of the system provided by the wavefunction by some three-dimensional entity. For reasons that have to do with gravity, GBG argued that the ontology of spontaneous localization theories is provided by a *mass density field in three-dimensional space* defined in terms of the wavefunction as we have seen earlier. This three-dimensional mass field is the ontology of this spontaneous localization theory, dubbed GRWm or CSLm.

Now that we have discussed how people have argued that spontaneous localization theories need to be specified an ontology different from the wavefunction, we can explore different possibilities. First there is an alternative proposal to have spontaneous localization theories with a discrete spatiotemporal ontology, called *'flashes'*: they are events in space-time that are identified by the space-time points at which the wavefunction collapses. In this theory, labelled GRWf, matter thus is in this case a galaxy of such events.[65] This theory has been proposed because it seems to be the most suitable to be extended to relativity (see Section 45.6).

Then, there are spontaneous localization theories with a *particle ontology*, GRWp.[66] Here particles evolve according to the same guidance law of the pilot-wave theory, and the wavefunction evolves according to the Schrödinger equation. Then randomly the wavefunction collapses in the actual position of the particle, and the particle's location jumps at random. This theory has the advantage of having the simplest ontology, particles. However, with its doubly stochastic evolution, both for the particle and for the wavefunction, it is difficult to see what advantage this theory could have when compared to the pilot-wave theory, where both the wavefunction and the particle evolve deterministically.

## 45.5.3 Mutual Constraints

It turns out that one can put different bounds on the parameters depending on the ontology of the spontaneous localization theory. First, let me notice that GRWp puts no constraints on the parameters. In fact, if the collapse frequency is zero, namely there are no wavefunction collapses, the theory reduces to the pilot-wave theory (which solves the measurement problem by adding particles' positions, not by suppressing macroscopic superpositions). Also, even if $f$ is large the theory does not need to suppress the macroscopic superpositions in order to solve the measurement problem. In fact, if $f$ is not zero, the particles evolve stochastically but no value of $f$ or $d$ prevents

---

[65] Bell (1987).

[66] For the very first hint at such a theory see Bohm and Hiley (1993). Also, see Allori (2020) for an analysis of the various spontaneous localization theories of particles.

the theory from successfully solving the measurement problem. This is because in this framework the solution of the measurement problem depends on the particle ontology, which has no superpositions, while the superpositions of the wavefunction have no physical meaning.[67]

In the case of GRWm, macroscopic objects are made of mass density fields, which, in contrast to the case of GRWp, inherit the superpositions of the wavefunction. So in order to keep macroscopic superpositions suppressed one needs that there are enough collapses per second (that is, the collapse frequency $f$ needs to be large enough) and that the localization distance is not too large (that is, the localization distance $d$ needs to be sharp enough). Notice that if there are no collapses ($f = 0$), then one would go back to orthodox quantum theory but now with a mass density field, so the theory would still solve the measurement problem but this time resorting to a many-worlds strategy.

If instead the world is described by GRWf, since one flash occurs at every collapse, if there are few collapses per second then the flashes could not be enough to possibly represent a macroscopic object. That would make the theory unsatisfactory for a different reason than GRWm: while for small $f$ GRWm would describe many-worlds, GRWf would represent no world at all. GRWf would also be unsatisfactory if $d$, the characteristic length, is too large because that would mean that the flashes would be spread out in too large a region. Therefore, this would not allow for an empirically appropriate description of the macroscopic world as we see it. In this way, if experiments can pose a bound from above, philosophical considerations like the ones above provide a bound from below: if there are very few collapses (small collapse frequency) and they are not too sharp (large localization distance) then they are inefficacious and do not suppress macroscopic superpositions.[68]

Interestingly, if future experiments end up falsifying quantum theory and confirming spontaneous localization theories for parameters in the philosophically unsatisfactory region one would have to conclude that a flash ontology could not be correct as a parameter in this region would make the theory describe no macroscopic world. The same would be true in the case of a mass density ontology, unless we embrace the idea that there are many worlds. So, this would leave us either with a particle, single-world ontology, or with a mass-density, many-worlds ontology.[69]

## 45.6 RELATIVISTIC EXTENSIONS

Let me conclude this chapter with the next challenge the spontaneous localization programme is facing, namely making the theories compatible with relativity theory. All

---

[67] See Allori (2020) and references therein.
[68] In fact, GRWm could now be interpreted as a theory representing a different world for each superposition term (Allori *et al.* 2008).
[69] Feldmann and Tumulka (2012).

the spontaneous localization theories discussed so far are non-relativistic. The problems in constructing a relativistic extension of this type of theories have to do with *reconciling the instantaneous collapse process with the nonlocality of quantum correlations*. We have seen in Section 45.3.4 that spontaneous localization theories we have considered (in virtue of being non-linear and stochastic) do not allow for superluminal signalling. However, even if a theory does not allow superluminal signalling that does not mean it is necessarily compatible with relativity: the pilot-wave theory does not allow superluminal signalling (there's no access to the particles' configuration) but it requires a preferred foliation, arguably violating the spirit of relativity.

As already anticipated, the tension between quantum theory and relativity is exemplified by the violation of Bell's inequality and therefore of locality. Some have argued that there are two ways in which a theory could violate locality: in an EPR-type experiment where there are two experimenters at opposite sides of a source of entangled particles, the outcomes of one side may depend on the type of experiment the other experimenter decides to perform (*parameter dependence*), or they can depend on the other experimenter's outcome (*outcome dependence*).[70] It is also argued that *outcome dependence is easier to reconcile with relativity*. The idea is roughly that in a parameter-dependent theory the settings on one side would have a causal influence on the result on the other side, suggesting action at a distance. Instead, in an outcome-dependent theory the correlations between the outcomes cannot be explained locally, and thus the theory does not necessarily conflict with relativity.[71] Deterministic theories such as the pilot-wave theory are parameter dependent, thus can be made relativistic invariant only by adding a preferred slicing of space-time (a foliation).[72] However, this seems to be contrary to the spirit of relativity, according to which there is no preferred frame. Instead, stochastic theories like *spontaneous localization theories are outcome dependent* (in virtue of the stochasticity of the law),[73] and thus people are being optimistic about obtaining a genuine relativistic invariant spontaneous localization theory.

The first attempt aimed at making CSL relativistic invariant, following Bell,[74] used a Tomonaga–Schwinger equation instead of the Schrödinger equation in the evolution of the theory.[75] However, this creates an infinite production of energy per unit of time and volume, due to the white noise. As already mentioned, proposals with a non-white noise have been put forward but they are mathematically very difficult to explore.[76]

[70] Suppes and Zanotti (1976); van Fraassen (1982); Jarrett (1984); Shimony (1984).
[71] Ghirardi (2010); Myrvold (2015). For a criticism, see Norsen (2009).
[72] Berndl *et al.* (1996).        [73] Ghirardi *et al.* (1993).        [74] Bell (1989, 1990).
[75] While the Schrödinger equation tells you what happens to the state from one time to the next as the system advances infinitesimally, this theory instead takes as fundamental an arbitrary space-like surface, and defines the evolution of the state from one such surface to the next.
[76] See Pearle (1989); Diósi (1990); Bassi and Ghirardi (2002) for nonrelativistic spontaneous localization models with non-white noise. See Nicrosini and Rimini (2003) for an attempt to reduce the energy increase in the relativistic theory. See Pearle (2009) and Myrvold (2017) for criticisms.

Pearle has proposed without success to avoid this infinite energy increase by considering a tachyonic noise in place of a white noise.[77] On a different front, Dowker and collaborators have put forward a spontaneous collapse model on a lattice.[78]

Important progress has been made by Tumulka, generalizing a previous idea of Bell based on the flash ontology.[79] He has proposed a relativistic discrete CSL theory dubbed rGRWf, which describes a system of distinguishable noninteracting 'particles', based on the multi-time formalism and one Dirac equation per 'particle'.[80] This theory has also been recently extended to interacting 'particles'.[81]

A relativistic GRWp theory needs to be fully developed.[82] Instead, a relativistic GRW theory with a mass density ontology has been proposed, disputing the claim that relativistic invariance requires a flash ontology.[83]

Clearly, more work needs to be done to reconcile spontaneous localization theories with the theory of relativity, but the prospects of succeeding seem much better than perceived in the past. To conclude, let me mention a completely different attitude toward the issue.[84] The idea is that spontaneous localization theories should be understood as phenomenological and thus there is *no need to require for them relativistic invariance*. If one assumes that the random field causing the collapse of the wavefunction is a physical field filling space and possibly having a cosmological origin, then the noise would select a privileged reference frame. The underlying theory, out of which these equations would emerge at an appropriate scale, should instead respect relativistic invariance. Such a theory would explain the origin of the collapse field which, because of the initial conditions, would break the relevant symmetry.

However, this attitude may undermine the main motivation to prefer spontaneous localization theories over, say, the pilot-wave theory. Among the non-relativistic solutions of the measurement problem, the simplest seems to be the pilot-wave theory because both the evolution for the particles' motion and that for the wavefunction are deterministic. Instead, when considering relativistic extensions, we cannot say the same. Rather one would lean towards spontaneous localization theories because they do not require a preferred foliation, in contrast with the pilot-wave theory. However, if we no longer require relativistic invariance for these theories because they are phenomenological, why would we ever want to consider non-linear, stochastic modifications of the wavefunction evolution if the measurement problem may be solved by the deterministic, linear equation of the pilot-wave theory? This question seems particularly pressing if one thinks that theories need an ontology in three-dimensional space (a primitive ontology), so that in both the pilot-wave theory and spontaneous localization theory one needs to add something over and above the wavefunction.

---

[77] See Pearle (1999) for the proposal, and for his admission the model is unsuccessful, Pearle (2009). Recently new proposals which involve the introduction of auxiliary fields have been put forward by Bedingham (2011).

[78] Dowker and Henson (2004); Dowker and Herbauts (2004).    [79] Bell (1987).

[80] Tumulka (2006a).    [81] Tumulka (2020).    [82] However, see Allori (2020).

[83] Bedingham *et al.* (2014).    [84] Adler (2004).

Nonetheless, I think this question arises also for wavefunction realists. In fact, on the one hand they would claim that spontaneous localization theories do not add anything, and thus they should be preferred to other deterministic theories with an inflated ontology like the pilot-wave theory, even if they are not relativistic invariant. Indeed, they could say that, given that in gravitation induced collapse theories the collapse is real and not merely effective like in the pilot-wave theory, spontaneous localization theories thought of as phenomenological allow us to unify all known forces by connecting the collapse with gravity. However, I think this line of thought ultimately backfires, as theories with gravitation induced collapse would look particularly cumbersome. In fact, there would be the fundamental quantum field living in a high dimensional space, represented by the wavefunction in configuration space, and then there would be the gravitational field in three-dimensional space. How would the two fields ultimately interact, if they belong to two different spaces? One may respond that the fundamental space is the 'largest' one, and that the gravitational field merely is confined in three dimensions. Still, why is that? What is special about the gravitational field with respect to the others?

These questions suggest that we are far from having a clear picture about quantum theory in general, and the nature and origin of the collapse in particular. Even so, tremendous progress has been made in recent years, both at the experimental and the theoretical level: new experimental settings have been proposed to test the boundaries of the spontaneous localization theories, several new models have been proposed to combine these theories with relativity, and more and more physicists and philosophers have taken the foundations of quantum theory more seriously. These encouraging results allow us to be cautiously optimistic that soon we will have more answers to our questions.

# REFERENCES

Adler, S. L. (2002). Environmental Influence on the Measurement Process in Stochastic Reduction Models. *Journal of Physics A*, 35(4), 841–58.

Adler, S. L. (2004). *Quantum Theory as an Emergent Phenomenon*. Cambridge: Cambridge University Press.

Adler, S. L. (2007). Lower and Upper Bounds on CSL Parameters from Latent Image Formation and IGM Heating. *Journal of Physics A*, 40, 2935–57.

Adler, S. L., and Bassi, A. (2007). Collapse Models with Non-white Noise. *Journal of Physics A*, 40(50), 15083–98.

Adler, Stephen, L., and Angelo Bassi: (2008). 'Collapse Models with Non-white Noises II: Particle-density Coupled Noises.' Journal of Physics A 41: 395308.

Adler, S. L., and Ramazanoglu, F. M. (2009). Photon-Emission Rate from Atomic Systems in the CSL Model. *Journal of Physics A*, 42(10), 109801.

Adler, S. L., Brody, D. C., Brun, T. A., and Hughston, L. P. (2001). Martingale Models for Quantum State Reduction. *Journal of Physics A*, 34(42), 8795–820.

Adler, S. L., Bassi, A., and Donadi, S. (2013). On Spontaneous Photon Emission in Collapse Models. *Journal of Physics A*, 46(24), 245304.

Albert, D. Z., and Ney, A. (eds) (2013). *The Wave Function: Essays in the Metaphysics of Quantum Mechanics*. Oxford: Oxford University Press.

Allori, V. (2013). Primitive Ontology and the Structure of Fundamental Physical Theories. In D. Albert and A. Ney (eds), *The Wave Function: Essays in the Metaphysics of Quantum Mechanics*, Oxford: Oxford University Press, pp. 58–75.

Allori, V. (2015). Primitive Ontology in a Nutshell. *International Journal of Quantum Foundations*, 1(3), 107–122.

Allori, V. (2019). Scientific Realism without the Wave-Function. In J. Saatsi and S. French (eds), *Scientific Realism and the Quantum*, Oxford: Oxford University Press, pp. 212–228.

Allori, V. (2020). Spontaneous Localization Theories with a Particle Ontology. In V. Allori, A. Bassi, D, Dürr, and N. Zanghì (eds), *Do Wave Functions Jump? Perspectives on the Work of GianCarlo Ghirardi*, Cham: Springer, pp. 73–94.

Allori, V., Goldstein, S., Tumulka, R., and Zanghì, N. (2008). On the Common Structure of Bohmian Mechanics and the Ghirardi–Rimini–Weber Theory. *The British Journal for the Philosophy of Science*, 59(3), 353–89.

Arndt, M., Nairz, O., Vos-Andreae, J. Keller, C., van der Zouw, G., and Zeilinger, A. (1999). Wave–Particle Duality of C60 Molecules. *Nature*, 401(6754), 680–682.

Aspect, A., Grangier, P., and Roger, G. (1981). Experimental Tests of Realistic Local Theories via Bell's Theorem. *Physical Review Letters*, 47(7), 460–466.

Aspect, A., Grangier, P., and Roger, G. (1982). Experimental Test of Bell's Inequalities Using Time-Varying Analyzers. *Physical Review Letters*, 49(25), 1804–07.

Bacciagaluppi, G., and Valentini, A. (2009). *Quantum Theory at the Crossroads: Reconsidering the 1927 Solvay Conference*. Cambridge: Cambridge University Press.

Bahrami, M., Grossardt, A., Donadi, S., and Bassi, A. (2014a). The Schrödinger – Newton Equation and its Foundations. *New Journal of Physics*, 16, 115007.

Bahrami, M., Paternostro, M., Bassi, A., and Ulbricht, H. (2014b). Proposal for a Noninterferometric Test of Collapse Models in Optomechanical Systems. *Physical Review Letters*, 112(21), 210404.

Bahrami, M., Smirne, A., and Bassi, A. (2014). Role of Gravity in the Collapse of the Wave Function: A Probe into the Diósi–Penrose Model. *Physical Review A*, 90, 062105.

Barchielli, A., Lanz, L., and Prosperi, G. M. (1982). A Model for the Macroscopic Description and Continual Observations in Quantum Mechanics. *Il Nuovo Cimento B*, 72, 79–121.

Barchielli, A., Lanz, L., and Prosperi, G. M. (1983a). Statistics of Continuous Trajectories in Quantum Mechanics: Operation-valued Stochastic Processes. *Foundations of Physics*, 13, 779–812.

Barchielli, A., Lanz, L., and Prosperi, G. M. (1983b). Statistic of Continuous Trajectories and the Interpretation of Quantum Mechanics. In S. Kamefuchi *et al.* (eds), *Proceedings of the International Symposium on the Foundations of Quantum Mechanics*, Tokyo: The Physical Society of Japan, pp. 165–179.

Bassi, A. (2005). Collapse Models: Analysis of the Free Particle Dynamics. *Journal of Physics A*, 38(14), 3173–92.

Bassi, A., Deckert, D.-A., and Ferialdi, L. (2010). Breaking Quantum Linearity: Constraints from Human Perception and Cosmological Implications. *Europhysics Letters*, 92, 50006.

Bassi, A., and Donadi, S. (2014). Spontaneous Photon Emission from a Non-Relativistic Free Charged Particle in Collapse Models: A Case Study. *Physics Letters A*, 378(10), 761–5.

Bassi, A., and Ferialdi, L. (2009a). The Non- Markovian Dynamics for a Free Quantum Particle Subject to Spontaneous Collapse in Space: General Solution and Main Properties. *Physical Review A*, **80**, 012116.

Bassi, A., and Ferialdi, L. (2009b). Non-Markovian Quantum Trajectories: An Exact Result. *Physical Review Letters*, **103**, 050403.

Bassi, A., and Ferialdi, L. (2012a). Dissipative Collapse Models with Nonwhite Noises. *Physical Review A*, **86**(2), 022108.

Bassi, A., and Ferialdi, L. (2012b). Exact Solution for a Non-Markovian Dissipative Quantum Dynamics. *Physical Review Letters*, **108**(17), 170404.

Bassi, A., and Ghirardi, G. (2002). Dynamical Reduction Models with General Gaussian Noises. *Physical Review A*, **65**, 042114.

Bassi, A., and Ghirardi, G. (2003). Dynamical Reduction Models. *Physics Reports*, **379**, 257.

Bassi, A., and Ghirardi, G. (2020). Collapse Theories. In E. N. Zalta (ed.), *The Stanford Encyclopedia of Philosophy*. URL = <https://plato.stanford.edu/entries/qm-collapse/>.

Bassi, A., and Hejazi, K. (2015). No-faster-than-light-signaling Implies Linear Evolutions. A Re-derivation. *European Journal of Physics*, **36**(5), 055027.

Bassi, A., and Ippoliti, E. (2004). Numerical Analysis of a Spontaneous Collapse Model for a Two-level System. *Physical Review A*, **69**, 012105.

Bassi, A., Lochan, K., Satin, S., Singh, T. P., and Ulbricht, H. (2013). Models of Wave-function Collapse, Underlying Theories, and Experimental Tests. *Review of Modern Physics*, **85**, 47141.

Bassi, A., Ippoliti, E., and Adler, S. L. (2005). Towards Quantum Superpositions of a Mirror: An Exact Open System Analysis. *Physical Review Letters*, **94**, 030401.

Bassi, A., Ippoliti, E., and Vacchini, B. (2005). On the Energy Increase in Space-collapse Models. *Journal of Physics A*, **38**(37), 8017–38.

Bedingham, D. J. (2011). Relativistic State Reduction Dynamics. *Foundations of Physics*, **41**, 686–704.

Bedingham, D., Dürr, D., Ghirardi, G., Goldstein, S., Tumulka, R., and Zanghì, N. (2014). Matter Density and Relativistic Models of Wave Function Collapse. *Journal of Statistical Physics*, **154**(1–2), 623–31.

Bell, J. S. (1964). On the Einstein–Podolsky–Rosen Paradox. *Physics*, **1**, 195–200.

Bell, J. S. (1987). Are There Quantum Jumps? In C.W. Kilmister (ed.), *Schrödinger: Centenary Celebration of a Polymath*, Cambridge: Cambridge University Press, pp. 172–85.

Bell, J. S. (1989). The Trieste Lecture of John Stewart Bell. Printed in: A. Bassi and G. C. Ghirardi (2007), *Journal of Physics A*, **40**(12), 2919–33.

Bell, J. S. (1990). Against 'Measurement'. In A. I. Miller (ed.), *Sixty-Two Years of Uncertainty: Historical, Philosophical, and Physical Inquiries into the Foundations of Quantum Mechanics*, New York: Springer, pp. 17–31.

Benatti, F., Ghirardi, G., Rimini, A., and Weber, T. (1988). Operations Involving Momentum Variables in non-Hamiltonian Evolution Equations. *Il Nuovo Cimento B*, **101**, 333–55.

Berndl, K., Dürr, D., Goldstein, S., and Zanghì, N. (1996). Nonlocality, Lorentz Invariance, and Bohmian Quantum Theory. *Physical Review A*, **53**, 2062–73.

Białynicki-Birula, I., and Mycielski, J. (1976). Nonlinear Wave Mechanics. *Annals of Physics*, **100**(1–2), 62–93.

Bilardello, M., Donadi, S., Vinante, A., and Bassi, A. (2016). Bounds on Collapse Models from Cold-atom Experiments. *Physica A*, **462**, 764–82.

Bohm, D. (1952). A Suggested Interpretation of the Quantum Theory in Terms of 'Hidden' Variables, I and II. *Physical Review*, 85(2), 166–193.

Bohm, D., and Hiley, B. J. (1993). *The Undivided Universe*. London: Routledge.

Bricmont, J. (2016). *Making Sense of Quantum Mechanics*. Cham: Springer.

Buffa, M., Nicrosini, O., and Rimini, A. (1995). Dissipation and Reduction Effects of Spontaneous Localization on Superconducting States. *Foundations of Physics Letters*, 8(2), 105–25.

Carlesso, M., Bassi, A., Falferi, P. and Vinante, A. (2016). Experimental Bounds on Collapse Models from Gravitational Wave Detectors. *Physical Review D*, 94(12), 124036.

Carlesso, M., Paternostro, M., Ulbricht, H., Vinante, A., and Bassi, A. (2018). Non-Interferometric Test of the Continuous Spontaneous Localization Model Based on Rotational Optomechanics. *New Journal of Physics*, 20(8), 083022.

Carlesso, M., Vinante, A., and Bassi, A. (2018). Multilayer Test Masses to Enhance the Collapse Noise. *Physical Review A*, 98(2), 022122.

Collett, B., and Pearle, P. (2003). Wavefunction Collapse and Random Walk. *Foundations of Physics*, 33(10), 1495–1541.

Curceanu, C., Hiesmayr, B. C., and Piscicchia, K. (2015). X-rays Help to Unfuzzy the Concept of Measurement. *Journal of Advanced Physics*, 4(3), 263–6.

Curceanu, C., Bartalucci, S., Bassi, A., Bazzi, M., Bertolucci, S., Berucci, C., Bragadireanu, A. M., Cargnelli, M., Clozza, A., De Paolis, L., Di Matteo, S., Donadi, S., D'Uffizi, A., Egger, J.-P., Guaraldo, C., Iliescu, M., Ishiwatari, T., Laubenstein, M., Marton, J., Milotti, E., Pichler, A., Pietreanu, D., Piscicchia, K., Ponta, T., Sbardella, E., Scordo, A., Shi, H., Sirghi, D. L., Sirghi, F., Sperandio, L., Vazquez Doce, O., and Zmeskal, J. (2016). Spontaneously Emitted X-Rays: An Experimental Signature of the Dynamical Reduction Models. *Foundations of Physics*, 46(3), 263–8.

Davies, E. B. (1976). *Quantum Theory of Open Systems*. London: Academic Press.

de Broglie, L. (1924). *Recherches sur la Theorie des Quanta (Researches on the Quantum Theory)*. Thesis, Paris (1924). Reprinted in de Broglie, L. (2015), *Annals of Physics*, 3, 22.

Diósi, L. (1984). Gravitation and Quantum Mechanical Localization of Macroobjects. *Physics Letters A*, 105, 199–202.

Diósi, L. (1988). Localized Solution of Simple Nonlinear Quantum Langevin-equation. *Physics Letters A*, 132(5), 233–6.

Diósi, L. (1988). Quantum-stochastic Processes as Models for State Vector Reduction. *Journal of Physics A*, 21, 2885–98.

Diósi, L. (1989). Models for Universal Reduction of Macroscopic Quantum Fluctuations. *Physical Review A*, 40(3), 1165–74.

Diósi, L. (1990). Relativistic Theory for Continuous Measurement of Quantum Fields. *Physical Review A*, 42(9), 5086–92.

Diósi, L. (2007). Notes on Certain Newton Gravity Mechanisms of Wavefunction Localization and Decoherence. *Journal of Physics A*, 40(12), 2989.

Diósi, L. (2015). Testing Spontaneous Wave-Function Collapse Models on Classical Mechanical Oscillators. *Physical Review Letters*, 114(5), 050403.

Diósi, L., Gisin, N., and Strunz, W. T. (1998). Non-Markovian Quantum State Diffusion. *Physical Review A*, 58, 1699.

Doebner, H. D., and Goldin, G. A. (1992). On a General Nonlinear Schrödinger Equation Admitting Diffusion Currents. *Physics Letters A*, 162(5), 397–401.

Donadi, S., and Bassi, A. (2014). The Emission of Electromagnetic Radiation from a Quantum System Interacting with an External Noise: A General Result. *Journal of Physics A*, **48**(3), 035305.

Donadi, S., Piscicchia, K., Curceanu, C., Diósi, L., Laubenstein, M., and Bassi, A. (2020). Underground Test of Gravity-related Wave Function Collapse. *Nature Physics*. https://doi.org/10.1038/s41567-020-1008-4

Donadi, S., Deckert, D.-A., and Bassi, A. (2014). On the Spontaneous Emission of Electromagnetic Radiation in the CSL Model. *Annals of Physics*, **340**(1), 70–86.

Dove, C., and Squires, E. J. (1995). Symmetric Versions of Explicit Wavefunction Collapse Models. *Foundations of Physics*, **25**, 1267–82.

Dowker, F., and Henson, J. (2004). Spontaneous Collapse Models on a Lattice. *Journal of Statistical Physics*, **115**(5/6), 1327–39.

Dowker, F., and Herbauts, I. (2004). Simulating Causal Wavefunction Collapse Models. *Classical and Quantum Gravity*, **21**(12), 2963–79.

Dürr, D., Goldstein, S., and Zanghì, N. (2011). *Quantum Physics without Quantum Philosophy*. Berlin, Heidelberg: Springer.

Dürr, D., Hinrichs, G., and Kolb, M. (2011). On a Stochastic Trotter Formula with Application to Spontaneous Localization Models. *Journal of Statistical Physics*, **143**, 1096–1119.

Eibenberger, S., Gerlich, S., Arndt, M., Mayor, M., and Tüxen, J. (2013). Matter–Wave Interference of Particles Selected from a Molecular Library with Masses Exceeding 10 000 Amu. *Physical Chemistry Chemical Physics*, **15**(35), 14696–700.

Einstein, A., Podolsky, B., and Rosen, N. (1935). Can Quantum-Mechanical Description of Physical Reality be Considered Complete? *Physical Review*, **47**, 777–80.

Ellis, J., Hagelinm, J. S., Nanopoulos, D. V., and Srednicki, M. A. (1984). Search for Violations of Quantum Mechanics. *Nuclear Physics B*, **241**(2), 381–405.

Ellis, J., Mohanty, S., and Nanopoulos, D. V. (1989). Quantum Gravity and the Collapse of the Wave Function. *Physics Letters B*, **221**, 113–19.

Everett, H. III (1957). Relative State Formulation of Quantum Mechanics. *Reviews of Modern Physics*, **29**(3), 454–62.

Fein, Y. Y., Geyer, P., Zwick, P., Kiałka, F., Pedalino, S., Mayor, M., Gerlich, S., and Arndt, M. (2019). Quantum Superposition of Molecules beyond 25 KDa. *Nature Physics*, **15**(12), 1242–5.

Feldmann, W., and Tumulka, R. (2012). Parameter Diagrams of the GRW and CSL Theories of Wavefunction Collapse. *Journal of Physics A*, **45**(6), 065304.

Fonda, L., Ghirardi, G., and Rimini, A. (1978). Decay Theory of Unstable Quantum Systems. *Reports on Progress in Physics*, **41**(4), 587–631.

Freedman, S. J., and Clauser, J. F. (1972). Experimental test of local hidden-variable theories. *Physical Review Letters*, **28**(938), 938–41.

Frenkel, A. (1990). Spontaneous Localizations of the Wave Function and Classical Behavior. *Foundation of Physics*, **20**, 159–88.

Fu, Q. (1997). Spontaneous Radiation of Free Electrons in a Nonrelativistic Collapse Model. *Physical Review A*, **56**(3), 1806–11.

Gasbarri, G., Toroš, M., Donadi, S., and Bassi, A. (2017). Gravity Induced Wave Function Collapse. *Physical Review D*, **96**(10), 104013.

Gatarek, D., and Gisin, N. (1991). Continuous Quantum Jumps and Infinite-dimensional Stochastic Equations. *Journal of Mathematical Physics*, **32**(8), 2152.

Ghirardi, G. (1999). Some Lessons from Relativistic Reduction Models. In H. P. Breuer and F. Petruccione (eds), *Open Systems and Measurement in Relativistic Quantum Theory*, Berlin: Springer-Verlag, pp. 117–152.

Ghirardi, G. (2010). Does Quantum Nonlocality Irremediably Conflict with Special Relativity? *Foundations of Physics*, 40(9), 1379–95.

Ghirardi, GianCarlo, and Renata Grassi: 1991. 'Dynamical Reduction Models: Some General Remarks.' In: D. Costantini *et al.* (eds), *Nuovi Problemi della Logica e della Filosofia della Scienza*. Bologna: Editrice Clueb.

Ghirardi, G., Grassi, R., Butterfield, J., and Fleming, G. N. (1993). Parameter Dependence and Outcome Dependence in Dynamical Models for State Vector Reduction. *Foundations of Physics*, 23(3), 341–64.

Ghirardi, G., Grassi, R., and Benatti, F. (1995). Describing the Macroscopic World: Closing the Circle within the Dynamical Reduction Program. *Foundation of Physics*, 25, 5–38.

Ghirardi, G., Grassi, R., and Rimini, A. (1990). Continuous-spontaneous-reduction Model Involving Gravity. *Physical Review A*, 42, 1057.

Ghirardi, G., Pearle, P., and Rimini, A. (1990). Markov-processes in Hilbert-space and Continuous Spontaneous Localization of Systems of Identical Particles. *Physical Review A*, 42, 78.

Ghirardi, G., Rimini, A., and Weber, T. (1980). A General Argument against Superluminal Transmission through the Quantum Mechanical Measurement Process. *Lettere al Nuovo Cimento*, 27(10), 293–8.

Ghirardi, G., Rimini, A., and Weber, T. (1986). Unified Dynamics for Microscopic and Macroscopic Systems. *Physical Review D*, 34, 470.

Ghirardi, G., Nicrosini, O., Rimini, A., and Weber, T. (1988). Spontaneous Localization of a System of Identical Particles. *Il Nuovo Cimento B*, 102(4), 383–96.

Gisin, N. (1984). Quantum Measurements and Stochastic Processes. *Physical Review Letters*, 52(19), 1657–60.

Gisin, N. (1989). Stochastic Quantum Dynamics and Relativity. *Helvetica Physica Acta*, 62(4), 363–71.

Goldstein, S., Norsen, T., Tausk, D. V., and Zanghì, N. (2011). Bell's Theorem. *Scholarpedia*, 6(10), 8378.

Hackermüller, L., Hornberger, K., Brezger, B., Zeilinger, A., and Arndt, M. (2004). Decoherence of Matter Waves by Thermal Emission of Radiation. *Nature*, 427(6976), 711–14.

Halliwell, J., and Zoupas, A. (1995). Quantum State Diffusion, Density Matrix Diagonalization, and Decoherent Histories: A Model. *Physical Review D*, 52, 7294.

Helou, B., Slagmolen, B. J. J., McClelland, D. E., and Chen, Y. (2017). LISA Pathfinder Appreciably Constrains Collapse Models. *Physical Review D*, 95(8), 084054.

Jarrett, J. P. (1984). On the Physical Significance of the Locality Conditions in the Bell Arguments. *Noûs*, 18(4), 569–89.

Károlyházy, F. (1976). Gravitation and Quantum Mechanics of Macroscopic Object. *Nuovo Cimento A*, 42, 390–402.

Károlyházy, F., Frenkel, A., and Lukács, B. (1986). On the Possible Role of Gravity in the Reduction of the Wave Function. In R. Penrose and C. J. Isham (eds), *Quantum Concepts in Space and Time*, Clarendon: Oxford, pp. 109–128.

Kent, A. (1989). 'Quantum Jumps' and Indistinguishability. *Modern Physics Letters A*, 4(19), 1839–45.

Komori, K., Enomoto, Y., Ooi, C. P., Miyazaki, Y., Matsumoto, N., Sudhir, V., Michimura, Y., and Ando, M. (2020). Attonewton-Meter Torque Sensing with a Macroscopic Optomechanical Torsion Pendulum. *Physical Review A*, **101**(1), 011802.

Laloë, F., Mullin, W. J., and Pearle, P. (2014). Heating of Trapped Ultracold Atoms by Collapse Dynamics. *Physical Review A*, **90**, 052119.

Leggett, A. J. (2002). Testing the Limits of Quantum Mechanics: Motivation, State of Play, Prospects. *Journal of Physics: Condensed Matter*, **14**, R415–R451.

Ludwig, G. (1967). Attempt of an Axiomatic Foundation of Quantum Mechanics and More General Theories, II. *Communications in Mathematical Physics*, **4**, 331–48.

Ludwig, G. (1968). Attempt of an Axiomatic Foundation of Quantum Mechanics and More General Theories, III. *Communications in Mathematical Physics*, **9**, 1–12.

Ludwig, G. (1970). Deutung des Begriffs 'physikalische Theorie' und axiomatische Grundlegung der Hilbertraumstruktur der Quantenmechanik durch Hauptsätze des Messens. *Lecture Notes in Physics* 4. Berlin: Springer.

Marshall, W., Simon, C., Penrose, R., and Bouwmeester, D. (2003). Towards Quantum Superpositions of a Mirror. *Physical Review Letters*, **91**(13), 130401.

Myrvold, W. C. (2015). Lessons of Bell's Theorem: Nonlocality, Yes; Action at a Distance, Not Necessarily. In M. Bell and S. Gao (eds), *Quantum Nonlocality and Reality: 50 Years of Bell's Theorem*, Cambridge: Cambridge University Press, pp. 238–260.

Myrvold, W. C. (2017). Relativistically Invariant Markovian Dynamical Collapse Theories Must Employ Nonstandard Degrees of Freedom. *Physical Review A*, **96**, 062116.

Nicrosini, O., and Rimini, A. (2003). Relativistic Spontaneous Localization: A Proposal. *Foundations of Physics*, **33**(7), 1061–84.

Nimmrichter, S., Hornberger, K., and Hammerer, K. (2014). Optomechanical Sensing of Spontaneous Wave-Function Collapse. *Physical Review Letters*, **113**(2), 020405.

Norsen, T. (2009). Local Causality and Completeness: Bell vs. Jarrett. *Foundations of Physics*, **39**, 273–94.

Norsen, T. (2017). *Foundations of Quantum Mechanics: An Exploration of the Physical Meaning of Quantum Theory*. Cham: Springer.

Pearle, P. (1976). Reduction of Statevector by a Nonlinear Schrödinger equation. *Physical Review D*, **13**, 857.

Pearle, P. (1979). Toward Explaining Why Events Occur. *International Journal of Theoretical Physics*, **18**, 489.

Pearle, P. (1989). Combining Stochastic Dynamical State-Vector Reduction with Spontaneous Localization. *Physical Review A*, **39**(5), 2277–89.

Pearle, P. (1993). Ways to Describe Dynamical State-vector Reduction. *Physical Review A*, **48**, 913.

Pearle, P. (1996). Wavefunction Collapse Models with Nonwhite Noise. In R. K. Clifton (ed.), *Perspectives on Quantum Reality*, Dordrecht: Springer, pp. 93–109.

Pearle, P. (1999). Relativistic Collapse Model with Tachyonic Features. *Physical Review A*, **59**(1), 80–101.

Pearle, P. (2009). How Stands Collapse II. In W. Myrvold and J. Christian (eds), *Quantum Reality, Relativistic Causality, and Closing the Epistemic Circle*, The Western Ontario Series in Philosophy, Dordrecht: Springer, pp. 257–292.

Pearle, P. (2012). Cosmogenesis and Collapse. *Foundations of Physics*, **42**, 4–18.

Pearle, P., and Squires, E. (1994). Bound State Excitation, Nucleon Decay Experiments and Models of Wave Function Collapse. *Physical Review Letters*, **73**, 1.

Penrose, R. (1986). Gravity and State Vector Reduction. In R. Penrose and C. J. Isham (eds), *Quantum Concepts in Space and Time*, Oxford: Clarendon, pp. 129–146.

Penrose, R. (1996). On Gravity's Role in Quantum State Reduction. *General Relativity and Gravitation*, **28**, 581–600.

Piscicchia, K., Bassi, A., Curceanu, C., Grande, R., Donadi, S., Hiesmayr, B., and Pichler, A. (2017). CSL Collapse Model Mapped with the Spontaneous Radiation. *Entropy*, **19**(7), 319.

Polchinski, J. (1991). Weinberg's Nonlinear Quantum Mechanics and the Einstein–Podolsky–Rosen Paradox. *Physical Review Letters*, **66**, 397.

Pontin, A., Bullier, N. P., Toroš, M., and Barker, P. F. (2020). An Ultra-narrow Line Width Levitated Nanooscillator for Testing Dissipative Wavefunction Collapse. *Physical Review Research*, **2**, 023349.

Rae, A. I. M. (1990). Can GRW Theory Be Tested by Experiments on SQUIDS? *Journal of Physics A*, **23**(2), L57–L60.

Schrinski, B., Stickler, B. A., and Hornberger, K. (2017). Collapse-Induced Orientational Localization of Rigid Rotors. *Journal of the Optical Society of America B*, **34**(6), C1–C7.

Schrödinger, E. (1928). *Collected Papers on Wave Mechanics*. Glasgow: Blackie and Son Limited.

Schrödinger, E. (1935). Die gegenwärtige Situation in der Quantenmechanik. *Naturwissenschaften*, **23**(48–50), 807–812, 823–828, 844–849.

Shimony, A. (1984). Controllable and Uncontrollable Non-Locality. In S. Kamefuchi *et al.* (eds), *Proceedings of the International Symposium: Foundations of Quantum Mechanics in the Light of New Technology*, Tokyo: Physical Society of Japan, pp. 225–230.

Suppes, P., and Zanotti, M. (1976). On the Determinism of Hidden Variable Theories with Strict Correlation and Conditional Statistical Independence of Observables. In P. Suppes (ed.), *Logic and Probability in Quantum Mechanics*, Synthese Library 78, Dordrecht: Springer Netherlands, pp. 445–455.

Toroš, M., and Bassi, A. (2018). Bounds on Quantum Collapse Models from Matter-Wave Interferometry: Calculational Details. *Journal of Physics A*, **51**(11), 115302.

Tumulka, R. (2006). A Relativistic Version of the Ghirardi–Rimini–Weber Model. *Journal of Statistical Physics*, **125**(4), 821–40.

Tumulka, R. (2006). On Spontaneous Wave Function Collapse and Quantum Field Theory. *Proceedings of the Royal Society A*, **462**(2070), 1897–1908.

Tumulka, R. (2006). Collapse and Relativity. In A. Bassi, D. Dürr, T. Weber, and N. Zanghì (eds), *Quantum Mechanics: Are there Quantum Jumps? and On the Present Status of Quantum Mechanics*, AIP Conference Proceedings 844, American Institute of Physics, pp. 340–351.

Tumulka, R. (2020). A Relativistic GRW Flash Process with Interaction. In V. Allori, A. Bassi, D. Dürr, and N. Zanghì (eds), *Do Wave Functions Jump? Perspectives on the Work of GianCarlo Ghirardi*, Cham: Springer, pp. 321–347.

van Fraassen, B. C. (1982). The Charybdis of Realism: Epistemological Implications of Bell's Inequality. *Synthese*, **52**(1), 25–38.

Vinante, A., Bahrami, M., Bassi, A., Usenko, O., Wijts, G., and Oosterkamp, T. H. (2016). Upper Bounds on Spontaneous Wave-Function Collapse Models Using Millikelvin-Cooled Nanocantilevers. *Physical Review Letters*, **116**(9), 090402-7.

Vinante, A., Mezzena, R., Falferi, P., Carlesso, M., and Bassi, A. (2017). Improved Noninterferometric Test of Collapse Models Using Ultracold Cantilevers. *Physical Review Letters*, **119**(11), 110401.

SPONTANEOUS LOCALIZATION THEORIES

von Neumann, J. (1932). *Mathematische Grundlagen der Quantenmechanik*. Berlin: Springer. Translated by R. T. Beyer as: *Mathematical Foundations of Quantum Mechanics*. Princeton, NJ: Princeton University Press (1955).

Wallace, D. M. W. (2012). *The Emergent Multiverse: Quantum Theory According to the Everett Interpretation*. Oxford: Oxford University Press.

Weinberg, S. (1989a). Precision Tests of Quantum Mechanics. *Physical Review Letters*, **62**, 485.

Weinberg, S. (1989b). Testing Quantum Mechanics. *Annals of Physics*, **194**, 336–86.

Weinberg, S. (1989c). Particle States as Realizations (Linear or Nonlinear) of Spacetime Symmetries. *Nuclear Physics B*, **6**, 67–75.

Zheng, D., Leng, Y., Kong, X., Li, R., Wang, Z., Luo, X., Zhao, J., Duan, C.-K., Huang, P., Du, J., Carlesso, M., and Bassi, A. (2020). Room Temperature Test of the Continuous Spontaneous Localization Model Using a Levitated Micro-Oscillator. *Physical Review Research*, **2**(1), 013057.

...........................................................................................

# CHAPTER 46

# THE NON-INDIVIDUALS INTERPRETATION OF QUANTUM MECHANICS

...........................................................................................

DÉCIO KRAUSE, JONAS R. B. ARENHART, AND OTÁVIO BUENO

## 46.1 INTRODUCTION

...........................................................................................

IN this chapter, we present a 'non-individuals' interpretation of quantum mechanics. This is not among the well-known interpretations of quantum theory, such as Bohmian mechanics, the many-worlds interpretation, or GRW. Talk of non-individuals in quantum mechanics concerns primarily the metaphysics of the theory (French and Krause (2006) and French (2019)). However, as we argue, there is a clear sense in which non-individuality, as a metaphysical concept, is associated with interpreting quantum mechanics.

Our arguments establish salient connections between interpreting quantum theory and the metaphysics of quantum individuality. We specify the meaning of non-individuality and relate it to current efforts regarding the interpretation of quantum theory. We begin by recalling that an interpretation of quantum mechanics, as currently understood, is related to solutions to the measurement problem. Such solutions, we claim, are directly connected to attempts at providing a picture of what the world looks like according to quantum mechanics. In particular, we claim that an interpretation, in the relevant sense, brings with it an ontology: an account of what the furniture of the world is like, according to quantum mechanics. This step takes us from interpretation to ontology. The shift from ontology to non-individuality is just another, albeit controversial, further step. Philosophical interest in ontology is typically accompanied by interest in articulating a deeper understanding of the nature of the relevant entities, and a proposal for a metaphysics of non-individuals finds its way at this point.

If quantum mechanics is understood as dealing with objects of a given kind, whether particles, fields, or something else, it may be asked: what are these objects metaphysically? This, in turn, leads to questions regarding whether they are individuals or not, and if they are, which principle of individuality determines that that is the case? The non-individuals interpretation of quantum mechanics takes the relevant entities as lacking individuality, adding a further metaphysical interpretative layer over the theory's bare entities. This is known as the *Received View* of quantum non-individuality (see (French and Krause 2006, Chapter 3) for a historical overview).

This only tells us that a metaphysical view is attached to the ontology of an interpretation, but it does not specify what non-individuals are, or how the relevant concepts should be understood. Once a metaphysical interpretation is connected with an ontology, the task of making the metaphysics explicit still remains. Considered at this level of generality, the Received View is only a scheme, rather than a specific thesis. It acquires metaphysical content only once it is specified what an individual is and how something may fail to be one—that is, once an articulation of the Received View in metaphysical terms is provided. We examine two ways of articulating the view: a metaphysically inflationary form, and a metaphysically deflationary way.

The Received View, as a thesis about the nature of quantum entities, has its historical roots in the work done by those who first formulated quantum mechanics. It originates from attempts to understand the nature of the entities described by quantum theory. In particular, quantum particles' lack of well-defined trajectories and the presence of permutation invariance contrasted strikingly with the behaviour of classical particles (French, and Krause, 2006, Chapters 2 and 3). In this way, the Received View, even if considered as adding a metaphysics over the entities posited by an interpretation of quantum mechanics, has a historical pedigree.

This chapter is structured as follows. We begin, in the next section, by connecting an interpretation of quantum mechanics—understood as aiming to address the measurement problem—and an ontology. The plan is to make room to interpret further some entities in metaphysical terms. In Section 46.3 we discuss how a metaphysical profile can be attached to such entities, and how a philosophical motivation is typically offered for such move. We then briefly examine, in Section 46.4, the main concepts involved in articulating a metaphysics of non-individuals. This is followed, in Section 46.5, by the presentation of what is perhaps the most well-known metaphysics of non-individuals: one in which the concept of identity has no place. In Section 46.6 we analyse a less revisionary alternative to the standard account. Finally, in Section 46.7, a conclusion is offered.

## 46.2 INTERPRETATION AND ONTOLOGY

The idea that a physical theory needs an interpretation is rather recent in the history of physics, appearing after the development of modern physics itself (see also the discussion in Mittelstaedt (2011)). Independently of interpretational issues in quantum mechanics, the very notions of theory and interpretation remain subjects of philo-sophical controversy, but we will not pursue this here (see Krause and Arenhart (2017)). Following Esfeld (2019), we take the fact that quantum theory needs an interpretation to be part of the very aim of that physical theory. As Esfeld puts it:

> [...] a physical theory has to (i) spell out an ontology of what there is in nature according to the theory, (ii) provide a dynamics for the elements of the ontology and (iii) deduce measurement outcome statistics from the ontology and dynamics by treating measurement interactions within the ontology and dynamics; in order to do so, the ontology and dynamics have to be linked with an appropriate probability measure. Thus, the question is: What is the law that describes the individual processes that occur in nature (dynamics) and what are the entities that make up these individual processes (ontology)?    (Esfeld, 2019, p. 222)

As is well known, to find an appropriate account of the dynamics has been a key task to ensure the proper formulation and understanding of quantum mechanics (dynamical laws are crucial to characterize the theory). This is also a source of difficulty in theory individuation: formulations of quantum theory that differ in their dynamics may turn out to be *distinct* theories. Basically, the problem of formulating a dynamics for quantum entities—whatever they are—is now known as the *measurement problem*.

According to the standard way of formulating the measurement problem (Maudlin, 1995, p. 7), the following three apparently reasonable propositions are inconsistent (despite being pairwise consistent):

1. A The wave function of a system is *complete*, i.e., the wave function specifies (directly or indirectly) all of the physical properties of a system.
2. A The wave function always evolves in accord with a linear dynamical equation (e.g., the Schrödinger equation).
3. A Measurements of the spin of an electron (typically) have determinate outcomes, i.e., at the end of the measurement, the measuring device is either in a state that indicates spin up (and not down) or spin down (and not up). (Maudlin, 1995, p. 7)

The major difficulty in ensuring consistency comes from having the Schrödinger equation (or something to the same effect) as the dynamics of quantum systems and having to account for specific and determinate measurement results by using that very dynamics. The problem derives from the linearity of the equation and the probabilistic character of quantum theory. When one attempts to describe a measurement situation composed by a quantum system plugged into a measurement apparatus through such an equation, one ends up with superpositions of distinct states of the system being measured and distinct measurement results. The entire system (quantum system + measuring apparatus) typically does not evolve, following the Schrödinger equation, to a definite eigenstate of the observable being measured and the measuring apparatus pointing to the correspondent result. Rather, the whole system evolves in a state of superposition of different states of the system and different states of the apparatus (associated with distinct measurement results). However, such superpositions are never observed: only determinate results are. Hence, to restore consistency, something needs to go.

To provide a solution to the measurement problem is to offer an interpretation of quantum theory. Some interpretations deny that the wave function is complete (hidden variables interpretations); others challenge that the Schrödinger equation is the only account of the dynamics; yet others question that we need determinate outcomes (non-collapse interpretations).

Relevant to us is that the measurement problem is closely connected to the problem of providing a dynamics to quantum theory. However one addresses this issue, a problem remains. As noted above, following Esfeld, a physical theory must provide not only the dynamics, but also an ontology, that is, a more or less explicit character-ization of the entities that are posited by the theory. How is the problem of ontology related to the measurement problem?

The two problems are closely connected. A dynamics is a dynamics for the entities postulated by the theory. Different interpretations populate the world with distinct kinds of entities. The many-worlds interpretation posits a plurality of worlds, whereas Bohmian mechanics puts forward particles endowed with trajectories and a pilot wave. Attempts to explain measurements as acts of conscious activity, in turn, postulate the existence of consciousness apart from matter. Clearly, interpretations contribute dir-ectly to an ontology. As Laura Ruetsche (2015, p. 3433) notes, when one believes in a theory, one believes in an interpretation of the theory, which provides an account of what the world is like according to it. Interpretations articulate ontologies for theories.

If ontology is the part of metaphysics concerned with determining the world's furniture, to solve the measurement problem involves describing the ontology of quantum mechanics. In this respect, other chapters in this book advance distinct ontologies as part of their account of the measurement problem. This makes ontology importantly related to scientific development, and not something of just a philosoph-ical concern, making room for a partially naturalistic approach to ontology (see Esfeld (2019), Sklar (2010)).

In addition to the particular entities postulated by each interpretation, there is much in common among various interpretations. Most of them posit the existence of atoms, electrons, protons, etc. (see the discussion in Ladyman (2019)). These entities are also put forward in each formulation of quantum theory; hence, they are part of the theory's ontology. But this is not the case for every interpretation of quantum mechanics. Wave function realism, for instance, populates the world with a wave function, from which everything else (including particles) should follow. However, even if these entities are a by-product of the fundamental ontology, they must be recovered somehow, and the question regarding their metaphysical status is still pertinent. A major problem con-cerns specifying precisely what kind of things the entities that are typically dealt with by quantum mechanics are: particles or fields? What is the picture of reality suggested by the theory?

Modern physics—including quantum mechanics—is formed, to a large extent, by field theories. Quantum physics is significantly concerned with the study of *quantum fields*. 'Particles' arise as second-order entities, emerging when fields interact and momentum and energy are transferred, so that certain field excitations arise. The excitations (the 'particles') come in units, namely, in 1, 2 . . ., *n* units of excitation. The interactions increase or decrease the number of excitations by one and they are supposed to happen in space and time, so that the excitations behave as particles, hence the term. However, when an experiment with cathode rays is conducted, one (supposedly) sees *electron* beams, rather than electrons, despite the way one describes them. When neutral hydrogen atoms are ionized, it is with electrons that one interacts empirically, not with a mathematical field. When something is accelerated, mathematical fields or wave functions are not accelerated, but 'particles' are. Although the mathematical formalism refers to a mathematical entity called 'field', a particle interpretation needs to be provided to accommodate these empirical circumstances. The vocabulary of experimental physics is mostly a vocabulary of particles. Even if we can dispense with the notion of particle when it comes to mathematics, talk of particles still remains on the experimental front. Michael Redhead talks about the 'particle grin', by comparing the situation with the Cheshire cat (French and Krause, 2006, p. 361).

In orthodox quantum mechanics, before the advent of quantum field theories, the systems were supposed to behave either as particles or as waves, but never as both. The terms 'particles' and 'waves' were taken from analogies with what was known from classical physics and ordinary experience. In quantum mechanics, there is limited knowledge of the *real* entities under investigation, except for what is postulated by the theory.

These are the *quantum objects* we will focus on. According to the Received View of quantum non-individuality, the behaviour of entities of this kind, independently of whether they are described by orthodox quantum mechanics or field theories, is not compatible with the traditional metaphysical conception of *individuals*. This chapter aims to make these points clear. But, first, we need to say a bit more about the key concepts involved in the metaphysics of individuality.

## 46.3 NON-INDIVIDUALITY AND METAPHYSICAL PROFILES

Once it is acknowledged that particles of some kind need to be accommodated, a further question arises: what kind of things are these objects? A philosophical theory about objects may be considered, and the question about whether quantum entities are individuals becomes central. Moreover, if the tradition is followed, and if particular entities must have an individuation principle, according to which principle of individuation should we individuate quantum entities?

To come to grips with these questions, a proper understanding of principles of individuation is called for. We will consider the issue in the next section. First, we motivate the need for engaging with metaphysics when it comes to quantum entities. Historically, such entities have been treated as lacking individuality, in an informal sense. Some of the founding fathers of quantum theory noted the odd behaviour of quantum entities, and thought that this required close consideration (French and Krause, 2006, Chapter 3). Erwin Schrödinger, in particular, examined the issue attentively. He starts an essay on the nature of quanta explicitly claiming that quantum entities have no individuality:

> This essay deals with the elementary particle, more particularly with a certain feature that this concept has acquired—or rather lost—in quantum mechanics. I mean this: that the elementary particle is not an individual; it cannot be identified, it lacks 'sameness'. [ ... ] In technical language it is covered by saying that the particles 'obey' a new-fangled statistics, either Bose–Einstein or Fermi–Dirac statistics. The implication, far from obvious, is that the unsuspected epithet 'this' is not quite properly applicable to, say, an electron, except with caution, in a restricted sense, and sometimes not at all.   (Schrödinger, 1998, p. 197)

Note that Schrödinger directly relates what he saw as a new metaphysical character of quanta (they are not individuals, cannot be identified, lack 'sameness') with quantum statistics. Quantum statistics is connected with permutation symmetry in quantum mechanics, which leads to quanta being indiscernible by any observable available in the theory. As will become clear, this relation between statistics and new metaphysical features of the entities is not as direct as Schrödinger expected, although any kind of metaphysical approach to quantum mechanics will need to accommodate the odd statistics and be compatible with it.

Crucial for us is that the lack of individuality, which emerges directly from quantum theory (at least from a historical point of view), is a metaphysical question about the metaphysical status of the particles regarding their individuality. While some may think that considering the nature of quantum entities may lead us too far from the theory itself—due to the fact that the problem of individuality is a metaphysics problem—Schrödinger (among others) saw the connection between non-individuality and physics as a pretty direct one. Furthermore, according to some philosophers, it is only after such questions about the nature of the relevant entities are properly addressed that one can be confident of having a *clear picture* of what one is accepting when the theory is accepted. Steven French, in particular, dubs this demand for a metaphysical picture *Chakravartty's challenge*, given Anjan Chakravartty's identification of the need for a metaphysical clarification for a reasonable articulation of realism (French (2014) and (French and Krause, 2006, p. 244) explicitly emphasize the need for a metaphysical articulation of the ontology). Whether Chakravartty himself agrees with this understanding of his own work is a matter of controversy (Chakravartty, 2019). French continues:

But how do we obtain this clear picture? A simple answer would be, through physics which gives us a certain picture of the world as including particles, for example. But is this clear enough? Consider the further, but apparently obvious, question, are these particles individual objects, like chairs, tables, or people are? In answering this question, we need to supply, I maintain, or at least allude to, an appropriate metaphysics of individuality, and this exemplifies the general claim that in order to obtain Chakravartty's clear picture and hence obtain an appropriate realist understanding we need to provide an appropriate metaphysics. Those who reject any such need are either closet empiricists or 'ersatz' realists.    (French, 2014, p. 48)

In other words, in addition to the ontology—the items posited by a theory—further metaphysical questions still need to be addressed (Arenhart, 2019). In particular, if the entities a theory posits are particulars, questions about identity and individuality arise quite directly.

This suggests that a further interpretational layer must be added to the posits of a theory, a layer involving something like the metaphysical profile of the relevant entities. This metaphysical layer does not perform any role in the solution to the measurement problem, or to fix the ontology that comes with such solution. Rather, it is presupposed that the ontology is already in place, and one works from there. This suggests that one cannot *extract* the metaphysical profile of the entities from the theory, as Schrödinger had suggested, but rather must impose it from above.

One could ask why this extra step is required. As French suggested in the quotation above, addressing these issues is the only way to articulate a completely realistic view of what the world is like according to the theory in question. Otherwise, something will be missing. But whether or not one takes this extra step, we take it that individuality and non-individuality are metaphysical issues that add an additional layer to the description of reality provided by quantum theory. In this sense, one shifts from a mere operational use of quantum theory to an interpreted version of the theory, and from there, to an interpretation plugged with a metaphysical characterization of its posits. This leads one, in principle, to a fuller version of the theory.

We can now examine what individuality is (and its absence), and some of the accompanying concepts of identity, indiscernibility, and identification.

## 46.4 INDIVIDUALS, INDIVIDUATION, AND IDENTITY

There is a lot of ambiguity in discussions of quantum identity and individuality. Unfortunately, such ambiguities also contaminate formulations of the Received View and, as a consequence, some of its criticisms. In this section, we specify more precisely the meaning of the terms we use. This will inform our examination of the approaches to the Received View that will be developed in the next sections (see also Krause and Arenhart (2018), and Arenhart *et al.* (2019)).

By *identity* we mean the *logical relation* of identity, typically symbolized by the binary predicate symbol '='. A sentence involving the identity symbol, such as '$a = b$' describes a relation that, intuitively speaking, is true when $a$ and $b$ are names of one and the same thing, and is false when $a$ names a distinct object from $b$. Obviously, this explanation of the meaning of identity is circular, but it is not supposed to work as a definition of identity. (The issue of whether identity can be defined or explained in more basic terms is delicate, and many argue that identity is fundamental; see Bueno (2014) and Shumener (2017) for further discussion.)

Typically, identity is presented by its first-order logical axioms:

**Reflexivity** $\forall x(x = x)$
**Substitution law** $x = y \rightarrow (\alpha \rightarrow \alpha[y/x])$, with the known restrictions.

The law of substitution, in its first-order version, states that terms that are related by identity in any formula can be replaced by one another without disturbing the truth value of such formula. In material terms, it becomes the law of the *indiscernibility of identicals*, which states that identical things are indiscernible by properties and relations. With regard to the relation between identity and indiscernibility, this is only half of the story, with identity implying indiscernibility. The other half is much more controversial, stating that indiscernibility implies identity. This is the famous *principle of identity of indiscernibles* (PII). In material terms, it states that items that share all properties and relations are numerically identical. In first-order languages, PII is stated as a schema:

**PII** $(\alpha(x) \leftrightarrow \alpha(x/y)) \rightarrow x = y$

PII has counterexamples in first-order languages (it is enough to use a poor language), not being a truth of first-order logic. The metaphysical thesis (the material reading), however, has proved more resistant. In quantum mechanics, controversy around the principle is notorious. While some have claimed that it is a false principle in quantum theory (French and Krause, 2006, Chapter 4), others have tried to save weaker versions of the principle, going as far as to claim that quantum mechanics proves weaker versions of PII (Muller and Saunders, 2008). We will not engage with this particular controversy in this chapter.

Finally, we need also to specify what is meant by the notion of *individual*. According to the *Online Etymology Dictionary*, 'individual' is a medieval term that means 'single object or thing'. In other usages of the term, for instance in Middle English, '*individuum* was used in [the] sense of "individual member of a species" (early 15c.)'. In current metaphysics, one of the tasks of metaphysicians consists in explaining what is it that confers individuality to an individual, that is, what is it that makes that individual precisely the individual that it is. As Jonathan Lowe (Lowe, 2003, p. 75) has put it, the role of an individuation principle with regard to an object is to account for 'whatever it is that makes it the single object that it is—whatever it is that makes it *one* object, distinct from others, and the very object that it is as opposed to any other thing'.

The notion of individual is related to the notion of numerical identity, but not, necessarily, with the notion of discernibility. In other words, individuality does not exclude the possibility that 'different' individuals have the same qualitative features. After all, an 'individual member of a species' need not be *discernible* from other members. The additional supposition that every individual should be a unity *discernible* from other individuals was a central piece in Leibniz's metaphysics, which adopts PII. This principle, in turn, is responsible for the connection between identity and indiscernibility. The idea that identity is grounded in qualitative features has its epistemic attractions, but it is not required by the notion of individuality. It remains an open issue whether to adopt an individuality principle that links identity with indiscernibility as articulated by PII, or else to leave open the possibility that individuals can be indiscernible.

# 46.5 Non-individuals, Lack of Identity

As just noted, the concepts of identity, individuality, and indiscernibility can be understood as being largely independent. These concepts can now be used to frame a metaphysics of non-individuality; one that contributes to the metaphysical profile of quantum entities. This is required for a more complete picture of what the world looks like according to quantum mechanics.

Schrödinger, it was noted above, declared that quantum entities lost their individuality. What emerges from this, as we take it, is the need to specify a connection between quantum theory and a metaphysics of non-individuality. To examine this issue, consider another passage in which Schrödinger stresses the point:

> I beg to emphasize this and I beg you to believe it: it is not a question of our being able to ascertain the identity in some instances and not being able to do so in others. It is beyond doubt that the question of 'sameness', of identity, really and truly has no meaning.   (Schroödinger, 1996, pp. 121–122)

Strictly speaking, Schrödinger is claiming that sameness and identity fail to make sense for quantum entities. This is not obviously the same as to claim that quantum entities are non-individuals in light of the distinctions we made above. How can we make sense of this claim metaphysically? In what follows, a particular approach is developed that puts some metaphysical flesh on the bones of the Received View, as described above, and which attempts to capture the literal reading of 'losing identity', as Schrödinger suggested.

Consider again the notion of individuality. There must be something that accounts for what an entity is, and makes it precisely that entity. Some have thought that an entity must be an individual in virtue of possessing a special non-qualitative property of being identical to itself, and that this is what explains its individuality. Socrates is an individual in virtue of being endowed with the property of 'being identical to Socrates'. Nothing else has this property, only Socrates. Hence, this accounts for the difference between Socrates and everything else, as a kind of individual essence. At the same time, this is a non-qualitative property: one cannot use it to *qualitatively* distinguish Socrates from anything else. That means that, in terms of qualities, two individuals may coincide, while still having different essences—their non-qualitative individuating properties. In terms of the concepts formulated in the previous section, two items may be numerically distinct individuals, but at the same time be qualitatively indiscernible. This non-qualitative property is called in the literature a 'haecceity'. Individuals are the individuals they are in virtue of their haecceity. Typically, a haecceity is expressed in terms of self-identity: Socrates's haecceity is expressed by the fact that 'Socrates = Socrates'. It is, in other terms, the property of 'being identical to Socrates'.

This approach to individuality accommodates the link between individuality and identity, and also, as French and Krause (2006) have observed, between lack of identity and lack of individuality, which is presumably what Schrödinger had in mind—although Schrödinger would not have made the point in terms of haecceity. French and Krause consider principles of individuality that rely on non-qualitative features to provide a form of transcendental individuality—and haecceitism is not the only one available, although it is our focus here (French and Krause, 2006, Chapter 1). If individuality is related to identity as just suggested, failure of identity amounts to failure of individuality. The case for non-individuality is then made metaphysically: a non-individual is something that has no haecceity, and given what haecceities are, non-individuals are entities for which self-identity fails. As French and Krause note:

> [ ... ] the idea is apparently simple: regarded in haecceistic terms, 'Transcendental Individuality' can be understood as the identity of an object with itself; that is, '$a = a$'. We shall then defend the claim that the notion of non-individuality can be captured in the quantum context by formal systems in which self-identity is not always well-defined, so that the reflexive law of identity, namely, $\forall x(x = x)$, is not valid in general.    (French and Krause, 2006, pp. 13–14)

We will sketch below a way of articulating the approach formally, preserving the close connection between identity and individuality that is central to the conception. As French and Krause point out:

> We are supposing a strong relationship between individuality and identity [ ... ] for we have characterized 'non-individuals' as those entities for which the relation of self-identity $a = a$ does not make sense.    (French and Krause, 2006, p. 248)

As a result, it is possible to advance a metaphysical conception of non-individuality along the lines suggested by Schrödinger. Understood in this way, non-individuals are indiscernible by their properties. This is a physical issue, rather than a logical or philosophical one. We may approach indiscernibility in the way physicists do: the relevant entities are characterized by physical laws. Electrons, for instance, are those entities specified by a law according to which a mass $m = 9.109 \times 10^{-31}$ kg and a charge $e = -4.80320451(10) \times 10^{-10}$ esu is always accompanied by spin $1/2\hbar$, a Bohr magneton, etc. (see (Toraldo di Francia, 1981, p. 342); 'esu' is the abbreviation for the electrostatic unity of charge). These entities are, on di Francia's suggestive terminology, *nomological* objects. In other words, one does not analyse these objects in order to obtain their properties; the properties are specified by the relevant theory. It then emerges that all electrons are absolutely indiscernible. (This idea can be questioned, though; see Dieks and Lubberdink (2011).)

### 46.5.1  The Mathematics

On the Received View, we noted, non-individuals are entities devoid of a haecceity and, hence, cannot figure in the relation of identity. They are also nomological entities, which are specified by physical laws, and do not obey the standard theory of identity. We will examine it now.

The considerations so far motivate the development of a formal system to accommodate non-individuals, along the lines suggested by French and Krause (2006) (see, in particular, the quotes above). The key idea is that these entities can be aggregated in collections, which are called *quasi-sets* and can have a cardinal, but not an ordinal. Non-individuals do not bear unambiguous names or identifications. Given the lack of a haecceity, all of them would have the same 'identity card', which is, thus, useless. But non-individuals of different kinds (e.g. electrons and positrons) can be distinguished from one another, such as, by their electric charge (the positron's charge is the opposite of the electron's). However, since they do not obey the theory of identity, it is not appropriate to claim that they are *different*. Talk of *discernible* and *indiscernible* is enough.

All of this conflicts with standard logic and mathematics. Let us start with the notion of identity. Identity can be either defined or taken as primitive. Suppose we have a language with just three primitive predicate symbols: $P$ and $Q$ are unary symbols, whereas $R$ is binary. It can then be stated that:

$$x = y := (P(x) \leftrightarrow P(y)) \wedge (Q(x) \leftrightarrow Q(y))$$
$$\wedge \forall z(R(x,z) \leftrightarrow R(y,z)) \wedge \forall z(R(z,x) \leftrightarrow R(z,y)). \tag{1}$$

This can be generalized to any finite quantity of predicates. Of course, this offers no definition of numerical identity, and provides only indiscernibility relative to the language's predicates. It would be better to indicate this fact with $x \sim y$. The same objection can be raised against any attempt to use equivalence relations or congruences to stand for identity. For reasons that will emerge shortly, these strategies only hide the problem, making individuals appear to be non-individuals, while they are not. But why are they not?

The reasons have to do with set theory, the typical framework in which to develop all standard mathematics. Take a system such as ZFC, Zermelo–Fraenkel set theory with the Axiom of Choice (we are not including ur-elements for now). Axiomatized as a first-order theory, as is now usual, ZFC has '=' as a primitive binary symbol, satisfying the following axiom schema:

($=_1$) $\forall x(x = x)$ (Reflexive Law of identity).
($=_2$) $\forall x \forall y(x = y \rightarrow (\alpha(x) \rightarrow \alpha(y)))$, in which $\alpha(x)$ is a formula (that may contain parameters), $x$ is free and $\alpha(y)$ is obtained from $\alpha(x)$ by the substitution of some free occurrences of $x$ by $y$ (Axiom of Substitution of Identicals).
($=_3$) $\forall x \forall y(\forall z(z \in x \leftrightarrow z \in y) \rightarrow x = y)$ (Axiom of Extensionality).

The converse of the extensionality axiom follows from ($=_2$). It is also easy to prove that identity is symmetric and transitive. For our purposes, the key consequence is that the universe of sets—the von Neumann cumulative hierarchy of well-founded sets $V = \langle V, \in \rangle$ (Jech, 2003)—does not have nontrivial automorphisms. This has important implications.

First, from a set-theoretic perspective, mathematics is conducted inside a mathematical structure. A group is a structure of the kind $\mathcal{G} = \langle G, \star \rangle$; a vector space is a structure of the form $\mathcal{V} = \langle V, K, +, \cdot \rangle$; a classical particle mechanics is a structure $\mathcal{CPM} = \langle P, \mathbf{s}, m, \mathbf{f}, \mathbf{g} \rangle$ (McKinsey et al., 1953), and so on. These structures are (or can be viewed as) sets of ZFC. Let $\mathcal{E} = \langle D, R_i \rangle$ be a structure, in which relations $R_i$ hold for the elements of the domain $D$ (all other structures can be reduced to this case). Two objects $a$ and $b$, which belong to $D$, are $\mathcal{E}$-indiscernible if there is an automorphism $h$ of $\mathcal{E}$ such that $h(a) = b$ (thus, $h^{-1}(b) = a$). As an illustration, $i$ and $-i$ are $\mathcal{C}$-indiscernible in the complex field structure $\mathcal{C}$, for the operation of taking the conjugate is an automorphism. Structures that have automorphisms other than the identity function are said to be *deformable*, or *non-rigid*; otherwise, they are *rigid*.

A significant theorem states that *every* structure can be extended to a rigid structure (da Costa and Rodrigues, 2007). This means that the apparently indiscernible objects of the domain of a given structure can be individuals with identity on the outside of it. Furthermore, recall that the whole universe is rigid. So, within a mathematical framework, such as ZF, ZFC, NBG, KM, and other 'standard' set theories, indiscernible things can only be taken to be $\mathcal{E}$-indiscernible for some structure $\mathcal{E}$. This manoeuvre is adopted when congruence relations are used to stand for identity. Good foundational work should avoid this trick.

To conclude this section, some remarks on set theories with ur-elements are in order (see Suppes, 1972). Ur-elements are entities that are not sets but can be members of sets. They do not have elements, relative to a membership relation, although they can obey mereological axioms. Set theories with ur-elements are useful for applications in the empirical sciences, in which many objects, such as eucalyptuses, elephants, and electrons, are not sets, but can be members of them.

Let $A$ be the set of ur-elements. It is known that every permutation of $A$ can be extended to an automorphism of the entire universe, which is not rigid, in contrast to the universe of 'pure' sets. This means that, in certain situations, two ur-elements apparently could not be discerned by any means and so they might be good candidates to represent quantum objects. However, this should be viewed with suspicion. Even ur-elements obey the axiom of pairing, which states that given sets $a$ and $b$ of ur-elements, there is a set $z$ that has only $a$ and $b$ as its elements. If $a = b$, the unitary of $a$ is denoted by $\{a\}$. The 'property' $I_a(x) \leftrightarrow x \in \{a\}$, which is called *the identity of $a$* and is satisfied only by $a$, can then be defined. This property can be used to establish, as a consequence of PII, that $a$ has something that is not shared by any other object in the universe. Hence, although ur-elements can be taken to be indiscernible, they are not. They do have identity: every ur-element is *distinct* from any other entity in the universe.

## 46.5.2 A More Appropriate Mathematics

We noted that if quantum objects are non-individuals, this does not entail that they cannot be discerned in some cases: quantum entities of different kinds, such as electrons and protons, can be distinguished. But the fact that these objects are discernible in certain instances does not grant them identity. Since standard set theories presuppose the identity of all objects in their domains, they are not adequate to study entities of this sort. To address this issue, a novel set theory has been articulated. We now provide some details about it.

Quasi-sets are collections of objects, some of which lack identity conditions, that is, they are non-individuals. Other elements satisfy the rules of standard logic, including those of identity. Those that do are *individuals*; they are either sets or ur-elements. When restricted to individuals, the theory yields a copy of ZFU, Zermelo–Fraenkel set theory with ur-elements. The latter are $M$-objects, whereas non-individuals are $m$-objects ($M$ for 'macro', $m$ for 'micro'). The theory allows for $m$-objects of several kinds, and can accommodate protons, electrons, positrons, and so on. Although $m$-objects of a given kind, e.g. electrons, may or may not be discerned, one cannot claim that they are *different*, for identity does not apply to them.

Operations on quasi-sets are similar to those on sets: intersection, union, and power set, for instance, are all specified for quasi-sets. A quasi-set may have a cardinal, but when indistinguishable $m$-objects are involved, no ordinal can be associated to it. So, a quasi-set cannot be ordered, named, or labelled without ambiguity (see French and Krause, 2006). Given these remarks, we can now examine how this theory can be used to do physics.

## 46.5.3 Quantum Physics

The main philosophical problem regarding quantum physics does not concern the results of the theory. These are well established and properly supported. The problem concerns interpretation. Our aim is to present a reasonable account of quantum theory that takes quantum objects, which are, in any case, involved in any interpretation, as non-individuals. We consider how a different mathematical theory can accommodate this point.

The application of quasi-set theory involves various steps, culminating in a formulation of quantum theory in terms of quasi-sets. On standard formulations, quantum objects themselves (or quantum systems) appear only in an indirect way. Despite Max Jammer's comment to the effect that the theory's primitive notions include *system* (Jammer, 1974, p. 5), the standard formalism is formulated in terms of *states* and *observables*. (There is no need to review the axioms here. We just note that there are plenty of different alternative ways of formulating quantum mechanics: Styer *et al.* (2002) present nine.) On the standard formulations, it is assumed that there is a *set* of quantum systems to begin with. However, *sets* from classical set theories, such as those mentioned above, are collections of *distinct* objects. The members of a set do have identity: they are individuals. Thus, something goes wrong if sets are assumed in this context. Our proposal is to assume that quantum systems form a quasi-set instead.

At least two ways are available to accommodate quantum mechanics in quasi-set theory. The first concerns non-relativistic quantum mechanics, QM. The core idea was advanced in (Krause and Arenhart, 2017, Section 5.8.1), and explicitly assumes, as in the case of classical particle mechanics mentioned earlier, that there is a quasi-set of quantum systems, $S$, such that some of them, depending on the particular model under consideration, can be indiscernible. The structure has the form:

$$\mathcal{QM} = \langle S, \{\mathcal{H}_i\}, \{\hat{A}_{ij}\}, \{\mathcal{U}_{ik}\}, \mathcal{B}(\mathbb{R})\rangle, i \in I, j \in J, k \in K \tag{2}$$

in which:

1. $S$ is a quasi-set whose elements are called *physical objects*, or *physical systems*.
2. $\{H_i\}$ is a family of complex separable Hilbert spaces whose cardinality is defined in the particular application of the theory.
3. $\{\hat{A}_{ij}\}$, for each Hilbert space $\mathcal{H}_i$, is a family of self-adjunct (or Hermitian) operators over $H_i$.
4. $\{U_{ik}\}$, again for each $\mathcal{H}_i$, is a family of unitary operators over $H_i$.
5. $\mathcal{B}(\mathbb{R})$ is the collection of Borel sets over the set of real numbers.

The Hilbert space formalism, whose postulates we will not review, does not involve space and time, something needed in order to apply it to the world. We will return to this point.

To each quantum system $s \in S$, we associate a 4-tuple of the form:

$$\sigma = \langle \mathbb{E}^4, \psi(\mathbf{x},t), \Delta, P \rangle, \tag{3}$$

in which $\mathbb{E}^4$ is the Galilean spacetime (see Penrose, 2007, Chapter 17), such that each point is denoted by a 4-tuple $\langle x,y,z,t \rangle$, in which $\mathbf{x} = \langle x,y,z \rangle$ denotes the coordinates of the system and $t$ is a parameter representing time. $\psi(\mathbf{x},t)$ is a function over $\mathbb{E}^4$, which is called the *wave function* of the system. $\Delta \in \mathcal{B}(\mathbb{R})$ is Borelian. $P$ is a function defined in $\mathcal{H}_i \times \{\hat{A}_{ij}\} \times \mathcal{B}(\mathbb{R})$, for some $i$ (determined by the physical system $s$), and having values in $[0, 1]$. Thus, $P(\psi, \hat{A}, \Delta) \in [0, 1]$ is the probability that the measurement of the observable $A$ (represented by the self-adjunct operator $\hat{A}$) for the system in state $\psi(\mathbf{x}, t)$ lies in $\Delta$.

The relation between the state vector and the wave function is given as follows: Let $(\mathbf{x},t)$ denote the position operator at time $t$. Then $\psi(\mathbf{x},t) = \langle (\mathbf{x},t) | \psi \rangle$, that is, the wave function is described by the Fourier coefficients of the expansion of the state vector in the orthonormal basis of the position operator.

Suitable postulates connecting all these concepts need not be reviewed here (Krause and Arenhart, 2017). The key shift is the adoption of quasi-set theory. This approach is able to resolve the difficulties raised by Dalla Chiara and di Francia (1993) about the interpretation of a language $L$ of quantum mechanics if it is formulated in a standard set theory, such as ZFC. We review some of their arguments and explain why these criticisms do not hold in quasi-set theory.

1. The standard interpretation assumes that its domain $D$ is a set of individuals. In contrast, in quasi-set theory, the domain includes a quasi-set of indiscernible entities, so that some quantum objects can be taken to be indiscernible without the usual tricks.
2. For any individual $d \in D$ in the standard interpretation, the language can be extended to a language $L'$, which contains a name $a$ and a denotation function $\rho$, such that $\rho(a) = d$. In quasi-set theory, this is not the case, for $m$-objects cannot be named. In fact, a suitable language for QM should not allow for proper names.
3. If $L$ is at least a second-order language, Leibniz's PII holds in standard semantics:

$$\forall x \forall y (x = y \rightarrow \exists F(F(x) \wedge \neg F(y))).$$

In contrast, in quasi-set theory, this principle fails for indiscernible $m$-objects, for no haecceity is assumed and they may share all their properties and relations.

The second approach to use quasi-set theory to accommodate quantum mechanics focuses on the Fock space formalism. This was started in Domenech *et al.* (2008) and Domenech *et al.* (2010) (see also French and Krause 2006, Chap. 9). The Fock space formalism includes creation and annihilation operators, making room for (non-interacting) fields. In the papers just mentioned, two kinds of vector spaces (Hilbert spaces) are defined; they are called *quasi*-spaces—one for bosons, another for fermions.

The main idea is that, by positing that some entities can be strongly indiscernible, it is possible to dispense with certain postulates, such as the indistinguishability postulate. Quantum results emerge naturally with this move.

Moreover, within quasi-set theory, the so-called quantum statistics (B–E and F–D) are obtained quite naturally too. Given that certain objects can be indiscernible in a strong sense, the formulas that provide the counting states in both B–E and F–D statistics can be obtained just by counting the states—without having to discharge some of them, as is usual in the standard approach, to mimic indiscernibility. In this respect, the quasi-set-theoretic framework seems to be much more natural for the quantum domain (French and Krause, 2006, Section 7.6).

In quasi-set theory, not only indiscernible *objects* can be considered, but also indiscernible *properties*. As is well known, to arrive at statistical predictions from the quantum formalism, the same experiments need to be repeated enough times to allow for the computation of mean values and probabilities, and the determination of all necessary stochastic properties of experimental outcomes. But what counts as 'the same experiment'? De Barros *et al.* (2019) elaborated on this notion to underscore how indistinguishability is deeply connected to contextuality. They realized that in order to perform *n* experiments, physicists must prepare *n* 'identical' copies of a same quantum system. But, given the indistinguishability postulate—according to which different mean values in permutations of indistinguishable objects is not observable—the particles involved cannot be identified. Interestingly, this is so even if we perform a thought experiment, provided that we want to respect what the logic of QM seems to suggest regarding identity, at least given an interpretation of quantum theory.

When considering indistinguishable properties, we neither perform 'the same' experiment twice nor measure two indistinguishable properties on 'the same' quantum system, but we measure indistinguishable properties—prepared in 'the same' way—over indistinguishable quantum systems. Thus, by seriously considering the notion of indistinguishability as something distinct from identity, something that quasi-set theory enables us to do, we can read the results by Kochen and Specker and realize that the core of their theorem (the 'paradox') can be avoided, for the contradiction assumes that 'the same' properties are measured in 'the same' particles in different contexts. However, if we assume the apparently obvious fact that we measure indistinguishable properties over indistinguishable particles, there will be no surprise in acknowledging that the obtained results may differ. In other words, within this context, the Kochen–Specker result can be put within parentheses (de Barros *et al.*, 2019).

## 46.6 NON-INDIVIDUALS WITH IDENTITY

Despite the developments in the Received View, some may argue that the loss of identity, as an expressive resource, is just too much to swallow. The relation between identity and individuality that was suggested leads to a relation of identity heavily burdened by the metaphysical role of individuation—a role that, according to some, identity should not have. But for those who want to keep the idea that quantum mechanics does not deal with individuals, some options are still available. Given the distinctions we provided regarding identity and individuality, loss of individuality need not entail loss of identity. So, even if one is not willing to embrace the version of the Received View above, there is no need to abandon the Received View altogether, formulated as the thesis that quantum entities are non-individuals. It is enough to apply distinct principles of individuality, and to consider those whose failure do not imply lack of identity (Arenhart (2017), and Krause and Arenhart (2018)).

Here we sketch an alternative version of the Received View that allows us to have non-individuals while identity is still maintained. Of course, in this context, identity and individuality must be separated: the fact that one may use identity to refer to some entities does not imply that they have a haecceity. To do so, identity should be deflated from the metaphysical content it had been burdened with in the previous approach, and individuality endowed with more epistemic features. Recall that an individual is a unity of some kind. The idea gains a twist with the addition that an individual is something that can be repeatedly identified as being *the same* individual, that is, it requires identification over time.

Bueno's approach to individuality in Bueno (2014) provides one such option. According to his approach, an individual must satisfy two conditions:

**Discernibility:**    The item is discernible from any other individual.
**Re-identification:**    The item is re-identifiable over time.

Although identity can be employed in the formulation of these conditions, the fact that certain entities have identity, by itself, does not confer them individuality. As Bueno (2014) remarks, this account of individuality is compatible with a view in which the relation of identity is devoid of metaphysical content. Interestingly, a notion of non-individuality also emerges: a non-individual is something that fails to be an individual. As a result, three kinds of non-individuals are obtained. Something is a non-individual provided that (i) it is indiscernible from others of its kind; or (ii) it lacks conditions of identification over time; or (iii) it is indiscernible from other entities of its kind and lacks conditions of identification over time. Quantum entities, on a reasonable interpretation, are non-individuals according to the third option.

As noted above, the typical version of the Received View states that non-individuals are entities without identity. However, if the Received View is merely the thesis that quantum entities are non-individuals, the proposal we have presented here also provides a formulation of the Received View. This approach has an interesting historical motivation. Consider again Schrödinger's quote that supports the previous version of the Received View:

I beg to emphasize this and I beg you to believe it: it is not a question of our being able to ascertain the identity in some instances and not being able to do so in others. It is beyond doubt that the question of 'sameness', of identity, really and truly has no meaning.   (Schrödinger, 1996, pp. 121–122)

In this passage, Schrödinger is not completely clear about what it is meant by 'identity'. Perhaps the failure is not of numerical identity, but rather of identity over time. As he notes, in a given experiment, one cannot claim, with full certainty, that a particle that appears at an instant $t_1$ in a bubble chamber is the same that appears at a later instant $t_2$. The particle's identity cannot be determined because it cannot be re-identified in distinct instants of time. So, what fails is not the particle's numerical identity, but its identity over time. This makes such a particle a non-individual according to proposal (ii) above.

If non-individuals are items lacking individuality in this second sense—that is, they cannot be re-identified over time—Hans Dehmelt's positron Priscilla as well as other quantum objects he has trapped are non-individuals. After all, as soon as they leave the apparatus that trapped them, these objects cannot be identified anymore. This is not just an epistemological feature but belongs to the very nature of these entities. (The same happens with more recent trapped quantum objects by Haroche and Wineland; for a discussion of Priscilla, see Krause (2011).)

One of the benefits of this approach is that it does not require revising the logic of identity. Quantum theory, however, does not favour one metaphysics over another. Thus, not only is quantum mechanics empirically underdetermined, given that its data are compatible with distinct interpretations, but it is also metaphysically underdetermined: there is no specific quantum mechanical support for a preferred metaphysical interpretation of its entities. This is as it should be: metaphysical interpretations were designed to provide answers to metaphysical questions rather than to problems connected to the applications of the theory.

## 46.7 CONCLUSIONS

In quantum mechanics, similarly to what happens in arithmetic, one starts with informal notions. A puzzle emerges when it is realized that quantum 'particles' have lost their individuality. To make sense of this, a more fine-grained framework, which does not presuppose the identity of the objects under consideration, is called for. Such framework, articulated in quasi-set theory, offers the basis for an interpretation of quantum mechanics with regard to the nature of the entities in question.

Of course, one can be an instrumentalist about quantum theory, refuse to engage with these issues, and insist that QM only provides tools for computing probabilities. But something would be missing. One can be legitimately interested in the world views that QM offers. One of them suggests that quantum objects are better thought of as non-individuals. We take this view seriously, and have provided a logic and a mathematics compatible with it. And a significant source of understanding regarding the foundations of quantum mechanics then results.

## ACKNOWLEDGEMENTS

Our thanks go to Olival Freire Jr for his help and support during the writing of this chapter.

## REFERENCES

Arenhart, J. R. B. (2017). The received view on quantum non-individuality: formal and metaphysical analysis. *Synthese*, **194**(4), 1323–47.

Arenhart, J. R. B. (2019). Bridging the gap between science and metaphysics, with a little help from quantum mechanics. In J. D. Dantas, E. Erickson, and S. Molick (eds), *Proceedings of the 3rd Filomena Workshop*, Natal: PPGFIL UFRN, pp. 9–33.

Arenhart, J. R. B., Bueno, O., and Krause, D. (2019). Making sense of non-individuals in quantum mechanics. In O. Lombardi, S. Fortin, C. López, and F. Holik (eds), *Quantum Worlds. Perspectives on the Ontology of Quantum Mechanics*, Cambridge: Cambridge University Press, pp. 185–204.

Bueno, O. (2014). Why identity is fundamental. *American Philosophical Quarterly*, **51**(4), 325–32.

Chakravartty, A. (2019). Physics, metaphysics, dispositions, and symmetries – À la French. *Studies in Hisory and Philosophy of Science*, **74**, 10–15.

da Costa, N. C. A., and Rodrigues, A. A. M. (2007). Definability and invariance. *Studia Logica: An International Journal for Symbolic Logic*, **86**(1), 1–30.

Dalla Chiara, M. L., and di Francia, T. G. (1993). Individuals, kinds and names in physics. In G. Corsi, M. L. Dalla Chiara, and G. C. Ghirardi (eds), *Bridging the Gap: Philosophy, Mathematics, and Physics – Lectures on the Foundations of Science*, Vol. 140 of Boston Studies in the Philosophy of Science, Dordrecht: Springer, pp. 261–83.

de Barros, J. A., Holik, F., and Krause, D. (2019). Indistinguishability and the origins of contextuality in physics. *Philosophical Transactions of the Royal Society A*, **377**(2157).

Dieks, D., and Lubberdink, A. (2011). How classical particles emerge from the quantum world. *Foundations of Physics*, **41**(6), 1051–64.

Domenech, G., Holik, F., Kniznik, L., and Krause, D. (2010). No labeling quantum mechanics of indiscernible particles. *International Journal of Theoretical Physics*, **49**, 3085–91.

Domenech, G., Holik, F., and Krause, D. (2008). Q-spaces and the foundations of quantum mechanics. *Foundations of Physics*, **38**, 969–94.

Esfeld, M. (2019). Individuality and the account of nonlocality: The case for the particle ontology in quantum physics. In O. Lombardi, S. Fortin, C. López, and F. Holik (eds), Quantum Worlds: Perspectives on the Ontology of Quantum Mechanics, Cambridge: Cambridge University Press, pp. 222–44.

French, S. (2014). *The structure of the world: Metaphysics and representation*. Oxford: Oxford University Press.

French, S. (2019). Identity and Individuality in Quantum Theory. In E. N. Zalta (ed.), *The Stanford Encyclopedia of Philosophy*, winter 2019 edn, Stanford, CA: Metaphysics Research Lab, Stanford University.

French, S., and Krause, D. (2006). *Identity in physics: A historical, philosophical, and formal analysis*. Oxford: Oxford University Press.

Jammer, M. (1974). *The Philosophy Of Quantum Mechanics: The Interpretations Of Quantum Mechanics In Historical Perspective*. New York: Wiley and Sons.

Jech, T. (2003). *Set theory. The Third Millennium Edition, Revised and Expanded*. Berlin: Springer.

Krause, D. (2011). Is Priscilla, the trapped positron, an individual? Quantum physics, the use of names, and individuation. *Arbor*, **187**, 61–66.

Krause, D., and Arenhart, J. R. B. (2017). *The Logical Foundations of Scientific Theories: Languages, Structures, and Models*. Abingdon: Routledge.

Krause, D., and Arenhart, J. R. B. (2018). Quantum non-individuality: Background concepts and possibilities. In S. Wuppuluri and F. Doria (eds), *The Map and the Territory*, Cham: Springer, pp. 281–305.

Ladyman, J. (2019). What is the quantum face of realism? In O. Lombardi, S. Fortin, C. López, and F. Holik (eds), Quantum Worlds. Perspectives on the Ontology of Quantum Mechanics, Cambridge: Cambridge University Press, pp. 121–32.

Lowe, J. E. (2003). Individuation. In M. J. Loux and D. W. Zimmerman (eds), *The Oxford Handbook of Metaphysics*, Oxford: Oxford University Press, pp. 75–95.

Maudlin, T. (1995). Three measurement problems. *Topoi*, **14**(1), 7–15.

McKinsey, J. C. C., Sugar, A. C., and Suppes, P. (1953). Axiomatic foundations of classical particle mechanics. *Journal of Rational Mechanics and Analysis*, **2**, 253–77.

Mittelstaedt, P. (2011). The problem of interpretation of modern physics. *Foundations of Physics*, **41**(11), 1667–76.

Muller, F. A., and Saunders, S. (2008). Discerning fermions. *British Journal for the Philosophy of Science*, **59**, 499–548.

Penrose, R. (2007). *The Road to Reality: A Complete Guide to the Laws of the Universe*. New York: Vintage Books.

Ruetsche, L. (2015). The Shaky Game+ 25, or: On locavoracity. *Synthese*, **192**(11), 3425–442.

Schrödinger, E. (1996). *Nature and the Greeks and Science and Humanism*. Cambridge: Cambridge University Press.

Schrödinger, E. (1998). What is an elementary particle? In E. Castellani (ed.), *Interpreting Bodies: Classical and Quantum Objects in Modern Physics*, Princeton, NJ: Princeton University Press, pp. 197–210.

Shumener, E. (2017). The metaphysics of identity: Is identity fundamental? *Philosophy Compass*, **12**(1).

Sklar, L. (2010). I'd love to be a naturalist—if only I knew what naturalism was. *Philosophy of Science*, **77**(5), 1121–137.

Styer, D. F., Balkin, M. S., Becker, K. M., Burns, M. R., Dudley, C. E., Forth, S. T., Gaumer, J. S., Kramer, M. A., Oertel, D. C., Park, L. H., Rinkoski, M. T., Smith, C. T., and Wotherspoon, T. D. (2002). Nine formulations of quantum mechanics. *American Journal of Physics*, **70**(3), 288–97.

Suppes, P. (1972). *Axiomatic Set Theory*. New York: Dover.

Toraldo di Francia, G. (1981). *The Investigation of the Physical World*. Cambridge: Cambridge University Press.

..........

# CHAPTER 47

# MODAL INTERPRETATIONS OF QUANTUM MECHANICS

..........

DENNIS DIEKS

## 47.1 INTRODUCTION: ORIGIN AND MOTIVATING IDEAS OF MODAL INTERPRETATIONS

..........

QUANTUM mechanics as it emerged in the 1920s and 1930s assigned an essential role to the concept of measurement. In von Neumann's synthesizing book (von Neumann, 1932) this was formalized by the distinction between ordinary physical evolution and what happens during a measurement intervention: states of isolated systems evolve unitarily, with an evolution operator determined by the system's Hamiltonian, whereas in a measurement the state is assumed to 'collapse' to one of the eigenstates of the measured observable.

This combination of unitary evolution and collapses has remained a central part of the standard account of quantum mechanics. Nevertheless, attributing a special status to measurements and collapses raises serious problems. For example, the question immediately arises when exactly the 'projection postulate' (effecting collapses) should be applied instead of unitary evolution determined by the Hamiltonian governing the system and its interaction with the environment. Even the consistency of the existence of two different evolution mechanisms is debatable: situations like those of the Wigner's friend paradox, in which an observer performs a measurement while being part of a larger experiment by a second observer, raise the question of whether the collapse caused by the first measurement is compatible with the unitary description used by the second observer as long as no second measurement has taken place (this thought experiment will be discussed in Section 47.3).

A theoretical scheme with only ordinary physical interaction and exclusively unitary evolution would be able to avoid such ambiguities and would also fulfil the desideratum that an anthropocentric notion like 'measurement' should not play a central role in fundamental physical theory. In the 1970s, 1980s, and 1990s this thought motivated a number of physicists and philosophers of physics to develop 'no-collapse' interpretations designed to reproduce the empirical consequences of standard quantum mechanics. The resulting interpretative schemes do away with collapses, and provide a physical interpretation of unitary quantum mechanics in terms of possible values of the physical quantities associated with quantum objects.

Since these no-collapse interpretations have to predict possibilities and their probabilities, they must contain *modal* statements—where the term 'modal' is meant in the sense of *modal logic*, the logic dealing with what is actual, possible, or necessary. For this reason van Fraassen (1972), who came to the foundations of quantum mechanics via logic, introduced the name 'modal interpretation'.

In order to link the mathematical formalism of unitary quantum mechanics to physical reality, interpretative decisions have to be made: it has to be specified in what way mathematical quantities refer to physical objects and processes. As a minimum requirement, a connection has to be established with observations and measurement results. Since one of the motivations for modal interpretations is to stop assigning a special status to 'measurement', it is natural to go beyond a strict empiricist attitude and to ascribe properties also to unobserved physical systems (at least hypothetically—see the discussion of the 'Copenhagen variant' of the modal interpretation in Section 47.4). There is a certain leeway here and different interpretative decisions can be made, with different modal interpretations as a result.

In this chapter we will sketch the history of modal interpretations and briefly review their current status. Even before van Fraassen coined the term 'modal interpretation', interpretative schemes along modal lines had been proposed: Bohm's 1952 theory (preceded by de Broglie's proposal from the 1920s) is an important example. Everett's 1957 *relative-state* interpretation may also be understood as a modal scheme, as we will discuss. We will dwell on Everett's theory in some detail, as this provides an opportunity to position modal interpretations with respect to their important no-collapse rival, the many-worlds interpretation.

After intensive discussions of the details of modal schemes in the 1990s and early 2000s, it became clear that all modal interpretations provide descriptions of the physical world that in one way or another are 'exotic'. For example, some interpretations introduce physical properties that are perspectival or hyperplane-dependent; others assume the existence of a preferred inertial frame of reference and a form of action at a distance. As a result, interest in modal interpretations of quantum mechanics waned. However, recent years have seen renewed interest in the possibilities of 'one-world interpretations of unitary quantum mechanics', which actually are modal schemes (e.g., Frauchiger and Renner (2018); Bub (2018); Conroy (2018); Healey (2018); Dieks (2019b)). This accords with the actual practice of physics, in which collapses are often taken to be not fundamental and in which unitary quantum mechanics is used as a universally applicable tool for generating probabilistic predictions.

# 47.2 BOHM'S THEORY AS A MODAL INTERPRETATION

Bohm's theory (Bohm, 1952) starts from unitary quantum mechanics in the position basis representation (unitary *wave* mechanics). This fits the physical interpretation proposed by Bohm, namely that the formalism describes point-like particles with always determinate positions, whose motion is guided by the wave function. Moreover, in Bohm's picture the square of the wave function gives us the probabilities of possible particle positions.

A single-particle wave function $\Psi(\mathbf{x}) = R(\mathbf{x})e^{iS(\mathbf{x})/\hbar}$ is thus associated with a particle that may find itself at position $\mathbf{x}_0$ with probability $|R(\mathbf{x}_0)|^2 = |\Psi(\mathbf{x}_0)|^2$. The wave function also determines the *momentum* of this particle, via the equation $\mathbf{p}_0 = \nabla S(\mathbf{x}_0)$.

The wave function $\Psi(\mathbf{x})$ undergoes deterministic Schrödinger evolution and defines, via $\mathbf{p} = \nabla S(\mathbf{x})$, for any initial position a unique particle trajectory starting from that position. But since $\Psi(\mathbf{x})$ only specifies a probability distribution for the initial localization of the particle, the theory does not fix *which* deterministic trajectory is in fact realized.

The Bohm theory thus deals with a range of physical possibilities (namely, a range of possible trajectories, each starting from one position within a range of possible initial positions) of which only one is actually realized; and it specifies the probability for each possibility to be realized. This makes it a modal interpretation.

Concerning the ontological status of $\Psi(\mathbf{x})$ different stances may be taken (Dürr, Goldstein, and Zanghì, 1997; Goldstein and Teufel, 2001; Goldstein, 2017). Bohm originally introduced $\Psi(\mathbf{x})$ as a real physical field, comparable to the electromagnetic field of Maxwell's theory. A different view, favoured by a number of later Bohmians,[1] is that $\Psi(\mathbf{x})$ and the 'guidance equation' $\mathbf{p} = \nabla S(\mathbf{x})$ are merely part of the nomological framework of the theory; that is, they are expressions figuring in the *laws* governing particle motion without themselves referring to something physical. In this case we arrive at a pure particle ontology, with the wave function in a role similar to that of, e.g., the Lagrangian in classical mechanics.[2]

---

[1] In the case of a many-particle system $\Psi(\mathbf{x})$ is defined in configuration space and not in ordinary three-dimensional space, in contrast to the electromagnetic or gravitational force fields. As a consequence, the forces derivable from $\Psi(\mathbf{x})$ also have an unusual character: they are many-body forces, so that the force on any particle depends on the simultaneous positions of all other particles. Another unusual feature, from a classical point of view, is that $\Psi(\mathbf{x})$ evolves autonomously, in accordance with the Schrödinger equation, without being influenced by the particles. These peculiarities may be seen as supporting the view that $\Psi$ figures centrally in the laws of the theory, but does not represent a material field.

[2] This interpretative strategy is similar to one adopted in some later modal interpretations, in which a distinction is made between the 'dynamical state' and the 'value state' so that the dynamical state is an element of the nomological framework governing evolution, whereas the value state represents values actually taken by physical quantities.

# 47.3 EVERETT'S RELATIVE-STATE INTERPRETATION

## 47.3.1 Relative States and Typicality

Like Bohm, Everett objected to the standard unitary-evolution-plus-collapses formalism (Everett, 1957; DeWitt and Graham, 1973; Barrett and Byrne, 2012); he even argued that this formalism is inconsistent. To demonstrate the inconsistency, Everett considered a situation of the type later made famous by Wigner: an observer $A$ performs a measurement on a system $S$ while a second observer $B$, isolated from the $A + S$ system, knows the initial states of $A$ and $S$ and uses quantum mechanics to predict the state of $A + S$ (and the states of $A$ and $S$ separately) after the measurement.

If $A$ measures the observable $\mathcal{O}$, while the initial state of $S$ is a superposition of eigenstates of $\mathcal{O}$, $c_1|o_1\rangle + c_2|o_2\rangle$ say, the collapse treatment predicts that $A$ will register either $o_1$ or $o_2$ and the state of $S$ will be projected correspondingly onto either $|o_1\rangle$ or $|o_2\rangle$. So after the measurement $A$ will attribute one of these pure vector states to $S$. However, observer $B$ will still describe the interaction between $A$ and $S$ by means of unitary evolution. This will lead to the entangled state $c_1|o_1\rangle|A_1\rangle + c_2|o_2\rangle|A_2\rangle$ for system $A + S$, with $|A_1\rangle$ and $|A_2\rangle$ states in which $A$ has registered the outcome $o_1$ or $o_2$, respectively. In this entangled state neither $A$ nor $S$ has a pure (vector) state of its own. Therefore, after the measurement $A$ and $B$ will describe system $S$ (and system $A + S$) in ways that are not consistent with each other.[3]

Everett removed this discrepancy by eliminating collapses, so that after a measurement object system and observer are generally in an entangled total state. The task then is to relate such entangled states, which are superpositions of eigenstates of the measured observable, to our experience. Unlike Bohm, Everett did not introduce a privileged determinate observable for this purpose; he deemed this 'superfluous' and in fact criticized Bohm's approach on these grounds.[4]

Everett modelled observers as physical systems with a memory register that will become correlated, during a measurement interaction, to the measured quantity. The state of observer $A$, in the above Wigner-friend scenario, illustrates the idea: in the post-measurement state $c_1|o_1\rangle|A_1\rangle + c_2|o_2\rangle|A_2\rangle$ the two post-measurement states of $A$ represent two different memory contents, correlated with two different values of the measured observable $\mathcal{O}$. The problem is of course that after the measurement neither $A$ nor $S$ are in an eigenstate representing a definite measurement result.

---

[3] By contrast, Wigner (1961) later used the same example to argue that collapses take place, as an objective physical event, as soon as the first *conscious* observer performs a measurement.

[4] Everett wrote about Bohm's theory 'our main criticism of this view is on the grounds of simplicity', and he stated that adding particles with determinate positions was 'superfluous since, as we have endeavored to illustrate, the pure wave theory is itself satisfactory' (Everett, 1956, p. 154).

Everett addressed this problem by introducing the notion of *relative states*: although subsystems of a composite system generally do not possess vector states of their own, they do have such states *relative to the rest of the total system*. In our example $|o_1\rangle$ is the relative state of $S$ with respect to the relative state $|A_1\rangle$ of $A$, and $|o_2\rangle$ is the relative state of $S$ with respect to the relative state $|A_2\rangle$ of $A$. Generally, if the total 'absolute' state of a composite system $X + Y$ is given by $\Sigma_i|x_i\rangle|y_i\rangle$, $X$ has the relative state $|x_i\rangle$ with respect to $Y$'s relative state $|y_i\rangle$.

Now, if the *relative state* of system $S$ is an eigenstate of an observable, then the interpretative rule introduced by Everett is that $S$ possesses the value given by the associated eigenvalue *in a relational way*. In the example, $S$ has the value $o_1$ of $\mathcal{O}$ with respect to $A$ having the memory content 1, and the value $o_2$ with respect to $A$ having the memory content 2. Generally, *all* possible outcomes, and *all* possible associated records, are represented in the post-measurement state; and for all values of $i$, the record indicating $i$ is realized relative to the definite outcome being $i$.

Everett emphasized that all terms in post-measurement superpositions are equally **real**, in the sense that their presence can be demonstrated, in principle, via experiments. For example, in the Wigner's friend case, $B$ could decide to measure an observable of which $c_1|o_1\rangle|A_1\rangle + c_2|o_2\rangle|A_2\rangle$ is an eigenstate. In this way $B$ could verify, without disturbing the system $A + S$, that there has been no collapse—that there still are two 'branches' after the measurement by $A$.

In order to reproduce the predictions of standard quantum mechanics, the interpretation has to specify probabilities. But since all branches remain in place during measurements, one cannot speak of the probability of one branch to be realized to the exclusion of all others. Everett responded to this problem by assigning 'weights' to the terms in a superposition, which measure their 'typicality'. The idea is that although all branches are there, some of them may be more 'typical' than others. If certain features enjoy a high degree of typicality, i.e. are instantiated in a set of branches with a large typicality value, an observer should expect to find them. The typicality measure adopted by Everett is given by the squares of the absolute values of the coefficients occurring in the superposition. So, in the state $c_1|o_1\rangle|A_1\rangle + c_2|o_2\rangle|A_2\rangle$ the two terms, and the relative properties represented by them, possess typicalities $|c_1|^2$ and $|c_2|^2$, respectively.

It is easy to show that the total typicality measure of those branches that conform to the usual quantum statistics (i.e., that exhibit the right relative frequencies of outcomes in repeated experiments, while passing tests of randomness) will tend to 1 in the limit when the number of consecutive measurements grows indefinitely. Everett takes this to mean that an observer is entitled to be confident that in the long run the probabilistic predictions of standard quantum mechanics will be verified.[5]

---

[5] It is important to note that simply *counting* the number of terms in the superposition with a certain relative frequency of outcomes would give results that conflict with quantum mechanics. For example, after $N$ consecutive measurements of observable $\mathcal{O}$ on systems all in the state $c_1|o_1\rangle + c_2|o_2\rangle$, the final state will contain $2^N$ terms, each one representing a different series of outcomes. The *number* of terms

## 47.3.2 Branches as Modalities

Bohm and Everett thus both accepted the same unitary quantum mechanics formalism, but added different interpretative principles to it. Bohm's interpretation is clearly modal: according to it, the theory specifies a range of physical possibilities, of which *one* is actually realized with a probability supplied by the theory. Everett, on the other hand, emphasized the reality of all terms of post-measurement superpositions, on the grounds that they can interfere. This suggests the picture of coexisting although different physical realities, a picture famously fleshed out by DeWitt (DeWitt, 1970; DeWitt and Graham, 1973). DeWitt's many-worlds interpretation assumed that in a measurement *worlds* are multiplied, so as to make a different measurement outcome determinate in each world.

However, Everett himself did not use the notion of many worlds in his writings but only emphasized the simultaneous existence of many *superposition terms* in the uncollapsed total state. Moreover, he asserted that for him the essential issue was only whether unitary quantum mechanics was *empirically faithful*, as he expressed it; that is, whether our actual observational records can be satisfactorily represented by one of the terms in the total state (Barrett, 2011a, 2011b; Barrett and Byrne, 2012; Barrett, 2015, 2016, 2018; Conroy, 2012, 2018). If unitary quantum mechanics in this way is able to achieve the same predictive success as the standard collapse theory, Everett maintained, it should be preferred since it is simpler, consistent, and not marred by the measurement problem.

Conroy (2012, 2018) therefore argues that Everett had the view that the wave function serves the purpose of providing us with a set of factual and counterfactual descriptions of only one world (our own). A term in a superposition provides a *factual* description if it corresponds to what occurs in our world, in which case the other terms describe *counterfactually* what our world could have been like if some different measurement result had materialized. This reconstruction appears to be compatible with the historical record, and if it is right Everett himself saw his relative-state interpretation as a theory about one single world. This would make his interpretation almost modal. The only missing element would be the replacement of the typicality measure by probabilities for physical possibilities.

Everett did not introduce such probabilities because no specific superposition term is picked out for actuality during measurement interactions: the whole superposition remains in place as can be verified by means of interference experiments. But, as

with $k$ occurrences of $o_1$ and $(N-k)$ occurrences of $o_2$ will be $\binom{N}{k}$, so their relative counting weight is $\binom{N}{k}/2^N$. But their weight in the Everett measure is different, namely $\binom{N}{k}|c_1|^{2k}|c_2|^{2(N-k)}$. Not coincidentally, the latter expression is numerically equal to the *probability* predicted by standard quantum mechanics for the realization of $k$ occurrences of $o_1$ and $(N-k)$ occurrences of $o_2$, regardless of their order.

stressed by Barrett (2011a), that all terms in the quantum state are relevant for future predictions does not imply that they correspond to separate physical actualities. To see this, it suffices to take our minds back to the situation in Bohm's theory: in that case there existed the interpretative option of *denying* that the (always uncollapsed) wave function is physically real, while still admitting that it fully determines the motion of particles and needs all its superposition terms to do so correctly. As we will see in the following sections, it is typical of modal interpretations to likewise always retain full superpositions and accept their importance for predicting interference, while not recognizing the physical reality of all branches. The conclusion that no probabilities can be assigned to physical possibilities because all branches remain in place, so that we have to take recourse to typicality as a surrogate, is therefore not compelling.

It may be added that the use of typicality as a replacement of probability in the case of coexisting states of affairs is problematic. Everett's motivation for introducing typicality was the assumption that typicality can guide our expectations: the more typical a branch is, the more justified we should be in our belief to be described by that branch. But it is not at all clear why typicality should be entitled to play this guiding role.

To see this, apply the branch picture to a fair quantum lottery with a massive number of lottery tickets: after the drawing the total quantum state will consist of a very large collection of coexisting branches with equal weights, each corresponding to a different ticket winning the prize. What should I, as a participant, expect about the outcome? In the total quantum state each individual lottery ticket has its prize-winning branch, but I am interested in my own chances. It seems intuitively plausible to respond that the number of branches in which my specific ticket does *not* win is overwhelming, so that there is little reason for me to feel confident about winning. However, it is difficult to see how this can be fleshed out into a logically valid argument without *assuming as a premise* that it is most reasonable to expect to find myself described by a branch of whose type there are many; without the help of probability considerations this premise would have to be a first principle. The problem becomes more pressing in a scenario in which we from the usual one-world perspective would say that the tickets do not possess equal winning chances; e.g., some cost more and have higher chances. This corresponds to a quantum description in terms of branches with unequal weights. In this case the typicality of branches of a certain sort cannot be decided by counting (compare footnote 5): the number of privileged tickets could be minute, and still it could be completely typical for one of these very few tickets to win. The intuitive appeal of typicality as a placeholder for abundance has now disappeared. Reasoning with the help of typicality in this situation is without problems if only one lottery ticket actually wins, so that we can define typicality as reflecting the *probability* of winning. But in the coexisting realities scenario this cannot be done, and typicality becomes an uninterpreted number.[6]

---

[6] There is hardly any literature analysing the meaning and status of typicality. A recent exception is Wilhelm (2019), who identifies two possible grounds for accepting typicality as a guide for rational expectation: 'typical' can mean occurring in *most* situations of a certain sort, of which we will encounter

It follows, then, that a connection between typicality and rational expectation must be *postulated* at some point in the argument. But such a postulate or axiom does not clarify the conceptual situation. If typicality is taken as a basis for expectation without further explanation, it remains a mysterious notion and our expectations would in this case seem less than rational.[7]

The alternative is to go in the opposite direction: to take probability as fundamental (and, if something like typicality would be needed after all, to reduce it to probability). This brings us to the central idea of modal interpretations, namely interpreting uncollapsed post-measurement superpositions in terms of possibilities of which only a single one is realized.

## 47.4 VAN FRAASSEN'S COPENHAGEN MODAL INTERPRETATION

The general strategy for interpreting unitary quantum mechanics in terms of possibilities and their probabilities was formulated by van Fraassen (1972) and elaborated into what he called 'the Copenhagen variant of the modal interpretation' van Fraassen (1981, 1991).[8] The leading idea was, according to (van Fraassen, 1981, p. 273):

> to deny neither the determinism of the total system evolution nor the indeterminism of outcomes, but to say that the two are different aspects of the total situation. Specifically, we can deny the identification of value-attributions to observables with attributions of states; the state can then develop deterministically, with only statistical constraints on changes in the values of the observables.

In order to sever the direct link between the quantum state and determinate value assignments to observables, van Fraassen made a distinction between two concepts of state: the *value state* and the *dynamical state*. The latter is the usual state of unitary quantum mechanics, which evolves according to a deterministic evolution equation (the Schrödinger equation in non-relativistic quantum mechanics) and which will become entangled during interactions. The value state, by contrast, is introduced to tell us which observables possess determinate values and what these values are.

---

one, and/or typicality is connected to causality. Both grounds are missing in the many-worlds case: typical branches will generally not be the most numerous ones, and the typicality measure is introduced independently of any causal and probabilistic background.

[7] Wallace (2012) has improved on the typicality account and has undertaken to introduce physical (objective) *probability* into a many-worlds picture, instead of typicality. The matter remains controversial, though: the meaning of probability in a scenario in which all possible measurement outcomes are actually realized stays contentious.

[8] See also the historical remarks by van Fraassen (1991), p. 494.

According to the standard interpretation of quantum mechanics an observable pertaining to a system has a definite value if and only if the system's state is an eigenstate of that observable. This so-called eigenstate–eigenvalue link is the background of the projection postulate: after a successful measurement the 'pointer' of a measuring device should indicate one determinate outcome, so that it should be in the associated eigenstate of the pointer observable. In no-collapse interpretations determinate measurement outcomes evidently have to be accommodated in a different way. At this point van Fraassen employs his distinction between two types of state: at the end of a measurement the total *dynamical state* is the uncollapsed result of unitary evolution, but the pointer observable nevertheless possesses one particular determinate value, represented by its *value state*.

After a measurement interaction,[9] for example of the von Neumann type

$$\Sigma_i c_i |a_i\rangle |b_0\rangle \rightarrow \Sigma_i c_i |a_i\rangle |b_i\rangle, \tag{1}$$

in which an observable $A$ is measured on an object system by a device with pointer observable $B$, the pointer observable must possess one definite value. It is true that the post-measurement state of the total system occurring in eq. (1) is a superposition (which yields, when restricted to a state of the measuring device alone, an improper mixture), but this is now not problematic because we are dealing with the dynamical state, which does not specify the actual physical state of affairs. The actually realized outcome is represented by the value state, and in this case this value state is stipulated to be one of the eigenstates of the pointer observable $B$, $\{|b_i\rangle\}$. The probability that $|b_i\rangle$ corresponds to the measurement result that actually occurs is postulated to be $|c_i|^2$.

Stated in this way, the interpretation only establishes a link between the total dynamical state and possible value states for the case of pointer observables after an interaction with a measuring device. It is natural to ask what the possible value states are in general, also outside of measurement situations. Concerning this question van Fraassen adopts a cautious attitude, loosely inspired by what he regards as the spirit of the Copenhagen interpretation (van Fraassen, 1981, pp. 280–282), namely not to attribute determinate properties to systems outside of measurement situations 'unless the system's state forces us to do otherwise'.

Elaborating this 'Copenhagen cue', van Fraassen proposes a minimalist general principle for attributing value states: Given that system $X$ is described by dynamical state $W$ (generally, a mixed state obtained by partial tracing from the total dynamical state), then the actual values of observables of $X$ correspond to one pure value state $|x\rangle$ in the image space of $W$. In general, we cannot know *which* pure state in $W$'s image space is the value state.

---

[9] Importantly, in van Fraassen's proposal, and in modal interpretations in general, measurements are ordinary physical interactions described by unitary dynamics. What distinguishes measurements from other physical interactions is only that measurement interactions result in a correlation between the pre-measurement state of the object and the post-measurement state of the measuring device (van Fraassen, 1981, Ch. 7.4).

This 'agnostic' Copenhagen principle is supplemented by a more precise rule that only applies to states resulting from measurement interactions. In the latter case the pointer's value state is one of the eigenstates of the pointer observable (fixed by the measurement interaction), and its probability is given by the Born rule (as in the example of eq. (1)). This more precise rule connects van Fraassen's Copenhagen modal interpretation to experiment.

Van Fraassen's interpretation is close to the one-world take on the Everett interpretation explained in Section 47.3.2. Van Fraassen views the entangled states resulting from interactions as playing an essential role in the dynamics of the theory, but not as qualifying for the predicate 'physically real'—only value states correspond to properties instantiated in physical reality. The possibility of interference between 'branches' surely exists, but it is an interference between components of the dynamical state, not an interference between worlds. In other words, the branches are only 'dynamically real'. Everett took the coexistence of all branches to exclude the application of the concept of probability and concluded that he had to take recourse to the notion of typicality. This manoeuvre now proves unnecessary: if only one *value state* corresponds to actuality, the concept of probability can be applied straightforwardly.

# 47.5 Non-Copenhagen Modal Interpretations

Van Fraassen's Copenhagen scheme can be criticized for being informative about value states only in the context of measurements (e.g., Monton, 1999). It is natural to inquire about the possibilities of going beyond van Fraassen's cautious empiricism in order to develop more encompassing interpretations that could define value states in general situations. Such interpretations would be *realist*, in the sense that they could describe the physical world also on the (sub)microscopic level, even if no macroscopic measurements are taking place.

A variety of proposals in this direction have been made, mostly independently of and partly even prior to van Fraassen's work. Bohm's theory is an early example, and as we have discussed Everett's interpretation might be seen as 'modal in disguise'. In the 1980s and 1990s new modal schemes were proposed by various authors: Kochen (1985); Krips (1987); Dieks (1988, 1989a, 1989b, 1994); Bub (1992, 1994).

The interpretations of Kochen, Healey, and Dieks employ the 'bi-orthogonal decomposition theorem' for the purpose of defining value states.[10] This theorem

---

[10] Dieks suggested using the bi-orthogonal decomposition theorem at the end of the 1970s, motivated by the form of the total post-measurement state as given in eq. (1). He reasoned that von Neumann's measurement problem could be avoided by giving a probabilistic interpretation of this state, and noted that an arbitrary two-particle state could be written in the same form by using the bi-orthogonal decomposition.

says that any vector in a tensor product Hilbert space $\mathcal{H}_1 \otimes \mathcal{H}_2$ can be written in the form

$$|\Psi\rangle = \Sigma_i c_i |a_i\rangle |b_i\rangle, \qquad (2)$$

where $\{|a_i\rangle\}$ and $\{|b_i\rangle\}$ are orthogonal bases of $\mathcal{H}_1$ and $\mathcal{H}_2$, respectively. If the values of $|c_i|^2$ are different for all $i$, these bases are uniquely determined by $|\Psi\rangle$. The interpretative idea now is that a dynamical state $|\Psi\rangle$ of a composite system defines possible value states $\{|a_i\rangle\}$ and $\{|b_i\rangle\}$ via eq. (2); and that *one* specific pair $(|a_k\rangle, |b_k\rangle)$ represents the actual properties of the two physical subsystems. That the pair $(|a_k\rangle, |b_k\rangle)$ is realized from among the members of the set $\{(|a_i\rangle, |b_i\rangle)\}$ is postulated to have the probability $|c_k|^2$.

This idea can be generalized to include degeneracy (occurring when not all values of $|c_i|^2$ are different) and cases in which the dynamical states are not vector states but mixtures (Vermaas and Dieks, 1995). In these cases the definite-valued properties of the subsystems, and their probabilities, are given by the spectral decomposition of the mixed states (obtained by 'partial tracing') of the subsystems: $\mathcal{W} = \Sigma_i p_i \mathcal{P}_i$, with $\{\mathcal{P}_i\}$ the (possibly multidimensional) mutually orthogonal projection operators occurring in the spectral decomposition and $\{p_i\}$ their associated probabilities. In this general case the value states need not be pure vector states, but may be represented by the higher-dimensional subspaces of Hilbert space onto which the operators $\{\mathcal{P}_i\}$ project.

For states resulting from an ideal von Neumann measurement, without degeneracy, application of the bi-orthogonal decomposition theorem leads to the same value states as assigned by van Fraassen's interpretation: after such an interaction a dynamical state of the form of Eq. (2) will obtain, and the one-dimensional projection operators $\{|b_i\rangle\langle b_i|\}$ define the spectral resolution of the pointer's density matrix. The pointer will therefore indicate a definite outcome corresponding to one of the value states $\{|b_i\rangle\}$, with the correct Born probability.[11]

Another modal strategy relies on the introduction of a preferred observable that is postulated to be always definite-valued. The paradigmatic example is Bohm's theory, in which the position observable plays this privileged role. The more recent 'modal-Hamiltonian interpretation' (Lombardi and Castagnino, 2008; Ardenghi, Castagnino, and Lombardi, 2009; Lombardi, Castagnino, and Ardenghi, 2010) takes the *Hamiltonian* of a total system to determine which observables are determinate, namely all observables commuting with the Hamiltonian and sharing its symmetries. As

---

[11] To describe real laboratory measurements more general schemes than those of idealized von Neumann measurements need to be considered; in particular, one has to take account of decoherence by the environment (Bacciagaluppi and Hemmo, 1996; Monton, 1999). As Bacciagaluppi and Hemmo (1996) show, the bi-orthogonal decomposition combined with decoherence leads to the assignment of empirically correct pointer-observable value states in situations with a finite number of degrees of freedom; usually position will become determinate. In the case of *continuous* decoherence models the situation is less clear; Bacciagaluppi (2000) shows that delocalized properties may occur. It remains to be seen how this works out in accurate models of real observations.

Lombardi and Castagnino (2008) argue, this scheme deals successfully with several important cases (free particle with spin, harmonic oscillator, free hydrogen atom, Zeeman effect, fine structure, and the Born–Oppenheimer approximation).

Bub and coworkers (Bub and Clifton (1996); Bub (1997); Bub, Clifton, and Goldstein (2000)) have proved that all these modal schemes can be characterized as variations on the same theme: all of them determine possible definite properties of physical systems (representable by value states) on the basis of the total dynamical state $|\Psi\rangle$ plus one chosen observable $\mathcal{R}$ that is postulated to be always definite-valued. As these authors show, only one unique set of possible value states can be defined from $|\Psi\rangle$ and any such given privileged observable $\mathcal{R}$. This unique set of value states is generated by the projections of $|\Psi\rangle$ onto the eigenspaces $R_i$ of $\mathcal{R}$, plus the 'null space' of vectors orthogonal to all these projections.

Appropriate choices for the determinate observable $\mathcal{R}$ lead to the various modal schemes we have mentioned. If the *position* observable is chosen, Bohm's interpretation results. Choosing the *Hamiltonian* leads to the modal-Hamiltonian interpretation.

If the *density operator* $\mathcal{W}$ (obtained by partial tracing) of a system is taken as the relevant observable representing determinate properties, the value states defined by the bi-orthogonal decomposition theorem (or its generalization) are found.

The Everettian empiricist modal scheme discussed in Section 47.3.2 follows from choosing as privileged the observable of which *memory states* are the eigenstates.

Van Fraassen's Copenhagen modal interpretation can similarly be characterized as following from the choice of the *pointer observable defined by measurement interactions* as privileged.[12]

In the last three cases the preferred observable is itself determined by the total state $|\Psi\rangle$ and its evolution in Hilbert space. It is therefore not fixed a priori what the determinate observables are, as is the case in the examples of Bohm or the modal-Hamiltonian interpretation. These modal schemes without *a priori* privileged observables may therefore be said to respect the symmetries of the Hilbert space formalism: they do not add additional mathematical elements to it and in this specific sense do not involve 'hidden variables'.[13]

## 47.6 PERSPECTIVAL PROPERTIES

One of the reasons for eliminating collapses was that nested measurements (in Wigner's friend-type experiments) raise the problem of *where* collapses occur (see

---

[12] The various modal interpretations all aim at making macroscopic measurement records determinate, and if macroscopic pointer positions are not made determinate by postulate, decoherence by the environment usually helps to achieve this aim. See, however, footnote 11 for a problem faced by the bi-orthogonal decomposition version.

[13] Detailed discussions of various modal schemes can be found in (Dieks and Vermaas, 1998; Vermaas, 1999; Lombardi and Dieks, 2021).

Section 47.3.1). However, in unitary quantum mechanics a variation on this problem returns: after observer $A$ (Wigner's friend, who performs a measurement inside a sealed laboratory while observer $B$, Wigner, is outside) has found a determinate outcome, observer $B$ can still decide to verify that $A$'s laboratory is described by a superposition containing terms corresponding to all possible outcomes.

Everett introduced *relational* properties to cope with this situation. The same manoeuvre is available to modal interpretations that refrain from fixing, in an *a priori* manner, observables that are always determinate (see Section 47.5). For example, if we model the Wigner's friend set-up in the way explained in Section 47.3.1, the final state will have the form

$$\{c_1|o_1\rangle|A_1\rangle + c_2|o_2\rangle|A_2\rangle\} \otimes |W_0\rangle, \tag{3}$$

in which $|W_0\rangle$ represents the outside observer (Wigner) in his 'neutral' state. According to modal interpretations of the bi-orthogonal, van Fraassen and Everettian types this state has the following physical meaning: for the internal observer *one* of the definite results $o_1$ or $o_2$ has come out, but for the outside observer the laboratory including internal observer and object system has the definite property corresponding to the value state $c_1|o_1\rangle|A_1\rangle + c_2|o_2\rangle|A_2\rangle$ (Dieks, 2009).

That different observers may thus have different 'perspectives', associated with incompatible property attributions, relates to the non-commutativity of the algebra of quantum observables. In the case at hand, Wigner's observable (with eigenstate $c_1|o_1\rangle|A_1\rangle + c_2|o_2\rangle|A_2\rangle$) does not commute with the observable measured by his friend, and it is therefore not surprising that their measurement reports fail to agree.[14] Perspectivalism and contextuality are anchored in the non-commutative structure of Hilbert space operators.

A further illustration of this is provided by the extension of the Wigner's friend experiment proposed by Frauchiger and Renner (2018).[15] In this new set-up there are two laboratories and two friends, one in each laboratory; and two external observers. Friend $A$ performs a quantum measurement (of observable $\mathcal{A}$) in his lab and on the basis of its result sends a qubit in one of two states to friend $B$, who performs a spin measurement on it (observable $\mathcal{B}$). According to unitary quantum mechanics, the two laboratories plus internal observers are in an entangled state after this procedure. The external observers $X$ and $Y$ subsequently measure global observables, $\mathcal{X}$ and $\mathcal{Y}$ on the labs of $A$ and $B$, respectively. Frauchiger and Renner (2018) show that combining all the resulting measurement results in one Boolean deductive scheme leads to a contradiction. Fortin and Lombardi (2020) argue that this contradiction arises from treating all results as existing independently of context, and combining them in one deductive

---

[14] Even if two non-commuting observables share eigenstates, the Kochen and Specker theorem tells us that in general we should not assume that these shared eigenstates encode values of physical quantities independently of the different contexts defined by the two non-commuting observables.

[15] Bub (2018) has presented a particularly transparent version of this thought experiment.

argument even though some quantities pertain to non-commuting observables. Just as in the original case of Wigner's friend, the observables representing measurements inside the two respective laboratories do not commute with the external observables $\mathcal{X}$ and $\mathcal{Y}$. As in the original Wigner's friend thought experiment, contradictions can be avoided by introducing perspectival (relational) properties (Dieks, 2019a).

For modal interpretations based on the bi-orthogonal decomposition of the total state $|\Psi\rangle$, or on the spectral resolution of the partial trace density matrices $\mathcal{W}_i$, a general framework for defining such perspectival states was formulated and discussed by Bene and Dieks (2002). Basically, their approach is to assign a value state to each system via the usual procedure (taking a projection operator from the spectral resolution of the system's density operator $\mathcal{W}$), and then to define value states of other systems, with respect to the first system, on the basis of the relative state (in the sense of Everett) of the rest of the universe. For the Wigner's friend case, this leads to the expected conclusion: relative to Wigner the laboratory plus friend has the property represented by the superposition of the two outcome states. Relative to Wigner's friend, however, the experiment has a definite outcome and the laboratory possesses the corresponding physical properties.

## 47.7 RELATIVISTIC COVARIANCE

Modal interpretations face difficulties in accommodating the principle of relativity. This was first noticed for the case of the Bohm theory: application of Bohmian dynamics in the same way in different Lorentz frames turns out to lead to contradictory explanations of the measurement results (Berndl, Dürr, Goldstein, and Zanghì, 1996). Bohmians have responded by introducing a preferred inertial frame of reference in which Bohm's laws of motion are exclusively valid. However, it turns out to be impossible to identify this preferred frame on the basis of the empirical predictions of the theory: these predictions for outcomes of measurements remain the same, regardless of which frame is assumed to be privileged. A consistent Bohmian story about how these empirical results are produced can be told in any frame, even though these stories contradict each other. Bohmian mechanics thus satisfies what could be called 'empirical Lorentz invariance'. Still, the situation is uncomfortable: the lack of detectability of absolute rest motivated Einstein in 1905 to reject the concept of absolute rest, and analogously the lack of detectability of the Bohmian privileged frame raises doubts about its existence.

The proof by Berndl, Dürr, Goldstein, and Zanghì (1996) that demonstrated this problem for Bohm's theory has been extended by Dickson and Clifton (1998) and Myrvold (2002) to cover other modal interpretations. The core difficulty laid bare by their theorems resides in the treatment of correlations between the properties of (approximately) localized systems that are spacelike separated from each other. The standard assumption is that for two such systems (e.g., in the EPR situation) these correlations should be given by $\langle\Psi|\mathcal{P}_1\otimes\mathcal{P}_2|\Psi\rangle$, with $\mathcal{P}_1$ and $\mathcal{P}_2$ the projection operators projecting on the value states of system 1 and 2, respectively. The no-go theorems show, however, that this correlation formula cannot be valid along all different hyperplanes on which the two systems can be compared; this applies both to modal interpretations with a fixed determinate observable and to modal interpretations that use only the quantum state for determining which observables are definite-valued (the distinction of Section 47.5).

That microscopic properties assignments along different hyperplanes do not always mesh may be taken as a signal that there is only one inertial frame in which the interpretational rules may be legitimately applied. This is the Bohmian way out; as noted, this introduces a tension with 'the spirit of relativity'. Another response is to assume that properties are hyperplane-dependent, and various authors have made concrete proposals in this direction (e.g. Aharonov and Albert (1980); Dieks (1985); Myrvold (2003); Bene and Dieks (2002); Dieks (2019b); some of these proposals are couched in the language of hyperplane-dependent *collapse*). We will here indicate yet another road to relativistic covariance, analogous to how proponents of the many-worlds interpretation have dealt with the issue (Timpson and Brown, 2002; Wallace and Timpson, 2010; Wallace, 2012).[16]

Suppose that Alice and Bob, who are spacelike separated from each other, share a pair of spin-1/2 particles in the singlet state

$$\frac{1}{\sqrt{2}}(|\uparrow\rangle_A|\downarrow\rangle_B - |\downarrow\rangle_A|\uparrow\rangle_B), \tag{4}$$

where the labels $A$ and $B$ refer to Alice's and Bob's position, respectively. The spin states at $A$ and $B$ are in this case given by the unity operator in the respective two-dimensional spin spaces. According to the modal interpretation based on $W$ and its spectral decomposition (Section 47.5) this means that the spin properties of the two particles are degenerate, so that no definite spin directions are singled out.[17] Now

---

[16] Gao (2018) has illustrated the tension between standard unitary quantum mechanics and relativity by means of a simple and elegant thought experiment. He uses the fact that in unitary quantum mechanics local measurements can also be *undone* locally (in principle). Now suppose Alice performs a spin measurement on her part of an EPR singlet pair, and that this measurement is undone a bit later. Bob, who is spacelike separated from Alice, measures the spin of his particle, in the same direction. Because of the spacelike separation, it depends on the frame of reference whether Bob's measurement takes place before or after Alice's experiment has been undone. Standard reasoning now implies that Bob's result has to be opposite to what Alice found or is independent of it, depending on the frame of reference that is used. Using this argument for repeated measurements one can construct a contradiction in terms of predicted relative frequencies of outcomes. In (Dieks, 2019b) it was proposed to solve this problem by assuming that Bob's results could be different on different hyperplanes. The approach adopted in the present chapter is different and probably preferable: it implies that it will not make any difference to the statistics of Bob's results whether Alice, at spacelike separation, does or does not perform a measurement—neither is it relevant whether or not her measurement is undone.

[17] The absence of definite spins in well-defined directions in this case is due to the rotational symmetry of the singlet state, which in turn is a consequence of the equality of the coefficients in state (4).

suppose that Alice measures the spin of her particle with a device $M_A$, designed to measure spin in the $z$-direction. The result is

$$\frac{1}{\sqrt{2}}(|M_A \uparrow\rangle|\uparrow\rangle_A|\downarrow\rangle_B - |M_A\downarrow\rangle|\downarrow\rangle_A|\uparrow\rangle_B), \tag{5}$$

with the two device states representing the outcomes $\uparrow$ and $\downarrow$, respectively. Alice's particle has now acquired a definite spin in the $z$-direction[18] and Bob's particle has acquired a definite spin in the opposite direction, *relative to Alice and her particle* (see Section 47.6). However, nothing has happened to the density matrix of Bob's particle itself,[19] and we are therefore entitled to assume that its spin property *as defined in itself* has remained untouched by Alice's measurement. So, a difference has been created between the spin of Bob's particle as defined in itself, and its spin *from the perspective of Alice*. This is comparable to the difference between the descriptions given by Wigner and his friend, respectively, concerning the outcome within the sealed laboratory.

That the situation at $B$, from Alice's perspective, has changed because of something that occurred locally at $B$ does not require superluminal influences. A perfect anti-correlation between $A$ and $B$ was present in eq. (4) from the start, and this correlation is still present after Alice's measurement; the only thing that has happened is that a local change at $A$ has affected Alice's perspective on the world. By contrast, if the spin at $B$ in itself had changed as a result of Alice's measurement, this would have implied an action at a distance. But this latter spin value has *not* changed.

When Bob also performs a $z$-spin measurement, while still spacelike separated from Alice, this results in the state

$$\frac{1}{\sqrt{2}}(|M_A\uparrow\rangle|\uparrow\rangle_A|\downarrow\rangle_B|M_B\downarrow\rangle - |M_A\downarrow\rangle|\downarrow\rangle_A|\uparrow\rangle_B|M_B\uparrow\rangle). \tag{6}$$

The same interpretative principles now tell us that Bob's particle has also acquired a definite spin in the $z$-direction, and from Bob's viewpoint the spin of Alice's particle has become definite in the opposite direction. Importantly, we do *not* suppose that the spin value found by Bob is opposite to that measured by Alice: the two measurements have taken place at spacelike separation, and we therefore assume that they cannot have had an influence on each other.

This account differs from the standard analysis, which says that a measurement by Alice, on her half of the singlet, immediately changes the situation at $B$. According to this standard account Bob's subsequent measurement at $B$ *must* have an outcome

---

[18] There is a complication here because in state (5) the coefficients are still equal so that there is still rotational symmetry; breaking this symmetry requires a less idealized description of the measurement, in which the many degrees of freedom of the macroscopic device should be taken into account.

[19] This is a consequence of the no-signalling theorem.

opposite to that obtained by Alice; by contrast, we assume here that Bob's outcome is unrestrained by Alice's result.

At first sight, this account seems to conflict with experience: in EPR-type experiments correlations are found between the measurements at the two wings, so the just-postulated independence of outcomes at spacelike separation appears to be refuted immediately. However, no such conflict with experience actually arises. This is because any *comparison* of Bob's and Alice's results will bear out the perfect anti-correlation encoded in states (4), (5), and (6). Such a comparison must take place in the intersection of the future light-cones of Alice's and Bob's measurements; let's say via a device $M_C$ (alternatively, we could bring Alice into contact with Bob, or vice versa). Any interaction representing such a comparison will lead to a state like

$$\frac{1}{\sqrt{2}}(|M_A \uparrow\rangle| \uparrow\rangle_A| \downarrow\rangle_B|M_B \downarrow\rangle|M_C \uparrow \downarrow\rangle - |M_A \downarrow\rangle| \downarrow\rangle_A| \uparrow\rangle_B|M_B \uparrow\rangle|M_C \downarrow \uparrow\rangle). \qquad (7)$$

So, the result of the comparison will always confirm what Alice and Bob expected all along: the two parties have obtained opposite spin values in their experiments.

If Bob measures in a direction that is not parallel to Alice's but at an angle $\theta$, the same reasoning leads us to conclude that again Bob's result (obtained at spacelike separation) is independent of Alice's: it is 'up' or 'down', both with probability $1/2$. But when causal contact is established between Alice and Bob and a comparison of results takes place, the four combinations up-up, up-down, down-up, and down-down are found with the respective probabilities $\sin^2(\frac{\theta}{2})$, $\cos^2(\frac{\theta}{2})$, $\cos^2(\frac{\theta}{2})$, and $\sin^2(\frac{\theta}{2})$. These are the standard probabilities responsible for violations of Bell inequalities. However, their derivation as given here does not involve non-local changes of properties.

Obviously, this account has counterintuitive aspects: it assumes that the properties of an object as defined in itself may be completely different from its properties as judged from another perspective. But any unitary interpretation of quantum mechanics must exhibit a similar peculiarity, given Wigner's friend and Schrödinger's cat situations.

Summing up, modal interpretations are interpretations of unitary quantum mechanics in terms of possibilities and their probabilities. They eliminate the projection postulate and always retain superpositions, in a spirit similar to that of the Everett interpretation; but they describe one single world. All modal interpretations possess features that are unexpected from a pre-quantum point of view (e.g., a preferred inertial system in the case of Bohmian schemes, and the occurrence of perspectival properties in other modal interpretations). But what we expect on the basis of pre-quantum physics is a bad guide to the question of whether interpretations of quantum mechanics draw a coherent and acceptable picture of the physical world.

## Acknowledgement

I thank Guido Bacciagaluppi for his suggestion to explore the affinity between Everett's 'relative-state formulation of quantum mechanics' and modal ideas.

## REFERENCES

Aharonov, Y., and Albert, D. Z. (1980). States and Observables in Relativistic Quantum Field Theories. *Physical Review D*, **21**, 3316–24.

Ardenghi, J. S., Castagnino, M., and Lombardi, O. (2009). Quantum mechanics: modal interpretation and Galilean transformations. *Foundations of Physics*, **39**, 1023–45.

Bacciagaluppi, G. (2000). Delocalized properties in the modal interpretation of a continuous model of decoherence. *Foundations of Physics*, **30**, 1431–44.

Bacciagaluppi, G., and Hemmo, M. (1996). Modal interpretations, decoherence and measurements. *Studies in History and Philosophy of Modern Physics*, **27**, 239–77.

Barrett, J. A. (2011a). Everett's pure wave mechanics and the notion of worlds. *European Journal for Philosophy of Science*, **1**, 277–302.

Barrett, J. A. (2011b). On the Faithful Interpretation of Pure Wave Mechanics. *The British Journal for the Philosophy of Science*, **62**, 693–709.

Barrett, J. A. (2015). Pure wave mechanics and the very idea of empirical adequacy. *Synthese*, **192**, 3071–104.

Barrett, J. A. (2016). Quantum Worlds. *Principia*, **20**, 45–60.

Barrett, J. A. (2018). Everett's Relative-State Formulation of Quantum Mechanics. In E. N. Zalta (ed.), *The Stanford Encyclopedia of Philosophy*, winter 2018 edition, URL = https://plato.stanford.edu/archives/win2018/entries/qm-everett/

Barrett, J. A., and Byrne, P. (eds) (2012). *The Everett Interpretation of Quantum Mechanics— Collected Works 1955–1980 with Commentary*. Princeton, NJ: Princeton University Press.

Bene, G., and Dieks, D. (2002). A perspectival version of the modal interpretation of quantum mechanics and the origin of macroscopic behavior. *Foundations of Physics*, **32**, 645–71.

Berndl, K., Dürr, D., Goldstein, S., and Zanghì, N. (1996). Nonlocality, Lorentz invariance, and Bohmian quantum theory. *Physical Review A*, **53**, 2062–73.

Bohm, D. (1952). A suggested interpretation of the quantum theory in terms of 'hidden' variables, I and II. *Physical Review*, **85**, 166–93.

Brown, H., and Wallace, D. (2005). Solving the Measurement Problem: De Broglie–Bohm Loses Out to Everett. *Foundations of Physics*, **35**, 517–40.

Bub, J. (1992). Quantum Mechanics Without the Projection Postulate. *Foundations of Physics*, **22**, 737–54.

Bub, J. (1994). How to Interpret Quantum Mechanics. *Erkenntnis*, **41**, 253–73.

Bub, J. (1997). *Interpreting the Quantum World*. Cambridge: Cambridge University Press.

Bub, J. (2018). In defense of a 'single-world' interpretation of quantum mechanics. *Studies in History and Philosophy of Modern Physics*. doi: 10.1016/j.shpsb.2018.03.002

Bub, J., and Clifton, R. (1996). A uniqueness theorem for 'no collapse' interpretations of quantum mechanics. *Studies in History and Philosophy of Modern Physics*, **27**, 181–219.

Bub, J., Clifton, R., and Goldstein, S. (2000). Revised proof of the uniqueness theorem for 'no collapse' interpretations of quantum mechanics. *Studies in History and Philosophy of Modern Physics*, **31**, 95–98.

Conroy, C. (2012). The relative facts interpretation and Everett's note added in proof. *Studies in History and Philosophy of Modern Physics*, **43**, 112–20.

Conroy, C. (2018). Everettian actualism. *Studies in History and Philosophy of Modern Physics*, **63**, 24–33.

DeWitt, B. S. (1970). Quantum mechanics and reality. *Physics Today*, **23**, 30–35.

DeWitt, B. S., and Neill Graham, R. (eds) (1973). *The many-worlds interpretation of quantum mechanics*. Princeton, NJ: Princeton University Press.

Dickson, M., and Clifton, R. (1998). Lorentz invariance in modal interpretations. In D. Dieks and P. Vermaas (eds), *The Modal Interpretation of Quantum Mechanics*, Dordrecht: Kluwer Academic Publishers, pp. 9–47.

Dieks, D. (1985). On the covariant description of wavefunction collapse. *Physics Letters A*, **108**, 379–83.

Dieks, D. (1988). The formalism of quantum theory: an objective description of reality? *Annalen der Physik*, 7, 174–90.

Dieks, D. (1989a). Quantum mechanics without the projection postulate and its realistic interpretation. *Foundations of Physics*, 38, 1397–423.

Dieks, D. (1989b). Resolution of the measurement problem through decoherence of the quantum state. *Physics Letters A*, **142**, 439–46.

Dieks, D. (1994). Modal Interpretation of Quantum Mechanics, Measurements, and Macroscopic Behavior. *Physical Review A*, **49**, 2290–2300.

Dieks, D. (2009). Objectivity in perspective: Relationism in the interpretation of quantum mechanics. *Foundations of Physics*, **39**, 760–75.

Dieks, D. (2017). Quantum Information and Locality. In O. Lombardi, S. Fortin, F. Holik, and C. López (eds), *What is Quantum Information?*, Cambridge: Cambridge University Press, pp. 93–112.

Dieks, D. (2019a). Quantum Mechanics and Perspectivalism. In O. Lombardi, S. Fortin, C. López, and F. Holik (eds), *Quantum Worlds. Perspectives on the Ontology of Quantum Mechanics*, Cambridge: Cambridge University Press, pp. 51–70.

Dieks, D. (2019b). Quantum Reality, Perspectivalism and Covariance. *Foundations of Physics*, **49**, 629–46.

Dieks, D., and Vermaas, P. (eds) (1998). *The Modal Interpretation of Quantum Mechanics*. Dordrecht: Kluwer Academic Publishers.

Dürr, D., Goldstein, S., and Zanghì, N. (1997). Bohmian Mechanics and the Meaning of the Wave Function. In R. S. Cohen, M. Horne, and J. Stachel (eds), *Experimental Metaphysics—Quantum Mechanical Studies for Abner Shimony*, Volume One (Boston Studies in the Philosophy of Science vol. 193), Boston: Kluwer Academic Publishers, pp. 25–38.

Everett, H. (1956). The theory of the universal wave function. In J. A. Barrett, and P. Byrne (eds) (2012), *The Everett Interpretation of Quantum Mechanics—Collected Works 1955–1980 with Commentary*, Princeton, NJ: Princeton University Press, pp. 72–172.

Everett, H. (1957). Relative state formulation of quantum mechanics. *Reviews of Modern Physics*, **29**, 454–62.

Fortin, S., and Lombardi, O. (2020). The Frauchiger–Renner argument: A new no-go result? *Studies in History and Philosophy of Modern Physics*. doi: 10.1016/j.shpsb.2019.12.002

Frauchiger, D., and Renner, R. (2018). Quantum theory cannot consistently describe the use of itself. *Nature Communications*, **9**, 3711.

Gao, S. (2018). Unitary quantum theories are incompatible with special relativity. http://philsci-archive.pitt.edu/15357/

Goldstein, S. (2017). Bohmian Mechanics. In E. N. Zalta (ed.), *The Stanford Encyclopedia of Philosophy* (summer 2017 edition). URL = https://plato.stanford.edu/archives/sum2017/entries/qm-bohm/

Goldstein, S., and Teufel, S. (2001). Quantum Spacetime without Observers: Ontological Clarity and the Conceptual Foundations of Quantum Gravity. In C. Callender and N. Huggett (eds), *Physics meets Philosophy at the Planck Scale*, Cambridge: Cambridge University Press, pp. 275–89.

Healey, R. (1989). *The Philosophy of Quantum Mechanics: An Interactive Interpretation*. Cambridge: Cambridge University Press.

Healey, R. (2018). Quantum theory and the limits of objectivity. *Foundations of Physics*, 48, 1568–89.

Kochen, S. (1985). A New Interpretation of Quantum Mechanics. In P. Lahti and P. Mittelstaedt (eds), *Symposium on the Foundations of Modern Physics*, Singapore: World Scientific, pp. 151–69.

Krips, H. (1987). *The Metaphysics of Quantum Theory*. Oxford: Clarendon Press.

Lombardi, O., and Castagnino, M. (2008). A modal-Hamiltonian interpretation of quantum mechanics. *Studies in History and Philosophy of Modern Physics*, 39, 380–443.

Lombardi, O., Castagnino, M., and Ardenghi, J. S. (2010). The modal-Hamiltonian interpretation and the Galilean covariance of quantum mechanics. *Studies in History and Philosophy of Modern Physics*, 41, 93–103.

Lombardi, O., and Dieks, D. (2021). Modal Interpretations of Quantum Mechanics. In E. N. Zalta (ed.), *The Stanford Encyclopedia of Philosophy* (Winter 2021 edition). URL = https://plato.stanford.edu/archives/win2021/entries/qm-modal/

Maudlin, T. (2010). Can the World Be Only Wavefunction? In S. Saunders, J. Barrett, A. Kent, and D. Wallace (eds), *Many Worlds?: Everett, Quantum Theory, and Reality*, Oxford: Oxford University Press, pp. 121–43.

Monton, B. (1999). Van Fraassen and Ruetsche on Preparation and Measurement. *Philosophy of Science*, 66, S82–S91.

Myrvold, W. (2002). Modal interpretations and relativity. *Foundations of Physics*, 32, 1773–84.

Myrvold, W. (2003). Relativistic Quantum Becoming. *The British Journal for the Philosophy of Science*, 54, 475–500.

Timpson, C. G., and Brown, H. R. (2002). Entanglement and Relativity. In R. Lupacchini and V. Fano (eds), *Understanding Physical Knowledge*, Bologna: CLUEB, pp. 147–66.

van Fraassen, B. C. (1972). A formal approach to the philosophy of science. In R. Colodny (ed.), *Paradigms and Paradoxes: The Philosophical Challenge of the Quantum Domain*, Pittsburgh: University of Pittsburgh Press, pp. 303–66.

van Fraassen, B. C. (1981). A Modal Interpretation of Quantum Mechanics. In E. Beltrametti and B. C. van Fraassen (eds), *Current Issues in Quantum Logic*, Singapore: World Scientific, pp. 229–58.

van Fraassen, B. C. (1991). *Quantum Mechanics*. Oxford: Clarendon Press.

Vermaas, P. (1999). *A Philosopher's Understanding of Quantum Mechanics: Possibilities and Impossibilities of a Modal Interpretation*. Cambridge: Cambridge University Press.

Vermaas, P., and Dieks, D. (1995). The modal interpretation of quantum mechanics and its generalization to density operators. *Foundations of Physics*, 25, 145–58.

von Neumann, J. (1932). *Mathematische Grundlagen der Quantentheorie*. Berlin: Springer. English translation: *Mathematical Foundations of Quantum Mechanics*. Princeton, NJ: Princeton University Press, 1955.

Wallace, D. (2012). *The emergent multiverse: Quantum theory according to the Everett interpretation*. Oxford: Oxford University Press.

Wallace, D., and Timpson, C.G. (2010). Quantum Mechanics on Spacetime I: Spacetime State Realism. *The British Journal for the Philosophy of Science*, 61, 697–727.

Wigner, E. (1961). Remarks on the mind–body problem. In I. J. Good (ed.), *The scientist speculates*, New York: Basic Books, pp. 284–302.

Wilhelm, I. (2019). Typical: A Theory of Typicality and Typicality Explanation. *The British Journal for the Philosophy of Science*. doi: 10.1093/bjps/axz016.

# CHAPTER 48

# A BRIEF HISTORICAL PERSPECTIVE ON THE CONSISTENT HISTORIES INTERPRETATION OF QUANTUM MECHANICS

GUSTAVO RODRIGUES ROCHA, DEAN RICKLES, AND FLORIAN J. BOGE

## 48.1 INTRODUCTION

DEVELOPED, with distinct flavours, between 1984 and the early 1990s, by Robert Griffiths (1984, 1986), Roland Omnès, (1987, 1988a,b,c), James Hartle (1991, 1993), and Murray Gell-Mann (see for instance Gell-Mann and Hartle (1990a)), the consistent (or decoherent) histories approach to quantum mechanics, according to its proponents, aims at introducing 'probabilities into quantum mechanics in a fully consistent and physically meaningful way', which can be applied to a closed system (such as the entire universe), dispensing the notion of measurement (and, consequently, dissolving its inevitable paradoxes).

The formalism of the consistent/decoherent histories approach will be presented in Section 48.2, reviewing it in historical context in Section 48.3. Robert Griffiths and Roland Omnès, whose seminal works will be explored in Sections 48.3 and 48.4, considered their contributions to be 'a clarification of what is, by now, a standard approach to quantum probabilities' (Freire, 2015, p. 321), namely, the Copenhagen perspective.

Gell-Mann and Hartle, on the other hand, who first applied the expression 'decoherence' to histories (Gell-Mann and Hartle, 1990a, 1994a,b), following in the footsteps of Zeh (1970), Zurek (1981), and Joos (see for instance Joos and Zeh (1985)), considered themselves to be post-Everettians, ascribing to Everett's 1957 work the merit of first suggesting 'how to generalize the Copenhagen framework so as to apply quantum mechanics to cosmology' (Gell-Mann and Hartle, 1990b, p. 323).

Indeed, as will be outlined in more detail in Section 48.5, Gell-Mann and Hartle had their sights set on quantum cosmology. However, instead of 'many worlds' (Everett), they make use of the term 'many alternative histories of the universe' (Gell-Mann and Hartle, 1994a,b), as 'branching' (Everett) is replaced by 'decoherent histories' in their interpretation. Gell-Mann and Hartle considered then decoherence history to be the right way to develop Everett's ideas.

Moreover, by introducing the role of 'information gathering and utilizing systems' (IGUSes) in the formalism, Gell-Mann's and Hartle's emphasis on the applications of decoherent histories to cosmology and 'complex adaptive systems' also added an 'evolutionary' basis for our existence in the universe and for our preference for quasi-classical descriptions of nature (Gell-Mann and Hartle, 1994b; Hartle, 1991).

# 48.2 THE FORMALISM OF THE CONSISTENT HISTORIES APPROACH

## 48.2.1 History Spaces

The formalism of the consistent histories approach rests fundamentally on the notion of a *history*, say $\alpha$, defined as a sequence of *events*, $E_1, \ldots, E_n$, occurring, respectively, at times $t_1 < \ldots < t_n$ (Griffiths, 1984). In a classical setting, the $E_i$ could be, say, the falling of positions $\mathbf{x}_i$ of a particle performing a random walk into cells $V_j$ of a partition of some bounded subset of $\mathbb{R}^3$ at times $t_i$. The set of all sequences $\langle \mathbf{x}_1 \in V_j, \ldots, \mathbf{x}_n \in V_k \rangle$ would then correspond to the set of possible histories, and at the same time define a *sample space* (Griffiths, 2003, p. 110) or *realm* (the quantum counterpart of *sample space* in ordinary probability theory).

In quantum settings, events are represented by *projectors* $\hat{P}_k$ that project to some (not necessarily one-dimensional) subspace of a suitable Hilbert space $\mathcal{H}$. To define an expression for the history of a system $S$ with Hilbert space $\mathcal{H}^{(S)}$, Griffiths (1996) introduces a *different-time* tensor product, $\odot$, so that $\breve{\mathcal{H}} = \mathcal{H}_1^{(S)} \odot \ldots \odot \mathcal{H}_n^{(S)}$ corresponds to an $n$-fold tensor product of $S$'s Hilbert space considered at times $t_1, \ldots, t_n$. Accordingly, a history $\alpha$ can be represented by a *history operator* $\hat{Y}_\alpha$, $\hat{Y}_\beta$ on $\breve{\mathcal{H}}$, where each $\hat{P}_i$ corresponds to some projection on the respective $\mathcal{H}_i^{(S)}$ (i.e., a projection on $\mathcal{H}^{(S)}$ considered at time $t_i$).

Defining histories in this way, one can also introduce logical operations on them (e.g. Griffiths, 2003, p. 113 ff.). However, this is straightforward only in the case where $\mathcal{H}^{(S)}$ is finite-dimensional (see Griffiths, 1996, Section VIII B). The intricacies of extending the formalism to infinite-dimensional spaces have been explored, e.g., by Isham (1994); Isham and Linden (1994), but shall not be covered here.

In the finite-dimensional case, the (negated) proposition '$\alpha$ did not occur', for instance, can be defined by, $\hat{Y}_\alpha^\perp := \check{1} - \hat{Y}_\alpha$ where $\check{1}$ is the identity on $\check{\mathcal{H}}$. Defining pendants for logical operations in this way, it is then possible to associate probabilities to histories, and to regard as a *sample space* any set of histories with operators $\hat{Y}_\alpha$ that satisfy, $\sum_\alpha \hat{Y}_\alpha = \check{1}$, $\hat{Y}_\alpha \hat{Y}_\beta = \delta_{\alpha\beta} \hat{Y}_\alpha$, i.e., where any two distinct histories (or rather, their operators) are mutually orthogonal, and where they jointly exhaust the entire space (their operators sum to unity).

In quantum theory, those probabilities are, of course, given by the Born rule. Application to histories is only possible, however, if the dynamics in between events is specified. This is most conveniently established by defining so-called *chain operators*. A chain operator $\hat{C}_\alpha$ for history $\alpha$ with operator $\hat{Y}_\alpha = \hat{P}_1 \odot \ldots \odot \hat{P}_n$ on $\check{\mathcal{H}}$ is given by

$$\hat{C}_\alpha := \hat{P}_n \hat{U}(t_n; t_{n-1}) \hat{P}_{n-1} \hat{U}(t_{n-1}; t_{n-2}) \ldots \hat{U}(t_3; t_2) \hat{P}_2 \hat{U}(t_2; t_1) \hat{P}_1, \qquad (1)$$

where the $\hat{U}(t_i; t_{i-1})$ terms are unitary operators representing the time evolution between events. For instance, if the dynamics is governed by a time-independent Hamiltonian $\hat{H}$, than $\hat{U}(t_i; t_{i-1}) = e^{\frac{i}{\hbar}\hat{H}(t_{i-1} - t_i)}$. It should be clear that the step from $\hat{Y}_\alpha$ to $\hat{C}_\alpha$ is thus non-trivial: different choices of dynamics will result in different chain operators.

It is useful to also introduce a second index $a$ on the projectors, so they may be distinguished not only by their application at points in time but also by their function to represent different events. For instance, $a_i$ could be the value an observable $A$ takes on at time $t_i$, so that $\sum_{a_i} \hat{P}_{a_i} = 1$ (being the identity on one copy of the space $\mathcal{H}^{(S)}$). We here use $a$ as a generic label, however, which need not be associated with the same observable $A$ in each occurrence. Using this notation, one may also define a Heisenberg operator

$$\hat{P}_{a_j}^i := e^{\frac{i}{\hbar}\hat{H}t_j} \hat{P}_{a_j} e^{-\frac{i}{\hbar}\hat{H}t_j}, \qquad (2)$$

so that $\hat{C}_\alpha$ reduces to

$$\hat{C}_\alpha := \hat{P}_{a_n}^n \hat{P}_{a_{n-1}}^{n-1} \ldots \hat{P}_{a_1}^1. \qquad (3)$$

Using chain operators, we retrieve the expression

$$p(\alpha) = p(\wedge_i E_i) = \mathrm{Tr}[\hat{C}_\alpha^\dagger \hat{C}_\alpha] \qquad (4)$$

for the probability of $\alpha$'s occurrence, where Tr denotes the trace. Moreover, if a (normalized) density operator $\hat{\rho}_S$ can be associated to $S$ at some initial time, we can extend this to

$$p(\alpha|\hat{\rho}_S) = \mathrm{Tr}[\hat{C}_\alpha^\dagger \hat{\rho}_S \hat{C}_\alpha] \qquad (5)$$

(Isham and Linden, 1994, p. 5455).

## 48.2.2 Consistency, Compatibility, and Grain

Understanding Griffiths' *consistency conditions* for histories requires us to think of the trace functional as a more general 'weight', $w$, not necessarily to be identified with a probability under all circumstances. If the weight $w(\alpha \wedge \beta) = \mathrm{Tr}(\hat{C}_\alpha \dagger \hat{C}_\beta)$ jointly assigned to histories $\alpha \neq \beta$ *vanishes*, history operators $\hat{Y}_\alpha, \hat{Y}_\beta$ satisfy a pendant of an additivity condition $p(A \vee B) = p(A) + p(B)$ for $A, B$ mutually exclusive events. Under these conditions, $\alpha$ and $\beta$ may be called *consistent*. A more straightforward way of putting this is 'that the chain operators associated with the different histories in the sample space be mutually orthogonal in terms of the inner product defined [by $\mathrm{Tr}(\hat{C}_\alpha \dagger \hat{C}_\beta)$ ]' (Griffiths, 2003, p. 141).[1]

Additionally, one may introduce the notion of a *family of histories*, $\mathcal{F}$, defined as a set of sequences $\alpha = \langle E_1, \ldots E_n \rangle, \beta = \langle \tilde{E}_1, \ldots \tilde{E}_n \rangle, \ldots$, such that at each point of any of the sequences that constitute the family one out of a set of mutually exclusive and jointly exhaustive events occurs. In other words: that the event set defines a sample space of events possible at that point in time. A family $\mathcal{F}$ of histories is said to be *consistent*, moreover, if any two $\alpha \neq \beta$ from $\mathcal{F}$ are, and both a consistent family and its operator-pendant are also called *framework* by Griffiths (2003, p. 141). The centrepiece of Griffiths' work is the suggestion that 'only those sequences, or histories, satisfying a consistency requirement are considered meaningful' (Griffiths, 1986, p. 516).

An equally important but weaker notion is that of *compatibility*, which will be further explored in Section 48.3. Two families, $\mathcal{F}$ and $\mathcal{G}$, are said to be compatible just in case they have a common *refinement* that is itself consistent (Griffiths, 2003, p. 147), and otherwise are called *mutually incompatible*. 'Refinement' here refers to the process of resolving a projector $\hat{P}$ in terms of projectors $\hat{P}', \hat{P}''$ that project onto subspaces of the space projected onto by $\hat{P}$. Conversely, if we sum $\hat{P}', \hat{P}''$ to get $\hat{P}$, we may call this a *coarsening* (or coarse-graining) of the event space.

## 48.2.3 Decoherence

As a final point, consider the connection to *decoherence*, made especially by Gell-Mann and Hartle (1990b). Decoherence, as originally discovered by Zeh (1970), refers to the dynamic vanishing of interference due to the interaction of a system with its environment. In brief, if $S$ couples to environment $E$ in a suitable way, and the dynamics is dominated by the interaction, it is possible to show that the effective (pure state) density matrix

---

[1] That this indeed defines an inner product can be seen by investigating its properties (positive definiteness, skew-symmetry, and anti-linearity in one argument); see Griffiths (2003), p. 139.

$$\hat{\rho}_S = \text{Tr}_E[\hat{\rho}_{SE}] = \sum_{i,j} \alpha_i \alpha_j^* |S_i\rangle \langle S_j| \text{Tr}[|\varepsilon_i\rangle\langle\varepsilon_j|] = \ldots = \sum_{i,j} \alpha_i \alpha_j^* |S_i\rangle \langle S_j| \langle \varepsilon_j||\varepsilon_i\rangle, \quad (6)$$

becomes $\hat{\rho}_S \approx \sum_j |\alpha_j|^2 |S_i\rangle\langle S_j|$, because $\langle \varepsilon_j || \varepsilon_i\rangle$ vanishes dynamically for $i \neq j$.

The problems with interpreting decoherence directly are well known: it does not lead to an 'or' from an 'and' (Bell, 1990), and the possibility of recoherence does not favour a straightforward epistemic reading either, even though $\hat{\rho}_S$ looks suspiciously like a statistical mixture. In any case, many-worlds theorists have counted decoherence in their favour for reasons touched on below.

As far as the consistent histories apporach is concerend, Gell-Mann and Hartle (1990b) realized, apparently independently of Griffiths and Omnès, that the expression $\text{Tr}[\hat{C}_\alpha{}^\dagger \hat{\rho}_0 \hat{C}_\alpha]$, with $\hat{\rho}_0$ 'the initial density matrix of the universe' (Gell-Mann and Hartle, 1990b, p. 326), corresponds to a complex functional $D(\alpha, \beta)$ that can be interpreted in terms of the decoherence between histories $\alpha, \beta$.

The reference to the initial state of the universe here indicates Gell-Mann and Hartle's motivation of providing a foundation for quantum cosmology. However, decoherence in the sense of Zeh and others originally meant the vanishing of coherence due to the fact that essentially any system is 'open'. Considerations of decoherence for the sake of cosmology not just require 'assuming [...] that the key features [...][of decoherence] transcend the Hilbert-space language' (Halliwell, 1989, pp. 2915–6), but to think of the universe as one *closed system*, whose internal dynamics provides consistency conditions for situated, reasoning agents (see Section 48.5 for details).

Intuitively, when $D$ vanishes for two histories $\alpha \neq \beta$, they have 'nothing in common' and thus do not interfere. This coincides with the notion of consistency adopted, at least since 1994, by Griffiths (see Griffiths, 1994). However, it is, first, important to note that this can only happen trivially for two *completely fine-grained* histories, i.e., where every projector is one-dimensional.

The sense of triviality here is that 'what happens at the last moment is uniquely correlated with the alternatives at all previous moments' (Gell-Mann and Hartle, 2007, p. 13). Hence, histories are always distinguished and decoherence does not do any real work. Using that $D(\alpha, \beta)$ reduces to $\langle \Psi| \hat{C}_\alpha{}^\dagger \hat{C}_\beta |\Psi\rangle$ for a pure initial state, it can be shown that there is only trivial decoherence of completely fine-grained histories in this sense.

Now, second, for *coarse-grained* histories $\alpha, \beta$, where $D(\alpha, \beta)$ can also vanish non-trivially, one can distinguish at least two 'levels' of decoherence:

$$D(\alpha, \beta) = p(\alpha)\delta_{\alpha\beta} \quad \text{(medium decoherence)}$$
$$Re[D(\alpha, \beta)] = p(\alpha)\delta_{\alpha\beta}. \quad \text{(weak decoherence)}$$

A good motivation for seeing that the weaker condition also quantifies decoherence is that, if $\bar{\alpha}$ is a coarsening of $\alpha$ and $\beta$, in the sense that $\hat{C}_{\bar{\alpha}} = \hat{C}_\alpha + \hat{C}_\beta$, and $\hat{\rho} = |\Psi\rangle\langle\Psi|$, we have

$$p(\bar{\alpha}|\hat{\rho}_S) = \langle\Psi| \hat{C}_{\bar{\alpha}}^\dagger \hat{C}_{\bar{\beta}}|\Psi\rangle = \ldots = p(\alpha) + p(\beta) + 2\mathrm{Re}\langle\Psi|\hat{C}_\alpha^\dagger \hat{C}_\beta|\Psi\rangle, \qquad (7)$$

which only gives back the classical sum over unresolved possibilities in case the final term vanishes. $\hat{C}_{\bar{\alpha}}$ here is not necessarily itself a chain of time-dependent projectors, and thus called a *class operator* (cf. Hartle, 1995). Furthermore, a family of histories of which any two are at least weakly decoherent is also called a *realm* by Gell-Mann and Hartle (1990a), which notion thus delivers a generalization of Griffiths' notion of a framework.

Joos *et al.* (2003, p. 242) have actually lamented that, in contrast to weak decoherence, medium decoherence

> can also be fulfilled for very non-classical situations [ ... ], [so] we prefer to call this a consistency condition to avoid confusion with the physical concept of decoherence [ ... ].

Nevertheless, Gell-Mann and Hartle have investigated even stronger notions of decoherence. In Gell-Mann and Hartle (1994a), an elaborate construction is presented, which, among other things, allowed them to establish a 'connection [ ... ] between the notion of strong decoherence and the work of many authors who investigate the evolution of a reduced density matrix for a system in the presence of an environment' (Gell-Mann and Hartle, 1994a, p. 24) or that 'strong decoherence implies the diagonalization of a variety of reduced density matrices' (Gell-Mann and Hartle, 1994a, p. 26).

# 48.3 Interpreting and Understanding the Formalism in Historical Context

## 48.3.1 Single-Framework Rule and Realm-Dependent Reality: From Method to Reality and Back

As described in Section 48.2 above, a *history* is a sequence of events, $E_1, \ldots, E_n$, represented by *projectors* $\hat{P}_k$ on the Hilbert space $\mathcal{H}$ at times $t_1 < \ldots < t_n$. Moreover, we have seen that it is only allowed to have probabilities assigned to a *family* (or set) of histories, $\mathcal{F}$, providing it fulfills Griffiths' consistency conditions, namely, (i) the sum of probabilities of the histories in the family equals one, and (ii) all pairs of histories within the family are orthogonal.

A family of histories that meets these conditions is called a *consistent framework* (or just *framework* for short) or *realm* (which constitute, as seen in Section 48.2, the quantum counterpart of *sample spaces* in ordinary probability theory). In other words, a *realm* is a decoherent set of alternative coarse-grained histories or, as Griffiths would phrase it, 'a *framework* is a Boolean algebra of commuting projectors based upon a suitable sample space' (Griffiths, 2003, p. 216). Finally, we have seen that *different realms can be mutually incompatible.*

Thus, an important feature of the consistent histories approach is this notion, suggesting a *realm-dependent reality*, that quantum systems possess *multiple incompatible frameworks*. Hartle claims that the reason we hardly realize why this is the only way to make meaningful statements is because human languages contain 'excess baggage' (Hartle, 2007).

Indeed, the so-called *single-framework rule* (which states that two or more incompatible quantum descriptions cannot be combined to form a meaningful quantum description) must be followed or enforced in order to avoid inconsistencies (Griffiths, 2009). 'Once this notion has been accepted (or "digested")', as P. C. Hohenberg put it well, 'the theory unfolds in a logical manner without further paradoxes or mysteries' (Hohenberg, 2010, p. 2843).

A historian of science might then wonder within what sort of historical context these premises would be more prone to be accepted (or 'digested'). What does it take for one to accept the idea (following from the formalism) of a *realm-dependent reality*? In what follows, it will be suggested that this context is twofold: i) that of an informational worldview based on previous attempts (going as far back as to the turn of the 20th century) to formalize the sciences (as will be seen in Section 48.3, followed by additional information about Griffiths' and Omnès' trajectories in Section 48.4); and ii) that of having the universe as a whole taken as an object of science (the ultimate closed system, as will be seen in Section 48.5, based on the works by Gell-Mann and Hartle).

To begin with the former (the importance of formalization to the epistemology behind the consistent histories interpretation), an exposition of Omnès' thoughts will be outlined in the following subsections.

Furthermore, as will be seen in Sections 48.3.3 and 48.5, as these two contexts come together in the last quarter of the 20th century, the importance of the study of complex adaptive systems becomes more relevant and apparent to the consistent histories approach, as epitomized by the introduction of IGUSes in the formalism.

The main argument in the following two subsections will be that the concept of a realm-dependent reality seems to rely on a sort of 'structural objectivity' (see Daston and Galison, 2007), as formalism should be taken as more trustful than experience and common sense in the making of our metaphysics and epistemology.

Omnès states very clearly that, when choosing the method one should follow to interpret quantum theory: 'we must not begin with the classical but with the quantum world, and deduce the former together with all its appearances' (Omnès, 1994, p. 164), i.e., 'to deduce common sense from the quantum premises', as best formulated by a formal logic/a formal physics—formalism being the utmost reach of modern science, as if we had achieved a certain historical 'stage of reason' (Foucault, 1969, p. 191).

Indeed, Omnès concludes that our current mature 'scientific reasoning' follows (or should follow) a four-stage method, i.e., 'empiricism, concept formation, development, and verification' (epitomized by the history of science itself), by which 'reality is summoned twice, at the beginning' (a more empiricist phase) and at the end of the process (when formalism gradually imposes itself) (i.e., from method to reality and back). Omnès' reference to the history of science encapsulates his worldview.

## 48.3.2 Information, Formalization, and the Emergence of Structural Objectivity

Behind the acceptance of an informational worldview underlies a generalized practice of formalism, which could be seen, following Hacking, as a *style of reasoning*, 'a matter not only of thought but of action', constituted by 'techniques of thinking' (Hacking, 1990). It is no accident that Omnès calls our time not the 'Information Age', but the 'Age of Formalism' (Omnès, 1994). The 'Information Age' was in certain sense the end result of the 'Age of Formalism', since, historically, the latter came first (paving the way to the former) and eventually became embodied in the technological advances of the second half of the 20th century.

In fact, just as Hacking (1990) claimed that it took from around 1660 to 1900 for us to move between two different *styles of reasoning*, the former characterized by universal Cartesian causation and determinism, as central concepts both of science and nature (at a time when chance was seen as superstition, vulgarity, and unreason), the latter, a 'universe of absolute irreducible chance' (where 'nature is at bottom irreducibly stochastic'), Omnès also claimed that our path to the *style of reasoning* of formalism took us 'approximately one hundred years, from 1850 to 1950. Several names stand out: Weierstrass, Dedekind, Cantor, Frege, Peano, Hilbert, Russel, and Whitehead' (Omnès, 1994, p. 102).

The concept of 'information' became in the fourth quarter of the 20th century what cultural historian Arthur Lovejoy used to call a *unit-idea*, a primary and recurrent dynamic unit which blends an aggregate of 'implicit or incompletely explicit assumptions, or more or less unconscious mental habits, operating in the thought of a generation' (Lovejoy, 1936, p. 7) or what the philosopher Michael Foucault once called an *episteme*, 'a worldview, a slice of history common to all branches of knowledge, which imposes on each one the same norms and postulates, a general stage of reason, a certain structure of thought that the men of a particular period cannot escape' (Foucault, 1969, p. 191).

The interplay between metaphysics and physical sciences were well explored by historians and philosophers of science such as Pierre Duhem (1906), Emile Meyerson (1908), and E. A. Burtt (1924), who somehow disclosed the metaphysics underlying our physical theories. The concept of 'information', from the late 1970s onwards, started playing the role of our constitutive metaphysics, just as the concepts of 'matter', 'force', and 'energy' did from the 17th to the 19th century. Meanwhile, inasmuch as 'information' was being consolidated as a *unit-idea*, quantum field theory was becoming our most basic theory to describe the physical world.

This historically constructed informational metaphysics eventually interwove with our foremost physical theories. A most revealing text in this regard is Omnès' *Quantum Philosophy: Understanding and Interpreting Contemporary Science*. As we have seen, formalization is a key idea behind the central concept of information underlying our most fundamental metaphysics. Omnès is aware that the founding fathers of quantum mechanics, such as Niels Bohr, did not have this powerful informational/ computational metaphor at hand when formulating the puzzles emerging from quantum theory: 'many of these conclusions, by the way, had already been guessed by Bohr thanks to a stroke of genius, unaided by the powerful mathematical techniques and the guide of a discursive construction available to latecomers' (Omnès, 1994, p. 233).

## 48.3.3 Cybernetics and Measurement: From von Neumann and the Cybernetics Group to the Santa Fe Institute

It has been pointed out that one important feature of the consistent histories approach is that it does not depend on an assumed separation of classical and quantum domain, on notions of measurement, or on collapse of the wave function. Indeed, there is no role assigned to observers (or measurement) in consistent histories. How does the project of the formalization of the sciences relates to these features?

In setting an observer-independent criterion, the consistent histories approach was following a historical trend (well noticed by Omnès) that had already been sweeping all branches of knowledge by the last quarter of the 20th century, i.e., structural objectivity— a radicalization in a certain sense of a former mechanical objectivity (based on the idea of representation and the ideal of truth as correspondence to nature).

Both mechanical and structural objectivity shared a common enemy, i.e., subjectivity. However, structural objectivity, as opposed to mechanical objectivity, gave up 'one's own sensations and ideas in favour of formal structures accessible to all thinking beings' (Daston and Galison, 2007, p. 257). Moreover, structural objectivity is an even stronger instantiation of a trend for objectivation in modern science: 'Structural objectivity was, in some senses, an intensification of mechanical objectivity, more royalist than the king' (Daston and Galison, 2007, p. 259).

If mechanical objectivity 'had been a solitary and paradoxically egotistical pursuit: the restraint of the self by the self', as instantiated by the transcendental subject in Kantianism, structural objectivity, in contrast, 'demanded self-effacement' (Daston and Galison, 2007, p. 300) or, to rephrase it, an even more radical desubjectivation. Although a plethora of names and research programmes across a wide range of disciplines could be mentioned, cybernetics and its offspring (e.g., system theories) will be the focus of this section.

A number of authors (Heims, 1991; Dupuy, 1994; Lafontaine, 2004) have emphasized the seminal importance of the Cybernetics Group as a crucial move towards this direction.

A series of multidisciplinary conferences, supported by the Macy Foundation, was held between 1946 and 1953 to discuss a variety of topics that came to be known as cybernetics. Members of the so-called Cybernetics Group included mathematicians John von Neumann and Norbert Wiener, anthropologists Margaret Mead and Gregory Bateson, psychologist Kurt Lewin, and neuropsychiatrist Warren McCulloch, among many others.

Lafontaine has convincedly argued that 'there is a certain paradigmatic unit underlying this imponent theorical diversity: from structuralism to systems theory, from postmodernism to posthumanism, from cyberspace to biotechnological reshaping of the bodies, in any case, it can be verified, over and over again, the same denial of the humanist heritage, the same logic of desubjectivation' (Lafontaine, 2004, p. 17).

It sounds like Lovejoy's *unit-idea* or Foucault's *episteme*. Dupuy has also called attention to the fact that although 'cybernetics was soon forgotten, apparently consigned to a dark corner of modern intellectual history', several decades later it re-emerged, and, after having undergone a mutation, now bears 'the features of the various disciplines that make up what today is known as cognitive science' (Dupuy, 1994, p. ix).

The formalization of the mind, processes, and behaviours, the coadunation of the understanding of animals, humans, and machines, led to a 'subjectless philosophy of mind'. As pinpointed by Dupuy: 'consider perhaps the most significant contribution of cybernetics to philosophy: mind minus the subject' (Dupuy, 1994, p. 107).

The informational metaphor was of fundamental importance for this *Weltanschauung*. The Cybernetics Group's research programme was inspired and informed by behaviourism, but it took its functionalism to the next level. The replacement of the concept of 'behaviour' by the concept of 'information' had enormous consequences, as these models can be indistinguishably applied to living systems and machines.

In the early 1950s, for instance, the father of information theory, Claude Shannon, who had studied under Norbert Wiener at MIT, demonstrated how 'Theseus', a life-sized magnetic mouse controlled by relay circuits (a perfect correlate to the paradigmatic mice experiments of behaviourism), could learn his way around a maze.

If the concept of 'behaviour' didn't distinguish humans from animals, the concept of 'information' pushes it even further, as it doesn't differentiate the living from the non-living, the subject being dissolved into a physical symbol system. It is no wonder that, within this context, as the mouse and the maze form one single systemic network of information exchange, it is nonsensical to separate them substantially.

As an incipient example of a complex adaptive system, Theseus is in structural coupling with the maze, adapting itself through (positive and negative) feedback loops to its environment (in the language of cybernetics this would be called a 'servomechanism'). The observer-independent criterion of the consistent histories approach, summarized by the single-framework rule, likewise, follows a similar cybernetic philosophy of nature.

As cognition in cybernetics and system theories is a process (and not a 'thing', such as 'mind'), extracting information about a complex adaptive system has nothing to do with an act of measurement in the usual sense of an observer and an observed. It is rather a quest to select a meaningful framework in an indefinite network of interactions. It is also the result of applying formal operations that, although drawn from the empirical inquiry of nature, cannot be reducible to any of our common intuitions.

Drawing the consequences of applying this *style of reasoning* to the consistent histories approach seems to be straightforward:

i) the need to divide the world into a system and an external observer becomes inoperative, and

ii) the concept of history gains a fresh meaning as the evolution of a complex adaptive system.

Jeffrey Barrett and Peter Byrne, in their comments on the letters exchange between Everett and Wheeler, as well as Everett and Norbert Wiener, have rightly pointed out a generation change taking place in interpreting the paradoxes of quantum mechanics:

> Information theory was a starting point for Everett, and it is not surprising that physicists of Bohr's and Stern's generation tended to think of information in terms of 'meaning', whereas Everett thought of information as a formal notion that might be represented in the state of almost any physical system—in keeping with his background in game theory and the new science of 'cybernetics'. That is perhaps why Everett could easily conceive of an observer as a servomechanism, whereas Bohr (a neo-Kantian) and Stern (a Bohrian) could not separate measurement from human agency.    (Barrett and Byrne, 2012, p. 214)

For someone who embraces formalism as the foremost realization of the progress of modern science, the notion, in a consistent quantum approach, that reality, or what is real, is meaningful only relative to a *framework* should sound just as natural as the notion, in special relativity, that simultaneity is relative to a reference frame. Both concepts, although counter-intuitive, are products of formalism (Einstein having provided an interpretation to Lorentz transformation). As Omnès would agree:

> With the advent of relativity, the theory of knowledge has forever ceased to be cast in an intuitive representation, to be based solely on concepts whose only authentic formulation involves a mathematical formalism.    (Omnès, 1994, p. 197)

In the same manner, for Omnès, the interpretation of quantum mechanics should follow from a small set of basic principles, not a host of ad hoc rules needed to deal with particular cases, as it should follow from the formalism to the common sense of experience (and not the other way around).

Omnès, indeed, brought to our attention, in his 1994 book, how the consistent histories approach mirrors and departs from Kant's philosophy which Omnès holds in high respect. However, he seems to have missed an important feature in his comparison.

Omnès summarized the measurement problem and the function of interpretation as following: 'the alpha of facts seems to contradict the omega of theory' and 'the purpose of interpretation is to reconcile these extremes', and found in Kant's unsolvable antinomies, the twelve categories of understanding, and the forms of sensible intuition (i.e., space and time) an unparalleled attempt at formalization of physical sciences. However, Omnès opposes the *synthetic* a priori character of Kant's proposal: 'the roots of logic, if not of reason, are not to be sought within the structure of our mind but outside it, in physical reality. This of course implies a complete reversal of Kant's approach' (Omnès, 1994, p. 76). For Omnès, even though it was not quite as Kant had in mind, formalism would have taken Kant's proposal to fruition in the 20th century.

However, Omnès' narrative missed an important aspect of Kantianism. Back in 1963, Ricoeur, when commenting about the structuralism of Claude Lévi-Strauss and its resemblances with Kantianism, got right to the point: 'Kantianism, yes, but without the transcendental subject' (Ricoeur, 1963, p. 618). As a matter of fact, the formalization of all branches of scientific knowledge culminates in the end of a long modern history of the 'metaphysics of subjectivity'—to use an expression coined by Heidegger (whose hostility towards cybernetics was well known at the time).

Similarly, by means of the use of appropriate frameworks, the separation of the system from the measurement apparatus becomes pointless in the consistent history approach—as 'measurement' is inextricably subsumed in what Griffiths calls 'quantum reasoning', which is nothing but the usual rules for manipulating probabilities providing the *single-framework rule* is enforced.

Would that quantum reasoning be so convincing in a different historical context (other than the Age of Formalism)? As Hacking suggested, *styles of reasoning* can be sometimes incommensurable in comparison with backgrounds so different in their historical contexts.

Lafontaine devotes two chapters in her monograph, 'Colonization 1: The Structural Subject' and 'Colonization 2: The Systemic Subject', to documenting the 'colonization' of cybernetics over, respectively, structuralism and system theories/complex theories, the latter being part of the so-called second cybernetics. Dupuy mentions three offspring coming out of the Cybernetics Group:

> one of these, an offspring of the second cybernetics, grew out of the various attempts made during the 1970s and 1980s to formalize self-organizing in biological systems using Boolean automata networks; in France this work was conducted chiefly by a team directed by Henri Atlan and including Françoise Fogelman-Soulié, Gérard Weisbuch, and Maurice Milgran; in the United States, by Stuart Kauffman, a former student of Warren McCulloch, and others at the Santa Fe Institute for the Study of Complex Systems. Another, also springing from the second cybernetic, was the Chilean school of autopoiesis.    (Dupuy, 1994, p. 103)

The Santa Fe Institute (SFI) began functioning as an independent institution in 1984 and included among its co-founders George Cowan, David Pines, and Murray Gell-Mann

(Pines, 1988). 'The Santa Fe Institute', described Gell-Mann, 'is devoted to the study of complex systems, including their relation to the simple laws that underlies them, but emphasizing the behaviour of the complex systems themselves' (Gell-Mann, 1990).

Stuart Kauffman, a former student of Warren McCulloch, the key figure of the Cybernetics Group, approached Cowan and joined the SFI. As a multidisciplinary research and teaching institute, remembered Cowan, 'we struggled to find a term that would embrace their commonalities and settled on "Complexity". Later our mantra became "Complex Adaptive Systems"' (Cowan, 2010, p. 144).

As will be seen in Section 48.5, concepts such as IGUSes ('information gathering and utilizing systems'), developed by Gell-Mann and Hartle in their consistent histories approach, clearly convey the episteme (in the sense given by Foucault) underlying the Santa Fe Institute.

# 48.4 Robert Griffiths and Roland Omnès: Context and Logic

Having reflected upon the philosophical implications drawn from consistent histories by Omnès in Section 48.3, in the following some extra important facts will be given about Omnès' and Griffiths' trajectories in formulating the consistent histories approach.

Robert Griffiths is a statistical physicist at Carnegie Mellon University, having studied Physics at Princeton and at Stanford University. After a productive career on statistical mechanics in the 1960s and 1970s, Griffiths turned his attention to foundational issues of quantum mechanics in the 1980s, working on the consistent histories approach thereafter, and publishing his major work on the subject in 2002.

When Griffiths first introduced the concept of 'consistent histories', as opposed to 'measurement', in his 1984 paper, 'Consistent Histories and Interpretation of Quantum Mechanics', the idea of 'history' had already been put forward, although with different meaning, by R. P. Feynman and P. A. M. Dirac. It is meant by 'consistent' that Griffiths' proposal contains no paradoxes—as it can be used in the study of closed systems (not requiring the artificial separation between the classical and quantum domains and without the need to divide the world into a system and an observer).

In a follow-up 1986 paper, 'Making Consistent Inferences from Quantum Measurements', Griffiths made it clearer as to what motivated him to seek for a way to reason consistently in quantum settings:

> It is widely acknowledged that there are conceptual difficulties (in particular, problems of logical consistency) in making inferences about microscopic phenomena from the results of experimental measurements. [ ... ] It seems to me that it is possible to speak sensibly about microscopic phenomena and how they are related to

measurements. However, the procedure for doing this, namely, the logical process of inference, must be appropriate to the domain of study and must be consistent with and indeed based on quantum theory itself.    (Griffiths, 1986, p. 512)

'Following essentially Griffiths' proposal', Omnès, afterwards, noticing the logical background behind the consistent histories, constructed a 'consistent Boolean logics describing the history of a system' (Omnès, 1988a, p. 893). Omnès claimed that to be a progress over Griffiths' proposal. Both proponents of the consistent histories approach, in the end, aimed at the same, i.e., a consistent and complete reformulation of Bohr's view. Niels Bohr presented the complementarity principle, as the essence of quantum mechanics, at the Solvay Conference in 1927. Having been supported by a number of physicists this view became known as the 'Copenhagen Interpretation'.

In fact, in his 1994 book, Omnès reminded his readers that consistent histories 'might be considered as being simply a modernized version of the interpretation first proposed by Bohr in the early days of quantum mechanics' (Omnès, 1994, p. 498) and Griffiths, in the same vein, claimed that the consistent histories approach could be seen as merely consisting of 'Copenhagen done right' (Hohenberg, 2010, p. 2835)— although, it should be mentioned, historians and philosophers alike often point out that there is no such thing as a monolithic view of the Copenhagen interpretation.

Roland Omnès is an Emeritus Professor at the University of Paris XI in Orsay, having begun to work on an axiomatization of quantum mechanics in his 1987 paper, 'Interpretation of Quantum Mechanics'. As one can read in his final note at the end of his paper, when asked by a referee how his approach differed from Griffiths' consistent histories, Omnès replied that, 'as far as mathematical techniques are concerned, Griffith's construction is used', however, 'the conceptual foundations are different because what is proposed is a revision of the logical foundations of quantum mechanics' (Omnès, 1987, p. 172).

As a matter of fact, in a series of papers in the following couple of years, titled 'Logical Reformulation of Quantum Mechanics', Omnès fully developed this idea (Omnès, 1988a, b,c, 1989). In the mid 1990s, Omnès developed even more completely the formalism, besides his reflections on historical, epistemological, and philosophical aspects of the consistent histories approach, in three major textbooks (Omnès, 1994, 1999; Omnès and Sangalli, 1999), his last word so far on the philosophical aspects of physics and mathematics having been expressed more recently in a 2005 book (Omnès, 2005).

# 48.5 DECOHERENT HISTORIES: CONSISTENT HISTORIES à LA GELL-MANN AND HARTLE

Murray Gell-Mann (1929–2019) was the quintessential particle theorist. James Hartle (born 1939), a product (as a postdoc as well as an undergraduate student) of John

Wheeler's Princeton relativity school, is the quintessential cosmologist, though he also possesses strong particle physics acumen, having been a Ph.D. student of Gell-Mann's in fact. Together, beginning in the late 1980s (see Gell-Mann and Hartle, 1994a), they introduced the so-called 'decoherent histories' approach to quantum mechanics, which can be viewed largely as an application of the consistent histories approach of Griffiths and Omnès to cosmology, as described by classical general relativity, with the lessons of Everett and decoherence thrown into the mix.

Though apparently discovered independently of these earlier approaches, Gell-Mann and Hartle do cite Griffiths' and Omnès' works in their initial paper, noting at least consistency with them, despite their own wider perspective. Whereas the consistent histories approaches of Griffiths and Omnès have their roots in making sense of any sequence of events in a subsystem, with every element described solely quantum mechanically (as described in Section 48.2 above), the approach of Gell-Mann and Hartle focuses on histories in a more global sense, corresponding to the possible histories of the universe understood as a system in its own right: the ultimate closed system. Griffiths and Omnès had no such grand aspirations. Expanding consistent histories to the universe brings in a host of additional conceptual and technical problems, especially involving the emergence of a seemingly classical world with space, time, and observers experiencing an apparently unique history. They sought a framework, as Hartle puts it, 'within which to erect a fundamental description of the universe which would encompass all scales—from the microscopic scales of the elementary particle interactions to the most distant reaches of the realm of the galaxies—from the moment of the big bang to the most distant future that one can contemplate' (Hartle, 1992, p. 1).

Obviously, to a quantum cosmologist, an approach which does minimal violence to quantum mechanics yet allows it to be applied to genuinely closed systems (with no reference to measurement or observers as having special importance) is almost a holy grail. Of course, there had already been the Everett interpretation which made the same claims, but that seemed either too extravagant or unable to explain its basic process of branching (and the related emergence of a classical world), as well as other problems which we turn to below.

It was precisely this emergence of classicality (as actually observed in our universe: the quasi-classical domain of ordinary experience) that concerned Gell-Mann and Hartle, with their histories amounting to what they called 'realms', defined by their mutual isolation from quantum interference (i.e. they are decoherent, enabling systems to exhibit apparently definite properties). Since there was no observer required, only processes endogenous to quantum mechanics, the emergence of a classical universe could be modelled using only quantum mechanics, with no additional collapse postulates or hidden variables. One simply has a theory that specifies probabilities for alternative consistent histories for the universe, with the decoherence condition allowing probabilities to be thus assigned. Hartle refers to this framework, with its merging of the ideas of consistent histories, decoherence theory, and cosmology, as 'Post-Everett' quantum mechanics (see Hartle, 1992, pp. 5–6, and Hartle, 1991).

Prior to his work with Gell-Mann, Hartle had been working with Stephen Hawking applying path-integral techniques to Hawking's only just discovered theory that black holes emit radiation. The difficulty they faced in this paper was applying the (flat-space) quantum field theoretic technique of path-integration in the context of a curved spacetime background, in which a black hole appears. This led to their classic paper on the wavefunction of the universe, in 1983. The basis of this work is the Wheeler–DeWitt equation:

$$\mathcal{H}\Psi[h_{ij}, \phi] = 0 \tag{8}$$

This is solved by appropriate universal wavefunctions, $\Psi[h_{ij}, \phi]$, of the 3-metric $h_{ij}$ on a 3-surface and the configuration of matter on that 3-surface. This is of course a Hamiltonian approach (the '$\mathcal{H}$' denotes the Hamiltonian): one develops the wave function from an initial time-slice to other times. The Wheeler–DeWitt equation is essentially just a kind of Schrödinger equation, describing the time-development of wave-functions, though here these functions depend on the metric on a hypersurface and the embedding of that hypersurface into spacetime, rather than position and time. This equation develops the wave function deterministically, with only the various components of the superposition having amplitudes assigned.

The decoherent histories paper, with Gell-Mann, needs to be understood against this quantum cosmological work, which was an attempt to make quantum mechanics apply to the universe as a whole. If it is indeed the case that quantum mechanics is not limited in its application, then there is no reason why the universe should not also be described by a wavefunction. But this then stands in need of an interpretation, and clearly the Copenhagen interpretation falls short here: what serves as the classical domain when the universe itself is subject to quantization? What counts as an external observer in this case, where, as Bryce DeWitt puts it, 'the external observer has no place to stand' (DeWitt, 1968, p. 319)?

However, the idea here is that quantum mechanics now assigns probabilities to possible, consistent histories of the universe rather than measurement outcomes by classical observers. As mentioned, Gell-Mann and Hartle applied the notion of decoherence to ground the classical probabilities of the consistent histories approach, hence their name 'decoherent histories' to describe those histories that can represent classical universes of the kind we observe (i.e. with no strange interfering between histories). The central object describing the non-interfering alternative histories (as well, of course, as the degree of interference) is the 'decoherence functional' which involves pairs of such (coarse-grained) alternatives as given in a sum-over-histories formulation (and so spits out a complex number when fed such pairs):

$$D(\alpha', \alpha) = N \int_{c_{\alpha'}} \delta x' \int_\alpha \delta x \rho_f(x_f, x'_f) \exp\{iS[x'(\tau)] - S[x(\tau)]\}/\hbar\}\rho_i(x'_0, x_0) \tag{9}$$

Here, the $c_\alpha$ are the exclusive classes of histories. This then grounds the probability sum rules of quantum mechanics since, with decoherence as ordinarily understood (e.g. in the sense of Zeh), the off-diagonal terms of the decoherence functional represent the amount of interference between the alternative histories. When this interference is absent (or zero for all practical purposes), the diagonal terms provide the probabilities, which can be given a classical interpretation because of the negligible interference between them.

Crucially, the selection of non-interfering alternatives is achieved not by some reduction process, resulting from measurement say, but by the specification of an appropriate initial condition and Hamiltonian. This is seemingly required in situations involving genuinely closed quantum systems. That is, in the case of closed system, decoherence replaces measurement as the deciding factor in the emergence of a classical world, as it must if both observer and observed are contained in the system. But interpretational issues are not entirely side-stepped here simply because measurement and observation have been dethroned.

Gell-Mann and Hartle appear to view both Copenhagen and Everett as approximations in some sense of their more fundamental interpretation of closed systems. A quasi-classical world emerges to provide a home for Copenhagen interpretations, and if one restricts the analysis of probabilities for correlations to a single moment of time (a Now), then Everettian interpretations are also provided for (see, e.g., Hartle, 1991, p. 154). Indeed, in acknowledgement of the ability to view decoherent histories as a kind of generalization of many worlds, Hartle proposes the title 'many histories' for his own approach (Hartle, 1991, p. 155).

But it is somewhat difficult to nail down exactly what the interpretation is in the case of decoherent histores. Certainly, Griffiths believes that he presents a realist interpretation, on account of the absence of observers. But to do this, it seems as though the reality of many histories must be included, otherwise, on measurement we face a situation in which all but one of the histories are wiped out. In other words, on these histories approaches the probabilities for the histories are not probabilities for 'getting' such a history on a measurement of a sequence of events, but probabilities *for* those properties themselves, irrespective of whether or not a measurement is made (that is, they are beables). But then, it seems, we face a problem of predictions, since there are no outcomes here only records *qua* correlations which appear classical thanks to decoherence, as in modern Everettian interpretations. In which case, realism about the consistent histories just is many worlds. When discussing the issue of 'reality', Gell-Mann and Hartle tend to allow only what is internal to the formalism to have any say. In particular, decoherence and the approximate emergence of classicality that it generates (for some suitable initial conditions) is all that matters.[2] Given this internalist

---

[2] Note that an observer (even as modelled endogenously) is not required for the emergence of a quasi-classical realm. Gell-Mann and Hartle speak more generally of 'measurement situations'. There are certain operators that 'habitually decohere' as a result, thus generating the quasi-classical realm (see Gell-Mann and Hartle, 1994a). However, it must still be admitted that any coarse-graining (which generates some level of complexity) is a matter of selection.

viewpoint, all that quantum mechanics can do, on the question of which alternative history is 'more real', is supply a probability. This naturally sends the position into many-worlds-type interpretive issues, since one can then ask about the status of the ensemble of all alternatives, the history space (by analogy with the multiverse).

The problem is that we do have a clear experience of one history happening. Gell-Mann and Hartle bite the many-worlds bullet, but add in a story about why we experience our specific universal history, again with a story that is supposed to be internal to quantum mechanics itself, employing Everett's innovation of including the observer as part of the overall quantum mechanical story (see Gell-Mann and Hartle, 1994a, p. 325). Fay Dowker (a Ph.D. student of Hartle's) and Adrian Kent argued that the decoherent histories programme is really a promissory note requiring that 'a to-be-found theory of experience will identify as the fundamentally correct ones among all the possibilities offered by the formalism' (Dowker and Kent, 1996, p. 1641). They suggest that quantum mechanics cannot supply everything, and must be supplemented by a 'selection principle' (added as an additional axiom) to adequately capture our experience. The reason is natural selection: while there are many co-existing histories each giving a 'way the world could be', we have evolved in such a way to pick out ours. This addition bears the hallmarks of the Santa Fe Institute, with its focus on the role of complexity in the universe. In this case, complexity is fundamental in an almost Kantian manner: without complexity there would be no organisms capable of experiencing the specific history we do in fact observe. However, it is not so very far from Griffiths' own position on the status of the histories, which is that we choose for our convenience, or even *aesthetic* reasons (see his comments after Bryce DeWitt's talk at the 1994 Mazagon conference: DeWitt, 1994, p. 231)! The concept of an 'IGUS' (or information gathering and utilizing system) grounds the explanation for Gell-Mann and Hartle. Our sense of classical, single-historied reality amounts to a pattern of agreement between a kind of ecosystem of IGUSes (that is, *us*, as modelled in physics) on the quasi-classical realm, which then seems robust and invariant (i.e. real). The IGUSes come with coarse-grainings of the world appropriate to their constitution (as localized entities with limited sensory capabilities, and so on (Hartle, 1993, p. 121)). They are also equipped with a memory, which involves correlations between the projection operators of the quasi-classical realm and of their internal states (e.g. brain-states). This near-isomorphic mapping between projection operators is what constitutes experience or perception of a highly classical world. Hence, the concept of reality is somewhat contorted here: it is not a 'world out there' (which, it seems, *must* be some kind of history-ensemble). Instead, it is a relational construct involving correlations between memory records (themselves contingent on the initial condition for the universe along with the dynamics) and a specific history that is then naturally selected. In other words: this is an anthropic notion, though not one that treats the anthropic components independently of the other basic physical principles: it is more of a bootstrapped conception reality.

In this way, the separate elements of Copenhagen approaches are dealt with endogenously. It is the *particular* coarse-grainings that the IGUSes settle upon that are to be described by evolution.[3] The response to Zurek's puzzlement is this: while quantum mechanics itself does not show any preference for one alternative history or set of such over another (pace probabilities), the IGUSes (or observers) in the world *do* show a preference, as only some decoherent sets of histories are suitable from the point of view of an IGUS. In this response, we find a curious blending of the complexity work of Gell-Mann's Santa Fe Institute, since an IGUS is just a kind of complex adaptive system concerned with its survival (Gell-Mann and Hartle, 1994a), with the cosmological approach of Hartle.

If anything, rather than using a classical world to get a handle on the quantum world, as in standard Copenhagen interpretations, here one generates an approximation of the Copenhagen's classical world through quantum mechanics itself (cf. Hartle, 1989, p. 179). In this case, no collapse and no interaction between the classical and quantum worlds are required, with the former essentially just *being* the latter, only from a different point of view, and one that emerges only given certain special initial conditions and only then emerges fairly late in the universe's evolution. That is, whereas Copenhagen involves a dualism, the decoherent histories approach, like Everettian approaches, is monist. The separate postulation of a classical world is, in Hartle's words, 'excess baggage' caused by the initial focus on laboratory experiments (Hartle, 1992, p. 6). Of course, what remains is to explain the initial condition of the universe at the root of the emergence of the reality we observe, which allows for records of an apparently unique classical history. This is largely dependent on the production of a quantum theory of gravity.

Ultimately, the links between the relative-state interpretation and decoherent histories became more apparent, with the latter simply supplying a solution to one of the key problems with the former (see, e.g., Wallace, 2003). In the many-worlds approach of David Wallace and Simon Saunders, the decoherent histories function to select those classical worlds that follow probability theory with the various alternatives summing to unity (this unitarity is of course just what it means to be consistent here). In other words, decoherence of a family of histories implies that they can be treated as separate classical, exclusive alternatives each with an associated probability. Hartle himself later referred to his approach with Gell-Mann as an 'extension and to some extent a completion of Everett's work' (Hartle, 2011, p. 986).

# 48.6 CONCLUSION

Common to all history-based interpretations of quantum mechanics is the notion that one should be able to extract from it (rather than put in by hand) the everyday classical world. The basic idea is that one can treat families of histories using probabilities more or less along the lines of classical stochastic theory, with quantum mechanics then

---

[3] See also Saunders (1993b,a) along these same lines.

assigning probabilities to each possible history in such a family, so long as certain consistency conditions are satisfied. While the original proposals of Griffiths (1984) and Omnès, 1987) utilized predominantly formal consistency conditions, the later approach of Gell-Mann and Hartle (1990a), which Griffiths would later also adopt, used decoherence to ground the proper assignment of probabilities to histories, which are then capable of describing the kind of world we see. There is also no special meaning assigned to measurement and observation (and, indeed observers) in such interpretations: these simply constitute just another process to be modelled within the formalism (we showed how cybernetic and complexity ideas provided a hospitable research landscape for modelling observers and their observations). This was the main motivation behind Gell-Mann and Hartle's approach since they desired an interpretation fit for cosmological applications, in which external measurement and observers make no sense. This link to quantum cosmology (and quantum gravity) has certainly led to the increased endurance of the consistent histories approach, and new developments and applications continue to appear as a result.

## REFERENCES

Barrett, J. A., and Byrne, P. (eds) (2012). *The Everett Interpretation of Quantum Mechanics: Collected Works 1955–1980 with Commentary*. Princeton, NJ: Princeton University Press.

Bell, J. S. (1990). Against 'measurement'. In A. I. Miller (ed.), *Sixty-Two Years of Uncertainty. Historical, Philosophical, and Physical Inquiries into the Foundations of Quantum Mechanics*, New York: Plenum Press, pp. 17–32.

Burtt, E. A. (2003 [1924]). *The Metaphysical Foundations of Modern Physical Science*. New York: Dover Publications Inc.

Cowan, G. (2010). *Manhattan Project to the Santa Fe Institute: The Memoirs of George A Cowan*. Albequerque, NM: University of New Mexico Press.

Daston, L., and Galison, P. (2007). *Objectivity*. Brooklyn, NY: Zone Books.

DeWitt, B. (1968). The Everett–Wheeler interpretation of quantum mechanics. In B. DeWitt and J. Wheeler (eds), *Batelle Rencontres*, New York: W. A. Benjamin, Inc., pp. 318–32.

DeWitt, B. (1994). Decoherence without complexity and without an arrow of time. In J. J. Halliwell, J. Pérez-Mercader, and W. H. Zurek (eds), *Physical Origins of Time Asymmetry*, Cambridge: Cambridge University Press, pp. 221–33.

Dowker, F., and Kent, A. (1996). On the consistent histories approach to quantum mechanics. *Journal of Statistical Physics*, 82(5/6), 1575–646.

Duhem, P. M. M. (1991 [1906]). *The Aim and Structure of Physical Theory*. Princeton, NJ: Princeton University Press.

Dupuy, J.-P. (1994). *On the Origins of Cognitive Science: The Mechanization of the Mind*. Cambridge, MA: The MIT Press.

Foucault, M. (1969). *The Archaeology of Knowledge and the Discourse on Language*. New York: Pantheon Books.

Freire, O. (2015). *The Quantum Dissidents: Rebuilding the Foundations of Quantum Mechanics (1950–1990)*. Berlin, Heidelberg: Springer.

Gell-Mann, M. (1990). *The Santa Fe Institute*. Santa Fe Institute, Santa Fe, New Mexico.

Gell-Mann, M., and Hartle, J. (1990a). Alternative decohering histories in quantum mechanics. In K. Phua and Y. Yamaguchi (eds), *Proceedings of the 25th International Conference on High Energy Physics*, Singapore: World Scientific.

Gell-Mann, M., and Hartle, J. B. (1990b). Quantum mechanics in the light of quantum cosmology. In W. H. Zurek (ed.), *Complexity, entropy and the physics of information*, Reading, MA: Addison Wesley, pp. 321–43.

Gell-Mann, M., and Hartle, J. B. (1994a). Strong decoherence. In D. H. Feng and B. Hu (eds), *Quantum Classical Correspondence: Proceedings of the 4th Drexel Symposium on Quantum Nonintegrability*, Boston, MA: International Press, pp. 3–35.

Gell-Mann, M., and Hartle, J. B. (1994b). Time symmetry and asymmetry in quantum mechanics and quantum cosmology. In J. J. Halliwell, J. Pérez-Mercader, and W. H. Zurek (eds), *Physical Origins of Time Asymmetry*, Cambridge: Cambridge University Press, pp. 311–45.

Gell-Mann, M., and Hartle, J. B. (2007). Quasiclassical coarse graining and thermodynamic entropy. *Physical Review A*, **76**(2), 1–16.

Griffiths, R. B. (1984). Consistent histories and the interpretation of quantum mechanics. *Journal of Statistical Physics*, **36**(1–2), 219–72.

Griffiths, R. B. (1986). Making consistent inferences from quantum measurements. In D. Greenberger (ed.), *New Techniques and Ideas in Quantum Measurement Theory*, New York: New York Academy of Sciences, pp. 512–17.

Griffiths, R. B. (1994). A consistent history approach to the logic of quantum mechanics. In *Symposium on the Foundations of Modern Physics 1994: 70 Years of Matter Waves, Helsinki, Finland, 13–16 June 1994*, Paris: Editions Frontières, pp. 85–98.

Griffiths, R. B. (1996). Consistent histories and quantum reasoning. *Physical Review A*, **54**(4), 2759.

Griffiths, R. B. (2003). *Consistent Quantum Theory*. Cambridge: Cambridge University Press.

Griffiths, R. B. (2009). Consistent histories. In D. Greenberger (ed.), *Compendium of Quantum Physics: Concepts, Experiments, History and Philosophy*, Berlin, Heidelberg: Springer, pp. 117–22.

Hacking, I. (1990). *The Taming of Change*. Cambridge: Cambridge University Press.

Halliwell, J. J. (1989). Decoherence in quantum cosmology. *Physical Review D*, **39**(10), 2912.

Hartle, J. (1989). Quantum cosmology and quantum mechanics. In M. Kafatos (ed.), *Bell's Theorem, Quantum Theory and Conceptions of the Universe*, Dordrecht: Kluwer Academic, pages 179–80.

Hartle, J. (1991). The quantum mechanics of cosmology. In S. Coleman, J. B. Hartle, T. Piran, and S. Weinberg (eds), *Quantum Cosmology and Baby Universes*, Singapore: World Scientific, pages 65–158.

Hartle, J. (1992). Excess baggage. In J. Schwarz (ed.), *Elementary Particles and the Universe: Essays in Honor of Murray Gell-Mann*, Cambridge: Cambridge University Press, pp. 1–16.

Hartle, J. (1993). The quantum mechanics of closed systems. In B. L. Hu, M. P. Ryan Jr, and C. V. Vishveshwara (eds), *Directions in General Relativity: Volume 1: Proceedings of the 1993 International Symposium, Maryland: Papers in Honor of Charles Misner*, Cambridge: Cambridge University Press, pp. 104–124.

Hartle, J. B. (1995). Spacetime quantum mechanics and the quantum mechanics of spacetime. In B. Julia and J. Zinn-Justin (eds), *Gravitation and Quantizations: Proceedings of the 1992 Les Houches Summer School*, Amsterdam: Elsevier, pp. 285–480.

Hartle, J. B. (2007). Quantum physics and human language. *Journal of Physics A: Mathematical and Theoretical*, **40**(12), 3101–121.

Hartle, J. B. (2011). The quasiclassical realms of this quantum universe. *Foundations of Physics*, 41(6), 982–1006.

Heims, S. J. (1991). *The Cybernetics Group*. Cambridge, MA: The MIT Press.

Hohenberg, P. C. (2010). Colloquium: An introduction to consistent quantum theory. *Reviews of Modern Physics*, 82(4), 2835.

Isham, C. J. (1994). Quantum logic and the histories approach to quantum theory. *Journal of Mathematical Physics*, 35(5), 2157–85.

Isham, C. J., and Linden, N. (1994). Quantum temporal logic and decoherence functionals in the histories approach to generalized quantum theory. *Journal of Mathematical Physics*, 35(10), 5452–76.

Joos, E., Zeh, H., Kiefer, C., Giulini, D., Kupsch, J., and Stamatescu, I.-O. (2003). *Decoherence and the Appearance of a Classical World in Quantum Theory*, 2nd edn. Berlin, Heidelberg: Springer.

Joos, E., and Zeh, H. D. (1985). The emergence of classical properties through interaction with the environment. *Zeitschrift für Physik B Condensed Matter*, 59(2), 223–43.

Lafontaine, C. (2004). *O Império Cibernético: Das Máquinas de Pensar ao Pensamento Máquina*. Chelas, Portugal: Instituto Piaget Divisão Editorial.

Lovejoy, A. O. (1936). *The Great Chain of Being: A Study of the History of an Idea*. Cambridge, MA: Harvard University Press.

Meyerson, E. (1908). *Identity and Reality*, 3rd edn. London: George Allen & Unwin Ltd,

Omnès, R. (1987). Interpretation of quantum mechanics. *Physics Letters A*, 125(4), 169–72.

Omnès, R. (1988a). Logical reformulation of quantum mechanics. I. Foundations. *Journal of Statistical Physics*, 53(3–4), 893–932.

Omnès, R. (1988b). Logical reformulation of quantum mechanics. II. Interferences and the Einstein–Podolsky–Rosen experiment. *Journal of Statistical Physics*, 53(3–4), 933–55.

Omnès, R. (1988c). Logical reformulation of quantum mechanics. III. Classical limit and irreversibility. *Journal of Statistical Physics*, 53(3–4), 957–75.

Omnès, R. (1989). Logical reformulation of quantum mechanics. IV. Projectors in semiclassical physics. *Journal of Statistical Physics*, 57(1–2), 357–82.

Omnès, R. (1994). *The Interpretation of Quantum Mechanics*. Princeton, NJ: Princeton University Press.

Omnès, R. (1999). *Understanding Quantum Mechanics*. Princeton, NJ: Princeton University Press.

Omnès, R. (2005). *Converging Realities: Toward a Common Philosophy of Physics and Mathematics*. Princeton, NJ: Princeton University Press.

Omnès, R., and Sangalli, A. (1999). *Quantum Philosophy: Understanding and Interpreting Contemporary Science*. Princeton, NJ: Princeton University Press.

Pines, D. (1988). Designing a University for the Millennium: A Santa Fe Institute perspective. A keynote address to the April 1998 Fred Emery Conference of Sabanci University, Istanbul, Turkey.

Ricoeur, P. (1963). Structure et Herméneutique. *Esprit*, 322(11), 596–627.

Saunders, S. (1993a). Decoherence and evolutionary adaptation. *Physics Letters A*, 184(1), 1–5.

Saunders, S. (1993b). Decoherence, relative states, and evolutionary adaptation. *Foundations of Physics*, 23(12), 1553–85.

Wallace, D. (2003). Everett and structure. *Studies in History and Philosophy of Modern Physics*, 34(1), 87–105.

Zeh, H. D. (1970). On the interpretation of measurement in quantum theory. *Foundations of Physics*, 1(1), 69–76.

Zurek, W. H. (1981). Pointer basis of quantum apparatus: Into what mixture does the wave packet collapse? *Physical Review D*, 24(6), 1516–25.

# CHAPTER 49

## EINSTEIN, BOHM, AND BELL

### *A Comedy of Errors*

JEAN BRICMONT

## 49.1 INTRODUCTION

THE history of quantum mechanics, as told in general to students, can be summarized as follows: Einstein did not like the indeterminism of ordinary quantum mechanics. Therefore, he invented several clever arguments in order to prove that quantum mechanics was wrong, but Niels Bohr successfully answered all his objections.

In 1952, Bohm, elaborating on earlier ideas of de Broglie, tried to introduce a 'hidden variables' theory. However, this theory was nonlocal, therefore non-relativistic, and could only deal (maybe) with the special case of non-relativistic quantum mechanics.

Finally, in 1964, John Bell proved that any theory reproducing certain quantum mechanical predictions must be either nonlocal or non-realistic. Given that nonlocality obviously conflicts with relativity, we must reject realism, since the predictions in question have been verified experimentally.

In order to show that this version of history is grossly misleading, we shall first explain what the expression 'hidden variables' means and why it is difficult to introduce those variables.

Then, we will see that what mostly worried Einstein about quantum mechanics was not its indeterminism but its nonlocal aspect (as well as the central role of the observer) and that is what led him to think that some hidden variables were necessary.

In 1952, following the work of de Broglie in 1927, Bohm did introduce hidden variables, but not those that Einstein hoped for. In fact, his theory was explicitly nonlocal.

We will also discuss what Bell proved about that issue of nonlocality and how the conclusions of his analysis vindicate rather than refute Bohm's theory.

Finally, we will briefly review how all this was misunderstood by most physicists, including some of the most famous ones.

# 49.2 WHAT ARE 'HIDDEN VARIABLES'?

According to ordinary quantum mechanics, the 'complete description' of a physical system is given by its quantum state $\Psi$, a vector of norm one. The only physical meaning of that state is given through the measurement of 'observables', represented by self-adjoint operators acting on that Hilbert space: if an operator $\mathcal{A}$ has a basis of eigenvectors:

$$\mathcal{A}\Psi_i = \lambda_i \Psi_i \qquad (2.1)$$

$i = 1, 2, ...$, one writes:

$$\Psi = \sum_i c_i \Psi_i, \qquad (2.2)$$

where $\sum_i |c_i|^2 = 1$.

Then, the probability that the result of the measurement of the observable represented by the self-adjoint operator $\mathcal{A}$ is equal to $\lambda_i$ equals $|c_i|^2$.

The meaning attributed to the quantum state $\Psi$ is purely statistical and does not say what goes on for an individual system having that quantum state.

Thus, one could naturally ask, what produces this result for an individual quantum system? There are two possibilities:

1. The measurement creates the result, at random, by some mechanism not described by the quantum theory.
2. The measurement reveals a property that the individual system had before the measurement.

The second possibility amounts to introducing what are called *hidden variables*: one assumes that, for each observable $\mathcal{A}$ and each system having the quantum state $\Psi$, there is a value $v(\mathcal{A})$ equal to one of the possible results of measurement $\lambda_i$ and that this value is a property that any given individual system has before the measurement. The map $v$ is called a value map.

This means that if one measures the position or the momentum or the spin of a particle, one simply discovers the corresponding property that the particle had before its measurement.

The state $\Psi$ would, according to that view, give the statistical distribution of value maps through the formula:

$$\text{Probability}(v(\mathcal{A}) = \lambda_i) = |c_i|^2. \tag{2.3}$$

which of course would then imply the quantum predictions.

This view is sometimes called the *statistical interpretation of quantum mechanics*. However, it encounters serious difficulties because of several theorems by Bell (Bell, 1966) and Kochen and Specker (Kochen and Specker, 1967), proving that the mere assumption that such a map $v$ exists leads to a contradiction:

**Theorem 2.1:** Let $\mathcal{H}$ be a Hilbert space of at least four dimensions, and let $\mathcal{O}$ be the set of self-adjoint operators on $\mathcal{H}$. There does not exist a map $v : \mathcal{O} \to \mathbb{R}$ such that:

1) $\forall \mathcal{A} \in \mathcal{O}$,

$$v(\mathcal{A}) \text{ is an eigenvalue of } \mathcal{A}. \tag{2.4}$$

2) $\forall \mathcal{A}, \mathcal{B} \in \mathcal{O}$ with $[\mathcal{A}, \mathcal{B}] = \mathcal{A}\mathcal{B} - \mathcal{B}\mathcal{A} = 0$, we have

$$v(\mathcal{A}\mathcal{B})) = v(\mathcal{A})v(\mathcal{B}), \tag{2.5}$$

or

$$v(\mathcal{A} + \mathcal{B}) = v(\mathcal{A}) + v(\mathcal{B}). \tag{2.6}$$

For simple proofs of parts of that theorem, see (Mermin, 1985) and (Peres, 1990, 1991). Note that the assumptions of this theorem do not use very much the quantum formalism, only properties (2.4–2.6), that hold for results of measurements. In particular, one does not assume that the map $v$ can be known or controlled by us, but simply that it exists.

Obviously, the quantum state cannot give a probability distribution on objects (the maps $v$) that do not exist! So, the statistical interpretation, at least as it is stated here, is ruled out.

It seems that we are left only with the first of the two possibilities above. But this also encounters difficulties.

# 49.3 FROM LOCALITY TO HIDDEN VARIABLES

Let us first consider an experimental situation that may, at first sight, seem baffling: one measures, in a given direction, the spin of a particle that is prepared in such a way that, according to quantum mechanics, the result will be 'random' with both outcomes, up and down, equally probable.

Yet, there is someone, call him Bob, who claims to be clairvoyant and who does predict, in each run of the experiment and with 100 per cent success, what the result will be.

How does Bob do it? Well, as with every paranormal claim, there is a trick: Bob has an accomplice, Alice, and there is a box between Bob and Alice that sends to each of them one particle belonging to an entangled pair of particles, which means that their joint quantum state $\Psi$ is of the form:

$$\Psi = \frac{1}{\sqrt{2}}(|1\uparrow\rangle|2\downarrow\rangle - |1\downarrow\rangle|2\uparrow\rangle), \tag{3.1}$$

where the factors labelled 1 refer to the particle sent to Alice and the one labelled 2 to the one sent to Bob. That state, according to ordinary quantum mechanics, means that the spin measured by Bob will have equal probability to be up or down, but is perfectly anti-correlated with the one measured by Alice: if the spin is up for Alice, it will be down for Bob and vice versa.

So the trick is simply that Alice measures the spin of her particle first and sends to Bob a message giving her result. Then Bob can, with certainty, predict what the result of the spin measurement on his side will be, namely the opposite of Alice's result.

So, there is no mystery here and, within ordinary quantum mechanics, this is 'explained' by the collapse rule: when Alice measures the spin of her particle, the state (3.1) collapses to $|1\uparrow\rangle|2\downarrow\rangle$ if Alice sees the spin up and to $|1\downarrow\rangle|2\uparrow\rangle$ if she sees the spin down.

And, after the collapse, the spin of the particle on Bob's side is no longer undetermined: it is, with certainty, up if the collapsed state is $|1\downarrow\rangle|2\uparrow\rangle$ and down if the collapsed state is $|1\uparrow\rangle|2\downarrow\rangle$.

But one may ask, what does this collapse mean? As in the previous section, there are two possibilities:

1. The measurement by Alice creates the result, at random, by some mechanism not described by the quantum theory, but it creates it also for Bob.
2. The measurement reveals a property that the individual system had before the measurement; in other words, because of the measurement by Alice, we simply *learn* something about the physical situation on Bob's side (and of course also on her side).

It is important, for the rest of the discussion, to keep that dilemma, which we will call *the Einstein–Podolsky–Rosen dilemma* or *EPR dilemma* (Einstein, Podolsky, and Rosen, 1935), in mind and, if one chooses one horn of that dilemma, to stick to it and to its meaning. A lot of confusion comes from the fact that some people jump from one horn of the dilemma to the other in the course of a discussion.

Consider the first horn of the dilemma. What could this physical operation be? One may think that, before Alice's measurement, the spin of the particle on Bob's side is genuinely undetermined or non-existent and that it acquires a given value only after Alice's measurement.

But this means that some sort of nonlocal action or action at a distance took place: the action of Alice measuring the spin of the particle sent to her affects the physical situation where Bob is, no matter how far apart Alice and Bob are.

According to the second horn of the dilemma, the spin of the particle on Bob's side was determined before Alice's measurement and we simply *learn* its value by doing that measurement. But that also means that there is a fact of the matter about the spin of the particle on Bob's side (it is either up or down) that is not described by the state (3.1). And this fact is what was called a 'hidden variable' in the previous section.

Of course, since Bob could measure the spin of the particle sent to him before Alice, the same argument shows that the spin of the particle on Alice's side was determined before Bob's measurement.

It is true that EPR did not formulate their argument the way we do (see Einstein, Podolsky, and Rosen, 1935), but for the rest of our paper, this dilemma is all we need.

Indeed, this dilemma raises a very serious problem: if we do not want to accept nonlocality, which is a consequence of the first horn of the dilemma, then we seem forced to introduce hidden variables. But we just saw in the previous section that there are impossibility theorems against the introduction of such variables.

Before turning to Bell's solution of this dilemma, let us state the properties that this nonlocal action would have to possess if we have to accept the first horn of the dilemma:

1. *That action is instantaneous*: the spin of Bob has to be determined immediately once Alice measures her spin.
2. *The action extends arbitrarily far and does not decrease with the distance*: the fact that the spin of Bob is determined once Alice measures her spin does not change with the distance between Alice and Bob.
3. *This effect is individuated*: suppose we have a thousand people called Alice and Bob, and that we send to each of them a particle belonging to a pair of particles whose quantum state is given by (3.1). Then, each measurement by an Alice of the particle of the pair sent to her will determine the spin of the particle of the pair sent to her corresponding Bob, but will not affect the spin of any particle sent to another Bob.
4. *That action cannot be used to transmit messages*: assume that many pairs of particles are sent to Alice and Bob; since, according to quantum theory, the results of repeated measurements by Alice are random and since those results cannot be controlled by her, it is impossible to use that kind of experiment to send a message from Alice to Bob.

Note that this nonlocality is very different from the nonlocality in Newtonian gravity: the latter is instantaneous but its effect decreases with distance (through the inverse square law), is not individuated (it affects equally all bodies at a given distance), but can, in principle, be used to send signals.

# 49.4 BELL'S SOLUTION OF EPR'S DILEMMA

In 1964, Bell proved a no hidden variable theorem (Bell, 1964), similar but not identical to Theorem 2.1 which, combined with the EPR dilemma of Section 49.3 proves nonlocality, since it rules out the other horn of that dilemma.

We will explain Bell's argument in a simplified form, after (Dürr *et al.*, 1992), and moreover start with an anthropomorphic thought experiment, but which is completely analogous to what happens in real experiments and could even, in principle, be realized in the anthropomorphic form presented here, as we will explain at the end of this section.

Two people, Alice and Bob are together in the middle of a room and go towards two different doors, located at $X$ and $Y$. At the doors, each of them is given a number, 1,2,3 (let's call them 'questions', although they do not have any particular meaning) and has to say 'Yes' or 'No' (the reason why we introduce 3 questions will become clear below). This experiment is repeated many times, with Alice and Bob meeting together each time in the middle of the room, and the questions and answers vary apparently at random. When Alice and Bob are together in the room, they can decide to follow whatever 'strategy' they want in order to answer the questions, but the statistics of their answers, although they look random, must nevertheless satisfy two properties. We impose these properties because they are analogous to what happens in real physical experiments, but translating the experiments into an anthropomorphic language may help us see how paradoxical our final conclusions are.

The *first property* is that, when the same question is asked at $X$ and $Y$, one always gets the same answer. How can that be realized? One obvious possibility is that Alice and Bob agree upon which answers they will give before moving toward the doors. They may decide, for example, that they will both say 'Yes' if the question is 1, 'No' if it is 2, and 'Yes' if it is 3. They can choose different strategies at each repetition of the experiment and choose those strategies 'at random' so that the answers will look random.

Another possibility is that, when Alice reaches door $X$, she calls Bob and tells him which question was asked and the answer she gave. Then, of course, Bob can just give the same answer as Alice if he is asked the same question and any other answer if the question is different (Figure 49.1).

But let us assume that the answers are given simultaneously, so that the second possibility is ruled out unless there exists some superluminal action at a distance between Alice at $X$ and Bob at $Y$. Maybe Alice and Bob communicate by telepathy!

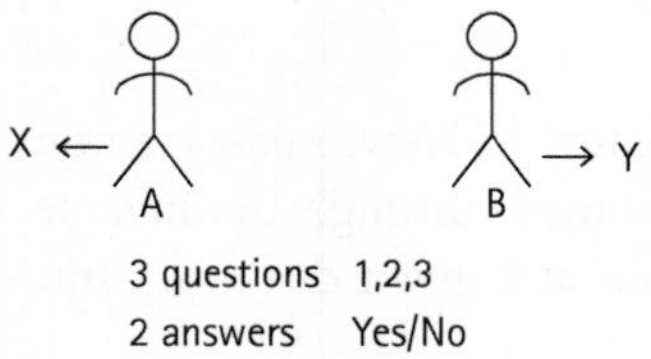

**FIGURE 49.1** The anthropomorphic experiment.

$X \leftarrow$  A    B  $\rightarrow Y$

3 questions   1,2,3
2 answers   Yes/No

Of course, this is not to be taken seriously, but that is the sort of interaction that Einstein had in mind when he spoke of God using 'telepathic methods' (letter from Albert Einstein to Cornel Lanczos, 21 March 1942, quoted in (Dukas and Hoffmann, 1979), p. 68) or of 'spooky actions at a distance' (Born, 1971, p. 158).

The question that the reader should ask at this point is whether there is *any other possibility*: either the answers are predetermined or (assuming simultaneity of the answers) there is a 'spooky action at a distance', namely a communication of some sort takes place between Alice and Bob *when* they are asked the questions.

These two possibilities are exactly those considered in Section 49.3, where, expressed through the analogy used here, there was only one 'question': is the spin measured by Alice and Bob up or down? But in both situations, we face the same dilemma: either the answers are predetermined or there is some form of action at a distance between Alice and Bob.

The reason we need three possible questions here is that there is a *second property* of the statistics of the answers: when the two questions addressed to Alice and Bob are *different*, then the answers must be the same in only a quarter of the cases. And this property, combined with the idea that the properties are predetermined, leads to a contradiction:

**Theorem (Bell).** We cannot have these two properties together:

1. The answers are determined before the questions are asked and are the same on both sides (for the same questions).
2. The frequency of having the same answers on both sides when the questions are different is $\frac{1}{4}$.

**Proof.** There are three questions numbered 1, 2, and 3, and two answers Yes and No. If the answers are given in advance, there are $2^3 = 8$ possibilities:

| 1 | 2 | 3 |
|---|---|---|
| Y | Y | Y |
| Y | Y | N |
| Y | N | Y |
| Y | N | N |
| N | Y | Y |
| N | Y | N |
| N | N | Y |
| N | N | N |

In *each case* there are at least *two questions* with the same answer. Therefore,

Frequency (answer to 1 = answer to 2)
+ Frequency (answer to 2 = answer to 3)
+ Frequency (answer to 3 = answer to 1) $\geq 1$.

But if

$$\begin{aligned}
\text{Frequency (answer to 1} &= \text{answer to 2)} \\
= \text{Frequency (answer to 2} &= \text{answer to 3)} \\
= \text{Frequency (answer to 3} &= \text{answer to 1)} = \tfrac{1}{4},
\end{aligned}$$

we get $\tfrac{3}{4} \geq 1$, which is a contradiction.∎

The inequality saying that the sum of the frequencies is greater than or equal to 1 is an example of what is called a *Bell inequality*, i.e., an inequality which is a logical consequence of the assumption of pre-existing values, and which is violated by the quantum predictions.

Before drawing conclusions from this theorem, let us briefly describe how Alice and Bob could realize these statistics. According to quantum mechanics, one can send two particles in opposite directions and with a quantum state such that, if one measures the spin of those particles along the same direction on both sides (at $X$ and $Y$), one always obtains the same answer.[1] And one can choose three different directions in such a way that the frequency of having the same answers on both sides when the directions along which the spin is measured are different is $\tfrac{1}{4}$ (see (Dürr *et al.*, 1992) or (Bricmont, 2016, Appendix 4.A) for the quantum calculation).

Then Alice and Bob can just arrange things so that such correlated particles are sent to them when they reach $X$ and $Y$. Then, they measure the spin in directions corresponding to the questions being asked to them, and give their answers according to the results of their measurements.

It is easy to check that the nonlocality proven here does have the four properties listed in Section 49.3: the results are in principle obtained instantaneously (or at least the time between them is much less that the one taken by the light to go from $X$ to $Y$); the correlations between the results obtained by Alice and Bob exist no matter how far Alice and Bob are from each other and do not depend on their distance; those correlations are between two people initially together (or two particles coming from a common source) and not between one object and the rest of the Universe (as would be the case for the gravitational force); and, since neither side can control what the results will be,[2] these strange correlations cannot be used to transmit messages.

---

[1] To be precise, the spins are anti-correlated: if one measures the spin of the particle at $X$ and one finds that it is up, the spin of the particle at $Y$ will be down. But one can decide that an up spin at $X$ means that the answer is yes on that side, and that an up spin at $Y$ means that the answer is no on that side. In that way, the perfect anti-correlations are translated into getting always the same answers when one measures the spin along the same direction on both sides.

[2] We will not give the proof of this fact, but is a well-established consequence of the quantum formalism; see (Eberhard, 1978); (Ghirardi, Rimini, and Weber, 1980); (Bell, 1976), (Bohm and Hiley, 1993), p. 139, or (Maudlin, 1994), Chapter 4.

## 49.5 CONCLUSIONS OF THE
## EPR–BELL ARGUMENT

Taken by itself and forgetting about the EPR argument, Bell's result can be called a 'no hidden variables theorem'. Indeed, Bell showed that the mere supposition that the values of the spin pre-exist to their 'measurement' (remember that those values are called hidden variables, since they are not included in the usual quantum description), combined with the perfect anti-correlation when the directions along which measurements are made are the same, and the $\frac{1}{4}$ result for the frequencies of correlations when measurements are made along different directions, leads to a contradiction. Since the last two claims are empirical predictions of quantum mechanics that have been amply verified (in a somewhat different form), this means that these hidden variables or pre-existing values cannot exist.

But Bell, of course, always presented his result *in combination with* the EPR argument (see Section 49.7.2), which shows that the assumption of locality, combined with the perfect correlations when the directions of measurement (or questions) are the same, implies the existence of those hidden variables shown by Bell to be impossible, because their mere existence leads to a contradiction. So, for Bell, his result, combined with the EPR argument, was not a 'no hidden variables theorem', but a nonlocality theorem, the result about the impossibility of hidden variables being only one step in a two-step argument.[3]

But what does this mean? It means that some action at a distance does exist in Nature, but it does not tell us what this action consists of. And we cannot answer that question without having a theory that goes beyond ordinary quantum mechanics. The de Broglie–Bohm theory is such a theory, and since nonlocality is explicit in that theory, we will briefly explain it.

## 49.6 WHAT HAPPENS IN DE
## BROGLIE–BOHM THEORY?

In the de Broglie–Bohm theory, or pilot-wave theory, also known as Bohmian mechanics, the complete state of a closed physical system composed of $N$ particles is a pair $(\Psi, X)$, where $\Psi$ is the usual quantum state (given by the tensor product of wave functions with some possible internal states), and $X = (X_1, ..., X_N)$ is the configuration

---

[3] Actually, one can give a proof of nonlocality without Bell's argument about measurements in different directions, using only the perfect correlations when the directions of the spin measurements are the same, and Theorem 2.1; see (Bricmont, Goldstein, and Hemmick, 2019a, 2019b), (Hemmick and Shakur, 2012), and references therein.

representing the positions of the particles (that exist, independently of whether one 'looks' at them or one measures them; each $X_i \in \mathbb{R}^3$).[4]

These positions are the 'hidden variables' of the theory, in the sense that they are not included in the purely quantum description $\Psi$, but they are not at all hidden: it is only the particles' positions that one detects directly, in any experiment (think, for example, of the impacts on the screen in the two-slit experiment). So the expression 'hidden variables' is really a misnomer, at least in the context of the de Broglie–Bohm theory.

Both objects, the quantum state and the particles' positions, evolve according to deterministic laws, the quantum state guiding the motion of the particles. Indeed, the time evolution of the complete physical state is composed of two laws (we consider, for simplicity, spinless particles, for which the quantum state is just the wave function):

1. The wave function evolves according to the usual Schrödinger equation.
2. The particle positions $\mathbf{X} = \mathbf{X}(t)$ evolve in time according to a guiding equation determined by the quantum state: their velocity is a function of the wave function. If one writes:[5]

$$\Psi(x_1,...,x_N) = R(x_1,..., x_N)e^{iS(x_1,...,x_N)},$$

then:

$$\frac{dX_k(t)}{dt} = \nabla_k S(X_1(t),...,X_N(t)), k = 1,...,N, \tag{6.1}$$

where $\nabla_k$ is the gradient with respect the coordinates of the $k$-th particle.[6]

While we cannot go into details, we will briefly explain what are measurements of observables in that theory and how quantum predictions are recovered, why this theory is not refuted by the no hidden variables theorems, and how nonlocality manifests itself here.

1. To understand what are 'measurements of observables' in the de Broglie–Bohm theory, we need first to observe that all measurements are eventually

---

[4] For elementary introductions to this theory, see (Bricmont, 2017), (Tumulka, 2004), and for more advanced ones, see (Bell, 2004), (Bohm, 1989), (Bohm and Hiley, 1993), (Bricmont, 2016), (Dürr et al., 2005), (Dürr et al., 2012), (Dürr and Teufel, 2009), (Goldstein, 2013), and (Norsen, 2017). There are also pedagogical videos made by students in Munich, available at: https://cast.itunes.uni-muenchen.de/vod/playlists/URqb5J7RBr.html.

[5] We use lower case letters for the generic arguments of the quantum state and upper case ones for the actual positions of the particles.

[6] Although equation (6.1) was used by de Broglie (Bacciagaluppi and Valentini, 2009) and also by John Bell (Bell, 2004), Bohm rewrote (6.1) as a second-order equation (in time); this made the theory look closer to a Newtonian dynamics, but it also introduced the *quantum potential*, which has caused a certain amount of perplexity and confusion.

measurements of positions: if one measures the spin of a particle, one checks whether it goes up or down in a Stern–Gerlach experiment. If one measures its velocity, one measures the distance travelled in a given amount of time, etc.

So, if one recovers the quantum predictions for measurements of positions, one will automatically recover those predictions for the measurements of all observables.

Since the de Broglie–Bohm theory is deterministic, any probabilistic statement must be justified by statistical assumptions on the initial conditions of the system.

That justification is rather easy, because of a fundamental consequence of the de Broglie–Bohm dynamics, *equivariance*: If the probability density $\rho_{t_0}(\mathbf{x})$ for the initial configuration $\mathbf{X}_{t_0}$ is given by $\rho_{t_0}(\mathbf{x}) = |\Psi(\mathbf{x},t_0)|^2$, then the probability density for the configuration $\mathbf{X}_t$ at any time t is given by

$$\rho_t(\mathbf{x}) = |\Psi(\mathbf{x},t)|^2, \tag{6.2}$$

where $\Psi(\mathbf{x},t)$ is the solution to Schrödinger's equation and the probability density $\rho_t(\mathbf{x})$ is determined by (6.1). Equation (6.2) follows easily from (6.1) and Schrödinger's equation.

Because of equivariance, the quantum predictions for the results of measurements of any quantum observable are obtained if one assumes that the initial density satisfies $\rho_{t_0}(\mathbf{x}) = |\Psi(\mathbf{x}, t_0)|^2$. The assertion that configurational probabilities at any time $t_0$ are given by this 'Born rule' is called the *quantum equilibrium hypothesis*. However, the justification of the quantum equilibrium hypothesis, although it is, in the end, rather natural, is too long to be discussed here (see Dürr *et al.*, 2005).

2. Next, an analysis of measurements in the de Broglie–Bohm theory shows that these 'measurements' do not reveal any intrinsic property of the particles but are rather interactions with an apparatus and that the result of a given measurement will depend not only on the complete state of particle $(\Psi, X)$, but also on the way the apparatus is set up.

For example, one could, in a Stern–Gerlach experiment, obtain opposite results for the measurement of the spin of a particle, starting with exactly the same wave function and the same position of the particle, but with different orientations of the gradient of the magnetic field in the apparatus, see e.g. (Bricmont, Goldstein, and Hemmick, 2019a), Section 6.

Because of this, measurements in the de Broglie–Bohm theory are called *contextual*.[7] The word measurement is quite misleading in the context of the de Broglie–Bohm theory, since they do not measure any property of the system undergoing a

---

[7] For an example of contextuality of the measurements of momentum, see (Bricmont, Goldstein, and Hemmick, 2019b, Section5).

'measurement'. A better expression would be 'interaction with a specific apparatus', whose statistical results are given by the quantum formalism.

3. In the de Broglie–Bohm theory, there are no hidden variables except the particle positions and the no hidden variables theorems do not rule out the introduction of those variables alone. And since, as we just saw, there are no hidden variables for any other observable such as spin, momentum, etc., the de Broglie–Bohm theory is not refuted by Theorem 2.1.
4. The theory is nonlocal because, in equation (6.1), we see that the evolution of each $X_k(t)$ depends on the positions of all the other coordinates $X_i(t)$, $i = 1, 2, ... N$
5. Take $N = 2$; then acting via a potential in Schrödinger's equation, on the quantum state in a neighbourhood of where particle 2 is located, will affect, via (6.1) for $k = 2$ the position of $X_2$, but then also, via the same equation for $k = 1$, the position of $X_1$ no matter how far apart $X_1$ and $X_2$ are.

The merits of the de Broglie–Bohm theory cannot be emphasized strongly enough:

1. It is a clear theory about what goes on in the world, with no particular role assigned to 'observers'. It is not an alternative theory to quantum mechanics, but rather its (natural) completion, by saying what goes on outside of 'laboratories' and how measurements work.
2. Measurements have no particular role either: they are just interactions between a particle and an apparatus, whose statistical results are accounted for through natural assumptions on the initial conditions of quantum systems.
3. The theory is nonlocal, a feature that might have been considered a (grave) defect before Bell's result but has become a quality since then: indeed, Bell (in combination with EPR) shows that *any* empirically adequate theory must be nonlocal.

Unfortunately the ideas of Einstein, EPR, and Bell have been very frequently misunderstood; this, in turn, has led to an enormous lack of appreciation of de Broglie and Bohm's ideas.

# 49.7 MISUNDERSTANDINGS OF EINSTEIN, BELL, AND BOHM

## 49.7.1 Misunderstandings of Einstein and of EPR

One rather common misconception is to think that the goal of EPR was to show that one could measure, say, the spin in one direction at $X$ and in another direction at $Y$ for pairs of particles with the quantum state described at the end of Section 49.4, and therefore *know* the values of the spin of both particles in two different directions.[8] Since the spin along two different axes cannot be measured simultaneously, according to ordinary quantum mechanics, this would mean that the goal of EPR would be to refute that limitation in our ability to perform measurements.

But, as we explained in Section 49.4, the point of the EPR paper was not to claim that one could *measure* quantities that are impossible to measure simultaneously according to quantum mechanics, but rather that, if one can learn something about a physical system by making a measurement on a distant system, then, barring actions at a distance, that 'something' must already be there before the measurement is made on the distant system.

In his 1949 *Reply to criticisms*, Einstein did actually pose the question in the form of a dilemma, implied by the EPR 'paradox':

> [ . . . ] the paradox forces us to relinquish one of the following two assertions:
> (1) the description by means of the $\psi$-function is complete,
> (2) the real states of spatially separated objects are independent of each other.
> (Einstein, 1949), in (Schilpp, 1949), p. 682

This is identical to our previous alternatives: (1) means that there are no hidden variables and (2) means locality. So, for Einstein, either we introduce hidden variables (of the type that were later proven to be impossible by Bell) or we give up locality.

Let us now see how Bohr and Born responded to Einstein.

### 49.7.1.1 *Bohr and Einstein*

The root of the difference between Einstein and Bohr is that Einstein was arguing at the level of ontology, of *what there is*, insisting that a complete description of physical systems *must* go beyond the description given by ordinary quantum mechanics, if the world is local. Bohr, on the other hand, was answering systematically at the level of epistemology, of *what we can know*. Bohr was not answering Einstein, he was simply not listening to his objections.

The 'onslaught' of the EPR paper 'came down upon us as a bolt from the blue', wrote Léon Rosenfeld, who was in Copenhagen at that time and who reported EPR's argument to Bohr. The latter then worked intensely 'week after week' on a reply.[9] There is a widespread belief among physicists that Bohr came up with an adequate answer to the EPR paper (Bohr, 1935). But Bohr's answer is hard to understand.

---

[8] That argument relies on the perfect correlations between spin measurements: measuring the spin in one direction at $X$ immediately tells you what the result of the measurement of the spin in the same direction at $Y$ would have been. And a measurement at $Y$ in another direction would likewise give you the value of the spin in that direction at both $X$ and $Y$. But this argument of course assumes locality, which is false. While it is true that EPR did assume locality and thus could have made that argument, it was not what they aimed at.

[9] See Rosenfeld's commentary in (Wheeler and Zurek, 1983), p. 142.

EPR had written that:

> If, without in any way disturbing a system, we can predict with certainty (i.e., with probability equal to unity) the value of a physical quantity, then there exists an element of physical reality corresponding to this physical quantity.
>
> (Einstein, Podolsky, and Rosen, 1935), reprinted in
> (Wheeler and Zurek, 1983, pp. 138–141)

This again means that if, by performing a measurement at $X$, we can learn something about the physical situation at $Y$, then what we learn must exist before we learn it, since $X$ and $Y$ can be far apart. Here of course, EPR assumed locality.

Bohr replied:

> [...] the wording of the above-mentioned criterion [...] contains an ambiguity as regards the meaning of the expression 'without in any way disturbing a system'. Of course there is in a case like that just considered no question of a mechanical disturbance of the system under investigation during the last critical stage of the measuring procedure. But even at this stage there is essentially the question of *an influence on the very conditions which define the possible types of predictions regarding the future behaviour of the system* [...] their argumentation does not justify their conclusion that quantum mechanical description is essentially incomplete [...] This description may be characterized as a rational utilization of all possibilities of unambiguous interpretation of measurements, compatible with the finite and uncontrollable interaction between the objects and the measuring instruments in the field of quantum theory.
>
> (Bohr, 1935, quoted in Bell, 2004, p. 155; italics in the original)

Bell dissects this passage as follows:

> Indeed I have very little idea what this means. I do not understand in what sense the word 'mechanical' is used, in characterizing the disturbances which Bohr does not contemplate, as distinct from those which he does. I do not know what the italicized passage means — 'an influence on the very conditions [...]'. Could it mean just that different experiments on the first system give different kinds of information about the second? But this was just one of the main points of EPR, who observed that one could learn *either* the position *or* the momentum of the second system.[10] And then I do not understand the final reference to 'uncontrollable interactions between measuring instruments and objects', it seems just to ignore the essential point of EPR that in the absence of action at a distance, only the first system could be supposed disturbed by the first measurement and yet definite predictions become possible for the second system. Is Bohr just rejecting the premise — 'no action at a distance' — rather than refuting the argument?
>
> (Bell, 2004, pp. 155–156; italics in the original)

---

[10]   Here Bell refers to the original EPR argument which used the position and momentum variables, instead of spin variables in different directions as we do here, following (Bohm, 1952) (Note by J.B.).

## 49.7.1.2 *Born and Einstein*

Einstein wrote an article in 1948 and sent it to Born (Einstein (1948); it is reproduced in the correspondence between them, edited by Born in 1971), begging him to read it as if he had just arrived 'as a visitor from Mars'. Einstein hoped that the article would help Born to 'understand my principal motives far better than anything else of mine you know' (Born, 1971, p. 168). In that article, Einstein restated his unchanging criticisms:

> If one asks what, irrespective of quantum mechanics, is characteristic of the world of ideas of physics, one is first of all struck by the following: the concepts of physics relate to a real outside world, that is, ideas are established relating to things such as bodies, fields, etc., which claim a 'real existence' that is independent of the perceiving subject—ideas which, on the other hand, have been brought into as secure a relationship as possible with the sense-data. It is further characteristic of these physical objects that they are thought of as arranged in a spacetime continuum. An essential aspect of this arrangement of things in physics is that they lay claim, at a certain time, to an existence independent of one another, provided these objects 'are situated in different parts of space'. [ ... ]
>
> The following idea characterizes the relative independence of objects far apart in space ($A$ and $B$): external influence on $A$ has no direct influence on $B$.[11]
>
> (Einstein, 1948), reproduced in (Born, 1971), pp. 170–171

Here is Born's reaction to that very same article:

> The root of the difference between Einstein and me was the axiom that events which happens in different places $A$ and $B$ are independent of one another, in the sense that an observation on the states of affairs at $B$ cannot teach us anything about the state of affairs at $A$.    (Born, 1971, p. 176)

As Bell says:

> Misunderstanding could hardly be more complete. Einstein had no difficulty accepting that affairs in different places could be correlated. What he could not accept was that an intervention at one place could *influence*, immediately, affairs at the other.
>
> These references to Born are not meant to diminish one of the towering figures of modern physics. They are meant to illustrate the difficulty of putting aside preconceptions and listening to what is actually being said. They are meant to encourage *you*, dear listener, to listen a little harder.    (Bell, 2004, p. 144; italics in the original)

What Born said was that making an experiment in one place *teaches* us something about what is happening at another place, which is unsurprising. If, in the anthropomorphic example given in Section 49.4, both people had agreed on a common strategy,

---

[11]  The places $A$ and $B$ are what we call $X$ and $Y$ in the rest of this paper (Note by J.B.).

one would learn what Bob would answer to question 1, 2, or 3, by asking Alice that same question. But, and that was Einstein's point, it would mean that the answers were predetermined and, thus, that quantum mechanics was incomplete.

So, it seems that Born did in fact agree with Einstein that quantum mechanics is incomplete, but simply did not understand what Einstein meant by that.

## 49.7.2  Misunderstandings of Bell

As we saw, if one forgets about the EPR argument, Bell's result, can be seen as a 'no hidden variables theorem'. But for Bell, his result, combined with the EPR result was not a 'no hidden variables theorem', but a nonlocality theorem. Bell was perfectly clear about this:

> Let me summarize once again the logic that leads to the impasse. The EPRB correlations[12] are such that the result of the experiment on one side immediately foretells that on the other, whenever the analyzers happen to be parallel. If we do not accept the intervention on one side as a causal influence on the other, we seem obliged to admit that the results on both sides are determined in advance anyway, independently of the intervention on the other side, by signals from the source and by the local magnet setting. But this has implications for non-parallel settings which conflict with those of quantum mechanics. So we *cannot* dismiss intervention on one side as a causal influence on the other.
>
> (Bell, 2004, pp. 149–150; italics in the original)

He was also conscious of the misunderstandings of his results:

> It is important to note that to the limited degree to which *determinism*[13] plays a role in the EPR argument, it is not assumed but *inferred*. What is held sacred is the principle of 'local causality'—or 'no action at a distance'. [ ... ]
>
> It is remarkably difficult to get this point across, that determinism is not a *presupposition* of the analysis.    (Bell, 2004, p. 143; italics in the original)

And he added, unfortunately only in a footnote:

> My own first paper on this subject (*Physics* 1, 195 (1964))[14] starts with a summary of the EPR argument *from locality to* deterministic hidden variables. But the commentators have almost universally reported that it begins with deterministic hidden variables.    (Bell, 2004, p. 157, footnote 10; italics in the original)

[12] Here EPRB means EPR and Bohm, who reformulated the EPR argument in term of spins (Bohm, 1952), which is the formulation used here and in most of the literature (Note by J.B.).

[13] Here, 'determinism' refers to the idea of pre-existing values (Note by J.B.).

[14] (Bell, 1964) (Note by J.B.).

An example of such a commentator is the famous physicist Murray Gell-Mann, Nobel Prize winner and discoverer of the quarks, who wrote:

> Some theoretical work of John Bell revealed that the EPRB experimental setup could be used to distinguish quantum mechanics from hypothetical hidden variable theories [ ... ] After the publication of Bell's work, various teams of experimental physicists carried out the EPRB experiment. The result was eagerly awaited, although virtually all physicists were betting on the corrections of quantum mechanics, which was, in fact, vindicated by the outcome.   (Gell-Mann, 1994, p. 172)

So Gell-Mann opposes hidden variables theories to quantum mechanics, but the only hidden variables that Bell considered were precisely those that were needed, because of the EPR argument, in order to 'save' locality. So if there is a contradiction between the existence of those hidden variables and experiments, it is not just quantum mechanics that is vindicated, but locality that is refuted.

Eugene Wigner, another Nobel Prize winner, also saw Bell's result solely as a no hidden variables result: 'In my opinion, the most convincing argument against the theory of hidden variables was presented by J.S. Bell' (Wigner, 1983, p. 291).[15]

Basically the same mistake was made in 1999 by one of the most famous physicists of our time, Stephen Hawking:

> Einstein's view was what would now be called a hidden variables theory. Hidden variables theories might seem to be the most obvious way to incorporate the Uncertainty Principle into physics.[16] They form the basis of the mental picture of the universe, held by many scientists, and almost all philosophers of science. But these hidden variable theories are wrong. The British physicist, John Bell, who died recently, devised an experimental test that would distinguish hidden variable theories. When the experiment was carried out carefully, the results were inconsistent with hidden variables.   (Hawking, 1999)

The misconceptions about Bell continue: in an article about a Symposium held in 2000 in Vienna, in honor of the 10th anniversary of John Bell's death, the physicist Gerhard Grössing wrote:

> In the 11 August 2000 issue of 'Science', D. Kleppner and R. Jackiw of MIT published a review article on 'One Hundred Years of Quantum Physics' [Kleppner and Jackiw, 2000]. They discuss Bell's inequalities and also briefly mention the possibility of 'hidden variables'. However, according to the authors, the corresponding experiments had shown the following: 'Their collective data came down

---

[15] See (Goldstein, 1996) for a further discussion of Wigner's views.

[16] By this, Hawking means that hidden variables would be a way to maintain determinism, despite the fact that precise measurements of both position and velocities are prohibited by Heisenberg's uncertainty principle (Note by J.B.).

decisively against the possibility of hidden variables. For most scientists this
resolved any doubt about the validity of quantum mechanics.'   (Grössing, 2000)

In the February 2001 issue of *Scientific American*, Max Tegmark and John Archibald
Wheeler published an article entitled '100 Years of the Quantum' (quoted by Grössing,
2000), where one reads:

> Could the apparent quantum randomness be replaced by some kind of unknown
> quantity carried out inside particles, so-called 'hidden variables'? CERN theorist
> John Bell showed that in this case, quantities that could be measured in certain
> difficult experiments would inevitably disagree with standard quantum predictions.
> After many years, technology allowed researchers to conduct these experiments and
> eliminate hidden variables as a possibility.   (Tegmark and Wheeler, 2001, p. 76)

More recently, the Nobel Prize winner Frank Wilczek wrote:

> In 1964, the physicist John Bell identified certain constraints that must apply to any
> physical theory that is both local—meaning that physical influences don't travel
> faster than light—and realistic, meaning that the physical properties of a system
> exist prior to measurement. But decades of experimental tests, [ ... ] show that the
> world we live in evades those constraints.   (Wilczek, 2015)

All the quotes above commit the same mistake: they ignore the fact that Bell's result,
combined with the EPR argument, refutes locality not merely the existence of 'hidden
variables'.

David Mermin summarized the situation described in this section in an amusing way:

> Contemporary physicists come in two varieties.
>
> Type 1 physicists are bothered by EPR and Bell's theorem.
>
> Type 2 (the majority) are not, but one has to distinguish two subvarieties.
>
> Type 2a physicists explain why they are not bothered. Their explanations tend
> either to miss the point entirely (like Born's to Einstein)[17] or to contain physical
> assertions that can be shown to be false.
>
> Type 2b are not bothered and refuse to explain why. Their position is unassailable.
> (There is a variant of type 2b who say that Bohr straightened out the whole
> business, but refuse to explain how.)   (Mermin, 1993)

To conclude, let us highlight some more positive reactions to Bell's result. Henry
Stapp, a particle theorist at Berkeley, wrote that 'Bell's theorem is the most profound
discovery of science' (Stapp, 1975, p. 271). And David Mermin mentions an unnamed

---

[17]  See Section 49.7.1.2 (Note by J.B.).

'distinguished Princeton physicist' who told him: 'Anybody who's not bothered by Bell's theorem has to have rocks in his head' (Mermin, 1993).

## 49.7.3  Lack of Appreciation for the de Broglie–Bohm Theory

Given the almost universal misunderstanding of the EPR–Bell result as refuting hidden variables theories or maybe only local hidden variables theories (but suggesting that nonlocal ones are too crazy to consider), it is not surprising that most people did not appreciate de Broglie and Bohm's ideas.

### 49.7.3.1  *Reactions to de Broglie*

In his 1924–1927 notes to the French *Académie des Sciences* (de Broglie, 1924b, 1926, 1927a), and in his Ph.D. thesis (de Broglie, 1924a), which pre-date all other major works on quantum physics, de Broglie not only associated waves with particles, but also introduced the idea that the motion of particles may be guided by waves. The French physicist Paul Langevin sent a copy of de Broglie's thesis to Einstein, who replied that de Broglie had 'lifted a corner of the great veil'[18] and Einstein circulated the thesis in Germany.

Near the end of his 1927 paper published in *Le Journal de Physique et le Radium* (de Broglie, 1927b), de Broglie wrote:

> [...] one will assume the existence, as distinct realities, of the material point and of the continuous wave represented by the function, and one will take it as a postulate that the motion of the point is determined as a function of the phase of the wave [...]. One then conceives the continuous wave as guiding the motion of the particle. It is a pilot wave.
>
> (de Broglie, 1927b, quoted in Bacciagaluppi and Valentini, 2009, p. 65,
> and reprinted in de Broglie, 1953, p. 52)

De Broglie was invited to present a report at the 5th Solvay conference in Brussels, in October 1927. There, he presented his 'pilot wave' idea, for a non-relativistic system of $N$ particles guided by a wave function in configuration space that determines the particle velocities according to the law of motion (6.1).

Moreover, de Broglie did derive the Born rule by proving the equivariance formula, which implies the Born rule for probabilities, as we explained briefly in Section 49.6.

As de Broglie said in his report to the Solvay Conference, in order to justify the introduction of real particles, 'it seems a little paradoxical to construct a configuration space with the coordinates of points that do not exist' (Bacciagaluppi and Valentini, 2009, p. 346).

---

[18] Letter from Einstein to Paul Langevin, 16 December 1924, Einstein Archives, Jerusalem, 15–376.

But the general reaction at the Solvay Conference was negative. In particular, Pauli raised an objection having to do with the account, within the de Broglie theory, of certain scattering phenomena that de Broglie did not answer satisfactorily. A complete answer to Pauli's objection was given in 1952 by Bohm, who should be credited for explaining how the de Broglie–Bohm theory accounts for all the quantum mechanical predictions.

De Broglie left the Solvay Conference discouraged. Not only because of Pauli's criticisms, or because of the generally negative reactions, but also because he could not take the wave function *defined on configuration space* seriously as a physical quantity, which is essential to understanding the de Broglie–Bohm theory. Already in his 1927 paper, written before the Solvay Conference (de Broglie, 1927b), de Broglie had said:

> Physically, there can be no question of a propagation in a configuration space whose existence is purely abstract: the wave picture of our system must include $N$ waves propagating in real space and not a single wave propagating in the config-uration space.    (de Broglie, 1953, p. 48)

For de Broglie, the wave function had a purely probabilistic meaning, hence a subjective and non-physical meaning, and only the existence of another solution, or rather $N$ solutions for $N$ particles, more or less localized in space, would have given a meaning to the whole scheme. Since he could not solve the mathematical problems linked with these other solutions, he gave up his alternative and followed the Copenhagen orthodoxy for the next 25 years. That is, until he received a manuscript from a 'young American physicist', David Bohm. He reacted by, of course, mentioning his priority, but also by restating his old objections towards his own theory (de Broglie, 1953, p. 67).

De Broglie stressed the obvious nonlocality of his and Bohm's theory, about which he wrote:

> If the wave $\Psi$ is a 'physical reality', how can one understand that a measure done in one region of space could affect this physical reality in other regions of space that can be very far from the first one?    (de Broglie, 1953, pp. 67–68)

Bell did put his finger on the root of the problem, but of course much later:

> *No one can understand this theory [the de Broglie–Bohm theory] until he is willing to think of $\psi$ as a real objective field rather than a 'probability amplitude'. Even though it propagates not in 3-space but in 3N-space.*
> (Bell, 2004, p. 128; italics in the original)

And this is something that nobody before Bohm was willing to think. De Broglie is a tragic figure in the history of physics. He was undoubtedly a precursor and he did influence his successors in an indirect way. However, his 1927 theory is to a large extent forgotten, and he has been rather neglected in his own country, despite his Nobel Prize and his having been the 'perpetual secretary' for the mathematical and physical sciences section of the French *Académie des Sciences* from 1942 until 1975.

It should be emphasized that de Broglie's theory had nothing to do with 'solving the measurement problem', unlike almost all the subsequent discussions about quantum mechanics. It was meant to be a physical theory, albeit a provisional one. Of course, de Broglie did not fully understand his own theory, and neither did he really believe in it. But who can blame him? De Broglie wrote 25 years before Bohm and 37 years before Bell and nobody could accept then the most revolutionary feature of the de Broglie theory, namely its nonlocality.

The tragedy of de Broglie is that he did 'lift a corner of the great veil' and saw far beyond anybody else at the time; so far, indeed, that he could not see clearly himself.

### 49.7.3.2  *Reactions to Bohm*

In 1952, David Bohm published two papers in *Physical Review* which give in complete detail what we call the de Broglie–Bohm theory (Bohm, 1989).

Let us now consider the reactions to Bohm's theory of some of the most famous people associated with the Copenhagen interpretation, Pauli and Heisenberg. In 1951, Pauli wrote to Bohm:

> I just received your long letter of 20 November, and I also have studied more thoroughly the details of your paper. I do not see any longer the possibility of any logical contradiction as long as your results agree completely with those of the usual wave mechanics and as long as no means is given to measure the values of your hidden parameters both in the measuring apparatus and in the observe [sic] system. As far as the whole matter stands now, your 'extra wave-mechanical predictions' are still a check, which cannot be cashed.[19]

But he also wrote, in 1951, to his friend Markus Fierz (Pauli, 1951, quoted in Peat, 1997, p. 128): 'Bohm keeps writing me letters "such as might have come from a sectarian cleric trying to convert me particularly to de Broglie's old theory of the pilot wave". In the last analysis Bohm's whole approach is "foolish simplicity", which "is of course beyond all help".'

Pauli's reaction misses the point of the de Broglie–Bohm theory, namely to complete ordinary quantum mechanics into a real physical theory, saying what happens in the world outside of laboratories instead of being simply a recipe for predicting 'results of measurements', which is what ordinary quantum mechanics is.

What about Heisenberg's reaction? He wrote a rather detailed defence of the orthodoxy in *Physics and Philosophy* (Heisenberg, 1959), where he discusses Bohm's ideas. One of Heisenberg's objections was based on his belief that the introduction of hidden variables would destroy interference effects. But that would only be true if those hidden variables would in effect collapse the quantum state, whereas, in the de Broglie–Bohm theory, they do nothing of the sort and only complete the usual quantum

---

[19]  Letter from Wolfgang Pauli to David Bohm, 3 December 1951, quoted in (Peat, 1997)., p. 127.

description. Another objection that Heisenberg raised against Bohm was that the wave function could not be 'real', because it was defined on configuration space (Heisenberg, 1959, p. 116). But of course, if even the wave function is not real, while supposedly constituting a 'complete description' of physical systems, then nothing is real, and that surely cannot be what Heisenberg intended.

Heisenberg also falls back on the familiar idea that Bohm's theory is just like the Copenhagen interpretation, but expressed in a different language, which 'from a strictly positivistic standpoint' amounts to an 'exact repetition' and not a counterproposal to the Copenhagen interpretation (Heisenberg, 1959, p. 115). This does not square with the fact that Heisenberg was also arguing that quantum mechanics had put an end to the 'ontology of materialism', namely 'to the idea of an objective real world whose smallest parts exist objectively in the same sense as stones or trees exist, independently of whether or not we observe them' (Heisenberg, 1959, p. 115). One cannot have it both ways: saying that Bohm's theory reintroduces an 'objective ontology' and, at the same time, saying that it is just the same theory (expressed in a different language) as one that denies the possibility of such an ontology.

Heisenberg's objections may be contradictory, but he certainly expressed his dismissal of Bohm's idea in a contemptuous way. Referring without citation to Bohr, he said:

> When such strange hopes[20] were expressed, Bohr used to say that they were similar in structure to the sentence 'We may hope that it will later turn out that sometimes $2 \times 2 = 5$, for this would be of great advantage for our finances'.
> (Heisenberg, 1959, p. 117)

However, the most vehement critique of Bohm was probably the Belgian physicist Léon Rosenfeld, who wrote to Bohm:

> I shall certainly not enter into any controversy with you or anybody else on the subject of complementarity, for the simple reason that there is not the slightest controversial point about it. [ ... ] It is just because we have undergone this process of purification through error that we feel so sure of our results. [ ... ] there is no truth in your suspicion that we may just be talking ourselves into complementarity by a kind of magical incantation. I am inclined to return that it is just among your Parisian admirers that I notice some disquieting signs of primitive mentality.[21]
> (Léon Rosenfeld, letter to Bohm, 20 May 1952, quoted in   (Peat, 1997), p. 130)

While Bohm was marginalized, partly for political reasons (see Freire (2015); Olwell (1999)), but also because of his heretical views on quantum mechanics, Rosenfeld was

---

[20] Here, Heisenberg attributes to Bohm the hope that in the future the quantum theory may be proved false, but this was not the main point raised by Bohm, who stressed that his theory provided at least a coherent and empirically adequate alternative to the orthodoxy (Note by J.B.).

[21] He was probably referring to people around de Broglie, such as Jean-Pierre Vigier (Note by J.B.).

in a position of power, which, according to the historian of science A. S. Jacobsen, he did not always use very generously with his opponents:

> In addition, [Rosenfeld] served as consultant or referee in matters of epistemology of physics and the like at several well-reputed publishing houses and at the influential journal *Nature*. In this capacity he used his influence effectively, and several books and papers, among them some by Frenkel,[22] Bohm, and de Broglie, were rejected on this account.    (Jacobsen, 2007, p. 23)

## 49.8 CONCLUSIONS

The lesson of this sad history is that we are unlikely to understand the real issues raised by quantum mechanics, whether it be the role of the observer or nonlocality, if we continue to adhere to a history of quantum mechanics written by the victors (at least until now) and that systematically misrepresents the views of Einstein, Bell, de Broglie, and Bohm. Reading those authors without prejudice would already constitute a big step in the right direction.

## REFERENCES

Bacciagaluppi, G., and Valentini, A. (2009). *Quantum Mechanics at the Crossroads. Reconsidering the 1927 Solvay Conference*. Cambridge: Cambridge University Press.

Bell, J. S. (1964). On the Einstein–Podolsky–Rosen paradox. *Physics* 1, 195–200. Reprinted as Chap. 2 in Bell (2004).

Bell, J. S. (1966). On the problem of hidden variables in quantum mechanics. *Reviews of Modern Physics*, 38, 447–52. Reprinted as Chap. 1 in Bell (2004).

Bell, J. S. (1976). The theory of local beables. *Epistemological Letters*, March 1976. Reprinted as Chap. 7 in Bell (2004).

Bell, J. S. (2004). *Speakable and Unspeakable in Quantum Mechanics. Collected Papers on Quantum Philosophy*, 2nd edn, with an introduction by Alain Aspect. (1st edn 1993.) Cambridge: Cambridge University Press.

Bohm, D. (1952). A suggested interpretation of the quantum theory in terms of 'hidden variables', Parts 1 and 2. *Physical Review*, 89, 166–93. Reprinted in Wheeler and Zurek (1983), pp. 369–390.

Bohm, D. (1989). *Quantum Theory*, new edition. New York: Dover Publications. First edition: 1951, Englewood Cliffs, NJ: Prentice Hall.

Bohm, D., and Hiley, B. J. (1993). *The Undivided Universe*. London: Routledge.

---

[22] Iakov Frenkel was a Soviet physicist who had submitted a paper critical of complementarity to *Nature* that Rosenfeld managed to have withdrawn by the translator, Pauline Yates (Freire, 2019, p. 38) (Note by J.B.).

Bohr, N. (1935). Can quantum mechanical description of reality be considered complete? *Physical Review*, 48, 696–702.

Born, M. (ed.) (1971). *The Born–Einstein Letters*. London: Macmillan.

Bricmont, J. (2016). *Making Sense of Quantum Mechanics*. Berlin: Springer.

Bricmont, J. (2017). *Quantum Sense and Nonsense*. Cham, Switzerland: Springer International Publishing.

Bricmont, J., Goldstein, S., and Hemmick, D. (2019a). EPR–Bell–Schrödinger proof of non-locality using position and momentum. arXiv:1906.06687

Bricmont, J., Goldstein, S., and Hemmick, D. (2019b). Schrödinger's paradox and proofs of nonlocality using only perfect correlations. *Journal of Statistical Physics*. https://doi.org/ 10.1007/s10955-019-02361-w

de Broglie, L. (1924a). *Recherches sur la théorie des quanta*. Ph.D. Thesis, Université de Paris.

de Broglie, L. (1924b). Sur la dynamique du quantum de lumière et les interférences. *C. R. Acad. Sci. Paris*, 179, 1039–41. Reprinted in de Broglie (1953), pp. 23–25.

de Broglie, L. (1926). Sur la possibilité de relier les phénomènes d'interférence et de diffraction à la théorie des quantas de lumières. *C. R. Acad. Sci. Paris*, 183, 447. Reprinted in de Broglie (1953), pp. 25–27.

de Broglie, L. (1927a). La structure atomique de la matière et du rayonnement en la mécanique ondulatoire. *C. R. Acad. Sci. Paris*, 184, 273–74. Reprinted in de Broglie (1953), pp. 27–29.

de Broglie, L. (1927b). La mécanique ondulatoire et la structure atomique de la matière et du rayonnement. *Le Journal de Physique et le Radium*, 8, 225–41. Reprinted, with additional comments, in de Broglie (1953), pp. 29–61.

de Broglie L. (1953). *La physique quantique restera-t-elle indéterministe?* Paris: Gauthier-Villars.

Dukas, H., and Hoffmann, B. (eds) (1979). *Albert Einstein: The Human Side*. Princeton, NJ: Princeton University Press.

Dürr, D., Goldstein, S., Tumulka, R., and Zanghì, N. (2005). John Bell and Bell's theorem. In D. M. Borchert (ed.), *Encyclopedia of Philosophy*, New York: Macmillan Reference.

Dürr, D., Goldstein, S., and Zanghì, N. (1992). Quantum equilibrium and the origin of absolute uncertainty. *Journal of Statistical Physics*, 67, 843–907.

Dürr, D., Goldstein, S., and Zanghì, N. (2012). *Quantum Physics Without Quantum Philosophy*. Berlin, Heidelberg: Springer.

Dürr, D., and Teufel, S. (2009). *Bohmian Mechanics. The Physics and Mathematics of Quantum Theory*. Berlin, Heidelberg: Springer.

Eberhard, P. H. (1978). Bell's theorem and the different concepts of locality. *Il Nuovo Cimento*, 46B, 392–419.

Einstein, A. (1948). Quantum mechanics and reality. *Dialectica*, 2, 320–24.

Einstein, A. (1949). Remarks concerning the essays brought together in this co-operative volume. In P. A. Schilpp (ed.), *Albert Einstein, Philosopher-Scientist*, Evanston, IL: The Library of Living Philosophers, pp. 665–88.

Einstein, A., Podolsky, B., and Rosen, N. (1935). Can quantum mechanical description of reality be considered complete? *Physical Review*, 47, 777–80.

Freire O. (2015). *The Quantum Dissidents—Rebuilding the Foundations of Quantum Mechanics (1950–1990)*. Berlin, Heidelberg: Springer-Verlag.

Freire, O. (2019). *David Bohm. A Life Dedicated to Understanding the Quantum World*. Cham, Switzerland: Springer Nature.

Gell-Mann, M. (1994). *The Quark and the Jaguar*. London: Little, Brown and Co.

Ghirardi, G. C., Rimini, A., and Weber, T. (1980). A general argument against superluminal transmission through the quantum mechanical measurement process. *Lettere Al Nuovo Cimento*, **27**, 293–8.

Goldstein, S. (2013). Bohmian mechanics. In E. N. Zalta (ed.), *The Stanford Encyclopedia of Philosophy* (spring 2013 edition). plato.stanford.edu/archives/spr2013/entries/qm-bohm/

Goldstein, S. (1996). Quantum philosophy: the flight from reason in science. In P. Gross, N. Levitt, and M.W. Lewis (eds), *The Flight from Science and Reason, Annals of the New York Academy of Sciences* **775**, 119–25. Reprinted in Dürr *et al.* (2005), Chapter 4.

Grössing, G. (2000). Serious Matter: The John Bell Scandal. Available from https://nonlinearstudies.at/49/a-serious-matter-the-john-bell-scandal/

Hawking, S. (1999). Does God play dice? Available online: www.hawking.org.uk/doesgod-play-dice.html

Heisenberg, W. (1959). *Physics and Philosophy. The Revolution in Modern Science*. London: Allen and Unwin. First edition, 1958, New York: Harper and Row.,

Hemmick, D. L., and Shakur, A. M. (2012). *Bell's theorem and quantum realism. Reassessment in light of the Schrödinger paradox*. Heidelberg: Springer.

Jacobsen, A. S. (2007). Léon Rosenfeld's Marxist defense of complementarity. *Historical Studies in the Physical and Biological Sciences*, **37**, Supplement, 3–34.

Kleppner, D., and Jackiw, R. (2000). One hundred years of quantum physics, *Science*, **289**, 893–8.

Kochen, S., and Specker, E. P. (1967). The problem of hidden variables in quantum mechanics. *Journal of Mathematics and Mechanics*, **17**, 59–87.

Maudlin, T. (2011). *Quantum Nonlocality and Relativity*, 3rd edn. Cambridge: Blackwell. 1st edn, 1994.

Mermin, D. (1993). Hidden variables and the two theorems of John Bell. *Reviews of Modern Physics*, **65**, 803–15.

Mermin D. (1985). Is the moon there when nobody looks? Reality and the quantum theory. *Physics Today*, **38**, 38–47.

Norsen, T. (2017). *Foundations of Quantum Mechanics: An Exploration of the Physical Meaning of Quantum Theory*. Cham, Switzerland: Springer International Publishing.

Olwell, R. (1999). Physical isolation and marginalization in physics—David Bohm's Cold War exile. *Isis*, **90**, 738–56.

Pauli, W (1951). Letter to Markus Fierz. In W. Pauli, *Wissenschaftlicher Briefwechsel mit Bohr, Einstein, Heisenberg u.a./Scientific Correspondence with Bohr, Einstein, Heisenberg a.o.: Band/Volume IV. History of Mathematics and Physical Sciences*, K. von Meyenn (ed.), Springer-Verlag, Berlin, 2005, pp. 131–32.

Peat, D. (1997). *Infinite Potential: The Life and Times of David Bohm*. New York:Basic Books.

Peres A. (1990). Incompatible results of quantum measurements. *Physics Letters A*, **151**, 107–8.

Peres A. (1991). Two simple proofs of the Kochen–Specker theorem. *Journal of Physics A: Math. Gen.*, **24**, L175–L178.

Schilpp, P. A. (ed.) (1949). *Albert Einstein, Philosopher-Scientist*. Evanston, IL: The Library of Living Philosophers.

Stapp, H. P. (1975). Bell's theorem and world process. *Il Nuovo Cimento*, **29B**, 270.

Tegmark, M., and Wheeler, J. A. (2001). 100 years of quantum mysteries. *Scientific American*, February 2001, 72–9.

Tumulka R. (2004). Understanding Bohmian mechanics—A dialogue. *American Journal of Physics*, 72, 1220–26.

Wheeler, J. A., and Zurek, W. H. (eds) (1983). *Quantum Theory and Measurement*. Princeton, NJ: Princeton University Press.

Wigner, E. P. (1983). Interpretation of quantum mechanics. In J. A. Wheeler and W. H. Zurek (eds), Quantum Theory and Measurement, Princeton, NJ: Princeton University Press, pp. 260–314.

Wilczek, F. (2015). Einstein's Parable of Quantum Insanity, *Quanta Magazine*, 23 September.

CHAPTER 50

....................................................................................................

# THE STATISTICAL (ENSEMBLE) INTERPRETATION OF QUANTUM MECHANICS

....................................................................................................

## ALEXANDER PECHENKIN

## 50.1 DEFINITIONS

....................................................................................................

THE standard empirical interpretation of quantum mechanics is already statistical (Michael Redhead (1987) referred to it as the 'minimal instrumentalist interpretation'). To establish a connection between the mathematical scheme of quantum mechanics and the empirical data, the wave function of a system (for example, an electron) is represented as a linear superposition of the eigenfunctions of some dynamic variables (either a spatial coordinate, or momentum, energy, spin, etc.). Square modules of the coefficients in this superposition then give the probabilities that the dynamic variable will have one or another eigenvalue. This example illustrates the simple case of a variable with a discrete spectrum.

The so-called 'statistical interpretation of quantum mechanics' (the name the 'ensemble interpretation' is also often used, which can help avoid a terminological confusion) is related to the empirical interpretation but goes further and represents a semantic understanding and interpretation of the minimal procedure described above. It claims that we can actually understand the standard statistical procedure, if 'the system' is meant not as an individual object, for example a single electron, but as a large ensemble of similarly prepared ones: 'the conceptual (infinite) set of single electrons which have been subjected to same state preparation technique generally by interaction with suitable apparatus . . . A momentum eigenstate represents the ensemble whose members are single electrons each having the same momentum, but distributed uniformly over all positions' (Ballentine, 1970, p. 361). Coefficients in the superposition of eigenfunctions can be interpreted therefore simply and straightforwardly as proportional numbers of electrons in the ensemble that belongs to one or another eigenstate.

There is a distinction between the ensemble interpretation of quantum mechanics and the Copenhagen interpretation about their reconstruction of the physical world. In explaining quantum mechanics the Copenhagen type of interpretation refers to thought experiments with a single physical system, whereas the ensemble interpretation basically uses the image of statistical collectives. The ensemble interpretation thus disagrees with the orthodox, Copenhagen interpretation, which insists that the super-position represents a single electron or object that could turn out in one or another eigenstate as a result of an (uncontrollable, unpredictable, and, non-transparent) interaction with the measuring apparatus.

The ensemble interpretation agrees that the standard quantum mechanical description imposes certain limits on what is known about the system: it cannot describe independently each single electron in the ensemble, but only properties of their entire collective. The interpretation leaves open, however, the question of whether another theory, fuller and deeper than quantum mechanics, may be developed someday, which would be capable of describing the properties and behaviour of individual electrons within the ensemble. Only in the quantum case have we already discovered quantum mechanics as a replacement of the classical statistical mechanics, but we do not know yet whether we will be able to invent the appropriate quantum analogue of the Newtonian dynamics.

The concept of complementarity plays a crucial role in the Copenhagen interpretation. Since an experiment is always an interaction between a quantum system and a measuring device, the quantum object does not exist independently from the experimental situation. In different situations, different, complementary results are obtained: while measuring the spatial coordinate, one loses information about the momentum of the system, and vice versa. Some proponents of the ensemble approach may also apply the idea of complementarity in their explanations of quantum phenomena, yet the idea plays an auxiliary, illustrative role and is not as essential for the ensemble approach as it is for the Copenhagen philosophy.

# 50.2 EARLY HISTORY

## 50.2.1 Karl Popper's Interpretation of Quantum Mechanics

In his classic book on the history of the philosophy of quantum mechanics, Max Jammer described the philosopher Karl Popper's argument with the Copenhagen philosophy as a prehistory of the ensemble interpretation. Calling Popper 'one of the great humane thinkers of our time', Jammer carefully analysed his critique of Heisenberg's 'subjective interpretation, which reads "The more precisely the position of a particle is measured the less is known about its momentum and vice versa"' (Jammer,

1974, pp. 174–176). Instead, Popper advocated a statistical objective interpretation: he 'always held that the problems to which quantum mechanics is applied are essentially statistical problems and as such require statistical answers. In fact, according to Popper, the vectors in Hilbert space provide statistical assertions from which no predictive inferences can be drawn about the behaviour of individual particles' (Jammer, 1974, p. 448). Jammer emphasized that Popper was influenced by Einstein's presentation at the Fifth Solvay Congress (1927): 'According to *viewpoint I*, Einstein declared, the de Broglie–Schrödinger waves do not represent one individual particle but rather an ensemble of particles distributed in space . . . According to *viewpoint II* quantum mechanics is considered as a complete theory of individual processes; each particle moving toward the screen is described as a wave packet which, after diffraction, arrives at a certain point $P$ on the screen, and $|\psi(r)|^2$ expresses the probability (probability density) that at a given moment one and the same particle shows its presence at $r$' (Jammer, 1974, pp. 115-6).

Popper published his own discussion of quantum mechanics seven years later, in his 1934 brief paper in *Naturwissenschaften* and in the 1935 book *Logik der Forschung*. The main target of Popper's criticism was Bohr's complementarity principle and the interpretation of the uncertainty principle as presented in Heisenberg's 'Chicago Lectures' (Heisenberg, 1930). Popper emphasized that while the uncertainty relations per se can be considered a mathematical theorem, one of the unavoidable conclusions of quantum theory, the same did not apply to Heisenberg's interpretation of uncertainty. 'Popper thus advocated what he called the *statistical objective interpretation*. Given an ensemble . . . of particles from which, at a certain moment and with a given precision $\Delta x$ those having a certain position $x$ are selected; the momenta $p$ of the latter will then show a random scattering with a range of scatter $\Delta p$ where $\Delta x \Delta p \geq h$, and vice versa' (Jammer, 1974, p. 176).

Popper called the uncertainty relations the 'statistical scatter relations'. The term 'uncertainty principle' emphasized a limitation of our knowledge and is connected with the 'subjective' interpretation of quantum mechanics. "Our statistical scatter relations comes to this, Popper writes. If one tries . . . to obtain as homogeneous an aggregate as possible, then this attempt will ancounter a definite barrier in these scatter relations. For example, we can obtain by means of physical selection a plane monochromatic ray—a ray of electrons of equal momentum. But if we attempt to make this aggregate of electrons still more homogeneous . . . so as to obtain electrons which not only have the same momentum, but have also passed through some narrow slit determining a positional range $\nabla x$, then we are bound to fall" (Popper, 1959, pp. 226–227, reprinted in 1992, 1995).

Later, the German-American applied physicist and philosopher Henry (Hans) Margenau, who also developed the ensemble approach to quantum mechanics, called Heisenberg's uncertainty principle 'the statistical dispersion principle' (Margenau, 1963). This terminology was also accepted by Leslie Ballentine, one the most influential proponents of the ensemble approach to quantum mechanics in the late 20th and early 21st centuries.

## 50.2.2 Ensemble Approach in the USA

As a young postdoc who had collaborated with Bohr and Kramers on the famous Bohr–Kramers–Slater theory of quantum radiation and subsequently made important contributions to the theory of multi-electron systems, John Clarke Slater put forward an ensemble conception of quantum mechanics in a paper delivered at the 1928 Symposium of the American Physical Society. His work even preceded that of Popper (Whitaker, 1996; Pechenkin, 2012). Slater did not write his paper as a polemical piece against the Copenhagen interpretation, but simply elucidated what he regarded as the correct physical meaning of quantum theory, 'the physics' which lay behind its mathematical apparatus. He stressed that quantum mechanics 'operates with ensembles . . . Just as in ordinary statistical mechanics, we must here choose an ensemble, by considering the sort of statistical distributions actually present in the repetitions of the experiment being performed' (Slater, 1929, p. 453). In his recollections he provides the following explanation: 'An ensemble . . . represents a collection of many repetitions of the same experiment, agreeing as concern the large scale or macroscopic properties which we can control, but taking different values of microscopic properties which are on such a small scale that we cannot experimentally determine or control them. It does not necessarily imply a system with a great many particles in it. The essence of the ensemble is the large number of repetitions of the experiment . . . The probability of finding certain coordinates in certain ranges . . . means simply the fraction of all systems in the ensemble which lie within the prescribed limits' (Slater, 1975, p. 44).

After having published his 1929 paper, Slater did not publish any more on the foundations of quantum mechanics. In some of his books on the theory of multi-electron systems, for example (Slater, 1960, p. 438), he advised readers to study the principles of quantum theory with the 1937 book by Edwin C. Kemble.

Kemble, who as a university teacher had earlier introduced Slater to the new quantum mechanics, also proposed the statistical conception as his preferred interpretation in 1935. He was encouraged to do so by the Einstein–Podolsky–Rosen argument of 1935 and, in turn, encouraged the Harvard Professor Wendell Furry's favourable remarks on the ensemble approach in 1936. Kemble generally followed Slater's argument, but he used a more refined terminology. He spoke of an ensemble as consisting of 'the preparation of state and measurement', meaning the 'Gibbsian assemblage of identical systems so prepared that the past histories of all its members are the same in all details that can affect the future behaviour of that of the original system . . . Ideally the predictions of quantum mechanics should be tested by a series of observations on a suitably prepared assemblage of completely independent systems each in its own separate box or laboratory' (Kemble, 1937, pp. 54–5).

The most comprehensive presentation of Kemble's philosophical interpretation is presented in his 1937 textbook. Like Slater, Kemble did not explicitly contradict the Copenhagen interpretation, but he certainly must have understood the differences.

Kemble distinguished between subjective and objective (or true) quantum states. The subjective state is the state of a single system that is not independent from the observer. Any measurement on this system involves an interaction between it and the measuring device, which inevitably modifies the future behaviour. The objective state is a Gibbsian ensemble, and the wave function that represents it can predict the statistical results of any future experiment on the ensemble. 'The objective state of an individual system is operationally undefined', he concluded (Kemble, 1937, p. 52).

The above-mentioned physicist and philosopher Margenau started to publish philosophical papers on the ensemble interpretation also in 1936. Unlike Slater and Kemble, he did criticize the 'subjectivist' point of view that the wave function 'describes our knowledge' of the system but not the system itself, and is 'as a conventional carrier, a symbol for the sum total of all that knowledge which the speculative observer can possibly accumulate'. Although the subjectivist point of view is consistent with positivism, to which Margenau was sympathetic, it still leads, according to Margenau, to 'converting physics into a highly expressionistic type of psychology' (Margenau, 1937, pp. 346–349).

Margenau's rejection of subjectivism did not mean, however, that for him the wave function represented reality. He viewed the belief in objective reality as psychological prejudice of the same kind as the belief that the wave function is a description of our knowledge. The real sense of the wave function is revealed in the ensemble interpretation: 'When a given state is realized . . . the experimental correlate of the state is a number of probability aggregates, one for every observable, with definitely assignable properties and probabilities. Experimental indeterminacy as an outcome of a given measurement is inherent in the fundamental manner of describing states' (Lindsay and Margenau, 1936, p. 421). He wrote that 'quantum mechanical state is synonymous with probability distribution' (Margenau, 1936, p. 242). Whereas the subjectivist interpretation of the wave function relied on the subjectivist definition of probability, Margenau formulated his objective understanding of probability with the help of Richard von Mises' definition of probability as empirical frequency.

## 50.2.3  Ensemble Approach in the USSR

Konstantin V. Nikolsky, the first Soviet proponent of the ensemble approach, worked at Lebedev Physical Institute of the Academy of Sciences in Moscow. According to existing recollections and archival documents, his research on the foundations of quantum theory was supported by the institute's director Sergei I. Vavilov (who after World War II also became the President of the USSR Academy of Sciences i.e. he reached the highest position in the Soviet hierarchy of scientists). However, his immediate inspiration came from the 1936 debate between Einstein and Bohr regarding the description of physical reality. The exchange was translated into Russian and followed with great attention by Soviet physicists. Einstein, like several other physicists on various occasions, mentioned that the EPR paradox could be resolved by the assumption that 'the ψ function does not, in any sense, describe the condition of *one single system* . . . it relates rather to many systems, to "an ensemble of systems" in the sense of statistical mechanics' (Einstein, 1936, p. 375; Ballentine, 1972). Einstein did not follow through with his suggestion, however; he instead took it to signify that quantum mechanics was incomplete, and he continued to look for a better, fuller, description at the individual level. Nikolsky took up the idea much more seriously and developed it as a way to sort out systematically the difficulties in the interpretation of quantum mechanics.

Quantum mechanics to him was a non-classical statistical theory. He formulated its laws as describing ensembles of particles and analysed mathematically the differences between the statistical description in the classical and the quantum version. In the latter case, the finiteness of the Planck quantum meant that the statistical properties of the ensemble changed in the process of measurement, through interaction with macro-scopic measuring devices, whereas at the classical level such interaction could be considered negligible. The statistical reformulation of quantum laws preserved this uncertainty in measurement, concluded Nikolsky, but it allowed physicists to 'charac-terize quantum processes as objective ones', whereby the probabilities of experimental outcomes corresponded not to indeterminacy in individual behaviour, but to objective probabilities—the relative number of particles within a quantum ensemble for which the measured variable had a particular value (Nikolsky, 1936, p. 540).

In a subsequent book that developed the approach further, Nikolsky spoke about two kinds of measurement: one is the measurement that formed (in Kemble's termin-ology, prepared) an ensemble. The latter is the measurement in the proper sense, one that sorted an existing ensemble according to the values of a physical variable and measured the latter. In this second sense, Nikolsky wrote about the ensemble of 'quantum processes', that is, the 'ensemble of experiments with single quantum particles that had initially been set into certain conditions' (Nikolsky, 1941, pp. 26–27) or the 'ensemble of passages of microparticles through a diffraction device' (Nikolsky, 1941, p. 148). These ensembles of quantum processes were ensembles of measurement operations that generated statistics.

Similar ideas were brewing within the circle of Moscow physicists, including the much more senior and authoritative Leonid I. Mandelstam. Educated before the revolution, Mandelstam was no Marxist either politically or philosophically. He tried to stay away from any ideological discourse, while his views in physics were strongly positivist. A personal friend of von Mises, Mandelstam was quite sympathetic to the latter's 'objective' treatment of probability. At the same time, he collaborated with the Marxist philosopher Boris M. Hessen, who also supported the ensemble approach. Hessen in 1930–1936 held influential administrative positions as the Dean of the Department of Physics at Moscow State University and as deputy director of the Lebedev Physical Institute. Mandelstam supervised Hessen's philosophical analysis of the foundations of statistical mechanics and probability (Hessen, 1929; Hessen, 1930a, b), and also very likely advised Nikolsky during the second half of the 1930s, when both worked at the Lebedev Physical Institute. In 1939 he applied the statistical ensemble idea to sort out interpretational difficulties in the famous thought experiments that lay at the foundation of quantum physics.

Mandelstam did not occupy administrative posts, but his expertise and authority on fundamental questions of physics was revered by his Moscow colleagues, and his judgements and intuitions influenced many of his well-established scientific colleagues by way of backroom consultations. On quantum physics, Mandelstam worked only episodically, and during the 1930s he focused primarily on establishing a completely new field of non-linear physics, but the debate between Einstein and Bohr drew his attention to the fundamental problem of quantum interpretation. In spring 1939 he presented his views on the Einstein–Podolsky–Rosen controversy in an advanced lecture course at Moscow University on 'The Foundations of Quantum Mechanics (The Theory of Indirect Measurement)' (Mandelstam, 1939). The course was attended by students as well as mature scientists who took detailed notes which allowed the posthumous publication of these lectures in 1950, in volume 5 of Mandelstam's *Collected Works*.

Like a number of early pioneers of the ensemble approach, Mandelstam was influenced by Johann von Neumann's 1932 mathematical analysis of the foundations of quantum mechanics which, as Max Jammer observed, was written 'with the general tenor of the statistical ensemble interpretation' (Jammer, 1974, p. 443; von Neumann, 1932). Von Neumann's influence would eventually make the very term 'ensembles' the typical word of choice for quantum physicists, whereas their mathematical colleagues, such as von Mises and his Russian opponent Khinchin, spoke of 'collectives'. The 'general tenor' refers only to the mathematical language of the book, perfectly adapted to analyse statistical ensembles, but not to its philosophical conclusions. Von Neumann believed that quantum mechanics described individual processes and that he had proven 'the impossibility of a causal atomic theory', whereas Mandelstam was not at all convinced by that proof.

In his analysis of the EPR paradox, Mandelstam was torn between his equal reverence towards Einstein and Bohr. He did not explicitly contradict the Copenhagen interpretation but did express dissatisfaction with 'vague' and 'mystical' formulations about the reduction of the wave packet in Bohr's debate against the EPR argument. In the end, one can say that he found a third way out of the challenging dilemma as formulated by Einstein: either quantum mechanics was incomplete, or a mysteriously non-local theory. If one systematically follows through with the suggestion that in quantum mechanics, 'physical parameters (observables) refer to populations, rather than individual cases', and that quantum measurement acts upon a statistical 'collective', explained Mandelstam, the EPR paradox ceases to be a paradox and does not require the two spatially separated systems I and II to interact instantaneously. The interaction had happened earlier, before the separation, whereas subsequent measurements selected different sub-populations from an ensemble of such separated and no longer interacting pairs of systems. 'Once I understood the mistake, I could no longer see and present the matter in any other way', confessed Mandelstam. 'The essence of the problem is that when we are performing different *measurements* on system II, we are selecting different subsets . . . For each subset separately, the uncertainty relation holds for system I . . . But nothing prevents us from measuring exactly the [two non-commuting observables] from two different subsets' (Mandelstam, 1939).

Even before the publication of Mandelstam's lectures in 1950, the essence of his argument was known, at least in Moscow, to the circle of interested physicists, including Dmitry I. Blokhintsev. Blokhintsev considered himself a pupil of Mandelstam and used a similar 'indirect measurement' approach to reformulate Heisenberg's famous experiment illustrating the uncertainty relation: 'Let's now consider the indirect determination of the coordinates of micro-particles . . . We shall show that in this case also ensembles which satisfy the uncertainty relation will be produced. An example of the indirect measurement is the determination of the position of particles by means of a microscope. A particle near $x=0$ is illuminated with light of wavelength $\lambda$. The beam of light is parallel to the axis $OX$. Scattering light will enter the microscope lens. It is known from the theory of the microscope that the position of the particles is determined with the accuracy $\Delta x = \lambda/sin\varepsilon$, where $2\varepsilon$ is the angle subtended by the objective at the position of the object. Thus, an ensemble of particles with $\Delta x \approx \lambda/sin\ \varepsilon$ can be selected. When $\lambda$ is sufficiently small the quantity $\Delta x$ may in principle be arbitrarily small. However, the momentum of the photon is changed in each scattering process . . . The correct mathematical theory of this experiment, proceeding from the statistical interpretation of the wave function has been given by L. I. Mandelstam in his *Lectures*. In the non-Russian literature this experiment is customarily considered as an experiment with one particle[1]. However, for a particle we can only get one scattering (after which this particle belongs to a different ensemble) and the position of the particle can not be judged from one scattered quantum (there being no image in the focal plane)' (Blokhintsev, 1964, p. 53).

## 50.2.4 Two Types of the Ensemble Interpretation

Home and Whitaker (1992) distinguish two types of ensemble interpretation: 1) pre–assigned initial values (PIV) and 2) minimal ensemble interpretations. According to the PIV interpretation, each particle in the ensemble has a definite, yet unknown, momentum and position. This assumption is so natural and pleasing, that for many realists it represents 'the' ensemble interpretation, a typical example of which is presented in Popper's *The Logic of Scientific Discovery* (Popper, 1959, Chapter 9). All the observables have predetermined values which could then be discovered

---

[1] In the Russian original text Blokhintsev writes "in Western writings" In the fifth edition this expression has been omitted.

during measurement. The minimal ensemble interpretation only assumes that all that is known about the ensemble is contained in its wave function. This view is closer to the Copenhagen approach; as a result some philosophers of science prefer not to demarcate these two types of interpretations. However, historically they were distinguished, because the minimal ensemble interpretation was usually proposed, intertwined, and sometimes mixed with ideas of the PIV interpretation, rather than with the Copenhagen school.

The minimal ensemble view is analogous to the concept of the Gibbs ensemble from classical thermodynamics, in which the microsystem $\mu$ is considered in relation to the macroscopic thermostat $M$ of fixed temperature $\Theta$. The probability $W_\Theta(P, Q)$ for the values of dynamic variables $(P,Q)$ refers to the ensemble of states of $\mu$ formed during unlimited interactions between the system and the thermostat, in other words, by a series of reproductions of the system $\mu$ in the same macroscopic situation determined by the temperature $\Theta$. The probability $W_\Theta(P,Q)$ thus depends on the characteristics of both parts: the microsystem $(P,Q)$ and the macroscopic situation characterized by the thermostat's temperature $\Theta$. The minimal quantum ensemble, similar to the Gibbs canonical ensemble, is formed by unlimited reproductions of the states of (more than one) microsystems $\mu$ in the macroscopic condition $M$.

The proponents of ensemble interpretations often wavered between or typically did not distinguish between the two types. Slater's and Kemble's views were closer to the minimal ensemble interpretation. Margenau at first described the measurement as related to the PIV 'real ensemble': 'numeral observations, or a single collective of observations, on a physical assemblage of many similar systems in the same state' (Margenau, 1937, p. 352). But then he expressed ideas closer to the Gibbsian ideal ensemble, though without using the term: 'Numerous repeated observations on the same system, state in question being re-prepared before each observation' (Margenau, 1937). He emphasized the latter ensemble's fundamental role in the foundations of quantum mechanics.

In his 1939 lectures Mandelstam also initially assumed 'real ensembles'. However, when he came to discussing major philosophical problems, such as the reduction of the wave packet and the Einstein–Podolsky–Rosen argument, he switched to ideal Gibbsian ensembles. Mandelstam also used Gibbsian ensembles in his 1942 manuscript 'On Energy in Wave Mechanics' which is historically and logically connected with his and Tamm's article 'The Energy–Time Uncertainty Relation in Nonrelativistic Quantum Mechanics' (Mandelstam and Tamm, 1945; Mandelstam, 1947–1955, vol. 2, pp. 306–315). Mandelstam discussed the measurement of energy in his manuscript: 'Let the wave function be $\psi(x,t)$. In order for statistics to make sense, reiteration must be performed, that is, the experiment must be repeated many times, where $t$ is the time elapsed from the beginning of the experiment in each of the experiments. The measurement at a "given moment of time" is the measurement in different experiments, but always at the same amount of time passed since the beginning of that experiment' (Mandelstam, 1947–1955, vol. 3, p. 402).

# 50.3 Philosophical Premises

## 50.3.1 Popper's Philosophy of Science

In contemporary literature on the philosophy of science, Popper is often characterized as a realist. However, the terms 'realism' and 'scientific realism' were not used in his 1934–1935 papers and book; they also did not appear in the English edition of his *The Logic of Scientific Discovery* (1959). Popper's goal then was to formulate the objective interpretation that would allow the treatment of quantum mechanics as a scientific theory not dependent upon metaphysics.

Popper turned to the concept of realism in volume 1 of his 1975 *Postscript to the Logic of Scientific Discovery*. While his treatment of the structure of the uncertainty relations did not essentially rely on realism, the latter provided an implicit background for Popper's criticism of indeterminist metaphysics of quantum mechanics. 'The task of science', Popper writes, 'which I have suggested is to find satisfactory explanation which can hardly be understood if we are not realists. For a satisfactory explanation is one which is not ad hoc, and this idea—the idea of independent evidence—can hardly be understood without the idea of discovery of progressing to deeper levels of explanation, without the idea that there is something for us to discover, and something for us to discuss critically' (Popper, 1975, p. 145). A further comment by Popper explained the motivation behind his 1934 attack on the Bohr–Heisenberg approach: 'It may perhaps . . . be mentioned that my interpretation of Heisenberg's indeterminacy formulae as scatter relations was both an attempt to criticize Heisenberg's Machian positivism ('observables') and to eliminate what I regarded as his metaphysical dogmatism: his theory that indeterminacy formulae indicated the limit of scientific knowledge' (Popper, 1975, p. 181). In volume 3 of his Postscript, *Quantum Theory and the Schism in Physics*, Popper developed the propensity interpretation of probability, which also helped him to criticize the Bohr–Heisenberg approach to quantum mechanics.

## 50.3.2 Operationalism

Both Slater and Kemble studied physics under Percy Bridgman, known not only as the prominent experimentalist, but also as a philosopher of physics who advanced the philosophy of operationalism. Kemble subscribed to this philosophy openly and also described Slater's 1928 paper on ensembles as providing the operationalist sense of the probability in quantum mechanics. In his 1938 paper 'Operational Reasoning, Reality and Quantum Mechanics,' Kemble traced his philosophy of science back to Ernst Mach and stated that the modern physicist needed to adopt 'the method of operational reasoning recently advocated by Bridgman' (Kemble, 1938, pp. 266–269). In a later 1963 interview he characterized Bridgman's point of view as 'heaven sent' for the foundations of modern science (*APS Archives*). Kemble was also influenced by John von Neumann and aspired his book to bridge a 'gap between exact technique of von Neumann and the usual less rigorous formulation of the [quantum] theory' (Kemble, 1937, VII).

In his subsequent writings, he used operationalism in a very generalized sense. For example, his operationalist analysis of the concept of probability meant a discussion 'in terms of mental operations involved in determining numerical probabilities' (Kemble, 1941, p. 204). Kemble combined operationalism with a search for objectivity, in particular with the idea of 'objective state', the key concept of his interpretation. He wrote that the definition of probability as frequency in his 1937 book had 'an air of objectivity appealing to the student of natural science' (Kemble, 1941). But his object-ivity was different from 'objective reality': 'The province of the physicist is the study of a portion of inner world of experience', he wrote. However, 'large scale experience fits the common-sense idea of an external world' (Kemble, 1938, pp. 274–5). 'No experiment is informative in scientific sense unless it is actually, or by implications, many times repeated' (Kemble, 1938, p. 270). In this sense, his primary concern was with 'statistical experiments'. In 1939 Kemble made the following generalization: 'the primary object of inquiry in physics ought *always* be identical with Gibbsian ensemble of systems similarly prepared' (Kemble, 1939, p. 1015).

Margenau is known as a critic of operationalism (although, according to Kemble, his criticism was based on a very narrow interpretation of Bridgman's writings (Kemble, 1938, p. 267)). He still admitted being 'strongly influenced by Bridgman's theory of operational definitions which were later recognized as one important class of rules of correspondence' (Margenau, 1978, XIV). By the 'rule of correspondence', the idea of which was influenced by Bridgman's operationalism, Margenau understood statements that connected the 'constructs' with sensual information. Operational definitions were regarded by him as a class of 'epistemic definitions' by means of which physical magnitudes acquired numerical values, whereas the epistemic definitions themselves were a class of the 'rules of correspondence'.

In the Soviet Union, the term 'operationalism' was used in a negative sense, to criticize Mandelstam's philosophy of physics. A famous mathematician, specialist in geometry, and a committed Marxist, Alexander D. Alexandrov presented a paper at Lebedev Physical Institute in 1952 with a politically inspired attack on Mandelstam's lectures on quantum mechanics: 'This is a prescriptional view of the definition of scientific concepts, the view which was developed by the idealists Percy Bridgman and Philip Frank. The trend to reduce concept to operations is a trend of subjective idealism which tends to eliminate the objective reality and to reduce everything to experimental data' (cited in Sonin, 1994, p. 181).

In his own works, Mandelstam never referred to Bridgman, but like many physicists of his generation, he read Mach, Duhem, and other positivists. He was also strongly influenced by von Mises, whose positivism also emphasized the operational meaning of the scientific concepts, even if von Mises also did not cite Bridgman and operationalism in his *Kleines Lehrbuch des Positivismus. Einführung in die empirische*

*Wissenschaftsauffassung* (1939). The reference to Bridgman's *Logic of Modern Physics* appeared in the 1951 book written by von Mises after he became a professor at Harvard University.

Operationalist ideas circulated widely in physics, with or without familiarity with Bridgman, but often inspired by Einstein's theory of relativity. In this sense, one also does find operationalism in Mandelstam's lectures on quantum mechanics: 'Quantum mechanics rightly abandons the prejudice that the laws of the macroscopic world remain valid in the micro-world. But only the mathematical part of the theory completely proceeds from this point of view. The textbooks still do not take sufficiently into account that prescriptions for the transaction [from the mathematical technique to the real objects] must differ from those in classics. If in classics I state that $x$ is the position of a material point, by this I mean a clear prescription: if I set properly a rigid rod graduated according to a definite prescription, then the material point will coincide with the marking $x$ on the rod. When we talk about molecules, this prescription cannot be performed . . . Thus, calling $x$ 'the position' I only pretend that I have an established relation to the nature. Without such a definition, the theory is hanging in the air . . .

The uncertainty relation is troubling, since by calling $x$ and $p$ position and momentum respectively, we imply the corresponding classical magnitudes . . . Why do we call $p$ the momentum? This is self-delusion again. Until we have a new prescription, it would be better not to use old terms' (Mandelstam, 1947–1955, Vol. 4, pp. 354, 358)[2].

## 50.3.3 Materialism

Unlike Mandelstam, Nikolsky declared his worldview openly as materialistic, the philosophy that presupposes the objective reality of things and natural phenomena. Materialism can be traced back to Marx and Engels and further to Feuerbach and to the French materialists, Denis Diderot and Baron d'Holbach.

Materialism should not be confused with scientific realism which had gained popularity by the end of XX. It was closer to metaphysical realism, from which scientific realists usually dissociate themselves. However, in contrast to metaphysical realism materialism believes that matter is substance and space and time are its attributes. True, in contrast to "metaphysics" (the Soviet Union philosophers declared themselves as "dialectics") materialists insisted that knowledge of what this substance is infinite and science will never come to the final picture of the world (Weltbild). Nevertheless, all the materialists dissociated themselves from positivists who admitted the reference to human being experience in the Weltbild.[3]

---

[2] For comments see (Pechenkin, 2000).

[3] It should be noted that in the Soviet Union dialectical materialism was a kind of religion and people often referred to it without any penetration to its ideas. It is not true with respect to Nikolsky who tried to treat materialism as the philosophy of physics.

Nikolsky professed materialism already in his 1934 book, where he insisted that 'all physical phenomena are processes that progress in time.' The Copenhagen approach to quantum mechanics he considered, on the contrary, idealistic: 'Heisenberg's approach leads to explicitly idealistic conclusions' (Nikolsky, 1934, p. 28).

In his 1937 response to Fock's criticism, Nikolsky drew an ideological line. He pointed out that there was more than one existing interpretation of quantum mechanics, including those by Einstein and Erwin Schrödinger, and he referred to his statistical proposal as representing Einstein's stance on the issue. 'In order to achieve a materialistic description', Nikolsky wrote, 'it is necessary to represent physical events objectively, in space and time, as existing independently from the observer . . . Quantum mechanics in Bohr's interpretation does not satisfy this condition' (Nikolsky, 1937, pp. 555–7). Nikolsky's quotations with examples of idealistic statements, including pronouncements on the influence of an observer on the behaviour of atomic particles, actually came from Werner Heisenberg's rather than Bohr's philosophizing. In proofs, he added a critique of yet another, more recent 'positivistic' representation of quantum mechanics by Pascual Jordan in Germany.

Nikolsky argued that proponents of the 'Soviet branch of the Copenhagen school' (which included Nikolsky's teacher Vladimir Fock, and also Landau, Bronstein, and a few others) should stop masquerading its interpretation as compatible with materialism and try, instead, to develop a 'truly materialistic theory of atomic phenomena'.[4] Still, in his description of quantum ensembles, Nikolsky combined his materialism with the frequency interpretation of probability which had an operationalist meaning.

# 50.4 POST-WORLD WAR II

## 50.4.1 Margenau's 1950 'Nature of Physical Reality'

In his book *Nature of Physical Reality* Margenau presented his philosophy of physics for a wide range of intelligentsia. He did not use the term 'the ensemble interpretation'. However, he claimed that 'as we have repeatedly emphasized the statistical character of quantum mechanics has been designed for the purpose enabling that science to deal with the situations in which a single measurement means very little with respect the state of the system'. 'The uncertainty principle correlates the spread on a collective position measurements made when a system is in a state specified in the most exact quantum mechanical matter (by means of a function $\psi$) with the spread of momentum measurements on the same state' (Margenau, 1950, p. 375). *Nature of Physical Reality* contains some critical comments concerning Kemble's interpretation of quantum

---

[4] On Matvey Bronstein and other representatives of the Soviet branch of the Copenhagen school, see Gorelik and Frenkel, 1990).

mechanics. Margenau wrote that 'Kemble mixes different ideas of measurement and different formulations of probability' (p. 387).

Margenau's *Nature of Physical Reality* had been written in the style of the 'standard' (or hypothetico-deductive) model of a scientific theory, the model which emphasized the theoretical statements and the rules of correspondence which connect theoretical terms with sense datum. According to Lenzen (1951, p. 69), 'like positivist Margenau finds physical reality among constructs which serve to introduce order into manifold data that are selected to constitute Nature. But he is not willing to restrict philosophy to this field. He considers that the method of physical science does not explain the origin of experience and he clearly acknowledges that the solution of the problem in the term of philosophy of critical realism has meaning'.

B. van Fraassen (1991, p. 257) developed his modal interpretation of quantum mechanics that is different from Margenau's ensemble approach. Nevertheless he referred to Margenau's interpretation by explaining some concepts. 'Margenau was right to point out that a particle subjected to measurement is usually absorbed at the end. This made repetition of a measurement on the same particle rather difficult'.

## 50.4.2  Blokhintsev's Interpretation

After World War II, Dmitry I. Blokhintsev became the main proponent of the ensemble approach to quantum mechanics. Almost certainly inspired by Mandelstam's 1939 lectures and Nikolsky's 1940 book, Blokhintsev first chose a somewhat cautious research path. Rather than addressing the general interpretation heads on, he attacked a series of mathematical and technical problems that were relevant for both quantum statistics and the eventual systematic development of the ensemble interpretation. Some calculations were done in collaboration, possibly as assignments for advanced students in connection with the quantum mechanics course he taught. The titles of his published papers speak for themselves: 'The Quantum Gibbs Ensemble and its Relation to the Classical Ensemble' (1940, 2 parts); 'On the Separation of the System into Two Parts – a Classical and a Quantum one' (1941); 'Atom as Observed in the Electron Microscope' (1947); 'The Principle of Detailed Equilibrium and Quantum Mechanics' (1947); 'The Relationship between the Mathematical Formalisms of Quantum Mechanics with that of Classical Mechanics' (1948).[5] The importance of these papers for the ensemble interpretation would become obvious when their results appeared in his textbook of 1949.

In 1944 he published a textbook, *Introduction to Quantum Mechanics*, which was written 'in the spirit of Heisenberg's interpretation according to which the wave function represents the man's knowledge of the state rather than the state of the system itself'. Five years later Blokhintsev published a revised edition entitled *Fundamentals of*

---

[5] All recently reprinted in volume 2 of his selected works (Blokhintsev, 2009b).

*Quantum Mechanics* 'which by virtue of its excellent didactic approach became one of the most popular textbooks ever on quantum mechanics in the Russian language' (Jammer, 1974, p. 445). This book went through five editions and was reprinted many times. In contrast to the earlier version, Blokhintsev's 1949 textbook relied on an articulated anti-Copenhagen approach: 'The chapter which concerns the concept of state in quantum mechanics has been drastically changed', Blokhintsev reported, 'and the idealistic conceptions of quantum mechanics which are now popular abroad have been subjected to criticism' (Blokhintsev, 1963, p. 9).

The main assumption of Blokhintsev's approach was spelled out upfront in the book's introduction: 'quantum mechanics is a statistical theory . . . but different from classical statistical mechanics . . . Unlike statistical mechanics, modern quantum mechanics is not based on a theory of individual processes. It operates right from the start with statistical collectives—ensembles . . . and studies these ensembles in their relationship with macroscopic measuring devices' (Blokhintsev, 1949, pp. 10–11). The ensemble interpretation appeared in his book, for the first time, as a consistently and systematically developed framework for the entire body of quantum mechanics. The main changes did not concern the theory's mathematical formalism and applications, but its basic definitions and the explanations of key experimental facts and thought experiments, including the wave function, the indeterminacy relation, measurement, reduction, and the EPR paradox. $\psi$ was defined as a description of a pure ensemble of identically prepared atomic systems, rather than of one individual system. Measuring devices acted on it like 'reference systems', or more precisely, like 'spectral analysers': 'the process of measurement transforms a pure ensemble into a mixed one . . . which is, in practice, a spectral resolution of the initial ensemble into components or sub-ensembles . . . determined by the specific type of measuring instrument' (Blokhintsev, 1949, p. 76).

Epistemological problems were further discussed in the conclusion, which dealt with the questions of causality in quantum mechanics and the role of the observer. Blokhintsev openly criticized some of major claims of the Copenhagen interpretation (Blokhintsev, 1964, p. 540). For him, the 'quantum ensembles' approach helped to restore materialism as the philosophy of physics: 'The quantum mechanics in fact deals with an objective nature of quantum ensemble as existing independently from the observer. The properties of a single micro phenomenon are examined through the statistical laws which are entirely objective' (Blokhintsev, 1964, p. 496).

Quantum ensembles allowed Blokhintsev to reformulate the interpretation of quantum mechanics: 'The predictions which result from the knowledge of the wave function are in general statistical one. If any single measurement is made, the result of it will therefore show only to what extent our expectations were justified—whether the probable or improbable event has occurred. The only objective result of the measurement is obtained when the measurements are repeatedly carried out in a large number of identical experiments . . . It is important to note that in quantum region we can not repeat an experiment on the same particle, since the measurement can in general change the state of the microparticle . . . In carrying out a large number ($N \geq 1$) of identical experiments, it is therefore necessary to imagine a large number of particles or systems which are independently in the same macroscopic condition. We call such a group of microparticles (systems) a quantum ensemble or simply an ensemble' (Blokhintsev, 1964, p. 7; 1976, p. 59).

In his subsequent book on the philosophy of quantum mechanics (1966, English translation, 1968) Blokhintsev formulated the following argument in favour of the ensemble interpretation: 'If the wave function is a characteristic of a single particle, it would be of interest to perform such a measurement on a single particle (say, an electron) which would allow us to determine its own individual wave function. No such measurement is possible' (Blokhintsev 1968, p. 50). This argument was later reproduced by one of the active proponents of the ensemble approach, Ballentine (Ballentine, 1970, p. 379). In the last, fifth edition of his textbook (1976), Blokhintsev omitted his attacks on the Copenhagen interpretation, noting that such criticisms were of historical importance only. However, the physics part of his ensemble interpretation remained the same.

Home and Whitaker refer to Blokhintsev as a proponent of the minimal ensemble interpretation which does not introduce the superstructure (does not mention the space coordinates and momentums that characterized all the particles belonging to the ensemble). They cite his discussion concerning Gibbsian ensembles that appeared in the 1976 edition of Blokhintsev's *Fundamentals of Quantum Mechanics* (Home and Whitaker, 1992, p. 285). The conception of quantum ensembles, Blokhintsev wrote, is very close to the conception of the Gibbs classical ensemble. It should be noted that the sophisticated terminology of the Gibbsian ensemble appeared in his philosophical treatment of the foundations of quantum mechanics. At the same time, his presentation of quantum mechanics in the core chapters of his textbook can be read as implying the concept of PIV ensembles. 'Let us suppose that we attempt to fix the value of one coordinate, for instance, $y$', wrote Blokhintsev in the section dedicated to the uncertainty relations. 'To do so we place a screen with its plane perpendicular to the direction of the waves, the screen having a slit. Let the half-width of the slit be $d$. If a particle passes through the slit, then at the instant when it does so, its coordinate is fixed by slit diffraction will occur. The waves are deflected from their original direction of propagation. The momentum of the particles is also changed when the screen is put in' (Blokhintsev, 1976, pp. 49–50).

## 50.4.3  Blokhintsev's Ideological Premises

Blokhintsev's *Fundamentals of Quantum Mechanics* of 1949 was more openly ideological than what was normally expected from Soviet physicists.[6] He actively discussed quantum philosophy, criticized idealistic versions, and connected the ensemble

---

[6] See also (Kojevnikov, 2012).

interpretation with dialectical materialism. 'The seemingly paradoxical nature of quantum mechanics emerges only if one attempts to understand its novel laws from the point of view of old classical mechanics', wrote Blokhintsev after explaining the EPR paradox along Mandelstam's lines. Overall, according to him, quantum mechanics demonstrated the restricted nature of classical atomistic concepts and uncovered qualitatively new statistical regularities in the microscopic world, which had been tested experimentally. It thus confirmed an important tenet of dialectical materialism, namely that every particular state of our knowledge about nature is only approximate and relative, while qualitatively new laws and regularities emerge at every fundamentally different level of material existence. Lenin's dictum from *Materialism and Empiriocriticism* that the main criterion of a materialistic epistemology is the assumption that nature and its laws exist objectively, independently of the observer, was equally satisfied by quantum mechanics' ensemble interpretation. 'Therefore, from the point of view of dialectical materialism, quantum mechanics should be regarded as the most important development of atomism in the 20th century', Blokhintsev happily concluded in his textbook (Blokhintsev, 1949, p. 555).

In 1951 he published a paper with an even stronger political pitch. He attacked the main philosophical claims of Bohr, Heisenberg, and Jordan: the principle of observability and the principle of complementarity, especially the former.

According to Blokhintsev this principle presupposed a positivist substitution of the study of objective reality, matter, for the description of human sensory data. Bohr's principle of complementarity, Blokhintsev insisted, is based on the principle of observability. 'The Copenhagen school overlooks the fact that quantum mechanics is applicable only to statistical ensembles and focuses on the interrelation of a single phenomenon and the device. This is a significant methodological error. In this interpretation, all quantum mechanics becomes instrumental in nature and the objective side of things is lost.' As his guiding philosophical inspiration, Blokhintsev referred to Lenin's 1909 book *Materialism and Empiriocriticism* that criticized Mach and Machists. Blokhintsev also cited an important ideological document of Soviet Communism *Kratkii Kurs Istorii KPSS* (Short Course on the History of the Communist Party of the Soviet Union) in particular, the chapter 'On Dialectical and Historical Materialism' written by Stalin.

In 1957 Blokhintsev published a paper 'Lenin's *Materialism and Empiriocriticism* and Modern Ideas on the Structure of Elementary Particles'. At first sight this paper looks like a review of some problems in physics. However, at the beginning of the paper and at its end Blokhintsev cited the famous dictum from Lenin's book that 'the electron is as inexhaustible as the atom', implying that Lenin foresaw the later development of the physics of elementary particles. Blokhintsev also referred to *Materialism and Empiriocriticism* in his textbook on quantum mechanics: 'Lenin emphasized in his book that the fundamental point of epistemology is the objectivity of nature's laws' (Blokhintsev, 1964, p. 516).

It is probable that Blokhintsev was honest in his attachment to Marxism-Leninism. By the end of his life he wrote in a synopsis of his own papers and books: 'My philosophical credo has been formed by Lenin's ideas brilliantly presented in his book *Materialism and Empiriocritisism.* I had to defend the ideas of the founders of dialectical materialism from both their opponents and the primitive defenders—dogmatists among the Soviet philosophers' (Blokhintsev, 2009a, p. 62).

## 50.4.4 The Criticism of the Ensemble Interpretation

The criticism of the ensemble approach to quantum mechanics started in 1950 with Vladimir A. Fock. The manuscript of Mandelstam's lectures on quantum mechanics was then under preparation for publication in *Collected Works* and the editors asked Fock to read some of its parts. Fock corrected a few technical details, but in private conversations he also expressed disagreement with Mandelstam's interpretation (Pechenkin, 2014; 2019, p. 223). In 1951 Fock published a review of Mandelstam's by then published lectures. While highly appreciating Mandelstam's writings on theoretical physics, Fock carefully expressed his reservations regarding the ensembles:

> The wave function allows us to find the probabilities of the results of measurement if the object had been in a given state before the measurement. In this sense quantum mechanics is a statistical theory. L. I. Mandelstam emphasized that every statistical theory presupposed the concept of an ensemble to which the statistics refers. This is true. However, there is ambiguity in Mandelstam's formulations. Let us measure some variable by observing an object in some state. We arrive at some statistical ensemble. Yet, by observing the same object we can produce a measurement of a different variable. This measurement would produce different statistics. It is thus not possible to connect the concept of statistical collective with the concept of the state. We came to different statistical ensembles by combining the wave function with one or another variable. Mandelstam tells us about a statistical collective to which the wave function refers. In fact, we have a sub-ensemble. When using different instruments, the measurement produces different sub-ensembles.   (Fock, 1951, p. 162)

> According to quantum mechanics, the most complete description of a state of a micro-object is provided by the wave function which allows us to find statistics of the measurement results over an object that was before the measurement in the state represented by the wave function. The statistical collective is determined not only by an assignment of the wave function, but also by an indication of the measured quantities (and the corresponding measuring settings, which make the interaction with the object necessary for the measurement) . . . Another ambiguity is associated with the first one. The author repeatedly speaks of the choice of sub-ensembles of some ensembles of measurement results. Meanwhile, in fact, we are talking about different collectives obtained by the use of different measuring devices, and one of the sub-collective of which the author speaks refers to one ensemble, the other one refers to a completely different ensemble.   (Fock, 1951, p. 162)

In 1957 Fock published a paper specially dedicated to the interpretation of quantum mechanics, in which he again addressed Mandelstam's work (Fock, 1957, p. 471). 'The question about a statistical collective to which the probabilities are designated needs to be discussed. One of the first who formulated this question was L. I. Mandelstam, but he gave a wrong answer to it. Mandelstam spoke about a micromechanical collective, to which the wave function refers, he also called it an electronic collective, emphasizing thereby that he meant a set of micro-objects prepared in a certain way. The elements of statistical ensembles discussed in quantum mechanics are not objects themselves but the results of experiments with these objects, any definite experiment corresponding to one definite ensemble. Since the probability distributions resulting from the wave function for different variables refer to different experiments, they refer to different ensembles. Thus, the wave function does not refer to any definite statistical ensemble. The elements of statistical collectives considered in quantum mechanics are not the objects themselves, but the results of experiments with them, and a specific experimental setup corresponds to a particular ensemble. Since the probability distributions obtained from the wave function for various quantities refer to different experimental designs, they refer to different ensembles.' Further, in the style of his own version of the interpretation of quantum mechanics Fock writes the following: 'More fundamental reason of the impossibility to attach any statistical collective to the wave function results from the fact that the conception of the wave function refers to the potentially possible (to yet non-produced experiments) and the conception of the statistical ensemble refers to the already produced ensemble' (Fock, 1957, p. 472).

Fock also criticized Blokhintsev's interpretation of quantum mechanics (Fock, 1952) by pointing to logical contradictions in it. He emphasized three points: 1) according to Blokhintsev, the quantum ensemble consists of a considerable number of particles (systems) which exist independently from one another in the same quantum state represented by the wave function; 2) if in such an ensemble all particles are characterized by the same quantum state represented by the wave function, then every particle belonging to the ensemble is found in this state; 3) the wave function does not represent an individual particle. What was the outcome of this? The first statement is not consistent with the third statement. The first statement and the second statement together constitute a vicious circle.

Fock's position stirred up a debate. For instance, in his paper entitled 'My reply to Academician Fock' Blokhintsev wrote that Fock proceeded from the point of view that the wave function represented the state of one particle and this point of view led him to idealism. To explain Blokhintsev wrote that by proceeding from Fock's position we should admit our permanent ignorance with respect to the objective state of a particle because in principle there is no way to measure the wave function of an individual particle.

Blokhintsev's reply contained a theoretical part which can be summed up as follows (he did not formulate the point by point reply): 1. Combination of particles into an ensemble is carried out not speculatively, but on the basis of the connection of the particles with the same macroscopic environment. The wave function represents the fact that a particle belongs to some macroscopic circumstance and a set of particles forms an ensemble. This is my definition of the quantum ensemble. There is no vicious circle in my definition; 2. I have emphasized that the quantum ensembles are objective. The state of particles in the sense of their belonging to an ensemble is definable, and it is not definable after the experiment (as Fock believes), but before the experiment, i.e. objectively, regardless of the observer. 3. Fock confusingly mixes up the representation of a quantum state with respect to a measurement apparatus with the action of the apparatus on the particles belonging to the ensemble. The representations of the wave function in relation to different apparatuses are equivalent, the actions of the apparatuses on the ensemble are different. 4. According to the conception of quantum ensembles the wave function (or density operator) is the basic characteristic of the quantum ensemble. Fock illegally treats the quantum ensembles as statistical collectives of measurement results on the same single particle (Blokhintsev, 1952, pp. 172–174).

Another example of this debate is M. A. Markov (2010). He noted that Blokhintsev summed up his philosophy of quantum mechanics by applying the vague concept of popularization of physics: 'from the methodological point of view the conception of quantum ensembles distinguishes from the Copenhagen school approach by prescribing a *more modest* role to the observer' (Blokhintsev, 1976, p. 616).

At the beginning of this chapter we referred to Michael Redhead's treatment of the empirical interpretation of quantum mechanics. We emphasized that the ensemble interpretation is close to the empirical interpretation. The ensemble point of view does not proceed from the object of nature. It proceeds from observations, from the statistical collective of observation results. Redhead's later approach to quantum mechanics can be seen as a further development of Fock's criticism. For Redhead, the statistical interpretation is an instrumental (empirical) interpretation (as explained in the beginning of this chapter). To formulate further interpretations that provide 'understanding', Redhead formulates the following question: If a system is not in an eigenstate of some dynamical variable, what can we say about the value of this magnitude for this system? The Copenhagen answer is that we can say nothing: the question about the value of dynamical variable is metaphysical. The theory of hidden variables provides a different answer: the variable has a value, but this value is unknown, generally speaking. The third answer is induced by the concept of propensity introduced by Popper, as Popper suggested that indeed we can talk about the tendency (the propensity) of this variable to have any value. It is interesting that Popper's concept of propensity is close to the concept of potentially possible suggested by Fock.

## 50.4.5 Later Decades: Ballentine

The Canadian physicist Leslie Ballentine became the main proponent of the statistical (ensemble) approach to quantum mechanics in recent decades (Ballentine, 1970, 1990). Unlike Blokhintsev, Ballentine refrained from using any ideological arguments or justifications. Ballentine's definition of the ensemble interpretation was cited in the opening section of this chapter. Home and Whitaker treat it as the PIV type and discuss it together with Popper's. However, Ballentine is more radical: 'The statistical interpretation which regards quantum states as being descriptive of ensembles of similarly prepared systems, is completely open with respect of hidden variables. It does not demand them, but makes the search for them entirely reasonable' (Ballentine, 1970, p. 374). In connection with the above declaration (which is often cited), Ballentine analysed the von Neumann no-go theorem and Bell's rebuttal of it. By describing von Neumann's proof, Ballentine pointed out the implicit assumptions used by von Neumann. He emphasized that Bell's demonstration refers to the mathematical observations to be taken into account in any future theory of hidden variables.

## 50.5 CONCLUSIONS

The present paper is historical. It does not argue either in favour of the ensemble approach or against it. Currently, the ensemble interpretation is not popular. For example, the *Stanford Encyclopedia* contains articles on the Copenhagen, Many Worlds, Modal, etc., interpretations of quantum theory, but no article on the statistical interpretation. In April 2001 the Russian journal *Uspekhi Fizicheskikh Nauk* published a series of papers on the theory of measurement in quantum mechanics but paid no attention to Blokhintsev's writings which used to be cited quite frequently during the Soviet Union era. Nevertheless, the ensemble interpretation remains present in important papers that discuss modern foundations of quantum mechanics. References to this interpretation continue to help physicists and philosophers to formulate their own views on quantum mechanics. The ensemble interpretation of quantum mechanics belongs to the methodology of physics. It was inspired by a variety of philosophical positions: falsificationism (Popper), operationalism (Slater, Kemble, Mandelstam), materialism (Nikolsky, Blokhintsev), and positivism combined with Kantianism (Margenau). Sometimes, ideological justifications were also used and philosophical conceptions provided the linguistic frameworks within which the interpretation had been formulated.

## REFERENCES

Ballentine, L. (1970). The statistical interpretation of quantum mechanics. *Reviews of Modern Physics*, **42**, 358–381.

Ballentine, L. (1972). Einstein's Interpretation of Quantum Mechanics. *American Journal of Physics*, **40**, 1763–1771.

Ballentine, L. (1990). *Quantum Mechanics*. New York: Prentice-Hall International.

Blokhintsev, D. I. (1944). *Vvedenie v kvantovoiu mekhaniku* (Quantum Mechanics. An Introduction). Moscow: GITTL.

Blokhintsev, D. I. (1949). *Osnovy Kvantovoi Mekhaniki.* (Foundations of Quantum Mechanics). The second revised edition. Izdanie vtoroe, pererabotannoe. Moscow-Leningrad: GITTL.

Blokhintsev, D. I. (1951). Kritika idealisticheskogo ponimania kvantovoi mekhaniki (The criticism of the idealistic understanding of quantum theory). *Uspeki Fizicheskikh Nauk*, **XLV**, issue 2.

Blokhintsev, D. I. (1952). D.I. Moi otvet akademiku Foku (My reply to Academician Fock). *Voprosy filosofii*, **N6**, 171–173.

Blokhintsev, D. I. (1959). Kniga Lenina 'Materialism i empiriokritizism' i sovremennye predstavlenia o structure elementarnykh chastitz (Lenin's book 'Materialism and empiriocriticism' and modern conceptions of the structure of elementary particles). *Uspekhi Fizicheskikh Nauk*, **69**, 3–12.

Blokhintsev, D. I. (1963). *Osnovy Kvantovoi Mekhaniki. Uchebnoie Posobie Dlia Universitetov* (Foundations of quantum mechanics. Textbook for universities). Moscow: Vyshaia Shkolar.

Blokhintsev, D. I. (1964). *Quantum Mechanics.* Dordrecht: Reidel (translation of the 1963 edition).

Blokhintsev, D. I. (1968). *The Philosophy of Quantum Mechanics.* Dordecht: Reidel (translation of the book *Printsipialnye Voprosy Kvantovoi Mekhaniki.* Moscow, 1966).

Blokhintsev, D. I. (1976). *Osnovy Kvantovoi Mekhaniki* (Foundations of quantum mechanics) Fifth edition. Moscow: Nauka.

Blokhintsev, D. I. (2009a). Synopsis of my writings. In D. I. Blokhintsev, *Izbrannye trudy* (Selected writings). Vol.1. Moscow: Fizmatlit. 18–72.

Blokhintsev, D. I. (2009b). *Izbrannye Trudy.* Moscow: Fizmatlit. 2 vols.

Einstein, A. (1936). Physics and Reality. *Journal of the Franklin Institute*, 221, 349–382.

Fock, V. A. (1951). Review of Mandelstam's 'Complete Works' (Vol.5). *Uspekhi Fizicheskikh Nauk*, **45**, 160–163.

Fock, V. A. (1952). O tak nazyvaemykh ansambliakh v kvantovoi mekhanike (On the so-called ensembles in quantum mechanics). *Voprosy Filosofii*, 4, 160–163.

Fock, V. A. (1957). Ob interpretatsii kvantovoi mekhaniki (On the interpretation of quantum mechanics). *Uspekhi Fizicheskikh Nauk*, 62, 461–471.

Gorelik, G. E., Frenkel V. Ia. (1990). *Matvei Petrovich Bronstein. 1906–1938.* Moscow: Nauka.

Gorelik, G. E., and Frenkel, V. Ya. (1990). *Matvei Petrovich Bronstein. 1906–1938.* Moscow: Nauka.

Heisenberg W. (1930). *Die physikalischen Prinzipien der Quantentheorie.* Suttgart: S. Hirzel.

Hessen, B. M. (1929). Statisticheskii metod v fizike (The statistical method in physics and the von Mises new foundations of the theory of probability). *Estesvoznanie i Marksism*, 1, 3–58.

Hessen, B. M. (1930a). Materialistic dialectics and modern physics. Abstract of the paper presented to the First All-Union conference on physics in Odessa (Archives of RAN. Fond 1515, list 2, item 17).

Hessen, B. M. (1930b). Foreword. Russian translation of A. Haas. *Waves of Matter and Quantum Mechanics.* Moscow: Gos. Izd.: I-XXXIII.

Home, D., and Whitaker, E. (1992). Ensembles interpretations of quantum mechanics. A modern perspective. *Physics Reports*, 210, 223–317.

Jammer, M. (1974). *The Philosophy of Quantum Mechanics. Interpretation of Quantum Mechanics in Historical Perspective.* New York: John Wiley and Sons.

Kemble, E. C. (1935). The correlation of the wave function with the states of physical systems. *Phys. Rev.*, 47, 263–275.

Kemble, E. C. (1937). *The Fundamental Principles of Quantum Mechanics.* New York: McGraw Hill.

Kemble, E. C. (1938). Operational reasoning. Reality and quantum mechanics. *J. Franklin Inst.*, **225**, 263.

Kemble, E. C. (1939). Fluctuations, thermodynamics, and entropy. *Physical Review*, **56**, 1013–1122.

Kemble, E. C. (1941). The Probability Concept. *Philosophy of Science*, 8(2), 204–232.

Kojevnikov, A. (2012). Probability, Marxism, and Quantum Ensembles. *Jahrbuch für Europäische Wissenschaftskultur/Yearbook for European Culture of Science*, 6, 211–235.

Lenzen, V. F. (1951). Review of Margenau's 'Nature of physical reality'. *Isis*, 42(1), 68–69.

Lindsay, R., and Margenau, H. (1936). *Foundations of physics.* New York: Dover.

Mandelstam, L. I. (1939). Lektsii po Osnovam Kvantovoi Mekhaniki (Teoriia Kosvennykh Izmerenii). In *Lektsii po Optike, Teorii Otnositel'nosti i Kvantoivoi Mekhanike*, Moscow, pp. 325–388.

Mandelstam, L. I. (1947–55). *Collected Works.* 5 volumes. Moscow: Izd. AN SSSR.

Mandelstam, L., and Tamm, I. (1945). The uncertainty relation between energy and time in nonrelativistic quantum mechanics. *J. Phys. USSR*, **9**, 249–254.

Margenau, H. (1936). Quantum mechanical description, *Physical Review*, **49**, 240–42.

Margenau, H. (1937). Critical points in modern physical theory. *Philosophy of Science*, **4**, 337–370.

Margenau, H. (1950). *The nature of physical reality. A philosophy of modern physics.* New York: McGraw Hill.

Margenau, H. (1963). Measurement and quantum states. *Philosophy of Science*, **30**, 1–16; 138–157.

Markov, M. (2010). *On three interpretations of quantum mechanics*, 2nd edn. Moscow: URSS.

Mises, R. von (1939). *Kleines Lehrbuch des Positivismus. Einführung in die empiristishe Wissenschaftauffassung.* Den Haag.

Mises, R. von (1951). *Positivism. A Study in Human Understanding.* Cambridge: Harvard Univ. Press.

Nikolsky, K. V. (1934). *Kvantovaia Mekhanika Molekuly* (Quantum mechanics of a molecule). Moscow: GTTI.

Nikolsky, K. V. (1936). *Printsipy Kvantovoi Mekhaniki* (Principles of quantum mechanics). *Uspekhi Fizicheskikh Nauk*, **16**, 537–565.

Nikolsky, K. V. (1937). Otvet Foku (A reply to V. Fock). *Uspekhi Fizicheskikh Nauk*, **17**, 554–560.

Nikolsky, K.V. (1941). *Kvantovye processy* (Quantum Processes). Moscow: GTTI.

Pechenkin, A. (2000). Operationalism as the philosophy of Soviet physics. The philosophical background of L. Mandelstam and his school. *Synthese*, **124**, 200, 407–432.

Pechenkin, A. (2002). Mandelstam's Interpretation of Quantum Mechanics in Comparative Perspective. *International Studies in the Philosophy of Science*, **16**, 265–284.

Pechenkin, A. (2012). The early ensemble interpretations of quantum mechanics in USA and USSR. *SHPMP*, **43**(1), 25–34.

Pechenkin, A. (2014). *Leonid Isaakovich Mandelstam. Research, teaching, life.* Cham: Springer.

Pechenkin, A. (2019). *L. I. Mandelstam and his school in physics,* 2nd edn. Cham: Springer.

Popper K. (1934). Zur Kritik der Ungenauigkeitsrelationen. *Naturwissenschaften.* Bd. 22, S.807

Popper K. (1935). *Logik der Forschung.* Vienna. Julius Springer Verlag.

Popper, K. (1959). *The Logic of Scientific Discovery.* London: Hutchinson.

Popper, K. (1975). *Postscript to the Logic of Scientific Discovery.* London: Hutchinson.

Popper, K. (1992, 1995). *The Logic of Scientific Discovery.* Reprint. London: Routledge.

Redhead, M. (1987). *Incompleteness, Nonlocality, and Realism: A Prolegomenon to the Philosophy of Quantum Mechanics.* Oxford: Clarendon Press.

Slater, J. C. (1929). Physical meaning of wave mechanics. *J. Franklin Institute,* **207**, 449–455.

Slater, J.C. (1961). *Quantum Theory of Atomic Structure.* N.Y.: Mc Graw Hill.

Slater, J. C. (1975). *Solid State and Molecular Theory. A Scientific Biography.* New York: Wiley.

Sonin, A.S. (1994). Fizicheskii idealism. Istoria odnoi ideologicheskoi kompanii (Idealism in Physics. The History of an Ideological Compain). Moscow: Izd. Fiz.Mat.Lit.

van Fraassen, B. C. (1991). *Quantum mechanics. An empiricist point of view.* Oxford: Oxford University Press.

von Neumann, J. (1932). *Mathematische Grundlagen der Quantenmechanik.* Berlin: Springer.

Whitaker, A. (1996). *Einstein, Bohr and the Quantum Dilemma.* Cambridge: Cambridge University Press.

# CHAPTER 51

# STOCHASTIC INTERPRETATIONS OF QUANTUM MECHANICS

EMILIO SANTOS

## 51.1 INTRODUCTION

A stochastic interpretation of quantum mechanics is the logical consequence of the statistical character of quantum mechanics if it is interpreted as incompleteness of the description. Unlike classical systems, measurement on a system in a well defined ('pure') state may give rise to several possible results, and quantum mechanics cannot predict the actual result but only the probabilities of them. This is similar to what happens in classical statistical mechanics, whence we might assume that the reason is the same, namely that the quantum mechanical description is incomplete. This was the view supported by several of the 'founding fathers' of quantum theory, the best known being Albert Einstein. However, for a number of reasons, that I will not detail here, the mainstream of the scientific community did not support that view, and the opinions ranged from the pragmatic, even instrumentalistic, interpretation of Bohr and his followers (the Copenhagen interpretation) to the assumption that there is not a strict causality in the microworld, that is, identical causes may led to different effects.

In classical physics we define as 'pure' a state such that all parameters describing the state have fixed values. For instance a pure state of a system of point particles corresponds to the positions and velocities of all particles being known, whilst a 'mixed' state is represented by a probability distribution in phase space. Supporters of the incompleteness of the quantum description might assume that also in quantum mechanics the state of a system of particles should correspond to some probability distribution in phase space. However it is the case that no distribution may be associated to every quantum wave function (or state vector in the Hilbert space) that is free from difficulties. In particular the most popular, the Wigner function, is not positive definite in general whence it cannot be a probability distribution. As a consequence of these facts, naive statistical interpretations are not possible and some people fond of an intuitive picture of the quantum world have replaced them by stochastic interpretations. The difference is that in stochastic interpretations it is usually assumed that particles' motions are very irregular so that the velocity of a particle is not well defined, whence we should not search for distributions in phase space.

In the following I will give a short review of the two most popular stochastic interpretations, known as 'stochastic mechanics' and 'stochastic (or random) electro-dynamics'. I will discuss also an extension of the latter to optics, named 'stochastic optics'.

# 51.2 STOCHASTIC MECHANICS

## 51.2.1 Hydrodynamical Interpretation of Schrödinger equation

The purpose of stochastic mechanics was, at least initially, to derive Schrödinger equation from stochastic assumptions about the motion of the particles. As is well known the complex, linear, time-dependent Schrödinger equation of one particle may be transformed into two nonlinear equations via writing the wavefunction in polar form, that is

$$\psi(\mathbf{r}, t) = \sqrt{\rho(\mathbf{r}, t)}\, exp\left[iS(\mathbf{r}, t)/\, \hbar\right]. \tag{1}$$

Separating real and imaginary parts, we get two equations involving only real quantities, that is

$$\frac{\partial \rho}{\partial t} - \nabla(\rho \mathbf{v}) = 0, \mathbf{v} \equiv -\nabla S/m \tag{2}$$

$$\frac{\partial S}{\partial t} = \frac{1}{2m}(\nabla S)^2 + V(\mathbf{r}) - \frac{\hbar^2}{2m}\left(\frac{\nabla^2 \sqrt{\rho}}{\sqrt{\rho}}\right). \tag{3}$$

The former may be interpreted as a continuity equation of a fluid with density $\rho(\mathbf{r},t)$ and velocity $\mathbf{v}(\mathbf{r}, t)$. Then the latter is a dynamical equation that may be written in terms of $\rho$ and $\mathbf{v}$. In fact if we apply the gradient operator, $\nabla$, to eq.(3) we get, taking the latter eq.(2) into account,

$$\frac{d\mathbf{v}}{dt} \equiv \frac{\partial \mathbf{v}}{\partial t} + (\mathbf{v}.\nabla)\mathbf{v} = -\frac{1}{m}\nabla[V(\mathbf{r}) + U(\mathbf{r})],$$

$$U(\mathbf{r}) = -\frac{\hbar^2}{2m^2}\frac{\nabla^2\sqrt{\rho}}{\sqrt{\rho}}. \tag{4}$$

This hydrodynamical interpretation was proposed by Madelung in 1926.[1] Indeed for $\hbar = 0$ eq. (4) becomes the Euler equation of motion of an inviscid fluid under the action of an external force. The modification introduced by quantum mechanics derives from the term $U(\mathbf{r},t)$, called 'quantum potential' in the de Broglie–Bohm theory, which will not be commented on here; see (Holland, 1993).

If we accept Born's interpretation of $\rho(\mathbf{r},t)$ as a probability density of the position of the particle, it is natural to look at the Madelung fluid as a fictitious fluid consisting of a statistical ensemble of particles that represents our information about the unique particles of the system. Thus $U(\mathbf{r},t)$ has the resemblance of the potential of a force produced by some field, which would provide an explanation for the quantum behaviour. Indeed many people have attempted to make a derivation of the Schrödinger equation, or to provide an interpretation of wave mechanics, via explaining the origin of the terms in eq.(4). A deterministic approach was proposed by Bohm, reformulating a previous theory of de Broglie,[2] and a stochastic one by E. Nelson,[3] with the name of 'stochastic mechanics'. It should not be confused with 'stochastic quantization',[4] which is a formal procedure consisting of the replacement of real time by an imaginary time, that is $t \rightarrow it$, in order to make convergent some relevant integrals, which allows the formulation of quantum field theory as similar to classical statistical mechanics.

## 51.2.2 Stochastic Interpretation of Schrödinger Equation

Here I start presenting the theory, which rests upon the assumption that the motion of a (quantum) particle may be treated as a stochastic process, that is a probability distribution on the set of possible paths. I shall consider the particle in one dimension for simplicity, and later I will comment on the modifications needed for a 3N-dimensional theory. The presentation here follows L. de la Peña.[5] We assume that the particle's path may be represented by a continuous, but nowhere differentiable, function of time, $x(t)$. Lack of differentiability means that it is not possible to get an instantaneous velocity as the time derivative of the position coordinate. However, if we consider a Markov stochastic process, it is possible to define two stochastic time derivatives, forward, $D_+$ and backward, $D_-$, as follows

---

[1] (Madelung, 1926).     [2] (Holland, 1993).
[3] (Nelson, 1966); (Nelson, 1967).     [4] (Masijuma, 2009).
[5] (de la Peña, 1969); (de la Peña, 1971); (de la Peña and Cetto, 1971).

$$v_+(t) \equiv D_+x(t) \equiv \lim_{\Delta t \to 0} \frac{\langle x(t+\Delta t) - x(t) \rangle}{\Delta t},$$

$$v_-(t) \equiv D_-x(t) \equiv \lim_{\Delta t \to 0} \frac{\langle x(t) - x(t-\Delta t) \rangle}{\Delta t},$$

(5)

where $\langle \rangle$ means ensemble average, and $v_+(t), v_-(t)$ are called forward and backward velocities, respectively. A stochastic process is a Markov process in case the past and future are conditionally independent given the present. The notion is time symmetric.

As there are two velocities there are four accelerations. Identifying a linear combination of the accelerations with the quotient of force by mass, it is possible to obtain different theories. For a particular choice of the linear combination the (Einstein–Smoluchowski) theory of Brownian motion is obtained. Another choice leads to

$$\frac{\partial v}{\partial t} + v\frac{\partial v}{\partial x} - 2D^2 \frac{\partial}{\partial x}\left(\frac{1}{\sqrt{\rho}}\frac{\partial^2 \sqrt{\rho}}{\partial x^2}\right) = f/m = -\frac{1}{m}\frac{\partial V(x)}{\partial x},$$

(6)

where $m$ is the mass of the particle. The latter equation corresponds to the force $f$ being conservative and it agrees with eq.(4) provided that we identify $2D = \hbar/m$ This is the basis for the interpretation of the Schrödinger equation as the equation for the evolution of the probability density of a particle moving under the action of a given conservative force plus a non-dissipative stochastic force.

## 51.2.3 Does Stochastic Mechanics Provide a Realistic Interpretation of Quantum Mechanics?

Stochastic mechanics has had successes but also failures which may be summarized as follows.[6] The main success is of course the derivation of Schrödinger equation with the correct relation between the wave function and the probability density of the particle positions (i. e. Born's rule). It is also possible to explain why identical particles satisfy either Bose–Einstein or Fermi–Dirac statistics (see (Nelson, 1985), Section 20) and why particles have integer or half-odd spin.[7] Nelson (1985, Section 17) has pointed out that it is also possible to get a stochastic picture of the two-slit experiment, explaining how particles have trajectories going through just one slit or the other, but nevertheless producing a probability density as for interfering waves. All these developments are rather formal and it is difficult to get an intuitive picture from them. The stochastic processes associated to these phenomena are rather strange, to say the least. In particular the existence of nodal surfaces in the stationary states of the Schrödiger equation is counterintuitive. It is possible to believe that there are stochastic processes which, for unexplained reasons, give rise to barriers which cannot be crossed (Nelson, 1985).

[6] (Nelson, 1985); (Nelson, 2012).
[7] (Dankel, 1977); (Dohrn and Guerra, 1978); (Wallstrom, 1988).

Understanding many-body (non-relativistic) quantum mechanics is more difficult. It is necessary to assume that every particle has an influence on all other particles via the quantum potential, that is the third term in the left side of eq.(6). In fact, in the many-body equation that term contains derivatives with respect to all the particle's positions. The result is that, if there are two uncoupled systems, an alteration of the second affects the first. This makes it unrealistic to regard the trajectories as physically real, as Nelson has pointed out.[8] In summary, stochastic mechanics does not provide a clear intuitive model of the behaviour of quantum systems.

# 51.3 STOCHASTIC ELECTRODYNAMICS

Stochastic electrodynamics (SED) is a theory that combines the laws of classical electrodynamics with the assumption that the vacuum is not empty but there is a random electromagnetic field (or zeropoint field, ZPF) filling the whole space. A review of the work done on this up until 1995 is the book by L. de la Peña and A. M. Cetto (de la Peña and Cetto, 1996), see also (Santos, 2012). The application of the theory to some simple systems provides a picture of several phenomena usually considered as purely quantum, like the stability of the classical (Rutherford) atom, Heisenberg uncertainty relations, entanglement, specific heats of solids, behaviour of atoms in cavities, paramagnetism in the vacuum, etc.

The origin of SED may be traced back to Walter Nernst, who extended to the electromagnetic field the zeropoint fluctuations of oscillators assumed by Planck in his second radiation theory of 1912. Nernst also suggested that the zeropoint fluctuations might explain some empirical facts, like the stability of atoms and the chemical bond. The proposal was forgotten due to the success of Bohr's model of 1913 and the subsequent development of the (old) quantum theory. Many years later the idea was put forward again several times, e. g. by Braffort et al. in 1954 and by Marshall in 1963.[9]

A fundamental hypothesis leading to SED is that the quantum vacuum fields are real stochastic fields, in particular the vacuum electromagnetic field, the so-called zeropoint field, ZPF. The spectrum $\rho_{ZPF}(\omega)$ (energy per unit volume and unit frequency interval) of the ZPF in free space is fully characterized, except for a constant, by the condition of Lorentz invariance.[10] This gives

$$\rho_{ZPF}(\omega) = \frac{1}{2\pi^2 c^3} \hbar\omega^3, \tag{7}$$

that corresponds to an average energy $\frac{1}{2}\hbar\omega$ per normal mode. The Planck constant $\hbar$ enters the theory via fixing the scale of the assumed vacuum radiation. The full characterization of the vacuum electromagnetic field in free space may be achieved

---

[8] (Nelson, 2012).    [9] (Braffort et al., 1954); (Marshall, 1963).
[10] (Milonni, 1994); (de la Peña and Cetto, 1996).

via an expansion in normal modes leading to a set of (complex) amplitudes $\{a_j\}$. The amplitudes have random phases and two possible probability distributions $W_0$ have been proposed, that is:

$$W_0 = \prod_j \frac{2}{\pi} exp\left(-\frac{2|a_j|^2}{\hbar\omega_j}\right), W_0 = \prod_j \delta\left(|a_j|^2 - \frac{1}{2}\hbar\omega_j\right), \tag{8}$$

where $\delta()$ is Dirac's delta. Both forms are consistent with eq.(7) and they may be proved equivalent, via the central limit theorem, in the limit where the normalization volume goes to infinity. The spectrum eq. (7) is appropriate for systems at zero Kelvin, but SED may be also studied at a finite temperature, where we should add to eq. (7) the thermal Planck spectrum. SED studies the motion of charged particles using classical electrodynamics but the back actions of the charged particles on the ZPF are neglected. In addition SED may provide an interpretation for phenomena in cavities, where the spectrum eq.(7) is modified but the average energy $\frac{1}{2}\hbar\omega$ per normal mode still holds true.

There are many examples where SED fails, predicting results in contradiction with quantum mechanics and with experiments. Therefore SED may be taken as an approximation to quantum mechanics in a limited domain, although this opinion has been questioned by some authors.[11] Indeed by introducing additional assumptions it is possible to extend the validity of the SED, for instance deriving the Schrödinger equation. There are at least two reasons for the failure of strict SED (as defined above), namely the neglect of the back action of the particles on the electromagnetic radiation and the fact that SED applies to charged particles. A natural generalization of SED would be taking into account other vacuum fields, in particular metric fluctuations of spacetime. Attempts in that direction have been made without too much success.[12]

We shall be concerned with the intepretation of nonrelativistic quantum mechanics (QM in the following), which deals with particles, like electrons, nuclei, atoms, or molecules, moving with low velocity compared to light, and ignoring spin. A short review of SED is presented in the following, where we comment on the analogies and differences between SED and QM treatments of some simple systems.

### 51.3.1 The Harmonic Oscillator

The harmonic oscillator in one dimension is the most simple system to be treated within SED (the free particle requires a more careful study in order to avoid divergences). I revisit a well known treatment of the oscillator in SED,[13] emphasizing the aspects that may provide a clue for a more complete stochastic interpretation of quantum mechanics.

---

[11] (de la Peña et al., 2015).    [12] (Santos, 2006).
[13] (Santos, 1974); (de la Peña and Cetto, 1996); (Santos, 2012).

If a charged particle moves in one dimension in a potential well and at the same time is immersed in electromagnetic noise (the ZPF), it may arrive at a dynamical equilibrium between absorption and emission of radiation. In order to study the equilibrium I shall write the differential equation for the one-dimensional motion of the particle in the non-relativistic approximation. The passage to more dimensions is straightforward. Neglecting magnetic effects of the ZPF and the dependence of the field on the position coordinate, which corresponds to the common electric dipole approximation, the differential equation of motion of the particle in a harmonic oscillator potential is

$$m \ddot{x} = -m\omega_0^2 x + m\tau \dddot{x} + eE(t), \tag{9}$$

where $m(e)$ is the particle mass (charge) and $E(t)$ is the $x$ component of the electric field of the radiation (the zeropoint field, ZPF). The equation of the mechanical classical oscillator is modified by the two latter terms of eq.(9). The second term on the right side is the damping force due to emission of radiation. In it we have introduced the parameter $\tau$ defined by

$$\tau = \frac{2e^2}{3mc^3} \Rightarrow \tau\omega_0 = \frac{2}{3} \frac{e^2}{\hbar c} \frac{\hbar\omega_0}{mc^2} << 1. \tag{10}$$

The dimensionless quantity $\tau\omega_0$ is very small, being the product of two small numbers, namely the fine structure constant, $\alpha \equiv e^2/\hbar c \sim 1/137$, and the nonrelativistic ratio $\hbar\omega_0/mc^2 \simeq v^2/c^2 << 1$. Thus the two latter terms of eq.(9) may be taken as small, which allows some useful approximations. Eq.(9) is a stochastic differential equation of Langevin type with coloured (non-white) noise. It is named the Braffort–Marshall equation after early workers on SED.[14] Solving an equation of this kind when the stochastic properties of $E(t)$ and eqs.(8) are known means finding the evolution of the probability distribution of the relevant quantities as functions of time, starting from given initial conditions. When the time goes to infinity the probability distributions become independent of the initial conditions, leading to the stationary (equilibrium) distribution.

Several solutions of eq.(9) have been published.[15] The most simple is the stationary solution, which may be found by Fourier transform of eq.(9). It allows getting the spectral density of the oscillator from the spectrum of the field, eq.(7). The resulting spectrum of $x(t)$ is

[14] (Braffort *et al.*, 1954); (Marshall, 1963).
[15] (Santos, 1974); (de la Peña and Cetto, 1996); (Santos, 2012).

$$S_x(\omega) = \frac{\hbar\tau\omega^3}{\pi m[(\omega_0^2 - \omega^2)^2 + \tau^2\omega^6]}, \tag{11}$$

whence it is trivial to get the quadratic means of the relevant variables. Calculating the integrals is lengthy but it becomes trivial in the limit $\tau \to 0$ where the integrand is highly peaked at $\omega \simeq \omega_0$. We get

$$\langle x^2 \rangle = \int_0^\infty S_x(\omega)d\omega \simeq \frac{\hbar}{2m\omega_0}, \langle v^2 \rangle = \langle \dot{x}^2 \rangle \simeq \frac{\hbar\omega_0}{2m}, \tag{12}$$

where a similar procedure has been used for the quadratic mean velocity. Hence the total mean energy to zeroth order in the small quantity $\tau\omega_0$, is

$$\langle E \rangle = \left\langle \frac{1}{2}m\omega_0^2 x^2 + \frac{1}{2}mv^2 \right\rangle = \frac{1}{2}\hbar\omega_0. \tag{13}$$

These results agree with the quantum prediction for the ground state of the oscillator, the unique stationary state. Taking eqs. (8) into account it is also possible to calculate the probability distributions of coordinate and velocity, that are Gaussian with zero mean. Hence is easy to get the probability distribution of the energy, see eq.(14) below.

Calculating the corrections due to the finite value of the parameter $\tau$ in eqs.(12) to (13) is straightforward, although it is also lengthy and will not be reproduced here. An interesting point is that the correction is not analytical in $\tau$ (or in the fine structure constant $\alpha$), but the leading, logarithmic, term agrees with the radiative corrections of quantum electrodynamics. Thus the radiative corrections (to the nonrelativistic oscillator) in SED may be obtained exactly, whilst in quantum electrodynamics the calculations (in more interesting relativistic systems) requires perturbative techniques.

## 51.3.2 Comparison between the Stationary State in SED and the Ground State in QM

The stationary state of the oscillator in SED is rather similar to the ground state of the oscillator in QM, as shown above. However the probability distribution of the energy does not agree. Indeed the SED prediction is an exponential distribution of energies whilst the quantum prediction is a sharp energy, that is

$$\rho_{SED}(E) = \frac{1}{2}\hbar\omega_0 exp\left(-\frac{2E}{\hbar\omega_0}\right), \rho_{QM}(E) = \delta\left(E - \frac{1}{2}\hbar\omega_0\right). \tag{14}$$

I will devote some space to explain the discrepancy because it clarifies the SED interpretation of QM.

Firstly I mention that the conflict between QM and SED in eq.(14) is an example of the general argument used by von Neumann in his celebrated theorem of 1932 proving that hidden variable theories are incompatible with QM.[16] That theorem prevented research in hidden variables theories until Bell's rebuttal in 1966.[17] J. von Neumann starts with the assumption that any linear relation between quantum observables should correspond to a similar linear relation between the possible (dispersion free) values in a hypothetical hidden variables theory. In our case the energy $E$ is a linear combination of $v^2$ and $x^2$. Thus as the energy predicted by quantum mechanics, $E = \hbar\omega_0/2$, is sharp, any pair of values of $v^2$ and $x^2$ in the hidden variables theory should fulfil, according to von Neumann's hypothesis,

$$mv^2 + m\omega_0^2 x^2 = \hbar\omega_0, \tag{15}$$

which is not compatible with the SED distributions. For instance the possible value $v^2 = 2\hbar\omega_0/m$ is incompatible with eq.(15) because it would imply $x^2 \leq 0$. Bell's rebuttal to von Neumann pointed out that the contradiction only arises when two of the quantum observables involved do not commute and in this case the measurement of the three observables should be made in, at least, two different experiments. Thus a contextual hidden variables theory is possible, that is a theory where it is assumed that the value obtained in the measurement depends on both the state of the observed system and the full experimental context.

In our case the apparent contradiction disappears if we take into account how the energy of a state is defined *operationally* (i. e. how it may be measured). In SED the stationary state corresponds to a dynamical equilibrium between the oscillator and the ZPF. Checking empirically whether a dynamical equilibrium exists requires a long time, ideally infinite. If we define the energy of the oscillator in equilibrium as the average over an infinite time, it would be obviously sharp. In fact the probability distribution of the 'mean energy over a time interval of size $\Delta t$' has a smaller dispersion the greater $\Delta t$ is, and will be dispersion free in the limit $\Delta t \to \infty$. Thus it is natural to assume that the ground state energy as defined by QM actually corresponds to measurements made over infinitely long times. This fits fairly well with the quantum energy–time (Heisenberg) uncertainty relation

$$\Delta E \Delta t \geq \hbar/2, \tag{16}$$

which predicts that the measured energy does possess a dispersion $\Delta E$ if the measurement involves a finite time $\Delta t$. Thus no contradiction exists between SED and QM for the energy in the ground state.

---

[16] (von Neumann, 1932).     [17] (Bell, 1966).

### 51.3.3 Other Applications to Linear Systems

The generalization of the harmonic oscillator in SED to many dimensions is straightforward using the appropriate extension of eq.(9). In the following I summarize the results of two simple examples of coupled oscillators. Firstly, a system of two one-dimensional oscillators at a long distance as an example of van der Waals forces. The system is interesting because it shows that a phenomenon similar to quantum entanglement appears also in SED. The second example is a system of an array of coupled three-dimensional oscillators at a finite temperature, that reproduces the Debye theory of the specific heat of solids.

#### 51.3.3.1 *A Model for Quantum Entanglement in the London–Van Der Waals Forces*

As is well known, entanglement is a quantum property of systems with several degrees of freedom, which appears when the total state vector (or wave function) cannot be written as a product of vectors associated to one degree of freedom each. It is most relevant if the degrees of freedom involved may be measured in two distant places. In recent times entanglement has been the subject of intense study, and a resource for many applications, especially in the domain of quantum information, where entanglement usually involves spin or polarization. Here I will illustrate, with a simple example, that entanglement might be understood as a correlation induced by quantum vacuum fluctuations acting in two different places.

I shall consider the London theory of the van der Waals forces in a simple model of two one-dimensional oscillating electric dipoles. Each dipole consists of a particle at rest and another particle (which we will name an electron) with mass $m$ and charge $e$. In the model it is assumed that every electron moves in a harmonic oscillator potential and there is an additional interaction between the electrons. Thus the Hamiltonian is

$$H = \frac{p_1^2}{2m} + \frac{1}{2}m\omega_0^2 x_1^2 + \frac{p_2^2}{2m} + \frac{1}{2}m\omega_0^2 x_2^2 - Kx_1 x_2, \qquad (17)$$

where $x_1 (x_2)$ is the position of the electron of the first (second) dipole with respect to the equilibrium position. The positive parameter $K$ depends on the distance bewteen the dipoles, but this dependence is irrelevant for our purposes.

The solution of the Schrödinger equation may be got analytically, that is

$$\Psi = \sqrt{\frac{m}{\pi\hbar\sqrt{\omega_+\omega_-}}}exp\left\{-\frac{m}{4\hbar}[(\omega_+ + \omega_-)(x_1^2 + x_2^2) + 2(\omega_+ - \omega_-)(x_1 x_2)]\right\}, \quad (18)$$

where

$$\omega_+ \equiv \sqrt{\omega_0^2 + K/m}, \omega_- \equiv \sqrt{\omega_0^2 - K/m}.$$

This wave function cannot be written as a product of two wave functions, associated to one electron each. Therefore the state is entangled and it may give rise to an EPR-type paradox.[18] For instance a measurement of the position of the second electron giving $x_2 = 1$ (alternatively $x_2 = -1$) leads to a wave function for the first electron,

$$\psi(x_1) \propto exp\left\{-\frac{m}{4\hbar}[(\omega_+ + \omega_-)(x_1^2 + 1) \pm 2(\omega_+ - \omega_-)x_1]\right\}.$$

If the wave function is assumed to represent a pure quantum state, then the state of the first electron is modified by a measurement performed in the second one, that may be far away.

In sharp contrast the interpretation offered by SED is transparent: there is a random motion of the electrons induced by the ZPF, and the correlation is produced by the interaction. The SED calculation starts from equations of motion, similar to eq.(9), that is

$$\begin{aligned}
m\,\ddot{x}_1 &= -m\omega_0^2 x_1 - Kx_2 + \frac{2e^2}{3c^3}\dddot{x}_1 + eE_1(t),\\
m\,\ddot{x}_2 &= -m\omega_0^2 x_2 - Kx_1 + \frac{2e^2}{3c^3}\dddot{x}_2 + eE_2(t).
\end{aligned} \tag{19}$$

We may neglect the $x$ dependence within each dipole but, as we assume that the distance between dipoles is large, we shall take the stochastic processes $E_1(t)$ and $E_2(t)$ as uncorrelated. The coupled eqs.(19) may be decoupled via writing new equations which are the sum and the difference of the former, and introducing the new variables $x_\pm(t) = \frac{1}{\sqrt{2}}[x_1(t) \pm x_2(t)]$. The solution is straightforward and the result agrees with the quantum prediction eq.(18). Entanglement appears as *a correlation between the (quantum) fluctuations* of the dipoles, these fluctuations induced by the vacuum fields (the ZPF) and the correlation due to the interaction between the dipoles.

### 51.3.3.2 *Specific Heats of Solids*

An application of SED at a finite temperature is the calculation of the specific heat of solids, which we summarize in the following.[19] We may consider a solid as a set of positive ions immersed in an electron gas. As is well known the electrons contribute only slightly to the specific heat at not too high temperatures. In SED we study the motion of the ions under the action of three forces. The first one derives from the interaction with the neighbour ions and the electron gas, that may be modelled by an oscillator potential which increases when the distance between the neighbour ions departs from the equilibrium configuration. The second is the random background radiation with the Planck spectrum (including the ZPF) and the third one is the radiation reaction. This gives rise to a discrete set of coupled third-order differential

---

[18] (Einstein, Podolsky, and Rosen, 1935).    [19] (Blanco *et al.*, 1991).

equations that may be decoupled by the introduction of normal mode coordinates. After that, every equation is similar to eq.(9) and may be solved in analogous form. The net result is that the mean (potential plus kinetic) energy in equilibrium becomes

$$E(\omega) = \frac{1}{2}\,\hbar\omega\,coth\left(\frac{\hbar\omega}{2kT}\right), \tag{20}$$

where $\omega$ is the frequency of the mode. With an appropriate distribution of modes this leads to the Debye quantum theory.

There are other interesting results of SED at a finite temperature, in particular relating to magnetic properties.[20]

### 51.3.3.3  *The Particle in a Homogeneous Magnetic Field*

Another linear problem that has been extensively studied within SED is the motion of a charged particle in a homogeneous magnetic field.[21] The most relevant result is the prediction of magnetic properties of a free charge, which departs from classical physics and agrees with QM. The SED calculation of the free charged particle in a homogeneous field is easy. Taking the $Z$ axis in the direction of the magnetic field, the motion in the $Z$ direction is uniform and what remains is a linear two-dimensional problem. Hence it is possible to derive and solve the appropriate stochastic equations, similar to eqs.(9). The result is that the motion is circular with the Larmor frequency $\omega_0$. The stationary values for the magnetic moment, mean energy, and mean angular momentum agree with the quantum ground state in the limit $\alpha \to 0$. As in the harmonic oscillator there is a discrepancy for the probability distribution.

## 51.3.4  Failure in Nonlinear Systems

SED reproduces quantum results, and agrees with experiments, in a limited domain. The domain includes those systems of charged particles interacting with external forces that may be treated linearly. However there are many phenomena, even in the non-relativistic domain, where SED gives wrong predictions.

The study of the stationary state of nonlinear systems in SED cannot made via Fourier transform as in the linear ones. The method used is to derive a Fokker–Planck equation for the classical constants of the motion. The diffusion coefficient is derived from the action of the ZPF and the drift from the radiation reaction. A simple example is the planar rigid rotor modelled by a charged particle compelled to move in a circle with given radius $R$. The system has a single classical motion constant, namely the rotation frequency $\omega$. The predictions of SED are the following probability density and mean energy (Santos, 2012):

---

[20] (de la Peña and Cetto, 1996).      [21] (de la Peña and Cetto, 1996); (Boyer, 1975).

$$\rho(\omega) = \frac{2I}{\hbar} exp\left(-\frac{2I\omega}{\hbar}\right) \Rightarrow \langle E \rangle = \frac{I}{2} \int_0^\infty \omega^2 \rho(\omega) d\omega = \frac{\hbar}{I},$$

where $I = mR^2$ is the moment of inertia. This result does not agree with the quantum prediction, namely $E = 0$. Similar disagreement is obtained for the three-dimensional rigid rotor.

The hydrogen atom is the most relevant nonlinear system within elementary quantum mechanics, and therefore a crucial test for the validity of SED. The result was that the atom is not stable but ionizes spontaneously due to the orbits passing close to the nucleus.[22] Of course this is the opposite to the classical result (without ZPF) where the atom collapses. Numerical solutions for the hydrogen atom in SED have been made[23] that explain the stability of the atom, but the numerical methods have uncertainties that make it difficult to know whether the agreement with quantum predictions is fair (e. g. for the stationary probability distribution).

### 51.3.5 The Equilibrium between Radiation and Matter

This is another problem treated by several authors,[24] and it has been claimed that Planck's law may be derived from classical postulates, usually within the framework of SED. For instance Boyer (2012) suggested that early derivations of the classical law by van Vleck (van Vleck and Huber, 1977) involved Newtonian dynamics and that a study with relativistic dynamics should led to Planck's law. However it has been shown that thermal equilibrium of relativistic particles also leads to the Rayleigh–Jeans law.[25] In my opinion all alleged classical derivations involve assumptions that do not derive from classical laws. Boyer (2013) has shown that the ZPF, when viewed in either an accelerated frame or a gravitational field, gives rise to Planck's law, thus providing a classical derivation of the Davies–Unruh effect. However, this is a different problem from the thermal equilibrium spectrum.

## 51.4 STOCHASTIC OPTICS

The hypothesis that the vacuum electromagnetic field (ZPF) is real allows a stochastic interpretation of several quantum optical phenomena, which is known as stochastic optics. It is briefly reviewed in the following.

---

[22] (Marshall and Claverie, 1980).     [23] (Cole and Zou, 2003).
[24] (de la Pena *et al.*, 2015); (Boyer, 1971).     [25] (Blanco *et al.*, 1983a, 1983b).

## 51.4.1 Vacuum Radiation Field in Confined Space

In confined space the normal modes of the vacuum electromagnetic field are no longer plane waves, although the mean energy per normal mode remains $\frac{1}{2}\hbar\omega$. This gives rise to phenomena that may be interpreted in terms of stochastic vacuum fields.

A well known example is the Casimir effect,[26] the attractive force between two metallic parallel plates at a small distance. Casimir derived it in 1946 from quantum electrodynamics. The stochastic derivation[27] provides a clear intuitive picture: The force is due to the fact that the radiation pressure of the ZPF is different on the two sides of every plate, smaller on the side oriented towards the other plate. Similar calculations have been made for other geometries, like a charged hollow sphere.

Many experiments have been performed with cavities. For instance, it has been shown that the lifetime of an atom changes when it is enclosed in a metallic cavity.[28] The emission of radiation is enhanced, and the lifetime shortened, when the radiation emitted corresponds to any of the normal modes inside the cavity, and inhibited if it is not. A stochastic interpretation is provided modelling an atom as a charged dipole oscillator placed between two metallic plates and immersed in ZPF.[29] It is shown that the rate of emission of the oscillator in a excited state is modified by the boundaries of the cavity, in good agreement with experiments.

## 51.4.2 Wave–Particle Behaviour of Light

Here I comment on an anticorrelation-recombination experiment exhibiting the wave–particle behaviour of light.[30] A radiation signal consisting of well separated photons is sent to one of the incoming channels of a beamsplitter $BS1$ and two photon detectors are placed in front of the outgoing channels. No coincidences are observed, which shows the corpuscular behaviour of light: a photon is not divided, but goes to one of the detectors. If the detectors are removed and the two outgoing radiation beams are recombined in another beamsplitter $BS2$, then the detection in the outgoing channel depends on the difference in lengths of the two paths from $BS1$ to $BS2$, a typically wave behaviour. The stochastic intepretation,[31] assuming that vacuum fields are real, is as follows. The field $E$ entering $BS1$ should give two outgoing fields $E/\sqrt{2}$. However there is another (vacuum) field $E_0$ entering via the second incoming channel of $BS1$, whence the outgoing fields will be

---

[26] (Milonni, 1994).     [27] (de la Peña and Cetto, 1996).

[28] (Jhe et al., 1987).    [29] (Franca et al., 1992).

[30] (Grangier et al., 1986).    [31] (Marshall and Santos, 1988).

$$E_1 = \frac{E + E_0}{\sqrt{2}}, \quad E_2 = \frac{E - E_0}{\sqrt{2}}. \tag{21}$$

Depending on the relative phases, one of the intensities outgoing from *BS1* may be large and the other one small, whence it is probable that the former is detected but not the latter, which explains the anticorrelation. In the recombination process the fields eq.(21) will enter *BS2* giving rise to the following intensity in one of the outgoing channels

$$\left| \frac{E_1}{\sqrt{2}} + exp(i\phi) \frac{E_2}{\sqrt{2}} \right|^2 = \frac{1}{2}(|E|^2 + |E_0|^2) + \frac{1}{2}(|E|^2 - |E_0|^2)cos\phi,$$

where $\phi$ is the relative phase due to the different path lengths. A quantitative treatment is able to explain the observed interference effect.[32]

## 51.4.3 Spontaneous Parametric Down-Conversion (SPDC)

SPDC consists of the production of correlated radiation beams (entangled photon pairs) when a laser inpinges on an appropriate crystal having nonlinear electric susceptibility. A stochastic interpretation of the phenomenon is possible[33] that actually closely parallels the standard quantum electrodynamical one, the difference being the use of either classical amplitudes of the vacuum or quantum field operators.

A large number of experiments have been performed using entangled photon pairs produced by SPDC. Stochastic interpretations of many of these experiments using the Weyl–Wigner formalism in the Heisenberg picture have been published, which we cannot detail here.[34] For an introduction to the Weyl–Wigner formalism in quantum optics with an application to the experimental tests of Bell inequalities see (Santos, 2019). This provides an appealing picture of many experiments in terms of stochastic processes.

The stochastic treatment is possible in spite of the Wigner function of quantum states not being positive definite in general. In fact in the Heisenberg picture only the initial state is involved, that is the vacuum state whose Wigner function, given by the former eq.(8), is positive definite. Actually the stochastic interpretation of SPDC experiments presents difficulties because sometimes either the assumed 'stochastic processes' have non-positive (pseudo) probabilities or the calculated values do not agree with the empirical results.

---

[32] (Marshall and Santos, 1988).      [33] (Dechoum *et al.*, 2000).
[34] See (Casado, Marshall, and Santos, 1997, 1998, 1999); (Casado, Fernández-Rueda, et al., 1997a, 1997b); (Casado, Fernández-Rueda, et al., 2000); (Casado, Marshall, and Santos, 2001); (Casado, Guerra, and Plácido, 2008, 2010, 2014, 2015, 2018).

# REFERENCES

Bell, J. S. (1966). *Rev. Modern Phys.*, **38**, 447.

Blanco, R., França, H. M., and Santos, E. (1991). *Phys. Rev. A*, **43**, 693.

Blanco, R., Pesquera, L., and Santos, E. (1983a). *Phys. Rev. D*, **27**, 1254.

Blanco, R., Pesquera, L., and Santos, E. (1983b). *Phys. Rev. D*, **29**, 2240–2254.

Boyer, T. H. (1971). *Phys. Rev. A*, **3**, 2035.

Boyer, T. H. (1975). *Phys. Rev. D*, **11**, 809.

Boyer, T. H. (2012). *Found. Phys.*, **42**, 595, and references therein.

Boyer, T. H. (2013). *Found. Phys.*, **43**, 923, and references therein.

Braffort, P., Spighel, M., and Tzara, C. (1954). *C. R. Acad. Sc. Paris*, **239**, 157; erratum: id **239**, 925.

Casado, A., Fernández-Rueda, A., Marshall, T. W., Risco-Delgado, R., and Santos, E. (1997a). *Phys. Rev. A*, **55**, 3879.

Casado, A., Fernández-Rueda, A., Marshall, T. W., Risco-Delgado, R., and Santos, E. (1997b). *Phys. Rev. A*, **56**, 2477.

Casado, A., Fernández-Rueda, A., Marshall, T. W., Martínez, J., Risco-Delgado, R., and Santos, E. (2000). *Eur. Phys. J. D*, **11**, 465.

Casado, A., Guerra, S., and Plácido, J. (2008). *J. Phys. B: At. Mol. Opt. Phys.*, **41**, 045501.

Casado, A., Guerra, S., and Plácido, J. (2010). *Adv. Math. Phys.*, **2010**, 501521.

Casado, A., Guerra, S., and Plácido, J. (2014). *Eur. Phys. J. D*, **68**, 338–348.

Casado, A., Guerra, S., and Plácido, J. (2015). *J. Mod. Opt.* **62**, 377–386.

Casado, A., Guerra, S., and Plácido, J. (2018). *J. Mod. Opt.*, **65**, 1960–1974.

Casado, A., Marshall, T. W., Risco-Delgado, R., and Santos, E. (2001). *Eur. Phys. J. D*, **13**, 109.

Casado, A., Marshall, T. W., and Santos, E. (1997). *J. Opt. Soc. Amer. B*, **14**, 494.

Casado, A., Marshall, T. W., and Santos, E. (1998). *J. Opt. Soc. Amer. B*, **15**, 1572.

Casado, A., Marshall, T. W., and Santos, E. (1999). *J. Mod. Opt.*, **47**, 1273.

Cole, D. C., and Zou, Y. (2003). arXiv: 0307154 [quant-ph]

Dankel, T. (1977). *J. Math. Phys.*, **18**, 255.

Dechoum, K., Marshall, T. W., and Santos, E. (2000). *J. Mod. Opt.*, **47**, 1273.

de la Peña, L. (1969). *J. Math. Phys.*, **10**, 1620.

de la Peña, L. (1971). *J. Math. Phys.*, **12**, 453.

de la Peña, L., and Cetto, A. M. (1971). *Phys. Rev. D*, **3**, 795.

de la Peña, L., and Cetto, A. M. (1996). *The quantum dice. An introduction to stochastic electrodynamics*. Dordrecht: Kluwer Academic Publishers.

de la Peña, L., Cetto, A. M., and Valdés Hernández, A. (2015). *The emerging quantum. The physics behind quantum mechanics*. Heidelberg: Springer.

Dohrn, D., and Guerra, F. (1978). *Lett. Nuovo Cimento*, **22**, 271.

Einstein, A., Podolsky, B., and Rosen, N. (1935). *Phys.Rev.*, **47**, 777.

França, H. M., Marshall T. W., and Santos, E. (1992). *Phys. Rev. A*, **45**, 6436.

Grangier, P., Roger, G., and Aspect, A. (1986). *Europhys. Lett.*, **1**, 173.

Holland, P. R. (1993). *The Quantum Theory of Motion*. Cambridge: Cambridge University Press.

Jhe, W., Anderson, A., Hinds, E. A., Meshed, D., Moi, L., and Haroche, S. (1987). *Phys. Rev. Lett.*, **58**, 666.

Madelung, E. (1926). *Z. Phys.*, **40**, 322.

Marshall, T. W. (1963). *Proc. Roy. Soc. A*, **276**, 475.

Marshall, T. W., and Claverie, P. (1980). *J. Math. Phys.*, **32**, 1819.

Marshall, T. W., and Santos, E. (1988). *Found. Phys.*, **18**, 185.

Masujima, M. (2009). *Path Integral Quantization and Stochastic Quantization*. Berlin: Springer.

Milonni, P. W. (1994). *The Quantum Vacuum. An introduction to quantum electrodynamics.* San Diego: Academic Press.

Nelson, E. (1966). *Phys. Rev.*, **150**, 1079.

Nelson, E. (1967). *Dynamical theories of Brownian motion.* Princeton, NJ: Princeton University Press.

Nelson, E. (1985). *Quantum fluctuations.* Princeton, NJ: Princeton University Press.

Nelson, E. (2012). Review of stochastic mechanics. *Journal of Physics: Conference Series*, **361**, 012011.

Santos, E. (1974). *Nuovo Cimento*, **19B**, 57. There is a misprint in Appendix A: sinh (cosh) should read si (ci) everywhere.

Santos, E. (2006). *Phys. Lett. A*, **352**, 49.

Santos, E. (2012). Stochastic electrodynamics and the interpretation of quantum physics. arXiv:1205.0916

Santos, E. (2019). Local realism weaker than Bell's compatible with quantum predictions for entangled photon pairs. arXiv:1908.03924

van Vleck, J. H., and Huber, D. L. (1977). *Rev. Mod. Phys.*, **49**, 939.

von Neumann, J. (1932). *Mathematische Grundlagen der Quantenmechanik.* Berlin: Springer-Verlag. English translation, Princeton University Press, 1955.

Wallstrom, T. (1988). *On the derivation of Schrödinger equation from stochastic mechanics.* Princeton, NJ: Princeton University Press.

# Index

Note: Tables and figures are indicated by an italic 't' and 'f', respectively, following the page number. QM = quantum mechanics/quantum theory

## A

α-particle tracks in cloud chamber 213
acausality
  and Forman Thesis 896
  and science journalism in Japan 691–3
acousto-optic modulators 598
action-angle formalism, Stark effect 83
adiabatic principle 83
Adler, SL 42
aether 911
Agar, J 871
agential realism 1031, 1041–4
  interpretation of quantum physics 1044–8
  worldly intra-actions 1032
agential separability 1053
Aharonov, Y 28, 672, 1087
Aharonov-Bohm
  correlation genesis of EPR 741
  effect 1087
  principal citations 743t
AHQP project see Archives for the History of Quantum Physics
Albareda, G 33
Albert, D 69, 70, 1120
Alekseev, IS 681
algebras 439, 442, 484
Allahverdyan, A 29
Allen, HS 923–4
Allori, V 77
Amaldi, E 643
America
  Cold War 497, 609, 670, 736, 756–7, 759, 847–8, 888, 896–7
  new experiments on Bell's theorem 830
  Vietnam war 644–5, 657, 830, 890

American Institute of Physics (AIP) 653, 864, 888
American Philosophical Society 889
American textbooks 722, 725
analysers
  for γ-rays 598
  for low-energy photons 595–7
  time-varying 597f
Andrade e Silva, E 236, 649, 653
angular correlation of Compton-scattered annihilation photons 745–6
angular momentum, diatomic molecules 82
annihilation radiation
  angular correlation between two photons 738, 745–6
  experiments 736
  photons 590
  positrons 589–90, 736
anthropic principle, Wheeler 406, 1192
applications
  articulation of the theory 178
  defined in QM 174
  extending range of validity of theory 177
  foundational 184–93
  'sharp' 178
  vs foundations 175–80
approximation 181
  Born–Oppenheimer 101, 1166
  classical mechanics's continuity 1062
  common electric dipole 1253
  Copenhagen interpretation of QM 1191
  dBB theory 33
  decoherence hides interference 1055
  justification for 290–1, 362
  linearity 1094
  non-relativistic 1253

approximation (*cont.*)
  and Stefan–Boltzmann law  100
  thermodynamic  815
  in treatment of a measurement process  14
  WKB  411
*Archive for the History of Quantum Physics*
      (AHQP project)  833, 888–90
Arrhenius, S  620
artificial radioactivity  590
Aspect, A  3–4, 424, 510, 593–8, 609, 772, 791,
      828–32
  Bell inequalities  175, 344, 743
  CHSH inequality violation  967
astronomical random number generators
      (ARNGs)  354, 360*f*
atom *see* hydrogen atom; physics of the atom
atomic beams  591
Auffèves, A  23
Auletta, G  14, 15
Austria-Hungary, post-World War 1  891–2
avalanche process, reduction  289
Axiom of Choice, with Zermelo–Fraenkel set
      theory  1146
axiomatic field theory  387
axiomatic reconstructions, quantum
      mechanics  418–19
axiomization of QT
  functional analysis  473–94
  in terms of Hilbert spaces (and
      operators)  482
axioms
  CDP  464–7
  five, operational approach  22, 422–4,
      452–60

**B**
Bacciagaluppi, G
  and Crull, E  220, 224, 273, 528, 571, 577,
      579, 852, 857, 1171
  and Valentini, A  60, 63, 203, 226, 271, 630,
      827, 958, 1104, 1110, 1206
Badino, M  4, 97
Balibar, F  27
Balian, R  29
Ballentine, L  28, 412, 516–17, 526, 1223,
      1225, 1228, 1238, 1242–3
Balmer lines  80

Banach algebras  484
Banach spaces  478
band theory of free electrons  669
Baracca, A  647–9, 656–7
Barad, K  1031
Barchielli, Lanz, Prosperi (BLP)  1112
Bari, Italy, Selleri's group  652–5
Barkla, CG  922, 924, 930
Barrett, JA  67
Bassi, A  42
Bauer, E  12, 287, 425
Bayesian probabilities  23–4, 1007–8
Bayesianism  1008
Bedingham, DJ  65
Bell, JS  14, 20, 343*f*
  conclusion of EPR–Bell argument  1205
  entanglement  3
  Erice School  659
  experimental metaphysics  791–2
  Forman Thesis  899
  'impossibility proofs'  965–9
  misunderstandings  1212–15
  non-locality of QM  1085
  quantum gravitation in QM  413
  solution of EPR dilemma  1202–4
Bell test  341*f*
Bell-Kochen-Specker theorem  1044
Bell's dilemma (measurement problem)
      63–6
Bell's inequalities  65, 175, 340–60, 516,
      779, 1078
  Cosmic Bell tests  359–61
  crowd-sourcing (Bellsters)  358
  experiments (BIEs)  588–601
  Fine and  764
  forbidden local hidden variables  1014–15
  freedom-of-choice loophole and
      tests  356–61
  Leggett and  514–15
  local hidden variable theories (LHVT)
      499, 588
  locality and fair-sampling loopholes  340,
      352–6, 358, 361, 758, 966
  non-classicality  417
  simple form  341*f*–2
  Tsirelson's bound  682
  violation  454, 772, 791, 1123, 1171

Bell–CHSH (Clauser, Horne, Shimony, Holt)
    parameter 343–58, 966
Beller, M 135, 157, 165, 184–6, 207–8,
    212, 629, 712–16, 785, 789,
    828, 835
Belot, G 69
Beltrametti, E 451, 660
Benatti, F 68, 1110, 1120–1
Bennett, C 831
Benseny, A 33
Bergia, S 656–7
Bethe, H 1, 373, 378, 716, 864, 900
bi-orthogonal decomposition
    theorem 1164–6
    degeneracy 1165
Bijker, WE 258
Birkhoff, G
    and Kreyszig, E 479–80
    and von Neumann 418, 423, 444–6,
        482–6, 660
BKS see Bohr–Kramers–Slater theory (BKS)
blackbody radiation 77, 79, 96, 99–101,
    235, 241
    spectral density 235
Bloch, F 668
    early QM 183–4, 190
Bloch wave 190
Blokhintsev, DL 675, 679–81, 782–3, 847,
    899, 1228–40
    Mandelstam controversy 677, 1230,
        1240–3
BLP see Barchielli, Lanz, Prosperi
Blum, A 137, 144, 146, 192, 206, 383, 395,
    400, 404, 844
Bohm, D 31, 672
    causal interpretation of QM 1076–7
    on causality violation 1082
    emigration from US and idealogical Soviet
        campaign 783
    entanglement 3
    EPR-type experiments 341, 1088
    on incommensurability 834
    indeterminism of QM 1082
    philosophy 1088–91
    principal citations 743t
    Rosenfeld contra 788
    two-state system 1081–2

Bohm's theory
    (1952) 1157
    complementarity, and
        instrumentalism 784
    and dialectical materialism 782–3
    emergence of scientific realism 783–5
    empty valleys 1086
    equivalence 1084
    existence theory 783
    guiding field 1083–4, 1086
    hidden variable theories 499, 778, 788,
        977–82
    denunciation 851
    holomovement 1087
    as modal interpretation 1157
    non-locality 1085, 1087
    quantum gravity 34–6
    relativistic covariance 1168–70
    responses
        impossibility proof, von
            Neumann 851–5
        Pauli and Heisenberg 850–3
    on Schrödinger equation 1083
    and scientific realism 783–5
    trajectories 33
    see also de Broglie–Bohm theory
Bohm–Aharonov hypothesis, and Furry's
    hypothesis 741–3t, 747, 1087
Bohm–Bub hidden variable theory 968, 973–82
Bohr, N 18–21, 988
    1913 classical description 915–19
    angular momentum quantum number 78
    aphorism 820
    atom modeling 77, 78, 80, 84
        and zero-point energy (ZPE) 120–1
    Bohr frequency condition 84
    classical background 813–18
    on communicability and
        reproducibility 425, 1037, 1043
    Como lecture on complementarity 424,
        535, 688–9, 812
    complementarity argument 21, 31, 145–65,
        273–4, 556–7, 658, 827–8
    complementarity and uncertainty 159,
        798–801, 809
    constitutive role of experimental
        contexts 811

Bohr, N (*cont.*)
  contexts, constitutive role 808–13
  Copenhagen interpretation 543–66,
    977, 988
  correspondence principle 78, 85, 124, 138,
    178, 438, 917–19, 919
  dialectical materialism 673
  early work 915–19
  Einstein–Bohr debate 43, 306, 321, 561,
    630, 634–7, 635, 1045
  epistemology
    lesson from QM 787–820, 827
    vs ontology 557–8
  EPR paradox, correlation genesis 741
  EPR semantics 802–8
  human presence as observer 537–8
  'inadequacy of classical
    electrodynamics' 916
  indefinability thesis, semantics 808
  indeterminacy of the measurement
    interaction 1034
  on information (two-slit interference) 1045
  instrumentalism 812
  measurement matters 1032
  non-epistemic interpretation, of
    uncertainty relations 798
  'ordinary electrodynamics 917
  philosophy-physics,
    reconstruction 1032–40
  planetary model of atom 945–8
  position and momentum, mutually
    exclusive 1033
  QED 395
  QFT, measurement 379–84, 395–7, 670
  QM, classical vs macroscopical 382–3
  quantization conditions 77–89
  quantum atom, and classical
    electrodynamics 915–19
  and Rosenfeld, on QED 382, 395
  stationary states 555–6
  on uncertainty relations of Landau and
    Peierls 381
  wave mechanics as a corrective vs
    uncertainty relation 165
  wave–particle duality 165–6
  ZP vacuum infinite energy density 376
Bohr–Balmer quantization, hydrogen 87

Bohr–Einstein debates 674
Bohr–Heisenberg–Copenhagen
    interpretation, burying of
    creativity 175
Bohr–Kramers–Slater theory (BKS) 137, 152,
    161, 260, 902, 1226–7
Bohr–Rosenfeld argument 121, 383, 399,
    401, 559
Bohr–Sommerfeld planetary theory of
    atom 945–8
Bohr–Sommerfeld quantum theory of
    atom 84–5, 241–2
  defects 138
  stationary states and their energies 137
Bologna, Bergia, S 656–7
Bolshevik regime 667
  *see also* communism
Bolshevik Revolution 673
Boltzmann, L
  combinatorial models, thermal
    equilibrium 97–100
  continuity principle 99
  Maxwell–Boltzmann statistics 99, 260
  quantum theory 82
  statistical mechanics 96–9, 104
  Stefan–Boltzmann law 100
Bonsack, F 763–4, 772
Boolean frame 978–81
Bopp, F 850, 858
Born, M 670
  $\alpha$-particle tracks in cloud chamber
    213–14
  commutation relation 79
  correspondence with Einstein 857–8
  correspondence principle 178
  deterministic completion of QM 855
  doubts on completeness of QM 862
  and Heisenberg, statistical interpretation of
    wavefunction, phase-1 205–9
  on merging of Heisenberg's and
    Schrödinger's ideas 211
  new quantum theory 78–9, 87
  probabilistic interpretation of
    transformation theory 147–60, 185
  $\psi$ interpretation 308
  radiation/light and matter 178–9
  on Schrödinger 857

wave mechanics extended to aperiodic
    phenomena 184
Waynflete Lectures in Oxford 857
Born probability, von Neumann 482
Born rule (1926) 13–18, 29, 33, 35, 36, 65, 79,
    813, 957, 1009, 1207
    probability assignments 810
    rational justification (Schrödinger) 802
Born–Oppenheimer approximation 101
boron oxides, half-integral quantum
    numbers 127
Borrelli, A 852
Bose, SN 243
    Planck's radiation law 259
Bose–Einstein condensation 244
Bose–Einstein statistics 99, 111–12, 186, 243,
    258–60, 275
bosons 243, 275
    and fermions, indistinguishability 256
Bragg, WH, 'paradox of quantity' 283
Brans, K, local-realist model 357
Brassard, G 831
Brazil, Marxist physicists 848
breakthroughs, experiment-led vs
    theory-led 1073
Brewster angle 595–6f
Bricmont, J 34
Brillouin, L 426
British perspectives, classical physics
    910–15
    ether, classical physics 912
    McLaren 910–12
Bronstein, M 671
Brooksby, C 33
Brownian movement 289
Brukner, C 14, 460–2
Bub, J 14, 957, 1166
    modal interpretations of QM 1156–7
Burgers, J 83
Busch, P 23

C
calculus 473
    history of functional analysis 473–4
Caldeira, A 496, 503–9, 527
Camilleri, K 136, 158, 163, 165, 268, 273, 496,
    635, 849, 884

canonical commutation relations 55
canonical genres of QM 709–22
Carlip, S 44
Carnap, R 779, 781
Carney, D 44
Carrier, M, on application-oriented
    research 180
Cartesian coordinates, order and 1089
Cartesian product assumption 68
Casimir energy 377
Casimir, H
    renormalization programme 378
    ZPE 131, 376–8
Cassinelli, G 660
causal interpretation 1076
causal locality, outcome and parameter
    independence 65
causal perturbation theory 485
causality
    causal order 429–30
    common cause principle 64–6
    principle of local causality 65
    see also acausality
Caves, CM 24, 1007
cavity quantum electrodynamics
    (QED) 607–8
cavity radiation, using Planck's law 105
CDP see Chiribella, D'Ariano, Perinotti (CDP)
Chapel Hill conference 396, 404, 407
'characterization' problem 287–9
chemical bond, Heisenberg's concept of
    resonance/exchange 189
Chen Ying Yang 852
Chicago, History of Science Society, Forman
    Thesis 895
Chien-Shiung Wu
    angular correlation of Compton-scattered
        annihilation photons 745–6
    career awards 735
    experimental philosophy 735–50
    Kasday–Ullman–Wu (KUW)
        experiment 735, 744–50
    letter to Bell 746f
Chiribella, D'Ariano, Perinotti (CDP),
    axioms 464–7
Chiribella, G 464–7
Christov, IP 33

CHSH theorem/inequality (Clauser, Horne, Shimony, Holt) 340, 343–58, 966–7
  apparatus used in proof of Bell's inequality 589f
  Bell tests 593, 963
  quantum archaeology 609
Cini, M 645, 648, 657–8
classical determinism, rescue by Bohm 740
classical 'distinguishable' vs quantum 'indistinguishable' 255–8
classical electrodynamics
  Germany 919–22
  and quantum atom 915–19
classical mechanics 77, 1062
classical phase space 77–8
classical physics 1075
  1911–20, Germany 919–22
  Bohr (1913) 915–19
  co-creation with modern 909–25
  communicability, concepts 425, 525–7
classical thermodynamics (1890s) 909
classical vs quantum mechanics, non-Booleanity 977–8
classical vs quantum probability, impossibility theorem 217, 227
Clauser, J 3, 343–58, 347f, 744, 748–9, 757, 829, 832
  see also CHSH
Clausius, R 82
Clifford algebra 1091, 1100
Clifton, R 65, 69
closed theory, defined 856
Colbeck, R and Renner, R 973, 982
Cold War 497, 609, 670, 736, 756–7, 759, 888, 896–7
  attack on 'Copenhagen school' by Soviet physicists 847
  'quantum dissidents' 562, 642, 648, 830, 847
  Zhdanov's speech 847–8
collapse postulate 1105–6
  theories 1075
  wavefunction collapse 1110
collapse process, strength of 1113–14
collapse of state vector 60–1
  dynamical collapse theory programme 63–4

Colston conference, Bristol (1957) 291, 789, 850, 858
common cause principle 64–5
common sense interpretations of QM 10–13, 44
communicability 425, 1037, 1043
communism 561, 658
  dialectical materialism 672, 782–3
  see also Marxism; Soviet Union
commutation relation, Heisenberg's quantization condition 86
commutator 55, 87
Como lecture see Bohr, N, Como lecture on complementarity
complementarity
  argument 21, 31, 145–65, 273–4, 556–7
  banishment 561
  Copenhagen interpretation of QM 523–5, 1224
  interpretation 306–11
  Nishina Yoshio on 688–91, 698–9, 706
  as paradigm 785–8, 835
  principle 556–7, 587
  tests 601–5
  and uncertainty 159, 798–801
  see also Bohr, N
completeness of QM 528–31
compositional role for systems 419–20, 423, 455
Compton, AH, on photons/'light bullets' 236, 271
Compton effect 239
Compton scattering
  annihilation photons, angular correlation 745–6
  distribution formula 598
  Klein–Nishina formula 739
configuration space problem 1120
Conroy, C, modal interpretations of QM 1156–7, 1160
consistent histories 24–7, 1188–93
  families of 24–7
consistent/decoherent histories approach to QM 1175–93
  formalism 1176–80
  Gell-Mann and Hartle 1188–93
  Griffiths and Omnès 1187–8

historical context 1180-7
information gathering and utilizing systems
	(IGUSes) 1176, 1192
contexts, constitutive role 808-13
contexts, systems, modalities (CSM)
	interpretations 23
continuity, defined 455
continuous spontaneous localization (CSL)
	theory 64-5, 1113
	identical particles 1114
	probability rule 39, 41
	relativistic invariant, Bell 1123
continuous wave function 31, 87
continuum wave theory, matter waves 87
convivial solipsism (ConSol), comparison
	with QBism 1023-5
Copenhagen interpretation of QM 16, 21-2,
	175, 400-7, 424-5, 521-42, 992
	approximation 1191
	attack on 847-8
	Bohm and 740, 977
	Bohr and 543-62, 977-82
	classical concepts 525-7
	communism and 561, 847
	complementarity 523-5
	completeness of QM 528-31
	defence of 848
	Dirac and 859
	Everett and 987-8
	first/original 531-3
	Göttingen–Copenhagen 184, 309
	Institute for Theoretical Physics 546-7
	Landau/Lifschitz 1012
	materialism and antireductionism 672
	measurement-induced wave-packet
		collapse 534-5
	myth of 848, 849
		NO unitary interpretation 539
	objective indeterminacy 531-2
	orthodoxy and heterodoxy, postwar
		era 847-60
	paradigm (Hanson) 778
	philosophical challenges to 784
	postwar response to new ideas 560-2
	quantum indeterminacy 535-6
	quantum postulate 553-5
	Russian branch 670-2, 671

	subject observer role 522, 536-9
	thought experiments 558-60
	wave–particle duality 532-3
	see also Heisenberg
'Copenhagen observer' 1061
Copenhagen perspective, quantum
	probabilities 1175
Copenhagen variant of modal interpretation,
	van Fraassen 1162
corpuscular theory of light 104-5, 237, 246,
	260, 263, 268, 271, 273, 283-4, 311,
	575, 647, 799, 1260
correlation function 341
correlation interpretation 13
correspondence principle 78, 85, 124, 138,
	178, 438-9, 919
Cosmic Bell tests 359-61
cosmological constant $\Lambda$ 131-2
cosmology,
	see also universe
Coulomb's law 946
coupled three-dimensional oscillators 1256-7
Cox, RT 23
Crull, E, and Bacciagaluppi 220, 224, 273,
	528, 571, 577, 579, 852, 857, 1171
CSL see continuous spontaneous localization
	(CSL) theory
CSM see contexts, systems, modalities (CSM)
	interpretations
Curie–Langevin law 123, 125
Cybernetics Group 1183-4, 1186
cybernetics and measurement 1193-7

D
D-ontology 96-9, 103, 111
Dakic, B 460-2
Dalla Chiara, ML 452, 660, 1149
Daneri, A 643
Daneri, Loinger, Prosperi 282, 290-1, 643,
	647, 815, 828
D'Ariano, GM 464-7
	Chiribella, D'Ariano, Perinotti (CDP)
		axioms 464-7
dark energy 131-2
Darrigol, O 8, 78-9, 373
	formalism of quantum mechanics 22
Davies-Unruh effect 1259

de Broglie, L  22, 31
  matter-wave theory  267, 283–4
  tragedy of  1216–17
de Broglie–Bohm theory  22, 31–5, 65, 77,
      1205–8
  contextual measurements  1207
  equivariance  1207
  lack of appreciation for  1215–19
  merits of  1208
  no hidden variables  1208
  preferred relation of distant
      simultaneity  65
  reactions to  1215–17
  reactions to Bohm  1217–19
  world wave function  412
  see also pilot-wave theory
de Finetti, B  23
de Witt see DeWitt, B
Debye, P  119, 123, 125, 266, 598, 629, 918,
      1256, 1258
decoherence  9–10, 12, 62, 304
  birth of  507–9
  concept described  830–1
  decohering-worlds  1013–14
  defined  496
  eation  605–8
  environment-induced  296, 605
  experiments  605–8
  facts are stable  1057–8
  grounding the proper assignment of
      probabilities to histories  1194
  Károlyházy on  413
  Schrödinger equation and  1012–13
decoherent/consistent histories approach to
      QM  1175–93
  information gathering and utilizing systems
      (IGUSes)  1176, 1192
definitions (Δ) and axioms (A)  455
  matrix density representation of states  456
degrees of freedom, defined  455
Dehmelt, H  606–7
Del Santo, F  641, 654–5, 830, 835
delayed-choice experiment  601f, 1082
Delbrück, M  259
Denmark
  International Education Board
      (IEB)  547–50

Rask-Ørsted Foundation  545
  see also Copenhagen
density matrix  1112
density operator  9, 1166
d'Espagnat, B  8, 21, 646, 648
  Erice School  659
detection sets, photomultipliers and electronic
      circuits  599
determinate observable, choices  1166
determinate record problem  993
determinism, classical, rescue by Bohm  740
deterministic linear dynamics  992
Deutsch, D  413, 516, 831
DeWitt, B  649, 828, 988
  on Everett interpretation of QM  394,
      407–13, 416, 988, 996, 1001–3, 1160
  'false problem'  30, 383, 1192
  incompatibility between QM and
      gravitation  828
  quantization of gravitational field  383,
      396–401, 404
  Wheeler–DeWitt equation  410, 1190
di Francia, TG  648, 651, 660, 1145, 1149
dialectical materialism  673, 678
  and Bohm's theory  782–3
Dickson, M  35, 65
Dieks, D  35, 36, 63, 258, 1164
  modal interpretations of QM  1156–7
Dieudonné, J  474–81, 485
differential equations, linear, Sturm–Liouville
      theory  474
differential-space theory, Wiener and Siegel
      (1955)  973
Dijksterhuis, EJ  473
Diósi, L  42
  QMUPL  1114
Diósi-Penrose (DP) theory  1116, 1119
Dirac, P  14, 79
  and Born, doubts on completeness of
      QM  862
  commutation relation, Poisson brackets  86
  delta-function  485
  deterministic completion of QM  859
  hole theory  668
  indistinguishable particles  272
  magnetic monopole paper  192
  nature of scientific progress  859

$q$-number theory 87
quantum phenomena, state
        probabilities 211
reductionist declaration 272
relativistic theory 190
to Bohr, renormalization of QM 859
Dirac–Jordan statistical transformation
        theory 87
Dirichlet problem 474
discernability 1151
disciplinarity 876–8
discreteness 1061
    defined 455
dispersion theory for elastic scattering of
        light 78, 79, 84–5, 88
displacement operators 1090
dissipative effects 1115–16
distinguishability, and individual identity 95
distribution theory 485–6
Doppler-shift, tuneable laser 606
Dorling, J 235, 872, 873, 879, 880
down-conversion, spontaneous parametric
        down-conversion (SPDC)
        594–5, 1261
Drude, P 83
Drude–Lorentz metal model 668
Duhem–Quine thesis of
        underdetermination 827, 833, 840
Duncan, A 77, 79
dynamical collapse mechanism 1110
dynamical collapse theory 63–4, 68
dynamical equations 10, 32
dynamical law, Schrödinger equation 535
dynamical state 35–6, 1162
dynamical variables of quantum
        theory 58–60
Dyson, F, renormalization
        programme 378–9, 387

**E**
$\epsilon = h\nu$ 78, 79
earliest quantum conditions 79–80
early debates on future of QM 135–72,
        181–93
Easlea, B, measurement problem 500
'Eastern thought' 687
Eckart, C 87

Eckert, M 82
Ehlers, J 255
Ehrenfest, P 79, 81, 83, 104–6, 110–11, 119,
        260–2, 265, 270–1, 627$f$, 874, 901,
        918–19, 922–3
    adiabatic invariants 83
    entropy and radiation law 79–80,
        104–5, 259
    Lorentz–Ehrenfest calculation 206
eigen-oscillations, crystal lattice 130, 205
eigenfunction, description of a 'pure
        ensemble' 675
eigenstate–eigenvalue link 56, 68, 218
eigenvalues, Born rule 16
Einstein, A
    1921 Nobel prize 925
    1922 lecture 'Geometry and
        Experience' 777
    analogy between gas and radiation 104
    Bose–Einstein statistics 99, 111–12, 186,
        256–66
    correspondence 855–8
    criticism of Bohm's theory 855
    on epistemological problems in atomic
        physics 1045
    'God does not play dice' 306–7, 317, 1223
    misunderstandings 1208–12
    modern/classical physics 920–1
    non-independence interpretation 269
    pilot-wave theory 633–4
    on Planck's quantum hypothesis 234, 241
    QM an incomplete theory of individual
        process 855–6, 859
    QM criticism 303–30, 914
    QM, dissatisfaction with 802
    QM, 'most complete possible theory
        compatible with experience' 855–6
    QM, subsequent advances 328–30
    on radiation
        corpuscularization 104–5
        emission and absorption of 911
    state vector *qua* set of prescriptions, and
        'physical' state of quantum
        systems 802
    on statistical interpretation of
        wavefunction 204
    terminology used 235$t$

Einstein, A (*cont.*)
  zero-point energy (ZPE) 122–4
  *see also* Bose–Einstein statistics
Einstein, Bohm and Bell
  'comedy of errors' 1197–219
  'history for students' 1197
  summary/conclusions 1219
Einstein, Podolsky, Rosen, (EPR) article/
      argument 14, 43, 190, 306–10, 320–5,
      341, 588, 597, 609, 742, 803–8, 827,
      828, 1077–8, 1212–13
  challenge 1044
  dilemma 1200
  discussion of experimental proof
      (Bohm) 740
  EPR singlet pair, spin measurement
      experiment 1169
  EPR-type paradox 1257
  falsifiying alternatives to QT 1110
  misunderstandings 1208–12
  and non-locality 1014, 1110
Einstein and Schrödinger, doubts on
      satisfactoriness of QM 862–3
Einstein–Bohr debate 43, 306, 321, 561, 630,
      634–7, 1045
Einstein–de Haas effect 123
'elastic quantum' (phonon) 669
electrodynamics
  conceptions of Maxwell and Hertz 940–4
  problem of interpretation 938
electromagnetic fields
  multi-fields 69
  quantizing 130, 235, 374–5
  quantum nature? 374
  ZPE 122, 376–8
electromagnetism
  bidirectionality 941
  Maxwell's study 940–4
  separation between formalism and physical
      content 942
electron impact spectroscopy 592
electrons
  anomalous magnetic moment *g* 371
  band theory 669
  bombardment, and optical pumping 591
  Fermi statistics 186
  free, band theory 669

indiscernibility 1145
infinite self-energy 395
microscope thought experiment 535
wavefunction 1117
electron–positron annihilation, Wu–Shaknov
      experiments 736
electrostatic unity of charge (esu) 1145
Ellis, Hagelin, and Nanopoulos 1117
empirical immediacy/veracity 438
empirical interpretation of QM 1223
empty waves 33
energy transfer Compton effect 239
enfolding, folding-unfolding 1088, 1091
Englert, B 15, 33
ensemble interpretation of QM 1227–9
  and Copenhagen interpretation
      of QM 1224
  criticism by Fock 1240–2
ensembles,
  *see also* statistical/ensemble interpretation
      of QM
entangled photons, sources 591, 595
entanglement *see* quantum entanglement
entropy 101–2
  competing definitions 266
  extensivity 106–8
episteme 1182, 1184
*Epistemological Letters*
  1970s 755–71
  association with Ferdinand Gonseth 762–4
  Bonsack's pamphlet 772
  historical context 756–61
  informality of 770–1
  list of recipients 765–6t
epistemology
  definitions 1233
  lesson from QM, Bohr 787–804
  vs ontology 557–8
EPR *see* Einstein, Podolsky, Rosen, (EPR)
      article/argument
Epstein, P 82, 920–1
equilibrium constant, defined 106
equivalence proof
  matrix and wave mechanics 181, 443
  pure states 481
equivariance 1084
  de Broglie–Bohm theory 1207

Erice School of 1976  659–60
ether  911
Eucken, A, ZPE  122
Euclidean geometry  1090
Everett, H  2, 29–30, 64, 848, 987–1107
    content of deduction  1009–11
    Copenhagen interpretation of QM
        and  987–8
    formulation of QM  987–8
    interpretation of QT, and Wheeler  400–7
    measurement problem  989–92
    observable, of which memory states are
        eigenstates  1166
    'Relative State' Formulation of Quantum
        Mechanics  987–8
    relative-state interpretation of QM  296,
        1156, 1158–62
    Theory of Universal Wave Function  988
    on von Neumann formulation of QM  988
    world wave function  412
evolution equation  1112
exchange symmetry  267–9, 271–2
exciton  669
exclusion principle, integration into
        QM  186–7
experimental metaphysics  791–2
experimental outcomes of quantum
        mechanics  53
extensionality axiom  1146

F
Fabry–Perot superconducting cavity  608
facts of QM, defined  1057–8
families of histories  24–7
    consistency condition  26
    evolution of isolated system  26–7
    using probabilities  1193
Farady, M  938–9
Fein, YY  9
Fermi, E, QED  378
fermions, Fermi–Dirac statistics  186, 257, 271
Ferretti, B  641
ferromagnetism
    collectivist alternative  669
    and QM  190
Feyerabend, PK  290–1, 522, 778, 781, 784,
        786–91, 833, 835, 849

Feynman diagrams  245, 378
    perturbation theory  245
Feynman, R, two-slit experiment  213
Fick and Kant  259, 260
Finetti, B  23
Fock, V
    antireductionism  672–5, 679
    and Blokhintsev controversy  675, 679–81
    and Bohr  680–1
    ensemble interpretation of QM  679,
        1240–2
    space formalism  658, 670–2, 1149–50
    visit to Copenhagen  561
form and meaning  709–22
    textbooks  709–10
formalism
    action-angle formalism, Stark effect  83
    consistent/decoherent histories
        approach  1176–80
    demarcation from interpretation  174,
        194–5
    Hilbert space  419, 1166
    historical context  1180–7
    information theory  1185
    as a style of reasoning  1182
Forman, P
    Archive for the History of Quantum Physics
        (AHQP)  833, 888–90
    on Heidelberg  273–4
    reference point for science study  905
Forman Thesis (1971)  891–900
    and acausality  896
    examples  901–4
    impact and controversy  895–900
    rebuttals by Hendry  872–4, 879, 880,
        897–8, 900, 902
    reception, modernity and
        postmodernity  871–86
    Weimar Culture, Causality, and Quantum
        Theory (1918–27)  828, 872–86, 891–5
foundational applications  184–93
foundations vs applications  175–80
Fourier analysis  474
Fowler, RH  269–71, 712, 925–6
France, Marxist physicists  848
Frauchiger and Renner, modal interpretations
        of QM  1156–7

Freedman, S  346–50, 745, 829, 967
  see also CHSH; Clauser
Freire Jr, O  4, 825–42
French, S  11, 55, 247, 256, 274, 1067,
    1135–47
Frenkel, Y  668–9
Friedrichs, K, new mathematical
    physics 385–7
Fries, JK  569
  reconceptualization of Kant  570
Frisch, O  744–5
Fuchs, C  15, 24, 1007
  positive operator-valued measure
    (POVM)  427–8
  on QBism  1009, 1013
functional analysis  473–5, 486
  axiomization of QT  473–94
  distribution theory  485–6
  Hilbert and  475
  history  473
  since 1932  483–7
  unexpected application to QT  475
  von Neumannn and  475
Furry, W  289, 1226
  hypothesis  741–2
    and Bohm–Aharonov hypothesis  747
fuzzy link  69

G
Gabrielse, G  44
Galilean realist tradition of physical
    theory  59, 784
Gamow, G, Thirty Years that Shook
    Physics  888
Gao, S, tension between standard unitary QM
    and relativity  1171
gas
  ideal gas, quantum theory  104, 106–12
  monoatomic gas, yearly energy
    increase  1113
  undulatory theory  105, 191, 263, 266,
    271, 283
Gasbarri, G  42
Gaussian distribution  362
Gaussian wave  32
Gavroglu, K  7
GBG see Ghirardi, Benatti, Grassi

Gedankenexperiment debates, Einstein and
    Bohr  617, 633, 1044, 1046
Geiger–Müller counter  600
Gell-Mann, M  24, 645, 1175–81
  consistent histories approach  1188–93
  and Hartle, J  645, 1175–81, 1187–93
general relativity  7–8, 34, 42–3
  Einstein's special theory  238
  quantization  401, 404
general states with density operators  482
geons, Wheeler on  402–3
Germany
  'Appeal to the Civilised World (1914)'  616
  classical physics (1911–20)  919–22
  post-World War 1  891–2
  quantum theory and classical
    electrodynamics  919–22
  see also Forman Thesis (1971); Weimar
    Culture, Causality, and Quantum
    Theory (1918–27), Paul Forman
γ-ray absorption experiments  590
Ghirardi, Benatti, Grassi  1120–1
Ghirardi, G  655–6, 1110, 1120–1
  measurement problem  831
  objective collapse model  642
Ghirardi, Rimini, Weber theory  37–41, 64–6,
    297, 656, 1106, 1111–12, 1121
  collapse rule  1112
  dynamical collapse mechanism  1112
  hittings  1112
  non-linear stochastic equation, density
    matrix  1112
  QMSL  1106
  wavefunction behaviour  1112
Gifford Lectures, Eddington  910
Gindensperger, E  33
Ginzburg, V  501, 677
Gisin, N  1110
  non-linear deterministic theory  1110
Giustina, M  355f
Gleason, AM  23
Gleason theorem, CSM interpretations
    (contexts, systems, modalities)  23
'God does not play dice' (Einstein)  306–7,
    317, 1223
Gonseth, F  760, 762–4
Gooday, G  716, 910, 912

Göttingen quantum mechanics 214, 475
  von Neumann on QM 214, 475
  ZPE 130
Göttingen–Copenhagen physicists 86,
    184, 309
Grangier, P 23
Grassi, R 41, 66, 1120–1
Gravitational effects of light quanta,
    Rosenfeld, L 394–400
gravitational field 43, 159, 376, 379, 394–402,
    413, 636, 778, 1066, 1116–17, 1125
  measurement 397
  quantization 394–402
gravitational wave detectors 1119
gravitation–induced collapse theories
    41–2, 1125
gravity 8, 34
Green functions, QFT 378, 1090–1
Green, HS 290
Greene, B 44
Griffiths, R 24, 27, 1175–81, 1194
  consistent histories approach 1186–91
Grinbaum, A 419, 423, 429–30, 464
GRW see Ghirardi, Rimini, Weber
guiding equation 1083, 1084f
guiding field 1083, 1084–5, 1084f

H
Haag, R, new mathematical physics 385–7
haecceity 1144
half-integral quantum numbers 126
Halpern, O 260, 264
Hamilton, WR 87, 941
Hamiltonian term
  modal–Hamiltonian interpretation 1166
  Poisson algebra of infinitesimal
      evolutions 439–43
  Schrödinger evolution and 1113
Hamilton–Jacobi theory 82, 87, 442, 480, 740,
    834, 1061
Hanson, complementarity as paradigm
    785–8
Hardy, L, reconstruction 452–60
  five axioms for probabilities 22, 422–4,
      467–8
    'five reasonable axioms' 458–60, 467
  one-half spin system 452–5

shut up and contemplate! 460
  variant 455
harmonic oscillator
  coupled oscillators 1256–60
  one-dimensional 78
  in SED 1252–4
Haroche, S 496, 503, 606–8, 831, 1152
Harrigan, N 66
Hartfield, B 43
Hartle, J 24
  and Gell-Mann, M, consistent histories
      approach 645, 1175–83
Hartz, T 199, 371–92, 407, 684
Hawking, S 408, 413, 1190, 1213
  evaporation of a black hole 413
  wave function collapse 413
Healey, R 24, 36
Heilbron, J 394, 590, 626, 695, 709–16, 722,
    878, 889–90, 895–6, 900
  AHQP project 889
Heisenberg picture, Weyl–Wigner formalism
    1260
Heisenberg, W 13, 16, 21, 25, 675
  Anschaulichkeit 137–47, 163, 379
  concept of resonance/exchange 189
  'Copenhagen interpretation' coined 848
  equation of motion 59
  ferromagnetism 190
    collectivist alternative 669
  half-integer quantum number 243
  indeterminacy principle 283–4, 531–2
  notion of 'closed theory' 856–7
  QM a 'closed theory' 856
  quantization condition, commutation
      relation 86
  quantum indeterminacy 535–6
  restrictions on reality of particles 273
  and Schrödinger 379
  spacetime 'pictures' 59, 137–47
    redefinition 163
  statistical interpretation of wavefunction
      phase-2 209–14
  Umdeutung reinterpretation of
      amplitudes 78, 85, 88, 137–8, 180
  uncertainty principle 127, 157–67, 211,
      798–9
  wave mechanics 949

Heisenberg, W (*cont.*)
    wave-packet collapse 142–3, 152, 159–61,
        166–7, 534–5
    wave–particle duality 532–3
    ZPE 376
Heisenberg–Pauli quantum
        electrodynamics 670
Heitler, W 189, 287–8, 548
helium
    liquefaction 126
    spectrum 187
helium-3, superfluidity of 500–1
Henderson, J, on orthodoxy 839, 848
Hendry, J, on Weimar (WCC) 872–4, 879,
        880, 897–8, 900–2
Hentschel, K, photon concept 236–7, 244,
        248, 592
Herbert, N, Flash experiment 652–3, 656
Hermann, G 567, 828
    Kantian interpretation of QM 567–83
    mathematics, philosophy, and
        physics 567–83
Hermitian operators 37, 39, 962
Hertz, H, theory of electrodynamics
        940–4
Hesse, M, on particles 257
Hessen, B 1228
    *Social and Economic Roots of Newton's
        Principia* 694, 887, 896, 897
hidden parameters 222–3, 226–7, 854,
        960, 1078
hidden variable theories 63, 65, 150, 221,
        788, 828, 957–80, 1078,
        1198–9, 1213
    alternatives to QM 1075–6
    Bell's model 499
    Bohmian mechanics 499, 778, 850–2,
        1084–5
    Bohm–Bub theory 968, 973–6
    contextual, of environmental type 968
    Italy 647
    Kasday–Ullman–Wu (KUW)
        experiment 735, 744–9
    stochastic hidden variables 969–73
    *see also* impossibility proofs
Hilbert
    axiomatic method 476–7

functional analysis 475–8
    *Sixth Problem* 475
Hilbert space 9, 41, 87
    associated probability distributions 437
    capacity 57
    defined 9
    formalism 174, 419, 827
        observables determined by total
            state 1166
    generalization 484
    of infinite dimension 449–52
    operators 423–4
        non-commutative structure 1167
    structure 468–70
    transformation theory 167
histories
    consistency 24–7
    families of histories 24–7, 1193
    functional analysis 473
    Karl Popper's interpretation of
        QM 1224–5
    Kuhn's model 905
    milestones 424–8
    overview of physics
        (1960s/1970s) 756
    quantum field theory (QFT) 373–9
    science and technology 887
    textbooks 175, 712–16, 727–8
    *see also* consistent/decoherent histories
        approach to QM
historiography, quantum, and cultural
        history 887–905
hittings 1112
    wavefunction evolution (Schrödinger
        equation) 1112
    wavefunction evolves linearly 1113
Hohenberg, P 24
Holevo, AS 682
Holland, P 31, 33
holomovement, hologram 1087
Holt, R 343
    *see also* CHSH
Hörmander, L, microlocal analysis 485
Hornberger, K 9
Horne, M 343–58
    *see also* CHSH
Howard, D 21

Hund, F, applications extending range of
     validity of theory 177
hydrogen atom
     application of matrix mechanics 186
     Bohr–Balmer quantization 87
     Einstein–Stern model 122
     Lamb shift 378
     spectrum 187
     Stark effect 82, 186, 945
     test for validity of SED 1259
hydrogen molecule, theories 189
hyperplane-dependent collapse 1169

**I**
ideal gas, quantum theory 106–12
identity
     discernibility and re-identification 1151
     first-order logical axioms 1142
     individual, vs distinguishability 95
     logical relation of identity 1142
     Schrödinger on 1151
impossibilities
     micro-causality 464
     quantum cryptography 464–5
'impossibility proofs'
     Bell 965–9
     Kochen and Specker 962–5
     von Neumann 958–62
impossibility theorem
     classical vs quantum probability
          217, 227
     proof, von Neumann, J 217, 227, 528,
          851, 958–62
incompatibility 27
     between QM and gravitation 828
indeterminacy, disturbance analysis
     535–6
indirect measurement, as third way 676–7
indiscernibility 1142, 1144–5
indistinguishability
     bosons and fermions 256
     defined 255
     particles 1114
indistinguishable properties 1150
individuality 1141–3
     defined by discernibility and
          re-identification 1151

individuation principle 1142
information, emphasizing 14–15, 418
'information' concept 1182
information gathering and utilizing systems
     (IGUSes) 1176, 1192
information theory 1185
information-theoretic postquantum
     models 428–31
informational quantum theory 15
instrumentalism 784, 1009
     and Bohm complementarity 784
instrumentation, complementarity
     principle 556–7
instrumentation and quantum
     foundations 587–610
integral equations 474
interference, as a many-particle effect
     9–10, 209
interferometry 9–10, 27
     experiments 604–6
internal model of measurement 993–4
International Solvay Institute for Physics
     (ISIP) 625
internationalization of science see Solvay
     Councils
interpretation
     debate and quantum gravity (QG)
          393–413
     pluralism 723
     problem of 1074
     of theory, defined 8
interpretation of QM 937–52
     the claim 937–8
     the argument 938–40
     electromagnetism (case 1) 940–4
     physics of the atom (case 2) 945–8
     Schrödinger's wave mechanics 948–53
     summary/conclusion 953
interpretation–ontology–non-individuality
     shifts 1135
ion trapping, laser cooling 606–7
Ishiwara, J 82
Italy, postwar FQM 641–61
     1960s 643–4, 830
     1970s–1980s 651–60
     1988 effect 644–51
     quantum dissidents 642, 648

## J

Jacobi, 82, 87, 442, 474, 480, 740, 834, 1061
James, RW, on ZPE 125
Jammer, M 1, 4, 8, 18, 826, 829, 842
    on applications 177
    on like particles 256
    *Philosophy of Quantum Mechanics, The* 829
Jánossy, L 848
Janssen, M 77, 79
Japanese reactions to QM 1927–1943
    687–702
    1920s 689–91
    acausality and science journalism 691–3
    Bohr's visit 697–701
    complementarity, Sakai on 690–1
    dialectical materialism and QM 693–6
    Kyoto School philosophy and
        Marxism 693–6
    QM in 1920s 689
    responses to EPR paper 696–7
    RIKEN 690
    summary 701–2
JASON group 645
Jeans, J 104, 266, 692, 910–14
    Rayleigh–Jeans law 128, 1239
Joffe, A 82, 236, 259, 903–4
Joliot-Curie's experiments 590
Joliot–Curie, I and F 590, 625
Jones and Clifton 982
Jordan, P
    on applications 79, 176
    on measurement 288, 670
    philosophical anxieties over QM 859–61
    Schrödinger function, state
        probabilities 211
Jost, R, new mathematical physics 385–7, 714
Julia, B 44

## K

Kaiser, DI 385–7, 589, 641–2, 652–6, 716,
    722–7, 756–60, 830, 898, 900
    Bell's inequalities, loopholes 758
    US quantum physics teaching 722
Kant, principle of causality 567
Kant, I 552
Kantian interpretation of QM 569–83, 700,
    800, 833, 1039, 1186, 1243

Károlyházy, F 1116
Kasday, LR 745
Kasday–Ullman–Wu (KUW)
    experiment 735, 744–9
Kato, T, Schrödinger operators 484
Keesom, W, ZPE and 125
Kelvin, Lord (aka W Thomson) 941–2
Kenji Ito, Japanese reactions to QM 1927–1943
    687–702
Khinchin, A, founder of Moscow probability
    school 676–7
Kiefer, C 44
kinetic energy, and temperature 117–20
Kirchhoff, G 100, 942, 952
Klein–Nishina formula, Compton
    scattering 739
Klyshko, D 594, 681
knowledge transfer/applications,
    bidirectionality 180
Kochen, S 36
Kochen–Specker theorem 1150
    'impossibility proofs' 962–5
    non-classicality 417
Kojevnikov, A 682, 887
Koyré, A 897
Kramers, H 78, 84–5
    BKS theory 137, 152, 161, 260
    Bohr–Kramers–Slater theory (BKS) 137,
        152, 161, 260, 902, 1226–7
Kramers–Heisenberg, dispersion theory for
    elastic scattering of light 78, 79,
    84–5, 88
Kreyszig, E 479–80
Kronecker delta 79
Krutkow, G 106
Krylov, N 671
Kubetsky, L 599
Kuhn, TS
    *Archives for the History of Quantum
        Physics* 833, 889–90
    *Black-Body Theory and the Quantum
        Discontinuity* 890
    complementarity as paradigm
        785–8, 835
    model 905
    *Structure of Scientific Revolutions, The*
        833, 889

# L

Ladenburg, R, classical dispersion
    theory 84–5
Ladyman, J 1067, 1138
Lafontaine, C 1183–4, 1186
Lagrange, L 180, 474, 941
Lakatos, I 785, 835
Laloë, F 2, 7, 33, 562, 825
Lamb, W
    Lamb shift, hydrogen atom 378
    on photons 234
Landau, L
    analyses of field measurements 380–4
    Lifshitz and 424, 669–70, 727, 1012
    solid-state physics 669
Landsman, K 473
Langevin, P 273
Lanz, L 1112
Laqueur, W 879–81, 885
laser
    continuous wave (CW) tuneable laser 593
    Maiman 241
    principle of 242f
    spontaneous parametric down-conversion
        (SPDC) 594–5, 1261
    tuneable 606
laser cooling 606–7
laser fields, quantum Monte Carlo
    methods 33
lattice
    definition 446
    irreducible vs complemented 446
law of identity 1146
law of indiscernibility of identicals 1142
law of total probability 356–7
Lebedev Institute, Moscow 674
Leggett, Sir A 30, 495–519, 842
    *The Problems of Physics* 514
Leggett–Garg inequalities 516
Leibniz
    metaphysics 837, 1143
    principle of indiscernibles 95
Lemaitre, G, cosmological constant $\Lambda$ 131–2
Letertre, L, operational reconstruction 422–4
Lewis, GN, on 'photon' use 237
Lewis, P 69
Lifshitz, EM 424, 679, 727, 1012

Lifshitz, IM 669
light
    Newton's corpuscular theory 237, 246
    wave–particle duality 31, 165–6, 241
light quantum/photon *see* photons
light/radiation
    corpuscular theory 104–5, 237, 246, 260,
        263, 268, 271, 273, 283–4, 311, 575,
        647, 799, 1260
    projective elements 1098
    scattering, inelastic 86
Lindblad form 40
Lindemann, F 124
linear deterministic 1109
linear dynamics of QM 992
linear operators on Hilbert space,
    diagonizability 217
linear space of states (Hilbert space) 9, 41, 57,
    87, 167
linearity condition 56, 62
local-realist theories/models 342, 344, 357
locality condition of relativity 1110
localization theories, spontaneous 1123
Loewer, B 69
logical positivism, and logical
    empiricism 777–9
Lombardi, O 35, 63
London, F 12, 287, 425
London theory of van der Waals forces 1256
loop quantum gravity 393
Lopreore, CL 33
Lorentz, HA 83
    Chair, Solvay Councils 622–5, 627f–9
    transformations 1097
    *see also* Drude–Lorentz metal model
Lorentz invariance 852
Ludwig, G 289, 295, 297, 452, 467
Luft, KF, electron impact spectroscopy 592

# M

McCormmach, R 720, 877, 890, 896
Mach, E 27, 403–4, 411, 952, 1232–3, 1239
    Mach's principle 411
        integration into quantum gravity 403–4
Machida, S 295
Mackey, G 419, 451–2, 486
    probabilities, states and evolution 451

McLaren, SB, on new physics 910–15
macromolecule, loss of coherence 1119
Madelung, E 173
Maiman, T, laser 241
Mandelstam, LI 671–2, 1228–36, 1239–40
    indirect measurement 676–7
    von Neumann's 1932 analysis
        of QM 1229
many-worlds interpretation of QM 848, 987,
        1002–4, 1062, 1106, 1169
    measurements as acts of conscious
        activity 1138
Margenau, H 281, 294, 761, 1231, 1233,
        1235–6
    ensemble approach to QM 1225, 1227
Markov, AA, Markov processes 39–41, 1115,
        1249–50
Markov, MA
    dismissal 678
    on 'modest role for observer' 1242
Marxism 897
    France 848
    Japan 693–6
    see also communism
Masanes, L 463–4
mass density field M, defined 1114
materialism and antireductionism 672
mathematical physics, new (1950s) 385–7
mathematics
    of nonindividuality 1145–8
    quasi-sets 1147–9
    set-theoretic perspective 1146
matrix and equivalence proof 87, 181
matrix mechanics 86, 140, 210
    Born, Heisenber, Jordan 86
    equivalence proof 181, 443
    transformation theory 147–57
    and wave mechanics 135
    Zeeman effect 83, 187
matter waves 603–4
    continuum wave theory 87
    wave–particle duality 112, 165–6
Maxwell, G 938
Maxwell, JC 839, 938–48, 953–4,
        1074, 1157
    distribution of particles 98
    electromagnetism 940–4

equations 253, 911, 921
    Hertz and 940–4, 953
Maxwell–Boltzmann statistics 99, 260
'meaning', demarcation from
        interpretation 194–5
measurement 1009
    Bohr on 1049
    and cybernetics 1193–7
    defined 1105
    human presence as observer 537–8
measurement problem 10, 57, 61–4, 281–98,
        989–94, 1059–60, 1074
    1950s and 1960s 282–3, 860–1
    determinate record and probability
        problems 992
    Easlea 500
    Everett 989–92
    insolubility proofs 292–3
    and interpretation problems 1074
    Jordan and Wigner on 859–61
    legitimising heterodoxy 861
    nested measurements 990
    nonlinearity in Schrödinger equation 831
    objectivist approaches 295–7
    Omnès on 1186
    'paradox of quantity' 283
    pre-1970s 293–5
    pre-von Neumann period 283–5
    providing a dynamics to QT 1138
    QBism 1011–14
    resolution (Barad) 1044, 1049–54
    subjectivist approach 287–8
        'characterization' problem 287–9
        'completeness' problem 287–8
    thermodynamic amplification (TDA)
        progamme 288–91
    von Neumann, wave–particle
        duality 285–6
Mehra, J 7, 8, 79
Meissner, KW, electron impact
        spectroscopy 592
Mermin, ND 20, 24, 1007
mesons 275
metaphysics
    according to Einstein 863
    experimental 791–2
    of individualism 1032

of non-individuality 1143
and physical sciences 1182
profile of elementary particles 1139–41
Michelson–Morley experiment, electronic
    theory of matter 912
micro-mechanics, Schrödinger 950
microlocal analysis 485
microscope, resolving power 535
microscope thought experiment 535
military applications, positron–electron
    studies 736
Millikan, R, on light as electromagnetic
    waves 236, 239
Minkowski, R, electron impact
    spectroscopy 592
Mitchell, D 716, 910, 912
Möbius transformation 1097
modal interpretations of QM 31, 35, 1155–75
    based on bi-orthogonal decomposition of
        total state ($\psi$) 169
    Copenhagen variant 1162–4
    non-Copenhagen variant 1164–6
    perspectival properties of
        observables 1166–8
    relatavistic covariance 1168–71
    summary 1171
modal logic 1156
momentum eigenstate 1223
Mompart, J 33
monoatomic gas, yearly energy increase 1113
Monton, B 65, 69
Moscow, Lebedev Institute 674
Moyal, J, Moyal algebra 442–3
Müller, M 463–4
Mulliken, RS, evidence for ZPE 127
multi-fields 69
multiplet intensity problem 137
Myrvold, WC 64, 65, 66, 68, 69, 77

N
Nagai Kazuo 695
Nakane, M 82
Namiki, M 295
Nath, P 44
natural language, principle/postulate 419, 431
natural reconstructions of QM 437–72
    empirical immediacy/veracity 438

Nauenberg, M 829
Needell, A 909
Nelson, L 569
neopositivist philosophy 1925–1950 777–82
Nernst, W 122
    heat theorem 619
    modern/classical physics 921
    Solvay Councils, vision 616f, 619
    zero-point energy 127–9, 376
neutron interferometry experiments 604
new physical field, filling space 1115
Newton, I 473
    corpuscular theory of light 237, 246
    velocity of light 237–8
Newtonian mechanics, Heisenberg on 856
Ney, A 70, 1120
Nieuwenhuizen, T 29
Nikolsky, KV 674–5, 1234–6, 1243
    ensembles 1227
    materialism 1234–6
Nishida Kitaro 695
Nishina Yoshio
    and complementarity 688–91, 698–9, 706
    Klein–Nishina formula 739
NMR, perturbating effect of measurement
    process 10–11
Nobel Foundation 1920, new physics 922–7
Noether, E 567–70
nomic view, quantum states 77
nomological objects 1145
non-Booleanity, classical to quantum
    mechanics 977–8
non-classicality 417, 419, 421t
non-commutative geometry 485
non-commutative structure, Hilbert space
    operators 1167
non-individuals
    interpretation of QM 1135–55
    lack of identity 1143–4
    wphyith identity 1150–1
non-linear deterministic theory 1110
non-linear stochastic evolution equation for
    wavefunction 1113
non-locality in quantum physics 325–8, 1100
non-statistical theory 214, 217
non-trivial complex manifold 1097f
Norris, C 18

nuclear coincidence methods 592
nuclear decay law 1111
nuclear physics 192

# O

objectification problem 285–7
objective collapse model 642
objective indeterminacy 531–2
observables
    determined by total state $\psi$ 1166
    perspectival properties 1166–8
    preferred 1079, 1165
observer
    consciousness of 283–4, 425–30
    human presence as 537
    intervention, wave-packet collapse 522
    privileged role for subjective
        observer 536–9
    and reality (Penrose) 1075
    subjective experience 992
old quantum theory 80–3
Omelyanovskij, M 658, 681
Omnès, R 24, 330, 816, 1175, 1179,
        1181–9, 1194
    consistent histories approach 1186–8
Onnes, K, ZPE 125
ontic state space 67, 205
ontological indeterminacy 1031
ontology 1135–6
    defined 96
    vs epistemology 557–8
operational definitions 147, 1233
operational reconstruction 422–4
operators 55
    adjoints 55, 1079
    algebras 484
    canonical variables of system 55
    normal 1079
    rings (von Neumann algebras) 484
Oppenheimer, JR, on QM 188–9
optical pumping, and electron
        bombardment 591
optical-mechanical analogy, equivalence
        proof 87
optics, stochastic optics 1259–61
orbits, Bohr (1922) 945
order, implicate/explicate 1089

Oriols, X 33
orthodox QM/QT 16–42, 1103
orthodoxy and heterodoxy, postwar
        era 847–65
    diachronicity 848
    rethinking 862–5
    see also 'Copenhagen interpretation'
outcome independence 65
Oxford workshops/symposia 46, 412–15, 857

# P

P $\psi$ function, statistical QM 674
    $\psi$ and $\varphi$, quantum states 67–8, 213
    $\psi(x)$, modulus squared 210
Pais, A 307, 310
paradigm interpretation 778, 785–8, 835
parameter independence 65
partial differential equations (PDEs) 474
particles
    classical 'distinguishable' vs quantum
        'indistinguishable' 255–8
    'like' 256
    wave–particle duality 31, 112, 165–6, 241,
        271, 283–5, 309, 1083, 1260–1
Paty, M 303–38
Pauli, W 87
    artificial asymmetry between position and
        momentum 851
    exclusion principle 186–7
    on 'final QM', (1954), Zurich 858
    on quantum numbers 243
    spin matrices 187, 243
    ZPE 376
Pauli–de Broglie exchange, on pilot wave
        theory 633
Pearle, P 39, 40, 64
    discrete spontaneous localization
        theory 1114
    non-linear equation, state vectors 1111
Pechenkin, A xi, 674, 1216, 1240
Peierls, R, analyses of field
        measurements 380–4, 670
Peil, S 44
Penrose, R 41
    angular momentum 1077
    on Beauty and Truth 1074
    collapse theories 1075

Diósi-Penrose (DP) theory 1116, 1119
entanglement in quantum
    measurement 1092
non-locality of QM 1095–8
observables 1079
Oxford symposium 412–13
quantum environmental decoherence 1092
quantum gravitation in QM 413
reduction of wave function 1092
twistors 1099
wave-packet reduction 1093–5
Peres, A 14, 21, 30, 1053
on experimental measurement 1011–14
Perinotti, Chiribella, D'Ariano, Perinotti
    (CDP) axioms 464–7
Perinotti, P 464–7
perspectival (relational) properties of
    observables 1168
perturbation theory, Feynman diagrams 245
Peter, P 34, 44
Petersen, A 18
phase space
    carving/splicing 77, 78, 81
    quantization rule 81
philosophy of science
    20C 777–90
    core concepts 1031
    on QT/QM 53–77, 305
phonon 669
photocathodes 600
photoelectric effect 235, 239
photomultiplier tubes by Kubetsky 600*f*
photons
    annihilation radiation 590
    Bose–Einstein statistics 271
    bunching Hong–Ou–Mandel dip 244
    cascade, source of photon pairs 591*f*
    defined 247
    emission and absorption/transfer of
        momentum 239
    experiments 233–54
    massless exchange particle of
        electromagnetic interaction 244
    pure radiation fields 396
    quanta, 'loose' 260
    reduction, transfer of momentum 239
    strictly massless energy quanta 239

terminology used by Einstein 235*t*–7
today's interpretation 246–8
twelve layers of meaning (QFD/
    QFT) 237–46
ultra-relativistic limit 240
wave–particle duality 31, 112, 165–6, 241
*see also* bosons; light quantum/photon
physics
    classical 915–19, 919–22, 921, 1075
    classical and modern, co-creation 909–25
    development 44
    formalism vs physical content 953
    future of 858–9
    historic milestones 424–8
    interpretation of QM 1136
    new mathematical physics (1950s) 385–7
    new physics, and concepts of the
        classical 922–7
    and philosophy 496–9
    post WWI 922–7
    *Problems of Physics, The*, Leggett, Sir A 514
    purpose of 14
    Soviet ideology 677–80
physics of the atom 945–8
    atomic nucleus in 1932 192
    nurturing intuition, Schrödinger 950–1
    planetary model (Bohr and
        Sommerfeld) 946, 948
    problem of interpretation 938
physics, new, Nobel Foundation 1920 922–7
pilot-wave theory 31, 33, 63, 1124
    parameter dependence 1123
    Pauli–de Broglie exchange on 633
    *see also* de Broglie–Bohm theory
Pinho, EJC 34, 44
Pinto-Neto, N 34, 44
Piron, C 419, 439, 448–51, 457, 469, 763
Pironio, S 358, 431
Pitowsky, L 14
Planck, M
    classical theory, contradictions 911
    combinatorics in radiation theory 96,
        101–4
    constant *h* 117
    electromagnetic entropy 101–2
    energy distribution law 102–3
    and Kuhn 890

Planck, M (*cont.*)
  new theories 834
  quantization conditions 77–89
  quantum hypothesis 234
  quantum theory of ideal gas 109–11
  radiation law $\epsilon = h\nu$ 79–80, 105, 259
  resonators and free field 102, 104
  second theory 80, 101, 118
  theory of blackbody radiation 79, 99–101,
    235, 241
  Third Law of Thermodynamics 619, 1259
Planck spectrum, includes ZPF 1257
Planck–Einstein condition 78
Planck–Sommerfeld rule, quantizing phase
    integrals 82
planetary model of atom (Bohr and
    Sommerfeld) 945–8
pluralism
  interpretative 723
  theoretical 779, 788–91, 833, 838–9
Pockels cell 603
Podolsky, B 14
  *see also* Einstein, Podolsky, Rosen, (EPR)
    article/argument
Poincaré, H 623, 912–14
  derivation of Planck's law 913–14
pointer observable, defined by measurement
    interactions as privileged 1166
Poisson algebra of infinitesimal evolutions,
    Hamiltonian mechanics 439–43
Poisson brackets
  commutation relation 86
  limit of large quantum numbers 86
polaron 669
Polchinski, J 44
Popescu–Rohrlich boxes 430
Popper, K 654, 782
  on Bohr's complementarity principle 1225
  on Copenhagen interpretation of
    QM 522–3, 1224–5
  insinuation of realist surrender 784
  on instrumentalism and Bohm
    complementarity 784
  statistical objective interpretation 1225
  uncertainty/'statistical scatter'
    relations 1225
  on validity of QM 835

positive operator-valued measure (POVM),
    Fuchs 427–8
positivity condition 56
positron annihilation 589–90, 736
positron–electron studies, military
    applications 736
postquantum models 428–31
  information-theoretic
    reconstructions 428–31
  toy models 428–31
Poynting, JH 953
$\Psi$, incomplete description of individual
    quantum system 204
Pranger, RJ 18
'pre-established harmony' 479
preferred basis problem 979, 994, 1012, 1024,
    1108, 1111–13, 1120
preparation independence postulate 67
preparation uninformativeness condition 68
Prezhdo, OV 33
primitive ontology approach 1120
Princeton interpretation of von Neumann and
    Wigner 285
principle of identity of indiscernibles
    (PII) 1142
principle/postulate, natural language
    419, 431
probabilities
  defined 422, 455
  physical, many-worlds picture 1162
  states and evolution 451
probability amplitude 309
probability assignments, Born rule 810
probability problem 993
*Problems of Physics, The*, Leggett, Sir A 514
projection postulate 281, 286, 1080
  and linear evolution 1059–60
projective geometry of finite dimension,
    defined 446
projector, observables 1079
Prosperi, GM 282, 643, 647, 649, 660, 1112
protective measurements 27
psychophysical parallelism 11, 61
pure conditioning, defined 455
pure states
  equivalence proof 481
  and mixed states 1107

pure wave mechanics, relative state
        formulation, Everett 987–8, 992–9
Pusey, MA 67

# Q

QBism 1007–30
  criticisms 1016–17
  instrumentalism 1009
  interpretation 1015, 1017–23
  no-go theorems 1015
  probabilities 1013–14
  subjective/objective 1010
  summary 1025
QM
  additional variables 31
  axiomization through functional
        analysis 473–94
  causal interpretation 1076
  commutation relation 79
  completeness of QM 855–6
  consistent (or decoherent) histories
        approach 1175–93
  constructing 54–60
  continuity/discontinuity 136
  continuum wave theory 87
  creation in 1925/6 243
  defined 54, 281
  Einstein' criticism 303–38
  empirically underdetermined 1152
  entangled and unentangled states 58
  finality, Pauli on 858
  foundations
  change in foundations: 1950s,
        1960s 499–500
  classical/modern 909–23
  fundamental equations, first 1055
  fundamental mistakes 1056
  informational 15, 418
  interpretation 8–9
  Italy, post-war 641–61
  Kantian interpretation by
        Hermann 569–83
  matrix and wave mechanics 135
  non-classical concepts 417
  noncommutativity or non-
        Booleanity 977–80
  observer, consciousness of 283–4, 425–30

old quantum theory 80–3
overview 7–15
philosophical anxieties over 859–61
post 1911
  Britain and Germany 909–27
  as centrality of 'modern physics' 927–30
predictions of, as relations/correlations 457
projection postulate 11–12, 17
'proving it wrong' 509–17
reconstructions 417–31
  from 5 assumptions 455–8
  natural 437–72
resource theories 417–31
scandal of modern science 1104
Schrödinger and Heisenberg
        pictures 58–60
with spontaneous localization (QMSL) 64
standard formulation 989
statistical/ensemble interpretation 27, 285,
        293–4, 675–9, 1224–44
subjective/objective 1010
textbook interpretations 16–17
as theory of ensembles 674–6
transformation theory 147–57
with universal position localization
        (QMUPL) 1114, 1117
wave mechanics see QM; quantum theory
        typicality and probability 1997–9
see also Born rule; hidden-variable theories
QM interpretations
20C philosophy of science 777–90
classes 28
contexts, systems, modalities (CSM) 23
Copenhagen 16, 21–2, 175, 400–7, 424–5
cut between formalism and
        interpretation 174, 194–5
dBB theory 31–2
demarcation from formalism 174, 194–5
domain of applicability 187
early debates 135–72, 184–90
ensemble/statistical 27–9
Everett/many-worlds 29–31, 69
informational 15, 418
meaning and 194–5
measurement problem and
        ontology 1135–6
precursor theories 193–4

QM interpretations (*cont.*)
    standard/orthodox 16–42, 1103
    veiled reality 21
    Wigner's 12
    *see also* complementarity
QMUPL (QM with universal position
    localization) 1114, 1117
quantization 77–89
    conditions, Bohr, N 77–89
    defined 54
    first/second 1114
    of general relativity 401–4
    of gravitational field 383, 396–401, 404
    old quantum theory 80–3
    procedure, Rosenfeld on 395–9
    radiation theory 396
    rule
       phase space 81
       Reiche, F 126, 129, 913, 921
       rotating molecule 126
    second quantization formalism 1114
    transition to quantum mechanics 83–8
quantum archaeology 609
quantum Bayesianism *see* QBism
quantum computing 742
quantum contextuality 420, 421*t*
quantum controversy 825–42
    first part 827–33
    second part 833–41
    uniqueness 841–2
quantum correlation bound 419–20
quantum cosmology 1176
quantum cryptography 329, 418, 431, 464–5,
    653, 742
quantum description, and relativity 34, 42–3
'quantum dissidents' 562, 642, 648, 830, 847
quantum electrodynamics (QED) 244, 372
    infinite self-energy of the electron 395
quantum encryption infrastructure 363
quantum entanglement 36, 190–1
    contextuality as non-classical
       resource 420*t*
    entangled and unentangled states 58, 967
    and EPR experiment 1077–8
    geometrics 1079
    includes Q cryptography, teleportation,
       computing 742

leak of 15
    model, London theory of van der Waals
       forces 1256
    projection postulate 1080
    wave function and 141
quantum equilibrium 32, 1207
    hypothesis, *see also* Born rule (1926)
quantum eraser experiment, Scully 1046
quantum field theory (QFT) 244, 371–87
    defined 54, 371
    distribution theory 485–6
    eliminating infinities 375
    history 373–9
    locality and speed of light limit 1086
    measurement of QF 379–84
    postwar pragmatism and its
       opponents 384–7
    regularization and renormalization 378
    'small war' 379–81
    ZPE 131
quantum foundations (QF) 587–610
quantum gravity (QG) 34, 393–413
quantum historiography and cultural
    history 887–905
    founding period (1960s) 888–91
    Forman Thesis 891–5
       controversy 895–900
       examples 901–4
quantum indeterminacy, disturbance
    analysis 535–6
quantum information 304
    dominant interest 831
    field 515–17
    historic milestones 424–8
    information-theoretic postquantum
       models 428–31
    published papers on 832*f*
    quest for reconstruction of QT 417–32
quantum logic 439, 444–54
    to resource theories 418–21
quantum Monte Carlo methods, laser
    fields 33
quantum numbers, angular momentum 81
quantum optics 592
    Weyl–Wigner formalism 1261
quantum particles, indistinguishable
    255–8

quantum physics
  Soviet ideology 677–80
  *see also* physics
quantum potential (Bohm) 1206
quantum probabilities, Copenhagen
      perspective 1175
quantum random number generators
      (QRNGs) 354, 360*f*
'quantum states' 1061
quantum states
  defined 56–7
  expectation values 65
  monism 68–9
  nomic view 77
  ontological status 68–5
  $\varphi$-ontic state 67
  preferred basis 1108
  $\psi$, deterministic evolution during an
      A-measurement 974–6
  $\psi$ and $\varphi$ 67–8
  $\psi$-epistemic model 67
  $\psi$-ontic state 67
  realism 66–8
  self-adjoint 56
quantum statistics interpretations 255–75
  early
      lack/loss of individuality of
          particles 273–4
      statistical interdependence of
          particles 258–64
      symmetry of quantum formalism pre
          1926 264–72
  fundamental tension 95–6, 186
  general formalism 269–71
  new form of exchange symmetry 267–9
  photons, exchange symmetry,
      electrons 271–2
  summary 274–5
  theoretical formalism 257–8
quantum superposition of macroscopically
      distinct states (QSMDS) 37, 40, 43
quantum system, $K$ and $N$ 22
quantum teleportation 742
quantum (unsolved) controversy (1974) 826
quantum velocity term 32–3
quantumness 417
quasi-sets 1145, 1147–9
Quine, WVO 781, 827, 833, 840

**R**
Rabi, I, probabilistic interpretation of
      QM 863–4
Radder, H
  affirming WCC's support of German
      physicists' ideology 882
  and Weimar (WCC) 880–2
radiation pressure 239, 240
radiation theory 103–6, 914
  natural radiation and entropy 101–2
  quantization 396
  spontaneous parametric down-conversion
      (SPDC) 594–5, 1261
  *see also* blackbody radiation; wave–particle
      duality
Raman light scattering 86
Raman spectra 243, 246
random collapse dynamics 992
Rask-Ørsted Foundation 545
Rayleigh, Lord 104, 910, 912
Rayleigh–Jeans law 104, 128, 1259
Rayleigh–Schrödinger perturbation
      theory 191–2
re-identification 1151
realism (Popper) 784
reality/relations 1032, 1075
Received View of quantum
      nonindividuality 1136, 1145
  alternative version 1151–2
  entities without identity 1151
Rechenberg, H 7, 8, 79
reconstructions, stages 422–4
Redhead, M 372, 872, 965, 1139,
      1223, 1242
reduction, avalanche process 289
reflexive law of identity 1146
reflexivity 1142
Reiche, F
  quantization rule 126, 129, 913, 921
  Thomas-(Reiche-)Kuhn sum rule 79,
      86, 88
Reichenbach, H 64, 568–70, 658, 777, 780,
      780–1
  common cause principle 64–5
'relata precede' relations 1031–2
relational interpretation of quantum physics
      (RQM) 64, 1031, 1055–71
relational properties 1167

relative state interpretation of QM 987, 1193
  pure wave mechanics 987-8, 992-9
relativistic covariance 1168-71
  hyperplane-dependent collapse 1169-71
relativistic invariant spontaneous localization
        theory 1123-4
relativistic quantum electrodynamics, von
        Neumann 858-9
relativity
  modal interpretations 1168
  physical limits 1109
Renn, J 7
renormalization programme 372, 378,
        385, 387
representationalism 1033
resource theories 418-21
response probability functions $fk(\lambda)$ 61
retrocausation in conventional formulation of
        QM 1082
Riemann, B 473, 1089
Riemann sphere 1081f, 1097
  Möbius transformation 1095f
Rimini, A 37, 40, 64
  (GRW) theory 37-41, 64-6, 297
rings of operators, von Neumann
        algebras 484
Rio de Janeiro and São Paulo conference 855
Rockefeller-funded International Education
        Board (IEB) 547
Rome, Cini, M 645, 648, 657-8
Rosen, N 14
Rosenfeld, L
  Bohr's interpretation of QM 395-9
  Bohr-Rosenfeld argument 121, 383, 399,
        401, 559
  contra Bohm, origins of theoretical
        pluralism 788-91
  'false problem' 174, 383
  gravitational effects of light
        quanta 394-400
  quantization procedure 395-9
  'quantum dissidents' and 562
  wave equation in five dimensions 394
rotating molecule, quantization rule 126
Rovelli, C 44, 64, 422, 1055
RQM (relational interpretation of quantum
        physics) 64, 1031, 1055-71

Rudolph, T 67
Russia see Soviet Union
Rutherford, E 546, 616f, 625, 912, 916-17,
        925-6, 1251
Ryazanov, GV 681
Rybakov, Y 681
Rydberg constant 80
Rydberg, J, atoms and cavities 608-9f

S
S-ontology 96-9, 103, 111
Sackur-Tetrode entropy 111
Sakai Takuzo 690-1
Santa Fe Institute 1183-7, 1192-3
Satosi Watanabe 426, 701
Schack, R 1007
Schaffer, S 887, 899
Schlosshauer, M 5, 10
Schmidt decomposition 36
Schrödinger dynamics 12
Schrödinger dynamics (modified) 37-42
Schrödinger, E 41, 44
  cat paradox 14, 19-20, 288, 503-6
  discontinuity issues 145-7
  evolution for wavefunction 1107
  and Heisenberg, Anschaulichkeit 379
  on identity 1151-2
  intuitive conception of physics 950-1
  'measurement problem' 1056
  micro-mechanics of the atom 949-50
  oscillatory states of a coupled system 191
  rational justification for Born rule 802
  transformation theory 147-57
  'wave function' and quantum state 1056
  wave mechanics 140-7, 184, 948-53, 957
     continuous scale transition 948-53, 957
  wave packet theory 160
Schrödinger equation 29-30, 59, 62,
        1104, 1109
  Bohm on 740
  dynamical law 535
  measurement processes 1111
  non-linear modifications 846, 1110
  non-linear stochastic modification 1111
  and the psi-function 905
  SL term 37-8
  solution 1256-7

state vector 17–20
statistical interpretation of
    wavefunction 203
stochastic interpretation 1249–50
Schrödinger function, state probabilities 211
Schrödinger operators 484
Schrödinger picture 58–9
Schrödinger-Newton (SN) equation 1116
Schwartz, L, distribution theory 485–6
Schwarzschild, K 78, 82
    Stark effect 82
Schweber, SS 1, 7, 373, 379, 384–7, 832
Schwinger, J, renormalization
    programme 378
science journalism, Japan 691–3
scientific internationalism, interwar 544–5
scientific realism 783–5
    and Bohm's theory 783–5
scintillation counters 600
Scully, MO 33
    quantum eraser experiments 33, 1046
second quantization formalism 1114
Segal, I, new mathematical physics 385–7
self-adjoint operator 55–6
Selleri, F 645–8, 652–5
separability of knower and known 1033
separability/nonseparability 58
set theories with ur-elements 1146–7
Shaknov, Wu-Shaknov experiment
    (1950) 735–43
Shapin, S 882, 887, 899
Shiffren, L 33
Shimony, A 1100
    see also CHSH
Shubin, S 669
Siegel and Wiener (1955), differential-space
    theory 973
simplicity (axiom) 422
Singh, TP 42
single user theory 1008
singularity 33
Slater, JC
    BKS theory 137, 152, 161, 260, 902
    ensemble approach in USA 1226–7
Solèr, MP 450–1, 486, 573, 583
Solvay Councils
    1911, British perspectives 910–15

1911–1930 f–9, 212, 213, 307, 386,
    615–37, 910
1913 625, 626–7
1927 618f–34
forced closure 1927 630–3
international boycotts 628–30
ZPE 60, 118, 122–3, 166
Solvay, E, and Solvay process 621f
Sommerfeld, A 78, 80
    on indeterminate [unbestimmt]
        physics 863
    phase integral 78
    planetary model of atom 945–8
    rephrasing Bohr's quantization
        condition 81
Sources for History of Quantum Physics
    (SHQP) project (Heilbron) 711
Soviet Union 658–9, 667–83
    attack on 'Copenhagen school' by Soviet
        physicists 847
    Bohr–Einstein debates 674
    Cold War 497, 609, 670, 736, 756–7, 759,
        847–8, 888, 896–7
    collectivism 668–9
    Copenhagen school/interpretation of
        QM 670–2
    death of Stalin 679
    democratic vs imperialist world camps 678
    dialectical materialism 673
    ensemble interpretation 679
    ideology and quantum physics 677–80
    indirect measurement as a third way 676–7
    Moscow, Lebedev Institute 674
    postwar QM, international context 680–2
    statistical/ensemble interpretation of
        QM 1227–30
    summary 682
space and time, competing definitions 397
spacetime 59
    Galilean realist tradition 59
    Minkowski 59, 479, 962, 968, 1096–7, 1099
    state realism 77–85
spacetime structure of atom 43
    abandoning 139
specific heat, Debye theory 1256, 1258
spectral theorem, for (possibly unbounded)
    self-adjoint operators 481

spectroscopy
    atomic beams 592
    doublet splitting spectra 241
    electron impact spectroscopy 592
    Zeeman effect 241
Spekkens, RW 66
    toy postquantum models 428–31
Spengler, O
    *Decline of the West* 892
    philosophy 896
    views on causality 871, 873, 900
spin, angular momentum 1077
spinors
    non-relativistic spinor 1100
    twistors 1077
spontaneous localization 37–8
spontaneous localization theories 1103–25
    compatiblity with relativity theory 1121–2
    dynamical reduction theories 1106
    empirical and ontological
        adequacy 1117–22
    experimental tests 1118
    Ghirardi, Rimini, Weber 37–41, 64–6, 297,
        656, 1106, 1111–12, 1121
    ideas driving 1106–11
    measurement problem 1104
    overview 1111–17
    precursors 1111
    relativistic extensions 1122–5
    tests 1113
    *see also* continuous spontaneous
        localization theory (CSL)
spontaneous parametric down-conversion
    (SPDC) 594–5, 1261
Squids (superconducting quantum
    interference devices) 504
Staley, R 178, 195, 715, 873, 899, 900, 909, 918
    refuting WCC 873
standard interpretation of states 989–92
Standard Model 304
Stapp, HP 21, 344–5
Stark effect 82–3, 186, 917, 920, 945
Stark, J, action-angle formalism 83
state, value state and dynamical state 1162
state reduction 281
state vector 37
    correlation interpretation 13

reduction 12–13, 17
    Schrödinger equation 17–20
    of universe 30
    and wave function 1149
stationary states 555–6
statistical interpretation of
    wavefunction 203–32
    indeterminism 221–7
    phase-1, Born on 205–9
    phase-2, Born to Heisenberg 209–14
    phase-3, von Neumann 214–21
statistical objective interpretation,
    Popper 1225
statistical/ensemble interpretation of QM 27,
    285, 293–4, 675–9, 1224–44
    interpretation 27, 285, 293–4, 675–9,
        1223–44
    $\psi$ function 674
    QM as theory of 674–6
    subensembles 29, 218
    USSR 1227–30
Stefan–Boltzmann law 100
Stern–Gerlach magnet 161, 213, 289,
    592, 1207
stochastic differential equations 682
stochastic electrodynamics (SED) 1251–9
    applications to linear systems 1256–9
    approximation to QM 1252
    equilibrium between radiation and
        matter 1259
    failure in nonlinear systems 1258–9
    harmonic oscillator 1252–4
    hydrogen atom, test for validity
        of SED 1259
    London theory of van der Waals forces,
        coupled oscillators 1256
    origin 1251–2
    particle in a homogeneous magnetic
        field 1258
    specific heat of solids 1257
    stationary state, vs ground state in
        QM 1254–5
    ZPF 1257
stochastic evolution equation,
    wavefunction 1113
stochastic interpretations, of Schrödinger
    equation 1249–50

stochastic interpretations of QM 1247–63
  realistic interpretation? 1250
stochastic mechanics, interpretation of
      QM 1250–1
stochastic optics
  vacuum electromagnetic field (ZPF) 1257,
      1259–61
  vacuum radiation field in confined
      space 1260
Stokes's rule, on photoluminescence 234
Stone, MH, theorem 483
Stone–von Neumann theorem 483
Struyve, W 42
Sturm–Liouville theory, linear second-order
      differential equations 474
subjective appearance of quantum
      probabilities 988
subjective observer 536–9, 992
  privileged role for 536–9
subspaces, defined 423, 455
substitution law 1142
superconducting ring, Cooper pairs 1119
superluminal signalling 33, 1109, 1123
  avoiding 1110
superposition principle 1104, 1160
superposition state 1075, 1160
  a statistical mixture 1107
superposition of state functions 304
  defined 57, 288
'superstring revolution' 393
surrealistic trajectories 33
Süssman, G 291
system–environment interaction, destroys
      quantum superposition 605
Szily, K 82

T
tachyonic noise, in place of white
      noise 1124
Tagliagambe, S 658
Tait, PG 941
Taketani Mituo 696
Takeuchi Tokio 693
Tamm, IE 669, 671–2
Tanabe Hajime 694–701
Tani Kazuo 696
Tastevin, G 33

temperature
  and kinetic energy 117–20
  see also zero-point energy (ZPE)
tension
  creative, early development of
      QM 173–202
  statistical mechanics 95
Terletsky, Ya. P. 658, 678, 680, 783, 847
textbooks 709–28
  American dominance 722, 725
  canonical genres of QM 709–22
  on Copenhagen interpretation 723
  foundations 722–6
  as historical sources 175, 712–16
  on quantum physics, history 727–8
  quantum textbook generations 717–21
  summary/conclusions 726–8
theoretical pluralism 779, 788–91, 833, 838–9
theory foundations
  range of validity 179
  vs applications 175–80
thermal equilibrium, combinatorial
      models 97–100
thermodynamic amplification (TDA)
      progamme 288–91
Thomas, W 86
  Thomas–Kuhn sum rule 79, 86, 88
Thomson, JJ, 'paradox of quantity' 283
Thomson, W 941–2
thought experiments see Copenhagen
      interpretation of QM
  see also Wigner
Tilloy, A 42
Timpson, GG 77
  on axioms 423
Tomonaga, S-I 187, 243, 378, 497, 690, 1123
Tomonaga–Schwinger
  equation 1123
  renormalization programme 378
trajectories, plotting 33
transformation theory 147–57
  Born, s probabilistic interpretation 147–60
  Dirac–Jordan statistical transformation
      theory 147–57
  transition probabilities, quantities other
      than energy 210
  von Neumann, Hilbert space 167

transition between ontologically different
modalities 1074
trigger problem 1108, 1113, 1120
Trimmer, JD 19
Tsirelson, B, bound, Bell–CHSH
inequalities 344, 419, 682
tunneling 513–14
twentieth century philosophy of
science 777–90
1925–1950 779–82
twistor
Clifford algebra 1100
Penrose on 1099
two-slit experiment 1075–6, 1086*f*

**U**

uncertainty, and complementarity 159,
798–801
uncertainty relations (Heisenberg) 127,
157–67, 211
Landau and Peierls 381
non-epistemic interpretation of Bohr,
N 798–800
wave mechanics as a corrective 165
undulatory theory 105, 191, 263, 266,
271, 283
Universe
big bang period 33, 44
'big bounce' 34
state vector 30
ur-elements 1146–7
USA *see* America
USSR *see* Soviet Union

**V**

vacuum electromagnetic field
(ZPF) 1259–61
vacuum radiation field in confined
space 1260
vacuum/dark energy 130–2, 131–2
Valentini, A 23, 60, 63, 273
and Bacciagaluppi, G 60, 63, 203, 212,
214, 226, 271, 630, 827, 958, 1104,
1110, 1206
value state 35–6
and dynamical state 1162

van der Waals forces, coupled oscillators 1256
van Fraassen, B 35
Copenhagen variant of modal
interpretation 1162–4, 1166
on modal interpretation of QM 1156
van Vleck, J 78, 86, 185, 927, 1259
Varenna school on FQM 648–50, 745
KUW results 745
Vey, J, Moyal algebra 442–3
Vietnam war 644–5, 657, 830, 890
Vigier, JP 603–4, 647, 654
von Helmholtz, H 83
von Neumann, J 11–12
algebras 484
Born probability 482
chain (infinite regression) 11–13, 15,
19–20
collapse postulate 1105
'ensembles' 215–16, 218
formulation of QM 988
Göttingen paper on QM 214, 475
hidden parameters 222–3, 226–7, 854,
960, 1078
Hilbert space, formalism 174, 419, 827
ideal measurements 1165
impossibility theorem 217, 227, 528, 851,
958–62
response to Bohm 851–5
mathematical abstraction 423
measurement problem 1164
non-statistical theory 214, 217
psycho-physical parallelism 61
QT foundations 479–83
quantum logic 439
standard formulation of QM 988
state vector reduction postulate 17
statistical interpretation of wavefunction,
phase-3 214–21
transformation theory/Hilbert space
167, 484
wave theory vs statistical theory in
QM 215–16
wave–particle duality 285–6
'von Neumann–Wigner interpretation' 61
von Weizsäcker, CF 852, 1082
Vonsovskiy, SI 669

# W

Wallace, D 77
Walther, H 608
Warburg, E, modern/classical physics 625, 629, 921
Watanabe Satosi 426, 701
wave mechanics
  continuous scale transition 948–53, 957
  Heisenberg's 949
  Schrödinger's 948–53
  without probability, Everett 987
  *see also* QM/QT; quantum theory
wave-packet collapse 142–3, 152, 159–61, 166–7, 281
  Heisenberg on 142–3, 152, 159–61, 166–7, 534–6
  observer intervention 522
wave-packet reduction 1093–5
  Penrose 1092–4
wavefunction 57
  defined 1104, 1108, 1223
  ensemble approach to QM 1227
  entanglement 141
  finite 87
  non-linear stochastic modification of Schrödinger equation 1111, 1113
  particle-like properties of matter and 284–5
  quantum-mechanical 69
  realism 69–77, 1120
  reduction, R element 1092
  represented physical systems? 1120
  role 63
  state vector 37–8, 1149
  status 34, 38
  trigger problem 1108
  universal 402, 404–6
  visualizable alternatives 827
  *see also* statistical interpretation of wave function
wavefunction $\psi$ 87, 1055
  Bohmian mechanics 1084
wave–particle duality 31, 112, 165–6, 241, 271, 283–5, 309, 1083, 1260–1
  behaviour of light 1260
  matter waves 112
  measurement problem 285–6

WCC *see Weimar Culture, Causality, and Quantum Theory* (1918–27), Paul Forman
Weart, S 890
Weber, T 37, 64
  authoritative refusals of assent 872
  culture 552, 724–6, 828, 871–6, 878–86
  (GRW) theory 37–41, 64–6, 297
  Norton Wise on 874, 877, 880–4, 900
  postmodernity 882–5
  Radder, H, affirming WCC's support of German physicists' ideology 882
  rebuttals by Hendry 872–4, 879, 880, 897–8, 900, 902
  reception of WCC by moderns 879–82
  summary/conclusions 886
  *Weimar Culture, Causality, and Quantum Theory* (1918–27), Paul Forman 872–86
Weinberg, S 2, 8
Western capitalism 782
Western philosophy 1031–2
Weyl, H
  on exchange symmetry 272
  quantization 440–1
  quantization formula 272, 440–1
  unbounded self-adjoint operators 481
Weyl–Wigner formalism, in quantum optics 1261
Wheeler, J 988
  anthropic principle 406, 1192
  delayed-choice experiments 601
  Everett interpretation of QT 400–7
  'It from Bit' 417
  pair theory, Wu-Shaknov experiment (1950) 737
  quantization of general relativity 401–2
Wheeler–DeWitt equation 410, 1190
which-path experiments 1044
wholeness 1076
Wiener, N 87, 148, 206–7, 478, 864, 1184
Wiener and Siegel (1955), differential-space theory 973
Wien's displacement law 100, 102, 104, 259

Wightman, A
  axiomatic quantum field theory 485
  new mathematical physics 385–7
Wigner, EP 295, 645
  formula 12–13, 15, 25
  friend thought-experiment 15, 282, 990–1,
      1016, 1058–9, 1166–70
    extension 1167–8
  interpretation of state vector reduction 12
  irreducible representations of Poincaré
      group 387
  on measurement problem 861
  perspectival properties of
      observables 1167–8
  philosophical anxieties over QM 859–61
  Weyl–Wigner formalism, Heisenberg
      picture 1260–1
Wilson, AH 669
Wilson, W 82
Wineland, D 606–7, 831, 1152
Wise, N 874, 877, 880–4, 900
Wolfke, M 106
World War I
  new physics and concepts of the
      classical 922–7
    Zhdanov's speech 847
  postwar era 384–7, 560–2, 641–61, 680–2,
      847–60
  postwar scientific internationalism 544–5
World War II, see also Cold War
World wave function 412
Wu see Chien-Shiung Wu
Wu-Shaknov experiment (WS, 1950) 735–43
  angular correlation between two
      photons 738f, 740t
  EPR paradox (Bohm) 740, 742
Wyatt, RE 33

X
x-ray bremsstrahlung, short-wave
      limit 235
x-ray diffraction, and ZPE 125

Y
Yukawa Hideki 687–8, 694, 701
  $\psi$ and $\varphi$, quantum states 67–8

Z
Zeeman effect 945
  matrix mechanics 83, 187
  splitting of spectral lines 241
Zeh, HD 10, 30, 310, 605
Zeilinger, A 2, 4, 14, 832
  Bell's inequality, locality loophole 351
  Erice School 659
zero-point energy (ZPE) 117–34
  Bohr atom 120–1
  Casimir energy 377–8
  'dead as doornail' 123
  Einstein's ambivalence 122–4
  half-integral quantum numbers 126–7
  physical meaning 376–8
  rotational 126
  of space 130
  vacuum/dark energy 130–2
ZFC, Zermelo–Fraenkel set theory 1146
Zhdanov, AA
  idealogical Soviet campaign 782
  speech (1947) 847–8
Zinn-Justin, J 44
ZPF, stochastic optics 1257
Zurek, WH 10, 310, 605
Zurich, International Congress of
      Philosophers (1954), Pauli on 'final
      QM' 858